McGraw-Hill

Dictionary of Bioscience

Second Edition

McGraw-Hill

New York Chicago San Francisco Lisbon London Madrid Mexico City Milan New Delhi San Juan Seoul Singapore Sydney Toronto

The McGraw-Hill Companies

1 2 3 4 5 6 7 8 9 0 DOC/DOC 0 9 8 7 6 5 4 3

ISBN 0-07-141043-0

This book is printed on recycled, acid-free paper containing a minimum of 50% recycled, de-inked fiber.

This book was set in Helvetica Bold and Novarese Book by the Clarinda Company, Clarinda, Iowa. It was printed and bound by RR Donnelley, The Lakeside Press.

Library of Congress Cataloging-in-Publication Data

McGraw-Hill dictionary of bioscience — 2nd. ed.
p. cm.
ISBN 0-07-141043-0 (alk. paper)
1. Biology—Dictionaries. 2. Life sciences—Dictionaries. I. Title: Dictionary of bioscience. II. McGraw-Hill dictionary of scientific and technical terms. 6th ed.

QH302.5.M382 2002
570′.3—dc21 2002033193

Contents

Staff

Mark D. Licker, Publisher—Science

Elizabeth Geller, Managing Editor
Jonathan Weil, Senior Staff Editor
David Blumel, Staff Editor
Alyssa Rappaport, Staff Editor
Charles Wagner, Digital Content Manager
Renee Taylor, Editorial Assistant

Roger Kasunic, Vice President—Editing, Design, and Production

Joe Faulk, Editing Manager
Frank Kotowski, Jr., Senior Editing Supervisor

Ron Lane, Art Director

Thomas G. Kowalczyk, Production Manager
Pamela A. Pelton, Senior Production Supervisor

Henry F. Beechhold, Pronunciation Editor
Professor Emeritus of English
Former Chairman, Linguistics Program
The College of New Jersey
Trenton, New Jersey

Preface

The *McGraw-Hill Dictionary of Bioscience* provides a compendium of 22,000 terms that are central to the life sciences and are encountered in many other fields of science and engineering. The coverage in this Second Edition has been substantially reorganized, with approximately 5000 new terms in areas such as forensic science, medicine, neuroscience, pathology, pharmacology, and veterinary medicine. The terminology previously classified as cytology and molecular biology now appears as cell and molecular biology. In some instances, terminology appearing in the previous edition has been moved to more appropriate companion dictionaries. It is hoped that this organization reflects the scope of modern life sciences and the needs of readers.

All told, the terms and definitions in the Dictionary represent 26 fields. In addition to those mentioned above, the reader will find coverage in: Anatomy, Biochemistry, Biology, Biophysics, Botany, Ecology, Embryology, Evolution, Genetics, Histology, Immunology, Invertebrate zoology, Microbiology, Mycology, Physiology, Systematics, Vertebrate zoology, Virology, and Zoology.

All of the definitions are drawn from the *McGraw-Hill Dictionary of Scientific and Technical Terms*, Sixth Edition (2003). Each definition is classified according to the field with which it is primarily associated; if a definition is used in more than one area, it is identified by the general label [BIOL]. The pronunciation of each term is provided along with synonyms, acronyms, and abbreviations where appropriate. A guide to the use of the Dictionary appears on pages vi-vii, explaining the alphabetical organization of terms, the format of the book, cross referencing, chemical formulas, and how synonyms, variant spellings, abbreviations, and similar information are handled. The Pronunciation Key is provided on page x. The Appendix provides conversion tables for commonly used scientific units as well as other listings of scientific data, including the universal (standard) genetic code, a classification of living organisms, and a table of the major groups of viruses.

It is the editors' hope that the Second Edition of the *McGraw-Hill Dictionary of Bioscience* will serve the needs of scientists, engineers, students, teachers, librarians, and writers for high-quality information, and that it will contribute to scientific literacy and communication.

Mark D. Licker
Publisher

How to Use the Dictionary

ALPHABETIZATION. The terms in the *McGraw-Hill Dictionary of Bioscience*, Second Edition, are alphabetized on a letter-by-letter basis; word spacing, hyphen, comma, solidus, and apostrophe in a term are ignored in the sequencing. For example, an ordering of terms would be:

allelotropism
Allen-Doisy unit
Allen's rule
all-or-none law
allosteric transition
allostery

FORMAT. The basic format for a defining entry provides the term in boldface, the field in small capitals, and the single definition in lightface:

term [FIELD] Definition

A term may be followed by multiple definitions, each introduced by a boldface number:

term [FIELD] **1.** Definition. **2.** Definition. **3.** Definition.

A term may have definitions in two or more fields:

term [ECOL] Definition. [GEN] Definition.

A simple cross-reference entry appears as:

term *See* another term.

A cross reference may also appear in combination with definitions:

term [ECOL] Definition. [GEN] *See* another term.

CROSS REFERENCING. A cross-reference entry directs the user to the defining entry. For example, the user looking up "aiophyllous" finds:

aiophyllous *See* evergreen.

The user then turns to the "E" terms for the definition. Cross references are also made from variant spellings, acronyms, abbreviations, and symbols.

aesthacyte *See* esthacyte.
AMP *See* adenylic acid.
D-loop *See* displacement loop.

ALSO KNOWN AS . . . , etc. A definition may conclude with a mention of a synonym of the term, a variant spelling, an abbreviation for the term, or other such information, introduced by "Also known as . . . ," "Also spelled . . . ," "Abbreviated . . . ," "Symbolized . . . ," "Derived from . . . ." When a term has more than one definition, the positioning of any of these phrases conveys the extent of applicability. For example:

term [ECOL] **1.** Definition. Also known as synonym. **2.** Definition. Symbolized T.

In the above arrangement, "Also known as . . ." applies only to the first definition; "Symbolized . . ." applies only to the second definition.

term [ECOL] **1.** Definition. **2.** Definition. [GEN] Definition. Also known as synonym.

In the above arrangement, "Also known as . . ." applies only to the second field.

term [ECOL] Also known as synonym. **1.** Definition. **2.** Definition. [GEN] Definition.

In the above arrangement, "Also known as . . ." applies only to both definitions in the first field.

term Also known as synonym. [ECOL] **1.** Definition. **2.** Definition. [GEN] Definition.

In the above arrangement, "Also known as . . ." applies to all definitions in both fields.

CHEMICAL FORMULAS. Chemistry definitions may include either an empirical formula (say, for abietic acid, $C_{20}H_{30}O_2$) or a line formula (for succinamide, $H_2NCOCH_2CONH_2$), whichever is appropriate.

Fields and Their Scope

[ANAT] **anatomy**—The branch of morphology concerned with the gross and microscopic structure of animals, especially humans.

[BIOCHEM] **biochemistry**—The study of the chemical substances that occur in living organisms, the processes by which these substances enter into or are formed in the organisms and react with each other and the environment, and the methods by which the substances and processes are identified, characterized, and measured.

[BIOL] **biology**—The science of living organisms.

[BIOPHYS] **biophysics**—The science that uses the experimental and theoretical approaches of physics to study the mechanisms of biological processes.

[BOT] **botany**—That branch of biology dealing with the structure, function, diversity, evolution, reproduction, and utilization of plants and their interactions within the environment.

[CELL MOL] **cell and molecular biology**—The study of the structures, functions, and molecular aspects (proteins, enzymes, nucleic acids) of the living cell.

[ECOL] **ecology**—The study of the interrelationships between organisms and their environment.

[EMBRYO] **embryology**—The study of the development of the organism from the zygote, or fertilized egg.

[EVOL] **evolution**—The processes of biological and organic change in organisms by which descendants come to differ from their ancestors, and a history of the sequence of such change.

[FOREN SCI] **forensic science**—The recognition, collection, identification, individualization, and interpretation of physical evidence, and the application of science and medicine for criminal and civil law or regulatory purposes.

[GEN] **genetics**—The science concerned with biological inheritance, that is, with the causes of the resemblances and differences among related individuals.

[HISTOL] **histology**—The study of the structure and chemical composition of animal tissues as related to their function.

[IMMUNOL] **immunology**—The division of biological science concerned with the native or acquired resistance of higher animal forms and humans to infection with microorganisms.

[INV ZOO] **invertebrate zoology**—The branch of zoology concerned with the taxonomy, behavior, and morphology of invertebrate animals.

[MED] **medicine**—The study of cause and treatment of human disease, including the healing arts dealing with diseases which are treated by a physician or a surgeon.

[MICROBIO] **microbiology**—The science and study of microorganisms and of antibiotic substances.

[MYCOL] **mycology**—The branch of biological science concerned with the study of fungi.

[NEUROSCI] **neuroscience**—The study of the brain and nervous system, including the anatomy and histology of the nervous system, development, sensation and perception, learning, memory, motor control, behavior, aging, and neurological and psychiatric disorders. Studies range from the molecular basis of nervous system development and function to attempts to understand the basis of consciousness and behavior.

[PATH] **pathology**—The branch of biological science which deals with the nature of disease, through study of its causes, its processes, and its effects, together with the associated alterations of structure and function; and the laboratory findings of disease, as distinguished from clinical signs and symptoms.

[PHARM] **pharmacology**—The science of detecting and measuring the therapeutic and toxic effects of drugs or other chemicals on biological systems, as well as the development and testing of new drugs and alternative uses of existing drugs.

[PHYSIO] **physiology**—The branch of biological science concerned with the basic activities that occur in cells and tissues of living organisms, and involving physical and chemical studies of these organisms.

[SYST] **systematics**—The science of animal and plant classification.

[VERT ZOO] **vertebrate zoology**—The branch of zoology concerned with the taxonomy, behavior, and morphology of vertebrate animals.

[VET MED] **veterinary medicine**—The branch of medical practice which treats the diseases and injuries of animals.

[VIROL] **virology**—The science that deals with the study of viruses.

[ZOO] **zoology**—The science that deals with the taxonomy, behavior, and morphology of animal life.

Pronunciation Key

Vowels

a as in b**a**t, th**a**t
ā as in b**ai**t, cr**a**te
ä as in b**o**ther, f**a**ther
e as in b**e**t, n**e**t
ē as in b**ee**t, tr**ea**t
i as in b**i**t, sk**i**t
ī as in b**i**te, l**igh**t
ō as in b**oa**t, n**o**te
ȯ as in b**ough**t, t**au**t
u̇ as in b**oo**k, p**u**ll
ü as in b**oo**t, p**oo**l
ə as in b**u**t, sof**a**
au̇ as in cr**ow**d, p**ow**er
ȯi as in b**oi**l, sp**oi**l
yə as in form**u**la, spectac**u**lar
yü as in f**ue**l, m**u**le

Semivowels/Semiconsonants

w as in **w**ind, t**w**in
y as in **y**et, on**ion**

Stress (Accent)

ˈ precedes syllable with primary stress

ˌ precedes syllable with secondary stress

¦ precedes syllable with variable or indeterminate primary/secondary stress

Consonants

b as in **b**i**b**, dri**bb**le
ch as in **ch**arge, stre**tch**
d as in **d**og, ba**d**
f as in **f**ix, sa**f**e
g as in **g**ood, si**g**nal
h as in **h**and, be**h**ind
j as in **j**oint, di**g**it
k as in **c**ast, bri**ck**
k̲ as in Ba**ch** (used rarely)
l as in **l**oud, be**ll**
m as in **m**ild, su**mm**er
n as in **n**ew, de**n**t
n̲ indicates nasalization of preceding vowel
ŋ as in ri**ng**, si**ng**le
p as in **p**ier, sli**p**
r as in **r**ed, sca**r**
s as in **s**ign, po**s**t
sh as in **su**gar, **sh**oe
t as in **t**imid, ca**t**
th as in **th**in, brea**th**
t̲h̲ as in **th**en, brea**the**
v as in **v**eil, wea**v**e
z as in **z**oo, crui**s**e
zh as in bei**g**e, trea**s**ure

Syllabication

· Indicates syllable boundary when following syllable is unstressed

aapamoor [ECOL] A moor with elevated areas or mounds supporting dwarf shrubs and sphagnum, interspersed with low areas containing sedges and sphagnum, thus forming a mosaic. { 'äp·ə,mür }

aardvark [VERT ZOO] A nocturnal, burrowing, insectivorous mammal of the genus *Orycteropus* in the order Tubulidentata. Also known as earth pig. { 'ärd,värk }

aardwolf [VERT ZOO] *Proteles cristatus.* A hyenalike African mammal of the family Hyaenidae. { 'ärd,wu̇lf }

ABA *See* abscisic acid.

abaca [BOT] *Musa textilis.* A plant of the banana family native to Borneo and the Philippines, valuable for its hard fiber. Also known as Manila hemp. { 'ä·bä,kä *or* 'ä·bə,kä }

abactinal [INV ZOO] In radially symmetrical animals, pertaining to the surface opposite the side where the mouth is located. { a'bak·tin·əl }

abalone [INV ZOO] A gastropod mollusk composing the single genus *Haliotis* of the family Haliotidae. Also known as ear shell; ormer; paua. { ,ab·ə'lō·nē }

abambulacral [INV ZOO] Pertaining to that part of the surface of an echinoderm that lacks tube feet. { ab,am·byə'lak·rəl }

A band [HISTOL] The region between two adjacent I bands in a sarcomere; characterized by partial overlapping of actin and myosin filaments. { 'ā band }

abapertural [INV ZOO] Away from the shell aperture, referring to mollusks. { ab'ap·ər,chu̇r·əl }

abapical [BIOL] On the opposite side to, or directed away from, the apex. { ab'ap·i·kəl }

abarognosis [MED] Lack of ability to estimate the weight of an object one is holding. { ā,bar·əg'nō·sis }

abasia [MED] Lack of muscular coordination in walking. { ā'bā·zhə }

abaxial [BIOL] On the opposite side to, or facing away from, the axis of an organ or organism. { ab'ak·sē·əl }

abb [VERT ZOO] A coarse wool from the fleece areas of lesser quality. { ab }

Abderhalden reaction [PATH] A chemical blood test for the identification of certain enzymes associated with pregnancy and a few diseases. { 'äp·dər,häl·dən rē'ak·shən }

abdomen [ANAT] The portion of the vertebrate body between the thorax and the pelvis. The cavity of this part of the body. [INV ZOO] The elongate region posterior to the thorax in arthropods. { ab'dōm·ən *or* 'ab·də,mən }

abdominal gestation [MED] Development of a fetus outside the uterus in the abdominal cavity. { ab'däm·ə·nəl je'stā·shən }

abdominal gills [INV ZOO] Paired, segmental, leaflike, filamentous expansions of the abdominal cuticle for respiration in the aquatic larvae of many insects. { ab'däm·ə·nəl 'gilz }

abdominal hernia *See* ventral hernia. { ab'däm·ə·nəl 'hər·nē·ə }

abdominal hysterectomy [MED] Surgical removal of all or part of the uterus through an incision in the abdomen. { ab'däm·ə·nəl ,his·tə'rek·tə·mē }

abdominal limb [INV ZOO] In most crustaceans, any of the segmented abdominal appendages. { ab'däm·ə·nəl 'lim }

abdominal pregnancy *See* abdominocyesis. { ab'däm·ə·nəl 'preg·nən·sē }

abdominal regions [ANAT] Nine theoretical areas delineated on the abdomen by two horizontal and two parasagittal lines: above, the right hypochondriac, epigastric, and left hypochondriac; in the middle, the right lateral, umbilical, and left lateral; and below, the right inguinal, hypogastric, and left inguinal. { ab'däm·ə·nəl 'rē·jənz }

abdominal vascular accident [MED] Vascular occlusion and hemorrhage in an abdominal organ, usually the small intestine, or in the peritoneal cavity. { ab'däm·ə·nəl 'vas·kyə·lər 'ak·sə·dənt }

abdominocyesis [MED] Implantation and development of the fertilized ovum in the peritoneal cavity. Also known as abdominal pregnancy. { ab,däm·ə·nō·sī'ē·səs }

abducens [NEUROSCI] The sixth cranial nerve in vertebrates; a paired, somatic motor nerve arising from the floor of the fourth ventricle of the brain and supplying the lateral rectus eye muscles. { ab'dyü·sənz }

abduction [PHYSIO] Movement of an extremity or other body part away from the axis of the body. { ab'dək·shən }

abductor [PHYSIO] Any muscle that draws a

part of the body or an extremity away from the body axis. { ab'dək·tər }

abenteric [MED] Involving abdominal organs and structures outside the intestine. { ˌab·en 'ter·ik }

aberrant [BIOL] An atypical group, individual, or structure, especially one with an aberrant chromosome number. { ə'ber·ənt }

Abies [BOT] The firs, a genus of trees in the pine family characterized by erect cones, absence of resin canals in the wood, and flattened needlelike leaves. { 'ā·bēˌēz }

abiocoen [ECOL] A nonbiotic habitat. { 'āˌbī·ōˌsēn }

abiogenesis [BIOL] The obsolete concept that plant and animal life arise from nonliving organic matter. Also known as autogenesis; spontaneous generation. { ¦āˌbī·ō'jen·ə·sis }

abiotic [BIOL] Referring to the absence of living organisms. { ¦aˌbī'äd·ik }

abiotic environment [ECOL] All physical and nonliving chemical factors, such as soil, water, and atmosphere, which influence living organisms. { ¦aˌbī'äd·ik in'vī·rənˌmənt }

abiotic substance [ECOL] Any fundamental chemical element or compound in the environment. { ¦aˌbī'äd·ik 'səb·stəns }

abiotrophy [MED] Disordered functioning of an organ or system, as in Huntington's chorea, due to an inherited pathologic trait, which trait, however, may remain latent in the individual rather than becoming apparent; this mechanism is still conceptual. { ¦a·bī'ä·trə·fē }

abjection [MYCOL] The discharge or casting off of spores by the spore-bearing structure of a fungus. { ab'jek·shən }

ablastin [IMMUNOL] An antibodylike substance elicited by *Trypanosoma lewisi* in the blood serum of infected rats that inhibits reproduction of the parasite. { ə'blas·tən }

ablation [MED] The removal of tissue or a part of the body by surgery, such as by excision or amputation. { ə'blā·shən }

ABO blood group [IMMUNOL] An immunologically distinct, genetically determined group of human erythrocyte antigens represented by two blood factors (A and B) and four blood types (A, B, AB, and O). { ā·bē'ō 'bləd ˌgrüp }

ABO blood group system [IMMUNOL] A set of multiple alleles found on a single locus on human chromosome 9 that specifies the presence or absence of certain red cell antigens, which determines the ABO blood group. { ˌāˌbē'ō 'bləd ˌgrüp ˌsis·təm }

abomasitis [VET MED] Inflammation of the abomasum in ruminants. { ˌa·bō·mə'sīd·əs }

abomasum [VERT ZOO] The final chamber of the complex stomach of ruminants; has a glandular wall and corresponds to a true stomach. { ˌab·ō'mā·səm }

aboral [INV ZOO] Opposite to the mouth. { a'bȯr·əl }

abortifacient [MED] Any agent that induces abortion. { əˌbȯrd·ə'fā·shənt }

abortion [MED] The spontaneous or induced expulsion of the fetus prior to the time of viability, most often during the first 20 weeks of the human gestation period. { ə'bȯr·shən }

abortive [BIOL] Imperfectly formed or developed. { ə'bȯrd·iv }

abortive infection [VIROL] The viral infection of a cell in which viral components may be synthesized without the production of infective viruses. Also known as nonproductive infection. { ə'bȯrd·iv in'fek·shən }

abortive transduction [MICROBIO] Failure of exogenous fragments that were introduced into a bacterial cell by viruses to become inserted into the bacterial chromosome. { ə'bȯrd·iv tranz'dək·shən }

abortus [MED] An aborted fetus. { ə'bȯrd·əs }

abrachiocephalia [MED] Congenital lack of arms and head. Also known as acephalobrachia. { ¦āˌbrāk·ē·ə·sə'fal·yə }

abranchiate [ZOO] Without gills. { ā'braŋk·ē·ət }

abrasion [MED] A spot denuded of skin, mucous membrane, or superficial epithelium by rubbing or scraping. { ə'brā·zhən }

abrin [BIOCHEM] A highly poisonous protein found in the seeds of *Abrus precatorius*, the rosary pea. { 'a·brin }

abrupt [BOT] Ending suddenly, as though broken off. { ə'brəpt }

abruptio placenta [MED] A pregnancy disorder in which the placenta separates prematurely from the uterus. { əˌbrəp·shē·o plə'sent·ə }

abscess [MED] A localized collection of pus surrounded by inflamed tissue. { 'ab·ses }

abscisic acid [BIOCHEM] $C_{16}H_{20}O_4$ A plant hormone produced by fruits and leaves that promotes abscission and dormancy and retards vegetative growth. Abbreviated ABA. Formerly known as abscisin. { ab'sis·ik 'as·əd }

abscisin *See* abscisic acid. { ab'sis·ən }

abscission [BOT] A physiological process promoted by abscisic acid whereby plants shed a part, such as a leaf, flower, seed, or fruit. { ab 'sizh·ən }

abscission layer [BOT] A zone of cells whose breakdown causes separation of a leaf or other structure from the stem. { ab'sizh·ən ˌlā·ər }

absolute plating efficiency [CELL MOL] The percentage of individual cells that give rise to colonies when the cells are inoculated into culture media. { 'ab·səˌlüt 'plād·iŋ iˌfish·ən·sē }

absolute refractory period [NEUROSCI] A period ranging from 0.5 to 2 milliseconds during which neural tissue is totally unresponsive. { 'ab·səˌlüt ri'frak·trē ˌpir·ē·əd }

absolute threshold [PHYSIO] The minimum stimulus energy that an organism can detect. { 'ab·səˌlüt 'thresh·hȯld }

absorbed dose [MED] The part of an administered medication which is not excreted by the recipient's body. { əb'sȯrbd 'dōs }

absorption [BIOL] The net movement (transport) of water and solutes from outside a cell or an organism to the interior. [IMMUNOL] **1.** Removal of antibodies from an antiserum by addition of antigen. **2.** Removal of antigens from a mixture by addition of antibodies. [PHYSIO] Passage of a chemical substance, a pathogen, or radiant energy through a body membrane. { əb'sȯrp·shən }

absorption atelectasis *See* obstructive atelectasis. { əb'sȯrp·shən ˌad·ə'lek·tə·sis }

absorption test [IMMUNOL] Analysis of the antigenic components of bacterial cells and large macromolecules by a series of precipitation or agglutination reactions with specific antibodies. { əb'sȯrp·shən ˌtest }

abstinence syndrome [MED] A disturbance of metabolic equilibrium that occurs when a narcotic drug is withdrawn from the user. { 'abz·tə·nəns 'sin·drōm }

abstriction [MYCOL] In fungi, the cutting off of spores in hyphae by formation of septa followed by abscission of the spores, especially by constriction. { ab'strik·shən }

abterminal [BIOL] Referring to movement from the end toward the middle; specifically, describing the mode of electric current flow in a muscle. { ab'tərm·ən·əl }

A-B toxin [BIOCHEM] A toxin found in some bacteria and plants that is composed of two functionally distinct parts termed A, the enzymatically active portion, and B, the receptor binding portion; it can catalyze chemical reactions inside animal cells. { ¦āˌ¦bē 'täk·sən }

abzyme *See* catalytic antibody. { 'abˌzīm }

Acala [BOT] A type of cotton indigenous to Mexico and cultivated in Texas, Oklahoma, and Arkansas. { ə'kal·ə }

acalculia [MED] A form of aphasia characterized by inability to do mathematical calculations. { ˌā·kal'kyü·lē·ə }

acalyculate [BOT] Lacking a calyx. { ¦ā·kə¦lik·yüˌlāt }

Acalyptratae *See* Acalyptreatae. { ¦ā·kə'lip·trədˌē }

Acalyptreatae [INV ZOO] A large group of small, two-winged flies in the suborder Cyclorrhapha characterized by small or rudimentary calypters. Also spelled Acalyptratae. { ¦ā·kə·lip'trē·əˌdē }

acantha [BIOL] A sharp spine; a spiny process, as on vertebrae. { ə'kan·thə }

Acanthaceae [BOT] A family of dicotyledonous plants in the order Scrophulariales distinguished by their usually herbaceous habit, irregular flowers, axile placentation, and dry, dehiscent fruits. { əˌkan'thās·ēˌē }

acanthaceous [BOT] Having sharp points or prickles; prickly. { əˌkan'thā·shəs }

Acantharia [INV ZOO] A subclass of essentially pelagic protozoans in the class Actinopodea characterized by skeletal rods constructed of strontium sulfate (celestite). { əˌkan'tha·rē·ə }

Acanthaster [INV ZOO] A genus of Indo-Pacific starfishes, including the crown-of-thorns, of the family Asteriidae; economically important as a destroyer of oysters in fisheries. { əˌkan'thas·tər }

acanthella [INV ZOO] A transitional larva of the phylum Acanthocephala in which rudiments of reproductive organs, lemnisci, a proboscis, and a proboscis receptacle are formed. { əˌkan'thel·ə }

acanthocarpous [BOT] Having spiny fruit. { əˌkan·thə'kär·pəs }

Acanthocephala [INV ZOO] The spiny-headed worms, a phylum of helminths; adults are parasitic in the alimentary canal of vertebrates. { əˌkan·thō'sef·ə·lə }

Acanthocheilonema perstans [INV ZOO] A tropical filarial worm, parasitic in humans. { əˌkan·thəˌkī·lə'nē·mə 'pərs·təns }

acanthocheilonemiasis [MED] A parasitic infection of humans caused by the filarial nematode *Acanthocheilonema perstans*. { əˌkan·thəˌkī·lə·ne'mī·ə·səs }

acanthocladous [BOT] Having spiny branches. { əˌkan·thə'klad·əs }

acanthocyte [PATH] A crenated red cell which has a distinctive spiky outline. { ə'kan·thōˌsīt }

acanthocytosis [MED] A disorder of erythrocytes in which spiny projections appear on the blood cells. { əˌkan·thəˌsī'tō·səs }

acanthoid [BIOL] Shaped like a spine. { ə'kan·thȯid }

acanthoma [MED] A tumor composed of epidermal cells. { ˌak·ən'thō·mə }

Acanthometrida [INV ZOO] An order of marine protozoans in the subclass Acantharia with 20 or less skeletal rods. { əˌkan·thə'met·rə·də }

Acanthophis antarcticus [VERT ZOO] The death adder, a venomous snake found in Australia and New Guinea; venom is neurotoxic. { ə'kan·thə·fəs ant'ärk·tə·kəs }

Acanthophractida [INV ZOO] An order of marine protozoans in the subclass Acantharia; skeleton includes a latticework shell and skeletal rods. { əˌkan·thə'frak·tə·də }

acanthopodia [INV ZOO] The long subpseudopodia of amoebas of the suborder Acanthopodina, order Amoebida. { əˌkan·thə'pōd·ē·ə }

acanthopodous [BOT] Having a spiny or prickly petiole or peduncle. { ¦āˌkan¦thä·pə·dəs }

Acanthopteri [VERT ZOO] An equivalent name for the Perciformes. { aˌkan'thäp·tə·rī }

Acanthopterygii [VERT ZOO] An equivalent name for the Perciformes. { aˌkanˌthäp·tə'rē·jē·ī }

acanthosis [MED] Any thickening of the prickle-cell layer of the epidermis; associated with many skin diseases. { aˌkan'thō·səs }

acanthosoma [INV ZOO] The last primitive larval stage, the mysis, in the family Sergestidae. { əˌkan·thə'sō·mə }

Acanthosomatidae [INV ZOO] A small family of insects in the order Hemiptera. { əˌkan·thə·sə'mad·əˌdē }

acanthosphere [BOT] A specialized ciliated body in *Nitella* cells. { ə'kan·thəˌsfir }

acanthostegous [INV ZOO] Being overlaid with two series of spines, as the ovicell or ooecium of certain bryozoans. { əˌkan·thə'steg·əs }

acanthozooid [INV ZOO] A specialized individual in a bryozoan colony that secretes tubules which project as spines above the colony's outer surface. { ə,kan·thə'zō·öid }

Acanthuridae [VERT ZOO] The surgeonfishes, a family of perciform fishes in the suborder Acanthuroidei. { ə,kan'thu̇·rə·dē }

Acanthuroidei [VERT ZOO] A suborder of chiefly herbivorous fishes in the order Perciformes. { ə,kan·thə'rȯid·ē,ī }

acapnia [MED] Absence of carbon dioxide in the blood and tissues. { ə'kap·nē·ə }

Acari [INV ZOO] The equivalent name for Acarina. { 'a·kə·rē }

acariasis [MED] Any skin disease resulting from infestation with acarids or mites. { a·kə'rī·ə·səs }

Acaridiae [INV ZOO] A group of pale, weakly sclerotized mites in the suborder Sarcoptiformes, including serious pests of stored food products and skin parasites of warm-blooded vertebrates. { ,a·kə'rid·ē,ē }

Acarina [INV ZOO] The ticks and mites, a large order of the class Arachnida, characterized by lack of body demarcation into cephalothorax and abdomen. { ,a·kə'rēn·ə }

acarology [INV ZOO] A branch of zoology dealing with the mites and ticks. { ,a·kə'rä·lə·je }

acarophily [ECOL] A symbiotic relationship between plants and mites. { ¦a·kə·rō¦fil·ē }

acarpellous [BOT] Lacking carpels. { ā'kär·pə·ləs }

acarpous [BOT] Not producing fruit. { ā'kär·pəs }

acatalasemia [MED] Lack of catalase in the blood. { ¦ā·ka·tə·lə'sē·mē·ə }

acatalasia [MED] Congenital absence of the enzyme catalase. { ¦ā·ka·tə'lā·zhē·ə *or* zhə }

acatamathesia [MED] **1.** Inability to understand conversation. **2.** Morbid blunting or deterioration of the senses, as in mental deafness and blindness. { ¦ā·ka·tə·mə'thē·zhē·ə *or* zhə }

acaulous [BOT] **1.** Lacking a stem. **2.** Being apparently stemless but having a short underground stem. { ¦ā'kȯl·əs }

accelerated hypertension *See* malignant hypertension. { ak'sel·ər,ā·dəd 'hī·pər,ten·shən }

acceleration globulin [BIOCHEM] A globulin that acts to accelerate the conversion of prothrombin to thrombin in blood clotting; found in blood plasma in an inactive form. { ak,sel·ə'rā·shən 'gläb·yə·lən }

acceleration phase [MICROBIO] A period of increasing growth which is followed by the log phase in a culture of microbes. { ak·sel·ə'rā·shən ,fāz }

acceleration stress [MED] The effect of an increase in gravitational force upon human physiology and behavior, particularly during takeoff and reentry in space flight. { ak,sel·ə'rā·shən ,stres }

acceleration tolerance [PHYSIO] The maximum *g* forces an individual can withstand without losing control or consciousness. { ak,sel·ə'rā·shən 'täl·ər·əns }

acceleratory reflex [PHYSIO] Any reflex originating in the labyrinth of the inner ear in response to a change in the rate of movement of the head. { ak'sel·ə·rə,tȯr·ē 'rē·fleks }

accessorius [ANAT] Any muscle that reinforces the action of another. { ak·sə'sȯr·ē·əs }

accessory bud [BOT] An embryonic shoot occurring above or to the side of an axillary bud. Also known as supernumerary bud. { ak'ses·ə·rē ,bəd }

accessory cell [BOT] A morphologically distinct epidermal cell adjacent to, and apparently functionally associated with, guard cells on the leaves of many plants. [IMMUNOL] Any nonlymphocytic cell that helps in the induction of the immune response by presenting antigen to a helper T lymphocyte. { ak'ses·ə·rē ,sel }

accessory chromosome *See* supernumerary chromosome. { ak'ses·ə·rē 'krōm·ə,sōm }

accessory fruit [BOT] A fruit derived from the ovary and its contents as well as other parts of the flower; it is usually derived from an inferior (inserted below other floral parts) ovary. Also known as inferior fruit. { ak'ses·ə·rē ,früt }

accessory gland [ANAT] A mass of glandular tissue separate from the main body of a gland. [INV ZOO] A gland associated with the male reproductive organs in insects. { ak'ses·ə·rē ,gland }

accessory movement *See* synkinesia. { ak'ses·ə·rē ,müv·mənt }

accessory nerve [NEUROSCI] The eleventh cranial nerve in tetrapods, a paired visceral motor nerve; the bulbar part innervates the larynx and pharynx, and the spinal part innervates the trapezius and sternocleidomastoid muscles. { ak'ses·ə·rē ,nərv }

accessory pigments [BIOCHEM] Light-absorbing pigments, including carotenoids and phycobilins, which complement chlorophyll in plants, algae, and bacteria by trapping light energy for photosynthesis. { ak¦ses·ə·rē 'pig·məns }

accessory sexual characters [ANAT] Those structures and organs (excluding the gonads) composing the genital tract and including accessory glands and external genitalia. { ak'ses·ə·rē 'seksh·ə·wəl 'kar·ik·tərz }

accessory species [ECOL] A species comprising 25–50% of a community. { ak'ses·ə·rē 'spē·shēz }

accidental species [ECOL] A species which constitutes less than one-fourth of the population of a stand. { ¦ak·sə¦den·təl 'spē·shēz }

accidental whorl [FOREN] A type of whorl fingerprint pattern which is a combination of two different types of pattern, with the exception of the plain arch, with two or more deltas; or a pattern which possesses some of the requirements for two or more different types; or a pattern which conforms to none of the definitions; in accidental whorl tracing three types appear: an outer (O), inner (I), or meeting (M). { ¦ak·sə¦den·təl 'wərl }

accident-prone [MED] Predisposed to sustain

more accidents than others exposed to the same hazard. { 'ak·sə,dent ,prōn }

Accipitridae [VERT ZOO] The diurnal birds of prey, the largest and most diverse family of the order Falconiformes, including hawks, eagles, and kites. { ,ak·sə'pi·trə,dē }

acclimated microorganism [ECOL] Any microorganism that is able to adapt to environmental changes such as a change in temperature, or a change in the quantity of oxygen or other gases. { ə'klīm·əd·əd ,mī·krō'ȯr·gə·niz·əm }

acclimation *See* acclimatization. { ,ak·lə'mā·shən }

acclimatization [BIOL] Physiological, emotional, and behavioral adjustment by an individual to changes in the environment. [EVOL] Adaptation of a species or population to a changed environment over several generations. Also known as acclimation. { ə,klī·mə·tə'zā·shən }

accommodation [ECOL] A population's location within a habitat. [PHYSIO] A process in most vertebrates whereby the focal length of the eye is changed by automatic adjustment of the lens to bring images of objects from various distances into focus on the retina. { ə,käm·ə'dā·shən }

accommodation reflex [PHYSIO] Changes occurring in the eyes when vision is focused from a distant to a near object; involves pupil contraction, increased lens convexity, and convergence of the eyes. { ə,käm·ə'dā·shən 'rē·fleks }

accrescent [BOT] Growing continuously with age, especially after flowering. { ə'krēs·ənt }

accretion line [HISTOL] A microscopic line on a tooth, marking the addition of a layer of enamel or dentin. { ə'krē·shən ,līn }

accumbent [BOT] Describing an organ that leans against another; specifically referring to cotyledons having their edges folded against the hypocotyl. { ə'kəm·bənt }

accumulated dose [PHYSIO] The total amount of radiation absorbed by an organism as a result of exposure to radiation. { ə'kyü·myə,lād·əd 'dōs }

accumulator plant [BOT] A plant or tree that grows in a metal-bearing soil and accumulates an abnormal content of the metal. { ə'kyü·myə ,lād·ər ,plant }

acellular [BIOL] Not composed of cells. { 'ā·sel·yə·lər }

acellular gland [PHYSIO] A gland, such as intestinal glands, the pancreas, and the parotid glands, that secretes a noncellular product. { 'ā·sel·yə·lər ,gland }

acellular slime mold [MYCOL] The common name for members of the Myxomycetes. { 'ā·sel·yə·lər 'slīm ,mōld }

acellular vaccine [IMMUNOL] A vaccine consisting of one or more parts of an infectious agent, rather than the whole cell. { ¦ā¦sel·yü·lər ,vak'sēn }

acentric [BIOL] Not oriented around a middle point. [GEN] A chromosome or chromosome fragment lacking a centromere. { ,ā'sen·trik }

acentrous [VERT ZOO] Lacking vertebral centra and having the notochord persistent throughout life, as in certain primitive fishes. { ,ā'sen·trəs }

Acephalina [INV ZOO] A suborder of invertebrate parasites in the protozoan order Eugregarinida characterized by nonseptate trophozoites. { ā,sef·ə'līn·ə }

acephalobrachia *See* abrachiocephalia. { ā,sef·ə·lə'brāk·ē·ə }

acephalocardia [MED] Congenital lack of a head and a heart. { ā,sef·ə·lə'kärd·ē·ə }

acephalochiria [MED] Congenital lack of a head and hands. { ā,sef·ə·lə'kir·ē·ə }

acephalocyst [INV ZOO] An abnormal cyst of the *Echinococcus granulosus* larva, lacking a head and brood capsules, found in human organs. { ā'sef·ə·lə,sist }

acephalopodia [MED] Congenital lack of a head and feet. { ā,sef·ə·lə'pōd·ē·ə }

acephalorrhachia [MED] Congenital lack of a head and vertebral column. { ā¦sef·ə·lə'rāk·ē·ə }

acephalostomia [MED] Congenital lack of a head, with a mouthlike orifice in the neck or chest. { ā¦sef·ə·lə'stōm·ē·ə }

acephalothoracica [MED] Congenital lack of a head and thorax. { ā¦sef·ə·lə·thə'ras·ə·kə }

acephalous [BOT] Having the style originate at the base instead of at the apex of the ovary. [ZOO] Lacking a head. { ā'sef·ə·ləs }

Acer [BOT] A genus of broad-leaved, deciduous trees of the order Sapindales, commonly known as the maples; the sugar or rock maple (*A. saccharum*) is the most important commercial species. { 'ā·sər *or* 'ä,kər }

acerate [BOT] Needle-shaped, specifically referring to leaves. { 'as·ə,rāt }

Acerentomidae [INV ZOO] A family of wingless insects belonging to the order Protura; the body lacks tracheae and spiracles. { ,a·sə·rən'tōm·ə·dē }

acervate [BIOL] Growing in heaps or dense clusters. { 'a·sər,vāt }

acervulus [MYCOL] A cushion- or disk-shaped mass of hyphae, peculiar to the Melanconiales, on which there are dense aggregates of conidiophores. { ə'sər·vyə·ləs }

acetabulectomy [MED] Excision of the acetabulum. { ə,sēd·ə·bə'lek·tə·mē }

acetabuloplasty [MED] Plastic surgery involving repair or enlargement of the cavity of the acetabulum to restore its normal state. { ə,sēd·ə'bu̇l·ə,plas·tē }

acetabulum [ANAT] A cup-shaped socket on the hipbone that receives the head of the femur. [INV ZOO] **1.** A cavity on an insect body into which a leg inserts for articulation. **2.** The sucker of certain invertebrates such as trematodes and tapeworms. { ,a·sə'tab·yə·ləm }

acetaldehydase [BIOCHEM] An enzyme that catalyzes the oxidation of acetaldehyde to acetic acid. { ¦as·əd,al·də'hī,dās }

3-acetamido-4-hydroxybenzenearsonic acid [PHARM] $(HO)(CH_3CONH)C_6H_3AsO(OH)_2$ An odorless, white to slightly yellow powder with a slightly acid taste; soluble in alkali and alkali carbonate solutions; used in veterinary medicine. Also known as acetarsone. { ˌthrē ˌas·əd'am·ə·dō ˌfȯr ˌhīˌdräk·sēˌbenˌzēn·är'sän·ik 'as·əd }

acetarsone *See* 3-acetamido-4-hydroxybenzenearsonic acid. { ˌa·səd'ärˌsōn }

acetazolamide [PHARM] $(CH_3CONH)C_2N_2S(SO_2NH_2)$ An odorless, white to faintly yellowish-white, crystalline powder with a melting point of 258°C; slightly soluble in water; used as a diuretic. { ˌas·ə·tə'zäl·əˌmīd }

acetic acid bacteria *See* Acetobacter. { ə'sēd·ik 'as·əd ˌbak'tir·ē·ə }

acetic fermentation [MICROBIO] Oxidation of alcohol to produce acetic acid by the action of bacteria of the genus *Acetobacter*. { ə'sēd·ik fər·mən'tā·shən }

acetic thiokinase [BIOCHEM] An enzyme that catalyzes the formation of acetyl coenzyme A from acetate and adenosine triphosphate. { ə'sēd·ik ˌthī·ə'kīnˌās }

acetoacetyl coenzyme A [BIOCHEM] $C_{25}H_{41}O_{18}N_7P_3S$ An intermediate product in the oxidation of fatty acids. { ¦as·əˌtō·ə'sēd·əl ˌkō'en·zīm 'ā }

Acetobacter [MICROBIO] A genus of gram-negative, aerobic bacteria of uncertain affiliation comprising ellipsoidal to rod-shaped cells as singles, pairs, or chains; they oxidize ethanol to acetic acid. Also known as acetic acid bacteria; vinegar bacteria. { ə'sēd·ōˌbak·tər }

Acetobacter aceti [MICROBIO] An aerobic, rod-shaped bacterium capable of efficient oxidation of glucose, ethyl alcohol, and acetic acid; found in vinegar, beer, and souring fruits and vegetables. { ə'sēd·ōˌbak·tər ə'sēd·ē }

Acetobacter suboxydans [MICROBIO] A short, nonmotile vinegar bacterium that can oxidize ethanol to acetic acid; useful for industrial production of ascorbic and tartaric acids. { ə'sēd·ōˌbak·tər səb'äks·ə·dəns }

acetoclastic bacteria [MICROBIO] Bacteria that utilize acetic acid only and produce methane during anaerobic fermentation. { ˌa·sə·tō¦klas·tik bak'tir·ē·ə }

acetoclastis [MICROBIO] The process, carried out by some methanogens, of splitting acetate into methane and carbon dioxide. { ˌa·sə·tō'klas·təs }

acetogenic bacteria [BIOCHEM] Anaerobic bacteria capable of reducing carbon dioxide to acetic acid or converting sugars into acetate. { ˌa·sə·tō¦jen·ik bak'tir·ē·ə }

acetolactic acid [BIOCHEM] $C_5H_8O_4$ A monocarboxylic acid formed as an intermediate in the synthesis of the amino acid valine. { ˌas·ə·tə'lak·tik 'as·əd }

acetomeroctol [PHARM] $CH_3COOHgC_6H_3(OH)C(CH_3)_2CH_3C(CH_3)_3$ A white solid with a melting point of 155–157°C; soluble in alcohol, ether, and chloroform; used as an antiseptic. { ˌas·əˌtō·mə'räk·tȯl }

Acetomonas [MICROBIO] A genus of aerobic, polarly flagellated vinegar bacteria in the family Pseudomonadaceae; used industrially to produce vinegar, gluconic acid, and L-sorbose. { əˌsed·ə'mōn·əs }

acetone body *See* ketone body. { 'as·əˌtōn ˌbäd·ē }

acetone fermentation [MICROBIO] Formation of acetone by the metabolic action of certain anaerobic bacteria on carbohydrates. { 'as·əˌtōn fər·mən'tā·shən }

acetonemia [MED] A condition characterized by large amounts of acetone bodies in the blood. Also known as ketonemia. { ˌas·ə·tə'nēm·ē·ə }

acetophenetidin [PHARM] $CH_3CONHC_6H_4OC_2H_5$ Odorless, white crystals or powder with a melting point of 135°C; soluble in alcohol, chloroform, and ether; used as an analgesic. { ˌas·əˌtä·fə'net·əˌdin }

acetovanillon [PHARM] $HOC_6H_3(OCH_3)COCH_3$ Fine, needlelike crystals with a melting point of 115°C and a faint vanilla odor; freely soluble in hot water, alcohol, benzene, chloroform, and ether; used as a cardiotonic drug. { əˌsē·dō·və'nil·ən }

21-acetoxypregnenolone [PHARM] $C_{23}H_{34}O_4$ Needlelike crystals which become opaque at about 80°C or on prolonged standing; melting point is 184–185°C; soluble in chloroform and toluene; used as an antiarthritic drug. { ˌtwen·tē'wən ˌas·əˌtäk·sēˌpreg'nen·əˌlōn }

***para*-acetylaminophenyl salicylate** [PHARM] $C_6H_4(NHCOCH_3)OOCC_6H_4OH$ Odorless and tasteless, fine, white, crystalline scales with a melting point of 187–188°C; soluble in alcohol, ether, and hot water; used as an analgesic. Also known as phenetsal. { ¦par·ə ¦as·ə·təl¦am·əˌnō 'fēn·əl sə'lis·əˌlāt }

acetylase [BIOCHEM] Any enzyme that catalyzes the formation of acetyl esters. { ə'sed·əlˌās }

acetylcarbromal [PHARM] $(C_2H_5)_2CBrCONHCONHCOCH_3$ Crystals with a slightly bitter taste and a melting point of 109°C; soluble in alcohol and ethyl acetate; used as a sedative. { əˌsed·əl'kär·bəˌməl }

acetylcholine [NEUROSCI] $C_7H_{17}O_3N$ A compound released from certain autonomic nerve endings which acts in the transmission of nerve impulses to excitable membranes. { əˌsed·əl'kōˌlēn }

acetylcholine receptor [CELL MOL] A receptor in the membranes of certain cell structures, such as synapses or the neuromuscular junction, to which the transmitter substance acetylcholine binds. Nicotinic acetylcholine receptors are gated ion channels that open in response to acetylcholine, leading to an increase in membrane conductance; muscarinic acetylcholine receptors are G-protein-linked receptors inducing membrane ion channel changes or intracellular processes such as smooth muscle contraction. { əˌsed·əl 'kōˌlēn riˌsep·tər }

acetylcholinesterase [BIOCHEM] An enzyme found in excitable membranes that inactivates acetylcholine. { əˌsed·əl·kō·lən'es·təˌrās }

acetyl coenzyme A [BIOCHEM] $C_{23}H_{39}O_{17}N_7P_3S$

A coenzyme, derived principally from the metabolism of glucose and fatty acids, that takes part in many biological acetylation reactions; oxidized in the Krebs cycle. { ə,sed·əl ,kō'en,zīm 'ā }

acetylcysteine [PHARM] $HSCH_2CH(NHCOCH_3)COOH$ Crystals with a melting point of 109–110°C; used as a mucolytic drug. { ə,sed·əl'sis·tē,ēn *or* ,tēn }

acetyl phosphate [BIOCHEM] $C_2H_5O_5P$ The anhydride of acetic and phosphoric acids occurring in the metabolism of pyruvic acid by some bacteria; phosphate is used by some microorganisms, in place of adenosine triphosphate, for the phosphorylation of hexose sugars. { ə'sed·əl 'fäs,fāt }

achaetous [INV ZOO] Without setae. Also known as asetigerous. { ə'kēd·əs }

A chain *See* heavy chain. { 'ā ,chān }

achalasia [MED] Inability of a hollow muscular organ or ring of muscle (sphincter) to relax. { ,ak·ə'lāzh·ē·ə }

achalasia of the cardia [MED] Enlargement of the esophagus as a result of cardiospasm. { ,ak·ə'lāzh·ē·ə əv thə 'kärd· ē·ə }

ache [MED] A constant dull or throbbing pain. { āk }

acheb [ECOL] Short-lived vegetation regions of the Sahara composed principally of mustards (Cruciferae) and grasses (Gramineae). { ə'cheb }

achene [BOT] A small, dry, indehiscent fruit formed from a simple ovary bearing a single seed. { ə'kēn }

achiasmate [CELL MOL] Pertaining to meiosis that lacks chiasma. { ,ā·kī'az,māt }

achilary [BOT] In flowers, having the lip (labellum) undeveloped or lacking. { ə'kil·ə·rē }

Achilles jerk [PHYSIO] A reflex action seen as plantar flection in response to a blow to the Achilles tendon. Also known as Achilles tendon reflex. { ə'kil·ēz ,jərk }

Achilles tendon [ANAT] The tendon formed by union of the tendons of the calf muscles, the soleus and gastrocnemius, and inserted into the heel bone. { ə'kil·ēz 'ten·dən }

Achilles tendon reflex *See* Achilles jerk. { ə'kil·ēz 'ten·dən 'rē·fleks }

achlamydeous [BOT] Lacking a perianth. { ¦ā·klə'mid·ē·əs }

achlorhydria [MED] Absence of hydrochloric acid in the stomach. { ¦ā·klȯr'hīd·rē·ə }

Acholeplasma [MICROBIO] The single genus of the family Acholeplasmataceae, comprising spherical and filamentous cells. { ə,kōl·ə'plaz·mə }

Acholeplasmataceae [MICROBIO] A family of the order Mycoplasmatales; characters same as for the order and class (Mollicutes); members do not require sterol for growth. { ə,kōl·ə,plaz·mə'tās·ē,ē }

acholia [MED] Suppression or absence of bile secretion into the small intestine. { ¦ā'kōl·ē·ə }

achondroplasia [MED] A hereditary deforming disease of the skeletal system, inherited in humans as an autosomal dominant trait and characterized by insufficient growth of the long bones, resulting in reduced length. Also known as chondrodystrophy fetalis. { ¦ā,kän·drə 'plāzh·ē·ə }

achondroplastic dwarf [MED] A human with short legs and arms due to achondroplasia. { ¦ā,kän·drə'plas·tik 'dwȯrf }

achordate [VERT ZOO] Lacking a notochord. { ¦ā'kȯr,dāt }

achreocythemia [MED] An anemia characterized by pale erythrocytes due to hemoglobin deficiency. { ¦ā,krē·ə,sī'thēm·ē·ə }

achroglobin [BIOCHEM] A colorless respiratory pigment present in some mollusks and urochordates. { ¦ak·rə'glōb·ən }

achromasia [MED] Absence of normal skin pigment. { ¦ā·krō'māzh·ē·ə }

Achromatiaceae [MICROBIO] A family of gliding bacteria of uncertain affiliation; cells are spherical to ovoid or cylindrical, movements are slow and jerky, and microcysts are not known. { a,krō·mə·dē'ās·ē,ē }

achromatic interval [PHYSIO] The difference between the achromatic threshold and the smallest light stimulus at which the hue is detectable. { ¦a·krə¦mad·ik 'int·ər·vəl }

achromatic threshold [PHYSIO] The smallest light stimulus that can be detected by a dark-adapted eye, so called because all colors lose their hue at this illumination. { ¦a·krə¦mad·ik 'thresh,hōld }

achromatin [CELL MOL] The portion of the cell nucleus which does not stain easily with basic dyes. { ,ā'krō·mə·tən }

Achromatium [MICROBIO] The type genus of the family Achromatiaceae. { ,a·krə'māsh·ē·əm }

achromatophilia [BIOL] The property of not staining readily. { ¦ā·krō,mad·ə'fil·ē·ə }

achromic [BIOL] Colorless; lacking normal pigmentation. { ¦ā'krō·mik }

Achromobacter [MICROBIO] A genus of motile and nonmotile, gram-negative, rod-shaped bacteria in the family Achromobacteraceae. { ,ā'krō·mə,bak·tər }

Achromobacteraceae [MICROBIO] Formerly a family of true bacteria, order Eubacteriales, characterized by aerobic metabolism. { ,ā,krō·mə,bak·tər'ās·ē,ē }

Achroonema [MICROBIO] A genus of bacteria in the family Pelonemataceae; cells have smooth, delicate, porous walls. { ,ak·rō'ōn·ə·mə }

achylia [MED] Absence of chyle. { ,ā'kil·ē·ə }

achylia gastrica [MED] Lack of secretion of hydrochloric acid and proteolytic enzymes by the stomach. { ,ā'kil·ē·ə 'gas·trik·ə }

acicula [BOT] A needle-shaped part, for example, of a plant or crystal. { ə'sik·yə·lə }

aciculate [BOT] Finely scored on the surface. { ə'sik·yə·lət }

aciculignosa [ECOL] Narrow sclerophyll or coniferous vegetation that is mostly subalpine, subarctic, or continental. { ə,sik·yə·lig'nōs·ə }

Acidaminococcus [MICROBIO] A genus of bacteria in the family Veillonellaceae; cells are often oval or kidney-shaped and occur in pairs; amino acids can supply the single energy source. { ˌas·əd'a·mə·nō'käk·əs }

acid-base balance [PHYSIO] Physiologically maintained equilibrium of acids and bases in the body. { 'as·əd 'bās 'bal·əns }

acid cell [HISTOL] A parietal cell of the stomach. { 'as·əd ˌsel }

acidemia [MED] A condition in which the pH of the blood falls below normal. { ˌas·ə'dēm·ē·ə }

acid-fast bacteria [MICROBIO] Bacteria, especially mycobacteria, that stain with basic dyes and fluorochromes and resist decoloration by acid solutions. { 'as·əd ¦fast bak'tir·ē·ə }

acid-fast stain [MICROBIO] A differential stain used in identifying species of *Mycobacterium* and one species of *Nocardia*. { 'as·əd ˌfast 'stān }

acid hydrolase [BIOCHEM] Any of a group lysosomal digestive enzymes that function optimally in an acidic environment and can sever (by hydrolysis) particular chemical bonds found in natural materials. { ˌas·əd 'hī·drəˌlās }

acidophil [BIOL] **1.** Any substance, tissue, or organism having an affinity for acid stains. **2.** An organism having a preference for an acid environment. [HISTOL] **1.** An alpha cell of the adenohypophysis. **2.** *See* eosinophil. { ə'sid·əˌfil }

acidophilia *See* eosinophilia. { əˌsid·ə'fil·ē·ə }

acidophilic erythroblast *See* normoblast. { ə'sid·əˌfil·ik ə'rith·rəˌblast }

acidosis [MED] A condition of decreased alkali reserve of the blood and other body fluids. { ˌas·ə'dō·səs }

acidotrophic [BIOL] Having an acid nutrient requirement. { əˌsid·ə'trōf·ik }

acid phosphatase [BIOCHEM] An enzyme in blood which catalyzes the release of phosphate from phosphate esters; optimum activity at pH 5. { 'as·əd 'fäs·fəˌtās }

acid tide [MED] A period of increased acidity of urine and body fluids. { 'as·əd ˌtīd }

acinar [ANAT] Pertaining to an acinus. { 'as·ə·nər }

acinar cell [ANAT] Any of the cells lining an acinous gland. { 'as·ə·nər 'sel }

Acinetobacter [MICROBIO] A genus of nonmotile, short, plump, almost spherical rods in the family Neisseriaceae; strictly aerobic; resistant to penicillin. { ˌas·əˌnēd·ə'bak·tər }

aciniform [ZOO] Shaped like a berry or a bunch of grapes. { ə'sin·əˌfȯrm }

acinotubular gland *See* tubuloalveolar gland. { ˌas·əˌnō'tü·byə·lər 'gland }

acinous [BIOL] Of or pertaining to acini. { 'as·əˌnəs }

acinous gland [ANAT] A multicellular gland with sac-shaped secreting units. Also known as alveolar gland. { 'as·əˌnəs ˌgland }

acinus [ANAT] The small terminal sac of an acinous gland, lined with secreting cells. [BOT] An individual drupelet of a multiple fruit. { 'as·ə·nəs }

Acipenser [VERT ZOO] A genus of actinopterygian fishes in the sturgeon family, Acipenseridae. { 'as·əˌpen·sər }

Acipenseridae [VERT ZOO] The sturgeons, a family of actinopterygian fishes in the order Acipenseriformes. { ˌas·əˌpen'ser·ə·dē }

Acipenseriformes [VERT ZOO] An order of the subclass Actinopterygii represented by the sturgeons and paddlefishes. { ˌas·ə·penˌser·ə'fȯrˌmēz }

Acmaeidae [INV ZOO] A family of gastropod mollusks in the order Archaeogastropoda; includes many limpets. { ak'mē·əˌdē }

acmic [ECOL] A phase or period in which an aquatic population undergoes seasonal changes. { 'ak·mik }

acne [MED] A pleomorphic, inflammatory skin disease involving sebaceous follicles of the face, back, and chest and characterized by blackheads, whiteheads, papules, pustules, and nodules. { 'ak·nē }

acne rosacea [MED] A form of acne occurring in older persons and seen as reddened inflamed areas on the forehead, nose, and cheeks. { 'ak·nē rō'zā·shə }

Acnidosporidia [INV ZOO] An equivalent name for the Haplosporea. { akˌnī·dəˌspō·'rid·ē·ə }

Acoela [INV ZOO] An order of marine flatworms in the class Turbellaria characterized by the lack of a digestive tract and coelomic cavity. { ā'sēl·ə }

Acoelea [INV ZOO] An order of gastropod mollusks in the subclass Opistobranchia; includes many sea slugs. { ˌā·sə'lē·ə }

Acoelomata [INV ZOO] A subdivision of the animal kingdom; individuals are characterized by lack of a true body cavity. { āˌsēl·ə'mäd·ə }

acoelomate [ZOO] Pertaining to an animal that lacks a coelom. { ā'sēl·əˌmāt }

acoelous [ZOO] **1.** Lacking a true body cavity or coelom. **2.** Lacking a true stomach or digestive tract. { ˌā'sēl·əs }

acolpate [BOT] Of pollen grains, lacking furrows or grooves. { ˌā'kōlˌpāt }

Aconchulinida [INV ZOO] An order of protozoans in the subclass Filosia comprising a small group of naked amebas having filopodia. { əˌkäŋ·kə'lin·ə·də }

aconitase [BIOCHEM] An enzyme involved in the Krebs citric acid cycle that catalyzes the breakdown of citric acid to *cis*-aconitic and isocitric acids. { ə'kän·əˌtās }

aconite [BOT] Any plant of the genus *Aconitum*. Also known as friar's cowl; monkshood; mousebane; wolfsbane. [PHARM] A toxic drug obtained from the dried tuberous root of *Aconitum napellus*; the principal alkaloid is aconitine. { 'ak·əˌnīt }

aconitine [PHARM] $C_{34}H_{47}O_{11}N$ A poisonous, white, crystalline alkaloid compound obtained from aconites such as monkshood (*Aconitum napellus*). { ə'kän·əˌtēn }

acorn [BOT] The nut of the oak tree, usually surrounded at the base by a woody involucre. { 'āˌkȯrn }

acorn barnacle [INV ZOO] Any of the sessile barnacles that are enclosed in conical, flat-bottomed shells and attach to ships and near-shore rocks and piles. { 'ā,kȯrn ,bär·nə·kəl }

acorn sugar *See* quercitol. { 'ā,kȯrn ,shu̇g·ər }

acorn worm [INV ZOO] Any member of the class Enteropneusta, free-living animals that usually burrow in sand or mud. Also known as tongue worm. { 'ā,kȯrn ,wərm }

acotyledon [BOT] A plant without cotyledons. { ā,käd·əl'ēd·ən }

acouchi [VERT ZOO] A hystricomorph rodent represented by two species in the family Dasyproctidae; believed to be a dwarf variety of the agouti. { ə'kü·shē }

acoustic nerve *See* auditory nerve. { ə'küs·tik ,nərv }

acoustic reflex [NEUROSCI] Brief, involuntary closure of the eyes due to stimulation of the acoustic nerve by a sudden sound. { ə'küs·tik 'rē,fleks }

ACP *See* acyl carrier protein.

acquired [BIOL] Not present at birth, but developed by an individual in response to the environment and not subject to hereditary transmission. { ə'kwīrd }

acquired immune deficiency syndrome [MED] A disease that is caused by the human immunodeficiency virus (HIV) and compromises the competency of the immune system; characterized by persistent lymphadenopathy and opportunistic infections such as *Pneumocystis carinii* pneumonia, cytomegalovirus, disseminated histoplasmosis, candidiasis, isosporiasis, and malignancies such as Kaposi's sarcoma. Abbreviated AIDS. { ə'kwīrd ə¦myün də¦fish·ən·sē 'sin,drōm }

acquired immunity [IMMUNOL] Resistance to a microbial or other antigenic substance taken on by a naturally susceptible individual; may be either active or passive. { ə'kwīrd ə'myün·ə·dē }

acquired immunological tolerance [IMMUNOL] Failure of immunological responsiveness, that is, inability of antigen-sensitive cells to synthesize antibodies; induced by exposure to large amounts of an antigen. Also known as immunological paralysis. { ə'kwīrd ,im·yü·nə'laj·ə·kəl 'täl·ə·rəns }

acrania [MED] Partial or complete absence of the cranium at birth. { ā'krān·ē·ə }

Acrania [ZOO] A group of lower chordates with no cranium, jaws, vertebrae, or paired appendages; includes the Tunicata and Cephalochordata. { ā'krān·ē·ə }

Acrasiales [BIOL] A group of microorganisms that have plant and animal characteristics; included in the phylum Myxomycophyta by botanists and Mycetozoia by zoologists. { ə'krāzh·ē'ā·lēz }

Acrasida [MYCOL] An order of Mycetozoia containing cellular slime molds. { ə'kras·ə·də }

Acrasieae [BIOL] An equivalent name for the Acrasiales. { ə,krāz·ē'ē,ē }

acrasin [BIOCHEM] The chemotactic substance thought to be secreted by, and to effect aggregation of, myxamebas during their fruiting phase. { ə'krāz·ən }

Acrasiomycota [MYCOL] The phylum containing the cellular slime molds. { ə¦krā·zē·ō·mī'käd·ə }

acraspedote [INV ZOO] Describing tapeworm segments which are not overlapping. { ə'kras·pə,dōt }

acroagnosis [MED] Loss or absence of sense perception in a limb. { ,ak·rō·ag'nō·səs }

acroblast [CELL MOL] A vesicular structure in the spermatid formed from Golgi material. { 'ak·rə,blast }

acrocarpous [BOT] In some mosses of the subclass Eubrya, having the sporophyte at the end of a stem and therefore exhibiting the erect habit. { ,ak·rə'kär·pəs }

acrocentric chromosome [CELL MOL] A chromosome having the centromere close to one end. { ¦ak·rə¦sen·trik 'krō·mə·sōm }

acrocephalosyndactylism [MED] A congenital malformation consisting of an enlarged, pointed skull and defective separation of fingers and toes. Also known as Apert's syndrome. { ,ak·rə,sef·ə·lō,sin'dak·tə,liz·əm }

acrocephaly *See* oxycephaly. { ,ak·rə'sef·ə·lē }

Acroceridae [INV ZOO] The humpbacked flies, a family of orthorrhaphous dipteran insects in the series Brachycera. { a,krä'ser·ə·de }

acrodermatitis enteropathica [MED] An often fatal inherited disease involving inefficient intestinal absorption of zinc; readily treated by adding zinc to the diet. { ,a·krō,dər·mə¦tī·təs ,en·tə·rə'pa·thə·kə }

acrodomatia [ECOL] Specialized structures on certain plants adapted to shelter mites; relationship is presumably symbiotic. { ,ak·rə·də 'māsh·ē·ə }

acrodont [ANAT] Having teeth fused to the edge of the supporting bone. { 'ak·rə,dänt }

acrodynia [MED] A childhood syndrome associated with mercury ingestion and characterized by periods of irritability alternating with apathy, anorexia, pink itching hands and feet, photophobia, sweating, tachycardia, hypertension, and hypotonia. { ,ak·rō'din·ē·ə }

acromegaly [MED] A chronic condition in adults caused by hypersecretion of the growth hormone and marked by enlarged jaws, extremities, and viscera, accompanied by certain physiological changes. { ,ak·rō'meg·ə·lē }

acromelalgia *See* erythromelalgia. { ,ak·rō·mə'läl·jē·ə }

acromere [HISTOL] The distal portion of a rod or cone in the retina. { 'ak·rō,mēr }

acromion [ANAT] The flat process on the outer end of the scapular spine that articulates with the clavicle and forms the outer angle of the shoulder. { ə'krō·me,än }

acron [EVOL] Unsegmented head of the ancestral arthropod. [INV ZOO] **1.** The preoral, nonsegmented portion of an arthropod embryo. **2.** The prostomial region of the trochophore larva of some mollusks. { 'ak,rän }

acronematic [BIOL] Referring to a flagellum without hairs. { ,ak·rō·nə'mad·ik }

acroparesthesia [MED] A chronic self-limited symptom complex associated with a variety of systemic diseases, characterized by tingling, pins-and-needles sensations, numbness or stiffness, and occasionally pains in the hands and feet. { ,ak·rō,par·ə'thēzh·ē·ə }

acropetal [BOT] From the base toward the apex, as seen in the formation of certain organs or the spread of a pathogen. { ə'krä·pəd·əl }

acroscopic [BOT] Facing, or on the side toward, the apex. { ,ak·rə'skäp·ik }

acrosin [BIOCHEM] A proteolytic enzyme located in the acrosome of a spermatozoon; thought to be involved in penetration of the egg. { 'ak·rə·sin }

acrosome [CELL MOL] The anterior, crescent-shaped body of spermatozoon, formed from Golgi material of the spermatid. Also known as perforatorium. { 'ak·rə,sōm }

acrosome reaction [CELL MOL] A form of cellular exocytosis that allows sperm to penetrate the zona pellucida of ovulated eggs. { 'ak·rə,sōm rē,ak·shən }

acrospore [MYCOL] In fungi, a spore formed at the outer tip of a hypha. { 'ak·rə,spȯr }

acrotarsium [ANAT] Instep of the foot. { ,ak·rō'tär·sē·əm }

Acrothoracica [INV ZOO] A small order of burrowing barnacles in the subclass Cirripedia that inhabit corals and the shells of mollusks and barnacles. { ,ak·rə·thə'ras·ik·ə }

Acrotretida [INV ZOO] An order of brachiopods in the class Inarticulata; representatives are known from Lower Cambrian to the present. { ,ak·rō'tred·ə·də }

Acrotretidina [INV ZOO] A suborder of inarticulate brachiopods of the order Acrotretida; includes only species with shells composed of calcium phosphate. { ,ak·rō·tre'tī·də·nə }

Actaeonidae [INV ZOO] A family of gastropod mollusks in the order Tectibranchia. { ,ak·tē'än·ə·dē }

Actaletidae [INV ZOO] A family of insects belonging to the order Collembola characterized by simple tracheal systems. { ,ak·tə'led·ə·dē }

ACTH *See* adrenocorticotropic hormone.

Actidione [MICROBIO] Trade name for the antibiotic cyclohexamide. { ,ak·tə'dī,ōn }

actin [BIOCHEM] A muscle protein that is the chief constituent of the Z-band myofilaments of each sarcomere. { 'ak·tən }

actinal [INV ZOO] In radially symmetrical animals, referring to the part from which the tentacles or arms radiate or to the side where the mouth is located. { 'ak·tə·nəl }

Actiniaria [INV ZOO] The sea anemones, an order of cnidarians in the subclass Zoantharia. { ak,tin·ē'a·rē·ə }

actinobacillosis [VET MED] A bacterial disease of domestic animals caused by *Actinobacillus lignieresii*. { ,ak·tə·nō,bas·ə'lō·səs }

Actinobacillus [MICROBIO] A species of gram-negative, oval, spherical, or rod-shaped bacteria that are of uncertain affiliation; coccal and bacillary cells are often interspersed, giving a "Morse code" form; species are pathogens of animals, occasionally of humans. { ,ak·tə·nō·bə'sil·əs }

Actinobacillus lignieresii [MICROBIO] The causative agent of actinobacillosis. { ,ak·tə·nō·bə,sil·əs ,lin·yir'ās·ē,ē }

Actinobacillus pleuropneumoniae [MICROBIO] The etiologic agent of pleuropneumoniae in swine. { ,ac·tə·nō·bə,sil·əs ,plu̇·rə·nəprmō·nē,ī }

Actinobacillus suis [MICROBIO] The etiologic agent of various lesions in piglets. { ,ak·tə·nō·bə,sil·əl 'sü·is }

Actinobifida [MICROBIO] A genus of bacteria in the family Micromonosporaceae with a dichotomously branched substrate; an aerial mycelium is formed which produces single spores. { ,ak·tə·nō'bī·fə·də }

actinocarpous [BOT] Having flowers and fruit radiating from one point. { ,ak·tə·nō'kär·pəs }

actinochitin [BIOCHEM] A form of birefringent or anisotropic chitin found in the seta of certain mites. { ,ak·tə·nō'kī·tən }

Actinochitinosi [INV ZOO] A group name for two closely related suborders of mites,the Trombidiformes and the Sarcoptiformes. { ,ak·tə·nō,kī·tə'nō·sē }

Actinolaimoidea [INV ZOO] A superfamily of nematodes in the order Dorylaimida, containing some species with remarkable elaborations of the stoma and the characteristic axial spear. { ¦ak·tə·nō·lə'mȯid·ē·ə }

actinomere [INV ZOO] One of the segments composing the body of a radially symmetrical animal. { ,ak'tin·ə,mir }

actinomorphic [BIOL] Descriptive of an organism, organ, or part that is radially symmetrical. { ,ak·tə·nō'mȯr·fik }

Actinomyces [MICROBIO] The type genus of the family Actinomycetaceae; anaerobic to facultatively anaerobic; includes human and animal pathogens. { ,ak·tə·nō'mī·sēs }

Actinomycetaceae [MICROBIO] A family of bacteria in the order Actinomycetales; gram-positive, diphtheroid cells which form filaments but not mycelia; chemoorganotrophs that ferment carbohydrates. { ,ak·tə·nō,mī·sə'tās·ē,ē }

Actinomycetales [MICROBIO] An order of bacteria; cells form branching filaments which develop into mycelia in some families. { ,ak·tə·nō,mī·sə'tā·lēz }

actinomycete [MICROBIO] Any member of the bacterial family Actinomycetaceae. { ,ak·tə·nō'mī,sēt }

actinomycin [MICROBIO] The collective name for a large number of red chromoprotein antibiotics elaborated by various strains of *Streptomyces*. { ,ak·tə·nō'mī·sən }

actinomycosis [MED] An infectious bacterial disease caused by *Actinomyces bovis* in cattle, hogs, and occasionally in humans. Also known as lumpy jaw. { ,ak·tə·nō,mī'kō·səs }

actinomyosin [BIOCHEM] A protein complex formed by the combination of actin and myosin during muscle contraction. { ˌak·tə·nō'mī·əs·ən }

Actinomyxida [INV ZOO] An order of protozoan invertebrate parasites of the class Myxosporidea characterized by trivalved spores with three polar capsules. { ˌak·tə·nō'mik·sə·də }

actinophage [MICROBIO] A bacteriophage that infects and lyses members of the order Actinomycetales. { ak'tin·əˌfāj }

Actinophryida [INV ZOO] An order of protozoans in the subclass Heliozoia; individuals lack an organized test, a centroplast, and a capsule. { ˌak·tə·nō'frī·ə·də }

Actinoplanaceae [MICROBIO] A family of bacteria in the order Actinomycetales with well-developed mycelia and spores formed on sporangia. { ˌak·tə·nō·plə'nās·ēˌē }

Actinoplanes [MICROBIO] A genus of bacteria in the family Actinoplanaceae having aerial mycelia and spherical to subspherical sporangia; spores are spherical and motile by means of a tuft of polar flagella. { ˌak·tə·nō'plā·nēz }

Actinopodea [INV ZOO] A class of protozoans belonging to the superclass Sarcodina; most are free-floating, with highly specialized pseudopodia. { ˌak·tə·nō'pōd·ē·ə }

Actinopteri [VERT ZOO] An equivalent name for the Actinopterygii. { ˌak·tə'näp·təˌrī }

Actinopterygii [VERT ZOO] The ray-fin fishes, a subclass of the Osteichthyes distinguished by the structure of the paired fins, which are supported by dermal rays. { ˌak·təˌnäp·tə'rij·ēˌī }

actinostele [BOT] A protostele characterized by xylem that is either star-shaped in cross section or has ribs radiating from the center. { ak'tin·əˌstēl }

actinostome [BIOL] **1.** The mouth of a radiate animal. **2.** The peristome of an echinoderm. { ak'tin·əˌsōm }

actinotherapy *See* radiation therapy. { ˌak·tə·nō'the·rə·pē }

actinotrocha [INV ZOO] The free-swimming larva of *Phoronis*, a genus of small, marine, tubicolous worms. { ˌak·tə·nō'trō·kə }

actinula [INV ZOO] A larval stage of some hydrozoans that has tentacles and a mouth; attaches and develops into a hydroid in some species, or metamorphoses into a medusa. { ak'tin·yə·lə }

action current [PHYSIO] The electric current accompanying membrane depolarization and repolarization in an excitable cell. { 'ak·shən ˌkə·rənt }

action potential [NEUROSCI] A transient change in electric potential at the surface of a nerve or muscle cell occurring at the moment of excitation. { 'ak·shən pəˌten·chəl }

action spectrum [PHYSIO] Graphic representation of the comparative effects of different wavelengths of light on living systems or their components. { 'ak·shən ˌspek·trəm }

actium [ECOL] A rocky seashore community. { 'ak·tē·əm }

activated macrophage [IMMUNOL] A macrophage whose ability to destroy microbes or other cells has been enhanced because of stimulation by a lymphokine. { 'ak·təˌvād·əd 'mak·rəˌfāj }

activating enzyme [BIOCHEM] An enzyme that catalyzes a reaction involving adenosine triphosphate and a specific amino acid to give a product that subsequently reacts with a specific transfer ribonucleic acid. { 'ak·təˌvād·iŋ 'enˌzīm }

activating receptor [NEUROSCI] A sense organ at the end of a nerve that triggers a specific response when it is stimulated. { 'ak·təˌvād·iŋ rə'sep·tər }

activation [CELL MOL] A change that is induced in an amino acid before it is utilized for protein synthesis. [PHYSIO] The designation for all changes in the ovum during fertilization, from sperm contact to the dissolution of nuclear membranes. { ˌak·tə'vā·shən }

activator [GEN] A molecule that modifies a repressor in a way that enables it to stimulate operon transcription. { 'ak·təˌvād·ər }

activator ribonucleic acid [GEN] Ribonucleic acid molecules which form a sequence-specific complex with receptor genes linked to producer genes. { 'ak·təˌvā·tər ¦rībō¦nü¦klē·ik 'as·əd }

active anaphylaxis [IMMUNOL] The allergic response following reintroduction of an antigen into a hypersensitive individual. { 'ak·tiv 'an·ə·fə'lak·səs }

active biomass [MICROBIO] The amount of a culture that is actively growing. { ¦ak·tiv 'bī·ōˌmas }

active center [BIOCHEM] **1.** A flexible portion of an enzyme that binds to the substrate and converts it into the reaction product. **2.** In carrier and receptor proteins, the portion of the molecule that interacts with the specific target compounds. { 'ak·tiv 'sen·tər }

active immunity [IMMUNOL] Disease resistance in an individual due to antibody production after exposure to a microbial antigen following disease, inapparent infection, or inoculation. { 'ak·tiv im'yü·nət·ē }

active immuno-gene therapy [MED] Immunogene therapy in which the tumor is directly altered with cytokine genes in order to induce an endogenous reaction against it. { ¦ak·tiv ˌim·yə·nō'jēn ˌther·ə·pē }

active immunotherapy [IMMUNOL] A type of immunotherapy that attempts to stimulate the host's intrinsic immune response to the tumor, either nonspecifically or specifically. { ¦ak·tiv ¦im·yə·nō'ther·ə·pē }

active site [CELL MOL] The region of an enzyme molecule at which binding with the substrate occurs. Also known as binding site; catalytic site. { 'ak·tiv 'sīt }

active transport [PHYSIO] The pumping of ions or other substances across a cell membrane against an osmotic gradient, that is, from a lower to a higher concentration. { 'ak·tiv 'tranzˌpȯrt }

actomyosin [BIOCHEM] A protein complex consisting of myosin and actin; the major constituent of a contracting muscle fibril. { ˌak·tə'mī·ə·sən }

actophilous [ECOL] Having a seashore growing habit. { ,ak'tä·fə·ləs }

acuate [BIOL] **1.** Having a sharp point. **2.** Needle-shaped. { 'ak·yə,wāt }

acuity [BIOL] Sharpness of sense perception, as of vision or hearing. { ə'kyü·ə·dē }

Aculeata [INV ZOO] A group of seven superfamilies that constitute the stinging forms of hymenopteran insects in the suborder Apocrita. { ə,kyü·lē'ä·də }

aculeate [BIOL] Pertaining to something that is prickly. { ə'kyü·lē·ət }

aculeus [INV ZOO] **1.** A sharp, hairlike spine, as on the wings of certain lepidopterans. **2.** An insect stinger modified from an ovipositor. { ə'kyü·lē·əs }

Aculognathidae [INV ZOO] The ant-sucking beetles, a family of coleopteran insects in the superfamily Cucujoidea. { ə,kyü·ləg'nath·ə,dē }

acuminate [BOT] Tapered to a slender point, especially referring to leaves. { ə'kyüm·ə·nət }

acupuncture [MED] The ancient Chinese art of puncturing the body with long, fine gold or silver needles to relieve pain and cure disease. { 'ak·yü,pəŋk·chər }

acute [BIOL] Ending in a sharp point. [MED] Referring to a disease or disorder of rapid onset, short duration, and pronounced symptoms. { ə'kyüt }

acute alcoholism [MED] Drunkenness accompanied by an acute, transient disturbance of physiological and mental functions. { ə'kyüt 'al·kə·hȯ,liz·əm }

acute appendicitis [MED] A sudden, severe attack of appendicitis characterized by abdominal pain, usually localized in the lower right quadrant, with nausea, vomiting, and constipation. { ə'kyüt ə,pen·də'sīd·əs }

acute arthritis [MED] A severe joint inflammation with a short course. { ə'kyüt ärth'rīd·əs }

acute ascending myelitis [MED] Severe inflammation of the spinal cord beginning in the lower segments and progressing toward the head. { ə'kyüt ə'send·iŋ ,mī·ə'līd·əs }

acute bacterial endocarditis [MED] Fulminant, rapidly progressive endocarditis, usually associated with a significant systemic infection. { ə'kyüt bak'tir·ē·əl ,en·dō,kär'dīd·əs }

acute benign lymphoblastosis *See* infectious mononucleosis. { ə'kyüt bə'nīn ,lim·fə,blas'tōs·əs }

acute berylliosis [MED] Severe chemical pneumoconiosis that is caused by inhalation of beryllium salts. { ə'kyüt bə,ril·ē'os·əs }

acute cerebellar ataxia [MED] A severe childhood syndrome of sudden onset characterized by muscular incoordination, impaired articulation, oscillations of the eyeballs, and decreased intraocular pressure. { ə'kyüt ,ser·ə'bel·ər ə'tak·sē·ə }

acute dermatitis [MED] Any severe inflammation of the skin. { ə'kyüt ,dər·mə'tīd·əs }

acute glomerulonephritis [MED] Severe kidney inflammation, usually following infection with group A hemolytic streptococci, particularly type 12. { ə'kyüt glä,mər·yə,lō·nə'frīd·əs }

acute granulocytic leukemia [MED] A severe blood disorder in which the abnormal white cells are immature forms of granulocytes. Also known as myeloblastic leukemia. { ə'kyüt gran·yə·lə'sid·ik lü'kē·mē·ə }

acute infective encephalomyelitis *See* epidemic neuromyasthenia. { ə'kyüt in'fek·təv en,sef·ə·lō,mī·ə'līd·əs }

acute inflammation [MED] Severe inflammation with rapid progress and pronounced symptoms. { ə'kyüt in·flə'mā·shən }

acute leukemia [MED] A severe blood disorder characterized by rapid onset and progress, with anemia and hemorrhagic manifestations; immature forms of leukocytes are predominant. { ə'kyüt lü'kē·mē·ə }

acute lymphocytic leukemia [MED] A severe blood disorder in which abnormal leukocytes are identified as immature forms of lymphocytes. Also known as lymphoblastic leukemia. { ə'kyüt lim·fə'sid·ik lü'kē·mē·ə }

acute monocytic leukemia [MED] A severe blood disorder in which abnormal leukocytes are identified as immature forms of monocytes. Also known as monoblastic leukemia. { ə'kyüt ,män·ə'sid·ik lü'kē·mē·ə }

acute necrotizing hemorrhagic encephalomyelitis [MED] A sudden, severe central nervous system disease with variable symptoms; pathology includes hemorrhages and necrosis of the white matter. { ə'kyüt ,nek·rə'tīz·iŋ hem·ə'raj·ik en ,sef·ə·lō,mī·ə'līd·əs }

acute nonsuppurative hepatitis *See* interstitial hepatitis. { ə'kyüt ,nän'səp·yə,rād·iv hep·ə'tīd·əs }

acute pancreatitis [MED] A disease of unknown etiology that causes sudden liberation of activated pancreatic enzymes that digest the pancreatic parenchyma, leading to dissolution of fat and production of calcium soaps, and rupture of pancreatic vessels with resultant hemorrhage and shock. { ə'kyüt 'pan·krē·ə'tīd·əs }

acute-phase protein [IMMUNOL] Any of a group of proteins that are produced by the liver and appear in the blood in increased amounts shortly after the onset of infection or tissue damage; they include C-reactive protein, fibrinogen, proteolytic enzyme inhibitors, and transferrin. { ə¦kyüt ¦fāz 'prō,tēn }

acute-phase reaction [IMMUNOL] During inflammation, change in the rates of synthesis of certain serum proteins that are important in nonspecific defense reactions. { ə¦kyüt ¦fāz rē'ak·shən }

acute radiation syndrome [MED] A complex of symptoms involving the intestinal tract, blood-forming organs, and skin following whole-body irradiation. { ə'kyüt 'rād·ē'a·shən 'sin,drōm }

acute respiratory disease [MED] Severe adenovirus infection of the respiratory tract characterized by fever, sore throat, and cough. { ə'kyüt 'res·prə,tȯr·ē di,zēz }

acute rheumatic fever [MED] A severe infectious process caused by beta hemolytic streptococci; characterized by fever and frequently accompanied by painful inflamed joints, endocarditis, chorea, or glomerulonephritis. { ə'kyüt rü'mad·ik 'fē·vər }
acute rhinitis [MED] Inflammation of the nasal mucous membrane due to either infection or allergy. { ə'kyüt rī'nīd·əs }
acute toxic encephalopathy [MED] A severe childhood syndrome characterized by sudden onset of coma or stupor, fever, convulsions, and impaired respiratory and cardiovascular functioning. { ə'kyüt 'täk·sik en,sef·ə'läp·ə·thē }
acute transfection [GEN] Short-term deoxyribonucleic acid infection of cells. { ə'kyüt tranz 'fek·shən }
acute tubular necrosis *See* lower nephron nephrosis. { ə'kyüt 'tüb·yə·lər nə'krō·səs }
acute yellow atrophy [MED] Rapid liver destruction following viral hepatitis, toxic chemicals, or other agents. { ə'kyüt 'yel·ō 'a·trə·fē }
acutifoliate [BOT] Possessing sharply pointed leaves. { ə,kyüd·ə'fō·lē·āt }
acutilobate [BOT] Possessing sharply pointed lobes. { ə,kyüd·ə'lō,bāt }
acyclic [BOT] Having flowers arranged in a spiral instead of a whorl. { ā'sik·lik }
acyl carnitine *See* fatty acyl carnitine. { 'a·səl 'kär·nə,tēn }
acyl carrier protein [BIOCHEM] A protein in fatty acid synthesis that picks up aceytl and malonyl groups from acetyl coenzyme A and malonyl coenzyme A and links them by condensation to form β-keto acid acyl carrier protein, releasing carbon dioxide and the sulfhydryl form of acyl carrier protein. Abbreviated ACP. { 'a·səl 'kar·ē·ər 'prō,tēn }
acyl-coenzyme A *See* fatty acyl-coenzyme A. { 'a·səl kō'en,zim 'ā }
AD *See* Alzheimer's disease.
adamantinoma *See* ameloblastoma. { ad·ə,man·tə'nōm·ə }
adambulacral [INV ZOO] Lying adjacent to the ambulacrum. { ¦ad·am·byə'lāk·rəl }
adapertural [INV ZOO] Near the aperture, specifically of a conch. { ,ad'ap·ə,chər·əl }
adapical [BOT] Near or toward the apex or tip. { ,ad'a·pi·kəl }
adaptation [GEN] Adjustment to new or altered environmental conditions by changes in genotype (natural selection) or phenotype. [PHYSIO] The occurrence of physiological changes in an individual exposed to changed conditions; for example, tanning of the skin in sunshine, or increased red blood cell counts at high altitudes. { ,a,dap'tā·shən }
adaptation syndrome [MED] Endocrine-mediated stress reaction of the body in response to systemic injury; involves an initial stage of shock, followed by resistance or adaptation and then healing or exhaustion. { ,a,dap'tā·shən ,sin ,drōm }
adaptins [CELL MOL] Peripheral membrane proteins that play an important role in the assembly of clathrin-coated vesicles in the trans-Golgi network and plasma membrane during receptor-mediated endocytosis. { ə'dap·tins }
adaptive colitis *See* irritable colon. { ə'dap·tiv kə'līd·əs }
adaptive disease [PHYSIO] The physiologic changes impairing an organism's health as the result of exposure to an unfamiliar environment. { ə'dap·tiv di,zēz }
adaptive divergence [EVOL] Divergence of new forms from a common ancestral form due to adaptation to different environmental conditions. { ə'dap·tiv də'vər·jəns }
adaptive enzyme [MICROBIO] Any bacterial enzyme formed in response to the presence of a substrate specific for that enzyme. { ə'dap·tiv 'en,zīm }
adaptive immune response [IMMUNOL] An immune response based on the principle of clonal recognition, such that upon first exposure to an antigen, primed lymphocytes either differentiate into immune effector cells or form an expanded pool of memory cells that respond to secondary exposure to the same antigen by mounting an amplified and more rapid response. { ə¦dap·tiv i'myün ri,späns }
adaptive mutations [GEN] Mutations conferring an advantage in a selective environment which arise after nongrowing or slowly growing cells are exposed to the selective environment. { ə¦dap·tiv myü'tā·shənz }
adaptive norm [GEN] The mix of genotypes of a well-adapted species or population. { ə'dap·tiv 'nȯrm }
adaptive radiation [EVOL] Diversification of a dominant evolutionary group into a large number of subsidiary types adapted to more restrictive modes of life (different adaptive zones) within the range of the larger group. { ə'dap·tiv ,rād·ē'ā·shən }
adaptive value [GEN] The property of a given genotype that confers fitness to an organism in a given environment. { ə'dap·tiv 'val·yü }
adaptor [GEN] Any of the short synthetic oligonucleotide strands that have one sticky end and one blunt end; the blunt end joins to the blunt end of a deoxyribonucleic acid fragment, forming a new fragment with two sticky ends that can be more easily spliced into a vector. { ə'dap·tər }
adaptor protein [CELL MOL] A specialized protein that links protein components of the signaling pathway, thereby aiding intracellular signal transduction. { ə'dap·tər ,prō,tēn }
adaxial [BIOL] On the same side as or facing toward the axis of an organ or organism. { ,ad'ak·sē·əl }
adder [VERT ZOO] Any of the venomous viperine snakes included in the family Viperidae. { 'ad·ər }
addiction [MED] Habituation to a specific practice, such as drinking alcoholic beverages or using drugs. { ə'dik·shən }
Addis count [PATH] A renal function test which

estimates the blood cell count in a 12-hour urine specimen. { 'ad·əs ˌkau̇nt }

Addison's disease [MED] A primary failure or insufficiency of the adrenal cortex to secrete hormones. { 'ad·ə·sənz diˌzēz }

additive factor [GEN] Any of a group of nonallelic genes that affect the same phenotypic characteristics. { 'ad·ə·div 'fak·tər }

additive gene action [GEN] **1.** A form of allelic interaction in which dominance is absent, resulting in a heterozygote that is intermediate in phenotype between homozygotes for the alternative alleles. **2.** The cumulative contribution made by all loci in a group of nonallelic genes to a polygenic trait. { ¦ad·ə·div ¦jēn ˌak·shən }

additive genetic variance [GEN] That part of the genetic variance of a quantitative character attributed to the average effects of substituting one allele for another at a given locus or at the multiple loci governing a polygenic trait. { ¦ad·ə·div jə¦ned·ik 'ver·ē·əns }

adduction [PHYSIO] Movement of one part of the body toward another or toward the median axis of the body. { ə'dək·shən }

adductor [ANAT] Any muscle that draws a part of the body toward the median axis. { ə'dək·tər }

Adeleina [INV ZOO] A suborder of protozoan invertebrate parasites in the order Eucoccida in which the sexual and asexual stages are in different hosts. { ˌad·ə'līn·ə }

adelphous [BOT] Having stamens fused together by their filaments. { ə'del·fəs }

adenase [BIOCHEM] An enzyme that catalyzes the hydrolysis of adenine to hypoxanthine and ammonia. { 'ad·ənˌās }

adenine [BIOCHEM] $C_5H_5N_5$ A purine base, 6-aminopurine, occurring in ribonucleic acid and deoxyribonucleic acid and as a component of adenosine triphosphate. { 'ad·ənˌēn }

adenitis [MED] Inflammation of a gland or lymph node. { ˌad·ən'īd·əs }

adenoacanthoma [MED] An adenocarcinoma, common in the endometrium, in which squamous cells replace the cylindrical epithelium. { ¦ad·ənˌō·əˌkan'thō·mə }

adeno-associated satellite virus [VIROL] A defective virus that is unable to reproduce without the help of an adenovirus. { ¦ad·ənˌō·ə'sō·shēˌād·əd 'sad·əˌlīt ˌvī·rəs }

adenocarcinoma [MED] A malignant tumor originating in glandular or ductal epithelium and tending to produce acinic structures. { ¦ad·ənˌōˌkär·sən'ō·mə }

adenohypophysis [ANAT] The glandular part of the pituitary gland, composing the anterior and intermediate lobes. { ¦ad·ənˌōˌhī'pä·fə·səs }

adenoid [ANAT] **1.** A mass of lymphoid tissue. **2.** Lymphoid tissue of the nasopharynx. Also known as pharyngeal tonsil. { 'adˌnȯid }

adenoma [MED] A benign tumor of glandular origin and structure. { ˌad·ən'ō·mə }

adenomatoid tumor [MED] A benign genital-tract tumor composed of stroma whose spaces are lined by cells that resemble epithelium, endothelium, and mesothelium. { ˌad·ən'ä·məˌtȯid 'tü·mər }

adenomatosis [MED] A condition characterized by multiple adenomas within an organ or in several related organs. { ˌad·ənˌō·mə'tō·səs }

adenomatous goiter [MED] An asymmetric goiter due to isolated nodular masses of thyroid tissue. Also known as multiple colloid goiter; nodular goiter. { ˌad·ənˌō'ma·təs 'gȯit·ər }

adenomere [EMBRYO] The embryonic structure which will become the functional portion of a gland. { ˌad·ən'ō·mir }

adenomyoma [MED] A benign tumor of glandular and muscular elements occurring principally in the uterus and rectum. { ˌad·ənˌōˌmī'ō·mə }

adenomyosis [MED] **1.** The invasion of muscular tissue, such as of the uterine wall or Fallopian tubes, by endometrial tissue. **2.** Any abnormal growth of muscle or glandular tissues. { ˌad·ənˌōˌmī'ō·səs }

adenopathy [MED] Any glandular disease; common usage limits the term to any abnormal swelling or enlargement of lymph nodes. { ˌad·ən'ä·pə·thē }

Adenophorea [INV ZOO] A class of unsegmented worms in the phylum Nematoda. { ˌad·ən·ə'fȯr·ē·ə }

adenophyllous [BOT] Having leaves with glands. { ˌad·ən'ä·fə·ləs }

adenosine [BIOCHEM] $C_{10}H_{13}N_5O_4$ A nucleoside composed of adenine and D-ribose. { ə'den·əˌsēn }

adenosine 3′,5′-cyclic monophosphate *See* cyclic adenylic acid. { ə¦den·ə·sēn ¦thrē¦prīm ¦fīv¦prīm 'sīk·lik 'mä·nō'fäs·fāt }

adenosine 3′,5′-cyclic phosphate *See* cyclic adenylic acid. { ə¦den·ə·sēn ¦thrē¦prīm ¦fīv¦prīm 'sīk·lik 'fäs·fāt }

adenosine diphosphatase [BIOCHEM] An enzyme that catalyzes the hydrolysis of adenosine diphosphate. Abbreviated ADPase. { ə¦den·əˌsēn ˌdī'fäs·fəˌtās }

adenosine diphosphate [BIOCHEM] $C_{10}H_{15}N_5O_{10}P_2$ A coenzyme composed of adenosine and two molecules of phosphoric acid that is important in intermediate cellular metabolism. Abbreviated ADP. { ə¦den·əˌsēn ˌdī'fäs·fāt }

adenosine monophosphate *See* adenylic acid. { ə¦den·əˌsēn 'mä·nō'fäs·fāt }

adenosine 3′,5′-monophosphate *See* cyclic adenylic acid. { ə¦den·əˌsēn ¦thrē¦prīm ¦fīv¦prīm 'mä·nō'fäs·fāt }

adenosine triphosphatase [BIOCHEM] An enzyme that catalyzes the hydrolysis of adenosine triphosphate. Abbreviated ATPase. { ə¦den·əˌsēn ˌtrī'fäs·fəˌtās }

adenosine triphosphate [BIOCHEM] $C_{10}H_{16}N_5O_{12}P_3$ A coenzyme composed of adenosine diphosphate with an additional phosphate group; an important energy compound in metabolism. Abbreviated ATP. { ə¦dēn·əˌsēn ˌtri·'fäsˌfāt }

adenosis [MED] Any nonneoplastic glandular disease, especially one involving the lymph nodes. { ˌad·ən'ō·səs }

adeno-SV40 hybrid virus [VIROL] A defective virus particle in which part of the genetic material of papovavirus SV40 is encased in an adenovirus protein coat. { ¦ad·ən,ō ¦es¦vē¦fȯr·tē 'hī·brəd 'vī·rəs }

Adenoviridae [VIROL] A family of double-stranded DNA viruses with icosahedral symmetry; usually found in the respiratory tract of the host species and often associated with respiratory diseases. Also known as adenovirus. { ,ad·ən·ō'vīr·ə,dē }

adenovirus *See* Adenoviridae. { ¦ad·ən,o'vī·rəs }

adenylcyclase [BIOCHEM] The catalyzing enzyme in the conversion of adenosine triphosphate to cyclic adenosine monophosphate during metabolism. { ¦ad·ən,il'sī,klās }

adenylic acid [BIOCHEM] **1.** A generic term for a group of isomeric nucleotides. **2.** The phosphoric acid ester of adenosine. Also known as adenosine monophosphate (AMP). { ¦ad·ən¦il·ik 'as·əd }

adeoniform [INV ZOO] **1.** A lobate, bilamellar zooarium. **2.** Resembling the fossil bryozoan *Adeona*. { ,ad·ē'ä·nə,fȯrm }

Adephaga [INV ZOO] A suborder of insects in the order Coleoptera characterized by fused hind coxae that are immovable. { ə'def·ə·gə }

adequate contact [MED] The degree of contact required between an infectious and a susceptible individual to cause infection of the latter. { 'ad·ə·kwət 'kän,takt }

adequate stimulus [NEUROSCI] The energy of any specific mode that is sufficient to elicit a response in an excitable tissue. { 'ad·ə·kwət 'stim·yə·ləs }

adfluvial [BIOL] Migrating between lakes and rivers or streams. { ad'flü·vē·əl }

ADH *See* vasopressin.

adherens junction [CELL MOL] An integrin-mediated anchoring junction that connects the cytoskeleton (actin filaments) of a cell to the extracellular matrix or to the cytoskeleton of surrounding cells. { ad'hir·enz jəŋk·shən }

adhering junction [CELL MOL] An intercellular junction that promotes adhesion between cells. Also known as desmosome. { ad¦hir·iŋ 'jəŋk·shən }

adhesion [BOT] Growing together of members of different and distinct whorls. [MED] The abnormal union of an organ or part with some other part by formation of fibrous tissue. { ad'hē·zhən }

adhesive cell [INV ZOO] Any of various glandular cells in ctenophores, turbellarians, and hydras used for adhesion to a substrate and for capture of prey. Also known as colloblast; glue cell; lasso cell. { ad'hēz·iv 'sel }

adiadochokinesis [MED] A type of motor incoordination associated with cerebellar damage in which repetitive movements controlled by antagonistic muscles cannot be performed without severe muscular incoordination. { ə,dē·ə,dō·kō·kə'nē·səs }

Adie's syndrome [MED] Impaired pupillary reaction to light and absent tendon reflexes. { 'a,dēz ,sin,drōm }

Adimeridae [INV ZOO] An equivalent name for the Colydiidae. { ,ad·ə'mer·ə·dē }

adiphenine [PHARM] $C_{20}H_{25}NO_2 \cdot HCl$ The compound 2-diethylaminoethyl diphenylacetate hydrochloride; a cholinergic blocking agent. { ə'dif·ə·nēn }

adipocellulose [BIOCHEM] A type of cellulose found in the cell walls of cork tissue. { ,ad·ə·pō'sel·yə,lōs }

adipocere [MED] A light-colored, waxy material formed by postmortem conversion of body fats to higher fatty acids. { 'ad·ə·pə,sir }

adipogenesis [PHYSIO] The formation of fat or fatty tissue. { ,ad·ə·pō'jen·ə·səs }

adiponecrosis neonatorum [MED] Localized fatty-tissue necrosis occurring in large, healthy infants born after difficult labor. { ¦ad·ə·pō·nə'krō·səs ,nē·ō·nə'tȯr·əm }

adipose [BIOL] Fatty; of or relating to fat. { 'ad·ə,pōs }

adipose fin [VERT ZOO] A modified posterior dorsal fin that is fleshy and lacks rays; found in salmon and typical catfishes. { 'ad·ə,pōs ,fin }

adipose tissue [HISTOL] A type of connective tissue specialized for lipid storage. { 'ad·ə,pōs 'tish·ü }

adiposis dolorosa [MED] An uncommon type of obesity in which the excess fat deposits are tender and painful. { ad·ə'pō·səs ,dō·lə'rōs·ə }

adiposogenital dystrophy [MED] A syndrome involving obesity, retarded gonad development, and sometimes diabetes insipidus resulting from impaired functioning of the pituitary and hypothalamus. Also known as Froehlich's syndrome. { ¦ad·ə,pō·sō'jen·əd·əl 'dis·trə·fē }

adipsia [MED] Absence of thirst or avoidance of drinking. { ā'dip·sē·ə }

adjustor neuron [NEUROSCI] Any of the interconnecting nerve cells between sensory and motor neurons of the central nervous system. { ə'jəs·tər 'nü,rän }

adjuvant [PHARM] A material that enhances the action of a drug or antigen. { 'aj·ə·vənt }

adjuvant chemotherapy [MED] Chemotherapy that is used to destroy suspected undetectable residual tumor after surgery or radiation treatment has eradicated all detectable tumor; effective in the treatment of breast and colon cancer. { ¦a·jə·vənt ,kē·mō'ther·ə·pē }

ad lib [BIOL] Shortened form for ad libitum; without limit or restraint. { ,ad 'lib }

adnate [BIOL] United through growth; used especially for unlike parts. [BOT] Pertaining to growth with one side adherent to a stem. { 'ad,nāt }

adnexa [BIOL] Subordinate or accessory parts, such as eyelids, Fallopian tubes, and extraembryonic membranes. { ad'neks·ə }

adonite *See* adonitol. { 'ad·ə,nīt }

adonitol [BIOCHEM] $C_5H_{12}O_5$ A pentitol from the dicotyledenous plant *Adonis vernalis*; large crystals that are optically inactive and melt at

102°C; it does not reduce Fehling's solution, and is freely soluble in water and hot alcohol. Also known as adonite; ribitol. { ə'dän·ə,tōl }

adont hinge [INV ZOO] A type of ostracod hinge articulation which either lacks teeth and has overlapping valves or has a ridge and groove. { 'ā,dänt ,hinj }

adoptive immunity [IMMUNOL] Immunity resulting from the transfer of an immune function from one organism to another through the transfer of immunologically competent cells. Also known as transfer immunity. { ə¦däp·təv ə'myü·nəd·ē }

adoptive immunotherapy [IMMUNOL] The transfer of immunologically competent white blood cells or their precursors into the host. { ə¦däp·tiv ,im·yə·nō'ther·ə·pē }

adoral [ZOO] Near the mouth. { ,a'dȯr·əl }

ADP *See* adenosine diphosphate.

ADPase *See* adenosine diphosphatase.

adrenal cortex [ANAT] The cortical moietie of the suprarenal glands which secretes glucocorticoids, mineralocorticoids, androgens, estrogens, and progestagens. { ə'drēn·əl 'kȯr·teks }

adrenal cortex hormone [BIOCHEM] Any of the steroids produced by the adrenal cortex. Also known as adrenocortical hormone; corticoid. { ə'drēn·əl 'kȯr·teks 'hȯr,mōn }

adrenal cortical insufficiency [MED] Failure of the adrenal cortex to secrete adequate hormones. { ə'drēn·əl 'kȯrt·i·kəl ,in·sə'fish·ən·sē }

adrenalectomy [MED] Surgical removal of an adrenal gland. { ə,drēn·əl'ek·tə·mē }

adrenal gland [ANAT] An endocrine organ located close to the kidneys of vertebrates and consisting of two morphologically distinct components, the cortex and medulla. Also known as suprarenal gland. { ə'drēn·əl ,gland }

adrenaline *See* epinephrine. { ə'dren·əl·ən }

adrenal medulla [ANAT] The hormone-secreting chromaffin cells of the adrenal gland that produce epinephrine and norepinephrine. { ə'drēn·əl mə'dəl·ə }

adrenal virilism [MED] **1.** The development of male characteristics in the female resulting from excessive production of adrenal hormones with androgenic activity. **2.** A rare form of pseudohermaphroditism. { ə'drēn·əl 'vir·ə,liz·əm }

adrenergic [PHYSIO] Describing the chemical activity of epinephrine or epinephrine-like substances. { ,ad·rə'nər·jik }

adrenergic blocking agent [BIOCHEM] Any substance that blocks the action of epinephrine or an epinephrine-like substance. { ,ad·rə'nər·jik 'bläk·iŋ ,ā·jənt }

adrenochrome [BIOCHEM] $C_9H_9O_3N$ A brick-red oxidation product of epinephrine which can convert hemoglobin into methemoglobin. { ə'dren·ə,krōm }

adrenocortical hormone *See* adrenal cortex hormone. { ə¦drēn·ō'kȯrd·ə·kəl 'hȯr,mōn }

adrenocorticosteroid [BIOCHEM] **1.** A steroid that is obtained from the adrenal cortex. **2.** A steroid that resembles adrenal cortex steroids or has physiological effects like them. { ə¦drē·nō,kȯrd·ə·kō'stir,ȯid }

adrenocorticotropic hormone [BIOCHEM] The chemical secretion of the adenohypophysis that stimulates the adrenal cortex. Abbreviated ACTH. Also known as adrenotropic hormone. { ə¦drēn·ō'kȯrd·ə·kō'träp·ik 'hȯr,mōn }

adrenogenital syndrome [MED] A group of symptoms associated with hypersecretion of adrenal cortex hormones; effects vary with sex and time of development. { ə¦drēn·ō'jen·ə·təl 'sin ,drōm }

adrenomedullary [PHYSIO] Pertaining to the adrenal gland medulla. { ə¦drē·nō·mə'dəl·ə·rē }

adrenotropic [PHYSIO] Of or pertaining to an effect on the adrenal cortex. { ə¦drēn·ə'träp·ik }

adrenotropic hormone *See* adrenocorticotropic hormone. { ə¦drēn·ə'träp·ik 'hȯr,mōn }

adret [ECOL] The sunny (usually south) face of a mountain featuring high timber and snow lines. { 'ad·rət }

adult polycystic kidney disease [MED] An autosomal dominant disease that is characterized by the formation of cysts along the length of the nephron that causes the kidneys to enlarge, resulting in kidney failure in midadulthood. { ə,dəlt ,päl·ə,sis·tik 'kid·nē diz,ēz }

adult rickets *See* osteomalacia. { ə'dəlt'rik·əts }

advanced [EVOL] Denoting a later stage within a lineage that demonstrates evolutionary progression. { əd'vanst }

adventitia [ANAT] The external, connective-tissue covering of an organ or blood vessel. Also known as tunica adventitia. { ,ad·ven'tish·ə }

adventitious [BIOL] **1.** Also known as adventive. Acquired spontaneously or accidentally, not by heredity. **2.** Arising, as a tissue or organ, in an unusual or abnormal place. { ,ad·ven'tish·əs }

adventitious bud [BOT] A bud that arises at points on the plant other than at the stem apex or a leaf axil. { ,ad·ven'tish·əs 'bəd }

adventitious deafness [MED] A type of deafness that occurs at any point during a lifetime and may have a course either of gradual, progressive development or of sudden onset. { ,ad·ven'tish·əs 'def·nəs }

adventitious embryo [MED] An embryo developing outside the uterus. { ,ad·ven'tish·əs 'em·brē·ō }

adventitious root [BOT] A root that arises from any plant part other than the primary root (radicle) or its branches. { ,ad·ven'tish·əs 'rüt }

adventitious vein [INV ZOO] The vessel between the intercalary and accessory veins on certain insect wings. { ,ad·ven'tish·əs 'vān }

adventitious virus [VIROL] A contaminant virus present by chance in a virus preparation. { ,ad·ven'tish·əs 'vī·rəs }

adventive [BIOL] **1.** An organism that is introduced accidentally and is imperfectly naturalized; not native. **2.** *See* adventitious. { ad 'ven·tiv }

advolution [BIOL] Development or growth with

increasing similarities; growth toward; the opposite of evolution. { ˌad·vəˈlü·shən }

aebi [BIOL] A unit for the standardization of a phosphatase. { ˌāˈē·bē }

aeciospore [MYCOL] A spore produced by an aecium. { ˈēsh·ē·əˌspȯr }

aecium [MYCOL] The fruiting body or sporocarp of rust fungi. { ˈēsh·ē·əm }

aedeagus [INV ZOO] The copulatory organ of a male insect. { ¦ē·dēˈā·gəs }

Aedes [INV ZOO] A genus of the dipterous subfamily Culicinae in the family Culicidae, with species that are vectors for many diseases of humans. { āˈē·dēz }

Aegeriidae [INV ZOO] The clearwing moths, a family of lepidopteran insects in the suborder Heteroneura characterized by the lack of wing scales. { ˌē·jəˈrē·əˌdē }

Aegialitidae [INV ZOO] An equivalent name for the Salpingidae. { ˌē·jyəˈlid·əˌdē }

Aegidae [INV ZOO] A family of isopod crustaceans in the suborder Flabellifera whose members are economically important as fish parasites. { ˈē·jə·dē }

aegithognathous [VERT ZOO] Referring to a bird palate in which the vomers are completely fused and truncate in appearance. { ˌē·gəˌthägˈnā·thəs }

Aegothelidae [VERT ZOO] A family of small Australo-Papuan owlet-nightjars in the avian order Caprimulgiformes. { ˌē·gəˈthel·əˌdē }

Aegypiinae [VERT ZOO] The Old World vultures, a subfamily of diurnal carrion feeders of the family Accipitridae. { ˌē·jəˈpī·ə·nē }

Aegyptianella [MICROBIO] A genus of the family Anaplasmataceae; organisms from inclusions in red blood cells of birds. { əˌjip·shəˈnel·ə }

aelophilous [BOT] Describing a plant whose disseminules are dispersed by wind. { ˌēˈlä·fə·ləs }

Aelosomatidae [INV ZOO] A family of microscopic fresh-water annelid worms in the class Oligochaeta characterized by a ventrally ciliated prostomium. { ˌeˌlä·səˈmad·əˌdē }

Aepophilidae [INV ZOO] A family of bugs in the hemipteran superfamily Saldoidea. { ˌē·pōˈfil·əˌdē }

aequorin [BIOCHEM] A bioluminescent protein that is produced by jellyfish of the genus *Aequorea* and emits light in the presence of calcium or strontium. { ˈē·kwəˌrin }

aerenchyma [BOT] A specialized tissue in some water plants characterized by thin-walled cells and large intercellular air spaces. { ˌaˈreŋk·ə·mə }

aerial [BIOL] Of, in, or belonging to the air or atmosphere. { ˈe·rē·əl }

aerial mycelium [MYCOL] A mass of hyphae that occurs above the surface of a substrate. { ˈe·rē·əl mīˈsē·lē·əm }

aerial root [BOT] A root exposed to the air, usually anchoring the plant to a tree, and often functioning in photosynthesis. { ˈe·rē·əl ˈrüt }

aerial stem [BOT] A stem with an erect or vertical growth habit above the ground. { ˈe·rē·əl ˈstem }

aeroallergen [MED] Any airborne particulate matter that can induce allergic responses in sensitive persons. { ˌe·rōˈal·ər·jən }

aerobe [BIOL] An organism that requires air or free oxygen to maintain its life processes. { ˈeˌrōb }

aerobic bacteria [MICROBIO] Any bacteria requiring free oxygen for the metabolic breakdown of materials. { eˈrōb·ik ˌbakˈtir·ē·ə }

aerobic process [BIOL] A process requiring the presence of oxygen. { eˈrōb·ik ˈpräs·əs }

aerobiology [BIOL] The study of the atmospheric dispersal of airborne fungus spores, pollen grains, and microorganisms; and, more broadly, of airborne propagules of algae and protozoans, minute insects such as aphids, and pollution gases and particles which exert specific biologic effects. { ˌe·rōˌbīˈäl·ə·jē }

aerobioscope [MICROBIO] An apparatus for collecting and determining the bacterial content of a sample of air. { ˌe·rōˈbi·əˌskōp }

aerobiosis [BIOL] Life existing in air or oxygen. { ˌe·rōˌbiˈō·səs }

Aerococcus [MICROBIO] A genus of bacteria in the family Streptococcaceae; spherical cells have the tendency to form tetrads; they ferment glucose with production of dextrorotatory lactic acid (homofermentative). { ˈe·rōˌkäk·əs }

aerocyst [BOT] An air vesicle in certain species of algae. { ˈe·rōˌsist }

aerodontalgia [MED] A toothache brought on by atmospheric decompression. { ˌeˌrōˌdän ˈtal·jē·ə }

aeroembolism [MED] A condition marked by the presence of nitrogen bubbles in the blood and other body tissues resulting from a sudden fall in atmospheric pressure. Also known as air embolism. { ˌe·rōˈem·bəˌliz·əm }

aeromedicine *See* aerospace medicine. { ¦e·rō¦med·ə·sin }

Aeromonas [MICROBIO] A genus of bacteria in the family Vibrionaceae; straight, motile rods with rounded ends; most species are pathogenic to marine and fresh-water animals. { e·rō ˈmōn·əs }

aerootitis *See* barotitis. { ¦e·rōˌōˈtīd·əs }

aerophyte *See* epiphyte. { ˈe·rōˌfīt }

aeroplankton [ECOL] Small airborne organisms such as insects. { ¦e·rōˈplaŋk·tən }

aerosinusitis *See* barosinusitis. { ˌe·rōˌsī·nə ˈsīd·əs }

aerospace medicine [MED] The branch of medicine dealing with the effects of flight in the atmosphere or space upon the human body and with the prevention or cure of physiological or psychological malfunctions arising from these effects. Also known as aeromedicine; aviation medicine. { ¦e·rō¦spās ˈmed·ə·sin }

Aerosporin [MICROBIO] Trade name for the antibiotic polymyxin B. { ¦e·rō¦spȯr·ən }

aerotaxis [BIOL] The movement of an organism, especially aerobic and anaerobic bacteria, with reference to the direction of oxygen or air. { ˌe·rō'tak·səs }

aerotolerant [MICROBIO] Able to survive in the presence of oxygen. { ˌe·rō'täl·ə·rənt }

aerotropism [BOT] A response in which the growth direction of a plant component changes due to modifications in oxygen tension. { ˌe·rō'trōˌpiz·əm }

aeschynomenous [BOT] Having sensitive leaves that droop when touched, such as members of the Leguminosae. { ˌes·kə'näm·ə·nəs }

Aesculus [BOT] A genus of deciduous trees or shrubs belonging to the order Sapindales. Commonly known as buckeye. { ˌes·kyə·ləs }

Aeshnidae [INV ZOO] A family of odonatan insects in the suborder Anisoptera distinguished by partially fused eyes. { 'esh·nə·dē }

aesthacyte *See* esthacyte. { 'es·thəˌsīt }

aesthesia *See* esthesia. { es'thē·zhə }

aesthete [BOT] A plant organ with the capacity to respond to definite physical stimuli. { 'esˌthēt }

aestidurilignosa [ECOL] A mixed woodland of evergreen and deciduous hardwoods. { ˌes·tə·dəˌril·əg'nōs·ə }

aestilignosa [ECOL] A woodland of tropophytic vegetation in temperate regions. { es·tə·lig'nōs·ə }

aestivation [BOT] The arrangement of floral parts in a bud. [PHYSIO] The condition of dormancy or torpidity. { ˌes·tə'vā·shən }

afebrile [MED] Without fever. { ¦ā¦fēbˌril }

affection [MED] Any pathology or diseased state of the body. { ə'fek·shən }

afferent [PHYSIO] Conducting or conveying inward or toward the center, specifically in reference to nerves and blood vessels. { 'af·ə·rənt }

afferent neuron [NEUROSCI] A nerve cell that conducts impulses toward a nerve center, such as the central nervous system. { 'af·ə·rənt 'nu̇ˌrän }

affinity [IMMUNOL] The strength of the attractive forces between an antigen and an antibody. { ə'fin·əd·ē }

affinity labeling [BIOCHEM] A method for introducing a label into the active site of an enzyme by relying on the tight binding between the enzyme and its substrate (or cofactors). { ə'fin·əd·ē 'lā·bə·liŋ }

afibrinogenemia [MED] Complete absence of fibrinogen in the blood. { ˌā·fī¦brin·ə·jə'nē·mē·ə }

aflatoxicosis [MED] Aflatoxin poisoning. { ə¦flād·ōˌtäk·sə'kō·səs }

aflatoxin [BIOCHEM] The toxin produced by some strains of the fungus *Aspergillus flavus*, the most potent carcinogen yet discovered. { ˌaf·lə'täk·sin }

African horse sickness [VET MED] An infectious, mosquito-borne virus disease of equines characterized by fever and edematous swelling. { 'af·ri·kən hȯrs 'sik·nəs }

African sleeping sickness [MED] A disease of humans confined to tropical Africa, caused by the protozoans *Trypanosoma gambiense* or *T. rhodesiense*; symptoms include local reaction at the site of the bite, fever, enlargement of adjacent lymph nodes, skin rash, edema, and during the late phase, somnolence and emaciation. Also known as African trypanosomiasis; maladie du sommeil; sleeping sickness. { 'af·ri·kən 'slēp·iŋ 'sik·nəs }

African swine fever *See* hog cholera. { 'af·ri·kən 'swīn ˌfēv·ər }

African trypanosomiasis *See* African sleeping sickness. { 'af·ri·kən trəˌpan·ə·sə'mī·ə·səs }

African violet [BOT] *Saintpaulia ionantha*. A flowering plant typical of the family Gesneriaceae. { 'af·ri·kən 'vī·ə·lət }

afterbirth [EMBRYO] The placenta and fetal membranes expelled from the uterus following birth of offspring in viviparous mammals. { 'af·tərˌbərth }

afterimage [NEUROSCI] A visual sensation occurring after the stimulus to which it is a response has been removed. { 'af·tərˌim·əj }

afterloading [MED] Placing nonradioactive holders in a patient during an operative procedure to provide radiographic information used for dosimetric evaluation, and to receive the radioactive sources during the course of postoperative treatment. { 'af·tərˌlōd·iŋ }

afterpain [MED] Lower abdominal pain after passage of the placenta secondary to uterine contractions. { 'af·tərˌpān }

afterpotential [NEUROSCI] A small positive or negative wave that follows and is dependent on the main spike potential, seen in the oscillograph tracing of an action potential passing along a nerve. { 'af·tərpə¦ten·chəl }

afterripening [BOT] A period of dormancy after a seed is shed during which the synthetic machinery of the seed is prepared for germination and growth. { 'af·tərˌrī·pən·iŋ }

aftershaft [VERT ZOO] An accessory, plumelike feather near the upper umbilicus on the feathers of some birds. { 'af·tərˌshaft }

agalactia [MED] Nonsecretion or imperfect secretion of milk after childbirth. { ˌā·gə'lak·shə }

agameon [BIOL] An organism which reproduces only by asexual means. Also known as agamospecies. { ā'gam·ē·ən }

agamete [BIOL] An asexual reproductive cell that develops into an adult individual. { ā'gaˌmēt }

agamic [BIOL] Referring to a species or generation which does not reproduce sexually. { ā'gam·ik }

Agamidae [VERT ZOO] A family of Old World lizards in the suborder Sauria that have acrodont dentition. { ə'gam·əˌdē }

agammaglobulinemia [MED] The condition characterized by lack of or extremely low levels of gamma globulin in the blood, together with defective antibody production and frequent infections; primary agammaglobulinemia occurs in three clinical forms: congenital, acquired, and transient. { āˌgam·ə'gläb·yə·lən·ē·mē·ə }

agamogony [BIOL] Asexual reproduction, specifically schizogony. { ˌā·gə'mäg·ə·nē }

agamospecies *See* agameon. { ˌa·gə·mō'spē·shēz }

agamospermy [BOT] Apogamy in which sexual union is incomplete because of abnormal development of the pollen and the embryo sac. { 'ā·gam·əˌspərm·ē }

Agaontidae [INV ZOO] A family of small hymenopteran insects in the superfamily Chalcidoidea; commonly called fig insects for their role in cross-pollination of figs. { ˌa·gā'än·təˌdē }

agar-gel reaction [IMMUNOL] A precipitin type of antigen-antibody reaction in which the reactants are introduced into different regions of an agar gel and allowed to diffuse toward each other. { 'äg·ər ˌjel ri'ak·shən }

Agaricales [MYCOL] An order of fungi in the class Basidiomycetes containing all forms of fleshy, gilled mushrooms. { əˌgar·ə'kā·lēz }

agarophyte [BOT] Any seaweed that yields agar. { ə'gar·əˌfīt }

agarose [BIOCHEM] The gelling component of agar; possesses a double-helical structure which forms a three-dimensional framework capable of holding water molecules in the interstices. { 'ag·əˌrōs }

agar plate count *See* plate count. { 'äg·ər 'plāt ˌkau̇nt }

Agavaceae [BOT] A family of flowering plants in the order Liliales characterized by parallel, narrow-veined leaves, a more or less corolloid perianth, and an agavaceous habit. { 'ag·ə'vās·ēˌē }

age [BIOL] Period of time from origin or birth to a later time designated or understood; length of existence. { āj }

age distribution [ECOL] The distribution of different age groups in a population. { 'āj dis·trə'byü·shən }

agenesis [BIOL] Absence of a tissue or organ due to lack of development. { 'ā·jen·ə·səs }

agglutination [CELL MOL] The joining of two organisms of the same species for the purpose of sexual reproduction. { əˌglü·tə'nā·shən }

agglutination reaction [IMMUNOL] Clumping of a particulate suspension of antigen by a reagent, usually an antibody. { əˌglüt·ən'ā·shən rē'ak·shən }

agglutinin [IMMUNOL] An antibody from normal or immune serum that causes clumping of its complementary particulate antigen, such as bacteria or erythrocytes. { ə'glüt·ən·ən }

agglutinogen [IMMUNOL] An antigen that stimulates production of a specific antibody (agglutinin) when introduced into an animal body. { əˌglü'tin·ə·jən }

agglutinoid [IMMUNOL] An agglutin that lacks the power to agglutinate but has the ability to unite with its agglutinogen. { ə'glüt·ənˌȯid }

aggregate [BOT] Referring to fruit formed in a cluster, from a single flower, such as raspberry, or from several flowers, such as pineapple. { 'ag·rə·gət }

aggregate fruit [BOT] A type of fruit composed of a number of small fruitlets all derived from the ovaries of a single flower. { 'ag·rə·gət 'früt }

aggregation [BIOL] A grouping or clustering of separate organisms. { ˌag·rə'gā·shən }

aggressin [BIOCHEM] A protein produced by a pathogenic microbe which inhibits the host's immune system. { ə'gre·sən }

aggressive mimicry [ZOO] Mimicry used to attract or deceive a species in order to prey upon it. { ə'gres·iv 'mim·ə·krē }

aging [BIOL] Growing older. { 'āj·iŋ }

aging-lung emphysema [MED] An asymptomatic pulmonary disease associated with aging, characterized by alveolar dilation due to loss of tissue elasticity. { ¦āj·iŋ ¦ləŋ em·fə'zē·mə }

aglomerular [HISTOL] Lacking glomeruli. { ¦ā·glə'mər·yə·lər }

Aglossa [VERT ZOO] A suborder of anuran amphibians represented by the single family Pipidea and characterized by the absence of a tongue. { ā'gläs·ə }

aglycon [BIOCHEM] The nonsugar compound resulting from the hydrolysis of glycosides; an example is 3,5,7,3′,4′-pentahydroxyflavylium, or cyanidin. { ə'glīˌkän }

aglyphous [VERT ZOO] Having solid teeth. { 'a·glə'fəs }

agmatine [BIOCHEM] $C_5H_{14}N_4$ Needlelike crystals with a melting point of 231°C; soluble in water; a product of the enzymatic decarboxylation of arginine. { 'ag·məˌtēn }

agnate [BIOL] Related exclusively through male descent. { 'agˌnāt }

Agnatha [VERT ZOO] The most primitive class of vertebrates, characterized by the lack of true jaws. { 'ag·nə'thə }

agnathia [MED] Lack of the jaws. { ag'nath·ē·ə }

agnosia [MED] Loss of the ability to recognize persons or objects and their meaning. { ag 'nozh·ē·ə }

Agonidae [VERT ZOO] The poachers, a small family of marine perciform fishes in the suborder Cottoidei. { ə'gän·ə·dē }

agonist [BIOCHEM] A chemical substance that can combine with a cell receptor and cause a reaction or create an active site. [PHYSIO] A contracting muscle that is resisted or counteracted by another muscle, called an antagonist, with which it is paired. { 'ag·əˌnist }

agouti [VERT ZOO] A hystricomorph rodent, *Dasyprocta*, in the family Dasyproctidae, with 13 species. { ə'güd·ē }

agranular leukocyte [HISTOL] A type of white blood cell, including lymphocytes and monocytes, characterized by the absence of cytoplasmic granules and by a relatively large spherical or indented nucleus. { ˌā'gran·yə·lər 'lü·kəˌsīt }

agranular reticulum [CELL MOL] Endoplasmic reticulum lacking ribosomes. { ā'gran·yə·lər ri'tik·yə·ləm }

agranulocytosis [MED] An acute febrile illness, usually resulting from drug hypersensitivity, manifested as severe leukopenia, often with complete disappearance of granulocytes. { ¦a,gran·yə·lō,sī'tō·səs }

agraphia [MED] Loss of the ability to write. { ,ā'graf·ē·ə }

agravic illusion [MED] An apparent movement of a target in the visual field due to otolith response in zero gravity. Also known as oculoagravic illusion. { ,ā'grav·ik il'ü·zhən }

agrestal [ECOL] Growing wild in the fields. { ə'grest·əl }

agretope [IMMUNOL] In antigen presentation, the part of an antigen that interacts with a class II histocompatibility molecule. { 'ag·rə,tōp }

agriculture [BIOL] The production of plants and animals useful to humans, involving soil cultivation and the breeding and management of crops and livestock. { 'ag·rə,kəl·chər }

agrioecology [ECOL] The ecology of cultivated plants. { ¦ag·rē·ō,ē'käl·ə·jē }

Agrionidae [INV ZOO] A family of odonatan insects in the suborder Zygoptera characterized by black or red markings on the wings. { ,ag·rē'än·ə,dē }

Agrobacterium [MICROBIO] A genus of bacteria in the family Rhizobiaceae; cells do not fix free nitrogen, and three of the four species are plant pathogens, producing galls and hairy root. { ¦ag·rō,bak'tir·e·əm }

agroecosystem [ECOL] A model for the functionings of an agricultural system with all its inputs and outputs. { ¦ag·rō'ek·ō,sis·təm }

Agromyzidae [INV ZOO] A family of myodarian cyclorrhaphous dipteran insects of the subsection Acalypteratae; commonly called leaf-miner flies because the larvae cut channels in leaves. { ¦ag·rō'mīz·ə,dē }

agrophilous [ECOL] Having a natural habitat in grain fields. { ə'gräf·ə·ləs }

agrostology [BOT] A division of systematic botany concerned with the study of grasses. { ¦ag·rə¦stä·lə·jē }

ahaptoglobinemia [MED] An inherited lack of haptoglobin, a blood serum protein. { ,ā¦hap·tō,glō·bə'nēm·ē·ə }

ahermatypic [INV ZOO] Non-reef-building, as applied to corals. { ¦ā,hər·mə¦tip·ik }

AIA *See* anti-immunoglobulin antibody.

AIDS *See* acquired immune deficiency syndrome. { ādz }

AIDS-related complex [MED] A set of symptoms, such as lymph node enlargement, fever, loss of weight, diarrhea, and minor opportunistic diseases, associated with a weakened immune system, indicating a less severe form of infection by the HIV virus than AIDS itself. Abbreviated ARC. { 'ādz rə,lād·əd ¦käm,pleks }

ainhum [MED] A tropical disease of unknown etiology that is peculiar to black males, in which a toe is slowly and spontaneously amputated by a fibrous ring. { 'ī·nyüm }

aiophyllous *See* evergreen. { ,ī·ō'fil·əs }

air cell [ZOO] A cavity or receptacle for air such as an alveolus, an air sac in birds, or a dilation of the trachea in insects. { 'er ,sel }

air embolism *See* aeroembolism. { ¦er 'em·bə,liz·əm }

air hunger [MED] The deep, gasping respiration characteristic of severe diabetic acidosis and coma. { 'er ,həŋ·gər }

air layering [BOT] A method of vegetative propagation, usually of a wounded part, in which the branch or shoot is enclosed in a moist medium until roots develop, and then it is severed and cultivated as an independent plant. { 'er ,lā·ər·iŋ }

air pollution [ECOL] The presence in the outdoor atmosphere of one or more contaminants such as dust, fumes, gas, mist, odor, smoke, or vapor in quantities and of characteristics and duration such as to be injurious to human, plant, or animal life or to property, or to interfere unreasonably with the comfortable enjoyment of life and property. { ¦er pə'lü·shən }

air sac [INV ZOO] One of the large, thin-walled structures associated with the tracheal system of some insects. [VERT ZOO] In birds, any of the small vesicles that are connected with the respiratory system and located in bones and muscles to increase buoyancy. { 'er ,sak }

airsickness [MED] Motion sickness associated with flying due to the effects of acceleration. { 'er,sik·nəs }

air spora [BIOL] Airborne fungus spores, pollen grains, and microorganisms. { 'er ,spȯr·ə }

Aizoaceae [BOT] A family of flowering plants in the order Caryophyllales; members are unarmed leaf-succulents, chiefly of Africa. { ā,īz·ə'wā·sē,ē }

akaryote [CELL MOL] A cell that lacks a nucleus. { ,ā'ka·rē,ōt }

akathisia [MED] Motor restlessness ranging from a feeling of inner disquiet, often localized in the muscles, to an inability to sit still or lie quietly. { ,a·kə'thiz·ē·ə }

akinesia [MED] **1.** Loss or impairment of motor function. **2.** Immobility from any cause. { ¦a·ki'nēzh·ə }

akinete [BOT] A thick-walled resting cell of unicellular and filamentous green algae. { ,a'kī,nēt }

akureyri disease *See* epidemic neuromyasthenia. { ,a·kyu̇'rā·ri di,zēz }

ala [BIOL] A wing or winglike structure. { 'ā·lə }

alalia [MED] Loss of speech. { ¦ā'lāl·yə }

alang-alang *See* cogon. { 'ä,läŋ'ä,läŋ }

alanine [BIOCHEM] $C_3H_7NO_2$ A white, crystalline, nonessential amino acid of the pyruvic acid family. { 'al·ə,nēn }

alar [BIOL] Winglike or pertaining to a wing. { 'ā·lər }

alarm reaction [BIOL] The sum of all nonspecific phenomena which are elicited by sudden exposure to stimuli, which affect large portions of the body, and to which the organism is quantitatively or qualitatively not adapted. { ə'lärm rē'ak·shən }

alarm song [INV ZOO] A stress signal occurring in many families of beetles. { ə'lärm ,sȯŋ }

alate [BIOL] Possessing wings or winglike structures. { 'ā,lāt }

Alaudidae [VERT ZOO] The larks, a family of Oscine birds in the order Passeriformes. { ə'lau̇·də,dē }

albatross [VERT ZOO] Any of the large, long-winged oceanic birds composing the family Diomedeidae of the order Procellariformes. { 'al·bə,trȯs }

albinism [BIOL] The state of having colorless chromatophores, which results in the absence of pigmentation in animals that are normally pigmented. [MED] A hereditary, metabolic disorder transmitted as an autosomal recessive and characterized by the inability to form melanin in the skin, hair, and eyes due to tyrosinase deficiency. { 'al·bə,niz·əm }

albino [BIOL] A human or animal with a congenital deficiency of pigment in the skin, hair, and eyes. [BOT] An abnormal plant with colorless chromatophores. { al'bī·nō }

albomaculatus [BOT] A variegation consisting of irregularly distributed white and green regions on plants due to the mitotic segregation of genes or plastids. { ¦al·bō,ma·kyə'läd·əs }

albomycin [MICROBIO] An antibiotic produced by *Actinomyces subtropicus*; effective against penicillin-resistant pneumococci and staphylococci. { ,al·bō'mīs·ən }

albuginea [HISTOL] A layer of white, fibrous connective tissue investing an organ or other body part. { ,al·byü'jin· ē·ə }

albumen [CELL MOL] The white of an egg, composed principally of albumin. { ,al'byü·mən }

albumin [BIOCHEM] Any of a group of plant and animal proteins which are soluble in water, dilute salt solutions, and 50% saturated ammonium sulfate. { ,al'byü·mən }

albumin-globulin ratio [BIOCHEM] The ratio of the concentrations of albumin to globulin in blood serum. { ,al'byü·mən 'gläb·yə·lən ,rā·shō }

albuminoid *See* scleroprotein. [BIOL] Having the characteristics of albumin. { ,al'byü·mə,nȯid }

albumin suspension test [PATH] A blood-grouping test in which the determination is made by suspending the red cells in diluted bovine albumin instead of in a saline solution. { ,al'byü·mən sə'spen·chən ,test }

albuminuria [MED] The presence of albumin in the urine; usually indicating renal disease. { al ,byü·mə'nu̇r·ē·ə }

albumose [BIOCHEM] A protein derivative formed by the action of a hydrolytic enzyme, such as pepsin. { 'al·byə,mōs }

alburnum *See* sapwood. { al'bər·nəm }

Alcaligenes [MICROBIO] A genus of gram-negative, aerobic rods and cocci of uncertain affiliation; cells are motile, and species are commonly found in the intestinal tract of vertebrates. { ,al·kə'lij·ə,nēz }

alcaptonuria *See* alkaptonuria. { al,kap·tə'nu̇r·ē·ə }

Alcedinidae [VERT ZOO] The kingfishers, a worldwide family of colorful birds in the order Coraciiformes; characterized by large heads, short necks, and heavy, pointed bills. { ,al·sə'din·ə,dē }

Alcidae [VERT ZOO] A family of shorebirds, predominantly of northern coasts, in the order Charadriiformes, including auks, puffins, murres, and guillemots. { 'al·sə,dē }

Alciopidae [INV ZOO] A pelagic family of errantian annelid worms in the class Polychaeta. { ,al·sē'äp·ə,dē }

alcohol dehydrogenase [BIOCHEM] The enzyme that catalyzes the oxidation of ethanol to acetaldehyde. { 'al·kə,hȯl ,dē·hī'drä·jə,nās }

alcoholic [MED] An individual who consumes excess amounts of alcoholic beverages to the extent of being addicted, habituated, or dependent. { ,al·kə'hȯl·ik }

alcoholic fermentation [MICROBIO] The process by which certain yeasts decompose sugars in the absence of oxygen to form alcohol and carbon dioxide; method for production of ethanol, wine, and beer. { ,al·kə'hȯl·ik ,fər·mən 'tā·shən }

alcoholic hepatitis [MED] A frequently occurring form of hepatitis that is caused by excessive ethyl alcohol intake and is characterized by fever, high white blood cell count, and jaundice. { ,al·kə,hȯl·ik ,hep·ə'tīd·əs }

alcoholic hyaline *See* Mallory bodies. { ,al·kə,hȯl·ik 'hī·ə·lən }

alcoholism [MED] Compulsive consumption of and dependence on alcoholic beverages, usually leading to pathology of the digestive and nervous systems. { 'al·kə,hȯ,liz·əm }

Alcyonacea [INV ZOO] The soft corals, an order of littoral anthozoans of the subclass Alcyonaria. { ,al·sī·ə'nās·ē·ə }

Alcyonaria [INV ZOO] A subclass of the Anthozoa; members are colonial cnidarians, most of which are sedentary and littoral. { ,al·sī·ə'ner·ē·ə }

aldehyde dehydrogenase [BIOCHEM] An enzyme that catalyzes the conversion of an aldehyde to its corresponding acid. { 'al·də,hīd ,dē'hī·drə·jə,nās }

aldehyde lyase [BIOCHEM] Any enzyme that catalyzes the nonhydrolytic cleavage of an aldehyde. { 'al·də,hīd 'lī,ās }

alder [BOT] The common name for several trees of the genus *Alnus*. { 'ȯl·dər }

aldolase [BIOCHEM] An enzyme in anaerobic glycolysis that catalyzes the cleavage of fructose 1,6-diphosphate to glyceraldehyde 3-phosphate; used also in the reverse reaction. { 'al·də,lās }

aldosterone [BIOCHEM] $C_{21}H_{28}O_5$ A steroid hormone extracted from the adrenal cortex that functions chiefly in regulating sodium and potassium metabolism. { al'däs·tə,rōn }

aldosteronism [MED] Hypertension induced by excessive secretion of aldosterone. { al'däs·tə·rə,niz·əm }

Aldrich syndrome [MED] A recessive, sex-linked disease characterized by a complex of symptoms, including eczematoid dermatitis, thrombocytopenia, black stool, and a deficiency of immune globulins. { 'ȯl·drich ˌsinˌdrōm }

alecithal [CELL MOL] Referring to an egg without yolk, such as the eggs of placental mammals. { ā'les·ə·thəl }

Alepocephaloidei [VERT ZOO] The slickheads, a suborder of deap-sea teleostean fishes of the order Salmoniformes. { əˌlep·ō·sə·fə'lȯi·deˌī }

aletophyte [ECOL] A weedy plant growing on the roadside or in fields where natural vegetation has been disrupted by humans. { ə'lēd·əˌfīt }

aleukemia [MED] Leukemia in which the white blood cell count is normal or low. { ¦ā·lü'kē·mē·ə }

aleuron [BOT] Protein in the form of grains stored in the embryo, endosperm, or perisperm of many seeds. { 'al·yəˌrän }

aleurospore [MYCOL] A simple terminal or lateral, thick-walled, nondeciduous spore produced by some fungi of the order Moniliales. { ə'lyür·əˌspȯr }

Aleutian disease [VET MED] A disease of mink characterized by accumulations of plasma cells in several organs, hyaline changes in the walls of small arteries, and interstitial fibrosis of the kidneys. { ə'lü·shən dizˌēz }

alewife [VERT ZOO] *Pomolobus pseudoharengus.* A food fish of the herring family that is very abundant on the Atlantic coast. { 'ālˌwīf }

alexia [MED] Loss of the ability to read. { ə'lek·sē·ə }

Alexinic unit [BIOL] A unit for the standardization of blood serum. { ¦a·lek¦sin·ik 'yü·nət }

Aleyrodidae [INV ZOO] The whiteflies, a family of homopteran insects included in the series Sternorrhyncha; economically important as plant pests. { ˌal·ə'räd·əˌdē }

alfalfa [BOT] *Medicago sativa.* A herbaceous perennial legume in the order Rosales, characterized by a deep taproot. Also known as lucerne. { al'fal·fə }

Alfalfa mosaic virus [VIROL] The type species of the genus *Alfamovirus* of the family Bromoviridae. Abbreviated AMV. { al¦fal·fə mō¦zā·ik 'vī·rəs }

Alfalfa mosaic virus group *See* Alfamovirus. { al ¦fal·fə mō¦zā·ik 'vī·rəs ˌgrüp }

Alfamovirus [VIROL] A genus of plant viruses in the family Bromoviridae that is characterized by virions which are either bacilliform or ellipsoidal and contain single-stranded ribonucleic acid genomes; alfalfa mosaic virus is the type species. Also known as Alfalfa mosaic virus group. { al 'fam·əˌvī·rəs }

algae [BOT] General name for the chlorophyll-bearing organisms in the plant subkingdom Thallobionta. { 'al·jē }

algae bloom [ECOL] A heavy growth of algae in and on a body of water as a result of high phosphate concentration from farm fertilizers and detergents. { 'al·jē ˌblüm }

algae wash [ECOL] A shoreline drift consisting almost entirely of filamentous algae. { 'al·jē ˌwash }

algal [BOT] Of or pertaining to algae. { 'al·gəl }

alged malaria *See* falciparum malaria. { 'al·jəd mə'ler·ē·ə }

algesia [PHYSIO] Sensitivity to pain. { al'jēz·ē·ə }

algesimeter [PHYSIO] A device used to determine pain thresholds. { ˌal·jə'sim·əd·ər }

algesiroreceptor [PHYSIO] A pain-sensitive cutaneous sense organ. { ˌal·jə¦si·rō·ri¦sep·tər }

alginate [BOT] An algal polysaccharide that is a major constituent of the cell walls of brown algae. { 'al·jəˌnāt }

algology [BOT] The study of algae. Also known as phycology. [MED] The science and study of phenomena associated with pain. { al 'gäl·ə·jē }

algometer [MED] An instrument for measuring pressure stimuli which produce pain. { al'gä·məd·ər }

algophage *See* cyanophage. { 'al·gəˌfāj }

algor mortis [PATH] Postmortem cooling of the body. { ¦al·gər ¦mȯr·təs }

alien substitution [GEN] The replacement of one or more chromosomes by those from a different species. { ¦āl·ē·ən ˌsəb·stə'tü·shən }

aliesterase [BIOCHEM] Any one of the lipases or nonspecific esterases. { al·ē'es·təˌrās }

alimentary [BIOL] Of or relating to food, nutrition, or diet. { ¦al·ə¦men·trē }

alimentary canal [ANAT] The tube through which food passes; in humans, includes the mouth, pharynx, esophagus, stomach, and intestine. { ¦al·ə¦men·trē kə'nal }

alimentation [BIOL] Providing nourishment by feeding. { ˌal·ə·mən'tā·shən }

aliquot [MED] A representative sample of a larger quantity. { 'al·əˌkwät }

Alismataceae [BOT] A family of flowering plants belonging to the order Alismatales characterized by schizogenous secretory cells, a horseshoe-shaped embryo, and one or two ovules. { əˌliz·mə'tās·ēˌē }

Alismatales [BOT] A small order of flowering plants in the subclass Alismatidae, including aquatic and semiaquatic herbs. { əˌliz·mə'tā·lēz }

Alismatidae [BOT] A relatively primitive subclass of aquatic or semiaquatic herbaceous flowering plants in the class Liliopsida, generally having apocarpous flowers, and trinucleate pollen and lacking endosperm. { əˌliz'mad·əˌdē }

alisphenoid [ANAT] **1.** The bone forming the greater wing of the sphenoid in adults. **2.** Of or pertaining to the sphenoid wing. { ¦al·ə¦sfēˌnȯid }

alivincular [INV ZOO] In some bivalves, having the long axis of the short ligament transverse to the hinge line. { ¦al·ə¦viŋ·kyə·lər }

alkalemia [MED] An increase in blood pH above normal levels. { ˌal·kə'lēm·ē·ə }

alkali denaturation test [PATH] A blood test for the measurement of fetal hemoglobin in terms of its resistance to alkali denaturation. { 'al·kəˌlī dəˌnach·ə'rā·shən ˌtest }

alkali disease [MED] Selenium poisoning.

[VET MED] **1.** Botulism of ducks. **2.** Trembles of cattle. { 'al·kə,lī diz,ēz }

alkaline phosphatase [BIOCHEM] A phosphatase active in alkaline media. { 'al·kə,līn 'fäs·fə,tās }

alkaline tide [PHYSIO] The temporary decrease in acidity of urine and body fluids after eating, attributed by some to the withdrawal of acid from the body due to gastric digestion. { 'al·kə,līn 'tīd }

alkaliphile [PHYSIO] An organism that prefers or is able to withstand an alkaline environment (pH value above 9). { 'al·kə·lə,fīl }

alkalosis [MED] A condition of high blood alkalinity caused either by high intake of sodium bicarbonate or by loss of hydrochloric acid or blood carbon dioxide. { ,al·kə'lō·səs }

alkaptonuria [MED] A hereditary metabolic disorder transmitted as an autosomal recessive in humans in which large amounts of homogentisic acid (alkapton) are excreted in the urine due to a deficiency of homogentisic acid oxidase. Also spelled alcaptonuria. { al,kap·tə'nür·ē·ə }

allachesthesia [MED] A tactile sensation experienced remote from the point of stimulation but on the same side of the body. { 'al·ək·əs'thēzh·ə }

allantoic acid [BIOCHEM] $C_4H_8N_4O_4$ A crystalline acid obtained by hydrolysis of allantoin; intermediate product in nucleic acid metabolism. { ¦al·ən¦tō·ik 'as·əd }

allantoin [BIOCHEM] $C_4H_6N_4O_3$ A crystallizable oxidation product of uric acid found in allantoic and amniotic fluids and in fetal urine. { ə'lan·tə'wən }

allantoinase [BIOCHEM] An enzyme, occurring in nonmammalian vertebrates, that catalyzes the hydrolysis of allantoin. { ə'lan·tə·wə,nās }

allantois [EMBRYO] A fluid-filled, saclike, extraembryonic membrane lying between the chorion and amnion of reptilian, bird, and mammalian embryos. { ə'lan·tə'wəs }

allantoxanic acid [BIOCHEM] $C_4H_3N_3O_4$ An acid formed by oxidation of uric acid or allantoin. { ,a,lan,täk'san·ik 'as·əd }

allanturic acid [BIOCHEM] $C_3H_4N_2O_3$ An acid formed principally by the oxidation of allantoin. { ¦al·ən¦tür·ik 'as·əd }

Alleculidae [INV ZOO] The comb claw beetles, a family of mostly tropical coleopteran insects in the superfamily Tenebrionoidea. { ,ól·ə'kyü·lə,dē }

Allee's principle [GEN] The concept of an intermediate optimal population density by which groups of organisms often flourish best if neither too few nor too many individuals are present. { a'lēz ,prin·sə·bəl }

Alleghenian life zone [ECOL] A biome that includes the eastern mixed coniferous and deciduous forests of New England. { ¦al·ə¦gān·ē·ən 'līf ,zōn }

allele [GEN] One of the alternate forms of a gene at a gene locus on a chromosome. Also known as allelomorph. { ə'lēl }

allele frequency [GEN] The fraction of all alleles at a given locus constituted by a particular allele in a population. Also known as gene frequency. { ə'lēl ¦frē·kwən·sē }

allelic exclusion [GEN] Expression of a single immunoglobulin allele by a B lymphocyte. { ə¦lē,lik iks'klü·zhən }

allelic mutant [GEN] A cell or organism with characters different from those of the parent due to alterations of one or more alleles. { ə¦lēl·ik ¦myüt·ənt }

allelochemic [PHYSIO] Pertaining to a semiochemical that acts as an interspecific agent of communication. { ə¦lē·lō¦kem·ik }

allelomorph *See* allele. { ə'lē·lə,mórf }

allelotropism [BIOL] A mutual attraction between two cells or organisms. { ə¦lē·lō¦trä,piz·əm }

Allen-Doisy unit [BIOL] A unit for the standardization of estrogens. { ¦al·ən ¦dóiz·ē ,yü·nət }

Allen's rule [VERT ZOO] The generalization that the protruding parts of a warm-blooded animal's body, such as the tail, ears, and limbs, are shorter in animals from cold parts of the species range than from warm parts. { 'al·ənz ,rül }

allergen [IMMUNOL] Any antigen, such as pollen, a drug, or food, that induces an allergic state in humans or animals. { 'al·ər,jen }

allergic arteritis [MED] Inflammation of the arterial walls resulting from an allergic state. { ə'lərj·ik ,är·tə'rīd·əs }

allergic dermatitis [MED] Inflammation of the skin following contact of an allergen with sensitized tissue. { ə'lərj·ik 'dər·mə'tīd·əs }

allergic granulomatosis [MED] A form of pulmonary vasculitis, frequently associated with asthma. { ə¦lər·jik ¦gran·yə·lō·mə'tō·səs }

allergic reaction *See* allergy. { ə'lərj·ik rē'ak·shən }

allergic rhinitis *See* hay fever. { ə'lərj·ik rī'nīd·əs }

allergic vasculitis syndrome [MED] A skin disease, possibly immunologic, characterized by ulcers which result from destructive inflammation of underlying blood vessels. { ə'lərj·ik vas·kyü'līd·əs ,sin,drōm }

allergy [MED] A type of antigen-antibody reaction marked by an exaggerated physiologic response to a substance that causes no symptoms in nonsensitive individuals. Also known as allergic reaction. { 'al·ər·jē }

alliance [SYST] A group of related families ranking between an order and a class. { ə'lī·əns }

alligator [VERT ZOO] Either of two species of archosaurian reptiles in the genus *Alligator* of the family Alligatoridae. { 'al·ə,gād·ər }

Alligatorinae [VERT ZOO] A subgroup of the crocodilian family Crocodylidae that includes alligators, caimans, *Melanosuchus*, and *Paleosuchus*. { ,al·ə·gə'tór·ə,nē }

Allium [BOT] A genus of bulbous herbs in the family Liliaceae including leeks, onions, and chives. { 'al·ē·əm }

alloantibody [IMMUNOL] Antibody that reacts with an antigen occurring in a genetically different member of the same species. { ¦a·lō¦ant·i,bäd·ē }

alloantigen *See* isoantigen. { ¦a·lō¦ant·i·jən }

allocheiria [MED] A form of allachesthesia in which the tactile sensation is experienced on the side opposite the one to which the stimulus was applied; seen in tabes dorsalis and other conditions. { ˌa·lō'kī·rē·ə }

allochoric [BOT] Describing a species that inhabits two or more closely related communities, such as forest and grassland, in the same region. { ˌa·lə'kȯr·ik }

allochthonous [ECOL] Pertaining to organisms or organic sediments in a given ecosystem that originated in another system. { ə'läk·thə·nəs }

allodynia [MED] Painful response to a stimulus that does not normally elicit pain. { ˌa·lə'dī·nē·ə }

Alloeocoela [INV ZOO] An order of platyhelminthic worms of the class Turbellaria distinguished by a simple pharynx and a diverticulated intestine. { əˌlē·ə'sēl·ə }

allogeneic [IMMUNOL] Referring to a transplant made to a different genotype within the same species. Also spelled allogenic. { ¦al·ə·jə¦nē·ik }

allogeneic graft *See* allograft. { ¦al·ə·jə¦nē·ik 'graft }

allogenic [ECOL] Caused by external factors, as in reference to the change in habitat of a natural community resulting from drought. [IMMUNOL] *See* allogeneic. { ¦a·lə¦jen·ik }

allograft [BIOL] Graft from a donor transplanted to a genetically dissimilar recipient of the same species. Also known as allogeneic graft. { 'a·lōˌgraft }

Allogromiidae [INV ZOO] A little-known family of protozoans in the order Foraminiferida; adults are characterized by a chitinous test. { ˌa·lə·grə'mī·əˌdē }

Allogromiina [INV ZOO] A suborder of marine and fresh-water protozoans in the order Foraminiferida characterized by an organic test of protein and acid mucopolysaccharide. { ˌa·lə·grə'mī·ə·nə }

Alloionematoidea [INV ZOO] A superfamily of parasitic nematodes belonging to the order Rhabditida, having either no lips or six small amalgamated lips, and a rhabditiform esophagus with a weakly developed valve in the posterior bulb. { əˌlȯi·ōˌnem·ə'tȯid·ē·ə }

allometry [BIOL] **1.** The quantitative relation between a part and the whole or another part as the organism increases in size. Also known as heterauxesis; heterogony. **2.** The quantitative relation between the size of a part and the whole or another part, in a series of related organisms that differ in size. { ə'läm·ə·trē }

allomone [PHYSIO] A chemical produced by an organism which induces in a member of another species a behavioral or physiological reaction favorable to the emitter; may be mutualistic or antagonistic. { 'a·ləˌmōn }

allomorphosis [EVOL] Allometry in phylogenetic development. { ¦al·ō·mȯr'fō·səs }

Allomyces [MYCOL] A genus of aquatic phycomycetous fungi in the order Blastocladiales characterized by basal rhizoids, terminal hyphae, and zoospores with a single posterior flagellum. { a·lō'mīˌsēz }

alloparapatric speciation [EVOL] A mode of gradual speciation in which new species originate through populations that are initially allopatric but eventually become parapatric before effective reproductive isolation has evolved. { ¦al·əˌpa·rə¦pa·trik ˌspē·sē'ā·shən }

allopathy [MED] A system of medicine that employs remedies whose effects are unlike those of the disease, in contrast to homeopathy. { ə'läp·ə·thē }

allopatric [ECOL] Referring to populations or species that occupy naturally exclusive, but usually adjacent, geographical areas. { ¦a·lō¦pa·trik }

allopatric speciation [ECOL] Differentiation of populations in geographical isolation to the point where they are recognized as separate species. { ¦al·ō¦pa·trik ˌspē·sē'ā·shən }

allopelagic [ECOL] Relating to organisms living at various depths in the sea in response to influences other than temperature. { ¦a·lō·pə'laj·ik }

allophene [GEN] A mutant phenotype that can revert to a normal phenotype if it is transplanted to a wild-type host. { 'al·əˌfēn }

allophore [HISTOL] A chromatophore which contains a red pigment that is soluble in alcohol; found in the skin of fishes, amphibians, and reptiles. { 'a·lōˌfȯr }

alloploid *See* allopolyploid. { 'al·əˌplȯid }

allopolyploid [GEN] An organism or strain arising from a combination of chromosome sets from two diploid organisms belonging to different species. Also known as alloploid. { ˌa·lə'päl·əˌplȯid }

allopregnancy [IMMUNOL] A pregnancy in which the male partner is allogeneic with respect to the female. { ¦al·ō'preg·nən·sē }

allopurinol [PHARM] $C_5H_4ON_4$ The compound 4-hydroxypyrazolo-3,4-*d*-pyrimidine; inhibits xanthine oxidase, an enzyme required for uric acid formation. { ˌa·lə'pyür·əˌnȯl }

all-or-none law [NEUROSCI] The principle that transmission of a nerve impulse is either at full strength or not at all. { ¦ȯl ər ¦nən ˌlȯ }

allosome [GEN] **1.** Sex chromosome. **2.** Any atypical chromosome. { 'a·lōˌsōm }

allosteric activation [BIOCHEM] The increase in an enzyme's activity that occurs when an allosteric effector binds to its specific regulatory site on the enzyme. Also known as positive allosteric control. { ¦a·lə¦stir·ik ˌak·tə'vāshən }

allosteric control *See* feedback inhibition. { ¦a·lə¦stir·ik kən'trōl }

allosteric effector [BIOCHEM] A small molecule that reacts either with a nonbinding site of an enzyme molecule, or with a protein molecule, and causes a change in the function of the molecule. Also known as allosteric modulator. { ¦a·lə¦stir·ik ə'fek·tər }

allosteric enzyme [BIOCHEM] Any of the regulatory bacterial enzymes, such as those involved in end-product inhibition. { ¦a·lə¦stir·ik 'en ˌzīm }

allosteric modulator *See* allosteric effector. { ¦a·lə¦stir·ik 'mäd·yəˌlād·ər }

allosteric site [BIOCHEM] The inactive (or less active) region of an enzyme molecule. { ¦a·lə¦ster·ik 'sīt }

allosteric transition [BIOCHEM] A reversible exchange of one base pair for another on a protein molecule that alters the properties of the active site and changes the biological activity of the protein. { ¦a·lə¦stir·ik tranz'ish·ən }

allostery [BIOCHEM] The property of an enzyme able to shift reversibly between an active and an inactive configuration. { 'a·lōˌstir·ē }

allotetraploid *See* amphidiploid. { ˌa·lō'te·trə ˌplȯid }

Allotriognathi [VERT ZOO] An equivalent name for the Lampridiformes. { ə'lä·trē'äg·nəˌthī }

allotype [IMMUNOL] Inherited variations in the genes coding for certain amino acid sequences in the constant region of the immunoglobulin molecules. [SYST] A paratype of the opposite sex to the holotype. { 'a·ləˌtīp }

alloxan [BIOCHEM] $C_4H_2N_2O_4$ Crystalline oxidation product of uric acid; induces diabetes experimentally by selective destruction of pancreatic beta cells. Also known as mesoxyalyurea. { ə'läk·sən }

allozygote [GEN] An individual who has different mutations in the two alleles at a given locus. { ˌal·ə'zīˌgōt }

allozyme [GEN] One of two or more forms of an enzyme that are specified by allelic genes. { 'al·əˌzīm }

allspice [BOT] The dried, unripe berries of a small, tropical evergreen tree, *Pimenta officinalis*, of the myrtle family; yields a pungent, aromatic spice. { 'ȯlˌspīs }

alm [ECOL] A meadow in alpine or subalpine mountain regions. { älm }

almond [BOT] *Prunus amygdalus*. A small deciduous tree of the order Rosales; it produces a drupaceous edible fruit with an ellipsoidal, slightly compressed nutlike seed. { 'ä·mənd }

Almquist unit [BIOL] A unit for the standardization of vitamin K. { 'ämˌkwist ˌyü·nət }

aloe [PHARM] The dried resinous juice extracted from the leaves of the genus *Aloe*, especially A. *vulgaris* of the West Indies and A. *perryi* of Africa; used in purgative mixtures. { 'a·lō }

alopecia [MED] Loss of hair; baldness. { ˌa·lə'pē·shə }

alopecia areata [MED] A type of alopecia that is characterized as an autoimmune disorder and usually presents with one or many oval, slightly erythematous, asymptomatic patches of hair loss. { ˌa·lə'pē·shə ˌa·rē'ad·ə }

Alopiidae [VERT ZOO] A family of pelagic isurid elasmobranchs commonly known as thresher sharks because of their long, whiplike tail. { ˌal·ə'pī·əˌdē }

alpaca [VERT ZOO] *Lama pacos*. An artiodactyl of the camel family (Camelidae); economically important for its long, fine wool. { al'pak·ə }

alpage [ECOL] A summer grazing area composed of natural plant pasturage in upland or mountainous regions. { 'al·pəj }

alpestrine [ECOL] Referring to organisms that live at high elevation but below the timberline. Also known as subalpine. { ˌal'pes·trən }

alpha-adrenergic receptor *See* alpha receptor. { ¦al·fə ˌad·rə¦nər·jik ri'sep·tər }

alpha cell [HISTOL] Any of the acidophilic chromophiles in the anterior lobe of the adenohypophysis. { 'al·fə ˌsel }

alpha fetoprotein [IMMUNOL] A serum protein that has become associated with the detection of certain types of cancer and fetal abnormalities. Abbreviated AFP. { 'al·fə ¦fēd·ō'prōˌtēn }

alpha globulin [BIOCHEM] A heterogeneous fraction of serum globulins containing the proteins of greatest electrophoretic mobility. { 'al·fə 'gläb·yə·lən }

alpha helix [CELL MOL] A spatial configuration of the polypeptide chains of proteins in which the chain assumes a helical form, 0.54 nanometer in pitch, 3.6 amino acids per turn, presenting the appearance of a hollow cylinder with radiating side groups. { 'al·fə 'hē·liks }

alpha hemolysis [MICROBIO] Partial hemolysis of red blood cells with green discoloration in a blood agar medium by certain hemolytic streptococci. { 'al·fə hi'mäl·ə·səs }

alpha receptor [CELL MOL] Any of a group of receptors on cell membranes that are thought to be associated with vasoconstriction, relaxation of intestinal muscle, and contraction of the nictitating membrane, iris dilator muscle, smooth muscle of the spleen, and muscular layer of the uterine wall. Also called alpha-adrenergic receptor. { ¦al·fə ri¦sep·tər }

alpha rhythm [PHYSIO] An electric current from the occipital region of the brain cortex having a pulse frequency of 8 to 13 per second; associated with a relaxed state in normal human adults. { 'al·fə ˌrith·əm }

alpha taxonomy [SYST] The level of taxonomic study concerned with characterizing and naming species. { 'al·fə tak·'sän·ə·mē }

Alpheidae [INV ZOO] The snapping shrimp, a family of decapod crustaceans that is included in the section Caridea. { ˌal'fē·əˌdē }

alpine [ECOL] Any plant native to mountain peaks or boreal regions. { 'alˌpīn }

alpine tundra [ECOL] Large, flat or gently sloping, treeless tracts of land above the timberline. { 'alˌpīn 'tənˌdrə }

Alport's syndrome [MED] A very rare genetic disease of the glomeruli that results in glomerular scarring and eventual renal failure within the second or third decade of life. { 'alˌpȯrts ˌsinˌdrōm }

alsike clover [BOT] *Trifolium hybridium*. A species of clover with pink or white flowers that is grown for forage; native to Alsike, Sweden. { 'alˌsik 'klō·ver }

alteration enzyme [IMMUNOL] A protein of bacteriophage T4 that is injected into a host bacterium along with the deoxyribonucleic acid of the bacteriophage and that modifies the ribonucleic acid polymerase of the host so that the ribonucleic acid is unable to initiate transcription at host promoters. { ȯl·tə'rā·shən ˌen‚zīm }

alternate [BOT] **1.** Of the arrangement of leaves on opposite sides of the stem at different levels. **2.** Of the arrangement of the parts of one whorl between members of another whorl. { 'ȯl·tər·nət }

alternate phyllotaxy [BOT] **1.** An arrangement of leaves that occur individually at nodes and on opposite sides of the stem. **2.** A spiral arrangement of leaves on a stem, with one leaf at a node. { 'ȯl·tər·nət 'fil·ə‚tak·sē }

alternation of generations *See* metagenesis. { ˌȯl·tər'nā·shən əv ˌjen·ə'rā·shənz }

alternative RNA splicing [CELL MOL] A process in gene expression that enables the production of multiple forms of messenger ribonucleic acid (mRNA) from a single RNA transcript, thus enabling the production of multiple forms of protein from one gene. { ȯl¦tər·nəd·iv ¦är¦en'ā ˌsplīs·iŋ }

alterne [ECOL] A community exhibiting alternating dominance with other communities in the same area. { 'ȯl‚tərn }

altherbosa [ECOL] Communities of tall herbs, usually succeeding where forests have been destroyed. { ¦al·thər¦bōs·ə }

altitude acclimatization [PHYSIO] A physiological adaptation to reduced atmospheric and oxygen pressure. { 'al·tə‚tüd ə‚klī·məd·ə'zā·shən }

altitude sickness [MED] In general, any sickness brought on by exposure to reduced oxygen tension and barometric pressure. { 'al·tə‚tüd ˌsik·nəs }

altitudinal vegetation zone [ECOL] A geographical band of physiognomically similar vegetation correlated with vertical and horizontal gradients of environmental conditions. { ¦al·tə¦tüd·ən·əl ˌvej·ə'tā·shən ˌzōn }

altricial [VERT ZOO] Pertaining to young that are born or hatched immature and helpless, thus requiring extended development and parental care. { ˌal'trish·əl }

alula [ZOO] **1.** Digit of a bird wing homologous to the thumb. **2.** *See* calypter. { 'al·yə·lə }

aluminosis [MED] A lung disorder caused by inhalation of alumina dust. { ə‚lüm·ə'nō·səs }

aluminum acetylsalicylate [PHARM] $[C_6H_4(OCOCH_3)\text{-}(COO)]_2AlOH$ White, odorless granules or powder; decomposes on melting and in alkali, hydroxides, and carbonates; used in medicine. Also known as aluminum aspirin. { ə'lüm·ə·nəm ə¦sēd·əl·sə¦lis·ə‚lāt }

aluminum therapy [MED] Therapy intended mainly for prevention rather than treatment of silicosis; provides for inhalation of powdered aluminum and alumina (Al_2O_3) dust by miners in the change house. { ə'lüm·ə·nəm 'ther·ə·pē }

alvar [ECOL] Dwarfed vegetation characteristic of certain Scandinavian steppelike communities with a limestone base. { 'al‚vär }

alveator [INV ZOO] A type of pedicellaria in echinoderms. { 'al·vē‚ād·ər }

alveolar [BIOL] Of or relating to an alveolus. { al'vē·ə·lər }

alveolar-capillary block syndrome [MED] Arterial oxygen deficiency due to improper functioning of the membranes between the alveoli and capillaries. { al'vē·ə·lər ¦kap·ə‚ler·ē ¦bläk ˌsin ˌdrōm }

alveolar gland *See* acinous gland. { al'vē·ə·lər ˌgland }

alveolar oxygen pressure [PHYSIO] The oxygen pressure in the alveoli; the value is about 105 mmHg. { al'vē·ə·lər 'äk·sə·jən ˌpresh·ər }

alveolar process [ANAT] The ridge of bone surrounding the alveoli of the teeth. { al¦vē·ə·lər ¦prä‚ses }

alveolar ridge [ANAT] The bony remains of the alveolar process of the maxilla or mandible. { al¦vē·ə·lər 'rij }

alveolated cell *See* epithelioid cell. { al'vē·ə‚lād·əd ˌsel }

alveolitoid [INV ZOO] A type of tabulate coral having a vaulted upper wall and a lower wall parallel to the surface of attachment. { al'vē·ə·lə‚tȯid }

alveoloplasty [MED] Surgical shaping of the alveolar ridges in preparation for dentures or after the removal of several teeth. { al'vē·ə‚lō‚plas·tē }

alveolus [ANAT] **1.** A tiny air sac of the lung. **2.** A tooth socket. **3.** A sac of a compound gland. { al'vē·ə·ləs }

Alydidae [INV ZOO] A family of hemipteran insects in the superfamily Coreoidea. { ə'lid·ə‚dē }

alymphocytic agammaglobulinemia [IMMUNOL] A type of immune globulin deficiency usually transmitted as an autosomal recessive that is characterized by a complete absence of lymphocytes; affected infants cannot produce a humoral or a cell-mediated immune response and are severely predisposed to infections, and usually die within the first year of life. { ˌā·lim·fə¦sid·ik ˌā‚gam·ə‚gläb·yü‚li'nē·mē·ə }

alymphocytosis [MED] Absence or deficiency of blood lymphocytes. { ¦ā‚lim·fō‚sī'tō·səs }

Alysiella [MICROBIO] A genus of bacteria in the family Simonsiellaceae; cells are arranged in pairs within the filaments. { ə‚lis·ē'el·ə }

Alzheimer's disease [MED] A progressive neurodegenerative disease characterized by loss of function and death of nerve cells in several areas of the brain, leading to loss of cognitive function such as memory and language; the most common cause of senile dementia. Abbreviated AD. { 'älts‚hī·mərz di‚zēz }

amacrine cell [ANAT] An interneuron located in the inner plexiform layer of the vertebrate retina that influences retinal signal processing in response to visual stimuli at the level of contact between the bipolar and ganglion cells. { 'am·ə‚krēn ˌsel }

amantadine [PHARM] $C_{10}H_{17}N$ A symmetrical amine used as a viral chemoprophylactic because it selectively inhibits certain myxoviruses; also of value in the treatment of parkinsonism. { ə'man·tə,dēn }

amanthophilous [BOT] Of plants having a habitat in sandy plains or hills. { ,a·mən'thä·fə·ləs }

Amaranthaceae [BOT] The characteristic family of flowering plants in the order Caryophyllales; they have a syncarpous gynoecium, a monochlamydeous perianth that is more or less scarious, and mostly perfect flowers. { ,a·mə·rə'thā·sē,ē }

amaranth [BOT] An annual plant (seldom perennial) of the genus *Amaranthus* that is distributed worldwide in warm and humid regions and is distinguished by small chaffy flowers (arranged in dense, green or red, monoecious or dioecious inflorescences) and by dry, membranous, indehiscent, one-seeded fruit. { 'am·ə·ranth }

Amaryllidaceae [BOT] The former designation for a family of plants now included in the Liliaceae. { ,a·mə,ri·lə'dā·sē,ē }

amatoxin [BIOCHEM] Any of a group of toxic peptides that selectively inhibit ribonucleic acid polymerase in mammalian cells; produced by the mushroom *Amanita phalloides*. { ¦am·ə¦täk·sən }

amaurosis [MED] Total or partial blindness. { ,a,mò'rō·səs }

amaurotic familial idiocy [MED] A hereditary condition, transmitted as an autosomal recessive, predominantly in Jewish children, characterized by blindness, muscular weakness, and subnormal mental development; when the onset is in infancy, the disease is known commonly as Tay-Sachs disease. { ¦a,mò¦räd·ik fə'mil·yəl 'id·ē·ə·sē }

amavadin [BIOCHEM] A vanadium coordination complex and natural product obtained from the poisonous mushroom *Amanita muscaria*. { ə'mav·ə,din }

amber codon [VIROL] The polypeptide chain-termination messenger-RNA codon UAG, which brings about the termination of protein translation. { 'am·bər 'kō,dän }

ambergris [PHYSIO] A fatty substance formed in the intestinal tract of the sperm whale; used in the manufacture of perfume. { 'am·bə,gris }

amber mutation [GEN] Alteration of a codon to UAG, one of the three codons that result in premature polypeptide chain termination in all living organism. { 'am·bər myü'tā·shən }

ambidextrous [PHYSIO] Capable of using both hands with equal skill. { ¦am·bə¦dek·strəs }

ambigenous [BOT] Of a perianth whose outer leaves resemble the calyx while the inner leaves resemble the corolla. { ,am'bi·jən·əs }

ambiguous codon [GEN] A codon capable of coding for more than one amino acid sequence. { am'big·yə·wəs 'kō,dän }

ambisexual [MED] An individual having undifferentiated primordia of both sexes. { ¦am·bi ¦sek·shə·wəl }

ambitus [BIOL] The periphery or external edge, as of a mollusk shell or leaf. { 'am·bə·təs }

amblyopia [MED] Dimness of vision, especially that not due to refractive errors or organic disease of the eye; may be congenital or acquired. { am'blē'ōp·ē·ə }

Amblyopsidae [VERT ZOO] The cave fishes, a family of actinopterygian fishes in the order Percopsiformes. { am·blē'äp·sə,dē }

Amblyopsiformes [VERT ZOO] An equivalent name for the Percopsiformes. { am·blē¦äp·sə'fòr·mēz }

Amblypygi [INV ZOO] An order of chelicerate arthropods in the class Arachnida, commonly known as the tailless whip scorpions. { am 'blip·i,jī }

amboceptor [IMMUNOL] According to P. Ehrlich, an antibody present in the blood of immunized animals which contains two specialized elements: a cytophil group that unites with a cellular antigen, and a complementophil group that joins with the complement. { 'am·bō ,sep·tər }

amboceptor unit [BIOL] A unit for the standardization of blood serum. { 'am·bō,sep·tər ,yü·nət }

ambulacrum [INV ZOO] In echinoderms, any of the radial series of plates along which the tube feet are arranged. { ,am·byə'lak·rəm }

ambulatorial [ZOO] **1.** Capable of walking. **2.** In reference to a forest animal, having adapted to walking, as opposed to running, crawling, or leaping. { ,am·byə·lə'tòr·ē·əl }

Ambystoma [VERT ZOO] A genus of common salamanders; the type genus of the family Ambystomatidae. { am'bis·tə·mə }

Ambystomatidae [VERT ZOO] A family of urodele amphibians in the suborder Salamandroidea; neoteny occurs frequently in this group. { am,bis·tə'mad·ə,dē }

Ambystomoidea [VERT ZOO] A suborder to which the family Ambystomatidae is sometimes elevated. { am,bis·tə'mòid·ē·ə }

AMCHA *See* tranexamic acid.

ameba [INV ZOO] The common name for a number of species of naked unicellular protozoans of the order Amoebida; an example is a member of the genus *Amoeba*. { ə'mē·bə }

amebiasis [MED] A parasitic disease of humans caused by the ameba *Entamoeba histolytica*, characterized by clinical-pathological intestinal manifestations, including an acute dysentery phase. Also known as amebic dysentery. { ,am·ē'bī·ə·səs }

amebic abscess [MED] Liquefactive necrosis of the brain and liver, without suppuration, caused by amebas, usually *Entamoeba histolytica*. { ə'mē·bik 'ab,ses }

amebic dysentery *See* amebiasis. { ə'mē·bik 'dis·ən,ter·ē }

amebocyte [INV ZOO] One of the wandering ameboid cells in the tissues and fluids of many invertebrates that function in assimilation and excretion. { ə'mēb·ə,sīt }

ameboid movement [CELL MOL] A type of cellular locomotion involving the formation of pseudopodia. { ə'mēb·ȯid 'müv·mənt }

ameiosis [GEN] Nonreduction of chromosome number due to suppression failure of one of the two meiotic divisions, resulting n failure to reduce the chromosome complement from diploid to haploid, as in parthenogenesis. { ,ā·mī'ō·səs }

Ameiuridae [VERT ZOO] A family of North American catfishes belonging to the suborder Siluroidei. { ,a·mī'yür·ə,dē }

ameloblast [EMBRYO] One of the columnar cells of the enamel organ that form dental enamel in developing teeth. { 'am·ə·lō,blast }

ameloblastic odontoma [MED] A neoplasm of epithelial and mesenchymal odontogenic tissue. Also known as odontoblastoma. { ¦am·ə·lō ¦blas·tik ,ō,dän'tō·mə }

ameloblastoma [MED] An epithelial tumor associated with the enamel organ; cells of basal layers resemble the ameloblast. Also known as adamantinoma. { ,am·ə·lō·bla'stō·mə }

amelogenesis imperfecta [MED] An inherited dental disorder that causes defective formation of tooth enamel. { ,am·ə·lō,jen·ə·səs ,im·pər'fek·tə }

amenorrhea [MED] Absence of menstruation due to either normal or abnormal conditions. { ¦ā,men·ə'rē·ə }

amensalism [ECOL] A type of interaction that is neutral to one species but harmful to a second species. { ā'men·sə,liz·əm }

ament [BOT] A catkin. [MED] A person with congenital mental deficiency; an idiot. { 'ā,ment }

amentia [MED] Congenital subnormal intellectual development. { ,ā'mensh·ə }

Amera [INV ZOO] One of the three divisions of the phylum Vermes proposed by O. Bütschli in 1910 and given the rank of a subphylum. { 'am·ə·rə }

American boreal faunal region [ECOL] A zoogeographic region comprising marine littoral animal communities of the coastal waters off east-central North America. { ə'mer·ə·kən 'bȯr·ē·əl 'fȯn·əl ,rē·jən }

American-Egyptian cotton [BOT] A type of cotton developed by hybridization of Egyptian and American plants. { ə'mer·ə·kən i¦jip·shən 'kät·ən }

American lion *See* puma. { ə'mer·ə·kən 'lī·ən }

American mucocutaneous leishmaniasis [MED] A form of leishmaniasis caused by *Leishmania braziliensis*, transmitted by sandflies of the genus *Phlebotomus*, and characterized by skin ulcers and ulceration and necrosis of the mucosa of the mouth and nose. Also known as South American leishmaniasis. { ə'mer·ə·kən ¦myü·kō,kyü'tān·e·əs ,lēsh·mən'ī·ə·səs }

American spotted fever *See* Rocky Mountain spotted fever. { ə'mer·ə·kən ,späd·əd 'fēv·ər }

Amerosporae [MYCOL] A spore group of the Fungi Imperfecti characterized by one-celled or threadlike spores. { ,am·ə'räs·pə,rē }

ametabolous metamorphosis [INV ZOO] A growth stage of certain insects characterized by an increase in size without distinct external changes. { ¦ā·mə¦tab·ə·ləs ,med·ə'mȯr·fə·səs }

amethopterin [PHARM] $C_{20}H_{22}N_8O_5 \cdot H_2O$ An antimetabolite effective as a folic acid antagonist and used for treatment of acute and subacute leukemia. Also known as methotrexate. { ,am·ə'thäp·tə,rin }

ametoecious [ECOL] Of a parasite that remains with the same host. { ,am·ə'tēsh·əs }

ametropia [MED] Any deficiency in the refractive ability of the eye that causes an unfocused image to fall on the retina. { ,a·mə'trōp·ē·ə }

amictic [INV ZOO] **1.** In rotifers, producing diploid eggs that are incapable of being fertilized. **2.** Pertaining to the egg produced by the amictic female. { ə'mik·tik }

amidase [BIOCHEM] Any enzyme that catalyzes the hydrolysis of nonpeptide C=N linkages. { 'am·ə,dās }

amidohydrolase [BIOCHEM] An enzyme that catalyzes deamination. { ə¦mē·dō¦nī·drə,lās }

Amiidae [VERT ZOO] A family of actinopterygian fishes in the order Amiiformes represented by a single living species, the bowfin (*Amia calva*). { ə'mī·ə,dē }

Amiiformes [VERT ZOO] An order of actinopterygian fishes characterized by an abbreviate heterocercal tail, fusiform body, and median fin rays. { ə,mī·ə'fȯr,mēz }

amine oxidase [BIOCHEM] An enzyme that catalyzes the oxidation of tyramine and tryptamine to aldehyde. { ə'mēn 'äk·sə,dās }

aminoacetic acid *See* glycine. { ə'mē,no,ə'sēd·ik 'as·əd }

amino acid [BIOCHEM] Any of the organic compounds that contain one or more basic amino groups and one or more acidic carboxyl groups and that are polymerized to form peptides and proteins; only 20 of the more than 80 amino acids found in nature serve as building blocks for proteins; examples are tyrosine and lysine. { ə'mē,nō 'as·əd }

aminoaciduria [MED] A group of disorders in which excess amounts of amino acids are excreted in the urine; caused by abnormal protein metabolism. { ə¦mē·nō,as·ə¦dür·ē·ə }

aminoacyl tRNA [CELL MOL] A molecular complex formed during protein synthesis by the linkage of a transfer ribonucleic acid molecule (tRNA) with its corresponding amino acid. { ə¦mē·nō¦as·əl ¦tē,är,en'ā }

aminoacyl-tRNA synthetase [CELL MOL] An enzyme catalyzing the linkage of a transfer ribonucleic acid (tRNA) molecule to its corresponding amino acid during protein synthesis. { ə¦mē·nō¦as·əl ¦tē,är,en¦ā 'sin·thə,tās }

4-aminoantipyrine [PHARM] $C_{11}H_{13}N_3O$ Pale yellow crystals with a melting point of 109°C; soluble in water, benzene, and ethanol; used as an antipyretic and analgesic. { ¦fȯr ə¦mē·nō,an·tē¦pī,rēn }

***para*-aminobenzoic acid** [BIOCHEM] $C_7H_7O_2N$ A yellow-red, crystalline compound that is part of the folic acid molecule; essential in metabolism of certain bacteria. Abbreviated PABA. { ¦par·ə ə¦mē·nō,ben¦zō·ik 'as·əd }

amino diabetes [MED] A congenital disorder characterized by excessive quantities of amino acids, glucose, and phosphate in the urine, resulting from deficient resorption in the proximal convoluted tubules of the kidney. Also known as Fanconi's syndrome. { ə'mē·nō ,dī·ə'bēd·ēz }

2-aminoethanethiol [PHARM] C_2H_7NS Crystals having a disagreeable odor and a melting point of 97–98.5°C; used in radiation sickness therapy. { ¦tü ə¦mē·nō,eth·ən¦eth·ē,ól }

α-aminohydrocinnamic acid *See* phenylalanine. { ¦al·fə ə¦mē·nō¦hī·drō·sə¦nam·ik 'as·əd }

4-amino-3-hydroxybutyric acid [PHARM] H_2N-$CH_2CH(OH)CH_2COOH$ Crystals that decompose at 218°C; soluble in water; used as an anticonvulsant. { ¦fór ə¦me·nō ¦thrē ,hī¦dräk·sē,byü¦tir·ik 'as·əd }

***para*-(aminomethyl)benzenesulfonamide** [PHARM] $C_7H_{10}N_2O_2S$ Crystals with a melting point of 151–152°C; soluble in dilute alkali and acid; used as an antibacterial agent. { ¦par·ə ə¦mē·nō¦meth·əl ¦ben,zēn,səl¦fän·ə,mīd }

2-amino-5-nitrothiazole [PHARM] $C_3H_3N_3O_2S$ A fluffy, greenish to orange-yellow powder with a slightly bitter taste; decomposes at 202°C; used as a veterinary drug in turkeys and chickens for histomonads and for trichomoniasis in pigeons. { ¦tü ə'mē·nō ¦fīv ,nī·trō'thī·ə,zōl }

6-aminopenicillanic acid [PHARM] $C_8H_{12}N_2O_3S$ Nonhygroscopic crystals that decompose at 209°C; used in the manufacture of synthetic penicillins. Abbreviated 6-APA. { ¦siks ə¦me·nō ,pen·ə·sil'an·ik 'as·əd }

aminopeptidase [BIOCHEM] An enzyme which catalyzes the liberation of an amino acid from the end of a peptide having a free amino group. { ə,mē·nō'pep·tə,dās }

α-amino-β-phenylpropionic acid *See* phenylalanine. { ¦al·fe ə¦mē·nō ¦bād·ə ¦fen·əl¦prō·pē¦an·ik 'as·əd }

aminophylline [PHARM] $C_{16}H_{24}N_{10}O_4$ A drug in the form of white or slightly yellowish, water-soluble granules or powder, used as a smooth-muscle relaxant, myocardial stimulant, and diuretic. { ə,mē·nō'fī,lēn }

aminopolypeptidase [BIOCHEM] A proteolytic enzyme that cleaves polypeptides containing either a free amino group or a basic nitrogen atom having at least one hydrogen atom. { ə¦mē·nō,pä·lə'pep·tə,dās }

aminoprotease [BIOCHEM] An enzyme that hydrolyzes a protein and unites with its free amino group. { ə¦mē·nō'prō·tē,ās }

aminopterin [PHARM] $C_{19}H_{20}N_8O_5 \cdot 2H_2O$ A yellow crystalline acid which is similar to folic acid and is used clinically as an antagonist of folic acid. { ,a·mə'näp·tə·rən }

aminopyrine [PHARM] $C_{13}H_{17}N_3O$ Leafletlike crystals with a melting point of 107–109°C; soluble in alcohol, benzene, and chloroform; used as an antipyretic and analgesic in humans and animals. { ə'mē·nō'pī,rēn }

***para*-aminosalicylic acid** [PHARM] $C_7H_7NO_3$ White, crystalline drug used with other drugs in the treatment of tuberculosis. Abbreviated PAS. { ¦par·ə ə¦me·nō¦sal·ə¦sil·ik 'as·əd }

amino sugar [BIOCHEM] A monosaccharide in which a non-glycosidic hydroxyl group is replaced by an amino or substituted amino group; an example is D-glucosamine. { ə¦mē·nō ¦shu̇g·ər }

amino terminal [BIOCHEM] The end part of a polypeptide chain which contains a free alpha-amino group. { ə¦mē·nō ¦tərm·ən·əl }

aminotransferase *See* transaminase. { ə,mē·nō 'tranz·fə,rās }

amitosis [CELL MOL] Cell division by simple fission of the nucleus and cytoplasm without chromosome differentiation. { ¦ā,mī'tō·səs }

ammocoete [ZOO] A protracted larval stage of lampreys. { 'a·mə,sēt }

ammocolous [ECOL] Describing plants having a habitat in dry sand. { ə'mä·kə·ləs }

Ammodiscacea [INV ZOO] A superfamily of foraminiferal protozoans in the suborder Textulariina, characterized by a simple to labyrinthic test wall. { ,a·mə,dis'kāsh·ə }

Ammodytoidei [VERT ZOO] The sand lances, a suborder of marine actinopterygian fishes in the order Perciformes, characterized by slender, eel-shaped bodies. { ə,mäd·i'tói·dē,ī }

ammonifiers [ECOL] Fungi, or actinomycetous bacteria, that participate in the ammonification part of the nitrogen cycle and release ammonia (NH_3) by decomposition of organic matter. { ə'män·ə,fī·ərz }

ammonium sulfate fractionation [BIOCHEM] A protein purification technique that utilizes the varying precipitation rates of different proteins in ammonium sulfate to obtain considerable separation. { ə¦mōn·ē·əm ¦səl,fāt ,frak·shə'nā·shən }

ammonotelic [BIOL] Pertaining to the excretion of nitrogen primarily as ammonium ion, $[NH_4^+]$. { ə¦mä·nō'tēl·ik }

Ammotheidae [INV ZOO] A family of marine arthropods in the subphylum Pycnogonida. { ,a·mə'thē·ə,dē }

amnesia [MED] The pathological loss or impairment of memory brought about by psychogenic or physiological disturbances. { am'nēzh·ə }

amnesic aphasia [MED] Loss of memory for the appropriate names of objects, conditions, or relations, accompanied by fragmented or hesitant speech. { am'nēz·ik ə'fāzh·ə }

amnicolous [ECOL] Describing plants having a habitat on sandy riverbanks. { ,am·nə'kä·ləs }

amniocentesis [MED] A procedure during pregnancy by which the abdominal wall and fetal membranes are punctured with a cannula to withdraw amniotic fluid. { ¦am·nē·ō,sen'tē·səs }

amniochorionic [EMBRYO] Relating to both amnion and chorion. { ¦am·nē·ō,kór·ē'än·ik }

amniogenesis [EMBRYO] The development or

formation of the amnion. { ¦am·nē·ō'jen·ə·səs }

amniography [MED] Radiography of the fetus after injection of radiopaque material into the amniotic sac. { ˌam·nē'äg·rə·fē }

amnioma [MED] A broad flat tumor of the skin resulting from adhesion of the amnion after birth. { ˌam·nē'ō·mə }

amnion [EMBRYO] A thin extraembryonic membrane forming a closed sac around the embryo in birds, reptiles, and mammals. { 'am·nēˌän }

amnionitis [MED] Inflammation of the amnion resulting from infection. { ˌam·nē·ō'nīd·əs }

amniorrhea [MED] The premature escape or discharge of the amniotic fluid. { ˌam·ne·ō'rē·ə }

amniorrhexis [MED] Rupture of the amnion. { ¦am·nē·ō'rek·səs }

amnioscope [MED] An endoscope for studying amniotic fluid in the intact amniotic sac. { 'am·ne·ōˌskōp }

amnioscopy [MED] Visual examination of the amniotic fluid in the lower part of the amniotic sac by means of an amnioscope. { 'am·ne·ōˌskōp·ē }

Amniota [VERT ZOO] A collective term for the Reptilia, Aves, and Mammalia, all of which have an amnion during development. { ˌam·nē'äd·ə }

amniote [ZOO] An animal that develops an amnion during its embryonic stage; includes birds, reptiles, and mammals. { 'am·nēˌōt }

amniotic fluid [PHYSIO] A substance that fills the amnion to protect the embryo from desiccation and shock. { ¦am·nē¦äd·ik 'flü·əd }

amniotome [MED] An instrument used to puncture the fetal membranes. { 'am·nē·əˌtōm }

amniotomy [MED] Artificial rupture of the fetal membranes by means of an amniotome to induce labor. { ˌam·nē'äd·ə·mē }

amobarbital [PHARM] $C_{11}H_{18}N_2O_3$ A white, crystalline powder with a bitter taste and a melting point of 156–161°C; soluble in alcohol; used in medicine. { ˌam·ō'bär·bə·təl }

amodiaquine [PHARM] $C_{20}H_{22}ClN_3O$ A crystalline compound that melts at 208°C; used as an antimalarial. { ˌam·ō'dī·əˌkwēn }

Amoeba [INV ZOO] A genus of naked, rhizopod protozoans in the order Amoebida characterized by a thin pellicle and thick, irregular pseudopodia. { ə'mē·bə }

Amoebida [INV ZOO] An order of rhizopod protozoans in the subclass Lobosia characterized by the absence of a protective covering (test). { ˌam·ə'bī·də }

Amoebobacter [MICROBIO] A genus of bacteria in the family Chromatiaceae; cells are spherical and nonmotile, have gas vacuoles, and contain bacteriochlorophyll *a* on vesicular photosynthetic membranes. { əˌmē·bə'bak·tər }

amorphic allele [GEN] An allele that lacks gene activity. { ə'mȯr·fik ə'lēl }

Amorphosporangium [MICROBIO] A genus of bacteria in the family Actinoplanaceae with irregular sporangia and rod-shaped spores; they are motile by means of polar flagella. { ə¦mȯr·fə·spə'ran·jəm }

AMP *See* adenylic acid.

3′,5′-AMP *See* cyclic adenylic acid.

Ampeliscidae [INV ZOO] A family of tube-dwelling amphipod crustaceans in the suborder Gammaridea. { ˌam·pə'lis·əˌdē }

Ampharetidae [INV ZOO] A large, deep-water family of polychaete annelids belonging to the Sedentaria. { ˌam·fə'red·əˌdē }

Ampharetinae [INV ZOO] A subfamily of annelids belonging to the family Ampharetidae. { ˌam·fə'ret·əˌnē }

amphetamine [PHARM] $C_6H_5CH_2CHNHCH_3$ A volatile, colorless liquid used as a central nervous system stimulant. { ˌam·fed·əˌmēn }

amphiarthrosis [ANAT] An articulation of limited movement in which bones are connected by fibrocartilage, such as that between vertebrae or that at the tibiofibular junction. { ˌam·fī·är'thrō·səs }

amphiaster [INV ZOO] Type of spicule found in some sponges. { 'am·fēˌas·tər }

Amphibia [VERT ZOO] A class of vertebrate animals in the superclass Tetrapoda characterized by a moist, glandular skin, gills at some stage of development, and no amnion during the embryonic stage. { am'fib·ē·ə }

Amphibicorisae [INV ZOO] A subdivision of the insect order Hemiptera containing surface water bugs with exposed antennae. { amˌfib·ə'kȯr·əˌsē }

Amphibioidei [INV ZOO] A family of tapeworms in the order Cyclophyllidea. { amˌfəˌbī'ȯid·ēˌī }

amphibiotic [ZOO] Being aquatic during the larval stage and terrestrial in the adult stage. { ¦am·fəˌbī¦äd·ik }

amphibious [BIOL] Capable of living both on dry or moist land and in water. { ˌam'fib·ē·əs }

amphiblastic cleavage [EMBRYO] The unequal but complete cleavage of telolecithal eggs. { ¦am·fə¦blas·tik 'klēv·ij }

amphiblastula [EMBRYO] A blastula resulting from amphiblastic cleavage. [INV ZOO] The free-swimming flagellated larva of many sponges. { ¦am·fə¦blas·chə·lə }

amphibolic [MED] Uncertain; wavering; refers to the stage of a disease when prognosis is uncertain. [ZOO] Possessing the ability to turn either backward or forward, as the outer toe of certain birds. { ¦am·fə¦bäl·ik }

amphibolic pathway [BIOCHEM] A microbial biosynthetic and energy-producing pathway, such as the glycolytic pathway. { ¦am·fə¦bäl·ik 'pathˌwā }

Amphibolidae [INV ZOO] A family of gastropod mollusks in the order Basommatophora. { ˌam·fə'bäl·əˌdē }

amphicarpic [BOT] Having two types of fruit, differing either in form or ripening time. { ¦am·fə¦kär·pik }

amphichrome [BOT] A plant that produces flowers of different colors on the same stalk. { 'am·fəˌkrōm }

Amphicoela [VERT ZOO] A small suborder of

amphibians in the order Anura characterized by amphicoelous vertebrae. { ¦am·fə¦sēl·ə }

amphicoelous [VERT ZOO] Describing vertebrae that have biconcave centra. { ¦am·fə ¦sēl·əs }

amphicribral [BOT] Having the phloem surrounded by the xylem, as seen in certain vascular bundles. { ¦am·fə¦krib·rəl }

amphicryptophyte [BOT] A marsh plant with amphibious vegetative organs. { ¦am·fə¦krip·tə,fīt }

amphicytula [EMBRYO] A zygote that is capable of holoblastic unequal cleavage. { ,am·fə'sich·ə·lə }

amphid [INV ZOO] Either of a pair of sensory receptors in nematodes, believed to be chemoreceptors and situated laterally on the anterior end of the body. { 'am·fəd }

amphidetic [INV ZOO] Of a bivalve ligament, extending both before and behind the beak. { ¦am·fə¦ded·ik }

amphidiploid [GEN] A tetraploid organism or species produced when chromosome of a hybrid between two species double, yielding a diploid set of chromosomes from each parent. Also known as allotetraploid. { ¦am·fə¦di,plòid }

Amphidiscophora [INV ZOO] A subclass of sponges in the class Hexactinellida characterized by an anchoring tuft of spicules and no hexasters. { ¦am·fə·di'skäf·ə·rə }

Amphidiscosa [INV ZOO] An order of hexactinellid sponges in the subclass Amphidiscophora characterized by amphidisc spicules, that is, spicules having a stellate disk at each end. { ¦am·fə·di'skō·sə }

amphigastrium [BOT] Any of the small appendages located ventrally on the stem of some liverworts. { ¦am·fə'gas·trē·əm }

amphigean [ECOL] An organism that is native to both Old and New Worlds. { ¦am·fə¦jē·ən }

Amphilinidea [INV ZOO] An order of tapeworms in the subclass Cestodaria characterized by a protrusible proboscis, anterior frontal glands, and no holdfast organ; they inhabit the coelom of sturgeon and other fishes. { ,am·fə·lə'nid·ē·ə }

amphimixis [PHYSIO] The union of egg and sperm in sexual reproduction. { ,am·fə'mik·səs }

Amphimonadidae [INV ZOO] A family of zoomastigophorean protozoans in the order Kinetoplastida. { ,am·fə·mə'näd·ə,dē }

Amphineura [INV ZOO] A class of the phylum Mollusca; members are bilaterally symmetrical, elongate marine animals, such as the chitons. { ,am·fə'nu̇r·ə }

Amphinomidae [INV ZOO] The stinging or fire worms, a family of amphinomorphan polychaetes belonging to the Errantia. { ,am·fə'näm·ə,dē }

Amphinomorpha [INV ZOO] Group name for three families of errantian polychaetes: Amphenomidae, Euphrosinidae, and Spintheridae. { ¦am·fə·nə¦mȯr·fə }

amphioxus [ZOO] Former designation for the lancelet, *Branchiostoma*. { ,am·fē'äk·səs }

amphiphilic [BIOCHEM] Describing a molecule having a polar region that is separated from the nonpolar region. { ,am·fə'fil·ək }

amphiphloic [BOT] Pertaining to the central vascular cylinder of stems having phloem on both sides of the xylem. { ,am·fə'flō·ik }

amphiphyte [ECOL] A plant growing on the boundary zone of wet land. { 'am·fə,fīt }

amphiplatyan [ANAT] Describing vertebrae having centra that are flat both anteriorly and posteriorly. { ¦am·fə¦plad·ē·ən }

amphipneustic [VERT ZOO] Having both gills and lungs through all life stages, as in some amphibians. { ¦am·fə¦nüs·tik }

Amphipoda [INV ZOO] An order of crustaceans in the subclass Malacostraca; individuals lack a carapace, bear unstalked eyes, and respire through thoracic branchiae or gills. { am'fip·ə·də }

amphipodous [INV ZOO] Having both walking and swimming legs. { am'fip·ə·dəs }

amphisarca [BOT] An indehiscent fruit characterized by many cells and seeds, pulpy flesh, and a hard rind; melon is an example. { ,am·fə'sär·kə }

Amphisbaenidae [VERT ZOO] A family of tropical snakelike lizards in the suborder Sauria. { ¦am·fəs¦bēn·ə,dē }

Amphisopidae [INV ZOO] A family of isopod crustaceans in the suborder Phreactoicoidea. { ¦am·fə¦säp·ə,dē }

amphispore [MYCOL] A specialized urediospore with a thick, colorful wall; a resting spore. { 'am·fə,spȯr }

Amphistaenidae [VERT ZOO] The worm lizards, a family of reptiles in the suborder Sauria; structural features are greatly reduced, particularly the limbs. { ¦am·fə¦stēn·ə,dē }

amphistome [INV ZOO] An adult type of digenetic trematode having a well-developed ventral sucker (acetabulum) on the posterior end. { 'am·fə,stōm }

amphistylic [VERT ZOO] Having the jaw suspended from the brain case and the hyomandibular cartilage, as in some sharks. { ¦am·fə¦stīl·ik }

amphitene *See* zygotene. { 'am·fə,tēn }

amphithecium [BOT] The external cell layer during development of the sporangium in mosses. { ,am·fə'thē·shē·əm }

amphitriaene [INV ZOO] A poriferan spicule having three divergent rays at each end. { ,am·fə'trī,ēn }

amphitrichous [BIOL] Having flagella at both ends, as in certain bacteria. { ,am'fī·trə·kəs }

Amphitritinae [INV ZOO] A subfamily of sedentary polychaete worms in the family Terebellidae. { ,am·fə'trī·tə,nē }

amphitropical distribution [ECOL] Distribution of mostly temperate organisms which are discontinuous between the Northern and Southern hemispheres. { ¦am·fə¦träp·ə·kəl ,dis·trə'byü·shən }

amphitropous [BOT] Having a half-inverted ovule with the funiculus attached near the middle. { ˌam'fi·trə·pəs }

Amphiumidae [VERT ZOO] A small family of urodele amphibians in the suborder Salamandroidea composed of three species of large, eel-like salamanders with tiny limbs. { ˌam·fē'yü·məˌdē }

amphivasal [BOT] Having the xylem surrounding the phloem, as seen in certain vascular bundles. { ¦am·fə¦vā·zəl }

Amphizoidae [INV ZOO] The trout stream beetles, a small family of coleopteran insects in the suborder Adephaga. { ˌam·fə'zō·əˌdē }

Amphoriscidae [INV ZOO] A family of calcareous sponges in the order Sycettida. { ¦am·fə¦ris·əˌdē }

amphotericin [MICROBIO] An amphoteric antifungal antibiotic produced by *Streptomyces nodosus* and having of two components, A and B. { ˌam·fə'ter·ə·sən }

amphotericin A [MICROBIO] The relatively inactive component of amphotericin. { ˌam·fə'ter·ə·sən 'ā }

amphotericin B [MICROBIO] $C_{46}H_{73}O_{20}N$ The active component of amphotericin, suitable for systemic therapy of deep or superficial mycotic infections. { ˌam·fə'ter·ə·sən 'bē }

amplexicaul [BOT] Pertaining to a sessile leaf with the base or stipules embracing the stem. { am'plek·səˌkȯl }

amplexus [BOT] Having the edges of a leaf overlap the edges of a leaf above it in vernation. [VERT ZOO] The copulatory embrace of frogs and toads. { am'plek·səs }

ampliate [BIOL] Widened or enlarged. { 'am·plēˌāt }

amplification [GEN] **1.** Treatment with an antibiotic or other agent to increase the relative proportion of plasmid to bacterial deoxyribonucleic acid. **2.** Bulk replication of a gene library. { ˌam·plə·fə'kā·shən }

amplitude of accommodation [PHYSIO] The range of focal powers of which the eye is capable, expressed in diopters. { 'am·pləˌtüd əv əˌkäm·ə'dā·shən }

ampule [MED] A small, hermetically sealed flask that usually contains medicine for parenteral administration. { 'amˌpyül }

Ampulicidae [INV ZOO] A small family of hymenopteran insects in the superfamily Sphecoidea. { ˌam·pyə'lis·əˌdē }

ampulla [ANAT] A dilated segment of a gland or tubule. [BOT] A small air bladder in some aquatic plants. [INV ZOO] The sac at the base of a tube foot in certain echinoderms. { am'pu̇l·ə }

ampulla of Lorenzini [VERT ZOO] Any of the cutaneous receptors in the head region of elasmobranchs; thought to have a thermosensory function. { am'pu̇l·ə əv ˌlȯ·rent'zēˌnē }

ampulla of Vater [ANAT] Dilation at the junction of the bile and pancreatic ducts and the duodenum in humans. Also known as papilla of Vater. { am'pu̇l·ə əv 'fät·ər }

Ampullariella [MICROBIO] A genus of bacteria in the family Actinoplanaceae having cylindrical or bottle-shaped sporangia and rod-shaped spores arranged in parallel chains; motile by means of a polar tuft of flagella. { ˌam·pyəˌlar·ē'el·ə }

ampullary organ [PHYSIO] An electroreceptor most sensitive to direct-current and low-frequency electric stimuli; found over the body surface of electric fish and also in certain other fish such as sharks, rays, and catfish. { ¦am·pyəˌler·ē 'ȯr·gən }

amputation [MED] The surgical, congenital, or spontaneous removal of a limb or projecting body part. { ˌam·pyə'tā·shən }

AMV *See* Alfalfa mosaic virus.

amygdalin [BIOCHEM] $C_6H_5CH(CN)OC_{12}H_{21}O_{10}$ A glucoside occurring in the kernels of certain plants of the genus *Prunus*. { ə'mig·dəˌlən }

amylase [BIOCHEM] An enzyme that hydrolyzes reserve carbohydrates, starch in plants and glycogen in animals. { 'am·əˌlās }

amyloid [PATH] An abnormal protein deposited in tissues, formed from the infiltration of an unknown substance, probably a carbohydrate. { 'am·əˌlȯid }

amyloid beta protein [BIOCHEM] A 42-amino-acid proteolytic product of the amyloid precursor protein that accumulates in the brains (frontal cortex and hippocampus) of persons with Alzheimer's disease. { ¦am·əˌlȯid ¦bād·ə 'prōˌtēn }

amyloid body [PATH] Any of the microscopic hyaline bodies that stain like amyloid with metachromatic aniline dyes. { 'am·əˌlȯid ˌbäd·ē }

amyloidosis [MED] Deposition of amyloid in one or more organs of the body. { ˌam·ə·loi'dō·səs }

amylolysis [BIOCHEM] The enzyme-catalyzed hydrolysis of starch to soluble products. { ˌam·ə'läl·ə·səs }

amylolytic enzyme [BIOCHEM] A type of enzyme capable of denaturing starch molecules; used in textile manufacture to remove starch added to slash sizing agents. { ˌam·ə'läd·ik 'enˌzīm }

amylopectin [BIOCHEM] A highly branched, high-molecular-weight carbohydrate polymer composed of about 80% corn starch. { ˌam·ə·lō'pek·tən }

amylopectinosis [MED] A hereditary disease arising from an enzyme deficiency and characterized by abnormal accumulation of glycogen in tissues. Also known as Andersen's disease. { ˌam·ə·lōˌpek·tə'nō·səs }

amyloplast [BOT] A colorless cell plastid packed with starch grains and occurring in cells of plant storage tissue. { 'am·ə·lōˌplast }

amylopsin [BIOCHEM] An enzyme in pancreatic juice that acts to hydrolyze starch into maltose. { ˌam·ə'läp·sən }

amylose [BIOCHEM] A linear starch polymer. { 'am·əˌlōs }

amyotonia [MED] Absence of muscle tone. { ¦ā'mī·ə'tō· nē·ə }

amyotonia congenita [MED] A congenital disease of the central nervous system characterized by absence of voluntary muscle tone and reflexes. Also known as Oppenheim's disease. { ¦ā'mī·ə'tō·nē·ə kən'jen·əd·ə }

amyotrophic lateral sclerosis [MED] A progressive neurological disorder characterized by loss of connection and death of motor neurons in the cortex and spinal cord. { ¦a'mī·ə¦träf·ik 'la·trəl sklə'rō·səs }

ANA *See* antinuclear antibody.

Anabaena [BOT] A genus of blue-green algae in the class Cyanophycaea; members fix atmospheric nitrogen. { ˌan·ə'bēn·ə }

Anabantidae [VERT ZOO] A fresh-water family of actinopterygian fishes in the order Perciformes, including climbing perches and gourami. { ˌan·ə'ban·təˌdē }

Anabantoidei [VERT ZOO] A suborder of fresh-water labyrinth fishes in the order Perciformes. { ˌan·əˌban'tȯi·dēˌī }

anabiosis [BIOL] State of suspended animation induced by desiccation and reversed by addition of moisture; can be achieved in rotifers. { ˌan·əˌbī'ō·səs }

anabolic [BIOCHEM] Pertaining to anabolism. [EVOL] Pertaining to anaboly. { ˌan·ə'bäl·ik }

anabolic steroid [BIOCHEM] Any of a group of steroid hormones that increase anabolism. { ˌan·ə¦bäl·ik 'stiˌrȯid }

anabolism [BIOCHEM] A part of metabolism involving the union of smaller molecules into larger molecules; the method of synthesis of tissue structure. { ˌan'ab·əˌliz·əm }

anaboly [EVOL] The addition, through evolutionary differentiation, of a new terminal stage to the morphogenetic pattern. { ə'nab·ə·lī }

Anacanthini [VERT ZOO] An equivalent name for the Gadiformes. { ˌan·ə'kan·thəˌnī }

Anacardiaceae [BOT] A family of flowering plants, the sumacs, in the order Sapindales; many species are allergenic to humans. { ˌan·əˌkärd·ē'ās·ēˌē }

anaconda [VERT ZOO] *Eunectes murinus*. The largest living snake, an arboreal-aquatic member of the boa family (Boidae). { ˌan·ə'kän·də }

Anactinochitinosi [INV ZOO] A group name for three closely related suborders of mites and ticks: Onychopalpida, Mesostigmata, and Ixodides. { ə¦nak·tə·nə¦kīt·ən'ōˌsī }

Anacystis [BOT] A genus of blue-green algae in the class Cyanophycea. { ˌan·ə'sis·təs }

anadromous [VERT ZOO] Said of a fish, such as the salmon and shad, that ascends fresh-water streams from the sea to spawn. { ə'na·drə·məs }

Anadyomenaceae [BOT] A family of green marine algae in the order Siphonocladales characterized by the expanded blades of the thallus. { əˌna·dyəˌmen'ās·ēˌē }

anaerobe [BIOL] An organism that does not require air or free oxygen to maintain its life processes. { 'an·əˌrōb }

anaerobic bacteria [MICROBIO] Any bacteria that can survive in the partial or complete absence of air; two types are facultative and obligate. { ¦an·ə¦rōb·ik ˌbak'tir·ē·ə }

anaerobic condition [BIOL] The absence of oxygen, preventing normal life for organisms that depend on oxygen. { ¦an·ə¦rōb·ik kən'dish·ən }

anaerobic glycolysis [BIOCHEM] A metabolic pathway in plants by which, in the absence of oxygen, hexose is broken down to lactic acid and ethanol with some adenosine triphosphate synthesis. { ¦an·ə¦rōb·ik glī'käl·ə·səs }

anaerobic petri dish [MICROBIO] A glass laboratory dish for plate cultures of anaerobic bacteria; a thioglycollate agar medium and restricted air space give proper conditions. { ¦an·ə¦rōb·ik 'pē·trē ˌdish }

anaerobiosis [BIOL] A mode of life carried on in the absence of molecular oxygen. { ˌan·əˌrō'bī·ə·səs }

anaerophyte [ECOL] A plant that does not need free oxygen for respiration. { ə'ner·əˌfīt }

anagen effluvium [MED] Acute hair loss that usually follows chemotherapy or radiotherapy. { ¦an·ə·jən ə¦flü·vē·əm }

anakinesis [BIOCHEM] A process in living organisms by which energy-rich molecules, such as adenosine triphosphate, are formed. { ˌan·ə·kə'nē·səs }

anal [ANAT] Relating to or located near the anus. { 'ān·əl }

analbuminemia [MED] A disorder transmitted as an autosomal recessive, characterized by drastic reduction or absence of serum albumin. { ¦anˌalˌbyü·mə¦nēm·ē·ə }

analeptic [PHARM] Any drug used to restore respiration and a wakeful state. { ¦an·ə¦lep·tik }

anal fin [VERT ZOO] An unpaired fin located medially on the posterior ventral part of the fish body. { 'ān·əl ˌfin }

analgesia [PHYSIO] Insensibility to pain with no loss of consciousness. { ˌan·əl'jēzh·ə }

analgesic [PHARM] Any drug, such as salicylates, morphine, or opiates, used primarily for the relief of pain. { ˌan·əl'jēz·ik }

anal gland [INV ZOO] A gland in certain mollusks that secretes a purple substance. [VERT ZOO] A gland located near the anus or opening into the rectum in many vertebrates. { 'ān·əl ˌgland }

analogous [BIOL] Referring to structures that are similar in function and general appearance but not in origin, such as the wing of an insect and the wing of a bird. { ə'nal·ə·gəs }

anal plate [EMBRYO] An embryonic plate formed of endoderm and ectoderm through which the anus later ruptures. [VERT ZOO] **1.** One of the plates on the posterior portion of the plastron in turtles. **2.** A large scale anterior to the anus of most snakes. { 'ān·əl ˌplāt }

anal sphincter [ANAT] Either of two muscles, one voluntary and the other involuntary, controlling closing of the anus in vertebrates. { 'ān·əl 'sfiŋk·tər }

analytical transmission electron microscopy [BIOPHYS] A type of electron microscopy that allows the simultaneous characterization of crystal morphology, structure, and composition at high spatial resolution. { ,an·ə¦lid·ə·kəl ,tranz ¦mish·ən i¦lek,trän mī'kräs·kə·pē }

anamnesis [MED] Information gained from the patient and others regarding the patient's medical history. { ,an·əm'nē·səs }

anamnestic response [IMMUNOL] A rapidly increased antibody level following renewed contact with a specific antigen, even after several years. Also known as booster response. { ,an·əm'nes·tik ri'späns }

Anamnia [VERT ZOO] Vertebrate animals which lack an amnion in development, including Agnatha, Chondrichthyes, Osteichthyes, and Amphibia. { a'nam·nē·ə }

Anamniota [VERT ZOO] The equivalent name for Anamnia. { ¦a,nam·nē'ōd·ə }

anamniote [ZOO] An animal that does not develop an amnion during its embryonic stage. { an'am·nē,ōt }

anamorphism *See* anamorphosis. { ,an·ə'mȯr·fiz·əm }

anamorphosis [EVOL] Gradual increase in complexity of form and function during evolution of a group of animals or plants. Also known as anamorphism. { ,an·ə'mȯr·fə·səs }

anaphase [CELL MOL] **1.** The stage in mitosis and in the second meiotic division when the centromere splits and the chromatids separate and move to opposite poles. **2.** The stage of the first meiotic division when the two halves of a bivalent chromosome separate and move to opposite poles. { 'an·ə,fāz }

anaphase-promoting complex [CELL MOL] A large protein complex activated during mitosis that promotes the metaphase to anaphase transition by effecting the proteolytic destruction of several key mitotic regulators. Abbreviated APC. { ¦an·ə·fāz prə,mōd·iŋ 'käm,pleks }

anaphoresis [MED] Deficient functioning of sweat glands. [PHYSIO] Movement of positively charged ions into tissues under the influence of an electric current. { ¦an·ə·fə'rē·səs }

anaphylactic shock [MED] A syndrome seen as one of the clinical manifestations of anaphylaxis. { ¦an·ə·fə¦lak·tik 'shäk }

anaphylactoid reaction [MED] A nonallergic reaction resembling anaphylaxis and depending on the toxicity of the inductant. { ¦an·ə·fə¦lak ,tȯid rē'ak·shən }

anaphylatoxin [IMMUNOL] The vasodilator principal, a toxic substance released by tissues of sensitized animals when antigen and antibody react. { ¦an·ə,fil·ə'täk·sən }

anaphylaxis [MED] Hypersensitivity following parenteral injection of an antigen; local or systemic allergenic reaction occurs when the antigen is reintroduced after a time lapse. { ¦an·ə·fə'lak·səs }

anaplasia [MED] Reversion of cells to an embryonic, immature, or undifferentiated state; degree usually corresponds to malignancy of a tumor. { ,an·ə'plāzh·ə }

Anaplasma [MICROBIO] A genus of the family Anaplasmataceae; organisms form inclusions in red blood cells of ruminants. { ,an·ə'plaz·mə }

Anaplasmataceae [MICROBIO] A family of the order Rickettsiales; obligate parasites, either in or on red blood cells or in the plasma of various vertebrates. { ,an·ə,plaz·mə'tās·ē,ē }

anapolysis [INV ZOO] Lifetime retention of ripe proglottids in some tapeworms. { ,an·ə'päl·ə·səs }

anapophysis [ANAT] An accessory process on the dorsal side of the transverse process of the lumbar vertebrae in humans and other mammals. { ,an·ə'päf·ə·səs }

Anapsida [VERT ZOO] A subclass of reptiles characterized by a roofed temporal region in which there are no temporal openings. { ə'nap·sə·də }

anasarca [MED] Generalized edema of the subcutaneous connective tissue and serous body cavities. { ,an·ə'sär·kə }

Anaspidacea [INV ZOO] An order of the crustacean superorder Syncarida. { ə¦nas·pə'dās·ē·ə }

Anaspididae [INV ZOO] A family of crustaceans in the order Anaspidacea. { ə¦nas·pə'dī,dē }

anastomosis [MED] **1.** A surgical communication made between blood vessels, for example, between the portal vein and the inferior vena cava. **2.** An opening created by surgery, trauma, or disease between two or more normally separate spaces or organs. { ə,nas·tə'mō·səs }

anastomotic operation [MED] A surgical procedure to create an anastomosis. { ,an·ə·stə¦mäd·ik ,äp·ə'rā·shən }

anastral [CELL MOL] Lacking asters. { a'nas·trəl }

anastral mitosis [CELL MOL] Mitosis in which a spindle forms but no centrioles or asters are observed; typically occurs in plants. { ,ā¦nas·trəl mī'tō·səs }

Anatidae [VERT ZOO] A family of waterfowl, including ducks, geese, mergansers, pochards, and swans, in the order Anseriformes. { ə'nad·ə,dē }

anatomical dead space *See* dead space. { ,an·ə'täm·ə·kəl 'ded ,spās }

anatomical landmark *See* anatomical reference point. { ,an·ə¦täm·ə·kəl 'lan,märk }

anatomical position [ANAT] A reference posture used in anatomical description in which the subject stands erect against a wall with feet parallel and touching, and arms adducted and supinated, with palms facing forward. { ,an·ə¦täm·ə·kəl pə'zi·shən }

anatomical reference point [ANAT] A prominent structure or feature of the human body that can be located and described by visual inspection or palpation at the body's surface; used to define movements and postures. Also known as anatomical landmark. { ,an·ə¦täm·ə·kəl 'ref·rəns ,pȯint }

anatomy [BIOL] A branch of morphology dealing with the structure of animals and plants. { ə'nad·ə·mē }

anatomy of function [PHYSIO] A description of

the changes in the configuration of limbs during the performance of a task; considered a subdiscipline of kinesiology. { ə¦nad·ə·mē əv 'fəŋk·shən }

anatropous [BOT] Having the ovule fully inverted so that the micropyle adjoins the funiculus. { ə'na·trə·pəs }

anautogenous insect [INV ZOO] Any insect in which the adult female must feed before producing eggs. { ¦ā,nȯ¦täj·ə·nəs 'in,sekt }

anaxial [BIOL] Lacking an axis, therefore being irregular in form. { a'nak·sē·əl }

Ancalomicrobium [MICROBIO] A genus of prosthecate bacteria; nonmotile, unicellar forms with two to eight prosthecae per cell; reproduction is by budding of cells. { ,an¦kal·ō,mī¦krob·ē·əm }

ancestroecium [INV ZOO] The tube that encloses an ancestrula. { ,an·səs'trēsh·əm }

ancestrula [INV ZOO] The first polyp of a bryozoan colony. { ,an'ses·trə·lə }

anchor [INV ZOO] **1.** An anchor-shaped spicule in the integument of sea cucumbers. **2.** An anchor-shaped ossicle in echinoderms. { 'aŋ·kər }

anchorage-dependent cell [CELL MOL] A cell that grows, survives, or maintains function only when attached to an inert surface, such as glass or plastic. { ¦aŋ·kə,rij di¦pen·dənt 'sel }

anchovy [VERT ZOO] Any member of the Engraulidae, a family of herringlike fishes harvested commercially for human consumption. { 'an ,chō·vē }

anchylosis *See* ankylosis. { aŋ·ki'lō·səs }

ancipital [BOT] Having two edges, specifically referring to flattened stems, as of certain grasses. { an'sip·əd·əl }

ancora [INV ZOO] The initial, anchor-shaped growth stage of graptolithinids. { 'aŋ·kə·rə }

ancyloid [INV ZOO] A limpet-shaped or patelliform shell with the apex directed anteriorly. { 'an·sə,lȯid }

Ancylostoma [INV ZOO] A genus of roundworms, commonly known as hookworms, in the order Ancylostomidae; parasites of humans, dogs, and cats. { ,aŋ·kə'läs·tə·mə }

Ancylostoma duodenale [INV ZOO] The Old World hookworm, a human intestinal parasite that causes microcytic hypochromic anemia. { ,aŋ·kə'läs·tə·mə ¦dü·ə·di¦näl }

Ancylostomidae [INV ZOO] A family of nematodes belonging to the group Strongyloidea. { ¦aŋ·kə·lō·stä'mad·ə,dē }

Andersen's disease *See* amylopectinosis. { 'an·dər-sənz di,zēz }

Andreaeales [BOT] The single order of mosses of the subclass Andreaeobrya. { ,an·drē·ē'ā·lēz }

Andreaeceae [BOT] The single family of the Andreaeales, an order of mosses. { ,an·drē'ē·sē,ē }

Andreaeobrya [BOT] The granite mosses, a subclass of the class Bryopsida. { ,an·drē·ē'äb·rē·ə }

Andreaeopsida [BOT] A class of the plant division Bryophyta distinguished by longitudinal splitting of the mature capsule into four valves; commonly known as granite mosses. { ,an·drē·ē'äp·səd·ə }

Andrenidae [INV ZOO] The mining or burrower bees, a family of hymenopteran insects in the superfamily Apoidea. { ,an'dren·ə,dē }

androecious [BOT] Pertaining to plants that have only male flowers. { an'drē·shəs }

androecium [BOT] The aggregate of stamens in a flower. { ,an'drēsh·ē·əm }

androgen [BIOCHEM] A class of steroid hormones produced in the testis and adrenal cortex which act to regulate masculine secondary sexual characteristics. { 'an·drə·jən }

androgenesis [EMBRYO] Development of an embryo from a fertilized irradiated egg, involving only the male nucleus. { ,an·drə'jen·ə·səs }

androgenetic alopecia [MED] The most common cause of hair loss, characterized by gradual progression, with miniaturization of genetically programmed hair follicles. { ,an·drə·jə¦ned·ik ,al·ə'pē·shə }

androgenetic merogony [EMBRYO] The fertilization of egg fragments that lack a nucleus. { ¦an·drə·jə¦ned·ik mə'rä·gə·nē }

androgenic gland [INV ZOO] A gland found in most malacostracan crustaceans and producing hormones that control the development of the testes and male sexual characteristics. { ¦an·drə¦jen·ik 'gland }

androgen insensitivity syndrome *See* testicular feminization. { ¦an·drə·jən in,sen·sə'tiv·ə·dē ,sin,drōm }

androgenital syndrome [MED] An inherited syndrome of humans in which there is masculinization of genitals in females with XX chromosomal constitution. { ¦an·drə¦gen·ə·dəl 'sin ,drōm }

androgen unit [BIOL] A unit for the standardization of male sex hormones. { 'an·drə·jən ,yü·nət }

androgyny [MED] A form of pseudohermaphroditism in humans in which the individual has female external sexual characteristics, but has undescended testes. Also known as male pseudohermaphroditism. { an'dräj·ə·nē }

android pelvis *See* masculine pelvis. { 'an,drȯid 'pel·vəs }

andromerogony [EMBRYO] Development of an egg fragment following cutting, shaking, or centrifugation of a fertilized or unfertilized egg. { ¦an,drō·mə'räg·ə·nē }

androphile [ECOL] An organism, such as a mosquito, showing a preference for humans as opposed to animals. { 'an·drō,fīl }

androphore [BOT] A stalk that supports stamens or antheridia. [INV ZOO] A gonophore in cnidarians in which only male elements develop. { 'an·drō,fȯr }

androsin [BIOCHEM] $C_{15}H_{20}O_8$ A glucoside found in the herb *Apocynum androsaemifolium*; yields glucose and acetovanillone on hydrolysis. { 'an·drə·sən }

androsperm [BIOL] A sperm cell carrying a Y chromosome. { 'an·drə,spərm }

androstane [BIOCHEM] $C_{19}H_{32}$ The parent steroid hydrocarbon for all androgen hormones. Also known as etioallocholane. { 'an·drə,stān }

androstenedione [BIOCHEM] $C_{19}H_{26}O_2$ Any one of three isomeric androgens produced by the adrenal cortex. { ¦an·drō·stə¦ned·ē,ōn }

androsterone [BIOCHEM] $C_{19}H_{30}O_2$ An androgenic hormone occurring as a hydroxy ketone in the urine of men and women. { ,an'dräs·tə,rōn }

Anelytropsidae [VERT ZOO] A family of lizards represented by a single Mexican species. { ə,nel·ə'träp·sə,dē }

anemia [MED] A condition marked by significant decreases in hemoglobin concentration and in the number of circulating red blood cells. Also known as oligochromemia. { ə'nēm·ē·ə }

anemic necrosis [MED] Tissue death following a critical decrease in blood flow or oxygen levels. { ə'nēm·ik nə'krō·səs }

anemochory [ECOL] Wind dispersal of plant and animal disseminules. { ə'nēm·ə,kȯr·ē }

anemogenic [MED] Causing anemia. { ə¦nēm·ə¦jen·ik }

anemophilous [BOT] Pollinated by wind-carried pollen. { ¦an·ə¦mäf·ə·ləs }

anemotaxis [BIOL] Orientation movement of a free-living organism in response to wind. { ,an·ə·mō'tak·səs }

anemotropism [BIOL] Orientation response of a sessile organism to air currents and wind. { ,an·ə'mä·trə,piz·əm }

anencephalia [MED] A congenital anomaly marked by severely defective development of the brain, together with absence of the bones of the cranial vault and the cerebral and cerebellar hemispheres, and with only a rudimentary brainstem and some traces of basal ganglia present. { ,an,en·sə'fāl·yə }

anencephaly [MED] A neural tube defect resulting in gross underdevelopment of the brain in fetal life. { ,an,en'sef·ə·lē }

anenterous [ZOO] Having no intestine, as a tapeworm. { a'nen·tə·rəs }

Anepitheliocystidia [INV ZOO] A superorder of digenetic trematodes proposed by G. LaRue. { ¦an·ə·pə¦thē·lī·ō,sis'tid·ē·ə }

anergy [IMMUNOL] The condition of exhibiting no response to an antigen or antibody. [MED] The condition of exhibiting a lack of energy. { 'a,nər·jē }

anesthesia [PHYSIO] **1.** Insensibility, general or local, induced by anesthetic agents. **2.** Loss of sensation, of neurogenic or psychogenic origin. { ,an·əs'thēzh·ə }

anesthesiology [MED] A branch of medicine dealing with the administration of anesthetics. { ,an·əs,thēz·ē'äl·ə·jē }

anesthetic [PHARM] A drug, such as ether, that produces loss of sensibility. { ¦an·əs¦thed·ik }

anestrus [VERT ZOO] A prolonged period of inactivity between two periods of heat in cyclically breeding female mammals. { a'nes·trəs }

aneucentric aberration [GEN] An aberration that results in a chromosome with more than one centromere. { ¦an·yə¦sen·trik ab·ə'rā·shən }

aneuploidy [GEN] Deviation from a normal haploid, diploid, or polyploid chromosome complement by the presence in excess of, or in defect of, one or more individual chromosomes. { 'a·nyü,plȯid·ē }

aneurine *See* thiamine. { 'an·yə,rēn }

aneurysm [MED] Localized abnormal dilation of an artery due to weakening of the vessel wall. { 'an·yə,riz·əm }

aneusomatic organism [GEN] An organism whose cells contain variable numbers of individual chromosomes. { ¦an·yə·sō¦mad·ik 'ȯr·gə,niz·əm }

angel dust *See* phencyclidine. { 'ān·jəl ,dəst }

Angelman syndrome [MED] A genetic disorder that is caused by defects on the maternally derived chromosome 15, causing severe mental retardation, absence of speech, microcephaly, facial dysmorphism, seizures, neonatal hypotonia, ataxic movements, and inappropriate laughter. { 'aŋ·gəl·mən ,sin,drōm }

angiectasia [MED] Abnormal blood vessel dilation. { ,an·jē·ek'tā·zē·ə }

angiitis *See* vasculitis. { ,an·jē'īd·əs }

angina [MED] **1.** A sore throat. **2.** Any intense, constricting pain. { 'an·jə·nə *or* an'jī·nə }

angina pectoris [MED] Constricting chest pain which may be accompanied by pain radiating down the arms, up into the jaw, or to other sites. { 'an·jə·nə 'pek·tə·rəs }

angioblast [EMBRYO] A mesenchyme cell derived from extraembryonic endoderm that differentiates into embryonic blood cells and endothelium. { 'an·jē·ə,blast }

angiocardiography [MED] Roentgenographic visualization of the heart chambers and thoracic vessels following injection of a radiopaque material. { ¦an·jē·ə,kärd·ē¦äg·rə·fē }

angioedema [MED] The development of giant hives beneath the surface of the skin, especially around the eyes and lips; usually due to an allergic response. { ,an·jē·ō·i'dē·mə }

angiogenesis [EMBRYO] The origin and development of blood vessels. { ¦an·jē·ō¦jen·ə·səs }

angiogram [MED] An x-ray photograph of blood vessels following injection of a radiopaque material. { 'an·jē·ə,gram }

angiography [MED] Roentgenographic visualization of blood vessels following injection of a radiopaque material. { ,an·jē'äg·rə·fē }

angiokinesis *See* vasomotion. { ¦an·jē·ō·kə'nē·səs }

angiology [MED] The branch of medicine concerned with the blood vessels and the lymphatic system. { ,an·jē'äl·ə·jē }

angioma [MED] A tumor composed of lymphatic vessels or blood. { ,an·jē'ō·mə }

angioneurotic edema [MED] Acute, localized accumulations of tissue fluid causing swellings around the face; condition may be due either to heredity or a food allergy. { ¦an·jē·ō·nə'räd·ik i'dē·mə }

angiopathy [MED] Any disease of the vascular system. { ˌan·je'äp·ə·thē }

angioplacentography [MED] Radiography of the blood vessels of placentas through the injection of radiopaque dye. { ¦an·jē·ō·pləˌsen'täg·rə·fē }

angioplasty [MED] A procedure for alleviating blockage of an artery in which a balloon-tipped catheter is threaded into an artery to a point of obstruction and inflated to push the vessel open. { 'an·jē·əˌplas·tē }

angiosarcoma [MED] A malignant soft-tissue tumor arising from vascular elements. { ˌan·jē·ōˌsär'kō·mə }

angioscotoma [MED] A visual-field disturbance caused by dilated blood-vessels in the retina. { ¦an·jē·ō·skə'tō·mə }

angiospasm *See* vasospasm. { 'an·jē·əˌspaz·əm }

angiosperm [BOT] The common name for members of the plant division Magnoliophyta. { 'an·jē·ōˌspərm }

Angiospermae [BOT] An equivalent name for the Magnoliophyta. { ¦an·jē·ə¦spər·mē }

angiotensin [BIOCHEM] A decapeptide hormone that influences blood vessel constriction and aldosterone secretion by the adrenal cortex. Also known as hypertensin. { ˌan·jē·ə'ten·sən }

angitis [MED] An inflammatory condition of the walls of blood or lymph vessels. { an'jīd·əs }

anglerfish [VERT ZOO] Any of several species of the order Lophiiformes characterized by remnants of a dorsal fin seen as a few rays on top of the head that are modified to bear a terminal bulb. { 'aŋ·glərˌfish }

Anguidae [VERT ZOO] A family of limbless, snakelike lizards in the suborder Sauria, commonly known as slowworms or glass snakes. { 'aŋ·gwəˌdē }

Anguilliformes [VERT ZOO] A large order of actinopterygian fishes containing the true eels. { aŋˌgwil·ə'fòrˌmēz }

anguilliform motion [VERT ZOO] A type of locomotion in which a fish such as an eel moves its entire body, from head to tail, with considerable amplitude. { aŋ¦gwil·əˌform 'mō·shən }

Anguilloidei [VERT ZOO] The typical eels, a suborder of actinopterygian fishes in the order Anguilliformes. { aŋ·gwə'lòid·ēˌī }

anhalonium alkaloid [PHARM] Any of the alkaloids found in mescal buttons, including hordenine and mescaline; used as a cerebral stimulant and motor depressant. Also known as cactus alkaloid. { ˌan·ə'lōn·ē·əm 'al·kə ˌlòid }

anhidrosis [MED] Absent or deficient secretion of sweat. { ˌan·hī'drō·səs }

Anhimidae [VERT ZOO] The screamers, a family of birds in the order Anseriformes characterized by stout bills, webbed feet, and spurred wings. { an'him·əˌdē }

Anhingidae [VERT ZOO] The anhingas or snakebirds, a family of swimming birds in the order Pelecaniformes. { anˌhin·jəˌdē }

anhydrase [BIOCHEM] Any enzyme that catalyzes the removal of water from a substrate. { an'hīˌdrās }

anhydremia [MED] A decreased amount of water in the plasma. { ˌanˌhī'drēm·yə }

anhydrobiosis [PHYSIO] A type of cryptobiosis induced by dehydration. { ¦anˌhī·drō ˌbī'ō·səs }

anileridine [PHARM] $C_{22}H_{28}N_2O_2$ A white, crystalline powder; soluble in alcohol and chloroform; used as a narcotic. { ˌan·əl'er·əˌdēn }

Aniliidae [VERT ZOO] A small family of nonvenomous, burrowing snakes in the order Squamata. { ˌan·əl'ī·əˌdē }

animal [ZOO] Any living organism distinguished from plants by the lack of chlorophyll, the requirement for complex organic nutrients, the lack of a cell wall, limited growth, mobility, and greater irritability. { 'an·ə·məl }

animal community [ECOL] An aggregation of animal species held together in a continuous or discontinuous geographic area by ties to the same physical environment, mainly vegetation. { 'an·ə·məl kə'myü·nəd·ē }

animal ecology [ECOL] A study of the relationships of animals to their environment. { 'an·ə·məl i'käl·ə·jē }

Animalia [SYST] The animal kingdom. { ˌan·ə'māl·yə }

animal kingdom [SYST] One of the two generally accepted major divisions of living organisms which live or have lived on earth (the other division being the plant kingdom). { 'an·ə·məl ˌkiŋ·dəm }

animal locomotion [ZOO] Progressive movement of an animal body from one point to another. { 'an·ə·məl ˌlō·kə'mō·shən }

animal pole [CELL MOL] The region of an ovum which contains the least yolk and where the nucleus gives off polar bodies during meiosis. { 'an·ə·məl ˌpōl }

animal virus [VIROL] A small infectious agent able to propagate only within living animal cells. { 'an·ə·məl 'vī·rəs }

Anisakidae [INV ZOO] A family of parasitic roundworms in the superfamily Ascaridoidea. { ˌan·ə'säk·əˌdē }

anise [BOT] The small fruit of the annual herb *Pimpinella anisum* in the family Umbelliferae; fruit is used for food flavoring, and oil is used in medicines, soaps, and cosmetics. { 'an·əs }

anisocarpous [BOT] Referring to a flower whose number of carpels is different from the number of stamens, petals, and sepals. { ¦aˌnis·ə'kär·pəs }

anisochela [INV ZOO] A chelate sponge spicule with dissimilar ends. { ¦aˌnis·ə'kēl·ə }

anisocytosis [MED] A condition in which the erythrocytes show a considerable variation in size due to excessive quantities of hemoglobin. { ¦aˌnis·əˌsī'tō·səs }

anisodactylous [VERT ZOO] Having unequal digits, especially referring to birds with three toes forward and one backward. { ¦aˌnis·ə'dak·tə·ləs }

anisogamete *See* heterogamete. { ¦a,nis·ə'ga ,mēt }

anisogamy *See* heterogamy. { ¦a,nis'äg·ə·mē }

anisomerous [BOT] Referring to flowers that do not have the same number of parts in each whorl. { ,a,nī'säm·ə·rəs }

anisometric particle [VIROL] Any unsymmetrical, rod-shaped plant virus. { ¦a,nī·sə¦me·trik 'pärd·ə·kəl }

Anisomyaria [INV ZOO] An order of mollusks in the class Bivalvia containing the oysters, scallops, and mussels. { ,a,nī·sə,mī'a·rē·ə }

anisophyllous [BOT] Having leaves of two or more shapes and sizes. { ¦a,nī·sə¦fil·əs }

Anisoptera [INV ZOO] The true dragonflies, a suborder of insects in the order Odonata. { ,a,nī'säp·tə·rə }

anisostemonous [BOT] Referring to a flower whose number of stamens is different from the number of carpels, petals, and sepals. { ¦a,nī·sä'stem·ə·nəs }

Anisotomidae [INV ZOO] An equivalent name for Leiodidae. { ¦a,nī·sə'täm·ə,dē }

anisotropy [BOT] The property of a plant that assumes a certain position in response to an external stimulus. [ZOO] The property of an egg that has a definite axis or axes. { ¦a,nī'sä·trə·pē }

ankle [ANAT] The joint formed by the articulation of the leg bones with the talus, one of the tarsal bones. { 'aŋ·kəl }

ankylosing spondylitis *See* spondylitis. { 'aŋ·kə,lōz·iŋ ,spän·də'līd·əs }

ankyrin [CELL MOL] A protein found in the cell membrane of erythrocytes that attaches the membrane to the cytoskeleton protein spectrin. { aŋ'kī·rən }

anlage [EMBRYO] Any group of embryonic cells when first identifiable as a future organ or body part. Also known as blastema; primordium. { 'än,läg·ə }

annatto [BOT] *Bixa orellana*. A tree found in tropical America, characterized by cordate leaves and spinose, seed-filled capsules; a yellowish-red dye obtained from the pulp around the seeds is used as a food coloring. { ə'näd·ō }

anneal [GEN] To recombine strands of complementary deoxyribonucleic acid that were separated by heating or other means of denaturation. { ə'nēl }

Annedidae [VERT ZOO] A small family of limbless, snakelike, burrowing lizards of the suborder Sauria. { ə'ned·ə,dē }

Annelida [INV ZOO] A diverse phylum comprising the multisegmented wormlike animals. { ə'nel·ə·də }

annidation [ECOL] The phenomenon whereby a mutant is maintained in a population because it can flourish in an available ecological niche that the parent organisms cannot utilize. { ,an·ə'dā·shən }

Anniellidae [VERT ZOO] A family of limbless, snakelike lizards in the order Squamata. { ,an·ē'el·ə,dē }

Annonaceae [BOT] A large family of woody flowering plants in the order Magnoliales, characterized by hypogynous flowers, exstipulate leaves, a trimerous perianth, and distinct stamens with a short, thick filament. { ,a·nə'nās·ē,ē }

annual growth ring *See* annual ring. { 'an·yə·wəl 'grōth ,riŋ }

annual plant [BOT] A plant that completes its growth in one growing season and therefore must be planted annually. { 'an·yə·wəl 'plant }

annual ring [BOT] A line appearing on tree cross sections marking the end of a growing season and showing the volume of wood added during the year. Also known as annual growth ring. { 'an·yə·wəl 'riŋ }

annuation [ECOL] The annual variation in the presence, absence, or abundance of a member of a plant community. { ,an·yə'wā·shən }

annular budding [BOT] Budding by replacement of a ring of bark on a stock with a ring bearing a bud from a selected species or variety. { 'an·yə·lər 'bəd·iŋ }

annular hernia *See* umbilical hernia. { 'an·yə·lər 'hərn·ē·ə }

annular vessel [BOT] A xylem tube or duct with internal lignified rings. { 'an·yə·lər 'ves·əl }

annulus [ANAT] Any ringlike anatomical part. [BOT] **1.** An elastic ring of cells between the operculum and the mouth of the capsule in mosses. **2.** A line of cells, partly or entirely surrounding the sporangium in ferns, which constricts, thus causing rupture of the sporangium to release spores. **3.** A whorl resembling a calyx at the base of the strobilus in certain horsetails. [MYCOL] A ring of tissue representing the remnant of the veil around the stipe of some agarics. { 'an·yə·ləs }

Anobiidae [INV ZOO] The deathwatch beetles, a family of coleopteran insects of the superfamily Bostrichoidea. { ,an·ə'bī·ə,dē }

anodontia *See* hypodontia. { ,an·ə'dän·chə }

anole [VERT ZOO] Any arboreal lizard of the genus *Anolis*, characterized by flattened adhesive digits and a prehensile outer toe. { ə'nō·lē }

Anomalinacea [INV ZOO] A superfamily of marine and benthic sarcodinian protozoans in the order Foraminiferida. { ə,näm·ə·lə'nās·ē·ə }

anomalous trichromatism [PHYSIO] A mild defect in red-green color vision in which the subject, when asked to mix red and green light to match a certain shade of yellow, produces a different shade than does someone with normal color vision. { ə'näm·ə·ləs trī'krōm·ə,tiz·əm }

Anomaluridae [VERT ZOO] The African flying squirrels, a small family in the order Rodentia characterized by the climbing organ, a series of scales at the root of the tail. { ə¦näm·ə¦lür·ə,dē }

anomaly [BIOL] An abnormal deviation from the characteristic form of a group. [MED] Any part of the body that is abnormal in its position, form, or structure. { ə'näm·ə·lē }

anomeric carbon [BIOCHEM] The carbon about which anomers rotate. { ¦an·ə,mir·ik 'kär·bən }

Anomocoela [VERT ZOO] A suborder of toadlike

amphibians in the order Anura characterized by a lack of free ribs. { ¦an·ə·mō¦sē·lə }

anomocoelous [ANAT] Describing a vertebra with a centrum that is concave anteriorly and flat or convex posteriorly. { ¦an·ə·mō¦sē·ləs }

Anomura [VERT ZOO] A section of the crustacean order Decapoda that includes lobsterlike and crablike forms. { ˌan·ə'mu̇r·ə }

anoopsia [MED] Strabismus in which the eye is turned upward. { ˌan·ō'äp·sē·ə }

Anopheles [INV ZOO] A genus of mosquitoes in the family Culicidae; members are vectors of malaria, dengue, and filariasis. { ə'näf·əˌlēz }

anopheline [INV ZOO] Pertaining to mosquitoes of the genus *Anopheles* or a closely related genus. { ə'näf·ə·lən }

Anopla [INV ZOO] A class or subclass of the phylum Rhynchocoela characterized by a simple tubular proboscis and by having the mouth opening posterior to the brain. { 'an·ə·plə }

Anoplocephalidae [INV ZOO] A family of tapeworms in the order Cyclophyllidea. { ¦an·əˌplä·sə'fal·əˌdē }

Anoplura [INV ZOO] The sucking lice, a small group of mammalian ectoparasites usually considered to constitute an order in the class Insecta. { ˌan·ə'plu̇r·ə }

anorexia [MED] Loss of appetite. { ˌan·ə'rek·sē·ə }

anorgasmy [MED] Inability, usually psychic, to reach a climax during coitus. { ˌan·ȯr'gaz·mē }

anorthopia [MED] A defect of vision in which straight lines do not seem straight, and parallelism or symmetry is not properly perceived. { ˌa·nȯr'thōp·ē·ə }

anoscope [MED] An instrument for examining the lower rectum and anal canal. { 'ā·nəˌskōp }

anosmia [MED] Absence of the sense of smell. { ə'näz·mē·ə }

Anostraca [INV ZOO] An order of shrimplike crustaceans generally referred to the subclass Branchiopoda. { ə'näs·trə·kə }

anovulation [MED] The absence of ovulation. { ˌanˌäv·yə'lā·shən }

anoxemia [MED] Condition of having an insufficient supply of oxygen in the bloodstream. { ˌa·näk'sem·ē·ə }

anoxia [MED] The failure of oxygen to gain access to, or to be utilized by, the body tissues. { a'näk·sē·ə }

anoxybiosis [PHYSIO] A type of cryptobiosis induced by lack of oxygen. { aˌnäk·səˌbī'ō·səs }

Ansbacher unit [BIOL] A unit for the standardization of vitamin K. { 'änzˌbäk·ər ˌyü·nət }

Anser [VERT ZOO] A genus of birds in the family Anatidae comprising the typical geese. { 'an·sər }

Anseranatini [VERT ZOO] A subfamily of aquatic birds in the family Anatidae represented by a single species, the magpie goose. { ˌan·sər·ə'nat·əˌnī }

Anseriformes [VERT ZOO] An order of birds, including ducks, geese, swans, and screamers, characterized by a broad, flat bill and webbed feet. { ˌan·sər·ə'fȯrˌmēz }

Anson unit [BIOL] A unit for the standardization of trypsin and proteinases. { 'an·sən ˌyü·nət }

ant [INV ZOO] The common name for insects in the hymenopteran family Formicidae; all are social, and colonies exhibit a highly complex organization. { ant }

antagonism [BIOL] **1.** Mutual opposition as seen between organisms, muscles, physiologic actions, and drugs. **2.** Opposing action between drugs and disease or drugs and functions. { an'tag·əˌniz·əm }

antagonist [BIOCHEM] A molecule that bears sufficient structural similarity to a second molecule to compete with that molecule for binding sites on a third molecule. [PHARM] A drug or other chemical substance capable of reducing the physiological activity of another chemical substance; refers especially to a drug that opposes the action of a drug or other chemical substance on the nervous system by combining with and blocking the nerve receptor. [PHYSIO] A muscle that contracts with, and limits the action of, another muscle, called an agonist, with which it is paired. { an'tag·əˌnist }

antarctic faunal region [ECOL] A zoogeographic region describing both the marine littoral and terrestrial animal communities on and around Antarctica. { ˌant'ärd·ik 'fȯn·əl ˌrē·jən }

antazoline [PHARM] $C_{17}H_{19}N_3$ A bitter, white, crystalline powder with a melting point of 237–241°C; used as an antihistamine. { an'taz·əˌlēn }

anteater [VERT ZOO] Any of several mammals, in five orders, which live on a diet of ants and termites. { 'antˌēd·ər }

antebrachium *See* forearm. { ˌan·tə'brāk·ē·əm }

antecosta [INV ZOO] The internal, anterior ridge of the tergum or sternum of many insects that provides a surface for attachment of the longitudinal muscles. { 'an·təˌkäs·tə }

anteflexion [MED] Forward bending of an organ on itself. { ˌan·tē'flek·shən }

antelope [VERT ZOO] Any of the hollow-horned, hoofed ruminants assigned to the artiodactyl subfamily Antilopinae; confined to Africa and Asia. { 'an·təlˌōp }

antemortem [MED] Before death. { ¦an·tē ¦mȯr·dəm }

antenatal [MED] Occurring or existing before birth. { ¦an·te¦nād·əl }

antenna [INV ZOO] Any one of the paired, segmented, and movable sensory appendages occurring on the heads of many arthropods. { an'ten·ə }

antenna chlorophyll [BIOCHEM] Chlorophyll molecules which collect light quanta. { an'ten·ə 'klȯr·əˌfil }

antenna complex [CELL MOL] A complex of protein molecules found in the thylakoid membrane of chloroplasts that captures and transfers light energy to the photochemical reaction center. { an'ten·ə ˌkämˌpleks }

antennal gland [INV ZOO] An excretory organ in

the cephalon of adult crustaceans and best developed in the Malacostraca. Also known as green gland. { an'ten·ə ˌgland }

Antennata [INV ZOO] An equivalent name for the Mandibulata. { ˌan·tə'näd·ə }

antenodal [INV ZOO] Before or in front of the nodus, a cross vein near the middle of the costal border of the wing of dragonflies. { ˌan·tē'nōd·əl }

antepartum [MED] Pertaining to the period before delivery or birth. { ˌan·tē'pard·əm }

anter [INV ZOO] Part of a bryozoan operculum which serves to close off a portion of the operculum. { 'an·tər }

anteriad [ZOO] Toward the anterior portion of the body. { an'tir·ēˌad }

anterior [ZOO] Situated near or toward the front or head of an animal body. { an'tir·ē·ər }

anterior commissure [NEUROSCI] A bundle of nerve fibers that cross the midline of the brain in front of the third ventricle and serve to connect parts of the cerebral hemispheres. { an'tir·ē·ər 'käm·ə·shu̇r }

anterior horn [NEUROSCI] The ventral column of gray matter in the spinal cord containing the cell bodies of motor (efferent) neurons. { an¦tir·ē·ər 'hȯrn }

anterior horn cell [NEUROSCI] A motor neuron in the anterior horn gray matter of the spinal cord; directly innervates skeletal muscle. { an¦tir·ē·ər 'hȯrn ˌsel }

anterograde amnesia [MED] Loss of memory for the period subsequent to a sudden trauma or a seizure. { 'an·tə·rəˌgrād am'nēzhə }

anteromedial [ZOO] Anterior and toward the middle of the body. { ¦an·tə·rə¦mēd·ē·əl }

anthelate [BOT] An open, paniculate cyme. { 'an·thəˌlāt }

anthelminthic [PHARM] A chemical substance used to destroy tapeworms in domestic animals. Also spelled anthelmintic. { ¦an·thel¦min·thik }

anthelmintic *See* anthelminthic. { ¦an·thel¦min·tik }

anther [BOT] The pollen-producing structure of a flower. { 'an·thər }

antheraxanthin [BIOCHEM] A neutral yellow plant pigment unique to the Euglenophyta. { ¦an·thərˌaks¦an·thən }

anther culture [BOT] A haploid tissue culture derived from anthers or pollen cells. { 'an·thər ˌkəl·chər }

antheridiophore [BOT] A specialized stemlike structure that supports an antheridium in some mosses and liverworts. { ˌan·thə'rid·ē·əˌfȯr }

antheridium [BOT] **1.** The sex organ that produces male gametes in cryptogams. **2.** A minute structure within the pollen grain of seed plants. { ˌan·thə'rid·ē·əm }

antheriferous [BOT] Anther-bearing. { ˌan·thə'rif·ə·rəs }

antherozoid [BOT] A motile male gamete produced by plants. { ˌan·thə·rə'zō·əd }

anther smut [MYCOL] *Ustilago violacea*. A smut fungus that attacks certain plants and forms spores in the anthers. { 'an·thər ˌsmət }

anthesis [BOT] The flowering period in plants. { an'thē·səs }

Anthicidae [INV ZOO] The antlike flower beetles, a family of coleopteran insects in the superfamily Tenebrionoidea. { an'this·əˌdē }

anthill [INV ZOO] A mound of earth deposited around the entrance to ant and termite nests in the ground. { 'antˌhil }

anthoblast [INV ZOO] A developmental stage of some corals; produced by budding. { 'an·thəˌblast }

anthocarpous [BOT] Describing fruits having accessory parts. { ¦an·thə¦kär·pəs }

anthocaulis [INV ZOO] The stemlike basal portion of some solitary corals; the oral portion becomes a new zooid. { ¦an·thə'kȯl·əs }

Anthocerotae [BOT] A small class of the plant division Bryophyta, commonly known as hornworts or horned liverworts. { ˌan·thə'ser·əˌtē }

anthocodium [INV ZOO] The free, oral end of an anthozoan polyp. { ˌan·thə'kōd·ē·əm }

Anthocoridae [INV ZOO] The flower bugs, a family of hemipteran insects in the superfamily Cimicimorpha. { ˌan·thə'kȯr·əˌdē }

anthocyanidin [BIOCHEM] Any of the colored aglycone plant pigments obtained by hydrolysis of anthocyanins. { ˌan·thəˌsī'an·ə·dən }

anthocyanin [BIOCHEM] Any of the intensely colored, sap-soluble glycoside plant pigments responsible for most scarlet, purple, mauve, and blue coloring in higher plants. { ˌan·thə'sī·ə·nən }

anthocyathus [INV ZOO] The oral portion of the anthocaulus of some solitary corals that becomes a new zooid. { an·thəˌsī'ā·thəs }

Anthomedusae [INV ZOO] A suborder of hydrozoan cnidarians in the order Hydroida characterized by athecate polyps. { ˌan·thō·mi'dü·sē }

Anthomyzidae [INV ZOO] A family of cyclorrhaphous myodarian dipteran insects belonging to the subsection Acalypteratae. { ˌan·thō'mīz·əˌdē }

anthophagous [ZOO] Feeding on flowers. { an'thä·fə·gəs }

anthophore [BOT] A stalklike extension of the receptacle bearing the pistil and corolla in certain plants. { 'an·thəˌfȯr }

Anthosomidae [INV ZOO] A family of fish ectoparasites in the crustacean suborder Caligoida. { ˌan·thə'säm·əˌdē }

anthostele [INV ZOO] A thick-walled, nonretractile aboral region of certain cnidarians. { 'an·thəˌstēl }

Anthozoa [INV ZOO] A class of marine organisms in the phylum Cnidaria including the soft, horny, stony, and black corals, the sea pens, and the sea anemones. { ˌan·thō'zō·ə }

anthozooid [INV ZOO] Any of the individual zooids of a compound anthozoan. { ˌan·thō'zōˌȯid }

anthracosilicosis [MED] Chronic lung inflammation caused by inhalation of carbon and silicon particles. { ˌan·thrə·kəˌsil·ə'kō·səs }

anthracosis [MED] The accumulation of inhaled black coal dust particles in the lung accompanied by chronic inflammation. Also known as blacklung. { 'an·thrə'kō·səs }

anthralin [PHARM] $C_{14}H_{10}O_3$ A yellow powder with a melting point of 176–181°C; soluble in chloroform, acetone, and benzene; used in treatment of psoriasis. { 'an·thrə·lən }

anthraquinone pigments [BIOCHEM] Coloring materials which occur in plants, fungi, lichens, and insects; consists of about 50 derivatives of the parent compound, anthraquinone. { ˌan·thrə·kwi'nōn 'pig·məns }

anthrax [VET MED] An acute, infectious bacterial disease of sheep and cattle caused by *Bacillus anthracis*; transmissible to humans. Also known as splenic fever; wool-sorter's disease. { 'an ˌthraks }

Anthribidae [INV ZOO] The fungus weevils, a family of coleopteran insects in the superfamily Curculionoidea. { an'thrib·əˌdē }

anthropochory [ECOL] Dispersal of plant and animal disseminules by humans. { ¦an·thrə·pə¦kȯr·ē }

anthropodesoxycholic acid *See* chenodeoxycholic acid. { ¦an·thrəˌpō·de¦zäk·sə'käl·ik 'as·əd }

anthropogenic [ECOL] Referring to environmental alterations resulting from the presence or activities of humans. { ¦an·thrə·pə¦jen·ik }

anthropoid [VERT ZOO] Pertaining to or resembling the Anthropoidea. { 'an·thrəˌpȯid }

Anthropoidea [VERT ZOO] A suborder of mammals in the order Primates including New and Old World monkeys. { ˌan·thrə'pȯid·ē·ə }

anthropology [BIOL] The study of the interrelations of biological, cultural, geographical, and historical aspects of humankind. { ˌan·thrə'päl·ə·jē }

anthroponoses [MED] Diseases transmitted from humans to animals. { ˌan·thrə·pə'nōˌsēz }

anthroposphere [ECOL] The biosphere of the great geological activities of humankind. Also known as noosphere. { an'thrä·pəˌsfir }

Anthuridea [INV ZOO] A suborder of crustaceans in the order Isopoda characterized by slender, elongate, subcylindrical bodies, and by the fact that the outer branch of the paired tail appendage (uropod) arches over the base of the terminal abdominal segment, the telson. { ˌan·thə'rīd·ē·ə }

antiagglutinin [IMMUNOL] A substance that neutralizes a corresponding agglutinin. { ˌan·tē·ə'glüt·ən·ən }

antiaggressin [IMMUNOL] An antibody that neutralizes aggressin, a substance produced by pathogenic microorganisms to enhance virulence. { ˌan·tē·ə'gres·ən }

antianaphylaxis [IMMUNOL] A condition in which a sensitized animal resists anaphylaxis. { ¦an·tēˌan·ə·fə'lak·səs }

antiarrhythmic [MED] An agent that prevents or alleviates cardiac arrhythmia. { ˌan·tēˌā'rith·mik }

antiauxin [BIOCHEM] A molecule that competes with an auxin for receptor sites. { ¦an·tē'ȯk·sən }

antibacterial agent [MICROBIO] A synthetic or natural compound which inhibits the growth and division of bacteria. { ¦an·tēˌbak'tir·ē·əl 'ā·jənt }

antibiosis [ECOL] Antagonistic association between two organisms in which one is adversely affected. { ¦an·tēˌbī¦ō·səs }

antibiotic [MICROBIO] A chemical substance, produced by microorganisms and synthetically, that has the capacity in dilute solutions to inhibit the growth of, and even to destroy, bacteria and other microorganisms. { ¦an·tēˌbī¦äd·ik }

antibiotic assay [MICROBIO] A method for quantitatively determining the concentration of an antibiotic by its effect in inhibiting the growth of a susceptible microorganism. { ¦an·tēˌbī¦äd·ik 'aˌsā }

antibody [IMMUNOL] A protein, found principally in blood serum, originating either normally or in response to an antigen and characterized by a specific reactivity with its complementary antigen. Also known as immune body. { 'an·təˌbäd·ē }

antibody binding site *See* antibody combining site. { ¦an·təˌbäd·ē 'bīnd·iŋ ˌsīt }

antibody combining site [CELL MOL] **1.** The portion of an antibody molecule that makes physical contact with the corresponding antigenic determinant. **2.** The portion of an antigen that makes physical contact with the corresponding antibody. Also known as antibody binding site. { ¦an·təˌbäd·ē kəm'bīn·iŋ ˌsīt }

antibody-deficiency syndrome [MED] Any of the human defects of antibody production, such as hypogammaglobulinemia, agammaglobulinemia, and dysgammaglobulinemia, usually associated with reduced serum concentrations of immunoglobulins. { 'an·təˌbäd·ē di'fish·ən·sē ˌsinˌdrōm }

antibody-dependent cell-mediated cytotoxicity [IMMUNOL] An immunologic response in which an immunologic effector cell binds to a target cell coated with antibodies, triggering a series of metabolic events that leads to lysis of the target cell. { ˌant·iˌbäd·ē di¦pen·dənt ˌsel¦mēd·ēˌād·əd ˌsī·tō·täk'sis·əd·ē }

antibody-mediated immunity *See* humoral immunity. { ¦an·təˌbäd·ē ¦mēd·ēˌād·əd i'myün·əd·ē }

antiboreal faunal region [ECOL] A zoogeographic region including marine littoral faunal communities at the southern end of South America. { 'an·tēˌbȯr·ē·əl 'fȯn·əl ˌrē·jən }

anticarcinogen [PHARM] Any substance which is antagonistic to the action of a carcinogen. { 'an·tēˌkär'sin·ə·jən }

anticholinesterase [BIOCHEM] Any agent, such as a nerve gas, that inhibits the action of cholinesterase and thereby destroys or interferes with nerve conduction. { ˌan·tiˌkō·lə'nes·təˌrās }

anticipation [GEN] The occurrence of a phenotype at a younger age or in a more severe form in succeeding generations of a family. { an,tis·ə'pā·shən }

anticlinal [BOT] Pertaining to a cell layer that runs at right angles across the circumference of a plant part. { ¦an·tē¦klīn·əl }

anticoagulant [PHARM] An agent, such as sodium citrate, that prevents coagulation of a colloid, especially blood. { ¦an·tē,kō'ag·yə·lənt }

anticoding strand *See* antisense strand. { ,an·tē'kōd·iŋ ,strand }

anticodon [GEN] A three-nucleotide sequence in transfer RNA that complements the codon in messenger RNA. { ¦an·tē¦kō,dän }

anticonvulsant [PHARM] An agent, such as Dilantin, that prevents or arrests a convulsion. { ,an·tē·kən'vəl·sənt }

anticryptic [ECOL] Pertaining to protective coloration that makes an animal resemble its surroundings so that it is inconspicuous to its prey. { ¦an·tē¦krip·tik }

anticusp [INV ZOO] An anterior, downward projection in conodonts. { 'an·tē,kəsp }

antidepressant [PHARM] A drug, such as imipramine and tranylcypromine, that relieves depression by increasing central sympathetic activity. { ,an·tē·di'pres·ənt }

antidiabetic [PHARM] An agent, such as insulin, that is effective in controlling diabetes. { ¦an·tē,dī·ə¦bed·ik }

antidiarrheal [PHARM] An agent, such as Kaopectate, that prevents or arrests diarrhea. { ¦an·tē,dī·ə¦rē·əl }

antidiuretic [PHARM] An agent, such as vasopressin, that prevents the excretion of urine. { ¦an·tē,dī·yə¦red·ik }

antidiuretic hormone *See* vasopressin. { ¦an·tē,dī·yə¦red·ik 'hȯr,mōn }

antidote [PHARM] An agent that relieves or counteracts the action of a poison. { 'an·tə,dōt }

antienzyme [BIOCHEM] An agent that selectively inhibits the action of an enzyme. { ,an·tē'en,zīm }

antifertility agent [PHARM] A drug that prevents the formation of a fertilized ovum upon sexual intercourse. { ,an·tē·fər'til·əd·ē 'ā·jənt }

antifertilizin [BIOCHEM] An immunologically specific substance produced by animal sperm to implement attraction by the egg before fertilization. { ¦an·tē·fər'til·ə·zən }

antifibrinolysin [BIOCHEM] Any substance that inhibits the proteolytic action of fibrinolysin. { ¦an·tē,fī·brə'näl·ə·sən }

antifreeze proteins [BIOCHEM] Proteins that decrease the nonequilibrium freezing point of water without significantly affecting the melting point by directly binding to the surface of an ice crystal, thereby disrupting its normal structure and growth pattern and inhibiting further ice growth; found in a number of fish, insects, and plants. { 'an·ti,frēz ¦prō,tēnz }

antigen [IMMUNOL] A substance which reacts with the products of specific humoral or cellular immunity, even those induced by related heterologous immunogens. { 'an·tə·jən }

antigenaemia [IMMUNOL] A condition in which viral antigen is present in the blood; occurs in viral hepatitis and may occur in smallpox, myxomatosis, and yellow fever. { ,an·tə·jə'nē·mē·ə }

antigen-antibody reaction [IMMUNOL] The combination of an antigen with its antibody. { 'an·tə·jən ¦an·tə¦bäd·ē rē'ak·shən }

antigenic competition [IMMUNOL] A decrease in immune response to one antigenic peptide due to a concurrent immune response to a different antigenic peptide. { ,an·tə¦jen·ik ,käm·pə'tish·ən }

antigenic determinant [IMMUNOL] The portion of an antigen molecule that determines the specificity of the antigen-antibody reaction. { ,an·tə¦jen·ik di'tər·mə·nənt }

antigenic drift [IMMUNOL] Minor change of an antigen on the surface of a pathogenic microorganism. { ,an·tə¦jen·ik 'drift }

antigenicity [IMMUNOL] Ability of an antigen to induce an immune response and combine with specific antibodies or T-cell receptors. { ,an·tə·jə'nis·əd·ē }

antigenic mimicry [IMMUNOL] Acquisition or production of host antigens by a parasite, enabling it to avoid detection by the host's immune system. { 'an·tə¦jen·ik 'mim·ə·krē }

antigenic modulation [IMMUNOL] Loss of detectable antigen from the surface of a cell after incubation with antibodies. { ,an·tə¦jen·ik ,mäj·ə'lā·shən }

antigenic shift [VIROL] An abrupt major change in the antigenicity of a virus; believed to result from recombination of genes. { ,an·tə¦jen·ik 'shift }

antigenic variation [IMMUNOL] Alteration of an antigen on the surface of a microorganism; may enable a pathogenic mocroorganism to evade destruction by the host's immune system. { ,an·tə¦jen·ik ,ver·ē'ā·shən }

antigen presentation [IMMUNOL] The process whereby a cell expresses antigen on its surface in a form that can be recognized by a T lymphocyte. { ¦an·tə·jən ,prē·zən'tā·shən }

antihemophilic factor [BIOCHEM] A soluble protein clotting factor in mammalian blood. Also known as factor VIII; thromboplastinogen. { ¦an·tē,hē·mə'fil·ik ,fak·tər }

antihemorrhagic vitamin *See* vitamin K. { ¦an·tē,hem·ə'raj·ik 'vīd·ə·mən }

antihistamine [PHARM] A drug that prevents or diminishes the effect of histamine; used in treating allergic reactions and common-cold symptoms. { ,an·tē'hist·ə,mēn }

antihypertensive agent [PHARM] A substance, such as reserpine, that reduces hypertension. { ,an·tē,hī·pər'ten·siv 'ā·jənt }

anti-idiotype antibody [IMMUNOL] An antibody that is the mirror image of the original antibody formed against a specific surface antigen. { ,an·tē¦id·ē·ə,tīp 'an·tə,bäd·ē }

anti-immunoglobulin antibody [IMMUNOL] An antibody produced in response to a foreign antibody introduced into an experimental animal.

Abbreviated AIA. { ¦an·tē,im·yə·nō¦gläb·yə·lən 'an·tē,bäd·ē }

anti-infective vitamin *See* vitamin A. { ,an·tē·in 'fek·div ,vīd·ə·mən }

anti-inflammatory agent [PHARM] A substance, such as cortisone, that counteracts inflammation. { ,an·tē·in'flam·ə·tȯr·ē ,ā·jənt }

Antilocapridae [VERT ZOO] A family of artiodactyl mammals in the superfamily Bovoidea; the pronghorn is the single living species. { ,an·tə,lō'kap·rə,dē }

Antilopinae [VERT ZOO] The antelopes, a subfamily of artiodactyl mammals in the family Bovidae. { ,an·tə'lōp·ə,nē }

antilymphocyte serum [IMMUNOL] An immunosuppressive agent effective in prolonging the lives of homografts in experimental animals by reducing the circulating lymphocytes. { ¦an·tē'lim·fə,sīt ,sir·əm }

antilysin [IMMUNOL] A substance antagonistic to the action of a lysin. { 'an·tē¦lī·sən }

antimalarial [PHARM] **1.** A drug, such as quinacrine, that prevents or suppresses malaria. **2.** Acting against malaria. { ,an·tē·mə'ler·ē·əl }

antimere [INV ZOO] Any one of the equivalent parts into which a radially symmetrical animal may be divided. { 'an·tē,mir }

antimetabolite [PHARM] A substance, such as sulfanilamide or amethopterin, that inhibits utilization of an essential metabolite because it is an analog of the metabolite. { ¦an·tē·mə'tab·ə,līt }

antimicrobial agent [MICROBIO] A chemical compound that either destroys or inhibits the growth of microscopic and submicroscopic organisms. { ,an·tē,mī'krōb·ē·əl ,ā·jənt }

antimitotic drug [PHARM] A substance, such as colchicine, vincristine, or vinblastine, that interferes with mitotic cellular division; used in the chemotherapy of leukemia. { ,an·tē,mī'täd·ik ,drəg }

antimutagen [GEN] A compound that is antagonistic to the action of mutagenic agents on bacteria. { ¦an·tē'myüd·ə·jən }

antineoplastic drug [PHARM] An agent, such as mercaptopurine compounds, that is antagonistic to the growth of a neoplasm. { ¦an·tē,nē·ō¦plas·tik 'drəg }

antinuclear antibody [IMMUNOL] Antibody to deoxyribonucleic acid, ribonucleic acid, histone, or nonhistone proteins found in the serum of individuals with certain autoimmune diseases. Abbreviated ANA. { ¦an·tē¦nü·klē·ər 'an·tə ,bäd·ē }

antioncogene [GEN] Any of a class of genes that are involved in the negative regulation of normal growth; the loss or mutation of these genes leads to malignant growth. More generally called tumor suppressor gene. { ¦an·tē'aŋ·kə,jēn }

antiparallel [GEN] The opposite orientation of the two complementary strands or deoxyribonucleic acid, 5′ to 3′ and 3′ to 5′. { ¦an·tē'par·ə,lel }

antiparasitic agent [PHARM] An agent, such as emetine or quinine, that destroys or suppresses human and animal parasites. { ,an·tē¦par·ə¦sid·ik 'ā·jənt }

Antipatharia [INV ZOO] The black or horny corals, an order of tropical and subtropical cnidarians in the subclass Zoantharia. { ,an·tē·pə'thar·ē·ə }

antipetalous [BOT] Having stamens positioned opposite to, rather than alternating with, the petals. { ¦an·ti¦ped·ə·ləs }

antipodal [BOT] Any of three cells grouped at the base of the embryo sac, that is, at the end farthest from the micropyle, in most angiosperms. { an'tip·əd·əl }

antiporter [CELL MOL] A channel protein that simultaneously or sequentially transports two different types of substrates across a cell membrane, one into the cell (for example sodium ion) and one out of the cell (for example calcium ion). { an·tē'pȯrd·ər }

antipruritic [PHARM] An agent, such as camphor, that relieves itching. { ,an·tē·prü'rid·ik }

antipyretic [PHARM] Any agent, such as aspirin, that reduces or prevents fever. { ,an·tē·pī'red·ik }

antipyrine [PHARM] $C_{11}H_{12}ON_2$ A compound used as an antipyretic, analgesic, and antirheumatic. { ,an·te'pī,rēn }

antirachitic vitamin *See* vitamin D. { ,an·tē·rə'kid·ik 'vid·ə·mən }

anti-Rh agglutinin [IMMUNOL] An antibody against any Rh antigen; it must be acquired and is never natural. { 'an·tē,är¦āch ə'glüt·ən·ən }

anti-Rh immunoglobulin [IMMUNOL] A serum protein that destroys Rh-positive fetal erythrocytes in an Rh-negative mother when administered after delivery. { ¦an·tē,är¦āch ,im·yə·nə'gläb·yə·lən }

anti-Rh serum [IMMUNOL] A blood serum containing anti-Rh antibodies. { ¦an·tē,är¦āch 'sir·əm }

antirostrum [INV ZOO] The terminal segment of the appendages of certain mites. { ,an·tē'räs·trəm }

antisense [GEN] A strand of deoxyribonucleic acid having a sequence identical to messenger ribonucleic acid. { ¦an·tē'sens }

antisense drug [MED] A gene-based drug containing material that inhibits the synthesis of abnormal protein (which is typically caused by a specific disease state) by specifically binding to the ribonucleic acid responsible for its formation. { ,an·tē¦sens 'drəg }

antisense ribonucleic acid [CELL MOL] A ribonucleic acid (RNA) transcript (or portion of one) that is complementary to another nucleic acid, usually another RNA molecule. { ,an·tē¦sens ,rī·bō¦nü¦klē·ik 'as·əd }

antisense strand [CELL MOL] The strand of a double-stranded deoxyribonucleic acid molecule from which ribonucleic acid is transcribed. Also known as anticoding strand. { ¦an·tē 'sens ,strand }

antiseptic [MICROBIO] A substance used to destroy or prevent the growth of infectious microorganisms on or in the human or animal body. { ¦an·tə¦sep·tik }

antiserum [IMMUNOL] The serum component of blood that contains antibodies specific to one or more antigens. { 'an·tē,sir·əm }

antishock agent [PHARM] A substance, such as a cesium salt, that relieves a state of shock. { ,an·tē'shäk ,ā·jənt }

antisideric [PHARM] A pharmaceutical that counteracts the physiological effects of iron. { ¦an·tē·sə¦der·ik }

antismallpox vaccine *See* smallpox vaccine. { ¦an·tē¦smȯl,päks ,vak'sēn }

antispasmodic [PHARM] An agent, such as benzyl benzoate, that relieves convulsions and the pain of muscular spasms. { ¦an·tē,spaz'mäd·ik }

antistreptolysin [IMMUNOL] The antibody that neutralizes the streptolysin of group A hemolytic streptococci. { ,an·tē,strep·tə'līs·ən }

antitermination factor [BIOCHEM] Protein that interferes with normal termination of ribonucleic acid synthesis. { ,an·tē,tər·mə'nā·shən 'fak·tər }

antithrombin [BIOCHEM] A substance in blood plasma that inactivates thrombin. { ,an·tē'thräm·bən }

antitoxin [IMMUNOL] An antibody elaborated by the body in response to a bacterial toxin that will combine with and generally neutralize the toxin. { ,an·tē'täk·sən }

antitubercular agent [PHARM] A substance, such as streptomycin or isoniazid, used in the treatment of tuberculosis. { ¦an·tē·tə¦bərk·yə·lər 'ā·jənt }

antitumor antibiotic [MICROBIO] A substance, such as actinomycin, luteomycin, or mitomycin C, which is produced by microorganisms and is effective against some forms of cancer. { ,an·tē'tüm·ər ,an·tē,bī'äd·ik }

antitussive [PHARM] An agent, such as benylin expectorants, that relieves coughing. { ¦an·tē¦təs·iv }

antivenin [IMMUNOL] An immune serum that neutralizes the venoms of certain poisonous snakes and black widow spiders. { ¦an·tē¦ven·ən }

antivernalization [BOT] Delayed flowering in plants due to treatment with heat. { ,an·tē,vərn·əl·ə'zā·shən }

antiviral agent [PHARM] A substance, such as interferon or amantadine, that decreases virus multiplication in the body. { ¦an·tē¦vī·rəl 'ā·jənt }

antivitamin [BIOCHEM] Any substance that prevents a vitamin from normal metabolic functioning. { ,an·tē'vīd·ə·mən }

antixerophthalmic vitamin *See* vitamin A. { ¦an·tē,zir,äf'thal·mik 'vīd·ə·mən }

antler [VERT ZOO] One of a pair of solid bony, usually branched outgrowths on the head of members of the deer family (Cervidae); shed annually. { 'ant·lər }

ant lion [INV ZOO] The common name for insects of the family Myrmeleontidae in the order Neuroptera; larvae are commonly called doodlebugs. { 'ant ,lī·ən }

antrorse [BIOL] Turned or directed forward or upward. { 'an,trȯrs }

antrum [ANAT] A cavity of a hollow organ or a sinus. { 'an·trəm }

Anura [VERT ZOO] An order of the class Amphibia comprising the frogs and toads. { ə'nu̇r·ə }

anuresis [MED] Retention of urine in the urinary bladder due to inability to void. { ,an·yə'rē·səs }

anuria [MED] Complete absence of urinary output. { ə'nyu̇r·ē·ə }

anus [ANAT] The posterior orifice of the alimentary canal. { 'ā·nəs }

anvil *See* incus. { 'an·vəl }

anxiolytic agent [PHARM] A drug that relieves anxiety. { ,aŋk·sē·ō¦lid·ik 'ā·jənt }

aorta [ANAT] The main vessel of systemic arterial circulation arising from the heart in vertebrates. [INV ZOO] The large dorsal or anterior vessel in many invertebrates. { ā'ȯrd·ə }

aortic aneurysm [MED] Dilation of the wall of the aorta, usually the ascending portion. { ā'ȯrd·ik 'an·yə,riz·əm }

aortic arch [ANAT] The portion of the aorta extending from the heart to the third thoracic vertebra; single in warm-blooded vertebrates and paired in fishes, amphibians, and reptiles. { ā'ȯrd·ik 'ärch }

aortic body *See* aortic paraganglion. { ā'ȯrd·ik 'bäd·ē }

aortic incompetence [MED] A condition in which blood from the aorta flows back into the left ventricle because of the incapacity of the aortic valve. { ā'ȯrd·ik in'käm·pə·təns }

aortic paraganglion [ANAT] A structure in vertebrates belonging to the chromaffin system and found on the front of the abdominal aorta near the mesenteric arteries. Also known as aortic body; organs of Zuckerkandl. { ā'ȯrd·ik ,pa·rə'gaŋ·glē,än }

aortic stenosis [MED] Abnormal narrowing of the aortic valve orifice; may be either congenital or acquired. { ā'ȯrd·ik stə'nō·səs }

aortic valve [ANAT] A heart valve comprising three flaps which guards the passage from the left ventricle to the aorta and prevents the backward flow of blood. { ā'ȯrd·ik 'valv }

aortitis [MED] Inflammation of the aorta. { ,ā,ȯr'tīd·əs }

aortography [MED] Radiography of the aorta through a radiopaque dye injection. { ,ā,ȯr'tä·grə·fē }

6-APA *See* 6-aminopenicillanic acid.

apandrous [BOT] Lacking male organs or having nonfunctional male organs. { ,a'pan·drəs }

apatetic [ECOL] Pertaining to the imitative protective coloration of an animal subject to being preyed upon. { ¦a·pə¦ted·ik }

APC *See* anaphase-promoting complex.

ape [VERT ZOO] Any of the tailless primates of

the families Hylobatidae and Pongidae in the same superfamily as humans. { āp }

Apert's syndrome *See* acrocephalosyndactylism. { 'a,pərts ,sin,drōm }

apetalous [BOT] Lacking petals. { ,ā'ped·əl·əs }

apex [ANAT] **1.** The upper portion of a lung extending into the root. **2.** The pointed end of the heart. **3.** The tip of the root of a tooth. [BOT] The pointed tip of a leaf. { 'ā,peks }

apex impulse [PHYSIO] The point of maximum outward movement of the left ventricle of the heart during systole, normally localized in the fifth left intercostal space in the midclavicular line. Also known as left ventricular thrust. { 'ā,peks ¦im,pəls }

Apgar score [MED] An index used to evaluate a newborn infant's physical condition based on a rating of 0–2 for each of five criteria: heart rate, respiratory effort, muscle tone, response to stimulation, and skin color. { 'ap·gär ,skòr }

aphagia [MED] Inability to swallow; may be organic or psychic in origin. { ə'fāj·ə }

aphakia [MED] Absence of the lens of the eye. { ə'fāk· ē·ə }

Aphanomyces [MYCOL] A genus of fungi in the phycomycetous order Saprolegniales; species cause root rot in plants. { 'af·ə·nə'mī,sēz }

aphasia [MED] Impairment in the use or comprehension of language that is caused by lesions of the cerebral cortex. { ə'fāzh·ə }

aphasic seizure [MED] A transient inability to speak due to an abnormal electrical discharge from the speech areas of the brain. { ə'fāz·ik 'sēzh·ər }

Aphasmidea [INV ZOO] An equivalent name for the Adenophorea. { ¦a,faz'mid·ē·ə }

Aphelenchoidea [INV ZOO] A superfamily of plant and insect-associated nematodes in the order Tylenchida. { ,af·ə,leŋ'kòid·ē·ə }

Aphelenchoidoidea [INV ZOO] A superfamily of parasitic nematodes containing only one family, characterized by the lack of an isthmus in the esophagus and, in males, thorn-shaped spicules. { ,af·ə,leŋ,kòi'dòid·ē·ə }

Aphelocheiridae [INV ZOO] A family of hemipteran insects belonging to the superfamily Naucoroidea. { ¦af·ə,läk·ə'rī,dē }

aphid [INV ZOO] The common name applied to the soft-bodied insects of the family Aphididae; they are phytophagous plant pests and vectors for plant viruses and fungal parasites. { ā·fəd }

Aphididae [INV ZOO] The true aphids, a family of homopteran insects in the superfamily Aphidoidea. { ə'fid·ə,dē }

Aphidoidea [INV ZOO] A superfamily of sternorrhynchan insects in the order Homoptera. { ,a·fə'dòid·ē·ə }

Aphis [INV ZOO] A genus of aphid, the type genus of the family Aphididae. { 'ā·fəs }

aphonia [MED] Loss of voice and power of speech. { ā'fōn·ē·ə }

Aphredoderidae [VERT ZOO] A family of actinopterygian fishes in the order Percopsiformes containing one species, the pirate perch. { ,a·frə·də'der·ə,dē }

aphrodisiac [PHYSIO] Any chemical agent or odor that stimulates sexual desires. { ,af·rə'dē·zē,ak }

Aphroditidae [INV ZOO] A family of scale-bearing polychaete worms belonging to the Errantia. { ,af·rə'did·ə,dē }

aphtha [MED] White, painful oral ulcer of unknown cause. { ,af·thə }

Aphylidae [INV ZOO] An Australian family of hemipteran insects composed of two species; not placed in any higher taxonomic group. { ə'fil·ə,dē }

aphyllous [BOT] Lacking foliage leaves. { ā 'fil·əs }

aphytic zone [ECOL] The part of a lake floor that lacks plants because it is too deep for adequate light penetration. { ā'fid·ik ,zōn }

apical [BOT] Relating to the apex or tip. { 'ap·i·kəl }

apical bud *See* terminal bud. { 'ap·i·kəl ,bəd }

apical dominance [BOT] Inhibition of lateral bud growth by the apical bud of a shoot, believed to be a response to auxins produced by the apical bud. { 'ap·i·kəl 'däm·ə·nəns }

apicalia [INV ZOO] Paired sensory cilia on the head of gnathostomulids. { ¦ap·ə¦kal·yə }

apical meristem [BOT] A region of embryonic tissue occurring at the tips of roots and stems. Also known as promeristem. { 'ap·i·kəl 'mer·ə,stem }

apical plate [INV ZOO] A group of cells at the anterior end of certain trochophore larvae; believed to have nervous and sensory functions. { 'ap·i·kəl 'plāt }

apiculate [BOT] Ending abruptly in a short, sharp point. { ə'pik·yə·lət }

Apidae [INV ZOO] A family of hymenopteran insects in the superfamily Apoidea including the honeybees, bumblebees, and carpenter bees. { 'a·pə,dē }

Apioceridae [INV ZOO] A family of orthorrhaphous dipteran insects in the series Brachycera. { ,ap·ē·ō'ser·ə,dē }

apiology [INV ZOO] The scientific study of bees, particularly honeybees. { ,ā·pē'äl·ə·jē }

Apis [INV ZOO] A genus of bees, the type genus of the Apidae. { 'ā·pəs }

Apistobranchidae [INV ZOO] A family of spioniform annelid worms belonging to the Sedentaria. { ə¦pis·tə¦braŋk·ə,dē }

Aplacophora [INV ZOO] A subclass of vermiform mollusks in the class Amphineura characterized by no shell and calcareous integumentary spicules. { ¦ā,pla'käf·ə·rə }

aplanogamete [BIOL] A gamete that lacks motility. { ā'plan·ə·gə,mēt }

aplanospore [MYCOL] A nonmotile, asexual spore, usually a sporangiospore, common in the Phycomycetes. { ā'plan·ə,spòr }

aplasia [MED] Defective development which results in the virtual absence of a tissue or an organ; only a remnant appears. { ā'plāzh·ə }

aplastic anemia [MED] A blood disorder in

which lymphocytes predominate while there is a deficiency of erythrocytes, hemoglobin, and granulocytes. { ā'plas·tik ə'nēm·yə }

aplysiatoxin [BIOCHEM] A bislactone toxin produced by the blue-green alga *Lyngbya majuscula*. { ə¦plīzh·ə¦tak·sən }

apnea [MED] A transient cessation of respiration. { 'ap·nē·ə }

Apneumonomorphae [INV ZOO] A suborder of arachnid arthropods in the order Araneida characterized by the lack of book lungs. { ā,nü·mə,nō'mȯr,fē }

apneusis [PHYSIO] In certain lower vertebrates, sustained tonic contraction of the respiratory muscles to allow prolonged inspiration. { ap'nü·səs }

apocarpous [BOT] Having carpels separate from each other. { ¦ap·ə¦kär·pəs }

apocodeine [PHARM] $C_{18}H_{19}NO_2$ White crystals with a melting point of 124°C; decomposes on melting; soluble in alcohol and ether; used in medicine. { ¦ap·ə¦kō,dēn }

apocrine gland [PHYSIO] A multicellular gland, such as a mammary gland or an axillary sweat gland, that extrudes part of the cytoplasm with the secretory product. { 'ap·ə·krən ,gland }

Apocynaceae [BOT] A family of tropical and subtropical flowering trees, shrubs, and vines in the order Gentianales, characterized by a well-developed latex system, granular pollen, a poorly developed corona, and the carpels often united by the style and stigma; well-known members are oleander and periwinkle. { ə,päs·ə'nās·ē,ē }

Apoda [VERT ZOO] The caecilians, a small order of wormlike, legless animals in the class Amphibia. { 'a·pəd·ə }

Apodacea [INV ZOO] A subclass of echinoderms in the class Holothuroidea characterized by simple or pinnate tentacles and reduced or absent tube feet. { ,a·pə'dās·ē·ə }

apodeme [INV ZOO] An internal ridge or process on an arthropod exoskeleton to which organs and muscles attach. { 'ap·ə,dēm }

Apodes [VERT ZOO] An equivalent name for the Anguilliformes. { 'ap·ə,dēz }

Apodi [VERT ZOO] The swifts, a suborder of birds in the order Apodiformes. { 'ap·ə,dī }

Apodida [INV ZOO] An order of worm-shaped holothurian echinoderms in the subclass Apodacea. { ə'päd·ə·də }

Apodidae [VERT ZOO] The true swifts, a family of apodiform birds belonging to the suborder Apodi. { ə'päd·ə,dē }

Apodiformes [VERT ZOO] An order of birds containing the hummingbirds and swifts. { ə,päd·ə'fȯr,mēz }

apoenzyme [BIOCHEM] The protein moiety of an enzyme; determines the specificity of the enzyme reaction. { ¦a·pō¦en,zīm }

apoferritin [BIOCHEM] A protein found in intestinal mucosa cells that has the ability to combine with ferric ion. { ¦ap·ə¦fer·ət·ən }

apogamy [BIOL] Asexual, parthenogenetic development of diploid cells, such as the development of a sporophyte from a gametophyte without fertilization. { ə'päg·ə·mē }

apogeny [BOT] Loss of the function of reproduction. { ə'päj·ə·nē }

apogeotropism [BOT] Negative geotropism; growth up or away from the soil. { ¦a·pō,jē·ō'trä,piz·əm }

Apogonidae [VERT ZOO] The cardinal fishes, a family of tropical marine fishes in the order Perciformes; males incubate eggs in the mouth. { ,ap·ə'gän·ə,dē }

Apoidea [INV ZOO] The bees, a superfamily of hymenopteran insects in the suborder Apocrita. { ə'pȯid·ē·ə }

apoinducer [BIOCHEM] A protein that, when bound to deoxyribonucleic acid, activates transcription by ribonucleic acid polymerase. { ¦a·pō·in¦dü·sər }

apolipoprotein [BIOCHEM] A protein that combines with a lipid to form a lipoprotein. { ¦a·pō¦li·pō'prō,tēn }

apolysis [INV ZOO] In most tapeworms, the shedding of ripe proglottids. { ə'päl·ə·səs }

apomeiosis [CELL MOL] Meiosis that is either suppressed or imperfect. { ¦ap·ə,mī¦ō·səs }

apomixis [EMBRYO] Parthenogenetic development of sex cells without fertilization. { ,ap·ə'mik·səs }

apomorph [SYST] Any derived character occurring at a branching point and carried through one descending group in a phyletic lineage. { 'ap·ə,mȯrf }

apomorphine [PHARM] $C_{17}H_{17}NO_2$ A crystalline alkaloid obtained by dehydration of morphine; acts as a powerful emetic. { ¦ap·ə¦mȯr,fēn }

apomyoglobin [BIOCHEM] Myoglobin that lacks its heme group. { ¦ap·ə¦mī·e,glōb·ən }

aponeurosis [ANAT] A broad sheet of regularly arranged connective tissue that covers a muscle or serves to connect a flat muscle to a bone. { ¦ap·ə,nu̇'rō·səs }

apophyllous [BOT] Having the parts of the perianth distinct. { ə'päf·ə·ləs }

apophysis [ANAT] An outgrowth or process on an organ or bone. [MYCOL] A swollen filament in fungi. { ə'päf·ə·səs }

apoplexy [MED] **1.** A symptom complex caused by an acute vascular lesion of the brain and characterized by unconsciousness with various degrees of paralysis and sensory impairment. **2.** Sudden, severe hemorrhage into any organ. { 'ap·ə,plek·sē }

apoprotein [BIOCHEM] The protein portion of a conjugated protein exclusive of the prosthetic group. { ¦ap·ə¦prō,tēn }

apoptosis [CELL MOL] Death of cells triggered by extracellular signals or genetically programmed events, carried out by processes within the cell, and characterized by systemic breakdown of cellular constituents, in particular chromosomal deoxyribonucleic acid; may be involved in normal development and aging, or may serve to eliminate defective or damaged cells.

Also known as programmed cell death. { ˌā·pō'tō·səs }

apopyle [INV ZOO] Any one of the large pores in a sponge by which water leaves a flagellated chamber to enter the exhalant system. { 'ap·əˌpīl }

Aporidea [INV ZOO] An order of tapeworms of uncertain composition and affinities; parasites of anseriform birds. { ˌap·ə'rīd·ē·ə }

aporogamy [BOT] Entry of the pollen tube into the embryo sac through an opening other than the micropyle. { ˌap·ə'räg·ə·mē }

aposematic [ECOL] Pertaining to colors or structures on an organism that provide a special means of defense against enemies. Also known as sematic. { ¦ap·ə·sə¦mad·ik }

apospory [MYCOL] Suppression of spore formation with development of the haploid (sexual) generation directly from the diploid (asexual) generation. { 'ap·əˌspȯr·ē }

apostatic selection [ECOL] Predation on the most abundant forms in a population, leading to balanced distribution of a variety of forms. { ¦ap·ə¦stad·ik sə'lek·shən }

Apostomatida [INV ZOO] An order of ciliated protozoans in the subclass Holotrichia; majority are commensals on marine crustaceans. { əˌpäs·tə'mad·ə·də }

apotele [ANAT] A scalloped ridge around the edge of an otolith. { 'ap·əˌtēl }

apothecaries' measure [PHARM] A system of units of volume, usually of liquid drugs, in which 16 fluid ounces equals 1 pint. { ə'päth·əˌker·ēz 'mezh·ər }

apothecaries' weight [PHARM] A system of units of mass, usually of drugs, in which 1 pound equals 5760 grains or 1 troy pound. { ə'päth·əˌker·ēz 'wāt }

apothecium [MYCOL] A spore-bearing structure in some Ascomycetes and lichens in which the fruiting surface or hymenium is exposed during spore maturation. { ¦ap·ə¦thēsh·əm }

apozymase [BIOCHEM] The protein component of a zymase. { ˌap·ə'zīˌmās }

appendage [BIOL] Any subordinate or nonessential structure associated with a major body part. [ZOO] Any jointed, peripheral extension, especially limbs, of arthropod and vertebrate bodies. { ə'pen·dij }

appendectomy [MED] Surgical removal of the vermiform appendix. { ˌap·ən'dek·tə·mē }

appendicitis [MED] Inflammation of the vermiform appendix. { əˌpen·də'sīd·əs }

appendicular skeleton [ANAT] The bones of the pectoral and pelvic girdles and the paired appendages in vertebrates. { ˌap·ən'dik·yə·lər 'skel·ə·tən }

appendiculate [BIOL] Having or forming appendages. { ˌap·ən'dik·yəˌlāt }

appendix [ANAT] **1.** Any appendage. **2.** *See* vermiform appendix. { ə'pen·diks }

appendix testis [MED] A remnant of the cranial part of the paramesonephric or Müllers duct, attached to the testis. Also known as hydatid of Morgagni. { ə'pen·diks 'tes·təs }

appestat [PHYSIO] The center for appetite regulation in the hypothalamus. { 'ap·əˌstat }

appetitive behavior [ZOO] Any behavior that increases the probability that an animal will be able to satisfy a need; for example, a hungry animal will move around to find food. { ə'ped·ə·tiv bi'hāv·yər }

apple [BOT] *Malus domestica.* A deciduous tree in the order Rosales which produces an edible, simple, fleshy, pome-type fruit. { 'ap·əl }

apple of Peru *See* jimsonweed. { 'ap·əl əv pə'rü }

applied anatomy [ANAT] **1.** A discipline that considers problems involving the biomechanical functions of a body. **2.** The application of anatomical principles to specific fields of human endeavor, for example, surgical anatomy. { ə'plīd ə'nad·ə·mē }

applied ecology [ECOL] Activities involved in the management of natural resources. { ə'plīd i'käl·ə·jē }

applied potential tomography [MED] A method of producing images of the electrical impedance of tissues, in which potentials are applied to the body through skin electrodes, and the resulting currents give rise to measurable potentials elsewhere on the body from which the impedance of organs and tissues can be determined. { ə'plīd pə'ten·chəl tə'mäg·rə·fē }

apposition eye [INV ZOO] A compound eye found in diurnal insects and crustaceans in which each ommatidium focuses on a small part of the whole field of light, producing a mosaic image. { ˌap·ə'zish·ən ˌī }

appressed [BIOL] Pressed close to or lying flat against something. { ə'prest }

apraxia [MED] The inability to perform purposeful acts as a result of brain lesions; characteristically, paralysis is absent and kinesthesia is unimpaired. { ā'prak·sē·ə }

apricot [BOT] *Prunus armeniaca.* A deciduous tree in the order Rosales which produces a simple fleshy stone fruit. { 'ap·rəˌkät }

aproctous [MED] Having an imperforate anus. [ZOO] Lacking an anus. { ā'präk·təs }

Apsidospondyli [VERT ZOO] A term used to include, as a subclass, amphibians in which the vertebral centra are formed from cartilaginous arches. { ¦ap·sə·də'spän·dəˌlī }

apterium [VERT ZOO] A bare space between feathers on a bird's skin. { ap'tir·ē·əm }

apterous [BIOL] Lacking wings, as in certain insects, or winglike expansions, as in certain seeds. { 'ap·tə·rəs }

Apterygidae [VERT ZOO] The kiwis, a family of nocturnal ratite birds in the order Apterygiformes. { ¦ap·tə¦rij·əˌdē }

Apterygiformes [VERT ZOO] An order of ratite birds containing three living species, the kiwis, characterized by small eyes, limited eyesight, and nostrils at the tip of the bill. { ˌap·təˌrij·ə'fȯrˌmēz }

Apterygota [INV ZOO] A subclass of the Insecta characterized by being primitively wingless. { ˌap·tə·rə'gōd·ə }

aptyalism [MED] Deficiency or absence of saliva. { ā'tī·ə,liz·əm }

Apus [VERT ZOO] A genus of birds comprising the Old World swifts. { 'ā·pəs }

apyrase [BIOCHEM] Any enzyme that hydrolyzes adenosine triphosphate, with liberation of phosphate and energy, and that is believed to be associated with actomyosin activity. { 'ap·ə,rās }

apyrexia [MED] Absence of fever. { ¦ā,pī'rek·sē·ə }

aquaculture *See* aquiculture. { 'ak·wə,kəl·chər }

aquatic [BIOL] Living or growing in, on, or near water; having a water habitat. { ə'kwäd·ik }

aqueous desert [ECOL] A marine bottom environment with little or no macroscopic invertebrate shelled life. { 'āk·wē·əs 'dez·ərt }

aqueous humor [PHYSIO] The transparent fluid filling the anterior chamber of the eye. { 'āk·wē·əs 'yü·mər }

aqueous micelle [BIOCHEM] A spherical aggregate, 4–8 nanometers in diameter, formed dynamically from surfactants in water above a characteristic concentration, the critical micelle concentration. { ¦āk·wē·əs mi'sel }

aquiculture [BIOL] Cultivation of natural faunal resources of water. Also spelled aquaculture. { 'ak·wə,kəl·chər }

Aquifoliaceae [BOT] A family of woody flowering plants in the order Celastrales characterized by pendulous ovules, alternate leaves, imbricate petals, and drupaceous fruit; common members include various species of holly (*Ilex*). { ,ak·wə,fōl·ē'ās·ē,ē }

aquiherbosa [ECOL] Herbaceous plant communities in wet areas, such as swamps and ponds. { ,ak·wē,hər'bōs·ə }

aquiprata [ECOL] Communities of plants which are found in areas such as wet meadows where groundwater is a factor. { ə'kwip·rəd·ə }

araban [BIOCHEM] A polysaccharide composed of the pentose sugar L-arabinose. { 'ar·ə,ban }

Arabellidae [INV ZOO] A family of polychaete worms belonging to the Errantia. { ,ar·ə'bel·ə,dē }

arabinose [BIOCHEM] $C_5H_{10}O_5$ A pentose sugar obtained in crystalline form from plant polysaccharides such as gums, hemicelluloses, and some glycosides. { ə'rab·ə,nōs }

Araceae [BOT] A family of herbaceous flowering plants in the order Arales; plants have stems, roots, and leaves, the inflorescence is a spadix, and the growth habit is terrestrial or sometimes more or less aquatic; well-known members include dumb cane (*Dieffenbachia*), jack-in-the-pulpit (*Arisaema*), and *Philodendron*. { ə'rās·ē,ē }

arachidonate [BIOCHEM] A salt or ester of arachidonic acid. { ,a¦rak·ə¦dän,āt }

arachidonic acid [BIOCHEM] $C_{20}H_{32}O_2$ An essential unsaturated fatty acid that is a precursor in the biosynthesis of prostaglandins, thromboxanes, and leukotrienes. { ə¦rak·ə¦dan·ik 'as·əd }

Arachnia [MICROBIO] A genus of bacteria in the family Actinomycetaceae; branched diphtheroid rods and branched filaments form filamentous microcolonies; facultatively anaerobic; the single species is a human pathogen. { ə'rak·nē·ə }

Arachnida [INV ZOO] A class of arthropods in the subphylum Chelicerata characterized by four pairs of thoracic appendages. { ə'rak·nəd·ə }

arachnodactyly [MED] A rare congenital defect of the skeletal system marked by abnormally long hand and foot bones. { ə¦rak·nə'dak·tə·lē }

arachnoid [ANAT] A membrane that covers the brain and spinal cord and lies between the pia mater and dura mater. [BOT] Of cobweblike appearance, caused by fine white hairs. Also known as araneose. [INV ZOO] Any invertebrate related to or resembling the Arachnida. { ə'rak,nȯid }

arachnoidal granulations [ANAT] Projections of the arachnoid layer of the cerebral meninges through the dura mater. Also known as arachnoid villi; Pacchionian bodies. { ¦a,rak¦nȯid·əl ,gran·yə'lā·shənz }

Arachnoidea [INV ZOO] The name used in some classification schemes to describe a class of primitive arthropods. { ¦a,rak'nȯid·ē·ə }

arachnoid villi *See* arachnoidal granulations. { ə'rak,nȯid'vil·ē }

arachnology [INV ZOO] The study of arachnids. { ,a,rak'näl·ə·jē }

Aradidae [INV ZOO] The flat bugs, a family of hemipteran insects in the superfamily Aradoidea. { ə'rad·ə,dē }

Aradoidea [INV ZOO] A small superfamily of hemipteran insects belonging to the subdivision Geocorisae. { ,a·rə'dȯid·ē·ə }

Arales [BOT] An order of monocotyledonous plants in the subclass Arecidae. { ə'rā,lēz }

Araliaceae [BOT] A family of dicotyledonous trees and shrubs in the order Umbellales; there are typically five carpels and the fruit, usually a berry, is fleshy or dry; well-known members are ginseng (*Panax*) and English ivy (*Hedera helix*). { ə,rāl·ē'ās·ē,ē }

Aramidae [VERT ZOO] The limpkins, a family of birds in the order Gruiformes. { ə'ram·ə,dē }

Aran-Duchenne atrophy [MED] A muscular system disorder of adults involving progressive spinal muscular atrophy. { ¦ä·rän ,dyü¦shen 'a·trə·fē }

Araneae [INV ZOO] An equivalent name for Araneida. { ə'rān·ē,ē }

Araneida [INV ZOO] The spiders, an order of arthropods in the class Arachnida. { ,a·rə'nē·ə·də }

araneology [INV ZOO] The study of spiders. { ə,rān·ē'äl·ə·jē }

araneose *See* arachnoid. { ə'rān·ē,ōs }

araucaria [BOT] A primitive conifer of the genus *Araucaria* with broad leathery leaves, large cones, and edible seeds that is indigenous to South America and Australia. { ,ar,ō'kä·rē·ə }

Arbacioida [INV ZOO] An order of echinoderms in the superorder Echinacea. { är,bās·ē'ȯid·ə }

arbitrarily primed polymerase chain reaction [CELL MOL] A deoxyribonucleic acid (DNA) fingerprinting technique in which one short arbitrary primer is used to amplify multiple DNA fragments of different length, which yield a fingerprint after separation in gel electrophoresis. Also known as random amplification. { ,ar·bə¦trer·ə·lē ¦prīmd pə¦lim·ə,rās 'chān rē,ak·shən }

arboreal Also known as arboreous. [BOT] Relating to or resembling a tree. [ZOO] Living in trees. { är'bȯr·ē·əl }

arboreous [BOT] **1.** Wooded. **2.** *See* arboreal. { är'bȯr·ē·əs }

arborescence [BIOL] The state of being treelike in form and appearance. { ¦är·bə¦res·əns }

arboretum [BOT] An area where trees and shrubs are cultivated for educational and scientific purposes. { ,är·bə'rēd·əm }

arboriculture [BOT] The cultivation of ornamental trees and shrubs. { ¦är·bə·rə¦kəl·chər }

arborization [BIOL] A treelike arrangement, such as a branched dendrite or axon. { ,är·bə·rə'zā·shən }

arborization block *See* intraventricular heart block. { ,är·bə·rə'zā·shən ,bläk }

arbor vitae [NEUROSCI] The treelike arrangement of white nerve tissue seen in a median section of the cerebellum. { 'är·bər 'vīd·ē }

arborvitae [BOT] Any of the ornamental trees, sometimes called the tree of life, in the genus *Thuja* of the order Pinales. { ¦ar·bər¦vīd·ē }

arboviral encephalitides [MED] Diseases which are caused by arthropod-borne viruses (arboviruses), such as the encephalitis infections. { ¦är·bə¦vī·rəl ,en·sef·ə'līd·ə,dēz }

arbovirus [VIROL] Small, arthropod-borne animal viruses that are unstable at room temperature and inactivated by sodium deoxycholate; cause several types of encephalitis. Also known as arthropod-borne virus. { 'är·bə,vī·rəs }

arbuscule [MYCOL] A treelike haustorial organ in certain mycorrhizal fungi. { är'bə·skyül }

arcade [INV ZOO] A type of cell associated with the pharyngeal region of nematodes and united with like cells by an arch. { är'kād }

Arcellinida [INV ZOO] An order of rhizopodous protozoans in the subclass Lobosia characterized by lobopodia and a well-defined aperture in the test. { ,är·sə'lin·ə·də }

archaebacteria [MICROBIO] A group of unusual prokaryotic organisms that microscopically resemble true bacteria but differ biochemically and genetically, and form a distinct evolutionary group; some occur widely in oxygen-free environments and produce methane, while others are found in extreme salty or acidic conditions or grow at high temperatures. { ,ar·kē·bak'tir·ē·ə }

Archaeogastropoda [INV ZOO] An order of gastropod mollusks that includes the most primitive snails. { ,ärk·ē·ə,gas'träp·ə·də }

archallaxis [BIOL] Deviation from an ancestral pattern early in development, eliminating duplication of the phylogenetic history. { ,ärk·ə'lak·səs }

Archangiaceae [MICROBIO] A family of bacteria in the order Myxobacterales; microcysts are rod-shaped, ovoid, or spherical and are not enclosed in sporangia, and fruiting bodies are irregular masses. { ¦ärk,an·jē'ās·ē,ē }

Archangium [MICROBIO] The single genus of the family Archangiaceae; sporangia are lacking, and there is no definite slime wall. { ärk'an·jē·əm }

archegoniophore [BOT] The stalk supporting the archegonium in liverworts and ferns. { ¦ärk·ə'gōn·ē·ə,fȯr }

archegonium [BOT] The multicellular female sex organ in all plants of the Embryobionta except the Pinophyta and Magnoliophyta. { ,ark·ə'gōn·ē·əm }

archencephalon [EMBRYO] The primitive embryonic forebrain from which the forebrain and midbrain develop. { ¦ärk,in'sef·ə·län }

archenteron [EMBRYO] The cavity of the gastrula formed by ingrowth of cells in vertebrate embryos. Also known as gastrocoele; primordial gut. { ¦ärk'en·tə,rän }

archeocyte [INV ZOO] A type of ovoid amebocyte in sponges, characterized by large nucleolate nuclei and blunt pseudopodia; gives rise to germ cells. { 'är·kē·ə,sīt }

archerfish [VERT ZOO] The common name for any member of the fresh-water family Toxotidae in the order Perciformes; individuals eject a stream of water from the mouth to capture insects. { 'är·chər,fish }

archespore [BOT] A cell from which the spore mother cell develops in either the pollen sac or the ovule of an angiosperm. { 'är·kə,spȯr }

archetype [EVOL] A hypothetical ancestral type conceptualized by eliminating all specialized character traits. { 'är·ki,tīp }

Archiacanthocephala [INV ZOO] An order of worms in the phylum Acanthocephala; adults are endoparasites of terrestrial vertebrates. { ¦är·kē·ə,kan·thə'sef·ə·lə }

Archiannelida [INV ZOO] A group name applied to three families of unrelated annelid worms: Nerillidae, Protodrilidae, and Dinophilidae. { ¦är·kē·ə'nel·ə·də }

Archichlamydeae [BOT] An artificial group of flowering plants, in the Englerian system of classification, consisting of those families of dicotyledons that lack petals or have petals separate from each other. { ¦är·kē·klə'mid·ē,ē }

archicoel [ZOO] The segmentation cavity persisting between the ectoderm and endoderm as a body cavity in certain lower forms. { 'är·kē,sēl }

Archidiidae [BOT] A subclass of the plant class Bryopsida; consists of a single genus, *Archidium*, unique in having spores scattered in a single layer of the endothecium and having no quadrant stage in the early ontogeny of the capsule. { ,är·kə'dī·ə,dē }

archigastrula [EMBRYO] A gastrula formed by invagination, as opposed to ingrowth of cells. { ¦är·kē'gas·trə·lə }

Archigregarinida [INV ZOO] An order of telosporean protozoans in the subclass Gregarinia; endoparasites of invertebrates and lower chordates. { ¦är·kē,greg·ə'rin·ə·də }

archinephridium [INV ZOO] One of a pair of primitive nephridia found in each segment of some annelid larvae. { ¦är·kē·nə¦frid·ē·əm }

archinephros [VERT ZOO] The paired excretory organ of primitive vertebrates and the larvae of hagfishes and caecilians. { ¦är·kē'ne,frōs }

archipallium [PHYSIO] The olfactory pallium or the olfactory cerebral cortex; phylogenetically, the oldest part of the cerebral cortex. { ¦är·ki ¦pal·ē·əm }

Archosauria [VERT ZOO] A subclass of reptiles composed of five orders: Thecodontia, Saurischia, Ornithschia, Pterosauria, and Crocodilia. { ,är·kə'sȯr·ē·ə }

Archostemata [INV ZOO] A suborder of insects in the order Coleoptera. { ,är·kə·stə'mäd·ə }

arch pattern [FOREN] A fingerprint pattern in which ridges enter on one side of the impression, form a wave or angular upthrust, and flow out the other side. { 'ärch ,pad·ərn }

arcocentrum [ANAT] A centrum formed of modified, fused mesial parts of the neural or hemal arches. { ¦ar·kō¦sen·trəm }

arctic-alpine [ECOL] Of or pertaining to areas above the timberline in mountainous regions. { ¦ärd·ik ¦al,pīn }

arctic tree line [ECOL] The northern limit of tree growth; the sinuous boundary between tundra and boreal forest. { 'ärd·ik ¦trē ,līn }

Arctiidae [INV ZOO] The tiger moths, a family of lepidopteran insects in the suborder Heteroneura. { ärk'tī·ə,dē }

Arcturidae [INV ZOO] A family of isopod crustaceans in the suborder Valvifera characterized by an almost cylindrical body and extremely long antennae. { ,ärk'tu̇r·ə,dē }

arcuale [EMBRYO] Any of the four pairs of primitive cartilages from which a vertebra is formed. { ,ärk·yə'wā·lē }

arcuate [ANAT] Arched or curved; bow-shaped. { 'ärk·yə·wət }

arc-welder's disease *See* siderosis. { 'ark¦weld·ərz di,zēz }

Ardeidae [VERT ZOO] The herons, a family of wading birds in the order Ciconiiformes. { är'dē·ə,dē }

area amniotica [EMBRYO] The transparent part of the blastodisc in mammals. { 'er·ē·ə ,am·nē'äd·ə·kə }

area opaca [EMBRYO] The opaque peripheral area of the blastoderm of birds and reptiles, continuous with the yolk. { 'er·ē·ə ō'päk·ə }

area pellucida [EMBRYO] The central transparent area of the blastoderm of birds and reptiles, overlying the subgerminal cavity. { 'er·ē·ə pə'lü·səd·ə }

area placentalis [EMBRYO] The part of the trophoblast in immediate contact with the uterine mucosa in the embryos of early placental vertebrates. { 'er·ē·ə pla·sən'tāl·əs }

area vitellina [EMBRYO] The outer nonvascular zone of the area opaca; consists of ectoderm and endoderm. { 'er·ē·ə ,vid·ə'līn·ə }

Arecaceae [BOT] The palms, the single family of the order Arecales. { ,ar·ə'ka·sē,ē }

Arecales [BOT] An order of flowering plants in the subclass Arecidae composed of the palms. { ,ar·ə'kā,lēz }

Arecidae [BOT] A subclass of flowering plants in the class Liliopsida characterized by numerous, small flowers subtended by a prominent spathe and often aggregated into a spadix, and broad, petiolate leaves without typical parallel venation. { ə'res·ə,dē }

areg [ECOL] A sand desert. { 'a,reg }

arena *See* lek. { ə'rēn·ə }

Arenaviridae [VIROL] A family of ribonucleic acid animal viruses consisting of a single genus, Arenavirus, having an enveloped, spherical pleomorphic form. { ə,rēn·ə'vī·rə,dē }

Arenicolidae [INV ZOO] The lugworms, a family of mud-swallowing worms belonging to the Sedentaria. { ə,ren·ə'käl·ə,dē }

arenicolous [ZOO] Living or burrowing in sand. { ,a·rə'nik·ə·ləs }

areography [ECOL] Descriptive biogeography. { ,ar·ē'äg·rə·fē }

areola [ANAT] **1.** The portion of the iris bordering the pupil of the eye. **2.** A pigmented ring surrounding a nipple, vesicle, or pustule. **3.** A small space, interval, or pore in a tissue. { ə'rē·ə·lə }

areola mammae [ANAT] The circular pigmented area surrounding the nipple of the breast. Also known as areola papillaris; mammary areola. { ə'rē·ə·lə 'mam·ē }

areola papillaris *See* areola mammae. { ə'rē·ə·lə pap·ə'lär·əs }

areolar tissue [HISTOL] A loose network of fibrous tissue and elastic fiber that connects the skin to the underlying structures. { ə'rē·ə·lər 'tish·ü }

Argasidae [INV ZOO] The soft ticks, a family of arachnids in the suborder Ixodides; several species are important as ectoparasites and disease vectors for humans and domestic animals. { är 'gas·ə,dē }

argentaffin cell [HISTOL] Any of the cells of the gastrointestinal tract that are thought to secrete serotonin. { är'jen·tə·fən ,sel }

argentaffin fiber *See* reticular fiber. { är'jen·tə·fən ,fī·bər }

Argentinoidei [VERT ZOO] A family of marine deepwater teleostean fishes, including deep-sea smelts, in the order Salmoniformes. { ¦är,jen·tə'nȯid·ē,ī }

argentophil [BIOL] Of cells, tissues, or other structures, having an affinity for silver. { är 'jen·tə,fil }

Argidae [INV ZOO] A small family of hymenopteran insects in the superfamily Tenthredinoidea. { 'är·jə,dē }

arginase [BIOCHEM] An enzyme that catalyzes the splitting of urea from the amino acid arginine. { 'ar·jə,nās }

arginine [BIOCHEM] $C_6H_{14}N_4O_2$ A colorless, crystalline, water-soluble, essential amino acid of the α-ketoglutaric acid family. { 'ar·jə,nēn }

Arguloida [INV ZOO] A group of crustaceans known as the fish lice; taxonomic status is uncertain. { ,är·gə'lȯid·ə }

argyria [MED] A dusky-gray or bluish discoloration of the skin and mucous membranes produced by the prolonged administration or application of silver preparations. { är'jir·ē·ə }

argyrophil lattice fiber *See* reticular fiber. { 'är·jə·rō,fil 'lad·əs 'fī·bər }

Arhynchobdellae [INV ZOO] An order of annelids in the class Hirudinea characterized by the lack of an eversible proboscis; includes most of the important leech parasites of human and warm-blooded animals. { ¦ā,riŋ'käb·də,lē }

Arhynchodina [INV ZOO] A suborder of ciliophoran protozoans in the order Thigmotrichida. { ¦ā,riŋ'kä·də·nə }

arhythmia *See* arrhythmia. { ā'rith·mē·ə }

arhythmicity [BIOL] A condition that is characterized by the absence of an expected behavioral or physiologic rhythm. { ¦ā,rith'mis·əd·ē }

ariboflavinosis [MED] Dietary deficiency of riboflavin, associated with the syndrome of angular cheilosis and stomatitis, corneal vascularity, nasolabial seborrhea, and genitorectal dermatitis. { ¦ā,rīb·ə,flav·ə'nō·səs }

arid biogeographic zone [ECOL] Any region of the world that supports relatively little vegetation due to lack of water. { 'ar·əd ¦bī·ō,gē·ō'graf·ik ,zōn }

Arid Transition life zone [ECOL] The zone of climate and biotic communities occurring in the chaparrals and steppes from the Rocky Mountain forest margin to California. { 'ar·əd trans'ish·ən 'līf ,zōn }

arietiform [VERT ZOO] Shaped like a ram's horns; specifically, describing the dark facial marking that extends across the nose of kangaroo rats. { ,ar·ē'ed·ə,fȯrm }

Ariidae [VERT ZOO] A family of tropical saltwater catfishes in the order Siluriformes. { ə'rī·ə,dē }

aril [BOT] An outgrowth of the funiculus in certain seeds that either remains as an appendage or envelops the seed. { 'ar·əl }

arilode [BOT] An aril originating from tissues in the micropyle region; a false aril. { 'ar·ə,lōd }

Arionidae [INV ZOO] A family of mollusks in the order Stylommatophora, including some of the pulmonate slugs. { ,ar·ē'än·ə,dē }

arista [INV ZOO] The bristlelike or hairlike structure in many organisms, especially at or near the tip of the antenna of many Diptera. { ə'ris·tə }

Aristolochiaceae [BOT] The single family of the plant order Aristolochiales. { ə,ris·tə,lō·kē'ās·ē,ē }

Aristolochiales [BOT] An order of dicotyledonous plants in the subclass Magnoliidae; species are herbaceous to woody, often climbing, with perigynous to epigynous, apetalous flowers, uniaperturate or nonaperturate pollen, and seeds with a small embryo and copious endosperm. { ə,ris·tə,lō·kē'ā,lēz }

aristopedia [INV ZOO] Replacement of the arista by a nearly perfect leg. { ə,ris·tə'pēd·ē·ə }

Aristotle's lantern [INV ZOO] A five-sided feeding and locomotor apparatus surrounding the esophagus of most sea urchins. { 'ar·ə,städ·əlz 'lant·ərn }

arm [ANAT] The upper or superior limb in humans which comprises the upper arm with one bone and the forearm with two bones. { ärm }

armadillo [VERT ZOO] Any of 21 species of edentate mammals in the family Dasypodidae. { ,är·mə'dil·ō }

Armilliferidae [INV ZOO] A family of pentastomid arthropods belonging to the suborder Porocephaloidea. { ,är·mə·lə'fer·ə,dē }

Armour unit [BIOL] A unit for the standardization of adrenal cortical hormones and trypsin. { 'är·mər ,yü·nət }

armyworm [INV ZOO] Any of the larvae of certain species of noctuid moths composing the family Phalaenidae; economically important pests of corn and other grasses. { 'är·mē ,wərm }

Arneth's classification *See* Arneth's index. { 'är ,nets ,klas·ə·fə'kā·shən }

Arneth's count *See* Arneth's index. { 'är,nets ,kau̇nt }

Arneth's formula *See* Arneth's index. { 'är,nets ,fȯr·myə·lə }

Arneth's index [HISTOL] A system for dividing peripheral blood granulocytes into five classes according to the number of nuclear lobes, the least mature cells being tabulated on the left, giving rise to the terms "shift to left" and "shift to right" as an indication of granulocytic immaturity or hypermaturity respectively. Also known as Arneth's classification; Arneth's count; Arneth's formula. { 'är,nets ,in,deks }

Arnold sterilizer [MICROBIO] An apparatus that employs steam under pressure at 212°F (100°C) for fractional sterilization of specialized bacteriological culture media. { ärn·əld 'ster·ə,liz·ər }

Arodoidea [INV ZOO] A superfamily of hemipteran insects belonging to the subdivision Geocorisae. { ,a·rə'dȯid·ē·ə }

arolium [INV ZOO] A pad projecting between the tarsal claws of some insects and arachnids. { ə'rōl·ē·əm }

aromatic amino acid [BIOCHEM] An organic acid containing at least one amino group and one or more aromatic groups; for example, phenylalanine, one of the essential amino acids. { ¦ar·ə¦mad·ik ə'mēn·ō 'as·əd }

aromatic spirits of ammonia [PHARM] A flavored, hydroalcoholic solution of ammonia and ammonium carbonate having an aromatic, pungent odor; used as a reflex stimulant. Also known as hartshorn salts; smelling salts. { ¦ar·ə¦mad·ik ¦spir·ət əv ə'mōn·yə }

aromatic sulfuric acid [PHARM] A preparation consisting of sulfuric acid, tincture of ginger, oil of cinnamon, and alcohol; formerly used as a tonic and astringent. { ¦ar·ə¦mad·ik ,səl'fyür·ik 'as·əd }

aromatic vinegar [PHARM] A flavored solution of acetic acid used as smelling salts. { ¦ar·ə¦mad·ik 'vin·ə·gər }

array-type microelectrode [NEUROSCI] A type

of microelectrode that monitors electrically active cells in culture and can potentially be used to explore electrical activity of neural networks during development and learning. { ə¦rā ¦tīp ˌmī·krō·i'lekˌtrōd }

arrested evolution [EVOL] Evolution that was extremely slow in comparison with that characteristic of most organic lineages. { ə'res·təd ˌev·ə'lü·shən }

arrhenoblastoma [MED] A solid, sometimes malignant, tumor of the ovary that usually produces male sex hormones, inducing virilism. { ˌa·rəˌnōˌbla'stō·mə }

arrhenotoky [BIOL] Production of only male offspring by a parthenogenetic female. { ˌa·rə'näd·ə·kē }

arrhinencephalia [MED] A congenital malformation in which part or all of the rhinencephalon is absent and the nose is malformed. { əˌrīn·en·sə'fal·yə }

arrhythmia [MED] Absence of rhythm, especially of heart beat or respiration. Also spelled arhythmia. { ā'rith·mē·ə }

Arridae [VERT ZOO] A family of catfishes in the suborder Siluroidei found from Cape Cod to Panama. { 'a·rəˌdē }

arrowhead [BOT] Any aquatic plant of the genus *Sagitarria* (water plantain family) that has arrowhead-shaped leaves and white flowers. { 'a·rōˌhed }

arrow of time [PHYSIO] The uniform and unique direction associated with the apparent inevitable flow of time into the future. { ¦ar·ō əv 'tīm }

arrowroot [BOT] Any of the tropical American plants belonging to the genus *Maranta* in the family Marantaceae. { 'ar·ōˌrüt }

arrowworm [INV ZOO] Any member of the phylum Chaetognatha; useful indicator organism for identifying displaced masses of water. { 'ar·ōˌwərm }

arsenotherapy [MED] Treatment of disease by means of arsenical drugs. { ˌärs·ən·ō'ther·ə·pē }

arsphenamine [PHARM] $C_{12}H_{12}As_2N_2O_2 \cdot 2HCl \cdot 2H_2O$ The antisyphilitic diaminodihydroxyarsenobenzene dihydrochloride, effective also on protozoan infections, first prepared by P. Ehrlich in 1909. Also known as Ehrlich's 606. { är 'sfen·əˌmēn }

Artacaminae [INV ZOO] A subfamily of polychaete annelids in the family Terebellidae of the Sedentaria. { ˌär·tə'kam·əˌnē }

arteriogram [MED] A roentgenogram of an artery after injection with radiopaque material. { är'tir·ē·əˌgram }

arteriography [MED] **1.** Graphic presentation of the pulse. **2.** Roentgenography of the arteries after the intravascular injection of a radiopaque substance. { ärˌtir·ē'äg·rə·fē }

arteriole [ANAT] An artery of small diameter that terminates in capillaries. { är'tir·ēˌōl }

arteriolopathy [MED] Disease of the arterioles. { ärˌtir·ē·ə'läp·ə·thē }

arteriolosclerosis [MED] Thickening of the lining of arterioles, usually due to hyalinization or fibromuscular hyperplasia. { är¦tir·ē·əˌlō·sklə'rō·səs }

arteriometer [MED] An instrument that measures arterial pulsations. { ärˌtir·ē'äm·əd·ər }

arteriosclerosis [MED] A degenerative arterial disease marked by hardening and thickening of the vessel walls. { ärˌtir·ē·ō·sklə'rō·səs }

arteriosclerosis obliterans [MED] Hardening of the artery walls with obstruction of the lumen due to proliferation of the innermost vessel layer. { ärˌtir·ē·ō·sklə'rō·səs ō'blid·ərˌänz }

arteriotomy [MED] Incision or opening of an artery. { ärˌtir·ē'äd·ə·mē }

arteriovenous anastomosis [ANAT] A blood vessel that connects an arteriole directly to a venule without capillary intervention. { ärˌtir·ē·ō'vē·nəs əˌnas·tə'mō·səs }

arteriovenous aneurysm [MED] **1.** Dilation of the walls of an artery and a vein via an abnormal canal (fistula) between the vessels. **2.** Dilation of an arteriovenous fistula. { ärˌtir·ē·ō'vē·nəs 'an·yəˌriz·əm }

arteritis [MED] Inflammation of an artery. { ˌärd·ə'rīd·əs }

artery [ANAT] A vascular tube that carries blood away from the heart. { 'ärd·ə·rē }

arthochromatic erythroblast *See* normoblast. { ¦är·thrōˌkrō ¦mad·ik ə'rith·rəˌblast }

Arthoniaceae [BOT] A family of lichens in the order Hysteriales. { ärˌthän·ē'ās·ēˌē }

arthritis [MED] Any inflammatory process affecting joints or their component tissues. { är'thrīd·əs }

arthritis urethritica *See* Reiter's syndrome. { är'thrīd·əs ˌyu̇·rē'thrid·ə·kə }

Arthrobacter [MICROBIO] A genus of gram-positive, aerobic rods in the coryneform group of bacteria; metabolism is respiratory, and cellulose is not attached. { 'är·thrōˌbak·tər }

arthrobranch [INV ZOO] In Malacostraca, the gill attached to the joint between the body and the first leg segment. { 'är·thrōˌbraŋk }

arthrodesis [MED] Fusion of a joint by removing the articular surfaces and securing bony union. Also known as operative ankylosis. { är'thräd·ə·səs }

arthrodia [ANAT] A diarthrosis permitting only restricted motion between a concave and a convex surface, as in some wrist and ankle articulations. Also known as gliding joint. { är'thrōd·ē·ə }

Arthrodonteae [BOT] A family of mosses in the subclass Eubrya characterized by thin, membranous peristome teeth composed of cell walls. { ˌär·thrō'dänt·ēˌē }

arthrogram [MED] A roentgenogram of a joint space after injection of radiopaque material. { 'är·thrəˌgram }

arthrography [MED] Roentgenography of a joint space after the injection of radiopaque material. { ˌär'thräg·rə·fē }

arthrogryposis [MED] Permanent fixation of a joint in a flexed position. { ˌär·thrōˌgrī'pō·səs }

Arthromitaceae [MICROBIO] Formerly a family

of nonmotile bacteria in the order Caryophanales found in the intestine of millipedes, cockroaches, and toads. { ˌär·thräm·əˈtās·ēˌē }

arthropathy [MED] **1.** Any joint disease. **2.** A neurotrophic disorder of a joint, usually due to lack of pain sensation, found in association with tabes dorsalis, leprosy, syringomyelia, diabetic polyneuropathy, and occasionally multiple sclerosis and myelodysplasias. { ˌärˈthräp·ə·thē }

arthroplasty [MED] **1.** The making of an artificial joint. **2.** Reconstruction of a new and functioning joint from an ankylosed one; a plastic operation upon a joint. { ˈär·thrōˌplas·tē }

arthropod [INV ZOO] Any invertebrate (of the phylum Arthropoda) with a hard exoskeleton, segmented body, and jointed legs (for example, insects, arachnids, myriapods, and crustaceans). { ˈarth·rōˌpäd }

Arthropoda [INV ZOO] The largest phylum in the animal kingdom; adults typically have a segmented body, a sclerotized integument, and many-jointed segmental limbs. { ärˈthräp·ə·də }

arthropod-borne virus *See* arbovirus. { ¦är·thrəˌpäd ¦bȯrn ˈvī·rəs }

arthropodin [BIOCHEM] A water-soluble protein which forms part of the endocuticle of insects. { ärˈthräp·ə·dən }

arthroscope [MED] An endoscope for examining the interior of a joint. { ˈärth·rəˌskōp }

arthroscopy [MED] Visual examination of the interior of a joint by means of an arthroscope. { ärˈthrä·skə·pē }

arthrosis [ANAT] An articulation or suture uniting two bones. [MED] Any degenerative joint disease. { ärˈthrō·səs }

arthrospore [BOT] A jointed, vegetative resting spore resulting from filament segmentation in some blue-green algae and hypha segmentation in many Basidiomycetes. { ˈär·thrōˈspȯr }

Arthrotardigrada [INV ZOO] A suborder of microscopic invertebrates in the order Heterotardigrada characterized by toelike terminations on the legs. { ¦är·thrōˌtard·əˈgräd·ə }

arthrotomy [MED] Surgical incision into a joint. { ärˈthräd·ə·mē }

Arthur unit [BIOL] A unit for the standardization of splenin A. { ˈär·thər ˌyü·nət }

Arthus reaction [IMMUNOL] An allergic reaction of the immediate hypersensitive type that results from the union of antigen and antibody, with complement present, in blood vessel walls. { ˈär·thəs rēˈak·shən }

artichoke [BOT] *Cynara scolymus*. A herbaceous perennial plant belonging to the order Asterales; the flower head is edible. { ˈärd·əˌchōk }

article [INV ZOO] A segment of an arthropod leg between two articulations. { ˈärd·ə·kəl }

articulamentum [INV ZOO] The innermost layer of a calcareous plate in a chiton. { ärˌtik·yə·ləˈmen·təm }

articular cartilage [ANAT] Cartilage that covers the articular surfaces of bones. { ärˈtik·yə·lər ˈkärt·lij }

articular disk [ANAT] A disk of fibrocartilage, dividing the cavity of certain joints. { ärˈtik·yə·lər ˈdisk }

articular membrane [INV ZOO] A flexible region of the cuticle between sclerotized areas of the exoskeleton of an arthropod; functions as a joint. { ärˈtik·yə·lər ˈmemˌbrān }

Articulata [INV ZOO] **1.** A class of the Brachiopoda having hinged valves that usually bear teeth. **2.** The only surviving subclass of the echinoderm class Crinoidea. { ärˌtik·yəˈläd·ə }

articulation [ANAT] *See* joint. [BOT] A joint between two parts of a plant that can separate spontaneously. [INV ZOO] A joint between rigid parts of an animal body, such as the segments of an appendage in insects. [PHYSIO] The act of enunciating speech. { ärˌtik·yəˈlā·shən }

artifact [HISTOL] A structure in a fixed cell or tissue formed by manipulation or by the reagent. [MED] Noise or spurious signals that occur during various radiological imaging techniques; can reach a level where they appear in the image with as much strength as the signals produced by real objects. { ˈärd·əˌfakt }

artificial chromosome [GEN] A functional chromosome constructed by genetic engineering, having a centromere (and a telomere at each end, if linear rather than circular) and thus transmissable in cell division after introduction into a cell. { ˌärd·ə¦fish·əl ˈkrō·məˌsōm }

artificial heart [MED] An endoprosthetic device used to replace or assist the heart. { ¦ärd·ə¦fish·əl ˈhärt }

artificial hypothermia [MED] A surgical technique used, for example, in heart surgery, in which blood is cooled by a heat exchanger; body temperature is lowered to approximately 85°F (29.4°C), reducing oxygen requirements of tissues, particularly brain cells, and permitting temporary cessation of circulation. { ¦ärd·ə¦fish·əl ˌhī·pōˈthər·mē·ə }

artificial insemination [MED] A process by which spermatozoa are collected from males and deposited in female genitalia by instruments rather than by natural service. { ¦ärd·ə¦fish·əl inˌsem·əˈnā·shən }

artificial kidney [MED] An apparatus that performs the work of the kidney in purifying blood; used only in cases of renal failure or shutdown. { ¦ärd·ə¦fish·əl ˈkid·nē }

artificial nerve graft [MED] Used to enhance peripheral nerve regeneration, a porous or resorbable tube containing matrix material that may lead axons to grow in the desired direction. { ˌärd·ə¦fish·əl ˈnərv ˌgraft }

artificial parthenogenesis [PHYSIO] Activation of an egg by chemical and physical stimuli in the absence of sperm. { ¦ärd·ə¦fish·əl ¦pär·thə·nō¦gen·ə·səs }

artificial respiration [MED] The maintenance of breathing by artificial ventilation, in the absence of normal spontaneous respiration; effective methods include mouth-to-mouth breathing and the use of a respirator. { ¦ärd·ə¦fish·əl ˌres·pə'rā·shən }

artificial selection [GEN] A breeding method whereby particular genetic traits are selected by human manipulation. { ¦ärd·ə¦fish·əl si'lek·shən }

Artiodactyla [VERT ZOO] An order of terrestrial, herbivorous mammals characterized by having an even number of toes and by having the main limb axes pass between the third and fourth toes. { ˌärd·ē·ō'dak·tə·lə }

arytenoid [ANAT] Relating to either of the paired, pyramid-shaped, pivoting cartilages on the dorsal aspect of the larynx, in humans and most other mammals, to which the vocal cords and arytenoid muscles are attached. { ˌar·ə'tēˌnȯid }

asbestosis [MED] A chronic lung inflammation caused by inhalation of asbestos dust. { as ˌbe'stō·səs }

Ascaphidae [VERT ZOO] A family of amphicoelous frogs in the order Anura, represented by four living species. { ə'skaf·əˌdē }

ascariasis [MED] Any parasitic infection of humans or domestic mammals caused by species of *Ascaris*. { ˌas·kə'rī·ə·səs }

ascarid [INV ZOO] The common name for any roundworm belonging to the superfamily Ascaridoidea. { 'as·kə·rəd }

Ascaridata [INV ZOO] An equivalent name for the Ascaridina. { əˌskar·ə'däd·ə }

Ascaridida [INV ZOO] An order of parasitic nematodes in the subclass Phasmidia. { əˌskar·ə'dī·də }

Ascarididae [INV ZOO] A family of parasitic nematodes in the superfamily Ascaridoidea. { ˌas·kə'rid·əˌdē }

Ascaridina [INV ZOO] A suborder of parasitic nematodes in the order Ascaridida. { əˌskar·ə'dī·nə }

Ascaridoidea [INV ZOO] A large superfamily of parasitic nematodes of the suborder Ascaridina. { əˌskar·ə'dȯid·ē·ə }

Ascaris [INV ZOO] A genus of roundworms that are intestinal parasites in mammals, including humans. { 'as·kə·rəs }

Ascaroidea [INV ZOO] An equivalent name for Ascaridoidea. { ˌas·kə'rȯid·ē·ə }

ascending aorta [ANAT] The first part of the aorta, extending from its origin in the heart to the aortic arch. { ə'send·iŋˌā'ȯrd·ə }

ascending colon [ANAT] The portion of the colon that extends from the cecum to the bend on the right side below the liver. { ə'send·iŋ 'kōl·ən }

Ascheim-Zondek test [PATH] A human pregnancy test that uses the reaction of ovaries in immature white mice to an injection of urine from a woman. { 'äshˌhīm 'tsänˌdek ˌtest }

Aschelmintha [INV ZOO] A theoretical grouping erected by B. G. Chitwood as a series that includes the phylum Nematoda. { ˌaskˌhel'min·thə }

Aschelminthes [INV ZOO] A heterogeneous phylum of small to microscopic wormlike animals; individuals are pseudocoelomate and mostly unsegmented and are covered with a cuticle. { ˌaskˌhel'minˌthēz }

Aschoff body [MED] The lesion of rheumatic fever found around blood vessels in the myocardium. { 'äˌshȯf ˌbäd·ē }

Ascidiacea [INV ZOO] A large class of the phylum Tunicata; adults are sessile and may be solitary or colonial. { əˌsid·ē'āsh·ē·ə }

ascidiform [BOT] Pitcher-shaped, as certain leaves. { ə'sid·əˌfȯrm }

ascidium [BOT] A pitcher-shaped plant organ or part. { ə'sid·ē·əm }

ascites [MED] An abnormal accumulation of serous fluid in the abdominal cavity. { ə'sīd·ēz }

Asclepiadaceae [BOT] A family of tropical and subtropical flowering plants in the order Gentianales characterized by a well-developed latex system; milkweed (*Asclepias*) is a well-known member. { əˌsklēp·ē·ə'dās·ēˌē }

ascocarp [MYCOL] The mature fruiting body bearing asci with ascospores in higher Ascomycetes. { 'as·kəˌkärp }

ascogenous [MYCOL] Pertaining to or producing asci. { a'skäj·ə·nəs }

ascogonium [MYCOL] The specialized female sexual organ in higher Ascomycetes. { ˌas·kə'gōn·ē·əm }

Ascolichenes [BOT] A class of the lichens characterized by the production of asci similar to those produced by Ascomycetes. { ¦as·kəˌlī'kē·nēz }

Ascomycetes [MYCOL] A class of fungi in the subdivision Eumycetes, distinguished by the ascus. { ˌas·kōˌmī'sēd·ēz }

ascon [INV ZOO] A sponge or sponge larva having incurrent canals leading directly to the paragaster. { 'aˌskän }

ascorbic acid [BIOCHEM] $C_6H_8O_6$ A white, crystalline, water-soluble vitamin found in many plant materials, especially citrus fruit. Also known as vitamin C. { ə'skȯr·bik 'as·əd }

ascospore [MYCOL] An asexual spore representing the final product of the sexual process, borne on an ascus in Ascomycetes. { 'as·kəˌspȯr }

Ascothoracica [INV ZOO] An order of marine crustaceans in the subclass Cirripedia occurring as endo- and ectoparasites of echinoderms and cnidarians. { ¦as·kə·thə'ras·ə·kə }

ascus [MYCOL] An oval or tubular spore sac bearing ascospores in members of the Ascomycetes. { 'as·kəs }

A selection [ECOL] Selection that favors species adapted to consistently adverse environments. { 'ā siˌlek·shən }

Aselloidea [INV ZOO] A group of free-living, fresh-water isopod crustaceans in the suborder Asellota. { ˌa·sə'lȯid·ē·ə }

Asellota [INV ZOO] A suborder of morphologically and ecologically diverse aquatic crustaceans in the order Isopoda. { ə'sel·ə·də }

asepsis [MED] The state of being free from pathogenic microorganisms. { ā'sep·səs }

aseptic meningitis [MED] A type of meningitis in which the raised cell count in the cerebrospinal fluid is predominantly lymophocytic; usually caused by a viral infection, but can result from other causes. { ā'sep·tik ˌmen·ən'jīd·əs }

asetigerous *See* achaetous. { ā·sə'tij·ə·rəs }

asexual [BIOL] **1.** Not involving sex. **2.** Exhibiting absence of sex or of functional sex organs. { ā'seksh·ə·wəl }

asexual reproduction [BIOL] Formation of new individuals from a single individual without the involvement of gametes. { ā'seksh·ə·wəl ˌrē·prə'dək·shən }

ash [BOT] **1.** A tree of the genus *Fraxinus*, deciduous trees of the olive family (Oleaceae) characterized by opposite, pinnate leaflets. **2.** Any of various Australian trees having wood of great toughness and strength; used for tool handles and in work requiring flexibility. { ash }

Asian flu [MED] An acute viral respiratory infection of humans caused by influenza A-2 virus. { 'āzh·ən 'flü }

Asilidae [INV ZOO] The robber flies, a family of predatory, orthorrhaphous, dipteran insects in the series Brachycera. { ə'sil·əˌdē }

asomatognosia [MED] Lacking awareness of paralysis because the brain is damaged. { āˌsōm·əˌtäg'nōzh·ə }

Asopinae [INV ZOO] A family of hemipteran insects in the superfamily Pentatomoidea including some predators of caterpillars. { ə'sōp·əˌnē }

asparaginase [BIOCHEM] An enzyme that catalyzes the hydrolysis of asparagine to asparaginic acid and ammonia. { ˌas·pə'raj·əˌnās }

asparagine [BIOCHEM] $C_4H_8N_2O_3$ A white, crystalline amino acid found in many plant seeds. { ə'spar·əˌjēn }

asparagus [BOT] *Asparagus officinalis*. A dioecious, perennial monocot belonging to the order Liliales; the shoot of the plant is edible. { ə'spar·ə·gəs }

aspartase [BIOCHEM] A bacterial enzyme that catalyzes the deamination of aspartic acid to fumaric acid and ammonia. { ə'spärˌtās }

aspartate [BIOCHEM] A compound that is an ester or salt of aspartic acid. { ə'spärˌtāt }

aspartic acid [BIOCHEM] $C_4H_7NO_4$ A nonessential, crystalline dicarboxylic amino acid found in plants and animals, especially in molasses from young sugarcane and sugarbeet. { ə'spärd·ik 'as·əd }

aspartokinase [BIOCHEM] An enzyme that catalyzes the reaction of aspartic acid with adenosinetriphosphate to give aspartyl phosphate. { əˌspärd·ō'kīˌnās }

aspartoyl [BIOCHEM] $-COCH_2CH(NH_2)CO-$ A bivalent radical derived from aspartic acid. { ə'spärd·əˌwil }

aspartyl phosphate [BIOCHEM] $H_2O_3POOCH_2$-$CHNH_2$- COOH An intermediate in the biosynthesis of pyrimidines. { ə'spärd·əl 'fäsˌfāt }

aspect [ECOL] Seasonal appearance. { 'a ˌspekt }

aspection [ECOL] Seasonal change in appearance or constitution of a plant community. { a'spek·shən }

aspen [BOT] Any of several species of poplars (*Populus*) characterized by weak, flattened leaf stalks which cause the leaves to flutter in the slightest breeze. { 'as·pən }

Aspergillaceae [MYCOL] Former name for the Eurotiaceae. { ˌas·pər·jə'lās·ēˌē }

Aspergillales [MYCOL] Former name for the Eurotiales. { ˌas·pər·jə'lā·lēz }

aspergillic acid [BIOCHEM] $C_{12}H_{20}O_2N_2$ A diketopiperazine-like antifungal antibiotic produced by certain strains of *Aspergillus flavus*. { ¦as·pər¦jil·ik 'as·əd }

aspergillin [MYCOL] **1.** A black pigment found in spores of some molds of the genus *Aspergillus*. **2.** A broad-spectrum antibacterial antibiotic produced by the molds *Aspergillus flavus* and *A. fumigatus*. { ˌas·pər'jil·ən }

aspergillosis [MED] A rare fungus infection of humans and animals caused by several species of *Aspergillus*. { ˌas·pər·jil'ō·səs }

Aspergillus [MYCOL] A genus of fungi including several species of common molds and some human and plant pathogens. { ˌas·pər'jil·əs }

asperifoliate [BOT] Rough-leaved. { ¦as·pər·ə¦fōl·ēˌāt }

aspermatism [MED] **1.** Failure to ejaculate or secrete semen. **2.** Defective or absent sperm formation. { ā'spərm·əˌtiz·əm }

asperulate [BOT] Delicately roughened. { a'sper·əˌlāt }

asphradium [INV ZOO] An organ, believed to be a chemoreceptor, in mollusks. Also spelled osphradium. { a'sfrād·ē·əm }

asphyxia [MED] Suffocation due to oxygen deprivation, resulting in anoxia and carbon dioxide accumulation in the body. { a'sfik·sē·ə }

asphyxiation [MED] Suffocation caused by lowering the oxygen supply through the blood to bodily organs to a level that is incapable of supporting life. { asˌfik·sē'ā·shən }

aspiculate [INV ZOO] Lacking spicules, referring to Porifera. { a'spik·yə·lət }

Aspidiotinae [INV ZOO] A subfamily of homopteran insects in the superfamily Coccoidea. { aˌspid·ē'ä·təˌnē }

Aspidiphoridae [INV ZOO] An equivalent name for the Sphindidae. { aˌspid·ə'fȯr·əˌdē }

Aspidobothria [INV ZOO] An equivalent name for the Aspidogastrea. { ¦as·pəˌdō'bäth·rē·ə }

Aspidobothroidea [INV ZOO] A group of trematodes accorded class rank by W. J. Hargis. { ¦as·pəˌdō·bə'thrȯid·ē·ə }

Aspidobranchia [INV ZOO] An equivalent name for the Archaeogastropoda. { ¦as·pəˌdō'braŋk·ē·ə }

Aspidochirotacea [INV ZOO] A subclass of bilaterally symmetrical echinoderms in the class Holothuroidea characterized by tube feet and 10–30 shield-shaped tentacles. { ¦as·pəˌdōˌkī·rə'tās·ē·ə }

Aspidochirotida [INV ZOO] An order of holothurioid echinoderms in the subclass Aspidochirotacea characterized by respiratory trees and dorsal tube feet converted into tactile warts. { ¦as·pə‚dō‚kī'räd·ə·də }

Aspidocotylea [INV ZOO] An equivalent name for the Aspidogastrea. { ¦as·pə‚dō‚käd·ə'lē·ə }

Aspidodiadematidae [INV ZOO] A small family of deep-sea echinoderms in the order Diadematoida. { ¦as·pə‚dō‚dī·ə·də'mad·ə‚dē }

Aspidogastrea [INV ZOO] An order of endoparasitic worms in the class Trematoda having strongly developed ventral holdfasts. { ¦as·pə‚dō'gas·trē·ə }

Aspidogastridae [INV ZOO] A family of trematode worms in the order Aspidogastrea occurring as endoparasites of mollusks. { ¦as·pə‚dō'gas·trə‚de }

aspidospermine [PHARM] $C_{22}H_{30}O_2N_2$ White to brownish-yellow crystals with a melting point of 132–136°C; soluble in water, alcohol, chloroform, and ether; used in medicine. { ¦as·pə‚dō'spər‚mēn }

aspidospondyly [VERT ZOO] The condition in which the vertebral centra and spines are separate. { ‚as·pi·dō'spän·də·lē }

aspiration [MED] The removal of fluids from a cavity by suction. [MICROBIO] The use of suction to draw up a sample in a pipette. { ‚as·pə'rā·shən }

asporogenic mutant [MICROBIO] A bacillus that is unable to form spores due to alterations at any of several gene loci. { ¦ā‚spȯr·ə¦jen·ik 'myüt·ənt }

asporogenous [BOT] Not producing spores, especially of certain yeasts. { ¦ā·spə'räj·ə·nəs }

Aspredinidae [VERT ZOO] A family of salt-water catfishes in the order Siluriformes found off the coast of South America. { ‚a·sprə'din·ə·dē }

ass [VERT ZOO] Any of several perissodactyl mammals in the family Equidae belonging to the genus *Equus*, especially *E. hemionus* and *E. asinus*. { as }

assemblage [ECOL] A group of organisms sharing a common habitat by chance. { ə'sem·blij }

assimilation [PHYSIO] The conversion of nutritive materials into protoplasm. { ə‚sim·ə'lā·shən }

assimilative nitrate reduction [MICROBIO] The reduction of nitrates by some aerobic bacteria for purposes of assimilation. { ə‚sim·ə'lād·iv 'nī‚trāt ri‚dək·shən }

assimilative sulfate reduction [MICROBIO] The reduction of sulfates by certain obligate anaerobic bacteria for purposes of assimilation. { ə‚sim·ə'lād·iv 'səl‚fāt ri‚dək·shən }

associated automatic movement *See* synkinesia. { ə'sō·sē‚ād·əd ¦ȯd·ə'mad·ik 'müv·mənt }

association [ECOL] Major segment of a biome formed by a climax community, such as an oak-hickory forest of the deciduous forest biome. { ə‚sō·sē'ā·shən }

association area [PHYSIO] An area of the cerebral cortex that is thought to link and coordinate activities of the projection areas. { ə‚sō·sē'ā·shən ‚er·ē·ə }

association center [INV ZOO] In invertebrates, a nervous center coordinating and distributing stimuli from sensory receptors. { ə‚sō·sē'ā·shən ‚sen·tər }

association constant [BIOCHEM] A quantitative description of the affinity of a ligand for a protein that binds to it. { ə‚sō·sē'ā·shən ¦kän·stənt }

association fiber [NEUROSCI] One of the white nerve fibers situated just beneath the cortical substance and connecting the adjacent cerebral gyri. { ə‚sō·sē'ā·shən ‚fi·bər }

association neuron [NEUROSCI] A neuron, usually within the central nervous system, between sensory and motor neurons. { ə‚sō·sē'ā·shən 'nü‚rän }

assortative mating [GEN] Nonrandom mating with respect to phenotypes. { ə'sȯrd·əd·iv ¦mād·iŋ }

Astacidae [INV ZOO] A family of fresh-water crayfishes belonging to the section Macrura in the order Decapoda, occurring in the temperate regions of the Northern Hemisphere. { ‚as·tə'sī‚dē }

astacin [BIOCHEM] $C_{40}H_{48}O_4$ A red carotenoid ketone pigment found in crustaceans, as in the shell of a boiled lobster. { 'as·tə·sən }

Astacinae [INV ZOO] A subfamily of crayfishes in the family Astacidae including all North American species west of the Rocky Mountains. { ‚as·tə'sī‚nē }

astasia [MED] Lack of muscular coordination in standing. { ə'stāzh·ə }

astaxanthin [BIOCHEM] $C_{40}H_{52}O_4$ A violet carotenoid pigment found in combined form in certain crustacean shells and bird feathers. { ¦as·tə'zan·thən }

Asteidae [INV ZOO] A small, obscure family of cyclorrhaphous myodarian dipteran insects in the subsection Acalypteratae. { ‚as·tē'ī‚dē }

astelic [BOT] Lacking a stele or having a discontinuous arrangement of vascular bundles. { ā 'stēl·ik }

aster [BOT] Any of the herbaceous ornamental plants of the genus *Aster* belonging to the family Compositae. [CELL MOL] The star-shaped structure that encloses the centrosome at the end of the spindle during mitosis. { 'as·tər }

Asteraceae [BOT] An equivalent name for the Compositae. { ‚as·tə'rās·ē‚ē }

Asterales [BOT] An order of dicotyledonous plants in the subclass Asteridae, including aster, sunflower, zinnia, lettuce, artichoke, and dandelion; the ovary is inferior, flowers are borne in involucrate, centripetally flowering heads, and the calyx, when present, is modified into a set of scale-, hair-, or bristlelike structures called the pappus. { ‚as·tə'rāl·ēz }

astereognosis [MED] Loss of recognition of objects by touch, although recognition occurs through another sense, usually vision. Also known as tactile agnosia. { ā‚ste·rē·äg'nō·səs }

Asteridae [BOT] A large subclass of dicotyledonous plants in the class Magnoliopsida; plants are sympetalous, with unitegmic, tenuinucellate ovules and with the stamens usually as many as, or fewer than, the corolla lobes and alternate with them. { ˌas·tə'rīˌdē }

Asteriidae [INV ZOO] A large family of echinoderms in the order Forcipulatida, including many predatory sea stars. { ˌas·tə'ri·əˌdē }

Asterinidae [INV ZOO] The starlets, a family of echinoderms in the order Spinulosida. { ˌas·tə'rin·əˌdē }

asternal [ANAT] **1.** Not attached to the sternum. **2.** Without a sternum. { ā'stərn·əl }

Asteroidea [INV ZOO] The starfishes, a subclass of echinoderms in the subphylum Asterozoa characterized by five radial arms. { ˌas·tə'rȯid·ē·ə }

Asteroschematidae [INV ZOO] A family of ophiuroid echinoderms in the order Phrynophiurida with individuals having a small disk and stout arms. { ˌas·tə·rōˌskē'mad·əˌdē }

Asterozoa [INV ZOO] A subphylum of echinoderms characterized by a star-shaped body and radially divergent axes of symmetry. { ˌas·tə·rə'zō·ə }

asthenia [MED] Loss or lack of strength. { as'thē·nē·ə }

asthenopia [MED] Weakness of the eye muscles or of visual acuity, sometimes accompanied by pain and headache. { as·thə'nōp·ē·ə }

asthma [MED] A pulmonary disease marked by labored breathing, wheezing, and coughing; cause may be emotional stress, chemical irritation, or exposure to an allergen. { 'az·mə }

Asticcacaulis [MICROBIO] A genus of prosthecate bacteria; cells are rod-shaped with an appendage (pseudostalk), and reproduction is by binary fission of cells. { əˌstik·ə'kȯl·əs }

astichous [BOT] Not arranged in rows. { 'as·tə·kəs }

astigmatism [MED] A defect of vision due to irregular curvatures of the refractive surfaces of the eye so that focal points of light are distorted. { ə'stig·məˌtiz·əm }

astogeny [INV ZOO] Morphological and size changes associated with zooids of aging colonial animals. { a'stäj·ə·nē }

astomatal [BOT] Lacking stomata. Also known as astomous. { ā'stōm·əd·əl }

Astomatida [INV ZOO] An order of mouthless protozoans in the subclass Holotrichia; all species are invertebrate parasites, typically in oligochaete annelids. { as·tō'mad·ə·də }

astomatous [INV ZOO] Lacking a mouth, especially a cytostome, as in certain ciliates. { ā'stäm·əd·əs }

astomocnidae nematocyst [INV ZOO] A stinging cell whose thread has a closed end and either is adhesive or acts as a lasso to entangle prey. { ˌas·tə'mäk·nəˌdē ni'mad·əˌsist }

astomous [BOT] **1.** Having a capsule that bursts irregularly and is not dehiscent by an operculum. **2.** *See* astomatal. { 'as·tə·məs }

astraeid [INV ZOO] Of a group of corals that are imperforate. { a'strē·əd }

astragalus [ANAT] The bone of the ankle which articulates with the bones of the leg. Also known as talus. { ə'strag·ə·ləs }

Astrea [INV ZOO] A genus of mollusks in the class Gastropoda. { 'as·trē·ə }

astringent [MED] A substance applied to produce local contraction of blood vessels, to shrink mucous membranes, or to check discharges such as serum or mucus. { ə'strin·jənt }

astrocyte [NEUROSCI] A star-shaped cell; specifically, a neuroglial cell. { 'as·trəˌsīt }

astrocytoma [MED] A slow-growing glial tumor made up of cells resembling astrocytes; often it will undergo malignant change and assume the appearance and growth characteristics of a glioblastoma. { ˌas·trəˌsī'tōm·ə }

astroglia [NEUROSCI] Neuroglia composed of astrocytes. { ə'sträg·lē·ə }

Astropectinidae [INV ZOO] A family of echinoderms in the suborder Paxillosina occurring in all seas from tidal level downward. { ˌas·trōˌpek'tin·əˌdē }

astropyle [INV ZOO] A small, rounded projection from the central capsule of some radiolarians. { 'as·trōˌpīl }

astrosclereid [BOT] A type of sclereid cell that tends to be radiately branched but is otherwise quite variable, occurring in the leaves of many plants and in the petioles of *Camellia*. { ˌas·trə'skler·ē·əd }

astrosphere [CELL MOL] The center of the aster exclusive of the rays. { 'as·trōˌsfir }

asty [INV ZOO] A bryozoan colony. { 'aˌstī }

asulcal [BIOL] Without a sulcus. { ā'səl·kəl }

asymbolia [MED] An aphasia in which there is an inability to understand or use acquired symbols, such as speech, writing, or gestures, as a means of communication. { ˌāˌsim'bōl·ē·ə }

asymmetric cell division [CELL MOL] A phenomenon in which a mother cell divides into daughter cells that are unequal in size or cytoplasmic content, usually resulting in a different developmental fate for each. { ¦ā·səˌme·trik 'sel də'vizh·ən }

asynapsis [CELL MOL] Absence of pairing of homologous chromosomes during meiosis. { ˌā·si'nap·səs }

asynergia [MED] Faulty coordination of groups of organs or muscles normally acting in unison; particularly, the abnormal state of muscle antagonism in cerebellar disease. { ˌā·si'nərj·ē·ə }

asystole [MED] The absence of cardiac contraction; cardiac arrest. { ā'sis·tə·lē }

atactostele [BOT] A type of monocotyledonous siphonostele in which the vascular bundles are dispersed irregularly throughout the center of the stem. { ə'tak·təˌstēl }

atavism [EVOL] Appearance of a distant ancestral form of an organism or one of its parts due to reactivation of ancestral genes. { 'ad·əˌviz·əm }

ataxia [MED] Lack of muscular coordination due to any of several nervous system diseases. { ə'tak·sē·ə }

atelectasis [MED] **1.** Total or partial collapsed state of the lung. **2.** Failure of the lung to expand at birth. { ˌad·əl'ek·stə·səs }

Ateleopoidei [VERT ZOO] A family of oceanic fishes in the order Cetomimiformes characterized by an elongate body, lack of a dorsal fin, and an anal fin continuous with the caudal fin. { əˌtel·ē·ə'pȯid·ēˌī }

ateliosis [MED] Infantilism or dwarfism characterized by general, but proportional, underdevelopment and normal intelligence; associated with anterior pituitary deficiencies. { əˌtel·ē'ō·səs }

Atelopodidae [VERT ZOO] A family of small, brilliantly colored South and Central American frogs in the suborder Procoela. { əˌtel·ə'päd·əˌdē }

Atelostomata [INV ZOO] A superorder of echinoderms in the subclass Euechinoidea characterized by a rigid, exocylic test and lacking a lantern, or jaw, apparatus. { əˌtel·ə'stōm·əd·ə }

Athalamida [INV ZOO] An order of naked amebas of the subclass Granuloreticulosia in which pseudopodia are branched and threadlike (reticulopodia). { ˌath·ə'läm·əd·ə }

Athecanephria [INV ZOO] An order of tube-dwelling, tentaculate animals in the class Pogonophora characterized by a saclike anterior coelom. { ˌath·ə·kə'nef·rē·ə }

athecate [INV ZOO] Lacking a theca. { 'ath·əˌkāt }

Atherinidae [VERT ZOO] The silversides, a family of actinopterygian fishes of the order Atheriniformes. { ˌath·ə'rin·əˌdē }

Atheriniformes [VERT ZOO] An order of actinopterygian fishes in the infraclass Teleostei, including flyingfishes, needlefishes, killifishes, silversides, and allied species. { ˌath·əˌrin·ə'fȯrˌmēz }

atheroma [MED] A lipid deposit in the inner wall of an artery; characteristic of atherosclerosis. { ˌath·ə'rōm·ə }

atherosclerosis [MED] Deposition of lipid with proliferation of fibrous connective tissue cells in the inner walls of the arteries. { ˌath·ə·rōˌsklə'rō·səs }

athetosis [MED] Slow, recurrent, involuntary wormlike movements of various parts of the body associated with lesions of the basal ganglia. { ˌath·ə'tō·səs }

athetotic speech [MED] Disorder of articulation rhythm involving a general jerkiness in speech production that interferes with the normal rate of speech; associated with athetosis. { ¦ath·əˌtäd·ik 'spēch }

Athiorhodaceae [MICROBIO] Formerly the nonsulfur photosynthetic bacteria, a family of small, gram-negative, nonsporeforming, motile bacteria in the suborder Rhodobacteriineae. { āˌthī·əˌrō'dās·ēˌē }

athlete's foot [MED] Dermatophytosis of the feet, usually affecting the skin between the toes. { 'athˌlēts 'fu̇t }

athrocyte [HISTOL] A cell that engulfs extraneous material and stores it as granules in the cytoplasm. { 'ath·rəˌsīt }

Atlantacea [INV ZOO] A superfamily of mollusks in the subclass Prosobranchia. { ¦atˌlan¦tās·ē·ə }

Atlantic salmon [INV ZOO] *Salmo salar.* A species of salmon that occurs throughout the North Atlantic Ocean and spawns in eastern North America and western Europe; it can complete more than one migratory cycle and thus can breed multiple times. { ət¦lan·tik 'sa·mən }

atlas [ANAT] The first cervical vertebra. { 'at·ləs }

atony [MED] Absence or extremely low degree of tonus. { 'at·ən·ē }

atopic allergy [MED] A type of immediate hypersensitivity in humans resulting from spontaneous sensitization, usually by inhaled or ingested antigens; for example, asthma, hay fever, or hives. { ā'täp·ik 'al·ər·jē }

atopic dermatitis [MED] A chronic eruption of red patches accompanied by intense itching that usually begins in infancy but may continue into adult life; the disease has a genetic predisposition, but its expression is modified by environmental factors. { ˌāˌtäp·ik ˌdər·mə'tīd·əs }

atopy [MED] Clinically evident hypersensitivity. { 'ad·ə·pē }

ATP *See* adenosine triphosphate.

ATPase [BIOCHEM] An enzyme that hydrolyzes adenosine triphosphate into adenosine diphosphate and phosphate. { ˌāˌtē'pēˌās }

ATP synthase [BIOCHEM] An enzyme that catalyzes the conversion of phosphate and adenosine diphosphate into adenosine triphosphate during oxidative phosphorylation in mitochondria and bacteria or phosphorylation in chloroplasts. { ¦ā¦tē¦pē 'sinˌthās }

Atractidae [INV ZOO] A family of parasitic nematodes in the superfamily Oxyuroidea. { a'trak·təˌdē }

atresia [MED] Imperforation or closure of a natural orifice or passage of the body. { ə'trē·zhə }

atrial flutter [MED] A cardiac arrhythmia characterized by rapid, irregular atrial impulses and ineffective atrial contractions; the heartbeat varies from 60 to 180 per minute and is grossly irregular in intensity and rhythm. Also known as auricular flutter. { 'ā·trē·əl 'fləd·ər }

atrial septum [ANAT] The muscular wall between the atria of the heart. Also known as interatrial septum. { 'ā·trē·əl 'sep·təm }

atrichia [MED] Any congenital or acquired condition in which hair is essentially absent. { ā'trik·ē·ə }

Atrichornithidae [VERT ZOO] The scrubbirds, a family of suboscine perching birds in the suborder Menurae. { aˌtrī·kȯr'nith·əˌdē }

atrichous [CELL MOL] Lacking flagella. { 'a·trə·kəs }

atriopore [ZOO] The opening of an atrium as seen in lancelets and tunicates. { 'ā·trē·əˌpȯr }

atrioventricular bundle *See* bundle of His. { ¦ā·trē·ōˌven'trik·yə·lər 'bən·dəl }

atrioventricular canal [EMBRYO] The common passage between the atria and ventricles in the heart of the mammalian embryo before division of the organ into right and left sides. { ¦ā·trē·ō,ven'trik·yə·lər kə'nal }

atrioventricularis communis [MED] A congenital malformation of the heart in which partitioning has not occurred. { ¦ā·trē·ō,ven,trik·yə'lär·əs kə'myü·nəs }

atrioventricular node [ANAT] A group of slow-conducting fibers in the atrium of the vertebrate heart that are stimulated by impulses originating in the sinoatrial node and conduct impulses to the bundle of His. { ¦ā·trē·o,ven¦trik·yə·lər 'nōd }

atrioventricular valve [ANAT] A structure located at the orifice between the atrium and ventricle which maintains a unidirectional blood flow through the heart. { ¦a·trē·ō,ven'trik·yə·lər 'valv }

atrium [ANAT] **1.** The heart chamber that receives blood from the veins. **2.** The main part of the tympanic cavity, below the malleus. **3.** The external chamber to receive water from the gills in lancelets and tunicates. { 'ā·trē·əm }

atrophic arthritis *See* rheumatoid arthritis. { ā'trōf·ik ar'thrīd·əs }

atrophic gastritis [MED] Chronic inflammation of the stomach with atrophy of the mucosa. { ā'trōf·ik gas'trīd·əs }

atrophy [MED] Diminution in the size of a cell, tissue, or organ that was once fully developed and of normal size. { 'a·trə·fē }

atropine [PHARM] $C_{17}H_{23}O_3N$ An alkaloid extracted from *Atropa belladonna* and related plants of the family Solanaceae; used to relieve muscle spasms and pain, and to dilate the pupil of the eye. { 'a·trə,pēn }

atropine sulfate [PHARM] $C_{34}H_{48}N_2O_{10}S$ In the hydrate form the compound is either granules or powder; melting point is 190–194°C; used as an anticholinergic drug for humans and as an anticholinergic drug and smooth muscle relaxant drug for animals. { 'a·trə,pēn 'səl,fāt }

atropinization [PHYSIO] The physiological condition of being under the influence of atropine. { ə¦trō·pə·nə'zā·shən }

atropous *See* orthotropous. { 'a·trə·pəs }

attachment [VIROL] The initial stage in the infection of a cell by a virus that follows a chance collision by the virus with a suitable receptor area on the cell. { ə'tach·mənt }

attenuated vaccine [IMMUNOL] A suspension of weakened bacteria, viruses, or fractions thereof used to produce active immunity. { ə'ten·yə,wād·əd ,vak'sēn }

attenuation [BOT] Tapering, sometimes to a long point. [MICROBIO] Weakening or reduction of the virulence of a microorganism. { ə,ten·yə'wā·shən }

Atyidae [INV ZOO] A family of decapod crustaceans belonging to the section Caridea. { a'tī·ə,dē }

A-type virus particles [VIROL] A morphologically defined group of double-shelled spherical ribonucleic acid virus particles, often found in tumor cells. { 'ā ¦tīp 'vī·rəs ,pard·ə·kəlz }

Auberger blood group system [IMMUNOL] An immunologically distinct, genetically determined human erythrocyte antigen, demonstrated by reaction with anti-Au^2 antibody. { ¦ō·bər,zhā 'bləd ,grüp ,sis·təm }

Auchenorrhyncha [INV ZOO] A group of homopteran families and one superfamily, in which the beak arises at the anteroventral extremity of the face and is not sheathed by the propleura. { ,ȯk·ə·nə'riŋ·kə }

audiogenic seizure [MED] A transient episode of muscular, sensory, or psychic dysfunction induced by sound. { ¦ȯd·ē·ō¦jēn·ik 'sē·zhər }

audition [PHYSIO] Ability to hear. { ȯ'dish·ən }

auditory [PHYSIO] Pertaining to the act or the organs of hearing. { 'ȯd·ə,tȯr·ē }

auditory association area [NEUROSCI] The cortical association area in the brain just inferior to the auditory projection area, related to it anatomically and functionally by association fibers. { 'ȯd·ə,tȯr·ē ə,sō·sē'ā·shən ,er·ē·ə }

auditory impedance [PHYSIO] The acoustic impedance of the ear. { 'ȯd·ə,tȯr·ē im'pēd·əns }

auditory nerve [NEUROSCI] The eighth cranial nerve in vertebrates; either of a pair of sensory nerves composed of two sets of nerve fibers, the cochlear nerve and the vestibular nerve. Also known as acoustic nerve; vestibulocochlear nerve. { 'ȯd·ə,tȯr·ē 'nərv }

auditory placode [EMBRYO] An ectodermal thickening from which the inner ear develops in vertebrates. { 'ȯd·ə,tȯr·ē 'pla,kōd }

Auerbach's plexus *See* myenteric plexus. { 'au̇r ,baks ¦plek·səs }

aufwuch [ECOL] A plant or animal organism which is attached or clings to surfaces of leaves or stems of rooted plants above the bottom stratum. { 'ȯf,wək }

auk [VERT ZOO] Any of several large, short-necked diving birds (*Alca*) of the family Alcidae found along North Atlantic coasts. { ȯk }

Aulodonta [INV ZOO] An order of echinoderms proposed by R. Jackson in 1921. { ,ȯl·ə'dän·tə }

aulodont dentition [INV ZOO] In echinoderms, grooved teeth with epiphyses that do not meet, so the foramen magnum of the jaw is open. { 'ȯl·ə,dänt ,den'tish·ən }

aulophyte [ECOL] A nonparasitic plant that lives in the cavity of another plant for shelter. { 'ȯl·ə,fīt }

aura [MED] An unusual sensation preceding the appearance of more definite symptoms; in epilepsy, auras frequently precede the convulsive seizure. { 'ȯr·ə }

aural [BIOL] Pertaining to the ear or the sense of hearing. { 'ȯr·əl }

aurelia [INV ZOO] A morphological grouping of paramecia, including the elongate, cigar-shaped species which appear to be nearly circular in cross section. { ȯ'rēl·yə }

Aurelia [INV ZOO] A genus of scyphozoans. { ȯ'rēl·yə }

aureofacin [MICROBIO] An antifungal antibiotic produced by a strain of *Streptomyces aureofaciens*. { ,ȯr·ē·ō'fās·ən }

aureothricin [MICROBIO] $C_9H_{10}O_2N_2S_2$ An antibacterial antibiotic produced by a strain of *Actinomyces*. { ,ȯr·ē·ō'thrīs·ən }

aureusidin [BIOCHEM] $C_{15}H_{11}O_5$ A yellow flavonoid pigment found typically in the yellow snapdragon. Also known as 4,6,3′,4′-tetrahydroxyaurone. { ,ȯr·e·ə'sīd·ən }

auricle [ANAT] **1.** An ear-shaped appendage to an atrium of the heart. **2.** Any ear-shaped structure. **3.** *See* pinna. { 'ȯr·ə·kəl }

auricular fibrillation [MED] Arrhythmic contractions of the auricles. { ȯ'rik·yə·lər fib·rə'lā·shən }

auricular flutter *See* atrial flutter. { ȯ'rik·yə·lər 'fləd·ər }

auricularia larva [INV ZOO] A barrel-shaped, food-gathering larval form with a winding ciliated band, common to holothurians and asteroids. { ȯ,rik·yə'lar·ē·ə 'lär·və }

auricularis [ANAT] Any of the three muscles attached to the cartilage of the external ear. { ȯ¦rik·yə'lär·əs }

aurophore [INV ZOO] A bell-shaped structure which is part of the float of certain cnidarians. { 'ȯr·ə,fȯr }

aurothioglucose [PHARM] $C_6H_{11}AuO_5S$ A compound of gold with thioglucose, used for the treatment of rheumatoid arthritis and nondisseminated lupus erythematosus; administered in oil suspension. { ,ȯr·ō,thī·ə'glü,kōs }

auscultation [MED] The act of listening to sounds from internal organs, especially the heart and lungs, to aid in diagnosing their physical state. { ,ȯs·kəl'tā·shən }

Australia antigen [IMMUNOL] An infectious agent that causes hepatitis in some people; similar to an inherited serum protein in being polymorphic. { ȯ'strāl·yə 'ant·i·jən }

Australian faunal region [ECOL] A zoogeographic region that includes the terrestrial animal communities of Australia and all surrounding islands except those of Asia. { ȯ'strāl·yən 'fȯn·əl ,rē·jən }

Australian X disease *See* Murray Valley encephalitis. { ȯ'strāl·yən 'eks di,zēz }

austral region [ECOL] A North American biogeographic region including the region between transitional and tropical zones. { 'ȯs·trəl ,rē·jən }

Austroastacidae [INV ZOO] A family of crayfish in the order Decapoda found in temperate regions of the Southern Hemisphere. { ¦ȯs·trō·ə'stās·ə,dē }

Austrodecidae [INV ZOO] A monogeneric family of marine arthropods in the subphylum Pycnogonida. { ,ȯs·trə'des·ə,dē }

Austroriparian life zone [ECOL] The zone in which occurs the climate and biotic communities of the southeastern coniferous forests of North America. { ¦ȯs·trō,rī'per·ē·ən 'līf ,zōn }

autecology *See* autoecology. { ,ȯd·i'käl·ə·jē }

autoagglutination [IMMUNOL] Agglutination of an individual's erythrocytes by his own serum. Also known as autohemagglutination. { ¦ȯd·ō·ə,glüt·ən'ā·sh; ppn }

autoagglutinin [IMMUNOL] An antibody in an individual's blood serum that causes agglutination of his own erythrocytes. { ¦ȯd·ō·ə'glüt·ən·ən }

autoantibody [IMMUNOL] An antibody formed by an individual against his own tissues; common in hemolytic anemias. { ¦ȯd·ō'ant·i ,bäd·ē }

autoantigen [IMMUNOL] A tissue within the body which acquires the ability to incite the formation of complementary antibodies. { ¦ȯd·ō'ant·i·jən }

autoasphyxiation [PHYSIO] Asphyxiation by the products of metabolic activity. { ¦ȯd·ō·as,fik·sē'ā·shən }

autobasidium [MYCOL] An undivided basidium typically found in higher Basidiomycetes. { ¦ȯd·ō·bə'sid·ē·əm }

autocarp [BOT] **1.** A fruit formed as the result of self-fertilization. **2.** A fruit consisting of the ripened pericarp without adnate parts. { 'ȯd·ō,kärp }

autocarpy [BOT] Production of fruit by self-fertilization. { 'ȯd·ō,kärp·ē }

autochory [ECOL] Active self-dispersal of individuals or their disseminules. { 'ȯd·ō,kȯr·ē }

autochthonous [ECOL] Pertaining to organisms or organic sediments that are indigenous to a given ecosystem. { ȯ'täk·thə·nas }

autochthonous microorganism [MICROBIO] An indigenous form of soil microorganisms, responsible for chemical processes that occur in the soil under normal conditions. { ȯ'täk·thə·nas ,mī·krō'ȯr·gə,niz·əm }

autocoenobium [INV ZOO] An asexually produced coenobium that is a miniature of the parent. { ¦ȯd·ō·sē'nō·bē·əm }

autocopulation [INV ZOO] Self-copulation; sometimes occurs in certain hermaphroditic worms. { ¦ȯd·ō,käp·yə'lā·shan }

autocrine signaling [PHYSIO] Signaling in which cells respond to substances that they themselves release. { ¦ȯd·ə,krīn 'sig·nəl·iŋ }

autodeme [ECOL] A plant population in which most individuals are self-fertilized. { 'ȯd·ō,dēm }

autoecious [BOT] *See* autoicous. [MYCOL] Referring to a parasitic fungus that completes its entire life cycle on a single host. { ȯ'tēsh·əs }

autoecology [ECOL] The study of how a particular species responds to the environment. Also spelled autecology. { ,ȯd·ō i'käl·ə·jē }

autogamy [BIOL] A process of self-fertilization that results in homozygosis; occurs in some flowering plants, fungi, and protozoans. { 'ȯd·ō,gam·ē }

autogenesis *See* abiogenesis. { ¦ȯd·ō·'jen·ə·səs }

autogenous [BIOL] Originating or derived from sources within the same individual. { ȯ'täj·ə·nəs }

autogenous control [CELL MOL] Regulation of gene expression by a product of the gene itself that either inhibits or enhances the gene's activity. { ȯ'täj·ə·nəs kən'trōl }

autogenous insect [INV ZOO] Any insect in which adult females can produce eggs without first feeding. { ȯ'täj·ə·nəs 'in,sekt }

autogenous vaccine [IMMUNOL] A vaccine prepared from a culture of microorganisms taken directly from the infected person. { ȯ'täj·ə·nəs ,vak'sēn }

autograft *See* autotransplant. { 'ȯd·ō,graft }

autohemagglutination *See* autoagglutination. { ¦ȯd·ō,hēm·ə,glüt·ən'ā·shən }

autohemolysis [PHYSIO] The spontaneous lysis of blood which occurs during an incubation in a hematological procedure. { ¦ȯd·ō·hə'mäl·ə·səs }

autohemorrhage [INV ZOO] Voluntary exudation or ejection of nauseous or poisonous blood by certain insects as a defense against predators. { ¦ȯd·ō'hem·rij }

autohemotherapy [MED] Treatment of disease with the patient's own blood, withdrawn by venipuncture and then injected intramuscularly. { ¦ȯd·ō,hēm·ə'ther·ə·pē }

autoicous [BOT] Having male and female organs on the same plant but on different branches. Also spelled autoecious. { ȯ'tȯi·kəs }

autoimmune disease [IMMUNOL] An illness involving the formation of autoantibodies which appear to cause pathological damage to the host. { ¦ȯd·ō·ə'myün di,zēz }

autoimmune hemolytic anemia [IMMUNOL] An autoimmune disease in which antibodies initiate complement lysis of red blood cells. { ¦ȯd·ō·i¦myün ,hē·mə¦lid·ik ə'nēm·yə }

autoimmunity [IMMUNOL] An immune state in which antibodies are formed against the person's own body tissues. { ¦ȯd·ō·ə'myün·əd·ē }

autoimmunization [IMMUNOL] Immunization obtained by natural processes within the body. { ¦ȯd·ō,i·myə·nə'zā·shən }

autoinfection [MED] Reinfection by an organism existing within the body or transferred from one part of the body to another. { ¦ȯd·ō,in 'fek·shən }

autoinoculation [MED] **1.** Spread of a disease from one part of the body to another. **2.** Injection of an autovaccine. { ¦ȯd·ō·in,äk·yə'lā·shən }

autointoxication [MED] Poisoning by metabolic products elaborated within the body; generally, toxemia of pathologic states. { ¦ȯd·ō·in,täk·sə'kā·shən }

autologous [BIOL] Derived from or produced by the individual in question, such as an autologous protein or an autologous graft. { ȯ'täl·ə·gəs }

autolysis [PATH] Self-digestion by body cells following somatic or organ death or ischemic injury. { ȯ'täl·ə·səs }

autolysosome *See* autophagic vacuole. { ¦ȯd·ō'lī·sə,sōm }

autolytic enzyme [BIOCHEM] A bacterial enzyme, located in the cell wall, that causes disintegration of the cell following injury or death. { ¦ȯd·əl¦id·ik 'en,zīm }

Autolytinae [INV ZOO] A subfamily of errantian polychaetes in the family Syllidae. { ,ȯd·ə'lid·ə·nē }

automated fingerprint identification system [FOREN] A searchable database of finger and palm print records used to verify the identity of users of systems or of criminals, or to link unsolved crimes. { ¦ȯd·ə,mād·əd ¦fiŋ·gər,print ī,den·tə·fə'kā·shən ,sis·təm }

automatism [BIOL] Spontaneous activity of tissues or cells. [MED] An act performed with no apparent exercise of will, as in sleepwalking and certain hysterical and epileptic states. { ȯ'täm·ə,tiz·əm }

automutagen [GEN] Any mutagenic chemical formed as a product of metabolism. { ¦ȯd·ō'myü·də·jən }

autonomic agent [NEUROSCI] A compound that reduces or enhances nerve-impulse transmission across synaptic junctions, especially in the autonomic nervous system. { ¦ȯd·ə¦näm·ik 'ā·jənt }

autonomic movement [BOT] A plant movement that results from internal growth changes and is independent of changes in the external environment. { ¦ȯd·ə¦näm·ik 'müv·mənt }

autonomic nervous system [NEUROSCI] The visceral or involuntary division of the nervous system in vertebrates, which enervates glands, viscera, and smooth, cardiac, and some striated muscles. { ¦ȯd·ə¦näm·ik 'nər·vəs ,sis·təm }

autonomic reflex system [PHYSIO] An involuntary biological control system characterized by the uncontrolled functioning of smooth muscles and glands to maintain an acceptable internal environment. { ¦ȯd·ə¦näm·ik 'rē,fleks ,sis·təm }

autophagic vacuole [CELL MOL] A membrane-bound cellular organelle that engulfs pieces of the substance of the cell itself. Also known as autolysosome. { ¦ȯd·ō¦fā·jik 'vak·yə·wōl }

autophagocytosis [CELL MOL] The cellular process of phagocytizing a portion of protoplasm by a vacuole within the cell. { ¦ȯd·ō,fag·ə·sī'tō·səs }

autophagy [CELL MOL] The cellular process of self-digestion. { 'ȯd·ə,fā·jē }

autopolyploid [GEN] A cell or organism having three or more sets of chromosomes all derived from the same species. { ¦ȯd·ō¦päl·i,plȯid }

autopsy [PATH] A postmortem examination of the body to determine cause of death. { 'ȯ,tap·sē }

autoserum [IMMUNOL] A serum obtained from a patient used for treatment of that patient. { 'ȯd·ō,sir·əm }

autosexing [BIOL] Displaying differential sex characters at birth, noted particularly in fowl bred for sex-specific colors and patterns. { 'ȯd·ō,seks·iŋ }

autoskeleton [INV ZOO] The endoskeleton of a sponge. { 'ȯd·ō,skel·ə·tən }

autosomal dominant hearing loss [MED] Typically, a progressive form of hearing loss in which one of an individual's two copies of the autosomal dominant hearing loss gene is mutated. { ˌȯd·ə¦sōm·əl ¦däm·ə·nənt ˈhēr·iŋ ˌlȯs }

autosomal recessive hearing loss [MED] Typically, a congenital, severe loss of hearing (up to and including complete deafness) that occurs only if an individual inherits a mutant copy of an autosomal recessive hearing loss gene from each parent. { ˌȯd·ə¦sōm·əl ri¦ses·iv ˈhēr·iŋ ˌlȯs }

autosomal trait [GEN] Any characteristic determined by autosomal genes. { ¦ȯd·ə¦sō·məl ˈtrāt }

autosome [GEN] Any chromosome other than a sex chromosome. { ˈȯd·ōˌsōm }

autospore [BOT] In algae, a nonmotile asexual reproductive cell or a nonmotile spore that is a miniature of the cell that produces it. { ˈȯd·ōˌspȯr }

autostylic [VERT ZOO] Having the jaws attached directly to the cranium, as in chimeras, amphibians, and higher vertebrates. { ¦ȯd·ō¦stīl·ik }

autosyndesis [CELL MOL] The act of pairing of homologous chromosomes from the same parent during meiosis in polyploids. { ¦ȯd·ō¦sin·də·səs }

autotomy [MED] Surgical removal of a part of one's own body. [ZOO] The process of self-amputation of appendages in crabs and other crustaceans and tails in some salamanders and lizards under stress. { ȯˈtäd·ə·mē }

autotransplant [BIOL] Tissue removed from an organism and grafted on another site of the same organism. Also known as autograft. { ¦ȯd·ōˈtranzˌplant }

autotroph [BIOL] An organism capable of synthesizing organic nutrients directly from simple inorganic substances, such as carbon dioxide and inorganic nitrogen. { ˈȯd·ōˌträf }

autotrophic ecosystem [ECOL] An ecosystem that has primary producers as a principal component, and sunlight as the major initial energy source. { ¦ȯd·əˌtrōf·ik ˈek·ōˌsis·təm }

autotrophic succession [ECOL] A type of ecological succession that involves organisms that can utilize renewable resources. { ¦ȯd·əˌtrō·fik səkˈsesh·ən }

autozooecium [INV ZOO] The tube enclosing an autozooid. { ¦ȯd·ōˌzōˈēsh·ē·əm }

autozooid [INV ZOO] An unspecialized feeding individual in a bryozoan colony, possessing fully developed organs and exoskeleton. { ¦ȯd·ō¦zōˌȯid }

auxanogram [MICROBIO] A plate culture provided with variable growth conditions to determine the effects of specific environmental factors. { ȯgˈzan·əˌgram }

auxanography [MICROBIO] The study of growth-inhibiting or growth-promoting agents by means of auxanograms. { ˌȯg·zəˈnäg·rə·fē }

auxesis [PHYSIO] Growth resulting from increase in cell size. { ˌȯgˈzē·səs }

auximone [BIOCHEM] Any of certain growth-promoting substances occurring principally in sphagnum peat decomposed by nitrogen bacteria. { ˈȯk·səˌmōn }

auxin [BIOCHEM] Any organic compound which promotes plant growth along the longitudinal axis when applied to shoots free from indigenous growth-promoting substances. { ˈȯk·sən }

auxoautotrophic [BIOL] Requiring no exogenous growth factors. { ¦ȯk·sō¦ȯd·ə¦trä·fik }

auxocyte [BIOL] A gamete-forming cell, such as an oocyte or spermatocyte, or a sporocyte during its growth period. Also known as gonotocont. { ˈȯk·səˌsīt }

auxoheterotrophic [BIOL] Requiring exogenous growth factors. { ¦ȯk·sō¦hed·ə·rə¦trä·fik }

auxospore [INV ZOO] A reproductive cell in diatoms formed in association with rejuvenescence by the union of two cells that have diminished in size through repeated divisions. { ˈȯk·səˌspȯr }

auxotonic [BOT] Induced by growth rather than by exogenous stimuli. { ˌȯk·səˈtän·ik }

auxotrophic mutant [GEN] An organism that requires a specific exogenous growth factor, such as an amino acid, for its growth. { ¦ȯk·sə¦trä·fik ˈmyüt·ənt }

available-chlorine method [MICROBIO] A technique for the standardization of chlorine disinfectants intended for use as germicidal rinses on cleaned surfaces; increments of bacterial inoculum are added to different disinfectant concentrations, and after incubation the results indicate the capacity of the disinfectant to handle an increasing bacterial load before exhaustion of available chlorine, the germicidal principle. { əˈvāl·ə·bəl ˈklȯrˌēn ˌmeth·əd }

avalanche conduction [NEUROSCI] Conduction of a nerve impulse through several neurons which converge, increasing the discharge intensity by summation. { ˈav·əˌlanch kənˈdək·shən }

Avena [BOT] A genus of grasses (family Gramineae), including oats, characterized by an inflorescence that is loosely paniculate, two-toothed lemmas, and deeply furrowed grains. { əˈvēn·ə }

avenin [BIOCHEM] The glutelin of oats. { əˈvēn·ən }

Aves [VERT ZOO] A class of animals composed of the birds, which are warm-blooded, egg-laying vertebrates primarily adapted for flying. { ˈāˌvēz }

avianize [VIROL] To attenuate a virus by repeated culture on chick embryos. { ˈav·ē·əˌnīz }

avian leukosis [VET MED] A disease complex in fowl probably caused by viruses and characterized by autonomous proliferation of blood-forming cells. { ˈav·ē·ən lüˈkō·səs }

avian pneumoencephalitis *See* Newcastle disease. { ˈav·ē·ən ¦nü·mō·inˌsef·əˈlīd·əs }

avian pseudoplague *See* Newcastle disease. { ˈav·ē·ən ˈsüd·ōˌplāg }

avian tuberculosis [VET MED] A tuberculosis-like mycobacterial disease of fowl caused by *Mycobacterium avium*. { 'av·ē·ən tə,bər·kyə'lō·səs }

aviation medicine *See* aerospace medicine. { ,ā·vē'ā·shən 'med·ə·sən }

avicolous [ECOL] Living on birds, as of certain insects. { ā'vik·ə·ləs }

avicularium [INV ZOO] A specialized individual in a bryozoan colony with a beak that keeps other animals from settling on the colony. { ə,vik·yə'lar·ē·ən }

aviculture [VERT ZOO] Care and breeding of birds, especially wild birds, in captivity. { 'ā·və,kəl·chər }

avidin [BIOCHEM] A protein constituting 0.2% of the total protein in egg white; molecular weight is 70,000; combines firmly with biotin but loses this ability when subjected to heat. { 'av·əd·ən }

avidity [IMMUNOL] The total binding strength of a polyvalent antibody (an antibody that has multiple binding sites for the same antigen) with a polyvalent antigen (an antigen with multiple identical antibody-binding sites); equivalent to the sum of the affinities at each of the binding sites. { ə'vid·əd·ē }

avifauna [VERT ZOO] **1.** Birds, collectively. **2.** Birds characterizing a period, region, or environment. { ¦ā·və¦fȯn·ə }

avitaminosis [MED] Any vitamin-deficiency disease. { ¦ā,vīd·ə·mə'nō·səs }

avocado [BOT] *Persea americana*. A subtropical evergreen tree of the order Magnoliales that bears a pulpy pear-shaped edible fruit. { ,av·ə'käd·ō }

avulsion [MED] Tearing one part away from the other, either by trauma or surgery. { ə'vəl·shən }

awn [BOT] Any of the bristles at the ends of glumes or bracts on the spikelets of oats, barley, and some wheat and grasses. Also known as beard. { ȯn }

axenic culture [BIOL] The growth of organisms of a single species in the absence of cells or living organisms of any other species. { ā'zen·ik 'kəl·chər }

axial filament [CELL MOL] The central microtubule elements of a cilium or flagellum. [INV ZOO] An organic fiber which serves as the core for deposition of mineral substance to form a ray of a sponge spicule. { 'ak·sē·əl 'fil·ə·mənt }

axial gland [INV ZOO] A structure enclosing the stone canal in certain echinoderms; its function is uncertain. { 'ak·sē·əl 'gland }

axial gradient [EMBRYO] In some invertebrates, a graded difference in metobolic activity along the anterior-posterior, dorsal-ventral, and medial-lateral embryonic axes. { 'ak·sē·əl 'grād·ē·ənt }

axial musculature [ANAT] The muscles that lie along the longitudinal axis of the vertebrate body. { 'ak·sē·əl 'məs·kyə·lə·chər }

axial skeleton [ANAT] The bones composing the skull, vertebral column, and associated structures of the vertebrate body. { 'ak·sē·əl 'skel·i·tən }

axiation [EMBRYO] The formation or development of axial structures, such as the neural tube. { ,ak·sē'ā·shən }

Axiidae [INV ZOO] A family of decapod crustaceans, including the hermit crabs, in the suborder Reptantia. { ,ak'sī·ə,dē }

axil [BIOL] The angle between a structure and the axis from which it arises, especially for branches and leaves. { 'ak·səl }

axilla [ANAT] The depression between the arm and the thoracic wall; the armpit. [BOT] An axil. { ak'sil·ə }

axillary [ANAT] Of, pertaining to, or near the axilla or armpit. [BOT] Placed or growing in the axis of a branch or leaf. { 'ak·sə,ler·ē }

axillary bud [BOT] A lateral bud borne in the axil of a leaf. { 'ak·sə,ler·ē 'bəd }

axillary sweat gland [ANAT] An apocrine gland located in the axilla. { 'ak·sə,ler·ē 'swet ,gland }

Axinellina [INV ZOO] A suborder of sponges in the order Clavaxinellida. { ,ak·sə'nə'lī·nə }

axis [ANAT] **1.** The second cervical vertebra in higher vertebrates; the first vertebra of amphibians. **2.** The center line of an organism, organ, or other body part. { 'ak·səs }

axis cylinder [NEUROSCI] **1.** The central mass of a nerve fiber. **2.** The core of protoplasm in a medullated nerve fiber. { 'ak·səs ¦sil·ən·dər }

axis of pelvis [ANAT] A curved line which forms right angles to the pelvic-cavity planes. { 'ak·səs əv 'pel·vəs }

axoblast [INV ZOO] **1.** The germ cell in mesozoans; cells are linearly arranged in the longitudinal axis and produce the primary nematogens. **2.** The individual scleroblast of the axis epithelium which produces spicules in octocorals. { 'ak·sə,blast }

axocoel [INV ZOO] The anterior pair of coelomic sacs in the dipleurula larval ancestral stage of echinoderms. { 'ak·sə,sēl }

axogamy [BOT] Having sex organs on a leafy stem. { ak'säg·ə·mē }

axolemma [NEUROSCI] The plasma membrane of an axon. { ,ak·sə'lem·ə }

axolotl [VERT ZOO] The neotenous larva of some salamanders in the family Ambystomidae. { ¦ak·sə¦läd·əl }

axolotl unit [BIOL] A unit for the standardization of thyroid extracts. { ¦ak·sə¦läd·əl 'yü·nət }

axon [NEUROSCI] The process or nerve fiber of a neuron that carries the unidirectional nerve impulse away from the cell body. Also known as neuraxon; neurite. { 'ak,sän }

axonal transport [NEUROSCI] The movement of organelles and molecules down a nerve cell's axon to its terminals along its cytoplasmic microtubule network. { ak¦sän·əl 'tranz,pȯrt }

axoneme [CELL MOL] A bundle of fibrils enclosed by a membrane that is continuous with the plasma membrane. { 'ak·sə,nēm }

Axonolaimoidea [INV ZOO] A superfamily of free-living nematodes with species inhabiting marine and brackish-water environments. { ¦ak·sə·nō·lə'mȯid·ē·ə }

axoplasm [NEUROSCI] The protoplasm of an axon. { 'ak·sə,plaz·əm }

axopodium [INV ZOO] A semipermanent pseudopodium composed of axial filaments surrounded by a cytoplasmic envelope. { ,ak·sə'pōd·ē·əm }

aye-aye [VERT ZOO] *Daubentonia madagascariensis*. A rare prosimian primate indigenous to eastern Madagascar; the single species of the family Daubentoniidae. { 'ī,ī }

azaserine [PHARM] $C_4H_7O_4N_3$ An antibiotic produced by a species of *Streptomyces* or by synthesis; used in treatment of acute leukemia. { ə'zas·ə,rēn }

6-azauradine [PHARM] $C_8H_{11}N_3O_6$ Crystals having a melting point of 160–161°C; the base is used as an antineoplastic agent, and the triacetate is used as a drug for psoriasis. { ¦siks ə'zȯr·ə,dēn }

2-azetidinecarboxylic acid [BIOCHEM] $C_4H_7NO_2$ Crystals which discolor at 200°C and darken until 310°C; soluble in water; a specific antagonist of L-proline; used in the production of abnormally high-molecular-weight polypeptides. { ¦tü ə,zed·ə,dēn·ə,kär,bäk'sil·ik 'as·əd }

Azomonas [MICROBIO] A genus of large, motile, oval to spherical bacteria in the family Axotobacteraceae; members produce no cysts and secrete large quantities of capsular slime. { ,az·ə'mō·nəs }

azomycin [MICROBIO] $C_3H_3O_2N_3$ An antimicrobial antibiotic produced by a strain of *Nocardia mesenterica*. { ā·zō'mis·ən }

azoospermia [PHYSIO] **1.** Absence of motile sperm in the semen. **2.** Failure of formation and development of sperm. { ¦ā,zō·ō'spər·mē·ə }

azotemia [MED] The presence of excessive amounts of nitrogenous compounds in the blood. { ,az·ə'tem·ē·ə }

Azotobacter [MICROBIO] A genus of large, usually motile, rod-shaped, oval, or spherical bacteria in the family Azotobacteraceae; form thick-walled cysts, and may produce large quantities of capsular slime. { ə'zōd·ə,bak·tər }

Azotobacteraceae [MICROBIO] A family of large, bluntly rod-shaped, gram-negative, aerobic bacteria capable of fixing molecular nitrogen. { ə¦zōd·ə,bak·tə'rās·ē,ē }

azoturia [MED] A condition characterized by excess amounts of urea or other nitrogenous substances in the urine. { 'az·ə'tu̇r·ē·ə }

azure B [CELL MOL] $C_{15}H_{16}ClN_3S$ A metachromatic basic dye that imparts a green color to chromosomes, a blue color to nucleoli and cytoplasmic ribosomes, and a red color to deposits containing mucopolysaccharides. { ¦azh·ər 'bē }

azygospore [MYCOL] A spore which is morphologically similar to a zygospore but is formed parthenogenetically. Also known as parthenospore. { ā'zī·gə,spȯr }

azygos vein [ANAT] A branch of the right precava which drains the intercostal muscles and empties into the superior vena cava. { ā'zī·gəs ,vān }

azygote [BIOL] An individual produced by haploid parthenogenesis. { ā'zī,gōt }

B

Babesia [INV ZOO] The type genus of the Babesiidae, a protozoan family containing red blood cell parasites. { bə'bezh·ə }

babesiasis [VET MED] A tick-borne protozoan disease of mammals other than humans caused by species of *Babesia*. { ˌbab·ə'zī·ə·səs }

Babesiidae [INV ZOO] A family of protozoans in the suborder Haemosporina containing parasites of vertebrate red blood cells. { ˌbab·ə'zī·əˌdē }

Babinski reflex [MED] An abnormal reflex after infancy associated with a disturbance of the pyramidal tract, characterized by extension of the great toe with fanning of the other toes on sharply stroking the lateral aspect of the sole. { bə'binz·kē 'riˌfleks }

baboon [VERT ZOO] Any of five species of large African and Asian terrestrial primates of the genus *Papio*, distinguished by a doglike muzzle, a short tail, and naked callosities on the buttocks. { ba'bün }

babuina [VERT ZOO] A female baboon. { ˌbab·ə'wēn·ə }

baby-pig disease [VET MED] Acute hypoglycemia of newborn pigs; usually fatal if untreated. { ¦bā·bē 'pig diˌzēz }

baccate [BOT] **1.** Bearing berries. **2.** Having pulp like a berry. { 'bakˌāt }

bacciferous [BOT] Bearing berries. { bak'sif·ə·rəs }

Bacillaceae [MICROBIO] The single family of endospore-forming rods and cocci. { ˌbas·ə'lās·ēˌē }

Bacillariophyceae [BOT] The diatoms, a class of algae in the division Chrysophyta. { ˌbas·əˌler·ē·ə'fīs·ēˌē }

Bacillariophyta [BOT] An equivalent name for the Bacillariophyceae. { ˌbas·əˌler·ē'ä·fəd·ə }

bacillary [MICROBIO] **1.** Rod-shaped. **2.** Produced by, pertaining to, or resembling bacilli. { 'bas·əˌler·ē }

bacillary dysentery [MED] A highly infectious bacterial disease of humans, localized in the bowels; caused by *Shigella*. { 'bas·əˌler·ē 'dis·ənˌter·ē }

bacillary white diarrhea *See* pullorum disease. { 'bas·əˌler·ē ¦wīt ˌdi·ə'rē·ə }

bacilluria [MED] The presence of bacilli in the urine. { ˌbas·ə'lu̇r·ē·ə }

bacillus [MICROBIO] Any rod-shaped bacterium. { bə'sil·əs }

Bacillus [MICROBIO] A genus of bacteria in the family Bacillaceae; rod-shaped cells are aerobes or facultative anaerobes and usually produce catalase. { bə'sil·əs }

Bacillus anthracis [MICROBIO] A gram-positive, rod-shaped, endospore-forming bacterium that is the causative agent of anthrax; its spores can remain viable for many years in soil, water, and animal hides and products. { bə¦sil·əs ˌan 'thrak·əs }

Bacillus Calmette-Guérin vaccine [IMMUNOL] A vaccine prepared from attenuated human tubercle bacilli and used to immunize humans against tuberculosis. Abbreviated BCG vaccine. { bə'sil·əs ˌkal'met ˌgā'ra͟n ˌvakˌsēn }

Bacillus cereus [MICROBIO] A spore-forming bacterium that often survives cooking and grows to large numbers in improperly refrigerated foods; it produces both a diarrheal toxin and an emetic toxin in the gastrointestinal tract following its ingestion via contaminated meats, dried foods, and rice. { bəˌsil·əs 'sir·ē·əs }

Bacillus sphaericus [MICROBIO] An aerobic, spore-producing bacterium that forms a protein complex during sporulation that is toxic to the larvae of certain mosquitoes but is apparently harmless to all other forms of life. { bəˌsil·əs 'sfī·ri·kəs }

bacitracin [MICROBIO] A group of polypeptide antibiotics produced by *Bacillus licheniformis*. { ˌbas·ə'trās·ən }

back [ANAT] The part of the human body extending from the neck to the base of the spine. { bak }

backbone *See* spine. { 'bakˌbōn }

back bulb [BOT] A pseudobulb on certain orchid plants that remains on the plant after removal of the terminal growth, and that is used for propagation. { 'bak ˌbəlb }

backcross [GEN] A cross between an F_1 heterozygote and an individual of P_1 genotype. { 'bakˌrȯs }

backcross parent *See* recurrent parent. { 'bakˌkrȯs ˌper·ənt }

background extinction [EVOL] Intervals of lower extinction intensity between mass extinctions. { 'bakˌgrau̇nd ikˌstiŋk·shən }

background genotype [GEN] The genotype of

the organism other than those genetic loci responsible for the phenotype. { ¦ba,kraủnd 'jē·nə,tīp }

backmarsh [ECOL] Marshland formed in poorly drained areas of an alluvial floodplain. { 'bak,märsh }

backswamp depression [ECOL] A low swamp found adjacent to river levees. { 'bak,swamp di'presh·ən }

bacteremia [MED] Presence of bacteria in the blood. { 'bak·tə'rē·mē·ə }

bacteremic shock [MED] A state of shock occurring during the course of bacteremia, especially if caused by gram-negative bacteria. { ¦bak·tə¦rē·mik 'shäk }

bacteria [MICROBIO] Extremely small, relatively simple prokaryotic microorganisms traditionally classified with the fungi as Schizomycetes. { bak'tir·ē·ə }

Bacteriaceae [MICROBIO] A former designation for Brevibacteriaceae. { bak,tir·ē'ās·ē,ē }

bacterial bronchopneumonia [MED] Bacterial infection of the lung which has spread from infected bronchi. { bak'tir·ē·əl ¦bräŋ·kō·nủ'mō·nyə }

bacterial capsule [MICROBIO] A thick, mucous envelope, composed of polypeptides or carbohydrates, surrounding some bacteria. { bak'tir·ē·əl 'kap·səl }

bacterial coenzyme [MICROBIO] Organic molecules that participate directly in a bacterial enzymatic reaction and may be chemically altered during the reaction. { bak'tir·ē·əl kō'en,zīm }

bacterial competence [MICROBIO] The ability of cells in a bacterial culture to accept and be transformed by a molecule of transforming deoxyribonucleic acid. { bak'tir·ē·əl 'käm·pə·təns }

bacterial encephalitis [MED] Inflammation of the brain caused by primary or secondary bacterial infection. { bak'tir·ē·əl in,sef·ə'līd·əs }

bacterial endocarditis [MED] Inflammation of the endocardium due to bacterial invasion. Also known as subacute bacterial endocarditis. { bak'tir·ē·əl ,en·dō,kär'dīd·əs }

bacterial endoenzyme [MICROBIO] An enzyme produced and active within the bacterial cell. { bak'tir·ē·əl ,en·dō'en,zīm }

bacterial endospore [MICROBIO] A body, resistant to extremes of temperature and to dehydration, produced within the cells of gram-positive, sporeforming rods of *Bacillus* and *Clostridium* and by the coccus *Sporosarcina*. { bak'tir·ē·əl 'en·dō,spȯr }

bacterial enzyme [MICROBIO] Any of the metabolic catalysts produced by bacteria. { bak'tir·ē·əl 'en,zīm }

bacterial infection [MED] Establishment of an infective bacterial agent in or on the body of a host. { bak'tir·ē·əl in'fek·shən }

bacterial luminescence [MICROBIO] A light-producing phenomenon exhibited by certain bacteria. { bak'tir·ē·əl lü·mə'nes·əns }

bacterial metabolism [MICROBIO] Total chemical changes carried out by living bacteria. { bak'tir·ē·əl mə'tab·ə,liz·əm }

bacterial motility [MICROBIO] Self-propulsion in bacteria, either by gliding on a solid surface or by moving the flagella. { bak'tir·ē·əl mō'til·əd·ē }

bacterial photosynthesis [MICROBIO] Use of light energy to synthesize organic compounds in green and purple bacteria. { bak'tir·ē·əl ,fōd·ō'sin·thə·səs }

bacterial pigmentation [MICROBIO] The organic compounds produced by certain bacteria which give color to both liquid cultures and colonies. { bak'tir·ē·əl ,pig·mən'tā·shən }

bacterial pneumonia [MED] Consolidation of the lung caused by inflammatory exudation due to bacterial infection. { bak'tir·ē·əl nə'mō·nyə }

bacterial vaccine [IMMUNOL] A preparation of living, attenuated, or killed bacteria used to enhance the immune reaction in an individual already infected with the same bacteria. { bak'tir·ē·əl vak'sēn }

bacterial vaginosis [MED] Inflammation of the vagina that causes a nonirritating white or gray vaginal discharge, often with a distinctive fishy odor; it results from overgrowth of various normal vaginal bacteria and by depletion of *Lactobacillus* species, especially strains that produce hydrogen peroxide. { bak,tir·ē·əl ,vaj·ə'nō·səs }

bactericidin [IMMUNOL] An antibody that kills bacteria in the presence of complement. { bak¦tir·ə¦sīd·ən }

bacterin [IMMUNOL] A suspension of killed or weakened bacteria used in artificial immunization. { 'bak·tə·rən }

bacteriochlorophyll [BIOCHEM] $C_{52}H_{70}O_6N_4Mg$ A tetrahydroporphyrin chlorophyll compound occurring in the forms *a* and *b* in photosynthetic bacteria; there is no evidence that *b* has the empirical formula given. { bak¦tir·ē·ə'klȯr·ə,fil }

bacteriocin [MICROBIO] Any of a group of proteins produced by various strains of gram-negative bacteria that may inhibit the growth of other strains of the same or related species. { bak'tir·ē·ə,sīn }

bacteriocinogen [GEN] A plasmid deoxyribonucleic acid found in some strains of bacteria which specifies production of a bacteriocin, an antibiotic for some other bacteria. { bak,tir·ē·ə'sin·ə·jən }

bacteriocyte [INV ZOO] A modified fat cell found in certain insects that contains bacterium-shaped rods believed to be symbiotic bacteria. { bak'tir·ē·ə,sīt }

bacteriogenic [MICROBIO] Caused by bacteria. { bak¦tir·ē·ə'jen·ik }

bacteriologist [MICROBIO] A specialist in the study of bacteria. { bak,tir·ē'äl·ə·jəst }

bacteriology [MICROBIO] The science and study of bacteria; a specialized branch of microbiology. { bak,tir·ē'äl·ə·jē }

bacteriolysin [MICROBIO] An antibody that is active against and causes lysis of specific bacterial cells. { bak,tir·ē·ə'līs·ən }

bacteriolysis [MICROBIO] Dissolution of bacterial cells. { bak,tir·ē'äl·ə·səs }

Bacterionema [MICROBIO] A genus of bacteria in the family Actinomycetaceae; characteristically cells are rods with filaments attached and produce filamentous microcolonies; facultative anaerobes; carbohydrates are fermented. { bak ,tir·ē'än·ə·mə }

bacteriophage [VIROL] Any of the viruses that infect bacterial cells; each has a narrow host range. Also known as phage. { bak'tir·ē·ə,fāj }

bacteriorhodopsin [BIOCHEM] A purple substance in the cell membranes of halobacteria (found in extremely saline environments) during conditions of low oxygen, and consisting of the protein bacteriopsin and retinal, the same carotenoid found in the visual pigments of animals; in response to light, the purple membrane pumps protons out of the cell, providing the energy gradient for synthesis of adeniosine triphosphate. { bak,tir·ē·ō·rō'däp·sən }

bacterioruberin [MICROBIO] A carotenoid pigment found in some halophilic aerobic archaebacteria that gives them a striking red color and seems to protect them from strong sunlight in their natural environments. { bak,tir·ē·ə'rüb·ə·rən }

bacteriostasis [MICROBIO] Inhibition of bacterial growth and metabolism. { bak,tir·ē·ō'stā·səs }

bacteriostatic agent [MICROBIO] A substance that inhibits the growth of bacteria. { ,bak¦tir·ē·ō¦stad·ik 'ā·jənt }

bacteriotoxin [MICROBIO] **1.** Any toxin that destroys or inhibits growth of bacteria. **2.** A toxin produced by bacteria. { bak,tir·ē·ō'täk·sən }

bacteriotropin [IMMUNOL] An antibody that is increased in amount during specific immunization and that renders the corresponding bacterium more susceptible to phagocytosis. { bak,tir·ē'ä·trə·pən }

bacterioviridin *See* chlorobium chlorophyll. { bak ,tir·ē·ə'vir·ə·dən }

bacteriuria [MED] The occurrence of bacteria in the urine. { bak,tir·ē'yu̇·rē·ə }

bacteroid [MICROBIO] A bacterial form of irregular shape, frequently associated with special conditions. { 'bak·tə,ro̊id }

Bacteroidaceae [MICROBIO] The single family of gram-negative anaerobic bacteria; cells are nonsporeforming rods; some species are pathogenic. { ,bak·tə,ro̊i'dās·ē,ē }

Bactrian camel [VERT ZOO] *Camelus bactrianus.* The two-humped camel. { 'bak·trē·ən 'kam·əl }

Baculoviridae [VIROL] A family of invertebrate viruses comprising the genus *Baculovirus* that is characterized by enveloped rod-shaped virions containing a molecule of supercoiled double-stranded deoxyribonucleic. { ,bak·yə·lə'vir·ə,dī }

Baculovirus [VIROL] A genus of invertebrate viruses that contains enveloped, double-stranded, supercoiled deoxyribonucleic, includes the subgroups nuclear polyhedrosis virus and granulosis virus. { ,bak·yə·lə'vī·rəs }

baculum [VERT ZOO] The penis bone in lower mammals. Also known as os priapi. { 'bak·yə·ləm }

badger [VERT ZOO] Any of eight species of carnivorous mammals in six genera comprising the subfamily Melinae of the weasel family (Mustelidae). { 'baj·ər }

baeocyte [INV ZOO] A motile or nonmotile blue-green bacterial endospore. { 'bē·ə,sīt }

bagasse disease *See* bagassosis. { bə'gas di,zēz }

bagassosis [MED] A pneumoconiosis caused by the inhalation of bagasse dust, a dry sugarcane residue. Also known as bagasse disease. { ,bag·ə'sō·səs }

Bagridae [VERT ZOO] A family of semitropical catfishes in the suborder Siluroidei. { 'bag·rə,dē }

Bainbridge reflex [PHYSIO] A poorly understood reflex acceleration of the heart rate due to rise of pressure in the right atrium and vena cavae, possibly mediated through afferent vagal fibers. { 'bān,brij 'rē,fləks }

Bairdiacea [INV ZOO] A superfamily of ostracod crustaceans in the suborder Podocopa. { ,ber·dē'ās·ē·ə }

BAL *See* dimercaprol.

Balaenicipitidae [VERT ZOO] A family of wading birds composed of a single species, the shoebill stork (*Balaeniceps rex*), in the order Ciconiiformes. { bə,lēn·ə·sə'pid·ə,dē }

Balaenidae [VERT ZOO] The right whales, a family of cetacean mammals composed of five species in the suborder Mysticeti. { bə'lēn·ə,dē }

balanced anesthesia [MED] Anesthesia produced by safe doses of two or more agents or methods of anesthesia, each of which contributes to the total desired effect. { 'bal·ənst an·əs'thē·zhə }

balanced polymorphism [GEN] Maintenance in a population of two or more alleles in equilibrium at frequencies too high to be explained, particularly for the rarer of them, by mutation; commonly due to the selective advantage of a heterozygote over both homozygotes. { 'bal·ənst ¦päl·i'mo̊r,fiz·əm }

balanced translocation [GEN] Positional change of one or more chromosome segments in cells or gametes without alteration of the normal diploid or haploid complement of genetic material. { 'bal·ənst tranz·lō'kā·shən }

balancer [INV ZOO] *See* haltere. [VERT ZOO] Either of a pair of rodlike lateral appendages on the heads of some larval salamanders. { 'bal·ən·sər }

Balanidae [INV ZOO] A family of littoral, sessile barnacles in the suborder Balanomorpha. { bə'lan·ə,dē }

balanitis [MED] Inflammation of the glans of the penis or of the clitoris. { bal·ə'nīd·əs }

Balanoglossus [INV ZOO] A cosmopolitan genus of tongue worms belonging to the class Enteropneusta. { ,bal·ə·nō'glä·səs }

Balanomorpha [INV ZOO] The symmetrical barnacles, a suborder of sessile crustaceans in the order Thoracica. { ,bal·ə·nō'mo̊r·fə }

Balanopaceae [BOT] A small family of dioecious dicotyledonous plants in the order Fagales characterized by exstipulate leaves, seeds with endosperm, and the pistillate flower solitary in a multibracteate involucre. { ˌbal·ə·nō'pās·ēˌē }

Balanopales [BOT] An ordinal name suggested for the Balanopaceae in some classifications. { ˌbal·ə'näp·əˌlēz }

Balanophoraceae [BOT] A family of dicotyledonous terrestrial plants in the order Santalales characterized by dry nutlike fruit, one to five ovules, unisexual flowers, attachment to the stem of the host, and the lack of chlorophyll. { ˌbal·əˌnäf·ə'rās·ēˌē }

balanoposthitis [MED] Inflammation of the glans penis and of the prepuce. { ¦bal·ə·nō·päs'thīd·əs }

Balanopsidales [BOT] An order in some systems of classification which includes only the Balanopaceae of the Fagales. { ˌbal·ənˌäp·sə'dā·lēz }

balantidiasis [MED] An intestinal infection of humans caused by the protozoan *Balantidium coli*. { ˌbal·ən·tə'dī·ə·səs }

Balantidium [INV ZOO] A genus of protozoans in the order Trichostomatida containing the only ciliated protozoan species parasitic in humans, *Balantidium coli*. { ˌbal·ən'tid·ē·əm }

Balanus [INV ZOO] A genus of barnacles composed of sessile acorn barnacles; the type genus of the family Balanidae. { 'bal·ə·nəs }

Balbiani chromosome *See* polytene chromosome. { ˌbäl·bē'än·ē 'krō·məˌsōm }

Balbiani rings [CELL MOL] Localized swellings of a polytene chromosome. { ˌbäl·bē'än·ē ˌriŋz }

baldness [MED] Loss or absence of hair. { 'bȯld·nəs }

baleen [VERT ZOO] A horny substance, growing as fringed filter plates suspended from the upper jaws of whalebone whales. Also known as whalebone. { be'lēn }

Balfour's law [EMBRYO] The law that the speed with which any part of the ovum segments is roughly proportional to the protoplasm's concentration in that area; the segment's size is inversely proportional to the protoplasm's concentration. { 'bal·fərz ˌlȯ }

ball-and-socket joint *See* enarthrosis. { ¦bȯl ən 'säk·ət ˌjȯint }

ballistocardiogram [MED] The recording made by a ballistocardiograph. { bə¦lis·tō'kärd·ē·əˌgram }

ballistocardiograph [MED] A device to measure the volume of blood passing through the heart in a given period of time. { bə¦lis·tō'kärd·ē·əˌgraf }

ballistospore [MYCOL] A type of fungal spore that is forcibly discharged at maturity. { bə 'lis·təˌspȯr }

ball joint *See* ball-and-socket joint. { 'bȯl ˌjȯint }

balm [MED] A soothing or healing medication. { bäm }

balsa [BOT] *Ochroma lagopus*. A tropical American tree in the order Malvales; its wood is strong and lighter than cork. { 'bȯl·sə }

Balsaminaceae [BOT] A family of flowering plants in the order Geraniales, including touch-me-not (*Impatiens*); flowers are irregular with five stamens and five carpels, leaves are simple with pinnate venation, and the fruit is an elastically dehiscent capsule. { ˌbȯl·sə·mə'nās·ēˌē }

bamboo [BOT] The common name of various tropical and subtropical, perennial, ornamental grasses in five genera of the family Gramineae characterized by hollow woody stems up to 6 inches (15 centimeters) in diameter. { bam'bü }

Bambusoideae [BOT] A subfamily of grasses, composed of bamboo species, in the family Gramineae. { ˌbam·bə'sȯid·ēˌē }

banana [BOT] Any of the treelike, perennial plants of the genus *Musa* in the family Musaceae; fruit is a berry characterized by soft, pulpy flesh and a thin rind. { bə'nan·ə }

Bancroft's filariasis *See* wuchereriasis. { 'baŋ·krȯfs ˌfil·ə'rī·ə·səs }

band [CELL MOL] Any of the characteristic transverse stripes exhibited by polytene or metaphase chromosomes that are stained. { band }

bandage [MED] A strip of gauze, muslin, flannel, or other material, usually in the form of a roll, but sometimes triangular or tailed, used to hold dressing in place, to apply pressure, to immobilize a part, to support a dependent or injured part, to obliterate tissue cavities, or to check hemorrhage. { 'ban·dij }

banded anteater *See* marsupial anteater. { 'ban·dəd 'antˌēd·ər }

bandicoot [VERT ZOO] **1.** Any of several large Indian rats of the genus *Nesokia* and related genera. **2.** Any of several small insectivorous and herbivorous marsupials comprising the family Peramelidae and found in Tasmania, Australia, and New Guinea. { 'ban·diˌküt }

Bangiophyceae [BOT] A class of red algae in the plant division Rhodophyta. { ˌbaŋ·ē·ə'fīs·ēˌē }

Bang's disease *See* contagious abortion. { 'baŋz diz'ēz }

banner [BOT] The fifth or posterior petal of a butterfly-shaped (papilionaceous) flower. { 'ban·ər }

Banti's disease [MED] Portal hypertension, congestive splenomegaly, and hypersplenism due to an obstructive lesion in the splenic vein, portal vein, or intrahepatic veins. { 'bän·tēz diˌzēz }

barb [VERT ZOO] A side branch on the shaft of a bird's feather. { bärb }

barbel [VERT ZOO] **1.** A slender, tactile process near the mouth in certain fishes, such as catfishes. **2.** Any European fresh-water fish in the genus *Barbus*. { 'bär·bəl }

barbellate [BIOL] Having short, stiff, hooked bristles. { 'bär·bəˌlāt }

barbicel [VERT ZOO] One of the small, hook-bearing processes on a barbule of the distal side of a barb or a feather. { 'bär·bəˌsel }

barbiturate [PHARM] Any of a group of ureides, such as phenobarbital, Amytal, or Seconal, that act as central nervous system depressants. { bär'bich·ə·rət }

barbiturism [MED] Intoxication following an overdose of barbiturates; characterized by delirium, coma, and sometimes death. { bär'bich·ə,riz·əm }

bariatrics [MED] A branch of medicine that deals with the treatment of obesity. { ,ba·rē'a,triks }

barium enema [MED] A suspension of barium sulfate administered as an enema into the lower bowel to render it radiopaque. { 'bar·ē·əm 'en·ə·mə }

barium hypophosphite [PHARM] $BaH_4(PO_2)_2$ A toxic, white, crystalline powder; soluble in water; used in medicine. { 'bar·ē·əm ,hī·pō'fäs,fāt }

barium meal [MED] A suspension of barium sulfate taken orally to render the upper gastrointestinal tract radiopaque. { 'bar·ē·əm ,mēl }

bark [BOT] The tissues external to the cambium in a stem or root. { bärk }

bark graft [BOT] A graft made by slipping the scion beneath a slit in the bark of the stock. { 'bärk ,graft }

bark pocket [BOT] An opening between tree annual rings which contains bark. { 'bärk ,päk·ət }

barley [BOT] A plant of the genus *Hordeum* in the order Cyperales that is cultivated as a grain crop; the seed is used to manufacture malt beverages and as a cereal. { 'bär·lē }

barley yellow dwarf virus [VIROL] The type species of the genus *Luteovirus*. { ¦bär·lē 'yel·ō ,dwȯrf ,vī·rəs }

barley yellow dwarf virus group *See* Luteovirus. { ¦bär·lē 'yel·ō ,dwȯrf ,vī·rəs ,grüp }

barnacle [INV ZOO] The common name for a number of species of crustaceans which compose the subclass Cirripedia. { 'bär·nə·kəl }

baroduric bacteria [MICROBIO] Bacteria that can tolerate conditions of high hydrostatic pressure. { ¦bar·ə¦dür·ik bak'tir·ē·ə }

barophile [MICROBIO] An organism that thrives under conditions of high hydrostatic pressure. { 'bar·ə,fīl }

barosinusitis [MED] Inflammation of the sinuses, characterized by edema and hemorrhage, due to expansion of air within the sinuses at decreased barometric pressure. Also known as aerosinusitis. { ¦bar·ō,sī·nə'sīd·əs }

barotaxis [BIOL] Orientation movement of an organism in response to pressure changes. { ¦bar·ə¦tak·səs }

barotitis [MED] Inflammation of the ear, or a part of it, caused by changes in atmospheric pressure. Also known as aerootitis. { ,bar·ə'tīd·əs }

barotrauma [MED] Injury to air-containing structures, such as the middle ears, sinuses, lungs, and gastrointestinal tract, due to unequal pressure differences across their walls. { ,bar·ə'trau̇·mə }

barracuda [VERT ZOO] The common name for about 20 species of fishes belonging to the genus *Sphyraena* in the order Perciformes. { ,bar·ə'küd·ə }

Barr body [CELL MOL] A condensed, inactivated X chromosome inside the nuclear membrane in interphase somatic cells of women and most female mammals. Also known as sex chromatin. { 'bär ,bäd·ē }

barrel chest *See* emphysematous chest. { 'bar·əl ,chest }

barrier [ECOL] Any physical or biological factor that restricts the migration or free movement of individuals or populations. { 'bar·ē·ər }

barrier marsh [ECOL] A type of marsh that restricts or prevents invasion of the area beyond it by new species of animals. { 'bar·ē·ər ,märsh }

barrier zone [BOT] In a tree, new tissue formed by the cambium after it has been wounded; serves as both an anatomical and a chemical wall. { 'bar·ē·ər ,zōn }

bartholinitis [MED] Inflammation of Bartholin's glands and/or their ducts which is caused by bacteria from feces or a sexually transmitted infection, such as gonorrhea. { ¦bär·tə·lə'nīd·əs }

Bartholin's glands [ANAT] Two pea-sized glands located on each side of the labia minora that secrete a lubricating fluid upon sexual arousal. { 'bärt·əl·ənz ,glanz }

Bartonella [MICROBIO] A genus of the family Bartonellaceae; parasites in or on red blood cells and within fixed tissue cells; found in humans and in the arthropod genus *Phlebotomus*. { ,bärt·ən'el·ə }

Bartonellaceae [MICROBIO] A family of the order Rickettsiales; rod-shaped, coccoid, ring- or disk-shaped cells; parasites of human and other vertebrate red blood cells. { ,bärt·ən,e'lās·ē,ē }

bartonellosis *See* Carrion's disease. { ,bärt·ən·e'lō·səs }

barysphere *See* centrosphere. { 'bar·ə,sfir }

basal [BIOL] Of, pertaining to, or located at the base. [PHYSIO] Being the minimal level for, or essential for maintenance of, vital activities of an organism, such as basal metabolism. { 'bā·səl }

basal body [CELL MOL] A cellular organelle that induces the formation of cilia and flagella and is similar to and sometimes derived from a centriole. Also known as kinetosome. { 'bā·səl ,bäd·ē }

basal-cell carcinoma [MED] A locally invasive, rarely metastatic nevoid tumor of the epidermis. Also known as basal-cell epithelioma. { 'bā·səl ,sel ,kärs·ən'ō·mə }

basal-cell epithelioma *See* basal-cell carcinoma. { 'bā·səl ,sel ,ep·ə,thē·lē'ō·mə }

basal disc [BIOL] The expanded basal portion of the stalk of certain sessile organisms, used for attachment to the substrate. { 'bā·səl 'disk }

basal ganglia [NEUROSCI] The corpus striatum, or the corpus striatum and the thalamus considered together as the important subcortical centers. { 'bā·səl 'gaŋ·glē·ə }

basalia [VERT ZOO] The cartilaginous rods that support the base of the pectoral and pelvic fins in elasmobranchs. { bə'sal·ē·ə }

basalis [HISTOL] The basal portion of the endometrium; it is not shed during menstruation. { bə'sal·əs }

basal lamina [EMBRYO] The portion of the gray matter of the embryonic neural tube from which motor nerve roots develop. { 'bā·səl 'lam·ə·nə }

basal membrane [ANAT] The tissue beneath the pigment layer of the retina that forms the outer layer of the choroid. { 'bā·səl 'mem,brān }

basal metabolic rate [PHYSIO] The amount of energy utilized per unit time under conditions of basal metabolism; expressed as calories per square meter of body surface or per kilogram of body weight per hour. Abbreviated BMR. { 'bā·səl med·ə'bäl·ik 'rāt }

basal metabolism [PHYSIO] The sum total of anabolic and catabolic activities of an organism in the resting state providing just enough energy to maintain vital functions. { 'bā·səl mə'tab·ə,liz·əm }

basal wall [BOT] The wall dividing the oospore into an anterior and a posterior half in plants bearing archegonia. { 'bā·səl 'wȯl }

base *See* nitrogenous base. { bās }

base analog [CELL MOL] A molecule similar enough to a purine or pyrimidine base to substitute for the normal bases, resulting in abnormal base pairing. { 'bās 'an·ə,läg }

Basedow's disease *See* exophthalmic goiter. { 'bäz·ə,dōz di,zēz }

base excision repair [CELL MOL] A deoxyribonucleic acid (DNA) repair system in which an altered base is removed from the sugar backbone by action of a specific DNA glycolase and then the abasic sugar is removed by apurinic/apyrimidic (AP) lyase and AP endonuclease, leaving a one-nucleotide gap that is then filled in and ligated. { 'bās ek'siz·zhən ri,per }

basement membrane [HISTOL] A delicate connective-tissue layer underlying the epithelium of many organs. { 'bās·mənt 'mem,brān }

basendite [INV ZOO] In crustaceans, either of a pair of lobes at the end of each specialized paired appendage. { 'bā'sen,dīt }

base pair [CELL MOL] Two nitrogenous bases, one purine and one pyrimidine, that pair in double-stranded deoxyribonucleic acid. { 'bās ,per }

base pairing [CELL MOL] The hydrogen bonding of complementary purine and pyrimidine bases-adenine with thymine, guanine with cytosine-in double-stranded deoxyribonucleic acids or ribonucleic acids or in DNA/RNA hybrid molecules. { 'bās 'per·iŋ }

base sequence [GEN] The specific order of purine and pyrimidine bases in a polynucleotide such as deoxyribonucleic acid or ribonucleic acid. { 'bās 'sē·kwəns }

base stacking [CELL MOL] The orientation of adjacent base pairs such that their planes are parallel and their surfaces almost touch, as occurs in double-stranded deoxyribonucleic acid molecules. { 'bās ,stak·iŋ }

basichromatic [BIOL] Staining readily with basic dyes. { ¦bā·si¦krō¦mad·ik }

basidiocarp [MYCOL] The fruiting body of a fungus in the class Basidiomycetes. { bə'sid·ē·ə,kärp }

Basidiolichenes [BOT] A class of the Lichenes characterized by the production of basidia. { bə,sid·ē·ō,lī'kē,nēz }

Basidiomycetes [MYCOL] A class of fungi in the subdivision Eumycetes; important as food and as causal agents of plant diseases. { bə,sid·ē·ō,mī'sēd,ēz }

basidiophore [MYCOL] A basidia-bearing sporophore. { bə'sid·ē·ə,fȯr }

basidiospore [MYCOL] A spore produced by a basidium. { bə'sid·ē·ə,spȯr }

basidium [MYCOL] A cell, usually terminal, occurring in Basidiomycetes and producing spores (basidiospores) by nuclear fusion followed by meiosis. { bə'sid·ē·əm }

basifixed [BOT] Attached at or near the base. { ¦bās·ə¦fikst }

basil [BOT] The common name for any of the aromatic plants in the genus *Ocimum* of the mint family; leaves of the plant are used for food flavoring. { 'bāz·əl *or* 'baz·əl }

basilar [BIOL] Of, pertaining to, or situated at the base. { 'bas·ə·lər }

basilar groove [ANAT] The cavity which is located on the upper surface of the basilar process of the brain and upon which the medulla rests. { 'bas·ə·lər ¦grüv }

basilar membrane [ANAT] A membrane of the mammalian inner ear supporting the organ of Corti and separating two cochlear channels, the scala media and scala tympani. { 'bas·ə·lər 'mem,brān }

basilar meningitis [MED] Inflammation of the meninges which affects chiefly the base of the brain, or in which exudate collects predominantly at the basal cisterns. { 'bas·ə·lər ,men·ən'jīd·əs }

basilar papilla [ANAT] **1.** A sensory structure in the lagenar portion of an amphibian's membranous labyrinth between the oval and round windows. **2.** The organ of Corti in mammals. { 'bas·ə·lər pə'pil·ə }

basilar plate [EMBRYO] An embryonic cartilaginous plate in vertebrates that is formed from the parachordals and anterior notochord and gives rise to the ethmoid and other bones of the skull. { 'bas·ə·lər ¦plāt }

basilar process [ANAT] A strong, quadrilateral plate of bone forming the anterior portion of the occipital bone, in front of the foramen magnum. { 'bas·ə·lər 'präs·əs }

basilic vein [ANAT] The large superficial vein of the arm on the medial side of the biceps brachii muscle. { bə'sil·ik 'vān }

basin swamp [ECOL] A fresh-water swamp at the margin of a small calm lake, or near a large lake protected by shallow water or a barrier. { 'bās·ən ,swämp }

basion [ANAT] In craniometry, the point on the anterior margin of the foramen magnum where the midsagittal plane of the skull intersects the plane of the foramen magnum. { 'bās·ē,än }

basipetal [BIOL] Movement or growth from the apex toward the base. { bā'sip·əd·əl }

basipodite [INV ZOO] The distal segment of the protopodite of a biramous appendage in arthropods. { bā'sip·ə,dīt }

basipterygium [VERT ZOO] A basal bone or cartilage supporting one of the paired fins in fishes. { bə,sip·tə'rij·ē·əm }

basisternum [INV ZOO] In insects, the anterior one of the two sternal skeletal plates. { ¦bās·ə¦stər·nəm }

basistyle [INV ZOO] Either of a pair of flexible processes on the hypopygium of certain male dipterans. { ¦bās·ə¦stīl }

basitarsus [INV ZOO] The basal segment of the tarsus in arthropods. { ¦bās·ə¦tär·səs }

basket cell [HISTOL] A type of cell in the cerebellum whose axis-cylinder processes terminate in a basketlike network around the cells of Purkinje. { 'bas·kət ,sel }

basket star [INV ZOO] The common name for ophiuroid echinoderms belonging to the family Gorgonocephalidae. { 'bas·kət ,stär }

Basommatophora [INV ZOO] An order of mollusks in the subclass Pulmonata containing many aquatic snails. { bə,säm·ə'täf·rə }

basonym [SYST] The original, validly published name of a taxon. { 'bās·ə,nim }

basophil [HISTOL] A white blood cell with granules that stain with basic dyes and are water-soluble. { 'bās·ə,fil }

basophilia [BIOL] An affinity for basic dyes. [MED] An increase in the number of basophils in the circulating blood. [PATH] Stippling of the red cells with basic staining granules, representing a degenerative condition as seen in severe anemia, leukemia, malaria, lead poisoning, and other toxic states. { ,bās·ə'fil·ē·ə }

basophilous [BIOL] Staining readily with the basic dyes. [ECOL] Of plants, growing best in alkaline soils. { bə'säf·ə·ləs }

bass [VERT ZOO] The common name for a number of fishes assigned to two families, Centrarchidae and Serranidae, in the order Perciformes. { bas }

basset [VERT ZOO] A French breed of short-legged hunting dogs with long ears and a typical hound coat. { 'bas·ət }

basswood [BOT] A common name for trees of the genus *Tilia* in the linden family of the order Malvales. Also known as linden. { 'bas,wu̇d }

bast *See* phloem. { bast }

bast fiber [BOT] Any fiber stripped from the inner bark of plants, such as flax, hemp, jute, and ramie; used in textile and paper manufacturing. { 'bast ,fī·bər }

bat [VERT ZOO] The common name for all members of the mammalian order Chiroptera. { bat }

Batales [BOT] A small order of dicotyledonous plants in the subclass Caryophillidae of the class Magnoliopsida containing a single family with only one genus, *Batis*. { bə'tā,lēz }

Batesian mimicry [ECOL] Resemblance of an innocuous species to one that is distasteful to predators. { 'bāt·sē·ən 'mim·ə·krē }

Bathyctenidae [INV ZOO] A family of bathypelagic coelenterates in the phylum Ctenophora. { ,ba·thik'ten·ə,dē }

Bathyergidae [VERT ZOO] A family of mammals, including the South African mole rats, in the order Rodentia. { ,bath·ē'ər·jə,dē }

Bathylaconoidei [VERT ZOO] A suborder of deep-sea fishes in the order Salmoniformes. { ¦bath·ə,la·kə'nȯi·dē,ī }

Bathynellacea [INV ZOO] An order of crustaceans in the superorder Syncarida found in subterranean waters in England and central Europe. { ¦bath·ə,nel'ās·ē·ə }

Bathynellidae [INV ZOO] The single family of the crustacean order Bathynellacea. { ,bath·ə'nel·ə,dē }

Bathyodontoidea [INV ZOO] A superfamily of nematodes of the order Mononchida containing the single family Bathyodontidae, characterized by a high, usually well-developed lip region, a hexaradiate oral opening with cuticularized walls, and a two-fold stoma; eight longitudinal rows of pores occur along the length of the body. { ¦bath·ē·ō·dän'tȯid·ē·ə }

Bathypteroidae [VERT ZOO] A family of benthic, deep-sea fishes in the order Salmoniformes. { bə,thip·tə'rȯi·dē }

Bathysquillidae [INV ZOO] A family of mantis shrimps, with one genus (*Bathysquilla*) and two species, in the order Stomatopoda. { ¦bath·ə'skwil·ə,dē }

Batoidea [VERT ZOO] The skates and rays, an order of the subclass Elasmobranchii. { bə'tȯid·ē·ə }

Batrachoididae [VERT ZOO] The single family of the order Batrachoidiformes. { ,ba·trə'kȯi·də,dē }

Batrachoidiformes [VERT ZOO] The toadfishes, an order of teleostean fishes in the subclass Actinopterygii. { ,ba·trə,kȯi·də'fȯr,mēz }

battered-child syndrome [MED] A clinical condition in young children due to serious physical abuse, generally from a parent or foster parent. { ¦bad·ərd ¦chīld ¦sin,drōm }

battery of genes [GEN] The set of producer genes which is activated when a particular sensor gene activates its set of integrator genes. { 'bad·ə·rē əv 'jēnz }

Battey disease [MED] A tuberculosislike disease of humans caused by *Mycobacterium intracellulare*. { 'bad·ē di'zēz }

bay [BOT] *Laurus nobilis*. An evergreen tree of the laurel family. { bā }

bayberry [BOT] **1.** *Pimenta acris*. A West Indian tree related to the allspice; a source of bay oil. Also known as bay-rum tree; Jamaica bayberry; wild cinnamon. **2.** Any tree of the genus *Myrica*. { 'bā,ber·ē }

bay-rum tree *See* bayberry. { ¦bā 'rəm ,trē }

B cell [IMMUNOL] One of a heterogeneous population of bone-marrow-derived lymphocytes which participates in the immune responses. Also known as B lymphocyte. { 'bē ˌsel }

BCG vaccine *See* Bacillus Calmette-Guérin vaccine. { ¦bē¦sē¦jē vak'sēn }

B chain *See* light chain. { 'bē ˌchān }

bcl-2 [BIOCHEM] A family of proteins that operate in the effector phase of apoptosis and may either promote or inhibit apoptosis.

Bdelloidea [INV ZOO] An order of the class Rotifera comprising animals which resemble leeches in body shape and manner of locomotion. { də'lȯid·ē·ə }

Bdellomorpha [INV ZOO] An order of ribbonlike worms in the class Enopla containing the single genus *Malacobdella*. { ˌdel·ə'mȯr·fə }

Bdellonemertini [INV ZOO] An equivalent name for the Bdellomorpha. { ˌdel·ə·nə'mer·təˌnī }

Bdellovibrio [MICROBIO] A genus of bacteria of uncertain affiliation; parasites of other bacteria; curved, motile rods with a polar flagellum in parasitic state; in nonparasitic state, cells are helical. { ˌdel·ə'vī·brēˌō }

beak [BOT] Any pointed projection, as on some fruits, that resembles a bird bill. [INV ZOO] The tip of the umbo in bivalves. [VERT ZOO] **1.** The bill of a bird or some other animal, such as the turtle. **2.** A projecting jawbone element of certain fishes, such as the sawfish and pike. { bēk }

bean [BOT] The common name for various leguminous plants used as food for humans and livestock; important commercial beans are true beans (*Phaseolus*) and California blackeye (*Vigna sinensis*). { bēn }

bear [VERT ZOO] The common name for a few species of mammals in the family Ursidae. { ber }

beard *See* awn. { bird }

beaver [VERT ZOO] The common name for two different and unrelated species of rodents, the mountain beaver (*Aplodontia rufa*) and the true or common beaver (*Castor canadensis*). { 'bē·vər }

Beckwith-Wiedemann syndrome [MED] A congenital, generalized overgrowth syndrome attributed to a relative deficiency of maternally derived genes that is characterized by visceromegaly and predisposition to childhood tumors, especially Wilms' tumor. { ¦bekˌwith 'wēd·ə·män ˌsin ˌdrōm }

bedbug [INV ZOO] The common name for a number of species of household pests in the insect family Cimicidae that infest bedding, and by biting humans obtain blood for nutrition. { 'bedˌbəg }

Bedsonia [MICROBIO] The psittacosis-lymphogranuloma-trachoma (PLT) group of bacteria belonging to the Chlamydozoaceae; all are obligatory intracellular parasites. { bed'sō·nē·ə }

bedsore *See* decubitus ulcer. { 'bedˌsȯr }

bee [INV ZOO] Any of the membranous-winged insects which compose the superfamily Apoidea in the order Hymenoptera characterized by a hairy body and by sucking and chewing mouthparts. { bē }

beech [BOT] Any of various deciduous trees of the genus *Fagus* in the beech family (Fagaceae) characterized by smooth gray bark, triangular nuts enclosed in burs, and hard wood with a fine grain. { bēch }

bee dance [INV ZOO] Circling and wagging movements exhibited by worker bees to give other bees in the hive information about the location of a new source of food. { 'bē ˌdans }

beehive [INV ZOO] A colony of bees. Also known as hive. { 'bēˌhīv }

beer drinkers' cardiomyopathy [MED] Congestive heart failure and nonspecific cardiomyopathy presumed due to cobalt added to beer. { 'bir ˌdriŋ·kərz ¦kärd·ē·ō·mī'äp·ə·thē }

beet [BOT] *Beta vulgaris*. The red or garden beet, a cool-season biennial of the order Caryophyllales grown for its edible, enlarged fleshy root. { bēt }

beetle [INV ZOO] The common name given to members of the insect order Coleoptera. { 'bēd·əl }

beet yellows virus [VIROL] The type species of the plant-virus genus *Closterovirus*. Abbreviated BYV. { ¦bēt 'yel·ōz ˌvī·rəs }

beet yellows virus group *See* Closterovirus. { ¦bēt 'yel·ōz ˌvī·rəs ˌgrüp }

Beggiatoa [MICROBIO] A genus of bacteria in the family Beggiatoaceae; filaments are individual, and cells contain sulfur granules when grown on media containing hydrogen sulfide. { bə'jad·ə·wə }

Beggiatoaceae [MICROBIO] A family of bacteria in the order Cytophagales; cells are in chains in colorless, flexible, motile filaments; microcysts are not known. { bə¦jad·ə'wās·ēˌē }

Beggiatoales [MICROBIO] Formerly an order of motile, filamentous bacteria in the class Schizomycetes. { bəˌjad·ə'wāˌlēz }

Begoniaceae [BOT] A family of dicotyledonous plants in the order Violales characterized by an inferior ovary, unisexual flowers, stipulate leaves, and two to five carpels. { bəˌgō·nē'ās·ēˌē }

behavioral ecology [ECOL] The branch of ecology that focuses on the evolutionary causes of variation in behavior among populations and species. { bi'hāv·yə·rəl ē'käl·ə·jē }

behavioral isolation [BIOL] An isolating mechanism in which two allopatric species do not mate because of differences in courtship behavior. Also known as ethological isolation. { bi'hāv·yə·rəl ī·sə'lā·shən }

behavioral toxicology [MED] The study of behavioral abnormalities induced by exogenous agents such as drugs, chemicals in the general environment, and chemicals encountered in the workplace. { bə¦hāv·yə·rəl ˌtäk·sə'käl·ə·jē }

Behçet's disease [MED] Chronic disease of young adult males characterized by recurrent painful ulcers of the mouth and genitalia, inflammation of the irises, and joint pains. { 'bā ˌchets diˌzēz }

Beijerinckia [MICROBIO] A genus of bacteria in the family Azotobacteraceae; slightly curved or pear-shaped rods with large, refractile lipoid bodies at the ends of the cell. { bī·zhə'riŋ·kyə }

bejel [MED] An infectious nonvenereal treponemal disease occurring principally in children in the Middle East. { 'be·jəl }

belladonna [BOT] *Atropa belladonna.* A perennial poisonous herb that belongs to the family Solanaceae; atropine is produced from the roots and leaves; used as an antispasmodic, as a cardiac and respiratory stimulant, and to check secretions. Also known as deadly nightshade. { ˌbel·ə'dän·ə }

Bell's law [NEUROSCI] **1.** The law that in the spinal cord the ventral roots are motor and the dorsal roots sensory in function. **2.** The law that in a reflex arc the nerve impulse can be conducted in one direction only. { 'belz ˌlȯ }

belly [ANAT] **1.** The abdominal cavity or the abdomen. **2.** The most prominent, fleshy, central portion of a muscle. { 'bel·ē }

Belondiroidea [INV ZOO] A diverse superfamily of nematodes belonging to the order Dorylaimida, consisting of a diverse group whose principal common characteristic is a thick sheath of spiral muscles enclosing the basal swollen portion of the esophagus. { bəˌlän·də'rȯid·ē·ə }

Beloniformes [VERT ZOO] The former ordinal name for a group of fishes now included in the order Atheriniformes. { ˌbe·lə·nə'fȯrˌmēz }

Belostomatidae [INV ZOO] The giant water bugs, a family of hemipteran insects in the subdivision Hydrocorisae. { ˌbel·ə·stō'mad·əˌdē }

belt [ECOL] **1.** Any altitudinal vegetation zone or band from the base to the summit of a mountain. **2.** Any benthic vegetation zone or band from sea level to the ocean depths. **3.** Any of the concentric vegetation zones around bodies of fresh water. { belt }

bemegride [PHARM] $C_8H_{13}NO_2$ Platelets which melt at 127°C and sublime at 100°C (2 mmHg pressure); soluble in water and in acetone; used as an analeptic drug in barbiturate poisoning for humans and as an analeptic drug and central nervous system stimulant in animals. Also known as methetharimide; β,β-methylethylglutarimide. { 'bem·əˌgrīd }

Bence-Jones protein [PATH] An abnormal group of globulins appearing in the serum and urine, usually in association with multiple myeloma and characterized by coagulation at 50–60°C. { ¦bens ¦jōnz ¦prōˌtēn }

Bence-Jones proteinuria [MED] The presence of Bence-Jones protein in the urine. { ¦bens ¦jōnz ˌprō·tə'nür·ē·ə }

bending [CELL MOL] A conformational change characterized by a localized bend or kink in deoxyribonucleic acid due to heterogeneities in local structural composition. { 'ben·diŋ }

bends *See* caisson disease. { benz }

benign [MED] Of no danger to life or health. { bə'nīn }

benign lymphoreticulosis *See* cat scratch disease. { bə'nīn ¦lim·fə·rəˌtik·yə'lō·səs }

benign myalgic encephalomyelitis *See* neuromyasthenia. { bə'nīn mī'al·jik ˌenˌsef·ə·lōˌmī·ə'līd·əs }

benign tertian malaria *See* vivax malaria. { bə'nīn 'tər·shən mə'ler·ē·ə }

benign tumor [MED] A nonmalignant neoplasm. { bə'nīn 'tü·mər }

benthos [ECOL] Bottom-dwelling forms of marine life. Also known as bottom fauna. { 'benˌthäs }

Benzedrine *See* amphetamine. { 'ben·zəˌdrēn }

benzestrol [PHARM] $C_{20}H_{26}O_2$ A white powder; melting point 162–166°C; soluble in ethanol and acetone, insoluble in water; used as a medicine. { ben'zeˌstrȯl }

benzethonium chloride [PHARM] $C_{27}H_{42}ClNO_2$ Crystallizes as thin, hexagonal plates; melting point is 164–166°C; soluble in water, alcohol, chloroform, and acetone; used as a topical antiseptic on humans and animals. { ˌben·zə'thōn·ē·əm 'klȯrˌīd }

benzodiazepine [MED] A group of tranquilizers that are used to combat anxiety and convulsions. { ˌben·zōˌdī'az·əˌpēn }

2,3-benzopyrrole *See* indole. { ¦tü ¦thrē ¦ben·zō¦pīˌrōl }

benzyl penicillin potassium [MICROBIO] $C_{16}H_{17}KN_2O_4S$ Moderately hygroscopic crystals; soluble in water; inactivated by acids and alkalies; obtained from fermentation of *Penicillium chrysogenum*; used as an antimicrobial drug in human and animal disease. { 'ben·zəl ˌpen·ə'sil·ən pə'tas·ē·əm }

benzyl penicillin sodium [MICROBIO] $C_{16}H_{17}N_2NaO_4S$ Crystals obtained from a methanol-ethyl acetate acidified extract of fermentation broth of *Penicillium chrysogenum*; used as an antimicrobial in human and animal disease. { 'ben·zəl ˌpen·ə'sil·ən 'sōd·ē·əm }

Berberidaceae [BOT] A family of dicotyledonous herbs and shrubs in the order Ranunculales characterized by alternate leaves, perfect, well-developed flowers, and a seemingly solitary carpel. { ˌbər·bə·rə'dās·ēˌē }

Bergmann's rule [ECOL] The principle that in a polytypic wide-ranging species of warm-blooded animals the average body size of members of each geographic race varies with the mean environmental temperature. { 'bərg·mənz ˌrül }

Bergman-Turner unit [BIOL] A unit for the standardization of thyroid extract. { ¦bərg·mən ¦tər·nər ˌyü·nət }

beriberi [MED] A disorder resulting from the deficiency of vitamin B_1 and characterized by neurologic symptoms, cardiovascular abnormalities, edema, and cerebral manifestations. { ¦ber·ē'ber·ē }

Berkefeld filter [MICROBIO] A diatomaceous-earth filter that is used for the sterilization of heat-labile liquids, such as blood serum, enzyme solutions, and antibiotics. { 'berk·əˌfeld ˌfil·tər }

Bermuda grass [BOT] *Cynodon dactylon.* A long-lived perennial in the order Cyperales. { bər'myüd·ə ˌgras }

Beroida [INV ZOO] The single order of the class Nuda in the phylum Ctenophora. { bə'rō·ə·də }

berry [BOT] A usually small, simple, fleshy or

pulpy fruit, such as a strawberry, grape, tomato, or banana. { 'ber·ē }

Bertrand's rule [MICROBIO] The rule stating that in those compounds, and only in those compounds, having cis secondary alcoholic groups containing at least one carbon atom of D configuration which is subtended by a primary alcohol group, or having a methyl-substituted primary alcohol group of D configuration, the D-carbon atom will be dehydrogenated by the vinegar bacteria *Acetobacter suboxydans*, yielding a ketone. { 'ber,tränz ,rül }

Beryciformes [VERT ZOO] An order of actinopterygian fishes in the infraclass Teleostei. { bə,ris·ə'fór,mēz }

Berycomorphi [VERT ZOO] An equivalent name for the Beryciformes. { bə,rik·ō'mór·fī }

berylliosis [MED] Chronic lung inflammation due to inhalation of beryllium oxide dust. { bə,ril·ē'ō·səs }

Berytidae [INV ZOO] The stilt bugs, a small family of hemipteran insects in the superfamily Pentatomorpha. { bə'rid·ə,dē }

Bessey unit [BIOL] A unit for the standardization of phosphatase. { 'bes·ē ,yü·nət }

beta blocker [NEUROSCI] An adrenergic blocking agent capable of blocking nerve impulses to special sites (beta receptors) in the cerebellum; reduces the rate of heartbeats and the force of heart contractions. { 'bād·ə¦bläk·ər }

beta carotene [BIOCHEM] $C_{40}H_{56}$ A carotenoid hydrocarbon pigment found widely in nature, always associated with chlorophylls; converted to vitamin A in the liver of many animals. { 'bād·ə 'kar·ə,tēn }

beta catenin [BIOCHEM] A multifunctional protein that is involved in Wnt signal transduction and plays an essential role in intercellular adhesion. { ,bād·ə 'kat·ən·in }

beta cell [HISTOL] **1.** Any of the basophilic chromophiles in the anterior lobe of the adenohypophysis. **2.** One of the cells of the islets of Langerhans which produce insulin. { 'bād·ə ,sel }

betacyanin [BIOCHEM] A group of purple plant pigments found in leaves, flowers, and roots of members of the order Caryophyllales. { ¦bād·ə¦sī·ə·nən }

beta globulin [BIOCHEM] A heterogeneous fraction of serum globulins containing transferrin and various complement components. { 'bād·ə 'gläb·yə·lən }

beta hemolysis [MICROBIO] A sharply defined, clear, colorless zone of hemolysis surrounding certain streptococci colonies growing on blood agar. { 'bād·ə hə'mäl·ə·səs }

Betaherpesvirinae [VIROL] A subfamily of animal, double-stranded linear deoxyribonucleic acid viruses of the family Herpesviridae, which are enveloped by a lipid bilayer and several glycoproteins. Also known as cytomegalovirus group. { ¦bād·ə¦hər·pēz'vī·rə,nē }

beta-lactamase [MICROBIO] A bacterial enzyme that catalyzes the hydrolysis of the lactam ring in some penicillin antibiotics, rendering them ineffective. { ¦bād·ə 'lak·tə,mās }

betalain [BIOCHEM] The name for a group of 35 red or yellow compounds found only in plants of the family Caryophyllales, including red beets, red chard, and cactus fruits. { 'bed·ə,lān }

betamethasone [PHARM] A white, crystalline powder with a melting point of 240°C; a corticosteroid hormone. { ¦bād·ə'meth·ə,zōn }

betanin [BIOCHEM] An anthocyanin that contains nitrogen and constitutes the principal pigment of garden beets. { 'bēd·ə·nən }

beta oxidation [BIOCHEM] Catabolism of fatty acids in which the fatty acid chain is shortened by successive removal of two carbon fragments from the carboxyl end of the chain. { 'bād·ə äks·ə'dā·shən }

beta rhythm [PHYSIO] An electric current of low voltage from the brain, with a pulse frequency of 13–30 per second, encountered in a person who is aroused and anxious. { 'bād·ə ,rith·əm }

beta taxonomy [SYST] The level of taxonomic study dealing with the arrangement of species into lower and higher taxa. { 'bād·ə tak'sän·ə·mē }

betaxanthin [BIOCHEM] The name given to any of the yellow pigments found only in plants of the family Caryophyllales; they always occur with betacyanins. { ¦bād·ə'zan·thən }

betel nut [BOT] A dried, ripe seed of the palm tree *Areca catechu* in the family Palmae; contains a narcotic. { 'bēd·əl ,nət }

Bethylidae [INV ZOO] A small family of hymenopteran insects in the superfamily Bethyloidea. { bə'thil·ə,dē }

Bethyloidea [INV ZOO] A superfamily of hymenopteran insects in the suborder Apocrita. { ,beth·ə'lói·dē·ə }

Betula [BOT] The birches, a genus of deciduous trees composing the family Betulaceae. { 'bech·ə·lə }

Betulaceae [BOT] A small family of dicotyledonous plants in the order Fagales characterized by stipulate leaves, seeds without endosperm, and by being monoecious with female flowers mostly in catkins. { ,bech·ə'lās·ē,ē }

Betz cell [HISTOL] Any of the large conical cells composing the major histological feature of the precentral motor cortex in humans. { 'bets ,sel }

Bezold-Abney phenomenon [PHYSIO] The perception of light at very high intensities as colorless. { 'bāt,zōlt 'ab·nē fə,näm·ə,nän }

Bial's test [PATH] A test for the presence of a pentose in urine, utilizing oracin, hydrochloric acid, and ferric chloride; a green color or green precipitate indicates pentose. { 'byälz ,test }

Bibionidae [INV ZOO] The March flies, a family of orthorrhaphous dipteran insects in the series Nematocera. { ,bib·ē'än·ə,dē }

bicameral [BIOL] Having two chambers, as the heart of a fish. { bī'kam·ə·rəl }

bicapsular [BIOL] **1.** Having two capsules. **2.** Having a capsule with two locules. { bī'kap·sə·lər }

bicarinate [BIOL] Having two keellike projections. { bī'kar·ə‚nāt }

bicaudal [ZOO] Having two tails. { bī'kȯd·əl }

bicellular [BIOL] Having two cells. { bī'sel·yə·lər }

bicephalous [ZOO] Having two heads. { bī'sef·ə·ləs }

biceps [ANAT] **1.** A bicipital muscle. **2.** The large muscle of the front of the upper arm that flexes the forearm; biceps brachii. The thigh muscle that flexes the knee joint and extends the hip joint; biceps femoris. { 'bī‚seps }

biciliate [BIOL] Having two cilia. { bī'sil·ē‚āt }

bicipital [ANAT] **1.** Pertaining to muscles having two origins. **2.** Pertaining to ribs having double articulation with the vertebrae. [BOT] Having two heads or two supports. { bī'sip·əd·əl }

bicipital groove *See* intertubercular sulcus. { bī'sip·əd·əl ‚grüv }

bicipital tuberosity [ANAT] An eminence on the anterior inner aspect of the neck of the radius; the tendon of the biceps muscle is inserted here. { bī'sip·əd·əl ‚tüb·ə'räs·əd·ē }

bicollateral bundle [BOT] A vascular bundle in which phloem is located both externally and internally with respect to the xylem, with all tissues lying on the same radius and with the internal phloem lying next to the pith. { ‚bī·kə'lad·ə·rəl 'bənd·əl }

bicornuate uterus [ANAT] A uterus with two horn-shaped processes on the superior aspect. { bī'kȯr·yə‚nāt 'yüd·ə·rəs }

Bicosoecida [INV ZOO] An order of colorless, free-living protozoans, each member having two flagella, in the class Zoomastigophorea. { bī·kō'se·shə·də }

bicostate [BOT] Of a leaf, having two principal longitudinal ribs. { 'bī·kō‚stāt }

bicuspid [ANAT] Any of the four double-pointed premolar teeth in humans. [BIOL] Having two points or prominences. { bī'kəs·pəd }

bicyclic [BOT] Having or arranged in two whorls, as in petals. { bī'sī·klik }

Bidder's organ [VERT ZOO] A structure in the males of some toad species that may develop into an ovary in older individuals. { 'bid·ərz ‚ȯr·gən }

bidentate [BIOL] Having two teeth or teethlike processes. { bī'den‚tāt }

bidirectional replication [CELL MOL] A mechanism of replication of deoxyribonucleic acid that involves two replicating forks moving in opposite directions away from the same origin. { ‚bī·də‚rek·shən·əl ‚rep·lə'kā·shən }

biennial plant [BOT] A plant that requires two growing seasons to complete its life cycle. { bī'en·ē·əl ¦plant }

bifacial [BOT] Of a leaf, having dissimilar tissues on the upper and lower surfaces. { bī'fā·shəl }

bifanged [ANAT] Of a tooth, having two roots. { bī'faŋgd }

bifid [BIOL] Divided into two equal parts by a median cleft. { 'bī‚fid }

Bifidobacterium [MICROBIO] A genus of bacteria in the family Actinomycetaceae; branched, bifurcated, club-shaped or spatulate rods forming smooth microcolonies; metabolism is saccharoclastic. { ‚bī·fə·dō·bak'tir·ē·əm }

biflabellate [INV ZOO] The shape of certain insect antennae, characterized by short joints with long, flattened processes on opposite sides. { ¦bī·flə'bel·ət }

biflagellate [BIOL] Having two flagella. { bī'flaj·ə‚lāt }

bifoliate [BOT] Two-leaved. { bī'fōl·ē·ət }

biforate [BIOL] Having two perforations. { 'bī·fə‚rāt }

bifunctional vector *See* shuttle vector. { ¦bī¦fəŋk·shən·əl 'vek·tər }

bigeminal pulse [PHYSIO] A pulse that is characterized by a double impulse produced by coupled heartbeats, that is, an extra heartbeat occurs just after the normal beat. { bī'jem·ə·nəl 'pəls }

Bignoniaceae [BOT] A family of dicotyledonous trees or shrubs in the order Scrophulariales characterized by a corolla with mostly five lobes, mature seeds with little or no endosperm and with wings, and opposite or whorled leaves. { ‚big·nō·nē'ās·ē‚ē }

bijugate [BOT] Of a pinnate leaf, having two pairs of leaflets. { 'bī·jə‚gāt }

bilabiate [BOT] Having two lips, such as certain corollas. { bī'lāb·ē·ət }

bilateral [BIOL] Of or relating to both right and left sides of an area, organ, or organism. { bī'lad·ə·rəl }

bilateral cleavage [EMBRYO] The division pattern of a zygote that results in a bilaterally symmetrical embryo. { bī'lad·ə·rəl 'klēv·ij }

bilateral hermaphroditism [ZOO] The presence of an ovary and a testis on each side of the animal body. { bī'lad·ə·rəl hər'maf·rə‚dīd‚iz·əm }

bilateral symmetry [BIOL] Symmetry such that the body can be divided by one median, or sagittal, dorsoventral plane into equivalent right and left halves, each a mirror image of the other. { bī'lad·ə·rəl 'sim·ə·trē }

Bilateria [ZOO] A major division of the animal kingdom embracing all forms with bilateral symmetry. { ‚bī·lə'tir·ē·ə }

bile [PHYSIO] An alkaline fluid secreted by the liver and delivered to the duodenum to aid in the emulsification, digestion, and absorption of fats. Also known as gall. { bīl }

bile acid [BIOCHEM] Any of the liver-produced steroid acids, such as taurocholic acid and glycocholic acid, that appear in the bile as sodium salts. { 'bīl 'as·əd }

bile duct [ANAT] Any of the major channels in the liver through which bile flows toward the hepatic duct. { 'bīl ‚dəkt }

bile pigment [BIOCHEM] Either of two colored organic compounds found in bile: bilirubin and biliverdin. { 'bīl 'pig·mənt }

bile salt [BIOCHEM] The sodium salt of glycocholic and taurocholic acids found in bile. { 'bīl ˌsȯlt }

bilharzias *See* schistosomiasis. { ˌbil'härz·ē·əs }

bilharziasis *See* schistosomiasis. { ˌbilˌhär'ze·ə·səs }

biliary atresia [MED] Failure of the bile ducts to develop in the embryo. { 'bil·ēˌer·ē ə'trēzh·ə }

biliary cirrhosis [MED] A progressive inflammatory disease of the liver due to obstruction of bile ducts. { 'bil·ēˌer·ē sə'rō·səs }

biliary colic [MED] Severe abdominal pain caused by passage of a gallstone through the bile ducts into the duodenum. { 'bil·ēˌer·ē 'käl·ik }

biliary diskinesia [MED] A functional spasticity of the sphincter of Oddi with disturbances in the speed of evacuation of the biliary tract. { 'bil·ēˌer·ē ˌdis·kə'nēzh·ə }

biliary system [ANAT] The complex of canaliculi, or microscopic bile ducts, that empty into the larger intrahepatic bile ducts. { 'bil·ēˌer·ē ˌsis·təm }

bilicyanin [BIOCHEM] A blue pigment found in gallstones; an oxidation product of biliverdin or bilirubin. { ¦bil·ə¦sī·ə·nən }

bilification [PHYSIO] Formation and excretion of bile. { ˌbil·ə·fə'kā·shən }

biliprotein [BIOCHEM] The generic name for the organic compounds in certain algae that are composed of phycobilin and a conjugated protein. { ˌbil·ə'prōˌtēn }

bilirubin [BIOCHEM] $C_{33}H_{36}N_4O_6$ An orange, crystalline pigment occurring in bile; the major metabolic breakdown product of heme. { ˌbil·ə'rü·bən }

bilirubinemia [MED] The presence of bilirubin in the blood. { ¦bil·əˌrü·bə'nē·mē·ə }

biliverdin [BIOCHEM] $C_{33}H_{34}N_4O_6$ A green, crystalline pigment occurring in the bile of amphibians, birds, and humans; oxidation product of bilirubin in humans. { ˌbil·ə'vərd·ən }

bill [INV ZOO] A flattened portion of the shell margin of the broad end of an oyster. [VERT ZOO] The jaws, together with the horny covering, of a bird. [ZOO] Any jawlike mouthpart. { bil }

bilobate [BIOL] Divided into two lobes. { 'bī'lōˌbāt }

bilobular [BIOL] Having two lobules. { ˌbī'läb·yə·lər }

bilocular [BIOL] Having two cells or compartments. { ˌbī'läk·yə·lər }

bilophodont [ZOO] Having two transverse ridges, as the molar teeth of certain animals. { bī'läf·əˌdänt }

bimanous [ANAT] Having the distal part of the two forelimbs modified as hands, as in primates. { bī'man·əs }

bimaxillary [ZOO] Pertaining to the two halves of the maxilla. { ¦bī'max·əˌler·ē }

binary fission [BIOL] A method of asexual reproduction accomplished by the splitting of a parent cell into two equal, or nearly equal, parts, each of which grows to parental size and form. { 'bīn·ə·rē 'fish·ən }

binate [BOT] Growing in pairs. { 'bīˌnāt }

binding site *See* active site. { 'bīn·diŋ ˌsīt }

binaural hearing [PHYSIO] The perception of sound by stimulation of two ears. { bī'nȯr·əl 'hir·iŋ }

binocular [BIOL] **1.** Of, pertaining to, or used by both eyes. **2.** Of a type of visual perception which provides depth-of-field focus due to angular difference between the two retinal images. { bī'näk·yə·lər }

binocular accommodation [PHYSIO] Automatic lens adjustment by both eyes simultaneously for focusing on distant objects. { bī'näk·yə·lər əˌkäm·ə'dā·shən }

binomen [SYST] A binomial name assigned to species, as *Canis familiaris* for the dog. { bī'nō·mən }

binomial nomenclature [SYST] The Linnean system of classification requiring the designation of a binomen, the genus and species name, for every species of plant and animal. { bī'nō·mē·əl ˌnō·mən'klā·chər }

binuclear [CELL MOL] Having two nuclei. { bī'nü·klē·ər }

binucleolate [CELL MOL] Having two nucleoli. { ¦bī·nü'klē·əˌlāt }

bioacoustics [BIOL] The study of the relation between living organisms and sound. { ¦bī·ō·ə'kü·stiks }

bioactivity [BIOL] The effect that a substance has on a living organism or tissue after interaction. { ˌbī·ō·ak'tiv·əd·ē }

bioartificial organs [MED] Devices, used for both short-term and long-term organ replacement, that are designed and manufactured for membrane biocompatibility, diffusion limitations, device retrieval in the event of failure, and mechanical stability. { ˌbī·ōˌard·ə¦fish·əl 'ȯr·gənz }

bioastronautics [BIOL] The study of biological, behavioral, and medical problems pertaining to astronautics. { ¦bī·ōˌas·trə'nȯd·iks }

bioavailability [PHYSIO] The extent and rate at which a substance, such as a drug, is absorbed into a living system or is made available at the site of physiological activity. { ˌbī·ō·əˌvāl·ə'bil·əd·ē }

biobubble [ECOL] A model concept of the ecosphere in which all living things are considered as particles held together by nonliving forces. { 'bī·ōˌbəb·əl }

biocalorimetry [BIOPHYS] The measurement of the energetics of biological processes such as biochemical reactions, association of ligands to biological macromolecules, folding of proteins into their native conformations, phase transitions in biomembranes, and enzymatic reactions, among others. { ˌbī·ōˌkal·ə'rim·ə·trē }

biocatalyst [BIOCHEM] A biochemical catalyst, especially an enzyme. { ˌbī·ō'kad·əl·ist }

biocenology [ECOL] The study of natural communities and of interactions among the members of these communities. { ˌbī·ō·sə'näl·ə·jē }

biocenose *See* biotic community. { ˌbī·ō'sēˌnōs }

biochemical [BIOCHEM] **1.** A chemical that produces an effect on living organisms after making contact. **2.** Referring to chemical substances found in or having an effect on living organisms. { ˌbī·ō'kem·i·kəl }

biochemical engineering [BIOCHEM] The application of chemical engineering principles to conceive, design, develop, operate, or utilize processes and products based on biological and biochemical phenomena; this field is included in a wide range of industries, such as health care, agriculture, food, enzymes, chemicals, waste treatment, and energy. { ˌbī·ō¦kem·i·kəl ˌen·jə'nir·iŋ }

biochemical oxygen demand [MICROBIO] The amount of dissolved oxygen required to meet the metabolic needs of aerobic microorganisms in water rich in organic matter, such as sewage. Abbreviated BOD. Also known as biological oxygen demand. { ¦bī·ō'kem·ə·kəl 'äk·sə·jən di'mand }

biochemical oxygen demand test [MICROBIO] A standard laboratory procedure for measuring biochemical oxygen demand; standard measurement is made for 5 days at 20°C. Abbreviated BOD test. { ¦bī·ō'kem·ə·kəl 'äk·sə·jən di'mand ˌtest }

biochemical pharmacology [MED] The study of the effects of chemicals on biochemical reactions in living systems and the effects of these systems on the chemicals, that is, their metabolism. { ˌbī·ōˌkem·ə·kəl ˌfär·mə'käl·ə·jē }

biochemorphology [BIOCHEM] The science dealing with the chemical structure of foods and drugs and their reactions on living organisms. { ¦bī·ō¦ke·mȯr¦fäl·ə·jē }

biochore [ECOL] A group of similar biotopes. { 'bī·ōˌkȯr }

biochrome [BIOCHEM] Any naturally occurring plant or animal pigment. { 'bī·ōˌkrōm }

biociation [ECOL] A subdivision of a biome distinguished by the predominant animal species. { bīˌäs·ē'ā·shən }

bioclimatic law [ECOL] The law which states that phenological events are altered by about 4 days for each 5° change of latitude northward or longitude eastward; events are accelerated in spring and retreat in autumn. { ¦bī·ōˌklī'mad·ik 'lȯ }

bioclimatograph [ECOL] A climatograph showing the relation between climatic conditions and some living organisms. { ¦bī·ōˌklī'mad·əˌgraf }

bioclimatology [ECOL] The study of the effects of the natural environment on living organisms. { ¦bī·ōˌklī·mə'täl·ə·jē }

biocoenosis [ECOL] A group of organisms that live closely together and form a natural ecologic unit. { ˌbī·ō·sə'nō·səs }

biocompatibility [PHYSIO] The condition of being compatible with living tissue by virtue of a lack of toxicity or ability to cause immunological rejection. { ˌbī·ō·kəmˌpad·ə'bil·əd·ē }

biocontainment *See* biological containment. { ˌbī·ō·kən'tān·mənt }

biocontrol *See* biological control. { ˌbī·ō·kən'trōl }

bioconversion [BIOPHYS] The process of converting biomass to a source of usable energy. { ˌbī·o·kən'vər·zhən }

biocycle [ECOL] A group of similar biotopes composing a major division of the biosphere; there are three biocycles: terrestrial, marine, and fresh-water. { 'bī·ōˌsī·kəl }

biocytin [BIOCHEM] $C_{16}H_{28}N_4O_4S$ Crystals with a melting point of 241–243°C; obtained from dilute methanol or acetone solutions; characterized by its utilization by *Lactobacillus casei* and L. *delbrückii* LD5 as a biotin source, and by its unavailability as a biotin source to L. *arabinosus*. Also known as biotin complex of yeast. { ˌbī·ō'sīt·ən }

biocytinase [BIOCHEM] An enzyme present in the blood and liver which hydrolyzes biocytin into biotin and lysine. { ˌbī·ō'sīt·ənˌās }

biodegradation [ECOL] The destruction of organic compounds by microorganisms. { ˌbī·ōˌdeg·rə'dā·shən }

biodistribution kinetics [BIOL] A mathematical description of the in vivo distribution of a radionuclide present in various organs as a function of time following its administration. { ˌbī·ōˌdis·trə¦byü·shən ki'ned·iks }

biodiversity [ECOL] The range of living organisms (such as plant and animal species) in an environment during a specific time period. { ˌbī·ō·di'vər·sə·dē }

biodynamic [BIOPHYS] Of or pertaining to the dynamic relation between an organism and its environment. { ˌbi·ō·dī'nam·ik }

biodynamics [BIOPHYS] The study of the effects of dynamic processes (motion, acceleration, weightlessness, and so on) on living organisms. { ˌbi·ō·dī'nam·iks }

bioelectric current [PHYSIO] A self-propagating electric current generated on the surface of nerve and muscle cells by potential differences across excitable cell membranes. { ¦bī·ō·i'lek·trik 'kər·ənt }

bioelectricity [PHYSIO] The generation by and flow of an electric current in living tissue. { ¦bī·ōˌiˌlek'tris·əd·ē }

bioelectric model [PHYSIO] A conceptual model for the study of animal electricity in terms of physical principles. { ¦bī·ō·i'lek·trik 'mäd·əl }

bioelectrochemistry [PHYSIO] The study of the control of biological growth and repair processes by electrical stimulation. { ¦bī·ō·i¦lek·trō'kem·ə·strē }

bioelectronics [BIOPHYS] **1.** The application of electronic theories and techniques to the problems of biology. **2.** The use of biotechnology in electronic devices such as biosensors, molecular electronics, and neuronal interfaces; more speculatively, the use of proteins in constructing circuits. { ¦bī·ōˌiˌlek'trän·iks }

bioenergetics [BIOCHEM] The branch of biology dealing with energy transformations in living organisms. { ¦bī·ōˌen·ər'jed·iks }

bioengineer *See* genetic engineer. { ˌbī·ōˌen·jə'nir }

bioethics [BIOL] A discipline concerned with the application of ethics to biological problems, especially in the field of medicine. { ˌbī·ō'eth·iks }

biofilm [MICROBIO] A microbial (bacterial, fungal, algal) community, enveloped by the extracellular biopolymer which these microbial cells produce, that adheres to the interface of a liquid and a surface. { 'bī·ōˌfilm }

bioflavonoid [BIOCHEM] A group of compounds obtained from the rinds of citrus fruits and involved with the homeostasis of the walls of small blood vessels; in guinea pigs a marked reduction of bioflavonoids results in increased fragility and permeability of the capillaries; used to decrease permeability and fragility in capillaries in certain conditions. Also known as citrus flavonoid compound; vitamin P complex. { ¦bī·ō'flav·əˌnȯid }

biogalvanic battery [MED] A battery that is implanted in the body and depends on interaction of metal electrodes with oxygen and fluids of the body to generate sufficient power for a heart pacemaker or other implanted medical device. { ˌbī·ō·gal'van·ik 'bad·ə·rē }

biogenesis [BIOL] Development of a living organism from a similar living organism. { ¦bī·ō'jen·ə·səs }

biogenetic law *See* recapitulation theory. { ¦bī·ō·jə'ned·ik 'lȯ }

biogenetics *See* genetic engineering. { ˌbī·ō·jə'ned·iks }

biogenic [BIOL] **1.** Essential to the maintenance of life. **2.** Produced by actions of living organisms. { ¦bī·ō¦jen·ik }

biogenic amine [BIOCHEM] Any of a group of organic compounds that contain one or more amine groups ($-NH_2$) and have a possible role in brain functioning, including catecholamines and indoles. { ¦bī·ō¦jen·ik 'a·mēn }

biogeographic realm [ECOL] Any of the divisions of the landmasses of the world according to their distinctive floras and faunas. { ˌbī·ōˌjē·əˌgraf·ik 'relm }

biogeography [ECOL] The science concerned with the geographical distribution of animal and plant life. { ¦bī·ō·jē'äg·rə·fē }

biogeosphere [ECOL] The region of the earth extending from the surface of the upper crust to the maximum depth at which organic life exists. { ˌbī·ō'jē·əˌsfir }

biohazard [BIOL] Any biological agent or condition that presents a hazard to life. { 'bī·ōˌhaz·ərd }

biohydrology [ECOL] Study of the interactions between water, plants, and animals, including the effects of water on biota as well as the physical and chemical changes in water or its environment produced by biota. { ¦bī·ōˌhī'dräl·ə·jē }

bioinorganic chemistry [BIOCHEM] The application of the principles of inorganic chemistry to problems of biology and biochemistry. Also known as inorganic biochemistry; metallobiochemistry. { ˌbī·ōˌin·ȯr¦gan·ik 'kem·ə·strē }

bioleaching [BIOCHEM] The dissolution of metals from their mineral source by naturally occurring microorganisms. Also known as biooxidation. { ¦bī·ō'lēch·iŋ }

biological [BIOL] Of or pertaining to life or living organisms. [IMMUNOL] A biological product used to induce immunity to various infectious diseases or noxious substances of biological origin. { ¦bī·ə¦läj·ə·kəl }

biological balance [ECOL] Dynamic equilibrium that exists among members of a stable natural community. { ¦bī·ə¦läj·ə·kəl 'bal·əns }

biological clock [PHYSIO] Any physiologic factor that functions in regulating body rhythms. { ¦bī·ə¦läj·ə·kəl 'kläk }

biological containment [GEN] A technique by which the genetic constitution of an organism is altered in order to minimize its ability to grow outside the laboratory. Also known as biocontainment. { ¦bī·ə¦läj·ə·kəl kən'tān·mənt }

biological control [ECOL] Natural or applied regulation of populations of pest organisms, especially insects, through the role or use of natural enemies. Also known as biocontrol. { ¦bī·ə¦läj·ə·kəl kən'trōl }

biological equilibrium [BIOPHYS] A state of body balance for an actively moving animal, when internal and external forces are in equilibrium. { ¦bī·ə¦läj·ə·kəl ˌē·kwə'lib·rē·əm }

biological half-life [PHYSIO] The time required by the body to eliminate half of the amount of an administered substance through normal channels of elimination. { ¦bī·ə¦läj·ə·kəl 'haf ˌlīf }

biological indicator [BIOL] An organism that can be used to determine the concentration of a chemical in the environment. { ˌbī·ə¦läj·ə·kəl 'in·dəˌkād·ər }

biological invasion [ECOL] The process by which species (or genetically distinct populations), with no historical record in an area, breach biogeographic barriers and extend their range. { ˌbi·ə¦läj·i·kəl in'vā·zhən }

biological magnification [ECOL] The increasing concentration of toxins from pesticides, herbicides, and various types of waste in living organisms that accompanies cycling of nutrients through the trophic levels of food webs. { ˌbī·ə¦läj·ə·kəl ˌmag·nə·fə'kā·shən }

biological oil-spill control [ECOL] The use of cultures of microorganisms capable of living on oil as a means of degrading an oil slick biologically. { ¦bī·ə¦läj·ə·kəl 'ȯil ˌspil kən'trōl }

biological oxidation [BIOCHEM] Energy-producing reactions in living cells involving the transfer of hydrogen atoms or electrons from one molecule to another. { ¦bī·ə¦läj·ə·kəl ˌäk·sə'dā·shən }

biological oxygen demand *See* biochemical oxygen demand. { ¦bī·ə¦läj·ə·kəl 'äk·sə·jən di'mand }

biological productivity [ECOL] The quantity of organic matter or its equivalent in dry matter, carbon, or energy content which is accumulated during a given period of time. { ¦bī·ə¦läj·ə·kəl prə,dək'tiv·əd·ē }

biological shield [MICROBIO] A structure designed to prevent the migration of living organisms from one part of a system to another; used on sterilized space probes. { ¦bī·ə¦läj·ə·kəl 'shēld }

biological specificity [BIOL] The principle that defines the orderly patterns of metabolic and developmental reactions giving rise to the unique characteristics of the individual and of its species. { ¦bī·ə¦läj·ə·kəl spes·ə'fis·əd·ē }

biological standardization [PHARM] The standardization of drugs or biological products that cannot be chemically analyzed by studying the drugs' pharmacologic action on animals. { ¦bī·ə¦läj·ə·kəl ,stan·dərd·ə'zā·shən }

biological value [BIOCHEM] A measurement of the efficiency of the protein content in a food for the maintenance and growth of the body tissues of an individual. { ¦bī·ə¦läj·ə·kəl 'val·yü }

bioluminescence [BIOL] The emission of visible light by living organisms. { ,bī·ō,lü·mə 'nes·əns }

biolysis [BIOL] **1.** Death and the following tissue disintegration. **2.** Decomposition of organic materials, such as sewage, by living organisms. { bī'äl·ə·səs }

biomagnetism [BIOPHYS] The production of a magnetic field by a living organism. { ¦bī·ō'mag·nə,tiz·əm }

biomass [ECOL] The dry weight of living matter, including stored food, present in a species population and expressed in terms of a given area or volume of the habitat. { 'bī·ō,mas }

biomaterial [MED] A natural or synthetic nondrug material that is compatible with living tissue and is suitable for surgical implanting; it can be used to enhance, treat, or replace organs, tissues, and functions in a living organism. { ¦bī·ō·mə¦tir·ē·əl }

biomathematics [BIOPHYS] Mathematical methods applied to the study of living organisms. { ,bī·ō,math·ə'mad·iks }

biome [ECOL] A complex biotic community covering a large geographic area and characterized by the distinctive life-forms of important climax species. { 'bī,ōm }

biomechanics [BIOPHYS] The study of the mechanics of living things. { ¦bī·ō·mə'kan·iks }

biomedical chemical engineering [MED] The application of chemical engineering principles to the solution of medical problems due to physiological impairment. { ,bī·ō¦med·i·kəl ¦kem·i·kəl ,en·jə'nir·iŋ }

biomedicine [MED] The science concerned with the study of the environment required for astronauts in space vehicles. { ,bī·ō'med·ə·sən }

biomere [ECOL] A biostratigraphic unit bounded by abrupt nonevolutionary changes in the dominant elements of a single phylum. { 'bī·ō,mir }

biometeorology [BIOL] The study of the relationship between living organisms and atmospheric phenomena. { ¦bī·ō,mēd·ē·ə'räl·ə·jē }

biometer [BIOL] An instrument which is used to measure minute amounts of carbon dioxide given off by the functioning tissue of an organism. { bī'äm·ə·tər }

biomimetics [BIOCHEM] A branch of science in which synthetic systems are developed by using information obtained from biological systems. { ¦bī·ō·mə¦med·iks }

biomineralization [PHYSIO] A mineralization process carried out within a living organism, such as formation of bone in vertebrates. { ,bī·ō,min·rəl·ə'zā·shən }

biomining [MICROBIO] The use of microorganisms to recover metals of value, such as gold, silver, and copper, from sulfide minerals. { 'bī·ō,mīn·iŋ }

biomolecular [BIOCHEM] Pertaining to organic molecules occurring in living organisms, especially macromolecules. { ,bī·ō·mə'le·kyə·lər }

bion [ECOL] An independent, individual organism. { 'bī,än }

bionavigation [VERT ZOO] The ability of animals such as birds to find their way back to their roost, even if the landmarks on the outward-bound trip were effectively concealed from them. { ,bī·ō,nav·ə'gā·shən }

bionomics *See* ecology. { ,bī·ō'näm·iks }

biophage *See* macroconsumer. { 'bī·ō,fāj }

biophagous [ZOO] Feeding on living organisms. { bī'äf·ə·gəs }

biopharmaceutics [PHARM] The study of the relationships between physical and chemical properties, dosage, and administration of a drug and its activity in humans and animals. { ,bī·ō,färm·ə'süd·iks }

biopharming [MED] The application of genetic engineering on living organisms to induce or increase the production of pharmacologically active substances. { 'bī·ō,färm·iŋ }

biophile [BIOCHEM] Any element concentrated or found in the bodies of living organisms and organic matter; examples are carbon, nitrogen, and oxygen. { 'bī·ə,fīl }

biopolymer [BIOCHEM] A biological macromolecule such as a protein or nucleic acid. { ¦bī·ō'päl·ə·mər }

biopotency [BIOCHEM] Capacity of a chemical substance, as a hormone, to function in a biological system. { ¦bī·ō'pōt·ən·sē }

biopotential [PHYSIO] Voltage difference measured between points in living cells, tissues, and organisms. { ¦bī·ō·pə'ten·chəl }

bioprocess [BIOCHEM] **1.** A technique used to produce a commercial substance (such as alcohol) by a biological process (that is, via microbial fermentation). **2.** A technique used to prepare a biological material (usually genetically engineered) for commercial use. { ¦bī·o¦prä·səs }

bioprospecting [PHARM] The search for new pharmaceutical (and sometimes nutritional or agricultural) products from natural sources, such as plants, microorganisms, and sometimes animals. { ,bī·ō'prä·spek·tiŋ }

biopsy [PATH] The removal and examination of

tissues, cells, or fluids from the living body for the purposes of diagnosis. { 'bī,äp·sē }

bioreactor [BIOCHEM] A vessel that is used for the fermentation and production of living organisms, such as bacteria or yeast. { 'bī·ō·rē,ak·tər }

biorefinery [BIOCHEM] A large, integrated processing facility that produces chemicals and biochemicals from plant matter, wood waste, and waste paper. { ¦bī·ō·ri¦fīn·rē }

bioregion [ECOL] A region with borders that are naturally defined by topographic systems (such as mountains, rivers, and oceans) and ecological systems (such as deserts, rainforests, and tundras). { 'bī·ō,rē·jən }

bioregionalism [ECOL] An environmentalist movement to make political boundaries coincide with bioregions. { ,bī·ō'rē·jən·əl,iz·əm }

bioremediation [ECOL] The use of a biological process (via plants or microorganisms) to clean up a polluted environmental area (such as an oil spill). { ,bī·ō·ri,mē·dē'ā·shən }

biorheology [BIOPHYS] The study of the deformation and flow of biological fluids, such as blood, mucus, and synovial fluid. { ¦bī·ō·rē'äl·ə·jē }

biorhythm [PHYSIO] A biologically inherent cyclic variation or recurrence of an event or state, such as a sleep cycle or circadian rhythm. { 'bī·ō,rith·əm }

biosafety [BIOL] The establishment and maintenance of safe conditions in a biological research laboratory to ensure that pathogenic microbes are contained (and not released to workers or the environment). { ¦bī·ō¦sāf·tē }

bioscience [BIOL] The study of the nature, behavior, and uses of living organisms as applied to biology. { ¦bī·ō¦sī·əns }

bioseparation [BIOCHEM] The recovery of a product from solutions of cells and media; the process used must avoid harsh conditions that could damage the product. { ,bī·ō,sep·ə'rā·shən }

bioseries [EVOL] A historical sequence produced by the changes in a single hereditary character. { ¦bī·ō,sir·ēz }

biosocial [ZOO] Pertaining to the interplay of biological and social influences. { ,bī·ō'sō·shəl }

biosonar [PHYSIO] A guidance system in certain animals, such as bats, utilizing the reflection of sounds that they produce as they move about. { ¦bī·ō¦sō,när }

biospeleology [BIOL] The study of cave-dwelling organisms. { ,bī·ō,spē·lē'äl·ə·jē }

biosphere [ECOL] The life zone of the earth, including the lower part of the atmosphere, the hydrosphere, soil, and the lithosphere to a depth of about 1.2 miles (2 kilometers). { 'bī·ə,sfir }

biostasy [ECOL] Maximum development of organisms when, during tectonic repose, residual soils form extensively on the land and calcium carbonate deposition is widespread in the sea. { bī'äs·tə·sē }

biostrome [INV ZOO] A flat-bedded, fossil, reeflike structure. { 'bī·ə,strōm }

biosynthesis [BIOCHEM] Production, by synthesis or degradation, of a chemical compound by a living organism. { ,bī·ō'sin·thə·səs }

biota [BIOL] **1.** Animal and plant life characterizing a given region. **2.** Flora and fauna, collectively. { bī'ōd·ə }

biotechnology [GEN] The use of advanced genetic techniques to construct novel microbial, plant, and animal strains or obtain site-directed mutants to improve the quantity or quality of products or obtain other desired phenotypes. { ¦bī·ō·tek'näl·ə·jē }

biotherapy [MED] Treatment of disease with biologicals, that is, materials produced by living organisms. { ¦bī·ō'ther·ə·pē }

biotic [BIOL] **1.** Of or pertaining to life and living organisms. **2.** Induced by the actions of living organisms. { bī'äd·ik }

biotic community [ECOL] An aggregation of organisms characterized by a distinctive combination of both animal and plant species in a particular habitat. Also known as biocenose. { bī'äd·ik kə'myün·əd·ē }

biotic district [ECOL] A subdivision of a biotic province. { bī'äd·ik 'dis·trikt }

biotic environment [ECOL] That environment comprising living organisms, which interact with each other and their abiotic environment. { bī'äd·ik in'vī·ərn·mənt }

biotic isolation [ECOL] The occurrence of organisms in isolation from others of their species. { bī'äd·ik ,i·sə'lā·shən }

biotic potential [ECOL] The maximum possible growth rate of living things under ideal conditions. { bī'äd·ik pə'ten·chəl }

biotic province [ECOL] A community, according to some systems of classification, occupying an area where similarity of climate, physiography, and soils leads to the recurrence of similar combinations of organisms. { bī'äd·ik 'präv·əns }

biotin [BIOCHEM] $C_{10}H_{16}N_2O_3S$ A colorless, crystalline vitamin of the vitamin B complex occurring widely in nature, mainly in bound form. { 'bī·ə·tən }

biotin carboxylase [BIOCHEM] An enzyme which condenses bicarbonate with biotin to form carboxybiotin. { 'bī·ə·tən kär'bäk·sə,lās }

biotin complex of yeast *See* biocytin. { 'bī·ə·tən 'käm,pleks əv 'yēst }

biotope [ECOL] An area of uniform environmental conditions and biota. { 'bī·ə,tōp }

biotransformation [BIOCHEM] The series of chemical reactions that occur in a compound, especially a drug, as a result of enzymatic or metabolic activities by a living organism. { ¦bī·ō,tranz·fər'mā·shən }

biotype [GEN] A group of organisms having the same genotype. { 'bī·ə,tīp }

biparasitic [ECOL] Parasitic upon or in a parasite. { ¦bī,par·ə'sid·ik }

biparous [BOT] Bearing branches on dichotomous axes. [VERT ZOO] Bringing forth two young at a birth. { 'bī·pə·rəs }

bipartite uterus [ANAT] A uterus divided into two parts almost to the base. { bī'pär,tīt 'yüd·ə·rəs }
bipectinate [INV ZOO] Of the antennae of certain moths, having two margins with comblike teeth. [ZOO] Branching like a feather on both sides of a main shaft. { bī'pek·tə,nāt }
biped [VERT ZOO] **1.** A two-footed animal. **2.** Any two legs of a quadruped. { 'bī,ped }
bipedal [BIOL] Having two feet. { bī'ped·əl }
bipeltate [BOT] Having two shield-shaped parts. [ZOO] Having a shell or other covering resembling a double shield. { bī'pel,tāt }
bipenniform [ANAT] Of the arrangement of muscle fibers, resembling a feather barbed on both sides. { bī'pen·ə,fȯrm }
biphasic [BOT] Possessing both a sporophyte and a gametophyte generation in the life cycle. { bī'fāz·ik }
biphyletic [EVOL] Descended in two branches from a common ancestry. { ¦bī·fī'led·ik }
Biphyllidae [INV ZOO] The false skin beetles, a family of coleopteran insects in the superfamily Cucujoidea. { ,bī'fil·ə,dē }
bipinnaria [INV ZOO] The complex, bilaterally symmetrical, free-swimming larval stage of most asteroid echinoderms. { ¦bī·pi'ner·ē·ə }
bipinnate [BOT] Pertaining to a leaf that is pinnate for both its primary and secondary divisions. { bī'pin,āt }
bipolar flagellation [MICROBIO] The presence of flagella at both poles in certain bacteria. { bī'pō·lər ,flaj·ə'lā·shən }
Bipolarina [INV ZOO] A suborder of protozoan parasites in the order Myxosporida. { ,bī·pō·lə'rī·nə }
bipotential [BIOL] Having the potential to develop in either of two mutually exclusive directions. { ¦bī·pə'ten·chəl }
bipotentiality [BIOL] **1.** Capacity to function either as male or female. **2.** Hermaphroditism. { ¦bī·pə,ten·chē'al·əd·ē }
biradial symmetry [BIOL] Symmetry both radial and bilateral. Also known as disymmetry. { bī'rād·ē·əl 'sim·ə·trē }
biramous [BIOL] Having two branches, such as an arthropod appendage. { bī'rā·məs }
birch [BOT] The common name for all deciduous trees of the genus *Betula* that compose the family Betulaceae in the order Fagales. { bərch }
bird [VERT ZOO] Any of the warm-blooded vertebrates which make up the class Aves. { bərd }
bird louse [INV ZOO] The common name for any insect of the order Mallophaga. Also known as biting louse. { 'bərd ,laùs }
bird of prey [VERT ZOO] Any of various carnivorous birds of the orders Falconiformes and Strigiformes which feed on meat taken by hunting. { 'bərd əv 'prā }
birimose [BOT] Opening by two slits, as an anther. { bī'rī,mōs }
birotulate [INV ZOO] A sponge spicule characterized by two wheel-shaped ends. { bī'räch·ə,lāt }
birth [BIOL] The emergence of a new individual from the body of its parent. { bərth }
birth canal [ANAT] The channel in mammals through which the fetus is expelled during parturition; consists of the cervix, vagina, and vulva. { 'bərth kə'nal }
birth control [MED] Limitation of the number of children born by preventing or reducing the frequency of impregnation. { bərth kən'trōl }
birth defect *See* congenital anomaly. { 'bərth di'fekt }
birthmark [MED] Any abnormal cellular or vascular benign nevus that is present at birth or that appears sometime later. { 'bərth,märk }
birth rate [BIOL] The ratio between the number of live births and a specified number of people in a population over a given period of time. { 'bərth ,rāt }
biserial [BIOL] Arranged in two rows or series. { ,bī'sir·ē·əl }
biserrate [BIOL] **1.** Having serrated serrations. **2.** Serrate on both sides. { ,bī'ser,āt }
bisexual [BIOL] Of or relating to two sexes. { ,bī'sek·shə·wəl }
bison [VERT ZOO] The common name for two species of the family Bovidae in the order Artiodactyla; the wisent or European bison (*Bison bonasus*), and the American species (*Bison bison*). { 'bīs·ən }
bisporangiate [BOT] Having two different types of sporangia. { ¦bī·spə'ran·jē,āt }
bispore [BOT] In certain red algae, an asexual spore that is produced in pairs. { 'bī,spȯr }
bite [BIOL] **1.** To seize with the teeth. **2.** Closure of the lower teeth against the upper teeth. [MED] Skin injury produced by an animal's teeth or the mouthparts of an insect. { bīt }
bitegmic [BOT] Having two integuments, especially in reference to ovules. { bī'teg·mik }
biternate [BOT] Of a ternate leaf, having each division ternate. { bī'tər,nāt }
bitewing [MED] A dental x-ray film having a central wing on which the teeth can close to hold it in place. { 'bīt,wiŋ }
biting louse *See* bird louse. { 'bīd·iŋ ,laùs }
Bitot's spots [MED] The silver-gray, shiny, triangular spots on the cornea characteristic of xerosis conjunctivae. { 'bē·tōs ,späts }
bittern [VERT ZOO] Any of various herons of the genus *Botaurus* characterized by streaked and speckled plumage. { 'bid·ərn }
Bittner milk factor *See* milk factor. { 'bit·nər 'milk ,fak·tər }
bivalent antibody [IMMUNOL] An antibody possessing two antibody combining sites. { bī'vāl·ənt 'ant·i,bäd·ē }
bivalent chromosome [CELL MOL] The structure formed following synapsis of a pair of homologous chromosomes from the zygotene stage of meiosis up to the beginning of anaphase. { bī'vā·lənt 'krō·mə,sōm }
bivalve [INV ZOO] The common name for a number of diverse, bilaterally symmetrical animals, including mollusks, ostracod crustaceans,

and brachiopods, having a soft body enclosed in a calcareous two-part shell. { 'bī,valv }

Bivalvia [INV ZOO] A large class of the phylum Mollusca containing the clams, oysters, and other bivalves. { bī'val·vē·ə }

biventer [ANAT] A muscle having two bellies. { bī'ven·tər }

bivittate [ZOO] Having a pair of longitudinal stripes. { bī'vi,tāt }

bivium [INV ZOO] The pair of starfish rays that extend on either side of the madreporite. { 'bī·vē·əm }

bivoltine [INV ZOO] **1.** Having two broods in a season, used especially of silkworms. **2.** Of insects, producing two generations a year. { bī'vōl,tēn }

black band disease [INV ZOO] A coral reef disease that is characterized by a thick black band of tissue that advances rapidly across infected corals, leaving empty coral skeletons behind. { ¦blak¦band di,zēz }

blackberry [BOT] Any of the upright or trailing shrubs of the genus *Rubus* in the order Rosales; an edible berry is produced by the plant. { 'blak,ber·ē }

blackbird [VERT ZOO] Any bird species in the family Icteridae, of which the males are predominantly or totally black. { 'blak,bərd }

black coral [INV ZOO] The common name for antipatharian cnidarians having black, horny axial skeletons. { ¦blak 'kär·əl }

black death *See* plague. { ¦blak 'deth }

black disease [VET MED] Necrotic hepatitis of sheep, resulting from infection with *Clostridium novyi* type B, with the necessary conditions for the growth of the clostridia provided by the damaged liver tissue produced by the fluke *Fasciola hepatica*. { 'blak di'zēz }

blackeye bean *See* cowpea. { 'blak,ī ¦bēn }

blackhead *See* comedo. { 'blak,hed }

blackleg [VET MED] An acute, usually fatal bacterial disease of cattle, and occasionally of sheep, goats, and swine, caused by *Clostridium chauvoei*. { 'blak,leg }

blacklung *See* anthracosis. { 'blak,ləng }

black measles *See* hemorrhagic measles. { ¦blak 'mē·zəlz }

black membrane [CELL MOL] An artificial planar membrane that forms over a hole in the partition between two aqueous compartments and is optically black when viewed in incident light; used to study the permeability of bilayer membranes and the mobility of bilayer components. { ¦blak 'mem,brān }

black mold [MYCOL] Any dark fungus belonging to the order Mucorales. { 'blak ,mōld }

black scour [VET MED] Hemorrhagic enteritis of sheep, swine, and cattle, usually associated with a heavy worm burden but sometimes caused by bacterial infection. { 'blak ,skaůr }

blacktongue [VET MED] A niacin-deficiency disease of dogs, with black discoloration of the tongue. { 'blak,təŋ }

black vomit [MED] Dark vomited matter, consisting of digested blood and gastric contents. { ¦blak 'väm·ət }

blackwater [MED] Any disease that is characterized by dark-colored urine. { 'blak,wȯd·ər }

blackwater fever [MED] A complication of falciparum malaria, characterized by intravascular hemolysis, hemoglobinuria, tachycardia, high fever, and poor prognosis. { 'blak,wȯd·ər 'fev·ər }

bladder [ANAT] Any saclike structure in humans and animals, such as a swimbladder or urinary bladder, that contains a gas or functions as a receptacle for fluid. { 'blad·ər }

bladder cell [INV ZOO] Any of the large vacuolated cells in the outer layers of the tunic in some tunicates. { 'blad·ər ,sel }

blade [BOT] The broad, flat portion of a leaf. Also known as lamina. [VERT ZOO] A single plate of baleen. { blād }

blast cell [HISTOL] An undifferentiated precursor of a human blood cell in the reticuloendothelial tissue. { 'blast ,sel }

blastema [EMBRYO] **1.** A mass of undifferentiated protoplasm capable of growth and differentiation. **2.** *See* anlage. { bla'stēma }

blasto- [EMBRYO] A germ or bud, with reference to early embryonic stages of development. { 'blas·tō }

Blastobacter [MICROBIO] A genus of budding bacteria; cells are rod-, wedge-, or club-shaped, and several attach at the narrow end to form a rosette; reproduce by budding at the rounded end. { ¦blas·tō¦bak·tər }

Blastobasidae [INV ZOO] A family of lepidopteran insects in the superfamily Tineoidea. { ¦blas·tō'bas·ə,dē }

blastocarpous [BOT] Germinating in the pericarp. { ¦blas·tō¦kär·pəs }

blastochyle [EMBRYO] The fluid filling the blastocoele. { 'blas·tə,kīl }

Blastocladiales [MYCOL] An order of aquatic fungi in the class Phycomycetes. { ¦blas·tō¦klā·dē'ā·lēz }

blastocoele [EMBRYO] The cavity of a blastula. Also known as segmentation cavity. { 'blas·tə,sēl }

blastocone [EMBRYO] An incomplete blastomere. { 'blas·tō,kōn }

blastocyst [EMBRYO] A modified blastula characteristic of placental mammals. { 'blas·tə,sist }

blastocyte [EMBRYO] An embryonic cell that is undifferentiated. [INV ZOO] An undifferentiated cell capable of replacing damaged tissue in certain lower animals. { 'blas·tə,sīt }

blastoderm [EMBRYO] The blastodisk of a fully formed vertebrate blastula. { 'blas·tə,dərm }

blastodisk [EMBRYO] The embryo-forming, protoplasmic disk on the surface of a yolk-filled egg, such as in reptiles, birds, and some fish. { 'blas·tə,disk }

blastokinesis [INV ZOO] Movement of the embryo into the yolk in some insect eggs. { ,blas·tō,kə'nē·səs }

blastoma [MED] **1.** A tumor whose parenchymal

cells have certain embryonal characteristics. **2.** A true tumor. { ˌbla'stōm·ə }

blastomere [EMBRYO] A cell of a blastula. { 'blas·təˌmir }

blastomycosis [MED] A term for two infectious, yeastlike fungus diseases of humans: North American blastomycosis, caused by *Blastomyces dermatitidis*, and South American (paracoccidioidomycosis) caused by *Blastomyces brasiliensis*. { ˌblas·təˌmī'kō·səs }

blastophore [CELL MOL] The cytoplasm that is detached from a spermatid during its transformation to a spermatozoon. [INV ZOO] An amorphose core of cytoplasm connecting cells of the male morula of developing germ cells in oligochaetes. { 'blas·təˌfór }

blastopore [EMBRYO] The opening of the archenteron. { 'blas·təˌpór }

blastospore [MYCOL] A fungal resting spore that arises by budding. { 'blas·təˌspór }

blastostyle [INV ZOO] A zooid on certain hydroids that lacks a mouth and tentacles and functions to produce medusoid buds. { 'blas·təˌstīl }

blastotomy [EMBRYO] Separation of cleavage cells during early embryogenesis. { ¦blas·'täd·ə·mē }

blastozooid [INV ZOO] A zooid produced by budding. { ˌblas·tə'zōˌóid }

blastula [EMBRYO] A hollow sphere of cells characteristic of the early metazoan embryo. { 'blas·chə·lə }

blastulation [EMBRYO] Formation of a blastula from a solid ball of cleaving cells. { ˌblas·chə'lā·shən }

Blattabacterium [MICROBIO] A genus of the tribe Wolbachiae; straight or slightly curved rod-shaped cells; symbiotic in cockroaches. { ˌblad·ə·bak'tir·ē·əm }

Blattidae [INV ZOO] The cockroaches, a family of insects in the order Orthoptera. { 'blad·ə·dē }

bleb [MED] A localized collection of fluid, as serum or blood, in the epidermis. { bleb }

bleed [MED] To exude blood from a wound. { blēd }

bleeder [MED] **1.** A person subject to frequent hemorrhages, as a hemophiliac. **2.** A blood vessel from which there is persistent uncontrolled bleeding. **3.** A blood vessel which has escaped closure by cautery or ligature during a surgical procedure. { 'blēd·ər }

bleeding time [PHYSIO] The time required for bleeding to stop after a small puncture wound. { 'blēd·iŋ ˌtīm }

blending inheritance [GEN] Inheritance in which the character of the offspring is a blend of those in the parents; a common feature for quantitative characters, such as stature, determined by large numbers of genes and affected by environmental variation. { 'blen·diŋ in'her·ə·təns }

Blenniidae [VERT ZOO] The blennies, a family of carnivorous marine fishes in the suborder Blennioidei. { ble'nī·əˌdē }

Blennioidei [VERT ZOO] A large suborder of small marine fishes in the order Perciformes that live principally in coral and rock reefs. { ˌble·nē'ói·dēˌī }

blennorrhagia [MED] Excessive discharge of mucus. Also known as blennorrhea. { ble·nə'rā·jē·ə }

blennorrhea *See* blennorrhagia. { ˌble·nə'rē·ə }

Blephariceridae [INV ZOO] A family of dipteran insects in the suborder Orthorrhapha. { ˌblef·ə·ri'ser·əˌdē }

blepharism [MED] Spasm of the eyelids causing rapid, repetitive involuntary winking. { 'blef·əˌriz·əm }

blepharitis [MED] Inflammation of the eyelids. { ˌblef·ə'rīd·əs }

blepharoconjunctivitis [MED] Inflammation of the eyelids and the conjunctiva. { ¦blef·ə·rō·kənˌjəŋk·tə'vīd·əs }

blepharoplast [MICROBIO] A cytoplasmic granule bearing a bacterial flagellum. { 'blef·ə·rōˌplast }

blepharoplasty [MED] Any plastic surgical operation on the eyelid. Also known as tarsoplasty. { 'blef·ə·rəˌplas·tē }

blepharospasm [MED] Spasmodic winking due to spasms of the orbicular muscle of the eyelid. { 'blef·ə·rōˌspaz·əm }

blindness [MED] **1.** Loss or absence of the ability to perceive visual images. **2.** The condition of a person having less than 1/10 (20/200 on the Snellen test) normal vision. { 'blīnd·nəs }

blind passage [MICROBIO] Transfer of some material from an inoculated animal or cell culture that does not exhibit evidence of infection, to a fresh animal or cell culture. { 'blīnd 'pas·ij }

blind spot [NEUROSCI] A place on the retina of the eye that is insensitive to light, where the optic nerve passes through the eyeball's inner surface. { 'blīnd ˌspät }

blister [MED] A local swelling of the skin resulting from the accumulation of serous fluid between the epidermis and true skin. { 'blis·tər }

bloat [VET MED] Distension of the rumen in cattle and other ruminants due to excessive gas formation following heavy fermentation of legumes eaten wet. { blōt }

blocked reading frame *See* closed reading frame. { ¦bläkt 'rēd·iŋ ˌfrām }

blocking [HISTOL] **1.** The process of embedding tissue in a solid medium, such as paraffin. **2.** A histochemical process in which a portion of a molecule is treated to prevent it from reacting with some other agent. { 'bläk·iŋ }

blocking antibody [IMMUNOL] Antibody that inhibits the reaction between antigen and other antibodies or sensitized T lymphocytes. { ¦bläk·iŋ ¦ant·iˌbäd·e }

blood [HISTOL] A fluid connective tissue consisting of the plasma and cells that circulate in the blood vessels. { bləd }

blood agar [MICROBIO] A nutrient microbiologic culture medium enriched with whole blood and used to detect hemolytic strains of bacteria. { 'bləd ,äg·ər }

blood blister [MED] A blister that is filled with blood. { 'bləd ,blis·tər }

blood-brain barrier [NEUROSCI] A barrier to the entry of substances from the blood into brain tissue; believed to be formed primarily by the endothelial cells of the brain vasculature. { ¦bləd ¦brān 'bar·ē·ər }

blood cell [HISTOL] An erythrocyte or a leukocyte. { 'bləd ,sel }

blood chimerism [GEN] Having red blood cells of two different genotypes stemming from two different fertilize eggs or individuals. { 'bləd ka'mir,iz·əm }

blood count [PATH] Determination of the number of white and red blood cells in a definite volume of blood. { 'bləd ,kau̇nt }

blood crisis [MED] The sudden appearance of large numbers of nucleated erythrocytes in the circulating blood. { 'bləd ,krī·səs }

blood dyscrasia [MED] Obsolete term. Any abnormal condition of the formed elements of blood or of the constituents required for clotting. { 'bləd də'skrāzh·ə }

blood group [IMMUNOL] An immunologically distinct, genetically determined class of human erythrocyte antigens, identified as A, B, AB, and O. Also known as blood type. { 'bləd ,grüp }

blood island [EMBRYO] One of the areas in the yolk sac of vertebrate embryos allocated to the production of the first blood cells. { 'bləd ,ī·lənd }

bloodline [BIOL] A line of direct ancestors, especially in a pedigree. { 'bləd,līne }

blood-plate hemolysis [MICROBIO] Destruction of red blood cells in a blood agar medium by a bacterial toxin. { 'bləd ,plāt hə'mäl·ə·səs }

blood platelet *See* thrombocyte. { 'bləd ,plāt·lət }

blood poisoning *See* septicemia. { 'bləd ,poiz·ən·iŋ }

blood pressure [PHYSIO] Pressure exerted by blood on the walls of the blood vessels. { 'bləd ,presh·ər }

bloodstream [PHYSIO] The flow of blood in its circulation through the body. { 'bləd ,strēm }

blood sugar [BIOCHEM] The carbohydrate, principally glucose, of the blood. { 'bləd ,shu̇g·ər }

blood test [PATH] **1.** A serologic test for syphilis. **2.** A blood count. **3.** A test for detection of blood, usually one based on the peroxidase activity of blood, such as the benzidine test or guaiac test. { 'bləd ,test }

blood type *See* blood group. { 'bləd ,tīp }

blood typing [IMMUNOL] Determination of an individual's blood group. { 'bləd ,tīp·iŋ }

blood vessel [ANAT] A tubular channel for blood transport. { 'bləd ,ves·əl }

bloom [BOT] **1.** An individual flower. Also known as blossom. **2.** To yield blossoms. **3.** The waxy coating that appears as a powder on certain fruits, such as plums, and leaves, such as cabbage. [ECOL] A colored area on the surface of bodies of water caused by heavy planktonic growth. { blüm }

blossom *See* bloom. { 'bläs·əm }

blotting [CELL MOL] The transfer of electrophoretically separated polypeptides onto a solid support medium, such as nitrocellulose paper or a nylon membrane. { 'bläd·iŋ }

blowball [BOT] A fluffy seed ball, as of the dandelion. { 'blō,bȯl }

blowhole [VERT ZOO] The nostril on top of the head of cetacean mammals. { 'blō,hōl }

blowpipe [BIOL] A small tube, tapering to a straight or slightly curved tip, used in anatomy and zoology to reveal or clean a cavity. { 'blō,pīp }

blubber [INV ZOO] A large sea nettle or medusa. [VERT ZOO] A thick insulating layer of fat beneath the skin of whales and other marine mammals. { 'bləb·ər }

blue baby [MED] An infant with congenital cyanosis due to cardiac or pulmonary defect, causing shunting of unoxygenated blood into the systemic circulation. { 'blü ,bā·bē }

blueberry [BOT] Any of several species of plants in the genus *Vaccinium* of the order Ericales; the fruit, a berry, occurs in clusters on the plant. { 'blü,ber·ē }

blue comb [VET MED] A disease of domestic fowl and other birds that resembles Bright's disease in humans; may be a viral disease or caused by excessive salt intake. { 'blü ,kōm }

bluefish [VERT ZOO] *Pomatomus saltatrix*. A predatory fish in the order Perciformes. Also known as skipjack. { 'blü,fish }

bluegrass [BOT] The common name for several species of perennial pasture and lawn grasses in the genus *Poa* of the order Cyperales. { 'blü,gras }

blue-green algae *See* cyanobacteria. { ¦blü¦grēn 'al·jē }

blue-green algal virus *See* cyanophage. { ¦blü ¦grēn ¦al·gəl 'vī·rəs }

blue mold [MYCOL] Any fungus of the genus *Penicillium*. { 'blü ,mōld }

blue nevus [MED] A nevus composed of spindle-shaped pigmented melanocytes in the middle and lower two-thirds of the dermis. { 'blü 'nē·vəs }

bluestem grass [BOT] The common name for several species of tall, perennial grasses in the genus *Andropogon* of the order Cyperales. { 'blü ,stem ¦gras }

bluetongue [VET MED] An arthropod-borne disease of ruminant species that is caused by a ribonucleic acid-containing virus in the genus *Orbivirus*, family Reoviridae; acute infection evokes high fever, excessive salivation, nasal discharge, hyperemia (buccal and nasal mucosa, skin, coronet band), and erosions and ulcerations of mucosal surfaces in the mouth. { 'blü,təŋ }

Blumeriella jaapii [MYCOL] Previously known as *Coccomyces hiemalis*; a fungal plant pathogen that causes cherry leaf spot. { ˌblü·mer·ēˌel·ə ˈjäp·ēˌē }

blunt dissection [MED] In surgery, the exposure of structures or separation of tissues without cutting. { ¦blənt diˈsek·shən }

B lymphocyte *See* B cell. { ¦bē ˈlim·fəˌsīt }

BMI *See* body mass index.

BMR *See* basal metabolic rate.

BMV *See* brome mosaic virus.

boa [VERT ZOO] Any large, nonvenomous snake of the family Boidae in the order Squamata. { ˈbō·ə }

boar *See* wild boar. { bȯr }

Boas' test [PATH] A test that uses an alcoholic solution of resorcinol and sucrose to determine the presence of free hydrochloric acid in gastric juices. { ˈbōˌas ˌtest }

BOD *See* biochemical oxygen demand.

Bodonidae [INV ZOO] A family of protozoans in the order Kinetoplastida characterized by two unequally long flagella, one of them trailing. { bəˈdän·əˌdē }

BOD test *See* biochemical oxygen demand test. { ¦bē¦ōˈdē ˌtest }

body cavity [ANAT] The peritoneal, pleural, or pericardial cavities, or the cavity of the tunica vaginalis testis. { ˈbäd·ē ˌkav·əd·ē }

body mass index [MED] An estimation of the amount of fat stored in adipose tissue that can be calculated by dividing the body weight in kilograms by the square of the height in meters. Abbreviated BMI. { ¦bäd·ē ¦mas ˌin·deks }

body rhythm [PHYSIO] Any bodily process having some degree of regular periodicity. { ˈbäd·ē ˌrith·əm }

body-righting reflex [PHYSIO] A postural reflex, initiated by the asymmetric stimulation of the body surface by the weight of the body, so that the head tends to assume a horizontal position. { ˈbäd·ē ˌrīd·iŋ ˈrēˌfleks }

bog [ECOL] A plant community that develops and grows in areas with permanently waterlogged peat substrates. Also known as moor; quagmire. { bäg }

bog moss [ECOL] Moss of the genus *Sphagnum* occurring as the characteristic vegetation of bogs. { ˈbäg ˌmȯs }

Bohr effect [BIOCHEM] The effect of carbon dioxide and pH on the oxygen equilibrium of hemoglobin; increase in carbon dioxide prevents an increase in the release of oxygen from oxyhemoglobin. { ˈbȯr iˈfekt }

Boidae [VERT ZOO] The boas, a family of nonvenomous reptiles of the order Squamata, having teeth on both jaws and hindlimb rudiments. { ˈbō·əˌdē }

boil *See* furuncle. { bȯil }

boil disease [VET MED] A protozoan disease of fish caused by *Myxobolus pfeiffer* that forms large tumorous masses in the muscles and connective tissues, finally causing death. { ˈbȯil diˌzēz }

boll [BOT] A pod or capsule (pericarp), as of cotton and flax. { bōl }

boll weevil [INV ZOO] A beetle, *Anthonomus grandis*, of the order Coleoptera; larvae destroy cotton plants and are the most important pests in agriculture. { ˈbōl ˌwē·vəl }

bolus [PHARM] A pill of large size. [PHYSIO] The mass of food prepared by the mouth for swallowing. { ˈbō·ləs }

Bombacaceae [BOT] A family of dicotyledonous tropical trees in the order Malvales with dry or fleshy fruit usually having woolly seeds. { ˌbäm·bəˈkās·ēˌē }

Bombay blood group system [IMMUNOL] A system comprising an immunologically distinct, genetically determined group of human erythrocytes characterized by the lack of A, B, or H antigens. { bämˈbā ˈbləd ˌgrüp ˌsis·təm }

Bombidae [INV ZOO] A family of relatively large, hairy, black and yellow bumblebees in the hymenopteran superfamily Apoidea. { ˈbäm·bəˌdē }

Bombycidae [INV ZOO] A family of lepidopteran insects of the superorder Heteroneura that includes only the silkworms. { bämˈbis·əˌdē }

bombykol [BIOL] The first pheromone to be characterized chemically; it is an unsaturated straight-chain alcohol secreted in microgram amounts by females of the silkworm moth (*Bombyx mori*) and is capable of attracting male silkworm moths at large distances. { ˈbäm·bəˌkȯl }

Bombyliidae [INV ZOO] The bee flies, a family of dipteran insects in the suborder Orthorrhapha. { ˌbäm·bəˈlī·əˌdē }

Bombyx [INV ZOO] The type genus of Bombycidae. { ˈbämˌbiks }

Bombyx mori [INV ZOO] The species name of the commercial silkworm. { ˈbämˌbiks ˈmȯr·ē }

bone [ANAT] One of the parts constituting a vertebrate skeleton. [HISTOL] A hard connective tissue that forms the major portion of the vertebrate skeleton. { bōn }

bone conduction [BIOPHYS] Transmission of sound vibrations to the internal ear via the bones of the skull. { ˈbōn kənˈdək·shən }

Bonellidae [INV ZOO] A family of wormlike animals belonging to the order Echiuroinea. { bō ˈnel·əˌdē }

bone marrow [HISTOL] A vascular modified connective tissue occurring in the long bones and certain flat bones of vertebrates. { ˈbōn ˌmar·ō }

bonsai [BOT] The production of a mature, very dwarfed tree in a relatively small container. { bōnˈsī }

bony fish [VERT ZOO] The name applied to all members of the class Osteichthyes. { ˈbō·nē ˌfish }

bony labyrinth [ANAT] The system of canals within the otic bones of vertebrates that houses the membranous labyrinth of the inner ear. { ˈbō·nē ˈlab·əˌrinth }

Boodleaceae [BOT] A family of green marine algae in the order Siphonocladales. { bōˌäd·lēˈās·ēˌē }

book gill [INV ZOO] A type of gill in king crabs consisting of folds of membranous tissue arranged like the leaves of a book. { ˈbu̇k ˌgil }

book louse [INV ZOO] A common name for a

number of insects belonging to the order Psocoptera; important pests in herbaria, museums, and libraries. { 'bu̇k ˌlau̇s }

book lung [INV ZOO] A saccular respiratory organ in many arachnids consisting of numerous membranous folds arranged like the pages of a book. { 'bu̇k ˌləŋ }

Boöpidae [INV ZOO] A family of lice in the order Mallophaga, parasitic on Australian marsupials. { bō'äp·əˌdē }

booster [IMMUNOL] The dose of an immunizing agent given to stimulate the effects of a previous dose of the same agent. { 'büs·tər }

booster response *See* anamnestic response. { 'büs·tər ri'späns }

Bopyridae [INV ZOO] A family of epicaridean isopods in the tribe Bopyrina known to parasitize decapod crustaceans. { bō'pī·rəˌdē }

Bopyrina [INV ZOO] A tribe of dioecious isopods in the suborder Epicaridea. { bō·pī'rə·nə }

Bopyroidea [INV ZOO] An equivalent name for Epicaridea. { ˌbō·pi'ròid·ē·ə }

Boraginaceae [BOT] A family of flowering plants in the order Lamiales comprising mainly herbs and some tropical trees. { bəˌra·jə'nās·ēˌē }

bordered [BOT] Having a margin with a distinctive color or texture; used especially of a leaf. { 'bȯrd·ərd }

bordered pit [BOT] A wood-cell pit having the secondary cell wall arched over the cavity of the pit. { 'bȯrd·ərd 'pit }

Bordetella [MICROBIO] A genus of gram-negative, aerobic bacteria of uncertain affiliation; minute coccobacilli, parasitic and pathogenic in the respiratory tract of mammals. { ˌbȯr·də'tel·ə }

Bordetella avium [MICROBIO] A nonsporulating, gram-negative coccobacillus that causes respiratory infections in birds. { ˌbȯr·dəˌtel·ə 'ā·vē·əm }

Bordetella bronchiseptica [MICROBIO] An aerobic, gram-negative bacterium that is a pathogen in many domestic and wild mammals, including horses, swine, dogs, and rodents, and may cause a variety of respiratory diseases in them. { ˌbȯr·dəˌtel·ə ˌbraŋ·ki'sep·ti·kə }

boreal [ECOL] Of or relating to northern geographic regions. { 'bȯr·ē·əl }

boreal forest *See* taiga. { 'bȯr·ē·əl 'fär·əst }

Boreal life zone [ECOL] The zone comprising the climate and biotic communities between the Arctic and Transitional zones. { 'bȯr·ē·əl 'līf ˌzōn }

borer [INV ZOO] Any insect or other invertebrate that burrows into wood, rock, or other substances. { 'bȯr·ər }

Borhyaenidae [VERT ZOO] A family of carnivorous mammals in the superfamily Borhyaenoidea. { ˌbȯr·ē'ēn·əˌdē }

Borhyaenoidea [VERT ZOO] A superfamily of carnivorous mammals in the order Marsupialia. { ˌbȯr·ē·ə'nȯid·ē·ə }

boring sponge [INV ZOO] Marine sponge of the family Clionidae represented by species which excavate galleries in mollusks, shells, corals, limestone, and other calcareous matter. { 'bȯr·iŋ ˌspənj }

Borrelia [MICROBIO] A genus of bacteria in the family Spirochaetaceae; helical cells with uneven coils and parallel fibrils coiled around the cell body for locomotion; many species cause relapsing fever in humans. { bə'rel·ē·ə }

Borrelia anserina [MICROBIO] A motile, helical bacterial pathogen propagated by ticks of the genus *Argas* that causes borreliosis in geese, ducks, turkeys, pheasants, chickens, and other birds. { bəˌrel·ē·ə an'ser·ə·nə }

Borrelia burgdorferi [MICROBIO] A gram-negative, helically shaped bacterium that is the causative agent of Lyme disease. { bəˌrēl·yə ˌbərg'dȯr·fə·rē }

bosque *See* temperate and cold scrub. { 'bäsk, 'bä·skā }

Bostrichidae [INV ZOO] The powder-post beetles, a family of coleopteran insects in the superfamily Bostrichoidea. { bä'strik·əˌdē }

Bostrichoidea [INV ZOO] A superfamily of beetles in the coleopteran suborder Polyphaga. { ˌbä·strə'kȯid·ē·ə }

botanical *See* crude drug. { bə'tan·ə·kəl }

botanical garden [BOT] An institution for the culture of plants collected chiefly for scientific and educational purposes. { bə'tan·ə·kəl 'gär·dən }

botany [BIOL] A branch of the biological sciences which embraces the study of plants and plant life. { 'bät·ən·ē }

bothridium [INV ZOO] A muscular holdfast organ, often with hooks, on the scolex of tetraphyllidean tapeworms. { bä'thrid·ē·əm }

Bothriocephaloidea [INV ZOO] The equivalent name for the Pseudophyllidea. { ˌbä·thrē·ōˌsef·ə'lȯid·ē·ə }

bothrium [INV ZOO] A suction groove on the scolex of pseudophyllidean tapeworms. { 'bäth·rē·əm }

botryomycosis [VET MED] A chronic infectious bacterial disease of horses caused by *Staphylococcus aureus* and characterized by localized fibromatous tumors. { ˌbä·trēˌmī'kō·səs }

bottle graft [BOT] A plant graft in which the scion is a detached branch and is protected from wilting by keeping the base of the branch in a bottle of water until union with the stock. { 'bäd·əl ˌgraft }

bottom break [BOT] A branch that arises from the base of a plant stem. { 'bäd·əm ˌbrāk }

bottom fauna *See* benthos. { 'bäd·əm ˌfȯn·ə }

botulin [MICROBIO] The neurogenic toxin which is produced by *Clostridium botulinum* and *C. parabotulinum* and causes botulism. Also known as botulinus toxin. { 'bäch·ə·lən }

botulinus [MICROBIO] A bacterium that causes botulism. { 'bäch·ə'lī·nəs }

botulinus toxin *See* botulin. { 'bäch·ə'lī·nəs 'täk·sən }

botulism [MED] Food poisoning due to intoxication by the exotoxin of *Clostridium botulinum* and *C. parabotulinum*. { 'bäch·ə,liz·əm }

bough [BOT] A main branch on a tree. { bau̇ }

Bournesville's disease *See* tuberous sclerosis. { 'bürn,vēlz di,zēz }

bouton [NEUROSCI] A club-shaped enlargement at the end of a nerve fiber. Also known as end bulb. { bü'tōn }

boutonneuse fever *See* fièvre boutonneuse. { 'büt·ən,üz ,fē·vər }

Bovidae [VERT ZOO] A family of pecoran ruminants in the superfamily Bovoidea containing the true antelopes, sheep, and goats. { 'bō·və,dē }

bovine [VERT ZOO] **1.** Any member of the genus *Bos*. **2.** Resembling or pertaining to a cow or ox. { 'bō,vīn }

bovine mastitis [VET MED] Inflammation of the udder of a cow; may result from injury or bacterial infection. { 'bō,vīn ma·'stīd·əs }

bovine staggers [VET MED] A disease of cattle in southern Africa caused by eating the poisonous herb *Matricaria nigellaefolia* and characterized by staggering, emaciation, and finally paralysis. { 'bō,vīn 'stag·ərz }

Bovoidea [VERT ZOO] A superfamily of pecoran ruminants in the order Artiodactyla comprising the pronghorns and bovids. { bō'vȯid·ē·ə }

bowel [ANAT] The intestine. { bau̇l }

Bowen's disease [MED] **1.** Intraepithelial squamous-cell carcinoma of the skin, forming distinctive plaques. **2.** A similar carcinoma occurring in mucous membranes. { 'bō·ənz di,zēz }

bowfin [VERT ZOO] *Amia calva*. A fish recognized as the only living species of the family Amiidae. Also known as dogfish; grindle; mudfish. { 'bō,fin }

Bowman's capsule [ANAT] A two-layered membranous sac surrounding the glomerulus and constituting the closed end of a nephron in the kidneys of all higher vertebrates. { ¦bō·mənz 'kap·səl }

brace root *See* prop root. { 'brās ,rüt }

brachial [ZOO] Of or relating to an arm or armlike process. { 'brā·kē·əl }

brachial artery [ANAT] An artery which originates at the axillary artery and branches into the radial and ulnar arteries; it distributes blood to the various muscles of the arm, the shaft of the humerus, the elbow joint, the forearm, and the hand. { 'brā·kē·əl 'ärd·ə·rē }

brachial cavity [INV ZOO] The anterior cavity which is located inside the valves of brachiopods and into which the brachia are withdrawn. { 'brā·kē·əl 'kav·əd·ē }

brachial plexus [NEUROSCI] A plexus of nerves located in the neck and axilla and composed of the anterior rami of the lower four cervical and first thoracic nerves. { 'brā·kē·əl 'plek·səs }

Brachiata [INV ZOO] A phylum of deuterostomous, sedentary bottom-dwelling marine animals that live encased in tubes. { ,bra·kē'ad·ə }

brachiate [BOT] Possessing widely divergent branches. [ZOO] Having arms. { 'bra·kē,āt }

brachiolaria [INV ZOO] A transitional larva in the development of certain starfishes that is distinguished by three anterior processes homologous with those of the adult. { ,bra·kē·ō'la·re·ə }

Brachiopoda [INV ZOO] A phylum of solitary, marine, bivalved coelomate animals. { ,brā·kē'ä·pə·də }

brachioradialis [ANAT] The muscle of the arm that flexes the elbow joint; origin is the lateral supracondylar ridge of the humerus, and insertion is the lower end of the radius. { ¦brā·kē·ō,rā·dē'äl·əs }

brachium [ANAT] The upper arm or forelimb, from the shoulder to the elbow. [INV ZOO] **1.** A ray of a crinoid. **2.** A tentacle of a cephalopod. **3.** Either of the paired appendages constituting the lophophore of a brachiopod. { 'brā·kē·əm }

Brachyarchus [MICROBIO] A genus of bacteria of uncertain affiliation; rod-shaped cells bent in a bowlike configuration and usually arranged in groups of two, four, or more cells. { ¦bra·kē'är·kəs }

brachyblast [BOT] A short shoot often bearing clusters of leaves. { 'bra·kə,blast }

brachydactylia [MED] Abnormal shortening of fingers or toes. { ,brak·i·dak'til·ē·ə }

brachydont [ANAT] Of teeth, having short crowns, well-developed roots, and narrow root canals; characteristic of humans. { 'brak·ə,dänt }

Brachygnatha [INV ZOO] A subsection of brachyuran crustaceans to which most of the crabs are assigned. { bra'kig·nə·thə }

Brachypsectridae [INV ZOO] A family of coleopteran insects in the superfamily Cantharoidea represented by a single species. { ,bra·kip 'sek·trə,dē }

Brachypteraciidae [VERT ZOO] The ground rollers, a family of colorful Madagascan birds in the order Coraciiformes. { brə,kip·ter·ə'sī·ə,dē }

brachypterous [INV ZOO] Having rudimentary or abnormally small wings, referring to certain insects. { brə'kip·tə·rəs }

brachysclereid [BOT] A sclereid that is more or less isodiametric and is found in certain fruits and in the pith, cortex, and bark of many stems. Also known as stone cell. { ,brak·i'sklir·ē·əd }

brachysm [BOT] Plant dwarfing in which there is shortening of the internodes only. { 'bra,kiz·əm }

brachytherapy [MED] Radiation treatment using a solid or enclosed radioisotopic source on the surface of the body or at a short distance from the area to be treated. { ¦brak·i'ther·ə·pē }

Brachyura [INV ZOO] The section of the crustacean order Decapoda containing the true crabs. { ,bra·kē'yu̇r·ə }

bracket fungus [MYCOL] A basidiomycete characterized by shelflike sporophores, sometimes seen on tree trunks. { 'brak·ət ,fəŋ·gəs }

Braconidae [INV ZOO] The braconid wasps, a family of hymenopteran insects in the superfamily Ichneumonoidea. { brə'kän·ə,dē }

bract [BOT] A modified leaf associated with plant reproductive structures. { brakt }

bracteolate [BOT] Having bracteoles. { brak 'tē·ə·lət }

bracteole [BOT] A small bract, especially if on the floral axis. Also known as bractlet. { 'brak·tē,ōl }

bractlet *See* bracteole. { 'brak·lət }

bradyauxesis [BIOL] Allometric growth in which a part lags behind the body as a whole in development. { ¦brā·dē·ȯg'zē·səs }

bradycardia [MED] Slow heart rate. { ¦brād·i¦kärd·ē·ə }

bradydiastole [MED] Prolongation of diastole beyond normal limits. { ,brā·dē·dī'as·tə·lē }

bradyesthesia [MED] Retardation in the transmission of sensory impressions. { ,brā·dē·es 'thēzh·yə }

bradykinesia [MED] Extreme slowness in movement. { ,brā·dē·kə'nēzh·yə }

bradykinetic syndrome [MED] A neurologic condition characterized by a generalized slowness of motor activity. { ¦brā·dē·ki'ned·ik 'sin,drōm }

bradykinin [BIOCHEM] $C_{50}H_{73}N_{15}O_{11}$ A polypeptide kinin; forms an amorphous precipitate in glacial acetic acid; released from plasma precursors by plasmin; acts as a vasodilator. Also known as callideic I; kallidin I. { ¦brād·i'kī·nən }

bradypnea [MED] Abnormal slowness of respiration. { ,brād·ē'nē·ə }

Bradypodidae [VERT ZOO] A family of mammals in the order Edentata comprising the true sloths. { ,brād·i'pä·də,dē }

bradytely [EVOL] Evolutionary change that is either arrested or occurring at a very slow rate over long geologic periods. { 'brād·ə,te·lē }

Brailsford-Morquio syndrome *See* Morquio's syndrome. { ¦brālz·fərd 'mȯr·ke·ō ¦sin,drōm }

brain [ANAT] The portion of the vertebrate central nervous system enclosed in the skull. [ZOO] The enlarged anterior portion of the central nervous system in most bilaterally symmetrical animals. { brān }

braincase *See* cranium. { 'brān,kās }

brain coral [INV ZOO] A reef-building coral resembling the human cerebrum in appearance. { 'brān ,kär·əl }

brain hormone [INV ZOO] A neurohormone secreted by the insect brain that regulates the release of ecdysone from the prothoracic glands. { 'brān ¦hȯr,mōn }

brainstem [ANAT] The portion of the brain remaining after the cerebral hemispheres and cerebellum have been removed. { 'brān,stem }

brain wave [PHYSIO] A rhythmic fluctuation of voltage between parts of the brain, ranging from about 1 to 60 hertz and 10 to 100 microvolts. { 'brān ,wāv }

bramble [BOT] **1.** A plant of the genus *Rubus*. **2.** A rough, prickly vine or shrub. { 'bram·bəl }

branch [BOT] A shoot or secondary stem on the trunk or a limb of a tree. { branch }

branched acinous gland [ANAT] A multicellular structure with saclike glandular portions connected to the surface of the containing organ or structure by a common duct. { ¦brancht 'as·ə·nəs ,gland }

branched-chain ketoaciduria *See* maple syrup urine disease. { ¦brancht ¦chān ,kēd·ō·as·ə 'dür·ē·ə }

branched tubular gland [ANAT] A multicellular structure with tube-shaped glandular portions connected to the surface of the containing organ or structure by a common secreting duct. { ¦brancht ¦tüb·yə·lər 'gland }

branchia *See* gill. { ,braŋ·kē·ə }

branchial [ZOO] Of or pertaining to gills. { 'braŋ·kē·əl }

branchial arch [VERT ZOO] One of the series of paired arches on the sides of the pharynx which support the gills in fishes and amphibians. { 'braŋ·kē·əl 'ärch }

branchial basket [ZOO] A cartilaginous structure that supports the gills in protochordates and certain lower vertebrates such as cyclostomes. { 'braŋ·kē·əl 'bas·kət }

branchial cleft [EMBRYO] A rudimentary groove in the neck region of air-breathing vertebrate embryos. [VERT ZOO] One of the openings between the branchial arches in fishes and amphibians. { 'braŋ·kē·əl 'kleft }

branchial heart [INV ZOO] A muscular enlarged portion of a vein of a cephalopod that contracts and forces the blood into the gills. { 'braŋ·kē·əl 'härt }

branchial plume [INV ZOO] An accessory respiratory organ that extends out under the mantle in certain Gastropoda. { 'braŋ·kē·əl 'plüm }

branchial pouch [ZOO] In cyclostomes and some sharks, one of the respiratory cavities occurring in the branchial clefts. { 'braŋ·kē·əl 'pau̇ch }

branchial sac [INV ZOO] In tunicates, the dilated pharyngeal portion of the alimentary canal that has vascular walls pierced with clefts and serves as a gill. { 'braŋ·kē·əl 'sak }

branchial segment [EMBRYO] Any of the paired pharyngeal segments indicating the visceral arches and clefts posterior to and including the third pair in air-breathing vertebrate embryos. { 'braŋ·kē·əl 'seg·mənt }

branchiate [VERT ZOO] Having gills. { 'braŋ·kē,āt }

branching adaptation *See* divergent adaptation. { 'branch·iŋ ,ad,ap'tā·shən }

branchiocranium [VERT ZOO] The division of the fish skull constituting the mandibular and hyal regions and the branchial arches. { ¦braŋ·kē·ō'krā·nē·əm }

branchiomere [EMBRYO] An embryonic metamere that will differentiate into a visceral arch and cleft; a branchial segment. { 'braŋ·kē·ə,mir }

branchiomeric musculature [VERT ZOO] Those muscles derived from branchial segments in vertebrates. { ¦braŋ·kē·ə¦mer·ik 'məs·kyə·lə,chər }

Branchiopoda [INV ZOO] A subclass of crustaceans containing small or moderate-sized animals commonly called fairy shrimps, clam shrimps, and water fleas. { ˌbraŋ·kē'äp·ə·də }

branchiostegite [INV ZOO] A gill cover and chamber in certain malacostracan crustaceans, formed by lateral expansion of the carapace. { ˌbraŋ·kē'äs·tə ˌjīt }

Branchiostoma [ZOO] A genus of lancelets formerly designated as amphioxus. { ˌbraŋ·kē'äs·tə·mə }

Branchiotremata [INV ZOO] The hemichordates, a branch of the subphylum Oligomera. { ¦braŋ·kē·ə'trem·əd·ə }

Branchiura [INV ZOO] The fish lice, a subclass of fish ectoparasites in the class Crustacea. { ˌbraŋ·kē'yu̇r·ə }

Branhamella [MICROBIO] A genus of bacteria in the family Neisseriaceae; cocci occur in pairs with flattened adjacent sides; parasites of mammalian mucous membranes. { ˌbran·ə'mel·ə }

brass chills *See* metal fume fever. { 'bras ˌchilz }

brass founder's ague *See* metal fume fever. { 'bras ˌfau̇n·dərz ˌāg }

Brassica [BOT] A large genus of herbs in the family Cruciferae of the order Capparales, including cabbage, watercress, and sweet alyssum. { 'bras·ə·kə }

Brassicaceae [BOT] An equivalent name for the Cruciferae. { ˌbras·ə'kās·ēˌē }

brassin [BIOCHEM] Any of a class of plant hormones characterized as long-chain fatty-acid esters; brassins act to induce both cell elongation and cell division in leaves and stems. { 'bra·sən }

Brathinidae [INV ZOO] The grass root beetles, a small family of coleopteran insects in the superfamily Staphylinoidea. { brə'thī·nəˌdē }

Braulidae [INV ZOO] The bee lice, a family of cyclorrhaphous dipteran insects in the section Pupipara. { 'brau̇l·əˌdē }

Braxton-Hicks contractions [MED] Intermittent painless contractions of the uterus that occur throughout pregnancy. { 'brak·stən 'hiks kənˌtrak·shənz }

Brazil nut [BOT] *Bertholletia excelsa*. A large broadleafed evergreen tree of the order Lecythedales; an edible seed is produced by the tree fruit. { brə'zil ˌnət }

breadfruit [BOT] *Artocarpus altilis*. An Indo-Malaysian tree, a species of the mulberry family (Moraceae). The tree produces a multiple fruit which is edible. { 'bredˌfrüt }

bread mold [MYCOL] Any fungus belonging to the family Mucoraceae in the order Mucorales. { 'bred ˌmōld }

breakage and reunion [CELL MOL] The classical model of crossing over by means of physical breakage and crossways reunion of completed chromatids during the process of meiosis. { 'brāk·ij ən rē'yün·yən }

breakbone fever *See* dengue. { 'brākˌbōn ˌfē·vər }

breast [ANAT] The human mammary gland. { brest }

breathing [PHYSIO] Inhaling and exhaling. { 'brē<u>th</u>·iŋ }

breeding [GEN] Controlled mating and selection, or hybridization of plants and animals in order to modify the species with respect to one or more phenotypic traits. { 'brēd·iŋ }

bregma [ANAT] The point at which the coronal and sagittal sutures of the skull meet. { 'breg·mə }

Brentidae [INV ZOO] The straight-snouted weevils, a family of coleopteran insects in the superfamily Curculionoidea. { 'bren·təˌdē }

Brevibacteriaceae [MICROBIO] Formerly a family of gram-positive, rod-shaped, schizomycetous bacteria in the order Eubacteriales. { ˌbrev·əˌbakˌtir·ē'ās·ēˌē }

Brevibacterium [MICROBIO] A genus of short, unbranched, rod-shaped bacteria in the coryneform group. { ˌbrev·əˌbak'tir·ē·əm }

brevitoxin [BIOCHEM] One of several ichthyotoxins produced by the dinoflagellate *Ptychodiscus brevis*. { ˌbrev·ə'täk·sən }

Brewer anaerobic jar [MICROBIO] A glass container in which petri dish cultures are stacked and maintained under anaerobic conditions. { 'brü·ər an·ə'rō·bik 'jär }

bridge graft [BOT] A plant graft in which each of several scions is grafted in two positions on the stock, one above and the other below an injury. { 'brij ˌgraft }

bright-field microscope [CELL MOL] A type of light microscope that produces a dark image against a brighter background; commonly used for the visualization of stained cells. { ¦brīt ˌfēld 'mī·krəˌskōp }

Bright's disease [MED] Any of several kidney diseases attended by glomerulonephritis. { 'brīts diˌzēz }

Brill's disease [MED] A mild recurrence of typhus some years after the initial infection. { ¦brilz diˌzēz }

Brisingidae [INV ZOO] A family of deep-water echinoderms with as many as 44 arms, belonging to the order Forcipulatida. { brə'sin·jəˌdē }

bristle [BIOL] A short stiff hair or hairlike structure on an animal or plant. { 'bris·əl }

bristlecone pine [BOT] A small slow-growing evergreen tree of the genus *Pinus* that grows at high altitudes in the western United States, having dense branches with rust-brown bark and short needles in bunches of five and thorn-tipped cone scales. The two types are *P. longaeva*, which lives longer than any other tree (over 4000 years), and *P. monophylla*, the single-leaf pinyon. { ˌbris·əlˌkōn 'pīn }

British antilewisite *See* dimercaprol. { 'brid·ish an·tē'lü·wəˌsīt }

brittle star [INV ZOO] The common name for all members of the echinoderm class Ophiuroidea. { ¦brid·əl 'stär }

broadleaf tree [BOT] Any deciduous or evergreen tree having broad, flat leaves. { 'bro̍d ˌlēf ˌtrē }

broad-sense heritability [GEN] The degree to which individual phenotypes are determined by their genotypes; expressed as the ratio of the total genetic variance to the total phenotypic variance. { ¦bród ˌsens ˌher·ə·tə'bil·ə·dē }

broad-spectrum antibiotic [MICROBIO] An antibiotic that is effective against both gram-negative and gram-positive bacterial species. { ¦bród ¦spek·trəm ˌant·i·bī'äd·ik }

Broca's area [NEUROSCI] In the human brain, an area in the inferior left frontal lobe—one of several areas believed to activate the fibers of the precentral gyrus concerned with movements necessary for speech production; injury to this area generally results in nonfluent aphasia, with effortful articulation, loss of syntax, but relatively well-preserved auditory comprehension. { 'brō·kəz ˌer·ē·ə }

broccoli [BOT] *Brassica oleracea* var. *italica*. A biennial crucifer of the order Capparales which is grown for its edible stalks and buds. { 'brak·ə·lē }

Brodmann's area 4 *See* motor area. { 'bräd·mənz ˌer·ē·ə 'fór }

Brodmann's area 17 *See* visual projection area. { 'bräd·mənz ˌer·ē·ə sev·ən'tēn }

Brodmann's areas [PHYSIO] Numbered regions of the cerebral cortex used to identify cortical functions. { 'bräd·mənz ˌer·ē·əz }

broken wind *See* heaves. { ¦brō·kən 'wind }

bromatium [ECOL] A swollen hyphal tip on fungi growing in ants nests that is eaten by the ants. { brō'māsh·əm }

bromegrass [BOT] The common name for a number of forage grasses of the genus *Bromus* in the order Cyperales. { 'brōm,gras }

bromelain [BIOCHEM] An enzyme that digests protein and clots milk; prepared by precipitation by acetone from pineapple juice; used to tenderize meat, to chill-proof beer, and to make protein hydrolysates. Also spelled bromelin. { 'brō·mə,lān }

Bromeliaceae [BOT] The single family of the flowering plant order Bromeliales. { brō,mel·ē'ās·ē,ē }

Bromeliales [BOT] An order of monocotyledonous plants in the subclass Commelinidae, including terrestrial xerophytes and some epiphytes. { brō,mel·ē'ā·lēz }

bromelin *See* bromelain. { 'brō·mə·lən }

brome mosaic virus [VIROL] The type species of the genus *Bromovirus*. Abbreviated BMV. { ¦brōm mō¦zā·ik ˌvī·rəs }

brome mosaic virus group *See* Bromovirus. { ˌbrōm mō,zā·ik 'vī·rəs ˌgrüp }

bromism [MED] A disease state produced by prolonged usage or overdosage of bromide compounds. { 'brō,miz·əm }

5-bromodeoxyruridine [BIOCHEM] $C_9H_{11}O_5NBr$ A thymidine analog that can be incorporated into deoxyribonucleic acid during its replication; induces chromosomal breakage in regions rich in heterochromatin. Abbreviated BUDR. { ¦fīv ¦brō·mō·dē,äk·sē'rur·ə,dēn }

bromodiethylacetylurea [PHARM] $C(C_2H_5)_2Br$-$CONHCONH_2$ A white, crystalline powder with a melting point of 116–117°C; soluble in chloroform, ether, and alcohol; used in medicine. { ¦brō·mō,dī¦eth·əl·ə,sēd·əl'y ü·rē·ə }

bromouracil [BIOCHEM] $C_4H_3N_2O_2Br$ 5-Bromouracil, an analog of thymine that can react with deoxyribonucleic acid to produce a polymer with increased susceptibility to mutation. { ˌbrō·mō'yůr·ə·səl }

Bromoviridae [VIROL] A family of ribonucleic acid (RNA)-containing plant viruses that are characterized by nonenveloped virions with three segments of linear, positive-sense, single-stranded RNA. It includes the genera *Bromovirus*, *Cucomovirus*, *Ilarvirus*, *Alfamovirus*, and *Oleavirus*. { ˌbrō·mə'vir·ə,dī }

Bromovirus [VIROL] A genus of plant viruses in the family Bromoviridae characterized by icosahedral particles with genomes consisting of four species of single-stranded ribonucleic acid separately encapsidated. Also known as Brome mosaic virus group. { 'brō·mə ˌvī·rəs }

bromuration [HISTOL] A process in which a tissue section is treated with a solution of bromine or a bromine compound. { ˌbräm·yə'rā·shən }

bronchial adenoma [MED] A low-grade malignant or potentially malignant tumor of bronchi. { 'bräŋ·kē·əl ˌad·ən'ō·mə }

bronchial asthma [MED] Asthma usually due to hypersensitivity to an inhaled or ingested allergen. { 'bräŋ·kē·əl 'az·mə }

bronchial tree [ANAT] The arborization of the bronchi of the lung, considered as a structural and functional unit. { 'bräŋ·kē·əl 'trē }

bronchiectasis [MED] Dilation of the bronchi and bronchioles following a chronic inflammatory process or an infection attended by pus formation. { ¦bräŋ·kē'ek·tə·səs }

bronchiolar carcinoma [MED] Adenocarcinoma of the lung characterized by mucus-producing cells which spread over the alveoli. { ¦bräŋ·kē¦ō·lər ˌkärs·ən'ō·mə }

bronchiole [ANAT] A small, thin-walled branch of a bronchus, usually terminating in alveoli. { 'bräŋ·kē,ōl }

bronchiolitis [MED] Inflammation of the bronchioles. { ˌbräŋ·kē·ō'līd·əs }

bronchiolitis obliterans [MED] Inflammation of the bronchioles with the formation of an exudate and fibrous tissue that obliterate the lumen. { ˌbräŋ·kē·ō'līd·əs ō'blid·ə,ranz }

bronchitis [MED] An inflammation of the bronchial tubes. { bräŋ'kīd·əs }

bronchoconstriction [PHYSIO] Narrowing of the air passages in bronchi and bronchioles. { ¦bräŋ·kō·kən¦strik·shən }

bronchoconstrictor [PHARM] Any agent that causes a narrowing of the air passages in bronchi and bronchioles. { ¦bräŋ·kō·kən¦strik·tər }

bronchodilation [PHYSIO] Widening of the air passages in bronchi and bronchioles. { ¦bräŋ·kō,dī'lā·shən }

bronchodilator [MED] An instrument used to

increase the caliber of the pulmonary air passages. [PHARM] Any agent that causes a widening of the air passages in bronchi and bronchioles. { ¦bräŋ·kō'dī,lād·ər }

bronchoesophagology [MED] The specialty concerned with endoscopic examination through the mouth of the esophagus and tracheobronchial tree. { ¦bräŋ·kō·ē,säf·ə'gäl·ə·jē }

bronchofiberscope [MED] A fiber-optic endoscope adapted for viewing the trachea and bronchi. { ¦bräŋ·kə'fī·bər,skōp }

bronchogram [MED] Radiography of the bronchial tree made after the introduction of a radiopaque substance. { 'bräŋ·kə,gram }

bronchography [MED] Roentgenographic visualization of the bronchial tree following injection of a radiopaque material. { ,bräŋ'käg·rə·fē }

bronchopneumonia [MED] Inflammation of the lungs which has spread from infected bronchi. Also known as lobular pneumonia. { ¦bräŋ·kō·nə'mō·nyə }

bronchorrhea [MED] Excessive discharge of mucus from the bronchial mucous membranes. { ,bräŋ·kə'rē·ə }

bronchoscope [MED] An instrument for the visual examination of the interior of the bronchi. { 'bräŋ·kə,skōp }

bronchospasm [MED] Temporary narrowing of the bronchi due to violent, involuntary contraction of the smooth muscle of the bronchi. { 'bräŋ·kō,spaz·əm }

bronchospirometry [MED] The determination of various aspects of the functional capacity of a single lung or lung segment. { ¦bräŋ·kō·spə'räm·ə·trē }

bronchus [ANAT] Either of the two primary branches of the trachea or any of the bronchi's pulmonary branches having cartilage in their walls. { 'bräŋ·kəs }

brood [BOT] Heavily infested by insects. [ZOO] **1.** The young of animals. **2.** To incubate eggs or cover the young for warmth. **3.** An animal kept for breeding. { brüd }

brood capsule [INV ZOO] A secondary scolex-containing cyst constituting the infective agent of a tapeworm. { 'brüd ,kap·səl }

brood parasitism [ECOL] A type of social parasitism among birds characterized by a bird of one species laying and abandoning its eggs in the nest of a bird of another species. { ¦brüd ,par·ə·sə,tiz·əm }

brood pouch [VERT ZOO] A pouch of an animal body where eggs or embryos undergo certain stages of development. { 'brüd ,paùch }

Brotulidae [VERT ZOO] A family of benthic teleosts in the order Perciformes. { brō'tü·lə,dē }

brown algae [BOT] The common name for members of the Phaeophyta. { ¦braùn ¦al·jē }

brown fat cell [HISTOL] A moderately large, generally spherical cell in adipose tissue that has small fat droplets scattered in the cytoplasm. { ¦braùn 'fat ,sel }

brown induration [MED] A pathologic condition marked by acute pulmonary congestion and edema with leakage of blood into the alveoli. { 'braùn in·də'rā·shən }

brown lung disease *See* byssinosis. { ,braùn 'ləŋ di,zēz }

brown seaweed [BOT] A common name for the larger algae of the division Phaeophyta. { ¦braùn 'sē,wēd }

browse [BIOL] **1.** Twigs, shoots, and leaves eaten by livestock and other grazing animals. **2.** To feed on this vegetation. { braùz }

Brucella [MICROBIO] A genus of gram-negative, aerobic bacteria of uncertain affiliation; single, nonmotile coccobacilli or short rods, all of which are parasites and pathogens of mammals. { brü'sel·ə }

Brucellaceae [MICROBIO] Formerly a family of small, coccoid to rod-shaped, gram-negative bacteria in the order Eubacteriales. { ,brü·sə'lās·ē,ē }

brucellergen [BIOCHEM] A nucleoprotein fraction of brucellae used in skin tests to detect the presence of *Brucella* infections. { ,brü'sel·ər·jen }

brucellergen test [IMMUNOL] A diagnostic skin test for detection of *Brucella* infections. { ,brü 'sel·ər·jen ,test }

brucellosis [MED] **1.** An infectious bacterial disease of humans caused by *Brucella* species acquired by contact with diseased animals. Also known as Malta fever; Mediterranean fever; undulant fever. **2.** *See* contagious abortion. { ,brü·sə'lō·səs }

Bruchidae [INV ZOO] The pea and bean weevils, a family of coleopteran insects in the superfamily Chrysomeloidea. { 'brü·kə,dē }

Bruch's membrane [ANAT] The membrane of the retina that separates the pigmented layer of the retina from the choroid coat of the eye. { 'brüks ,mem,brān }

Brücke-Abney phenomenon [PHYSIO] The inability to distinguish colors other than blue-violet, green, and red at very low light levels. { 'brük·ə 'ab·nē fə,näm·ə,nän }

Brunner's glands [ANAT] Simple, branched, tubular mucus-secreting glands in the submucosa of the duodenum in mammals. Also known as duodenal glands; glands of Brunner. { 'brən·ərz ,glanz }

brush *See* tropical scrub. { brəsh }

brush border [CELL MOL] A superficial protoplasic modification in the form of filiform processes or microvilli; present on certain absorptive cells in the intestinal epithelium and the proximal convolutions of nephrons. { 'brəsh ,bòr·dər }

brussels sprouts [BOT] *Brassica oleracea* var. *gemmifera*. A biennial crucifer of the order Capparales cultivated for its small, edible, headlike buds. { ¦brəs·əlz 'spraùts }

Bruton's disease [IMMUNOL] A hereditary type of agammaglobulinemia that is a sex-linked recessive disorder characterized by a deficiency of all types of immunoglobulins, reflecting a failure of the entire humoral antibody marrow system; the thymus may be normal, but the lymph nodes and spleen lack lymph cell follicles. { 'brüt·ənz di,zēz }

bruxism [MED] A clenching and grinding of the teeth that occurs unconsciously while the individual is awake or sleeping. { 'brək,siz·əm }

Bryales [BOT] An order of the subclass Bryidae; consists of mosses which often grow in disturbed places. { brī'ā·lēz }

Bryatae *See* Bryopsida. { 'brī·ə,tē }

Bryidae [BOT] A subclass of the class Bryopsida; includes most genera of the true mosses. { 'brī·ə,dē }

bryology [BOT] The study of bryophytes. { brī'äl·ə·jē }

Bryophyta [BOT] A small phylum of the plant kingdom, including mosses, liverworts, and hornworts, characterized by the lack of true roots, stems, and leaves. { brī'ä·fə·də }

Bryopsida [BOT] The mosses, a class of small green plants in the phylum Bryophyta. Also known as Musci. { brī'äp·sə·də }

Bryopsidaceae [BOT] A family of green algae in the order Siphonales. { brī,äp·sə'dās·ē,ē }

Bryopsidales *See* Siphonales. { brī,äp·sə'dā,lēz }

bryostatin-1 [PHARM] A polyketide isolated from the bryozoan *Bugula neritina* that has both anticancer and immune modulating activity. { ¦brī·ə,stat·ən 'wən }

Bryoxiphiales [BOT] An order of the class Bryopsida in the subclass Bryidae; consists of a single genus and species, *Bryoxiphium norvegicum*, the sword moss. { brī,äk·sə·fē'ā·lēz }

Bryozoa [INV ZOO] The moss animals, a major phylum of sessile aquatic invertebrates occurring in colonies with hardened exoskeleton. { ,brī·ə'zō·ə }

bubo [MED] An inflammatory enlargement of lymph nodes, usually of the groin or axilla; commonly associated with chancroid, lymphogranuloma venereum, and plague. { 'bü,bō }

bubonic plague *See* plague. { bü¦ban·ik 'plāg }

buccal cavity [ANAT] The space anterior to the teeth and gums in the mouths of all vertebrates having lips and cheeks. Also known as vestibule. { 'bək·əl ¦kav·əd·ē }

buccal gland [ANAT] Any of the mucous glands in the membrane lining the cheeks of mammals, except aquatic forms. { 'bək·əl ,gland }

Buccinacea [INV ZOO] A superfamily of gastropod mollusks in the order Prosobranchia. { ,bək·sin'ās·ē·ə }

Buccinidae [INV ZOO] A family of marine gastropod mollusks in the order Neogastropoda containing the whelks in the genus *Buccinum*. { bək'sin·ə,dē }

Bucconidae [VERT ZOO] The puffbirds, a family of neotropical birds in the order Piciformes. { bə'kän·ə,dē }

Bucerotidae [VERT ZOO] The hornbills, a family of Old World tropical birds in the order Coraciiformes. { ,byü·sə'räd·ə,dē }

buck [VERT ZOO] A male deer. { bək }

bucket *See* calyx. { 'bək·ət }

buckeye [BOT] The common name for deciduous trees composing the genus *Aesculus* in the order Sapindales; leaves are opposite and palmately compound, and the seed is large with a firm outer coat. { 'bək,ī }

Bucky diaphragm *See* Potter-Bucky grid. { 'bək·ē ¦dī·ə,fram }

bud [BOT] An embryonic shoot containing the growing stem tip surrounded by young leaves or flowers or both and frequently enclosed by bud scales. { bəd }

budbreak [BOT] Initiation of growth from a bud. { 'bəd,brāk }

budding [BIOL] A form of asexual reproduction in which a new individual arises as an outgrowth of an older individual. Also known as gemmation. [BOT] A method of vegetative propagation in which a single bud is grafted laterally onto a stock. [VIROL] A form of virus release from the cell in which replication has occurred, common to all enveloped animal viruses; the cell membrane closes around the virus and the particle exits from the cell. { 'bəd·iŋ }

budding bacteria [MICROBIO] Bacteria that reproduce by budding. { 'bəd·iŋ bak'tir·ē·ə }

bud grafting [BOT] Grafting a plant by budding. { 'bəd ,graf·tiŋ }

budling [BOT] The shoot that develops from the bud which was the scion of a bud graft. { 'bəd·liŋ }

bud scale [BOT] One of the modified leaves enclosing and protecting buds in perennial plants. { 'bəd ,skāl }

bud scale scar [BOT] A characteristic marking left on a stem when a bud falls off. { 'bəd ,skāl ,skär }

Buerger's disease *See* thromboangitis obliterans. { 'bər·gərz di,zēz }

buffalo [VERT ZOO] The common name for several species of artiodactyl mammals in the family Bovidae, including the water buffalo and bison. { 'bəf·ə,lō }

buffer [ECOL] An animal that is introduced to serve as food for other animals to reduce the losses of more desirable animals. { 'bəf·ər }

Bufonidae [VERT ZOO] A family of toothless frogs in the suborder Procoela including the true toads (*Bufo*). { byü'fän·ə,dē }

bug [INV ZOO] Any insect in the order Hemiptera. { bəg }

buildup [ECOL] A significant increase in a natural population, usually as a result of progressive changes in ecological relations. { 'bil,dəp }

bulb [BOT] A short, subterranean stem with many overlapping fleshy leaf bases or scales, such as in the onion and tulip. { bəlb }

bulbar paralysis [MED] A clinical syndrome due to involvement of the nuclei of the last four or five cranial nerves, characterized principally by paralysis or weakness of the muscles which control swallowing, talking, movement of the tongue and lips, and sometimes respiratory paralysis. { 'bəl·bər pə'ral·ə·səs }

bulbil [BOT] A secondary bulb usually produced on the aerial part of a plant. { 'bəl·bəl }

bulbocavernosus *See* bulbospongiosus. { ¦bəl·bō,ka·vər'nō·səs }

bulb of the penis [ANAT] The expanded proximal portion of the corpus spongiosum of the penis. { ¦bəlb əv t͟hə 'pē·nəs }

bulbospongiosus [ANAT] A muscle encircling the bulb and adjacent proximal parts of the penis in the male, and encircling the orifice of the vagina and covering the lateral parts of the vestibular bulbs in the female. Also known as bulbocavernosus. { ¦bəl·bō,spən·jē'ō·səs }

bulbourethral gland [ANAT] Either of a pair of compound tubular glands, anterior to the prostate gland in males, which discharge into the urethra. Also known as Cowper's gland. { ¦bəl·bō·yu̇'rēth·rəl ,gland }

bulimia [MED] Excessive, insatiable appetite, seen in psychotic states; a symptom of diabetes mellitus and of certain cerebral lesions. Also known as hyperphagia. { bə'lēm·ē·ə }

Buliminacea [INV ZOO] A superfamily of benthic, marine foraminiferans in the suborder Rotaliina. { ,byü,lim·ə'nās·ē·ə }

bullate [BIOL] Appearing blistered or puckered, especially of certain leaves. { 'bə,lāt }

bulliform [BOT] Type of plant cell involved in tissue contraction or water storage, or of uncertain function. { 'bu̇l·ə,fȯrm }

bulliform cell [BOT] One of the large, highly vacuolated cells occurring in the epidermis of grass leaves. Also known as motor cell. { 'bu̇l·ə,fȯrm ,sel }

bullous emphysema [MED] Acute overinflation of the lungs due to extreme efforts to inhale to overcome a bronchial obstruction. { 'bu̇l·əs ,em·fə'zē·mə }

bumblebee [INV ZOO] The common name for several large, hairy social bees of the genus *Bombus* in the family Apidae. { 'bəm·bəl,bē }

bundle branch [ANAT] Either of the components of the atrioventricular bundle passing to the right and left ventricles of the heart. { 'bənd·əl ,branch }

bundle of His [ANAT] A small region of heart muscle located in the right auricle and specialized to relay contraction impulses from the right auricle to the ventricles. Also known as atrioventricular bundle. { 'bənd·əl əv ,äch,ī'es }

bundle scar [BOT] A mark within a leaf scar that shows the point of an abscised vascular bundle. { 'bənd·əl ,skär }

bundle sheath [BOT] A sheath around a vascular bundle that consists of a layer of parenchyma. { 'bənd·əl ,shēth }

α-bungarotoxin [NEUROSCI] A neurotoxin found in snake venom which blocks neuromuscular transmission by binding with acetylcholine receptors on motor end plates. { ¦al·fə ¦bəŋ·gə·rə'täk·sən }

bunion [MED] A swelling of a bursa of the foot, especially of the metatarsophalangeal joint of the great toe; associated with thickening of the adjacent skin and a forcing of the great toe into adduction. { 'bən·yən }

bunodont [ANAT] Having tubercles or rounded cusps on the molar teeth, as in humans. { ¦byü·nə,dänt }

bunolophodont [VERT ZOO] **1.** Of teeth, having the outer cusps in the form of blunt cones and the inner cusps as transverse ridges. **2.** Having such teeth, as in tapirs. { ¦byü·nə¦läf·ə,dänt }

Bunonematoidea [INV ZOO] A superfamily of nematodes in the order Rhabditida, characterized by asymmetrical bodies in both the labia and the distribution of sensory organs. { ,bü·nō,nem·ə'tȯid·ē·ə }

bunoselenodont [VERT ZOO] **1.** Of teeth, having the inner cusps in the form of blunt cones and the outer ones as longitudinal crescents. **2.** Having such teeth, as in the extinct titanotheres. { ¦byü·nō·sə¦lē·nə,dänt }

Bunyaviridae [VIROL] A family of enveloped spherical viruses whose lipid envelopes contain at least one virus-specific glycopeptide; members develop in the cytoplasm and mature by budding. { ¦bən·yə'vī·rə,dē }

Bunyavirus [VIROL] A genus of viruses in the family Bunyaviridae; contains a minimum of 87 species which exhibit some degree of antigenic relationship. { 'bən·yə,vī·rəs }

Buprestidae [INV ZOO] The metallic wood-boring beetles, the large, single family of the coleopteran superfamily Buprestoidea. { byü 'pres·tə,dē }

Buprestoidea [INV ZOO] A superfamily of coleopteran insects in the suborder Polyphaga including many serious pests of fruit trees. { ,byü·pres'tȯid·ē·ə }

Burhinidae [VERT ZOO] The thick-knees or stone curlews, a family of the avian order Charadriiformes. { byü'rin·ə,dē }

Burkett's lymphoma [MED] A malignant lymphoma of children, typically involving the retroperitoneal area and the mandible, but sparing the peripheral lymph nodes, bone marrow, and spleen. { 'bər·kəts lim'fō·mə }

burl [BOT] A hard, woody outgrowth on a tree, usually resulting from the entwined growth of a cluster of adventitious buds. { bərl }

Burmanniaceae [BOT] A family of monocotyledonous plants in the order Orchidales characterized by regular flowers, three or six stamens opposite the petals, and ovarian nectaries. { bər,man·ē'ās·ē,ē }

burn [MED] An injury to tissues caused by heat, chemicals, electricity, or irradiation effects. { bərn }

Burnett's syndrome *See* milk-alkali syndrome. { bər'nets ,sin,drōm }

burr [BOT] **1.** A rough or prickly envelope on a fruit. **2.** A fruit so characterized. { bər }

burr ball *See* lake ball. { 'bər ,bȯl }

burro [VERT ZOO] A small donkey used as a pack animal. { 'bu̇r·ō }

bursa [ANAT] A simple sac or cavity with smooth walls containing a clear, slightly sticky fluid and interposed between two moving surfaces of the body to reduce friction. { 'bər·sə }

bursa of Fabricius [VERT ZOO] A thymus-like organ in the form of a diverticulum at the lower end of the alimentary canal in birds. { 'bər·sə əv fə'brēsh·əs }

Burseraceae [BOT] A family of dicotyledonous plants in the order Sapindales characterized by an ovary of two to five cells, prominent resin ducts in the bark and wood, and an intrastaminal disk. { ˌbər·sə'rās·ēˌē }

bursicle [BOT] A purse or pouchlike receptacle. { ˌbər·sə·kəl }

bursitis [MED] Inflammation of a bursa. { ˌbər'sīd·əs }

burster [BOT] An abnormally double flower having the calyx split or fragmented. { 'bər·stər }

bushbaby [VERT ZOO] Any of six species of African arboreal primates in two genera (*Galago* and *Euoticus*) of the family Lorisidae. Also known as galago; night ape. { 'bu̇shˌbā·bē }

busulfan [PHARM] $CH_3SO_2O(CH_2)_4OSO_2CH_3$ Crystals with a melting point of 114–118°C; soluble in acetone at 25°C; used as an antineoplastic drug. { ˌbyü'səl·fən }

butanol fermentation [MICROBIO] Butanol production as a result of the fermentation of corn and molasses by the anaerobic bacterium *Clostridium acetobutylicum*. { 'byüt·ənˌȯl ˌfər·mən'tā·shən }

Butomaceae [BOT] A family of monocotyledonous plants in the order Alismatales characterized by secretory canals, linear leaves, and a straight embryo. { ˌbyüd·ə'mās·ēˌē }

butt [BOT] The portion of a plant from which the roots extend, for example, the base of a tree trunk. { bət }

butterfly [INV ZOO] Any insect belonging to the lepidopteran suborder Rhopalocera, characterized by a slender body, broad colorful wings, and club-shaped antennae. { 'bəd·ərˌflī }

buttocks [ANAT] The two fleshy parts of the body posterior to the hip joints. { 'bəd·əks }

buttress [BOT] A ridge of wood developed in the angle between a lateral root and the butt of a tree. { 'bə·trəs }

***n*-butyl-*para*-aminobenzoate** [PHARM] $H_2NC_6H_4COOC_4H_9$ A white, crystalline powder with a melting point of 57–59°C; soluble in dilute acids, alcohol, chloroform, ether, and fatty oils; used as a local anesthetic, in burn treatment and ointments, and in suntan preparations to absorb ultraviolet light. { ¦en ¦byüd·əl ¦par·ə ˌam·ə·nō 'ben·zəˌwāt }

butyric fermentation [BIOCHEM] Fermentation in which butyric acid is produced by certain anaerobic bacteria acting on organic substances, such as butter; occurs in putrefaction and in digestion in herbivorous mammals. { byü'tir·ik fər·mən'tā·shən }

butyrinase [BIOCHEM] An enzyme that hydrolyzes butyrin, found in the blood serum. { 'byüd·ə·rəˌnās }

Butyrivibrio [MICROBIO] A genus of gram-negative, strictly anaerobic bacteria of uncertain affiliation; motile curved rods occur singly, in chains, or in filaments; ferment glucose to produce butyrate. { ¦byüd·ə·rə'vī·brēˌō }

Buxbaumiales [BOT] An order of very small, atypical mosses (Bryopsida) composed of three genera and found on soil, rock, and rotten wood. { ˌbəksˌbōm·ē'ā·lēz }

B virus [VIROL] An animal virus belonging to subgroup A of the herpesvirus group. { 'bē ˌvī·rəs }

Byrrhidae [INV ZOO] The pill beetles, the single family of the coleopteran insect superfamily Byrrhoidea. { 'bir·əˌdē }

Byrrhoidea [INV ZOO] A superfamily of coleopteran insects in the superorder Polyphaga. { bə'rȯid·ē·ə }

byssinosis [MED] A pneumoconiosis caused by the inhalation of cotton dust. Also known as brown lung disease. { ¦bis·ə'nō·səs }

Byturidae [INV ZOO] The raspberry fruitworms, a small family of coleopteran insects in the superfamily Cucujoidea. { bī'tu̇r·əˌdē }

BYV *See* beet yellows virus.

C

caatinga [ECOL] A sparse, stunted forest in areas of little rainfall in northeastern Brazil; trees are leafless in the dry season. { kä'tiŋ·gə }

Ca^{2+} ATPase *See* calcium pump. { ¦sē,ā¦tü¦pləs ¦ā¦tē'pās *or* ¦kal·sē·əm ¦tü¦pləs ¦a¦te'pās }

cabbage [BOT] *Brassica oleracea* var. *capitata*. A biennial crucifer of the order Capparales grown for its head of edible leaves. { 'kab·ij }

cabezon [VERT ZOO] *Scorpaenichthys marmoratus*. A fish that is the largest of the sculpin species, weighing as much as 25 pounds (11.3 kilograms) and reaching a length of 30 inches (76 centimeters). { 'kab·ə,zōn, *or* ,ka·bə'zōn }

Cabot's ring [CELL MOL] A ringlike body in immature erythrocytes that may represent the remains of the nuclear membrane. { 'kab·əts 'riŋ }

cacao [BOT] *Theobroma cacao*. A small tropical tree of the order Theales that bears capsular fruits which are a source of cocoa powder and chocolate. Also known as chocolate tree. { kə'kaů }

cachexia [MED] Weight loss, weakness, and wasting of the body encountered in certain diseases or in terminal illnesses. { ka'kek·sē·ə }

cacomistle [VERT ZOO] *Bassariscus astutus*. A raccoonlike mammal that inhabits the southern and southwestern United States; distinguished by a bushy black-and-white ringed tail. Also known as civet cat; ringtail. { 'kak·ə,mis·əl }

caconym [SYST] A taxonomic name that is linguistically unacceptable. { 'kak·ə,nim }

Cactaceae [BOT] The cactus family of the order Caryophyllales; represented by the American stem succulents, which are mostly spiny with reduced leaves. { kak'tās·ē,ē }

cactus [BOT] The common name for any member of the family Cactaceae, a group characterized by a fleshy habit, spines and bristles, and large, brightly colored, solitary flowers. { 'kak·təs }

cactus alkaloid *See* anhalonium alkaloid. { 'kak·təs 'al·kə,lȯid }

cadaver [MED] A dead animal or human body to be studied by dissection. { kə'dav·ər }

cadaverine [BIOCHEM] $C_5H_{14}N_2$ A nontoxic, organic base produced as a result of the decarboxylation of lysine by the action of putrefactive bacteria on flesh. { kə'dav·ə,rēn }

caddis fly [INV ZOO] The common name for all members of the insect order Trichoptera; adults are mothlike and the immature stages are aquatic. { 'kad·əs ,flī }

cadherin [CELL MOL] Any of a family of calcium-dependent cell adhesion glycoproteins that play a fundamental role in tissue differentiation and structure. { kad'hir·en }

caducicorn [VERT ZOO] Having deciduous horns, as certain deer. { kə'dü·sə,kȯrn }

caducous [BOT] Lasting on a plant only a short time before falling off. { 'kad·ə·kəs }

caecilian [VERT ZOO] The common name for members of the amphibian order Apoda. { sē'sil·yən }

caecum *See* cecum. { 'sē·kəm }

Caenolestoidea [VERT ZOO] A superfamily of marsupial mammals represented by the single living family Caenolestidae. { ,sē·nə·le'stȯid·ē·ə }

caenostylic [VERT ZOO] Having the first two visceral arches attached to the cranium and functioning in food intake; a condition found in sharks, amphibians, and chimaeras. { ¦sē·nə¦stīl·ik }

Caesalpinoidea [BOT] A subfamily of dicotyledonous plants in the legume family, Leguminosae. { ,sē,zal·pə'nȯid·ē·ə }

cage compound *See* clathrate. { ¦kāj ¦käm,paůnd }

caiman [VERT ZOO] Any of five species of reptiles of the genus *Caiman* in the family Alligatoridae, differing from alligators principally in having ventral armor and a sharper snout. { 'kā·mən }

caisson disease [MED] A condition resulting from a rapid change in atmospheric pressure from high to normal, causing nitrogen bubbles to form in the blood and body tissues. Also known as bends; compressed-air illness. { 'kā ,sän di,zēz }

Calabar swelling [MED] Edematous, painful, subcutaneous swelling occurring in the body of natives of Calabar and of other parts of West Africa, probably due to an allergic reaction to *Loa loa* infection. { 'kal·ə,bär ,swel·iŋ }

calamine [PHARM] A powder mixture of zinc oxide and ferric oxide, used in skin lotions and ointments. { 'kal·ə,mīn }

Calanoida [INV ZOO] A suborder of the crustacean order Copepoda, including the larger and more abundant of the pelagic species. { ,kal·ə'nȯid·ə }

Calappidae [INV ZOO] The box crabs, a family of reptantian decapods in the subsection Oxystomata of the section Brachyura. { kə'lap·ə,dē }

calathiform [BIOL] Cup-shaped, being almost hemispherical. { kə'lath·ə,fȯrm }

calcaneal exostosis *See* heel spur. { kal'kā·nē·əl ,ek·sə'stō·səs }

calcaneocuboid ligament [ANAT] The ligament that joins the calcaneus and the cuboid bones. { kal¦kan·ē·ō¦kyü,bȯid ¦lig·ə·mənt }

calcaneum [ANAT] **1.** A bony projection of the metatarsus in birds. **2.** *See* calcaneus. { kal 'kan·ē·əm }

calcaneus [ANAT] A bone of the tarsus, forming the heel bone in humans. Also known as calcaneum. { kal'kan·ē·əs }

calcar [ZOO] A spur or spurlike process, especially on an appendage or digit. { 'kal,kär }

calcarate [BOT] Having spurs. { 'kal·kə,rāt }

Calcarea [INV ZOO] A class of the phylum Porifera, including sponges with a skeleton composed of calcium carbonate spicules. { kal'kar·ē·ə }

calcareous algae [BOT] Algae that grow on limestone or in soil impregnated with lime. { kal'ker·ē·əs 'al·jē }

calcarine fissure *See* calcarine sulcus. { 'kal·kə,rēn 'fish·ər }

calcarine sulcus [ANAT] A sulcus on the medial aspect of the occipital lobe of the cerebrum, between the lingual gyrus and the cuneus. Also known as calcarine fissure. { 'kal·kə,rēn 'səl·kəs }

Calcaronea [INV ZOO] A subclass of sponges in the class Calcarea in which the larva are amphiblastulae. { kal·kə'rō·nē·ə }

calcemia *See* hypercalcemia. { kal'sē·mē·ə }

calcicole [BOT] Requiring soil rich in calcium carbonate for optimum growth. { 'kal·sə,kōl }

calcicosis [MED] A form of pneumoconiosis caused by the inhalation of marble (calcium carbonate) dust. { ,kal·sə'kō·səs }

calciferol [BIOCHEM] A synthetic form of vitamin D that is prepared by ultraviolet irradiation of ergosterol, a vitamin D precursor found in plants. { kal'sif·ə,rōl }

calciferous [BIOL] Containing or producing calcium or calcium carbonate. { kal'sif·ə·rəs }

calciferous gland [INV ZOO] One of a series of glands that secrete calcium carbonate into the esophagus of certain oligochaetes. { kal'sif·ə·rəs ,gland }

calcification [PHYSIO] The deposit of calcareous matter within the tissues of the body. { ,kal·sə·fə'kā·shən }

calcifuge [ECOL] A plant that grows in an acid medium that is poor in calcareous matter. { 'kal·sə,fyüj }

Calcinea [INV ZOO] A subclass of sponges in the class Calcarea in which the larvae are parenchymulae. { kal'sin·ē·ə }

calcinosis [MED] Deposition of calcium salts in the skin, subcutaneous tissue, or other part of the body in certain pathologic conditions. { ,kal·sə'nō·səs }

calciphylaxis [IMMUNOL] A sudden local calcification in tissues in response to induced hypersensitivity following systemic sensitization by a calcifying factor. { ,kal·sə·fə'lak·səs }

calcitonin [BIOCHEM] A polypeptide, calcium-regulating hormone produced by the ultimobranchial bodies in vertebrates. Also known as thyrocalcitonin. { kal·sə'tō·nən }

calcitriol [BIOCHEM] 1,25-dihydroxycholecalciferol, the form of vitamin D that is involved in intestinal absorption of Ca^{2+} and Ca^{2+} resorption by bone. { ,kal·sə'trī,ōl }

calcium metabolism [BIOCHEM] Biochemical and physiological processes involved in maintaining the concentration of calcium in plasma at a constant level and providing a sufficient supply of calcium for bone mineralization. { 'kal·sē·əm mə'tab·ə,liz·əm }

calcium pump [CELL MOL] An enzyme that uses the energy generated by the hydrolysis of adenosine triphosphate (ATP) to move calcium ions out of the cytoplasm or, in the case of muscle cells, from the cytoplasm to the sarcoplasmic reticulum. Also known as Ca^{2+} ATPase. { ¦cal·se·əm ,pəmp }

calculus [ANAT] A small and cuplike structure. [PATH] An abnormal, solid concretion of minerals and salts formed around organic materials and found chiefly in ducts, hollow organs, and cysts. { 'kal·kyə·ləs }

calf [VERT ZOO] The young of the domestic cow, elephant, rhinoceros, hippopotamus, moose, whale, and others. { kaf }

calf diphtheria [VET MED] Disease of calves caused by species of *Sphaerophorus* which affects the mouth and pharynx and develops into pneumonia and septicemia, resulting in death. { 'kaf dif'thir·ē·ə }

Caliciaceae [BOT] A family of lichens in the order Caliciales in which the disk of the apothecium is borne on a short stalk. { kə,lē·sē'ās·ē,ē }

Caliciales [BOT] An order of lichens in the class Ascolichenes characterized by an unusual apothecium. { kə,lē·se'ā,lēz }

Caliciviridae [VIROL] A family of nonenveloped ribonucleic acid viruses with characteristic hollow surface structures that resemble cups. { kə,lē·sē'vī·rə,dē }

Caligidae [INV ZOO] A family of fish ectoparasites belonging to the crustacean suborder Caligoida. { kə'lij·ə,dē }

Caligoida [INV ZOO] A suborder of the crustacean order Copepoda, including only fish ectoparasites and characterized by a sucking mouth with styliform mandibles. { kal·ə'gȯid·ə }

caliper splint [MED] A splint designed for the leg, consisting of two metal rods from a posterior-thigh band or a padded ischial ring to a metal plate attached to the sole of the shoe at the instep. { 'kal·ə·pər ,splint }

Callichthyidae [VERT ZOO] A family of tropical

catfishes in the suborder Siluroidei. { ˌka ˌlik'thī·əˌdē }

callideic I *See* bradykinin. { ¦kal·i¦dē·ik ¦wən }

calling song [INV ZOO] A high-intensity insect sound which may play a role in habitat selection among certain species. { 'kȯl·iŋ ˌsȯŋ }

Callionymoidei [VERT ZOO] A suborder of fishes in the order Perciformes, including two families of colorful marine bottom fishes known as dragonets. { kə¦län·ē¦mȯid·ēˌī }

Callipallenidae [INV ZOO] A family of marine arthropods in the subphylum Pycnogonida lacking palpi and having chelifores and 10-jointed ovigers. { ˌkal·ə·pə'len·əˌdē }

Calliphoridae [INV ZOO] The blow flies, a family of myodarian cyclorrhaphous dipteran insects in the subsection Calypteratae. { ˌkal·ə'fȯr·əˌdē }

Callithricidae [VERT ZOO] The marmosets, a family of South American mammals in the order Primates. { kal·ə'thris·əˌdē }

Callitrichales [BOT] An order of flowering plants, division Magnoliophyta (Angiospermae), in the subclass Asteridae of the class Magnoliopsida (dicotyledons); consists of three small families with about 50 species, most of which are aquatics or small herbs of wet places. { kəˌlit·rə'kā·lēz }

Callorhinchidae [VERT ZOO] A family of ratfishes of the chondrichthyan order Chimaeriformes. { kal·ə'riŋk·əˌdē }

callose [BIOCHEM] A carbohydrate component of plant cell walls; associated with sieve plates where calluses are formed. [BIOL] Having hardened protuberances, as on the skin or on leaves and stems. { 'kaˌlōs }

callus [BOT] **1.** A thickened callose deposit on sieve plates. **2.** A hard tissue that forms over a damaged plant surface. [MED] Hard, thick area on the surface of the skin. { 'kal·əs }

calmodulin [BIOCHEM] A calcium-modulated protein consisting of a single polypeptide with 148 amino acids and a molecular weight of 16,700, found in all eukaryotes. { kal'mäj·ə·lən }

Calobryales [BOT] An order of liverworts; characterized by prostrate, simple or branched, leafless stems and erect, leafy branches of a radial organization. { ˌkal·ō·brī'ā·lēz }

caloreceptor [PHYSIO] A cutaneous sense organ that is stimulated by heat. { ¦kal·ō·ri'sep·tər }

calorimetric-respirometric ratio [PHYSIO] The ratio expressing that the oxidation of all carbon compounds produces approximately the same amount of heat per amount of oxygen. Abbreviated CR ratio. { kə¦lȯr·ə¦me·trik ri¦spī·rə¦me·trik 'rā·shō }

calotte [BIOL] A cap or caplike structure. [INV ZOO] **1.** The four-celled polar cap in Dicyemidae. **2.** A ciliated, retractile disc in certain bryozoan larva. **3.** A dark-colored anterior area in certain nematomorphs. { kə'lät }

calpain [CELL MOL] A calcium-dependent cysteine protease in the cytoplasm that is central to most processes in cell biology (plasma membrane-associated signaling events; cell proliferation, differentiation, activation, and communication; and programmed cell death). Its overactivation has been observed in muscular dystrophy, Alzheimer's disease, AIDS, cataract formation, multiple sclerosis, and arthritis. { 'kalˌpān }

calpastatin [CELL MOL] A protein found in all cells that is both the specific inhibitor and substrate of calpains. { ¦kal·pə'stat·ən }

calsequestrin [CELL MOL] An acidic protein, with a molecular weight of 44,000, which binds calcium inside the sarcoplasmic reticulum. { ¦kal·sə¦kwes·trən }

calthrop [INV ZOO] A sponge spicule having four axes in which the rays are equal or almost equal in length. { 'kal·thrəp }

calvarium [ANAT] A skull lacking facial parts and the lower jaw. { kal'ver·ē·əm }

Calvé's disease *See* osteochondrosis. { kal'vāz diˌzēz }

Calvin-Benson cycle *See* Calvin cycle. { ¦kal·vən 'ben·sən ˌsī·kəl }

Calvin cycle [CELL MOL] A metabolic process during photosynthesis that uses light indirectly to convert carbon dioxide to sugar in the stroma of chloroplasts. Also known as Calvin-Benson cycle; carbon fixation cycle. { 'kal·vən ˌsī·kəl }

calving [VERT ZOO] Giving birth to a calf. { 'kav·iŋ }

Calycerales [BOT] An order of flowering plants, division Magnoliophyta (Angiospermae), in the subclass Asteridae of the class Magnoliopsida (dicotyledons); consists of a single family with about 60 species native to tropical America. { 'kal·ə·sə'rā·lēz }

calyculate [BOT] Having bracts that imitate a second, external calyx. { kə'lik·yə·lət }

calymma [INV ZOO] The outer, vacuolated protoplasmic layer of certain radiolarians. { kə'lim·ə }

Calymmatobacterium [MICROBIO] A genus of gram-negative, usually encapsulated, pleomorphic rods of uncertain affiliation; the single species causes granuloma inguinale in humans. { kə¦lim·əd·ōˌbak'tir·ē·əm }

Calymnidae [INV ZOO] A family of echinoderms in the order Holasteroida characterized by an ovoid test with a marginal fasciole. { kə'lim·nəˌdē }

calypter [INV ZOO] A scalelike or lobelike structure above the haltere of certain two-winged flies. Also known as alula; squama. { kə'lip·tər }

Calyptoblastea [INV ZOO] A suborder of cnidarians in the order Hydroida, including the hydroids with protective cups around the hydranths and gonozooids. { kə¦lip·tō'blas·tē·ə }

calyptra [BOT] **1.** A membranous cap or hoodlike covering, especially the remains of the archegonium over the capsule of a moss. **2.** Tissue surrounding the archegonium of a liverwort. **3.** Root cap. { kə'lip·trə }

Calypteratae [INV ZOO] A subsection of dipteran insects in the suborder Cyclorrhapha characterized by calypters associated with the wings. { kə'lip·tə·rə,tē }

calyptrate [BOT] Having a calyptra. { kə'lip ,trāt }

calyptrogen [BOT] The specialized cell layer from which a root cap originates. { kə'lip·trə·jən }

Calyssozoa [INV ZOO] The single class of the bryozoan subphylum Entoprocta. { kə,lis·ə 'zō·ə }

calyx [BOT] The outermost whorl of a flower; composed of sepals. [INV ZOO] A cup-shaped structure to which the arms are attached in crinoids. [MED] **1.** A cuplike structure. **2.** In the kidney, a collecting structure extending from the renal pelvis. { 'kā,liks }

calyx tube [BOT] A tube formation resulting from fusion of the lateral edges of a group of sepals. { 'kā·liks ,tüb }

Camacolaimoidea [INV ZOO] A superfamily of nematodes consisting of a single family, the Camacolaimidae; they occur in marine or brackish-water environments. { ,kam·ə·kō·lə'mȯid·ē·ə }

Camallanida [INV ZOO] An order of phasmid nematodes in the subclass Spiruria, including parasites of domestic animals. { kam·ə'lan·ə·də }

Camallanoidea [INV ZOO] A superfamily of parasitic nematodes in the subclass Spiruria. { kə,mal·ə'nȯid·ē·ə }

Camarodonta [INV ZOO] An order of Euechinoidea proposed by R. Jackson and abandoned in 1957. { kam·ə·rə'dän·tə }

camarodont dentition [INV ZOO] In echinoderms, keeled teeth meeting the epiphyses so that the foramen magnum of the jaw is closed. { 'kam·ə·rə,dänt den'tish·ən }

Cambaridae [INV ZOO] A family of crayfishes belonging to the section Macrura in the crustacean order Decapoda. { kam'bär·ə,dē }

Cambarinae [INV ZOO] A subfamily of crayfishes in the family Astacidae, including all North American species east of the Rocky Mountains. { kam'bär·ə,nē }

cambium [BOT] A layer of cells between the phloem and xylem of most vascular plants that is responsible for secondary growth and for generating new cells. { 'kam·be·əm }

camel [VERT ZOO] The common name for two species of artiodactyl mammals, the bactrian camel (*Camelus bactrianus*) and the dromedary camel (*C. dromedarius*), in the family Camelidae. { 'kam·əl }

Camelidae [VERT ZOO] A family of tylopod ruminants in the superfamily Cameloidea of the order Artiodactyla, including four species of camels and llamas. { ka'mel·ə,dē }

Cameloidea [VERT ZOO] A superfamily of tylopod ruminants in the order Artiodiodactyla. { kam·ə'lȯid·ē·ə }

cAMP *See* cyclic adenylic acid.

Campanulaceae [BOT] A family of dicotyledonous plants in the order Campanulales characterized by a style without an indusium but with well-developed collecting hairs below the stigmas, and by a well-developed latex system. { kam,pan·yə'lās·ē,ē }

Campanulales [BOT] An order of dicotyledonous plants in the subclass Asteridae distinguished by a chiefly herbaceous habit, alternate leaves, and inferior ovary. { kam,pan·yə'lā,lēz }

campanulate [BOT] Bell-shaped; applied particularly to the corolla. { kam'pan·yə·lət }

campestrian [ECOL] Of or pertaining to the northern Great Plains area. { kam'pes·trē·ən }

camphoric acid [PHARM] $C_{10}H_{16}O_4$ A compound crystallizing in leaflets from water and in monoclinic prisms from alcohol; melting point is 186–188°C; used as a central respiratory stimulant. { kam'fȯr·ik 'as·əd }

camphor tree [BOT] *Cinnamomum camphora*. A plant of the laurel family (Lauraceae) in the order Magnoliales from which camphor is extracted. { 'kam·fər ,trē }

Campodeidae [INV ZOO] A family of primarily wingless insects in the order Diplura which are most numerous in the Temperate Zone of the Northern Hemisphere. { kam·pə'dē·ə,dē }

campodeiform [INV ZOO] Elongate, flattened, and narrowed posteriorly. { kam'pō·dē·ə ,fȯrm }

campos [ECOL] The savanna of South America. { 'käm,pōs }

camptothecin [PHARM] An alkaloid belonging to the family of drugs called topoisomerase inhibitors which is used as an anticancer drug, it is the first known naturally produced DNA topoisomerase inhibitor and was originally isolated from the Chinese tree *Camptothecin acuminate*. { ,kam·tō'thē·sən }

Campylobacter [MICROBIO] A genus of bacteria in the family Spirillaceae; spirally curved rods that are motile by means of a polar flagellum at one or both poles. { ,kam·pə·lə'bak·tər }

Campylobacter enteritis [MED] A water-borne gastroenteritis caused by *Campylobacter jejune*. { kam¦pī·lə,bak·tər ,en·tə'rīd·əs }

Campylobacter jejune [MICROBIO] A microaerophilic pathogen associated with raw meats and unpasteurized milk; ingestion of a small amount can cause diarrhea, cramps, and nausea. { kam¦pī·lə,bak·tər jə'jü·nē }

campylotropous [BOT] Having the ovule symmetrical but half inverted, with the micropyle and funiculus at right angles to each other. { ¦kam·pə¦lä·trə·pəs }

Canaceidae [INV ZOO] The seashore flies, a family of myodarian cyclorrhaphous dipteran insects in the subsection Acalypteratae. { ,kan·ə'sē·ə,dē }

Canadian life zone [ECOL] The zone comprising the climate and biotic communities of the portion of the Boreal life zone exclusive of the Hudsonian and Arctic-Alpine zones. { kə'nād·ē·ən 'līf ,zōn }

canal [BIOL] A tubular duct or passage in bone or soft tissues. { kə'nal }

canal cell [BOT] One of the row of cells that make up the axial row within the neck of an archegonium. { kə'nal ˌsel }

canaliculate [BIOL] Having small channels, canals, or grooves. { ˌkan·əl'ik·yəˌlāt }

canaliculus [HISTOL] **1.** One of the minute channels in bone radiating from a Haversian canal and connecting lacunae with each other and with the canal. **2.** A passage between the cells of the cell cords in the liver. { ˌkan·əl'ik·yə·ləs }

canalization [EVOL] The effect of natural selection on development to produce pathways that are insensitive to minor genetic or environmental variation; results in the phenotypic norm of the species. [MED] Surgical method of wound drainage without tubes by forming channels. [PHYSIO] The formation of new channels in tissues, such as the formation of new blood vessels through a thrombus. { ˌkan·əl·ə'zā·shən }

canalized character [GEN] A trait whose variability is restricted within narrow boundaries even when individual members of the species are subjected to enhanced environmental pressures or mutations. { ¦kan·əˌlīzd 'kar·ik·tər }

canal of Schlemm [ANAT] An irregular channel at the junction of the sclera and cornea in the eye that drains aqueous humor from the anterior chamber. { kə'nal əv 'shlem }

canal valve [ANAT] The semilunar valve in the right atrium of the heart between the orifice of the inferior vena cava and the right atrioventricular orifice. Also known as eustachian valve. { kə'nal ˌvalv }

canary-pox virus [VIROL] An avian poxvirus that causes canary pox, a disease closely related to fowl pox. { kə'ner·ē ˌpäks ˌvī·rəs }

canavanine [BIOCHEM] $C_5H_{12}O_3N_4$ An amino acid found in the jack bean. { kə'nav·əˌnēn }

cancellate [BIOL] Lattice-shaped. Also known as clathrate. { 'kan·səˌlāt }

cancellous [BIOL] Having a reticular or spongy structure. { kan'sel·əs }

cancellous bone [HISTOL] A form of bone near the ends of long bones having a cancellous matrix composed of rods, plates, or tubes; spaces are filled with marrow. { kan'sel·əs ˌbōn }

cancer [MED] Any malignant neoplasm, including carcinoma and sarcoma. { 'kan·sər }

cancer eye [VET MED] A malignant epithelioma of the eye of cattle, common in regions of intense sunlight. { 'kan· sər ˌī }

cancroid [MED] A squamous-cell carcinoma. { 'kaŋˌkrȯid }

Candida [MYCOL] A genus of yeastlike, pathogenic imperfect fungi that produce very small mycelia. { 'kan·də·də }

Candida utilis [MICROBIO] An asexual yeast species used industrially in the production of single-cell protein for food and fodder. { ¦kan·də·də 'yü·də·lis }

candidiasis [MED] A fungus infection of the skin, lungs, mucous membranes, and viscera of humans caused by a species of *Candida*, usually *C. albicans*. Also known as moniliasis. { ˌkan·də'dī·ə·səs }

cane [BOT] **1.** A hollow, usually slender, jointed stem, such as in sugarcane or the bamboo grasses. **2.** A stem growing directly from the base of the plant, as in most Rosaceae, such as blackberry and roses. { kān }

canescent [BOT] Having a grayish epidermal covering of short hairs. { kə'nes·ənt }

Canidae [VERT ZOO] A family of carnivorous mammals in the superfamily Canoidea, including dogs and their allies. { 'kan·əˌdē }

canine [ANAT] A conical tooth, such as one located between the lateral incisor and first premolar in humans and many other mammals. Also known as cuspid. [VERT ZOO] Pertaining or related to dogs or to the family Canidae. { 'kāˌnīn }

canine distemper [VET MED] A pantropic virus disease occurring among animals of the family Canidae. { 'kāˌnīn dis'tem·pər }

Canis [VERT ZOO] The type genus of the dog family (Canidae), including dogs, wolves, and jackals. { 'kā·nəs }

canker [VET MED] A localized chronic inflammation of the ear in cats, dogs, foxes, ferrets, and others caused by the mite *Otodectes cynotis*. { 'kaŋ·kər }

canker sore [MED] Small ulceration of the mucous membrane of the mouth, sometimes caused by a food allergy. { 'kaŋ·kər ˌsȯr }

cankerworm [INV ZOO] Any of several lepidopteran insect larvae in the family Geometridae which cause severe plant damage by feeding on buds and foliage. { 'kaŋ·kərˌwərm }

Cannabaceae [BOT] A family of dicotyledonous herbs in the order Urticales, including Indian hemp (*Cannabis sativa*) and characterized by erect anthers, two styles or style branches, and the lack of milky juice. { kan·ə'bās·ēˌē }

Cannabis [BOT] A genus of tall annual herbs in the family Cannabaceae having erect stems, leaves with three to seven elongate leaflets, and pistillate flowers in spikes along the stem. { 'kan·ə·bəs }

cannabism [MED] Poisoning resulting from excessive or habitual use of cannabis. { 'kan·əˌbiz·əm }

Cannaceae [BOT] A family of monocotyledonous plants in the order Zingiberales characterized by one functional stamen, a single functional pollen sac in the stamen, mucilage canals in the stem, and numerous ovules in each of the one to three locules. { kə'nās·ēˌē }

cannula [MED] A small tube that can be inserted into a body cavity, duct, or vessel. { 'kan·yə·lə }

Canoidea [VERT ZOO] A superfamily belonging to the mammalian order Carnivora, including all dogs and doglike species such as seals, bears, and weasels. { kə'nȯid·ē·ə }

canonical sequence [CELL MOL] An archetypical nucleotide or amino acid sequence to which all variants are compared. { kə'nän·ə·kəl 'sē·kwəns }

cantala [BOT] A fiber produced from agave (*Agave cantala*) leaves; used to make twine. Also known as Cebu maguey; maguey; Manila maguey. { kan'täl·ə }

cantaloupe [BOT] The fruit (pepo) of *Cucumis malo*, a small, distinctly netted, round to oval muskmelon of the family Cucurbitaceae in the order Violales. { 'kant·əl,ōp }

canthariasis [MED] Infection or disease caused by coleopteran insects or their larvae. { kan·thə'rī·ə·səs }

Cantharidae [INV ZOO] The soldier beetles, a family of coleopteran insects in the superfamily Cantharoidea. { kan'thar·ə,dē }

Cantharoidea [INV ZOO] A superfamily of coleopteran insects in the suborder Polyphaga. { kan·thə'ròid·ē·ə }

canthus [ANAT] Either of the two angles formed by the junction of the eyelids, designated outer or lateral, and inner or medial. { 'kan·thəs }

cap [GEN] In many eukaryotic messenger ribonucleic acids, the structure at the 5′ end consisting of 7′-methyl-guanosine-pppX, where X is the first nucleotide encoded in the deoxyribonucleic acid; it is added posttranscriptionally. { kap }

capacitation [PHYSIO] The process of physiological change occurring in mammalian spermatozoa during passage through the female reproductive tract that enables them to penetrate the egg membrane. { kə,pas·ə'tā·shən }

cap-binding protein [CELL MOL] A protein that specifically recognizes the methylated cap of eukaryotic messenger ribonucleic acid (mRNA) and is essential in the regulation of mRNA translation. { ¦kap ,bīnd·iŋ 'prō,tēn }

capillaroscope [MED] A microscope used for diagnostic examination of the cutaneous capillaries, as in the nail beds and conjunctiva. { ,kap·ə'lar·ə,skōp }

capillary [ANAT] The smallest vessel of both the circulatory and lymphatic systems; the walls are composed of a single cell layer. { 'kap·ə,ler·ē }

capillary angioma *See* hemangioma. { 'kap·ə,ler·ē ,an·jē'ō·mə }

capillary bed [ANAT] The capillaries, collectively, of a given area or organ. { 'kap·ə,ler·ē ,bed }

capillary pressure [PHYSIO] Pressure exerted by blood against capillary walls. { 'kap·ə,ler·ē ,presh·ər }

capillitium [MYCOL] A network of threadlike tubes or filaments in which spores are embedded within sporangia of certain fungi, such as the slime molds. { ,kap·ə'lish·ē·əm }

capitate [BIOL] Enlarged and swollen at the tip. [BOT] Forming a head, as certain flowers of the Compositae. { 'kap·ə,tāt }

capitellate [BOT] **1.** Having a small knoblike termination. **2.** Grouped to form a capitulum. { ¦kap·ə¦te,lāt }

Capitellidae [INV ZOO] A family of mud-swallowing annelid worms, sometimes called bloodworms, belonging to the Sedentaria. { ,kap·ə'te·lə,dē }

capitellum [ANAT] A small head or rounded process of a bone. { ,kap·ə'te·ləm }

Capitonidae [VERT ZOO] The barbets, a family of pantropical birds in the order Piciformes. { ,kap·ə'tä·nə,dē }

capitulum [BIOL] A rounded, knoblike, usually terminal proturberance on a structure. [BOT] One of the rounded cells on the manubrium in the antheridia of lichens belonging to the Caliciales. { kə'pich·ə·ləm }

Capnocytophaga [MICROBIO] A genus of bacteria comprising fusiform, fermentative, gram-negative rods which require carbon dioxide for growth and show gliding motility. { ,kap·nō,sī'täf·ə·gə }

Caponidae [INV ZOO] A family of arachnid arthropods in the order Araneida characterized by having tracheae instead of book lungs. { kə'pä·nə,dē }

Capparaceae [BOT] A family of dicotyledonous herbs, shrubs, and trees in the order Capparales characterized by parietal placentation; hypogynous, mostly regular flowers; four to many stamens; and simple to trifoliate or palmately compound leaves. { ,kap·ə'rās·ē,ē }

Capparales [BOT] An order of dicotyledonous plants in the subclass Dilleniidae. { ,kap·ə'rā,lēz }

capping [CELL MOL] Addition of a methylated cap to eukaryotic messenger ribonucleic acid molecules. { 'kap·iŋ }

Caprellidae [INV ZOO] The skeleton shrimps, a family of slender, cylindrical amphipod crustaceans in the suborder Caprellidea. { kə'prel·ə,dē }

Caprellidea [INV ZOO] A suborder of marine and brackish-water animals of the crustacean order Amphipoda. { ,kap·rə'lid·ē·ə }

Caprifoliaceae [BOT] A family of dicotyledonous, mostly woody plants in the order Dipsacales, including elderberry and honeysuckle; characterized by distinct filaments and anthers, typically five stamens and five corolla lobes, more than one ovule per locule, and well-developed endosperm. { ,kap·rə,fōl·ē'ās·e,ē }

Caprimulgidae [VERT ZOO] A family of birds in the order Caprimulgiformes, including the nightjars, or goatsuckers. { ,kap·rə'məl·jə,dē }

Caprimulgiformes [VERT ZOO] An order of nocturnal and crepuscular birds, including nightjars, potoos, and frog-mouths. { ,kap·rə,məl·jə'fòr,mēz }

Capripoxvirus [VIROL] A genus of viruses belonging to the subfamily Chordopoxvirinae; natural hosts are ungulates; these viruses can be mechanically transmitted by arthropods. { 'kap·ri,päks,vī·rəs }

Capsaloidea [INV ZOO] A superfamily of ectoparasitic trematodes in the subclass Monogenea characterized by a sucker-shaped holdfast with anchors and hooks. { ,kap·sə'lòid·ē·ə }

capsanthin [BIOCHEM] $C_{40}H_{58}IO_3$ Carmine-red carotenoid pigment occurring in paprika. { kap'san·thən }

capsicum [BOT] The fruit of a plant of the genus

Capsicum, especially C. *frutescens*, cultivated in southern India and the tropics; a strong irritant to mucous membranes and eyes. { 'kap·sə·kəm }

capsid [INV ZOO] The name applied to all members of the family Miridae. [VIROL] In a virus, the protein shell surrounding the nucleic acid and its associated protein core. Also known as protein coat. { 'kap·səd }

capsomere [VIROL] An individual protein subunit of a capsid. { 'kap·sə,mir }

capsular ligament [ANAT] A saclike ligament surrounding the articular cavity of a freely movable joint and attached to the bones. { 'kap·sə·lər 'lig·ə·mənt }

capsulate [BIOL] Enclosed in a capsule. { 'kap·sə,lāt }

capsule [ANAT] A membranous structure enclosing a body part or organ. [BOT] A closed structure bearing seeds or spores; it is dehiscent at maturity. [MICROBIO] A thick, mucous envelope, composed of polypeptide or carbohydrate, surrounding certain microorganisms. [PHARM] A soluble shell in which drugs are enclosed for oral administration. { 'kap·səl }

capybara [VERT ZOO] *Hydrochoerus capybara*. An aquatic rodent (largest rodent in existence) found in South America and characterized by partly webbed feet, no tail, and coarse hair. { ,kap·ə'bar·ə }

Cap Z protein [BIOCHEM] A microfilament capping protein located in the Z line of skeletal muscle, made up of two subunits that selectively bind to the positive ends of actin filaments, stabilizing them and preventing depolymerization. { ¦kap ¦zē 'prō,tēn }

Carabidae [INV ZOO] The ground beetles, a family of predatory coleopteran insects in the suborder Adephaga. { kə'rab·ə,dē }

caraboid larva [INV ZOO] The morphologically distinct larva of certain beetles, characterized by a narrow, elongate body with long legs, a head that occasionally bears a single ocellus on each side, and three-segmented antennae with a well-developed sensorium. { 'kar·ə,bȯid 'lar·və }

Caracarinae [VERT ZOO] The caracaras, a subfamily of carrion-feeding birds in the order Falconiformes. { ,karə'kar·ə,nē }

Carangidae [VERT ZOO] A family of perciform fishes in the suborder Percoidei, including jacks, scads, and pompanos. { kə'ran·jə,dē }

carangiform motion [VERT ZOO] A type of fish locomotion in which the fish moves its head slightly but builds considerable amplitude of motion toward the tail. { kə'ran·jə,fȯrm ,mō·shən }

carapace [INV ZOO] A dorsolateral, chitinous case covering the cephalothorax of many arthropods. [VERT ZOO] The bony, dorsal part of a turtle shell. { 'kar·ə,pās }

Carapidae [VERT ZOO] The pearlfishes, a family of sinuous, marine shore fishes in the order Gadiformes that live as commensals in the body cavity of holothurians. { kə'rap·ə,dē }

carate *See* pinta. { kə'räd·ē }

caraway [BOT] *Carum carvi*. A white-flowered perennial herb of the family Umbelliferae; the fruit is used as a spice and flavoring agent. { 'kar·ə,wā }

carbachol [PHARM] $C_6H_{15}ClN_2O_2$ Hygroscopic, hard, prismatic crystals with a melting point of 200–203°C; soluble in water, methanol, and alcohol; used as a cholinergic drug in humans and parasympathomimetic drug in larger animals. { 'kär·bə,kȯl }

carbamazepine [PHARM] $C_{15}H_{12}N_2O$ An anticonvulsant that is used in the treatment of epilepsy, and is also useful in the treatment of periods of mania associated with bipolar depression. { ,kär·bə'maz·ə,pēn }

carbamic acid [BIOCHEM] NH_2COOH An amino acid known for its salts, such as urea and carbamide. { kar'bam·ik 'as·əd }

carbamino [BIOCHEM] A compound formed by the combination of carbon dioxide with a free amino group in an amino acid or a protein. { kär'bam·ə,nō }

carbamyl phosphate [BIOCHEM] $NH_2COPO_4H_2$ The ester formed from reaction of phosphoric acid and carbamyl acid. { 'kär·bə,mil 'fäs,fāt }

carbohydrase [BIOCHEM] Any enzyme that catalyzes the hydrolysis of disaccharides and more complex carbohydrates. { ,kär·bō'hī,drās }

carbohydrate [BIOCHEM] Any of the group of organic compounds composed of carbon, hydrogen, and oxygen, including sugars, starches, and celluloses. { ,kär·bō'hī,drāt }

carbohydrate metabolism [BIOCHEM] The sum of the biochemical and physiological processes involved in the breakdown and synthesis of simple sugars, oligosaccharides, and polysaccharides and in the transport of sugar across cell membranes. { ,kär·bō'hī,drāt me'tab·ə,liz·əm }

carbomycin [MICROBIO] $C_{42}H_{67}O_{16}N$ Colorless, crystalline antibiotic produced by *Streptomyces halstedii*; principally active against gram-positive bacteria. { kär·bō'mīs·ən }

carbomycin B [MICROBIO] $C_{42}H_{67}O_{15}N$ A colorless, crystalline antibiotic differing from carbomycin only in having one less oxygen atom in its molecule. { kär·bō'mīs·ən 'bē }

carbon fixation [CELL MOL] During photosynthesis, the process by which plants convert carbon dioxide from the air into organic molecules. { 'kär·bən fik¦sā·shən }

carbon fixation cycle *See* Calvin cycle. { ¦kär·bən fik'sā·shən ,sī·kəl }

carbonic anhydrase [BIOCHEM] An enzyme which aids carbon dioxide transport and release by catalyzing the synthesis, and the dehydration, of carbonic acid from, and to, carbon dioxide and water. { kär'bän·ik an'hī,drās }

carbonmonoxyhemoglobin [BIOCHEM] A stable combination of carbon monoxide and hemoglobin formed in the blood when carbon monoxide is inhaled. Also known as carbonylhemoglobin; carboxyhemoglobin. { ¦kär·bən·mə¦näk·sē'hē·mə,glō·bən }

carbonylhemoglobin *See* carbonmonoxyhemoglobin. { 'kär·bə,nil¦hēm·ə'glō·bən }

carboplatin [MED] A platinum coordination compound and anticancer agent used for advanced gynecologic malignancies, especially ovarian tumors, and for head and neck cancers and lung cancers. { ¦kär·bō'plat·ən }

carboxyhemoglobin *See* carbonmonoxyhemoglobin. { kär¦bäk·sē¦hē·mə,glō·bən }

carboxylase [BIOCHEM] Any enzyme that catalyzes a carboxylation or decarboxylation reaction. { kär'bäk·sə,lās }

carboxyl terminal [BIOCHEM] The end of a polypeptide chain with a free carboxyl group. { kär 'bäk·səl 'tər·mən·əl }

carboxypeptidase [BIOCHEM] Any enzyme that catalyzes the hydrolysis of a peptide at the end containing the free carboxyl group. { kär,bäk·sē'pep·tə,dās }

carbuncle [MED] A bacterial infection of subcutaneous tissue caused by *Staphylococcus aureus*; multiple sinuses are created by tissue destruction. { 'kär,bəŋ·kəl }

Carcharhinidae [VERT ZOO] A large family of sharks belonging to the charcharinid group of galeoids, including the tiger sharks and blue sharks. { ,kär·kə'rīn·ə,dē }

Carchariidae [VERT ZOO] A family of shallow-water predatory sharks belonging to the isurid group of galeoids. { ,kär·kə'rī·ə,dē }

carcharodont [VERT ZOO] Possessing sharp, flat, triangular teeth with serrated margins, like those of the human-eating sharks. { kär'kar·ə,dänt }

carcinoembryonic antigen [IMMUNOL] A glycoprotein found in tissues of the fetal gut during the first two trimesters of pregnancy and in the peripheral blood of individuals with some forms of cancer, such as digestive-system or breast cancer. { ¦kärs·ən·ō,em·brē¦än·ik 'ant·i·jən }

carcinogen [MED] Any agent that incites development of a carcinoma or any other sort of malignancy. { kär'sin·ə·jən }

carcinogenesis [CELL MOL] The processes of tumor development. { ,kärs·ən·ō'jen·ə·səs }

carcinoid [MED] A potentially malignant tumor of the argentaffin cells of the stomach and intestine. { 'kärs·ən,ȯid }

carcinoid syndrome [MED] A complex of symptoms arising from the metastasis of a carcinoid tumor to the liver. { 'kärs·ən,ȯid 'sin,drōm }

carcinoma [MED] A malignant epithelial tumor. { ,kärs·ən'ō·mə }

carcinoma in situ [MED] A malignant tumor in the premetastatic stage, when cells are at the site of origin. { ,kärs·ən'ō·mə in 'si·chü }

carcinomatosis [MED] Metastasis of a primary carcinoma to many sites throughout the body. { ,kärs·ən,äm·ə'tō·səs }

Carcinus maenas [INV ZOO] A decapod crustacean commonly found on the coasts of northwest Europe and the northeast United States that feeds on invertebrates such as mollusks, polychaete worms, and other crustaceans, and periodically sheds its exoskeleton in order to grow. Also known as shore crab. { ¦kär·sən·əs 'mī·nəs }

cardamom *See* cardamon. { 'kärd·ə·məm }

cardamon [BOT] *Elettaria cardamomum*. A perennial herbaceous plant in the family Zingiberaceae; the seed of the plant is used as a spice. Also spelled cardamom. { 'kärd·ə·mən }

cardia [ANAT] **1.** The orifice where the esophagus enters the stomach. **2.** The large, blind diverticulum of the stomach adjoining the orifice. [INV ZOO] Anterior enlargement of the ventriculus in some insects. { 'kärd·ē·ə }

cardiac [ANAT] **1.** Of, pertaining to, or situated near the heart. **2.** Of or pertaining to the cardia of the stomach. { 'kärd·ē,ak }

cardiac arrest [MED] Cessation of the heartbeat. { 'kärd·ē,ak ə'rest }

cardiac arrhythmia [MED] Any disturbance or irregularity of the heartbeat. { ¦kärd·ē,ak ā'rith·mē·ə }

cardiac cirrhosis [MED] Progressive fibrosis of the liver due to prolonged venous blood retention as a result of prolonged and severe heart failure. { 'kärd·ē,ak sə'rō·səs }

cardiac cycle [PHYSIO] The sequence of events in the heart between the start of one contraction and the start of the next. { 'kärd·ē,ak ,sī·kəl }

cardiac edema [MED] Accumulation of fluids throughout the body, as a function of cardiac failure. { 'kärd·ē,ak i'dē·mə }

cardiac electrophysiology [PHYSIO] The science that is concerned with the mechanism, spread, and interpretation of the electric currents which arise within heart muscle tissue and initiate each heart muscle contraction. { 'kärd·ē,ak i,lek·trō,fiz·ē'äl·ə·jē }

cardiac failure [MED] A complex of symptoms resulting from failure of the heart to pump sufficient quantities of blood. Also known as heart failure. { 'kärd·ē,ak 'fāl·yər }

cardiac gland [ANAT] Any of the mucus-secreting, compound tubular structures near the esophagus or in the cardia of the stomach of vertebrates; capable of secreting digestive enzymes. { 'kärd·ē,ak ,gland }

cardiac glycoside [BIOCHEM] A class of naturally occurring glycosides that exhibit the ability to strengthen the contraction of heart muscles. { ,kärd·ē,ak 'glī·kə,sīd }

cardiac input [PHYSIO] The amount of venous blood returned to the heart during a specified period of time. { 'kärd·ē,ak 'in,pu̇t }

cardiac loop [EMBRYO] The embryonic heart formed by bending and twisting of the cardiac tube. { 'kärd·ē,ak 'lüp }

cardiac massage [MED] Rhythmic compression of the heart by a physician or other person in the effort to maintain effective circulation following heart failure. { 'kärd·ē,ak mə'säzh }

cardiac murmur [MED] Any adventitious sound heard in the region of the heart. Also known as heart murmur. { 'kärd·ē,ak ¦mər·mər }

cardiac muscle [HISTOL] The principal tissue of the vertebrate heart; composed of a syncytium of striated muscle fibers. { 'kärd·ē,ak ¦məs·əl }

cardiac output [PHYSIO] The total blood flow

from the heart during a specified period of time. { 'kärd·ē,ak 'au̇t,pu̇t }

cardiac pacemaker *See* pacemaker. { 'kärd·ē,ak 'pā,smā·kər }

cardiac plexus [NEUROSCI] A network of visceral nerves situated at the base of the heart; contains both sympathetic and vagal nerve fibers. { 'kärd·ē,ak 'plek·səs }

cardiac sphincter [ANAT] The muscular ring at the orifice between the esophagus and stomach. { 'kärd·ē,ak 'sfiŋk·tər }

cardiac tamponade [MED] Cardiac compression caused by an accumulation of fluid within the pericardium. { 'kärd·ē,ak ¦tam·pə¦nād }

cardiac valve [ANAT] Any of the structures located within the orifices of the heart that maintain unidirectional blood flow. { 'kärd·ē,ak 'valv }

cardiectomy [MED] Excision of the cardiac end of the stomach. { ,kärd·ē'ek·tə·mē }

cardinal teeth [INV ZOO] Ridges and grooves on the inner surfaces of both valves of a bivalve mollusk near the anterior end of the hinge. { 'kärd·nəl ,tēth }

cardinal vein [EMBRYO] Any of four veins in the vertebrate embryo which run along each side of the vertebral column; the paired veins on each side discharge blood to the heart through the duct of Cuvier. { 'kärd·nəl ,vān }

Cardiobacterium [MICROBIO] A genus of gram-negative, rod-shaped bacteria of uncertain affiliation; facultative anaerobes that ferment fructose, glucose, mannose, sorbitol, and sucrose; the single species causes endocarditis in humans. { ¦kärd·ē·ō,bak'tir·ē·əm }

cardioblast [INV ZOO] Any of certain early embryonic cells in insects from which the heart develops. { 'kärd·ē·ə,blast }

cardiocirculatory *See* cardiovascular. { ¦kärd·ē·ō'sər·kyə·lə,tȯr·ē }

cardiodynamics [PHYSIO] The dynamics of the heart's action in pumping blood. { ¦kärd·ē·ō·dī'nam·iks }

cardiogenic plate [EMBRYO] An area of splanchnic mesoderm in the early mammalian embryo from which the heart develops. { ¦kärd·ē·ō¦jen·ik 'plāt }

cardiogenic shock [MED] Shock due to inadequate arterial blood flow following left ventricular failure or pulmonary embolism. { ¦kärd·ē·ō¦jen·ik 'shäk }

cardiography [MED] Analysis of heart movements in the cardiac cycle by means of electronic instruments, especially by tracings. { ,kärd·ē'äg·rə·fē }

cardiolipin [BIOCHEM] A complex phospholipid found in the ether alcohol extract of powdered beef heart; mixed with leutin and cholesterol, it functions as the antigen in the Wassermann complement-fixation test for syphilis. Also known as diphosphatidyl glycerol. { ,kärd·ē·ō'lip·ən }

cardiology [MED] The study of the heart. { ,kärd·ē'äl·ə·jē }

cardiomyopathy *See* myocardiopathy. { ¦kärd·ē·ō,mī'äp·ə·thē }

cardiopathy [MED] Any disorder or disease of the heart. { ,kärd·ē'äp·ə·thē }

cardiopulmonary [PHYSIO] Pertaining to the heart and lungs. { ¦kärd·ē·ō'pu̇l·mə,ner·ē }

cardiopulmonary bypass [MED] A surgical technique for avoiding circulation of the blood through the heart and lungs by directing the flow from the entrance of the right atrium, conditioning it by mechanical means, and reentering it at the aorta. { ¦kärd·ē·ō'pu̇l·mə,ner·ē 'bī·pas }

cardiopulmonary resuscitation [MED] The simultaneous forced ventilation of the lungs and squeezing of the heart ventricles to sustain the flow of oxygenated blood throughout the system; often applied in cases of cardiac arrest. Abbreviated CPR. { ¦kärd·ē·ō'pu̇l·mə,ner·ē ri,səs·ə'tā·shən }

cardiorrhaphy [MED] Suturing of the heart muscle. { ,kärd·ē'ȯr·ə·fē }

cardioscope [MED] **1.** An instrument for the examination or visualization of the interior of the cardiac chambers. **2.** An instrument which, by means of a cathode-ray oscillograph, projects an electrocardiographic record on a luminous screen. { 'kärd·ē·ə,skōp }

cardiospasm [MED] Failure of the cardiac sphincter to relax; associated with spasm of the cardiac portion of the stomach and dilation of the esophagus. { 'kärd·ē·ō,spaz·əm }

cardiotachometer [MED] An electronic amplifier that times and records pulse rates of the heart. { ¦kärd·ē·ō·tə'käm·əd·ər }

cardiotomy [MED] Dissection or incision of the heart or the cardia of the stomach. { ,kärd·ē'äd·ə·mē }

cardiotonic drug [PHARM] Any agent, such as digitalis, that increases cardiac muscle tonus. { ¦kärd·ē·ō¦tän·ik 'drəg }

cardiovascular [PHYSIO] Pertaining to the heart and circulatory system. Also known as cardiocirculatory. { ¦kärd·ē·ō'vas·kyə·lər }

cardiovascular system [ANAT] Those structures, including the heart and blood vessels, which provide channels for the flow of blood. { ¦kärd·ē·ō'vas·kyə·lər ,sis·təm }

cardiovascular toxicity [MED] The adverse effects on the heart or blood systems which result from exposure to toxic chemicals. { ¦kärd·ē·ō¦vas·kyə·lər tak'sis·əd·ē }

Cardiovirus [VIROL] A genus of viruses of the family Picornaviridae; consists of strains of encephalomyocarditis virus and mouse encephalomyelitis. { 'kär·dē·ō,vī·rəs }

carditis [MED] Inflammation of the heart tissues. { kär'dīd·əs }

Carettochelyidae [VERT ZOO] A family of reptiles in the order Chelonia containing only one species, the New Guinea plateless turtle (*Carettochelys insculpta*). { kə¦red·ō·kə'lī·ə,dē }

Cariamidae [VERT ZOO] The long-legged cariamas, a family of birds in the order Gruiformes. { ,kär·ē'am·ə,dē }

Caribosireninae [VERT ZOO] A subfamily of trichechiform sirenean mammals in the family Dugongidae. { ˌkar·əˌbäs·ə'ren·əˌnē }

Caridea [INV ZOO] A large section of decapod crustaceans in the suborder Natantia including many diverse forms of shrimps and prawns. { kə'rid·ē·ə }

caries [MED] **1.** Bone decay. **2.** Tooth decay. Also known as dental caries. { 'kar·ēz }

carina [BIOL] **1.** A ridge or a keel-shaped anatomical structure. **2.** *See* keel. { kə'rī·nə }

carinate [BIOL] Having a ridge or keel, as the breastbone of certain birds. { 'kar·əˌnāt }

Carinomidae [INV ZOO] A monogeneric family of littoral ribbonlike worms in the order Palaeonemertini. { ˌkär·ə'näm·əˌdē }

cariostatic [PHYSIO] Acting to halt bone or tooth decay. { ¦kar·ē·ə¦stad·ik }

carnassial [ANAT] Of or pertaining to molar or premolar teeth specialized for cutting and shearing. { kär'nas·ē·əl }

carnation ringspot virus [VIROL] The type species of the plant-virus genus *Dianthovirus*; symptoms of infection include leaf mottling and ringspotting, plant stunting and distortion, and flower distortion. Abbreviated CRSV. { kär¦nā·shən 'riŋˌspät ˌvī·rəs }

carnitine [BIOCHEM] $C_7H_{15}NO_3$ α-Amino-β-hydroxybutyric acid trimethylbetaine; a constituent of striated muscle and liver, identical with vitamin B_T. { 'kär·nəˌtēn }

Carnivora [VERT ZOO] A large order of placental mammals, including dogs, bears, and cats, that is primarily adapted for predation as evidenced by dentition and jaw articulation. { kär'niv·ə·rə }

carnivorous [BIOL] Eating flesh or, as in plants, subsisting on nutrients obtained from animal protoplasm. { kär'niv·ə·rəs }

carnivorous plant *See* insectivorous plant. { kär'niv·ə·rəs 'plant }

carnosine [BIOCHEM] $C_9H_{14}N_4O_3$ A colorless, crystalline dipeptide occurring in the muscle tissue of vertebrates. { 'kär·nəˌsēn }

Carolinian life zone [ECOL] A zone comprising the climate and biotic communities of the oak savannas of eastern North America. { ¦kar·ə¦lin·ē·ən 'līf ˌzōn }

carotenase [BIOCHEM] An enzyme that effects the hydrolysis of carotenoid compounds, used in bleaching of flour. { kə'rät·ənˌās }

carotene [BIOCHEM] $C_{40}H_{56}$ Any of several red, crystalline, carotenoid hydrocarbon pigments occurring widely in nature, convertible in the animal body to vitamin A, and characterized by preferential solubility in petroleum ether. Also known as carotin. { 'kar·əˌtēn }

carotenemia [MED] The presence of carotene in the blood; may cause yellowing of the skin. { ˌka'rä·tə'nēm·ē·ə }

carotenoid [BIOCHEM] A class of labile, easily oxidizable, yellow, orange, red, or purple pigments that are widely distributed in plants and animals and are preferentially soluble in fats and fat solvents. { kə'rät·ənˌȯid }

carotenol *See* xanthophyll. { kə'rät·ənˌȯl }

carotid artery [ANAT] Either of the two principal arteries on both sides of the neck that supply blood to the head and neck. Also known as common carotid artery. { kə'räd·əd 'ärd·ə·rē }

carotid body [ANAT] Either of two chemoreceptors sensitive to changes in blood chemistry which lie near the bifurcations of the carotid arteries. Also known as glomus caroticum. { kə'räd·əd ˌbäd·ē }

carotid ganglion [NEUROSCI] A group of nerve cell bodies associated with each carotid artery. { kə'räd·əd 'gaŋ·glēˌän }

carotid sinus [ANAT] An enlargement at the bifurcation of each carotid artery that is supplied with sensory nerve endings and plays a role in reflex control of blood pressure. { kə'räd·əd 'sī·nəs }

carotin *See* carotene. { 'kar·əˌtin }

carotol [BIOCHEM] $C_{15}H_{25}OH$ A sesquiterpenoid alcohol in carrots. { 'kar·əˌtȯl }

carp [VERT ZOO] The common name for a number of fresh-water, cypriniform fishes in the family Cyprinidae, characterized by soft fins, pharyngeal teeth, and a suckerlike mouth. { kärp }

carpal tunnel syndrome [MED] A condition caused by compression of the median nerve in the passage between the wrist and carpal bones; characterized by nocturnal pain, numbness, and tingling in the hand. { ¦kär·pəl ¦tən·əl 'sin ˌdrōm }

carpel [BOT] The basic specialized leaf of the female reproductive structure in angiosperms; a megasporophyll. { 'kär·pəl }

carpogonium [BOT] The basal, egg-bearing portion of the female reproductive organ in some thallophytes, especially red algae. { ˌkär·pə'gō·nē·əm }

carpology [BOT] The study of the morphology of fruit and seeds. { kär'päl·ə·jē }

carpophagous [ZOO] Feeding on fruits. { kär'pa·fə·gəs }

carpophore [BOT] The portion of a flower receptacle that extends between and attaches to the carpels. [MYCOL] The stalk of a fruiting body in fungi. { 'kär·pəˌfȯr }

carpophyte [BOT] A thallophyte that forms a sporocarp following fertilization. { 'kär·pəˌfīt }

carposporangium [BOT] In red algae, a sporangium that forms the cystocarp and contains carpospores. { ˌkär·pō·spə'ran·jē·əm }

carpospore [BOT] In red algae, a diploid spore produced terminally by a gonimoblast, giving rise to the diploid tetrasporic plant. { 'kär·pəˌspȯr }

carpus [ANAT] **1.** The wrist in humans or the corresponding part in other vertebrates. **2.** The eight bones of the human wrist. [INV ZOO] The fifth segment from the base of a generalized crustacean appendage. { 'kär·pəs }

carrageen [BOT] *Chondrus crispus*. A cartilaginous red algae harvested in the northern Atlantic as a source of carrageenan. Also known as Irish moss; pearl moss. { 'kar·əˌgēn }

carrier [GEN] An individual who is heterozygous for a recessive gene. [IMMUNOL] A protein to which a hapten becomes attached, thereby rendering the hapten immunogenic. [MED] A person who harbors and eliminates an infectious agent and so transmits it to others, but who may not show signs of the disease. { 'kar·ē·ər }

carrier culture [VIROL] A cell culture exhibiting a persistent infection; only a small fraction of the cell population is infected, but the viruses released when the infected cells die infect a small number of other cells. { 'kar·ē·ər ˌkəl·chər }

carrier molecule [IMMUNOL] An immunogenic molecule such as a foreign protein to which a hapten is coupled, thus enabling the hapten to induce an immune response. { 'kar·ē·ər 'mäl·əˌkyül }

carrier protein [CELL MOL] A membrane protein that transports a specific solute across the cell membrane by binding to the solute on one side of the membrane and then releasing it on the other. { 'kar·ē·ər ˌprōˌtēn }

Carrion's disease [MED] A bacterial infection of humans endemic in the Andes which is caused by *Bartonella bacilliformis* and attacks red blood cells and blood-forming organs. Also known as bartonellosis. { ¦kar·ē¦ōnz diˌzēz }

carrot [BOT] *Daucus carota.* A biennial umbellifer of the order Umbellales with a yellow or orange-red edible root. { 'kar·ət }

carrying capacity [ECOL] The maximum population size that the environment can support without deterioration. { 'kar·ē·iŋ kə'pas·əd·ē }

carsickness [MED] Motion sickness resulting from acceleratory movements of a train or automobile. { 'kärˌsik·nəs }

Carterinacea [INV ZOO] A monogeneric superfamily of marine, benthic foraminiferans in the suborder Rotaliina characterized by a test with monocrystal calcite spicules in a granular groundmass. { ˌkärd·ər·ə'nās·ē·ə }

cartilage [HISTOL] A specialized connective tissue which is bluish, translucent, and hard but yielding. { 'kärd·əl·ij }

cartilage bone [HISTOL] Bone formed by ossification of cartilage. { 'kärd·əl·ij 'bōn }

cartilaginous fish [VERT ZOO] The common name for all members of the class Chondrichthyes. { ¦kärd·əl¦aj·ə·nas 'fish }

caruncle [ANAT] Any normal or abnormal fleshy outgrowth, such as the comb and wattles of fowl or the mass in the inner canthus of the eye. [BOT] A fleshy outgrowth developed from the seed coat near the hilum in some seeds, such as the castor bean. { 'kaˌrəŋ·kəl }

Caryophanaceae [MICROBIO] Formerly a family of large, gram-negative bacteria belonging to the order Caryophanales and having disklike cells arranged in chains. { ¦kar·ē·ō·fə'nās·ēˌē }

Caryophanales [MICROBIO] Formerly an order of bacteria in the class Schizomycetes occurring as trichomes which produce short structures that function as reproductive units. { ˌkar·ēˌäf·ə'nā·lēz }

Caryophanon [MICROBIO] A genus of gram-positive, large, rod-shaped or filamentous bacteria of uncertain affiliation. { ˌkar·ē'äf·əˌnän }

Caryophyllaceae [BOT] A family of dicotyledonous plants in the order Caryophyllales differing from the other families in lacking betalains. { ¦kar·ē·ō·fə'lās·ēˌē }

Caryophyllales [BOT] An order of dicotyledonous plants in the subclass Caryophyllidae characterized by free-central or basal placentation. { ¦kar·ē·ō·fə'lā·lēz }

Caryophyllidae [BOT] A relatively small subclass of plants in the class Magnoliopsida characterized by trinucleate pollen, ovules with two integuments, and a multilayered nucellus. { ˌkar·ē·ō'fil·əˌdē }

caryopsis [BOT] A small, dry, indehiscent fruit having a single seed with such a thin, closely adherent pericarp that a single body, a grain, is formed. { ˌkar·ē'äp·səs }

casaba melon [BOT] *Cucumis melo.* A winter muskmelon with a yellow rind and sweet flesh belonging to the family Cucurbitaceae of the order Violales. { kəˌsäb·ə ˌmel·ən }

cascade [CELL MOL] A molecular system that is capable of self-propagation or amplification. { ka'skād }

cascade regulation [CELL MOL] **1.** In prokaryotes, a form of genetic regulation in which one operon codes for the production of an internal inducer that turns on one or more operons. **2.** In eukaryotes, a multistep model of genetic regulation involving mechanisms that interface with messenger ribonucleic acid formation, transport, and translation. { ka'skād ˌreg·yə'lā·shən }

caseation necrosis [PATH] Tissue death involving loss of cellular integrity with the consequent conversion to a cheeselike substance; typical in tuberculosis. { ˌkas·ē'ā·shən nə'krō·səs }

caseous lymphadenitis [VET MED] A chronic bacterial disease of sheep and goats caused by *Corynebacterium pseudotuberculosis*, characterized by caseation of the lymph glands and sometimes the lungs, liver, spleen, and kidneys. { 'kā·shəs limˌfad·ən'ī·dəs }

cashew [BOT] *Anacardium occidentale.* An evergreen tree of the order Sapindales grown for its kidney-shaped edible nuts and resinous oil. { 'kash·ü }

Casparian strip [BOT] A thin band of suberin- or lignin-like deposition in the radial and transverse walls of certain plant cells during the primary development phase of the endodermis. { ka'spar·ē·ən ˌstrip }

caspase [CELL MOL] Any of a family of intracellular proteins that mediate apoptosis. { 'kasˌpās }

cassava [BOT] *Manihot esculenta.* A shrubby perennial plant grown for its starchy, edible tuberous roots. Also known as manihot; manioc. { kə'sav·ə }

cassette [MYCOL] In yeast, any of the sites lying

in tandem that contain nucleotide sequences that can be substituted for one another. { kə'set }

Cassidulinacea [INV ZOO] A superfamily of marine, benthic foraminiferans in the suborder Rotaliina, characterized by a test of granular calcite with monolamellar septa. { ˌkas·əˌdü·lə'nās·ē·ə }

Cassiduloida [INV ZOO] An order of exocyclic Euechinoidea possessing five similar ambulacra which form petal-shaped areas (phyllodes) around the mouth. { ˌkas·ə·də'lȯid·ē·ə }

cassowary [VERT ZOO] Any of three species of large, heavy, flightless birds composing the family Casuariidae in the order Casuariiformes. { 'kas·əˌwer·ē }

cast [MED] **1.** A rigid dressing used to immobilize a part of the body. **2.** *See* strabismus. [PHYSIO] A mass of fibrous material or exudate having the form of the body cavity in which it has been molded; classified from its source, such as bronchial, renal, or tracheal. { kast }

castaneous [BIOL] Chestnut-colored. { ka 'stān·ē·əs }

caste [INV ZOO] One of the levels of mature social insects in a colony that carry out a specific function; examples are workers and soldiers. { kast }

Castle's intrinsic factor *See* intrinsic factor. { 'kas·əlz in¦trin·zik 'fak·tər }

Castniidae [INV ZOO] The castniids; large diurnal, butterflylike moths composing the single, small family of the lepidopteran superfamily Castnioidea. { ˌkast'nī·əˌdē }

Castnioidea [INV ZOO] A superfamily of neotropical and Indo-Australian lepidopteran insects in the suborder Heteroneura. { ˌkast·nī'ȯid·ē·ə }

castor bean [BOT] The seed of the castor oil plant (*Ricinus communis*), a coarse, erect annual herb in the spurge family (Euphorbiaceae) of the order Geraniales. { 'kas·tər ˌbēn }

castoreum gland [VERT ZOO] A preputial scent gland in the beaver. { ka'stȯr·ē·əm ˌgland }

castration [MED] Removing, or inhibiting the function or development of, the ovaries or testes. { ka'strā·shən }

casual carrier [MED] A person who carries an infectious microorganism but never manifests the disease. { 'kazh·ə·wəl 'kar·ē·ər }

Casuariidae [VERT ZOO] The cassowaries, a family of flightless birds in the order Casuariiformes lacking head and neck feathers and having bony casques on the head. { ˌkazh·əˌwa'rē·əˌdē }

Casuariiformes [VERT ZOO] An order of large, flightless, ostrichlike birds of Australia and New Guinea. { ˌkazh·əˌwa·rē·ə'fȯrˌmēz }

Casuarinaceae [BOT] The single, monogeneric family of the plant order Casuarinales characterized by reduced flowers and green twigs bearing whorls of scalelike, reduced leaves. { ˌkazh·əˌwa·rə'nās·ēˌē }

Casuarinales [BOT] A monofamilial order of dicotyledonous plants in the subclass Hamamelidae. { ˌkazh·əˌwa·rə'nā·lēz }

cat [VERT ZOO] The common name for all members of the carnivoran mammalian family Felidae, especially breeds of the domestic species, *Felis domestica*. { kat }

CAT *See* computerized tomography.

catabiosis [PHYSIO] Degenerative changes accompanying cellular senescence. { ˌkad·əˌbī'ō·səs }

catabolism [BIOCHEM] That part of metabolism concerned with the breakdown of large protoplasmic molecules and tissues, often with the liberation of energy. { kə'tab·əˌliz·əm }

catabolite [BIOCHEM] Any product of catabolism. { kə'tab·əˌlīt }

catabolite repression [BIOCHEM] An intracellular regulatory mechanism in bacteria whereby glucose, or any other carbon source that is an intermediate in catabolism, prevents formation of inducible enzymes. { kə'tab·əˌlīt ri'presh·ən }

catadromous [VERT ZOO] Pertaining to fishes which live in fresh water and migrate to spawn in salt water. { kə'ta·drə·məs }

catalase [BIOCHEM] An enzyme that catalyzes the decomposition of hydrogen peroxide into molecular oxygen and water. { 'kad·əlˌās }

catalectrotonus [PHYSIO] The negative electric potential during the passage of a current on the surface of a nerve or muscle in the region of the cathode. { ˌkad·əlˌek·'träd·ən·əs }

catalytic antibody [IMMUNOL] A large protein that is naturally produced by the immune system and has the capability of catalyzing a chemical reaction similarly to enzymes. Also known as abzyme. { ¦kad·əˌlid·ik 'an·tiˌbäd·ē }

catalytic site *See* active site. { ¦kad·əl¦id·ik ˌsīt }

catamount *See* puma. { 'kad·ə·maůnt }

cataplexy [MED] **1.** Sudden loss of muscle tone provoked by exaggerated emotion, such as excessive anger or laughter, often associated with a profound desire for sleep. **2.** Prostration by the sudden onset of disease. **3.** Hypnotic sleep. { 'kad·əˌplek·sē }

Catapochrotidae [INV ZOO] A monospecific family of coleopteran insects in the superfamily Cucujoidea. { ˌkad·ə·pə'kräd·əˌdē }

cataract [MED] An opacity in the crystalline lens or the lens capsule of the eye. { 'kad·əˌrakt }

catarobic [ECOL] Pertaining to a body of water characterized by the slow decomposition of organic matter, and oxygen utilization which is insufficient to prevent the activity of aerobic organisms. { ¦kad·ə¦rō·bik }

catarrh [MED] An old term for an inflammation of mucous membranes, particularly of the respiratory tract. { kə'tär }

catarrhal conjunctivitis [MED] A usually acute inflammation of the conjunctiva with smarting of the eyes, heaviness of the lids, photophobia, and excessive mucous or mucopurulent secretion, caused by a variety of contagious organisms, but sometimes becoming chronic as a sequela of the acute form or because of irritation from polluted atmosphere or allergic factors. Also known as pinkeye. { kə'tär·əl kən,jəŋk·ti'vīd·əs }

catarrhal jaundice *See* infectious hepatitis. { kə'tär·əl 'jȯn·dəs }

catecholamine [BIOCHEM] Any one of a group of sympathomimetic amines containing a catechol moiety, including especially epinephrine, norepinephrine (levarterenol), and dopamine. { ¦kad·ə'käl·ə,mīn }

category [SYST] In a hierarchical classification system, the level at which a particular group is ranked. { 'kad·ə,gȯr·ē }

catenin [BIOCHEM] Any of a family of 80–102-kilodalton proteins that are thought to have a major role in regulation of cell-to-cell adhesion, which is related to their interaction with E-cadherin and the actin cytoskeleton. { 'kat·ə·nən }

catenulate [BIOL] Having a chainlike form. { kə'ten·yə,lāt }

Catenulida [INV ZOO] An order of threadlike, colorless fresh-water rhabdocoeles with a simple pharynx and a single, median protonephridium. { kə,ten·yə'lī·də }

caterpillar [INV ZOO] **1.** The wormlike larval stage of a butterfly or moth. **2.** The larva of certain insects, such as scorpion flies and sawflies. { 'kad·ər,pil·ər }

caterpillar fungus *See* Cordyceps sinensis. { 'kat·ər,pil·ər ,fəŋ·gəs }

catfish [VERT ZOO] The common name for a number of fishes which constitute the suborder Siluroidei in the order Cypriniformes, all of which have barbels around the mouth. { 'kat,fish }

Catharanthus roseus [BOT] The Madagascar periwinkle plant from which the anticancer compounds vinblastine and vincristine are derived. { ,kath·ə,ran·thəs 'rō·zē·əs }

cathartic [PHARM] Any drug, such as castor oil, mineral oil, or a laxative, that causes defecation. { kə'thär·dik }

Cathartidae [VERT ZOO] The New World vultures, a family of large, diurnal predatory birds in the order Falconiformes that lack a voice and have slightly webbed feet. { kə'thär·də,dē }

cathepsin [BIOCHEM] Any of several proteolytic enzymes occurring in animal tissue that hydrolyze high-molecular-weight proteins to proteoses and peptones. { kə'thep·sən }

catheter [MED] A hollow, tubular device for insertion into a cavity, duct, or vessel to permit injection or withdrawal of fluids or to establish patency of the passageway. { 'kath·ə·dər }

catheterization [MED] Insertion or use of a catheter. { ,kath·əd·ə·rə'zā·shən }

catkin [BOT] An indeterminate type of inflorescence that resembles a scaly spike and sometimes is pendant. { 'kat·kən }

Catostomidae [VERT ZOO] The suckers, a family of cypriniform fishes in the suborder Cyprinoidei. { ,kad·ə'stäm·ə,dē }

CAT scan [MED] An image of a sectional view of a portion of the body made by computerized tomography. { 'kat ,skan }

CAT scanner [MED] An instrument consisting of integrated x-ray and computing equipment and used for computerized tomography. { 'kat ,skan·ər }

cat scratch disease [MED] A benign systematic illness in humans characterized by malaise, fever, and a granulomatous lymphadenitis; the causative organism has not been identified. Also known as benign lymphoreticulosis. { 'kat ¦skrach di,zēz }

cauda equina [NEUROSCI] The roots of the sacral and coccygeal nerves, collectively; so called because of their resemblance to a horse's tail. { 'kaủd·ə i'kwīn·ə }

caudal [ZOO] Toward, belonging to, or pertaining to the tail or posterior end. { 'kȯd·əl }

caudal artery [VERT ZOO] The extension of the dorsal aorta in the tail of a vertebrate. { 'kȯd·əl 'ärd·ə·rē }

caudal vertebra [ANAT] Any of the small bones of the vertebral column that support the tail in vertebrates; in humans, three to five are fused to form the coccyx. { 'kȯd·əl 'vər·tə·brə }

Caudata [VERT ZOO] An equivalent name for Urodela. { kaủ·dad·ə }

caudate [ZOO] **1.** Having a tail or taillike appendage. **2.** Any member of the Caudata. { 'kȯ,dāt }

caudate lobe [ANAT] The tailed lobe of the liver that separates the right extremity of the transverse fissure from the commencement of the fissure for the inferior vena cava. { 'kȯ,dāt 'lōb }

caudate nucleus [ANAT] An elongated arched gray mass which projects into and forms part of the lateral wall of the lateral ventricle. { 'kȯ,dāt 'nü·klē·əs }

caudex [BOT] The main axis of a plant, including stem and roots. { 'kȯ,deks }

caudicle [BOT] A slender appendage attaching pollen masses to the stigma in orchids. { 'kȯd·ə·kəl }

Caulerpaceae [BOT] A family of green algae in the order Siphonales. { ,kȯ·lər'pās·ē,ē }

Caulerpales *See* Siphonales. { ,kȯ·lər'pa,lēz }

caulescent [BOT] Having an aboveground stem. { kȯ'les·ənt }

cauliflorous [BOT] Producing flowers on the older branches or main stem. { ¦kȯl·ə¦flȯr·əs }

cauliflory [BOT] Of flowers, growth on the main stem of limbs of a tree. { 'kol·ə,flȯr·ē }

cauliflower [BOT] *Brassica oleracea* var. *botrytis*. A biennial crucifer of the order Capparales grown for its edible white head or curd, which is a tight mass of flower stalks. { 'kȯl·ə,flaủ·ər }

cauline [BOT] Belonging to or arising from the stem, particularly if on the upper portion. { 'kȯ,līn }

Caulobacter [MICROBIO] A genus of prosthecate bacteria; cells are rod-shaped, fusiform, or vibrioid and stalked, and reproduction is by binary fission of cells. { ,kȯl·ō'bak·tər }

Caulobacteraceae [MICROBIO] Formerly a family of aquatic, stalked, gram-negative bacteria in the order Pseudomonadales. { ¦kȯl·ə,bak·tə 'rās·ē,ē }

caulocarpic [BOT] Having stems that bear flowers and fruit every year. { ¦kȯl·ō¦kär·pik }

Caulococcus [MICROBIO] A genus of bacteria of uncertain affiliation; coccoid cells may be connected by threads; reproduces by budding. { ¦kȯl·ō¦käk·əs }

caulome [BOT] The stem structure or stem axis of a plant as a whole. { 'kȯ,lōm }

causalgia [MED] A sensation of burning pain, especially of the palms and soles, which may be of psychic or organic origin. { kȯ'zal·jē·ə }

cauterization [MED] Use of a device or chemical agent to coagulate or destroy tissue. { ,kȯd·ə·rə'zā·shən }

cautery [MED] Any agent or device used to coagulate or destroy tissue by means of heat, cold, electric current, or caustic chemicals. { 'kȯd·ə·rē }

caveolae [CELL MOL] Tiny indentations in the cell surface membrane which trap fluids during the process of micropinocytosis. { kə'vē·ə·lē }

cavernicolous [BIOL] Inhabiting caverns. { ¦kav·ər¦nik·ə·ləs }

cavernous sinus [ANAT] Either of a pair of venous sinuses of the dura mater located on the side of the body of the sphenoid bone. { 'kav·ər·nəs 'sī·nəs }

Caviidae [VERT ZOO] A family of large, hystricomorph rodents distinguished by a reduced number of toes and a rudimentary tail. { kə'vī·ə,dē }

CA virus *See* croup-associated virus. { sē'ā ,vī·rəs }

cavitation [PATH] The formation of one or more cavities in an organ or tissue, especially as the result of disease. { ,kav·ə'tā·shən }

cavity [BIOL] A hole or hollow space in an organ, tissue, or other body part. { 'kav·əd·ē }

cavy [VERT ZOO] Any of the rodents composing the family Caviidae, which includes the guinea pig, rock cavies, mountain cavies, capybara, salt desert cavy, and mara. { 'kā·vē }

CBP *See* CREB-binding protein.

C cells [HISTOL] Calcitonin-secreting cells located in the thyroid gland in mammals and in the ultimobranchial body in lower animals. { 'sē ,selz }

CCR5 [MED] Belonging to the seven-transmembrane chemokine receptor family, the major cofactor for primary macrophage-tropic human immunodeficiency virus-1 strains.

CDC gene *See* cell-division-cycle gene. { ¦sē ¦dē'sē jēn }

CD1 [IMMUNOL] Glycoproteins that are expressed on the surface of immature thymocytes, Langerhans cells, and certain B cells, similar in structure to class I major histocompatibility complex molecules.

CD4 [IMMUNOL] A T-cell signaling and/or costimulatory monomeric transmembrane glycoprotein involved in major histocompatibility complex II adhesion.

Cdk *See* cyclin-dependent kinase.

cDNA *See* complementary deoxyribonucleic acid.

Cebidae [VERT ZOO] The New World monkeys, a family of primates in the suborder Anthropoidea including the capuchins and howler monkeys. { 'seb·ə,dē }

Cebrionidae [INV ZOO] The robust click beetles, a family of cosmopolitan coleopteran insects in the superfamily Elateroidea. { ,seb·rē'än·ə,dē }

Cecropidae [INV ZOO] A family of crustaceans in the suborder Caligoida which are external parasites on fish. { sə'kräp·ə,dē }

cecum [ANAT] The blind end of a cavity, duct, or tube, especially the sac at the beginning of the large intestine. Also spelled caecum. { 'sē·kəm }

cedar [BOT] The common name for a large number of evergreen trees in the order Pinales having fragrant, durable wood. { 'sē·dər }

ceiba *See* kapok. { 'sā·bə }

Celastraceae [BOT] A family of dicotyledonous plants in the order Celastrales characterized by erect and basal ovules, a flower disk that surrounds the ovary at the base, and opposite or sometimes alternate leaves. { ,sel·ə'strās·ē,ē }

Celastrales [BOT] An order of dicotyledonous plants in the subclass Rosidae marked by simple leaves and regular flowers. { ,sel·ə'strā·lēz }

celery [BOT] *Apium graveolens* var. *dulce*. A biennial umbellifer of the order Umbellales with edible petioles or leaf stalks. { 'sel·rē }

celiac [ANAT] Of, in, or pertaining to the abdominal cavity. { 'sēl·ē,ak }

celiac syndrome [MED] A complex of symptoms produced by intestinal malabsorption of fat and marked by bulky, loose, foul-smelling stools, high in fatty acid content. Also known as idiopathic steatorrhea. { 'sēl·e,ək ,sin ,drōm }

cell [BIOL] The microscopic functional and structural unit of all living organisms, consisting of a nucleus, cytoplasm, and a limiting membrane. { sel }

cell adhesion molecule [CELL MOL] A class of membrane proteins comprising the outer surfaces of cell membranes in the developing nervous system that is thought to be intimately involved in guiding development during embryonic life. Abbreviated CAM. { ¦sel ad,hē·zhən 'mäl·ə·kyül }

cell-associated virus [VIROL] Virus particles that remain attached to or within the host cell after replication. { 'sel ə¦sō·shē,ād·əd 'vī·rəs }

cell body [NEUROSCI] The part of a nerve cell (neuron) containing the nucleus and several cytoplasmic structures such as mitochondria and neurofibrils. { 'sel ¦bäd·ē }

cell constancy [BIOL] The condition in which the entire body, or a part thereof, consists of a fixed number of cells that is the same for all adults of the species. { 'sel 'kän·stən·sē }

cell culture *See* tissue culture. { 'sel ,kəl·chər }

cell cycle [CELL MOL] In eukaryotic cells, the

cycle of events consisting of cell division, including mitosis and cytokinesis, and interphase. { 'sel ˌsī·kəl }

cell determination [PHYSIO] The process by which multipotential cells become committed to a particular development pathway. { ¦sel ˌdiˌtər·mə¦nā·shən }

cell differentiation [CELL MOL] The series of events involved in the development of a specialized cell having specific structural, functional, and biochemical properties. { 'sel dif·əˌren·chē'ā·shən }

cell division [CELL MOL] The process by which living cells multiply; may be mitotic or amitotic. { 'sel di'vizh·ən }

cell-division-cycle gene [CELL MOL] A gene that regulates the cell cycle. Also known as CDC gene. { 'sel dəˌvizh·ən ¦sī·kəl jēn }

cell fractionation [CELL MOL] A laboratory technique that uses differential centrifugation to separate the different components of the cell, resulting in nuclear, mitochondrial, microsomal, and soluble fractions. { ¦sel ˌfrak·shə'nā·shən }

cell-free extract [CELL MOL] A fluid obtained by breaking open cells; contains most of the soluble molecules of a cell. { 'sel ˌfrē 'ekˌstrakt }

cell inclusion [CELL MOL] A small, nonliving intracellular particle, usually representing a form of stored food, not immediately vital to life processes. { 'sel in'klü·zhan }

cell junction [CELL MOL] A specialized site on a cell at which it is attached to another cell or to the extracellular matrix. { 'sel ¦jəŋk·shen }

cell lineage [EMBRYO] The developmental history of individual blastomeres from their first cleavage division to their ultimate differentiation into cells of tissues and organs. { 'sel 'lin·yəj }

cell-lineage mutant [CELL MOL] Any mutation that affects the division of cells or the fates of their progeny. { ¦sel ˌlin·ē·əj 'myüt·ənt }

cell-mediated immunity [IMMUNOL] Immune responses produced by the activities of T cells rather than by immunoglobulins. { ¦sel ¦mē·dēˌād·əd i'myü·nəd·ē }

cell membrane [CELL MOL] A thin layer of protoplasm, consisting mainly of lipids and proteins, which is present on the surface of all cells. Also known as plasmalemma; plasma membrane. { 'sel ¦memˌbrān }

cell movement [CELL MOL] **1.** Intracellular movement of cellular components. **2.** The movement of a cell relative to its environment. { 'sel ˌmüv·mənt }

cellobiase [BIOCHEM] An enzyme that participates in the hydrolysis of cellobiose into glucose. { ˌsel·ō'bīˌās }

cell pathology [PATH] Abnormalities of the events taking place within cells. { 'sel pə'thäl·ə·jē }

cell permeability [CELL MOL] The permitting or activating of the passage of substances into, out of, or through cells. { 'sel pər·mē·ə'bil·əd·ē }

cell plate [CELL MOL] A membrane-bound disk formed during cytokinesis in plant cells which eventually becomes the middle lamella of the wall formed between daughter cells. { 'sel ˌplāt }

cell recognition [CELL MOL] The mutual recognition of cells, as expressed by specific cellular adhesion, due to a specific complementary interaction between molecules on adjacent cell surfaces. { 'sel ˌrek·əgˌnish·ən }

cell sap [CELL MOL] The liquid content of a plant cell vacuole. { 'sel ˌsap }

cells of Paneth [HISTOL] Coarsely granular secretory cells found in the crypts of Lieberkühn in the small intestine. Also known as Paneth cells. { ¦selz əv 'pän·ət }

cell-surface differentiation [CELL MOL] The specialization or modification of the cell surface. { 'sel ˌsər·fəs dif·əˌren·chē'ā·shən }

cell-surface ionization [PHYSIO] The presence of a negative charge on the surface of all living cells suspended in aqueous salt solutions at neutral pH. { 'sel ˌsər·fəs ˌī·ən·ə'zā·shən }

cell theory [BIOL] **1.** A principle that describes the cell as the fundamental unit of all living organisms. **2.** A principle that describes the properties of an organism as the sum of the properties of its component cells. { 'sel ˌthē·ə·rē }

cellular [BIOL] Characterized by, consisting of, or pertaining to cells. { 'sel·yə·lər }

cellular affinity [BIOL] The phenomenon of selective adhesiveness observed among the cells of certain sponges, slime molds, and vertebrates. { 'sel·yə·lər ə'fin·əd·ē }

cellular immunity [IMMUNOL] Immune responses carried out by active cells rather than by antibodies. { ¦sel·yə·lər i'myü·nəd·ē }

cellular immunology [IMMUNOL] The study of the cells of the lymphoid organs, which are the main agents of immune reactions in all vertebrates. { ¦sel·yə·lər ˌim·yə'näl·ə·jē }

cellular infiltration [MED] **1.** Passage of cells into tissues in the course of inflammation. **2.** Migration of or invasion by cells of neoplasms. { 'sel·yə·lər in·fil'trā·shən }

cellular slime molds [BIOL] A group of funguslike protozoa that form slimy aggregations on decaying organic matter; they differ from true slime molds in that the pseudoplasmodium is made of a group of separate cells rather than an amebalike mass. { 'sel·yə·lər 'slīm ˌmōlz }

cellulase [BIOCHEM] Any of a group of extracellular enzymes, produced by various fungi, bacteria, insects, and other lower animals, that hydrolyze cellulose. { 'sel·yəˌlās }

cellulitis [MED] Inflammation of connective tissue, especially the loose subcutaneous tissue. { ˌsel·yə'līd·əs }

cellulolytic [BIOL] Having the ability to hydrolyze cellulose; applied to certain bacteria and protozoans. { ¦sel·yə¦lid·ik }

Cellulomonas [MICROBIO] A genus of gram-positive, irregular rods in the coryneform group of bacteria; metabolism is respiratory; most strains produce acid from glucose, and cellulose is attacked by all strains. { ˌsel·yə'läm·ə·nəs }

cellulose [BIOCHEM] $(C_6H_{10}O_5)_n$ The main polysaccharide in living plants, forming the skeletal structure of the plant cell wall; a polymer of β-D-glucose units linked together, with the elimination of water, to form chains comprising 2000–4000 units. { 'sel·yə,lōs }

cell wall [CELL MOL] A semirigid, permeable structure that is composed of cellulose, lignin, or other substances and that envelops most plant cells. { ¦sel ¦wȯl }

Celyphidae [INV ZOO] A family of myodarian cyclorrhaphous dipteran insects in the subsection Acalypteratae. { se'lif·ə,dē }

cement [HISTOL] Calcified tissue which fastens the roots of teeth to the alveolus. Also known as cementum. [INV ZOO] Any of the various adhesive secretions, produced by certain invertebrates, that harden on exposure to air or water and are used to bind objects. { si'ment }

cement gland [INV ZOO] A structure in many invertebrates that produces cement. { si'ment ,gland }

cementum [HISTOL] *See* cement. [MED] A tissue closely resembling bone which covers the root of a tooth. { si'men·təm }

centesis [MED] Surgical puncture or perforation, as of a tumor or membrane. { ,sen'tē·səs }

centimorgan [GEN] A unit of genetic map distance, equal to the distance along a chromosome that gives a recombination frequency of 1%. { 'sent·ə,mȯrg·ən }

centipede [INV ZOO] The common name for an arthropod of the class Chilopoda. { 'sent·ə,pēd }

central apnea [MED] A pause in breathing lasting more than 10 seconds that is caused by a failure of commands from the brain. Also known as central sleep apnea. { ¦sen·trəl 'ap·nē·ə }

central apparatus [CELL MOL] The centrosome or centrosomes together with the surrounding cytoplasm. Also known as cytocentrum. { 'sen·trəl ,ap·ə'rad·əs }

central canal [ANAT] The small canal running through the center of the spinal cord from the conus medullaris to the lower part of the fourth ventricle; represents the embryonic neural tube. { 'sen·trəl kə'nal }

central deafness [MED] Deafness that results from some injury or failure to function in the central nervous system. { 'sen·trəl 'def·nəs }

central dogma [GEN] The concept, subject to several exceptions, that genetic information is coded in self-replicating deoxyribonucleic acid and undergoes unidirectional transfer to messenger ribonucleic acids in transcription that act as templates for protein synthesis in translation. { 'sen·trəl 'dȯg·mə }

Centrales [BOT] An order of diatoms (Bacillariophyceae) in which the form is often circular and the markings on the valves are radial. { sen 'trā·lēz }

central nervous system [NEUROSCI] The division of the vertebrate nervous system comprising the brain and spinal cord. { 'sen·trəl 'nər·vəs ,sis·təm }

central paralysis [MED] Paralysis due to a lesion of the brain or spinal cord. { 'sen·trəl pa'ral·ə·səs }

central placentation [BOT] Having the ovules located in the center of the ovary. { 'sen·trəl ,pla·sən'tā·shən }

central pocket loop [ANAT] A whorl type of fingerprint pattern having two deltas and at least one ridge that make a complete circuit. { ¦sen·trəl 'päk·ət ,lüp }

central sulcus [ANAT] A groove situated about the middle of the lateral surface of the cerebral hemisphere, separating the frontal from the parietal lobe. { 'sen·trəl 'səl·kəs }

Centrarchidae [VERT ZOO] A family of fishes in the order Perciformes, including the fresh-water or black basses and several sunfishes. { ,sen 'trär·kə,dē }

centric [ANAT] Having all teeth of both jaws meet normally with perfect distribution of forces in the dental arch. { 'sen·trik }

centrilobular emphysema [MED] A disorder marked by pulmonary inflation, primarily affecting the respiratory bronchioles and usually more severe in the upper lobes. { ¦sen·trə'lä·byə·lər ,em·fə'sē·mə }

centriole [CELL MOL] A complex cellular organelle forming the center of the centrosome in most cells; usually found near the nucleus in interphase cells and at the spindle poles during mitosis. { 'sen·trē,ōl }

Centrohelida [INV ZOO] An order of protozoans in the subclass Heliozoia lacking a central capsule and having axopodia or filopodia, and siliceous scales and spines. { ¦sen·trō'hel·ə·də }

centrolecithal ovum [CELL MOL] An egg cell having the yolk centrally located; occurs in arthropods. { ¦sen·tro'les·ə·thəl 'ō·vəm }

Centrolenidae [VERT ZOO] A family of arboreal frogs in the suborder Procoela characterized by green bones. { ,sen·trə'len·ə,dē }

centromere [CELL MOL] A specialized chromomere to which the spindle fibers are attached during mitosis. Also known as kinetochore; kinomere; primary constriction. { 'sen·trə,mir }

centromere distance [GEN] The distance of a gene from a centromere, measured in terms of recombination frequency. { 'sen·trə,mir ,dis·təns }

centromere effect [GEN] The reduced level of genetic recombination shown by genetic loci close to the centromere. { 'sen·trə,mir i,fekt }

centromere shift [GEN] A type of chromosomal defect in which the centromere changes position during chromosomal rearrangement in the G1 phase of the cell cycle. { 'sen·trə,mir ,shift }

centrosome [CELL MOL] A spherical hyaline region of the cytoplasm surrounding the centriole in many cells; plays a dynamic part in mitosis as the focus of the spindle pole. { 'sen·trə,sōm }

centrosome cycle [CELL MOL] Duplication of

the centrosome during interphase (S phase) of the animal cell cycle followed by separation of the resulting centrioles and associated microtubles at the beginning of mitosis to form the poles of the mitotic spindle. Following mitosis, each daughter cell has a new centrosome in association with its chromosomes. { 'sen·trə,sōm ,sī·kəl }

Centrospermae [BOT] An equivalent name for the Caryophyllales. { ,sen·trō'spər,mē }

Centrospermales [BOT] An equivalent name for the Caryophyllales. { ,sen·trō·spər'mā·lēz }

centrosphere [CELL MOL] The differentiated layer of cytoplasm immediately surrounding the centriole. { 'sen·trə,sfir }

centrum [ANAT] The main body of a vertebra. [BOT] The central space in hollow-stemmed plants. { 'sen·trəm }

cephalalgia [MED] Headache or head pain. { ,sef·ə'lal·jə }

Cephalaspidomorphi [VERT ZOO] An equivalent name for Monorhina. { ,sef·ə¦las·pə·də'mȯr·fī }

cephalic [ZOO] Of or pertaining to the head or anterior end. { sə'fal·ik }

cephalic vein [ANAT] A superficial vein located on the lateral side of the arm which drains blood from the radial side of the hand and forearm into the axillary vein. { sə'fal·ik 'vān }

cephalin [BIOCHEM] Any of several acidic phosphatides whose composition is similar to that of lecithin but having ethanolamine, serine, and inositol instead of choline; found in many living tissues, especially nervous tissue of the brain. { 'sef·ə·lən }

Cephalina [INV ZOO] A suborder of protozoans in the order Eugregarinida that are parasites of certain invertebrates. { sef·ə'lī·nə }

cephalization [ZOO] Anterior specialization resulting in the concentration of sensory and neural organs in the head. { ,sef·ə·lə'zā·shən }

Cephalobaenida [INV ZOO] An order of the arthropod class Pentastomida composed of primitive forms with six-legged larvae. { ,sef·ə·lō 'bēn·ə·də }

Cephaloboidea [INV ZOO] A superfamily of free-living nematodes in the order Rhabditida distinguished by cephalic elaborations or ornamentations. { 'sef·ə·lə'bȯid·ē·ə }

Cephalocarida [INV ZOO] A subclass of Crustacea erected to include the primitive crustacean *Hutchinsoniella macracantha*. { ,sef·ə·lō'kar·ə·də }

Cephalochordata [VERT ZOO] A subphylum of the Chordata comprising the lancelets, including *Branchiostoma*. { ¦sef·ə·lō,kȯr'däd·ə }

cephalodium [BOT] A small wart-like growth containing nitrogen-fixing cyanobacteria that is found on or in the thallus of some lichens with photobionts. { ,sef·ə'lō·dē·əm }

Cephaloidae [INV ZOO] The false longhorn beetles, a small family of coleopteran insects in the superfamily Tenebrionoidea. { ,sef·ə'lȯid·ē }

cephalomere [INV ZOO] One of the somites that make up the head of an arthropod. { sə'fal·ə,mir }

cephalont [INV ZOO] A sporozoan just prior to spore formation. { 'sef·ə,länt }

cephalopelvic disproportion [MED] A condition in which the fetus is unable to pass safely through the pelvis during labor because of pelvic contraction, an unfavorable fetal position, or a large fetal head in relation to pelvic size. Abbreviated CPD. { ¦sef·ə·lō¦pel·vik ,dis·prə'pȯr·shən }

Cephalopoda [INV ZOO] Exclusively marine animals constituting the most advanced class of the Mollusca, including squids, octopuses, and *Nautilus*. { ,sef·ə'läp·ə·də }

cephalosporin [MICROBIO] Any of a group of antibiotics produced by strains of the imperfect fungus *Cephalosporium*. { ,sef·ə·lə'spȯr·ən }

cephalothin [MICROBIO] An antibiotic derived from the fungus *Cephalosporium*, resembling penicillin units in structure and activity, and effective against many gram-positive cocci that are resistant to penicillin. { 'sef·ə·lə·thən }

cephalothorax [INV ZOO] The body division comprising the united head and thorax of arachnids and higher crustaceans. { ¦sef·ə·lə'thȯr ,aks }

Cephalothrididae [INV ZOO] A family of ribbonlike worms in the order Palaeonemertini. { ,sef·ə·lō'thrī·də,dē }

cephalotrichous flagellation [CELL MOL] Insertion of flagella in polar tufts. { ,sef·ə·lō'trī·kəs ,flaj·ə'lā·shən }

Cephidae [INV ZOO] The stem sawflies, composing the single family of the hymenopteran superfamily Cephoidea. { 'sef·ə,dē }

Cephoidea [INV ZOO] A superfamily of hymenopteran insects in the suborder Symphyta. { sa'fȯid·ē·ə }

Ceractinomorpha [INV ZOO] A subclass of sponges belonging to the class Demospongiae. { sə¦rak·tə·nə'mȯr·fə }

Cerambycidae [INV ZOO] The longhorn beetles, a family of coleopteran insects in the superfamily Chrysomeloidea. { se·rəm'bī·sə,dē }

cerambycoid larva [INV ZOO] A beetle larva that is morphologically similar to a caraboid larva except for the former's absence of legs. { sə'ram·bə,kȯid 'lar·və }

Ceramonematoidea [INV ZOO] A superfamily of marine nematodes in the order Desmodorida characterized by distinctive cuticular ornamentation, giving the appearance of a body covered with rings of crested tilelike plates. { ,ser·ə·mō ,nem·ə'tȯid·ē·ə }

Cerapachyinae [INV ZOO] A subfamily of predacious ants in the family Formicidae, including the army ant. { ,ser·ə·pə'kī·ə,nē }

Ceraphronidae [INV ZOO] A superfamily of hymenopteran insects in the superfamily Proctotrupoidea. { ,ser·ə'frän·ə,dē }

cerata [INV ZOO] Respiratory papillae of the mantle in certain nudibranchs. { sa'räd·ə }

cerate [PHARM] An unguent made of wax, resin, or spermaceti mixed with oil, lard, and medicinal ingredients; used for external application. { 'sir,āt }

ceratine [INV ZOO] A hornlike material secreted by some anthozoans. { 'ser·ə,tēn }

Ceratiomyxaceae [MYCOL] The single family of the fungal order Ceratiomyxales. { sə,rāsh·ē·ō,mik'sās·ē,ē }

Ceratiomyxales [MYCOL] An order of myxomycetous fungi in the subclass Ceratiomyxomycetidae. { sə,rāsh·ē·ō,mik 'sā·lēz }

Ceratiomyxomycetidae [MYCOL] A subclass of fungi belonging to the Myxomycetes. { sə,rāsh·ē·ō,mik·sō,mī'sed·ə, dē }

Ceratomorpha [VERT ZOO] A suborder of the mammalian order Perissodactyla including the tapiroids and the rhinoceratoids. { ,ser·ə·tō 'mȯr·fə }

Ceratophyllaceae [BOT] A family of rootless, free-floating dicotyledons in the order Nymphaeales characterized by unisexual flowers and whorled, cleft leaves with slender segments. { ,ser·ə·tō·fə'lās·ē,ē }

cercaria [INV ZOO] The larval generation which terminates development of a digenetic trematode in the intermediate host. { sər'kar·ē·ə }

cerci [ZOO] A pair of posterior abdominal appendages with delicate hairs found on many insects, including cockroaches and crickets, that are very sensitive to the air currents generated by a moving object. { 'sər·sē }

Cercopidae [INV ZOO] A family of homopteran insects belonging to the series Auchenorrhyncha. { sar'käp·ə,dē }

Cercopithecidae [VERT ZOO] The Old World monkeys, a family of primates in the suborder Anthropoidea. { ¦sər·kō·pə'thē·sə,dē }

cercopod [INV ZOO] **1.** Either of two filamentous projections on the posterior end of notostracan crustaceans. **2.** *See* cercus. { 'sər·kə ,päd }

Cercospora [MYCOL] A genus of imperfect fungi having dark, elongate, multiseptate spores. { sər'käs·pə·rə }

cercus [INV ZOO] Either of a pair of segmented sensory appendages on the last abdominal segment of many insects and certain other anthropods. Also known as cercopod. { 'sər·kəs }

cere [VERT ZOO] A soft, swollen mass of tissue at the base of the upper mandible through which the nostrils open in certain birds, such as parrots and birds of prey. { sir }

cerea flexibilitas [MED] The flexibility often present in catatonia in which the person's arm or leg remains in the position in which it is placed. { 'ser·ē·ə ,flek·sə'bil·ə,täs }

cereal [BOT] Any member of the grass family (Graminae) which produces edible, starchy grains usable as food by humans and livestock. Also known as grain. { 'sir·ē·əl }

cerebellar ataxia [MED] Incoordination of muscles due to disease of the cerebellum. { ,ser·ə'bel·ər ā'tak·sē·ə }

cerebellum [NEUROSCI] The part of the vertebrate brain lying below the cerebrum and above the pons, consisting of three lobes and concerned with muscular coordination and the maintenance of equilibrium. { ,ser·ə'bel·əm }

cerebral arteriosclerosis [MED] Hardening of the arteries of the brain, sometimes resulting in an organic mental disorder that may be primarily neurologic, primarily psychologic, or a combination of both. Also known as multiple infarct dementia. { sə'rē·brəl är,tir·ē·ō·sklə'rō·səs }

cerebral cortex [NEUROSCI] The superficial layer of the cerebral hemispheres, composed of gray matter and concerned with coordination of higher nervous activity. { sə'rē·brəl 'kȯr,teks }

cerebral hemisphere [NEUROSCI] Either of the two lateral halves of the cerebrum. { sə'rē·brəl 'hem·ə,sfir }

cerebral localization [NEUROSCI] Designation of a specific region of the brain as the area controlling a specific physiologic function or as the site of a lesion. { sə'rē·brəl lō·kə·lə'zā·shən }

cerebral palsy [MED] Any nonprogressive motor disorder in humans caused by brain damage incurred during fetal development. { sə'rē·brəl 'pȯl·zē }

cerebral peduncle [NEUROSCI] One of two large bands of white matter (containing descending axons of upper motor neurons) which emerge from the underside of the cerebral hemispheres and approach each other as they enter the rostral border of the pons. { sə'rē·brəl pi'dəŋ·kəl }

cerebrose *See* galactose. { 'ser·ə,brōs }

cerebroside [BIOCHEM] Any of a complex group of glycosides found in nerve tissue, consisting of a hexose, a nitrogenous base, and a fatty acid. Also known as galactolipid. { 'ser·ə·brō,sīd }

cerebroside lipoidosis *See* Gaucher's disease. { 'ser·ə·brō,sīd ,li'pȯi'dō·səs }

cerebrospinal axis [ANAT] The axis of the body composed of the brain and spinal cord. { sə¦rē·brō'spīn·əl 'ak·səs }

cerebrospinal fluid [PHYSIO] A clear liquid that fills the ventricles of the brain and the spaces between the arachnoid mater and pia mater. { sə¦rē·brō'spīn·əl 'flü·əd }

cerebrospinal meningitis [MED] Inflammation of the meninges of the brain and spinal cord. { sə¦rē·brō'spīn·əl ,men·ən'jīd·əs }

cerebrovascular accident [MED] A symptom complex resulting from cerebral hemorrhage, embolism, or thrombosis of the cerebral vessels, characterized by sudden loss of consciousness. { sa¦rē·brō'vas·kyə·lər 'ak·sə·dənt }

cerebrum [NEUROSCI] The enlarged anterior or upper part of the vertebrate brain consisting of two lateral hemispheres. { sə'rē·brəm }

Cerelasmidae [INV ZOO] A family of Psamminida with a soft test composed principally of organic cement; xenophyae, when present, are not systematically arranged. { ,ser·ə'las·mə,dē }

cerelose *See* glucose. { 'sir·ə,lōs }

Ceriantharia [INV ZOO] An order of the Zoantharia distinguished by the elongate form of the anemone-like body. { ¦ser·ē·ən'thar·ē·ə }

Ceriantipatharia [INV ZOO] A subclass proposed by some authorities to include the anthozoan orders Antipatharia and Ceriantharia. { ,sir·ē,an·tə·pə'thar·ē·ə }

Cerithiacea [INV ZOO] A superfamily of gastropod mollusks in the order Prosobranchia. { ˌser·əˌthī'ās·ē·ə }

cernuous [BOT] Drooping or inclining. { 'sərn·yə·wəs }

Cerophytidae [INV ZOO] A small family of coleopteran insects in the superfamily Elateroidea. { ˌser·ə'fīd·əˌdē }

certation [BOT] Competition in growth rate between pollen tubes of different genotypes resulting in unequal chances of accomplishing fertilization. { sər'tā·shən }

ceruloplasmin [BIOCHEM] The copper-binding serum protein in human blood. { sə¦rül·ō¦plaz·mən }

cerumen [PHYSIO] The waxy secretion of the ceruminous glands of the external ear. Also known as earwax. { sə'rü·mən }

ceruminous gland [ANAT] A modified sweat gland in the external ear that produces earwax. { sə'rü·mə·nəs ˌgland }

cervical [ANAT] Of or relating to the neck, a necklike part, or the cervix of an organ. { 'sər·və·kəl }

cervical canal [ANAT] Canal of the cervix of the uterus. { 'sər·və·kəl kə'nal }

cervical dysplasia [MED] Abnormal growth of the epithelial cells lining the cervix. Also known as cervical intraepithelial neoplasia. { ¦sər·və·kəl di'splāzh·yə }

cervical flexure [EMBRYO] A ventrally concave flexure of the embryonic brain occurring at the junction of hindbrain and spinal cord. { 'sər·və·kəl 'flek·shər }

cervical ganglion [NEUROSCI] Any ganglion of the sympathetic nervous system located in the neck. { 'sər·və·kəl 'gaŋ·glēˌän }

cervical plexus [NEUROSCI] A plexus in the neck formed by the anterior branches of the upper four cervical nerves. { 'sər·və·kəl 'plek·səs }

cervical sinus [EMBRYO] A triangular depression caudal to the hyoid arch containing the posterior visceral arches and grooves. { 'sər·və·kəl 'sī·nəs }

cervical vertebra [ANAT] Any of the bones in the neck region of the vertebral column; the transverse process has a characteristic perforation by a transverse foramen. { 'sər·və·kəl 'vərd·ə·brə }

cervicitis [MED] Inflammation of the cervix uteri. { ˌsər·və'sīd·əs }

Cervidae [VERT ZOO] A family of pecoran ruminants in the superfamily Cervoidea, characterized by solid, deciduous antlers; includes deer and elk. { 'sər·vəˌdē }

cervix [ANAT] A constricted or necklike portion of a structure. { 'sər·viks }

Cervoidea [VERT ZOO] A superfamily of tylopod ruminants in infraorder Pecora, including deer, giraffes, and related species. { sər'vȯid·ē·ə }

cesarean section [MED] Delivery of the fetus through an abdominal incision. { sə'zer·ē·ən 'sek·shən }

cespitose [BOT] **1.** Tufted; growing in tufts, as grass. **2.** Having short stems forming a dense turf. { 'ses·pəˌtōs }

Cestida [INV ZOO] An order of ribbon-shaped ctenophores having a very short tentacular axis and an elongated pharyngeal axis. { 'ses·tə·də }

Cestoda [INV ZOO] A subclass of tapeworms including most members of the class Cestoidea; all are endoparasites of vertebrates. { se'stō·də }

Cestodaria [INV ZOO] A small subclass of worms belonging to the class Cestoidea; all are endoparasites of primitive fishes. { ˌses·tə'dar·ē·ə }

Cestoidea [INV ZOO] The tapeworms, endoparasites composing a class of the phylum Platyhelminthes. { se'stȯid·ē·ə }

Cetacea [VERT ZOO] An order of aquatic mammals, including the whales, dolphins, and porpoises. { sē'tā·shə }

cetology [VERT ZOO] The study of whales. { sē'täl·ə·jē }

Cetomimiformes [VERT ZOO] An order of rare oceanic, deepwater, soft-rayed fishes that are structurally diverse. { ˌsēd·əˌmim·ə'fȯrˌmēz }

Cetomimoidei [VERT ZOO] The whalefishes, a suborder of the Cetomimiformes, including bioluminescent, deep-sea species. { ˌsēd·ə·mə 'mȯid·ēˌī }

Cetorhinidae [VERT ZOO] The basking sharks, a family of large, galeoid elasmobranchs of the isurid line. { ˌsēd·ə'rīn·əˌdē }

chaeta *See* seta. { 'kēd·ə }

Chaetodontidae [VERT ZOO] The butterflyfishes, a family of perciform fishes in the suborder Percoidei. { ˌkēd·ō'dän·təˌdē }

Chaetognatha [INV ZOO] A phylum of abundant planktonic arrowworms. { ˌkē'täg·nə·thə }

Chaetonotoidea [INV ZOO] An order of the class Gastrotricha characterized by two adhesive tubes connected with the distinctive paired, posterior tail forks. { ˌkēˌtän·ō'tȯid·ē·ə }

Chaetophoraceae [BOT] A family of algae in the order Ulotrichales characterized as branched filaments which taper toward the apices, sometimes bearing terminal setae. { ˌkēd·ō·fə'rās·ēˌē }

Chaetopteridea [INV ZOO] A family of spioniform polychaete annelids belonging to the Sedentaria. { ˌkēˌtäp·tə'rīd·ē·ə }

Chagas' disease [MED] An acute and chronic protozoan disease of humans caused by the hemoflagellate *Trypanosoma (Schizotrypanum) cruzi.* Also known as South American trypanosomiasis. { 'shäg·əs diˌzēz }

chainette [INV ZOO] In some cestodes, a longitudinal row of similar spines, usually with lateral winglike expansions, found near the base of the tentacles. { chā'net }

chalaza [BOT] The region at the base of the nucellus of an ovule; gives rise to the integuments. [CELL MOL] One of the paired, spiral, albuminous bands in a bird's egg that attach the yolk to the shell lining membrane at the ends of the egg. { kə'lāz·ə }

chalazion [MED] A small tumor of the eyelid

formed by retention of tarsal gland secretions. Also known as a Meibomian cyst. { kə'lāz·ē·ən }

chalazogamy [BOT] A process of fertilization in which the pollen tube passes through the chalaza to reach the embryo sac. { ˌkal·ə'zäg·ə·mē }

Chalcididae [INV ZOO] The chalcids, a family of hymenopteran insects in the superfamily Chalcidoidea. { kal'sid·əˌdē }

Chalcidoidea [INV ZOO] A superfamily of hymenopteran insects in the suborder Apocrita, including primarily insect parasites. { ˌkal·sə'dȯid·ē·ə }

chalice cell *See* goblet cell. { 'chal·əs ˌsel }

chalicosis [MED] A pulmonary affection caused by inhalation of stone dust. { ˌkal·ə'kō·səs }

chalkstone [PATH] A gouty deposit, usually of sodium urate, in the hands or feet. { 'chȯk ˌstōn }

challenge [IMMUNOL] Administration of an antigen to ascertain state of immunity. { 'chal·ənj }

chalones [BIOCHEM] Substances thought to be molecules of the protein-polypeptide class that are produced as part of the growth-control systems of tissues; known to inhibit cell division by acting on several phases in the mitotic cycle. { 'kaˌlōnz }

Chamaeleontidae [VERT ZOO] The chameleons, a family of reptiles in the suborder Sauria. { kəˌmēl·ē'än·təˌdē }

Chamaemyidae [INV ZOO] The aphid flies, a family of myodarian cyclorrhaphous dipteran insects of the subsection Acalypteratae. { ˌkam·ə'mī·əˌdē }

chamaephyte [ECOL] Any perennial plant whose winter buds are within 10 inches (25 centimeters) of the soil surface. { 'kam·əˌfīt }

Chamaesiphonales [BOT] An order of blue-green algae of the class Cyanophyceae; reproduce by cell division, colony fragmentation, and endospores. { ˌkam·ēˌsī·fə'nā·lēz }

chambered pith [BOT] A form of pith in which the parenchyma collapses or is torn during development, leaving the sclerenchyma plates to alternate with hollow zones. { ˌchām·bərd 'pith }

Chambersielloidea [INV ZOO] A superfamily of nematodes in the order Rhabdita characterized by highly unusual elaborations of the oral opening in the form of six mandibles. { chāmˌbər·sē·ə'lȯid·ē·ə }

chameleon [VERT ZOO] The common name for about 80 species of small to medium-size lizards composing the family Chamaeleontidae. { kə'mēl·yən }

chamois [VERT ZOO] *Rupicapra rupicapra.* A goatlike mammal included in the tribe Rupicaprini of the family Bovi-dae. { 'sham·ē }

chancre [MED] **1.** A lesion or ulcer at the site of primary inoculation by an infecting organism. **2.** The initial lesion of syphilis. { 'shaŋ·kər }

chancroid [MED] A lesion of the genitalia, usually of venereal origin, caused by *Hemophilus ducreyi.* Also known as soft chancre. { 'chaŋ ˌkrȯid }

Chanidae [VERT ZOO] A monospecific family of teleost fishes in the order Gonorynchiformes which contain the milkfish (*Chanos chanos*), distinguished by the lack of teeth. { 'kan·əˌdē }

channel protein [CELL MOL] Protein forming an aqueous pore spanning the lipid bilayer of the cell membrane which when open allows certain solutes to traverse the membrane. { 'chan·el ˌprōˌtēn }

Channidae [VERT ZOO] The snakeheads, a family of fresh-water perciform fishes in the suborder Anabantoidei. { 'kan·əˌdē }

Chaoboridae [INV ZOO] The phantom midges, a family of dipteran insects in the suborder Orthorrhapha. { ˌkā·ə'bȯr·əˌdē }

chaotropic [BIOCHEM] Having the ability to destabilize hydrogen bonding and hydrophobic interactions. { ˌkā·ə'träp·ik }

chaparral [ECOL] A vegetation formation characterized by woody plants of low stature, impenetrable because of tough, rigid, interlacing branches, which have simple, waxy, evergreen, thick leaves. { ¦shap·ə¦ral }

Characeae [BOT] The single family of the order Charales. { kə'rās·ēˌē }

Characidae [VERT ZOO] The characins, the single family of the suborder Characoidei. { kə'ras·əˌdē }

Characoidei [VERT ZOO] A suborder of the order Cypriniformes including fresh-water fishes with toothed jaws and an adipose fin. { ˌkar·ə'kȯid·ēˌī }

character convergence [ECOL] A process whereby two relatively evolved species interact so that one converges toward the other with respect to one or more traits. { 'kar·ik·tər kənˌvər·jəns }

character displacement [ECOL] An outcome of competition in which two species living in the same area have evolved differences in morphology or other characteristics that lessen competition for food resources. { 'kar·ik·tər dis'plās·mənt }

character progression [ECOL] The geographic gradation of expression of specific characters over the range of distribution of a race or species. { 'kar·ik·tər prəˌgresh·ən }

character stasis [GEN] Long-term constancy in a phenotypic character within a lineage. { 'kar·ik·tər ˌstā·səs }

Charadrii [VERT ZOO] The shore birds, a suborder of the order Charadriiformes. { kə'rad·rēˌī }

Charadriidae [VERT ZOO] The plovers, a family of birds in the superfamily Charadrioidea. { kə·rə'drī·əˌdē }

Charadriiformes [VERT ZOO] An order of cosmopolitan birds, most of which live near water. { kəˌrad·rē·ə'fȯrˌmēz }

Charadrioidea [VERT ZOO] A superfamily of the suborder Charadrii, including plovers, sandpipers, and phalaropes. { kəˌrad·rē'ȯid·ē·ə }

Charales [BOT] Green algae composing the single order of the class Charophyceae. { kə'rā·lēz }

Chareae [BOT] A tribe of green algae belonging to the family Characeae. { 'kar·ē,ē }

Charophyceae [BOT] A class of green algae in the division Chlorophyta. { ,kar·ə'fīs·ē,ē }

Charophyta [BOT] A group of aquatic plants, ranging in size from a few inches to several feet in height, that live entirely submerged in water. { kə'räf·əd·ə }

chartaceous [BOT] Resembling paper. { chär 'tā·shəs }

chartreusin [MICROBIO] $C_{18}H_{18}O_{18}$ Crystalline, greenish-yellow antibiotic produced by a strain of *Streptomyces chartreusis*; active against gram-positive microorganisms, acid-fast bacilli, and phage of *Staphylococcus pyogenes*. { shär'trü·zən }

chasmogamy [BOT] The production of a hermaphroditic floral type that opens at anthesis and may be visited by an insect vector, providing a means of cross-pollination. { kaz'mäg·ə·mē }

chasmophyte [ECOL] A plant that grows in rock crevices. { 'kaz·mə,fīt }

Chauffard-Still disease *See* Still's disease. { shō 'fär 'stil di,zēz }

chE *See* cholinesterase.

Cheadle's disease *See* infantile scurvy. { chēd·əlz diz'ēz }

check cross [GEN] The crossing of an unknown genotype with a phenotypically similar individual of known genotype. { 'chek ,krȯs }

check ligament [ANAT] A thickening of the orbital fascia running from the insertion of the lateral rectus muscle to the medial orbital wall (medial check ligament) or from the insertion of the lateral rectus muscle to the lateral orbital wall (lateral check ligament). { 'chek ,lig·ə·mənt }

checkpoint [CELL MOL] A point in the eukaryotic cell cycle at which the cycle may continue if specific conditions are present or will stop if conditions are not right. { 'chek,pȯint }

Chediak-Higashi anomaly [PATH] Deeply staining, coarse, peroxidase-positive granules in the cytoplasm of neutrophils and eosinophils in certain disease states. { 'ched·ē·ak ,hi'gä,shē ə,näm·ə·lē }

cheek [ANAT] The wall of the mouth in humans and other mammals. [ZOO] The lateral side of the head in submammalian vertebrates and in invertebrates. { chēk }

cheek pouch [VERT ZOO] A saclike dilation of the cheeks in certain animals, such as rodents, in which food is held. { 'chēk ,paủch }

cheetah [VERT ZOO] *Acinonyx jubatus*. A doglike carnivoran mammal belonging to the cat family, having nonretractile claws and long legs. { 'chēd·ə }

cheiloplasty [MED] Any plastic operation upon the lip. { 'kī·lō,plas·tē }

cheilosis [MED] Cracking at the corners of the mouth and scaling of the lips, usually associated with riboflavin deficiency. { kī'lō·səs }

Cheilostomata [INV ZOO] An order of ectoproct bryozoans in the class Gymnolaemata possessing delicate erect or encrusting colonies composed of loosely grouped zooecia. { ,kī·lə'stō·məd·ə }

cheiromegaly [MED] Enlargement of one or both hands that is not attributable to disease of the hypophysis. Also spelled chiromegaly. { ,kī·rə'meg·ə·lē }

cheiroplasty [MED] Any plastic operation performed on the hand. Also spelled chiroplasty. { 'kī·rə,plas·tē }

chela [INV ZOO] **1.** A claw or pincer on the limbs of certain crustaceans and arachnids. **2.** A sponge spicule with talonlike terminal processes. { 'kē·lə }

chelate [INV ZOO] Pertaining to an appendage with a pincerlike organ or claw. { 'kē,lāt }

chelicera [INV ZOO] Either appendage of the first pair in arachnids, usually modified for seizing, crushing, or piercing. { kə'lis·ə·rə }

Chelicerata [INV ZOO] A subphylum of the phylum Arthropoda; chelicerae are characteristically modified as pincers. { kə,lis·ə'räd·ə }

Chelidae [VERT ZOO] The side-necked turtles, a family of reptiles in the suborder Pleurodira. { 'kel·ə,dē }

chelifore [INV ZOO] Either of the first pair of appendages on the cephalic segment of pycnogonids. { 'kel·ə,fȯr }

cheliform [INV ZOO] Having a forcepslike organ formed by a movable joint closing against an adjacent segment; referring especially to a crab's claw. { 'kel·ə,fȯrm }

cheliped [INV ZOO] Either of the paired appendages bearing chelae in decapod crustaceans. { 'kel·ə,ped }

chellin *See* khellin. { 'kel·ən }

Chelonariidae [INV ZOO] A family of coleopteran insects in the superfamily Dryopoidea. { ke,län·ə'rī·ə,dē }

Chelonethida [INV ZOO] An equivalent name for the Pseudoscorpionida. { ,kel·ə'neth·ə·də }

Chelonia [VERT ZOO] An order of the Reptilia, subclass Anapsida, including the turtles, terrapins, and tortoises. { ke'lōn·ē·ə }

Cheloniidae [VERT ZOO] A family of reptiles in the order Chelonia including the hawksbill, loggerhead, and green sea turtles. { ,kel·ə'nī·ə,dē }

Cheluridae [INV ZOO] A family of amphipod crustaceans in the suborder Gammaridea. { kə'lür·ə,dē }

Chelydridae [VERT ZOO] The snapping turtles, a small family of reptiles in the order Chelonia. { kə'lid·rə,dē }

chemical burn [MED] Tissue destruction caused by caustic agents, irritant gases, or other chemical agents. { 'kem·i·kəl ,bərn }

chemical carcinogenesis [MED] The chemical-induced cancerous transformation of normal cells via a multistep process in which the genetic code of the cells is altered and then the altered cells are promoted to form tumors. { ¦kem·i·kəl ,kär·sə·nə'gen·ə·səs }

chemical ecology [ECOL] The study of ecological interactions mediated by the chemicals that organisms produce. { ¦kem·i·kəl ē'käl·ə·jē }

chemical meningitis [MED] Meningeal inflammation brought about by foreign irritants such as alcohol, detergents, chemotherapeutic agents, or contrast agents used in some radiologic imaging procedures. { ˌkem·ə·kəl ˌmen·in'jīd·əs }

chemical pathology [PATH] The study of disease by using chemical methods. { 'kem·i·kəl pə'thäl·ə·jē }

chemical sense [NEUROSCI] A process of the nervous system for reception of and response to chemical stimulation by excitation of specialized receptors. { 'kem·i·kəl 'sens }

chemiosmotic coupling [BIOCHEM] The mechanism by which adenosine diphosphate is phosphorylated to adenosine triphosphate in mitochondria and chloroplasts. { ¦kem·ē‚äs¦mäd·ik 'kəp·liŋ }

chemoautotroph [MICROBIO] Any of a number of autotrophic bacteria and protozoans which do not carry out photosynthesis. { ˌkē·mō‚ȯd·ə'träf·ik }

chemodectoma [MED] A benign tumor of the carotid body. { ¦kē·mō‚dek'tō·mə }

chemodifferentiation [EMBRYO] The process of cellular differentiation at the molecular level by which embryonic cells become specialized as tissues and organs. { ¦kē·mō‚dif·ə‚ren·chē'ā·shən }

chemoheterotroph [BIOL] An organism that derives energy and carbon from the oxidation of preformed organic compounds. { ¦kē·mō'hed·ə·rə‚träf }

chemokine [CELL MOL] A small (7–14 kilodaltons of soluble protein) chemoattractant cytokine produced by cells and tissues at the beginning of an immune system response to infection, allergen, injury, and so forth that controls the nature and magnitude of immune cell infiltration and inflammation at the affected site. { 'kē·mə‚kīn }

chemoorganotroph [BIOL] An organism that requires an organic source of carbon and metabolic energy. { ¦kē·mō‚ȯr'gan·ə‚träf }

chemoprevention [MED] Prevention of illness through pharmaceutical means; for example, use of drugs to arrest or reverse development of premalignant neoplasia. { ¦kē·mō·pri¦ven·shən }

chemoprophylaxis [MED] Use of drugs to prevent the development of infectious diseases. { ¦kē·mō‚prō·fə'lak·səs }

chemoreception [PHYSIO] Reception of a chemical stimulus by an organism. { ˌkē·mō·ri'sep·shən }

chemoreceptor [PHYSIO] Any sense organ that responds to chemical stimuli. { ˌkē·mō·ri'sep·tər }

chemosis [MED] An eye disorder characterized by swelling of the mucous membrane that covers the eyeball and lines the inner surface of the eyelids. { ke'mō·səs }

chemostat [MICROBIO] An apparatus, and a principle, for the continuous culture of bacterial populations in a steady state. { 'kē·mə‚stat }

chemosurgery [MED] Surgical removal of diseased or unwanted tissue by the application of chemicals. { ¦kē·mō¦sərj·ə·rē }

chemosynthesis [BIOCHEM] The synthesis of organic compounds from carbon dioxide by microorganisms using energy derived from chemical reactions. { ˌkē·mō'sin·thə·səs }

chemotaxin [BIOCHEM] A chemical that promotes movement of a cell or microorganism in the process of chemotaxis. { 'kē·mō‚tak·sən }

chemotaxis [BIOL] The orientation or movement of a motile organism with reference to a chemical agent. { ˌkē·mō'tak·səs }

chemotaxonomy [BOT] The classification of plants based on natural products. { ˌkē·mō‚tak'sän·ə·mē }

chemotherapeutic [PHARM] Any agent used for chemotherapy. { ˌkē·mō‚ther·ə‚pyüd·ik }

chemotherapeutic index [PHARM] The relationship between toxicity of a compound for the body and the toxicity for parasites. { ˌkē·mō‚ther·ə‚pyüd·ik 'in‚deks }

chemotherapy [MED] Administering chemical substances for treatment of disease, especially cancer and diseases caused by parasites. { ˌkē·mō'ther·ə·pē }

chemotropism [BIOL] Orientation response of a sessile organism with reference to chemical stimuli. { ˌkē·mō'trō‚piz·əm }

chemotypes [BOT] Plants of the same species that are chemically different but otherwise indistinguishable. { 'ē·mə‚tīps }

chenic acid *See* chenodeoxycholic acid. { 'kēn·ik 'as·əd }

chenodeoxycholic acid [BIOCHEM] $C_{24}H_{40}O_4$ A constituent of bile; needlelike crystals with a melting point of 119°C; soluble in alcohol, methanol, and acetic acid; used on an experimental basis to prevent and dissolve gallstones. Also known as anthropodesoxycholic acid; chenic acid; gallodesoxycholic acid. { ¦kē·nō‚dē‚äk·sē¦kō·lik 'as·əd }

Chenopodiaceae [BOT] A family of dicotyledonous plants in the order Caryophyllales having reduced, mostly greenish flowers. { ˌkē·nə‚pō·dē'ās·ē‚ē }

Chermidae [INV ZOO] A small family of minute homopteran insects in the superfamily Aphidoidea. { 'kər·mə·dē }

Cherminae [INV ZOO] A subfamily of homopteran insects in the family Chermidae; all forms have a beak and an open digestive tract. { 'kər·mə·nē }

cherry [BOT] **1.** Any trees or shrub of the genus *Prunus* in the order Rosales. **2.** The simple, fleshy, edible drupe or stone fruit of the plant. { 'cher·ē }

chersophyte [ECOL] A plant that grows in dry waste lands. { 'kərz·ə‚fīt }

chestnut [BOT] The common name for several species of large, deciduous trees of the genus *Castanea* in the order Fagales, which bear sweet, edible nuts. { 'ches‚nət }

chevron [VERT ZOO] The bone forming the hemal arch of a caudal vertebra. { 'shev·rən }

chevrotain [VERT ZOO] The common name for four species of mammals constituting the family Tragulidae in the order Artiodactyla. Also known as mouse deer. { 'shev·rə,tān }

Cheyne-Stokes respiration [MED] Breathing characterized by periods of hyperpnea alternating with periods of apnea; rhythmic waxing and waning of respiration; occurs most commonly in older patients with heart failure and cerebrovascular disease. { 'chān·ē 'stōks res·pə'rā·shən }

chiasma [ANAT] A cross-shaped point of intersection of two parts, especially of the optic nerves. [CELL MOL] The point of junction and fusion between paired chromatids or chromosomes, first seen during diplotene of meiosis. { kī'az·mə }

Chiasmodontidae [VERT ZOO] A family of deep-sea fishes in the order Perciformes. { kī,az·mə'dän·tə,dē }

chicken [VERT ZOO] *Galus galus*. The common domestic fowl belonging to the order Galliformes. { 'chik·ən }

chickenpox [MED] A mild, highly infectious viral disease of humans caused by a herpesvirus and characterized by vesicular rash. Also known as varicella. { 'chik·ən,päks }

chick unit [BIOL] A unit for the standardization of pantothenic acid. { 'chick ,yü·nət }

chicory [BOT] *Cichorium intybus*. A perennial herb of the order Campanulales grown for its edible green leaves. { 'chik·ə·rē }

chief cell [HISTOL] **1.** A parenchymal, secretory cell of the parathyroid gland. **2.** A cell in the lumen of the gastric fundic glands. { ¦chēf 'sel }

chi element [GEN] Any of the special sites in bacterial deoxyribonucleic acid near which enhancement of genetic recombination occurs. { 'kī ,el·ə·mənt }

chigger [INV ZOO] The common name for bloodsucking larval mites of the Trombiculidae which parasitize vertebrates. { 'chig·ər }

chilarium [INV ZOO] One of a pair of processes between the bases of the fourth pair of walking legs in the king crab. { kī'lar·ē·əm }

Chilopoda [INV ZOO] The centipedes, a class of the Myriapoda that is exclusively carnivorous and predatory. { ,kī'läp·ə·də }

Chimaeridae [VERT ZOO] A family of the order Chimaeriformes. { kī'mir·ə,dē }

Chimaeriformes [VERT ZOO] The single order of the chondrichthyan subclass Holocephali comprising the ratfishes, marine bottom-feeders of the Atlantic and Pacific oceans. { kī·mir·ə'fór,mēz }

chimera [BIOL] An organism or a part made up of tissues or cells exhibiting chimerism. { kī'mir·ə }

chimeric deoxyribonucleic acid [GEN] A recombinant deoxyribonucleic acid (DNA) molecule that contains sequences from more than one organism. [CELL MOL] A deoxyribonucleic acid (DNA) molecule that has resulted from recombination or from the splicing together of DNA from two sources. { kī'mir·ik ,dē¦äk·se¦rī·bō¦nü¦klē·ik 'as·əd }

chimerism [BIOL] The admixture of cell populations from more than one zygote. { 'kī·mə,riz·əm }

chimopelagic [ECOL] Pertaining to, belonging to, or being marine organisms living at great depths throughout most of the year; during the winter they move to the surface. { ¦kī·mō·pə'laj·ik }

chimpanzee [VERT ZOO] Either of two species of Primates of the genus *Pan* indigenous to central-west Africa. { ,chim,pan'zē }

chin [ANAT] The lower part of the face, at or near the symphysis of the lower jaw. { chin }

China grass *See* ramie. { 'chī·nə ,gras }

chinchilla [VERT ZOO] The common name for two species of rodents in the genus *Chinchilla* belonging to the family Chinchillidae. { chin 'chil·ə }

Chinchillidae [VERT ZOO] A family of rodents comprising the chinchillas and viscachas. { chin'chil·ə,dē }

chinook salmon [VERT ZOO] *Oncorhynchus tshawytscha*. The Pacific's largest salmon, possibly exceeding 46 kilograms (100 pounds) at maturity, often spawns in tributaries located a considerable distance from the ocean. Also known as king salmon. { shə¦nu̇ 'sa·mən }

Chionididae [VERT ZOO] The white sheathbills, a family of birds in the order Charadriiformes. { ,kī·ə'nid·ə,dē }

chionophile [ECOL] Having a preference for snow. { ,kī'än·ə,fīl }

chip blower [MED] A dental instrument used to blow drilling debris from a tooth cavity that is being prepared for filling. { 'chip ,blō·ər }

chipmunk [VERT ZOO] The common name for 18 species of rodents belonging to the tribe Marmotini in the family Sciuridae. { 'chip,məŋk }

Chiridotidae [INV ZOO] A family of holothurians in the order Apodida having six-spoked, wheel-shaped spicules. { ,kī·rə'däd·ə,dē }

chiromegaly *See* cheiromegaly. { ,kī·rə'meg·ə·lē }

chiroplasty *See* cheiroplasty. { 'kī·rə,plas·tē }

chiropodist [MED] One who treats minor ailments of the feet. Also known as podiatrist. { kə'räp·əd·əst }

chiropractic [MED] A system of therapeutics based upon the theory that disease is caused by abnormal function of the nervous system; attempts to restore normal function are made through manipulation and treatment of the structures of the body, especially those of the spinal column. { 'kī·rə,prak·tik }

chiropractor [MED] One who practices the chiropractic arts. { 'kī·rə,prak·tər }

Chiroptera [VERT ZOO] The bats, an order of mammals having the front limbs modified as wings. { kī'räp·tə·rə }

chiropterophilous [BIOL] Pollinated by bats. { kī¦räp·tə¦räf·ə·ləs }

chiropterygium [VERT ZOO] A typical vertebrate limb, thought to have evolved from a finlike appendage. { ¦kī·räp·tə¦rij·ē·əm }

Chirotheuthidae [INV ZOO] A family of mollusks comprising several deep-sea species of squids. { ˌkī·rō'thyüd·əˌdē }

chitin [BIOCHEM] A white or colorless amorphous polysaccharide that forms a base for the hard outer integuments of crustaceans, insects, and other invertebrates. { 'kīt·ən }

chitinase [BIOCHEM] An externally secreted digestive enzyme produced by certain microorganisms and invertebrates that hydrolyzes chitin. { 'kīt·ənˌās }

chitinivorous bacterium [MICROBIO] Any bacterium which secretes chitinase and can digest chitin; organisms extract chitin from lobster exoskeletons, causing an infection called soft-shell disease. { ˌkīt·ən'iv·ə·rəs bak'tir·ē·əm }

chiton [INV ZOO] The common name for over 600 extant species of mollusks which are members of the class Polyplacophora. { 'kīt·ən }

Chitral fever *See* phlebotomus fever. { 'chi·trəl ˌfē·vər }

chlamydeous [BOT] **1.** Pertaining to the floral envelope. **2.** Having a perianth. { klə'mid·ē·əs }

Chlamydia [MICROBIO] The single genus of the family Chlamydiaceae. { klə'mid·ē·ə }

Chlamydiaceae [MICROBIO] The single family of the order Chlamydiales; characterized by a developmental cycle from a small elementary body to a larger initial body which divides, with daughter cells becoming elementary bodies. { kləˌmid·ē'ās·ēˌē }

Chlamydiales [MICROBIO] An order of coccoid rickettsias; gram-negative, obligate, intracellular parasites of vertebrates. { kləˌmid·ē'ā·lēz }

Chlamydobacteriaceae [MICROBIO] Formerly a family of gram-negative bacteria in the order Chlamydobacteriales possessing trichomes in which false branching may occur. { ¦klam·əˌdō ˌbak·tir·ē'ās·ēˌē }

Chlamydobacteriales [MICROBIO] Formerly an order comprising colorless, gram-negative, algae-like bacteria of the class Schizomycetes. { ¦klam·əˌdōˌbak·tir·ē'ā·lēz }

Chlamydomonadidae [INV ZOO] A family of colorless, flagellated protozoans in the order Volvocida considered to be close relatives of protozoans that possess chloroplasts. { ¦klam·əˌdō·mə'näd·əˌdē }

Chlamydoselachidae [VERT ZOO] The frilled sharks, a family of rare deep-water forms having a combination of hybodont and modern shark characteristics. { ¦klam·əˌdō·se'lak·əˌdē }

chlamydospore [MYCOL] A thick-walled, unicellar resting spore developed from vegetative hyphae in almost all parasitic fungi. { klə 'mid·əˌspȯr }

Chlamydozoaceae [MICROBIO] A family of small, gram-negative, coccoid bacteria in the order Rickettsiales; members are obligate intracytoplasmic parasites, or saprophytes. { ¦klam·əˌdō·zō'ās·ēˌē }

chloasma [MED] Patchy tan, brown, and black hyperpigmentation, especially on the brow and cheek; of unknown cause, but may be due to the action of sunshine upon perfume or to endocrinopathy. { klō'az·mə }

Chloracea [MICROBIO] The green sulfur bacteria, a family of photosynthetic bacteria in the suborder Rhodobacteriineae. { klȯr'ās·ē·ə }

chloracne [MED] An acnelike eruption caused by chlorinated hydrocarbons. { klȯr'ak·nē }

chloragogen [INV ZOO] Of or pertaining to certain specialized cells forming the outer layer of the alimentary tract in earthworms and other annelids. { ¦klȯr·ə¦gō·jən }

chloramphenicol [MICROBIO] $C_{11}H_{12}O_2N_2Cl_2$ A colorless, crystalline, broad-spectrum antibiotic produced by *Streptomyces venezuelae*; industrial production is by chemical synthesis. Also known as chloromycetin. { ˌklȯrˌam'fen·əˌkȯl }

Chlorangiaceae [BOT] A primitive family of colonial green algae belonging to the Tetrasporales in which the cells are directly attached to each other. { ˌklȯrˌan·jē'ās·ēˌē }

chloranthy [BOT] A reverting of normally colored floral leaves or bracts to green foliage leaves. { 'klȯrˌan·thē }

chlorcyclizine hydrochloride [PHARM] $ClC_6H_4CH(C_6H_5)C_4H_8N_2CH_3 \cdot HCl$ A white, crystalline solid with a melting point of 222–227°C; soluble in water, chloroform, and alcohol; used as an antihistamine. { klȯr'sī·kləˌzēn hī·drə 'klȯrˌīd }

chlordiazepoxide hydrochloride [PHARM] $C_{16}H_{14}ON_3Cl$ A white crystalline conpound, soluble in water; the hydrochloride salt is used as a tranquilizer. { ¦klȯr·dīˌaz·ə'pakˌsīd ˌhī·drə 'klȯrˌīd }

chlorenchyma [BOT] Chlorophyll-containing tissue in parts of higher plants, as in leaves. { klȯr'eŋ·kə·mə }

chloride shift [PHYSIO] The reversible exchange of chloride and bicarbonate ions between erythrocytes and plasma to effect transport of carbon dioxide and maintain ionic equilibrium during respiration. { 'klȯrˌīd ˌshift }

chlorin [BIOCHEM] A saturated porphyrin for which one double bond at a single pyrrole ring has been reduced. { 'klȯr·ən }

Chlorobacteriaceae [MICROBIO] The equivalent name for Chlorobiaceae. { ¦klȯr·ōˌbak·tir·ē'ās·ēˌē }

Chlorobiaceae [MICROBIO] A family of bacteria in the suborder Chlorobiineae; cells are nonmotile and contain bacteriochlorophylls *c*, *d*, or *e* in chlorobium vesicles attached to the cytoplasmic membrane. { ˌklȯr·ōˌbī'ās·ēˌē }

Chlorobiineae [MICROBIO] The green sulfur bacteria, a suborder of the order Rhodospirillales; contains the families Chlorobiaceae and Chloroflexaceae. { ˌklȯr·ōˌbī'i·əˌnē }

Chlorobium [MICROBIO] A genus of bacteria in the family Chlorobiaceae; cells are ovoid, rod- or vibrio-shaped, and nonmotile, do not have gas vacuoles, contain bacteriochlorophyll *c* or *d*, and are free-living. { klȯr'ō·bē·əm }

chlorobium chlorophyll [BIOCHEM] $C_{51}H_{67}O_4N_4Mg$ Either of two spectral forms of chlorophyll occurring as esters of farnesol in certain (*Chlorobium*) photosynthetic bacteria. Also known as bacterioviridin. { klȯr'ō·bē·əm 'klȯr·ə,fil }

Chlorococcales [BOT] A large, highly diverse order of unicellular or colonial, mostly freshwater green algae in the class Chlorophyceae. { ,klȯr·ō·kä'kā·lēz }

chlorocruorin [BIOCHEM] A green metalloprotein respiratory pigment found in body fluids or tissues of certain sessile marine annelids. { ¦klȯr·ə'krȯr·ən }

Chlorodendrineae [BOT] A suborder of colonial green algae in the order Volvocales comprising some genera with individuals capable of detachment and motility. { ,klȯr·ō,den'drin·ē,ē }

Chloroflexaceae [MICROBIO] A family of phototrophic bacteria in the suborder Chlorobiineae; cells possess chlorobium vesicles and bacteriochlorophyll *a* and *c*, are filamentous, and show gliding motility; capable of anaerobic, phototrophic growth or aerobic chemotrophic growth. { ¦klȯr·ō·fleks'as·ē,ē }

Chloroflexus [MICROBIO] The single genus of the family Chloroflexaceae. { ,klȯr·ō'flek,sēz }

chlorogenic acid [BIOCHEM] $C_{16}H_{18}O_9$ An important factor in plant metabolism; isolated from green coffee beans; the hemihydrate crystallizes in needlelike crystals from water. { ¦klȯr·ə¦jen·ik 'as·əd }

chloroguanide hydrochloride [PHARM] $C_{11}H_{16}N_5Cl$ A very effective suppressive drug in low doses, against the three kinds of malaria. { ¦klȯr·ə¦guän,īd ,hī·drə'klȯr,īd }

chloroma [MED] A focal tumorous proliferation of granulocytes, with or without the blood findings of granulocytic leukemia; the sectioned surfaces of the mass are green. { klə'rō·mə }

chloromethapyrilene citrate *See* chlorothen citrate. { ¦klȯr·ō,meth·ə'pī·rə,lēn 'sī,trāt }

Chloromonadida [INV ZOO] An order of flattened, grass-green or colorless, flagellated protozoans of the class Phytamastigophorea. { ¦klȯr·ō·mə'näd·ə·də }

Chloromonadina [INV ZOO] The equivalent name for Chloromonadida. { ¦klȯr·ō·mə'näd·ə·nə }

Chloromonadophyceae [BOT] A group of algae considered by some to be a class of the division Chrysophyta. { ¦klȯr·ō·mə,näd·ə'fīs·ē,ē }

Chloromonadophyta [BOT] A division of algae in the plant kingdom considered by some to be a class, Chloromondophyceae. { ¦klȯr·ō,mä·nə'dä·fə·də }

chloromycetin *See* chloramphenicol. { ,klȯr·ō,mī'sēt·ən }

***para*-chlorophenol** [PHARM] C_6H_5ClO Crystals with a typical phenol odor and a melting point of 43.2–43.7°C; soluble in alcohol, ether, glycerol, and chloroform; used as a topical antiseptic. { ¦par·ə ¦klȯr·ə'fen,ōl }

Chlorophyceae [BOT] A class of microscopic or macroscopic green algae, division Chlorophyta, composed of fresh- or salt-water, unicellular or multicellular, colonial, filamentous or sheetlike forms. { ,klȯr·ō'fīs·ē,ē }

chlorophyll [BIOCHEM] The generic name for any of several oil-soluble green tetrapyrrole plant pigments which function as photoreceptors of light energy for photosynthesis. { 'klȯr·ə,fil }

chlorophyll a [BIOCHEM] $C_{55}H_{72}O_5N_4Mg$ A magnesium chelate of dihydroporphyrin that is esterified with phytol and has a cyclopentanone ring; occurs in all higher plants and algae. { 'klȯr·ə,fil 'ā }

chlorophyllase [BIOCHEM] An enzyme that splits or hydrolyzes chlorophyll. { 'klȯr·ə·fə,lās }

chlorophyll b [BIOCHEM] $C_{55}H_{70}O_6N_4Mg$ An ester similar to chlorophyll *a* but with a $-CHO$ substituted for a $-CH_3$; occurs in small amounts in all green plants and algae. { 'klȯr·ə,fil 'bē }

Chlorophyta [BOT] The green algae, a highly diversified plant division characterized by chloroplasts, having chlorophyll *a* and *b* as the predominating pigments. { klō'räf·ə·də }

Chloropidae [INV ZOO] The chloropid flies, a family of myodarian cyclorrhaphous dipteran insects in the subsection Acalypteratae. { klō'räp·ə,dē }

chloroplast [BOT] A type of cell plastid occurring in the green parts of plants, containing chlorophyll pigments, and functioning in photosynthesis and protein synthesis. { 'klȯr·ə,plast }

chloroplast deoxyribonucleic acid [BIOCHEM] The circular deoxyribonucleic acid duplex, generally 40 to 80 copies, contained within a chloroplast. Abbreviated ctDNA. Also known as chloroplast genome. { ¦klȯr·ə,plast dē,äk·sē,rī·bō·nü¦klē·ik 'as·əd }

chloroplast genome *See* chloroplast deoxyribonucleic acid. { 'klȯr·ə,plast 'jē,nōm }

Chloropseudomonas [MICROBIO] An invalid genus of bacteria; originally described as a member of the family Chlorobiaceae, with motile rod-shaped cells, without gas vacuoles, containing bacteriochlorophyll *c*, and capable of photoheterotrophic growth; cultures now known to be symbiotic of one of a number of typical Chlorobiaceae and a chemoorganotrophic bacterium capable of reducing elemental sulfur to sulfide. { ,klȯr·ō,süd·ə'mō·nəs }

chloropsia [MED] A defect of vision in which all objects appear green. { klō'räp·sē·ə }

6-chloropurine [PHARM] $C_5H_3ClN_4$ A compound crystallizing as blunt needles from water; crystals decompose at 175–177°C; soluble in dimethylformamide and ether; used as an antineoplastic drug. { ¦siks ,klȯr·ō'pyu̇r,ēn }

chlorosis [MED] A form of macrocytic anemia in young females characterized by marked reduction in hemoglobin and a greenish skin color. { klə'rō·səs }

chlorothen citrate [PHARM] $C_{14}H_{18}ClN_3S \cdot C_6H_8O_7$ A white, crystalline powder with a melting range of 112–116°C; used in medicine. Also known as chloromethapyrilene citrate. { 'klȯr·ə·thən 'sī,trāt }

chlorpromazine [PHARM] $C_{17}H_{19}ClN_2S$ A gray-white, crystalline compound used as a sedative and in preventing or relieving nausea and vomiting. { klȯr'prō·mə‚zēn }

chlorpropamide [PHARM] $C_3H_7NHCONHSO_2$-C_6H_4Cl A crystalline compound with a melting point of 127–129°C; soluble in alcohol; used in the treatment of diabetes. { klȯr'prō·prə‚mīd }

chlortetracycline [MICROBIO] $C_{22}H_{23}O_8N_2Cl$ Yellow, crystalline, broad-spectrum antibiotic produced by a strain of *Streptomyces aureofaciens*. { ¦klȯr‚te·trə'sī‚klēn }

chlorzoxazone [PHARM] $C_7H_4ClNO_2$ Crystals with a melting point of 191–191.5°C; soluble in aqueous solutions of alkali hydroxides and in ammonia; used as a skeletal muscle relaxant. { klȯr'zäks·ə‚zōn }

choana [ANAT] A funnel-shaped opening, especially the posterior nares. [INV ZOO] A protoplasmic collar surrounding the basal ends of the flagella in certain flagellates and in the choanocytes of sponges. { 'kō·ə·nə }

Choanichthyes [VERT ZOO] An equivalent name for the Sarcopterygii. { ‚kō·ə'nik·thē‚ēz }

choanocyte [INV ZOO] Any of the choanate, flagellate cells lining the cavities of a sponge. Also known as collar cell. { kō'an·ə‚sīt }

Choanoflagellida [INV ZOO] An order of single-celled or colonial, colorless flagellates in the class Zoomastigophorea; distinguished by a thin protoplasmic collar at the anterior end. { ¦kō·ə·nō·flə'jel·ə·də }

Choanolaimoidea [INV ZOO] A superfamily of marine nematodes in the order Chromadoria distinguished by a complex stoma in two parts. { ¦kō·ə·nō·lə'mȯid·ē·ə }

choanosome [INV ZOO] The inner layer of a sponge; composed of choanocytes. { kō'an·ə‚sōm }

chocolate tree *See* cacao. { 'chäk·lət ‚trē }

choked disk *See* papilledema. { ¦chōkt 'disk }

cholagogic [PHYSIO] Inducing the flow of bile. { ¦käl·ə¦gäj·ik }

cholagogue [PHARM] Any agent that causes an increased flow of bile into the intestine. { 'kō·lə‚gäg }

cholaic acid *See* taurocholic acid. { kō'lā·ik 'as·əd }

cholane [BIOCHEM] $C_{24}H_{42}$ A tetracyclic hydrocarbon which may be considered as the parent substance of sterols, hormones, bile acids, and digitalis aglycons. { 'kō‚lān }

cholangiogram [MED] The x-ray film produced by means of cholangiography. { kə'lan·jē·ə‚gram }

cholangiography [MED] Roentgenography of the bile ducts. { kə‚lan·jē'äg·rə·fē }

cholangiolitis [MED] Inflammation of the bile capillaries. { kə‚lan·jē·ə'līd·əs }

cholangioma [MED] Adenocarcinoma of the bile ducts. { kə‚lan·jē'ō·mə }

cholangitis [MED] Inflammation of the bile ducts. { ‚kō· lən'jīd·as }

cholate [BIOCHEM] Any salt of cholic acid. { 'kō‚lāt }

cholecalciferol [PHARM] $C_{27}H_{44}O$ Colorless crystals with a melting range of 84–88°C; soluble in alcohol, chloroform, and fatty oils; derived from the vitamin D_3 of tuna liver oil and used as an antirachitic vitamin. Also known as vitamin D_3. { ‚kō·lə‚kal'sif·ə‚rȯl }

cholecystectomy [MED] Surgical removal of the gallbladder and cystic duct. { ‚kō·lə‚sis'tek·tə·mē }

cholecystitis [MED] Inflammation of the gallbladder. { ‚kō·lə‚sis'tīd·əs }

cholecystography [MED] Radiography of the gallbladder following injection or ingestion of a radiopaque substance excreted in bile. Also known as Graham-Cole test. { ‚kō·lə‚si'stäg·rə·fē }

cholecystokinin [BIOCHEM] A hormone produced by the mucosa of the upper intestine which stimulates contraction of the gallbladder. { ‚kō·lə‚sis·tə'kī·nən }

cholecystostomy [MED] The establishment of an opening into the gallbladder, usually for external drainage of its contents. { ‚kō·lə‚si'stä·stə·mē }

choledochoduodenal junction [ANAT] The point where the common bile duct enters the duodenum. { ¦kō·lə¦däk·ə‚dü'wäd·ən·əl 'jəŋk·shən }

choledocholithiasis [MED] The presence of calculi in the common bile duct. { ‚kō·lə‚däk·ə‚li'thī·ə·səs }

choledochostomy [MED] The draining of the common bile duct through the abdominal wall. { ‚kō·lə‚dä'kä·stə·mē }

choleglobin [BIOCHEM] Combined native protein (globin) and open-ring iron-porphyrin, which is bile pigment hemoglobin; a precursor of biliverdin. { ¦kō·lə¦glō·bən }

cholelithiasis [MED] The production of or the condition associated with gallstones in the gallbladder or bile ducts. { ‚kō·lə‚li'thī·ə·səs }

cholera [MED] **1.** An acute, infectious bacterial disease of humans caused by *Vibrio comma*; characterized by diarrhea, delirium, stupor, and coma. **2.** Any condition characterized by profuse vomiting and diarrhea. { 'käl·ə·rə }

cholera [MED] An acute, severe gastroenteritis. { 'käl·ə·rə }

cholera vibrio [MICROBIO] *Vibrio comma*, the bacterium that causes cholera. { 'käl·ə·rə 'vib·rē‚ō }

cholesteatoma [MED] An epidermal inclusion cyst of the middle ear, or mastoid bone, sometimes in the external ear canal, brain, or spinal cord. Also known as pearly tumor. { kə‚les·tē·ə'tō·mə }

cholesterol [BIOCHEM] $C_{27}H_{46}O$ A sterol produced by all vertebrate cells, particularly in the liver, skin, and intestine, and found most abundantly in nerve tissue. { kə'les·tə‚rȯl }

cholic acid [BIOCHEM] $C_{24}H_{40}O_5$ An unconjugated, crystalline bile acid. { 'kō·lik 'as·əd }

choline [BIOCHEM] $C_5H_{15}O_2N$ A basic hygroscopic substance constituting a vitamin of the B complex; used by most animals as a precursor of acetylcholine and a source of methyl groups. { 'kō,lēn }

choline acetyltransferase [BIOCHEM] An enzyme that transfers the acetyl group to choline in the synthesis of acetylcholine from acetyl coenzyme A and choline. { ¦kō,lēn ə,sed·əl'tranz·fə,rās }

cholinergic [PHYSIO] Liberating, activated by, or resembling the physiologic action of acetylcholine. { ¦kō·lə¦nər·jik }

cholinergic nerve [NEUROSCI] Any nerve, such as autonomic preganglionic nerves and somatic motor nerves, that releases a cholinergic substance at its terminal points. { ¦kō·lə¦nər·jik 'nərv }

cholinesterase [BIOCHEM] An enzyme found in blood and in various other tissues that catalyzes hydrolysis of choline esters, including acetylcholine. Abbreviated chE. { 'kō·lə'nes·tə,rās }

choluria [MED] The presence of bile in the urine. { kō'lu̇·rē·ə }

cholytaurine *See* taurocholic acid. { ,käl·ə'tȯ,rēn }

Chondrichthyes [VERT ZOO] A class of vertebrates comprising the cartilaginous, jawed fishes characterized by the absence of true bone. { kän'drik·thē,ēz }

chondrification [PHYSIO] Formation of or conversion into cartilage. { ,kän·drə·fə'kā·shən }

chondrin [BIOCHEM] A horny gelatinous protein substance obtainable from the collagen component of cartilage. { 'kän·drən }

chondrioid [MICROBIO] A cell organelle in bacteria that is functionally equivalent to the mitochondrion of eukaryotes. { 'kän·drē,ȯid }

chondriome [CELL MOL] Referring collectively to the chondriosomes (mitochondria) of a cell as a functional unit. { 'kän·drē,ōm }

chondriosome [CELL MOL] Any of a class of self-perpetuating lipoprotein complexes in the form of grains, rods, or threads in the cytoplasm of most cells; thought to function in cellular metabolism and secretion. { 'kän·drē·ə,sōm }

chondroblast [HISTOL] A cell that produces cartilage. { 'kän·drō,blast }

Chondrobrachii [VERT ZOO] The equivalent name for Ateleopoidei. { ¦kän·drō'brā·kē,ī }

chondroclast [HISTOL] A cell that absorbs cartilage. { 'kän·drō,klast }

chondrocranium [ANAT] The part of the adult cranium derived from the cartilaginous cranium. [EMBRYO] The cartilaginous, embryonic cranium of vertebrates. { ¦kän·drō'krā·nē·əm }

chondrocyte [HISTOL] A cartilage cell. { 'kän·drō,sīt }

chondrodysplasia *See* enchondromatosis. { ¦kän·drō·də 'splā·zhə }

chondrodystrophy fetalis *See* achondroplasia. { ¦kän·drō ¦dis·trə·fē fə'tal·əs }

chondrogenesis [EMBRYO] The development of cartilage. { ¦kän·drō·jə'nē·səs }

chondroitin [BIOCHEM] A nitrogenous polysaccharide occurring in cartilage in the form of condroitinsulfuric acid. { kän'drō·ə·tən }

chondrology [ANAT] The anatomical study of cartilage. { kän'dräl·ə·jē }

chondroma [MED] A benign tumor of bone, cartilage, or other tissue which simulates the structure of cartilage in its growth. { kän'drō·mə }

chondromalacia [MED] Softening of a cartilage. { ¦kän·drō·mə'lā·shə }

chondromucoid [BIOCHEM] A mucoid found in cartilage; a glycoprotein in which chondroitinsulfuric acid is the prosthetic group. { ¦kän·drō'myü,kȯid }

Chondromyces [MICROBIO] A genus of bacteria in the family Polyangiaceae; sporangia are stalked, and vegetative cells are short rods or spheres. { ,kän·drō'mī,sēz }

chondrophone [INV ZOO] In bivalve mollusks, a structure or cavity supporting the internal hinge cartilage. { 'kän·drō,fōn }

Chondrophora [INV ZOO] A suborder of polymorphic, colonial, free-floating cnidarians of the class Hydrozoa. { kän'drä·fə·rə }

chondroprotein [BIOCHEM] A protein (glycoprotein) occurring normally in cartilage. { ¦kän·drō'prō,tēn }

chondrosarcoma [MED] A malignant tumor of cartilage. { ¦kän·drō·sär'kō·mə }

chondroskeleton [ANAT] **1.** The parts of the bony skeleton which are formed from cartilage. **2.** Cartilaginous parts of a skeleton. [VERT ZOO] A cartilaginous skeleton, as in Chondrostei. { ¦kän·drō'skel·ə·tən }

Chonotrichida [INV ZOO] A small order of vase-shaped ciliates in the subclass Holotrichia; commonly found as ectocommensals on marine crustaceans. { ,kän·ə'trik·ə·də }

chordamesoderm [EMBRYO] The portion of the mesoderm in the chordate embryo from which the notochord and related structures arise, and which induces formation of ectodermal neural structures. { ,kȯrd·ə'mes·ə,dərm }

Chordata [ZOO] The highest phylum in the animal kingdom, characterized by a notochord, nerve cord, and gill slits; includes the urochordates, lancelets, and vertebrates. { kȯr 'däd·ə }

Chordodidae [INV ZOO] A family of worms in the order Gordioidea distinguished by a rough cuticle containing thickenings called areoles. { kȯr'däd·ə,dē }

chordoma [MED] A rarely malignant tumor derived from persistent remnants of the notochord. { kȯr'dō·mə }

Chordopoxivirinae [VIROL] A subfamily of vertebrate deoxyribonucleic acid viruses of the Poxviridae family whose members replicate within the cytoplasm. { ¦kȯr·dō,päk·sə'vī·rə,nē }

chordotomy [MED] Surgical division of a spinal nerve tract to relieve severe intractable pain. { kȯr'däd·ə·mē }

chorea [MED] A nervous disorder seen as part of a syndrome following an organic dysfunction or an infection and characterized by irregular, involuntary movements of the body, especially of the face and extremities. { kə'rē·ə }

choreiform syndrome [MED] A complex of symptoms representing a form or component of minimal brain dysfunction in children, manifested by twitching movements of the face, trunk, and extremities. { kə'rē·ə,fȯrm ,sin,drōm }

chorioadenoma [MED] A tumor intermediate in malignancy between a hydatidiform mole and choriocarcinoma. { ¦kȯr·ē·ō,ad·ən'ō·mə }

chorioallantois [EMBRYO] A vascular fetal membrane that is formed by the close association or fusion of the chorion and allantois. { ¦kȯr·ē·ō·ə'lant·ə·wəs }

choriocarcinoma [MED] A highly malignant tumor derived from chorionic tissue; found most commonly in the uterus and testis. Also known as chorioepithelioma. { ¦kȯr·ē·ō,kärs·ən'ō·mə }

chorioepithelioma *See* choriocarcinoma. { ¦kȯr·ē·ō,ep·ə,thē·lē'ō·mə }

chorion [EMBRYO] The outermost of the extraembryonic membranes of amniotes, enclosing the embryo and all of its other membranes. { 'kȯr·ē·än }

chorionic adenoma [MED] A benign tumor of the placenta. { ,kȯr·ē'än·ik ad·ən'ō·mə }

chorionic gonadotropin *See* human chorionic gonadotropin. { ,kȯr·ē'än·ik gō,nad·ə'trō·pən }

chorionic villus sampling [MED] A technique in which samples of chorionic villi are taken from the placenta for the purpose of genetic testing; usually performed at the end of the second month of pregnancy. { ¦kör·ē,an·ik 'vil·əs ,sam·pliŋ }

chorionitis *See* scleroderma. { ,kȯr·ē·ə'nīd·əs }

chorioretinal [ANAT] Pertaining to the choroid and retina of the eye. { ,kȯr·ē·ə'ret·ən·əl }

chorioretinitis [MED] Inflammation of the choroid and retina of the eye. { ¦kȯr·ē·ō,ret·ən'īd·əs }

choripetalous *See* polypetalous. { ¦kȯr·ə'ped·əl·əs }

chorisepalous *See* polysepalous. { ¦kȯr·ə'sep·əl·əs }

chorisis [BOT] Separation of a leaf or floral part into two or more parts during development. { 'kȯr·ə·səs }

Choristida [INV ZOO] An order of sponges in the class Demospongiae in which at least some of the megascleres are tetraxons. { kə'ris·tə·də }

choroid [ANAT] The highly vascular layer of the vertebrate eye, lying between the sclera and the retina. { 'kȯr,ȯid }

choroiditis [MED] Inflammation of the choroid. { ,kȯr ,ȯi'dīd·əs }

choroid plexus [ANAT] Any of the highly vascular, folded processes that project into the third, fourth, and lateral ventricles of the brain. { ¦kȯr ,ȯid 'plek·səs }

chorology [ECOL] The study of how organisms are distributed geographically. { kə'räl·ə·jē }

Christmas disease [MED] A hereditary, sex-linked, hemophilia-like disease involving failure of the clotting mechanism due to a deficiency of Christmas factor. Also known as Factor IX deficiency. { 'kris·məs di,zēz }

Christmas factor [BIOCHEM] A soluble protein blood factor involved in blood coagulation. Also known as factor IX; plasma thromboplastin component (PTC). { 'kris·məs ,fak·tər }

Chromadoria [INV ZOO] A subclass of nematode worms in the class Adenophorea. { ,krō·mə'dȯr·ē·ə }

Chromadorida [INV ZOO] An order of principally aquatic nematode worms in the subclass Chromadoria. { ,krō·mə'dȯr·ə·də }

Chromadoridae [INV ZOO] A family of soil and fresh-water, free-living nematodes in the superfamily Chromadoroidea; generally associated with algal substances. { ,krō·mə'dȯr·ə,dē }

Chromadoroidea [INV ZOO] A superfamily of small to moderate-sized, free-living nematodes with spiral, transversely ellipsoidal amphids and a striated cuticle. { ,krō·mə·də'rȯid·ē·ə }

chromaffin [BIOL] Staining with chromium salts. { krō'ma·fən }

chromaffin body *See* paraganglion. { krō'ma·fən ,bäd·ē }

chromaffin cell [HISTOL] Any cell of the suprarenal organs in lower vertebrates, of the adrenal medulla in mammals, of the paraganglia, or of the carotid bodies that stains with chromium salts. { krō'ma·fən ,sel }

chromaffin system [PHYSIO] The endocrine organs and tissues of the body that secrete epinephrine; characterized by an affinity for chromium salts. { krō'ma·fən ,sis·təm }

Chromatiaceae [MICROBIO] A family of bacteria in the suborder Rhodospirillineae; motile cells have polar flagella, photosynthetic membranes are continuous with the cytoplasmic membrane, all except one species are anaerobic, and bacteriochlorophyll *a* or *b* is present. { ,krō·mad·ē'as·ē,ī }

chromatic adaptation [PHYSIO] A decrease in sensitivity to a color stimulus with prolonged exposure. { krō'mad·ik ,ad,ap'tā·shən }

chromatic valence [PHYSIO] A relative measure of the hue-producing effectiveness of a chromatic stimulus. { krō'mad·ik 'vāl·əns }

chromatic vision [PHYSIO] Vision pertaining to the color sense, that is, the perception and evaluation of the colors of the spectrum. { krō'mad·ik 'vizh·ən }

chromatid [CELL MOL] **1.** One of the pair of strands formed by longitudinal splitting of a chromosome which are joined by a single centromere in somatic cells during mitosis. **2.** One of a tetrad of strands formed by longitudinal splitting of paired chromosomes during diplotene of meiosis. { 'krō·mə·təd }

chromatin [BIOCHEM] The deoxyribonucleoprotein complex forming the major portion of the nuclear material and of the chromosomes. { 'krō·mə·tən }

chromatin diminution [GEN] Elimination during development of one or more deoxyribonucleic acid sequences from the genome of somatic cells. { ¦krō·mə·ten ,dim·yə'nü·shən }

Chromatium [MICROBIO] A genus of bacteria in the family Chromatiaceae; cells are ovoid to rod-shaped, are motile, do not have gas vacuoles, and contain bacteriochlorophyll *a* on vesicular photosynthetic membranes. { krō'mash·ē·əm }

chromatophore [CELL MOL] A type of pigment cell found in the integument and certain deeper tissues of lower animals that contains color granules capable of being dispersed and concentrated. { krō'mad·ə,fȯr }

chromatophorotrophin [INV ZOO] Any crustacean neurohormone which controls the movement of pigment granules within chromatophores. { krō¦mad·ə¦fȯr·ə'trō·fən }

chromatoplasm [BOT] The peripheral protoplasm in blue-green algae containing chlorophyll, accessory pigments, and stored materials. { krō'mad·ə,plaz·əm }

chromatopsia [MED] A disorder of visual sensation in which color impressions are disturbed or arise subjectively, with objects appearing as unnaturally colored or colorless objects as colored; may be caused by a disturbance of the optic centers, psychic disturbance, or drugs. { ,krō·mə'täp·sē·ə }

chromatosis [MED] A pathologic process or pigmentary disease in which there is a deposit of coloring matter in a normally unpigmented site, or an excessive deposit in a normally pigmented area. { ,krō·mə'tō·səs }

Chromobacterium [MICROBIO] A genus of gram-negative, aerobic or facultatively anaerobic, motile, rod-shaped bacteria of uncertain affiliation; they produce violet colonies and violacein, a violet pigment with antibiotic properties. { ¦krō·mō,bak'tir·ē·əm }

chromoblastomycosis [MED] A granulomatous skin disease caused by any of several fungi, usually *Hormodendrum pedrosoi*, and characterized by warty nodules which may ulcerate. Also known as chromomycosis. { ¦krō·mō¦blas·tō·mī'kō·səs }

chromocenter [CELL MOL] An irregular, densely staining mass of heterochromatin in the chromosomes, with six armlike extensions of euchromatin, in the salivary glands of *Drosophila*. { 'krō·mō,sen·tər }

chromocyte [HISTOL] A pigmented cell. { 'krō·mə,sīt }

chromogen [BIOCHEM] A pigment precursor. [MICROBIO] A microorganism capable of producing color under suitable conditions. { 'krō·mə,jen }

chromogenesis [BIOCHEM] Production of colored substances as a result of metabolic activity; characteristic of certain bacteria and fungi. { ¦krō·mō'jen·ə·səs }

chromolipid *See* lipochrome. { ¦krō·mō'lip·id }

chromomere [CELL MOL] Any of the linearly arranged chromatin granules in leptotene and pachytene chromosomes and in polytene nuclei. { 'krō·mō,mir }

chromomycin [MICROBIO] Any of five components of an antibiotic complex produced by a strain of *Streptomyces griseus*; components are designated A_1 to A_5, of which A_3 ($C_{51}H_{72}O_{32}$) is biologically active. { ¦krō·mō'mī·sən }

chromomycosis *See* chromoblastomycosis. { ¦krō·mō·mī'kō·səs }

chromonema [CELL MOL] The coiled core of a chromatid; it is thought to contain the genes. { ,krō·mō'nē·mə }

chromophile [BIOL] Staining readily. { 'krō·mō,fīl }

chromophobe [BIOL] Not readily absorbing a stain. { 'krō·mə,fōb }

Chromophycota [BOT] A division of the plant kingdom comprising nine classes of algae ranging in size and complexity from unicellular flagellates to gigantic kelps; distinguished by the presence (in almost all) of chlorophyll *c* to complement chlorophyll *a*, and usually having brownish or yellowish chloroplasts. Also known as Chromophyta. { ¦krō·mō'fī·kəd·ə }

chromophyll [BIOCHEM] Any plant pigment. { 'krō·mə,fil }

Chromophyta *See* Chromophycota. { krō'mäf·əd·ə }

chromoplasm [BOT] The pigmented, peripheral protoplasm of blue-green algae cells; contains chlorophyll, carotenoids, and phycobilins. { 'krō·mō,plaz·əm }

chromoplast [CELL MOL] Any colored cell plastid, excluding chloroplasts. { 'krō·mō,plast }

chromoprotein [BIOCHEM] Any protein, such as hemoglobin, with a metal-containing pigment. { ¦krō·mō'prō,tēn }

chromosomal hybrid sterility [GEN] Sterility caused by inability of homologous chromosomes to pair during meiosis due to a chromosome aberration. { ,krō·mə'sō·məl 'hī·brəd stə'ril·əd·ē }

chromosomal mosaic [CELL MOL] An individual showing at least two cell lines with different karyotypes. { ¦krō·mə,sō·məl mō'zā·ik }

chromosome [GEN] A linear (usually) or circular structure containing deoxyribonucleic acid (DNA) complexed with histone and nonhistone proteins, a centromere, and a telomere at each end, if linear. Chromosomes are seen in animals, plants, and other eukaryotes during mitotic and meiotic cell divisions. The single DNA molecule in each chromosome carries a unique complement of linearly arranged genes. { 'krō·mə ,sōm }

chromosome aberration [GEN] Modification of the normal chromosome complement due to deletion, duplication, or rearrangement of genetic material. { 'krō·mə,sōm ab·ə'rā·shən }

chromosome arm [CELL MOL] One of the two main segments of the chromosome that are separated by the centromere. { 'krō·mə,sōm 'ärm }

chromosome banding [GEN] **1.** Unique patterns of cross striations seen on polytene chromosomes, as in *Drosophila*. **2.** Unique pattern of stain intensities along each chromosome in mammals and some other species, using any one of various staining procedures; examples are C-, Q-, R-, and G-banding. { 'krō·mə,sōm ,ban·diŋ }

chromosome breakage syndrome [MED] Any of a number of human genetic disorders characterized by increased frequencies of broken and rearranged chromosomes. { 'krō·mə,sōm 'brā·kij ,sin,drōm }

chromosome complement [GEN] The species-specific, normal diploid set of chromosomes in somatic cells. { 'krō·mə,sōm 'käm·plə,mənt }

chromosome condensation [CELL MOL] The process whereby chromosomes become shorter and thicker during prophase as a consequence of coiling and supercoiling of chromatic strands. { ¦krō·mə,sōm ,kän·dən'sā·shən }

chromosome congression *See* congression. { ¦krō·mə,sōm kəŋ'gresh·ən }

chromosome diminution [EMBRYO] During embryogenesis, the elimination of certain chromosomes from cells that form somatic tissues. Also known as chromosome elimination. { ¦krō·mə,sōm ,dim·ə'nyü·shən }

chromosome elimination *See* chromosome diminution. { ¦krō·mə,sōm i,lim·ə'nā·shən }

chromosome instability syndrome [GEN] Disorders due to defective DNA repair; in humans, these are marked by chromosome changes, increased risk of cancer, and other phenotypic changes. { ¦krō·mə,sōm ,in·stə'bil·əd·ē ,sin ,drōm }

chromosome loss [CELL MOL] Failure of a chromosome to become incorporated into a daughter nucleus at cell division. { ¦krō·mə,sōm 'lȯs }

chromosome map *See* genetic map. { 'krō·mə,sōm ,map }

chromosome painting [GEN] Rendering a specific chromosome or chromosome segment distinguishable by deoxyribonucleic acid hybridization with a pool of many fluorescence-labeled DNA fragments derived from the full length of a chromosome or segment. { 'krō·mə,sōm ,pānt·iŋ }

chromosome puff [CELL MOL] Chromatic material accumulating at a restricted site on a chromosome; thought to reflect functional activity of the gene at that site during differentiation. { 'krō·mə,sōm ,pəf }

chromosome transfer [GEN] The transfer of isolated metaphase chromosomes into cultured mammalian cells. { ¦krō·mə,sōm ,tranz·fər }

chromosome walking [GEN] Sequential isolation of overlapping molecular clones in order to span large intervals on the chromosome. { 'krō·mə,sōm ,wȯk·iŋ }

chronaxie [PHYSIO] The time interval required to excite a tissue by an electric current of twice the galvanic threshold. { 'krä,nak·sē }

chronic [MED] Long-continued; of long duration. { 'krän·ik }

chronic alcoholism [MED] Excessive consumption of alcohol over a prolonged period of time. { 'krän·ik 'al·kə,hȯ,liz·əm }

chronic appendicitis [MED] Inflammation of the vermiform appendix characterized by recurring attacks of right-sided abdominal pain over an extended period of time. { 'krän·ik ə,pen·də'sīd·əs }

chronic carrier [MED] A person who harbors and transmits an infectious agent for an indefinite period. { 'krän·ik 'kar·ē·ər }

chronic catarrhal enteritis [MED] Inflammation of the intestinal tract associated with vascular congestion, mucosal edema, and increased outpouring of mucus in the intestinal lumen. { 'krän·ik kə'tär·əl en·tə'rīd·əs }

chronic glomerulonephritis [MED] Diffuse inflammation of the glomeruli over a prolonged period of time; characterized by progressive fibrosis and associated with hypertension and uremia. { 'krän·ik glə,mer·ə·lō·nə'frīd·əs }

chronic granulomatosis [MED] A disorder of the phagocytes in which they ingest bacteria normally but fail to kill them; disease is usually fatal due to overwhelming bacterial infection. { 'krän·ik ,gran·yə,lō·mə'tō·səs }

chronic hepatitis [MED] A syndrome that is defined clinically by evidence of liver disease for at least six consecutive months. { ,krän·ik ,hep·ə'tīd·əs }

chronic hyperplastic perihepatitis *See* polyserositis. { 'krän·ik hī·pər'plas·tik ,per·ē,hep·ə 'tīd·əs }

chronic infectious arthritis *See* rheumatoid arthritis. { 'krän·ik in'fek·shəs är'thrīd·əs }

chronic leukemia [MED] A leukemia in which the life expectancy is prolonged; leukemias are classified as to acute or chronic, and according to the predominant cell type; the life expectancy is highly variable depending on the latter. { 'krän·ik lü'kēm·e·ə }

chronic myeloid leukemia [MED] A form of leukemia in which immature granulocytes are predominant and the life expectancy is 1–20 years or more. { 'krän·ik 'mī·ə,loid lü'kem·ē·ə }

chronic rhinitis [MED] Inflammation of the nasal mucous membrane due to repeated attacks of acute rhinitis; associated with membrane hypertrophy and later atrophy. { 'krän·ik rī'nīd·əs }

chronotherapy [PHARM] Pharmacological treatment timed to the biological rhythms of the person or organism being treated in order to enhance the effect of the drugs used or to reduce undesirable side effects. { ¦krä·nō'ther·ə·pē }

Chroococcales [BOT] An order of blue-green algae (Cyanophyceae) that reproduce by cell division and colony fragmentation only. { ,krō·ō·kə'kā·lēz }

Chryomyidae [INV ZOO] A family of myodarian cyclorrhaphous dipteran insects in the subsection Acalypteratae. { ,krī·ō'mī·ə,dē }

Chrysididae [INV ZOO] The cuckoo wasps, a family of hymenopteran insects in the superfamily Bethyloidea having brilliant metallic blue and green bodies. { krə'sid·ə,dē }

chrysocarpous [BOT] Bearing yellow fruits. { ¦kris·ə¦kär·pəs }

Chrysomelidae [INV ZOO] The leaf beetles, a

family of coleopteran insects in the superfamily Chrysomeloidea. { ˌkris·ə'mel·əˌdē }

Chrysomeloidea [INV ZOO] A superfamily of coleopteran insects in the suborder Polyphaga. { ¦kris·ə·mə'lȯid·ē·ə }

Chrysomonadida [INV ZOO] An order of yellow to brown, flagellated colonial protozoans of the class Phytamastigophorea. { ¦kris·ə·mə'näd·ə·də }

Chrysomonadina [INV ZOO] The equivalent name for the Chrysomonadida. { ¦kris·əˌmän·ə'dī·nə }

Chrysopetalidae [INV ZOO] A small family of scale-bearing polychaete worms belonging to the Errantia. { ˌkris·ō·pə'tal·əˌdē }

Chrysophyceae [BOT] Golden-brown algae making up a class of fresh- and salt-water unicellular forms in the division Chrysophyta. { ˌkris·ō'fīs·ēˌē }

Chrysophyta [BOT] The golden-brown algae, a division of plants with a predominance of carotene and xanthophyll pigments in addition to chlorophyll. { krə'säf·ə·də }

chrysotherapy [MED] The use of gold compounds in the treatment of disease. { ¦kris·ō'ther·ə·pē }

Chthamalidae [INV ZOO] A small family of barnacles in the suborder Thoracica. { thə'mal·əˌdē }

chum salmon [INV ZOO] *Oncorhynchus keta*. The salmon with the broadest geographic distribution, from the south coast of Korea to the central coast of California. Also known as dog salmon. { 'chəm ˌsa·mən }

chyle [PHYSIO] Lymph containing emulsified fat, present in the lacteals of the intestine during digestion of ingested fats. { kīl }

chylomicron [BIOCHEM] One of the extremely small lipid droplets, consisting chiefly of triglycerides, found in blood after ingestion of fat. { ˌkīl·ə'mīˌkrän }

chylophyllous [BOT] Having succulent or fleshy leaves. { ¦kīl·ō¦fil·əs }

chylothorax [MED] An accumulation of chyle in the pleural cavity. { ¦kīl·ō¦thȯrˌaks }

chyluria [MED] The presence of chyle or lymph in the urine, usually caused by a fistulous communication between the urinary and lymphatic tracts or by lymphatic obstruction. { kī'lu̇r·ē·ə }

chyme [PHYSIO] The semifluid, partially digested food mass that is expelled into the duodenum by the stomach. { kīm }

chymopapain [BIOCHEM] Any one of several proteolytic enzymes obtained from papaya. { ˌkī·mō·pə'pī·ən }

chymosin *See* rennin. { 'kī·mə·sən }

chymotrypsin [BIOCHEM] A proteinase in the pancreatic juice that clots milk and hydrolyzes casein and gelatin. { ˌkī·mə'trip·sən }

chymotrypsinogen [BIOCHEM] An inactive proteolytic enzyme of pancreatic juice; converted to the active form, chymotrypsin, by trypsin. { ¦kī·mōˌtrip'sin·ə·jən }

Chytridiales [MYCOL] An order of mainly aquatic fungi of the class Phycomycetes having a saclike to rhizoidal thallus and zoospores with a single posterior flagellum. { kīˌtrid·ē'ā·lēz }

Chytridiomycetes [MYCOL] A class of true fungi. { kīˌtrid·ē·ōˌmī'sēd·ēz }

cibarium [INV ZOO] In insects, the space anterior to the mouth cavity in which food is chewed. { sə'bar·ē·əm }

Cicadellidae [INV ZOO] Large family of homopteran insects belonging to the series Auchenorrhyncha; includes leaf hoppers. { ˌsik·ə'del·əˌdē }

Cicadidae [INV ZOO] A family of large homopteran insects belonging to the series Auchenorrhyncha; includes the cicadas. { sə'kad·əˌdē }

cicatrix [BIOL] A scarlike mark, usually caused by previous attachment of a part or organ. [MED] The connective-tissue scar formed at the site of a healing wound. { 'sik·əˌtriks }

Cichlidae [VERT ZOO] The cichlids, a family of perciform fishes in the suborder Percoidei. { 'sik·ləˌdē }

Cicindelidae [INV ZOO] The tiger beetles, a family of coleopteran insects in the suborder Adephaga. { ˌsi·sən'del·əˌdē }

Ciconiidae [VERT ZOO] The tree storks, a family of wading birds in the order Ciconiiformes. { ˌsi·kə'nī·əˌdē }

Ciconiiformes [VERT ZOO] An order of predominantly long-legged, long-necked birds, including herons, storks, ibises, spoonbills, and their relatives. { səˌkōn·ē·ə'fȯrˌmēz }

Cidaroida [INV ZOO] An order of echinoderms in the subclass Perischoechinoidea in which the ambulacra comprise two columns of simple plates. { ˌsid·ə'rȯi·də }

CID disease *See* combined immunological deficiency disease. { ¦sē¦ī¦dē diˌzēz }

ciguatoxin [BIOCHEM] A toxin produced by the benthic dinoflagellate *Gambierdiscus toxicus*. { ¦sēg·wə¦täk·sən }

Ciidae [INV ZOO] The minute, tree-fungus beetles, a family of coleopteran insects in the superfamily Cucujoidea. { 'sī·əˌdē }

cilia [ANAT] Eyelashes. [CELL MOL] Relatively short, centriole-based, hairlike processes on certain anatomical cells and motile organisms. { 'sil·ē·ə }

ciliary body [ANAT] A ring of tissue lying just anterior to the retinal margin of the eye. { ¦sil·ēˌer·ē ¦bäd·ē }

ciliary movement [BIOL] A type of cellular locomotion accomplished by the rhythmical beat of cilia. { ¦sil·ēˌer·ē 'müv·mənt }

ciliary muscle [ANAT] The smooth muscle of the ciliary body. { ¦sil·ēˌer·ē 'məs·əl }

ciliary process [ANAT] Circularly arranged choroid folds continuous with the iris in front. { ¦sil·ēˌer·ē 'präs·əs }

Ciliatea [INV ZOO] The single class of the protozoan subphylum Ciliophora. { ˌsil·ē'ad·ē·ə }

ciliated epithelium [HISTOL] Epithelium composed of cells bearing cilia on their free surfaces. { 'sil·ēˌād·əd ep·ə'thēl·ē·əm }

ciliolate [BIOL] Ciliated to a very minute degree. { 'sil·ē·əˌlāt }

Ciliophora [INV ZOO] The ciliated protozoans, a homogeneous subphylum of the Protozoa distinguished principally by a mouth, ciliation, and infraciliature. { ,sil·ē'äf·ə·rə }

Cimbicidae [INV ZOO] The cimbicid sawflies, a family of hymenopteran insects in the superfamily Tenthredinoidea. { ,sim'bis·ə,dē }

Cimex [INV ZOO] The type genus of Cimicidae, including bedbugs and related forms. { 'sī,meks }

Cimicidae [INV ZOO] The bat, bed, and bird bugs, a family of flattened, wingless, parasitic hemipteran insects in the superfamily Cimicimorpha. { sī'mis·ə,dē }

Cimicimorpha [INV ZOO] A superfamily, or group according to some authorities, of hemipteran insects in the subdivision Geocorisae. { ,sī·mə·sə'mȯr·fə }

Cimicoidea [INV ZOO] A superfamily of the Cimicimorpha in some systems of classification. { ,sī·mə'kȯid·ē·ə }

cinchona [BOT] The dried, alkaloid-containing bark of trees of the genus *Cinchona*. { siŋ'kō·nə }

Cinclidae [VERT ZOO] The dippers, a family of insect-eating songbirds in the order Passeriformes. { 'siŋ·klə,dē }

cinclides [INV ZOO] Pores in the body wall of some sea anemones for the release of water and stinging cells. { siŋ'klī,dēz }

cinclis [INV ZOO] Singular of cinclides. { 'siŋ·kləs }

cineplasty *See* kineplasty. { 'sin·ə,plas·tē }

cinereous [BIOL] **1.** Ashen in color. **2.** Having the inert and powdery quality of ashes. { sə'nir·ē·əs }

Cingulata [VERT ZOO] A group of xenarthran mammals in the order Edentata, including the armadillos. { ,siŋ·gyə'läd·ə }

cingulate [BIOL] Having a girdle of bands or markings. { 'siŋ·gyə·lət }

cingulum [ANAT] The ridge around the base of the crown of a tooth. [BOT] The part of a plant between stem and root. [INV ZOO] **1.** Any girdlelike structure. **2.** A band of color or a raised line on certain bivalve shells. **3.** The outer zone of cilia on discs of certain rotifers. **4.** The clitellum in annelids. [NEUROSCI] The tract of association nerve fibers in the brain, connecting the callosal and the hippocampal convolutions. { 'sin·gyə·ləm }

cinnamon [BOT] *Cinnamomum zeylanicum*. An evergreen shrub of the laurel family (Lauraceae) in the order Magnoliales; a spice is made from the bark. { 'sin·ə·mən }

circadian rhythm [PHYSIO] A self-sustained cycle of physiological changes that occurs over an approximately 24-hour cycle, generally synchronized to light-dark cycles in an organism's environment. { sər'kād·ē·ən 'rith·əm }

circinate [BIOL] Having the form of a flat coil with the apex at the center. { 'sərs·ən,āt }

circinate vernation [BOT] Uncoiling of new leaves from the base toward the apex, as in ferns. { 'sərs·ən,āt vər'nā·shən }

circle of Willis [ANAT] A ring of arteries at the base of the cerebrum. { 'sər·kəl əv 'wil·əs }

circular deoxyribonucleic acid [BIOCHEM] A single- or double-stranded ring of deoxyribonucleic acid found in certain bacteriophages and in human wart virus. Also known as ring deoxyribonucleic acid. [CELL MOL] A deoxyribonucleic acid molecule that has no free 5′ or 3′ ends; characteristic of prokaryotes but also found in mitochondria, chloroplasts, and some viral genomes. { 'sər·kyə·lər dē¦äk·sē¦rī,bō¦nü¦klē·ik 'as·əd }

circulation [PHYSIO] The movement of blood through defined channels and tissue spaces; movement is through a closed circuit in vertebrates and certain invertebrates. { ,sər·kyə'lā·shən }

circulatory system [ANAT] The vessels and organs composing the lymphatic and cardiovascular systems. { 'sər·kyə·lə,tȯr·ē ,sis·təm }

circulin [MICROBIO] Any of a group of peptide antibiotics produced by *Bacillus circulans* which are related to polymixin and are active against both gram-negative and gram-positive bacteria. { 'sər·kyə·lən }

circulus [BIOL] Any of various ringlike structures, such as the vascular circle of Willis or the concentric ridges on fish scales. { 'sər·kyə·ləs }

circumboreal distribution [ECOL] The distribution of a Northern Hemisphere organism whose habitat includes North American, European, and Asian stations. { ¦sər·kəm'bȯr·ē·əl ,dis·trə'byü·shən }

circumcision [MED] Surgical excision of the foreskin. { ,sər·kəm'sizh·ən }

circumduction [ANAT] Movement of the distal end of a body part in the form of an arc; performed at ball-and-socket and saddle joints. { ,sər·kəm'dək·shən }

circumflex artery [ANAT] Any artery that follows a curving or winding course. { 'sər·kəm,fleks 'ärd·ə·rē }

circumnutation [BOT] The bending or turning of a growing stem tip that occurs as a result of unequal rates of growth along the stem. { ¦sər·kəm·nü'tā·shən }

circumpharyngeal connective [INV ZOO] One of a pair of nerve strands passing around the esophagus in annelids and anthropods, connecting the brain and subesophageal ganglia. { ¦sər·kəm·fə'rin·jē·əl kə'nek·tiv }

circumscissile [BOT] Dehiscing along the line of a circumference, as exhibited by a pyxidium. { ¦sər·kəm'sis·əl }

circumvallate papilla *See* vallate papilla. { ¦sər·kəm¦va,lāt pa'pil·ə }

Cirolanidae [INV ZOO] A family of isopod crustaceans in the suborder Flabellifera composed of actively swimming predators and scavengers with biting mouthparts. { ,sir·ə'lan·ə,dē }

Cirratulidae [INV ZOO] A family of fringe worms belonging to the Sedentaria which are important detritus feeders in coastal waters. { ,sir·ə'tül·ə,dē }

cirrhosis [MED] A progressive, inflammatory

disease of the liver characterized by a real or apparent increase in the proportion of hepatic connective tissue. { sə'rō·səs }

cirriform [ZOO] Having the form of a cirrus; generally applied to a prolonged, slender process. { 'sir·ə,fȯrm }

Cirripedia [INV ZOO] A subclass of the Crustacea, including the barnacles and goose barnacles; individuals are free-swimming in the larval stages but permanently fixed in the adult stage. { ,sir·ə'pēd·ē·ə }

Cirromorpha [INV ZOO] A suborder of cephalopod mollusks in the order Octopoda. { ¦sir·ō¦mȯr·fə }

cirrus [INV ZOO] **1.** The conical locomotor structure composed of fused cilia in hypotrich protozoans. **2.** Any of the jointed thoracic appendages of barnacles. **3.** Any hairlike tuft on insect appendages. **4.** The male copulatory organ in some mollusks and trematodes. [VERT ZOO] Any of the tactile barbels of certain fishes. [ZOO] A tendrillike animal appendage. { 'sir·əs }

cirrus sac [INV ZOO] A pouch or channel containing the copulatory organ (cirrus) in certain invertebrates. { 'sir·əs ,sak }

cis-active [CELL MOL] Describing a genetic element (such as a promoter or other regulatory locus) that promotes or suppresses two unrelated targets (such as genes) on the same chromosome as a result of their relative positions on the chromosome. { ,sis 'ak·tiv }

cisco [VERT ZOO] Any of several North American freshwater whitefishes of the genus *Coregonus*. { 'sis,kō }

cisplatin [MED] A transition-metal complex that is used in the treatment of cancer, with particular effectiveness in the treatment of testicular and ovarian cancers; second- and third-generation variants of this drug have been developed to mitigate undesirable side effects. { ¦sis'plat·ən }

cistern [ANAT] A closed, fluid-filled sac or vesicle, such as the subarachnoid spaces or the vesicles comprising the dictyosomes of a Golgi apparatus. { 'sis·tərn }

cistron [CELL MOL] The genetic unit (deoxyribonucleic acid fragment) that codes for a particular polypeptide; mutants do not complement each other within a cistron. Also known as structural gene. { 'sis,trän }

Citheroniinae [INV ZOO] Subfamily of lepidopteran insects in the family Saturniidae, including the regal moth and the imperial moth. { ,sith·ə·rō'nī·ə,nē }

citramalase [BIOCHEM] An enzyme that is involved in the fermentation of glutamate by *Clostridium tetanomorphum*; catalyzes the breakdown of citramalic acid to acetate and pyruvate. { ,si·trə'ma,lās }

citrate [BIOCHEM] A salt or ester of citric acid. { 'si,trāt }

citrate test [MICROBIO] A differential cultural test to identify genera within the bacterial family Enterobacteriaceae that are able to utilize sodium citrate as a sole source of carbon. { 'si ,trāt ,test }

citric acid [BIOCHEM] $C_6H_8O_7 \cdot H_2O$ A colorless crystalline or white powdery organic, tricarboxylic acid occurring in plants, especially citrus fruits, and used as a flavoring agent, as an antioxidant in foods, and as a sequestering agent; the commercially produced form melts at 153°C. { 'si,trik 'as·əd }

citric acid cycle *See* Krebs cycle. { 'si,trik 'as·əd 'sī·kəl }

citriculture [BOT] The cultivation of citrus fruits. { 'si·trə,kəl·chər }

Citrobacter [MICROBIO] A genus of bacteria in the family Enterobacteriaceae; motile rods that utilize citrate as the only carbon source. { ,si·trō'bak·tər }

citron [BOT] *Citrus medica*. A shrubby, evergreen citrus tree in the order Sapindales cultivated for its edible, large, lemonlike fruit. { 'si·trən }

citronella [BOT] *Cymbopogon nardus*. A tropical grass; the source of citronella oil. { ,si·trə'nel·ə }

citrulline [BIOCHEM] $C_6H_{13}O_3N_3$ An amino acid formed in the synthesis of arginine from ornithine. { 'si·trə,lēn }

citrus flavonoid compound *See* bioflavonoid. { 'si·trəs 'flav·ə,nȯid ,käm,paůnd }

citrus fruit [BOT] Any of the edible fruits having a pulpy endocarp and a firm exocarp that are produced by plants of the genus *Citrus* and related genera. { 'si·trəs ,früt }

civet [PHYSIO] A fatty substance secreted by the civet gland; used as a fixative in perfumes. [VERT ZOO] Any of 18 species of catlike, nocturnal carnivores assigned to the family Viverridae, having a long head, pointed muzzle, and short limbs with nonretractile claws. { 'siv·ət }

civet cat *See* cacomistle. { 'siv·ət ,kat }

civet gland [VERT ZOO] A large anal scent gland in civet cats that secretes civet. { 'siv·ət ,gland }

civetone [BIOCHEM] $C_{17}H_{30}O$ 9-Cycloheptadecen-1-one, a macrocyclic ketone component of civet used in perfumes because of its pleasant odor and lasting quality; believed to function as a sex attractant among civet cats. { 'siv·ə,tōn }

clade [SYST] A taxonomic group containing a common ancestor and its descendants. { klād }

cladism [SYST] A theory of taxonomy by which organisms are grouped and ranked on the basis of the most recent phylogenetic branching point. { 'kla,diz·əm }

Cladistia [VERT ZOO] The equivalent name for Polypteriformes. { klə'dis·tē·ə }

Cladocera [INV ZOO] An order of small, freshwater branchiopod crustaceans, commonly known as water fleas, characterized by a transparent bivalve shell. { klə'däs·ə·rə }

Cladocopa [INV ZOO] A suborder of the order Myodocopida including marine animals having a carapace that lacks a permanent aperture when the two valves are closed. { klə'däk·ə·pə }

Cladocopina [INV ZOO] The equivalent name for Cladocopa. { ¦klad·ə¦käp·ə·nə }

cladode *See* cladophyll. { 'kla,dōd }

cladogenesis [EVOL] Evolution associated with altered habit and habitat, usually in isolated species populations. { ,klad·ə'jen·ə·səs }

cladogenic adaptation *See* divergent adaptation. { ¦klad·ə¦jen·ik ,ad,ap'tā·shən }

cladogram [EVOL] A dendritic diagram which shows the evolution and descent of a group of organisms. { 'klad·ə,gram }

Cladoniaceae [BOT] A family of lichens in the order Lecanorales, including the reindeer mosses and cup lichens, in which the main thallus is hollow. { kla¦dō·nē'as·ē,ē }

Cladophorales [BOT] An order of coarse, wiry, filamentous, branched and unbranched algae in the class Chlorophyceae. { kla,däf·ə'rā·lēz }

cladophyll [BOT] A branch arising from the axil of a true leaf and resembling a foliage leaf. Also known as cladode. { 'klad·ə,fil }

cladoptosis [BOT] The annual abscission of twigs or branches instead of leaves. { ,kla,däp 'tō·səs }

cladus [BOT] A branch of a ramose spicule. { 'klā·dəs }

clam [INV ZOO] The common name for a number of species of bivalve mollusks, many of which are important as food. { klam }

Clambidae [INV ZOO] The minute beetles, a family of coleopteran insects in the superfamily Dascilloidea. { 'klam·bə,dē }

clammy [BIOL] Moist and sticky, as the skin or a stem. { 'klam·ē }

clamp connection [MYCOL] In the Basidiomycetes, a lateral connection formed between two adjoining cells of a filament and arching over the septum between them and permitting a type of pseudosexual activity. { 'klamp kə,nek· shən }

clam worm [INV ZOO] The common name for a number of species of dorsoventrally flattened annelid worms composing the large family Nereidae in the class Polychaeta; all have a distinct head, with numerous appendages. { 'klam ,wərm }

clan [ECOL] A very small community, perhaps a few square yards in area, in climax formation, and dominated by one species. { klan }

Clariidae [VERT ZOO] A family of Asian and African catfishes in the suborder Siluroidei. { kla'rī·ə,dē }

clasper [VERT ZOO] A modified pelvic fin of male elasmobranchs and holocephalians used for the transmission of sperm. { 'klasp·ər }

class [SYST] A taxonomic category ranking above the order and below the phylum or division. { klas }

classical pathway [IMMUNOL] The pathway by which antigen-antibody complex activates the complement system. { ¦klas·i·kəl 'path,wā }

classic botulism [MED] Botulism typically due to ingestion of preformed toxin. { ,klas·ik 'bäch·ə,liz·əm }

classic epidemic typhus [MED] An epidemic disease caused by *Rickettsia prowazeki* var. *prowazekii*, and characterized by violent headache, a rash, neurological symptoms, and high fever. Also known as epidemic typhus. { 'klas·ik ,ep· ə'dem·ik 'tī,fəs }

classification [SYST] A systematic arrangement of plants and animals into categories based on a definite plan, considering evolutionary, physiologic, cytogenetic, and other relationships. { ,klas·ə·fə'kā·shən }

class switch [IMMUNOL] A switch of B-lymphocyte expression from one antibody class to another. { 'klas ,swich }

class II-associated invariant chain peptide [IMMUNOL] A residual fragment of the invariant chain that is essential for proper loading of exogenous peptide on the class II major histocompatibility complex. { ¦klas ¦tü ə¦sō·sē,ād·əd in¦ver· ē·ənt ¦chān 'pep,tīd }

clastogenesis [CELL MOL] The loss, addition, or rearrangement of chromosomes. { ,klas· tə'jen·ə·səs }

clathrate *See* cancellate. { 'klath,rāt }

clathrin [CELL MOL] A protein that forms a lattice-shaped coating, through the assembly of subunits called triskelions, on the cytosolic side of membrane regions called coated pits during the initial stages of receptor-mediated endocytosis. Invagination of the pit results in a clathrin-coated vesicle. { 'klath·rən }

clathrin-coated pit [CELL MOL] A partially invaginated membrane structure (bud or pit) involved in receptor-mediated endocytosis consisting of a cluster of receptor proteins attached on the cytosolic side by means of adaptin molecules to the protein clathrin, which forms a lattice-shaped coating. Complete invagination of the pit and release from the membrane results in a clathrin-coated vesicle containing cargo molecules. { ¦klath·rən ,kōd·əd 'pit }

clathrin-coated vesicle *See* clathrin-coated pit. { 'klath·rən ,kōd·əd 'ves·ə·kəl }

Clathrinida [INV ZOO] A monofamilial order of sponges in the subclass Calcinea having an asconoid structure and lacking a true dermal membrane or cortex. { kla'thrin·ə·də }

Clathrinidae [INV ZOO] The single family of the order Clathrinida. { kla'thrin·ə,dē }

Clathrochloris [MICROBIO] A genus of bacteria in the family Chlorobiaceae; cells are spherical to ovoid and arranged in chains united in trellislike aggregates, are nonmotile, contain gas vacuoles, and are free-living. { ¦klath·rō¦klȯr·əs }

claustrum [ANAT] A thin layer of gray matter in each cerebral hemisphere between the lenticular nucleus and the island of Reil. { 'klȯ,strəm }

clava [BIOL] A club-shaped structure, as the tip on the antennae of certain insects or the fruiting body of certain fungi. { 'klā·va }

clavate [BIOL] Club-shaped. Also known as claviform. { 'klā,vāt }

Clavaxinellida [INV ZOO] An order of sponges in the class Demospongiae; members have monaxonid megascleres arranged in radial or plumose tracts. { kla¦vak·sə'nel·ə·də }

clavicle [ANAT] A bone in the pectoral girdle

of vertebrates with articulation occurring at the sternum and scapula. { 'klav·ə·kal }

claviculate [ANAT] Having a clavicle. { kla'vik·yə·lət }

claviform *See* clavate. { 'klav·ə,fórm }

clavus [INV ZOO] Any of several rounded or fingerlike processes, such as the club of an insect antenna or the pointed anal portion of the hemelytron in hemipteran insects. { 'klāv·əs }

claw [ANAT] A sharp, slender, curved nail on the toe of an animal, such as a bird. [INV ZOO] A sharp-curved process on the tip of the limb of an insect. { klò }

clearance test [PATH] The use of a substance such as urea or creatinine, or an injected foreign substance such as inulin to measure renal excretory activity; the ratio of the amount of these excreted substances in two 1-hour periods contrasted with the level of these substances in the blood is calculated in terms of the amount of blood cleared of these substances in a given unit time. { 'klir·əns ,test }

clear-cell carcinoma *See* renal-cell carcinoma. { ¦klir ¦sel ,kärs·ən'ō·mə }

cleavage [EMBRYO] The subdivision of activated eggs into blastomeres. { 'klēv·ij }

cleavage nucleus [EMBRYO] The nucleus of a zygote formed by fusion of male and female pronuclei. { 'klēv·ij ,nü·klē·əs }

cleft grafting [BOT] A top-grafting method in which the scion is inserted into a cleft cut into the top of the stock. { 'kleft ,graft·iŋ }

cleft lip *See* harelip. { ¦kleft 'lip }

cleft palate [MED] A birth defect resulting from incomplete closure of the palate during embryogenesis. { ¦kleft 'pal·ət }

cleidocranial dysostosis [MED] A congenital defect in which there is deficient formation of bone in the skull and clavicle. { ¦klī·dō¦krān·ē·əl ¦dis·ä'stō·səs }

cleistocarp *See* cleistothecium. { 'klī·stə,kärp }

cleistocarpous [BOT] Of mosses, having the capsule opening irregularly without an operculum. [MYCOL] Forming or having cleistothecia. { ¦klī·stə¦kär·pəs }

cleistogamy [BOT] The production of small closed flowers that are self-pollinating and contain numerous seeds. { ,klī'stäg·ə·mē }

cleistothecium [MYCOL] A closed sporebearing structure in Ascomycetes; asci and spores are freed of the fruiting body by decay or desiccation. Also known as cleistocarp. { ¦klī·stə'thē·sē·əm }

cleithrum [VERT ZOO] A bone external and adjacent to the clavicle in certain fishes, stegocephalians, and primitive reptiles. { 'klī·thrəm }

Cleridae [INV ZOO] The checkered beetles, a family of coleopteran insects in the superfamily Cleroidea. { 'kler·ə,dē }

Cleroidea [INV ZOO] A superfamily of coleopteran insects in the suborder Polyphaga. { klə'ròid·ē·ə }

climacteric *See* menopause. { klī'mak·tə·rik }

climate therapy *See* climatotherapy. { 'klīm·ət ,ther·ə·pē }

climatic climax [ECOL] A climax community viewed, by some authorities, as controlled by climate. { klī'mad·ik 'klī,maks }

climatopathology [MED] The study of disease in relation to the effects of the natural environment. { ¦klī·mə·tō·pə'thäl·ə·jē }

climatophysiology [PHYSIO] The study of the interaction of the natural environment with physiologic factors. { ¦klī·mə·tō,fiz·ē'äl·ə·jē }

climatotherapy [MED] Placing a person in a suitable climate to treat a certain disease. Also known as climate therapy; climotherapy. { ¦klī·mə·tō'ther·ə·pē }

climax [ECOL] A mature, relatively stable community in an area, which community will undergo no further change under the prevailing climate; represents the culmination of ecological succession. { 'klī,maks }

climax community [ECOL] The final stage in ecological succession in which a relatively constant environment is reached and species composition no longer changes in a directional fashion, but fluctuates about some mean, or average, community composition. { 'klī,maks kə,myü·nə·dē }

climax plant formation [ECOL] A mature, stable plant population in a climax community. { 'klī,maks 'plant fór'mā·shən }

climbing bog [ECOL] An elevated boggy area on a swamp margin, usually occurring where there is a short summer and considerable rainfall. { 'klīm·iŋ 'bäg }

climbing stem [BOT] A long, slender stem that climbs up a support or along the tops of other plants by using spines, adventitious roots, or tendrils for attachment. { 'klīm·iŋ 'stem }

climotherapy *See* climatotherapy. { 'klīm·ə,ther·ə·pē }

cline [BIOL] A graded series of morphological or physiological characters exhibited by a natural group (as a species) of related organisms, generally along a line of environmental or geographic transition. { klīn }

clinical chemistry [PATH] The science involving chemical analysis of body tissues to diagnose disease. { 'klin·ə·kəl 'kem·ə·strē }

clinical genetics [GEN] The study of the role of genetic factors in human disease susceptibility by observation of patients and their families. { 'klin·ə·kəl jə'ned·iks }

clinical microbiology [MED] The adaptation of microbiological techniques to the study of the etiological agents of infectious disease. { ,klin·ə·kəl ,mī·krō·bī'äl·ə·jē }

clinical pathology [PATH] A medical specialty encompassing the diagnostic study of disease by means of laboratory tests of material from the living patient. { 'klin·ə·kəl pə'thäl·ə·jē }

clinical pharmacology [PHARM] The study and evaluation of the effects of drugs in humans. { 'klin·ə·kəl ,fär·mə'käl·ə·jē }

clinical sports medicine [MED] Sports medicine that focuses on the prevention and treatment of athletic injuries and the design of exercise and nutrition programs for maintaining peak physical performance. { ˌklin·ə·kəl 'spȯrts ˌmed·ə·sən }

clinical teleconferencing [MED] The use of teleconferencing in medical diagnosis and treatment, allowing rural health-care facilities to perform diagnosis and treatment that would otherwise be available only in metropolitan areas.

clinical trial [MED] A research study used to find better ways to treat individuals with a specific disease, patients are evaluated after being administered a new treatment or drug. { ¦klin·i·kəl 'trīl }

Clionidae [INV ZOO] The boring sponges, a family of marine sponges in the class Demospongiae. { ˌklī'än·əˌdē }

clisere [ECOL] The succession of ecological communities, especially climax formations, as a consequence of intense climatic changes. { klīˌsir }

clitellum [INV ZOO] The thickened, glandular, saddlelike portion of the body wall of some annelid worms. { klə'tel·əm }

clitoris [ANAT] The homolog of the penis in females, located in the anterior portion of the vulva. { 'klid·ə·rəs }

cloaca [INV ZOO] The chamber which functions as a respiratory, excretory, and reproductive duct in certain invertebrates. [VERT ZOO] The chamber which receives the discharges of the intestine, urinary tract, and reproductive canals in monotremes, amphibians, birds, reptiles, and many fish. { klō'ā·kə }

cloacal bladder [VERT ZOO] A diverticulum of the cloacal wall in monotremes, amphibians, and some fish, into which urine is forced from the cloaca. { klō'ā·kəl 'blad·ər }

cloacal gland [VERT ZOO] Any of the sweat glands in the cloaca of lower vertebrates, as snakes or amphibians. { klō'ā·kəl 'gland }

clock gene [GEN] Any of a number of genes that interact to determine the duration of development and lifespan. { 'kläk ˌjēn }

clonal selection theory [IMMUNOL] Theory to explain the specificity of the adaptive immune response according to which there is a large pool of lymphocytes, each having genetically predetermined specificity for only one of a vast array of possible antigens. Upon encountering an antigen, the lymphocytes sensitive to it reproduce much more rapidly than the others, thus leading to a build-up of antigen-specific cells large enough to mount the response. { ¦klōn·əl si'lek·shən ˌthē·ə·rē }

clone [BIOL] All individuals, considered collectively, produced asexually or by parthenogenesis from a single individual. [GEN] **1.** An organism whose diploid nuclear genome was derived from a somatic cell of another organism of the same species using biotechnolog. **2.** A copy of a genetically engineered DNA sequence. { klōn }

cloning vector [CELL MOL] A carrier, such as a bacterial plasmid or bacteriophage, used to insert a genetic sequence, such as a deoxyribonucleic acid fragment or a complete gene, into a host cell such that the foreign genetic material is capable of replication. { 'klōn·iŋ ˌvek·tər }

clonorchiasis [MED] A parasitic infection of humans and other fish-eating mammals which is caused by the trematode *Opisthorchis* (*Clonorchis*) *sinensis*, which is usually found in the bile ducts. { ˌklōn·ȯr'kī·ə·səs }

Clonothrix [MICROBIO] A genus of sheathed bacteria; cells are attached and encrusted with iron and manganese oxides, and filaments are tapered. { 'klän·əˌthriks }

clonus [PHYSIO] Irregular, alternating muscular contractions and relaxations. { 'klō·nəs }

closed ecological system [ECOL] A community into which a new species cannot enter due to crowding and competition. { ¦klōzd ek·ə'läj·ə·kəl ˌsis·təm }

closed reading frame [CELL MOL] A reading frame containing terminator codons that prevent the translation of subsequent nucleotides into protein. Also known as blocked reading frame. { ¦klōzd 'rēd·iŋ ˌfrām }

closed reduction [MED] Reduction of fractures or dislocations by manipulation without surgical intervention. { ¦klōzd ri'dək·shən }

Closteroviridae [VIROL] A family of plant viruses characterized by flexuous rod-shaped particles containing one molecule of single-stranded positive-sense ribonucleic acid; includes the genera *Closterovirus* and *Crinivirus*. { ˌklä·stə·rə'vir·əˌdī }

Closterovirus [VIROL] A genus of plant viruses belonging to the family Closteroviridae that has a wide host range and is transmitted primarily by aphids; beet yellows virus is the type species. { ¦klä·stə·rə'vī·rəs }

Clostridium [MICROBIO] A genus of bacteria in the family Bacillaceae; usually motile rods which form large spores that distend the cell; anaerobic and do not reduce sulfate. { klä'strid·ē·əm }

Clostridium perfringens [MICROBIO] A spore-forming, toxin-producing bacterium that can contaminate meat left at room temperature. The ingested cells release toxin in the digestive tract, resulting in cramps and diarrhea. { kläˌstrid·ē·əm pər'frin·jənz }

Clostridium tetani [MED] A spore-forming bacterium that produces a powerful toxin, tetanospasmin, that blocks inhibitory synapses in the central nervous system and thus causes the severe muscle spasms characteristic of tetanus. { kläˌstrid·ē·əm 'tet·ənˌī }

clot [PHYSIO] A semisolid coagulum of blood or lymph. { klät }

clot retraction time [PATH] The length of time required for the appearance or completion of the contraction or shrinkage of a blood clot, resulting in the extrusion of serum. { ¦klät ri'trak·shən ˌtīm }

clotting factor [PHYSIO] Any of several plasma components that are involved in the clotting of blood, such as fibrinogen, prothrombin, and thromboplastin. { 'kläd·iŋ ˌfak·tər }

clotting time [PHYSIO] The length of time required for shed blood to coagulate under standard conditions. Also known as coagulation time. { 'kläd·iŋ ,tīm }

cloud forest *See* temperate rainforest. { 'klau̇d ,fär·əst }

cloudy swelling [PATH] A retrogressive change in the cytoplasm of parenchymatous cells, whereby the cell enlarges and its outline becomes irregular, with resultant swelling of the organ. { 'klau̇d·ē 'swel·iŋ }

clove [BOT] **1.** The unopened flower bud of a small, conical, symmetrical evergreen tree, *Eugenia caryophyllata*, of the myrtle family (Myrtaceae); the dried buds are used as a pungent, strongly aromatic spice. **2.** A small bulb developed within a larger bulb, as in garlic. { klōv }

clover [BOT] **1.** A common name designating the true clovers, sweet clovers, and other members of the Leguminosa. **2.** A herb of the genus *Trifolium*. { 'klō·vər }

clubfoot [MED] Congenital malpositioning of a foot such that the forefoot is inverted and rotated with a shortened Achilles tendon. { 'kləb,fu̇t }

club fungi [MYCOL] The common name for members of the class Basidiomycetes. { 'kləb ,fən,jī }

clubhead fungus *See* Cordyceps ophioglossoides. { ¦kləb,hed 'fəŋ·gəs }

club moss [BOT] The common name for members of the class Lycopodiatae. { 'kləb ,mȯs }

Clupeidae [VERT ZOO] The herrings, a family of fishes in the suborder Clupoidea composing the most primitive group of higher bony fishes. { klü'pē·ə,dē }

Clupeiformes [VERT ZOO] An order of teleost fishes in the subclass Actinopterygii, generally having a silvery, compressed body. { ,klü·pē·ə'fȯr,mēz }

clupeine [BIOCHEM] A protamine found in salmon sperm, mainly composed of arginine (74.1%) and small percentages of threonine, serine, proline, alanine, valine, and isoleucine. { 'klü·pē,ēn }

Clupoidea [VERT ZOO] A suborder of fishes in the order Clupeiformes comprising the herrings and anchovies. { ,klü'pȯid·ē·ə }

Clusiidae [INV ZOO] A family of myodarian cyclorrhaphous dipteran insects in the subsection Acalypteratae. { ,klü'sī·ə,dē }

cluster gene [GEN] A gene that codes for a multifunctional enzyme. { 'kləs·tər ,jēn }

cluster headache [MED] A type of migraine that is characterized by severe unilateral pain in the eye or temple and tends to recur in a series of attacks. { 'kləs·tər ,hed,āk }

clutch [VERT ZOO] A nest of eggs or a brood of chicks. { kləch }

Clypeasteroida [INV ZOO] An order of exocyclic Euechinoidea having a monobasal apical system in which all the genital plates fuse together. { ,klip·ē,as·tə'rȯid·ə }

clypeus [INV ZOO] An anterior medial plate on the head of an insect, commonly bearing the labrum on its anterior margin. [MYCOL] A disk of black tissue about the mouth of the perithecia in certain ascomycetes. { 'klip·ē·əs }

clysis [MED] **1.** Administration of an enema. **2.** Subcutaneous or intravenous administration of fluids. { 'klī·səs }

Clythiidae [INV ZOO] The flat-footed flies, a family of cyclorrhaphous dipteran insects in the series Aschiza characterized by a flattened distal end on the hind tarsus. { klə'thī·ə,dē }

CMV *See* cucumber mosaic virus.

Cnidaria [INV ZOO] A phylum of the Radiata whose members typically bear tentacles and possess intrinsic nematocysts. Also known as Coelenterata. { nī'dar·ē·ə }

cnidoblast [INV ZOO] A cell that produces nematocysts. Also known as cnidocyte; nettle cell; stinging cell. { 'nīd·ə,blast }

cnidocil [INV ZOO] The trigger on a cnidoblast that activates discharge of the nematocyst when touched. { 'nīd·ə,sil }

cnidocyte *See* cnidoblast. { 'knīd·ə,sīt }

cnidophore [INV ZOO] A modified structure bearing nematocysts in certain cnidarians. { 'nīd·ə,fȯr }

Cnidospora [INV ZOO] A subphylum of spore-producing protozoans that are parasites in cells and tissues of invertebrates, fishes, a few amphibians, and turtles. { nī'däs·pə·rə }

CoA *See* coenzyme A.

coadaptation [EVOL] The selection process that tends to accumulate favorably interacting genes in the gene pool of a population. { ,kō,ad·əp'tā·shən }

coagulability test [PATH] Any of several clinical tests of the ability of blood to coagulate, such as clot retraction time and quantification, prothrombin time, partial thromboplastin time, and platelet enumeration. { kō,ag·yə·lə'bil·əd·ē test }

coagulase [BIOCHEM] Any enzyme that causes coagulation of blood plasma. { kō'ag·yə,lās }

coagulation time *See* clotting time. { kō,ag·yə'lā·shən ,tīm }

coalescence [BOT] The union of plant parts of the same kind such as the united sepals of flowering plants. { ,kō·ə'les·əns }

coancestry [GEN] The degree of relationship between two parents of a diploid individual. { ,kō'an,səs·trē }

coarctation [MED] **1.** A compression of the wall of a vessel, narrowing the lumen and reducing the volume (or flow). **2.** A stricture or occlusion resulting from an outside force deforming a vessel. { ,kō·ärk'tā·shən }

coated pit [CELL MOL] A cell surface depression that is coated with clathrin on its cytoplasmic surface and functions in receptor-mediated endocytosis. { 'kōd·əd 'pit }

coated vesicle [CELL MOL] An intracellular structure formed by an invagination of the membrane surrounded by a cagelike protein coating. { ¦kōd·əd 'ves·ə·kəl }

coati [VERT ZOO] The common name for three species of carnivorous mammals assigned to the raccoon family (Procyonidae) characterized by their elongated snout, body, and tail. { kə 'wäd·ē }

cobalamin *See* vitamin B_{12}. { kə'bȯl·ə·mən }

cobalt-beam therapy [MED] Therapy involving the use of gamma radiation from a cobalt-60 source mounted in a cobalt bomb. Also known as cobalt therapy. { ¦kō,bȯlt ¦bēm 'ther·ə·pē }

cobalt therapy *See* cobalt-beam therapy. { 'kō ,bȯlt ,ther·ə·pē }

Cobitidae [VERT ZOO] The loaches, a family of small fishes, many eel-shaped, in the suborder Cyprinoidei, characterized by barbels around the mouth. { kə'bid·ə,dē }

cobra [VERT ZOO] Any of several species of venomous snakes in the reptilian family Elaphidae characterized by a hoodlike expansion of skin on the anterior neck that is supported by a series of ribs. { 'kō·brə }

coca [BOT] *Erythroxylon coca.* A shrub in the family Erythroxylaceae; its leaves are the source of cocaine. { 'kō·kə }

cocaine [PHARM] $C_{17}H_{21}O_4N$ An alkaloid obtained from coca leaves that is used for local anesthesia and as a tonic in digestive and nervous disorders. { kō'kān }

cocarboxylase *See* thiamine pyrophosphate. { ¦kō·kär'bäk·sə,lās }

cocarcinogen [MED] A noncarcinogenic agent which augments the carcinogenic process. { ¦kō·kär'sin·ə·jən }

Coccidia [INV ZOO] A subclass of protozoans in the class Telosporea; typically intracellular parasites of epithelial tissue in vertebrates and invertebrates. { käk'sid·ē·ə }

Coccidioides immitis [MED] A mold primarily found in desert soil that converts into spherules containing endospores when growing within the body and that causes coccidioidomycosis or San Joaquin valley fever. { ,käk·sid·ē¦ȯi,dēz i'mīd·əs *or* i'mēd·əs }

coccidioidomycosis [MED] An infectious fungus disease of humans and animals of either a pulmonary or a cutaneous nature; caused by *Coccidioides immitis.* Also known as San Joaquin Valley fever. { käk¦sid·ē¦ȯid·ō·mī'kō·səs }

coccidiosis [MED] The state of or the conditions associated with being infected by coccidia. { käk¦sid·ē¦ō·səs }

coccine [INV ZOO] For protozoa, denoting the sessile state during which reproduction does not occur. { 'käk,sēn }

Coccinellidae [INV ZOO] The ladybird beetles, a family of coleopteran insects in the superfamily Cucujoidea. { käk·sə'nel·ə,dē }

coccobacillus [MICROBIO] A short, thick, oval bacillus, midway between the coccus and the bacillus in appearance. { ¦kä·kō·ba'sil·əs }

coccoid [MICROBIO] A spherical bacterial cell. { 'kä,kȯid }

Coccoidea [INV ZOO] A superfamily of homopteran insects belonging to the Sternorrhyncha; includes scale insects and mealy bugs. { kä 'kȯid·ē·ə }

coccolith [BOT] One of the small, interlocking calcite plates covering members of the Coccolithophorida. { 'käk·ə,lith }

Coccolithophora [INV ZOO] An order of phytoflagellates in the protozoan class Phytamastigophorea. { ,käk·ō·li'thäf·ə·rə }

Coccolithophorida [BOT] A group of unicellular, biflagellate, golden-brown algae characterized by a covering of coccoliths. { ,käk·ō,lith·ə'fór·ə·də }

Coccomyces hiemalis *See* Blumeriella jaapii. { ,käk·ə,mī,sēz hē'mäl·əs }

Coccomyxaceae [BOT] A family of algae belonging to the Tetrasporales composed of elongate cells which reproduce only by vegetative means. { ,käk·ō,mik'sās·ē,ē }

cocculin *See* picrotoxin. { 'käk·yə·lən }

coccus [MICROBIO] A form of eubacteria which are more or less spherical in shape. { 'käk·əs }

coccygeal body [ANAT] A small mass of vascular tissue near the tip of the coccyx. { käk'sij·ē·əl 'bä,dē }

coccygectomy [MED] Surgical excision of the coccyx. { käk·sə'jek·tə·mē }

coccyx [ANAT] The fused vestige of caudal vertebrae forming the last bone of the vertebral column in humans and certain other primates. { 'käk,siks }

cochlea [ANAT] The snail-shaped canal of the mammalian inner ear; it is divided into three channels and contains the essential organs of hearing. { 'kōk·lē·ə }

cochlear duct *See* scala media. { 'kōk·lē·ər 'dəkt }

Cochleariidae [VERT ZOO] A family of birds in the order Ciconiiformes composed of a single species, the boatbill. { ,kōk·lē·ə'rī·ə,dē }

cochlear implant [NEUROSCI] A sensory prosthesis that restores some hearing to deaf people by electrically stimulating the auditory nerve. { ¦kō·klē·ər 'im,plant }

cochlear nerve [NEUROSCI] A sensory branch of the auditory nerve which receives impulses from the organ of Corti. { 'kok·lē·ər 'nərv }

cochlear nucleus [NEUROSCI] One of the two nuclear masses in which the fibers of the cochlear nerve terminate; located ventrad and dorsad to the inferior cerebellar peduncle. { 'kok·lē·ər 'nük·lē·əs }

cochleate [BIOL] Spiral; shaped like a snail shell. { 'kōk·lē,āt }

cock [VERT ZOO] The adult male of the domestic fowl and of gallinaceous birds. { käk }

cockle [INV ZOO] The common name for a number of species of marine mollusks in the class Bivalvia characterized by a shell having convex radial ribs. { 'käk·əl }

cockroach *See* roach. { 'käk,rōch }

coconut [BOT] *Cocos nucifera.* A large palm in the order Arecales grown for its fiber and fruit, a large, ovoid, edible drupe with a fibrous exocarp and a hard, bony endocarp containing fleshy meat (endosperm). { 'kō·kə,nət }

cocoon [INV ZOO] **1.** A protective case formed by the larvae of many insects, in which they pass the pupa stage. **2.** Any of the various protective egg cases formed by invertebrates. { kə'kün }

cod [VERT ZOO] The common name for fishes of the subfamily Gadidae, especially the Atlantic cod (*Gadus morrhua*). { käd }

codecarboxylase [BIOCHEM] The prosthetic component of the enzyme carboxylase which catalyzes decarboxylation of D-amino acids. Also known as pyridoxal phosphate. { ¦kō·də·kär 'bäk·sə,lās }

codehydrogenase I *See* diphosphopyridine nucleotide. { ¦kō·dē'hī·drə·jə,nās ¦wən }

codehydrogenase II *See* triphosphopyridine nucleotide. { ¦kō·dē'hī·drə·jə,nās ¦tü }

codeine [PHARM] $C_{18}H_{21}NO_3$ An alkaloid prepared from morphine; used as mild analgesic and cough suppressant. { 'kō,dēn }

Codiaceae [BOT] A family of green algae in the order Siphonales having macroscopic thalli composed of aggregates of tubes. { ,kō·dē'as·ē,ē }

coding ratio [BIOCHEM] The number of bases in nucleic acids divided by the number of amino acids whose sequence the bases determine in a particular polypeptide. { 'kōd·iŋ ,rā·shō }

coding strand *See* sense strand. { 'kōd·iŋ ,strand }

codominance [GEN] A condition in which each allele of a heterozygous pair expresses itself fully, as in human blood group AB individuals. { kō 'däm·ə·nəns }

codon [GEN] The basic unit of the genetic code, comprising sequential, nonoverlapping three-nucleotide sequences in messenger ribonucleic acid, each of which is translated into one amino acid; 61 of the 64 codons code for a specific protein synthesis; the other 3 are stop codons that specify termination of the growing polypeptide or protein chain. { 'kō,dän }

codon family [GEN] A group of codons that code for the same amino acid and differ only in the nucleotide that occupies the third codon position. { 'kō,dän ,fam·lē }

codon fidelity [CELL MOL] The constancy of the genetic coding process as maintained during deoxyribonucleic acid replication and the synthesis of proteins by a series of proofreading reactions that remove errors. { 'kō,dän fi,del·əd·ē }

codon misreading [CELL MOL] The mistranslation of a codon in messenger ribonucleic acid that increases errors in protein synthesis by generation of amino acid substitutions. { 'kō,dän mis,rēd·iŋ }

coelacanth [VERT ZOO] Any member of the Coelacanthiformes, an order of lobefin fishes represented by a single living genus, *Latimeria*. { 'sē·lə,kanth }

Coelacanthiformes [VERT ZOO] An order of lobefin fishes in the subclass Crossopterygii which were common fresh-water animals of the Carboniferous and Permian; one genus, *Latimeria*, exists today. { ¦sē·lə¦kan·thə'fȯr,mēz }

Coelacanthini [VERT ZOO] The equivalent name for Coelacanthiformes. { ,sē·lə'kan·thə,nī }

Coelenterata *See* Cnidaria. { sə,len·tə'räd·ə }

coelenteron [INV ZOO] The internal cavity of cnidarians. { sə'len·tə,rän }

coeloblastula [EMBRYO] A simple, hollow blastula with a single-layered wall. { ,sē·lō'blas·chə·lə }

coelom [ZOO] The mesodermally lined body cavity of most animals higher on the evolutionary scale than flatworms and nonsegmented roundworms. { 'sē·ləm }

Coelomata [ZOO] The equivalent name for Eucoelomata. { ,sē·lə'mad·ə }

coelomocyte [INV ZOO] A corpuscle, including amebocytes and eleocytes, in the coelom of certain animals, especially annelids. { 'sē·lō·mə,sīt }

coelomoduct [INV ZOO] Either of a pair of ciliated excretory and reproductive channels passing from the coelom to the exterior in certain invertebrates, including annelids and mollusks. { 'sē·lō·mə,dəkt }

Coelomomycetaceae [MYCOL] A family of entomophilic fungi in the order Blastocladiales which parasitize primarily mosquito larvae. { ¦sē·lō·mə,mī·sə'tās·ē,ē }

coelomostome [INV ZOO] The opening of a coelomoduct into the coelom. { 'sē·lō·mə,stōm }

Coelomycetes [MYCOL] A group set up by some authorities to include the Sphaerioidaceae and the Melanconiales. { ¦sē·lō·mī'sēd,ēz }

Coelopidae [INV ZOO] The seaweed flies, a family of myodarian cyclorrhaphous dipteran insects in the subsection Acalypteratae whose larvae breed on decomposing seaweed. { sə'lō·pə,dē }

coeloplanula [INV ZOO] A hollow planula having a wall of two layers of cells. { ,sē·lə'plan·yə·lə }

Coenagrionidae [INV ZOO] A family of zygopteran insects in the order Odonata. { ,sē,nag·rē'än·ə,dē }

coencytic [MYCOL] Pertaining to filaments or mycelia that lack septa. Also known as nonseptate. { ,kō·in'sī·tik }

coenenchyme [INV ZOO] The mesagloea surrounding and uniting the polyps in compound anthozoans. Also known as coenosarc. { sə 'neŋ,kīm }

Coenobitidae [INV ZOO] A family of terrestrial decapod crustaceans belonging to the Anomura. { ,sē·nə'bid·ə,dē }

coenobium [INV ZOO] A colony of protozoans having a constant size, shape, and cell number, but with undifferentiated cells. { sə'nō·bē·əm }

coenocyte [BIOL] A multinucleate mass of protoplasm formed by repeated nucleus divisions without cell fission. { 'sē·nə,sīt }

Coenomyidae [INV ZOO] A family of orthorrhaphous dipteran insects in the series Brachycera. { ,sē·nə'mī·ə,dē }

coenosarc [INV ZOO] **1.** The living axial part of a hydroid colony. **2.** *See* coenenchyme. { 'sē·nə,särk }

coenosteum [INV ZOO] The calcareous skeleton of a compound coral or bryozoan colony. { sə'näs·tē·əm }

Coenothecalia [INV ZOO] An order of the class Alcyonaria that forms colonies; lacks spicules but has a skeleton composed of fibrocrystalline argonite. { ˌsē·nō·thə'kāl·ē·ə }

coenotype [BIOL] An organism having the characteristic structure of the group to which it belongs. { 'sē·nəˌtīp }

coenurosis [VET MED] An infestation by a coenurus, the metacestode of *Taenia* species; most common in sheep, rabbits, and other herbivores. { ˌsē·nyə'rō·səs }

coenzyme [BIOCHEM] The nonprotein portion of an enzyme; a prosthetic group which functions as an acceptor of electrons or functional groups. { kō'enˌzīm }

coenzyme I *See* diphosphopyridine nucleotide. { kō'enˌzīm 'wən }

coenzyme II *See* triphosphopyridine nucleotide. { kō'enˌzīm 'tü }

coenzyme A [BIOCHEM] $C_{21}H_{36}O_{16}N_7P_3S$ A coenzyme in all living cells; required by certain condensing enzymes to act in acetyl or other acyl-group transfer and in fatty-acid metabolism. Abbreviated CoA. { kō'enˌzīm 'ā }

coevolution [EVOL] An evolutionary pattern based on the interaction among major groups or organisms with an obvious ecological relationship; for example, plant and plant-eater, flower and pollinator. { ¦kōˌev·ə'lü·shən }

cofactor [BIOCHEM] A specific substance required for the activity of an enzyme, such as a coenzyme or metal ion. { 'kōˌfak·tər }

coffee [BOT] Any of various shrubs or small trees of the genus *Coffea* (family Rubiaceae) cultivated for the seeds (coffee beans) of its fruit; most coffee beans are obtained from the Arabian species, *C. arabica*. { kȯf·ē }

cogon [BOT] *Imperate cylindrica*. A grass found in rainforests. Also known as alang-alang. { kō 'gōn }

cog region [BIOCHEM] Any group of similar sequences of nucleotides that occurs in deoxyribonucleic acid molecules and may specifically be recognized by endonucleases or other enzymes. { 'käg ˌrē·jən }

cohesion [BOT] The union of similar plant parts or organs, as of the petals to form a corolla. { kō'hē·zhən }

cohesive end *See* sticky end. { kō'hē·siv ˌend }

cohesive terminus [CELL MOL] Either of the ends of single-stranded deoxyribonucleic acid that are complementary in the nucleotide sequences and can join, by base pairing, to form circular molecules. { kō'hē·siv 'tər·mə·nəs }

cohort selection [EVOL] A type of natural selection due to interactions among groups of similar ages in a population. { 'kōˌhȯrt siˌlek·shən }

coho salmon [VERT ZOO] *Oncorhynchus kisutch*. A species that is widespread across both the Asian and western North American coasts, but is the rarest salmon throughout much of its range. Also known as silver salmon. { 'kōˌhō ˌsa·mən }

coiled tubular gland [ANAT] A structure having a duct interposed between the surface opening and the coiled glandular portion; an example is a sweat gland. { ¦kȯild ¦tü·byə·lər 'gland }

coincidence [GEN] A numerical value equal to the number of double crossovers observed, divided by the number expected; a measure of interference. { kō'in·sə·dəns }

coincidental evolution [EVOL] The maintenance of sequence homology among nonallelic members of a multigene family within a species. Also known as concerted evolution; horizontal evolution. { ˌkōˌin·sə¦dent·əl ˌev·ə'lü·shən }

co-inducer [BIOCHEM] A molecule that interacts with a repressor to free the operon from restraints on its transcription into messenger ribonucleic acid. { ˌkō·in'dü·sər }

cointegrate structure [CELL MOL] The circular molecule formed by fusing two replicons, one possessing a transposon, the other lacking it. { kō¦in·tə·grət 'strək·chər }

coisogenic strain [GEN] An animal strain known to differ from the inbred partner strain at a single locus. { ¦kō'i·səˌjen·ik 'strān }

coitus [ZOO] The act of copulation. Also known as intercourse. { 'kō·əd·əs }

cola [BOT] *Cola acuminata*. A tree of the sterculia family (Sterculiaceae) cultivated for cola nuts, the seeds of the fruit; extract of cola nuts is used in the manufacture of soft drinks. { 'kō·lə }

colchicum [BOT] Any plant of the genus *Colchicum*, a part of the lily family (mainly the autumn crocus). { 'käl·chi·kəm }

cold agglutination phenomenon [IMMUNOL] Clumping of human blood group O erythrocytes at 0–4°C, but not at body temperature; occurs in primary atypical pneumonia, trypanosomiasis, and other unidentifiable states. { 'kōld əˌglüt·ən'ā·shən fə'näm·əˌnän }

cold agglutinin [IMMUNOL] A nonspecific panagglutinin found in many normal human serums which produce maximum clumping of erythrocytes at 4°C and none at 37°C. { 'kōld ə'glüt·ən·ən }

cold-blooded [PHYSIO] Having body temperature approximating that of the environment and not internally regulated. { 'kōld ¦bləd·əd }

cold desert *See* tundra. { ¦kōld 'dez·ərt }

cold hemagglutination [IMMUNOL] A phenomenon caused by the presence of cold agglutinin. { 'kōld ˌhēm·əˌglüt·ən'ā·shən }

cold-induced vasodilation [MED] A sequence of vasoconstriction followed by vasodilation that acts as a protective mechanism to prevent cold weather injury to the extremities. { ¦kōld in ˌdüst ˌvā·zō·di'lā·shən }

cold injury [MED] Physical trauma following exposure to very low temperatures. { 'kōld ˌin·jə·rē }

cold-sensitive mutation [GEN] An alteration that causes a gene to be inactive at low temperature. { 'kōld ¦sen·səd·iv myü'tā·shən }

cold torpor [PHYSIO] Condition of reduced body temperature in poikilotherms. { 'kōld ˌtȯr·pər }

colectomy [MED] Excision of all or a portion of the colon. { kə'lek·tə·mē }

Coleochaetaceae [BOT] A family of green algae in the suborder Ulotrichineae; all occur as attached, disklike, or parenchymatous thalli. { ˌkō·lēˌō·kē'tās·ēˌē }

Coleoidea [INV ZOO] A subclass of cephalopod mollusks including all cephalopods except *Nautilus*, according to certain systems of classification. { ˌkō·lē'óid·ē·ə }

Coleophoridae [INV ZOO] The case bearers, moths with narrow wings composing a family of lepidopteran insects in the suborder Heteroneura; named for the silk-and-leaf shell carried by larvae. { ˌkō·lē·ō'fór·əˌdē }

Coleoptera [INV ZOO] The beetles, holometabolous insects making up the largest order of the animal kingdom; general features of the Insecta are found in this group. { ˌkō·lē'äp·tə·rə }

coleoptile [BOT] The first leaf of a monocotyledon seedling. { ˌkō·lē'äp·təl }

coleorhiza [BOT] The sheath surrounding the radicle in monocotyledons. { ˌkō·lē·ə'rīz·ə }

Coleorrhyncha [INV ZOO] A monofamilial group of homopteran insects in which the beak is formed at the anteroventral extremity of the face and the propleura form a shield for the base of the beak. { ˌkō·lē·ə'riŋ·kə }

Coleosporaceae [MYCOL] A family of parasitic fungi in the order Uredinales. { ˌkō·lē·ō·spə'rās·ēˌē }

colic [MED] **1.** Acute paroxysmal abdominal pain usually caused by smooth muscle spasm, obstruction, or twisting. **2.** In early infancy, paroxysms of pain, crying, and irritability caused by swallowing air, overfeeding, intestinal allergy, and emotional factors. { 'käl·ik }

colic artery [ANAT] Any of the three arteries that supply the colon. { 'käl·ik 'ärd·ə·rē }

colicin [MICROBIO] A bacteriocin produced by coliform bacteria, such as *Escherichia coli*. { 'kä·lə·sən }

coliform bacteria [MICROBIO] Colon bacilli, or forms which resemble or are related to them. { 'kä·ləˌfórm bak'tir·ē·ə }

Coliidae [VERT ZOO] The colies or mousebirds, composing the single family of the avian order Coliiformes. { kə'lī·əˌdē }

Coliiformes [VERT ZOO] A monofamilial order of birds distinguished by long tails, short legs, and long toes, all four of which are directed forward. { kəˌlī·ə'fórˌmēz }

colinearity [CELL MOL] The relationship between the linear sequence of codons in deoxyribonucleic acid and the order of amino acids in the polypeptide product that it specifies. Also spelled collinearity. { kōˌlin·ē'ar·əd·ē }

coliphage [VIROL] Any bacteriophage able to infect *Escherichia coli*. { 'kä·ləˌfāj }

colistin [MICROBIO] $C_{45}H_{85}O_{10}N_{13}$ A basic polypeptide antibiotic produced by *Bacillus colistinus*; consists of an A and B component, active against a broad spectrum of gram-positive microorganisms and some gram-negative microorganisms. { kə'lis·tən }

colitis [MED] Inflammation of the large bowel, or colon. { kə'līd·əs }

collagen [BIOCHEM] A fibrous protein found in all multicellular animals, especially in connective tissue. { 'kä·lə·jən }

collagenase [BIOCHEM] Any proteinase that decomposes collagen and gelatin. { 'kä·lə·jəˌnās }

collagen disease [MED] Any of various clinical syndromes characterized by widespread alterations of connective tissue, including inflammation and fibrinoid degeneration. { 'kä·lə·jən diˌzēz }

collar cell *See* choanocyte. { 'käl·ər ˌsel }

collard [BOT] *Brassica oleracea* var. *acephala*. A biennial crucifer of the order Capparales grown for its rosette of edible leaves. { 'käl·ərd }

collateral [ANAT] A side branch of a blood vessel or nerve. { kə'lad·ə·rəl }

collateral bud [BOT] An accessory bud produced beside an axillary bud. { kə¦lad·ə·rəl 'bəd }

collateral bundle [BOT] A vascular bundle in which the phloem and xylem lie on the same radius, with the phloem located toward the periphery of the stem and the xylem toward the center. { kə¦lad·ə·rəl 'bənd·əl }

collateral circulation [PHYSIO] The circulation established for an organ or a part of an organ through the intercommunication of blood vessels when the original direct blood supply is obstructed or abolished. { kə¦lad·ə·rəl ˌsər·kyə'lā·shən }

collateral fiber [NEUROSCI] A lateral branch of an axon. { kə¦lad·ə·rəl 'fīb·ər }

collateral ligament [ANAT] Any of various stabilizing ligaments on either side of a hinge joint such as the knee or elbow. { kə¦lad·ə·rəl 'lig·ə·mənt }

collateral respiration [PHYSIO] The passage of air between lobules within the same lobe of a lung, enabling ventilation of a lobule whose branchiole is obstructed. { ke¦lad·ə·rəl ˌres·pə'rā·shən }

collecting tubule [ANAT] One of the ducts conveying urine from the renal tubules (nephrons) to the minor calyces of the renal pelvis. { kə'lek·tiŋ ˌtüˌbyül }

Collembola [INV ZOO] The springtails, an order of primitive insects in the subclass Apterygota having six abdominal segments. { kə'lem·bə·lə }

collenchyma [BOT] A primary, or early-differentiated, subepidermal supporting tissue in leaf petioles and vein ribs formed before vascular differentiation. { kə'leŋ·kə·mə }

collenchyme [INV ZOO] A loose mesenchyme that fills the space between ectoderm and endoderm in the body wall of many lower invertebrates, such as sponges. { 'kä·lənˌkīm }

Colles' fracture [MED] A fracture of the radius about 1 inch (2.5 centimeters) above the wrist with dorsal displacement of the distal fragment. { 'kä·lə·səz ˌfrak·chər }

Colletidae [INV ZOO] The colletid bees, a family of hymenopteran insects in the superfamily Apoidea. { kə'led·əˌdē }

colliculus [ANAT] **1.** Any of the four prominences of the corpora quadrigemina. **2.** The anterolateral, apical elevation of the arytenoid cartilages. [NEUROSCI] The elevation of the optic nerve where it enters the retina. { kə'lik·yə·ləs }

collinearity *See* colinearity. { kō,lin·ē'ar·əd·ē }

colloblast [INV ZOO] An adhesive cell on the tentacles of ctenophores. { 'käl·ə,blast }

colloidal osmotic pressure *See* oncotic pressure. { kə'lȯid·əl äz'mäd·ik ,presh·ər }

colloid goiter [MED] A soft, diffuse enlargement of the thyroid gland in which colloid fills the acinar spaces. { ¦käl,ȯid ¦gȯid·ər }

Collothecacea [INV ZOO] A monofamilial suborder of mostly sessile rotifers in the order Monogonata; many species are encased in gelatinous tubes. { ,käl·ə·thə'kās·ē·ə }

Collothecidae [INV ZOO] The single family of the Collothecacea. { ,käl·ə'thes·ə,dē }

coloboma [MED] A congenital, pathologic, or operative fissure, especially of the eye or eyelid. { ,käl·ə'bō·mə }

colon [ANAT] The portion of the human intestine extending from the cecum to the rectum; it is divided into four sections: ascending, transverse, descending, and sigmoid. Also known as large intestine. { 'kō·lən }

colon bacillus *See* Escherichia coli. { 'kō·lən bə'sil·əs }

colonization [ECOL] The establishment of an immigrant species in a peripherally unsuitable ecological area; occasional gene exchange with the parental population occurs, but generally the colony evolves in relative isolation and in time may form a distinct unit. { ,käl·ə·nə'zā·shən }

colonoscopy [MED] Visual examination of the inner surface of the colon by means of an endoscope. { ,kō·lə'näs·kə·pē }

colony [BIOL] A localized population of individuals of the same species which are living either attached or separately. [MICROBIO] A cluster of microorganisms growing on the surface of or within a solid medium; usually cultured from a single cell. { 'käl·ə·nē }

colony count [MICROBIO] The number of colonies of bacteria growing on the surface of a solid medium. { 'käl·ə·nē ,kau̇nt }

colony-stimulating factor [IMMUNOL] A group of lymphocytes that induce the maturation and proliferation of leukocyte, macrophage, and monocyte lines present in bone marrow. { 'käl·ə·nē ¦stim·yə,lād·iŋ ,fak·tər }

Colorado tick fever [MED] A nonexanthematous acute viral disease of humans occurring in the western United States and transmitted by a bite of the tick *Dermacentor andersoni*; characterized by a short course, intermittent fever, leukopenia, and occasionally meningoencephalitis. Also known as mountain tick fever. { ,kal·ə'rad·ō 'tik ,fē·vər }

color blindness [MED] Inability to perceive one or more colors. { 'kəl·ər ,blīnd·nəs }

color index [PATH] The amount of hemoglobin per erythrocyte relative to normal, equal to the percent normal hemoglobin concentration divided by percent normal erythrocyte count. Abbreviated CI. { 'kəl·ər ,in,deks }

color vision [PHYSIO] The ability to discriminate light on the basis of wavelength composition. { 'kəl·ər ,vizh·ən }

Colossendeidae [INV ZOO] A family of deep-water marine arthropods in the subphylum Pycnogonida, having long palpi and lacking chelifores, except in polymerous forms. { ,käl·ə·sen'dā·ə,dē }

colostomy [MED] Surgical formation of an artificial anus by joining the colon to an opening in the anterior abdominal wall. { kə'läs·tə·mē }

colostrum [PHYSIO] The first milk secreted by the mammary gland during the first days following parturition. { kə'läs·trəm }

colotomy [MED] Incision of the colon; may be abdominal, lateral, lumbar, or iliac, according to the region of entrance. { kə'läd·ə·mē }

colposcope [MED] An instrument for the visual examination of the vagina and cervix; a vaginal speculum. { 'käl·pə,skōp }

colposcopy [MED] Visual examination of cells of the vagina and cervix by means of an endoscope. { käl'päs·kə·pē }

colpotomy [MED] Incision of the vagina. { käl 'päd·ə·mē }

Colubridae [VERT ZOO] A family of cosmopolitan snakes in the order Squamata. { kə'lü·brə,dē }

Columbidae [VERT ZOO] A family of birds in the order Columbiformes composed of the pigeons and doves. { kə'ləm·bə,dē }

Columbiformes [VERT ZOO] An order of birds distinguished by a short, pointed bill, imperforate nostrils, and short legs. { kə,ləm·bə'fȯr,mēz }

columella *See* stapes. [BIOL] Any part shaped like a column. [BOT] A sterile axial body within the capsules of certain mosses, liverworts, and many fungi. { ,käl·yə'mel·ə }

columnar epithelium [HISTOL] Epithelium distinguished by elongated, columnar, or prismatic cells. { kə'ləm·nər ep·ə'thēl·ē·əm }

columnar stem [BOT] An unbranched, cylindrical stem bearing a set of large leaves at its summit, as in palms, or no leaves, as in cacti. { kə'ləm·nər ,stem }

Colydiidae [INV ZOO] The cylindrical bark beetles, a large family of coleopteran insects in the superfamily Cucujoidea. { käl·ə'dī·ə,dē }

coma [MED] Unconsciousness from which the patient cannot be aroused. { 'kō·mə }

Comasteridae [INV ZOO] A family of radially symmetrical Crinozoa in the order Comatulida. { ,kō·mə'ster·ə,dē }

comatose [MED] In a condition of coma; resembling coma. { 'käm·ə,tōs }

Comatulida [INV ZOO] The feather stars, an order of free-living echinoderms in the subclass Articulata. { ,kō·mə'tül·ə·də }

comb [INV ZOO] **1.** A system of hexagonal cells constructed of beeswax by a colony of bees. **2.** A comblike swimming plate in ctenophores.

[VERT ZOO] A crest of naked tissue on the head of many male fowl. { kōm }

comb growth unit [BIOL] A unit for the standardization of male sex hormones. { 'kōm ˌgrōth ˌyü·nət }

combinatorial control [CELL MOL] Control of gene expression requiring presence or absence of a particular combination of regulatory proteins. { ˌkäm·bə·nəˈtȯr·ē·əl kənˈtrōl }

combined immunological deficiency disease [MED] A severe and usually fatal disease in which the individual lacks not only the T (thymus-derived) cells, which are responsible for graft rejection and for defense against viruses, but also the B (bone-marrow-derived) cells, which are responsible for production of globulins and antibodies. Also known as CID disease. { kəmˈbīnd ˌim·yə·nəˈläj·ə·kəl dəˈfish·ən·sē diˌzēz }

comedo [MED] A collection of sebaceous material and keratin retained in the hair follicle and excretory duct of the sebaceous gland, whose surface is covered with a black dot caused by oxidation of sebum at the follicular orifice. Also known as blackhead. { 'kä·məˌdō }

comedocarcinoma [MED] A type of adenocarcinoma of the breast in which the ducts are filled with cells which, when expressed from the cut surface, resemble comedos. { ¦kä·mə·dōˌkärs·ənˈō·mə }

Comesomatidae [INV ZOO] A family of free-living nematodes in the superfamily Comesomatoidea found as deposit feeders on soft bottom sediments. { ˌkō·mə·sōˈmad·əˌdē }

Comesomatoidea [INV ZOO] A superfamily of marine nematodes in the order Chromadorida distinguished by their wide multispiral amphids that make at least two complete turns. { ˌkō·mə·sō·məˈtȯid·ē·ə }

cometabolism [ECOL] A process in which compounds not utilized for growth or energy are transformed to other products by microorganisms. { ˌkō·məˈtab·əˌliz·əm }

Commelinaceae [BOT] A family of monocotyledonous plants in the order Commelinales characterized by differentiation of the leaves into a closed sheath and a well-defined, commonly somewhat succulent blade. { ˌkä·mə·ləˈnās·ēˌē }

Commelinales [BOT] An order of monocotyledonous plants in the subclass Commelinidae marked by having differentiated sepals and petals but lacking nectaries and nectar. { ˌkä·mə·ləˈnā·lēz }

Commelinidae [BOT] A subclass of flowering plants in the class Liliopsida. { ˌkä·məˈlī·nəˌdē }

commensal [ECOL] An organism living in a state of commensalism. { kəˈmen·səl }

commensalism [ECOL] An interspecific, symbiotic relationship in which two different species are associated, wherein one is benefited and the other neither benefited nor harmed. { kəˈmen·səˌliz·əm }

commercial lecithin *See* lecithin. { kəˈmər·shəl 'les·ə·thən }

commissure [BIOL] A joint, seam, or closure line where two structures unite. { 'käm·əˌshu̇r }

commissurotomy [MED] The surgical destruction of a commissure, usually the anterior commissure, particularly in the treatment of certain psychiatric disorders. { ˌkäm·əˌshu̇ˈräd·ə·mē }

commitment [CELL MOL] The establishment of a unique developmental sequence in a cell that differs from the prior state. { kəˈmit·mənt }

common bile duct [ANAT] The duct formed by the union of the hepatic and cystic ducts. { ¦käm·ən 'bīl ˌdəkt }

common carotid artery *See* carotid artery. { ¦käm·ən kəˈräd·əd 'ärd·ə·rē }

common cold [MED] A viral disease of humans most frequently caused by the rhinovirus and accompanied by inflammation of the mucous membranes of the nose, throat, and eyes. { ¦käm·ən 'kōld }

common hepatic duct *See* hepatic duct. { ¦käm·ən heˈpad·ik 'dəkt }

common iliac artery *See* iliac artery. { ¦käm·ən ˌil·ēˌak 'ärd·ə·rē }

communicable disease [MED] An infectious disease that can be transmitted from one individual to another either directly by contact or indirectly by fomites and vectors. { kəˈmyü·nə·kə·bəl diˌzēz }

communicating hydrocephaly [MED] A form of hydrocephaly in which there is normal communication between the ventricles and the subarachnoid space of the brain. { kəˈmyü·nəˌkād·iŋ ˌhī·drōˈsef·ə·lē }

communicating junction *See* gap junction. { kəˈmyü·nəˌkād·iŋ ˌjəŋk·shən }

communication disorder [MED] An interference with an individual's ability to comprehend or express ideas, experiences, knowledge, and feelings. { kəˌmyü·nəˈkā·shən ˌdisˈȯrd·ər }

community [ECOL] Aggregation of organisms characterized by a distinctive combination of two or more ecologically related species; an example is a deciduous forest. Also known as ecological community. { kəˈmyü·nə·dē }

community classification [ECOL] Arrangement of communities into classes with respect to their complexity and extent, their stage of ecological succession, or their primary production. { kəˈmyü·nə·dē ˌklas·ə·fəˈkā·shən }

comose [BOT] Having a tuft of soft hairs. { 'kōˌmōs }

Comoviridae [VIROL] A family of plant viruses that are characterized by icosahedral particles containing a bipartite genome consisting of two single-stranded positive-sense polyadenylated ribonucleic acid molecules; includes the genus *Nepovirus*. { ˌkō·məˈvir·əˌdī }

Comovirus [VIROL] A genus of plant viruses belonging to the family Comoviridae; the type species is cowpea mosaic virus. Also known as cowpea mosaic virus group. { 'kō·məˌvī·rəs }

companion cell [BOT] A specialized parenchyma cell occurring in close developmental and physiologic association with a sieve-tube member. { kəm'pan·yən ˌsel }

comparative embryology [EMBRYO] A branch of embryology that deals with the similarities and differences in the development of animals or plants of different orders. { kəm'par·əd·iv ˌem·brē'äl·ə·jē }

comparative genomic hybridization [GEN] A method that uses fluorescence in situ hybridization and comparison of the strength of hybridization signal to determine any differences in copy number of deoxyribonucleic acid sequences anywhere in the nuclear genome. { kəm¦par·əd·iv jəˌnō·mik ˌhī·brəd·ə'zā·shen }

comparative pathology [MED] Investigation and comparison of disease in various animals, including humans, to arrive at resemblances and differences which may clarify disease as a phenomenon of nature. { kəm'par·əd·iv pə'thäl·ə·jē }

compartment [CELL MOL] Any of the membrane-bound organelles within cells. { kəm 'pärt·mənt }

compatibility [IMMUNOL] Ability of two bloods or other tissues to unite and function together. [PHARM] The capacity of two or more ingredients in a medicine to mix without chemical change or loss of therapeutic effectiveness. { kəmˌpad·ə'bil·ə·dē }

compensation point [BOT] The light intensity at which the amount of carbon dioxide released in respiration equals the amount used in photosynthesis, and the amount of oxygen released in photosynthesis equals the amount used in respiration. { ˌkäm·pən'sā·shən ˌpȯint }

compensatory emphysema [MED] Simple, nonobstructive overdistension of lung segments or an entire lung in intrathoracic adaptation to collapse, destruction, or removal of portions of the lung or the opposite lung. { kəm'pen·səˌtȯr·ē ˌem·fə'sē·mə }

compensatory hypertrophy [MED] An increase in the size of an organ following injury or removal of the opposite paired organ or of part of the same organ. { kəm'pen·səˌtȯr·ē hī'pər·trə·fē }

competence [EMBRYO] The ability of a reacting system to respond to the inductive stimulus during early developmental stages. { 'käm·pəd·əns }

competent cell [CELL MOL] A cell that is able to incorporate exogenous deoxyribonucleic acid and undergo genetic transformation. [EMBRYO] A cell that can respond to an inducer during embryonic development. { ¦käm·pəd·ənt 'sel }

competition [ECOL] The inter- or intraspecific interaction resulting when several individuals share an environmental necessity. { ˌkäm·pə'tish·ən }

competitive displacement [ECOL] The inability of a species to successfully live in an area because a second species dominates local resources. { kəmˌped·əd·iv di'splās·mənt }

competitive enzyme inhibition [BIOCHEM] Prevention of an enzymatic process resulting from the reversible interaction of an inhibitor with a free enzyme. { kəm'ped·əd·iv 'enˌzīm ˌin·ə 'bish·ən }

competitive exclusion [ECOL] The result of a competition in which one species is forced out of part of the available habitat by a more efficient species. { kəm'ped·əd·iv iks'klüzh·ən }

competitive-exclusion principle *See* Gause's principle. { kəm'ped·əd·iv iks'klüzh·ən ˌprin·sə·pəl }

competitive inhibition [BIOCHEM] Enzyme inhibition in which the inhibitor competes with the natural substrate for the active site of the enzyme; may be overcome by increasing substrate concentration. { kəm'ped·əd·iv ˌin·ə'bish·ən }

compital [BOT] **1.** Of the vein of a leaf, intersecting at a wide angle. **2.** Of a fern, bearing sori at the intersection of two veins. { 'käm·pəd·əl }

complement [IMMUNOL] A heat-sensitive, complex system in fresh human and other sera which, in combination with antibodies, is important in the host defense mechanism against invading microorganisms. { 'käm·plə·mənt }

complemental air [PHYSIO] The amount of air that can still be inhaled after a normal inspiration. { ˌkäm·plə'ment·əl 'er }

complementary base pairing [CELL MOL] The formation of weak hydrogen bonds between complementary nitrogenous bases (for example, guanine and cytosine) on opposite strands of a double-stranded nucleic acid molecule (such as deoxyribonucleic acid), contributing to the overall stability of the double-stranded structure. { ˌkäm·plə¦men·tə·rē ¦bās 'per·iŋ }

complementary deoxyribonucleic acid [CELL MOL] A deoxyribonucleic acid molecule that is synthesized by reverse transcriptase from a ribonucleic acid template. Abbreviated cDNA. Also known as copy DNA. { ˌkäm·plə¦men·trē ˌdē¦äk·sē¦rī·bōˌnü¦klē·ik 'as·əd }

complementary deoxyribonucleic acid library [CELL MOL] A collection of complementary deoxyribonucleic acid molecules, representative of all the various messenger ribonucleic acid molecules produced by a specific type of cell of a given species, spliced into a corresponding collection of DNA vectors. { ˌkäm·plə¦men·trē dē¦äk·sē¦rīb·ōˌnü¦klē·ik 'as·əd }

complementary genes [GEN] Nonallelic genes for which a wild type allele of one abolishes the phenotypic effect of a mutation of the other. { ˌkäm·plə'men·trē 'jēnz }

complementation group [GEN] Different mutations in the same cistron or gene. { ˌkäm·plə·mən'tā·shən ˌgrüp }

complementation test [GEN] An analytic procedure for determining whether two non expressed mutants are in the same cistron or gene. { ˌkäm·plə·mən'tā·shən ˌtest }

complement cascade [IMMUNOL] The sequential activation of complement proteins resulting in lysis of a target cell. { 'käm·plə·mənt kas 'kād }

complement fixation [IMMUNOL] The binding of complement to an antigen-antibody complex so that the complement is unavailable for subsequent reaction. { 'käm·plə·mənt ˌfik'sā·shən }

complement-fixation test [IMMUNOL] A diagnostic test to determine the presence of antigen or antibody in the blood by adding complement to the test system; used especially in diagnosing syphilis. { 'käm·plə·mənt ˌfik'sā·shən ˌtest }

complete antibody [IMMUNOL] An antibody that can directly agglutinate saline-suspended red blood cells. { kəm¦plēt 'ant·iˌbäd·ē }

complete blood count [PATH] Differential and absolute determinations of the numbers of each type of blood cell in a sample and, by extrapolation, in the general circulation. { kəm'plēt 'bləd ˌkau̇nt }

complete flower [BOT] A flower having all four floral parts, that is, having sepals, petals, stamens, and carpels. { kəm'plēt 'flau̇·ər }

complete leaf [BOT] A dicotyledon leaf consisting of three parts: blade, petiole, and a pair of stipules. { kəm'plēt 'lēf }

complex *See* syndrome. { 'kämˌpleks }

complicate [INV ZOO] Folded lengthwise several times, as applied to insect wings. { 'käm·pləˌkāt }

Compositae [BOT] The single family of the order Asterales; perhaps the largest family of flowering plants, it contains about 19,000 species. { kəm'päz·əˌtē }

composite gene [GEN] Any gene arising by recombination between two nonallelic genes and containing portions of both genes. { kəm'päz·ət 'jēn }

composite nerve [NEUROSCI] A nerve containing both sensory and motor fibers. { kəm'päz·ət 'nərv }

compound acinous gland [ANAT] A structure with spherical secreting units connected to many ducts that empty into a common duct. { 'käm ˌpau̇nd 'as·ə·nəs ˌgland }

compound eye [INV ZOO] An eye typical of crustaceans, insects, centipedes, and horseshoe crabs, constructed of many functionally independent photoreceptor units (ommatidia) separated by pigment cells. { 'kämˌpau̇nd 'ī }

compound gland [ANAT] A secretory structure with many ducts. { 'kämˌpau̇nd 'gland }

compound layering [BOT] A plant propagation technique in which more than one portion of the same stem is buried. { ¦kämˌpau̇nd 'lā·ər·iŋ }

compound leaf [BOT] A type of leaf with the blade divided into two or more separate parts called leaflets. { 'kämˌpau̇nd 'lēf }

compound pistil [BOT] A pistil composed of two or more united carpels. { 'kämˌpau̇nd 'pis·təl }

compound sugar *See* oligosaccharide. { 'käm ˌpau̇nd 'shu̇g·ər }

compound tubular-acinous gland [ANAT] A structure in which the secreting units are simple tubes with acinous side chambers and all are connected to a common duct. { 'kämˌpau̇nd ¦tüb·yə·lər ¦as·ə·nəs ˌgland }

compound tubular gland [ANAT] A structure having branched ducts between the surface opening and the secreting portion. { 'käm ˌpau̇nd 'tüb·yə·lər ˌgland }

compressed-air illness *See* caisson disease. { kəm¦prest ¦er 'il·nəs }

compression syndrome *See* crush syndrome. { kəm'presh·ən ˌsinˌdrōm }

compression wood [BOT] Dense wood found at the base of some tree trunks and on the undersides of branches. { kəm'presh·ən ˌwu̇d }

computed tomography *See* computerized tomography. { kəm'pyüd·əd tə'mä·grə·fē }

computer forensics [FOREN] The study of evidence from attacks on computer systems in order to learn what has occurred, how to prevent it from recurring, and the extent of the damage. { kəm¦pyüd·ər fə'ren·ziks }

computerized axial tomography *See* computerized tomography. { kəm¦pyüd·əˌrīzd ¦ak·sē·əl tə'mä·grə·fē }

computerized tomography [MED] The process of producing a picture showing human body organs in cross section by first electronically detecting the variation in x-ray transmission through the body section at different angles, and then using this information in a digital computer to reconstruct the x-ray absorption of the tissues at an array of points representing the cross section. Abbreviated CT. Also known as computed tomography; computerized axial tomography (CAT). { kəm'pyüd·əˌrīzd tə'mäg·rə·fē }

conarium *See* pineal body. { kō'nar·ē·əm }

concatamer [CELL MOL] Two or more identical linear molecules (for example, deoxyribonucleic acid and ribonucleic acid) covalently linked in tandem. { kən'kad·ə·mər }

Concato's disease *See* polyserositis. { kōŋ'käd·ōz diz'ēz }

concentration-dilution test [PATH] A renal function test to measure the ability of the kidney to concentrate and dilute urine under stress; specific gravity for urine from a normal kidney fluctuates from 1.030 on a restricted fluid intake to 1.003 on a high water intake. { ˌkän·sən'trā·shən də'lü·shən ˌtest }

concentric bundle [BOT] A vascular bundle in which xylem surrounds phloem, or phloem surrounds xylem. { kən'sen·trik 'bən·dəl }

conceptacle [BOT] A cavity which is shaped like a flask with a pore opening to the outside, contains reproductive structures, and is bound in a thallus such as in the brown algae. { kən'sep·tə·kəl }

conception [BIOL] Fertilization of an ovum by the sperm resulting in the formation of a viable zygote. { kən'sep·shən }

conceptus [BIOL] The product of a conception, including the embryo or fetus and extraembryonic membranes, at any stage of development from fertilization to birth. { kən'sep·təs }

concerted evolution *See* coincidental evolution. { kən'sərd·əd ev·ə'lü·shən }

conch [INV ZOO] The common name for several species of large, colorful gastropod mollusks of the family Strombidae; the shell is used to make cameos and porcelain. { käŋk }

conchiolin [BIOCHEM] A nitrogenous substance that is the organic basis of many molluscan shells. { käŋ'kī·ə·lən }

Conchorhagae [INV ZOO] A suborder of benthonic wormlike animals in the class Kinorhyncha. { käŋ'kȯr·ə,gē }

Conchostraca [INV ZOO] An order of mussellike crustaceans of moderate size belonging to the subclass Branchiopoda. { käŋ'käs·trə·kə }

concordance [GEN] Similarity in appearance of members of a twin pair with respect to one or more specific traits. { kən'kȯrd·əns }

concordant segregation [GEN] The simultaneous appearance or disappearance of gene markers in hybrid cells undergoing chromosome loss. { kən'kȯrd·ənt ,seg·rə'gā·shən }

concrescence [BIOL] Convergence and fusion of parts originally separate, as the lips of the blastopore in embryogenesis. { kən'krēs·əns }

concretionary [PATH] A compact mass of inorganic material formed in a body cavity or in tissue. { kən'krē·shə,ner·ē }

concurrent infection [MED] Two or more forms of an infection existing simultaneously. { kən'kər·ənt in'fek·shən }

concussion [MED] A state of shock following traumatic injury, especially cerebral trauma, in which there is temporary functional impairment without physical evidence of damage to impaired tissues. { kən'kəsh·ən }

condensin [CELL MOL] The class of proteins that form condensin complexes, which drive the coiling of interphase chromosomes, leading to their condensation prior to mitosis. { kən 'den·sən }

condensin complex [CELL MOL] One of the protein complexes that drive the coiling of interphase chromosomes, leading to their condensation prior to mitosis. { kən'den·sən ¦käm,pleks }

conditional lethal mutant [GEN] A lethal mutant that expresses characteristics of the wild type when grown under certain conditions, as at a particular temperature, and mutant characteristics under other conditions. { kən'dish·ən·əl 'lē·thəl 'myüt·ənt }

condor [VERT ZOO] *Vultur gryphus.* A large American vulture having a bare head and neck, dull black plumage, and a white neck ruff. { 'kän ,dȯr }

conduction deafness [MED] Deafness involving an impairment of the mechanism that conducts sound to the sense organ. { kən'dək·shən ,def·nəs }

conductive hearing loss [PHYSIO] Failure of sound to be transmitted properly to the receptors in the inner ear so that sounds must be made louder to be heard. { kən,dək·tiv 'hir·iŋ ,lȯs }

conductive paste [MED] **1.** A substance applied to the skin to lower its electrical resistance in an area to which electrodes will be applied. **2.** A gel applied to the skin to lower its acoustical impedance and to accommodate an ultrasonic probe; enables ultrasonic energy to penetrate to underlying tissues without severe attenuation at the skin interface. { kən'dək·tiv 'päst }

conduplicate [BOT] Folded lengthwise and in half with the upper faces together, applied to leaves and petals in the bud. { kən'düp·lə·kət }

condyle [ANAT] A rounded bone prominence that functions in articulation. [BOT] The antheridium of certain stoneworts. [INV ZOO] A rounded, articular process on arthropod appendages. { 'kän,dīl }

condyloid articulation [ANAT] A joint, such as the wrist, formed by an ovoid surface that fits into an elliptical cavity, permitting all movement except rotation. { 'kän·də,lȯid är,tik·yə'lā·shən }

condyloma acuminata [MED] A venereal disease characterized by wartlike growths on the genital organs; thought to be of viral origin. { ,kän·də'lō·mə ə,kyü·mə'näd·ə }

cone [BOT] The ovulate or staminate strobilus of a gymnosperm. [HISTOL] A photoceptor of the vertebrate retina that responds differentially to light across the visible spectrum, providing both color vision and visual acuity in bright light. { kōn }

confocal laser-scanning fluorescence microscopy [BIOPHYS] A technique that allows three-dimensional microscopic image sectioning of a specimen such as a cell by scanning the object step by step with a laser spot, instead of illuminating it as a whole. From each point the fluorescence is measured and, after analog-to-digital conversion, stored as a matrix in computer memory. { ,kan¦fō·kəl ,lā·zər ,skan·iŋ flə¦res·ənt mī'kräs·kə·pē }

confocal microscope [BIOPHYS] A microscope that creates high-resolution images of very small objects by using a condenser lens to focus the illuminating light from a point source into a very small diffraction-limited spot within the specimen, and an objective lens to focus the light emitted from that spot onto a small pinhole in an opaque screen. { ¦kän,fō·kəl 'mī·krə,skōp }

congeneric [SYST] Referring to the species of a given genus. { ¦kän·jə'ner·ik }

congenic [GEN] Describing organisms that differ in genotype at a specified locus. { kən 'jen·ik }

congenic strain [GEN] An animal line that differs from its inbred partner strain by only a short chromosomal segment that includes the differential locus (the locus to be studied). { kən¦jen·ik 'strān }

congenital [MED] Dating from or existing before birth. { kən'jen·əd·əl }

congenital agammaglobulinemia [MED] A congenital deficiency (in serum) of immunoglobulins, characterized clinically by increased susceptibility to bacterial infections; may be a sex-linked recessive tract, affecting male infants, or sporadic, affecting both sexes. { kən'jen·əd·əl ā¦gam·ə,gläb·yə·lə'nē·mē·ə }

congenital anomaly [MED] A structural or functional abnormality of the human body that develops before birth. Also known as birth defect. { kən'jen·əd·əl ə'näm·ə·lē }

congenital disease [MED] Any disorder or disease state that is present at birth. { kən'jen·əd·əl di'zēz }

congenital pathology [MED] The study of diseases and defects existing at birth. { kən'jen·əd·əl pə'thäl·ə·jē }

congestin [BIOCHEM] A toxin produced by certain sea anemones. { kən'jes·tən }

congestion [MED] An abnormal accumulation of fluid, usually blood, but occasionally bile or mucus, within the vessels of an organ or part. { kən'jes·chən }

congestive heart failure [MED] A state in which circulatory congestion exists as a result of heart failure. { kən'jes·tiv 'härt ,fāl·yər }

conglutination [IMMUNOL] The completion of an agglutinating system, or the enhancement of an incomplete one, by the addition of certain substances. [MED] Abnormal union of two contiguous surfaces or bodies. { kən,glüt·ən'ā·shən }

conglutination phenomenon [IMMUNOL] Clumping of cells or particles, such as red cells or bacteria, when treated with conglutinin in the presence of antibody and nonhemolytic complement. { kən,glüt·ən'ā·shən fə'näm·ə·nən }

conglutinin [IMMUNOL] A heat-stable substance in bovine and other serums that aids or causes agglomeration or lysis of certain sensitized cells or particles. { kən'glüt·ən·ən }

congo red test [PATH] Diagnostic test for amyloidosis, in which congo red is injected intravenously; 30% disappears within 1 hour in normal individuals, but in amyloidosis 40–100% disappears. { 'käŋ·gō 'red ,test }

congression [CELL MOL] The movement of chromosomes to the spindle equator during mitosis. Also known as chromosome congression. { kəŋ'gresh·ən }

Conidae [INV ZOO] A family of marine gastropod mollusks in the order Neogastropoda containing the poisonous cone shells. { 'kän·ə,dē }

conidiophore [MYCOL] A specialized aerial hypha that produces conidia in certain ascomycetes and imperfect fungi. { kə'nid·ē·ə,fȯr }

conidiospore *See* conidium. { kə'nid·ē·ə,spȯr }

conidium [MYCOL] Unicellular, asexual reproductive spore produced externally upon a conidiophore. Also known as conidiospore. { kə 'nid·ē·əm }

conifer [BOT] The common name for plants of the order Pinales. { 'kän·ə·fər }

Coniferales [BOT] The equivalent name for Pinales. { kə,nif·ə'rā·lēz }

Coniferophyta [BOT] The equivalent name for Pinicae. { kə,nif·ə'räf·əd·ə }

coniferous forest [ECOL] An area of wooded land predominated by conifers. { kə'nif·ə·rəs 'fär·əst }

coniine [PHARM] $C_5H_{10}NC_3H_7$ A colorless, oily liquid with a mousy odor and a boiling point of 166°C; soluble in alcohol, ether, and oils; used as a sedative. Also known as propylpiperidine. { 'kō·nē·ən }

coniine hydrobromide [PHARM] $C_8H_{18}BrN$ Prismatic crystals melting at 211°C; soluble in chloroform, water, and alcohol; used as an antispasmodic drug. { 'kō·nē·ən ,hī·drə'brō,mīd }

conjoint tendon [ANAT] The common tendon of the transverse and internal oblique muscles of the abdomen. { kən'jȯint 'ten·dən }

Conjugales [BOT] An order of fresh-water green algae in the class Chlorophyceae distinguished by the lack of flagellated cells, and conjugation being the method of sexual reproduction. { ,kän·jə'gā·lēz }

conjugase [BIOCHEM] Any of a group of enzymes which catalyze the breakdown of pteroylglutamic acid. { 'kän·jə,gās }

conjugate division [MYCOL] Division of dikaryotic cells in certain fungi in which the two haploid nuclei divide independently, each daughter cell receiving one product of each nuclear division. { 'kän·jə·gət də'vizh·ən }

conjugated protein [BIOCHEM] A protein combined with a nonprotein group, other than a salt or a simple protein. { 'kän·jə,gād·əd 'prō,tēn }

conjugation [BOT] Sexual reproduction by fusion of two protoplasts in certain thallophytes to form a zygote. [INV ZOO] Sexual reproduction by temporary union of cells with exchange of nuclear material between two individuals, principally ciliate protozoans. [MICROBIO] A process involving contact between two bacterial cells during which genetic material is passed from one cell to the other. { ,kän·jə'gā·shən }

conjugon [GEN] Any of a number of different genetic elements in bacterial deoxyribonucleic acid that promote bacterial conjugation and gene transfer. { 'kän·jə,gän }

conjunctiva [ANAT] The mucous membrane covering the eyeball and lining the eyelids. { kən'jəŋk·tə·və }

conjunctivitis [MED] Inflammation of the conjunctiva. { kən,jəŋk·tə'vīd·əs }

connate leaf [BOT] A leaf shaped as though the bases of two opposite leaves had fused around the stem. { kə'nāt 'lēf }

connective tissue [HISTOL] A primary tissue, distinguished by an abundance of fibrillar and nonfibrillar extracellular components. { kə'nek·tiv 'tish·ü }

connexins [CELL MOL] A group of transmembrane proteins that form the intermembrane channels of gap junctions; they are used by inorganic ions and most small organic molecules to pass through cell interiors. { kə'nek·sənz }

connexon [CELL MOL] Any of the cylindrical channels associated with gap junctions. { kə'nek,sän }

connivent [BIOL] Converging so as to meet, but not fused into a single part. { kə'nīv·ənt }

Conopidae [INV ZOO] The wasp flies, a family of dipteran insects in the suborder Cyclorrhapha. { kə'näp·ə,dē }

conotheca [INV ZOO] The thin integument of

the phragmocone in certain mollusks. { ¦kō·nə'thē·kə }

consanguineous [GEN] Pertaining to two or more individuals that have a common recent ancestor. { ¦kän·saŋ¦gwin·ē·əs }

consanguinity [GEN] Genetic blood relationship arising from a common ancestor. { ¦kän·saŋ¦gwin·əd·ē }

consensual eye reflex *See* consensual light reflex. { kən¦sench·yə·wəl ¦ī 'rē,fleks }

consensual light reflex [NEUROSCI] The reaction of both pupils when only one eye is exposed to a change in light intensity. Also known as consensual eye reflex. { kən¦sench·yə·wəl ¦līt 'rē,fleks }

consensus sequence [GEN] An average nucleotide sequence; each nucleotide is the most frequent at its position in the sequence. { kən'sen·səs ,sē·kwəns }

conservation [ECOL] Those measures concerned with the preservation, restoration, beneficiation, maximization, reutilization, substitution, allocation, and integration of natural resources. { ,kän·sər'vā·shən }

conservative replication [CELL MOL] Replication of a molecule of deoxyribonucleic acid (DNA) such that one DNA molecule would consist of both the original parent strands and the replicated molecule would contain two newly synthesized strands. { kən¦sər·vəd·iv ,rep·lə'kā·shən }

conservative substitution [CELL MOL] Replacement of an amino acid in a polypeptide by one with similar characteristics. { kən'sər·vəd·iv ,səb·stə'tü·shən }

conserved sequence [EVOL] A sequence of nucleotides in genetic material or of amino acids in a polypeptide chain that has changed only slightly or not at all during an evolutionary period of time. { kən'sərvd 'sē·kwəns }

consociation [ECOL] A climax community of plants which is dominated by a single species. { kən,sō·sē'ā·shən }

consortism *See* symbiosis. { 'kän,sȯrd,iz·əm }

conspecific [SYST] Referring to individuals or populations of a single species. { ¦kän·spə'sif·ik }

constant guidance [CELL MOL] The oriented response of isolated tissue cells in culture according to the topography of their substratum. { 'kän,takt ,gīd·əns }

constipation [MED] The passage of hard, dry stools. { kän·stə'pā·shən }

constitutive enzyme [BIOCHEM] An enzyme whose concentration in a cell is constant and is not influenced by substrate concentration. { 'kän·stə,tüd·iv 'en,zīm }

constitutive gene [GEN] A gene that encodes a product required in the maintenance of basic cellular processes or cell architecture. Also known as housekeeping gene. { 'kän·stə,tüd·iv ,jēn }

constitutive heterochromatin [GEN] A type of heterochromatin that is always condensed and is often centered on either side of the centromere, and that stains to give a C band. { 'kän·stə,tüd·iv ¦hed·ə·rō'krō·məd·ən }

constitutive mutation [GEN] A mutation that modifies an operator gene or a regulator gene, resulting in unregulated expression of structural genes that are normally regulated. { 'kän·stə,tüd·iv myü'tā·shən }

constitutive promoter [CELL MOL] An unregulated promoter segment of deoxyribonucleic acid that allows continuous transcription of its cognate gene. { ¦kän·stə,tüd·iv prə'mōd·ər }

constitutive secretory pathway [CELL MOL] A secretory pathway found in all cells by which transport vesicles continuously leave the Golgi apparatus and fuse with the plasma membrane, and their contents are exported to the extracellular space or used as components of the plasma membrane. { ¦kän·stə,tüd·iv ¦sek·rə,tȯr·ē 'path ,wā }

constriction *See* stricture. { kən'strik·shən }

constrictive pericarditis [MED] Inflammation and fibrosis of the pericardium resulting in constriction of the heart and restriction of contraction and blood flow. Also known as Pick's disease. { kən'strik·div per·i,kär'dīd·əs }

constructional apraxia *See* optic apraxia. { kən'strək·shən·əl ā'prak·sē·ə }

consumer [ECOL] A nutritional grouping in the food chain of an ecosystem, composed of heterotrophic organisms, chiefly animals, which ingest other organisms or particulate organic matter. { kən'süm·ər }

consumption *See* tuberculosis. { kən'səm·shən }

consumptive coagulopathy [MED] Reduction in one or more of the blood elements involved in coagulation as the result of marked blood clotting. { kən'səm·div ,kō,ag·yə'läp·ə·thē }

contact-dependent signaling [CELL MOL] A type of intracellular communication whereby a signal molecule remains bound to the signaling cell surface, rather than being released into the extracellular space, and influences only cells that come into contact with it. { ¦kän,tak di,pen·dənt 'sig·nəl·iŋ }

contact dermatitis [MED] An acute or chronic inflammation of the skin resulting from irritation by or sensitization to some substance coming in contact with the skin. { 'kän,takt dər·mə'tīd·əs }

contact inhibition [CELL MOL] Cessation of cell division when cultured cells are in physical contact with each other. { 'kän,takt in·ə'bish·ən }

contact paralysis [CELL MOL] The cessation of forward extension of the pseudopods of a cell as a result of its collision with another cell. { 'kän,takt pə,ral·ə·səs }

contagion [MED] **1.** The process whereby disease spreads from one person to another, by direct or indirect contact. **2.** The bacterium or virus which transmits disease. { kən'tā·jən }

contagious abortion [VET MED] Brucellosis in cattle caused by *Brucella abortus* and inducing abortion. Also known as Bang's disease; infectious abortion. { kən'tā·jəs ə'bȯr·shən }

contagious disease [MED] An infectious disease communicable by contact with a person suffering from it, with the bodily discharge, or with an object touched by the person. { kən'tā·jəs di'zēz }

contagious polyarthritis [VET MED] An infectious bacterial disease of mice caused by members of the genus *Bacteroides*; characterized by inflammation and abcess formation in the joints. { kən'tā·jəs ¦päl·e·är'thrīd·əs }

contagious pustular dermatitis [VET MED] An infectious disease of sheep and goats characterized by vesicles on the skin which are transformed into pustules. { kən'tā·jəs 'pəs·chə·lər ˌdər·mə'tīd·əs }

containment [CELL MOL] Prevention of the replication of the products of recombinant deoxyribonucleic acid technology outside the laboratory. { kən'tān·mənt }

contamination [MICROBIO] The process or act of soiling with bacteria. { kən,tam·ə'nā·shən }

contig [GEN] A region of chromosome defined by its hybridization to one or more cloned deoxyribonucleic acid fragments from an overlapping array of clones. [CELL MOL] A group of cloned nucleotide sequences that are contiguous. { kən'tig }

contiguous gene syndrome [GEN] A characteristic complex phenotype produced by deletion of a short chromosome segment, resulting from haplo-insufficiency of several genes in the deleted segment. { kən¦tig·yə·wəs 'jēn 'sin'drōm }

continuous cell line [CELL MOL] A group of morphologically uniform cells that can be propagated in vitro for an indefinite time. { kən¦tin·yə·wəs 'sel ˌlīn }

contorted [BOT] Twisted; applied to proximate leaves whose margins overlap. { kən'tȯrd·əd }

contour feather [VERT ZOO] Any of the large flight feathers or long tail feathers of a bird. Also known as penna; vane feather. { 'kän,tůr ˌfet͟h·ər }

contraception [MED] Prevention of impregnation. { ¦kän·trə¦sep·shən }

contraceptive [MED] Any mechanical device or chemical agent used to prevent conception. { ¦kän·trə¦sep·tiv }

contracted pelvis [MED] A pelvis having one or more major diameters reduced in size, interfering with parturition. { kən'trak·təd 'pel·vəs }

contractile [BIOL] Displaying contraction; having the property of contracting. { kən'trak·təl }

contractile ring [CELL MOL] A cytoskeletal structure that forms in animal cells and many unicellular eukaryotes during cell division, contraction of which causes the plasma membrane to pinch inward and the cell to divide. { kən'trak·təl ˌriŋ }

contractile vacuole [CELL MOL] A tiny, intracellular, membranous bladder that functions in maintaining intra- and extracellular osmotic pressures in equilibrium, as well as excretion of water, such as occurs in protozoans. { kən'trak·təl 'vak·yə·wōl }

contraction [PHYSIO] Shortening of the fibers of muscle tissue. { kən'trak·shən }

contracture [MED] **1.** Shortening, as of muscle or scar tissue, producing distortion or deformity or abnormal limitation of movement of a joint. **2.** Retarded relaxation of muscle, as when it is injected with veratrine. { kən'trak·chər }

contraindication [MED] A symptom, indication, or condition in which a remedy or a method of treatment is inadvisable or improper. { ¦kän·trə,in·də'kā·shən }

contralateral [PHYSIO] Opposite; acting in unison with a similar part on the opposite side of the body. { ¦kän·trə'lad·ə·rəl }

contransformation [GEN] The incorporation of two or more linked genes on the same fragment of foreign deoxyribonucleic acid into a bacterial genome. { ˌkän·tranz·fər,mā·shən }

co-transport [CELL MOL] The simultaneous transport of two substrates across a cell membrane, either in the same direction (symport) or in opposite directions (antiport). { ˌkō 'tranz,pȯrt }

control element [CELL MOL] A site within a gene or operon that acts to control gene expression. { kən'trōl ˌel·ə·mənt }

control gene [GEN] A gene that regulates the time and rate at which neighboring structural genes are transcribed in messenger ribonucleic acid. { kən'trōl ˌjēn }

controller node [GEN] A genetic unit of regulation consisting of a set of regulators, effectors, and receptors located near the gene that the unit controls. { kən'trō·lər ˌnōd }

controlling element [GEN] Any of a class of transposable genetic elements that have the capacity to control gene expression at several loci, as well as to render target genes extremely likely to mutate. { kən'trōl·iŋ 'el·ə·mənt }

contusion [MED] A subcutaneous bruise caused by an injury in which the skin is not broken. { kən'tü·zhən }

Conulata [INV ZOO] A subclass of free-living cnidarians in the class Scyphozoa; individuals are described as tetraramous cones to elongate pyramids having tentacles on the oral margin. { ˌkän·əl'äd·ə }

conus arteriosus [EMBRYO] The cone-shaped projection from which the pulmonary artery arises on the right ventricle of the heart in man and mammals. { 'kō·nəs är,tir·ē'ō·səs }

convalescence [MED] The period and process of recovery after an illness or injury. { ˌkän·və'les·əns }

convalescent carrier [MED] A person who harbors an infectious agent after recovery from a clinical attack of a disease. { ¦kän·və¦les·ənt 'kar·ē·ər }

convalescent serum [IMMUNOL] The serum of the blood of one or more patients recovering from an infectious disease; used for prophylaxis of the particular infection. { ¦kän·və¦les·ənt 'sir·əm }

convergence [EVOL] Development of similarities between animals or plants of different groups resulting from adaptation to similar habitats. [NEUROSCI] The coming together of a group of afferent nerves upon a motoneuron of the ventral horn of the spinal cord. { kən'vər·jəns }

conversion polarity [GEN] A gradient in the frequency of gene conversion from one end of a gene to the other. { kən'vər·zhən pə,lar·əd·ē }

conversive heating [MED] The conversion of some form of energy, especially radio waves, into heat for use in thermotherapy. { kən'vər·siv 'hēd·iŋ }

convertase [BIOCHEM] An enzyme that cleaves inactive protein precursors into smaller biologically active molecules. { 'än·vər,tās }

convivium [ECOL] A population exhibiting differentiation within the species and isolated geographically, generally a subspecies or ecotype. { kən'viv·ē·əm }

convolute [BIOL] Twisted or rolled together, specifically referring to leaves, mollusk shells, and renal tubules. { 'kän·və,lüt }

convolution [ANAT] A fold, twist, or coil of any organ, especially any one of the prominent convex parts of the brain, separated from each other by depressions or sulci. { ,kän·və'lü·shən }

Convolvulaceae [BOT] A large family of dicotyledonous plants in the order Polemoniales characterized by internal phloem, the presence of chlorophyll, two ovules per carpel, and plicate cotyledons. { kən,väl·və'lās·ē,ē }

convulsion [MED] An episode of involuntary, generally violent muscular contractions, rhythmically alternated with periods of relaxation; associated with many systematic and neurological diseases. { kən'vəl·shən }

convulsive disorder [MED] Any pathologic condition in which convulsions are a common symptom and characteristic electroencephalogram patterns are displayed. { kən'vəl·siv dis,ȯr·dər }

convulsive equivalent *See* epileptic equivalent. { kən'vəl·siv i'kwiv·ə·lənt }

Cooke unit [BIOL] A unit for the standardization of pollen antigenicity. { 'ku̇k ,yü·nət }

Coombs serum [IMMUNOL] An immune serum containing antiglobulin that is used in testing for Rh and other sensitizations. { 'kümz ,sir·əm }

Copeognatha [INV ZOO] An equivalent name for Psocoptera. { ,kō·pē'äg·nə·thə }

Copepoda [INV ZOO] An order of Crustacea commonly included in the Entomostraca; contains free-living, parasitic, and symbiotic forms. { kō'pep·ə·də }

copperhead [VERT ZOO] *Agkistrodon contortrix*. A pit viper of the eastern United States; grows to about 3 feet (90 centimeters) in length and is distinguished by its coppery-brown skin with dark transverse blotches. { 'käp·ər,hed }

coppice [ECOL] A growth of small trees that are repeatedly cut down at short intervals; the new shoots are produced by the old stumps. { 'käp·əs }

coproantibody [IMMUNOL] An antibody whose presence in the intestinal tract can be demonstrated by its presence in an extract of the feces. { ¦käp·rō'ant·i,bäd·ē }

coprolalia [MED] Uncontrollable barking or grunting of profane language that is commonly associated with Tourette's syndrome. { ,käp·rə'lāl·yə }

coprophagy [ZOO] Feeding on dung or excrement. { kə'präf·ə·jē }

coprophilous [ECOL] Living in dung. { kə 'präf·ə·ləs }

copulation [ZOO] The sexual union of two individuals, resulting in insemination or deposition of the male gametes in proximity to the female gametes. { ,käp·yə'lā·shən }

copulatory bursa [INV ZOO] **1.** A sac that receives the sperm during copulation in certain insects. **2.** The caudal expansion of certain male nematodes that functions as a clasper during copulation. { 'käp·yə·lə,tȯr·ē 'bər·sə }

copulatory organ [ANAT] An organ employed by certain male animals for insemination. { 'käp·yə·lə,tȯr·ē ,ȯr·gən }

copulatory spicule *See* spiculum. { 'käp·yə·lə,tȯr·ē 'spik·yəl }

copy deoxyribonucleic acid *See* complementary deoxyribonucleic acid. { ¦käp·ē de¦äk·sē,rī·bō·nü¦klē·ik 'as·əd }

copy error [CELL MOL] A mutation that occurs during deoxyribonucleic acid replication as a result of an error in base pairing. { 'käp·ē ,er·ər }

coquina [INV ZOO] A small marine clam of the genus *Donax*. { kō'kē·nə }

Coraciidae [VERT ZOO] The rollers, a family of Old World birds in the order Coraciiformes. { ,kȯr·ə'sī·ə,dē }

Coraciiformes [VERT ZOO] An order of predominantly tropical and frequently brightly colored birds. { ,kȯr·ə,sī·ə'fȯr,mēz }

coracoid [ANAT] One of the paired bones on the posterior-ventral aspect of the pectoral girdle in vertebrates. { 'kȯr·ə,kȯid }

coracoid ligament [ANAT] The transverse ligament of the scapula which crosses over the suprascapular notch. { 'kȯr·ə,kȯid 'lig·ə·mənt }

coracoid process [ANAT] The beak-shaped process of the scapula. { 'kȯr·ə,kȯid ,präs·əs }

coral [INV ZOO] The skeleton of certain solitary and colonial anthozoan cnidarians; composed chiefly of calcium carbonate. { 'kä·rəl }

Corallanidae [INV ZOO] A family of sometimes parasitic, but often free-living, isopod crustaceans in the suborder Flabellifera. { ,kä·rə'lan·ə,dē }

Corallidae [INV ZOO] A family of dimorphic cnidarians in the order Gorgonacea. { kə'ral·ə,dē }

Corallimorpharia [INV ZOO] An order of solitary sea anemones in the subclass Zoantharia resembling coral in many aspects. { kə¦ral·ə,mȯr'far·ē·ə }

Corallinaceae [BOT] A family of red algae, division Rhodophyta, having compact tissue with lime deposits within and between the cell walls. { kə,ral·ə'nās·ē,ē }

coralline [INV ZOO] Any animal that resembles coral, such as a bryozoan or hydroid. { 'kär·ə,lēn }

coralline algae [BOT] Red algae belonging to the family Corallinaceae. { 'kär·ə,lēn 'al·jē }

corallite [INV ZOO] Skeleton of an individual polyp in a compound coral. { 'kär·ə,līt }

coralloid [BIOL] Resembling coral, or branching like certain coral. { 'kär·ə,lòid }

corallum [INV ZOO] Skeleton of a compound coral. { kə'ral·əm }

Corbiculidae [INV ZOO] A family of fresh-water bivalve mollusks in the subclass Eulamellibranchia; an important food in the Orient. { kòr·bə'kyül·ə,dē }

cordate [BOT] Heart-shaped; generally refers to a leaf base. { 'kòr,dāt }

cord factor [MICROBIO] A toxic glycolipid found as a surface component of tubercle bacilli that is responsible for virulence and serpentine growth. { 'kòrd ,fak·tər }

cordon [BOT] A plant trained to grow flat against a vertical structure, in a single horizontal shoot or two opposed horizontal shoots. { 'kòrd·ən }

Cordulegasteridae [INV ZOO] A family of anisopteran insects in the order Odonata. { ¦kòrd·yə·lə,ga'ster·ə,dē }

Cordyceps ophioglossoides [MYCOL] A mushroom that is a parasite on the fruiting bodies of the truffle found in the soil of bamboo, oak, and pine woods that has antitumor properties and is an immune booster. Also known as clubhead fungus. { ¦kòrd·ə,seps ,ō·fē·ə·glə'sòi,dēz }

Cordyceps sinensis [MYCOL] A type of mushroom found on the cold mountain tops and snowy grass marshlands of China that infects insect larvae with spores that germinate before the cocoons are formed; it has been successfully used in clinical trials to treat liver diseases, high cholesterol, and loss of sexual drive. Also known as caterpillar fungus. { ¦kòrd·ə,seps sī 'nen·sis }

core [ANAT] A fingerprint focal point which is the point on a ridge that is located in the approximate center of the finger impression. { kòr }

Coreidae [INV ZOO] The squash bugs and leaf-footed bugs, a family of hemipteran insects belonging to the superfamily Coreoidea. { kə'rē·ə,dē }

coremium [MYCOL] A small bundle of conidiophores in certain imperfect fungi. { kə'rē·mē·əm }

corepressor [CELL MOL] A certain metabolite which, through combination with a repressor apoprotein produced by a regulator gene, can cause the binding of the protein to the operator gene region of a deoxyribonucleic acid chain. { ¦kō·ri,pres·ər }

coriaceous [BIOL] Leathery, applied to leaves and certain insects. { ¦kòr·ē¦ā·shəs }

coriander [BOT] *Coriandrum sativum*. A strong-scented perennial herb in the order Umbellales; the dried fruit is used as a flavoring. { ,kòr·ē'an·dər }

Cori ester *See* glucose-1-phosphate. { 'kòr·ē ,es·tər }

Coriolis effect [PHYSIO] The physiological effects (nausea, vertigo, dizziness, and so on) felt by a person moving radially in a rotating system, as a rotating space station. { kòr·ē'ō·ləs i'fekt }

corium [ANAT] *See* dermis. [INV ZOO] Middle portion of the forewing of hemipteran insects. { 'kòr·ē·əm }

Corixidae [INV ZOO] The water boatmen, the single family of the hemipteran superfamily Corixoidea. { kə'rik·sə,dē }

Corixoidea [INV ZOO] A superfamily of hemipteran insects belonging to the subdivision Hydrocorisae that lack ocelli. { kə,rik'sòid·ē·ə }

cork [BOT] A protective layer of cells that replaces the epidermis in older plant stems. { kòrk }

corm [BOT] A short, erect, fleshy underground stem, usually broader than high and covered with membranous scales. { kòrm }

cormatose [BOT] Having or producing a corm. { 'kòr·mə,tōs }

cormidium [INV ZOO] The assemblage of individuals dangling in clusters from the main stem of pelagic siphonophores. { kòr'mid·ē·əm }

corn [BOT] *Zea mays*. A grain crop of the grass order Cyperales grown for its edible seeds (technically fruits). [MED] A small, sharply circumscribed, conically shaped deep-seated area of thickened skin composed of the fibrous protein keratin. Also known as heloma. { kòrn }

Cornaceae [BOT] A family of dicotyledonous plants in the order Cornales characterized by perfect or unisexual flowers, a single ovule in each locule, as many stamens as petals, and opposite leaves. { kòr'nās·ē,ē }

Cornales [BOT] An order of dicotyledonous plants in the subclass Rosidae marked by a woody habit, simple leaves, well-developed endosperm, and fleshy fruits. { kòr'nā·lēz }

cornea [ANAT] The transparent anterior portion of the outer coat of the vertebrate eye covering the iris and the pupil. [INV ZOO] The outer transparent portion of each ommatidium of a compound eye. { 'kòr·nē·ə }

corneal reflex [MED] Automatic closing of the eyelids as a result of irritation of the cornea. { 'kòr·nē·əl 'rē,fleks }

cornicle [INV ZOO] Either of two protruding horn-shaped dorsal tubes in aphids which secrete a waxy fluid. { 'kòr·nə·kəl }

corniculate [BIOL] Possessing small horns or hornlike processes. { kòr'nik·yə·lət }

corniculate cartilage [ANAT] The cartilaginous nodule on the tip of the arytenoid cartilage. { kòr'nik·yə·lət ¦kärt·lij }

cornification [PHYSIO] Conversion of stratified squamous epithelial cells into a horny layer and into derivatives such as nails, hair, and feathers. { ,kòr·nə·fə'kā·shən }

corn sugar *See* dextrose. { 'kòrn ,shu̇g·ər }

cornu [ANAT] A horn or hornlike structure. { 'kòr·nü }

corolla [BOT] Collectively, the petals of a flower. { kə'räl·ə }
corollate [BOT] Having a corolla. { kə'rä,lāt }
corolline [BOT] Relating to, resembling, or being borne on a corolla. { kȯr·ə,līn }
corona [BOT] **1.** An appendage or series of fused appendages between the corolla and stamens of some flowers. **2.** The region where stem and root of a seed plant merge. Also known as crown. [INV ZOO] **1.** The anterior ring of cilia in rotifers. **2.** A sea urchin test. **3.** The calyx and arms of a crinoid. { kə'rō·nə }
coronal suture [ANAT] The union of the frontal with the parietal bones transversely across the vertex of the skull. { kə'rō·nəl 'sü·chər }
corona radiata [HISTOL] The layer of cells immediately surrounding a mammalian ovum. { kə'rō·nə ,rā·dē'äd·ə }
coronary artery [ANAT] Either of two arteries arising in the aortic sinuses that supply the heart tissue with blood. { 'kär·ə,ner·ē 'ärd·ə·rē }
coronary disease [MED] Any condition that reduces the flow of blood to the heart muscles. { 'kär·ə,ner·ē di'zēz }
coronary failure [MED] Prolonged precordial pain or discomfort without conventional evidence of myocardial infarction; subendocardial ischemia caused by a disparity between coronary blood flow and myocardial needs; this condition is more commonly referred to as coronary artery insufficiency. { 'kär·ə,ner·ē 'fāl·yər }
coronary occlusion [MED] Complete blockage of a coronary artery. { 'kär·ə,ner·ē ə'klü·zhən }
coronary sinus [ANAT] A venous sinus opening into the heart's right atrium which drains the cardiac veins. { 'kär·ə,ner·ē 'sī·nəs }
coronary stenosis [MED] Narrowing of the lumen of a coronary artery. { 'kär·ə,ner·ē stə [PHYSIO] 'nō·səs }
coronary sulcus [ANAT] A groove in the external surface of the heart separating the atria from the ventricles, containing the trunks of the nutrient vessels of the heart. { 'kär·ə,ner·ē 'səl·kəs }
coronary thrombosis [MED] Formation of a thrombus in a coronary artery. { 'kär·ə,ner·ē thräm'bō·səs }
coronary valve [ANAT] A semicircular fold of the endocardium of the right atrium at the orifice of the coronary sinus. { 'kär·ə,ner·ē 'valv }
coronary vein [ANAT] **1.** Any of the blood vessels that bring blood from the heart and empty into the coronary sinus. **2.** A vein along the lesser curvature of the stomach. { 'kär·ə,ner·ē 'vān }
Coronatae [INV ZOO] An order of the class Scyphozoa which includes mainly abyssal species having the exumbrella divided into two parts by a coronal furrow. { ,kȯr·ə'näd·ē }
Coronaviridae [VIROL] A family of vertebrate viruses consisting of the single genus Coronavirus; the prototype, avian infectious virus, has an enveloped spherical form with large spikes and a helical nucleocapsid with single-stranded ribonucleic acid. { kə,rō·nə'vī·rə,dē }
coronavirus [VIROL] A major group of animal viruses including avian infectious bronchitis virus and mouse hepatitis virus. { kə¦rō·nə [PHYSIO] ¦vī·rəs }
coronet band [ZOO] The area above the hoof containing the germinal cells from which hoof tissue is formed. { 'kär·ə,net ¦band }
coronoid [BIOL] Shaped like a beak. { 'kȯr·ə,nȯid }
coronoid fossa [ANAT] A depression in the humerus into which the apex of the coronoid process of the ulna fits in extreme flexion of the forearm. { 'kȯr·ə,nȯid ¦fäs·ə }
coronoid process [ANAT] **1.** A thin, flattened process projecting from the anterior portion of the upper border of the ramus of the mandible, and serving for the insertion of the temporal muscle. **2.** A triangular projection from the upper end of the ulna, forming the lower part of the radial notch. { 'kȯr·ə,nȯid ¦präs·əs }
coronule [INV ZOO] A peripheral ring of spines on some diatom shells. { 'kȯr·ə,nyül }
Corophiidae [INV ZOO] A family of amphipod crustaceans in the suborder Gammaridea. { kȯr·ə'fī·ə,dē }
corpora quadrigemina [ANAT] The inferior and superior colliculi collectively. Also known as quadrigeminal body. { 'kȯr·pə·rə ,kwäd·rə'jem·ə·nə }
cor pulmonale [MED] Hypertrophy and dilation of the right ventricle that is secondary to obstruction to the pulmonary blood flow and consequent pulmonary hypertension. { ¦kȯr ,pu̇l·mə'näl·ē }
corpus albicans [HISTOL] The white fibrous scar in an ovary; produced by the involution of the corpus luteum. { 'kȯr·pəs 'al·bə,kanz }
corpus allatum [INV ZOO] An endocrine structure near the brain of immature arthropods that secretes a juvenile hormone, neotenin. { 'kȯr·pəs ə'lād·əm }
corpus callosum [NEUROSCI] A band of nerve tissue connecting the cerebral hemispheres in humans and higher mammals. { 'kȯr·pəs kə'lō·səm }
corpus cardiacum [INV ZOO] One of a pair of separate or fused bodies of nervous tissue in many insects that lie posterior to the brain and dorsal to the esophagus and that function in the storage and secretion of brain hormone. { 'kȯr·pəs ,kärd·ē'ak·əm }
corpus cavernosum [ANAT] The cylinder of erectile tissue forming the clitoris in the female and the penis in the male. { 'kȯr·pəs ,ka·vər'nō·səm }
corpus cerebelli [NEUROSCI] The central lobe or zone of the cerebellum; regulates reflex tonus of postural muscles in mammals. { 'kȯr·pəs ,ser·ə'bel·ē }
corpuscle [ANAT] A small, rounded body. [NEUROSCI] An encapsulated sensory-nerve end organ. { 'kȯr·pəs·əl }
corpus luteum [HISTOL] The yellow endocrine body formed in the ovary at the site of a ruptured Graafian follicle. { 'kȯr·pəs 'lüd·ē·əm }

corpus striatum [ANAT] The caudate and lenticular nuclei, together with the internal capsule which separates them. { 'kȯr·pəs ˌstrī'ād·əm }

corrective therapy [MED] A program, and the techniques, designed to improve or maintain the health of a patient by improving neuromuscular activities and personal health habits and promoting relaxation by adjustment to stresses. { kə'rek·tiv 'ther·ə·pē }

corresponding points [PHYSIO] Any two retinal areas in the respective eyes so that the area in one eye has an identical direction in the opposite retina. { ˌkär·ə'spänd·iŋ 'pȯins }

corridor [ECOL] A land bridge that allows free migration of fauna in both directions. { 'kär·ə·dər }

Corrigan's pulse [MED] A pulse characterized by a rapid, forceful ascent (water-hammer quality) and rapid downstroke or descent (collapsing quality); seen with aortic regurgitation and hyperkinetic circulatory states. { 'kär·ə·gənz 'pəls }

corrins [BIOCHEM] Cobalt-containing compounds (porphyrin-like macrocycles) that act in concert with enzymes to catalyze essential reactions in humans. { 'kȯ·rinz }

Corrodentia [INV ZOO] The equivalent name for Psocoptera. { ˌkȯr·ə'dench·ə }

cortex [ANAT] The outer portion of an organ or structure, such as of the brain and adrenal glands. [BOT] A primary tissue in roots and stems of vascular plants that extends inward from the epidermis to the phloem. [CELL MOL] A peripheral layer in many cells that includes the plasma membrane and associated cytoskeletal and extracellular components. [INV ZOO] The peripheral layer of certain protozoans. { 'kȯrˌteks }

cortical granule [CELL MOL] Any of the round to elliptical membrane-bound bodies that occur in the cortex of animal oocytes, contain mucopolysaccharides, and participate in formation of the fertilization membrane. { 'kȯrd·ə·kəl 'gran·yəl }

cortical stimulator [MED] An electronic instrument used in nerve and mental therapy to deliver an electric shock of prescribed strength by means of a pulsating current. { 'kȯrd·ə·kəl 'stim·yəˌlād·ər }

corticoid *See* adrenal cortex hormone. { 'kȯrd·əˌkȯid }

corticosteroid [BIOCHEM] **1.** Any steroid hormone secreted by the adrenal cortex of vertebrates. **2.** Any steroid with properties of an adrenal cortex steroid. { ¦kȯrd·əˌkō'stirˌȯid }

corticosterone [BIOCHEM] $C_{21}H_{30}O_4$ A steroid hormone produced by the adrenal cortex of vertebrates that stimulates carbohydrate synthesis and protein breakdown and is antagonistic to the action of insulin. { ˌkȯrd·ə'käs·təˌrōn }

corticotrophic [PHYSIO] Having an effect on the adrenal cortex. { ¦kȯrd·əˌkō¦trä·fik }

corticotropin [BIOCHEM] A hormonal preparation having adrenocorticotropic activity, derived from the adenohypophysis of certain domesticated animals. { ˌkȯrd·əˌkō'trō·pən }

corticotropin-releasing hormone [BIOCHEM] A substance produced by the hypothalamus that stimulates the pituitary gland to produce adrenocorticotropic hormone (ACTH). Abbreviated CRH. { ˌkȯrd·ə·kō'trō·pən ri¦lēs·iŋ ˌhȯrˌmōn }

Corticoviridae [VIROL] A family of nontailed bacterial viruses (bacteriophages) characterized by a nonenveloped icosahedral particle containing a circular double-stranded deoxyribonucleic acid genome. { ˌkȯrd·i kō'vir·əˌdī }

Corticovirus [VIROL] The only genus of the family Corticoviridae. { ˌkȯrd·i·kō 'vī·rəs }

cortin unit [BIOL] A unit for the standardization of adrenal cortical hormones. { 'kȯrt·ən ˌyü·nət }

cortisol *See* hydrocortisone. { 'kȯrd·əˌsȯl }

cortisone [BIOCHEM] $C_{21}H_{28}O_5$ A steroid hormone produced by the adrenal cortex of vertebrates that acts principally in carbohydrate metabolism. { 'kȯrd·əˌsōn }

Corvidae [VERT ZOO] A family of large birds in the order Passeriformes having stout, long beaks; includes the crows, jays, and magpies. { 'kȯr·vəˌdē }

Corylophidae [INV ZOO] The equivalent name for Orthoperidae. { ˌkȯr·ə'läf·əˌdē }

corymb [BOT] An inflorescence in which the flower stalks arise at different levels but reach the same height, resulting in a flat-topped cluster. { 'kȯˌrim }

corymbose [BOT] Resembling or pertaining to a corymb. { kə'rimˌbōs }

Corynebacteriaceae [MICROBIO] Formerly a family of nonsporeforming, usually nonmotile rod-shaped bacteria in the order Eubacteriales including animal and plant parasites and pathogens. { ¦kȯr·əˌnēˌbakˌtir·ē'ās·ēˌē }

corynebacteriophage [VIROL] Any bacteriophage able to infect *Corynebacterium* species. { ¦kȯr·əˌnē·bak'tir·ē·əˌfāzh }

Corynebacterium [MICROBIO] A genus of gram-positive, straight or slightly curved rods in the coryneform group of bacteria; club-shaped swellings are common; includes human and animal parasites and pathogens, and plant pathogens. { ¦kȯr·əˌnē·bak'tir·ē·əm }

Corynebacterium diphtheriae [MICROBIO] A facultatively aerobic, nonmotile species of bacteria that causes diphtheria in humans. Also known as Klebs-Loeffler bacillus. { ¦kȯr·əˌnē·bak'tir·ē·əm dif'thir·ēˌī }

Coryphaenidae [VERT ZOO] A family of pelagic fishes in the order Perciformes characterized by a blunt nose and deeply forked tail. { ˌkȯr·ə'fēn·əˌdē }

coryza [MED] Inflammation of the mucous membranes of the nose, usually marked by sneezing and discharge of watery mucous. { kə'rī·zə }

Cosmocercidae [INV ZOO] A group of nematodes assigned to the suborder Oxyurina by some authorities and to the suborder Ascaridina by others. { ˌkäz·mə'sər·səˌdē }

Cosmocercoidea [INV ZOO] A superfamily of parasitic nematodes having either three or six lips surrounding a weakly developed stoma. { ˌkäz·mō·sər'kȯid·ē·ə }

cosmoid scale [VERT ZOO] A structure in the skin of primitive rhipidistians and dipnoans that is composed of enamel, a dentine layer (cosmine), and laminated bone. { 'käzˌmȯid 'skāl }

cosmopolitan [ECOL] Having a worldwide distribution wherever the habitat is suitable, with reference to the geographical distribution of a taxon. { ¦käz·mə¦päl·ət·ən }

Cossidae [INV ZOO] The goat or carpenter moths, a family of heavy- bodied lepidopteran insects in the superfamily Cossoidea having the abdomen extending well beyond the hindwings. { 'käs·əˌdē }

Cossoidea [INV ZOO] A monofamilial superfamily of lepidopteran insects belonging to suborder Heteroneura. { kə'sȯid·ē·ə }

Cossuridae [INV ZOO] A family of fringe worms belonging to the Sedentaria. { kə'syu̇r·əˌdē }

Cossyphodidae [INV ZOO] The lively ant guest beetles, a small family of coleopteran insects in the superfamily Tenebrionoidea. { ˌkäs·ə'fä·dəˌdē }

costa [BIOL] A rib or riblike structure. [BOT] The midrib of a leaf. [INV ZOO] The anterior vein of an insect's wing. { 'käs·tə }

Costaceae [BOT] A family of monocotyledonous plants in the order Zingiberales distinguished by having one functional stamen with two pollen sacs and spirally arranged leaves and bracts. { kȯs'tās·ēˌē }

costal cartilage [ANAT] The cartilage occupying the interval between the ribs and the sternum or adjacent cartilages. { 'käst·əl ¦kärd·əl·ij }

costal process [ANAT] An anterior or ventral projection on the lateral part of a cervical vertebra. [EMBRYO] An embryonic rib primordium, the ventrolateral outgrowth of the caudal, denser half of a sclerotome. { 'käst·əl ¦präs·əs }

costate [BIOL] Having ribs or ridges. { 'käˌstāt }

Cotingidae [VERT ZOO] The cotingas, a family of neotropical suboscine birds in the order Passeriformes. { kō'tin·jəˌdē }

Cottidae [VERT ZOO] The sculpins, a family of perciform fishes in the suborder Cottoidei. { 'käd·əˌdē }

Cottiformes [VERT ZOO] An order set up in some classification schemes to include the Cottoidei. { ˌkäd·ə'fȯrˌmēz }

Cottoidei [VERT ZOO] The mail-cheeked fishes, a suborder of the order Perciformes characterized by the expanded third infraorbital bone. { kə'tȯid·ēˌī }

cotton [BOT] Any plant of the genus *Gossypium* in the order Malvales; cultivated for the fibers obtained from its encapsulated fruits or bolls. { 'kät·ən }

cottonmouth *See* water moccasin. { 'kät·ən ˌmau̇th }

cottonwood [BOT] Any of several poplar trees (*Populus*) having hairy, encapsulated fruit. { 'kät·ənˌwu̇d }

cotyledon [BOT] The first leaf of the embryo of seed plants. { ˌkäd·əl'ēd·ən }

cotylocercous cercaria [INV ZOO] A digenetic trematode larva characterized by a sucker or adhesive gland on the tail. { käd·əl'äs·ə·rəs ˌsər'kar·ē·ə }

cotype *See* syntype. { 'kōˌtīp }

cougar *See* puma. { 'kü·gər }

cough [MED] A sudden, violent expulsion of air after deep inspiration and closure of the glottis. { kȯf }

Couinae [VERT ZOO] The couas, a subfamily of Madagascan birds in the family Cuculidae. { 'kü·əˌnē }

Coulter counter [MICROBIO] An electronic device for counting the number of cells in a liquid culture. { 'kōl·tər ¦kau̇nt·ər }

coumarin glycoside [BIOCHEM] Any of several glycosidic aromatic principles in many plants; contains coumaric acid as the aglycon group. { 'kü·mə·rən 'glī·kəˌsīd }

counterimmunoelectrophoresis [IMMUNOL] Immunoelectrophoresis which uses two wells of application, one above the other, along the electrical axis—the anodal well filled with antibody and the cathodal with a negatively charged antigen; electrophoresis results in the antigen and antibody migrating cathodally and anodally, respectively, and a line of precipitation appears where the two meet. { ¦kau̇nt·ər¦im·yə·nō·iˌlek·trō·fə'r ē·səs }

counterstain [BIOL] A second stain applied to a biological specimen to color elements other than those demonstrated by the principal stain. { 'kau̇nt·ərˌstān }

counting chamber [MICROBIO] An accurately dimensioned chamber in a microslide which can hold a specific volume of fluid and which is usually ruled into units to facilitate the counting under the microscope of cells, bacteria, or other structures in the fluid. { 'kau̇nt·iŋ ˌchām·bər }

courtship [ECOL] A sequence of behavioral patterns that eventually may lead to completed mating. { 'kȯrtˌship }

covert [ECOL] A refuge or shelter, such as a coppice, for game animals. { 'kō·vərt }

covey [VERT ZOO] **1.** A brood of birds. **2.** A small flock of birds of one kind, used typically of partridge and quail. { 'kəv·ē }

cow [VERT ZOO] A mature female cattle of the genus *Bos*. { kau̇ }

Cowdria [MICROBIO] A genus of the tribe Ehrlichieae; coccoid to ellipsoidal, pleomorphic, or rod-shaped cells; intracellular parasites in cytoplasm and vacuoles of vascular endothelium of ruminants. { 'kau̇·drē·ə }

cowpea [BOT] *Vigna sinensis*. An annual legume in the order Rosales cultivated for its edible seeds. Also known as blackeye bean. { 'kau̇ˌpē }

cowpea mosaic virus group *See* Comovirus. { ¦kau̇ˌpē mōˌzā·ik 'vī·rəs ˌgrüp }

Cowper's gland *See* bulbourethral gland. { 'ku̇p·ərz ˌgland }

cowpox *See* vaccinia. { 'kau̇ˌpäks }

cowpox virus [VIROL] The causative agent of cowpox in cattle. { 'kau̇ˌpäks ˌvī·rəs }

coxa [INV ZOO] The proximal or basal segment of the leg of insects and certain other arthropods which articulates with the body. { 'käk·sə }

coxal cavity [INV ZOO] A cavity in which the coxa of an arthropod limb articulates. { 'käk·səl 'kav·əd·ē }

coxal gland [INV ZOO] One of certain paired glands with ducts opening in the coxal region of arthropods. { 'käk·səl 'gland }

Coxiella [MICROBIO] A genus of the tribe Rickettsieae; short rods which grow preferentially in host cell vacuoles. { ˌkäk·sē'el·ə }

coxitis [MED] Inflammation of the hip joint. { käk'sīd·əs }

coxopodite [INV ZOO] The basal joint of a crustacean limb. { käk'säp·əˌdīt }

coxsackie disease [MED] A variety of syndromes resulting from a coxsackievirus infection. { ku̇k'säk·ē di'zēz }

coxsackievirus [VIROL] A large subgroup of the enteroviruses in the picornavirus group including various human pathogens. { ku̇k'säk·ēˌvī·rəs }

coyote [VERT ZOO] *Canis latrans*. A small wolf native to western North America but found as far eastward as New York State. Also known as prairie wolf. { 'kīˌōd·ē }

CPD *See* cephalopelvic disproportion.

C-peptide [BIOCHEM] A metabolically inactive polypeptide chain that is a by-product of normal insulin production by the beta cells in the pancreas. { ¦sē 'pepˌtīd }

C_3 plant [BOT] A plant that produces the 3-carbon compound phosphoglyceric acid as the first stage of photosynthesis. { ¦sē'thrē ˌplant }

C_4 plant [BOT] A plant that produces the 4-carbon compound oxalocethanoic (oxaloacetic) acid as the first stage of photosynthesis. { ¦sē'fȯr ˌplant }

CPR *See* cardiopulmonary resuscitation.

crab [INV ZOO] **1.** The common name for a number of crustaceans in the order Decapoda having five pairs of walking legs, with the first pair modified as chelipeds. **2.** The common name for members of the Merostoma. { krab }

crabapple [BOT] Any of several trees of the genus *Malus*, order Rosales, cultivated for their small, edible pomes. { 'krabˌap·əl }

Cracidae [VERT ZOO] A family of New World tropical upland game birds in the order Galliformes; includes the chachalacas, guans, and curassows. { 'kra·səˌdē }

cradle cap [MED] Heavy, greasy crusts on the scalp of an infant; seborrheic dermatitis of infants. { 'krād·əl ˌkap }

Crambiinae [INV ZOO] The snout moths, a subfamily of lepidopteran insects in the family Pyralididae containing small marshland and grassland forms. { kram'bī·əˌnē }

cramp [MED] **1.** Painful, involuntary contraction of a muscle, such as a leg or foot cramp that may occur in normal individuals at night or in swimming. **2.** Any cramplike pain, as of the intestine, or that accompanying dysmenorrhea. **3.** Spasm of certain muscles, which may be intermittent or constant, from excessive use. { kramp }

cranberry [BOT] Any of several plants of the genus *Vaccinium*, especially *V. macrocarpon*, in the order Ericales, cultivated for its small, edible berries. { 'kranˌber·ē }

Cranchiidae [INV ZOO] A family of cephalopod mollusks in the subclass Dibranchia. { ˌkraŋ'kī·əˌdē }

crane [VERT ZOO] The common name for the long-legged wading birds composing the family Gruidae of the order Gruiformes. { krān }

Craniacea [INV ZOO] A family of inarticulate branchiopods in the suborder Craniidina. { ˌkrā·nē'ās·ē·ə }

cranial capacity [ANAT] The volume of the cranial cavity. { 'krān·ē·əl kə'pas·əd·ē }

cranial flexure [EMBRYO] A flexure of the embryonic brain. { 'krān·ē·əl 'flek·shər }

cranial fossa [ANAT] Any of the three depressions in the floor of the interior of the skull. { 'krān·ē·əl 'fäs·ə }

cranial nerve [NEUROSCI] Any of the paired nerves which arise in the brainstem of vertebrates and pass to peripheral structures through openings in the skull. { 'krān·ē·əl 'nərv }

Craniata [VERT ZOO] A major subdivision of the phylum Chordata comprising the vertebrates, from cyclostomes to mammals, distinguished by a cranium. { ˌkrā·nē'ād·ə }

craniectomy [MED] Surgical removal of strips or pieces of the cranial bones. { krā·nē·'ek·tə·mē }

Craniidina [INV ZOO] A subdivision of inarticulate branchiopods in the order Acrotretida known to possess a pedicle; all forms are attached by cementation. { krā·ne'īd·ən·ə }

craniobuccal pouch [EMBRYO] A diverticulum from the buccal cavity in the embryo from which the anterior lobe of the hypophysis is developed. Also known as Rathke's pouch. { ¦krā·nē·ō'bək·əl 'pau̇ch }

cranioclasis [MED] The operation of breaking the fetal head by means of a cranioclast. { ˌkrā·nē·ō'klā·səs }

cranioclast [MED] A heavy forceps for crushing the fetal head. { 'krā·nē·ōˌklast }

craniofacial dysmorphism [MED] Malformation of the cranium and the low face. { ¦krā·nē·ō¦fā·shəl dis'mor·fiz·əm }

craniopharyngioma [MED] An epithelial tumor of the craniopharyngeal canal, usually in children. { ¦krā·nē·ō·fəˌrin·jē'ō·mə }

cranioplasty [MED] Surgical correction of defects in the cranial bones, usually by implants of metal, plastic material, or bone. { 'krā·nē·ōˌplas·tē }

cranioscopy [MED] Examination of the human skull. { ˌkra·nē'as·kə·pē }

craniosynostosis [MED] The union of separate cranial bones into a single bone structure. { ˌkrā·nē·ōˌsin·ə'stō·səs }

cranium [ANAT] That portion of the skull enclosing the brain. Also known as braincase. { 'krā·nē·əm }

craspedon [INV ZOO] A cnidarian medusa stage possessing a velum. { 'kras·pəˌdän }

craspedote [INV ZOO] Having a velum, used specifically for velate hydroid medusae. { 'kras·pəˌdōt }

Crassulaceae [BOT] A family of dicotyledonous plants in the order Rosales notable for their succulent leaves and resistance to desiccation. { ˌkras·ə'lās·ēˌē }

crassulacean acid metabolism [BOT] A type of photosynthesis exhibited by many succulent plants in which carbon dioxide is taken up and stored during the night to allow the stomata to remain closed during the daytime, decreasing water loss. Abbreviated CAM. { ˌkras·əˌlā·shən ¦as·əd mə'tab·əˌliz·əm }

craterization [MED] Surgical excision of part of a bone so that a crater remains. { ˌkrād·ə·rə'zā·shən }

craw [ZOO] **1.** The crop of a bird or insect. **2.** The stomach of a lower animal. { kró }

crayfish [INV ZOO] The common name for a number of lobsterlike fresh-water decapod crustaceans in the section Astacura. { 'krāˌfish }

C-reactive protein [IMMUNOL] A plasma protein that is present normally in low concentration, and after trauma or infection in much higher concentration; the biological function is unknown. { ¦sē rēˌak·tiv 'prōˌtēn }

creatine [BIOCHEM] $C_4H_9O_2N_3$ α-Methylguanidine-acetic acid; a compound present in vertebrate muscle tissue, principally as phosphocreatine. { 'krē·əˌtēn }

creatine kinase [BIOCHEM] An enzyme of vertebrate skeletal and myocardial muscle that catalyzes the transfer of a high-energy phosphate group from phosphocreatine to adenosine diphosphate with the formation of adenosine triphosphate and creatine. { ¦krē·əˌtēn 'kīˌnās }

creatinine [BIOCHEM] $C_4H_7ON_3$ A compound present in urine, blood, and muscle that is formed from the dehydration of creatine. { ˌkrē'at·ənˌēn }

creatinuria [MED] The occurrence of creatine in the urine. { ˌkrē·ə·tə'nu̇r·ē·ə }

CREB *See* cyclic AMP-responsive element binding protein. { ¦sē¦är¦ē'bē *or* kreb }

CREB-binding protein [CELL MOL] A transcriptional coactivator that binds a phosphorylated domain of cyclic AMP-responsive element binding protein (CREB) to facilitate transcription initiation. Abbreviated CBP. { ¦kreb ¦bin·diŋ 'prōˌtēn }

Credé procedure [MED] Instillation of silver nitrate drops into the eyes of a newborn infant to prevent ophthalmia neonatorum. { krə'dā prə'sē·jər }

creeping disk [INV ZOO] The smooth and adhesive undersurface of the foot or body of a mollusk or of certain other invertebrates, on which the animal creeps. { 'krē·piŋ ˌdisk }

creeping eruption [MED] A red line of eruption on the skin produced by larva burrowing in the dermis; characterizes the condition of larva migrans. { 'krē·piŋ i'rəp·shən }

C region [IMMUNOL] The parts of heavy or light chains of immunoglobulin molecules within the same class that have the same amino acid sequence regardless of the molecule. { 'sē ˌrē·jən }

cremocarp [BOT] A dry dehiscent fruit consisting of two indehiscent one-seeded mericarps which separate at maturity and remain pendant from the carpophore. { 'krem·əˌkärp }

crenate [BIOL] Having a scalloped margin; used specifically for foliar structures, shrunken erythrocytes, and shells of certain mollusks. { 'krēˌnāt }

crenation [PHYSIO] A notched appearance of shrunken erythrocytes; seen when they are exposed to the air or to strong saline solutions. { krə'nā·shən }

Crenothrix [MICROBIO] A genus of sheathed bacteria; cells are nonmotile, sheaths are attached and encrusted with iron and manganese oxides, and filaments may have swollen tips. { 'kren·əˌthriks }

Crenotrichaceae [MICROBIO] Formerly a family of bacteria in the order Chlamydobacteriales having trichomes that are differentiated at the base and tip and attached to a firm substrate. { ¦kren·ə·trə'kās·ēˌē }

crenulate [BIOL] Having a minutely crenate margin. { 'kren·əlˌāt }

creosote bush [BOT] *Larrea divaricata.* A bronze-green, xerophytic shrub characteristic of all the American warm deserts. { 'krē·əˌsōt ˌbu̇sh }

crepitation [MED] A noise produced by the rubbing of fractured ends of bones, by cracking joints, and by pressure upon tissues containing abnormal amounts of air, as in cellular emphysema. { ˌkrep·ə'tā·shən }

crepuscular [ZOO] Active during the hours of twilight or preceding dawn. { krə'pəs·kyə·lər }

cress [BOT] Any of several prostrate crucifers belonging to the order Capparales and grown for their flavorful leaves; includes watercress (*Nasturtium officinale*), garden cress (*Lepidium sativum*), and upland or spring cress (*Barbarea verna*). { kres }

cretin [MED] An individual afflicted with cretinism. { 'krēt·ən *or* 'kret·ən }

cretinism [MED] A type of dwarfism caused by hypothyroidism and associated with generalized body changes, including mental deficiency. { krēt·ənˌiz·əm }

CRH *See* corticotropin-releasing hormone.

crib death *See* sudden infant death syndrome. { 'krib ˌdeth }

cribellum [INV ZOO] **1.** A small accessory spinning organ located in front of the ordinary spinning organ in certain spiders. **2.** A chitinous plate perforated with the openings of certain gland ducts in insects. { krə'bel·əm }

Cribrariaceae [BOT] A family of true slime molds in the order Liceales. { krə,brer·ē'ās·ē,ē }

cribriform [BIOL] Perforated, like a sieve. { 'krib·rə,fórm }

cribriform fascia [ANAT] The sievelike covering of the fossa ovalis of the thigh. { 'krib·rə,fórm ¦fā·shə }

cribriform plate [ANAT] **1.** The horizontal plate of the ethmoid bone, part of the floor of the anterior cranial fossa. **2.** The bone lining a dental alveolus. { 'krib·rə,fórm ¦plāt }

Cricetidae [VERT ZOO] A family of the order Rodentia including hamsters, voles, and some mice. { krə'sed·ə,dē }

Cricetinae [VERT ZOO] A subfamily of mice in the family Cricetidae. { krə'set·ən,ē }

cricket [INV ZOO] **1.** The common name for members of the insect family Gryllidae. **2.** The common name for any of several related species of orthopteran insects in the families Tettigoniidae, Gryllotalpidae, and Tridactylidae. { 'krik·ət }

cricoid [ANAT] The signet-ring-shaped cartilage forming the base of the larynx in humans and most other mammals. { 'krī,kóid }

Criconematoidea [INV ZOO] A superfamily of plant parasitic nematodes of the order Diplogasterida distinguished by their ectoparasitic habit and males that have atrophied mouthparts and do not feed. { ,krī·kō,nem·ə'tóid·ē·ə }

cri du chat syndrome [MED] An inherited condition characterized by mental subnormality, physical abnormalities, and the emitting of a flat, toneless, catlike cry in infancy. { ,krē dü 'shä ,sin,drōm }

criminal abortion [MED] Illegal interruption of pregnancy. { 'krim·ən·əl ə'bór·shən }

Crinoidea [INV ZOO] A class of radially symmetrical crinozoans in which the adult body is flower-shaped and is either carried on an anchored stem or is free-living. { krə'nóid·ē·ə }

Crinozoa [INV ZOO] A subphylum of the Echinodermata comprising radially symmetrical forms that show a partly meridional pattern of growth. { ,krī·nə,zō·ə }

crisis [MED] The turning point in the course of a disease. { 'krī·səs }

crissum [VERT ZOO] **1.** The region surrounding the cloacal opening in birds. **2.** The vent feathers covering the circumcloacal region. { 'kris·əm }

crista [BIOL] A ridge or crest. [CELL MOL] A fold on the inner membrane of a mitochondrion. { 'kris·tə }

cristate [BIOL] Having a crista. { 'kri,stāt }

Cristispira [MICROBIO] A genus of bacteria in the family Spirochaetaceae; helical cells with 3–10 complete turns; they have ovoid inclusion bodies and bundles of axial fibrils; commensals in mollusks. { ,kris·tə'spī·rə }

critical care medicine [MED] The treatment of acute, life-threatening disorders, usually in intensive care units. { ,krid·ə·kəl 'ker ¦med·ə·sən }

critical frequency of fusion [NEUROSCI] A sufficiently high flash rate at which the eye fails to detect the flicker of a light; that is, the light pulses seem to fuse to form a steady light indistinguishable from a continuous light that has the same total energy per unit time. { ¦krid·ə·kəl ¦frē·kwən·sē əv 'fyü·zhən }

critical region [GEN] The shortest segment of a chromosome whose gain or loss results in a particular complex phenotype, such as Down syndrome. { 'krid·ə·kəl 'rē·jən }

crocodile [VERT ZOO] The common name for about 12 species of aquatic reptiles included in the family Crocodylidae. { 'kräk·ə,dīl }

Crocodilia [VERT ZOO] An order of the class Reptilia which is composed of large, voracious, aquatic species, including the alligators, caimans, crocodiles, and gavials. { 'kräk·ə¦dil·ē·ə }

Crocodylidae [VERT ZOO] A family of reptiles in the order Crocodilia including the true crocodiles, false gavial, alligators, and caimans. { ,kräk·ə'dil·ə,dē }

Crocodylinae [VERT ZOO] A subfamily of reptiles in the family Crocodylidae containing the crocodiles, *Osteolaemus*, and the false gavial. { ,kräk·ə'dil·ə,nē }

crocus [BOT] A plant of the genus *Crocus*, comprising perennial herbs cultivated for their flowers. { 'krō·kəs }

Crohn's disease [MED] Chronic inflammation of the colon and stomach of unknown etiology that involves the full thickness of the intestinal wall, often with bowel narrowing and obstruction of the lumen. It is usually accompanied by granulomas; and abdominal cramps, alteration of bowel function, and diminished food intake are common. { 'krōnz diz,ēz }

cromolyn sodium [PHARM] $C_{23}H_{14}Na_2O_{11}$ An inhaled anti-inflammatory agent that prevents the degranulation of mast cells; effective in preventing bronchoconstriction. { 'krō·mə·lin ¦sōd·ē·əm }

cron [EVOL] A time unit equal to 10^6 years; used in reference to evolutionary processes. { krän }

Cronartium ribicola [MYCOL] A heteroecious rust fungus that causes white pine blister rust; it produces pycnia and aecia on pine stems, and uredinia and telia on currants and gooseberries. { krə¦närd·ē·əm ,rib·i'kō·lə }

crop [VERT ZOO] A distensible saccular diverticulum near the lower end of the esophagus of birds which serves to hold and soften food before passage into the stomach. { kräp }

crossbreed [BIOL] To propagate new individuals by breeding two distinctive varieties of a species. Also known as outbreed. { 'krós,brēd }

crossed electrophoresis [IMMUNOL] Immunoelectrophoresis that uses an initial separation along one axis of the plate, after which only a strip of medium along the axis is preserved; new medium containing antisera is poured beside the strip, the plate is turned 90°, and electrophoresis is resumed. { 'króst i,lek·trō·fə'rē·səs }

crossed paralysis [MED] Paralysis of the arm

and leg on one side, associated with contralateral cranial nerve palsies caused by a brainstem lesion involving cranial nerve nuclei and the ipsilateral pyramidal tract. { 'kröst pə'ral·ə·səs }

cross-eye *See* esotropia. { 'krös ,ī }

cross-fertilization [BOT] Fertilization between two separate plants. [ZOO] Fertilization between different kinds of individuals. { 'krös ,fərd·əl·ə'zā·shən }

crossing barrier [GEN] Any of the genetically controlled mechanisms that either prevent or significantly reduce the ability of individuals in a population to hybridize with individuals of other populations. { 'krös·iŋ ,bar·ē·ər }

crossing-over [GEN] The exchange of genetic material between paired homologous chromosomes during meiosis. Also known as crossover. { ¦krös·iŋ 'ō·vər }

crossing-over map [GEN] A genetic map made by utilizing the frequency of crossing-over as a measure of the relative distances between genes in one linkage group. { ¦krös·iŋ 'ō·vər ,map }

crossing-over value [GEN] The frequency of crossing-over between two linked genes. { ¦krös·iŋ 'ō·vər ,val·yü }

crosslink [CELL MOL] A covalent linkage between the complementary strands of deoxyribonucleic acid (DNA) duplex or between bases of a single strand of DNA. { 'krös ,liŋk }

cross matching [IMMUNOL] Determination of blood compatibility for transfusion by mixing donor cells with recipient serum, and recipient cells with donor serum, and examining for an agglutination reaction. { 'krös ,mach·iŋ }

Crossosomataceae [BOT] A monogeneric family of xerophytic shrubs in the order Dilleniales characterized by perigynous flowers, seeds with thin endosperm, and small, entire leaves. { ¦krä·sə,sō·mə¦tās·ē,ē }

crossover *See* crossing-over. { 'krös,ō·vər }

crossover experiment [MED] An experiment or clinical investigation in which subjects are divided randomly into at least as many groups as there are kinds of treatment to be given, and then the groups are interchanged until every subject has received each treatment. { 'krös,ō·vər ik'sper·ə·mənt }

cross-pollination [BOT] Transfer of pollen from the anthers of one plant to the stigmata of another plant. { ¦krös ,pä·lə,nā·shən }

cross-reacting antibody [IMMUNOL] Antibody that reacts with an antigen that did not stimulate the production of that antibody. { 'krös rē,ak·tiŋ 'ant·i,bäd·ē }

cross-reacting antigen [IMMUNOL] Antigen that reacts with an antibody whose production was induced by a different antigen. { 'krös rē,ak·tiŋ 'ant·i·jən }

cross-reacting material [BIOCHEM] A protein produced by a mutant gene that is enzymatically inactive but shows serological properties similar to the protein of the wild-type gene. { 'krös rē,ak·tiŋ mə'tir·ē·əl }

cross-reaction [IMMUNOL] Reaction between an antibody and a closely related, but not complementary, antigen. { 'krös rē'ak·shən }

cross-tolerance [MED] Tolerance or resistance to the action of a drug brought about by continued use of another drug of similar pharmacologic action. { ¦krös 'täl·ə·rəns }

Crotalidae [VERT ZOO] A family of proglyphodont venomous snakes in the reptilian suborder Serpentes. { krō'tal·ə,dē }

croup [MED] Any condition of upper-respiratory pathway obstruction in children, especially acute inflammation of the pharynx, larynx, and trachea, characterized by a hoarse, brassy, and stridulent cough and difficulties in breathing. { krüp }

croup-associated virus [VIROL] A virus belonging to subgroup 2 of the parainfluenza viruses and found in children with croup. Also known as CA virus; laryngotracheobronchitis virus. { ¦krüp ə¦sō·sē,ād·əd 'vī·rəs }

crow [VERT ZOO] The common name for a number of predominantly black birds in the genus *Corvus* comprising the most advanced members of the family Corvidae. { krō }

crown [ANAT] **1.** The top of the skull. **2.** The portion of a tooth above the gum. [BOT] **1.** The topmost part of a plant or plant part. **2.** *See* corona. { kraún }

crown grafting [BOT] A method of vegetative propagation whereby a scion 3–6 inches (8–15 centimeters) long is grafted at the root crown, just below ground level. { 'kraún ,graf·tiŋ }

CR ratio *See* calorimetric-respirometric ratio. { ¦sē'är ,rā·shō }

CRSV *See* carnation ringspot virus.

cruciate [ANAT] Resembling a cross. { 'krü·shē,āt }

Cruciferae [BOT] A large family of dicotyledonous herbs in the order Capparales characterized by parietal placentation; hypogynous, mostly regular flowers; and a two-celled ovary with the ovules attached to the margins of the partition. { krü'sif·ə,rē }

cruciform structure [CELL MOL] A cross-shaped configuration of deoxyribonucleic acid produced by intrastrand base pairing of complementary inverted repeats. { 'krü·sə,fórm ,strək·chər }

crude drug [PHARM] **1.** A plant or animal drug containing all principles characteristic of the drug. **2.** The dried leaves, bark, or rhizome of a plant containing therapeutically active principles. Also known as botanical; plant extract. { ¦krüd 'drəg }

crude lecithin *See* lecithin. { ¦krüd 'les·ə·thən }

crura [ANAT] Plural of crus. { 'krúr·ə }

crus [ANAT] **1.** The shank of the hindleg, that portion between the femur and the ankle. **2.** Any of various parts of the body resembling a leg or root. { krüs }

crush kidney *See* lower nephron nephrosis. { ¦krəsh 'kid·nē }

crush syndrome [MED] A severe, often fatal condition that follows a severe crushing injury, particularly involving large muscle masses, characterized by fluid and blood loss, shock, hematuria, and renal failure. Also known as compression syndrome. { 'krəsh ˌsinˌdrōm }

Crustacea [INV ZOO] A class of arthropod animals in the subphylum Mandibulata having jointed feet and mandibles, two pairs of antennae, and segmented, chitin-encased bodies. { krə'stā·shə }

crustacyanin [BIOCHEM] A carotenoprotein that determines the color of lobster shells. { ˌkrəs·tə'sī·ə·nən }

crustecdysone [BIOCHEM] $C_{27}H_{44}O_7$ 20-Hydroxyecdysone, the molting hormone produced by Y organs in crustaceans. { krəs'tek·dəˌsōn }

crustose [BOT] Of a lichen, forming a thin crustlike thallus which adheres closely to the substratum of rock, bark, or soil. { 'krəsˌtōs }

crust vegetation [ECOL] Zonal growths of algae, mosses, lichens, or liverworts having variable coverage and a thickness of only a few centimeters. { 'krəst ˌvej·ə'tā·shən }

cryalgesia [MED] Pain caused by cold. Also known as crymodynia. { ˌkrī·al'jē·zhə }

cryanesthesia [MED] Loss of sensation of cold. { ˌkrīˌan·əs'thē·zhə }

cryesthesia [PHYSIO] **1.** The sensation of coldness. **2.** Exceptional sensitivity to low temperatures. { ˌkrī·əs'thē·zhə }

crymodynia *See* cryalgesia. { ˌkrī·mə'dīn·ē·ə }

cryoanalgesia [MED] Loss of sensation of pain resulting from the use of a cryoprobe. { ˌkrī·ōˌan·əl'jē·zhə }

cryoanesthesia [MED] Regional anesthesia produced by localized application of cold. { ˌkrī·ōˌan·əs'thē·zhə }

cryobiology [BIOL] The use of low-temperature environments in the study of living plants and animals. { ¦krī·ō·bī'äl·ə·jē }

cryobiosis [PHYSIO] A type of cryptobiosis induced by low temperatures. { ¦krī·ō·bī'ō·səs }

cryocautery [MED] A substance or instrument that causes destruction of tissue by freezing. { ˌkrī·ō'kȯd·ə·rē }

cryoextraction [MED] Removal of a cataract by means of a cryoprobe. { ˌkrī·ō·ik'strak·shən }

cryofibrinogen [PATH] An abnormal fibrinogen that precipitates upon cooling but redissolves when warmed to room temperature; rarely found in human plasma. { ˌkrī·ō·fī'brin·ə·jən }

cryoglobulin [PATH] An abnormal protein, usually an immunoglobulin, which precipitates from plasma between 40 and 70°F (4.4 and 21°C). { ¦krī·ō'gläb·yə·lən }

cryophilic *See* cryophilous. { ˌkrī·ə'fil·ik }

cryophilous [ECOL] Having a preference for low temperatures. Also known as cryophilic. { krī'äf·ə·ləs }

cryophyte [ECOL] A plant that forms winter buds below the soil surface. { 'krī·əˌfīt }

cryoprecipitate [BIOCHEM] The precipitate of a cryoglobulin. { ¦krī·ō·prə'sip·əˌtāt }

cryoprobe [MED] A blunt instrument that can be chilled to a temperature of −162°F (−108°C); used to freeze tissues in cryosurgery. { 'krī·ōˌprōb }

cryosurgery [MED] Selective destruction of tissue by freezing, as the use of a liquid nitrogen probe to the brain in parkinsonism. { ¦krī·ō'sərj·ə·rē }

cryotherapy [MED] A form of therapy which consists of local or general use of cold. { ¦krī·ō'ther·əˌpē }

Cryphaeaceae [BOT] A family of mosses in the order Isobryales distinguished by a rough calyptra. { ˌkrī·fē'ās·ēˌē }

crypt [ANAT] **1.** A follicle or pitlike depression. **2.** A simple glandular cavity. { kript }

cryptic behavior [ZOO] A behavior pattern that maximizes an organism's ability to conceal itself. { 'krip·tik bə'hāv·yər }

cryptic coloration [ZOO] A phenomenon of protective coloration by which an animal blends with the background through color matching or countershading. { 'krip·tik kəl·ə'rā·shən }

Cryptobiidae [INV ZOO] A family of flagellate protozoans in the order Kinetoplastida including organisms with two flagella, one free and one with an undulating membrane. { ˌkrip·tō'bī·əˌdē }

cryptobiosis [PHYSIO] A state in which metabolic rate of the organism is reduced to an imperceptible level. { ¦krip·tō·bī'ō·səs }

cryptobiotic [ECOL] Living in concealed or secluded situations. { ¦krip·tō·bī'äd·ik }

Cryptobranchidae [VERT ZOO] The giant salamanders and hellbenders, a family of tailed amphibians in the suborder Cryptobranchoidea. { ˌkrip·tə'braŋ·kəˌdē }

Cryptobranchoidea [VERT ZOO] A primitive suborder of amphibians in the order Urodela distinguished by external fertilization and aquatic larvae. { ˌkrip·təˌbraŋ'kȯid·ē·ə }

Cryptocerata [INV ZOO] A division of hemipteran insects in some systems of classification that includes the water bugs (Hydrocorisae). { ˌkrip·tō·sə'räd·ə }

Cryptochaetidae [INV ZOO] A family of myodarian cyclorrhaphous dipteran insects in the subsection Acalypteratae. { ˌkrip·tə'kēd·əˌdē }

cryptochromes [CELL MOL] Light-sensitive proteins found in both plants and animals that detect and change conformation in response to blue light; in animals, they play an important role in circadian rhythm. { 'krip·təˌkrōm }

Cryptococcaceae [MYCOL] A family of imperfect fungi in the order Moniliales in some systems of classification; equivalent to the Cryptococcales in other systems. { ˌkrip·tə·käk'sā·sēˌē }

Cryptococcales [MYCOL] An order of imperfect fungi, in some systems of classification, set up to include the yeasts or yeastlike organisms whose perfect or sexual stage is not known. { ˌkrip·tə·kä'kāˌlēz }

cryptococcal meningitis [MED] Inflammation of the meninges due to yeasts of the genus *Cryptococcus.* { ˌkrip·tə'käk·əl ˌmen·ən'jīd·əs }

cryptococcosis [MED] A yeast infection of humans, primarily of the central nervous system, caused by *Cryptococcus neoformans*. Also known as torulosis. { ˌkrip·tə·käˈkō·səs }

Cryptococcus [MYCOL] A genus of encapsulated pathogenic yeasts in the order Moniliales. { ˌkrip·təˈkäk·əs }

Cryptodira [VERT ZOO] A suborder of the reptilian order Chelonia including all turtles in which the cervical spines are uniformly reduced and the head folds directly back into the shell. { ˌkrip·təˈdī·rə }

cryptogam [BOT] An old term for nonflowering plants. { ˈkrip·təˌgam }

cryptomedusa [INV ZOO] The final stage in the reduction of a hydroid medusa to a rudiment having sex cells within the gonophore. { ¦krip·tō·məˈdü·sə }

cryptomitosis [INV ZOO] Cell division in certain protozoans in which a modified spindle forms, and chromatin assembles with no apparent chromosome differentiation. { ¦krip·tō·mīˈtō·səs }

Cryptomonadida [BIOL] An order of the class Phytamastigophorea considered to be protozoans by biologists and algae by botanists. { ˌkrip·tō·məˈnäd·ə·də }

Cryptomonadina [BIOL] The equivalent name for Cryptomonadida. { ˌkrip·tō·män·əˈdī·nə }

cryptonephridic [INV ZOO] In certain insects, referring to Malpighian tubules independently attached to the hindgut (in contrast to being free). { ¦krip·tō·ne¦frid·ik }

Cryptophagidae [INV ZOO] The silken fungus beetles, a family of coleopteran insects in the superfamily Cucujoidea. { ˌkrip·təˈfāj·əˌdē }

Cryptophyceae [BOT] A class of algae of the Pyrrhophyta in some systems of classification; equivalent to the division Cryptophyta. { ˌkrip·təˈfīs·ēˌē }

Cryptophyta [BOT] A division of the algae in some classification schemes; equivalent to the Cryptophyceae. { ˌkrip·təˈfīd·ə }

cryptophyte [BOT] A plant that produces buds either underwater or underground on corms, bulbs, or rhizomes. { ˈkrip·təˌfīt }

Cryptopidae [INV ZOO] A family of epimorphic centipedes in the order Scolopendromorpha. { kripˈtäp·əˌdē }

cryptorchidism *See* cryptorchism. { kripˈtȯr·kə ˌdiz·əm }

cryptorchism [MED] Failure of the testes to descend into the scrotum from the abdomen or inguinal canals. Also known as cryptorchidism. { kripˈtȯrˌkiz·əm }

cryptotope [IMMUNOL] A determinant (or epitope) of an immunological antigen or immunogen which is initially hidden and becomes functional only when the molecule is broken or degraded. { ˈkrip·tōˌtōp }

cryptoviolin *See* phycobiliviolin. { ˌkrip·təˈvī·ə·lən }

cryptovirogenic [VIROL] Possessing the ability to produce infective virus particles after derepression of the viral genome within the cell. { ¦krip·tōˌvī·rəˈjen·ik }

cryptoxanthin [BIOCHEM] $C_{40}H_{57}O$ A xanthophyll carotenoid pigment found in plants; convertible to vitamin A by many animal livers. { ¦krip·tōˈzan·thən }

crypts of Lieberkühn [ANAT] Simple, tubular glands which arise as evaginations into the mucosa of the small intestine. { ˈkrips əv ˈlē·bərˌkyün }

crystalliferous bacteria [MICROBIO] Bacteria, especially *Bacillus thuringiensis*, that are characterized by the formation of a protein crystal in the sporangium at the time of spore formation. { ¦kris·tə¦lif·ə·rəs bakˈtir·ē·ə }

crystalline lens *See* lens. { ˈkris·tə·lən ˈlenz }

crystallin protein [BIOCHEM] Any of a group of stable structural components distributed nonuniformly in the lens of the eye of vertebrates. { ˈkrist·əl·ən ˈprōˌtēn }

CT *See* computerized tomography.

ctDNA *See* chloroplast deoxyribonucleic acid.

ctenidium [INV ZOO] **1.** The comb- or featherlike respiratory apparatus of certain mollusks. **2.** A row of spines on the head or thorax of some fleas. { təˈnid·ē·əm }

Ctenodrilidae [INV ZOO] A family of fringe worms belonging to the Sedentaria. { ˌten·əˈdrī·ləˌdē }

ctenoid scale [VERT ZOO] A thin, acellular structure composed of bonelike material and characterized by a serrated margin; found in the skin of advanced teleosts. { ˈtenˌȯid ˌskāl }

Ctenophora [INV ZOO] The comb jellies, a phylum of marine organisms having eight rows of comblike plates as the main locomotory structure. { təˈnäf·ə·rə }

Ctenostomata [INV ZOO] An order of bryozoans in the class Gymnolaemata recognized as inconspicuous, delicate colonies made up of relatively isolated, short, tubular zooecia with chitinous walls. { ˌten·əˈstäm·ə·də }

Ctenostomatida [INV ZOO] The equivalent name for Odontostomatida. { ˌten·ə·stəˈmad·ə·də }

C-type virus particle [VIROL] One of a morphologically similar group of enveloped virus particles having a central, spherical ribonucleic acid-containing nucleoid; associated with certain cancers, as sarcomas and leukemias. { ˈsē ¦tīp ˈvī·rəs ˌpärd·ə·kəl }

cubeb [BOT] The dried, nearly ripe fruit (berries) of a climbing vine, *Piper cubeba*, of the pepper family (Piperaceae). { ˈkyüˌbeb }

cuboid [ANAT] The outermost distal tarsal bone in vertebrates. [INV ZOO] Main vein of the wing in many insects, particularly the flies (Diptera). { ˈkyüˌbȯid }

cuboidal epithelium [HISTOL] A single-layered epithelium made up of cubelike cells. { kyü ˈbȯid·əl ˌep·əˈthēl·ē·əm }

Cubomedusae [INV ZOO] An order of cnidarians in the class Scyphozoa distinguished by a cubic umbrella. { ¦kyü·bo·məˈdü·sē }

cuckoo [VERT ZOO] The common name for about 130 species of primarily arboreal birds in the family Cuculidae; some are social parasites. { 'ku̇,kü }

Cucujidae [INV ZOO] The flat-back beetles, a family of predatory coleopteran insects in the superfamily Cucujoidea. { kə'kü·yə,dē }

Cucujoidea [INV ZOO] A large superfamily of coleopteran insects in the suborder Polyphaga. { ,kü·kə'yȯid·ē·ə }

Cuculidae [VERT ZOO] A family of perching birds in the order Cuculiformes, including the cuckoos and the roadrunner, characterized by long tails, heavy beaks and conspicuous lashes. { kə'kyü·lə,dē }

Cuculiformes [VERT ZOO] An order of birds containing the cuckoos and allies, characterized by the zygodactyl arrangement of the toes. { kə,kyü·lə'fȯr,mēz }

cucullus [INV ZOO] A transverse flap at the anterior edge of the carapace that hangs over the mouthparts of certain arachnids. { kyü'kəl·əs }

Cucumariidae [INV ZOO] A family of dendrochirotacean holothurian echinoderms in the order Dendrochirotida. { ,kü·kə·mə'rī·ə,dē }

cucumber [BOT] *Cucumis sativus.* An annual cucurbit in the family Cucurbitaceae grown for its edible, immature fleshy fruit. { 'kyü·kəm·bər }

cucumber mosaic virus [VIROL] The type species of the genus *Cucumovirus.* Abbreviated CMV. { ¦kyü·kəm·bər mō¦zāik 'vī·rəs }

cucumber mosaic virus group *See* Cucumovirus. { ¦kyü·kəm·bər mō¦zā·ik 'vī·rəs ,grüp }

Cucumovirus [VIROL] A genus of the family Bromoviridae that is characterized by icosahedral particles containing one molecule of one of the four single-stranded ribonucleic acid species; the type species is cucumber mosaic virus. Also known as cucumber mosaic virus group. { 'kyü·kəm,o'vī·rəs }

cul-de-sac [ANAT] Blind pouch or diverticulum. { 'kəl·də,sak }

culdoscope [MED] An instrument used to visualize female pelvic organs, introduced through the vagina or a perforation into the retrouterine pouch. { 'kəl·də,skōp }

Culex [INV ZOO] A genus of mosquitoes important as vectors for malaria and several filarial parasites. { 'kyü,leks }

Culicidae [INV ZOO] The mosquitoes, a family of slender, orthorrhaphous dipteran insects in the series Nematocera having long legs and piercing mouthparts. { kyü'lis·ə,dē }

Culicinae [INV ZOO] A subfamily of the dipteran family Culicidae. { kyü'lis·ə,nē }

culm [BOT] **1.** A jointed and usually hollow grass stem. **2.** The solid stem of certain monocotyledons, such as the sedges. { kəlm }

culmen [VERT ZOO] The edge of the upper bill in birds. { 'kəl·mən }

cultigen [BIOL] A cultivated variety or species of organism for which there is no known wild ancestor. Also known as cultivar. { 'kəl·tə·jən }

cultivar *See* cultigen. { 'kəl·tə,vär }

cultural ecology [ECOL] The branch of ecology that involves the study of the interaction of human societies with one another and with the natural environment. { ¦kəl·chər·əl ē'käl·ə·jē }

culture [BIOL] A growth of living cells or microorganisms in a controlled artificial environment. { 'kəl·chər }

culture alteration [CELL MOL] A persistent change in the properties of cultured cells, such as altered morphology, virus susceptibility, nutritional requirements, or proliferative capacity. { 'kəl·chər ,ȯl·tə,rā·shən }

culture community [ECOL] A plant community which is established or modified through human intervention; for example, a fencerow, hedgerow, or windbreak. { 'kəl·chər kə'myü·nəd·ē }

cultured pearl [INV ZOO] A natural pearl grown by means of controlled stimulation of the oyster. { ¦kəl·chərd 'pərl }

culture medium [MICROBIO] The nutrients and other organic and inorganic materials used for the growth of microorganisms and plant and animal tissue in culture. { ¦kəl·chər ,mēd·ē·əm }

Cumacea [INV ZOO] An order of the class Crustacea characterized by a well-developed carapace which is fused dorsally with at least the first three thoracic somites and overhangs the sides. { kyü'mās·ē·ə }

cumatophyte [ECOL] A plant that grows under surf conditions. { kyü'mad·ə,fīt }

cumin [BOT] *Cuminum cyminum* An annual herb in the family Umbelliferae; the fruit is valuable for its edible, aromatic seeds. { 'kyü·mən }

cumulative trauma [MED] An injury or work strain that results from the repeated or continuous application of a work stress that would not ordinarily be harmful in single applications or in multiple applications of short duration. { 'kyü·myə·ləd·iv 'trau̇·mə }

cumulus oophorus [HISTOL] The layer of gelatinous, follicle cells surrounding the ovum in a Graafian follicle. { 'kyü·myə·ləs ¦ō·ə'fȯr·əs }

cuneate [BIOL] Wedge-shaped with the acute angle near the base, as in certain insect wings and the leaves of various plants. { 'kyü·nē,āt }

cuneiform [ANAT] **1.** Any of three wedge-shaped tarsal bones. **2.** Either of a pair of cartilages lying dorsal to the thyroid cartilage of the larynx. **3.** Wedge-shaped, chiefly referring to skeletal elements. { 'kyü·nē·ə,fȯrm }

cunnus [ANAT] The vulva. { 'kən·əs }

Cupedidae [INV ZOO] The reticulated beetles, the single family of the coleopteran suborder Archostemata. { kyü'ped·ə,dē }

cupule [BOT] **1.** The cup-shaped involucre characteristic of oaks. **2.** A cup-shaped corolla. **3.** The gemmae cup of the Marchantiales. [INV ZOO] A small sucker on the feet of certain male flies. { 'kyü,pyül }

Curculionidae [INV ZOO] The true weevils or snout beetles, a family of coleopteran insects in the superfamily Curculionoidea. { kər,kyü·lē'än·ə,dē }

Curculionoidea [INV ZOO] A superfamily of coleopteran insects in the suborder Polyphaga. { kər,kyü·lē·ə'nȯid·ē·ə }

curculionoid larva [INV ZOO] A kind of beetle larva having a highly reduced and grublike body. { kər'kyü·lē·ə,nȯid 'lär·və }

Curcurbitaceae [BOT] A family of dicotyledonous herbs or herbaceous vines in the order Violales characterized by an inferior ovary, unisexual flowers, one to five stamens but typically three, and a sympetalous corolla. { kər,kər·bə'tās·ē,ē }

Curcurbitales [BOT] The ordinal name assigned to the Curcurbitaceae in some systems of classification. { kər,kər·bə'tā·lēz }

curd [BOT] The edible flower heads of members of the mustard family such as broccoli. { kərd }

curet [MED] An instrument, shaped like a spoon or scoop, for scraping away tissue. { kyů'ret }

curettage [MED] Scraping of the inside of a body cavity or the hollow of an organ with a curet. { ,kyů·rə'täzh }

curling factor *See* griseofulvin. { 'kərl·iŋ ,fak·tər }

Curling's ulcer [MED] An acute gastric ulcer associated with severe skin burns. { 'kərl·iŋz ,əl·sər }

currant [BOT] A shrubby, deciduous plant of the genus *Ribes* in the order Rosales; the edible fruit, a berry, is borne in clusters on the plant. { 'kər·ənt }

cursorial [VERT ZOO] Adapted for running. { kər'sȯr·ē·əl }

Cuscutaceae [BOT] A family of parasitic dicotyledonous plants in the order Polemoniales which lack internal phloem and chlorophyll, have capsular fruit, and are not rooted to the ground at maturity. { kə,skyü'tās·ē,ē }

Cushing's syndrome [MED] A complex of symptoms including facial and truncal obesity, hypertension, edema, and osteoporosis, resulting from oversecretion of adrenocortical hormones. { 'kůsh·iŋz ,sin,drōm }

cusp [ANAT] **1.** A pointed or rounded projection on the masticating surface of a tooth. **2.** One of the flaps of a heart valve. { kəsp }

cuspid *See* canine. { 'kəs·pəd }

cuspidate [BIOL] Having a cusp; terminating in a point. { 'kəs·pə,dāt }

cut [BIOCHEM] A double-strand incision in a duplex deoxyribonucleic acid molecule. [CELL MOL] A double-strand incision in a duplex deoxyribonucleic acid molecule. { kət }

cut-and-paste transposition [CELL MOL] A form of deoxyribonucleic acid (DNA) transposition in which the transposed segment is cut from the donor DNA and inserted in the receptor DNA location. { ¦kət ən ¦pāst ,tranz·pə'zish·en }

cutaneous anaphylaxis [IMMUNOL] Hypersensitivity that is marked by an intense skin reaction following parenteral contact with a sensitizing agent. { kyü'tā·nē·əs an·ə·fə'lak·səs }

cutaneous anthrax *See* malignant pustule. { kyü 'tā·nē·əs 'an,thraks }

cutaneous appendage [ANAT] Any of the epidermal derivatives, including the nails, hair, sebaceous glands, mammary glands, and sweat glands. { kyü'tā·nē·əs ə'pen·dij }

cutaneous blastomycosis [MED] A form of North American blastomycosis considered by some to be a clinical manifestation of the systemic form. { kyü'tā·nē·əs ¦blas·tō·mī'kō·səs }

cutaneous coccidioidomycosis [MED] A primary skin infection by the fungus *Coccidioides immitis*; a skin infection secondary to a pulmonary lesion. { kyü'tā·nē·əs käk¦sid·ē¦ȯid·ō·mī'kō·səs }

cutaneous leishmaniasis [MED] A parasitic skin infection by *Leishmania tropica* characterized by deep ulcers of the skin and subcutaneous tissue. { kyü'tā·nē·əs lēsh·mə'nī·ə·səs }

cutaneous mycosis [MED] Any of a group of infections (collectively known as dermatophytoses, ringworms, or tineas) that are caused by keratinophilic fungi (dermatophytes). In general, the infections are limited to the nonliving keratinized layers of skin, hair, and nails, but a variety of pathologic changes can occur depending on the etiologic agent, site of infection, and immune status of the host. { kyü¦tān·ē·əs mī'kō·səs }

cutaneous pain [PHYSIO] A sensation of pain arising from the skin. { kyü'tā·nē·əs 'pān }

cutaneous reaction [MED] **1.** Any change in the outer layers of the skin, as in sunburn or the rash in measles. **2.** Any immediate or delayed immune reaction in the skin resulting from antigen-antibody interaction. { kyü'tā·nē·əs rē'ak·shən }

cutaneous sensation [PHYSIO] Any feeling originating in sensory nerve endings of the skin, including pressure, warmth, cold, and pain. { kyü'tā·nē·əs sen'sā·shən }

Cuterebridae [INV ZOO] The robust botflies, a family of myodarian cyclorrhaphous dipteran insects in the subsection Calypteratae. { kyü·də'reb·rə,dē }

cuticle [ANAT] The horny layer of the nail fold attached to the nail plate at its margin. [BIOL] A noncellular, hardened or membranous secretion of an epithelial sheet, such as the integument of nematodes and annelids, the exoskeleton of arthropods, and the continuous film of cutin on certain plant parts. { 'kyüd·ə·kəl }

cutin [BIOCHEM] A mixture of fatty substances characteristically found in epidermal cell walls and in the cuticle of plant leaves and stems. { 'kyüt·ən }

cutis *See* dermis. { 'kyüd·əs }

cutting [BOT] A piece of plant stem with one or more nodes, which, when placed under suitable conditions, will produce roots and shoots resulting in a complete plant. { 'kəd·iŋ }

cuttlefish [INV ZOO] An Old World decapod mollusk of the genus *Sepia*; shells are used to manufacture dentifrices and cosmetics. { 'kəd·əl,fish }

C value [CELL MOL] The amount (mass or molecular weight) of deoxyribonucleic acid per haploid cell. { 'sē ,val·yü }

C value paradox [CELL MOL] The observation that the amount of deoxyribonucleic acid in the haploid genome is not related to its evolutionary complexity. { ¦sē ˌval·yü 'par·əˌdäks }

Cyamidae [INV ZOO] The whale lice, a family of amphipod crustaceans in the suborder Caprellidea that bear a resemblance to insect lice. { sī'am·əˌdē }

cyanobacteria [MICROBIO] A group of one-celled to many-celled aquatic organisms. Also known as blue-green algae. { ¦sī·ə·noˌbak'tir·ē·ə }

cyanocobalamin *See* vitamin B_{12}. { ¦sī·ə·nō·kō 'bal·ə·mən }

cyanophage [VIROL] A virus that replicates in blue-green algae. Also known as algophage; blue-green algal virus. { sī'an·əˌfāj }

cyanophilous [BIOL] Having an affinity for blue or green dyes. { ¦sī·ə¦näf·ə·ləs }

Cyanophyceae [BOT] A class of photosynthetic monerans distinguished by their algalike biology and bacteriumlike cell organization. { ˌsī·ə·nō'fīs·ēˌē }

cyanophycin [BIOCHEM] A granular protein food reserve in the cells of blue-green algae, especially in the peripheral cytoplasm. { ˌsī·ə·nō'fīs·ən }

Cyanophyta [BOT] An equivalent name for the Cyanophyceae. { ˌsī·ə'näf·ə·də }

cyanosis [MED] A bluish coloration in the skin and mucous membranes due to deficient levels of oxygen in the blood. { ˌsī·ə'nō·səs }

Cyatheaceae [BOT] A family of tropical and pantropical tree ferns distinguished by the location of sori along the veins. { sī·ˌath·ə'ās·ēˌē }

cyathium [BOT] An inflorescence in which the flowers arise from the base of a cuplike involucre. { sī'ath·ē·əm }

Cyathoceridae [INV ZOO] The equivalent name for the Lepiceridae. { sīˌath·ə'räs·əˌdē }

Cyatholaimoidea [INV ZOO] A superfamily of nematodes of the order Chromadorida, distinguished by tightly coiled multispiral amphids located a short distance posterior to the cephalic sensilla. { ˌsī·ə·thō·lə'mȯid·ē·ə }

cybrid [GEN] A hybrid cell produced by fusing a cell nucleus with a cell of the same or a different species whose nucleus has been removed; some are able to proliferate. In plants, an individual produced following fusion of protoplasts from different species with complete elimination of the chromosomes of one of the species. { 'sī·brəd }

Cycadales [BOT] An ancient order of plants in the class Cycadopsida characterized by tuberous or columnar stems that bear a crown of large, usually pinnate leaves. { ˌsī·kə'dā·lēz }

Cycadatae *See* Cycadopsida. { sī'kad·əˌdē }

Cycadicae [BOT] A subdivision of large-leaved gymnosperms with stout stems in the plant division Pinophyta; only a few species are extant. { sī'kad·əˌsē }

Cycadophyta [BOT] An equivalent name for Cycadecae elevated to the level of a division. { sī·kə'däf·əd·ə }

Cycadophytae [BOT] An equivalent name for Cycadicae. { sī·kə'däf·əˌtē }

Cycadopsida [BOT] A class of gymnosperms in the plant subdivision Cycadicae. { sī·kə'däp·sə·də }

Cyclanthaceae [BOT] The single family of the order Cyclanthales. { ˌsīˌklan'thās·ēˌē }

Cyclanthales [BOT] An order of monocotyledonous plants composed of herbs; or, seldom, composed of more or less woody plants with leaves that usually have a bifid, expanded blade. { ˌsīˌklan'thā·lēz }

cyclase [BIOCHEM] An enzyme that catalyzes cyclization of a compound. { 'sīˌklās }

cyclic adenylic acid [BIOCHEM] $C_{10}H_{12}N_5O_6P$ An isomer of adenylic acid; crystal platelets with a melting point of 219–220°C; a key regulator which acts to control the rate of a number of cellular processes in bacteria, most animals, and some higher plants. Abbreviated cAMP. Also known as adenosine 3′,5′-cyclic monophosphate; adenosine 3′,5′-cyclic phosphate; adenosine 3′,5′-monophosphate; 3′,5′-AMP; cyclic AMP. { 'sīk·lik ¦ad·ən¦il·ik 'as·əd }

cyclic AMP *See* cyclic adenylic acid. { 'sīk·lik ¦āˌem¦pē }

cyclic AMP-dependent protein kinase [BIOCHEM] A serine/threonine protein kinase that phosphorylates a variety of substrates and regulates many important processes such as cell growth and differentiation and the flow of ions across the cell membrane. Also known as protein kinase A; PKA. { ¦sī·klik ¦ā¦em¦pē di¦pen·dənt ˌproˌtēn 'kīˌnās }

cyclic AMP-responsive element binding protein [CELL MOL] A deoxyribonucleic acid-binding transcription factor that becomes modified in response to an extracellular signal. Abbreviated CREB. { ¦sī·klik ¦ā¦em¦pē ri¦spän·siv ¦el·ə·mənt ¦bīnd·iŋ 'prōˌtēn }

cyclic GMP [BIOCHEM] A 3′,5′-cyclic ester of guanosine monophosphate that is involved in vision transduction through its direct effects on Na^{+} and Ca^{2+} channels in the plasma membrane of rod cells. { ¦sī·klik ¦jē¦em'pē }

cyclic nucleotide phosphodiesterase [BIOCHEM] Any of a group of enzymes that degrade cyclic nucleotides. { ¦sīk·lik ˌnü·klē·əˌtīd ˌfäs·fə·dī'es·təˌrās }

cyclin [CELL MOL] Any member of a family of proteins that regulate the cell cycle and whose cellular levels rise steadily until mitosis, then fall abruptly to zero. As cyclins reach a threshold level, they are thought to drive cells into G2 phase and thus toward mitosis. { 'sī·klin }

cyclin-Cdk complex [CELL MOL] A complex of the regulatory protein cyclin with the catalytic protein Cdk that regulates the cell cycle by selectively phosphorylating various protein substrates at different stages of the cycle. { ¦sīk·lin ¦sē¦dē¦kā 'kämˌpleks }

cyclin-dependent kinase [CELL MOL] A family of kinases that, once activated by cyclin, regulate the cell cycle by adding phosphate groups to a variety of protein substrates that control processes in the cycle. Abbreviated Cdk. { ¦sī·klin di¦pen·dənt 'kīˌnās }

cyclin-dependent kinase activating kinase [CELL MOL] A kinase that activates cyclin-dependent kinases via phosphorylation. { ¦sī·klin di¦pen·dənt ¦kī,nās ¦ak·tə,vād·iŋ 'kī,nās }
cyclitis [MED] Inflammation of the ciliary body of the eye. { sə'klīd·əs }
cyclizine hydrochloride [PHARM] $(C_6H_5)_2CHC_4H_8N_2CH_3 \cdot HCl$ A white, crystalline powder with a melting point of 285°C; used in medicine. { 'sī·klə,zēn ,hī·drə'klȯr,īd }
cycloamylose [BIOCHEM] A member of a group of cyclic oligomers of glucose in which the individual glucose units are connected by 1,4 bonds. Also known as cyclodextrin; Schardinger dextrin. { ¦sī·klō'am·ə,lōs }
cyclobarbital [PHARM] $C_{12}H_{16}N_2O_3$ A hypnotic and sedative of short duration. { ¦sī·klō'bär·bə,tȯl }
cyclodextrin *See* cycloamylose. { ¦sī·klō'dek·strən }
cyclodialysis [MED] Detaching the ciliary body from the sclera in order to effect reduction of intraocular tension in certain cases of glaucoma, especially in aphakia. { ¦sī·klō·dī'al·ə·səs }
cyclodiathermy [MED] Destruction, by diathermy, of the ciliary body. { ¦sī·klō'dī·ə,thər·mē }
cycloheptaamylose [BIOCHEM] A cycloamylose with seven glucose units in a cyclic array. { ,sī·klə,hep·tə'am·ə,lōs }
cyclohexaamylose [BIOCHEM] A cycloamylose with six glucose units in a cyclic array. { ,sī·klə,hek·sə'am·ə,lōs }
cycloheximide [MICROBIO] $C_{15}H_{23}NO_4$ Colorless crystals with a melting point of 119.5–121°C; soluble in water, in amyl acetate, and in common organic solvents such as ether, acetone, and chloroform; used as an agricultural fungicide. { ¦sī·klō'hek·sə,mīd }
cycloid scale [VERT ZOO] A thin, acellular structure which is composed of a bonelike substance and shows annual growth rings; found in the skin of soft-rayed fishes. { 'sī,klȯid ,skāl }
cyclomorphosis [ECOL] Cyclic recurrent polymorphism in certain planktonic fauna in response to seasonal temperature or salinity changes. { ,sī·klō'mȯr·fə·səs }
cyclooxygenase [BIOCHEM] An enzyme that catalyzes the conversion of arachidonic acid into prostaglandins. { ,sī·klō'äks·ə·jə,nās }
cyclophilin [BIOCHEM] An abundant cytoplasmic protein that catalyzes cis-trans isomerizations; it has a high affinity for the immunosuppressive drug cyclosporin A. { ,sī·kə'fil·ən }
Cyclophoracea [INV ZOO] A superfamily of gastropod mollusks in the order Prosobranchia. { ,sī·klō·fə'rās·ē·ə }
Cyclophoridae [INV ZOO] A family of land snails in the order Pectinibranchia. { ,sī·klō'fȯr·ə,dē }
Cyclophyllidea [INV ZOO] An order of platyhelminthic worms comprising most tapeworms of warm-blooded vertebrates. { ,sī·klō·fə'lid·ē·ə }
cyclopia [MED] A congenital anomaly characterized by fusion of the eye sockets with various degrees of fusion of the eyes, to the occurrence of a single median eye. { sī'klō·pē·ə }
Cyclopinidae [INV ZOO] A family of copepod crustaceans in the suborder Cyclopoida, section Gnathostoma. { ,sī·klō'pin·ə,dē }
Cyclopoida [INV ZOO] A suborder of small copepod crustaceans. { ,sī·klō'pȯid·ē·ə }
Cyclopteridae [VERT ZOO] The lumpfishes and snailfishes, a family of deep-sea forms in the suborder Cottoidei of the order Perciformes. { ,sī,kläp'ter·ə,dē }
Cyclorhagae [INV ZOO] A suborder of benthonic, microscopic marine animals in the class Kinorhyncha of the phylum Aschelminthes. { sī'klȯr·ə,gē }
Cyclorrhapha [INV ZOO] A suborder of true flies, order Diptera, in which developing adults are always formed in a puparium from which they emerge through a circular opening. { sī'klȯr·ə·fə }
cycloserine [MICROBIO] $C_3H_6O_2N_2$ Broad-spectrum, crystalline antibiotic produced by several species of *Streptomyces*; useful in the treatment of tuberculosis and urinary-tract infections caused by resistant gram-negative bacteria. { ¦sī·klō'se,rēn }
cyclosis [CELL MOL] Massive rotational streaming of cytoplasm in certain vacuolated cells, such as the stonewort *Nitella* and *Paramecium*. { sī'klō·səs }
Cyclosporeae [BOT] A class of brown algae, division Phaeophyta, in which there is only a free-living diploid generation. { ¦sī·klō'spȯr·ē,ē }
cyclosporin A [BIOCHEM] A cyclic peptide produced by some fungi. It inactivates helper T cells, making it useful as an immunosuppressive drug, especially in the prevention of graft rejection in transplantation surgery. { ¦sī·klə,spȯr·ən 'ā }
Cyclostomata [INV ZOO] An order of bryozoans in the class Stenolaemata. [VERT ZOO] A subclass comprising the simplest and most primitive of living vertebrates characterized by the absence of jaws and the presence of a single median nostril and an uncalcified cartilaginous skeleton. { ¦sī·klō'stō·mə·də }
cyclotron cataract *See* irradiation cataract. { 'sī·klə,trän 'kad·ə,rakt }
Cydippida [INV ZOO] An order of the pelagic ctenophores; members retain the cydippid state (resemble the cydippid larva) until the adult stage is reached in development. { sī'dip·ə·də }
Cydippidea [INV ZOO] An order of the Ctenophora having well-developed tentacles. { ,sī·də'pid·ē·ə }
Cydnidae [INV ZOO] The ground or burrower bugs, a family of hemipteran insects in the superfamily Pentatomorpha. { 'sid·nə,dē }
cylindrarthrosis [ANAT] A joint characterized by rounded articular surfaces. { ¦sil·ən,drär'thrō·səs }
Cylindrocapsaceae [BOT] A family of green algae in the suborder Ulotrichineae comprising thick-walled, sheathed cells having massive chloroplasts. { sə,lin·drō,kap'sās·ē,ē }
Cylindrocorporoidea [INV ZOO] A superfamily

of both free-living and parasitic nematodes of the order Diplogasterida having well-developed lips surrounding the oral opening and lateral lips bearing small amphids. { sə,lin·drə,kȯr·pə 'rȯid·ē·ə }

cyme [BOT] An inflorescence in which each main axis terminates in a single flower; secondary and tertiary axes may also have flowers, but with shorter flower stalks. { sīm }

cymose [BOT] Of, pertaining to, or resembling a cyme. { 'sī,mōs }

Cymothoidae [INV ZOO] A family of isopod crustaceans in the suborder Flabellifera; members are fish parasites with reduced maxillipeds ending in hooks. { ,sī·mə'thȯi,dē }

Cynipidae [INV ZOO] A family of hymenopteran insects in the superfamily Cynipoidea. { sə'nip·ə,dē }

Cynipoidea [INV ZOO] A superfamily of hymenopteran insects in the suborder Apocrita. { ,sin·ə'pȯid·ē·ə }

Cynoglossidae [VERT ZOO] The tonguefishes, a family of Asiatic flatfishes in the order Pleuronectiformes. { ,sin·ə'gläs·ə,dē }

Cyperaceae [BOT] The sedges, a family of monocotyledonous plants in the order Cyperales characterized by spirally arranged flowers on a spike or spikelet; a usually solid, often triangular stem; and three carpels. { ,sip·ə'rās·ē,ē }

Cyperales [BOT] An order of monocotyledonous plants in the subclass Commelinidae with reduced, mostly wind-pollinated or self-pollinated flowers that have a unilocular, two-or three-carpellate ovary bearing a single ovule. { sip·ə'rā·lēz }

Cypheliaceae [BOT] A family of typically crustose lichens with sessile apothecia in the order Caliciales. { sə,fel·ē'ās·ē,ē }

cyphonautes [INV ZOO] The free-swimming bivalve larva of certain bryozoans. { ,sī·fə'nōd·ēz }

Cyphophthalmi [INV ZOO] A family of small, mitelike arachnids in the order Phalangida. { ,sī·fə'thal,mī }

Cypovirus [VIROL] A genus in the family Reoviridae that infects invertebrates; the type species is cytoplasmic polyhedrosis virus. { 'sip,o'vī·rəs }

Cypraecea [INV ZOO] A superfamily of gastropod mollusks in the order Prosobranchia. { sī'prēsh·ē·ə }

Cypraeidae [INV ZOO] A family of colorful marine snails in the order Pectinibranchia. { sī'prē·ə,dē }

cypress [BOT] The common name for members of the genus *Cupressus* and several related species in the order Pinales. { 'sī·prəs }

Cypridacea [INV ZOO] A superfamily of mostly fresh-water ostracods in the suborder Podocopa. { ,sī·prə'dās·ē·ə }

Cypridinacea [INV ZOO] A superfamily of ostracods in the suborder Myodocopa characterized by a calcified carapace and having a round back with a downward-curving rostrum. { ,sī,prid·ə'nās·ē·ə }

Cyprinidae [VERT ZOO] The largest family of fishes, including minnows and carps in the order Cypriniformes. { sī'prin·ə,dē }

Cypriniformes [VERT ZOO] An order of actinopterygian fishes in the suborder Ostariophysi. { sī,prin·ə'fōr,mēz }

Cyprinodontidae [VERT ZOO] The killifishes, a family of actinopterygian fishes in the order Atheriniformes that inhabit ephemeral tropical ponds. { sī,prin·ə'dän·tə,dē }

Cyprinoidei [VERT ZOO] A suborder of primarily fresh-water actinopterygian fishes in the order Cypriniformes having toothless jaws, no adipose fin, and faliciform lower pharyngeal bones. { ,sī·prə'nȯi·dē,ī }

cypris [INV ZOO] An ostracod-like, free-swimming larval stage in the development of Cirripedia. { 'sī·prəs }

Cyrtophorina [INV ZOO] The equivalent name for Gymnostomatida. { ,sərd·ə'fə'rī·nə }

cyrtopia [INV ZOO] A type of crustacean larva (Ostracoda) characterized by an elongation of the first pair of antennae and loss of swimming action in the second pair. { sər'tō·pē·ə }

cyst [MED] A normal or pathologic sac with a distinct wall, containing fluid or other material. { sist }

cystacanth [INV ZOO] The infective larva of the Acanthocephala; lies in the hemocele of the intermediate host. { 'sis·tə,kanth }

L-cystathionine [BIOCHEM] $C_7H_{14}N_2O_4S$ An amino acid formed by condensation of homocysteine with serine, catalyzed by an enzyme transsulfurase; found in high concentration in the brain of primates. { ¦el ,sis·tə'thī·ə,nēn }

cystectomy [MED] **1.** Excision of the gallbladder, or part of the urinary bladder. **2.** Removal of a cyst. **3.** Removal of a piece of the anterior capsule of the lens for the extraction of a cataract. { si'stek·tə·mē }

cysteine [BIOCHEM] $C_3H_7O_2NS$ A crystalline amino acid occurring as a constituent of glutathione and cystine. { 'si,stēn }

cystic disease [MED] A disorder of women, usually at or near menopause, characterized by the development of large cysts in the breast. { 'sis·tik di'zēz }

cystic duct [ANAT] The duct of the gallbladder. { 'sis·tik 'dəkt }

cysticercosis [MED] The infestation in humans by cysticerci of the genus *Taenia*. { ¦sis·tə·sər'kō·səs }

cysticercus [INV ZOO] A larva of tapeworms in the order Cyclophyllidea that has a bladder with a single invaginated scolex. { ¦sis·tə'sər·kəs }

cystic fibrosis [MED] A hereditary disease of the pancreas transmitted as an autosomal recessive; involves obstructive lesions, atrophy, and fibrosis of the pancreas and lungs, and the production of mucus of high viscosity. Also known as mucoviscidosis. { ¦sis·tik fī'brō·səs }

cystic fibrosis transmembrane conductance regulator [CELL MOL] A specialized chloride channel that is regulated by cyclic adenosine monophosphate; its disruption has been implicated in cystic fibrosis. { ¦sis·tik fī¦brō·səs tranz ¦mem,brān kən'dək·təns ,reg·yə,lād·ər }

cystic kidney [MED] A kidney with one or more cysts. { 'sis·tik 'kid·nē }

cystine [BIOCHEM] $C_6H_{12}N_2S_2$ A white, crystalline amino acid formed biosynthetically from cysteine. { 'si,stēn }

cystinosis [MED] A congenital metabolic disorder involving sulfur-containing amino acids, usually cystine; characterized by deposits of cystine crystals in the body organs. { ,sis·tə'nō·səs }

cystinuria [MED] The presence in the urine of crystals of cystine together with some lysine, arginine, and ornithine. { ,sis·tə'nu̇r·ē·ə }

cystitis [MED] Inflammation of a fluid-filled organ, especially the urinary bladder. { si'stīd·əs }

Cystobacter [MICROBIO] A genus of bacteria in the family Cystobacteraceae; vegetative cells are tapered, sporangia are sessile, and microcysts are rigid rods. { ¦sis·tə'bak·tər }

Cystobacteraceae [MICROBIO] A family of bacteria in the order Myxobacterales; vegetative cells are tapered, and microcysts are rod-shaped and enclosed in sporangia. { ¦sis·tə,bak·tə 'rās·ē,ē }

cystocarp [BOT] A fruiting structure with a special protective envelope, produced after fertilization in red algae. { 'sis·tə,kärp }

cystocele [MED] Herniation of the urinary bladder into the vagina. { 'sis·tə,sēl }

cystocercous cercaria [INV ZOO] A digenetic trematode larva that can withdraw the body into the tail. { ¦sis·tə¦sər·kəs sər'kar·ē,ə }

cystography [MED] Radiography of the urinary bladder after the injection of a contrast medium. { si'stäg·rə·fē }

cystolith [BOT] A concretion of calcium carbonate arising from the cell walls of modified epidermal cells in some flowering plants. { 'sis·tə,lith }

cystoma [MED] A cystic mass, especially in or near the ovary. { si'stō·mə }

cystometer [MED] An instrument used to determine pressure in the urinary bladder under standard conditions. { si'stäm·əd·ər }

cystopyelitis [MED] Inflammation of the urinary bladder and the renal pelvis. { ¦sis·tə,pī·ə'līd·əs }

cystopyelography [MED] Radiography of the urinary bladder, ureter, and renal pelvis after injection of a radiopaque material. { ¦sis·tə,pī·ə'läg·rə·fē }

cystopyelonephritis [MED] Inflammation of the urinary bladder, renal pelvis, and renal parenchyma. { ¦sis·tə,pī·ə,lō·nə'frīd·əs }

cystoscope [MED] An optical instrument for visual examination of the urinary bladder, ureters, and kidneys. { 'sis·tə,skōp }

cystoureteritis [MED] Inflammation of the urinary bladder and the ureters. { ¦sis·tə·yu̇,rēd·ə'rīd·əs }

Cystoviridae [VIROL] A family of enveloped ribonucleic acid (RNA)-containing bacteriophages characterized by a spherical virion containing three molecules of linear double-stranded RNA. { ,sis·tə'vir·ə,dī }

Cystovirus [VIROL] The sole genus of the family Cystoviridae containing the type species *Pseudomonas* Phage phi6. { 'sis·tə,vī·rəs }

cytase [BIOCHEM] Any of several enzymes in the seeds of cereals and other plants, which hydrolyze the cell-wall material. { 'sī,tās }

Cytheracea [INV ZOO] A superfamily of ostracodes in the suborder Podocopa comprising principally crawling and digging marine forms. { ,sith·ə'rās·ē·ə }

Cytherellidae [INV ZOO] The family comprising all living members of the ostracod suborder Platycopa. { ,sith·ə'rel·ə,dē }

cytidine [BIOCHEM] $C_9H_{13}N_3O_5$ Cytosine riboside, a nucleoside composed of one molecule each of cytosine and D-ribose. { 'sid·ə,dēn }

cytidylic acid [BIOCHEM] $C_9H_{14}O_8N_3P$ A nucleotide synthesized from the base cytosine and obtained by hydrolysis of nucleic acid. { ¦sid·ə¦dil·ik 'as·əd }

cytocentrum *See* central apparatus. { ¦sīd·ō'sen·trəm }

cytochalasin [BIOCHEM] One of a series of structurally related fungal metabolic products which, among other effects on biological systems, selectively and reversibly block cytokinesis while not affecting karyokinesis; the molecule with minor variations consists of a benzyl-substituted hydroaromatic isoindolone system, which in turn is fused to a small macrolide-like cyclic ring. { ¦sīd·ō·kə'lā·sən }

cytochemistry [CELL MOL] The science concerned with the chemistry of cells and cell components, primarily with the location of chemical constituents and enzymes. { ¦sīd·ō'kem·ə·strē }

cytochrome [BIOCHEM] Any of the complex protein respiratory pigments occurring within plant and animal cells, usually in mitochondria, that function as electron carriers in biological oxidation. { 'sīd·ə,krōm }

cytochrome a_3 *See* cytochrome oxidase. { 'sīd·ə,krōm ¦ā səb'thrē }

cytochrome oxidase [BIOCHEM] Any of a family of respiratory pigments that react directly with oxygen in the reduced state. Also known as cytochrome a_3. { 'sīd·ə,krōm 'äk·sə,dās }

cytocidal [CELL MOL] Causing cell death. { sī'täs·əd·əl }

cytocidal unit [BIOL] A unit for the standardization of adrenal cortical hormones. { sī'täs·əd·əl ,yü·nət }

cytocrine gland [CELL MOL] A cell, especially a melanocyte, that passes its secretion directly to another cell. { 'sīd·ə·krən ,gland }

cytodiagnosis [PATH] The determination of the nature of an abnormal liquid by the study of cells it contains. { ¦sīd·ō,dī·ig'nō·səs }

cytoduction [MYCOL] In yeast, the production of cells with mixed cytoplasm but with the nucleus of one or the other parent. { 'sīd·ə,dək·shən }

cytogamy [CELL MOL] Fusion or conjugation of cells. { sī'täg·ə·mē }

cytogenetics [CELL MOL] The comparative study of the mechanisms and behavior of chromosomes in populations and taxa, and the effect of chromosomes on inheritance and evolution. { ¦sīd·ō·jə'ned·iks }

cytogenous gland [PHYSIO] A structure that secretes living cells; an example is the testis. { sī'tä·jə·nəs ,gland }

cytohet [CELL MOL] A cell containing two genetically distinct types of a specific organelle. { 'sīd·ə,het }

cytokine [CELL MOL] Any of a group of peptides that are released by some cells and affect the behavior of other cells, serving as intercellular signals. { 'sīd·ə,kīn }

cytokine receptor [CELL MOL] A type of cell-surface receptor that binds to cytokines, initiating the Jak-STAT signaling pathway within the cell. { 'sīd·ə,kīn ri,sep·tər }

cytokinesis [CELL MOL] Division of the cytoplasm following nuclear division. { ¦sīd·ō·kə'nē·səs }

cytokinin [BIOCHEM] Any of a group of plant hormones which elicit certain plant growth and development responses, especially by promoting cell division. { ¦sīd·ō'kī·nən }

cell and molecular biology [BIOL] A branch of the biological sciences which deals with the structure, behavior, growth, and reproduction of cells and the function and chemistry of cell components. { sī'täl·ə·jē }

cytolysin *See* perforin. { ,sīd·ə'līs·ən }

cytolysis [PATH] Disintegration or dissolution of cells, usually associated with a pathologic process. { sī'täl·ə·səs }

cytolysosome [CELL MOL] An enlarged lysosome that contains organelles such as mitochondria. { ¦sīd·ō'lī·sə,sōm }

cytomegalic [MED] Of, pertaining to, or characterizing the greatly enlarged cells with enlarged nuclei and inclusion bodies found in tissues in cytomegalic inclusion disease. { ¦sīd·ō·mə¦gal·ik }

cytomegalic inclusion disease [MED] A virus infection primarily of infants characterized by jaundice, liver enlargement, and circulatory disturbances. { ¦sīd·ō·mə¦gäl·ik in'klü·zhən di'zēz }

cytomegalovirus [VIROL] An animal virus belonging to subgroup B of the herpesvirus group; causes cytomegalic inclusion disease and pneumonia. { ¦sīd·ō¦meg·ə·lō'vī·rəs }

cytomegalovirus group *See* Betaherpesvirinae. { ¦sīd·ō¦meg·ə·lō'vī·rəs ,grüp }

cytomegalovirus infection [MED] A common asymptomatic infection caused by cytomegalovirus, which can produce life-threatening illnesses in the immature fetus and in immunologically deficient subjects. { ,sīd·ō¦meg·ə·lō,vī·rəs in'fek·shən }

cytomegalovirus mononucleosis [MED] A self-limited illness such as infectious mononucleosis, the main manifestation of which is fever; it is the only cytomegalovirus illness clearly described in mature, immunologically normal subjects. { ,sīd·ō¦meg·ə·lō,vī·rəs ,män·ə,nü·klē'ō·səs }

cytomixis [CELL MOL] Extrusion of chromatin from one cell into the cytoplasm of an adjoining cell. { ,sīd·ə'mik·səs }

cytomorphosis [CELL MOL] All the structural alterations which cells or successive generations of cells undergo from the earliest undifferentiated stage to their final destruction. { ,sīd·ə'mȯr·fə·səs }

cyton [NEUROSCI] The central body of a neuron containing the nucleus and excluding its processes. { 'sī,tän }

cytopathic effect [CELL MOL] A change in the microscopic appearance of cells in a culture after being infected with a virus. { ¦sīd·ə¦path·ik i,fekt }

cytopathology [PATH] A branch of pathology concerned with abnormalities within cells. { ¦sīd·ō·pə'thäl·ə·jē }

cytopenia [PATH] A blood cell count below normal. { ,sīd·ō'pē·nē·ə }

Cytophaga [MICROBIO] A genus of bacteria in the family Cytophagaceae; cells are unsheathed, unbranched rods or filaments and are motile; microcysts are not known; decompose agar, cellulose, and chitin. { sī'täf·ə·gə }

Cytophagaceae [MICROBIO] A family of bacteria in the order Cytophagales; cells are rods or filaments, unsheathed cells are motile, filaments are not attached, and carotenoids are present. { ,sīd·ō·fə'gās·ē,ē }

Cytophagales [MICROBIO] An order of gliding bacteria; cells are rods or filaments and motile by gliding, and fruiting bodies are not produced. { ,sīd·ō·fə'gā·lēz }

cytopharynx [INV ZOO] A channel connecting the surface with the protoplasm in certain protozoans; functions as a gullet in ciliates. { ¦sīd·ō'far·iŋks }

cytophilic antibody [IMMUNOL] A substance capable of combining directly with the receptors of a corresponding antigenic cell. Also known as cytotropic antibody. { ¦sid·ə¦fil·ik 'ant·i,bäd·ē }

cytoplasm [CELL MOL] The protoplasm of an animal or plant cell external to the nucleus. { 'sīd·ə,plaz·əm }

cytoplasmic inheritance [GEN] The control of genetic difference by genes carried in cytoplasmic organelles such as mitochondria or chloroplasts. Also known as extrachromosomal inheritance. { ,sīd·ə'plaz·mik in'her·ə·təns }

cytoplasmic male sterility [BOT] The maternally inherited inability of a higher plant to produce viable pollen. { ¦sīd·ə,plaz·mik ¦māl stə'ril·əd·ē }

cytoplasmic streaming [CELL MOL] Intracellular movement involving irreversible deformation of the cytoplasm produced by endogenous forces. { ,sīd·ə'plaz·mik 'strem·iŋ }

cytoplast [CELL MOL] The cytoplasmic substance of eukaryotic cells, including a network of proteins forming an internal skeleton and the attached nucleus and organelles. { 'sīd·ə,plast }

cytopyge [INV ZOO] A fixed point for waste discharge in the body of a protozoan, especially a ciliate. { 'sīd·ə,pīj }

cytosine [BIOCHEM] $C_4H_5ON_3$ A pyrimidine occurring as a fundamental unit or base of nucleic acids. { 'sīd·ə,sēn }

cytoskeleton [CELL MOL] Protein fibers composing the structural framework of a cell. { ¦sīd·ō'skel·ə·tən }

cytosol [CELL MOL] The fluid portion of the cytoplasm, that is, the cytoplasm exclusive of organelles and membranes. { 'sīd·ə,säl *or* 'sīd·ə,sōl }

cytosome [CELL MOL] The cytoplasm of the cell, as distinct from the nucleus. { 'sīd·ə,sōm }

cytostasis [CELL MOL] Inhibition of the ability of cells to continue growing. { ,sīd·ə'stā·səs }

cytostatic [CELL MOL] Inhibiting cell development. { ,sīd·ō 'stad·ik }

cytostome [INV ZOO] The mouth-like opening in many unicellular organisms, particularly Ciliophora. { 'sīd·ə,stōm }

cytotaxis [PHYSIO] Attraction of motile cells by specific diffusible stimuli emitted by other cells. { ¦sīd·ō'tak·səs }

cytotechnologist [PATH] A person trained to prepare smears of and examine exfoliated cells, referring abnormalities to a physician. { ,sīd·ō,tek'näl·ə·jəst }

cytotoxic [CELL MOL] Pertaining to an agent, such as a drug or virus, that exerts a toxic effect on cells. { ¦sīd·ə'täk·sik }

cytotoxic T cell [IMMUNOL] A type of T cell which protects against pathogens that invade host cell cytoplasm, where they cannot be bound by antibodies, by recognizing and killing the host cell before the pathogens can proliferate and escape. { ¦sīd·ə,täk·sik 'tē ,sel }

cytotrophoblast [EMBRYO] The inner, cellular layer of a trophoblast, covering the chorion and the chorionic villi during the first half of pregnancy. { ¦sīd·ō'trō·fə,blast }

cytotropic antibody *See* cytophilic antibody. { ¦sīd·ō'trä·pik 'an·tə,bäd·ē }

cytotropism [BIOL] The tendency of individual cells and groups of cells to move toward or away from each other. { sī'tä·trə,piz·əm }

Czapek's agar [MICROBIO] A nutrient culture medium consisting of salt, sugar, water, and agar; used for certain mold cultures. { 'chä·peks ,äg·ər }

D

Dacromycetales [MYCOL] An order of jelly fungi in the subclass Heterobasidiomycetidae having branched basidia with the appearance of a tuning fork. { ˌdak·rəˌmī·səˈtā·lēz }

dacryoblennorrhea [MED] Chronic inflammation of the lacrimal sac of the eye accompanied by discharge of mucus. { ˈdak·rəˌblen·əˈrē·ə }

dacryocyst *See* lacrimal sac. { ˈdak·rəˌsist }

dacryocystitis [MED] Inflammation of the lacrimal sac. { ˌdak·rəˌsisˈtīd·əs }

dacryon [ANAT] The point of the face where the frontomaxillary, the maxillolacrimal, and frontolacrimal sutures meet. { ˈdak·rēˌän }

Dactylochirotida [INV ZOO] An order of dendrochirotacean holothurians in which there are 8–30 digitate or digitiform tentacles, which sometimes bifurcate. { ˌdak·tə·lō·kəˈräd·ə·də }

dactylognathite [INV ZOO] The distal segment of a maxilliped in crustaceans. { ˌdak·tə·lōˈnaˌthīt }

dactylography [FOREN] The scientific study of fingerprints as a device for identifying people. { ˌdak·təˈläg·rə·fē }

Dactylogyroidea [INV ZOO] A superfamily of trematodes in the subclass Monogenea; all are fish ectoparasites. { ¦dak·tə·loˌjiˈrȯid·ē·ə }

dactylopodite [INV ZOO] The distal segment of ambulatory limbs in decapods and of certain limbs in other arthropods. { ˌdak·təˈläp·əˌdīt }

dactylopore [INV ZOO] Any of the small openings on the surface of Milleporina through which the bodies of the polyps are extended. { dak ˈtil·əˌpȯr }

Dactylopteridae [VERT ZOO] The flying gurnards, the single family of the perciform suborder Dactylopteroidei. { ˌdak·tə·lōˈter·əˌdē }

Dactylopteroidei [VERT ZOO] A suborder of marine shore fishes in the order Perciformes, characterized by tremendously expansive pectoral fins. { ˌdak·tə·lō·təˈrȯid·ēˌī }

Dactyloscopidae [VERT ZOO] The sand stargazers, a family of small tropical and subtropical perciform fishes in the suborder Blennioidei. { ˌdak·tə·ˈskäp·əˌdē }

Dactylosporangium [MICROBIO] A genus of bacteria in the family Actunoplanaceae; fingerlike sporangia are formed in clusters, each containing a single row of three or four motile spores. { ˌdak·tə·lō·spəˈrän·jē·əm }

dactylosternal [VERT ZOO] Of turtles, having marginal fingerlike processes in joining the plastron to the carapace. { ¦dak·tə·lōˈstərn·əl }

dactylozooid [INV ZOO] One of the long defensive polyps of the Milleporina, armed with stinging cells. { ¦dak·tə·ləˈzōˌȯid }

dactylus [INV ZOO] The structure of the tarsus of certain insects which follows the first joint; usually consists of one or more joints. { ˈdak·tə·ləs }

DAF *See* decay accelerating factor.

Da Fano bodies [VIROL] Minute basophilic areas of abnormal staining found within cells infected with human herpesvirus 1 or 2. { däˈfän·ō ˌbäd·ēz }

Dalatiidae [VERT ZOO] The spineless dogfishes, a family of modern sharks belonging to the squaloid group. { ˌdal·əˈtī·əˌdē }

Dallis grass [BOT] The common name for the tall perennial forage grasses composing the genus *Paspalum* in the order Cyperales. { ˈda·ləs ˌgras }

Danaidae [INV ZOO] A family of large tropical butterflies, order Lepidoptera, having the first pair of legs degenerate. { dəˈnā·əˌdē }

dandruff [MED] Scales of dry sebum formed on the scalp in seborrhea. { ˈdan·drəf }

dandy fever *See* dengue. { ˈdan·dē ˌfē·vər }

Dane particle [VIROL] The causative virus of type B viral hepatitis visualized ultrastructurally in its complete form. { ˈdān ˌpard·ə·kəl }

Danysz reaction [IMMUNOL] A toxin-antitoxin reaction that occurs when an exact equivalence of toxin is added to antitoxin, not in one portion but in successive increments. { ˈdä·nish rē ˈak·shən }

DAP *See* diaminopimelate.

Daphniphyllales [BOT] An order of dicotyledonous plants in the subclass Hamamelidae, consisting of a single family with one genus, *Daphniphyllum*, containing about 35 species; dioecious trees or shrubs native to eastern Asia and the Malay region, they produce a unique type of alkaloid and often accumulate aluminum and sometimes produce iridoid compounds. { ˌdaf·ni·fəˈlā·lēz }

dapsone *See* 4,4′-sulfonyldianiline. { ˈdapˌsōn }

Darrier's disease [MED] A genetically determined disease characterized by patches of papules of the horny layer of skin. Also known as keratosis follicularis. { ˈdar·ēˌāz dizˌēz }

dart [INV ZOO] A small sclerotized structure ejected from the dart sac of certain snails into the body of another individual as a stimulant before copulation. { därt }

dart sac [INV ZOO] A dart-forming pouch associated with the reproductive system of certain snails. { 'därt ˌsak }

darwin [EVOL] A unit of evolutionary rate of change; if some dimension of a part of an animal or plant, or of the whole animal or plant, changes from l_0 to l_t over a time of t years according to the formula $l_t = l_0 \exp (Et/10^6)$, its evolutionary rate of change is equal to E darwins. { 'där·wən }

Darwinism [BIOL] The theory of the origin and perpetuation of new species based on natural selection of those offspring best adapted to their environment because of genetic variation and consequent vigor. Also known as Darwin's theory. { 'där·wəˌniz·əm }

Darwin's finch [VERT ZOO] A bird of the subfamily Fringillidae; Darwin studied the variation of these birds and used his data as evidence for his theory of evolution by natural selection. { ¦där·winz 'finch }

Darwin's theory *See* Darwinism. { ¦där·winz 'thē·ə·rē }

Darwinulacea [INV ZOO] A small superfamily of nonmarine, parthenogenetic ostracods in the suborder Podocopa. { därˌwin·ə'lās·ē·ə }

Dasayatidae [VERT ZOO] The stingrays, a family of modern sharks in the batoid group having a narrow tail with a single poisonous spine. { ˌda·sā'ad·əˌdē }

Dascillidae [INV ZOO] The soft-bodied plant beetles, a family of coleopteran insects in the superfamily Dascilloidea. { ˌdə'sil·əˌdē }

Dascilloidea [INV ZOO] Superfamily of coleopteran insects in the suborder Polyphaga. { ˌdas·ə'lȯid·ē·ə }

dasheen [BOT] *Colocasia esculenta*. A plant in the order Arales, grown for its edible corm. { da'shēn }

Dasycladaceae [BOT] A family of green algae in the order Dasycladales comprising plants formed of a central stem from which whorls of branches develop. { ˌdas·ə·klə'dās·ēˌē }

Dasycladales [BOT] An order of lime-encrusted marine algae in the division Chlorophyta, characterized by a thallus composed of nonseptate, highly branched tubes. { ˌdas·ə·klə'dā·lēz }

Dasyonygidae [INV ZOO] A family of biting lice, order Mallophaga, that are confined to rodents of the family Procaviidae. { ˌdas·ē·ə'nij·əˌdē }

Dasypodidae [VERT ZOO] The armadillos, a family of edentate mammals in the infraorder Cingulata. { ˌdas·ə'päd·əˌdē }

Dasytidae [INV ZOO] An equivalent name for Melyridae. { də'sid·əˌdē }

Dasyuridae [VERT ZOO] A family of mammals in the order Marsupialia characterized by five toes on each hindfoot. { das·ē'yu̇r·əˌdē }

Dasyuroidea [VERT ZOO] A superfamily of marsupial mammals. { ˌdas·ē·yə'rȯid·ē·ə }

Daubentoniidae [VERT ZOO] A family of Madagascan prosimian primates containing a single species, the aye-aye. { ˌdō·bən·tō'nī·əˌdē }

daughter nucleus [CELL MOL] One of the two cell nuclei resulting from a nuclear division. { 'dȯd·ər ˌnü·klē·əs }

Dawsoniales [BOT] An order of mosses comprising rigid plants with erect stems rising from a rhizomelike base. { dȯˌsō·nē'ā·lēz }

day neutral [BOT] Reaching maturity regardless of relative length of light and dark periods. { 'dā ˌnü·trəl }

day-neutral response [PHYSIO] A photoperiodic response that is independent or nearly independent of day length. { ¦dā ˌnü·trəl ri'späns }

deadly nightshade *See* belladonna. { ¦ded·lē 'nītˌshād }

dead space [ANAT] The space in the trachea, bronchi, and other air passages which contains air that does not reach the alveoli during respiration, the amount of air being about 140 milliliters. Also known as anatomical dead space. [MED] A cavity left after closure of a wound. [PHYSIO] A calculated expression of the anatomical dead space plus whatever degree of overventilation or underperfusion is present; it is alleged to reflect the relationship of ventilation to pulmonary capillary perfusion. Also known as physiological dead space. { 'ded ˌspās }

deafness [MED] Temporary or permanent impairment or loss of hearing. { 'def·nəs }

deamidase [BIOCHEM] An enzyme that catalyzes the removal of an amido group from a compound. { dē'am·əˌdās }

deaminase [BIOCHEM] An enzyme that catalyzes the hydrolysis of amino compounds, removing the amino group. { de'am·əˌnās }

death [MED] Cessation of all life functions; can involve the whole organism, an organ, individual cells, or cell parts. { deth }

death point [PHYSIO] The limit (as of extremes of temperature) beyond which an organism cannot survive. { 'deth ˌpȯint }

death rate *See* mortality rate. { 'deth ˌrāt }

de Beurmann-Gougerot disease *See* sporotrichosis. { dē 'bu̇r·män 'güzh·rō di'zēz }

debridement [MED] A surgical procedure for removing lacerated, morbid, or contaminated tissue. { dē'brīd·mənt }

debromoaplysiatoxin [BIOCHEM] A bislactone toxin related to aplysiatoxin and produced by the blue-green alga *Lyngbya majuscula*. { dēˌbrō·mō·ə'plizh·əˌtäk·sən }

decanth larva *See* lycophore larva. { ˌdē'kanth ˌlär·və }

Decapoda [INV ZOO] **1.** A diverse order of the class Crustacea including the shrimps, lobsters, hermit crabs, and true crabs; all members have a carapace, well-developed gills, and the first three pairs of thoracic appendages specialized as maxillipeds. **2.** An order of dibranchiate cephalopod mollusks containing the squids and cuttle fishes, characterized by eight arms and two long tentacles. { də'kap·əd·ə }

decarboxylase [BIOCHEM] An enzyme that

hydrolyzes the carboxyl radical, COOH. { ˌdē·kär'bäk·səˌlās }

decay accelerating factor [IMMUNOL] A plasma protein involved with complement regulation on the red blood cell surface; it accelerates the breakdown of C3 and C5 convertases. Abbreviated DAF. { di¦kā ik'sel·əˌrād·iŋ ˌfak·tər }

decerebellate [MED] Lacking the cerebellum, generally by experimental removal. { ¦dēˌser·ə'bel·ət }

decerebrate [MED] Lacking the cerebrum either by experimental removal or by disconnection. { dē'ser·əˌbrət }

decerebrate rigidity [MED] Exaggerated postural tone in the antigravity muscles due to release of vestibular nuclei from cerebral control. { dē'ser·əˌbrət ri'jid·əd·ē }

decidua [MED] The endometrium of pregnancy and associated fetal membranes which are cast off at parturition. { di'sij·ə·wə }

deciduitis [MED] Inflammation of the decidua. { dəˌsid·yə'wīd·əs }

deciduoma [MED] **1.** A mass of tissue formed in the uterus following pregnancy as the result of hyperplasia of chorionic or decidual cells. **2.** Decidual tissue induced in the uterus, usually by physical trauma. { ˌdes·i'dwō·mə }

deciduous [BIOL] Falling off or being shed at the end of the growing period or season. { di'sij·ə·wəs }

deciduous teeth [ANAT] Teeth of a young mammal which are shed and replaced by permanent teeth. Also known as milk teeth. { di'sij·ə·wəs 'tēth }

deckzelle [INV ZOO] In certain hydroids, one of the supporting or epithelial cells which are usually columnar or cuboidal. { ¦dek¦zel }

declinate [BIOL] Curved toward one side or downward. { 'dek·ləˌnāt }

declining population [ECOL] A population in which old individuals outnumber young individuals. { də'klin·iŋ ˌpäp·yə'lā·shən }

decomposer [ECOL] A heterotrophic organism (including bacteria and fungi) which breaks down the complex compounds of dead protoplasm, absorbs some decomposition products, and releases substances usable by consumers. Also known as microcomposer; microconsumer; reducer. { de·kəm'pō·zər }

decompound [BOT] Divided or compounded several times, with each division being compound. { dē'kämˌpau̇nd }

decompression illness *See* aeroembolism. { dē·kəm'presh·ən ˌil·nəs }

decorticate [BIOL] Lacking a cortical layer. { dē'kȯrd·əˌkāt }

decubitus ulcer [MED] An ulcer of the skin and subcutaneous tissues following prolonged lying down, due to pressure on bony protuberances. Also known as bedsore; pressure ulcer. { də'kyüb·əd·əs 'əl·sər }

decumbent [BOT] Lying down on the ground but with an ascending tip, specifically referring to a stem. { di'kəm·bənt }

decurrent [BOT] Running downward, especially of a leaf base extended past its insertion in the form of a winged expansion. { di'kər·ənt }

decussate [BOT] Of the arrangement of leaves, occurring in alternating pairs at right angles. { 'dek·əˌsāt }

dedifferentiation [BIOL] Disintegration of a specialized habit or adaptation. [CELL MOL] Loss of recognizable specializations that define a differentiated cell. [PHYSIO] Return of a specialized cell or structure to a more general or primitive condition. { dēˌdif·əˌren·chē'ā·shən }

deep fascia [ANAT] The fibrous tissue between muscles and forming the sheaths of muscles, or investing other deep, definitive structures, as nerves and blood vessels. { 'dēp 'fā·shə }

deep hibernation [PHYSIO] Profound decrease in metabolic rate and physiological function during winter, with a body temperature near 0°C, in certain warm-blooded vertebrates. Also known as hibernation. { 'dēp ˌhī·bər'nā·shən }

deep pain [NEUROSCI] Pattern of somesthetic sensation of pain, usually indefinitely localized, originating in the viscera, muscles, and other deep tissues. { 'dēp 'pān }

deep palmar arch [ANAT] The anastomosis between the terminal part of the radial artery and the deep palmar branch of the ulnar artery. Also known as deep volar arch; palmar arch. { 'dēp ¦pä·mər ˌärch }

deep volar arch *See* deep palmar arch. { ¦dēp ¦vō·lər ˌärch }

deer [VERT ZOO] The common name for 41 species of even-toed ungulates that compose the family Cervidae in the order Artiodactyla; males have antlers. { dir }

defecation [PHYSIO] The process by which fecal wastes that reach the lower colon and rectum are evacuated from the body. { ˌdef·ə'kā·shən }

defective interfering virus [VIROL] A virus generated at the peak of an infection that can interfere with replication of the normal virus and may modify the outcome of the disease. { di'fek·tiv ˌin·tərˌfir·iŋ 'vī·rəs }

defective virus [VIROL] A virus, such as adeno-associated satellite virus, that can grow and reproduce only in the presence of another virus. { di'fek·tiv 'vī·rəs }

defeminization [PHYSIO] Loss or reduction of feminine attributes, usually caused by ovarian dysfunction or removal. { dēˌfem·ə·nə'zā·shən }

defervescence *See* lysis. { def·ər'ves·əns }

defibrillation [MED] Stopping a local quivering of muscle fibers, especially of the heart. { dē ˌfib·rə'lā·shən }

defibrillator [MED] An electronic instrument used for stopping fibrillation during a heart attack by applying controlled electric pulses to the heart muscles. { dē'fib·rəˌlād·ər }

deficiency disease [MED] Any disease resulting from a dietary deficiency of minerals, vitamins, or essential nutrients. { də'fish·ən·sē diˌzēz }

definitive host [BIOL] The host in which a parasite reproduces sexually. { də'fin·əd·iv 'hōst }

deflexed [BIOL] Turned sharply downward. { dē'flekst }

defluvium [PATH] The pathological loss of a part of an animal or plant, as nails or bark. { dē'flü·vē·əm }

defoliate [BOT] To remove leaves or cause leaves to fall, especially prematurely. { dē'fō·lē,āt }

degenerate code [GEN] The observation that more than one triplet sequence of nucleotides (codon) can specify the insertion of the same amino acid into a polypeptide chain. { di'jen·ə·rət 'kōd }

degeneration [MED] **1.** Deterioration of cellular integrity with no sign of response to injury or disease. **2.** General deterioration of a physical, mental, or moral state. { di,jen·ə'rā·shən }

degenerative arthritis *See* degenerative joint disease. { di'jen·ə·rəd·iv är'thrīd·əs }

degenerative disease [MED] General debility and diseases associated with advancing age. { di'jen·ə·rəd·iv di'zēz }

degenerative joint disease [MED] A chronic joint disease characterized pathologically by degeneration of articular cartilage and hypertrophy of bone, clinically by pain on activity which subsides with rest. Also known as degenerative arthritis; hypertrophic arthritis; osteoarthritis; senescent arthritis. { di'jen·ə·rəd·iv 'jȯint di,zēz }

Degeneriaceae [BOT] A family of dicotyledonous plants in the order Magnoliales characterized by laminar stamens; a solitary, pluriovulate, unsealed carpel; and ruminate endosperm. { ,dē·jen·ə,rī'ās·ē,ē }

deglutition [PHYSIO] Act of swallowing. { ,dē ,glü'tish·ən }

degradative plasmid [GEN] A type of plasmid that specifies a set of genes involved in biodegradation of an organic compound. { ¦deg·rə,dād·iv 'plaz·mid }

dehiscence [BOT] Spontaneous bursting open of a mature plant structure, such as fruit, anther, or sporangium, to discharge its contents. [MED] A defect in the boundary of a bony canal or cavity. { də'his·əns }

dehiscent [BOT] Becoming open at maturity to release seeds, as certain fruits. { di'his·ənt }

dehydrase [BIOCHEM] An enzyme which catalyzes the removal of water from a substrate. { dē'hī,drās }

dehydrochlorinase [BIOCHEM] An enzyme that dechlorinates a chlorinated hydrocarbon such as the insecticide DDT; found in some insects that are resistant to DDT. { dē,hī·drō'klȯr·ə,nās }

dehydrochlorination [BIOCHEM] Removal of hydrogen and chlorine or hydrogen chloride from a compound. { dē,hī·drō,klȯr·ə'nā·shən }

dehydrocholesterol [BIOCHEM] $C_{27}H_{43}OH$ A provitamin of animal origin found in the skin of humans, in milk, and elsewhere, which upon irradiation with ultraviolet rays becomes vitamin D. { dē¦hī·drō·kə'les·tə,rȯl }

dehydrogenase [BIOCHEM] An enzyme which removes hydrogen atoms from a substrate and transfers it to an acceptor other than oxygen. { dē'hī·drə·jə,nās }

delamination [BIOL] The separation of cells into layers. [EMBRYO] Gastrulation in which the endodermal layer splits off from the inner surface of the blastoderm and the space between this layer and the yolk represents the archenteron. { dē,lam·ə'nā·shən }

delayed hypersensitivity [IMMUNOL] Abnormal reactivity in a sensitized individual beginning several hours after contact with the allergen. { di'lād ,hī·pər,sen·sə'tiv·əd·ē }

delayed speech [MED] A speech disorder characterized by a complete absence of vocalization or vocalization with no communicative value; speech is considered delayed when it fails to develop by the second year, caused by impaired hearing, severe childhood illness, or emotional disturbance. { di'lād 'spēch }

deletion [GEN] Loss of a chromosome segment of any size, down to a part of a single gene. { di'lē·shən }

deliquescence [BOT] The condition of repeated divisions ending in fine divisions; seen especially in venation and stem branching. { del·ə'kwes·əns }

delirium [MED] Severely disordered mental state associated with fever, intoxication, head trauma, and other encephalopathies. { di'lir·ē·əm }

delirium tremens [MED] Delirium associated with tremors, insomnia, and other physical and neurological symptoms frequently following chronic alcoholism. { di'lir·ē·əm 'trem·ənz }

delomorphous cell *See* parietal cell. { ¦dē·lō¦mȯr·fəs 'sel }

Delphinidae [VERT ZOO] A family of aquatic mammals in the order Cetacea; includes the dolphins. { del'fin·ə,dē }

delphinidin [BIOCHEM] $C_{15}H_{11}O_7Cl$ A purple or brownish-red anthocyanin compound occurring widely in plants. { del'fin·ə·dən }

Delphinus [VERT ZOO] A genus of cetacean mammals, including the dolphin. { del'fē·nəs }

delta [ANAT] A fingerprint focal point which is the point on a ridge at or in front of and nearest the center of the divergence of the type lines. { 'del·tə }

delta hepatitis [MED] A type of viral hepatitis caused by the delta agent hepatitis D virus, a defective ribonucleic acid virus that requires the helper function of hepatitis B virus for its replication and expression. { ,del·tə ,hep·ə'tīd·əs }

delta rhythm [PHYSIO] An electric current generated in slow waves with frequencies of 0.5–3 per second from the forward portion of the brain of normal subjects when asleep. { 'del·tə ,rith·əm }

deltoid [ANAT] The large triangular shoulder muscle; originates on the pectoral girdle and inserts on the humerus. [BIOL] Triangular in shape. { 'del,tȯid }

deltoid ligament [ANAT] The ligament on the medial side of the ankle joint; the fibers radiate from the medial malleolus to the talus, calcaneus, and navicular bones. { 'del,tȯid 'lig·ə·mənt }

demarcation potential *See* injury potential. { dē ,mär'kā·shən pə,ten·chəl }

Dematiaceae [MYCOL] A family of fungi in the order Moniliales; sporophores are not grouped, hyphae are always dark, and the spores are hyaline or dark. { də,mad·ē'ās·ē,ē }

deme [ECOL] A local population in which the individuals freely interbreed among themselves but not with those of other demes. { dēm }

demersal [BIOL] Living at or near the bottom of the sea. { də'mər·səl }

demethylchlortetracycline [MICROBIO] $C_{21}H_{21}O_8N_2Cl$ A broad-spectrum tetracycline antibiotic produced by a mutant strain of *Streptomyces aureofaciens*. { dē¦meth·əl,klȯr,te·trə'sī,klēn }

demilune [BIOL] Crescent-shaped. { 'dem·i ,lün }

demineralization [MED] **1.** Removal or loss of minerals and salts from the body, especially by disease. **2.** In particular, the continual dissolving of tooth mineral that occurs at the surface of teeth as the result of the action of weak acids created by plaque-forming bacteria. { dē,min·rə·lə'zā·shən }

Demodicidae [INV ZOO] The pore mites, a family of arachnids in the suborder Trombidiformes. { ,dem·ə'dis·ə,dē }

demographic genetics [BIOL] A branch of population genetics and ecology concerned with genetic differences related to age, population size, genetic alteration in competitive ability, and viability. { ¦dem·ə,graf·ik jə¦ned·iks }

demography [ECOL] The statistical study of populations with reference to natality, mortality, migratory movements, age, and sex, among other social, ethnic, and economic factors. { də'mäg·rə·fē }

Demospongiae [INV ZOO] A class of the phylum Porifera, including sponges with a skeleton of one-to four-rayed siliceous spicules, or of spongin fibers, or both. { dem·ə'spən·jē,ē }

demyelinating disease [MED] Any disease associated with the destruction or removal of myelin from nerves. { dē'mī·ə·lə,nād·iŋ di,zēz }

demyelination [PATH] Destruction of myelin; loss of myelin from nerve sheaths or nerve tracts. { dē,mī·ə·lə'nā·shən }

denaturation map [CELL MOL] A map that shows the positions of denaturation loops of deoxyribonucleic acid and that provides a unique way to distinguish different molecules of deoxyribonucleic acid. { di'nā·chə,rā·shən ,map }

dendrite [NEUROSCI] The part of a neuron that carries the unidirectional nerve impulse toward the cell body. Also known as dendron. { 'den,drīt }

dendritic cell [CELL MOL] A specialized cell of the lymphoid reticuloendothelial system that presents antigens for detection by lymphocytes. { den'drid·ik ¦sel }

Dendrobatinae [VERT ZOO] A subfamily of anuran amphibians in the family Ranidae, including the colorful poisonous frogs of Central and South America. { ,den·drō'bat·ən·ē }

dendrobranchiate gill [INV ZOO] A respiratory structure of certain decapod crustaceans, characterized by extensive branching of the two primary series. { ¦den·drō'braŋ·kē,āt 'gil }

Dendroceratida [INV ZOO] A small order of sponges of the class Demospongiae; members have a skeleton of spongin fibers or lack a skeleton. { ,den·drō·sə'räd·əd·ə }

Dendrochirotacea [INV ZOO] A subclass of echinoderms in the class Holothuroidea. { ,den·drō,kī·rō'tās·ē·ə }

Dendrochirotida [INV ZOO] An order of dendrochirotacean holothurian echinoderms with 10–30 richly branched tentacles. { ,den·drō,kī 'räd·əd·ə }

Dendrocolaptidae [VERT ZOO] The woodcreepers, a family of passeriform birds belonging to the suboscine group. { ,den·drō·kə'lap·tə,dē }

dendroecology [ECOL] The use of tree rings to study changes in ecological processes over time such as defoliation by insect outbreaks; the effects of air, water, and soil pollution on tree growth and forest health; the age, maturity, and successional status of forest stands; and the effects of human disturbances and management on forest vitality. { ,den·drō·ē'käl·ə·jē }

dendrogram [BIOL] A genealogical tree; the trunk represents the oldest common ancestor, and the branches indicate successively more recent divisions of a lineage for a group. { 'den·drə,gram }

dendroid [BIOL] Branched or treelike in form. { 'den,drȯid }

Dendromurinae [VERT ZOO] The African tree mice and related species, a subfamily of rodents in the family Muridae. { ¦den·drō'myůr·ə·nē }

dendron *See* dendrite. { 'den,drän }

dendrophagous [ZOO] Feeding on trees, referring to insects. { den'dräf·ə·gəs }

dendrophysis [MYCOL] A hyphal thread with arboreal branching in certain fungi. { den'dräf·ə·səs }

denervate [MED] To interfere with or cut off the nerve supply to a part of the body, or to remove a nerve; may occur by excision, drugs, or a disease process. { dē'nər,vāt }

denervation hypersensitivity [NEUROSCI] Extreme sensitivity of an organ that has recovered from the removal or interruption of its nerve supply. { ,dē·nər'vā·shən ,hīp·ər,sen·sə'tiv·əd·ē }

dengue [MED] An acute viral disease of humans characterized by fever, rash, prostration, and lymphadenopathy; transmitted by the mosquito *Aedes aegypti*. Also known as breakbone fever; dandy fever. { 'deŋ·gē }

Dengue fever [MED] An infection borne by the *Aedes* female mosquito, and caused by one of four closely related but antigenically distinct Dengue virus serotypes (DEN-1, DEN-2, DEN-3, and DEN-4). It starts abruptly after an incubation period of 2–7 days with high fever, severe headache, myalgia, and rash. It is found throughout the tropical and subtropical zones. Also known as break-bone fever. { 'deŋ·gē ˌfēv·ər }

Dengue hemmorhagic fever [MED] A severe and potentially fatal form of Dengue fever that is characterized by loss of appetite, vomiting, high fever, headache, and abdominal pain; shock and circulatory failure may occur. { ¦deŋ·gē ˌhem·əˌraj·ik 'fē·vər }

denitrification [MICROBIO] The reduction of nitrate or nitrite to gaseous products such as nitrogen, nitrous oxide, and nitric oxide; brought about by denitrifying bacteria. { dēˌnī·trə·fə'kā·shən }

denitrifying bacteria [MICROBIO] Bacteria that reduce nitrates to nitrites or nitrogen gas; most are found in soil. { dē'nī·trəˌfī·iŋ bak'tir·ē·ə }

denitrogenate [PHYSIO] To remove nitrogen from the body by breathing nitrogen-free gas. { dē'nī·trə·jəˌnāt }

dense connective tissue [HISTOL] A fibrous connective tissue with an abundance of enlarged collagenous fibers which tend to crowd out the cells and ground substance. { ¦dens kə¦nek·tiv 'tish·yü }

dense fibrillar component [CELL MOL] A component of the nucleolus that lacks granules and stains more intensely than other nucleolar components. { ¦dens ¦fi·brə·lər kəm¦pōn·ənt }

density-dependent factor [ECOL] A factor that affects the birth rate or mortality rate of a population in ways varying with the population density. { ¦den·səd·ē ˌdi¦pen·dənt ˌfak·tər }

density-independent factor [ECOL] A factor that affects the birth rate or mortality rate of a population in ways that are independent of the population density. { ¦den·səd·ē ˌin·də¦pen·dənt ˌfak·tər }

Densovirus [VIROL] A genus of the animal virus family Parvoviridae whose virion is nonenveloped, with deoxyribonucleic acid single-stranded; replicates autonomously. { ˌden·sō'vī·rəs }

dental [ANAT] Pertaining to the teeth. { 'dent·əl }

dental arch [ANAT] The parabolic curve formed by the cutting edges and masticating surfaces of the teeth. { ¦dent·əl 'ärch }

dental bridge [MED] A prosthetic device used to replace missing teeth. { 'dent·əl ˌbrij }

dental calculi [MED] Calcareous deposits of organic and mineral matter on the teeth. Also known as tartar. { ¦dent·əl 'kal·kyə·lē }

dental caries *See* caries. { ¦dent·əl 'kar·ēz }

dental epithelium [HISTOL] The cells forming the boundary of the enamel organ. { 'dent·əl ep·ə'thē·lē·əm }

dental follicle *See* dental sac. { ¦dent·əl 'fäl·ə·kəl }

dental formula [VERT ZOO] An expression of the number and kind of teeth in each half jaw, both upper and lower, of mammals. { ¦dent·əl 'fór·myə·lə }

Dentaliidae [INV ZOO] A family of mollusks in the class Scaphopoda; members have pointed feet. { ˌdent·əl'ī·əˌdē }

dental materials science [MED] An interdisciplinary area that applies biology, chemistry, and physics to the development, understanding, and evaluation of materials used in the practice of dentistry; principally involved in restorative dentistry, prosthodontics, pedodontics, and orthodontics. { 'dent·əl məˌtir·ē·əlz 'sī·əns }

dental pad [VERT ZOO] A firm ridge that replaces incisors in the maxilla of cud-chewing herbivores. { 'dent·əl ˌpad }

dental papilla [EMBRYO] The mass of connective tissue located inside the enamel organ of a developing tooth, and forming the dentin and dental pulp of the tooth. { 'dent·əl pə'pil·ə }

dental plate [INV ZOO] A flat plate that replaces teeth in certain invertebrates, such as some worms. [VERT ZOO] A flattened plate that represents fused teeth in parrot fishes and related forms. { 'dent·əl ˌplāt }

dental pulp [HISTOL] The vascular connective tissue of the roots and pulp cavity of a tooth. { 'dent·əl ˌpəlp }

dental ridge [EMBRYO] An elevation of the embryonic jaw that forms a cusp or margin of a tooth. { 'dent·əl ˌrij }

dental sac [EMBRYO] The connective tissue that encloses the developing tooth. Also known as dental follicle. { 'dent·əl ˌsak }

dentate [BIOL] **1.** Having teeth. **2.** Having toothlike or conical marginal projections. { 'denˌtāt }

dentate fissure *See* hippocampal sulcus. { 'den ˌtāt 'fish·ər }

dentate nucleus [NEUROSCI] An ovoid mass of nerve cells located in the center of each cerebellar hemisphere, which give rise to fibers found in the superior cerebellar peduncle. { 'denˌtāt 'nü·klē·əs }

denticle [ZOO] A small tooth or toothlike projection, as the type of scale of certain elasmobranchs. { 'dent·ə·kəl }

denticulate [ZOO] Having denticles; serrate. { den'tik·yə·lət }

dentigerous [BIOL] Having teeth or toothlike structures. { den'tij·ə·rəs }

dentin [HISTOL] A bonelike tissue composing the bulk of a vertebrate tooth; consists of 70% inorganic materials and 30% water and organic matter. { 'dent·ən }

dentinoblast [HISTOL] A mesenchymal cell that forms dentin. { den'tēn·əˌblast }

dentinogenesis [PHYSIO] The formation of dentin. { den¦tēn·ə¦jen·ə·səs }

dentinogenesis imperfecta [MED] An inherited dental disorder that causes defective formation of dentin. { ˌden·tə·nōˌjen·ə·səs ˌim·pər 'fek·tə }

dentinoma [MED] A benign odontogenic tumor made up of dentin. { ˌden·tə'nō·mə }

dentistry [MED] A branch of medical science concerned with the prevention, diagnosis, and treatment of diseases of the teeth and adjacent tissues and the restoration of missing dental structures. { 'dent·ə·strē }

dentition [VERT ZOO] The arrangement, type, and number of teeth which are variously located in the oral or in the pharyngeal cavities, or in both, in vertebrates. { den'tish·ən }

denture [MED] A partial or complete prosthetic appliance to replace one or more missing teeth. { 'den·chər }

deoperculate [BOT] Of mosses and liverworts, to shed the operculum. { dē·ō'pər·kyə,lāt }

deoxycholate [BIOCHEM] A salt or ester of deoxycholic acid. { dē¦äk·sə'kō,lāt }

deoxycholic acid [BIOCHEM] $C_{24}H_{40}O_4$ One of the unconjugated bile acids; in bile it is largely conjugated with glycine or taurine. { dē¦äk·sə'käl·ik 'as·əd }

deoxycorticosterone [BIOCHEM] $C_{21}H_{30}O_3$ A steroid hormone secreted in small amounts by the adrenal cortex. { dē¦äk·sē,kȯrd·ə'kä·stə,rōn }

deoxyribonuclease [BIOCHEM] An enzyme that catalyzes the hydrolysis of deoxyribonucleic acid to nucleotides. Abbreviated DNase. { dē¦äk·sē,rī·bō'nü·klē,ās }

deoxyribonucleic acid [BIOCHEM] A linear polymer made up of deoxyribonucleotide repeating units (composed of the sugar 2-deoxyribose, phosphate, and a purine or pyrimidine base) linked by the phosphate group joining the 3′ position of one sugar to the 5′ position of the next; most molecules are double-stranded and antiparallel, resulting in a right-handed helix structure kept together by hydrogen bonds between a purine on one chain and a pyrimidine on another; carrier of genetic information, which is encoded in the sequence of bases; present in chromosomes and chromosomal material of cell organelles such as mitochondria and chloroplasts, and also present in some viruses. Abbreviated DNA. { dē¦äk·sē,rī·bō·nü¦klē·ik 'as·əd }

deoxyribonucleic acid clone [CELL MOL] A deoxyribonucleic acid segment inserted via a vector into a host cell and replicated along with the vector to form many copies per cell. { dē¦äk·sē,rī·bō·nü¦klē·ik ¦as·əd 'klōn }

deoxyribonucleic acid complexity [CELL MOL] A measure of the fraction of nonrepetitive deoxyribonucleic acid that is characteristic of a given sample. { dē¦äk·sē,rī·bō·nü¦klē·ik ¦as·əd kəm'plek·səd·ē }

deoxyribonucleic acid-directed ribonucleic acid polymerase [BIOCHEM] An enzyme which transcribes a ribonucleic acid (RNA) molecule complementary to deoxyribonucleic acid (DNA); required for initiation of DNA replication as well as transcription of RNA. { dē¦äk·sē,rī·bō·nü¦klē·ik ¦as·əd də'rek·təd ¦rī·bō·nü¦kle·ik ¦as·əd pə'lim·ə,rās }

deoxyribonucleic acid footprinting [CELL MOL] A method for determining the sequence of deoxyribonucleic acid-binding proteins. { dē¦äk·sē,rī·bō·nü¦klē·ik ¦as·əd 'fu̇t,print·iŋ }

deoxyribonucleic acid hybridization [CELL MOL] A technique for selectively binding specific segments of single-stranded deoxyribonucleic acid (DNA) or ribonucleic acid by base pairing to complementary sequences on single-stranded DNA molecules that are trapped on a nitrocellulose filter. { dē¦äk·sē,rī·bō·nü¦klē·ik ¦as·əd ,hī·brə·də'zā·shən }

deoxyribonucleic acid lesion [CELL MOL] Deoxyribonucleic acid deformations that may result in gene mutation or changes in chromosome structure. { de¦ak·sē,rī·bō·nü¦klē·ik ¦as·əd 'lē·zhən }

deoxyribonucleic acid ligase [BIOCHEM] An enzyme which joins the ends of two deoxyribonucleic acid chains by catalyzing the synthesis of a phosphodiester bond between a 3′-hydroxyl group at the end of one chain and a 5′-phosphate at the end of the other. { dē¦äk·sē,rī·bō·nü¦klē·ik ¦as·əd 'lī,gās }

deoxyribonucleic acid polymerase I [BIOCHEM] An enzyme which catalyzes the addition of deoxyribonucleotide residues to the end of a deoxyribonucleic acid (DNA) strand; generally considered to function in the repair of damaged DNA. { dē¦äk·sē,rī·bō·nü¦klē·ik ¦as·əd pə'lim·ə,rās ¦wən }

deoxyribonucleic acid polymerase II [BIOCHEM] An enzyme similar in action to DNA polymerase I but with lower activity. { dē¦äk·sē,rī·bō·nü¦klē·ik ¦as·əd pə'lim·ə,rās ¦tü }

deoxyribonucleic acid polymerase III [BIOCHEM] An enzyme thought to be the primary enzyme involved in deoxyribonucleic acid replication. { dē¦äk·sē,rī·bō·nü¦klē·ik ¦as·əd pə'lim·ə,rās ¦thrē }

deoxyribonucleoprotein [BIOCHEM] A protein containing molecules of deoxyribonucleic acid in close association with protein molecules. { dē¦äk·sē,rī·bō,nü·klē·ō 'prō,tēn }

deoxyribonucleotide [BIOCHEM] A nucleotide that contains deoxyribose and is a constituent of deoxyribonucleic acid. { dē¦äk·sē,rī·bō'nü·klē·ə,tīd }

deoxyribose [BIOCHEM] $C_5H_{10}O_4$ A pentose sugar in which the hydrogen replaces the hydroxyl groups of ribose; a major constituent of deoxyribonucleic acid. { dē¦äk·sē'rī,bōs }

deoxyribovirus [VIROL] Any virus that contains deoxyribonucleic acid. { dē¦äk·sē,rī·bō'vī·rəs }

deoxy sugar [BIOCHEM] A substance which has the characteristics of a sugar, but which shows a deviation from the required hydrogen-to-oxygen ratio. { dē¦äk·sē 'shu̇g·ər }

depauperate [BIOL] Inferiority of natural development or size. { dē'pȯ·pə·rət }

dependence [MED] Habituation to, abuse of, or addiction to a substance. { di'pen·dəns }

Dependovirus [VIROL] A genus of the animal-virus family Parvoviridae that is characterized by defective viruses that require a helper virus (usually an adenovirus) for their replication. { di'pen·də,vī·rəs }

dephosphorylate [BIOCHEM] To remove a phosphate group. { ,dē·fäs'fȯr·ə,lāt }

depletion [ECOL] Using a resource, such as water or timber, faster than it is replenished. { də'plē·shən }

deposit feeder [INV ZOO] Any animal that feeds on the detritus that collects on the substratum at the bottom of water. Also known as detritus feeder. { də'päz·ət ˌfēd·ər }

depressed fracture [MED] A fracture of the skull in which the fractured part is depressed below the normal level. { di'prest 'frak·chər }

depressive neurosis *See* dysthymia. { di¦pres·iv nü'rō·səs }

depressor [ANAT] A muscle that draws a part down. { di'pres·ər }

depressor nerve [NEUROSCI] A nerve which, upon stimulation, lowers the blood pressure either in a local part or throughout the body. { di 'pres·ər ˌnərv }

depth perception [PHYSIO] Ability to judge spatial relationships. { 'depth pər'sep·shən }

depurination [BIOCHEM] Detachment of guanine from sugar in a deoxyribonucleic acid molecule. { dēˌpyu̇r·ə'nā·shən }

de Quervain's disease [MED] Inflammation of tendons and their sheaths at the styloid process of the radius, often causing pain in the inner side of the wrist. { də'ker·vənz dizˌēz }

derelict land [ECOL] Land that, because of mining, drilling, or other industrial processes, or by serious neglect, is unsightly and cannot be beneficially utilized without treatment. { 'der·əˌlikt ˌland }

derepression [MICROBIO] Transfer of microbial cells from an enzyme-repressing medium to a nonrepressing medium. [CELL MOL] Increased production of a gene product due to interference with the action of a repressor on the operator portion of the operon. { dē·ri 'presh·ən }

dermabrasion [MED] The surgical removal of scarred or tatooed skin by mechanical means such as rotating wire brushes or sandpaper. { ¦dər·mə¦brā·zhən }

Dermacentor [INV ZOO] A genus of ticks, important as vectors of disease. { 'dər·məˌsen·tər }

Dermacentor andersoni [INV ZOO] The wood tick, which is the vector of Rocky Mountain spotted fever and tularemia. { 'dər·məˌsen·tər an·dər'sō·nē }

Dermacentor variabilis [INV ZOO] A North American tick which is parasitic primarily on dogs but may attack humans and other mammals. { 'dər·məˌsen·tər ver·ē'ab·ə·ləs }

dermal [ANAT] Pertaining to the dermis. { 'dər·məl }

dermal bone [ANAT] A type of bone that ossifies directly from membrane without a cartilaginous predecessor; occurs only in the skull and shoulder region. Also known as investing bone; membrane bone. { ¦dər·məl 'bōn }

dermal denticle [VERT ZOO] A toothlike scale composed mostly of dentine with a large central pulp cavity, found in the skin of sharks. { ¦dər·məl 'dent·i·kəl }

dermalia [INV ZOO] Dermal microscleres in sponges. { dər'mal·yə }

dermal pore [INV ZOO] One of the minute openings on the surface of poriferans leading to the incurrent canals. { ¦dər·məl 'pōr }

Dermaptera [INV ZOO] An order of small or medium-sized, slender insects having incomplete metamorphosis, chewing mouthparts, short forewings, andcerci. { dər'map·tə·rə }

Dermatemydinae [VERT ZOO] A family of reptiles in the order Chelonia; includes the river turtles. { ˌdər·mə·tə'mī·dəˌnē }

dermatitis [MED] Inflammation of the skin. { ˌdər·mə'tīd·əs }

Dermatocarpaceae [BOT] A family of lichens in the order Pyrenulales having an umbilicate or squamulose growth form; most members grow on limestone or calcareous soils. { dər¦mad·ōˌkär'pās·ēˌē }

dermatocranium [ANAT] Bony parts of the skull derived from ossifications in the dermis of the skin. { dər¦mad·ə'krā·nē·əm }

dermatogen [BOT] The outer layer of primary meristem or the primordial epidermis in embryonic plants. Also known as protoderm. { dər'mad·əˌjen }

dermatoglyphics [ANAT] **1.** The integumentary patterns on the surface of the fingertips, palms, and soles. **2.** The study of these patterns. { dər¦mad·ə¦glif·iks }

dermatograph [MED] A crayonlike or similar instrument, used to mark the skin before surgery to outline positions of organs. { dər'mad·əˌgraf }

dermatologist [MED] A physician who specializes in diseases of the skin. { ˌdər·mə'täl·ə·jəst }

dermatology [MED] The science of the structure, function, and diseases of the skin. { ˌdər·mə'täl·ə·jē }

dermatome [ANAT] An area of skin delimited by the supply of sensory fibers from a single spinal nerve. [EMBRYO] Lateral portion of an embryonic somite from which the dermis will develop. [MED] Instrument for cutting skin for grafting. { 'dər·məˌtōm }

dermatomyositis [MED] An inflammatory reaction of unknown cause involving degenerative changes of skin and muscle. { dər¦mad·ōˌmī·ə'sīd·əs }

dermatopathic lymphadenitis *See* lipomelanotic reticulosis. { dər¦mad·ə¦path·ik ˌlim¦fad·ən'īd·əs }

dermatopathology [MED] A branch of pathology concerned with diseases of the skin. { dər¦mad·ō·pə'thäl·ə·jē }

Dermatophilaceae [MICROBIO] A family of bacteria in the order Actinomycetales; cells produce mycelial filaments or muriform thalli; includes human and mammalian pathogens. { dər¦mad·ō·fə'lās·ēˌē }

Dermatophilus [MICROBIO] A genus of bacteria in the family Dermatophilaceae; mycelial filaments are long, tapering, and branched, and are divided transversely and longitudinally (in two planes) by septa; spherical spores are motile. { dər¦mad·ə'fil·əs }

dermatophyte [MYCOL] A fungus parasitic on skin or its derivatives. { dər'mad·əˌfīt }

dermatophytosis *See* cutaneous mycosis. { dər'mad·ōˌfī'tō·səs }

dermatoplast [BOT] In angiosperms, a cell with a cell wall. { dər'mad·əˌplast }

dermatosclerosis *See* scleroderma. { dər¦mad·ō·sklə'rō·səs }

dermatosis [MED] Any skin disease. { ˌdər·mə'tō·səs }

Dermestidae [INV ZOO] The skin beetles, a family of coleopteran insects in the superfamily Dermestoidea, including serious pests of stored agricultural grain products. { dər'mes·tə·dē }

Dermestoidea [INV ZOO] A superfamily of coleopteran insects in the suborder Polyphaga. { ˌdər·mə'stȯid·ē·ə }

dermis [ANAT] The deep layer of the skin, a dense connective tissue richly supplied with blood vessels, nerves, and sensory organs. Also known as corium; cutis. { 'dər·məs }

Dermochelidae [VERT ZOO] A family of reptiles in the order Chelonia composed of a single species, the leatherback turtle. { ˌdər·mə'kel·əˌdē }

dermoepidermal junction [HISTOL] The area of separation between the stratum basale of the epidermis and the papillary layer of the dermis. { ¦dər·mōˌep·ə¦dər·məl 'jəŋk·shən }

dermographia [MED] A condition in which the skin is peculiarly susceptible to irritation, characterized by elevations or wheals with surrounding erythematous axon reflex flare, caused by tracing a fingernail or a blunt instrument over the skin. { 'dər·mə'graf·ē·ə }

dermoid cyst [MED] A benign cystic teratoma with skin, skin appendages, and their products as the most prominent components, usually involving the ovary or the skin. { ¦dərˌmȯid 'sist }

dermometer [PHYSIO] An instrument used to measure the electrical resistance of the skin. { dər'mäm·əd·ər }

dermonecrotic [MED] Pertaining to or causing necrosis of the skin. { ˌdər·mō·nə'kräd·ik }

Dermoptera [VERT ZOO] The flying lemurs, an ancient order of primatelike herbivorous and frugivorous gliding mammals confined to southeastern Asia and eastern India. { dər'mäp·tə·rə }

Derodontidae [INV ZOO] The tooth-necked fungus beetles, a small family of coleopteran insects in the superfamily Dermestoidea. { ˌder·ə'dän·təˌdē }

derris [BOT] Any of certain tropical shrubs in the genus *Derris* in the legume family (Leguminosae), having long climbing branches. { 'der·əs }

Derxia [MICROBIO] A genus of bacteria in the family Azotobacteraceae; rod-shaped, pleomorphic, motile cells; older cells contain large refractive bodies. { 'dərk·sē·ə }

DES *See* diethylstilbesterol.

Descemet's membrane [HISTOL] A layer of the cornea between the posterior surface of the stroma and the anterior surface of the endothelium which contains collagen arranged on a crystalline lattice. { des'māz ˌmemˌbrān }

descending [ANAT] Extending or directed downward or caudally, as the descending aorta. [NEUROSCI] In the nervous system, efferent; conducting impulses or progressing down the spinal cord or from central to peripheral. { di'sen·diŋ }

descending colon [ANAT] The portion of the colon on the left side, extending from the bend below the spleen to the sigmoid flexure. { di 'send·iŋ 'kōl·ən }

descriptive anatomy [ANAT] Study of the separate and individual portions of the body, with regard to form, size, character, and position. { di'skrip·tiv ə'nad·ə·mē }

descriptive botany [BOT] The branch of botany that deals with diagnostic characters or systematic description of plants. { di'skrip·tiv 'bät·ən·ē }

desensitization [IMMUNOL] Loss or reduction of sensitivity to infection or an allergen accomplished by means of frequent, small doses of the antigen. Also known as hyposensitization. { dēˌsen·sə·tə'zā·shən }

deserticolous [ECOL] Living in a desert. { ¦dez·ər¦tik·ə·ləs }

desertification [ECOL] The creation of desiccated, barren, desertlike conditions due to natural changes in climate or possibly through mismanagement of the semiarid zone. { dəˌzərd·ə·fə'kā·shən }

desexualization [PHYSIO] Depriving an organism of sexual characters or power, as by spaying or castration. { dēˌseksh·ə·lə'zā·shən }

design feature [ECOL] An organismal trait that can influence rates of death and reproduction, and hence Darwinian fitness. { di'zīn ˌfē·chər }

desma [INV ZOO] A branched, knobby spicule in some Demospongiae. { 'dez·mə }

desmacyte [INV ZOO] A bipolar collencyte found in the cortex of certain sponges. { 'dez·məˌsīt }

Desmanthos [MICROBIO] A genus of bacteria in the family Pelonemataceae; unbranched, relatively straight filaments with a thickened base which are arranged in bundles partially enclosed in a sheath. { dez'man·thȯs }

desmid [BOT] Any member of a group of microscopic, unicellular green algae of the family Desmidiaceae, having cells of varying shapes but always composed of mirror-image semicells, often demarcated by a median constriction or incision, and a cell wall that has pores, which are frequently ornamented. { 'dez·məd }

Desmidiaceae [BOT] A family of desmids, mostly unicellular algae in the order Conjugales. { dezˌmid·ē'ās·ēˌē }

desmin [BIOCHEM] The muscle protein forming the Z lines in striated muscle. { 'dez·mən }

desmochore [ECOL] A plant having sticky or barbed disseminules. { 'dez·məˌkȯr }

Desmodontidae [VERT ZOO] A small family of

chiropteran mammals comprising the true vampire bats. { ˌdez·mə'dän·tə̩dē }

Desmodoroidea [INV ZOO] A superfamily of marine-and brackish-water-inhabiting nematodes with an annulated, usually smooth cuticle. { 'dez·mə·də'rȯid·ē·ə }

Desmokontae [BOT] The equivalent name for Desmophyceae. { ˌdez·mə'kän·tē }

desmolase [BIOCHEM] Any of a group of enzymes which catalyze rupture of atomic linkages that are not cleaved through hydrolysis, such as the bonds in the carbon chain of D-glucose. { 'dez·mə̩lās }

desmoneme [INV ZOO] A nematocyst having a long coiled tube which is extruded and wrapped around the prey. { ˌdez·mə'nēm }

desmopelmous [VERT ZOO] A type of bird foot in which the hindtoe cannot be bent independently because planter tendons are united. { ˌdez·mə'pel·məs }

Desmophyceae [BOT] A class of rare, mostly marine algae in the division Pyrrhophyta. { ˌdez·mə'fīs·ē̩ē }

desmoplasia [MED] **1.** The formation and proliferation of connective tissue, frequently in the growth of tumors. **2.** The formation of adhesions. { ˌdez·mə'plā·zhə }

Desmoscolecida [INV ZOO] An order of the class Nematoda. { ˌdez·mə·skə'les·ə·də }

Desmoscolecidae [INV ZOO] A family of nematodes in the superfamily Desmoscolecoidea; individuals resemble annelids in having coarse-annulation. { ˌdez·mə·skə'les·ə̩dē }

Desmoscolecoidea [INV ZOO] A small superfamily of free-living nematodes characterized by a ringed body, an armored head set, and hemispherical amphids. { ˌdez·mə̩skō·lə'kȯid·ē·ə }

desmose [INV ZOO] A fibril connecting the centrioles during mitosis in certain protozoans. { 'dez̩mōs }

desmosome *See* adhering junction. { 'dez·mə ̩sōm }

Desmothoracida [INV ZOO] An order of sessile and free-living protozoans in the subclass Heliozoia having a spherical body with a perforate, chitinous test. { ˌdez·mə·thə'ras·ə·də }

Desor's larva [INV ZOO] An oval, ciliated larva of certain nemertineans in which the gastrula remains inside the egg membrane. { də'zȯrz ̩lär·va }

desquamation [PHYSIO] Shedding; a peeling and casting off, as of the superficial epithelium, mucous membranes, renal tubules, and the skin. { dē·skwə'mā·shən }

Desulfotomaculum [MICROBIO] A genus of bacteria in the family Bacillaceae; motile, straight or curved rods with terminal to subterminal spores; anaerobic and reduce sulfate. { dē̩səl·fəd·ə'mak·yə·ləm }

Desulfovibrio [MICROBIO] A genus of gram-negative, strictly anaerobic bacteria of uncertain affiliation; motile, curved rods reduce sulfates and other sulfur compounds to hydrogen sulfide. { dē̩səl·fə'vib·rē̩ō }

detached meristem [BOT] A meristematic region originating from apical meristem but becoming discontinuous with it because of differentiation of intervening tissue. { di'tacht 'mer·ə̩stem }

determinate cleavage [EMBRYO] A type of cleavage which separates portions of the zygote with specific and distinct potencies for development as specific parts of the body. { də'tər·mə·nət 'klē·vij }

determinate growth [BOT] Growth in which the axis, or central stem, being limited by the development of the floral reproductive structure, does not grow or lengthen indefinitely. { də'tər·mə·nət 'grōth }

detorsion [INV ZOO] Untwisting of the 180° visceral twist imposed by embryonic torsion on many gastropod mollusks. [MED] Untwisting of an abnormal torsion, as of a ureter or intestine. { dē'tȯr·shən }

detoxification [BIOCHEM] The act or process of removing a poison or the toxic properties of a substance in the body. { dē̩täk·sə·fə'kā ·shən }

detritivore [ECOL] An organism that consumes dead organic matter. { di'trid·ə̩vȯr }

detritus [ECOL] Dead plants and corpses or cast-off parts of various organisms. { də'trīd·əs }

detritus feeder *See* deposit feeder. { də'trīd·əs 'fēd·ər }

detritus food web [ECOL] A trophic web that is based on the consumption of dead organic material. { di̩trīd·əs 'füd ̩web }

detrivorous [BIOL] Referring to an organism that feeds on dead animals or partially decomposed organic matter. { də'triv·ə·rəs }

deutencephalon *See* epichordal brain. { 'düt 'en'sef·ə̩län }

deuteranomaly [MED] A partial deuteranopia. { ˌdüt̩er·ə'näm·ə·lē }

deuteranopia [MED] Defective vision consisting of red-green color confusion, with no marked reduction in the brightness of any color. { ¦düd·ər·ə'nō·pē·ə }

deuterogamy [BOT] Secondary pairing of sexual cells or nuclei replacing direct copulation in many fungi, algae, and higher plants. { ˌdüd·ə'räg·ə·mē }

Deuteromycetes [MYCOL] The equivalent name for Fungi Imperfecti. { ¦düd·ə·rō̩mī'sēd·ēz }

Deuterophlebiidae [INV ZOO] A family of dipteran insects in the suborder Cyclorrhapha. { ˌdüd·ə·rō·flə'bī·ə̩dē }

Deuterostomia [ZOO] A division of the animal kingdom which includes the phyla Echinodermata, Chaetognatha, Hemichordata, and Chordata. { ˌdüd·ə·rō'stō·mē·ə }

deutocerebrum [INV ZOO] The median lobes of the insect brain. { ¦düd·ō'ser·ə·brəm }

deutoplasm [EMBRYO] The nutritive yolk granules in egg cells. { 'düd·ə̩plaz·əm }

development [BIOL] A process of regulated growth and differentiation that results from interaction of the genome with the cytoplasm, the internal cellular environment, and the external environment. { də'vel·əp·mənt }

developmental control gene [GEN] A gene whose primary function is the regulation of cell fates during development. { di,vel·əp¦men·təl kən'trōl ,jēn }

developmental disability [MED] A substantial handicap or impairment originating before the age of 18 that may be expected to continue indefinitely. { də¦vel·əp,ment·əl ,dis·ə,bil·əd·ē }

developmental genetics [GEN] A branch of genetics primarily concerned with the manner in which genes control or regulate development. { də¦vel·əp,ment·əl jə'ned·iks }

developmental instability [GEN] Variation of development within a genotype due to local fluctuations in internal or external environmental conditions. { di,vel·əp¦men·təl ,in·stə'bil·əd·ē }

developmental noise [GEN] Any uncontrollable variation in phenotype due to random events during development. { di,vel·əp¦men·təl 'nȯiz }

developmental toxicity [MED] Adverse effects on the developing child which result from exposure to toxic chemicals or other toxic substances, can include birth defects, low birth weight, and functional or behavioral weaknesses that show up as the child develops. { di,vel·əp¦ment·əl tak'sis·ə·dē }

devernalization [BOT] Annulment of the vernalization effect. { dē,vərn·əl·ə'zā·shən }

deviation [EVOL] Evolutionary differentiation involving interpolation of new stages in the ancestral pattern of morphogenesis. { ,dēv·ē'ā·shən }

dewclaw [VERT ZOO] **1.** A vestigial digit on the foot of a mammal which does not reach the ground. **2.** A claw or hoof terminating such a digit. { 'dü,klȯ }

dewlap [ANAT] A fleshy or fatty fold of skin on the throat of some humans. [BOT] One of a pair of hinges at the joint of a sugarcane leaf blade. [VERT ZOO] A fold of skin hanging from the neck of some reptiles and bovines. { 'dü,lap }

dew retting [MICROBIO] A type of retting process in which the stems of fiber plants are spread out in moist meadows, and the pectin decomposition is accomplished by molds and aerobic bacteria with the formation of CO_2 and H_2. { 'dü ,red·iŋ }

Dexaminidae [INV ZOO] A family of amphipod crustaceans in the suborder Gammeridea. { dek'sam·in·ə,dē }

dexterotropic [BIOL] Turning toward the right; applied to cleavage, shell formation, and whorl patterns. { ¦dek·stə·rō ¦träp·ik }

dextran [BIOCHEM] Any of the several polysaccharides, $(C_5H_{10}O_5)_n$, that yield glucose units on hydrolysis. { 'dek ,stran }

dextranase [BIOCHEM] An enzyme that hydrolyzes 1,6-α-glucosidic linkages in dextran. { 'dek·strə,nās }

dextrin [BIOCHEM] A polymer of D-glucose which is intermediate in complexity between starch and maltose. { 'dek·strən }

dextrocardia [MED] The presence of the heart in the right hemithorax, with the cardiac apex directed to the right. { ,dek·strō'kär·dē·ə }

dextromethorphan hydrobromide [PHARM] $C_{18}H_{25}NO \cdot HBr \cdot H_2O$ White crystals or a crystalline powder; soluble in alcohol and chloroform; used in medicine. { ¦dek·strō·mə¦thȯr·fən,hī·drō'brō,mīd }

dextrorse [BOT] Twining toward the right. { 'dek,strȯrs }

dextrose [BIOCHEM] $C_6H_{12}O_6 \cdot H_2O$ A dextrorotatory monosaccharide obtained as a white, crystalline, odorless, sweet powder, which is soluble in about one part of water; an important intermediate in carbohydrate metabolism; used for nutritional purposes, for the temporary increase of blood volume, and as a diuretic. Also known as corn sugar; grape sugar. { 'dek,strōs }

dextrotopic cleavage [EMBRYO] A clockwise spiral cleavage pattern. { ¦dek·strə¦täp·ik 'klē·vij }

DFP *See* isoflurophate.

diabetes [MED] Any of various abnormal conditions characterized by excessive urinary output, thirst, and hunger; usually refers to diabetes mellitus. { ,dī·ə'bēd·ēz }

diabetes insipidus [MED] A form of diabetes due to a disfunction of the hypothalamus. { ,dī·ə'bēd·ēz in'sip·ə·dəs }

diabetes mellitus [MED] A metabolic disorder arising from a defect in carbohydrate utilization by the body, related to inadequate or abnormal insulin production by the pancreas. { ,dī·ə'bēd·ēz 'mel·ə·dəs }

diabetic acidosis [MED] Metabolic acidosis seen in diabetes mellitus, due to an excess of ketone bodies. { ¦dī·ə'bed·ik ,as·ə'dō·səs }

diabetic gangrene [MED] A moist form of gangrene occurring in persons with diabetes mellitus, often following a minor injury. { ¦dī·ə¦bed·ik ,gaŋ'grēn }

diabetic glomerulosclerosis *See* intercapillary glomerulosclerosis. { ¦dī·ə¦bed·ik glə¦mər·yə·lō·sklə'rō·səs }

diactine [INV ZOO] A type of sponge spicule which develops in two directions from a central point. { dī'ak,tēn }

diacylglycerol [BIOCHEM] A product of the cleavage of membrane-associated phospholipid by phospholipase C that activates isozymes of protein kinase C, an important regulator of cell function. { ,dī·ə·səl'glis·ə,rōl }

diadelphous stamen [BOT] A stamen that has its filaments united into two sets. { ¦dī·ə¦del·fəs 'stā·mən }

Diadematacea [INV ZOO] A superorder of Euchinoidea having a rigid or flexible test, perforate tubercles, and branchial slits. { ,dī·ə,dē·mə'tās·ē·ə }

Diadematidae [INV ZOO] A family of large euechinoid echinoderms in the order Diadematoida having crenulate tubercles and long spines. { ,dī·ə·də'mad·ə,dē }

Diadematoida [INV ZOO] An order of echinoderms in the superorder Diadematacea with hollow primary radioles and crenulate tubercles. { ˌdī·ə·dē·mə'tȯid·ə }

diadromous [BOT] Having venation in the form of fanlike radiations. [VERT ZOO] Of fish, migrating between salt and fresh waters. { dī'ad·rə·məs }

Diadumenidae [INV ZOO] A family of anthozoans in the order Actiniaria. { dī·ə·dü'men·əˌdē }

diageotropism [BIOL] Growth orientation of a sessile organism or structure perpendicular to the line of gravity. { ¦dī·ə·jē'ä·trəˌpiz·əm }

diagnosis [MED] Identification of a disease from its signs and symptoms. [SYST] In taxonomic study, a statement of the characters that distinguish a taxon from coordinate taxa. { ˌdī·əg'nō·səs }

diagnostic bacteriology [MICROBIO] A branch of medical bacteriology that focuses on the identification of bacteria by their ability to grow on various selective media and by the characteristic appearance of their colonies on test media. { ˌdī·əgˌnäs·tik bakˌtir·ē'äl·ə·jē }

diaheliotropism [BOT] Movement of plant leaves which follow the sun such that they remain perpendicular to the sun's rays throughout the day. { ˌdī·əˌhē·lē·ə'träˌpiz·əm }

diakinesis [CELL MOL] The last stage of meiotic prophase, when the chromatids attain maximum contraction and the bivalents move apart and position themselves against the nuclear membrane. { dī·ə·kə'nē·səs }

diamine oxidase [BIOCHEM] A flavoprotein which catalyzes the aerobic oxidation of amines to the corresponding aldehyde and ammonia. { 'dī·əˌmēn 'äk·səˌdās }

diaminopimelate [BIOCHEM] $C_7H_{14}O_4N_2$ A compound that serves as a component of cell wall mucopeptide in some bacteria and as a source of lysine in all bacteria. Abbreviated DAP. { dī¦am·əˌnō'pim·əˌlāt }

diandrous [BOT] Having two stamens. { dī 'an·drəs }

Dianemaceae [MICROBIO] A family of slime molds in the order Trichales. { ˌdī·ə·nə'mās·ēˌē }

Dianthovirus [VIROL] A genus of plant viruses within the family Tombusviridae that are characterized by icosahedral particles containing two single-stranded positive-strand ribonucleic acid molecules; Carnation ringspot virus is the type species. Also known as Carnation ringspot virus group. { dī 'an·thəˌvī·rəs }

diapause [PHYSIO] A period of spontaneously suspended growth or development in certain insects, mites, crustaceans, and snails. { 'dī·əˌpȯz }

diapedesis [MED] Hemorrhage of blood cells, especially erythrocytes, through an intact vessel wall into the tissues. { ˌdī·ə·pə'dē·səs }

Diapensiaceae [BOT] The single family of the Diapensiales, an order of flowering plants. { ˌdī·əˌpen·sē'ās·ēˌē }

Diapensiales [BOT] A monofamilial order of dicotyledonous plants in the subclass Dilleniidae comprising certain herbs and dwarf shrubs in temperate and arctic regions of the Northern Hemisphere. { ˌdī·əˌpen·sē'ā·lēz }

Diaphanocephalidae [INV ZOO] A family of parasitic roundworms belonging to the Strongyloidea; snakes are the principal host. { di¦af·ə·nō·sə'fal·əˌdē }

Diaphanocephaloidea [INV ZOO] A superfamily of nematodes represented by a single family, Diaphanocephalidae, distinguished by the modification of the stoma into two massive lateral jaws and the absence of a corona radiata or lips. { dī¦af·ə·nōˌsef·ə'lȯid·ē· pp }

diaphorase [BIOCHEM] Mitochondrial flavoprotein enzymes which catalyze the reduction of dyes, such as methylene blue, by reduced pyridine nucleotides such as reduced diphosphopyridine nucleotide. { dī'af·əˌrās }

diaphragm [ANAT] The dome-shaped partition composed of muscle and connective tissue that separates the abdominal and thoracic cavities in mammals. { 'dī·əˌfram }

diaphragmatic hernia [MED] Protrusion of an abdominal organ through the diaphragm into the thoracic cavity. { ¦di·əˌfrag¦mad·ik 'hər·nē·ə }

diaphragmatic respiration [PHYSIO] Respiration effected primarily by movement of the diaphragm, changing the intrathoracic pressure. { ¦di·əˌfrag¦mad·ik ˌres·pə'rā·shən }

diaphragmed pith [BOT] Pith in which plates or nests of sclerenchyma may be interspersed with the parenchyma. { ˌdī·əˌframd 'pith }

diaphyseal aclasis *See* multiple hereditary exostoses. { dī'af·sē·əl 'ak·lə·səs }

diaphysis [ANAT] The shaft of a longbone. { dī'af·ə·səs }

diapophysis [ANAT] The articular portion of a transverse process of a vertebra. { ˌdī·ə'päf·ə·səs }

Diapriidae [INV ZOO] A family of hymenopteran insects in the superfamily Proctotrupoidea. { ˌdī·ə'prī·əˌdē }

diarch [BOT] Of a plant, having two protoxylem points or groups. { 'dīˌärch }

diarrhea [MED] The passage of loose or watery stools, usually at more frequent than normal intervals. { ˌdī·ə'rē·ə }

diarthrosis [ANAT] A freely moving articulation, characterized by a synovial cavity between the bones. { ¦dī·är'thrō·səs }

diastase [BIOCHEM] An enzyme that catalyzes the hydrolysis of starch to maltose. Also known as vegetable diastase. { 'di·əˌstās }

diastasis [MED] Any simple separation of parts normally joined together, as the separation of an epiphysis from the body of a bone without true fracture, or the dislocation of an amphiarthrosis. [PHYSIO] The final phase of diastole, the phase of slow ventricular filling. { dī'as·tə·səs }

diastema [ANAT] A space between two types of teeth, as between an incisor and premolar.

[CELL MOL] Modified cytoplasm of the equatorial plane prior to cell division. { ˌdī·ə'stē·mə }

diastole [PHYSIO] The rhythmic relaxation and dilation of a heart chamber, especially a ventricle. { dī'as·tə·lē }

diastolic pressure [PHYSIO] The lowest arterial blood pressure during the cardiac cycle; reflects relaxation and dilation of a heart chamber. { ¦dī·ə¦stäl·ik 'presh·ər }

diathermy [MED] The therapeutic use of high-frequency electric currents to produce localized heat in body tissues. { 'dī·əˌthər·mē }

diatom [INV ZOO] The common name for algae composing the class Bacillariophyceae; noted for the symmetry and sculpturing of the siliceous cell walls. { 'dī·əˌtäm }

diatropism [BOT] Growth orientation of certain plant organs that is transverse to the line of action of a stimulus. { dī'a·trəˌpiz·əm }

diauxic growth [MICROBIO] The diphasic response of a culture of microorganisms based on a phenotypic adaptation to the addition of a second substrate; characterized by a growth phase followed by a lag after which growth is resumed. { dī'ökˌsik 'grōth }

diazepam [PHARM] $C_{16}H_{13}ClN_2O$ A benzodiazepine with a melting point of 125–126°C; used as a minor tranquilizer to relieve muscle spasms, anxiety, and tension. Also known by trade name Valium. { dī'az·əˌpam }

diazotroph [MICROBIO] An organism that carries out nitrogen fixation; examples are *Clostridium* and *Azotobacter*. { dī'az·əˌträf }

Dibamidae [VERT ZOO] The flap-legged skinks, a small family of lizards in the suborder Sauria comprising three species confined to southeastern Asia. { dī'bäm·əˌdē }

***N,N'*-dibenzylethylenediamine** [PHARM] $C_6H_5CH_2NHCH_2CH_2NHCH_2C_6H_5$ An oily liquid, soluble in most organic solvents; used in the manufacture of a repository form of penicillin. { ¦en ¦enˌprīm dī¦ben·zil¦eth·əˌlēn¦dī·əˌmēn }

Dibothriocephalus *See* Diphyllobothrium. { ¦dī·bəˌthrī·ə'sef·ə·ləs }

Dibranchia [INV ZOO] A subclass of the Cephalopoda containing all living cephalopods except *Nautilus*; members possess two gills and, when present, an internal shell. { dī'braŋ·kē·ə }

dicarpellate [BOT] Having two carpels. { dī 'kär·pəˌlāt }

dicaryon *See* dikaryon. { dī'kar·ēˌän }

dicentric [CELL MOL] Having two centromeres. { dī'sen·trik }

dicephaly [MED] A severe congenital anomaly in which the infant is born with two distinct heads. { dī'sef·ə·lē }

dicerous [INV ZOO] Having two tentacles or two antennae. { 'dī·sə·rəs }

Dice's life zones [ECOL] Biomes proposed by L.R. Dice based on the concept of the biotic province. { 'dīs·əz 'līf ˌzōnz }

dichasium [BOT] A cyme producing two main axes from the primary axis or shoot. { dī'kā·zhē·əm }

Dichelesthiidae [INV ZOO] A family of parasitic copepods in the suborder Caligoida; individuals attach to the gills of various fishes. { dī¦ke·ləs'thī·əˌdē }

dichlamydeous [BOT] Having both calyx and corolla. { dī·klə'mid·ē·əs }

dichogamous [BOT] Referring to a type of flower in which the pistils and stamens reach maturity at different times. { dī'käg·ə·məs }

dichogamy [BIOL] Producing mature male and female reproductive structures at different times. { dī'käg·ə·mē }

dichoptous [INV ZOO] Having the margins of the compound eyes separate. { dī'käp·təs }

dichotomous venation [BOT] A vascular arrangement in leaves such that the veins are forked, with each vein dividing at intervals into smaller veins of approximately equal size. { dīˌkäd·ə·məs ve'nā·shən }

dichotomy [BIOL] **1.** Divided in two parts. **2.** Repeated branching or forking. { dī'käd·ə·mē }

dichotriaene [INV ZOO] A type of sponge spicule with three rays. { ¦di·kō'trīˌēn }

dichromatic [BIOL] Having or exhibiting two color phases independently of age or sex. { dī·krə'mād·ik }

dichromatism [MED] Partial color blindness in which vision is apparently based on two primary colors rather than the normal three. { dī'krō·məˌtiz·əm }

Dicksoniaceae [BOT] A family of tree ferns characterized by marginal sori which are terminal on the veins and protected by a bivalved indusium. { ˌdik·sə·nē'ās·ēˌē }

Dick test [IMMUNOL] A skin test to determine immunity to scarlet fever; *Streptococcus pyogenes* toxin is injected intracutaneously and produces a reaction if there is no circulating antitoxin. { 'dik ˌtest }

diclinous [BOT] Having stamens and pistils on different flowers. { dī'klī·nəs }

dicoccous [BOT] Composed of two adherent one-seeded carpels. { dī'käk·əs }

dicotyledon [BOT] Any plant of the class Magnoliopsida, all having two cotyledons. { ˌdīˌkäd·əl'ēd·ən }

Dicotyledoneae [BOT] The equivalent name for Magnoliopsida. { dīˌkäd·əl·ə'dän·ēˌē }

Dicranales [BOT] An order of mosses having erect stems, dichotomous branching, and dense foliation. { ˌdī·krə'nā·lēz }

dicrotic [MED] Pertaining to a secondary pressure wave in an artery on the descending limb of a main wave during diastole of the heart. { dī'kräd·ik }

dictyoblastospore [MYCOL] A blastospore with both cross and longitudinal septa. { ¦dik·tē·ō¦blas·təˌspȯr }

Dictyoceratida [INV ZOO] An order of sponges of the class Demospongiae; includes the bath sponges of commerce. { ¦dik·tē·ō·sə'rad·əd·ə }

Dictyonema pavonium [BOT] A common tropical basidiolichen with lobed thalli and the blue-green Scytonema as photobiont. { ˌdik·tē·əˌnē·mə pə'vō·nē·əm }

dictyosome [CELL MOL] A stack of two or more cisternae; a component of the Golgi apparatus. { 'dik·tē·ə,sōm }

Dictyosporae [MYCOL] A spore group of the imperfect fungi characterized by multicelled spores with cross and longitudinal septae. { ,dik·tē·ə'spȯr·ē }

dictyospore [MYCOL] A multicellular spore in certain fungi characterized by longitudinal walls and cross septa. { 'dik·tē·ə,spȯr }

dictyostele [BOT] A modified siphonostele in which the vascular tissue is dissected into a network of distinct strands; found in certain fern stems. { 'dik·tē·ə,stēl }

Dictyosteliaceae [MICROBIO] A family of microorganisms belonging to the Acrasiales and characterized by strongly differentiated fructifications. { ¦dik·tē·ō,stel·ē'ās·ē,ē }

Dicyemida [INV ZOO] An order of mesozoans comprising minute, wormlike parasites of the renal organs of cephalopod mollusks. { ,dī ,sī'em·ə·də }

didelphic [ANAT] Having a double uterus or genital tract. { dī'del·fik }

Didelphidae [VERT ZOO] The opossums, a family of arboreal mammals in the order Marsupialia. { dī'del·fə,dē }

dideoxy method [CELL MOL] A method of DNA sequencing utilizing chain-terminating (dideoxy) nucleotides. { ,dī·dē'äk·sē ,meth·əd }

Didymelales [BOT] An order of dicotyledonous plants in the subclass Hamamelidae, characterized by the primitive nature of the wood, which has vessels with scalariform perforations, and a pistil, which has one carpel; dioecious, evergreen trees restricted to Madagascar. { ,dī·də·mə'lā·lēz }

Didymiaceae [MICROBIO] A family of slime molds in the order Physarales. { ,dī·də·mī'ās·ē,ē }

Didymosporae [MYCOL] A spore group of the imperfect fungi characterized by two-celled spores. { dī·də·mə'spȯr·ē }

didymous [BIOL] Occurring in pairs. { 'did·ə·məs }

didynamous [BOT] Having four stamens occurring in two pairs, one pair long and the other short. { dī'din·ə·məs }

die [MED] To pass from physical life. { dī }

dieback [ECOL] A large area of exposed, unprotected swamp or marsh deposits resulting from the salinity of a coastal lagoon. { 'dī,bak }

die down [BOT] Normal seasonal death of aboveground parts of herbaceous perennials. { 'dī ,dau̇n }

Diego blood group [IMMUNOL] A genetically determined, immunologically distinct group of human erythrocyte antigens recognized by reaction with a specific antibody. { dē'ā·gō 'bləd ,grüp }

diencephalon [EMBRYO] The posterior division of the embryonic forebrain in vertebrates. { ¦dī·en'sef·ə,län }

diesterase [BIOCHEM] An enzyme such as a nuclease which splits the linkages binding individual nucleotides of a nucleic acid. { dī'es·tə,rās }

diestrus [PHYSIO] The long, quiescent period following ovulation in the estrous cycle in mammals; the stage in which the uterus prepares for the reception of a fertilized ovum. { dī'es·trəs }

diet [BIOL] The food or drink regularly consumed. [MED] Food prescribed, regulated, or restricted as to kind and amount, for therapeutic or other purpose. { 'dī·ət }

dietetics [MED] The science concerned with applying the principle of nutrition to the feeding of people under various economic conditions or for therapeutic purposes. { ,dī·ə'ted·iks }

2,2-diethyl-1,3-propanediol [PHARM] $HOCH_2C(C_2H_5)_2CH_2OH$ Crystals with a melting point of 61–61.6°C; used as a skeletal muscle relaxant. { ¦tü ¦tü ,dī¦eth·əl ¦wən ¦thrē ¦prō,pān'dī,ȯl }

diethylstilbesterol [BIOCHEM] $C_{18}H_{20}O_2$ A white, crystalline, nonsteroid estrogen that is used therapeutically as a substitute for natural estrogenic hormones. Also known as stilbestrol. Abbreviated DES. { ,dī¦eth·əl·stil'bes·tə,rȯl }

dietician [MED] A person trained in dietetics, or the scientific management of meals for individuals or groups. { ,dī·ə'tish·ən }

Dietl's crisis [MED] Recurrent attacks of radiating pain in the costovertebral angle, accompanied by nausea, vomiting, tachycardia, and hypotension, caused by kinking or twisting of the ureter with intermittent obstructive dilation. { 'dēd·əlz ,krī·səs }

differential blood count *See* differential leukocyte count. { ,dif·ə'ren·chəl 'bləd ,kau̇nt }

differential centrifugation [CELL MOL] The separation of mixtures such as cellular particles in a medium at various centrifugal forces to separate particles of different density, size, and shape from each other. { ,dif·ə'ren·chəl ,sen·trə·fə'gā·shən }

differential diagnosis [MED] Distinguishing between diseases of similar character by comparing their signs and symptoms. [SYST] In taxonomic study, a statement of the characters that distinguish a given taxon from other, specifically mentioned equivalent taxa. { ,dif·ə'ren·chəl ,dī·əg'nō·səs }

differential leukocyte count [PATH] The percentage of each variety of leukocytes in the blood, usually based on counting 100 leukocytes. Also known as differential blood count. { ,dif·ə'ren·chəl 'lü·kə,sīt ,kau̇nt }

differentiation antigen [IMMUNOL] A cell surface antigen that is expressed only during a specific period of embryological differentiation. { dif·ə,ren·chē¦ā·shən 'ant·i·jən }

diffuse hypergammaglobulinemia [MED] General increase in serum immunoglobulins due to infection, hepatic disease, collagen diseases, and advanced sarcoidosis. { də'fyüs ¦hī·pər ¦gam·ə¦gläb·yə·lə'nēm·ē·ə }

diffuse placenta [EMBRYO] A placenta having villi diffusely scattered over most of the surface of the chorion; found in whales, horses, and other mammals. { də'fyüs plə'sent·ə }

diffusion respiration [PHYSIO] Exchange of gases through the cell membrane, between the cells of unicellular or other simple organisms and the environment. { də'fyü·zhən res·pə'rā·shən }

digastric [ANAT] Of a muscle, having a fleshy part at each end and a tendinous part in the middle. { dī'gas·trik }

Digenea [INV ZOO] A group of parasitic flatworms or flukes constituting a subclass or order of the class Trematoda and having two types of generations in the life cycle. { dī'jē·nē·ə }

digenesis [BIOL] Sexual and asexual reproduction in succession. { dī'jen·ə·səs }

Di George's syndrome *See* thymic aplasia. { də'jȯrj·əz ˌsinˌdrōm }

digestion [PHYSIO] The process of converting food to an absorbable form by breaking it down to simpler chemical compounds. { də'jes·chən }

digestive efficiency [ECOL] A measure of the amount of ingested chemical energy actually absorbed by an animal. { dī¦jes·tiv i'fish·ən·sē }

digestive enzyme [BIOCHEM] Any enzyme that causes or aids in digestion. { də'jes·tiv 'enˌzīm }

digestive gland [PHYSIO] Any structure that secretes digestive enzymes. { də'jes·tiv ˌgland }

digestive system [ANAT] A system of structures in which food substances are digested. { də'jes·tiv ˌsis·təm }

digestive tract [ANAT] The alimentary canal. { də'jes·tiv ˌtrakt }

digicitrin [BIOCHEM] $C_{21}H_{21}O_{10}$ A flavone compound that is found in foxglove leaves. { ˌdij·ə'si·trən }

digitalis [PHARM] The dried leaf of the purple foxglove plant (*Digitalis purpurea*), containing digitoxin and gitoxin; constitutes a powerful cardiac stimulant and diuretic. { dij·ə'tal·əs }

Digitalis [BOT] A genus of herbs in the figwort family, Scrophulariaceae. { dij·ə'tal·əs }

digital subtraction angiography [MED] A form of digital radiography that delineates blood vessels by subtracting a digitized tissue background image from an image of tissue injected with an intravascular contrast material with a high content of iodine that attenuates x-rays. Abbreviated DSA. { 'dij·əd·əl səb'trak·shən ˌan·jē'äg·rə·fē }

digitate [ANAT] Having digits or digitlike processes. { 'dij·əˌtāt }

digitellum [INV ZOO] A tentacle-like gastric filament in scyphozoans. { ˌdij·ə'tel·əm }

digitigrade [VERT ZOO] Pertaining to animals, such as dogs and cats, which walk on the digits with the posterior part of the foot raised from the ground. { 'dij·ə·dəˌgrād }

digitinervate [BOT] Having straight veins extending from the petiole like fingers. { ¦dij·ə·də'nərˌvāt }

digitipinnate [BOT] Having digitate leaves with pinnate leaflets. { ¦dij·ə·də'pinˌāt }

digitonin [PHARM] $C_{41}H_{64}O_{13}$ A glycoside derived from the purple foxglove plant (*Digitalis purpurea*); a white powder melting at 255–256°C; used as a medicine for cardiac conditions. { ˌdij·ə'tō·nən }

digitus [INV ZOO] In insects, the claw-bearing terminal segment of the tarsus. { 'dij·əd·əs }

diglucoside [BIOCHEM] A compound containing two glucose molecules. { dī'glü·kəˌsīd }

dihydrostreptomycin [MICROBIO] $C_{21}H_{41}O_{12}N_7$ A hydrogenated derivative of streptomycin having the same action as streptomycin. { dī¦hī·drōˌstrep·tə'mī·sən }

dihydroxyacetonephosphoric acid [BIOCHEM] $C_3H_7O_6P$ A phosphoric acid ester of dehydroxyacetone, produced as an intermediate substance in the conversion of glycogen to lactic acid during muscular contraction. { ¦dīˌhī¦dräk·sē¦as·əˌtōn·fäs'fȯr·ik 'as·əd }

dihydroxyphenylalanine [BIOCHEM] $C_9H_{11}NO_4$ An amino acid that can be formed by oxidation of tyrosine; it is converted by a series of biochemical transformations, utilizing the enzyme dopa oxidase, to melanins. Also known as dopa. { ¦dīˌhī¦dräk·sēˌfen·əl'al·əˌnēn }

diisopropyl phosphorofluoridate [PHARM] $[(CH_3)_2CHO]_2FPO$ A colorless, oily liquid that inhibits cholinesterase, prolongs meiosis, and is effective in treating glaucoma. { dī¦īˌsō'prō·pəl ˌfäs·fəˌrō'flu̇r·əˌdāt }

dikaryon [MYCOL] **1.** Also spelled dicaryon. **2.** A pair of distinct, unfused nuclei in the same cell brought together by union of plus and minus hyphae in certain mycelia. **3.** Mycelium containing dikaryotic cells. { dī'kar·ēˌän }

diktoma *See* neuroepithelioma. { dik'tō·mə }

dilator [PHYSIO] Any muscle, instrument, or drug causing dilation of an organ or part. { dī'lād·ər }

dill [BOT] *Anethum graveolens*. A small annual or biennial herb in the family Umbelliferae; the aromatic leaves and seeds are used for food flavoring. { dil }

Dilleniaceae [BOT] A family of dicotyledonous trees, woody vines, and shrubs in the order Dilleniales having hypogynous flowers and mostly entire leaves. { diˌlen·ē'ās·ēˌē }

Dilleniales [BOT] An order of dicotyledonous plants in the subclass Dilleniidae characterized by separate carpels and numerous stamens. { diˌlen·ē'ā·lēz }

Dilleniidae [BOT] A subclass of plants in the class Magnoliopsida distinguished by being syncarpous, having centrifugal stamens, and usually having bitegmic ovules and binucleate pollen. { dil·ə'nī·əˌdē }

dilution gene [GEN] Any modifier gene that acts to reduce the effect of another gene. { də'lü·shən ˌjēn }

dilution method [MICROBIO] A technique in which a series of cultures is tested with various concentrations of an antibiotic to determine the minimum inhibiting concentration of antibiotic. { də'lü·shən ˌmeth·əd }

dimercaprol [PHARM] $C_3H_8OS_2$ 2,3-Dimercapto-1-propanol, a colorless, water-soluble oily liquid with a mercaptanlike odor; used as an antidote for arsenic, gold, and mercury poisoning. Also known as British antilewisite (BAL). { ¦dī·mər'ka,prōl }

dimerous [BIOL] Composed of two parts. { 'di·mər·əs }

dimethicone [PHARM] $CH_3[Si(CH_3)_2O]Si(CH_3)_3$ A colorless oil consisting of dimethylsiloxane polymers; used in ointments and topical drugs. { dī'meth·ə,kōn }

diminution [BOT] Increasing simplification of inflorescences on successive branches. { ,dim·ə'nü·shən }

Dinidoridae [INV ZOO] A family of hemipteran insects in the superfamily Pentatomoidea. { ,dī·nə'dȯr·ə,dē }

dinitrogen fixation *See* nitrogen fixation. { dī'nī·trə·jən fik'sā·shən }

Dinoflagellata [INV ZOO] The equivalent name for Dinoflagellida. { ¦dī·nō,flaj·ə'läd·ə }

Dinoflagellida [INV ZOO] An order of flagellate protozoans in the class Phytamastigophorea; most members have fixed shapes determined by thick covering plates. { ¦dī·nō,flə'jel·ə·də }

dinokaryon [CELL MOL] Nuclear organization peculiar to dinoflagellates and characterized by the absence of a chromosome coiling cycle. { ,dī·nə'kar·ē,än }

Dinophilidae [INV ZOO] A family of annelid worms belonging to the Archiannelida. { ,dī·nō'fil·ə,dē }

Dinophyceae [BOT] The dinoflagellates, a class of thallophytes in the division Pyrrhophyta. { ,dī·nō'fīs·ē,ē }

diocoel [EMBRYO] The cavity of the diencephalon, which becomes the third brain ventricle. { 'dī·ə,sēl }

Dioctophymatida [INV ZOO] An order of parasitic nematode worms in the subclass Enoplia. { dī,äk·tə·fə'mad·ə·də }

Dioctophymoidea [INV ZOO] An order or superfamily of parasitic nematodes characterized by the peculiar structure of the copulatory bursa of the male. { dī,äk·tə·fə'mȯid·ē·ə }

dioecious [BIOL] Having the male and female reproductive organs on different individuals. Also known as dioic. { dī'ē·shəs }

dioic *See* dioecious. { dī'ō·ik }

Diomedeidae [VERT ZOO] The albatrosses, a family of birds in the order Procellariiformes. { ,dī·ə·mə'dī·ə,dē }

Diopsidae [INV ZOO] The stalk-eyed flies, a family of myodarian cyclorrhaphous dipteran insects in the subsection Acalypteratae. { dī'äp·sə,dē }

diorchism [ANAT] Having two testes. { dī'ȯr ,kiz·əm }

Dioscoreaceae [BOT] A family of monocotyledonous, leafy-stemmed, mostly twining plants in the order Liliales, having an inferior ovary and septal nectaries and lacking tendrils. { ,dī·ə,skȯr·ē'ās·ē,ē }

dioxygenase [BIOCHEM] Any of a group of enzymes which catalyze the insertion of both atoms of an oxygen molecule into an organic substrate according to the generalized formula $AH_2 + O_2 \rightarrow A(OH)_2$. { dī'äk·sə·jə,nās }

dipeptidase [BIOCHEM] An enzyme that hydrolyzes a dipeptide. { dī'pep·tə,dās }

diphosphatidyl glycerol *See* cardiolipin. { dī,fäs 'fad·əd·əl 'glis·ə,rȯl }

diphosphopyridine nucleotide [BIOCHEM] $C_{21}H_{27}O_{14}N_7P_2$ An organic coenzyme that functions in enzymatic systems concerned with oxidation-reduction reactions. Abbreviated DPN. Also known as codehydrogenase I; coenzyme I; nicotinamide adenine dinucleotide (NAD). { dī,fäs·fə'pir·ə,dēn 'nü·klē·ə,tīd }

diphtheria [MED] A communicable bacterial disease of humans caused by the growth of *Corynebacterium diphtheriae* on any mucous membrane, especially of the throat. { dif'thir·ē·ə }

diphtheria antitoxin [IMMUNOL] An antibody produced in animals or in humans after contact with diphtheria toxin or toxoid. { dif,thir·ē·ə ,ant·i'täk·sən }

diphtheritic myocarditis [MED] Inflammation of the cardiac muscle arising from local or generalized diphtheria. { ,dif·thə'rid·ik ,mī·ə·kär 'dīd·əs }

diphycercal [VERT ZOO] Pertaining to a tail fin, having symmetrical upper and lower parts, and with the vertebral column extending to the tip without upturning. { ¦dī·fə'ser·kəl }

diphyletic [EVOL] Originating from two lines of descent. { dī·fə'led·ik }

Diphyllidea [INV ZOO] A monogeneric order of tapeworms in the subclass Cestoda; all species live in the intestine of elasmobranch fishes. { dī·fə'lid·ē·ə }

Diphyllobothrium [INV ZOO] A genus of tapeworms; including parasites of humans, dogs, and cats. Formerly known as *Dibothriocephalus*. { dī,fil·ō'bäth·rē·əm }

Diphyllobothrium latum [INV ZOO] A large tapeworm that infects humans, dogs, and cats; causes anemia and disorders of the nervous and digestive systems in humans. { dī,fil·ō'bäth·rē·əm 'lād·əm }

diphyllous [BOT] Having two leaves. { dī'fil·əs }

diphyodont [ANAT] Having two successive sets of teeth, deciduous followed by permanent, as in humans. { dī'fī·ə,dänt }

dipicolinic acid [BIOCHEM] $C_7H_5O_4N \cdot 1\frac{1}{2}H_2O$ A chelating agent composing 5–15% of the dry weight of bacterial spores. { dī¦pik·ə¦lin·ik 'as·əd }

diplacusis [MED] A difference in the pitch perceptions of the two ears when stimulated by the same sound frequency. { dī·plə'kyü·səs }

Diplasiocoela [VERT ZOO] A suborder of amphibians in the order Anura typically having the eighth vertebra biconcave. { di,plā·zē·ō'sē·lə }

diplegia [MED] Paralysis of similar parts on the two sides of the body. { dī'plē·jə }

dipleurula [INV ZOO] **1.** A hypothetical bilaterally symmetrical larva postulated to be an ancestral form of echinoderms and chordates. **2.** Any bilaterally symmetrical, ciliated echinoderm larva. { dī'plu̇r·ə·lə }

diplobiont [BIOL] An organism characterized by alternating, morphologically dissimilar haploid and diploid generations. { ¦dip·lō¦bī,änt }

diploblastic [ZOO] Having two germ layers, referring to embryos and certain lower invertebrates. { ¦dip·lō¦blas·tik }

diploblastula [INV ZOO] A two-layered, flagellated larva of certain ceractinomorph sponges. Also known as parenchymella. { ¦dip·lō¦blas·chə·lə }

diplococci [MICROBIO] A pair of micrococci. { ,dīp·lō'kä·kē *or* 'käk·sī }

Diplogasteroidea [INV ZOO] A superfamily of nematodes in the subclass Diplogasteria, having a stoma of variable and often complex shape, a very distinctive esophagus, and a muscular corpus with well-developed valve. { ,dip·lō,gas·tə'ro̊id·ē·ə }

diploglossate [VERT ZOO] Pertaining to certain lizards, having the ability to retract the end of the tongue into the basal portion. { ¦dip·lō'glä,sāt }

diplohaplont [BIOL] An organism characterized by alternating, morphologically similar haploid and diploid generations. { ¦dip·lō¦hap,länt }

diploid [GEN] Having two complete chromosome pairs in a nucleus (2N). { 'di,plo̊id }

diploidization [GEN] The process by which a tetraploid organism attains the diploid state, involving repeated chromosome loss. { ,di,plo̊id·ə'zā·shən }

diploid merogony [EMBRYO] Development of a part of an egg in which the nucleus is the normal diploid fusion product of egg and sperm nuclei. { 'di,plo̊id mə'räg·ə·nē }

diplolepidious [BOT] Double-scaled, specifically referring to the peristome of mosses with two rows of scales on the outside and one row on the inner. { ¦dip·lō·lə'pid·ē·əs }

Diplomonadida [INV ZOO] An order of small, colorless protozoans in the class Zoomastigophorea, having a bilaterally symmetrical body with four flagella on each side. { ,dip·lō·mə'nad·ə·də }

Diplomystidae [VERT ZOO] A family of catfishes in the suborder Siluroidei confined to the waters of Chile and Argentina. { ,dip·lō'mis·tə,dē }

diplont [BIOL] An organism with diploid somatic cells and haploid gametes. { 'dip,länt }

diplopia [MED] A disorder characterized by double vision. { də'plō·pē·ə }

Diplopoda [INV ZOO] The millipeds, a class of terrestrial tracheate, oviparous arthropods; each body segment except the first few bears two pairs of walking legs. { də'plä·pə·də }

Diplorhina [VERT ZOO] The subclass of the class Agnatha that includes the jawless vertebrates with paired nostrils. { ,dip·lə'rī·nə }

diplosome [CELL MOL] A double centriole. { 'dip·lə,sōm }

diplospondyly [ANAT] Having two centra in one vertebra. { ,dip·lə'spän·də·lē }

diplotene [CELL MOL] The stage of meiotic prophase during which pairs of nonsister chromatids of each bivalent repel each other and are kept from falling apart by the chiasmata. { 'dip·lō,tēn }

Diplura [INV ZOO] An order of small, primarily wingless insects of worldwide distribution. { də'plu̇r·ə }

Dipneumonomorphae [INV ZOO] A suborder of the order Araneida comprising the spiders common in the United States, including grass spiders, hunting spiders, and black widows. { dī,nü·mən·ō'mo̊r·fē }

Dipneusti [VERT ZOO] The equivalent name for Dipnoi. { dip'nü,stī }

Dipnoi [VERT ZOO] The lungfishes, a subclass of the Osteichthyes having lungs that arise from a ventral connection in the gut. { 'dip,no̊i }

Dipodidae [VERT ZOO] The Old World jerboas, a family of mammals in the order Rodentia. { də'päd·ə,dē }

diporpa larva [INV ZOO] A developmental stage of a monogenean trematode. { 'di,po̊r·pə 'lär·və }

Diprionidae [INV ZOO] The conifer sawflies, a family of hymenopteran insects in the superfamily Tenthredinoidea. { ,dip·rē'än·ə,dē }

Diprotodonta [VERT ZOO] A proposed order of marsupial mammals to include the phalangers, wombats, koalas, and kangaroos. { dī,prōd·ə'dän·tə }

Dipsacales [BOT] An order of dicotyledonous herbs and shrubs in the subclass Asteridae characterized by an inferior ovary and usually opposite leaves. { ,dip·sə'kā·lēz }

Dipsocoridae [INV ZOO] A family of hemipteran insects in the superfamily Dipsocoroidea; members are predators on small insects under bark or in rotten wood. { ,dip·sə'ko̊r·ə,dē }

Dipsocoroidea [INV ZOO] A superfamily of minute, ground-inhabiting hemipteran insects belonging to the subdivision Geocorisae. { ,dip·sə·kə'ro̊id·ē·ə }

Diptera [INV ZOO] The true flies, an order of the class Insecta characterized by possessing only two wings and a pair of balancers. { 'dip·tə·rə }

Dipteriformes [VERT ZOO] The single order of the subclass Dipnoi, the lungfishes. { ,dip·tə·rə'fo̊r·mēz }

Dipterocarpaceae [BOT] A family of dicotyledonous plants in the order Theales having mostly stipulate, alternate leaves, a prominently exserted connective, and a calyx that is mostly winged in fruit. { ¦dip·tə·rō,kär'pās·ē,ē }

dipterous [BIOL] **1.** Of, related to, or characteristic of Diptera. **2.** Having two wings or winglike structures. { 'dip·tə·rəs }

direct immunofluorescence [IMMUNOL] The use of labeled reactant to reveal the presence of an unlabeled one. { də'rekt ¦im·yə,nō·flu̇r'es·əns }

directive [INV ZOO] Any of the dorsal and ventral paired mesenteries of certain anthozoan cnidarians. { də'rek·tiv }

direct repeat [CELL MOL] Identical or closely related nucleotide sequences present in two or more copies in the same orientation within the same molecule. { di¦rekt ri'pēt }

disaccharide [BIOCHEM] Any of the class of compound sugars which yield two monosaccharide units upon hydrolysis. { dī'sak·ə,rīd }

disc *See* disk. { disk }

Discellaceae [MYCOL] A family of fungi of the order Sphaeropsidales, including saprophytes and some plant pathogens. { ,dis·ə'lās·ē,ē }

discifloral [BOT] Having flowers with enlarged, disklike receptacles. { ¦dis·kə¦flȯr·əl }

disciform [BIOL] Disk-shaped. { 'dis·kə,fȯrm }

Discinacea [INV ZOO] A family of inarticulate brachiopods in the suborder Acrotretidina. { ,dis·kə'nās·ē·ə }

disclimax [ECOL] A climax community that includes foreign species following a disturbance of the natural climax by humans or domestic animals. Also known as disturbance climax. { dis'klī·maks }

discoaster [BOT] A star-shaped coccolith. { dis'kō·ə·stər }

discoblastula [EMBRYO] A blastula formed by cleavage of a meroblastic egg; the blastoderm is disk-shaped. { ¦dis·kō¦blas·chə·lə }

discocephalous [INV ZOO] Having a sucker on the head. { ¦dis·kō¦sef·ə·ləs }

discoctaster [INV ZOO] A type of spicule with eight rays terminating in discs in hexactinellid sponges. { dis'käk·tə·stər }

discodactylous [VERT ZOO] Having sucking disks on the toes. { ¦dis·kō¦dak·tə·ləs }

discodermolide [PHARM] A polyketide isolated from deep-water sponges of the genus *Discodermia* that is a potent antitumor agent which inhibits the proliferation of cancer cells by interfering with the cell's microtubule network. { ,disk·ə'dər·mə,līd }

discogastrula [EMBRYO] A gastrula formed from a blastoderm. { ¦dis·kō¦gas·trə·lə }

Discoglossidae [VERT ZOO] A family of anuran amphibians in and typical of the suborder Opisthocoela. { ¦dis·kō¦gläs·ə,dē }

discoid [BIOL] **1.** Being flat and circular in form. **2.** Any structure shaped like a disc. { 'dis,kȯid }

discoidal cleavage [EMBRYO] A type of cleavage producing a disc of cells at the animal pole. { dis'kȯid·əl 'klē·vij }

Discolichenes [BOT] The equivalent name for Lecanorales. { ¦dis·kō·lī¦kē·nēz }

Discolomidae [INV ZOO] The tropical log beetles, a family of coleopteran insects in the superfamily Cucujoidea. { ,dis·kō'läm·ə,dē }

Discomycetes [MYCOL] A group of fungi in the class Ascomycetes in which the surface of the fruiting body is exposed during maturation of the spores. { ,dis·kō,mī'sēd·ēz }

discontinuous coding sequence [CELL MOL] The coding sequence in deoxyribonucleic acid of eukaryotic split genes consisting of exons and introns. { ,dis·kən'tin·yə·wəs 'kōd·iŋ ,sē·kwəns }

discopodous [INV ZOO] Having a disk-shaped foot. { di'skäp·ə·dəs }

Discorbacea [INV ZOO] A superfamily of foraminiferan protozoans in the suborder Rotaliina characterized by a radial, perforate, calcite test and a monolamellar septa. { ,dis·kər'bās·ē·ə }

disease [MED] An alteration of the dynamic interaction between an individual and his or her environment which is sufficient to be deleterious to the well-being of the individual and produces signs and symptoms. { di'zēz }

disjunction [CELL MOL] Separation of chromatids or homologous chromosomes during anaphase. { dis'jəŋk·shən }

disjunctor [MYCOL] A small cellulose body between the conidia of certain fungi, which eventually breaks down and thus frees the conidia. { dis'jəŋk·tər }

disk [BIOL] Any of various rounded and flattened animal and plant structures. Also spelled disc. { disk }

disk flower [BOT] One of the flowers on the disk of a composite plant. { 'disk ,flau̇·ər }

dislocation [MED] Displacement of one or more bones of a joint. { ,dis·lō'kā·shən }

disomaty [CELL MOL] Duplication of chromosomes unaccompanied by nuclear division. { dī'sō·məd·ē }

Disomidae [INV ZOO] A family of spioniform annelid worms belonging to the Sedentaria. { də'säm·ə,dē }

disophenol [PHARM] $I_2C_6H_2(NO_2)OH$ Light yellow, feathery crystals with a melting point of 157°C; soluble in alcohol; used as an antihelminthic drug in animals. { də'sä·fə,nȯl }

disorientation [MED] Mental confusion as to one's normal relationship to his or her environment, especially time, place, and people; associated with organic brain disorders. { dis,ȯr·ē·ən'tā·shən }

dispermy [PHYSIO] Entrance of two spermatozoa into an ovum. { 'dī,spər·mē }

dispersal barrier [ECOL] A physical structure that prevents organisms from crossing into new space. { də'spər·səl ,bar·ē·ər }

displacement loop [CELL MOL] In circular deoxyribonucleic acid (DNA), a small region in which ribonucleic acid is paired with one strand of DNA, effectively displacing the other DNA strand. Also known as D-loop. { dis'plās·mənt ,lüp }

dissect [BIOL] To divide, cut, and separate into different parts. { də'sekt }

disseminated necrotizing periarteritis *See* polyarteritis nodosa. { də'sem·ə,nād·əd 'nek·rə,tiz·iŋ ¦per·ē,ärd·ə'rīd·əs }

disseminule [BIOL] An individual organism or part of an individual adapted for the dispersal of a population of the same organisms. { də'sem·ə,nyül }

dissepiment [BOT] A partition which divides a fruit or an ovary into chambers. { də'sep·ə·mənt }

dissociation [MED] Independent, uncoordinated functioning of the atria and ventricles.

[MICROBIO] The appearance of a novel colony type on solid media after one or more subcultures of the microorganism in liquid media. { də,sō·sē'ā·shən }

dissogeny [ZOO] Having two sexually mature stages, larva and adult, in the life of an individual. { də'sä·jə·nē }

distal [BIOL] Located away from the point of origin or attachment. { 'dist·əl }

distal convoluted tubule [ANAT] The portion of the nephron in the vertebrate kidney lying between the loop of Henle and the collecting tubules. { 'dist·əl ,kän·və'lüd·əd 'tü·byül }

distemper [VET MED] Any of several contagious virus diseases of mammals, especially the form occurring in dogs, marked by fever, respiratory inflammation, and destruction of myelinated nerve tissue. { dis'tem·pər }

distichous [BIOL] Occurring in two vertical rows. { 'dis·tə·kəs }

distoclusion [MED] Malocclusion of the teeth in which those of the lower jaw are in distal relation to the upper teeth. { ¦dis·tə¦klü·zhən }

distome [INV ZOO] A digenetic trematode characterized by possession of an oral and a ventral sucker. { 'dī,stōm }

disturbance climax *See* disclimax. { də'stər·bəns 'klī,maks }

disulfiram [PHARM] $C_{10}H_{20}N_2S_4$ A drug used to treat alcohol abuse that blocks the metabolism of acetaldehyde, the major metabolite of ethanol, causing a rapid buildup of acetaldehyde and a severe physiological syndrome intended to prevent or modify further immediate drinking behavior. Also known as Antabuse. { di'səl·fə,ram }

ditokous [VERT ZOO] Producing two eggs or giving birth to two young at one time. { 'did·ə·kəs }

diuresis [MED] Increased excretion of urine. { ,dī·yü'rē·səs }

diuretic [PHARM] Any agent that increases the volume and flow of urine. { ,dī·yü'red·ik }

diuretic hormone [BIOCHEM] A neurohormone that promotes water loss in insects by increasing the volume of fluid secreted into the Malpighian tubules. { ,dī·yü'red·ik 'hȯr,mōn }

diurnal [BIOL] Active during daylight hours. { dī'ərn·əl }

diurnal migration [BIOL] The daily rhythmic movements of organisms in the sea from deeper water to the surface at the approach of darkness and their return to deeper water before dawn. { dī'ərn·əl mī'grā·shən }

divaricate [BIOL] Broadly divergent and spread apart. { dī'var·ə,kāt }

divaricator [ZOO] A muscle that causes separation of parts, as of brachipod shells. { dī'var·ə,kād·ər }

divergent adaptation [EVOL] Adaptation to different kinds of environment that results in divergence from a common ancestral form. Also known as branching adaptation; cladogenic adaptation. { də'vər·jənt ,ad,ap'tā·shən }

divergent transcription [CELL MOL] The initiation of genetic transcription at two promoters that are facing in opposite directions. { də'vər·jənt ,tran'skrip·shən }

diverticulitis [MED] Inflammation of a diverticulum. { ,dī·vər,tik·yə'līd·əs }

diverticulosis [MED] Presence of many diverticula in the intestine. { ,dī·vər,tik·yə'lō·səs }

diverticulum [MED] An abnormal outpocketing or sac on the wall of a hollow organ. { ,dī·vər'tik·yə·ləm }

diving bird [VERT ZOO] Any bird adapted for diving and swimming, including loons, grebes, and divers. { 'dīv·iŋ ,bərd }

division I of meiosis [CELL MOL] The first nuclear division of meiosis, which results in two daughter cells, each containing either the maternal or paternal chromosome from each homologous pair; it is divided into four stages: prophase I (homologous chromosome pairing, synapsis, and crossing over), metaphase I (interlocked homologous chromosomes line up on the middle of the meiotic spindle), anaphase I (homologous chromosome pairs separate and move to opposite poles), and telephase I (nuclear envelope reforms around each daughter nucleus). { də,vizh·ən ¦wən əv mī'ō·səs }

division II of meiosis [CELL MOL] The second division stage of meiosis, in which the daughter cells resulting from division I divide themselves to produce a total of four gametes, each with a haploid number of cells; it is divided into four stages: prophase II (nuclear envelope breaks down and new meiotic spindle forms), metaphase II (chromosomes recondense and align themselves on a new pair of spindles), anaphase II (separation of sister centromeres and movement of the two sister chromatids to opposite poles), and telophase II (the nuclei begin to re-form and the second cell division occurs). { də,vizh·ən ¦tü əv mī'ō·səs }

Dixidae [INV ZOO] A family of orthorrhaphous dipteran insects in the series Nematocera. { 'dik·sə,dē }

dizygotic twins [BIOL] Twins derived from two eggs. Also known as fraternal twins. { ¦dī·zī'gäd·ik 'twinz }

D-loop *See* displacement loop. { 'de ,lüp }

DMC *See* penicillamine.

DMRT 1 gene [GEN] A gene widely required for male sex differentiation, for example in *Drosophila melanogaster* (called dsx), the roundworm *Caenorhabditis elegans* (called mab-3), chickens, and probably humans. { ¦dē¦em¦är¦tē 'wən jēn }

DNA *See* deoxyribonucleic acid.

DNA fingerprint [GEN] Each human individual's virtually unique pattern of deoxyribonucleic acid (DNA) fragment sites produced by restriction enzyme digestion, separated by gel electrophoresis, and hybridized to labeled DNA. { ¦dē¦en¦ā 'fiŋ·gər,print }

DNA fingerprinting *See* genetic fingerprinting. { ,dē,en'ā 'fiŋ·gər,print·iŋ }

DNA microarray [CELL MOL] A microscopic

spot containing identical single-stranded polymeric molecules of deoxyribonucleotides (DNAs), usually oligonucleotides or complementary DNAs, attached to a solid support (such as a membrane, a polymer, or glass) used to simultaneously analyze the expression levels of the corresponding genes. { ¦dē¦en¦ā 'mī·krō·ə,rā }

DNA primase [CELL MOL] An enzyme involved in the initiation of deoxyribonucleic acid (DNA) replication that catalyzes the polymerization of short ribonucleic acid (RNA) primers on the template DNA. { ¦dē¦en¦ā 'prī,mās }

DNase *See* deoxyribonuclease.

DNA sequencing [CELL MOL] The determination of the sequence of nucleotides in deoxyribonucleic acid (DNA) molecules. { ¦dē¦en¦ā 'sē·kwən·siŋ }

DNA vaccine [IMMUNOL] A type of noninfectious vaccine that directly injects deoxyribonucleic plasmids that express antigens of interest, resulting in foreign protein expression within the cells of the vaccine; however, the vaccine itself is unable to replicate. { ¦de¦en¦ā vak'sēn }

Doctrine of Signatures [MED] An archaic concept that a medicinal plant was often stamped with some clear indication (signature) of its specific remedial power; for example, plants with yellow sap were said to cure jaundice. { ¦däk·trən əv 'sig·nə·chərz }

dodecamerous [BOT] Having the whorls of floral parts in multiples of 12. { ¦dō·də¦kam·ə·rəs }

dodo [VERT ZOO] *Raphus calcullatus.* A large, flightless, extinct bird of the family Raphidae. { 'dō,dō }

doe [VERT ZOO] The adult female deer, antelope, goat, rabbit, or any other mammal of which the male is referred to as buck. { dō }

dog [VERT ZOO] Any of various wild and domestic animals identified as *Canis familiaris* in the family Canidae; all are carnivorous and digitigrade, are adapted to running, and have four toes with nonretractable claws on each foot. { dȯg }

dogfish *See* bowfin. { 'dȯg,fish }

Doisy unit [BIOL] A unit for standardization of vitamin K. { 'dȯi·zē ,yü·nət }

dolabriform [BIOL] Shaped like an ax head. { dō'lab·rə,fȯrm }

dolichol [BIOCHEM] Any of a group of long-chain unsaturated isoprenoid alcohols containing up to 84 carbon atoms; found free or phosphorylated in membranes of the endoplasmic reticulum and Golgi apparatus. { 'däl·ə,kȯl }

Dolichopodidae [INV ZOO] The long-legged flies, a family of orthorrhaphous dipteran insects in the series Brachycera. { ,däl·ə·kō'päd·ə,dē }

dolioform [BIOL] Barrel-shaped. { 'dō·lē·ə ,fȯrm }

doliolaria larva [INV ZOO] A free-swimming larval stage of crinoids and holothurians having an apical tuft and four or five bands of cilia. { ,dō·lē·ə'lar·ē·ə 'lär·və }

Doliolida [INV ZOO] An order of pelagic tunicates in the class Thaliacea; transparent forms, partly or wholly ringed by muscular bands. { ,dō·lē'ä·lə·də }

dolphin [VERT ZOO] The common name for about 33 species of cetacean mammals included in the family Delphinidae and characterized by the pronounced beak-shaped mouth. { 'däl·fən }

domestication [BIOL] The adaptation of an animal or plant through breeding in captivity to a life intimately associated with and advantageous to humans. { də,mes·tə'kā·shən }

dominance [ECOL] The influence that a controlling organism has on numerical composition or internal energy dynamics in a community. [GEN] The expression of a heritable trait in the heterozygote such as to make it phenotypically indistinguishable from the homozygote. { 'däm·ə·nəns }

dominant allele [GEN] The member of a pair of alleles which is phenotypically indistinguishable in both the homozygous and heterozygous condition. { 'däm·ə·nənt ə'lēl }

dominant hemisphere [PHYSIO] The cerebral hemisphere which controls certain motor activities; usually the left hemisphere in right-handed individuals. { 'däm·ə·nənt 'hem·ə,sfir }

dominant negative mutation [CELL MOL] Mutation resulting in a gene product that can interfere with the function of the normal gene product in heterozygotes. { ¦däm·ə·nənt ¦neg·əd·iv myü'tā·shən }

dominant species [ECOL] A species of plant or animal that is particularly abundant or controls a major portion of the energy flow in a community. { 'däm·ə·nənt 'spē,shēz }

donation [MYCOL] In conjugation, a process involving a nonconjugative plasmid and a conjugative plasmid in which the latter provides the missing conjugative function to the former so that the former may be transferred. { dō'nā·shən }

donkey [VERT ZOO] A domestic ass (*Equus asinus*); a perissodactyl mammal in the family Equidae. { 'dəŋ·kē }

donor splicing site [CELL MOL] The boundary between the left (5′) end of an intron and the right (3′) end of an exon in messenger ribonucleic acid. Also known as left splicing junction. { 'dō·nər ,splīs·iŋ ,sīt }

Donovan body [MED] The causative microorganism of granuloma inguinale, demonstrated in stained mononuclear cells and characterized by one or two opposite polar chromatin masses. { 'dän·ə·vən ,bäd·ē }

doodlebug [INV ZOO] The larva of an ant lion. { 'düd·əl,bəg }

dopa *See* dihydroxyphenylalanine. { 'dō·pə }

L-dopa *See* levodopa. { ¦el 'dō·pə }

dopamine [BIOCHEM] $C_8H_{11}O_2N$ An intermediate in epinephrine and norepinephrine biosynthesis; the decarboxylation product of dopa. { 'dō·pə,mēn }

dopamine hypothesis [MED] A theory that explains the pathogenesis of schizophrenia and other psychotic states as due to excesses in dopamine activity in various brain areas. { 'dōp·ə,mēn hī,päth·ə·səs }

dopa oxidase [BIOCHEM] An enzyme that catalyzes the oxidation of dihydroxyphenylalanine to melanin; occurs in the skin. { 'dō·pə 'äk·sə,dās }

Doradidae [VERT ZOO] A family of South American catfishes in the suborder Siluroidei. { də'ra·də,dē }

Dorilaidae [INV ZOO] The big-headed flies, a family of cyclorrhaphous dipteran insects in the series Aschiza. { dȯr·ə'lā·ə,dē }

Dorippidae [INV ZOO] The mask crabs, a family of brachyuran decapods in the subsection Oxystomata. { də'rip·ə,dē }

dormancy [BOT] A state of quiescence during the development of many plants characterized by their inability to grow, though continuing their morphological and physiological activities. { 'dȯr·mən·sē }

dormant bud *See* latent bud. { 'dȯr·mənt 'bəd }

dormouse [VERT ZOO] The common name applied to members of the family Gliridae; they are Old World arboreal rodents intermediate between squirrels and rats. { 'dȯr,maùs }

Dorngeholz *See* thornbush. { 'dȯrn·gə,hōlts }

Dorngestrauch *See* thornbush. { 'dorn·gə ,straùk }

dornveld *See* thornbush. { 'dȯrn,felt }

dorsal [ANAT] Located near or on the back of an animal or one of its parts. { 'dȯr·səl }

dorsal aorta [ANAT] The portion of the aorta extending from the left ventricle to the first branch. [INV ZOO] The large, dorsal blood vessel in many invertebrates. { 'dȯr·səl ā'ȯrd·ə }

dorsal column [NEUROSCI] A column situated dorsally in each lateral half of the spinal cord which receives the terminals of some afferent fibers from the dorsal roots of the spinal nerves. { 'dȯr·səl 'käl·əm }

dorsal fin [VERT ZOO] A median longitudinal vertical fin on the dorsal aspect of a fish or other aquatic vertebrate. { 'dȯr·səl 'fin }

dorsalia [INV ZOO] Paired sensory bristles on the dorsal aspect of the head of gnathostomalids. { dȯr'sal·yə }

dorsal lip [EMBRYO] In an amphibian embryo, the margin or lip of the fold of blastula wall marking the dorsal limit of the blastopore during gastrulation and constituting the primary organizer, is necessary to the development of neural tissue, and forms the originating point of chordamesoderm. { 'dȯr·səl 'lip }

dorsiferous [BOT] Of ferns, bearing sori on the back of the frond. [ZOO] Bearing the eggs or young on the back. { dȯr'sif·ə·rəs }

dorsiflex [ZOO] To flex or cause to flex in a dorsal direction. { 'dȯr·sə,fleks }

dorsiflexion sign *See* Homan's sign. { ,dȯr·sə'flek·shən ,sīn }

dorsigrade [VERT ZOO] Walking on the back of the toes. { 'dȯr·sə,grād }

dorsocaudad [ANAT] To or toward the dorsal surface and caudal end of the body. { ¦dȯr·sō¦kȯ,dad }

dorsomedial [ANAT] Located on the back, toward the midline. { ¦dȯr·sō¦mēd·ē·əl }

dorsoposteriad [ANAT] To or toward the dorsal surface and posterior end of the body. { ¦dȯr·sō·pō'stir·ē,ad }

dorsum [ANAT] The entire dorsal surface of the animal body. The upper part of the tongue, opposite the velum. { 'dȯr·səm }

Dorvilleidae [INV ZOO] A family of minute errantian annelids in the superfamily Eunicea. { ,dȯr·və'lē·ə,dē }

Dorylaimoidea [INV ZOO] An order or superfamily of nematodes inhabiting soil and fresh water. { ,dor·ə·lə'mȯid·ē·ə }

Dorylinae [INV ZOO] A subfamily of predacious ants in the family Formicidae, including the army ant (*Eciton hamatum*). { dȯ'rī·lə·nē }

dosage [GEN] The number of copies of a particular gene. [MED] The prescribed or correct amount of medicine or other therapeutic agent administered to treat a given illness. Also known as dose. { 'dō·sij }

dosage compensation [GEN] **1.** A mechanism that equalizes the expression in males and females of genes located on the X chromosome, despite their presence in two doses in the homogametic sex and a single dose in the heterogametic sex. **2.** A mechanism that equalizes the expression of X-linked and autosomal genes by doubling the expression level of X-linked genes in male *Drosophila* and in both male and female mammals with their single active X. { 'dō·sij ,käm·pən,sā·shən }

dose [MED] **1.** The measure, expressed in number of roentgens, of a property of x-rays at a particular place; used in radiology. **2.** *See* dosage. { dōs }

dose fractionation [BIOPHYS] The application of a radiation dose in two or more fractions separated by a certain minimal time interval. { 'dōs ,frak·shə,nā·shən }

double circulation [PHYSIO] A circulatory system in which blood flows through two separate circuits, as pulmonary and systemic. { ¦dəb·əl sər·kyə'lā·shən }

double fertilization [BOT] In most seed plants, fertilization involving fusion between the egg nucleus and one sperm nucleus, and fusion between the other sperm nucleus and the polar nuclei. { ¦dəb·əl ,fərd·əl·ə'zā·shən }

double-loop pattern [FOREN] A whorl type of fingerprint pattern consisting of two separate loop formations and two deltas. { ¦dəb·əl ,lüp 'pad·ərn }

double minute chromosomes [GEN] Chromatin circles that vary in size from tiny dots (common) to the size of a large chromosome, consisting of multiple copies of a short rearranged DNA segment that has undergone amplification. { ¦dəb·əl mī,nyüt 'krō·mə,sōmz }

double-work [BOT] In plant propagation, to graft or bud a scion to an intermediate variety that is itself grafted on a stock of still another variety. { ¦dəb·əl ¦wərk }

doubling dose [GEN] The radiation dose that would double the rate of spontaneous mutation. { ¦dəb·liŋ ¦dōs }

Douglas-fir [BOT] *Pseudotsuga menziesii.* A large coniferous tree in the order Pinales; cones are characterized by bracts extending beyond the scales. Also known as red fir. { ¦dəg·ləs 'fər }

Dounce homogenizer [BIOL] An apparatus consisting of a glass tube with a tight-fitting glass pestle used manually to disrupt tissue suspensions to obtain single cells or subcellular fractions. { 'daůns hə'mäj·ə,nīz·ər }

dove [VERT ZOO] The common name for a number of small birds of the family Columbidae. { dəv }

Down's syndrome [MED] A syndrome of congenital defects, especially mental retardation, typical facies responsible for the term mongolism, and cytogenetic abnormality consisting of trisomy 21 or its equivalent in the form of an unbalanced translocation. Also known as mongolism; trisomy 21 syndrome. { 'daůnz ,sin ,drōm }

downstream [GEN] Further along in the direction of transcription on one strand of the deoxyribonucleic acid sequence of a gene; for a linked gene, downstream may be in the opposite direction along the chromosome. { 'daůn,strēm }

DPN *See* diphosphopyridine nucleotide.

Draconematoidea [INV ZOO] A superfamily of marine nematodes in the order Desmodorida distinguished by a body that, when relaxed, is dorsally and then ventrally arched into a shallow sigmoid shape. { ,drā·kō,nem·ə'tȯid·ē·ə }

Dracunculoidea [INV ZOO] An order or superfamily of parasitic nematodes characterized by their habitat in host tissues and by the way larvae leave the host through a skin lesion. { drə,kəŋ·kyə'lȯid·ē·ə }

dragonfly [INV ZOO] Any of the insects composing six families of the suborder Anisoptera and having four large, membranous wings and compound eyes that provide keen vision. { 'drag·ən,flī }

Drepanidae [INV ZOO] The hooktips, a small family of lepidopteran insects in the suborder Heteroneura. { dre'pan·ə,dē }

dressing [MED] **1.** Application of various materials for protecting a wound and encouraging healing. **2.** Material so applied. { 'dres·iŋ }

Drilidae [INV ZOO] The false firefly beetles, a family of coleopteran insects in the superfamily Cantharoidea. { 'dril·ə,dē }

Drilonematoidea [INV ZOO] A superfamily of parasitic nematodes in the subclass Spiruria. { ,drī·lō,nem·ə'tȯid·ē·ə }

Dromadidae [VERT ZOO] A family of the avian order Charadriiformes containing a single species, the crab plover (*Dromas ardeola*). { drō 'mad·ə,dē }

dromedary [VERT ZOO] *Camelus dromedarius* The Arabian camel, distinguished by a single hump. { 'dräm·ə,der·ē }

Dromiacea [INV ZOO] The dromiid crabs, a subsection of the Brachyura in the crustacean order Decapoda. { ,drō·mē'ā·shē·ə }

Dromiceidae [VERT ZOO] The emus, a monospecific family of flightless birds in the order Casuariiformes. { ,drō·mə'sē·ə,dē }

drone [INV ZOO] A haploid male bee or ant; one of the three castes in a colony. { drōn }

dropfoot [MED] A condition in which the foot drags along the ground and gives the person a characteristic shuffling gait, caused by the failure of the muscle responsible for raising the foot during walking as the leg is swung forward. { 'dräp,fůt }

droplet [MED] A tiny drop of matter consisting of water, mucus, and bacterial products, released through the nasal passages or expectorated. { 'dräp·lət }

droplet infection [MED] Infection by contact with airborne droplets of sputum carrying infectious agents. { 'dräp·let in,fek·shən }

dropsy *See* edema. { 'dräp·sē }

Droseraceae [BOT] A family of dicotyledonous plants in the order Sarraceniales, distinguished by leaves that do not form pitchers, parietal placentation, and several styles. { ,drä·sə'rās·ē,ē }

Drosophilidae [INV ZOO] The vinegar flies, a family of myodarian cyclorrhaphous dipteran insects in the subsection Acalypteratae, including the fruit fly (*Drosophila melanogaster*). { ,drä·sə'fil·ə,dē }

drug [PHARM] **1.** Any substance used internally or externally as a medicine for the treatment, cure, or prevention of a disease. **2.** A narcotic preparation. { drəg }

drug idiosyncrasy [MED] A peculiarity of constitution that makes an individual respond differently to a drug or treatment than do most people. { ¦drəg ,id·ī·ō'siŋ·krə·sē }

drug-induced parkinsonism *See* pseudoparkinsonism. { ¦drəg in¦düst 'pärk·ən·sən,iz·əm }

drug pacemaker [PHARM] A pharmaceutical agent capable of increasing the ventricular rate in a diseased heart. { ¦drəg 'pās,māk·ər }

drug resistance [MICROBIO] A decreased reactivity of living organisms to the injurious actions of certain drugs and chemicals. { ¦drəg ri'zis·təns }

drug sensitivity gene [MED] A gene that encodes enzymes which catalyze the conversion of a prodrug into active anticancer metabolites. { ¦drəg 'sen·si¦tiv·ə,tē ¦gen }

drug tolerance [MED] Condition that may follow repeated ingestion of a drug in so that the effect produced by the original dose no longer occurs. { 'drəg ,tä·lə·rəns }

drupaceous [BOT] Of, pertaining to, or characteristic of a drupe. { drü'pā·shəs }

drupe [BOT] A fruit, such as a cherry, having a thin or leathery exocarp, a fleshy mesocarp, and a single seed with a stony endocarp. Also known as stone fruit. { drüp }

drupelet [BOT] An individual drupe of an aggregate fruit. Also known as grain. { 'drüp·lət }

dry cough [MED] A cough not accompanied by

expectoration. Also known as nonproductive cough; unproductive cough. { ¦drī 'kȯf }

dry gangrene [MED] Local death of a part caused by arterial obstruction without associated venous obstruction or infection. { ¦drī 'gaŋ,grēn }

Dryinidae [INV ZOO] A family of hymenopteran insects in the superfamily Bethyloidea. { drī'in·ə,dē }

Dryomyzidae [INV ZOO] A family of myodarian cyclorrhaphous dipteran insects in the subsection Acalypteratae. { ,drī·ō'mīz·ə,dē }

Dryopidae [INV ZOO] The long-toed water beetles, a family of coleopteran insects in the superfamily Dryopoidea. { drī'äp·ə,dē }

Dryopoidea [INV ZOO] A superfamily of coleopteran insects in the suborder Polyphaga, including the nonpredatory aquatic beetles. { drī·ə'pȯid·ē·ə }

dry rot [MICROBIO] A rapid decay of seasoned timber caused by certain fungi which cause the wood to be reduced to a dry, friable texture. { 'drī ,rät }

dry socket [MED] Inflammation of the dental alveolus, especially the inflamed condition following the removal of a tooth. Also known as alveolitis. { ¦drī 'säk·ət }

DSA *See* digital subtraction angiography.

D_1 trisomy [MED] A syndrome resulting from the presence in triplicate of chromosomes 13–15; manifested in severe congenital anomalies and usually resulting in death in infancy. Also known as trisomy 13–15. { ¦dē ,səb,wən 'trī,sō·mē }

Duchenne's dystrophy [MED] A sex-linked or autosomal recessive form of muscular dystrophy, which is progressive with pseudohypertrophy. { dü'shenz 'dis·trə·fē }

duck [VERT ZOO] The common name for a number of small waterfowl in the family Anatidae, having short legs, a broad flat bill, and a dorsoventrally flattened body. { dək }

duckbill platypus *See* platypus. { ¦dək,bil 'plad·ə·pu̇s }

duck wheat *See* tartary buckwheat. { 'dək ,wēt }

Ducrey test [IMMUNOL] A skin test to determine past or present infection with *Hemophilus ducreyi*. { dü'krā ,test }

duct [ANAT] An enclosed tubular channel for conducting a glandular secretion or other body fluid. { dəkt }

ductless gland *See* endocrine gland. { ¦dək·ləs 'gland }

duct of Cuvier [EMBRYO] Either of the paired common cardinal veins in a vertebrate embryo. { ¦dəkt əv küv'yā }

duct of Santorini [ANAT] The dorsal pancreatic duct in a vertebrate embryo; persists in adult life in some species and serves as the pancreatic duct in the adult elasmobranch, pig, and ox. { ¦dəkt əv san·tə'rē·nē }

ductus arteriosus [EMBRYO] Blood shunt between the pulmonary artery and the aorta of the mammalian embryo. { 'dək·təs ,är·tir·ē'ō·səs }

ductus deferens *See* vas deferens. { 'dək·təs 'def·ə,renz }

ductus venosus [EMBRYO] Blood shunt between the left umbilical vein and the right sinus venosus of the heart in the mammalian embryo. { 'dək·təs ve'nō·səs }

Duffy blood group [IMMUNOL] A genetically determined, immunologically distinct group of human erythrocyte antigens defined by their reaction with anti-Fy* serum. { 'dəf·ē 'bləd ,grüp }

Dugongidae [VERT ZOO] A family of aquatic mammals in the order Sirenia comprising two species, the dugong and the sea cow. { dü 'gän·jə,dē }

Dugonginae [VERT ZOO] The dugongs, a subfamily of sirenian mammals in the family Dugongidae characterized by enlarged, sharply deflected premaxillae and the absence of nasal bones. { dü'gän·jə,nē }

dulse [BOT] Any of several species of red algae of the genus *Rhodymenia* found below the intertidal zone in northern latitudes; an important food plant. { dəls }

dumping syndrome [MED] An imperfectly understood symptom complex of disagreeable or painful epigastric fullness, nausea, weakness, giddiness, sweating, palpitations, and diarrhea, occurring after meals in patients who have gastric surgery which interferes with the function of the pylorus. { 'dəmp·iŋ ,sin,drōm }

duodenal glands *See* Brunner's glands. { dü¦äd·ən·əl 'glanz }

duodenal ulcer [MED] A peptic ulcer occurring in the wall of the duodenum, the first portion of the small intestine. { dü¦äd·ən·əl 'əl·sər }

duodenum [ANAT] The first section of the small intestine of mammals, extending from the pylorus to the jejunum. { dü'äd·ən·əm *or* dü·ə'dē·nəm }

duplex deoxyribonucleic acid [CELL MOL] The deoxyribonucleic acid double helix. { ¦dü,pleks dē¦äk·se,rī·bō·nü¦klē·ik 'as·əd }

duplex uterus [ANAT] A condition in certain primitive mammals, such as rodents and bats, that have two distinct uteri opening separately into the vagina. { ¦dü,pleks 'yüd·ə·rəs }

dura mater [ANAT] The fibrous membrane forming the outermost covering of the brain and spinal cord. Also known as endocranium. { 'du̇r·ə ,mā·dər }

Durham fermentation tube [MICROBIO] A test tube containing lactose or lauryl tryptose and an inverted vial for gas collection; used to test for the presence of coliform bacteria. { 'du̇r·əm ,fər·mən'tā·shən ,tüb }

dwarf [BIOL] Being an atypically small form or variety of something. [MED] An abnormally small individual; especially one whose bodily proportions are altered. { dwȯrf }

dwarfism [MED] Underdevelopment of the body due to surgical removal of the pituitary gland or hyposecretion of growth hormone. { 'dwȯr,fiz·əm }

dwarf mouse unit [BIOL] A unit for the standardization of somatotropin. { ¦dwȯrf 'mau̇s ,yü·nət }

dyad [CELL MOL] Either of the two pair of chromatids produced by separation of a tetrad during the first meiotic division. { 'dī,ad }

dynamic ileus *See* spastic ileus. { dī¦nam·ik 'il·ē·əs }

dynamic work [BIOPHYS] Performance of work by a muscle in which one end of the muscle moves with respect to the other, resulting in external movement. { dī'nam·ik ,wərk }

dynein [CELL MOL] A large enzyme complex that hydrolyzes adenosine triphosphate to provide energy to power retrograde [from (+) to (−)] transport along microtubules. { dī'nē·ən }

dysarthria [MED] Impairment of articulation caused by any disorder or lesion affecting the tongue or speech muscles. { di'sär·thrē·ə }

dysarthrosis [MED] Deformity, dislocation, or disease of a joint. A false joint. { ¦dis·är'thrō·səs }

dysautonomia [MED] **1.** Abnormal functioning of the autonomic nervous system. **2.** A congenital syndrome with aberrations in the autonomic nervous system function, including indifference to pain, diminished secretion of tears, poor vasomotor control, motor incoordination, labile cardiovascular reactions, frequent attacks of bronchial pneumonia, and hypersalivation with aspiration and trouble in swallowing. Also known as Riley-Day syndrome. { ¦dis¦ȯd·ə¦näm·ē·ə }

dysbarism [MED] A condition of the body resulting from the existence of a pressure differential between the total ambient pressure and the total pressure of dissolved and free gases within the body tissues, fluids, and cavities; characteristic symptoms include aeroembolism and abdominal gas pains. { 'dis·bə,riz·əm }

dyschondroplasia *See* enchondromatosis. { di,skän·drō'plā·zhə }

dysentery [MED] Inflammation of the intestine characterized by pain, intense diarrhea, and the passage of mucus and blood. { 'dis·ən,ter·ē }

dysfibrinogenemia [MED] The presence of abnormal fibrinogens in the blood. { ,dis·fī,brin·ō·jə'nē·mē·ə }

dysfunction [MED] Impaired or abnormal functioning of a body part. { ,dis'fəŋk·shən }

dysgammaglobulinemia [MED] A quantitative or qualitative abnormality of serum globulins. { ¦dis,gam·ə,gläb·yə·lə'nē·mē·ə }

dysgerminoma [MED] An ovarian tumor composed of large polygonal cells of germ-cell origin, resembling seminoma of the testis, but less malignant. Also known as embryoma of the ovary. { ¦dis·jər·mə'nō·mə }

dyshidrosis [MED] Any disturbance in sweat production or excretion. { ¦dis·hī'drō·səs }

dyshistogenesis [MED] The morphological result of an abnormal organization of cells into tissues. { ,dis,his·tə'jen·ə·səs }

Dysideidae [INV ZOO] A family of sponges in the order Dictyoceratida. { ,dis·ə'dē·ə,dē }

dyskaryosis [PATH] Any abnormality of the nuclei of exfoliated cells, without significant change in cell integrity. { di,skar·ē'ō·səs }

dyskeratosis [MED] **1.** Imperfect keratinization of individual epidermal cells. **2.** Keratinization of corneal epithelium. { di,sker·ə'tō·səs }

dyskinesia [MED] **1.** Disordered movements of voluntary or involuntary muscles, particularly those seen in disorders of the extrapyramidal system. **2.** Impaired voluntary movements. { ,dis·kə'nē·zhə }

dyslexia [MED] Impairment of the ability to read. { dis'lek·sē·ə }

dyslogia [MED] **1.** Difficulty in the expression of ideas by speech. **2.** Impairment of reasoning or the faculty to think logically. { dis'lō·jē·ə }

dysmenorrhea [MED] Difficult or painful menstruation. { dis,men·ə'rē·ə }

dysostosis [MED] Defective formation of bone. { ¦dis·ä'stō·səs }

dyspareunia [MED] The occurrence of pain during sexual intercourse. { ,dis·pə'rün·ē·ə }

dyspepsia [MED] Disturbed digestion. { dis'pep·sē·ə }

dysphagia [MED] Difficulty in swallowing, or inability to swallow, of organic or psychic causation. { dis'fā·jə }

dysphasia [MED] Partial aphasia due to a brain lesion. { dis'fā·zhə }

dysphonia [MED] An impairment of the voice. { dis'fō·nē·ə }

dysphoria [MED] **1.** The condition of not feeling well or of being ill at ease. **2.** Morbid impatience and restlessness, anxiety, or fidgetiness. { dis'fȯr·ē·ə }

dysplasia [PATH] Abnormal development or growth, especially of cells. { di'splā·zhə }

dyspnea [MED] Difficult or labored breathing. { 'dis·nē·ə }

dysrhythmia [MED] Disordered rhythm of the brain waves. { dis'rith·mē·ə }

dyssebacia [MED] Plugging of the sebaceous glands, especially around the nose, mouth, and forehead, with a dry, yellowish material. { ,di·sə'bā·shə }

dyssomnia [MED] A group of disorders characterized by difficulty in going to sleep or staying asleep or excessive daytime sleepiness. { di'säm·nē·ə }

dysthymia [MED] Any childhood condition caused by malfunction of the thymus. { ,dis'thī·mē·ə }

dystonia [PHYSIO] Disorder or lack of muscle tonicity. { di'stōn·ē·ə }

dystophic [BIOL] Pertaining to an environment that does not supply adequate nutrition. { di'stäf·ik }

dystrophy [MED] **1.** Defective nutrition. **2.** Defective or abnormal development or degeneration. { 'dis·trə·fē }

dysuria [MED] Painful urination. { dis'yu̇r·ē·ə }

Dytiscidae [INV ZOO] The predacious diving beetles, a family of coleopteran insects in the suborder Adephaga. { dī'tis·ə,dē }

E

eager *See* bore. { 'ē·gər }

eagle [VERT ZOO] Any of several large, strong diurnal birds of prey in the family Accipitridae. { 'ē·gəl }

Eagle's medium [MICROBIO] A tissue-culture medium, developed by H. Eagle, containing vitamins, amino acids, inorganic salts and serous enrichments, and dextrose. { 'ē·gəlz ˌmēd·ē·əm }

ear [ANAT] The receptor organ that sends both auditory information and space orientation information to the brain in vertebrates. { ir }

eardrum *See* tympanic membrane. { 'irˌdrəm }

earlobe [ANAT] The pendulous, fleshy lower portion of the auricle or external ear. { 'irˌlōb }

early enzyme [BIOCHEM] Any of the enzymes that are synthesized in a bacterial cell under the direction of an invading bacteriophage. { ¦ər·lē 'enˌzīm }

early gene [GEN] Any gene expressed very soon after a growth stimulus that initiates cell proliferation; it is divided into immediate early and delayed early classes. { 'ər·lē ˌjēn }

earlywood [BOT] The portion of the annual ring that is formed during the early part of a tree's growing season. { 'ər·lēˌwu̇d }

ear shell *See* abalone. { 'ir ˌshel }

earth pig *See* aardvark. { 'ərth ˌpig }

earthstar [MYCOL] A fungus of the genus *Geastrum* that resembles a puffball with a double peridium, the outer layer of which splits into the shape of a star. { 'ərthˌstär }

earthworm [INV ZOO] The common name for certain terrestrial members of the class Oligochaeta, especially forms belonging to the family Lumbricidae. { 'ərthˌwərm }

earwax *See* cerumen. { 'irˌwaks }

earwig [INV ZOO] The common name for members of the insect order Dermaptera. { 'irˌwig }

East African sleeping sickness *See* Rhodesian trypanosomiasis. { ¦ēst ¦af·rə·kən 'slēp·iŋ ˌsik·nəs }

eastern equine encephalitis [MED] A mosquito-borne virus infection of horses and mules in the eastern and southern United States caused by a member of arbovirus group A. { ¦ē·stərn ¦ēˌkwīn enˌsef·ə'līd·əs }

Eaton agent [MICROBIO] The name applied to *Mycoplasma pneumoniae* when it was regarded as a virus. { 'ēt·ən ˌā·jənt }

Eaton agent pneumonia [MED] Pneumonitis in humans, caused by *Mycoplasma pneumoniae*. Also known as primary atypical pneumonia. { 'ēt·ən ˌā·jənt nəˌmōn·yə }

Ebenaceae [BOT] A family of dicotyledonous plants in the order Ebenales, in which a latex system is absent and flowers are mostly unisexual with the styles separate, at least distally. { ˌeb·ə'nās·ēˌē }

Ebenales [BOT] An order of woody, sympetalous dicotyledonous plants in the subclass Dilleniidae, woody having axile placentation and usually twice as many stamens as corolla lobes. { ˌeb·ə'nā·lēz }

ebony [BOT] Any of several African and Asian trees of the genus *Diospyros*, providing a hard, durable wood. { 'eb·ə·nē }

ebracteate [BOT] Without bracts, or much reduced leaves. { ē'brak·tēˌāt }

ebracteolate [BOT] Without bracteoles. { ē 'brak·tē·əˌlāt }

Ebriida [INV ZOO] An order of flagellate protozoans in the class Phytamastigophorea characterized by a solid siliceous skeleton. { ē'brī·ə·də }

ebullism [PHYSIO] The formation of bubbles, especially of water vapor bubbles in biological fluids, owing to reduced ambient pressure. { 'eb·yəˌliz·əm }

ecad [ECOL] A type of plant which is altered by its habitat and possesses nonheritable characteristics. { 'ēˌkad }

eccentric contraction [BIOPHYS] The increase in tension that occurs in a muscle as it lengthens. { ek¦sen·trik kən'trak·shən }

ecchymosis [MED] A subcutaneous hemorrhage marked by purple discoloration of the skin. { ¦ek·ə'mō·səs }

eccrine gland [PHYSIO] One of the small sweat glands distributed all over the human body surface; they are tubular coiled merocrine glands that secrete clear aqueous sweat. { 'ek·rən ˌgland }

ecdysis [INV ZOO] Molting of the outer cuticular layer of the body, as in insects and crustaceans. { 'ek·də·səs }

ecdysone [BIOCHEM] The molting hormone of insects. { 'ek·dəˌsōn }

ecesis [ECOL] Successful naturalization of a

plant or animal population in a new environment. { ə'sē·səs }

ECG *See* electrocardiogram.

Echeneidae [VERT ZOO] The remoras, a family of perciform fishes in the suborder Percoidei. { ,ek·ə'nā·ə,dē }

echidna [VERT ZOO] A spiny anteater; any member of the family Tachyglossidae. { ə'kid·nə }

Echinacea [INV ZOO] A suborder of echinoderms in the order Euechinoidea; individuals have a rigid test, keeled teeth, and branchial slits. { ,ek·ə'nā·shə }

echinate [ZOO] Having a dense covering of spines or bristles. { ə'kī,nāt }

Echinidae [INV ZOO] A family of echinacean echinoderms in the order Echinoida possessing trigeminate or polyporous plates with the pores in a narrow vertical zone. { ə'kī·nə,dē }

Echiniscoidea [INV ZOO] A suborder of tardigrades in the order Heterotardigrada characterized by terminal claws on the legs. { ə,kī·nə'skoid·ē·ə }

echinococcosis [MED] Infestation by the larva (hydatid) of *Echinococcus granulosis* in humans, and in some canines and herbivores. Also known as hydatid disease; hydatidosis. { ə,kī·nə·kä'kō·səs }

Echinococcus [INV ZOO] A genus of tapeworms. { ə,kī·nə'kä·kəs }

echinococcus cyst [INV ZOO] A cyst formed in host tissues by the larva of *Echinococcus granulosus*. { ə,kī·nə'kä·kəs 'sist }

Echinodera [INV ZOO] The equivalent name for Kinorhyncha. { ek·ə'nä·də·rə }

Echinodermata [INV ZOO] A phylum of exclusively marine coelomate animals distinguished from all others by an internal skeleton composed of calcite plates, and a water-vascular system to serve the needs of locomotion, respiration, nutrition, or perception. { ,ek·ə·nə'dər·məd·ə }

Echinoida [INV ZOO] An order of Echinacea with a camarodont lantern, smooth test, and imperforate noncrenulate tubercles. { ,ek·ə'noid·ə }

Echinoidea [INV ZOO] The sea urchins, a class of Echinozoa having a compact body enclosed in a hard shell, or test, formed by regularly arranged plates which bear movable spines. { ,ek·ə'noid·e·ə }

Echinometridae [INV ZOO] A family of echinoderms in the order Echinoida, including polyporous types with either an oblong or a sphericaltest. { ,ek·ə·nō'me·trə,dē }

echinomycin [MICROBIO] $C_{50}H_{60}O_{12}N_{12}S_2$ A toxic polypeptide antibiotic produced by species of *Streptomyces*. { ,ek·ə·nō'mīs·ən }

echinopluteus [INV ZOO] The bilaterally symmetrical larva of sea urchins. { ,ek·ə·nō'plüd·ē·əs }

Echinosteliaceae [MYCOL] A family of slime molds in the order Echinosteliales. { ,ek·ə·nō,ste·lē'ās·ē,ē }

Echinosteliales [MYCOL] An order of slime molds in the subclass Myxogastromycetidae. { ,ek·ə·nō,ste·lē'ā·lēz }

echinostome cercaria [INV ZOO] A digenetic trematode larva characterized by the large anterior acetabulum and a collar with spines. { ə'kī·nə,stōm sər'kar·ē·ə }

Echinothuriidae [INV ZOO] A family of deep-water echinoderms in the order Echinothurioida in which the large, flexible test collapses into a disk at atmospheric pressure. { ,ek·ə·nō·thə'rī·ə,dē }

Echinothurioida [INV ZOO] An order of echinoderms in the superorder Diadematacea with solid or hollow primary radioles, diademoid ambulacral plates, noncrenulate tubercles, and the anus within the apical system. { ,ek·ə·nō·thu̇·rē'ȯid·ə }

Echinozoa [INV ZOO] A subphylum of free-living echinoderms having the body essentially globoid with meridional symmetry and lacking appendages. { ,ek·ə·nō'zō·ə }

Echiurida [INV ZOO] A small group of wormlike organisms regarded as a separate phylum of the animal kingdom; members have a saclike or sausage-shaped body with an anterior, detachable prostomium. { ,ek·ē'yu̇r·ə·də }

Echiuridae [INV ZOO] A small family of the order Echiuroinea characterized by a flaplike prostomium. { ,ek·ē'yu̇r·ə,dē }

Echiuroidea [INV ZOO] A phylum of schizocoelous animals. { ,ek·ē·yə'rȯid·ē·ə }

Echiuroinea [INV ZOO] An order of the Echiurida. { ,ek·ē·yə'rȯi·nē·ə }

echocardiography [MED] A diagnostic technique for the heart that uses a transducer held against the chest to send high-frequency sound waves which pass harmlessly into the heart; as they strike structures within the heart, they are reflected back to the transducer and recorded on an oscilloscope. { ,ek·ō,kärd·ē'äg·rə·fē }

echoencephalograph [MED] An instrument that uses ultrasonic pulses and echo-ranging techniques to give a pictorial representation of intracranial structure. Also known as sonoencephalograph. { ¦ek·ō,en'sef·lə,graf }

echogram [MED] The pictorial display of anatomical structures using pulse-echo techniques. { 'ek·ō,gram }

echolalia [MED] The purposeless, often seemingly involuntary repetition of words spoken by another person; a disorder seen in certain psychotic states and in certain organic brain syndromes. Also known as echophrasia. { ,ek·ō'lā·lē·ə }

echolic [MED] Producing abortion or accelerating labor. { e'käl·ik }

echolocation [BIOPHYS] An animal's use of sound reflections to localize objects and to orient in the environment. { 'ek·ō·lō,kā·shən }

echophrasia *See* echolalia. { ,ek·ō'frā·zhə }

echo ranging [VERT ZOO] An auditory feedback mechanism in bats, porpoises, seals, and certain other animals whereby reflected ultrasonic sounds are utilized in orientation. { 'ek·ō ,rānj·iŋ }

echouterograph [MED] An instrument that

uses ultrasonic pulses and echo-ranging techniques to give a pictorial representation of the uterus. { ˌek·ō'yüd·ə·rəˌgraf }

echovirus [VIROL] Any member of the Picornaviridae family, genus *Enterovirus*; the name is derived from the group designation enteric cytopathogenic human orphan virus. { 'ek·ōˌvī·rəs }

echylosis [CELL MOL] The release of nonparticulate material from the cell through an apparently intact cell membrane. { ē'kī·lə·səs }

eclampsia [MED] A disorder occurring during the latter half of pregnancy, characterized by elevated blood pressure, edema, proteinuria, and convulsions or coma. { i'klam·sē·ə }

eclipsed antigen [IMMUNOL] An antigenic determinant of parasitic origin resembling an antigenic determinant of the parasite's host to such a degree that it does not elicit the formation of antibody by the host. { i¦klipst 'ant·i·jen }

eclipse period [VIROL] A phase in the proliferation of viral particles during which the virus cannot be detected in the host cell. { i'klips ˌpir·ē·əd }

eclosion [INV ZOO] The process of an insect hatching from its egg. { ē'klō·zhən }

ECM *See* extracellular matrix.

ecmnesia [MED] A lapse in memory of recent events, with normal memory for earlier events. { ek'nē·zhə }

ecocline [ECOL] A genetic gradient of adaptability to an environmental gradient; formed by the merger of ecotypes. { 'ek·ōˌklīn }

ecological association [ECOL] A complex of communities, such as an elm-hackberry association, which develops in accord with variations in physiography, soil, and successional history within the major subdivision of a biotic realm. { ek·ə'läj·ə·kəl əˌsō·shē'ā·shən }

ecological climatology [BIOL] A branch of bioclimatology, including the physiological adaptation of plants and animals to their climate, and the geographical distribution of plants and animals in relation to climate. { ek·ə'läj·ə·kəl klī·mə'täl·ə·jē }

ecological community *See* community. { ek·ə'läj·ə·kəl kə'myün·əd·ē }

ecological energetics [ECOL] The study of the flow of energy within an ecological system from the time the energy enters the living system until it is ultimately degraded to heat and irretrievably lost from the system. Also known as production ecology. { ˌek·əˌläj·ə·kəl ˌen·ər'jed·iks }

ecological interaction [ECOL] The relation between species that live together in a community; specifically, the effect an individual of one species may exert on an individual of another species. { ek·ə'läj·ə·kəl in·tər'ak·shən }

ecological modeling [ECOL] The conceptualization and implementation of computer simulations of the behavior of living systems. { ˌek·əˌläj·ə·kəl 'mäd·əl·iŋ }

ecological physiology [BIOL] The science of the interrelationships between the physiology of organisms and their environment. { ˌē·kə¦läj·ə·kəl fiz·ē'äl·ə·jē }

ecological pyramid [ECOL] A pyramid-shaped diagram representing quantitatively the numbers of organisms, energy relationships, and biomass of an ecosystem; numbers are high for the lowest trophic levels (plants) and low for the highest trophic level (carnivores). { ek·ə'läj·ə·kəl 'pir·ə·mid }

ecological succession [ECOL] A gradual process incurred by the change in the number of individuals of each species of a community and by establishment of new species populations that may gradually replace the original inhabitants. { ek·ə'läj·ə·kəl sək'sesh·ən }

ecological system *See* ecosystem. { ek·ə'läj·ə·kəl 'sis·təm }

ecological zoogeography [ECOL] The study of animal distributions in terms of their environments. { ˌek·əˌläj·ə·kəl ˌzō·ō·jē'äg·rə·fē }

ecology [BIOL] A study of the interrelationships which exist between organisms and their environment. Also known as bionomics; environmental biology. { ē'käl·ə·jē }

economic entomology [BIOL] A branch of entomology concerned with the study of economic losses of commercially important animals and plants due to insect predation. [ECOL] The study of insects that have a direct influence on humanity, with an emphasis on pest management. { ˌek·ə'näm·ik ˌen·tə'mäl·ə·jē }

ecophene [GEN] The range of phenotypic modifications produced by one genotype within the limits of the habitat under which the genotype is found in nature. { 'ē·kəˌfēn }

ecophenotype [ECOL] A nongenetic phenotypic modification in response to environmental conditions. { ˌē·kō'phēn·əˌtīp }

ecospecies [ECOL] A group of ecotypes capable of interbreeding without loss of fertility or vigor in the offspring. { 'ē·kōˌspē·shēz }

ecosystem [ECOL] A functional system which includes the organisms of a natural community together with their environment. Derived from ecological system. { 'ek·oˌsis·təm *or* 'ē·kōˌsis·təm }

ecosystem mapping [ECOL] The drawing of maps that locate different ecosystems in a geographic area. { 'ek·oˌsis·təm ˌmap·iŋ }

ecotone [ECOL] A zone of intergradation between ecological communities. { 'ek·əˌtōn }

ecotrine [ECOL] A metabolite produced by one kind of organism and utilized by another. { 'ek·əˌtrēn }

ecotype [ECOL] A subunit, race, or variety of a plant ecospecies that is restricted to one habitat; equivalent to a taxonomic subspecies. { 'ek·əˌtīp }

ecstasy [MED] A trancelike state with loss of sensory perception and voluntary control. { 'ek·stə·sē }

ectasia [MED] Dilation, especially of a hollow organ. { ek'tā·zhə }

Ecterocoelia [INV ZOO] The equivalent name for Protostomia. { ˌek·tə·rō'sēl·yə }

ectethmoid [ANAT] Either one of the lateral cellular masses of the ethmoid bone. { ek 'teth,mȯid }

ecthyma [MED] An inflammatory skin disease characterized by large flat pustules that ulcerate and become crusted, and are surrounded by a distinct inflammatory areola. { 'ek·thə·mə }

ectocardia [MED] An abnormal position of the heart; it may be outside the thoracic cavity or misplaced within the thorax. { ¦ek·tō'kärd·ē·ə }

ectocommensal [ECOL] An organism living on the outer surface of the body of another organism, without affecting its host. { ¦ek·tō·kə'men·səl }

ectocornea [ANAT] The outer layer of the cornea. { ,ek·tō'kōr·nē·ə }

ectocyst [INV ZOO] **1.** The outer layer of the wall of a zooecium. **2.** *See* epicyst. { 'ek·tə,sist }

ectoderm [EMBRYO] The outer germ layer of an animal embryo. Also known as epiblast. [INV ZOO] The outer layer of a diploblastic animal. { 'ek·tə,dərm }

ectoenzyme [BIOCHEM] An enzyme which is located on the external surface of a cell. { ¦ek·tō'en,zīm }

ectogenesis [EMBRYO] Development of an embryo or of embryonic tissue outside the body in an artificial environment. { ,ek·tō'jen·ə·səs }

ectogony [BOT] The influence of pollination and fertilization on structures outside the embryo and endosperm; effect may be on color, chemical composition, ripening, or abscission. { ek'täg·ə·nē }

ectomere [EMBRYO] A blastomere that will differentiate into ectoderm. { 'ek·tə,mir }

ectomesoblast [EMBRYO] An undifferentiated layer of embryonic cells from which arises the epiblast and mesoblast. { ¦ek·tō'me·zō,blast }

ectomesoderm [EMBRYO] Mesoderm which is derived from ectoderm and is always mesenchymal; a type of primitive connective tissue. { ¦ek·tō'me·zō,dərm }

ectomycorrhizae [ECOL] A type of mycorrhizae composed of a fungus sheath around the outside of root tips, with individual hyphae penetrating between the cortical cells of the root to absorb photosynthates. { ,ek·tō·mī'kȯr·ə,zī }

ectoparasite [ECOL] A parasite that lives on the exterior of its host. { ¦ek·tō'par·ə,sīt }

ectophagous [INV ZOO] The larval stage of a parasitic insect which is in the process of development externally on a host. { ek'täf·ə·gəs }

ectophloic siphonostele [BOT] A type of stele with pith that has the phloem only on the outside of the xylem. { ,ek·tə'flō·ək sī'fän·ə,stēl }

ectophyte [ECOL] A plant which lives externally on another organism. { 'ek·tə,fīt }

ectopia [MED] A congenital or acquired positional abnormality of an organ or other part of thebody. { ek'tōp·ē·ə }

ectopic expression [GEN] Phenotypic expression of a gene in a type of cell or tissue in which it is usually inactive. { ek¦täp·ik iks'presh·ən }

ectopic pairing [CELL MOL] Pairing between nonhomologous segments of the salivary gland chromosomes in *Drosophila*, presumably involving mainly heterochromatic regions. { ek'täp·ik 'per·iŋ }

ectopic pregnancy [MED] Embryonic development outside the uterus, usually within the Fallopian tube. { ek¦täp·ik 'preg·nən·sē }

ectoplasm [CELL MOL] The outer, gelled zone of the cytoplasmic ground substance in many cells. Also known as ectosarc. { 'ek·tə,plaz·əm }

Ectoprocta [INV ZOO] A subphylum of colonial bryozoans having eucoelomate visceral cavities and the anus opening outside the circlet of tentacles. { ek·tō'präk·tə }

ectopterygoid [VERT ZOO] A membrane bone located ventrally on the skull, situated behind the palate and extending to the quadrate; found in some fishes and reptiles. { ¦ek·tō'ter·ə,gȯid }

ectosarc *See* ectoplasm. { 'ek·tə,särk }

ectosome [INV ZOO] The outer, cortical layer of a sponge. { 'ek·tə,sōm }

ectostosis [PHYSIO] Formation of bone immediately beneath the perichondrium and surrounding and replacing underlying cartilage. { ¦ek·tə'stō·səs }

ectosymbiont [ECOL] A symbiont that lives on the surface of or is physically separated from its host. { ¦ek·tō'sim·bē,änt }

ectotherm [PHYSIO] An animal that obtains most of its heat from the environment and therefore has a body temperature very close to that of its environment. { 'ek·tə,thərm }

Ectothiorhodospira [MICROBIO] A genus of bacteria in the family Chromatiaceae; cells are spiral to slightly bent rods, are motile, contain bacteriochlorophyll *a* on lamellar stock membranes, and produce and deposit sulfur as globules outside the cells. { ¦ek·tō,thī·ə,rō'däs·pə·rə }

ectotrophic [BIOL] Obtaining nourishment from outside; applied to certain parasitic fungi that live on and surround the roots of the host plant. { ¦ek·tə'träf·ik }

ectozoa [ECOL] Animals which live externally on other organisms. { ,ek·tə'zō·ə }

Ectrephidae [INV ZOO] An equivalent name for Ptinidae. { ek'tref·ə,dē }

ectrodactylia [MED] Congenital absence of any of the fingers or toes or parts of them. { ,ek·trō·dak'til·ē·ə }

ectromelia [MED] A congenital absence or an anomaly of one or more limbs. { ,ek·tra'mē·lē·ə }

ectromelia virus [VIROL] A member of subgroup I of the poxvirus group; causes mousepox. { ,ek·tra'mē·lē·ə 'vī·rəs }

eczema [MED] Any skin disorder characterized by redness, thickening, oozing from blisters or papules, and occasional formation of fissures and crusts. { 'ek·sə·mə }

eczematoid reaction [MED] A dermal and epidermal inflammatory response characterized by erythema, edema, vesiculation, and exudation in the acute stage, and by erythema, edema, thickening of the epidermis, and scaling in the chronic stage. { ek'zē·mə,tȯid rē,ak·shən }

ED_{50} *See* effective dose 50.

edaphic community [ECOL] A plant community that results from or is influenced by soil factors such as salinity and drainage. { ē'daf·ik kə'myün·əd·ē }

edaphon [BIOL] Flora and fauna in soils. { 'ed·ə,fän }

edema [MED] An excessive accumulation of fluid in the cells, tissue spaces, or body cavities due to a disturbance in the fluid exchange mechanism. Also known as dropsy. { ə'dē·mə }

Edentata [VERT ZOO] An order of mammals characterized by the absence of teeth or the presence of simple prismatic, unspecialized teeth with no enamel. { ,ē,den'tä·də }

edentate [VERT ZOO] **1.** Lacking teeth. **2.** Any member of the Edentata. { ē'den,tāt }

edentulous [VERT ZOO] Having no teeth; especially, having lost teeth that were present. { ē'den·chə·ləs }

edge effect [ECOL] The influence of adjacent plant communities on the number of animal species present in the direct vicinity. { 'ej i,fekt }

editing *See* proofreading. { 'ed·əd·iŋ }

Edman degradation technique [BIOCHEM] In protein analysis, an approach to amino-end-group determination involving the use of a reagent, phenylisothiocyanate, that can be applied to the liberation of a derivative of the amino-terminal residue without hydrolysis of the remainder of the peptide chain. { 'ed·mən ,deg·rə'dā·shən tek,nēk }

eduction [CELL MOL] Loss of host genetic material when the plasmid that had been integrated into the host chromosome exits. { ē'dək·shən }

Edwardsiella [MICROBIO] A genus of bacteria in the family Enterobacteriaceae; motile rods that produce hydrogen sulfide from TSI agar. { e¦dwärd·zē'el·ə }

Edwards' syndrome *See* trisomy 18 syndrome. { 'ed·wərdz,sin,drōm }

EEG *See* electroencephalogram.

eel [VERT ZOO] The common name for a number of unrelated fishes included in the orders Anguilliformes and Cypriniformes; all have an elongate, serpentine body. { ēl }

eel grass *See* tape grass. { ēl ,gras }

EEP *See* electroencephalophone.

effective dose 50 [PHARM] The amount of a drug required to produce a response in 50% of the subjects to whom the drug is given. Abbreviated ED_{50}. Also known as median effective dose. { ə¦fek·tiv ¦dōs 'fif·tē }

effective lethal phase [GEN] The developmental stage at which a lethal gene generally causes death of the organism carrying it. { ə¦fek·tiv ¦lē·thəl 'fāz }

effector [BIOCHEM] Anactivator of an allosteric enzyme. [PHYSIO] A structure that is sensitive to a stimulus and causes an organism or part of an organism to react to the stimulus, either positively or negatively. { ə'fek·tər }

effector organ [PHYSIO] Any muscle or gland that mediates overt behavior, that is, movement or secretion. { ə'fek·tər ,ȯr·gən }

effector system [PHYSIO] A system of effector organs in the animal body. { ə,fek·tər ,sis·təm }

efferent [PHYSIO] Carrying or conducting away, as the duct of an exocrine gland or a nerve. { 'ef·ə·rənt }

efflorescence [BOT] The period or process of flowering. { ,ef·lə'res·əns }

efflux pump [CELL MOL] An active transport system for the removal of some antibiotics (such as tetracyclines, macrolides, and quinolones) from bacterial cells. { 'ē·fləks ,pəmp }

effuse [BOT] Expanded; spread out in a definite form. { e'fyüz }

effusion [MED] A pouring out of any fluid into a body cavity or tissue. { e'fyü·zhən }

egest [PHYSIO] **1.** To discharge indigestible matter from the digestive tract. **2.** To rid the body of waste. { ē'jest }

egg [CELL MOL] **1.** A large, female sex cell enclosed in a porous, calcareous or leathery shell, produced by birds and reptiles. **2.** *See* ovum. { eg }

egg apparatus [BOT] A group of three cells, consisting of the egg and two synergid cells, in the micropylar end of the embryo sac in seed plants. { 'eg ,ap·ə,rad·əs }

egg capsule *See* egg case. { 'eg ,kap·səl }

egg case [INV ZOO] **1.** A protective capsule containing the eggs of certain insects and mollusks. Also known as egg capsule. **2.** A silk pouch in which certain spiders carry their eggs. Also known as egg sac. [VERT ZOO] A soft, gelatinous (amphibians) or strong, horny (skates) envelope containing the egg of certain vertebrates. { 'eg ,kās }

eggplant [BOT] *Solanum melongena.* A plant of the order Polemoniales grown for its edible egg-shaped, fleshy fruit. { 'eg,plant }

egg raft [ZOO] A floating mass of eggs; produced by a variety of aquatic organisms. { 'eg ,raft }

egg sac [ZOO] **1.** The structure containing the eggs of certain microcrustaceans. **2.** *See* egg case. { 'eg ,sak }

egg tooth [VERT ZOO] A toothlike prominence on the tip of the beak of a bird embryo and the tip of the nose of an oviparous reptile, which is used to break the eggshell. { 'eg ,tüth }

eglandular [BIOL] Without glands. { ē'glan·dyə·lər }

Egyptian cotton [BOT] Long-staple, high-quality cotton grown in Egypt. { i'jip·shən 'kät·ən }

Egyptian henna *See* henna. { i'jip·shən 'hen·ə }

Ehrlichia [MICROBIO] A genus of the tribe Ehrlichieae; coccoid to ellipsoidal or pleomorphic cells; intracellular parasites in cytoplasm of host leukocytes. { er'lik·ē·ə }

Ehrlichieae [MICROBIO] A tribe of the family Rickettsiaceae; spherical and occasionally pleomorphic cells; pathogenic for some mammals, not including humans. { ,er·lə'kī·ē,ē }

ehrlichiosis [MED] A tick-borne bacterial infection caused by two distinct *Ehrlichia* species that infect white blood cells; the infection may be asymptomatic, but it also can produce illness ranging from a few mild symptoms to an overwhelming multisystem disease. { är,lik·ē'ō·səs }

Ehrlich's 606 *See* arsphenamine. { 'er·liks ¦sik·sō'siks }

eicosanoid [BIOCHEM] Any member of a group of naturally occurring substances composed of prostaglandins, thromboxanes, and leukotrienes that are derived from polyunsaturated fatty acids, particularly arachidonic acid, and exhibit various types of biological activity. { ī'käs·ə,nȯid }

Eimeriina [INV ZOO] A suborder of coccidian protozoans in the order Eucoccida in which there is no syzygy and the microgametocytes produce a large number of microgametes. { ,ī·mə'rī·ə·nə }

ejaculation [PHYSIO] The act or process of suddenly discharging a fluid from the body; specifically, the ejection of semen during orgasm. { i,jak·yə'lā·shən }

ejaculatory duct [ANAT] The terminal part of the ductus deferens after junction with the duct of a seminal vesicle, embedded in the prostate gland and opening into the urethra. { i'jak·yə·lə,tȯr·ē 'dəkt }

ejaculatory incompetence [MED] Inability of a male to reach orgasm and ejaculate during sexual intercourse despite adequacy of erection. { i¦jak·yə·lə,tȯr·ē in'käm·pəd·əns }

ejecta [PHYSIO] Excrement. { ē'jek·tə }

EKG *See* electrocardiogram.

Elaeagnaceae [BOT] A family of dicotyledonous plants in the order Proteales, noted for peltate leaf scales which often give the leaves a silvery-gray appearance. { ,el·ē·ag'nās·ē,ē }

elaioplast [HISTOL] An oil-secreting leucoplast. { ə'lī·ə,plast }

Elaphomycetaceae [MYCOL] A family of underground, saprophytic or mycorrhiza-forming fungi in the order Eurotiales characterized by ascocarps with thick, usually woody walls. { ,el·ə·fō,mī·sə'tās·ē,ē }

Elapidae [VERT ZOO] A family of poisonous reptiles, including cobras, kraits, mambas, and coral snakes; all have a pteroglyph fang arrangement. { ə'lap·ə,dē }

Elasipodida [INV ZOO] An order of deep-sea aspidochirotacean holothurians in which there are no respiratory trees and bilateral symmetry is often quite conspicuous. { ə,laz·ə'päd·ə·də }

Elasmidae [INV ZOO] A family of hymenopteran insects in the superfamily Chalcidoidea. { ə'laz·mə,dē }

Elasmobranchii [VERT ZOO] The sharks and rays, a subclass of the class Chondrichthyes distinguished by separate gill openings, amphistylic or hyostylic jaw suspension, and ampullae of Lorenzini in the head region. { ə,laz·mə'braŋ·kē,ī }

Elassomatidae [VERT ZOO] The pygmy sunfishes, a family of the order Perciformes. { ə,las·ō'mad·ə,dē }

elastase [BIOCHEM] An enzyme which acts on elastin to change it chemically and render it soluble. { i'la,stās }

elastic cartilage [HISTOL] A type of cartilage containing elastic fibers in the matrix. { i'las·tik 'kärt·lij }

elastic fiber [HISTOL] A homogeneous, fibrillar connective tissue component that is highly refractile and appears yellowish when arranged in bundles. { i'las·tik 'fī·bər }

elastic tissue [HISTOL] A type of connective tissue having a predominance of elastic fibers, bands, or lamellae. { i'las·tik 'tish·ü }

elastin [BIOCHEM] An elastic protein composing the principal component of elastic fibers. { i'las·tən }

elastosis [MED] **1.** Retrogressive change in elastic tissue. **2.** Retrogressive change in cutaneous connective tissue resulting in excessive amounts of material which give the staining reactions for elastin. { i,la'stō·səs }

elater [BOT] A spiral, filamentous structure that functions in the dispersion of spores in certain plants, such as liverworts and slime molds. { 'el·ə·tər }

Elateridae [INV ZOO] The click beetles, a large family of coleopteran insects in the superfamily Elateroidea; many have light-producing organs. { ,el·ə'ter·ə,dē }

Elateroidea [INV ZOO] A superfamily of coleopteran insects in the suborder Polyphaga. { i,lad·ə'rȯid·ē·ə }

elaterophore [BOT] A tissue bearing elaters, found in some liverworts. { i'lad·ə·rə,fȯr }

elbow [ANAT] The arm joint formed at the junction of the humerus, radius, and ulna. { 'el,bō }

elective culture [MICROBIO] A type of microorganism grown selectively from a mixed culture by culturing in a medium and under conditions selective for only one type of organism. { i¦lek·tiv 'kəl·chər }

electrical synapse [NEUROSCI] An anatomically specialized junction between two nerve cells at which one cell influences the other by means of electrical current from one flowing directly into the other. { i'lek·trə·kəl 'si,naps }

electric anesthesia [MED] Anesthesia produced by electrical means, as with interrupted direct current. { i'lek·trik ,an·əs'thē·zhə }

electric burn [MED] A burn caused by electric current. { i¦lek·trik 'bərn }

electric eel [VERT ZOO] *Electrophorus electricus.* An eellike cypriniform electric fish of the family Gymnotidae. { i¦lek·trik 'ēl }

electric fish [VERT ZOO] Any of several fishes capable of producing electric discharges from an electric organ. { i¦lek·trik 'fish }

electric organ [VERT ZOO] An organ consisting of rows of electroplaques which produce an electric discharge. { i¦lek·trik 'ȯr·gən }

electric shock [PHYSIO] The sudden pain, convulsion, unconsciousness, or death produced by the passage of electric current through the body. { i¦lek·trik 'shäk }

electrocardiogram [MED] A graphic recording

of the electrical manifestations of the heart action as obtained from the body surfaces. Abbreviated ECG; EKG. { i,lek·trō'kärd·ē·ə,gram }

electrocardiograph [MED] The instrument used to obtain anelectrocardiogram. { i,lek·trō'kärd·ē·ə,graf }

electrocardiography [MED] The medical specialty concerned with the production and interpretation of electrocardiograms. { i¦lek·trō ,kärd·ē'äg·rə·fē }

electrocardiophonograph [MED] An instrument that records graphic traces of heart sounds to fix precisely the times at which valve action occurs and to reveal valvular defects which affect blood flow. { i¦lek·trō¦kärd·ē·ō'fō·nə,graf }

electrocauterization [MED] The application of a direct galvanic current to tissues to cause destruction or coagulation. { i,lek·trō,kȯd·ə·rə 'zā·shən }

electrochemical gradient [CELL MOL] The combined effect of a solute's concentration gradient across a membrane and the electric charge gradient across the membrane (the membrane potential), which drives the solute to cross the membrane. { i,lek·trō,kem·i·kəl 'grād·ē·ənt }

electrochemical proton gradient [CELL MOL] The combined effect of the electric charge gradient across a membrane and the pH gradient across the membrane (the membrane potential), which drives H^+ and OH^- ions to cross the membrane. { i,lek·trō,kem·i·kəl 'prō,tän ,grād·ē·ənt }

electrocoagulation [MED] The coagulation of tissue by means of a high-frequency electric current. { i¦lek·trō·kō,ag·yə'lā·shən }

electroconvulsive shock [MED] The technique of eliciting convulsions by applying an electric current through the brain for a brief period. { i¦lek·trō·kən¦vəl·siv 'shäk }

electrocorticogram [MED] The record obtained by electrocorticography. { i¦lek·trō'kȯrd·ə·kə ,gram }

electrocorticography [MED] The technique of surveying the electrical activity of the cerebral cortex. { i¦lek·trə,kȯrd·ə'käg·rə·fē }

electrodermal response *See* galvanic skin response. { i,lek·trə¦dərm·əl ri'späns }

electrodermography [MED] The recording of the electrical resistance of the skin. { i¦lek·trō·dər'mäg·rə·fē }

electrodesiccation [MED] The use of a single terminal electrode with a small sparking distance to destroy lesions or seal off blood vessels. { i¦lek·trō,des·ə'kā·shən }

electrodiagnosis [MED] Diagnosis of disease states by recording the spontaneous electrical activity of tissue or organs, or by the response to stimulation of electrically excitable tissue. { i¦lek·trō,dī·əg'nō·səs }

electroencephalogram [MED] A graphic recording of the electric discharges of the cerebral cortex as detected by electrodes on the surface of the scalp. Abbreviated EEG. { i,lek·trō·en 'sef·ə·lə,gram }

electroencephalograph [MED] An instrument used to make electroencephalograms. { i,lek·trō·en'sef·ə·lə,graf }

electroencephalography [MED] The medical specialty concerned with the production and interpretation of electroencephalograms. { i¦lek·trō·en,sef·ə'läg·rə·fē }

electroencephalophone [MED] An instrument that provides an audible presentation of brain waves. Abbreviated EEP. { i,lek·trō·en'sef·ə·lə,fōn }

electroencephaloscope [MED] An instrument for displaying on a cathode-ray screen the waveforms of voltages generated by various sections of the brain. { i¦lek·trō·en'sef·ə·lə,skōp }

electrogenesis [PHYSIO] The generation of electric current by living tissue. { i¦lek·trə'jen·ə·səs }

electrogram [PHYSIO] The graphic representation of electric events in living tissues; commonly, an electrocardiogram or electroencephalogram. { i'lek·trə,gram }

electroinjection [BIOL] The use of electric-field impulses to introduce foreign deoxyribonucleic acid directly into intact cells. { i,lek·trō·in'jek·shən }

electrokymograph [MED] An instrument that provides a continuous recording of the movements of an internal organ such as the heart, generally by recording the movements or the changes in density of the shadow of the organ as presented on a fluoroscope. { i,lek·trō'kī·mə,graf }

electromyogram [MED] **1.** A graphic recording of the electrical response of a muscle to electrical stimulation. **2.** A graphic recording of eye movements during reading. { i¦lek·trō'mī·ə ,gram }

electromyograph [MED] An instrument used for making electromyograms. { i¦lek·trō'mī·ə ,graf }

electromyography [MED] A medical specialty concerned with the production and study of electromyograms. { i¦lek·trō·mī'äg·rə·fē }

electronarcosis [MED] Profound stupor or unconsciousness produced by passing an electric current through the brain. { i¦lek·trō·när'kō·səs }

electron carrier [CELL MOL] A molecule that accepts electrons from electron donors and donates them to electron acceptors, creating an energy-producing electron transport chain such as occurs in respiration and photosynthesis. { i'lek,trän ,kar·ē·ər }

electronic tonometer *See* tonometer. { i,lek'trän·ik tō'näm·əd·ər }

electron transport system [BIOCHEM] The components of the final sequence of reactions in biological oxidations; composed of a series of oxidizing agents arranged in order of increasing strength and terminating in oxygen. { i'lek,trän 'trans,pȯrt ,sis·təm }

electrooculogram [PHYSIO] A record of the standing voltage between the front and back of the eye that is correlated with eyeball movement and obtained by electrodes placed on the skin near the eye. { i,lek·trə·'ok·yül·ə,gram }

electrooculography [PHYSIO] The production and study of electroculograms. { i¦lek·trə·okyü 'läg·rə·fē }

electrophonic effect [BIOPHYS] The sensation of hearing produced when an alternating current of suitable frequency and magnitude is passed through a person. { i,lek·trə'fän·ik i'fekt }

electrophoretic mobility [BIOCHEM] A characteristic of living cells in suspension and biological compounds (proteins) in solution to travel in an electric field to the positive or negative electrode, because of the charge on these substances. { i¦lek·trō·fə'red·ik mō'bil·əd·ē }

electrophoretic variants [BIOCHEM] Phenotypically different proteins that are separable into distinct electrophoretic components due to differences in mobilities; an example is erythrocyte acid phosphatase. { i¦lek·trō·fə'red·ik 'ver·ē·əns }

electrophrenic respiration [MED] Artificial respiration in which the nerves that control breathing are stimulated electrically through appropriately placed electrodes. { i,lek·trə'fren·ik ,res·pə'rā·shən }

electrophysiology [PHYSIO] The branch of physiology concerned with determining the basic mechanisms by which electric currents are generated within living organisms. { i,lek·trō,fiz·ē'ä·lə·jē }

electroplax [VERT ZOO] One of the structural units of an electric organ of some fishes, composed of thin, flattened plates of modified muscle that appear as two large, waferlike, roughly circular or rectangular surfaces. { i'lek·trō,plaks }

electroporation [BIOL] The application of electric pulses to increase the permeability of cell membranes. [CELL MOL] The application of electric pulses to animal cells or plant protoplasts to increase membrane permeability. { i,lek·trō·pə'rā·shən }

electroretinogram [MED] A graphic recording of the electric discharges of the retina. Abbreviated ERG. { i¦lek·trō'ret·ən·ə,gram }

electroshock therapy [MED] Treatment of mental patients by passing an electric current of 85–110 volts through the brain. { i'lek·trō,shäk 'ther·ə·pē }

electrostethophone [MED] A stethoscope consisting of a microphone and audio amplifier feeding headphones, used for detection and study of sounds arising within the body. { i¦lek·trō'steth·ə,fōn }

electrosurgery [MED] The use of electricity to perform surgical procedures, as the use of electricity to simultaneously cut tissue and arrest bleeding. { i¦lek·trō'sərj·ə·rē }

electrotaxis [BIOL] Movement of an organism in response to stimulation by electric charges. { i¦lek·trō'tak·səs }

electrotherapy [MED] The therapeutic use of electricity. { i¦lek·trō'ther·ə·pē }

electrotonus [PHYSIO] The change of condition in a nerve or a muscle during the passage of a current of electricity. { i,lek'trät·ən·əs }

electrotropism [BIOL] Orientation response of a sessile organism to stimulation by electric charges. { i,lek'trä·trə,piz·əm }

elephant [VERT ZOO] The common name for two living species of proboscidean mammals in the family Elephantidae; distinguished by the elongation of the nostrils and upper lip into a sensitive, prehensile proboscis. { 'el·ə·fənt }

elephantiasis [MED] A parasitic disease of humans caused by the filarial nematode *Wuchereria bancrofti*; characterized by cutaneous and subcutaneous tissue enlargement due to lymphatic obstruction. { ,el·ə·fan'tī·ə·səs }

Elephantidae [VERT ZOO] A family of mammals in the order Proboscidea containing the modern elephants and extinct mammoths. { el·ə'fan·tə,dē }

elfinwood *See* krummholz. { 'el·fən,wu̇d }

elimination coefficient [GEN] The frequency with which certain genotypes die prematurely or are hindered during reproduction and are genetically eliminated as a consequence. { ə¦lim·ə'nā·shən ,kō·ə,fish·ənt }

ELISA *See* enzyme-linked immunosorbent assay. { ə'līz·ə *or* ¦ē¦el¦ī¦es'ā }

elixir [PHARM] A sweetened, aromatic solution, usually hydroalcoholic, sometimes containing soluble medicants; intended for use only as a flavor or vehicle. { ə'lik·sər }

elk [VERT ZOO] *Alces alces*. A mammal (family Cervidae) in Europe and Asia that resembles the North American moose but is smaller; it is the largest living deer. { elk }

elliptocytosis [MED] A rare hereditary disease of man characterized by the presence of large numbers of oval or elliptic erythrocytes in the circulating blood. { ə,lip·tə,sī'tō·səs }

elm [BOT] The common name for hardwood trees composing the genus *Ulmus*, characterized by simple, serrate, deciduous leaves. { elm }

Elmidae [INV ZOO] The drive beetles, a small family of coleopteran insects in the superfamily Dryopoidea. { 'el·mə,dē }

elongation factor [BIOCHEM] Any of several proteins required for elongation of growing polypeptide chains during protein synthesis. { ē,loŋ'gā·shən ,fak·tər }

Elopidae [VERT ZOO] A family of fishes in the order Elopiformes, including the tarpon, ladyfish, and machete. { e'läp·ə,dē }

Elopiformes [VERT ZOO] A primitive order of actinopterygian fishes characterized by a single dorsal fin composed of soft rays only, cycloid scales, and toothed maxillae. { e,läp·ə 'fōr,mēz }

El Tor vibrio [MICROBIO] Any of the rod-shaped paracholera vibrios; many strains can be agglutinated with anticholera serum. { el 'tȯr 'vib·rē·ō }

elusive ulcer *See* Hunner's ulcer. { i'lü·siv 'əl·sər }

elytron [INV ZOO] **1.** One of the two sclerotized or leathery anterior wings of beetles which serve to cover and protect the membranous hindwings.

2. A dorsal scale of certain Polychaeta. { 'el·ə,trän }

emaciation [MED] A wasted condition of the body; the process of losing flesh so as to become extremely lean. { i,mā·sē'ā·shən }

emarginate [BIOL] Having a margin that is notched or slightly forked. { ē'mär·jə,nāt }

EMB agar [MICROBIO] A culture medium containing sugar, eosin, and methylene blue, used in the confirming test for coliform bacteria. { ¦ē¦em¦bē 'äg·ər }

Emballonuridae [VERT ZOO] The sheath-tailed bats, a family of mammals in the order Chiroptera. { em,bal·ə'nu̇r·ə,dē }

embalm [MED] To treat a cadaver with antiseptics and preservatives to prevent decay, before burial or dissection. { em'bäm }

Embden-Meyerhof pathway *See* glycolytic pathway. { ¦em·dən ¦mī·ər,hȯf 'path,wā }

Embiidina [INV ZOO] An equivalent name for Embioptera. { em,bē·ə'dī·nə }

Embioptera [INV ZOO] An order of silk-spinning, orthopteroid insects resembling the grasshoppers; commonly called the embiids or webspinners. { ,em·bē'äp·tə·rə }

Embiotocidae [VERT ZOO] The surfperches, a family of perciform fishes in the suborder Percoidei. { ,em·bē·ə'täs·ə,dē }

embolectomy [MED] Surgical removal of an embolus. { ,em·bə'lek·tə·mē }

embolism [MED] The blocking of a blood vessel by an embolus. { 'em·bə,liz·əm }

embolus [MED] A clot or other mass of particulate matter foreign to the bloodstream which lodges in a blood vessel and causes obstruction. { 'em·bə·ləs }

emboly [EMBRYO] Formation of a gastrula by the process of invagination. { 'em·bə·lē }

embryo [BOT] The young sporophyte of a seed plant. [EMBRYO] **1.** An early stage of development in multicellular organisms. **2.** The product of conception up to the third month of human pregnancy. { 'em·brē·ō }

Embryobionta [BOT] The land plants, a subkingdom of the Plantae characterized by having specialized conducting tissue in the sporophyte (except bryophytes), having multicellular sex organs, and producing an embryo. { ¦em·brē·ō·bī'än·tə }

embryogenesis [EMBRYO] The formation and development of an embryo from an egg. { ,em·brē·ō'jen·ə·səs }

embryoid [BOT] An embryolike structure originating from somatic cells, such as immature plant embryos, inflorescences, or leaves cultivated in culture. { 'em·brē,ȯid }

embryology [BIOL] The study of the development of the organism from the zygote, or fertilized egg. { em·brē'äl·ə·jē }

embryoma of the ovary *See* dysgerminoma. { ,em·brē'ō·mə əv t͟hē 'ō·və·rē }

embryonal-cell lipoma *See* liposarcoma. { em 'brī·ən·əl ,sel li'pō·mə }

embryonate [EMBRYO] **1.** To differentiate into a zygote. **2.** Containing an embryo. { 'em·brē·ə'nāt }

embryonated egg culture [VIROL] Embryonated hen's eggs inoculated with animal viruses for the purpose of identification, isolation, titration, or for quantity cultivation in the production of viral vaccines. { 'em·brē·ə,nād·əd 'eg ,kəl·chər }

embryonic differentiation [EMBRYO] The process by which specialized and diversified structures arise during embryogenesis. { ,em·brē'än·ik ,dif·ə,ren·chē'ā·shən }

embryonic inducer [EMBRYO] The acting system in embryos, which contributes to the formation of specialized tissues by controlling the mode of development of the reacting system. { ,em·brē'än·ik in'dü·sər }

embryonic induction [EMBRYO] The influence of one cell group (inducer) over a neighboring cell group (induced) during embryogenesis. Also known as induction. { ,em·brē'än·ik in'dək·shən }

embryonic stem cell [EMBRYO] Undifferentiated cell derived from the inner cell mass of the blastocyst that can give rise to any of the three embryonic germ layers, and thus can form any cell or tissue type of the body, but cannot give rise to the full spectrum of cells required to complete fetal development. { ,em·brē¦än·ik 'stem ,sel }

embryopathy [MED] Any abnormal development of an embryo, either morphological or biochemical. { ,em·brē'äp·ə·thē }

Embryophyta [BOT] The equivalent name for Embryobionta. { ,em·brē'äf·əd·ə }

embryo rescue [GEN] A technique for crossing wild and domestic species of plants in which the wild species is used as the male parent, and the embryos are excised approximately one month after pollination and placed on an artificial medium, where a small fraction survive. { 'em·brē·ō 'res·kyü }

embryo sac [BOT] The female gametophyte of a seed plant, containing the egg, synergids, and polar and antipodal nuclei; fusion of the antipodals and a pollen generative nucleus forms the endosperm. { 'em·brē·ō ,sak }

embryotomy [MED] Any mutilation of the fetus in the uterus to aid in its removal when natural delivery is impossible. { 'em·brē'äd·ə·mē }

emerged bog [ECOL] A bog which grows vertically above the water table by drawing water up through the mass of plants. { ə¦mərjd 'bäg }

emergency medicine [MED] The medical specialty that comprises the immediate decision making and action necessary to prevent death or further disability under emergency conditions. { ə,mər·jən·sē 'med·ə·sən }

emesis [MED] The act of vomiting. { 'em·ə·səs }

emetic [PHARM] Any agent that induces emesis. { i'med·ik }

emetine [PHARM] $C_{29}H_{40}N_2O_4$ Cephaeline methyl ether, the principal alkaloid of ipecac; a white powder, sparingly soluble in water; it is emetic, diaphoretic, and expectorant, but its chief utility is as an amebicide. { 'em·ə,tēn }

emigration [ECOL] The movement of individuals or their disseminules out of a population or population area. { ,em·ə'grā·shən }

emiocytosis [CELL MOL] Fusion of intracellular granules with the cell membrane, followed by discharge of the granules outside of the cell; applied chiefly to the mechanism of insulin secretion. Also known as reverse pinocytosis. { ,em·ē,sī'tō·səs }

emission inventory [ECOL] A quantitative detailed compilation of pollutants emitted into the atmosphere of a given community. { i'mish·ən 'in·vən,tȯr·ē }

emission tomography [MED] A technique which uses the emission of gamma-ray photons from radioactive tracers to construct images of the distribution of the tracers in the human body. { i'mish·ən tō'mäg·rə·fē }

emmetropia [MED] Normal vision. { ,em·ə 'trō·pē·ə }

emollient [PHARM] A softening agent, especially for use on skin and mucous membranes; lanolin is widely used as a base. { ə'mäl·yənt }

emphysema [MED] A pulmonary disorder characterized by overdistention and destruction of the air spaces in the lungs. { ,em·fə'sē·mə }

emphysematous chest [MED] The altered contour of the chest seen in pulmonary emphysema, with increased anteroposterior diameter, flaring at the lower rib margins, low position of the diaphragm, and minimal respiratory motion. Also known as barrel chest. { ,em·fə·sə'mad·əs 'chest }

Empididae [INV ZOO] The dance flies, a family of orthorrhaphous dipteran insects in the series Nematocera. { em'pid·ə,dē }

empodium [INV ZOO] A small peripheral part located between the claws of the tarsi of many insects and arachnids. { em'pōd·ē·əm }

empyema [MED] The presence of pus in the body cavity, hollow organ, or tissue space; when the term is used without qualification, it generally refers to pus in the pleural space. { ,em,pī'ē·mə }

emu [VERT ZOO] *Dromiceius novae-hollandiae.* An Australian ratite bird, the second largest living bird, characterized by rudimentary wings and a feathered head and neck without wattles. { 'ē,myü }

emulsan [BIOCHEM] A lipopolysaccharide produced by a strain of *Acinetobacter calcoaceticus*, used to stabilize oil-in-water emulsions. { i'məl·sən }

Emydidae [VERT ZOO] A family of aquatic and semiaquatic turtles in the suborder Cryptodira. { e'mid·ə,dē }

enamel organ [EMBRYO] The epithelial ingrowth from the dental lamina which covers the dental papilla, furnishes a mold for the shape of a developing tooth, and forms the dental enamel. { i'nam·əl ,ȯr·gən }

Enantiozoa [INV ZOO] The equivalent name for Parazoa. { ə¦nan·te·ə'zō·ə }

enarthrosis [ANAT] A freely movable joint that allows a wide range of motion on all planes. Also known as ball-and-socket joint. { ,e,när 'thrō·səs }

Enantiornithines [VERT ZOO] Opposite birds, so called because their foot bones fuse in the opposite direction of modern birds, from the subclass Sauriuvae. { i,nan·tē'ȯrn·ə,thēnz }

Encalyptales [BOT] An order of true mosses (subclass Bryidae) characterized by broad papillose leaves and erect capsules covered by very long calyptrae. { en,ka·lip'tā·lēz }

encephalitis [MED] Inflammation of the brain. { en,sef·ə'līd·əs }

encephalitis lethargica [MED] Epidemic encephalitis, probably of viral etiology, characterized by lethargy, ophthalmoplegia, hyperkinesia, and at times residual neurologic disability, particularly parkinsonism with oculogyric crisis. Also known as epidemic encephalitis; sleeping sickness; von Economo's disease. { en,sef·ə'līd·əs lə'thär·jə·kə }

encephalocele [MED] Hernia of the brain through a congenital or traumatic opening in the cranium. { en'sef·ə·lō,sēl }

encephalogram [MED] A roentgenogram of the brain made in encephalography. { en'sef·ə·lə,gram }

encephalography [MED] Roentgenography of the brain following removal of cerebrospinal fluid, by lumbar or cisternal puncture, and its replacement by air or other gas. { en,sef·ə'läg·rə·fē }

encephaloid carcinoma *See* medullary carcinoma. { en'sef·ə,lȯid ,kärs·ən'ō·mə }

encephalomalacia [MED] **1.** Infarction of the brain. **2.** Any softening or fragmentation of the brain. { en¦sef·ə·lō·mə'lā·shə }

encephalomyelitis [MED] Inflammation of the brain and spinal cord. { en¦sef·ə·lō,mī·ə'līd·əs }

encephalomyocarditis [MED] An acute febrile RNA virus disease accompanied by pharyngitis, stiff neck, and hyperactive deep reflexes; certain species of wild rats are the reservoir; human infections range from a mild febrile illness to a severe encephalomyelitis. { en¦sef·ə·lō,mī·ə,kär'dīd·əs }

encephalopathy [MED] Any disease of the brain. { en,sef·ə'läp·ə·thē }

Encholaimoidea [INV ZOO] A superfamily of nematodes of the order Dorylaimida, characterized by two circlets of cephalic sense organs on the lips, pouchlike amphids with slitlike openings, a stoma armed with an axial spear, and a body cuticle marked by widely spaced annulations giving a platelike appearance. { ,en·kō·lə'mȯid·ē·ə }

enchondroma [MED] A benign tumor composed of dysplastic cartilage cells, occurring in the metaphysis of cylindric bones, especially of the hands and feet. { ,en,kän'drō·mə }

enchondromatosis [MED] A rare disorder principally involving tubular bones, especially those of the feet and hands, characterized by hamartomatous proliferation of cartilage in the metaphysis, indistinguishable in single lesions from enchondromas. Also known as chondrodysplasia; dyschondroplasia; Ollier's disease. { ˌenˌkän·drō·mə'tō·səs }

enchymatous [PHYSIO] Of gland cells, distended with secreted material. { en'kim·əd·əs }

enclosure compound *See* clathrate. { in'klō·zhər ˌkämˌpau̇nd }

Encyrtidae [INV ZOO] A family of hymenopteran insects in the superfamily Chalcidoidea. { en'sərd·əˌdē }

encystment [BIOL] The process of forming or becoming enclosed in a cyst or capsule. { en 'sist·mənt }

Endamoeba [INV ZOO] The type genus of the Endamoebidae comprising insect parasites and, in some systems of classification, certain vertebrate parasites. { ¦end·ə'mē·bə }

endarch [BOT] Formed outward from the center, referring to xylem or its development. { 'enˌdärk }

endarteritis [MED] Inflammation of the lining (tunica intima) of an artery. { ¦endˌärt·ə'rīd·əs }

endarteritis obliterans [MED] Endarteritis, particularly of small arteries, accompanied by degeneration of the intima, leading to occlusion of the blood vessel. Also known as obliterating endarteritis. { ¦endˌärt·ə'rīd·əs ō'blit·əˌränz }

end bulb *See* bouton. { 'end ˌbəlb }

end bulb of Krause *See* Krause's corpuscle. { 'end ˌbəlb əv 'krau̇s }

Endeidae [INV ZOO] A family of marine arthropods in the subphylum Pycnogonida. { en'dē·əˌdē }

endemic [MED] Peculiar to a certain region, specifically referring to a disease which occurs more or less constantly in any locality. { en 'dem·ik }

endemic goiter [MED] Goiter peculiar to areas that are iodine-poor in food, water, or soil. { en 'dem·ik 'gȯid·ər }

endemic rural plague *See* sylvatic plague. { en 'dem·ik ¦ru̇r·əl 'plāg }

endemic typhus *See* murine typhus. { en'dem·ik 'tī·fəs }

endemism [MED] The state or quality of being endemic. { 'en·dəˌmiz·əm }

endergonic [BIOCHEM] Of or pertaining to a biochemical reaction in which the final products possess more free energy than the starting materials; usually associated with anabolism. { ¦en·dər¦gän·ik }

endermic [MED] Acting through the skin by absorption, such as medication applied to the skin. { en'dər·mik }

endexine [BOT] An inner membranous layer of the exosporium. { en'dekˌsēn }

endite [INV ZOO] **1.** One of the appendages on the inner aspect of an arthropod limb. **2.** A ridgelike chewing surface on the inner part of the pedipalpus or maxilla of many arachnids. { 'enˌdīt }

end labeling [BIOCHEM] The addition of a radioactively labeled group to one end of a deoxyribonucleic acid strand. { 'end ˌlab·əl·iŋ }

endobasion [ANAT] The anteriormost point of the margin of the foramen magnum at the level of its smallest diameter. { ¦en·dō'bā·sēˌän }

endobiotic [ECOL] Referring to an organism living in the cells or tissues of ahost. { ¦en·dō·bī'äd·ik }

endobranchiate [ZOO] Animal form with endodermal gills. { ¦en·dō'braŋ·kēˌāt }

endocardial fibroelastosis [MED] Fibrous or fibroelastic thickening of the endocardium, of unknown cause. { ¦en·dō¦kärd·ē·əl ¦fī·brō·ə ˌlas'tō·səs }

endocarditis [MED] Inflammation of the endocardium. { ¦en·do·kär'dīd·əs }

endocardium [ANAT] The membrane lining the heart. { ˌen·dō'kärd·ē·əm }

endocarp [BOT] The inner layer of the wall of a fruit or pericarp. { 'en·dōˌkärp }

endocervicitis [MED] Inflammation of the mucous membrane of the uterine cervix. { ¦en·dōˌsər·və'sīd·əs }

endocervix [ANAT] The glandular mucous membrane of the cervix uteri. { ¦en·dō'sər·viks }

endochondral ossification [PHYSIO] The conversion of cartilage into bone. Also known as intracartilaginous ossification. { ¦en·dō'kän·drəl ˌäs·ə·fə'kā·shən }

endocommensal [ECOL] A commensal that lives within the body of its host. { ¦en·dō·kə'men·səl }

endocorpuscular [CELL MOL] Located within an erythrocyte. { ¦en·dō·kȯr'pəs·kyə·lər }

endocranium [ANAT] **1.** The inner surface of the cranium. **2.** *See* dura mater. [INV ZOO] The processes on the inner surface of the head capsule of certain insects. { ¦en·dō'krā·nē·əm }

endocrine gland [PHYSIO] A ductless structure whose secretion (hormone) is passed into adjacent tissue and then to the bloodstream either directly or by way of the lymphatics. Also known as ductless gland. { 'en·də·krən ˌgland }

endocrine signaling [PHYSIO] Signaling in which endocrine cells release hormones that act on distant target cells. { 'en·də·krən ˌsig·nəl·iŋ }

endocrine system [PHYSIO] The chemical coordinating system in animals, that is, the endocrine glands that produce hormones. { 'en·də·krən ˌsis·təm }

endocrine toxicity [MED] Any adverse structural and/or functional changes to the endocrine system which may result from exposure to chemicals; can harm human and animal reproduction and development. { ¦en·də·krən täk'sis·əd·ē }

endocrinology [PHYSIO] The study of the endocrine glands and the hormones that they synthesize and secrete. { ˌen·də·krə'näl·ə·jē }

endocuticle [INV ZOO] The inner, elastic layer of an insect cuticle. { ¦en·dō'kyüd·i·kəl }

endocyst [INV ZOO] The soft layer consisting of ectoderm and mesoderm, lining the ectocyst of bryozoans. { 'en·dəˌsist }

endocytic vacuole [CELL MOL] A membrane-bound cellular organelle containing extracellular particles engulfed by the mechanisms of endocytosis. { ¦en·də¦sīd·ik ′vak·yə,wōl }

endocytobiosis [ECOL] Symbiosis in which the symbionts live within host cells. { ,en·dō,sī·tō·bī′ō·səs }

endocytosis [CELL MOL] **1.** An active process in which extracellular materials are introduced into the cytoplasm of cells by either phagocytosis or pinocytosis. **2.** The process by which animal cells internalize large molecules and large collections of fluid. { ¦en·do·sī′tō·səs }

endodeoxyribonuclease *See* restriction endonuclease. { ¦en·dō·dē¦äk·sē,rī·bō′nü·klē,ās }

endoderm [EMBRYO] The inner, primary germ layer of an animal embryo; sometimes referred to as the hypoblast. Also known as entoderm; hypoblast. { ′en·dō,dərm }

endodermis [BOT] The innermost tissue of the cortex of most plant roots and certain stems consisting of a single layer of at least partly suberized or cutinized cells; functions to control the movement of water and other substances into and out of the stele. { ¦en·dō¦dər·məs }

endodontics [MED] A branch of dentistry that treats diseases and injuries affecting the root tips or nerves of teeth; root canal is a common procedure. { ,en·dō′dän·tiks }

endoenzyme [BIOCHEM] An intracellular enzyme, retained and utilized by the secreting cell. { ¦en·dō¦en,zīm }

endogamy [BIOL] Sexual reproduction between organisms which are closely related. [BOT] Pollination of a flower by another flower of the same plant. { en′däg·ə·mē }

endogenote [MICROBIO] The genetic complement of the partial zygote formed as a result of gene transfer during the process of recombination in bacteria. { en′däj·ə,nōt }

endogenous [BIOCHEM] Relating to the metabolism of nitrogenous tissue elements. [MED] Pertaining to diseases resulting from internal causes. { en′däj·ə·nəs }

endogenous pyrogen [BIOCHEM] A fever-inducing substance (protein) produced by cells of the host body, such as leukocytes and macrophages. { en′däj·ə·nəs ′pī·rə·jən }

endogenous virus [GEN] An inactive virus that is integrated into the chromosome of its host cell and can, therefore, exhibit vertical transmission. { en¦däj·ən·əs ′vī,rəs }

endoglycosidase [BIOCHEM] An enzyme which releases intact glycans from their linkages with amino acids. { ,en·dō·glī′kō·sə,dās }

endognath [INV ZOO] The inner and main branch of a crustacean's oral appendage. { ′en·dəg,nath }

endolecithal [INV ZOO] A type of egg found in turbellarians with yolk granules in the cytoplasm of the egg. Also spelled entolecithal. { ¦en·dō′les·ə·thəl }

endolithic [ECOL] Living within rocks, as certain algae and coral. { ¦en·də¦lith·ik }

endolymph [PHYSIO] The lymph fluid found in the membranous labyrinth of the ear. { ′en·də,limf }

endolymphatic stromomyosis *See* interstitial endometriosis. { ¦en·də,lim¦fad·ik ¦strō·mō,mī′ō·səs }

endomembrane system [CELL MOL] In eukaryotes, the functional continuum of membraneous cell components consisting of the nuclear envelope, endoplastic reticulum, and Golgi apparatus as well as vesicles and other structures derived from these major components. { ,en·dō′mem ,brān ,sis·təm }

endomeninx [EMBRYO] The internal part of the meninx primitiva that differentiates into the pia mater and arachnoid membrane. { ¦en·dō¦mē·niŋks }

endomere [EMBRYO] A blastomere that forms endoderm. { ′en·də,mir }

endometrioma [MED] Endometriosis in which there is a discrete tumor mass. { ¦en·dō,mē·trē′ō·mə }

endometriosis [MED] The presence of endometrial tissue in abnormal locations, including the uterine wall, ovaries, or extragenital sites. { ¦en·dō,mē·trē′ō·səs }

endometritis [MED] Inflammation of the endometrium. { ¦en·dō·mə′trīd·əs }

endometrium [ANAT] The mucous membrane lining the uterus. { ¦en·dō¦mē·trē·əm }

endomitosis [CELL MOL] Division of the chromosomes without dissolution of the nuclear membrane; results in polyploidy or polyteny. { ¦en·dō,mī′tō·səs }

endomixis [INV ZOO] Periodic division and reorganization of the nucleus in certain ciliated protozoans. { ¦en·dō¦mik·səs }

Endomycetales [MICROBIO] Former designation for Saccharomycetales. { ,en·də,mī·sə′tā·lēz }

Endomycetoideae [MICROBIO] A subfamily of ascosporogenous yeasts in the family Saccharomycetaceae. { ,en·də,mī·sə′tȯid·ē,ē }

Endomychidae [INV ZOO] The handsome fungus beetles, a family of coleopteran insects in the superfamily Cucujoidea. { ,en·də′mīk·ə,dē }

endomysium [HISTOL] The connective tissue layer surrounding an individual skeletal muscle fiber. { ,en·də′miz·ē·əm }

endoneural fibroma *See* neurofibroma. { ¦en·dō ¦nu̇r·əl fī′brō·mə }

endoneurium [HISTOL] Connective tissue fibers surrounding and joining the individual fibers of a nerve trunk. { ,en·dō′nu̇r·ē·əm }

endonuclease [BIOCHEM] Any of a group of enzymes which degrade deoxyribonucleic acid or ribonucleic acid molecules by attaching nucleotide linkages within the polynucleotide chain. { ¦en·dō′nü·klē,ās }

endoparasite [ECOL] A parasite that lives inside its host. { ¦en·dō′par·ə,sīt }

endopeptidase [BIOCHEM] An enzyme that acts upon the centrally located peptide bonds of a protein molecule. { ¦en·dō′pep·tə,dās }

endoperoxide [BIOCHEM] Any of various intermediates in the biosynthesis of prostaglandins. { ˌen·dō·pəˈräkˌsīd }

endophagous [INV ZOO] Of an insect larva, living within and feeding upon the host tissues. { enˈdäf·ə·gəs }

endophallus [INV ZOO] Inner wall of the phallus of insects. { ¦en·dōˈfal·əs }

endophyte [ECOL] A plant that lives within, but is not necessarily parasitic on, another plant. { ˈen·dəˌfīt }

endoplasm [CELL MOL] The inner, semifluid portion of the cytoplasm. { ˈen·dəˌplaz·əm }

endoplasmic reticulum [CELL MOL] A vacuolar system of the cytoplasm in differentiated cells that functions in protein synthesis and sequestration. Abbreviated ER. { ¦en·də¦plaz·mik rə ˈtik·yə·ləm }

endopleurite [INV ZOO] **1.** The portion of a crustacean apodeme which develops from the interepimeral membrane. **2.** One of the laterally located parts on the thorax of an insect which fold inward, extending into the body cavity. { ˌen·dəˈplu̇rˌīt }

endopodite [INV ZOO] The inner branch of a biramous crustacean appendage. { enˈdäp·əˌdīt }

endopolyploid cell [CELL MOL] Any cell whose chromosome number has been increased by endomitosis and for which the degree of ploidy is proportional to the number of times that endomitosis has taken place. { ˌen·dōˈpäl·əˌplȯid ˌsel }

Endoprocta [INV ZOO] The equivalent name for Entoprocta. { ˌen·dəˈpräk·tə }

endoprosthesis [MED] A prosthesis that is used internally. { ¦en·dō·präsˈthē·səs }

endopterygoid [VERT ZOO] A paired dermal bone of the roof of the mouth in fishes. { ¦en·dōˈter·əˌgȯid }

Endopterygota [INV ZOO] A division of the insects in the subclass Pterygota, including those orders which undergo a holometabolous metamorphosis. { ¦en·dōˌter·əˈgäd·ə }

endoreduplication [CELL MOL] Appearance in mitotic cells of certain chromosomes or chromosome sets in the form of multiples. [GEN] Two to twelve or more rounds of replication and chromosome duplication without mitotic cell division, as in the production of polytene chromosomes. { ¦en·dō·rēˌdü·pləˈkā·shən }

end organ [NEUROSCI] The expanded termination of a nerve fiber in muscle, skin, mucous membrane, or other structure. { ˈend ˌȯr·gən }

β-endorphin [BIOCHEM] A 31-amino acid peptide fragment of pituitary β-lipotropic hormone having morphinelike activity. { ¦bād·ə enˈdȯr·fən }

endosalpingioma *See* serous cystadenoma. { ¦en·dō·salˌpin·jēˈō·mə }

endosalpinx [ANAT] The mucous membrane that lines the fallopian tube. { ¦en·dō¦sal ˌpiŋks }

endosarc [INV ZOO] The inner, relatively fluid part of the protoplasm of certain unicellular organisms. { ˈen·dəˌsärk }

endoscope [MED] An instrument used to visualize the interior of a body cavity or hollow organ. { ˈen·dəˌskōp }

endoskeleton [ZOO] An internal skeleton or supporting framework in an animal. { ¦en·dōˈskel·ə·tən }

endosmosis [PHYSIO] The passage of a liquid inward through a cell membrane. { ¦en·dō·äsˈmō·səs }

endosome [CELL MOL] A mass of chromatin near the center of a vesicular nucleus. [INV ZOO] The inner layer of certain sponges. { ˈen·dəˌsōm }

endosperm [BOT] **1.** The nutritive protein material within the embryo sac of seed plants. **2.** Storage tissue in the seeds of gymnosperms. { ˈen·dəˌspərm }

endosperm nucleus [BOT] The triploid nucleus formed within the embryo sac of most seed plants by fusion of the polar nuclei with one sperm nucleus. { ˈen·dəˌspərm ˈnü·klē·əs }

endospore [BIOL] An asexual spore formed within a cell. { ˈen·dəˌspȯr }

endosteum [ANAT] The membrane lining of bone marrow cavities. { enˈdäs·tē·əm }

endostome [BOT] The opening in the inner integument of a bitegmic ovule. { ˈen·dəˌstōm }

endostyle [INV ZOO] A ciliated groove or pair of grooves in the pharynx of lower chordates. { ˈen·dəˌstīl }

endosymbiont theory [CELL MOL] A theory that the mitochondria of eukaryotes and the chloroplasts of green plants and flagellates originated as free-living prokaryotes that invaded primitive eukaryotic cells and become established as permanent symbionts in the cytoplasm. { ˌen·dōˈsim·bēˌänt ˌthē·ə·rē }

endosymbiosis [ECOL] A mutually beneficial relationship in which one organism lives inside the other. { ˌen·dōˌsim·bēˈō·səs }

endosymbiotic infection [VIROL] A virus infection in which virus replication occurs in cells without a cytopathic effect. { ¦en·dōˌsim·bēˈäd·ik inˈfek·shən }

endotergite [INV ZOO] A dorsal plate to which muscles are attached in the insect skeleton. { ¦en·dōˈtərˌjīt }

endotesta [BOT] An inner layer of the testa in various seeds. { ˈen·dōˌtes·tə }

endothecium [BOT] The middle of three layers that make up an immature anther; becomes the inner layer of a mature anther. { ˌen·dəˈthē·shē·əm }

endothelial cell [HISTOL] A type of squamous epithelial cell composing the endothelium. { ˌen·də¦thē·lē·əl ˈsel }

endotheliochorial placenta [EMBRYO] A type of placenta in which the maternal blood is separated from the chorion by the maternal capillary endothelium; occurs in dogs. { ¦en·dəˌthē·lē·əˈkȯr·ē·əl pləˈsen·tə }

endothelioma [MED] Any tumor arising from, or resembling, endothelium; usually a benign growth, but occasionally a malignant tumor. { ˌen·dōˌthē·lēˈō·mə }

endothelium [HISTOL] The epithelial layer of cells lining the heart and vessels of the circulatory system. { ,en·də'thē·lē·əm }

endotherm [PHYSIO] An animal that produces enough heat from its own metabolism and employs devices to retard heat loss so that it is able to keep its body temperature higher than that of its environment. { 'en·də,thərm }

endothermy [PHYSIO] The utilization of metabolic heat for thermoregulation. { 'en·dō ,thər·mē }

endotoxin [MICROBIO] A biologically active substance produced by gram-negative bacteria and consisting of lipopolysaccharide, a complex macromolecule containing a polysaccharide covalently linked to a unique lipid structure, termed lipid A. { ,en·dō'täk·sən }

endotracheal [ANAT] Within the trachea. { ¦en·dō'trā·kē·əl }

endotrophic [BIOL] Obtaining nourishment from within; applied to certain parasitic fungi that live in the root cortex of the host plant. { ¦en·də¦trä·fik }

end-plate potential [NEUROSCI] Depolarization of the postsynaptic membrane at the neuromuscular junction, mediated by acetylcholine, in response to action potentials arriving at the endings of presynaptic motor neurons. { ¦end ,plāt 'pə'ten·chəl }

end-product inhibition [BIOCHEM] In sequential enzyme systems, a control mechanism in which accumulation of final product from a metabolic reaction causes inhibition of product formation. { ¦end ,präd·əkt ,in·ə'bish·ən }

enema [MED] A rectal injection of liquid for therapeutic, diagnostic, or nutritive purposes. { 'en·ə·mə }

energy balance [PHYSIO] The relation of the amount of utilizable energy taken into the body to that which is employed for internal work, external work, and the growth and repair of tissues. { 'en·ər·jē ,bal·əns }

energy metabolism [BIOCHEM] The chemical reactions involved in energy transformations within cells. { 'en·ər·jē mə'tab·ə,liz·əm }

energy pyramid [ECOL] An ecological pyramid illustrating the energy flow within an ecosystem. { 'en·ər·jē ,pir·ə·mid }

Engel-Recklinghausen disease *See* osteitis fibrosa cystica. { ¦eŋ·gəl ¦rek·liŋ,haùz·ən di,zēz }

engram [NEUROSCI] A memory imprint; the alteration that has occurred in nervous tissue as a result of an excitation from a stimulus, which hypothetically accounts for retention of that experience. Also known as memory trace. { 'en,gram }

Engraulidae [VERT ZOO] The anchovies, a family of herringlike fishes in the suborder Clupoidea. { ,en'gròl·ə,dē }

enhancer gene [GEN] Any modifier gene that acts to enhance the action of a nonallelic gene. { en'han·sər ,jēn }

Enicocephalidae [INV ZOO] The gnat bugs, a family of hemipteran insects in the superfamily Enicocephaloidea. { ,en·ə·kō·sə'fal·ə,dē }

Enicocephaloidea [INV ZOO] A superfamily of the Hemiptera in the subdivision Geocorisae containing a single family. { ,en·ə·kō·sef·ə'lòid·ē·ə }

enkephalin [BIOCHEM] A mixture of two polypeptides isolated from the brain; central mode of action is an inhibition of neurotransmitter release. { en'kef·ə·lən }

enolase [BIOCHEM] An enzyme that catalyzes the reversible dehydration of phosphoglyceric acid to phosphopyruvic acid. { 'ē·nə,lās }

enone [BIOCHEM] An alpha-, beta-unsaturated ketone. { 'ē,nōn }

enophthalmos [MED] Recession of the eyeball into the orbital cavity. { ,e,näf'thal·məs }

Enopla [INV ZOO] A class or subclass of ribbonlike worms of the phylum Rhynchocoela. { e'näp·ē·ə }

Enoplia [INV ZOO] A subclass of nematodes in the class Adenophorea. { e'näp·lē·ə }

Enoplida [INV ZOO] An order of nematodes in the subclass Enoplia. { e'näp·lə·də }

Enoplidae [INV ZOO] A family of free-living marine nematodes in the superfamily Enoploidea, characterized by a complex arrangement of teeth and mandibles. { e'näp·lə,dē }

Enoploidea [INV ZOO] A superfamily of small to very large free-living marine nematodes having pocketlike amphids opening to the exterior via slitlike apertures. { e·nə'plòid·ē·ə }

Enoploteuthidae [INV ZOO] A molluscan family of deep-sea squids in the class Cephalopoda. { e,näp·lə'tü·thə,dē }

enrichment culture [MICROBIO] A medium of known composition and specific conditions of incubation which favors the growth of a particular type or species of bacterium. { in'rich·mənt ,kəl·chər }

enrichment medium [MICROBIO] A liquid cultural medium of a given composition which permits preferential emergence of certain organisms that initially may have made up a relatively minute proportion of a mixed inoculum. { in 'rich·mənt ,mē·dē·əm }

ensiform [BIOL] Sword-shaped. { 'en·sə,fòrm }

Entamoeba [INV ZOO] A genus of parasite amebas in the family Endamoebidae, including some species of the genus *Endamoeba* which are parasites of humans and other vertebrates. { ¦ent·ə'mē·bə }

enteralgia [MED] Pain in the intestine. { ,en·tə'ral·jē·ə }

enterectomy [MED] Excision of a part of the intestine. { ,en·tə'rek·tə·mē }

enteric bacilli [MICROBIO] Microorganisms, especially the gram-negative rods, found in the intestinal tract of humans and animals. { en 'ter·ik bə'sil·ī }

enteric cytopathogenic human orphan virus *See* echovirus. { en'ter·ik ¦sī·dō,path·ə'jen·ik ¦yü·mən 'òr·fən ,vī·rəs }

enteritis [MED] Inflammation of the intestinal tract. { ent·ə'rīd·əs }

Enterobacter [MICROBIO] A genus of bacteria in

the family Enterobacteriaceae; motile rods found in the intestine of humans and other animals; some strains are encapsulated. { ˌent·ə·rō ˈbak·tər }

Enterobacteriaceae [MICROBIO] A family of gram-negative, facultatively anaerobic rods; cells are nonsporeforming and may be nonmotile or motile with peritrichous flagella; includes important human and plant pathogens. { ˌent·ə·rōˌbak·tir·ēˈās·ēˌē }

enterobiasis [MED] Infestation of the intestinal tract of humans with the nematode *Enterobius vermacularis* (pinworm); characterized by mild enteritis. { ˌent·ə·rōˈbī·ə·səs }

enterococci [MICROBIO] Spherical bacteria in short chains. { ˌen·tə·rəˈkäk·ē }

enterocoel [ZOO] A coelom that arises by mesodermal outpocketing of the archenteron. { ˈent·ə·rōˌsēl }

Enterocoela [SYST] A section of the animal kingdom that includes the Echinodermata, Chaetognatha, Hemichordata, and Chordata. { ˌent·ə·rōˈsēl·ə }

enterocolitis [MED] Inflammation of the small intestine and colon. { ˌent·ə·rō·kəˈlīd·əs }

enterocyte [HISTOL] A cell that lines the intestinal wall. { ˈen·tə·rəˌsit }

enterohydrocoel [INV ZOO] In crinoids, an anterior cavity derived from the archenteron. { ˌent·ə·rōˈhī·drəˌsēl }

enterokinase [BIOCHEM] An enzyme which catalyzes the conversion of trypsinogen to trypsin. { ˌent·ə·rōˈkīˌnās }

enterolith [PATH] A concretion formed in the intestine. { ˈent·ə·rōˌlith }

enteron [ANAT] The alimentary canal. { ˈent·əˌrän }

enteropathy [MED] Disease of the intestine. { ˌen·təˈräp·ə·thē }

Enteropneusta [INV ZOO] The acorn worms or tongue worms, a class of the Hemichordata; free-living solitary animals with no exoskeleton and with numerous gill slits and a straight gut. { ˌent·ə·rəˈnüs·tə }

enteroptosis *See* visceroptosis. { ˌent·ə·räpˈtō·səs }

enterorrhagia [MED] Intestinal hemorrhage. { ˌent·ə·rōˈrāj·ē·ə }

enterotoxin [MICROBIO] A toxin produced by *Micrococcus pyogenes* var. *aureus* (*Staphylococcus aureus*) which gives rise to symptoms of food poisoning in humans and monkeys. { ˌent·ə·rōˈtäk·sən }

enterovirus [VIROL] One of the two subgroups of human picornaviruses; includes the polioviruses, the coxsackieviruses, and the echoviruses. { ˌent·ə·rōˈvī·rəs }

Enterozoa [ZOO] Animals with a digestive tract or cavity; includes all animals except Protozoa, Mesozoa, and Parazoa. { ˌent·ə·rəˈzō·ə }

entire [BIOL] Having a continuous, unimpaired margin. { enˈtīr }

Entner-Doudoroff pathway [BIOCHEM] A sequence of reactions for glucose degradation, with the liberation of energy; the distinguishing feature is the formation of 2-keto-3-deoxy-6-phosphogluconate from 6-phosphogluconate and the cleaving of this compound to yield pyruvate and glyceraldehyde-3-phosphate. { ¦ent·nər ¦dō·dəˌrȯf ˌpathˌwā }

entoblast [EMBRYO] A blastomere that differentiates into endoderm. { ˈent·əˌblast }

entoderm *See* endoderm. { ˈent·əˌdərm }

Entodiniomorphida [INV ZOO] An order of highly evolved ciliated protozoans in the subclass Spirotrichia, characterized by a smooth, firm pellicle and the lack of external ciliature. { ˌent·əˌdī·nē·əˈmȯr·fə·də }

entolecithal *See* endolecithal. { ˌent·əˈles·ə·thəl }

entomogenous [BIOL] Growing on or in an insect body, as certain fungi. { ˌent·əˈmäj·ə·nəs }

entomology [INV ZOO] A branch of the biological sciences that deals with the study of insects. { ˌent·əˈmäl·ə·jē }

entomophagous [ZOO] Feeding on insects. { ˌent·əˈmäf·ə·gəs }

entomophilic fungi [MYCOL] Fungi that parasitize insects. { ˌen·tə·məˌfil·ik ˈfən·jī }

entomophilous [ECOL] Pollinated by insects. { ˌent·əˈmäf·ə·ləs }

Entomophthoraceae [MYCOL] The single family of the order Entomophthorales. { ˌent·əˌmäf·thəˈrās·ēˌ; ame }

Entomophthorales [MYCOL] An order of mainly terrestrial fungi in the class Phycomycetes having a hyphal thallus and nonmotile sporangiospores, or conidia. { ˌent·əˌmäf·thəˈrā·lēz }

Entomostraca [INV ZOO] A group of Crustacea comprising the orders Cephalocarida, Branchiopoda, Ostracoda, Copepoda, Branchiura, and Cirripedia. { ˌent·əˈmä·strə·kə }

Entoniscidae [INV ZOO] A family of isopod crustaceans in the tribe Bopyrina that are parasitic in the visceral cavity of crabs and porcellanids. { ˌent·əˈnis·əˌdē }

entoplastron [VERT ZOO] The anterior median bony plate of the plastron of chelonians. { ¦en·tō¦plasˌträn }

Entoprocta [INV ZOO] A group of bryozoans, sometimes considered to be a subphylum, having a pseudocoelomate visceral cavity and the anus opening inside the circlet of tentacles. { ˌent·əˈpräk·tə }

entry site [CELL MOL] The ribosome site available for initial binding of transfer ribonucleic acid during genetic translation. { ˈen·trē ˌsīt }

entypy [EMBRYO] The formation of the amnion in certain mammals by the invagination of the embryonic knob into the yolk sac, without the formation of any amniotic folds. { ˈent·ə·pē }

enucleate [CELL MOL] To remove the nucleus from a cell. [MED] To remove an organ or a tumor in its entirety, as an eye from its socket. { ēˈnü·klēˌāt }

enuresis [MED] Urinary incontinence, especially in the absence of organic cause. { ˌen·yəˈrē·səs }

envelope [CELL MOL] The sum of all cell-surface elements that are located outside the plasma membrane. [VIROL] The outer membranous lipoprotein coat of certain viruses. Also known as bulb. { 'en·və,lōp }

environment [ECOL] The sum of all external conditions and influences affecting the development and life of organisms. { in'vī·ərn·mənt *or* in'vī·rən·ment }

environmental biology *See* ecology. { in¦vī·ərn ¦ment·əl bī'äl·ə·jē }

environmental pathology [MED] A branch of pathology concerned with abiotic environmental agents that influence human health. { in,vī·ərn¦ment·əl pa'thäl·ə·jē }

environmental resistance [ECOL] The effect of physical and biological factors in preventing a species from reproducing at its maximum rate. { in,vī·ərn¦men·təl ri'zis·təns }

environmental toxicology [MED] A broad field of study encompassing the production, fate, and effects of natural and synthetic pollutants in the environment. { in,vī·ərn,ment·əl ,täk·sə'käl·ə·jē }

environmental variance [GEN] That portion of the phenotypic variance caused by differences in the environments to which the individuals in a population have been exposed. { in¦vī·ərn,ment·əl 'ver·ē·əns }

enzootic [VET MED] **1.** A disease affecting animals in a limited geographic region. **2.** Pertaining to such a disease. { ¦en·zō¦äd·ik }

enzyme [BIOCHEM] Any of a group of catalytic proteins that are produced by living cells and that mediate and promote the chemical processes of life without themselves being altered or destroyed. { 'en,zīm }

enzyme induction [MICROBIO] The process by which a microbial cell synthesizes an enzyme in response to the presence of a substrate or of a substance closely related to a substrate in the medium. { 'en,zīm in'dək·shən }

enzyme inhibition [BIOCHEM] Prevention of an enzymic process as a result of the interaction of some substance with the enzyme so as to decrease the rate of reaction. { 'en,zīm ,in·ə'bish·ən }

enzyme-linked immunosorbent assay [MED] A laboratory technique in which a monoclonal antibody conjugated to an enzyme is used to rapidly detect and quantify the presence of an antigen in a sample. Abbreviated ELISA. { ¦en·zīm ¦liŋkt ,im·yə·nə¦sȯr·bənt 'a,sā }

enzyme-linked receptor [CELL MOL] The type of cell-surface receptor having an intracellular domain that either functions as an enzyme or is enzyme-associated, which is stimulated upon binding of a ligand to the receptor. { 'en,zīm ,liŋkt ri¦sep·tər }

enzyme repression [BIOCHEM] The process by which the rate of synthesis of an enzyme is reduced in the presence of a metabolite, often the end product of a chain of reactions in which the enzyme in question operates near the beginning. { 'en,zīm ri'presh·ən }

enzyme unit [BIOCHEM] The amount of an enzyme that will catalyze the transformation of 10^{-6} mole of substrate per minute or, when more than one bond of each substrate is attacked, 10^{-6} of 1 gram equivalent of the group concerned, under specified conditions of temperature, substrate concentration, and pH number. { 'en,zīm ,yü·nət }

enzymology [BIOCHEM] A branch of science dealing with the chemical nature, biological activity, and biological significance of enzymes. { ,en·zə'mäl·ə·jē }

Eocanthocephala [INV ZOO] An order of the Acanthocephala characterized by the presence of a small number of giant subcuticular nuclei. { ¦ē·ō¦kan·thō'sef·ə·lə }

Eosentomidae [INV ZOO] A family of primitive wingless insects in the order Protura that possess spiracles and tracheae. { ¦ē·ō,sen¦täm·ə,dē }

eosinophil [HISTOL] A granular leukocyte having cytoplasmic granules that stain with acid dyes and a nucleus with two lobes connected by a thin thread of chromatin. { ,ē·ə'sin·ə,fil }

eosinophil chemotactic factor [IMMUNOL] A peptide released from mast cell granules that stimulates chemotaxis of eosinophils; may be responsible for accumulation of eosinophils at sites of inflammation and allergic reactions. Abbreviated ECF. { ,ē·ə¦sin·ə·fil ¦kē·mō¦tak·tik 'fak·tər }

eosinophilia [MED] A greater than average number of circulating eosinophils. Also known as acidophilia; oxyphilia. { ,e·ə,sin·ə'fil·ē·ə }

eosinophilic erythroblast *See* normoblast. { ¦ē·ə¦sin·ə¦fil·ik ə'rith·rə,blast }

eosinophilic granuloma [MED] A disease, principally of childhood, characterized by foci of bone inflammation and granulation containing lipids, mononuclear cells, and eosinophils. { ¦ē·ə¦sin·ə¦fil·ik gran·yə'lō·mə }

eosinophilic pneumonitis *See* Loeffler's syndrome. { ¦ē·ə¦sin·ə¦fil·ik nü·mə'nīd·əs }

Epacridaceae [BOT] A family of dicotyledonous plants in the order Ericales, distinguished by palmately veined leaves, and stamens equal in number with the corolla lobes. { ,ep·ə·krə'dās·ē,ē }

epaulette [INV ZOO] **1.** Any of the branched or knobbed processes on the oral arms of many Scyphozoa. **2.** The first haired scale at the base of the costal vein in Diptera. { ¦ep·ə¦let }

epaxial [BIOL] Above or dorsal to an axis. { e'pak·sē·əl }

epaxial muscle [ANAT] Any of the dorsal trunk muscles of vertebrates. { e¦pak·sē·əl ¦məs·əl }

ependyma [HISTOL] The layer of epithelial cells lining the cavities of the brain and spinal cord. Also known as ependymal layer. { e'pen·də·mə }

ependymal layer *See* ependyma. { e'pen·də·məl ,lā·ər }

ependymoma [MED] A tumor of the central nervous system whose essential portion consists of cells derived from and resembling ependymal cells. Also known as medulloepithelioma. { e,pen·də'mō·mə }

Eperythrozoon [MICROBIO] A genus of the family Anaplasmataceae; rings and coccoids occur on erythrocytes and in the plasma of various vertebrates. { ˌep·əˌrith·rə'zō·ən }

ephapse [NEUROSCI] A false synapse between neighboring neurons where current current from one affects the other. { e'faps }

ephaptic transmission [PHYSIO] Electrical transfer of activity to a postephaptic unit by the action current of a preephaptic cell. { e'fap·tik tranz'mish·ən }

Ephedra [BOT] A genus of low, leafless, green-stemmed shrubs belonging to the order Ephedrales; source of the drug ephedrine. { ə'fed·rə }

Ephedrales [BOT] A monogeneric order of gymnosperms in the subdivision Gneticae. { ˌe·fə'drā·lēz }

ephemeral plant [BOT] An annual plant that completes its life cycle in one short moist season; desert plants are examples. { ə'fem·ə·rəl 'plant }

Ephemerida [INV ZOO] An equivalent name for the Ephemeroptera. { ˌe·fə'mer·ə·də }

Ephemeroptera [INV ZOO] The mayflies, an order of exopterygote insects in the subclass Pterygota. { əˌfem·ə'räp·tə·rə }

ephidrosis *See* hyperhidrosis. { ə'fid·rə·səs }

Ephydridae [INV ZOO] The shore flies, a family of myodarian cyclorrhaphous dipteran insects in the subsection Acalypteratae. { ə'fid·rəˌdē }

ephyra [INV ZOO] A larval, free-swimming medusoid stage of scyphozoans; arises from the scyphistoma by transverse fission. Also known as ephyrula. { 'e·fə·rə }

ephyrula *See* ephyra. { e'fir·ə·lə }

epiandrum [INV ZOO] The genital orifice of a male arachnid. { ˌep·ē'an·drəm }

epibasidium [MYCOL] A lengthening of the upper part of each cell of the basidium of various heterobasidiomycetes. { ˌep·ə·bə'sid·ē·əm }

epibiosis [ECOL] The arrangement in which organisms live on top of each other. { ˌep·ə·bī'ō·səs }

epibiotic [ECOL] Living, usually parasitically, on the surface of plants or animals; used especially of fungi. { ˌep·ə·bī'äd·ik }

epiblast *See* ectoderm. { 'ep·əˌblast }

epiblem [BOT] A tissue that replaces the epidermis in most roots and in stems of submerged aquatic plants. { 'ep·əˌblem }

epiboly [EMBRYO] The growing or extending of one part, such as the upper hemisphere of a blastula, over and around another part, such as the lower hemisphere, in embryogenesis. { ə'pib·ə·lē }

epibranchial [ANAT] Of or pertaining to the segment below the pharyngobranchial region in a branchial arch. { ¦ep·ə'bräŋ·kē·əl }

epicalyx [BOT] A ring of fused bracts below the calyx forming a structure that resembles the calyx. { ¦ep·ə'kā·liks }

epicardium [ANAT] The inner, serous portion of the pericardium that is in contact with the heart. [INV ZOO] A tubular prolongation of the branchial sac in certain ascidians which takes part in the process of budding. { ˌep·ə'kärd·ē·əm }

Epicaridea [INV ZOO] A suborder of the Isopoda whose members are parasitic on various marine crustaceans. { ˌep·ə·kə'rid·ē·ə }

epicarp [BOT] The outer layer of the pericarp. Also known as exocarp. { 'ep·əˌkärp }

epichordal [VERT ZOO] Located upon or above the notochord. { ¦ep·ə'kȯrd·əl }

epichordal brain [EMBRYO] The area of origin of the hindbrain or rhombencephalon, located on the dorsal side of the notochord. Also known as deutencephalon. { ¦ep·ə'kȯrd·əl 'brān }

epicnemial [ANAT] Of or pertaining to the anterior portion of the tibia. { ¦ep·ək¦nē·mē·əl }

epicondyle [ANAT] An eminence on the condyle of a bone. { ¦ep·ə'känˌdīl }

epicondylitis [MED] Infection or inflammation of an epicondyle. { ˌep·əˌkän·də'līd·əs }

epicone [INV ZOO] The part anterior to the equatorial groove in a dinoflagellate. { 'ep·əˌkōn }

epicotyl [BOT] The embryonic plant stem above the cotyledons. { ˌep·ə'käd·əl }

epicranium [INV ZOO] The dorsal wall of an insect head. [VERT ZOO] The structures covering the cranium in vertebrates. { ¦ep·ə'krā·nē·əm }

epicuticle [INV ZOO] The outer, waxy layer of an insect cuticle or exoskeleton. { ¦ep·ə'kyüd·i·kəl }

epicyst [INV ZOO] The outer layer of a cyst wall in encysted protozoans. Also known as ectocyst. { 'ep·əˌsist }

epidemic [MED] A sudden increase in the incidence rate of a disease to a value above normal, affecting large numbers of people and spread over a wide area. { ¦ep·ə¦dem·ik }

epidemic diarrhea of the newborn [MED] Contagious, fulminating diarrhea with high mortality, seen in newborns; caused by enteropathogenic strains of *Escherichia coli*, strains of *Staphylococcus*, other bacteria, and possibly viruses. { ¦ep·ə¦dem·ik ˌdī·ə'rē·ə əv t͟hə 'nüˌbȯrn }

epidemic encephalitis *See* encephalitis lethargica. { ¦ep·ə¦dem·ik inˌsef·ə'līd·əs }

epidemic gastroenteritis [MED] Inflammation of the stomach and intestine, of viral origin; considered to be epidemic when symptoms are manifested by a member of the patient's family within 10 days of the patient's recovery. { ¦ep·ə¦dem·ik ¦gas·trōˌent·ə'rīd·əs }

epidemic hepatitis *See* infectious hepatitis. { ¦ep·ə¦dem·ik ˌhep·ə'tīd·əs }

epidemic jaundice *See* infectious hepatitis. { ¦ep·ə¦dem·ik 'jȯn·dəs }

epidemic keratoconjunctivitis [MED] Inflammation of the cornea and conjunctiva, caused by a virus; epidemic by nature. { ¦ep·ə¦dem·ik ¦ker·əd·ō·kənˌjəŋk·tə'vīd·əs }

epidemic neuromyasthenia [MED] A prolonged, debilitating disease of the nervous system of adults; characterized by fatigue, headache, muscle pain, paresis, and emotional and mental disturbances; no etiologic agent has been isolated. Also known as acute infective encephalomyelitis; Akureyri disease; benign myalgic encephalomyelitis; epidemic vegetative neuritis; Iceland disease. { ¦ep·ə¦dem·ik ¦nü·rō·mī·əs'thē·nē·ə }

epidemic pleurodynia [MED] An acute epidemic disease of humans, caused by coxsackie B virus; characterized by severe pain in the lower thorax and upper abdomen, and associated with fever and malaise. { ¦ep·ə¦dem·ik ˌplür·ə'din·ē·ə }

epidemic roseola *See* rubella. { ¦ep·ə¦dem·ik ˌrō·zē'ō·lə }

epidemic typhus *See* classic epidemic typhus. { ¦ep·ə¦dem·ik 'tī·fəs }

epidemic vegetative neuritis *See* epidemic neuromyasthenia. { ¦ep·ə¦dem·ik 'vej·əˌtād·iv nə'rīd·əs }

epidemiological study [MED] A population study designed to examine associations (commonly, hypothesized causal relations) between personal characteristics and environmental exposures that increase the risk of disease. { ˌep·əˌdē·mē·ə¦läj·ə·kəl 'stəd·ē }

epidemiology [MED] The study of the mass aspects of disease. { ˌep·əˌdē·mē'äl·ə·jē }

epidermal growth factor [PHYSIO] A polypeptide produced in animals that stimulates and sustains the replication of epidermal cells (of ectodermal or endodermal origin); its human equivalent is urogastrone. { ˌep·ə¦dərm·əl 'grōth ˌfak·tər }

epidermal ridge [ANAT] Any of the minute corrugations of the skin on the palmar and plantar surfaces of humans and other primates. { ˌep·ə'dər·məl 'rij }

epidermis [BOT] The outermost layer (sometimes several layers) of cells on the primary plant body. [HISTOL] The outer nonsensitive, nonvascular portion of the skin comprising two strata of cells, the stratum corneum and the stratum germinativum. { ˌep·ə'dər·məs }

epidermoid carcinoma *See* squamous-cell carcinoma. { ˌep·ə'dərˌmȯid ˌkärs·ən'ō·mə }

epidermoid cyst [MED] A cyst lined by stratified squamous epithelium without associated cutaneous glands. { ˌep·ə'dərˌmȯid 'sist }

epidermolysis [MED] The easy separation of various layers of skin, primarily of the epidermis from the corium, observed in certain pathological conditions. { ˌep·ə·dər'mäl·ə·səs }

epidermolysis bullosa [MED] A congenital skin disease characterized by the development of vesicles and bullae upon slight, or even without, trauma. { ˌep·ə·dər'mäl·ə·səs bů'lō·sə }

epididymis [ANAT] The convoluted efferent duct lying posterior to the testis and connected to it by the efferent ductules of the testis. { ˌep·ə'did·ə·məs }

epididymitis [MED] Inflammation of the epididymis. { ˌep·əˌdid·ə'mīd·əs }

epidural [ANAT] Located on or over the dura mater. { ¦ep·ə¦dür·əl }

epifauna [ZOO] Benthic fauna that live on a surface, such as the sea floor, other organisms, or objects. { 'ep·əˌfȯn·ə }

epigaster [EMBRYO] The portion of the intestine in vertebrate embryos which gives rise to the colon. { 'ep·əˌgas·tər }

epigastric region [ANAT] The upper and middle part of the abdominal surface between the two hypochondriac regions. Also known as epigastrium. { ¦ep·ə¦gas·trik ˌrē·jən }

epigastrium [ANAT] *See* epigastric region. [INV ZOO] The ventral side of mesothorax and metathorax in insects. { ˌep·ə'gas·trē·əm }

epigean [BOT] Pertaining to a plant or plant part that grows above the ground surface. [ZOO] Living near or on the ground surface, applied especially to insects. { ¦ep·ə¦jē·ən }

epigenesis [EMBRYO] Development in gradual stages of differentiation. { ˌep·ə'jen·ə·səs }

epigenetics [GEN] The study of those processes by which genetic information ultimately results in distinctive physical and behavioral characteristics. { ¦ep·ə·jə¦ned·iks }

epigenotype [GEN] The total developmental system through which the adult form of an organism is realized, comprising the interactions among genes and between genes and the nongenetic environment. { ˌep·ə'jēn·əˌtīp }

epigenous [BOT] Developing or growing on a surface, especially of a plant or plant part. { ə'pij·ə·nəs }

epiglottis [ANAT] A flap of elastic cartilage covered by mucous membrane that protects the glottis during swallowing. { ˌep·ə'gläd·əs }

epigynous [BOT] Having the perianth and stamens attached near the top of the ovary; that is, the ovary is inferior. { ə'pij·ə·nəs }

epigynum [INV ZOO] **1.** The genital pore of female arachnids. **2.** The plate covering this opening. { ə'pij·ə·nəm }

epilation [MED] Removal of the hair by the roots by the use of forceps, chemical means, or roentgenotherapy. { ˌep·ə'lā·shən }

epilemma [HISTOL] The perineurium of very small nerves. { ˌep·ə'lem·ə }

epilepsy [MED] A condition characterized by the paroxysmal recurrence of transient, uncontrollable episodes of abnormal neurological or mental function, or both. { 'ep·əˌlep·sē }

epileptic equivalent [MED] Episodes of sensory or motor phenomena experienced by an epileptic instead of convulsions. Also known as convulsive equivalent. { ¦ep·əˌlep·tik i'kwiv·ə·lənt }

epimerase [BIOCHEM] A type of enzyme that catalyzes the rearrangement of hydroxyl groups on a substrate. { ə'pim·əˌrās }

epimere [ANAT] The dorsal muscle plate of the lining of a coelomic cavity. [EMBRYO] The dorsal part of a mesodermal segment in the embryo of chordates. { 'ep·əˌmir }

epimeron [INV ZOO] **1.** The posterior plate of

the pleuron in insects. **2.** The portion of a somite between the tergum and the insertion of a limb in arthropods. { ˌep·əˈmirˌän }

epimorphosis [PHYSIO] Regeneration in which cell proliferation precedes differentiation. { ˌep·əˈmȯr·fə·səs }

epimyocardium [EMBRYO] The layer of the embryonic heart from which both the myocardium and epicardium develop. { ˌep·əˌmī·əˈkärd·ē·əm }

epimysium [ANAT] The connective-tissue sheath surrounding a skeletal muscle. { ˌep·əˈmī·sē·əm }

epinasty [BOT] Growth changes in which the upper surface of a leaf grows, thus bending the leaf downward. { ˈep·əˌnas·tē }

epinephrine [BIOCHEM] $C_9H_{13}O_3N$ A hormone secreted by the adrenal medulla that acts to increase blood pressure due to stimulation of heart action and constriction of peripheral blood vessels. Also known as adrenaline. { ˌep·əˈne·frən }

epineural [ANAT] Arising from or on the outside of a nerve trunk. [INV ZOO] The nervous tissue dorsal to the ventral nerve cord in arthropods. { ˌep·əˈnu̇r·əl }

epineural canal [INV ZOO] A canal that runs between the radial nerve and the epithelium in echinoids and ophiuroids. { ˌep·əˈnu̇r·əl kəˈnal }

epineurium [ANAT] The connective-tissue sheath of a nerve trunk. { ˌep·əˈnu̇r·ē·əm }

epipetalous [BOT] Having stamens located on the corolla. { ¦ep·əˈped·əl·əs }

epiphallus [INV ZOO] A sclerite in some orthopterans in the floor of the genital chamber. { ¦ep·əˈfal·əs }

epipharynx [INV ZOO] An organ attached beneath the labrium of many insects. { ¦ep·əˈfar·iŋks }

epiphora [MED] An abnormal increase in tearing of one or both eyes. { ˌep·ə ˈfȯr·ə }

epiphragm [BOT] A membrane covering the aperture of the capsule in certain mosses. [INV ZOO] A membranous or calcareous partition that covers the aperture of certain hibernating land snails. { ˈep·əˌfram }

epiphyll [ECOL] A plant that grows on the surface of leaves. { ˈep·əˌfil }

epiphyseal arch [EMBRYO] The arched structure in the third ventricle of the embryonic brain, which marks the site of development of the pineal body. { ə¦pif·ə¦sē·əl ˈärch }

epiphyseal plate [ANAT] **1.** The broad, articular surface on each end of a vertebral centrum. **2.** The thin layer of cartilage between the epiphysis and the shaft of a long bone. Also known as metaphysis. { ə¦pif·ə¦sē·əl ˈplāt }

epiphysiolysis [MED] The separation of an epiphysis from the shaft of abone. { əˌpif·ə·sēˈäl·ə·səs }

epiphysis [ANAT] **1.** The end portion of a long bone in vertebrates. **2.** *See* pineal body. { əˈpif·ə·səs }

epiphyte [ECOL] A plant which grows nonparasitically on another plant or on some nonliving structure, such as a building or telephone pole, deriving moisture and nutrients from the air. Also known as aerophyte. { ˈep·əˌfīt }

epiplankton [BIOL] Plankton occurring in the sea from the surface to a depth of about 100 fathoms (180 meters). { ¦ep·əˈplaŋk·tən }

epipleural [ANAT] Arising from a rib. [VERT ZOO] An intramuscular bone arising from and extending between some of the ribs in certain fishes. { ¦ep·əˈplu̇r·əl }

epiploic foramen [ANAT] An aperture of the peritoneal cavity, formed by folds of the peritoneum and located between the liver and the stomach. Also known as foramen of Winslow. { ¦ep·ə¦plō·ik fəˈrā·mən }

epipodite [INV ZOO] A branch of the basal joint of the protopodite of thoracic limbs of many arthropods. { əˈpip·əˌdīt }

epipodium [BOT] The apical portion of an embryonic phyllopodium. [INV ZOO] **1.** A ridge or fold on the lateral edges of each side of the foot of certain gastropod mollusks. **2.** The elevated ring on an ambulacral plate in Echinoidea. { ˌep·əˈpōd·ē·əm }

Epipolasina [INV ZOO] A suborder of sponges in the order Clavaxinellida having radially arranged monactinal or diactinal megascleres. { ˌep·ə·pəˈlaz·ə·nə }

epiproct [INV ZOO] A plate above the anus forming the dorsal part of the tenth or eleventh somite of certain insects. { ˈep·əˌpräkt }

epipubis [VERT ZOO] A single cartilage or bone located in front of the pubis in some vertebrates, particularly in some amphibians. { ¦ep·əˈpyü·bəs }

episclera [ANAT] The loose connective tissue lying between the conjunctiva and the sclera. { ¦ep·əˈskler·ə }

episepalous [BOT] Having stamens growing on or adnate to the sepals. { ¦ep·əˈsep·ə·ləs }

episiotomy [MED] Medial or lateral incision of the vulva during childbirth, to avoid undue laceration. { əˌpēz·ēˈäd·ə·mē }

episome [GEN] A circular genetic element in bacteria, presumably a deoxyribonucleic acid fragment, which is not necessary for survival of the organism and which can be integrated in the bacterial chromosome or remain free. { ˈep·əˌsōm }

epispadias [MED] A congenital defect of the anterior urethra in which the canal terminates on the dorsum of the penis and posterior to its normal opening. { ˌep·əˈspād·ē·əs }

episperm *See* testa. { ˈep·əˌspərm }

epistasis [GEN] The suppression of the effect of one gene by another. [MED] A checking or stoppage of a hemorrhage or other discharge. [PATH] A scum or film of substance floating on the surface of urine. { əˈpis·tə·səs }

episternum [VERT ZOO] A dermal bone or pair of bones ventral to the sternum of certain fishes and reptiles. { ¦ep·əˈstər·nəm }

epistome [INV ZOO] **1.** The area between the

mouth and the second antennae in crustaceans. **2.** The plate covering this region. **3.** The area between the labrum and the epicranium in many insects. **4.** A flap covering the mouth of certain bryozoans. **5.** The area just above the labrum in certain dipterans. { 'ep·ə‚stōm }

epithalamus [ANAT] A division of the vertebrate diencephalon including the habenula, the pineal body, and the posterior commissure. { ¦ep·ə'thal·ə·məs }

epitheca [INV ZOO] **1.** An external, calcareous layer around the basal portion of the theca of many corals. **2.** A protective covering of the epicone. **3.** The outer portion of a diatom frustule. { ¦ep·ə'thē·kə }

epitheliochorial placenta [EMBRYO] A type of placenta in which the maternal epithelium and fetal epithelium are in contact. Also known as villous placenta. { ‚ep·ə¦thē·lē·ō'kȯr·ē·əl plə 'sen·tə }

epithelioid cell [HISTOL] A macrophage that resembles an epithelial cell. Also known as alveolated cell. { ‚ep·ə'thē·lē‚ȯid ‚sel }

epithelioma [MED] A tumor derived from epithelium; usually a skin cancer, occasionally cancer of a mucous membrane. { ‚ep·ə‚thē·lē'ō·mə }

epitheliomuscular cell [INV ZOO] An epithelial cell with an elongate base that contains contractile fibrils; common among cnidarians. { ‚ep·ə¦thē·lē·ō'məs·kyə· lər 'sel }

epithelium [HISTOL] A primary animal tissue, distinguished by cells being close together with little intercellular substance; covers free surfaces and lines body cavities and ducts. { ‚ep·ə'thē·lē·əm }

epithema [VERT ZOO] A horny outgrowth on the beak of certain birds. { ‚ep·ə'thē·mə }

epitoke [INV ZOO] The posterior portion of marine polychaetes; contains the gonads. { 'ep·ə‚tōk }

epitoky [INV ZOO] In certain polychaetes, development of the posterior sexual part from the anterior sexless part. { 'ep·ə‚täk·ē }

epitope [IMMUNOL] The portion of the antigen molecule that determines its capacity to combine with the specific combining site of its corresponding antibody in an antigen-antibody interaction. { 'ep·ə‚tōp }

epitrichium [EMBRYO] The outer layer of the fetal epidermis of many mammals. { ¦ep·ə'trik·ē·əm }

epitrochlear [ANAT] Of or pertaining to a lymph node that lies above the trochlea of the elbow joint. { ¦ep·ə'trō·klē·ər }

epituberculosis [MED] A massive pulmonary shadow seen in x-ray films in active juvenile tuberculosis, probably caused by bronchial obstruction. { ¦ep·ə·tə‚bər·kyə'lō·s/əs }

epitympanum [ANAT] The attic of the middle ear, or tympanic cavity. { ¦ep·ə'tim·pə·nəm }

epivalve [INV ZOO] **1.** The upper or apical shell of certain dinoflagellates. **2.** The upper shell of a diatom. { 'ep·ə‚valv }

epixylous [ECOL] Growing on wood; used especially of fungi. { ¦ep·ə¦zī·ləs }

epizoic [BIOL] Living on the body of an animal. { ¦ep·ə¦zō·ik }

epizootic [VET MED] **1.** Affecting many animals of one kind in one region simultaneously; widely diffuse and rapidly spreading. **2.** An extensive outbreak of an epizootic disease. { ¦ep·ə·zō¦äd·ik }

epizootiology [VET MED] The study of epizootics. { ‚ep·ə·zō‚äd·ē'äl·ə·jē }

eponychium [ANAT] The horny layer of the nail fold attached to the nail plate at its margin; represents the remnant of the embryonic condition. [EMBRYO] A horny condition of the epidermis from the second to the eighth month of fetal life, indicating the position of the future nail. { ‚ep·ə'nik·ē·əm }

epoophoron [ANAT] A blind longitudinal duct and 10–15 transverse ductules in the mesosalpinx near the ovary which represent remnants of the reproductive part of the mesonephros in the female; homolog of the head of the epididymis in the male. Also known as parovarium; Rosenmueller's organ. { ¦ep·ō'äf·ə‚rän }

Epsilonematoidea [INV ZOO] A superfamily of small (0.5-millimeter) marine nematodes in the order Desmodorida; the body is strongly arched in a sigmoid manner when relaxed. { ¦ep·si‚län·ə·mə'tȯid·ē·ə }

Epstein-Barr virus [VIROL] Herpeslike virus particles first identified in cultures of cells from Burkett's malignant lymphoma. { ¦ep·stīn ¦bär ‚vī·rəs }

epulis [MED] A benign tumorlike lesion of the gingiva. { ə'pyü·ləs }

equatorial plane [CELL MOL] The plane in a cell undergoing mitosis that is midway between the centrosomes and perpendicular to the spindle fibers. { ‚e·kwə'tȯr·ē·əl 'plān }

Equidae [VERT ZOO] A family of perissodactyl mammals in the superfamily Equoidea, including the horses, zebras, and donkeys. { 'ēk·wə‚dē }

equilibrium population [EVOL] A population in which the gene frequencies have reached an equilibrium between mutation pressure and selection pressure. { ‚ē·kwə¦lib·rē·əm ‚päp·yə¦lā·shən }

equine [VERT ZOO] **1.** Resembling a horse. **2.** Of or related to the Equidae. { 'ē‚kwīn }

equine encephalitis [MED] A disease of equines and humans caused by one of three viral strains: eastern, western, and Venezuelan equine viruses. Also known as equine encephalomyelitis. { 'ē‚kwīn en‚sef·ə'līd·əs }

equine encephalomyelitis *See* equine encephalitis. { 'ē‚kwīn en¦sef·ə·lō‚mī·ə'līd·əs }

Equisetales [BOT] The horsetails, a monogeneric order of the class Equisetopsida; the only living genus is *Equisetum*. { ‚e·kwə·sə'tā·lēz }

Equisetatae *See* Equisetopsida. { ‚e·kwə·sə 'tā‚tē }

Equisetineae [BOT] The equivalent name for the Equisetophyta. { ‚e·kwə·sə'tin·ē‚ē }

Equisetophyta [BOT] A division of the subkingdom Embryobionta represented by a single living genus, *Equisetum*. { ˌe·kwə·sə'täf·ə·də }

Equisetopsida [BOT] A class of the division Equisetophyta whose members made up a major part of the flora, especially in moist or swampy places, during the Carboniferous Period. { ˌe·kwə·sə'täp·sə·də }

equitant [BOT] Of leaves, overlapping transversely at the base. { 'e·kwəd·ənt }

equivalence zone *See* zone of optimal proportion. { i'kwiv·ə·ləns ˌzōn }

Equoidea [VERT ZOO] A superfamily of perissodactyl mammals in the suborder Hippomorpha comprising the living and extinct horses and their relatives. { ē'kwȯīd·ē·ə }

Equus [VERT ZOO] The genus comprising the large, one-toed modern horses, including donkeys and zebras. { 'e·kwəs }

ER *See* endoplasmic reticulum.

erection [PHYSIO] The enlarged state of erectile tissue when engorged with blood, as of the penis or clitoris. { i'rek·shən }

erector [PHYSIO] Any muscle that produces erection of a part. { i'rek·tər }

erect stem [BOT] A stem that stands, having a vertical or upright habit. { i'rekt 'stem }

Eremascoideae [BOT] A monogeneric subfamily of ascosporogenous yeasts characterized by mostly septate mycelia, and spherical asci with eight oval to round ascospores. { er·ə·mə'skȯid·ē,ē }

Erethizontidae [VERT ZOO] The New World porcupines, a family of rodents characterized by sharply pointed, erectile hairs and four functional digits. { ˌer·ə·thə'zänt·əˌdē }

ERG *See* electroretinogram.

Ergasilidae [INV ZOO] A family of copepod crustaceans in the suborder Cyclopoida in which the females are parasitic on aquatic animals, while the males are free-swimming. { ər·gə'sil·əˌdē }

ergastic [CELL MOL] Pertaining to the nonliving components of protoplasm. { ər'gas·tik }

ergastoplasm [CELL MOL] A cytoplasm component which shows an affinity for basic dyes; a form of the endoplasmic reticulum. { ˌər'gas·təˌplaz·əm }

ergosterin *See* ergosterol. { ər'gäs·tə·rən }

ergosterol [BIOCHEM] $C_{28}H_{44}O$ A crystalline, water-insoluble, unsaturated sterol found in ergot, yeast, and other fungi, and which may be converted to vitamin D_2 on irradiation with ultraviolet light or activation with electrons. Also known as ergosterin. { ər'gäs·təˌrȯl }

ergot [MYCOL] The dark purple or black sclerotium of the fungus *Claviceps purpurea*. { 'ər·gət }

ergotism [MED] Acute or chronic intoxication resulting from ingestion of grain infected with ergot fungus, or from chronic use of drugs containing ergot. { 'ər·gəˌtiz·əm }

Ericaceae [BOT] A large family of dicotyledonous plants in the order Ericales distinguished by having twice as many stamens as corolla lobes. { ˌer·ə'kās·ē,ē }

ericaceous mycorrhizal fungi [MYCOL] Mycorrhizal fungi that form symbiotic relationships with plants in the Ericaceae family; they are divided into three subgroups based on the presence or absence of a Hartig net, a fungal sheath, and fungal hyphae within the root cells. { er·əˈkā·shəs ˌmī·kəˈrīz·əl 'fən·jī }

Ericales [BOT] An order of dicotyledonous plants in the subclass Dilleniidae; plants are generally sympetalous with unitegmic ovules and they have twice as many stamens as petals. { ˌer·ə'kā·lēz }

ericophyte [ECOL] A plant that grows on a heath or moor. { 'er·ək·əˌfīt }

Erinaceidae [VERT ZOO] The hedgehogs, a family of mammals in the order Insectivora characterized by dorsal and lateral body spines. { ˌer·ə·nə'sē·əˌdē }

Erinnidae [INV ZOO] A family of orthorrhaphous dipteran insects in the series Brachycera. { ə'rin·əˌdē }

Eriocaulaceae [BOT] The single family of the order Eriocaulales. { ˌer·ē·ōˌkȯ'lās·ēˌē }

Eriocaulales [BOT] An order of monocotyledonous plants in the subclass Commelinidae, having a perianth reduced or lacking and having unisexual flowers aggregated on a long peduncle. { ˌer·ē·ōˌkȯ'lā·lēz }

Eriococcinae [INV ZOO] A family of homopteran insects in the superfamily Coccoidea; adult females and late instar nymphs have an anal ring. { ˌer·ē·ō'käk·səˌnē }

Eriocraniidae [INV ZOO] A small family of lepidopteran insects in the superfamily Eriocranioidea. { ˌer·ē·ō·krə'nī·əˌdē }

Eriocranioidea [INV ZOO] A superfamily of lepidopteran insects in the suborder Homoneura comprising tiny moths with reduced, untoothed mandibles. { ˌer·ē·ōˌkrā·nē'ȯid·ē·ə }

Eriophyidae [INV ZOO] The bud mites or gall mites, a family of economically important plant-feeding mites in the suborder Trombidiformes. { ˌer·ē·ō'fī·əˌdē }

eriophyllous [BOT] Having leaves covered by a cottony pubescence. { er·ē'äf·ə·ləs }

erose [BIOL] Having an irregular margin. { ē'rōs }

erosion [MED] **1.** Surgical removal of tissues by scraping. **2.** Excision of a joint. { ə'rō·zhən }

Erotylidae [INV ZOO] The pleasing fungus beetles, a family of coleopteran insects in the superfamily Cucujoidea. { ˌer·ə'tī·ləˌdē }

Errantia [INV ZOO] A group of 34 families of polychaete annelids in which the anterior region is exposed and the linear body is often long and is dorsoventrally flattened. { ə'ran·chə }

eruciform [INV ZOO] In certain insect larvae, having a soft cylindrical body with a well-defined head and usually short thoracic legs. { ə'rüs·əˌfȯrm }

Erwinia [MICROBIO] A genus of motile, rod-shaped bacteria in the family Enterobacteriaceae; these organisms invade living plant tissues and cause dry necroses, galls, wilts, and soft rots. { ər'win·ē·ə }

Erwinieae [MICROBIO] Formerly a tribe of phytopathogenic bacteria in the family Enterobacteriaceae, including the single genus *Erwinia*. { ər'win·ē,ē }

erysipelas [MED] An acute, infectious bacterial disease caused by *Streptococcus pyogenes* and characterized by inflammation of the skin and subcutaneous tissues. { ,er·ə'sip·ə·ləs }

erysipeloid [MED] A bacterial infection caused by *Erysipelothrix rhuscopathiae* and occurring on the hands of people who handle infected meat or fish. { ,er·ə'sip·ə,lòid }

Erysipelothrix [MICROBIO] A genus of gram-positive, rod-shaped bacteria of uncertain affiliation; cells have a tendency to form long filaments. { ,er·ə'sip·ə·lō,thriks }

Erysiphaceae [MYCOL] The powdery mildews, a family of ascomycetous fungi in the order Erysiphales with light-colored mycelia and conidia. { ,er·ə·sə'fās·ē,ē }

Erysiphales [MYCOL] An order of ascomycetous fungi which are obligate parasites of seed plants, causing powdery mildew and sooty mold. { ,er·ə·sə'fā·lēz }

erythema [MED] Localized redness of the skin in areas of variable size. { ,er·ə'thē·mə }

erythema migrans [MED] An expanding skin lesion characterized by a small red papule or macule that is a unique clinical marker for Lyme disease. { ,er·ə¦thē·mə 'mī,granz }

erythema multiforme [MED] An acute inflammatory skin disease characterized by red macules, papules, or tubercles on the extremities, neck, and face. { ,er·ə'thē·mə ,məl·tə'fòr,mē }

erythema nodosum [MED] The occurrence of pink to blue, tender nodules on the anterior surfaces of the lower legs; more frequent in women than men. { ,er·ə'thē·mə nō'dō·səm }

erythremia *See* erythrocytosis; polycythemia vera. { ,er·ə'thrē·mē·ə }

erythroblast [HISTOL] A nucleated cell occurring in bone marrow as the earliest recognizable cell of the erythrocytic series. { ə'rith·rə,blast }

erythroblastosis [MED] The abnormal presence of erythroblasts in the blood. [VET MED] A virus disease of birds; considered to be part of the avian leukosis complex in which there is an abnormal number of erythroblasts in the blood. { ə,rith·rə,bla'stō·səs }

erythroblastosis fetalis [MED] A form of hemolytic anemia affecting the fetus and newborn infant when a mother is Rh-negative and has developed antibodies against an Rh-positive fetus. Also known as hemolytic disease of newborn. { ə,rith·rə,bla'stō·səs fē'tal·əs }

erythrocruorin [BIOCHEM] Any of the iron-porphyrin protein respiratory pigments found in the blood and tissue fluids of certain invertebrates; corresponds to hemoglobin in vertebrates. { ə,rith·rə'krü·ə·rən }

erythrocyte [HISTOL] A type of blood cell that contains a nucleus in all vertebrates but humans and that has hemoglobin in the cytoplasm. Also known as red blood cell. { ə'rith·rə,sīt }

erythrocytopoiesis *See* erythropoiesis. { ə,rith·rə,sīd·ə,pòi'ē·səs }

erythrocytosis [MED] An increase in the number of circulating erythrocytes of more than two standard deviations above the mean normal, usually occurring secondary to hypoxia. Also known as erythremia; polycythemia. { ə,rith·rə,sī'tō·səs }

erythrodermia [MED] A skin condition in which the whole body surface is marked by an inflammatory blood vessel dilation. { ə,rith·rə'dər·mē·ə }

erythromelalgia [MED] A cutaneous vasodilation of the feet or, more rarely, of the hands; characterized by redness, mottling, changes in skin temperature, and neuralgic pains. Also known as acromelalgia; Mitchell's disease. { ə,rith·rō·mə'lal·jē·ə }

erythromycin [MICROBIO] A crystalline antibiotic produced by *Streptomyces erythreus* and used in the treatment of gram-positive bacterial infections. { ə,rith·rə'mīs·ən }

D-erythropentose *See* ribulose. { ¦dē ə,rith·rə'pen,tōs }

erythrophilous [BIOL] Having an affinity for red dyes and other coloring matter. { ¦er·ə¦thräf·ə·ləs }

erythrophore [ZOO] A chromatophore containing a red pigment, especially a carotenoid. { ə'rith·rə,fòr }

erythropia *See* erythropsia. { ,er·ə'thrō·pē·ə }

erythropoiesis [PHYSIO] The process by which erythrocytes are formed. Also known as erythrocytopoiesis. { ə,rith·rə,pòi'ē·səs }

erythropoietin [BIOCHEM] A hormone, thought to be produced by the kidneys, that regulates erythropoiesis, at least in higher vertebrates. { ə,rith·rə'pòi·ət·ən }

erythropsia [MED] An abnormality of vision in which all objects appear red; red vision. Also known as erythropia. { ,er·ə'thräp·sē·ə }

erythrosis [MED] **1.** Overproliferation of erythropoietic tissue, as found in polycythemia. **2.** The unusual red skin color of individuals with polycythemia. { ə'rith·rə·səs }

Erythroxylaceae [BOT] A homogeneous family of dicotyledonous woody plants in the order Linales characterized by petals that are internally appendiculate, three carpels, and flowers without a disk. { ,er·ə,thräk·sə'lās·ē,ē }

erythrulose [BIOCHEM] $C_4H_8O_4$ A ketose sugar occurring as an oxidation product of erythritol due to the action of certain bacteria. { ə'rith·rə,lōs }

Esbach's reagent [PATH] A solution of 1 gram of trinitrophenol and 2 grains of citric acid in 100 milliliters of water; used in determining albumin in urine. { 'es,bäks rē,ā·jənt }

eschar [MED] A dry crust or slough, especially one formed after a thermal or chemical burn. { 'es·kər }

escharotic [MED] **1.** Caustic. **2.** Producing an eschar. { ¦es·kə¦räd·ik }

Escherichia [MICROBIO] A genus of bacteria in

the family Enterobacteriaceae; straight rods occurring singly or in pairs. { ˌesh·əˈrik·ē·ə }

Escherichia coli [MICROBIO] The type species of the genus, occurring as part of the normal intestinal flora in vertebrates. Also known as colon bacillus. { ˌesh·əˈrik·ē·ə ˈkōˌlī }

Escherichia coli O157:H7 [MICROBIO] An unusually virulent food-borne pathogen that is found primarily in cattle and causes severe, sometimes life-threatening illness; symptoms include hemorrhagic colitis, hemolytic uremic syndrome, and thrombotic thrombocytopenic purpura. { ˌes·kəˈrēk·ē·ə ˈkōˌlī ¦ō¦wən¦fīv¦sev·ən ˈāch¦sev·ən }

Escherichieae [MICROBIO] Formerly a tribe of bacteria in the family Enterobacteriaceae defined by the ability to ferment lactose, with the rapid production of acid and visible gas. { ˌesh·ə·rəˌkī·ēˌē }

esculin [PHARM] $C_{15}H_{16}O_9$ A substance extracted from the leaves and bark of the horse chestnut tree; used as a skin protectant. { ˈes·kyə·lən }

esculoside *See* esculin. { esˈkyü·ləˌsīd }

Esocidae [VERT ZOO] The pikes, a family of fishes in the order Clupeiformes characterized by an elongated beaklike snout and sharp teeth. { əˈsäs·əˌdē }

Esocoidei [VERT ZOO] A small suborder of fresh-water fishes in the order Salmoniformes; includes the pikes, mudminnows, and pickerels. { ˌes·əˈkȯid·ēˌī }

esophageal diverticulum [MED] An outpocketing of the wall of the esophagus, or of the pharynx just above the opening to the esophagus. { ə¦säf·ə¦jē·əl ˌdī·vərˈtik·yə·ləm }

esophageal fistula [MED] Congenitally, an abnormal tube communicating between the esophagus and an internal organ, usually the trachea; an acquired esophageal fistula usually communicates between the esophagus and the skin through an external opening, or may communicate with internal organs. { ə¦säf·ə¦jē·əl ˈfis·chə·lə }

esophageal gland [ANAT] Any of the digestive glands within the submucosa of the esophagus; secretions are chiefly mucus and serve to lubricate the esophagus. { ə¦säf·ə¦jē·əl ˈgland }

esophageal hiatus [ANAT] The opening in the diaphragm for passage of the esophagus. { ə¦säf·ə¦jē·əl hīˈād·əs }

esophageal teeth [VERT ZOO] The enamel-tipped hypapophyses of the posterior cervical vertebrae of certain snakes, which penetrate the esophagus and function to break eggshells. { ə¦säf·ə¦jē·əl ˈtēth }

esophagitis [MED] Inflammation of the esophagus. { əˌsäf·əˈjīd·əs }

esophagogastrostomy [MED] Establishment, by surgery, of an anastomosis between the esophagus and the stomach; may be performed by the abdominal route or by transpleural operation. { ə¦säf·ə·gōˌgaˈsträs·tə·mē }

esophagoscopy [MED] Endoscopic examination of the interior of the esophagus. { eˌsäf·əˈgäs·kə·pē }

esophagus [ANAT] The tubular portion of the alimentary canal interposed between the pharynx and the stomach. Also known as gullet. { əˈsäf·ə·gəs }

esotropia [MED] Convergent strabismus, occurring when one eye fixes upon an object and the other deviates inward. Also known as cross-eye. { ˌes·əˈtrō·pē·ə }

essential amino acid [BIOCHEM] Any of eight of the 20 naturally occurring amino acids that are indispensable for optimum animal growth but cannot be formed in the body and must be supplied in the diet. { iˈsen·chəl əˈmē·nō ˌas·əd }

essential fatty acid [BIOCHEM] Any of the polyunsaturated fatty acids which are required in the diet of mammals; they are probably precursors of prostaglandins. { iˈsen·chəl ˈfad·ē ˌas·əd }

essential hypertension [MED] Elevation of the systemic blood pressure, of unknown origin. Also known as primary hypertension. { iˈsen·chəl ˌhī·pərˈten·chən }

esterase [BIOCHEM] Any of a group of enzymes that catalyze the synthesis and hydrolysis of esters. { ˈes·təˌrās }

esthacyte [INV ZOO] A simple sensory cell occurring in certain lower animals, such as sponges. Also spelled aesthacyte. { ˈes·thə ˌsīt }

esthesia [PHYSIO] The capacity for sensation, perception, or feeling. Also spelled aesthesia. { esˈthē·zhə }

esthesioneuroblastoma *See* neuroepithelioma. { es¦thē·zē·ō¦nu̇r·ōˌblaˈstō·mə }

esthesioneuroepithelioma *See* neuroepithelioma. { es¦thē·zē·ō¦nu̇r·ōˌep·əˌthē·lēˈō·mə }

esthiomene [MED] The chronic ulcerative lesion of the vulva in lymphogranuloma venereum. { esˈthī·əˌmēn }

estimated exposure concentration [MED] Measured or calculated amount or mass concentration of a substance to which an organism is likely to be exposed. { ˌes·təˌmād·əd ikˈspō·zhər ˌkəns·ən¦trā·shən }

estivation [PHYSIO] **1.** The adaptation of certain animals to the conditions of summer, or the taking on of certain modifications, which enables them to survive a hot, dry summer. **2.** The dormant condition of an organism during the summer. { ˌes·təˈvā·shən }

estivoautumnal malaria *See* falciparum malaria. { ¦es·tə·vōˌȯ¦təm·nəl məˈler·ē·ə }

estradiol [BIOCHEM] $C_{18}H_{24}O_2$ An estrogenic hormone produced by follicle cells of the vertebrate ovary; provokes estrus and proliferation of the human endometrium, and stimulates ICSH (interstitial-cell-stimulating hormone) secretion. { ˌes·trəˈdīˌȯl }

estriol [BIOCHEM] $C_{18}H_{24}O_3$ A crystalline estrogenic hormone obtained from human pregnancy urine. { ˈeˌstrīˌȯl }

estrogen [BIOCHEM] Any of various natural or

synthetic substances possessing the biologic activity of estrus-producing hormones. { 'es·trə·jən }

estrogenic hormone [BIOCHEM] A hormone, found principally in ovaries and also in the placenta, which stimulates the accessory sex structures and the secondary sex characteristics in the female. { ¦es·trə¦jen·ik 'hȯr,mōn }

estrone [BIOCHEM] $C_{18}H_{22}O_2$ An estrogenic hormone produced by follicle cells of the vertebrate ovary; functions the same as estradiol. { 'e,strōn }

estrous cycle [PHYSIO] The physiological changes that take place between periods of estrus in the female mammal. { 'es·trəs ,sī·kəl }

estrus [PHYSIO] The period in female mammals during which ovulation occurs and the animal is receptive to mating. { 'es·trəs }

etherize [MED] To produce anesthesia by administration of ether. { 'ē·thə,rīz }

ethionine [BIOCHEM] $C_5H_{13}O_2N$ An amino acid that is the ethyl analog of and the biological antagonist of methionine. { e'thī·ə,nēn }

Ethiopian zoogeographic region [ECOL] A geographic unit of faunal homogeneity including all of Africa south of the Sahara. { ,ē·thē'ō·pē·ən ¦zō·ō,jē·ə·¦graf·ik 'rē·jən }

ethmoid bone [ANAT] An irregularly shaped cartilage bone of the skull, forming the medial wall of each orbit and part of the roof and lateral walls of the nasal cavities. { 'eth,mȯid ,bōn }

ethmoturbinate [ANAT] Of or pertaining to the masses of ethmoid bone which form the lateral and superior portions of the turbinate bones in mammals. { ¦eth·mō'tər·bə,nāt }

ethogram [ECOL] An extensive list, inventory, or description of the behavior of an organism. { 'ē·thə,gram }

ethological isolation *See* behavioral isolation. { ,ē·thə'läj·ə·kəl ī·sə'lā·shən }

ethology [VERT ZOO] The study of animal behavior in a natural context. { ē'thäl·ə·jē }

ethyl chlorophyllide [BIOCHEM] A compound formed by replacing the phytyl ($C_{20}H_{39}$) tail of the chlorophyll molecule with a short ethyl (C_2H_5) tail; crystallizes easily and has an absorption spectrum and electrochemical properties similar to those of chlorophyll. { ¦eth·əl ,klȯr·ō'fil,īd }

1-ethyl-3-hydroxypiperidine [PHARM] $C_7H_{15}NO$ A liquid with a boiling point of 93–95°C; used as an antispasmodic drug. Also known as 1-ethyl-3-piperidinol. { ¦wan ¦eth·əl ¦thrē hī,dräk·sē·pə'per·ə,dēn }

etioallocholane *See* androstane. { ,ēd·ē·ō,al·ə'kō,lān }

etioblast [BOT] An immature chloroplast, containing prolamellar bodies. { 'ēd·ē·ō,blast }

etiolation [BOT] The yellowing or whitening of green plant parts grown in darkness. { ,ed·ē·ə'lā·shən }

etiology [MED] Any factors which cause disease. { ,ēd·ē'äl·ə·jē }

etiopathogenesis [MED] The cause and development of a disease or abnormal condition. { ,ē·tē·ō,path·ə'jen·ə·səs }

etioplast [BOT] The plastid of a dark-grown plant that contains crystalline prolamellar bodies. { 'ēd·ē·ō,plast }

E trisomy *See* trisomy 18 syndrome. { ¦ē 'trī,sō·mē }

Eubacteriales [MICROBIO] Formerly an order of the class Schizomycetes; considered the true bacteria and characterized by simple, undifferentiated, rigid cells of either spherical or straight, rod-shaped form. { ,yü·bak,tir·ē'ā·lēz }

Eubacterium [MICROBIO] A genus of bacteria in the family Propionibacteriaceae; obligate anaerobes producing a mixture of organic acids (butyric, acetic, formic, and lactic) from carbohydrates and peptone. { ,yü·bak'tir·ē·əm }

Eubasidiomycetes [MYCOL] An equivalent name for Homobasidiomycetidae. { ,yü·bə,sid·ē·ō,mī'sēd·ēz }

Eubrya [BOT] A subclass of the mosses (Bryopsida); the leafy gametophytes arise from buds on the protonemata, which are nearly always filamentous or branched green threads attached to the substratum by rhizoids. { yü'brī·ə }

Eubryales [BOT] An order of mosses (Bryatae); plants have the sporophyte at the end of a stem, vary in size from small to robust, and generally grow in tufts. { ¦yü,brī'ā·lēz }

Eucalyptus [BOT] A large genus of evergreen trees belonging to the myrtle family (Myrtaceae) and occurring in Australia and New Guinea. { ,yü·kə'lip·təs }

eucalyptus gum [PHARM] The dried gummy exudate from *Eucalyptus longirostris* of Australia, composed of kinotannic acid, kino red, glucoside, catechol, and pyrocatechol; used in medicine as an astringent and antidiarrheal agent. Also known as eucalyptus kino; red gum. { ,yü·kə'lip·təs ,gəm }

eucalyptus kino *See* eucalyptus gum. { ,yü·kə'lip·təs 'kē·nō }

Eucarida [INV ZOO] A large superorder of the decapod crustaceans, subclass Malacostraca, including shrimps, lobsters, hermit crabs, and crabs; characterized by having the shell and thoracic segments fused dorsally and the eyes on movable stalks. { ,yü'kar·ə·də }

Eucaryota [BIOL] Primitive, unicellular organisms having a well-defined nuclear membrane, chromosomes, and mitotic cell division. { yü·kar·ē'ōd·ə }

eucaryote *See* eukaryote. { yü'kar·ē,ōt }

eucatropine hydrochloride [PHARM] $C_{17}H_{25}O_3N \cdot HCl$ A white, odorless powder that melts at 183–186°C, is soluble in alcohol and chloroform, and insoluble in ether; used in medicine as a mydriatic. { yü'ka·trə,pēn ,hī·drə'klȯr,īd }

Eucestoda [INV ZOO] The true tapeworms, a subclass of the class Cestoda. { yü'ses·tə·də }

Eucharitidae [INV ZOO] A family of hymenopteran insects in the superfamily Chalcidoidea. { yü·kə'rid·ə,dē }

euchromatin [CELL MOL] The portion of the

chromosomes that stains with low intensity, uncoils during interphase, and condenses during cell division. { yü'krō·mə·tən }

Eucinetidae [INV ZOO] The plate thigh beetles, a family of coleopteran insects in the superfamily Dascilloidea. { yü·sə'ned·ə,dē }

Euclasterida [INV ZOO] An order of asteroid echinoderms in which the arms are sharply distinguished from a small, central disk-shaped body. { ,yü·klə'ster·ə·də }

Eucleidae [INV ZOO] The slug moths, a family of lepidopteran insects in the suborder Heteroneura. { yü'klē·ə,dē }

Euclymeninae [INV ZOO] A subfamily of annelids in the family Maldonidae of the Sedentaria, having well-developed plaques and an anal pore within the plaque. { ,yü·klə'men·ə,nē }

Eucnemidae [INV ZOO] The false click beetles, a family of coleopteran insects in the superfamily Elateroidea. { yük'nem·ə,dē }

Eucoccida [INV ZOO] An order of parasitic protozoans in the subclass Coccidia characterized by alternating sexual and asexual phases; stages of the life cycle occur intracellularly in vertebrates and invertebrates. { yü'käk·sə·də }

Eucoelomata [ZOO] A large sector of the animal kingdom including all forms in which there is a true coelom or body cavity; includes all phyla above Aschelminthes. { yü,sē·lə'mäd·ə }

Eucommiales [BOT] A monotypic order of dicotyledonous plants in the subclass Hamamelidae; plants have two, unitegmic ovules and lack stipules. { yü,käm·ē'ā·lēz }

Eudactylinidae [INV ZOO] A family of parasitic copepod crustaceans in the suborder Caligoida; found as ectoparasites on the gills of sharks. { yü,dak·tə'lin·ə,dē }

eudoxid *See* eudoxome. { yü'däk,sīd }

eudoxome [INV ZOO] Cormidium of most calycophoran siphonophores which lead a free existence. Also known as eudoxid. { yü'däk,sōm }

Euechinoidea [INV ZOO] A subclass of echinoderms in the class Echinoidea; distinguished by the relative stability of ambulacra and interambulacra. { yü,ek·ə'nȯid·ē·ə }

eugenics [GEN] The attempt to improve the phenotypes of future generations of the human population by fostering the reproduction of those with favorable phenotypes and genotypes and hampering or preventing breeding by those with "undesirable" phenotypes and genotypes. The concept is largely discredited. { yü'jen·iks }

Euglena [BIOL] A genus of organisms with one or two flagella, chromatophores in most species, and a generally elongate, spindle-shaped body; classified as algae by botanists (Euglenophyta) and as protozoans by zoologists (Euglenida). { yü'glē·nə }

Euglenida [INV ZOO] An order of protozoans in the class Phytamastigophorea, including the largest green, noncolonial flagellates. { yü 'glen·ə·də }

Euglenidae [INV ZOO] The antlike leaf beetles, a family of coleopteran insects in the superfamily Tenebrionoidea. { yü'glen·ə,dē }

Euglenoidina [INV ZOO] The equivalent name for Euglenida. { ,yü·glə,nȯi'dī·nə }

Euglenophyceae [BOT] The single class of the plant division Euglenophyta. { yü,glē·nə'fīs·ē,ē }

Euglenophyta [BOT] A division of the plant kingdom including one-celled, chiefly aquatic flagellate organisms having a spindle-shaped or flattened body, naked or with a pellicle. { ,yü·glə'näf·əd·ə }

euglobulin [BIOCHEM] True globulin; a simple protein that is soluble in distilled water and dilute salt solutions. { yü'gläb·yə·lən }

Eugregarinida [INV ZOO] An order of protozoans in the subclass Gregarinia; parasites of certain invertebrates. { ,yü·grə,gar·ə'nīd·ə }

Eukaryotae [BIOL] A superkingdom that includes living and fossil organisms comprising all taxonomic groups above the primitive unicellular prokaryotic level. { yü,kar·ē'ō,tē }

eukaryote [BIOL] A cell with a definitive nucleus. Also spelled eucaryote. { yü'kar·ē,ōt }

Eulamellibranchia [INV ZOO] The largest subclass of the molluscan class Bivalvia, having a heterodont shell hinge, leaflike gills, and well-developed siphons. { ¦yü·lə,mel·ə'braŋ·kē·ə }

Eulenburg's disease *See* paramyotonia congenita. { 'ȯi·lən,bərkz di,zēz }

Eulophidae [INV ZOO] A family of hymenopteran insects in the superfamily Chalcidoidea including species that are parasitic on the larvae of other insects. { yü'läf·ə,dē }

Eumalacostraca [INV ZOO] A series of the class Crustacea comprising shrimplike crustaceans having eight thoracic segments, six abdominal segments, and a telson. { yü,mal·ə'käs·trə·kə }

Eumetazoa [ZOO] A section of the animal kingdom that includes the phyla above the Porifera; contains those animals which have tissues or show some tissue formation and organ systems. { yü,med·ə'zō·ə }

eumitosis [CELL MOL] Typical mitosis. { ¦yü ,mī'tō·səs }

Eumycetes [MYCOL] The true fungi, a large group of microorganisms characterized by cell walls, lack of chlorophyll, and mycelia in most species; includes the unicellular yeasts. { ,yü,mī'sēd·ēz }

Eumycetozoida [INV ZOO] An order of protozoans in the subclass Mycetozoia; includes slime molds which form a plasmodium. { ¦yü,mī,sed·ə'zȯid·ə }

Eumycophyta [MYCOL] An equivalent name for the Eumycetes. { ,yü,mī'käf·əd·ə }

Eumycota [MYCOL] An equivalent name for Eumycetes. { ,yü,mī'kōd·ə }

Eunicea [INV ZOO] A superfamily of polychaete annelids belonging to the Errantia. { yü'nis·ē·ə }

Eunicidae [INV ZOO] A family of polychaete annelids in the superfamily Eunicea having characteristic pharyngeal armature consisting of maxillae and mandibles. { yü'nis·əd·ē }

eunuch [MED] An individual who has undergone complete loss of testicular function. { 'yü·nik }

eupavarine [PHARM] $C_{20}H_{21}NO_2$ A crystalline alkaloid that melts at 214°C and is soluble in many organic solvents and hot water; used in medicine as an antispasmodic. { yü'pav·ə,rēn }

Eupelmidae [INV ZOO] A family of hymenopteran insects in the superfamily Chalcidoidea. { yü'pel·mə,dē }

Euphausiacea [INV ZOO] An order of planktonic malacostracans in the class Crustacea possessing photophores which emit a brilliant blue-green light. { yü,fö·zē'ās·ē·ə }

euphenics [GEN] The production of a satisfactory phenotype by means other than eugenics. { yü'fen·iks }

Eupheterochlorina [INV ZOO] A suborder of flagellate protozoans in the order Heterochlorida. { yü¦fed·ə·rō,klə'rī·nə }

Euphorbiaceae [BOT] A family of dicotyledonous plants in the order Euphorbiales characterized by dehiscent fruit having more than one seed and by epitropous ovules. { yü,för·bē'ās·ē,ē }

Euphorbiales [BOT] An order of dicotyledonous plants in the subclass Rosidae having simple leaves and unisexual flowers that are aggregated and reduced. { yü,för·bē'ā·lēz }

Euphrosinidae [INV ZOO] A family of amphinomorphan polychaete annelids with short, dorsolaterally flattened bodies. { yü·frə'zin·ə,dē }

Euplexoptera [INV ZOO] The equivalent name for Dermaptera. { yü,plek'säp·tə·rə }

euploid [GEN] Having a chromosome complement that is an exact multiple of the haploid complement. { 'yü,plöid }

eupnea [PHYSIO] Normal or easy respiration rhythm. { 'yüp·nē·ə }

Eupodidae [INV ZOO] A family of mites in the suborder Trombidiformes. { yü'päd·ə,dē }

European boreal faunal region [ECOL] A zoogeographic region describing marine littoral faunal regions of the northern Atlantic Ocean between Greenland and the northwestern coast of Europe. { ¦yür·ə¦pē·ən ¦bör·ē·əl 'fön·əl 'rē·jən }

Eurotiaceae [MYCOL] A family of ascomycetous fungi of the order Eurotiales in which the asci are borne in cleistothecia or closed fruiting bodies. { yə,rōd·ē'ās·ē,ē }

Eurotiales [MYCOL] An order of fungi in the class Ascomycetes bearing ascospores in globose or broadly oval, delicate asci which lack a pore. { yə,rōd·ē'ā·lēz }

Euryalae [INV ZOO] The basket fishes, a family of echinoderms in the subclass Ophiuroidea. { yə'rī·ə,lē }

Euryalina [INV ZOO] A suborder of ophiuroid echinoderms in the order Phrynophiurida characterized by a leathery integument. { yür·ē'a·lə·nə }

eurybathic [ECOL] Living at the bottom of a body of water. { ¦yür·ə¦bath·ik }

eurygamous [INV ZOO] Mating in flight, as in many insect species. { yü'rig·ə·məs }

euryhaline [ECOL] Pertaining to the ability of marine organisms to tolerate a wide range of saline conditions, and therefore a wide variation of osmotic pressure, in the environment. { ¦yür·ə¦ha,līn }

Eurylaimi [VERT ZOO] A monofamilial suborder of suboscine birds in the order Passeriformes. { ,yür·ə'lā,mī }

Eurylaimidae [VERT ZOO] The broadbills, the single family of the avian suborder Eurylaimi. { ,yür·ə'lā·mə,dē }

euryon [ANAT] One of the two lateral points functioning as end points to measure the greatest transverse diameter of the skull. { 'yür·ē,än }

Euryphoridae [INV ZOO] A family of copepod crustaceans in the order Caligoida; members are fish ectoparasites. { ,yür·ə'fōr·ə,dē }

euryplastic [BIOL] Referring to an organism with a marked ability to change and adapt to a wide spectrum of environmental conditions. { ¦yür·ə¦plas·tik }

Eurypygidae [VERT ZOO] The sun bitterns, a family of tropical and subtropical New World birds belonging to the order Gruiformes. { ,yür·ə'pij·ə,dē }

eurypylous [INV ZOO] Having a wide opening; applied to sponges with wide apopyles opening directly into excurrent canals, and wide prosopyles opening directly from incurrent canals. { ¦yür·ə¦pī·ləs }

eurytherm [BIOL] An organism that is tolerant of a wide range of temperatures. { 'yür·ə,thərm }

Eurytomidae [INV ZOO] The seed and stem chalcids, a family of hymenopteran insects in the superfamily Chalcidoidea. { ,yür·ə'täm·ə,dē }

eurytopic [ECOL] Referring to organisms which are widely distributed. { ,yür·ə'täp·ik }

Eusiridae [INV ZOO] A family of pelagic amphipod crustaceans in the suborder Gammaridea. { yü'sir·ə,dē }

eusocial [ZOO] Pertaining to animal societies, such as those of certain insects, in which sterile individuals work on behalf of reproductive individuals. { ,yü'sō·shəl }

eusporangiate [BOT] Having sporogenous tissue derived from a group of epidermal cells. { ¦yü·spə¦ran·jē,āt }

eustachian tube [ANAT] A tube composed of bone and cartilage that connects the nasopharynx with the middle ear cavity. { yü'stā·shən ,tüb }

eustachian valve *See* canal valve. { yü'stā·shən ,valv }

eustele [BOT] A modified siphonostele containing collateral or bicollateral vascular bundles; found in most gymnosperm and angiosperm stems. { yü'stēl }

eusternum [INV ZOO] The anterior sternal plate in insects. { yü'stər·nəm }

Eustigmatophyceae [BOT] A small class of mostly nonmotile, photosynthetic, unicellular algae in the division Chromophycota, characterized by the unique organization of motile cells, photosynthetic pigments including chlorophyll *a*, beta-carotene, and violaxanthin, and a single parietal yellow-green chloroplast; live chiefly in fresh water but also in marine and soilhabitats. { ˌyüs·tigˌma·də'fī·sē·ē }

Eusuchia [VERT ZOO] The modern crocodiles, a suborder of the order Crocodilia characterized by a fully developed secondary palate and procoelous vertebrae. { yü'sü·kē·ə }

Eusyllinae [INV ZOO] A subfamily of polychaete annelids in the family Syllidae having a thick body and unsegmented cirri. { yü'sil·əˌnē }

Eutardigrada [INV ZOO] An order of tardigrades which lack both a sensory cephalic appendage and a club-shaped appendage. { yüˌtärd·ə 'grād·ə }

eutely [BIOL] Having the body composed of a constant number of cells, as in certain rotifers. { 'yüd·əl·ē }

euthanasia [MED] The act or practice of putting to death or allowing the death, in a relatively painless way, of persons or animals with incurable or painful disease. { ˌyü·thə'nā·zhə }

euthenics [BIOL] The science that deals with the improvement of the future of humanity by changing the environment. { yü'then·iks }

Eutheria [VERT ZOO] An infraclass of therian mammals including all living forms except the monotremes and marsupials. { yü'thir·ē·ə }

Eutrichosomatidae [INV ZOO] Small family of hymenopteran insects in the superfamily Chalcidoidea. { yü¦trik·əˌsō'mad·əˌdē }

eutrophication [ECOL] The process by which a body of water becomes, either by natural means or by pollution, excessively rich in dissolved nutrients, resulting in increased primary productivity that often leads to a seasonal deficiency in dissolved oxygen. { yü·trə·fə'kā·shən }

Evaniidae [INV ZOO] The ensign flies, a family of hymenopteran insects in the superfamily Proctotrupoidea. { ˌev·ə'nī·əˌdē }

evaporative heat regulation [PHYSIO] The composite process by which an animal body is cooled by evaporation of sensible perspiration; this avenue of heat loss serves as a physical means of regulating the body temperature. { i'vap·əˌrād·iv 'hēt ˌreg·yəˌlā·shən }

Eventognathi [VERT ZOO] The equivalent name for Cypriniformes. { ˌeˌven'täg·nəˌthī }

event-related potential [NEUROSCI] Electrical activity produced by the brain in response to a sensory stimulus or associated with the execution of a motor, cognitive, or psychophysiologic task. { i¦vent riˌlād·əd pə'ten·chəl }

everglade [ECOL] A type of wetland in southern Florida usually containing sedges and at least seasonally covered by slowly moving water. { 'ev·ərˌglād }

evergreen [BOT] Pertaining to a perennially green plant. Also known as aiophyllous. { 'ev·ərˌgrēn }

evoked potential [NEUROSCI] Electrical response of any neuron to stimuli. { ē'vōkt pə'ten·chəl }

evolution [BIOL] The processes of biological and organic change in organisms by which descendants come to differ from their ancestors. { ˌev·ə'lü·shən }

evolutionarily significant unit [ECOL] A distinct local population within a species that has very different behavioral and phenological traits and thus harbors enough genetic uniqueness to warrant its own management and conservation agenda. Abbreviated ESU. { ˌev·əˌlü·shə¦ner·ə·lē sig¦nif·i·kənt 'yü·nət }

evolutionary distance [EVOL] The number of base substitutions per homologous site that have occurred since the divergence of two deoxyribonucleic acid sequences. { ˌev·ə¦lü·s·həˌner·ē 'dis·təns }

evolutionary divergence [EVOL] The degree of divergence, at the intra- and interspecific levels, of two or more populations, which presumably have evolved from a common ancestor. { ˌev·ə¦lü·shəˌner·ē də'vər·jəns }

evolutionary force [EVOL] Any factor that brings about changes in gene frequencies or chromosome frequencies in a population and is thus capable of causing evolutionary change. { ˌev·ə¦lü·shəˌner·ē 'fȯrs }

evolutionary plasticity [EVOL] The genetic adaptibility of populations or lines of descent. { ˌev·ə¦lü·shəˌner·ē plas'tis·əd·ē }

evolutionary progress [EVOL] The acquisition of new macromolecular and metabolic processes by which competitive superiority is achieved. { ˌev·ə¦lü·shəˌner·ē 'präˌgres }

evolutionary rate [EVOL] The amount of evolutionary change per unit of time. { ˌev·ə¦lü·shəˌner·ē 'rāt }

evolutionary tree [EVOL] **1.** A diagram that portrays the hypothesized genealogical ties and sequence of evolutionary relationships linking individual organisms, populations, or taxa. Also known as phylogenetic tree. **2.** A diagram that depicts the evolutionary relationship of protein or nucleic acid sequences. { ˌev·ə¦lü·shəˌner·ē 'trē }

evolutionary trend [EVOL] Any trend in the evolution of phyletic lines that is a consequence of genotypic cohesion. { ˌev·ə¦lü·shəˌner·ē 'trend }

evolution pressure [EVOL] The result of the combined action of mutation pressure, immigration and hybridization pressure, and selection pressure, giving rise to systematic changes in the gene frequency of a population. { ˌev·ə¦lü·shən ¦presh·ər }

evolvon [EVOL] The operational unit in evolution, assumed to consist of a deoxyribonucleic acid master sequence with a series of redundant sequences that constitute a repository of genetic information. { 'ev·əˌlän }

ewe [VERT ZOO] A mature female sheep, goat, or related animal, as the smaller antelopes. { yü }

Ewing's sarcoma [MED] A primary malignant tumor of bone, usually arising as a central tumor in long bone. { 'yü·iŋz sär'kō·mə }

exalate [BOT] Being without winglike appendages. { 'ek·sə,lāt }
exalbuminous *See* exendospermous. { ¦eks,al 'byü·mə·nəs }
exanthema [MED] **1.** An eruption on the skin. **2.** Any disease or fever accompanied by a skin eruption. { ,eg,zan'thē·mə }
exanthem subitum [MED] A mild, sometimes epidemic viral disease of young children, with abrupt onset, high fever, and rash. Also known as roseola infantum. { eg'zan·thəm 'sü·bəd·əm }
exarch [BOT] A vascular bundle in which the primary wood is centripetal. { 'ek,särk }
exasperate [BIOL] Having a surface roughened by stiff elevations or bristles. { ig'zas·pə·rət }
exaspidean [VERT ZOO] Of the tarsal envelope of birds, being continuous around the outer edge of the tarsus. { ,eg·zə'spid·ē·ən }
exchange transfusion [MED] The replacement of most or all of the recipient's blood in small amounts at a time by blood from a donor, a technique used particularly in cases of erythroblastosis fetalis, in certain types of poisoning such as salicylism, and occasionally in liver failure. Also known as replacement transfusion. { iks'chānj tranz,fyü·zhən }
excipient [PHARM] Any inert substance combined with an active drug for preparing an agreeable or convenient dosage form. { ek'sip·ē·ənt }
Excipulaceae [MYCOL] The equivalent name for Discellaceae. { ,ek·sə·pə'lās·ē,ē }
excision [GEN] Recombination involving removal of a genetic element. [MED] The cutting out of a part; removal of a foreign body or growth from a part, organ, or tissue. { ek'sizh·ən }
excision enzyme [BIOCHEM] A bacterial enzyme that removes damaged dimers from the deoxyribonucleic acid molecule of a bacterial cell following light or ultraviolet radiation or nitrogen mustard damage. { ek'sizh·ən 'en,zīm }
excitable [BIOL] Referring to a tissue or organism that exhibits irritability. { ek'sīd·ə·bəl }
exclusion principle [ECOL] The principle according to which two species cannot coexist in the same locality if they have identical ecological requirements. { ik'sklü·zhən ,prin·sə·pəl }
exclusive species [ECOL] A species which is completely or nearly limited to one community. { ik'sklü·siv 'spē·shēz }
Excorallanidae [INV ZOO] A family of free-living and parasitic isopod crustaceans in the suborder Flabellifera which have mandibles and first maxillae modified as hooklike piercing organs. { ek ,skȯr·ə'lan·ə,dē }
excoriation [MED] Abrasion of a portion of the skin. { ek,skȯr·ē'ā·shən }
excrement [PHYSIO] An excreted substance; the feces. { 'ek·skrə·mənt }
excrescence [BIOL] **1.** Abnormal or excessive increase in growth. **2.** An abnormal outgrowth. { ek'skrē·səns }
excretion [PHYSIO] The removal of unusable or excess material from a cell or a living organism. { ek'skrē·shən }
excretory system [ANAT] Those organs concerned with solid, fluid, or gaseous excretion. { 'ek·skrə,tȯr·ē ,sis·təm }
excurrent [BIOL] Flowing out. [BOT] **1.** Having an undivided main stem or trunk. **2.** Having the midrib extending beyond the apex. { eks'kə·rənt }
exendospermous [BOT] Lacking endosperm. Also known as exalbuminous. { eks¦en·də ¦spər·məs }
exercise physiology [MED] A field of sports medicine that involves the study of the body's response to physical stress; comprises the science of fitness, the preservation of fitness, and the role of fitness in the prevention and treatment of disease. { ,ek·sər,sīz ,fiz·ē'äl·ə·jē }
exergonic [BIOCHEM] Of or pertaining to a biochemical reaction in which the end products possess less free energy than the starting materials; usually associated with catabolism. { ,ek·sər'gän·ik }
exfoliation [MED] **1.** The separation of bone or other tissue in thin layers. **2.** A peeling and shedding of the horny layer of the skin. { eks,fō·lē'ā·shən }
exfoliative cytology [PATH] The study of cells shed spontaneously from the body surfaces; used principally in the diagnosis of cancer. { eks'fō·lē,ād·iv sī'täl·ə·jē }
exfoliative erythroderma [MED] A type of dermatitis that is characterized by widespread warm redness and scaling; nail degeneration and loss, hair loss, fever, chills, and enlargement of the lymph nodes may also occur. { eks,fōl·ē,ād·iv ə,rith·rə'dər·mə }
exhalation [PHYSIO] The giving off or sending forth in the form of vapor; expiration. { ,eks·ə'lā·shən }
exhaustion delirium [MED] Acute, confusional, delirious reactions brought about by extreme fatigue, long wasting illness, or prolonged insomnia. { ig'zȯs·chən də'lir·ē·əm }
exine *See* exosporium. { 'ek,sēn }
exite [INV ZOO] A movable appendage or lobe located on the external side of the limb of a generalized arthropod. { 'ek,sīt }
exobiology [BIOL] The search for and study of extraterrestrial life. { ¦ek·sō·bī'äl·ə·jē }
exocarp *See* epicarp. { 'ek·sō,kärp }
exocellular [CELL MOL] Referring to reactions or processes that are initiated inside a cell and take place outside it. { ,ek·sə'sel·yə·lər }
exoccipital [ANAT] Lying to the side of the foramen magnum, as the exoccipital bone. { ¦eks·äk'sip·əd·əl }
exochorion [INV ZOO] The outer of two layers forming the covering of an insect egg. { ¦ek·sō'kȯr·ē,än }
exocoel [INV ZOO] The space between pairs of adjacent mesenteries in anthozoan polyps. { 'ek·sə,sēl }

Exocoetidae [VERT ZOO] The halfbeaks, a family of actinopterygian fishes in the order Atheriniformes. { ˌek·səˈsēd·əˌdē }

exocrine gland [PHYSIO] A structure whose secretion is passed directly or by ducts to its exterior surface, or to another surface which is continuous with the external surface of the gland. { ˈek·sə·krən ˌgland }

exocuticle [INV ZOO] The middle layer of the cuticle of insects. { ¦ek·sōˈkyüd·ə·kəl }

exocyclic [CELL MOL] Pertaining to the outside of a ring structure. { ˌek·soˈsī·klik }

exocytosis [CELL MOL] The extrusion of material from a cell. { ¦ek·sō·sīˈtō·səs }

exodermis *See* hypodermis. { ˌek·sōˈdər·məs }

exoenzyme [BIOCHEM] An enzyme that functions outside the cell in which it was synthesized. { ¦ek·sō¦wenˌzīm }

exogamy [GEN] Union of gametes from organisms that are not closely related. Also known as outbreeding. { ekˈsäg·ə·mē }

exogastrula [EMBRYO] An abnormal gastrula that is unable to undergo invagination or further development because of a quantitative increase of presumptive endoderm. { ¦ek·sōˈgas·trə·lə }

exogenote [GEN] The genetic fragment transferred from the donor to the recipient cell during the process of recombination in bacteria. { ˌek ˈsäj·əˌnōt }

exogenous [BIOL] **1.** Due to an external cause; not arising within the organism. **2.** Growing by addition to the outer surfaces. [PHYSIO] Pertaining to those factors in the metabolism of nitrogenous substances obtained from food. { ˌekˈsäj·ə·nəs }

exognathite [INV ZOO] The external branch of an oral appendage of a crustacean. { ekˈsäg·nəˌthīt }

Exogoninae [INV ZOO] A subfamily of polychaete annelids in the family Syllidae having a short, small body of few segments. { ¦ek·sōˈgä·nəˌnē }

exogynous [BOT] Having the style longer than and exserted beyond the corolla. { ekˈsäj·ə·nəs }

exon [GEN] The segment or segments of a gene which code for its final messenger ribonucleic acid. { ˈekˌsän }

exonephric [INV ZOO] Having the excretory organs discharge through the body wall. { ¦ek·sōˈneˌfrik }

exon shuffling [GEN] In eukaryotic split genes, the creation of new genes by the addition or removal of exons through unequal crossing over within introns intervening between the exons of a split gene. { ˈekˌsän ˌshəf·liŋ }

exonuclease [BIOCHEM] Any of a group of enzymes which catalyze hydrolysis of single nucleotide residues from the end of a deoxyribonucleic acid chain. { ¦ek·sōˈnü·klēˌās }

exopeptidase [BIOCHEM] An enzyme that acts on the terminal peptide bonds of a protein chain. { ¦ek·sōˈpep·təˌdās }

exophoria [MED] A type of heterophoria in which the visual lines tend outward. { ˌek·səˈfór·ē·ə }

exophthalmic goiter *See* hyperthyroidism. { ¦ek ˌsäf¦thal·mik ˈgȯid·ər }

exophthalmos [MED] Abnormal protrusion of the eyeball from the orbit. { ˌekˌsäfˈthal·məs }

exopodite [INV ZOO] The outer branch of a biramous crustacean appendage. { ekˈsäp·əˌdīt }

exoprosthesis [MED] An externally applied prosthesis. { ¦ek·sō·prəsˈthē·səs }

Exopterygota [INV ZOO] A division of the insect subclass Pterygota including those insects which undergo a hemimetabolous metamorphosis. { ¦ekˌsäpˌter·ə¦gōd·ə }

exoskeleton [INV ZOO] The external supportive covering of certain invertebrates, such as arthropods. [VERT ZOO] Bony or horny epidermal derivatives, such as nails, hoofs, and scales. { ¦ek·sōˈskel·ə·tən }

exosmosis [PHYSIO] Passage of a liquid outward through a cell membrane. { ¦ek·sō·äsˈmō·səs }

exospore [MYCOL] An asexual spore formed by abstriction, as in certain Phycomycetes. { ˈek·sōˌspór }

exosporium [BOT] The outer of two layers forming the wall of spores such as pollen and bacterial spores. Also known as exine. { ˌek·səˈspór·ē·əm }

exostome [BOT] The opening through the outer integument of a bitegmic ovule. { ˈek·səˌstōm }

exostosis [MED] A benign cartilage-capped protuberance from the surface of long bones but also seen on flat bones, caused by chronic irritation as from infection, trauma, or osteoarthritis. { ˌek·səˈtō·səs }

exotheca [INV ZOO] The tissue external to the theca of corals. { ˌek·səˈthē·kə }

exotic [ECOL] Not endemic to an area. { ig ˈzäd·ik }

exotic viral disease [MED] A viral disease that occurs only rarely in human populations of developed countries. { ig¦zäd·ik ¦vī·rəl dizˌēz }

exotoxin [MICROBIO] A toxin that is excreted by a microorganism. { ¦ek·sə¦täk·sən }

expanding population [ECOL] A population containing a large proportion of young individuals. { ikˈspand·iŋ ˌpäp·yəˈlā·shən }

expectorant [PHARM] **1.** Tending to promote expectoration. **2.** An agent that promotes expectoration. { ikˈspek·tə·rənt }

expectorate [PHYSIO] To eject phlegm or other material from the throat or lungs. { ikˈspek·təˌrāt }

experimental ecology [ECOL] The manipulation of organisms or their environments to discover the underlying mechanisms governing distribution and abundance. { ikˌsper·əˌment·əl ēˈkäl·ə·jē }

expiratory reserve volume [PHYSIO] At the end of a normal expiration, the quantity of air that can be expelled by forcible expiration. { ekˈspī·rəˌtór·ē ri¦zərv ˌväl·yəm }

expiratory standstill [PHYSIO] Suspension of

action at the end of expiration. { ek'spī·rə,tȯr·ē 'stand,stil }

explant [CELL MOL] An excised fragment of a tissue or an organ used to start a cell culture. { 'eks,plant }

explosive evolution [EVOL] Rapid diversification of a group of fossil organisms in a short geological time. { ik'splō·siv ev·ə'lü·shən }

exponential growth [MICROBIO] The period of bacterial growth during which cells divide at a constant rate. Also known as logarithmic growth. { ,ek·spə'nen·chəl 'grōth }

exposure [MED] The state of being open to some action or influence that may affect detrimentally, as cold, disease, or wetness. { ik 'spō·zhər }

exposure dose [MED] A measure of the radiation in a certain place based upon its ability to produce ionization in air. { ik'spō·zhər ,dōs }

exposure limit [MED] The maximum radiation dose equivalent permitted under specified conditions. { ik'spō·zhər ,lim·ət }

exposure pathway [MED] The mode of intake of a substance; for example, inhalation, ingestion, or absorption. { ik'spō·zhər ,path·wā }

exposure rate [MED] Exposure dose per unit time. { ik'spō·zhər ,rāt }

expression vector [CELL MOL] A cloning vector that promotes the expression of foreign gene inserts. { ik'spresh·ən ,vek·tər }

exserted [BIOL] Protruding beyond the enclosing structure, such as stamens extending beyond the margin of the corolla. { ek'sərd·əd }

exsheath [INV ZOO] To escape from the residual membrane of a previous developmental stage, as pertaining to the larva of certain nematodes, microfilaria, and so on. { ek'shēth }

exstipulate [BOT] Lacking stipules. { ek'stip·yə,lāt }

exstrophy [MED] Eversion; the turning inside out of a part. { 'ek·strə·fē }

extension [PHYSIO] A movement which has the effect of straightening a limb. { ik'sten·chən }

external auditory meatus [ANAT] The external passage of the ear, leading to the tympanic membrane in reptiles, birds, and mammals. { ek'stərn·əl 'ȯd·ə,tȯr·ē mē'ād·əs }

external carotid artery [ANAT] An artery which originates at the common carotid and distributes blood to the anterior part of the neck, face, scalp, side of the head, ear, and dura mater. { ek'stərn·əl kə'räd·əd 'ärd·ə·rē }

external ear [ANAT] The portion of the ear that receives sound waves, including the pinna and external auditory meatus. { ek¦stərn·əl 'ēr }

external fertilization [PHYSIO] Those processes involved in the union of male and female sex cells outside the body of the female. { ek¦stərn·əl ,fərd·əl·ə'zā·shən }

external gill [ZOO] A gill that is external to the body wall, as in certain larval fishes and amphibians, and in many aquatic insects. { ek¦stərn·əl 'gil }

external respiration [PHYSIO] The processes by which oxygen is carried into living cells from the outside environment and by which carbon dioxide is carried in the reverse direction. { ek¦stərn·əl ,res·pə'rā·shən }

exteroceptor [PHYSIO] Any sense receptor at the surface of the body that transmits information about the external environment. { ¦ek·stə·rō¦sep·tər }

extinction [EVOL] The worldwide death and disappearance of a specific organism or group of organisms. { ek'stiŋk·shən }

extirpate [BIOL] To uproot, destroy, make extinct, or exterminate. { 'ek·stər,pāt }

extracellular [BIOL] Outside the cell. { ¦ek·strə'sel·yə·lər }

extracellular matrix [CELL MOL] A filamentous structure that is attached to the outer cell surface and provides anchorage, traction, and positional recognition to the cell. [HISTOL] A filamentous structure of glycoproteins and proteoglycans that is attached to the cell surface and provides cells with anchorage, traction for movement, and positional recognition. Abbreviated ECM. { ¦ek·strə¦sel·yə·lər 'mā·triks }

extrachromosomal inheritance *See* cytoplasmic inheritance. { ¦ek·strə,krō·mə'sō·məl in'her·ət·əns }

extracorporeal shock-wave lithotripsy [MED] A treatment for renal calculi (kidney stones) in which powerful ultrasonic shock waves are focused on the stones, thereby breaking them into small fragments that can be excreted, thus avoiding surgery. Also known as lithotripsy. { ,ek·strə·kȯr¦pȯr·ē·əl ¦shäk,wāv 'lith·ə,trip·sē }

extract [PHARM] **1.** A pharmaceutical preparation obtained by dissolving the active constituents of a drug with a suitable menstruum, evaporating the solvent, and adjusting to prescribed standards. **2.** A preparation, usually in a concentrated form, obtained by treating plant or animal tissue with a solvent to remove desired odiferous, flavorful, or nutritive components of the tissue. { 'ek,strakt }

extraction [MED] The act or process of pulling out a tooth. { ik'strak·shən }

extraembryonic coelom [EMBRYO] The cavity in the extraembryonic mesoderm; it is continuous with the embryonic coelom in the region of the umbilicus, and is obliterated by growth of the amnion. { ¦ek·strə,em·brē'än·ik 'sē·ləm }

extraembryonic membrane *See* fetal membrane. { ¦ek·strə,em·brē'än·ik 'mem,brān }

extrajunctional receptor [PHYSIO] An acetylcholine receptor which occurs randomly over a muscle fiber surface outside the area of the neuromuscular junction. { ¦ek·strə¦jənk·shən·əl ri'sep·tər }

extrapyramidal system [NEUROSCI] Descending tracts of nerve fibers arising in the cortex and subcortical motor areas of the brain. { ¦ek·strə,pir·ə'mid·əl 'sis·təm }

extrasystole [MED] Premature beat of the heart. { ¦ek·strə'sis·tə·lē }

extrauterine pregnancy [MED] Gestation outside the uterus. { ¦ek·trə'yüd·ə,rēn 'preg·nən·sē }

extravasation [MED] The pouring out or eruption of a body fluid from its proper channel or vessel into the surrounding tissue. { ik,strav·ə'sā·shən }

extremophiles [PHYSIO] Microorganisms belonging to the domains Bacteria and Archaea that can live and thrive in environments with extreme conditions such as high or low temperatures and pH levels, high salt concentrations, and high pressure. { ek'trem·ə,fīlz }

extrinsic factor *See* vitamin B_{12}. { ek¦strinz·ik ¦fak·tər }

extrinsic protein *See* peripheral membrane protein. { ¦ek¦strin·sik 'prō,tēn }

extrophy [MED] Malformation of an organ. { 'ek·strə·fē }

extrorse [BIOL] Directed outward or away from the axis of growth. { ek'strȯrs }

extroversion [BIOL] A turning outward. { ¦ek·strə¦vər·zhən }

exudate [MED] **1.** A proteinaceous material that passes through blood vessel walls into the surrounding tissue in inflammation or a superficial lesion. **2.** Any substance that is exuded. { 'ek·syü,dāt }

exumbrella [INV ZOO] The outer, convex surface of the umbrella of jellyfishes. { ,ek·səm'brel·ə }

eye [ZOO] A photoreceptive sense organ that is capable of forming an image in vertebrates and in some invertebrates such as the squids and crayfishes. { ī }

eyeball [ANAT] The globe of the eye. { 'ī,bȯl }

eyeball potential [PHYSIO] Very small electrical potentials at the eyeball surface resulting from depolarization of muscles controlling eye position. { 'ī,bȯl pə,ten·chəl }

eyelid [ANAT] A movable, protective section of skin that covers and uncovers the eyeball of many terrestrial animals. { 'ī,lid }

eye socket *See* orbit. { 'ī ,säk·ət }

eyespot [BOT] **1.** A small photosensitive pigment body in certain unicellular algae. **2.** A dark area around the hilum of certain seeds, as some beans. [INV ZOO] A simple organ of vision in many invertebrates consisting of pigmented cells overlying a sensory termination. { 'ī,spät }

eyestalk [INV ZOO] A movable peduncle bearing a terminal eye in decapod crustaceans. { 'ī,stȯk }

F

F_1 *See* first filial generation.

F_2 [GEN] Notation for the progeny produced by intercrossing members of the first filial generation. Also known as second generation.

Fabales [BOT] An order of dicotyledonous plants whose members typically have stipulate, compound leaves, ten to many stamens which are often united by the filaments, and a single carpel which gives rise to a legume; many harbor symbiotic nitrogen-fixing bacteria in the roots. { fə'bā·lēz }

Fab region [IMMUNOL] Region of an antibody molecule that contains the antigen binding site; Fab is derived from the term antigen binding fragment. { 'fab ˌrē·jən }

Fabriciinae [INV ZOO] A subfamily of small to minute, colonial, sedentary polychaete annelids in the family Sabellidae. { ˌfa·brə'sī·əˌnē }

face [ANAT] The anterior portion of the head, including the forehead and jaws. { fās }

facet [ANAT] A small plane surface, especially on a bone or a hard body; may be produced by wear, as a worn spot on the surface of a tooth. [INV ZOO] The surface of a simple eye in the compound eye of arthropods and certain other invertebrates. { 'fas·ət }

facial artery [ANAT] The external branch of the external carotid artery. { ¦fā·shəl 'ärd·ə·rē }

facial bone [ANAT] The bone comprising the nose and jaws, formed by the maxilla, zygoma, nasal, lacrimal, palatine, inferior nasal concha, vomer, mandible, and parts of the ethmoid and sphenoid. { 'fā·shəl ˌbōn }

facial nerve [NEUROSCI] The seventh cranial nerve in vertebrates; a paired composite nerve, with motor elements supplying muscles of facial expression and with sensory fibers from the taste buds of the anterior two-thirds of the tongue and from other sensory endings in the anterior part of the throat. { 'fā·shəl ˌnərv }

facies [ANAT] Characteristic appearance of the face in association with a disease or abnormality. [ECOL] The makeup or appearance of a community or species population. { 'fā·shēz }

facilitated glucose transport [BIOCHEM] The movement of glucose across cell membranes that is driven by the glucose concentration gradient but assisted (facilitated) by carrier proteins. { fə¦sil·əˌtād·əd 'glü·kōs ˌtranzˌpȯrt }

facilitated transport [PHYSIO] The transport of certain materials across a cell membrane, down a concentration gradient, assisted by enzymelike carrier proteins embedded in the membrane and without the explicit provision of energy. { fə'sil·əˌtād·əd 'transˌpȯrt }

factor I *See* fibrinogen. { 'fak·tər 'wən }

factor II *See* prothrombin. { 'fak·tər 'tü }

factor III *See* thromboplastin. { 'fak·tər 'thrē }

factor IV [BIOCHEM] Calcium ions involved in the mechanism of blood coagulation. { 'fak·tər 'fōr }

factor V *See* proaccelerin. { 'fak·tər 'fīv }

factor VII [BIOCHEM] A procoagulant, related to prothrombin, that is involved in the formation of a prothrombin-converting principle which transforms prothrombin to thrombin. Also known as stable factor. { 'fak·tər 'se·vən }

factor VIII *See* antihemophilic factor. { 'fak·tər 'āt }

factor IX *See* Christmas factor. { 'fak·tər 'nīn }

factor IX deficiency [MED] *See* Christmas disease.

factor X *See* Stuart factor. { 'fak·tər 'ten }

factor XI [BIOCHEM] A procoagulant present in normal blood but deficient in hemophiliacs. Also known as plasma thromboplastin antecedent (PTA). { 'fak·tər ə'le·vən }

factor XII [BIOCHEM] A blood clotting factor effective experimentally only in vitro; deficient in hemophiliacs. Also known as Hageman factor. { 'fak·tər 'twelv }

facultative aerobe [MICROBIO] An anaerobic microorganism which can grow under aerobic conditions. { 'fa·kəlˌtād·iv 'erˌōb }

facultative anaerobe [MICROBIO] A microorganism that grows equally well under aerobic and anaerobic conditions. { 'fak·əlˌtād·iv 'an·əˌrōb }

facultative heterochromatin [GEN] Chromosomal material that may alternate in form and function between euchromatin and heterochromatin. { 'fak·əlˌtād·iv ¦ hed·əˌrō'krō·mə·tən }

facultative parasite [ECOL] An organism that can exist independently but may be parasitic on certain occasions, such as the flea. { 'fak·əlˌtād·iv 'par·əˌsīt }

facultative photoheterotroph [MICROBIO] Any bacterium that usually grows anaerobically in light but can also grow aerobically in the dark. { 'fak·əlˌtād·iv ¦fōd·ō¦hed·ə·rəˌträf }

FAD *See* flavin adenine dinucleotide.

Fagaceae [BOT] A family of dicotyledonous plants in the order Fagales characterized by stipulate leaves, seeds without endosperm, female flowers generally not in catkins, and mostly three styles and locules. { fə'gās·ē,ē }

Fagales [BOT] An order of dicotyledonous woody plants in the subclass Hamamelidae having simple leaves and much reduced, mostly unisexual flowers. { fə'gā·lēz }

fagopyrism [VET MED] Photosensitization of the skin and mucous membranes, accompanied by convulsions; produced especially in sheep and swine by feeding on the buckwheat plant, *Fagopyrum sagittatum*, or clovers and grasses containing flavin or carotene and xanthophyll. { ,fa·gō'pī,riz·əm }

Fahrenholz's rule [ECOL] The rule that in groups of permanent parasites the classification of the parasites usually corresponds directly to the natural relationships of the hosts. { 'fär·ən,hōlt·səz ,rül }

falcate [BIOL] Shaped like a sickle. { 'fal,kat }

falciform [BIOL] Sickle-shaped. { 'fal·sə,fȯrm }

falciform ligament [ANAT] The ventral mesentery of the liver; its peripheral attachment extends from the diaphragm to the umbilicus and contains the round ligament of the liver. { 'fal·sə,fȯrm 'lig·ə·mənt }

falciger [INV ZOO] Seta with a distally blunt and curved tip. { 'fal·sə·gər }

falciparum malaria [MED] A severe form of malaria caused by *Plasmodium falciparum* and characterized by sudden attacks of chills, fever, and sweating at irregular intervals; the infecting organism usually localizes in a specific organ, causing capillary blockage. Also known as alged malaria; estivoautumnal malaria; malignant malaria; pernicious malaria. { fal'sip·ə·rəm mə 'ler·ē·ə }

falcon [VERT ZOO] Any of the highly specialized diurnal birds of prey composing the family Falconidae; these birds have been captured and trained for hunting. { 'fal·kən }

Falconidae [VERT ZOO] The falcons, a family consisting of long-winged predacious birds in the order Falconiformes. { fal'kän·ə,dē }

Falconiformes [VERT ZOO] An order of birds containing the diurnal birds of prey, including falcons, hawks, vultures, and eagles. { fal,kän·ə'fōr·mēz }

falculate [ZOO] Curved and with a sharp point. { 'fal· kyə,lāt }

falling disease [VET MED] A terminal manifestation of copper deficiency in which the animal collapses and dies because of heart failure. { 'fȯl·iŋ di,zēz }

Fallopian tube [ANAT] Either of the paired oviducts that extend from the ovary to the uterus for conduction of the ovum in mammals. { fə'lō·pē·ən 'tüb }

false ligament [ANAT] Any peritoneal fold which is not a true supporting ligament. { ¦fȯls 'lig·ə·mənt }

false rib [ANAT] A rib that is not attached to the sternum directly; any of the five lower ribs on each side in humans. { ¦fȯls 'rib }

false ring [BOT] A layer of wood that is less than a full season's growth and often does not form a complete ring. { ¦fȯls 'riŋ }

falx [ANAT] A sickle-shaped structure. { falks }

familial [BIOL] Of, pertaining to, or occurring among the members of a family. { fə'mil·yəl }

familial aldosterone deficiency [MED] A hereditary metabolic disorder, probably due to a defect in the enzyme involved in dehydrogenation of 18-hydroxycorticosterone to aldosterone, characterized by growth retardation and hypoaldosteronism. { fə'mil·yəl al'däs·tə,rōn di,fish·ən·sē }

familial dysautonomia [MED] A hereditary disease transmitted as an autosomal recessive and characterized from infancy by evidence of autonomic nervous system dysfunction, including feeding difficulties, absence of overflow tears, indifference to pain, absent corneal reflexes and deep tendon reflexes, and absence of fungiform papillae on the tongue; most common in Jewish children. { fə'mil·yəl ,dis,ȯd·ə'nō·mē·ə }

familial Mediterranean fever [MED] A hereditary disease of unknown cause characterized by recurrent fever, abdominal and chest pain, arthralgia, and rash, sometimes terminating in renal failure. Abbreviated FMF. Also known as familial recurring polyserositis; periodic disease; periodic peritonitis. { fə'mil·yəl ¦med·ə·tə¦rā·nē·ən ¦ fē·vər }

familial osteochondrodystrophy *See* Morquio's syndrome. { fə'mil·yəl ¦ äs·tē·ō,kän·drə'dis·trə·fē }

familial polyposis [MED] A hereditary condition transmitted as an autosomal dominant and characterized by the appearance of polyps in the small intestine and colon; malignant degeneration is common. { fə'mil·yəl ,päl·ə'pō·səs }

familial recurring polyserositis *See* familial Mediterranean fever. { fə'mil·yəl ri'kər·iŋ ,päl·ē,ser·ə'sīd·əs }

familial splenic anemia *See* Gaucher's disease. { fə'mil·yəl 'splēn·ik ə'nē·myə }

family [SYST] A taxonomic category based on the grouping of related genera. { 'fam·lē }

fan [BIOL] Any structure, such as a leaf or the tail of a bird, resembling an open fan. { fan }

Fanconi's anemia [MED] An infantile anemia that resembles pernicious anemia; related to excessive chromosomal breakage and associated with the risk of developing leukemia. { 'fäŋ· kō·nēz ə'nē·myə }

Fanconi's syndrome *See* amino diabetes. { 'fäŋ·kō·nēz 'sin,drōm }

fang [ANAT] The root of a tooth. [VERT ZOO] A long, pointed tooth, especially one of a venomous serpent. { faŋ }

Fantl unit [BIOL] A unit for the standardization of thrombin. { 'fant·əl ,yü·nət }

faradization [BIOPHYS] Use of a faradic current to stimulate muscles and nerves. { ,far·əd·ə'zā·shən }

farcy *See* glanders. { 'fär·sē }

farinaceous [BIOL] Having a mealy surface covering. { ¦far·ə¦nā·shəs }

Farinales [BOT] An order that includes several groups regarded as orders of the Commelinidae in other systems of classification. { ˌfar·ə'nā·lēz }

Farinosae [BOT] The equivalent name for Farinales. { ˌfar·ə'nō·sē }

farinose [BIOL] Covered with a white powdery substance. { 'far·əˌnōs }

farmer's lung [MED] An acute pulmonary disorder caused by the inhalation of spores from moldy hay or straw. { 'fär·mərz ˌləŋ }

farnesol [BIOCHEM] $C_{15}H_{25}OH$ A colorless liquid extracted from oils of plants such as citronella, neroli, cyclamen, and tuberose; it has a delicate floral odor, and is an intermediate step in the biological synthesis of cholesterol from mevalonic acid in vertebrates; used in perfumery. { 'fär·nəˌsȯl }

farsightedness *See* hypermetropia. { 'fär¦sīd·əd·nəs }

fascia [HISTOL] Layers of areolar connective tissue under the skin and between muscles, nerves, and blood vessels. { 'fā·shə }

fasciate [BOT] Having bands or stripes. { 'fa·sheˌāt }

fascicle [BOT] A small bundle, as of fibers or leaves. { fas·i·kəl }

fasciculate [BOT] Arranged in tufts or fascicles. { fə'sik·yə·lət }

fasciculation potential [PHYSIO] An action potential which is quantitatively comparable to that of a motor unit and which represents spontaneous contraction of a bundle of muscle fibers. { fəˌsik·yə'lā·shən pəˌten·chəl }

fasciculus [ANAT] A bundle or tract of nerve, muscle, or tendon fibers isolated by a sheath of connective tissues and having common origins, innervation, and functions. { fə'sik·yə·ləs }

Fasciola hepatica [INV ZOO] A digenetic trematode which parasitizes sheep, cattle, and occasionally humans. { fə·'sē·ə·lə he'pad·ə·kə }

fasciole [INV ZOO] A band of cilia on the test of certain sea urchins. { 'fas·ēˌōl }

fascioliasis [MED] The infection of humans with *Fasciola hepatica*. { fəˌsē·ə'lī·ə·səs }

fasciolopsiasis [MED] The presence of the parasite *Fasciolopsis buski* in a person's small intestine. { fəˌsē·əˌläp'sī·ə·səs }

Fasciolopsis buski [INV ZOO] A large, fleshy trematode, native to eastern Asia and the southwestern Pacific, which parasitizes humans. { fəˌsē·ə'läp·səs 'bəs·kē }

fascioscapulohumeral dystrophy [MED] A progressive hereditary form of muscular dystrophy involving atrophy of the muscles of the face, pectoral girdle, and upper arm. { ¦fa·sē·ō¦skap·yə·lō¦hyü·mə·rəl 'di·strə·fē }

Fas protein [CELL MOL] A cell-surface protein receptor expressed on essentially all cells of the body that when bound to its ligand (FasL) signals a caspase cascade, ultimately resulting in apoptosis (programmed cell death). { ¦ef¦ā'es ˌprōˌtēn }

fastigiate [BOT] **1.** Having erect branches that are close to the stem. **2.** Becoming narrower at the top. [ZOO] Arranged in a conical bundle. { fa'stij·e·āt }

fat [ANAT] Pertaining to an obese person. [BIOCHEM] Any of the glyceryl esters of fatty acids which form a class of neutral organic compounds. [PHYSIO] The chief component of fat cells and other animal and plant tissues. { fat }

fat body [INV ZOO] A nutritional reservoir of fatty tissue surrounding the viscera or forming a layer beneath the integument in the immature larval stages of many insects. [VERT ZOO] A mass of adipose tissue attached to each genital gland in amphibians. { 'fat ˌbäd·ē }

fat cell [HISTOL] The principal component of adipose connective tissue; two types are yellow fat cells and brown fat cells. { 'fat ˌsel }

fate map [EMBRYO] A graphic scheme indicating the definite spatial arrangement of undifferentiated embryonic cells in accordance with their destination to become specific tissues. { 'fāt ˌmap }

fat embolus [MED] An embolus composed principally of fatty substances. { 'fat ˌem·bə·ləs }

fatigue [PHYSIO] Exhaustion of strength or reduced capacity to respond to stimulation following a period of activity. { fə'tēg }

fat-metabolizing hormone *See* ketogenic hormone. { 'fat mə¦tab·əˌlīz·iŋ ˌhȯrˌmōn }

fat necrosis [MED] Pathologic death of adipose tissue often accompanied by soap production from the hydrolyzed fat; associated with pancreatitis. { ¦fat nə'krō·səs }

fatty acid peroxidase [BIOCHEM] An enzyme present in germinating plant seeds which catalyzes the oxidation of the carboxyl carbon of fatty acids to carbon dioxide. { ¦fad·ē 'as·əd pə'räk·səˌdās }

fatty acyl carnitine [BIOCHEM] Transport form of fatty acids which allows them to cross the mitochondrial membrane; formed by reaction of fatty acyl-coenzyme A with carnitine by employing the enzyme carnitine acyltransferase. Also known as acyl carnitine. { 'fad·ē 'as·əl 'kär·nəˌtēn }

fatty acyl-coenzyme A [BIOCHEM] Activated form of fatty acids formed by the enzyme acyl-coenzyme A synthetase at the expense of adenosinetriphosphate. Also known as acyl-coenzyme A. { 'fad·ē 'as·əl kō¦enˌzīm 'ā }

fatty infiltration [PHYSIO] Infiltration of an organ or tissue with excessive amounts of fats. { 'fad·ē ˌin·fil'trā·shən }

fatty metamorphosis [MED] Fatty degeneration, fatty infiltration, or both. { 'fad·ē ˌmed·ə'mȯr·fə·səs }

faucal [BIOL] Of or pertaining to the fauces. [INV ZOO] The opening of a spiral shell. { 'fȯ·kəl }

fauces [ANAT] The passage in the throat between the soft palate and the base of the tongue.

[BOT] The throat of a calyx, corolla, or similar part. { 'fȯ,sēz }

faucial tonsil *See* palatine tonsil. { 'fȯ·shəl 'tän·səl }

fauna [ZOO] **1.** Animals. **2.** The animal life characteristic of a particular region or environment. { 'fȯn·ə }

faunal extinction [EVOL] The worldwide death and disappearance of diverse animal groups under circumstances that suggest common and related causes. Also known as mass extinction. { 'fȯn·əl ik'stiŋk·shən }

faunal region [ECOL] A division of the zoosphere, defined by geographic and environmental barriers, to which certain animal communities are bound. { 'fȯn·əl ,rē·jən }

favism [MED] An acute hemolytic anemia, usually in persons of Mediterranean area descent, occurring when an individual with glucose-6-phosphate dehydrogenase deficiency of erythrocytes eats the beans or inhales the pollen of *Vicia faba*. { 'fä,viz·əm }

favus [MED] A fungal infection of the scalp, usually caused by *Trichophyton schoenleini*, characterized by round, yellow, cup-shaped crusts having a peculiar mousy odor. Also known as tinea favosa. { 'fā·vəs }

Fc region [IMMUNOL] Region of an antibody molecule that binds to antibody receptors on the surface of cells such as macrophages and mast cells, and to complement protein; F_c is derived from the term crystallizable fragment. { ¦ef'sē ,rē·jən }

feather [VERT ZOO] An ectodermal derivative which is a specialized keratinous outgrowth of the epidermis of birds; functions in flight and in providing insulation and protection. { 'feth·ər }

feathering [VERT ZOO] Plumage. { 'feth·ə·riŋ }

febrile convulsion [MED] A type of convulsion that occurs in infants and young children in association with fever. { ¦fē,brīl kən'vəl·shən }

febrile disease [MED] Any disease associated with or characterized by fever. { 'feb·rəl di,zēz }

fecalith [MED] A hardened piece of fecal matter formed in the intestine or vermiform appendix. { 'fek·ə,lith }

feces [PHYSIO] The waste material eliminated by the gastrointestinal tract. { 'fē·sēz }

Fechner fraction [PHYSIO] The smallest difference in the brightness of two sources that can be detected by the human eye divided by the brightness of one of them. { 'fek·ner ,frak·shən }

Fechner law [PHYSIO] The intensity of a sensation produced by a stimulus varies directly as the logarithm of the numerical value of that stimulus. { 'fek·nər ,lȯ }

fecundity [BIOL] The innate potential reproductive capacity of the individual organism, as denoted by its ability to form and separate from the body the mature germ cells. { fə'kən·dəd·ē }

feedback inhibition [BIOCHEM] A cellular control mechanism by which the end product of a series of metabolic reactions inhibits the activity of the first enzyme in the sequence. Also known as allosteric control. { 'fēd,bak ,in·ə,bish·ən }

feeding mechanism [ZOO] A mechanism by which an animal obtains and utilizes food materials. { 'fēd·iŋ ,mek·ə,niz·əm }

Felidae [VERT ZOO] The cats and saber-toothed cats, a family of mammals in the superfamily Feloidea. { 'fel·ə,dē }

feline [VERT ZOO] **1.** Of or relating to the genus *Felis*. **2.** Catlike. { 'fē,līn }

Felis [VERT ZOO] The type genus of the Felidae, comprising the true or typical cats, both wild and domestic. { 'fē·ləs }

fell-field [ECOL] A culture community of dwarfed, scattered plants or grasses above the timberline. { 'fel ,fēld }

Feloidea [VERT ZOO] A superfamily of catlike mammals in the order Carnivora. { fə'lȯid·ē·ə }

Felon's unit [BIOL] A unit for the standardization of antipneumococcic serum. { 'fel·ənz ,yü·nət }

Felty's syndrome [MED] A complex of symptoms involving rheumatoid arthritis, splenomegaly, lymphadenopathy, and anemia. { 'fel·tēz ,sin,drōm }

female [BOT] A flower lacking stamens. [ZOO] An individual that bears young or produces eggs. { 'fē,māl }

female heterogamety [GEN] The production by females of two kinds of gametes differing in sex chromosome complement, as in birds and Lepidoptera. { ¦fē,māl ¦hed·ə·rō·gə'mēd·ē }

female homogamety [GEN] The production by females of a single type of gamete with respect to sex chromosome complement, as in mammals and *Drosophila*. { ¦fē,māl ¦hō·mō·gə'mēd·ē }

female pseudohermaphroditism *See* gynandry. { ¦fē,māl ¦sü·dō·hər¦maf·rə'dīd,iz·əm }

feminizing syndrome [MED] Any of a number of symptom complexes in which males tend to take on feminine characteristics due to alterations of adrenocorticotropin output. { 'fem·ə,niz·iŋ ,sin,drōm }

femoral artery [ANAT] The principal artery of the thigh; originates as a continuation of the external iliac artery. { ¦fem·ə·rəl 'ärd·ə·rē }

femoral hernia [MED] A hernia that occurs at the passage of the arteries and veins from the abdomen into the legs below the inguinal ligament. { ¦fem·ə·rəl 'hər·nē·ə }

femoral nerve [NEUROSCI] A mixed nerve of the leg; the motor portion innervates muscles of the thigh, and the sensory portion innervates portions of the skin of the thigh, leg, hip, and knee. { ¦fem·ə·rəl 'nərv }

femoral ring [ANAT] The abdominal opening of the femoral canal. { ¦fem·ə·rəl 'riŋ }

femoral vein [ANAT] A vein accompanying the femoral artery. { ¦fem·ə·rəl 'vān }

femur [ANAT] **1.** The proximal bone of the hind or lower limb in vertebrates. **2.** The thigh bone in humans, articulating with the acetabulum and tibia. { 'fē·mər }

fenestra [ANAT] An opening in the medial wall

of the middle ear. [MED] An opening in a bandage or plaster splint for examination or drainage. { fə'nes·trə }

fenestrated membrane [HISTOL] One of the layers of elastic tissue in the tunica media and tunica intima of large arteries. { 'fen·ə,strād·əd 'mem,brān }

fenestration [BIOL] **1.** A transparent or windowlike break or opening in the surface. **2.** The presence of windowlike openings. { ,fen·ə'strā·shən }

fennel [BOT] *Foeniculum vulgare.* A tall perennial herb of the family Umbelliferae; a spice is derived from the fruit. { 'fen·əl }

ferment [BIOCHEM] An agent that can initiate fermentation and other metabolic processes. { ¦fər¦ment }

fermentation [MICROBIO] An enzymatic transformation of organic substrates, especially carbohydrates, generally accompanied by the evolution of gas; a physiological counterpart of oxidation, permitting certain organisms to live and grow in the absence of air; used in various industrial processes for the manufacture of products such as alcohols, acids, and cheese by the action of yeasts, molds, and bacteria; alcoholic fermentation is the best-known example. Also known as zymosis. { ,fər·mən'tā·shən }

fermentation tube [MICROBIO] A culture tube with a vertical closed arm to collect gas formed in a broth culture by microorganisms. { ,fər·mən'tā·shən ,tüb }

fern [BOT] Any of a large number of vascular plants composing the division Polypodiophyta. { fərn }

ferredoxins [BIOCHEM] Iron-containing proteins that transfer electrons, usually at a low potential, to flavoproteins; the iron is not present as a heme. { ,fer·ə'däk·sənz }

ferret [VERT ZOO] *Mustela nigripes.* The largest member of the weasel family, Mustelidae, and a relative of the European polecat; has yellowish fur with black feet, tail, and mask. { 'fer·ət }

ferrichrome [MICROBIO] A cyclic hexapeptide that is a microbial hydroxamic acid and is involved in iron transport and metabolism in microorganisms. { 'fer·ə,krōm }

ferrihemoglobin [BIOCHEM] Hemoglobin in the oxidized state. Also known as methemoglobin. { ¦fe·ri,hē·mə'glō·bən }

ferrimycin [MICROBIO] The representative antibiotic of the sideromycin group; a hydroxamic acid compound. { ,fe·ri'mīs·ən }

ferriporphyrin [BIOCHEM] A red-brown to black complex of iron and porphyrin in which the iron is in the 3+ oxidation state. { ¦fe·ri'pȯr·fə·rəŋ }

ferritin [BIOCHEM] An iron-protein complex occurring in tissues, probably as a storage form of iron. { 'fer·ət·ən }

ferrochelatase [BIOCHEM] A mitochondrial enzyme which catalyzes the incorporation of iron into the protoporphyrin molecule. { ¦fe·rō'kel·ə,tās }

ferroporphyrin [BIOCHEM] A red complex of porphyrin and iron in which the iron is in the 2+ oxidation state. { ¦fe·rō'pȯr·fə·rən }

fertility [BIOL] The state of or capacity for abundant productivity. { fər'til·əd·ē }

fertility factor [GEN] An episomal bacterial sex factor which determines the role of a bacterium as either a male donor or as a female recipient of genetic material. Also known as F factor; sex factor. { fər'til·əd·ē ,fak·tər }

fertilization [PHYSIO] The physicochemical processes involved in the union of the male and female gametes to form the zygote. { 'fərd·əl·ə'zā·shən }

fertilization membrane [CELL MOL] A membrane that separates from the surface of and surrounds many eggs following activation by the sperm; prevents multiple fertilization. { 'fərd·əl·ə'zā·shən ,mem,brān }

fertilizin [BIOCHEM] A mucopolysaccharide, derived from the jelly coat of an egg, that plays a role in sperm recognition and the stimulation of sperm motility and metabolic activity. { fər'til·ə·zən }

fervenulin [MICROBIO] $C_7H_7N_5O_2$ An antibiotic from culture filtrates of *Streptomyces fervens*; yellow, orthorhombic crystals can be formed; melting point is 178–179°C. Also known as planomycin. { fər'ven·ə·lin }

fescue [BOT] A group of grasses of the genus *Festuca*, used for both hay and pasture. { 'fes,kyü }

fescue foot [VET MED] A gangrenous condition of cattle feet caused by grazing tall fescue infected with the endophytic symbiotic fungus *Acremonium coenophialum.* { 'fes·kyü ,fu̇t }

fetal alcohol syndrome [MED] A spectrum of changes in the offspring of women who consume alcoholic beverages during pregnancy, ranging from mild mental changes to severe growth deficiency, mental retardation, and abnormal facial features. { ,fēd·əl 'al·kə,hȯl ,sin,drōm }

fetal asphyxia [MED] Deprivation of oxygen to the fetus due to interference with its blood supply. { 'fēd·əl əs'fik·sē·ə }

fetal fat-cell lipoma *See* liposarcoma. { 'fēd·əl ¦fat ¦sel li'pō·mə }

fetal hemoglobin [BIOCHEM] A normal embryonic hemoglobin having alpha chains identical to those of normal adult human hemoglobin, and gamma chains similar to adult beta chains. { 'fēd·əl 'hē·mə,glō·bən }

fetal membrane [EMBRYO] Any one of the membranous structures which surround the embryo during its development period. Also known as extraembryonic membrane. { 'fēd·əl 'mem,brān }

fetometamorphism [INV ZOO] A life cycle variation in the Cantharidae (Coleoptera); the larvae hatch prematurely as legless, immature prelarvae. { ¦fē·dō,med·ə'mȯr,fiz·əm }

fetus [EMBRYO] **1.** The unborn offspring of viviparous mammals in the later stages of development. **2.** In human beings, the developing body in utero from the beginning of the ninth week after fertilization through the fortieth week of intrauterine gestation, or until birth. { 'fēd·əs }

fever [MED] An elevation in the central body temperature of warm-blooded animals caused by abnormal functioning of the thermoregulatory mechanisms. { 'fē·vər }

Feyliniidae [VERT ZOO] The limbless skinks, a family of reptiles in the suborder Sauria represented by four species in tropical Africa. { ,fā·lə'nī·ə,dē }

F factor *See* fertility factor. { 'ef ,fak·tər }

fiber [BOT] **1.** An elongate, thick-walled, tapering plant cell that lacks protoplasm and has a small lumen. **2.** A very slender root. { 'fī·bər }

fiber flax [BOT] The flax plant grown in fertile, well-drained, well-prepared soil and cool, humid climate; planted in the early spring and harvested when half the seed pods turn yellow; used in the manufacture of linen. { 'fī·bər ,flaks }

fibril [BIOL] A small thread or fiber, as a root hair or one of the structural units of a striated muscle. { 'fī·brəl }

fibrillation [PHYSIO] An independent, spontaneous, local twitching of muscle fibers. { ,fib·rə'lā·shən }

fibrillose [BIOL] Having fibrils. { 'fīb·rə,lōs }

fibrin [BIOCHEM] The fibrous, insoluble protein that forms the structure of a blood clot; formed by the action of thrombin. { 'fī·brən }

fibrinase [BIOCHEM] An enzyme that catalyzes the formation of covalent bonds between fibrin molecules. Also known as fibrin-stabilizing factor. { 'fī·brə,nās }

fibrinogen [BIOCHEM] A plasma protein synthesized by the parenchymal cells of the liver; the precursor of fibrin. Also known as factor I. { fī'brin·ə·jən }

fibrinogenopenia [MED] A congenital hemorrhagic diathesis in which there is a decrease in fibrinogen in the plasma. { fī,brin·ə,jēn·ə'pē·nē·ə }

fibrinoid [BIOCHEM] A homogeneous, refractile, oxyphilic substance occurring in degenerating connective tissue, as in term placentas, rheumatoid nodules, and Aschoff bodies, and in pulmonary alveoli in some prolonged pneumonitides. { 'fī·brə,nȯid }

fibrinolysin *See* plasmin. { ,fī·brə'näl·ə·sən }

fibrinolysis [PHYSIO] Liquefaction of coagulated blood by the action of plasmin on fibrin. { ,fī·brə'näl·ə·səs }

fibrinous pericarditis [MED] Inflammation of the pericardium involving the deposition of fibrin and leukocytes between the layers of the pericardium; seen in rheumatic carditis and acute infectious diseases. { 'fī·brə·nəs ,per·ə,kär'dīd·əs }

fibrin-stabilizing factor *See* fibrinase. { ¦fī·brən 'stā·bə,līz·iŋ ,fak·tər }

fibroadenoma [MED] A benign tumor containing both fibrous and glandular elements. { ¦fī·brō,ad·ən'ō·mə }

fibroblast [HISTOL] A stellate connective tissue cell found in fibrous tissue. Also known as a fibrocyte. { 'fī·brə,blast }

fibroblast growth factor [PHYSIO] A family of proteins important in the development of the nervous and skeletal systems. { ¦fī·brō,blast 'grōth ,fak·tər }

fibrocartilage [HISTOL] A form of cartilage rich in dense, closely opposed bundles of collagen fibers; occurs in intervertebral disks, in the symphysis pubis, and in certain tendons. { ¦fī·brō 'kärd·əl·ij }

fibrocyte *See* fibroblast. { 'fī·brə,sīt }

fibroid [HISTOL] Composed of fibrous tissue. { 'fī,brȯid }

fibroid deoxyribonucleic acid [CELL MOL] Sections of relatively uncoiled double-stranded deoxyribonucleic acid thought to be regions of specific base sequences for coding rather than gene control. { 'fī,brȯid dē¦äk·sē,rī·bō·nü¦klē·ik 'as·əd }

fibroid tumor *See* fibroma. { 'fī,brȯid 'tü·mər }

fibroin [BIOCHEM] A protein secreted by spiders and silkworms which rapidly solidifies into strong, insoluble thread that is used to form webs or cocoons. { 'fī·brə·wən }

fibroma [MED] A benign tumor composed primarily of fibrous connective tissue. Also known as fibroid tumor. { fī'brō·mə }

fibroma molluscum *See* neurofibromatosis. { fī'brō·mə mə'ləs·kəm }

fibromatosis [MED] **1.** The occurrence of multiple fibromas. **2.** Localized proliferation of fibroblasts without apparent cause. { ¦fī·brō·mə'tō·səs }

fibromyoma [MED] A benign tumor, usually of smooth muscle, with a prominent fibrous stroma; commonly a uterine leiomyoma. { ¦fī·brō·mī'ō·mə }

fibromyosis *See* interstitial endometrosis. { ¦fī·bro·mī'ō·səs }

fibromyositis *See* myositis. { ¦fī·brō,mī·ə'sīd·əs }

fibronectin [BIOCHEM] A type of large glycoprotein that is found on the surface of cells and mediates cellular adhesion, control of cell shape, and cell migration. { 'fī·brə'nek·tən }

fibroplasia [MED] The growth of fibrous tissue, as in the second phase of wound healing. { ,fī·brə'plā·zhə }

fibrosarcoma [MED] A sarcoma composed of spindle cells that produce collagenous fibrils. { ¦fī·brō·sär'kō·mə }

fibrosing adenomatosis *See* sclerosing adenomatosis. { fī'brōs·iŋ ,ad·ən,ō·mə'tō·səs }

fibrosis [MED] Growth of fibrous connective tissue in an organ or part in excess of that naturally present. { fī'brō·səs }

fibrositis [MED] Inflammation of white fibrous connective tissue, usually in a joint region. { ,fī·brə'sīd·əs }

fibrous dysplasia [MED] **1.** Extensive formation of fibrous tissue and transformation of bony tissue in one or more bones. **2.** Development of abnormal amounts of fibrous tissue in the mammary glands. { 'fī·brəs dis'plā·zhə }

fibrous osteoma *See* ossifying fibroma. { 'fī·brəs ,äs·tē'ō·mə }

fibrous protein [BIOCHEM] Any of a class of

highly insoluble proteins representing the principal structural elements of many animal tissues. { 'fī·brəs 'prō,tēn }

fibula [ANAT] The outer and usually slender bone of the hind or lower limb below the knee in vertebrates; it articulates with the tibia and astragalus in humans, and is ankylosed with the tibia in birds and some mammals. { 'fib·yə·lə }

Ficus [BOT] A genus of tropical trees in the family Moraceae including the rubber tree and the fig tree. [INV ZOO] A genus of gastropod mollusks having pear-shaped, spirally ribbed sculptured shells. { 'fī·kəs }

Fiedler's myocarditis *See* interstitial myocarditis. { 'fēd·lərz ,mī·ə,kär'dīd·əs }

field [MED] The area in which surgery is taking place, bounded on all sides by sterilized tissue or drapes. Also known as sterile field. { fēld }

fièvre boutonneuse [MED] A mild febrile rickettsial disease of humans caused by *Rickettsia conori*; characterized by a rash, tache noire (primary ulcer), and swollen lymph glands. Also known as boutonneuse fever; Marseilles fever. { 'fyev·rə bü·tə'nāz }

fig [BOT] *Ficus carica*. A deciduous tree of the family Moraceae cultivated for its edible fruit, which is a syconium, consisting of a fleshy hollow receptacle lined with pistillate flowers. { fig }

Figitidae [INV ZOO] A family of hymenopteran insects in the superfamily Cynipoidea. { fə'jid·ə,dē }

Fijivirus [VIROL] A genus in the viral family Reoviridae that is the causative agent of Fiji disease in plants and insects. { 'fē·jē,vī·rəs }

filament [BOT] **1.** The stalk of a stamen which supports the anther. **2.** A chain of cells joined end to end, as in certain algae. [INV ZOO] A single silk fiber in the cocoon of a silkworm. { 'fil·ə·mənt }

filamentous bacteria [MICROBIO] Bacteria, especially in the order Actinomycetales, whose cells resemble filaments and are often branched. { ,fil·ə'men·təs bak'tir·ē·ə }

filamentous bacteriophage [VIROL] Threadlike bacterial viruses that use pili to attach to the host. { ,fil·ə¦men·təs bak'tir·ē·ə,fāj }

filaria [INV ZOO] A parasitic filamentous nematode belonging to the order Filaroidea. { fə'lar·ē·ə }

filariasis [MED] A disease due to the presence of hairlike nematodes (filariae) in humans, including *Wuchereria bancrofti*, *W. pacifica*, and *Onchocerca volvulus*. { ,fil·ə'rī·ə·səs }

Filarioidea [INV ZOO] An order of the class Nematoda comprising highly specialized parasites of humans and domestic animals. { ,fil·ə'rōid·ē·ə }

filbert [BOT] Either of two European plants belonging to the genus *Corylus* and producing a thick-shelled, edible nut. Also known as hazelnut. { 'fil·bərt }

filial generation [GEN] Any generation following the parental generation. { ¦fil·ē·əl ,jen·ə¦rā·shən }

Filicales [BOT] The equivalent name for Polypodiales. { ,fil·ə'kā·lēz }

Filicineae [BOT] The equivalent name for Polypodiatae. { ,fil·ə'sin·ē,ē }

Filicornia [INV ZOO] A group of hyperiid amphipod crustaceans in the suborder Genuina having the first antennae inserted anteriorly. { ,fil·ə'kȯr·nē·ə }

filiform [BIOL] Threadlike or filamentous. { 'fil·ə,fȯrm }

filiform papilla [ANAT] Any one of the papillae occurring on the dorsum and margins of the oral part of the tongue, consisting of an elevation of connective tissue covered by a layer of epithelium. { 'fil·ə,fȯrm pə'pil·ə }

film [BIOL] A thin, membranous skin, such as a pellicle. [MED] A pathological opacity, as of the cornea. { film }

filoplume [VERT ZOO] A specialized feather that may be decorative, sensory, or both; it is always associated with papillae of contour feathers. { 'fil·ə,plüm }

filopodia [INV ZOO] Filamentous pseudopodia. { ,fil·ə'pōd·ē·ə }

filoreticulopodia [INV ZOO] Branched, filamentous pseudopodia. { ,fil·ə,red·ə,kyül·ə'pōd·ē·ə }

Filosia [INV ZOO] A subclass of the class Rhizopodea characterized by slender filopodia which rarely anastomose. { fī'lō·shə }

filterable virus [VIROL] Virus particles that remain in a fluid after passing through a diatomite or glazed porcelain filter with pores too minute to allow the passage of bacterial cells. { 'fil·trə·bəl 'vī·rəs }

filter feeder [INV ZOO] A microphagous organism that uses complex filtering mechanisms to trap particles suspended in water. { 'fil·tər ,fēd·ər }

filtration sterilization [MICROBIO] The physical removal of microorganisms from liquid that may be destroyed by heat (such as blood serum, enzyme solutions, antibiotics, and some bacteriological media and medium constituents) by filtering through materials having relatively small pores. { fil'trā·shən ,ster·ə·lə,zā·shən }

fimbria *See* pilus. { 'fim·brē·ə }

fimbriate [BIOL] Having a fringe along the edge. { 'fim·brē,āt }

fin [VERT ZOO] A paddle-shaped appendage on fish and other aquatic animals that is used for propulsion, balance, and guidance. { fin }

final common pathway *See* lower motor neuron. { ¦fīn·əl ¦käm·ən 'path,wā }

finch [VERT ZOO] The common name for birds composing the family Fringillidae. { finch }

fin fold [EMBRYO] A median integumentary fold extending along the body of a fish embryo which gives rise to the dorsal, caudal, and anal fins. { 'fin ,fōld }

finger [ANAT] Any of the four digits on the hand other than the thumb. { 'fiŋ·gər }

fingerprint [FOREN] **1.** A pattern of distinctive epidermal ridges on the bulbs of the inside of the end joints of fingers and thumbs. **2.** An impression of a human fingerprint. { 'fiŋ·gər,print }

fingerprint pattern area [FOREN] The part of a fingerprint that contains the cores, deltas, and ridges that are used for classification. { ¦fiŋ·gər,print 'pad·ərn ,er·ē·ə }

fin rot [VET MED] A bacterial disease of hatchery fishes characterized by necrosis and erosion of the fin tissue. { 'fin ,rät }

fin spine [VERT ZOO] A bony process that supports the fins of certain fishes. { 'fin ,spīn }

fir [BOT] The common name for any tree of the genus *Abies* in the pine family; needles are characteristically flat. { fər }

fire disclimax [ECOL] A community that is perpetually maintained at an early stage of succession through recurrent destruction by fire followed by regeneration. { 'fīr dis'klī,maks }

firefly [INV ZOO] Any of various flying insects which produce light by bioluminescence. { 'fīr,flī }

first-degree burn [MED] A mild burn characterized by pain and reddening of the skin. { ¦fərst də¦grē 'bərn }

first filial generation [GEN] The first generation resulting from a cross between parents homozygous for different alleles at a locus; all members are heterozygous at the locus. Symbolized F_1. { ¦fərst ¦fil·ē·əl jen·ə'rā·shən }

fish [VERT ZOO] The common name for the cold-blooded aquatic vertebrates belonging to the groups Cyclostomata, Chondrichthyes, and Osteichthyes. { fish }

FISH *See* fluorescent in situ hybridization. { fish *or* ¦ef¦ī¦es'āch }

fisher [VERT ZOO] *Martes pennanti*. An arboreal, carnivorous mammal of the family Mustelidae; a relatively large weasellike animal with dark fur, found in northern North America. { 'fish·ər }

fisheries conservation [ECOL] Those measures concerned with the protection and preservation of fish and other aquatic life, particularly in sea waters. { 'fish·ə·rēz ,kän·sər'vā·shən }

fishery [ECOL] A place for harvesting fish or other aquatic life, particularly in sea waters. { 'fish·ə·rē }

fish lice [INV ZOO] The common name for all members of the crustacean group Arguloida. { 'fish ,līs }

fish stock [ECOL] Any natural population of fish which is an isolated and self-perpetuating group of the same species. { 'fish ,stäk }

Fissidentales [BOT] An order of the Bryopsida having erect to procumbent, simple or branching stems and two rows of leaves arranged in one plane. { ,fis·ə,den'tā·lēz }

fission [BIOL] A method of asexual reproduction among bacteria, algae, and protozoans by which the organism splits into two or more parts, each part becoming a complete organism. { 'fish·ən }

fission fungi [MICROBIO] A misnomer once used to describe the Schizomycetes. { ¦fish·ən 'fən,jī }

fissiped [VERT ZOO] **1.** Having the toes separated to the base. **2.** Of or relating to the Fissipeda. { 'fis·ə,ped }

Fissipeda [VERT ZOO] Former designation for a suborder of the Carnivora. { fə'sip·ə·də }

Fissurellidae [INV ZOO] The keyhole limpets, a family of gastropod mollusks in the order Archeogastropoda. { ,fis·ə'rel·ə,dē }

fistula [MED] An abnormal congenital or acquired communication between two surfaces or between a viscus or other hollow structure and the exterior. { 'fis·chə·lə }

fistulous withers [VET MED] A chronic inflammation of the withers of a horse accompanied by fluid discharge, which may be initiated by mechanical injury but depends on bacterial (*Brucella abortus*) infection for development. { 'fis·chə·ləs 'wit͟h·ərz }

fitness [GEN] A measure of reproductive success for a genotype, based on the average number of surviving progeny of this genotype as compared to the average number of other, competing genotypes. { 'fit·nəs }

fix [BIOL] To kill, harden, or preserve a tissue, organ, or organism by immersion in dilute acids, alcohol, or solutions of coagulants. { fiks }

fixator [PHYSIO] A muscle whose action tends to hold a body part in a certain position or limit its movement. { 'fik,sād·ər }

fixed allele [GEN] An allele that is homozygous in all members of a population. { ,fikst ə'lēl }

fixed prosthodontics [MED] A subdivision of prosthodontics that focuses on the replacement of missing teeth by dental bridges. { ,fikst ,präs·thə'dän·tiks }

flabellate [BIOL] Fan-shaped. { flə'bel·ət }

Flabellifera [INV ZOO] The largest and morphologically most generalized suborder of isopod crustaceans; the biramous uropods are attached to the sides of the abdomen and may form, with the last abdominal fragment, a caudal fan. { ,flab·ə'lif·ə·rə }

Flabelligeridae [INV ZOO] The cage worms, a family of spioniform worms belonging to the Sedentaria; the anterior part of the body is often concealed by a cage of setae arising from the first few segments. { flə,bel·ə'jer·ə,dē }

flabellum [INV ZOO] Any structure resembling a fan, as the epipodite of certain crustacean limbs. { flə'bel·əm }

flaccid [BOT] Deficient in turgor. [PHYSIO] Soft, flabby, or relaxed. { 'flas·əd *or* 'flak·səd }

flacherie [INV ZOO] A fatal bacterial disease of caterpillars, especially silkworms, marked by loss of appetite, dysentery, and flaccidity of the body; after death the body darkens and liquefies. { ,flash·ə'rē }

Flacourtiaceae [BOT] A family of dicotyledonous plants in the order Violales having the characteristics of the more primitive members of the order. { flə,kůrd·ē'ās·ē,ē }

flagella [BIOL] Relatively long, whiplike, centriole-based locomotor organelles on some motile cells. { flə'jel·ə }

Flagellata [INV ZOO] The equivalent name for Mastigophora. { ˌflaj·əˈläd·ə }

flagellate [BIOL] **1.** Having flagella. **2.** An organism that propels itself by means of flagella. **3.** Resembling a flagellum. [INV ZOO] Any member of the protozoan superclass Mastigophora. { ˈflaj·əˌlāt }

flagellated chamber [INV ZOO] An outpouching of the wall of the central cavity in Porifera that is lined with choanocytes; connects with incurrent canals through prosophyles. { ¦flaj·əˌlād·əd ˈchām·bər }

flagellation [BIOL] The arrangement of flagella on an organism. { ˌflaj·əˈlā·shən }

flagellin [MICROBIO] The protein component of bacterial flagella. { fləˈjel·ən }

flagilliflory [BOT] Of flowers, hanging down freely from ropelike twigs. { fləˈjil·əˌflȯr·ē }

flame bulb [INV ZOO] The enlarged terminal part of the flame cell of a protonephridium, consisting of a tuft of cilia. { ˈflām ˌbəlb }

flame cell [INV ZOO] A hollow cell that contains the terminal branches of excretory vessels in certain flatworms and rotifers and some other invertebrates. { ˈflām ˌsel }

flamingo [VERT ZOO] Any of various long-legged and long-necked aquatic birds of the family Phoenicopteridae characterized by a broad bill resembling that of a duck but abruptly bent downward and rosy-white plumage with scarlet coverts. { fləˈmiŋ·gō }

flank [VERT ZOO] The part of a quadruped mammal between the ribs and the pelvic girdle. { flaŋk }

flash burn [MED] Tissue injury resulting from exposure to high-intensity radiant heat. { ˈflash ˌbərn }

flash coloration [ECOL] A type of protective coloration in which the prey is cryptic when at rest but reveals brilliantly colored parts while escaping. { ˌflash kəl·əˈrā·shən }

flash pasteurization [MICROBIO] A pasteurization method in which a heat-labile liquid, such as milk, is briefly subjected to temperatures of 230°F (110°C). { ¦flash pas·chə·rəˈzā·shən }

flatfish [VERT ZOO] Any of a number of asymmetrical fishes which compose the order Pleuronectiformes; the body is laterally compressed, and both eyes are on the same side of the head. { ˈflatˌfish }

flatulence [MED] Excessive intestinal gas. { ˈflach·ə·ləns }

flatus [MED] Gas in the intestinal tract. { ˈflād·əs }

flatwood [ECOL] An almost-level zone containing mostly imperfectly drained, acid soils and vegetation consisting of wiregrass and saw palmetto at ground level, shrubs such as gallberry and waxmyrtle, and trees such as longleaf and slash pines. { ˈflatˌwu̇d }

flatworm [INV ZOO] The common name for members of the phylum Platyhelminthes; individuals are dorsoventrally flattened. { ˈflatˌwərm }

flavan [BIOCHEM] $C_{15}H_{14}O$ 2-Phenylbenzopyran, an aromatic heterocyclic compound from which all flavonoids are derived. { ˈfla·vən }

flavanol [BIOCHEM] Yellow needles with a melting point of 169°C, derived from flavanone; a flavanoid pigment used as a dye. Also known as 3-hydroxyflavone. { ˈfla·vəˌnȯl }

flavanone [BIOCHEM] $C_{15}H_{12}O_2$ A colorless crystalline ketone, that often occurs in plants in the form of a glycoside. { ˈfla·vəˌnōn }

flavin [BIOCHEM] Any of several water-soluble yellow pigments occurring as coenzymes of flavoproteins. { ˈfla·vən }

flavin adenine dinucleotide [BIOCHEM] $C_{27}H_{33}N_9O_{15}P_2$ A coenzyme that functions as a hydrogen acceptor in aerobic dehydrogenases (flavoproteins). Abbreviated FAD. { ˈfla·vən ˈad·ənˌēn dīˈnü·klē·əˌtīd }

flavin mononucleotide *See* riboflavin 5′-phosphate. { ˈfla·vən ¦mä·nōˈnü·klē·əˌtīd }

flavin phosphate *See* riboflavin 5′-phosphate. { ˈfla·vən ˈfäsˌfāt }

Flavobacterium [MICROBIO] A genus of bacterium of uncertain affiliation; gram-negative coccobacilli or slender rods producing pigmented (yellow, red, orange, or brown) growth on solid media. { ¦fla·vō·bakˈtir·ē·əm }

flavone [BIOCHEM] **1.** Any of a number of ketones composing a class of flavonoid compounds. **2.** $C_{15}H_{10}O_2$ A colorless crystalline compound occurring as dust on the surface of many primrose plants. { ˈflaˌvōn }

flavonoid [BIOCHEM] Any of a series of widely distributed plant constituents related to the aromatic heterocyclic skeleton of flavan. { ˈfla·vəˌnȯid }

flavonol [BIOCHEM] **1.** Any of a class of flavonoid compounds that are hydroxy derivatives of flavone. **2.** $C_{16}H_{10}O_2$ A colorless, crystalline compound from which many yellow plant pigments are derived. { ˈfla·vəˌnȯl }

flavoprotein [BIOCHEM] Any of a number of conjugated protein dehydrogenases containing flavin that play a role in biological oxidations in both plants and animals; a yellow enzyme. { ¦fla·vō¦prōˌtēn }

flax [BOT] *Linum usitatissimum.* An erect annual plant with linear leaves and blue flowers; cultivated as a source of flaxseed and fiber. { flaks }

flaxseed [BOT] The seed obtained from the seed flax plant; a source of linseed oil. Also known as linseed. { ˈflakˌsēd }

flea [INV ZOO] Any of the wingless insects composing the order Siphonaptera; most are ectoparasites of mammals and birds. { flē }

flea allergy dermatitis [VET MED] Inflammation of the skin of small pets (dogs, cats, and ferrets) that results from an allergic reaction to protein substances deposited on or under the surface of the skin at the time of flea feeding. Also known as flea-bite dermatitis. { ¦flēˌal·ər·jē ˌdər·məˈtīd·əs }

flea-bite dermatitis *See* flea allergy dermatitis. { ¦flēˌbīt ˌdər·məˈtīd·əs }

flea-borne typhus *See* murine typhus. { 'flē ,bȯrn 'tī·fəs }

fleece [VERT ZOO] Coat of wool shorn from sheep; usually taken off the animal in one piece. { flēs }

Flehmen response [ECOL] A courtship behavior displayed by the males of some mammalian species in which the upper lip is curled and the neck is extended, facilitating the reception of olfactory cues. { 'flā·mən ri,späns }

flesh [ANAT] The soft parts of the body of a vertebrate, especially the skeletal muscle and associated connective tissue and fat. { flesh }

fleshy fruit [BOT] A fruit having a fleshy pericarp that is usually soft and juicy, but sometimes hard and tough. { ¦flesh·ē ¦früt }

Flexibacter [MICROBIO] A genus of bacteria in the family Cytophagaceae; cells are unsheathed rods or filaments and are motile; microcysts are not known. { ¦flek·sə¦bak·tər }

flexion [BIOL] Act of bending, especially of a joint. { 'flek·shən }

flexion reflex [PHYSIO] An unconditioned, segmental reflex elicited by noxious stimulation and consisting of contraction of the flexor muscles of all joints on the same side. Also known as the nocioceptive reflex. { 'flek·shən ,rē,fleks }

Flexithrix [MICROBIO] A genus of bacteria in the family Cytophagaceae; cells are usually sheathed filaments, and unsheathed cells are motile; microcysts are not known. { 'flek·sə,thriks }

flexor [PHYSIO] A muscle that bends or flexes a limb or a part. { 'flek·sər }

flexuous [BIOL] **1.** Flexible. **2.** Bending in a zigzag manner. **3.** Wavy. { 'flek·shə·wəs }

flexure [EMBRYO] A sharp bend of the anterior part of the primary axis of the vertebrate embryo. [VERT ZOO] The last joint of a bird's wing. { 'flek·shər }

flicker fusion [PHYSIO] The tendency to perceive an oscillating or flickering sensory input signal as continuous when the frequency is above a specific threshold frequency. { 'flik·ər ,fyü·zhən }

flight feather [VERT ZOO] Any of the long contour feathers on the wing of a bird. Also known as remex. { 'flīt ,feth·ər }

flipper [VERT ZOO] A broad, flat appendage used for locomotion by aquatic mammals and sea turtles. { 'flip·ər }

float [BIOL] An air-filled sac in many pelagic flora and fauna that serves to buoy up the body of the organism. { flōt }

floating rib [ANAT] One of the last two ribs in humans which have the anterior end free. { ¦flōd·iŋ 'rib }

floatoblast [INV ZOO] A free-floating statoblast having a float of air cells. { 'flōd·ə,blast }

floccose [BOT] Covered with tufts of woollike hairs. { 'flä,kōs }

flocculate [BIOL] Having small tufts of hairs. { 'fläk·yə,lət (adjective) *or* 'fläk·yə,lāt (verb) }

flocculonodular lobes [ANAT] The pair of lateral cerebellar lobes in vertebrates which function to regulate vestibular reflexes underlying posture; referred to functionally as the vestibulocerebellum. { ¦fläk·yə·lō¦näj·ə·lər 'lōbz }

flocculus [ANAT] A prominent lobe of the cerebellum situated behind and below the middle cerebellar peduncle on each side of the median fissure. { 'fläk·yə·ləs }

floccus [BOT] A tuft of woolly hairs. { 'fläk·əs }

flora [BOT] **1.** Plants. **2.** The plant life characterizing a specific geographic region or environment. { 'flȯr·ə }

floral axis [BOT] A flower stalk. { ¦flȯr·əl 'ak·səs }

floral diagram [BOT] A diagram of a flower in cross section showing the number and arrangement of floral parts. { ¦flȯr·əl 'dī·ə,gram }

floret [BOT] A small individual flower that is part of a compact group of flowers, such as the head of a composite plant or inflorescence. { 'flȯr·ət }

Florey unit [BIOL] A unit for the standardization of penicillin. { 'flȯr·ē ,yü·nət }

floricome [INV ZOO] A type of branched hexaster spicule. { 'flȯr·ə,kōm }

Florideophyceae [BOT] A class of red algae, division Rhodophyta, having prominent pit connections between cells. { flə¦rid·ē·ō¦fīs·ē,ē }

floriferous [BOT] Blooming freely, used principally of ornamental plants. { flȯ'rif·ə·rəs }

florigen [BIOCHEM] A plant hormone that stimulates buds to flower. { 'flȯr·ə·jen }

florivorous [ZOO] Feeding on flowers. { flȯ 'riv·ə·rəs }

florula [ECOL] Plants which grow in a small, confined habitat, for example, a pond. { 'flȯr·yə·lə }

floscelle [INV ZOO] A flowerlike structure around the mouth of some echinoids. { flȯ'sel }

Flosculariacea [INV ZOO] A suborder of rotifers in the order Monogononta having a malleoramate mastax. { ¦fläs·kyə,lar·ē'ās·ē·ə }

Flosculariidae [INV ZOO] A family of sessile rotifers in the suborder Flosculariacea. { ,fläs·kyə·lə'rī·ə,dē }

flosculous [BOT] **1.** Composed of florets. **2.** Of a floret, tubular in form. { 'fläs·kyə·ləs }

flosculus [BOT] A floret. { 'fläs·kyə·ləs }

flounder [VERT ZOO] Any of a number of flatfishes in the families Pleuronectidae and Bothidae of the order Pleuronectiformes. { 'flau̇n·dər }

flow bog [ECOL] A peat bog with a surface level that fluctuates in accordance with rain and tides. { 'flō ,bäg }

flow cytometry [CELL MOL] A technique for optical analysis and separation of cells and metaphase chromosomes based on light scattering and fluorescence. { ¦flō sī¦täm·ə·trē }

flower [BOT] The characteristic reproductive structure of a seed plant, particularly if some or all of the parts are brightly colored. { 'flau̇·ər }

flowers of sulfur [PHARM] One of three forms of pharmaceutical sulfur, made by sublimation; the other two forms are precipitated sulfur and washed sulfur. Also known as sublimed sulfur. { 'flau̇·ərz əv 'səl·fər }

flow karyotype [CELL MOL] A karyotype that is based on flow cytometry measurements. { ¦flō 'kar·ē·ə‚tīp }

flow of variability [GEN] The movement of genetic variability within a population as a result of hybridization and segregation. { ¦flō əv ‚ver·ē·ə'bil·əd·ē }

floxuridine [PHARM] $C_9H_{11}FN_2O_5$ Crystals that melt at 150–151°C; used as an antiviral drug and as an inhibitor of deoxyribonucleic acid synthesis. Abbreviated FUDR. { ‚fläks'yůr·ə‚dēn }

fluctuation test [MICROBIO] A method of demonstrating that bacterial mutations preexist in a population before they are selected; a large parent population is divided into small parts which are grown independently and the number of mutants in each subculture determined; the number of mutants in the subculture will fluctuate because in some a mutant arises early (giving a large number of progeny), while in others the mutant arises late and gives few progeny. { ‚flək·chə'wā·shən ‚test }

flufenamic acid [PHARM] $C_{14}H_{10}F_3NO_2$ Pale yellow needles with a melting point of 125°C; used as an anti-inflammatory drug or analgesic. { ¦flü·fə¦nam·ik 'as·əd }

fluke [INV ZOO] The common name for more than 40,000 species of parasitic flatworms that form the class Trematoda. [VERT ZOO] A flatfish, especially summer flounder. { flük }

fluorescent antibody [IMMUNOL] An antibody that is labeled by a fluorescent dye, such as fluorescein. { flů'res·ənt 'an·tē‚bäd·ē }

fluorescent antibody test [IMMUNOL] A clinical laboratory test based on the antigen used in the diagnosis of syphilis and lupus erythematosus and for identification of certain bacteria and fungi, including the tubercle bacillus. { flů'res·ənt 'an·tē‚bäd·ē ‚test }

fluorescent in situ hybridization [GEN] A technique in which a deoxyribonucleic acid (DNA) probe is labeled with a fluorescent dye (that can be visualized under a fluorescent microscope) and then hybridized with target DNA, usually chromosome preparations on a microscopic slide. It is used to precisely map genes to a specific region of a chromosome in prepared karyotype, or can enumerate chromosomes, or can detect chromosomal deletions, translocations, or gene amplifications in cancer cells. Abbreviated FISH. { flə¦res·ənt in¦sit‚chü ‚hī·brə·də'zā·shən }

fluorescent staining [CELL MOL] The use of fluorescent dyes to mark specific cell structures, such as chromosomes. { flů¦res·ənt 'stān·iŋ }

fluorochromasia [CELL MOL] The immediate appearance of fluorescence inside viable cells on exposure to a fluorogenic substrate. { ¦flůr·ō·krə'mā·zhə }

fluorouracil [PHARM] $C_4H_3FN_2O_2$ Crystals that decompose at 282–283°C; used as an antineoplastic drug and as an inhibitor of deoxyribonucleic acid synthesis. Also known as 2,4-dioxo-5-fluoropyrimidine; 5-FU. { ¦flůr·ō'yůr·ə‚sil }

flush [ECOL] An evergreen herbaceous or nonflowering vegetation growing in habitats where seepage water causes the surface to be constantly wet but rarely flooded. { fləsh }

flutter [MED] Rapid, regular contraction of the atrial muscle of the heart. { 'fləd·ər }

fly [INV ZOO] The common name for a number of species of the insect order Diptera characterized by a single pair of wings, antennae, compound eyes, and hindwings modified to form knoblike balancing organs, the halters. { flī }

flying fish [VERT ZOO] Any of about 65 species of marine fishes which form the family Exocoetidae in the order Atheriniformes; characteristic enlarged pectoral fins are used for gliding. { ¦flī·iŋ ¦fish }

flyway [VERT ZOO] A geographic migration route for birds, including the breeding and wintering areas that it connects. { 'flī‚wā }

FMF *See* familial Mediterranean fever.

FMN *See* riboflavin 5′-phosphate.

foal [VERT ZOO] A young horse, especially one under 1 year of age. { fōl }

focal adhesion plaque [CELL MOL] Points of attachment that form when cells attach to a substrate. { ¦fō·kəl əd'hē·zhən ‚plak }

focal infection [MED] Infection in a limited area, such as the tonsils, teeth, sinuses, or prostate. { ¦fō·kəl in¦fek·shən }

focal seizure [MED] An epileptic manifestation of a restricted nature, usually without loss of consciousness, due to irritation of a localized area of the brain. { ¦fō·kəl 'sē·zhər }

foehn sickness [MED] A phenomenon in humans in alpine regions, marked by adverse psychological and physiological effects during prolonged periods of foehn wind. { 'fān ‚sik·nəs }

fog climax [ECOL] A community that deviates from a climatic climax because of the persistent occurrence of a controlling fog blanket. { ¦fäg 'klī‚maks }

fog forest [ECOL] The dense, rich forest growth which is found at high or medium-high altitudes on tropical mountains; occurs when the tropical rain forest penetrates altitudes of cloud formation, and the climate is excessively moist and not too cold to prevent plant growth. { 'fäg ‚fär·əst }

fold [ANAT] A plication or doubling, as of various parts of the body such as membranes and other flat surfaces. { fōld }

foliaceous [BOT] Consisting of or having the form or texture of a foliage leaf. [ZOO] Resembling a leaf in growth form or mode. { ‚fō·lē'ā·shəs }

foliage [BOT] The leaves of a plant. { 'fō·lē·ij }

foliage leaf [BOT] The chief photosynthetic organ of most vascular plants. { 'fō·lē·ij ‚lēf }

foliar [BOT] Of, pertaining to, or consisting of leaves. { 'fō·lē·ər }

foliate papilla [VERT ZOO] One of the papillae found on the posterolateral margin of the tongue of many mammals, but vestigial or absent in humans. { 'fō·lē·ət pə'pil·ə }

foliation [BOT] **1.** The process of developing

into a leaf. **2.** The state of being in leaf. { ˌfō·lē'ā·shən }

folic acid [BIOCHEM] $C_{19}H_{19}N_7O_6$ A yellow, crystalline vitamin of the B complex; it is slightly soluble in water, usually occurs in conjugates containing glutamic acid residues, and is found especially in plant leaves and vertebrate livers. Also known as pteroylglutamic acid (PGA). { 'fō·lik 'as·əd }

foliferous [BOT] Producing leaves. { fə'lif·ə·rəs }

foliicolous [BIOL] Growing or parasitic upon leaves, as certain fungi. { 'fä·lē·ə'kə·ləs }

foliobranchiate [VERT ZOO] Having leaflike gills. { ¦fō·lē·ō¦braŋ·kē‚āt }

foliolate [BOT] Having leaflets. { 'fō·lē·ə‚lāt }

follicle [BIOL] A deep, narrow sheath or a small cavity. [BOT] A type of dehiscent fruit composed of one carpel opening along a single suture. { 'fäl·ə·kəl }

follicle cell [CELL MOL] A cell through which the developing ovum receives material for growth. { 'fäl·ə·kəl ˌsel }

follicle-stimulating hormone [BIOCHEM] A protein hormone released by the anterior pituitary of vertebrates which stimulates growth and secretion of the Graafian follicle and also promotes spermatogenesis. Abbreviated FSH. { ¦fäl·ə·kəl ¦stim·yə‚lād·iŋ 'hȯr‚mōn }

follicular lymphoma [MED] A premalignant lymphoma in which the lymph nodes show enlarged follicles composed predominantly of closely packed, large reticuloendothelial cells. { fə'lik·yə·lər lim'fō·mə }

folliculate [BIOL] Having or composed of follicles. { fə'lik·yə‚lāt }

folliculitis [MED] Inflammation of a follicle or group of follicles. { fə‚lik·yə'līd·əs }

folliculitis keloidalis *See* keloid acne. { fə‚lik·yə'līd·əs ˌkē·lȯi'dal·əs }

fomite [MED] An inanimate object contaminated with an infectious organism (for example, a dish, clothing, towel, needle, or dust). { 'fō‚mīt }

fontanelle [ANAT] A membrane-covered space between the bones of a fetal or young skull. [INV ZOO] A depression on the head of termites. { fänt·ən'el }

food [BIOL] A material that can be ingested and utilized by the organism as a source of nutrition and energy. { füd }

food allergy [IMMUNOL] A hypersensitivity to certain foods. { 'füd ˌal·ər·jē }

food-borne disease [MED] Any disease transmitted by contaminated foods. { ¦füd ¦bȯrn di'zēz }

food chain [ECOL] The scheme of feeding relationships by trophic levels which unites the member species of a biological community. { 'füd ˌchān }

food infection [MED] A type of bacterial food poisoning in which the host is infected by organisms carried by food. { 'füd in‚fek·shən }

food poisoning [MED] Poisoning due to intake of food contaminated by bacteria or poisonous substances produced by bacteria. { 'füd ˌpȯiz·ən·iŋ }

food pyramid [ECOL] An ecological pyramid representing the food relationship among the animals in a community. { 'füd ˌpir·ə‚mid }

food vacuole [CELL MOL] A membrane-bound organelle in which digestion occurs in cells capable of phagocytosis. Also known as heterophagic vacuole; phagocytic vacuole. { 'füd ˌvak·yə‚wōl }

food web [ECOL] A modified food chain that expresses feeding relationships at various, changing trophic levels. { 'füd ˌweb }

foot [ANAT] Terminal portion of a vertebrate leg. [BOT] In a fern, moss, or liverwort, the basal part of the young sporophyte that attaches it to the gametophyte. [INV ZOO] An organ for locomotion or attachment. { fu̇t }

foot-and-mouth disease [VET MED] A highly contagious virus disease of cattle, pigs, sheep, and goats that is transmissible to humans; characterized by fever, salivation, and formation of vesicles in the mouth and pharynx and on the feet. Also known as hoof-and-mouth disease. { ¦fu̇t ən 'mau̇th di‚zēz }

foot gland [INV ZOO] A glandular structure which secretes an adhesive substance in many animals. Also known as pedal gland. { 'fu̇t ˌgland }

foot rot *See* foul foot. { 'fu̇t ˌrät }

foramen [BIOL] A small opening, orifice, pore, or perforation. { fə'rā·mən }

foramen magnum [ANAT] A large oval opening in the occipital bone at the base of the cranium that allows passage of the spinal cord, accessory nerves, and vertebral arteries. { fə'rā·mən 'mag·nəm }

foramen of Magendie [ANAT] The median aperture of the fourth ventricle of the brain. { fə'rā·mən əv mə‚zhän'dē }

foramen of Monro *See* interventricular foramen. { fə'rā·mən əv mən'rō }

foramen of Winslow *See* epiploic foramen. { fə'rā·mən əv 'winz·lō }

foramen ovale [ANAT] An opening in the sphenoid for the passage of nerves and blood vessels. [EMBRYO] An opening in the fetal heart partition between the two atria. { fə'rā·mən ō'vä·lē }

foramen primum [EMBRYO] A temporary embryonic interatrial opening. { fə'rā·mən 'prī·məm }

Foraminiferida [INV ZOO] An order of dominantly marine protozoans in the subclass Granuloreticulosia having a secreted or agglutinated shell enclosing the ameboid body. { fə¦ram·ə·nə'fer·ə·də }

forb [BOT] A weed or broadleaf herb. { fȯrb }

forced expiratory volume [MED] During the performance of a forced vital capacity measurement, the volume of exhaled gas over a specific time interval. { ¦fȯrst ik¦spī·rə‚tȯr·ē 'väl·yəm }

forced vital capacity [MED] Maximum gas volume which can be expired, as quickly and forcibly as possible, after a maximum inspiration. { ¦fȯrst 'vīd·əl kə'pas·əd·ē }

forceps [INV ZOO] A pair of curved, hard, movable appendages at the end of the abdomen of certain insects, for example, the earwig. [MED] A device with two blades or limbs opposite each other which is operated by handles or by direct force on the blades; used in surgery to grasp, compress, and hold tissue, a body part, or surgical substances. { 'fȯr·səps }

forcipate [BIOL] Shaped like forceps; deeply forked. { 'fȯr·sə,pāt }

forcipate trophus [INV ZOO] A type of masticatory apparatus in certain predatory rotifers which resembles forceps and is used for grasping. { 'fȯr·sə,pāt 'trō·fəs }

Forcipulatida [INV ZOO] An order of echinoderms in the subclass Asteroidea characterized by crossed pedicellariae. { fȯr¦sip·ə'lad·ə·də }

forearm [ANAT] The part of the upper extremity between the wrist and the elbow. Also known as antebrachium. { 'fȯr,ärm }

forebrain [EMBRYO] The most anterior expansion of the neural tube of a vertebrate embryo. [VERT ZOO] The part of the adult brain derived from the embryonic forebrain; includes the cerebrum, thalamus, and hypothalamus. { 'fȯr ,brān }

forefinger [ANAT] The index finger; the first finger next to the thumb. { 'fȯr,fiŋ·gər }

forefoot [VERT ZOO] An anterior foot of a quadruped. { 'fȯr,fu̇t }

foregut [EMBRYO] The anterior alimentary canal in a vertebrate embryo, including those parts which will develop into the pharynx, esophagus, stomach, and anterior intestine. { 'fȯr ,gət }

forehead [ANAT] The part of the face above the eyes. { 'fär·əd }

forelimb [ANAT] An appendage (as a wing, fin, or arm) of a vertebrate that is, or is homologous to, the foreleg of a quadruped. { 'fȯr,lim }

forensic anthropology [FOREN] The application of physical anthropology theory and techniques to answering legal questions involving human skeletal identification and analysis. { fə¦ren·sik ,an·thrə'päl·ə·jē }

forensic biology [FOREN] The analysis of the biological or genetic properties of evidence. { fə'ren·sik bī'äl·ə·jē }

forensic chemistry [FOREN] The application of chemistry to the study of materials or problems in cases where the findings may be presented as technical evidence in a court of law. { fə'ren·sik 'kem·ə·strē }

forensic engineering [FOREN] The application of accepted engineering practices and principles for discussion, debate, argumentative, or legal purposes. { fə'ren·sik ,en·jə'nir·iŋ }

forensic entomology [FOREN] The application of insect evidence to criminal investigations and civil cases. { fə¦ren·sik ,en·tə'mä·lə·jē }

forensic medicine [FOREN] Application of medical evidence or medical opinion for purposes of civil or criminal law. { fə'ren·sik 'med·ə·sən }

forensic odontology [FOREN] A subspecialty of forensic medicine which focuses on the identification of deceased persons by dental examination, or of perpetrators by bite marks. { fə'ren·sik ,ō·dän'täl·ə·jē }

forensic pathology [FOREN] A subspecialty of forensic medicine which deals with the cause and manner of death. { fə'ren·sik pə'thäl·ə·jē }

forensic physics [FOREN] The application of physics for discussion, debate, argumentative, or legal purposes. { fə'ren·sik 'fiz·iks }

forensic psychiatry [FOREN] A branch of psychiatry dealing with legal issues related to mental disorders. { fə¦renz·ik sī'kī·ə·trē }

forensic science [FOREN] The recognition, collection, identification, individualization, and interpretation of physical evidence, and the application of science and medicine for criminal and civil law, or regulatory purposes. { fə'ren·sik 'sī·əns }

forensic toxicology [FOREN] An interdisciplinary field applying the methods of analytical chemistry, pharmacology, and toxicology to the analysis and interpretation of drugs and chemicals in biological samples for legal purposes. { fə'ren·sik ,täk·sə'käl·ə·jē }

forest [ECOL] An ecosystem consisting of plants and animals and their environment, with trees as the dominant form of vegetation. { 'fär·əst }

forest conservation [ECOL] Those measures concerned with the protection and preservation of forest lands and resources. { 'fär·əst ,kän·sər'vā·shən }

forest ecology [ECOL] The science that deals with the relationship of forest trees to their environment, to one another, and to other plants and to animals in the forest. { 'fär·əst i,käl·ə·jē }

forest ecosystem [ECOL] The entire assemblage of forest organisms (trees, shrubs, herbs, bacteria, fungi, and animals, including people) together with their environmental substrate (the surrounding air, soil, water, organic debris, and rocks), interacting inside a defined boundary. { ¦fär·əst 'ek·ō,sis·təm }

forestry [ECOL] The management of forest lands for wood, forages, water, wildlife, and recreation. { 'fär·ə·strē }

forest-tundra [ECOL] A temperate and cold savanna which occurs at high altitudes and consists of scattered or clumped trees and a shrub layer of varying coverage. { ¦fär·əst ¦tən·drə }

formamidase [BIOCHEM] An enzyme involved in tryptophane catabolism; catalyzes the conversion of N-formylkynurenine to kynurenine and formate. { fȯr'mam·ə,dās }

Formicariidae [VERT ZOO] The antbirds, a family of suboscine birds in the order Passeriformes. { ,fȯr·mə·kə'rī·ə,dē }

Formicidae [INV ZOO] The ants, social insects composing the single family of the hymenopteran superfamily Formicoidea. { fȯr'mis·ə,dē }

formicivorous [ZOO] Feeding on ants. { ,fȯr·mə'siv·ə·rəs }

Formicoidea [INV ZOO] A monofamilial superfamily of hymenopteran insects in the suborder

Apocrita, containing the ants. { ˌfȯr·mə'kȯid·ē·ə }

formyl methionine [BIOCHEM] Formylated methionine; initiates peptide chain synthesis in bacteria. { 'fȯrˌmil mə'thī·əˌnēn }

fornix [ANAT] A structure that is folded or arched. [BOT] A small scale, especially in the corolla tube of some plants. { 'fȯrˌniks }

Forssman antibody [IMMUNOL] A heterophile antibody that reacts with Forssman antigen. { 'fȯrs·mən 'an·təˌbäd·ē }

Forssman antigen [IMMUNOL] A heterophile antigen, occurring in a variety of unrelated animals, which elicits production of hemolysin (Forssman antibody) for sheep red blood cells. { 'fȯrs·mən 'an·tə·jən }

fossa [ANAT] A pit or depression. [VERT ZOO] *Cryptoprocta ferox*. A Madagascan carnivore related to the civets. { 'fäs·ə }

fossorial [VERT ZOO] Adapted for digging. { fä'sȯr·ē·əl }

foulbrood [INV ZOO] The common name for three destructive bacterial diseases of honeybee larvae. { 'fau̇lˌbrüd }

foul foot [VET MED] A feedlot disease of cattle and sheep marked by inflammation and ulceration of the feet; common in wet feedlots. Also called foot rot. { 'fau̇l ˌfu̇t }

fouling organism [ECOL] Any aquatic organism with a sessile adult stage that attaches to and fouls underwater structures of ships. { 'fau̇l·iŋ ˌȯr·gəˌniz·əm }

founder effect [GEN] The overrepresentation of a specific allele at one or more loci in a new population that arises from a small number of individuals whose small gene pool may be unrepresentative of the parental population initially or as a result of the ensuing genetic drift. { 'fau̇n·dər iˌfekt }

fourré *See* temperate and cold scrub; tropical scrub. { fü'rā }

fovea [BIOL] A small depression or pit. { 'fō·vē·ə }

fovea centralis [ANAT] A small, rodless depression of the retina in line with the visual axis, which affords acute vision. { 'fō·vē·ə sen'tral·əs }

foveal vision [PHYSIO] Vision achieved by looking directly at objects in the daylight so that the image falls on or near the fovea centralis. Also known as photopic vision. { ¦fō·vē·əl 'vizh·ən }

foveola [BIOL] A small pit, especially one in the embryonic gastric mucosa from which gastric glands develop. { fō'vē·ə·lə }

foveolate [BIOL] Having small depressions; pitted. { 'fō·vē·əˌlāt }

fowl pox [VET MED] A disease of birds caused by a virus and characterized by wartlike nodules on the skin, particularly on the head. { 'fau̇l ˌpäks }

fox [VERT ZOO] The common name for certain members of the dog family (Canidae) having relatively short legs, long bodies, large erect ears, pointed snouts, and long bushy tails. { fäks }

fraction kill hypothesis [MED] A principle of chemotherapy that a uniform dose of a drug will destroy a constant fraction rather than a constant number of tumor cells regardless of the size of the tumor or the number of cells present. { 'frak·shən ˌkil hīˌpäth·ə·səs }

fracture [MED] The breaking of bone, cartilage, or teeth. { 'frak·shər }

fragile site [GEN] The chromosomal position of a deoxyribonucleic acid sequence predisposed to spontaneous or induced breakage; sometimes contains short repetitive sequences. { 'fraj·əl ¦sīt }

fragile X syndrome [MED] A hereditary condition resulting from a trinucleotide repeat at an inherited fragile site on the long arm of the X chromosome. Affected males usually have some characteristic facial features, enlarged testes, and mental retardation. Females with one fragile X chromosome and one normal X chromosome may have a lesser degree of mental retardation. { ˌfraj·əl 'eks ˌsinˌdrōm }

fragility test [PATH] A measure of the resistance of red blood cells to osmotic hemolysis in hypotonic salt solutions of graded dilutions. { frə'jil·əd·ē ˌtest }

fragmentation [CELL MOL] Amitotic division; a type of asexual reproduction. { ˌfrag·mən'tā·shən }

frambesia *See* yaws. { fram'bē·zhə }

frameshift mutation [GEN] The addition or deletion of nucleotides to anexon in numbers other than three, which shifts the translation reading frame so a new set of codons beyond the point of abnormality in the messenger ribonucleic acid is read. Also known as phase-shift mutation. { ¦frāmˌshift myü'tā·shən }

frameshift suppression [GEN] Reversion of a frameshift mutation by a second frameshift mutation in the same gene. { 'frāmˌshift sə ˌpresh·ən }

Francisella [MICROBIO] A genus of gram-negative, aerobic bacteria of uncertain affiliation; cells are small, coccoid to ellipsoidal, pleomorphic rods and can be parasitic on mammals, birds, and arthropods. { ˌfran·si'sel·ə }

Frankia [MICROBIO] The single genus of the family Frankiaceae. { ˌfraŋ'kē·ə }

Frankiaceae [MICROBIO] A family of bacteria in the order Actinomycetales; filamentous cells form true mycelia; they are symbiotic and found in active, nitrogen-fixing root nodules. { ˌfraŋ·kē'ās·ēˌē }

fraternal twins *See* dizygotic twins. { frə¦tərn·əl 'twinz }

freckle [MED] A pigmented macule resulting from focal increase in melanin, usually associated with exposure to sunlight, commonly on the face. { 'frek·əl }

freemartin [VERT ZOO] An intersexual, usually sterile female calf twin born with a male. { 'frēˌm-ärt·ən }

free ribosome [CELL MOL] A ribosome located by itself or in a group (known as a polysome or polyribosome) in the cytosol, rather than bound to the endoplasmic reticulum; synthesizes soluble cytosolic proteins and most extrinsic membrane proteins. { ¦frē 'rī·bə,sōm }

freestone [BOT] A fruit stone to which the fruit does not cling, as in certain varieties of peach. { 'frē,stōn }

Free test [PATH] Demonstration of a delayed hypersensitivity reaction to Bedsonia antigen in the diagnosis of lymphogranuloma venereum. { 'frē ,test }

freeze-fracture electron microscopy [CELL MOL] A technique used to visualize the inside of cellular membranes. Rapidly frozen cells are ruptured (fractured) so as to split open the membrane and expose the interior surfaces. A thin layer of carbon together with a metal (usually platinum) is then evaporated over the specimen to produce a replica of the surface, which is removed and examined in the electron microscope. { ¦frēz,frak·chər i,lek,trän mi'kräs·kə·pē }

Fregatidae [VERT ZOO] Frigate birds or man-o'-war birds, a family of fish-eating birds in the order Pelecaniformes. { fre'gad·ə,dē }

F_c region [IMMUNOL] Region of an antibody molecule that binds to antibody receptors on the surface of cells such as macrophages and mast cells, and to complement protein; F_c is derived from the term crystallizable fragment. { ¦ef'sē ,rē·jən }

Frenatae [INV ZOO] The equivalent name for Heteroneura. { 'frē·nə,tē }

French measles *See* rubella. { ¦french 'mē·zəlz }

frenulum [ANAT] **1.** A small fold of integument or mucous membrane. **2.** A small ridge on the upper part of the anterior medullary velum. [INV ZOO] A spine on most moths that projects from the hindwings and is held to the forewings by a clasp, thus coupling the wings together. { 'fren·yə·ləm }

frenum [ANAT] A fold of tissue that restricts the movements of an organ. { 'frē·nəm }

frequency coding [NEUROSCI] A means by which the central nervous system analyzes the content of a receptor message; the frequency of discharged impulses is a function of the rate of rise of the generator current and indirectly of the stimulus strength; the greater the stimulus intensity, the higher the impulse frequency of the message. { 'frē·kwən·sē ,kōd·iŋ }

frequency-dependent selection [EVOL] A type of natural selection that decreases the frequency of more common phenotypes in a population and increases the frequency of less common phenotypes. { ¦frē·kwən·sē di¦pen·dənt si'lek·shən }

frequency theory [NEUROSCI] A theory of human hearing according to which every specific frequency of sound energy is represented by nerve impulses of the same frequency, and pitch differentiation and analysis are carried out by the brain centers. Also known as telephone theory. { 'frē·kwən·sē ,thē·ə·rē }

fresh-water ecosystem [ECOL] The living organisms and nonliving materials of an inland aquatic environment. { 'fresh ,wōd·ər 'ek·ō,sis·təm }

Freund's adjuvant [IMMUNOL] A water-oil emulsion containing a killed microorganism (usually *Mycobacterium tuberculosis*) which enhances antigenicity. { 'frȯinz 'a·jə·vənt }

friar's cowl *See* aconite. { 'frī·ərz ,kau̇l }

friction ridge [ANAT] One of the integumentary ridges on the plantar and palmar surfaces of primates. { 'frik·shən ,rij }

Friedlander's bacillus *See* Klebsiella pneumoniae. { 'frēd,lan·dərz bə,sil·əs }

Friedlander's pneumonia [MED] Inflammation of the lungs caused by *Klebsiella pneumoniae*. { 'frēd,lan·dərz nə'mō·nyə }

Friedman test [PATH] A pregnancy test in which a female rabbit is given an intravenous injection of urine from the patient; formation of corpora lutea in the ovaries indicates a positive test. { 'frēd·mən ,test }

Friedrich's ataxia [MED] A hereditary sclerosis of the spine with speech impairment, lateral curvature of the spine, and palsy of the lower limbs. Also known as hereditary spinal ataxia. { 'frēd·riks ā'tak·sē·ə }

frigidoreceptor [PHYSIO] A cutaneous sense organ which is sensitive to cold. { ¦frij·ə·dō·ri ¦sep·tər }

Fringillidae [VERT ZOO] The finches, a family of oscine birds in the order Passeriformes. { frin 'jil·ə,dē }

fritillary [BOT] The common name for plants of the genus Fritillaria. [INV ZOO] The common name for butterflies in several genera of the subfamily Nymphalinae. { frə'til·ə·rē }

Froehlich's syndrome *See* adiposogenital dystrophy. { 'frā·liks ,sin,drōm }

frog [VERT ZOO] The common name for a number of tailless amphibians in the order Anura; most have hindlegs adapted for jumping, scaleless skin, and large eyes. { fräg }

frond [BOT] **1.** The leaf of a palm or fern. **2.** A foliaceous thallus or thalloid shoot. { fränd }

frontal bone [ANAT] Either of a pair of flat membrane bones in vertebrates, and a single bone in humans, forming the upper frontal portion of the cranium; the forehead bone. { 'frənt·əl ,bōn }

frontal crest [ANAT] A median ridge on the internal surface of the frontal bone in humans. { 'frənt·əl ,krest }

frontal eminence [ANAT] The prominence of the frontal bone above each superciliary ridge in humans. { ¦frənt·əl 'em·ə·nəns }

frontalia [INV ZOO] Paired sensory bristles on the frontal aspect of the head of gnathostomulids. { frən'tal·yə }

frontal lobe [NEUROSCI] The anterior portion of a cerebral hemisphere, bounded behind by the central sulcus and below by the lateral cerebral sulcus. { ¦frənt·əl ¦lōb }

frontal nerve [NEUROSCI] A somatic sensory nerve, attached to the ophthalmic nerve, which innervates the skin of the upper eyelid, the forehead, and the scalp. { ¦frənt·əl ¦nərv }

frontal plane [ANAT] Any plane parallel with the long axis of the body and perpendicular to the sagittal plane. [MED] In electrocardiography and vectorcardiography, the projection of the vertical axis. { ¦frənt·əl ¦plān }

frontal sinus [ANAT] Either of a pair of air spaces within the frontal bone above the nasal bridge. { ¦frənt·əl 'sī·nəs }

frostbite [MED] Injury to skin and subcutaneous tissues, and in severe cases to deeper tissues also, from exposure to extreme cold. { 'fróst,bīt }

frost cracks [BOT] Cracks in wood that have split outward from ray shakes. { 'fróst ,kraks }

frost ring [BOT] A false annual growth ring in the trunk of a tree due to out-of-season defoliation by frost and subsequent regrowth of foliage. { 'fróst ,riŋ }

frozen section [BIOL] A thin slice of material cut from a frozen sample of tissue or organ. { ¦frōz·ən 'sek·shən }

fructescence [BOT] The period of fruit maturation. { ,frək'tes·əns }

fructification [BOT] **1.** The process of producing fruit. **2.** A fruit and its appendages. [MYCOL] A sporogenous structure. { 'frək·tə·fə'kā·shən }

fructivorous *See* frugivorous. { ¦frək¦tiv·ə·rəs }

D-fructopyranose *See* fructose. { ¦dē ¦frək·tō'pī·rə,nōs }

fructose [BIOCHEM] $C_6H_{12}O_5$ The commonest of ketoses and the sweetest of sugars, found in the free state in fruit juices, honey, and nectar of plant glands. Also known as D-fructopyranose. { 'frük,tōs }

frugivore [ZOO] A fruit-eating animal. { 'frü·ji,vór }

frugivorous [ZOO] Fruit-eating. Also known as fructivorous. { frü'jiv·ə·rəs }

fruit [BOT] A fully matured plant ovary with or without other floral or shoot parts united with it at maturity. { früt }

fruit bud [BOT] A fertilized flower bud that matures into a fruit. { 'früt ,bəd }

fruit fly [INV ZOO] **1.** The common name for those acalypterate insects composing the family Tephritidae. **2.** Any insect whose larvae feed on fruit or decaying vegetable matter. { 'früt ,flī }

fruiting body [BOT] A specialized, spore-producing organ. { 'früd·iŋ ,bäd·ē }

fruiting myxobacteria *See* Myxobacterales. { 'früd·iŋ ,mik·sō·bak'tir·ē·ə }

frustule [INV ZOO] **1.** The shell and protoplast of a diatom. **2.** A nonciliated planulalike bud in some hydrozoans. { 'frəs,chül }

frutescent [BIOL] *See* fruticose. [BOT] Shrublike in habit. { frü'tes·ənt }

fruticose [BIOL] Resembling a shrub; applied especially to lichens. Also known as frutescent. { 'früd·ə,kōs }

FSH *See* follicle-stimulating hormone.

ft *See* foot.

5-FU *See* fluorouracil.

Fucales [BOT] An order of brown algae composing the class Cyclosporeae. { fyü'kā·lēz }

fuchsinophile [BIOL] Having an affinity for the dye fuchsin. { fyük'sin·ə,fil }

fucoidin [BIOCHEM] A gum composed of L-fucose and sulfate acid ester groups obtained from *Fucus* species and other brown algae. { fyü 'kȯid·ən }

Fucophyceae [BOT] A class of brown algae. { ,fyü·kə'fīs·ē,ē }

L-fucopyranose *See* L-fucose. { ¦el ,fyü·kō'pī·rə,nōs }

L-fucose [BIOCHEM] $C_6H_{12}O_5$ A methyl pentose present in some algae and a number of gums and identified in the polysaccharides of blood groups and certain bacteria. Also known as 6-deoxy-L-galactose; L-fucopyranose; L-galactomethylose; L-rhodeose. { ¦el 'fyü,kōs }

fucoxanthin [BIOCHEM] $C_{40}H_{60}O_6$ A carotenoid pigment; a partial xanthophyll ester found in diatoms and brown algae. { ¦fyü·kō¦zan·thən }

Fucus [BOT] A genus of dichotomously branched brown algae; it is harvested in the kelp industry as a source of algin. { 'fyü·kəs }

FUDR *See* floxuridine.

fugacious [BOT] Lasting a short time; used principally to describe plant parts that fall soon after being formed. { fyü'gā·shəs }

fulcrate [BIOL] Having a fulcrum. { 'fu̇l,krāt }

fulcrate trophus [INV ZOO] A type of masticatory apparatus in certain rotifers characterized by an elongate fulcrum. { 'fu̇l,krāt 'trō·fəs }

Fuld-Gross unit [BIOL] A unit for the standardization of trypsin. { ¦fu̇ld ¦grōs ,yü·nət }

Fulgoroidea [INV ZOO] The lantern flies, a superfamily of homopteran insects in the series Auchenorrhyncha distinguished by the anterior and middle coxae being of equal length and joined to the body at some distance from the median line. { ,fu̇l·gə'rȯid·ē·ə }

fulmar [VERT ZOO] Any of the oceanic birds composing the family Procellariidae; sometimes referred to as foul gulls because of the foul-smelling substance spat at intruders upon their nests. { 'fu̇l·mər }

fulminate [MED] Of a disease, to come suddenly and follow a severe, intense, and rapid course. { 'fu̇l·mə,nāt }

fumagillin [MICROBIO] $C_{26}H_{34}O_7$ An insoluble, crystalline antibiotic produced by a strain of the fungus *Aspergillus fumigatus*. { ,fyü·mə'jil·ən }

fumarase [BIOCHEM] An enzyme that catalyzes the hydration of fumaric acid to malic acid, and the reverse dehydration. { 'fyü·mə,rās }

Fumariaceae [BOT] A family of dicotyledonous plants in the order Papaverales having four or six stamens, irregular flowers, and no latex system. { fyü,ma·rē'ās·ē,ē }

Funariales [BOT] An order of mosses; plants are usually annual, are terrestrial, and have stems that are erect, short, simple, or sparingly branched. { fyü,nar·ē'ā·lēz }

functional anatomy [ANAT] The study of the human body and its parts with emphasis on those features that are directly involved in physiological function. { 'fəŋk·shə·nəl ə'nad·ə·mē }

functional disorder [MED] A disorder in which

the performance of an organ or organ system is abnormal, but not as a result of known changes in structure. { ¦fəŋk·shən·əl dis'ȯrd·ər }

functional electrical stimulation [MED] Therapeutic application of controlled amounts of electric current to muscles to make them contract in a nearly normal manner. { 'fəŋk·shən·əl i¦lek·trə·kəl ˌstim·yə'lā·shən }

functional paralysis *See* hysterical paralysis. { ¦fəŋk·shən·əl pə'ral·ə·səs }

functional residual capacity [PHYSIO] The volume of gas which remains within the lungs at expiratory standstill. { ¦fəŋk·shən·əl ri¦zij·ə·wəl kə'pas·əd·ē }

fundamental frequency of the voice [PHYSIO] The rate of vibration of the vocal folds. Abbreviated F0. { ¦fən·dəˌment·əl ¦frē·kwən·sē əv thə 'vȯis }

fundamental number [GEN] The number of chromosome arms of a karyotype. { ¦fən·də¦men·təl 'nəm·bər }

fundic gland [ANAT] Any of the glands of the corpus and fundus of the stomach. { 'fən·dik ˌgland }

fundus [ANAT] The bottom of a hollow organ. { 'fən·dəs }

fungal ecology [ECOL] The subdiscipline in mycology and ecology that examines fungal community composition and structure; responses, activities, and interactions of single fungus species; and the functions of fungi in ecosystems. { ˌfəŋ·gəl i'käl·ə·jē }

fungal sheath [MYCOL] A compact layer of fungal hyphae that surrounds the young root surface of the host plant and prevents direct contact between the root and the soil. { 'fəŋ·gəl ˌshēth }

fungi [MYCOL] Nucleated, usually filamentous, sporebearing organisms devoid of chlorophyll. { 'fənˌjī }

fungiform [BIOL] Mushroom-shaped. { 'fən·jəˌfȯrm }

fungiform papilla [ANAT] One of the low, broad papillae scattered over the dorsum and margins of the tongue. { 'fən·jəˌfȯrm pə'pil·ə }

Fungi Imperfecti [MYCOL] A class of the subdivision Eumycetes; the name is derived from the lack of a sexual stage. { 'fənˌjī ˌim·pər'fekˌtī }

Fungivoridae [INV ZOO] The fungus gnats, a family of orthorrhaphous dipteran insects in the series Nematocera; the larvae feed on fungi. { ˌfən·jə'vȯr·əˌdē }

fungivorous [ZOO] Feeding on or in fungi. { fən'jiv·ə·rəs }

fungus [MYCOL] Singular of fungi. { 'fəŋ·gəs }

funicle *See* funiculus. { 'fyün·ə·kəl }

funiculus [ANAT] **1.** Also known as funicle. **2.** Any structure in the form of a chord. **3.** A column of white matter in the spinal cord. [BOT] The stalk of an ovule. [INV ZOO] A band of tissue extending from the adoral end of the coelom to the adoral body wall in bryozoans. { fə'nik·yə·ləs }

funnel chest [MED] A developmental deformity in which the sternum is depressed and the ribs and costal cartilages curve inward. { 'fən·əl ˌchest }

fur [VERT ZOO] The coat of a mammal. { fər }

furanones [BIOCHEM] Analogs of homoserine lactones that appear to interfere with the development of typical biofilm structure, leaving these organisms more susceptible to treatment with natural biocides. { 'fyür·əˌnōnz }

furanose [BIOCHEM] A sugar whose cyclic or ring structure resembles that of furan. { 'fyu̇r·əˌnōs }

furca [INV ZOO] A forked process as the last abdominal segment of certain crustaceans, and as part of the spring in collembolans. { 'fər·kə }

furcate [BIOL] Forked. { 'fərˌkāt }

furcocercous cercaria [INV ZOO] A free-swimming, digenetic trematode larva with a forked tail. { ¦fər·kō¦sər·kəs sər'kar·ē·ə }

furcula [ZOO] A forked structure, especially the wishbone of fowl. { 'fər·kyə·lə }

Furipteridae [VERT ZOO] The smoky bats, a family of mammals in the order Chiroptera having a vestigial thumb and small ears. { ˌfu̇·rip'ter·əˌdē }

Furnariidae [VERT ZOO] The oven birds, a family of perching birds in the superfamily Furnarioidea. { ˌfər·nə'rī·əˌdē }

Furnarioidea [VERT ZOO] A superfamily of birds in the order Passeriformes characterized by a predominance of gray, brown, and black plumage. { fərˌnar·ē'ȯid·ē·ə }

furuncle [MED] A small cutaneous abscess, usually resulting from infection of a hair follicle by *Staphylococcus aureus*. Also known as boil. { 'fyu̇rˌəŋ·kəl }

furunculosis [MED] A condition marked by numerous furuncles, or the recurrence of furuncles following healing of a preceding crop. { fyu̇ˌrəŋ·kyə'lō·səs }

Fusarium [MYCOL] A genus of fungi in the family Tuberculariaceae having sickle-shaped, multicelled conidia; includes many important plant pathogens. { fyü'za·rē·əm }

Fusarium oxysporum [MYCOL] A pathogenic fungus causing a variety of plant diseases, including cabbage yellows and wilt of tomato, flax, cotton, peas, and muskmelon. { fyü'za·rē·əm ˌäk·sə'spȯr·əm }

Fusarium solani [MYCOL] A pathogenic fungus implicated in root rot and wilt diseases of several plants, including sisal and squash. { fyü'za·rē·əm sō'lan·ē }

fusee [VET MED] A bony growth occurring on a horse's leg. Also spelled fuzee. { fyü'zē }

Fusicoccum amygdali [MYCOL] A fungal pathogen that produces fusicoccin, the cause of wilt disease in peach and almond trees. { ˌfyüz·iˌkäk·əm ə'mig·dəˌlē }

fusiform [BIOL] Spindle-shaped; tapering toward the ends. { 'fyü·zəˌfȯrm }

fusiform bacillus [MICROBIO] A bacillus having one blunt and one pointed end, as *Fusobacterium fusiforme*. { 'fyü·zəˌfȯrm bə'sil·əs }

G

GABOB *See* 4-amino-3-hydroxybutyric acid. { 'ga ˌbäb *or* ˌjēˌāˌbēˌō'bē }

Gadidae [VERT ZOO] A family of fishes in the order Gadiformes, including cod, haddock, pollock, and hake. { 'ga·dəˌdē }

Gadiformes [VERT ZOO] An order of actinopterygian fishes that lack fin spines and a swim bladder duct and have cycloid scales and many-rayed pelvic fins. { ˌgad·ə'fȯrˌmēz }

gait analysis [PHYSIO] An aspect of kinesiology that involves the study of walking or other types of ambulation. { 'gāt əˌnal·ə·səs }

galactan [BIOCHEM] Any of a number of polysaccharides composed of galactose units. Also known as galactosan. { gə'lak·tən }

galactocele [MED] **1.** A retention cyst caused by obstruction of one or more of the mammary ducts. **2.** A hydrocele with milky contents. { gə'lak·təˌsēl }

galactogen [BIOCHEM] A polysaccharide, in snails, that yields galactose on hydrolysis. { gə'lak·tə·jən }

galactoglucomannan [BIOCHEM] Any of a group of polysaccharides which are prominent components of coniferous woods; they are soluble in alkali and consist of D-glucopyranose and D-mannopyranose units. { gə¦lak·tōˌglü·kə 'man·ən }

galactokinase [BIOCHEM] An enzyme which reacts D-galactose with adenosine triphosphate to give D-galactose-1-phosphate and adenosine diphosphate. { gə¦lak·tə'kīˌnās }

galactolipid *See* cerebroside. { gə¦lak·tə'lip·id }

galactomannan [BIOCHEM] Any of a group of polysaccharides which are composed of D-galactose and D-mannose units, are soluble in water, and form highly viscous solutions; they are plant mucilages existing as reserve carbohydrates in the endosperm of leguminous seeds. { gə¦lak·tō'man·ən }

galactonic acid [BIOCHEM] $C_6H_{12}O_7$ A monobasic acid derived from galactose, occurring in three optically different forms, and melting at 97°C. Also known as pentahydroxyhexoic acid. { ¦gaˌlak¦tän·ik 'as·əd }

galactophore [ANAT] A duct that carries milk. { gə'lak·təˌfȯr }

galactopoiesis [PHYSIO] Formation of the components of milk by the cells composing the lobuloalveolar glandular structure. { gəˌlak·tə ˌpȯi'ē·səs }

galactorrhea [MED] Excessive flow of milk. { gəˌlak·tə'rē·ə }

galactosamine [BIOCHEM] $C_6H_{14}O_5N$ A crystalline amino acid derivative of galactose; found in bacterial cell walls. { gəˌlak'tō·səˌmēn }

galactosan *See* galactan. { gə'lak·təˌsan }

galactose [BIOCHEM] $C_6H_{12}O_6$ A monosaccharide occurring in both levo and dextro forms as a constituent of plant and animal oligosaccharides (lactose and raffinose) and polysaccharides (agar and pectin). Also known as cerebrose. { gə 'lakˌtōs }

galactosemia [MED] A congenital metabolic disorder caused by an enzyme deficiency and marked by high blood levels of galactose. { gəˌlak·tō'sē·mē·ə }

galactosidase [BIOCHEM] An enzyme that hydrolyzes galactosides. { gəˌlak·tə'sīˌdās }

galactoside [BIOCHEM] A glycoside formed by the reaction of galactose with an alcohol; yields galactose on hydrolysis. { gə'lak·təˌsīd }

galactosuria [MED] Passage of urine containing galactose. { gəˌlak·tə'su̇r·ē·ə }

galactosyl ceramide [BIOCHEM] A type of glycolipid that enriches brain tissue and is a major component of the myelin sheaths around nerves. { gəˌlak·təˌsil 'ser·əˌmīd }

galacturonic acid [BIOCHEM] The monobasic acid resulting from oxidation of the primary alcohol group of D-galactose to carboxyl; it is widely distributed as a constituent of pectins and many plant gums and mucilages. { gə¦lakt·yə¦rän·ik 'as·əd }

galago *See* bushbaby. { gə'lä·gō }

Galatheidea [INV ZOO] A group of decapod crustaceans belonging to the Anomura and having a symmetrical abdomen bent upon itself and a well-developed tail fan. { ˌgal·ə·thē'ī·dē·ə }

Galaxioidei [VERT ZOO] A suborder of mostly small, fresh-water fishes in the order Salmoniformes. { gəˌlak·sē'ȯid·ēˌī }

Galbulidae [VERT ZOO] The jacamars, a family of highly iridescent birds of the order Piciformes that resemble giant hummingbirds. { ˌgal'bu̇l·əˌdē }

galea [ANAT] The epicranial aponeurosis linking the occipital and frontal muscles. [BIOL] A helmet-shaped structure. [BOT] A helmet-shaped petal near the axis. [INV ZOO] **1.** The endopodite of the maxilla of certain insects. **2.** A spinning organ on the movable digit of chelicerae of pseudoscorpions. { 'gā·lē·ə }

galeate [BIOL] **1.** Shaped like a helmet. **2.** Having a galea. { 'ga·lē,āt }

galenical [PHARM] A medicinal preparation containing one or several active plant ingredients and produced so that inert constituents and other undesirable contents of the plant remain undissolved. { gə'len·i·kəl }

Galen's vein [ANAT] One of the two veins running along the roof the third ventricle that drain the interior of the brain. { 'gā·lənz 'vān }

gall [MED] A sore on the skin that is caused by chafing. [PHYSIO] *See* bile. { gȯl }

gallbladder [ANAT] A hollow, muscular organ in humans and most vertebrates which receives dilute bile from the liver, concentrates it, and discharges it into the duodenum. { 'gȯl,blad·ər }

galleria forest [ECOL] A modified tropical deciduous forest occurring along stream banks. { ,gal·ə'rē·ə ,fär·əst }

Galleriinae [INV ZOO] A monotypic subfamily of lepidopteran insects in the family Pyralididae; contains the bee moth or wax worm (*Galleria mellonella*), which lives in beehives and whose larvae feed on beeswax. { ,gal·ə'rī·ə,nē }

gallicolous [BIOL] Producing or inhabiting galls. { gə'lik·ə·ləs }

Galliformes [VERT ZOO] An order of birds that includes important domestic and game birds, such as turkeys, pheasants, and quails. { ,gal·ə'fȯr,mēz }

gallinaceous [VERT ZOO] Of, pertaining to, or resembling birds of the order Galliformes. { ,gal·ə'nā·shəs }

Gallionella [MICROBIO] A genus of appendaged bacteria; cells are kidney-shaped or rounded and occur on stalks; reproduce by binary fission. { ,gal·yə'nel·ə }

gallivorous [ZOO] Feeding on the tissues of galls, especially certain insect larvae. { gȯ'liv·ə·rəs }

gallodesoxycholic acid *See* chenodeoxycholic acid. { ¦ga·lō·de,zäk·sē'käl·ik 'as·əd }

gallop rhythm [MED] A three-sound sequence resulting from the intensification of the normal third or fourth heart sounds, occurring usually with a rapid ventricular rate. { 'gal·əp ,rith·əm }

gallstone [PATH] A nodule formed in the gallbladder or biliary tubes and composed of calcium, cholesterol, or bilirubin, or a combination of these. { 'gȯl,stōn }

Galumnidae [INV ZOO] A family of oribatid mites in the suborder Sarcoptiformes. { gə 'ləm·nə,dē }

galvanic skin response [PHYSIO] The electrical reactions of the skin to any stimulus as detected by a sensitive galvanometer; most often used experimentally to measure the resistance of the skin to the passage of a weak electric current. Also known as electrodermal response. { gal 'van·ik 'skin ri,späns }

galvanism [BIOL] The use of a galvanic current for medical or biological purposes. { 'gal·və,niz·əm }

galvanotaxis [BIOL] Movement of a free-living organism in response to an electrical stimulus. { ¦gal·və·nō¦tak·səs }

galvanotropism [BIOL] Response of an organism to electrical stimulation. { ,gal·və'nä·trə,piz·əm }

galvo *See* metal fume fever. { 'gal,vō }

game bird [BIOL] A bird that is legal quarry for hunters. { 'gām ,bərd }

Gamella [MICROBIO] A genus of bacteria in the family Streptococcaceae; spherical cells occurring singly or in pairs with flattened adjacent sides; ferment glucose. { gə'mel·ə }

gametangial copulation [MYCOL] Direct fusion of certain fungal gametangia without differentiation of the gametes. { ,ga·mə'tan·jē·əl ,käp·yə'la·shən }

gametangium [BIOL] A cell or organ that produces sex cells; occurs in algae, fungi, and plants. { ,gam·ə'tan·jē·əm }

gamete [BIOL] A cell which participates in fertilization and development of a new organism. Also known as germ cell; sex cell. { 'ga,mēt }

gamete intrafallopian transfer [MED] A variation of in vitro fertilization in which the spermatozoa and oocytes are placed directly into the fimbriated end of the Fallopin tube during the laparoscopy. Abbreviated GIFT. { ¦ga,mēt ,in·trə·fə¦lō·pē·ən 'tranz·fər }

gametic copulation [MYCOL] The fusion of pairs of differentiated, uninucleate sexual cells or gametes formed in specialized gametangia. { gə'med·ik ,käp·yə'lā·shən }

gametoblast [BOT] An archespore that has not yet undergone differentiation. { gə'mēd·ə ,blast }

gametocyte [HISTOL] An undifferentiated cell from which gametes are produced. { gə'mēd·ə,sīt }

gametogenesis [BIOL] The formation of gametes, or reproductive cells such as ova or sperm. { gə,mēd·ə'jen·ə·səs }

gametophore [BOT] A branch that bears gametangia. { gə'mēd·ə,fȯr }

gametophyte [BOT] **1.** The haploid generation producing gametes in plants exhibiting metagenesis. **2.** An individual plant of this generation. { gə'mēd·ə,fīt }

gamma globulin [IMMUNOL] Any of the serum proteins with antibody activity. { 'gam·ə 'gläb·yə·lən }

gamma hemolysis [MICROBIO] The absence of activity in the area surrounding a bacterial colony growing on blood agar. Also known as nonhemolysis. { ,gam·ə hi'mäl·ə·səs }

Gammaherpesvirinae [VIROL] A subfamily of animal, double-stranded linear DNA viruses of the family Herpesviridae, which are enveloped by a lipid bilayer and several glycoproteins. Also known as lymphoproliferative virus group. { ¦gam·ə¦hər,pēz'vī·rə,nē }

Gammaridea [INV ZOO] The scuds or sand hoppers, a suborder of amphipod crustaceans; individuals are usually compressed laterally, are poor walkers, and lack a carapace. { ,gam·ə'rid·ē·ə }

gamma taxonomy [SYST] The level of taxonomic study concerned with biological aspects of taxa, including intraspecific populations, speciation, and evolutionary rates and trends. { 'gam·ə tak'sän·ə·mē }

gamodeme [ECOL] An isolated breeding community. { 'gam·ə,dēm }

gamogony [INV ZOO] Spore formation by multiple fission in sporozoans. [ZOO] Sexual reproduction. { gə'mäg·ə·nē }

gamone [PHYSIO] Any substance released by a gamete that facilitates fertilization processes. { ga'mōn }

gamont [INV ZOO] The gametocyte of sporozoans. { 'ga,mänt }

gamopetalous [BOT] Having petals united at their edges. Also known as sympetalous. { ¦gam·ə¦ped·əl·əs }

gamophyllous [BOT] Having the leaves of the perianth united. { ¦gam·ə¦fil·əs }

gamosepalous [BOT] Having sepals united at their edges. Also known as synsepalous. { ¦ga·mō¦sep·ə·ləs }

gangliated cord [NEUROSCI] One of the two main trunks of the sympathetic nervous system, one trunk running along each side of the spinal column. { 'gaŋ·glē,ād·əd 'kȯrd }

ganglioma [MED] A form of ganglioneuroma in which neuronal and glial elements appear in about equal proportions. { ,gaŋ·glē'ō·mə }

ganglion [NEUROSCI] A group of nerve cell bodies, usually located outside the brain and spinal cord. { 'gaŋ·glē·ən }

ganglioneuroma [MED] A tumor composed of sympathetic ganglion cells and sheathed nerve fibers. { ¦gaŋ·glē·ō·nu̇'rō·mə }

ganglionitis [MED] Inflammation of a ganglion. { gaŋ·glē·ə'nīd·əs }

ganglioside [BIOCHEM] One of a group of glycosphingolipids found in neuronal surface membranes and spleen; they contain an N-acyl fatty acid derivative of sphingosine linked to a carbohydrate (glucose or galactose); they also contain N-acetylglucosamine or N-acetylgalactosamine, and N-acetylneuraminic acid. { 'gaŋ·glē·ō,sīd }

gangosa [MED] Destructive lesions of the nose and hard palate, sometimes more extensive, considered to be the tertiary stage of yaws. { gaŋ'gō·sə }

gangrene [MED] A form of tissue death usually occurring in an extremity due to insufficient blood supply. { gaŋ'grēn }

gangrenous stomatitis *See* noma. { 'gaŋ·grə·nəs ,stō·mə'tīd·əs }

Ganoderma lucidum [MYCOL] A mushroom found throughout the United States, Europe, South America, and Asia that appears to have antiallergic, anti-inflammatory, antibacterial, antioxidant, antitumor, and immunostimulating activity. Also known as ling-zhi; reishi mushroom. { ¦gen·ə,dər·mə 'lüs·ə·dəm }

ganoid scale [VERT ZOO] A structure having several layers of enamellike material (ganoin) on the upper surface and laminated bone below. { ¦ga,nȯid 'skāl }

ganoin [VERT ZOO] The enamellike covering of a ganoid scale. { 'gan·ə·wən }

gap [GEN] A short region that is missing in one strand of a double-stranded deoxyribonucleic acid. { gap }

gape [ANAT] The margin to margin distance between open jaws. [INV ZOO] The space between the margins of a closed mollusk valve. { gāp }

gap junction [NEUROSCI] An intercellular junction composed of cylindrical channels connecting adjacent cells; considered to be a low-resistance pathway for intercellular communication. Also known as communicating junction; nexus. { 'gap ,jəŋk·shən }

gar [VERT ZOO] The common name for about seven species of bony fishes in the order Semionotiformes having a slim form, an elongate snout, and close-set ganoid scales. { gar }

Gardner's syndrome [MED] A hereditary disorder transmitted as an autosomal dominant; manifested in childhood by multiple neoplasms, including bony and mesenteric tumors, fatty and fibrous skin, and intestinal polyps. { 'gärd·nərz ,sin,drōm }

gargoylism *See* Hurler's syndrome. { 'gär·gȯi,liz·əm }

garigue [ECOL] A low, open scrubland restricted to limestone sites in the Mediterranean area; characterized by small evergreen shrubs and low trees. { gə'rēg }

garlic [BOT] *Allium sativum*. A perennial plant of the order Liliales grown for its pungent, edible bulbs. { 'gär·lik }

Gartner's duct [ANAT] The remnant of the embryonic Wolffian duct in the adult female mammal. { 'gärt·nərz ,dəkt }

gas embolus [MED] An embolus composed of a gas resulting from trauma or other causes. { 'gas ,em·bə·ləs }

gas gangrene [MED] A localized, but rapidly spreading, necrotizing bacterial wound infection characterized by edema, gas production, and discoloration; caused by several species of *Clostridium*. { ¦gas 'gaŋ,grēn }

gas gland [VERT ZOO] A structure inside the swim bladder of many teleosts which secretes gas into the bladder. { 'gas ,gland }

Gasserian ganglion [NEUROSCI] A group of nerve cells of the sensory root of the trigeminal nerve. Also known as semilunar ganglion. { ga'ser·ē·ən 'gaŋ·glē·ən }

gas sterilization [MICROBIO] Sterilization of heat and liquid-labile materials by means of gaseous agents, such as formaldehyde, ethylene oxide, and β-propiolactone. { ¦gas ,ster·ə·lə'zā·shən }

Gasteromycetes [MYCOL] A group of basidiomycetous fungi in the subclass Homobasidiomycetidae with enclosed basidia and with basidiospores borne symmetrically on long sterigmata and not forcibly discharged. { ¦gas·tə·rō,mī 'sēd·ēz }

Gasterophilidae [INV ZOO] The horse bots, a family of myodarian cyclorrhaphous dipteran insects in the subsection Calypteratae, including individuals that cause myiasis in horses. { ¦gas·tə·rō'fil·ə,dē }

Gasterophilus [INV ZOO] A large genus of botflies in the family Gasterophilidae. { ,gas·tə'rä·fə·ləs }

Gasterosteidae [VERT ZOO] The sticklebacks, a family of actinopterygian fishes in the order Gasterosteiformes. { ,gas·tə·rō'stē·ə,dē }

Gasterosteiformes [VERT ZOO] An order of actinopterygian fishes characterized by a ductless swim bladder, a pelvic fin that is abdominal to subthoracic in position, and an elongate snout. { ,gas·tə,rä·stē·ə'fòr,mēz }

Gasteruptiidae [INV ZOO] A family of hymenopteran insects in the superfamily Proctotrupoidea. { ,gas·tə,rəp'tī·ə,dē }

gastralium [INV ZOO] A microsclere located just beneath the inner cell layer of hexactinellid sponges. [VERT ZOO] One of the riblike structures in the abdomen of certain reptiles. { ga 'strā·lē·əm }

gastrectomy [MED] Surgical removal of all or part of the stomach. { ga'strek·tə·mē }

gastric acid [BIOCHEM] Hydrochloric acid secreted by parietal cells in the fundus of the stomach. { 'gas·trik 'as·əd }

gastric cecum [INV ZOO] One of the elongated pouchlike projections of the upper end of the stomach in insects. { 'gas·trik 'sē·kəm }

gastric enzyme [BIOCHEM] Any digestive enzyme secreted by cells lining the stomach. { 'gas·trik 'en,zīm }

gastric filament [INV ZOO] In scyphozoans, a row of filaments on the surface of the gastric cavity which function to kill or paralyze live prey taken into the stomach. Also known as phacella. { 'gas·trik 'fil·ə·mənt }

gastric gland [ANAT] Any of the glands in the wall of the stomach that secrete components of the gastric juice. { 'gas·trik ,gland }

gastric hypothermia [MED] Cooling of the upper digestive tract; useful in the management of bleeding disorders. { 'gas·trik ,hī·pō'thər·mē·ə }

gastric juice [PHYSIO] The digestive fluid secreted by gastric glands; contains gastric acid and enzymes. { 'gas·trik ,jüs }

gastric mill [INV ZOO] A grinding apparatus consisting of calcareous or chitinous plates in the pharynx or stomach of certain invertebrates. { 'gas·trik ,mil }

gastric ostium [INV ZOO] The opening into the gastric pouch in scyphozoans. { 'gas·trik 'äs·tē·əm }

gastric pouch [INV ZOO] One of the pouchlike diversions of a scyphozoan stomach. { 'gas·trik 'paùch }

gastric shield [INV ZOO] A thickening of the stomach wall in some mollusks for mixing the contents. { 'gas·trik 'shēld }

gastric ulcer [MED] An ulcer of the mucous membrane of the stomach. { 'gas·trik 'əl·sər }

gastrin [BIOCHEM] A polypeptide hormone secreted by the pyloric mucosa which stimulates the pancreas to release pancreatic fluid and the stomach to release gastric acid. { 'gas·trən }

gastritis [MED] Inflammation of the stomach. { ga'strīd·əs }

gastroanastomosis [MED] The surgical formation of a communication between the two pouches of the stomach. Also known as gastrogastrostomy. { ¦ga·strō·ə,nas·tə'mō·səs }

gastroblast [INV ZOO] A feeding zooid of a tunicate colony. { 'ga·strə,blast }

gastrocnemius [ANAT] A large muscle of the posterior aspect of the leg, arising by two heads from the posterior surfaces of the lateral and medial condyles of the femur, and inserted with the soleus muscle into the calcaneal tendon, and through this into the back of the calcaneus. { ga,sträk'nē·mē·əs }

gastrocoele *See* archenteron. { 'ga·strə,sēl }

gastrodermis [INV ZOO] The cellular lining of the digestive cavity of certain invertebrates. { ¦ga·strō¦dər·məs }

gastroduodenitis [MED] Inflammation of the stomach and duodenum. { ¦ga·strō·dü,äd·ən 'īd·əs }

gastroenteritis [MED] Inflammation of the mucosa of the stomach and intestine. { ¦ga·strō,ent·ə'rīd·əs }

gastroenterology [MED] The branch of medicine concerned with study of the stomach and intestine. { ¦ga·strō,ent·ə'räl·ə·jē }

gastroenterostomy [MED] Surgical formation of a connection between the stomach and small intestine. { ¦ga·strō,ent·ə'räs·tə·mē }

gastroepiploic artery [ANAT] Either of two arteries arising from the gastroduodenal and splenic arteries respectively and forming an anastomosis along the greater curvature of the stomach. { ¦ga·strō¦ep·ə¦plōik 'ärd·ə·rē }

gastrogastrostomy *See* gastroanastomosis. { ¦ga·strō ,ga'sträs·tə·mē }

gastrointestinal hormone [BIOCHEM] Any hormone secreted by the gastrointestinal system. { ¦ga·strō,in'tes·tən·əl 'hòr,mōn }

gastrointestinal system [ANAT] The portion of the digestive system including the stomach, intestine, and all accessory organs. { ¦ga·strō ,in'tes·tən·əl 'sis·təm }

gastrointestinal tract [ANAT] The stomach and intestine. { ¦ga·strō,in'tes·tən·əl 'trakt }

gastrojejunostomy [MED] Surgical establishment of an anastomosis between the jejunum and the anterior or posterior wall of the stomach. { ¦ga·strō,jē·jə'nä·stə·mē }

gastrolith [VERT ZOO] A pebble swallowed by certain animals and retained in the gizzard or stomach, where it serves to grind food. { 'ga·strə,lith }

gastrolysis [MED] The breaking up of adhesions between the stomach and adjacent organs. { ga'sträl·ə·səs }

Gastromyzontidae [VERT ZOO] A small family of actinopterygian fishes of the suborder Cyprinoidei found in southeastern Asia. { ¦ga·strō ,mī'zän·tə,dē }

gastropexy [MED] The fixation of a prolapsed stomach in its normal position by suturing it to the abdominal wall or other structure. { 'ga·strə,pek·sē }

gastroplication [MED] An operation for relief of chronic dilation of the stomach by suturing a large horizontal fold in the stomach wall. { ¦ga·strō·plə'kā·shən }

Gastropoda [INV ZOO] A large, morphologically diverse class of the phylum Mollusca, containing the snails, slugs, limpets, and conchs. { ga 'sträp·ə·də }

gastropore [INV ZOO] A pore containing a gastrozooid in hydrozoan corals. { 'ga·strə,pȯr }

gastroptosis [MED] Prolapse or downward displacement of the stomach. { ,ga,sträp'tō·səs }

gastroscope [MED] A hollow, tubular instrument used to examine the inside of the stomach by passage through the mouth and esophagus. { 'ga·strə,skōp }

gastrosplenic ligament [ANAT] The fold of peritoneum passing from the stomach to the spleen. Also known as gastrosplenic omentum. { ¦ga·strō¦splen·ik 'lig·ə·mənt }

gastrosplenic omentum *See* gastrosplenic ligament. { ¦ga·strō¦splen·ik ō'men·təm }

gastrostome [INV ZOO] The opening of a gastropore. { 'ga·strə,stōm }

gastrostomy [MED] The establishment of a fistulous opening into the stomach, with an external opening in the skin; used for artificial feeding. { ga'sträs·tə·mē }

gastrostyle [INV ZOO] A spiculated projection that extends into the gastrozooid from the base of the gastropore. { 'ga·strə,stīl }

Gastrotricha [INV ZOO] A group of microscopic, pseudocoelomate animals considered either to be a class of the Aschelminthes or to constitute a separate phylum. { ga'strä·trə·kə }

gastrozooid [INV ZOO] A nutritive polyp of colonial coelenterates, characterized by having tentacles and a mouth. { ¦ga·strə'zō,ȯid }

gastrula [EMBRYO] The stage of development in animals in which the endoderm is formed and invagination of the blastula has occurred. { 'ga·strə·lə }

gastrulation [EMBRYO] The process by which the endoderm is formed during development. { ,ga·strə'lā·shən }

gas vacuole [BIOL] A membrane-bound, gas-filled cavity in some algae and protozoans; thought to control buoyancy. { ¦gas 'vak·yə,wōl }

Gaucher's cells [PATH] Abnormal macrophages associated with Gaucher's disease. { gō'shāz ,selz }

Gaucher's disease [MED] A rare chronic, probably hereditary disease in which cells loaded with cerebrosides become localized in reticuloendothelial tissue and eventually cause tissue destruction; manifestations include enlargement of the spleen, bronzing of the skin, and anemia. Also known as cerebroside lipoidosis; familial splenic anemia. { gō'shāz di,zēz }

Gaultheria [BOT] A genus of upright or creeping evergreen shrubs (Ericaceae). { gȯl'thir·ē·ə }

Gause's principle [ECOL] A statement that two species cannot occupy the same niche simultaneously. Also known as competitive-exclusion principle. { 'gaüz·əz ,prin·sə·pəl }

gavage [MED] The administration of nourishment through a stomach tube. { gə'väzh }

gavial [VERT ZOO] The name for two species of reptiles composing the family Gavialidae. { 'gā·vē·əl }

Gavialidae [VERT ZOO] The gavials, a family of reptiles in the order Crocodilia distinguished by an extremely long, slender snout with an enlarged tip. { ,gā·vē'al·ə,dē }

Gaviidae [VERT ZOO] The single, monogeneric family of the order Gaviiformes. { gə'vī·ə,dē }

Gaviiformes [VERT ZOO] The loons, a monofamilial order of diving birds characterized by webbed feet, compressed, bladelike tarsi, and a heavy, pointed bill. { ,gā·vē·ə'fȯr,mēz }

gDNA *See* genomic deoxyribonucleic acid.

Gecarcinidae [INV ZOO] The true land crabs, a family of decapod crustaceans belonging to the Brachygnatha. { ,jē·kär'sin·ə,dē }

gecko [VERT ZOO] The common name for more than 300 species of arboreal and nocturnal reptiles composing the family Gekkonidae. { 'gek·ō }

geitonogamy [BOT] Pollination and fertilization of one flower by another on the same plant. { ,gīt·ən'äg·ə·mē }

Gekkonidae [VERT ZOO] The geckos, a family of small lizards in the order Squamata distinguished by a flattened body, a long sensitive tongue, and adhesive pads on the toes of many species. { ge'kän·ə,dē }

Gelastocoridae [INV ZOO] The toad bugs, a family of tropical and subtropical hemipteran insects in the subdivision Hydrocorisae. { je,la·stō'kȯr·ə,dē }

gelatinase [BIOCHEM] An enzyme, found in some yeasts and molds, that liquefies gelatin. { 'jel·ə·tə,nās }

gelatin liquefaction [MICROBIO] Reduction of a gelatin culture medium to the liquid state by enzymes produced by bacteria in a stab culture; used in identifying bacteria. { 'jel·ət·ən ,lik·wə'fak·shən }

Gelechiidae [INV ZOO] A large family of minute to small moths in the lepidopteran superfamily Tineoidea, generally having forewings and trapezoidal hindwings. { ,jel·ə'kī·ə,dē }

gelfoam [BIOCHEM] Absorbable gelatin sponge partially insolubilized by crosslinking, used for arresting hemorrhage during surgery. { 'jel ,fōm }

gelsemine [PHARM] $C_{20}H_{22}O_2N_2$ A white, crystalline alkaloid with a melting point of 178°C; derived from *Gelsemium sempervirens*; soluble in alcohol, ether, and dilute acids; used in medicine as a central nervous system stimulant. { 'jel·sə,mēn }

geminate [BIOL] Growing in pairs or couples. { 'jem·ə·nət }

geminiflorous [BOT] Having flowers in pairs. { ,jem·ə'nif·lə·rəs }

gemma [BOT] A small, multicellular, asexual reproductive body of some liverworts and mosses. { 'jem·ə }

gemmation *See* budding. { je'mā·shən }

gemmiform [BOT] Resembling a gemma or bud. { 'jem·ə,fórm }

gemmiparous [BIOL] Producing a bud or reproducing by a bud. { je'mip·ə·rəs }

gemmule [BIOL] Any bud formed by gemmation. [INV ZOO] A cystlike, asexual reproductive structure of many Porifera that germinates when proper environmental conditions exist; it is a protective, overwintering structure which germinates the following spring. [NEUROSCI] A minute dendritic process functioning as a synaptic contact point. { 'je·myül }

Gempylidae [VERT ZOO] The snake mackerels, a family of the suborder Scombroidei comprising compressed, elongate, or eel-shaped spiny-rayed fishes with caniniform teeth. { jem'pil·ə,dē }

gena [ANAT] Cheek, or side of the head. { 'jē·nə }

gene [GEN] The basic unit of inheritance; composed of a deoxyribonucleic acid (DNA) sequence that contains the elements required for transcription of a complementary ribonucleic acid (RNA) which is sometimes the functional gene product but more often is converted into messenger RNAs that specify the amino acid sequence of a protein product. { jēn }

gene action [GEN] The functioning of a gene in determining the phenotype of an individual. { 'jēn ,ak·shən }

genealogy [GEN] A record of the descent of a family, group, or person from an ancestor or ancestors. { ,jē·nē'äl·ə·jē }

gene amplification [CELL MOL] Any process by which a deoxyribonucleic acid sequence is disproportionately duplicated in comparison with the parent genome. [GEN] Repeated replication in a single cell cycle of the deoxyribonucleic acid (DNA) in a limited portion of the genome, resulting in an increase in the number of copies of a particular gene or DNA segment. { 'jēn ,am·plə·fə,kā·shən }

gene assignment [CELL MOL] The physical or functional localization of specific genes to individual chromosomes. { 'jēn ə,sīn·mənt }

gene bank *See* gene library. { 'jēn ,baŋk }

gene cluster [GEN] Any group of two or more closely linked genes that encode for the same or similar products. { 'jēn ,kləs·tər }

genecology [BIOL] The study of species and their genetic subdivisions, their place in nature, and the genetic and ecological factors controlling speciation. { ¦jēn·ə'käl·ə·jē }

gene control region [GEN] The portion of a eukaryotic gene containing promoter and regulatory sequences of deoxyribonucleic acid that control transcription. { 'jēn kən,trōl ,rē·jən }

gene conversion [GEN] A situation in which gametocytes of an individual that is heterozygous for a pair of alleles undergo meiosis, and the gametes produced are in a 3:1 ratio rather than the expected 2:2 ratio, implying that one allele was converted to the other. { 'jēn kən,vər·zhən }

gene duplication [GEN] The reduction in fitness of a diploid population due to new mutant genes and those already in the gene pool. { 'jēn düp·lə'kā·shən }

gene escape [CELL MOL] The movement of genetic material from a genetically engineered organism to another population or another species. { 'jēn ə,skāp }

gene expression [GEN] The transcription, translation, and phenotypic manifestation of a gene. { 'jēn ik¦spresh·ən }

gene family [GEN] A group of related genes of similar sequence and function resulting from multiple duplications and subsequent mutational variation among the different copies. { 'jēn ,fam·lē }

gene flow [GEN] The passage and establishment of alleles characteristic of one breeding population into the gene pool of another population through hybridization and backcrossing. { 'jēn ,flō }

gene frequency *See* allele frequency. { 'jēn ,frē·kwən·sē }

gene library [GEN] A random collection of cloned deoxyribonucleic acid fragments in a vector; includes all the genetic information of the species. Also known as gene bank. { 'jēn ,lī·brēr·ē }

gene loss [GEN] Gene elimination from differentiating cells in some protozoans, insects, and crustaceans. { 'jēn ,lós }

gene overlap [GEN] The ability of a sequence of deoxyribonucleic acid to code for more than one protein by use of different reading frames. { ¦jēn 'ō·vər,lap }

gene penetrance *See* penetrance. { 'jēn ,pen·ə·trəns }

gene pool [GEN] The totality of the genes of a specific population at a given time. { 'jēn ,pül }

general anesthesia [MED] Loss of sensation with loss of consciousness, produced by administration of anesthetic drugs. { ¦jen·rəl ,an·əs'thē·zhə }

general anesthetic [PHARM] An agent that produces general anesthesia. { ¦jen·rəl ,an·əs 'thed·ik }

general paralysis [MED] A chronic, progressive form of syphilis caused by inflammatory and degenerative changes in the brain and meninges and characterized by physical, emotional, and intellectual deterioration; occurs at least 10 to

15 years after initial infection. { ¦jen·rəl pə'ral·ə·səs }

general paresis [MED] An inflammatory and degenerative disease of the brain caused by infection with *Treponema pallidum*. Also known as syphilitic meningoencephalitis. { ¦jen·rəl pə'rē·səs }

generation [BIOL] A group of organisms having a common parent or parents and comprising a single level in line of descent. { ˌjen·ə'rā·shən }

generation time [MICROBIO] The time interval required for a bacterial cell to divide or for the population to double. { ˌjen·ə'rā·shən ˌtīm }

generative nucleus [BOT] A haploid nucleus in a pollen grain that produces two sperm nuclei by mitosis. { 'jen·rəd·iv 'nü·klē·əs }

gene redundancy [GEN] The presence of additional copies of a gene within a cell. { 'jēn ri'dən·dən·sē }

gene regulatory protein [CELL MOL] The general name for a protein that controls gene expression via its ability to bind to specific deoxyribonucleic acid sequences during transcription. { ¦jēn ¦reg·yə·ləˌtòr·ē 'prōˌtēn }

general transcription factor [CELL MOL] In eukaryotes, a protein that binds to ribonucleic acid polymerase to form a preinitiation complex that is necessary to begin transcription. { jen·rəl tranz'kript·shən ˌfak·tər }

generic [BIOL] Pertaining to or having the rank of a biological genus. { jə'ner·ik }

gene scanning [CELL MOL] A method by which mutations are inserted at specific sites on a deoxyribonucleic acid (DNA) segment to determine those DNA sequences needed for gene activity. { ¦jēn ˌskan·iŋ }

gene sharing [GEN] The acquisition and maintenance of a second function for a gene without duplication and without loss of primary function. { ¦jēn ˌsher·iŋ }

gene splicing *See* recombinant technology. { 'jēn ˌsplīs·iŋ }

gene substitution [GEN] The replacement of an allele with a mutant allele. { ¦jēn ˌsəb·stəˌtü·shən }

gene suppression [GEN] The development of a normal phenotype in a mutant individual or cell due to a second mutation either in the same gene or in a different gene. { 'jēn səˌpresh·ən }

genet [VERT ZOO] The common name for nine species of small, arboreal African carnivores in the family Viverridae. { 'jen·ət }

gene targeting [GEN] Replacement, by genetic engineering and homologous recombination, of a mutant (or wild-type) gene by a wild-type (or mutant) copy that may also contain a reporter gene. { 'jēn ˌtär·gəd·iŋ }

gene therapy [GEN] An experimental technique in which a normal gene is inserted into an organism to correct a genetic defect. { ¦jēn ˌther·ə·pē }

genetic block [GEN] The reduction in enzyme activity due to a gene mutation. { jə¦ned·ik 'bläk }

genetic carrier [GEN] An individual who is heterozygous for a recessive gene that predisposes for a hereditary disease. { jə¦ned·ik kar·ē·ər }

genetic code [CELL MOL] The genetic information in the nucleotide sequences in deoxyribonucleic acid represented by a four-letter alphabet that makes up a vocabulary of 64 three-nucleotide sequences, or codons; a sequence of such codons (averaging about 100 codons) constructs a message for a polypeptide chain. { jə'ned·ik 'kōd }

genetic colonization [GEN] Natural introduction of genetic material into the deoxyribonucleic acid of a host cell; for example, transmission of a tumor-inducing plasmid into a plant cell by the bacterium *Agrobacterium tumefaciens*. { jə¦ned·ik ˌkäl·ə·nə'zā·shən }

genetic death [GEN] **1.** Preferential elimination of genotypes that are carriers of alleles that reduce the adaptive value or fitness of those genotypes. **2.** The death of an individual without reproducing. { jə¦ned·ik 'deth }

genetic differentiation [GEN] The accumulation of differences in allelic frequencies between completely or partially isolated populations due to evolutionary forces such as selection or genetic drift. { jə¦ned·ik ˌdif·əˌren·chē'ā·shən }

genetic distance [GEN] **1.** A measure of the allelic substitutions per locus that have occurred during the separate evolution of two populations or species. **2.** The distance between linked genes in terms of recombination frequency or map units. { jə¦ned·ik 'dis·təns }

genetic drift [GEN] The random fluctuation of gene frequencies from generation to generation that occurs in small populations. { jə¦ned·ik 'drift }

genetic engineering [GEN] The intentional production of new genes and alteration of genomes by the substitution or addition of new genetic material. Also known as biogenetics. { jə¦ned·ik en·jə'nir·iŋ }

genetic equilibrium [GEN] In a population, the condition in which the frequencies of allelic genes are maintained at the same values from generation to generation. { jə¦ned·ik ˌē·kwə'lib·rē·əm }

genetic fingerprinting [CELL MOL] Identification of chemical entities in animal tissues as indicative of the presence of specific genes. [FOREN] A forensic identification technique that enables virtually 100% discrimination between individuals from small samples of blood or semen, using probes for hypervariable minisatellite deoxyribonucleic acid. Also known as DNA fingerprinting. { jə¦ned·ik 'fiŋ·gərˌprint·iŋ }

genetic homeostasis [GEN] The tendency of Mendelian populations to maintain a constant genetic composition. { jə¦ned·ik ˌhō·mē·ō'stā·səs }

genetic identity [GEN] A measure of the proportion of genes that are identical in two populations. { jə¦ned·ik ī'den·əd·ē }

genetic induction [CELL MOL] Gene activation by a chemical inducer; results in transcription of structural genes. [GEN] Gene activation by a molecule that inactivates a repressor protein and thereby activates transcription of one or more structural genes. { jə¦ned·ik in'dək·shən }

genetic isolation [GEN] The absence of genetic exchange between populations or species as a result of geographic separation or of mechanisms that prevent reproduction. { jə¦ned·ik ī·səl'ā·shən }

genetic load [GEN] The reduction in fitness of a diploid population due to new mutant genes and those already in the gene pool. { jə¦ned·ik 'lōd }

genetic map [GEN] A graphic presentation of the linear arrangement of genes on a chromosome; gene positions are determined by percentages of recombination in linkage experiments. Also known as chromosome map. { jə¦ned·ik 'map }

genetic marker [GEN] A gene whose phenotypic expression is easily discerned and thereby can be used to identify an individual or a cell that carries it, or as a probe to mark a nucleus, chromosome, or locus. { jə¦ned·ik 'märk·ər }

genetic material [GEN] The nuclear (chromosomal) and cytoplasmic (mitochondrial and chloroplast) material that plays a fundamental role in determining the nature of all cell substances, cell structures, and cell effects; the genes have properties of self-propagation and variation. { jə¦ned·ik mə'tir·ē·əl }

genetics [BIOL] The science that is concerned with the study of biological inheritance. { jə'ned·iks }

genetic system [GEN] For a given species, the organization of genetic material and the ways in which the genetic material is transmitted. { jə¦ned·ik 'sis·təm }

genetic transformation *See* transformation. { jə¦ned·ik ˌtranz·fər'mā·shən }

genetic variance [GEN] The phenotypic variance in a population that is due to genetic heterogeneity. { jə¦ned·ik ˌver·ē·əns }

genial *See* mental. { 'jēn·yəl }

genic hybrid sterility [GEN] Sterility resulting from the interaction of genes in a hybrid to cause disturbances of sex-cell formation or meiosis. { 'jēn·ik 'hī·brəd stə'ril·əd·ē }

geniculate body [NEUROSCI] Any of the four oval, flattened prominences on the posterior inferior aspect of the thalamus; functions as the synaptic center for fibers leading to the cerebral cortex. { jə'nik·yə·lət ˌbäd·ē }

geniculate ganglion [NEUROSCI] A mass of sensory and sympathetic nerve cells located along the facial nerve. { jə'nik·yə·lət 'gaŋ·glē·ən }

geniculum [ANAT] **1.** A small, kneelike, anatomical structure. **2.** A sharp bend in any small organ. { je'nik·yə·ləm }

genioglossus [ANAT] An extrinsic muscle of the tongue, arising from the superior mental spine of the mandible. { ¦jē·nē·ō¦glä·səs }

genital atrium [ZOO] A common chamber receiving openings of male, female, and accessory organs. { 'jen·ət·əl 'ā·trē·əm }

genital coelom [INV ZOO] In mollusks, the lumina of the gonads. { 'jen·ət·əl 'sē·ləm }

genital cord [EMBRYO] A mesenchymal shelf bridging the coeloms in mammalian embryos, produced by fusion of the caudal part of the urogenital folds; fuses with the urinary bladder in the male, and is the primordium of the broad ligament and the uterine walls in the female. [INV ZOO] Strands of cells located in the genital canal which are primordial sex cells in crinoids. Also known as genital rachis. { 'jen·ət·əl 'kȯrd }

genitalia [ANAT] The organs of reproduction, especially those which are external. { ˌjen·ə'tāl·yə }

genital orifice *See* genital pore. { 'jen·ət·əl 'ȯr·ə·fəs }

genital pore [INV ZOO] A small opening on the side of the head in some gastropods through which the penis is protruded. Also known as genital orifice. { 'jen·ət·əl 'pȯr }

genital rachis *See* genital cord. { 'jen·ət·əl 'rā·kəs }

genital recess [INV ZOO] A depression between the calyx surface and anal cone in entoprocts which serves as a brood chamber. { 'jen·ət·əl 'rēˌses }

genital ridge [EMBRYO] A medial ridge or fold on the ventromedial surface of the mesonephros in the embryo, produced by growth of the peritoneum; the primordium of the gonads and their ligaments. { 'jen·ət·əl 'rij }

genital scale [INV ZOO] Any of the small calcareous plates in ophiuroids associated with the buccal shields. { 'jen·ət·əl 'skāl }

genital segment *See* gonosomite. { 'jen·ət·əl 'seg·mənt }

genital shield [INV ZOO] In ophiuroids, a support of a bursal slit in the arms located near the arm base. { 'jen·ət·əl 'shēld }

genital stolon [INV ZOO] Part of the axial complex in ophiuroids. { 'jen·ət·əl 'stō·lən }

genital sucker [INV ZOO] In some trematodes, a suckerlike structure surrounding the gonopore. { 'jen·ət·əl 'sək·ər }

genital tract [ANAT] The ducts of the reproductive system. { 'jen·ət·əl ˌtrakt }

genital tube [INV ZOO] A blood lacuna in crinoids, connected with the subtegminal plexus and suspended in the genital canal. { 'jen·ət·əl 'tüb }

genitourinary system *See* urogenital system. { ¦jen·ə·tō'yür·əˌner·ē ˌsis·təm }

genome [GEN] **1.** The genetic endowment of a species. **2.** The haploid set of chromosomes. { 'jēˌnōm }

genomic deoxyribonucleic acid [CELL MOL] Fragments of deoxyribonucleic acid (DNA) that are produced by the action of restriction enzymes on the DNA of a cell or organism. Abbreviated gDNA. { jə¦nōm·ik dēˌäk·sēˌrī·bō·nü¦klē·ik'as·əd }

genomic imprinting *See* parental imprinting. { jə¦nō·mik imˌprint·iŋ }

genomics [GEN] The study of the entire deoxyribonucleic sequence of an organism. { jə'nō·miks }

genomic stress [GEN] Any influence that may disrupt the stability of the genome by fostering chromosome damage or mutation, such as environmental factors or altered genetic background. { jə¦nōm·ik 'stres }

genotoxant [BIOCHEM] An agent that induces toxic, lethal, or heritable effects to nuclear and extranuclear genetic material in cells. { ˌjēn·ə'täk·sənt }

genotype [GEN] The genetic constitution of an organism, usually in respect to one gene or a few genes relevant in a particular context. [SYST] The type species of a genus. { 'jē·nəˌtīp }

genotype frequency [GEN] The proportion or frequency of any particular genotype among the individuals of a population. { 'jēn·əˌtīp ˌfrē·kwən·sē }

genotypic cohesion [EVOL] The process whereby balanced and superior gene combinations are held together under the force of genetic recombination, thus reducing the frequency of deleterious recombinants, and with it the genetic load. { ¦jēn·əˌtip·ik kō'hē·zhən }

genotypic distance [EVOL] For two individuals A and B, the probability that the genotype of A is not the same as that of B for a given locus; the distance is zero when the genotype is the same at the particular locus. { ¦jen·əˌtip·ik 'dis·təns }

genotypic structure [GEN] The set of the genotype frequencies of a population. { ¦jēn·əˌtip·ik 'strək·chər }

gentamicin [MICROBIO] A broad-spectrum antibiotic produced by a species of *Micromonospora*. { ¦jent·ə¦mīs·ən }

Gentianaceae [BOT] A family of dicotyledonous herbaceous plants in the order Gentianales distinguished by lacking stipules and having parietal placentation. { ˌjen·chə'nās·ēˌē }

Gentianales [BOT] A family of dicotyledonous plants in the subclass Asteridae having well-developed internal phloem and opposite, simple, mostly entire leaves. { ˌjen·chə'nā·lēz }

genu *See* knee. { 'ge·nü }

genus [SYST] A taxonomic category that includes groups of closely related species; the principal subdivision of a family. { 'jē·nəs }

genu valgum [MED] Inward or medial curving of the knee; knock-knee. { 'ge·nü 'val·gəm }

geobotany [BOT] The study of plants as related to their geologic environment. { ¦jē·ō'bät·ən·ē }

Geocorisae [INV ZOO] A subdivision of hemipteran insects containing those land bugs with conspicuous antennae and an ejaculatory bulb in the male. { ˌjē·ə'kȯr·əˌsē }

Geodermatophilus [MICROBIO] A genus of bacteria in the family Dermatophilaceae; the mycelium is rudimentary and a muriform thallus is produced; motile spores are elliptical to lanceolate. { ˌjē·ōˌdər·mə'täf·ə·ləs }

geographical botany *See* plant geography. { ¦jē·ə¦graf·ə·kəl 'bät·ən·ē }

geographic speciation [EVOL] Evolution of two or more species from a single species following geographic isolation. { ¦jē·ə¦graf·ik ˌspē·shē'ā·shən }

Geometridae [INV ZOO] A large family of lepidopteran insects in the superfamily Geometroidea that have slender bodies and relatively broad wings; includes measuring worms, loopers, and cankerworms. { ˌjē·ə'me·trəˌdē }

Geometroidea [INV ZOO] A superfamily of lepidopteran insects in the suborder Heteroneura comprising small to large moths with reduced maxillary palpi and tympanal organs at the base of the abdomen. { ˌjē·ə·mə'trȯid·ē·ə }

Geomyidae [VERT ZOO] The pocket gophers, a family of rodents characterized by fur-lined cheek pouches which open outward, a stout body with short legs, and a broad, blunt head. { ˌjē·ō'mī·əˌdē }

geophagia [ZOO] Soil ingestion by animals. { ˌjē·ə'fā·jə }

geophagous [ZOO] Feeding on soil, as certain worms. { jē'äf·ə·gəs }

Geophilomorpha [INV ZOO] An order of centipedes in the class Chilopoda including specialized forms that are blind, epimorphic, and dorsoventrally flattened. { ˌjē·ōˌfil·ə'mȯr·fə }

geophilous [ECOL] Living or growing in or on the ground. { jē'äf·ə·ləs }

geophyte [ECOL] A perennial plant that is deeply embedded in the soil substrata. { 'jē·əˌfīt }

Georyssidae [INV ZOO] The minute mud-loving beetles, a family of coleopteran insects belonging to the Polyphaga. { ˌjē·ə'ris·əˌdē }

geosensing [BOT] The sensing or detecting of gravity by a plant relative to its longitudinal axis. { 'jē·ōˌsens·iŋ }

Geosiridaceae [BOT] A monotypic family of monocotyledonous plants in the order Orchidales characterized by regular flowers with three stamens that are opposite the sepals. { ˌjē·əˌsir·ə'dās·ēˌē }

Geospizinae [VERT ZOO] Darwin finches, a subfamily of perching birds in the family Fringillidae. { ˌjē·ō'spiz·əˌnē }

geotaxis [PHYSIO] Movement of a free-living organism in response to the stimulus of gravity. { ¦jē·ō¦tak·səs }

geotropism [BOT] Response of a plant to the force of gravity. Also known as gravitropism. { jē'ä·trəˌpiz·əm }

Gephyrea [INV ZOO] A class of burrowing worms in the phylum Annelida. { jə'fir·ē·ə }

gephyrocercal [VERT ZOO] Having the dorsal and anal fins coming together smoothly at the aborted end of the vertebral column of a fish's tail. { ¦jef·ə·rō¦sər·kəl }

Geraniaceae [BOT] A family of dicotyledonous plants in the order Geraniales in which the fruit is beaked, styles are usually united, and the leaves have stipules. { jəˌrā·nē'ās·ēˌē }

Geraniales [BOT] An order of dicotyledonous

plants in the subclass Rosidae comprising herbs or soft shrubs with a superior ovary and with compound or deeply cleft leaves. { jə,rā·nē'ā·lēz }

Geranium [BOT] A genus of plants in the family Geraniaceae characterized by regular flowers, and glands alternating with the petals. { jə'rā·nē·əm }

Gerardiidae [INV ZOO] A family of anthozoans in the order Zoanthidea. { ,jər·är'dī·ə,dē }

gerbil [VERT ZOO] The common name for about 100 species of African and Asian rodents composing the subfamily Gerbillinae. { 'jər·bəl }

Gerbillinae [VERT ZOO] The gerbils, a subfamily of rodents in the family Muridae characterized by hindlegs that are longer than the front ones, and a long, slightly haired, usually tufted tail. { jər'bil·ə,nē }

Gerhardt's test [PATH] A test for acetoacetic acid in urine. { 'ger,härts ,test }

geriatrics [MED] The study of the biological and physical changes and the diseases of old age. { ,jer·ē'a·triks }

germ [BIOL] A primary source, especially one from which growth and development are expected. [MICROBIO] General designation for a microorganism. { jərm }

German measles *See* rubella. { ¦jər·mən 'mē·zəlz }

germarium [INV ZOO] The egg-producing portion of an ovary and the sperm-producing portion of a testis in Platyhelminthes and Rotifera. { jər'mar·ē·əm }

germ ball [INV ZOO] A group of cells in digenetic trematode miracidial larvae which are embryos. { 'jərm ,bȯl }

germ cell *See* gamete. { 'jərm ,sel }

germfree animal [MICROBIO] An animal having no demonstrable, viable microorganisms living in intimate association with it. { 'jərm,frē 'an·ə·məl }

germfree isolator [MICROBIO] An apparatus that provides a mechanical barrier surrounding the area in which germfree vertebrates and accessory equipment are housed. { 'jərm,frē 'ī·sə,lād·ər }

germinal epithelium [EMBRYO] The region of the dorsal coelomic epithelium lying between the dorsal mesentery and the mesonephros. { 'jər·mən·əl ,ep·ə'thē·lē·əm }

germinal spot [CELL MOL] The nucleolus of an egg cell. { 'jər·mən·əl ,spät }

germinal vesicle [CELL MOL] The enlarged nucleus of the primary oocyte before reduction divisions are complete. { 'jər·mən·əl 'ves·ə·kəl }

germination [BOT] The beginning or the process of development of a spore or seed. { ,jer·mə'nā·shən }

germ layer [EMBRYO] One of the primitive cell layers which appear in the early animal embryo and from which the embryo body is constructed. { 'jərm ,lā·ər }

germ-layer theory [EMBRYO] The theory that three primary germ layers, ectoderm, mesoderm, and endoderm, are established in the early embryo and all organs and structures are derived from a specific germ layer. { 'jərm ,lā·ər ,thē·ə·rē }

germline [BIOL] A lineage of cells from which gametes are derived. Also known as germ track. { 'jərm ,līn }

germline mutation [GEN] A mutation within a lineage of cells that form gametes. { ¦jərm,līn myü'tā·shən }

germovitellarium [INV ZOO] A sex gland which differentiates into a yolk-producing or egg-producing region. { ¦jər·mō,vīd·əl'a·rē·əm }

germ plasm [BIOL] The genetic material contained within a germ cell. { 'jərm ,plaz·əm }

germ theory [MED] The theory that contagious and infectious diseases are caused by microorganisms. { 'jərm ,thē·ə·rē }

germ track *See* germ line. { 'jərm ,trak }

geroderma [MED] The skin of old age, showing atrophy, loss of fat, and loss of elasticity. { ,jer·ō'dər·mə }

gerontology [PHYSIO] The scientific study of aging processes in biological systems, particularly in humans. { ,jer·ən'täl·ə·jē }

Gerrhosauridae [VERT ZOO] A small family of lizards in the suborder Sauria confined to Africa and Madagascar. { ,jer·ō'sȯr·ə,dē }

Gerridae [INV ZOO] The water striders, a family of hemipteran insects in the subdivision Amphibicorisae having long middle and hind legs and a median scent gland opening on the metasternum. { 'jer·ə,dē }

Gerroidea [INV ZOO] The single superfamily of the hemipteran subdivision Amphibicorisae; all members have conspicuous antennae and hydrofuge hairs covering the body. { jə'rȯid·ē·ə }

Gesneriaceae [BOT] A family of dicotyledonous plants in the order Scrophulariales characterized by parietal placentation, mostly opposite or whorled leaves, and a well-developed embryo. { ge,snir·ē'ās·ē,ē }

gestalt vision [PHYSIO] The visual perception and retention in the memory of an object in terms of its geometric shape. { gə¦shtält ,vizh·ən }

gestate [EMBRYO] To carry the young in the uterus from conception to delivery. { 'je,stāt }

gestation period [EMBRYO] The period in mammals from fertilization to birth. { jə'stā·shən ,pir·ē·əd }

GFP *See* green fluorescent protein.

GH *See* growth hormone.

Ghon complex [MED] The combination of a focus of subpleural tuberculosis with associated hilar and mediastinal lymph node tuberculosis. { 'gōn ,käm,pleks }

ghost layer [CELL MOL] A single layer of cultivated animal cells that has been treated with a nonionic detergent in order to disrupt the membranes. { 'gōst ,lā·ər }

giant-cell arteritis [MED] Inflammation of the arteries, particularly the carotid branches, characterized by the appearance of multinucleate giant cells in the exudate. Also known as temporal arteritis. { 'jī·ənt ˌsel ˌärd·ə'rīd·əs }

Giardia [INV ZOO] A genus of zooflagellates that inhabit the intestine of numerous vertebrates, and may cause diarrhea in humans. { je'ärd·ē·ə }

giardiasis [MED] Presence of the protozoon *Giardia lamblia* in the human small intestine. { jēˌär'dī·ə·səs }

Gibberella fujikuroi [MYCOL] A fungal pathogen that causes bakanae disease, a seed-borne disease of rice that is characterized by the growth of excessively long internodes, through its production of plant growth hormones called gibberellins. { ˌjib·əˌrel·ə ˌfü·jē'küˌroi }

gibberellic acid [BIOCHEM] $C_{18}H_{22}O_6$ A crystalline acid occurring in plants that is similar to the gibberellins in its growth-promoting effects. { ¦jib·ə¦rel·ik 'as·əd }

gibberellin [BIOCHEM] Any member of a family of naturally derived compounds which have a gibbane skeleton and a broad spectrum of biological activity but are noted as plant growth regulators. { ˌjib·ə'rel·ən }

gibbon [VERT ZOO] The common name for seven species of large, tailless primates belonging to the genus *Hylobates*; the face and ears are hairless, and the arms are longer than the legs. { 'gib·ən }

gid [VET MED] A chronic brain disease of sheep, less frequently of cattle, characterized by forced movements of circling or rolling, caused by the larval form of the tapeworm *Multiceps multiceps*. { gid }

GIFT *See* gamete intrafallopian transfer. { gift *or* ¦jē¦ī¦ef'tē }

gigantism [MED] Abnormal largeness of the body due to hypersecretion of growth hormone. { jī'ganˌtiz·əm }

Giganturoidei [VERT ZOO] A suborder of small, mesopelagic actinopterygian fishes in the order Cetomimiformes having large mouths and strong teeth. { jīˌgan·tə'ròid·ēˌī }

Gila monster [VERT ZOO] The common name for two species of reptiles in the genus *Heloderma* (Helodermatidae) distinguished by a rounded body that is covered with multicolored beaded tubercles, and a bifid protrusible tongue. { 'hē·lə ˌmän·stər }

gill [MYCOL] A structure consisting of radially arranged rows of tissue that hang from the underside of the mushroom cap of certain basidiomycetes. [VERT ZOO] The respiratory organ of water-breathing animals. Also known as branchia. { gil }

gill cover [VERT ZOO] The fold of skin providing external protection for the gill apparatus of most fishes; it may be stiffened by bony plates and covered with scales. { 'gil kəv·ər }

Gilles de la Tourette syndrome *See* Tourette's syndrome. { ˌzhēl də lä tü'ret ˌsinˌdrōm }

gill raker [VERT ZOO] One of the bony processes on the inside of the branchial arches of fishes which prevents the passage of solid substances through the branchial clefts. { 'gil ˌrāk·ər }

ginger [BOT] *Zingiber officinale.* An erect perennial herb of the family Zingiberaceae having thick, scaly branched rhizomes; a spice oleoresin is made by an organic solvent extraction of the ground dried rhizome. { 'jin·jər }

gingiva [ANAT] The mucous membrane surrounding the teeth sockets. { 'jin·jə·və }

gingival crevice [ANAT] The space between the free margin of the gingiva and the surface of a tooth. Also known as gingival sulcus. { 'jin·jə·vəl 'krev·əs }

gingival sulcus *See* gingival crevice. { 'jin·jə·vəl 'səl·kəs }

gingivectomy [MED] Excision of a portion of the gingiva. { ˌjin·jə'vek·tə·mē }

gingivitis [MED] Inflammation of the gingiva. { ˌjin·jə'vīd·əs }

gingivostomatitis [MED] An inflammation of the gingiva and oral mucosa. { ¦jin·jə·vōˌstō·mə'tīd·əs }

ginglymoarthrodia [ANAT] A composite joint consisting of one hinged and one gliding element. { ¦jiŋ·glə·mōˌär'thrō·dē·ə }

Ginglymodi [VERT ZOO] An equivalent name for Semionotiformes. { ˌjiŋ·glə'mōˌdī }

ginglymus [ANAT] A type of diarthrosis permitting motion only in one plane. Also known as hinge joint. { 'jiŋ·glə·məs }

ginkgo [BOT] A dioecious tree, commonly known as the maidenhair tree (*Ginkgo biloba*), that is native to China and is cultivated as a shade tree, it is the only surviving species of the class Ginkgoatae and is considered a living fossil. Also known as gingko tree. { 'giŋ·kō }

Ginkgoales [BOT] An order of gymnosperms composing the class Ginkgoopsida with one living species, the dioecious maidenhair tree (*Ginkgo biloba*). { ˌgiŋ·kō'ā·lēz }

Ginkgoatae *See* Ginkgoopsida. { ˌgiŋ'kō·əˌtē }

Ginkgoopsida [BOT] A class of the subdivision Pinicae containing the single, monotypic order Ginkgoales. { ˌgiŋ·kō'äp·sə·də }

Ginkgophyta [BOT] The equivalent name for Ginkgoopsida. { ˌgiŋ'käf·əd·ə }

gingko tree *See* ginkgo. { 'giŋ·kō ˌtrē }

ginseng [BOT] The common name for plants of the genus *Panax*, a group of perennial herbs in the family Araliaceae; the aromatic root of the plant has been used medicinally in China. { 'jinˌseŋ }

giraffe [VERT ZOO] *Giraffa camelopardalis.* An artiodactyl mammal in the family Giraffidae characterized by extreme elongation of the neck vertebrae, and two prominent horns on the head. { jə'raf }

Giraffidae [VERT ZOO] A family of pecoran ruminants in the superfamily Bovoidea including giraffe, okapi, and relatives. { jə'raf·əˌdē }

girdle [ANAT] Either of the ringlike groups of bones supporting the forelimbs (arms) and hindlimbs (legs) in vertebrates. [INV ZOO] **1.** Either of the hooplike bands constituting the sides of the two valves of a diatom. **2.** The peripheral portion of the mantle in chitons. { 'gərd·əl }

gitogenin [PHARM] $C_{27}H_{44}O_4$ A crystalline compound obtained from gitonin by heating with dilute hydrochloric acid, forms leaflets from benzene solutions, soluble in organic solvents such as chloroform, hot alcohol, and ether; used in medicine for treating coronary disease. { jə'täj·ə·nən }

gitoxigenin [PHARM] $C_{23}H_{34}O_5$ Platelike crystals, used as a cardiotonic drug. { jə,täk·sə'jen·ən }

gitoxin [PHARM] $C_{41}H_{64}O_{14}$ A secondary glycoside derived from *Digitalis purpurea*, crystallizes in stout prisms from chloroform methanol solution, soluble in a mixture of chloroform and alcohol; used in medicine for coronary disease. { jə'täk·sən }

gitter cell [PATH] A compound granule cell that is characteristic of certain brain lesions. { 'gid·ər ,sel }

gizzard [VERT ZOO] The muscular portion of the stomach of most birds where food is ground with the aid of ingested pebbles. { 'giz·ərd }

glabella [ANAT] The bony prominence on the frontal bone joining the supraorbital ridges. { glə'bel·ə }

glabrous [BIOL] Having a smooth surface; specifically, having the epidermis devoid of hair or down. { 'glab·rəs }

gladiate [BOT] Sword-shaped. { 'glad·ē,āt }

gladiolus *See* mesosternum. { ,glad·ē'ō·ləs }

Gladiolus [BOT] A genus of chiefly African plants in the family Iridaceae having erect, sword-shaped leaves and spikes of brightly colored irregular flowers. { ,glad·ē'ō·ləs }

gladius *See* pen. { 'glād·ē·əs }

gland [ANAT] A structure which produces a substance essential and vital to the existence of the organism. { gland }

glanders [VET MED] A bacterial disease of equines caused by *Actinobacillus mallei*; involves the respiratory system, skin, and lymphatics. Also known as farcy. { 'glan·dərz }

glands of Brunner *See* Brunner's glands. { ¦glanz əv 'brün·ər }

glands of Leydig [VERT ZOO] Unicellular, epidermal structures of urodele larvae and the adult *Necturus* that secrete a substance which digests the egg capsule and permits hatching. { ¦glanz əv 'lī·dig }

glandular fever *See* infectious mononucleosis. { ¦glan·jə·lər 'fē·vər }

glandulomuscular [ANAT] Of or pertaining to glands and muscles. { 'glan·jə·lə'məs·kyə·lər }

glans [ANAT] The conical body forming the distal end of the clitoris or penis. { glanz }

Glareolidae [VERT ZOO] A family of birds in the order Charadriiformes including the ploverlike coursers and the swallowlike pratincoles. { ,gla·rē'ä·lə,dē }

glareous [ECOL] Growing in gravelly soil; refers specifically to plants. { 'gla·rē·əs }

Glasser's disease [VET MED] A generalized bacterial infection of swine caused by *Mycoplasma hyorhinis*. { 'glas·ərz di,zēz }

glass ionomer [MED] The only dental restorative material that forms a durable chemical bond to dentin; it is formed by the reaction of aluminosilicate glass with polyacrylic acid. { ,glas ī'än·ə·mər }

glass sponge [INV ZOO] A siliceous sponge belonging to the class Hyalospongiae. { 'glas ,spənj }

glaucoma [MED] A disease of the eye characterized by increased fluid pressure within the eyeball. { glaů'kō·mə }

glaucous [BOT] Having a white or grayish powdery coating that gives a frosty appearance and rubs off easily. { 'glȯ·kəs }

gleba [MYCOL] The central, sporogenous tissue of the sporophore in certain basidiomycetous fungi. { 'glē·bə }

gleet [MED] The chronic stage of gonorrheal urethritis, characterized by a slight mucopurulent discharge. { glēt }

glenoid [ANAT] A smooth, shallow, socketlike depression, particularly of the skeleton. { 'gle,nȯid }

glenoid cavity [ANAT] The articular surface on the scapula for articulation with the head of the humerus. { ¦gle,nȯid 'kav·əd·ē }

glial cell *See* neuroglia. { 'glē·əl ,sel }

gliding bacteria [MICROBIO] The descriptive term for members of the orders Beggiatoales and Myxobacterales; they are motile by means of creeping movements. { ¦glīd·iŋ bak'tir·ē·ə }

gliding joint *See* arthrodia. { ¦glīd·iŋ 'jȯint }

gliding motility [MICROBIO] A means of bacterial self-propulsion by slow gliding or creeping movements on the surface of a substrate. { ¦glīd·iŋ mō'til·əd·ē }

glioma [MED] A malignant tumor derived from the supporting tissue of the central nervous system. { glī'ō·mə }

gliosis [MED] Proliferation of neuroglia in the brain or spinal cord, either as a replacement process or in response to a low-grade inflammation. { glī'ō·səs }

gliotoxin [MICROBIO] $C_{13}H_{14}O_4N_2S_2$ A heat-labile, bacteriostatic antibiotic produced by species of *Trichoderma* and *Cliocladium* and by *Aspergillus fumigatus*. { ¦glī·ō,täk·sən }

Gliridae [VERT ZOO] The dormice, a family of mammals in the order Rodentia. { 'glir·ə,dē }

Glisson's capsule [ANAT] The membranous sheet of collagenous and elastic fibers covering the liver. { 'glis·ənz ,kap·səl }

Globigerinacea [INV ZOO] A superfamily of foraminiferan protozoans in the suborder Rotaliina characterized by a radial calcite test with bilamellar septa and a large aperture. { glō,bij·ə·rə'nās·ē·ə }

globin [BIOCHEM] Any of a class of soluble histone proteins obtained from animal hemoglobins. { 'glō·bən }

globin zinc insulin [PHARM] A preparation of insulin modified by the addition of globin (derived from the hemoglobin of beef blood) and zinc chloride; it has intermediate duration of action. { ¦glō·bən ¦ziŋk 'in·sə·lən }

globoside [BIOCHEM] A glycoside of ceramide containing several sugar residues, but not neuraminic acid; obtained from human, sheep, and hog erythrocytes. { 'glō·bə,sīd }

globular deoxyribonucleic acid [CELL MOL] A compact arrangement of double-stranded deoxyribonucleic acid formed by additional twisting of the fiber into a helical double helix. { 'gläb·yə·lər dē¦äk·sē,rī·bō,nü¦klē·ik 'as·əd }

globular protein [BIOCHEM] Any protein that is readily soluble in aqueous solvents. { 'gläb·yə·lər 'prō,tēn }

globulin [BIOCHEM] A heat-labile serum protein precipitated by 50% saturated ammonium sulfate and soluble in dilute salt solutions. { 'gläb·yə·lən }

globulomaxillary cyst [MED] A cyst in the alveolar process between the upper lateral incisor and canine teeth. { ¦gläb·yə·lō¦mak·sə,ler·ē 'sist }

glochid *See* glochidium. { 'glō,kòid }

glochidium [BOT] A barbed hair. Also known as glochid. [INV ZOO] The larva of fresh-water mussels in the family Unionidae. { glō'kid·ē·əm }

gloea [INV ZOO] An adhesive mucoid substance secreted by certain protozoans and other lower organisms. { 'glē·ə }

glomerulonephritis [MED] Inflammation of the kidney, primarily involving the glomeruli. { glə¦mər·yə·lō·nə¦frīd·əs }

glomerulosclerosis [MED] Fibrosis of the renal glomeruli. { glə¦mər·yə·lō·sklə¦rō·səs }

glomerulus [ANAT] A tuft of capillary loops projecting into the lumen of a renal corpuscle. { glə'mər·yə·ləs }

glomus [ANAT] **1.** A fold of the mesothelium arising near the base of the mesentery in the pronephros and containing a ball of blood vessels. **2.** A prominent portion of the choroid plexus of the lateral ventricle of the brain. { 'glō·məs }

glomus aorticum *See* paraaortic body. { 'glō·məs ā'òrd·ə·kəm }

glomus caroticum *See* carotid body. { 'glō·məs kə'räd·ə·kəm }

glossa [INV ZOO] A tongue or tonguelike structure in insects, especially the median projection of the labium. { 'gläs·ə }

glossalgia [MED] Pain in the tongue. { glä 'sal·jə }

glossate [INV ZOO] Having a glossa or tonguelike structure. { 'glä,sāt }

Glossinidae [INV ZOO] The tsetse flies, a family of cyclorrhaphous dipteran insects in the section Pupipara. { glä'sin·ə,dē }

Glossiphoniidae [INV ZOO] A family of small leeches with flattened bodies in the order Rhynchobdellae. { ,glä·sə·fə'nī·ə,dē }

glossitis [MED] Inflammation of the tongue. { glä'sīd·əs }

glossopalatine nerve [NEUROSCI] The intermediate branch of the facial nerve. { ¦gläs·ō'pal·ə,tēn ,nərv }

glossopharyngeal nerve [NEUROSCI] The ninth cranial nerve in vertebrates; a paired mixed nerve that supplies autonomic innervation to the parotid gland and contains sensory fibers from the posterior one-third of the tongue and the anterior pharynx. { ¦gläs·ō·fə'rin·jē·əl ,nərv }

glossopyrosis [MED] Burning sensation of the tongue. { ,gläs·ō,pī'rō·səs }

glottis [ANAT] The opening between the margins of the vocal folds. { 'gläd·əs }

glove-and-stocking anesthesia [MED] Loss or diminution of sensation in the hands and feet, corresponding to the areas covered by gloves and stockings. { ¦gləv ən 'stäk·iŋ ,an·əs,thē·zhə }

glove anesthesia [MED] Loss or diminution of sensation in the hands, corresponding to the area covered by gloves. { 'gləv ,an·əs,thē·zhə }

glucagon [BIOCHEM] The protein hormone secreted by α-cells of the pancreas which plays a role in carbohydrate metabolism. Also known as hyperglycemic factor; hyperglycemic glycogenolytic factor. { 'glü·kə,gän }

glucamine [BIOCHEM] $C_6H_{15}O_4N$ An amine formed by reduction of glucosylamine or of glucose oxime. { 'glü·kə,mēn }

glucan [BIOCHEM] A polysaccharide composed of the hexose sugar D-glucose. { 'glü,kan }

glucocerebrosidase [BIOCHEM] An enzyme that removes the glucose from glycosyl ceramide and is defective or missing in Gaucher's disease. { ,glü·kō,ser·ə'bräs·i,dās }

glucocerebroside [BIOCHEM] A glycoside of ceramide that contains glucose. { ¦glü·kō·sə'rēb·rə,sīd }

glucocorticoid [BIOCHEM] A corticoid that affects glucose metabolism; secreted principally by the adrenal cortex. { ¦glü·kō¦kòrd·ə,kòid }

glucogenesis [BIOCHEM] Formation of glucose within the animal body from products of glycolysis. { ,glü·kō'jen·ə·səs }

glucokinase [BIOCHEM] An enzyme that catalyzes the phosphorylation of D-glucose to glucose-6-phosphate. { ¦glü·kō'kī,nās }

glucolipid [BIOCHEM] A glycolipid that yields glucose on hydrolysis. { ¦glü·kō'lip·əd }

glucomannan [BIOCHEM] A polysaccharide composed of D-glucose and D-mannose; a prominent component of coniferous trees. { ¦glü·kō ¦man·ən }

gluconeogenesis [BIOCHEM] Formation of glucose within the animal body from substances other than carbohydrates, particularly proteins and fats. { ¦glü·kō¦nē·ō'jen·ə·səs }

Gluconobacter [MICROBIO] A genus of bacteria in the family Pseudomonadaceae; ellipsoidal to rod-shaped cells having three to eight polar flagella and occurring singly, in pairs, or in chains. { glü¦kän·ə¦bak·tər }

D-glucopyranose *See* glucose. { ¦dē ¦glü·kə'pir·ə,nōs }

glucopyranoside [BIOCHEM] Any glucoside that contains a six-membered ring. { ¦glü·kō·pir'an·ə,sīd }

glucosamine [BIOCHEM] $C_6H_{13}O_5$ An amino

sugar; the most abundant in nature, occurring in glycoproteins and chitin. { 'glü,kōs·ə·mēn }

glucose [BIOCHEM] $C_6H_{12}O_6$ A monosaccharide; occurs free or combined and is the most common sugar. Also known as cerelose; D-glucopyranose. { 'glü,kōs }

glucose-6-phosphatase [BIOCHEM] An enzyme found in liver which catalyzes the hydrolysis of glucose-6-phosphate to free glucose and inorganic phosphate. { 'glü,kōs ¦siks 'fäs·fə,tās }

glucose phosphate [BIOCHEM] A phosphoric derivative of glucose, as glucose-1-phosphate. { 'glü,kōs 'fäs,fāt }

glucose-1-phosphate [BIOCHEM] $C_6H_{12}O_8P$ An ester of glucopyranose in which a phosphate group is attached to carbon atom 1; there are two types: α-D- and β-D-glucose-1-phosphates. Also known as Cori ester. { 'glü,kōs ¦wən 'fäs,fāt }

glucose-6-phosphate [BIOCHEM] $C_6H_{13}O_9P$ An ester of glucose with phosphate attached to carbon atom 6. Also known as Robisonester. { 'glü,kōs ¦siks 'fäs,fāt }

glucose-6-phosphate dehydrogenase [BIOCHEM] The mammalian enzyme that catalyzes the oxidation of glucose-6-phosphate by TPN^+ (triphosphopyridine nucleotide). { 'glü,kōs ¦siks 'fäs,fāt dē,hī'drä·jə,nās }

glucose tolerance test [PATH] A test to measure the ability of the liver to convert glucose to glycogen. { 'glü,kōs ¦täl·ə·rəns ,test }

glucosidase [BIOCHEM] An enzyme that hydrolyzes glucosides. { glü'kō·sə,dās }

glucoside [BIOCHEM] One of a group of compounds containing the cyclic forms of glucose, in which the hydrogen of the hemiacetal hydroxyl has been replaced by an alkyl or aryl group. { 'glü·kə,sīd }

glucosulfone sodium *See* sodium glucosulfone. { 'glü,kō ¦səl,fōn 'sōd·ē·əm }

glucosyl ceramide [BIOCHEM] A type of glycolipid that is present in the cell membranes of many cell types and is abundant in serum. { ,glü·kə,sil 'ser·ə,mīd }

glucosyltransferase [BIOCHEM] An enzyme that catalyzes the glucosylation of hydroxymethyl cytosine; a constituent of bacteriophage deoxyribonucleic acid. { ¦glü·kə,sil'-tranz·fə,rās }

glucuronic acid [BIOCHEM] $C_6H_{10}O_7$ An acid resulting from oxidation of the CH_2OH radical of D-glucose to COOH; a component of many polysaccharides and certain vegetable gums. Also known as glycuronic acid. { ¦glü·kyə¦rän·ik 'as·əd }

glucuronidase [BIOCHEM] An enzyme that catalyzes hydrolysis of glucuronides. Also known as glycuronidase. { ,glü·kyə'rän·ə,dās }

glucuronide [BIOCHEM] A compound resulting from the interaction of glucuronic acid with a phenol, an alcohol, or an acid containing a carboxyl group. Also known as glycuronide. { glü'kyür·ə,nīd }

D-glucuronolactone [BIOCHEM] $C_6H_8O_6$ A water-soluble crystalline compound found in plant gums in polymers with other carbohydrates, and an important structural component of almost all fibrous and connective tissues in animals; used in medicine as an antiarthritic. { ¦de glü¦kyür·ə·nō'lak,tōn }

glue cell *See* adhesive cell. { 'glü ,sel }

glume [BOT] One of two bracts at the base of a spikelet of grass. { glüm }

glumiferous [BOT] Bearing glumes. { glü'mif·ə·rəs }

Glumiflorae [BOT] An equivalent name for Cyperales. { ,glü·mə'flör,ē }

glutamate [BIOCHEM] A salt or ester of glutamic acid. { 'glüd·ə,māt }

glutamic acid [BIOCHEM] $C_5H_9O_4N$ A dicarboxylic amino acid of the α-ketoglutaric acid family occurring widely in proteins. { glü'tam·ik 'as·əd }

glutaminase [BIOCHEM] The enzyme which catalyzes the conversion of glutamine to glutamic acid and ammonia. { glü'tam·ə,nās }

glutamine [BIOCHEM] $C_5H_{10}O_3N_2$ An amino acid; the monamide of glutamic acid; found in the juice of many plants and essential to the development of certain bacteria. { 'glüd·ə,mēn }

glutamine synthetase [BIOCHEM] An enzyme which catalyzes the formation of glutamine from glutamic acid and ammonia, using adenosine triphosphate as a source of energy. { 'glüd·ə,mēn 'sin·thə,tās }

glutarate [BIOCHEM] The salt or ester of glutaric acid. { 'glüd·ə,rāt }

glutaric acid [BIOCHEM] $C_5H_5O_4$ A water-soluble, crystalline acid that occurs in green sugarbeets and in water extracts of crude wool. { glü'tar·ik 'as·əd }

glutathione [BIOCHEM] $C_{10}H_{17}O_6N_3S$ A widely distributed tripeptide that is important in plant and animal tissue oxidation reactions. { ¦glüd·ə'thī,ōn }

glutelin [BIOCHEM] A class of simple, heat-labile proteins occurring in seeds of cereals; soluble in dilute acids and alkalies. { 'glüd·əl·ən }

gluten [BIOCHEM] **1.** A mixture of proteins found in the seeds of cereals; gives dough elasticity and cohesiveness. **2.** An albuminous element of animal tissue. { 'glüt·ən }

gluten enteropathy [MED] A malabsorption disease characterized by inflammation and loss of the normal architecture of the small intestine following ingestion of some proteins. Also known as celiac sprue, nontropical sprue. { ,glüt·ən ,ent·ə'räp·ə·thē }

glutenin [BIOCHEM] A glutelin of wheat. { 'glüt·ən·ən }

glutethimide [PHARM] $C_{13}H_{15}NO_2$ A minor or sedative antianxiety tranquilizer that acts as a central nervous system depressant. { glü'teth·ə·məd }

gluteus maximus [ANAT] The largest and most superficial muscle of the buttocks. { 'glüd·ē·əs 'mak·sə·məs }

gluteus medius [ANAT] The muscle of the buttocks lying between the gluteus maximus and gluteus minimus. { 'glüd·ē·əs 'mēd·ē·əs }

gluteus minimus [ANAT] The smallest and deepest muscle of the buttocks. { 'glüd·ē·əs 'min·ə·məs }

glutinant nematocyst [INV ZOO] A nematocyst characterized by an open, sticky tube used for anchoring the cnidarian when walking on its tentacles. { 'glüt·ən·ənt nə'mad·ə,sist }

glutinous [BOT] Having a sticky surface. { 'glüt·ən·əs }

glycal [BIOCHEM] A monosaccharide having a double bond between carbons 1 and 2, and no hydroxyl group on either of those carbons. { 'glī,kal }

glycan *See* polysaccharide. { 'glī,kan }

glycemia [PHYSIO] The presence of glucose in the blood. { glī'sē·mē·ə }

glycemic index [MED] A ranking of foods based on how they affect blood glucose (sugar) levels in the 2–3 hours after eating, foods with carbohydrates that break down quickly during digestion have the highest glycemic indexes. { glī¦sēm·ik 'in,deks }

glyceraldehyde [BIOCHEM] $CH_2OHCOHCHO$ A colorless solid, isomeric with dehydroxyacetone; soluble in water and insoluble in organic solvents; an important intermediate in carbohydrate metabolism; used as a chemical intermediate in biochemical research and nutrition. { ,glis·ə'ral·də,hīd }

glycerate [BIOCHEM] A salt or ester of glyceric acid. { 'glis·ə,rāt }

glyceric acid [BIOCHEM] $C_3H_6O_4$ A hydroxy acid obtained by oxidation of glycerin. { glə'ser·ik 'as·əd }

Glyceridae [INV ZOO] A family of polychaete annelids belonging to the Errantia and characterized by an enormous eversible proboscis. { glə'ser·ə,dē }

glyceride [BIOCHEM] An ester of glycerin and an organic acid radical; fats are glycerides of certain long-chain fatty acids. { 'glis·ə,rīd }

glycerinated vaccine virus *See* smallpox vaccine. { 'glis·ə·rə,nād·əd 'vak,sēn ,vī·rəs }

glycerokinase [BIOCHEM] An enzyme that catalyzes the phosphorylation of glycerol to glycerophosphate during microbial fermentation of propionic acid. { ¦glis·ə·rō'kī,nās }

glycerophosphate [BIOCHEM] Any salt of glycerophosphoric acid. { ,glis·ə·rō'fäs,fāt }

glycerophosphoric acid [BIOCHEM] $C_3H_5(OH)_2OPO_3H_2$ Either of two pale-yellow, water-soluble, isomeric dibasic acids occurring in nature in combined form as cephalin and lecithin. { ¦glis·ə·rō,fäs'fȯr·ik 'as·əd }

glycine [BIOCHEM] $C_2H_5O_2N$ A white, crystalline amino acid found as a constituent of many proteins. Also known as aminoacetic acid. { 'glī,sēn }

glycocalyx [CELL MOL] The outer component of a cell surface, outside the plasmalemma; usually contains strongly acidic sugars, hence it carries a negative electric charge. { 'glī·kō'kā,liks }

glycocholic acid [BIOCHEM] $C_{26}H_{43}NO_6$ A bile obtained by the conjugation of cholic acid with glycine. { ¦gli·kō¦käl·ik 'as·əd }

glycocyamine [BIOCHEM] $C_3H_7N_3O_2$ A product of interaction of aminocetic acid and arginine, which on transmethylation with methionine is converted to creatine. Also known as guanidine-acetic acid. { 'glī·kō'sī·ə,mēn }

glycogen [BIOCHEM] A nonreducing, white, amorphous polysaccharide found as a reserve carbohydrate stored in muscle and liver cells of all higher animals, as well as in cells of lower animals. { 'glī·kə·jən }

glycogenesis [BIOCHEM] The metabolic formation of glycogen from glucose. { 'glī·kə'jen·ə·səs }

glycogenolysis [BIOCHEM] The metabolic breakdown of glycogen. { 'glī·kə·jə'näl·ə·səs }

glycogenosis [MED] One of several inborn errors in the metabolism of glycogen, classified on the basis of the enzyme deficiency and clinical findings as von Gierke's disease, Pompe's disease, limit dextrinosis, amylopectinosis, McArdle's disease, or Hers' disease. { 'glī·kō·jə'nō·səs }

glycogen storage disease *See* von Gierke's disease. { 'glī·kə·jən 'stȯr·ij di,zēz }

glycogen synthetase [BIOCHEM] An enzyme that catalyzes the synthesis of the amylose chain of glycogen. { 'glī·kə·jən 'sin·thə,tās }

glycolipid [BIOCHEM] Any of a class of complex lipids which contain carbohydrate residues. { ¦glī·kō¦lip·əd }

glycolysis [BIOCHEM] The enzymatic breakdown of glucose or other carbohydrate, with the formation of lactic acid or pyruvic acid and the release of energy in the form of adenosinetriphosphate. { ¦glī'käl·ə·səs }

glycolytic pathway [BIOCHEM] The principal series of phosphorylative reactions involved in pyruvic acid production in phosphorylative fermentations. Also known as Embden-Meyerhof pathway; hexose diphosphate pathway. { ¦glī·kə¦lid·ik 'path,wā }

glyconeogenesis [BIOCHEM] The metabolic process of glycogen formation from noncarbohydrate precursors. { ¦glī·kō¦nē·ō'jen·ə·səs }

glycopeptide *See* glycoprotein. { ,glī·kō'pep,tīd }

glycophyte [BOT] A plant requiring more than 0.5% sodium chloride solution in the substratum. { 'gli·kə,fīt }

glycoprotein [BIOCHEM] Any of a class of conjugated proteins containing both carbohydrate and protein units. Also known as glycopeptide. { ¦glī·kō'prō,tēn }

glycose [BIOCHEM] A simple sugar whose structure is in the form either of an open-chain aldehyde or ketone or of a cyclic hemiacetal. { 'glī,kōs }

glycosidase [BIOCHEM] An enzyme that hydrolyzes a glycoside. { glī'kō·sə,dās }

glycoside [BIOCHEM] A compound that yields on hydrolysis a sugar (glucose, galactose) and an aglycon; many of the glycosides are therapeutically valuable. { 'glī·kə,sīd }

glycosphingolipid [BIOCHEM] A glycoside of ceramide that is the most abundant and structurally diverse type of glycolipid in animals. { ¦glī·kō,sfiŋ·gō'lip·id }

glycosuria [MED] Presence of sugar in the urine. { ,glī·kō'shu̇r·ē·ə }

glycosyl [BIOCHEM] A univalent functional group derived from the cyclic form of glycose by removal of the hemiacetal hydroxyl group. { 'glī·kə,sil }

glycosylation [BIOCHEM] A chemical reaction in which glycosyl groups are added to a protein to produce a glycoprotein. { glī,käs·ə'lā·shən }

glycosyl glyceride [BIOCHEM] A class of glycolipids structurally analogous to phospholipids; they are the major glycolipids of plants and microorganisms but are rare in animals. { glī,kōs·əl 'glis·ə,rīd }

glycosyl phosphatidylinositol [BIOCHEM] Any of a class of glycolipids that serve as membrane anchors for a multitude of proteins in organisms ranging from yeast to protozoa to humans. { glī¦kōs·əl fäs,fäd·ə,dil·i'nos·ə,tȯl }

glycosyltransferase [BIOCHEM] Any of a group of enzymes that participate in the biosynthesis of glycoproteins by transferring one sugar at a time from suitable donors to particular acceptors. { ,glī·kō·sil'tranz·fə,rās }

glycotropic [BIOCHEM] Acting to antagonize the action of insulin. { ¦glī·kō¦träp·ik }

glycuresis [PHYSIO] Excretion of sugar seen normally in urine. { ,glik·yə'rē·səs }

glycuronic acid *See* glucuronic acid. { ¦glik·yə¦rän·ik 'as·əd }

glycuronidase *See* glucuronidase. { ,glik·yə'rän·ə,dās }

glycuronide *See* glucuronide. { gli'kyu̇r·ə,nīd }

glyoxalase [BIOCHEM] An enzyme present in various body tissues which catalyzes the conversion of methylglyoxal into lactic acid. { glī'äk·sə,lās }

glyoxylate cycle [BIOCHEM] A sequence of biochemical reactions related to respiration in germinating fatty seeds by which acetyl coenzyme A is converted to succinic acid and then to hexose. { glī'äk·sə,lāt ,sī·kəl }

glyoxylic acid [BIOCHEM] $CH(OH)_2COOH$ An aldehyde acid found in many plant and animal tissues, especially unripe fruit. { ¦glī,äk¦sil·ik 'as·əd }

glyoxysome [BOT] A specialized type of peroxisome found in plant tissues that is bounded by a single membrane and contains a broad spectrum of enzymes, including those of the glyoxylate cycle and the β-oxidation cycle in addition to catalase and oxidase. { glī'äk·sə,sōm }

Gmelin's test [PATH] A qualitative test for the pigments in bile; test solution is mixed with nitric acid containing nitrous acid; reaction is positive if color appears at the acid-solution junction. { gə'mäl·ənz ,test }

gnat [INV ZOO] The common name for a large variety of biting insects in the order Diptera. { nat }

Gnathiidea [INV ZOO] A suborder of isopod crustaceans characterized by a much reduced second thoracomere, short antennules and antennae, and a straight pleon. { nā'thī·ə·də }

gnathion [VERT ZOO] The most anterior point of the premaxillae on or near the middle line in certain lower mammals. { 'nā·thē,än }

gnathite [INV ZOO] A mouth appendage in arthropods. { 'nā,thīt }

Gnathobdellae [INV ZOO] A suborder of leeches in the order Arhynchobdellae having jaws and a conspicuous posterior sucker; it contains most of the important blood-sucking leeches of humans and other warm-blooded animals. { ¦nā ,thäb'del·ē }

gnathocephalon [INV ZOO] The part of the insect head lying behind the protocephalon; bears the maxillae and mandibles. { ,nā·thō'sef·ə,län }

gnathochilarium [INV ZOO] The lower lip of certain arthropods; thought to be fused maxillae. { ,nā·thō,kī'lar·ē·əm }

gnathopod [INV ZOO] Any of the crustacean paired thoracic appendages modified for manipulation of food but sometimes functioning in copulatory amplexion. { 'nā·thə,päd }

gnathopodite [INV ZOO] A segmental, modified appendage which serves as a jaw in arthropods. { nā'thä·pə,dīt }

gnathos [INV ZOO] A mid-ventral plate on the ninth tergum in lepidopterans. { 'nā,thōs }

gnathostegite [INV ZOO] One of a pair of broad plates formed from the outer maxillipeds of some crustaceans, which function to cover other mouthparts. { nə'thäs·tə,jīt }

Gnathostomata [INV ZOO] A suborder of echinoderms in the order Echinoidea characterized by a rigid, exocyclic test and a lantern or jaw apparatus. [VERT ZOO] A group of the subphylum Vertebrata which possess jaws and usually have paired appendages. { ,nā·thə'stō·məd·ə }

Gnathostomidae [INV ZOO] A family of parasitic nematodes in the superfamily Spiruroidea; sometimes placed in the superfamily Physalopteroidea. { ,nā·thə'stō·məd,ē }

Gnathostomulida [INV ZOO] Microscopic marine worms of uncertain systematic relationship, mainly characterized by cuticular structures in the pharynx and a monociliated skin epithelium. { nə¦thäs·tə'myül·ə·də }

gnathothorax [INV ZOO] The thorax and part of the head bearing feeding organs in arthropods, regarded as a primary region of the body. { ¦nā·thō'thȯr,aks }

Gnetales [BOT] A monogeneric order of the subdivision Gneticae; most species are lianas with opposite, oval, entire-margined leaves. { nə'tā·lēz }

Gnetatae *See* Gnetopsida. { nə'tād,ē }

Gneticae [BOT] A subdivision of the division Pinophyta characterized by vessels in the secondary wood, ovules with two integuments, opposite leaves, and an embryo with two cotyledons. { 'ned·ə,sē }

Gnetophyta [BOT] The equivalent name for Gnetopsida. { nə'täf·əd·ə }

Gnetopsida [BOT] A class of gymnosperms comprising the subdivision Gneticae. { nə'täp·səd·ə }

Gnostidae [INV ZOO] An equivalent name for Ptinidae. { 'näs·tə,dē }

gnotobiology [BIOL] That branch of biology dealing with known living forms; the study of higher organisms in the absence of all demonstrable, viable microorganisms except those known to be present. { ¦nō·dō·bī'äl·ə·jē }

gnotobiote [MICROBIO] **1.** An individual (host) living in intimate association with another known species (microorganism). **2.** The known microorganism living on a host. { ¦nō·dō'bī,ōt }

gnotobiotics [MICROBIO] The science involved with maintaining a microbiologically controlled environment, and with the knowledge necessary to obtain and use biological specimens in this environment. { ,nōd·ə·bī'äd·iks }

gnu [VERT ZOO] Any of several large African antelopes of the genera *Connochaetes* and *Gorgon* having a large oxlike head with horns that characteristically curve downward and outward and then up, with the bases forming a frontal shield in older individuals. { nü }

goat [VERT ZOO] The common name for a number of artiodactyl mammals in the genus *Capra*; closely related to sheep but differing in having a lighter build and hollow, swept-back, sometimes spiral or twisted horns. { gōt }

Gobiesocidae [VERT ZOO] The single family of the order Gobiesociformes. { gō¦bī·ə¦säs·ə,dē }

Gobiesociformes [VERT ZOO] The clingfishes, a monofamilial order of scaleless bony fishes equipped with a thoracic sucking disk which serves for attachment. { gō¦bī·ə,säs·ə'fōr,mēz }

Gobiidae [VERT ZOO] A family of perciform fishes in the suborder Gobioidei characterized by pelvic fins united to form a sucking disk on the breast. { gō'bī·ə,dē }

Gobioidei [VERT ZOO] The gobies, a suborder of morphologically diverse actinopterygian fishes in the order Perciformes; all lack a lateral line. { ,gō·bē'ȯid·ē,ī }

goblet cell [HISTOL] A unicellular, mucus-secreting intraepithelial gland that is distended on the free surface. Also known as chalice cell. [INV ZOO] Any of the unicellular choanocytes of the genus *Monosiga*. { 'gäb·lət ,sel }

goiter [MED] An enlargement of all or part of the thyroid gland; may be accompanied by a hormonal dysfunction. { 'gȯid·ər }

golden algae [BOT] The common name for members of the class Chrysophyceae. { 'gol·dən 'al·jē }

golden-brown algae [BOT] The common name for members of the division Chrysophyta. { 'gol·dən ,braùn 'al·jē }

goldenseal *See* Hydrastis canadensis. { 'gōl·dən ,sēl }

goldfish [VERT ZOO] *Crassius auratus*. An orange cypriniform fish of the family Cyprinidae that can grow to over 18 inches (46 centimeters); closely related to the carps. { 'gōl,fish }

gold sodium thiomalate [PHARM] $C_4H_3AuNa_2O_4S$ A crystalline compound, soluble in water; used as an antirheumatic in medicine. Also known as sodium aurothiomalate. { ¦gōld 'sōd·ē·əm ,thī·ō'ma,lāt }

gold sodium thiosulfate [PHARM] $Na_3Au(S_2O_3)_2\cdot 2H_2O$ A white crystalline compound that is freely soluble in water; used for treatment of rheumatoid arthritis and lupus erythematosus. { ¦gōld 'sōd·ē·əm ,thī·ō'səl,fāt }

Golgi apparatus [CELL MOL] A cellular organelle that is part of the cytoplasmic membrane system; it is composed of regions of stacked cisternae and it functions in secretory processes. { 'gȯl,jē ,ap·ə,rad·əs }

Golgi cell [NEUROSCI] **1.** A nerve cell with long axons. **2.** A nerve cell with short axons that branch repeatedly and terminate near the cell body. { 'gȯl·jē ,sel }

Golgi-Mazzoni's corpuscle [ANAT] A small sensory lamellar corpuscle located in the parietal pleura. { 'gȯl·jē mät'sō·nēz ,kȯr·pə·səl }

Golgi stack [CELL MOL] The central structure of the Golgi apparatus consisting of flattened membrane-bounded cisternae. Formerly known as dictyosome. { 'gōl·jē ,stak }

Golgi tendon organ [PHYSIO] Any of the kinesthetic receptors situated near the junction of muscle fibers and a tendon which act as muscle-tension recorders. { 'gȯl·jē 'ten·dən ,ȯr·gən }

Gomphidae [INV ZOO] A family of dragonflies belonging to the Anisoptera. { 'gäm·fə,dē }

gomphosis [ANAT] An immovable articulation, as that formed by the insertion of teeth into the bony sockets. { gäm'fō·səs }

gonad [ANAT] A primary sex gland; an ovary or a testis. { 'gō,nad }

gonadal agenesis [MED] Failure of the gonad to develop, or retrogression of the gonad at very early stages. Also known as gonadal dysgenesis. { gō'nad·əl ā'jēn·ə·səs }

gonadal dysgenesis *See* gonadal agenesis. { gō 'nad·əl dis'jen·ə·səs }

gonadectomy [MED] Surgical removal of a gonad. { ,gō·na'dek·tə·mē }

gonadotropic hormone [BIOCHEM] Either of two adenohypophyseal hormones, FSH (follicle-stimulating hormone) or ICSH (interstitial-cell-stimulating hormone), that act to stimulate the gonads. { gō¦nad·ə¦träp·ik 'hȯr,mōn }

gonadotropin [BIOCHEM] A substance that acts to stimulate the gonads. { gō,nad·ə'trō·pən }

gonapophysis [INV ZOO] A paired, modified appendage of the anal region in insects that functions in copulation, oviposition, or stinging. { gän·ə'päf·ə·səs }

gonia [HISTOL] Primordial sex cells, such as oogonia and spermatogonia. { 'gō·nē·ə }

Goniadidae [INV ZOO] A family of marine polychaete annelids belonging to the Errantia. { ,gō·nē'ad·ə,dē }

gonidium [BIOL] An asexual reproductive cell

or group of cells arising in a special organ on or in a gametophyte. { gō'nid·ē·əm }

gonimoblast [BOT] A filament arising from the fertilized carpogonium of most red algae. { 'gän·ə·mō,blast }

gonioscope [MED] A special optical instrument for studying in detail the angle of the anterior chamber of the eye. { 'gō·nē·ə,skōp }

gonitis [MED] Inflammation of the knee joint. { gō'nīd·əs }

gonococcal arthritis [MED] A blood-borne joint infection by *Neisseria gonorrhoeae* occurring as a manifestation of gonorrhea. { ¦gän·ə¦käk·əl är'thrīd·əs }

gonococcal epididymitis [MED] Inflammation of the epididymitis due to infection by *Neisseria gonorrhoeae*; a secondary manifestation of gonorrhea. { ¦gän·ə¦käk·əl ,ep·ə,did·ə'mīd·əs }

gonococcus *See* Neisseria gonorrhoeae. { ,gän·ō'käk·əs }

Gonodactylidae [INV ZOO] A family of mantis shrimp in the order Stomatopoda. { ,gän·ō·dak'til·ə,dē }

gonodendrum [INV ZOO] A branched structure which bears clusters of gonophores. { ,gän·ə'den·drəm }

gonomery [EMBRYO] In some insect embryos, grouping of maternal and paternal chromosomes separately during the first couple of mitotic divisions after fertilization. { gō'näm·ə·rē }

gonopalpon [INV ZOO] Tentaclelike, sensitive structures associated with cnidarian gonophores. { ,gän·ə'pal·pən }

gonophore [BOT] An elongation of the receptacle extending between the stamens and corolla. [INV ZOO] Reproductive zooid of a hydroid colony. { 'gän·ə,fȯr }

gonopodium [VERT ZOO] Anal fin modified as a copulatory organ in certain fishes. { ,gän·ə'pōd·ē·əm }

Gonorhynchiformes [VERT ZOO] A small order of soft-rayed teleost fishes having fusiform or moderately compressed bodies, single short dorsal and anal fins, a forked caudal fin, and weak toothless jaws. { ¦gän·ə,riŋ·kə'fȯr,mēz }

gonorrhea [MED] A bacterial infection of humans caused by the gonococcus (*Neisseria gonorrhoeae*) which invades the mucous membrane of the urogenital tract. { ,gän·ə'rē·ə }

gonorrheal urethritis [MED] Inflammation of the urethra, particularly in males, as the result of gonorrhea. { ,gän·ə'rē·əl ,yu̇r·ə'thrīd·əs }

gonorrheal vulvovaginitis [MED] Inflammation of the vulva and vagina as the result of gonorrhea. { ,gän·ə'rē·əl ¦vəl·vō,va·jə'nīd·əs }

gonosome [INV ZOO] Aggregate of gonophores in a hydroid colony. { 'gän·ə,sōm }

gonosomite [INV ZOO] The ninth segment of the abdomen of the male insect. Also known as genital segment. { ,gän·ə'sō,mīt }

gonostome [INV ZOO] The part of the genital duct of a coelomate invertebrate known as the coelomic funnel. Also known as coelomostome. { 'gän·ə,stōm }

gonostyle [INV ZOO] Gonapophysis of dipteran insects. { 'gän·ə,stīl }

gonotocont *See* auxocyte. { gə'näd·ə,känt }

gonotome [EMBRYO] The part of an embryonic somite involved in gonad formation. { 'gän·ə,tōm }

gonozooid [INV ZOO] A zooid of bryozoans and tunicates which produces gametes. { ¦gän·ə'zō,ȯid }

gonyautoxin [BIOCHEM] One of a group of saxitoxin-related compounds that are produced by the dinoflagellates *Gonyaulax catenella* and *G. tamarensis*. { ¦gō·nē·ō¦täk·sən }

gonys [VERT ZOO] A ridge along the mid-ventral line of the lower mandible of certain birds. { 'gō·nəs }

Goodpasture's syndrome [MED] A complex of symptoms associated with acute glomerulonephritis and pulmonary hemorrhage. { 'gu̇d,pas·chərz ,sin,drōm }

goose [VERT ZOO] The common name for a number of waterfowl in the subfamily Anatinae; they are intermediate in size and features between ducks and swans. { güs }

gooseberry [BOT] The common name for about six species of thorny, spreading bushes of the genus *Ribes* in the order Rosales, producing small, acidic, edible fruit. { 'güs,ber·ē }

gooseneck barnacle [INV ZOO] Any stalked barnacle, especially of the genus *Lepas*. { ¦güs,nek 'bär·ni·kəl }

gopher [VERT ZOO] The common name for North American rodents composing the family Geomyidae. Also known as pocket gopher. { 'gō·fər }

Gordiidae [INV ZOO] A monogeneric family of worms in the order Gordioidea distinguished by a smooth cuticle. { gȯr'dī·ə,dē }

Gordioidea [INV ZOO] An order of worms belonging to the Nematomorpha in which there is one ventral epidermal cord, a body cavity filled with mesenchymal tissue, and paired gonads. { ,gȯr·dē'ȯid·ē·ə }

gordioid larva [INV ZOO] The developmental stage of nematomorphs, free-living for a short time. { 'gȯr·dē,ȯid 'lär·və }

Gorgonacea [INV ZOO] The horny corals, an order of the cnidarian subclass Alcyonaria; colonies are fanlike or featherlike with branches spread radially or oppositely in one plane. { ,gȯr·gə'nās·ē·ə }

gorgonin [BIOCHEM] The protein, frequently containing iodine and bromine, composing the horny skeleton of members of the Gorgonacea; contains iodine and bromine. { 'gȯr·gə·nən }

Gorgonocephalidae [INV ZOO] A family of ophiuroid echinoderms in the order Phrynophiurida in which the individuals often have branched arms. { ,gȯr·gə,nō·sə'fal·ə,dē }

gorilla [VERT ZOO] *Gorilla gorilla*. An anthropoid ape, the largest living primate; the two African subspecies are the lowland gorilla and the mountain gorilla. { gə'ril·ə }

gossypose *See* raffinose. { 'gäs·ə,pōs }

gout [MED] A condition of purine metabolism

resulting in increased blood levels of uric acid with ultimate deposition as urates in soft tissues around joints. { gau̇t }

G protein *See* GTP-binding protein. { 'jē ˌprōˌtēn }

G-protein-linked receptor [CELL MOL] A cell surface receptor that consists of a polypeptide chain threaded across the membrane seven times and that, when activated by the binding of a ligand, in turn activates a cytosolic G-protein molecule, which then initiates a cascade of reactions effecting the intracellular response to the extracellular signal (the ligand). { ¦jē ˌprōˌtēn ˌliŋkt ri'sep·tər }

gr *See* grain.

Graafian follicle [HISTOL] The mature mammalian ovum with its surrounding epithelial cells. { 'gräf·ē·ən 'fäl·ə·kəl }

Gracilariidae [INV ZOO] A family of small moths in the superfamily Tineoidea; both pairs of wings are lanceolate and widely fringed. { ˌgra·sil·ə'rī·əˌdē }

gracilis [ANAT] A long slender muscle on the medial aspect of the thigh. { 'gras·ə·ləs }

grade [EVOL] A stage of evolution in which a similar level of organization is reached by one or more species in the development of a structure, physiological process, or behavioral character. { grād }

graded topocline [ECOL] A topocline having a wide range, or ranging into different kinds of environment, thus subjecting its members to differential selection so that divergence between local races may become sufficient to warrant creation of varietal, or even specific, names. { ¦grād·əd 'täp·əˌklīn }

gradualism [EVOL] A model of evolution in which change is slow, steady, and on the whole ameliorative, resulting in a gradual and continuous increase in biological diversity. Also known as phyletic gradualism. { 'graj·ə·wəˌliz·əm }

graft [BIOL] **1.** To unite to form a graft. **2.** A piece of tissue transplanted from one individual to another or to a different place on the same individual. **3.** An individual resulting from the grafting of parts. [BOT] To unite a scion to an understock in such manner that the two grow together and continue development as a single plant without change in scion or stock. { graft }

Graham-Cole test *See* cholecystography. { ¦grā·əm 'kōl ˌtest }

Grahamella [MICROBIO] A genus of the family Bartonellaceae; intracellular parasites in red blood cells of rodents and other mammals. { ˌgrā·ə'mel·ə }

grain [BOT] **1.** A rounded, granular prominence on the back of a sepal. **2.** *See* cereal. **3.** *See* drupelet. { grān }

grain sorghum [BOT] *Sorghum bicolor.* A grass plant cultivated for its grain and to a lesser extent for forage. Also known as nonsaccharine sorghum. { ¦grān 'sȯr·gəm }

gramagrass [BOT] Any grass of the genus *Bouteloua*; pasture grass. { 'gram·əˌgras }

gramicidin [MICROBIO] A polypeptide antibacterial antibiotic produced by *Bacillus brevis*; active locally against gram-positive bacteria. { ˌgram·ə'sīd·ən }

Graminales [BOT] The equivalent name for Cyperales. { ˌgram·ə'nā·lēz }

Gramineae [BOT] The grasses, a family of monocotyledonous plants in the order Cyperales characterized by distichously arranged flowers on the axis of the spikelet. { grə'min·ēˌē }

graminicolous [ECOL] Living upon grass. { ¦gram·ə¦nik·ə·ləs }

graminivorous [ZOO] Feeding on grasses. { ¦gram·ə¦niv·ə·rəs }

graminoid [BOT] Of or resembling the grasses. { 'gram·əˌnȯid }

gram-negative [MICROBIO] Of bacteria, decolorizing and staining with the counterstain when treated with Gram's stain. { 'gram¦neg·əd·iv }

gram-negative diplococci [MICROBIO] Three bacteriologic genera composing the family Neisseriaceae: *Gemella, Veillonella,* and *Neisseria.* { 'gram¦neg·əd·iv ˌdip·lə'käk·sē }

gram-positive [MICROBIO] Of bacteria, holding the color of the primary stain when treated with Gram's stain. { 'gram ¦päs·əd·iv }

Gram's stain [MICROBIO] A differential bacteriological stain; a fixed smear is stained with a slightly alkaline solution of basic dye, treated with a solution of iodine in potassium iodide, and then with a neutral decolorizing agent, and usually counterstained; bacteria stain either blue (gram-positive) or red (gram-negative). { 'gramz ˌstān }

gram-variable [MICROBIO] Pertaining to staining inconsistently with Gram's stain. { 'gram ¦ver·ē·ə·bəl }

grana [CELL MOL] A multilayered membrane unit formed by stacks of the lobes or branches of a chloroplast thylakoid. { 'grän·ə }

grand mal [MED] A complete epileptic seizure involving sudden loss of consciousness and tonic convulsion of the skeletal musculature followed by clonic muscular spasms. { 'gran ¦mäl }

granellare [INV ZOO] In xenophyophores, that portion of the body consisting of the multinucleate plasmodium and its enclosing, branching organic tube. { 'gran·əlˌär }

granin [BIOCHEM] Any of a group of proteins that localize to secretory vesicles, are secreted by a regulated pathway, are posttranslationally glycosylated, and are typically responsive to hormones. { 'gran·ən }

granite moss [BOT] The common name for a group of the class Bryatae represented by two Arctic genera and distinguished by longitudinal splitting of the mature capsule into four valves. { 'gran·ət ¦mȯs }

Grantiidae [INV ZOO] A family of calcareous sponges in the order Sycettida. { gran'tī·əˌdē }

granular component [CELL MOL] The component of the nucleolus that contains the cleaved preribosomal particles. { ¦gran·yə·lər kəm¦pō·nənt }

granular gland [PHYSIO] A gland that produces and secretes a granular material. { 'gran·yə·lər 'gland }

granular leukocyte *See* granulocyte. { 'gran·yə·lər 'lü·kə,sīt }

granulation [MED] **1.** Tiny red granules made of capillary loops in the base of an ulcer. **2.** Process of granular tissue formation around a focus of inflammation. { ,gran·yə'lā·shən }

granuloblastosis [VET MED] An avian leukosis characterized by the presence of excessive numbers of immature granulocytes in the blood of affected birds. { ¦gran·yə·lō,bla'stō·səs }

granulocyte [HISTOL] A leukocyte containing granules in the cytoplasm. Also known as granular leukocyte; polymorph; polymorphonuclear leukocyte. { 'gran·yə·lō,sīt }

granulocytic leukemia [MED] A blood disease involving neoplastic transformation of granulocytes, principally the neutrophilic series. Also known as myelogenous leukemia; myeloid leukemia. { ¦gran·yə·lō¦sid·ik lü'kē·mē·ə }

granulocytopenia [MED] A deficiency of granulocytes in circulating blood. Also known as granulopenia. { ¦gran·yə·lō,sīd·ə'pēn·yə }

granulocytosis [MED] An increase in the number of granulocytes in the circulation. { ,gran·yə·lō,sī'tō·səs }

granuloma [MED] A discrete nodular lesion of inflammatory tissue in which granulation is significant. { ,gran·yə'lō·mə }

granuloma inguinale [MED] An infectious, chronic, destructive granulomatous lesion of humans most frequently localized in the genital and inguinal regions; caused by Donovan bodies (*Donovania granulomatis*). { ,gran·yə'lō·mə ,iŋ·gwə'nä·lē }

granuloma pyogenicum [MED] A hemangioma with superimposed inflammation on the skin or other epithelial surfaces. { ,gran·yə'lō·mə ,pī·ō'jen·ə·kəm }

granulomatosis [MED] Any disease characterized by multiple granulomas. { ,gran·yə,lō·mə 'tō·səs }

granulopenia *See* granulocytopenia. { ,gran·yə·lō'pēn·yə }

Granuloreticulosia [INV ZOO] A subclass of the protozoan class Rhizopodea characterized by reticulopodia which often fuse into networks. { ¦gran·yə·lō·re,tik·yə'lō·shə }

granulosis [INV ZOO] A virus disease of lepidopteran larvae characterized by the accumulation of small granular inclusion bodies (capsules) in the infected cells. { ,gran·yə'lō·səs }

grape [BOT] The common name for plants of the genus *Vitis* characterized by climbing stems with cylindrical-tapering tendrils and polygamodioecious flowers; grown for the edible, pulpy berries. { grāp }

grapefruit [BOT] *Citrus paradisi.* An evergreen tree with a well-rounded top cultivated for its edible fruit, a large, globose citrus fruit characterized by a yellow rind and white, pink, or red pulp. { 'grāp,früt }

grape sugar *See* dextrose. { 'grāp ,shu̇g·ər }

Graphidaceae [BOT] A family of mosses formerly grouped with lichenized Hysteriales but now included in the order Lecanorales; individuals have true paraphyses. { ,graf·ə'dās·ē,ē }

Graphiolaceae [MYCOL] A family of parasitic fungi in the order Ustilaginales in which teleutospores are produced in a cuplike fruiting body. { ,graf·ē·ō'lās·ē,ē }

Grapsidae [INV ZOO] The square-backed crabs, a family of decapod crustaceans in the section Brachyura. { 'grap·sə,dē }

grass [BOT] The common name for all members of the family Gramineae; moncotyledonous plants having leaves that consist of a sheath which fits around the stem like a split tube, and a long, narrow blade. { gras }

grasserie [INV ZOO] A polyhedrosis disease of silkworms characterized by spotty yellowing of the skin and internal liquefaction. Also known as jaundice. { 'gras·ə·rē }

grasshopper [INV ZOO] The common name for a number of plant-eating orthopteran insects composing the subfamily Saltatoria; individuals have hindlegs adapted for jumping, and mouthparts adapted for biting and chewing. { 'gras ,häp·ər }

grassland [ECOL] Any area of herbaceous terrestrial vegetation dominated by grasses and graminoid species. { 'gras,land }

grass sickness [VET MED] A disease of horses occurring mainly in Scotland; thought to be caused by a virus similar to the one that causes poliomyelitis in humans. { 'gras ,sik·nəs }

grass tetany [VET MED] A magnesium-deficiency disease of cows. Also known as hypomagnesemia. { 'gras ,tet·ən·ē }

Grave's disease *See* hyperthyroidism. { 'grāvz di,zēz }

graviceptor *See* gravireceptor. { 'grav·i,sep·tər }

gravid [ZOO] **1.** Of the uterus, containing a fetus. **2.** Pertaining to female animals when carrying young or eggs. { 'grav·əd }

Gravigrada [VERT ZOO] The sloths, a group of herbivorous xenarthran mammals in the order Edentata; members are completely hairy and have five upper and four lower prismatic teeth without enamel. { grə'vig·rə·də }

graviperception [BOT] The magnitude of acceleration required to induce a directed growth response in plants carried aboard spacecraft. { 'grav·ə·pər,sep·shən }

graviportal [BIOL] Weight-bearing. { ¦grav·ə ¦pȯrd·əl }

gravireceptor [NEUROSCI] One of the highly specialized nerve endings and receptor organs located in the skeletal muscles, tendons, joints, and inner ear that furnishes information to the brain with respect to body position, equilibrium, and the direction of gravitational forces. Also known as graviceptor. { 'grav·ə·ri,sep·tər }

gravitropism *See* geotropism. { grə'vi·trə,piz·əm }

Grawitz's tumor *See* renal-cell carcinoma. { 'grä·vits·əz ,tü·mər }

gray matter [NEUROSCI] The part of the central nervous system composed of nerve cell bodies, their dendrites, and the proximal and terminal unmyelinated portions of axons. { 'grā ˌmad·ər }

graze [VERT ZOO] To feed by browsing on, cropping, and eating grass. { grāz }

grazing food web [ECOL] A trophic web that is based on the consumption of the tissues of living organisms. { 'graz·iŋ ˌfüd ˌweb }

greasewood [BOT] Any plant of the genus *Sarcobatus*, especially *S. vermiculatus*, which is a low shrub that grows in alkali soils of the western United States. { 'grēsˌwu̇d }

greater omentum [ANAT] A fold of peritoneum that is attached to the greater curvature of the stomach and hangs down over the intestine and fuses with the mesocolon. { 'grād·ər ō'men·təm }

great galago *See* thick-tailed bushbaby. { ¦grāt gə'lā·gō }

grebe [VERT ZOO] The common name for members of the family Podicipedidae; these birds have legs set far posteriorly, compressed bladelike tarsi, individually broadened and lobed toes, and a rudimentary tail. { grēb }

Greeffiellidae [INV ZOO] A family of free-living nematodes in the superfamily Desmoscolecoidea. { grē·fē'el·əˌdē }

Greeffielloidea [INV ZOO] A superfamily of primarily marine, free-living nematodes in the order Desmoscolecida, distinguished by a prominent nonencrusted annulation that bears a ring of elongate spines or short scales, and large subdorsal and subventral tubular setae along the body. { grē·fə'lȯid·ē·ə }

green algae [BOT] The common name for members of the plant division Chlorophyta. { 'grēn ¦al·jē }

green fluorescent protein [CELL MOL] A protein that is produced by the bioluminescent jellyfish *Aequorea victoria*; used to trace the synthesis, location, and movement of proteins of interest in cell biology research. Abbreviated GFP. { ¦grēn flə¦res·ənt 'prōˌtēn }

green gland *See* antennal gland. { 'grēn ¦gland }

greenhouse [BOT] Glass-enclosed, climate-controlled structure in which young or out-of-season plants are cultivated and protected. { 'grēnˌhau̇s }

green mold [MYCOL] Any fungus, especially *Penicillium* and *Aspergillus* species, that is green or has green spores. { ¦grēn ¦mōld }

green muscardine [INV ZOO] A disease of the European corn borer, the wheat cockchafer, and other insects caused by the fungus *Metarhizium anisopliae*. { ˌgrēn 'məs·kərˌdēn }

greenstick fracture [MED] An incomplete fracture of a long bone, seen in children; the bone is bent but splintered only on the convex side. { 'grēnˌstik ˌfrak·chər }

green sulfur bacteria [MICROBIO] A physiologic group of green photosynthetic bacteria of the Chloraceae that are capable of using hydrogen sulfide (H_2S) and other inorganic electron donors. { 'grēn ¦səl·fər bak'tir·ē·ə }

Gregarinia [INV ZOO] A subclass of the protozoan class Telosporea occurring principally as extracellular parasites of invertebrates. { 'greg·ə'rin·ē·ə }

gressorial [VERT ZOO] Adapted for walking, as certain birds' feet. { gre'sȯr·ē·əl }

grid *See* Potter-Bucky grid. { grid }

Grifola frondosa [MYCOL] A type of mushroom found in parts of the eastern United States, Europe, and Asia, growing in masses at the base of stumps and on roots; has an anticancer effect in patients with lung and stomach cancers or leukemia. Also known as maitake mushroom. { grəˌfō·lə frän'dōs·ə }

Grimmiales [BOT] An order of mosses commonly growing upon rock in dense tufts or cushions and having hygroscopic, costate, usually lanceolate leaves arranged in many rows on the stem. { ˌgrim·ē'ā·lēz }

grindle *See* bowfin. { 'grind·əl }

grisein [MICROBIO] $C_{40}H_{61}O_{20}N_{10}SFe$ A red, crystalline, water-soluble antibiotic produced by strains of *Streptomyces griseus*. { grə'zēn }

griseofulvin [MICROBIO] $C_{17}H_{17}O_6Cl$ A colorless, crystalline antifungal antibiotic produced by several species of *Penicillium*. Also known as curling factor. { ˌgriz·ē·ō'fu̇l·vən }

griseolutein [MICROBIO] Either of two fractions, A or B, of broad-spectrum antibiotics produced by *Streptomyces griseoluteus*; more active against gram-positive than gram-negative microorganisms. { ˌgriz·ē·ō'lüd·ē·ən }

griseomycin [MICROBIO] A white, crystalline antibiotic produced by an actinomycete resembling *Streptomyces griseolus*. { ˌgriz·ē·ō'mīs·ən }

grizzly bear [VERT ZOO] The common name for a number of species of large carnivorous mammals in the genus *Ursus*, family Ursidae. { 'griz·lē ˌber }

groin [ANAT] Depression between the abdomen and the thigh. { grȯin }

Gromiida [INV ZOO] An order of protozoans in the subclass Filosia; the test, which is chitinous in some species and thin and somewhat flexible in others, is reinforced with sand grains or siliceous particles. { grə'mī·ə·də }

groove [BIOCHEM] Any of a group of depressions in the double helix of deoxyribonucleic acid that are believed to be sites occupied by nuclear proteins. { grüv }

gross anatomy [ANAT] Anatomy that deals with the naked-eye appearance of tissues. { 'grōs ə'nad·ə·mē }

gross primary production [ECOL] The incorporation of organic matter or biocontent by a grassland community over a given period of time. { ¦grōs 'prīˌmer·ē prə'dək·shən }

gross production rate [ECOL] The speed of assimilation of organisms belonging to a specific trophic level. { ¦grōs prə'dək·shən ˌrāt }

ground cover [BOT] Prostrate or low plants that cover the ground instead of grass. { 'grau̇nd ˌkəv·ər }

ground meristem [BOT] Partially differentiated meristematic tissue derived from the apical meristem that gives rise to ground tissue. { ¦grau̇nd 'mer·əˌstem }

groundplasm [CELL MOL] A polyphasic system in which the resolvable elements of the cytoplasm are suspended, including the larger organelles, enzymes of intermediate cell metabolism, contractile protein molecules, and the main cellular pool of soluble precursors. { 'graůnd ˌplaz·əm }

ground substance *See* matrix. { 'graůnd ˌsəb·stəns }

ground tissue [BOT] In leaves and young roots and stems, any tissue other than the epidermis and vascular tissues. { 'graůnd ˌtish·ü }

groundwater level *See* water table. { 'graůnd ˌwȯd·ər ˌlev·əl }

groundwater surface *See* water table. { 'graůnd ˌwȯd·ər ¦sər·fəs }

groundwater table *See* water table. { 'graůnd ˌwȯd·ər ¦tā·bəl }

group selection [EVOL] Selection in which changes in gene frequency are brought about by the differential extinction and proliferation of the local population. { ¦grüp si¦lek·shən }

grouse [VERT ZOO] Any of a number of game birds in the family Tetraonidae having a plump body and strong, feathered legs. { graůs }

growth [MED] Any abnormal, localized increase in cells, such as a tumor. [PHYSIO] Increase in the quantity of metabolically active protoplasm, accompanied by an increase in cell number or cell size, or both. { grōth }

growth cone [NEUROSCI] A specialized structure at the end of a growing nerve fiber that guides the fiber to its destination during the development of the nervous system by means of interaction with signaling molecules in its surroundings and its own motile mechanism. { 'grōth ˌkōn }

growth curve [MICROBIO] A graphic representation of the growth of a bacterial population in which the log of the number of bacteria or the actual number of bacteria is plotted against time. { 'grōth ˌkərv }

growth factor [PHYSIO] Any factor, genetic or extrinsic, which affects growth. { 'grōth ˌfak·tər }

growth form [ECOL] The habit of a plant determined by its appearance of branching and periodicity. { 'grōth ˌfȯrm }

growth hormone [BIOCHEM] **1.** A polypeptide hormone secreted by the anterior pituitary which promotes an increase in body size. Abbreviated GH. **2.** Any hormone that regulates growth in plants and animals. { 'grōth ¦hȯrˌmōn }

growth rate [MICROBIO] Increase in the number of bacteria in a population per unit time. { 'grōth ˌrāt }

growth regulator [BIOCHEM] A synthetic substance that produces the effect of a naturally occurring hormone in stimulating plant growth; an example is dichlorophenoxyacetic acid. { 'grōth ¦reg·yəˌlād·ər }

groyne *See* groin. { grȯin }

grub [INV ZOO] The larva of certain insects; commonly used in reference to beetles. { grəb }

Gruidae [VERT ZOO] The cranes, a family of large, tall, cosmopolitan wading birds in the order Gruiformes. { 'grü·əˌdē }

Gruiformes [VERT ZOO] A heterogeneous order of generally cosmopolitan birds including the rails, coots, limpkins, button quails, sun grebes, and cranes. { 'grü·ə'fȯrˌmēz }

Gryllidae [INV ZOO] The true crickets, a family of orthopteran insects in which individuals are dark-colored and chunky with long antennae and long, cylindrical ovipositors. { 'gril·əˌdē }

Grylloblattidae [INV ZOO] A monogeneric family of crickets in the order Orthoptera; members are small, slender, wingless insects with hindlegs not adapted for jumping. { ˌgril·ō'blad·əˌdē }

Gryllotalpidae [INV ZOO] A family of North American insects in the order Orthoptera which live in sand or mud; they eat the roots of seedlings growing in moist, light soils. { 'gril·ō'tal·pəˌdē }

Gs *See* stimulatory G protein.

GTP *See* guanosine 5′-triphosphate.

GTPase [CELL MOL] One of a family of monomeric GTP-binding proteins. { 'jēˌtē¦pās }

GTP-binding protein [CELL MOL] One of a large family of heterotrimeric or monomeric proteins that bind GTP (guanosine 5′-triphosphate) as intermediaries in intracellular signaling pathways. Also known as G protein. { ¦jēˌtē¦pē 'bīn·diŋ ˌprōˌtēn }

guaiazulene [PHARM] $C_{15}H_{18}$ A blue oil with a boiling point of 165–170°C; used as an anti-inflammatory drug. { ˌgwī'az·əˌlēn }

guanidine [BIOCHEM] CH_5N_2 Aminomethanamidine, a product of protein metabolism found in urine. { 'gwän·əˌdēn }

guanidine-acetic acid *See* glycocyamine. { 'gwän·əˌdēn ə¦sēd·ik 'as·əd }

guanine [BIOCHEM] $C_5H_5ON_5$ A purine base; occurs naturally as a fundamental component of nucleic acids. { 'gwänˌēn }

guanophore *See* iridocyte. { 'gwän·əˌfȯr }

guanosine [BIOCHEM] $C_{10}H_{13}O_5N_5$ Guanine riboside, a nucleoside composed of guanine and ribose. Also known as vernine. { 'gwän·əˌsēn }

guanosine 5′-triphosphate [CELL MOL] A nucleoside triphosphate that is instrumental in many cellular processes, including microtubule assembly, protein synthesis, and cell signaling, due to the energy it releases upon removal of its terminal phosphate group (producing guanosine 5′-diphosphate). Abbreviated GTP. { ¦gwän·əˌsēn ¦fīvˌprīm trī'fäsˌfāt }

guanosine tetraphosphate [BIOCHEM] A nucleotide which participates in the regulation of gene transcription in bacteria by turning off the synthesis of ribosomal ribonucleic acid. { 'gwän·əˌsēn te·trə'fäsˌfāt }

guanylic acid [BIOCHEM] A nucleotide composed of guanine, a pentose sugar, and phosphoric acid and formed during the hydrolysis of nucleic acid. Abbreviated GMP. Also known as guanosine monophosphate; guanosine phosphoric acid. { gwə'nil·ik 'as·əd }

guard cell [BOT] Either of two specialized cells

surrounding each stoma in the epidermis of plants; functions in regulating stoma size. { 'gärd ˌsel }

Guarnieri body [PATH] Eosinophilic cytoplasmic inclusion bodies found in the epidermal cells of patients with smallpox or chickenpox. { gwär'nyer·ē ˌbäd·ē }

guava [BOT] *Psidium guajava.* A shrub or low tree of tropical America belonging to the family Myrtaceae; produces an edible, aromatic, sweet, juicy berry. { 'gwäv·ə }

guayule [BOT] *Parthenium argentatum.* A subshrub of the family Compositae that is native to Mexico and the southwestern United States; it has been cultivated as a source of rubber. { wī'yü·lē }

gubernaculum [ANAT] A guiding structure, as the fibrous cord extending from the fetal testes to the scrotal swellings. [INV ZOO] **1.** A posterior flagellum of certain protozoans. **2.** A sclerotized structure associated with the copulatory spicules of certain nematodes. { ˌgü·bər'nak·yə·ləm }

Guest unit [BIOL] A unit for the standardization of plasmin. { 'gest ˌyü·nət }

guide ribonucleic acid [CELL MOL] A ribonucleic acid sequence that provides a template for the alignment of splice junctions. { ¦gīd ˌrī·bō·nü¦klē·ik 'as·əd }

guild [ECOL] A group of species that utilize the same kinds of resources, such as food, nesting sites, or places to live, in a similar manner. { gild }

Guillain-Barré syndrome *See* Landry-Guillain-Barré syndrome. { ¦gē·yan bə'rā ˌsinˌdrōm }

guinea fowl [VERT ZOO] The common name for plump African game birds composing the family Numididae; individuals have few feathers on the head and neck, but may have a crest of feathers and various fleshy appendages. { 'gin·ē ˌfau̇l }

guinea pig [VERT ZOO] The common name for several species of wild and domestic hystricomorph rodents in the genus *Cavia,* family Caviidae; individuals are stocky, short-eared, short-legged, and nearly tailless. { 'gin·ē ˌpig }

guinea worm [INV ZOO] *Dracunculus medinensis.* A parasitic nematode that infects the subcutaneous tissues of humans and other mammals. { 'gin·ē ˌwərm }

guitarfish [VERT ZOO] The common name for fishes composing the family Rhinobatidae. { gə'tärˌfish }

gular [ANAT] Of, pertaining to, or situated in the gula or upper throat. [VERT ZOO] A horny shield on the plastron of turtles. { 'gyü·lər }

gulfweed [BOT] Brown algae of the genus *Sargassum.* { 'gəlfˌwēd }

gull [VERT ZOO] The common name for a number of long-winged swimming birds in the family Laridae having a stout build, a thick, somewhat hooked bill, a short tail, and webbed feet. { gəl }

gullet [ANAT] *See* esophagus. [INV ZOO] A canal between the cytostome and reservoir that functions in food intake in ciliates. { 'gəl·ət }

gumbo *See* okra. { 'gəm·bō }

gumme [PATH] A mass of rubberlike necrotic tissue found in any of various organs and tissues in tertiary syphilis. { 'gəm·ə }

gum vein [BOT] Local accumulation of resin occurring as a wide streak in certain hardwoods. { 'gəm ˌvān }

Gunneraceae [BOT] A family of dicotyledonous terrestrial herbs in the order Haloragales, distinguished by two to four styles, a unilocular bitegmic ovule, large inflorescences with no petals, and drupaceous fruit. { ˌgən·ə'rās·ēˌē }

gustation [PHYSIO] The act or the sensation of tasting. { gə'stā·shən }

gustatoreceptor [ANAT] A taste bud. [PHYSIO] Any sense organ that functions as a receptor for the sense of taste. { ¦gəs·tə·tō ri'sep·tər }

gut [ANAT] The intestine. [EMBRYO] The embryonic digestive tube. { gət }

Guthrie test [PATH] A screening test for the detection of phenylketonuria in which the inhibition of growth of a strain of *Bacillus subtilis* by a phenylalanine analog is reversed by L-phenylalanine, as found in elevated concentration in the plasma of patients with phenylketonuria. { 'gəth·rē ˌtest }

guttation [BOT] The discharge of water from a plant surface, especially from a hydathode. { ˌgə'tā·shən }

Guttiferae [BOT] A family of dicotyledonous plants in the order Theales characterized by extipulate leaves and conspicuous secretory canals or cavities in all organs. { gə'tif·əˌrē }

Guttulinaceae [MICROBIO] A family of microorganisms in the Acrasiales characterized by simple fruiting structures with only slightly differentiated component cells containing little or no cellulose. { ˌgu̇d·əl·ə'nās·ēˌē }

Gymnarchidae [VERT ZOO] A monotypic family of electrogenic fishes in the order Osteoglossiformes in which individuals lack pelvic, anal, and caudal fins. { ˌjim'närk·əˌdē }

Gymnoascaceae [MYCOL] A family of ascomycetous fungi in the order Eurotiales including dermatophytes and forms that grow on dung, soil, and feathers. { ¦jim·nō·ə'skās·ēˌē }

Gymnoblastea [INV ZOO] A suborder of cnidarians in the order Hydroida comprising hydroids without protective cups around the hydranths and gonozooids. { ˌjim·nə'blas·tē·ə }

gymnoblastic [INV ZOO] Having naked medusa buds, referring to anthomedusan hydroids. { ¦jim·nə¦bla·stik }

gymnocarpous [BOT] Having the hymenium uncovered on the surface of the thallus or fruiting body of lichens or fungi. { ¦jim·nə¦kär·pəs }

gymnocephalous cercaria [INV ZOO] A type of digenetic trematode larva. { ¦jim·nə¦sef·ə·ləs sər'kar·ē·ə }

Gymnocerata [INV ZOO] An equivalent name for Hydrocorisae. { ˌjim·nō'ser·əd·ə }

Gymnodinia [INV ZOO] A suborder of flagellate protozoans in the order Dinoflagellida that are naked or have thin pellicles. { ˌjim·nə'din·ē·ə }

gymnogynous [BOT] Having a naked ovary. { ˌjimˈnäj·ə·nəs }

Gymnolaemata [INV ZOO] A class of ectoproct bryozoans possessing lophophores which are circular in basal outline and zooecia which are short, wide, and vaselike or boxlike. { ˌjim·nəˈlē·məd·ə }

Gymnonoti [VERT ZOO] An equivalent name for Cypriniformes. { ˌjim·nəˈnōd·ī }

Gymnophiona [VERT ZOO] An equivalent name for Apoda. { ˌjim·nəˈfī·ə·nə }

gymnoplast [BOT] In angiosperms, a cell without a cell wall. { ˈjim·nəˌplast }

Gymnopleura [INV ZOO] A subsection of brachyuran decapod crustaceans including the primitive burrowing crabs with trapezoidal or elongate carapaces, the first pereiopods subchelate, and some or all of the remaining pereiopods flattened and expanded. { ˌjim·nəˈplu̇r·ə }

gymnosperm [BOT] The common name for members of the division Pinophyta; seed plants having naked ovules at the time of pollination. { ˈjim·nəˌspərm }

Gymnospermae [BOT] The equivalent name for Pinophyta. { ˌjim·nəˈspər·mē }

Gymnosporangium juniperi-virginianae [MYCOL] A heteroecious fungal pathogen that is the cause of apple-cedar rust. { ˌjim·nō·spə¦ran·jē·əm ˌjü·niˌper·ē ˌvərˌjin·ēˈaˌnī }

Gymnostomatida [INV ZOO] An order of the protozoan subclass Holotrichia containing the most primitive ciliates, distinguished by the lack of ciliature in the oral area. { ˌjim·nəˈstō·məd·ə }

Gymnotidae [VERT ZOO] The single family of the suborder Gymnotoidei; eel-shaped fishes having numerous vertebrae, and anus located far forward, and lacking pelvic and developed dorsal fins. { ˌjimˈnäd·əˌdē }

Gymnotoidei [VERT ZOO] A monofamilial suborder of actinopterygian fishes in the order Cypriniformes. { ˌjim·nəˈtȯid·ēˌī }

gynaecandrous [BOT] Having staminate and pistillate flowers on the same spike. { ¦gī·nə¦kan·drəs }

gynander [BIOL] A mosaic individual composed of diploid female portions derived from both parents and haploid male portions derived from an extra egg or sperm nucleus. { gīnˈan·dər }

gynandromorph [BIOL] An individual of a dioecious species made up of a mosaic of tissues of male and female genotypes. { gīˈnan·drə ˌmȯrf }

gynandry [PHYSIO] A form of pseudohermaphroditism in which the external sexual characteristics are partly or wholly of the male aspect, but internal female genitalia are present. Also known as female pseudohermaphroditism; virilism. { gīnˈan·drē }

gynecology [MED] The branch of medicine dealing with diseases of women, particularly those affecting the sex organs. { ˌgīn·əˈkäl·ə·jē }

gynecomastia [MED] Abnormal enlargement of the mammary glands in the male. { ˌgīn·ə·kō ˈmas·tē·ə }

gynobase [BOT] A gynoecium-bearing elongation of the receptacle in certain plants. { ˈgīn·ōˌbās }

gynodioecious [BOT] Dioecious but with some perfect flowers on a plant bearing pistillate flowers. { ¦gīn·ō·dīˈē·shəs }

gynoecious [BOT] Pertaining to plants that have only female flowers. { gīˈnē·shəs }

gynoecium [BOT] The aggregate of carpels in a flower. { gīˈnē·sē·əm }

gynogenesis [EMBRYO] Development of a fertilized egg through the action of the egg nucleus, without participation of the sperm nucleus. { ¦gīn·ōˈjen·ə·səs }

gynogenetic diploid [GEN] A diploid organism having two maternal haploid chromosome sets; lethal in mammals because of imprinting. { ˌgī·nō·jə¦ned·ik ˈdipˌlȯid }

gynomerogony [EMBRYO] Development of a fragment of a fertilized egg containing the haploid egg nucleus. { ¦gīn·ō·məˈräg·ə·nē }

gynomonoecious [BOT] Having complete and pistillate flowers on the same plant. { ¦gīn·ō·mäˈnē·shəs }

gynophore [BOT] **1.** A stalk that bears the gynoecium. **2.** An elongation of the receptacle between pistil and stamens. { ˈgīn·əˌfȯr }

gynostemium [BOT] The column composed of the united gynoecia and androecium. { ¦gīn·ōˈstē·mē·əm }

gypsophilous [ECOL] Flourishing on a gypsum-rich substratum. { ˌjipˈsäf·ə·ləs }

gypsy moth [INV ZOO] *Porthetria dispar.* A large lepidopteran insect of the family Lymantriidae that was accidentally imported into New England from Europe in the late 19th century; larvae are economically important as pests of deciduous trees. { ˈjip·sē ¦mȯth }

Gyrinidae [INV ZOO] The whirligig beetles, a family of large coleopteran insects in the suborder Adephaga. { jəˈrin·əˌdē }

Gyrinocheilidae [VERT ZOO] A monogeneric family of cypriniform fishes in the suborder Cyprinoidei. { ¦jir·ə·nō·kīˈlī·əˌdē }

Gyrocotylidae [INV ZOO] An order of tapeworms of the subclass Cestodaria; species are intestinal parasites of chimaeroid fishes and are characterized by an anterior eversible proboscis and a posterior ruffled adhesive organ. { ¦jī·rō·käˈtil·əd·ē }

Gyrocotyloidea [INV ZOO] A class of trematode worms according to some systems of classification. { ¦jī·rōˌkäd·əlˈȯid·ē·ə }

Gyrodactyloidea [INV ZOO] A superfamily of ectoparasitic trematodes in the subclass Monogenea; the posterior holdfast is solid and is armed with central anchors and marginal hooks. { ¦jī·rōˌdak·təˈlȯid·ē·ə }

Gyropidae [INV ZOO] A family of biting lice in the order Mallophaga; members are ectoparasites of South American rodents. { jəˈräp·əˌdē }

gyrus [ANAT] One of the convolutions (ridges) on the surface of the cerebrum. { ˈjī·rəs }

H

HΔ [MED] A symbol used in electrocardiography for the longitudinal axis of the heart as projected on the frontal plane.

habenula [ANAT] **1.** Stalk of the pineal body. **2.** A ribbonlike structure. { hə'ben·yə·lə }

habenular commissure [ANAT] The commissure connecting the habenular ganglia in the roof of the diencephalon. { hə'ben·yə·lər 'käm·ə,shu̇r }

habenular ganglia [NEUROSCI] Olfactory centers anterior to the pineal body. { hə'ben·yə·lər 'gaŋ·glē·ə }

habenular nucleus [NEUROSCI] Either of a pair of nerve centers that are located at the base of the pineal body on either side and serve as an olfactory correlation center. { hə'ben·yə·lər 'nü·klē·əs }

habitat [ECOL] The part of the physical environment in which a plant or animal lives. { 'hab·ə,tat }

habitual abortion [MED] Recurring, successive spontaneous abortion. { hə'bich·ə·wəl ə,bȯr·shən }

habituation [MED] **1.** A condition of tolerance to the effects of a drug or a poison, acquired by its continued use; marked by a psychic craving for it when the drug is withdrawn. **2.** Mild drug addiction in which withdrawal symptoms are not severe. { hə,bich·ə'wā·shən }

habitus [BIOL] General appearance or constitution of an organism. { 'hab·ə·təs }

hackberry [BOT] **1.** *Celtis occidentalis.* A tree of the eastern United States characterized by corky or warty bark, and by alternate, long-pointed serrate leaves unequal at the base; produces small, sweet, edible drupaceous fruit. **2.** Any of several other trees of the genus *Celtis.* { 'hak,ber·ē }

hackmarack *See* tamarack. { 'hak·mə,rak }

haddock [VERT ZOO] *Melanogrammus aeglefinus.* A fish of the family Gadidae characterized by a black lateral line and a dark spot behind the gills. { 'had·ək }

Hadromerina [INV ZOO] A suborder of sponges in the class Clavaxinellida having monactinal megascleres, usually with a terminal knob at one end. { ,had·rō·mə'rī·nə }

Haeckel's law *See* recapitulation theory. { 'hek·əlz ,lȯ }

haematodocha [INV ZOO] A sac in the palpus of male spiders that fills with hemolymph and becomes distended during pairing. { ,hē·məd·ō'dō·kə }

Haemobartonella [MICROBIO] A genus of bacteria in the family Anaplasmataceae; parasites in or on red blood cells of many vertebrates. { ¦hē·mō,bart·ən'el·ə }

Haemophilus [MICROBIO] A genus of gram-negative coccobacilli or rod-shaped bacteria of uncertain affiliation; cells may form threads and filaments and are aerobic or facultatively anaerobic; strictly blood parasites. { hē'mä·fə·ləs }

Haemophilus aegyptius [MICROBIO] A pathogenic bacterium associated with acute contagious forms of conjunctivitis and Brazilian purpuric fever. { hē,mäf·ə·ləs ə'jip·tē·əs }

Haemophilus ducreyi [MICROBIO] A bacterial pathogen that causes the sexually transmitted disease soft chancre, or chancroid. { hē¦mäf·ə·ləs dü'krā,ī }

Haemophilus (para) gallinarum [MICROBIO] A bacterial pathogen that causes infectious coryza in chickens and some birds. { hē,mäf·ə·ləs ,par·ə ,gal·ə'när·əm }

Haemophilus parasuis [MICROBIO] A bacterial pathogen that frequently inhabits the normal upper respiratory tract, can cause secondary pneumonias and, in young or otherwise susceptible animals, generalized illness with arthritis, meningitis, pleuritis, and peritonitis. { hē,mäf·ə·ləs ,par·ə'sü·is }

Haemosporina [INV ZOO] A suborder of sporozoan protozoans in the subclass Coccidia; all are parasites of vertebrates, and human malarial parasites are included. { ,hē·mō'spȯr·ə·nə }

haerangium [MYCOL] The fruiting body of *Fugascus* and *Ceratostomella.* { hē'ran·je·əm }

Hafnia [MICROBIO] A genus of bacteria in the family Enterobacteriaceae; motile rods that can utilize citrate as the only source of carbon. { 'haf·ne·ə }

Haftplatte [INV ZOO] An adhesive plate or disk in some turbellarians; it is a glanduloepidermal organ. { 'häft,pläd·ə }

Hageman factor *See* factor XII. { 'häg·ə·män ,fak·tər }

hagfish [VERT ZOO] The common name for the jawless fishes composing the order Myxinoidea. { 'hag,fish }

H agglutinin [IMMUNOL] An antibody that is

type-specific for the flagella of cells or microorganisms. { ¦āch ə'glüt·ən·ən }

Hahnemannism *See* homeopathy. { 'hän·ə·mä ˌniz·əm }

hair [ZOO] **1.** A threadlike outgrowth of the epidermis of animals, especially a keratinized structure in mammalian skin. **2.** The hairy coat of a mammal, or of a part of the animal. { her }

hair ball *See* lake ball. { 'her ˌbȯl }

hair cell [HISTOL] The basic sensory unit of the inner ear of vertebrates; a columnar, polarized structure with specialized cilia on the free surface. { 'her ˌsel }

hair cycle [PHYSIO] The formation and growth of a new hair, followed by a resting stage, and ending with growth of another new hair from the same follicle. { 'her ˌsī·kəl }

hair follicle [ANAT] An epithelial ingrowth of the dermis that surrounds a hair. { 'her ˌfäl·ə·kəl }

hair gland [ANAT] Sebaceous gland associated with hair follicles. { 'her ˌgland }

hairpin loop [CELL MOL] Any double-stranded region of single-stranded deoxyribonucleic acid or ribonucleic acid formed by base-pairing between complementary base sequences on the same strand. { 'herˌpin ˌlüp }

hairworm [INV ZOO] The common name for about 80 species of worms composing the class Nematomorpha. { 'herˌwərm }

hairy cell leukemia *See* leukemic reticuloendotheliosis. { ¦her·ē ˌsel lü'kē·mē·ə }

hairy tongue [MED] Hyperplasia of the papillae forming hairlike projections on the tongue. { ¦her·ē 'təŋ }

Halacaridae [INV ZOO] A family of marine arachnids in the order Acarina. { ˌhal·ə'kar·əˌdē }

Haldane's rule [GEN] The rule that if one sex in a first generation of hybrids between species is rare, absent, or sterile, then it is the heterogametic sex. { 'hȯlˌdānz ˌrül }

Halecomorphi [VERT ZOO] The equivalent name for Amiiformes. { ˌhal·ə·kō'mȯr·fī }

Halecostomi [VERT ZOO] The equivalent name for Pholidophoriformes. { ˌhal·ə'käst·ə·mē }

half-hardy plant [BOT] A plant that can withstand relatively low temperatures but cannot survive severe freezing in cold climates unless carefully protected. { ¦haf ˌhär·dē 'plant }

halibut [VERT ZOO] Either of two large species of flatfishes in the genus *Hippoglossus*; commonly known as a right-eye flounder. { 'hal·ə·bət }

Halichondrida [INV ZOO] A small order of sponges in the class Demospongiae with a skeleton of diactinal or monactinal, siliceous megascleres (or both), a skinlike dermis, and small amounts of spongin. { ˌhal·ə'kän·drē·də }

Halictidae [INV ZOO] The halictid and sweat bees, a family of hymenopteran insects in the superfamily Apoidea. { hə'lik·təˌdē }

Halimeda [BOT] A genus of small, bushy green algae in the family Codiaceae composed of thick, leaflike segments; important as a fossil and as a limestone builder. { ˌhal·ə'mē·də }

Haliotis [INV ZOO] A genus of gastropod mollusks commonly known as the abalones. { ˌhal·ē'ōd·əs }

Haliplidae [INV ZOO] The crawling water beetles, a family of coleopteran insects in the suborder Adephaga. { hə'lip·ləˌdē }

Haller's organ [INV ZOO] A chemoreceptor on the tarsus of certain ticks. { 'hal·ərz ˌȯr·gən }

hallucinogen [PHARM] A substance, such as LSD, that induces hallucinations. { hə'lüs·ən·ə·jən }

hallucinogenic [PHARM] Of or pertaining to a hallucinogen. { həˌlüs·ən·ə'jen·ik }

hallux [ANAT] The first digit of the hindlimb; the big toe of a human. { 'hal·əks }

hallux valgus [MED] A deformity of the great toe, in which the head of the first metatarsal deviates away from the second metatarsal, that is, toward the outside of the foot, and the phalanges are deviated toward the second toe, causing prominence of the metatarsophalangeal joint. { 'hal·əks 'val·gəs }

halmophagous [ZOO] Pertaining to organisms which infest and eat stalks or culms of plants. { hal'mäf·ə·gəs }

Halobacteriaceae [MICROBIO] A family of gram-negative, aerobic rods and cocci which require high salt (sodium chloride) concentrations for maintenance and growth. { ˌhal·əˌbak·tir·ē'ās·ēˌē }

Halobacterium [MICROBIO] A genus of bacteria in the family Halobacteriaceae; single, rod-shaped cells which may be pleomorphic when media are deficient. { ˌhal·ə·bak'tir·ē·əm }

Halococcus [MICROBIO] A genus of bacteria in the family Halobacteriaceae; nonmotile cocci which occur in pairs, tetrads, or clusters of tetrads. { ˌhal·ə'käk·əs }

Halocypridacea [INV ZOO] A superfamily of ostracods in the suborder Myodocopa; individuals are straight-backed with a very thin, usually calcified carapace. { ˌhal·əˌsip·rə'dās· ē·ə }

halonate [MYCOL] Pertaining to a spore surrounded by a colored circle. { 'hal·əˌnāt }

halophile [BIOL] An organism that requires high salt concentrations for growth and maintenance. { 'hal·əˌfīl }

halophilism [BIOL] The phenomenon of demand for high salt concentrations for growth and maintenance. { ¦hal·ə¦fil·iz·əm }

halophyte [ECOL] A plant or microorganism that grows well in soils having a high salt content. { 'hal·əˌfīt }

Haloragaceae [BOT] A family of dicotyledonous plants in the order Haloragales distinguished by an apical ovary of 2–4 loculi, small inflorescences, and small, alternate or opposite or whorled, exstipulate leaves. { ˌhal·ə·rə'gās·ēˌē }

Haloragales [BOT] An order of dicotyledonous plants in the subclass Rosidae containing herbs with perfect or often unisexual, more or less reduced flowers, and a minute or vestigial perianth. { ˌhal·ə·rə'gā·lēz }

Halosauridae [VERT ZOO] A family of mostly extinct deep-sea teleost fishes in the order Notacanthiformes. { ˌhal·ə'sȯr·əˌdē }

halosere [ECOL] The series of communities succeeding one another, from the pioneer stage to the climax, and commencing in salt water or on saline soil. { 'hal·əˌsir }

halothane [PHARM] $C_2HBrClF_3$ A colorless, nonflammable liquid used as a general anesthetic, by inhalation. { 'hal·əˌthān }

haltere [INV ZOO] Either of a pair of capitate filaments representing rudimentary hindwings in Diptera. Also known as balancer. { 'hȯlˌtir }

Hamamelidaceae [BOT] A family of dicotyledonous trees or shrubs in the order Hamamelidales characterized by united carpels, alternate leaves, perfect or unisexual flowers, and free filaments. { ˌha·məˌmel·ə'dās·ēˌē }

Hamamelidae [BOT] A small subclass of plants in the class Magnoliopsida having strongly reduced, often unisexual flowers with poorly developed or no perianth. { ˌha·mə'mel·əˌdē }

Hamamelidales [BOT] A small order of dicotyledonous plants in the subclass Hamamelidae characterized by vessels in the wood and a gynoecium consisting either of separate carpels or of united carpels that open at maturity. { ˌha·məˌmel·ə'dā·lēz }

hamartoma [MED] An abnormal condition resulting in the formation of a mass of tissue of disproportionate size and distribution but composed of the normal tissue of the region. { ˌha·mər'tō·mə }

hamate [BIOL] Hook-shaped or hooked. { 'hā ˌmāt }

hammertoe [MED] A condition of the toe, usually the second, in which the proximal phalanx is extremely extended while the two distal phalanges are flexed. { 'ham·ərˌtō }

hammock *See* hummock. { 'ham·ək }

Hamoproteidae [INV ZOO] A family of parasitic protozoans in the suborder Haemosporina; only the gametocytes occur in blood cells. { ˌham·ō·prə'tē·əˌdē }

hamster [VERT ZOO] The common name for any of 14 species of rodents in the family Cricetidae characterized by scent glands in the flanks, large cheek pouches, and a short tail. { 'ham·stər }

hamstring muscles [ANAT] The biceps femoris, semitendinosus, and semimembranosus collectively. { 'hamˌstriŋ ˌməs·əlz }

hamulus [VERT ZOO] A hooklike process, especially a small terminal hook on the barbicel of a feather. { 'ham·yə·ləs }

hamus [BIOL] A hook or a curved process. { 'hā·məs }

hand [ANAT] The terminal part of the upper extremity modified for grasping. { hand }

hand-foot-and-mouth disease [MED] An infectious disease of humans caused by a coxsackie virus and characterized by maculopapular and vesicular eruptions in the mouth and on the hands and feet. { ¦hand ¦fu̇t ən 'mau̇th diˌzēz }

Hand-Schüller-Christian disease [MED] A childhood syndrome characterized by exopthalmos, diabetes insipidus, and softened or punched-out areas in the bones. { ¦hänt ¦shil·ər 'kris·chən diˌzēz }

hand-shoulder syndrome *See* shoulder-hand syndrome. { ¦hand ¦shōl·dər 'sinˌdrōm }

hanging-drop preparation [MICROBIO] A technique used in microscopy in which a specimen is placed in a drop of a suitable fluid on a cover slip and the cover slip is inverted over a concavity on a slide. { 'haŋ·iŋ ˌdräp ˌprep·ə'rā·shən }

hangover [MED] Aftereffect of excessive intake of alcohol or certain drugs, such as barbiturates. { 'haŋˌō·vər }

Hansen's disease [MED] An infectious disease of humans thought to be caused by *Mycobacterium leprae*; common manifestations are cutaneous and neural lesions. Also known as leprosy. { 'han·sənz diˌzēz }

H antigen [MICROBIO] A general term for microbial flagellar antigens; former designation for species-specific flagellar antigens of *Salmonella*. { 'āch ¦ant·i·jən }

haplobiont [BOT] A plant that produces only sexual haploid individuals. { ¦ha·plō¦bīˌänt }

haplocaulescent [BOT] Having a simple axis with the reproductive organs on the principal axis. { ¦ha·plō·kȯ'les·ənt }

Haplodoci [VERT ZOO] The equivalent name for Batrachoidiformes. { hə'pläd·əˌsī }

haploid [GEN] Having half of the diploid or full complement of chromosomes, that is, one complete set, as in mature gametes. { 'haˌplȯid }

haploidization [MYCOL] In certain fungi, the transformation of a diploid into a haploid cell by progressive loss of chromosomes due to nondisjunction. { ˌhapˌlȯid·i'zā·shən }

haplo-insufficiency [GEN] In a diploid organism, the presence of an abnormal phenotype in which only one of the two copies of a particular gene yields its normal product, and a reduced amount rather than a qualitative change in the gene product leads to the phenotype. It is responsible for some, but not all, autosomal dominant disorders. { ¦hapˌlō 'in·səˌfish·ən·sē }

Haplomi [VERT ZOO] An equivalent name for Salmoniformes. { ha'plōˌmī }

haplomitosis [CELL MOL] Type of primitive mitosis in which the nuclear granules form into threadlike masses rather than clearly differentiated chromosomes. { ˌha·plōˌmī'tō·səs }

haplont [BOT] A plant with only haploid somatic cells; the zygote is diploid. { 'haˌplänt }

haplophase [BIOL] Haploid stage in the life cycle of an organism. { 'ha·plōˌfāz }

Haplosclerida [INV ZOO] An order of sponges in the class Demospongiae including species with a skeleton made up of siliceous megascleres embedded in spongin fibers or spongin cement. { ˌha·plō'skler·ə·də }

haplosis [CELL MOL] Reduction of the chromosome number to half during meiosis. { ha 'plō·səs }

Haplosporea [INV ZOO] A class of Protozoa in the subphylum Sporozoa distinguished by the production of spores lacking polar filaments. { ¦ha·plō¦spȯr·ē·ə }

Haplosporida [INV ZOO] An order of Protozoa in the class Haplosporea distinguished by the production of uninucleate spores that lack both polar capsules and filaments. { ¦ha·plō¦spȯr·ə·də }

haplostele [BOT] A type of protostele with the core of xylem characterized by a smooth outline. { 'ha·plō,stēl }

haplo-sufficient gene [GEN] Any gene that allows the production of viable adults even when one copy of the gene in diploids is mutant or deleted from one of the homologous chromosomes. { ¦hap,lō sə,fish·ənt 'jēn }

haplotype [GEN] A set of alleles of closely linked loci on a chromosome that tend to be inherited together; commonly used in reference to the linked genes of the major histocompatibility complex. { 'hap·lə,tīp }

hapten [IMMUNOL] An incomplete antigen that cannot induce antibody formation by itself but can do so by coupling with a larger carrier molecule.

hapteron [BOT] A disklike holdfast on the stem of certain algae. { 'hap·tə,rän }

haptochlamydeous [BOT] Having the sporophylls protected by rudimentary perianth leaves. { ,hap·tō·klə'mid·ē·əs }

haptoglobin [BIOCHEM] An alpha globulin that constitutes 1–2% of normal blood serum; contains about 5% carbohydrate. { 'hap·tə,glō·bən }

Haptophyceae [BOT] A class of the phylum Chrysophyta that contains the Coccolithophorida. { ,hap·tə'fīs·ē,ē }

haptor [INV ZOO] The posterior organ of attachment in certain monogenetic trematodes characterized by multiple suckers and the presence of hooks. { 'hap·tər }

haptotropism [BIOL] Movement of sessile organisms in response to contact, especially in plants. { hap'tä·trə,piz·əm }

hardening [BOT] Treatment of plants designed to increase their resistance to extremes in temperature or drought. { 'hard·ən·iŋ }

Harderian gland [VERT ZOO] An accessory lacrimal gland associated with lower eyelid structures in all vertebrates except land mammals. { här'dir·ē·ən ,gland }

hard fiber [BOT] A heavily lignified leaf fiber used in making cordage, twine, and textiles. { 'härd ,fī·bər }

hard pad [VET MED] A disease of dogs, probably associated with the canine distemper virus; often characterized by encephalitis and hardening of the foot pads. { 'härd ,pad }

hard palate [ANAT] The anterior portion of the roof of the mouth formed by paired palatine processes of the maxillary bones and by the horizontal part of each palate bone. { 'härd ¦pal·ət }

hardwood forest [ECOL] **1.** An ecosystem having deciduous trees as the dominant form of vegetation. **2.** An ecosystem consisting principally of trees that yield hardwood. { 'härd,wu̇d ¦fär·əst }

hardy plant [BOT] A plant able to withstand low temperatures without artificial protection. { ¦här·dē ¦plant }

Hardy-Weinberg law [GEN] The concept that frequencies of both genes and genotypes will remain constant from generation to generation in an idealized population where mating is random and evolutionary forces (such as mutation, migration, selection, or genetic drift) are absent. { ¦här·dē 'wīn,bərg ,lȯ }

hare [VERT ZOO] The common name for a number of lagomorphs in the family Leporidae; they differ from rabbits in being larger with longer ears, legs, and tails. { her }

harelip [MED] A congenital defect, sometimes hereditary, marked by an abnormal cleft between the upper lip and the base of the nose. Also known as cleft lip. { 'her,lip }

Harpacticoida [INV ZOO] An order of minute copepod crustaceans of variable form, but generally being linear and more or less cylindrical. { här,pak·tə'kȯid·ə }

harpago [INV ZOO] Part of the clasper on the copulatory organ of certain male insects. { 'här·pə,gō }

Harpidae [INV ZOO] A family of marine gastropod mollusks in the order Neogastropoda. { 'här·pə,dē }

Hartig net [MYCOL] A complex network of fungal hyphae that is the site of nutrient exchange between the fungus and the host plant. { 'här·tig ,net }

Hart-Park virus [VIROL] A ribonucleic acid-containing animal virus of the rhabdovirus group. { 'härt 'pärk ,vī·rəs }

hartshorn salts *See* aromatic spirits of ammonia. { 'härts,hȯrn ,sȯls }

Hashimoto's disease *See* struma lymphomatosa. { ,ha·shi'mō·dōz di,zēz }

Hashimoto's struma *See* struma lymphomatosa. { ,ha·shi'mō·dōz 'strü·mə }

hashish [PHARM] A narcotic drug derived from the plant *Cannabis sativa*; can be smoked, chewed, or drunk. { 'hash,ēsh }

Hassal's body *See* thymic corpuscle. { 'has·əlz ,bäd·ē }

hastate [BIOL] Shaped like an arrowhead with divergent barbs. { 'ha,stāt }

Hatch-Slack pathway [BIOCHEM] A metabolic cycle involved in the non-light-requiring phase of photosynthesis in certain plants having specific metabolic and anatomical modifications in their mesophyll and bundle sheath cells which facilitate the temporary fixation of carbon dioxide (CO_2) into four-carbon organic acid; these acids are next broken down to three-carbon organic acids plus CO_2 in bundle sheath cells, where this freed CO_2 is then fixed into carbohydrates in a normal Calvin cycle pathway. { 'hach 'slak 'path,wā }

haustellum [INV ZOO] A proboscis modified for sucking. { hȯ'stel·əm }

haustoria [MYCOL] Specialized branches of hyphae that penetrate host cells and absorb nutrients from them. { hau̇'stȯr·ē·ə }

haustorial [MYCOL] Pertaining to fungi that have food-absorbing cells in the host. { hau̇ 'stȯr·ē·əl }

haustorium [BOT] **1.** An outgrowth of certain parasitic plants which serves to absorb food from the host. **2.** Food-absorbing cell of the embryo sac in nonparasitic plants. { hȯ'stȯr·ē·əm }

haustrum [ANAT] An outpocketing or pouch of the colon. { 'hȯ·strəm }

Haverhill fever [MED] An acute bacterial infection caused by *Streptobacillus moniliformis*, usually acquired by rat bite, and characterized by acute onset, intermittent fever, erythematous rash, and polyarthritis. Also known as streptobacillary fever. { 'hāv·ə·rəl ˌfē·vər }

Haversian canal [HISTOL] The central, longitudinal channel of an osteon containing blood vessels and connective tissue. { hə'vər·zhən kəˌnal }

Haversian lamella [HISTOL] One of the concentric layers of bone composing a Haversian system. { hə'vər·zhən lə'mel·ə }

Haversian system *See* osteon. { hə'vər·zhən ˌsis·təm }

HA virus *See* hemadsorption virus. { ¦āch¦ā ˌvī·rəs }

hawk [VERT ZOO] Any of the various smaller diurnal birds of prey in the family Accipitridae; some species are used for hunting hare and partridge in India and other parts of Asia. { hȯk }

hay fever [MED] An allergic disorder of the nasal membranes and related structures due to sensitization by certain plant pollens. Also known as allergic rhinitis; pollinosis. { 'hā ˌfē·vər }

Hayflick limit [PHYSIO] The finite replicative capacity of normal somatic cells. { 'hāˌflik ˌlim·ət }

Hay's test [PATH] A test for bile salts; the salts lower the surface tension of water, and therefore a light powder such as flowers of sulfur will not float in a solution containing a high concentration of the salts. { 'hāz ˌtest }

hazelnut *See* filbert. { 'ha·zəlˌnət }

H-B virus [VIROL] A subgroup-A picornavirus associated with diseases in rodents. { ¦āch ¦bē ˌvī·rəs }

HCG *See* human chorionic gonadotropin.

H chain *See* heavy chain. { 'āch ˌchān }

HDL *See* high-density lipoprotein.

head [ANAT] **1.** The region of the body consisting of the skull, its contents, and related structures. **2.** Proximal end of a long bone. [BOT] A dense cluster of nearly sessile flowers on a very short stem. { hed }

headache [MED] A deep form of pain, with a characteristic aching quality, localized in the head. { 'hedˌāk }

head bulb [INV ZOO] A structure armed with spines behind the lips of spiruroid nematodes. { 'hed ˌbəlb }

head fold [EMBRYO] A ventral fold formed by rapid growth of the head of the embryo over the embryonic disk, resulting in the formation of the foregut accompanied by anteroposterior reversal of the anterior part of the embryonic disk. { 'hed ˌfōld }

head organ [INV ZOO] One of the bulbous structures in the prohaptor of monogenetic trematodes which are openings for adhesive glands. { 'hed ˌȯr·gən }

head process [EMBRYO] The notochord or notochordal plate formed as an axial outgrowth of the primitive node. { 'hed ˌprä·səs }

head shield [INV ZOO] A conspicuous structure arching over the lips of certain nematodes. { 'hed ˌshēld }

health [MED] A state of dynamic equilibrium between an organism and its environment in which all functions of mind and body are normal. { helth }

hearing [PHYSIO] The general perceptual behavior and the specific responses made in relation to sound stimuli. { 'hir·iŋ }

heart [ANAT] The hollow muscular pumping organ of the cardiovascular system in vertebrates. { härt }

heartbeat [PHYSIO] Pulsation of the heart coincident with ventricular systole. { 'härtˌbēt }

heart block [MED] The cardiac condition resulting from defective transmission of impulses from atrium to ventricle. { 'härt ˌbläk }

heartburn [MED] A burning sensation emanating from the esophagus below the sternum. { 'härtˌbərn }

heart failure *See* cardiac failure. { 'härt ˌfāl·yər }

heart-failure cells [PATH] Macrophages containing hemosiderin granules found in the lung in certain heart disorders. { 'härt ˌfāl·yər ˌselz }

heart-lung machine [MED] A machine through which blood is shunted to maintain circulation during heart surgery. { 'härt 'ləng məˌshēn }

heart murmur *See* cardiac murmur. { 'härt 'mər·mər }

heart pacemaker *See* pacemaker. { 'härt 'pās ˌmāk·ər }

heart rate [PHYSIO] The number of heartbeats per minute. { 'härt ˌrāt }

heart valve [ANAT] Flaps of tissue that prevent reflux of blood from the ventricles to the atria or from the pulmonary arteries or aorta to the atria. { 'härt ˌvalv }

heartwater disease [VET MED] A septicemic infectious disease of cattle, sheep, and goats in Africa caused by the rickettsial microorganism *Cowdria ruminantium*. { 'härtˌwȯd·ər diˌzēz }

heartwood [BOT] Xylem of an angiosperm. { 'härtˌwu̇d }

heart worm [INV ZOO] *Dirofilaria immitis*. A filarial nematode parasitic on dogs and other carnivores. { 'härt ˌwərm }

heat cramps [MED] Painful voluntary-muscle spasm and cramps following strenuous exercise, usually in persons in good physical condition, due to loss of sodium chloride and water from excessive sweating. { 'hēt ˌkramps }

heat exhaustion [MED] A heat-exposure syndrome characterized by weakness, vertigo, headache, nausea, and peripheral vascular collapse, usually precipitated by physical exertion in a hot environment. { 'hēt ig,zós·chən }

heath *See* temperate and cold scrub. { hēth }

heather [BOT] *Calluna vulgaris.* An evergreen heath of northern and alpine regions distinguished by racemes of small purple-pink flowers. { 'heth·ər }

heat rash *See* miliaria. { 'hēt ,rash }

heat shock protein [CELL MOL] Any of a group of proteins that are synthesized in the cytoplasm of cells as part of the heat shock response and act to protect the chromosomes from damage. { ¦hēt ,shäk 'prō,tēn }

heat shock response [CELL MOL] A cellular reaction to a stimulus such as elevated temperatures or abrupt environmental changes, in which there is cessation or slowdown of normal protein synthesis and activation of previously inactive genes, resulting in the production of heat shock proteins. { 'hēt ,shäk ri,späns }

heat stress index [PHYSIO] Relation of the amount of evaporation or perspiration required for particular job conditions as related to the maximum evaporative capacity of an average person. Abbreviated HSI. { 'hēt ,stres ¦in ,deks }

heatstroke [MED] A heat-exposure syndrome characterized by hyperpyrexia and prostration due to diminution or cessation of sweating, occurring most commonly in persons with underlying disease. { 'hēt,strōk }

heaves [VET MED] Chronic emphysema in horses marked by labored breathing due to overdistension of the alveoli. Also known as broken wind. { hēvz }

heavy chain [IMMUNOL] The heavier of the two types of polypeptide chains occurring in immunoglobulin molecules, its molecular weight range being 50,000–70,000. Also known as A chain; H chain. { 'hev·ē 'chān }

heavy meromyosin [BIOCHEM] The larger of two fragments obtained from the muscle protein myosin following limited proteolysis by trypsin or chymotrypsin. { 'hev·ē ¦mer·ə¦mī·ə,sin }

Heberden-Rosenbach node *See* Heberden's node. { 'heb·ərd·ən 'rōz·ən·bäk ,nōd }

Heberden's arthritis (Obsolete) [MED] Degenerative joint disease of the terminal joints of the fingers, producing enlargement (Heberden's nodes) and flexion deformities. { 'heb·ərd·ənz är'thrīd·əs }

Heberden's node [MED] Nodose deformity of the fingers in degenerative joint disease. Also known as Heberden-Rosenbach node. { 'heb·ərd·ənz ,nōd }

Hebrovellidae [INV ZOO] A family of hemipteran insects in the subdivision Amphibicorisae. { ,heb·rō'vel·ə,dē }

hectocotylus [INV ZOO] A specialized appendage of male cephalopods adapted for the transference of sperm. { ,hek·tō'käd·əl·əs }

hedgehog [VERT ZOO] The common name for members of the insectivorous family Erinaceidae characterized by spines on their back and sides. { 'hej,häg }

hedonic gland [VERT ZOO] One of the mucus-secreting scent glands in many urodeles; functions in courtship. { hē'dän·ik ,gland }

Hedwigiaceae [BOT] A family of mosses in the order Isobryales. { ,hed·vig·ē'ās·ē,ē }

heel spur [MED] A bony growth produced by excessive musculoskeletal tension at the heel. Also known as calcaneal exostosis. { 'hēl ,spər }

heifer [VERT ZOO] A female cow less than 3 years of age that has not borne a calf. { 'hef·ər }

Heine-Medin disease *See* poliomyelitis. { 'hī·nə 'med·ən di,zēz }

Heinz bodies [PATH] Refractile spots seen in erythrocytes in hemolytic anemia that may represent denatured globulin. { 'hīnts ,bäd·ēz }

hekistotherm [ECOL] Plant adapted for conditions of minimal heat; can withstand long dark periods. { he'kis·tō,thərm }

HeLa cells [PATH] Human cancer cells maintained in tissue culture since 1953, originally excised from the cervical carcinoma of a patient named Helen Lane. { 'hel·ə ,selz }

Heleidae [INV ZOO] The biting midges, a family of orthorrhaphous dipteran insects in the series Nematocera. { hə'lē·ə,dē }

Heliasteridae [INV ZOO] A family of echinoderms in the subclass Asteroidea lacking pentameral symmetry but structurally resembling common asteroids. { ,hēl·ē·ə'ster·ə,dē }

helical repeat [CELL MOL] The number of base pairs in one turn of a deoxyribonucleic acid helix. { 'hel·ə·kəl ri'pēt }

helicase [BIOCHEM] An enzyme that is capable of unwinding the deoxyribonucleic acid double helix at a replication fork. { 'hel·ə,kās }

Helicinidae [INV ZOO] A family of gastropod mollusks in the order Archeogastropoda containing tropical terrestrial snails. { ,hel·ə'sin·ə,dē }

helicoid [INV ZOO] Of a gastropod shell, shaped like a flat coil or flattened spiral. { 'hel·ə,kȯid }

helicoid cyme [BOT] A type of determinate inflorescence having a coiled cluster, with flowers on only one side of the axis. { 'hel·ə,kȯid ¦sīm }

Heliconiaceae [BOT] A family of monocotyledonous plants in the order Zingiberales characterized by perfect flowers with a solitary ovule in each locule, schizocarpic fruit, and capitate stigma. { ,hel·ə,kän·ē'ās·ē,ē }

Helicosporae [MYCOL] A spore group of the Fungi Imperfecti characterized by spirally coiled, septate spores. { ,hel·ə'kä·spə,rē }

helicospore [INV ZOO] Mature spore of the Helicosporida characterized by a peripheral spiral filament. { 'hel·ə·kə,spȯr }

Helicosporida [INV ZOO] An order of protozoans in the class Myxosporidea characterized by production of spores with a relatively thick, single intrasporal filament and three uninucleate sporoplasms. { ¦hel·ə·kō¦spȯr·ə·də }

helicotrema [ANAT] The opening at the apex of

the cochlea through which the scala tympani and the scala vestibuli communicate with each other. { ˌhel·ə·kō'trē·mə }

Heligmosomidae [INV ZOO] A family of parasitic roundworms belonging to the Strongyloidea. { həˌlig·mō'sō·məˌdē }

Heliodinidae [INV ZOO] A family of lepidopteran insects in the suborder Heteroneura. { ˌhē·lē·ə'din·əˌdē }

heliophilous [ECOL] Attracted by and adapted for a high intensity of sunlight. { ¦hē·lē¦äf·ə·ləs }

heliophobe [MED] An individual who is extremely sensitive to the sun's rays. { 'hē·lē·əˌfōb }

heliophyte [ECOL] A plant that thrives in full sunlight. { 'hē·lē·əˌfīt }

Heliornithidae [VERT ZOO] The lobed-toed sun grebes, a family of pantropical birds in the order Gruiformes. { ¦hē·lēˌȯr'nith·əˌdē }

heliotaxis [BIOL] Orientation movement of an organism in response to the stimulus of sunlight. { ¦hē·lē·ō¦tak·səs }

heliotrope [BOT] A plant whose flower or stem turns toward the sun. { 'hē·lē·əˌtrōp }

heliotropism [BIOL] Growth or orientation movement of a sessile organism or part, such as a plant, in response to the stimulus of sunlight. { ˌhē·lē'ä·trəˌpiz·əm }

Heliozoia [INV ZOO] A subclass of the protozoan class Actinopodea; individuals lack a central capsule and have either axopodia or filopodia. { ˌhē·lē·ə'zȯi·ə }

heliozooid [BIOL] Ameboid, but with distinct filamentous pseudopodia. { ˌhē·lē·ə'zōˌȯid }

helix [CELL MOL] A spiral structure with a repeating pattern that characterizes many biological polymers, for example, double-stranded nucleic acids and proteins. { 'hēˌliks }

Helix [INV ZOO] A genus of pulmonate land mollusks including many of the edible snails; individuals have a coiled shell with a low conical spire. { 'hēˌliks }

helix-destabilizing protein [BIOCHEM] Any of a group of proteins that bind to single-stranded regions of duplex deoxyribonucleic acid and cause unwinding of the helix. { ¦hēˌliks di'stā·bəˌlīz·iŋ ˌprō·tēn }

hellbender [VERT ZOO] *Cryptobranchus alleganiensis*. A large amphibian of the order Urodela which is the most primitive of the living salamanders, retaining some larval characteristics. { 'hel¦ben·der }

Heller's test [PATH] A test for albumin in urine; presence of albumin is indicated by formation of a white ring at the junction of the solution and a concentrate solution of nitric acid. { 'hel·ərz ˌtest }

Helmholtz theory *See* Young-Helmholtz theory. { 'helmˌhōlts ˌthē·ə·rē }

helminth [INV ZOO] Any parasitic worm. { 'hel ˌminth }

helminthemesis [MED] Vomiting of worms. { ¦hel·mən'them·ə·səs }

helminthiasis [MED] Any disease caused by the presence of parasitic worms in the body. { ˌhel·mən'thī·ə·səs }

helminthic abscess [MED] An abscess caused by worms. { hel'min·thik 'abˌses }

helminthogogue [PHARM] An anthelminthic. { hel'min·thəˌgäg }

helminthoid [BIOL] Resembling a helminth. { hel'minˌthȯid }

helminthologist [BIOL] An individual who studies helminths. { ˌhelˌmən'thäl·ə·jəst }

helminthosporin [BIOCHEM] $C_{15}H_{10}O_5$ A maroon, crystalline pigment formed by certain fungi growing on a sugar substrate. { helˌmin·thə 'spȯr·ən }

Helminthosporium [MYCOL] A genus of parasitic fungi of the family Dematiaceae having conidiophores which are more or less irregular or bent and bear conidia successively on new growing tips. { helˌmin·thə'spȯr·ē·əm }

Helminthosporium victoriae [MYCOL] A fungal pathogen that produces victorin, the cause of Victoria blight of oats. { helˌmin·thəˌspȯr·ē·əm vik'tȯr·ēˌī }

Helobiae [BOT] The equivalent name for Helobiales. { he'lō·bēˌē }

Helobiales [BOT] An order embracing most of the Alismatidae in certain systems of classification. { heˌlō·bē'ā·lēz }

Heloderma [VERT ZOO] The single genus in the reptilian family Helodermatidae; contains the only known poisonous lizards, the Gila monster (H. *suspectum*) and the beaded lizard (H. *horridum*). { ˌhē·lō'dər·mə }

Helodermatidae [VERT ZOO] A family of lizards in the suborder Sauria. { ˌhē·lō·dər'mad·əˌdē }

Helodidae [INV ZOO] The marsh beetles, a family of coleopteran insects in the superfamily Dascilloidea. { hə'lōd·əˌdē }

heloma *See* corn. { hē'lōm·ə }

Helomyzidae [INV ZOO] The sun flies, a family of myodarian cyclorrhaphous dipteran insects in the subsection Acalypteratae. { ˌhe·lō'mīz·əˌdē }

helophyte [ECOL] A marsh plant; buds overwinter underwater. { 'he·ləˌfīt }

helophytia [ECOL] Differences in ecological control by fluctuations in water level such as in marshes. { ˌhe·lə'fī·shə }

Heloridae [INV ZOO] A family of hymenopteran insects in the superfamily Proctotrupoidea. { hə'lȯr·əˌdē }

Helotiales [MYCOL] An order of fungi in the class Ascomycetes. { həˌlō·shē'ā·lēz }

Helotidae [INV ZOO] The metallic sap beetles, a family of coleopteran insects in the superfamily Cucujoidea. { hə'läd·əˌdē }

helotism [ECOL] Symbiosis in which one organism is a slave to the other, as between certain species of ants. { 'hel·əˌtiz·əm }

Helotrephidae [INV ZOO] A family of true aquatic, tropical hemipteran insects in the subdivision Hydrocorisae. { ˌhe·lə'tref·əˌdē }

helper virus [VIROL] A virus that, by its infection of a cell, enables a defective virus to multiply by supplying one or more functions that the defective virus lacks. { 'hel·pər ,vī·rəs }

hem-, hema-, hemo-, haem- [HISTOL] Combining form for blood. { hēm, 'hēm·ə, 'hēm·o, hēm }

hemacytometer *See* hemocytometer. { ,hē·mə·sī'täm·əd·ər }

hemadsorption virus [VIROL] A descriptive term for myxoviruses that agglutinate red blood cells and cause the cells to adsorb to each other. Abbreviated HA virus. { ,hēm·ad'sȯrp·shən ,vī·rəs }

hemagglutination [IMMUNOL] Agglutination of red blood cells. { ,hē·mə,glüd·ən'ā·shən }

hemagglutination-inhibition test [IMMUNOL] A test to identify a virus antigen or to quantitate an antibody by adding virus-specific antibody to a mixture of agglutinating virus and red blood cells. { ,hē·mə,glüd·ən'ā·shən ,in·ə'bish·ən ,test }

hemagglutinin [IMMUNOL] An erythrocyte-agglutinating antibody. { ,hē·mə'glüd·ən·ən }

hemal arch [ANAT] **1.** A ventral loop on the body of vertebrate caudal vertebrae surrounding the blood vessels. **2.** In humans, the ventral vertebral process formed by the centrum together with the ribs. { 'hē·məl 'ärch }

hemal ring [INV ZOO] A vessel in certain echinoderms, variously located, associated with the coelom and axial gland. { 'hē·məl 'riŋ }

hemal sinus [INV ZOO] The two principal lacunae along the digestive tube in certain echinoderms. { 'hē·məl 'sī·nəs }

hemal tuft [INV ZOO] Series of fine vessels in echioderms arising from the axial gland. { 'hē·məl 'təft }

hemangioendothelioma [MED] A malignant tumor composed of anoplastic endothelial cells. Also known as hemangiosarcoma. { hē¦mān·jē·ō,en·dō,thē·lē 'ō·mə }

hemangioma [MED] A tumor composed of blood vessels. Also known as capillary angioma. { hē,mān·jē'ō·mə }

hemangiopericytoma [MED] A tumor composed of endothelium-lined tubes or cords of cells surrounded by spherical cells with supporting reticulin network. { hē¦mān·jē·ō,per·ə·sī'tō·mə }

hemangiosarcoma *See* hemangioendothelioma. { hē¦mān·jē·ō·sär'kō·mə }

hemapodium [INV ZOO] The dorsal lobe of a parapodium. { ,hē·mə'pō·dē·əm }

hemarthrosis [MED] Passage of blood into a joint. { ,hē·mär'thrō·səs }

hematein [BIOCHEM] $C_{16}H_{12}O_6$ A brownish stain and chemical indicator obtained by oxidation of hematoxylin. { ,hē·mə'tē·ən }

hematemesis [MED] Vomiting of blood. { ,hē·mə'tem·ə·səs }

hematidrosis [MED] The appearance of blood or blood products in sweat gland secretions. { hē,mad·ə'drō·səs }

hematoblast [HISTOL] An immature erythrocyte. { 'he·məd·ō,blast }

hematocele [MED] Collection of blood in a body part. { 'he·məd·ō,sēl }

hematochrome [BIOCHEM] A red pigment occurring in green algae, especially when plants are exposed to intense light on subaerial habitats. { hi'mad·ə,krōm }

hematocrit [PATH] The volume, after centrifugation, occupied by the cellular elements of blood, in relation to the total volume. { hi'mad·ə,krit }

hematodocha [INV ZOO] In some spiders, a thin sac on the male that is distended during copulation. { hi,mad·ə'dō·kə }

hematogenous [PHYSIO] **1.** Pertaining to the production of blood or of its fractions. **2.** Carried by way of the bloodstream. **3.** Originating in blood. { ¦hēm·ə¦täj·ə·nəs }

hematoidin crystals [PATH] Yellow to brown crystals in the feces following gastrointestinal hemorrhage. { ,hēm·ə'tȯid·ən ,krist·əlz }

hematologic disorder [MED] A disorder marked by aberrations in structure or function of the blood cells or the blood-clotting mechanism. { ,hē·mə·tə,läj·ik dis'ȯrd·ər }

hematologist [MED] A specialist in the study of blood. { ,hē·mə'täl·ə·jəst }

hematology [MED] The science of the blood, its nature, functions, and diseases. { ,hē·mə'täl·ə·jē }

hematoma [MED] A localized mass of blood in tissue; usually it clots and becomes encapsulated by connective tissue. { ,hē·mə'tō·mə }

hematometra [MED] Accumulation of blood in the uterus. { ,he·məd·ō'me·trə }

hematomyelia [MED] Hemorrhage into the spinal cord. { ,he·məd·ō,mī'ē·lē·ə }

hematopathology *See* hemopathology. { ,he·məd·ō·pə'thäl·ə·jē }

hematophagous [ZOO] Feeding on blood. { ¦hē·mə¦täf·ə·gəs }

hematopoiesis [PHYSIO] The process by which the cellular elements of the blood are formed. Also known as hemopoiesis. { ,he·məd·ō·pȯi'ē·səs }

hematopoietic system *See* reticuloendothelial system. { ¦he·məd·ō·pȯi¦ed·ik 'sis·təm }

hematopoietic tissue [HISTOL] Blood-forming tissue, consisting of reticular fibers and cells. Also known as hemopoietic tissue. { ¦he·məd·ō·pȯi¦ed·ik 'tish·ü }

hematopoietin [BIOCHEM] A substance which is produced by the juxtaglomerular apparatus in the kidney and controls the rate of red cell production. Also known as hemopoietin. { ,he·məd·ō'pȯi·ət·ən }

hematoporphyrin [BIOCHEM] $C_{34}H_{38}O_6N$ Iron-free heme, a porphyrin obtained by treating hemoglobin with sulfuric acid in vitro. Also known as hemoporphyrin. { ¦he·məd·ō¦pȯr·fə·rən }

hematorrhachis [MED] Hemorrhage into the spinal meninges, producing irritative phenomena. { ,he·mə'tȯr·ə·kəs }

hematosalpinx [MED] Accumulation of blood in a Fallopian tube. Also known as hemosalpinx. { ¦he·məd·ō'sal,piŋks }

hematoxylon [BOT] The heartwood of *Haematoxylon campechianum*. Also known as logwood. { ˌhē·mə'täk·səˌlän }

hematuria [MED] A pathological condition in which the urine contains blood. { ˌhe·mə'tu̇r·ē·ə }

heme [BIOCHEM] $C_{34}H_{32}O_4N_4Fe$ An iron-protoporphyrin complex associated with each polypeptide unit of hemoglobin. { hēm }

hemeralopia [MED] Day blindness. { ˌhem·ə·rə'lō·pē·ə }

hemerythrin *See* hemoerythrin. { ˌhem·ə'rith·rən }

heme synthetase [BIOCHEM] An enzyme which combines protoporphyrin IX, ferrous iron, and globin to form the intact hemoglobin molecule. { ˌhēm 'sin·thəˌtās }

hemi- [BIOL] **1.** Prefix for half. **2.** Prefix denoting one side of the body. { 'he·mē }

hemianesthesia [MED] Loss of sensation on one side of the body. { ¦he·mēˌan·əs'thē·zha }

hemianopsia [MED] Bilateral or unilateral blindness in one-half of the field of vision. { ˌhe·mē·ə'näp·sē·ə }

Hemiascomycetes [MYCOL] The equivalent name for Hemiascomycetidae. { ¦he·mēˌas·kōˌmī'sēd·ēz }

Hemiascomycetidae [MYCOL] A subclass of fungi in the class Ascomycetes. { ¦he·mēˌas·kōˌmī'sed·əˌdē }

hemiazygous vein [ANAT] A vein on the left side of the vertebral column which drains blood from the left ascending lumbar vein to the azygos vein. { ¦he·mē'az·ə·gəs 'vān }

hemiballismus [MED] Sudden, violent, spasmodic movements involving particularly the proximal portions of the extremities of one side of the body; caused by a destructive lesion of the contralateral subthalamic nucleus or its neighboring structures or pathways. { ¦he·mē·bə'liz·məs }

Hemibasidiomycetes [MYCOL] The equivalent name for Heterobasidiomycetidae. { ¦he·mē·bəˌsid·ē·ōˌmī's ēd·ēz }

hemicellulose [BIOCHEM] $(C_6H_{10}O_5)_n$ A type of polysaccharide found in plant cell walls in association with cellulose and lignin; it is soluble in and extractable by dilute alkaline solutions. Also known as hexosan. { ¦he·mē'sel·yəˌlōs }

hemicephaly [MED] Congenital absence of the cerebrum. { ¦he·mē¦sef·ə·lē }

Hemichordata [SYST] A group of marine animals categorized as either a phylum of deuterostomes or a subphylum of chordates; includes the Enteropneusta, Pterobranchia, and Graptolithina. { ¦he·mē·kȯr'däd·ə }

hemic murmur [MED] Blowing or rasping sound heard in the heart or vessels, usually in association with systole, in abnormal conditions of increased velocity of blood flow. { 'hē·mik 'mər·mər }

hemicolectomy [MED] Surgical removal of a portion of the colon. { ¦he·mē·kə'lek·tə·mē }

hemicryptophyte [ECOL] A plant having buds at the soil surface and protected by scales, snow, or litter. { ¦he·mē'krip·təˌfīt }

hemicydic [BOT] Of flowers, having the floral leaves arranged partly in whorls and partly in spirals. { 'he·mē'sī·dik }

hemidesmosome [HISTOL] A structure similar to a desmosome that joins a cell to its basilar membrane rather than to another cell. { ˌhem·ē'dez·məˌsōm }

hemidiaphragm [ANAT] A lateral half of a diaphragm. [MED] Diaphragm with normal muscle development only on one side. { ¦he·mē 'dī·əˌfram }

Hemidiscosa [INV ZOO] An order of sponges in the subclass Amphidiscophora distinguished by birotulates that are hemidiscs with asymmetrical ends. { ¦he·mē·dis'kō·sə }

Hemileia vastatrix [MYCOL] A fungus of the order Uredinales which is the causative agent of orange coffee rust. { ˌhem·əˌlē·yə 'vas·təˌtriks }

Hemileucinae [INV ZOO] A subfamily of lepidopteran insects in the family Saturnidae consisting of the buck moths and relatives. { ¦he·mē'lüs·ənˌē }

Hemimetabola [INV ZOO] A division of the insect subclass Pterygota; members are characterized by hemimetabolous metamorphosis. { ¦he·mē·me'tab·ə·lə }

hemimetabolous metamorphosis [INV ZOO] An incomplete metamorphosis; gills are present in aquatic larvae, or naiads. { ¦he·mē·me'tab·ə·ləs ˌmed·ə'mȯr·fə·səs }

hemin [BIOCHEM] $C_{34}H_{32}O_4N_4FeCl$ The crystalline salt of ferriheme, containing iron in the ferric state. { 'hē·mən }

hemiparasite [ECOL] A parasite capable of a saprophytic existence, especially certain parasitic plants containing some chlorophyll. Also known as semiparasite. { ¦he·mē'par·əˌsīt }

hemiparesis [MED] Muscle weakness on one side of the body. { ¦he·mē·pə'rē·səs }

hemipelagic [ECOL] Of the biogeographic environment of the hemipelagic region with both neritic and pelagic qualities. { ¦he·mē·pə'laj·ik }

hemipenis [VERT ZOO] Either of a pair of nonerectile, evertible sacs that lie on the floor of the cloaca in snakes and lizards; used as intromittent organs. { ¦he·mē'pē·nəs }

Hemipeplidae [INV ZOO] An equivalent name for Cucujidae. { ¦he·mē'pep·ləˌdē }

hemiplegia [MED] Unilateral paralysis of the body. { ˌhē·mə'plē·jə }

Hemiprocnidae [VERT ZOO] The crested swifts, a family comprising three species of perching birds found only in southeastern Asia. { ¦he·mē'präk·nəˌdē }

Hemiptera [INV ZOO] The true bugs, an order of the class Insecta characterized by forewings differentiated into a basal area and a membranous apical region. { he'mip·tə·rə }

Hemisphaeriales [MYCOL] A group of ascomycetous fungi characterized by the wall of the fruit body being a stroma. { ¦he·mēˌsfir·ē'ā·lēz }

hemithorax [ANAT] One side of the chest. { ¦he·mē'thȯr,aks }

hemizygous [GEN] In diploid organisms, the presence of single copy of a gene; it may be a result of deletion or chromosome loss, or simply may reflect the presence of a single copy of a sex chromosome, such as the X in male mammals. { ¦he·mē¦zī·gəs }

hemlock [BOT] The common name for members of the genus *Tsuga* in the pine family characterized by two white lines beneath the flattened, needlelike leaves. { 'hem,läk }

hemobilirubin [BIOCHEM] Bilirubin in normal blood serum before passage through the liver. { ¦hē·mō,bil·i'rü·bən }

hemoblast *See* hemocytoblast. { 'hē·mə,blast }

hemochorial placenta [EMBRYO] A type of placenta having the maternal blood in direct contact with the chorionic trophoblast. Also known as labyrinthine placenta. { ¦hē·mō'kȯr·ē·əl plə 'sen·tə }

hemochromatosis [MED] A disorder of iron metabolism characterized by excessive accumulation of iron in the liver and other tissues and by development of severe cirrhosis. { ¦hē·mō,krō·mə'tō·səs }

hemocoel [INV ZOO] An expanded portion of the blood system in arthropods that replaces a portion of the coelom. { 'hē·mə,sēl }

hemoconcentration [MED] An increase in the concentration of blood cells resulting from the loss of plasma or water from the bloodstream. { ¦hē·mō,käns·ən'trā·shən }

hemoconia [BIOCHEM] Round or dumbbell-shaped, refractile, colorless particles found in blood plasma. { ,hē·mə'kō·nē·ə }

hemoconiosis [MED] Condition of having an abnormal amount of hemoconia in the blood. { ,hē·mō,kō·nē'ō·səs }

hemocyanin [BIOCHEM] A blue respiratory pigment found only in mollusks and in arthropods other than insects. { ,hē·mō'sī·ə·nən }

hemocyte [INV ZOO] A cellular element of blood, especially in invertebrates. { 'hē·mə,sīt }

hemocytoblast [HISTOL] A pluripotential blast cell thought to be capable of giving rise to all other blood cells. Also known as hemoblast; stem cell. { ,hē·mə'sīd·ə,blast }

hemocytolysis [PHYSIO] The dissolution of blood cells. { ,hē·mə,sī'täl·ə·səs }

hemocytometer [PATH] A specifically designed, ruled and calibrated glass slide used with a microscope to count red and white blood cells. Also spelled hemacytometer. { ,hē·mə,sī'täm·əd·ər }

hemodialysis [MED] The filtering of toxic solutes and excess fluid from the blood via an external membranous coil placed between the blood and a rinsing solution. { ,hē·mō·di'al·ə·səs }

hemodichorial placenta [EMBRYO] A placenta with a double trophoblastic layer. { ,hē·mə·də'kȯr·ē·əl plə'sen·tə }

hemodynamics [PHYSIO] A branch of physiology concerned with circulatory movements of the blood and the forces involved in circulation. { ¦hē·mō·dī'nam·iks }

hemoendothelial placenta [EMBRYO] A placenta having the endothelium of vessels of chorionic villi in direct contact with the maternal blood. { ¦hē·mō,en·də'thē·lē·əl plə'sen·tə }

hemoerythrin [BIOCHEM] A red respiratory pigment found in a few annelid and sipunculid worms and in the brachiopod *Lingula*. Also known as hemerythrin. { ,hē·mō·ə'rith·rən }

hemoflagellate [INV ZOO] A parasitic, flagellate protozoan that lives in the blood of the host. { ,hē·mə'flaj·ə·lət }

hemoglobin [BIOCHEM] The iron-containing, oxygen-carrying molecule of the red blood cells of vertebrates comprising four polypeptide subunits in a heme group. { 'hē·mə,glō·bən }

hemoglobin A [BIOCHEM] The type of hemoglobin found in normal adults, which moves as a single component in an electrophoretic field, is rapidly denatured by highly alkaline solutions, and contains two titratable sulfhydryl groups per molecule. { 'hē·mə,glō·bən 'ā }

hemoglobin C [PATH] A slow-moving abnormal hemoglobin associated with intraerythrocytic crystal formation, target cells, and chronic hemolytic anemia. { 'hē·mə,glō·bən 'sē }

hemoglobin E [PATH] An abnormal hemoglobin found in people of Southeast Asia, migrating slightly faster than hemoglobin C; in the homozygous form it causes a mild hemolytic anemia with normochromic target cells. { 'hē·mə,glō·bən 'ē }

hemoglobinemia [MED] The presence of hemoglobin in the blood plasma. { ,hē·mə,glō·bə'nē·mē·ə }

hemoglobin H [PATH] An abnormal hemoglobin migrating more rapidly than normal hemoglobin on electrophoresis, and usually associated with thalassemia. { 'hē·mə,glō·bən 'āch }

hemoglobin M [PATH] An abnormal hemoglobin associated with hereditary methemoglobinemia, differing from normal hemoglobin in its electrophoretic mobility by the starch-block method. { 'hē·mə,glō·bən 'em }

hemoglobinopathy [MED] Any blood dyscrasia resulting from the genetically determined alteration of the chemical nature of hemoglobin. { ,hē·mə,glō·bə'näp·ə·thē }

hemoglobin S *See* sickle-cell hemoglobin. { 'hē·mə,glō·bən 'es }

hemoglobinuria [MED] A pathological condition in which the urine contains hemoglobin. { ,hē·mə,glō·bə'nu̇r·ē·ə }

hemoglobinuric nephrosis *See* lower nephron nephrosis. { ,he·mə,glō·bə'nu̇r·ik nə'frō·səs }

hemogram [PATH] **1.** Erythrocyte and leukocyte count per cubic millimeter of blood plus the differential leukocyte count and hemoglobin level in grams per 100 milliliters of blood. **2.** The differential leukocyte count. { 'hē·mə,gram }

hemohistioblast [HISTOL] The hypothetical reticuloendothelial cell from which all the cells of the blood are eventually differentiated. { ˌhē·mō'his·tē·ōˌblast }

hemolymph [INV ZOO] The circulating fluid of the open circulatory systems of many invertebrates. { 'hē·məˌlimf }

hemolysin [IMMUNOL] A substance that lyses erythrocytes. { ˌhē·mə'līs·ən }

hemolysis [PHYSIO] The lysis, or destruction, of erythrocytes with the release of hemoglobin. { hē'mäl·ə·səs }

hemolytic anemia [MED] A decrease in the blood concentration of hemoglobin and the number of erythrocytes, due to the inability of the mature erythrocytes to survive in the circulating blood. { ¦hē·mə¦lid·ik ə'nē·mē·ə }

hemolytic disease of newborn *See* erythroblastosis fetalis. { ¦hē·mə¦lid·ik di¦zēz əv 'nüˌbȯrn }

hemolytic jaundice [MED] Accumulation of bile pigments in the plasma as a result of excessive hemolysis. { ¦hē·mə¦lid·ik 'jȯn·dəs }

hemolytic uremic syndrome [MED] An illness characterized by the abrupt onset of decreased urine production, loss of kidney function, and anemia. It may be accompanied by edema, hypertension, blood-clotting disorders, and seizures. It is often caused by infection with *Escherichia coli* O157:H7 but has also been associated with *Salmonella* and *Shigella*. { hē·mə¦lid·ik yə¦rēm·ik 'sinˌdrōm }

hemomonochorial placenta [EMBRYO] A placenta with a single trophoblastic layer. { ¦hē·mōˌmän·ə'kȯr·ē·əl plə'sen·tə }

hemoparasite [INV ZOO] A parasitic animal that lives in the blood of a vertebrate. { ¦hē·mō 'par·əˌsīt }

hemopathology [MED] A branch of medicine dealing with blood diseases. Also known as hematopathology. { ¦hē·mō·pə'thäl·ə·jē }

hemopathy [MED] Any disease of the blood. { hē'mäp·ə·thē }

hemopericardium [MED] The presence of blood or bloody effusion in the pericardial sac. { ¦hē·mōˌper·ə'kärd·ē·əm }

hemoperitoneum [MED] An effusion of blood in the peritoneal cavity. { ¦hē·mōˌper·ə·tə'nē·əm }

hemopexin [BIOCHEM] A heme-binding protein in human plasma that may be a regulator of heme and drug metabolism, and a distributor of heme. { ˌhē·mə'pek·sən }

hemophilia [MED] A rare, hereditary blood disorder marked by a tendency toward bleeding and hemorrhages due to a deficiency of factor VIII. { ˌhē·mə'fil·ē·ə }

hemophilic bacteria [MICROBIO] Bacteria of the genera *Hemophilus*, *Bordetella*, and *Moraxella*; all are small, gram-negative, nonmotile, parasitic rods, dependent upon blood factors for growth. { ˌhē·mə'fil·ik bak'tir·ē·ə }

hemophilioid disease [MED] Any hemophilialike disease; it is the same as hemophilia clinically, but caused by a deficiency of factors IX, X, and XII. { ¦hē·mə¦fil·ēˌȯid diˌzēz }

Hemophilus [MICROBIO] A genus of hemophilic bacteria in the family Brucellaceae requiring hemin and nicotinamide nucleoside for growth. { hə'mäf·ə·ləs }

hemopoiesis *See* hematopoiesis. { ˌhē·mōˌpȯi'ē·səs }

hemopoietic tissue *See* hematopoietic tissue. { ˌhē·mōˌpȯi'ed·ik 'tish·ü }

hemopoietin *See* hematopoietin. { ˌhē·mōˌpȯi'ēt·ən }

hemoporphyrin *See* hematoporphyrin. { ˌhē·mō 'pȯr·fə·rən }

hemoptysis [MED] Discharge of blood from the larynx, trachea, bronchi, or lungs. { hē'mäp·tə·səs }

hemorrhage [MED] The escape of blood from the vascular system. { 'hem·rij }

hemorrhagic colitis [MED] An acute disease characterized by overtly bloody diarrhea that is caused by infection with the enterohemorrhagic strain of *Escherichia coli* (EC O157:H7). { ˌhem·ə¦raj·ik kə'līd·əs }

hemorrhagic diathesis [MED] Any condition marked by abnormal bleeding tendency. { ˌhem·ə'raj·ik dī'ath·ə·səs }

hemorrhagic fever virus [VIROL] Any of several arboviruses causing acute infectious human diseases characterized by fever, prostration, vomiting, and hemorrhage. { ˌhem·ə'raj·ik ¦fē·vər ˌvī·rəs }

hemorrhagic measles [MED] A grave variety of measles with a hemorrhagic eruption and severe constitutional symptoms. Also known as black measles. { ˌhem·ə'raj·ik 'mēz·əlz }

hemorrhagic pericarditis [MED] Inflammation of the pericardium accompanied by the hemorrhagic appearance of the exudate imparted by the presence of red cells. { ˌhem·ə'raj·ik ˌper·əˌkär'dīd·əs }

hemorrhagic pleuritis [MED] Inflammation of the pleura characterized by the presence of bloody fluid in the pleural cavity. { ˌhem·ə'raj·ik plə'rīd·əs }

hemorrhagic septicemia [VET MED] An infectious bacterial disease of fowl, rabbit, buffalo, and other animals caused by *Pasteurella mulfocida*. Also known as pasteurellosis. { ˌhem·ə'raj·ik ˌsep·tə'sē·mē·ə }

hemorrhagic unit [BIOL] A unit for the standardization of snake venom. { ˌhem·ə'raj·ik 'yü·nət }

hemorrhoid [MED] A varicosity of the external hemorrhoidal veins, causing painful swelling in the anal region. { 'hemˌrȯid }

hemorrhoidectomy [MED] Surgical removal of hemorrhoids. { ˌhemˌrȯi'dek·tə·mē }

hemosalpinx *See* hematosalpinx. { ˌhē·mō 'salˌpiŋks }

hemosiderin [BIOCHEM] An iron-containing glycoprotein found in most tissues and especially in liver. { ˌhē·mō'sid·ə·rən }

hemosiderosis [PHYSIO] Deposition of hemosiderin in body tissues without tissue damage, reflecting an increase in body iron stores. { ˌhē·mōˌsid·ə'rō·səs }

hemostasis [MED] **1.** The arrest of a flow of

blood or hemorrhage. **2.** The stopping or slowing of circulation. { ˌhē·mə'stā·səs }

hemostat [MED] An instrument to compress a bleeding vessel. { 'hē·məˌstat }

hemostatic [MED] An agent that arrests or checks bleeding, especially by shortening clotting time. { ¦hē·mə¦stad·ik }

hemothorax [MED] Accumulation of blood in the pleural cavity. { ¦hē·mō'thȯrˌaks }

hemotrichorial placenta [EMBRYO] A placenta with a triple trophoblastic layer. { ˌhē·mō·trə'kȯr·ē·əl plə'sen·tə }

hemotrophe [BIOCHEM] The nutritive substance supplied via the placenta to embryos of viviparous animals. { hē'mä·trə·fē }

hen [VERT ZOO] The female of several bird species, especially gallinaceous species. { hen }

henbane [BOT] *Hyoscyamus niger.* A poisonous herb containing the toxic alkaloids hyoscyamine and hyoscine; extracts have properties similar to belladonna. { 'henˌbān }

Henicocephalidae [INV ZOO] A family of hemopteran insects of uncertain affinities. { ˌhen·ə·kō·sə'fal·əˌdē }

Henle's loop *See* loop of Henle. { 'hen·lēz ˌlüp }

henna [BOT] *Lawsonia inermis.* An Old World plant having small opposite leaves and axillary panicles of white flowers; a reddish-brown dye extracted from the leaves is used in hair dyes. Also known as Egyptian henna. { 'hen·ə }

Hensen's node [EMBRYO] Thickening formed by a group of cells at the anterior end of the primitive streak in vertebrate gastrulas. { 'hen·səns ˌnōd }

heparin [BIOCHEM] An acid mucopolysaccharide acting as an antithrombin, antithromboplastin, and antiplatelet factor to prolong the clotting time of whole blood; occurs in a variety of tissues, most abundantly in liver. { 'hep·ə·rən }

hepar lobatum [MED] The liver in syphilitic cirrhosis, having a nodular lobulated appearance. { 'hēˌpär lō'bäd·əm }

hepatectomy [MED] Surgical removal of the liver or a part of it. { ˌhep·ə'tek·tə·mē }

Hepaticae [BOT] The equivalent name for Marchantiatae. { he'pad·əˌsē }

hepatic artery [ANAT] A branch of the celiac artery that carries blood to the stomach, pancreas, great omentum, liver, and gallbladder. { he'pad·ik 'ärd·ə·rē }

hepatic cecum [INV ZOO] A hollow outpocketing of the foregut of *Branchiostoma*; receives veins from the intestine. { he'pad·ik 'sē·kəm }

hepatic coma [MED] Unconscious state associated with advanced liver disease. { he'pad·ik 'kō·mə }

hepatic duct [ANAT] The common duct draining the liver. Also known as common hepatic duct. { he'pad·ik 'dəkt }

hepatic duct system [ANAT] The biliary tract including the hepatic ducts, gallbladder, cystic duct, and common bile duct. { he'pad·ik 'dəkt ˌsis·təm }

hepatic encephalopathy [MED] Behavioral, psychological, and neurological changes associated with advanced liver disease. { he'pad·ik enˌsef·ə'läp·ə·thē }

hepatic glycogenosis *See* von Gierke's disease. { he'pad·ik ˌglī·kə·jə'nō·səs }

hepatic plexus [NEUROSCI] Nerve network accompanying the hepatic artery to the liver. { he'pad·ik 'plek·səs }

hepatic portal system [ANAT] A system of veins in vertebrates which collect blood from the digestive tract and spleen and pass it through capillaries in the liver. { he'pad·ik 'pȯrd·əl ˌsis·təm }

hepatic vein [ANAT] A blood vessel that drains blood from the liver into the inferior vena cava. { he'pad·ik 'vān }

hepatitis [MED] Inflammation of the liver; commonly of viral origin but also occurring in association with syphilis, typhoid fever, malaria, toxemias, and parasitic infestations. { ˌhep·ə'tīd·əs }

hepatitis virus [VIROL] Any of several viruses causing hepatitis in humans and lower mammals. { 'hep·ə'tīd·əs 'vī·rəs }

hepatization [PATH] The conversion of tissue into a liverlike substance, as of the lungs during the exudative stage of pneumonia. { ˌhep·əd·ə'zā·shən }

hepatocyte [HISTOL] An epithelial cell constituting the major cell type in the liver. { hə'pad·əˌsīt }

hepatolenticular degeneration *See* Wilson's disease. { he¦pad·ōˌlen¦tik·yə·lər diˌjen·ə'rā·shən }

hepatoma [MED] A usually malignant neoplasm arising from parenchymal cells of the liver. { ˌhep·ə'tō·mə }

hepatomegaly [MED] Enlargement of the liver. { ˌhep·əd·ō'meg·ə·lē }

hepatopancreas [INV ZOO] A gland in crustaceans and certain other invertebrates that combines the digestive functions of the liver and pancreas of vertebrates. { ¦hep·əd·ō'paŋ·krē·əs }

hepatorenal syndrome [MED] A complex of syndromes due to hepatic and renal failure, including hyperpyrexia, oliguria, and coma. Also known as Heyd's syndrome. { ¦hep·əd·ō'rēn·əl 'sinˌdrōm }

hepatoscopy [MED] Inspection of the liver, as by laparotomy or peritoneoscopy. { ˌhep·ə'täs·kə·pē }

hepatotoxin [PHARM] An agent capable of damaging the liver. { ¦hep·əd·ō'täk·sən }

Hepialidae [INV ZOO] A family of lepidopteran insects in the superfamily Hepialoidea. { ˌhep·ē'al·əˌdē }

Hepialoidea [INV ZOO] A superfamily of lepidopteran insects in the suborder Homoneura including medium- to large-sized moths which possess rudimentary mouthparts. { ˌhep·ē·ə'lȯid·ē·ə }

Hepsogastridae [INV ZOO] A family of parasitic insects in the order Mallophaga. { ˌhep·sə'gas·trəˌdē }

heptose [BIOCHEM] Any member of the group

of monosaccharides containing seven carbon atoms. { 'hep,tōs }

heptulose [BIOCHEM] The generic term for a ketose formed from a seven-carbon monosaccharide. { 'hep·tə,lōs }

herb [BOT] **1.** A seed plant that lacks a persistent, woody stem aboveground and dies at the end of the season. **2.** An aromatic plant or plant part used medicinally or for food flavoring. { hərb }

herbaceous [BOT] **1.** Resembling or pertaining to a herb. **2.** Pertaining to a stem with little or no woody tissue. { hər'bā·shəs }

herbaceous dicotyledon [BOT] A type of dicotyledon in which the primary vascular cylinder forms an ectophloic siphonostele with widely separated vascular strands. { hər'bā·shəs ,dī,käd·əl'ēd·ən }

herbaceous monocotyledon [BOT] A type of monocotyledon with a vascular system composed of widely spaced strands arranged in one of four ways. { hər'bā·shəs ,män·ə,käd·əl'ēd·ən }

herbarium [BOT] **1.** A collection of plant specimens, pressed and mounted on paper or placed in liquid preservatives, and systematically arranged. **2.** A building where a herbarium is housed. { hər'ber·ē·əm }

herbicolous [ECOL] Living on herbs. { hər 'bik·əl·əs }

herbivore [VERT ZOO] An animal that eats only vegetation. { 'hər·bə,vȯr }

herbivory [ECOL] The consumption of plants without killing them. { hər'biv·ə·rē }

Herbst corpuscle [VERT ZOO] A cutaneous sense organ found in the mucous membrane of the tongue of the duck. { 'hərbst ¦kȯr·pə·səl }

herd [VERT ZOO] A number of one kind of wild, semidomesticated, or domesticated animals grouped or kept together under human control. { hərd }

herd immunity [IMMUNOL] Immunity of a sufficient number of individuals in a population such that infection of one individual will not result in an epidemic. { 'hərd i,myü·nəd·ē }

hereditary deforming chondrodysplasia *See* multiple hereditary exostoses. { hə'red·ə,ter·ē di 'fȯr·miŋ ¦kän·drō,dis'plā·zhə }

hereditary determinant [CELL MOL] A nuclear or extranuclear genetically functional unit that is replicated with conservation of specificity. { hə¦red·ə,ter·ē di'tər·mə·nənt }

hereditary disease [MED] A genetically determined illness transmitted from parent to child. { hə'red·ə,ter·ē di,zēz }

hereditary hemorrhagic telangiectasia [MED] An inherited disease characterized by dilatation of groups of capillaries and a tendency to hemorrhage. Also known as Osler-Rendu-Weber disease. { hə'red·ə,ter·ē ,hem·ə'raj·ik tə¦lan·jē·ek'tā·zhə }

hereditary hypophosphatemic rickets [MED] Sex-linked syndrome involving defective bone growth and decreased concentrations of phosphates in the serum. { hə'red·ə,ter·ē ¦hī·pō,fäs·fə¦tē·mik 'rik·əts }

hereditary nephritis [MED] A familial disease characterized by recurrent attacks of interstitial inflammation of the kidneys and discharge of blood in the urine. { hə'red·ə,ter·ē ne'frīd·əs }

hereditary spherocytosis [MED] A chronic congenital disorder of the erythrocytopoietic system characterized by a preponderance of spherical erythrocytes, increased osmotic fragility, hemolytic anemia, and splenomegaly. { hə'red·ə,ter·ē ,sfer·ō,sī'tō·səs }

hereditary spinal ataxia *See* Friedrich's ataxia. { hə'red·ə,ter·ē ¦spīn·əl ə'tak·sē·ə }

heredity [GEN] The transmission of phenotypes and alleles from one generation to the next. The sum of genetic endowment obtained from the parents. { hə'red·əd·ē }

heredofamilial [MED] Referring to a disease or disorder having a familial pattern of occurrence and thought to be hereditary. { hə¦red·ō·fə ¦mil·yəl }

Hering theory [PHYSIO] A theory of color vision which assumes that three qualitatively different processes are present in the visual system, and that each of the three is capable of responding in two opposite ways. { 'her·iŋ ,thē·ə·rē }

heritability [GEN] In a population, the ratio of the total genetic variance to the total phenotypic variance. { ,her·əd·ə'bil·əd·ē }

heritable change [GEN] A nonlethal genetic change that is passed on to the descendants. { ¦her·əd·ə·bəl 'chānj }

hermaphrodite [BIOL] An individual animal or plant exhibiting hermaphroditism. { hər'maf·rə,dīt }

hermaphroditic *See* monoecious. { hər,ma·frə 'did·ik }

hermaphroditism [PHYSIO] An abnormal condition, especially in humans and other higher vertebrates, in which both male and female reproductive organs are present in the individual. { hər'ma·frə·dīd,iz·əm }

hermatype *See* hermatypic coral. { 'hər·mə,tīp }

hermatypic coral [INV ZOO] Reef-building coral characterized by the presence of symbiotic algae within their endodermal tissue. Also known as hermatype. { ,hər·mə'tip·ik 'kär·əl }

hermit crab [INV ZOO] The common name for a number of marine decapod crustaceans of the families Paguridae and Parapaguridae; all lack right-sided appendages and have a large, soft, coiled abdomen. { 'hər·mət ,krab }

hernia [MED] Abnormal protrusion of an organ or other body part through its containing wall. Also called rupture. { 'hər·nē·ə }

hernial sac [MED] A pouch of peritoneum containing a herniated organ or other body part. { 'hər·nē·əl 'sak }

herniated disk [MED] An intervertebral disk in which the pulpy center has pushed through the fibrocartilage. Also known as slipped disk. { 'hər·nē,ad·əd 'disk }

herniation of the nucleus pulposus [MED] Occurrence of a herniated disk. { ˌhər·nē'ā·shən əv thə 'nü·klē·əs pəl'pō·səs }

herniorrhaphy [MED] Operation for repair and suturing of a hernia. { ˌhər·nē'ór·ə·fē }

herniotomy [MED] An operation for the relief of irreducible hernia, by cutting through the neck of the sac. { ˌhər·nē'äd·ə·mē }

heroin [PHARM] $C_{21}H_{23}O_5N$ A white, crystalline powder made from morphine; the hydrochloride compound is used as a sedative and narcotic. { 'her·ə·wən }

heron [VERT ZOO] The common name for wading birds composing the family Ardeidae characterized by long legs and neck, a long tapered bill, large wings, and soft plumage. { 'her·ən }

herpangia [MED] A mild viral disease of humans caused by a coxsackie virus and characterized by fever, anorexia, and grayish papules surrounded by a red areola in the mouth. { hər'pan·jē·ə }

herpes [MED] An acute inflammation of the skin or mucous membranes, characterized by the development of groups of vesicles on an inflammatory base. { 'hərˌpēz }

herpes simplex [MED] An acute vesicular eruption of the skin or mucous membranes caused by a virus, commonly seen as cold sores or fever blisters. { ¦hərˌpēz 'simˌpleks }

herpes simplex encephalitis *See* herpetic encephalitis. { ¦hərˌpēz 'simˌpleks inˌsef·ə'līd·əs }

herpes simplex virus [VIROL] Either of two types of subgroup A herpesviruses that are specific for humans; given the binomial designation *Herpesvirus hominis*. { ¦hərˌpēz 'simˌpleks 'vī·rəs }

Herpesviridae [VIROL] A family of deoxyribonucleic acid (DNA)-containing viruses characterized by enveloped virions containing one molecule of double-stranded linear DNA wrapped around an associated spool-shaped protein inside an icosahedron. It includes subfamilies Alphaherpesvirinae (herpes simplex virus group), Betaherpesvirinae (cytomegalovirus group), and Gammaherpesvirinae (lymphoproliferative virus group). { ˌhər·pēz'vir·əˌdī }

herpesvirus [VIROL] A major group of deoxyribonucleic acid-containing animal viruses, distinguished by a cubic capsid, enveloped virion, and affinity for the host nucleus as a site of maturation. { 'hərˌpēzˌvī·rəs }

herpes zoster [MED] A systemic virus infection affecting spinal nerve roots, characterized by vesicular eruptions distributed along the course of a cutaneous nerve. Also known as shingles; zoster. { ¦hərˌpēz 'zäs·tər }

herpetic encephalitis [MED] A type of meningoencephalitis characterized by large intranuclear inclusion bodies in the brain. Also known as herpes simplex encephalitis. { hər'ped·ik inˌsef·ə'līd·əs }

herpetic stomatitis [MED] Inflammation of the soft tissues of the mouth characterized by fever blisters. { hər'ped·ik ˌstō·mə'tīd·əs }

herpetic tonsillitis [MED] Acute inflammation of the tonsils characterized by fever and vesicles caused by a herpesvirus. { hər'ped·ik tän·sə'līd·əs }

Herpetosiphon [MICROBIO] A genus of bacteria in the family Cytophagaceae; cells are unbranched, sheathed rods or filaments; unsheathed segments are motile; microcysts are not known. { ¦hər·pəd·ō'sī·fən }

herring [VERT ZOO] The common name for fishes composing the family Clupeidae; fins are soft-rayed and have no supporting spines, there are usually four gill clefts, and scales are on the body but absent on the head. { 'her·iŋ }

Herring body [HISTOL] Any of the distinct colloid masses in the vertebrate pituitary gland, possibly representing greatly dilated endings of nerve fibers. { 'her·iŋ ˌbäd·ē }

Hesionidae [INV ZOO] A family of small polychaete worms belonging to the Errantia. { 'hes·ē'än·əˌdē }

hesperidium [BOT] A modified berry, with few seeds, a leathery rind, and membranous extensions of the endocarp dividing the pulp into chambers; an example is the orange. { 'hes·pə'rid·ē·əm }

Hesperiidae [INV ZOO] The single family of the superfamily Hesperioidea comprising butterflies known as skippers because of their rapid, erratic flight. { hes·pə'rī·əˌdē }

Hesperioidea [INV ZOO] A monofamilial superfamily of lepidopteran insects in the suborder Heteroneura including heavy-bodied, mostly diurnal insects with clubbed antennae that are bent, curved, or reflexed at the tip. { heˌspir·ē'óid·ē·ə }

Heterakidae [INV ZOO] A group of nematodes assigned either to the suborder Oxyurina or the suborder Ascaridina. { ˌhed·ə'rak·əˌdē }

Heterakoidea [INV ZOO] A superfamily of parasitic nematodes of the order Ascaridida, characterized by small, well-developed lips with paired sensilla in the labial region, an infundibular stoma, and a rarely cylindrical esophagus divided into three parts. { ˌhed·ə·rə'kóid·ē·ə }

heterandrous [BOT] Having stamens differing from each other in length or form. { ¦hed·ə¦ran·drəs }

heterauxesis *See* allometry. { ¦hed·ərˌóg'zē·səs }

heteroagglutinin [IMMUNOL] An antibody in normal blood serum capable of agglutinating foreign particles and erythrocytes of other species. { ¦hed·ə·rō·ə'glüd·ən·ən }

heteroallele [GEN] One of two or more alternative forms of a gene that differ at nonidentical mutation sites. { ˌhed·ə·rō·ə'lēl }

heteroauxin [BIOCHEM] $C_{10}H_9O_2N$ A plant growth hormone with an indole skeleton. { ¦hed·ə·rō'ók·sən }

Heterobasidiomycetidae [MYCOL] A class of fungi in which the basidium either is branched or is divided into cells by cross walls. { ¦hed·ə·rō·bəˌsid·ē·ōˌmī¦sed·əˌdē }

heteroblastic [EMBRYO] Arising from different tissues or germ layers, in referring to similar organs in different species. { ¦hed·ə·rō¦blas·tik }

Heterocapsina [BOT] An order of green algae in the class Xanthophyceae. [INV ZOO] A suborder of yellow-green to green flagellate protozoans in the order Heterochlorida. { ¦hed·ə·rō 'kap·sə·nə }

heterocarpous [BOT] Producing two distinct types of fruit. { ˌhed·ə·rō'kär·pəs }

heterocentric chromosome [GEN] A dicentric chromosome whose centromeres are of unequal strength. { ¦hed·ə·rəˌsen·trik 'krō·məˌsōm }

Heterocera [INV ZOO] A formerly recognized suborder of Lepidoptera including all forms without clubbed antennae. { ˌhed·ə'räs·ə·rə }

heterocercal [VERT ZOO] Pertaining to the caudal fin of certain fishes and indicating that the upper lobe is larger, with the vertebral column terminating in this lobe. { ¦hed·ə·rō¦sər·kəl }

Heteroceridae [INV ZOO] The variegated mud-loving beetles, a family of coleopteran insects in the superfamily Dryopoidea. { ¦hed·ə·rō¦ser·əˌdē }

Heterocheilidae [INV ZOO] A family of parasitic roundworms in the superfamily Ascaridoidea. { ¦hed·ə·rō¦kī·ləˌdē }

heterochlamydeous [BOT] Having the perianth differentiated into a distinct calyx and a corolla. { ¦hed·ə·rō·klə'mid·ē·əs }

Heterochlorida [INV ZOO] An order of yellow-green to green flagellate oraganisms of the class Phytamastigophorea. { ¦hed·ə·rō'klȯr·ə·də }

heterochromatin [CELL MOL] Specialized chromosome material which remains tightly coiled even in the nondividing nucleus and stains darkly in interphase. { ¦hed·ə·rō'krō·məd·ən }

heterochromia [PHYSIO] A condition in which the two irises of an individual have different colors, or in which one iris has two colors. { ˌhed·ə·rō'krō·mē·ə }

heterochronic mutation [GEN] A mutation that perturbs the relative timing of events during postembryonic development. { ¦hed·ə·rəˌkrä·nik myü'tā·shən }

heterochronism [EMBRYO] Deviation from the normal sequence of organ formation; a factor in evolution. { ˌhed·ə'räk·rəˌniz·əm }

heterochrony [EVOL] An evolutionary phenomenon that involves changes in the rate and timing of species development. { ˌhed·ə'rä·krə·nē }

heterococcolith [BIOL] A coccolith with crystals arranged into boat, trumpet, or basket shapes. { ˌhed·ə·rō'käk·əˌlith }

heterocoelous [ANAT] Pertaining to vertebrae with centra having saddle-shaped articulations. { ¦hed·ə·rō¦sē·ləs }

Heterocotylea [INV ZOO] The equivalent name for Monogenea. { ˌhed·ə·rō·kə'til·ē·ə }

heterocyst [BOT] Clear, thick-walled cell occurring at intervals along the filament of certain blue-green algae. { 'hed·ə·rəˌsist }

heterodactylous [VERT ZOO] Having the first two toes turned backward. { ¦hed·ə·rō¦dakt·əl·əs }

Heterodera [INV ZOO] The cyst nematodes, a genus of phytoparasitic worms that live in the internal root systems of many plants. { ˌhed·ə'räd·ə·rə }

heterodimer [BIOCHEM] A protein made of paired polypeptides that differ in their amino acid sequences. { ˌhed·ə·rō'dī·mər }

heterodisomy [GEN] A type of uniparental disomy in which two different homologous chromosomes and their frequently different alleles are inherited from the same parent. { ˌhed·əˌrō·dī'sō·mē }

heterodont [ANAT] Having teeth that are variable in shape and differentiated into incisors, canines, and molars. [INV ZOO] In bivalves, having two types of teeth on one valve which fit into depressions on the other valve. { 'hed·ə·rəˌdänt }

Heterodonta [INV ZOO] An order of bivalve mollusks in some systems of classification; hinge teeth are few in number and variable in form. { ˌhed·ə·rə'dän·tə }

Heterodontoidea [VERT ZOO] A suborder of sharks in the order Selachii which is represented by the single living genus *Heterodontus*. { ¦hed·ə·rōˌdän'tȯid·ē·ə }

heteroduplex [CELL MOL] A double-stranded molecule of deoxyribonucleic acid in which the two strands show noncomplementary sections. [GEN] A double-stranded deoxyribonucleic molecule comprising strands from different individuals. { ˌhed·ə·rō'düˌpleks }

heteroecious [BIOL] Pertaining to forms that pass through different stages of a life cycle in different hosts. { ˌhed·ə'rē·shəs }

heterogamete [BIOL] A gamete that differs in size, appearance, structure, or sex chromosome content from the gamete of the opposite sex. Also known as anisogamete. { ¦hed·ə·rō 'gaˌmēt }

heterogametic sex [GEN] That sex of some species which produces two or more different kinds of gametes that differ in their sex chromosome content. { ¦hed·ə·rō·gə'med·ik 'seks }

heterogamety [GEN] The production of different kinds of gametes by one sex of a species. { ¦hed·ə·rō'gam·əd·ē }

heterogamous [BIOL] Of or pertaining to heterogamy. { ˌhed·ə'räg·ə·məs }

heterogamy [BIOL] **1.** Alternation of a true sexual generation with a parthenogenetic generation. **2.** Sexual reproduction by fusion of unlike gametes. Also known as anisogamy. [BOT] Condition of producing two kinds of flowers. { ˌhed·ə'räg·ə·mē }

heterogeneity [BIOL] The condition or state of being different in kind or nature. { ¦hed·ə·rə·jə'nē·əd·ē }

heterogeneous nuclear ribonuclear protein [CELL MOL] Any of a large family of proteins involved in the processing of pre-messenger ribonucleic acid (mRNA). { ¦hed·ə·rəˌjēn·ē·əs ¦nü·klē·ər ¦rī·bōˌnü·klē·ər 'prōˌtēn }

heterogeneous ribonucleic acid [CELL MOL] A

large molecule of ribonucleic acid that is believed to be the precursor of messenger ribonucleic acid. Abbreviated H-RNA. { ˌhed·əˈrə'jē·nē·əs ¦rī·bō·nü¦klē·ik 'as·əd }

Heterogeneratae [BOT] A class of brown algae distinguished by a heteromorphic alteration of generations. { ¦hed·ə·rō·ji'ner·əd·ē }

heterogenesis [BIOL] Alternation of generations in a complete life cycle, especially the alternation of a dioecious generation with one or more parthenogenetic generations. { ¦hed·ə·rō'jen·ə·səs }

heterogenetic antigen *See* heterophile antigen. { ¦hed·ə·rō·jə¦ned·ik 'ant·i·jən }

heterogenous [BIOL] Not originating within the body of the organism. { ˌhed·ə'räj·ə·nəs }

heterogenous vaccine [IMMUNOL] A vaccine derived from a source other than the patient. { ˌhed·ə'räj·ə·nəs vak'sēn }

Heterognathi [VERT ZOO] An equivalent name for Cypriniformes. { ˌhed·ə'räg·nəˌthī }

heterogony [BIOL] **1.** Alteration of generations in a complete life cycle, especially of a dioecious and hermaphroditic generation. **2.** *See* allometry. [BOT] Having heteromorphic perfect flowers with respect to lengths of the stamens or styles. { ˌhed·ə'räg·ə·nē }

heterograft [IMMUNOL] A tissue or organ obtained from an animal of one species and transplanted to the body of an animal of another species. Also known as heterologous graft. { 'hed·ə·rōˌgraft }

heterohemolysin [IMMUNOL] Hemolytic amboceptor against the erythrocytes of a species different from that used to obtain the amboceptor. { ¦hed·ə·rō·hə'mäl·ə·sən }

heterokaryon [GEN] Cell with two or more nuclei originating from different cell types or species. [MYCOL] A bi- or multinucleate cell having genetically different kinds of nuclei. { ¦hed·ə·rō'kar·ēˌän }

heterokaryosis [MYCOL] The condition of a bi- or multinucleate cell having nuclei of genetically different kinds. { ˌhed·ə·rōˌkar·ē'ō·səs }

heterokaryotype [GEN] A karyotype that is heterozygous for a chromosome mutation. { ˌhed·ə·rə'kar·ē·əˌtīp }

heterokont [BIOL] An individual, especially among certain algae, having unequal flagella. { 'hed·ə·rōˌkänt }

heterolactic fermentation [MICROBIO] A type of lactic acid fermentation by which small yields of lactic acid are produced and much of the sugar is converted to carbon dioxide and other products. { ¦hed·ə·rō'lak·tik ˌfər·mən'tā·shən }

heterolateral [ANAT] Of, pertaining to, or located on the opposite side. { ˌhed·ə·rō'lad·ə·rəl }

heterolecithal [CELL MOL] Of an egg, having the yolk distributed unevenly throughout the cytoplasm. { ˌhed·ə·rō'les·ə·thəl }

heterologous graft *See* heterograft. { ˌhed·ə'räl·ə·gəs 'graft }

heterologous stimulus [PHYSIO] A form of energy capable of exciting any sensory receptor or form of nervous tissue. { ˌhed·ə'räl·ə·gəs 'stim·yə·ləs }

heterologous tumor [MED] A neoplasm composed of tissues that differ from those of the organ at the site of the tumor. { ˌhed·ə'räl·ə·gəs 'tü·mər }

heteromedusoid [INV ZOO] A styloid type of sessile gonophore. { ¦hed·ə·rō'med·yəˌsȯid }

Heteromera [INV ZOO] The equivalent name for Tenebrionoidea. { ¦hed·ə¦räm·ə·rə }

heteromerous [BOT] Of a flower, having one or more whorls made up of a different number of members than the remaining whorls. { ¦hed·ə¦räm·ə·rəs }

heterometaplasia [MED] Change in the character of an autograft. { ¦hed·ə·rō¦med·ə'plā·zhə }

Heteromi [VERT ZOO] An equivalent name for Notacanthiformes. { ˌhed·ə'rōˌmī }

heteromixis [MYCOL] In Fungi, sexual reproduction which involves the fusion of genetically different nuclei, each from a different thallus. { ¦hed·ə·rō'mik·səs }

heteromorphic [CELL MOL] Having synoptic or sex chromosomes that differ in size or form. [MED] Differing from the normal in size or morphology. [ZOO] Having a different form at each stage of the life history. { ¦hed·ə·rō'mȯr·fik }

heteromorphosis [BIOL] Regeneration of an organ or part that differs from the original structure at the site. [EMBRYO] Formation of an organ at an abnormal site. Also known as homoeosis. { ˌhed·ə·rō'mȯr·fə·səs }

Heteromyidae [VERT ZOO] A family of the mammalian order Rodentia containing the North American kangaroo mice and the pocket mice. { ˌhed·ə'räm·əˌdē }

Heteromyinae [VERT ZOO] The spiny pocket mice, a subfamily of the rodent family Heteromyidae. { ˌhed·ə·rō'mī·əˌnē }

Heteromyota [INV ZOO] A monospecific order of wormlike animals in the phylum Echiurida. { ˌhed·ə·rō'mī·ə·tə }

Heterondontidae [VERT ZOO] The Port Jackson sharks, a family of aberrant modern elasmobranchs in the suborder Heterodontoidea. { ˌhed·ə·rän'dänt·əˌdē }

Heteronemertini [INV ZOO] An order of the class Anopla; individuals have a middorsal blood vessel and a body wall composed of three muscular layers. { ¦hed·ə·rōˌnem·ər'tī·nī }

Heteroneura [INV ZOO] A suborder of Lepidoptera; individuals are characterized by fore- and hindwings that differ in shape and venation and by sucking mouthparts. { ˌhed·ə·rō'nůr·ə }

heteronuclear culture [CELL MOL] A cell culture showing a marked variation in chromosome complement among the cells. { ¦hed·ə·rəˌnü·klē·ər ¦kəl·chər }

heteropelmous [VERT ZOO] Having bifid flexor tendons of the toes. { ¦hed·ə·rō¦pel·məs }

heterophagic vacuole *See* food vacuole. { ˌhed·ə·rō'faj·ik 'vak·yə·wōl }

heterophile agglutination test [PATH] A test for the presence of heterophile antibodies in the serum produced in infectious mononucleosis; agglutination of sheep red cells is a positive test. Also known as heterophile antibody test; Paul-Bunnell test. { 'hed·ə·rə,fīl ə,glüd·ən'ā·shən ,test }

heterophile antibody [IMMUNOL] Substance that will react with heterophile antigen; found in the serum of patients with infectious mononucleosis. { 'hed·ə·rə,fīl 'an·ti·bäd·ē }

heterophile antibody test *See* heterophile agglutination test. { 'hed·ə·rə,fīl 'an·ti,bäd·ē ,test }

heterophile antigen [IMMUNOL] A substance that occurs in unrelated species of animals but has similar serologic properties among them. Also known as heterogenetic antigen. { 'hed·ə·rə,fīl 'ant·i·jən }

heterophile leukocyte [HISTOL] A neutrophile of vertebrates other than humans. { 'hed·ə·rə,fīl 'lü·kə,sīt }

heterophyiasis [MED] Presence of the minute intestinal fluke *Heterophyes heterophyes* in the small intestine of humans. { ¦hed·ə·rō'fī·ə·səs }

heterophyllous [BOT] Having more than one form of foliage leaves on the same plant or stem. { ,hed·ə'räf·ə·ləs }

heterophyte [BOT] A plant that depends upon living or dead plants or their products for food materials. { 'hed·ə·rə,fīt }

Heteropiidae [INV ZOO] A family of calcareous sponges in the order Sycettida. { ¦hed·ə·rō'pī·ə,dē }

heteroplasia [MED] **1.** The presence of a tissue in an abnormal location. **2.** A process whereby tissues are displaced to or developed in locations foreign to their normal habitats. { ,hed·ə·rō'plā·zhə }

heteroplastic [BIOL] Pertaining to transplantation between individuals of different species within the same genus. { ,hed·ə·rō'plas·tik }

heteroplastidy [BOT] The condition of having two kinds of plastids, chloroplasts and leukoplasts. { ,hed·ə·rō'plas·təd·ē }

heteroploidy [GEN] A chromosome complement in which one or more chromosomes, or parts of chromosomes, are present in number different from the numbers of the rest. { ¦hed·ə·rō¦plȯid·ē }

heteropolysaccharide [BIOCHEM] A polysaccharide which is a polymer consisting of two or more different monosaccharides. { ¦hed·ə·rō ,päl·ē'sak·ə,rīd }

Heteroporidae [INV ZOO] A family of trepostomatous bryozoans in the order Cyclostomata. { ¦hed·ə·rō'pȯr·ə,dē }

Heteroptera [INV ZOO] The equivalent name for Hemiptera. { ,hed·ə'räp·tə·rə }

heteropycnosis [CELL MOL] Differential condensation of certain chromosomes, such as sex chromosomes, or chromosome parts. { ¦hed·ə·rō·pik'nō·səs }

heterosis [GEN] The increase in size, yield, and performance found in some hybrid plants and aninals, especially if the parents are from inbred stocks. Also known as hybrid vigor. { ,hed·ə'rō·səs }

Heterosomata [VERT ZOO] The equivalent name for Pleuronectiformes. { ,hed·ə·rō'sō·məd·ə }

Heterospionidae [INV ZOO] A monogeneric family of spioniform worms found in shallow and abyssal depths of the Atlantic and Pacific oceans. { ¦hed·ə·rō,spī'än·ə,dē }

heterospory [BOT] Development of more than one type of spores, especially relating to the microspores and megaspores in ferns and seed plants. { ,hed·ə'räs·pə·rē }

heterostemony [BOT] Presence of two or more different types of stamens in the same flower. { ,hed·ə·rō'stem·ə·nē }

heterostyly [BOT] Condition or state of flowers having unequal styles. { ,hed·ə'räst·əl·ē }

Heterotardigrada [INV ZOO] An order of the tardigrades exhibiting wide morphologic variations. { ¦hed·ə·rō,tär'dig·rə·də }

heterotaxia [ANAT] The reversed polarity of one or more individual organs with respect to the left-right axis. { ,hed·ə·rə'tak·sē·ə }

heterothallic [BOT] Pertaining to a mycelium with genetically incompatible hyphae, therefore requiring different hyphae to form a zygospore; refers to fungi and some algae. { ¦hed·ə·rō ¦thal·ik }

heterotherm [ECOL] An animal that is endothermic part of the time but can reduce metabolic heat production and lower body temperature when conservation of food energy supplies is necessary. { 'hed·ə·rə,thərm }

heterotopia [ECOL] An abnormal habitat. [MED] Displacement of an organ or other body part from its natural position. { ,hed·ə·rō'tō·pē·ə }

heterotopic [BIOL] Pertaining to transplantation of tissue from one site to another on the same organism. [MED] Occurring in an abnormal anatomic location. { ¦hed·ə·rō¦täp·ik }

heterotopic epithelium [MED] Intestinal epithelium and goblet cells occurring in the stomach. { ¦hed·ə·rō¦täp·ik ,ep·ə'thē·lē·əm }

heterotopic pregnancy [MED] Double pregnancy with one fetus within and the other outside the uterus. { ¦hed·ə·rō¦täp·ik 'preg·nən·sē }

heterotopic transplantation [BIOL] A graft transplanted to an abnormal anatomical location on the host. { ¦hed·ə·rō¦täp·ik ,tranz·plən'tā·shən }

Heterotrichida [INV ZOO] A large order of large ciliates in the protozoan subclass Spirotrichia; buccal ciliature is well developed and some species are pigmented. { ,hed·ə·rō'trik·ə·də }

Heterotrichina [INV ZOO] A suborder of the protozoan order Heterotrichida. { ,hed·ə·rō'trik·ə·nə }

heterotrichous [BOT] In certain algae, a body that is divided into both prostrate and erect parts. { ¦hed·ə¦rä·trə·kəs }

heterotroph [BIOL] An organism that obtains nourishment from the ingestion and breakdown of organic matter. { 'hed·ə·rō,träf }

heterotrophic ecosystem [ECOL] An ecosystem that depends upon preformed organic matter that is imported from autotrophic ecosystems elsewhere. { ¦hed·ə·rə,träf·ik 'ek·ō,sis·təm }

heterotrophic effect [BIOCHEM] The interaction between different ligands, such as the effect of an inhibitor or activator on the binding of a substrate by an enzyme. { ¦hed·ə·rō¦träf·ik i'fekt }

heterotrophic succession [ECOL] A type of ecological succession that involves decomposer organisms. { ¦hed·ə·rə,träf·ik sək'sesh·ən }

heterotropia *See* strabismus. { ,hed·ə·rō'trō·pē·ə }

heterotropic enzyme [BIOCHEM] A type of allosteric enzyme in which a small molecule other than the substrate serves as the allosteric reflector. { ¦hed·ə·rō¦träp·ik 'en,zīm }

heteroxenous [BIOL] Requiring more than one host to complete a life cycle. { ¦hed·ə¦räk·sə·nəs }

heterozooid [INV ZOO] Any of the specialized, nonfeeding zooids in a bryozoan colony. { ,hed·ə·rō'zō,ȯid }

heterozygote [GEN] An individual that has different alleles at one or more loci and therefore produces gametes of two or more different kinds with respect to their loci. { ¦hed·ə·rō'zī,gōt }

heterozygote advantage *See* heterozygote superiority. { ,hed·ə·rō¦zī,gōt ad'van·tij }

heterozygote superiority [GEN] The greater fitness of an organism that is heterozygous at a given genetic locus as compared with either homozygote. Also known as heterozygote advantage. { ,hed·ə·rō¦zī,gōt sü,pir·ē'ȯr·əd·ē }

heterozygous [GEN] Of or pertaining to a heterozygote. { ¦hed·ə·rō¦zī·gəs }

Hevea [BOT] The rubber tree genus of the order Euphoriales from which the largest volumes of latex are harvested for use in the manufacture of natural rubber. { 'hē·vē·ə }

hexacanth [INV ZOO] Having six hooks; refers specifically to the embryo of certain tapeworms. { 'hek·sə,kanth }

Hexacorallia [INV ZOO] The equivalent name for Zoantharia. { ,hek·sə·kə'ral·ē·ə }

hexactin [INV ZOO] A spicule, especially in Porifera, having six equal rays at right angles to each other. { hek'sak·tən }

Hexactinellida [INV ZOO] A class of the phylum Porifera which includes sponges with a skeleton made up basically of hexactinal siliceous spicules. { hek,sak·tə'nel·ə·də }

Hexactinosa [INV ZOO] An order of sponges in the subclass Hexasterophora; parenchymal megascleres form a rigid framework and consist of simple hexactins. { hek,sak·tə'nō·sə }

hexamethonium [PHARM] One of a homologous series of polymethylene bis(trimethylammonium) ions, of the general formula $[(CH_3)_3N(CH_2)_6N(CH_3)_3]^{2+}$, in which *n* is 6; possesses potent ganglion-blocking action, effecting reduction in blood pressure; used clinically as a salt, commonly bromide or iodide. { ,hek·sə·mə'thō·nē·əm 'klȯr,īd }

Hexanchidae [VERT ZOO] The six- and seven-gill sharks, a group of aberrant modern elasmobranchs in the suborder Notidanoidea. { ,hek'saŋ·kə,dē }

hexapetalous [BOT] Having or being a perianth comprising six petaloid divisions. { ¦hek·sə'ped·əl·əs }

Hexapoda [INV ZOO] An equivalent name for Insecta. { hek'säp·əd·ə }

hexaster [INV ZOO] A type of hexactin with branching rays that form star-shaped figures. { 'hek,sas·tər }

Hexasterophora [INV ZOO] A subclass of sponges of the class Hexactinellida in which parenchymal microscleres are typically hexasters. { ,hek,sas·tə'räf·ə·rə }

hexobarbital [PHARM] $C_{12}H_{16}N_2O_3$ A sedative and hypnotic of short duration of action; used also, as the sodium derivative, to induce surgical anesthesia. { ¦hek·sə'bär·bə·tȯl }

hexoglycan [BIOCHEM] A polysaccharide that yields hexose monosaccharides on hydrolysis. { ,hek·sə'glī,kan }

hexokinase [BIOCHEM] Any enzyme that catalyzes the phosphorylation of hexoses. { ¦hek·sō'kī,nās }

hexosamine [BIOCHEM] A primary amine derived from a hexose by replacing the hydroxyl with an amine group. { hek'säs·ə,mēn }

hexosaminidase A [BIOCHEM] An enzyme which catalyzes the hydrolysis of the N-acetylgalactosamine residue from certain gangliosides. { hek,säs·ə'min·ə,dās 'ā }

hexosan *See* hemicellulose. { 'hek·sō,san }

hexose [BIOCHEM] Any monosaccharide that contains six carbon atoms in the molecule. { 'hek,sōs }

hexose diphosphate pathway *See* glycolytic pathway. { 'hek,sōs dī'fäs,fāt 'path,wā }

hexose monophosphate cycle [BIOCHEM] A pathway for carbohydrate metabolism in which one molecule of hexose monophosphate is completely oxidized. { 'hek,sōs ¦män·ō¦fäs,fāt 'sī·kəl }

hexose phosphate [BIOCHEM] Any one of the phosphoric acid esters of a hexose, notably glucose, formed during the metabolism of carbohydrates by living organisms. { 'hek,sōs 'fäs,fāt }

hexulose [BIOCHEM] A ketose made from a six-carbon-chain monosaccharide. { 'heks·yə,lōs }

Heyd's syndrome *See* hepatorenal syndrome. { 'hīdz ,sin,drōm }

Hfr *See* high-frequency recombination.

HG [MED] A symbol used in electrocardiography for the longitudinal axis of the heart as projected on the frontal plane.

hiascent [BIOL] Gaping. { hī'ā·shənt }

hiatus [ANAT] A space or a passage through an organ. { hī'ād·əs }

hiatus hernia [MED] Hernia through the esophageal hiatus, usually of a portion of the stomach. { hī'ād·əs 'hər·nē·ə }

hibernaculum [BIOL] A winter shelter for plants or dormant animals. [BOT] A winter bud or other winter plant part. [INV ZOO] A winter resting bud produced by a few fresh-water bryozoans which grows into a new colony in the spring. { ˌhī·bər'nak·yə·ləm }

hibernation [PHYSIO] **1.** Condition of dormancy and torpor found in cold-blooded vertebrates and invertebrates. **2.** *See* deep hibernation. { ˌhī·bər'nā·shən }

hiccup *See* singultus. { 'hik·əp }

hickory [BOT] The common name for species of the genus *Carya* in the order Fagales; tall deciduous tree with pinnately compound leaves, solid pith, and terminal, scaly winter buds. { 'hik·ə·rē }

hide [VERT ZOO] Outer covering of an animal. { hīd }

hidradenitis [MED] Inflammation of sweat glands. { ˌhīˌdrad·ən'īd·əs }

hidradenoma papilliferum [MED] Benign tumor of sweat glands, usually on the vulva or perineum. { ˌhīˌdrad·ən'ō·mə ˌpap·ə'lif·ə·rəm }

hidrosis [MED] Abnormally profuse sweat. [PHYSIO] The formation and excretion of sweat. { hī'drō·səs }

high-altitude disease *See* mountain sickness. { 'hī ¦al·təˌtüd diˌzēz }

high-altitude erythremia *See* mountain sickness. { 'hī ¦al·təˌtüd ˌer·ə'thrē·mē·ə }

high-density lipoprotein [BIOCHEM] A lipoprotein containing more proteins than lipids that transports excess cholesterol from tissues to the liver for excretion. Abbreviated HDL. { ¦hī ˌden·səd·ē ˌlī·pō'prōˌtēn }

high enema [MED] An enema injected into the colon. { 'hī 'en·ə·mə }

high-frequency recombination [MICROBIO] A bacterial cell type, especially *Escherichia coli*, having an integrated F factor and characterized by a high frequency of recombination. Abbreviated Hfr. { 'hī ¦frē·kwən·sē ˌrēˌkäm·bə'nā·shən }

highmoor bog [ECOL] A bog whose surface is covered by sphagnum mosses and is not dependent upon the water table. { 'hīˌmür 'bäg }

Hikojima serotype [IMMUNOL] An immunologically distinct group of *Vibrio* somatic O antigens. { ˌhē·kō'jē·mə 'ser·əˌtīp }

Hill plot [BIOCHEM] A graphic representation of the Hill reaction. { 'hil ˌplät }

Hill reaction [BIOCHEM] The release of molecular oxygen by isolated chloroplasts in the presence of a suitable electron receptor, such as ferricyanide. { 'hil rēˌak·shən }

hilum [ANAT] *See* hilus. [BOT] Scar on a seed marking the point of detachment from the funiculus. { 'hī·ləm }

hilus [ANAT] An opening or recess in an organ, usually for passage of a vessel or duct. Also known as hilum. { 'hī·ləs }

Himantandraceae [BOT] A family of dicotyledonous plants in the order Magnoliales characterized by several, uniovulate carpels and laminar stamens. { hə¦mant·ən'drās·ēˌē }

himantioid [MYCOL] Pertaining to a mycelium arranged in spreading fanlike cords. { hə 'man·tēˌȯid }

Himantopterinae [INV ZOO] A subfamily of lepidopteran insects in the family Zygaenidae including small, brightly colored moths with narrow hindwings, ribbonlike tails, and long hairs covering the body and wings. { hə¦man·tō'ter·əˌnē }

hindbrain *See* rhombencephalon. { 'hīnˌbrān }

hindgut [EMBRYO] The caudal portion of the embryonic digestive tube in vertebrates. { 'hīnˌgət }

hinge joint *See* ginglymus. { 'hinj ˌjȯint }

hinge plate [INV ZOO] **1.** In bivalve mollusks, the portion of a valve that supports the hinge teeth. **2.** The socket-bearing part of the dorsal valve in brachiopods. { 'hinj ˌplāt }

hinge tooth [INV ZOO] A projection on a valve of a bivalve mollusk near the hinge line. { 'hinj ˌtüth }

Hiodontidae [VERT ZOO] A family of tropical, fresh-water actinopterygian fishes in the order Osteoglossiformes containing the mooneyes of North America. { ˌhī·ə'dänt·əˌdē }

hip [ANAT] **1.** The region of the junction of thigh and trunk. **2.** The hip joint, formed by articulation of the femur and hipbone. { hip }

hipbone [ANAT] A large broad bone consisting of three parts, the ilium, ischium, and pubis; makes up a lateral half of the pelvis in mammals. Also known as innominate. { 'hipˌbōn }

Hippidea [INV ZOO] A group of decapod crustaceans belonging to the Anomura and including cylindrical or squarish burrowing crustaceans in which the abdomen is symmetrical and bent under the thorax. { hi'pid·ē·ə }

Hippoboscidae [INV ZOO] The louse flies, a family of cyclorrhaphous dipteran insects in the section Pupipara. { ˌhip·ə'bäs·kəˌdē }

hippocampal sulcus [ANAT] A fissure on the brain situated between the para hippocampal gyrus and the fimbria hippocampi. Also known as dentate fissure. { ¦hip·ə¦kam·pəl 'səl·kəs }

hippocampus [ANAT] A ridge that extends over the floor of the descending horn of each lateral ventricle of the brain. { ˌhip·ə'kam·pəs }

Hippocampus [VERT ZOO] A genus of marine fishes in the order Gasterosteiformes which contains the sea horses. { ˌhip·ə'kam·pəs }

Hippocrateaceae [BOT] A family of dicotyledonous plants in the order Celastrales distinguished by an extrastaminal disk, mostly opposite leaves, seeds without endosperm, and a well-developed latex system. { ˌhip·əˌkrād·ē'ās·ēˌē }

hippocrepiform [BIOL] Horseshoe-shaped. { ¦hip·ə¦krep·əˌfȯrm }

Hippoglossidae [VERT ZOO] A family of actinopterygian fishes in the order Pleuronectiformes composed of the flounders and plaice. { ˌhip·ə'gläs·əˌdē }

Hippomorpha [VERT ZOO] A suborder of the mammalian order Perissodactyla containing horses, zebras, and related forms. { ˌhip·ə'mȯr·fə }

Hippopotamidae [VERT ZOO] The hippopotamuses, a family of palaeodont mammals in the superfamily Anthracotherioidea. { ˌhip·ə·pəd'am·əˌdē }

hippopotamus [VERT ZOO] The common name for two species of artiodactyl ungulates composing the family Hippopotamidae. { ˌhip·ə'päd·ə·məs }

Hipposideridae [VERT ZOO] The Old World leaf-nosed bats, a family of mammals in the order Chiroptera. { ˌhiˌpō·sə'der·əˌdē }

Hirschsprung's disease [MED] A disease caused by absence of the myenteric ganglion cells in a segment of rectum or distal colon, resulting in spasm of the affected part and dilation of the bowel proximal to the defect. { 'hərshˌprůŋz diˌzēz }

hirsute [BIOL] Shaggy; hairy. { 'hərˌsüt }

hirsutism [MED] An abnormal condition characterized by growth of hair in unusual places and in unusual amounts. { 'hər·səˌtiz·əm }

Hirudinea [INV ZOO] A class of parasitic or predatory annelid worms commonly known as leeches; all have 34 body segments and terminal suckers for attachment and locomotion. { ˌhi·rə'din·ē·ə }

Hirudinidae [VERT ZOO] The swallows, a family of passeriform birds in the suborder Oscines. { ˌhi·rə'din·əˌdē }

***his* operon** [GEN] A sequence of nine contagious genes in the bacterial chromosome invariious species, these code for all the enzymes of histidine biosynthesis. { 'his 'äp·əˌrän }

hispid [BIOL] Having a covering of bristles or minute spines. { 'his·pəd }

hispidulous [BIOL] Hispid to a minute degree. { his'pij·ə·ləs }

histamine [BIOCHEM] $C_5H_9N_3$ An amine derivative of histadine which is widely distributed in human tissues. { 'his·təˌmēn }

Histeridae [INV ZOO] The clown beetles, a large family of coleopteran insects in the superfamily Histeroidea. { hi'ster·əˌdē }

Histeroidea [INV ZOO] A superfamily of coleopteran insects in the suborder Polyphaga. { ˌhis·tə'rȯid·ē·ə }

histidase [BIOCHEM] An enzyme found in the liver of higher vertebrates that catalyzes the deamination of histidine to urocanic acid. { 'his·təˌdās }

histidine [BIOCHEM] $C_6H_9O_2N_3$ A crystalline basic amino acid present in large amounts in hemoglobin and resulting from the hydrolysis of most proteins. { 'his·təˌdēn }

histidine-kinase-associated receptor [CELL MOL] A class of enzyme-linked cell-surface receptors found in bacteria, yeast, and plant cells that transduce intracellular signals via a two-component pathway that results in the phosphorylation of an intracellular messenger protein. { ¦his·təˌdēn ¦kīˌnās əˌsō·sēˌād·əd ri'sep·tər }

histidinemia [MED] An asymptomatic, hereditary metabolic disorder involving a deficiency of histidase with high blood and urine levels of histidine, urocanic acid, and sometimes alanine. { ¦his·tə·də'nē·ē·ə }

histiocyte *See* macrophage. { 'his·tē·əˌsīt }

histiocytoma [MED] **1.** Benign tumor composed of histiocytes. **2.** Dermatofibroma. { ˌhis·tē·ōˌsī'tō·mə }

histiocytosis [MED] Abnormal proliferation of histiocytes, especially in hematopoietic tissues. { ˌhis·tē·ōˌsī'tō·səs }

Histioteuthidae [INV ZOO] A family of cephalopod mollusks containing several species of squids. { ˌhis·tē·ō'tü·thəˌdē }

histochemistry [BIOCHEM] A science that deals with the distribution and activities of chemical components in tissues. { ¦hi·stō'kem·ə·strē }

histocompatibility [IMMUNOL] The capacity to accept or reject a tissue graft. { ¦hi·stō·kəm 'pad·ə'bil·əd·ē }

histocompatibility gene [GEN] In mammals, any of the genes, specially those of the major histocompatibility complex, that influence immunological properties of cellular antigens. { ˌhis·tə·kəm'pad·əˌbil·əd·ē ˌjēn }

histodifferentiation [EMBRYO] Differentiation of cell groups into tissues. { ¦hi·stō·dif·əˌren·chē'ā·shən }

histogen [BOT] A clearly delimited region or primary tissue from which the specific parts of a plant organ are thought to be produced. { 'his·tə·jən }

histogenesis [EMBRYO] The developmental process by which the definite cells and tissues which make up the body of an organism arise from embryonic cells. { ˌhis·tə'jen·ə·səs }

histoincompatibility [IMMUNOL] The condition in which a recipient rejects a tissue graft. { ¦hi·stōˌin·kəmˌpad·ə'bil·əd·ē }

histologist [ANAT] An individual who specializes in histology. { hi'stäl·ə·jəst }

histology [ANAT] The study of the structure and chemical composition of animal and plant tissues as related to their function. { hi'stäl·ə·jē }

histolysis [PATH] Disintegration of organic tissue. { hi'stäl·ə·səs }

histomycosis [MED] Infection of deep tissues by a fungus. { ˌhis·təˌmī'kō·səs }

histone [BIOCHEM] Any of the strong, soluble basic proteins of cell nuclei that are precipitated by ammonium hydroxide. { 'hiˌstōn }

histopathology [PATH] A branch of pathology that deals with tissue changes associated with disease. { ˌhi·stō·pə'thäl·ə·jē }

histophysiology [PHYSIO] The science of tissue functions. { ˌhi·stōˌfiz·ē'äl·ə·jē }

Histoplasma [MYCOL] A genus of parasitic fungi. { ˌhis·tə'plaz·mə }

Histoplasma capsulatum [MYCOL] The parasitic fungus that causes histoplasmosis in humans. { ˌhis·tə'plaz·mə ˌkap·sə'läd·əm }

histoplasmin [PHARM] A standardized liquid concentrate of soluble growth factors developed by the fungus *Histoplasma capsulatum*. { ˌhis·tə'plaz·mən }

histoplasmin test [IMMUNOL] Skin test for hypersensitivity reaction to *Histoplasma capsulatum* products in the diagnosis of histoplasmosis. { ˌhis·tə'plaz·mən ˌtest }

histoplasmoma [MED] A tumorlike swelling caused by an inflammatory reaction to *Histoplasma capsulatum*. { ˌhis·təˌplaz'mō·mə }

histoplasmosis [MED] An infectious fungus disease of the lungs of humans caused by *Histoplasma capsulatum*. { ˌhis·təˌplaz'mō·səs }

historadiography [BIOPHYS] A technique for taking x-ray pictures of cells, tissues, or sometimes whole small organisms. { ˌhis·təˌrād·ē'äg·rə·fē }

historical biogeography [ECOL] The study of how species' distributions have changed over time in relationship to the history of landforms, ocean basins, and climate, as well as how those changes have contributed to the evolution of biotas. { hisˌtär·i·kəl ˌbī·ō·jē'ag·rə·fē }

historical zoogeography [ECOL] The study of animal distributions in terms of evolutionary history. { hisˌtär·i·kəl ˌzō·ō·jē'äg·rə·fē }

histotome [BIOL] A microtome used to cut tissue sections for microscopic examination. { 'his·təˌtōm }

Histriobdellidae [INV ZOO] A small family of errantian polychaete worms that live as ectoparasites on crayfishes. { ˌhis·trē·əb'del·əˌdē }

hitchhiking effect [GEN] The increase in frequency of a neutral allele at a locus closely linked to a selectively favored allele at a different locus. { 'hichˌhīk·iŋ iˌfekt }

HIV *See* human immunodeficiency virus.

hives *See* urticaria. { hīvz }

HLA *See* human leukocyte antigen.

HLA complex [IMMUNOL] The major histocompatibility complex of humans. { ¦āch¦el'ā ˌkäm ˌpleks }

hoarding behavior [VERT ZOO] The carrying of food to the home nest for storage, in quantities exceeding daily need. { 'hȯrd·iŋ biˌhāv·yər }

hoarhound *See* marrubiumu. { 'hȯrˌhau̇nd }

hoarse [MED] Having a harsh, discordant voice, caused by an abnormal condition of the larynx or throat. { hȯrs }

hoary [BOT] Having grayish or whitish color, referring to leaves. { hȯr·ē }

Hodgkin's disease [MED] A disease characterized by a neoplastic proliferation of atypical histiocytes in one or several lymph nodes. Also known as lymphogranulomatosis. { 'häj·kənz diˌzēz }

Hodotermitidae [INV ZOO] A family of lower (primitive) termites in the order Isoptera. { ˌhäd·ō·tər'mid·əˌdē }

Hofbauer cell [HISTOL] A large, possibly phagocytic cell found in chorionic villi. { 'hōf·bau̇r ˌsel }

hog cholera [VET MED] A fatal infectious virus disease of swine characterized by fever, diarrhea, and inflammation and ulceration of the intestine; secondary infection by *Salmonella cholerae suis* is common. Also known as African swine fever. { 'häg ˌkäl·ə·rə }

holandric trait [GEN] Any trait appearing only in males. { hə¦lanˌdrik 'trāt }

holarctic zoogeographic region [ECOL] A major unit of the earth's surface extending from the North Pole to 30–45°N latitude and characterized by faunal homogeneity. { hō'lärd·ik ˌzō·ōˌjē·ə¦graf·ik 'rē·jən }

Holasteridae [INV ZOO] A family of exocyclic Euechinoidea in the order Holasteroida; individuals are oval or heart-shaped, with fully developed pore pairs. { ˌhäl·ə'ster·əˌdē }

Holasteroida [INV ZOO] An order of exocyclic Euechinoidea in which the apical system is elongated along the anteroposterior axis and teeth occur only in juvenile stages. { ˌhäl·ə·stə 'rȯid·ə }

holcodont [VERT ZOO] Having the teeth fixed in a long, continuous groove. { 'häl·kəˌdänt }

holdfast [BOT] **1.** A suckerlike base which attaches the thallus of certain algae to the support. **2.** A disklike terminal structure on the tendrils of various plants used for attachment to a flat surface. [INV ZOO] An organ by which parasites such as tapeworms attach themselves to the host. { 'hōlˌfast }

Holectypoida [INV ZOO] An order of exocyclic Euechinoidea with keeled, flanged teeth, with distinct genital plates, and with the ambulacra narrower than the interambulacra on the adoral side. { hōˌlek·tə'pȯid·ə }

holism [BIOL] The view that the whole of a complex system, such as a cell or organism, is functionally greater than the sum of its parts. Also known as organicism. { 'hōˌliz·əm }

holly [BOT] The common name for the trees and shrubs composing the genus *Ilex*; distinguished by spiny leaves and small berries. { 'häl·ē }

hollywood lignumvitae *See* lignumvitae. { 'häl·ēˌwu̇d ˌlig·nəm'vī·dē }

Holobasidiomycetes [MYCOL] An equivalent name for Homobasidiomycetidae. { ¦häl·ō·bəˌsid·ē·ō·mī' sēd·ēz }

holoblastic [EMBRYO] Pertaining to eggs that undergo total cleavage due to the absence of a mass of yolk. { ¦häl·ə¦blas·tik }

holobranch [VERT ZOO] A gill with a row of filaments on each side of the branchial arch. { 'häl·əˌbraŋk }

holocarpic [BOT] **1.** Having the entire thallus developed into a fruiting body or sporangium. **2.** Lacking rhizoids and haustoria. { ¦häl·ō¦kär·pik }

holocellulose [BIOCHEM] The total polysaccharide fraction of wood, straw, and so on, that is composed of cellulose and all of the hemicelluloses and that is obtained when the extractives and the lignin are removed from the natural material. { ¦häl·ō'sel·yəˌlōs }

holocentric chromosome [GEN] A chromosome with centromer; activity/spread along its entire length. { ˌhō·ləˌsen·trik 'krō·məˌsōm }

Holocentridae [VERT ZOO] A family of nocturnal beryciform fishes found in shallow tropical and subtropical reefs; contains the squirrelfishes and soldierfishes. { ˌhäl·ə'sen·trəˌdē }

Holocephali [VERT ZOO] The chimaeras, a subclass of the Chondrichthyes, distinguished by four pairs of gills and gill arches, an erectile

dorsal fin and spine, and naked skin. { ¦häl·ō¦sef·ə,lī }
holochrome [BIOCHEM] A colored chromophore bound to an apoprotein. { 'hōl·ə,krōm }
holococcolith [BIOL] A coccolith with simple rhombic or hexagonal crystals arranged like a mosaic. { ¦häl·ō'käk·ə,lith }
holocoenosis [ECOL] The nature of the action of the environment on living organisms. { ¦häl·ō·sə¦nō·səs }
holocrine gland [PHYSIO] A structure whose cells undergo dissolution and are entirely extruded, together with the secretory product. { 'häl·ə·krən ,gland }
holoechinate [INV ZOO] Having the whole body covered with spines. { ¦hō·lō·ē'kī,nāt }
holoenzyme [BIOCHEM] A complex, fully active enzyme, containing an apoenzyme and a coenzyme. { ¦häl·ō'en,zīm }
hologamy [BIOL] Condition of having gametes similar in size and form to somatic cells. [BOT] Condition of having the whole thallus develop into a gametangium. { hə'läg·ə·mē }
hologony [INV ZOO] Condition of having the germinal area extend the full length of a gonad; refers specifically to certain nematodes. { hə'läg·ə·nē }
hologynic trait [GEN] Any trait appearing only in females. { ,hō·lə,gī·nik 'trāt }
Holometabola [INV ZOO] A division of the insect subclass Pterygota whose members undergo holometabolous metamorphosis during development. { ¦häl·ō·mə'tab·ə·lə }
holometabolous metamorphosis [INV ZOO] Complete metamorphosis, during which there are four stages; the egg, larva, pupa, and imago or adult. { ¦häl·ō·mə'tab·ə·ləs ,med·ə'mör·fə·səs }
holomorphosis [BIOL] Complete regeneration of a lost body structure. { ¦häl·ō'mör·fə·səs }
holomyarian [INV ZOO] Having zones of muscle layers but no muscle cells; refers specifically to certain nematodes. { ¦häl·ō·mī¦a·rē·ən }
holonephros [VERT ZOO] Type of kidney having one nephron beside each somite along the entire length of the coelom; seen in larvae of myxinoid cyclostomes. { ¦häl·ō'ne,frōs }
holophyte [BIOL] An organism that obtains food in the manner of a green plant, that is, by synthesis of organic substances from inorganic substances using the energy of light. { 'häl·ə,fīt }
holoplankton [ZOO] Organisms that live their complete life cycle in the floating state. { ,häl·ō'plaŋk·tən }
holorhinal [VERT ZOO] Among birds, having a rounded anterior margin on the nasal bones. { ¦häl·ō'rīn·əl }
holospondyly [VERT ZOO] The condition in which the vertebral centra and spines are single-pieced or fused. { ,hōl·ə'span·də·lē }
Holostei [VERT ZOO] An infraclass of fishes in the subclass Actinopterygii descended from the Chondrostei and ancestral to the Teleostei. { hə'läs·tē,ī }
holostome [INV ZOO] A type of adult digenetic trematode having a portion of the ventral surface modified as a complex adhesive organ. { 'häl·ə,stōm }
Holothuriidae [VERT ZOO] A family of aspidochirotacean echinoderms in the order Aspidochirotida possessing tentacular ampullae and only the left gonad. { ,häl·ō·thə'rī·ə,dē }
Holothuroidea [INV ZOO] The sea cucumbers, a class of the subphylum Echinozoa characterized by a cylindrical body and smooth, leathery skin. { ,häl·ō,thu̇·rē'ȯid·ē·ə }
Holothyridae [INV ZOO] The single family of the acarine suborder Holothyrina. { ,häl·ō'thī·rə,dē }
Holothyrina [INV ZOO] A suborder of mites in the order Acarina which are large and hemispherical with a deep-brown, smooth, heavily sclerotized cuticle. { ,häl·ō'thī·rə·nə }
Holotrichia [INV ZOO] A major subclass of the protozoan class Ciliatea; body ciliation is uniform with cilia arranged in longitudinal rows. { ,häl·ō'trik·ē·ə }
holotype [SYST] A nomenclatural type for the single specimen designated as "the type" by the original author at the time of publication of the original description. { 'häl·ə,tīp }
holozoic [ZOO] Obtaining food in the manner of most animals, by ingesting complex organic matter. { ¦häl·ə¦zō·ik }
Homalopteridae [VERT ZOO] A small family of cypriniform fishes in the suborder Cyprinoidei. { ¦häm·ə·läp'ter·ə,dē }
Homalorhagae [INV ZOO] A suborder of the class Kinorhyncha having a single dorsal plate covering the neck and three ventral plates on the third zonite. { ,häm·ə'lȯr·ə,gē }
Homalozoa [INV ZOO] A subphylum of echinoderms characterized by the complete absence of radial symmetry. { hə¦mal·ə¦zō·ə }
Homan's sign [MED] Pain in the calf and popliteal area on passive dorsiflexion of the foot, indicating deep venous thrombosis of the calf. Also known as dorsiflexion sign. { 'hō·mənz ,sīn }
Homaridae [INV ZOO] A family of marine decapod crustaceans containing the lobsters. { hō 'mar·ə,dē }
Homarus [INV ZOO] A genus of the family Homaridae comprising most species of lobsters. { 'häm·ə·rəs }
homaxial [BIOL] Having all axes equal. { hōm 'ak·sē·əl }
homeobox [CELL MOL] A highly conserved sequence of deoxyribonucleic acid (DNA) that occurs in the coding region of development-controlling regulatory genes and codes for a protein domain that is similar in structure to certain DNA-binding proteins and is thought to be involved in the control of gene expression during morphogenesis and development. { 'hō·mē·ə,bäks }
homeohydric [PHYSIO] Pertaining to the ability to restrict cellular water loss regardless of environmental conditions. { ,hō·mē·o 'hī·drik }

homeomorph [BIOL] An organism which exhibits a superficial resemblance to another organism even though they have different ancestors. { 'hō·mē·ə,mȯrf }

homeopathy [MED] A system of medicine expounded by Samuel Hahnemann that treats disease by administering to the patient small doses of drugs which produce the signs and symptoms of the disease in a healthy person. Also known as Hahnemannism. { ,hō·mē'äp·ə·thē }

homeosis *See* heteromorphosis. { ,hō·mē'ō·səs }

homeostasis [BIOL] In higher animals, the maintenance of an internal constancy and an independence of the environment. { ,hō·mē·ō'stā·səs }

homeotherm [PHYSIO] An endotherm that maintains a constant body temperature, as do most mammals and birds. { 'hō·mē·ə,thərm }

homeothermia [PHYSIO] The condition of being warm-blooded. { ¦hō·mē·ə¦thər·mē·ə }

homeotic mutation [GEN] A gene whose mutation in *Drosophila* can lead to the substitution of one body part for another, such as a leg for an antenna. { 'hō·mē,äd·ik myü'tā·shən }

home range [ECOL] The physical area of an organism's normal activity. { 'hōm ¦rānj }

Hominidae [VERT ZOO] A family of primates in the superfamily Hominoidea containing one living species, *Homo sapiens*. { hä'min·ə,dē }

Hominoidea [VERT ZOO] A superfamily of the order Primates comprising apes and humans. { hä·mə'nȯid·ē·ə }

Homo [VERT ZOO] The genus of human beings, including modern humans and many extinct species. { 'hō·mō }

homoallele [GEN] One of two or more alternative forms of a gene that differ at identical mutation sites. { ,hō·mō·ə'lēl }

Homobasidiomycetidae [MYCOL] A subclass of basidiomycetous fungi in which the basidium is not divided by cross walls. { ¦hä·mō·bə,sid·ē·ō,mī'sed·ə,dē }

homocercal [VERT ZOO] Pertaining to the caudal fin of certain fishes which has almost equal lobes, with the vertebral column terminating near the middle of the base. { ,häm·ə'sər·kəl }

homochlamydeous [BIOL] Having all members of the perianth similar or not differentiated into calyx or corolla. { ¦hä·mō·klə'mid·ē·əs }

homochromy [ZOO] A form of protective coloration whereby the individual blends into the background. { hə'mäk·rə·mē }

homocysteine [BIOCHEM] $C_4H_9O_2NS$ An amino acid formed in animals by demethylation of methionine. { ¦hä·mə'sis·tēn }

homocystinuria [MED] A hereditary disease characterized by a deficiency of the enzyme serine dehydratase causing incompletely dislocated lenses after the age of 10, thromboembolisms, and usually mental retardation. { ,hō·mō,sis·tə'nu̇r·ē·ə }

homodimer [BIOCHEM] A protein made of paired identical polypeptides. { ,hō·mō'dī·mər }

homodont [VERT ZOO] Having all teeth similar in form; characteristic of nonmammalian vertebrates. { 'hä·mə,dänt }

homoduplex [CELL MOL] A deoxyribonucleic acid duplex in which the nitrogenous bases of the two strands are precisely complementary. { ,hō·mō'dü,pleks }

homodynamic [INV ZOO] Developing through continuous successive generations without a diapause; applied to insects. { ,hä·mə,dī'näm·ik }

homoecious [BIOL] Having one host for all stages of the life cycle. { hō'mē·shəs }

homoeomerous [BOT] Having algae distributed uniformly throughout the thallus of a lichen. { ¦hō·mē¦äm·ə·rəs }

homofermentative lactobacilli [MICROBIO] Bacteria that produce a single end product, lactic acid, from fermentation of carbohydrates. { ¦hō·mō·fər'men·tə·tiv ¦lak·tō·bə¦sil·ē }

homogametic sex [GEN] The sex that produces one type of gamete with respect to sex chromosome content, such as female mammals and male birds. { ¦hä·mō·gə¦med·ik 'seks }

homogamous [BIOL] Of or pertaining to homogamy. { hə'mäg·ə·məs }

homogamy [BIOL] Inbreeding due to isolation. [BOT] Condition of having all flowers alike. { hə'mäg·ə·mē }

homogenate [BIOL] A tissue that has been finely divided and mixed. { hə'mäj·ə·nət }

homogeneously staining region [CELL MOL] In human chromosomes, an extended chromosomal segment that has a banding pattern and represents a site of gene amplification; found mostly in cancer cells. { ,hō·mə¦jēn·ē·əs·lē ¦stān·iŋ ,rē·jən }

homogentisase [BIOCHEM] The enzyme that catalyzes the conversion of homogentisic acid to fumaryl acetoacetic acid. { ¦hä·mə¦jen·tə,sās }

homogentisic acid [BIOCHEM] $C_8H_8O_4$ An intermediate product in the metabolism of phenylalanine and tyrosine; found in excess in persons with phenylketonuria and alkaptonuria. { ¦hä·mə,jen¦tiz·ik 'as·əd }

homogony [BOT] Condition of having one type of flower, with stamens and pistil of uniform length. { hə'mäg·ə·nē }

homograft rejection [IMMUNOL] An immunologic process by which an individual destroys and casts off a tissue transplanted from a donor of the same species. { 'hä·mə,graft ri'jek·shən }

homoiochlamydeous [BOT] Having perianth leaves alike, not differentiated into sepals and petals. { hō¦mȯi·ō·klə'mid·ē·əs }

homoiogenetic [EMBRYO] Of a determined part of an embryo, capable of inducing formation of a similar part when grafted into an undetermined field. { hō¦mȯi·ō·jə'ned·ik }

homoiothermal [PHYSIO] Referring to an organism which maintains a constant internal temperature which is often higher than that of the environment; common among birds and mammals. Also known as warm-blooded. { hō¦mȯi·ō ¦thər·məl }

homokaryon [MYCOL] A bi- or multinucleate

cell having nuclei all of the same kind. { ˌhä·mə'kar·ēˌän }

homokaryosis [MYCOL] The condition of a bi- or multinucleate cell having nuclei all of the same kind. { ˌhä·məˌkar·ē'ō·səs }

homokaryotype [GEN] A karyotype that is homozygous for a chromosome mutation. { ˌhō·mə'kar·ē·əˌtīp }

homolateral [MED] Situated on the same side. Also known as ipsilateral. { ˌhä·mə'lad·ə·rəl }

homolecithal [CELL MOL] Referring to eggs having small amounts of evenly distributed yolk. Also known as isolecithal. { ˌhä·mə'les·ə·thəl }

homolog [GEN] One of a pair of homologous chromosomes. Also spelled homologue. { 'häm·əˌläg }

homologous [BIOL] Pertaining to a structural relation between parts of different organisms due to evolutionary development from the same or a corresponding part, such as the wing of a bird and the pectoral fin of a fish. { hə'mäl·ə·gəs }

homologous chromosomes [GEN] A pair of chromosomes, one inherited from each parent, that have corresponding gene sequences and that pair during meiosis. { hə¦mäl·ə·gəs 'krō·məˌsōmz }

homologous recombination [GEN] Deoxyribonucleic acid exchange between identical chromosome regions on homologous chromosomes that occurs naturally during meiosis. { hə¦mäl·ə·gə ˌre·käm·bə'nā·shən }

homologous serum jaundice [MED] A type of hepatitis caused by a filtrable virus that exists in the blood plasma and may be passed to another person through blood transfusion. { hə'mäl·ə·gəs ¦sir·əm 'jȯn·dəs }

homologous stimulus [PHYSIO] A form of energy to which a specific sensory receptor is most sensitive. { hə'mäl·ə·gəs 'stim·yə·ləs }

homologous tumor [MED] A neoplasm composed of tissue identical with those of the organ at the site of the tumor. { hə'mäl·ə·gəs 'tü·mər }

homology [BIOL] A fundamental similarity between structures or processes in different organisms that usually results from their having descended from a common ancestor. { hə'mäl·ə·jē }

homomorphism [BOT] Having perfect flowers consisting of only one type. { ˌhä·mə'mȯrˌfiz·əm }

Homoneura [INV ZOO] A suborder of the Lepidoptera with mandibulate mouthparts, and fore- and hindwings that are similar in shape and venation. { ˌhä·mə'nu̇r·ə }

homonomous hemianopsia [MED] Partial blindness affecting the inner half of one field of vision or the outer half of the other; caused by optic nerve lesions posterior to the chiasma. { hə'män·ə·məs ¦hē·mē·ə'näp·sē·ə }

homopetalous [BOT] Having all petals identical. { ¦hä·mə¦ped·əl·əs }

homoplastic [BIOL] Pertaining to transplantation between individuals of the same species. { ˌhō·mō'plas·tik }

homoplastidy [BOT] The condition of having one kind of plastid. { ˌhō·mō'plas·təd·ē }

homoplasy [BIOL] Correspondence between organs or structures in different organisms acquired as a result of evolutionary convergence or of parallel evolution. { 'hä·məˌplā·sē }

homopolymer tail [CELL MOL] A segment that contains only one sort of nucleotide at the 3′ end of a deoxyribonucleic acid or ribonucleic acid molecule. { ˌhō·mə¦päl·i·mər 'tāl }

homopolysaccharide [BIOCHEM] A polysaccharide which is a polymer of one kind of monosaccharide. { ¦hä·mōˌpäl·ē'sak·əˌrīd }

Homoptera [INV ZOO] An order of the class Insecta including a large number of sucking insects of diverse forms. { hō'mäp·tə·rə }

Homo sapiens [VERT ZOO] Modern human species; a large, erect, omnivorous terrestrial biped of the primate family Hominidae. { ¦hō·mō 'sap·ē·ənz }

Homosclerophorida [INV ZOO] An order of primitive sponges of the class Demospongiae with a skeleton consisting of equirayed, tetraxonid, siliceous spicules. { ¦hä·mō¦skler·ə 'fȯr·ə·də }

homoserine [BIOCHEM] $C_4H_9O_3N$ An amino acid formed as an intermediate product in animals in the metabolic breakdown of cystathionine to cysteine. { hə'mäs·əˌrēn }

homosexual [BIOL] Of, pertaining to, or being the same sex. { ¦hō·mə'sek·shə·wəl }

homospory [BOT] Production of only one kind of asexual spore. { hə'mäs·pə·rē }

homothallic [MYCOL] Having genetically compatible hyphae and therefore forming zygospores from two branches of the same mycelium. { ˌhä·mə'thal·ik }

homotropic enzyme [BIOCHEM] A type of allosteric enzyme in which the substrate serves as the allosteric reflector. { ¦hä·mə¦träp·ik 'en ˌzīm }

homotropous [BOT] Having the radicle directed toward the hilum. { hō'mä·trə·pəs }

homotype [SYST] A taxonomic type for a specimen which has been compared with the holotype by another than the author of the species and determined by him to be conspecific with it. { 'hä·məˌtīp }

homotypy [BIOL] Protective device based on resemblance of shape to the background. { ¦hä·mə¦tī·pē }

homovanillic acid [BIOCHEM] $HOC_6H_3(OCH_3)$-CH_2CO_2H A major metabolite of 3-O-methyl dopa; used in enzyme determination. Abbreviated HVA. { ¦hä·mə·və¦nil·ik 'as·əd }

homozygote [GEN] An individual who has identical alleles at one or more loci and therefore produces identical gametes with respect to these loci. { ¦hō·mə'zīˌgōt }

homozygous [GEN] Of or pertaining to a homozygote. { ˌhō·mə'zī·gəs }

honeybee [INV ZOO] *Apis mellifera*. The bee kept

for the commercial production of honey; a member of the dipterous family Apidae. { 'hən·ē,bē }

honeycomb [INV ZOO] A mass of wax cells in the form of hexagonal prisms constructed by honeybees for their brood and honey. { 'hən·ē,kōm }

honeycomb lung [MED] **1.** Condition of the lung in emphysema. **2.** A lung containing small pus-filled cavities. { 'hən·ē,kōm ,ləŋ }

honeydew [INV ZOO] The viscous secretion deposited on leaves by many aphids and scale insects; an attractant for ants. { 'hən·ē,dü }

Honey Dew melon [BOT] A variety of muskmelon (*Cucumis melo*) belonging to the Violales; fruit is large, oval, smooth, and creamy yellow to ivory, without surface markings. { 'hən·ē ,dü ,mel·ən }

honey tube [INV ZOO] Either of a pair of cornicles on the dorsal aspect of one abdominal segment in certain aphids. { 'hən·ē ,tüb }

hoof [VERT ZOO] **1.** Horny covering for terminal portions of the digits of ungulate mammals. **2.** A hoofed foot, as of a horse. { hu̇f }

hoof-and-mouth disease *See* foot-and-mouth disease. { ¦hu̇f ən 'mau̇th di,zēz }

Hookeriales [BOT] An order of the mosses with irregularly branched stems and leaves that appear to be in one plane. { hu̇,kir·ē'ā·lēz }

hookworm [INV ZOO] The common name for parasitic roundworms composing the family Ancylostomidae. { 'hu̇k,wərm }

hookworm disease [MED] Microcytic hypochromic anemia in humans produced by the nematodes *Necator americanus* or *Ancylostoma duodenale* in the the intestine. { 'hu̇k,wərm di,zēz }

hop [BOT] *Humulus lupulus*. A dioecious liana of the order Urticales distinguished by herbaceous vines produced from a perennial crown; the inflorescence, a catkin, of the female plant is used commercially for beer production. { häp }

hophornbeam [BOT] Any tree of the genus *Ostrya* in the birch family recognized by its very scaly bark and the fruit which closely resembles that of the hopvine. { ¦häp 'hȯrn,bēm }

Hoplocarida [INV ZOO] A superorder of the class Crustacea with the single order Stomatopoda. { ,häp·lō'kar·ə·də }

Hoplonemertini [INV ZOO] An order of unsegmented, ribbonlike worms in the class Enopla; all species have an armed proboscis. { ¦häp·lō,ne·mər'tī,nī }

Hoplophoridae [INV ZOO] A family of prawns containing numerous bathypelagic representatives. { ,häp·lə'fȯr·ə,dē }

hordeolum [MED] A furuncular inflammation of the connective tissue of the eyelids near a hair follicle. Also known as sty. { hȯr'dē·ə·ləm }

Hordeum [BOT] A genus of the order Cyperales containing all species of barley. { 'hȯr·dē·əm }

horehound *See* marrubium. { 'hȯr,hau̇nd }

horizontal cells [NEUROSCI] Interneurons located in the outer plexiform layer of the vertebrate retina that influence retinal signal processing in response to visual stimuli at the level of contact between the photoreceptor cells and the bipolar cells. { ,här·ə,zänt·əl 'sels }

horizontal evolution *See* coincidental evolution. { ,hä·ri¦zänt·əl ,ev·ə'lü·shən }

horizontal plane [ANAT] A transverse plane at right angles to the longitudinal axis of the body. { ,här·ə'zänt·əl 'plān }

horizontal transmission [GEN] Passage of genetic information by invasive processes between individual organisms or cells of different species. { ,här·ə¦zänt·əl tranz'mish·ən }

hormesis [BIOL] Providing stimulus by nontoxic amounts of a toxic agent. { 'hȯr·mə·səs }

hormogonium [BOT] Portion of a filament between heterocysts in certain algae; detaches as a reproductive body. { ,hȯr·mə'gō·nē·əm }

hormone [BIOCHEM] **1.** A chemical messenger produced by endocrine glands and secreted directly into the bloodstream to exert a specific effect on a distant part of the body. **2.** An organic compound that is synthesized in minute quantities in one part of a plant and translocated to another part, where it influences a specific physiological process. { 'hȯr,mōn }

hormone-responsive element [CELL MOL] Specific regulatory nucleotide sequences in most hormonally regulated genes that are located near the promoter and mediate the action of steroid hormones. { 'hȯr,mōn ri,spän·siv ,el·ə·mənt }

hornbeam [BOT] Any tree of the genus *Carpinus* in the birch family distinguished by doubly serrate leaves and by small, pointed, angular winter buds with scales in four rows. { 'hȯrn,bēm }

horned liverwort [BOT] The common name for bryophytes of the class Anthocerotae. Also known as hornwort. { ¦hȯrnd 'liv·ər,wȯrt }

horned toad [VERT ZOO] The common name for any of the lizards of the genus *Phrynosoma*; a reptile that resembles a toad but is less bulky. { ¦hȯrnd 'tōd }

Horner's syndrome [MED] A complex of symptoms due to unilateral destruction of the cervical sympathetics. { 'hȯrn·ərz ,sin,drōm }

hornet [INV ZOO] The common name for a number of large wasps in the hymenopteran family Vespidae. { 'hȯr·nət }

hornwort *See* horned liverwort. { 'hȯrn,wȯrt }

horny coral [INV ZOO] The common name for cnidarian members of the order Gorgonacea. { 'hȯr·nē 'kär·əl }

horse [VERT ZOO] *Equus caballus*. A herbivorous mammal in the family Equidae; the feet are characterized by a single functional digit. { hȯrs }

horse chestnut [BOT] *Aesculus hippocastanum*. An ornamental buckeye tree in the order Sapindales, usually with seven leaflets per leaf and resinous buds. { 'hȯrs ¦ches·nət }

horsepox [VET MED] A disease of horses such as pseudotuberculosis, contagious pustular stomatitis, or a vesicular exanthema. { 'hȯrs ,päks }

horseradish [BOT] *Armoracia rusticana*. A perennial crucifer belonging to the order Capparales and grown for its pungent roots, used as a condiment. { 'hȯrs¦rad·ish }

horse serum [IMMUNOL] Immune serum obtained from the blood of horses. { 'hȯrs ˌsir·əm }

horseshoe crab [INV ZOO] The common name for arthropods composing the subclass Xiphosurida, especially the subgroup Limulida. { 'hȯrˌshü ˌkrab }

horseshoe kidney [MED] Congenital fusion of two kidneys at one pole. { 'hȯrˌshü ˌkid·nē }

horsetail [BOT] The common name for plants of the genus *Equisetum* composing the order Equisetales. Also known as scouring rush. { 'hȯrsˌtāl }

horticulture [BOT] The art and science of growing plants. { 'hȯrd·əˌkəl·chər }

host [BIOL] **1.** An organism on or in which a parasite lives. **2.** The dominant partner of a symbiotic or commensal pair. { hōst }

host structure *See* host. { ¦hōst 'strək·chər }

hot flash [PHYSIO] A sudden transitory sensation of heat, often involving the whole body, due to cessation of ovarian function; a symptom of the climacteric. { 'hät ˌflash }

hot spot [CELL MOL] A site in a gene at which there is an unusually high frequency of mutation. { 'hät ˌspät }

hourglass stomach [MED] A stomach with an equatorial constriction usually caused by formation of scar tissue around an ulcer. { 'au̇rˌglas 'stəm·ək }

housefly [INV ZOO] *Musca domestica.* A dipteran insect with lapping mouthparts commonly found near human habitations; a vector in the transmission of many disease pathogens. { 'hau̇sˌflī }

housekeeping gene *See* constitutive gene. { 'hau̇sˌkēp·iŋ ˌjēn }

house physician [MED] A physician employed by a hospital. { 'hau̇s fə'zish·ən }

Howell-Jolly bodies [PATH] Small, round basophilic inclusions of nuclear material in erythrocytes of splenectomized persons. { 'hau̇·əl 'jäl·ē ˌbäd·ēz }

Howship's lacuna [HISTOL] Minute depressions in the surface of a bone undergoing resorption. { 'hau̇·shəps lə'kü·nə }

HPV *See* human papillomavirus.

H-RNA *See* heterogeneous ribonucleic acid.

HSI *See* heat stress index.

H substance [BIOCHEM] An agent similar to histamine and believed to play a role in local blood vessel response in tissue damage. { 'āch ˌsəb·stəns }

HTLV *See* human T-lymphotropic virus.

huarizo [VERT ZOO] A hybrid offspring of a male llama and a female alpaca, bred for its fine fleece. { wä'rē·zō }

Hubbard tank [MED] A large, specially designed tank in which a patient may be immersed for various therapeutic underwater exercises. { 'həb·ərd ˌtaŋk }

huckleberry [BOT] The common name for shrubs of the genus *Gaylussacia* in the family Ericaceae distinguished by an ovary with 10 locules and 10 ovules; the dark-blue berries are edible. { 'hək·əlˌber·ē }

Hudsonian life zone [ECOL] A zone comprising the climate and biotic communities of the northern portions of North American coniferous forests and the peaks of high mountains. { ¦həd¦sō·nē·ən 'līf ˌzōn }

Huhner's test [PATH] An examination of seminal fluid obtained from the vaginal fornix and cervical canal after a specified interval following coitus; used in fertility studies to evaluate spermatozoal survival and activity in the female lower genital tract. { 'hyü·nərz ˌtest }

hull [BOT] The outer, usually hard, covering of a fruit or seed. { həl }

hulless buckwheat *See* tartary buckwheat. { 'həl·əs 'bəkˌwēt }

human biogeography [ECOL] The science concerned with the distribution of human populations on the earth. { 'hyü·mən ¦bī·ō·jē'äg·rə·fē }

human chorionic gonadotropin [BIOCHEM] A gonadotropic and luteotropic hormone secreted by the chorionic vesicle. Abbreviated HCG. Also known as chorionic gonadotropin. { 'hyü·mən kȯr·ē'än·ik gōˌnad·ə'trō·pən }

human community [ECOL] That portion of a human ecosystem composed of human beings and associated plant and animal species. { 'hyü·mən kə'myün·əd·ē }

human ecology [ECOL] The branch of ecology that considers the relations of individual persons and of human communities with their particular environment. { 'hyü·mən ē'käl·ə·jē }

human immunodeficiency virus [VIROL] The retrovirus that causes acquired immune deficiency syndrome. Abbreviated HIV. { 'hyü·mən ¦im·yə·nō·di'fish·ən·sē ˌvī·rəs }

human leukocyte antigen [IMMUNOL] Any of a group of antigens present on the surface of nucleated body cells that are coded for by the major histocompatibility complex of humans and thus allow the immune system to distinguish self and nonself. Abbreviated HLA. { ¦yü·mən ¦lü·kə ˌsīt 'ant·i·jən }

human measles immune serum [IMMUNOL] Serum from the blood of a person who has recovered from measles. { 'hyü·mən ¦mē·zəlz i'myün ˌsir·əm }

human myeloma protein [IMMUNOL] Any of a group of the first structurally discrete immunoglobulins known, resulting from the malignant expansion of a normal clone of antibody-producing cells. { ˌhyü·mən ˌmī·əˌlō·mə 'prōˌtēn }

human papillomavirus [MED] One of a family of more than 100 different viruses, most commonly spread via sexual contact, that cause warts on the hands and feet and in the genital area; several types are associated with premalignant and malignant changes in the cervix. Abbreviated HPV. { ¦yü·mən ˌpap·ə'lō·mə ˌvī·rəs }

human threadworm *See* pinworm. { 'hyü·mən 'thredˌwərm }

human T-lymphotropic virus type 1 [VIROL] A retrovirus associated with a rare form of T-cell leukemia occurring primarily in the Caribbean area, Africa, and Japan. Also known as human T-cell lymphotropic virus type I. { 'hyü·mən 'tē ¦lim·fə¦träp·ik ,vī·rəs ¦tīp 'wən }

human T-lymphotropic virus type 2 [VIROL] A retrovirus associated with at least one form of leukemia. Also known as human T-cell lymphotropic virus type 2. { 'hyü·mən 'tē ¦lim·fə¦träp·ik ,vī·rəs ¦tīp 'tü }

human T-lymphotropic virus type 3 [VIROL] A former designation for the human immunodeficiency virus. Also known as human T-cell lymphotropic virus type 3. { 'hyü·mən 'tē ¦lim·fə¦träp·ik ,vī·rəs ¦tīp 'thrē }

humeral [ANAT] Of or pertaining to the humerus or the shoulder region. { 'hyüm·ə·rəl }

humeroglandular effector system [PHYSIO] Glandular effector system in which the activating agent is a blood-borne chemical. { ¦hyüm·ə·rō ¦glan·jə·lər i'fek·tər ,sis·təm }

humeromuscular effector system [PHYSIO] Muscular effector system in which the activating agent is a blood-borne chemical. { ¦hyüm·ə·rō¦məs·kyə·lər i'fek·tər ,sis·təm }

humerus [ANAT] The proximal bone of the forelimb in vertebrates; the bone of the upper arm in humans, articulating with the glenoid fossa and the radius and ulna. { 'hyüm·ə·rəs }

humicolous [ECOL] Of or pertaining to plant species inhabiting medium-dry ground. { hyü 'mik·ə·ləs }

humid transition life zone [ECOL] A zone comprising the climate and biotic communities of the northwest moist coniferous forest of the north-central United States. { 'hyü·məd tran ¦zish·ən 'līf ,zōn }

humifuse [BIOL] Spread over the ground surface. { 'hyü·mə,fyüs }

Humiriaceae [BOT] A family of dicotyledonous plants in the order Linales characterized by exappendiculate petals, usually five petals, flowers with an intrastaminal disk, and leaves lacking stipules. { hyü,mir·ē'ās·ē,ē }

humivore [ECOL] An organism that feeds on humus. { 'hyü·mə,vór }

hummingbird [VERT ZOO] The common name for members of the family Trochilidae; fast-flying, short-legged, weak-footed insectivorous birds with a tubular, pointed bill and a fringed tongue. { 'həm·iŋ,bərd }

hummock [ECOL] A rounded or conical knoll frequently formed of earth and covered with vegetation. { 'həm·ək }

humor [PHYSIO] A fluid or semifluid part of the body. { 'hyü·mər }

humoral immunity [IMMUNOL] Immunity in which immune responses are mediated by immunoglobulins. Also known as antibody-mediated immunity; immunoglobulin-mediated immunity. { ¦hyüm·ə·rəl i'myü·nəd·ē }

humpback *See* kyphosis. { 'həmp,bak }

Hunner's ulcer [MED] A chronic ulcer of the urinary bladder, frequently in association with interstitial cystitis. Also known as elusive ulcer. { 'hən·ərz 'əl·sər }

Hunter syndrome [MED] An X-linked recessive disease in which a deficiency of the enzyme iduronate sulfatase leads to the accumulation of mucopolysaccharides in various body tissues, resulting in developmental abnormalities, skeletal deformations, mental retardation, and, in severe cases, early death. Also known as mucopolysaccharidosis II. { 'hənt·ər ,sin,drōm }

Huntington's chorea [MED] A rare hereditary disease of the basal ganglia and cerebral cortex resulting in choreiform (dancelike) movements, intellectual deterioration, and psychosis. { 'hənt·iŋ·tənz kə'rē·ə }

Hurler's syndrome [MED] Mucopolysaccharidosis I, a hereditary condition transmitted as an autosomal recessive in which there is excessive chondroitin sulfate B and heparin sulfate in the urine and tissues, and which is marked clinically by a complex of symptoms including grotesque skeletal and facial deformities, skin and cardiac changes, clouding of the cornea, and mental deficiency. Also known as gargoylism; lipochondrodystrophy. { 'hər·lərz ,sin,drōm }

Hürthle cells [PATH] Enlarged epithelial cells of the thyroid follicles containing acidophilic cytoplasm, seen most frequently in adenomas. { 'hirt·lē ,selz }

Hurtley's test [PATH] A test for the presence of acetoacetic acid in urine: take 10 cubic centimeters of a urine specimen and add 2.5 cm^3 of concentrated hydrochloric acid and 1 cm^3 of a freshly prepared 1% solution of sodium nitrite, then shake and let stand for 2 minutes; add 15 cm^3 of 0.880 ammonia and 5 cm^3 of 10% ferrous sulfate, then shake and pour into a large boiling tube, allowing it to stand; a violet or purple color confirms the presence of acetoacetic acid. { 'hərt·lēz ,test }

husk [BOT] The outer coat of certain seeds, particularly if it is a dry, membranous structure. { həsk }

Hutchinson-Gilford syndrome *See* progeria. { 'həch·ən·sən 'gil·fərd 'sin,drōm }

Hutchinson's freckle *See* melanotic freckle. { 'həch·ən·sənz ,frek·əl }

Hutchinson's teeth [MED] Deformity of permanent incisor teeth associated with congenital syphilis; the crown of the incisor is wider in the cervical portion than at the incisal edge, and the incisal edge has a characteristic crescent-shaped notch. { 'həch·ən·sənz ,tēth }

Huxley's anastomosis [HISTOL] Polyhedral cells forming the middle layer of a hair root sheath. { 'həks·lēz ə,nas·tə'mō·səs }

HVA *See* homovanillic acid.

Hyaenidae [VERT ZOO] A family of catlike carnivores in the superfamily Feloidea including the hyenas and aardwolf. { hī'e·nə,dē }

Hyalellidae [INV ZOO] A family of amphipod crustaceans in the suborder Gammaridea. { ,hī·ə'lel·ə,dē }

hyaline [BIOCHEM] A clear, homogeneous, structureless material found in the matrix of cartilage, vitreous body, mucin, and glycogen. { 'hī·ə·lən }

hyaline cartilage [HISTOL] A translucent connective tissue comprising about two-thirds clear, homogeneous matrix with few or no collagen fibrils. { 'hī·ə·lən 'kärt·lij }

hyaline cast [PATH] A clear, structureless mass of proteinaceous material found in the urine in association with certain kidney diseases. { 'hī·ə·lən ¦kast }

hyaline degeneration [PATH] Degenerative change involving tissues and cells so that they become clear, structureless, and homogeneous. { 'hī·ə·lən dē,jen·ə'rā·shən }

hyaline membrane [HISTOL] **1.** A basement membrane. **2.** A membrane of a hair follicle between the inner fibrous layer and the outer root sheath. { 'hī·ə·lən 'mem,brān }

hyaline membrane disease [MED] A disease occurring during the first few days of neonatal life, characterized by respiratory distress due to formation of a hyalinelike membrane within the alveoli. { 'hī·ə·lən 'mem,brān di,zēz }

hyaline test [INV ZOO] A translucent wall or shell of certain foraminiferans composed of layers of calcite interspersed with separating membranes. { 'hī·ə·lən ,test }

hyalinization [PATH] Replacement of the normal components of cells or tissues by a hyaline material. { ,hī·ə·lə·nə'zā·shən }

Hyalodictyae [MYCOL] A subdivision of the spore group Dictyosporae characterized by hyaline spores. { ,hi·ə·lō'dik·tē,ē }

Hyalodidymae [MYCOL] A subdivision of the spore group Didymosporae characterized by hyaline spores. { ,hi·ə·lō'did·ə,mē }

Hyalohelicosporae [MYCOL] A subdivision of the spore group Helicosporae characterized by hyaline spores. { ¦hī·ə·lō,hel·ə'käs·pə,rē }

hyaloid membrane [ANAT] The limiting membrane surrounding the vitreous body of the eyeball, and forming the suspensory ligament. { 'hī·ə,lȯid'mem,brān }

Hyalophragmiae [MYCOL] A subdivision of the spore group Phragmosporae characterized by hyaline spores. { ,hī·ə·lō'frag·mē,ē }

hyaloplasm [CELL MOL] The optically clear, viscous to gelatinous ground substance of cytoplasm in which formed bodies are suspended. { hī'al·ə,plaz·əm }

Hyaloscolecosporae [MYCOL] A subdivision of the spore group Scalecosporae characterized by hyaline spores. { ¦hī·ə·lō,skäl·ə'käs·pə,rē }

Hyalosporae [MYCOL] A subdivision of the spore group Amerosporae characterized by hyaline spores. { ,hī·ə'läs·pə·rē }

Hyalostaurosporae [MYCOL] A subdivision of the spore group Staurosporae characterized by hyaline spores. { ¦hī·ə·lō·stȯ'räs·pə,rē }

hyaluronate [BIOCHEM] A salt or ester of hyaluronic acid. { ,hī·ə'lu̇r·ə,nāt }

hyaluronate lyase *See* hyaluronidase. { ,hī·ə'lu̇r·ə,nāt 'lī,ās }

hyaluronic acid [BIOCHEM] A polysaccharide found as an integral part of the gellike substance of animal connective tissue. { ¦hī·ə·lu̇¦rän·ik 'as·əd }

hyaluronidase [BIOCHEM] Any one of a family of enzymes which catalyze the breakdown of hyaluronic acid. Also known as hyaluronate lyase; spreading factor. { ,hī·ə·lu̇'rän·ə,dās }

hybrid [GEN] The offspring of parents of different species or varieties. { 'hī·brəd }

hybrid-arrested translation [CELL MOL] A method for identifying the proteins coded for by a cloned deoxyribonucleic acid (DNA) sequence by depending on the ability of the cloned DNA to form a base pair with its messenger ribonucleic acid and thereby inhibit its translation. { ¦hī·brəd ə,res·təd tranz'lā·shən }

hybrid dysgenesis [GEN] A syndrome of abnormal traits that appears in the hybrids between certain strains of the fruit fly *Drosophila melanogaster*, and includes such traits as partial sterility and greatly elevated genetic mutations and chromosome rearrangements. Its cause is mobilization of a transposable element. { 'hī,brid dis 'jen·ə·səs }

hybrid enzyme [BIOCHEM] A form of polymeric enzyme occurring in heterozygous individuals that shows a hybrid molecular form made up of subunits differing in one or more amino acids. { 'hī·brəd 'en,zīm }

hybrid gene [CELL MOL] Any gene constructed by recombinant deoxyribonucleic acid technology that contains segments derived from different parents. { 'hī·brəd ,jēn }

hybridization [CELL MOL] The production of viable hybrid somatic cells following experimentally induced cell fusion. [GEN] **1.** Production of a hybrid by pairing complementary ribonucleic acid and deoxyribonucleic acid (DNA) strands. **2.** Production of a hybrid by pairing complementary DNA single strands. { ,hī·brəd·ə'zā·shən }

hybridization probe [CELL MOL] A small molecule of deoxyribonucleic acid or ribonucleic acid that is radioactively labeled and used to identify complementary nucleic acid sequences by hybridization. { ,hī·brəd·ə'zā·shən 'prōb }

hybrid merogony [EMBRYO] The fertilization of cytoplasmic fragments of the egg of one species by the sperm of a related species. { 'hī·brəd mə'räj·ə·nē }

hybrid molecule [BIOCHEM] A single molecule, usually protein, peculiar to heterozygotes and containing two structurally different polypeptide chains determined by two different alleles. { 'hī·brəd 'mäl·ə,kyül }

hybridoma [IMMUNOL] A hybrid myeloma formed by fusing myeloma cells with lymphocytes that produce a specific antibody; the individual cells can be cloned, and each clone produces large amounts of identical (monoclonal) antibody. { ,hī·brə'dō·mə }

hybrid plasmid [CELL MOL] In recombinant deoxyribonucleic acid (DNA) technology, any plasmid chimera containing inserted DNA sequences. { 'plaz·mid }

hybrid sterility [GEN] Inability to form functional gametes in a hybrid due to disturbances in sex-cell development or in meiosis, caused by incompatible genetic constitution. { 'hī·brəd stə'ril·əd·ē }

hybrid swarm [GEN] A collection of hybrids produced when there is a breakdown of isolating barriers between two species whose areas of distribution overlap. { 'hī·brəd 'swȯrm }

hybrid vigor *See* heterosis. { 'hī·brəd 'vig·ər }

hybrid zone [ECOL] A geographic zone in which two populations hybridize after the breakdown of the geographic barrier that separated them. { ¦hī·brəd ¦zōn }

Hydatellales [BOT] An order of monocotyledonous flowering plants, division Magnoliophyta, of the class Liliopsida, characterized by small, submersed or partly submersed aquatic annuals with greatly simplified internal anatomy; consists of a single family with five species native to Australia, New Zealand, and Tasmania. { ˌhī·dad·ə'lā·lēz }

hydathode [BOT] An opening of the epidermis of higher plants specialized for exudation of water. { 'hīd·əˌthōd }

hydatid [MED] **1.** A cyst formed in tissues due to growth of the larval stage of *Echinococcus granulosus*. **2.** A cystic remnant of an embryonal structure. { 'hīd·ə·dəd }

hydatid disease *See* echinococcosis. { 'hīd·ə·dəd diˌzēz }

hydatid of Morgagni *See* appendix testis. { 'hīd·ə·dəd əv mȯr'gan·yē }

hydatidosis *See* echinococcosis. { ˌhīd·əˌti'dō·səs }

hydatiform mole [MED] A benign placental tumor formed as a cystic growth of the chorionic villi. Also known as hydatiform tumor. { hī 'dad·əˌfȯrm 'mōl }

hydatiform tumor *See* hydatiform mole. { hī'dad·əˌfȯrm 'tü·mər }

hydatosis [MED] Multiple hydatid cysts. { ˌhīd·ə'tō·səs }

hydra [INV ZOO] Any species of *Hydra* or related genera, consisting of a simple, tubular body with a mouth at one end surrounded by tentacles, and a foot at the other end for attachment. { 'hī·drə }

Hydra [INV ZOO] A common genus of cnidarians in the suborder Anthomedusae. { 'hī·drə }

Hydrachnellae [INV ZOO] A family of generally fresh-water predacious mites in the suborder Trombidiformes, including some parasitic forms. { ˌhīˌdrak'ne·lē }

Hydraenidae [INV ZOO] The equivalent name for Limnebiidae. { hī'drē·nəˌdē }

hydragogue [MED] Causing the discharge of watery fluid, especially from the bowel. { 'hī·drəˌgäg }

hydralazine [PHARM] $C_8H_8N_4$ An antihypertensive drug; used as the hydrochloride salt. { hī'dral·əˌzēn }

hydramnios *See* polyhydramnios. { hī'dram·nēˌäs }

hydranencephaly [MED] A congenital anomaly in which there are vestiges of a cerebellum, occipital lobes, and basal nuclei; frontal and parietal lobes are replaced by a cyst; and the neurocranium is undeveloped. { ¦hī·drəˌen'sef·ə·lē }

hydranth [INV ZOO] Nutritive individual in a hydroid colony. { 'hīˌdranth }

hydrargyrism *See* mercurialism. { hī'drär·jəˌriz·əm }

hydrarthrosis [MED] An accumulation of fluid in a joint. { ˌhī·drär'thrō·səs }

hydrase [BIOCHEM] An enzyme that catalyzes removal or addition of water to a substrate without hydrolyzing it. { 'hīˌdrās }

Hydrastis canadensis [BOT] A perennial medicinal herb in the family Ranunculaceae from which the alkaloid hydrastine is derived. Also known as goldenseal. { hīˌdras·tis ˌkan·ə'den·sis }

hydremia [MED] An excessive amount of water in the blood; disproportionate increase in plasma volume as compared with red blood cell volume. { hī'drē·mē·ə }

hydric [ECOL] Characterized by or thriving in abundance of moisture. { 'hī·drik }

Hydrobatidae [VERT ZOO] The storm petrels, a family of oceanic birds in the order Procellariiformes. { 'hī·drō'bad·əˌdē }

hydrocaulus [INV ZOO] Branched, upright stem of a hydroid colony. { 'hī·drō'kȯl·əs }

hydrocele [MED] Accumulation of fluid in the membranes surrounding the testis. { 'hī·drəˌsēl }

hydrocephaly [MED] Increased volume of cerebrospinal fluid in the skull. { ˌhī·drə'sef·ə·lē }

Hydrocharitaceae [BOT] The single family of the order Hydrocharitales, characterized by an inferior, compound ovary with laminar placentation. { ˌhī·drōˌkar·ə'tās·ēˌē }

Hydrocharitales [BOT] A monofamilial order of aquatic monocotyledonous plants in the subclass Alismatidae. { ˌhī·drōˌkar·ə'tā·lēz }

hydrochlorthiazide [PHARM] $C_7H_8ClN_3O_4S_2$ An orally effective diuretic and antihypertensive drug. { ¦hī·drōˌklȯr'thī·əˌzīd }

hydrocholeresis [MED] Choleresis characterized by an increase of water output, or of a bile relatively low in specific gravity, viscosity, and content of total solids. { ¦hī·drōˌkō·lə'rē·səs }

hydrochory [BIOL] Dispersal of disseminules by water. { 'hī·drəˌkȯr·ē }

hydrocladium [INV ZOO] Branchlet of a hydrocaulus. { ˌhī·drə'klād·ē·əm }

hydrocoele [INV ZOO] **1.** Water vascular system in Echinodermata. **2.** Embryonic precursor of the system. { 'hī·drəˌsel }

Hydrocorallina [INV ZOO] An order in some systems of classification set up to include the cnidarian groups Milleporina and Stylasterina. { ˌhī·drōˌkȯr·ə'lī·nə }

Hydrocorisae [INV ZOO] A subdivision of the Hemiptera containing water bugs with concealed antennae and without a bulbus ejaculatorius in the male. { ˌhī·drō'kȯr·əˌsē }

hydrocortisone [BIOCHEM] $C_{21}H_{30}O_5$ The generic name for 17-hydroxycorticosterone; an adrenocortical steroid occurring naturally and prepared synthetically; its effects are similar to cortisone, but it is more active. Also known as cortisol. { ˌhī·drə'kȯrd·əˌzōn }

hydrocystoma [MED] A group of clear vesicles, usually located around the eyes, composed of cystic sweat glands. { ˌhī·drə·si'stō·mə }

Hydrodamalinae [VERT ZOO] A monogeneric subfamily of sirenian mammals in the family Dugongidae. { ˌhī·drō·də'mal·əˌnē }

hydroecium [INV ZOO] The closed, funnel-shaped tube at the upper end of cnidarians belonging to the Siphonophora. { hī'drē·shəm }

hydrofuge [ZOO] Of a structure, shedding water, as the hair on certain animals. { 'hī·drəˌfyüj }

hydrogenase [BIOCHEM] Enzyme that catalyzes the oxidation of hydrogen. { hī'dräj·əˌnās }

hydrogen bacteria [MICROBIO] Bacteria capable of obtaining energy from the oxidation of molecular hydrogen. { 'hī·drə·jən bak'tir·ē·ə }

hydrogenosome [MICROBIO] A membrane-bound organelle found in some anaerobic fungi and protozoa that is involved in the production of hydrogen and carbon dioxide. { ˌhī·drə'jen·əˌsōm }

hydroid [INV ZOO] **1.** The polyp form of a hydrozoan cnidarian. Also known as hydroid polyp; hydropolyp. **2.** Any member of the Hydroida. { 'hīˌdrȯid }

Hydroida [INV ZOO] An order of cnidarians in the class Hydrozoa including usually colonial forms with a well-developed polyp stage. { hī'drȯi·də }

hydroid polyp *See* hydroid. { 'hīˌdrȯid 'päl·əp }

hydrolase [BIOCHEM] Any of a class of enzymes which catalyze the hydrolysis of proteins, nucleic acids, starch, fats, phosphate esters, and other macromolecular substances. { 'hī·drəˌlās }

hydrolytic enzyme [BIOCHEM] A catalyst that acts like a hydrolase. { ¦hī·drə¦lid·ik 'enˌzīm }

Hydrometridae [INV ZOO] The marsh treaders, a family of hemipteran insects in the subdivision Amphibicorisae. { ˌhī·drə'me·trəˌdē }

hydronephrosis [MED] Accumulation of urine in and distension of the renal pelvis and calyces due to obstructed outflow. { ˌhī·drə·nə'frō·səs }

hydropathy [MED] The system of internal and external use of water in attempting to cure disease. { hī'drä·pə·thē }

hydropericardium [MED] Accumulation of serous fluid in the pericardial sac. { ¦hī·drōˌper·ə'kärd·ē·əm }

Hydrophiidae [VERT ZOO] A family of proglyphodont snakes in the suborder Serpentes found in Indian-Pacific oceans. { ˌhī·drō'fī·əˌdē }

Hydrophilidae [INV ZOO] The water scavenger beetles, a large family of coleopteran insects in the superfamily Hydrophiloidea. { ˌhī·drə'fil·əˌdē }

Hydrophiloidea [INV ZOO] A superfamily of coleopteran insects in the suborder Polyphaga. { ˌhī·drə·fə'lȯid·ē·ə }

hydrophilous [ECOL] Inhabiting moist places. { hī'dräf·ə·ləs }

hydrophobia *See* rabies. { 'hī·drə'fō·bē·ə }

hydrophobic [MED] Of, pertaining to, or suffering from hydrophobia. { ¦hī·drə'fō·bik }

Hydrophyllaceae [BOT] A family of dicotyledonous plants in the order Polemoniales distinguished by two carpels, parietal placentation, and generally imbricate corolla lobes in the bud. { ˌhī·drō·fə'lās·ēˌē }

hydrophyllium [INV ZOO] A transparent body partly covering the spore sacs of siphonophoran cnidarians. { ˌhī·drə'fil·ē·əm }

hydrophyte [BOT] **1.** A plant that grows in a moist habitat. **2.** A plant requiring large amounts of water for growth. Also known as hygrophyte. { 'hī·drəˌfīt }

hydroplanula [INV ZOO] A cnidarian larval stage between the planula and actinula stages. { ¦hī·drə'plan·yə·lə }

hydropolyp *See* hydroid. { ¦hī·drō'päl·əp }

hydroponics [BOT] Growing of plants in a nutrient solution with the mechanical support of an inert medium such as sand. { ˌhī·drə'pän·iks }

hydropore [INV ZOO] In certain asteroids and echinoids, an opening on the aboral surface of a canal which extends from the ring canal in one of the interradii. { 'hī·drəˌpȯr }

hydrorhiza [INV ZOO] Rootlike structure of a hydroid colony. { ˌhī·drō'rī·zə }

hydrosalpinx [MED] A distension of a Fallopian tube with fluid. { ¦hī·drō'salˌpiŋks }

Hydroscaphidae [INV ZOO] The skiff beetles, a small family of coleopteran insects in the suborder Myxophaga. { ˌhī·drə'skaf·əˌdē }

hydrosere [ECOL] Community in which the pioneer plants invade open water, eventually forming some kind of soil such as peat or muck. { 'hī·drəˌsir }

hydroskeleton [INV ZOO] Water contained within the coelenteron and serving a skeletal function in most cnidarian polyps. { ¦hī·drō 'skel·ət·ən }

hydrospire [INV ZOO] Either of a pair of flattened tubes composing part of the respiratory system in blastoids. { 'hī·drəˌspīr }

hydrospore [INV ZOO] Opening into the hydrocoele on the right side in echinoderm larvae. { 'hī·drəˌspȯr }

hydrotheca [INV ZOO] Cup-shaped portion of the perisarc in some cnidarians that serves to hold and protect a withdrawn hydranth. { ˌhī·drō'thē·kə }

hydrotherapy [MED] Treatment of disease by external application of water. { ˌhī·drə'ther·ə·pē }

hydrothorax [MED] Collection of serous fluid in the pleural spaces. { ˌhī·drə'thȯrˌaks }

hydrotropism [BIOL] Orientation involving growth or movement of a sessile organism or part, especially plant roots, in response to the presence of water. { hī'drä·trəˌpiz·əm }

hydroureter [MED] Accumulation of urine in and distention of the ureter due to obstructed outflow. { ¦hī·drō'yůr·əd·ər }

hydroxyamphetamine [PHARM] $C_9H_{13}ON$ A sympathetic amine used as the hydrobromide salt orally as a drug and locally as a mydriatic and nasal decongestant. { hī¦drak·sə·am¦fed·ə,mēn }

β-hydroxybutyric dehydrogenase [BIOCHEM] The enzyme that catalyzes the conversion of L-β-hydroxybutyric acid to acetoacetic acid by dehydrogenation. { ¦bād·ə hī¦dräk·sē·byü'tir·ik dē·hī'dräj·ə·nās }

hydroxycarbamide *See* hydroxyurea. { hī¦dräk·sē'kär·bə,mīd }

hydroxychloroquine [PHARM] $C_{18}H_{26}ClN_3O$ A drug used as the sulfate salt for the treatment of malaria, lupus erythematosus, and rheumatoid arthritis. { hī¦dräk·sē'klȯr·ə,kwīn }

3-hydroxyflavone *See* flavanol. { ¦thrē hī¦dräk·sē'fla,vōn }

hydroxyine [PHARM] $C_{21}H_{27}ClN_2O_2$ A tranquilizer, also possessing antiemetic and antihistaminic effects; used as the hydrochloride salt. { hī'dräk·sə,lēn }

hydroxylase [BIOCHEM] Any of several enzymes that catalyze certain hydroxylation reactions involving atomic oxygen. { hī'dräk·sə,lās }

hydroxyproline [BIOCHEM] $C_5H_9O_3N$ An amino acid that is essentially limited to structural proteins of the collagen type. { hī¦dräk·sə'prō,lēn }

***para*-hydroxypropiophenone** [PHARM] $HOC_6H_4COC_2H_5$ A crystalline substance with a melting point of 149°C; soluble in alcohol and ether; used as an inhibitor of pituitary gonadotropic hormone. { ¦par·ə hī¦dräk·sē,prō·pē'ä·fə,nōn }

8-hydroxyquinoline sulfate [PHARM] $C_{18}H_{16}N_2O_6S$ A pale yellow, crystalline powder with a melting point of 175–178°C; soluble in water; used as an antiseptic, deodorant, and antiperspirant. { ¦āt hī¦dräk·sē'kwin·ə·lən 'səl,fāt }

5-hydroxytryptamine *See* serotonin. { ¦fīv hī¦dräk·sē'trip·tə,mēn }

5-hydroxytryptophan [BIOCHEM] $C_{11}H_{12}N_2O_3$ Minute rods or needlelike crystals; the biological precursor of serotonin. { ¦fīv hī¦dräk·sē'trip·tə,fan }

hydroxyurea [PHARM] $HONHCONH_2$ Needle-like crystals with a melting point of 133–136°C; used as an antineoplastic agent. Also known as hydroxycarbamide. { ,hī¦dräk·sē·yů'rē·ə }

Hydrozoa [INV ZOO] A class of the phylum Cnidaria which includes the fresh-water hydras, the marine hydroids, many small jellyfish, a few corals, and the Portuguese man-of-war. { ,hī·drə'zō·ə }

hyena [VERT ZOO] An African carnivore represented by three species of the family Hyaenidae that resemble dogs but are more closely related to cats. { hī'ē·nə }

hygiene [MED] The science that deals with the principles and practices of good health. { 'hī,jēn }

Hygrobiidae [INV ZOO] The squeaker beetles, a small family of coleopteran insects in the suborder Adephaga. { hī·grə'bī·ə,dē }

hygroma [MED] A congenital disorder in which a lymph-filled cystic cavity is formed from distended lymphatics. { hī'grō·mə }

hygromycin [MICROBIO] $C_{25}H_{33}O_{12}N$ A weakly acidic, soluble antibiotic with a fairly broad spectrum, produced by a strain of *Streptomyces hygroscopicus*. { ,hī·grə'mīs·ən }

hygrophyte *See* hydrophyte. { 'hī·grə,fīt }

hygroscopic [BOT] Being sensitive to moisture, such as certain tissues. { ¦hī·grə¦skäp·ik }

hylaea *See* tropical rainforest. { hī·lē·ə }

Hylidae [VERT ZOO] The tree frogs, a large amphibian family in the suborder Procoela; many are adapted to arboreal life, having expanded digital disks. { 'hī·lə,dē }

Hylobatidae [VERT ZOO] A family of anthropoid primates in the superfamily Hominoidea including the gibbon and the siamang of southeastern Asia. { ,hī·lō'bad·ə,dē }

hylophagous [ZOO] Feeding on wood, as termites. { hi¦läf·ə·gəs }

hylotomous [ZOO] Cutting wood, as wood-boring insects. { hī¦läd·ə·məs }

hymen [ANAT] A mucous membrane partly closing off the vaginal orifice. Also known as maidenhead. { 'hī·mən }

hymenium [MYCOL] The outer, sporebearing layer of certain fungi or their fruiting bodies. { hī'mē·nē·əm }

hymenolepiasis [MED] Intestinal infection by tapeworms of the genus *Hymenolepis*. { ¦hī·mə·nō·lə'pī·ə·səs }

Hymenolepis [INV ZOO] A genus of tapeworms parasitic in humans, birds, and mammals. { ,hī·mə'näl·ə·pəs }

Hymenomycetes [MYCOL] A group of the Homobasidiomycetidae including forms such as mushrooms and pore fungi in which basidia are formed in an exposed layer (hymenium) and basidiospores are borne asymmetrically on slender stalks. { ¦hī·mə·nō,mī¦sēd·ēz }

hymenophore [MYCOL] Portion of a sporophore that bears the hymenium. { hī'men·ə,fȯr }

hymenopodium [MYCOL] **1.** Tissue beneath the hymenium in certain fungi. **2.** A genus of the Moniliales. { ,hī·mə·nō'pōd·ē·əm }

Hymenoptera [INV ZOO] A large order of insects including ants, wasps, bees, sawflies, and related forms; head, thorax and abdomen are clearly differentiated; wings, when present, and legs are attached to the thorax. { ,hī·mə'näp·trə }

Hymenostomatida [INV ZOO] An order of ciliated protozoans in the subclass Holotrichia having fairly uniform ciliation and a definite buccal ciliature. { ¦hī·mə·nō·stō'mad·ə·də }

hymenotomy [MED] Surgical incision of the hymen. { hī·mə'näd·ə·mē }

Hynobiidae [VERT ZOO] A family of salamanders in the suborder Cryptobranchoidea. { ,hī·nō'bī·ə,dē }

hyobranchium [VERT ZOO] A Y-shaped bone

supporting the tongue and tongue muscles in a snake. { ¦hī·ō'braŋ·kē·əm }

Hyocephalidae [INV ZOO] A monospecific family of hemipteran insects in the superfamily Pentatomorpha. { ˌhī·ō·sə'fal·əˌdē }

hyoglossus [ANAT] An extrinsic muscle of the tongue arising from the hyoid bone. { ˌhī·ō'glä·səs }

hyoid [ANAT] **1.** A bone or complex of bones at the base of the tongue supporting the tongue and its muscles. **2.** Of or pertaining to structures derived from the hyoid arch. { 'hīˌȯid }

hyoid arch [EMBRYO] Either of the second pair of pharyngeal segments or gill arches in vertebrate embryos. { 'hīˌȯid ˌärch }

hyoid tooth [VERT ZOO] One of a number of teeth on the tongue of fishes. { 'hīˌȯid ˌtüth }

hyomandibular [VERT ZOO] The upper portion of the hyoid arch in fishes. { ¦hī·ōˌman'dib·yə·lər }

hyomandibular cleft [EMBRYO] The space between the hyoid arch and the mandibular arch in the vertebrate embryo. { ¦hī·ōˌman'dib·yə·lər 'kleft }

hyomandibular pouch [EMBRYO] A portion of the endodermal lining of the pharyngeal cavity which separates the paired hyoid and mandibular arches in vertebrate embryos. { ¦hī·ōˌman 'dib·yə·lər 'pau̇ch }

hyostylic [VERT ZOO] Having the jaws and cranium connected by the hyomandibular, as certain fishes. { ¦hī·ō'stī·lik }

hypacusia [MED] Impairment of hearing. { ¦hip·ə¦kyü·zhə }

hypalgesia [PHYSIO] Diminished sensitivity to pain. { ¦hip·al'jē·zhə }

hypandrium [INV ZOO] A plate covering the genitalia on the ninth abdominal segment of certain male insects. { hi'pan·drē·əm }

hypanthium [BOT] Expanded receptacle margin to which the sepals, petals, and stamens are attached in some flowers. { hi'pan·thē·əm }

hypantrum [VERT ZOO] In reptiles, a notch on the anterior portion of the neural arch that articulates with the hyposphene. { hi'pan·trəm }

hypaxial musculature [ANAT] The ventral portion of the axial musculature of vertebrates including subvertebral flank and ventral abdominal muscle groups. { hi'pak·sē·əl 'məs·kyə·ləˌchər }

hyperacid [PHYSIO] Containing more than the normal concentration of acid in the gastric juice. { ¦hī·pər'as·əd }

hyperactivity [PHYSIO] Excessive or pathologic activity. { ¦hī·pər·ak'tiv·əd·ē }

hyperacusia [MED] A hearing impairment characterized by an acute sense of hearing. { ˌhī·pər·ə'kyü·zhə }

hyperadrenalism [MED] Hypersecretion of adrenal hormones marked by increased basal metabolism, decreased sugar tolerance, and glycosuria. Also known as hypercorticism. { ¦hī·pər·ə'dren·əlˌiz·əm }

hyperadrenocorticism [MED] Hypersecretion of adrenocortical hormones resulting in Cushing's syndrome, or virilism. { 'hī·pər·əˌdrē·nō 'kȯrd·əˌsiz·əm }

hyperaldosteronism [MED] Hypersecretion of aldosterone by the adrenal cortex. { ¦hī·pərˌal·dō'ster·əˌniz·əm }

hyperalgesia [PHYSIO] Increased or heightened sensitivity to pain stimulation. { ¦hī·pər·al'jē·zhə }

hyperbaric [MED] Pertaining to an anesthetic solution with a specific gravity greater than that of the cerebrospinal fluid. { ¦hī·pər¦bar·ik }

hyperbaric medicine [PHARM] Any agent having a specific gravity greater than that of spinal fluid; used for spinal anesthesia. { ¦hī·pər¦bar·ik 'med·ə·sən }

hyperbaric oxygenation [MED] The administration of oxygen, under greater than atmospheric pressure, by placing the patient in a room or chamber especially designed for the purpose. { ¦hī·pər¦bar·ik ˌäk·sə·jə'nā·shən }

hyperbilirubinemia [MED] **1.** Excessive amounts of bilirubin in the blood. **2.** A severe, prolonged physiologic jaundice. { ¦hī·pərˌbil·əˌrü·bə'nē·mē·ə }

hypercalcemia [MED] Excessive amounts of calcium in the blood. Also known as calcemia. { ˌhī·pərˌkal'sē·mē·ə }

hypercapnia [MED] Excessive amount of carbon dioxide in the blood. { ˌhī·pər'kap·nē·ə }

hyperchlorhydria [MED] Excessive secretion of hydrochloric acid in the stomach. { ¦hī·pərˌklȯr'hī·drē·ə }

hypercholesteremia [MED] Elevated cholesterol levels in the blood. { ¦hī·pər·kəˌles·tə'rē·mē·ə }

hyperchromatic [BIOL] Staining more intensely than normal. { ¦hī·pər·krō'mad·ik }

hyperchromatism [PATH] **1.** Excessive pigment formation in the skin. **2.** A condition in which cells or parts of cells stain more intensely than is normal. { ˌhī·pər'krō·məˌtiz·əm }

hyperchromic [PATH] Pertaining to increased hemoglobin content in erythrocytes due to increased cell thickness, not increased hemoglobin concentration. { ¦hī·pər¦krō·mik }

hyperchromic anemia [MED] Any of several blood disorders in which erythrocytes show an increase in hemoglobin and a reduction in number. { ¦hī·pər¦krō·mik ə'nē·mē·ə }

hypercoagulability [MED] Coagulation of blood more rapidly than normal. { ˌhī·pər·kōˌag·yə·lə'bil·əd·ē }

hypercoracoid [VERT ZOO] The upper of two bones at the base of the pectoral fin in teleosts. { ¦hī·pər'kȯr·əˌkȯid }

hypercorticism *See* hyperadrenalism. { ˌhī·pər 'kȯrd·əˌsiz·əm }

hyperdynamic ileus *See* spastic ileus. { ¦hī·pər·dī'nam·ik 'il·ē·əs }

hyperemesis [MED] Excessive vomiting. { hī·pər'em·ə·səs }

hyperemesis gravedorium [MED] Pernicious vomiting in pregnancy. { hī·pər'em·ə·səs ˌgrav·ə'dȯr·ē·əm }

hyperemia [MED] An excess of blood within an organ or tissue caused by blood vessel dilation or impaired drainage, especially of the skin. { ˌhī·pə'rē·mē·ə }

hyperergia [IMMUNOL] An altered state of reactivity to antigenic materials, in which the response is more marked than usual; one form of allery or pathergy. { ˌhī·pər'ər·jē·ə }

hyperesthesia [PHYSIO] Increased sensitivity or sensation. { ˌhī·pər·əs'thē·zhə }

hypergammaglobulinemia [MED] Increased blood levels of gamma globulin, usually associated with hepatic disease. { ¦hī·pərˌgam·ə¦gläb·yə·lə'nē·mē·ə }

hypergene *See* supergene. { 'hī·pərˌjēn }

hyperglobulinemia [MED] Increased blood levels of globulin. { ˌhī·pərˌgläb·yə·lə'nē·mē·ə }

hyperglycemia [MED] Excessive amounts of sugar in the blood. { ¦hī·pərˌglī'sē·mē·ə }

hyperglycemic factor *See* glucagon. { ˌhī·pərˌglī'sē·mik 'fak·tər }

hyperglycemic glycogenolytic factor *See* glucagon. { ˌhī·pərˌglī'sē·mik ¦glī·kōˌjen·ə¦lid·ik 'fak·tər }

hyperglycinenemia [MED] A hereditary metabolic disorder of males in which blood levels of glycine are excessive, resulting in vomiting, dehydration, osteoporosis, and mental retardation. { ¦hī·pərˌglīs·ən'ē·mē·ə }

hyperhidrosis [MED] Excessive sweating, which may be localized or generalized, chronic or acute, and often accumulating in visible drops on the skin. Also known as ephidrosis; polyhidrosis; sudatoria. { ¦hī·pərˌhī'drō·səs }

Hyperiidea [INV ZOO] A suborder of amphipod crustaceans distinguished by large eyes which cover nearly the entire head. { ˌhī·pə·rī'id·ē·ə }

hyperimmune antibody [IMMUNOL] An antibody having the characteristics of a blocking antibody. { ¦hī·pər·ə'myün 'ant·iˌbäd·ē }

hyperimmune serum [IMMUNOL] An antiserum that provides a very high degree of immunity due to a high antibody titer. { ¦hī·pər·ə'myün 'sir·əm }

hyperinsulinism [MED] Condition caused by abnormally high levels of insulin in the blood. { ˌhī·pər'in·sə·ləˌniz·əm }

hyperkalemia *See* hyperpotassemia. { ˌhī·pər·kə'lē·mē·ə }

hyperkeratosis [MED] **1.** Hypertrophy of the cornea. **2.** Hypertrophy of the horny layer of the skin. { ¦hī·pər·ker·ə'tō·səs }

hyperkinesia [MED] Excessive and usually uncontrollable muscle movement. { ¦hī·pər·kə'nē·zhə }

hyperleptinemia [MED] Increased serum leptin level. { ˌhī·pərˌlep·tə'nē·mē·ə }

hyperlipemia [MED] Excessive amounts of fat in the blood. { ¦hī·pər·lə'pē·mē·ə }

Hypermastigida [INV ZOO] An order of the multiflagellate protozoans in the class Zoomastigophorea; all inhabit the alimentary canal of termites, cockroaches, and woodroaches. { ˌhī·pərˌma'stij·ə·də }

hypermenorrhea *See* menorrhagia. { ¦hī·pərˌmen·ə'rē·ə }

hypermetabolism [MED] Any state in which there is an increase in basal metabolic rate. { ¦hī·pər·mə'tab·əˌliz·əm }

hypermetamorphism [INV ZOO] Type of embryological development in certain insects in which one or more stages have been interpolated between the full-grown larva and the adult. { ¦hī·pərˌmed·ə'mȯrˌfiz·əm }

hypermetropia [MED] A defect of vision resulting from too short an eyeball so that unaccommodated rays focus behind the retina. Also known as farsightedness; hyperopia. { ˌhī·pər·mə'trō·pē·ə }

hypermotility [MED] Increased motility, as of the stomach or intestine. { ¦hī·pərˌmō'til·əd·ē }

hypernatremia [MED] Excessive amounts of sodium in the blood. { ¦hī·pər·nə'trē·mē·ə }

hypernephroma *See* renal-cell carcinoma. { ˌhī·pər·nə'frō·mə }

Hyperoartii [VERT ZOO] A superorder in the subclass Monorhina distinguished by the single median dorsal nasal opening leading into a blind hypophyseal sac. { ˌhī·pə·rō'är·shēˌī }

hyperopia *See* hypermetropia. { ˌhī·pə'rō·pē·ə }

hyperosmia [MED] An abnormally acute sense of smell. { ˌhī·pə'räz·mē·ə }

hyperostosis [MED] Hypertrophy of bony tissue. { ˌhī·pəˌrä'stō·səs }

Hyperotreti [VERT ZOO] A suborder in the subclass Monorhina distinguished by the nasal opening which is located at the tip of the snout and communicates with the pharynx by a long duct. { ˌhī·pə·rō'trēdˌī }

hyperoxemia [MED] Extreme acidity of the blood. { ˌhī·pərˌäk'sē·mē·ə }

hyperparasite [ECOL] An organism that is parasitic on other parasites. { ¦hī·pər'par·əˌsīt }

hyperparathyroidism [MED] Condition caused by increased functioning of the parathyroid glands. { ˌhī·pərˌpar·ə'thīˌrȯidˌiz·əm }

hyperpathia [MED] An exaggerated or excessive perception of or response to any stimulus as being disagreeable or painful. { ˌhī·pər'path·ē·ə }

hyperperistalsis [MED] An increase in the rate and depth of the peristaltic waves. { ¦hī·pərˌper·ə'stäl·səs }

hyperphagia *See* bulimia. { ˌhī·pər'fā·jə }

hyperphosphaturia [MED] An excess of phosphates in the urine. Also known as phosphaturia. { ¦hī·pərˌfäs·fə'tu̇r·ē·ə }

hyperpigmentation [MED] Increased pigmentation. { ¦hī·pərˌpig·mən'tā·shən }

hyperpituitarism [MED] Any abnormal condition resulting from overactivity of the anterior pituitary. { ˌhī·pər·pi'tü·ə·təˌriz·əm }

hyperplasia [MED] Increase in cell number causing an increase in the size of a tissue or organ. { ˌhī·pər'plā·zhə }

hyperploid [GEN] Having one or more chromosomes or parts of chromosomes in excess of the haploid number, or of a whole multiple of the haploid number. { 'hī·pərˌplȯid }

hyperpnea [MED] Increase in depth and rate of respiration. { ,hī·pər'nē·ə }

hyperpotassemia [MED] Excessive amounts of potassium in the blood. Also known as hyperkalemia. { ¦hī·pə,päd·ə'sē·mē·ə }

hyperproteinemia [MED] Excessive protein levels in the blood. { ¦hī·pər,prōd·ə'nē·mē·ə }

hyperpyrexia [MED] Extremely high fever. { ¦hī·pər·pī'rek·sē·ə }

hyperreflexia [MED] A condition of abnormally increased reflex action. { ¦hī·pər·ri'flek·sē·ə }

hyperresonance [MED] Exaggeration of normal resonance on percussion of the chest; heard chiefly in pulmonary emphysema and pneumothorax. { ¦hī·pə'rez·ən·əns }

hypersensitivity [IMMUNOL] The state of being abnormally sensitive, especially to allergens; responsible for allergic reactions. { ¦hī·pər,sen·sə'tiv·əd·ē }

hypersensitization [IMMUNOL] The process of producing hypersensitivity. { ¦hī·pər,sen·sə·tə'zā·shən }

hypersomnia [MED] Excessive sleepiness. { ,hī·pər'säm·nē·ə }

hypersplenism [MED] Condition caused by abnormal spleen activity. { ,hī·pər'sple,niz·əm }

hypertelorism [ANAT] An unusually large distance between paired body parts or organs. { ¦hī·pər'tel·ə,riz·əm }

hypertely [EVOL] An extreme overdevelopment of an organ or body part during evolution that is disadvantageous to the organism. [ZOO] An extreme degree of imitative coloration, beyond the aspect of utility. { hī'pərd·əl·ē }

hypertensin *See* angiotensin. { ,hī·pər'ten·sən }

hypertension [MED] Abnormal elevation of blood pressure, generally regarded to be levels of 165 systolic and 95 diastolic. { ¦hī·pər'ten·chən }

hyperthecosis [PATH] Abnormal thickening of the inner layer of the Graafian follicle with increased leutein formation. { ¦hī·pər·thə'kō·səs }

hyperthermia [PHYSIO] A condition of elevated body temperature. { ,hī·pər'thər·mē·ə }

hyperthermophile [MICROBIO] An extremophile that thrives in high-temperature (above 60°C or 140°F) environments. { ,hī·pər'thər·mə,fīl }

hyperthyroidism [MED] The constellation of signs and symptoms caused by excessive thyroid hormone in the blood, either from exaggerated functional activity of the thyroid gland or from excessive administration of thyroid hormone, and manifested by thyroid enlargement, emaciation, sweating, tachycardia, exophthalmos, and tremor. Also known as exophthalmic goiter; Grave's disease; thyrotoxicosis; toxic goiter. { ¦hī·pər'thī,rȯid,iz·əm }

hyperthyrotropinism [MED] Excessive thyrotropic hormone secretion by the adenohypophysis. { ¦hī·pər¦thī·rō'träp·ə,niz·əm }

hypertonia [MED] Abnormal increase in muscle tonicity. { ,hī·pər'tō·nē·ə }

hypertonic [PHYSIO] **1.** Excessive or above normal in tone or tension, as a muscle. **2.** Having an osmotic pressure greater than that of physiologic salt solution or of any other solution taken as a standard. { ,hī·pər'tän·ik }

hypertonic bladder [MED] Hypertonia of the urinary bladder. { ,hī·pər'tän·ik 'blad·ər }

hypertonic contracture [MED] Prolonged muscular spasms in spastic paralysis. { ,hī·pər'tän·ik kən'trak·chər }

hypertrophic arthritis *See* degenerative joint disease. { ¦hī·pər¦trä·fik är'thrīd·əs }

hypertrophic gastritis [MED] Chronic inflammation of the stomach with hypertrophy of the mucosa and rugae. { ¦hī·pər¦trä·fik ga'strīd·əs }

hypertrophy [PATH] Increase in cell size causing an increase in the size of an organ or tissue. { hī'pər·trə·fē }

hyperuricemia [MED] Abnormally high level of uric acid in the blood. Also known as lithemia. { ¦hī·pər,yu̇r·ə'sē·mē·ə }

hypervariable minisatellite [GEN] One of a number of tandem repetitive regions of deoxyribonucleic acid dispersed throughout the human and other genomes that display polymorphism associated with the allelic variation in the number of repetitive copies of each minisatellite. { ,hī·pər'ver·ē·ə·bəl ¦min·ē'sad·əl,īt }

hyperventilation [MED] Increase in air intake or of the rate or depth of respiration. { ¦hī·pər,vent·əl'ā·shən }

hypervitaminosis [MED] Condition caused by intake of toxic amounts of a vitamin. { ,hī·pər,vīd·ə·mə'nō·səs }

hypesthesia [MED] Reduced or subnormal tactile sensibility. { ,hī·pə'sthē·zhə }

hypha [MYCOL] One of the filaments composing the mycelium of a fungus. { 'hī·fə }

hyphidium [MYCOL] A sterile hymenial structure of hyphal origin. { hī'fid·ē·əm }

Hyphochytriales [MYCOL] An order of aquatic fungi in the class Phycomycetes having a saclike to limited hyphal thallus and zoospores with two flagella. { ,hī·fō·ki,trī'ā·lēz }

Hyphochytridiomycetes [MYCOL] A class of the true fungi; usually grouped with other classes under the general term Phycomycetes. { ¦hī·fō·ki¦trid·ē·ō,mī'sēd,ēz }

hyphoid [MYCOL] Hypha-like. { 'hī,fȯid }

Hyphomicrobiaceae [MICROBIO] Formerly a family of bacteria in the order Hyphomicrobiales; cells occurring in free-floating groups with individual cells attached to each other by a slender filament. { ¦hī·fō·mī,krō·bē'ās·ē,ē }

Hyphomicrobiales [MICROBIO] Formerly an order of bacteria in the class Schizomycetes containing forms that multiply by budding. { ,hī·fō·mī,krō·bē'ā·lēz }

Hyphomicrobium [MICROBIO] A genus of prosthecate bacteria; they reproduce by budding at hyphal tips; cells are rod-shaped with pointed ends, ovoid, or bean-shaped; hyphae are not septate. { ,hī·fō·mī'krō·bē·əm }

Hyphomonas [MICROBIO] A genus of prosthecate bacteria; cells are oval or pear-shaped, and reproduction is by budding of the hyphae or by direct budding of cells. { ˌhi·fō'mō·nəs }

hyphopodium [MYCOL] Hypha with a haustorium in certain ectoparasitic fungi. { ˌhī·fō'pō·dē·əm }

Hypnineae [BOT] A suborder of the Hypnobryales characterized by complanate, glossy plants with ecostate or costate leaves and paraphyllia rarely present. { hip'nī·nēˌē }

Hypnobryales [BOT] An order of mosses composed of procumbent and pleurocumbent plants with usually symmetrical leaves arranged in more than two rows. { ¦hip·nōˌbrī'ā·lēz }

hypnotherapy [MED] Treatment of disease by means of hypnotism. { ¦hip·nō'ther·ə·pē }

hypnotic [PHARM] A drug which induces sleep. Also known as somnificant; soporific. { hip 'näd·ik }

hypnotoxin [BIOCHEM] A supposed hormone produced by brain tissue and inducing sleep. { ¦hip·nə'täk·sən }

hypoadrenalism *See* hypoadrenia. { ¦hī·pō·ə 'dren·əlˌiz·əm }

hypoadrenia [MED] Reduced functioning of the adrenal glands. Also known as hypoadrenalism. { ˌhī·pō·ə'drē·nē·ə }

hypoadrenocorticism [MED] Lowered or subnormal adrenal cortex activity. { ¦hī·pō·ə¦drē·nō'kȯrd·əˌsiz·əm }

hypoalbuminemia [MED] Abnormally low levels of albumin in the blood. { ¦hī·pō·alˌbyü·mə'nē·mē·ə }

hypoallergenic [PHARM] Having a low tendency to induce allergic reactions; used particularly for formulated dermatologic preparations. { ¦hī·poˌal·ər'jen·ik }

hypobaric [MED] Pertaining to an anesthetic solution of specific gravity lower than the cerebrospinal fluid. { ¦hī·pō¦bar·ik }

hypobasal [BOT] Located posterior to the basal wall. { ¦hī·pō¦bā·səl }

hypoblast *See* endoderm. { 'hī·pōˌblast }

hypobranchial musculature [ANAT] The ventral musculature in vertebrates extending from the pectoral girdle forward to the hyoid arch, chin, and tongue. { ¦hī·po'braŋ·kē·əl 'məs·kyə·lə ˌchər }

hypocalcemia [MED] Condition in which there are reduced levels of calcium in the blood. { ˌhī·pōˌkal'sē·mē·ə }

hypocalcification [MED] Reduction of normal amounts of mineral salts in calcified tissue. { ˌhī·pōˌkal·sə·fə'kā·shən }

hypocalciuria [MED] Decreased excretion of calcium in the urine. { ¦hī·pōˌkal·sē'yu̇r·ē·ə }

hypocapnia [MED] Reduced or subnormal blood levels of carbon dioxide. { ¦hī·po'kap·nē·ə }

hypochil [BOT] Lower portion of the lip in orchids. Also known as hypochillium. { 'hī·pəˌkil }

Hypochilidae [INV ZOO] A family of true spiders in the order Araneida. { ˌhī·pə'kil·əˌdē }

hypochillium *See* hypochil. { ˌhī·pə'kil·ē·əm }

Hypochilomorphae [INV ZOO] A monofamilial suborder of spiders in the order Araneida. { ˌhī·pəˌkil·ə'mȯr·fē }

hypochloremia [MED] Reduction in the amount of blood chlorides. { ¦hī·pō·klȯ'rē·mē·ə }

hypochlorhydria [MED] Reduction in the hydrochloric acid content of gastric juice. { ˌhī·pōˌklȯr'hī·drē·ə }

hypochlorization [MED] Reduction of the dietary intake of sodium chloride. { ¦hī·pə¦klȯr·ə'zā·shən }

hypochnoid [MYCOL] Having generally compacted hyphae. { hī'päkˌnȯid }

hypocholesterolemia [MED] Subnormal levels of serum cholesterol. { ¦hī·pō·kəˌles·tə·rə'lē·mē·ə }

hypochondriac region [ANAT] The upper, lateral abdominal region just below the ribs on each side of the body. { ¦hī·pə¦kän·drēˌak ˌrē·jən }

hypochromia [PATH] Lack of complete saturation of the erythrocyte stroma with hemoglobin, as judged by pallor of the unstained or stained erythrocytes when examined microscopically. { ˌhī·pə'krō·mē·ə }

hypochromic microcytic anemia [MED] An anemia associated with erythrocytes of reduced size and hemoglobin content. { ˌhī·pə'krō·mik ˌmī·krə¦sid·ik əˌnē·mē·ə }

hypocleidium [ANAT] The median, ventral bone between clavicles. [VERT ZOO] Median process on the wishbone of birds. { ˌhī·pə'klīd·ē·əm }

hypocone [ANAT] The posterior inner cusp of an upper molar. [INV ZOO] Region of a dinoflagellate posterior to the girdle. { 'hī·pəˌkōn }

Hypocopridae [INV ZOO] A small family of coleopteran insects in the superfamily Cucujoidea. { ˌhī·pə'käp·rəˌdē }

hypocoracoid [VERT ZOO] The lower of two bones at the base of the pectoral fin in teleosts. { ¦hī·pō'kȯr·əˌkȯid }

hypocotyl [BOT] The portion of the embryonic plant axis below the cotyledon. { 'hī·pəˌkäd·əl }

hypocrateriform [BIOL] Saucer-shaped. { ˌhī·pə·krə'ter·əˌfȯrm }

Hypocreales [MYCOL] An order of fungi belonging to the Ascomycetes and including several entomophilic fungi. { ˌhī·pōˌkrē'ā·lēz }

Hypodermatidae [INV ZOO] The warble flies, a family of myodarian cyclorrhaphous dipteran insects in the subsection Calypteratae. { ¦hī·pō·dər'mad·əˌdē }

hypodermic needle [MED] A hollow needle, with a slanted open point, used for subcutaneous and intramuscular injections of fluid. { ¦hī·pə¦dər·mik 'nēd·əl }

hypodermic syringe *See* syringe. { ¦hī·pə¦dər·mik sə'rinj }

hypodermis [BOT] The outermost cell layer of the cortex of plants. Also known as exodermis. [INV ZOO] The layer of cells that underlies and secretes the cuticle in arthropods and some other invertebrates. { ˌhī·pə'dər·mis }

hypodermoclysis [MED] Subcutaneous injections of large quantities of fluid for therapeutic purposes. { ˌhī·pə·dərˈmäk·lə·səs }

hypodontia [MED] The congenital absence of teeth. Also known as anodontia; oligodontia. { ˌhī·pəˈdän·chə }

hypoergia [IMMUNOL] A state of less than normal reactivity to antigenic materials, in which the response is less marked than usual; one form of allergy or pathergy. { ¦hī·pō¦er·jē·ə }

hypofibrinogenemia [MED] A decrease in plasma fibrogen level. { ¦hī·po·fī¦brin·ə·jəˈnē·mē·ə }

hypogammaglobulinemia [MED] Reduced blood levels of gamma globulin. { ¦hī·pōˌgam·ə¦gläb·yə·ləˈnē·mē·ə }

hypogeal *See* hypogeous. { ¦hī·pə¦jē·əl }

hypogeous [BIOL] Living or maturing below the surface of the ground. Also known as hypogeal. { ˌhī·pəˈjē·əs }

hypoglossal nerve [NEUROSCI] The twelfth cranial nerve; a paired motor nerve in tetrapod vertebrates innervating tongue muscles; corresponds to the hypobranchial nerve in fishes. { ¦hī·pə¦gläs·əl ˈnərv }

hypoglossal nucleus [NEUROSCI] A long nerve nucleus throughout most of the length of the medulla oblongata; cells give rise to the hypoglossal nerve fibers. { ¦hī·pə¦gläs·əl ˈnü·klē·əs }

hypoglycemia [MED] Condition caused by low levels of sugar in the blood. { ¦hī·pōˌglīˈsē·mē·ə }

hypogonadism [MED] Reduced hormonal secretion by the testes or ovaries. { ˌhī·pōˈgō ˌnaˌdiz·əm }

hypogynium [BOT] Structure that supports the ovary in plants such as sedges. { ˌhī·pəˌjīn·ē·əm }

hypogynous [BOT] Having all flower parts attached to the receptacle below the pistil and free from it. { ˌhīˈpäj·ə·nəs }

hypohidrosis [MED] Deficient perspiration. { ¦hī·pōˌhīˈdrō·səs }

hypokalemia [PHYSIO] A reduction in the normal amount of potassium in the blood. { ˌhī·pō·kəˈlē·mē·ə }

hypokinesia [MED] Subnormal muscular movements. { ˌhī·pō·kiˈnē·zhə }

hypokinetic syndrome [MED] General decrease in motor functions due to a form of minimal brain dysfunction. { ¦hī·pō·kiˈned·ik ˈsinˌdrōm }

hypomagnesemia *See* grass tetany. { ˌhī·pōˌmag·nəˈsē·mē·yə }

hypomenorrhea [MED] A deficient amount of menstrual flow at the regular period. { ¦hī·pō ˌmen·əˈrē·ə }

hypomere [EMBRYO] The lateral or lower mesodermal plate zone in vertebrate embryos. [INV ZOO] The basal portion of certain sponges that contain no flagellated chambers. { ˈhī·pəˌmir }

hypometabolism [MED] Metabolism below the normal rate. { ¦hī·pō·məˈtab·əˌliz·əm }

hypomorphic allele [GEN] An allele that has reduced levels of gene activity. { ¦hī·pə¦mȯr·fik əˈlēl }

hypomotility [MED] Decreased motility, especially of the gastrointestinal tract. { ¦hī·pō·mōˈtil·əd·ē }

hyponasty [BOT] A nastic movement involving inward and upward bending of a plant part. { ˈhī·pəˌnas·tē }

hyponatremia [MED] Subnormal or reduced blood sodium levels. { ˌhī·pə·nəˈtrē·mē·ə }

hyponeural sinus [INV ZOO] Tubular portion of the coelom containing hemal vessels and motor nerves in certain echinoderms. { ˌhī·pəˈnu̇r·əl ˈsī·nəs }

hyponychium [HISTOL] The thickened stratum corneum of the epidermis, which lies under the free edge of the nail. { hīˈpäŋ·kē·əm }

hypoovarianism [MED] Decrease in ovarian endocrine activity. { ¦hī·pōˌōˈver·ē·əˌniz·əm }

hypoparathyroidism [MED] Condition caused by insufficient functioning of the parathyroid gland. { ˌhī·pōˌpar·əˈthīˌrȯiˌdiz·əm }

hypophagia [MED] Undereating. { ˌhī·pəˈfā·jē·ə }

hypophalangism [MED] Congenital absence of one or more phalanges in a finger or toe. { ˌhī·pō·fəˈlan·jiz·əm }

hypopharynx [ANAT] *See* laryngopharynx. [INV ZOO] A sensory, tonguelike structure on the floor of the mouth of many insects; sometimes modified for piercing. { ¦hī·pōˈfar·iŋks }

hypophosphatasia [MED] **1.** Alkaline phosphatase deficiency. **2.** A hereditary metabolic disorder characterized by subnormal amounts of alkaline phosphatase in the tissues. { ˌhī·pō ˌfas·fəˈtā·zhə }

hypophyseal cachexia *See* Simmonds' disease. { hī¦päf·ə¦sē·əl kəˈkek·sē·ə }

hypophyseal duct tumor [MED] Any tumor derived from epithelial remnants of Rathke's pouch. { hī¦päf·ə¦sē·əl ¦dəkt ˈtü·mər }

hypophysectomy [MED] Surgical removal of the pituitary gland. { hī¦päf·ə¦sek·təˌmē }

hypophysis [ANAT] A small rounded endocrine gland which lies in the sella turcica of the sphenoid bone and is attached to the floor of the third ventricle of the brain in all craniate vertebrates. Also known as pituitary gland. { hīˈpäf·ə·səs }

hypopituitarism [MED] Condition caused by insufficient secretion of pituitary hormones, especially of the adenohypophysis. Also known as panhypopituitarism. { ˌhī·pō·pəˈtü·ə·təˌriz·əm }

hypopituitary cachexia *See* Simmonds' disease. { hī·pō·pəˈtü·əˌter·ē kəˈkek·sē·ə }

hypoplankton [BIOL] Forms of marine life whose swimming ability lies somewhere between that of the plankton and the nekton; includes some mysids, amphipods, and cumacids. { ¦hī·pō¦plaŋk·tən }

hypoplasia [MED] Failure of a tissue or organ to achieve complete development. { ˌhī·pōˈplā·zhə }

hypoplastic dwarf [MED] A normally proportioned individual of subnormal size. { ¦hī·pō¦plas·tik ˈdwȯrf }

hypoplastron [VERT ZOO] Either of the third

pair of lateral bony plates in the plastron of most turtles. { ¦hī·pō¦plas·trən }

hypopleura [INV ZOO] Sclerite above and in front of the hind coxa in Diptera. { ¦hī·pō¦plu̇r·ə }

hypoploid [GEN] Having a deficit of one or more chromosomes, or parts of chromosomes, from a whole multiple of the haploid number. { 'hī·pō,plȯid }

hypoploidy [GEN] The condition or state of being hypoploid. { 'hī·pə,plȯid·ē }

hypoproliferative anemia [MED] Decreased concentration of hemoglobin and number of red blood cells due to subnormal numbers of erythrocyte primordial cells in relation to the stimulus of anemia. { ¦hī·pō·prə'lif·rəd·iv ə'nē·mē·ə }

hypoproteinemia [MED] Abnormally low levels of protein in the blood. { ¦hī·pō,prō·də'nē·mē·ə }

hypoprothrombinemia [MED] Deficiency of prothrombin in the blood. { ¦hī·pō,prō¦thräm·bə'nē·mē·ə }

hypopus [INV ZOO] The resting larval stage of certain mites. { 'hī·pə·pəs }

hypopygium [INV ZOO] A modified ninth abdominal segment together with the copulatory apparatus in Diptera. { ,hī·pə'pij·ē·əm }

hypopyon [MED] A collection of pus in the anterior chamber of the eye. { hə'pō·pē,än }

hyporeactive [MED] Characterized by decreased responsiveness to stimuli. { ¦hī·pō·rē'ak·tiv }

hyporeflexia [MED] A condition in which reflexes are below normal, due to a variety of causes. { ¦hī·pō·ri'flek·sē·ə }

hyporheic zone [ECOL] The saturated sediment environment below a stream that exchanges water, nutrients, and fauna with surface flowing waters. { ,hī·pə'rē·ik ,zōn }

hyposensitivity [MED] Condition marked by diminished sensitivity to stimuli. { ,hī·pō,sen·sə'tiv·əd·ē }

hyposmia [MED] Decreased olfactory sensitivity. { hī'päz·mē·ə }

hypospadias [MED] **1.** Congenital anomaly in which the urethra opens on the ventral surface of the penis or in the perineum. **2.** Congenital anomaly in which the urethra opens into the vagina. { ,hī·pō'spād·ē·əs }

hypospermatogenesis [MED] Decreased sperm production. { ¦hī·pō·spər,mad·ə'jen·ə·səs }

hypostasis [MED] A condition involving settling of blood in dependent parts of an organ. { hī'päs·tə·səs }

hypostatic [GEN] Subject to being suppressed, as a gene that can be suppressed by a nonallelic gene. { ¦hī·pō'stad·ik }

hyposthenia [MED] Weakness; subnormal strength. { ¦hī·pə'sthē·nē·ə }

hyposthenuria [MED] The secretion of urine of low specific gravity. { hī,päs·thə'nu̇r·ē·ə }

hypostome [INV ZOO] **1.** Projection surrounding the oral aperture in many cnidarian polyps. **2.** Anteroventral part of the head in Diptera. **3.** Median ventral mouthpart in ticks. **4.** Raised area on the posterior oral margin in crustaceans. { 'hī·pə,stōm }

hypostracum [INV ZOO] The innermost layer of the cuticle of ticks lying above the hypodermis. { ¦hī·pō¦strak·əm }

hyposulculus [INV ZOO] A groove of the siphonoglyph below the pharynx in anthozoans. { ,hī·pō'səl·kyə·ləs }

hyposynergia [MED] Defective coordination. { ¦hī·pō·sə'nər·jē·ə }

hypotarsus [VERT ZOO] A process on the tarsometatarsal bone in birds. { ¦hī·pō¦tär·səs }

hypotelorism [MED] Decrease in distance between two organs or body parts. { ,hī·pō'tel·ə,riz·əm }

hypotension [MED] Abnormally low blood pressure, commonly considered to be levels below 100 diastolic and 40 systolic. { ¦hī·pō'ten·chən }

hypothalamic center [NEUROSCI] Any of the neural centers which regulate autonomic functions. { ¦hī·pō·thə'lam·ik ¦sen·tər }

hypothalamic releasing factor [NEUROSCI] Any of the hormones secreted by the hypothalamus which travel by way of nerve fibers to the anterior pituitary, where they cause selective release of specific pituitary hormones. { ¦hī·pō·thə'lam·ik ri'lēs·iŋ ,fak·tər }

hypothalamoneurohypophyseal system [PHYSIO] The hormones and neurosecretory structures involved in the endocrine activity of the adenohypophysis, neurohypophysis, and hypothalamus. { ¦hī·pō·thə¦lam·ə¦nu̇r·ō,hī¦päf·ə'sē·əl 'sis·təm }

hypothalamoneurohypophyseal tract [NEUROSCI] A bundle of nerve fibers connecting the supraoptic and paraventricular neurons of the hypothalamus with the infundibular stem and neurohypophysis. { ¦hī·pō·thə¦lam·ə¦nu̇r·ō,hī¦päf·ə'sē·əl 'trakt }

hypothalamus [NEUROSCI] The floor of the third brain ventricle; site of production of several substances that act on the adenohypophysis. { ¦hī·pō'thal·ə·məs }

hypotheca [INV ZOO] **1.** The lower valve of a diatom frustule. **2.** Covering on the hypocone in dinoflagellates. { ¦hī·pō'thē·kə }

hypothenar [ANAT] Of or pertaining to the prominent portion of the palm above the base of the little finger. { hī'päth·ə,när }

hypothermia [PHYSIO] Condition of reduced body temperature in homeotherms. { ,hī·pō'thər·mē·ə }

hypothyroidism [MED] Condition caused by deficient secretion of the thyroid hormone. { ¦hī·pō'thī,rȯi,diz·əm }

hypotonia [MED] Decrease of normal tonicity or tension, especially diminution of intraocular pressure or of muscle tone. { ,hī·pə'tō·nē·ə }

hypotonic [PHYSIO] **1.** Pertaining to subnormal muscle strength or tension. **2.** Referring to a solution with a lower osmotic pressure than physiological saline. { ¦hī·pə'tän·ik }

Hypotrichida [INV ZOO] An order of highly specialized protozoans in the subclass Spirotrichia characterized by cirri on the ventral surface and a lack of ciliature on the dorsal surface. { ˌhī·pə'trik·ə·də }

hypotype [SYST] A specimen of a species, which, though not a member of the original type series, is known from a published description or listing. { 'hī·pəˌtīp }

hypovitaminosis [MED] Condition due to deficiency of an essential vitamin. { ˌhī·pəˌvīd·ə·mə'nō·səs }

hypovolemia [MED] Low blood volume. { ˌhī·pōˌvä'lē·mē·ə }

hypovolemic shock [MED] Shock caused by reduced blood volume which may be due to loss of blood or plasma as in burns, the crush syndrome, perforating gastrointestinal wounds, or other trauma. Also known as wound shock. { ¦hī·pōˌvä¦lē·mik 'shäk }

hypoxanthine [BIOCHEM] $C_5H_4ON_4$ An intermediate product derived from adenine in the hydrolysis of nucleic acid. { ¦hī·pō'zanˌthēn }

hypoxemia *See* hypoxia. { ˌhīˌpäk'sē·mē·ə }

hypoxia [ECOL] A condition characterized by a low level of dissolved oxygen in an aquatic environment. [MED] Oxygen deficiency; any state wherein a physiologically inadequate amount of oxygen is available to or is utilized by tissue, without respect to cause or degree. Also known as hypoxemia. { hī'päk·sē·ə }

hypoxic encephalopathy [MED] Brain damage syndrome caused by hypoxia. { hī'päk·sik enˌsef·ə'läp·əˌthē }

hypozygal [INV ZOO] In comatulids, the proximal member of adjacent brachials in an articulation. { ˌhī·pō'zīg·əl }

hypsodont [VERT ZOO] Of teeth, having crowns that are high or deep and roots that are short. { 'hip·səˌdänt }

hypural [VERT ZOO] Of or pertaining to the bony structure formed by fusion of the hemal spines of the last few vertebrae in most teleost fishes. { hī'pyůr·əl }

Hyracoidea [VERT ZOO] An order of ungulate mammals represented only by the conies of Africa, Arabia, and Syria. { ˌhī·rə'kȯid·ē·ə }

hyster- [MED] A combining form that denotes a relation to or a connection with the uterus. { 'his·tər }

hysterectomy [MED] Surgical removal of all or part of the uterus. { ˌhis·tə'rek·tə·mē }

hysteriaceous [MYCOL] Of, belonging to, or resembling the Hysteriales. { hiˌster·ē'ā·shəs }

Hysteriales [BOT] An order of lichens in the class Ascolichenes including those species with an ascolocular structure. { hiˌster·ē'ā·lēz }

hysterical anesthesia [MED] A loss of cutaneous pain sensation accompanying hysteria. { hi'ster·ə·kəl ˌan·əs'thē·zhə }

hysterical paralysis [MED] Muscle weakness or paralysis without loss of reflex activity, in which no organic nerve lesion can be demonstrated, but which is due to psychogenic factors. Also known as functional paralysis. { hi'ster·ə·kəl pə'ral·ə·səs }

hysterochroic [MYCOL] Having fruiting bodies which discolor progressively from base to apex with age. { ¦his·tə·rō¦krō·ik }

hysterogram [MED] A roentgenogram with opacification of the cavity of the uterus by the injection of contrast medium. { 'his·tə·rōˌgram }

hysterography [MED] Roentgenologic examination of the uterus after the introduction of a contrast medium. { ˌhis·tə'räg·rə·fē }

hystero-oophorectomy [MED] Surgical removal of the uterus and ovaries. { ¦his·tə·rōˌōˌō·fə'rek·tə·mē }

hysteropexy [MED] Fixation of the uterus by a surgical operation to correct displacement. { hisˌte·rō'pek·sē }

hysterorrhaphy [MED] The closure of a uterine incision by suture. { ˌhis·ter'ȯr·ə·fē }

hysterorrhexis [MED] Rupture of the uterus. { ˌhis·tə·rō'rek·səs }

hysterosalpingectomy [MED] Surgical removal of the uterus and oviducts. { ˌhis·tə·rōˌsal·pin'jek·tə·mē }

hysterosalpingography [MED] Roentgenographic examination of the uterus and oviducts after injection of a radiopaque substance. { ˌhis·tə·rōˌsal·piŋ'gäg·rə·fē }

hysterosalpingo-oophorectomy [MED] The excision of the uterus, oviducts, and ovaries. { ¦his·tə·rō·sal¦piŋ·gōˌōˌō·fə'rek·tə·mē }

hysteroscope [MED] A uterine speculum with a reflector. { 'his·tə·rəˌskōp }

hysterosoma [INV ZOO] A body division of an acarid mite composed of the metapodosoma and opisthosoma. { ¦his·tə·rə'sō·mə }

hysterotomy [MED] **1.** Incision into the uterine wall. **2.** Cesarean section. { ˌhis·tə'räd·ə·mē }

Hystricidae [VERT ZOO] The Old World porcupines, a family of Rodentia ranging from southern Europe to Africa and eastern Asia and into the Philippines. { hi'stris·əˌdē }

Hystricomorpha [VERT ZOO] A superorder of the class Rodentia. { ˌhis·trə·kō'mȯr·fə }

H zone [HISTOL] The central portion of an A band in a sarcomere; characterized by the presence of myosin filaments. { 'āch ˌzōn }

I

IAA *See* indoleacetic acid.

iatrogenic [MED] Pertaining to an abnormal state or condition produced by a physician in a patient by inadvertent or incorrect treatment. { ī,a·trə'jen·ik }

Iballidae [INV ZOO] A small family of hymenopteran insects in the superfamily Cynipoidea. { ī'bal·ə,dē }

I band [HISTOL] The band on either side of a Z line; encompasses portions of two adjacent sarcomeres and is characterized by the presence of actin filaments. { 'ī ,band }

IBD *See* inflammatory bowel disease.

ibis [VERT ZOO] The common name for wading birds making up the family Threskiornithidae and distinguished by a long, slender, downward-curving bill. { 'ī·bəs }

I blood group [IMMUNOL] The erythrocyte antigens defined by reactions with anti-I and anti-i antibodies, which occur both in acquired hemolytic anemia and naturally in normal persons of the rare phenotype i. { ¦ī 'bləd ,grüp }

IBS *See* irritable bowel syndrome.

IC_{50} *See* incapacitating concentration 50. { ¦ī¦sē 'fif·tē }

Icacinaceae [BOT] A family of dicotyledonous plants in the order Celastrales characterized by haplostemonous flowers, pendulous ovules, stipules wanting or vestigial, a polypetalous corolla, valvate petals, and usually one (sometimes three) locules. { i,ka·sə,nās·ē,ē }

Iceland disease *See* epidemic neuromyasthenia. { 'īs·lənd di,zēz }

I-cell disease *See* inclusion-cell disease. { 'ī,sel diz,ēz }

ich [VET MED] A dermatitis of fresh-water fishes caused by the invasion of the skin by the ciliated protozoan *Ichthyophthirius multifiliis*. { ik }

ichneumon [INV ZOO] The common name for members of the family Ichneumonidae. { ik'nü·mən }

Ichneumonidae [INV ZOO] The ichneumon flies, a large family of cosmopolitan, parasitic wasps in the superfamily Ichneumonoidea. { ik·nü'män·ə,dē }

Ichneumonoidea [INV ZOO] A superfamily of hymenopteran insects; members are parasites of other insects. { ik,nü·mə'nöid·ē·ə }

ichor [MED] An acrid, watery or blood-tinged discharge from an ulcer or wound. { 'ī,kör }

ichthammol [PHARM] A viscid fluid obtained by the destructive distillation of certain bituminous schists, followed by sulfonation of the distillate and neutralization with ammonia; used as a weak antiseptic and stimulant in skin diseases, and occasionally as an expectorant. { 'ik·thə·mól }

ichthyism [MED] Food poisoning caused by eating spoiled fish. { 'ik·thē,iz·əm }

Ichthyobdellidae [INV ZOO] A family of leeches in the order Rhynchobdellae distinguished by cylindrical bodies with conspicuous, powerful suckers. { ,ik·thē·äb'del·ə,dē }

ichthyology [VERT ZOO] A branch of vertebrate zoology that deals with the study of fishes. { ,ik·thē'äl·ə·jē }

ichthyophagous [ZOO] Subsisting on a diet of fish. { ,ik·thē'äf·ə·gəs }

ichthyosarcotoxism [MED] Poisoning caused by eating the flesh of fish containing toxic substances. { ,ik·thē·ō,sär·kə'täk,siz·əm }

ichthyosis [MED] A congenital skin disease characterized by dryness and scales, especially on the extensor surfaces of the extremities. { ,ik·thē'ō·səs }

ichthyosis congenita [MED] A severe form of ichthyosis characterized by cracked, thickened skin and mucous membranes. { ,ik·thē'ō·səs kən'jen·əd·ə }

ichthyosis simplex [MED] A common, childhood form of ichthyosis characterized by large, papery scales on the skin. { ,ik·thē'ō·səs 'sim,pleks }

Ichthyotomidae [INV ZOO] A monotypic family of errantian annelids in the superfamily Eunicea. { ,ik·thē·ō'täm·ə,dē }

icosahedral virus [VIROL] A virion in the form of an icosahedron. { ī¦kä·sə¦hē·drəl 'vī·rəs }

Icosteidae [VERT ZOO] The ragfishes, a family of perciform fishes in the suborder Stromateoidei found in high seas. { ,ī,kä'stē·ə,dē }

icotype [SYST] A typical, accurately identified specimen of a species, but not the basis for a published description. { 'ī·kə,tīp }

ICSH *See* luteinizing hormone.

icteric [MED] Pertaining to or characterized by jaundice. { ik'ter·ik }

Icteridae [VERT ZOO] The troupials, a family of New World perching birds in the suborder Oscines. { ik'ter·ə,dē }

icteroanemia [VET MED] A disease of swine

characterized by jaundice, anemia, and erythrocytolysis. { ¦ik·tə·rō·ə'nē·mē·ə }

icterogenic [MED] Causing icterus, or jaundice. { ¦ik·tə·rō¦jen·ik }

icterohematuria [VET MED] A disease of sheep caused by the protozoan *Babesia ovis* and characterized by hemolysis of erythrocytes accompanied by jaundice. { ¦ik·tə·rō,hē·mə'tu̇r·ē·ə }

icterohemorrhagic fever *See* Weil's disease. { ¦ik·tə·rō,hem·ə'raj·ik 'fē·vər }

icterus *See* jaundice. { 'ik·tə·rəs }

icterus gravis [MED] Acute yellow atrophy of the liver marked by jaundice and nervous system dysfunctions. { 'ik·tə·rəs 'grav·əs }

icterus gravis neonatorum [MED] Icterus gravis in the newborn caused by physiologic jaundice, erythroblastosis fetalis, and severe jaundice. { 'ik·tə·rəs 'grav·əs nē,än·ə'tȯr·əm }

icterus index [PATH] Measure of serum bilirubin levels by comparing the yellow blood serum from a jaundiced patient with the colors of standard potassium dichromate solutions. { 'ik·tə·rəs ,in,deks }

ID_{50} *See* infective dose 50.

ideational apraxia [MED] Inability to perform meaningful motor functions or to use objects properly due to mental confusion caused by diffuse brain disease. { ,id·ē'ā·shən·əl a'prak·sē·ə }

identical twins *See* monozygotic twins. { ī'dent·ə·kəl 'twinz }

ideokinetic apraxia *See* ideomotor apraxia. { ¦id·ē·ō·kə'ned·ik a'prak·sē·ə }

ideomotor [PHYSIO] **1.** Pertaining to involuntary movement resulting from or accompanying some mental activity, as moving the lips while reading. **2.** Pertaining to both ideation and motor activity. { 'id·ē·ə,mōd·ər }

ideomotor apraxia [MED] A nervous system disorder caused by lesions of the cerebral cortex in which simple single acts can be performed but not a sequence of associated acts. Also known as ideokinetic apraxia. { 'id·ē·ə,mōd·ər a'pak·sē·ə }

ideotype [SYST] A specimen identified as belonging to a specific taxon but collected from other than the type locality. { 'id·ē·o,tīp }

idioblast [BOT] A plant cell that differs markedly in shape or function from neighboring cells within the same tissue. { 'īd·ē·ō,blast }

idiochromatin [CELL MOL] The portion of the nuclear chromatin thought to function as the physical carrier of genes. { ¦id·ē·ō'krō·mə·tən }

idiogram [GEN] A diagram of chromosome morphology, especially the banding pattern, that is used to compare karyotypes of different cells, individuals, or species. { 'id·ē·ə,gram }

idiomuscular [PHYSIO] Pertaining to any phenomenon occurring in a muscle which is independent of outside stimuli. { ¦id·ē·ō'məs·kyə·lər }

idiopathic arthritis *See* Reiter's syndrome. { ¦id·ē·ə¦path·ik är'thrīd·əs }

idiopathic colitis [MED] Any form of colitis for which the causative agent is not identified. { ¦id·ē·ə¦path·ik kə'līd·əs }

idiopathic eunuchoidism [MED] A primary type of eunuchoidism in which there is no mammary gland enlargement and testes remain at a prepubertal stage of development. { ¦id·ē·ə¦path·ik ,yü·nə'kȯi,diz·əm }

idiopathic familial jaundice [MED] A familial form of obstructive jaundice of unknown cause in which there is decreased ability to excrete conjugated bilirubin into the bile ducts. { ¦id·ē·ə¦path·ik fə'mil·yəl 'jȯn·dəs }

idiopathic hypercholesterolemia [MED] A genetic derangement of fat metabolism characterized by high blood, cell, and plasma levels of cholesterol. { ¦id·ē·ə¦path·ik ¦hī·pər·kə,les·tə·rə'lē·mē·ə }

idiopathic megacolon [MED] Hypertrophy and dilation of the colon. { ¦id·ē·ə¦path·ik ¦meg·ə¦kō·lən }

idiopathic pulmonary hemisiderosis [MED] A disease of unknown etiology characterized by recurrent hemorrhaging from pulmonary capillaries. { ¦id·ē·ə¦path·ik 'pu̇l·mə,ner·ē ¦hem·ə,sid·ə'rō·səs }

idiopathic steatorrhea *See* celiac syndrome. { ¦id·ē·ə¦path·ik ,stē·əd·ə'rē·ə }

idiopathic thrombocytopenic purpura [MED] Thrombocytopenic purpura of unknown causes. { ¦id·ē·ə¦path·ik ¦thräm·bō,sīd·ə'pē·nik 'pər·pyə·rə }

idiopathic ulcerative colitis [MED] A form of ulcerative colitis of unknown cause. { ¦id·ē·ə¦path·ik ,əl·sə'rād·iv kə'līd·əs }

idiopathy [MED] **1.** A primary disease; one not a result of any other disease, but of spontaneous origin. **2.** Disease for which no cause is known. { ,id·ē'äp·ə·thē }

idiophase [MICROBIO] A period in culture growth during which secondary metabolites are produced. { 'id·ē·ə,fāz }

idiosome [CELL MOL] **1.** A hypothetical unit of a cell, such as the region of modified cytoplasm surrounding the centriole or centrosome. **2.** A sex chromosome. { 'id·ē·ə,sōm }

Idiostolidae [INV ZOO] A small family of hemipteran insects in the superfamily Pentatomorpha. { ,id·ē·ə'stäl·ə,dē }

idiosyncrasy [MED] A peculiarity of constitution that makes an individual react differently from most persons to drugs, diet, treatment, or other situations. { ,id·ē·ə'siŋ·krə·sē }

idiotype [IMMUNOL] The unique amino acid sequence and corresponding three-dimensional structure of the variable region of an immunoglobulin molecule that determines its antigenic specificity. { 'id·ē·ə,tīp }

Idoteidae [INV ZOO] A family of isopod crustaceans in the suborder Valvifera having a flattened body and seven pairs of similar walking legs. { ,ī·dō'tē·ə,dē }

id reaction [MED] Papular and vesicular eruptions on the skin, occurring suddenly following exacerbation of foci of some cutaneous fungus infections. { 'id rē'ak·shən }

IDU *See* 5-iodo-2′-deoxyuridine.

IDUR *See* 5-iodo-2′-deoxyuridine.

Ig *See* immunoglobulin entries.

iguana [VERT ZOO] The common name for a number of species of herbivorous, arboreal reptiles in the family Iguanidae characterized by a dorsal crest of soft spines and a dewlap; there are only two species of true iguanas. { i'gwän·ə }

iguanid [VERT ZOO] The common name for members of the reptilian family Iguanidae. { i'gwän·əd }

Iguanidae [VERT ZOO] A family of reptiles in the order Squamata having teeth fixed to the inner edge of the jaws, a nonretractile tongue, a compressed body, five clawed toes, and a long but rarely prehensile tail. { i'gwän·ə,dē }

IL *See* interleukin entries.

Ilarvirus [VIROL] A genus of the family Bromoviridae that is characterized by three types of particles, which are quasi-isometric, different in diameter, and contain four types of ribonucleic acid (1.1, 0.9, 0.7, 0.3 $\times 10^6$) separately encapsidated, except for the 3 $\times 10^5$ species which may be found in combination with others; the type species is tobacco streak virus. Also known as tobacco streak virus group. { 'ī·lə,vī·rəs }

ileitis [MED] Inflammation of the ileum. { ,il·ē'īd·əs }

ileocecal valve [ANAT] A muscular structure at the junction of the ileum and cecum which prevents reflex of the cecal contents. { ¦il·ē·ō¦sē·kəl 'valv }

ileocolic artery [ANAT] A branch of the superior mesenteric artery that supplies blood to the terminal part of the ileum and the beginning of the colon. { ¦il·ē·ō'käl·ik 'ärd·ə·rē }

ileocolic intussusception [MED] Slipping of the ileum through the ileocecal valve into the colon. { ¦il·ē·ō'käl·ik ,in·tə·sə'sep·shən }

ileocolitis [MED] Inflammation of both the ileum and the colon. { ,il·ē·ō·kə'līd·əs }

ileocolostomy [MED] Surgical formation of a bypass channel between the ileum and the colon. { ,il·ē·ō·kə'lä·stə·mē }

ileocutaneous [MED] Pertaining to a joining of the ileum and the skin resulting from a fistulous connection. { ,il·ē·ō,kyü'tā·nē·əs }

ileostomy [MED] Surgical formation of an artificial anus through the abdominal wall into the ileum. { ,il·ē'ä·stə·mē }

ileum [ANAT] The last portion of the small intestine, extending from the jejunum to the large intestine. { 'il·ē·əm }

ileus [MED] Acute intestinal obstruction of neurogenic origin. { 'il·ē·əs }

iliac artery [ANAT] Either of the two large arteries arising by bifurcation of the abdominal aorta and supplying blood to the lower trunk and legs (or hind limbs in quadrupeds). Also known as common iliac artery. { 'il·ē,ak 'ärd·ə·rē }

iliac fascia [ANAT] The fascia covering the pelvic surface of the iliacus muscle. { 'il·ē,ak 'fā·shə }

iliac region *See* inguinal region. { 'il·ē,ak ,rē·jən }

iliacus [ANAT] The portion of the iliopsoas muscle arising from the iliac fossa and sacrum. { i'lī·ə·kəs }

iliac vein [ANAT] Any of the three veins on each side of the body which correspond to and accompany the iliac artery. { 'il·ē,ak ¦vān }

iliocostalis [ANAT] The lateral portion of the erector spinal muscle that extends the vertebral column and assists in lateral movements of the trunk. { ,il·ē·ō,kä'sta·ləs }

iliofemoral ligament [ANAT] A strong band of dense fibrous tissue extending from the anterior inferior iliac spine to the lesser trochanter and the intertrochanteric line. Also known as Y ligament. { ,il·ē·ō'fem·ə·rəl ,lig·ə·mənt }

iliolumbar ligament [ANAT] A fibrous band that radiates laterally from the transverse processes of the fourth and fifth lumbar vertebrae and attaches to the pelvis by two main bands. { ,il·ē·ō'ləm,bär ,lig·ə·mənt }

iliopsoas [ANAT] The combined iliacus and psoas muscles. { ,il·ē·ō'sō·əs }

iliotibial tract [ANAT] A thickened portion of the fascia lata extending from the lateral condyle of the tibia to the iliac crest. { ,il·ē·ō'tib·ē·əl ,trakt }

ilium [ANAT] Either of a pair of bones forming the superior portion of the pelvis bone in vertebrates. { 'il·ē·əm }

Illiciales [BOT] An order of dicotyledonous flowering plants, division Magnoliophyta, of the class Magnoliopsida, characterized by having woody plants with scattered spherical cells containing volatile oils. { i,lis·ē'ā·lēz }

illness [MED] **1.** The state of being sick. **2.** A sickness, disease, or disorder. { 'il·nəs }

imaginal disk [INV ZOO] Any of the thickened areas within the sac of the body wall in holometabolous insects which give rise to specific organs in the adult. { ə'maj·ən·əl ,disk }

imago [INV ZOO] The sexually mature, usually winged stage of insect development. { ə'mä·gō }

imbed *See* embed. { im'bed }

imbricate [BIOL] Having overlapping edges, such as scales, or the petals of a flower. { 'im·brə·kət }

immarginate [BIOL] Lacking a clearly defined margin. { i'mär·jə,nāt }

immediate hypersensitivity [IMMUNOL] A type of hypersensitivity in which the response rapidly occurs following exposure of a sensitized individual to the antigen. { i'mē·dē·ət ,hī·pər·sen·sə'tiv·əd·ē }

immersion foot [MED] A serious and disabling condition of the feet due to prolonged immersion in seawater at 60°F (15.6°C) or lower, but not at freezing temperature. { ə'mər·zhən ,fu̇t }

immigrant [ECOL] An organism that settles in a zone where it was previously unknown. { 'im·ə·grənt }

immigration [ECOL] The one-way inward movement of individuals or their disseminules into a population or population area. [GEN] Gene flow from one population into another by interbreeding between members of the populations. { ,im·ə'grā·shən }

immobilize [MED] To render motionless or to fix in place, as by splints or surgery. { i'mō·bə,līz }

immortalization [CELL MOL] The process whereby a cell line gains the ability to undergo continuous cell division. { i,mȯrd·əl·ə'zā·shən }

immune [IMMUNOL] **1.** Safe from attack; protected against a disease by an innate or an acquired immunity. **2.** Pertaining to or conferring immunity. { i'myün }

immune body *See* antibody. { i'myün ,bäd·ē }

immune complex disease [MED] A disease that results from deposition of antigen-antibody complexes in tissues. { i¦myün ¦käm,pleks di ,zēz }

immune hemolysin [IMMUNOL] A substance formed in blood in response to an injection of erythrocytes from another species. { i'myün hē·mə'līs·ən }

immune horse serum [IMMUNOL] Serum obtained from the blood of an immunized horse. { i'myün 'hȯrs ,sir·əm }

immune lysin [IMMUNOL] An antibody that will disrupt a particular type of cell in the presence of complement and cofactors, such as magnesium or calcium ions. { i'myün 'līs·ən }

immune opsonin [IMMUNOL] A substance produced in blood in response to an infection or to inoculation with dead cells of the infecting species of bacteria. { i'myün 'äp·sə·nən }

immune precipitation [IMMUNOL] A method of isolating a protein from mixtures by using a specific antibody as the precipitating agent. { i'myün prə,sip·ə'tā·shən }

immune protein [IMMUNOL] Any antibody. { i'myün 'prō,tēn }

immune response [IMMUNOL] The physiological responses stemming from activation of the immune system by antigens, consisting of a primary response in which the antigen is recognized as foreign and eliminated, and a secondary response to subsequent contact with the same antigen. { i'myün ri,späns }

immune response gene [IMMUNOL] Any of a group of genes in the major histocompatibility complex that determines the degree of immune response. Abbreviated IR gene. { i¦myün ri 'späns ,jēn }

immune serum [IMMUNOL] Blood serum obtained from an immunized individual and carrying antibodies. { i'myün ,sir·əm }

immunity [IMMUNOL] The condition of a living organism whereby it resists and overcomes an infection or a disease. { i'myü·nəd·ē }

immunization [IMMUNOL] Rendering an organism immune to a specific communicable disease. { ,im·yə·nə'zā·shən }

immunization therapy [MED] The use of vaccines or anti-serums to produce immunity against a specific disease. { ,im·yə·nə'zā·shən ,ther·ə·pē }

immunoassay [IMMUNOL] A laboratory detection method that uses antibodies to react with specific substances. { ,im·yə·nō'a,sāy }

immunochemistry [IMMUNOL] A branch of science dealing with the chemical changes associated with immunity factors. { ¦im·yə·nō'kem·ə·strē }

immunocompromised [IMMUNOL] Having an impaired or weakened immune system (usually due to drugs or illness). { ,im·yə·nō'käm·pre ,mīzd }

immunodeficiency [IMMUNOL] Any defect of antibody function or cell-mediated immunity. { ¦im·yə,nō·də'fish·ən·sē }

immunodiffusion [IMMUNOL] A serological procedure in which antigen and antibody solutions are permitted to diffuse toward each other through a gel matrix; interaction is manifested by a precipitin line for each system. { ¦im·yə·nō·də'fyü·zhən }

immunoelectrophoresis [IMMUNOL] A serological procedure in which the components of an antigen are separated by electrophoretic migration and then made visible by immunodiffusion of specific antibodies. { ¦im·yə·nō·i,lek·trə·fə'rē·səs }

immunofluorescence [IMMUNOL] Fluorescence as the result of, or identifying, an immune response; a specifically stained antigen fluoresces in ultraviolet light and can thus be easily identified with a homologous antigen. { ¦im·yə·nō·flə'res·əns }

immunogen [IMMUNOL] A substance which stimulates production of specific antibody or of cellular immunity, and which can react with these products. { ə'myü·nə·jən }

immuno-gene therapy [MED] A type of gene therapy used in cancer treatment that aims to enhance the immune response against tumor cells. { 'im·yə·nō,jēn ¦ther·ə·pē }

immunogenetics [MED] A branch of immunology dealing with the relationships between immunity and genetic factors in disease. { ¦im·yə·nō·jə'ned·iks }

immunogenic [IMMUNOL] Producing immunity. { ¦im·yə·nō¦jen·ik }

immunogenic peptides [IMMUNOL] Peptides that are recognized by T cells primed by immunization or transplantation. { ,im·yə·nō,jen·ik 'pep,tīdz }

immunoglobulin [IMMUNOL] Any of a set of serum glycoproteins which have the ability to bind other molecules with a high degree of specificity. Abbreviated Ig. { ¦im·yə·nō'glä·byə·lən }

immunoglobulin A [IMMUNOL] A class of immunoglobulins that inhibits the binding of microorganisms to mucosal surfaces; large amounts are found in breast milk, saliva, and gastrointestinal secretions. Abbreviated IgA. { ,im·yə·nō,gläb·yə·lən 'ā }

immunoglobulin A deficiency [IMMUNOL] An immune system disorder in which lower than normal amounts of immunoglobulin A are produced, resulting in increased susceptibility to infections such as chronic sinusitis, chronic pulmonary infections, and digestive problems. { ˌim·yə·nōˌgläb·yə·lən ¦ā di'fish·ən·sē }

immunoglobulin D [IMMUNOL] A class of immunoglobulins that are found on the surface of B cells and in minute amounts in normal human serum. { ˌim·yə·nōˌgläb·yə·lən 'dē }

immunoglobulin domain [IMMUNOL] The basic structural unit of immunoglobulins. { ˌim·yə·nōˌgläb·yə·lən də'mān }

immunoglobulin E [IMMUNOL] A class of immunoglobulins present in minute amounts in normal human serum that is active against parasites and acts as a mediator of immediate hypersensitivity. Abbreviated IgE. Also known as reagin. { ˌim·yə·nōˌgläb·yə·lən 'ē }

immunoglobulin G [IMMUNOL] The most abundant immunoglobulin class in human serum; it is associated with complement fixation, opsonization, fixation to macrophages, and membrane transport. Abbreviated IgG. { ˌim·yə·nōˌgläb·yə·lən 'jē }

immunoglobulin M [IMMUNOL] The first immunoglobulin class to appear during the primary immune response; it is mainly contained in the bloodstream, where it can readily neutralize agents attempting to gain entrance through the blood. Abbreviated IgM. { ˌim·yə·nōˌgläb·yə·lən 'em }

immunoglobulin-mediated immunity *See* humoral immunity. { ˌim·yə·nō¦gläb·yə·lən ¦mē·dēˌād·əd i'myü·nəd·ē }

immunogold electron microscopy [CELL MOL] A technique in which cellular components are visualized with an electron microscope by using gold particles as antibody/protein labels. { ¦im·yə·nəˌgold iˌlekˌträn mī'kräs·kə·pē }

immunogranulomatous disease [MED] A condition in which a deviation from the standard immune mechanisms is considered to be associated with widespread granulomatosis. { ¦im·yə·nōˌgran·yə'läm·əd·əs diˌzēz }

immunological deficiency [IMMUNOL] A state wherein the immune mechanisms are inadequate in their ability to perform their normal function, that is, the elimination of foreign materials (usually infectious agents such as bacteria, viruses, and fungi). { ˌim·yə·nəˌläj·ə·kəl di'fish·ən·sē }

immunological memory [IMMUNOL] The capacity of the immune system to respond more rapidly and vigorously to the second contact with a specific antigen than to the primary contact. { ˌim·yə·nə¦läj·ə·kəl 'mem·rē }

immunological ontogeny [IMMUNOL] The origin and development of the lymphocyte system, from its earliest stages to the two major populations of mature lymphocytes: the thymus-dependent or T lymphocytes, and the thymus-independent or B lymphocytes. { ˌim·yə·nəˌläj·ə·kəl än 'täj·ə·nē }

immunological paralysis *See* acquired immunological tolerance. { ¦im·yə·nə¦läj·ə·kəl pə'ral·ə·səs }

immunological phylogeny [IMMUNOL] The study of immunology and the immune system in evolution. { ¦im·yə·nəˌläj·i·kəl fī'läj·ə·nē }

immunologic cytotoxicity [IMMUNOL] The mechanism by which the immune system destroys or damages foreign or abnormal cells. { ¦im·yə·nə¦läj·ik ˌsīd·ō·täk'sis·əd·ē }

immunologic suppression [IMMUNOL] The use of x-irradiation, chemicals, corticosteroid hormones, or antilymphocyte antisera to suppress antibody production, particularly in graft transplants. Also known as immunosuppression. { ¦im·yə·nə¦läj·ik sə'presh·ən }

immunologic tolerance [IMMUNOL] **1.** A condition in which an animal will accept a homograft without rejection. **2.** A state of specific unresponsiveness to an antigen or antigens in adult life as a consequence of exposure to the antigen in utero or in the neonatal period. { ¦im·yə·nə¦läj·ik 'täl·ə·rəns }

immunologist [IMMUNOL] A person who specializes in immunology. { ˌim·yə'näl·ə·jəst }

immunology [BIOL] A branch of biological science concerned with the native or acquired resistance of higher animal forms and humans to infection with microorganisms. { ˌim·yə'näl·ə·jē }

immunonephelometry [IMMUNOL] The application of nephelometry to the quantification of antigen or antibody. { ¦im·yə·nōˌnef·ə'läm·ə·trē }

immunopathology [MED] The study of various human and animal diseases in which humoral and cellular immune factors seem important in causing pathological damage to cells, tissues, and the host. { ¦im·yə·nō·pə'thäl·ə·jē }

immunopotentiation [IMMUNOL] Enhancement of an immune response by a variety of adjuvants. { ˌim·yə·nō·pəˌten·chē'ā·shən }

immunoprecipitation [IMMUNOL] A protein purification method which involves the formation of an antibody-protein complex to separate out the protein of interest. { ˌim·yə·nō·prəˌsip·ə'tā·shən }

immunoselective adsorption [IMMUNOL] A process that exploits biospecific interactions between antibodies and their corresponding antigens in order to separate pure high-value bioactive products from biological sources (for example, blood, food materials, and products of genetically engineered organisms). { ˌim·yə·nō·siˌlek·tiv ad'sörp·shən }

immunosuppression *See* immunologic suppression. { ¦im·yə·nō·sə'presh·ən }

immunosuppressive [PHARM] Any drug or agent used to suppress antibody production. { ¦im·yə·nō·sə'pres·iv }

immunotherapy [MED] **1.** Therapy utilizing immunosuppressives. **2.** *See* serotherapy. { ¦im·yə·nō'ther·ə·pē }

immunotoxicity [IMMUNOL] Adverse effects on the normal functioning of the immune system, caused by exposure to a toxic chemical. The result can be higher rates of infectious diseases or cancer, more severe cases of such autoimmune disease, or allergic reactions. { ˌim·yə·nō·täk'sis·əd·ē }

immunotoxin [IMMUNOL] Conjugate of antibody and toxic protein such that the specificity of the antibody molecule is combined with the cytotoxic property of the toxin. { ˌim·yə·nō 'täk·sən }

impaction [MED] **1.** The state of being lodged and retained in a body part. **2.** Confinement of a tooth in the jaw so that its eruption is prevented. **3.** A condition in which one fragment of bone is firmly driven into another fragment so that neither can move against the other. { im'pak·shən }

imparipinnate *See* odd-pinnate. { ˌim·par·ə'pi ˌnāt }

impedance plethysmography [MED] A technique by which changes in the volume of certain segments of the body can be determined by impedance measurements. These changes are related to factors involving the mechanical activity of the heart, to conditions of the circulatory system, or to respiratory flow and other physiological functions. { imˌpēd·əns ˌpleth·ə'smäg·rə·fē }

Impennes [VERT ZOO] A superorder of birds for the order Sphenisciformes in some systems of classification. { im'pen·ēz }

imperfect flower [BOT] A flower lacking either stamens or carpels. { im'pər·fikt 'flaù·ər }

imperforate [BIOL] Lacking a normal opening. { im'pər·fə·rət }

imperforate anus [MED] A congenital malformation in which the large intestine ends blindly. { im'pər·fə·rət 'ā·nəs }

impermeable junction *See* tight junction. { im¦pər·mē·ə·bəl 'jəŋk·shən }

impetigo [MED] An acute, contagious, inflammatory skin disease caused by streptococcal or staphylococcal infections and characterized by vesicular or pustular lesions. { ˌim·pə'tīˌgō }

implant [MED] **1.** A quantity of radioactive material in a suitable container, intended to be embedded in a tissue or tumor for therapeutic purposes. **2.** A tissue graft placed in depth in the body. { 'imˌplant }

implantation [MED] **1.** Placement of a tissue transplant in depth in the body. **2.** Placement in the body of a device for mechanical repair, such as for a ventral hernia or a fracture. **3.** Embedding of an embryo into the endometrium. { ˌimˌplan'tā·shən }

implanted device [MED] A heart pacemaker or other medical electronic device that is surgically placed in the body. { im'plant·əd di'vīs }

implexed [INV ZOO] In insects, having the integument infolded for muscle attachment. { 'imˌplekst }

impotence [MED] **1.** Inability in the male to perform the sexual act. **2.** Lack of sexual vigor. { 'im·pəd·əns }

impregnate [MED] To fertilize or cause to become pregnant. { im'pregˌnāt }

impunctate [BIOL] Lacking pores. { im'pəŋk ˌtāt }

IMViC test [MICROBIO] A group of four cultural tests used to differentiate genera of bacteria in the family Enterobacteriaceae and to distinguish them from other bacteria; tests are indole, methyl red, Voges-Proskauer, and citrate. { 'im ˌvik ˌtest }

inactivated vaccine *See* killed vaccine. { in¦ak·təˌvād·əd vakˌsēn }

inadunate [BIOL] Not united. [INV ZOO] In crinoids, having the arms free from the calyx. { i'näj·ə·nət }

inanimate [MED] Lacking consciousness or life. { in'an·ə·mət }

inanition [MED] The exhausted, pathologic condition resulting from starvation. { ˌin·ə'nish·ən }

inaperturate [BIOL] Lacking apertures. { in'ap·əˌchùr·ət }

inarching [BOT] A kind of repair grafting in which two plants growing on their own roots are grafted together and one plant is severed from its roots after the graft union is established. { 'in'ärch·iŋ }

Inarticulata [INV ZOO] A class of the phylum Brachiopoda; valves are typically not articulated and are held together only by soft tissue of the living animal. { ¦in·ärˌtik·yə'läd·ə }

inborn [BIOL] Of or pertaining to a congenital or hereditary characteristic. { 'inˌbòrn }

inbreeding [GEN] Reproduction behavior between closely related individuals; self-fertilization, as in some plants, is the most extreme form. { 'inˌbrēd·iŋ }

inbreeding coefficient [GEN] A measure of the rate of inbreeding or the degree to which an individual is inbred. Also known as Wright's inbreeding coefficient. { 'inˌbrēd·iŋ ˌkō·əˌfish·ənt }

inbreeding depression [GEN] A decrease in fitness and vigor as a result of inbreeding. { 'in ˌbrēd·iŋ diˌpresh·ən }

inbred strain [GEN] Animal strain that results when individuals that are more closely related to each other than randomly chosen individuals mate together for many generations. { 'in ˌbred ˌstrān }

incapacitating concentration 50 [PHYSIO] The concentration of gas or smoke that will incapacitate 50% of the test animals within a given time of exposure. Abbreviated IC_{50}. { ˌin·kə¦pas·əˌtād·iŋ ˌkans·ən¦trā·shən 'fif·tē }

incapsidation [VIROL] The construction of a capsid around the genetic material of a virus. { inˌkap·sə'dā·shən }

incarcerated hernia [MED] A hernia in which the intestinal loop is permanently trapped in the hernia sac. { in'kär·səˌrād·əd 'hər·nē·ə }

incertae sedis [SYST] Placed in an uncertain taxonomic position. { iŋ¦kərˌtī 'sā·dəs }

incipient species [EVOL] Populations that are in the process of diverging to the point of speciation but still have the potential to interbreed. { in'sip·ē·ənt ¦spē·shēz }

Incirrata [INV ZOO] A suborder of cephalopod mollusks in the order Octopoda. { ¦in·sə¦räd·ə }

incised [BIOL] Having a deeply and irregularly notched margin. [MED] Made by cutting, as a wound. { in'sīzd }

incision [MED] A cut or wound of the body tissue, as an abdominal incision or a vertical or oblique incision. { in'sizh·ən }

incisional hernia [MED] Abnormal protrusion of an organ through an operative or accidental incision. Also known as postoperative hernia; posttraumatic hernia. { in'sizh·ən·əl 'hər·nē·ə }

incisive canal [ANAT] The bifurcated bony passage from the floor of the nasal cavity to the incisive fossa. { in'sī·siv kə¦nal }

incisive foramen [ANAT] One of the two to four openings of the incisive canal on the floor of the incisive fossa. { in'sī·siv fə'rā·mən }

incisive fossa [ANAT] **1.** A bony pit behind the upper incisors into which the incisive canals open. **2.** A depression on the maxilla at the origin of the depressor muscle of the nose. **3.** A depression of the mandible at the origin of the mentalis muscle. { in'sī·siv 'fäs·ə }

incisor [ANAT] A tooth specialized for cutting, especially those in front of the canines on the upper jaw of mammals. { in'sīz·ər }

inclusion [CELL MOL] A visible product of cellular metabolism within the protoplasm. { in 'klü·zhən }

inclusion blennorrhea *See* inclusion conjunctivitis. { in'klü·zhən ,blen·ə'rē·ə }

inclusion body [VIROL] Any of the abnormal structures appearing within the cell nucleus or the cytoplasm during the course of virus multiplication. { in'klü·zhən ,bäd·ē }

inclusion-cell disease [MED] A rare genetic disorder in which lysosomal hydrolases are transported out of the cell into the blood, rather than into the lysosome, resulting in the accumulation of undigested macromolecules within the lysosome. Abbreviated I-cell disease. { iŋ'klü·zhən ,sel ,diz,ēz }

inclusion compound *See* clathrate. { in'klü·zhən 'käm,paùnd }

inclusion conjunctivitis [MED] An acute inflammation of the conjunctiva with pus formation caused by a virus and identified from epithelial-cell inclusion bodies in conjunctival scrapings. Also known as inclusion blennorrhea; paratrachoma; swimmer's conjunctivitis; swimming-pool conjunctivitis. { in'klü·zhən kən,jəŋk·tə'vīd·əs }

inclusion cyst [MED] A cyst formed by the implantation of epithelial tissue into another structure. { in'klü·zhən ,sist }

inclusion encephalitis [MED] A chronic inflammation of the brain in which large inclusion bodies occur in the nuclei of oligodendria and sometimes in nerve cells. { in'klü·zhən en,sef·ə 'līd·əs }

incoherence [MED] Lack of coherence, relevance, or continuity of ideas or language. { ,in·kō'hir·əns }

incompatibility [IMMUNOL] Genetic or antigenic differences between donor and recipient tissues that result in a rejection response. { ,in·kəm,pad·ə'bil·əd·ē }

incompetence [FOREN] Inability to function within the law, as the incompetence of an individual to drive when under the influence of alcohol. [MED] Insufficiency or inadequacy in performing natural functions. { in'käm·pəd·əns }

incomplete abortion [MED] Expulsion of only part of the product of conception, with some of the membranes or placenta remaining in the uterus. { ,in·kəm'plēt ə'bȯr·shən }

incomplete antibody [IMMUNOL] An antibody that cannot directly agglutinate saline-suspended red blood cells, it needs an additive to complete the agglutination. { ,in·kəm,plēt 'ant·i,bäd·ē }

incomplete dominance [GEN] Expression of alleles such that the phenotype of the heterozygote is intermediate between that of the two homozygotes. { ¦iŋ·kəm,plēt 'däm·ən·əns }

incomplete flower [BOT] A flower lacking one or more modified leaves, such as petals, sepals, pistils, or stamens. { ,in·kəm'plēt 'flaù·ər }

incontinence [MED] Inability to control the natural evacuations, as the feces or the urine; specifically, involuntary evacuation from organic causes. { in'känt·ən·əns }

incrassate [BIOL] State of being swollen or thickened. { in'kra,sāt }

incretion [PHYSIO] An internal secretion. { in 'krē·shən }

incross [GEN] Mating between individuals from the same inbred line. { 'in,krȯs }

incubation [MED] The phase of an infectious disease process between infection by the pathogen and appearance of symptoms. [VERT ZOO] The act or process of brooding. { ,iŋ·kyə'bā·shən }

incubation period [MED] The period of time required for the development of symptoms of a disease after infection, or of altered reactivity after exposure to an allergen. [VERT ZOO] The brooding period required to bring an egg to hatching. { ,iŋ·kyə'bā·shən ,pir·ē·əd }

incubator [MED] A small chamber with controlled oxygen, temperature, and humidity for newborn infants requiring special care. [MICROBIO] A laboratory cabinet with controlled temperature for the cultivation of bacteria, or for facilitating biologic tests. { 'iŋ·kyə,bād·ər }

incubatory carrier [MED] A person infected with a certain microorganism but in such an early stage of disease that clinical manifestations are not apparent. { 'iŋ·kyə·bə,tȯr·ē ¦kar·ē·ər }

incubous [BOT] The juxtaposition of leaves such that the anterior margins of older leaves overlap the posterior margins of younger leaves. { 'iŋ·kyə·bəs }

incudate [BIOL] Of, pertaining to, or having an incus. { 'iŋ·kyə,dāt }

incumbent [BIOL] Lying on or down. [ECOL] Referring to the occupation and utilization of resources to the exclusion of other species. { in'kəm·bənt }

incurrent canal [INV ZOO] A canal through which water enters a sponge. { in'kər·ənt kə'nal }

incurrent siphon *See* inhalant siphon. { in'kə·rənt 'sī·fən }

Incurvariidae [INV ZOO] A family of lepidopteran insects in the superfamily Incurvarioidea; includes yucca moths and relatives. { ,in,kər·və'rī·ə,dē }

Incurvarioidea [INV ZOO] A monofamilial superfamily of lepidopteran insects in the suborder Heteroneura having wings covered with microscopic spines, a single genital opening in the female, and venation that is almost complete. { ,in·kər,var·ē'ŏīd·ē·ə }

incus [ANAT] The middle one of three ossicles in the middle ear. Also known as anvil. { 'iŋ·kəs }

indeciduate placenta [EMBRYO] A placenta having the maternal and fetal elements associated but not fused. { ¦in·də¦sij·ə·wət plə 'sent·ə }

indehiscent [BOT] **1.** Remaining closed at maturity, as certain fruits. **2.** Not splitting along regular lines. { ¦in·də'his·ənt }

independent assortment [GEN] In meiosis, the random assortment of the alleles at two or more loci because they are on different chromosome pairs or far apart on the same chromosome pair. { ,in·də'pen·dənt ə'sȯrt·mənt }

indeterminate cleavage [EMBRYO] Cleavage in which all the early cells have the same potencies with respect to development of the entire zygote. { ,in·də'tərm·ə·nət 'klē·vij }

indeterminate growth [BOT] Growth of a plant in which the axis is not limited by development of a reproductive structure, and therefore growth continues indefinitely. { ,in·də'tərm·ə·nət 'grōth }

indican [BIOCHEM] C_8H_6NOSOK The potassium salt of indoxylsulfate found in urine as a result of bacterial action on tryptophan in the bowel. { 'in·də,kan }

indicator medium [MICROBIO] A usually solid culture medium containing substances capable of undergoing a color change in the vicinity of a colony which has effected a particular chemical change, such as fermenting a certain sugar. { 'in·də,kād·ər ,mē·dē·əm }

indicator plant [BOT] A plant used in geobotanical prospecting as an indicator of a certain geological phenomenon. { 'in·də,kād·ər ,plant }

indicator species [ECOL] A species whose presence is directly related to a particular quality in its environment at a given location. { 'in·də,kād·ər ,spē·shēz }

indirect Coombs test *See* Rh blocking test. { ,in·də'rekt 'kümz ,test }

indirect developing test *See* Rh blocking test. { ,in·də'rekt di'vel·ə·piŋ ,test }

indirect hernia [MED] A form of inguinal hernia that passes out of the abdomen through the inguinal canal. { ,in·də'rekt 'hər·nē·ə }

indirect immunofluorescence [IMMUNOL] The use of a labeled indicator antibody which reacts with an unlabeled detector antibody that has previously reacted with an antigen. { ,in·də'rekt ¦im·yə·nō·flü'res·əns }

indirect vision *See* peripheral vision. { ,in·də'rekt 'vizh·ən }

individuation [EMBRYO] The process whereby, through induction, a spatially organized tissue, organ, or embryo develops. { ,in·di,vij·ə'wā·shən }

indoleacetic acid [BIOCHEM] $C_{10}H_9O_2N$ A decomposition product of tryptophan produced by bacteria and occurring in urine and feces; used as a hormone to promote plant growth. Abbreviated IAA. { ¦in,dōl·ə¦sēd·ik 'as·əd }

indolent [MED] **1.** Of or relating to a slow-growing, nonpainful neoplasm. **2.** Slowness in the process of healing. { 'in·də·lənt }

indole test [MICROBIO] A test for the production of indole from tryptophan by microorganisms; a solution of *para*-dimethylaminobenzaldehyde, amyl alcohol, and hydrochloric acid added to the incubated culture of bacteria shows a red color in the alcoholic layer if indole is present. { 'in,dōl ,test }

Indo-Pacific faunal region [ECOL] A marine littoral faunal region extending eastward from the east coast of Africa, passing north of Australia and south of Japan, and ending in the east Pacific south of Alaska. { ¦in·dō·pə'sif·ik 'fȯn·əl ,rē·jən }

Indriidae [VERT ZOO] A family of Madagascan prosimians containing wholly arboreal vertical clingers and leapers. { in'drī·ə,dē }

inducer [EMBRYO] The cell group that functions as the acting system in embryonic induction by controlling the mode of development of the reacting system. Also known as inductor. { in'dü·sər }

inducible enzyme [BIOCHEM] An enzyme which is present in trace quantities within a cell but whose concentration increases dramatically in the presence of substrate molecules. { in'dü·sə·bəl 'en,zīm }

induction [EMBRYO] *See* embryonic induction. [MED] The period from administration of the anesthetic to loss of consciousness by the patient. { in'dək·shən }

inductor *See* inducer. { in'dək·tər }

inductura [INV ZOO] A layer of lamellar shell material along the inner lip of the aperture in gastropods. { in'dək·chə·rə }

indumentum [BOT] A covering, such as one that is woolly. [MYCOL] A covering of hairs. [VERT ZOO] The plumage covering a bird. { ,in·də'men·təm }

induplicate [BOT] Having the edges turned or rolled inward without twisting or overlapping; applied to the leaves of a bud. { in'dü·plə·kət }

induration [BIOL] The process of hardening, especially by increasing the fibrous elements. [MED] Hardening of a tissue or organ due to an accumulation of blood, inflammation, or neoplastic growth. { ,in·də'rā·shən }

indusium [ANAT] A covering membrane such as the amnion. [BOT] An epidermal outgrowth covering the sori in many ferns. [MYCOL] The annulus of certain fungi. { in'dü·zē·əm }

industrial hygiene [MED] The science that deals with the anticipation and control of unhealthy conditions in workplaces in order to prevent illness among employees. { in'dəs·trē·əl 'hī,jēn }

industrial microbiology [MICROBIO] The study, utilization, and manipulation of those microorganisms capable of economically producing desirable substances or changes in substances, and the control of undesirable microorganisms. { in'dəs·trē·əl ,mī·krō·bī'äl·ə·jē }

industrial microorganism [MICROBIO] Any microorganism utilized for industrial microbiology. { in'dəs·trē·əl ,mī·krō'ór·gə,niz·əm }

industrial yeast [MICROBIO] Any yeast used for the production of fermented foods and beverages, for baking, or for the production of vitamins, proteins, alcohol, glycerol, and enzymes. { in'dəs·trē·əl ¦yēst }

indwelling catheter [MED] A thin tube communicating to the body surface, inserted into the vascular system and positioned to permit pressure measurements or blood sampling over a long period of time. { 'in,dwel·iŋ 'kath·əd·ər }

inequilateral [BIOL] Having the two sides or ends unequal, as the ends of a bivalve mollusk on either side of a line from umbo to gape. { in¦ē·kwə'lad·ə·rəl }

inermous [BIOL] Lacking mechanisms for defense or offense, especially spines. { i'nər·məs }

inertia [MED] Sluggishness, especially of muscular activity. { i'nər·shə }

inevitable abortion [MED] An abortion that has progressed to a stage where termination of the pregnancy cannot be avoided. { i'nev·əd·ə·bəl ə'bór·shən }

in extremis [MED] At the point of death. { ,in ik'strā·məs }

infant botulism [MED] Botulism that involves ingestion of *Clostridium botulinum* spores with subsequent germination and toxin production in the gastrointestinal tract, found mostly in children aged 6 months or younger. { ,in·fənt 'bäch·ə,liz·əm }

infantile amaurotic familial idiocy *See* Tay-Sachs disease. { 'in·fən,tīl ¦a,mó¦räd·ik fə¦mil·yəl 'id·ē·ə·sē }

infantile celiac disease [MED] Celiac syndrome of infants and young children. { 'in·fən,tīl 'sē·lē,ak di,zēz }

infantile cortical hyperostosis [MED] A condition occurring during the first 3 months of life in which there is fever and painful swelling of the soft tissue of the lower jaw, characterized by periosteal proliferation of the mandible. { 'in·fən,tīl 'kórd·ə·kəl ,hī·pə,rä'stō·səs }

infantile diarrhea [MED] An acute gastrointestinal disease in infants resulting from damage of the intestinal mucosa by an infectious organism. { 'in·fən,tīl ,dī·ə'rē·ə }

infantile eczema [MED] An allergic inflammation of the skin in young children, usually due to common antigens such as food or inhalants. { 'in·fən,tīl 'ek·sə·mə }

infantile genitalia [ANAT] The genital organs of an infant. [MED] Underdeveloped genitals in an adult. { 'in·fən,tīl ,jen·ə'tāl·yə }

infantile neuroaxonal dystrophy [MED] A familial disease of the central nervous system occurring early in life and characterized by axonal swellings, arrested development, atrophy of the optic nerves, and eventual blindness. { 'in·fən,tīl ¦nu̇·rō¦ak·sən·əl 'dis·trə·fē }

infantile paralysis *See* poliomyelitis. { 'in·fən,tīl pə'ral·ə·səs }

infantile scurvy [MED] Acute scurvy of infants and young children characterized by periosteal hemorrhage and swelling, especially of long bones. Also known as Cheadle's disease; Moeller-Barlow disease. { 'in·fən,tīl 'skər·vē }

infantile spasm [MED] A type of seizure seen in infants and young children, characterized by a sudden, brief, massive myoclonic jerk. { 'in·fən,tīl 'spaz·əm }

infantilism [MED] Persistence of physical, behavioral, or mental infantile characteristics into childhood, adolescence, or adult life. { 'in·fən·tə,liz·əm }

infant respiratory distress syndrome [MED] A disorder usually affecting prematurely born infants and characterized by a rapid breathing rate, respiratory muscle retraction during expiration, and blood gas values reflecting oxygen deficiency, excessive carbon dioxide, and acidosis. { 'in·fənt 'res·prə,tór·ē di'stres ,sin,drōm }

infarct [MED] Localized death of tissue that is caused by obstructed inflow of arterial blood. Also known as infarction. { 'in,färkt }

infarction [MED] **1.** Condition or process leading to the development of an infarct. **2.** *See* infarct. { in'färk·shən }

infauna [ZOO] Aquatic animals which live in the bottom sediment of a body of water. { in'fón·ə }

infect [MED] To cause an infection, as by contamination with or invasion by a pathogen. [MICROBIO] To cause a phage infection of bacteria. { in'fekt }

infection [MED] **1.** Invasion of the body by a pathogenic organism, with or without disease manifestation. **2.** Pathologic condition resulting from invasion of a pathogen. { in'fek·shən }

infectious [MED] Caused by infection. { in 'fek·shəs }

infectious abortion *See* contagious abortion. { in 'fek·shəs ə'bór·shən }

infectious anemia [VET MED] A virus disease of horses and mules characterized by intermittent fever, weakness, jaundice, and hemorrhages of mucous membranes; it is often fatal. Also known as swamp fever. { in'fek·shəs ə'nē·mē·ə }

infectious arthritis [MED] An inflammatory joint disease caused by microbial invasion of the articular tissue. { in'fek·shəs är'thrīd·əs }

infectious bronchitis [VET MED] A highly contagious respiratory viral disease of chickens. { in 'fek·shəs bräŋ'kīd·əs }

infectious canine hepatitis [VET MED] An acute inflammatory liver disease of dogs caused by a virus. { in'fek·shəs 'kā,nīn ,hep·ə'tīd·əs }

infectious conjunctivitis [MED] Conjunctivitis due to invasion by a microorganism. { in'fek·shəs kən,jəŋk·tə'vīd·əs }

infectious disease [MED] Any disease caused by invasion by a pathogen which subsequently grows and multiplies in the body. { in'fek·shəs di'zēz }

infectious drug resistance [MICROBIO] A type of drug resistance that is transmissible from one bacterium to another by infectivelike agents referred to as resistance factors. { in'fek·shəs 'drəg ri,zis·təns }

infectious endocarditis [MED] Inflammation of the endocardium due to an infectious microorganism. { in'fek·shəs ,en·dō,kär'dīd·əs }

infectious hepatitis [MED] Type A viral hepatitis, an acute infectious virus disease of the liver associated with hepatic inflammation and characterized by fever, liver enlargement, and jaundice. Also known as catarrhal jaundice; epidemic hepatitis; epidemic jaundice; virus hepatitis. { in'fek·shəs ,hep·ə'tīd·əs }

infectious laryngotracheitis [VET MED] A highly contagious respiratory disease of viral etiology affecting chickens. { in'fek·shəs lə¦rin·jō,trā·kē'īd·əs }

infectious mononucleosis [MED] A disorder of unknown etiology characterized by irregular fever, pathology of lymph nodes, lymphocytosis, and high serum levels of heterophil antibodies against sheep erythrocytes. Also known as acute benign lymphoblastosis; glandular fever; kissing disease; lymphocytic angina; monocytic angina; Pfeiffer's disease. { in'fek·shəs ,män·ō,nü·klē'ō·səs }

infectious myocarditis [MED] Inflammation of the myocardium due to an infectious microorganism. { in'fek·shəs ,mī·ō,kär'dīd·əs }

infectious myxomatosis [VET MED] An infectious virus disease of rabbits characterized by myxomatous lesions. { in'fek·shəs ,mik·sō·mə'tō·səs }

infectious nucleic acid [VIROL] Purified viral nucleic acid capable of infecting a host cell and causing the production of viral progeny. { in ¦fek·shəs nü¦klē·ik 'as·əd }

infectious papillomatosis [VET MED] A virus disease of cattle characterized by the appearance of warts on the body. { in'fek·shəs ,pap·ə,lō·mə'tō·səs }

infectious rhinitis [MED] Inflammation of the nasal mucous membrane due to an infectious microorganism. { in'fek·shəs rī'nīd·əs }

infectious transfer [GEN] The rapid spread of extrachromosomal episomes from donor to recipient cells in a bacterial population. { in¦fek·shən 'tranz,fər }

infectious unit [VIROL] The smallest number of virus particles that will cause a lytic infection in a susceptible cell. { in'fek·shəs ,yü·nət }

infectious uroarthritis *See* Reiter's syndrome. { in'fek·shəs ¦yu̇r·ō·är'thrīd·əs }

infective dose 50 [MICROBIO] The dose of microorganisms required to cause infection in 50% of the experimental animals; a special case of the median effective dose. Abbreviated ID_{50}. Also known as median infective dose. { in'fek·tiv ¦dōs 'fif·tē }

inferior [BIOL] The lower of two structures. { in'fir·ē·ər }

inferior alveolar artery [ANAT] A branch of the internal maxillary artery supplying the mucous membrane of the mouth and teeth of the lower jaw. { in'fir·ē·ər al¦vē·ə·lər 'ärd·ə·rē }

inferior alveolar nerve [NEUROSCI] A branch of the mandibular nerve that innervates the teeth of the lower jaw. { in'fir·ē·ər al¦vē·ə·lər 'nərv }

inferior cerebellar peduncle [NEUROSCI] A large bundle of nerve fibers running from the medulla oblongata to the cerebellum. Also known as restiform body. { in'fir·ē·ər ¦ser·ə¦bel·ər 'pē,dəŋ·kəl }

inferior colliculus [NEUROSCI] One of the posterior pair of rounded eminences arising from the dorsal portion of the mesencephalon. { in'fir·ē·ər kə'lik·yə·ləs }

inferior fruit *See* accessory fruit. { in,fir·ē·ər 'früt }

inferior ganglion [NEUROSCI] **1.** The lower sensory ganglion in the glossopharyngeal nerve. **2.** The lower sensory ganglion on the vagus. { in'fir·ē·ər 'gaŋ·glē·ən }

inferior hypogastric plexus [NEUROSCI] A network of nerves in the pelvic fascia containing autonomic nerve elements. { in'fir·ē·ər ,hī·pə'gas·trik 'plek·səs }

inferiority [MED] An organic or psychic state or condition of being inferior or less adequate. { in,fir·ē'är·əd·ē }

inferior mesenteric ganglion [NEUROSCI] A sympathetic ganglion within the inferior mesenteric plexus at the origin of the inferior mesenteric artery. { in'fir·ē·ər ,mez·ən'ter·ik 'gaŋ·glē·ən }

inferior temporal gyrus [NEUROSCI] A convolution on the temporal lobe of the cerebral hemispheres lying below the middle temporal sulcus and extending to the inferior sulcus. { in'fir·ē·ər ¦tem·pə·rəl 'jī·rəs }

inferior vena cava [ANAT] A large vein which drains blood from the iliac veins, lower extremities, and abdomen into the right atrium. { in'fir·ē·ər ,vē·nə 'kā·və }

inferior vena cava syndrome [MED] Edema and venous distention of the abdomen and legs due to obstruction of the inferior vena cava. { in'fir·ē·ər ¦vē·nə ¦kā·və ,sin,drōm }

inferior vermis [NEUROSCI] The inferior portion of the vermis of the brain. { in'fir·ē·ər 'vər·məs }

inferior vestibular nucleus [NEUROSCI] The terminal nucleus for the spinal vestibular nerve tract. { in'fir·ē·ər və'stib·yə·lər 'nü·klē·əs }

infertility [MED] Involuntary reduction in reproductive ability. { ˌin·fərˈtil·əd·ē }

infest [MED] To live on or within the host's body. { inˈfest }

infestation [MED] The state or condition of having animal parasites on or in the body. { ˌin·feˈstā·shən }

infiltrating lipoma *See* liposarcoma. { inˈfilˌtrād·iŋ liˈpō·mə }

inflammation [MED] Local tissue response to injury characterized by redness, swelling, pain, and heat. { ˌin·fləˈmā·shən }

inflammatory arthritis [MED] A type of arthritis that is characterized by inflammation of tissues associated with joints; rheumatoid arthritis is the most common variety. { in¦flam·əˌtȯr·ē ärthˈrīd·əs }

inflammatory bowel disease [MED] A general term for two closely related conditions, ulcerative colitis and regional enteritis or Crohn's disease. { inˌflam·əˌtȯr·ē ˈbau̇l dizˌēz }

inflammatory carcinoma [MED] A carcinoma, usually of the breast, associated with inflammation and characterized by rapid metastasis. { inˈflam·əˌtȯr·ē ˌkärs·ənˈō·mə }

inflammatory response [IMMUNOL] A nonspecific defensive reaction of the body to invasion by a foreign substance or organism that involves phagocytosis by white blood cells and is often accompanied by accumulation of pus and an increase in the local temperature. { in¦flam·əˌtȯrē riˈspäns }

inflammatory tissue [MED] Tissue characterized by exudation or cell proliferation caused by trauma. { inˈflam·əˌtȯr·ē ˈtish·ü }

inflated [BIOL] **1.** Distended, applied to a hollow structure. **2.** Open and enlarged. { in ˈflād·əd }

inflected [BOT] Curved or bent sharply inward, downward, or toward the axis. Also known as inflexed. { inˈflek·təd }

inflexed *See* inflected. { inˈflekst }

inflorescence [BOT] A flower cluster segregated from any other flowers on the same plant, together with the stems and bracts (reduced leaves) associated with it. { ˌin·fləˈres·əns }

inflorescence structure [BOT] The way that the flowers are clustered or arranged on a flowering branch. { ˌin·flə¦res·əns ˈstrək·chər }

influent [ECOL] An organism that disturbs the ecological balance of a community. { ˈinˌflü·ənt }

influenza [MED] An acute virus disease of the respiratory system characterized by headache, muscle pain, fever, and prostration. { ˌin·flüˈen·zə }

influenzal meningitis [MED] Inflammation of the meninges caused by *Hemophilus influenzae*. { ˌin·flüˈenz·əl ˌmen·ənˈjīd·əs }

influenzal pneumonia [MED] Pneumonia resulting from infection by *Hemophilus influenzae*. { ˌin·flüˈenz·əl nəˈmōn·yə }

influenza vaccine [IMMUNOL] A vaccine prepared from formaldehyde-attenuated mixtures of strains of influenza virus. { ˌin·flüˈen·zə vakˈsēn }

influenza virus [VIROL] Any of three immunological types, designated A, B, and C, belonging to the myxovirus group which cause influenza. { ˌin·flüˈen·zə ˌvī·rəs }

influx *See* mouth. { ˈinˌfləks }

informosome [CELL MOL] A type of cellular particle that is thought to be a complex of messenger ribonucleic acid with ribonucleoprotein. { inˈfōr·məˌsōm }

infrabasal [BIOL] Inferior to a basal structure. { ¦in·frəˈbā·səl }

infrabranchial [VERT ZOO] Situated below the gills. { ¦in·frəˈbraŋ·kē·əl }

infracentral [ANAT] Located below the centrum. { ¦in·frəˈsen·trəl }

infracerebral gland [INV ZOO] A structure lying ventral to the brain in annelids which is thought to produce a hormone that inhibits maturation of the gametes. { ¦in·frə·səˈrē·brəl ¦gland }

infraciliature [INV ZOO] The neuromotor apparatus, silverline system, or neuroneme system of ciliates. { ˌin·frəˈsil·yə·chər }

infraclass [SYST] A subdivision of a subclass; equivalent to a superorder. { ˈin·frəˌklas }

infraclavicle [VERT ZOO] A bony element of the shoulder girdle located below the cleithrum in some ganoid and crossopterygian fishes. { ¦in·frəˈklav·ə·kəl }

infrafoliar [BOT] Located below the leaves. { ˌin·frəˈfō·lē·ər }

infraglenoid [ANAT] Below the glenoid cavity of the scapula. { ¦in·frəˈgleˌnȯid }

infraglenoid tubercle [ANAT] A rough impression below the glenoid cavity, from which the long head of the triceps muscle arises. { ¦in·frəˈgleˌnȯid ˈtü·bər·kəl }

infraorbital [ANAT] Located beneath the orbit. { ¦in·frəˈȯr·bəd·əl }

infraspinous [ANAT] Below the spine of the scapula. { ¦in·frəˈspī·nəs }

infraspinous fossa [ANAT] The recess on the posterior surface of the scapula occupied by the infraspinatus muscle. { ¦in·frəˈspī·nəs ˈfäs·ə }

infratemporal [ANAT] Situated below the temporal fossa. { ¦in·frəˈtem·prəl }

infratemporal fossa [ANAT] An irregular space situated below and medial to the zygomatic arch, behind the maxilla and medial to the upper part of the ramus of the mandible. { ¦in·frəˈtem·prəl ˈfäs·ə }

infructescence [BOT] An inflorescence's fruiting stage. { ˌinˌfrəkˈtes·əns }

infundibular canal [INV ZOO] A pathway from the mantle cavity through the funnel for water in cephalopods. { ˌin·fənˈdib·yə·lər kəˈnal }

infundibular ganglion [INV ZOO] A branch of pedal ganglion which supplies the funnel in cephalopods. { ˌin·fənˈdib·yə·lər ˈgaŋ·glē·ən }

infundibular process [ANAT] The distal portion of the neural lobe of the pituitary. { ˌin·fənˈdib·yə·lər ˈprä·səs }

infundibulum [ANAT] **1.** A funnel-shaped passage or part. **2.** The stalk of the neurohypophysis. { ˌin·fən'dib·yə·ləm }

infusion [MED] The slow injection of a solution into a vein or into subcutaneous or other tissue of the body. { in'fyü·zhən }

infusoriform larva [INV ZOO] The final larval stage, arising from germ cells within the infusorigen, in the life cycle of dicyemid mesozoans. { ¦in·fyə¦zȯr·əˌfȯrm 'lär·və }

infusorigen [INV ZOO] An individual that gives rise to the infusoriform larva in dicyemid mesozoans. { ˌin·fyə'zȯr·ə·jən }

ingesta [BIOL] Food and other substances taken into an animal body. { in'jes·tə }

ingestion [BIOL] The act or process of taking food and other substances into the animal body. { in'jes·chən }

Ingolfiellidea [INV ZOO] A suborder of amphipod crustaceans in which both abdomen and maxilliped are well developed and the head often bears a separate ocular lobe lacking eyes. { in ¦gäl·fē·ə¦lid·ē·ə }

ingrown [MED] Of a hair or nail, grown inward so that the normally free end is embedded in or under the skin. { 'inˌgrōn }

inguinal canal [ANAT] A short, narrow passage between the abdominal ring and the inguinal ring in which lies the spermatic cord in males and the round ligament in females. { 'iŋ·gwən·əl kə'nal }

inguinal fold [EMBRYO] A fold of embryonic tissue on the urogenital ridge in which the gubernaculum testis develops. { 'iŋ·gwən·əl 'fōld }

inguinal gland [ANAT] Any of the superficial lymphatic glands in the groin. { 'iŋ·gwən·əl ¦gland }

inguinal hernia [MED] Protrusion of the abdominal viscera through the inguinal canal. { 'iŋ·gwən·əl 'her·nē·ə }

inguinal ligament [ANAT] The thickened lower portion of the aponeurosis of the external oblique muscle extending from the anterior superior spine of the ileum to the tubercle of the pubis and the pectineal line. Also known as Poupart's ligament. { 'iŋ·gwən·əl 'lig·ə·mənt }

inguinal lymphadenopathy [MED] Tenderness and swelling of the lymph nodes of the groin. { ˌiŋ·gwən·əl limˌfad·ən'ä·pə·thē }

inguinal region [ANAT] The abdominal region occurring on each side of the body as a depression between the abdomen and the thigh. Also known as iliac region. { 'iŋ·gwən·əl ˌrē·jən }

inhalant canal [INV ZOO] The incurrent canal in sponges and mollusks. { in'hā·lənt kəˌnal }

inhalant siphon [INV ZOO] A channel for water intake in the mantle of bivalve mollusks. Also known as incurrent siphon. { in'hā·lənt 'sī·fən }

inhalation [PHYSIO] The process of breathing in. { ˌin·ə'lā·shən }

inhalator [MED] A device for facilitating the inhalation of a gas or spray, as for providing oxygen or oxygen-carbon dioxide mixtures for respiration in resuscitation. { 'in·əˌlād·ər }

inhaler [MED] **1.** A device containing a solid medication through which air is drawn into the air passages. **2.** An atomizer containing a liquid medication. { in'hā·lər }

inheritance [GEN] **1.** The acquisition of characteristics by transmission of particular alleles from ancestor to descendant. **2.** The sum total of characteristics dependent upon the constitution of the sperm-fertilized ovum. { in'her·əd·əns }

inhibiting antibody [IMMUNOL] A substance sometimes produced in the blood of immunized persons which is thought to prevent the expected antigen-reagin reaction. { in'hib·əd·iŋ 'ant·iˌbäd·ē }

inhibition immunonephelometry [IMMUNOL] A technique that uses a constant amount of antibody and a predetermined quantity of the same reactant as that assayed, chemically coupled to a macromolecular carrier, used for the assay of substances of low molecular weight. { ˌin·əˌbish·ən ˌim·yə·nōˌne·fə'lam·ə·trē }

inhibition index [BIOCHEM] The amount of antimetabolite that will overcome the biological effect of a unit weight of metabolite. { ˌin·ə'bish·ən ˌinˌdeks }

inhibitory G protein [CELL MOL] A guanine nucleotide-binding protein that lowers cellular levels of the second messenger cyclic adenosine monophosphate by inhibiting adenyl cyclase, the enzyme that catalyzes its conversion from adenosine triphosphate. { in¦hib·eˌtȯr·ē 'jē ˌprōˌtēn }

inhibitory postsynaptic potential [NEUROSCI] A transient, graded hyperpolarization of the postsynaptic membrane, mediated by a chemical neurotransmitter, in response to action potentials arriving at the endings of the presynaptic neurons. { in'hib·ə·tȯr·ē pōst·sə'nap·tik pə 'ten·chəl }

Iniomi [VERT ZOO] An equivalent name for Salmoniformes. { ˌin·ē'ōˌmī }

initiation codon [GEN] A codon that signals the first amino acid in a protein sequence; usually AUG, but sometimes GUG. Also known as start codon. { iˌnish·ē'ā·shən 'kōˌdän }

initiation complex [CELL MOL] An intermediate of protein synthesis consisting of messenger ribonucleic acid, initiator codons, initiation factors, and initiator transfer ribonucleic acid. { iˌnish·ē'ā·shən ˌkämˌpleks }

initiation factor [CELL MOL] Any protein required for the initiation of protein synthesis. { iˌnish·ē'ā·shən ˌfak·tər }

initiator codon [CELL MOL] A codon that acts as a start signal for the synthesis of a polypeptide. { i'nish·ēˌād·ər 'kōˌdän }

initiator ribonucleic acid [CELL MOL] An oligoribonucleotide that primes the initiation of Okazaki fragments during deoxyribonucleic acid synthesis. { iˌnish·ē¦ād·ər ˌrī·bō·nü¦klē·ik 'as·əd }

initiator tRNA [CELL MOL] A special type of transfer ribonucleic acid (RNA) that initiates protein synthesis by binding to the amino acid methionine and delivering it to the small ribosomal subunit. { i¦nish·ēˌād·ər ¦tēˌärˌen'ā }

injection [MED] **1.** Introduction of a fluid into the skin, vessels, muscle, subcutaneous tissue, or any cavity of the body. **2.** The substance injected. { in'jek·shən }

injection chimera [BIOL] A chimera produced experimentally by inserting embryonic cells of different genetic makeup into the preimplantation blastocyst. { in'jek·shən kī,mir·ə }

injury [MED] **1.** A structural or functional stress or trauma that induces a pathologic process. **2.** Damage resulting from the stress. { 'in·jə·rē }

injury current *See* injury potential. { 'in·jə·rē ,kə·rənt }

injury potential [PHYSIO] The potential difference observed between the injured and the noninjured regions of an injured tissue or cell. Also known as demarcation potential; injury current. { 'in·jə·rē pə,ten·chəl }

ink sac [INV ZOO] An organ attached to the rectum in many cephalopods which secretes and ejects an inky fluid. { 'iŋk ,sak }

inlet of the pelvis [ANAT] The space within the brim of the pelvis. { 'in,let əv thə 'pel·vəs }

innate [BIOL] Pertaining to a natural or inborn character dependent on genetic constitution. [BOT] Positioned at the apex of a supporting structure. [MYCOL] Embedded in, especially of an organ such as the fruiting body embedded in the thallus of some fungi. { i'nāt }

inner cell mass [EMBRYO] The cells at the animal pole of a blastocyst which give rise to the embryo and certain extraembryonic membranes. { ¦in·ər 'sel ,mas }

inner ear [ANAT] The part of the vertebrate ear concerned with labyrinthine sense and sound reception; consists generally of a bony and a membranous labyrinth, made up of the vestibular apparatus, three semicircular canals, and the cochlea. Also known as internal ear. { ¦in·ər 'ir }

innervation [ANAT] The distribution of nerves to a part. [PHYSIO] The amount of nerve stimulation received by a part. { ,in·ər'vā·shən }

innominate *See* hipbone. { i¦näm·ə·nət }

innominate artery [ANAT] The first artery branching from the aortic arch; distributes blood to the head, neck, shoulder, and arm on the right side of the body. { i¦näm·ə·nət 'ärd·ə·rē }

inoculation [BIOL] Introduction of a disease agent into an animal or plant to produce a mild form of disease and render the individual immune. [MICROBIO] Introduction of microorganisms onto or into a culture medium. { i,näk·yə'lā·shən }

inoculum [MICROBIO] A small amount of substance containing bacteria from a pure culture which is used to start a new culture or to infect an experimental animal. { i'näk·yə·ləm }

inoperculate [BIOL] Lacking an operculum. { ¦in·ä'pər·kyə·lət }

inorganic biochemistry *See* bioinorganic chemistry. { ,in·ȯr,gan·ik ,bī·ō'kem·ə·strē }

inosculation *See* anastomosis. { in,äs·kyə'lā·shən }

inosine [BIOCHEM] $C_{10}H_{12}N_4O_5$ A compound occurring in muscle; a hydrolysis product of inosinic acid. { 'in·ə,sēn }

inosinic acid [BIOCHEM] $C_{10}H_{13}N_4O_8P$ A nucleotide constituent of muscle, formed by deamination of adenylic acid; on hydrolysis it yields hypoxanthine and D-ribose-5-phosphoric acid. { ¦in·ə¦sin·ik 'as·əd }

Inoviridae [VIROL] A family of nontailed bacterial viruses (bacteriophages) characterized by a nonenveloped rod-shaped virion containing a single-stranded circular deoxyribonucleic acid genome. { ē·nō'vir·ə,dē }

Inovirus [VIROL] A genus of bacterial viruses of the family Inoviridae that are characterized by semiflexible filamentous virions with helical symmetry. { 'ī·nə,vī·rəs }

inquiline [ZOO] An animal that inhabits the nest of another species. { 'in·kwə,līn }

inrolling [BOT] Inward rolling of the corolla of a flower, a physical process associated with senescence. { 'in,rōl·iŋ }

insect [INV ZOO] **1.** A member of the Insecta. **2.** An invertebrate that resembles an insect, such as a spider, mite, or centipede. { 'in,sekt }

Insecta [INV ZOO] A class of the Arthropoda typically having a segmented body with an external, chitinous covering, a pair of compound eyes, a pair of antennae, three pairs of mouthparts, and two pairs of wings. { in'sek·tə }

insect control [ECOL] Regulation of insect populations by biological or chemical means. { 'in ,sekt kən,trōl }

insectistasis [ECOL] The use of pheromones to trap, confuse, or inhibit insects in order to hold populations below a level where they can cause significant economic damage. { in¦sek·tə¦stā·səs }

Insectivora [VERT ZOO] An order of mammals including hedgehogs, shrews, moles, and other forms, most of which have spines. { in,sek'tiv·ə·rə }

insectivorous [BIOL] Feeding on a diet of insects. { in,sek'tiv·ə·rəs }

insectivorous plant [BOT] A plant that captures and digests insects as a source of nutrients by using specialized leaves. Also known as carnivorous plant. { in,sek'tiv·ə·rəs 'plant }

insect pathology [INV ZOO] A biological discipline embracing the general principles of pathology as applied to insects. { 'in,sekt pə'thäl·ə·jē }

insect physiology [INV ZOO] The study of the functional properties of insect tissues and organs. { 'in,sekt ,fiz·ē'äl·ə·je }

insemination [BIOL] The planting of seed. [PHYSIO] **1.** The introduction of sperm into the vagina. **2.** Impregnation. { in,sem·ə'nā·shən }

inserted [BIOL] United or attached to the supporting structure by natural growth. { in'sərd·əd }

insertion [ANAT] The point at which a muscle is attached to a bone that moves when the muscle contracts; it is the distal end of the muscle.

[CELL MOL] The addition of an extranumerary base pair to double-stranded deoxyribonucleic acid; causes errors in transcription. { in'sər·shən }

insertion mutagenesis [CELL MOL] Gene alteration due to insertion of unusual nucleotide sequences from sources such as transposons, viruses, or synthetic deoxyribonucleic acid. { in'sər·shən ,myüd·ə,jen·ə·səs }

insertion site [CELL MOL] **1.** In a cloning vector molecule of deoxyribonucleic acid (DNA), a restriction site into which foreign DNA can be inserted. **2.** The position at which a transposable genetic element is integrated. { in'sər·shən ,sīt }

in situ hybridization [CELL MOL] A technique permitting identification of particular deoxyribonucleic acid or ribonucleic acid sequences while these sequences remain in their original location in the cell. A cell or tissue is treated with a fixative and then exposed to a labeled (by radioactivity or fluorescence) single-stranded nucleic acid probe that hybridizes with the targeted nucleic acid sequence, revealing its location on a chromosome band when the hybridized sequence is analyzed microscopically. { ,in ,si·chü ,hī·brəd·ə·zā·shən }

insomnia [MED] Sleeplessness; disturbed sleep; prolonged inability to sleep. { in'säm·nē·ə }

insomniac [MED] A person who is susceptible to insomnia. { in'säm·nē,ak }

inspiration [PHYSIO] The drawing in of the breath. { ,in·spə'rā·shən }

inspiratory capacity [PHYSIO] Commencing from expiratory standstill, the maximum volume of gas which can be drawn into the lungs. { in'spī·rə,tȯr·ē kə'pas·əd·ē }

inspiratory reserve volume [PHYSIO] The amount of air that can be inhaled by forcible inspiration after completion of a normal inspiration. { in'spī·rə,tȯr·ē ri¦zərv ,väl·yəm }

inspirometer [MED] An instrument for measuring the amount of air inspired. { ,in·spə'räm·əd·ər }

instar [INV ZOO] A stage between molts in the life of arthropods, especially insects. { 'in,stär }

instep [ANAT] The arch on the medial side of the foot. { 'in,step }

instinct [ZOO] A precise form of behavior in which there is an invariable association of a particular series of responses with specific stimuli; an unconditioned compound reflex. { 'in ,stiŋkt }

instinctive behavior [ZOO] Any species-typical pattern of responses not clearly acquired through training. { in'stiŋk·tiv bi'hā·vyər }

instinctual [ZOO] Of or pertaining to instincts. { in'stiŋk·chə·wəl }

insulin [BIOCHEM] A protein hormone produced by the beta cells of the islets of Langerhans which participates in carbohydrate and fat metabolism. { 'in·sə·lən }

insulinase [BIOCHEM] An enzyme produced by the liver which is able to inactivate insulin. { 'in·sə·lə,nās }

insulin-like growth factor *See* somatomedin. { ¦in·sə·lən ,līk 'grōth ,fak·tər }

insulinoma *See* islet-cell tumor. { ,in·sə·lə'nō·mə }

insulin shock [MED] Clinical manifestation of hypoglycemia due to excess amounts of insulin in the blood. { 'in·sə·lən ¦shäk }

insulin shock therapy [MED] Administration of large doses of insulin to induce hypoglycemic comas, followed by administration of glucose, in the treatment of certain psychotic disorders. { 'in·sə·lən ¦shäk 'ther·ə·pē }

insuloma *See* islet-cell tumor. { ,in·sə'lō·mə }

integral membrane protein [CELL MOL] A protein that is firmly anchored in the plasma membrane via interactions between its hydrophobic domains and the membrane phospholipids. Also known as intrinsic protein. { ¦int·i·grəl 'mem,brān ,prō,tēn }

integrase [BIOCHEM] An enzyme that facilitates prophage integration into or excision from a bacterial chromosome. { 'int·ə,grās }

integration [GEN] Recombination involving insertion of a genetic element. { ,int·ə'grā·shən }

integration efficiency [CELL MOL] The frequency with which a segment of foreign deoxyribonucleic acid is incorporated into a host bacterial genome. { int·ə'grā·shən i,fish·ən·sē }

integrin [CELL MOL] A heterodimeric transmembrane receptor protein of animal cells that binds to components of the extracellular matrix on the outside of a cell and to the cytoskeleton on the inside of the cell, functionally connecting the cell interior to its exterior; in blood cells, integrins are also involved in cell-cell adhesion. { in'teg·rən }

integument [ANAT] An outer covering, especially the skin, together with its various derivatives. { in'teg·yə·mənt }

integumentary musculature [VERT ZOO] Superficial skeletal muscles which are spread out beneath the skin and are inserted into it in some terrestrial vertebrates. { in¦teg·yə¦men·trē 'məs·kyə·lə·chər }

integumentary pattern [ANAT] Any of the features of the skin and its derivatives that are arranged in designs, such as scales, epidermal ridges, feathers coloration, or hair. { in¦teg·yə¦men·trē 'pad·ərn }

integumentary system [ANAT] A system encompassing the integument and its derivatives. { in¦teg·yə¦men·trē 'sis·təm }

intention tremor [MED] A clinical manifestation of certain diseases of the nervous system characterized by involuntary trembling of the limbs brought on by voluntary movements, and ceasing on rest. { in'ten·chən ,trem·ər }

interambulacrum [INV ZOO] In echinoderms, an area between two ambulacra. { ¦in·tər,am·byə'la·krəm }

interarticular [ANAT] Situated between articulating surfaces. { ¦in·tər,är'tik·yə·lər }

interatrial [ANAT] Located between the atria of the heart. { ¦in·tər'ā·trē·əl }

interatrial septal defect [MED] A congenital malformation of the septum between the atria of the heart. { ¦in·tər'ā·trē·əl ¦sept·əl 'dē,fekt }

interatrial septum *See* atrial septum. { ¦in·tər'ā·trē·əl 'sep·təm }

interaxillary [BOT] Located within or between the axils of leaves. { ,in·tər'ak·sə,ler·ē }

intercalary [BOT] Referring to growth occurring between the apex and leaf. { in'tər·kə,ler·ē }

intercalary meristem [BOT] A meristem that is forming between regions of permanent or mature meristem. { in'tər·kə,ler·ē 'mer·ə,stem }

intercalated disc [HISTOL] A dense region at the junction of cellular units in cardiac muscle. { in'tər·kə,lād·əd 'disk }

intercalated nucleus [ANAT] A nucleus of the medulla oblongata in the central gray matter of the ventricular floor located between the hypoglossal nucleus and the dorsal motor nucleus of the vagus. { in'tər·kə,lād·əd 'nü·klē·əs }

intercalating agent [CELL MOL] A chemical substance that can insert itself between base pairs in a deoxyribonucleic acid molecule. { ,in·tər'ka,lād·iŋ ,ā·jənt }

intercapillary [ANAT] Located between capillaries. { ¦in·tər'kap·ə,ler·ē }

intercapillary glomerulosclerosis [PATH] Nodular eosinophilic hyalin deposits on the periphery of glomeruli in individuals with diabetes. Also known as diabetic glomerulosclerosis; Kimmelstiel-Wilson disease. { ¦in·tər'kap·ə,ler·ē glə¦mer·yə·lō·sklə'rō·səs }

intercarpal [ANAT] Located between the carpal bones. { ¦in·tər'kärp·əl }

intercavernous sinuses [ANAT] Venous sinuses located on the median line of the dura mater, connecting the cavernous sinuses of each side. { ¦in·tər'kav·ər·nəs 'sī·nə·səz }

intercellular [HISTOL] Of or pertaining to the region between cells. { ¦in·tər'sel·yə·lər }

intercellular cement [HISTOL] A substance bonding epithelial cells together. { ¦in·tər'sel·yə·lər si'ment }

intercellular junction [CELL MOL] Any specialized region of contact between the membranes of adjacent cells. { ¦in·tər'səl·yə·lər 'jəŋk·shən }

intercellular plexus [NEUROSCI] A network of neuronal processes surrounding the cells in a sympathetic ganglion. { ¦in·tər'sel·yə·lər 'plek·səs }

intercellular space [HISTOL] A space between adjacent cells. { ¦in·tər'sel·yə·lər ¦spās }

intercellular substance [HISTOL] Tissue component that lies between cells. { ¦in·tər'sel·yə·lər ¦səb·stəns }

intercentrum [VERT ZOO] A type of crescentic intervertebral structure between successive centra in certain reptilian and mammalian tails. { ¦in·tər'sen·trəm }

interclavicle [VERT ZOO] A membrane bone in front of the sternum and between the clavicles in monotremes and most reptiles. { ¦in·tər'klav·ə·kəl }

intercostal [ANAT] Situated or occurring between adjacent ribs. { ¦in·tər¦käst·əl }

intercostal muscles [ANAT] Voluntary muscles between adjacent ribs. { ¦in·tər¦käst·əl ¦məs·əlz }

intercostal nerve [NEUROSCI] Any of the branches of the thoracic nerves in the intercostal spaces. { ¦in·tər¦käst·əl ¦nerv }

intercourse *See* coitus. { 'in·tər,kȯrs }

intercrescence [BIOL] A growing together of tissues. { ,in·tər'krēs·əns }

interfascicular cambium [BOT] The vascular cambium that develops between vascular bundles. { ¦in·tər·fa'sik·yə·lər 'kam·bē·əm }

interference phenomenon [VIROL] Inhibition by a virus of the simultaneous infection of host cells by some other virus. { ,in·tər'fir·əns fə'näm·ə,nən }

interference range [GEN] The smallest genetic distance that is large enough for two crossing-over events not to interfere with each other. { ,in·tər'fir·əns ,rānj }

interferon [BIOCHEM] A protein produced by intact animal cells when infected with viruses; acts to inhibit viral reproduction and to induce resistance in host cells. { ,in·tər'fir,än }

interferon-alpha [IMMUNOL] A low-molecular-weight protein produced by leukocytes in response to viral infection. { ,in·tər¦fir·än 'al·fə }

interferon-beta [IMMUNOL] An interferon produced by fibroblasts in response to viruses or foreign nucleic acids. { ,in·tər¦fir,än 'bād·ə }

interferon-gamma [IMMUNOL] An interferon produced by T lymphocytes and large granular lymphocytes in response to foreign macromolecules. { ,in·tər¦fir,än 'gam·ə }

interferonogen [VIROL] A preparation made of inactivated virus particles used as an inoculant to stimulate formation of interferon. { ¦in·tə·fə'rän·ə·jən }

interfoliaceous [BOT] Between a pair of leaves, such as between those which are opposite or verticillate. { ¦in·tər,fō·lē'ā·shəs }

intergenic crossing-over [CELL MOL] Recombination between distinct genes or cistrons. { ¦in·tər¦jen·ik ¦krȯs·iŋ'ō·vər }

intergenic suppression [GEN] The restoration of a suppressed function or character by a second mutation that is located in a different gene than the original or first mutation. { ¦in·tər¦jen·ik sə'presh·ən }

intergenote [CELL MOL] In hybrid bacteria, a chromosome with integrated deoxyribonucleic acid of foreign origin. { ,in·tər'jē,nōt }

interkinesis *See* interphase. { ¦in·tər·kə'nē·səs }

interlabium [INV ZOO] A small lobe situated between the lips in certain nematodes. { ¦in·tər'lā·bē·əm }

interleukin [IMMUNOL] Any of a class of proteins that are secreted mostly by macrophages and T lymphocytes and induce growth and differentiation of lymphocytes and hematopoietic stem cells. { ,in·tər'lü·kən }

interleukin-1 [IMMUNOL] A cytokine produced by macrophages, endothelial cells, lymphocytes, and epidermal cells that plays roles in the inflammatory process and in the immune response. Abbreviated IL-1. { ˌin·tərˌlük·ən 'wən }

interleukin-2 [IMMUNOL] A lymphokine secreted mostly by helper T lymphocytes that promotes the growth of T lymphocytes. Abbreviated IL-2. { ˌin·tərˌlük·ən 'tü }

interleukin-3 [IMMUNOL] A cytokine produced by a subset of helper T cells as well as by nonlymphoid cells; it is a growth factor for multiple lineages of hematopoietic cells and can act as an immunoregulatory factor. Abbreviated IL-3. { ˌin·tərˌlük·ən 'thrē }

interleukin-4 [IMMUNOL] A cytokine that is capable of a variety of activities, such as induction of proliferation by T cells, mast cells, megakarocytes, and erythroid precursors; induction of antibody secretion by B cells; and potentiation of the proliferation of mast cells. Abbreviated IL-4. { ˌin·tər'lük·ən 'fȯr }

interleukin-5 [IMMUNOL] A cytokine produced by helper T cells that promotes the development of B cells and stimulates them to produce IgA antibodies, induces proliferation and differentiation of eosinophils, and acts with IL-4 to enhance the production of IgE antibodies. Abbreviated IL-5. { ˌin·tərˌlük·ən 'fīv }

interleukin-6 [IMMUNOL] A cytokine derived from activated T lymphocytes that has many functions, including induction of B-cell growth; induction of B-cell differentiation and antibody production; induction of differentiation and proliferation of T cells; synergistic induction with IL-3 of hematopoietic cell growth; and induction of hepatocyte secretion of acute-phase inflammatory proteins. Abbreviated IL-6. { ˌin·tərˌlük·ən 'siks }

interleukin-7 [IMMUNOL] A cytokine that acts as a growth factor for precursors of B cells and T cells and also enhances the generation of interleukin-2-activated nonspecific killer cells and interleukin-2-activated antigen-specific cytotoxic T lymphocytes. Abbreviated IL-7. { ˌin·tərˌlük·ən 'sev·ən }

interleukin-8 [IMMUNOL] A group of peptides produced by a variety of cell types, which activate and recruit polymorphonuclear leukocytes in the inflammatory process, and are probably involved in initiation of labor and delivery in pregnant women. Abbreviated IL-8. { ˌin·tərˌlük·ən 'āt }

interleukin-9 [IMMUNOL] A cytokine produced by a type of activated helper T cells, it stimulates hemopoietic cells that develop into red blood cells. Abbreviated IL-9. { ˌin·tərˌlük·ən 'nīn }

interleukin-10 [IMMUNOL] An immunoregulatory cytokine produced by a subset of helper T cells as well as by B lymphocytes and some cells of the uterus during pregnancy, it inhibits secretion (and function) of cytokine by macrophages and the second population of helper T cells called Th1. Abbreviated IL-10. { ˌin·tərˌlük 'ten }

interleukin-11 [IMMUNOL] A pleiotropic cytokine produced by bone marrow-derived fibroblasts which supports the growth of certain cell types, such as B cells, neutrophils, and platelet-producing megakaryocytes. Abbreviated IL-11. { ˌin·tərˌlük·ən i'lev·ən }

interleukin-12 [IMMUNOL] A heterodimeric cytokine that stimulates nonspecific cytotoxic natural killer-type cells to produce gamma interferon. Abbreviated IL-12. { ˌin·tərˌlük·ən 'twelv }

interleukin-13 [IMMUNOL] A cytokine produced by activated T lymphocytes that inhibits inflammatory cytokine production induced by bacterial endotoxin and stimulates gamma-interferon production by natural killer cells, enhancing the effect of interleukin-2. Abbreviated IL-13. { ˌin·tərˌlük·ən 'thərˌtēn }

intermediary metabolism [BIOCHEM] Intermediate steps in the chemical synthesis and breakdown of foodstuffs within body cells. { ¦in·tərˌmēd·ēˌer·ē me'tab·əˌliz·əm }

intermediate filament [CELL MOL] Any of several classes of cell-specific cytoplasmic filaments of 8–12 nanometers in diameter; protein composition varies from one cell type to another. { ˌint·ər'mēd·ē·ət 'fil·ə·mənt }

intermediate ganglion [NEUROSCI] Any of certain small groups of nerve cells found along communicating branches of spinal nerves. { ˌin·tər'mēd·ē·ət 'gaŋ·glē·ən }

intermediate host [BIOL] The host in which a parasite multiplies asexually. { ˌin·tər'mēd·ē·ət 'hōst }

intermediate lobe [ANAT] The intermediate portion of the adenohypophysis. { ˌin·tər'mēd·ē·ət ¦lōb }

intermedin [BIOCHEM] A hormonal substance produced by the intermediate portion of the hypophysis of certain animal species which influences pigmentation; similar to melanocyte-stimulating hormone in humans. { ˌin·tər'mēd·ən }

intermembranous ossification [HISTOL] Ossification within connective tissue with no prior formation of cartilage. { ¦in·tər'mem·brə·nəs ˌäs·ə·fə'kā·shən }

intermeningeal [ANAT] Between any two of the three meninges covering the brain and spinal cord. { ˌin·tər·men·ən'jē·əl }

intermenstrual [PHYSIO] Between periods of menstruation. { ˌin·tər'men·strəl }

intermenstrual flow *See* metrorrhagia. { ˌin·tər'men·strəl 'flō }

intermetameric [ANAT] Between adjacent metameres. { ¦in·tər¦med·ə¦mer·ik }

intermetatarsal [ANAT] Between adjacent bones of the metatarsus. { ˌin·tərˌmed·ə'tär·səl }

intermitotic [CELL MOL] Of or pertaining to a stage of the cell cycle between two successive mitoses. { ˌin·tərˌmī'täd·ik }

intermittent claudication [MED] Cramping pain or weakness in the lower extremities during exercise, caused by occlusion of the arteries. { ˌin·tər¦mit·ənt klȯ·də'kā·shən }

intermural [ANAT] Between the walls of an organ. { ¦in·tər'myu̇r·əl }

intermuscular [ANAT] Between muscles. { ¦in·tər'məs·kyə·lər }

intermuscular hernia *See* interstitial hernia. { ¦in·tər'məs·kyə·lər 'hər·nē·ə }

intermuscular septum [ANAT] A connective-tissue partition between muscles. { ¦in·tər'məs·kyə·lər 'sep·təm }

internal acoustic meatus [ANAT] An opening in the hard portion of the temporal bone for passage of the facial and acoustic nerves and internal auditory vessels. { in'tərn·əl ə'kü·stik mē'ād·əs }

internal capsule [NEUROSCI] A layer of nerve fibers on the outer side of the thalamus and caudate nucleus, which it separates from the lenticular nucleus. { in'tərn·əl 'kap·səl }

internal carotid [ANAT] A main division of the common carotid artery, distributing blood through three sets of branches to the cerebrum, eye, forehead, nose, internal ear, trigeminal nerve, dura mater, and hypophysis. Also known as internal carotid artery. { in'tərn·əl kə'räd·əd }

internal carotid artery *See* internal carotid. { in'tərn·əl kə'räd·əd 'ärd·ə·rē }

internal carotid nerve [NEUROSCI] A sympathetic nerve which forms networks of branches around the internal carotid artery and its branches. { in'tərn·əl kə'räd·əd 'nərv }

internal ear *See* inner ear. { in'tərn·əl 'ir }

internal elastic membrane [HISTOL] A sheet of elastin found between the tunica intima and the tunica media in medium- and small-caliber arteries. { in'tərn·əl i'las·tik 'mem,brān }

internal fertilization [PHYSIO] Fertilization of the egg within the body of the female. { in'tərn·əl ,fərd·əl·ə'zā·shən }

internal fistula [ANAT] A fistula which has no opening through the skin. { in'tərn·əl 'fis·chə·lə }

internal granular layer [HISTOL] The fourth layer of the cerebral cortex. { in'tərn·əl 'gran·yə·lər ,lā·ər }

internal hemorrhage [MED] Bleeding within a body cavity or organ that is concealed from an observer. { in'tərn·əl 'hem·rij }

internal hernia [MED] A hernia of intraabdominal contents occurring within the abdominal cavity. { in'tərn·əl 'hər·nē·ə }

internal hydrocephaly *See* obstructive hydrocephaly. { in'tərn·əl ¦hī·drə'sef·ə·lē }

internal iliac artery [ANAT] The medial terminal division of the common iliac artery. { in'tərn·əl ¦il·ē,ak 'ärd·ə·rē }

internal respiration [PHYSIO] The gas exchange which occurs between the blood and tissues of an organism. { in'tərn·əl ,res·pə'rā·shən }

internal secretion [PHYSIO] A secreted substance that is absorbed directly into the blood. { in'tərn·əl si'krē·shən }

international unit [BIOL] A quantity of a vitamin, hormone, antibiotic, or other biological that produces a specific internationally accepted biological effect. { ¦in·tər¦nash·ən·əl 'yü·nət }

internode [BIOL] The interval between two nodes, as on a stem or along a nerve fiber. { 'in·tər,nōd }

internuncial neuron [NEUROSCI] A neuron located in the spinal cord which connects motor and sensory neurons. { ¦in·tər¦nən·chəl 'nu̇r,än }

interoceptor [PHYSIO] A sense receptor located in visceral organs and yielding diffuse sensations. { ,in·tə·rō'sep·tər }

interocular distance [ANAT] The distance between the centers of rotation of the human eyes. { ¦in·tər'äk·yə·lər 'dis·təns }

interorbital [ANAT] Between the orbits of the eyes. { ¦in·tər'ȯr·bəd·əl }

interparietal [ANAT] Between the parietal bones. { ¦in·tər·pə'rī·əd·əl }

interparietal hernia *See* interstitial hernia. { ¦in·tər·pə'rī·əd·əl 'hər·nē·ə }

interphase [CELL MOL] Also known as interkinesis. **1.** The period between succeeding mitotic divisions. **2.** The period between the first and second meiotic divisions in those organisms where nuclei are reconstituted at the end of the first division. { 'in·tər,fāz }

interproglottid gland [INV ZOO] Any of a number of cell clusters or glands arranged transversely along the posterior margin of the proglottids of certain tapeworms. { ¦in·tər,prō'gläd·əd ,gland }

interpterygoid [ZOO] A space between palatal plates in certain chordates. { ¦in·tər'ter·ə,gȯid }

interpulmonary [ANAT] Located between the lungs. { ¦in·tər'pu̇l·mə,ner·ē }

interradial canal [INV ZOO] Any of the radially arranged gastrovascular canals in certain jellyfishes and ctenophores. { ¦in·tər'rād·ē·əl kə ,nal }

interradius [INV ZOO] The area between two adjacent arms in echinoderms. { ,in·tər'rād·ē·əs }

interray [INV ZOO] A division of the radiate body of echinoderms. { ¦in·tər'rā }

interrenal [ANAT] Located between the kidneys. { ¦in·tər'rēn·əl }

interrupted gene *See* split gene. { 'in·tə,rəp·təd 'jēn }

interrupted mating [GEN] A technique for mapping bacterial genes by determining the sequence of gene transfer between conjugating bacteria. { 'in·tə,rəp·təd 'mād·iŋ }

interscapular [ANAT] Between the shoulders or shoulder blades. { ¦in·tər'skap·yə·lər }

intersegmental [BIOL] Situated between or involving segments. [EMBRYO] Situated between the primordial segments of the embryo. { ¦in·tər·seg'ment·əl }

intersegmental reflex [NEUROSCI] An unconditioned reflex arc connecting input and output by means of afferent pathways in the dorsal spinal roots and efferent pathways in the ventral spinal roots. { ¦in·tər·seg'ment·əl 'rē,fleks }

intersex [PHYSIO] An individual who is intermediate in sexual constitution between male and female. { 'in·tər,seks }

interspace [ANAT] An interval between the ribs or the fibers or lobules of a tissue or organ. { 'in·tər,spās }

interspersed repeats [CELL MOL] Repetitive deoxyribonucleic acid sequences. { ˌin·tər ˌspərsd ri'pēts }

interspersion [CELL MOL] A regular pattern of alternating sequences of repetitious and nonrepetitious deoxyribonucleic acid in the genome of eukaryotes. [ECOL] **1.** An intermingling of different organisms within a community. **2.** The level or degree of intermingling of one kind of organism with others in the community. { ˌin·tər'spər·zhən }

interspinal [ANAT] Situated between or connecting spinous processes; interspinous. { ¦in·tər'spīn·əl }

intersternite [INV ZOO] An intersegmental plate on the ventral surface of the abdomen in insects. { ¦in·tər'stər,nīt }

interstitial cell [HISTOL] A cell that is not peculiar to or characteristic of a particular organ or tissue but which comprises fibrous tissue binding other cells and tissue elements; examples are neuroglial cells and Leydig cells. { ¦in·tər¦stish·əl ¦sel }

interstitial-cell-stimulating hormone *See* luteinizing hormone. { ¦in·tər¦stish·əl ¦sel 'stim·yə,lād·iŋ ˌhȯr,mōn }

interstitial-cell tumor [MED] A benign tumor of the testes composed of Leydig cells. Also known as interstitioma; Leydig-cell tumor. { ¦in·tər¦stish·əl ¦sel 'tü·mər }

interstitial emphysema [MED] Escape of air from the alveoli into the interstices of the lung, commonly due to trauma or violent cough. { ¦in·tər¦stish·əl ˌem·fə'sē·mə }

interstitial endometriosis [MED] The presence of endometrial tissue in the form of the stroma thoughout the myometrium. Also known as endolymphatic stromomyosis; fibromyosis; parathelioma; stromal endometriosis; stromal myosis; stromatosis. { ¦in·tər¦stish·əl ¦en·dō·mə 'trē'ō·səs }

interstitial gland [HISTOL] **1.** Groups of Leydig cells which secrete angiogens. **2.** Groups of epithelioid cells in the ovarian medulla of some lower animals. { ¦in·tər¦stish·əl ¦gland }

interstitial hepatitis [MED] Pathologic deterioration and death of parenchymal cells in the liver associated with infiltration of lymphocytes and monocytes in the portal canals. Also known as acute nonsuppurative hepatitis; nonspecific hepatitis. { ¦in·tər¦stish·əl ˌhep·ə'tīd·əs }

interstitial hernia [MED] Protrusion of the intestine between the muscular planes of the abdominal wall. Also known as intermuscular hernia; interparietal hernia. { ¦in·tər¦stish·əl 'hər·nē·ə }

interstitial implants [MED] Solid or encapsulated radiation sources, made in the form of seeds, wires, or other shapes, to be inserted directly into tissue that is to be irradiated. { ¦in·tər¦stish·əl 'im,plans }

interstitial inflammation [MED] Inflammation of the interstitial tissues of an organ. { ¦in·tər¦stish·əl ˌin·flə'mā·shən }

interstitial keratitis [MED] Inflammation of the cornea in which the iris is almost completely hidden by the diffuse haziness of the corneal tissue. { ¦in·tər¦stish·əl ˌker·ə'tīd·əs }

interstitial lamella [HISTOL] Any of the layers of bone between adjacent Haversian systems. { ¦in·tər¦stish·əl lə'mel·ə }

interstitial myocarditis [MED] Inflammation of the myocardium accompanied by cellular infiltration of interstitial tissues. Also known as Fiedler's myocarditis. { ¦in·tər¦stish·əl ˌmī·ə·kär'dīd·əs }

interstitial nephritis *See* pyelonephritis. { ¦in·tər¦stish·əl nə'frīd·əs }

interstitial plasma-cell pneumonia *See* Pneumocystis carinii pneumonia. { ¦in·tər¦stish·əl 'plaz·mə ˌsel nə'mō·nē·ə }

interstitial pneumonia [MED] Inflammation of the lungs, particularly the stroma, including peribronchial tissue and interalveolar septa. { ¦in·tər¦stish·əl nə'mō·nē·ə }

interstitial radiation [MED] Radiation of tissues by implantation of a radioactive source material. { ¦in·tər¦stish·əl 'rād·ē'ā·shən }

interstitioma *See* interstitial-cell tumor. { 'in·tər'stich·ē'ō·mə }

intertergite [INV ZOO] One of the small plates between the tergites of certain insects. { ¦in·tər'tər,jīt }

intertubercular sulcus [ANAT] A deep groove on the anterior surface of the upper end of the humerus, separating the greater and lesser tubercles; contains the tendon of the long head of the biceps brachii muscle. Also known as bicipital groove. { ¦in·tər·tə'bər·kyə·lər 'səl·kəs }

intervallum [INV ZOO] The space between the walls of pleosponges. { ¦in·tər¦val·əm }

intervascular [ANAT] Located between or surrounded by blood vessels. { ¦in·tər'vas·kyə·lər }

intervening sequence [GEN] The one or more segments of a split gene that are transcribed but not included in the final messenger ribonucleic acid; each is flanked by two exons. Also known as intron. { ¦in·tər¦vēn·iŋ 'sē·kwəns }

interventricular foramen [ANAT] Either one of the two foramens that connect the third ventricle of the brain with each lateral ventricle. Also known as foramen of Monro. { ¦in·tər·ven'trik·yə·lər fə'rā·mən }

interventricular septal defect [MED] A congenital malformation of the septum between the ventricles of the heart. { ¦in·tər·ven'trik·yə·lər ¦sep·təl 'dē,fekt }

interventricular septum [ANAT] The muscular wall between the heart ventricles. Also known as ventricular septum. { ¦in·tər·ven'trik·yə·lər 'sep·təm }

intervertebral [ANAT] Being or located between the vertebrae. { ¦in·tər'vərd·ə·bəl }

intervillous spaces [HISTOL] Spaces in the placenta which communicate with the maternal blood vessels. { ¦in·tər'vil·əs 'spās·əz }

intestinal crura [INV ZOO] The main intestinal branches in certain trematodes. { in'tes·tən·əl 'krůr·ə }

intestinal digestion [PHYSIO] Conversion of

food to an assimilable form by the action of intestinal juices. { in'tes·tən·əl di'jes·chən }

intestinal dyspepsia [MED] Disturbed digestion due to diminished secretion of intestinal juices or lack of tonus in the wall of the intestine. { in'tes·tən·əl dis'pep·sē·ə }

intestinal hormone [BIOCHEM] Either of two hormones, secretin and cholecystokinin, secreted by the intestine. { in'tes·tən·əl 'hȯr,mōn }

intestinal juice [PHYSIO] An alkaline fluid composed of the combined secretions of all intestinal glands. { in'tes·tən·əl ¦jüs }

intestinal lipodystrophy *See* Whipple's disease. { in'tes·tən·əl ¦lip·ə'dis·trə·fē }

intestinal villi [ANAT] Fingerlike projections of the small intestine, composed of a core of vascular tissue covered by epithelium and containing smooth muscle cells and an efferent lacteal end capillary. { in'tes·tən·əl 'vil,ī }

intestine [ANAT] The tubular portion of the vertebrate digestive tract, usually between the stomach and the cloaca or anus. { in'tes·tən }

intima [HISTOL] The innermost coat of a blood vessel. Also known as tunica intima. { 'in·tə·mə }

intorsion [BIOL] Inward rotation of a structure about a fixed point or axis. { in'tȯr·shən }

intoxication [MED] **1.** Poisoning. **2.** The state produced by overindulgence in alcohol. { in ,täk·sə'kā·shən }

intraabdominal [ANAT] Occurring or being within the cavity of the abdomen. { ¦in·trə·ab 'däm·ən·əl }

intraatrial heart block [MED] A type of heart block which shows a broad, notched P wave of longer than normal duration on the electrocardiographic record. { ¦in·trə'ā·trē·əl 'härt ,bläk }

intracartilaginous ossification *See* endochondral ossification. { ¦in·trə,kärd·əl'aj·ə·nəs ,äs·ə·fə 'kā·shən }

intracavernous aneurysm [MED] A dilation of the wall of the internal carotid artery within the cavernous sinus. { ¦in·tər'kav·ər·nəs 'an·yə,riz·əm }

intracellular [CELL MOL] Within a cell. { ¦in·trə'sel·yə·lər }

intracellular canaliculi [CELL MOL] A system of minute canals within certain gland cells which are thought to drain the glandular secretions. { ¦in·trə'sel·yə·lər ,kan·əl'ik·yə,lī }

intracellular digestion [PHYSIO] Digestion which takes place within the cytoplasm of the organism, as in many unicellular protozoans. { ¦in·trə'sel·yə·lər di'jes·chən }

intracellular enzyme [BIOCHEM] An enzyme that remains active only within the cell in which it is formed. Also known as organized ferment. { ¦in·trə'sel·yə·lər 'en,zīm }

intracellular signaling protein [CELL MOL] One of a series of intracellular proteins that mediates a cell's response when an extracellular signal molecule binds to a receptor protein on the cell's surface, either by changing the cell's metabolism or shape, or movement, or gene expression. { ,in·trə¦sel·yə·lər 'sig·nəl·iŋ ,prō,tēn }

intracellular symbiosis [CELL MOL] Existence of a self-duplicating unit within the cytoplasm of a cell, such as a kappa particle in *Paramecium*, which seems to be an infectious agent and may influence cell metabolism. { ¦in·trə'sel·yə·lər ,sim·bē'ō·səs }

intracervical [ANAT] Located within the cervix of the uterus. { ¦in·trə'ser·və·kəl }

intracistron complementation [GEN] The process whereby two different mutant alleles, each of which determines in homozygotes an inactive enzyme, determine the formation of the active enzyme when present in the same nucleus. { ¦in·trə'sis·trən ,käm·plə·mən'tā·shən }

intracortical [ANAT] Occurring or located within the cortex. { ¦in·trə'kȯrd·ə·kəl }

intracranial [ANAT] Within the cranium. { ¦in·trə'krā·nē·əl }

intracranial aneurysm [MED] Dilation of a cerebral artery. { ¦in·trə'krā·nē·əl 'an·yə,riz·əm }

intracranial angiography [MED] Roentgenography of the blood vessels within the cranial cavity following the intravascular injection of a radiopaque material. { ¦in·trə'krā·nē·əl ,an·jē'äg·rə·fē }

intracytoplasmic [CELL MOL] Being or occurring within the cytoplasm of a cell. { ¦in·trə ,sīd·ə'plaz·mik }

intradermal [ANAT] Within the skin. { ¦in·trə 'dər·məl }

intradermal nevus [MED] A lesion containing melanocytes and located principally or completely within the derma. { ¦in·trə'dər·məl 'nē·vəs }

intradermopalpebral [ANAT] Within the skin of the eyelid. { ,in·trə,dərm·ō·pal'peb·rəl }

intraductal [ANAT] Within a duct. { ¦in·trə 'dək·təl }

intradural [ANAT] Within the dura mater. { ¦in·trə'dür·əl }

intraembryonic [EMBRYO] Within the embryo. { ¦in·trə,em·brē'än·ik }

intraepidermal [ANAT] Within the epidermis. { ¦in·trə,ep·ə'dər·məl }

intraepidermal epithelioma [MED] Carcinoma in situ, of either the squamous-cell or basal-cell type. { ¦in·trə,ep·ə'dər·məl ,ep·ə,thē·lē'ō·mə }

intraepithelial [ANAT] Within the epithelium. { ¦in·trə,ep·ə'thē·lē·əl }

intraesophageal [ANAT] Within the esophagus. { ¦in·trə·ə,säf·ə'jē·əl }

intrafascicular [BOT] Located or occurring within a vascular bundle. { ¦in·trə·fə'sik·yə·lər }

intrafusal fiber [HISTOL] Any of the striated muscle fibers contained in a muscle spindle. { ¦in·trə'fyüz·əl 'fī·bər }

intragenic [CELL MOL] Within a gene, in referring to certain events. { ¦in·trə¦jen·ik }

intragenic recombination [CELL MOL] Recombination occurring between the mutons of one cistron. { ¦in·trə¦jen·ik rē,käm·bə'nā·shən }

intragenic suppression [GEN] The restoration

of a suppressed function or character as a consequence of a second mutation located within the same gene as the original or first mutation. { ¦in·trə¦jen·ik sə'presh·ən }

intrahepatic [ANAT] Within the liver. { ¦in·trə·he'pad·ik }

intrajugular process [ANAT] **1.** A small, curved process on some occipital bones which partially or completely divides the jugular foramen into lateral and medial parts. **2.** A small process on the hard portion of the temporal bone which completely or partly separates the jugular foramen into medial and lateral parts. { ¦in·trə'jəg·yə·lər ˌprä·səs }

intralaminar nuclei [NEUROSCI] A diffuse group of nuclei located in the internal medullary lamina of the thalamus. { ˌin·trə'la·mən·ər 'nü·klēˌī }

intraluminal [ANAT] Within the lumen of a structure. { ˌin·trə'lü·mən·əl }

intramarginal [BIOL] Within a margin. { ˌin·trə'mär· jən·əl }

intramedullary [ANAT] Within the bone marrow. [NEUROSCI] **1.** Within the tissues of the spinal cord or medulla oblongata. **2.** Within the adrenal medulla. { ˌin·trə·mə'dəl·ə·rē }

intramembranous [HISTOL] Formed or occurring within a membrane. { ˌin·trə'mem·brə·nəs }

intramembranous ossification [HISTOL] Formation of bone tissue directly from connective tissue without a preliminary cartilage stage. { ˌin·trə'mem·brə·nəs ˌäs·ə·fə'kā·shən }

intramural [ANAT] Within the substance of the walls of an organ. { ¦in·trə'myu̇r·əl }

intramuscular [ANAT] Lying within or going into the substance of a muscle. { ¦in·trə'məs·kyə·lər }

intraocular [ANAT] Occurring within the globe of the eye. { ˌin·trə'äk·yə·lər }

intraocular pressure [PHYSIO] The hydrostatic pressure within the eyeball. { ˌin·trə'äk·yə·lər 'presh·ər }

intraparietal [ANAT] **1.** Within the wall of an organ or cavity. **2.** Within the parietal region of the cerebrum. **3.** Within the body wall. { ˌin·trə·pə'rī·əd·əl }

intrapartum [MED] Occurring during parturition. { ¦in·trə¦pär·dəm }

intraperitoneal [ANAT] **1.** Within the peritoneum. **2.** Within the peritoneal cavity. { ¦in·trəˌper·ə·tə'nē·əl }

intrapetiolar [BOT] **1.** Enclosed by the base of the petiole. **2.** Between the petiole and the stem. { ¦in·trə¦ped·ē¦äl·yə·lər }

intrapulmonic [ANAT] Being or occurring within the lungs. { ¦in·trə·pu̇l'män·ik }

intraspecific [BIOL] Being within or occurring among the members of the same species. { ¦in·trə·spə·'sif·ik }

intraspinal block [MED] Anesthesia of the spinal column by injection of an anesthetic into the spinal canal. { ¦in·trə'spīn·əl 'bläk }

intrathecal [ANAT] Within the subarachnoid space. { ¦in·trə'thē·kəl }

intrathoracic [ANAT] Within the thoracic cavity. { ¦in·trə·thə'ras·ik }

intratracheal [ANAT] Being or occurring within the trachea. { ¦in·trə'trā·kē·əl }

intrauterine [ANAT] Being or occurring within the uterus. { ¦in·trə'yüd·ə·rən }

intravaginal [ANAT] **1.** Being or occurring within the vagina. **2.** Located within a tendon sheath. [BOT] Located within a sheath, referring to branches of grass. { ¦in·trə'vaj·ə·nəl }

intravascular [ANAT] Within blood vessels or within a blood vessel. { ˌin·trə'vas·kyə·lər }

intravenous [ANAT] Located within, or going into, the veins. { ˌin·trə'vē·nəs }

intraventricular heart block [MED] Prolongation of the process of ventricular excitation, measured by the QRS complex of the electrocardiogram. Also known as arborization block; parietal block; peri-infarction block. { ˌin·trə·ven 'trik·yə·lər 'härt ˌbläk }

intravesical [ANAT] Within the urinary bladder. { ˌin·trə'ves·ə·kəl }

intravital [BIOL] Occurring while the cell or organism is alive. { ˌin·trə'vīd·əl }

intravital stain [BIOL] A nontoxic dye injected into the body to selectively stain certain cells or tissues. { ˌin·trə'vīd·əl 'stān }

intrinsic asthma [MED] Asthma caused by a respiratory tract infection. { in'trin·sik 'az·mə }

intrinsic factor [BIOCHEM] A substance, produced by the stomach, which combines with the extrinsic factor (vitamin B_{12}) in food to yield an antianemic principle; lack of the intrinsic factor is believed to be a cause of pernicious anemia. Also known as Castle's intrinsic factor. { in'trin·sik ¦fak·tər }

intrinsic nerve supply [NEUROSCI] The nerves contained entirely within an organ or structure. { in'trin·sik 'nərv səˌplī }

intrinsic protein *See* integral membrane protein. { in¦trin·zik 'prōˌtēn }

introgressive hybridization [GEN] The spreading of genes of a species into the gene complex of another due to hybridization between numerically dissimilar populations and the extensive backcrossing. { ¦in·trə¦gres·iv ˌhī·brəd·ə'zā·shən }

introitus [ANAT] An opening or entryway, especially the opening into the vagina. { in'trō·ə·dəs }

intromission [ZOO] The act or process of inserting one body into another, specifically, of the penis into the vagina. { ˌin·trə'mish·ən }

intromittent [ZOO] Adapted for intromission; applied to a copulatory organ. { ¦in·trə¦mit·ənt }

intron *See* intervening sequence. { 'inˌträn }

introrse [BIOL] Turned inward or toward the axis. { 'inˌtrȯrs }

introspective diplopia *See* physiologic diplopia. { ¦in·trə¦spek·tiv də'plō·pē·ə }

introversion [MED] The act or process of turning in upon itself, as a hollow organ. { ¦in·trə¦vər·zhən }

introvert [ZOO] **1.** A structure capable of introversion. **2.** To turn inward. { 'in·trə,vərt }

intubation [MED] The introduction of a tube into a hollow organ to keep it open, especially into the larynx to ensure the passage of air. { ,in,tü'bā·shən }

intussusception [MED] Passing of a portion of a structure into another part of the same structure, such as the invagination of a part of the intestine. { ,in·tə·sə'sep·shən }

inulase [BIOCHEM] An enzyme produced by certain molds that catalyzes the conversion of inulin to levulose. { 'in·yə,lās }

inulin [BIOCHEM] A polysaccharide made up of polymerized fructofuranose units. { 'in·yə·lən }

inunction [MED] Act of applying an oil or fatty material, especially rubbing an ointment into the skin as a therapeutic measure. { i'nəŋk·shən }

in utero [EMBRYO] Within the uterus, referring to the fetus. { in 'yüd·ə,rō }

invagination [EMBRYO] The enfolding of a part of the wall of the blastula to form a gastrula. [PHYSIO] **1.** The act of ensheathing or becoming ensheathed. **2.** The process of burrowing or enfolding to form a hollow space within a previously solid structure, as the invagination of the nasal mucosa within a bone of the skull to form a paranasal sinus. { in,vaj·ə'nā·shən }

invasion [MED] **1.** The phase of an infectious disease during which the pathogen multiplies and is distributed; precedes signs and symptoms. **2.** The process by which microorganisms enter the body. { in'vā·zhən }

inversion [GEN] A type of chromosomal rearrangement in which two breaks take place in a chromosome and the orientation of the fragment between breaks rotates 180° before rejoining. [MED] The act or process of turning inward or upside down. { in'vər·zhən }

inversion heterozygote [GEN] A diploid organism in which one member of a pair of homologous chromosomes has an inverted gene sequence and the other has the normal gene sequence. { in'vər·zhən ,hed·ə·rə'zī,gōt }

invertase *See* saccharase. { 'in·vər,tās }

Invertebrata [INV ZOO] A division of the animal kingdom including all animals which lack a spinal column; has no taxonomic status. { in,vərd·ə'bräd·ə }

invertebrate [INV ZOO] An animal lacking a backbone and internal skeleton. { in'vərd·ə,brət }

invertebrate pathology [INV ZOO] All studies having to do with the principles of pathology as applied to invertebrates. { in'vərd·ə,brət pə'thäl·ə·jē }

invertebrate zoology [ZOO] A branch of biology that deals with the study of Invertebrata. { in'vərd·ə,brət zō'äl·ə·jē }

inverted repeats [GEN] Two copies of the same nucleotide sequence oriented in opposite directions on the same molecule. Also known as IR sequences. { in'vərd·əd ri'pēts }

inverted terminal repeats [CELL MOL] Related or identical sequences of deoxyribonucleic acid in inverted form occurring at opposite ends of some transposons. { in¦vərd·əd ¦tər·mə·nəl ri'pēts }

invertin *See* saccharase. { in'vərt·ən }

investing bone *See* dermal bone. { in'vest·iŋ ,bōn }

in vitro [BIOL] Pertaining to a biological reaction taking place in an artificial apparatus. { in 'vē·trō }

in vivo [BIOL] Pertaining to a biological reaction taking place in a living cell or organism. { in 'vē·vō }

involucrate [BOT] Having an involucre. { ¦in·və¦lü·krət }

involucre [BOT] Bracts forming one or more whorls at the base of an inflorescence or fruit in certain plants. { 'in·və,lü·kər }

involucrum [ANAT] **1.** The covering of a part. **2.** New bone laid down by periosteum around a sequestrum in osteomyelitis. { ,in·və'lü·krəm }

involuntary fiber *See* smooth muscle fiber. { in 'väl·ən,ter·ē 'fī·bər }

involuntary muscle [PHYSIO] Muscle not under the control of the will; usually consists of smooth muscle tissue and lies within a vescus, or is associated with skin. { in'väl·ən,ter·ē 'məs·əl }

involute [BIOL] Being coiled, curled, or rolled in at the edge. { ¦in·və¦lüt }

involution [BIOL] A turning or rolling in. [EMBRYO] Gastrulation by ingrowth of blastomeres around the dorsal lip. [MED] **1.** The retrogressive change to their normal condition that organs undergo after fulfilling their functional purposes, as the uterus after pregnancy. **2.** The period of regression or the processes of decline or decay which occur in the human constitution after middle life. { ,in·və'lü·shən }

involution form [CELL MOL] A cell with a bizarre configuration caused by abnormal culture conditions. { ,in·və'lü·shən ,fȯrm }

involvucel [BOT] A secondary involucre. { in 'väl·və·səl }

involvucellate [BOT] Possessing an involvucel. { in¦väl·və¦sel,āt }

iodine solution *See* tincture of iodine. { 'ī·ə,dīn sə,lü·shən }

iodine tincture *See* tincture of iodine. { 'ī·ə,dīn ,tiŋk·chər }

5-iodo-2′-deoxyuridine [PHARM] $C_9H_{11}IN_2O_5$ A crystalline substance, used as an antiviral agent for the eye. Abbreviated IDU; IDUR; IUDR. { ¦fīv ī¦ō·dō ¦tü¦prīm dē,äk·sē'yu̇r·ə,dēn }

iodopsin [BIOCHEM] The visual pigment found in the retinal cones, consisting of retinene, combined with photopsin. { ,ī·ə'däp·sən }

ion channel *See* ionic channel. { 'ī,än ¦chan·əl }

ionic channel [CELL MOL] A transmembrane protein structure that forms an aqueous pore that allows only certain ion species to pass through the membrane. Also known as ion channel. { ī,än·ik 'chan·əl }

ionophore [BIOCHEM] Any of a class of compounds, generally cyclic, having the ability to carry ions across lipid barriers due to the property of cation selectivity; examples are valinomycin and nonactin. { ī'än·ə,fȯr }

iontophoresis [MED] A medical treatment used to drive positive or negative ions into a tissue, in which two electrodes are placed in contact with tissue, one of the electrodes being a pad of absorbent material soaked with a solution of the material to be administered, and a voltage is applied between the electrodes. { ī,än·tə·fə'rē·səs }

Iospilidae [INV ZOO] A small family of pelagic polychaetes assigned to the Errantia. { ,ī·ə'spil·ə,dē }

ipecac [BOT] Any of several low, perennial, tropical South American shrubs or half shrubs in the genus *Cephaelis* of the family Rubiaceae; the dried rhizome and root, containing emetine, cephaeline, and other alkaloids, is used as an emetic and expectorant. { 'ip·ə,kak }

Ipidae [INV ZOO] The equivalent name for Scolytidae. { 'ip·ə,dē }

ipsilateral *See* homolateral. { ¦ip·sə'lad·ə·rəl }

IR gene *See* immune response gene. { ¦ī'är ,jēn }

Iridaceae [BOT] A family of monocotyledonous herbs in the order Liliales distinguished by three stamens and an inferior ovary. { ,ir·ə'dās·ē,ē }

iridectomy [MED] Surgical removal of part of the iris. { ,ir·ə'dek·tə·mē }

iridocele [MED] A rupture of the cornea through which a portion of the iris protrudes. { i'rid·ə,sēl }

iridocyclitis [MED] Inflammation of the iris and the ciliary body. { ¦ir·ə·dō·sī'klīd·əs }

iridocyclochoroiditis [MED] Inflammation of the iris, the ciliary body, and the choroid. Also known as uveitis. { ¦ir·ə·dō¦sī·klō,kȯ·rȯi'dīd·əs }

iridocyte [HISTOL] A specialized cell in the integument of certain animal species which is filled with iridescent crystals of guanine and a variety of lipophores. Also known as guanophore; iridophore. { i'rid·ə,sīt }

iridophore *See* iridocyte. { i'rid·ə,fȯr }

Iridoviridae [VIROL] A family of double-stranded deoxyribonucleic acid-containing animal viruses that infects invertebrates and is characterized by an icosahedral virion that has a yellow-green glow in centrifuged pellets. { i,rid·ə'vir·ə,dē }

Iridovirus [VIROL] A genus of the animal-virus family Iridoviridae; characterized by the blue-to-purple iridescence emitted by purified viral particles and infected larvae. { i'rid·ə,vī·rəs }

iris [ANAT] A pigmented diaphragm perforated centrally by an adjustable pupil which regulates the amount of light reaching the retina in vertebrate eyes. [BOT] Any plant of the genus *Iris*, the type genus of the family Iridaceae, characterized by linear or sword-shaped leaves, erect stalks, and bright-colored flowers with the three inner perianth segments erect and the outer perianth segments drooping. { 'ī·rəs }

iris diaphragm *See* iris. { 'ī·rəs 'dī·ə,fram }

Irish moss *See* carrageen. { 'ī,rish 'mȯs }

Irish potato *See* potato. { 'ī·rish pə'tād·ō }

iritis [MED] Inflammation of the iris. { ə'rīd·əs }

iron bacteria [MICROBIO] The common name for bacteria capable of oxidizing ferrous iron to the ferric state. { 'ī·ərn bak'tir·ē·ə }

iron-binding protein [BIOCHEM] A serum protein, such as hemoglobin, for the transport of iron ions. { 'ī·ərn ,bīnd·iŋ 'prō,tēn }

iron-deficiency anemia [MED] Hypochromic microcytic anemia due to excessive loss, deficient intake, or poor absorption of iron. Also known as nutritional hypochromic anemia. { 'ī·ərn də,fish·ən·sē ə'nē·mē·ə }

iron metabolism [BIOCHEM] The chemical and physiological processes involved in absorption of iron from the intestine and in its role in erythrocytes. { 'ī·ərn mə'tab·ə,liz·əm }

Ironoidea [INV ZOO] A superfamily of presumably carnivorous, fresh-water and soil-dwelling nematodes in the order Enoplida, having two circlets of cephalic sensory organs, a cylindrical, elongate stoma armed either anteriorly with three large teeth or posteriorly with small teeth, and cuticularized stomatal walls. { ,i·rə'nȯid·ē·ə }

iron-porphyrin protein [BIOCHEM] Any protein containing iron and porphyrin; examples are hemoglobin, the cytochromes, and catalase. { 'ī·ərn'fȯr·fə·rən 'prō,tēn }

ironwood [BOT] Any of a number of hardwood trees in the United States, including the American hornbeam, the buckwheat, and the eastern hophornbeam. { 'ī·ərn ,wu̇d }

irradiation [BIOPHYS] Subjection of a biological system to sound waves of sufficient intensity to modify their structure or function. { i,rād·ē'ā·shən }

irradiation cataract [MED] A cataract that develops slowly following prolonged or intense irradiation, as by radium or roentgen rays. Also known as cyclotron cataract; radiation cataract. { i,rād·ē'ā·shən 'kad·ə,rakt }

irradiation cystitis [MED] Inflammation of the urinary bladder following radiation therapy of pelvic organs. { i,rād·ē'ā·shən sis'tīd·əs }

irregular [BOT] Lacking symmetry, as of a flower having petals unlike in size or shape. { i'reg·yə·lər }

irregular cleavage [EMBRYO] Division of a zygote into random masses of cells, as in certain cnidarians. { i'reg·yə·lər 'klē·vij }

irregular connective tissue [HISTOL] A loose or dense connective tissue with fibers irregularly woven and irregularly distributed; collagen is the dominant fiber type. { i'reg·yə·lər kə'nek·tiv ,tish·ü }

Irregularia [INV ZOO] An artificial assemblage of echinoderms in which the anus and periproct lie outside the apical system, the ambulacral plates remain simple, the primary radioles are hollow, and the rigid test shows some degree of bilateral symmetry. { iˌreg·yə'lar·ē·ə }

irrigation [MED] Therapeutic washing out by means of a continuous stream of water. { ˌir·ə'gā·shən }

irritability [PHYSIO] **1.** A condition or quality of being excitable; the power of responding to a stimulus. **2.** A condition of abnormal excitability of an organism, organ, or part, when it reacts excessively to a slight stimulation. { ˌir·əd·ə'bil·əd·ē }

irritable bowel syndrome [MED] A gastrointestinal disorder of unknown cause that is characterized by increased intestinal motility, causing recurrent abdominal pain, constipation or diarrhea that may alternate, and sensation of gaseousness and bloating. { ˌir·id·ə·bəl 'bau̇l ˌsin ˌdrōm }

irritable colon [MED] Any of several disturbed colonic functions associated with anxiety or emotional stress. Also known as adaptive colitis; mucous colitis; spastic colon; unstable colon. { 'ir·əd·ə·bəl 'kō·lən }

IR sequences *See* inverted repeats. { ¦ī'är ˌsē·kwən·səs }

isanthous [BOT] Having regular flowers. { ī 'san·thəs }

ischemia [MED] Localized tissue anemia as a result of obstruction of the blood supply or to vasoconstriction. { i'skē·mē·ə }

ischemic necrosis [MED] Local tissue death due to impaired blood supply. { i'skē·mik ne 'krō·səs }

ischemic neuropathy [MED] Nerve lesions characterized by numbness, tingling, and pain with loss of sensory and motor functions of the parts involved, due to obstruction of the blood supply to the nerves. { i'skē·mik nu̇'räp·ə·thē }

ischemic paralysis [MED] Impaired motor function due to obstructed circulation to the area. { i'skē·mik pə'ral·ə·səs }

ischemic tubulorrhexis *See* lower nephron nephrosis. { i'skē·mik ˌtü·byə·lə'rek·səs }

ischiopodite [INV ZOO] The segment nearest the basipodite of walking legs in certain crustaceans. Also known as ischium. { ˌis·kē'äp·əˌdīt }

ischiorectal region [ANAT] The region between the ischium and the rectum. { ¦is·kē·ə'rek·təl ˌrē·jən }

ischium [ANAT] Either of a pair of bones forming the dorsoposterior portion of the vertebrate pelvis; the inferior part of the human pelvis upon which the body rests in sitting. [INV ZOO] *See* ischiopodite. { 'is·kē·əm }

ischium-pubis index [ANAT] The ratio (length of pubis × 100/length of ischium) by which the sex of an adult pelvis may usually be determined; the index is greater than 90 in females, and less than 90 in males. { 'is·kē·əm 'pyü·bəs ˌinˌdeks }

island of Langerhans *See* islet of Langerhans. { 'ī·lənd əv 'läŋ·gərˌhänz }

island of Reil [ANAT] The insula of the cerebral hemisphere. { 'ī·lənd əv 'rīl }

islet-cell carcinoma [MED] A metastatic tumor of pancreatic cells of the islet of Langerhans in the pancreas. { ¦ī·lət ¦sel ˌkärs·ən'ō·mə }

islet-cell tumor [MED] A benign tumor of the pancreatic islet cells. Also known as insulinoma; insuloma; Langerhansian adenoma. { ¦ī·lət ¦sel 'tü·mər }

islet of Langerhans [HISTOL] A mass of cell cords in the pancreas that is of an endocrine nature, secreting insulin and a minor hormone like lipocaic. Also known as island of Langerhans; islet of the pancreas. { 'ī·lət əv 'läŋ·gərˌhänz }

islet of the pancreas *See* islet of Langerhans. { 'ī·lət əv t͟hə 'pan·krē·əs }

isoacceptor [CELL MOL] Any of several species of transfer ribonucleic acid that can accept the same amino acid. { ¦ī·sō·ak'sep·tər }

isoagglutinin [IMMUNOL] An agglutinin which acts upon the red blood cells of members of the same species. Also known as isohemagglutinin. { ¦ī·sō·ə'glut·ən·ən }

isoallele [GEN] An allele whose phenotype is indistinguishable from that of a different mutant allele at the same locus. { ¦ī·so·ə'lēl }

isoalloxazine mononucleotide *See* riboflavin 5′-phosphate. { ˌī·sō·ə'läk·səˌzēn ˌmänō'nü·klē·əˌtīd }

isoantibody [IMMUNOL] An antibody formed in response to immunization with tissue constituents derived from an individual of the same species. { ¦īˌsō'ant·iˌbäd·ē }

isoantigen [IMMUNOL] An antigen in an individual capable of stimulating production of a specific antibody in another member of the same species. Also known as alloantigen. { ¦ī·sō'ant·ə·jən }

isobiochore [ECOL] A boundary line on a map connecting world environments that have similar floral and faunal constituents. { ˌī·sə'bī·əˌkȯr }

Isobryales [BOT] An order of mosses in which the plants are slender to robust and up to 36 inches (90 centimeters) in length. { ¦ī·sō·brī'ā·lēz }

isocarpic [BOT] Having the same number of carpels and perianth divisions. { ˌī·sə'kär·pik }

isocercal [VERT ZOO] Of the tail fin of a fish, having the upper and lower lobes symmetrical and the vertebral column gradually tapering. { ¦ī·sə¦sər·kəl }

isochela [INV ZOO] **1.** A chela having two equally developed parts. **2.** A chelate spicule with both ends identical. { ¦ī·sə'kē·lə }

isochromosome [CELL MOL] An abnormal chromosome with a medial centromere and identical arms formed as a result of transverse, rather than longitudinal, splitting of the centromere. { ¦ī·sō'krō·məˌsōm }

isocitric acid [BIOCHEM] $HOOCCH_2CH(COOH)$-$CH(OH)COOH$ An isomer of citric acid that is involved in the Krebs tricarboxylic acid cycle in bacteria and plants. { ¦ī·sə¦si·trik 'as·əd }

isocoding mutation [GEN] A point mutation that changes a codon's nucleotide sequence to one that codes for the same amino acid specified by the initial codon. { ¦ī·sə,kōd·iŋ myü'tā·shən }

isochore [GEN] In mammals and birds, an organizational unit of chromosome DNA that is usually characterized by a fairly constant proportion of guanine-cytosine base pairs throughout its length (usually 200–1000 kilobase pairs long); genes are concentrated in the G-C rich isochores. { 'īs·ə,kȯr }

Isocrinida [INV ZOO] An order of stalked articulate echinoderms with nodal rings of cirri. { ¦ī·sō'krī·nə·də }

isodiametric [BIOL] Having equal diameters or dimensions. { ¦ī·sō,dī·ə'me·trik }

isodisomy [GEN] A type of uniparental disomy in which two copies of the same chromosome are inherited from one parent, with resultant homozygosity at all gene loci on the chromosome. { ,ī·sə,dī'sō·mē }

isodont [VERT ZOO] **1.** Having all teeth alike. **2.** Of a snake, having the maxillary teeth of equal length. { 'ī·sə,dänt }

isodulcitol *See* rhamnose. { ,ī·sə'dəl·sə,tȯl }

isoenzyme [BIOCHEM] Any of the electrophoretically distinct forms of an enzyme, representing different polymeric states but having the same function. Also known as isozyme. { ¦ī·sō'en,zīm }

Isoetaceae [BOT] The single family assigned to the order Isoetales in some systems of classification. { ,ī·sō·ə'tās·ē,ē }

Isoetales [BOT] A monotypic order of the class Isoetopsida containing the single genus *Isoetes*, characterized by long, narrow leaves with a spoonlike base, spirally arranged on an underground cormlike structure. { ,ī·sō·ə'tā·lēz }

Isoetatae *See* Isoetopsida. { ,ī·sō'ed·ə,tē }

Isoetopsida [BOT] A class of the division Lycopodiophyta; members are heterosporous and have a distinctive appendage, the ligule, on the upper side of the leaf near the base. { ,ī·sō·ə'täp·sə·də }

isoflavone [BIOCHEM] $C_{15}H_{10}O_2$ A colorless, crystalline ketone, occurring in many plants, generally in the form of a hydroxy derivative. { ¦ī·sō'fla,vōn }

isoflurophate [PHARM] $[(CH_3)_2CHO]_2P(O)F$ A liquid that forms hydrofluoric acid in the presence of moisture; used as a cholinergic drug in eye diseases in humans and as a miotic drug in animals. { ,ī·sə'flu̇r·ə,fāt }

isogamete [BIOL] A reproductive cell that is morphologically similar in both male and female and cannot be distinguished on form alone. { ¦ī·sō'ga,mēt }

isogamy [BIOL] Sexual reproduction by union of gametes or individuals of similar form or structure. { ī'säg·ə·mē }

isogenic [GEN] Having the same genotype, as all organisms of an inbred strain. [IMMUNOL] Referring to cells, tissues, or organs used in transplantation that originate in identical species. { ,ī·sə·jə'nē·ik }

Isogeneratae [BOT] A class of brown algae distinguished by having an isomorphic alternation of generations. { ,ī·sō,jen·ə'rä,tē }

isogony [BIOL] Growth of parts at such a rate as to maintain relative size differences. { 'ī'säg·ə·nē }

isograft [BIOL] A tissue transplant from one organism to another organism which is genetically identical. { 'ī·sə,graft }

isohemagglutinin *See* isoagglutinin. { ¦ī·sō,hē·mə'glüt·ən·ən }

isohemolysin [IMMUNOL] A hemolysin produced by an individual injected with erythrocytes from another individual of the same species. { ¦ī·sō·hə'mäl·ə·sən }

isohemolysis [IMMUNOL] Hemolysis induced by the action of an isohemolysin. { ¦ī·sō·hə'mäl·ə·səs }

isoimmunization [IMMUNOL] Immunization of an individual by the introduction of antigens from another individual of the same species. { ¦ī·sō,im·yə·nə'zā·shən }

isoinertial [BIOPHYS] Pertaining to the force of a human muscle that is applied to a constant mass in motion. { ¦ī·sō·i'nərsh·əl }

isokinetic [BIOPHYS] Pertaining to the force of a human muscle that is applied during constant velocity of motion. { ,i·sə·ki'ned·ik }

Isolaimoidea [INV ZOO] A superfamily of rather large, free-living soil nematodes in the order Isolaimida, characterized by six hollow tubes around the oral opening, two circlets of six circumoral sensilla, the absence of amphids, and an elongated triradiate stoma with thickened anterior walls. { ,ī·sō·lə'mȯid·ē·ə }

isolate [GEN] A population so cut off from others that mating occurs only within the group. { 'ī·sə,lāt }

isolating mechanism [GEN] A geographic barrier or biological difference that prevents mating or genetic exchange between individuals of different populations or species. { 'ī·sə,lād·iŋ mek·ə,niz·əm }

isolation [EVOL] The restriction or limitation of gene flow between distinct populations due to barriers to interbreeding. [MED] Separation of an individual with a communicable disease from other, healthy individuals. [MICROBIO] Separation of an individual or strain from a natural, mixed population. [PHYSIO] Separation of a tissue, organ, system, or other part of the body for purposes of study. { ,ī·sə'lā·shən }

isolecithal *See* homolecithal. { ¦ī·sō'les·ə·thəl }

isoleucine [BIOCHEM] $C_6H_{13}O_2$ An essential monocarboxylic amino acid occurring in most dietary proteins. { ¦ī·sō'lü,sēn }

isomerase [BIOCHEM] An enzyme that catalyzes isomerization reactions. { ī'säm·ə,rās }

isomerism [BIOL] The condition of having two or more comparable parts made up of identical numbers of similar segments. { ī'säm·ə,riz·əm }

isomerous [BIOL] Characterized by isomerism. { ī'säm·ə·rəs }

Isometopidae [INV ZOO] A family of hemipteran insects in the superfamily Cimicimorpha. { ¦ī·sō·mə'täp·ə,dē }

isometric contraction [PHYSIO] A contraction in which muscle tension is increased, but the muscle is not shortened because the resistance cannot be overcome. Also known as static contraction. { ¦ī·sə'me·trik kən'trak·shən }

isometric particle [VIROL] A plant virus particle that appears at first sight to be spherical when viewed in the electron microscope, but which is actually an icosahedron, possessing 20 sides. { ¦ī·sə'me·trik 'pärd·ə·kəl }

isometric work [BIOPHYS] Physiologic work that is performed by muscles in terms of energy utilization and heat production and involves muscular contraction that is not accompanied by movement. Also known as static work. { ¦ī·sə'me·trik 'wərk }

isoniazid [PHARM] $C_6H_7N_3O$ A drug used as a tuberculostatic. Also known as isonicotinic acid hydrazide. { ,ī·sə'nī·ə·zəd }

isonicotinic acid hydrazide *See* isoniazid. { ¦ī·sə,nik·ə'tin·ik ¦as·əd 'hī·drə,zīd }

isonymous substitution [GEN] Any deoxyribonucleic acid base-pair substitution within an exon that does not result in change in the amino acid sequence for which it codes. Also known as synonymous substitution. { ī¦sän·ə·məs ,səb·stə'tü·shən }

isophene [BIOL] A line on a chart connecting those places within a given region where a particular biological phenomenon (as the flowering of a certain plant) occurs at the same time. { 'ī·sə,fēn }

isophyllous [BOT] Having foliage leaves of similar form on a plant or stem. { ¦ī·sə¦fil·əs }

Isopoda [INV ZOO] An order of malacostracan crustaceans characterized by a cephalon bearing one pair of maxillipeds in addition to the antennae, mandibles, and maxillae. { ī'säp·ə·də }

isoprecipitin [IMMUNOL] A precipitin effective only against the serum of individuals of the same species from which it is derived. { ¦ī·sō·prə'sip·ə·tən }

isoprostanes [BIOCHEM] A class of natural products that are isomeric with prostaglandins but are formed in vivo by the nonenzymatic, free-radical oxygenation of arachidonic acid. { ,ī·sō'prä,stānz }

isoproterenol [PHARM] $C_{11}H_{17}NO_3$ A sympathomimetic amine used as a bronchodilator. { ,ī·sə,prōd·ə'rē,nól }

Isoptera [INV ZOO] An order of Insecta containing morphologically primitive forms characterized by gradual metamorphosis, lack of true larval and pupal stages, biting and prognathous mouthparts, two pairs of subequal wings, and the abdomen joined broadly to the thorax. { ī'säp·tə·rə }

isopygous [INV ZOO] Having a pygidium and a cephalon of equal size, as in certain trilobites. { ¦ī'säp·ə·gəs }

isoschizomer [BIOCHEM] One of two or more restriction endonucleases that cleave a deoxyribonucleic acid molecule at the same site. { ,ī·sə'siz·ə·mər }

Isospondyli [VERT ZOO] A former equivalent name for Clupeiformes. { ,ī·sə'spän·də,lī }

isospore [BIOL] A spore that does not display sexual dimorphism. { 'ī·sə,spór }

isosporiasis [MED] Infection with coccidia of the genus *Isospora*. { ,ī·sə·spə'rī·ə·səs }

isostemonous [BOT] Having the number of stamens of a flower equal to the number of perianth divisions. { ¦ī·sə¦stē·mə·nəs }

isothermal titration calorimeter [BIOPHYS] An instrument that directly measures the energetics (through heat effects) associated with biochemical reactions or processes occurring at constant temperatures. { ¦ī·sə¦thər·məl tī¦trā·shən ,kal·ə'rim·əd·ər }

isotonic [PHYSIO] **1.** Having uniform tension, as the fibers of a contracted muscle. **2.** Of a solution, having the same osmotic pressure as the fluid phase of a cell or tissue. { ¦ī·sə¦tän·ik }

isotonic sodium chloride solution *See* normal saline. { ¦ī·sə¦tän·ik ¦sōd·ē·əm 'klór,īd sə,lü·shən }

isotope farm [BOT] A carbon-14 (^{14}C) growth chamber, or greenhouse, arranged as a closed system in which plants can be grown in an atmosphere of carbon dioxide (CO_2) containing ^{14}C and thus become labeled with ^{14}C; isotope farms also can be used with other materials, such as heavy water (D_2O), phosphorus-35 (^{35}P), and so forth, to produce biochemically labeled compounds. { 'ī·sə,tōp ,färm }

isotropic [BIOL] Having a tendency for equal growth in all directions. [CELL MOL] An ovum lacking any predetermined axis. { ¦ī·sə¦trä·pik }

isotypes [IMMUNOL] **1.** A series of antigens, for example, blood types, common to all members of a species but differentiating classes and subclasses within the species. **2.** Different classes of immunoglobulins that have the same antigenic specificity. { 'ī·sə,tīps }

isozyme *See* isoenzyme. { 'ī·sə,zīm }

isthmus [BIOL] A passage or constricted part connecting two parts of an organ. { 'is·məs }

Istiophoridae [VERT ZOO] The billfishes, a family of oceanic perciform fishes in the suborder Scombroidei. { ,is·tē·ə'fór·ə,dē }

Isuridae [VERT ZOO] The mackerel sharks, a family of pelagic, predacious galeoids distinguished by a heavy body, nearly symmetrical tail, and sharp, awllike teeth. { i'súr·ə,dē }

itch [PHYSIO] An irritating cutaneous sensation allied to pain. { ich }

iteroparity [BIOL] Reproduction that occurs repeatedly over the life of the individual. { ,īd·ə·rə'par·əd·ē }

iteroparous [ZOO] Capable of breeding or reproducing multiple times. { ,īd·ə·rə'par·əs }

Ithomiinae [INV ZOO] The glossy-wings, a subfamily of weak-flying lepidopteran insects having

J

jaagsiekte [VET MED] A contagious disease of sheep, sometimes of goats and guinea pigs, resembling the more benign and diffuse forms of bronchiolar carcinoma in humans. { 'yäg ˌsēk·tə }

Jacanidae [VERT ZOO] The jacanas or lily-trotters, constituting the single family of the avian superfamily Jacanoidea. { jə'kan·əˌdē }

Jacanoidea [VERT ZOO] A monofamilial superfamily of colorful marshbirds distinguished by greatly elongated toes and claws, long legs, a short tail, and a straight bill. { ˌjak·ə'nȯid·ē·ə }

jackal [VERT ZOO] **1.** *Canis aureus.* A wild dog found in southeastern Europe, southern Asia, and northern Africa. **2.** Any of various similar Old World wild dogs; they resemble wolves but are smaller and more yellowish. { 'jak·əl }

jacket crown [MED] An artificial crown of a tooth consisting of a covering of porcelain or resin. { 'jak·ət ˌkrau̇n }

Jacksonian epilepsy [MED] Recurrent Jacksonian seizures. { jak'sō·nē·ən 'ep·əˌlep·sē }

Jacksonian seizure [MED] A focal seizure originating in one part of the motor or sensorimotor cortex and manifested usually by spasmodic contractions of or crawling or burning sensations of the skin; may become generalized, leading to loss of consciousness. { jak'sō·nē·ən 'sē·zhər }

Jacobsoniidae [INV ZOO] The false snout beetles, a small family of coleopteran insects in the superfamily Dermestoidea. { ˌjā·kəb·sə'nī·əˌdē }

Jacobson radical *See* radical. { 'jā·kəb·sən ˌrad·ə·kəl }

Jacobson's cartilage *See* vomeronasal cartilage. { 'jā·kəb·sənz ¦kärt·lij }

Jacobson's organ [VERT ZOO] An olfactory canal in the nasal mucosa which ends in a blind pouch; it is highly developed in reptiles and vestigial in humans. { 'jā·kəb·sənz ¦ȯr·gən }

Jadassohn's nevus *See* nevus sebaceus. { 'yä·dəˌzōnz 'nē·vəs }

jaguar [VERT ZOO] *Felis onca.* A large, wild cat indigenous to Central and South America; it is distinguished by a buff-colored coat with black spots, and has a relatively large head and short legs. { 'jagˌwär }

Jak-STAT pathway [CELL MOL] A rapid signal transduction pathway used by a variety of cytokines and growth factors to alter gene expression. Binding of a cytokine or growth factor to its receptor activates Jak (janus kinase—a cytoplasmic tyrosine kinase) and triggers it to phosphorylate and stimulate STAT (signal transducers and activators of transcription—a gene regulatory protein) to travel to the nucleus and induce transcription of a specific gene. { 'jakˌstat 'pathˌwā }

Jamaica bayberry *See* bayberry. { jə'mā·kə 'bāˌber·ē }

Japanese encephalitis [MED] A human viral infection epidemic in Japan, transmitted by the common house mosquito (*Culex pipiens*) and characterized by severe inflammation of the brain. { ¦jap·ə¦nēz inˌsef·ə'līd·əs }

Japygidae [INV ZOO] A family of wingless insects in the order Diplura with forcepslike anal appendages; members attack and devour small soil arthropods. { jə'pij·əˌdē }

jaundice [INV ZOO] *See* grasserie. [MED] Yellow coloration of the skin, mucous membranes, and secretions resulting from hyperbile-rubinemia. Also known as icterus. { 'jȯn·dəs }

jaundice of newborn [MED] Jaundice in infants during the first few days after birth, due to various causes. { 'jȯn·dəs əv 'nüˌbȯrn }

Java cotton *See* kapok. { 'jäv·ə 'kat·ən }

jaw [ANAT] Either of two bones forming the skeleton of the mouth of vertebrates: the upper jaw or maxilla, and the lower jaw or mandible. { jȯ }

jawless vertebrate [VERT ZOO] The common name for members of the Agnatha. { 'jȯ·ləs 'vərd·ə·brət }

jejunitis [MED] Inflammation of the jejunum. { ˌjē·jə'nīd·əs }

jejunostomy [MED] The making of an artificial opening through the abdominal wall into the jejunum. { ˌjē·jə'näs·tə·mē }

jejunum [ANAT] The middle portion of the small intestine, extending between the duodenum and the ileum. { jə'jü·nəm }

jellyfish [INV ZOO] Any of various free-swimming marine cnidarians belonging to the Hydrozoa or Scyphozoa and having a bell- or bowl-shaped body. Also known as medusa. { 'jel·ēˌfish }

jelly fungus [MYCOL] The common name for

many members of the Heterobasidiomycetidae, especially the orders Tremallales and Dacromycetales, distinguished by a jellylike appearance or consistency. { 'jel·ē ,fəŋ·gəs }

Jennerian vaccine *See* smallpox vaccine. { jə'nir·ē·ən vak'sēn }

jenny [VERT ZOO] **1.** A female animal, as a jenny wren. **2.** A female donkey. { 'jen·ē }

Jensen's sarcoma [VET MED] A transmissible malignant tumor originally observed in a rat inoculated with acid-fast bacteria from a cow with pseudotuberculous enteritis. { 'jen·sənz sär 'kō·mə }

jerboa [VERT ZOO] The common name for 25 species of rodents composing the family Dipodidae; all are adapted for jumping, having extremely long hindlegs and feet. { jər 'bō·ə }

jetty *See* groin. { 'jed·ē }

jimsonweed [BOT] *Datura stramonium.* A tall, poisonous annual weed having large white or violet trumpet-shaped flowers and globose prickly fruits. Also known as apple of Peru. { 'jim·sən,wēd }

jird [VERT ZOO] Any one of the diminutive rodents composing related species of the genus *Meriones* which are inhabitants of northern Africa and southwestern Asia; they serve as experimental hosts for studies of schistosomiasis. { jərd }

Johne's disease [VET MED] A chronic inflammation of the intestinal tract of sheep, cattle, and deer caused by *Mycobacterium paratuberculosis.* { 'yō·nəz di,zēz }

joint [ANAT] A contact surface between two individual bones. Also known as articulation. { jȯint }

joint body *See* joint mouse. { 'jȯint ,bäd·ē }

joint capsule [ANAT] A sheet of fibrous connective tissue enclosing a synovial joint. { 'jȯint ,kap·səl }

joint mouse [PATH] A small, loose body within a synovial joint, frequently calcified, derived from synovial membrane, organized fibrin fragments of articular cartilage, or arthritic osteophytes. Also known as joint body. { 'jȯint ,mau̇s }

Joppeicidae [INV ZOO] A monospecific family of hemipteran insects included in the Pentatomorpha; found in the Mediterranean regions. { ,jäp·ə'ī·sə,dē }

jordanon *See* microspecies. { 'jȯrd·ən,än }

Jordan's rule [EVOL] The rule that organisms which are closely related tend to occupy adjacent rather than identical or distant ranges. [VERT ZOO] The rule that fishes in areas of low temperatures tend to have more vertebrae than those in warmer waters. { 'jȯrd·ənz ,rül }

jugal [ANAT] Pertaining to the zygomatic bone. [VERT ZOO] In lower vertebrates, a bone lying below the orbit of the eye. { 'jü·gəl }

Jugatae [INV ZOO] The equivalent name for Homoneura. { 'jü·gə,tē }

jugate [BIOL] Structures which are joined together. { 'jü,gāt }

Juglandaceae [BOT] A family of dicotyledonous plants in the order Juglandales having unisexual flowers, a solitary basal ovule in a unilocular inferior ovary, and pinnately compound, exstipulate leaves. { ,jü,glan'dās·ē,ē }

Juglandales [BOT] An order of dicotyledonous plants in the subclass Hamamelidae distinguished by compound leaves; includes hickory, walnut, and butternut. { ,jü,glan'dā·lēz }

jugular [ANAT] Pertaining to the region of the neck above the clavicle. { 'jəg·yə·lər }

jugular compression [PATH] A test for a spinal subarachnoid block by noting the rate of rise and fall of the spinal fluid pressure following compression and release of the jugular veins. { 'jəg·yə·lər kəm'presh·ən }

jugular foramen [ANAT] An opening in the cranium formed by the jugular notches of the occipital and temporal bones for passage of an internal jugular vein, the ninth, tenth, and eleventh cranial nerves, and the inferior petrosal sinus. { 'jəg·yə·lər fə'rā·mən }

jugular foramen syndrome [MED] Injury to the jugular foramen, usually due to a basilar skull fracture, with resultant paralysis of ninth, tenth, and eleventh cranial nerves. { 'jəg·yə·lər fə'rā·mən ,sin,drōm }

jugular process [ANAT] A rough process external to the condyle of the occipital bone. { 'jəg·yə·lər ,prä·səs }

jugular vein [ANAT] The vein in the neck which drains the brain, face, and neck into the innominate. { 'jəg·yə·lər ,vān }

jugum [BOT] One pair of opposite leaflets of a pinnate leaf. [INV ZOO] **1.** The most posterior and basal portion of the wing of an insect. **2.** A crossbar connecting the two arms of the brachidium in some brachiopods. { 'jü·gəm }

jumping gene [GEN] A mobile genetic entity, such as a transposon. { ¦jəmp·iŋ ¦jēn }

Juncaceae [BOT] A family of monocotyledonous plants in the order Juncales characterized by an inflorescence of diverse sorts, vascular bundles with abaxial phloem, and cells without silica bodies. { ,jəŋ'kās·ē,ē }

Juncales [BOT] An order of monocotyledonous plants in the subclass Commelinidae marked by reduced flowers and capsular fruits with one too many anatropous ovules per carpel. { ,jəŋ'kā·lēz }

junctional complex [CELL MOL] Any specialized area of intercellular adhesion. { 'jəŋk·shən·əl ¦käm,pleks }

junctional nevus [MED] A skin lesion containing nevus cells at the junction of the epidermis and dermis. { 'jəŋk·shən·əl 'nē·vəs }

junctional receptor [PHYSIO] An acetylcholine receptor which occurs in clusters in a muscle membrane at the nerve-muscle junction. { 'jəŋk·shən·əl ri'sep·tər }

junction sequence [CELL MOL] Either of the two terminal regions of the intron in ribonucleic acid precursors. { 'jəŋk·shən ,sē·kwəns }

Jungermanniales [BOT] The leafy liverworts, an order of bryophytes in the class Marchantiatae characterized by chlorophyll-containing, ribbonlike or leaflike bodies and an undifferentiated thallus. { ˌjəŋ·gərˌman·ē'ā·lēz }

Jungermanniidae [BOT] A subclass of liverworts of the class Hepticopsida, division Bryophyta, distinguished by little or no tissue differentiation or organized into erect or prostrate stems with leafy appendages, leaves generally one cell in thickness and mostly arranged in three rows, with the third row of under-leaves commonly reduced or even lacking, and oil bodies usually present in all cells. { ˌjəŋ·gər'man·əˌdē }

jungle [ECOL] An impenetrable thicket of second-growth vegetation replacing tropical rain forest that has been disturbed; lower growth layers are dense. { 'jəŋ·gəl }

jungle yellow fever [MED] A form of yellow fever endemic in forested areas of Brazil. { 'jəŋ·gəl 'yel·ō 'fē·vər }

junk deoxyribonucleic acid *See* selfish deoxyribonucleic acid. { ¦jəŋk dē¦äk·sēˌrī·bō·nü¦klē·ik ˌas·əd }

jute [BOT] Either of two Asiatic species of tall, slender, half-shrubby annual plants, *Corchorus capsularis* and C. *olitorius*, in the family Malvaceae, useful for their fiber. { jüt }

juvenile cell *See* metamyelocyte. { 'jü·vən·əl 'sel }

juvenile hormone *See* neotenin. { 'jü·vən·əl 'hȯrˌmōn }

juvenile melanoma [MED] A benign compound nevus in young people; resembles a malignant melanoma histologically. { 'jü·vən·əl ˌmel·ə'nō·mə }

juvenile-onset diabetes [MED] A form of diabetes mellitus which develops early in life and presents much more severe symptoms than the more common maturity-onset diabetes. { 'jü·vən·əl ¦ȯnˌset ˌdī·ə'bēd·ēz }

Jynginae [VERT ZOO] The wrynecks, a family of Old World birds in the order Piciformes; a subfamily of the Picidae in some systems of classification. { jin'jīˌnē }

K

kaempferol [BIOCHEM] A flavonoid with a structure similar to that of quercetin but with only one hydroxyl in the B ring; acts as an enzyme cofactor and causes growth inhibition in plants. { 'kemp·fə,ròl }

Kahler's disease *See* multiple myeloma. { 'kä·lərz di,zēz }

Kahn flocculation test [PATH] A macroscopic precipitin test for identification of the antibody resulting from syphilitic infection, by using an antigen prepared from normal beef heart. { 'kän ,fläk·yə'lä·shən ,test }

kairomone [PHYSIO] A chemical produced by an organism that benefits the recipient, which is an individual of a different species. { 'kī·rə,mōn }

Kaiserling's method [BIOL] A method for preserving organ specimens and retaining their color by fixing in a solution of formalin, water, potassium nitrate, and potassium acetate, immersing in ethyl alcohol to restore color, and preserving in a solution of glycerin, aqueous arsenious acid, water, potassium acetate, and thymol. { 'kī·zər·liŋz ,meth·əd }

kaki [BOT] *Diospyros kaki.* The Japanese persimmon; it provides a type of ebony wood that is black with gray, yellow, and brown streaks, has a close, even grain, and is very hard. { 'kä·kē }

kakidrosis [MED] Secretion of sweat having a disagreeable odor. { ,kä·kə'drō·səs }

kala azar [MED] Visceral leishmaniasis due to the protozoan *Leishmania donovani*, transmitted by certain sandflies (*Phlebotomus*); characterized by chronic, irregular fever, enlargement of the spleen and liver, emaciation, anemia, and leukopenia. { ,käl·ə ə'zär }

kale [BOT] Either of two biennial crucifers, *Brassica oleracea* var. *acephala* and *B. fimbriata*, in the order Capparales, grown for the nutritious curled green leaves. { kāl }

kallidin I *See* bradykinin. { kə'līd·ən 'wən }

Kalotermitidae [INV ZOO] A family of relatively primitive, lower termites in the order Isoptera. { ,kal·ə,tər'mid·ə,dē }

Kamptozoa [INV ZOO] An equivalent name for Entoprocta. { ,kam·tə'zō·ə }

kanamycin [MICROBIO] $C_{18}H_{36}O_{11}N_4$ A water-soluble, basic antibiotic produced by strains of *Streptomyces kanamyceticus*; the sulfate salt is effective in infections caused by gram-negative bacteria. { ,kan·ə'mīs·ən }

kangaroo [VERT ZOO] Any of various Australian marsupials in the family Macropodidae generally characterized by a long, thick tail that is used as a balancing organ, short forelimbs, and enlarged hindlegs adapted for jumping. { ,kaŋ·gə'rü }

Kansasii disease [MED] A mycobacterial tuberculosislike infection caused by *Mycobacterium kansasii*, an orange-yellow acid-fast bacterium. { kan'zas·ē,ī di,zēz }

kapok [BOT] Silky fibers that surround the seeds of the kapok or ceiba tree. Also known as ceiba; Java cotton; silk cotton. { 'kā,päk }

kapok tree [BOT] *Ceiba pentandra.* A tree of the family Bombacaceae which produces pods containing seeds covered with silk cotton. Also known as silk cotton tree. { 'kā,päk ,trē }

kappa particle [CELL MOL] A self-duplicating nucleoprotein particle found in various strains of *Paramecium* and thought to function as an infectious agent; classed as an intracellular symbiont, occupying a position between the viruses and the bacteria and organelles. { 'kap·ə ,pard·ə·kəl }

Kaposi's sarcoma [MED] A multifocal malignant or benign neoplasm that occurs in the skin and sometimes in lymph nodes or viscera and is composed of primitive tissue that is involved in blood vessel formation. { kə'pō·sēz sär 'kōm·ə }

Karumiidae [INV ZOO] The termitelike beetles, a small family of coleopteran insects in the superfamily Cantharoidea distinguished by having a tenth tergum. { ,kar·ə'mī·ə,dē }

karyocyte *See* normoblast. { 'kar·ē·ə,sīt }

karyodesma *See* nucleodesma. { ,kar·ē·ə'dez·mə }

karyogamy [CELL MOL] Fusion of gametic nuclei, as in fertilization. { ,kar·ē'äg·ə·mē }

karyokinesis [CELL MOL] Nuclear division characteristic of mitosis. { ¦kar·ē·ō·kə'nē·səs }

karyolymph [CELL MOL] The clear material composing the ground substance of a cell nucleus. { 'kar·ē·ə,limf }

karyolysis [CELL MOL] Dissolution of a cell nucleus. { ,kar·ē'äl·ə·səs }

karyomastigont [INV ZOO] Pertaining to members of the protozoan order Oxymonadida; individuals can be uni- or multinucleate, and unattached forms give rise to two pairs of flagella. { ¦kar·ē·ō'mas·tə,gänt }

karyoplasm *See* nucleoplasm. { 'kar·ē·ə,plaz·əm }

karyoplasmic ratio *See* nucleocytoplasmic ratio. { ,kar·ē·ə'plaz·mik 'rā·shō }

karyorrhexis [CELL MOL] Fragmentation of a nucleus with scattering of the pieces in the cytoplasm. { ,kar·ē·ə'rek·səs }

karyosphere [CELL MOL] The fraction of nuclear volume to which the chromosomes are confined in nuclei that are rich in karyolymph. { 'kar·ē·ə,sfir }

karyotype [CELL MOL] **1.** The complement of chromosomes characteristic of an individual, species, genus, or other grouping. **2.** An organized array of the chromosomes from a single cell, grouped according to size, centromere position, and banding pattern, if any. { 'kar·ē·ə,tīp }

kasugamycin [MICROBIO] $C_{14}H_{28}ClN_3O_{10}$ A white, crystalline antibiotic used as a fungicide for rice crops. Also known as kasugamycin hydrochloride; kasumin. { kə,sü·gə'mīs·ən }

kasugamycin hydrochloride *See* kasugamycin. { kə,sü·gə'mīs·ən ,hī·drə'klȯr,īd }

kasumin *See* kasugamycin. { kə'sü·mən }

Kathlaniidae [INV ZOO] A family of nematodes assigned to the Ascaridina by some authorities and to the Oxyurina by others. { ,kath·lə'nī·ə,dē }

Kayser-Fleischer ring [PATH] A ring of golden-brown or brownish-green pigment behind the limbic border of the cornea, due to the deposition of copper. { 'kī·zər 'flī·shər ,riŋ }

kb *See* kilobase.

kbp *See* kilobase.

keel [VERT ZOO] The median ridge on the breastbone in certain birds. Also known as carina. { kēl }

Kell blood group system [IMMUNOL] A family of antigens found in erythrocytes and designated K, k, Kp^a, Kp^b, and Ku; antibodies to the K antigen, which occurs in about 10% of the population of England, have been associated with hemolytic transfusion reactions and with hemolytic disease. { 'kel 'bləd ,grüp ,sis·təm }

keloid [MED] A firm, elevated fibrous formation of tissue at the site of a scar. { 'kē,lȯid }

keloid acne [MED] Acnelike infection of hair follicles, especially on the nape of the neck, resulting in hard, white or reddish keloids and scarring. Also known as folliculitis keloidalis. { 'kē,lȯid 'ak·nē }

kelp [BOT] The common name for brown seaweed belonging to the Laminariales and Fucales. { kelp }

kenozooecium [INV ZOO] The outer, nonliving, hardened portion of a kenozooid. { ¦kēn·ə·zō'ē·sē·əm }

kenozooid [INV ZOO] A type of bryozoan heterozooid possessing a slender tubular or boxlike chamber, completely enclosed and lacking an aperture. { ¦kēn·ə¦zō,ȯid }

Kentucky coffee tree [BOT] *Gymnocladus dioica.* An extremely tall, dioecious tree of the order Rosales readily recognized when in fruit by its leguminous pods containing heavy seeds, once used as a coffee substitute. { kən'tək·ē 'kȯf·ē ,trē }

keratectomy [MED] Surgical removal of a portion of the cornea. { ,ker·ə'tek·tə·mē }

keratin [BIOCHEM] Any of various albuminoids characteristic of epidermal derivatives, such as nails and feathers, which are insoluble in protein solvents, have a high sulfur content, and generally contain cystine and arginine as the predominating amino acids. { 'ker·əd·ən }

keratinized tissue [HISTOL] Any tissue with a high keratin content, such as the epidermis or its derivatives. { 'ker·əd·ə,nīzd 'tish·ü }

keratinocyte [HISTOL] A specialized epidermal cell that synthesizes keratin. { kə'rat·ən·ə,sīt }

keratinous degeneration [CELL MOL] The occurrence of keratin granules in the cytoplasm of a cell, other than a keratinocyte. { kə'rat·ən·əs di,jen·ə'rā·shən }

keratitis [MED] Inflammation of the cornea. { ,ker·ə'tīd·əs }

keratitis rosacea [MED] The occurrence of small, sterile infiltrates at the periphery of the cornea. { ,ker·ə'īd·əs rō'zā·shə }

keratoconjunctivitis [MED] Concurrent inflammation of the cornea and the conjunctiva. Also known as shipyard eye. { ¦ker·əd·ō·kən,jəŋk·tə'vīd·əs }

keratohyalin [HISTOL] Granules in the stratum granulosum of keratinized stratified squamous epithelium which become keratin. { ,ker·əd·ō'hī·ə·lən }

keratomalacia [MED] Degeneration of the cornea characterized by infiltration and keratinization of the epithelium, eventually leading to thinning and perforation of the cornea; generally occurs in vitamin A deficiency. { ,ker·əd·ō·mə'lā·shə }

keratoplasty [MED] A plastic operation on the cornea, especially the transplantation of a portion of the cornea. { 'ker·əd·ō,plas·tē }

keratosis [MED] Any disease of the skin characterized by an overgrowth of the cornified epithelium. { ,ker·ə'tō·səs }

keratosis follicularis *See* Darrier's disease. { ,ker·ə¦tōs·əs fə,lik·yə'lär·əs }

Kerguelen faunal region [ECOL] A marine littoral faunal region comprising a large area surrounding Kerguelen Island in the southern Indian Ocean. { 'kər·gə·lən 'fȯn·əl ,rē·jən }

kernel [BOT] **1.** The inner portion of a seed. **2.** A whole grain or seed of a cereal plant, such as corn or barley. { 'kərn·əl }

kernicterus [PATH] Deposition of bilirubin in the gray matter of the brain and spinal cord, especially the basal ganglia, accompanied by nerve cell degeneration. { kər'nik·tə·rəs }

Kernig's sign [MED] In meningeal irritation, with the patient lying face up and the thigh flexed at the hip, the pain and spasm of the hamstring muscles when an attempt is made to completely extend the leg at the knee. { 'kər·nigz ˌsin }

ketoacidosis *See* ketosis. { ˌkēd·ōˌas·ə'dō·səs }

ketoadipic acid [BIOCHEM] $C_6H_8O_5$ An intermediate product in the metabolism of lysine to glutaric acid. { ¦kēd·ō·ə¦dip·ik 'as·əd }

ketogenesis [BIOCHEM] Production of ketone bodies. { ¦kēd·ō'jen·ə·səs }

ketogenic hormone [BIOCHEM] A factor originally derived from crude anterior hypophysis extract which stimulated fatty-acid metabolism; now known as a combination of adrenocorticotropin and the growth hormone. Also known as fat-metabolizing hormone. { ¦kēd·ə¦jen·ik 'hȯrˌmōn }

ketogenic substance [BIOCHEM] Any foodstuff which provides a source of ketone bodies. { ¦kēd·ə¦jen·ik 'səb·stəns }

ketoglutaric acid [BIOCHEM] $C_5H_6O_5$ A dibasic keto acid occurring as an intermediate product in carbohydrate and protein metabolism. { ¦kēd·ō·glü¦tar·ik 'as·əd }

ketohexose [BIOCHEM] Any monosaccharide composed of a six-carbon chain and containing one ketone group. { ¦kēd·ō'hekˌsōs }

ketolase [BIOCHEM] A type of enzyme that catalyzes cleavage of carbohydrates at the carbonyl carbon position. { 'kēd·əlˌās }

ketolysis [BIOCHEM] Dissolution of ketone bodies. { kē'täl·ə·səs }

ketone body [BIOCHEM] Any of various ketones which increase in blood and urine in certain conditions, such as diabetic acidosis, starvation, and pregnancy. Also known as acetone body. { 'kēˌtōn ˌbäd·ē }

ketonemia *See* acetonemia. { ˌkēd·ə'nē·mē·ə }

ketonuria [MED] Presence of ketone bodies in the urine. { ˌkēd·ə'nu̇r·ē·ə }

ketose [BIOCHEM] A carbohydrate that has a ketone group. { 'kēˌtōs }

ketosis [MED] Excess amounts of ketones in the body, especially associated with diabetes mellitus. Also known as ketoacidosis. { kē 'tō·səs }

ketosteroid [BIOCHEM] One of a group of neutral steroids possessing keto substitution, which produces a characteristic red color with *m*-dinitrobenzene in an alkaline solution; these compounds are principally metabolites of adrenal cortical and gonadal steroids. { ¦kēd·ō'stirˌȯid }

key [SYST] An arrangement of the distinguishing features of a taxonomic group to serve as a guide for establishing relationships and names of unidentified members of the group. { kē }

khelin *See* khellin. { 'kel·ən }

khellin [PHARM] $C_{14}H_{12}O_5$ A synthetic compound that crystallizes from methanol solution, has a bitter taste, melts at 154–155°C, and is more soluble in water than in organic solvents; used in medicine as an antispasmodic, a coronary vasodilator, and a bronchodilator. Also spelled chellin; khelin. Also known as visammin. { 'kel·ən }

Kidd blood group system [IMMUNOL] The erythrocyte antigens defined by reactions to anti-Jk[a] antibodies, originally found in the mother (Mrs. Kidd) of the erythroblastotic infant, and to anti-Jk[b] antibodies. { 'kid 'bləd ˌgrüp ˌsis·təm }

kidney [ANAT] Either of a pair of organs involved with the elimination of water and waste products from the body of vertebrates; in humans they are bean-shaped, about 5 inches (12.7 centimeters) long, and are located in the posterior part of the abdomen behind the peritoneum. { 'kid·nē }

Kienböck's disease *See* osteochondrosis. { kēn ˌbeks diˌzēz }

killed vaccine [IMMUNOL] A suspension of killed microorganisms used as antigens to produce immunity. Also known as inactivated vaccine. { 'kild vak'sēn }

killer whale [VERT ZOO] *Orcinus orca*. A predatory, cosmopolitan cetacean mammal, about 30 feet (9 meters) long, found only in cold waters. { 'kil·ər ˌwāl }

kilobase [GEN] Unit of length equal to 1000 base pairs in deoxyribonucleic acid or 1000 nitrogenous bases in ribonucleic acid. Abbreviated kb; kbp. { 'kil·ōˌbās }

Kimmelstiel-Wilson disease *See* intercapillary glomerulosclerosis. { 'kim·əlˌstēl 'wil·sən diˌzēz }

kinase [BIOCHEM] Any enzyme that catalyzes phosphorylation reactions. { 'kīˌnās }

kinematic chain [ANAT] A group of body segments that are connected by joints so that the segments operate together to provide a wide range of motion for a limb. { ¦kin·əˌmad·ik 'chān }

kineplasty [MED] An amputation of a limb in which tendons are arranged in the stump to permit their use in moving parts of the prosthetic appliance. Also spelled cineplasty. { 'kin· əˌplas·tē }

kinesiatrics [MED] The treatment of disease by systematic active or passive movements. Also known as kinesitherapy; kinetotherapy. { kə ˌnēz·ē'a·triks }

kinesin [CELL MOL] An enzyme that hydrolyzes adenosine triphosphate to provide energy to power anterograde [from (-) to (+)] movement along microtubules. { kī'nēs·ən }

kinesiology [PHYSIO] The study of human motion through anatomical and mechanical principles. { kəˌnēz·ē'äl·ə·jē }

kinesis [PHYSIO] The general term for physical movement, including that induced by stimulation, for example, light. { ki'nē·səs }

kinesitherapy *See* kinesiatrics. { kəˌnēz·ə'ther·ə· pē }

kinesthesis [PHYSIO] The system of sensitivity present in the muscles and their attachments. { ˌkin·əs'thē·səs }

kinesthetic apraxia *See* motor apraxia. { ˌkin· əs'thed·ik ā'prak·sē·ə }

kinetic intermediate [CELL MOL] A structural form that occurs transiently during protein folding. { kə¦ned·ik ˌin·tər'mēd·ē·ət }

kinetid [CELL MOL] In eukaryotic cells, any locomotory structure, that is, a cilium or flagellum. { ki'ned·əd }

kinetin [BIOCHEM] $C_{10}H_9ON_5$ A cytokinin formed in many plants which has a stimulating effect on cell division. { 'kin·ə·tən }

kinetochore [CELL MOL] Within the centromere, the granule upon which the spindle fibers attach. { kə'ned·ə,kȯr }

kinetoplast [CELL MOL] A genetically autonomous, membrane-bound organelle associated with the basal body at the base of flagella in certain flagellates, such as the trypanosomes. Also known as parabasal body. { kə'ned·ə,plast }

Kinetoplastida [INV ZOO] An order of colorless protozoans in the class Zoomastigophorea having pliable bodies and possessing one or two flagella in some stage of their life. { kə,ned·ə'plas·tə·də }

kinetosome *See* basal body. { kə'ned·ə,sōm }

kinetotherapy *See* kinesiatrics. { kə,ned·ə'ther·ə·pē }

kingdom [SYST] One of the primary divisions that include all living organisms: most authorities recognize two, the animal kingdom and the plant kingdom, while others recognize three or more, such as Protista, Plantae, Animalia, and Mycota. { 'kiŋ·dəm }

kingfisher [VERT ZOO] The common name for members of the avian family Alcedinidae; most are tropical Old World species characterized by short legs, long bills, bright plumage, and short wings. { 'kiŋ,fish·ər }

kinin [PHARM] Any of several pharmacologically active polypeptides that act as hypotensives, contracting isolated smooth muscles and increasing capillary permeability; an example is bradykinin. { 'kī·nən }

kink [CELL MOL] A bend between two helical segments of deoxyribonucleic acid achieved by unstacking one base pair and twisting the polynucleotide backbone. { kiŋk }

kinocilium [CELL MOL] A type of cilium containing one central pair of microfibrils and nine peripheral pairs; they extend from the apex of hair cells in all vertebrate ears except mammals. { ¦kin·ə'sil·ē·əm }

kinomere *See* centromere. { 'kin·ə,mir }

kinoplasm [CELL MOL] The substance of the protoplasm that is thought to form astral rays and spindle fibers. { 'kin·ə,plaz·əm }

Kinorhyncha [INV ZOO] A class of the phylum Aschelminthes consisting of superficially segmented microscopic marine animals lacking external ciliation. { ,kin·ə'riŋ·kē·ə }

Kinosternidae [VERT ZOO] The mud and musk turtles, a family of chelonian reptiles in the suborder Cryptodira found in North, Central, and South America. { ,kin·ə'stər·nə,dē }

kissing disease *See* infectious mononucleosis. { 'kis·iŋ di,zēz }

Kitasatoa [MICROBIO] A genus of bacteria in the family Actinoplanaceae; club-shaped sporangia, each containing a single chain of diplococcuslike uniflagellate spores. { kə¦ta·sə 'tō·ə }

kiwi [VERT ZOO] The common name for three species of nocturnal ratites of New Zealand composing the family Apterygidae; all have small eyes, vestigial wings, a long slender bill, and short powerful legs. { 'kē,wē }

Klebsiella [MICROBIO] A genus of bacteria in the family Enterobacteriaceae; nonmotile, encapsulated rods arranged singly, in pairs, or in chains; some species are human pathogens. { ,kleb·zē'el·ə }

Klebsiella pneumoniae [MICROBIO] An encapsulated pathogenic bacterium that causes severe pneumonitis in humans. Formerly known as Friedlander's bacillus; pneumobacillus. { ,kleb·zē'el·ə nə'mō·nē,ī }

Klebsiella rhinoscleromatis [MICROBIO] A gram-negative, nonmotile, pathogenic species of bacteria that causes the upper respiratory disease rhinoscleroma. { ,kleb·zē,el·ə ,rī·nō,skler·ə'mäd·əs }

Klebs-Loeffler bacillus *See* Corynebacterium diphtheriae. { 'klāps 'lef·lər bə,sil·əs }

Kleinschmidt spread [CELL MOL] A visualization technique for electron microscopy in which molecules are mounted in a positively charged protein monolayer, which is spread on the surface of water, and then are transferred to a hydrophobic grid. { 'klīn,shmit ,spred }

klendusity [BOT] The tendency of a plant to resist disease due to a protective covering, such as a thick cuticle, that prevents inoculation. { klen'dü·səd·ē }

Klinefelter's syndrome [MED] A complex of symptoms associated with hypogonadism in males as an accompaniment of an anomaly of the sex chromosomes; somatic cells are found to have a Y chromosome and more than one X chromosome. { 'klīn,fel·tərz ,sin,drōm }

Kline flocculation test [PATH] A microscopic precipitin test for identification of the antibody resulting from syphilitic infection. { 'klīn ,fläk·yə'lā·shən ,test }

klinorhynchy [BIOL] The property of a downwardly bent face. { 'klīn·ə,riŋ·kē }

klinotaxis [BIOL] Positive orientation movement of a motile organism induced by a stimulus. { ¦klī·nə'tak·səs }

knee [ANAT] **1.** The articulation between the femur and the tibia in humans. Also known as genu. **2.** The corresponding articulation in the hindlimb of a quadrupedal vertebrate. { nē }

kneecap *See* patella. { 'nē,kap }

Kneriidae [VERT ZOO] A small family of tropical African fresh-water fishes in the order Gonorynchiformes. { nə'rī·ə,dē }

knockout [GEN] Inactivation of a specific gene in a laboratory organism in order to study gene function. { 'näk,au̇t }

koala [VERT ZOO] *Phascolarctos cinereus.* An arboreal marsupial mammal of the family Phalangeridae having large hairy ears, gray fir, and two clawed toes opposing three others on each limb. { kō'äl·ə }

Koch's postulates [MICROBIO] A set of laws elucidated by Robert Koch: the microorganism identified as the etiologic agent must be present in every case of the disease; the etiologic agent must be isolated and cultivated in pure culture; the organism must produce the disease when inoculated in pure culture into susceptible animals; a microorganism must be observed in and recovered from the experimentally diseased animal. Also known as law of specificity of bacteria. { 'kōks 'päs·chə·ləts }

Kohler's disease *See* osteochondrosis. { 'kō·lərz di,zēz }

kohlrabi [BOT] A biennial crucifer, designated *Brassica caulorapa* and *B. oleracea* var. *caulo-rapa*, of the order Capparales grown for its edible turniplike, enlarged stem. { kōl'rä·bē }

koilonychia [MED] Spoon-shaped deformity of the fingernails, which may be familial or associated with a disease, such as iron-deficiency anemia. Also known as spoon nail. { ,kȯil·ō'nik·ē·ə }

koilorachic [MED] Having the lumbar spinal region concave ventrally. { ¦kȯil·ə¦rak·ik }

Kolmer test [PATH] A complement-fixation test for syphilis and other diseases. { 'kōl·mər ,test }

Komodo dragon [VERT ZOO] *Varanus komodoensis.* A predatory reptile of the family Varanidae found only on the island of Komodo; it is the largest living lizard and may grow to 10 feet (3 meters). { kə'mō·dō 'drag·ən }

Koonungidae [INV ZOO] A family of Australian crustaceans in the order Anaspidacea with sessile eyes and the first thoracic limb modified for digging. { kü'nən·jə,dē }

Koplik's sign [MED] Small red spots surrounded by white areas seen in the mucous membrane of the mouth in the prodromal stage of measles. Also known as Koplik's spots. { 'käp·liks ,sīn }

Koplik's spots *See* Koplik's sign. { 'käp·liks ,späts }

Kozak sequence [CELL MOL] A nucleotide sequence in the 5′ untranslated messenger ribonucleic acid region that allows ribosomes to recognize the initiator codon. { 'kō,zak ,sē·kwəns }

kraurosis [MED] A progressive, sclerosing, shriveling process of the skin, due to glandular atrophy. { krȯ'rō·səs }

Krause's corpuscle [NEUROSCI] One of the spheroid nerve-end organs resembling lamellar corpuscles, but having a more delicate capsule; found especially in the conjunctiva, the mucosa of the tongue, and the external genitalia; they are believed to be cold receptors. Also known as end bulb of Krause. { 'krau̇s·əz ,kȯr·pə·səl }

Krebs cycle [BIOCHEM] A sequence of enzymatic reactions involving oxidation of a two-carbon acetyl unit to carbon dioxide and water to provide energy for storage in the form of high-energy phosphate bonds. Also known as citric acid cycle; tricarboxylic acid cycle. { 'krebz ,sī·kəl }

Krebs-Henseleit cycle [BIOCHEM] A cyclic reaction pathway involving the breakdown of arginine to urea in the presence of arginase. { 'krebz 'hen·sə,līt ,sī·kəl }

Krebspest [INV ZOO] A fatal fungus disease of crayfish caused by *Aphanomyces mystaci.* { 'krebs,pest }

krill [INV ZOO] A name applied to planktonic crustaceans that constitute the diet of many whales, particularly whalebone whales. { kril }

kringle domain *See* kringle region. { 'kriŋ·gəl dō,mān }

kringle region [BIOCHEM] A unique protein structural configuration composed of three disulfide bonds. Also known as kringle domain. { 'kriŋ·gəl ,rē·jən }

Krukenberg's tumor [MED] Bilateral carcinoma of the ovaries; originally described as primary, but now denoting a metastatic form usually of gastric origin. { 'krü·kən,bərgz ,tü·mər }

krummholz [ECOL] Stunted alpine forest vegetation. Also known as elfinwood. { 'krüm ,hōlts }

K selection [ECOL] Selection favoring species that reproduce slowly where a resource is constant but available in limited quantities; population is maintained at or near the carrying capacity (K) of the habitat. { 'kā si,lek·shən }

kudzu [BOT] Any of various perennial vine legumes of the genus *Pueraria* in the order Rosales cultivated principally as a forage crop. { 'ku̇d,zü }

kumquat [BOT] A citrus shrub or tree of the genus *Fortunella* in the order Sapindales grown for its small, flame- to orange-colored edible fruit having three to five locules filled with an acid pulp, and a sweet, pulpy rind. { 'kəm,kwät }

Kupffer cell [HISTOL] One of the fixed macrophages lining the hepatic sinusoids. { 'ku̇p·fər ,sel }

Kurthia [MICROBIO] A genus of gram-positive, aerobic, rod-shaped to coccoid bacteria in the coryneform group; metabolism is respiratory. { 'kər·thē·ə }

Kurtoidei [VERT ZOO] A monogeneric suborder of perciform fishes having a unique ossification that encloses the upper part of the swim bladder, and an occipital hook in the male for holding eggs during brooding. { kər'tȯid·ē,ī }

kurtorachic [MED] Having the lumbar spinal region concave dorsally. { ¦kərd·ə¦rak·ik }

Kusnezovia [MICROBIO] A genus of bacteria of uncertain affiliation; coccoid, nonmotile cells attach to substrate and reproduce by budding. { ,küz·nə'zō·vē·ə }

K virus [VIROL] A group 2 papovavirus affecting rats and mice. { 'kā ¦vī·rəs }

kwashiorkor [MED] A nutritional deficiency disease in infants and young children, mainly in the tropics, caused primarily by a diet low in proteins and rich in carbohydrates. Also known as nutritional dystrophy. { ¦kwä·shē¦ór·kər }

Kyasanur Forest virus [VIROL] A group B arbovirus recognized as an agent that causes hemorrhagic fever. { kī'az·ə·nür ¦fär·əst 'vī·rəs }

kymograph [MED] A device for recording internal body movements by making tracings with a stylus on a revolving smoked drum. { 'kī·mə ˌgraf }

kymography [MED] Recording of the movements of internal organs by a kymograph. { kī'mäg·rə·fē }

kynurenic acid [BIOCHEM] $C_{10}H_7O_3N$ A product of tryptophan metabolism found in the urine of mammals. { ¦kin·yə¦ren·ik 'as·əd }

kynurenine [BIOCHEM] $C_{10}H_{12}O_3N_2$ An intermediate product of tryptophan metabolism occurring in the urine of mammals. { ˌkin·yə'reˌnēn }

kyphos [MED] The anteroposterior hump in the spine occurring in kyphoscoliosis. { 'kīˌfōs }

kyphoscoliosis [MED] Lateral curvature of the spine accompanied by rotation of the vertebrae. { ¦kī·fōˌskō·lē'ō·səs }

kyphosis [MED] Angular curvature of the spine, usually in the thoracic region. Also known as humpback; hunchback. { kī'fō·səs }

L

labeling index [CELL MOL] A measure of the mitotic activity of a cell population, defined as the number of cells in the S phase of the growth cycle divided by the total cells in the population. { 'lā·bəl·iŋ ˌin‚deks }

labellate [BIOL] Having a labellum. { lə'be ˌlāt }

labellum [BOT] The median membrane of the corolla of an orchid often differing in size and morphology from the other two petals. [INV ZOO] **1.** A prolongation of the labrum in certain beetles and true bugs. **2.** In Diptera, either of a pair of sensitive fleshy lobes consisting of the expanded end of the labium. { lə'bel·əm }

labial gland [ANAT] Any of the small, tubular mucous and serous glands underneath the mucous membrane of mammalian lips. [INV ZOO] A salivary gland, or modification thereof, opening at the base of the labium in certain insects. { 'lā·bē·əl ¦gland }

labial palp [INV ZOO] **1.** Either of a pair of fleshy appendages on either side of the mouth of certain bivalve mollusks. **2.** A jointed appendage attached to the labium of certain insects. { 'lā·bē·əl ¦palp }

labial papilla [INV ZOO] Any of the sensory bristles around the mouth of many nematodes; they are jointed projections of the cuticle. { 'lā·bē·əl pə'pil·ə }

labia majora [ANAT] A pair of outer skin folds forming the lateral border of the vulva in the female. { ˌlā·bē·ə mə'jȯr·ə }

labia minora [ANAT] A pair of inner skin folds, at the inner surfaces of the labia majora, surrounding the vulva in the female. { ˌlā·bē·ə mə'nȯr·ə }

Labiatae [BOT] A large family of dicotyledonous plants in the order Lamiales; members are typically aromatic and usually herbaceous or merely shrubby. { ˌlā·bē'ā‚tē }

labiate [ANAT] Having liplike margins that are thick and fleshy. [BIOL] Having lips. [BOT] Having the limb of a tubular calyx or corolla divided into two unequal overlapping parts. { 'lā·bē·ət }

labile factor *See* proaccelerin. { 'lā‚bīl ˌfak·tər }

labium [BIOL] **1.** A liplike structure. **2.** The lower lip, as of a labiate corolla or of an insect. { 'lā·bē·əm }

labium majus [ANAT] Either of the two outerfolds of skin that forms the lateral border of the vulva. { 'lā·bē·əm 'mā·jəs }

labium minus [ANAT] Either of the two inner folds, at the inner surfaces of the labia majora, surrounding the vulva in the female. { 'lā·bē·əm 'mē·nəs }

Laboulbeniales [MYCOL] An order of ascomycetous fungi made up of species that live primarily on the external surfaces of insects. { lə‚bül·ben·ē'ā·lēz }

Labridae [VERT ZOO] The wrasses, a family of perciform fishes in the suborder Percoidei. { 'lab·rə‚dē }

labrum [INV ZOO] **1.** The upper lip of certain arthropods, lying in front of or above the mandibles. **2.** The outer edge of a gastropod shell. { 'lā·brəm }

labyrinth [ANAT] **1.** Any body structure full of intricate cavities and canals. **2.** The inner ear. { 'lab·ə‚rinth }

labyrinthine placenta *See* hemochorial placenta. { ˌlab·ə'rin‚thēn plə'sen·tə }

labyrinthine reflex [PHYSIO] The involuntary response to stimulation of the vestibular apparatus in the inner ear. { ˌlab·ə'rin‚thēn 'rē‚fleks }

labyrinthine syndrome *See* Ménière's syndrome. { ˌlab·ə'rin‚thēn 'sin‚drōm }

labyrinthitis [MED] Inflammation of the labyrinth of the inner ear. { ˌlab·ə·rən'thīd·əs }

Labyrinthulia [INV ZOO] A subclass of the protozoan class Rhizopoda containing mostly marine, ovoid to spindle-shaped, uninucleate organisms that secrete a network of filaments (slime tubes) along which they glide. { ˌlab·ə·rən'thül·ē·ə }

Labyrinthulida [INV ZOO] The single order of the protozoan subclass Labyrinthulia. { ˌlab·ə·rən'thül·ə·də }

laccal [BIOCHEM] $C_{17}H_{31}C_6H_3(OH)_2$ A phenol compound which is found in the sap of lacquer trees, and which can be isolated in crystalline form. { 'la‚kȯl }

laccase [BIOCHEM] Any of a class of plant oxidases which catalyze the oxidation of phenols. { 'la‚kās }

laccate [BIOL] Having a lacquered appearance. { 'la‚kāt }

Lacciferinae [INV ZOO] A subfamily of scale insects in the superfamily Coccoidea in which the male lacks compound eyes, the abdomen is without spiracles in all stages, and the apical abdominal segments of nymphs and females do not form a pygidium. { lak·sə'fer·ə,nē }

lacerate [MED] To inflict a wound by tearing. { 'las·ə,rāt }

lacerated [BIOL] Having a deeply and irregularly incised margin or apex. { 'las·ə,rād·əd }

laceration [MED] A wound made by tearing. { ,las·ə'rā·shən }

Lacertidae [VERT ZOO] A family of reptiles in the suborder Sauria, including all typical lizards, characterized by movable eyelids, a fused lower jaw, homodont dentition, and epidermal scales. { lə'sərd·ə,dē }

Lachnospira [MICROBIO] A genus of weakly gram-positive, anaerobic bacteria of uncertain affiliation; motile, curved rods that ferment glucose and are found in the rumen of bovine animals. { lak'näs·pə·rə }

laciniate [BIOL] **1.** Having a fringed border. **2.** Narrowly and deeply incised to form irregular lobes, which may be pointed. { lə'sin·ē,āt }

lac operon [GEN] Three adjacent linked genes which code for the enzymes that act sequentially in lactose utilization in many bacteria. { 'lak 'äp·ə,rän }

lacquer tree *See* varnish tree. { 'lak·ər ,trē }

lacrimal [ANAT] Pertaining to tears, tear ducts, or tear-secreting organs. { 'lak·rə·məl }

lacrimal apparatus [ANAT] The functional and structural mechanisms for secreting and draining tears; includes the lacrimal gland, lake, puncta, canaliculi, sac, and nasolacrimal duct. { 'lak·rə·məl ,ap·ə'rad·əs }

lacrimal bone [ANAT] A small bone located in the anterior medial wall of the orbit, articulating with the frontal, ethmoid, maxilla, and inferior nasal concha. { 'lak·rə·məl ,bōn }

lacrimal canal *See* nasolacrimal canal. { 'lak·rə·məl kə,nal }

lacrimal canaliculus [ANAT] A small tube lined with stratified squamous epithelium which runs vertically a short distance from the punctum of each eyelid and then turns horizontally in the lacrimal part of the lid margin to the lacrimal sac. Also known as lacrimal duct. { 'lak·rə·məl ,kan·ə'lik·yə·ləs }

lacrimal duct *See* lacrimal canaliculus. { 'lak·rə·məl ,dəkt }

lacrimal gland [ANAT] A compound tubuloalveolar gland that secretes tears. Also known as tear gland. { 'lak·rə·məl ,gland }

lacrimal sac [ANAT] The dilation at the upper end of the nasolacrimal duct within the medial canthus of the eye. Also known as dacryocyst. { 'lak·rə·məl ,sak }

lacrimation [PHYSIO] **1.** Normal secretion of tears. **2.** Excessive secretion of tears, as in weeping. { ,lak·rə'mā·shən }

lactalbumin [BIOCHEM] A simple protein contained in milk which resembles serum albumin and is of high nutritional quality. { ,lak,tal 'byü·mən }

lactase [BIOCHEM] An enzyme that catalyzes the hydrolysis of lactose to dextrose and galactose. { 'lak,tās }

lactase deficiency syndrome [MED] Diarrhea induced by ingestion of a lactose-containing food such as milk, secondary to a congenital or acquired deficiency of lactase. { 'lak,tās də ,fish·ən·sē ,sin,drōm }

lactate [PHYSIO] To secrete milk. { 'lak,tāt }

lactate dehydrogenase [BIOCHEM] A zinc-containing enzyme which catalyzes the oxidation of several α-hydroxy acids to corresponding α-keto acids. { 'lak,tāt dē'hī·drə·jə,nās }

lactation [PHYSIO] Secretion of milk by the mammary glands. { lak'tā·shən }

lacteal [ANAT] One of the intestinal lymphatics that absorb chyle. [PHYSIO] Pertaining to or resembling milk. { 'lak·tē·əl }

lactescent [BIOL] Having a milky appearance. [PHYSIO] Secreting milk or a milklike substance. { lak'tes·ənt }

lactic acid [BIOCHEM] $C_3H_6O_3$ A hygroscopic α-hydroxy acid, occurring in three optically isomeric forms: L form, in blood and muscle tissue as a product of glucose and glycogen metabolism; D form, obtained by fermentation of sucrose; and DL form, a racemic mixture present in foods prepared by bacterial fermentation, and also made synthetically. { 'lak·tik 'as·əd }

lactic dehydrogenase [BIOCHEM] An enzyme that catalyzes the dehydrogenation of L-lactic acid to pyruvic acid. Abbreviated LDH. { 'lak·tik dē'hī·drə·jə,nās }

lactic dehydrogenase virus [VIROL] A virus of the rubella group which infects mice. { 'lak·tik dē'hī·drə·jə,nās ,vī·rəs }

lactiferous duct [BOT] A tubular channel consisting of latex vessels or latex cells; carries the latex produced by the plant. { lak'tif·ə·rəs 'dəkt }

lactin *See* lactose. { 'lak·tən }

lactivorous [ZOO] Feeding on milk. { lak'tiv·ə·rəs }

Lactobacillaceae [MICROBIO] The single family of gram-positive, asporogenous, rod-shaped bacteria; they are saccharoclastic, and produce lactate from carbohydrate metabolism. { ,lak·tō·bə,sil·ē'ās·ē,ē }

Lactobacilleae [MICROBIO] Formerly a tribe of rod-shaped bacteria in the family Lactobacillaceae. { ¦lak·tō·bə'sil·ē,ē }

Lactobacillus [MICROBIO] Lactic acid bacteria, the single genus of the family Lactobacillaceae; found in dairy products, meat products, fruits, beer, wine, and other food products. { ¦lak·tō·bə'sil·əs }

lactoferrin [BIOCHEM] An iron-binding protein found in milk, saliva, tears, and intestinal and respiratory secretions that interferes with the iron metabolism of bacteria; in conjunction with antibodies, it plays an important role in resistance to certain infectious diseases. { ¦lak·tō'fer·ən }

lactoflavin *See* riboflavin. { ¦lak·tō'flav·ən }

lactogenic hormone *See* prolactin. { ¦lak·tə¦jen·ik 'hòr,mōn }

lactoglobulin [BIOCHEM] A crystalline protein fraction of milk, which is soluble in half-saturated ammonium sulfate solution and insoluble in pure water. { ¦lak·tō'gläb·yə·lən }

lactonase [BIOCHEM] The enzyme that catalyzes the hydrolysis of 6-phosphoglucono-Δ-lactone to 6-phosphogluconic acid in the pentose phosphate pathway. { 'lak·tə,nās }

lactophenene *See para*-lactophenetide. { lak'täf·ə,nēn }

***para*-lactophenetide** [PHARM] $C_{11}H_{15}NO_3$ A water-soluble compound, crystallizing from ethyl acetate and hexane solution, and melting at 117–118°C; used in medicine as an analgesic and antipyretic. { ¦par·ə ,lak·tō'fen·ə,tīd }

lactose [BIOCHEM] $C_{12}H_{22}O_{11}$ A disaccharide composed of D-glucose and D-galactose which occurs in milk. Also known as lactin; milk sugar. { 'lak,tōs }

lactosuria [MED] Presence of lactose in the urine. { ,lak·tōs'yu̇r·ē·ə }

lactosyl ceramide [BIOCHEM] A neutral glycosphingolipid that is abundant in leukocyte membranes. { ,lak·tə,sil 'ser·ə,mīd }

lacuna [BIOL] A small space or depression. [HISTOL] A cavity in the matrix of bone or cartilage which is occupied by the cell body. { lə'kü·nə }

lacunar system [INV ZOO] A series of intercommunicating spaces branching from two longitudinal vessels in the hypodermis of many acanthocephalins. { lə'kü·nər ,sis·təm }

Lacydonidae [INV ZOO] A benthic family of pelagic errantian polychaetes. { ,las·ə'dän·ə,dē }

Laemobothridae [INV ZOO] A family of lice in the order Mallophaga including parasites of aquatic birds, especially geese and coots. { ,lē·mə'bäth·rə,dē }

Laennec's cirrhosis *See* portal cirrhosis. { lā'neks sə,rō·səs }

Lagenidiales [MYCOL] An order of aquatic fungi belonging to the class Phycomycetales characterized by a saclike to limited hyphal thallus and zoospores having two flagella. { ,la·jə,nid·ē'ā·lēz }

lageniform [BIOL] Flask-shaped. { lə'jen·ə,fo̊rm }

lagging [CELL MOL] Pertaining to chromosomes that show little or no movement during metaphase and anaphase of meiosis or mitosis. { 'lag·iŋ }

lagging strand [CELL MOL] In deoxyribonucleic acid (DNA) replication, the 3′ to 5′ DNA strand that is discontinuously synthesized as a series of Okazaki fragments in the 5′ to 3′ direction. { 'lag·iŋ ,strand }

Lagomorpha [VERT ZOO] The order of mammals including rabbits, hares, and pikas; differentiated from rodents by two pairs of upper incisors covered by enamel, vertical or transverse jaw motion, three upper and two lower premolars, fused tibia and fibula, and a spiral valve in the cecum. { ,lag·ə'mo̊r·fə }

lag phase [MICROBIO] The period of physiological activity and diminished cell division following the addition of inoculum of bacteria to a new culture medium. { 'lag ,fāz }

Lagriidae [INV ZOO] The long-jointed bark beetles, a family of coleopteran insects in the superfamily Tenebrionoidea. { lə'grī·ə,dē }

Lagynacea [INV ZOO] A superfamily of foraminiferan protozoans in the suborder Allogromiina having a free or attached test that has a membranous to tectinous wall and a single, ovoid, tubular, or irregular chamber. { ,lag·ə'nās·ē·ə }

lake ball [ECOL] A spherical mass of tangled, waterlogged fibers and other filamentous material of living or dead vegetation, produced mechanically along a lake bottom by wave action, and usually impregnated with sand and fine-grained mineral fragments. Also known as burr ball; hair ball. { 'lāk ,bo̊l }

lalopathy [MED] Any disorder of speech or disturbance of language. { la'läp·ə·thē }

laloplegia [MED] Inability to speak, caused by paralysis of the muscles concerned in speech, except those of the tongue. { ,lal·ə'plē·jə }

Lamarckism [EVOL] The theory that organic evolution takes place through the inheritance of modifications caused by the environment, and by the effects of use and disuse of organs. { lə'mär,kiz·əm }

lamb [VERT ZOO] A young sheep. { lam }

lambda bacteriophage [MICROBIO] A temperate phage that infects *Escherichia coli* and then undergoes one of two life cycles: (1) lytic, in which it infects the host cell, replicates, and causes the cell to lyse (burst) as new phage progeny emerge; or (2) lysogenic, in which it infects the bacterial host cell, integrates its deoxyribonucleic acid into the host's genome, and goes into a dormant phase during which it replicates along with the host chromosome until it is induced to undergo lytic growth. { 'lam·də bak¦tir·ē·ə,fāj }

lambda chain [IMMUNOL] A 22-kilodalton protein that is one of the two forms of smaller polypeptide chains (known as light chains) that occur in immunoglobulins. { 'lam·də ,chān }

lamb dysentery [VET MED] A bacterial infection and inflammation of the intestinal tract of lambs caused by *Clostridium perfringens*, chiefly along the English-Scottish border. { 'lam 'dis·ən,ter·ē }

lamb's wool [VERT ZOO] The first fleece taken from a sheep up to 7 months old, having natural tapered fiber tip and spinning qualities superior to those of wool taken from previously shorn sheep. { 'lamz ,wu̇l }

lamella [ANAT] A thin scale or plate. { lə'mel·ə }

lamellar bone [HISTOL] Any bone with a microscopic structure consisting of thin layers or plates. { lə'mel·ər ¦bōn }

lamellar chloroplast [CELL MOL] A type of chloroplast in which the layered structure extends more or less uniformly through the whole chloroplast body. { lə'mel·ər 'klȯr·ə,plast }

Lamellibranchiata [INV ZOO] An equivalent name for Bivalvia. { lə‚mel·ə‚braŋ·kē'äd·ə }

Lamellisabellidae [INV ZOO] A family of marine animals in the order Thecanephria. { lə‚mel·ə·sə'bel·ə‚dē }

Lamiaceae [BOT] An equivalent name for Labiatae. { ‚lā·mē'ās·ē‚ē }

Lamiales [BOT] An order of dicotyledonous plants in the subclass Asteridae marked by its characteristic gynoecium, consisting of usually two biovulate carpels, with each carpel divided between the ovules by a false partition, or with the two halves of the carpel seemingly wholly separate. { ‚lā·mē'ā·lēz }

lamina [ANAT] A thin sheet or layer of tissue; a scalelike structure. [BOT] *See* blade. { 'lam·ə·nə }

lamina cribrosa [ANAT] **1.** The portion of the sclera which is perforated for the passage of the optic nerve. **2.** The fascia covering the saphenous opening in the thigh. **3.** The anterior or posterior perforated space of the brain. **4.** The perforated plates of bone through which pass branches of the cochlear part of the vestibulocochlear nerve. { 'lam·ə·nə krə'brō·sə }

laminal placentation [BOT] Condition in which the ovules occur on the inner surface of the carpels. { 'lam·ən·əl ‚plas·ən'tā·shən }

Laminariales [BOT] An order of brown, large, structurally complicated, often highly differentiated members, commonly called kelps, of the algal class Phaeophyceae; distinctive features include a life history in which microscopic, filamentous, dioecious gametophytes alternate with a massive, parenchymatous sporophyte, and a mature sporophyte typically consisting of a holdfast, stipe, and one or more blades. { ‚lam·ə‚nar·ē'ā·lēz }

Laminariophyceae [BOT] A class of algae belonging to the division Phaeophyta. { ‚lam·i‚nar·ē·ō'fīs·ē‚ē }

lamina terminalis [ANAT] The layer of gray matter in the brain connecting the optic chiasma and the anterior commissure where the latter becomes continuous with the rostral lamina. { 'lam·ə·nə ‚tər·mə'nāl·is }

lamination [MED] An operation in embryotomy in which the skull is cut in slices. { 'lam·ə‚nā·shən }

laminectomy [MED] Surgical removal of the lateral portion of the neural arch from one or more vertebrae. { ‚lam·ə'nek·tə·mē }

lampbrush chromosome [CELL MOL] An exceptionally large chromosome characterized by fine lateral projections which are associated with active ribonucleic acid and protein synthesis. { 'lamp ‚brəsh 'krō·mə‚sōm }

lamprey [VERT ZOO] The common name for all members of the order Petromyzonida. { 'lam·prē }

Lampridiformes [VERT ZOO] An order of teleost fishes characterized by a compressed, often ribbonlike body, fins composed of soft rays, a ductless swim bladder, and protractile maxillae among other distinguishing features. { ‚lam·prid·ə'fōr‚mēz }

Lamprocystis [MICROBIO] A genus of bacteria in the family Chromatiaceae; cells are spherical and motile, have gas vacuoles, and contain bacteriochlorophyll *a* on vesicular photosynthetic membranes. { ‚lam·prə'sis·təs }

Lampropedia [MICROBIO] A genus of gram-negative, obligately anaerobic cocci of uncertain affiliation; cells form pairs, tetrads, or flat squared tablets. { ‚lam·prə'pēd·ē·ə }

Lampyridae [INV ZOO] The firefly beetles, a large cosmopolitan family of coleopteran insects in the superfamily Cantharoidea. { lam'pir·ə‚dē }

lanatoside [BIOCHEM] Any of three natural glycosides from the leaves of *Digitalis lanata*; on hydrolysis with acid, it yields one molecule of D-glucose, three molecules of digitoxose, and one molecule of acetic acid; all three glycosides are cardioactive. { lə'nad·ə‚sīd }

lance [MED] To cut or open, as with a lancet. { lans }

Lancefield groups [MICROBIO] Antigenically determined categories for classification of β-hemolytic streptococci. { 'lans‚fēld ‚grüps }

lancelet [ZOO] The common name for members of the subphylum Cephalochordata. { 'lans·lət }

lanceolate [BIOL] Shaped like the head of a lance. { 'lan·sē·ə‚lāt }

Lanceolidae [INV ZOO] A family of bathypelagic amphipod crustaceans in the suborder Hyperiidea. { ‚lan·sē'äl·ə‚dē }

lancet [MED] A sharp-pointed, double-edged cutting instrument used to make small incisions. { 'lan·sət }

landmark [CELL MOL] Any distinctive feature that can be used to identify a chromosome. { 'lan‚märk }

Landry-Guillain-Barré syndrome [MED] A diffuse motor-neuron paresis, rapid in onset, and usually ascending and symmetrical in distribution, with proximal involvement greater than distal, and motor deficits greater than sensory. Also known as Guillain-Barré syndrome; Landry's paralysis. { lan'drē ¦gē‚yan bə'rā ‚sin‚drōm }

Landry's paralysis *See* Landry-Guillain-Barré syndrome. { lan'drēz pə'ral·ə·səs }

landscape ecology [ECOL] The study of the ecological effects of spatial patterning of ecosystems. { ¦lan‚skāp ē'käl·ə·jē }

Langerhans cell [HISTOL] **1.** A type of cytotrophoblast in the human chorionic vesicle which is thought to secrete chorionic gonadotropin. **2.** A highly branched dendritic cell of the mammalian epidermis showing a lobulated nucleus and a diagnostic organelle resembling a tennis racket. { 'läŋ·ər‚hänz ‚sel }

Langerhansian adenoma *See* islet-cell tumor. { ¦läŋ·ər¦han·zē·ən ‚ad·ən'ō·mə }

Lange's nerve [INV ZOO] One of the paired cords of nervous tissue lying in the wall of the radial perihemal canal of asteroids. { 'läŋ·əz ‚nərv }

Lang's vesicle [INV ZOO] A seminal bursa in many polyclad flatworms. { 'laŋz ¦ves·ə·kəl }

Languriidae [INV ZOO] The lizard beetles, a cosmopolitan family of coleopteran insects in the superfamily Cucujoidea. { ˌlaŋ·gə'rī·əˌdē }

Laniatores [INV ZOO] A suborder of arachnids in the order Phalangida having flattened, often colorful bodies and found chiefly in tropical areas. { ˌlan·ē·ə'tȯr·ēz }

lanosterol [BIOCHEM] $C_{30}H_{50}O$ An unsaturated sterol occurring in wool fat and yeast. { lə'näs·təˌrȯl }

lantern fish [VERT ZOO] The common name for the deep-sea teleost fishes composing the family Myctophidae and distinguished by luminous glands that are widely distributed upon the body surface. { 'lan·tərn ˌfish }

Lanthonotidae [VERT ZOO] A family of lizards (Sauria) belonging to the Anguimorpha line; restricted to North Borneo. { lan·thə'näd·əˌdē }

lanugo [ANAT] A downy covering of hair, especially that seen on the fetus or persisting on the adult body. { lə'nü·gō }

lapachoic acid *See* lapachol. { lə'päch·ə·wik 'as·əd }

lapachol [BIOCHEM] $C_{15}H_{14}O_3$ A yellow crystalline compound obtained from lapacho, a hardwood in Argentina and Paraguay. Also known as lapachoic acid; targusic acid. { lə'päˌchȯl }

laparoscopy [MED] A method of visually examining the peritoneal cavity by means of a long slender endoscope equipped with sheath, obturator, biopsy forceps, a sphygmomanometer bulb and tubing, scissors, and a syringe; the endoscope is introduced into the peritoneal cavity through a small incision in the abdominal wall. Also known as peritoneoscopy. { ˌlap·ə'räs·kə·pē }

laparotomy [MED] A surgical incision through the abdominal wall into the abdominal cavity. { ˌlap·ə'räd·ə·mē }

lapidicolous [ECOL] Living under a stone. { ˌlap·ə'dik·ə·ləs }

lappet [ZOO] A lobe or flaplike projection, such as on the margin of a jellyfish or the wattle of a bird. { 'lap·ət }

larch [BOT] The common name for members of the genus *Larix* of the pine family, having deciduous needles and short, spurlike branches which annually bear a crown of needles. { lärch }

large intestine *See* colon. { 'lärj in'tes·tən }

Largidae [INV ZOO] A family of hemipteran insects in the superfamily Pentatomorpha. { 'lär·jəˌdē }

Laridae [VERT ZOO] A family of birds in the order Charadriiformes composed of the gulls and terns. { 'lar·əˌdē }

Larinae [VERT ZOO] A subfamily of birds in the family Laridae containing the gulls and characterized by a thick, slightly hooked beak, a square tail, and a stout white body, with shades of gray on the back and the upper wing surface. { 'lar·əˌnē }

larva [INV ZOO] An independent, immature, often vermiform stage that develops from the fertilized egg and must usually undergo a series of form and size changes before assuming characteristic features of the parent. { 'lär·va }

Larvacea [INV ZOO] A class of the subphylum Tunicata consisting of minute planktonic animals in which the tail, with dorsal nerve cord and notochord, persists throughout life. { lär'vā·shē·ə }

Larvaevoridae [INV ZOO] The tachina flies, a large family of dipteran insects in the suborder Cyclorrhapha distinguished by a thick covering of bristles on the body; most are parasites of arthropods. { ˌlär·və'vȯr·əˌdē }

larva migrans [INV ZOO] Fly larva, *Hypoderma* or *Gastrophilus*, that produces a creeping eruption in the dermis. [MED] Infestation of the dermis by various burrowing nematode larvae, producing a creeping eruption that may become contaminated with bacteria. { 'lär·və 'mīˌgranz }

larviporous [INV ZOO] Feeding on larva, referring especially to insects. { lär'vip·ə·rəs }

laryngeal pouch [VERT ZOO] A lateral saclike expansion of the cavity of the larynx that is greatly developed in certain monkeys. { lə'rin·jē·əl 'pau̇ch }

laryngectomy [MED] Surgical removal of all or part of the larynx. { ˌlar·ən'jek·tə·mē }

laryngismus stridulus [MED] **1.** Spasmodic croup. **2.** The laryngeal spasm sometimes seen in hypocalcemic states. { ˌlar·ən'jiz·məs 'strij·ə·ləs }

laryngitis [MED] Inflammation of the larynx. { ˌlar·ən'jīd·əs }

laryngology [MED] The science of anatomy, physiology, and diseases of the larynx. { ˌlar·əŋ'gäl·ə·jē }

laryngopharynx [ANAT] The lower portion of the pharynx, lying adjacent to the larynx. Also known as hypopharynx. { lə¦riŋ·gō'far·iŋks }

laryngoscope [MED] A tubular instrument, combining a light system and a telescopic system, used in the visualization of the interior larynx and adaptable for diagnostic, therapeutic, and surgical procedures. { lə'riŋ·gəˌskōp }

laryngospasm [MED] Sudden and uncontrollable closure of the larynx; often seen in anaphylactic reactions. { lə'riŋ·gəˌspaz·əm }

laryngotracheal groove [EMBRYO] A channel in the floor of the pharynx serving as the anlage of the respiratory system. { lə¦riŋ·gō'trā·kē·əl ˌgrüv }

laryngotracheal diphtheria [MED] A type of diphtheria that causes hoarseness, croupy cough, difficulty in breathing, and pallor. { ləˌriŋ·gōˌtrā·kē·əl dif'thir·ē·ə }

laryngotracheitis [MED] Inflammation of the larynx and trachea. { lə¦riŋ·gōˌtrā·kē'īd·əs }

laryngotracheobronchitis [MED] Acute inflammation of the mucosa of the larynx, trachea, and bronchi. { lə¦riŋ·gō¦trā·kē·ōˌbraŋ'kīd· əs }

laryngotracheobronchitis virus *See* croup-associated virus. { lə¦riŋ·gō¦trā·kē·ōˌbraŋ'kīd·əs 'vī·rəs }

larynx [ANAT] The complex of cartilages and related structures at the opening of the trachea into the pharynx in vertebrates; functions in protecting the entrance of the trachea, and in phonation in higher forms. { 'lar,iŋks }

laser photobiology [BIOPHYS] The interaction of laser light with biological molecules, and the applications to biology and medicine. { ¦lā·zər ,fōd·ō·bī'äl·ə·jē }

laser photocoagulator [MED] A laser combined with an ophthalmoscope for directing bursts of coherent light through a human eye to burn selected points on a detached retina; subsequent healing of the burns causes scars that weld the retina back into position. { 'lā·zər ,fōd·ō·kə'wag·yə,lād·ər }

Lasiocampidae [INV ZOO] The tent caterpillars and lappet moths, a family of cosmopolitan (except New Zealand) lepidopteran insects in the suborder Heteroneura. { ,las·ē·ō'kam·pə,dē }

Lassa fever [MED] An acute, highly communicable exotic infection that is endemic in western Africa. Caused by an arenavirus (the Lassa virus), it is characterized by high fever, weakness, headaches, mouth ulcers, hemorrhages under the skin, heart and kidney failure, and a high mortality rate. { 'läs·ə ,fē·vər }

lasso cell *See* adhesive cell. { 'las·ō ,sel }

late gene [GEN] Any gene expressed after the onset of deoxyribonucleic acid replication in viruses or eukaryotes. { 'lāt ,jēn }

latency [MED] The stage of an infectious disease, other than the incubation period, in which there are neither clinical signs nor symptoms. [PHYSIO] The period between the introduction of and the response to a stimulus. { 'lat·ən·sē }

latent bud [BOT] An axillary bud whose development is inhibited, sometimes for many years, due to the influence of apical and other buds. Also known as dormant bud. { 'lāt·ənt 'bəd }

latent period [MED] Any stage of an infectious disease in which there are no clinical signs of symptoms of the infection. [PHYSIO] The period between the introduction of a stimulus and the response to it. [VIROL] The initial period of phage growth after infection during which time virus nucleic acid is manufactured by the host cell. { 'lāt·ənt ¦pir·ē·əd }

latent virus [VIROL] A virus that remains dormant within body cells but can be reactivated by conditions such as reduced host defenses, toxins, or irradiation, to cause disease. { ¦lāt·ənt 'vī·rəs }

latent-virus infection [MED] A chronic, inapparent virus infection in which a virus-host equilibrium is established. [VIROL] A persistent viral infection in which there is little or no demonstrable presence of the virus and disease symptoms for a long time between episodes of recurrent outbreaks. { 'lāt·ənt ¦vī·rəs in,fek·shən }

laterad [ANAT] Toward the lateral aspect. { 'lad·ə,rad }

lateral [ANAT] At, pertaining to, or in the direction of the side; on either side of the medial vertical plane. { 'lad·ə·rəl }

lateral bud [BOT] Any bud that develops on the side of a stem. { 'lad·ə·rəl 'bəd }

lateral hermaphroditism [MED] A form of human hermaphroditism in which there is an ovary on one side and a testis on the other. { 'lad·ə·rəl hər'maf·rə,dīd,iz·əm }

lateralia [INV ZOO] Paired sensory bristles on the lateral aspect of the head of gnathostomulids. { ,lad·ə'rāl·yə }

lateral lemniscus [ANAT] The secondary auditory pathway arising in the cochlear nuclei and terminating in the inferior colliculus and medial geniculate body. { 'lad·ə·rəl lem'nis·kəs }

lateral line [INV ZOO] A longitudinal lateral line along the sides of certain oligochaetes consisting of cell bodies of the layer of circular muscle. [VERT ZOO] A line along the sides of the body of most fishes, often distinguished by differently colored scales, which marks the lateral line organ. { 'lad·ə·rəl 'līn }

lateral-line organ [VERT ZOO] A small, pear-shaped sense organ in the skin of many fishes and amphibians that is sensitive to pressure changes in the surrounding water. { 'lad·ə·rəl ,līn ,ȯr·gən }

lateral-line system [VERT ZOO] The complex of lateral-line end organs and nerves in skin on the sides of many fishes and amphibians. { 'lad·ə·rəl ,līn ,sis·təm }

lateral meristem [BOT] Strips or cylinders of dividing cells located parallel to the long axis of the organ in which they occur; the lateral meristem functions to increase the diameter of the organ. { 'lad·ə·rəl 'mer·ə,stem }

lateral root [BOT] A root branch arising from the main axis. { 'lad·ə·rəl 'rüt }

lateral sclerosis *See* amyotrophic lateral sclerosis; primary lateral sclerosis. { 'lad·ə·rəl sklə'rō·səs }

lateral ventricle [ANAT] The cavity of a cerebral hemisphere; communicates with the third ventricle by way of the interventricular foramen. { 'lad·ə·rəl 'ven·trə·kəl }

latewood [BOT] The portion of the annual ring that is formed after formation of earlywood has ceased. { 'lāt,wu̇d }

latex cell [BIOL] A coenocytic cell of a lactiferous duct in a latex-producing plant. { 'lā,teks ,sel }

latex vessel [BOT] An elongated cell joined end to end with other like cells to form a type of lactiferous duct. { 'lā,teks ,ves·əl }

Lathridiidae [INV ZOO] The minute brown scavenger beetles, a large cosmopolitan family of coleopteran insects in the superfamily Cucujoidea. { ,lath·rə'dī·ə,dē }

lathyrism [MED] Poisoning produced by ingestion of vetch (*Lathyrus*) and characterized by spastic paraplegia and decreased connective-tissue tensile strength. Also known as neurolathyrism. { 'lath·ə,riz·əm }

laticifer [BOT] A latex duct found in the midcortex of certain plants. { lā'tis·ə·fər }

latiferous [BOT] Containing or secreting latex. { lā'tif·ə·rəs }

Latimeridae [VERT ZOO] A family of deep-sea lobefin fishes (Coelacanthiformes) known from a single living species, *Latimeria chalumnae*. { ˌlad·ə'mer·əˌdē }

lattissimus dorsi [ANAT] The widest muscle of the back; a broad, flat muscle of the lower back that adducts and extends the humerus, is used to pull the body upward in climbing, and is an accessory muscle of respiration. { lə'tis·ə·məs 'dȯr·sē }

Lauraceae [BOT] The laurel family of the order Magnoliales distinguished by definite stamens in series of three, a single pistil, and the lack of petals. { lȯ'rās·ēˌē }

Laurales [BOT] An order of dicotyledonous flowering, mostly woody plants of the class Magnoliopsida, division Magnoliophyta; commonly have scattered spherical cells containing volatile oils, leaves usually simple and mostly entire, and flowers often pollinated by beetles. { lȯ'rā·lēz }

Lauratonematidae [INV ZOO] A family of marine nematodes of the superfamily Enoploidea; many females possess a cloaca. { ˌlȯr·ə·tō·nē'mad·əˌdē }

laurel forest *See* temperate rainforest. { 'lȯr·əl 'fär·əst }

Laurence-Moon-Bieldl syndrome [MED] A hereditary endocrine disorder of the pituitary or other hypothalamic structures transmitted as a dominant mutant gene and characterized principally by girdle-type obesity, mental retardation, and hypogenitalism. { 'lȯr·əns 'mün 'bēld·əl ˌsinˌdrōm }

Laurer's canal [INV ZOO] In certain flukes, a canal which passes from the oviduct to the ventral surface of the body. { 'lȯr·ərz kəˌnal }

laurisilva *See* temperate rainforest. { ¦lȯr·ə¦sil·və }

Lauxaniidae [INV ZOO] A family of myodarian cyclorrhaphous dipteran insects in the subsection Acalypteratae; larvae are leaf miners. { lȯk·sə'nī·əˌdē }

LAV *See* lymphadenopathy-associated virus.

lavage [MED] The therapeutic washing out of an organ. { lə'väzh }

law of minimum [BIOL] The law that those essential elements for which the ratio of supply to demand (A/N) reaches a minimum will be the first to be removed from the environment by life processes; it was proposed by J. von Liebig, who recognized phosphorus, nitrogen, and potassium as minimum in the soil; in the ocean the corresponding elements are phosphorus, nitrogen, and silicon. Also known as Liebig's law of minimum. { 'lȯ əv 'min·ə·məm }

law of specificity of bacteria *See* Koch's postulates. { 'lȯ əv ¦spes·ə¦fis·əd·ē əv bak'tir·ē·ə }

laxative [PHARM] An agent that stimulates bowel movement and relieves constipation. { 'lak·səd·iv }

layering [BOT] A propagation method by which root formation is induced on a branch or a shoot attached to the parent stem by covering the part with soil. [ECOL] A stratum of plant forms in a community, such as mosses, shrubs, or trees in a bog area. { 'lā·ə·riŋ }

LC_{50} *See* lethal concentration 50. { ¦el¦sē 'fif·tē }

L chain *See* light chain. { 'el ˌchān }

LD_{50} *See* lethal dose 50.

LDH *See* lactic dehydrogenase.

LDL *See* low-density lipoprotein.

lead encephalopathy [MED] Degeneration of the neurons of the brain accompanied by cerebral edema, due to lead poisoning. { 'led ˌen·sef·ə'läp·ə·thē }

leading strand [CELL MOL] In deoxyribonucleic acid (DNA) replication, the 5′ to 3′ DNA strand that is synthesized with few or no interruptions. { 'lēd·iŋ ˌstrand }

lead palsy *See* lead polyneuropathy. { 'led ¦pȯl·zē }

lead poisoning [MED] Poisoning due to ingestion or absorption of lead over a prolonged period of time; characterized by colic, brain disease, anemia, and inflammation of peripheral nerves. { 'led 'pȯiz·ən·iŋ }

lead polyneuropathy [MED] A distal polyneuropathy, affecting mainly the wrist and hand, seen principally in adults with chronic lead poisoning; characterized by weakness, paresthesias, pain, and glove-and-stocking anesthesia. Also known as lead palsy. { 'led ¦päl·ē·nu̇'räp·ə·thē }

leaf [BOT] A modified aerial appendage which develops from a plant stem at a node, usually contains chlorophyll, and is the principal organ in which photosynthesis and transpiration occur. { lēf }

leaf bud [BOT] A bud that produces a leafy shoot. { 'lēf ˌbəd }

leaf cushion [BOT] The small part of the thickened leaf base that remains after abscission in various conifers, and also in some extinct plants. { 'lēf ˌku̇sh·ən }

leaf fiber [BOT] A long, multiple-celled fiber extracted from the leaves of many plants that is used for cordage, such as sisal for binder, and abaca for manila hemp. { 'lēf ˌfī·bər }

leaf gap [BOT] The place where the vascular bundle of the stem interrupts above a leaf trace as a result of the diversion of vascular tissue from the stem into a leaf, occurring in many vascular plants. { 'lēf ˌgap }

leafhopper [INV ZOO] The common name for members of the homopteran family Cicadellidae. { 'lēfˌhäp·ər }

leaflet [BOT] **1.** A division of a compound leaf. **2.** A small or young foliage leaf. { 'lēf·lət }

leaf miner [INV ZOO] Any of the larvae of various insects which burrow into and eat the parenchyma of leaves. { 'lēf ˌmīn·ər }

leaf-nosed [VERT ZOO] Having a leaflike membrane on the nose, as certain bats. { 'lēf ˌnōzd }

leaf primordium [BOT] An immature leaf that arises as an emergence on the flanks of the apical meristem of the shoot tip. { 'lēf prī'mȯr·dē·əm }

leaf scar [BOT] A mark on a stem, formed by

secretion of suberin and a gumlike substance, showing where a leaf has abscised. { 'lēf ˌskär }

leaf scorch [BOT] Any of several disorders and fungus diseases marked by a burned appearance of the leaves; for example, caused by the fungus *Diplocarpon earliana* in strawberry. { 'lēf ˌskȯrch }

leaf trace [BOT] A section of the vascular bundle that leads from the stele to the base of the leaf. { 'lēf ˌtrās }

leaky [CELL MOL] Pertaining to a protein coded for by a mutant gene that shows subnormal activity. [GEN] Pertaining to a genetic block that is incomplete. { 'lēk·ē }

leaky mutant gene [GEN] An allele with reduced activity relative to that of the normal allele. { 'lēk·ē ¦myüt·ənt 'jēn }

Lecanicephaloidea [INV ZOO] An order of tapeworms of the subclass Cestoda distinguished by having the scolex divided into two portions; all species are intestinal parasites of elasmobranch fishes. { leˌkan·əˌsef·ə'lȯid·ē·ə }

Lecanoraceae [BOT] A temperate and boreal family of lichens in the order Lecanorales characterized by a crustose thallus and a distinct thalloid rim on the apothecia. { le·kə·nō'rās·ēˌē }

Lecanorales [BOT] An order of the Ascolichenes having open, discoid apothecia with a typical hymenium and hypothecium. { le·kə·nō'rā·lēz }

Lecideaceae [BOT] A temperate and boreal family of lichens in the order Lecanorales; members lack a thalloid rim around the apothecia. { ləˌsid·ē'ās·ēˌē }

lecithin [BIOCHEM] Any of a group of phospholipids having the general composition CH_2OR_1·$CHOR_2$·$CH_2OPO_2OHR_3$, in which R_1 and R_2 are fatty acids and R_3 is choline, and with emulsifying, wetting, and antioxidant properties. { 'les·ə·thən }

lecithinase [BIOCHEM] An enzyme that catalyzes the breakdown of a lecithin into its constituents. { 'les·ə·thəˌnās }

lecithinase A [BIOCHEM] An enzyme that catalyzes the removal of only one fatty acid from lecithin, yielding lipolecithin. { 'les·ə·thə ˌnās 'ā }

lecithinase C [BIOCHEM] An enzyme that catalyzes the removal of the nitrogenous base of lecithin to produce the base and a phosphatidic acid. { 'les·ə·thəˌnās 'sē }

lecithinase D [BIOCHEM] An enzyme that catalyzes the removal of the phosphorylated base from lecithins, producing α-β-diglyceride. { 'les·ə·thəˌnās 'dē }

le cri du chat syndrome [MED] A complex of congenital malformations resulting from a deletion in chromosome 4 or 5 and characterized by mental retardation and the production of a catlike cry. { lə ¦krē dyü 'shä ˌsinˌdrōm }

lectin [BIOCHEM] Any of various proteins that agglutinate erythrocytes and other types of cells and also have other properties, including mitogenesis, agglutination of tumor cells, and toxicity toward animals; found widely in plants, predominantly in legumes, and also occurring in bacteria, fish, and invertebrates. { 'lek·tən }

lectotype [SYST] A specimen selected as the type of a species or subspecies if the type was not designated by the author of the classification. { 'lek·təˌtīp }

Lecythidaceae [BOT] The single family of the order Lecythidales. { ¦les·ə·thə'dās·ēˌē }

Lecythidales [BOT] A monofamilial order of dicotyledonous tropical woody plants in the subclass Dilleniidae; distinguished by entire leaves, valvate sepals, separate petals, numerous centrifugal stamens, and a syncarpous, inferior ovary with axile placentation. { ¦les·ə·thə'dā·lēz }

Lederberg technique [MICROBIO] A method for rapid isolation of individual bacterial cells for demonstrating the spontaneous origin of bacterial mutants. { 'lā·dəˌberk tekˌnēk }

Leeaceae [BOT] A family of dicotyledonous plants in the order Rhamnales distinguished by solitary ovules in each locule, simple to compound leaves, a small embryo, and hypogynous flowers. { lē'ās·ēˌē }

leech [INV ZOO] The common name for members of the annelid class Hirudinea. { lēch }

leek [BOT] *Allium porrum*. A biennial herb known only by cultivation; grown for its mildly pungent succulent leaves and thick cylindrical stalk. { lēk }

LE factor [PATH] A substance in body fluids, especially the blood, of patients with systemic lupus erythematosus, and sometimes of those with other diseases. Derived from lupus erythematosus factor. { ¦el¦e ˌfak·tər }

left splicing junction *See* donor splicing site. { 'left ˌsplīs·iŋ ˌjəŋk·shən }

left ventricular thrust *See* apex impulse. { 'left ven¦trik·yə·lər 'thrəst }

leg [ANAT] The lower extremity of a human limb, between the knee and the ankle. [ZOO] An appendage or limb used for support and locomotion. { leg }

legena [VERT ZOO] An appendage of the sacculus containing sensory areas in the inner ear of tetrapods; termed the cochlea in humans. { lə'jē·nə }

Legionella pneumonia *See* Legionnaire's disease. { ˌlē·jəˌnel·ə nə'mō·nyə }

Legionnaire's disease [MED] A type of pneumonia usually caused by infection with the bacterium *Legionella pneumophila* that was first observed at an American Legion convention in Philadelphia, Pennsylvania, in 1976. Symptoms include headache, fever reaching 102–105°F (32–41°C), muscle aches, a generalized feeling of discomfort, cough, shortness of breath, chest pains, and sometimes abdominal pain and diarrhea. Also known as Legionella pneumonia. { ˌlē·jə'nerz diˌzēz }

legume [BOT] A dry, dehiscent fruit derived

from a single simple pistil; common examples are alfalfa, beans, peanuts, and vetch. { lə'gyüm }

Leguminosae [BOT] The legume family of the plant order Rosales characterized by stipulate, compound leaves, 10 or more stamens, and a single carpel; many members harbor symbiotic nitrogen-fixing bacteria in their roots. { lə ˌgyüm·ə'nō·sē }

Leiodidae [INV ZOO] The round carrion beetles, a cosmopolitan family of coleopteran insects in the superfamily Staphylinoidea; commonly found under decaying bark. { lī'äd·əˌdē }

leiomyofibroma [MED] A benign tumor composed of smooth muscle cells and fibrocytes. { ˌlī·oˌmī·ō·fī'brō·mə }

leiomyoma [MED] A benign tumor composed of smooth muscle cells. { ˌlī·ō·mī'ō·mə }

leiomyosarcoma [MED] A malignant tumor composed of anaplastic smooth muscle cells. { ˌlī·ōˌmī·ō·sär'kō·mə }

leiosporous [MYCOL] Having smooth spores. { lī'äs·pə·rəs }

Leishman-Donovan bodies [PATH] Small, oval protozoans lacking flagella and undulating membranes, found within macrophages of the skin, liver, and spleen in leishmanial infections such as kala-azar and mucocutaneous leishmaniasis. { 'lēsh·mən 'dän·ə·vən ˌbäd·ēz }

Leishmania [INV ZOO] A genus of flagellated protozoan parasites that are the etiologic agents of several diseases of humans, such as leishmaniasis. { lēsh'man·ē·ə }

Leishmania donovani [INV ZOO] The protozoan parasite that causes kala-azar. { lēsh'man·ē·ə ¦dan·ō¦vän·ē }

Leishmania infantum [INV ZOO] The protozoan parasite that causes infantile leishmaniasis. { lēsh'man·ē·ə in'fan·təm }

leishmaniasis [MED] Any of several infections caused by *Leishmania* species. { ˌlēsh·mə'nī·ə·səs }

Leitneriales [BOT] A monofamilial order of flowering plants in the subclass Hamamelidae; members are simple-leaved, dioecious shrubs with flowers in catkins, and have a superior, pseudomonomerous ovary with a single ovule. { ˌlīt·nir·ē'ā·lēz }

lek [ZOO] A gathering place for courtship. Also known as arena. { lek }

Lelapiidae [INV ZOO] A family of calcaronean sponges in the order Sycettida characterized by a rigid skeleton composed of tracts or bundles of modified triradiates. { le·lə'pī·əˌdē }

lemma [BOT] Either of the pair of bracts that are borne above the glumes and enclose the flower of a grass spikelet. { 'lem·ə }

lemming [VERT ZOO] The common name for the small burrowing rodents composing the subfamily Microtinae. { 'lem·iŋ }

Lemnaceae [BOT] The duckweeds, a family of monocotyledonous plants in the order Arales; members are small, free-floating, thalloid aquatics with much reduced flowers that form a miniature spadix. { lem'nās·ēˌē }

lemniscus [NEUROSCI] A secondary sensory pathway of the central nervous system, usually terminating in the thalamus. { lem'nis·kəs }

lemon [BOT] *Citrus limon*. A small evergreen tree belonging to the order Sapindales cultivated for its acid citrus fruit which is a modified berry called a hesperidium. { 'lem·ən }

lemur [VERT ZOO] The common name for members of the primate family Lemuridae; characterized by long tails, foxlike faces, and scent glands on the shoulder region and wrists. { 'lē·mər }

Lemuridae [VERT ZOO] A family of prosimian primates of Madagascar belonging to the Lemuroidea; all members are arboreal forest dwellers. { lə'myu̇r·əˌdē }

Lemuroidea [VERT ZOO] A suborder or superfamily of Primates including the lemurs, tarsiers, and lorises, or sometimes simply the lemurs. { ˌlem·ə'ròid·ē·ə }

lengthening reaction [PHYSIO] Sudden inhibition of the stretch reflex when extensor muscles are subjected to an excessive degree of stretching by forceful flexion of a limb. { 'leŋk·thəˌniŋ rēˌak·shən }

lenitic *See* lentic. { lə'nid·ik }

lens [ANAT] A transparent, encapsulated, nearly spherical structure located behind the pupil of vertebrate eyes, and in the complex eyes of many invertebrates, that focuses light rays on the retina. Also known as crystalline lens. { lenz }

lens crystallins [CELL MOL] A diverse family of water-soluble proteins that constitute up to 90% of the proteins found in the eye lens and together play a structural role by orienting themselves to facilitate refraction of light, some of the individual proteins exhibit other distinct functions when expressed elsewhere in the organism. { ¦lenz 'kris·təˌlinz }

lens placode [EMBRYO] The ectodermal anlage of the lens of the eye; its formation is induced by the presence of the underlying optic vesicle. { 'lenz 'plaˌkōd }

lentic [ECOL] Of or pertaining to still waters such as lakes, reservoirs, ponds, and bogs. Also spelled lenitic. { 'len·tik }

lenticel [BOT] A loose-structured opening in the periderm beneath the stomata in the stem of many woody plants that facilitates gas transport. { 'len·tə·səl }

lentiginose [ANAT] Of or pertaining to pigment spots in the skin; freckled. { len'tij·əˌnōs }

lentil [BOT] *Lens esculenta*. A semiviny annual legume having pinnately compound, vetchlike leaves; cultivated for its thin, lens-shaped, edible seed. { 'lent·əl }

leopard [VERT ZOO] *Felis pardus*. A species of wildcat in the family Felidae found in Africa and Asia; the coat is characteristically buff-colored with black spots. { 'lep·ərd }

Leotichidae [INV ZOO] A small Oriental family

of hemipteran insects in the superfamily Leptopodoidea. { ˌlē·ə'tik·əˌdē }

Lepadomorpha [INV ZOO] A suborder of barnacles in the order Thoracica having a peduncle and a capitulum which is usually protected by calcareous plates. { ˌlep·ə·də'mȯr·fə }

leper [MED] A person afflicted with Hansen's disease. { 'lep·ər }

Lepiceridae [INV ZOO] Horn's beetle, a family of Central American coleopteran insects composed of two species. { ˌlep·ə'ser·əˌdē }

Lepidocentroida [INV ZOO] The name applied to a polyphyletic assemblage of echinoids that are now regarded as members of the Echinocystitoida and Echinothurioida. { ˌlep·ə·dōˌsen 'trȯid·ə }

lepidophyllous [BOT] Having scaly leaves. { ¦lep·ə·dō¦fil·əs }

Lepidoptera [INV ZOO] Large order of scaly-winged insects, including the butterflies, skippers, and moths; adults are characterized by two pairs of membranous wings and sucking mouthparts, featuring a prominent, coiled proboscis. { ˌlep·ə'däp·tə·rə }

Lepidorsirenidae [VERT ZOO] A family of slender, obligate air-breathing, eellike fishes in the order Dipteriformes having small thin scales, slender ribbonlike paired fins, and paired ventral lungs. { ˌlep·ə·dō·sə'ren·əˌdē }

Lepidosaphinae [INV ZOO] A family of homopteran insects in the superfamily Coccoidea having dark-colored, noncircular scales. { ˌlep·ə·dō'saf·əˌnē }

Lepidosauria [VERT ZOO] A subclass of reptiles in which the skull structure is characterized by two temporal openings on each side which have reduced bony arcades, and by the lack of an antorbital opening in front of the orbit. { ˌlep·ə·dō'sȯr·ē·ə }

Lepidotrichidae [INV ZOO] A family of wingless insects in the order Thysanura. { ˌlep·ə·dō 'trik·əˌdē }

Lepismatidae [INV ZOO] A family of silverfish in the order Thysanura characterized by small or missing compound eyes. { ˌlep·əz'mad·əˌdē }

Lepisostei [VERT ZOO] An equivalent name for Semionotiformes. { ˌlep·ə'säs·tēˌī }

Lepisosteidae [VERT ZOO] A family of fishes in the order Semionotiformes. { ˌlep·ə·säs'tē·əˌdē }

Lepisosteiformes [VERT ZOO] An equivalent name for Semionotiformes. { ˌlep·əˌsäs·tē·ə'fȯrˌmēz }

Leporidae [VERT ZOO] A family of mammals in the order Lagomorpha including the rabbits and hares. { lə'pȯr·əˌdē }

lepospondylous [VERT ZOO] Having the notochord enclosed by cylindrical vertebrae shaped like an hourglass in longitudinal section. { ˌlep·ə'spänd·əl·əs }

leproma [MED] The cutaneous nodular lesion of leprosy. { le'prō·mə }

lepromatous leprosy [MED] A severely debilitating form of Hansen's disease characterized by the presence of multiple nodular lesions (lepromata) on the skin. { lə¦prä·məd·əs 'lep·rə·sē }

lepromin [IMMUNOL] An emulsion of ground lepromata containing the leprosy bacillus; used for intradermal skin tests in Hansen's disease. { lə'prō·mən }

leprosarium [MED] An institution for the treatment of lepers. { ˌlep·rə'sar·ē·əm }

leprosy *See* Hansen's disease. { 'lep·rə·sē }

Leptaleinae [INV ZOO] A subfamily of the Formicidae including largely arboreal ant forms which inhabit plants in tropical and subtropical regions. { ˌlep·tə'līˌnē }

Leptinidae [INV ZOO] The mammal nest beetles, a small European and North American family of coleopteran insects in the superfamily Staphylinoidea. { lep'tin·əˌdē }

Leptocardii [ZOO] The equivalent name for Cephalochordata. { ˌlep·tə'kärd·ēˌī }

leptocephalous larva [VERT ZOO] The marine larva of the fresh-water European eel *Anguilla vulgaris*. { ˌlep·tə'sef·ə·ləs 'lär·və }

leptocercal [VERT ZOO] Of the tail of a fish, tapering to a long, slender point. { ¦lep·tə¦sər·kəl }

Leptodactylidae [VERT ZOO] A large family of frogs in the suborder Procoela found principally in the American tropics and Australia. { ˌlep·təˌdak'til·əˌdē }

leptodactylous [VERT ZOO] Having slender toes, as certain birds. { ¦lep·tə¦dak·tə·ləs }

Leptodiridae [INV ZOO] The small carrion beetles, a cosmopolitan family of coleopteran insects in the superfamily Staphylinoidea. { ˌlep·tə'dir·əˌdē }

Leptomedusae [INV ZOO] A suborder of hydrozoan cnidarians in the order Hydroida characterized by the presence of a hydrotheca. { ¦lep·tō·mə'dü·sē }

leptomeninges [ANAT] The pia mater and arachnoid considered together. { ¦lep·tō·mə'nin·jēz }

leptomeningitis [MED] Inflammation of the leptomeninges of the brain or the spinal cord. { lep·tōˌmen·ən'jīd·əs }

Leptomitales [MYCOL] An order of aquatic Phycomycetes characterized by a hyphal thallus, or basal rhizoids and terminal hyphae, and zoospores with two flagella. { ˌlep·tō·mī'tā·lēz }

leptophyll [ECOL] A growth-form class of plants having a leaf surface area of 0.04 square inch (25 square millimeters) or less; common in alpine and desert habitats. { 'lep·təˌfil }

Leptopodidae [INV ZOO] A tropical and subtropical family of hemipteran insects in the superfamily Leptopodoidea distinguished by the spiny body and appendages. { ˌlep·tə'päd·əˌdē }

Leptopodoidea [INV ZOO] A superfamily of hemipteran insects in the subdivision Geocorisae. { ˌlep·tə·pə'dȯid·ē·ə }

Leptosomatidae [VERT ZOO] The cuckoo rollers, a family of Madagascan birds in the order

Coraciiformes composed of a single species distinguished by the downy covering on the newly hatched young. { ˌlep·tə·sə'mad·əˌdē }

Leptospira [MICROBIO] A genus of bacteria in the family Spirochaetaceae; thin, helical cells with bent or hooked ends. { ˌlep·tə'spī·rə }

leptospirosis [MED] Infection with spirochetes of the genus *Leptospira*. { ˌlep·tə·spə'rō·səs }

leptospirosis icterohemorrhagia *See* Weil's disease. { ˌlep·tə·spə'rō·səs ¦ik·tə·rōˌhem·ə'raj·ēˌə }

leptosporangium [BOT] A sporangium derived from a single actively dividing cell in a meristem. { ¦lep·tō·spə'ran·jē·əm }

Leptostraca [INV ZOO] A primitive group of crustaceans considered as one of a series of Malacostraca distinguished by an additional abdominal somite that lacks appendages, and a telson bearing two movable articulated prongs. { lep'täs·trə·kə }

Leptostromataceae [MYCOL] A family of fungi of the order Sphaeropsidales; pycnidia are black and shield-shaped, circular or oblong, and slightly asymmetrical; included are some fruit-tree pathogens. { ˌlep·təˌstrō·mə'tās·ēˌē }

leptotene [CELL MOL] The first stage of meiotic prophase, when the chromosomes appear as thin threads having well-defined chromomeres. { 'lep·təˌtēn }

Leptothrix [MICROBIO] A genus of sheathed bacteria; single cells are motile by means of polar or subpolar flagella, and sheaths are encrusted with iron or manganese oxides. { 'lep·tə ˌthriks }

Leptotrichia [MICROBIO] A genus of bacteria in the family Bacteroidaceae; straight or slightly curved rods with pointed or rounded ends arranged in filaments. { ˌlep·tə'trik·ē·ə }

Leptotyphlopidae [VERT ZOO] A family of small, harmless, burrowing circumtropical snakes (Serpentes) in the order Squamata; teeth are present only on the lower jaw and are few in number. { ˌlep·tō·ti'fläp·əˌdē }

Lepus [VERT ZOO] The type genus of the family Leporidae, comprising the typical hares. { 'lē·pəs }

Lernaeidae [INV ZOO] A family of copepod crustaceans in the suborder Caligoida; all are fixed ectoparasites, that is, they penetrate the skin of fresh-water fish. { lər'nē·əˌdē }

Lernaeopodidae [INV ZOO] A family of ectoparasitic crustaceans belonging to the Lernaeopodoida; individuals are attached to the walls of the fishes' gill chambers by modified second maxillae. { ˌlər·nē·ə'päd·əˌdē }

Lernaeopodoida [INV ZOO] The fish maggots, a group of ectoparasitic crustaceans characterized by a modified postembryonic development reduced to two or three stages, a free-swimming larva, and the lack of external signs of physical maturity in adults. { ˌlər·nē·ə'pȯid·ē·ə }

Lesch-Nyhan syndrome [MED] A hereditary disease of male children, transmitted as an X-linked recessive, characterized by hyperuricemia, deficiency of hypoxanthine-guanine phosphoribosyl transferase, mental retardation, spastic cerebral palsy, choreathetosis, and self-mutilating biting. { 'lesh 'nīˌhan ˌsinˌdrōm }

lesion [BIOL] A structural or functional alteration due to injury or disease. [CELL MOL] A damaged site in a gene, chromosome, or protein molecule. { 'lē·zhən }

Leskeineae [BOT] A suborder of mosses in the order Hypnobryales; plants are not complanate, paraphyllia are frequently present, leaves are costate, and alar cells are not generally differentiated. { ˌles'kī·nēˌē }

lespedeza [BOT] Any of various legumes of the genus *Lespedeza* having trifoliate leaves, small purple pea-shaped blossoms, and one seed per pod. { ˌles·pə'dē·zə }

lesser circulation *See* pulmonary circulation. { 'les·ər ˌsər·kyə'lā·shən }

lesser omentum [ANAT] A fold of the peritoneum extending from the lesser curvature of the stomach to the transverse hepatic fissure. { 'les·ər ō'men·təm }

Lestidae [INV ZOO] A family of odonatan insects belonging to the Zygoptera; distinguished by the wings being held in a V position while at rest. { 'les·təˌdē }

Lestoniidae [INV ZOO] A monospecific family of hemipteran insects in the superfamily Pentatomorpha found only in Australia. { ˌles·tə'nī·əˌdē }

lethal concentration 50 [PHYSIO] In a fire, the concentration of a gas or smoke that will kill 50% of the test animals within a given time of exposure. Abbreviated LC_{50}. { ¦leth·əl ˌkäns·ən¦trā·shən 'fif·tē }

lethal dose 50 [PHARM] The dose of a substance which is fatal to 50% of the test animals. Abbreviated LD_{50}. Also known as median lethal dose. { 'lē·thəl ¦dōs 'fif·tē }

lethal equivalent value [GEN] The product of the mean number of deleterious genes carried by each member of a population and the mean probability that each gene will cause premature death when homozygous. { ¦lē·thəl iˌkwiv·ə·lənt 'val·yü }

lethal gene [GEN] A gene mutation that causes premature death in heterozygotes if dominant, and in homozygotes if recessive. Also known as lethal mutation. { 'lē·thəl 'jēn }

lethal mutation *See* lethal gene. { 'lē·thəl myü 'tā·shən }

lethargy [MED] A morbid condition of drowsiness or stupor; mental torpor. { 'leth·ər·jē }

Letinula edodes [MYCOL] The second most widely cultivated mushroom in the world, it is native to Asia and is touted for its medicinal properties (cholesterol reduction and antitumor and immunostimulating activities) and flavorful addition to foods. Also known as the shiitake mushroom. { lə¦tin·yə·lə ē'dōˌdēz }

Letterer-Siwe disease [MED] A fatal disease of infants and young children, of unknown etiology, characterized by hyperplasia of the reticuloendothelial system without lipid storage. { 'led·ə·rər 'zē·və di,zēz }

lettuce [BOT] *Lactuca sativa.* An annual plant of the order Asterales cultivated for its succulent leaves; common varieties are head lettuce, leaf or curled lettuce, romaine lettuce, and iceberg lettuce. { 'led·əs }

Leucaltidae [INV ZOO] A family of calcinean sponges in the order Leucettida having numerous small, interstitial, flagellated chambers. { lü'kal·tə,dē }

Leucascidae [INV ZOO] A family of calcinean sponges in the order Leucettida having a radiate arrangement of flagellated chambers. { lü 'kas·ə,dē }

Leucettida [INV ZOO] An order of calcareous sponges in the subclass Calcinea having a leuconoid structure and a distinct dermal membrane or cortex. { lü'sed·ə·də }

leucine [BIOCHEM] $C_6H_{13}O_2N$ A monocarboxylic essential amino acid obtained by hydrolysis of protein-containing substances such as milk. { 'lü,sēn }

leucine amino peptidase [BIOCHEM] An enzyme that acts on peptides to catalyze the release of terminal amino acids, especially leucine residues, having free α-amino groups. { 'lü,sēn ə¦mē·nō 'pep·tə,dās }

Leucodontineae [BOT] A family of mosses in the order Isobryales with foliated branches, often bearing catkins. { ,lü·kə·dän'tin·ē,ē }

leucon [INV ZOO] A type of sponge having the choanocytes restricted to flagellated chambers inserted between the incurrent and excurrent canals, and a reduced or absent paragastric cavity. { 'lü,kän }

Leuconostoc [MICROBIO] A genus of bacteria in the family Streptococcaceae; spherical or lenticular cells occurring in pairs or chains; ferment glucose with production of levorotatory lactic acid (heterofermentative), ethanol, and carbon dioxide. { ¦lü·kə'näs,täk }

leucophore [HISTOL] A white reflecting chromatophore. { 'lü·kə,fór }

leucoplast [BOT] A nonpigmented plastid; capable of developing into a chromoplast. { 'lü·kə,plast }

Leucosiidae [INV ZOO] The purse crabs, a family of true crabs belonging to the Oxystomata. { ,lü·kə'sī·ə,dē }

leucosin [BIOCHEM] A simple protein of the albumin type found in wheat and other cereals. { 'lü·kə·sən }

Leucosoleniida [INV ZOO] An order of calcareous sponges in the subclass Calcaronea characterized by an asconoid structure and the lack of a true dermal membrane or cortex. { ,lü·kə,sō·lə'nī·ə·də }

Leucospidae [INV ZOO] A small family of hymenopteran insects in the superfamily Chalcidoidea distinguished by a longitudinal fold in the forewings. { lü'käs·pə,dē }

leucosporous [MYCOL] Having white spores. { lü'käs·pə·rəs }

Leucothoidae [INV ZOO] A family of amphipod crustaceans in the suborder Gammaridea including semiparasitic and commensal species. { lü'kä·thói,dē }

Leucothrix [MICROBIO] The type genus of the family Leucotrichaceae; cells do not form sulfur deposits. { 'lü·kə,thriks }

Leucotrichaceae [MICROBIO] A family of bacteria in the order Cytophagales; long, colorless, unbranched filaments having conspicuous crosswalls and containing cylindrical or ovoid cells; filaments attach to substrate. { ,lü·kə·tri'kās·ē,ē }

leucovorin [PHARM] Folinic acid used as a calcium salt to counteract the toxic effects of folic acid antagonists and for treatment of megaloblastic anemias. { lü'käv·ə·rən }

leukemia [MED] Any of several diseases of the hemopoietic system characterized by uncontrolled leukocyte proliferation. Also known as leukocythemia. { lü'kē·mē·ə }

leukemia virus *See* leukovirus. { lü'kē·mē·ə ¦vī·rəs }

leukemic reticuloendotheliosis [MED] A rare, usually chronic disorder characterized by proliferation of hairy cells, probably B lymphocytes, in reticuloendothelial organs and blood. Also known as hairy cell leukemia. { lü¦kēm·ik re,tik·yə·lō,en·dō,thē·lē'ō·səs }

leukemoid [MED] Similar to leukemia, that is, characterized by the presence of immature leukocytes in the blood, but of a different etiology. { 'lü·kə,móid }

leukocidin [BIOCHEM] A toxic substance released by certain bacteria which destroys leukocytes. { ,lü·kə'sīd·ən }

leukocyte [HISTOL] A colorless, ameboid blood cell having a nucleus and granular or nongranular cytoplasm. Also known as white blood cell; white corpuscle. { 'lü·kə,sīt }

leukocythemia *See* leukemia. { ,lü·kə,sī'thē·mē·ə }

leukocytolysin [BIOCHEM] A leukocyte-disintegrating lysin. { ,lü·kə,sī'täl·ə·sən }

leukocytopenia *See* leukopenia. { ,lü·kə,sīd·ə'pē·nē·ə }

leukocytopoiesis [PHYSIO] Formation of leukocytes. { ,lü·kə,sīd·ə,pói'ē·səs }

leukocytosis [MED] Elevation of the leukocyte count to values above the normal limit. { ,lü·kə,sī'tō·səs }

leukoderma [MED] Defective skin pigmentation, especially the congenital absence of pigment in patches or bands. { ,lü·kə'dər·mə }

leukodystrophy [MED] A condition thought to result from an inborn error of metabolism and characterized by progressive degeneration of the white matter of the cerebrum, or by defective buildup of myelin. { ,lü·kə'dis·trə·fē }

leukoencephalitis [MED] Inflammation of the white matter of the cerebrum. { ¦lü·kō,in,sef·ə'līd·əs }

leukoerythroblastic anemia *See* myelophthisic anemia. { ,lü·kō·ə,rith·rə'blas·tik ə'nē·mē·ə }

leukoerythroblastosis *See* myelophthisic anemia. { ¦lü·kō·ə,rith·rə,bla'stō·səs }

leukolymphosarcoma *See* leukosarcoma. { ¦lü·kō,lim·fō·sär'kō·mə }

leukoma [MED] A large and dense opacity of the cornea as a result of an ulcer, wound, or inflammation, which presents an appearance of ground glass. { lü'kō·mə }

leukonychia [MED] Whitish discoloration or spotting of the fingernails. { ,lü·kō'nik·ē·ə }

leukopenia [MED] A reduction in the leukocyte count to values below the normal limit. Also known as leukocytopenia. { ,lü·kō'pē·nē·ə }

leukophoresis [MED] The selective removal of large quantities of white blood cells from a donor's blood while returning other cells to the donor. { ¦lü·kō·fə'rē·səs }

leukoplakia [MED] Formation of thickened white patches on mucous membranes, particularly of the mouth and vulva. { ,lü·kō'plā·kē·ə }

leukorrhea [MED] A whitish, mucopurulent discharge from the female genital canal. { ,lü·kə'rē·ə }

leukosarcoma [MED] Lymphosarcoma accompanied by a small number of anaplastic lymphoid cells in the peripheral blood. Also known as leukolymphosarcoma; sarcoleukemia. { ¦lü·kə·sär'kō·mə }

leukosis [MED] An excess of white blood cells. { lü'kō·səs }

leukotomy *See* lobotomy. { lü'käd·ə·mē }

leukotriene [BIOCHEM] Any of a family of oxidized metabolites of certain polyunsaturated fatty acids, predominantly arachidonic acid, that mediate responses in allergic reactions and inflammations, produced in specific cells upon stimulation. { ,lü·kō'trī,ēn }

leukovirus [VIROL] A major group of animal viruses including those causing leukemia in birds, mice, and rats. Also known as leukemia virus. { ¦lü·kō¦vī·rəs }

leurocristine *See* vincristine. { ,lu̇·rə'kris,tēn }

levan [BIOCHEM] $(C_6H_{10}O_5)_n$ A polysaccharide consisting of repeating units of D-fructose and produced by a range of microorganisms, such as *Bacillus mesentericus*. { 'le,van }

levator [MED] An instrument used for raising a depressed portion of the skull. [PHYSIO] Any muscle that raises or elevates a part. { lə'vād·ər }

level above threshold [PHYSIO] **1.** Also known as sensation level. **2.** The pressure level of a sound in decibels above its threshold of audibility for the individual observer.In general, the level of any psychophysical stimulus, such as light, above its threshold of perception. { 'lev·əl ə,bəv 'thresh,hōld }

leveret [VERT ZOO] A young hare. { 'lev·rət }

levigate *See* glabrous. { 'lev·ə,gāt }

Leviviridae [VIROL] A family of single-stranded ribonucleic acid phages (including MS2 virus) characterized by icosahedral particles. { ,lev·ē'vir·ə,dī }

Levivirus [VIROL] The only genus of the family Leviviridae. { 'lev·i,vī·rəs }

levocardia [MED] Location of the heart on the left side associated with visceral situs inversus and congenital cardiac disease. { ¦lēv·ə¦kär·dē·ə }

levodopa [PHARM] $C_9H_{11}NO_4$ Crystals or crystalline powder, soluble in dilute hydrochloric acid and in formic acid; used as an anticholinergic drug and in the treatment of Parkinson's disease. Abbreviated L-dopa. { ¦lev·ə¦dō·pə }

levotropic cleavage [EMBRYO] Spiral cleavage with the cells displaced counterclockwise. { ¦lē·və¦träp·ik 'klēv·ij }

levulose [BIOCHEM] Levorotatory D-fructose. { 'lev·yə,lōs }

levulose tolerance test [PATH] A liver function test based on the observation that blood sugar increases in cases of hepatic disease following oral administration of levulose. { 'lev·yə,lōs 'täl·ə·rəns ,test }

Lewis blood group system [IMMUNOL] An antigen, designated by Le[a], first recognized in a Mrs. Lewis, occurring in about 22% of the population, detected by anti-Le[a] antibodies; primarily composed of soluble antigens of serum and body fluids like saliva, with secondary absorption by erythrocytes. { 'lü·əs 'bləd ,grüp ,sis·təm }

Leydig cell [HISTOL] One of the interstitial cells of the testes; thought to produce androgen. { 'lī·dig ,sel }

Leydig-cell tumor *See* interstitial-cell tumor. { 'lī·dig ,sel ,tü·mər }

L form [MICROBIO] A variant form of bacterial cells that has lost its cell wall because of the action of penicillin. { 'el ,fȯrm }

LGV *See* lymphogranuloma venereum.

LH *See* luteinizing hormone.

LH-RH *See* luteinizing-hormone releasing hormone.

liana [BOT] A woody or herbaceous climbing plant with roots in the ground. { lē'än·ə }

Libellulidae [INV ZOO] A large family of odonatan insects belonging to the Anisoptera and distinguished by a notch on the posterior margin of the eyes and a foot-shaped anal loop in the hindwing. { ,lī·bə'lü·lə,dē }

Libman-Sacks endocarditis [MED] Inflammation of, accompanied by the presence of watery vegetations on, the endocardium; complicates systemic lupus erythematosus. { 'lib·mən 'saks ,en·dō,kär'dīd·əs }

libriform [BOT] Elongated or thick-walled. { 'lib·rə,fȯrm }

Libytheidae [INV ZOO] The snout butterflies, a family of cosmopolitan lepidopteran insects in the suborder Heteroneura distinguished by long labial palps; represented in North America by a single species. { ,lib·ə'thē·ə,dē }

Liceaceae [MYCOL] A family of plasmodial slime molds in the order Liceales. { ,lī·sē'ās·ē,ē }

Liceales [MYCOL] An order of plasmodial slime molds in the subclass Myxogastromycetidae. { ,lī·sē'ā·lēz }

lichen [BOT] The common name for members of the Lichenes. { 'lī·kən }

Lichenes [BOT] A group of organisms consisting of fungi and algae growing together symbiotically. { lī'kē·nēz }

Lichenes Imperfecti [BOT] A class of the Lichenes containing species with no known method of sexual reproduction. { lī'kē·nēz ˌim·pər'fek·tī }

lichenification [MED] The process whereby the skin becomes leathery and hardened; often the result of chronic pruritis and the irritation produced by scratching or rubbing eruptions. { līˌken·ə·fə'kā·shən }

lichenology [BOT] The study of lichens. { ˌlī·kə'näl·ə·jē }

lichenophagous [ZOO] Feeding on lichens. { ˌlī·kə'näf·ə·gəs }

lichen planus [MED] A dermatologic disease of unknown etiology that also occurs in the mouth, on the tongue, or on the lips as smooth lacy networks of white lines or, less commonly, as white patches that may become ulcerative. { ˌlīk·ən 'plan·əs }

Lichnophorina [INV ZOO] A suborder of ciliophoran protozoans belonging to the Heterotrichida. { ˌlik·nə'fór·ə·nə }

licorice [BOT] *Glycyrrhiza glabra*. A perennial herb of the legume family (Leguminosae) cultivated for its roots, which when dried provide a product used as a flavoring in medicine, candy, and tobacco and in the manufacture of shoe polish. { 'lik·rəs }

Liebig's law of the minimum *See* law of minimum. { 'lē·bigz ¦lò əv t͟hə 'min·ə·məm }

Lieskeela [MICROBIO] A genus of sheathed bacteria; cells are nonmotile, and sheaths are not attached and may be encrusted with iron and manganese oxides. { ¦lēs·kē¦el·ə }

life cycle [BIOL] The functional and morphological stages through which an organism passes between two successive primary stages. { 'līf ˌsī·kəl }

life expectancy [BIOL] The expected number of years that an organism will live based on statistical probability. { 'līf ik'spek·tən·sē }

life form [ECOL] The form characteristically taken by a plant at maturity. { 'līf ˌfórm }

life zone [ECOL] A portion of the earth's land area having a generally uniform climate and soil, and a biota showing a high degree of uniformity in species composition and adaptation. { 'līf ˌzōn }

lifting torque [BIOPHYS] A measure of the stress arising from the performance of a task that requires lifting; it is the product of the weight of the load and the load's distance from a point within the vertebral column that serves as a fulcrum. { 'lift·iŋ ˌtórk }

ligament [HISTOL] A flexible, dense white fibrous connective tissue joining, and sometimes encapsulating, the articular surfaces of bones. { 'lig·ə·mənt }

ligamentum nuchae *See* nuchal ligament. { ˌlig·ə'men·təm 'nüˌkē }

ligase [BIOCHEM] An enzyme that catalyzes the union of two molecules, involving the participation of a nucleoside triphosphate which is converted to a nucleoside diphosphate or monophosphate. Also known as synthetase. { 'lī ˌgās }

ligation [CELL MOL] **1.** The process of joining two adjacent nitrogenous bases separated by a nick in one strand of a deoxyribonucleic acid duplex or of two linear nucleic acid molecules. **2.** Formation of a phosphodiester bond to join adjacent nucleotides in the same nucleic acid chain. [MED] Surgical tying of vessels or ducts with a ligature. { lī'gā·shən }

ligature [MED] A cord or thread used for tying vessels and ducts. { 'lig·ə·chər }

light adaptation [PHYSIO] The disappearance of dark adaptation; the chemical processes by which the eyes, after exposure to a dim environment, become accustomed to bright illumination, which initially is perceived as quite intense and uncomfortable. { 'līt ˌadˌap'tā·shən }

light chain [IMMUNOL] The smaller of the two types of chains found in immunoglobulin molecules, molecular weight 23,500. Also known as B chain; L chain. { 'līt 'chān }

light meromyosin [BIOCHEM] The smaller of two fragments obtained from the muscle protein myosin following limited proteolysis by trypsin or chymotrypsin. { 'līt ˌmer·ə'mī·ə·sən }

light reflex [PHYSIO] **1.** The postural orientation response of certain aquatic forms stimulated by the source of light; receptors may be on the ventral or dorsal surface. **2.** The response in which the pupil dilates when light levels are lowered, and constricts when light levels are raised. { 'līt ˌrēˌfleks }

Ligiidae [INV ZOO] A family of primitive terrestrial isopods in the suborder Oniscoidea. { lə'jī·əˌdē }

ligneous [BIOL] Of, pertaining to, or resembling wood. { 'lig·nē·əs }

lignify [BOT] To convert cell wall constituents into wood or woody tissue by chemical and physical changes. { 'lig·nəˌfī }

lignin [BIOCHEM] A substance that together with cellulose forms the woody cell walls of plants and cements them together. { 'lig·nən }

ligninase [BIOCHEM] Any of a group of enzymes that breaks down lignin. { 'lig·nəˌnās }

lignocellulose [BIOCHEM] Any of a group of substances in woody plant cells consisting of cellulose and lignin. { ¦lig·nō'sel·yəˌlōs }

lignosa [BOT] Woody vegetation. { lig'nō·sə }

lignumvitae [BOT] *Guaiacum sanctum*. A medium-sized evergreen tree in the order Sapindales that yields a resin or gum known as gum guaiac or resin of guaiac. Also known as hollywood lignumvitae. { 'lig·nəm'vīd·ē }

ligulate [BOT] **1.** Strap-shaped. **2.** Having ligules. { 'lig·yə·lət }

ligule [BOT] **1.** A small outgrowth in the axis of the leaves in Selaginellales. **2.** A thin outgrowth of a foliage leaf or leaf sheath. [INV ZOO] A small lobe on the parapodium of certain polychaetes. { 'lig·yül }

Liliaceae [BOT] A family of the order Liliales distinguished by six stamens, typically narrow, parallel-veined leaves, and a superior ovary. { ˌlil·ē'ās·ēˌē }

Liliales [BOT] An order of monocotyledonous plants in the subclass Liliidae having the typical characteristics of the subclass. { ˌlil·ē'ā·lēz }

Liliatae *See* Liliopsida. { lə'lī·əˌtē }

Liliidae [BOT] A subclass of the Liliopsida; all plants are syncarpous and have a six-membered perianth, with all members petaloid. { lə'lī·əˌdē }

Liliopsida [BOT] The monocotyledons, making up a class of the Magnoliophyta; characterized generally by a single cotyledon, parallel-veined leaves, and stems and roots lacking a well-defined pith and cortex. { ˌlil·ē'äp·səd·ə }

lily [BOT] **1.** Any of the perennial bulbous herbs with showy unscented flowers constituting the genus *Lilium*. **2.** Any of various other plants having similar flowers. { 'lil·ē }

Limacidae [INV ZOO] A family of gastropod mollusks containing the slugs. { lī'mas·əˌdē }

limb [ANAT] An extremity or appendage used for locomotion or prehension, such as an arm or a leg. [BOT] A large primary tree branch. { limb }

limbate [BIOL] Having a part of one color bordered with a different color. { 'limˌbāt }

limb bud [EMBRYO] A mound-shaped lateral proliferation of the embryonic trunk; the anlage of a limb. { 'lim ˌbəd }

limbic system [ANAT] The inner edge of the cerebral cortex in the medial and temporal regions of the cerebral hemispheres. { 'lim·bik ˌsis·təm }

limb-kinetic apraxia *See* motor apraxia. { 'lim kə¦ned·ik ā'prak·sē·ə }

limbus [BIOL] A border clearly defined by its color or structure, as the margin of a bivalve shell or of the cornea of the eye. { 'lim·bəs }

lime [BOT] *Citrus aurantifolia*. A tropical tree with elliptic oblong leaves cultivated for its acid citrus fruit which is a hesperidium. { līm }

lime water [PHARM] An alkaline aqueous solution of calcium hydroxide; used in medicine as an antacid. { 'līm ˌwȯd·ər }

limicolous [ECOL] Living in mud. { lī'mik·ə·ləs }

liminal length [PHYSIO] The amount of tissue that must be raised above threshold of excitability such that an action potential will propagate. { ˌlim·ən·əl 'leŋkth }

limited chromosome [CELL MOL] A chromosome that occurs only in germ cell nuclei. { ¦lim·əd·əd ¦krō·məˌsōm }

limivorous [ZOO] Feeding on mud, as certain annelids, for the organic matter it contains. { li'miv·ə·rəs }

Limnebiidae [INV ZOO] The minute moss beetles, a family of coleopteran insects in the superfamily Hydrophiloidea. { ˌlim·nə'bī·əˌdē }

limnetic [ECOL] Of, pertaining to, or inhabiting the pelagic region of a body of fresh water. { lim'ned·ik }

Limnichidae [INV ZOO] The minute false water beetles, a cosmopolitan family of coleopteran insects in the superfamily Dryopoidea. { lim 'nik·əˌdē }

Limnocharitaceae [BOT] A family of monocotyledonous plants in the order Alismatales characterized by schizogenous secretory canals, multiaperturate pollen, several or many ovules, and a horseshoe-shaped embryo. { ¦lim·nōˌkar·ə'tās·ēˌē }

limnology [ECOL] The science of the life and conditions for life in lakes, ponds, and streams. { lim'näl·ə·jē }

Limnomedusae [INV ZOO] A suborder of hydrozoan cnidarians in the order Hydroida characterized by naked hydroids. { ¦lim·nō·mə'dü·sē }

limnoplankton [BIOL] Plankton found in fresh water, especially in lakes. { ¦lim·nō'plaŋk·tən }

Limnoriidae [INV ZOO] The gribbles, a family of isopod crustaceans in the suborder Flabellifera that burrow into submerged marine timbers. { ˌlim·nə'rī·əˌdē }

limoniform [BOT] Lemon-shaped. { li'män·ə ˌfȯrm }

limpet [INV ZOO] Any of several species of marine gastropod mollusks composing the families Patellidae and Acmaeidae which have a conical and tentlike shell with ridges extending from the apex to the border. { 'lim·pət }

Limulacea [INV ZOO] A group of horseshoe crabs belonging to the Limulida. { ˌlim·yə'lās·ē·ə }

Limulida [INV ZOO] A subgroup of Xiphosurida including all living members of the subclass. { lə'myül·ə·də }

Limulodidae [INV ZOO] The horseshoe crab beetles, a family of coleopteran insects in the superfamily Staphylinoidea. { ˌlim·yə'läd·əˌdē }

Limulus [INV ZOO] The horseshoe crab; the type genus of the Limulacea. { 'lim·yə·ləs }

Linaceae [BOT] A family of herbaceous or shrubby dicotyledonous plants in the order Linales characterized by mostly capsular fruit, stipulate leaves, and exappendiculate petals. { lī'nās·ēˌē }

Linales [BOT] An order of dicotyledonous plants in the subclass Orsidae containing simple-leaved herbs or woody plants with hypogynous, regular, syncarpous flowers having five to many stamens which are connate at the base. { lī'nā·lēz }

lincomycin [MICROBIO] $C_{18}H_{34}O_6N_2S{\cdot}HCl$ A monobasic crystalline antibiotic, produced by *Streptomyces lincolnensis*, that is active as lincomycin hydrochloride mainly toward gram-positive microorganisms. { ˌliŋ·kə'mīs·ən }

linden *See* basswood. { 'lin·dən }

line [BOT] A unit of length, equal to 1/12 inch, or approximately 2.117 millimeters; it is most frequently used by botanists in describing the size of plants. { līn }

LINE *See* long interspersed nucleotide element.

linea alba [ANAT] A tendinous ridge extending in the median line of the abdomen from the pubis to the tiphoid process and formed by the blending of aponeuroses of the oblique and transverse muscles of the abdomen. { 'lin·ē·ə 'al·bə }

lineage [GEN] Individual descended from a common progenitor. { 'lin·ē·ij }

linearization [CELL MOL] Conversion of a circular deoxyribonucleic acid molecule into a linear molecule. { ˌlin·ē·ər·ə'zā·shən }

Lineidae [INV ZOO] A family of the Heteronemertini. { li'nē·əˌdē }

linellae [INV ZOO] Thin organic threads in the tests of some xenophyophores. { lə'nel·ē }

Lineolaceae [MICROBIO] A family of bacteria in some systems of classification that includes coenocytic members (*Lineola*) of the Caryophanales. { ˌlin·ē·ə'lās·ēˌē }

lineolate [BIOL] Marked with fine lines. { 'lin·ē·əˌlāt }

LINES *See* long interspersed elements. { līnz }

Lineweaver-Burk equation [BIOCHEM] A double reciprocal form of the Michaelis-Menten equation, written as $1/V_0 = K_m/V_{max}[S] + 1/V_{max}$, that yields a straight-line plot for reactions obeying basic Michaelis-Menten kinetics. It provides a more accurate estimate of V_{max} and is helpful in analyzing enzyme inhibition. { ¦līnˌwēv·ər 'bərk iˌkwā·zhən }

lingual artery [ANAT] An artery originating in the external carotid and supplying the tongue. { 'liŋ·gwəl 'ärd·ə·rē }

lingual gland [ANAT] A serous, mucous, or mucoserous gland lying deep in the mucous membrane of the mammalian tongue. { 'liŋ·gwəl 'gland }

lingual nerve [NEUROSCI] A branch of the mandibular nerve having somatic sensory components and innervating the mucosa of the floor of the mouth and the anterior two-thirds of the tongue. { 'liŋ·gwəl 'nərv }

lingual tonsil [ANAT] An aggregation of lymphoid tissue composed of 35–100 separate tonsillar units occupying the posterior part of the tongue surface. { 'liŋ·gwəl 'tän·səl }

Linguatulida [INV ZOO] The equivalent name for Pentastomida. { ˌliŋ·gwə'tül·ə·də }

Linguatuloidea [INV ZOO] A suborder of pentastomid arthropods in the order Porocephalida; characterized by an elongate, ventrally flattened, annulate, posteriorly attenuated body, simple hooks on the adult, and binate hooks in the larvae. { liŋˌgwach·ə'lȯid·ē·ə }

lingula [ANAT] A tongue-shaped organ, structure, or part thereof. { 'liŋ·gyə·lə }

Lingulacea [INV ZOO] A superfamily of inarticulate brachiopods in the order Lingulida characterized by an elongate, biconvex calcium phosphate shell, with the majority having a pedicle. { ˌliŋ·gyə'lās·ē·ə }

lingulate [BIOL] Tongue- or strap-shaped. { 'liŋ·gyəˌlāt }

Lingulida [INV ZOO] An order of inarticulate brachiopods represented by two living genera, *Lingula* and *Glottidia*. { liŋ'gyül·ə·də }

ling-zhi *See* Ganoderma lucidum. { ¦liŋ¦tsē }

liniment [PHARM] A heat-generating liquid that is thinner than ointment and is applied to the skin with friction. { 'lin·ə·mənt }

linin net [CELL MOL] The reticulum composed of chromatinic or oxyphilic substances in a cell nucleus. { 'lī·nən ¦net }

linkage [GEN] Failure of nonallelic genes to recombine at random in meiosis as a result of their being located within the same chromosome. { 'liŋ·kij }

linkage disequilibrium [GEN] The occurrence in a population of certain combinations of linked alleles in greater proportion than expected from the allele frequencies at the loci. { ¦liŋ·kij disˌē·kwə'lib·rē·əm }

linkage group [GEN] The set of gene loci that show significantly less than 50% recombination with one or more other genes within the set; given enough mapped genes, it corresponds to one complete chromosome of the complement. The genes located on a single chromosome. { 'liŋ·kij ˌgrüp }

linkage map [GEN] A diagrammatic representation of the linear order and genetic distance between genes in a linkage group. { 'liŋ·kij ˌmap }

linker *See* linker deoxyribonucleic acid. { 'liŋ·kər }

linker deoxyribonucleic acid [CELL MOL] **1.** A short, synthetic deoxyribonucleic acid (DNA) molecule that contains the recognition site for a specific restriction endonuclease. Also known as linker. **2.** A segment of DNA to which lysine-rich histone is bound and which connects the adjacent nucleosomes of a chromosome. { ¦liŋ·kər dēˌäk·sēˌrī·bō·nü¦klē·ik 'as·əd }

linoleic acid [BIOCHEM] $C_{17}H_{31}COOH$ A yellow unsaturated fatty acid, boiling at 229°C (14 mmHg), occurring as a glyceride in drying oils; obtained from linseed, safflower, and tall oils; a principal fatty acid in plants, and considered essential in animal nutrition; used in medicine, feeds, paints, and margarine. Also known as linolic acid; 9,12-octadecadienoic acid. { ¦lin·e¦lē·ik 'as·əd }

linolenate [BIOCHEM] A salt or ester of linolenic acid. { ˌlin·ə'lēˌnāt }

linolenic acid [BIOCHEM] $C_{17}H_{29}COOH$ One of the principal unsaturated fatty acids in plants and an essential fatty acid in animal nutrition; a colorless liquid that boils at 230°C (17 mmHg or 2266 pascals), soluble in many organic solvents; used in medicine and drying oils. Also known as 9,12,15-octadecatrienoic acid. { ¦lin·ə¦lin·ik 'as·əd }

linolic acid *See* linoleic acid. { lə'nō·lik 'as·əd }

linseed *See* flaxseed. { 'linˌsēd }

lion [VERT ZOO] *Felis leo*. A large carnivorous mammal of the family Felidae distinguished by a tawny coat and blackish tufted tail, with a heavy blackish or dark-brown mane in the male. { 'lī·ən }

Liopteridae [INV ZOO] A small family of hymenopteran insects in the superfamily Cynipoidea. { ˌlī·əp'ter·əˌdē }

lip [ANAT] A fleshy fold above and below the

entrance to the mouth of mammals. [MED] The margin of an open wound. { lip }

Liparidae [INV ZOO] The equivalent name for Lymantriidae. { lə'par·ə,dē }

lipase [BIOCHEM] An enzyme that catalyzes the hydrolysis of fats or the breakdown of lipoproteins. { 'lī,pās }

lipemia [MED] The presence of a fine emulsion of fatty substance in the blood. Also known as lipidemia; lipoidemia. { li'pē·mē·ə }

Liphistiidae [INV ZOO] A family of spiders in the suborder Liphistiomorphae in which the abdomen shows evidence of true segmentation by the presence of tergal and sternal plates. { ,lif·ə'stī·ə,dē }

Liphistiomorphae [INV ZOO] A suborder of arachnids in the order Araneida containing families with a primitively segmented abdomen. { lə,fis·tē·ə'mȯr·fē }

lipid [BIOCHEM] One of a class of compounds which contain long-chain aliphatic hydrocarbons and their derivatives, such as fatty acids, alcohols, amines, amino alcohols, and aldehydes; includes waxes, fats, and derived compounds. Also known as lipin; lipoid. { 'lip·əd }

lipid bilayer [CELL MOL] The foundational structure of plasma membranes; it is composed of two layers of phospholipids positioned such that their polar hydrophilic heads face outward and their nonpolar hydrophobic tails are directed inward, blocking entry of water and water-soluble material into the cell. { ¦lip·id 'bī,lā·ər }

lipidemia *See* lipemia. { ,lip·ə'dē·mē·ə }

lipid histiocytosis [MED] **1.** Any collection of histiocytes containing lipids. **2.** *See* Niemann-Pick disease. { 'lip·əd ¦his·tē·ō·sī'tō·səs }

lipid metabolism [BIOCHEM] The physiologic and metabolic processes involved in the assimilation of dietary lipids and the synthesis and degradation of lipids. { 'lip·əd me'tab·ə,liz·əm }

lipid nephrosis [MED] A chronic kidney disease of children associated with thickening of the basement membranes of glomeruli and characterized by edema, presence of protein in the urine, and abnormally high blood levels of albumin and cholesterol. { 'lip·əd ne'frō·səs }

lipidosis [MED] The generalized deposition of fat or fatty substances in reticuloendothelial cells. Also known as lipoidosis. { ,lip·ə'dō·səs }

lipid pneumonia [MED] **1.** Pneumonia resulting from aspiration of oily substances, such as nose drops. **2.** Deposition of lipids in tissues of chronically inflamed lungs. Also known as lipoid pneumonia. { 'lip·əd nu̇'mō·nyə }

lipid proteinosis [MED] A hereditary disorder characterized by extracellular deposits of phospholipid-protein conjugate involving various areas of the body, including the skin and air passages. { 'lip·əd ,prō·dē·ə'nō·səs }

lipid storage disease [MED] Any of various rare diseases characterized by the accumulation of large histiocytes containing lipids throughout reticuloendothelial tissues; examples are Goucher's disease, Niemann-Pick disease, and amaurotic familial idiocy. { 'lip·əd ,stȯr·ij di,zēz }

lipin [BIOCHEM] **1.** A compound lipid, such as a cerebroside. **2.** *See* lipid. { 'lip·ən }

lipoblastoma *See* liposarcoma. { lip·ə·bla'stō·mə }

lipochondrodystrophy *See* Hurler's syndrome. { ,lip·ə¦kän·drō'dis·trə·fē }

lipochrome [BIOCHEM] Any of various fat-soluble pigments, such as carotenoid, occurring in natural fats. Also known as chromolipid. { 'lip·ə,krōm }

lipodystrophy [MED] A disturbance of fat metabolism in which the subcutaneous fat disappears over some regions of the body, but is unaffected in others. { ,lip·ə'dis·trə·fē }

lipofuscin [BIOCHEM] Any of a group of lipid pigments found in cardiac and smooth muscle cells, in macrophages, and in parenchyma and interstitial cells; differential reactions include sudanophilia, Nile blue staining, fatty acid, glycol, and ethylene. { ,lip·ə'fyüs·ən }

lipogranuloma [MED] A small mass of fatty tissue associated with granulomatous inflammation. { ,lip·ə,gran·yə'lō·mə }

lipoic acid [BIOCHEM] $C_8H_{14}O_2S_2$ A compound which participates in the enzymatic oxidative decarboxylation of α-keto acids in a stage between thiamine pyrophosphate and coenzyme A. { lī'pō·ik 'as·əd }

lipoid [BIOCHEM] **1.** A fatlike substance. **2.** *See* lipid. { 'lī,pȯid }

lipoidemia *See* lipemia. { ,lī,pȯi'dē·mē·ə }

lipoidosis *See* lipidosis. { ,lī,pȯi'dō·səs }

lipoid pneumonia *See* lipid pneumonia. { 'lī,pȯid nu̇'mō·nyə }

lipolysis [PHYSIO] The release of fat from adipose tissue. { lə'päl·ə·səs }

lipoma [MED] A benign tumor composed of fat cells. { lī'pō·mə }

lipomatosis [MED] **1.** Multiple lipomas. **2.** Obesity. { lī,pō·mə'tō·səs }

lipomelanotic reticulosis [MED] A form of lymph node hyperplasia characterized by preservation of the architectural structure, inflammatory exudate, and hyperplasia of the reticulum cells which show phagocytosis of hemosiderin, melanin, and occasionally fat. Also known as dermatopathic lymphadenitis. { ,lip·ə,mel·ə'näd·ik ri,tik·yə'lō·səs }

Lipomycetoideae [MICROBIO] A subfamily of oxidative yeasts in the family Saccharomycetaceae characterized by budding cells and a saclike appendage which develops into an ascus. { ,lip·ə,mī·sə'tȯid·ē,ē }

lipomyxoma *See* liposarcoma. { ,lip·ə·mik'sō·mə }

lipophore [HISTOL] A chromatophore which contains lipochrome. { 'lip·ə,fȯr }

lipophosphoglycan [BIOCHEM] A class of glycosyl phosphatidylinositol attached to a large polysaccharide structure that coats the surfaces of many parasitic protozoa, such as *Leishmania donovani*. { ˌlip·ōˌfäs·fə'glī·kən }

lipoplex [MED] A deoxyribonucleic acid-liposome complex used as a gene delivery vehicle in nonviral gene therapy. { 'līp·əˌpleks }

lipopolysaccharide [BIOCHEM] Any of a class of conjugated polysaccharides consisting of a polysaccharide combined with a lipid. { ˌlip·ōˌpäl·ē'sak·əˌrīd }

lipoprotein [BIOCHEM] Any of a class of conjugated proteins consisting of a protein combined with a lipid. { ˌlip·ə'prōˌtēn }

liposarcoma [MED] A sarcoma originating in adipose tissue. Also known as embryonal-cell lipoma; fetal fat-cell lipoma; infiltrating lipoma; lipoblastoma; lipomyxoma; myxolipoma; myxoma lipomatodes. { ˌlip·ə·sär'kō·mə }

liposome [CELL MOL] One of the fatty droplets occurring in the cytoplasm, particularly of an egg. { 'lip·əˌsōm }

lipostat [MED] The set point of body weight. { 'līp·əˌstat }

lipotropic [BIOCHEM] Having an affinity for lipid compounds. [PHARM] Having a preventive or curative effect on the deposition of excessive fat in abnormal sites. { ˌlip·ə'trä·pik }

lipotropic hormone [BIOCHEM] Any hormone having lipolytic activity on adipose tissue. { ˌlip·ə'trä·pik 'hȯrˌmōn }

Lipotyphla [VERT ZOO] A group of insectivoran mammals composed of insectivores which lack an intestinal cecum and in which the stapedial artery is the major blood supply to the brain. { ˌlip·ə'tī·fē·ə }

lipoxidase [BIOCHEM] An enzyme catalyzing the oxidation of the double bonds of an unsaturated fatty acid. { li'päk·səˌdās }

liquor [PHARM] A solution of a medicinal substance in water. { 'lik·ər }

lirella [BOT] A long, narrow apothecium with a medial longitudinal furrow, occurring in certain lichens. { lə'rel·ə }

Liriopeidae [INV ZOO] The phantom craneflies, a family of dipteran insects in the suborder Orthorrhapha distinguished by black and white banded legs. { ˌlir·ē·ə'pē·əˌdē }

Lissamphibia [VERT ZOO] A subclass of Amphibia including all living amphibians; distinguished by pedicellate teeth and an operculum-plectrum complex of the middle ear. { ˌliˌsam 'fib·ē·ə }

Listeria [MICROBIO] A genus of small, gram-positive, motile coccoid rods of uncertain affiliation; found in animal and human feces. { li'stir·ē·ə }

listeriosis [MED] A bacterial disease of humans and some animals caused by *Listeria monocytogenes*; occurs primarily as meningitis or granulomatosis infantiseptica in humans, and takes many forms, such as meningoencephalitis, distemperlike disease, or generalized infection, in animals. { liˌstir·ē'ō·səs }

litchi *See* lychee. { 'līˌchē }

lithemia *See* hyperuricemia. { lə'thē·mē·ə }

lithiasis [MED] The formation of calculi in the body. { lə'thī·ə·səs }

Lithobiomorpha [INV ZOO] An order of chilopods in the subclass Pleurostigmophora; members are anamorphic and have 15 leg-bearing trunk segments, and when eyes are present, they are ocellar. { ˌlith·ōˌbī·ə'mȯrˌfə }

lithocholic acid [BIOCHEM] $C_{24}H_{40}O_3$ A crystalline substance with a melting point of 184–186°C; soluble in hot alcohol; found in ox, human, and rabbit bile. { ¦lith·ə¦kä·lik 'as·əd }

lithocyst [BOT] Epidermal plant cell in which cytoliths are formed. [INV ZOO] One of the minute sacs containing lithites in many invertebrates; thought to function in audition and orientation. { 'lith·əˌsist }

lithocyte [INV ZOO] A special cell in anthomedusae containing a statolith. { 'lith·əˌsīt }

Lithodidae [INV ZOO] The king crabs, a family of anomuran decapods in the superfamily Paguridea distinguished by reduced last pereiopods and by the asymmetrical arrangement of the abdominal plates in the female. { lə'thäd·əˌdē }

lithodomous [ZOO] Burrowing in rock. { lə 'thäd·ə·məs }

lithogenesis [PATH] The process of formation of calculi or stones. { ˌlith·ə'jen·ə·səs }

lithopedion [MED] A retained fetus that has become calcified. { ˌlith·ə'pē·dēˌän }

lithophagous [ZOO] Feeding on stone, as certain mollusks. { lə'thäf·ə·gəs }

lithophyte [ECOL] A plant that grows on rock. { 'lith·əˌfīt }

lithosere [ECOL] A succession of plant communities that originate on rock. { 'lith·əˌsir }

lithostyle [INV ZOO] A static organ in Narcomedusae. Also known as tentaculocyst. { 'lith·əˌstīl }

lithotomy [MED] Surgical removal of a calculus. { lə'thäd·ə·mē }

lithotripsy *See* extracorporeal shock-wave lithotripsy. { 'lith·əˌtrip·sē }

lithuria [MED] A condition marked by excess of uric (lithic) acid or its salts in the urine. { li 'thyu̇r·ē·ə }

little-drop technique [BIOL] A method for isolating single cells in which a drop of a cellular suspension containing a single cell, as determined by microscopic examination, is transferred with a capillary pipet to an appropriate culture medium. { 'lid·əl 'dräp ˌtekˌnēk }

Little's disease [MED] Spastic diplegia of infants which is characterized by spasticity of the lower extremities; involves degenerative and atrophic cerebral changes as well as congenital malformation. { 'lid·əlz diˌzēz }

littoral zone [ECOL] Of or pertaining to the biogeographic zone between the high- and low-water marks. { 'lit·ə·rəl ˌzōn }

Littorinidae [INV ZOO] The periwinkles, a family of marine gastropod mollusks in the order Pectinibranchia distinguished by their spiral, globular shells. { ˌlid·ə'rin·əˌdē }

lituate [BOT] Having a forked member or part

with the ends turned slightly outward, as in certain fungi. { 'lich·ə·wət }

Lituolacea [INV ZOO] A superfamily of benthic marine foraminiferans in the suborder Textulariina having a multilocular, rectilinear, enrolled or uncoiled test with a simple to labyrinthic wall. { ˌlich·ə'lās·ē·ə }

liver [ANAT] A large vascular gland in the body of vertebrates, consisting of a continuous parenchymal mass covered by a capsule; secretes bile, manufactures certain blood proteins and enzymes, and removes toxins from the systemic circulation. { 'liv·ər }

liver failure [MED] Severe functional disability of the liver marked clinically by a variety of signs and symptoms, including jaundice, coma, and abnormal blood levels of such things as ammonia, bilirubin, and alkaline phosphatase. { 'liv·ər ˌfāl·yər }

liver fluke [INV ZOO] Any trematode, especially *Clonorchis sinensis*, that lodges in the biliary passages within the liver. { 'liv·ər ˌflük }

liver phosphorylase [BIOCHEM] An enzyme that catalyzes the breakdown of liver glycogen to glucose-1-phosphate. { 'liv·ər ˌfäs'fȯr·əˌlās }

liverwort [BOT] The common name for members of the Marchantiatae. { 'liv·ərˌwȯrt }

live-virus vaccine [IMMUNOL] A suspension of attenuated live viruses injected to produce immunity. { 'līv ¦vī·rəs vak'sēn }

living fossil [BIOL] A living species belonging to an ancient stock otherwise known only as fossils. { 'liv·iŋ 'fäs·əl }

livor mortis [PATH] The reddish-blue discoloration of the cadaver that occurs in the dependent portions of the body due to gradual gravitational flow of unclotted blood. { 'līˌvȯr 'mȯrd·əs }

lizard [VERT ZOO] Any reptile of the suborder Sauria. { 'liz·ərd }

LK virus [VIROL] A type of equine herpesvirus. { ¦el¦kā 'vī·rəs }

llama [VERT ZOO] Any of three species of South American artiodactyl mammals of the genus *Lama* in the camel family; differs from the camel in being smaller and lacking a hump. { 'yäm·ə }

llano [ECOL] A savannah of Spanish America and the southwestern United States generally having few trees. { 'yä·nō }

lm *See* lumen.

loach [VERT ZOO] The common name for fishes composing the family Cobitidae; most are small and many are eel-shaped. { lōch }

lobar pneumonia [MED] An acute febrile disease involving one or more lobes of the lung, usually following pneumococcal infection. { 'lōˌbär nu̇'mō·nyə }

lobar sclerosis [MED] Neuroglial proliferation accompanied by atrophy of a cerebral lobe leading to mental and neurological deficits; most common in infants and children who have suffered prolonged hypoxia. { 'lōˌbär sklə'rō·səs }

Lobata [INV ZOO] An order of the Ctenophora in which the body is helmet-shaped. { lō'bäd·ə }

lobate [BIOL] Having lobes. [VERT ZOO] Of a fish, having the skin of the fin extend onto the bases of the fin rays. { 'lōˌbāt }

lobe [BIOL] A rounded projection on an organ or body part. { lōb }

lobectomy [MED] Surgical removal of a lobe of an organ, particularly of a lung. { lō'bek·tə·mē }

lobefin fish [VERT ZOO] The common name for members composing the subclass Crossopterygii. { 'lōbˌfin ˌfish }

lobeline [PHARM] $C_{22}H_{27}NO_2$ A crystalline compound isolated from the herb and seeds of Indian tobacco (*Labelia inflata*); melting point is 130–131°C; soluble in hot alcohol, chloroform, and benzene; used in medicine as a respiratory stimulant. { 'lō·bəˌlēn }

loblolly pine [BOT] *Pinus taeda*. A hard yellow pine of the central and southeastern United States having a reddish-brown fissured bark, needles in groups of three, and a full bushy top. { 'läbˌläl·ē 'pīn }

lobopodia [INV ZOO] Broad, thick pseudopodia. { ˌlō·bə'päd·ē·ə }

Lobosia [INV ZOO] A subclass of the protozoan class Rhizopodea generally characterized by lobopodia. { lō'bō·sē·ə }

lobotomy [MED] An operative section of the fibers between the frontal lobes of the brain. Also known as leukotomy; prefrontal lobotomy. { lō'bäd·ə·mē }

lobster [INV ZOO] The common name for several bottom-dwelling decapod crustaceans making up the family Homaridae which are commercially important as a food item. { 'läb·stər }

lobular pneumonia *See* bronchopneumonia. { 'läb·yə·lər nu̇'mō·nyə }

lobule [BIOL] **1.** A small lobe. **2.** A division of a lobe. { 'läb·yül }

local anesthetic [PHARM] A drug which induces loss of sensation only in the region to which it is applied. { 'lō·kəl ˌan·əs'thed·ik }

local immunity [IMMUNOL] Immunity localized in a specific tissue or region of the body. { 'lō·kəl i'myü·nəd·ē }

locellus [BOT] **1.** In some legumes, a secondary compartment of a unilocular ovary that is formed by a false partition. **2.** One of the two cavities of a pollen sac. { lō'sel·əs }

lochia [MED] The discharge from the uterus and vagina during the first few weeks after labor. { 'lō·kē·ə }

lociation [ECOL] One of the subunits of a faciation, distinguished by the relative abundance of a dominant species. { lō·sē'ā·shən }

lockjaw *See* tetanus. { 'läk ˌjȯ }

loco disease [VET MED] Poisoning in livestock resulting from ingestion of selenium-containing plants (loco weed); characterized by atrophy, delirium, convulsions, and stupor, often terminating in death. { 'lō·kō diˌzēz }

locomotor ataxia *See* tabes dorsalis. { ˌlō·kə'mōd·ər ə'tak·sē·ə }

locomotor system [ZOO] Appendages and associated parts, such as muscles, joints, and bones, concerned with motor activities and locomotion of the animal body. { ˌlō·kə'mōd·ər ˌsis·təm }

loco weed [BOT] Any species of *Astragalus* containing selenium taken up from the soil. { 'lō·kō ˌwēd }

loculate [BIOL] Having, or divided into, loculi. { 'läk·yə·lət }

locule [BOT] A small chamber in plant tissue within which specialized structures may develop, such as within an ovary or anther. { 'läˌkyül }

loculicidal [BOT] Pertaining to dehiscence that extends along the dorsal midline of a carpel. { ¦lak·yə¦lis·əd·əl }

Loculoascomycetes [MYCOL] A class in the subdivision Ascomycotina composed of organisms that form a well-developed mycelium which bears the sexual (ascus) and asexual (conidium) states, and that are distinguished from other ascomycetes by their method of ascocarp formation and their ascus structure. { ˌlō·kə·lōˌas·kō·mī'sēdˌēz }

loculus [BIOL] A small cavity or chamber. { 'läk·yə·ləs }

locus [GEN] The fixed position of a gene in a chromosome, occupied by allele. { 'lō·kəs }

locus control region [GEN] A segment of DNA that controls the chromatin structure and thus the potential for replication and transcription of an entire gene cluster, such as the beta-globin cluster in vertebrates. { ¦lō·kəs kən'trōl ˌrē·jən }

locust [BOT] Either of two species of commercially important trees, black locust (*Robinia pseudoacacia*) and honey locust (*Gladitsia triacanthos*), in the family Leguminosae. [INV ZOO] The common name for various migratory grasshoppers of the family Locustidae. { 'lō·kəst }

Locustidae [INV ZOO] A family of insects in the order Orthoptera; antennae are usually less than half the body length, hindlegs are adapted for jumping, and the ovipositor is multipartite. { lō'kəs·təˌdē }

LOD score *See* logarithm of the odds score. { ˌelˌō'dē ˌskȯr }

lodicule [BOT] One of the minute membranous bodies found at the base of the carpel in most flowering grasses; usually occurs in pairs. { 'läd·əˌkyül }

Loeffler's syndrome [MED] Extensive infiltration of the lung by eosinophils, and eosinophilia of the peripheral circulation. Also known as eosinophilic pneumonitis. { 'lef·lərz ˌsinˌdrōm }

Loganiaceae [BOT] A family of mostly woody dicotyledonous plants in the order Gentianales; members lack a latex system and have fully united carpels and axile placentation. { lōˌgān·ē'ā·sēˌē }

logarithmic growth *See* exponential growth. { 'läg·əˌrith·mik 'grōth }

logarithm of the odds score [GEN] A measure of the likelihood that genes are linked, expressed as the logarithm of the odds that an observed data set from the families is due to linkage at a specific map distance rather than to independent assortment on nonlinked genes. Abbreviated LOD score. { ¦läg·əˌrith·əm əv thē 'ädz ˌskȯr }

logistic growth [BIOL] Population growth in which the growth rate decreases with increasing number of individuals until it becomes zero when the population reaches a maximum. { lə'jis·tik 'grōth }

logomania [MED] Logorrhea so excessive as to be a form of a manic state; new words may be invented to keep up the garrulity. { ¦lō·gə'mā·nē·ə }

logoplegia [MED] Loss of ability to articulate, usually due to paralysis of the speech organs. { ˌlō·gə'plē·jē·ə }

logorrhea [MED] Excessive, usually rapid, incoherent, and uncontrollable talkativeness. { ˌlō·gə'rē·ə }

logotype [SYST] The selection or designation of a genotype after the generic name was published. { 'lō·gəˌtīp }

logwood *See* hematoxylon. { 'lägˌwu̇d }

LOH *See* loss of heterozygosity.

loiasis [MED] A filariasis of tropical Africa, caused by the filaria *Loa*, and characterized by diurnal periodicity of microfilariae in the blood, and transient cutaneous swelling caused by migrating adult worms. { ˌlȯi'ā·səs }

loment [BOT] A dry, indehiscent single-celled fruit that is formed from a single superior ovary; splits transversely in numerous segments at maturity. { 'lōˌment }

Lonchaeidae [INV ZOO] A family of minute myodarian cyclorrhaphous dipteran insects in the subsection Acalyptreatae. { läŋ'kē·əˌdē }

long bone [ANAT] A bone in which the length markedly exceeds the width, as the femur or the humerus. { 'lȯŋ ¦bōn }

long-day response [PHYSIO] A photoperiodic response that is evoked by increasing day lengths and decreasing night lengths. { ¦lȯŋ ˌdā ri'späns }

long-day plant [BOT] A plant that flowers in response to a long photoperiod. { 'lȯŋ ¦dā ˌplant }

longicollous [BIOL] Having a long beak or neck. { ¦län·jə¦käl·əs }

Longidorinae [INV ZOO] A subfamily of nematodes belonging to the Dorylaimoidea including economically important plant parasites. { ˌlän·jə'dȯr·əˌnē }

long interspersed elements [GEN] Families of deoxyribonucleic acid sequences, many truncated, that are inserted singly but in large numbers throughout the genome in many mammals, constituting about 15% of the human genome; some full-length copies are transposons. Abbreviated LINES. { ¦lȯŋ ¦in·tərˌspərst 'el·ə·məns }

long interspersed nucleotide element [CELL MOL] In mammalian deoxyribonucleic acid, any of the 5–10-kilobase repeated sequences that are grouped with the nonviral retroposons. Abbreviated LINE. { ¦loŋ ¦in·tərˌspərst 'n ü·klē·əˌtīd ˌel·ə·mənt }

Longipennes [VERT ZOO] An equivalent name for Charadriiformes. { ¦län·jə'pen·ēz }

Long's coefficient [PATH] The number 2.6, multiplied by the last two figures of the urine specific gravity, determined at 25°C, to derive the number of grams of solids per liter of urine. { 'lòŋz ,kō·i'fish·ənt }

long-term potentiation [NEUROSCI] A long-lasting increase in synaptic efficacy, believed to be involved in information storage in the brain. { ¦lòŋ ,tərm pə,ten·chē'ā·shən }

loon [VERT ZOO] The common name for birds composing the family Gaviidae, all of which are fish-eating diving birds. { lün }

loop of Henle [ANAT] The U-shaped portion of a renal tubule formed by a descending and an ascending tubule. Also known as Henle's loop. { 'lüp əv 'hen·lē }

loop pattern [FOREN] A type of fingerprint pattern in which one or more of the ridges enter on either side of the impression, recurve, touch, or pass an imaginary line from the delta to the core and terminate on the entering side. { 'lüp ,pad·ərn }

loose connective tissue [HISTOL] A type of irregularly arranged connective tissue having a relatively large amount of ground substance. { 'lüs kə'nek·tiv 'tish·ü }

Looser-Milkman syndrome *See* Milkman's syndrome. { 'lü·sər 'milk·mən ,sin,drōm }

Lopadorrhynchidae [INV ZOO] A small family of pelagic polychaete annelids belonging to the Errantia. { ,lō·pə·dō'riŋk·ə,dē }

Lophiiformes [VERT ZOO] A modified order of actinopterygian fishes distinguished by the reduction of the first dorsal fin to a few flexible rays, the first of which is on the head and bears a terminal bulb; includes anglerfish and allies. { ,lä·fē·ə'fór,mēz }

lophocercous [VERT ZOO] Having a ridgelike caudal fin that lacks rays. { ¦lä·fə¦sər·kəs }

lophocyte [INV ZOO] A specialized cell of uncertain function beneath the dermal membrane of certain Demospongiae which bears a process terminating in a tuft of fibrils. { 'läf·ə,sīt }

lophodont [VERT ZOO] Having molar teeth whose grinding surfaces have transverse ridges. { 'läf·ə,dänt }

Lophogastrida [INV ZOO] A suborder of free-swimming marine crustaceans in the order Mysidacea characterized by imperfect fusion of the sixth and seventh abdominal somites, seven pairs of gills and brood lamellae, and natatory, biramous pleopods. { ,läf·ə'gas·trə·də }

lophophore [INV ZOO] A food-gathering organ consisting of a fleshy basal ridge or lobe, from which numerous ciliated tentacles arise; found in Bryozoa, Phoronida, and Brachiopoda. { 'läf·ə,fór }

lophotrichous [CELL MOL] Having a polar tuft of flagella. { lə'fä·trə·kəs }

Loranthaceae [BOT] A family of dicotyledonous plants in the order Santalales in which the ovules have no integument and are embedded in a large, central placenta. { 'ló,ran 'thās·ē,ē }

lorate [BIOL] Strap-shaped. { 'lòr,āt }

lordosis [MED] Exaggerated forward curvature of the lumbar spine. { lòr,dō·səs }

L organisms [MICROBIO] Pleomorphic forms of bacteria occurring spontaneously, or favored by agents such as penicillin, which lack cell walls and grow in minute colonies; transition may be reversible under certain conditions. { 'el ,òr·gə,niz·əmz }

lorica [INV ZOO] A hard shell or case in certain invertebrates, as in many rotifers and protozoans; functions as an exoskeleton. { lə'rī·kə }

Loricaridae [VERT ZOO] A family of catfishes in the suborder Siluroidei found in the Andes. { ,lòr·ə·kə'rī·ə,dē }

Loricata [INV ZOO] An equivalent name for Polyplacophora. { ,lòr·ə'käd·ə }

loricate [INV ZOO] Of, pertaining to, or having a lorica. { 'lòr·ə,kāt }

loris [VERT ZOO] Either of two slow-moving, nocturnal, arboreal primates included in the family Lorisidae: the slender loris (*Loris tardigradus*) and slow loris (*Nycticebus coucang*). { 'lòr·əs }

Lorisidae [VERT ZOO] A family of prosimian primates comprising the lorises of Asia and the galagos of Africa. { lə'ris·ə,dē }

loss of heterozygosity [GEN] In a heterozygote, the loss of one of the two alleles at one or more loci in a cell lineage or cancer cell population due to chromosome loss, deletion, or mitotic crossing-over. Abbreviated LOH. { ¦lòs əv ,hed·e·rō·zī'gäs·əd·ē }

lotic [ECOL] Of or pertaining to swiftly moving waters. { 'lōd·ik }

louping ill [VET MED] A virus disease of sheep, similar to encephalomyelitis, transmitted by the tick *Ixodes racinus*. Also known as ovine encephalomyelitis; trembling ill. { 'lüp·iŋ ,il }

louping-ill virus [VIROL] A group B arbovirus that is infectious in sheep, monkeys, mice, horses, and cattle. { 'lüp·iŋ ,il ,vī·rəs }

louse [INV ZOO] The common name for the apterous ectoparasites composing the orders Anoplura and Mallophaga. { laùs }

low-density lipoprotein [BIOCHEM] A lipoprotein containing more lipids than protein that transports cholesterol from the liver to various tissues throughout the body. Abbreviated LDL. { ¦lō ,den·səd·ē ,lī·pō'prō,tēn }

Lower Austral life zone [ECOL] A term used by C.H. Merriam to describe the southern portion of the Austral life zone, characterized by accumulated temperatures of 18,000°F (10,000°C). { 'lō·ər ¦òs·trəl 'līf ,zōn }

lower esophageal ring *See* Schatzki's ring. { ,lō·ər i,saf·ə,jē·əl 'riŋ }

lower motor neuron [NEUROSCI] An efferent neuron which has its body located in the anterior gray column of the spinal cord or in the brainstem nuclei, and its axon passing by way of a peripheral nerve to skeletal muscle. Also known as final common pathway. { 'lō·ər 'mōd·ər ,nü,rän }

lower motor neuron disease [MED] An injury to

any part of a lower motor neuron, characterized by flaccid paralysis of the muscle, diminished or absent reflexes, and progressive atrophy of the muscle. { 'lō·ər ¦mōd·ər ¦nü,rän di,zēz }

lower nephron nephrosis [MED] Retrogressive kidney changes following traumatic injury and other conditions producing shock; sometimes accompanied by distal and collecting tubule necrosis. Also known as acute tubular necrosis; crush kidney; hemoglobinuric nephrosis; ischemic tubulorrhexis. { 'lō·ər 'ne,frän nə'frō·səs }

Lower Sonoran life zone [ECOL] A term used by C.H. Merriam to describe the climate and biotic communities of subtropical deserts and thorn savannas in the southwestern United States. { 'lō·ər sə'nór·ən 'līf ,zōn }

low-zone tolerance [IMMUNOL] Immunologic tolerance induced by repeated administration of very low doses of a protein antigen. { ¦lō ¦zōn 'täl·ə·rəns }

loxodont [VERT ZOO] Having molar teeth with shallow hollows between the ridges. { 'läk·sə,dänt }

loxolophodont [VERT ZOO] Having crests on the molar teeth that connect three of the tubercles and with the fourth or posterior inner tubercle being rudimentary or absent. { ¦läk·sə¦läf·ə,dänt }

Lucanidae [INV ZOO] The stag beetles, a cosmopolitan family of coleopteran insects in the superfamily Scarabaeoidea. { lü'kan·ə,dē }

lucerne *See* alfalfa. { lü'sərn }

Lucibacterium [MICROBIO] A genus of light-emitting bacteria in the family Vibrionaceae; motile, asporogenous rods with peritrichous flagella. { ,lü·si,bak'tir·ē·əm }

luciferase [BIOCHEM] An enzyme that catalyzes the oxidation of luciferin. { lü'sif·ə,rās }

luciferin [BIOCHEM] A species-specific pigment in many luminous organisms that emits heatless light when combined with oxygen. { lü'sif·ə·rən }

Luciocephalidae [VERT ZOO] A family of freshwater fishes in the suborder Anabantoidei. { ,lü·sē·ō·sə'fal·ə,dē }

Ludwig's angina [MED] Acute streptococcal cellulitis of the floor of the mouth. { 'ləd,wigz 'an·jə·nə }

Luidiidae [INV ZOO] A family of echinoderms in the suborder Paxillosina. { lü·ə'dī·ə,dē }

lumbago [MED] Backache in the lumbar or lumbosacral region. { ,ləm'bā,gō }

lumbar artery [ANAT] Any of the four or five pairs of branches of the abdominal aorta opposite the lumbar region of the spine; supplies blood to loin muscles, skin on the sides of the abdomen, and the spinal cord. { 'ləm,bär 'ärd·ə·rē }

lumbar nerve [NEUROSCI] Any of five pairs of nerves arising from lumbar segments of the spinal cord; characterized by motor, visceral sensory, somatic sensory, and sympathetic components; they innervate the skin and deep muscles of the lower back and the lumbar plexus. { 'ləm,bär 'nərv }

lumbar vertebrae [ANAT] Those vertebrae located between the lowest ribs and the pelvic girdle in all vertebrates. { 'ləm,bär 'vərd·ə,brā }

lumbodorsal fascia [ANAT] The sheath of the erector spinae muscle alone, or the sheaths of the erector spinae and the quadratus lumborum muscles. { ¦ləm·bō'dór·səl 'fā·shə }

lumbosacral plexus [NEUROSCI] A network formed by the anterior branches of lumbar, sacral, and coccygeal nerves which for descriptive purposes are divided into the lumbar, sacral, and pudendal plexuses. { ¦ləm·bō'sak·rəl 'plek·səs }

Lumbricidae [INV ZOO] A family of annelid worms in the order Oligochaeta; includes the earthworm. { ləm'bris·ə,dē }

Lumbriclymeninae [INV ZOO] A subfamily of mud-swallowing sedentary worms in the family Maldanidae. { ,ləm·bri·klī'men·ə,nē }

Lumbriculidae [INV ZOO] A family of aquatic annelids in the order Oligochaeta. { ,ləm·bri 'kyül·ə,dē }

Lumbricus [INV ZOO] A genus of earthworms recognized as the type genus of the family Lumbricidae. { 'ləm·brə·kəs }

Lumbrineridae [INV ZOO] A family of errantian polychaetes in the superfamily Eunicea. { ,ləm·bri'ner·ə,dē }

lumen [ANAT] The interior space within a tubular structure, such as within a blood vessel, a duct, or the intestine. { 'lü·mən }

lumpectomy [MED] Surgical removal of a tumor in the breast along with a small amount of surrounding tissue. { ləm'pek·tə·mē }

lumper [SYST] A taxonomist who tends to recognize large taxa. { 'ləm·pər }

lumpy jaw *See* actinomycosis. { 'ləm·pē ,jó }

lunate [BIOL] Crescent-shaped { 'lü,nāt }

lung [ANAT] Either of the paired air-filled sacs, usually in the anterior or anteroventral part of the trunk of most tetrapods, which function as organs of respiration. { ləŋ }

lung bud [EMBRYO] A primary outgrowth of the embryonic trachea; the anlage of a primary bronchus and all its branches. { 'ləŋ ,bəd }

lungfish [VERT ZOO] The common name for members of the Dipnoi; all have lungs that arise from a ventral connection with the gut. { 'ləŋ,fish }

lungworm [INV ZOO] Any of the nematodes that are parasites of terrestrial and marine nematodes, most commonly found in the respiratory tract, characterized by a reduced or absent stoma capsule, and an oral opening surrounded by six well-developed lips. { 'ləŋ,wərm }

lunule [BIOL] A crescent-shaped organ, structure, or mark. { 'lün,yül }

lupine [BOT] A leguminous plant of the genus *Lupinus* with an upright stem, leaves divided into several digitate leaflets, and terminal racemes of pea-shaped blossoms. { 'lü,pən }

lupus erythematosus [MED] An acute or subacute febrile collagen disease characterized by a butterfly-shaped rash over the cheeks and perilingual erythema. { 'lü·pəs ,er·ə,thē·mə'tō·səs }

lupus erythematosus factor *See* LE factor. { 'lü·pəs ,er·ə,thē·mə'tō·səs 'fak·tər }

lupus vulgaris [MED] True tuberculosis of the skin; a slow-developing, scarring, and deforming disease, often asymptomatic, frequently involving the face, and occurring in a wide variety of appearances. { 'lü·pəs vəl'gar·əs }

lutein [BIOCHEM] **1.** A dried, powdered preparation of corpus luteum. **2.** *See* xanthophyll. { 'lüd·ē·ən }

luteinization [PHYSIO] Acquisition of characteristics of cells of the corpus luteum by ovarian follicle cells following ovulation. { ,lüd·ē·ə·nə'zā·shən }

luteinizing hormone [BIOCHEM] A glycoprotein hormone secreted by the adenohypophysis of vertebrates which stimulates hormone production by interstitial cells of gonads. Abbreviated LH. Also known as interstitial-cell-stimulating hormone (ICSH). { 'lüd·ē·ə,nīz·iŋ 'hȯr,mōn }

luteinizing-hormone releasing hormone [BIOCHEM] A small peptide hormone released from the hypothalamus which acts on the pituitary gland to cause release of luteinizing hormone. Abbreviated LH-RH. { 'lüd·ē·ə,nīz·iŋ ¦hȯr,mōn ri'lēs·iŋ ¦hȯr,mōn }

luteoma [MED] A tumor of the ovary composed of cells resembling those of the corpus luteum. { ,lüd·ē'ō·mə }

luteotropic hormone *See* prolactin. { ¦lüd·ē·ə ¦träp·ik 'hȯr,mōn }

Luteoviridae [VIROL] A family of plant viruses containing positive-sense, single-stranded ribonucleic acid, includes the genera *Luteovirus*, *Polerovirus*, and *Enamovirus*. { ,lüd·ē·o'vir·ə,dī }

Luteovirus [VIROL] A genus of plant viruses in the family Luteoviridae that is characterized by icosahedral particles containing one molecule of linear, positive-sense, single-stranded ribonucleic acid, barley yellow dwarf virus is the type species. Also known as barley yellow dwarf virus group. { 'lüd·ē·ō,vī·rəs }

Lutheran blood group [IMMUNOL] The erythrocyte antigens defined by reactions with an antibody designated anti-Lu^a, initially detected in the serum of a multiply transfused patient with lupus erythematosus, who developed antibodies against erythrocytes of a donor named Lutheran, and by anti-Lu^b. { 'lüth·rən 'bləd ,grüp }

Lutjanidae [VERT ZOO] The snappers, a family of perciform fishes in the suborder Percoidei. { lü'chan·ə,dē }

lyase [BIOCHEM] An enzyme that catalyzes the nonhydrolytic cleavage of its substrate with the formation of a double bond; examples are decarboxylases. { 'lī,ās }

Lycaenidae [INV ZOO] A family of heteroneuran lepidopteran insects in the superfamily Papilionoidea including blue, gossamer, hairstreak, copper, and metalmark butterflies. { lī'sēn·ə,dē }

Lycaeninae [INV ZOO] A subfamily of the Lycaenidae distinguished by functional prothoracic legs in the male. { lī'sēn·ə,nē }

lychee [BOT] A tree of the genus *Litchi* in the family Sapindaceae, especially *L. chinensis* which is cultivated for its edible fruit, a one-seeded berry distinguished by the thin, leathery, rough pericarp that is red in most varieties. Also spelled litchi. { 'lī,chē }

lychnisc [INV ZOO] A hexactin in which the central part of the spicule is surrounded by a system of 12 struts. { 'lik·nisk }

Lychniscosa [INV ZOO] An order of sponges in the subclass Hexasterophora in which parenchymal megascleres form a rigid framework and are all or in part lychniscs. { ,lik·ni'skō·sə }

lycopene [BIOCHEM] $C_{40}H_{50}$ A red, crystalline hydrocarbon that is the coloring matter of certain fruits, as tomatoes; it is isomeric with carotene. { 'lī·kə,pēn }

lycoperdonosis [MED] A respiratory disease caused by inhalation of spores from the puffball mushroom, *Lycoperdon*. { ,lī·kō,pər·də'nō·səs }

lycophore larva [INV ZOO] A larva of certain cestodes characterized by cilia, large frontal glands, and 10 hooks. Also known as decanth larva. { 'lī·kə,fȯr ,lär·və }

Lycopodiales [BOT] The type order of Lycopodiopsida. { ,lī·kə·pō·dī'ā·lēz }

Lycopodiatae *See* Lycopodiopsida. { ,lī·kə·pō'dī·ə,tē }

Lycopodineae [BOT] The equivalent name for Lycopodiopsida. { ,lī·kə·pō'din·ē,ē }

Lycopodiophyta [BOT] A division of the subkingdom Embryobionta characterized by a dominant independent sporophyte, dichotomously branching roots and stems, a single vascular bundle, and small, simple, spirally arranged leaves. { ,lī·kə·pō·dī'äf·əd·ə }

Lycopodiopsida [BOT] The lycopods, the type class of Lycophodiophyta. { ,lī·kə·pō·dī'äp·səd·ə }

Lycopsida [BOT] Former subphylum of the Embryophyta now designated as the division Lycopodiophyta. { lī'käp·sə·də }

Lycoriidae [INV ZOO] A family of small, dark-winged dipteran insects in the suborder Orthorrhapha. { ,lī·kə'rī·ə,dē }

Lycosidae [INV ZOO] A family of hunting spiders in the suborder Dipneumonomorphae that actively pursue their prey. { lī'käs·ə,dē }

Lycoteuthidae [INV ZOO] A family of squids. { ,lī·kə'tü·thə,dē }

Lyctidae [INV ZOO] The large-winged beetles, a large cosmopolitan family of coleopteran insects in the superfamily Bostrichoidea. { 'lik·tə,de }

Lygaeidae [INV ZOO] The lygaeid bugs, a large family of phytophagous hemipteran insects in the superfamily Lygaeoidea. { lī'jē·ə,dē }

Lygaeoidea [INV ZOO] A superfamily of pentatomorphan insects having four-segmented antennae and ocelli. { ,lī·jē'ȯid·ē·ə }

Lymantriidae [INV ZOO] The tussock moths, a family of heteroneuran lepidopteran insects in the superfamily Noctuoidea; the antennae of males is broadly pectinate and there is a tuft of hairs on the end of the female abdomen. { ¦lī·mən'trī·ə,dē }

Lyme borreliosis *See* Lyme disease. { ,līm bə,rel·ē'ō·səs }

Lyme disease [MED] A complex multisystem human illness caused by the tick-borne spirochete *Borrelia burgdorferi*. Also known as Lyme borreliosis. { 'līm di,zēz }

Lymexylonidae [INV ZOO] The ship timber beetles composing the single family of the coleopteran superfamily Lymexylonoidea. { lə,mek·sə'län·ə,dē }

Lymexylonoidea [INV ZOO] A monofamilial superfamily of wood-boring coleopteran insects in the suborder Polyphaga characterized by a short neck and serrate antennae. { lə,mek·sə·lə'nȯid·ē·ə }

lymph [HISTOL] The colorless fluid which circulates through the vessels of the lymphatic system. { limf }

lymphadenitis [MED] Inflammation of lymph nodes. { ,lim,fad·ən'īd·əs }

lymphadenoid goiter *See* struma lymphomatosa. { ,lim'fad·ən,ȯid 'gȯid·ər }

lymphadenopathy [MED] Enlargement or disease of lymph nodes. { ,lim,fad·ən'äp·ə·thē }

lymphadenopathy-associated virus [VIROL] A former designation for the human immunodeficiency virus. Abbreviated LAV. { ,lim,fad·ən'äp·ə·thē ə¦sō·shē,ād·əd 'vī·res }

lymphadenosis [MED] Neoplasia or hyperplasia of lymph nodes. { ,lim,fad·ən'ō·səs }

lymphagogue [PHARM] An agent that stimulates lymph flow. { 'lim·fə,gäg }

lymphangiectasis [MED] Dilation in the wall of a lymphatic vessel. { ,lim,fan·jē'ek·tə·səs }

lymphangiectomy [MED] Surgical removal of a pathologic lymphatic channel, as for cancer. { ,lim,fan·jē'ek·tə·mē }

lymphangioendothelial sarcoma *See* lymphangiosarcoma. { ,lim,fan·jē·ō,en·dō'thē·lē·əl sär'kō·mə }

lymphangioendothelioma [MED] A tumor composed of aggregations of lymphatic vessels, between which are large mononuclear cells presumed to be endothelial cells. { ,lim,fan·jē·ō,en·dō,thē·lē'ō·mə }

lymphangiofibroma [MED] A benign tumor composed of lymphangiomatous and fibromatous elements. { ,limi,fan·jē·ō,fī'brō·mə }

lymphangioma [MED] An abnormal mass of lymphatic vessels. { ,lim,fan·jē'ō·mə }

lymphangiosarcoma [MED] A sarcoma whose parenchymal cells form vascular channels resembling lymphatics. Also known as lymphangioendothelial sarcoma. { ,lim,fan·jē·ō·sär'kō·mə }

lymphangitis [MED] Inflammation of lymphatic vessels. { ,lim,fan'jīd·əs }

lymphatic *See* lymph vessel. { lim'fad·ik }

lymphatic system [ANAT] A system of vessels and nodes conveying lymph in the vertebrate body, beginning with capillaries in tissue spaces and eventually forming the thoracic ducts which empty into the subclavian veins. { lim'fad·ik ,sis·təm }

lymphatic tissue [HISTOL] Tissue consisting of networks of lymphocytes and reticular and collagenous fibers. Also known as lymphoid tissue. { lim'fad·ik ,tish·ü }

lymphedema [MED] Edema resulting from lymph vessel obstruction. { ¦lim·fə¦dē·mə }

lymph gland *See* lymph node. { 'limf ,gland }

lymph heart [VERT ZOO] A muscular expansion of a lymphatic vessel which contracts, driving lymph to the veins, as in amphibians. { 'limf ,härt }

lymph node [ANAT] An aggregation of lymphoid tissue surrounded by a fibrous capsule; found along the course of lymphatic vessels. Also known as lymph gland. { 'limf ,nōd }

lymphoblast [HISTOL] Precursor of a lymphocyte. { 'lim·fə,blast }

lymphoblastic leukemia *See* acute lymphocytic leukemia. { ¦lim·fə¦blas·tik lü'kē·mē·ə }

lymphoblastosis [MED] An excessive number of lymphoblasts in peripheral blood; occasionally found in tissues. { ,lim·fō,bla'stō·səs }

lymphocyte [HISTOL] An agranular leukocyte formed primarily in lymphoid tissue; occurs as the principal cell type of lymph and composes 20–30% of the blood leukocytes. { 'lim·fə,sīt }

lymphocyte transformation *See* transformation. { ¦lim·fə,sīt ,tranz·fər'mā·shən }

lymphocytic angina *See* infectious mononucleosis. { ¦lim·fə¦sid·ik 'an·jə·nə }

lymphocytic choriomeningitis [MED] An acute viral meningitis caused by a specific virus endemic in mice; characterized clinically by rapid onset of symptoms of meningeal irritation, pleocytosis and often a rise in protein in the cerebrospinal fluid, and a short, benign course with recovery. { ¦lim·fə¦sid·ik ,kȯr·ē·ō,men·ən'jīd·əs }

lymphocytic leukemia [MED] A type of leukemia in which lymphocytic cells predominate. { ¦lim·fə¦sid·ik lü'kē·mē·ə }

lymphocytic lymphoma [MED] A malignant neoplasm of lymphoid tissue composed predominantly of lymphocytic cells. { ¦lim·fə¦sid·ik lim'fō·mə }

lymphocytic sarcoma *See* lymphosarcoma. { ¦lim·fə¦sid·ik sär'kō·mə }

lymphocytopenia [MED] Reduction of the absolute number of lymphocytes per unit volume of peripheral blood. Also known as lymphopenia. { ,lim·fə,sīd·ə'pē·nē·ə }

lymphocytosis [MED] An abnormally high lymphocyte count in the blood. { ,lim·fə,sī'tō·səs }

lymphocytotropic [IMMUNOL] Having an affinity for lymphocytes. { ,lim·fō,sīd·ə'träp·ik }

lymphoepithelioma [MED] A squamous-cell carcinoma of the nasopharynx whose parenchymal cells resemble elements of the reticuloendothelial system. { ,lim·fō,ep·ə,thē·lē'ō·mə }

lymphogranuloma inguinale *See* lymphogranuloma venereum. { ,lim·fə,gran·yə'lō·mə ,iŋ·gwə'nä·lē }

lymphogranulomatosis *See* Hodgkin's disease. { ˌlim·fəˌgran·yəˌlō·mə'tō·səs }

lymphogranuloma venereum [MED] A systemic infectious venereal disease caused by a microorganism belonging to the PLT-Bedsonia group, characterized by enlargement of inguinal lymph nodes and genital ulceration. Abbreviated LGV. Also known as lymphogranuloma inguinale; lymphopathia venereum; venereal bubo. { ˌlim·fəˌgran·yə'lō·mə və'nir·ē·əm }

lymphoid cell [HISTOL] A mononucleocyte that resembles a leukocyte. { 'limˌfȯid ˌsel }

lymphoid hemoblast of Pappenheim *See* pronormoblast. { 'limˌfȯid 'hē·məˌblast əv 'päp·ənˌhīm }

lymphoid organ [ANAT] An organ that produces lymphocytes or is associated with lymphocyte function, for example, the lymph nodes, spleen, and thymus. { 'limˌfȯid ˌȯr·gən }

lymphoid tissue *See* lymphatic tissue. { 'limˌfȯid ˌtish·ü }

lymphokine [IMMUNOL] A cytokine released from T lymphocytes after contact with an antigen. { 'lim·fəˌkīn }

lymphokine-activated killer cell [IMMUNOL] A cytotoxic cell that is able to lyse certain cell lines resistant to natural killer cells. { ¦lim·fəˌkīn ¦ak·təˌvād·əd 'kil·ər ˌsel }

lymphoma [MED] Any neoplasm, usually malignant, of the lymphoid tissues. { lim'fō·mə }

lymphopathia venereum *See* lymphogranuloma venereum. { ˌlim·fə'path·ē·ə və'nir·ē·əm }

lymphopenia *See* lymphocytopenia. { ˌlim·fə'pē·nē·ə }

lymphopoiesis [PHYSIO] The production of lymph. { ˌlim·fəˌpȯi'ē·səs }

lymphoproliferative virus group *See* Gammaherpesvirinae. { ¦lim·fō·prō'lif·rəd·iv 'vī·rəs ˌgrüp }

lymph sinus [ANAT] One of the tracts of diffuse lymphatic tissue between the cords and nodules, and between the septa and capsule of a lymph node. { 'limf ˌsī·nəs }

lymph vessel [ANAT] A tubular passage for conveying lymph. Also known as lymphatic. { 'limf ˌves·əl }

lyngbyatoxin A [BIOCHEM] An indole alkaloid toxin produced by *Lyngbya majuscula*. { 'liŋ·bē·əˌtäk·sən 'ā }

lynx [VERT ZOO] Any of several wildcats of the genus *Lynx* having long legs, short tails, and usually tufted ears; differs from other felids in having 28 instead of 30 teeth. { liŋks }

Lyomeri [VERT ZOO] The equivalent name for Saccopharyngiformes. { lī'äm·əˌrī }

Lyon hypothesis [GEN] The concept that mammalian females are X-chromosome mosaics as a result of the random inactivation of one X chromosome in some embryonic cells and their descendant and of the other X chromosome in the rest. { 'lī·ən hīˌpäth·ə·səs }

Lyon phenomenon [GEN] In the normal human female, the rendering of two X chromosomes inactive, or at least largely so, at an early stage in embryogenesis. { 'lī·ən fəˌnäm·əˌnän }

Lyopomi [VERT ZOO] An equivalent name for Notacanthiformes. { lī'äp·əˌmī }

Lysaretidae [INV ZOO] A family of errantian polychaete worms in the superfamily Eunicea. { ˌlī·sə'red·əˌdē }

lyse [CELL MOL] To undergo lysis. { līz }

Lysenkoism [BIOL] A pseudoscientific theory that flourished in the Soviet Union from the early 1930s to the mid-1960s; advocated by T. D. Lysenko, who called it agrobiology, it was claimed to be a revolutionary fusion of agronomy and biological science, and it opposed traditional biology and the gene concept but supported the inheritance of acquired characteristics. { lī'seŋ·kōˌiz·əm }

Lysianassidae [INV ZOO] A family of pelagic amphipod crustaceans in the suborder Gammaridea. { ˌlī·sē·ə'nas·əˌdē }

lysigenous [BIOL] Of or pertaining to the space formed following lysis of cells. { 'lī'sij·ə·nəs }

lysin [IMMUNOL] A substance, particularly antibodies, capable of lysing a cell. { 'līs·ən }

lysine [BIOCHEM] $C_6H_{14}O_2N_2$ An essential, basic amino acid obtained from many proteins by hydrolysis. { 'līˌsēn }

Lysiosquillidae [INV ZOO] A family of crustaceans in the order Stomatopoda. { ˌlī·sē·ō 'skwil·əˌdē }

lysis [CELL MOL] Dissolution of a cell or tissue by the action of a lysin. [MED] **1.** Gradual decline in the manifestations of a disease, especially an infectious disease. Also known as defervescence. **2.** Gradual fall of fever. { 'lī·səs }

lysogen [MICROBIO] A bacterial or prokaryotic cell whose genome contains integrated viral deoxyribonucleic acid. { 'līs·əˌjen }

lysogeny [MICROBIO] Lysis of bacteria, with the liberation of bacteriophage particles. { lī'säj·ə·nē }

lysosome [CELL MOL] A specialized cell organelle surrounded by a single membrane and containing a mixture of hydrolytic (digestive) enzymes. { 'lī·səˌsōm }

lysosomotropic drug [PHARM] A drug which can enter selectively the lysosomes of certain cell types; may be useful in chemotherapy for destruction of tumor cells. { ˌlī·səˌsō·mə'träp·ik 'drəg }

lysozyme [BIOCHEM] An enzyme present in certain body secretions, principally tears, which acts to hydrolyze certain bacterial cell walls. { 'lī·səˌzīm }

lyssacine [INV ZOO] An early stage of the skeletal network in hexactinellid sponges. { 'lī·səˌsēn }

Lyssacinosa [INV ZOO] An order of sponges in the subclass Hexasterophora in which parenchymal megascleres are typically free and unconnected but are sometimes secondarily united. { ¦lī·sə·sə¦nō·sə }

Lyssavirus [VIROL] A genus of the viral family Rhabdoviridae that is characterized by a bullet-shaped enveloped virion covered with projections that contains one molecule of linear, nega-

M

M *See* Morgan.
manikin [MED] **1.** A model of a term fetus; used in the teaching of obstetrics. **2.** A model of an adult human used to teach first aid and basic nursing skills. { 'man·ə·kən }
macadamia nut [BOT] The hard-shelled seed obtained from the fruit of a tropical evergreen tree, *Macadamia ternifolia*. { ˌmak·ə'dā·mē·ə ˌnət }
macaque [VERT ZOO] The common name for 12 species of Old World monkeys composing the genus *Macaca*, including the Barbary ape and the rhesus monkey. { mə'kak }
macaw [VERT ZOO] The common name for large South and Central American parrots of the genus *Ara* and related genera; individuals are brilliantly colored with a long tail, a hooked bill, and a naked area around the eyes. { mə'kȯ }
Macchiavello stain [MICROBIO] A differential staining procedure for rickettsiae, in which the fixed tissue smear is stained with basic fuchsin, differentiated with a 0.5% solution of citric acid, and counterstained with a 1% solution of methylene blue; rickettsiae stain red, and tissue cells stain blue. { ˌmäk·ē·ə'vel·ō ˌstān }
Machaeridea [INV ZOO] A class of homolozoan echinoderms in older systems of classification. { ˌmak·ə'rid·ē·ə }
Machilidae [INV ZOO] A family of insects belonging to the Thysanura having large compound eyes and ocelli and a monocondylous mandible of the scraping type. { mə'kil·əˌdē }
machopolyp *See* machozooid. { ¦ma·kō'päl·əp }
machozooid [INV ZOO] A defensive polyp equipped with stinging organs in certain hydroid colonies. Also known as machopolyp. { ¦ma·kō'zōˌȯid }
mackerel [VERT ZOO] The common name for perciform fishes composing the subfamily Scombroidei of the family Scombridae, characterized by a long slender body, pointed head, and large mouth. { 'mak·rəl }
mackerel shark [VERT ZOO] The common name for isurid galeoid elasmobranchs making up the family Isuridae; heavy-bodied fish with sharp-edged, awllike teeth and a nearly symmetrical tail. { 'mak·rəl 'shärk }
macrandrous [BOT] Having both antheridia and oogonia on the same plant; used especially for certain green algae. { ma'kran·drəs }
macrencephaly [MED] The condition of having an abnormally large brain. { ˌma·kren'sef·ə·lē }
Macristiidae [VERT ZOO] A family of oceanic teleostean fishes assigned by some zoologists to the order Ctenothrissiformes. { ˌmak·rə'stī·əˌdē }
macroblast of Naegeli *See* pronormoblast. { 'mak·rəˌblast əv 'neg·ə·lē }
macroblepharia [MED] The condition of having abnormally large eyelids. { ˌmak·rə·blə'far·ē·ə }
macrobrachia [MED] The condition of having excessive arm development. { ˌmak·rə'brāk·ē·ə }
macrocephalus [MED] An individual with an abnormally large head. { ˌmak·rə 'sef·ə·ləs }
macrocephaly [MED] The condition of having an abnormally large head. { 'mak·rə'sef·ə·lē }
macrocheiria [MED] The condition of having abnormal hand enlargement. { 'mak·rō 'kī·rə }
macroconsumer [ECOL] A large consumer which ingests other organisms or particulate organic matter. Also known as biophage. { ¦mak·rō·kən'sü·mər }
macrocrania [MED] The condition of having abnormally large skull size compared with face size. { ¦mak·rō'krā·nē·ə }
macrocyclic [MYCOL] Of a rust fungus, having binuclear spores as well as teliospores and sporidia, or having a life cycle that is long or complex. { ¦mak·rō'sī·klik }
Macrocypracea [INV ZOO] A superfamily of marine ostracodes in the suborder Podocopa having all thoracic legs different from each other, greatly reduced furcae, and long, thin Zenker's organs. { 'mak·rō·sə'prās·ē·ə }
macrocyst [MED] A large cyst, visible to the naked eye. { 'mak·rəˌsist }
macrocyte [HISTOL] An erythrocyte whose diameter or mean corpuscular volume exceed that of the mean normal by more than two standard deviations. Also known as macronormocyte. { 'mak·rəˌsīt }
macrocytic anemia [MED] A form of anemia characterized by the presence of macrocytes in the blood. { ¦mak·rə¦sid·ik ə'nē·mē·ə }
macrocytosis [MED] Presence of macrocytes in the blood. { ˌmak·rəˌsī'tō·səs }
macrodactyly [MED] The condition of having

abnormally large fingers or toes. { ¦mak·rō'dak·tə·lē }

Macrodasyoidea [INV ZOO] An order of wormlike invertebrates of the class Gastrotricha characterized by distinctive, cylindrical adhesive tubes in the cuticle which are moved by delicate muscle strands. { ˌmak·rōˌda·sē'ȯid·ē·ə }

macrodontia [MED] The condition of having abnormally large teeth. { ˌmak·rə'dän·chə }

macroencapsulation [CELL MOL] The envelopment of a large mass of xenotransplanted cells or tissue in planar membranes, hollow fibers, or diffusion chambers to isolate the cells from the body, thereby avoiding the immune responses that the foreign cells could initiate, and also to allow the desired metabolites (such as insulin and glucose for pancreatic islet cells) to diffuse in and out of the membrane. { ˌmak·rō·inˌkap·sə'lā·shən }

macroevolution [EVOL] The larger course of evolution by which the categories of animal and plant classification above the species level have been evolved from each other and have differentiated into the forms within each. { ¦mak·rōˌev·ə'lü·shən }

macrofauna [ECOL] **1.** Widely distributed fauna. **2.** Fauna of a macrohabitat. [ZOO] Animals visible to the naked eye. { ¦mak·rō 'fȯn·ə }

macroflora [BOT] Plants which are visible to the naked eye. [ECOL] **1.** Widely distributed flora. **2.** Flora of a macrohabitat. { ¦mak·rō'flȯr·ə }

macrofollicular adenoma [MED] **1.** A benign tumor of the thyroid with enlarged follicles. **2.** A type of malignant lymphoma. { ¦mak·rō·fə'lik·yə·lər ˌad·ən'ō·mə }

macrogamete [BIOL] The larger, usually female gamete produced by a heterogamous organism. { ¦mak·rō'gaˌmēt }

macroglia [NEUROSCI] That portion of the neuroglia composed of astrocytes. { mə'kräg·lē·ə }

macroglobulin [BIOCHEM] Any gamma globulin with a sedimentation constant of 19S. { ¦mak·rə'gläb·yə·lən }

macroglobulinemia [MED] **1.** Abnormal increase in macroglobulins in the blood. **2.** A disease characterized by proliferation of lymphocytes and plasmocytes and abnormally high macroglobulin blood levels. { ˌmak·rəˌgläb·yə·lə'nē·mē·ə }

macroglossia [MED] Enlargement of the tongue. { ˌmak·rō'gläs·ē·ə }

macrogyria [MED] The condition of having congenitally enlarged brain convolutions. { ˌmak·rə'jī·rē·ə }

macrohabitat [ECOL] An extensive habitat presenting considerable variation of the environment, containing a variety of ecological niches, and supporting a large number and variety of complex flora and fauna. { ¦mak·rō'hab·əˌtat }

Macrolepidoptera [INV ZOO] A former division of Lepidoptera that included the larger moths and butterflies. { ¦mak·rōˌlep·ə'däp·trə }

macrolide antibiotic [MICROBIO] A basic antibiotic characterized by a macrocyclic ring structure. { 'mak·rəˌlīd ˌant·iˌbī'äd·ik }

macrolymphocyte [HISTOL] A large lymphocyte. { ˌmak·rō'lim·fəˌsīt }

macromastia [MED] The condition of having abnormally enlarged breasts. { ˌmak·rō'mas·tē·ə }

macromelia [MED] The condition of having abnormally large arms or legs. { ¦mak·rō'mēl·yə }

macromere [EMBRYO] Any of the large blastomeres composing the vegetative hemisphere of telolecithal morulas and blastulas. { 'mak·rōˌmir }

Macromonas [MICROBIO] A genus of gram-negative, chemolithotrophic bacteria; large, motile, cylindrical to bean-shaped cells containing sulfur granules and calcium carbonate inclusions. { ¦mak·rə'mō·nəs }

macromutation [GEN] Any genetic change that leads to a pronounced phenotypic alteration. { ˌmak·rə·myü'tā·shən }

macronormocyte *See* macrocyte. { ˌmak·rə'nȯr·məˌsīt }

macronucleus [INV ZOO] A large, densely staining nucleus of most ciliated protozoans, believed to influence nutritional activities of the cell. { ¦mak·rō'nü·klē·əs }

macronutrient [BIOCHEM] An element required by animals or plants in large amounts. { ¦mak·rō'nü·trē·ənt }

macrophage [HISTOL] A large phagocyte of the reticuloendothelial system. Also known as a histiocyte. { 'mak·rəˌfāj }

macrophage inflammatory protein-1 [IMMUNOL] A protein produced by macrophages which has inflammatory and chemoattractant properties, it exists in two forms, MIP-1α and MIP-1β. { ¦mak·rəˌfāj in¦flam·əˌtȯr·ē 'prōˌtēn ¦wən }

macrophagy [BIOL] Feeding on large particulate matter. { 'mak·rəˌfā·jē }

macrophreate [INV ZOO] A comatulid with a large, deep cavity in the calyx. { ¦mak·rō'frēˌāt }

macrophyllous [BOT] Having large or long leaves. { ¦mak·rō'fil·əs }

macrophyte [ECOL] A macroscopic plant, especially one in an aquatic habitat. { 'mak·rəˌfīt }

macropinocytosis [CELL MOL] A mechanism of endocytosis in which large droplets of fluid are trapped underneath extensions (ruffles) of the cell surface. { ˌmak·rōˌpin·əˌsī'tō·səs }

Macropodidae [VERT ZOO] The kangaroos, a family of Australian herbivorous mammals in the order Marsupialia. { ˌmak·rə'päd·əˌdē }

macropodous [BOT] **1.** Having a large or long hypocotyl. **2.** Having a long stem or stalk. { mak'kräp·ə·dəs }

macroprosopus [MED] An individual with an abnormally large face. { ¦mak·rō·prə'sō·pəs }

macropsia [MED] A disturbance of vision in which objects seem larger than they are. Also known as megalopia. { ma'kräp·sē·ə }

macropterous [ZOO] Having large or long wings or fins. { ma'kräp·tə·rəs }

Macroscelidea [VERT ZOO] A monofamilial order of mammals containing the elephant shrews and their allies. { ˌmak·rō·sə'lid·ē·ə }

Macroscelididae [VERT ZOO] The single, African family of the mammalian order Macroscelidea. { ˌmak·rō·sə'lid·əˌdē }

macrosclereid [BOT] A sclereid cell that is rodlike and formed from the embryonic epidermis of certain seed coats. { ˌmak·rō 'skler·ē·əd }

macroseptum [INV ZOO] A primary septum in certain anthozoans. { ¦mak·rō'sep·təm }

macrosporangium [BOT] A spore case in which macrospores are produced. Also known as megasporangium. { ¦mak·rə·spə'ran·jē·əm }

macrospore [BOT] The larger of two spore types produced by heterosporous plants; the female gamete. Also known as megaspore. [INV ZOO] The larger gamete produced by certain radiolarians; the female gamete. { 'mak·rəˌspór }

macrosporogenesis [BOT] In angiosperms, the formation of macrospores and the production of the embryo sac from one or occasionally several cells of the subepidermal cell layer within the ovule of a closed ovary. Also known as megasporogenesis. { ¦mak·rō¦spór·ō'jen·ə·səs }

Macrostomida [INV ZOO] An order of rhabdocoels having a simple pharynx, paired protonephridia, and a single pair of longitudinal nerves. { ˌmak·rə'stäm·ə·də }

Macrostomidae [INV ZOO] A family of rhabdocoels in the order Macrostomida; members are broad and flattened in shape and have paired sex organs. { ˌmak·rə'stäm·əˌdē }

macrostylous [BOT] **1.** Having long styles. **2.** Having long styles and short stamens. { ¦mak·rō¦stī·ləs }

Macrotermitinae [INV ZOO] A subfamily of termites in the family Termitidae. { ˌmak·rōˌtər'mid·əˌnē }

macrothermophyte *See* megathermophyte. { ˌmak·rə'thər·məˌfīt }

Macrouridae [VERT ZOO] The grenadiers, a family of actinopterygian fishes in the order Gadiformes in which the body tapers to a point, and the dorsal, caudal, and anal fins are continuous. { mə'krür·əˌdē }

Macroveliidae [INV ZOO] A family of hemipteran insects in the subdivision Amphibicorisae. { ˌmak·rō·və'lī·əˌdē }

Macrura [INV ZOO] A group of decapod crustaceans in the suborder Reptantia including eryonids, spiny lobsters, and true lobsters; the abdomen is extended and bears a well-developed tail fan. { mə'krür·ə }

macrurous [ZOO] Having a long tail. { mə 'krür·əs }

macula [ANAT] Any anatomical structure having the form of a spot or stain. { 'mak·yə·lə }

macula densa [MED] A thickening of the epithelium of the ascending limb of the loop of Henle. { 'mak·yə·lə 'den·sə }

macula lutea [ANAT] A yellow spot on the retina; the area of maximum visual acuity, being made up almost entirely of retinal cones. { 'mak·yə·lə 'lüd·ē·ə }

maculate [BOT] Marked with speckles or spots. { 'mak·yə·lət }

maculopapule [MED] A small, circumscribed, discolored elevation of the skin; a macule and papule combined. { ¦mak·yə·lō'pap·yül }

madarosis [MED] Loss of the eyelashes or eyebrows. { ˌmad·ə'rō·səs }

Madreporaria [INV ZOO] The equivalent name for Scleractinia. { ˌmad·rə·pə'rar·ē·ə }

madreporite [INV ZOO] A delicately perforated sieve plate at the distal end of the stone canal in echinoderms. { 'mad·rəˌpórˌīt }

madura foot *See* mycetoma. { 'maj·ə·rə ˌfüt }

maduromycosis *See* mycetoma. { ¦maj·ə·rō ˌmī'kō·səs }

Magelonidae [INV ZOO] A monogeneric family of spioniform annelid worms belonging to Sedentaria. { ˌmaj·ə'län·əˌdē }

magenstrasse [ANAT] Gastric canal. { 'mäg·ənˌshträ·sə }

maggot [INV ZOO] Larva of a dipterous insect. { 'mag·ət }

maggot therapy [MED] Implantation of sterile cultivated maggots of the bluebottle fly into wounds in the treatment of chronic soft tissue infections and chronic osteomyelitis. { 'mag·ət ˌther·ə·pē }

magnesia magma *See* milk of magnesia. { mag'nē·zhə 'mag·mə }

magnetocardiograph [MED] An instrument that records the intensity of the magnetic field of the heart as produced by currents sent through the body by measuring the voltages associated with each heart beat. Abbreviated MCG. { mag¦nēd·ō'kär·dē·əˌgraf }

magnetoencephalogram [BIOPHYS] A measurement of the brain's magnetic field. Abbreviated MEG. { magˌned·ō·in'sef·ə·ləˌgram }

magnetoencephalography [MED] A method to detect the brain's electrical activity with an array of SQUID magnetometers positioned over the head. { magˌnēd·ō·inˌsef·ə'läg·rə·fē }

magnetosomes [MICROBIO] Intracellular, membrane-bound iron mineral crystals, often magnetite, that enable magnetotactic bacteria to orient and move in the direction of the earth's magnetic field; in marine environments, magnetosomes may contain the iron sulfide mineral greigite. { mag'ned·əˌsōmz }

magnetotactic bacteria [MICROBIO] A group of bacteria containing iron mineral crystals in intracellular structures, called magnetosomes, which enable the bacteria to orient and migrate along magnetic field lines. { magˌned·əˌtak·tik bak'tir·ē·ə }

magnocellular [CELL MOL] Having large cell bodies; said of various nuclei of the central nervous system. { ˌmag·nə'sel·yə·lər }

Magnolia [BOT] A genus of trees, the type genus of the Magnoliaceae, with large, chiefly white, yellow, or pinkish flowers, and simple, entire, usually large evergreen or deciduous alternate leaves. { mag'nōl·yə }

Magnoliaceae [BOT] A family of dicotyledonous plants of the order Magnoliales characterized by hypogynous flowers with few to numerous stamens, stipulate leaves, and uniaperturate pollen. { mag,nō·lē'ās·ē,ē }

Magnoliales [BOT] The type order of the subclass Magnoliidae; members are woody plants distinguished by the presence of spherical ethereal oil cells and by a well-developed perianth of separate tepals. { mag,nō·lē'ā·lēz }

Magnoliatae *See* Magnoliopsida. { ,mag·nō'lī·ə,tē }

Magnoliidae [BOT] A primitive subclass of flowering plants in the class Magnoliopsida generally having a well-developed perianth, numerous centripetal stamens, and bitegmic, crassinucellate ovules. { ,mag·nō'lī·ə,dē }

Magnoliophyta [BOT] The angiosperms, a division of vascular seed plants having the ovules enclosed in an ovary and well-developed vessels in the xylem. { mag,nō·lē'äf·əd·ə }

Magnoliopsida [BOT] The dicotyledons, a class of flowering plants in the division Magnoliophyta generally characterized by having two cotyledons and net-veined leaves, with vascular bundles borne in a ring enclosing a pith. { mag,nō·lē'äp·sə·də }

magnum [ANAT] Large, as in foramen magnum. { 'mag·nəm }

mahogany [BOT] Any of several tropical trees in the family Meliaceae of the Geraniales. { mə'häg·ə·nē }

maidenhead *See* hymen. { 'mād·ən,hed }

Maillard reaction [BIOCHEM] A reaction in which the amino group in an amino acid tends to form condensation products with aldehydes; believed to cause the Browning reaction when an amino acid and a sugar coexist, evolving a characteristic flavor useful in food preparations. { mī'yär rē,ak·shən }

main-band deoxyribonucleic acid [CELL MOL] A peak band of deoxyribonucleic acid obtained by density gradient centrifugation. { ¦mān ¦band ,dē,äk·sē,rī·bō·nü'klē·ik 'as·əd }

Maindroniidae [INV ZOO] A family of wingless insects belonging to the Thysanura proper. { ,mān·drō'nī·ə,dē }

maitake mushroom *See* Grifola frondosa. { ,mä·i,tä·ke 'məsh,rüm }

maize [BOT] *Zea mays*. Indian corn, a tall cereal grass characterized by large ears. { māz }

Majidae [INV ZOO] The spider, or decorator, crabs, a family of decapod crustaceans included in the Brachyura; members are slow-moving animals that often conceal themselves by attaching seaweed and sessile animals to their carapace. { 'maj·ə,dē }

major gene [GEN] Any gene individually associated with pronounced phenotypic effects. { 'mā·jər ,jēn }

major histocompatibility complex [IMMUNOL] In vertebrates, a family of genes that encode cell surface glycoproteins that regulate interactions among cells of the immune system, some components of the complement system, and perhaps other related functions connected with intercell recognition. Abbreviated MHC. { 'mā·jər ¦hi·stō·kəm'pad·ə'bil·əd·ē 'käm,pleks }

major histocompatibility molecule [IMMUNOL] Any of two classes of immunoregulatory cell-surface glycoproteins that are encoded by the major histocompatibility gene; class I molecules are found on almost every nucleated body cell and present antigens from the cytoplasm to cytotoxic T cells, whereas class II molecules are found only on macrophages, B cells, and CD4+ T cells and present antigens from outside the cell to helper T cells. { ¦mā·jər ,his·tō·kəm,pad·ə'bil·əd·ē 'mäl·ə,kyül }

major immunogene complex [IMMUNOL] A genetic region containing loci that code for lymphocyte surface antigens, histocompatibility antigens, immune response gene products, and proteins of the complement system. Abbreviated MIC. { 'mā·jar ,ə'myü·nə·jēn 'käm,pleks }

major operation [MED] An extensive, difficult, and potentially dangerous surgical procedure, usually requiring general anesthesia. { 'mā·jər ,äp·ə'rā·shən }

Malachiidae [INV ZOO] An equivalent name for Malyridae. { ,mal·ə'kī·ə,dē }

malacia [MED] Abnormal softening of tissues of an organ or other body structure. { mə'lāsh·ə }

Malacobothridia [INV ZOO] A subclass of worms in the class Trematoda; members typically have one or two soft, flexible suckers and are endoparasitic in vertebrates and invertebrates. { ,mal·ə·kō·bə'thrid·ē·ə }

Malacocotylea [INV ZOO] The equivalent name for Digenea. { ,mal·ə·kō,käd·əl'ē·ə }

malacology [INV ZOO] The study of mollusks. { ,mal·ə'käl·ə·jē }

malacoplakia [MED] The accumulation of modified histiocytes (malacoplakia cells) to produce soft, pale, elevated plaques, usually in the urinary bladder of middle-aged women. { ,mal·ə·kə'plā·kē·ə }

malacoplakia cell [PATH] A large histiocyte, occasionally multinucleate and containing Michaelis-Gutmann calcospherules in the cytoplasm; seen in malacoplakia. { ,mal·ə·kə'plā·kē·ə ,sel }

Malacopoda [INV ZOO] A subphylum of invertebrates in the phylum Oncopoda. { ,mal·ə'käp·ə·də }

Malacopterygii [VERT ZOO] An equivalent name for Clupeiformes in older classifications. { ¦mal·ə·kō·tə¦rij·ē,ī }

Malacostraca [INV ZOO] A large, diversified subclass of Crustacea including shrimps, lobsters, crabs, sow bugs, and their allies; generally characterized by having a maximum of 19 pairs of appendages and trunk limbs which are sharply differentiated into thoracic and abdominal series. { ,mal·ə'kä·strə·kə }

maladie du sommeil *See* African sleeping sickness. { ,mäl·ə'dē dyü sò'mā }

malady [MED] A disorder, disease, or illness. { 'mal·əd·ē }

malaise [MED] A general state of ill-being or the feeling of poor health. { mə'lāz }

Malapteruridae [VERT ZOO] A family of African catfishes in the suborder Siluroidei. { mə,lap·tə'rür·ə,dē }

malar [ANAT] Of or pertaining to the cheek or to the zygomatic bone. { 'mā·lər }

malar bone *See* zygomatic bone. { 'mā·lər ,bōn }

malaria [MED] A group of human febrile diseases with a chronic relapsing course caused by hemosporidian blood parasites of the genus *Plasmodium*, transmitted by the bite of the *Anopheles* mosquito. { mə'ler·ē·ə }

malaria pigment [PATH] Dark-brown, amorphous, microcrystalline and birefringent pigment found in parasitized erythrocytes, especially with malarial parasites, and in littoral phagocytes of spleen, liver, and bone marrow. { mə'ler·ē·ə ,pig·mənt }

malar stripe [VERT ZOO] **1.** The area extending from the corner of the mouth backward and down in birds. **2.** Area on side of throat below the base of the lower mandible. { 'mā·lər ,strīp }

malassimilation [MED] Faulty or inadequate assimilation. { ,mal·ə,sim·ə'lā·shən }

malate dehydrogenase *See* malic enzyme. { 'ma ,lāt dē'hī·drə·jə,nās }

Malayan filariasis [MED] Filariasis of humans caused by *Brugia malayi*, transmitted by *Mansonia* and *Anopheles* mosquitoes. { mə'lā·ən ,fil·ə'rī·ə·səs }

Malcidae [INV ZOO] A small family of Ethiopian and Oriental hemipternan insects in the superfamily Pentatomorpha. { 'mal·sə,dē }

Malcodermata [INV ZOO] The equivalent name for Cantharoidea. { ,mal·kō'dər·məd·ə }

Maldanidae [INV ZOO] The bamboo worms, a family of mud-swallowing annelids belonging to the Sedentaria. { mal'dan·ə,dē }

Maldaninae [INV ZOO] A subfamily of the Maldanidae distinguished by cephalic and anal plaques with the anal aperture located dorsally. { mal'dan·ə,nē }

mal de pinto *See* pinta. { ,mal de 'pin·tō }

male [BOT] A flower lacking pistils. [ZOO] **1.** Of or pertaining to the sex that produces spermatozoa. **2.** An individual of this sex. { māl }

male climacteric [PHYSIO] A condition presumably due to loss of testicular function, associated with an elevated urinary excretion of gonadotropins and symptoms of loss of sexual desire and potency, hot flashes, and vasomotor instability. { 'māl klə'mak·tə·rik }

male heterogamety [GEN] The production by males of some species, such as humans and *Drosophila*, of two types of gametes that differ in their sex chromosome content. { 'māl ¦hed·ə·rō¦gam·əd·ē }

male homogamety [GEN] The production by males of some species, as in birds, of a single type of gamete with respect to sex chromosome content. { 'māl ¦hō·mō¦gam·əd·ē }

male pseudohermaphroditism *See* androgyny. { māl ,süd·ō·hər'maf·rə,did,iz·əm }

male sterility [PHYSIO] The condition in which male gametes are absent, deficient in number, or nonfunctional. { māl stə'ril·əd·ē }

male Turner's syndrome *See* Ullrich-Turner syndrome. { māl 'tər·nərz ,sin,drōm }

malezal swamp [ECOL] A swamp resulting from drainage of water over an extensive plain with a slight, almost imperceptible slope. { mə'lēz·əl ,swämp }

malformation [MED] A deformity of a part of the body resulting from abnormal development. { ,mal·fər'mā·shən }

malic acid [BIOCHEM] $COOH \cdot CH_2 \cdot CHOH \cdot COOH$ Hydroxysuccinic acid: a dibasic hydroxy acid existing in two optically active isomers and a racemate form; found in apples and many other fruits. { 'mal·ik 'as·əd }

malic dehydrogenase [BIOCHEM] An enzyme in the Krebs cycle that catalyzes the conversion of L-malic acid to oxaloacetic acid. { 'mal·ik dē'hī·drə·jə,nās }

malic enzyme [BIOCHEM] An enzyme which utilizes nicotinamide-adenine dinucleotide phosphate (NADP) to catalyze the oxidative decarboxylation of malic acid to pyruvic acid and carbon dioxide. Also known as malate dehydrogenase. { 'mal·ik 'en,zīm }

malignant [CELL MOL] Pertaining to cells that have undergone phenotypic transformation by oncogenes or protooncogenes. [MED] **1.** Endangering the life or health of an individual. **2.** Of or pertaining to the growth and proliferation of certain neoplasms which terminate in death if not checked by treatment. { mə'lig·nənt }

malignant catarrh [VET MED] A catarrhal fever of cattle caused by a virus and characterized by acute inflammation and edema of the respiratory and digestive systems. { mə'lig·nənt kə'tär }

malignant disease [MED] Any disease that endangers the life of an individual over a short period of time, especially cancer. { mə'lig·nənt di'zēz }

malignant edema [VET MED] Inflammatory edema associated with certain infections, especially an acute wound infection in wild and domestic animals. { mə'lig·nənt i'dē·mə }

malignant embolus [MED] A blood-borne mass of malignant cells which have become dislodged from the parent neoplasm. { mə'lig·nənt 'em·bə·ləs }

malignant glaucoma [MED] A form of glaucoma associated with severe pain and rapidly leading to blindness. { mə'lig·nənt glau'kō·mə }

malignant hypertension [MED] A severe form of hypertension with a rapid course leading to progressive cardiac and renal vascular disease. Also known as accelerated hypertension. { mə'lig·nənt ,hī·pər'ten·chən }

malignant malaria *See* falciparum malaria. { mə'lig·nənt mə'ler·ē·ə }

malignant pustule [MED] The commonest form of anthrax in humans, resulting from contamination of the skin; characterized by a necrotic pustule surrounded by an area of edema and vesicles containing yellow fluid. Also known as cutaneous anthrax. { mə'lĭg·nənt 'pəs·chül }

malignant rhabdomyoma *See* rhabdomyosarcoma. { mə'lĭg·nənt ¦rab·də·mī'ō·mə }

malinger [MED] To pretend or exaggerate illness in order to avoid responsibilities. { mə'liŋ·gər }

malleate trophus [INV ZOO] A type of crushing masticatory apparatus in rotifers that are incidentally predatory, such as brachionids. { 'mal·ē,āt ,trō·fəs }

mallee *See* tropical scrub. { 'mä·lē }

malleolus [ANAT] A projection on the distal end of the tibia and fibula at the ankle. { mə'lē·ə·ləs }

malleoramate trophus [INV ZOO] An intermediate type of rotiferan masticatory apparatus having a looped manubrium and teeth on the incus (comprising the fulcrum and rami); developed for grinding. { ¦mal·ē·ə¦ra,māt ,trō·fəs }

malleus [ANAT] The outermost, hammer-shaped ossicle of the middle ear; attaches to the tympanic membrane and articulates with the incus. { 'mal·ē·əs }

Mallophaga [INV ZOO] Biting lice, a comparatively small order of wingless insects characterized by five-segmented antennae, distinctly developed mandibles, one or two terminal claws on each leg, and a prothorax developed as a distinct segment. { mə'läf·ə·gə }

Mallory bodies [PATH] Oval acidophilic hyalin inclusion bodies seen in the cytoplasm of hepatic cells in Laennec's cirrhosis. Also known as alcoholic hyaline. { 'mal·ə·rē ,bäd·ēz }

Mallory-Weiss syndrome [MED] Painless vomiting of blood secondary to lacerations of the distal esophagus and esophagogastric junction, usually a result of prolonged violent vomiting, coughing, or hiccuping. { 'mla·ə·rē 'wīs ,sin ,drōm }

malnutrition [MED] Defective nutrition due to inadequate intake of nutrients or to their faulty digestion, assimilation, or metabolism. { ¦mal·nü'trish·ən }

malocclusion [MED] Faulty occlusion of the teeth. { ¦mal·ə'klü·zhən }

malodor [PHYSIO] A bad or foul odor. { mal 'ōd·ər }

Malpighiaceae [BOT] A family of dicotyledonous plants in the order Polygalales distinguished by having three carpels, several fertile stamens, five petals that are commonly fringed or toothed, and indehiscent fruit. { mal,pig·ē'ās·ē,ē }

Malpighian corpuscle [ANAT] **1.** A lymph nodule of the spleen. **2.** *See* renal corpuscle. { mal'pig·ē·ən 'kȯr·pə·səl }

Malpighian layer [HISTOL] The germinative layer of the epidermis. { mal'pig·ē·ən ,lā·ər }

Malpighian pyramid *See* renal pyramid. { mal'pig·ē·ən 'pir·ə·məd }

Malpighian tubule [INV ZOO] Any of the blind tubes that open into the posterior portion of the gut in most insects and certain other arthropods and excrete matter or secrete substances such as silk. { mal'pig·ē·ən 'tü,byül }

malposition [MED] Abnormal position of an organ or other body structure, or of the fetus. { ¦mal·pə'zish·ən }

malpractice [MED] Improper or injurious medical or surgical treatment, through carelessness, ignorance, or intent. { mal'prak·təs }

malpresentation [MED] Abnormal position of the child at birth, making normal delivery difficult or impossible. { ¦mal,prē·zən'tā·shən }

Malta fever *See* brucellosis. { 'mȯl·tə 'fē·vər }

maltase [BIOCHEM] An enzyme that catalyzes the conversion of maltose to dextrose. { 'mȯl,tās }

Malthusianism [BIOL] The theory that population increases more rapidly than the food supply unless held in check by epidemics, wars, or similar phenomena. { mal'thü·zhə,niz·əm }

maltobiose *See* maltose. { ¦mȯl·tō'bī,ōs }

maltose [BIOCHEM] $C_{12}H_{22}O_{11}$ A crystalline disaccharide that is a product of the enzymatic hydrolysis of starch, dextrin, and glycogen; does not appear to exist free in nature. Also known as maltobiose; malt sugar. { 'mȯl,tōs }

maltose phosphorylase [BIOCHEM] An enzyme which reacts maltose with inorganic phosphate to yield glucose and glucose-1-phosphate. { 'mȯl,tōs ,fäs'fȯr·ə,lās }

maltosuria [MED] Presence of maltose in the urine. { ,mȯl·tə'syu̇r·ē·ə }

malt sugar *See* maltose. { 'mȯlt 'shu̇g·ər }

malunion [MED] Faulty union of the pieces of a fractured bone. { mal'yün·yən }

Malus *See* Pyxis. { 'mä·ləs }

Malvaceae [BOT] A family of herbaceous dicotyledons in the order Malvales characterized by imbricate or contorted petals, mostly unilocular anthers, and minutely spiny, multiporate pollen. { mal'vās·ē,ē }

Malvales [BOT] An order of flowering plants in the subclass Dilleniidae having hypogynous flowers with valvate calyx, mostly separate petals, numerous centrifugal stamens, and a syncarpous pistil. { mal'vā·lēz }

mamelon [BIOL] Any dome-shaped protrusion or elevation. { 'mam·ə·lən }

mamm-, mammo- [ANAT] A combining form meaning breast. { 'mam·ō }

mamma [ANAT] A milk-secreting organ characterizing all mammals. { 'mam·ə }

mamma aberrans [MED] A supernumerary breast. { 'mam·ə ,ab·ə'ränz }

mammal [VERT ZOO] A member of Mammalia. { 'mam·əl }

Mammalia [VERT ZOO] A large class of warm-blooded vertebrates containing animals characterized by mammary glands, a body covering of hair, three ossicles in the middle ear, a muscular diaphragm separating the thoracic and abdominal cavities, red blood cells without nuclei, and embryonic development in the allantois and amnion. { mə'māl·yə }

mammary [ANAT] Of or pertaining to the mamma, or breast. { 'mam·ə·rē }

mammary gland [PHYSIO] A highly modified sebaceous gland that secretes milk; a unique anatomical feature of mammals. { 'mam·ə·rē ˌgland }

mammary lymphatic plexus [ANAT] A network of anastomosing lymphatic vessels in the walls of the ducts and between the lobules of the mamma; also functions to drain skin, areola, and nipple. { 'mam·ə·rē lim'fad·ik 'plek·səs }

mammary region [ANAT] The space on the anterior surface of the chest between a line drawn through the lower border of the third rib and one drawn through the upper border of the xiphoid cartilage. { 'mam·ə·rē ˌrē·jən }

mammary ridge [EMBRYO] An ectodermal thickening forming a longitudinal elevation on the chest between the limbs from which the mammary glands develop. { 'mam·ə·rē ˌrij }

mammary-stimulating hormone [BIOCHEM] **1.** Estrogen and progesterone considered together as the hormones which induce proliferation of the mammary ductile and acinous elements respectively. **2.** *See* prolactin. { 'mam·ə·rē ¦stim·yəˌlād·iŋ 'hȯrˌmōn }

mammary tumor agent [VIROL] A milk-borne virus that produces mammary cancer in mice with the appropriate genotype. { 'mam·ə·rē ¦tü·mər ˌāj·ənt }

mammectomy *See* mastectomy. { ma'mek·tə·mē }

mammillary [ANAT] **1.** Of or pertaining to the nipple. **2.** Breast- or nipple-shaped. { 'ma·məˌler·ē }

mammillary body [NEUROSCI] Either of two small, spherical masses of gray matter at the base of the brain in the space between the hypophysis and oculomotor nerve, which receive and relay olfactory impulses. { 'ma·məˌler·ē ¦bäd·ē }

mammillary line [ANAT] A vertical line passing through the center of the nipple. { 'ma·məˌler·ē ¦līn }

mammillary process [ANAT] One of the tubercles on the posterior part of the superior articular processes of the lumbar vertebrae. { 'ma·məˌler·ē 'präs·əs }

mammillitis [MED] Inflammation of the nipple. { ˌmam·ə'līd·əs }

mammogen *See* prolactin. { ˌmam·ə'jen }

mammogenic hormone [BIOCHEM] **1.** Any hormone that stimulates or induces development of the mammary gland. **2.** *See* prolactin. { ¦mam·ə¦jen·ik 'hȯrˌmōn }

mammography [MED] Radiographic examination of the breast, performed with or without injection of the glandular ducts with a contrast medium. { ma'mäg·rə·fē }

mammoplasia [PHYSIO] Development of breast tissue. { ˌmam·ə'plā·zhə }

mammoplasty [MED] Plastic surgery performed to alter the shape of the breast. { 'mam·əˌplas·tē }

mammotropin *See* prolactin. { ¦mam·ə¦trä·pən }

Mancini method *See* single radial immunodiffusion. { man'sē·nē ˌmeth·əd }

mandarin [BOT] A large and variable group of citrus fruits in the species *Citrus reticulata* and some of its hybrids; many varieties of the trees are compact with willowy twigs and small, narrow, pointed leaves; includes tangerines, King oranges, Temple oranges, and tangelos. { 'man·də·rən }

mandible [ANAT] **1.** The bone of the lower jaw. **2.** The lower jaw. [INV ZOO] Any of various mouthparts in many invertebrates designed to hold or bite into food. { 'man·də·bəl }

mandibular arch [EMBRYO] The first visceral arch in vertebrates. { man'dib·yə·lər 'ärch }

mandibular cartilage [EMBRYO] The bar of cartilage supporting the mandibular arch. { man 'dib·yə·lər 'kärt·lij }

mandibular fossa [ANAT] The depression in the temporal bone that articulates with the mandibular condyle. { man'dib·yə·lər 'fäs·ə }

mandibular gland *See* submandibular gland. { man'dib·yə·lər 'gland }

mandibular nerve [ANAT] A mixed nerve branch of the trigeminal nerve; innervates various structures of the lower jaw and face. { man'dib·yə·lər 'nərv }

Mandibulata [INV ZOO] A subphylum of Arthropoda; members possess a pair of mandibles which characterize the group. { manˌdib·yə 'läd·ə }

mandrill [VERT ZOO] *Mandrillus sphinx*. An Old World cercopithecoid monkey found in west-central Africa and characterized by large red callosities near the ischium and by blue ridges on each side of the nose in males. { 'man·drəl }

mange [VET MED] Infestation of the skin of mammals by certain mites (Sarcoptoidea) which burrow into the epidermis; characterized by multiple lesions accompanied by severe itching. { mānj }

mango [BOT] *Mangifera indica*. A large evergreen tree of the sumac family (Anacardiaceae), native to southeastern Asia, but now cultivated in Africa, tropical America, Florida, and California for its edible fruit, a thick-skinned, yellowish-red, fleshy drupe. { 'maŋ·gō }

mangrove [BOT] A tropical tree or shrub of the genus *Rhizophora* characterized by an extensive, impenetrable system of prop roots which contribute to land building. { 'maŋˌgrōv }

mangrove swamp [ECOL] A tropical or subtropical marine swamp distinguished by the abundance of low to tall trees, especially mangrove trees. { 'maŋˌgrōv ˌswämp }

Manidae [VERT ZOO] The pangolins, a family of mammals comprising the order Pholidota. { 'man·əˌdē }

manihot *See* cassava. { 'man·əˌhät }

manikin [ENG] A correctly proportioned doll-like figure that is jointed and will assume any human position and hold it; useful in art to draw a human figure in action, or in medicine to show the relations of organs by means of movable parts. [MED] **1.** A model of a term fetus; used in the teaching of obstetrics. **2.** A model of an adult human used to teach first aid and basic nursing skills. { 'man·ə·kən }

Manila hemp *See* abaca. { mə'nil·ə 'hemp }

Manila maguey *See* cantala. { mə'nil·ə mə'gā }

manioc *See* cassava. { 'man·ē,äk }

manipulation [MED] Skillful use of the hands in moving body parts, as reducing a dislocation, or changing the position of a fetus. { mə,nip·yə'lā·shən }

mannan [BIOCHEM] Any of a group of polysaccharides composed chiefly or entirely of D-mannose units. { 'ma,nan }

mannose [BIOCHEM] $C_6H_{12}O_6$ A fermentable monosaccharide obtained from manna. { 'ma,nōs }

mansonelliasis [MED] A parasitic infection of humans by the filarioid nematode *Mansonella ozzardi*. { ,man·sən·ə'lī·ə·səs }

Mantidae [INV ZOO] A family of predacious orthopteran insects characterized by a long, slender prothorax bearing a pair of large, grasping legs, and a freely moving head with large eyes. { 'man·tə,dē }

mantis [INV ZOO] The common name for insects composing the family Mantidae. { 'man·təs }

mantle [ANAT] Collectively, the convolutions, corpus callosum, and fornix of the brain. [BIOL] An enveloping layer, as the external body wall lining the shell of many invertebrates, or the external meristematic layers in a stem apex. [VERT ZOO] The back and wing plumage of a bird if distinguished from the rest of the plumage by a uniform color. { 'mant·əl }

mantle cavity [INV ZOO] The space between mantle and body proper in bivalve mollusks. { 'mant·əl ,kav·əd·ē }

mantle lobe [INV ZOO] Either of the flaps on the dorsal and ventral sides of the mantle in bivalve mollusks. { 'mant·əl ,lōb }

Mantodea [INV ZOO] An order equivalent to the family Mantidae in some systems of classification. { man'tō·dē·ə }

Mantoux test [IMMUNOL] An intradermal test for tuberculin sensitivity, that is, for past or present infection with tubercle bacilli. { man'tü ,test }

manubrium [ANAT] **1.** The triangular cephalic portion of the sternum in humans and certain other mammals. **2.** The median anterior portion of the sternum in birds. **3.** The process of the malleus. [BOT] A cylindrical cell that projects inward from the middle of each shield composing the antheridium in stoneworts. [INV ZOO] The elevation bearing the mouth in hydrozoan polyps. { mə'nü·brē·əm }

manus [ANAT] The hand of a human or the forefoot of a quadruped. [INV ZOO] The proximal enlargement of the propodus of the chela of arthropods. { 'mä·nəs }

manus valga [MED] Clubhand with deviation of the ulna. { 'mä·nəs 'väl·gə }

manyplies *See* omasum. { 'men·ē,plīz }

MAP *See* microtubule-associated protein.

map distance [GEN] The frequency of meiotic recombination between linked genes, usually expressed in map units. { 'map ,dis·təns }

maple [BOT] Any of various broad-leaved, deciduous trees of the genus *Acer* in the order Sapindales characterized by simple, opposite, usually palmately lobed leaves and a fruit consisting of two long-winged samaras. { 'mā·pəl }

maple syrup urine disease [MED] A hereditary metabolic disorder caused by deficiency of branched-chain keto acid decarboxylase; characterized by the maple-syrup-like odor of urine. Also known as branched-chain ketoaciduria. { 'mā·pəl ¦sir·əp 'yu̇r·ən di,zēz }

map unit [GEN] A measure of genetic distance corresponding to a recombination frequency of 1% or 1 centriMorgan (cM). { 'map ,yü·nət }

maquis [ECOL] A type of vegetation composed of shrubs, or scrub, usually not exceeding 10 feet (3 meters) in height, the majority having small, hard, leathery, often spiny or needlelike drought-resistant leaves and occurring in areas with a Mediterranean climate. { mä'kē }

Marantaceae [BOT] A family of monocotyledonous plants in the order Zingiberales characterized by one functional stamen with a single functional pollen sac, solitary ovules in each locule, and mostly arillate seeds. { ,mar·ən'tās·ē,ē }

marantic [MED] **1.** Of or pertaining to marasmus. **2.** Of or pertaining to slowed circulation. { mə'ran·tik }

marantic endocarditis [MED] Nonbacterial thrombic endocarditis, usually associated with neoplasm or other debilitating disease. { mə'ran·tik ,en·dō,kär'dīd·əs }

marasmus [MED] Chronic severe wasting of the tissues of the body, particularly in children, due to malnutrition. { mə'raz·məs }

Marattiaceae [BOT] A family of ferns coextensive with the order Marattiales. { mə,rad·ē'ās·ē,ē }

Marattiales [BOT] An ancient order of ferns having massive eusporangiate sporangia in sori on the lower side of the circinate leaves. { mə,rad·ē'ā·lēz }

marble bone disease *See* osteopetrosis. { 'mär·bəl ¦bōn di,zēz }

Marburg virus [VIROL] A large virus transmitted to humans by the grivet monkey (*Cercopithecus aethiops*). { 'mär,bu̇rg ,vī·rəs }

marcescent [BOT] Withering without falling off. { mär'ses·ənt }

Marcgraviaceae [BOT] A family of dicotyledonous shrubs or vines in the order Theales having exstipulate leaves with scanty or no endosperm, two integuments, and highly modified bracts. { märk,grā·vē'ās·ē,ē }

Marchantiales [BOT] The thallose liverworts, an order of the class Marchantiopsida having a flat body composed of several distinct tissue layers, smooth-walled and tuberculate-walled rhizoids, and male and female sex organs borne on stalks on separate plants. { mär,shan·tē'ā·lēz }

Marchantiatae *See* Marchantiopsida. { mär¦shan·tē¦ā·tē }

Marchantiidae [BOT] A subclass of liverworts of the class Hepaticopsida, having gametophytes with ribbonlike or rosette-shaped thalli and generally reduced sporophytes. { ˌmär·shən'tī·əˌdē }

Marchantiopsida [BOT] The liverworts, a class of lower green plants; the plant body is usually a thin, prostrate thallus with rhizoids on the lower surface. { märˌshan·tē'äp·sə·də }

mare [VERT ZOO] A mature female horse or other equine. { 'mär·ā, mer }

Marfan's syndrome [MED] A hereditary connective-tissue disorder transmitted as an autosomal dominant; manifested by skeletal and ocular changes and by congenital heart disease. { 'märˌfanz ˌsinˌdrōm }

Margaritiferidae [INV ZOO] A family of gastropod mollusks with nacreous shells that provide an important source of commercial pearls. { märˌgär·ə·də'fer·əˌdē }

Margarodinae [INV ZOO] A subfamily of homopteran insects in the superfamily Coccoidea in which abdominal spiracles are present in all stages of development. { ˌmär·gə'räd·ənˌē }

marginal placentation [BOT] Arrangement of ovules near the margins of carpels. { 'mär·jən·əl ˌplas·ən'tā·shən }

marginal sinus [ANAT] **1.** One of the small, bilateral sinuses of the dura mater which skirt the edge of the foramen magnum, usually uniting posteriorly to form the occipital sinus. **2.** *See* terminal sinus. [EMBRYO] An enlarged venous sinus incompletely encircling the margin of the placenta. { 'mär·jən·əl 'sī·nəs }

marginal ulcer [MED] A peptic ulcer of the jejunum on the efferent margin of a gastrojejunostomy. { 'mär·jən·əl 'əl·sər }

marginate [BOT] Having a distinct margin or border. { 'mär·jəˌnāt }

maricolous [ECOL] Living in the sea. { mə'rik·ə·ləs }

Marie's ataxia [MED] A hereditary ataxia combining features of cerebellar, posterior column, and pyramidal tract lesions, with onset after age 20, normal or exaggerated deep tendon reflexes, and frequently optic atrophy and oculomotor palsies, but no clubfeet or scoliosis. { mə'rēz ā'tak·sē·ə }

Marie's disease [MED] Rheumatic spondylitis involving the spine only, or invading the shoulders and hips. { mə'rēz diˌzēz }

Marie-Strümpell disease *See* rheumatoid spondylitis. { mə'rē 'strim·pəl diˌzēz }

marihuana *See* marijuana. { ˌmar·ə'wän·ə }

marijuana [BOT] The Spanish name for the dried leaves and flowering tops of the hemp plant (*Cannabis sativa*), which have narcotic ingredients and are smoked in cigarettes. Also spelled marihuana. { ˌmar·ə'wän·ə }

marine biocycle [ECOL] A major division of the biosphere composed of all biochores of the sea. { mə'rēn 'bī·ōˌsī·kəl }

marine biology [BIOL] A branch of biology that deals with those living organisms which inhabit the sea. { mə'rēn bī'äl·ə·jē }

marine ecology [ECOL] An integrative science that studies the basic structural and functional relationships within and among living populations and their physical-chemical environments in marine ecosystems. { məˌrēn ē'käl·ə·jē }

marine littoral faunal region [ECOL] A geographically determined division of that portion of the zoosphere composed of marine animals. { mə'rēn 'lit·ə·rəl 'fön·əl ˌrē·jən }

marine marsh [ECOL] A flat, savannalike land expanse at the edge of a sea; usually covered by water during high tide. { mə'rēn 'märsh }

marine microbiology [MICROBIO] The study of the microorganisms living in the sea. { mə'rēn ˌmī·krō·bī'äl·ə·jē }

Marinesco-Sjögren-Garland syndrome [MED] A hereditary, congenital form of cerebral ataxia transmitted as an autosomal recessive; characterized by mental retardation, cataracts, minor skeletal anomalies, and hypertension. { mar·ə'nes·kō 'shō·grān 'gär·lənd ˌsinˌdrōm }

marita [INV ZOO] An adult trematode. { mə 'rīd·ə }

marjoram [BOT] Any of several perennial plants of the genera *Origanum* and *Majorana* in the mint family, Labiatae; the leaves are used as a food seasoning. { 'mär·jə·rəm }

marker [IMMUNOL] Any antigen that serves to distinguish cell types. { 'märk·ər }

marker gene [GEN] A gene with a known location on a chromosome and a clear-cut phenotype. { 'märk·ər ˌjēn }

marmoset [VERT ZOO] Any of 10 species of South American primates belonging to the family Callithricidae; individuals are primitive in that they have claws rather than nails and a nonprehensile tail. { 'mär·məˌset }

marmot [VERT ZOO] Any of several species of stout-bodied, short-legged burrowing rodents of the genus *Marmota* in the squirrel family Sciuridae. { 'mär·mət }

marrubium [BOT] *Marrubium vulgari*. An aromatic plant of the mint family, Labiatae; leaves have a bitter taste and are used as a tonic and anthelmintic. Also known as hoarhound; horehound. { mə'rü·bē·əm }

Marseilles fever *See* fièvre boutonneuse. { mär'sā ˌfē·vər }

marsh [ECOL] A transitional land-water area, covered at least part of the time by estuarine or coastal waters, and characterized by aquatic and grasslike vegetation, especially without peatlike accumulation. { märsh }

Marsileales [BOT] A small monofamilial order of heterosporous, leptosporangiate ferns (Polypodiophyta); leaves arise on long stalks from the rhizome, and sporangia are enclosed in modified folded leaves or leaf segments called sporocarps. { märˌsil·ē'ā·lēz }

Marsipobranchii [VERT ZOO] An equivalent name for Cyclostomata. { ˌmär·sə·pō'braŋ·kēˌī }

marsupial [VERT ZOO] **1.** A member of the Marsupialia. **2.** Having a marsupium. **3.** Of, pertaining to, or constituting a marsupium. { mär'sü·pē·əl }

marsupial anteater [VERT ZOO] *Myrmecobius fasciatus.* An anteater belonging to the Marsupialia. Also known as banded anteater. { mär'sü·pē·əl 'ant,ēd·ər }

marsupial frog [VERT ZOO] Any of various South American tree frogs which carry the eggs in a pouch on the back. { mär'sü·pē·əl 'fräg }

Marsupialia [VERT ZOO] The single order of the mammalian infraclass Metatheria, characterized by the presence of a marsupium in the female. { mär,sü·pē'ā·lē·ə }

marsupialization [MED] Surgical evacuation of pancreatic, hydatid, and other cysts when complete removal and closure are not possible. { mär,sü·pē·ə·lə'zā·shən }

marsupial mole [VERT ZOO] *Notoryctes typhlops.* A marsupial of Australia that resembles the euterian mole. { mär,sü·pē·əl 'mōl }

Marsupicarnivora [VERT ZOO] An order proposed to include the polydactylous and polyprotodont carnivorous superfamilies of Marsupialia. { mär,sü·pə·kär'niv·ə·rə }

marsupium [VERT ZOO] A fold of skin that forms a pouch enclosing the mammary glands on the abdomen of most marsupials. { mär'sü·pē·əm }

marten [VERT ZOO] Any of seven species of carnivores of the genus *Martes* in the family Mustelidae which resemble the weasel but are larger and of a semiarboreal habit. { 'märt·ən }

masculine [BIOL] Having an appearance or qualities distinctive for a male. { 'mas·kyə·lən }

masculine pelvis [ANAT] A female pelvis similar to the normal male pelvis in having a deeper cavity and more conical shape. Also known as android pelvis. { 'mas·kyə·lən 'pel·vəs }

masculinize [PHYSIO] To cause a female or a sexually immature animal to take on male secondary sex characteristics. { 'mas·kyə·lə,nīz }

masculinoma [MED] Adrenocorticoid adenoma of the ovary. { ,mas·kyə·lə'nō·mə }

masked messenger ribonucleic acid *See* maternal messenger ribonucleic acid. { ¦maskd 'mes·ən·jer ¦rī·bo·nü'klē·ik 'as·əd }

mask face [MED] An expressionless face seen in certain degenerative and inflammatory diseases of the basal ganglia and the extrapyramidal system; voluntary movements are near normal while involuntary movements are infrequent. { 'mask ,fās }

massage [MED] The act of rubbing, kneading, or stroking the superficial parts of the body with the hand or with an instrument, for therapeutic purposes. { mə'säzh }

mass bleaching [INV ZOO] A disease affecting coral reefs in which a reduction in the number of zooxanthellae (symbiotic plants) causes corals to lose their characteristic brown color over a period of several weeks and take on a brilliant white appearance. { ¦mas 'blēch·iŋ }

masseter [ANAT] The masticatory muscle, arising from the zygomatic arch and inserted into the lower jaw. { mə'sēd·ər }

mass extinction *See* faunal extinction. { 'mas ik 'stiŋk·shən }

mass reflex [NEUROSCI] A spread of reflexes suggesting lack of control by higher cortical centers; seen in normal newborns, in persons under the influence of drugs or in severe emotional states, and in encephalopathy or high spinal cord transections. { 'mas 'rē,fleks }

mast-, masto- [ANAT] A combining form denoting breast; denoting mastoid. { mast, 'mas·dō }

Mastacembeloidei [VERT ZOO] The spiny eels, a suborder of perciform fishes that are eellike in shape and have the pectoral girdle suspended from the vertebral column. { ,mas·tə,sem·bə'lȯid·ē,ī }

mastalgia [MED] Pain in the breast. { ma 'stal·jə }

mastatrophy [MED] Atrophy of the breast. { mast'a·trə·fē }

mastax [INV ZOO] The muscular pharynx in rotifers. { 'ma,staks }

mast cell [HISTOL] A connective-tissue cell with numerous large, basophilic, metachromatic granules in the cytoplasm. { 'mast ,sel }

mast-cell disease *See* mastocytosis. { 'mast ,sel di,zēz }

mastectomy [MED] Surgical removal of the breast. Also known as mammectomy. { ma 'stek·tə·mē }

Master's two-step test *See* two-step test. { 'mas·tərz 'tü ,step ,test }

masticate [PHYSIO] To chew. { 'mas·tə,kāt }

masticatory [PHARM] A medicine to be chewed but not swallowed. { 'mas·tə·kə,tȯr·ē }

Mastigamoebidae [INV ZOO] A family of ameboid protozoans possessing one or two flagella, belonging to the order Rhizomastigida. { ,mas·tə·gə'mē·bə,dē }

mastigoneme [CELL MOL] Any of the fine hairlike appendages that extend from the shaft of the flagellum in certain motile cells. { ,mas·tə'gō,nēm }

Mastigophora [INV ZOO] A superclass of the Protozoa characterized by possession of flagella. { ,mas·tə'gäf·ə·rə }

mastitis [MED] Inflammation of the breast. { ma'stīd·əs }

mastocytoma [VET MED] A local proliferation of mast cells forming a tumorous nodule; seen most frequently in dogs, but occasionally noted in humans. { ¦mas·tə,sī'tō·mə }

mastocytosis [MED] Excessive mast cell proliferation. Also known as mast-cell disease. { ¦mas·tə,sī'tō·səs }

mastodynia [MED] A type of cystic hyperplasia of the breast marked by an increase of connective tissue in the breast, without a proportionate increase in glandular epithelium. { ,mas·tə'din·ē·ə }

mastoid [ANAT] **1.** Breast-shaped. **2.** The portion of the temporal bone where the mastoid process is located. { 'ma,stȯid }

mastoid air cell *See* mastoid cell. { 'ma,stȯid 'er ,sel }

mastoid antrum [ANAT] An air-filled space between the upper portion of the middle ear and the mastoid cells. { 'ma,stȯid 'an·trəm }

mastoid canaliculus [ANAT] A small canal opening just above the stylomastoid foramen; gives passage to the auricular branch of the vagus nerve. { 'ma,stȯid ,kan·ə'lik·yə·ləs }

mastoid cell [ANAT] One of the compartments in the mastoid portion of the temporal bone, connected with the mastoid antrum and lined with a mucous membrane. Also known as mastoid air cell; mastoid sinus. { 'ma,stȯid ,sel }

mastoid foramen [ANAT] A small opening behind the mastoid process. { 'ma,stȯid fə'rā·mən }

mastoid fossa [ANAT] The depression behind the suprameatal spine on the lateral surface of the temporal bone. { 'ma,stȯid 'fäs·ə }

mastoiditis [MED] Inflammation of the mastoid cells. { ,ma,stȯi'dīd·əs }

mastoid process [ANAT] A nipple-shaped, inferior projection of the mastoid portion of the temporal bone. { 'ma,stȯid 'prä·səs }

mastoid sinus *See* mastoid cell. { 'ma,stȯid 'sī·nəs }

mastopathy [MED] Any disease or pain of the mammary gland. Also known as mazopathy. { ma'stäp·ə·thē }

mastopexy [MED] Surgical fixation of a pendulous breast. Also known as mazopexy. { 'mas·tə,pek·sē }

mastoplasia [MED] Hypertrophy of breast tissue. Also known as mastoplastia. { ,mas·tə'plā·zhə }

mastoplastia *See* mastoplasia. { ,mas·tə'plas·tē·ə }

mastoplasty [MED] Plastic surgery on the breast. { 'mas·tə,plas·tē }

mastorrhagia [MED] Hemorrhage from the breast. { ,mas·tə'ra·jē·ə }

Mastotermitidae [INV ZOO] A family of lower termites in the order Isoptera with a single living species, in Australia. { ,mas·tō·tər'mid·ə,dē }

mastotomy [MED] Incision of the breast. { ma'städ·ə·mē }

match [IMMUNOL] To select blood donors whose erythrocytes are compatible with those of the recipient. { mach }

mate [BIOL] **1.** To pair for breeding. **2.** To copulate. { māt }

mate killers [VIROL] Paramecia that contain the mu phage and cause their sensitive partners to die after conjugation. { 'māt ,kil·ərz }

materia medica [PHARM] **1.** The science that treats of the sources, properties, and preparation of medicinal substances. **2.** The materials from which medicinal substances are prepared. **3.** A treatise on the subject. { mə'tir·ē·ə 'med·ə·kə }

maternal [BIOL] Of, pertaining to, or related to a mother. { mə'tərn·əl }

maternal effect [GEN] Determination of characters of the progeny by the maternal parent; mediated by the genetic constitution of the mother. { mə'tərn·əl i,fekt }

maternal inheritance [GEN] The acquisition of characters transmitted through the cytoplasm of the egg. { mə'tərn·əl in'her·əd·əns }

maternal messenger ribonucleic acid [CELL MOL] In certain oocytes, messenger ribonucleic acid that is stored during oogenesis for translation during early embryogenesis. Also known as masked messenger ribonucleic acid. { mə¦tərn·əl ¦mes·ən·jər ,rī·bō·nü¦klē·ik 'as·əd }

maternal mortality rate [MED] The number of deaths reported as due to puerperal causes in a calendar year per 100 live births reported in the same year and place. { mə'tərn·əl mȯr'tal·əd·ē ,rāt }

maternal placenta [EMBRYO] The outer placental layer, developed from the decidua basalis. { mə'tərn·əl plə'sen·tə }

maternity [BIOL] **1.** Motherhood. **2.** The state of being pregnant. { mə'tərn·əd·ē }

mathematical biology [BIOL] A discipline that encompasses all applications of mathematics, computer technology, and quantitative theorizing to biological systems, and the underlying processes within the systems. { ¦math·ə¦mad·ə·kəl bī'äl·ə·jē }

mathematical biophysics [BIOL] A discipline which attempts to utilize mathematics to explain biophysical processes. { ¦math·ə¦mad·ə·kəl ,bī·ō'fiz·iks }

mathematical ecology [ECOL] The application of mathematical theory and technique to ecology. { ¦math·ə¦mad·ə·kəl ē'käl·ə·jē }

matico [BOT] *Piper angustifolium.* An aromatic wild pepper found in tropical America whose leaves are rich in volatile oil, gums, and tannins; leaves were used medicinally as a stimulant and hemostatic. { mə'tē·kō }

mating [BIOL] The meeting of individuals for sexual reproduction. { 'mād·iŋ }

mating type [MICROBIO] The genetically determined mating behavior characteristic of certain species of microorganisms; only different mating types can conjugate. { 'mad·iŋ ,tīp }

Matricaria [BOT] A genus of weedy herbs having a strong odor and white and yellow disk flowers; chamomile oil is obtained from certain species. { ,ma·trə'kar·ē·ə }

matrix [HISTOL] **1.** The intercellular substance of a tissue. Also known as ground substance. **2.** The epithelial tissue from which a toenail or fingernail develops. [MYCOL] The substrate on or in which fungus grows. { 'mā·triks }

matrix association region [CELL MOL] In eukaryotic interphase nuclei, any of the specific sites to which the chromatin loop domains are anchored. { ¦ma·triks ə,sō·sē'ā·shən ,rē·jən }

matrix metalloproteinase [BIOCHEM] Any member of a family of at least 19 structurally related zinc-dependent neutral endopeptidases collectively capable of degrading essentially all components of extracellular matrix. { ¦mā·triks mə,tal·ō'prō·tē·ə,nās }

matroclinous inheritance [GEN] Inheritance in which the offspring more closely resemble the female parent than the male parent. { ¦ma·trə,klī·nəs in'her·əd·əns }

matromycin *See* oleandomycin. { ¦ma·trə¦mīs·ən }

maturase [CELL MOL] Any enzyme encoded by self-splicing introns that catalyzes excision of the intron from its own primary transcript. { 'mach·ə,rās }

maturation [BIOL] **1.** The process of coming to full development. **2.** The final series of changes in the growth and formation of germ cells. [VIROL] The process that leads to incorporation of viral genomes into capsids and complete virions. { ,mach·ə'rā·shən }

mature [BIOL] **1.** Being fully grown and developed. **2.** Ripe. { mə'chu̇r }

maturity-onset diabetes [MED] A type of diabetes mellitus which develops later in life; characterized by more gradual development and less severe symptoms than juvenile-onset diabetes. { mə'chu̇r·əd·ē ¦ȯn,set ,dī·ə'bēd·əs }

Maunoir's hydrocele [MED] A congenital lymphatic cyst of the neck. { mōn'wärz 'hī·drə,sēl }

Mauriac syndrome [MED] A complex of symptoms associated with diabetes mellitus in children including retarded growth, obesity, and enlargement of the liver, probably related to inadequate control of the condition. { 'mȯ·rē·ak ,sin,drōm }

Mauritius hemp [BOT] A hard fiber obtained from the leaves of the cabuya, grown on the island of Mauritius; not a true hemp. { mȯ'rish·əs 'hemp }

maxilla [ANAT] **1.** The upper jawbone. **2.** The upper jaw. [INV ZOO] Either of the first two pairs of mouthparts posterior to the mandibles in certain arthropods. { mak'sil·ə }

maxillary air sinus *See* maxillary sinus. { 'mak·sə,ler·ē ¦er ,sī·nəs }

maxillary antrum *See* maxillary sinus. { 'mak·sə,ler·ē 'an·trəm }

maxillary arch *See* palatomaxillary arch. { 'mak·sə,ler·ē 'ärch }

maxillary artery [ANAT] A branch of the external carotid artery which supplies the deep structures of the face (internal maxillary) and the side of the face and nose (external maxillary). { 'mak·sə,ler·ē 'ärd·ə·rē }

maxillary hiatus [ANAT] An opening in the maxilla connecting the nasal cavity with the maxillary sinus. { 'mak·sə,ler·ē hī'ād·əs }

maxillary nerve [NEUROSCI] A somatic sensory branch of the trigeminal nerve; innervates the meninges, the skin of the upper portion of the face, upper teeth, and mucosa of the nose, palate, and cheeks. { 'mak·sə,ler·ē 'nərv }

maxillary process of the embryo [EMBRYO] An outgrowth of the dorsal part of the mandibular arch that forms the lateral part of the upper lip, the upper cheek region, and the upper jaw except the premaxilla. { 'mak·sə,ler·ē ¦prä·səs əv t͟hē 'em·brē·ō }

maxillary sinus [ANAT] A paranasal air cavity in the body of the maxilla. Also known as maxillary air sinus; maxillary antrum. { 'mak·sə,ler·ē 'sī·nəs }

maxilliped [INV ZOO] One of the three pairs of crustacean appendages immediately posterior to the maxillae. { mak'sil·ə,ped }

maxillofacial [MED] Pertaining to the jaws and face. { ,mak·sə·lō 'fā·shəl }

maximal breathing capacity [PHYSIO] The greatest respiratory minute volume which an individual can produce during a given period of extremely forceful breathing. { 'mak·sə·məl 'brēt͟h·iŋ kə,pas·əd·ē }

maximum breathing capacity [PHYSIO] The greatest volume of air an individual can breathe voluntarily in 10–30 seconds; expressed as liters per minute. Abbreviated MBC. { 'mak·sə·məm 'bret͟h·iŋ kə,pas·əd·ē }

maximum permissible concentration [MED] The maximum quantity/unit volume of radioactive material in air, water, and foodstuffs that is not considered an undue risk to human health. { 'mak·sə·məm pər'mis·ə·bəl ,kän·sən'trā·shən }

maximum permissible dose [MED] The dose of ionizing radiation that a person may receive in his lifetime without appreciable bodily injury. { 'mak·sə·məm pər'mis·ə·bəl 'dōs }

maxoplasia [MED] Degenerative disease of the breast. { ,mak·sō'plā·zhə }

mayfly [INV ZOO] The common name for insects composing the order Ephemeroptera. { 'mā,flī }

maz-, mazo- [EMBRYO] A combining form denoting placenta. { māz,'mā·zō }

mazaedium [BOT] The fruiting body of certain lichens, with the spores lying in a powdery mass in the capitulum. [MYCOL] A slimy layer on the hymenial surface of some ascomycetous fungi. { mə'zē·dē·əm }

mazopathy [MED] **1.** Disease of the placenta. **2.** *See* mastopathy. { mā'zäp·ə·thē }

mazopexy *See* mastopexy. { 'mā·zə,pek·sē }

Mazzoni's corpuscle [NEUROSCI] A specialized encapsulated sensory nerve end organ on a tendon in series with muscle fibers. { mät'zō·nēz ,kȯr·pə·səl }

MBC *See* maximum breathing capacity.

McArdle's syndrome [MED] A hereditary metabolic disorder caused by deficiency of muscle phosphorylase, with abnormal glycogen deposition in skeletal muscle leading to muscle fiber destruction. Also known as myophosphorylase deficiency glycogenosis. { mə'kärd·əlz ,sin ,drōm }

McBurney's incision [MED] A short diagonal incision in the lower right quadrant in which the muscle fibers are separated rather than cut; used for appendectomy. { mək'bər·nēz in,sizh·ən }

McBurney's point [ANAT] A point halfway between the umbilicus and the anterior superior iliac spine; a point of extreme tenderness in appendicitis. { mək'bər·nēz ,pȯint }

MCG *See* magnetocardiograph.

McGurk effect [PHYSIO] An auditory illusion

discovered by H. McGurk that demonstrates the important contribution made by visible face movements to normal speech perception. { mə'gərk i,fekt }

M-component hypergammaglobulinemia [MED] A form of hypergammaglobulinemia characterized by a single, prominent, more or less narrow band occurring from the slow gamma to the fast alpha-1 region of the electrophoretic strip. { 'em kəm,pō·nənt ¦hī·pər,gam·ə,gläb·yə·lə'nē·mē·ə }

M-cyclin [CELL MOL] A regulatory protein that binds with M-kinase, a cyclin-dependent kinase, activating the cell's entry into mitosis from the G2 stage of the cell cycle. { 'em¦sīk·lən }

MEA *See* 2-aminoethanethiol.

meadow [ECOL] A vegetation zone which is a low grassland, dense and continuous, variously interspersed with forbs but few if any shrubs. Also known as pelouse; Wiesen. { 'med·ō }

mealybug [INV ZOO] Any of various scale insects of the family Pseudococcidae which have a powdery substance covering the dorsal surface; all are serious plant pests. { 'mē·lē,bəg }

Meantes [VERT ZOO] The mud eels, a small suborder of the Urodela including three species of aquatic eellike salamanders with only anterior limbs. { mē'an,tēz }

measles [MED] An acute, highly infectious viral disease with cough, fever, and maculopapular rash; the appearance of Koplik spots on the oral mucous membranes marks the onset. Also known as rubeola. { 'mē·zəlz }

measles encephalitis [MED] Acute disseminated encephalitis following measles. { 'mē·zəlz in,sef·ə'līd·əs }

measles immune globulin [IMMUNOL] Sterile human globulin used to provide passive immunization against measles. { 'mē·zəlz i'myün 'gläb·yə·lən }

measles virus vaccine [IMMUNOL] A suspension of live attenuated or inactivated measles virus used for active immunization against measles. { 'mē·zəlz 'vī·rəs ,vak,sēn }

measly [MED] Infected with measles. { 'mēz·lē }

meatal plate [EMBRYO] A mass of ectodermal cells on the bottom of the branchial groove in a 2-month embryo. { mē'ād·əl 'plāt }

meatotomy [MED] Incision into and enlargement of a meatus. { ,mē·ə'täd·ə·mē }

meatus [ANAT] A natural opening or passage in the body. { mē'ād·əs }

mechanical dysmenorrhea [MED] Painful menstruation due to mechanical obstruction of the discharge of menstrual fluids. Also known as obstructive dysmenorrhea. { mi'kan·ə·kəl di ,smen·ə'rē·ə }

mechanoreceptor [PHYSIO] A receptor that provides the organism with information about such mechanical changes in the environment as movement, tension, and pressure. { ¦mek·ə·nō·ri'sep·tər }

Meckel's cartilage [EMBRYO] The cartilaginous axis of the mandibular arch in the embryo and fetus. { 'mek·əlz ,kärt·lij }

Meckel's diverticulum [EMBRYO] The persistent blind end of the yolk stalk forming a tube connected with the lower ileum. { 'mek·əlz ,dī·vər'tik·yə·ləm }

meconium [EMBRYO] A greenish mass of mucous, desquamated epithelial cells, lanugo, and vernix caseosa that collects in the fetal intestine, becoming the first fecal discharge of the newborn. { mə'kō·nē·əm }

meconium ileus [MED] Intestinal obstruction in the newborn with cystic fibrosis due to trypsin deficiency. { mə'kō·nē·əm 'il·ē·əs }

Mecoptera [INV ZOO] The scorpion flies, a small order of insects; adults are distinguished by the peculiar prolongation of the head into a beak, which bears chewing mouthparts. { me'käp·tə·rə }

mecystasis [PHYSIO] Increase in muscle length with maintenance of the original degree of tension. { me'sis·tə·səs }

media [HISTOL] The middle, muscular layer in the wall of a vein, artery, or lymph vessel. { 'mē·dē·ə }

mediad [ANAT] Toward the median line or plane of the body or of a part of the body. { 'mē·dē,ad }

medial [ANAT] **1.** Being internal as opposed to external (lateral). **2.** Toward the midline of the body. { 'mē·dē·əl }

medial arteriosclerosis [MED] Calcification of the tunica media of small and medium-sized muscular arteries. Also known as medial calcinosis; Mönckeberg's arteriosclerosis. { 'mē·dē·əl ,är·tir·ē·ō·sklə'rō·səs }

medial calcinosis *See* medial arteriosclerosis. { 'mē·dē·əl ,kal·sə'nō·səs }

medial lemniscus [ANAT] A lemniscus arising in the nucleus gracilis and nucleus cuneatus of the brain, crossing immediately as internal arcuate fibers, and terminating in the posterolateral ventral nucleus of the thalamus. { 'mē·dē·əl lem 'nis·kəs }

medial necrosis [MED] Death of cells in the tunica media of arteries. Also known as medionecrosis. { 'mē·dē·əl ne'krō·səs }

medial sclerosis [MED] An uncommon degenerative arterial disease that is characterized by ring-like calcifications that occur within the media (middle layer) of muscular arteries of the arms, legs, or genital tract, it rarely occurs in men or women under the age of 50, and its cause is obscure. { ,mēd·ē·əl sklə'rō·səs }

median effective dose *See* effective dose 50. { 'mē·dē·ən i'fek·tiv 'dōs }

median infective dose *See* infective dose 50. { 'mē·dē·ən in'fek·tiv 'dōs }

median lethal dose *See* lethal dose 50. { 'mē·dē·ən 'lēth·əl 'dōs }

median lethal time [MICROBIO] The period of time required for 50% of a large group of organisms to die following a specific dose of an injurious agent, such as a drug or radiation. { 'mē·dē·ən 'lēth·əl ,tīm }

median maxillary cyst [MED] Cystic dilation of embryonal inclusions in the incisive fossa or between the roots of the central incisors. Also known as nasopalatine cyst. { 'mē·dē·ən 'mak·sə,ler·ē ,sist }

median nasal process [EMBRYO] The region below the frontonasal sulcus between the olfactory sacs; forms the bridge and mobile septum of the nose and various parts of the upper jaw and lip. { 'mē·dē·ən 'nāz·əl ,prä·səs }

median nerve test [MED] A test for loss of function of the median nerve by having the patient abduct the thumb at right angles to the palm with fingertips in contact and forming a pyramid. { 'mē·dē·ən 'nərv ,test }

mediastinitis [MED] Inflammation of the mediastinum. { ,mē·dē,as·tə'nīd·əs }

mediastinum [ANAT] **1.** A partition separating adjacent parts. **2.** The space in the middle of the chest between the two pleurae. { ,mē·dē·ə'stī·nəm }

medical bacteriology [MED] A branch of medical microbiology that deals with the study of bacteria which affect human health, especially those which produce disease. { 'med·ə·kəl bak,tir·ē'äl·ə·jē }

medical climatology [MED] The study of the relation between climate and disease. { 'med·ə·kəl ,klī·mə'täl·ə·jē }

medical control systems [MED] Physiological and artificial systems that control one or more physiological variables or functions of the human body. { ,med·ə·kəl kən'trōl ,sis·təmz }

medical entomology [MED] The study of insects that are vectors for diseases and parasitic infestations in humans and domestic animals. { 'med·ə·kəl ,en·tə'mäl·ə·jē }

medical ethics [MED] Principles and moral values of proper medical conduct. { 'med·ə·kəl 'eth·iks }

medical examiner [FOREN] A professionally qualified physician duly authorized and charged by a governmental unit to determine facts concerning causes of death, particularly deaths not occurring under natural circumstances, and to testify thereto in courts of law. { 'med·ə·kəl ig 'zam·ə·nər }

medical geography [MED] The study of the relation between geographic factors and disease. { 'med·ə·kəl jē'äg·rə·fē }

medical history [MED] An account of a patient's past and present state of health obtained from the patient or relatives. { 'med·ə·kəl 'his·trē }

medical imaging [MED] The production of visual representations of body parts, tissues, or organs, for use in clinical diagnosis; encompasses x-ray methods, magnetic resonance imaging, single-photon-emission and positron-emission tomography, and ultrasound. { 'med·ə·kəl 'im·ij·iŋ }

medical information systems [MED] Standardized methods of collection, evaluation or verification, storage, and retrieval of data about a patient. { ,med·ə·kəl ,in·fər'mā·shən ,sis·təmz }

medical microbiology [MED] The study of microorganisms which affect human health. { 'med·ə·kəl ¦mī·krō·bī'äl·ə·jē }

medical mycology [MED] A branch of medical microbiology that deals with fungi that are pathogenic to humans. { 'med·ə·kəl mī'käl·ə·jē }

medical parasitology [MED] A branch of medical microbiology which deals with the relationship between humans and those animals which live in or on them. { 'med·ə·kəl ,par·ə·si'täl·ə·jē }

medical protozoology [MED] A branch of medical microbiology that deals with the study of Protozoa which are parasites of humans. { 'med·ə·kəl ,prō·dō·zō'äl·ə·jē }

medical radiography [MED] The use of x-rays to produce photographic images for visualizing internal anatomy as an aid in diagnosis. { 'med·ə·kəl ,rād·ē·'äg·rə·fē }

medication [MED] **1.** A medicinal substance. **2.** Treatment by or administration of a medicine. { ,med·ə'kā·shən }

medicinal [MED] Of, pertaining to, or having the nature of medicine. { mə'dis·ən·əl }

medicine [MED] **1.** Any agent administered for the treatment of disease. **2.** The science and art of treating and healing. { 'med·ə·sən }

medionecrosis [MED] Necrosis occurring in the tunica media of an artery. { ¦mē·dē·ō·nə'krō·səs }

Mediterranean anemia *See* thalassemia. { ,med·ə·tə'rā·nē·ən ə'nē·mē·ə }

Mediterranean faunal region [ECOL] A marine littoral faunal region including that offshore portion of the Atlantic Ocean from northern France to near the Equator. { ,med·ə·tə'rā·nē·ən 'fȯn·əl ,rē·jən }

Mediterranean fever *See* brucellosis. { ,med·ə·tə'rā·nē·ən 'fē·vər }

medius [ANAT] The middle finger. { 'mē·dē·əs }

medulla [ANAT] **1.** The central part of certain organs and structures such as the adrenal glands and hair. **2.** Marrow, such as of bone or the spinal cord. [BOT] **1.** Pith. **2.** The central spongy portion of some fungi. [NEUROSCI] *See* medulla oblongata. { mə'dəl·ə }

medulla oblongata [NEUROSCI] The somewhat pyramidal, caudal portion of the vertebrate brain which extends from the pons to the spinal cord. Also known as medulla. { mə'dəl·ə ,äb,lȯŋ 'gäd·ə }

medullary carcinoma [MED] A form of poorly differentiated adenocarcinoma, usually of the breast, grossly well circumscribed, gray-pink, and firm. Also known as encephaloid carcinoma. { mə'dəl·ə·rē ,kärs·ən'ō·mə }

medullary cord [ANAT] Dense lymphatic tissue separated by sinuses in the medulla of a lymph node. [EMBRYO] A primary invagination of the germinal epithelium of the embryonic gonad that differentiates into rete testis and seminiferous tubules or into rete ovarii. { mə'dəl·ə·rē 'kȯrd }

medullary ray [BOT] An extension of pith between vascular bundles in the plant stem. Also known as pith ray. { me'dəl·ə·rē ˌrā }

medulloblastoma [MED] A malignant neoplasm of the brain with a tendency to metastasize in the meninges. { məˌdəl·ə·bla'stō·mə }

medulloepithelioma [MED] **1.** A locally invasive tumor of the eye arising from the ciliary epithelium of the iris. **2.** *See* ependymoma. { mə ˌdəl·ōˌep·əˌthē·lē'ō·mə }

medusa *See* jellyfish. { mə'düs·ə }

MEG *See* magnetoencephalogram.

megacanthopore [INV ZOO] A large prominent tube, commonly projecting as a spine in a mature region of a bryozoan colony. { ˌmeg·ə'kan·thəˌpȯr }

Megachilidae [INV ZOO] The leaf-cutting bees, a family of hymenopteran insects in the superfamily Apoidea. { ˌmeg·ə'kil·əˌdē }

Megachiroptera [VERT ZOO] The fruit bats, a group of Chiroptera restricted to the Old World; most species lack a tail, but when present it is free of the interfemoral membrane. { ¦meg·ə·kī'räp·tə·rə }

megacine [MICROBIO] The bacteriocidin produced by *Bacillus megaterium*. { 'meg·ə·sən }

megacolon [MED] Hypertrophy and dilation of the colon associated with prolonged constipation. { 'meg·əˌkō·lən }

Megadermatidae [VERT ZOO] The false vampires, a family of tailless bats with large ears and a nose leaf; found in Africa, Australia, and the Malay Archipelago. { ˌmeg·ə·dər'mad·əˌdē }

megagametophyte [BOT] The female gametophyte in plants having two types of spores. { ¦meg·ə·gə'mēd·əˌfīt }

megakaryocyte [HISTOL] A giant bone-marrow cell characterized by a large, irregularly lobulated nucleus; precursor to blood platelets. { ¦meg·ə'kar·ē·əˌsīt }

megakaryocytopenia *See* megakaryophthisis. { ¦meg·əˌkar·ē·əˌsīd·ə'pē·nē·ə }

megakaryophthisis [MED] A scarcity of megakaryocytes in the bone marrow. Also known as megakaryocytopenia. { ¦meg·əˌkar·ē'äf·thə·səs }

megaloblast [PATH] A large nucleated erythroblast appearing in bone marrow in vitamin B_{12} or folic acid deficiency. { 'meg·ə·lōˌblast }

megaloblastic anemia [MED] Anemia characterized by the occurrence of megaloblasts in the bone marrow and blood. { ¦meg·ə·lō¦blas·tik ə'nē·mē·ə }

megaloblast of Sabin *See* pronormoblast. { 'meg·ə·lōˌblast əv 'sā·bən }

megalocardia [MED] Abnormal enlargement of the heart. { ¦meg·ə·lō'kär·dē·ə }

megalocephaly [MED] The condition of having a head whose maximum fronto-occipital circumference is greater than two standard deviations above the mean for age and sex. { ¦meg·ə·lō 'sef·ə·lē }

Megalodontoidea [INV ZOO] A superfamily of hymenopteran insects in the suborder Symphyta. { ¦meg·ə·lōˌdän'tȯid·ē·ə }

Megalomycteroidei [VERT ZOO] The mosaic-scaled fishes, a monofamilial suborder of the Cetomimiformes; members are rare species of small, elongate deep-sea fishes with degenerate eyes and irregularly disposed scales. { ˌmeg·ə·lōˌmik·tə'rȯid·ēˌī }

megalopia *See* macropsia. { ˌmeg·ə'lä·pē·ə }

megalops larva [INV ZOO] A preimago stage of certain crabs having prominent eyes and chelae. { 'meg·əˌläps ˌlär·və }

Megaloptera [INV ZOO] A suborder included in the order Neuroptera by some authorities. { ˌmeg·ə'läp·tə·rə }

Megalopygidae [INV ZOO] The flannel moths, a small family of lepidopteran insects in the suborder Heteroneura. { ¦meg·ə·lō'pij·əˌdē }

megalosphere [INV ZOO] The initial, large-chambered shell of sexual individuals of certain dimorphic species of Foraminifera. { 'meg·ə·lōˌsfir }

megaloureter [MED] Abnormal enlargement of a ureter. { ˌmeg·ə·lō'yür·əd·ər }

-megaly [MED] A combining form denoting abnormal enlargement. { 'meg·ə·lē }

Megamerinidae [INV ZOO] A family of myodarian cyclorrhaphous dipteran insects in the subsection Acalypteratae. { ˌmeg·ə·mə'rin·əˌdē }

megaphenic [GEN] Pertaining to genetic or environmental factors that are individually of large effect relative to the phenotypic standard deviation. { ˌmeg·ə'fē·nik }

megaphyllous [BOT] Having large leaves or leaflike extensions. { ¦meg·ə'fil·əs }

Megapodiidae [VERT ZOO] The mound birds and brush turkeys, a family of birds in the order Galliformes; distinguished by their method of incubating eggs in mounds of dirt or in decomposing vegetation. { ˌmeg·ə·pə'dī·əˌdē }

megarectum [MED] Abnormal enlargement of the rectum. { ¦meg·ə¦rek·təm }

megasclere [INV ZOO] A large sclerite. { 'meg·əˌsklir }

Megasphaera [MICROBIO] A genus of bacteria in the family Veillonellaceae; relatively large cells occurring in pairs arranged in chains. { mə'gas·fə·rə }

megasporangium *See* macrosporangium. { ¦meg·ə·spə'ran·jē·əm }

megaspore *See* macrospore. { 'meg·əˌspȯr }

megaspore mother cell *See* megasporocyte. { 'meg·əˌspȯr 'məth·ər ˌsel }

megasporocyte [BOT] A diploid cell from which four megaspores are produced by meiosis. Also known as megaspore mother cell. { ¦meg·ə'spȯr·əˌsīt }

megasporogenesis *See* macrosporogenesis. { ˌmeg·əˌspȯr·ə'jen·ə·səs }

megasporophyll [BOT] A leaf bearing megasporangia. { ˌmeg·ə'spȯr·əˌfil }

megathermophyte [ECOL] A plant that requires great heat and abundant moisture for normal growth. Also known as macrothermophyte. { ˌmeg·ə'thər·məˌfīt }

Megathymiinae [INV ZOO] The giant skippers, a

subfamily of lepidopteran insects in the family Hesperiidae. { ˌmeg·ə·thəˈmī·əˌnē }

meglumine *See N*-methyl glucamine. { ˈme·glə ˌmīn }

Mehlis' gland [INV ZOO] One of the large unicellular glands around the ootype of flatworms. { ˈmā·ləs ˌgland }

Meibomian cyst *See* chalazion. { mīˈbō·mē·ən ˈsist }

Meibomian gland *See* tarsal gland. { mīˈbō·mē·ən ˈgland }

meibomianitis [MED] Inflammation of the tarsal glands. { mīˌbō·mē·əˈnīd·əs }

Meig's syndrome [MED] A complex of symptoms associated with ovarian fibroma including abnormal accumulation of serous fluid in the pleural and peritoneal cavities. { ˈmegz ˌsin ˌdrōm }

Meinertellidae [INV ZOO] A family of wingless insects belonging to the Microcoryphia. { ˌmī·nərˈtel·əˌdē }

meiocyte [CELL MOL] A cell undergoing meiotic division. { ˈmī·əˌsīt }

meiofauna [ECOL] Small benthic animals ranging in size between macrofauna and microfauna; includes interstitial animals. { ¦mī·əˈfȯn·ə }

meioflora [ECOL] Small benthic plants ranging in size between macroflora and microflora; includes interstitial plants. { ¦mī·əˈflȯr·ə }

meiosis [CELL MOL] A type of cell division occurring in diploid or polyploid tissues that results in a reduction in chromosome number, usually by half. { mīˈō·səs }

meiospore [BIOL] A spore produced as the result of meiosis. { ˈmī·əˌspȯr }

meiotic drive [GEN] Preferential meiotic segregation favoring one chromosome over its homologue. { mē¦äd·ik ˈdrīv }

Meissner's corpuscle [NEUROSCI] An ovoid, encapsulated cutaneous sense organ presumed to function in touch sensation in hairless portions of the skin. { ˈmīs·nərz ˈkȯr·pə·səl }

Meissner's plexus *See* submucous plexus. { ˈmīs·nərz ˈplek·səs }

Melamphaidae [VERT ZOO] A family of bathypelagic fishes in the order Beryciformes. { ˌmel·əmˈfā·əˌdē }

Melampsoraceae [MYCOL] A family of parasitic fungi in the order Uredinales in which the teleutospores are laterally united to form crusts or columns. { ˌmel·əm·səˈrās·ēˌē }

Melanconiaceae [MYCOL] The single family of the order Melanconiales. { ˌmel·ənˌkō·nēˈās·ēˌē }

Melanconiales [MYCOL] An order of the class Fungi Imperfecti including many plant pathogens commonly causing anthracnose; characterized by densely aggregated cnidophores on an acervulus. { ˌmel·ənˌkō·nēˈā·lēz }

Melandryidae [INV ZOO] The false darkling beetles, a family of coleopteran insects in the superfamily Tenebrionoidea. { ˌmel·ənˈdrī·əˌdē }

melanin [BIOCHEM] Any of a group of brown or black pigments occurring in plants and animals. { ˈmel·ə·nən }

melanoblast [HISTOL] **1.** Precursor cell of melanocytes and melanophores. **2.** An immature pigment cell in certain vertebrates. **3.** A mature cell that elaborates melanin. { məˈlan·əˌblast }

melanoblastoma [MED] A malignant tumor composed principally of melanoblasts. { ¦mel·ə·nō·blaˈstō·mə }

melanocarcinoma [MED] A malignant melanoma derived from epithelial tissue. { ¦mel·ə·nōˌkärs·ənˈōˌmə }

melanocyte [HISTOL] A cell containing dark pigments. { məˈlan·əˌsīt }

melanocyte-stimulating hormone [BIOCHEM] A protein substance secreted by the intermediate lobe of the pituitary of humans which causes dispersion of pigment granules in the skin; similar to intermedins in other vertebrates. Abbreviated MSH. Also known as melanophore-dilating principle; melanophore hormone. { məˈlan·əˌsīt ¦stim·yəˌlād·iŋ ˈhȯrˌmōn }

melanocytoma [MED] A benign tumor composed principally of melanocytes. { məˌlan·əˌsīˈtō·mə }

melanocytosis [MED] An excessive number of melanocytes. { məˌlan·əˌsīˈtō·səs }

melanoderma [MED] Abnormal darkening of the skin. { ¦mel·ə·nōˈdər·mə }

melanogen [BIOCHEM] A colorless precursor of melanin. { məˈlan·ə·jən }

melanogenesis [BIOCHEM] The formation of melanin. { ˌmel·ə·nōˈjen·ə·səs }

melanoglosia [MED] Any blackening of the tongue associated with certain disorders and diseases in animals and humans. { ˌmel·ə·nō ˈgläs·ē·ə }

melanoid [MED] Dark-colored; resembling melanin. { ˈmel·əˌnȯid }

melanoma [MED] **1.** A malignant tumor composed of anaplastic melanocytes. **2.** A benign or malignant tumor composed of melanocytes. { mel·əˈnō·mə }

melanomatosis [MED] **1.** Widespread distribution of melanoma. **2.** Diffuse melanotic pigmentation of the meninges. { ˌmel·əˌnō·mə ˈtō·səs }

melanophage [HISTOL] A phagocytic cell which engulfs and contains melanin. { məˈlan·əˌfāj }

melanophore [HISTOL] A type of chromatophore containing melanin. { məˈlan·əˌfȯr }

melanophore-dilating principle *See* melanocyte-stimulating hormone. { məˈlan·əˌfȯr ˈdīˌlād·iŋ ˌprin·sə·pəl }

melanophore hormone *See* melanocyte-stimulating hormone. { məˈlan·əˌfȯr ˈhȯrˌmōn }

melanoprotein [BIOCHEM] A conjugated protein in which melanin is the associated chromagen. { ˌmel·ə·nōˈprōˌtēn }

melanosarcoma [MED] (Obsolete) A malignant melanoma. { ¦mel·ə·nō·särˈkō·mə }

melanosis coli [PATH] Melanotic pigmentation of the mucosa of the colon in large numbers of minute foci. { ˌmel·əˈnō·səs ˈkōˌlī }

melanosis iridis [PATH] Abnormal melanotic pigmentation of the iris. { ˌmel·əˈnō·səs iˈrīd·əs }

melanosome [CELL MOL] An organelle which contains melanin and in which tyrosinase activity is not demonstrable. { mə'lan·ə,sōm }

melanotic freckle [MED] An unevenly pigmented macule that sometimes develops on the skin, usually on the face, of an individual beyond middle life and enlarges progressively. Also known as Hutchinson's freckle. { ,mel·ə'näd·ik 'frek·əl }

melanuria [MED] The presence of black pigment in the urine. { ,mel·ə'nůr·ē·ə }

Melasidae [INV ZOO] The equivalent name for Eucnemidae. { mə'las·ə,dē }

Melastomataceae [BOT] A large family of dicotyledonous plants in the order Myrtales characterized by an inferior ovary, axile placentation, up to twice as many stamens as petals (or sepals), anthers opening by terminal pores, and leaves with prominent, subparallel longitudinal ribs. { ,mel·ə,stō·mə'tās·ē,ē }

melatonin [BIOCHEM] A hormone secreted by the pineal gland that acts on melanophores in the skins of amphibians and reptiles to concentrate the melanin in the center of the cells, lightening the body surface; in higher vertebrates it conveys information about time that influences reproduction and circadian physiology. { ,mel·ə'tōn·ən }

Meleagrididae [VERT ZOO] The turkeys, a family of birds in the order Galliformes characterized by a bare head and neck. { ,mel·ē·ə'grid·ə,dē }

melena [MED] The discharge of stools colored black by altered blood. { mə'lē·nə }

Meliaceae [BOT] A family of dicotyledonous plants in the order Sapindales characterized by mostly exstipulate, alternate leaves, stamens mostly connate by their filaments, and syncarpous flowers. { ,mel·ē'ās·ē,ē }

Melinae [VERT ZOO] The badgers, a subfamily of carnivorous mammals in the family Mustelidae. { 'mel·ə,nē }

Melinninae [INV ZOO] A subfamily of sedentary annelids belonging to the Ampharetidae which have a conspicuous dorsal membrane, with or without dorsal spines. { mə'lin·ə,nē }

melioidosis [VET MED] An endemic bacterial disease, primarily of rodents but occasionally communicable to humans, caused by *Pseudomonas pseudomallei* and characterized by infectious granulomas. { ,mel·ē,ȯi'dō·səs }

Meliolaceae [MYCOL] The sooty molds, a family of ascomycetous fungi in the order Erysiphales, with dark mycelia and conidia. { ,mel·ē·ə'lās·ē,ē }

melitose *See* raffinose. { 'mel·ə,tōs }

melitriose *See* raffinose. { mə'lī·trē,ōs }

Melittidae [INV ZOO] A family of hymenopteran insects in the superfamily Apoidea. { mə'lid·ə,dē }

melituria [MED] The presence of sugar in the urine. { ,mel·ə'tůr·ē·ə }

Meloidae [INV ZOO] The blister beetles, a large cosmopolitan family of coleopteran insects in the superfamily Meloidea; characterized by soft, flexible elytra and the strongly vesicant properties of the body fluids. { mə'lō·ə,dē }

Meloidea [INV ZOO] A superfamily of coleopteran insects in the suborder Polyphaga. { mə'lȯid·ē·ə }

melon [BOT] Either of two soft-fleshed edible fruits, muskmelon or watermelon, or varieties of these. [VERT ZOO] Around mass of fat on the forehead of some cetaceans between the blowhole and nose. { 'mel·ən }

melting profile [BIOCHEM] A plot of the degree of denaturation of the strands in a nucleic acid duplex in a specified time as a function of temperature. { 'melt·iŋ ,prō,fīl }

melting temperature [BIOCHEM] The temperature at which denaturing occurs for half of the double helices of deoxyribonucleic acid. { 'melt·iŋ ,tem·prə·chər }

Melusinidae [INV ZOO] A family of orthorrhaphous dipteran insects in the series Nematocera. { ,mel·ə'sin·ə,dē }

Melyridae [INV ZOO] The soft-winged flower beetles, a large family of cosmopolitan coleopteran insects in the superfamily Cleroidea. { mə'lir·ə,dē }

Membracidae [INV ZOO] The treehoppers, a family of homopteran insects included in the series Auchenorrhyncha having a pronotum that extends backward over the abdomen, and a vertical upper portion of the head. { mem'bras·əd·ē }

membrane [HISTOL] A thin layer of tissue surrounding a part of the body, separating adjacent cavities, lining cavities, or connecting adjacent structures. { 'mem,brān }

membrane bone *See* dermal bone. { 'mem,brān ,bōn }

membrane carrier [CELL MOL] Any protein that facilitates the movement of small molecules across cell membranes. { 'mem,brān ,kar·ē·ər }

membrane potential [PHYSIO] A potential difference across a living cell membrane. { 'mem ,brān pə,ten·chəl }

membranous glomerulonephritis [MED] A type of glomerulonephritis characterized by thickening of the basement membrane due to deposition of electron-dense material. { 'mem·brə·nəs glə¦mer·yə·lō·ne'frīd·əs }

membranous labyrinth [ANAT] The membranous portion of the inner ear of vertebrates. { 'mem·brə·nəs 'lab·ə,rinth }

membranous pregnancy [MED] Gestation in which there has been a rupture of the amniotic sac and the fetus is in direct contact with the wall of the uterus. { 'mem·brə·nəs 'preg·nən·sē }

membranous urethra [ANAT] The part of the urethra between the two facial layers of the urogenital diaphragm. { 'mem,brə·nəs yů'rē·thrə }

memory trace *See* engram. { 'mem·rē 'trās }

men-, meno- [PHYSIO] A combining form denoting menses. { men, 'men·ō }

menacme [PHYSIO] The period of a woman's life during which menstruation persists. { mə'nak·mē }

menarche [PHYSIO] The onset of menstruation. { mə'när·kē }

Mendelian genetics [GEN] Scientific study of the role of the nuclear genome in heredity, as opposed to cytoplasmic inheritance. { men 'dēl·yən jə'ned·iks }

Mendelian population [GEN] A group of interbreeding individuals; the total allelic gene content of the group is called their gene pool. { men'dēl·yən ,päp·yə'lā·shən }

Mendelian ratio [GEN] The ratio of occurrence of various phenotypes in F_1 and F_2 generations in any cross involving characters controlled by nuclear genes. { men'dēl·yən 'rā·shō }

Mendelism [GEN] The basic laws of inheritance as formulated by Mendel. { 'men·də'līz·əm }

Mendel's laws [GEN] Two basic principles of genetics formulated by Mendel: the law of segregation of alleles of a unit factor (gene), and the law of independent assortment of alleles of different unit factors. { 'men·dəlz ,lȯz }

Menetrier's disease [MED] Benign, diffuse hypertrophic gastritis; symptoms include vomiting, diarrhea, weight loss, and excessive secretion of mucus { 'men·ə,trirz di,zēz }

Ménière's syndrome [MED] A disease of the inner ear characterized by deafness, vertigo, and tinnitus; possibly an allergic process. Also known as labyrinthine syndrome { ¦mā·nē¦erz ,sin,drōm }

meninges [ANAT] The membranes that cover the brain and spinal cord; there are three in mammals and one or two in submammalian forms { mə'nin·jēz }

meninginococcemia [MED] **1.** The presence of meningococci in the blood **2.** A clinical disorder consisting of fever, skin hemorrhages, varying degrees of shock, and meningococci in the blood { mə,niŋ·gō,käk'sē·mē·ə }

meningioma [MED] A localized tumor composed of meningeal cells, involving the meninges and other central nervous system structures. Also known as meningothelioma. { mə,nin·jē'ō·mə }

meningism [MED] A condition in which signs and symptoms suggest meningitis, but clinical evidence for the disease is absent. Also known as meningismus. { mə'nin,jiz·əm }

meningismus *See* meningism. { ,men·ən'jiz·məs }

meningitis [MED] Inflammation of the meninges of the brain and spinal cord, caused by viral, bacterial, and protozoan agents. { ,men·ən'jīd·əs }

meningocele [MED] Hernia of the meninges through a defect in the skull or vertebral column, forming a cyst filled with cerebrospinal fluid. { mə'niŋ·gə,sēl }

meningococcal meningitis [MED] Inflammation of the meninges caused by the bacterium *Neisseria meningitidis* (meningococcus). { mə¦niŋ·gə¦käk·əl ,men·ən'jīd·əs }

meningococcus [MICROBIO] Common name for *Neisseria meningitidis*. { mə¦niŋ·gə¦käk·əs }

meningoencephalitis [MED] Inflammation of the brain and its meninges. { mə¦niŋ·gō·in,sef·ə'līd·əs }

meningoencephalocele [MED] A protrusion of the brain and its membranes through a defect in the skull. { mə¦niŋ·gō·in'sef·ə·lō,sēl }

meningoencephalomyelitis [MED] Combined inflammation of the meninges, brain, and spinal cord. { mə¦niŋ·gō·in,sef·ə·lō,mī·ə'līd·əs }

meningoencephalopathy [MED] Disease of the brain and its meninges. { mə¦niŋ·gō·in,sef·ə'läp·ə,thē }

meningomyelitis [MED] Inflammation of the spinal cord and its membranes. { mə,niŋ·gō,mī·ə'līd·əs }

meningomyocele [MED] Hernia of the spinal cord and its meninges through a defect in the vertebral column. { mə¦niŋ·gō'mī·ə,sēl }

meningothelioma *See* meningioma. { mə'niŋ·gə,thē·lē'ō·mə }

meningothelium [HISTOL] Epithelium of the arachnoid which envelops the brain. { mə,niŋ·gə'thē·lē·əm }

meningovascular [MED] Involving both the meninges and the cerebral blood vessels. { mə ,niŋ·gə'vas·kyə·lər }

meningovascular syphilis [MED] Syphilis of the central nervous system involving the formation of gummas of the leptomeninges and endarteritis of cerebral vessels. { mə,niŋ·gə'vas·kyə·lər 'sif·ə·ləs }

meninx [ANAT] Any one of the three membranes covering the brain and spinal cord. { 'me ,niŋks }

meninx primitiva [VERT ZOO] The single membrane covering the brain and spinal cord of certain submammalian vertebrates. { 'me,niŋks ,prim·ə'tī·və }

meniscectomy [MED] Surgical removal of a meniscus or semilunar cartilage. { ,men·ə'sek·tə,mē }

meniscitis [MED] Inflammation of the semilunar cartilages. { ,men·ə'sīd·əs }

meniscus [ANAT] A crescent-shaped body, especially an interarticular cartilage. { mə'nis·kəs }

Menispermaceae [BOT] A family of dicotyledonous woody vines in the order Ranunculales distinguished by mostly alternate, simple leaves, unisexual flowers, and a dioecious habit. { ,men·ə·spər'mās·ē,ē }

menometrorrhagia [MED] Excessive uterine bleeding during menstruation, plus irregular uterine bleeding at other times. { ,men·ə,me·trə'rā·jē·ə }

menopause [PHYSIO] The natural physiologic cessation of menstruation, usually occurring in the last half of the fourth decade. Also known as climacteric. { 'men·ə,pȯz }

menoplania [MED] Bleeding during menstruation from a part of the body other than the uterus. { ,men·ə'plā·nē·ə }

Menoponidae [INV ZOO] A family of biting lice (Mallophaga) adapted to life only upon domestic and sea birds. { ,men·ə'pän·ə,dē }

menorrhagia [MED] Excessive bleeding during

menstruation. Also known as hypermenorrhea. { ˌmen·əˈrā·jē·ə }

menorrhalgia [MED] Pelvic pain occurring at the menstrual period. { ˌmen·əˈral·jē·ə }

menorrhea [MED] Excessive menstruation. [PHYSIO] The normal flow of the menses. { ˌmen·əˈrē·ə }

menostasis [MED] Suppression of the menstrual flow. { məˈnäs·tə·səs }

menostaxis [MED] Prolonged menstruation. { ˌmen·əˈstak·səs }

menses *See* menstruation. { ˈmenˌsēz }

menstrual age [EMBRYO] The age of an embryo or fetus calculated from the first day of the mother's last normal menstruation preceding pregnancy. { ˈmen·strə·wəl ˈāj }

menstrual cycle [PHYSIO] The periodic series of changes associated with menstruation and the intermenstrual cycle; menstrual bleeding indicates onset of the cycle. { ˈmen·strə·wəl ˈsī·kəl }

menstrual period [PHYSIO] The time of menstruation. { ˈmen·strə·wəl ˈpir·ē·əd }

menstruate [PHYSIO] To discharge the products of menstruation. { ˈmen·strəˌwāt }

menstruation [PHYSIO] The periodic discharge of sanguineous fluid and sloughing of the uterine lining in women from puberty to the menopause. Also known as menses. { ˌmen·strəˈwā·shən }

mental [ANAT] Pertaining to the chin. Also known as genial. { ˈmen·təl }

Menthaceae [BOT] An equivalent name for Labiatae. { menˈthās·ēˌē }

mentum [ANAT] The chin. [BOT] A projection formed by union of the sepals at the base of the column in some orchids. [INV ZOO] **1.** A projection between the mouth and foot in certain gastropods. **2.** The median or basal portion of the labium in insects. { ˈmen·təm }

Menurae [VERT ZOO] A small suborder of suboscine perching birds restricted to Australia, including the lyrebirds and scrubbirds. { məˈnyu̇r·ē }

Menuridae [VERT ZOO] The lyrebirds, a family of birds in the suborder Menurae notable for their vocal mimicry. { məˈnyu̇r·əˌdē }

meperidine hydrochloride [PHARM] $C_{15}H_{21}NO_2$ A narcotic compound that is used medicinally as an analgesic and sedative. { məˌpir·əˌdēn ˌhī·drəˈklȯr·īd }

meprobromate [PHARM] $C_9H_{18}N_2O_4$ The compound 2-methyl-2-*n*-propyl-1,3-propanediol dicarbamate, a tranquilizer with anticonvulsant, muscle relaxant, and sedative actions. { ¦me·prōˈbrōˌmāt }

meralluride [PHARM] $C_9H_{16}HgN_2O_6$ A diuretic consisting of succinamic acid and theophylline, in approximately molecular proportions, administered as the sodium derivative. { məˈral·yəˌrīd }

mercurialism [MED] Chronic type of mercury poisoning. Also known as hydrargyrism. { mərˈkyu̇r·ē·əˌliz·əm }

mercurial nephrosis [MED] Nephrosis caused by poisoning with mercury bichloride. { mərˈkyu̇r·ē·əl neˈfrō·səs }

mercurial tremor [MED] A fine muscular tremor observed in persons with mercurialism or poisoning by other heavy metals. { mərˈkyu̇r·ē·əl ˈtrem·ər }

merganser [VERT ZOO] Any of several species of diving water fowl composing a distinct subfamily of Anatidae and characterized by a serrate bill adapted for catching fish. { mərˈgan·sər }

mericarp [BOT] An individual, one-seeded carpel of a schizocarp. { ˈmer·əˌkärp }

mericlinal chimera [BIOL] An organism or organ composed of two genetically different tissues, one of which partly surrounds the other. { ¦mer·əˌklīn·əl ˈkim·ə·rə }

Meridosternata [INV ZOO] A suborder of echinoderms including various deep-sea forms of sea urchins. { ¦mer·ə·dō·stərˈnäd·ə }

meristem [BOT] Formative plant tissue composed of undifferentiated cells capable of dividing and giving rise to other meristematic cells as well as to specialized cell types; found in growth areas. { ˈmer·əˌstem }

meristic [BIOL] Pertaining to a change in number or in geometric relation of parts of an organism. [ZOO] Of, pertaining to, or divided into segments. { məˈris·tik }

Merkel's corpuscles [NEUROSCI] Touch receptors consisting of flattened platelets at the tips of certain cutaneous nerves. { ˈmər·kəlz ˌkər·pə·səlz }

Mermithidae [INV ZOO] A family of filiform nematodes in the superfamily Mermithoidea; only juveniles are parasitic. { mərˈmith·əˌdē }

Mermithoidea [INV ZOO] A superfamily of nematodes composed of two families, both of which are invertebrate parasites. { mər·məˈthȯid·ē·ə }

meroblastic [EMBRYO] Of or pertaining to an ovum that undergoes incomplete cleavage due to large amounts of yolk. { ¦mer·ə¦blas·tik }

merocrine [PHYSIO] Pertaining to glands in which the secretory cells undergo cytological changes without loss of cytoplasm during secretion. { ˈmer·ə·krən }

merogony [EMBRYO] The normal or abnormal development of a part of an egg following cutting, shaking, or centrifugation of the egg before or after fertilization. { məˈräj·ə·nē }

meromixis [GEN] Genetic exchange in bacteria involving a unidirectional transfer of a partial genome. { ˌmer·əˈmik·səs }

meromyarian [INV ZOO] Having few muscle cells in each quadrant as seen in cross section; applied especially to nematodes. { ¦mer·ə·mī¦ar·ē·ən }

meromyosin [BIOCHEM] Protein fragments of a myosin molecule, produced by limited proteolysis. { ˌmer·əˈmī·ə·sən }

Meropidae [VERT ZOO] The bee-eaters, a family of brightly colored Old World birds in the order Coraciiformes. { məˈräp·əˌdē }

meroplankton [BIOL] Plankton composed of floating developmental stages (that is, eggs and larvae) of the benthos and nekton organisms. Also known as temporary plankton. { ¦mer·ə'plaŋk·tən }

merospermy [CELL MOL] Fusion of an egg with an anucleate sperm. { mi'räs·pər·mē }

merosporangium [MYCOL] A cylindrical sporangium containing sporangiospores in a row or chain-like formation. { ˌmer·ə·spə'ran·jē·əm }

Merostomata [INV ZOO] A class of primitive arthropods of the subphylum Chelicerata distinguished by their aquatic mode of life and the possession of abdominal appendages which bear respiratory organs; only three living species are known. { ˌmer·ə'stō·mə·də }

-merous [BIOL] Combining form that denotes having such parts or so many parts. { mər·əs }

Merozoa [INV ZOO] The equivalent name for Cestoda. { ˌmer·ə'zō·ə }

merozoite [INV ZOO] An ameboid trophozoite in some sporozoans produced from a schizont by schizogony. { ˌmer·ə'zōˌīt }

merozygote [MICROBIO] In bacteria, a zygote that has some diploid and some haploid genetic material because a chromosomal fragment was transferred by the F+ mate. { ˌmir·ə'zīˌgōt }

merycism *See* rumination. { 'mer·əˌsiz·əm }

mesappendix [ANAT] The mesentery of the vermiform appendix. { ˌmes·ə'pen·diks }

mesarch [BOT] Having metaxylem on both sides of the protoxylem in a siphonostele. [ECOL] Originating in a mesic environment. { 'meˌzärk }

mesarteritis [MED] Inflammation of the tunica media of an artery. { ¦mezˌärd·ə'rīd·əs }

mescal buttons [BOT] The dried tops from the cactus *Lophophora williamsii*; capable of producing inebriation and hallucinations. { me'skal ˌbət·ənz }

mesectoderm [EMBRYO] The portion of the mesenchyme arising from ectoderm. { mə 'zek·təˌdərm }

mesencephalon [EMBRYO] The middle portion of the embryonic vertebrate brain; gives rise to the cerebral peduncles and the tectum. Also known as midbrain. { ¦mez·ən'sef·əˌlän }

mesenchymal cell [HISTOL] An undifferentiated cell found in mesenchyme and capable of differentiating into various specialized connective tissues. { ¦mez·ən¦kī·məl ˌsel }

mesenchymal epithelium [HISTOL] A layer of squamous epithelial cells lining subdural, subarachnoid, and perilymphatic spaces, and the chambers of the eyeball. { ¦mez·ən¦kī·məl ˌep·ə'thē·lē·əm }

mesenchymal hyalin [PATH] A form of hyalin which results from degeneration or necrosis of nonepithelial tissue, usually of muscle, as in Zenker's hyaline necrosis, or of blood vessels. { ¦mez·ən¦kī·məl 'hī·ə·lən }

mesenchymal tissue [EMBRYO] Undifferentiated tissue composed of branching cells embedded in a coagulable fluid matrix. { ¦mez·ən¦kī·məl 'tish·ü }

mesenchyme [EMBRYO] That part of the mesoderm from which all connective tissues, blood vessels, blood, lymphatic system proper, and the heart are derived. { 'mez·ənˌkīm }

mesenchymoma [MED] A tumor composed of cells resembling those of embryonic mesenchyme, or of mesenchyme with its derivatives. { ¦mez·ən·kī'mō·mə }

mesendoderm [EMBRYO] Embryonic tissue which differentiates into mesoderm and endoderm. { mə'zen·dəˌdərm }

mesenteric [ANAT] Of or pertaining to the mesentery. { ¦mez·ən¦ter·ik }

mesenteric artery [ANAT] Either of two main arterial branches arising from the abdominal aorta: the inferior, supplying the descending colon and the rectum, and the superior, supplying the small intestine, the cecum, and the ascending and transverse colon. { ¦mez·ən¦ter·ik 'ärd·ə·rē }

mesenteric lymphadenitis [MED] Inflammation of the lymph nodes in the mesentery. { ¦mez·ən¦ter·ik limˌfad·ən'īd·əs }

mesenteron [EMBRYO] *See* midgut. [INV ZOO] Central gastric cavity in an actinozoan. { me'zen·təˌrän }

mesentery [ANAT] A fold of the peritoneum that connects the intestine with the posterior abdominal wall. { 'mez·ənˌter·ē }

mesentoderm [EMBRYO] **1.** The entodermal portion of the mesoderm. **2.** Undifferentiated tissue from which entoderm and mesoderm are derived. **3.** That part of the mesoderm which gives rise to certain structures of the digestive tract. { mə'zen·təˌdərm }

mesethmoid [ANAT] A bone or cartilage in the center of the ethmoid region of the vertebrate skull; usually constitutes the greater portion of the nasal septum. { me'zethˌmȯid }

mesh texture *See* reticulate. { 'mesh ˌteks·chər }

mesic [ECOL] **1.** Of or pertaining to a habitat characterized by a moderate amount of water. **2.** Of or pertaining to a mesophyte. { 'me·zik }

mesoappendix [ANAT] The mesentery of the vermiform appendix. { ¦me·zō·ə¦pen·diks }

mesobilirubin [BIOCHEM] $C_{33}H_{40}O_6N_4$ Yellow, crystalline by-product of bilirubin reduction. { ¦me·zōˌbil·i'rü·bən }

mesobilirubinogen [BIOCHEM] $C_{33}H_{44}O_6N_4$ Colorless, crystalline by-product of bilirubin reduction; may be converted to urobilin, stercobilinogen, or stercobilin. { ¦me·zōˌbil·i·rə'bin·ə·jən }

mesobiliverdin [BIOCHEM] $C_{28}H_{38}O_6N_4$ A structural isomer of phycoerythrin and phycocyanobilin released by certain biliprotein by treatment with alkali. { ¦me·zōˌbil·i'vərd·ən }

mesoblast [EMBRYO] Undifferentiated mesoderm of the middle layer of the embryo. { 'me·zōˌblast }

mesoblastema *See* mesoderm. { ˌme·zō·bla'stē·mə }

mesocardium [ANAT] Epicardium covering the blood vessels which enter and leave the heart. [EMBRYO] The mesentery supporting the embryonic heart. { ˌme·zō'kärd·ē·əm }

mesocarp [BOT] The middle layer of the pericarp. { 'mez·əˌkärp }

mesocercaria [INV ZOO] The developmental stage in the second intermediate host of *Alaria*, a digenetic trematode. { ˌme·zō·sər'kar·ē·ə }

mesocolon [ANAT] The part of the mesentery that is attached to the colon. { ¦me·zō'kō·lən }

mesoderm [EMBRYO] The third germ layer, lying between the ectoderm and endoderm; gives rise to the connective tissues, muscles, urogenital system, vascular system, and the epithelial lining of the coelom. Also known as mesoblastema. { 'mez·əˌdərm }

mesodermal tumor [MED] A tumor composed of cells normally derived from the mesoderm. { ¦mez·ə¦dər·məl 'tü·mər }

mesogaster [ANAT] The mesentery of the stomach. { 'mez·əˌgas·tər }

Mesogastropoda [INV ZOO] The equivalent name for Pectinibranchia. { ˌmez·ə·ga'sträp·ə·də }

mesoglea [INV ZOO] The gelatinous layer between the ectoderm and endoderm in cnidarians and certain sponges. { ¦me·zō'glē·ə }

mesoinositol *See* myoinositol. { ¦me·zō·i'näs·əˌtȯl }

mesokaryotic [INV ZOO] Pertaining to an organism that shares characteristics of both prokaryotic and eukaryotic organisms. { ˌmez·əˌkar·ē'äd·ik }

mesolamella [INV ZOO] A thin gelatinous membrane between the epidermis and gastrodermis in hydrozoans. { ¦me·zō·lə'mel·ə }

mesomere [EMBRYO] The muscle-plate region between the epimere and hypomere in vertebrates. { 'me·zōˌmir }

mesometrium [ANAT] The part of the broad ligament attached directly to the uterus. { ˌmez·ə'mē·trē·əm }

mesonephric duct [EMBRYO] The efferent duct of the mesonephros. Also known as Wolffian duct. { ¦me·zō¦nef·rik 'dəkt }

mesonephric fold *See* mesonephric ridge. { ¦me·zō¦nef·rik 'fōld }

mesonephric ridge [EMBRYO] A fold of the dorsal wall of the coelom lateral to the mesentery formed by development of the mesonephros. Also known as mesonephric fold. { ¦me·zō¦nef·rik 'rij }

mesonephros [EMBRYO] One of the middle of three pairs of embryonic renal structures in vertebrates; persists in adult fish and is replaced by the metanephros in higher forms. { ¦mez·ə'neˌfrōs }

mesophile [BIOL] An organism, as certain bacteria, that grows at moderate temperature. { 'mez·əˌfīl }

mesophily [ECOL] Physiological response of organisms living in environments with moderate temperatures and a fairly high, constant amount of moisture. { 'mez·əˌfil·ē }

mesophyll [BOT] Parenchymatous tissue between the upper and lower epidermal layers in foliage leaves. { 'mez·əˌfil }

mesophyte [ECOL] A plant requiring moderate amounts of moisture for optimum growth. { 'mez·əˌfīt }

mesopterygium [VERT ZOO] The middle one of three basal cartilages in the pectoral fin of sharks and rays. { ¦meˌzäp·tə'rij·ē·əm }

mesoptic vision [PHYSIO] Vision in which the human eye's spectral sensitivity is changing from the photoptic state to the scotoptic state. { me'zäp·tik 'vizh·ən }

mesorchium [EMBRYO] The mesentery that supports the embryonic testis in vertebrates. { me'zȯr·kē·əm }

mesosalpinx [ANAT] The portion of the broad ligament forming the mesentery of the uterine tube. { ¦me·zō'salˌpiŋks }

mesosere [ECOL] A sere originating in a mesic habitat and characterized by mesophytes. { 'me·zōˌsir }

mesosoma [INV ZOO] **1.** The anterior portion of the abdomen in certain arthropods. **2.** The middle of the body of some invertebrates, especially when the phylogenetic segmentation pattern cannot be determined. { ¦me·zō'sō·mə }

mesosome [MICROBIO] An extension of the cell membrane within a bacterial cell; possibly involved in cross-wall formation, cell division, and the attachment of daughter chromosomes following deoxyribonucleic acid replication. { 'mez·əˌsōm }

mesosternum [ANAT] The middle portion of the sternum in vertebrates. Also known as gladiolus. [INV ZOO] The ventral portion of the mesothorax in insects. { ¦me·zō'stər·nəm }

Mesostigmata [INV ZOO] The mites, a suborder of the Acarina characterized by a single pair of breathing pores (stigmata) that are located laterally in the middle of the idiosoma between the second and third, or third and fourth, legs. { ¦me·zōˌstig'mäd·ə }

Mesotaeniaceae [BOT] The saccoderm desmids, a family of fresh-water algae in the order Conjugales; cells are oval, cylindrical, or rectangular and have simple, undecorated walls in one piece. { ˌme·zōˌtē·nē'ās·ēˌē }

Mesotardigrada [INV ZOO] An order of tardigrades which combines certain echiniscoidean features with eutardigradan characters. { ¦me·zōˌtär'dig·rə·də }

mesotheca [INV ZOO] The middle lamina of bifoliate bryozoan colonies. { ¦me·zō'thē·kə }

mesothelioma [MED] A primary benign or malignant tumor composed of cells resembling the mesothelium. { ˌme·zōˌthē·lē'ō·mə }

mesothelium [ANAT] The simple squamous-cell epithelium lining the pleural, pericardial, peritoneal, and scrotal cavities. [EMBRYO] The lining of the wall of the primitive body cavity situated between the somatopleure and splanchnopleure. { ˌme·zō'thē·lē·əm }

mesotherm [ECOL] A plant that grows successfully at moderate temperatures. { 'mez·əˌthərm }

mesothorax [INV ZOO] The middle of three somites composing the thorax in insects. { ¦me·zō'thȯˌraks }

mesovarium [ANAT] A fold of the peritoneum

that connects the ovary with the broad ligament. { ˌme·zō'var·ē·əm }

Mesoveliidae [INV ZOO] The water treaders, a small family of hemipteran insects in the subdivision Amphibicorisae having well-developed ocelli. { ˌme·zō·və'lī·əˌdē }

mesoxyalyurea *See* alloxan. { məˌzäk·sē·al'yůr·ē·ə }

Mesozoa [INV ZOO] A division of the animal kingdom sometimes ranked intermediate between the Protozoa and the Metazoa; composed of two orders of small parasitic, wormlike organisms. { ˌmez·ə'zō·ə }

mesozooid [INV ZOO] A type of bryozoan heterozooid that produces slender tubes (mesozooecia or mesopores), internally subdivided by many closely spaced diaphragms, that open as tiny polygonal apertures. { ¦mez·ə¦zōˌȯid }

mesquite [BOT] Any plant of the genus *Prosopis*, especially *P. juliflora*, a spiny tree or shrub bearing sugar-rich pods; an important livestock feed. { mə'skēt }

messenger ribonucleic acid [NEUROSCI] A linear sequence of nucleotides which is transcribed from and complementary to a single strand of deoxyribonucleic acid and which carries the information for protein synthesis to the ribosomes. Abbreviated mRNA. { 'mes·ən·jər ¦rī·bō·nü'klē·ik 'as·əd }

metabiosis [ECOL] An ecological association in which one organism precedes and prepares a suitable environment for a second organism. { ˌmed·ə·bī'ō·səs }

metabolic [PHYSIO] Of or pertaining to metabolism. { ˌmed·ə'bäl·ik }

metabolic block [BIOCHEM] A nonfunctional reaction in a metabolic pathway due to a defective enzyme whose normal counterpart catalyzes the reaction. { ˌmed·ə¦bäl·ik 'bläk }

metabolic disorder [MED] Any disorder that involves an alteration in the normal metabolism of carbohydrates, lipids, proteins, water, and nucleic acids; evidenced by various syndromes and diseases. { ˌmed·ə'bäl·ik dis'ȯrd·ər }

metabolism [PHYSIO] The physical and chemical processes by which foodstuffs are synthesized into complex elements (assimilation, anabolism), complex substances are transformed into simple ones (disassimilation, catabolism), and energy is made available for use by an organism. { mə'tab·əˌliz·əm }

metabolite [BIOCHEM] A product of intermediary metabolism. { mə'tab·əˌlīt }

metabolize [PHYSIO] To transform by metabolism; to subject to metabolism. { mə'tab·əˌlīz }

metacarpus [ANAT] The portion of a hand or forefoot between the carpus and the phalanges. { ¦med·ə'kär·pəs }

metacentric [CELL MOL] Having the centromere near the middle of the chromosome. { ¦med·ə'sen·trik }

metacercaria [INV ZOO] Encysted cercaria of digenetic trematodes; the infective form. { ¦med·ə·sər'kar·ē·ə }

metacestode [INV ZOO] Encysted larva of a tapeworm; occurs in the intermediate host. { ¦med·ə'sesˌtōd }

Metachlamydeae [BOT] An artificial group of flowering plants, division Magnoliophyta, recognized in the Englerian system of classification; consists of families of dicotyledons in which petals are characteristically fused, forming a sympetalous corolla. { ˌmed·ə·klə'mid·ēˌē }

metachromatic granules [CELL MOL] Granules which assume a color different from that of the dye used to stain them. { ¦med·ə·krō'mad·ik 'gran·yülz }

metachromatic leukodystrophy [MED] A hereditary degenerative disease transmitted as an autosomal recessive, due to sulfatase A deficiency, with excess accumulation of sulfated lipids responsible for metachromasia in various tissues. Abbreviated MLD. Also known as sulfatide lipidosis. { ¦med·ə·krō'mad·ik ˌlü·kə'dis·trə·fē }

metachrosis [VERT ZOO] The ability of some animals to change color by the expansion and contraction of chromatophores. { ˌmed·ə'krō·səs }

metacneme [INV ZOO] A secondary mesentery in many zoantharians. { 'meˌtakˌnēm }

metacone [VERT ZOO] **1.** The posterior of three cusps of primitive upper molars. **2.** The posteroexternal cusp of an upper molar in higher vertebrates, especially mammals. { 'med·əˌkōn }

metaconid [VERT ZOO] The posteroexternal cusp of a lower molar in mammals; corresponds with the metacone. { ˌmed·ə'kō·nəd }

metagenesis [BIOL] The phenomenon in which one generation of certain plants and animals reproduces asexually, followed by a sexually reproducing generation. Also known as alternation of generations. { ¦med·ə'jen·ə·səs }

metagranulocyte *See* metamyelocyte. { ˌmed·ə'gran·yə·ləˌsīt }

metakaryocyte *See* normoblast. { ˌmed·ə'kar·ē·əˌsīt }

metal ague *See* metal fume fever. { 'med·əl ˌāg }

metal fume fever [MED] A febrile influenzalike occupational disorder following the inhalation of finely divided particles and fumes of metallic oxides. Also known as brass chills; brass founder's ague; galvo; Monday fever; metal ague; polymer fume fever; spelter shakes; teflon shakes; zinc chills. { 'med·əl ˌfyüm 'fē·vər }

metallobiochemistry *See* bioinorganic chemistry. { məˌtal·ōˌbī·ō'kem·ə·strē }

metallochaperones [BIOCHEM] **1.** Metalloproteins that aid in the insertion of the appropriate metal ion into a metalloenzyme. **2.** A family of proteins that shuttles metal ions to specific intracellular locations where metalloenzymes bind to the metal ions and use them as cofactors to carry out essential biochemical reactions. { məˌtal·ō'shap·əˌrōnz }

metalloenzymes [BIOCHEM] Metalloproteins that catalyze important cellular reactions. { məˌtal·ō 'enˌzīmz }

Metallogenium [MICROBIO] A genus of bacteria of uncertain affiliation; coccoid cells that attach to substrate; they germinate directly or form groups of elementary bodies by budding, and filaments form from these bodies. { mə,tal·ə'jē·nē·əm }

metalloporphyrin [BIOCHEM] A compound, such as heme, consisting of a porphyrin combined with a metal such as iron, copper, silver, zinc, or magnesium. { mə,tal·ō'pȯr·fə·rən }

metalloprotein [BIOCHEM] A protein enzyme containing a metallic atom as an inherent portion of its molecule. { mə,tal·ō'prō,tēn }

metallothionein [BIOCHEM] A group of vertebrate and invertebrate proteins that bind heavy metals; it may be involved in zinc homeostasis and resistance to heavy-metal toxicity. { mə,tal·ō'thī·ə,nēn }

metamere [ZOO] One of the linearly arranged similar segments of the body of metameric animals. Also known as somite. { 'med·ə,mir }

metamerism [ZOO] The condition of an animal body characterized by the repetition of similar segments (metameres), exhibited especially by arthropods, annelids, and vertebrates in early embryonic stages and in certain specialized adult structures. Also known as segmentation. { mə'tam·ə,riz·əm }

metamorphosis [BIOL] **1.** A structural transformation. **2.** A marked structural change in an animal during postembryonic development. [MED] A degenerative change in tissue or organ structure. { ,med·ə'mȯr·fə·səs }

metamyelocyte [HISTOL] A granulocytic cell intermediate in development between the myelocyte and granular leukocyte; characterized by a full complement of cytoplasmic granules and a bean-shaped nucleus. Also known as juvenile cell; metagranulocyte. { ,med·ə'mī·ə·lō,sīt }

metanauplius [INV ZOO] A primitive larval stage of certain decapod crustaceans characterized by seven pairs of appendages; follows the nauplius stage. { ,med·ə'nȯ·plē·əs }

metanephridium [INV ZOO] A type of nephridium consisting of a tubular structure lined with cilia which opens into the coelomic cavity. { ¦med·ə·ne'frid·ē·əm }

metanephrine [BIOCHEM] An inactive metabolite of epinephrine (3-O-methylepinephrine) that is excreted in the urine; it is recovered and measured as a test for pheochromocytoma. { ,med·ə'ne,frən }

metanephros [EMBRYO] One of the posterior of three pairs of vertebrate renal structures; persists as the definitive or permanent kidney in adult reptiles, birds, and mammals. { ,med·ə'ne ,fräs }

metanitricyte *See* normoblast. { ,med·ə'nī·trə ,sīt }

metaphase [CELL MOL] **1.** The phase of mitosis during which centromeres are arranged on the equator of the spindle. **2.** The phase of the first meiotic division when centromeric regions of homologous chromosomes come to lie equidistant on either side of the equator. { 'med·ə,fāz }

metaphase plate [CELL MOL] At metaphase, the imaginary plane located half way between the spindle poles along which the kinetochore microtubules align their chromosomes. { 'med·ə,fāz ,plāt }

metaphloem [BOT] The primary phloem that forms after differentiation of the protophloem. { ¦med·ə'flō·əm }

metaphyseal aclasis *See* multiple hereditary exostoses. { mə¦taf·ə¦sē·əl a'klas·əs }

metaphysis *See* epiphyseal plate. { mə'ta·fə·səs }

Metaphyta [BIOL] A kingdom set up to include mosses, ferns, and other plants in some systems of classification. { mə'taf·əd·ə }

metaplasia [PATH] Transformation of one form of tissue to another. { ¦med·ə'plā·zhə }

metaplasm [CELL MOL] The ergastic substance of protoplasm. { 'med·ə,plaz·əm }

metapodium [ANAT] **1.** The metatarsus in bipeds. **2.** The metatarsus and metacarpus in quadrupeds. [INV ZOO] Posterior portion of the foot of a mollusk. { ¦med·ə'pō·dē·əm }

metapodosoma [INV ZOO] Portion of the body bearing the third and fourth pairs of legs in Acarina. { ,med·ə,päd·ə'sō·mə }

metapterygium [VERT ZOO] The posterior one of three basal cartilages in the pectoral fin of sharks and rays. { mə¦tap·tə¦rij·ē·əm }

metarubricyte *See* normoblast. { ,med·ə'rü·brə,sīt }

metascutellum [INV ZOO] The scutellum of the metathorax in insects. { ¦med·ə·skü'tel·əm }

metasicula [INV ZOO] The succeeding part of the sicula or colonial tube of graptolites. { ¦med·ə'sik·yə·lə }

metasoma [INV ZOO] The posterior region of the body of certain invertebrates, a term used especially when the phylogenetic segmentation pattern cannot be identified. { ,med·ə'sō·mə }

metastasis [MED] Transfer of the causal agent (cell or microorganism) of a disease from a primary focus to a distant one through the blood or lymphatic vessels. { mə'tas·tə·səs }

metastasize [MED] To be transferred by metastasis. { mə'tas·tə,sīz }

metastatic anemia *See* myelophthisic anemia. { ,med·ə'stad·ik ə'nē·mē·ə }

metastatic capacity [MED] The malignancy of a tumor. { ,med·ə,stad·ik kə'pas·əd·ē }

metasternum [INV ZOO] The ventral portion of the metathorax in insects. { ¦med·ə'stər·nəm }

metastoma [INV ZOO] Median plate posterior to the mouth in certain crustaceans and related arthropods. { mə'tas·tə·mə }

Metastrongylidae [INV ZOO] A family of roundworms belonging to the Strongyloidea; species are parasitic in sheep, cattle, horses, dogs, and other domestic animals. { ¦med·ə,strän'jil·ə,dē }

Metastrongyloidea [INV ZOO] A superfamily of parasitic nematodes, characterized by a reduced or absent stoma capsule and an oral opening surrounded by six well-developed lips. { ¦med·ə,strän·jə'lȯid·ē·ə }

metatarsal [ANAT] Of or pertaining to the metatarsus. { ¦med·ə'tär·səl }

metatarsalgia [MED] Tenderness and burning

pain in the metatarsal region. { ˌmed·əˌtär'sal·jē·ə }

metatarsus [ANAT] The part of a foot or hindfoot between the tarsus and the phalanges. { ˌmed·ə'tär·səs }

Metatheria [VERT ZOO] An infraclass of therian mammals including a single order, the Marsupialia; distinguished by a small braincase, a total of 50 teeth, the inflected angular process of the mandible, and a pair of marsupial bones articulating with the pelvis. { ˌmed·ə'thir·ē·ə }

metathorax [INV ZOO] Posterior segment of the thorax in insects. { ¦med·ə'thórˌaks }

metatroch [INV ZOO] A segmented larval form following the trochophore in annelids. { 'med·əˌträk }

metaxylem [BOT] Primary xylem differentiated after and distinguished from protoxylem by thicker tracheids and vessels with pitted or reticulated walls. { ¦med·ə'zī·ləm }

Metazoa [ZOO] The multicellular animals that make up the major portion of the animal kingdom; cells are organized in layers or groups as specialized tissues or organ systems. { ˌmed·ə'zō·ə }

metazoea [INV ZOO] The last zoea of certain decapod crustaceans; metamorphoses into a megalopa. { ˌmed·ə·zō'ē·ə }

metencephalon [EMBRYO] The cephalic portion of the rhombencephalon; gives rise to the cerebellum and pons. { ˌmed·in'sef·əˌlän }

Meteoriaceae [BOT] A family of mosses in the order Isobryales in which the calyptra is frequently hairy. { ˌmēd·ēˌór·ē'ās·ēˌē }

meteorism [MED] Presence of abdominal gas causing severe distention. { 'mēd·ē·əˌriz·əm }

metestrus [PHYSIO] The beginning of the luteal phase following estrus. { med'es·trəs }

methadone [PHARM] $C_{21}H_{27}NO$ The compound 6-(dimethylamino)-4,4-diphenyl-3-heptanone, a narcotic analgesic, administered in the hydrochloride form for maintenance treatment of heroin addiction. { 'meth·əˌdän }

methanamide *See* formamide. { meth'an·ə·mid }

methane monooxygenase [BIOCHEM] A nonheme iron enzyme that catalyzes the oxidation of methane to methanol. { ˌmethˌān ˌmän·ō'ak·sə·jəˌnās }

methane-oxidizing bacteria [MICROBIO] Bacteria that derive energy from oxidation of methane. { 'methˌān ¦äk·səˌdīz·iŋ bak'tir·ē·ə }

Methanobacteriaceae [MICROBIO] The single family of methane-producing bacteria; anaerobes which obtain energy via formation of methane. { ¦meth·ə·nōˌbak·tir·ē'ās·ēˌē }

methanogen [BIOL] An organism carrying out methanogenesis, requiring completely anaerobic conditions for growth; considered by some authorities to be distinct from bacteria. { mə'than·ə·jən }

methanogenesis [BIOCHEM] The biosynthesis of the hydrocarbon methane; common in certain bacteria. Also known as bacterial methanogenesis. { ¦meth·ə·nō'jen·ə·səs }

Methanomonadaceae [MICROBIO] Formerly a family of bacteria in the suborder Pseudomonadineae; members identified as gram-negative rods are able to use carbon monoxide (*Carboxydomonas*), methane (*Methanomonas*), and hydrogen (*Hydrogenomonas*) as their sole source of energy for growth. { ¦meth·ə·nōˌmän·ə'dās·ēˌē }

methanotroph [MICROBIO] A bacterial organism that can use methane as its only source of carbon and energy. { mə'than·əˌträf }

methemoglobin *See* ferrihemoglobin. { metˌhē·mə'glō·bən }

methemoglobinemia [MED] The presence of methemoglobin in the blood. { ¦metˌhē·mə·glō·bə'nē·mē·ə }

methemoglobinuria [MED] The presence of methemoglobin in the urine. { ¦metˌhe·mə·glō·bə'nyůr·ē·ə }

methionine [BIOCHEM] $C_5H_{11}O_2NS$ An essential amino acid; furnishes both labile methyl groups and sulfur necessary for normal metabolism. { mə'thī·əˌnēn }

methotrexate *See* amethopterin. { ˌmeth·ə 'trekˌsāt }

methylated cap [CELL MOL] A modified guanine nucleotide that terminates a messenger ribonucleic acid molecule. { ¦meth·əˌlād·əd 'kap }

***N*-methyl glucamine** [PHARM] $C_7H_{17}NO_5$ A crystalline compound with a melting point of 128–129°C; used in medicine as a drug for leishmaniasis. Also known as meglumine. { ¦en ¦meth·əl 'glü·kəˌmēn }

methyl jasmonate [BOT] An enzymatic product of lipid metabolism produced via activation of lipoxygenase enzymes in most plant species when tissues are damaged or infected by pathogens; it acts as an external signal effecting communication among plants. { ˌmeth·əl 'jaz·məˌnāt }

Methylomonadaceae [MICROBIO] A family of gram-negative, aerobic bacteria which utilize only one-carbon compounds as a source of carbon. { ˌmeth·ə·lōˌmän·ə'dās·ēˌē }

methylotrophic bacteria [MICROBIO] Bacteria that are capable of growing on methane derivatives as their sole source of carbon and metabolic energy. { ¦meth·ə·lə¦trä·fik bak'tir·ē·ə }

methyl red test [MICROBIO] A cultural test for the ability of bacteria to ferment carbohydrate to form acid; uses methyl red as the indicator. { 'meth·əl ¦red ˌtest }

methylthiouracil [PHARM] $C_5H_6N_2OS$ A crystalline compound; used as a medicine for overactivity of the thyroid. Also known as 6-methyl-2-thiouracil. { ¦meth·əl¦thī·ō'yůr·ə·səl }

6-methyl-2-thiouracil *See* methylthiouracil. { ¦siks 'meth·əl ¦tü ¦thī·ō'yůr·ə·səl }

methyl transferase [BIOCHEM] Any of a group of enzymes which catalyze the reaction of S-adenosyl methionine with a suitable acceptor to yield the methylated acceptor molecule and S-adenosyl homocysteine. { 'meth·əl 'tranz·fəˌrās }

metopan hydrochloride [PHARM] $C_{18}H_{21}O_3N\cdot HCl$ A morphine derivative; a white, crystalline powder soluble in water; used as a sedative. { 'med·ə,pan ,hī·drə'klȯr,īd }

metraterm [INV ZOO] The distal portion of the uterus in trematodes. { 'mē·trə,tərm }

metria [MED] **1.** Any pathologic condition of the uterus. **2.** Any uterine inflammation occurring within the 6 weeks following childbirth. { 'mē·trē·ə }

Metridiidae [INV ZOO] A family of zoantharian cnidarians in the order Actiniaria. { ,me·trə'dī·ə,dē }

metritis [MED] Inflammation of the uterus, usually involving both the endometrium and myometrium. { mə'trīd·əs }

metrorrhagia [MED] Uterine bleeding during the intermenstrual cycle. Also known as intermenstrual flow; polymenorrhea. { ,mē·trə'rā·jē·ə }

metrorrhea [MED] Any pathologic discharge from the uterus. { ,mē·trə'rē·ə }

metrorrhexis [MED] Rupture of the uterine wall. { ¦mē·trə¦rek·səs }

metrosalpingitis [MED] Inflammation of the uterus and oviducts. { ,mē·trə,sal·pən'jīd·əs }

metrostaxis [MED] Slight, chronic bleeding from the uterus. { ,me·trə'stak·səs }

Metzgeriales [BOT] An order of liverworts in the subclass Jungermannidae, class Hepaticopsida, distinguished by archegonia produced behind a growing apex, a flat elongated gametophyte with no tissue differentiation or surface pores and, less commonly, a stem with two rows of leaves. { ,mets·gə·rē'ā·lēz }

Meyliidae [INV ZOO] A family of free-living nematodes in the superfamily Desmoscolecoidea. { ma'lī·ə,dē }

Meziridae [INV ZOO] A family of hemipteran insects in the superfamily Aradoidea. { mə'zir·ə,dē }

MHC *See* major histocompatibility complex.

MIC *See* major immunogene complex.

micelle [CELL MOL] A submicroscopic structural unit of protoplasm built up from polymeric molecules. { mī'sel }

Michaelis constant [BIOCHEM] A constant K_m such that the initial rate of reaction V, produced by an enzyme when the substrate concentration is high enough to saturate the enzyme, is related to the rate of reaction v at a lower substrate concentration c by the formula $V = v(1 + K_m/c)$. { mi'kā·ləs ,kän·stənt }

Michaelis-Menten equation [BIOCHEM] A mathematical equation expressing the hyperbolic relationship between the initial velocity, V_o, and the substrate concentration, [S], in a number of enzyme-catalyzed reactions such that $V_o = V_{max}[S]/K_m + [S]$, where V_{max} is the maximum velocity and K_m is the Michaelis constant. { ,mik·ä¦ā·ləs 'men·tən i,kwā·zhən }

micracanthopore [INV ZOO] Small, minute tubes projecting from the surface of bryozoan colonies. { ,mī·krə'kan·thrə,pȯr }

micrencephaly [MED] The condition of having an abnormally small brain. Also spelled microencephaly. { ¦mī·kren¦sef·ə·lē }

microabscess [MED] A small abscess. { ¦mī·krō'ab,ses }

microaerophilic [MICROBIO] Pertaining to those microorganisms requiring free oxygen but in very low concentration for optimum growth. { ¦mī·krō¦er·ə¦fil·ik }

microanatomy [ANAT] Anatomical study of microscopic tissue structures. { ¦mī·krō·ə'nad·ə·mē }

microaneurysm [MED] Dilation of the wall of a capillary, characteristic of certain disease entities. { ¦mī·krō'an·yə,riz·əm }

microangiopathy [MED] The development of lesions in small blood vessels throughout the body. { ,mī·krō,an·jē'äp·ə·thē }

microbe [MICROBIO] A microorganism, especially a bacterium of a pathogenic nature. { ¦mī,krōb }

microbial ecology [ECOL] The study of interrelationships between microorganisms and their living and nonliving environments. { mī,krōb·ē·əl ē'käl·ə·jē }

microbial insecticide [MICROBIO] Species-specific bacteria which are pathogenic for and used against injurious insects. { mī'krō·bē·əl in 'sek·tə,sīd }

microbiologist [MICROBIO] A scientist that studies a wide range of microorganisms in various subdisciplines of biology, such as bacteriology, mycology, parasitology, and virology. { ,mī·krō·bī'äl·ə·jist }

microbiology [MICROBIO] The science and study of microorganisms, including protozoans, algae, fungi, bacteria, viruses, and rickettsiae. { ¦mī·krō·bī'äl·ə·jē }

microbody [CELL MOL] Any of three distinct classes (peroxisomes, glyoxysomes, and microperoxisomes) of cytoplasmic organelles that are bounded by a single membrane and contain a variety of enzymes. { 'mī·krə,bäd·ē }

microcell [CELL MOL] A micronucleus within a layer of cytoplasm and a membrane. { 'mī·krə,sel }

microcentrum [CELL MOL] The centrosome, or a group of centrosomes, functioning as the dynamic center of a cell. { 'mī·krō,sen·trəm }

microcephalus [MED] An individual with microcephaly. { ,mī·krō'sef·ə·ləs }

microcephaly [MED] The condition of having an abnormally small head, with a circumference less than two standard deviations below the mean. { ,mī·krō'sef·ə·lē }

microceratous [INV ZOO] Having short antennae. { ,mī·krō'ser·ə·təs }

microcercous cercaria [INV ZOO] A cercaria with a very short broad tail. { ¦mi·krō¦sər·kəs sər'kar·ē·ə }

microchemistry [BIOCHEM] The chemistry of individual cells and minute organisms. { ¦mī·krō'kem·ə·strē }

Microchiroptera [VERT ZOO] A suborder of the mammalian order Chiroptera composed of the insectivorous bats. { ¦mī·krō·kī'räp·tə·rə }

microcirculation [PHYSIO] The flow of blood or

lymph in the vessels of the microcirculatory system. { ¦mī·krō,sər·kyə'lā·shən }

microcirculatory system [ANAT] Those vessels of the blood and lymphatic systems which are visible only with a microscope. { ¦mī·krō'sər·kyə·lə,tór·ē ,sis·təm }

microcneme [INV ZOO] Microsepta in certain anemones. { 'mī·krə,nēm }

Micrococcaceae [MICROBIO] A family of gram-positive cocci; chemoorganotrophic organisms with respiratory or fermentative metabolism. { ¦mī·krō·käk'sās·ē,ē }

microconsumer *See* decomposer. { ¦mī·krō·kən'sü·mər }

Microcotyloidea [INV ZOO] A superfamily of ectoparasitic trematodes in the subclass Monogenea. { ¦mī·krō,käd·əl'óid·ē·ə }

Microcyprini [VERT ZOO] The equivalent name for Cyprinodontiformes. { ,mī·krō·sə'prē,nē }

microcyst [MED] A very small cyst. { 'mī·krə ,sist }

microcyte [MED] A red blood cell whose diameter or mean corpuscular volume or both are more than two standard deviations below the normal mean. Also known as microerythrocyte. { 'mī·krə,sīt }

microcythemia [MED] Blood characterized by the presence of small red blood cells. { ¦mī·krō·sī'thē·mē·ə }

microcytic anemia [MED] Any form of anemia in which small erythrocytes occur in the blood. { ¦mī·krə¦sid·ik ə'nē·mē·ə }

microcytosis [MED] A blood disorder characterized by a preponderance of microcytes. { ,mī·krə·sī'tō·səs }

microdactyly [MED] A condition of abnormal smallness of fingers or toes. { ,mī·krō'dak·tə·lē }

microdissection [BIOL] Dissection under a microscope. { ¦mī·krō·di'sek·shən }

microencephaly *See* micrencephaly. { ¦mī·krō·en'sef·ə·lē }

microenvironment [ECOL] The specific environmental factors in a microhabitat. { ¦mī·krō·in'vī·ərn·mənt }

microerythrocyte *See* microcyte. { ¦mī·krō·ə'rith·rə,sīt }

microevolution [EVOL] **1.** Evolutionary processes resulting from the accumulation of minor changes over a relatively short period of time; evolutionary changes due to gene mutation. **2.** Evolution of species. { ¦mī·krō,ev·ə'lü·shən }

microfauna [ECOL] Microscopic animals such as protozoa and nematodes. { 'mī·krə,fón·ə }

microfibril [CELL MOL] The submicroscopic unit of a microscopic cellular fiber. { ¦mī·krō'fi·brəl }

microfilament [CELL MOL] One of the cytoplasmic fibrous structures, about 5 nanometers in diameter, virtually identical to actin; thought to be important in the processes of phagocytosis and pinocytosis. { ¦mī·krō'fil·ə·mənt }

microfilaria [INV ZOO] Slender, motile prelarval forms of filarial nematodes measuring 150–300 micrometers in length; adult filaria are mammalian parasites. { ¦mī·krō·fə'lar·ē·ə }

microflora [BOT] Microscopic plants. [ECOL] The flora of a microhabitat. { ¦mī·krō'flór·ə }

microgamete [BIOL] The smaller, or male gamete produced by heterogametic species. { ¦mī·krō'ga,mēt }

microgametocyte [BIOL] A cell that gives rise to microgametes. { ¦mī·krō·gə'mēd·ə,sīt }

microgametophyte [BOT] The male gametophyte in plants having two types of spores. { ¦mī·krō·gə'mēd·ə,fīt }

microgamy [BIOL] Sexual reproduction by fusion of the small male and female gametes in certain protozoans and algae. { mī'kräg·ə·mē }

microgastria [MED] A condition of abnormal smallness of the stomach. { ¦mī·krō'gas·trē·ə }

microgenesis [BIOL] Abnormally small development of a part. { ¦mī·krō'jen·ə·səs }

microgenitalism [MED] Having abnormally small genitalia. { ,mī·krō'jen·əd·əl,iz·əm }

microglia [NEUROSCI] Small neuroglia cells of the central nervous system having long processes and exhibiting ameboid and phagocytic activity under certain pathologic conditions. { mī'kräg·lē·ə }

microglossia [MED] A condition of abnormal smallness of the tongue. { ¦mī·krō'gläs·ē·ə }

micrognathia [MED] A condition of abnormal smallness of the jaws, particularly the mandible. { ,mī·krō'nā·thē·ə }

microgyria [MED] A condition of abnormal smallness of the gyri of the brain. { ¦mī·krə¦jī·rē·ə }

microhabitat [ECOL] A small, specialized, and effectively isolated location. { ¦mī·krō'hab·ə,tat }

Microhylidae [VERT ZOO] A family of anuran amphibians in the suborder Diplasiocoela including many heavy-bodied forms with a pointed head and tiny mouth. { ,mī·krō'hī·lə,dē }

microinfarct [MED] A very small infarct. { ¦mī·krō'in,färkt }

microinjection [CELL MOL] Injection of cells with solutions by using a micropipet. { ¦mī·krō·in'jek·shən }

microinvasion [MED] Invasion by tumor, especially a squamous-cell carcinoma of the uterine cervix, a very short distance into the tissues beneath the point of origin. { ¦mī·krō·in'vā·zhən }

microirradiation [BIOPHYS] A technique in which a laser beam is focused through the objective of a microscope onto a single cell to study the photosensitivity of its various parts. { ,mī·krō·i,rād·ē'ā·shən }

Microlepidoptera [INV ZOO] A former division of Lepidoptera. { ¦mī·krō,lep·ə'däp·tə·rə }

microlith [MED] A calculus of microscopic size. { 'mī·krə,lith }

microlithiasis [MED] The presence of numerous microliths. { ¦mī·krō·li'thī·ə·səs }

microlithiasis alveolaris pulmonum [MED] A

rare form of pulmonary calcification of unidentified etiology in which microliths, and larger osseous nodules, are found. { ¦mī·krō·li'thī·ə·səs ˌal·vē·ə'lar·əs pu̇l'mō·nəm }

Micromalthidae [INV ZOO] A family of coleopteran insects in the superfamily Cantharoidea; the single species is the telephone pole beetle. { ˌmī·krō'mȯl·thəˌdē }

micromanipulation [BIOL] The techniques and practice of microdissection, microvivisection, microisolation, and microinjection. { ¦mī·krō·məˌnip·yə'lā·shən }

micromere [EMBRYO] A small blastomere of the upper or animal hemisphere in eggs that undergo uneven cleavage. { 'mī·krəˌmir }

Micromonospora [MICROBIO] A genus of bacteria in the family Micromonosporaceae; the mycelium is well developed, branched, and septate; single spores are formed on hyphae.

Micromonosporaceae [MICROBIO] A family of bacteria in the order Actinomycetales; aerial hyphae are formed in all genera except *Micromonospora*; saprophytic soil organisms.

Micromonospora purpurea [MICROBIO] The bacterium that produces the antibiotic gentamycin.

micronekton [ECOL] Active pelagic crustaceans and other forms intermediate between thrusting nekton and feebler-swimming plankton. { ˌmī·krə'nekˌtän }

micronucleus [INV ZOO] The smaller, reproductive nucleus in multinucleate protozoans. { ¦mī·krō'nü·klē·əs }

micronutrient [BIOCHEM] An element required by animals or plants in small amounts. { ¦mī·krō'nü·trē·ənt }

microorganism [MICROBIO] A microscopic organism, including bacteria, protozoans, yeast, viruses, and algae. { ¦mī·krō'ȯr·gəˌniz·əm }

Micropezidae [INV ZOO] A family of myodarian cyclorrhaphous dipteran insects in the subsection Acalypteratae. { ˌmī·krō'pez·əˌdē }

microphage [HISTOL] A small phagocyte, especially a neutrophil. { 'mī·krəˌfāj }

microphagy [BIOL] Feeding on minute organisms or particles. { mī'kräf·ə·jē }

microphakia [MED] Congenital condition of abnormal smallness of the crystalline lens. { ˌmī·krō'fā·kē·ə }

microphenic [GEN] Pertaining to genetic or environmental factors that are numerous but individually of small effect relative to the phenotypic standard deviation. { ˌmī·krə'fen·ik }

microphthalmus [MED] A condition characterized by an abnormally small eyeball. Also known as nanophthalmus. { ˌmīˌkräf'thal·məs }

microphyllous [BOT] **1.** Having small leaves. **2.** Having leaves with a single, unbranched vein. { ¦mī·krō¦fil·əs }

Microphysidae [INV ZOO] A palearctic family of hemipteran insects in the subfamily Cimicimorpha. { ˌmī·krə'fīs·əˌdē }

microphyte [ECOL] **1.** A microscopic plant. **2.** A dwarfed plant due to unfavorable environmental conditions. { 'mī·krəˌfīt }

micropinocytosis [CELL MOL] A mechanism of endocytosis in which fluid droplets are internalized by indentations (caveolae) on the surface membrane which pinch off as tiny internal vesicles (micropinosomes). { ¦mī·krōˌpin·ə·sī'tō·səs }

micropinosome [CELL MOL] A very tiny vesicle that is pinched off from the plasma membrane of a cell during micropinocytosis. { ¦mī·krō'pin·əˌsōm }

micropipette [MED] An extremely fine glass capillary tube with a narrow tip on the order of micrometers that is used to extract or deliver minute quantities of fluid. { ¦mī·krō·pī'pet }

microplankton [ECOL] Zooplankton between 20 and 200 micrometers in size. { 'mī·krə ˌplaŋk·tən }

micropsia [MED] A visual disturbance in which objects appear undersized. { mī'kräp·sē·ə }

Micropterygidae [INV ZOO] The single family of the lepidopteran superfamily Micropterygoidea; members are minute moths possessing toothed, functional mandibles and lacking a proboscis. { mīˌkräp·tə'rij·əˌdē }

Micropterygoidea [INV ZOO] A monofamilial superfamily of lepidopteran insects in the suborder Homoneura. { mīˌkräp·tə·rə'gȯid·ē·ə }

Micropygidae [INV ZOO] A family of echinoderms in the order Diadematoida that includes only one genus, *Micropyga*, which has noncrenulate tubercles and umbrellalike outer tube feet. { ˌmī·krə'pij·əˌdē }

micropyle [BOT] A minute opening in the integument at the tip of an ovule through which the pollen tube commonly enters; persists in the seed as an opening or a scar between the hilum and point of radicle. { 'mī·krəˌpīl }

microsatellite deoxyribonucleic acid [CELL MOL] Any of a series of repeated motifs of a two to six base-pair sequence that are scattered throughout the nuclear genome and used as landmarks during physical mapping. Abbreviated msDNA. { ˌmī·krō¦sad·əlˌīt dēˌäk·sēˌrī·bō·nüˌklē·ik 'as·əd }

microsclere [INV ZOO] A minute sclerite in Porifera. { 'mī·krəˌsklir }

microscopical diagnosis [PATH] Identification of a disease by microscopic examination of specimens taken from the patient. { ˌmi·krə'skäp·ə·kəl ˌdī·əg'nō·səs }

microseptum [INV ZOO] An incomplete or imperfect mesentery in zoantharians. { ¦mī·krō 'sep·təm }

microsome [CELL MOL] **1.** A fragment of the endoplasmic reticulum. **2.** A minute granule of protoplasm. { 'mī·krəˌsōm }

microspecies [ECOL] A small, localized species population that is clearly differentiated from related forms. Also known as jordanon. { ¦mī·krō'spē·shēz }

microspike [CELL MOL] Any of the narrow cytoplasmic projections that extend or retract from the surface of a cell and may have a sensory function. { 'mi·krəˌspīk }

Microsporaceae [BOT] A monogeneric family

of green algae in the suborder Ulotrichineae; the chloroplast is a parietal network. { ˌmī·krō·spə'rās·ēˌē }

microsporangium [BOT] A sporangium bearing microspores. { ¦mī·krō·spə'ran·jē·əm }

microspore [BOT] The smaller spore of heterosporous plants; gives rise to the male gametophyte. { 'mī·krəˌspȯr }

microspore mother cell *See* microsporocyte. { 'mī·krəˌspȯr 'məth·ər ˌsel }

Microsporida [INV ZOO] The single order of the class Microsporidea. { ˌmī·krə'spȯr·ə·də }

Microsporidae [INV ZOO] The equivalent name for Sphaeriidae. { ˌmī·krə'spȯr·əˌdē }

Microsporidea [INV ZOO] A class of Cnidospora characterized by the production of minute spores with a single intrasporal filament or one or two intracapsular filaments and a single sporoplasm; mainly intracellular parasites of arthropods and fishes. { ˌmī·krə·spə'rid·ē·ə }

microsporidiosis [VET MED] Infection with microsporidians. { ˌmī·krō·spəˌrid·ē'ō·səs }

microsporocyte [BOT] A diploid cell from which four microspores are produced by meiosis. Also known as microspore mother cell. { ¦mī·krō'spȯr·əˌsīt }

microsporogenesis [BOT] In angiosperms, formation of microspores and production of the male gametophyte. { ˌmī·krəˌspȯr·ə'jen·ə·səs }

microsporophyll [BOT] A sporophyll bearing microsporangia. { ¦mī·krō'spȯr·əˌfil }

microsurgery [BIOL] Surgery on single cells by micromanipulation. { ¦mī·krō'sər·jə·rē }

Microtatobiotes [BIOL] An artificial taxonomic category, comprising two unrelated groups of biological entities, the rickettsiae and the viruses. { mī¦kräd·ə·dōˌbī'ōd·ēz }

microtherm [ECOL] A plant requiring a mean annual temperature range of 0–14°C for optimum growth. { 'mī·krəˌthərm }

Microtinae [VERT ZOO] A subfamily of rodents in the family Muridae that includes lemmings and muskrats. { mī'krät·ənˌē }

microtomy [BIOL] Cutting of thin sections of specimens with a microtome. { mī'kräd·ə·mē }

microtrabecular lattice [CELL MOL] A network of thin filaments that interconnect the cytoplasmic filaments. { ˌmī·krə·trə¦bek·yə·lər 'lad·əs }

microtrichia [INV ZOO] Small hairs on the integument of various insects, especially on the wings. { ˌmī·krō'trik·ē·ə }

microtubule [CELL MOL] One of the hollow tubelike filaments found in certain cell components, such as cilia and the mitotic spindle, and composed of repeating subunits of the protein tubulin. { ¦mī·krō'tüb·yül }

microtubule-associated protein [CELL MOL] A protein that enhances the rate and extent of the polymerization of intracellular microtubules or that modifies their properties once formed. Abbreviated MAP. { ¦mī·krōˌtüb·yül a¦sō·sēˌād·ed 'prōˌtēn }

microvillus [CELL MOL] One of the filiform processes that form a brush border on the surfaces of certain specialized cells, such as intestinal epithelium. { ¦mī·krō'vil·əs }

Microviridae [VIROL] A family of nontailed bacterial viruses (bacteriophages) that is characterized by icosahedral particles containing a single-stranded circular deoxyribonucleic acid genome, members infect enterobacteria. { ˌmī·krō'vir·əˌdī }

Microvirus [VIROL] The only genus of the family Microviridae. { 'mī·krōˌvī·rəs }

microwave thermography [MED] A method of measuring temperature through the detection of microwave radiation emitted from heated tissue during a therapeutic procedure. { 'mī·krəˌwāv thər'mäg·rə·fē }

microzoospermia [MED] A condition of abnormal smallness of sperm in the semen. { ¦mī·krō¦zō·ō'spər·mē·ə }

mictic [BIOL] **1.** Requiring or produced by sexual reproduction. **2.** Of or pertaining to eggs which without fertilization develop into males and with fertilization develop into amictic females, as occurs in rotifers. { 'mik·tik }

micturition *See* urination. { ˌmik·chə'ri·shən }

midaxillary line [ANAT] A perpendicular line drawn downward from the apex of the axilla. { ˌmid'ak·səˌler·ē 'līn }

midbrain [ANAT] Those portions of the adult brain derived from the embryonic midbrain. [EMBRYO] The middle portion of the embryonic vertebrate brain. Also known as mesencephalon. { 'midˌbrān }

midclavicular line [ANAT] A vertical line parallel to and midway between the midsternal line and a vertical line drawn downward through the outer end of the clavicle. { 'mid·klə¦vik·yə·lər 'līn }

middle ear [ANAT] The middle portion of the ear in higher vertebrates; in mammals it contains three ossicles and is separated from the external ear by the tympanic membrane and from the inner ear by the oval and round windows. { 'mid·əl 'ir }

middle lamella [CELL MOL] The layer of a cell wall that is derived from the phragmoplast. { 'mid·əl lə'mel·ə }

middle lobe syndrome [MED] A complex of symptoms due to enlarged lymph nodes compressing the bronchus of the right middle lobe, causing atelectasis, bronchiectasis, or chronic pneumonitis of the lobe. { 'mid·əl ¦lōb 'sin ˌdrōm }

midge [INV ZOO] Any of various dipteran insects, principally of the families Ceratopogonidae, Cecidomyiidae, and Chironomidae; many are biting forms and are vectors of parasites of man and other vertebrates. { mij }

midget [MED] An individual who is abnormally small, but otherwise normal. { 'mij·ət }

midgut [EMBRYO] The middle portion of the digestive tube in vertebrate embryos. Also known as mesenteron. [INV ZOO] The mesodermal intermediate part of an invertebrate intestine. { 'midˌgət }

midrib [BOT] The large central vein of a leaf. { 'mid,rib }

migraine [MED] Recurrent paroxysmal vascular headache, commonly having unilateral onset and often associated with nausea and vomiting. { 'mī,grān }

migrant [ZOO] An animal that moves from one habitat to another. { 'mī·grənt }

migration [GEN] The transfer of genetic information among populations by the movement of individuals or groups of individuals from one population into another. [VERT ZOO] Periodic movement of animals to new areas or habitats. { mī'grā·shən }

Mikulicz's disease [MED] Enlargement of salivary and lacrimal glands from any of various causes. { 'mi·kü,lich·əz dī,zēz }

mildew [MYCOL] **1.** A whitish growth on plants, organic matter, and other materials caused by a parasitic fungus. **2.** Any fungus producing such growth. { 'mil,dü }

miliaria [MED] An acute inflammatory skin disease, the lesions consisting of vesicles and papules, which may be accompanied by a prickling or tingling sensation. Also known as heat rash; prickly heat. { ,mil·ē'ar·ē·ə }

Milichiidae [INV ZOO] A family of myodarian cyclorrhaphous dipteran insects in the subsection Acalypteratae. { ,mil·ə'kī·ə,dē }

milieu interieur [PHYSIO] The fundamental concept that the living organism exists in an aqueous internal environment which bathes all tissues and provides a medium for the elementary exchange of nutrients and waste. { mēl'yü in ¦tir·ē·ər }

Miliolacea [INV ZOO] A superfamily of marine or brackish foraminiferans in the suborder Miliolina characterized by an imperforate test wall of tiny, disordered calcite rhombs. { ,mil·ē·ə'lās·ē·ə }

Miliolidae [INV ZOO] A family of foraminiferans in the superfamily Miliolacea. { ,mil·ē'äl·ə,dē }

Miliolina [INV ZOO] A suborder of the Foraminiferida characterized by a porcelaneous, imperforate calcite wall. { ,mil·ē'äl·ə·nə }

milk [PHYSIO] **1.** The whitish fluid secreted by the mammary gland for the nourishment of the young; composed of carbohydrates, proteins, fats, mineral salts, vitamins, and antibodies. **2.** Any whitish fluid in nature resembling milk, as coconut milk. { milk }

milk-alkali syndrome [MED] A complex of symptoms associated with prolonged excessive intake of milk and soluble alkali, including hypercalcemia, renal insufficiency, milk alkalosis, conjunctivitis, and calcinosis. Also known as Burnett's syndrome; milk-drinker's syndrome. { 'milk 'al·kə,lī ,sin,drōm }

milk-drinker's syndrome *See* milk-alkali syndrome. { 'milk ,driŋ·kərz 'sin,drōm }

milk factor [BIOCHEM] A filtrable, noncellular agent in the milk and tissues of certain strains of inbred mice; transmitted from the mother to the offspring by nursing. Also known as Bittner milk factor. { 'milk ,fak·tər }

milk fever [MED] A fever occurring during the first six weeks after childbirth, believed to be caused by puerperal infection. { 'milk ,fē·vər }

milk intolerance [MED] Extreme sensitivity to milk due to allergy to milk protein or lactose deficiency; characterized by diarrhea, abdominal cramps, and vomiting. { 'milk in'täl·ə·rəns }

milk leg *See* phlegmasia alba dolens. { 'milk ,leg }

milk line *See* mammary ridge. { 'milk ,līn }

Milkman's syndrome [MED] Decreased tubular reabsorption of phosphate, resulting in osteomalacia which gives a peculiar striped appearance (multiple pseudofractures) to the bones in roentgenograms. Also known as Looser-Milkman syndrome. { 'milk,manz ,sin,drōm }

milk of magnesia [PHARM] A white suspension of magnesium hydroxide in water; used as a cathartic. Also known as magnesia magma. { 'milk əv mag'nē·zhə }

milk sugar *See* lactose. { 'milk ,shu̇g·ər }

milk teeth *See* deciduous teeth. { 'milk ,tēth }

milkweed [BOT] Any of several latex-secreting plants of the genus *Asclepias* in the family Asclepiadaceae. { 'milk,wēd }

milky disease [INV ZOO] A bacterial disease of Japanese beetle larvae or related grubs caused by *Bacillus papilliae* and *B. lentimorbus* that penetrate the intestine and sporulate in the body cavity; blood of the grub eventually turns milky white. { 'mil·kē di,zēz }

Milleporina [INV ZOO] An order of the class Hydrozoa known as the stinging corals; they resemble true corals because of a calcareous exoskeleton. { mil·ə·pə'rī·nə }

Miller-Abbott tube [MED] A double-lumen rubber tube having a balloon at the end, inserted through the nasal passage and passed through the pylorus to locate and treat intestinal obstructions. { 'mil·ər 'ab·ət ,tüb }

millet [BOT] A common name applied to at least five related members of the grass family grown for their edible seeds. { 'mil·ət }

millipede [INV ZOO] The common name for members of the arthropod class Diplopoda. { 'mil·ə,pēd }

Millipore filter [MICROBIO] A filter capable of ultrafine separation, used for purification and analyses of fluids, among other applications. { 'mil·ə,pȯr ,fil·tər }

Milroy's disease [MED] Familial chronic lymphedema of the lower extremities. { 'mil,rȯiz di,zēz }

mimetic [ZOO] Pertaining to or exhibiting mimicry. { mə'med·ik }

mimetic camouflage [ECOL] Protective coloration that is achieved by a resemblance to some other existing object, which is recognized by the predator but not associated in its mind with feeding. { mə,med·ik 'kam·ə,fläzh }

mimicry [ZOO] Assumption of color, form, or behavior patterns by one species of another species, for camouflage and protection. { 'mim·ə·krē }

Mimidae [VERT ZOO] The mockingbirds, a family of the Oscines in the order Passeriformes. { 'mim·ə,dē }

Mimosoideae [BOT] A subfamily of the legume family, Leguminosae; members are largely woody and tropical or subtropical with regular flowers and usually numerous stamens. { ˌmim·əˈsȯid·ēˌē }

mineralocorticoid [BIOCHEM] A steroid hormone secreted by the adrenal cortex that regulates mineral metabolism and, secondarily, fluid balance. { ¦min·rə·lōˈkȯrd·əˌkȯid }

minicell [MICROBIO] A small anucleate bacterial cell produced by abnormal and unequal division of a parent cell. { ˈmin·ēˌsel }

minichromosome [CELL MOL] A eukaryotic chromosome reduced in size by deletion of a segment of deoxyribonucleic acid. [VIROL] Viral deoxyribonucleic acid combined with histone to form a chromatin-like structure. { ˌmin·ēˈkrō·məˌsōm }

minimal brain dysfunction syndrome [MED] A complex of learning and behavioral disabilities seen primarily in children of near-average or above-average intelligence exhibiting also deviations of function of the central nervous system. { ˈmin·ə·məl ¦brān disˈfəŋk·shən ˌsinˌdrōm }

minimal recognition length [CELL MOL] The shortest length of base pairs that will form a stable deoxyribonucleic acid duplex in genetic recombination. { ¦min·ə·məl ˌrek·igˈnish·ən ˌleŋkth }

Minimata disease [MED] A disorder resulting from methyl mercury poisoning, which occurred in epidemic proportions in 1956 in Minimata Bay, a Japanese coastal town, where the inhabitants ate fish contaminated by industrial pollution; the most obvious symptoms are tremors and involuntary movements. { ˌmin·ēˈmäd·ə diˌzēz }

minimum lethal dose [PHARM] The amount of an injurious agent which is the average of the smallest dose that kills and the largest dose that fails to kill. { ˈmin·ə·məm ¦lēth·əl ˈdōs }

miniplasmid [CELL MOL] Any plasmid that has been reduced in size by means of recombinant deoxyribonucleic acid technology. { ˌmin·ē ˈplaz·mid }

mink [VERT ZOO] Any of three species of slender-bodied aquatic carnivorous mammals in the genus *Mustela* of the family Mustelidae. { miŋk }

minnow [VERT ZOO] The common name for any fresh-water fish composing the family Cyprinidae, order Cypriniformes. { ˈmin·ō }

minor surgery [MED] Any superficial surgical procedure involving little hazard to the life of the patient and not requiring general anesthesia. { ˈmīn·ər ˈsər·jə·rē }

minus strand [CELL MOL] A polynucleotide strand that is complementary to, and formed by transcription from, another specific polynucleotide (plus) strand, with which it produces the double-stranded (double-helix) ribonucleic acid. { ˈmī·nəs ˌstrand }

miosis [PHYSIO] Contraction of the pupil of the eye. { mīˈō·səs }

miotic [PHARM] **1.** Causing miosis. **2.** Any agent that causes miosis. [PHYSIO] Of or pertaining to miosis. { mīˈäd·ik }

miracidium [INV ZOO] The ciliated first larva of a digenetic trematode; forms a sporocyst after penetrating intermediate host tissues. { ˌmī·rəˈsid·ē·əm }

Mirapinnatoidei [VERT ZOO] A suborder of tiny oceanic fishes in the order Cetomimiformes. { ˌmir·əˌpin·əˈtȯid·ēˌī }

Miridae [INV ZOO] The largest family of the Hemiptera; included in the Cimicomorpha, it contains herbivorous and predacious plant bugs which lack ocelli and have a cuneus and four-segmented antennae. { ˈmir·əˌdē }

Miripinnati [VERT ZOO] The equivalent name for Marapinnatoidei. { ¦mir·ə·pəˈnäd·ē }

miscarriage *See* spontaneous abortion. { ˈmis ˌkar·ij }

mispairing [CELL MOL] Pairing of a nucleotide in one chain of a deoxyribonucleic acid molecule that is not complementary to the nucleotide occupying the corresponding position in the other chain. { misˈper·iŋ }

misrepair [CELL MOL] Repair of deoxyribonucleic acid that gives rise to gene mutations or changes in chromosome structure. { ˈmis·riˌper }

missed abortion [MED] A condition in which a fetus weighing less than 500 grams dies and remains in the uterus for an extended period of time. { ˈmist əˈbȯr·shən }

missense codon [GEN] A mutant codon that directs the incorporation of a different amino acid and results in the synthesis of a protein with a sequence in which one amino acid has been replaced by a different one; in some cases the mutant protein may be unstable or less active. { ˈmis·əns ˈkōˌdän }

missense mutation [CELL MOL] A mutation that converts a codon coding for one amino acid to a codon coding for another amino acid. { ˈmis·əns myüˈtā·shən }

missense suppressor [CELL MOL] A suppressor that incorporates the correct amino acid at the site of a codon that has been altered because of a missense mutation. { ˈmis·əns səˌpres·ər }

mistletoe [BOT] **1.** *Viscum album.* The true, Old World mistletoe having dichotomously branching stems, thick leathery leaves, and waxy-white berries. **2.** Any of several species of green hemiparasitic plants of the family Loranthaceae. { ˈmis·əlˌtō }

mistranslation [CELL MOL] Incorporation of the wrong amino acid into a protein due to misreading of a codon. { ˌmis·tranzˈlā·shən }

Mitchell's disease *See* erythromelalgia. { ˈmich·əlz diˌzēz }

mite [INV ZOO] The common name for the acarine arthropods composing the diverse suborders Onychopalpida, Mesostigmata, Trombidiformes, and Sarcoptiformes. { mīt }

mitochondria [CELL MOL] Minute cytoplasmic organelles in the form of spherical granules, short rods, or long filaments found in almost all living cells; submicroscopic structure consists of an external membrane system. { ˌmīd·ə'kän·drē·ə }

mitochondrial crest [CELL MOL] Any of the infoldings of the mitochondrial inner membrane that extend into the matrix. { ˌmīd·əˈkän·drē·əl 'kres }

mitochondrial deoxyribonucleic acid [BIOCHEM] The circular deoxyribonucleic acid duplex, generally 5 to 10 copies, contained within a mitochondrion and maternally inherited since only the egg cell contributes significant numbers of mitochondria to the zygote. Abbreviated mtDNA. Also known as mitochondrial genome. { ˌmīd·əˈkän·drē·əl dēˈäk·sēˌrī·bō·nü'klē·ik 'as·əd }

mitochondrial genome *See* mitochondrial deoxyribonucleic acid. { ˌmīd·əˈkän·drē·əl 'jeˌnōm }

mitochondrial plasmid [CELL MOL] Plasmid-like deoxyribonucleic acid molecules in mitochondria of certain higher plants and some fungi. { ˌmīd·əˈkän·drē·əl 'plaz·mid }

mitogen [CELL MOL] A compound that stimulates cells to undergo mitosis. { 'mīd·əˌjen }

mitogen-activated protein kinases [CELL MOL] A large kinase network in which upstream kinases activate downstream kinases that, in response to phosphorylation, translocate to the nucleus and activate transcription factors. Abbreviated MAPKs. { ˌmīd·ə·jən ˈak·təˌvād·əd ˈprōˌtēn 'kīˌnās·əs }

mitogenesis [CELL MOL] **1.** Induction of mitosis. **2.** Formation as a result of mitosis. { ˌmīd·ə'jen·ə·səs }

mitomycin [MICROBIO] A complex of three antibiotics (mitomycin A, B, and C) produced by *Streptomyces caespitosus*. { ˈmīd·əˈmīs·ən }

mitoplast [CELL MOL] **1.** A mitochondrion that has had its outer membrane removed. **2.** The cytoplast of a mitotic cell after the chromosomes are extruded. { 'mīd·əˌplast }

mitosis [CELL MOL] Nuclear division involving exact duplication and separation of the chromosome threads so that each of the two daughter nuclei carries a chromosome complement identical to that of the parent nucleus. { mī'tō·səs }

mitotic apparatus [CELL MOL] A transitory organelle-like formation that is seen during mitosis and meiosis and consists of the asters, the spindle, and the traction fibers. { mīˈtäd·ik ˌap·ə'rad·əs }

mitotic center [CELL MOL] A structure that defines the poles toward which chromosomes move during mitosis and meiosis. { mīˈtäd·ik 'sen·tər }

mitotic chromosomes [CELL MOL] Condensed, replicated chromosomes held together as sister chromatids by the centromere as part of the process ensuring accurate segregation and transmission of genetic material to the daughter cells during cell division. { mīˈtäd·ik 'krō·məˌsōmz }

mitotic index [CELL MOL] The number of cells undergoing mitosis per thousand cells. { mī'täd·ik 'inˌdeks }

mitotic inhibitor [CELL MOL] A compound that inhibits mitosis. { mī'täd·ik in'hib·əd·ər }

mitotic poison [CELL MOL] A compound that prevents or affects the completion of mitosis. { mī'täd·ik 'pȯiz·ən }

mitral commissurotomy [MED] Any of several surgical procedures performed to relieve mitral stenosis. { 'mī·trəl ˌkäm·ə·sə'räd·ə·mē }

mitral stenosis [MED] Obstruction of the mitral valve, usually due to narrowing of the orifice. { 'mī·trəl stə'nō·səs }

mitral valve [ANAT] The atrioventricular valve on the left side of the heart. { 'mī·trəl 'valv }

mitriform [BIOL] Shaped like a miter. { 'mī·trəˌfȯrm }

mittelschmerz [MED] Pain or discomfort in the lower abdomen in women occurring midway in the intermenstrual interval, thought to be secondary to the irritation of the pelvic peritoneum by fluid or blood escaping from the point of ovulation in the ovary. { 'mid·əlˌshmerts }

Mitteniales [BOT] An order of true mosses, class Bryopsida, characterized by branches of protonema consisting of spherical cells which reflect light from a backing of chloroplasts, thus providing a glow. { ˌmit·ən'ā·lēz }

mixed bud [BOT] A bud that contains both rudimentary leaves and rudimentary flowers. { 'mikst 'bəd }

mixed cryoglobulin [BIOCHEM] A cryoglobulin with a monoclonal component made of immunoglobulin belonging to two different classes, one of which is monoclonal. { 'mikst ˌkrī·ō'gläb·yə·lən }

mixed deafness [MED] Combined conductive and sensorineural impairment. { 'mikst 'def·nəs }

mixed gland [PHYSIO] A gland that secretes more than one substance, especially a gland containing both mucous and serous components. { 'mikst 'gland }

mixed nerve [NEUROSCI] A nerve containing both sensory and motor components. { 'mikst 'nərv }

mixoploidy [CELL MOL] The presence of cells having different chromosome numbers in the same cell population. { ˈmik·səˌplȯid·ē }

mixotrophic [BIOL] Obtaining nutrition by combining autotrophic and heterotrophic mechanisms. { ˈmik·səˈträf·ik }

mixture [PHARM] A liquid medicine prepared by adding insoluble substances to a liquid medium, usually with a suspending agent. { 'miks·chər }

MLD *See* metachromatic leukodystrophy.

Mnesarchaeidae [INV ZOO] A family of lepidopteran insects in the suborder Homoneura; members are confined to New Zealand. { ˌnē·sär'kē·əˌdē }

Mobilina [INV ZOO] A suborder of ciliophoran protozoans in the order Peritrichida. { ˌmō·bə'lī·nə }

Mobulidae [VERT ZOO] The devil rays, a family of batoids that are surface feeders and live mostly on plankton. { mə'byül·əˌdē }

modal number [GEN] **1.** The typical chromosome number of a taxonomic group. **2.** The common chromosome number of a tumor cell population. { 'mōd·əl ˌnəm·bər }

modification [CELL MOL] In nucleic acid metabolism, any changes made to deoxyribonucleic acid or ribonucleic acid after their original incorporation into a polynucleotide chain. { ˌmäd·ə·fə'kā·shən }

modified base [CELL MOL] A nucleotide that is an altered form of the usual four nucleic acid bases. { ¦mäd·əˌfīd 'bās }

modifier gene [GEN] A gene that alters the phenotypic expression of a nonallelic gene. { 'mäd·əˌfī·ər ˌjēn }

modiolus [ANAT] The central axis of the cochlea. { mə'dī·ə·ləs }

modulating codon [CELL MOL] A codon that controls the frequency of transcription of a cistron. { ¦mäj·əˌlād·iŋ 'kōˌdän }

Moeller-Barlow disease *See* infantile scurvy. { 'məl·ər 'bär·lō diˌzēz }

molar [ANAT] **1.** A tooth adapted for grinding. **2.** Any of the three pairs of cheek teeth behind the premolars on each side of the jaws in humans. { 'mō·lər }

mold [MYCOL] Any of various woolly fungus growths. { mōld }

mole [MED] **1.** A mass formed in the uterus by the maldevelopment of all or part of the embryo or of the placenta and membranes. **2.** A fleshy, pigmented nevus. [VERT ZOO] Any of 19 species of insectivorous mammals composing the family Talpidae; the body is stout and cylindrical, with a short neck, small or vestigial eyes and ears, a long naked muzzle, and forelimbs adapted for digging. { mōl }

molecular biology [BIOL] That part of biology which attempts to interpret biological events in terms of the physicochemical properties of molecules in a cell. { mə'lek·yə·lərbī'äl·ə·jē }

molecular biomarker [MED] A biological indicator that signals a changed physiological state, stress, or injury due to disease or the environment, for example, an elevated serum level of prostate-specific antigen is a biomarker for prostate cancer. { me¦lek·yə·lər 'bī·ōˌmärk·ər }

molecular biophysics [BIOPHYS] The study of the physical properties and interactions of large molecules and particles of comparable size which play important roles in biology. { məˌlek·yə·lər ˌbi·ō'fiz·iks }

molecular chaperone [CELL MOL] Any of a class of cellular proteins involved in correct folding of certain polypeptide chains and their assembage into an oligomer. { mə'lək·yə·lər 'shap·əˌrän }

molecular genetics [CELL MOL] The approach which deals with the physics and chemistry of the processes of inheritance. { mə'lek·yə·lər jə'ned·iks }

molecular mimicry [IMMUNOL] The sharing, by two organisms closely related ecologically but not phylogenetically, of common macromolecular structures that are not attributable to evolutionary conservation of these structures. { mə'lek·yə·lər 'mim·iˌkrē }

molecular pathology [PATH] The study of the bases and mechanisms of disease on a molecular or chemical level. { mə'lek·yə·lər pə'thäl·ə·jē }

molecular recognition [CELL MOL] The ability of biological and chemical systems to distinguish between molecules and regulate behavior accordingly. { mə¦lek·yə·lər ˌrek·ig'nish·ən }

Molidae [VERT ZOO] A family of marine fishes, including some species of sunfishes, in the order Perciformes. { 'mäl·əˌdē }

Molisch test [BIOCHEM] A test used for the general detection of carbohydrates; a purple color is produced when the solution containing carbohydrate is treated with strong sulfuric acid in the presence of α-naphthol. { 'mȯl·ish ˌtest }

Mollicutes [MICROBIO] The mycoplasmas, a class of prokaryotic organisms lacking a true cell wall; cells are very small to submicroscopic. { mə'lik·yəˌdēz }

Mollusca [INV ZOO] One of the divisions of phyla of the animal kingdom containing snails, slugs, octopuses, squids, clams, mussels, and oysters; characterized by a shell-secreting organ, the mantle, and a radula, a food-rasping organ located in the forward area of the mouth. { mə'ləs·kə }

molluscum contagiosum [MED] A viral disease of the skin, characterized by one or more discrete, waxy, dome-shaped nodules with frequent umbilication. { mə'ləs·kəm kənˌtā·jē'ō·səm }

mollusk [INV ZOO] Any member of the Mollusca. { 'mäl·əsk }

Molossidae [VERT ZOO] The free-tailed bats, a family of tropical and subtropical insectivorous mammals in the order Chiroptera. { mə'läs·əˌdē }

Molpadida [INV ZOO] An order of sea cucumbers belonging to the Apodacea and characterized by a short, plump body bearing a taillike prolongation. { mäl'pā·də·də }

Molpadidae [INV ZOO] The single family of the echinoderm order Molpadida. { mäl'pā·dəˌdē }

molt [PHYSIO] To shed an outer covering as part of a periodic process of growth. { mōlt }

molting hormone [BIOCHEM] Any of several hormones which activate molting in arthropods. { 'mōl·tiŋ 'hȯrˌmōn }

Momotidae [VERT ZOO] The motmots, a family of colorful New World birds in the order Coraciiformes. { mə'mäd·əˌdē }

monactine [INV ZOO] A single-rayed spicule in the sponges. { mä'nak·tən }

monadelphous [BOT] Having the filaments of the stamens united into one set. { ¦män·ə¦del·fəs }

Monadidae [INV ZOO] A family of flagellated protozoans in the order Kinetoplastida having two flagella of uneven length. { mə'nad·əˌdē }

monandrous [BOT] Having one stamen. { mə'nan·drəs }

monaxon [INV ZOO] A spicule formed by growth along a single axis. { mä'nakˌsän }

Mönckeberg's arteriosclerosis *See* medial arteriosclerosis. { 'mən·kə,bərgz är¦tir·ē·ō·sklə'rō·səs }

Monday fever *See* metal fume fever. { 'mən,dā ,fē·vər }

monellin [BIOCHEM] A sweet protein obtained from the African plant *Dioscorephyllum cumminisii*, in its natural form, it consists of two polypeptide chains and is 3000 times sweeter than sucrose on a weight basis. { mə'nel·ən }

Monera [BIOL] A kingdom that includes the bacteria and blue-green algae in some classification schemes. { mə'nir·ə }

monestrous [PHYSIO] Having a single estrous cycle per year. { män'es·trəs }

Monge's disease *See* mountain sickness. { 'mōnzh·əz di,zēz }

Mongolian spot [MED] A focal bluish-gray discoloration of the skin of the lower back, also aberrantly on the face, present at birth and fading gradually. { mäŋ'gō·lē·ən 'spät }

mongolism *See* Down's syndrome. { 'mäŋ·gə,liz·əm }

mongoloid [MED] Having physical characteristics associated with Down's syndrome. { 'mäŋ·gə,lȯid }

mongoose [VERT ZOO] The common name for 39 species of carnivorous mammals which are members of the family Viveridae; they are plantigrade animals and have a long slender body, short legs, nonretractile claws, and scent glands. { 'mäŋ,güs }

Monhysterida [INV ZOO] An order of aquatic nematodes in the subclass Chromadoria. { ,män·hi'ster·ə·də }

Monhysteroidea [INV ZOO] A superfamily of free-living nematodes in the order Monhysterida characterized by single or paired outstretched ovaries, circular to cryptospiral amphids, and a stoma which is usually shallow and unarmed. { ,män ,hi·stə'rȯid·ē·ə }

Moniliaceae [MYCOL] A family of fungi in the order Moniliales; sporophores are usually lacking, but when present they are aggregated into fascicles, and hyphae and spores are hyaline or brightly colored. { mə,nil·ē'ās·ē,ē }

Moniliales [MYCOL] An order of fungi of the Fungi Imperfecti containing many plant pathogens; asexual spores are always formed free on the surface of the material on which the organism is living, and never occur in either pycnidia or acervuli. { mə,nil·ē'ā·lēz }

monilial vaginitis [MED] Inflammation of the vagina caused by a fungus of the genus *Monilia*. { mə'nil·ē·əl ,vaj·ə'nīd·əs }

moniliasis *See* candidiasis. { mə,nil·ē'ī·ə·səs }

moniliform [BIOL] Constructed with contractions and expansions at regular alternating intervals, giving the appearance of a string of beads. { mə'nil·ə,fȯrm }

Monilinia fructicola [MYCOL] A fungal pathogen in the class Discomycetes that causes brown rot of stone fruits. { ,mō·nə,lin·ē·ə ,frük·ti 'kō·lə }

monitor [VERT ZOO] Any of 27 carnivorous, voracious species of the reptilian family Varanidae characterized by a long, slender forked tongue and a dorsal covering of small, rounded scales containing pointed granules. { 'män·əd·ər }

monkey [VERT ZOO] Any of several species of frugivorous and carnivorous primates which compose the families Cercopithecidae and Cebidae in the suborder Anthropoidea; the face is typically flattened and hairless, all species are pentadactyl, and the mammary glands are always in the pectoral region. { 'məŋ·kē }

monkeypox [VET MED] An animal virus that causes a smallpox-like eruption but only rarely infects humans and has little potential for interhuman spread. { 'məŋ·kē,päks }

monkshood *See* aconite. { 'məŋks,hu̇d }

monoamine oxidase [BIOCHEM] A mitochondrial enzyme which oxidatively deaminates intraneuronal biogenic amines, some of which are important neurotransmitters in the peripheral and central nervous system. { män·ō'am,ēn 'äk·sə,dās }

monoamine oxidase inhibitor [PHARM] Any drug, such as isocarboxazid and tranylcypromine, that inhibits monoamine oxidase and thereby leads to an accumulation of the amines on which the enzyme normally acts. { män·ō 'am,ēn 'äk·sə,dās in'hib·əd·ər }

monoblast [HISTOL] A motile cell of the spleen and bone marrow from which monocytes are derived. { 'män·ō,blast }

monoblastic leukemia *See* acute monocytic leukemia. { ¦män·ō'blas·tik lü'kē·mē·ə }

Monoblepharidales [MYCOL] An order of aquatic fungi in the class Phycomycetes; distinguished by a mostly hyphal thallus and zoospores with one posterior flagellum. { ¦män·ō,blef·ə·rə'dā·lēz }

monocardiogram *See* vectorcardiogram. { ¦män·ō'kär·dē·ə,gram }

monocarpic [BOT] Bearing fruit once and then dying. { ¦män·ō¦kär·pik }

monocarpous [BOT] Having a single ovary. { ¦män·ō¦kär·pəs }

monochlamydous [BOT] Referring to flowers having only one set of floral envelopes, that is, either a calyx or a corolla. { ¦män·ō'klam·ə·dəs }

Monochoidea [INV ZOO] A superfamily of free-living, nonparasitic nematodes in the order Mononchida, characterized by angular, distinct lips bearing papilliform cephalic sensilla, an expanded lip region that is flattened anteriorly, and a heavily cuticularized, barrel or globular stoma, with one or more teeth or denticles. { ,män·ə'kȯid·ē·ə }

monochromasia [MED] Complete color blindness in which all colors appear as shades of gray. Also known as monochromatism. { ,män·ə·krə'mā·zhə }

monochromat [MED] An individual who suffers from total color blindness even at high light levels; such persons are typically deficient or lacking in cone receptors, so that their form vision is also poor. { ¦män·ə¦krō,mat }

monochromatism *See* monochromasia. { ¦män·ə¦krō·mə,tiz·əm }

Monochuloidea [INV ZOO] A superfamily of nonparasitic nematodes in the order Mononchida, distinguished by papilliform cephalic sense organs, an inconspicuous, small, slitlike amphid aperture, and a stoma with a thick-walled, slightly tapered anterior and an elongate, thin-walled posterior. { ,män·ə·kə'lȯid·ē·ə }

monocistronic messenger [BIOCHEM] A messenger ribonucleic acid molecule that contains the amino acid sequence for a single polypeptide chain. { ,män·ə·sis¦trän·ik 'mes·ən·jər }

Monocleales [BOT] An order of liverworts of the subclass Marchantiidae consisting of a single genus, *Monoclea*, which has the largest gametophyte of all liverworts, and lobed spore mother cells. { ¦män·ō·klē'ā·lēz }

monoclimax [ECOL] A climax community controlled primarily by one factor, as climate. { ¦män·ō¦klī,maks }

monoclinic [BOT] Having both stamens and pistils in the same flower. { ¦män·ə'klin·ik }

monoclonal antibody [IMMUNOL] A highly specific antibody produced by hybridoma cells; the antibody binds with a single antigenic determinant. { ¦män·ə¦klō·nəl 'ant·i,bäd·ē }

monoclonal cryoglobulin [BIOCHEM] A cryoglobulin composed of immunoglobin with only one class or subclass of heavy and light chain. { ¦män·ə¦klō·nəl ¦krī·ō'gläb·yə·lən }

monocolpate pollen [BOT] Pollen grains having a single furrow. { ¦män·ə'kōl,pāt ,päl·ən }

monocotyledon [BOT] Any plant of the class Liliopsida; all have a single cotyledon. { ¦män·ə,käd·əl'ēd·ən }

Monocotyledoneae [BOT] The equivalent name for Liliopsida. { ,män·ə,käd·əl·ə'dō·nē,ē }

monocrepid [INV ZOO] A desma formed by secondary deposits of silica on a monaxon. { ¦män·ə¦krep·əd }

monocular vision [MED] Sight with one eye. { mə'näk·yə·lər 'vizh·ən }

monocyclic [BOT] Referring to flower parts arranged in a single whorl. { ,män·ə'sī·klik }

monocyte [HISTOL] A large (about 12 micrometers), agranular leukocyte with a relatively small, eccentric, oval or kidney-shaped nucleus. { 'män·ə,sīt }

monocytic angina *See* infectious mononucleosis. { ¦män·ə¦sid·ik 'an·jə·nə }

monocytic leukemia [MED] A form of leukemia in which monocytic cells are predominant in the blood. Also known as myelomonocytic leukemia. { ¦män·ə¦sid·ik lü'kē·mē·ə }

monocytoma [MED] A neoplasm composed principally of monocytes, usually anaplastic. { ¦män·ō,sī'tō·mə }

monocytopenia [MED] Reduction in the number of circulating monocytes per unit volume of blood to below the minimum normal levels. { ,män·ō,sīd·ə'pē·nē·ə }

monocytosis [MED] Increase in the number of circulating monocytes per unit volume of blood to above the maximum normal levels. { ,män·ō,sī'tō·səs }

monodactylous [ZOO] Having a single digit or claw. { ¦män·ə'dak·tə·ləs }

Monodellidae [INV ZOO] A monogeneric family of crustaceans in the order Thermosbaenacea distinguished by seven pairs of biramous pereiopods on thoracomeres 2–8, and by not having the telson united to the last pleonite. { ¦män·ə'del·ə,dē }

monodelphic [VERT ZOO] **1.** Having a single genital tract, in the female. **2.** Having a single uterus. { ¦män·ō¦del·fik }

monoecious [BOT] **1.** Having both staminate and pistillate flowers on the same plant. **2.** Having archegonia and antheridia on different branches. [ZOO] Having male and female reproductive organs in the same individual. Also known as hermaphroditic. { mə'nē·shəs }

Monoedidae [INV ZOO] An equivalent name for Colydiidae. { mə'nē·də,dē }

monogamous bivalent [IMMUNOL] Antigen-antibody complex in which each bivalent antibody combines with two determinant groups on a single antigen molecule. { mə¦näg·ə·məs bī'vā·lənt }

monogastric [VERT ZOO] Having only one digestive cavity. { ¦män·ō'gas·trik }

Monogenea [INV ZOO] A diverse subclass of the Trematoda which are principally ectoparasites of fishes; individuals have enlarged anterior and posterior holdfasts with paired suckers anteriorly and opisthaptors posteriorly. { ,män·ə'jē·nē·ə }

monogenic [GEN] Relating to or controlled by one gene. { ,män·ə'jen·ik }

Monogenoidea [INV ZOO] A class of the Trematoda in some systems of classification; equivalent to the Monogenea of other systems. { ,män·ə·jə'nȯid·ē·ə }

Monogonota [INV ZOO] An order of the class Rotifera, characterized by the presence of a single gonad in both males and females. { ,män·ō·gō'näd·ə }

monogony [BIOL] Asexual reproduction. { mə'näg·ə·nē }

monogynous [BOT] Having only one pistil. [VERT ZOO] **1.** Having only one female in a colony. **2.** Consorting with only one female. { mə'näj·ə·nəs }

monohybrid [GEN] An individual heterozygous for alleles at one gene locus. { ¦män·ō'hī·brəd }

monokine [BIOCHEM] A cytokine released from macrophages. { 'män·ə,kīn }

monomeric [CELL MOL] Having a single polypeptide chain. { ,män·ə'mer·ik }

Monommidae [INV ZOO] A family of coleopteran insects in the superfamily Tenebrionoidea. { mə'nam·ə,dē }

monomorphic [BIOL] Having or exhibiting only a single form. { ¦män·ə¦mȯr·fik }

mononuclear [CELL MOL] Having only one nucleus. { ¦män·ō'nü·klē·ər }

mononucleosis [MED] Any of various conditions marked by an abnormal increase in monocytes in the peripheral blood. { ˌmän·əˌnü·klē'ō·səs }

monophagous [ZOO] Subsisting on a single kind of food. Also known as monotrophic. { mə'näf·ə·gəs }

Monophisthocotylea [INV ZOO] An order of the Monogenea in which the posthaptor is without discrete multiple suckers or clamps. { ˌmän·əˌfis·thəˌkäd·əl'ē·ə }

Monophlebinae [INV ZOO] A subfamily of the homopteran superfamily Coccoidea distinguished by a dorsal anus. { ˌmän·ə'fleb·əˌnē }

monophyletic [EVOL] Pertaining to any form evolved from a single interbreeding population. { ˌmän·ə·fə'led·ik }

monophyodont [VERT ZOO] Having only one set of teeth throughout life. { ¦män·ō'fī·əˌdänt }

Monopisthocotylea [INV ZOO] An order of trematode worms in the subclass Pectobothridia. { ˌmän·əˌfis·thəˌkäd·əl'ē·ə }

Monoplacophora [INV ZOO] A group of shell-bearing mollusks represented by few living forms; considered to be a sixth class of mollusks. { ˌmän·ō·plə'käf·ə·rə }

monoplegia [MED] Paralysis involving a single limb, muscle, or group of muscles. { ˌmän·ə'plē·jē·ə }

monoploid [GEN] **1.** Having only one set of chromosomes. **2.** Having the lowest haploid number of chromosomes in a polyploid series, in plants. { 'män·əˌplȯid }

monopodial [BOT] Stem branching in which there are lateral shoots on a primary axis. { ˌmän·ə'pōd·ē·əl }

monopodium [BOT] A primary axis that continues to grow while giving off successive lateral branches. { ˌmän·ə'pōd·ē·əm }

Monoposthioidea [INV ZOO] A superfamily of chiefly marine nematodes in the order Desmodorida, represented by the single family Monoposthiidae; distingushed by an annulate cuticle with spikelike ornamentation and a stoma that may or may not possess a well-developed tooth opposed by small subventral teeth. { ˌmän·ōˌpäs·thē'ȯid·ē·ə }

Monopylina [INV ZOO] A suborder of radiolarian protozoans in the order Oculosida in which pores lie at one pole of a single-layered capsule. { ˌmän·ō·pə'lī·nə }

monorchid [ANAT] **1.** Having one testis. **2.** Having one testis descended into the scrotum. { mä'nȯr·kəd }

Monorhina [VERT ZOO] The subclass of Agnatha that includes the jawless vertebrates with a single median nostril. { ˌmän·ə'rī·nə }

monorhinal [ANAT] Having only one nostril. { ¦män·ə¦rī·nəl }

monosaccharide [BIOCHEM] A carbohydrate which cannot be hydrolyzed to a simpler carbohydrate; a polyhedric alcohol having reducing properties associated with an actual or potential aldehyde or ketone group; classified on the basis of the number of carbon atoms, as triose (3C), tetrose (4C), pentose (5C), and so on. { ¦män·ō¦sak·əˌrīd }

Monosigales [BOT] A botanical order equivalent to the Choanoflagellida in some systems of classification. { ˌmän·ō·si'gā·lēz }

monosiphonous [BIOL] Having a single central tube, as in the thallus of certain filamentous algae or the hydrocaulus of some hydrozoans. { ¦män·ō'sī·fə·nəs }

monosome [CELL MOL] **1.** A single ribosome attached to messenger ribonucleic acid. **2.** A chromosome in the diploid chromosome complement that lacks a homolog. { 'män·əˌsōm }

monosomy [GEN] The condition in which one chromosome of a pair is missing a diploid organism. { 'män·əˌsōm·ē }

monospermous [BOT] Having or producing one seed. { ¦män·ō¦spər·məs }

monosporangium [BOT] A sporangium producing monospores. { ¦män·ō·spə'ran·jē·əm }

monospore [BOT] A simple or undivided nonmotile asexual spore; produced by the diploid generation of some algae. { 'män·əˌspȯr }

monostome [INV ZOO] A cercaria having only one mouth or sucker. { 'män·əˌstōm }

Monostylifera [INV ZOO] A suborder of the Hoplonemertini characterized by a single stylet. { ˌmän·ō·stī'lif·ə·rə }

Monotomidae [INV ZOO] The equivalent name for Rhizophagidae. { ˌmän·ə'täm·əˌdē }

Monotremata [VERT ZOO] The single order of the mammalian subclass Prototheria containing unusual mammallike reptiles, or quasi-mammals. { ¦män·ō·trə'mad·ə }

monotrichous [MICROBIO] Of bacteria, having an individual flagellum at one pole. { mə'nä·trə·kəs }

Monotropaceae [BOT] A family of dicotyledonous herbs or half shrubs in the order Ericales distinguished by a small, scarcely differentiated embryo without cotyledons, lack of chlorophyll, leaves reduced to scales, and anthers opening by longitudinal slits. { ˌmän·ō·trə'pās·ēˌē }

monotrophic *See* monophagous. { ¦män·ə¦träf·ik }

monotype [BIOL] A single type of organism that constitutes a species or genus. { 'män·əˌtīp }

monotypic [SYST] Pertaining to a taxon that constains only one immediately subordinate taxon. { ¦män·ə¦tip·ik }

monozygotic twins [BIOL] Twins which develop from a single fertilized ovum. Also known as identical twins. { ¦män·ōˌzī'gäd·ik 'twinz }

mons [ANAT] An eminence. { mänz }

monsoon forest [ECOL] A tropical forest occurring in regions where a marked dry season is followed by torrential rain; characterized by vegetation adapted to withstand drought. { ¦mänˌsün 'fär·əst }

mons pubis [ANAT] The eminence of the lower anterior abdominal wall above the superior rami of the pubic bones. { 'mänz 'pyü·bəs }

monster [MED] A congenitally malformed fetus which is incapable of properly performing the vital functions, or which exhibits marked structural differences from the normal. { 'män·stər }

Monstrilloida [INV ZOO] A suborder or order of microscopic crustaceans in the subclass Copepoda; adults lack a second antenna and mouthparts, and the digestive tract is vestigial. { ˌmän·strə'lȯid·ə }

mons veneris [ANAT] The mons pubis of the female. { 'mänz 'ven·ə·rəs }

montane [ECOL] Of, pertaining to, or being the biogeographic zone composed of moist, cool slopes below the timberline and having evergreen trees as the dominant life-form. { män'tān }

Montgomery's tubercles [ANAT] Elevations in the areola of the nipple due to apocrine sweat glands; most prominent during pregnancy and lactation. { mənt'gəm·rēz 'tü·bər·kəlz }

Monticellidae [INV ZOO] A family of tapeworms in the order Proteocephaloidea, in which some or all of the organs are in the cortical mesenchyme; catfish parasites. { ˌmän·tə'sel·əˌdē }

monticulus [ANAT] The median dorsal portion of the cerebellum. { män'tik·yə·ləs }

moor *See* bog. { mu̇r }

moose [VERT ZOO] An even-toed ungulate of the genus *Alces* in the family Cervidae; characterized by spatulate antlers, long legs, a short tail, and a large head with prominent overhanging snout. { müs }

Moraceae [BOT] A family of dicotyledonous woody plants in the order Urticales characterized by two styles or style branches, anthers inflexed in the bud, and secretion of a milky juice. { mə'rās·ēˌē }

Moraxella [MICROBIO] A genus of bacteria that are parasites of mucous membranes. { mə'rak·sə·lə }

morbid anatomy *See* pathologic anatomy. { 'mȯr·bəd ə'nad·ə·mē }

morbidity [MED] **1.** The quantity or state of being diseased. **2.** The conditions inducing disease. **3.** The ratio of the number of sick individuals to the total population of a community. { mȯr'bid·əd·ē }

Mordellidae [INV ZOO] The tumbling flower beetles, a family of coleopteran insects in the superfamily Meloidea. { mȯr'del·əˌdē }

morel [MYCOL] Any fungus belonging to the genus *Morchella*, distinguished by a large, pitted, spongelike cap; it is a highly prized food, but may be poisonous when taken with alcohol. { mə'rel }

mores [ECOL] Groups of organisms preferring the same physical environment and having the same reproductive season. { 'mȯrˌāz }

Morgan [GEN] The unit of genetic map distance (1 Morgan) between two loci that show one crossover per meiosis; 1 Morgan = 100 centimorgans. { 'mȯr·gən }

morgue [MED] A place where dead bodies are held pending identification and disposition. { mȯrg }

moribund [BIOL] **1.** In a dying or deathlike state. **2.** In a state of suspended life functions; dormant. { 'mȯr·ə·bənd }

Moridae [VERT ZOO] A family of actinopterygian fishes in the order Gadiformes. { 'mȯr·əˌdē }

Morinae [VERT ZOO] The deep-sea cods, a subfamily of the Moridae. { 'mȯr·əˌnē }

Mormyridae [VERT ZOO] A large family of electrogenic fishes belonging to the Osteoglossiformes; African river and lake fishes characterized by small eyes, a slim caudal peduncle, and approximately equal dorsal and anal fins in most. { mȯr'mir·əˌdē }

Mormyriformes [VERT ZOO] Formerly an order of fishes which are now assigned to the Osteoglossiformes. { ˌmȯr·mə·rə'fȯrˌmēz }

morning sickness [MED] Morning nausea associated with early pregnancy. { 'mȯrn·iŋ ˌsik·nəs }

Moro reflex [PHYSIO] The startle reflex observed in normal infants from birth through the first few months, consisting of abduction and extension of all extremities, followed by flexion and abduction of the extremities. { 'mȯr·ō ˌrēˌfleks }

morph [GEN] An individual variant in a polymorphic population. { mȯrf }

morphallaxis [PHYSIO] Regeneration whereby one part is transformed into another by reorganization of tissue fragments rather than by cell proliferation. { ˌmȯr·fə'lak·səs }

Morphinae [INV ZOO] A subfamily of large tropical butterflies in the family Nymphalidae. { 'mȯr·fəˌnē }

morphine [PHARM] $C_{17}H_{19}NO_3 \cdot H_2O$ A white crystalline narcotic powder, melting point 254°C, an alkaloid obtained from opium; used in medicine in the form of a hydrochloride or sulfate salt. { 'mȯrˌfēn }

***para*-morphine** *See* thebaine. { ¦par·ə 'mȯrˌfēn }

morphinism [MED] **1.** The condition caused by the habitual use of morphine. **2.** The morphine habit. { 'mȯr·fəˌniz·əm }

morphogen [BIOCHEM] Any compound that exerts a morphogenetic effect at low concentrations. { 'mȯr·fə·jən }

morphogene [GEN] Any gene involved directly or indirectly in the control of growth and morphogenesis. { 'mȯr·fəˌjēn }

morphogenesis [EMBRYO] The transformation involved in the growth and differentiation of cells and tissue. Also known as topogenesis. { ˌmȯr·fə'jen·ə·səs }

morphogenetic movement [EMBRYO] Any movement of or within a cell that changes the shape of differentiating cells or tissues. { ˌmȯr·fə·jə¦ned·ik müv·mənt }

morphogenetic stimulus [EMBRYO] A stimulus exerted by one part of the developing embryo on another, leading to morphogenesis in the reacting part. { ˌmȯr·fə·jə¦ned·ik 'stim·yə·ləs }

morphology [BIOL] A branch of biology that deals with structure and form of an organism at any stage of its life history. { mȯr'fäl·ə·jē }

morphospecies [SYST] A typological species

distinguished solely on the basis of morphology. { ¦mȯr·fō′spē,shēz }

Morquio's syndrome [MED] A hereditary disease transmitted as an autosomal recessive and characterized by large quantities of keratosulfate in urine, dwarfism, and a typical facies with broad mouth, prominent maxilla, short nose, and widely spaced teeth. Also known as Brailsford-Morquio syndrome; familial osteochondrodystrophy; mucopolysaccharidosis IV. { ′mȯr·kwē,ōz ,sin,drōm }

mortality rate [MED] For a given period of time, the ratio of the number of deaths occurring per 1000 population. Also known as death rate. { mȯr′tal·əd·ē ,rāt }

Morton's neuroma [MED] The thickening of the third intermetatarsal nerve over a period of years, causing a small benign fusiform tumor to eventually form in the space between the third and fourth toes, symptoms include painful burning, tingling, and numbness between the third and fourth toes, often accompanied by radiating electric-like shocks. { ¦mȯrt·ənz nə′rō·mə }

morula [EMBRYO] A solid mass of blastomeres formed by cleavage of the eggs of many animals; precedes the blastula or gastrula, depending on the type of egg. [INV ZOO] A cluster of immature male gametes in which differentiation occurs outside the gonad; common in certain annelids. { ′mȯr·ə·lə }

Moruloidea [INV ZOO] The only class of the phylum Mesozoa; embryonic development in the organisms proceeds as far as the morula or stereoblastula stage. { ,mȯr·ə′lȯid·ē·ə }

mosaic [BIOL] An organism or part made up of tissues or cells exhibiting mosaicism. [EMBRYO] An egg in which the cytoplasm of early cleavage cells is of the type which determines its later fate. { mō′zā·ik }

mosaic evolution [EVOL] The tendency of one or more characters to undergo evolutionary change at different rates than other characters in a lineage. { mō¦zā·ik ,ev·ə′lü·shən }

mosaicism [GEN] The coexistence in an individual of somatic cells of two or more genotypes or karyotypes; it is caused by gene or chromosome mutations, especially mitotic nondisjunction, after fertilization. { mō′zā·ə,siz·əm }

Moschcowitz's disease *See* thrombotic thrombocytopenic purpura. { ′mȯsh·kə,vit·səz di,zēz }

mosquito [INV ZOO] Any member of the dipterous subfamily Culicinae; a slender fragile insect, with long legs, a long slender abdomen, and narrow wings. { mə′skēd·ō }

moss [BOT] Any plant of the class Bryatae, occurring in nearly all damp habitats except the ocean. { mȯs }

moss forest *See* temperate rainforest. { ′mȯs ′fär·əst }

moss land [ECOL] An area which contains abundant moss but is not wet enough to be a bog. { ′mȯs ,land }

Motacillidae [VERT ZOO] The pipits, a family of passeriform birds in the suborder Oscines. { ,mōd·ə′sil·ə,dē }

moth [INV ZOO] Any of various nocturnal or crepuscular insects belonging to the lepidopteran suborder Heteroneura; typically they differ from butterflies in having the antennae feathery and rarely clubbed, a stouter body, less brilliant coloration, and proportionately smaller wings. { mȯth }

mother-of-pearl [INV ZOO] The pearly iridescent internal layer of the shell of various pearl-bearing bivalve mollusks. { ′mət͟h·ər əv ′pərl }

motile [BIOL] Being capable of spontaneous movement. { mōd·əl }

motility symbiosis [ECOL] A symbiotic relationship in which motility is conferred upon an organism by its symbiont. { mō¦til·əd·ē ,sim·bē′ō·səs }

motion sickness [MED] A complex of symptoms, including nausea, vertigo, and vomiting, occurring as the result of random multidirectional accelerations of a vehicle. { ′mō·shən ,sik·nəs }

motoneuron *See* motor neuron. { ¦mōd·ə′nu̇r,än }

motor [NEUROSCI] Pertaining to efferent nerves which innervate muscles and glands. [PHYSIO] That which causes action or movement. { ′mōd·ər }

motor alexia [MED] Inability to read aloud, while comprehension of the written word is preserved. { ′mōd·ər ā′lek·sē·ə }

motor aphasia [MED] A form of aphasia in which the patient knows what he wishes to say but is unable to get the words out, and is able to perceive and comprehend both spoken and written language but is unable to repeat what he sees or hears; due principally to a brain lesion. { ′mōd·ər ə′fā·zhə }

motor apraxia [MED] Inability to carry out, on command, a complex or skilled movement, though the purpose thereof is clear to the patient. Also known as kinesthetic apraxia; limb-kinetic apraxia. { ′mōd·ər ā′prak·sē·ə }

motor area [NEUROSCI] The ascending frontal gyrus containing nerve centers for voluntary movement; characterized by the presence of Betz cells. Also known as Broadman's area 4; motor cortex; pyramidal area. { ′mōd·ər ,er·ē·ə }

motor ataxia [MED] Inability to coordinate the muscles, which becomes apparent only on body movement. { ′mōd·ər ə′tak·sē·ə }

motor cell [BOT] *See* bulliform cell. [PHYSIO] An efferent nerve cell in the anterior horn of the spinal cord. { ′mōd·ər ,sel }

motor cortex *See* motor area. { ′mōd·ər ′kȯr,teks }

motor end plate [ANAT] A specialized area beneath the sarcolemma where functional contact is made between motor nerve fibers and muscle fibers. { ′mōd·ər ′end ,plāt }

motor nerve [NEUROSCI] A nerve composed wholly or principally of motor fibers. { ′mōd·ər ,nərv }

motor neuron [NEUROSCI] An efferent nerve cell. Also known as motoneuron. { ′mōd·ər ′nu̇r,än }

motor speech area [ANAT] The cortical area located in the triangular and opercular portions

of the inferior frontal gyrus; in right-handed people it is more developed on the left side. { 'mōd·ər 'spēch ˌer·ē·ə }
motor system [PHYSIO] Any portion of the nervous system that regulates and controls the contractile activity of muscle and the secretory activity of glands. { 'mōd·ər ˌsis·təm }
motor unit [ANAT] The axon of an anterior horn cell, or the motor fiber of a cranial nerve, together with the striated muscle fibers innervated by its terminal branches. { 'mōd·ər ˌyü·nət }
mottle [MED] An effect that occurs during radiological imaging when the dose of radiation is reduced to a level where quantum effects can be observed. { 'mäd·əl }
mountain lion *See* puma. { 'mau̇nt·ən ˌlī·ən }
mountain sickness [MED] A disease occurring in persons living at high altitudes when homeostatic adjustments to the lowered atmospheric oxygen tension fail or develop disproportionately. Also known as high-altitude disease; high-altitude erythremia; Monge's disease; seroche. { 'mau̇nt·ən ˌsik·nəs }
mountain tick fever *See* Colorado tick fever. { 'mau̇nt·ən 'tik ˌfē·vər }
mouse [VERT ZOO] Any of various rodents which are members of the families Muridae, Heteromyidae, Cricetidae, and Zapodidae; characterized by a pointed snout, short ears, and an elongated body with a long, slender, sparsely haired tail. { mau̇s }
mousebane *See* aconite. { 'mau̇sˌbān }
mouse deer *See* chevrotain. { 'mau̇s ˌdir }
mouth [ANAT] The oral or buccal cavity and its related structures. { mau̇th }
mouth-to-mouth resuscitation [MED] A method of artificial respiration in which the rescuer's mouth is placed over the victim's mouth and air is blown forcefully into the victim's lungs every few seconds to inflate them. { 'mau̇th tə 'mau̇th riˌsəs·ə'tā·shən }
mRNA *See* messenger ribonucleic acid.
msDNA *See* microsatellite deoxyribonucleic acid.
MSH *See* melanocyte-stimulating hormone.
mtDNA *See* mitochondrial deoxyribonucleic acid.
muc-, muci-, muco- [ZOO] A combining form denoting pertaining to mucus, mucin, mucosa. { myük,'myü·sē, 'myü·kō }
Mucedinaceae [MYCOL] The equivalent name for Moniliaceae. { myüˌsed·ən'ās·ēˌē }
mucigel [BIOCHEM] A complex polysaccharide material that is composed of root mucilage and bacterial slime and acts to control aggregation of soil particles in the rhizosphere in the vicinity of older portions of plant roots. { 'myü·səˌjel }
mucigen [BIOCHEM] A substance from which mucin is derived; contained in mucus-secreting epithelial cells. { 'myü·sə·jən }
mucin [BIOCHEM] A glycoprotein constituent of mucus and various other secretions of humans and lower animals. { 'myü·sən }
mucinosis [MED] Accumulations of materials containing mucin or mucinous substances in the skin; sometimes accompanied by papule and nodule formation. { ˌmyü·sə'nō·səs }
mucinous cyst *See* mucous cyst. { 'myüs·ən·əs 'sist }
mucocele [MED] **1.** Dilatation, particularly of a cavity, with mucus secretion. **2.** A polypoid lesion consisting of mucus and mucus-secreting tissue. { 'myü·kəˌsēl }
mucocutaneous [ANAT] Pertaining to a mucous membrane and the skin, and to the line where these join. { ¦myü·kō·kyü'tā·nē·əs }
mucoid [BIOCHEM] **1.** Any of various glycoproteins, similar to mucins but differing in solubilities and precipitation properties and found in cartilage, in the crystalline lens, and in white of egg. **2.** Resembling mucus. [MICROBIO] Pertaining to large colonies of bacteria characterized by being moist and sticky. { 'myüˌkȯid }
mucolytic [BIOCHEM] Effecting the dissolution, liquefaction, or dispersion of mucus and mucopolysaccharides. { ˌmyü·kə'lid·ik }
mucopolysaccharide [BIOCHEM] Any of a group of polysaccharides containing an amino sugar and uronic acid; a constituent of mucoproteins, glycoproteins, and blood-group substances. { ¦myü·kōˌpäl·ē'sak·əˌrīd }
mucopolysaccharidosis [MED] Any of several inborn metabolic disorders involving mucopolysaccharides; the six types are MPS I, Hurler's syndrome; MPS II, Hunter's syndrome; MPS III, Sanfillipo's syndrome; MPS IV, Morquio's syndrome; MPS V, Scheil's syndrome; and MPS VI, Maroteaux-Lamy's syndrome. { ¦myü·kōˌpäl·ēˌsak·ə·rə'dō·səs }
mucoprotein [BIOCHEM] Any of a group of glycoproteins containing a sugar, usually chondroitinsulfuric or mucoitinsulfuric acid, combined with amino acids or polypeptides. { ¦myü·kō'prōˌtēn }
mucopurulent [MED] Containing mucus and pus. { ¦myü·kō'pyu̇r·ə·lənt }
Mucorales [MYCOL] An order of terrestrial fungi in the class Phycomycetes, characterized by a hyphal thallus and nonmotile sporangiospores, or conidiospores. { ˌmyü·kə'rā·lēz }
mucormycosis [MED] An acute, usually fulminating fungus infection of humans caused by several genera of Mucorales, including *Absidia*, *Rhizopus*, and *Mucor*. { ¦myü·kȯrˌmī'kō·səs }
mucosa [HISTOL] A mucous membrane. { myü 'kō·sə }
mucosanguineous [MED] Containing mucus and blood. { ¦myü·kō·saŋ'gwin·ē·əs }
mucous [PHYSIO] Of or pertaining to mucus; secreting mucus. { 'myü·kəs }
mucous cell [PHYSIO] A mucus-secreting cell. { 'myü·kəs ˌsel }
mucous colitis *See* irritable colon. { 'myü·kəs kə'līd·əs }
mucous connective tissue [HISTOL] A type of loose connective tissue in which the ground substance is especially prominent and soft; occurs in the umbilical cord. { 'myü·kəs kə'nek·tiv 'tish·ü }
mucous cyst [MED] A retention cyst of a gland, containing a secretion rich in mucin. Also known as mucinous cyst. { 'myü·kəs ˌsist }

mucous degeneration [MED] Any retrogressive change associated with abnormal production of mucus. { 'myü·kəs dē,jen·ə'rā·shən }
mucous epithelium [EMBRYO] The epidermis of an embryo, excluding the epitrichium. [HISTOL] The germinative layer of a stratified squamous epithelium. { 'myü·kəs ,ep·ə'thē·lē·əm }
mucous gland [PHYSIO] A gland that secretes mucus. { 'myü·kəs ,gland }
mucous membrane [HISTOL] The type of membrane lining cavities and canals which have communication with air; it is kept moist by glandular secretions. Also known as tunica mucosa. { 'myü·kəs 'mem,brān }
mucoviscidosis *See* cystic fibrosis. { ,myü·kō,vis·ə'dō·səs }
mucro [BIOL] An abrupt, sharp terminal tip or process. { 'myü·krō }
mucronate [BIOL] Terminated abruptly by a sharp terminal tip or process. { 'myü·krə,nāt }
mucus [PHYSIO] A viscid fluid secreted by mucous glands, consisting of mucin, water, inorganic salts, epithelial cells, and leukocytes, held in suspension. { 'myü·kəs }
mudfish *See* bowfin. { 'məd,fish }
mud puppy [VERT ZOO] Any of several American salamanders of the genera *Necturus* and *Proteus* making up the family Proteidae; distinguished by having both lungs and gills as an adult. { 'məd ,pəp·ē }
Mugilidae [VERT ZOO] The mullets, a family of perciform fishes in the suborder Mugiloidei. { myü'jil·ə,dē }
Mugiloidei [VERT ZOO] A suborder of fishes in the order Perciformes; individuals are rather elongate, terete fishes with a short spinous dorsal fin that is well separated from the soft dorsal fin. { ,myü·jə'lȯid·ē,ī }
mulberry [BOT] Any of various trees of the genus *Morus* (family Moraceae), characterized by milky sap and simple, often lobed alternate leaves. { 'məl,ber·ē }
mule [VERT ZOO] The sterile hybrid offspring of the male ass and the mare, or female horse. { myül }
Müllerian duct *See* paramesonephric duct. { mi'ler·ē·ən 'dəkt }
Müllerian duct cyst [MED] A congenital cyst arising from vestiges of the Müllerian ducts. { mi'ler·ē·ən ¦dəkt 'sist }
Müllerian mimicry [ZOO] Mimicry between two aposematic species. { mi'ler·ē·ən 'mim·ə·krē }
Müller's larva [INV ZOO] The ciliated larva characteristic of various members of the Polycladida; resembles a modified ctenophore. { 'mil·ərz ,lär·və }
multicellular [BIOL] Consisting of many cells. { ¦məl·tē'sel·yə·lər }
multicipital [BIOL] Having many heads or branches arising from one point. { ,məl·tə'sip·əd·əl }
multicompartmental genome [VIROL] In certain viruses, separation of genetic information into encapsulated nucleic acid molecules. { ,məl·tə·kəm,pärt'ment·əl 'jē,nōm }
multifid [BIOL] Divided into many lobes. { 'məl·tə,fid }
multigene family [GEN] A set of genes that arose from duplications of a single ancestral gene and variation due to independent mutations and selection acting on individual members of the duplicate genes. { 'məl·tə,jēn ,fam·lē }
multiglandular [ANAT] Of or pertaining to several glands. { ,məl·tə'glan·jə·lər }
Multillidae [INV ZOO] An economically important family of Hymenoptera; includes the cow killer, a parasite of bumblebee pupae. { məl'til·ə,dē }
multilocular [BIOL] Having many small chambers or vesicles. { ¦məl·tē'äk·yə·lər }
multimer [BIOCHEM] A protein molecule composed of two or more monomers. { 'məl·tə·mər }
multiphasic personality inventory *See* Minnesota multiphasic personality inventory. { ¦məl·tə 'fāz·ik ,pər·sə'nal·əd·ē 'in·vən,tȯr·ē }
multiple-anomaly syndrome [MED] Any syndrome associated with numerous congenital abnormalities. { 'məl·tə·pəl ə¦näm·ə·lē ,sin ,drōm }
multiple cancellous exostoses *See* multiple hereditary exostoses. { 'məl·tə·pəl 'kan·sə·ləs ,ek·səs'tō·sēz }
multiple cartilaginous exostoses *See* multiple hereditary exostoses. { 'məl·tə·pəl ,kard·əl'aj·ə·nəs ,ek·səs'tō·sēz }
multiple colloid goiter *See* adenomatous goiter. { 'məl·tə·pəl 'kä,lȯid ,gȯid·ər }
multiple epidermis [BOT] Epidermis that is several layers thick, occurring in many species of *Ficus*, *Begonia*, and *Peperomia*. { 'məl·tə·pəl ,ep·ə'dər·məs }
multiple fruit [BOT] Any fruit derived from the ovaries and accessory structures of several flowers consolidated into one mass, such as a pineapple and mulberry. { 'məl·tə·pəl 'früt }
multiple hereditary exostoses [MED] An inherited form of exostosis, revealing itself at several sites in childhood or adolescence. Also known as diaphyseal aclasis; hereditary deforming chondrodysplasia; metaphyseal aclasis; multiple cancellous exostoses; multiple cartilaginous exostoses. { 'məl·tə·pəl hə'red·ə,ter·ē ,ek·səs 'tō·sēz }
multiple infarct dementia *See* cerebral arteriosclerosis. { ¦məl·tə·pəl ¦in,färkt di'men·chə }
multiple myeloma [MED] A primary bone malignancy characterized by diffuse osteoporosis, anemia, hyperglobulinemia, and other clinical features. Also known as Kahler's disease. { 'məl·tə·pəl ,mī·ə'lō·mə }
multiple neuritis *See* polyneuritis. { 'məl·tə·pəl nu̇'rīd·əs }
multiple neurofibroma *See* neurofibromatosis. { 'məl·tə·pəl ,nu̇r·ō·fī'brō·mə }
multiple neurofibromatosis *See* neurofibromatosis. { 'məl·tə·pəl ,nu̇r·ō·fī¦brō·mə'tō·səs }
multiple pregnancy [MED] Being pregnant with

more than one fetus. { 'məl·tə·pəl 'preg·nən·sē }

multiple sclerosis [MED] A degenerative disease of the nervous system of unknown cause in which there is demyelination followed by gliosis. { 'məl·tə·pəl sklə'rō·səs }

multiple serositis *See* polyserositis. { 'məl·tə·pəl ˌsir·ō'sīd·əs }

multiple sleep latency test [MED] A test used to document pathologic sleepiness and diagnose narcolepsy in which recordings of brain waves, muscle activities, and eye movements are taken while a person spends the day in a sleep laboratory, taking naps at intervals. { ¦məl·tə·pəl 'slēp ˌlāt·ən·sē ˌtest }

multipotent cell [EMBRYO] A cell capable of giving rise to only a limited number of cell types. { ¦məl·təˌpōt·ent 'sel }

multituberculate [VERT ZOO] Of teeth, having several or many simple conical cusps. { ˌməl·tē·təˌbər·kyə'lāt }

mummification [MED] **1.** Drying of a part of the body into a hard mass. **2.** Dry gangrene. { ˌməm·ə·fə'kā·shən }

mumps [MED] An acute contagious viral disease characterized chiefly by painful enlargement of a parotid gland. { məmps }

mumps orchitis [MED] Inflammation of the testis due to the mumps virus. { 'məmps ȯr'kīd·əs }

mu phage [VIROL] A temperate phage with properties similar to those of transposable genetic elements. { 'myü ˌfāj }

mural thrombus [MED] A thrombus attached to the wall of a blood vessel or mural endocardium. Also known as lateral thrombus. { 'myu̇r·əl 'thräm·bəs }

muramic acid [BIOCHEM] An organic acid found in the mucopeptide (murein) in the cell walls of bacteria and blue-green algae. { myu̇ 'ram·ik 'as·əd }

muramidase [BIOCHEM] Lysozyme when acting as an enzyme on the hydrolysis of the muramic acid-containing mucopeptide in the cell walls of some bacteria. { myu̇'ram·əˌdās }

murein [BIOCHEM] The peptidoglycan of bacterial cell walls. { myu̇r·ē·ən }

Muricacea [INV ZOO] A superfamily of gastropod mollusks in the order Prosobranchia. { ˌmyu̇r·ə'ka·shē·ə }

muricate [ZOO] Covered with sharp, hard points. { 'myu̇r·əˌkāt }

Muricidae [INV ZOO] A family of predatory gastropod mollusks in the order Neogastropoda; contains the rock snails. { ˌmyu̇'ris·əˌdē }

Muridae [VERT ZOO] A large diverse family of relatively small cosmopolitan rodents; distinguished from closely related forms by the absence of cheek pouches. { 'myu̇r·əˌdē }

muriform [BIOL] **1.** Resembling the arrangement of courses in a brick wall, especially having both horizontal and vertical septa. **2.** Pertaining to or resembling a rat or mouse. { 'myu̇r·əˌfȯrm }

Murinae [VERT ZOO] A subfamily of the Muridae which contains such forms as the striped mouse, house mouse, harvest mouse, and field mouse. { myu̇'rīˌnē }

murine plague [VET MED] Infection of the rat by the bacterium *Pasteurella pestis*; transmitted from rat to rat and from rat to human by a flea. { 'myu̇ˌrīn ¦plāg }

murine typhus [MED] A relatively mild, acute, febrile illness of worldwide distribution caused by *Rickettsia mooseri*, transmitted from rats to humans by the flea and characterized by headache, macular rash, and myalgia. Also known as endemic typhus; flea-borne typhus; rat typhus; shop typhus; urban typhus. { 'myu̇ˌrīn 'tī·fəs }

murmur [MED] A blowing or roaring heart sound heard through the wall of the chest; caused by blood flow through a defective valve. { 'mər·mər }

Murray Valley encephalitis [MED] An acute inflammation of the brain and spinal cord caused by a virus; confined to Australia and New Guinea. Also known as Australian X disease. { 'mər·ē ¦val·ē inˌsef·ə'līd·əs }

Musaceae [BOT] A family of monocotyledonous plants in the order Zingiberales characterized by five functional stamens, unisexual flowers, spirally arranged leaves and bracts, and fleshy, indehiscent fruit. { myü'zā·sēˌē }

muscardine diseases [INV ZOO] A group of insect diseases caused by the muscardine fungi, in which the fungal pathogen emerges from the body of the insect and covers the animal with a characteristic fungus mat. { 'məs·kərˌdēn di ˌzēz·əs }

muscarinism [MED] Poisoning due to ingestion of certain mushrooms. { məs'kar·əˌniz·əm }

Musci *See* Bryopsida. { 'məˌsī }

Muscicapidae [VERT ZOO] A family of passeriform birds assigned to the Oscines; includes the Old World flycatchers or fantails. { ˌməs·ə'kap·əˌdē }

Muscidae [INV ZOO] A family of myodarian cyclorrhaphous dipteran insects in the subsection Calypteratae; includes the houseflies, stable flies, and allies. { 'məs·əˌdē }

muscle [ANAT] A contractile organ composed of muscle tissue that changes in length and effects movement when stimulated. [HISTOL] A tissue composed of cells containing contractile fibers; three types are smooth, cardiac, and skeletal. { 'məs·əl }

muscle-contraction headache [MED] A type of headache characterized by dull constricting pain that can either occur intermittently or continue for days, months, or years. Also known as a tension headache. { ¦məs·əl kənˌtrak·shən ˌhedˌāk }

muscle fiber [HISTOL] The contractile cell or unit of which muscle is composed. { 'məs·əl ˌfī·bər }

muscle hemoglobin *See* myoglobin. { 'məs·əl 'hē·məˌglō·bən }

muscle tone *See* tonus. { 'məs·əl ˌtōn }

muscul-, musculo- [ZOO] A combining form denoting muscle, muscular. { 'məs·kə·lō }

muscular atrophy [MED] Degenerative reduction of muscle size, especially skeletal muscles, due to a lesion involving either the cell body or axon of the lower motor neuron. { 'məs·kyə·lər 'a·trə·fē }

muscular dystrophy [MED] A group of diseases characterized by degeneration of or injury to individual muscle cells, not primarily involving the nerve supply; the most common form is Duchenne-Greisinger disease. { 'məs·kyə·lər 'dis·trə·fē }

muscularis externa [HISTOL] The layer of the digestive tube consisting of smooth muscles. { ˌməs·kyə'lar·əs ek'stər·nə }

muscularis mucosae [HISTOL] Thin, deep layer of smooth muscle in some mucous membranes, as in the digestive tract. { ˌməs·kyə'lar·əs myü'kō·sē }

muscular system [ANAT] The muscle cells, tissues, and organs that effect movement in all vertebrates. { 'məs·kyə·lər ˌsis·təm }

musculoaponeurotic [HISTOL] Composed of muscle and of fibrous connective tissue in the form of a membrane. { ¦məs·kyə·lō¦ap·ə·nu̇¦räd·ik }

musculocutaneous [ANAT] Of or pertaining to muscles and skin. { ¦məs·kyə·lō·kyü'tā·nē·əs }

musculocutaneous nerve [NEUROSCI] A branch of the brachial plexus with both motor and somatic sensory components; innervates flexor muscles of the upper arm, and skin of the lateral aspect of the forearm. { ¦məs·kyə·lō·kyü'tā·nē·əs 'nərv }

musculoskeletal system [ANAT] The muscular and skeletal elements of vertebrates, considered as a functional unit. { ¦məs·kyə·lō'skel·ə·təl ˌsis·təm }

musculoskeletal toxicity [MED] Adverse effects to the structure and/or function of the muscles, bones, and joints caused by exposure to a toxic chemical, such as coal dust or cadmium. Also, the bone disorders arthritis, fluorosis, and osteomalacia can result. { ˌməs·kyə·lō¦skel·ət·əl ˌtak'sis·əd·ē }

mushroom [MYCOL] **1.** A fungus belonging to the basidiomycetous order Agaricales. **2.** The fruiting body (basidiocarp) of such a fungus. { 'məshˌrüm }

Musidoridae [INV ZOO] A family of orthorrhaphous dipteran insects in the series Brachycera distinguished by spear-shaped wings. { ˌmyü·zə'dȯr·əˌdē }

musk [PHYSIO] Any of various strong-smelling substances obtained from the musk glands of musk deer or similar animals; used in the form of a tincture as a fixative for perfume. { məsk }

musk bag *See* musk gland. { 'məsk ˌbag }

muskeg [ECOL] A peat bog or tussock meadow, with variably woody vegetation. { 'məˌskeg }

musk gland [VERT ZOO] A large preputial scent gland of the musk deer and various other animals, including skunk and musk-ox. Also known as musk bag. { 'məsk ˌgland }

muskmelon [BOT] *Cucumis melo.* The edible, fleshy, globular to long-tapered fruit of a trailing annual plant of the order Violales; surface is uniform to broadly sutured to wrinkled, and smooth to heavily netted, and flesh is pale green to orange; varieties include cantaloupe, Honey Dew, Casaba, and Persian melons. { 'məsk ˌmel·ən }

musk-ox [VERT ZOO] *Ovibos moschatus.* An even-toed ungulate which is a member of the mammalian family Bovidae; a heavy-set animal with a shag pilage, splayed feet, and flattened horns set low on the head. { 'məsˌkäks }

muskrat [VERT ZOO] *Ondatra zibethica.* The largest member of the rodent subfamily Microtinae; essentially a water rat with a laterally flattened, long, naked tail, a broad blunt head with short ears, and short limbs. { 'məˌskrat }

Musophagidae [VERT ZOO] The turacos, an African family of birds of uncertain affinities usually included in the order Cuculiformes; resemble the cuckoos anatomically but have two unique pigments, turacin and turacoverdin. { ˌmyü·zə'faj·əˌdē }

Muspiceoidea [INV ZOO] A superfamily of parasitic nematodes in the order Dioctophymatida, distinguished by a greatly reduced neurosensory structure, the absence of (except in one species) amphids and cephalic papillae, and, in females, a reduced digestive tube; males have never been reported. { myüsˌpī·sē'ȯid·ē·ə }

mustard [BOT] Any of several annual crucifers belonging to the genus *Brassica* of the order Capparales; leaves are lyrately lobed, flowers are yellow, and pods have linear beaks; the mustards are cultivated for their pungent seed and edible foliage, and the seeds of *B. niger* are used as a condiment, prepared as a powder, paste, or oil. { 'məs·tərd }

Mustilidae [VERT ZOO] A large, diverse family of low-slung, long-bodied carnivorous mammals including minks, weasels, and badgers; distinguished by having only one molar in each upper jaw, and two at the most in the lower jaw. { mə'stil·əˌdē }

mutable gene [GEN] Any of a class of unstable genes that spontaneously mutate at a sufficiently high rate to produce mosaicism. { ¦myüd·ə·bəl 'jēn }

mutagen [GEN] An agent that raises the frequency of mutation above the spontaneous or background rate. { 'myüd·ə·jən }

mutagen persistence [GEN] The stability of a mutagen in the environment or in the human body. { ¦myüd·ə·jən pər'sis·təns }

mutagen specificity [GEN] The tendency of a mutagen to induce only one type of mutation. { ¦myüd·ə·jən ˌspes·ə'fis·əd·ē }

mutant [GEN] An individual bearing an allele that has undergone mutation and is expressed in the phenotype. { 'myüt·ənt }

mutase [BIOCHEM] An enzyme able to catalyze a dismutation or a molecular rearrangement. { 'myüˌtās }

mutation [GEN] An abrupt change in the genotype of an organism, not resulting from recombination; genetic material may undergo qualitative or quantitative alteration, or rearrangement. { myü'tā·shən }

mutation fixation [CELL MOL] The condition of changing a premutational deoxyribonucleic acid lesion to mutation. { myü'tā·shən fik,sā·shən }

mutation hot spot [CELL MOL] Any locus in the deoxyribonucleic acid sequence or on a chromosome where mutations or aberrations occur preferentially. { myü¦tā·shən 'hät ,spöt }

mutator phenotype [GEN] The loss of function of one gene, such as one for the repair of damaged deoxyribonucleic acid, that greatly increases the mutation rates at other loci. { ¦myü ,tād·ər 'fēn·ə,tīp }

Mutillidae [INV ZOO] The velvet ants, a family of hymenopteran insects in the superfamily Scolioidea. { myü'til·ə,dē }

mutism [MED] Inability or refusal to speak. { 'myü,tiz·əm }

muton [CELL MOL] The smallest unit of genetic material capable of undergoing mutation. { 'myü,tän }

mutualism [ECOL] Mutual interactions between two species that are beneficial to both species. { 'myü·chə·wə,liz·əm }

muzzle [VERT ZOO] The snout of an animal, as a dog or horse. { 'məz·əl }

myalgia [MED] Pain in the muscles. { mī'al·jē·ə }

myasthenia [MED] Muscular weakness. { ,mī·əs·thēn·ē·ə }

myasthenia gravis [MED] A muscle disorder of unknown etiology characterized by varying degrees of weakness and excessive fatigability of voluntary muscle. { ,mī·əs·thēn·ē·ə 'grav·əs }

myasthenia reaction [MED] The electromyographic reaction observed in myasthenia gravis in which there is a gradual loss of intensity and duration for the tetanic contraction, and a gradual diminution in amplitude and frequency of motor unit discharges until the muscle is fatigued. { ,mī·əs·thēn·ē·ə ri'ak·shən }

myasthenic crisis [MED] Profound myasthenia and respiratory paralysis associated with myasthenia gravis. { ¦mī·əs¦thēn·ik 'krī·səs }

myatonia [MED] Lack of muscle tone. { ,mī·ə'tō·nē·ə }

Mycelia Sterilia [MYCOL] An order of fungi of the class Fungi Imperfecti distinguished by the lack of spores; certain members are plant pathogens. { mī'sel·yə stə'ril·yə }

mycelium [BIOL] A mass of filaments, or hyphae, composing the vegetative body of many fungi and some bacteria. { mī'sē·lē·əm }

Mycetaeidae [INV ZOO] The equivalent name for Endomychidae. { ,mī·sə'tē·ə,dē }

mycetocyte [INV ZOO] **1.** One of the cells clustered together to form a mycetome. **2.** An individual cell functioning like a mycetome. { mī'sēd·ə,sīt }

mycetoma [MED] A chronic fungus or bacterial infection, usually of the feet, resulting in swelling. Also known as madura foot; maduromycosis. { ,mī·sə'tō·mə }

mycetome [INV ZOO] One of the specialized structures in the body of certain insects for holding endosymbionts. { 'mī·sə,tōm }

Mycetophagidae [INV ZOO] The hairy fungus beetles, a cosmopolitan family of coleopteran insects in the superfamily Cucujoidea. { mī ,sēd·ə'faj·ə,dē }

Mycetozoa [BIOL] A zoological designation for organisms that exhibit both plant and animal characters during their life history (Myxomycetes); equivalent to the botanical Myxomycophyta. { mī,sēd·ə'zō·ə }

Mycetozoia [INV ZOO] A subclass of the protozoan class Rhizopodea. { mī,sēd·ə'zȯi·ə }

Mycobacteriaceae [MICROBIO] A family of bacteria in the order Actinomycetales; acid-fast, aerobic rods form a filamentous or myceliumlike growth. { ,mī·kō,bak·tir·ē'ās·ē,ē }

mycobacterial disease [MED] Any disease caused by species of *Mycobacterium*. { ¦mī·kō· bak¦tir·ē·əl di,zēz }

mycobactin [BIOCHEM] Any compound produced by some strains, and required for growth by other strains, of *Mycobacteria*. { ¦mī·kō¦bak· tən }

mycobiont [BOT] The fungal component of a lichen, commonly an ascomycete. { ¦mī·kə 'bī,änt }

mycology [BOT] The branch of botany that deals with the study of fungi. { mī'käl·ə·jē }

mycomycin [MICROBIO] $C_{13}H_{10}O_2$ An antibiotic produced by *Nocardia acidophilus* and a species of *Actinomyces*; characterized as a highly unsaturated aliphatic acid that shows strong activity against *Mycobacterium tuberculosis*. { ,mīk·ə'mīs·ən }

mycophagous [ZOO] Feeding on fungi. { mī 'käf·ə·gəs }

Mycophiformes [VERT ZOO] An equivalent name for Salmoniformes. { mī,käf·ə'fȯr·mēz }

Mycoplasmataceae [MICROBIO] A family of the order Mycoplasmatales; distinguished by sterol requirement for growth. { ,mī·kō,plaz·mə'tās· ē,ē }

Mycoplasmatales [MICROBIO] The single order of the class Mollicutes; organisms are gram-negative, generally nonmotile, nonsporing bacteria which lack a true cell wall. { ,mī·kō,plaz·mə 'tā·lēz }

mycorrhiza [BOT] A mutual association in which the mycelium of a fungus invades the roots of a seed plant. { ,mīk·ə'rīz·ə }

mycorrhizal fungi [MYCOL] Fungi that form symbiotic relationships in and on the roots of host plants. { ,mī·kə,rīz·əl 'fən,jī }

mycosis [MED] An infection with or a disease caused by a fungus. { mī'kō·səs }

mycosis fungoides [MED] A lymphoma of the skin, usually present in several sites when first diagnosed, that may remain confined to the skin for 10 or more years before eventually spreading to internal organs and causing death. { mī¦kōs· əs fəŋ'gȯi,dēz }

Mycosphaerella sentina [MYCOL] A fungal plant pathogen that causes leaf blight of pears, a disease that destroys the leaves of pear trees. { ,mī·kō·sfī,rel·ə sen'tē·nə }

Mycota [MYCOL] An equivalent name for Eumycetes. { mī'käd·ə }

mycotic stomatitis *See* thrush. { mī'käd·ik ˌstō·mə'tīd·əs }

mycotoxicosis [MED] Any of a group of diseases caused by accidental or recreational ingestion of toxic fungal metabolites, such as mushroom poisoning. { ˌmī·kōˌtäk·sə'kō·səs }

mycotoxin [MYCOL] A toxin produced by a fungus. { 'mī·kəˌtäk·sən }

Myctophidae [VERT ZOO] The lantern fishes, a family of deep-sea forms of the suborder Myctophoidei. { mik'täf·əˌdē }

Myctophoidei [VERT ZOO] A large suborder of marine salmoniform fishes characterized by having the upper jaw bordered only by premaxillae, and lacking a mesocoracoid arch in the pectoral girdle. { mik·tə'fȯid·ēˌī }

Mydaidae [INV ZOO] The mydas flies, a family of orthorrhaphous dipteran insects in the series Brachycera. { mī'dā·əˌdē }

mydriasis [MED] Prolonged dilation of the pupil of the eye. { mə'drī·ə·səs }

mydriatic [PHARM] An agent which produces dilation of the pupil, such as eucatropine hydrochloride. { ¦mid·rē¦ad·ik }

myel-, myelo- [ANAT] A combining form indicating relationship to marrow, often in specific reference to the spinal cord. { 'mī·əl, 'mī·ə·lō }

myelencephalon [EMBRYO] The caudal portion of the hindbrain; gives rise to the medulla oblongata. { ˌmī·ə·lən'sef·əˌlän }

myelin [NEUROSCI] A soft, white fatty substance that forms a sheath around certain nerve fibers. { 'mī·ə·lən }

myelin sheath [NEUROSCI] An investing cover of myelin around the axis cylinder of certain nerve fibers. { 'mī·ə·lən 'shēth }

myelitis [MED] **1.** Inflammation of the spinal cord. **2.** Inflammation of the bone marrow. { ˌmī·ə'līd·əs }

myeloblast [HISTOL] The youngest precursor cell for blood granulocytes, having a nucleus with finely granular chromatin and nucleoli and intensely basophilic cytoplasm. { 'mī·ə·lōˌblast }

myeloblastemia [MED] The presence of myeloblasts in the peripheral circulation. { ˌmī·ə·lō·bla'stē·mē·ə }

myeloblastic [HISTOL] Of, pertaining to, or characterized by the presence of myeloblasts. { ˌmī·ə·lə'blas·tik }

myeloblastic leukemia *See* acute granulocytic leukemia. { ˌmī·ə·lə'blas·tik lü'kē·mē·ə }

myeloblastoma [MED] A malignant tumor composed of myeloblasts. { ˌmī·ə·lō·bla'stō·mə }

myeloblastosis [MED] Diffuse proliferation of myeloblasts, with involvement of blood, bone marrow, and other tissues and organs. { ˌmī·ə·lō·bla'stō·səs }

myelocele [ANAT] The canal of the spinal cord. [MED] Spina bifida, with protrusion of the spinal cord. { 'mī·ə·lōˌsēl }

myelocyte [HISTOL] A motile precursor cell of blood granulocytes found in bone marrow. { 'mī·ə·ləˌsīt }

myelocytoma [MED] A malignant plasmacytoma. { ˌmī·ə·lə·sī'tō·mə }

myelocytosis [MED] The presence of myelocytes in the blood. { ˌmī·ə·lə·sī'tō·səs }

myelodysplasia [MED] Abnormal spinal cord development, especially the lumbosacral portion. { ˌmī·ə·lō·dis'plā·zhə }

myeloencephalitis [MED] Inflammation of the brain and spinal cord. { ˌmī·ə·lō·inˌsef·ə'līd·əs }

myelofibrosis [PATH] Growth of white, fibrous connective tissue in the bone marrow. { ˌmī·ə·lō·fī'brō·səs }

myelogenous leukemia *See* granulocytic leukemia. { ˌmī·ə'läj·ə·nəs lü'kē·mē·ə }

myelogram [MED] Roentgenogram of the spinal cord, made by myelography. [PATH] Differential cell study of material extracted from bone marrow. { 'mī·ə·ləˌgram }

myelography [MED] Roentgenographic visualization of the subarachnoid space, after the injection of air or an opaque medium. { ˌmī·ə'läg·rə·fē }

myeloid [ANAT] **1.** Of or pertaining to bone marrow. **2.** Of or pertaining to the spinal cord. { 'mī·əˌlȯid }

myeloid cell [HISTOL] Any of the white blood cell (leukocyte) types that do not fall into the lymphocyte category. { 'mī·əˌlȯid ˌsel }

myeloid leukemia *See* granulocytic leukemia. { 'mī·əˌlȯid lü'kē·mē·ə }

myeloid metaplasia [MED] The occurrence of hemopoietic tissue in abnormal places in the body. { 'mī·əˌlȯid ˌmed·ə'plā·zhə }

myeloid myeloma [MED] A malignant plasmacytoma. { 'mī·əˌlȯid ˌmī·ə'lō·mə }

myeloid reaction [MED] Increased numbers of granulocytes in the bone marrow and peripheral circulation, often with the appearance of immature granulocytes in the blood. { 'mī·əˌlȯid rēˌak·shən }

myeloid tissue [HISTOL] Red bone marrow attached to argyrophile fibers which form wide meshes containing scattered fat cells, erythroblasts, myelocytes, and mature myeloid elements. { 'mī·əˌlȯid ˌtish·ü }

myeloma [MED] A primary tumor of the bone marrow composed of any of the bone marrow cell types. { ˌmī·ə'lō·mə }

myelomalacia [MED] Softening of the spinal cord. { ˌmī·ə·lō'mā·shə }

myelomeningitis [MED] Inflammation of the spinal cord and its meninges. { ˌmī·ə·lōˌmen·ən'jīd·əs }

myelomeningocele [MED] Spina bifida with protrusion of the spinal meninges. { ˌmī·ə·lō·mə'niŋ·gōˌsēl }

myelomonocyte [HISTOL] **1.** A monocyte developing in bone marrow. **2.** A blood cell intermediate between monocytes and granulocytes. { ¦mī·ə·lō'män·əˌsīt }

myelomonocytic leukemia *See* monocytic leukemia. { ¦mī·ə·lōˌmän·ə¦sid·ik lü'kē·mē·ə }

myelopathic anemia *See* myelophthisic anemia. { ¦mī·ə·lō¦path·ik ə'nē·mē·ə }

myelophthisic anemia [MED] An anemia associated with space-occupying disorders of the bone marrow. Also known as leukoerythroblastic anemia; leukoerythroblastosis; metastatic anemia; myelopathic anemia; myelosclerotic anemia; osteosclerotic anemia. { ¦mī·ə·lō'this·ik ə'nē·mē·ə }

myelophthisis [MED] **1.** Loss of bone marrow. **2.** Atrophy of the spinal cord. { ¦mī·ə·lō'this·əs }

myeloplegia [MED] Spinal paralysis. { ˌmī·ə·lə'plē·jē·ə }

myelopoiesis [PHYSIO] The process by which blood cells form in the bone marrow. { ˌmī·ə·lōˌpȯi'ē·səs }

myelosclerosis [MED] **1.** Multiple sclerosis of the spinal cord. **2.** Hardening of the bone marrow. { ˌmī·ə·lō·sklə'rō·səs }

myelosclerotic anemia *See* myelophthisic anemia. { ¦mī·ə·lō·sklə¦räd·ik ə'nē·mē·ə }

myenteric [HISTOL] Of or pertaining to the muscular coat of the intestine. { ¦mī·ən¦ter·ik }

myenteric plexus [NEUROSCI] A network of nerves between the circular and longitudinal layers of the muscular coat of the digestive tract. Also known as Auerbach's plexus. { ¦mī·ən¦ter·ik 'plek·səs }

myenteron [HISTOL] The muscular coat of the intestine. { mī'ent·əˌrän }

Mygalomorphae [INV ZOO] A suborder of spiders (Araneida) including American tarantulas, trap-door spiders, and purse-web spiders; the tarantulas may attain a leg span of 10 inches (25 centimeters). { ˌmig·ə·lō'mȯrˌfē }

myiasis [MED] Infestation of vertebrates by the larvae, or maggots, of flies. { 'mī·ə·səs }

Mylabridae [INV ZOO] The equivalent name for Bruchidae. { mə'lab·rəˌdē }

Myliogatidae [VERT ZOO] The eagle rays, a family of batoids which may reach a length of 15 feet (4.6 meters). { ˌmil·ē·ə'gad·əˌdē }

Mymaridae [INV ZOO] The fairy flies, a family of hymenopteran insects in the superfamily Chalcidoidea. { mī'mar·əˌdē }

myoblast [EMBRYO] A precursor cell of a muscle fiber. { 'mī·əˌblast }

myocardial infarct [MED] An infarct in heart muscle. { ¦mī·ə¦kärd·ē·əl 'inˌfärkt }

myocardiopathy [MED] Disease of the myocardium. Also known as cardiomyopathy. { ¦mī·ōˌkärd·ē'äp·ə·thē }

myocarditis [MED] Inflammation of the myocardium. { ˌmī·əˌkär'dīd·əs }

myocardium [HISTOL] The muscular tissue of the heart wall. { ˌmī·ə'kärd·ē·əm }

myocardosis [MED] Any noninflammatory disease of the myocardium. { ˌmī·əˌkär'dō·səs }

myoclonic epilepsy [MED] Recurrent irregular, arrhythmic clonic muscle spasms, usually occurring more frequently in the morning or on going to sleep and often associated with other types of seizures. { ¦mī·ə¦klän·ik 'ep·əˌlep·sē }

myoclonic status [MED] Continual clonic spasms lasting an hour or more. { ¦mī·ə¦klän·ik 'stad·əs }

myoclonus [MED] **1.** Clonic muscle spasm. **2.** Any disorder characterized by scattered, irregular, arrhythmic muscle spasms. { mī'äk·lə·nəs }

myocoel [EMBRYO] Portion of the coelom enclosed in a myotome. { 'mī·əˌsēl }

myocomma [HISTOL] A ligamentous connection between successive myomeres. Also known as myoseptum. { ˌmī·ə'käm·ə }

myocyte [HISTOL] **1.** A contractile cell. **2.** A muscle cell. { 'mī·əˌsīt }

Myodaria [INV ZOO] A section of the Schizophora series of cyclorrhaphous dipterans; in this group adult antennae consist of three segments, and all families except the Conopidae have the second cubitus and the second anal veins united for almost their entire length. { ˌmī·ə'dar·ē·ə }

Myodocopa [INV ZOO] A suborder of the order Myodocopida; includes exclusively marine ostracodes distinguished by possession of a heart. { ˌmī·ə'däk·ə·pə }

Myodocopida [INV ZOO] An order of the subclass Ostracoda. { ¦mī·ə·də¦käp·ə·də }

Myodopina [INV ZOO] The equivalent name for Myodocopa. { ˌmī·ə'däp·ə·nə }

myodystrophy [MED] Muscle degeneration. { ˌmī·ə'dis·trəˌfē }

myoelastic fiber [HISTOL] An elastic fiber associated with the smooth muscles in bronchi and bronchioles. { ¦mī·ō·i'las·tik 'fī·bər }

myoelectric potential [PHYSIO] The electrical potential created by muscle action. { ¦mī·ō·i'lek·trik pə'ten·chəl }

myoelectric prosthesis [MED] A replacement device for lost limbs that uses the electromyographic activity of a contracting muscle as a control signal; it is most commonly used for below-elbow amputees in whom elbow function is retained. { ˌmī·ō·iˌlek·trik präs'thē·səs }

myoepithelial cells [HISTOL] Contractile epithelial cells resembling smooth muscle cells that are present in glands, notably the mammary gland, and aid in secretion. { ˌmī·ōˌep·ə'thē·lē·əl ˌsel }

myofascitis [MED] Muscular pain of obscure nature and origin in the lower back. { ¦mī·ō·fə'sīd·əs }

myofibril [CELL MOL] A contractile fibril in a muscle cell. [INV ZOO] *See* myoneme. { ¦mī·ō'fī·brəl }

myofilament [CELL MOL] The structural unit of muscle proteins in a muscle cell. { ¦mī·ō'fil·ə·mənt }

myofrisk [INV ZOO] A contractile structure surrounding the spines of certain radiolarians. { 'mī·ōˌfrisk }

myoglobin [BIOCHEM] A hemoglobinlike iron-containing protein pigment occurring in muscle fibers. Also known as muscle hemoglobin; myohemoglobin. { ¦mī·ə¦glō·bən }

myoglobinuria [MED] The presence of myoglobin in the urine. { ˌmī·əˌglō·bə'nu̇r·ē·ə }

myohematin [BIOCHEM] A cytochrome respiratory enzyme allied to hematin. { ¦mī·ō'hē·mə·tən }

myohemoglobin *See* myoglobin. { ¦mī·ō'hē·mə ˌglō·bən }

myoinositol [BIOCHEM] The commonest isomer of inositol. Also known as mesionositol. { ¦mī·ō·i'näs·əˌtȯl }

myokinase [BIOCHEM] An enzyme that catalyzes the reversible transfer of phosphate groups in adenosine diphosphate; occurs in muscle and other tissues. { ¦mī·ō'kīˌnās }

myolipoma [MED] A benign tumor composed of adipose and smooth muscle cells. { ¦mī·ō 'līˌpō·mə }

myology [MED] The study of muscles in both the normal and diseased states. { mī'äl·ə·jē }

myoma [MED] **1.** A benign uterine tumor composed principally of smooth muscle cells. **2.** Any neoplasm originating in muscle. { mī'ō·mə }

myomalacia [MED] Degeneration, with softening, of muscle tissue. { ¦mī·ō·mə'lā·shə }

myomere [EMBRYO] A muscle segment differentiated from the myotome, which divides to form the epimere and hypomere. { 'mī·əˌmir }

myometritis [MED] Inflammation of the myometrium. { ¦mī·ō·mə'trīd·əs }

myometrium [HISTOL] The muscular tissue of the uterus. { ˌmī·ə'mē·trē·əm }

Myomorpha [VERT ZOO] A suborder of rodents recognized in some systems of classification. { ˌmī·ə'mȯr·fə }

myoneme [INV ZOO] A contractile fibril in a protozoan. Also known as myofibril. { 'mī·ə ˌnēm }

myoneural junction [NEUROSCI] The point of junction of a motor nerve with the muscle which it innervates. Also known as neuromuscular junction. { ¦mī·ō'nu̇r·əl 'jəŋk·shən }

myopathia *See* myopathy. { ¦mī·ə'path·ē·ə }

myopathic facies [MED] An expressionless face with sunken cheeks and a drooping lower lip characteristic of patients with myopathies, especially myotonic dystrophy. { ¦mī·ə¦path·ik 'fā·shēz }

myopathy [MED] Any disease of the muscles. Also known as myopathia. { mī'äp·ə·thē }

myopericarditis [MED] A combination of myocarditis and pericarditis. { ¦mī·ōˌper·əˌkär'dīd·əs }

myophagia [PATH] The invasion of degenerated muscle sarcoplasm by histiocytes. { ¦mī·ə'fā·jē·ə }

myophosphorylase deficiency glycogenosis *See* McArdle's syndrome. { ˌmī·əˌfäs'fȯr·əˌlās di 'fish·ən·sē ¦glī·kō·jə'nō·səs }

myopia [MED] A condition in which the focal image is formed in front of the retina of the eye. Also known as nearsightedness. { mī'ō·pē·ə }

myoplasty [MED] Plastic surgery performed on muscle tissue. { 'mī·əˌplas·tē }

Myopsida [INV ZOO] A natural assemblage of cephalopod mollusks considered as a suborder in the order Teuthoida according to some systems of classification, and a group of the Decapoda according to other systems; the eye is covered by the skin of the head in all species. { mī'äp·sə·də }

myopsychopathy [MED] Any disease of the muscles associated with mental retardation or loss of intellect. { ¦mī·ōˌsī'käp·ə·thē }

myorhythmia [MED] Muscle tremor with a rate of 2–4 per second, and irregular intervals between cycles. { ¦mī·ō'ri<u>th</u>·mē·ə }

myosarcoma [MED] A sarcoma derived from muscle. { ¦mī·ō·sär'kō·mə }

myoseptum *See* myocomma. { ¦mī·ō'sep·təm }

myosin [BIOCHEM] A muscle protein, comprising up to 50% of the total muscle proteins; combines with actin to form actomycin. { 'mī·ə·sən }

myositis [MED] Inflammation of muscle. Also known as fibromyositis. { ˌmī·ə'sīd·əs }

myositis ossificans [MED] Muscle inflammation with bone formation in muscle, tendons, or ligaments. { ˌmī·ə'sīd·əs ä'sif·əˌkanz }

myostatic reflex *See* stretch reflex. { ¦mī·ə¦stad·ik 'rēˌfleks }

myosynovitis [MED] Inflammation of synovial membranes and surrounding musculature. { ¦mī·əˌsīn·ə·'vīd·əs }

myotasis [PHYSIO] Stretching of a muscle. { mī'äd·ə·səs }

myotome [ANAT] A group of muscles innervated by a single spinal nerve. [EMBRYO] The muscle plate that differentiates into myomeres. { 'mī·əˌtōm }

myotonia [MED] Tonic muscular spasm occurring after injury or infection. { ˌmī·ə'tō·nē·ə }

myotonia congenita intermittens *See* paramyotonia congenita. { ˌmī·ə'tō·nē·ə kən'jen·əd·ə ˌin·tər'mit·ənz }

myotonic [MED] Of, pertaining to, or characterized by myotonia. { ¦mī·ə¦tän·ik }

myotonic dystrophy [MED] A hereditary disease, transmitted as an autosomal dominant, characterized by lack of normal relaxation of muscles after contraction, slowly progressive muscular weakness and atrophy, especially of the face and neck, cataract formation, early baldness, gonadal atrophy, abnormal glucose tolerance curve, and, frequently, mental deficiency. { ¦mī·ə¦tän·ik 'dis·trə·fē }

Myoviridae [VIROL] A family of linear double-stranded deoxyribonucleic acid-containing bacterial viruses (bacteriophages) characterized by a contractile tail and an elongated (or icosahedral) head, a well-known genus is the T even phage, which contains coliphage T2. { ˌmī·ə'vir·əˌdī }

Myriangiales [MYCOL] An order of parasitic fungi of the class Ascomycetes which produce asci at various levels in uniascal locules within stromata. { ˌmir·ēˌan·jē'ā·lēz }

Myriapoda [INV ZOO] Informal designation for those mandibulate arthropods having two body tagmata, one pair of antennae, and more than three pairs of adult ambulatory appendages. { ˌmir·ē'äp·ə·də }

Myricaceae [BOT] The single family of the plant order Myricales. { ˌmir·əˈkās·ēˌē }

Myricales [BOT] An order of dicotyledonous plants in the subclass Hamamelidae, marked by its simple, resinous-dotted, aromatic leaves, and a unilocular ovary with two styles and a single ovule. { ˌmir·əˈkā·lēz }

Myrientomata [INV ZOO] The equivalent name for the Protura. { ˌmir·ē·ənˈtäm·əd·ə }

myrigoplasty [MED] A surgical procedure involving the simple repair of a persistent perforation of the tympanic membrane. { məˈrig·əˌplas·tē }

myringitis [MED] Inflammation of the tympanic membrane. { ˌmir·ənˈjīd·əs }

Myriotrochidae [INV ZOO] A family of holothurian echinoderms in the order Apodida, distinguished by eight or more spokes in each wheel-shaped spicule. { ˌmir·ē·əˈtrō·kəˌdē }

Myrmecophagidae [VERT ZOO] A small family of arboreal anteaters in the order Edentata. { ˌmər·mə·kōˈfaj·əˌdē }

myrmecophagous [ZOO] Feeding on ants. { ˌmər·məˈkäf·ə·gəs }

myrmecophile [ECOL] An organism, usually a beetle, that habitually inhabits the nest of ants. { mərˈmek·əˌfīl }

myrmecophyte [ECOL] A plant that houses and benefits from the habitation of ants. { mərˈmek·əˌfīt }

Myrmeleontidae [INV ZOO] The ant lions, a family of insects in the order Neuroptera; larvae are commonly known as doodlebugs. { ˌmər·mə·lēˈän·təˌdē }

Myrmicinae [INV ZOO] A large diverse subfamily of ants (Formicidae); some members are inquilines and have no worker caste. { mərˈmis·əˌnē }

Myrsinaceae [BOT] A family of mostly woody dicotyledonous plants in the order Primulales characterized by flowers without staminodes, a schizogenous secretory system, and gland-dotted leaves. { ˌmər·səˈnās·ēˌē }

Myrtaceae [BOT] A family of dicotyledonous plants in the order Myrtales characterized by an inferior ovary, numerous stamens, anthers usually opening by slits, and fruit in the form of a berry, drupe, or capsule. { mərˈtās·ēˌē }

Myrtales [BOT] An order of dicotyledonous plants in the subclass Rosidae characterized by opposite, simple, entire leaves and perigynous to epigynous flowers with a compound pistil. { mərˈtā·lēz }

Mysida [INV ZOO] A suborder of the crustacean order Mysidacea characterized by fusion of the sixth and seventh abdominal somites in the adult, lack of gills, and other specializations. { ˈmī·sə·də }

Mysidacea [INV ZOO] An order of free-swimming Crustacea included in the division Pericarida; adult consists of 19 somites, each bearing one pair of functionally modified, biramous appendages, and the carapace envelops most of the thorax and is fused dorsally with up to four of the anterior thoracic segments. { ˌmī·səˈdās·ē·ə }

mysis [INV ZOO] A larva of certain higher crustaceans, characterized by biramous thoracic appendages. { ˈmī·səs }

Mystacinidae [VERT ZOO] A monospecific family of insectivorous bats (Chiroptera) containing the New Zealand short-tailed bat; hindlegs and body are stout, and fur is thick. { ˌmis·təˈsin·əˌdē }

Mystacocarida [INV ZOO] An order of primitive Crustacea; the body is wormlike and the cephalothorax bears first and second antennae, mandibles, and first and second maxillae. { ˌmis·tə·kōˈkar·ə·də }

Mysticeti [VERT ZOO] The whalebone whales, a suborder of the mammalian order Cetacea, distinguished by horny filter plates of suspended from the upper jaws. { ˌmis·təˈsēˌtī }

Mytilacea [INV ZOO] A suborder of bivalve mollusks in the order Filibranchia. { ˌmid·əlˈā·shē·ə }

Mytilidae [INV ZOO] A family of mussels in the bivalve order Anisomyaria. { mīˈtil·əˌdē }

myx-, myxo- [ZOO] A combining form denoting mucus, mucous, mucin, mucinous. { ˈmik·sō }

myxadenitis [MED] Inflammation of mucous glands. { miksˌad·ənˈīd·əs }

myxadenoma [MED] An adenoma of a mucous gland. { miksˌad·ənˈō·mə }

myxameba [BIOL] An independent ameboid cell of the vegetative phase of Acrasiales. { ˌmiks·əˈmē·bə }

myxedema [MED] A condition caused by hypothyroidism characterized by a subnormal basal metabolic rate, dry coarse hair, loss of hair, mental dullness, anemia, and slowed reflexes. { ˌmiks·əˈdē·mə }

Myxicolinae [INV ZOO] A subfamily of sedentary polychaete annelids in the family Sabellidae. { ˌmik·səˈkäl·əˌnē }

Myxiniformes [VERT ZOO] The equivalent name for the Myxinoidea. { ˌmik·sə·nəˈfȯr·mēz }

Myxinoidea [VERT ZOO] The hagfishes, an order of eellike, jawless vertebrates (Agnatha) distinguished by having the nasal opening at the tip of the snout and leading to the pharynx, with barbels around the mouth and 6–15 pairs of gill pouches. { ˌmik·səˈnȯid·ē·ə }

myxoadenoma [MED] An adenoma of a mucous gland. { ˌmik·sōˌad·ənˈō·mə }

Myxobacterales [MICROBIO] An order of gliding bacteria; unicellular, gram-negative rods embedded in a layer of slime and capable of gliding movement; form fruiting bodies containing resting cells (myxospores) under certain environmental conditions. { ˌmik·səˌbak·təˈrā·lēz }

myxochondrofibrosarcoma [MED] A sarcoma composed of anaplastic myxoid, chondroid, and fibrous cells. { ¦mik·sə¦kän·drōˌfī·brəˌsärˈkō·mə }

Myxococcaceae [MICROBIO] A family of the order Myxobacterales; vegetative cells are straight to slightly tapered, and spherical to ovoid microcysts (myxospores) are produced. { ˌmik·səˌkäk 'sās·ēˌē }

myxofibroma of nerve sheath *See* neurofibroma. { ˌmik·səˌfī'brō·mə əv 'nərv ˌshēth }

Myxogastromycetidae [MYCOL] A large subclass of plasmodial slime molds (Myxomycetes). { ¦mik·səˌgas·trōˌmī'sed·əˌdē }

myxolipoma *See* liposarcoma. { ˌmik·sō·lī'pō·mə }

myxoma [MED] A benign tumor composed of mucinous connective tissue. { mik'sō·mə }

myxoma lipomatodes *See* liposarcoma. { mik'sō·mə līˌpō·mə'tō·dēz }

myxomatosis [VET MED] A virus disease of rabbits producing fever, skin lesions resembling myxomas, and mucoid swelling of mucous membranes. { mikˌsō·mə'tō·səs }

Myxomycetes [BIOL] Plasmodial (acellular or true) slime molds, a class of microorganisms of the division Mycota; they are on the borderline of the plant and animal kingdoms and have a noncellular, multinucleate, jellylike, creeping, assimilative stage (the plasmodium) which alternates with a myxameba stage. { ˌmik·sə ˌmī'sēd·ēz }

Myxomycophyta [BOT] An order of microorganisms, equivalent to the Mycetozoia of zoological classification. { ˌmik·səˌmī'käf·əd·ə }

Myxophaga [INV ZOO] A suborder of the Coleoptera. { mik'säf·ə·gə }

Myxophyceae [BOT] An equivalent name for the Cyanophyceae. { ˌmik·sə'fī·sēˌē }

myxosarcoma [MED] A sarcoma whose parenchyma is composed of anaplastic myxoid cells. { ˌmik·sō·sär'kō·mə }

Myxosporida [INV ZOO] An order of the protozoan class Myxosporidea characterized by the production of spores with one or more valves and polar capsules, and by possession of a single sporoplasm. { ˌmik·sə'spȯr·ə·də }

Myxosporidea [INV ZOO] A class of the protozoan subphylum Cnidospora; members are parasitic in some fish, a few amphibians, and certain invertebrates. { ˌmik·sə·spə'rid·ē·ə }

myxovirus [VIROL] A group of ribonucleic-acid animal viruses characterized by hemagglutination and hemadsorption; includes influenza and fowl plague viruses and the paramyxoviruses. { 'mik·səˌvī·rəs }

Myzopodidae [VERT ZOO] A monospecific order of insectivorous bats (Chiroptera) containing the Old World disk-winged bat of Madagascar; characterized by long ears and by a vestigial thumb with a monostalked sucking disk. { ˌmī·zə 'päd·əˌdē }

myzorhynchus [INV ZOO] An apical sucker on the scolex of certain tapeworms. { ˌmī·zə'riŋ·kəs }

Myzostomaria [INV ZOO] An aberrant group of Polychaeta; most are greatly depressed, broad, and very small, and true segmentation is delayed or absent in the adult; all are parasites of echinoderms. { ˌmī·zə·stə'mar·ē·ə }

Myzostomidae [INV ZOO] A monogeneric family of the Myzostomaria. { ˌmī·zə'stäm·əˌdē }

N

Nabidae [INV ZOO] The damsel bugs, a family of hemipteran insects in the superfamily Cimicimorpha. { 'nab·ə,dē }

Nabothian cyst [MED] Cystic distention of the Nabothian glands of the uterine cervix. { nə'bō·thē·ən 'sist }

Nabothian glands [ANAT] Mucous glands of the uterine cervix. { nə'bō·thē·ən 'glanz }

nacre [INV ZOO] An iridescent inner layer of many mollusk shells. { 'nā·kər }

NAD *See* diphosphopyridine nucleotide.

NADP+ *See* nicotinamide adenine dinucleotide phosphate. { ¦en,ā,dē,pē'pləs }

Naegeli-type leukemia [MED] A type of monocytic leukemia in which the leukoyctes resemble cells of the granulocytic series. { 'nā·gə·lē ¦tīp lü'kē·mē·ə }

nail [ANAT] The horny epidermal derivative covering the dorsal aspect of the terminal phalanx of each finger and toe. [MED] A metallic rod with one blunt end and one sharp end, used surgically to anchor bone fragments. { nāl }

Nairobi sheep disease [VET MED] A tick-borne viral disease of sheep and goats that is caused by a ribonucleic acid-containing virus of the genus *Nairovirus*, characterized by hemorrhagic gastroenteritis and high mortality. { nī,rō·bē 'shēp di,zēz }

Nairovirus [VIROL] A genus of the viral family Bunyaviridae that causes Nairobi sheep disease. { 'nī·rə,vī·rəs }

Najadaceae [BOT] A family of monocotyledonous, submerged aquatic plants in the order Najadales distinguished by branching stems and opposite or whorled leaves. { ,nāj·ə'dās·ē,ē }

Najadales [BOT] An order of aquatic and semiaquatic flowering plants in the subclass Alismatidae; the perianth, when present, is not differentiated into sepals and petals, and the flowers are usually not individually subtended by bracts. { ,nāj·ə'dā·lēz }

naked bud [BOT] A bud covered only by rudimentary foliage leaves. { 'nā·kəd 'bəd }

nalidixic acid [PHARM] $C_{12}H_{12}N_2O_3$ Pale buff, crystalline powder; melting point is 229–230°C; soluble in chloroform and in potassium hydroxide and sodium hydroxide solutions; used as an antibacterial drug in humans and animals. { ¦nal·ə¦dik·sik 'as·əd }

naltrexone [PHARM] $C_{20}H_{23}NO_4$ An opiate receptor antagonist that blocks the effects of endogenous opioids in the brain; used to treat alcoholism. { nal'trek,zōn }

Namanereinae [INV ZOO] A subfamily of largely fresh-water errantian annelids in the family Nereidae. { ,nā·mə,nə'rē·ə,nē }

nanism [MED] Dwarfed stature due to arrested development. { 'nā,niz·əm }

nannandrous [BOT] Pertaining to species of plants in which male members are markedly smaller than females, such as in some algal species of *Oedogonium* that have antheridia produced in special dwarf filaments. { na'nan·drəs }

nannoplankton [BIOL] Minute plankton; the smallest (usually from 2 to 20 nanometers) plankton, including algae, bacteria, and protozoans. Also spelled nanoplankton. { ¦nan·ō'plaŋk·tən }

nano- [BIOL] A prefix meaning dwarfed. { 'nan·ō }

nanocephalus [MED] A fetus with an undersized head. { ,nan·ə'sef·ə·ləs }

nanophanerophyte [ECOL] A shrub not exceeding 6.6 feet (2 meters) in height. { ¦nan·ō 'fan·ə·rə,fīt }

nanophthalmus *See* microphthalmus. { 'nan·ə ,thal·məs }

nanoplankton *See* nannoplankton. { ¦nan·ō 'plaŋk·tən }

nanozooid [INV ZOO] Dwarf zooid; bryozoan heterozooid possessing only a single tentacle. { ¦nan·ō¦zō,ȯid }

nape [ANAT] The back of the neck. { nāp }

napex [ANAT] That portion of the scalp just below the occipital protuberance. { 'nā,peks }

napiform [BOT] Turnip-shaped, referring to roots. { 'nāp·ə,fȯrm }

narco- [MED] Combining form meaning numbness, narcosis, or stupor. { 'nar·kō }

narcolepsy [MED] A disorder of sleep mechanism characterized by two or more of four distinct symptoms: uncontrollable periods of daytime drowsiness, cataleptic attacks of muscular weakness, sleep paralysis, and vivid nocturnal or hypnogogic hallucinations. { 'när·kə,lep·sē }

narcomania [MED] Morbid physiologic or psychologic craving for narcotics to avoid painful stimuli. { ,när·kə'mā·nē·ə }

Narcomedusae [INV ZOO] A suborder of hydrozoan cnidarians in the order Trachylina; the hydroid generation is represented by an actinula larva. { ,när·kə·mə'dü,sē }

narcosis [MED] Drug-produced state of profound stupor, unconsciousness, or arrested activity. { när'kō·səs }

narcosis therapy [MED] Prolonged, drug-induced sleep as treatment for certain mental disorders. Also known as sleep therapy. { när'kō·səs ,ther·ə·pē }

narcospasm [MED] Spasm accompanied by stupor. { 'när·kō,spaz·əm }

narcosynthesis [MED] Psychotherapeutic treatment under partial anesthesia, in which abreaction is a significant factor in obtaining positive results. { ¦när·kō'sin·thə·səs }

narcotic [PHARM] A drug which in therapeutic doses diminishes awareness of sensory impulses, especially pain, by the brain; in large doses, it causes stupor, coma, or convulsions. { när'käd·ik }

narcotine *See* noscapine. { 'när·kə,tēn }

naris *See* nostril. { 'nar·əs }

narrow-angle glaucoma [MED] Increased intraocular tension due to a block of the angle of the anterior chamber from contact of the iris by the trabecula. Also known as obstructive glaucoma. { 'nar·ō ¦aŋ·gəl glaü'kō·mə }

narrow-sense heritability [GEN] The degree to which individual phenotypes are determined by the genes transmitted from the parents; expressed as the ratio of the additive genetic variance to the total phenotypic variance. { ¦nar·ō ,sens ,her·ə·tə'bil·əd·ē }

narrow-spectrum antibiotic [MICROBIO] An antibiotic effective against a limited number of microorganisms. { 'nar·ō ¦spek·trəm ,ant·i,bī'äd·ik }

narwhal [VERT ZOO] *Monodon monoceros.* An arctic whale characterized by lack of a dorsal fin, and by possession in the male of a long, twisted, pointed tusk (or rarely, two tusks) which is a source of ivory. { 'när,wäl }

nasal [ANAT] Of or pertaining to the nose. { 'nā·zəl }

nasal bone [ANAT] Either of two rectangular bone plates forming the bridge of the nose; they articulate with the frontal, ethmoid, and maxilla bones. { 'nā·zəl ,bōn }

nasal cavity [ANAT] Either of a pair of cavities separated by a septum and located between the nasopharynx and anterior nares. { 'nā·zəl ¦kav·əd·ē }

nasal crest [ANAT] **1.** The linear prominence on the medial border of the palatal process of the maxilla. **2.** The linear prominence on the medial border of the palatine bone. **3.** The linear prominence on the internal border of the nasal bone and forming part of the nasal septum. { 'nā·zəl ¦krest }

nasal pit *See* olfactory pit. { 'nā·zəl ,pit }

nasal process of the frontal bone [ANAT] The downward projection of the nasal part of the frontal bone which terminates as the nasal spine. { 'nā·zəl ¦prä·səs əv thə 'frənt·əl ,bōn }

nasal process of the maxilla [ANAT] Frontal process of the maxilla. { 'nā·zəl ¦prä·səs əv thə mak'sil·ə }

nasal septum [ANAT] The partition separating the two nasal cavities. { 'nā·zəl ¦sep·təm }

nascent ribonucleic acid [CELL MOL] **1.** A ribonucleic acid (RNA) molecule in the process of being synthesized. **2.** A complete, newly synthesized RNA molecule before any alterations have been made. { ¦nās·ənt ¦rī·bō·nü,klē·ik 'as·əd }

Nasellina [INV ZOO] The equivalent name for Monopylina. { ,nas·ə'lī·nə }

nasolacrimal canal [ANAT] The bony canal that lodges the nasolacrimal duct. Also known as lacrimal canal. { ¦nā·zō'lak·rə·məl kə'nal }

nasolacrimal duct [ANAT] The membranous duct lodged within the nasolacrimal canal; it gives passage to the tears from the lacrimal sac to the inferior meatus of the nose. { ¦nā·zō'lak·rə·məl 'dəkt }

nasolacrimal groove [EMBRYO] The furrow, the maxillary, and the lateral nasal processes of the embryo. { ¦nā·zō'lak·rə·məl 'grüv }

nasopalatine cyst *See* median maxillary cyst. { ¦nā·zō'pal·ə,tēn 'sist }

nasopalatine duct [EMBRYO] A canal between the oral and nasal cavities of the embryo at the point of fusion of the maxillary and palatine processes. { ¦nā·zō'pal·ə,tēn 'dəkt }

nasopharynx [ANAT] The space behind the posterior nasal orifices, above a horizontal plane through the lower margin of the palate. { ¦nā·zō'far·iŋks }

nastic movement [BOT] Movement of a flat plant part, oriented relative to the plant body and produced by diffuse stimuli causing disproportionate growth or increased turgor pressure in the tissues of one surface. { 'nas·tik 'müv·mənt }

Nasutitermitinae [INV ZOO] A subfamily of termites in the family Termitidae, characterized by having the cephalic glands open at the tip of an elongated tube which projects anteriorly. { nə¦süd·ə·tər'mit·ən,ē }

Natalidae [VERT ZOO] The funnel-eared bats, a monogeneric family of small, tropical American insectivorous bats (Chiroptera) with large, funnellike ears. { nə'tal·ə,dē }

Natantia [INV ZOO] A suborder of decapod crustaceans comprising shrimp and related forms characterized by a long rostrum and a ventrally flexed abdomen. { nə'tan·chə }

Naticacea [INV ZOO] A superfamily of gastropod mollusks in the order Prosobranchia. { ,nad·ə'kā·shē·ə }

Naticidae [INV ZOO] A family of gastropod mollusks in the order Pectinibranchia comprising the moon-shell snails. { nə'tis·ə,dē }

native [BIOL] Grown, produced or originating in a specific region or country. { 'nād·iv }

native state [CELL MOL] The folded protein

configuration that is maintained through noncovalent interactions such as hydrophobic interactions, electrostatic interactions, and hydrogen bonds. { ¦nād·iv ¦stāt }

natremia [MED] Excessive amounts of sodium in the blood. { nə'trē·mē·ə }

natriuresis [PHYSIO] Excretion of sodium in the urine. { ¦na·trē·yu̇'rē·səs }

natriuretic [PHARM] A medicinal agent which inhibits reabsorption of cations, particularly sodium, from urine. { ¦nā·trē·yu̇'red·ik }

natural immunity [IMMUNOL] Native immunity possessed by the individuals of a race, strain, or species. { 'nach·rəl i'myü·nəd·ē }

naturalized [ECOL] Of a species, having become permanently established after being introduced. { 'nach·rə‚līzd }

natural killer cell [HISTOL] A large, granular lymphocyte that can lyse a variety of target cells when it is activated by interferon. Abbreviated NK. { ¦nach·rəl 'kil·ər ‚sel }

natural selection [EVOL] Darwin's theory of evolution, according to which organisms tend to produce progeny far above the means of subsistence; in the struggle for existence that ensues, only those progeny with favorable variations survive; the favorable variations accumulate through subsequent generations, and descendants diverge from their ancestors. { 'nach·rəl si'lek·shən }

Naucoridae [INV ZOO] A family of hemipteran insects in the superfamily Naucoroidea. { nȯ 'kȯr·ə‚dē }

Naucoroidea [INV ZOO] The creeping water bugs, a superfamily of hemipteran insects in the subdivision Hydrocorisae; they are suboval in form, with chelate front legs. { nȯ·kə'rȯid·ē·ə }

nauplius [INV ZOO] A larval stage characteristic of many groups of Crustacea; the oval, unsegmented body has three pairs of appendages: uniramous antennules, biramous antennae, and mandibles. { 'nȯ·plē·əs }

nausea [MED] Feeling of discomfort in the stomach region, accompanied by aversion to food and a tendency to vomit. { 'nȯ·zē·ə }

Nautilidae [INV ZOO] A monogeneric family of cephalopod mollusks in the order Nautiloidea; *Nautilus pompilius* is the only well-known living species. { nȯ'til·ə‚dē }

Nautiloidea [INV ZOO] A primitive order of tetrabranchiate cephalopods; shells are external and smooth, being straight or coiled and chambered with curved transverse septa. { ‚nȯd·əl'ȯid·ē·ə }

Nautilus [INV ZOO] The only living genus of the molluscan subclass Nautiloidea, containing the only living cephalopods with an external chambered shell and numerous cephalic tentacles, six species live in the western Pacific and around the East Indies. { 'nȯd·ə·ləs }

navel [ANAT] The umbilicus. { 'nā·vəl }

navicular [ANAT] A boat-shaped bone, especially the lateral bone on the radial side of the proximal row of the carpus. [BIOL] Resembling or having the shape of a boat. { nə'vik·yə·lər }

navicular cells [PATH] Boat-shaped squamous epithelial cells filled with glycogen and prominent in the exfoliated cells of the uterine cervix of pregnant women. { nə'vik·yə·lər ‚selz }

naviculoid [BIOL] Referring to a diatom, boat-shaped. { nə'vik·yə‚lȯid }

neallotype [SYST] A type specimen that, compared with the holotype, is of the opposite sex, and was collected and described later. { nē 'al·ə‚tīp }

Nearctic fauna [ECOL] The indigenous animal communities of the Nearctic zoogeographic region. { nē'ärd·ik 'fȯn·ə }

Nearctic zoogeographic region [ECOL] The zoogeographic region that includes all of North America to the edge of the Mexican Plateau. { nē'ärd·ik ¦zō·ō‚jē·ə'graf·ik ‚rē·jən }

near point [PHYSIO] The smallest distance from the eye at which a small object can be seen without blurring. { 'nir ‚pȯint }

nearsightedness *See* myopia. { 'nir¦sīd·əd·nəs }

nearthrosis [MED] A type of nonunion of broken ends of bones in which a cystic space resembling a joint cavity develops between poorly joined ends. { ¦nē·är'thrō·səs }

Nebaliacea [INV ZOO] A small, marine order of Crustacea in the subclass Leptostraca distinguished by a large bivalve shell, without a definite hinge line, an anterior articulated rostrum, eight thoracic and seven abdominal somites, a pair of articulated furcal rami, and the telson. { nə‚bā·lē'ā·shə }

neck [ANAT] The usually constricted communicating column between the head and trunk of the vertebrate body. { nek }

Neckeraceae [BOT] A family of mosses in the order Isobryales distinguished by undulate leaves. { ‚nek·ə'rās·ē‚ē }

necr-, necro- [MED] Combining form denoting death. { ¦ne·krō }

necrobiosis [MED] Death of a cell or group of cells under either normal or pathologic conditions. { ¦ne·krō‚bī'ō·səs }

necrophagous [ZOO] Feeding on dead bodies. { ne'kräf·ə·gəs }

necropsy [MED] To perform an autopsy. { 'ne ‚kräp·sē }

necrosis [MED] Death of a cell or group of cells as a result of injury, disease, or other pathologic state. { nə'krō·səs }

necrotic [MED] Pertaining to, causing, or undergoing necrosis. { nə'kräd·ik }

necrotic enteritis [VET MED] A bacterial infection of young swine caused by *Salmonella suipestifer* or *S. choleraesuis* and characterized by fever and necrotic and ulcerative inflammation of the intestine. { nə'kräd·ik ‚ent·ə'rīd·əs }

necrotize [MED] To undergo necrosis; to become necrotic. { 'nek·rə‚tīz }

necrozoospermia [MED] A condition in which spermatozoa are immobile. { ¦ne·krō‚zō·ō'spər·mē·ə }

nectar [BOT] A sugar-containing liquid secretion of the nectaries of many flowers. { 'nek·tər }

nectarine [BOT] A smooth-skinned, fuzzless fruit originating as a spontaneous somatic mutation of the peach, *Prunus persica* and *P. persica* var. *nectarina*. { ¦nek·tə¦rēn }

nectary [BOT] A secretory organ or surface modification of a floral organ in many flowers, occurring on the receptacle, in and around ovaries, on stamens, or on the perianth; secretes nectar. { 'nek·tə·rē }

nectocalyx [INV ZOO] A swimming bell of a siphonophore. Also known as nectophore. { ¦nek·tō'kā·liks }

Nectonematoidea [INV ZOO] A monogeneric order of worms belonging to the class Nematomorpha, characterized by dorsal and ventral epidermal chords, a pseudocoele, and dorsal and ventral rows of bristles; adults are parasites of true crabs and hermit crabs. { ¦nek·tō,ne·mə'tȯid·ē·ə }

nectophore *See* nectocalyx. { 'nek·tə,fȯr }

nectosome [INV ZOO] The part of a complex siphonophore that bears swimming bells. { 'nek·tə,sōm }

Nectrioidaceae [MYCOL] The equivalent name for Zythiaceae. { ¦nek·trē,ȯid·ē'ās·ē,ē }

needle [BOT] A slender-pointed leaf, as of the firs and other evergreens. { 'nēd·əl }

neencephalon [ANAT] The neopallium and the phylogenetically new acquisitions of the cerebellum and thalamus collectively. Also spelled neoencephalon. { ¦nē·in'sef·ə,län }

negative afterimage [PHYSIO] An afterimage that is seen on a bright background and is complementary in color to the initial stimulus. { 'neg·əd·iv ¦af·tər¦im·ij }

negative gene control [CELL MOL] Prevention of gene expression by the binding of specific repressor molecules to operator sites. { ¦neg·əd·iv 'jēn kən,trōl }

negative interference [GEN] A crossover exchange between homologous chromosomes which increases the likelihood of another in the same vicinity. { 'neg·əd·iv ,in·tər'fir·əns }

negative phase [IMMUNOL] The temporary quantitative reduction of serum antibodies immediately following a second inoculation of antigen. { 'neg·əd·iv 'fāz }

negative photokinesis [PHYSIO] The slower movement of an organism upon entering an illuminated area relative to its velocity of movement in the dark or in dim light. { ¦neg·əd·iv ,fōd·ō·kə'nē·səs }

negative phototaxis [PHYSIO] The orientation and movement of an organism away from the source of a light stimulus. { ¦neg·əd·iv ,fōd·ō'tak·səs }

negative regulator [GEN] Any regulator that acts to prevent transcription or translation. { 'neg·əd·iv ,reg·yə'lād·ər }

negative selection [IMMUNOL] The death of autoimmune lymphocytes shortly after they develop. Also known as clonal deletion. { ¦neg·əd·iv si'lek·shən }

negative staining [BIOL] A method in microscopy for demonstrating the form of cells, bacteria, and other small objects by staining the ground rather than the objects. { 'neg·əd·iv 'stān·iŋ }

Negri bodies [PATH] Acidophil cytoplasmic inclusion bodies in neurons, considered diagnostic of rabies. { 'nā·grē ,bäd·ēz }

Neididae [INV ZOO] A small family of thread-legged hemipteran insects in the superfamily Lygaeoidea. { nē'id·ə,dē }

Neisseriaceae [MICROBIO] The single family of gram-negative aerobic cocci and coccobacilli; some species are human parasites and pathogens. { 'nī·sər·ē'ās·ē,ē }

Neisseria gonorrhoeae [MICROBIO] A gram-negative coccus pathogen that causes the sexually transmitted disease gonorrhea. Also known as gonococcus. { nī·sə,rē·ə ,gän·ə'rē,ī }

nektobenthos [ECOL] Those forms of marine life that exist just above the ocean bottom and occasionally rest on it. { ¦nek·tə'ben,thȯs }

nekton [INV ZOO] Free-swimming aquatic animals, essentially independent of water movements. { 'nek·tən }

Nelumbonaceae [BOT] A family of flowering aquatic herbs in the order Nymphaeales characterized by having roots, perfect flowers, alternate leaves, and triaperturate pollen. { nə,ləm·bə'nās·ē,ē }

Nemata [INV ZOO] An equivalent name for Nematoda. { nə'mad·ə }

Nemataceae [BOT] A family of mosses in the order Hookeriales distinguished by having perichaetial leaves only. { ,nem·ə'tās·ē,ē }

Nemathelminthes [INV ZOO] A subdivision of the Amera which comprised the classes Rotatoria, Gastrotrichia, Kinorhyncha, Nematoda, Nematomorpha, and Acanthocephala. { ,nem·ə,thel'min·thēz }

Nematocera [INV ZOO] A series of dipteran insects in the suborder Orthorrhapha; adults have antennae that are usually longer than the head, and the flagellum consists of 10–65 similar segments. { ,nem·ə'täs·ə·rə }

nematocyst [INV ZOO] An intracellular effector organelle in the form of a coiled tube which may be rapidly everted in food gathering or defense by cnidarians. { nə'mad·ə,sist }

Nematoda [INV ZOO] A group of unsegmented worms which have been variously recognized as an order, class, and phylum. { ,nem·ə'tō·də }

nematode [INV ZOO] **1.** Any member of the Nematoda. **2.** Of or pertaining to the Nematoda. { 'nem·ə,tōd }

Nematodonteae [BOT] A group of mosses included in the subclass Eubrya in which there may be faint transverse bars on the peristome teeth. { nə¦mad·ə'dänt·ē,ē }

nematogen [INV ZOO] A reproductive phase of the Dicyemida during which vermiform larvae are formed asexually from the germ cells in the axial cells. { nə'mad·ə·jən }

Nematognathi [VERT ZOO] The equivalent name for Siluriformes. { ˌnem·əˈtäg·nəˌthī }

Nematoidea [INV ZOO] An equivalent name for Nematoda. { ˌnem·əˈtȯid·ē·ə }

nematology [INV ZOO] The study of nematodes. { ˌnem·əˈtäl·ə·jē }

Nematomorpha [INV ZOO] A group of the Aschelminthes or a separate phylum that includes the horsehair worms. { ˌnem·əd·əˈmȯr·fə }

Nematospora coryli [MICROBIO] A mycelial species with needle-shaped ascospores that causes yeast spot disease of various crops. { nəˌmad·əˌspȯr·ə ˈkȯr·əˌlē }

Nematosporoideae [BOT] A subfamily of the Saccharomycetaceae containing parasitic yeasts; two genera have been studied in culture: *Nematospora* with asci that contain eight spindle-shaped ascospores, and *Metschnikowia* whose asci contain one or two needle-shaped ascospores. { ¦nem·əd·ō·spəˈrȯid·ēˌē }

nematozooid [INV ZOO] A zooid bearing organs of defense, in hydroids and siphonophores. { ¦nem·əd·əˈzōˌȯid }

Nemertea [INV ZOO] An equivalent name for Rhynchocoela. { nəˈmərd·ē·ə }

Nemertina [INV ZOO] An equivalent name for Rhynchocoela. { ˌne·mərˈtī·nə }

Nemertinea [INV ZOO] An equivalent name for Rhynchocoela. { ˌne·mərˈtin·ē·ə }

Nemestrinidae [INV ZOO] The hairy flies, a family of dipteran insects in the series Brachycera of the suborder Orthorrhapha. { ¦nem·ə¦strin·əˌdē }

Nemichthyidae [VERT ZOO] A family of bathypelagic, eellike amphibians in the order Apoda. { ˌnem·ikˈthī·əˌdē }

Nemognathinae [INV ZOO] A subfamily of the coleopteran family Meloidae; members have greatly elongate maxillae that form a poorly fitted tube. { ˌnem·əgˈnath·əˌnē }

nemoral [ECOL] Pertaining to or inhabiting a grove or wooded area. { ˈnem·rəl }

neoadjuvant chemotherapy [MED] A type of chemotherapy that is used to shrink a tumor prior to surgery or radiation. { ˌnē·ō¦aj·ə·vənt ˌkē·mōˈther·ə·pē }

neoblast [INV ZOO] Any of various undifferentiated cells in annelids which migrate to and proliferate at sites of repair and regeneration. { ˈnē·əˌblast }

neocentric activity [CELL MOL] In plants, an aberrant behavior during meiosis in which specific chromosome regions act as secondary sites of attachment for spindle fibers. { ˌnē·əˈsen·trik ak¦tiv·əd·ē }

neocentromere [GEN] A functional centromere in a novel location; may lack specific classes of deoxyribonucleic acid usually present in a centromere. { ˌnē·ōˈsen·trəˌmir }

neocerebellum [ANAT] Phylogenetically, the most recent part of the cerebellum; receives cerebral cortex impulses via the corticopontocerebellar tract. { ¦nē·ōˌser·əˈbel·əm }

neocortex [ANAT] Phylogenetically the most recent part of the cerebral cortex; includes all but the olfactory, hippocampal, and piriform regions of the cortex. { ¦nē·ōˈkȯrˌteks }

neoencephalon *See* neencephalon. { ¦nē·ō·in ˈsef·əˌlän }

Neogastropoda [INV ZOO] An order of gastropods which contains the most highly developed snails; respiration is by means of ctenidia, the nervous system is concentrated, an operculum is present, and the sexes are separate. { ¦nē·ō·gaˈsträp·ə·də }

Neognathae [VERT ZOO] A superorder of the avian subclass Neornithes, characterized as flying birds with fully developed wings and sternum with a keel, fused caudal vertebrae, and absence of teeth. { nēˈäg·nəˌthē }

Neogregarinida [INV ZOO] An order of sporozoan protozoans in the subclass Gregarinia which are insect parasites. { ¦nē·ōˌgreg·əˈrin·ə·də }

neomorph [GEN] A mutant allele that produces an effect different from that produced by the normal allele. { ˈnē·əˌmȯrf }

neomycin [MICROBIO] The collective name for several colorless antibiotics produced by a strain of *Streptomyces fradiae*; the commercial fraction ($C_{23}H_{46}N_6O_{13}$) has a broad spectrum of activity. { ¦nē·ə¦mīs·ən }

neonatal [MED] Pertaining to a newborn infant. { ¦nē·əˈnād·əl }

neonatal impetigo [MED] A type of impetigo occurring in the newborn, characterized by bullae and caused by staphylococci or sometimes streptococci. { ¦nē·əˈnād·əl ˌim·pəˈtīˌgō }

neonatal line [ANAT] A prominent incremental line formed in the neonatal period in the enamel and dentin of a deciduous tooth or of a first permanent molar. { ¦nē·əˈnād·əl ˈlīn }

neonatal mortality rate [MED] The number of deaths reported among infants under 1 month of age in a calendar year per 1000 live births reported in the same year and place. { ¦nē·əˈnād·əl mȯrˈtal·əd·ē ˌrāt }

neonatal myasthenia [MED] Muscle weakness and ineffective motor activities in infants born of myasthenic mothers. { ¦nē·əˈnād·əl ˌmī·əsˈthē·nē·ə }

neonate [MED] A newborn infant. { ˈnē·əˌnāt }

neonatology [MED] The study of the newborn up to 2 months of age. { ¦nē·əˈnāˈtäl·ə·jē }

neopallium [ANAT] Phylogenetically, the new part of the cerebral cortex; formed from the region between the pyriform lobe and the hippocampus, it comprises the nonolfactory region. { ¦nē·ōˈpal·ē·əm }

neopalynology [BOT] A field of palynology concerned with extant microorganisms and disassociated microscopic parts of megaorganisms. { ¦nē·ōˌpal·əˈnäl·ə·jē }

neoplasia [MED] **1.** Formation of a neoplasm or tumor. **2.** Formation of new tissue. { ˌnē·əˈplā·zhə }

neoplasm [MED] An aberrant new growth of

abnormal cells or tissues; a tumor. { 'nē·ə,plaz·əm }

neoplastoid [CELL MOL] Pertaining to immortal mammalian cell lines that may behave like neoplasms. { ,nē·ə'plas,tȯid }

Neopseustidae [INV ZOO] A family of Lepidoptera in the superfamily Eriocranioidea. { ,nē·əp'sü·stə,dē }

Neoptera [INV ZOO] A section of the insect subclass Pterygota; members have a muscular and articular mechanism allowing the wings to be flexed over the abdomen when at rest. { nē'äp·tə·rə }

Neopterygii [VERT ZOO] An equivalent name for Actinopterygii. { nē,äp·tə'rij·ē,ī }

Neorhabdocoela [INV ZOO] A group of the Rhabdocoela comprising fresh-water, marine, or terrestrial forms, with a bulbous pharynx, paired protonephridia, sexual reproduction, and ventral gonopores. { ,nē·ō,rab·də'sē·lə }

Neornithes [VERT ZOO] A subclass of the class Aves containing all known birds except the fossil *Archaeopteryx*. { nē'ȯr·nə,thēz }

neossoptile [VERT ZOO] A downy feather on most newly hatched birds. { ,nē·ə'säp·təl }

neostigmine [PHARM] A quaternary ammonium cation that is used as the bromide ($C_{12}H_{19}BrN_{20}O_2$) and methylsulfate ($C_{13}H_{22}N_2O_6S$) salts; has anticholinesterase activity. { ¦nē·ō¦stig ,mēn }

neotenin [BIOCHEM] A hormone secreted by cells of the corpus allatum in arthropod larvae and nymphs; inhibits the development of adult characters. Also known as juvenile hormone. { nē'ät·ən·ən }

neoteny [VERT ZOO] A phenomenon peculiar to some salamanders, in which large larvae become sexually mature while still retaining gills and other larval features. { 'nē·ə,tē·nē *or* nē'ät·ən·ē }

Neotropical zoogeographic region [ECOL] A zoogeographic region that includes Mexico south of the Mexican Plateau, the West Indies, Central America, and South America. { ¦nē·ō'träp·ə·kəl ¦zō·ō,jē·ə'graf·ik 'rē·jən }

neotype [SYST] A specimen selected as type subsequent to the original description when the primary types are known to be destroyed; a nomenclatural type. { 'nē·ə,tīp }

neounitarian theory of hematopoiesis [HISTOL] A theory that under certain conditions, such as in tissue culture or in pathologic states, lymphocytes or cells resembling lymphocytes can become multipotent. { ¦nē·ō·yü·nə'tar·ē·ən ¦thē·ə·re əv ,hem·əd·ō·pȯi'ē·səs }

Nepenthaceae [BOT] A family of dicotyledonous plants in the order Sarraceniales; includes many of the pitcher plants. { ,nep·ən'thās·ē,ē }

nephr-, nephro- [ANAT] Combining form denoting kidney. { 'nef·rō }

nephrectomy [MED] Surgical removal of a kidney. { nə'frek·tə·mē }

nephric tubule *See* uriniferous tubule. { 'nef·rik 'tüb,yül }

nephridioblast [INV ZOO] An ecodermal precursor cell of a nephridium in certain animals. { nə'frid·ē·ə,blast }

nephridioduct [INV ZOO] The duct of a nephridium, sometimes serving as a common excretory and genital outlet. { nə'frid·ē·ə,dəkt }

nephridiopore [INV ZOO] The external opening of a nephridium. { nə'frid·ē·ə,pȯr }

nephridium [INV ZOO] Any of various paired excretory structures present in the Platyhelminthes, Rotifera, Rhynchocoela, Acanthocephala, Priapuloidea, Entoprocta, Gastrotricha, Kinorhyncha, Cephalochorda, and some Archiannelida and Polychaeta. { nə'frid·ē·əm }

nephritic [MED] **1.** Pertaining to or affected with nephritis. **2.** Pertaining to or affecting the kidney. { nə'frid·ik }

nephritis [MED] Inflammation of the kidney. { nə'frīd·əs }

nephroabdominal [ANAT] Of or pertaining to the kidneys and abdomen. { ¦nef·rō·ab'däm·ə·nəl }

nephroblastoma *See* Wilms' tumor. { ¦nef·rə·bla 'stō·mə }

nephrocalcinosis [PATH] Deposition of calcium salts in the kidney tubules. { ¦nef·rō,kal·sə 'nō·səs }

nephrocoel [ANAT] The cavity of a nephrotome. { ¦nef·rə,sēl }

nephrodystrophy *See* nephrosis. { ¦nef·rə'dis·trə·fē }

nephrogenic [EMBRYO] **1.** Having the potential to develop into kidney tissue. **2.** Of renal origin. { ¦nef·rə¦jen·ik }

nephrogenic cord [EMBRYO] The longitudinal cordlike mass of mesenchyme derived from the mesomere or nephrostomal plate of the mesoderm, from which develop the functional parts of the pronephros, mesonephros, and metanephros. { ¦nef·rə¦jen·ik 'kȯrd }

nephrogenic tissue [EMBRYO] The tissue of the nephrogenic cord derived from the nephrotome plate that forms the blastema or primordium from which the embryonic and definitive kidneys develop. { ¦nef·rə¦jen·ik 'tish·ü }

nephrolithiasis [PATH] Formation of renal calculi. { ¦nef·rō·li'thī·ə·səs }

nephrolithotomy [MED] Excision of renal calculi from the kidney. { ¦nef·rō·li'thäd·ə·mē }

nephrology [MED] The study of the kidney, including diseases. { nə'fräl·ə·jē }

nephrolysin [BIOCHEM] A toxic substance capable of disintegrating kidney cells. { nə'fräl·ə·sən }

nephrolysis [MED] **1.** Dissolution of kidney tissue by the action of a nephrolysin. **2.** Surgical detachment of a kidney from surrounding adhesions. { nə'fräl·ə·səs }

nephroma [MED] A tumor of the kidney. { nə'frō·mə }

nephromegaly [MED] Enlargement of the kidney. { ¦nef·rō'meg·ə·lē }

nephromixium [INV ZOO] A compound nephridium composed of flame cells and the coelomic funnel; functions as both an excretory organ and a genital duct. { ¦nef·rō'mik·sē·əm }

nephron [ANAT] The functional unit of a kidney, consisting of the glomerulus with its capsule and attached uriniferous tubule. { 'nef,rän }

nephropathy [MED] **1.** Any disease of the kidney. **2.** *See* nephrosis. { nə'fräp·ə·thē }

nephropexy [MED] Fixation of a floating kidney by means of surgery. { 'nef·rə,pek·sē }

Nephropidae [INV ZOO] The true lobsters, a family of decapod crustaceans in the superfamily Nephropidea. { nə'fräp·ə,dē }

Nephropidea [INV ZOO] A superfamily of the decapod section Macrura including the true lobsters and crayfishes, characterized by a rostrum and by chelae on the first three pairs of pereiopods, with the first pair being noticeably larger. { ,nef·rə'pid·ē·ə }

nephroptosis [MED] Prolapse of the kidney. { ,ne,fräp'tō·səs }

nephrorrhaphy [MED] **1.** The stitching of a floating kidney to the posterior wall of the abdomen or to the loin. **2.** Suturing a wound in the kidney. { nə'frȯr·ə·fē }

nephros [ANAT] The kidney. { 'nef·rəs }

nephrosclerosis [MED] Sclerosis of the renal arteries and arterioles. { ¦nef·rō·sklə'rō·səs }

nephrosis [PATH] Degenerative or retrogressive renal lesions, distinct from inflammation (nephritis) or vascular involvement (nephrosclerosis), especially as applied to tubular lesions (tubular nephritis). Also known as nephrodystrophy; nephropathy. { nə'frō·səs }

nephrostome [INV ZOO] The funnel-shaped opening of a nephridium into the coelom. { 'nef·rə,stōm }

nephrotic [MED] Pertaining to or affected by nephroses. { nə'fräd·ik }

nephrotic edema [MED] A type of edema occurring in persons with chronic lepoid nephrosis or the nephrotic stage of glomerular nephritis. { nə'fräd·ik i'dē·mə }

nephrotic syndrome [MED] A complex of symptoms, including proteinuria, hyperalbuminemia, and hyperlipemia, resulting from damage to the basement membrane of glomeruli. { nə'fräd·ik 'sin,drōm }

nephrotome [EMBRYO] The narrow mass of embryonic mesoderm connecting somites and lateral mesoderm, from which the pronephros, mesonephros, metanephros, and their ducts develop. { 'nef·rə,tōm }

nephrotomy [MED] Incision of the kidney. { nə'fräd·ə·mē }

Nephtyidae [INV ZOO] A family of errantian annelids of highly opalescent colors, distinguished by an eversible pharynx. { nef'tī·ə,dē }

Nepidae [INV ZOO] The water scorpions, a family of hemipteran insects in the superfamily Nepoidea, characterized by a long breathing tube at the tip of the abdomen, chelate front legs, and a short stout beak. { 'nep·ə,dē }

Nepoidea [INV ZOO] A superfamily of hemipteran insects in the subdivision Hydrocorisae. { nə'pȯid·ē·ə }

Nepovirus [VIROL] A genus of plant viruses in the family Comoviridae; tobacco ring spot virus is the type species. Also known as tobacco ring spot virus group. { 'nep·ə,vī·rəs }

Nepticulidae [INV ZOO] The single family of the lepidopteran superfamily Nepticuloidea. { ,nep·tə'kyül·ə,dē }

Nepticuloidea [INV ZOO] A monofamilial superfamily of heteroneuran Lepidoptera; members are tiny moths with wing spines, and the females have a single genital opening. { ,nep·tə·kyə'lȯid·ē·ə }

Nereidae [INV ZOO] A large family of mostly marine errantian annelids that have a well-defined head, elongated body with many segments, and large complex parapodia on most segments. { nə'rē·ə,dē }

Nerillidae [INV ZOO] A family of archiannelids characterized by well-developed parapodia and setae. { nə'ril·ə,dē }

Neritacea [INV ZOO] A superfamily of gastropod mollusks in the order Aspidobranchia. { ,ner·ə'tās·ē·ə }

Neritidae [INV ZOO] A family of primitive marine, fresh-water, and terrestrial snails in the order Archaeogastropoda. { nə'rid·ə,dē }

nerve [NEUROSCI] A bundle of nerve fibers or processes held together by connective tissue. { nərv }

nerve block [NEUROSCI] Interruption of impulse transmission through a nerve. { 'nərv ,bläk }

nerve cell *See* neuron. { 'nərv ,sel }

nerve cord [INV ZOO] Paired, ventral cords of nervous tissue in certain invertebrates, such as insects or the earthworm. [ZOO] Dorsal, hollow tubular cord of nervous tissue in chordates. { 'nərv ,kȯrd }

nerve deafness [MED] Deafness due to an abnormality of the sense organs or of the nerves involved in hearing. { 'nərv ,def·nəs }

nerve ending [NEUROSCI] **1.** The structure on the distal end of an axon. **2.** The termination of a nerve. { 'nərv ,end·iŋ }

nerve fiber [NEUROSCI] The long process of a neuron, usually the axon. { 'nərv ,fī·bər }

nerve growth factor [NEUROSCI] A multimeric protein that promotes nerve cell growth and may protect some types of nerve cells from damage, including nerve cells in the cholinergic system. { 'nərv ,grōth ,fak·tər }

nerve impulse [NEUROSCI] The transient physicochemical change in the membrane of a nerve fiber which sweeps rapidly along the fiber to its termination, where it causes excitation of other nerves, muscle, or gland cells, depending on the connections and functions of the nerve. { 'nərv ,im,pəls }

nerve net [INV ZOO] A network of continuous nerve cells characterized by diffuse spread of excitation, local and equipotential autonomy, spatial attenuation of conduction, and facilitation; occurs in cnidarians and certain other invertebrates. { 'nərv ,net }

nerve tracing [MED] A method used by chiropractors by which nerves are located and their pathologies are studied. { 'nərv ˌtrās·iŋ }

nerve tract [NEUROSCI] A bundle of nerve fibers having the same general origin and destination. { 'nərv ˌtrakt }

nervous [NEUROSCI] **1.** Of or pertaining to nerves. **2.** Originating in or affected by nerves. **3.** Affecting or involving nerves. { 'nər·vəs }

nervous system [ANAT] A coordinating and integrating system which functions in the adaptation of an organism to its environment; in vertebrates, the system consists of the brain, brainstem, spinal cord, cranial and peripheral nerves, and ganglia. { 'nər·vəs ˌsis·təm }

nervous tissue [HISTOL] The nerve cells and neuroglia of the nervous system. { 'nər·vəs ¦tish·ü }

Nesiotinidae [INV ZOO] A family of bird-infesting biting lice (Mallophaga) that are restricted to penguins. { ¦nes·ē·ō'tin·əˌdē }

nest [VERT ZOO] A bed, receptacle, or location in which the eggs of animals are laid and hatched. { nest }

net aerial production [ECOL] The biomass or biocontent which is incorporated into the aerial parts, that is, the leaf, stem, seed, and associated organs, of a plant community. { 'net 'er·ē·əl prə'dək·shən }

net plankton [ECOL] Plankton that can be removed from sea water by the process of filtration through a fine net. { 'net 'plaŋk·tən }

net primary production [ECOL] Over a specified period of time, the biomass or biocontent which is incorporated into a plant community. { 'net 'prīm·ə·rē prə'dək·shən }

net production rate [ECOL] The assimilation rate (gross production rate) minus the amount of matter lost through predation, respiration, and decomposition. { 'net prə'dək·shən ˌrāt }

nettle [BOT] A prickly or stinging plant of the family Urticaceae, especially in the genus *Urtica*. { 'ned·əl }

nettle cell *See* cnidoblast. { 'ned·əl ˌsel }

net-veined [BIOL] Having a network of veins, as a leaf or an insect wing. { 'net ˌvānd }

neural arc [NEUROSCI] A nerve circuit consisting of effector and receptor with intercalated neurons between them. { 'nu̇r·əl 'ärk }

neural arch *See* vertebral arch. { 'nu̇r·əl 'ärch }

neural canal [EMBRYO] The embryonic vertebral canal. { 'nu̇r·əl kə'nal }

neural cell adhesion molecule [NEUROSCI] A calcium-independent cell adhesion molecule expressed by migrating neurons that mediates intercellular binding via homophilic mechanisms (by binding to other neural cell adhesion molecules); important in neuronal aggregation. { ¦nu̇r·əl ˌsel ad'hē·zhen ˌmäl·əˌkyül }

neural crest [EMBRYO] Ectoderm composing the primordium of the cranial, spinal, and autonomic ganglia and adrenal medulla, located on either side of the neural tube. { 'nu̇r·əl 'krest }

neural ectoderm [EMBRYO] Embryonic ectoderm which will form the neural tube and neural crest. { 'nu̇r·əl 'ek·təˌdərm }

neural fold [EMBRYO] Either of a pair of dorsal longitudinal folds of the neural plate which unite along the midline, forming the neural tube. { 'nu̇r·əl 'fōld }

neuralgia [MED] Pain in or along the course of one or more nerves. Also known as neurodynia. { nu̇'ral·jē·ə }

neural groove [EMBRYO] A longitudinal groove between the neural folds of the vertebrate embryo before the neural tube is completed. { 'nu̇r·əl 'grüv }

neural lymphomatosis [VET MED] A form of the avian leukosis complex affecting primarily the sciatic nerve. { 'nu̇r·əl ˌlim·fə·mə'tō·səs }

neural plate [EMBRYO] The thickened dorsal plate of ectoderm that differentiates into the neural tube. { 'nu̇r·əl 'plāt }

neural prosthesis [MED] A prosthesis that bypasses a portion of the nervous system and provides electrical stimulation to existing muscles in situations where paralysis has interrupted the natural control pathways. { ˌnu̇r·əl präs'thē·səs }

neural spine [ANAT] The spinous process of a vertebra. { 'nu̇r·əl 'spīn }

neural tube [EMBRYO] The embryonic tube that differentiates into brain and spinal cord. { 'nu̇r·əl 'tüb }

neural tube defects [MED] Congenital defects resulting from the incomplete closure of the neural tube during embryogenesis. { ˌnu̇r·əl 'tüb ˌdēˌfeks }

neuraminic acid [BIOCHEM] $C_9H_{17}NO_8$ An amino acid, the aldol condensation product of pyruvic acid and N-acetyl-D-mannosamine, regarded as the parent acid of a family of widely distributed acyl derivatives known as sialic acids. { ¦nu̇r·ə¦min·ik 'as·əd }

neuraminidase [MICROBIO] A bacterial enzyme that acts to split salic acid from neuraminic acid glycosides. { ˌnu̇r·ə'min·əˌdās }

neurapophysis [EMBRYO] Either of two projections on each embryonic vertebra which unite to form the neural arch. { ¦nu̇r·ə'päf·ə·səs }

neurapraxia [MED] Injury to a nerve in which there is localized degeneration of the myelin sheath with transient nerve block. { ¦nu̇r·ə'prak·sē·ə }

neurasthenia [MED] A group of symptoms, now generally subsumed in the neurasthenic neurosis, formerly ascribed to debility or exhaustion of the nerve centers. { ˌnu̇r·əs'thē·nē·ə }

neurectomy [MED] Surgical removal of a portion of a nerve. { nu̇'rek·tə·mē }

neurenteric canal [EMBRYO] A temporary duct connecting the neural tube and primitive gut in certain vertebrate and tunicate embryos. { ¦nu̇r·ən'ter·ik kə'nal }

neurilemma [HISTOL] A thin tissue covering the axon directly, or covering the myelin sheath when present, of peripheral nerve fibers. { ˌnu̇r·ə'lem·ə }

neurilemmoma [MED] A solitary, encapsulated benign tumor originating in the neurilemma of peripheral, cranial, and sympathetic nerves. Also known as schwannoma. { ˌnu̇r·ə·lə'mō·mə }

neurine [BIOCHEM] $CH_2{=}CHN(CH_3)_3OH$ A very poisonous, syrupy liquid with fishy aroma; soluble in water and alcohol; a product of putrefaction of choline in brain tissue and bile, and in cadavers. Also known as trimethylvinylammonium hydroxide. { 'nüˌrēn }

neurinomatosis *See* neurofibromatosis. { ˌnu̇r·əˌnō·mə'tō·səs }

neurite *See* axon. { 'nu̇ˌrīt }

neuritic plaque [PATH] Abnormal clumps of degenerating neurons surrounding a core of amyloid protein; a characteristic pathological feature found in the brains of Alzheimer's disease patients. { nu̇ˌrid·ik 'plak }

neuritis [MED] Degenerative or inflammatory nerve lesions associated with pain, hypersensitivity, anesthesia or paresthesia, paralysis, muscular atrophy, and loss of reflexes in the innervated part of the body. { nu̇'rīd·əs }

neuroanatomy [ANAT] The study of the anatomy of the nervous system and nerve tissue. { ¦nu̇r·ō·ə'nad·ə·mē }

neuroarthropathy [MED] Joint disease associated with disease of the nervous system. { ¦nu̇r·ō·är'thräp·ə·thē }

neuroastrocytoma [MED] Ganglioneuroma, especially when on the floor of the third brain ventricle and in the temporal lobes and exhibiting neuronal elements within predominant astrocytic elements. { ¦nu̇r·ōˌas·trəˌsī'tō·mə }

neurobiotaxis [EVOL] Hypothetical migration of nerve cells and ganglia toward regions of maximum stimulation during phylogenetic development. { ¦nu̇r·ōˌbī·ō'tak·səs }

neuroblast [EMBRYO] Embryonic, undifferentiated neuron, derived from neural plate ectoderm. { 'nu̇r·ōˌblast }

neuroblastoma [MED] A malignant neoplasm composed of anaplastic sympathicoblasts; occurs usually in the adrenal medulla of children. { ˌnu̇r·ō·bla'stō·mə }

neuroblastomatosis *See* neurofibromatosis. { ˌnu̇r·ō·blaˌstō·mə'tō·səs }

neurobrucellosis [MED] Brucellosis with neurologic involvement, manifested by signs and symptoms of meningitis, encephalitis, radiculitis, or neuritis. { ˌnu̇r·ōˌbrü·sə'lō·səs }

neurochemistry [BIOCHEM] Chemistry of the nervous system. { ¦nu̇r·ō'kem·ə·strē }

neurochorioretinitis [MED] Chorioretinitis combined with optic neuritis. { ¦nu̇r·ōˌkȯr·ē·ōˌret·ən'īd·əs }

neurochoroiditis [MED] Choroiditis combined with optic neuritis. { ¦nu̇r·ōˌkȯ·rȯi'dīd·əs }

neurocirculatory asthenia [MED] A syndrome characterized by dyspnea, palpitation, chest pain, fatigue, and faintness. { ¦nu̇r·ō'sər·kyə·ləˌtȯr·ē as'thē·nē·ə }

neurocoele [ANAT] The system of cavities and ventricles in the brain and spinal cord. { 'nu̇r·ōˌsēl }

neurocranium [ANAT] The portion of the cranium which forms the braincase. { ¦nu̇r·ə'krā·nē·əm }

neurocutaneous [ANAT] **1.** Concerned with both the nerves and skin. **2.** Pertaining to innervation of the skin. { ¦nu̇r·ō·kyə'tā·nē·əs }

neurocyte [NEUROSCI] The body of a nerve cell. { 'nu̇r·əˌsīt }

neurodegeneration [NEUROSCI] The process in which neurons die. { ˌnu̇·rō·diˌjen·ə'rā·shən }

neurodegenerative [PATH] Characterized by the gradual and progressive loss of nerve cells. { ˌnu̇·rō·di'jen·rəd·iv }

neurodermatitis [MED] A skin disorder characterized by localized, often symmetrical, patches of pruritic dermatitis with lichenification, occurring in persons of nervous temperament. { ¦nu̇r·ōˌdər·mə'tīd·əs }

neurodermatosis [MED] A skin disease which is presumed to have a psychogenic component or basis. { ¦nu̇r·ōˌdər·mə'tō·səs }

neurodynia *See* neuralgia. { ¦nu̇r·ō'dī·nē·ə }

neuroelectricity [PHYSIO] A current or voltage generated in the nervous system. { ¦nu̇r·ōˌiˌlek'tris·əd·ē }

neuroendocrine [BIOL] Pertaining to both the nervous and endocrine systems, structurally and functionally. { ¦nu̇r·ō'en·də·krən }

neuroendocrinology [BIOL] The study of the structural and functional interrelationships between the nervous and endocrine systems. { ¦nu̇r·ōˌen·də·krə'näl·ə·jē }

neuroepidermal [BIOL] Pertaining to both the nerves and epidermis, structurally and functionally. { ¦nu̇r·ōˌep·ə'dər·məl }

neuroepithelioma [MED] A tumor resembling primitive medullary epithelium, containing cells of small cuboidal or columnar form with a tendency to form true rosettes, occurring in the retina, central nervous system, and occasionally in peripheral nerves. Also known as diktoma; esthesioneuroblastoma; esthesioneuroepithelioma. { ¦nu̇r·ōˌep·əˌthē·lē'ō·mə }

neuroethology [ZOO] The study of the neural basis of animal behavior. { ˌnu̇·rō·ē'thäl·ə·jē }

neurofibril [NEUROSCI] A fibril of a neuron, usually extending from the processes and traversing the cell body. { ¦nu̇r·ō'fī·brəl }

neurofibrillary tangle [PATH] The accumulation of twisted protein filaments (neurofibrils) within neurons of the cerebral cortex; a characteristic pathological feature found in the brains of Alzheimer's disease patients. { ˌnu̇·rōˌfīb·rəˌler·ē 'taŋ·gəl }

neurofibroma [MED] A tumor characterized by the diffuse proliferation of peripheral nerve elements. Also known as endoneural fibroma; myxofibroma of nerve sheath; neurofibromyxoma; perineural fibroblastoma; perineural fibroma. { ¦nu̇r·ō·fī'brō·mə }

neurofibromatosis [MED] A hereditary disease characterized by the presence of neurofibromas in the skin or along the pathway of peripheral nerves. Also known as fibroma molluscum; multiple neurofibroma; multiple neurofibromatosis; neurinomatosis; neuroblastomatosis; Smith-Recklinghausen's disease. { ¦nu̇r·ō·fī,brō·mə'tō·səs }

neurofibromyxoma *See* neurofibroma. { ¦nu̇r·ō,fī·brō·mik'sō·mə }

neurofibrosarcoma [MED] A malignant tumor composed of interlacing bundles of anaplastic spindle-shaped cells which resemble those of nerve sheaths. { ¦nu̇r·ō,fī·brō·sär'kō·mə }

neurofilament [PHYSIO] A type of intermediate filament, composed of three polypeptides (NF-L, NF-M, and NF-H), that helps support and strengthen the axons of nerve cells. { ,nu̇·rō'fil·ə·mənt }

neurogenesis [EMBRYO] The formation of nerves. { ¦nu̇r·ō'jen·ə·səs }

neurogenic [MED] Caused or affected by a trauma, dysfunction, or disease of the nervous system. [NEUROSCI] **1.** Originating in nervous tissue. **2.** Innervated by nerves. { ¦nu̇r·ō¦jen·ik }

neurogenic bladder [MED] A urinary bladder disorder due to lesions of the central or peripheral nervous system. { ¦nu̇r·ō¦jen·ik 'blad·ər }

neurogenic shock [MED] Shock caused by vasodilation leading to low blood pressure and serious reduction in venous return and in cardiac output; due to such causes as injury to the central nervous system, spinal anesthesia, or reflex. { ¦nu̇r·ō¦jen·ik 'shäk }

neuroglia [NEUROSCI] The nonnervous, supporting elements of the nervous system. { nu̇'räg·lē·ə }

neurohemal organ [ZOO] Any of various structures in vertebrates and some invertebrates that consist of clusters of bulbous, secretion-filled axon terminals of neurosecretory cells which function as storage-and-release centers for neurohormones. { ¦nu̇r·ō'hē·məl ,ȯr·gən }

neurohormone [NEUROSCI] A hormone produced by nervous tissue. { ¦nu̇r·ō'hȯr,mōn }

neurohumor [NEUROSCI] A hormonal transmitter substance, such as acetylcholine, released by nerve endings in the transmission of impulses. { ¦nu̇r·ō'hyü·mər }

neurohypophysis [NEUROSCI] The neural portion or posterior lobe of the hypophysis. { ¦nu̇r·ō·hī'päf·ə·səs }

neuroimmunomodulation [PHYSIO] The influences of the nervous system upon the immune system via neural and hormonal actions. { ,nu̇·rō,im·yə·no,mä·jə'lā·shən }

neurolathyrism *See* lathyrism. { ¦nu̇r·ō'lath·ə,riz·əm }

neuroleptic [PHARM] **1.** A drug that is useful in the treatment of mental disorders, especially psychoses. **2.** Pertaining to the actions of such a drug. { ,nu̇r·ō'lep·tik }

neuroleptoanalgesia [MED] A state of analgesic consciousness produced by the administration of neuroleptic drugs, allowing painless surgery to be performed on a wakeful subject. { ,nu̇r·ō,lep·tō,an·ə'jē·zhə }

neurologist [MED] A person versed in neurology, usually a physician who specializes in the diagnosis and treatment of disorders of the nervous system and the study of its functioning. { nu̇'räl·ə·jəst }

neurology [MED] The study of the anatomy, physiology, and disorders of the nervous system. { nu̇'räl·ə·jē }

neuroma [MED] A tumor of the nervous system. { nu̇'rō·mə }

neuromast [VERT ZOO] A lateral-line sensory organ in fishes and other lower vertebrates consisting of a cluster of receptor cells connected with nerve fibers. { 'nu̇r·ō,mast }

neuromere [EMBRYO] An embryonic segment of the central nervous system in vertebrates. { 'nu̇r·ō,mir }

neuromodulator [NEUROSCI] A chemical agent that is released by a neurosecretory cell and acts on other neurons in a local region of the central nervous system by modulating their response to neurotransmitters. { ,nu̇r·ō'mäj·ə,lād·ər }

neuromuscular [BIOL] Pertaining to both nerves and muscles, functionally and structurally. { ¦nu̇r·ō'məs·kyə·lər }

neuromuscular junction *See* myoneural junction. { ¦nu̇r·ō'məs·kyə·lər 'jəŋk·shən }

neuromyasthenia [MED] Fatigue, headache, intense muscle pain, slight or transient muscle weakness, mental disturbances, objective signs in neurologic examination but usually normal cerebrospinal fluid findings, occurring in epidemics and thought to be viral in origin. Also known as benign myalgic encephalomyelitis. { ¦nu̇r·ō,mī·əs'thē·nē·ə }

neuromyelitis [MED] Inflammation of the spinal cord and of nerves. { ¦nu̇r·ō,mī·ə'līd·əs }

neuron [NEUROSCI] A nerve cell, including the cell body, axon, and dendrites. { 'nu̇,rän }

neuron doctrine [NEUROSCI] A doctrine that the neuron is the basic structural and functional unit of the nervous system, and that it acts upon another neuron through the synapse. { 'nu̇,rän ,däk·trən }

neuronitis [MED] Inflammation of a neuron; particularly, neuritis involving the cells and roots of spinal nerves. { ,nu̇r·ō'nīd·əs }

neuropathy [MED] Any disease affecting neurons. { nu̇'räp·ə·thē }

neuropeptide [NEUROSCI] A polypeptide released by axons at the synapse; it may act as a neurotransmitter and have a direct effect on synapse function or as a neuromodulator, having a long-term effect on postsynaptic neurons. { ,nu̇r·ō'pep,tīd }

neuropharmacology [MED] The science dealing with the action of drugs on the nervous system. { ¦nu̇r·ō,fär·mə'käl·ə·jē }

neurophysiology [NEUROSCI] The study of the functions of the nervous system. { ¦nu̇r·ō,fiz·ē'äl·ə·jē }

neuropil [NEUROSCI] Nervous tissue consisting

of a fibrous network of nonmyelinated nerve fibers; gray matter with few nerve cell bodies; usually a region of synapses between axons and dendrites. { 'nůr·ō,pil }

neuroplasm [NEUROSCI] Protoplasm of nerve cells. { 'nůr·ō,plaz·əm }

neuropodium [NEUROSCI] A terminal branch of an axon. { ,nůr·ō'pō·dē·əm }

neuropore [EMBRYO] A terminal aperture of the neural tube before complete closure at the 20–25 somite stage. { 'nůr·ō,pȯr }

neuropsychopathy [MED] A mental disease based upon or manifesting itself in disorders or symptoms of the nervous system. { ¦nůr·ō·sī'käp·ə·thē }

Neuroptera [INV ZOO] An order of delicate insects having endopterygote development, chewing mouthparts, and soft bodies. { nů'räp·tə·rə }

neuroradiology [MED] The roentgenology of neurologic disease. { ¦nůr·ō,rād·ē'äl·ə·jē }

neurorrhexis [MED] Surgical tearing away of a nerve from its origin, as in the treatment of neuralgia. { ¦nůr·ō'rek·səs }

neurosarcoma [MED] A sarcoma composed of elements resembling those of the nervous system, or thought to be neurogenic. { ¦nůr·ō·sär'kō·mə }

neurosclerosis [MED] Hardening of nervous tissue. { ¦nůr·ō·sklə'rō·səs }

neurosecretion [NEUROSCI] The synthesis and release of hormones by nerve cells. { ¦nůr·ō·si'krē·shən }

neurosecretory cell [NEUROSCI] A neuron that releases one or more hormones into the circulatory system. { ,nůr·ō·si'krēd·ə·rē ,sel }

neurosurgery [MED] Surgery of the nervous system. { ¦nůr·ō'sər·jə·rē }

neurosyphilis [MED] Syphilitic infection of the nervous system. { ¦nůr·ō'sif·ə·ləs }

neurotoxicity [MED] Adverse effects on the structure or function of the central and/or peripheral nervous system caused by exposure to a toxic chemical, symptoms include muscle weakness, loss of sensation and motor control, tremors, cognitive alterations, and autonomic nervous system dysfunction. { ,nů·ro·täk'sis·əd·ē }

neurotoxin [BIOCHEM] A poisonous substance in snake venom that acts as a nervous system depressant by blocking neuromuscular transmission by binding acetylcholine receptors on motor end plates, or on the innervated face of an electroplax. { ¦nůr·ō'täk·sən }

neurotransmitter [NEUROSCI] A chemical agent that is released by a neuron at a synapse, diffuses across the synapse, and acts upon a postsynaptic neuron, a muscle, or a gland cell. { ,nůr·ō,tranz 'mid·ər }

neurotrophic ulcer [MED] A decubitus ulcer due to trophic disturbances following interruption or disease of afferent nerve fibers plus the factor of external trauma. { ¦nůr·ō¦träf·ik 'əl·sər }

neurotropic [BIOL] Having an affinity for nerve tissue. { ¦nůr·ō¦träp·ik }

neurovaricosis [MED] The formation or the presence of a varicosity on a nerve fiber. { ¦nůr·ō,var·ə'kō·səs }

neurovascular [BIOL] Pertaining structurally and functionally to both the nervous and vascular structures. { ¦nůr·ō'vas·kyə·lər }

neurulation [EMBRYO] Differentiation of nerve tissue and formation of the neural tube. { ,nůr·ə'lā·shən }

neuston [BIOL] Minute organisms that float or swim on surface water or on a surface film of water. { 'nü,stän }

neutralism [ECOL] A neutral interaction between two species, that is, one having no evident effect on either species. { 'nü·trə,liz·əm }

neutralizing antibody [IMMUNOL] An antibody that reduces or abolishes some biological activity of a soluble antigen or of a living microorganism. { 'nü·trə,līz·iŋ 'ant·i,bäd·ē }

neutral mutation [GEN] A mutation that has no phenotypic effect or adaptive significance. { 'nü·trəl myü,tā·shən }

neutron therapy [MED] Medical therapy involving irradiation with neutrons. { 'nü,trän ,ther·ə·pē }

neutropenia [MED] Abnormally low number of neutrophils in the peripheral circulation. { 'nü·trə,pē·nē·ə }

neutrophil [HISTOL] A large granular leukocyte with a highly variable nucleus, consisting of three to five lobes, and cytoplasmic granules which stain with neutral dyes and eosin. { 'nü·trə,fil }

neutrophilia [BIOL] Affinity for neutral dyes. [MED] An abnormal increase in leukocytes in the tissues or peripheral circulation. { ,nü·trə'fil·ē·ə }

neutrophilic leukemia [MED] Granulocytic leukemia in which the leukocytes resemble cells of the neutrophilic series. { ¦nü·trə¦fil·ik lü'kē·mē·ə }

neutrophilous [BIOL] Preferring an environment free of excess acid or base. { nü'träf·ə·ləs }

nevus [MED] A lesion containing melanocytes. { 'nē·vəs }

nevus sebaceus [MED] A nevus formed by an aggregate of sebaceous glands. Also known as Jadassohn's nevus. { 'nē·vəs si'bā·shəs }

newborn [MED] Born recently; said of human infants less than a month old, especially of those only a few days old. { 'nü,bȯrn }

Newcastle disease [VET MED] An acute viral disease of fowls, with respiratory, gastrointestinal, and central nervous system involvement; may be transmitted to human beings as a mild conjunctivitis. Also known as avian pneumoencephalitis; avian pseudoplague; Philippine fowl disease. { 'nü,kas·əl di,zēz }

Newcastle virus [VIROL] A ribonucleic acid hemagglutinating myxovirus responsible for Newcastle disease. { 'nü,kas·əl ,vī·rəs }

newt [VERT ZOO] Any of the small, semiaquatic salamanders of the genus *Triturus* in the family

Salamandridae; all have an aquatic larval stage. { nüt }

nexus *See* gap junction. { 'nek·səs }

Nezelof's syndrome [MED] A congenital immunodeficiency disease in which T and sometimes B cells are absent, causing a lack in immune function leading to recurrent infections. { 'nez·ə,lȯfs ,sin,drōm }

NGF *See* nerve growth factor.

NGU *See* nongonococcal urethritis.

niacin *See* nicotinic acid. { 'nī·ə·sən }

niacinamide *See* nicotinamide. { ,nī·ə'sin·ə·mīd }

niche [ECOL] The unique role or way of life of a plant or animal species. { nich }

nick [BIOCHEM] The absence of a phosphodiester bond between adjacent nucleotides in one strand of duplex deoxyribonucleic acid. { nik }

nickase [BIOCHEM] An enzyme that causes single-stranded breaks in duplex deoxyribonucleic acid, allowing it to unwind. { 'ni,kās }

Nicoletiidae [INV ZOO] A family of the insect order Thysanura proper. { ,nik·ə·lə'tī·ə,dē }

Nicomachinae [INV ZOO] A subfamily of the limnivorous sedentary annelids in the family Maldanidae. { ,nik·ə'mak·ə,nē }

nicotinamide [BIOCHEM] $C_6H_6ON_2$ Crystalline basic amide of the vitamin B complex that is interconvertible with nicotinic acid in the living organism; the amide of nicotinic acid. Also known as niacinamide. { ,nik·ə'tin·ə,mīd }

nicotinamide adenine dinucleotide *See* diphosphopyridine nucleotide. { ,nik·ə'tin·ə,mīd 'ad·ən,ēn dī'nü·klē·ə,tīd }

nicotinamide adenine dinucleotide phosphate [CELL MOL] A coenzyme and important component of the enzymatic systems concerned with biological oxidation-reduction systems. Abbreviated NADP+ (the reduced form is abbreviated NADPH). { ,nik·ə¦tin·e,mīd 'ad·ən,ēn dī ¦nü·klē·ə,tīd 'fäs,fāt }

nicotinamide adenine dinucleotide oxidase [BIOCHEM] Any of a group of proteins located on the cell surface that are responsible for functions which, in concert with other membrane proteins, allow cells to enlarge following cell division. Abbreviated NOX. { ,nik·ə,tin·ə,mīd ¦ad·ən·ēn di,nü·klē·ə,tīd 'äk·sə,dās }

nicotinic acid [BIOCHEM] $C_6H_5NO_2$ A component of the vitamin B complex; a white, water-soluble powder stable to heat, acid, and alkali; used for the treatment of pellagra. Also known as niacin. { ¦nik·ə¦tin·ik 'as·əd }

nictitating membrane [VERT ZOO] A membrane of the inner angle of the eye or below the eyelid in many vertebrates, and capable of extending over the eyeball. { 'nik·ə,tād·iŋ 'mem,brān }

nidamental gland [ZOO] Any of various structures that secrete covering material for eggs or egg masses. { ¦nīd·ə¦ment·əl 'gland }

nidicolous [ZOO] **1.** Spending a short time in the nest after hatching. **2.** Sharing the nest of another species. { nī'dik·ə·ləs }

nidus [MED] A focus of infection. [ZOO] A nest or breeding place. { 'nīd·əs }

Niemann-Pick disease [MED] A hereditary sphingolipidosis due to an enzyme deficiency resulting in abnormal accumulation of sphingomyelin; symptoms include anemia, enlargement of the liver, spleen, and lymph nodes, gastrointestinal disturbances, and various neurologic deficits. Also known as lipid hystiocytosis. { 'nē,män 'pik di,zēz }

night ape *See* bushbaby. { 'nīt ,āp }

night blindness [MED] Reduced dark adaptation resulting from vitamin A deficiency or from retinitis pigmentosa or other peripheral retinal disease. Also known as nyctalopia. { 'nīt ,blīnd·nəs }

night sweat [MED] Drenching perspiration occurring at night or during sleep in the course of certain febrile diseases. { 'nīt ,swet }

night vision *See* scotopic vision. { 'nīt ,vizh·ən }

nigrescent [BIOL] Blackish. { nī'gres·ənt }

nihilism [MED] Pessimism in regard to the efficacy of treatment, particularly the use of drugs. { 'nī·ə,liz·əm }

Nilionidae [INV ZOO] The false ladybird beetles, a family of coleopteran insects in the superfamily Tenebrionoidea. { ,nil·ē'än·ə,dē }

nipple [ANAT] The conical projection in the center of the mamma, containing the outlets of the milk ducts. { 'nip·əl }

Nippotaeniidea [INV ZOO] An order of tapeworms of the subclass Cestoda including some internal parasites of certain fresh-water fishes; the head bears a single terminal sucker. { ,nip·ō,tē·nē'ī·dē·ə }

Nissl bodies [CELL MOL] Chromophil granules of nerve cells which ultrastructurally are composed of large ribosomes. { 'nis·əl ,bäd·ēz }

Nitelleae [BOT] A tribe of stoneworts, order Charales, characterized by 10 cells in two tiers of five each composing the apical crown. { ni 'tel·ē,ē }

Nitidulidae [INV ZOO] The sap-feeding beetles, a large family of coleopteran insects in the superfamily Cucujoidea; individuals have five-jointed tarsi and antennae with a terminal three-jointed clavate expansion. { ,nid·ə'dyül·ə,dē }

nitric oxide synthase [BIOCHEM] An enzyme that catalyzes the stepwise conversion of the amino acid L-arginine to nitric oxide and L-citrulline. There are three types: the brain (or neuronal) and epithelial nitric oxide synthases, which are always present in cells, and inducible nitric oxide synthase, which is produced as needed. { ,nī·trik ¦äk,sīd ¦sin,thās }

nitrification [MICROBIO] Formation of nitrous and nitric acids or salts by oxidation of the nitrogen in ammonia; specifically, oxidation of ammonium salts to nitrites and oxidation of nitrites to nitrates by certain bacteria. { ,nī·trə·fə'kā·shən }

nitrifying bacteria [MICROBIO] Members of the family Nitrobacteraceae. { 'nī·trə,fī·iŋ bak'tir·ē·ə }

Nitrobacteraceae [MICROBIO] The nitrifying bacteria, a family of gram-negative, chemolithotrophic bacteria; autotrophs which derive energy from nitrification of ammonia or nitrite, and obtain carbon for growth by fixation of carbon dioxide. { ¦nī·trō‚bak·tə'rās·ē‚ē }

nitrocellulose filter [CELL MOL] A very thin filter whose fibers selectively bind single-stranded deoxyribonucleic acid (DNA) but not double-stranded DNA or ribonucleic acid. { ¦nī·trō'sel·yə‚lōs ‚fil·tər }

nitrogenase [BIOCHEM] An enzyme that catalyzes a six-electron reduction of N_2 in the process of nitrogen fixation. { nī'trä·jə‚nās }

nitrogen balance [PHYSIO] The difference between nitrogen intake (as protein) and total nitrogen excretion for an individual. { 'nī·trə·jən ‚bal·əns }

nitrogen fixation [MICROBIO] Assimilation of atmospheric nitrogen by heterotrophic bacteria. Also known as dinitrogen fixation. { 'nī·trə·jən ‚fik¦sā·shən }

nitrogen narcosis [MED] Narcosis caused by gaseous nitrogen at high pressure in the blood; produced in divers breathing air at depths of 100 feet (30 meters) or more. Also known as rapture of the deep. { 'nī·trə·jən när'kō·səs }

nitrogenous base [BIOCHEM] A purine or a pyrimidine derivative which is one of the three components of a nucleotide of nucleic acids. Also known as base. { nī'trä·jə·nəs 'bās }

nitrophenide [PHARM] $(NO_2C_6H_4)_2S$ Yellow crystals with a melting point of 83°C; soluble in ether; used in veterinary medicine and pharmaceuticals manufacture. { ¦nī·trō'fe‚nīd }

nitrophyte [BOT] A plant that requires nitrogen-rich soil for growth. { ¦nī·trə‚fīt }

nival [ECOL] **1.** Characterized by or living in or under the snow. **2.** Of or pertaining to a snowy environment. { 'nī·vəl }

niveous *See* niveus. { 'niv·ē·əs }

niveus [BIOL] Snow-white in color. Also spelled niveous. { 'niv·ē·əs }

NK *See* natural killer cell.

NMR tomography *See* zeugmatography. { ¦en ¦em'är tə'mäg·rə·fē }

Nocardiaceae [MICROBIO] A family of aerobic bacteria in the order Actinomycetales; mycelium and spore production is variable. { nō‚kär·dē'ās·ē‚ē }

nocardiosis [MED] Infection by species of the fungus *Nocardia* characterized by spreading granulomatous lesions. { nō‚kär·dē'ō·səs }

nociceptive reflex *See* flexion reflex. { ¦nō·sē·ō·ri'sep·tiv 'rē‚fleks }

nociceptor [PHYSIO] A sensory nerve ending that is particularly sensitive to noxious stimuli such as chemical changes in surrounding tissue evoked by injury. { 'nō·sə‚sep·tər }

noctalbuminuria [MED] Excretion of protein in night urine only. { ‚näkt·al‚byü·mə'nyu̇r·ē·ə }

Noctilionidae [VERT ZOO] The fish-eating bats, a tropical American monogeneric family of the Chiroptera having small eyes and long, narrow wings. { näk‚til·ē'än·ə‚dē }

Noctuidae [INV ZOO] A large family of dull-colored, medium-sized moths in the superfamily Noctuoidea; larva are mostly exposed foliage feeders, representing an important group of agricultural pests. { näk'tü·ə‚dē }

Noctuoidea [INV ZOO] A large superfamily of lepidopteran insects in the suborder Heteroneura; most are moderately large moths with reduced maxillary palpi. { ‚näk·tü'ȯid·ē·ə }

nocturia [MED] Excessive urination at night. { nak'tür·ē·ə }

nocturnal [BIOL] Active during the nighttime. { näk'tərn·əl }

nocturnal emission [PHYSIO] Normal, involuntary seminal discharge occurring during sleep in males after puberty. { näk'tərn·əl i'mish·ən }

nocturnal enuresis [MED] Involuntary nocturnal urination during sleep. { näk'tərn·əl ‚en·yu̇'rē·səs }

nodal rhythm [PHYSIO] A cardiac rhythm characterized by pacemaker function originating in the atrioventricular node, with a heart rate of 40–70 per minute. { 'nōd·əl 'rith·əm }

nodal tachycardia [MED] A cardiac arrhythmia characterized by a heart rate of 140–220 per minute. { 'nōd·əl ‚tak·ə'kärd·ē·ə }

nodal tissue [HISTOL] **1.** Tissue from the sinoatrial node, and the atrioventricular node and bundle and its branches, composed of a dense network of Purkinje fibers. **2.** Tissue from a lymph node. { 'nōd·əl ‚tish·ü }

Nodamura virus [VIROL] The type species of the *Nodavirus* genus, it infects mammals and insects. Abbreviated NOV. { ¦nō·də¦mu̇r·ə ‚vī·rəs }

Nodaviridae [VIROL] A family of single-strand ribonucleic acid-containing viruses that infect insects. { ‚nō·də'vir·ə‚dī }

Nodavirus [VIROL] The genus comprising the Nodaviridae family of insect viruses. { 'nō·də‚vī·rəs }

node [ANAT] **1.** A knob or protuberance. **2.** A small, rounded mass of tissue, such as a lymph node. [BOT] A site on a plant stem at which leaves and axillary buds arise. [NEUROSCI] A point of constriction along a nerve. { nōd }

node of Ranvier [NEUROSCI] The region of a local constriction in a myelinated nerve; formed at the junction of two Schwann cells. { 'nōd əv rän'vyā }

Nodosariacea [INV ZOO] A superfamily of Foraminiferida in the suborder Rotaliina characterized by a radial calcite test wall with monolamellar septa, and a test that is coiled, uncoiled, or spiral about the long axis. { ‚nōd·ə‚sar·ē'ās·ē·ə }

nodose [BIOL] Having many or noticeable protuberances; knobby. { 'nō‚dōs }

nodular goiter *See* adenomatous goiter. { 'näj·ə·lər 'gȯid·ər }

nodule [ANAT] **1.** A small node. **2.** A small aggregation of cells. [BOT] A bulbous enlargement found on roots of legumes and certain other plants, whose formation is stimulated by symbiotic, nitrogen-fixing bacteria that colonize the roots. [MED] A primary skin lesion, seen as a circumscribed solid elevation. { 'näj·ül }

nodules of the semilunar valves [ANAT] Small nodes in the midregion of the pulmonary and aortic semilunar valves. { 'näj·ülz əv thə 'sem·i,lü·nər 'valvz }

nodulose [BIOL] Having minute nodules or fine knobs. { 'näj·ə,lōs }

NOEL *See* no observed effect level.

nogalamycin [MICROBIO] $C_{39}H_{45}NO_{16}$ An antineoplastic antibiotic produced by *Streptomyces nogalaster.* { nō,gal·ə'mīs·ən }

noma [MED] Spreading gangrene beginning in a mucous membrane; considered to be a malignant form of infection by fusospirochetal organisms. Also known as gangrenous stomatitis. { 'nō·mə }

nomen dubium [SYST] A proposed taxonomic name invalid because it is not accompanied by a definition or description of the taxon to which it applies. { 'nō·mən 'dü·bē·əm }

nomen nudum [SYST] A proposed taxonomic name invalid because the accompanying definition or description of the taxon cannot be interpreted satisfactorily. { 'nō·mən 'nü·dəm }

non-A, non-B hepatitis [MED] A type of viral hepatitis that is most common among people who have received transfused blood and whose serologic tests show no evidence of hepatitis A, hepatitis B, or other types of virus such as Epstein-Barr. { ,nän ¦ā ,nän ¦bē ,hep·ə'tīd·əs }

noncommunicating hydrocephaly *See* obstructive hydrocephaly. { ¦nän·kə'myü·nə,kād·iŋ ,hī·drō 'sef·ə·lē }

noncompetitive inhibition [BIOCHEM] Enzyme inhibition in which the inhibitor can combine with either the free enzyme or the enzyme-substrate complex so that the inhibitor does not compete with the substrate for the enzyme. { ¦nän·kəm'ped·əd·iv ,in·ə'bish·ən }

noncongression [CELL MOL] The failure of pairing chromosomes, in certain stages of mitosis and meiosis, to orient in an orderly arrangement on the spindle equator. { ,nän·kən'gre·shən }

nonconjugative plasmid [GEN] Any plasmid that prevents conjugation of its bacterial host. { nän¦kän·jə,gād·iv 'plaz·mid }

nondisjunction [GEN] Failure of homologous chromosomes to separate symmetrically during cell division, with both ending up in the same daughter cell instead of in each daughter cell. { ¦nän'dis,jaŋk·shən }

nondisjunction mosaic [GEN] A population of cells with different chromosome numbers produced when one chromosome is lost during mitosis or when both members of a pair of chromosomes are included in the same daughter nucleus; can occur during embryogenesis or adulthood. { ¦nän·dis'jəŋk·shən mō'zā·ik }

nonessential amino acid [BIOCHEM] An amino acid which can be synthesized by an organism and thus need not be supplied in the diet. { ¦nän·i'sen·chəl ə'mē·nō 'as·əd }

nongonococcal urethritis [MED] Human urethral inflammation not associated with common bacterial pathogens; thought to be caused by bacteria of the Bedsonia group. Abbreviated NGU. { ¦nän,gän·ə'kä·kəl ,yür·ə'thrīd·əs }

nongranular leukocyte [HISTOL] A white blood cell, such as a lymphocyte or monocyte, with clear homogeneous cytoplasm. { ¦nän'gran·yə·lər 'lü·kə,sīt }

nonhemolysis *See* gamma hemolysis. { ,nän·hē 'mäl·ə·səs }

nonhistone protein [BIOCHEM] A class of acidic proteins in the cell nucleus associated with deoxyribonucleic acid. { nän¦hi,stōn 'prō,tēn }

Nonionacea [INV ZOO] A superfamily of Foraminiferida in the suborder Orbitoidacea, characterized by a granular calcite test wall with monolamellar septa, and a planispiral to trochospiral test. { ,nō·nē·ə'nās·ē·ə }

nonodontogenic cyst [MED] Any oral cyst that develops from epithelium which has been sequestered in bony or soft-tissue suture lines during embryonic development. { ,nän·ə¦dant·ə¦jen·ik 'sist }

nonosteogenic fibroma [MED] A tumor of bone, usually in the shaft of long bones; characterized by whorls of spindle-shaped connective tissue cells. { ¦nän¦äs·tē·ō¦jen·ik fī'brō·mə }

nonparalytic poliomyelitis [MED] Infection by poliomyelitis virus accompanied by upper respiratory or gastrointestinal symptoms, muscular pain and stiffness, and mild fever. { ,nän,par·ə'lid·ik ,pō·lē·ō,mī·ə'līd·əs }

nonpermissive cell [VIROL] A cell that does not support replication of a virus. { ,nän·pər¦mis·iv 'sel }

nonpositive tactile stimulus [PHYSIO] A cessation of a feedback signal from tactile sensors to the brain, such as when an object held in the hand falls. { ¦nän¦päz·əd·iv ¦tak,tīl 'stim·yə·ləs }

nonproductive cough *See* dry cough. { ,nän·prə¦dək·tiv 'kȯf }

nonproductive infection *See* abortive infection. { ¦nän·prə'dək·tiv in'fek·shən }

nonprotein nitrogen [BIOCHEM] The nitrogen fraction in the body tissues, excretions, and secretions, not precipitated by protein precipitants. { ¦nän¦prō,tēn 'nī·trə·jən }

nonpsychotic organic brain syndrome [MED] Organic brain syndrome in which there is no apparent psychosis. { ,nän·sī'käd·ik ȯr'gan·ik 'brān ,sin,drōm }

nonrhegmatogenous retinal detachment [MED] Retinal detachment that is not caused by a retinal hole or tear, but occurs as a final stage of such pathologic conditions as retinopathy of prematurity or diabetic retinopathy. { ,nän,reg·mə¦täj·ən·əs ¦ret·ən·əl di'tach·mənt }

nonsaccharine sorghum *See* grain sorghum. { ¦nän'sak·ə·rən 'sȯr·gəm }

nonselective medium [MICROBIO] A culture medium that supports the growth of all genotypes. { ¦nän·si,lek·tiv 'mē·dē·əm }

nonsense mutation [CELL MOL] A mutation that changes a codon that codes for one amino acid into a codon that does not specify any amino acid (a nonsense codon). { 'nän,səns myü,tā·shən }

nonsense suppression [GEN] Suppression of the termination effect of a nonsense codon on a growing polypeptide chain, allowing return to the normal phenotype. { 'nän,sens sə,presh·ən }

nonseptate *See* coencytic. { ,nän'sep,tāts }

nonspecific [MED] Not attributable to any one definite cause, as a disease not caused by one particular microorganism, or an immunity not conferred by a specific antibody. [PHARM] Of medicines or therapy, not counteracting any one causative agent. { ¦nän·spə'sif·ik }

nonspecific active immunotherapy [IMMUNOL] Active immunotherapy that utilizes materials that have no apparent antigenic relationship to the tumor but have modulatory effects on the immune system.

nonspecific hepatitis *See* interstitial hepatitis. { ¦nän·spə¦sif·ik ,hep·ə'tīd·əs }

nonspecific immunity [IMMUNOL] Resistance attributable to factors other than specific antibodies, including genetic, age, or hormonal factors. { ¦nän·spə¦sif·ik i'myün·əd·ē }

nonstriated muscle fiber *See* smooth muscle fiber. { ¦nän·strī'ād·əd 'məs·əl ,fī·bər }

nonsyndromic hearing loss [MED] A type of hearing loss in which the individual has no other symptoms except hearing loss. { ,hän·sin¦drō·mik 'hir·iŋ ,lòs }

nontropical sprue *See* gluten enteropathy. { ,nän ,träp·ə·kəl 'sprü }

nonvenereal syphilis [MED] Syphilis not acquired during sexual intercourse. { 'nän·və,nir·ē·əl 'sif·ə·ləs }

no observed effect level [MED] The highest dose at which no effects can be observed; used as a measure of chronic toxicity. { ¦nō əb¦zərvd i'fekt ,lev·əl }

noosphere *See* anthroposphere. { 'nō·ə,sfir }

noradrenaline *See* norepinephrine. { ,nòr·ə'dren·ə·lən }

noradrenergic system [NEUROSCI] A system of neurons that is responsible for the synthesis, storage, and release of the neurotransmitter norepinephrine. { nòr¦ad·rə¦nər·jik 'sis·təm }

norepinephrine [BIOCHEM] $C_8H_{11}O_3N$ A hormone produced by chromaffin cells of the adrenal medulla; acts as a vasoconstrictor and mediates transmission of sympathetic nerve impulses. Also known as noradrenaline. { ,nòr·ep·ə'ne·frən }

normalizing selection [GEN] Removal of alleles that produce divergence from the average phenotype in a population by selecting against deviant individuals. { 'nòr·mə,līz·iŋ si,lek·shən }

normal saline [PHYSIO] U.S. Pharmacopoeia title for a sterile solution of sodium chloride in purified water, containing 0.9 gram of sodium chloride in 100 milliliters; isotonic with body fluids. Also known as isotonic sodium chloride solution; normal salt solution; physiological saline; physiological salt solution; physiological sodium chloride solution; sodium chloride solution. { 'nòr·məl 'sā,lēn }

normal salt solution *See* normal saline. { 'nòr·məl 'sòlt sə,lü·shən }

normoblast [HISTOL] The smallest of the nucleated precursors of the erythrocyte; slightly larger than a mature adult erythrocyte. Also known as acidophilic erythroblast; arthochromatic erythroblast; eosinophilic erythroblast; karyocyte; metakaryocyte; metanitricyte; metarubricyte. { 'nòr·mə,blast }

normochromatic [CELL MOL] Pertaining to cells of the erythrocytic series which have a normal staining color; attributed to the presence of a full complement of hemoglobin. { ¦nòr·mə·krə'mad·ik }

normochromic [CELL MOL] Pertaining to erythrocytes which have a mean corpuscular hemoglobin (MCH), or color, index and a mean corpuscular hemoglobin concentration (MCHC), or saturation, index within two standard deviations above or below the mean normal. { ¦nòr·mə'krō·mik }

normocyte [HISTOL] An erythrocyte having both a diameter and a mean corpuscular volume (MCV) within two standard deviations above or below the mean normal. { 'nòr·mə,sīt }

normothermia [PHYSIO] A state of normal body temperature. { ,nòr·mə'thər·mē·ə }

North American blastomycosis [MED] A type of blastomycosis caused by the diphasic fungus *Blastomyces dermatitidis*; two recognized forms are cutaneous and systemic. { 'nòrth ə'mer·i·kən ,bla·stō,mī'kō·səs }

Northern blotting [GEN] Method of ribonucleic acid (RNA) detection and identification in which the intact RNA is separated by size via gel electrophoresis, transferred (blotted) onto nitrocellulose or nylon paper, and then hybridized with labeled DNA probes. { ¦nòr<u>th</u>·ern 'bläd·iŋ }

noscapine [PHARM] $C_{22}H_{23}NO_7$ A white, colorless, tasteless alkaloid obtained from opium; used as a nonaddicting antitussive.Formerly known as narcotine. { 'näs·kə,pēn }

nose [ANAT] The nasal cavities and the structures surrounding and associated with them in all vertebrates. { nōz }

nose leaf [VERT ZOO] A leaflike expansion of skin on the nose of certain bats; believed to have a tactile function. { 'nōz ,lēf }

nosocomial [MED] **1.** Pertaining to a hospital. **2.** Of disease, caused or aggravated by hospital life. { ¦näz·ə¦kō·mē·əl }

Nosodendridae [INV ZOO] The wounded-tree beetles, a small family of coleopteran insects in the superfamily Dermestoidea. { ,näz·ə'den·drə,dē }

nosology [MED] The science of classification of diseases. { nō'säl·ə·jē }

nostril [ANAT] One of the external orifices of the nose. Also known as naris. { 'näs·trəl }

nostrum [PHARM] A quack medicine. { 'näs·trəm }

Notacanthidae [VERT ZOO] A family of benthic, deep-sea teleosts in the order Notacanthiformes, including the spiny eel. { ,nōd·ə'kan·thə,dē }

Notacanthiformes [VERT ZOO] An order of actinopterygian fishes whose body is elongated, tapers posteriorly, and has no caudal fin. { ˌnōd·əˌkan·thə'fȯr·mēz }

notancephalia [MED] A congenital anomaly characterized by a deficiency in the occipital region of the skull. { ˌnō·tan·sə'fal·ē·ə }

notanencephalia [MED] A congenital anomaly marked by defective development or absence of the cerebellum. { ˌnō·tanˌen·sə'fal·ē·ə }

notch graft [BOT] A plant graft in which the scion is inserted in a narrow slit in the stock. { 'näch ˌgraft }

Notch signaling [EMBRYO] An evolutionarily conserved developmental pathway utilized during the differentiation of a plethora of tissue types, in organisms as diverse as nematodes and humans. { 'näch ˌsig·nəl·iŋ }

Noteridae [INV ZOO] The burrowing water beetles, a small family of coleopteran insects in the suborder Adephaga. { nō'ter·əˌdē }

Notidanoidea [VERT ZOO] A suborder of rare sharks in the order Selachii; all retain the primitive jaw suspension of the order. { ˌnōd·ə·də'nȯid·ē·ə }

notochord [VERT ZOO] An elongated dorsal cord of cells which is the primitive axial skeleton in all chordates; persists in adults in the lowest forms (*Branchiostoma* and lampreys) and as the nuclei pulposi of the intervertebral disks in adult vertebrates. { 'nōd·əˌkȯrd }

notochordal canal [EMBRYO] A canal formed by a continuation of the primary pit into the head process of mammalian embryos; provides a temporary connection between the yolk sac and amnion. { ¦nōd·ə¦kord·əl kə'nal }

notochordal plate [EMBRYO] A plate of cells representing the root of the head process of the embryo after the embryo becomes vesiculated. { ¦nōd·ə¦kord·əl 'plāt }

Notodelphyidiformes [INV ZOO] A tribe of the Gnathostoma in some systems of classification. { ¦nōd·ə·delˌfid·ə'fȯr·mēz }

Notodelphyoida [INV ZOO] A small group of crustaceans bearing a superficial resemblance to many insect larvae as a result of uniform segmentation, comparatively small trunk appendages, and crowding of inconspicuous oral appendages into the anterior portion of the head. { ˌnōd·əˌdel·fē'ȯid·ə }

Notodontidae [INV ZOO] The puss moths, a family of lepidopteran insects in the superfamily Noctuoidea, distinguished by the apparently three-branched cubitus. { ˌnōd·ə'dän·təˌdē }

Notogaean [ECOL] Pertaining to or being a biogeographic region including Australia, New Zealand, and the southwestern Pacific islands. { ¦nōd·ə¦jē·ən }

Notommatidae [INV ZOO] A family of rotifers in the order Monogonota including forms with a cylindrical body covered by a nonchitinous cuticle and with a slender posterior foot. { ˌnōd·ə'mat·əˌdē }

Notomyotina [INV ZOO] A suborder of echinoderms in the order Phanerozonida in which the upper marginals alternate in position with the lower marginals, and each tube foot has a terminal sucking disk. { ˌnōd·ə·mī'ät·ən·ə }

Notonectidae [INV ZOO] The backswimmers, a family of aquatic, carnivorous hemipteran insects in the superfamily Notonectoidea; individuals swim ventral side up, aided in breathing by an air bubble. { ˌnōd·ə'nek·təˌdē }

Notonectoidea [INV ZOO] A superfamily of Hemiptera in the subdivision Hydrocorisae. { ˌnōd·ə·nek'tȯid·ē·ə }

notopodium [INV ZOO] The dorsal branch of a parapodium in certain annelids. { ˌnōd·ə'pō·dē·əm }

Notopteridae [VERT ZOO] The featherbacks, a family of actinopterygian fishes in the order Osteoglossiformes; bodies are tapered and compressed, with long anal fins that are continuous with the caudal fin. { ˌnōˌtäp'ter·əˌdē }

Notostigmata [INV ZOO] The single suborder of the Opilioacriformes, an order of mites. { ˌnōd·ə·stig'mäd·ə }

Notostigmophora [INV ZOO] A subclass or suborder of the Chilopoda, including those centipedes embodying primitive as well as highly advanced characters, distinguished by dorsal respiratory openings. { ˌnōd·ə·stig'mäf·ə·rə }

Notostraca [INV ZOO] The tadpole shrimps, an order of crustaceans generally referred to the Branchiopoda, having a cylindrical trunk that consists of 25–44 segments, a dorsoventrally flattened dorsal shield, and two narrow, cylindrical cercopods on the telson. { nə'täs·trə·kə }

Nototheniidae [VERT ZOO] A family of perciform fishes in the suborder Blennioidei, including most of the fishes of the permanently frigid waters surrounding Antarctica. { ˌnōd·ə·thə'nī·əˌdē }

NOV *See* Nodamura virus.

novobiocin [MICROBIO] $C_{30}H_{36}O_{11}N_2$ A moderately broad-spectrum antibiotic produced by strains of *Streptomyces niveus* and *S. spheroides*; it is a dibasic acid and is converted either to the monosodium salt or to the calcium acid salt for pharmaceutical use. { ˌnō·və'bī·ə·sən }

NOX *See* nicotinamide adenine dinucleotide oxidase. { näks *or* ¦en¦ō'eks }

NPV *See* nuclear polyhedrosis virus.

nucellus [BOT] The oval central mass of tissue in the ovule; contains the embryo sac. { ˌnü'sel·əs }

nucha [ANAT] The nape of the neck. { 'nü·kə }

nuchal ligament [ANAT] An elastic ligament extending from the external occipital protuberance and middle nuchal line to the spinous process of the seventh cervical vertebra. Also known as ligamentum nuchae. { 'nü·kəl 'lig·ə·mənt }

nuchal organ [INV ZOO] Any of various sense organs on the prostomium of many annelids, which are sensitive to changes in the immediate environment of the individual. { 'nü·kəl ˌȯr·gən }

nuchal rigidity [MED] Stiffness in the nape of the neck, often accompanied by pain and spasm on attempts to move the head; the most common sign of meningitis. { 'nü·kəl ri'jid·əd·ē }

nuchal tentacle [INV ZOO] Any of various filiform or thick, fleshy tactoreceptors on anterior segments of many annelids. { 'nü·kəl ˌten·tə·kəl }

nuclear dimorphism [INV ZOO] In ciliated protozoas, the occurrence of two types of nuclei in a cell, each with different genetic functions. { ¦nü·klē·ər dī'mȯrˌfiz·əm }

nuclear envelope [CELL MOL] A structure consisting of two membranes that surrounds the nucleus; the outermost membrane is continuous with the rough endoplasmic reticulum. { 'nü·klē·ər 'en·vəˌlōp }

nuclear hormone receptor [CELL MOL] Any of a superfamily of proteins that directly regulate transcription in response to hormones and other ligands. { ¦nü·klē·ər 'hȯrˌmōn riˌsep·tər }

nuclear lamina [CELL MOL] A protein meshwork lining the inner surface of the nuclear envelope. { 'nü·klē·ər 'lam·ə·nə }

nuclear magnetic resonance tomography *See* zeugmatography. { 'nü·klē·ər mag'ned·ik 'rez·ən·əns tə'mäg·rə·fē }

nuclear medicine [MED] A branch of medicine in which radioactive pharmaceuticals are used for imaging or other diagnostic studies. { 'nü·klē·ər 'med·ə·sən }

nuclear medicine imaging [MED] A technique for producing images of the distribution of radioactive tracers in the human body by recording with a scintillation camera the gamma rays that are emitted by the trace. { 'nü·klē·ər ¦med·ə·sən 'im·ij·iŋ }

nuclear membrane [CELL MOL] The envelope surrounding the cell nucleus, separating the nucleoplasm from the cytoplasm; composed of two membranes and contains numerous pores. { 'nü·klē·ər 'memˌbrān }

nuclear plaque [MYCOL] In yeast, any region on the nuclear envelope from which the spindle originates. { 'nü·klē·ər 'plak }

nuclear polyhedrosis virus [VIROL] A Baculovirus subgroup characterized by the multiplication and formation of polyhedron-shaped inclusion bodies in the nuclei of infected host cells, used in the control of agriculture and forest insects. Abbreviated NPV. { ¦nü·klē·ər ˌpäl·ē·hē'drōs·əs ˌvī·rəs }

nuclear pore complex [CELL MOL] Any of the nonrandomly distributed, octagonal orifices in the nuclear envelope. { ¦nü·klē·ər ¦pȯr 'käm ˌpleks }

nuclear receptor superfamily [CELL MOL] A large family of intracellular receptor proteins that bind to hydrophobic signal molecules (such as steroid and thyroid hormones) or intracellular metabolites and are thus activated to bind to specific DNA sequences, affecting transcription. { ¦nü·klē·ər ri¦sep·tər 'sü·pərˌfam·lē }

nuclear sclerosis [MED] Hardening of the ocular lens nucleus. { 'nü·klē·ər sklə'rō·səs }

nuclear transfer [CELL MOL] Insertion of a diploid somatic nucleus into an egg from which the nucleus has been removed. { 'nü·klē·ər 'tranz·fər }

nuclease [BIOCHEM] An enzyme that catalyzes the splitting of nucleic acids to nucleotides, nucleosides, or the components of the latter. { 'nü·klēˌās }

nucleic acid [BIOCHEM] A large, acidic, chainlike molecule containing phosphoric acid, sugar, and purine and pyrimidine bases; two types are ribonucleic acid and deoxyribonucleic acid. { nü'klē·ik 'as·əd }

nuclein [BIOCHEM] Any of a poorly defined group of nucleic acid protein complexes occurring in cell nuclei. { 'nü·klē·ən }

nucleocapsid [VIROL] The nucleic acid of a virus and its surrounding capsid. { ¦nü·klē·ō'kap·səd }

nucleocytoplasmic ratio [CELL MOL] The ratio between the measured cross-sectional area or estimated volume of the nucleus of a cell to the volume of its cytoplasm. Also known as karyoplasmic ratio. { ¦nü·klē·ōˌsīd·ə'plaz·mik 'rā·shō }

nucleodesma [CELL MOL] A connection composed of fibrils between the nucleus and cytoplasm. Also know as karyodesma. { ˌnü·klē·ə'dez·mə }

nucleoid [CELL MOL] A discrete region within mitochondria, chloroplasts, and prokaryotes that contain molecules of deoxyribonucleic acid. [VIROL] The ribonucleic acid (RNA) core that is enveloped by a protein capsid in RNA tumor viruses. { 'nü·klēˌȯid }

nucleolar [CELL MOL] Of or pertaining to the nucleolus. { nü'klē·ə·lər }

nucleolus [CELL MOL] A small, spherical body composed principally of protein and located in the metabolic nucleus. Also known as plasmosome. { nü'klē·ə·ləs }

nucleoplasm [CELL MOL] The protoplasm of a nucleus. Also known as karyoplasm. { 'nü·klē·əˌplaz·əm }

nucleoprotein [BIOCHEM] Any member of a class of conjugated proteins in which molecules of nucleic acid are closely associated with molecules of protein. { ¦nü·klē·ō'prōˌtēn }

nucleoreticulum [CELL MOL] Any type of network found within a nucleus. { ¦nü·klē·ō·rə'tik·yə·ləm }

nucleosidase [BIOCHEM] Any enzyme involved in splitting a nucleoside into its component base and pentose. { ˌnü·klē·ə'sīˌdās }

nucleoside [BIOCHEM] The glycoside resulting from removal of the phosphate group from a nucleotide; consists of a pentose sugar linked to a purine or pyrimidine base. { 'nü·klē·əˌsīd }

nucleosome [CELL MOL] A morphologically repeating unit of deoxyribonucleic acid (DNA) containing 190 base pairs of DNA folded together with eight histone molecules. Also known as v-body. { 'nü·klē·ə‚sōm }

nucleospindle [CELL MOL] A mitotic spindle derived from nuclear material. { ¦nü·klē·ō 'spind·əl }

nucleotidase [BIOCHEM] Any of a group of enzymes which split phosphoric acid from nucleotides, leaving nucleosides. { ‚nü·klē·ə'tī‚dās }

nucleotide [BIOCHEM] An ester of a nucleoside and phosphoric acid; the structural unit of a nucleic acid. { 'nü·klē·ə‚tīd }

nucleotide excision repair [CELL MOL] A five-step pathway that is the major repair system for removing bulky lesions in deoxyribonucleic acid. { ¦nü·klē·ə‚tīd ik'sizh·ən rē‚per }

nucleus [CELL MOL] A small mass of differentiated protoplasm rich in nucleoproteins and surrounded by a membrane; found in most animal and plant cells, contains chromosomes, and functions in metabolism, growth, and reproduction. [NEUROSCI] A mass of nerve cells in the central nervous system. { 'nü·klē·əs }

nucleus pulposus [ANAT] The soft, fibrocartilaginous central portion of the intervertebral disk. { ¦nü·klē·əs pəl'pō·səs }

Nuda [INV ZOO] A class of the phylum Ctenophora distinguished by the lack of tentacles. { 'nüd·ə }

Nudechiniscidae [INV ZOO] A family of heterotardigrades in the suborder Echiniscoidea characterized by a uniform cuticle. { ‚nüd·ə·ki'nis·kə‚dē }

Nudibranchia [INV ZOO] A suborder of the Opisthobranchia containing the sea slugs; these mollusks lack a shell and a mantle cavity, and the gills are variable in size and shape. { ‚nüd·ə'braŋ·kē·ə }

null allele [GEN] An allele that does not produce a functional product and behaves as a recessive. { 'nəl ə‚lēl }

null cell [IMMUNOL] A lymphocyte without T- or B-cell markers on its surface. { 'nəl ‚sel }

numerical taxonomy [SYST] The numerical evaluation of the affinity or similarity between taxonomic units and the ordering of these units into taxa on the basis of their affinities. { nü'mer·i·kəl tak'sän·ə·mē }

Numididae [VERT ZOO] A family of birds in the order Galliformes commonly known as guinea fowl; there are few if any feathers on the neck or head, but there may be a crest of feathers and various fleshy appendages. { nü'mid·ə‚dē }

Nummulites [INV ZOO] A genus of unicellular shelled protozoa of the order Foraminiferida (superfamily Nummulitacea, family Nummulitidae). The discoidal, lenticular, or globular test or shell can reach a diameter of about 5 inches (12 centimeters) and is composed of finely perforate calcium carbonate, it consists of planispirally enrolled whorls of many tiny undivided chambers. { 'nəm·yə‚līts }

nurse cell [HISTOL] A cell type of the ovary of many animals which nourishes the developing egg cell. { 'nərs ‚sel }

nurse graft [BOT] A plant graft in which the scion remains united with the stock only until roots develop on the scion. { 'nərs ‚graft }

nursing [MED] The application of the principles of physical, biological, and social sciences in the physical and mental care of people. { 'nərs·iŋ }

nut [BOT] **1.** A fruit which has at maturity a hard, dry shell enclosing a kernel consisting of an embryo and nutritive tissue. **2.** An indehiscent, one-celled, one-seeded, hard fruit derived from a single, simple, or compound ovary. { nət }

nutation [BOT] The rhythmic change in the position of growing plant organs caused by variation in the growth rates on different sides of the growing apex. { nü'tā·shən }

nutmeg [BOT] *Myristica fragrans.* A dark-leafed evergreen tree of the family Myristicaceae cultivated for the golden-yellow fruits which resemble apricots; a delicately flavored spice is obtained from the kernels inside the seeds. { 'nət‚meg }

nutmeg liver [MED] Chronic passive hyperemia of the liver; the cut surface of the diseased organ resembles the cut surface of a nutmeg. { 'nət‚meg ¦liv·ər }

nutraceutical [BIOCHEM] Any food or food ingredient that is medically beneficial to an organism, including preventing disease. { ‚nü·trə 'süd·ə·kəl }

nutrient [BIOL] Providing nourishment. { 'nü·trē·ənt }

nutrient biopurification [ECOL] A process taking place within a nutrient cycle that maintains the pools of nutrient substances at optimum concentrations, to the exclusion of nonnutrient substances. { 'nü·trē·ənt ¦bī·ō‚pyu̇r·ə·fə'kā·shən }

nutrient foramen [ANAT] The opening into the canal which gives passage to the blood vessels of the medullary cavity of a bone. { 'nü·trē·ənt fə'rā·mən }

nutrilite [BIOCHEM] A nourishing compound. { 'nü·trə‚līt }

nutrition [BIOL] The science of nourishment, including the study of nutrients that each organism must obtain from its environment in order to maintain life and reproduce. { nü'trish·ən }

nutritional anemia [MED] Anemia resulting from certain nutritional deficiencies. { nü'trish·ən·əl ə'nē·mē·ə }

nutritional dystrophy *See* kwashiorkor. { nü'trish·ən·əl 'dis·trə·fē }

nutritional edema [MED] Edema resulting from starvation or malnutrition. { nü'trish·ən·əl i'dē·mə }

nutritional hypochromic anemia *See* iron-deficiency anemia. { nü'trish·ən·əl ¦hī·pō¦krō·mik ə'nē·mē·ə }

Nuttalliellidae [INV ZOO] A family of ticks (Ixodides) containing one rare African species, *Nuttalliella namaqua*, morphologically intermediate between the families Argasidae and Ixodidae. { nə‚tal·ē'el·ə‚dē }

nux vomica [BOT] The seed of *Strychnos nux-vomica*, an Indian tree of the family Loganiaceae;

contains the alkaloid strychnine, and was formerly used in medicine. { 'nəks 'väm·ə·kə }

Nyctaginaceae [BOT] A family of dicotyledonous plants in the order Caryophyllales characterized by an apocarpous, monocarpous, or syncarpous gynoecium, sepals joined to a tube, a single carpel, and a cymose inflorescence. { ˌnik·tə·jə'nās·ē,ē }

nyctalopia *See* night blindness. { ˌnik·tə'lō·pē·ə }

Nycteridae [VERT ZOO] The slit-faced bats, a monogeneric family of insectivorous chiropterans having a simple, well-developed nose leaf, and large ears joined together across the forehead. { nik'ter·əˌdē }

Nyctibiidae [VERT ZOO] A family of birds in the order Caprimulgiformes including the neotropical potoos. { ˌnik·tə'bī·əˌdē }

nyctinasty [BOT] A nastic movement in higher plants associated with diurnal light and temperature changes. { 'nik·təˌnas·tē }

Nyctribiidae [INV ZOO] The bat tick flies, a family of myodarian cyclorrhaphous dipteran insects in the subsection Acalyptratae. { ˌnik·trə'bī·əˌdē }

nygmata [INV ZOO] Sensory spots on wings of certain insects, such as some neuropterans. { nig'mäd·ə }

Nygolaimoidea [INV ZOO] A superfamily of predaceous nematodes of the order Dorylaimida, distinguished by an eversible stoma with a protrusible subventral mural tooth, and a bottle-shaped esophagus. { ˌnī·gō·lə'mȯid·ē·ə }

nymph [INV ZOO] Any immature larval stage of various hemimetabolic insects. { nimf }

Nymphaeaceae [BOT] A family of dicotyledonous plants in the order Nymphaeales distinguished by the presence of roots, perfect flowers, alternate leaves, and uniaperturate pollen. { ˌnim·fē'ās·ēˌē }

Nymphaeales [BOT] An order of flowering aquatic herbs in the subclass Magnoliidae; all lack cambium and vessels and have laminar placentation. { ˌnim·fē'ā·lēz }

Nymphalidae [INV ZOO] The four-footed butterflies, a family of lepidopteran insects in the superfamily Papilionoidea; prothoracic legs are atrophied, and the well-developed patagia are heavily sclerotized. { nim'fal·əˌdē }

Nymphalinae [INV ZOO] A subfamily of the lepidopteran family Nymphalidae. { nim'fal·əˌnē }

Nymphonidae [INV ZOO] A family of marine arthropods in the subphylum Pycnogonida; members have chelifores, five-jointed palpi, and ten-jointed ovigers. { nim'fän·əˌdē }

Nymphulinae [INV ZOO] A subfamily of the lepidopteran family Pyralididae which is notable because some species are aquatic. { nim'fyül·əˌnē }

Nysmyth's membrane [ANAT] The primary enamel cuticle which is the transitory remnants of the enamel organ and oral epithelium covering the enamel of a tooth after eruption. { 'nīˌsmiths ˌmemˌbrān }

Nyssaceae [BOT] A family of dicotyledonous plants in the order Cornales characterized by perfect or unisexual flowers with imbricate petals, a solitary ovule in each locule, a unilocular ovary, and more stamens than petals. { nə'sās·ēˌē }

nystagmus [MED] Involuntary oscillatory movement of the eyeballs. { nə'stag·məs }

nystatin [MICROBIO] $C_{46}H_{77}NO_{19}$ An antifungal antibiotic produced by *Streptomyces noursei*; used for the treatment of infections caused by *Candida* (*Monilia*) *albicans*. { 'nis·təd·ən }

O

oak [BOT] Any tree of the genus *Quercus* in the order Fagales, characterized by simple, usually lobed leaves, scaly winter buds, a star-shaped pith, and its fruit, the acorn, which is a nut; the wood is tough, hard, and durable, generally having a distinct pattern. { ōk }

O antigen [MICROBIO] A somatic antigen of certain flagellated microorganisms. { 'ō 'ant·i·jən }

oat [BOT] Any plant of the genus *Avena* in the family Graminae, cultivated as an agricultural crop for its seed, a cereal grain, and for straw. { ōt }

obclavate [BIOL] Inversely clavate. { äb'kla ˌvāt }

obcordate [BOT] Referring to a leaf, heart-shaped with the notch apical. { äb'kȯrˌdāt }

obdiplastemonous [BOT] Having the stamens arranged in two whorls, with members of the outer whorl positioned opposite the petals. { äb¦dip·lō¦stē·mə·nəs }

obduction [MED] The act or instance of performing a postmortem examination. { äb'dək·shən }

obese [ANAT] Extremely fat. { ō'bēs }

obesity [MED] An excessive accumulation of body fat which confers health risks such as diabetes, cardiovascular diseases, arthritis, and some types of tumors. { ō'bē·səd·ē }

objective sign [MED] A sign which can be detectable by someone other than the patient. { äb'jek·tiv 'sīn }

obligate [BIOL] Restricted to a specified condition of life, as an obligate parasite. { 'äb·lə·gət }

obligate aerobe [MICROBIO] A microorganism that uses oxygen for cellular respiration and requires some free molecular oxygen in its surroundings to support growth. { ¦äb·ləˌgāt 'er·ōb }

obligate anaerobe [MICROBIO] A microorganism that cannot use oxygen and can grow only in the absence of free oxygen. { ¦äb·liˌgāt 'an·əˌrōb }

oblique [ANAT] Referring to a muscle, positioned obliquely and having one end that is not attached to bone. [BOT] Referring to a leaf, having the two sides of a blade unequal. { ə'blēk }

obliterating endarteritis *See* endarteritis obliterans. { ə'blid·əˌrād·iŋ ¦endˌärd·ə'rīd·əs }

obliteration [MED] **1.** Complete removal of an organ or other body part by disease or surgical excision. **2.** Closure of a lumen. **3.** Loss of memory or consciousness of specific events. { əˌblid·ə'rā·shən }

obliterative appendicitis [MED] Obliteration of the lumen of the appendix by fibrofatty tissue. { ə'blid·əˌrād·iv əˌpen·də'sīd·əs }

obovate [BIOL] Inversely ovate. { 'äb·əˌvāt }

obsolete [BIOL] A part of an organism that is imperfect or indistinct, compared with a corresponding part of similar organisms. { ˌäb·sə'lēt }

obstetric [MED] **1.** Of or pertaining to obstetrics. **2.** Of or pertaining to pregnancy and childbirth. { äb'ste·trik }

obstetrical analgesia [MED] Analgesia induced to diminish or obliterate the pain of childbirth. { äb'ste·trə·kəl ˌan·əl'jē·zē·ə }

obstetric forceps [MED] A perforated, double-bladed traction forceps which can be applied to the fetal head in cases of difficult labor. { äb 'ste·trik 'fȯrˌseps }

obstetrician [MED] One who practices obstetrics. { ˌäb·stə'trish·ən }

obstetrics [MED] The branch of medicine that deals with pregnancy, labor, and the puerperium. { äb'ste·triks }

obstipation [MED] Constipation that is difficult to relieve. { ˌäb·stə'pā·shən }

obstruction [MED] Occlusion or stenosis of hollow viscera, ducts, and vessels. { əb'strək·shən }

obstructive apnea [MED] A pause in breathing while sleeping that lasts more than 10 seconds and is caused by a collapse of the upper airways. { əb¦stək·tiv 'ap·nē·ə }

obstructive atelectasis [MED] Collapse of all or a portion of the lung due to bronchial obstruction or occlusion. Also known as absorption atelectasis. { əb'strək·tiv ˌad·əl'ek·tə·səs }

obstructive dysmenorrhea *See* mechanical dysmenorrhea. { əb'strək·tiv ˌdis·men·ə'rē·ə }

obstructive emphysema [MED] Overdistension of the lung due to partial obstruction of the air passages, which permits air to enter the alveoli but which resists expiration of the air. { əb 'strək·tiv ˌem·fə'sē·mə }

obstructive glaucoma *See* narrow-angle glaucoma. { əb'strək·tiv glaů'kō·mə }

obstructive hydrocephaly [MED] Accumulation of cerebrospinal fluid in the brain ventricles caused by obstruction of the passage of the fluid from the ventricles to the subarachnoid space. Also known as internal hydrocephaly; noncommunicating hydrocephaly. { əb'strək·tiv ˌhī·drə'sef·ə·lē }

obstructive jaundice [MED] Jaundice caused by mechanical obstruction of the biliary passages, preventing the outflow of bile. { əb'strək·tiv 'jȯn·dəs }

obtund [MED] To make dull or reduce, as to obtund sensibility. { äb'tənd }

obturation [MED] **1.** The closing of an opening or passage. **2.** A form of intestinal obstruction in which the lumen is occupied by its normal contents or by foreign bodies. { ˌäb·tə'rā·shən }

obturator [ANAT] **1.** Pertaining to that which closes or stops up, as an obturator membrane. **2.** Either of two muscles, originating at the pubis and ischium, which rotate the femur laterally. [MED] A solid wire or rod contained within a hollow needle or cannula. { 'äb·təˌrād·ər }

obturator artery [ANAT] A branch of the internal iliac; it branches into the pubic and acetabular arteries. { 'äb·təˌrād·ər ¦ärd·ə·rē }

obturator foramen [ANAT] A large opening in the pelvis, between the ischium and the pubis, that gives passage to vessels and nerves; it is partly closed by a fibrous obturator membrane. { 'äb·təˌrād·ər fə'rā·mən }

obturator membrane [ANAT] **1.** A fibrous membrane closing the obturator foramen of the pelvis. **2.** A thin membrane between the crura and foot plates of the stapes. { 'äb·təˌrād·ər 'memˌbrān }

obturator nerve [NEUROSCI] A mixed nerve arising in the lumbar plexus; innervates the adductor, gracilis, and obturator externus muscles, and the skin of the medial aspect of the thigh, hip, and knee joints. { 'äb·təˌrād·ər ¦nərv }

obtuse [BOT] Of a leaf, having a blunt or rounded free end. { äb'tüs }

obvallate [BIOL] Surrounded by or as if by a wall. { äb'vaˌlāt }

obvolute [BIOL] Overlapping. { 'äb·vəˌlüt }

occipital arch [INV ZOO] A part of an insect cranium lying between the occipital suture and postoccipital suture. { äk'sip·əd·əl ¦ärch }

occipital artery [ANAT] A branch of the external carotid which branches into the mastoid, auricular, sternocleidomastoid, and meningeal arteries. { äk'sip·əd·əl ¦ärd·ə·rē }

occipital bone [ANAT] The bone which forms the posterior portion of the skull, surrounding the foramen magnum. { äk'sip·əd·əl ¦bōn }

occipital condyle [ANAT] An articular surface on the occipital bone which articulates with the atlas. [INV ZOO] A projection on the posterior border of an insect head which articulates with the lateral neck plates. { äk'sip·əd·əl 'känˌdīl }

occipital crest [ANAT] Either of two transverse ridges connecting the occipital protuberances with the foramen magnum. { äk'sip·əd·əl ¦krest }

occipital ganglion [INV ZOO] One of a pair of ganglia located just posterior to the brain in insects. { äk'sip·əd·əl 'gaŋ·glēˌän }

occipitalia [INV ZOO] An unpaired row of dorsal cilia on the head of gnathostomulids. { äkˌsip·ə'tal·ē·ə }

occipital lobe [ANAT] The posterior lobe of the cerebrum having the form of a three-sided pyramid. { äk'sip·əd·əl ¦lōb }

occipital pole [ANAT] The tip of the occipital lobe of the brain. { äk'sip·əd·əl ¦pōl }

occipital protuberance [ANAT] A prominence on the surface of the occipital bone to which the ligamentum nuchae is attached. { äk'sip·əd·əl prə'tü·bə·rəns }

occipitofrontalis [ANAT] A muscle in two parts, the frontal (inserting in the skin of the forehead) and the occipital (inserting in the galea sponeurotica). { äk¦sip·əd·ōˌfrən'tal·is }

occiput [ZOO] The back of the head of an insert or vertebrate. { 'äk·səˌpůt }

occluding junction *See* tight junction. { ə'klüd·iŋ ˌjəŋk·shən }

occlusal disharmony [MED] Increased or maldirected occlusal force on individual teeth or groups of teeth causing a malposition or functional aberration. { ə'klüs·əl dis'här·mə·nē }

occlusion [ANAT] The relationship of the masticatory surfaces of the maxillary teeth to the masticatory surfaces of the mandibular teeth when the jaws are closed. [MED] A closing or shutting up. [PHYSIO] The deficit in muscular tension when two afferent nerves that share certain motor neurons in the central nervous system are stimulated simultaneously, as compared to the sum of tensions when the two nerves are stimulated separately. { ə'klü·zhən }

occult blood [PATH] Blood in body products such as feces, not detectable on gross examination. { ə'kəlt 'bləd }

occult hydrocephaly [MED] A syndrome in which the brain ventricular system is enlarged while cerebrospinal fluid pressure remains normal, causing dementia, disturbances of equilibrium, and disorders of sphincter control. { ə'kəlt ˌhī·drə'sef·ə·lē }

occult virus [VIROL] A virus whose presence is assumed but which cannot be recovered. { ə'kəlt 'vī·rəs }

occupational acne [MED] Acne acquired from regular exposure to acnegenic materials in certain industries; disappears when the cause is removed. { ˌä·kyə'pā·shən·əl 'ak·nē }

occupational disease [MED] A functional or organic disease caused by factors arising from the operations or materials of an individual's industry, trade, or occupation. { ˌä·kyə'pā·shən·əl di'zēz }

occupational medicine [MED] The branch of medicine which deals with the relationship of humans to their occupations, for the purpose of the prevention of disease and injury and the promotion of optimal health, productivity, and social adjustment. { ˌä·kyə'pā·shən·əl 'med·i·sən }

occupational therapy [MED] The teaching of skills or the use of selected occupations for therapeutic or rehabilitation purposes. { ˌä·kyə'pā·shən·əl 'ther·ə·pē }

Oceanian [ECOL] Of or pertaining to the zoogeographic region that includes the archipelagos and islands of the central and south Pacific. { ˌō·shē'an·ē·ən }

oceanodromous [VERT ZOO] Of a fish, migratory in salt water. { ¦ō·shə¦nä·drə·məs }

ocellus [INV ZOO] A small, simple invertebrate eye composed of photoreceptor cells and pigment cells. { ō'sel·əs }

ocelot [VERT ZOO] *Felis pardalis.* A small arboreal wild cat, of the family Felidae, characterized by a golden head and back, silvery flanks, and rows of somewhat metallic spots on the body. { 'äs·əˌlät }

Ochnaceae [BOT] A family of dicotyledonous plants in the order Theales, characterized by simple, stipulate leaves, a mostly gynobasic style, and anthers that generally open by terminal pores. { äk'nās·ēˌē }

Ochotonidae [VERT ZOO] A family of the mammalian order Lagomorpha; members are relatively small, and all four legs are about equally long. { ˌäk·ə'tän·əˌdē }

ochratoxin A [BIOCHEM] $C_{20}H_{18}ClNO_6$ A toxic metabolite from *Aspergillus ochraceus*; a crystalline compound which exhibits green fluorescence; melting point is 169°C; inhibits phosphorylase and mitochondrial respiration in rat liver. { ¦äk·rə¦täk·sən 'ā }

ochre mutation [GEN] Alteration of a codon to UAA, a stop codon that results in premature termination of the polypeptide chain in prokaryotes and eukaryotes. { 'ō·kər myü'tā·shən }

Ochrobium [MICROBIO] A genus of bacteria in the family Siderocapsaceae; cells are ellipsoidal to rod-shaped and are surrounded by a delicate sheath, resembling a horseshoe, that is heavily embedded with iron oxides. { ō'krō·bē·əm }

ochroleucous [BIOL] Pale ocher or buff colored. { ˌō·krə'lü·kəs }

ochronosis [MED] A blue or brownish-blue discoloration of cartilage and connective tissue, especially around joints, caused by melanotic pigment. { ˌō·krə'nō·səs }

Ochteridae [INV ZOO] The velvety shorebugs, the single family of the hemipteran superfamily Ochteroidea. { äk'ter·əˌdē }

Ochteroidea [INV ZOO] A monofamilial tropical and subtropical superfamily of hemipteran insects in the subdivision Hydrocorisae; individuals are black with a silky sheen, and the antennae are visible from above. { äk·tə'rȯid·ē·ə }

ocrea [BOT] A tubular stipule or pair of coherent stipules. { 'ä·krē·ə }

Octocorallia [INV ZOO] The equivalent name for Alcyonaria. { ˌäk·tō·kə'ril·yə }

octopamine [PHARM] $C_8H_{11}NO_2$ A crystalline compound, used as an adrenergic drug. { äk'tō·pəˌmēn }

Octopoda [INV ZOO] An order of the dibranchiate cephalopods, characterized by having eight arms equipped with one to three rows of suckers. { äk'täp·ə·də }

Octopodidae [INV ZOO] The octopuses, in family of cephalopod mollusks in the order Octopoda. { äk·tə'päd·əˌdē }

octopus [INV ZOO] Any member of the genus *Octopus* in the family Octopodidae; the body is round with a large head and eight partially webbed arms, each bearing two rows of suckers, and there is no shell. { 'äk·təˌpu̇s }

ocular [BIOL] Of or pertaining to the eye. { 'äk·yə·lər }

ocular myasthenia [MED] A form of myasthenia gravis that is restricted to the eye muscles, symptoms may include droopy eyelids and blurred or double vision. { ˌäk·yə·lər ˌmī·əs'thē·nē·ə }

ocular skeleton [VERT ZOO] A rigid structure in most submammalian vertebrates consisting of a cup of hyaline cartilage enclosing the posterior part of the eye, and a thin-walled ring of intramembranous bones in the edge of the sclera at its junction with the cornea. { 'äk·yə·lər 'skel·ə·tən }

oculist *See* ophthalmologist. { 'äk·yə·ləst }

oculoagravic illusion *See* agravic illusion. { ¦äk·yə·lō·ā'grav·ik i'lü·zhən }

oculoglandular tularemia [MED] Infection by *Pasteurella tularensis* which, in addition to the usual symptoms of tularemia, causes swollen eyelids, conjunctivitis, swollen lymph nodes, and ulcers on the conjunctivae. { ¦äk·yə·lō'glan·jə·lər ˌtü·lə'rē·mē·ə }

oculomotor [PHYSIO] Pertaining to eye movement. { ¦äk·yə·lō'mōd·ər }

oculomotor nerve [NEUROSCI] The third cranial nerve; a paired somatic motor nerve arising in the floor of the midbrain, which innervates all extrinsic eye muscles except the lateral rectus and superior oblique, and furnishes autonomic fibers to the ciliary and pupillary sphincter muscles within the eye. { ¦äk·yə·lō'mōd·ər 'nərv }

oculomotor nucleus [NEUROSCI] A nucleus in the floor of the midbrain that gives rise to motor fibers of the oculomotor nerve. { ¦äk·yə·lō 'mōd·ər 'nü·klē·əs }

oculomotor paralysis [MED] Paralysis of the oculomotor nerve. { ¦äk·yə·lō'mōd·ər pə'ral·ə·səs }

Oculosida [INV ZOO] An order of the protozoan subclass Radiolaria; pores are restricted to certain areas in the central capsule, and an olive-colored material is always present near the astropyle. { ˌäk·yə'läs·ə·də }

odd-pinnate [BOT] Of a compound leaf, having a single leaflet at the tip of the petiole with leaflets on both sides of the petiole. Also known as imparipinnate. { 'äd 'piˌnāt }

Odiniidae [INV ZOO] A family of cyclorrhaphous myodarian dipteran insects in the subsection Acalyptratae. { ˌōd·ən'ī·əˌdē }

Odobenidae [VERT ZOO] A family of carnivorous mammals in the suborder Pinnipedia; contains a single species, the walrus (*Odobenus rosmarus*). { ˌō·dō'ben·əˌdē }

Odonata [INV ZOO] The dragonflies, an order of the class Insecta, characterized by a head with large compound eyes, and wings with clear or transparent membranes traversed by networks of veins. { ˌōd·ən'ad·ə }

odontalgia *See* toothache. { ˌōˌdän'täl·jē·ə }

odontectomy [MED] Surgical excision of a tooth. { ˌōˌdän'tek·tə·mē }

odontexesis [MED] Removal of deposits from the surface of teeth. { ˌōˌdän·tek'sē·səs }

odontoblast [HISTOL] One of the elongated, dentin-forming cells covering the dental papilla. { ō'dänt·əˌblast }

odontoblastoma *See* ameloblastic odontoma. { ō¦dänt·ə·bla'stō·mə }

Odontoceti [VERT ZOO] The toothed whales, a suborder of cetacean mammals distinguished by a single blowhole. { ōˌdänt·ə'sēˌtī }

odontoclast [HISTOL] A multinuclear cell concerned with resorption of the roots of milk teeth. { ō'dänt·əˌklast }

odont-, odonto- [VERT ZOO] A combining form meaning tooth. { ˌōˌdän·tō }

odontogenesis [EMBRYO] Formation of teeth. { ōˌdänt·ə'jen·ə·səs }

odontogenic [HISTOL] **1.** Pertaining to the origin and development of teeth. **2.** Originating in tissues associated with teeth. { ō¦dänt·ə¦jen·ik }

odontogenic cyst [MED] A cyst originating in tissues associated with teeth. { ō¦dänt·ə¦jen·ik 'sist }

odontogenic fibroma [MED] A benign tumor originating in the mesenchymal derivatives of the tooth germ. { ō¦dänt·ə¦jen·ik fī'brō·mə }

odontoid [BIOL] Toothlike. { ō'dänˌtȯid }

odontoid process [ANAT] A toothlike projection on the anterior surface of the axis vertebra with which the atlas articulates. { ō'dänˌtȯid ˌprä·səs }

odontology [VERT ZOO] A branch of science that deals with the formation, development, and abnormalities of teeth. { ˌōˌdän'täl·ə·jē }

odontoma [MED] A benign tumor representing a developmental excess, composed of mesodermal or octodermal tooth-forming tissue, alone or in association with the calcified derivatives of these structures. { ˌōˌdän'tō·mə }

Odontostomatida [INV ZOO] An order of the protozoan subclass Spirotrichia; individuals are compressed laterally and possess very little ciliature. { ō¦dänt·ə·stō'mad·ə·də }

Oecophoridae [INV ZOO] A family of small to moderately small moths in the lepidopteran superfamily Tineoidea, characterized by a comb of bristles, the pecten, on the scape of the antennae. { ˌēk·ə'fȯr·əˌdē }

Oedemeridae [INV ZOO] The false blister beetles, a large family of coleopteran insects in the superfamily Tenebrionoidea. { ˌēd·ə'mer·əˌdē }

Oedogoniales [BOT] An order of fresh-water algae in the division Chlorophyta; characterized as branched or unbranched microscopic filaments with a basal holdfast cell. { ˌēd·əˌgō·nē'ā·lēz }

Oegophiurida [INV ZOO] An order of echinoderms in the subclass Ophiuroidea, represented by a single living genus; members have few external skeletal plates and lack genital bursae, dorsal and ventral arm plates, and certain jaw plates. { ˌēg·ə·fi'yu̇r·ə·də }

Oegopsida [INV ZOO] A suborder of cephalopod mollusks in the order Decapoda of one classification system, and in the order Teuthoidea of another system. { ē'gäp·sə·də }

Oestridae [INV ZOO] A family of cyclorrhaphous myodarian dipteran insects in the subsection Calypteratae. { 'es·trəˌdē }

oidiophore [MYCOL] A hypha that produces oidia. { ō'id·əˌfȯr }

oidium [MYCOL] One of the small, thin-walled spores with flat ends produced by autofragmentation of the vegetative hyphae in certain Eumycetes. { ō'id·ē·əm }

Oikomonadidae [INV ZOO] A family of protozoans in the order Kinetoplastida containing organisms that have a single flagellum. { ˌȯik·ə·mə'nad·əˌdē }

oil gland *See* uropygial gland. { 'ȯil ˌgland }

ointment [PHARM] A semisolid preparation used for a protective and emollient effect or as a vehicle for the local or endermic administration of medicaments; ointment bases are composed of various mixtures of fats, waxes, animal and vegetable oils, and solid and liquid hydrocarbons. { 'ȯint·mənt }

okapi [VERT ZOO] *Okapia johnstoni.* An artiodactylous mammal in the family Giraffidae; has a hazel coat with striped hindquarters, and the head shape, lips, and tongue are the same as those of the giraffe, but the neck is not elongate. { ō'kä·pē }

Okazaki fragment [CELL MOL] In deoxyribonucleic acid replication, a discontinuous segment in which the lagging strand is synthesized. { ˌō·kə¦zä·kē ˌfrag·mənt }

okra [BOT] *Hibiscus esculentus.* A tall annual plant grown for its edible immature pods. Also known as gumbo. { 'ō·krə }

Olacaceae [BOT] A family of dicotyledonous plants in the order Santalales characterized by dry or fleshy indehiscent fruit, the presence of petals, stamens, and chlorophyll, and a 2–5-celled ovary. { ˌō·lə'sās·ēˌē }

Oleaceae [BOT] A family of dicotyledonous plants in the order Scrophulariales characterized generally by perfect flowers, two stamens, axile to parietal or apical placentation, a four-lobed corolla, and two ovules in each locule. { ˌō·lē'ās·ēˌē }

oleandomycin [MICROBIO] $C_{35}H_{61}O_{12}N$ A macrolide antibotic produced by *Streptomyces antibioticus*; active mainly against gram-positive microorganisms. Also known as matromycin. { ˌō·lēˌan·də'mīs·ən }

olecranon [ANAT] The large process at the distal end of the ulna that forms the bony prominence of the elbow and receives the insertion of the triceps muscle. { ō'lek·rə,nän }

Olethreutidae [INV ZOO] A family of moths in the superfamily Tortricoidea whose hindwings usually have a fringe of long hairs along the basal part of the cubitus. { ,ō·lə'thrüd·ə,dē }

olfaction [PHYSIO] **1.** The function of smelling. **2.** The sense of smell. { äl'fak·shən }

olfactoreceptor [PHYSIO] A structure which is a receptor for the sense of smell. { äl¦fak·tō·ri¦sep·tər }

olfactory aura [MED] Prodromal disagreeable olfactory sensation preceding or characterizing an epileptic attack. { äl'fak·trē 'ȯr·ə }

olfactory bulb [VERT ZOO] The bulbous distal end of the olfactory tract located beneath each anterior lobe of the cerebrum; well developed in lower vertebrates. { äl'fak·trē ,bəlb }

olfactory cell [NEUROSCI] One of the sensory nerve cells in the olfactory epithelium. { äl'fak·trē ,sel }

olfactory foramen [ANAT] Any of the openings in the cribriform plate of the ethmoid bone through which pass the fila olfactoria of the olfactory nerves. { äl'fak·trē fə'rā·mən }

olfactory gland [PHYSIO] A type of serous gland in the nasal mucous membrane. { äl'fak·trē ,gland }

olfactory lobe [VERT ZOO] A lobe projecting forward from the inferior surface of the frontal lobe of each cerebral hemisphere, including the olfactory bulb, tracts, and trigone; well developed in most vertebrates, but reduced in humans. { äl 'fak·trē ,lōb }

olfactory nerve [NEUROSCI] The first cranial nerve; a paired sensory nerve with its origin in the olfactory lobe and formed by processes of the olfactory cells which lie in the nasal mucosa; greatly reduced in humans. { äl'fak·trē ,nərv }

olfactory organ [PHYSIO] Any of the small chemoreceptors in the mucous membrane lining the upper part of the nasal gravity which receive stimuli interpreted as odors. { äl'fak·trē ,ȯr·gən }

olfactory pit [EMBRYO] A depression near the olfactory placode in the embryo that develops into part of the nasal cavity. Also known as nasal pit. { äl'fak·trē ,pit }

olfactory region [ANAT] The area on and above the superior conchae and on the adjoining nasal septum where the mucous membrane has olfactory epithelium and olfactory glands. { äl'fak·trē ,rē·jən }

olfactory stalk [ANAT] The structure that connects the olfactory bulb to the cerebrum of the vertebrate brain. { äl'fak·trē ,stȯk }

olfactory tract [NEUROSCI] A narrow tract of white nerve fibers originating in the olfactory bulb and extending posteriorly to the anterior perforated substance, where it enlarges to form a lateral root (olfactory trigone). { äl'fak·trē ,trakt }

oligemia [MED] A state in which the total blood volume is reduced. { ,äl·ə'gē·mē·ə }

Oligobrachiidae [INV ZOO] A monotypic family of the order Athecanephria. { ,äl·ə·gō·brə'kī·ə,dē }

Oligochaeta [INV ZOO] A class of the phylum Annelida including worms that exhibit both external and internal segmentation, and setae which are not borne on parapodia. { ,äl·ə·gō'kēd·ə }

oligochromemia *See* anemia. { ,äl·ə·gō·krə'mē·mē·ə }

oligocythemia [MED] A reduction in the total number of red blood cells in the body. { ,äl·ə·gō,sī'thē·mē·ə }

oligodendrocyte [NEUROSCI] Glial cell responsible for elaborating myelin in the central nervous system. { äl·i·gō'den·drə,sīt }

oligodendroglia [NEUROSCI] Small neuroglial cells with spheroidal or ovoid nuclei and fine cytoplasmic processes with secondary divisions. { ,äl·ə·gō·den'dräg·lē·ə }

oligodendroglioma [MED] A slowly growing, large, well-defined cerebral glioma, composed of small cells with richly chromatic nuclei and scanty, poorly staining cytoplasm. { ,äl·ə·gō·den,dräg·lē'ō·mə }

oligodontia *See* hypodontia. { ō,lig·ə'dän·chə }

oligodynamic action [MICROBIO] The inhibiting or killing of microorganisms by use of very small amounts of a chemical substance. { ¦äl·ə·gō·dī'nam·ik 'ak·shən }

oligogene [GEN] A gene that encodes segments of multiple structural genes and causes major phenotypic alterations when mutated.

oligomenorrhea [MED] Abnormally infrequent menstruation. { ,äl·ə·gō,men·ə'rē·ə }

Oligomera [INV ZOO] A subphylum of the phylum Vermes comprising groups with two or three coelomic divisions. { ,äl·ə'gäm·ə·rə }

oligomeric protein [BIOCHEM] A protein composed of two or more polypeptide chains. { ə,lig·ə'mer·ik 'prō,tēn }

oligomerous [BOT] Having one or more whorls with fewer members than other whorls of the flower. { ,äl·ə'gäm·ə·rəs }

oligomycin [MICROBIO] Any of a group of antifungal antibiotics produced by an actinomycete resembling *Streptomyces diastachromogenes*; the colorless, hexagonal crystals are soluble in many organic solvents. { ə,lig·ə'mīs·ən }

oligonucleotide [BIOCHEM] A polynucleotide of low molecular weight, consisting of less than 20 nucleotide polymers. { ,äl·ə·gō'nü·klē·ə,tīd }

oligophagous [ZOO] Eating only a limited variety of foods. { ¦äl·ə¦gäf·ə·gəs }

oligosaccharide [BIOCHEM] A sugar composed of two to eight monosaccharide units joined by glycosidic bonds. Also known as compound sugar. { ,äl·ə·gō'sak·ə,rīd }

oligospermia [MED] Scarcity of spermatozoa in the semen. { ¦äl·ə·gō'spər·mē·ə }

Oligotrichida [INV ZOO] A minor order of the Spirotrichia; the body is round in cross section, and the adoral zone of membranelles is often highly developed at the oral end of the organism. { ¦äl·ə·gō'trik·ə·də }

oliguria [MED] Diminished excretion of urine. { 'äl·ə,gyůr·ē·ə }

olivaceous [BIOL] **1.** Resembling an olive. **2.** Olive colored. { ,äl·ə'vā·shəs }

olivary nucleus [ANAT] A prominent, convoluted gray band that opens medially and occupies the upper two-thirds of the medulla oblongata. { 'äl·ə,ver·ē 'nü·klē·əs }

olive [BOT] Any plant of the genus *Olea* in the order Schrophulariales, especially the evergreen olive tree (*O. europea*) cultivated for its drupaceous fruit, which is eaten ripe (black olives) and unripe (green), and is of high oil content. { 'äl·əv }

Olividae [INV ZOO] A family of snails in the gastropod order Neogastropoda. { ō'liv·ə,dē }

Ollier's disease *See* enchondromatosis. { ōl'yāz di,zēz }

omasum [VERT ZOO] The third chamber of the ruminant stomach where the contents are mixed to a more or less homogeneous state. Also known as manyplies; psalterium. { ō'mā·səm }

ombrophilous [ECOL] Able to thrive in areas of abundant rainfall. { äm'bräf·ə·ləs }

ombrophobous [ECOL] Unable to live in the presence of long, continuous rain. { äm'bräf·ə·bəs }

omentum [ANAT] A fold of the peritoneum connecting or supporting abdominal viscera. { ō'ment·əm }

ommatidium [INV ZOO] The structural unit of a compound eye, composed of a cornea, a crystalline cone, and a receptor element connected to the optic nerve. { ,äm·ə'tid·ē·əm }

ommatophore [INV ZOO] A movable peduncle that bears an eye, as in snails. { ə'mad·ə,fȯr }

omnivore [ZOO] An organism that eats both animal and vegetable matter. { 'äm·nə,vȯr }

omohyoid [ANAT] **1.** Pertaining conjointly to the scapula and the hyoid bone. **2.** A muscle attached to the scapula and the hyoid bone. { ¦ō·mō'hī,ȯid }

Omophronidae [INV ZOO] The savage beetles, a small family of coleopteran insects in the suborder Adephaga. { ,o·mə'frän·ə,dē }

omphalitis [MED] Inflammation of the umbilicus. { ,äm·fə'līd·əs }

omphalomesenteric artery *See* vitelline artery. { ¦äm·fə·lō,mez·ən'ter·ik 'ärd·ə·rē }

omphalomesenteric duct *See* vitelline duct. { ¦äm·fə·lō,mez·ən'ter·ik 'dəkt }

omphalomesenteric vein *See* vitelline vein. { ¦äm·fə·lō,mez·ən'ter·ik 'vān }

omphaloproptosis [MED] Abnormal protrusion of the navel. { ¦äm·fə·lō·präp'tō·səs }

Omphralidae [INV ZOO] A family of orthorrhaphous dipteran insects in the series Nematocera. { äm'fral·ə,dē }

Onagraceae [BOT] A family of dicotyledonous plants in the order Myrtales characterized generally by an inferior ovary, axile placentation, twice as many stamens as petals, a four-nucleate embryo sac, and many ovules. { ,än·ə'grās·ē,ē }

Onchidiidae [INV ZOO] An intertidal family of sluglike pulmonate mollusks of the order Systellommatophora in which the body is oval or lengthened, with the convex dorsal integument lacking a mantle cavity or shell. { ,äŋ·kə'dī·ə,dē }

onchocerciasis [MED] Infection with the filaria *Onchocerca volvulus*; results in skin tumors, papular dermatitis, and ocular complications. { ,äŋ·kō·sər'kī·ə·səs }

onchogryposis [MED] A thickened, ridged, and curved condition of a nail. { ,äŋ·kō·grə'pō·səs }

Oncholaimoidea [INV ZOO] A superfamily of nematodes in the order Enoplida, characterized by a stoma armed with one dorsal tooth and two subventral teeth and sometimes with transverse rows of small denticles from its walls, and two whorls of cephalic sensilla. { ,äŋ·kō·lə'mȯid·ē·ə }

oncholysis [MED] A slow process of loosening of a nail from its bed, beginning at the free edge and progressing gradually toward the root. { äŋ'käl·ə·səs }

onchomycosis [MED] Any fungus disease of the nail. { ,äŋ·kō·mī'kō·səs }

onchosphere [INV ZOO] The hexacanth embryo identified as the earliest differentiated stage of cyclophyllidean tapeworms. { 'äŋ·kō,sfir }

oncocyte [HISTOL] A columnar-shaped cell with finely granular eosinophilic cytoplasm, found in salivary and certain endocrine glands, nasal mucosa, and other locations. { 'äŋ·kō,sīt }

oncocytoma [MED] A benign tumor composed principally of oncocytes; usually occurs in salivary glands. { ,äŋ·kō·sī'tō·mə }

oncofetal antigen [IMMUNOL] Any of a group of antigens that are commonly present both in fetal tissue during early development of life and in adult tissue when cancer occurs. { ¦aŋk·ə,fēd·əl 'ant·i·jən }

oncogene [GEN] A gene whose mutation can lead to cancer in experimental animals and humans. { 'äŋ·kō,jēn }

oncogenesis [MED] Processes of tumor formation. { ¦äŋ·kō'jen·ə·səs }

oncogenic virus [VIROL] A virus that transforms the infected cells so that they undergo uncontrolled proliferation. { ¦aŋ·kə,jen·ik 'vī·rəs }

oncology [MED] The study of the causes, development, characteristics, and treatment of tumors. { äŋ'käl·ə·jē }

oncolytic [MED] Pertaining to destruction of cancer cells. { ,äŋ·kə'lid·ik }

oncomouse [BIOL] A laboratory mouse that carries activated human cancer genes. { 'äŋk·ə,maůs }

Oncopoda [INV ZOO] A phylum of the superphylum Articulata. { äŋ'käp·ə·də }

Oncorhynchus [VERT ZOO] A genus of seven semelparous salmon species that occur naturally in the North Pacific Ocean and spawn in western

North America and coastal Asia. { ˌäŋ·kə'riŋ·kəs }

oncotic pressure [PHYSIO] Also known as colloidal osmotic pressure. **1.** The osmotic pressure exerted by colloids in a solution. **2.** The pressure exerted by plasma proteins. { äŋ'käd·ik 'presh·ər }

oncotomy [MED] Surgical incision of a tumor, abscess, or other swelling. { äŋ'käd·ə·mē }

Oncovirinae [VIROL] A subfamily of the Retroviridae family. { ˌäŋ·kō'vir·əˌnī }

onion [BOT] **1.** *Allium cepa*. A biennial plant in the order Liliales cultivated for its edible bulb. **2.** Any plant of the genus *Allium*. { 'ən·yən }

onisciform [INV ZOO] Ovate and slightly flattened. { ä'nis·əˌfórm }

Oniscoidea [INV ZOO] A terrestrial suborder of the Isopoda; the body is either dorsoventrally flattened or highly vaulted, and the head, thorax, and abdomen are broadly joined. { ˌän·ə'skóid·ē·ə }

ontogeny [EMBRYO] The origin and development of an organism from conception to adulthood. { än'täj·ə·nē }

Onuphidae [INV ZOO] A family of tubicolous, herbivorous, scavenging errantian annelids in the superfamily Eunicea. { än'yüf·əˌdē }

onych-, onycho- [ZOO] A combining form denoting claw or nail. { 'än·ə·kō }

onychia [MED] Inflammation of the nail matrix. { ō'nik·ē·ə }

onychomycosis [MED] A fungus disease of the nails. { ¦än·ə·kō·mī'kō·səs }

Onychopalpida [INV ZOO] A suborder of mites in the order Acarina. { ¦an·ə·kō'pal·pə·də }

Onychophora [INV ZOO] A phylum of wormlike animals that combine features of both the annelids and the arthropods. { ˌän·ə'käf·ə·rə }

onychorrhexis [MED] Longitudinal striation of the nail plate, with or without the formation of fissures. { ¦än·ə·kə'rek·səs }

Onygenaceae [MYCOL] A family of ascomycetous fungi in the order Eurotiales comprising forms that inhabit various animal substrata, such as horns and hoofs. { ˌän·ə·jə'nās·ēˌē }

oocyst [INV ZOO] The encysted zygote of some Sporozoa. { 'ō·əˌsist }

oocyte [HISTOL] An egg before the completion of maturation. { 'ō·əˌsīt }

oogamete [BIOL] A large, nonmotile female gamete containing reserve materials. { ¦ō·ə'gaˌmēt }

oogamous [BIOL] Of sexual reproduction, characterized by fusion of a motile sperm with an oogamete. { ō'äg·ə·məs }

oogenesis [PHYSIO] Processes involved in the growth and maturation of the ovum in preparation for fertilization. { ˌō·ə'jen·ə·səs }

oogonium [BOT] The unisexual female sex organ in oogamous algae and fungi. [HISTOL] A descendant of a primary germ cell which develops into an oocyte. { ˌō·ə'gō·nē·əm }

ookinete [INV ZOO] The elongated, mobile zygote of certain Sporozoa, as that of the malaria parasite. { ˌō·ə'kīˌnēt }

oolemma *See* zona pellucida. { ¦ō·ə'lem·ə }

oology [VERT ZOO] A branch of zoology concerned with the study of eggs, especially bird eggs. { ō'äl·ə·jē }

Oomycetes [MYCOL] A class of the Phycomycetes comprising the biflagellate water molds and downy mildews. { ˌō·ə·mī'sēd·ēz }

oophagous [ZOO] Feeding or living on eggs. { ō'äf·ə·gəs }

oophoritis [MED] Inflammation of the ovaries. { ˌō·ə·fə'rīd·əs }

ooplasm [CELL MOL] Cytoplasm of an egg. { 'ō·əˌplaz·əm }

oospore [BOT] A spore which is produced by heterogamous fertilization and from which the sporophytic generation develops. { 'ō·əˌspór }

oostegite [INV ZOO] In many crustaceans, a platelike expansion of the basal segment of a thoracic appendage that aids in forming an egg receptacle. { ō'äs·təˌjīt }

ootype [INV ZOO] In trematodes and tapeworms, a thickening of the oviduct near the ovaries. { 'ō·əˌtīp }

Opalinata [INV ZOO] A superclass of the subphylum Sarcomastigophora containing highly specialized forms which resemble ciliates. { o¦pal·ə¦näd·ə }

Opegraphaceae [BOT] A family of the Hysteriales characterized by elongated ascocarps; members are crustose on bark and rocks. { ˌō·pə·grə'fās·ēˌē }

open-angle glaucoma [MED] Bilateral, increased intraocular tension due to reduced aqueous outflow but with the angle open and the aqueous in free contact with the trabecula. { 'ō·pən ˌaŋ·gəl glau̇'kō·mə }

open bundle [BOT] A vascular bundle containing cambium. { 'ō·pən 'bən·dəl }

open-circle deoxyribonucleic acid *See* relaxed circular deoxyribonucleic acid. { 'ō·pən ˌsər·kəl dēˌäk·sēˌrī·bō·nü'klē·ik 'as·əd }

open community [ECOL] A community which other organisms readily colonize because some niches are unoccupied. { 'ō·pən kə'myü·nəd·ē }

open reading frame [CELL MOL] A stretch of triplets contained between an initator codon and a terminator codon. Abbreviated ORF. { ¦ō·pən ¦rēd·iŋ ¦frām }

open tuberculosis [MED] Tuberculosis in which tubercle bacilli are being discharged from the body; tuberculosis capable of transmission to other persons. { 'ō·pən təˌbər·kyə'lō·səs }

open water [ECOL] Lake water that is free from emergent vegetation, artificial obstructions, or tangled masses of underwater vegetation at very shallow depths. { 'ō·pən 'wód·ər }

operative ankylosis *See* arthrodesis. { 'äp·rəd·iv ˌaŋ·kə'lō·səs }

operator [GEN] A sequence at one end of an operon on which a repressor acts, thus regulating the transcription of the operon. { 'äp·əˌrād·ər }

operculum [ANAT] **1.** The soft tissue partially covering the crown of an erupting tooth. **2.** That part of the cerebrum which borders the lateral fissure. [BIOL] **1.** A lid, flap, or valve. **2.** A lidlike body process. { ō'pər·kyə·ləm }

operon [GEN] A functional unit composed of a number of adjacent cistrons on the chromosome; its transcription is regulated by a receptor sequence, the operator, and a repressor. { 'äp·ə,rän }

operon network [GEN] A group of operons and their associated regulator genes that interact such that the products of one operon activate or suppress another operon. { 'äp·ə,rän ,net,wərk }

Opheliidae [INV ZOO] A family of limivorous worms belonging to the annelid group Sedentaria. { ,äf·ə'lī·ə,dē }

Ophiacodonta [VERT ZOO] A suborder of extinct reptiles in the order Pelycosauria, including primitive, partially aquatic carnivores. { ¦af·ē·ə·kə'dän·tə }

Ophidiidae [VERT ZOO] A family of small actinopterygian fishes in the order Gadiformes, comprising the cusk eels and brotulas. { ,äf·ə'dī·ə,dē }

Ophiocanopidae [INV ZOO] A family of asterozoan echinoderms in the subclass Ophiuroidea. { ,äf·ē·ō·kə'näp·ə,dē }

Ophioglossales [BOT] An order of ferns in the subclass Ophioglossidae. { ,äf·ē·ō·glä'sā·lēz }

Ophioglossidae [BOT] The adder's-tongue ferns, a small subclass of the class Polypodiopsida; the plants are homosporous and eusporangiate and are distinguished by the arrangement of the sporogenous tissue in the characteristic fertile spike of the sporophyte. { ,äf·ē·ō'gläs·ə,dē }

Ophiomyxidae [INV ZOO] The single family of the echinoderm suborder Ophiomyxina distinguished by a soft, unprotected integument. { ,äf·ē·ō'mik·sə,dē }

Ophiomyxina [INV ZOO] A monofamilial suborder of ophiuroid echinoderms in the order Phrynophiurida. { ,äf·ē·ō'mik·sə·nə }

ophiopluteus [INV ZOO] The pluteus larva of brittle stars. { ¦äf·ē·ō'plüd·ē·əs }

Ophiurida [INV ZOO] An order of echinoderms in the subclass Ophiuroidea in which the vertebrae articulate by means of ball-and-socket joints, and the arms, which do not branch, move mainly from side to side. { ,äf·ə'yür·ə·də }

Ophiuroidea [INV ZOO] The brittle stars, a subclass of the Asterozoa in which the arms are usually clearly demarcated from the central disk and perform whiplike locomotor movements. { äf·ə·yə'röid·ē·ə }

ophthalmectomy [MED] Excision, or enucleation, of the eye. { ,äf·thəl'mek·tə·mē }

ophthalmia [MED] Inflammation of the eye, especially involving the conjunctiva. { äf'thal·mē·ə }

ophthalmia neonatorum [MED] Inflammation of the eyes in the newborn contracted during passage through the birth canal; may be gonorrheal or purulent. { äf'thal·mē·ə ,nē·ə·nə'tör·əm }

ophthalmia nodosa [MED] Inflammation of the eye due to lodging of caterpillar hairs in the conjunctiva, cornea, or iris. { äf'thal·mē·ə nō'dō·sə }

ophthalmic [ANAT] Of or pertaining to the eye. { äf'thal·mik }

ophthalmic nerve [NEUROSCI] A sensory branch of the trigeminal nerve which supplies the lacrimal glands, upper eyelids, skin of the forehead, and anterior portion of the scalp, meninges, nasal mucosa, and frontal, ethmoid, and sphenoid air sinuses. { äf'thal·mik 'nərv }

ophthalmodynamometer [MED] An instrument which measures the pressure necessary to collapse the retinal arteries. { äf¦thal·mō,dī·nə'mäm·əd·ər }

ophthalmological [PHARM] Any drugs used in the treatment of eye disease. { äf¦thal·mə¦läj·ə·kəl }

ophthalmologist [MED] A physician who specializes in ophthalmology. Also known as oculist. { ,äf,thal'mäl·ə·jəst }

ophthalmology [MED] The study of the anatomy, physiology, and diseases of the eye. { ,äf ,thal'mäl·ə·jē }

ophthalmomalacia [MED] Abnormal softness or subnormal tension of the eye. { äf¦thal·mə¦lā·shə }

ophthalmorrhexis [MED] Rupture of the eyeball. { äf¦thal·mə¦rek·səs }

opiate [PHARM] **1.** A sleep-inducing drug. **2.** Any narcotic. **3.** An opium preparation. **4.** Any tranquilizing agent. { 'ō·pē·ət }

Opilioacaridae [INV ZOO] The single family of moderately large mites of the suborder Notostigmata which comprises the Opiliocariformes. { ō¦pil·ē·ō'kar·ə,dē }

Opilioacariformes [INV ZOO] A small monofamilial order of the Acari comprising large mites characterized by long legs and by the possession of a pretarsus on the pedipalp, with prominent claws. { ō¦pil·ē·ō,kar·ə'fór·mēz }

opine [BIOCHEM] A type of amino acid usually not found in nature, such as that secreted by a crown gall. { 'ō,pīn }

opisthaptor [INV ZOO] A posterior adhesive organ in monogenetic trematodes. { ¦äp·əs¦thap·tər }

Opisthobranchia [INV ZOO] A subclass of the class Gastropoda containing the sea hares, sea butterflies, and sea slugs; generally characterized by having gills, a small external or internal shell, and two pairs of tentacles. { ə,pis·thə'braŋ·kē·ə }

Opisthocoela [VERT ZOO] A suborder of the order Anura; members have opisthocoelous trunk vertebrae, and the adults typically have free ribs. { ə,pis·thə'sē·lə }

opisthocoelous [ANAT] Of, related to, or being a vertebra with the centrum convex anteriorly and concave posteriorly. { ə¦pis·thə¦sē·ləs }

Opisthocomidae [VERT ZOO] A family of birds in the order Galliformes, including the hoatzins. { ə,pis·thə'käm·ə,dē }

opisthognathous [INV ZOO] Having the mouthparts ventral and posterior to the cranium.

[VERT ZOO] Having retreating jaws. { ¦ä·pəs¦thäg·nə·thəs }

opisthonephros [VERT ZOO] The fundamental adult kidney in amphibians and fishes. { ə‚pis·thə'ne‚fräs }

Opisthopora [INV ZOO] An order of the class Oligochaeta distinguished by meganephridiostomal, male pores opening posteriorly to the last testicular segment. { ¦ä·pəs¦thäp·ə·rə }

opisthotic [ANAT] Of, relating to, or being the posterior and inferior portions of the bony elements in the inner ear capsule. { ¦ä·pəs ¦thäd·ik }

opisthotonus [MED] A condition, caused by a tetanic spasm of the back muscles, in which the trunk is arched forward while the head and lower limbs are bent backward. { ‚ap·əs'thät·ən·əs }

opium [PHARM] A narcotic obtained from the unripe capsules of the opium poppy (*Papaver somniferum*); crude extract contains alkaloids such as morphine (5–15%), narcotine (2–8%), and codeine (0.1–2.5%). { 'ō·pē·əm }

Opomyzidae [INV ZOO] A family of cyclorrhaphous myodarian dipteran insects in the subsection Acalypteratae. { ‚äp·ə'mī·zə‚dē }

opossum [VERT ZOO] Any member of the family Didelphidae in the order Marsupialia; these mammals are arboreal and mainly omnivorous, and have many incisors, with all teeth pointed and sharp. { ə'päs·əm }

Oppenheim's disease *See* amyotonia congenita. { 'äp·ən‚hīmz di‚zēz }

opportunistic microorganism [MICROBIO] A normally harmless endogenous microorganism that produces disease due to fortuitous events that affect the host. { ¦äp·ər‚tü¦nis·tik ¦mī·krō'ór·gə‚niz·əm }

opportunistic species [ECOL] Species characterized by high reproduction rates, rapid development, early reproduction, small body size, and uncertain adult survival. { ¦äp·ər‚tü¦nis·tik 'spē·shēz }

opposite [BOT] **1.** Located side by side. **2.** Of leaves, being in pairs on an axis with each member separated from the other of the pair by half the circumference of the axis. { 'äp·ə·zət }

opsonic action [IMMUNOL] The effect produced upon susceptible microorganisms and other cells by opsonins, which renders them vulnerable to phagocytes. { äp'sän·ik 'ak·shən }

opsonic index [IMMUNOL] A numerical measure of the opsonic activity of sera, expressed as the ratio of the average number of bacteria engulfed per phagocytic cell in immune serum compared with the corresponding value for normal serum. { äp'sän·ik 'in‚deks }

opsonin [IMMUNOL] A substance in blood serum that renders bacteria more susceptible to phagocytosis by leukocytes. { 'äp·sə·nən }

opsonize [IMMUNOL] To render microorganisms susceptible to phagocytosis. { 'äp·sə‚nīz }

-opsy [MED] A combining form denoting examination, or denoting a condition of vision. { 'äp·sē }

optic [BIOL] Pertaining to the eye. { 'äp·tik }

optical axis [ANAT] An imaginary straight line passing through the midpoint of the cornea (anterior pole) and the midpoint of the retina (posterior pole). { 'äp·tə·kəl 'ak·səs }

optical-righting reflex *See* visual-righting reflex. { 'äp·tə·kəl ¦rīd·iŋ ‚rē‚fleks }

optic apraxia [MED] A form of apraxia in which the individual fails to represent spatial relations correctly in drawing or construction by other means. Also known as constructional apraxia. { 'äp·tik ā'prak·sē·ə }

optic canal [ANAT] The channel at the apex of the orbit, the anterior termination of the optic groove, just beneath the lesser wing of the sphenoid bone; it gives passage to the optic nerve and ophthalmic artery. { 'äp·tik kə'nal }

optic capsule [VERT ZOO] A cartilaginous capsule that develops around the eye in elasmobranchs and higher vertebrate embryos. { 'äp·tik 'kap·səl }

optic chiasma [NEUROSCI] The partial decussation of the optic nerves on the undersurface of the hypothalamus. { 'äp·tik kī'az·mə }

optic cup [EMBRYO] A two-layered depression formed by invagination of the optic vesicle from which the pigmented and sensory layers of the retina will develop. { 'äp·tik ‚kəp }

optic disk [NEUROSCI] The circular area in the retina that is the site of the convergence of fibers from the ganglion cells of the retina to form the optic nerve. { 'äp·tik ‚disk }

optic gland [INV ZOO] Either of a pair of endocrine glands in the octopus and squid which are found near the brain and produce a substance which causes gonadal maturation. { 'äp·tik ‚gland }

optic lobe [INV ZOO] A lateral lobe of the forebrain in certain arthropods. [NEUROSCI] One of the anterior pair of colliculi of the mammalian corpora quadrigemina. [VERT ZOO] Either of the corpora bigemina of lower vertebrates. { 'äp·tik ‚lōb }

optic nerve [NEUROSCI] The second cranial nerve; a paired sensory nerve technically consisting of three layers of special nerve cells in the retina of the eye; fibers converge to form the optic tracts. { 'äp·tik ‚nərv }

optic stalk [EMBRYO] The constriction of the optic vesicle which connects the embryonic eye and forebrain in vertebrates. { 'äp·tik ‚stók }

optic tectum [VERT ZOO] The roof of the mesencephalon constituting a major visual center and association area of the brain of premature vertebrates. { 'äp·tik 'tek·təm }

optic tract [NEUROSCI] The band of optic nerve fibers running from the optic chiasma to the lateral geniculate body and midbrain. { 'äp·tik ‚trakt }

optic vesicle [EMBRYO] An evagination of the lateral wall of the forebrain in vertebrate embryos which precedes formation of the optic cup. { 'äp·tik 'ves·ə·kəl }

optometrist [MED] One who measures the degrees of visual powers, without the aid of cycloplegic or mydriatic agents. { äp'täm·ə,trist }

optometry [MED] Measurement of visual powers. { äp'täm·ə·trē }

oral [ANAT] Of or pertaining to the mouth. { 'ȯr·əl }

oral and maxillofacial surgery [MED] A branch of dentistry that treats diseases and abnormalities of the maxillofacial region by surgical means. { ¦ȯr·əl and ,ak·sə·lō¦fā·shəl 'sər·jə·rē }

oral arm [INV ZOO] In a jellyfish, any of the prolongations of the distal end of the manubrium. { 'ȯr·əl ¦ärm }

oral cavity [ANAT] The cavity of the mouth. { 'ȯr·əl 'kav·əd·ē }

oral contraceptive [PHARM] Any medication taken by mouth that renders a woman nonfertile as long as the medication is continued. { 'ȯr·əl ,kän·trə'sep·tiv }

oral disc [INV ZOO] The flattened upper or free end of the body of a polyp that has the mouth in the center and tentacles around the margin. { 'ȯr·əl ¦disk }

oral groove [INV ZOO] A depressed, groovelike peristome. { 'ȯr·əl ¦grüv }

oral pathology [MED] A branch of dentistry that is concerned with the diseases of the teeth, oral cavity, and jaws, and with the oral manifestations of systemic diseases. { ,ȯr·əl pə'thäl·ə·jē }

orange [BOT] Any of various evergreen trees of the genus *Citrus*, cultivated for the edible fruit, a berry with an aromatic, leathery rind containing numerous oil glands. { 'är·inj }

orangeophile [HISTOL] A type of acidophile cell of the anterior lobe of the adenohypophysis, presumed to elaborate growth hormone. { ə'ran·jē·ə,fīl }

orangutan [VERT ZOO] *Pongo pygmaeus*. The largest of the great apes, a long-armed primate distinguished by long sparse reddish-brown hair, naked face and hands and feet, and a large laryngeal cavity which appears as a pouch below the chin. { ə'raŋ·ü,tan }

Orbiniidae [INV ZOO] A family of polychaete annelids belonging to the Sedentaria; the prostomium is exposed, and the thorax and abdomen are weakly separated. { ,ȯr·bə'nī·ə,dē }

Orbiniinae [INV ZOO] A subfamily of sedentary polychaete annelids in the family Orbiniidae. { ȯr·bə'nī·ə,nē }

orbit [ANAT] The bony cavity in the lateral front of the skull beneath the frontal bone which contains the eyeball. Also known as eye socket. { 'ȯr·bət }

orbital fossa [INV ZOO] A depression from which the eyestalk arises on the front of the carapace of crustaceans. { 'ȯr·bəd·əl 'fäs·ə }

Orbitoidacea [INV ZOO] A superfamily of foraminiferan protozoans in the suborder Rotaliina characterized by a low trochospire or a planispiral, uncoiled or branching test composed of radial calcite with bilamellar septa. { ,ȯr·bə·tȯi'dās·ē,ə }

Orbivirus [VIROL] A genus in the family Reoviridae that is the causative agent of bluetongue. { 'ȯrb·ə,vī·rəs }

orch-, orchi-, orchid-, orchido-, orchio- [ZOO] A combining form denoting testis. { 'ȯr·kē, 'ȯr·kəd·ō, 'ȯr·kē·ō }

orchid [BOT] Any member of the family Orchidaceae; plants have complex, specialized irregular flowers usually with only one or two stamens. { 'ȯr·kəd }

Orchidaceae [BOT] A family of monocotyledonous plants in the order Orchidales characterized by irregular flowers with only one or two stamens which are adnate to the style, and pollen grains which cohere in large masses called pollinia. { ,ȯr·kə'dās·ē,ē }

Orchidales [BOT] An order of monocotyledonous plants in the subclass Liliidae; plants are mycotropic and sometimes nongreen with numerous tiny seeds that have an undifferentiated embryo and little or no endosperm. { ,ȯr·kə'dā·lēz }

orchid mycorrhizal fungi [MYCOL] Mycorrhizal fungi that are characterized by the absence of a Hartig net and a fungal sheath and the presence of hyphal coils in the root cells. { ¦ȯr·kid ,mī·kə,rīz·əl 'fən,jī }

orchiectomy [MED] Surgical removal of one or both testes. { ,ȯr·kē'ek·tə·mē }

orchiopexy [MED] Surgical fixation of a testis. { 'ȯr·kē·ō,pek·sē }

order [SYST] A taxonomic category ranked below the class and above the family, made up either of families, subfamilies, or suborders. { 'ȯrd·ər }

ordered octad [MYCOL] A linear sequence of pairs of each of four haploid cells produced by a postmeiotic division within a fungal ascus. { 'ȯrd·ərd 'äk,tad }

ordered tetrad [MYCOL] A linear sequence of four haploid meiotic cells within a fungal ascus. { 'ȯrd·ərd 'te,trad }

Orectolobidae [VERT ZOO] An ancient isurid family of galeoid sharks, including the carpet and nurse sharks, which are primarily bottom feeders with small teeth and a blunt rostrum with barbels near the mouth. { ȯ,rek·tə'läb·ə,dē }

ORF *See* open reading frame. { ȯrf *or* ¦ō¦är'ef }

organ [ANAT] A differentiated structure of an organism composed of various cells or tissues and adapted for a specific function. { 'ȯr·gən }

organelle [CELL MOL] A specialized subcellular structure, such as a mitochondrion, having a special function; a condensed system showing a high degree of internal order and definite limits of size and shape. { ¦ȯr·gə¦nel }

organic brain syndrome [MED] A mental condition of multiple etiologies, resulting in diffuse impairment of brain tissue function and manifested by a complex of symptoms including impaired judgment and intellectual function, and often somatic and motor dysfunctions. { ȯr 'gan·ik 'brān ,sin,drōm }

organic evolution [EVOL] The processes of change in organisms by which descendants come to differ from their ancestors, and a history of the sequence of such changes. { ȯr'gan·ik ,ev·ə'lü·shən }

organicism *See* holism. { ȯr'gan·ə,siz·əm }

organism [BIOL] An individual constituted to carry out all life functions. { 'ȯr·gə,niz·əm }

organized ferment *See* intracellular enzyme. { ,ȯr·gə·nə'zā·shən 'fər·ment }

organizer [EMBRYO] Any part of the embryo which exerts a morphogenetic stimulus on an adjacent part or parts, as in the induction of the medullary plate by the dorsal lip of the blastopore. { 'ȯr·gə,niz·ər }

organizing pneumonia [MED] Pneumonia in which the healing process is characterized by organization and cicatrization of the exudate rather than by resolution and resorption. Also known as unresolved pneumonia. { 'ȯr·gə,nīz·iŋ nu'mō·nyə }

organ of Corti [NEUROSCI] A specialized structure located on the basilar membrane of the mammalian cochlea, which contains rods of Corti and hair cells connected to ganglia of the cochlear nerve. Also known as spiral organ. { 'ȯr·gən əv 'kȯrd·ē }

organ of Leydig [VERT ZOO] Two large accumulations of lymphoid tissues which run longitudinally the length of the esophagus in selachian fishes. { 'ȯr·gən əv 'lī,dig }

organogenesis [EMBRYO] The formation of an organ. { ȯr,gan·ə'jen·ə·səs }

organoleptic [PHYSIO] Having an effect or making an impression on sense organs; usually used in connection with subjective testing of food and drug products. { ȯr¦gan·ə¦lep·tik }

organotropic [MICROBIO] Of microorganisms, localizing in or entering the body by way of the viscera or, occasionally, somatic tissue. { ȯr ¦gan·ə¦träp·ik }

organs of Zuckerkandl *See* aortic paraganglion. { 'ȯr·gənz əv 'tsu̇k·ər,känd·əl }

orgasm [PHYSIO] The intense, diffuse, and subjectively pleasurable sensation experienced during sexual intercourse or genital manipulation, culminating in the male with seminal ejaculation and in the female with uterine contractions, warm suffusion, and pelvic throbbing sensations. { 'ȯr,gaz·əm }

orgasmolepsy [MED] Sudden loss of muscle tone during orgasm, accompanied by a transitory loss of consciousness. { ȯr¦gaz·mə¦lep·sē }

Oribatei [INV ZOO] A heavily sclerotized group of free-living mites in the suborder Sarcoptiformes which serve as intermediate hosts of tapeworms. { ,ȯr·ə'bad·ē,ī }

Oribatulidae [INV ZOO] A family of oribatid mites in the suborder Sarcoptiformes. { ,ȯr·ə·bə'tül·ə,dē }

Oriental zoogeographic region [ECOL] A zoogeographic region which encompasses tropical Asia from the Iranian Peninsula eastward through the East Indies to, and including, Borneo and the Philippines. { ,ȯr·ē'ent·əl ,zō·ə,jē·ə'graf·ik ,rē·jən }

origin [ANAT] The point at which the nonmoving end of a muscle is attached to a bone; it is at the proximal end of the muscle. { 'är·ə·jən }

origin of replication [CELL MOL] The nucleotide sequence from which deoxyribonucleic acid replication begins. { ¦är·ə·jən əv ,rep·lə'kā·shən }

Orleans process [MICROBIO] An older commercial method of vinegar production in which fermentation is carried out in a large cask in which holes have been drilled to permit the introduction of air. The cask has a spigot for the withdrawal of finished vinegar, and the bung (stopper) contains a tube so that fresh wine or other substrate can be added without disturbing the film of vinegar bacteria. { ¦ȯr·lē·ənz ,prä,ses }

ormer *See* abalone. { 'ȯr·mər }

Ormyridae [INV ZOO] A small family of hemipteran insects in the superfamily Chalcidoidea. { ȯr'mī·rə,dē }

Orneodidae [INV ZOO] A small family of lepidopteran insects in the superfamily Tineoidea; adults have each wing divided into six featherlike plumes. { ,ȯr·nē'äd·ə,dē }

ornithine [BIOCHEM] $C_5H_{12}O_2N_2$ An amino acid occurring in the urine of some birds, but not found in native proteins. { 'ȯr·nə,thēn }

ornithine cycle [BIOCHEM] A sequence of cyclic reactions in which potentially toxic products of protein catabolism are converted to nontoxic urea. { 'ȯr·nə,thīn ,sī·kəl }

ornithology [VERT ZOO] The study of birds. { ,ȯr·nə'thäl·ə·jē }

Ornithorhynchidae [VERT ZOO] A monospecific order of monotremes containing the semiaquatic platypus; characterized by a duck-billed snout, horny plates instead of teeth in the adult, and a flattened, well-developed tail. { ,ȯr·nə·thō'riŋ·kə,dē }

ornithosis [MED] Any form of psittacosis originating in birds other than psittacines. { ,ȯr·nə'thō·səs }

Oromericidae [VERT ZOO] An extinct family of camellike tylopod ruminants in the superfamily Cameloidea. { ,ȯr·ə·mə'ris·ə,dē }

oropharynx [ANAT] The oral pharynx, located between the lower border of the soft palate and the larynx. { ¦ȯr·ō'far,iŋks }

orophyte [ECOL] Any plant that grows in the subalpine region. { 'ȯr·ə,fīt }

orotic acid [BIOCHEM] $C_4H_4O_4N_2$ A crystalline acid which is a growth factor for certain bacteria and is also a pyrimidine precursor. { ə'räd·ik 'as·əd }

Oroya fever [MED] The severe form of Carrion's disease, characterized by a sudden, severe, and rapid course, often fatal anemia, and remittent fever. { ə'rȯi·ə ,fē·vər }

orphan drug [PHARM] A pharmaceutical developed to treat a disease that afflicts relatively few people. { ¦ȯr·fən 'drəg }

orphan virus [VIROL] Any nonpathogenic virus found in the human digestive and respiratory tracts. { ¦ȯr·fən 'vī·rəs }

Ortheziinae [INV ZOO] A subfamily of homopteran insects in the superfamily Coccoidea having abdominal spiracles present in all stages and a flat anal ring bearing pores and setae in immature forms and adult females. { ˌȯr·thə'zī·əˌnē }

orthoceratite [INV ZOO] Any nautiloid belonging to the genus *Orthoceras*, characterized by the presence of three longitudinal furrows on the body chamber. { ˌȯr·thə'ser·əˌtīt }

orthochromatic [BIOL] Having normal staining characteristics. { ˌȯr·thə·krə'mad·ik }

orthodontics [MED] A branch of dentistry that deals with the prevention and treatment of malocclusion. { ˌȯr·thə'dän·tiks }

orthogenesis [EVOL] A unidirectional evolutionary change among a related group of animals. { ˌȯr·thə'jen·ə·səs }

orthokeratology [MED] A procedure designed to reduce or eliminate refractive anomalies and binocular dysfunctions of the eye by the programmed application of contact lenses. { ¦ȯr·thōˌker·ə'täl·ə·jē }

orthokinesis [BIOL] Random movement of a motile cell or organism in response to a stimulus. { ¦ȯr·thə·ki'nē·səs }

orthologous locus [GEN] A gene that evolved without diverging from an ancestral locus. { ȯr'thäl·ə·gəs ˌlō·kəs }

Orthomyxoviridae [VIROL] A family of negative-strand ribonucleic acid viruses characterized by enveloped, spherical, pleomorphic virions with a helical nucleocapsid containing a fragmented genome; it includes the genus *Influenzavirus* (human influenza type A). { ˌȯr·thəˌmik·sə'vir·əˌdī }

Orthonectida [INV ZOO] An order of Mesozoa; orthonectids parasitize various marine invertebrates as multinucleate plasmodia, and sexually mature forms are ciliated organisms. { ˌȯr·thə'nek·tə·də }

orthopedics [MED] The branch of surgery concerned with corrective treatment of musculoskeletal deformities, diseases, and ailments by manual and instrumental measures. { ˌȯr·thə'pēd·iks }

Orthoperidae [INV ZOO] The minute fungus beetles, a family of coleopteran insects in the superfamily Cucujoidea. { ˌȯr·thə'per·əˌdē }

orthopnea [MED] A condition in which there is difficulty in breathing except when sitting or standing upright. { ȯr'thäp·nē·ə }

Orthopsida [INV ZOO] An order of echinoderms in the subclass Euechinoidea. { ȯr'thäp·sə·də }

Orthoptera [INV ZOO] A heterogeneous order of generalized insects with gradual metamorphosis, chewing mouthparts, and four wings. { ȯr'thäp·tə·rə }

Orthorrhapha [INV ZOO] A suborder of the Diptera; in this group of flies, the adult escapes from the puparium through a T-shaped opening. { or'thȯr·ə·fə }

orthoscope [MED] **1.** An instrument for examination of the eye through a layer of water, whereby the curvature and hence the refraction of the cornea is neutralized. **2.** An instrument used in drawing the projections of skulls. { 'ȯr·thəˌskōp }

orthosis [MED] A device applied to a human limb to control or enhance movement or to prevent bone movement or deformity, for example, a splint or an arch support. { ȯr'thōs·əs }

orthostatic [MED] Pertaining to or caused by standing upright. { ¦ȯr·thə¦stad·ik }

orthotonus [MED] Tetanic muscle spasm in which the body assumes a posture of rigid straightness. { ȯr'thät·ən·əs }

Orthotrichales [BOT] An order of true mosses in the subclass Bryidae, characterized by dull, tuft- or mat-forming plants that are probably heterogeneous, making a generalized description difficult. { ˌȯr·thō·trə'kā·lēz }

orthotropism [BOT] The tendency of a plant to grow with the longer axis oriented vertically. { ȯr'thä·trəˌpiz·əm }

orthotropous [BOT] Having a straight ovule with the micropyle at the end opposite the stalk. { ȯr'thä·trə·pəs }

Orussidae [INV ZOO] A small family of hymenopteran insects in the superfamily Siricoidea. { ȯ'rüs·əˌdē }

osage orange [BOT] *Maclura pomifera*. A tree in the mulberry family of the Urticales characterized by yellowish bark, milky sap, simple entire leaves, strong axillary thorns, and aggregate green fruit about the size and shape of an orange. { 'ōˌsāj 'är·inj }

osazone [BIOCHEM] Any of the compounds that contain two phenylhydrazine residues and are produced by a reaction between a reducing sugar and phenylhydrazine. { 'ō·səˌzōn }

Oscillatoriales [BOT] An order of blue-green algae (Cyanophyceae) which are filamentous and truly multicellular. { ¦äs·ə·ləˌtȯr·ē'ā·lēz }

Oscillospiraceae [MICROBIO] Formerly a family of large, gram-negative, motile bacteria of the order Caryophanales which lose motility on exposure to oxygen. { ˌäs·ə·lō·spə'rās·ēˌē }

Oscines [VERT ZOO] The songbirds, a suborder of the order Passeriformes. { 'äs·əˌnēz }

osculum [INV ZOO] An excurrent orifice in Porifera. { 'äs·kyə·ləm }

Osgood-Schlatter disease *See* osteochondrosis. { 'äzˌgu̇d 'shlad·ər diˌzēz }

Osler-Rendu-Weber disease *See* hereditary hemorrhagic telangiectasia. { 'ōs·lər 'rän·dü 'web·ər diˌzēz }

osmophile [MICROBIO] A microorganism adapted to media with high osmotic pressure. { 'äz·məˌfīl }

osmophore [BOT] A particular flower part specialized for odor production. { 'äz·məˌfȯr }

osmoreceptor [PHYSIO] One of a group of structures in the hypothalamus which respond to changes in osmotic pressure of the blood by regulating the secretion of the neurohypophyseal antidiuretic hormone. { ¦äz·mō·ri'sep·tər }

osmoregulatory mechanism [PHYSIO] Any physiological mechanism for the maintenance of an optimal and constant level of osmotic activity of the fluid in and around the cells. { ¦äz·mō'reg·yə·lə,tȯr·ē 'mek·ə,niz·əm }

osmotic diarrhea [MED] Diarrhea resulting from a rise in the osmotic pressure of fecal contents, diminishing the absorption of water by the intestine, it abates with fasting. { äz,mäd·ik ,dī·ə'rē·ə }

osmotic fragility [PHYSIO] Susceptibility of red blood cells to lyses when placed in dilute (hypotonic) salt solutions. { äz'mäd·ik frə'jil·əd·ē }

osmotic shock [PHYSIO] The bursting of cells suspended in a dilute salt solution. { äz'mäd·ik 'shäk }

osmotolerance [PHYSIO] The ability to withstand high solute concentrations. { ,äz·mō'täl·ə·rəns }

osphradium *See* asphradium. { äs'frād·ē·əm }

os priapi *See* baculum. { 'äs prī'ä·pē }

osseous [ANAT] Bony; composed of or resembling bone. { 'äs·ē·əs }

osseous system [ANAT] The skeletal system of the body. { 'äs·ē·əs ¦sis·təm }

osseous tissue [HISTOL] Bone tissue. { 'äs·ē·əs ¦tish·ü }

ossicle [ANAT] Any of certain small bones, as those of the middle ear. [INV ZOO] Any of various calcareous bodies. { 'äs·ə·kəl }

ossify [PHYSIO] To form or turn into bone. { 'äs·ə,fī }

ossifying fibroma [MED] A benign bone tumor derived from ossiferous connective tissue. Also known as fibrous osteoma; osteogenic fibroma. { 'äs·ə,fī·iŋ fī'brō·mə }

ost-, oste-, osteo- [ANAT] A combining form meaning bone. { äst, 'ä·stē, 'ä·stē·ō }

Ostariophysi [VERT ZOO] A superorder of actinopterygian fishes distinguished by the structure of the anterior four or five vertebrae which are modified as an encasement for the bony ossicles connecting the inner ear and swim bladder. { ä¦stär·ē·ō'fī,sī }

Osteichthyes [VERT ZOO] The bony fishes, a class of fishlike vertebrates distinguished by having a bony skeleton, a swim bladder, a true gill cover, and mesodermal ganoid, cycloid, or ctenoid scales. { ,ä·stē'ik·thē,ēz }

osteitis [MED] Inflammation of bone. { ,ä·stē'īd·əs }

osteitis fibrosa cystica [MED] Generalized skeletal demineralization due to an increased rate of bone destruction resulting from hyperparathyroidism. Also known as Engel-Recklinghausen disease; osteitis fibrosa generalisata. { ,ä·stē'īd·əs 'fī·brō·sə 'sis·tə·kə }

osteitis fibrosa generalisata *See* osteitis fibrosa cystica. { ,ä·stē'īd·əs 'fī·brō·sə ¦jen·ə·rə'lis·əd·ə }

osteoarthritis *See* degenerative joint disease. { ¦äs·tē·ō,är¦thrīd·əs }

osteoarthropathy [MED] Any disease of bony articulations. { ¦äs·tē·ō,är¦thräp·ə·thē }

osteoblast [HISTOL] A bone-forming cell of mesenchymal origin. { 'äs·tē·ə,blast }

osteochondritis [MED] Inflammation of both bone and cartilage. { ,äs·tē·ō,kän'drīd·əs }

osteochondroma [MED] A benign hamartomatous tumor originating in bone or cartilage. { ,äs·tē·ō,kän'drō·mə }

osteochondrosis [MED] A disease characterized by avascular necrosis of ossification centers followed by regeneration. Also known as Calvé's disease; Kienböck's disease; Köhler's disease; Osgood-Schlatter disease; Scheuermann's disease. { ,äs·tē·ō,kän'drō·səs }

osteoclasis [MED] Forcible fracture of a long bone without open operation, to correct a deformity. [PHYSIO] **1.** Destruction of bony tissue. **2.** Bone resorption. { ,äs·tē'äk·lə·səs }

osteoclast [HISTOL] A large multinuclear cell associated with bone resorption. [MED] A large surgical apparatus through which leverage can be exerted to effect osteoclasis. { 'äs·tē·ə,klast }

osteoclast differentiation factor [CELL MOL] A protein on the surface of osteoblasts that binds to receptors on osteoclast precursor cells and induces progression to osteoclasts. { ,äs·tē·ə,klast ,dif·ə,ren·chē'ā·shən ,fak·tər }

osteocyte [HISTOL] A bone cell. { 'äs·tē·ə,sīt }

osteodermia [MED] A condition characterized by ossification within the skin. { ¦äs·tē·ə¦dər·mē·ə }

osteodystrophy [MED] Any defective bone formation, as in rickets or dwarfism. { ,äs·tē·ō'di·strə·fē }

osteofibrosis [MED] Fibrosis of bone. { ¦äs·tē·ō,fī'brō·səs }

osteogenesis [PHYSIO] Formation or histogenesis of bone. { ¦äs·tē·ō'jen·ə·səs }

osteogenesis imperfecta [MED] A disease inherited as an autosomal dominant and characterized by hypoplasia of osteoid tissue and collagen, resulting in bone fractures. { ¦äs·tē·ō'jen·ə·səs ,im·pər'fek·tə }

osteogenesis imperfecta congenita [MED] A form of osteogenesis imperfecta in which fractures occur at or before birth. { ¦äs·tē·ō'jen·ə·səs ,im·pər'fek·tə kən'jen·əd·ə }

osteogenic fibroma *See* ossifying fibroma. { ¦äs·tē·ə¦jen·ik fī'brō·mə }

osteogenic sarcoma *See* osteosarcoma. { ¦äs·tē·ə¦jen·ik sär'kō·mə }

Osteoglossidae [VERT ZOO] The bony tongues, a family of actinopterygian fishes in the order Osteoglossiformes. { ,äs·tē·ō'gläs·ə,dē }

Osteoglossiformes [VERT ZOO] An order of soft-rayed, actinopterygian fishes distinguished by paired, usually bony rods at the base of the second gill arch, a single dorsal fin, no adipose fin, and a usually abdominal pelvic fin. { ,äs·tē·ō,gläs·ə'fȯr·mēz }

osteoid [HISTOL] The young hyaline matrix of true bone in which the calcium salts are deposited. { 'äs·tē,ȯid }

osteolathyrism [MED] Degeneration of bone collagen resulting from experimental administration of β-aminoproprionitrile. { ,äs·tē·ō'lath·ə,riz·əm }

osteology [ANAT] The study of anatomy and structure of bone. { ˌäs·tēˈäl·ə·jē }

osteolysis [MED] Degeneration of bone tissue. [PHYSIO] Resorption of bone. { ˌäs·tēˈäl·ə·səs }

osteoma [MED] A benign bone tumor, especially in membrane bones of the skull. { ˌäs·tēˈō·mə }

osteomalacia [MED] Failure of bone to ossify due to a reduced amount of available calcium. Also known as adult rickets. { ¦äs·tē·ō·məˈlā·shə }

osteometry [ANAT] The study of the size and proportions of the osseous system. { ˌäs·tēˈäm·ə·trē }

osteomyelitis [MED] Inflammation of bone tissue and bone marrow. { ¦äs·tē·ōˌmī·əˈlīd·əs }

osteon [HISTOL] A microscopic unit of mature bone composed of layers of osteocytes and bone surrounding a central canal. Also known as Haversian system. { ˈäs·tēˌän }

osteonephropathy [MED] Any syndrome involving bone changes accompanying kidney disease. { ¦äs·tē·ō·nəˈfräp·ə·thē }

osteopath [MED] A physician who specializes in osteopathy. { ˈäs·tē·əˌpath }

osteopathy [MED] **1.** A school of healing which teaches that the body is a vital mechanical organism whose structural and functional integrity are coordinate and interdependent, the abnormality of either constituting disease. **2.** Any disease of bone. { ˌäs·tēˈäp·ə·thē }

osteopenia [MED] The reduction in bone volume and bone structural quality. { ˌäs·tē·əˈpē·nē·ə }

osteopetrosis [MED] A rare developmental error of unknown cause but of familial tendency, characterized chiefly by excessive radiographic density of most or all of the bones. Also known as marble bone disease. { ¦äs·tē·ō·pəˈrō·səs }

osteophony [PHYSIO] Conduction of sound by bone. { ˌäs·tēˈäf·ə·nē }

osteoplasty [MED] Plastic surgery performed on bone, particularly bone tissue replacement or reconstruction. { ˈäs·tē·əˌplas·tē }

osteopoikilosis [MED] A bone affection of unknown cause and no symptoms, characterized by ellipsoidal, dense foci in all bones. { ¦äs·tē·ōˌpȯi·kəˈlō·səs }

osteoporosis [MED] Deossification with absolute decrease in bone tissue, resulting in enlargement of marrow and Haversian spaces, decreased thickness of cortex and trabeculae, and structural weakness. { ¦äs·tē·ō·pəˈrō·səs }

osteoprotegerin [BIOCHEM] A protein that plays a central role in regulating bone mass. { ¦äs·tē·prōˈteg·ərin }

osteosarcoma [MED] A malignant tumor principally composed of anaplastic cells of mesenchymal derivation. Also known as osteogenic sarcoma. { ¦äs·tē·ō·särˈkō·mə }

osteosclereid [BOT] A sclereid cell that is rodlike with swollen ends, occurring in seed coats and in some leaves. Also known as bone cell. { ˌä·tē·əˈskler·ē·əd }

osteosclerotic anemia *See* myelophthisic anemia. { ¦äs·tē·ō·skləˈräd·ik əˈnē·mē·ə }

osteotomy [MED] **1.** Surgical division of a bone. **2.** Making a section of a bone for the purpose of correcting a deformity. { ˌäs·tēˈäd·ə·mē }

ostiole [BIOL] A small orifice or pore. { ˈäs·tēˌōl }

ostium [BIOL] A mouth, entrance, or aperture. { ˈäs·tē·əm }

Ostomidae [INV ZOO] The bark-gnawing beetles, a family of coleopteran insects in the superfamily Cleroidea. { äˈstäm·əˌdē }

Ostracoda [INV ZOO] A subclass of the class Crustacea containing small, bivalved aquatic forms; the body is unsegmented and there is no true abdominal region. { äˈsträk·ə·də }

Ostreidae [INV ZOO] A family of bivalve mollusks in the order Anisomyaria containing the oysters. { äˈstrē·əˌdē }

ostrich [VERT ZOO] *Struthio camelus.* A large running bird with soft plumage, naked head, neck and legs, small wings, and thick powerful legs with two toes on each leg; the only living species of the Struthioniformes. { ˈöˌstrich }

ot-, oto- [ANAT] A combining form meaning ear. { ōt, ˈō·dō }

otalgia [MED] Pain in the ear. { ōˈtal·jə }

Otariidae [VERT ZOO] The sea lions, a family of carnivorous mammals in the superfamily Canoidea. { ˌōd·əˈrī·əˌdē }

Othniidae [INV ZOO] The false tiger beetles, a small family of coleopteran insects in the superfamily Tenebrionoidea. { ˌōthˈnī·əˌdē }

otic [ANAT] Of or pertaining to the ear or a part thereof. { ˈōd·ik }

otic capsule [EMBRYO] A cartilaginous capsule surrounding the auditory vesicle during development, later fusing with the spheroid and occipital cartilages. { ˈōd·ik ˈkap·səl }

otic ganglion [NEUROSCI] The nerve ganglion located immediately below the foramen ovale of the sphenoid bone. { ˈōd·ik ˈgaŋ·glēˌän }

Otitidae [INV ZOO] A family of cyclorrhaphous myodarian dipteran insects in the subsection Acalyptratae. { ōˈtid·əˌdē }

otitis [MED] Inflammation of the ear. { ōˈtīd·əs }

otitis externa [MED] Inflammation of the external ear. { ōˈtīd·əs ekˈstər·nə }

otitis media [MED] Inflammation of the middle ear. { ōˈtīd·əs ˈmē·dē·ə }

otocyst [EMBRYO] The auditory vesicle of vertebrate embryos. [INV ZOO] An auditory vesicle, otocell, or otidium in some invertebrates. { ˈōd·əˌsist }

otolaryngology [MED] A branch of medicine that deals with the ear, nose, and throat. Also known as otorhinolaryngology. { ¦ōd·ōˌlar·ənˈgäl·ə·jē }

otolith [ANAT] A calcareous concretion on the end of a sensory hair cell in the vertebrate ear and in some invertebrates. { ˈōd·əˌlith }

otology [MED] A branch of medicine that deals with the ear and its diseases. { ō'täl·ə·je }

otomycosis [MED] Fungus infection of the external ear, usually caused by *Aspergillus niger* and A. *fumigatus*. { ˌōd·əˌmī'kō·səs }

Otopheidomenidae [INV ZOO] A family of parasitic mites in the suborder Mesostigmata. { ˌōd·əˌfē·dō'men·əˌdē }

otorhinolaryngology *See* otolaryngology. { ¦ōd·ə¦rīn·ōˌlar·ən'gäl·ə ·jē }

otosclerosis [MED] Sclerosis of the inner ear, causing a progressive increase in deafness. { ¦ōd·ō·sklə'rō·səs }

otoscope [MED] An apparatus designed for examination of the ear and for rendering the tympanic membrane visible. { 'ōd·əˌskōp }

ototoxicity [MED] Drug- or chemical-induced damage to the ear resulting in high-frequency hearing loss and tinnitus or disequilibrium. { ˌō·tō·täk'sis·əd·ē }

otter [VERT ZOO] Any of various members of the family Mustelidae, having a long thin body, short legs, a somewhat flattened head, webbed toes, and a broad flattened tail; all are adapted to aquatic life. { 'äd·ər }

Ouchterlony test [IMMUNOL] A technique used to analyze an antigen-antibody mixture, in which the components are placed in multiple wells cut into agar on a flat slide and allowed to diffuse toward one another. { 'au̇ch·tərˌlȯn·ē ˌtest }

Oudin test [IMMUNOL] A technique used to measure antigen concentration, in which an antigen and an antibody are held in an agar matrix in a test tube and allowed to diffuse toward one another. { 'ü·dan ˌtest }

outbreed *See* crossbreed. { 'au̇tˌbrēd }

outbreeding *See* exogamy. { 'au̇tˌbrēd·iŋ }

outgroup [SYST] A monophyletic taxon that is used in a phylogenetic study to resolve which of two homologous character states are apomorphous. { 'au̇tˌgrüp }

outpatient [MED] A patient who comes to the hospital or clinic for diagnosis and treatment but who does not occupy a bed in the institution. { 'au̇tˌpā·shənt }

ovalbumin [BIOCHEM] The major, conjugated protein of eggwhite. { ¦ov·al'byü·mən }

oval window [ANAT] The membrane-covered opening into the inner ear of tetrapods, to which the ossicles of the middle ear are connected. { 'ō·vəl 'win·dō }

ovarian [ANAT] Of or pertaining to the ovaries. { ō'ver·ē·ən }

ovarian agenesis [MED] Failure of the ovaries to develop. { ō'ver·ē·ən ā'jen·ə·səs }

ovarian dysmenorrhea [MED] Dysmenorrhea caused by ovarian disease. { ō'ver·ē·ən dis ˌmen·ə'rē·ə }

ovarian follicle [HISTOL] An ovum and its surrounding follicular cells, found in the ovarian cortex. { ō'ver·ē·ən 'fäl·ə·kəl }

ovarian insufficiency [MED] Deficient functioning of the ovaries, leading to amenorrhea, oligomenorrhea, or abnormal dysfunctional uterine bleeding. { ō'ver·ē·ən ˌin·sə'fish·ən·sē }

ovariectomy [MED] Excision of an ovary. { ōˌvar·ē'ek·tə·mē }

ovariole [INV ZOO] The tubular structural unit of an insect ovary. { ō'var·ēˌōl }

ovary [ANAT] A glandular organ that produces hormones and gives rise to ova in female vertebrates. [BOT] The enlarged basal portion of a pistil that bears the ovules in angiosperms. { 'ōv·ə·rē }

overdominance [GEN] Monohybrid heterosis, that is, the phenotype is being more pronounced in the heterozygote than in either homozygote with respect to a specified pair of alleles. { ¦ō·vər'däm·ə·nəns }

overdose [MED] An excessive dose of medicine. { 'ō·vərˌdōs }

overlapping genes [GEN] Genes having nucleotide sequences that may overlap in a way that involves control genes or structural genes. { ¦ō·vərˌlap·iŋ 'jēnz }

overnight polysomnography [MED] The overnight sleep recording of brain waves, muscle activities, eye movements, heart activity, airflow at the nose and mouth, respiratory effort, and oxygen saturation that is used to diagnose abnormalities in sleep and/or wakefulness. { ˌō·vər¦nīt ˌpäl·ē·səm'näg·rə·fē }

overt infection [MED] A host-parasite interaction that results in some injury to the tissues of the host. { ˌō·vərt in'fek·shən }

overwinding [CELL MOL] Supercoiling of a deoxyribonucleic acid molecule in the same direction as that of the winding of the double helix, resulting in increased tension in the two strands of the molecule. { ¦ō·vər¦wīn·diŋ }

ovicell [INV ZOO] A broad chamber in certain bryozoans. { 'ō·vəˌsel }

ovicyst [INV ZOO] The pouch of a tunicate in which the eggs develop. { 'ō·vəˌsist }

oviduct [ANAT] A tube that serves to conduct ova from the ovary to the exterior or to an intermediate organ such as the uterus. Also known in mammals as Fallopian tube; uterine tube. { 'ō·vəˌdəkt }

oviger [INV ZOO] A modified leg used for carrying eggs in some pycnogonids. { 'ō·və·jər }

ovine encephalomyelitis *See* louping ill. { 'ōˌvīn in¦sef·ə·lōˌmī·ə'līd·əs }

oviparous [VERT ZOO] Producing eggs that develop and hatch externally. { ō'vip·ə·rəs }

oviposit [ZOO] To lay or deposit eggs, especially by means of a specialized organ, as found in certain insects and fishes. { 'ō·vəˌpäz·ət }

ovipositor [INV ZOO] A specialized structure in many insects for depositing eggs. [VERT ZOO] A tubular extension of the genital orifice in most fishes. { 'ō·vəˌpäz·əd·ər }

ovotesticular hermaphroditism [MED] A rare form of hermaphroditism in which an ovotestis is present on one or both sides. { ˌō·vōˌtes'tik·yə·lər hər'maf·rə·dəˌdiz·əm }

ovoviviparous [VERT ZOO] Producing eggs that develop internally and hatch before or soon after extrusion. { ¦ō·vōˌvī'vip·ə·rəs }

ovulation [PHYSIO] Discharge of an ovum or ovule from the ovary. { ˌäv·yəˈlā·shən }

ovule [BOT] A structure in the ovary of a seed plant that develops into a seed following fertilization. { ˈävˌyül }

ovum [CELL MOL] A female gamete. Also known as egg. { ˈō·vəm }

Oweniidae [INV ZOO] A family of limivorous polychaete annelids of the Sedentaria. { ˌō·wəˈnī·əˌdē }

owl [VERT ZOO] Any of a number of diurnal and nocturnal birds of prey composing the order Strigiformes; characterized by a large head, more or less forward-directed large eyes, a short hooked bill, and strong talons. { aůl }

Oxalidaceae [BOT] A family of dicotyledonous plants in the order Geraniales, generally characterized by regular flowers, two or three times as many stamens as sepals or petals, a style which is not gynobasic, and the fruit which is a beakless, loculicidal capsule. { äkˌsal·əˈdās·ēˌē }

oxalosis [MED] A rare hereditary metabolic disorder, inherited as an autosomal recessive, in which glyoxylic acid metabolism is impaired, resulting in overproduction of oxalic acid and deposition of calcium oxalate in body tissues. { ¦ak·sə¦lō·səs }

oxaluria [MED] The presence of oxalic acid or oxalates in the urine. { ˌäk·səlˈyůr·ē·ə }

oxidase [BIOCHEM] An enzyme that catalyzes oxidation reactions by the utilization of molecular oxygen as an electron acceptor. { ˈäk·sə ˌdās }

oxidative phosphorylation [BIOCHEM] Conversion of inorganic phosphate to the energy-rich phosphate of adenosinetriphosphatase by reactions associated with the electron transfer system. { ˌäk·səˌdād·iv ˌfäs·fə·rəˈlā·shən }

oxidoreductase [BIOCHEM] An enzyme catalyzing a reaction in which two molecules of a compound interact so that one molecule is oxidized and the other reduced, with a molecule of water entering the reaction. { ¦äk·sə·dō·riˈdəkˌtās }

oximeter [MED] A photoelectric photometer used to measure the oxygenated fraction of the hemoglobin in blood which is either circulating in a particular tissue of an intact animal or human being, or during, or shortly after, its withdrawal from the vascular system, by observation of the absorption of light transmitted through or reflected from the blood. { äkˈsim· əd·ər }

oximetry [PHYSIO] Optical measurement of the degree of oxygen saturation of the blood hemoglobin by determining the variation in the color of the blood. { äkˈsim·ə·trē }

oxycephaly [MED] A condition in which the head assumes a roughly conical shape due to premature closure of the coronal or lambdoid sutures, or to artificial pressure on the frontal and occipital regions of the infant's head. Also known as acrocephaly. { ˌäk·sēˈsef·ə·lē }

oxygenase [BIOCHEM] An oxidoreductase that catalyzes the direct incorporation of oxygen into its substrate. { ˈäk·sə·jəˌnās }

oxygenator [MED] An apparatus for introducing oxygen into the blood during extracorporal circulation. { ˈäk·sə·jəˌnād·ər }

oxygen debt [PHYSIO] **1.** A bodily condition in which oxygen demand is greater than oxygen supply. **2.** The amount of oxygen needed to restore the body to a steady state after a muscular exertion. { ˈäk·sə·jən ˌdet }

oxygen toxicity [PHYSIO] **1.** Harmful effects of breathing oxygen at pressures greater than atmospheric. **2.** A toxic effect in a living organism caused by a species of oxygen-containing reactive intermediate produced during the reduction of dioxygen. { ˈäk·sə·jən täkˈsis·əd·ē }

oxyhemoglobin [BIOCHEM] The red crystalline pigment formed in blood by the combination of oxygen and hemoglobin, without the oxidation of iron. { ¦äk·sēˈhē·məˌglō·bən }

oxylophyte [ECOL] A plant that thrives in or is restricted to acid soil. { äkˈsil·əˌfīt }

Oxymonadida [INV ZOO] An order of xylophagous protozoans in the class Zoomastigophorea; colorless flagellate symbionts in the digestive tract of the roach *Cryptocercus* and of certain termites. { ˌäk·sē·məˈnäd·ə·də }

oxyntic cell *See* parietal cell. { äkˈsin·tik ˈsel }

oxypetalous [BOT] Having sharp-pointed petals. { ¦äk·sēˈped·əl·əs }

oxyphilia *See* eosinophilia. { ˌäk·səˈfil·ē·ə }

oxyphytia [ECOL] Discordant habitat control due to an excessively acidic substratum. { ˌäk·səˈfīd·ē·ə }

oxyreductase [BIOCHEM] Any of a class of enzymes that catalyze electron-transfer reactions. { ˌäk·sē·riˈdəkˌtās }

oxysterol [BIOCHEM] Oxidized derivative of cholesterol. { ˌäk·sēˈstirˌȯl }

Oxystomata [INV ZOO] A subsection of the Brachyura, including those true crabs in which the first pair of pereiopods is chelate, and the mouth frame is triangular and forward. { ˌäk·sēˈstō·mə·də }

Oxystomatidae [INV ZOO] A family of free-living marine nematodes in the superfamily Enoploidea, distinguished by amphids that are elongated longitudinally. { ˌäk·sē·stōˈmad·əˌdē }

oxytetracycline [MICROBIO] $C_{22}H_{24}O_9N_2$ A crystalline, amphoteric, broad-spectrum antibiotic produced by *Streptomyces rimosus*; produced commercially by fermentation. { ¦äk·sēˌte·trəˈsī ˌklēn }

oxytocic [MED] Hastening parturition. [PHARM] A drug that hastens parturition. { ¦äk·sē¦tō·sik }

oxytocin [BIOCHEM] $C_{43}H_{66}O_{12}N_{12}S_2$ A polypeptide hormone secreted by the neurohypophysis that stimulates contraction of the uterine muscles. { ¦äk·sē¦tō·sən }

Oxyurata [INV ZOO] The equivalent name for Oxyurina. { ˌäk·sē·yůˈräd·ə }

Oxyuridae [INV ZOO] A family of the nematode superfamily Oxyuroidea. { ˌäk·sē'yůr·əˌdē }

Oxyurina [INV ZOO] A suborder of nematodes in the order Ascaridida. { ˌäk·sē'yů'rī·nə }

Oxyuroidea [INV ZOO] A superfamily of marine nematodes in the order Enoplida; contains species that maintain the most ancestral characters known in the phylum, such as a stoma composed entirely of esophastome, which is the ancestral primary blastocoel invagination. { ˌäk·sē·yů 'rȯid·ē·ə }

oyster [INV ZOO] Any of various bivalve mollusks of the family Ostreidae; the irregular shell is closed by a single adductor muscle, the foot is small or absent, and there is no siphon. { 'ȯi·stər }

P

P_1 [GEN] The parental generation; parents of the F_1 generation.

PABA *See para*-aminobenzoic acid.

PABA sodium *See* sodium *para*-aminobenzoate. { ¦pē¦ā¦bē¦ā ¦sōd·ē·əm }

paca [VERT ZOO] Any of several rodents of the genus *Cuniculus*, especially *C. paca*, with a white-spotted brown coat, found in South and Central America. { 'päk·ə }

Pacchionian bodies *See* arachnoidal granulations. { ¦pak·ē¦ō·nē·ən ˌbäd·ēz }

pacemaker [MED] A pulsed battery-operated oscillator implanted in the body to deliver electric impulses to the muscles of the lower heart, either at a fixed rate or in response to a sensor that detects when the patient's pulse rate slows or ceases. Also known as cardiac pacemaker; heart pacer. { 'pās,māk·ər }

pachyderm [VERT ZOO] Any of various nonruminant hooved mammals characterized by thick skin, including the elephants, hippopotamuses, rhinoceroses, and others. { 'pak·əˌdərm }

pachydermatous [MED] Abnormally thick-skinned. { ¦pak·ə¦dər·məd·əs }

pachydermia [MED] Abnormal thickening of the skin. { ˌpak·ə'dər·mē·ə }

pachyglossal [VERT ZOO] Of lizards, having a thick tongue. { ¦pak·ə'gläs·əl }

pachyglossia [MED] Abnormal thickness of the tongue. { ¦pak·ə'glä·sē·ə }

pachygyria [MED] A malformation of the brain characterized by its being too broad in form. { ¦pak·ə'jī·rē·ə }

pachymeningitis [MED] Inflammation of the dura mater. { ¦pak·ē,men·ən'jīd·əs }

pachynema *See* pachytene. { pə'kin·ə·mə }

pachyostosis [VERT ZOO] A bony thickening of vertebrae and ribs. { ˌak·ō·'stō·səs }

pachytene [CELL MOL] The third stage of meiotic prophase during which paired chromosomes thicken, each chromosome splits into chromatids, and breakage and crossing over between nonsister chromatids occur. Also known as pachynema. { 'pak·əˌtēn }

Pacific faunal region [ECOL] A marine littoral faunal region including offshore waters west of Central America, running from the coast of South America at about 5° south latitude to the southern tip of California. { pə'sif·ik 'fȯn·əl ˌrē·jən }

Pacific temperate faunal region [ECOL] A marine littoral faunal region including a narrow zone in the North Pacific Ocean, from Indochina to Alaska and along the west coast of the United States to about 40° north latitude. { pə'sif·ik 'tem·prət 'fȯn·əl ˌrē·jən }

Pacinian corpuscle [NEUROSCI] An encapsulated lamellar sensory nerve ending that functions as a kinesthetic receptor. { pə'chin·ē·ən 'kȯr·pə·səl }

packet [BIOL] A cluster of organisms in the form of a cube resulting from cell division in three planes. { 'pak·ət }

packet gland [INV ZOO] A cluster of gland cells opening through the epidermis of nemertines. { 'pak·ət ˌgland }

pactamycin [MICROBIO] An antitumor and antibacterial antibiotic produced by *Streptomyces pactum* var. *pactum*. { ˌpak·tə'mīs·ən }

pad [ANAT] A small circumscribed mass of fatty tissue, as in terminal phalanges of the fingers or the underside of the toes of an animal, such as a dog. { pad }

paedogamy [INV ZOO] A type of autogamy in certain protozoans whereby there is mutual fertilization of gametes derived from a single cell. { pē'däg·ə·mē }

paedomorphosis [EVOL] Phylogenetic change in which adults retain juvenile characters, accompanied by an increased capacity for further change; indicates potential for further evolution. { ˌpēd·ə'mȯr·fə·səs }

Paenungulata [VERT ZOO] A superorder of mammals, including proboscideans, xenungulates, and others. { pēn,əŋ·gyə'läd·ə }

Paeoniaceae [BOT] A monogeneric family of dicotyledonous plants in the order Dilleniales; members are mesophyllic shrubs characterized by cleft leaves, flowers with an intrastaminal disk, and seeds having copious endosperm. { ˌpē·ə·nē'ās·ē,ē }

PAF *See* platelet-activating factor. { ¦pē¦ā'ef }

Paget's cells [PATH] Large, epithelial cells with clear cytoplasm found in certain breast and skin cancers. { 'paj·əts ˌselz }

Paget's disease [MED] **1.** A type of carcinoma of the breast that involves the nipple or areola and the larger ducts, characterized by the presence of Paget's cells. **2.** Osseous hyperplasia simultaneous with accelerated deossification.

3. An apocrine gland skin cancer, composed principally of Paget's cells. { 'paj·əts di,zēz }

Paguridae [INV ZOO] The hermit crabs, a family of decapod crustaceans belonging to the Paguridea. { pə'gyůr·ə,dē }

Paguridea [INV ZOO] A group of anomuran decapod crustaceans in which the abdomen is nearly always asymmetrical, being either soft and twisted or bent under the thorax. { ,pag·yə'rid·ē·ə }

pain [PHYSIO] Patterns of somesthetic sensation, generally unpleasant, or causing suffering or distress. { pān }

pain spot [PHYSIO] Any of the small areas of skin overlying the endings of either very small myelinated (delta) or unmyelinated (C) nerve fibers whose stimulation, depending on the intensity and duration, results in the sensation of either pain or itching. { 'pān ,spät }

pain threshold [PHYSIO] The lowest limit for the perception of pain sensations. { 'pān ,thresh,hōld }

Palaeacanthocephala [INV ZOO] An order of the Acanthocephala including parasitic worms characterized by fragmented nuclei in the hypodermis, lateral placement of the chief lacunar vessels, and proboscis hooks arranged in long rows. { ¦pāl·ē·ə,kan·thə'sef·ə·lə }

Palaemonidae [INV ZOO] A family of decapod crustaceans in the group Caridea. { ¦pāl·ē'män·ə,dē }

Palaeocaridacea [INV ZOO] An order of crustaceans in the superorder Syncarida. { ¦pāl·ē·ō ,kar·ə'dās·ē·ə }

Palaeocaridae [INV ZOO] A family of the crustacean order Palaeocaridacea. { ¦pāl·ē·ō'kar·ə,dē }

Palaeodonta [VERT ZOO] A suborder of artiodactylous mammals including piglike forms such as the extinct "giant pigs" and the hippopotami. { ,pāl·ē·ə'dän·tə }

Palaeognathae [VERT ZOO] The ratites, making up a superorder of birds in the subclass Neornithes; merged with the Neognathae in some systems of classification. { ,pāl·ē'äg·nə,thē }

Palaeonemertini [INV ZOO] A family of the class Anopla distinguished by the two- or three-layered nature of the body-wall musculature. { ¦pāl·ē·ō,ne·mər'tī,nī }

Palaeopneustidae [INV ZOO] A family of deep-sea echinoderms in the order Spatangoida characterized by an oval test, long spines, and weakly developed fascioles and petals. { ¦pāl·ē·ō'nü·stə,dē }

Palaeopterygii [VERT ZOO] An equivalent name for the Actinopterygii. { ,pāl·ē,äp·tə'rij·ē,ī }

palaeotheriodont [VERT ZOO] Being or having lophodont teeth with longitudinal external tubercles that are connected with inner tubercles by transverse oblique crests. { ¦pāl·ē·ō¦ther·ē·ə,dänt }

palaeotropical *See* paleotropical. { ¦pāl·ē·ō'träp·ə·kəl }

palama [VERT ZOO] The membranous web on the feet of aquatic birds. { 'pal·ə·mə }

palate [ANAT] The roof of the mouth. { 'pal·ət }

palatine bone [ANAT] Either of a pair of irregularly L-shaped bones forming portions of the hard palate, orbits, and nasal cavities. { 'pal·ə ,tīn 'bōn }

palatine canal [ANAT] One of the canals in the palatine bone, giving passage to branches of the descending palatine nerve and artery. { 'pal·ə ,tīn kə'nal }

palatine gland [ANAT] Any of numerous small oral glands on the palate of mammals. { 'pal·ə,tīn 'gland }

palatine process [ANAT] A thick process that projects horizontally mediad from the medial aspect of the maxilla. [EMBRYO] An outgrowth on the ventromedial aspect of the maxillary process that develops into the definite palate. { 'pal·ə,tīn 'prä·səs }

palatine suture [ANAT] The median suture joining the bones of the palate. { 'pal·ə,tīn 'sü·chər }

palatine tonsil [ANAT] Either of a pair of almond-shaped aggregations of lymphoid tissue embedded between folds of tissue connecting the pharynx and posterior part of the tongue with the soft palate. Also known as faucial tonsil; tonsil. { 'pal·ə,tīn 'tän·səl }

palatomaxillary arch [ANAT] An arch formed by the palatine, maxillary, and premaxillary bones. Also known as maxillary arch. { ¦pal·ə·dō'mak·sə,ler·ē 'ärch }

palatoquadrate [VERT ZOO] A series of bones or a cartilaginous rod constituting part of the roof of the mouth or upper jaw of most nonmammalian vertebrates. { ¦pal·ə·dō'kwä,drāt }

palea [BOT] **1.** The upper, enclosing bract of a grass flower. **2.** A chaffy scale found on the receptacle of the disk flowers of some composite plants. [INV ZOO] One of the enlarged flattened setae forming the operculum of the tube of certain polychaete worms. { 'pā·lē·ə }

Palearctic [ECOL] Pertaining to a biogeographic region including Europe, northern Asia and Arabia, and Africa north of the Sahara. { ¦pāl·ē'ärd·ik }

paleate [BOT] Having a covering of chaffy scales, as some rhizomes. { 'pā·lē,āt }

paleontology [BIOL] The study of life of the past as recorded by fossil remains. { ,pāl·ē·ən'täl·ə·jē }

Paleoptera [INV ZOO] A section of the insect subclass Pterygota including primitive forms that are unable to flex their wings over the abdomen when at rest. { ,pāl·ē'äp·tə·rə }

paleosere [ECOL] A series of ecologic communities that have led to a climax community. { 'pāl·ē·ə,sir }

paleotropical [ECOL] Of or pertaining to a biogeographic region that includes the Oriental and Ethiopian regions. Also spelled palaeotropical. { ¦pāl·ē·ō'träp·ə·kəl }

palilalia [MED] Pathologic repetition of words or phrases. { ,pal·ə'lal·yə }

palindrome [GEN] A nucleic acid sequence that is self-complementary. { 'pal·ən,drōm }

palingenesis [EMBRYO] Unaltered recapitulation of ancestral features by the developing stages of an organism. { ,pal·ən'jen·ə·səs }

Palinuridae [INV ZOO] The spiny lobsters or langoustes, a family of macruran decapod crustaceans belonging to the Scyllaridea. { ,pal·ə 'nyür·ə,dē }

palisade cell [BOT] One of the columnar cells of the palisade mesophyll which contain numerous chloroplasts. { ,pal·ə'sād ,sel }

palisade mesophyll [BOT] A tissue system of the chlorenchyma in well-differentiated broad leaves composed of closely spaced palisade cells oriented parallel to one another, but with their long axes perpendicular to the surface of the blade. { ,pal·ə'sād 'mez·ə,fil }

pallanesthesia [MED] Absence of pallesthesia, or vibration sense. { pə¦lan·əs¦thē·zhə }

pallet [INV ZOO] One of a pair of plates on the siphon tubes of certain Bivalvia. { 'pal·ət }

pallial artery [INV ZOO] The artery that supplies blood to the mantle of a mollusk. { 'pal·ē·əl ,ärd·ə·rē }

pallial chamber [INV ZOO] The mantle cavity in mollusks. { 'pal·ē·əl ,chām·bər }

pallial line [INV ZOO] A mark on the inner surface of a bivalve shell caused by attachment of the mantle. { 'pal·ē·əl ,līn }

pallial nerve [INV ZOO] One of the pair of dorsal nerves that innervate the mantle in mollusks. { 'pal·ē·əl ,nərv }

pallial sinus [INV ZOO] An inward bend in the posterior portion of the pallial line in bivalve mollusks. { 'pal·ē·əl ,sī·nəs }

palliative [PHARM] **1.** Having a soothing or relieving quality. **2.** A drug that soothes or relieves symptoms of a disease. { 'pal·ē·əd·iv }

pallium [ANAT] The cerebral cortex. [INV ZOO] The mantle of a mollusk or brachiopod. { 'pal·ē·əm }

Pallopteridae [INV ZOO] A family of myodarian cyclorrhaphous dipteran insects in the subsection Acalypteratae. { ,pal·əp'ter·ə,dē }

pallor [MED] Paleness, especially of the skin and mucous membranes. { 'pal·ər }

palm [ANAT] The flexor or volar surface of the hand. [BOT] Any member of the monocotyledonous family Arecaceae; most are trees with a slender, unbranched trunk and a terminal crown of large leaves that are folded between the veins. { päm }

Palmales [BOT] An equivalent name for Arecales. { pä'mā·lēz }

palmar [ANAT] Of or pertaining to the palm of the hand. { 'päm·ər }

palmar aponeurosis [ANAT] Bundles of fibrous connective tissue which radiate from the tendons of the deep fascia of the forearm toward the proximal ends of the fingers. { 'päm·ər ,ap·ə·nə'rōs·əs }

palmar arch *See* deep palmar arch; superficial palmar arch. { 'päm·ər 'ärch }

palmar reflex [PHYSIO] Flexion of the fingers when the palm of the hand is irritated. { 'päm·ər 'rē,fleks }

palmate [BOT] Having lobes, such as on leaves, that radiate from a common point. [VERT ZOO] Having webbed toes. [ZOO] Having the distal portion broad and lobed, resembling a hand with the fingers spread. { 'pä,māt }

palmately compound leaf [BOT] A leaf with leaflets that originate from a common point at the end of the petiole. { ¦pä,māt·lē ,käm,paủnd 'lēf }

palmella stage [BOT] A stage in the life history of some unicellular flagellate algae in which the cells lose their flagella and form a gelatinous aggregation. { päl'mel·ə ,stāj }

palmelloid [BOT] Pertaining to a colony of cells that aggregates in a gelatinous matrix, as is characteristic of blue-green algae. { päl'me,lȯid }

palmetto fiber [BOT] Brush or broom fiber obtained from young leafstalks of the cabbage palm tree (*Sabal palmetto*). { päl'med·ō ,fī·bər }

palm nut [BOT] The edible seed of the African oil palm (*Elaeis guineensis*). { 'päm ,nət }

Palmyridae [INV ZOO] A mongeneric family of errantian polychaete annelids. { pal'mir·ə,dē }

palp [INV ZOO] Any of various sensory, usually fleshy appendages near the oral aperture of certain invertebrates. { palp }

palpable [MED] **1.** Capable of being felt or touched. **2.** Evident. { 'pal·pə·bəl }

palpal organ [INV ZOO] An organ on the terminal joint of each pedipalp of a male spider which functions to convey sperm to the female genital orifice. { 'pal·pəl ,ȯr·gən }

palpation [MED] Diagnostic examination by touch. { pal'pā·shən }

Palpatores [INV ZOO] A suborder of long-legged arachnids in the order Phalangida. { ,pal·pə'tȯr·ēz }

palpebra [ANAT] The eyelid. { 'pal·pə·brə }

palpebral disk [VERT ZOO] A scale, often transparent, covering the eyelid of certain lizards. { 'pal·pə·brəl 'disk }

palpebral fissure [ANAT] The opening between the eyelids. { 'pal·pə·brəl 'fish·ər }

palpebral fold [ANAT] A fold formed by the reflection of the conjunctiva from the eyelid onto the eye. { 'pal·pə·brəl 'fōld }

Palpicornia [INV ZOO] The equivalent name for Hydrophiloidea. { ,pal·pə'kȯr·nē·ə }

palpiger [INV ZOO] The palpi-bearing portion of an insect labium. { 'pal·pə·jər }

Palpigradida [INV ZOO] An order of rare tropical and warm-temperate arachnids; all are minute, whitish, eyeless animals with an elongate body that terminates in a slender, multisegmented flagellum set with setae. { ,pal·pə'grad·əd·ə }

palpitate [MED] To flutter, or beat abnormally fast; applied especially to the rate of the heartbeat. { 'pal·pə,tāt }

palpocil [INV ZOO] A fine, filamentous tactile hair. { 'pal·pə,sil }

palpus [INV ZOO] **1.** A process on a mouthpart of an arthropod that has a tactile or gustatory function. **2.** Any similar process on other invertebrates. { 'pal·pəs }

palsy [MED] Any of various special types of paralysis, such as cerebral palsy. { 'pȯl·zē }

paludal [ECOL] Relating to swamps or marshes and to material that is deposited in a swamp environment. { pə'lüd·əl }

paludification [ECOL] Bog expansion resulting from the gradual rising of the water table as accumulation of peat impedes water drainage. { pə,lüd·ə·fə'kā·shən }

palustrine [ECOL] Being, living, or thriving in a marsh. { pə'ləs·trən }

palytoxin [BIOCHEM] A water-soluble toxin produced by several species of *Palythoa*; considered to be one of the most poisonous substances known. { 'pal·ə,täk·sən }

pamabrom [PHARM] $C_{11}H_{18}BrN_5O_3$ A water-soluble, fine white powder, decomposing at 300°C; used in medicine as a diuretic. { 'pam·ə,brōm }

pamaquine naphthoate [PHARM] $C_{42}H_{45}N_3O_7$ A yellow to orange-yellow powder, soluble in alcohol and acetone; used as an antimalarial drug. { 'pam·ə,kwēn 'naf·thə,wāt }

pampa [ECOL] An extensive plain in South America, usually covered with grass. { 'päm·pə }

Pamphiliidae [INV ZOO] The web-spinning sawflies, a family of hymenopteran insects in the superfamily Megalodontoidea. { ,pam·fə'lī·ə,dē }

pampiniform [ANAT] Of the network of veins in the spermatic cord and in the broad ligament, having the form of a tendril. { pam'pin·ə,fȯrm }

pampiniform plexus [ANAT] A venous network in the spermatic cord in the male, and in the broad ligament in the female. { pam'pin·ə,fȯrm 'plek·səs }

pamprodactylous [VERT ZOO] Having the toes turned forward, as of certain birds. { ,pam·prə'dak·təl·əs }

panacinar emphysema [MED] Emphysema characterized by diffuse destruction of one lung. { ,pan·ə'sin·ər ,em·fə'sē·mə }

panagglutinin [IMMUNOL] An agglutinin lacking specificity, which agglutinates erythrocytes of various types. { ,pan·ə'glüt·ən·ən }

Panagrolaimoidea [INV ZOO] A superfamily of free-living nematodes in the order Rhabditida, characterized by a broad, open, thick-walled stoma that forms a chamber as long as its breadth. { pə,na·grō·lə'mȯid·ē·ə }

panarteritis [MED] **1.** Arteritis involving all the coats of an artery. **2.** *See* polyarteritis. { pan ,ärd·ə'rīd·əs }

panarthritis [MED] Inflammation of several joints. { ¦pan·är'thrīd·əs }

pancarditis [MED] Carditis involving the endocardium, myocardium, and pericardium. { ¦pan·kär'dīd·əs }

Pancarida [INV ZOO] A superorder of the subclass Malacostraca; the cylindrical, cruciform body lacks an external division between the thorax and pleon and has the cephalon united with the first thoracomere. { pan'kar·ə·də }

panclimax [ECOL] Two or more related climax communities or formations having similar climate, life forms, and genera or dominants. Also known as panformation. { pan'klī,maks }

pancreas [ANAT] A composite gland in most vertebrates that produces and secretes digestive enzymes, as well as at least two hormones, insulin and glucagon. { 'pan·krē·əs }

pancreatectomy [MED] Surgical removal of the pancreas. { ,pan·krē·ə'tek·tə·mē }

pancreatic diarrhea [MED] Diarrhea due to deficiency of pancreatic digestive enzymes, characterized by the passage of large, greasy stools having a high fat and nitrogen content. { ,pan·krē'ad·ik ,dī·ə'rē·ə }

pancreatic diverticulum [EMBRYO] One of two diverticula (dorsal and ventral) from the embryonic duodenum or hepatic diverticulum that form the pancreas or its ducts. { ¦pan·krē¦ad·ik ,dī·vər'tik·yə·ləm }

pancreatic duct [ANAT] The main duct of the pancreas formed from the dorsal and ventral pancreatic ducts of the embryo. { ¦pan·krē¦ad·ik ,dəkt }

pancreatic juice [PHYSIO] The thick, transparent, colorless secretion of the pancreas. { ¦pan·krē¦ad·ik 'jüs }

pancreatic lipase *See* steapsin. { ¦pan·krē¦ad·ik 'lī,pās }

pancreatin [BIOCHEM] A cream-colored, amorphous powder obtained from the fresh pancreas of a hog; contains amylopsin, trypsin, steapsin, and other enzymes. { pan'krē·əd·ən }

pancreatitis [MED] Inflammation of the pancreas. { ,pan·krē·ə'tīd·əs }

pancreozymin [BIOCHEM] A crude extract of the intestinal mucosa that stimulates secretion of pancreatic juice. { ,pan·krē·ō'zī·mən }

pancytopenia [MED] Abnormally low numbers of all formed elements in the blood. { ,pan·sīd·ə'pē·nē·ə }

panda [VERT ZOO] Either of two Asian species of carnivores in the family Procyonidae; the red panda (*Ailurus fulgens*) has long, thick, red fur, with black legs; the giant panda (*Ailuropoda melanoleuca*) is white, with black legs and black patches around the eyes. { 'pan·də }

Pandanaceae [BOT] The single, pantropical family of the plant order Pandanales. { ,pan·də'nās·ē,ē }

Pandanales [BOT] A monofamilial order of monocotyledonous plants; members are more or less arborescent and sparingly branched, with numerous long, firm, narrow, parallel-veined leaves that usually have spiny margins. { ,pan·də'nā·lēz }

pandanus [BOT] Any tree of the genus *Pandanus*, which contains more than 500 species. It is a characteristic component of the vegetation in the tropics of the Old World, especially on the Pacific islands and along continental coasts. Also known as screw pine. { pan'dan·əs }

Pandaridae [INV ZOO] A family of dimorphic crustaceans in the suborder Caligoida; members are external parasites of sharks. { pan'dar·ə,dē }

pandemic [MED] Epidemic occurring over a widespread geographic area. { pan'dem·ik }

Pandionidae [VERT ZOO] A monospecific family of birds in the order Falconiformes; includes the osprey (*Pandion haliaetus*), characterized by a reversible hindtoe, well-developed claws, and spicules on the scales of the feet. { pan·dē'än·ə,dē }

pandurate [BOT] Of a leaf, having the outline of a fiddle. { 'pan·dyůr·ət }

panendoscope [MED] A modification of the cystoscope, utilizing a Foroblique lens system, permitting adequate visualization of both the urinary bladder and the urethra. { pan'en·də,skōp }

Paneth cells *See* cells of Paneth. { 'pan·əth ,selz }

panformation *See* panclimax. { 'pan·fər,mā·shən }

pangene [CELL MOL] A hypothetical heredity-controlling protoplasmic particle proposed by Darwin. { 'pan,jēn }

pangenesis [BIOL] Darwin's comprehensive theory of heredity and development, according to which all parts of the body give off gemmules which aggregate in the germ cells; during development, they are sorted out from one another and give rise to parts similar to those of their origin. { pan'jen·ə·səs }

pangolin [VERT ZOO] Any of seven species composing the mammalian family Manidae; the entire dorsal surface of the body is covered with broad, horny scales, the small head is elongate, and the mouth is terminal in the snout. { 'paŋ·gə·lən }

panhypopituitarism *See* hypopituitarism. { pan ,hī·pō·pə'tü·ə·tə,riz·əm }

panicle [BOT] A branched or compound raceme in which the secondary branches are often racemose as well. { 'pan·ə·kəl }

panmixis [BIOL] Random mating within a breeding population; in a closed population this results in a high degree of uniformity. { pan 'mik·səs }

Panmycin [MICROBIO] A trade name for tetracycline. { pan'mīs·ən }

panniculitis [MED] Inflammation of the layer of subcutaneous fat, especially in the abdomen. { pə,nik·yə'līd·əs }

panniculus [ANAT] A membrane or layer. { pə'nik·yə·ləs }

pannose [BIOL] Having a felty or woolly texture. { 'pa,nōs }

pannus [MED] **1.** Vascularization accompanied by deposition of connective tissue beneath the cornal epithelium. **2.** Overgrowth of connective tissue on the articular surface of a diarthrodial joint. { 'pan·əs }

panophthalmitis [MED] Inflammation of all the tissues of the eyeball. { pan,äf,thal'mīd·əs }

panspermia [BIOL] The theoretical ability of life to travel from body to body within the solar system. { pan'spər·mē·ə }

pansporoblast [INV ZOO] A sporont of cnidosporan protozoans that contains two sporoblasts. { pan'spȯr·ə,blast }

Pantodontidae [VERT ZOO] A family of fishes in the order Osteoglossiformes; the single, small species is known as African butterflyfish because of its expansive pectoral fins. { ,pan·tə'dän·tə,dē }

pantophagous [ZOO] Feeding on a variety of foods. { pan'täf·ə·gəs }

Pantophthalmidae [INV ZOO] The wood-boring flies, a family of orthorrhaphous dipteran insects in the series Brachycera. { ,pan,täf'thal·mə,dē }

Pantopoda [INV ZOO] The equivalent name for Pycnogonida. { pan'täp·ə·də }

pantothenate [BIOCHEM] A salt or ester of pantothenic acid. { ,pan·tə'the,nāt }

pantothenic acid [BIOCHEM] $C_9H_{17}O_5N$ A member of the vitamin B complex that is essential for nutrition of some animal species. Also known as vitamin B_3. { ¦pan·tə¦then·ik 'as·əd }

panuveitis [MED] Inflammation of the entire uveal tract. { pan,yü·vē'īd·əs }

panzootic [VET MED] Affecting many animals of different species. { ,pan·zō'äd·ik }

papain [BIOCHEM] An enzyme preparation obtained from the juice of the fruit and leaves of the papaya (*Carica papaya*); contains proteolytic enzymes. { pə'pī·ən }

Papanicolaou test [PATH] A technique for the detection of precancerous and early noninvasive cancer by the staining and examination of exfoliated cells; used especially in the diagnosis of uterine cervical and endometrial cancer. Also known as Pap test. { ,pä·pə'nēk·ə,laů ,test }

Papaveraceae [BOT] A family of dicotyledonous plants in the order Papaverales, with regular flowers, numerous stamens, and a well-developed latex system. { pə,pav·ə'rās·ē,ē }

Papaverales [BOT] An order of dicotyledonous plants in the subclass Magnoliidae, marked by a syncarpous gynoecium, parietal placentation, and only two sepals. { pə,pav·ə'rā·lēz }

Papaver somniferum *See* poppy. { ,päp·ə,ver säm 'nif·ə·rəm }

Papilionidae [INV ZOO] A family of lepidopteran insects in the superfamily Papilionoidea; members are the only butterflies with fully developed forelegs bearing an epiphysis. { pə,pil·ē'än·ə,dē }

Papilionoidea [INV ZOO] A superfamily of diurnal butterflies (Lepidoptera) with clubbed antennae, which are rounded at the tip, and forewings that always have two or more veins. { pə,pil·ē·ə'nȯid·ē·ə }

Papilionoideae [BOT] A subfamily of the family Leguminosae with characteristic irregular flowers that have a banner, two wing petals, and two lower petals united to form a boat-shaped keel. { pə,pil·ē·ə'nȯid·ē,ē }

papilla [BIOL] A small, nipplelike eminence. { pə'pil·ə }

papilla of Vater *See* ampulla of Vater. { pə'pil·ə əv 'fät·ər }

papillary adenoma of ovary *See* serous cystadenoma. { 'pap·ə,ler·ē ,ad·ən'ō·mə əv 'ō·və·rē }

papillary carcinoma [MED] A carcinoma characterized by fingerlike outgrowths. { 'pap·ə,ler·ē ,kärs·ən'ō·mə }

papillary muscle [ANAT] Any of the muscular eminences in the ventricles of the heart from which the chordae tendineae arise. { 'pap·ə,ler·ē 'məs·əl }

papillate [BIOL] **1.** Having or covered with papillae. **2.** Resembling a papilla. Also known as papillose. { 'pap·ə,lāt }

papilledema [MED] Edema of the optic disk. Also known as choked disk. { ¦pap·əl·ə'dē·mə }

papillocystoma *See* serous cystadenoma. { ¦pap·ə·lō·si'stō·mə }

papilloma [MED] A growth pattern of epithelial tumors in which the proliferating epithelial cells grow outward from a surface, accompanied by vascularized cores of connective tissue, to form a branching structure. { ,pap·ə'lō·mə }

papillomatosis [MED] Widespread formation of papillomas. { ,pap·ə,lō·mə'tō·səs }

papillomatous [MED] Characterized by or pertaining to a papilloma. { ¦pap·ə¦läm·əd·əs }

papillose *See* papillate. { 'pap·ə,lōs }

Papovaviridae [VIROL] A family of deoxyribonucleic acid (DNA)-containing viruses characterized by a nonenveloped icosahedral virion containing double-stranded circular DNA that is complexed inside the nucleocapsid to histone proteins of host cell origin. { ,pä·pə·və'vir·ə,dī }

papovavirus [VIROL] A deoxyribonucleic acid-containing group of animal viruses, including papilloma and vacuolating viruses. { ¦pap·ə·və'vī·rəs }

pappataci fever *See* phlebotomus fever. { ¦päp·ə¦tä·chē ,fē·vər }

pappus [BOT] An appendage or group of appendages consisting of a modified perianth on the ovary or fruit of various seed plants; adapted to dispersal by wind and other means. { 'pap·əs }

paprika [BOT] *Capsicum annuum*. A type of pepper with nonpungent flesh, grown for its long red fruit from which a dried, ground condiment is prepared. { pə'prē·kə }

Pap test *See* Papanicolaou test. { 'pap ,test }

papula [BIOL] A small papilla. { 'pap·yə·lə }

papule [MED] A solid circumscribed elevation of the skin varying from less than 0.1 to 1 centimeter in diameter. { 'pap·yül }

papulonecrotic [MED] Pertaining to papule formation with a tendency to central necrosis; applied especially to a variety of skin tuberculosis. { ,pap·yə·lō·ne'kräd·ik }

paraaortic body [ANAT] One of the small masses of chromaffin tissue lying along the abdominal aorta. Also known as glomus aorticum. { ¦par·ə·ā¦ȯrd·ik 'bäd·ē }

parabasal body *See* kinetoplast. { ¦par·ə¦bā·səl 'bäd·ē }

parabiosis [BIOL] Experimental joining of two individuals to study the effects of one partner upon the other. { ¦par·ə·bī'ō·səs }

parabiotic [MED] Physiologically and anatomically associated. { ,par·ə·bī'äd·ik }

Paracanthopterygii [VERT ZOO] A superorder of teleost fishes, including the codfishes and allied groups. { ,par·ə,kan,thäp·tə'rij·ē,ī }

paracentesis [MED] Puncture of the wall of a fluid-filled cavity by means of a hollow needle to draw off the contents. { ,par·ə·sen'tē·səs }

paracentric inversion [GEN] A type of chromosomal alteration that occurs within one arm of a chromosome and does not span the centromere. { ¦par·ə¦sen·trik in'vər·zhən }

paracoccidioidomycosis *See* South American blastomycosis. { ¦par·ə·käk,sid·ē¦ȯid·ō·mī,kō·səs }

paracolon bacteria [MICROBIO] A group of bacteria intermediate between the *Escherichia-Aerobacter* genera and the *Salmonella-Shigella* group. { ¦par·ə'kō·lən bak'tir·ē·ə }

paracondyloid [VERT ZOO] A process on the outer side of each condyle of the occipital bone in the skull of certain mammals. { ¦par·ə'känd·əl,ȯid }

paracone [VERT ZOO] **1.** The anterior cusp of a primitive tricuspid upper molar. **2.** The principal anterior, external cusp of an upper molar in higher forms. { 'par·ə,kōn }

paraconid [VERT ZOO] **1.** The cusp of a primitive lower molar corresponding to the paracone. **2.** The anterior, internal cusp of a lower molar in higher forms. { ,par·ə'kän·əd }

paracrine signaling [PHYSIO] Signaling in which the target cell is close to the signaling cell and the signal molecule affects only adjacent target cells. { 'par·ə,krēn ,sig·nəl·iŋ }

Paracucumidae [INV ZOO] A family of holothurian echinoderms in the order Dendrochirotida; the body is invested with plates and has a simplified calcareous ring. { ¦par·ə·kə'kyüm·ə,dē }

paracystitis [MED] Inflammation of the connective tissue surrounding the urinary bladder. { ¦par·ə·si'stīd·əs }

paradidymis [ANAT] Atrophic remains of the paragenital tubules of the mesonephros, occurring near the convolutions of the epididymal duct. { ¦par·ə¦did·ə·məs }

paradoxical cold [PHYSIO] The arousal of a cold sensation by application of a hot probe to a cold point (a skin receptor that normally responds only to cold). { ,par·ə,däk·sə·əl 'kōld }

paradoxical embolus [MED] An embolus which is transported to the circulation in peripheral arteries through septal defect in the heart, usually a patent foramen ovale. { ,par·ə'däk·sə·kəl em'bə·ləs }

paradoxical warmth [PHYSIO] The arousal of a warm sensation by application of a cold probe to a warm point (a skin receptor that normally responds only to warmth). { ,par·ə,däk·sə·kəl 'wȯrmth }

paraesophageal cyst [MED] A bronchogenic cyst intimately connected with the esophageal wall, containing cartilage, and usually filled with a mucoid material and desquamated epithelial cells. { ¦par·ə·i,säf·ə'jē·əl 'sist }

paraganglion [NEUROSCI] Small structure associated with the sympathetic nervous system containing chromaffin tissue (hormonally active tissue related to the adrenal medulla). Also known as chromaffin body. { ¦par·ə'gaŋ·glē,än }

paragastric [ANAT] Located near the stomach. [INV ZOO] A cavity in Porifera into which radial canals open, and which opens to the outside through the cloaca. { ¦par·ə¦gas·trik }

paragenetic [GEN] Pertaining to chromosome changes that alter gene expression but not makeup. { ,par·ə·jə'ned·ik }

paragglutination [IMMUNOL] Agglutination of colon bacteria with the serum of a patient infected, or recovering from an infection, with dysentery bacilli. { ,par·ə,glüt·ən'ā·shən }

paragnath [INV ZOO] **1.** One of the paired leaflike lobes of the metastoma situated behind the mandibles in most crustaceans. **2.** One of the paired lobes of the hypopharynx in certain insects. **3.** One of the small, sharp and hard jaws of certain annelids. { 'par·əg,nath }

paragonimiasis [MED] Presence of the fluke *Paragonimus westermani* in the lungs or other tissues of humans. { ,par·ə,gän·ə'mī·ə·səs }

parainfluenza [MED] A viral condition similar to or resulting from influenza. [MICROBIO] An organism exhibiting growth characteristics of *Hemophilus influenzae*. { ¦par·ə,in·flü'en·zə }

parakeet [VERT ZOO] Any of various small, slender species of parrots with long tails in the family Psittacidae. { 'par·ə,kēt }

parakeratosis [PATH] Incomplete keratinization of epidermal cells characterized by retention of nuclei of cells attaining the level of the stratum corneum. { ¦par·ə,ker·ə'tō·səs }

paralalia [MED] Disturbance of the faculty of speech, characterized by distortion of sounds, or the habitual substitution of one sound for another. { ,par·ə'lāl·ē·ə }

paralexia [MED] Transposition or substitution of words or syllables in reading. { ,par·ə'lek·sē·ə }

paralgesia [MED] **1.** Paresthesia characterized by pain. **2.** Any perverted and disagreeable cutaneous sensation, as of formication, cold, or burning. { ,par·əl'jē·zē·ə }

parallel evolution [EVOL] Evolution of similar characteristics in different groups of organisms. { 'par·ə,lel ,ev·ə'lü·shən }

parallel muscle [ANAT] Any muscle having the long fibers arranged parallel to each other. { 'par·ə,lel 'məs·əl }

parallel-veined [BOT] Of a leaf, having the veins parallel, or nearly parallel, to each other. { 'par·ə,lel ¦vānd }

parallel venation [BOT] A vascular arrangement in leaves characterized by the longitudinal (or nearly so) orientation of veins of relatively uniform size. { ,par·ə,lel və'nā·shən }

paralogous locus [GEN] A gene that arose by duplication and later diverged in sequence or location from the parent gene. { pə'ral·ə·gəs ,lō·kəs }

paralutein cells [HISTOL] Epithelioid cells of the corpus luteum. { ¦par·ə'lüd·ē·ən ,selz }

paralysis [MED] Complete or partial loss of motor or sensory function. { pə'ral·ə·səs }

paralysis agitans *See* parkinsonism. { pə'ral·ə·səs 'aj·ə,tanz }

paralytic secretion [PHYSIO] Glandular secretion occurring in a denervated gland. { ¦par·ə¦lid·ik si'krē·shən }

paralytic spinal poliomyelitis [MED] An acute inflammatory virus disease chiefly involving the anterior horns of the gray matter of the spinal cord. { ¦par·ə¦lid·ik 'spīn·əl ¦pō·lē·ō,mī·ə'līd·əs }

Parameciidae [INV ZOO] A family of ciliated protozoans in the order Holotrichia; the body has differentiated anterior and posterior ends and is bounded by a hard but elastic pellicle. { ,par·ə·mə'sī·ə,dē }

paramecium [INV ZOO] A single-celled protozoan belonging to the family Parameciidae. { ,par·ə'mē·sē·əm }

Paramecium [INV ZOO] The genus of protozoans composing the family Parameciidae. { ,par·ə'mē·sē·əm }

paramedical [MED] Having a supplementary or secondary relation to medicine. { ¦par·ə'med·ə·kəl }

Paramelina [VERT ZOO] An order of marsupials that includes the bandicoots in some systems of classification. { ,par·ə'mel·ə·nə }

paramere [BIOL] One half of a bilaterally symmetrical animal or somite. [INV ZOO] Any of several paired structures of an insect, especially those on the ninth abdominal segment. { 'par·ə,mir }

paramesonephric duct [EMBRYO] An embryonic genital duct; in the female, it is the anlage of the oviducts, uterus, and vagina; in the male, it degenerates, leaving the appendix testes. Also known as Müllerian duct. { ¦par·ə¦me·zə¦nef·rik 'dəkt }

paramethadione [PHARM] $C_7H_{11}NO_3$ An anticonvulsant primarily useful in the treatment of petit mal epilepsy. { ¦par·ə¦meth·ə'dī,ōn }

paramo [ECOL] A biological community, essentially a grassland, covering extensive high areas in equatorial mountains of the Western Hemisphere. { 'pär·ə,mō }

paramutation [GEN] A mutation in which one member of a heterozygous pair of alleles permanently changes its partner allele. { ¦par·ə·myu 'tā·shən }

paramylum [BIOCHEM] A reserve, starchlike carbohydrate of various protozoans and algae. { pə'ram·ə·ləm }

paramyosin [BIOCHEM] A type of fibrous protein found in the adductor muscles of bivalves and thought to form the core of a filament with myosin molecules at the surface. { ,par·ə'mī·ə·sən }

paramyotonia congenita [MED] A heredofamilial condition characterized by recurrent muscular stiffness and weakness (myotonia) on exposure to cold, as well as on mechanical irritation; transmitted as an autosomal dominant and considered to be a variety of the hyperkalemic form of periodic paralysis. Also known as Eulenburg's disease; myotonia congenita intermittens. { ˌpar·əˌmī·ə'tō·nē·ə kən'jen·əd·ə }

Paramyxoviridae [VIROL] A family of negative-strand ribonucleic acid (RNA) viruses characterized by an enveloped spherical virion containing a single-stranded, nonfragmented molecule of RNA, contains the genera *Paramyxovirus* (sendai, mumps), *Morbillivirus* (measles), and *Pneumovirus* (respiratory syncytial virus). { ˌpar·əˌmik·sə 'vir·əˌdī }

paramyxovirus [VIROL] A subgroup of myxoviruses, including the viruses of mumps, measles, parainfluenza, and Newcastle disease; all are ribonucleic acid-containing viruses and possess an ether-sensitive lipoprotein envelope. { ¦par·əˌmik·sō'vī·rəs }

paranasal sinus [ANAT] Any of the paired sinus cavities of the human face; includes the frontal, ethmoid, and sphenoid sinuses. { ¦par·ə¦nā·zəl 'sī·nəs }

paranephritis [MED] **1.** Inflammation of the adrenal gland. **2.** Inflammation of the connective tissue adjacent to the kidney. { ¦par·ə·nə'frīd·əs }

paranesthesia [MED] Anesthesia of the body below the waist. { pə¦ran·əs'thē·zhə }

paranosmia [MED] A deviation in odor sensitivity involving change in odor quality. { ˌpar·ə'näz·mē·ə }

paranthropophytia [ECOL] Discrepant control of regions or areas due to immediate and continuous or periodic interference, as by certain cultivation practices. { ˌpar·anˌthräp·ə'fī·shə }

Paraonidae [INV ZOO] A family of small, slender polychaete annelids belonging to the Sedentaria. { ¦par·ə'än·əˌdē }

paraparesis [MED] Partial paralysis of the lower extremities. { ¦par·ə·pə'rē·səs }

parapatric [ECOL] Referring to populations or species that occupy nonoverlapping but adjacent geographical areas without interbreeding. { ¦par·ə¦pa·trik }

parapatric speciation [EVOL] Gradual speciation whereby new species are created from populations that maintain overlapping geographic zones of genetic contact. { ¦par·ə¦pa·trik ˌspe·shē'ā·shən }

parapertussis [MED] An acute bacterial respiratory infection similar to mild pertussis and caused by *Bordetella pertussis*. { ¦par·ə·pər'təs·əs }

paraphimosis [MED] **1.** Retraction and constriction, especially of the prepuce behind the glans penis. Also known as Spanish collar. **2.** Retraction of the eyelid behind the eyeball. { ˌpar·ə·fə'mō·səs }

paraphyletic [SYST] Describing a taxonomic group that does not contain all of the descendants of its most recent common ancestor. { ˌpar·ə·fī'led·ik }

paraphysis [BOT] A sterile filament borne among the sporogenous or gametogenous organs in many cryptogams. [VERT ZOO] A median evagination of the roof of the telencephalon of some lower vertebrates. { pə'raf·ə·səs }

paraplasm *See* hyaloplasm. { 'par·əˌplaz·əm }

paraplegia [MED] Paralysis of the lower limbs. { ˌpar·ə'plē·jə }

parapodium [INV ZOO] **1.** One of the short, paired processes on the sides of the body segments in certain annelids. **2.** A lateral expansion of the foot in gastropod mollusks. { ˌpar·ə'pōd·ē·əm }

parapolar cell [INV ZOO] Either of the first two trunk cells in the development of certain Mesozoa. { ¦par·ə¦pō·lər 'sel }

Parasaleniidae [INV ZOO] A family of echinacean echinoderms in the order Echinoida composed of oblong forms with trigeminate ambulacral plates. { ¦par·əˌsal·ə'nī·əˌdē }

Paraselloidea [INV ZOO] A group of the Asellota that contains forms in which the first pleopods of the male are coupled along the midline, and are lacking in the female. { ˌpar·ə·sə'lȯid·ē·ə }

parasexual cycle [GEN] A series of events leading to genetic recombination in vegetative or somatic cells even in mammals; it was first described in filamentous fungi; there are three essential steps: heterokaryosis; fusion of unlike haploid nuclei in the heterokaryon to yield heterozygous diploid nuclei; and recombination and segregation at mitosis by two independent processes, mitotic crossing-over and loss of chromosomes. { ¦par·ə'sek·shə·wəl 'sī·kəl }

parasite [BIOL] An organism that lives in or on another organism of different species from which it derives nutrients and shelter. { 'par·əˌsīt }

parasitemia [MED] The presence of parasites in the blood. { ˌpar·ə·si'tē·mē·ə }

Parasitica [INV ZOO] A group of hymenopteran insects that includes four superfamilies of the Apocrita: Ichneumonoidea, Chalcidoidea, Cynipoidea, and Proctotrupoidea; some are phytophagous, while others are parasites of other insects. { ˌpar·ə'sid·ə·kə }

parasitic castration [BIOL] Destruction of the reproductive organs by parasites. { ¦par·ə¦sid·ik ka'strā·shən }

parasitic stomatitis *See* thrush. { ¦par·ə¦sid·ik ˌstō·mə'tīd·əs }

parasitism [ECOL] A symbiotic relationship in which the host is harmed, but not killed immediately, and the species feeding on it is benefited. { 'par·ə·səˌtiz·əm }

parasitoidism [BIOL] Systematic feeding by an insect larva on living host tissues so that the host will live until completion of larval development. { ¦par·ə·sə'tȯidˌiz·əm }

parasitology [BIOL] A branch of biology which deals with those organisms, plant or animal, which have become dependent on other living creatures. { ˌpar·ə·sə'täl·ə·jē }

parasomnia [MED] A group of disorders characterized by abnormal movements or behavior intruding into sleep (for example, sleep walking, sleep terrors, sleep talking, nightmares, sleep paralysis, tooth grinding, and bed wetting). { ˌpar·ə'säm·nē·ə }
parasphenoid [VERT ZOO] A bone in the base of the skull of many vertebrates. { ¦par·ə'sfēˌnȯid }
parastacidae [INV ZOO] A family of crayfishes assigned to the Nephropoidea. { ˌpar·ə'stās·əˌdē }
parastyle [VERT ZOO] A small cusp anterior to the paracone of an upper molar. { 'par·əˌstīl }
parasympathetic [NEUROSCI] Of or pertaining to the craniosacral division of the autonomic nervous system. { ¦par·əˌsim·pə'thed·ik }
parasympathetic nervous system [NEUROSCI] The craniosacral portion of the autonomic nervous system, consisting of preganglionic nerve fibers in certain sacral and cranial nerves, outlying ganglia, and postganglionic fibers. { ¦par·əˌsim·pə'thed·ik 'nər·vəs ˌsis·təm }
parasympatholytic [PHARM] Blocking the action of parasympathetic nerve fibers. { ¦par·əsimˌpath·ə'lid·ik }
parasympathomimetic [PHARM] Of drugs, having an effect similar to that produced when the parasympathetic nerves are stimulated. { ˌpar·ə¦sim·pə·thō·mi'med·ik }
parathelioma *See* interstitial endometriosis. { ¦par·əˌthē·lē'ō·mə }
parathormone [BIOCHEM] A polypeptide hormone that functions in regulating calcium and phosphate metabolism. Also known as parathyroid hormone. { ˌpar·ə'thȯrˌmōn }
parathyroidectomy [MED] Excision of a parathyroid gland. { ¦par·əˌthīˌrȯi'dek·tə·mē }
parathyroid gland [ANAT] A paired endocrine organ located within, on, or near the thyroid gland in the neck region of all vertebrates except fishes. { ¦par·ə'thīˌrȯid ˌgland }
parathyroid hormone *See* parathormone. { ¦par·ə'thīˌrȯid 'hȯrˌmōn }
paratonic movement [BOT] The movement of the whole or parts of a plant due to the influence of an external stimulus, such as gravity, chemicals, heat, light, or electricity. { ¦par·ə¦tän·ik 'müv·mənt }
paratrachoma *See* inclusion conjunctivitis. { ¦par·ə·tra'kō·mə }
paratrichosis [MED] A condition in which the hair is either imperfect in growth or develops in abnormal places. { ¦par·ə·tri'kō·səs }
paratroch [INV ZOO] The ciliated band encircling the anus in certain trichophore larvae. { 'par·əˌtrōk }
paratuberculosis *See* Johne's disease. { ¦par·ə·tə ˌbər·kyə'lō·səs }
paratype [SYST] A specimen other than the holotype which is before the author at the time of original description and which is designated as such or is clearly indicated as being one of the specimens upon which the original description was based. { 'par·əˌtīp }
paratyphoid fever [MED] A bacterial disease of humans resembling typhoid fever and caused by *Salmonella paratyphi*. { ¦par'tīˌfȯid 'fē·vər }
parauterine organ [INV ZOO] In certain tapeworms, a pouchlike sac which receives and retains the embryos. { ¦par·ə'yüd·ə·rən ˌȯr·gən }
paraxonic [VERT ZOO] Pertaining to a state or condition wherein the axis of the foot lies between the third and fourth digits. { ¦par·ak 'sän·ik }
Parazoa [INV ZOO] A name proposed for a subkingdom of animals which includes the sponges (Porifera). { 'par·ə'zō·ə }
paregoric [PHARM] Camphorated opium tincture, a preparation of opium, camphor, benzoic acid, anise oil, glycerin, and diluted alcohol; used mainly as an antiperistaltic. { ¦par·ə¦gȯr·ik }
parenchyma [BOT] A tissue of higher plants consisting of living cells with thin walls that are agents of photosynthesis and storage; abundant in leaves, roots, and the pulp of fruit, and found also in leaves and stems. [HISTOL] The specialized epithelial portion of an organ, as contrasted with the supporting connective tissue and nutritive framework. { pə'reŋ·kə·mə }
parenchymal jaundice [MED] Any of various forms of jaundice in which the disease is due in part to damaged liver cells. { pə'reŋ·kə·məl 'jȯn·dəs }
parenchymella *See* diploblastula. { pə¦reŋ·kə 'mel·ə }
parenchymula [INV ZOO] The flagellate larva of calcinean sponges in which there is a cavity filled with gelatinous connective tissue. { ˌpar·əŋ 'kim·yə·lə }
parental imprinting [GEN] The condition whereby the extent of gene expression depends upon the sex of the parent that transmits the gene. Also known as genomic imprinting. { pə¦ren·təl 'imˌprint·iŋ }
parenteral [MED] Outside the intestine; not via the alimentary tract. { pər'ent·ə·rəl }
paresis [MED] **1.** A slight paralysis. **2.** Incomplete loss of muscular power. **3.** Weakness of a limb. { pə'rē·səs }
paresthesia [MED] Tingling, crawling, or burning sensation of the skin. { ˌpar·əs'thē·zhə }
Pareulepidae [INV ZOO] A monogeneric family of errantian polychaete annelids. { ˌpar·yü 'lep·əˌdē }
parietal [ANAT] Of or situated on the wall of an organ or other body structure. [BOT] Of a plant part, having a peripheral location or orientation; in particular, attached to the main wall of an ovary. { pə'rī·əd·əl }
parietal block *See* intraventricular heart block. { pə'rī·əd·əl ˌbläk }
parietal bone [ANAT] The bone that forms the side and roof of the cranium. { pə'rī·əd·əl ˌbōn }
parietal cell [HISTOL] One of the peripheral, hydrochloric acid-secreting cells in the gastric fundic glands. Also known as acid cell; delomorphous cell; oxyntic cell. { pə'rī·əd·əl ˌsel }
Parietales [BOT] An order of plants in the

Englerian system; families are placed in the order Violales in other systems. { pə,rī·ə'tā·lēz }

parietal lobe [ANAT] The cerebral lobe of the brain above the lateral cerebral sulcus and behind the central sulcus. { pə'rī·əd·əl ,lōb }

parietal peritoneum [ANAT] The portion of the peritoneum lining the interior of the body wall. { pə'rī·əd·əl ,per·ə·tə'nē·əm }

parietal pleura [ANAT] The pleura lining the inner surface of the thoracic cavity. { pə'rī·əd·əl 'plůr·ə }

Parinaud's oculoglandular syndrome [MED] Enlargement of the lymph node around the eye and conjunctivitis. { ¦pär·ə,nōz ,äk·yə·lō,glan·jə·lər 'sin,drōm }

Parkinje effect *See* Purkinje effect. { pər'kin·jē i,fekt }

parkinsonism [MED] A clinical state characterized by tremor at a rate of three to eight tremors per second, with "pill-rolling" movements of the thumb common, muscular rigidity, dyskinesia, hypokinesia, and reduction in number of spontaneous and autonomic movements; produces a masked facies, disturbances of posture, gait, balance, speech, swallowing, and muscular strength. Also known as paralysis agitans; Parkinson's disease. { 'pär·kən·sə,niz·əm }

Parkinson's disease *See* parkinsonism. { 'pär·kən·sənz di,zēz }

parkland *See* temperate woodland; tropical woodland. { 'pärk,land }

Parmeliaceae [BOT] The foliose shield lichens, a family of the order Lecanorales. { ,pär·mel·ē'ās·ē,ē }

Parnidae [INV ZOO] The equivalent name for Dryopidae. { 'par·nə,dē }

paromomycin [MICROBIO] A broad-spectrum antibiotic produced by *Streptomyces rimosus forma paromomycinus*; it is effective in the treatment of intestinal amebiasis in humans. { ,par·ə·me 'mīs·ən }

paronychia [MED] A suppurative inflammation about the margin of a nail. { ,par·ə'nik·ē·ə }

paronychium *See* perionychium. { ,par·ə'nik·ē·əm }

parotid duct [ANAT] The duct of the parotid gland. Also known as Stensen's duct. { pə 'räd·əd 'dəkt }

parotidectomy [MED] Excision of a parotid gland. { pə,räd·ə'dek·tə·mē }

parotid gland [ANAT] The salivary gland in front of and below the external ear; the largest salivary gland in humans; a compound racemose serous gland that communicates with the mouth by Steno's duct. { pə'räd·əd 'gland }

parotitis [MED] Inflammation of the parotid glands. { ,par·ə'tīd·əs }

parous [MED] Pertaining to an organism that has produced offspring. { 'par·əs }

parovarian cyst [MED] A cyst of mesonephric origin arising between the layers of the mesosalpinx, adjacent to the ovary. { ,par·ō'ver·ē·ən 'sist }

parovarium *See* epoophoron. { ,par·ə'var·ē·əm }

paroxysm [MED] **1.** A sudden attack, or the periodic crisis in the progress of a disease. **2.** A spasm, convulsion, or seizure. **3.** A burst of electrical activity during electroencephalography in the form of spikes, or spikes and waves, which indicates cerebral dysrhythmia or epileptic discharges. { 'par·ək,siz·əm }

parrot [VERT ZOO] Any member of the avian family Psittacidae, distinguished by the short, stout, strongly hooked beak. { 'par·ət }

pars anterior [ANAT] The major secretory portion of the anterior lobe of the adenohypophysis. Also known as pars distalis. { ¦pärz ,an'tir·ē·ər }

pars distalis *See* pars anterior. { ¦pärz 'dis·tə·ləs }

pars intermedia [ANAT] The intermediate lobe of the adenohypophysis. { ¦pärz ,in·tər'mē·dē·ə }

parsley [BOT] *Petroselinum crispum*. A biennial herb of European origin belonging to the order Umbellales; grown for its edible foliage. { 'pär·slē }

pars nervosa [ANAT] The inferior subdivision of the neurohypophysis. Also known as pars neuralis. { ¦pärz nər'vō·sə }

pars neuralis *See* pars nervosa. { ¦pärz nü'räl·əs }

parsnip [BOT] *Pastinaca sativa*. A biennial herb of Mediterranean origin belonging to the order Umbellales; grown for its edible thickened taproot. { 'pär·snəp }

pars tuberalis [ANAT] A pair of processes that grow forward or upward along the stalk of the adenohypophysis. { ¦pärz ,tü·bə'ral·əs }

parthenita [INV ZOO] A stage, such as the sporocyst, redia, or cercaria, in the development of a fluke which reproduces parthenogenetically. { pär'then·əd·ə }

parthenocarpy [BOT] Production of fruit without fertilization. { 'pär·thə·nō,kär·pē }

parthenogenesis [INV ZOO] A special type of sexual reproduction in which an egg develops without entrance of a sperm; common among rotifers, aphids, thrips, ants, bees, and wasps. { ¦pär·thə·nō'jen·ə·səs }

parthenomerogony [EMBRYO] Development of a nucleated fragment of an unfertilized egg following parthenogenetic stimulation. { ¦pär·thə·nō·mə'räg·ə·nē }

parthenospore *See* azygospore. { 'pär·thə·nə ,spȯr }

partial cleavage [EMBRYO] Cleavage in which only part of the egg divides into blastomeres. { 'pär·shəl 'klē·vij }

partridge [VERT ZOO] Any of the game birds comprising the genera *Alectoris* and *Perdix* in the family Phasianidae. { 'pär·trij }

parturient [MED] **1.** In labor; giving birth. **2.** Of or pertaining to parturition. { pär'tůr·ē·ənt }

parturifacient [MED] **1.** Inducing labor. **2.** An agent that induces labor. { pär¦tůr·ə¦fā·shənt }

parturiometer [MED] An instrument to determine the progress of labor by measuring the expulsive force of the uterus. { pär,tůr·ē'äm·əd·ər }

parturition [MED] The process of giving birth. { ,pär·chə'rish·ən }

parulis [MED] A subperiosteal abscess arising from dental structures. { pə'rü·ləs }

Parvoviridae [VIROL] A family of deoxyribonucleic acid (DNA)-containing animal viruses characterized by an icosahedral virion containing three or four structural proteins and a single-stranded DNA genome; includes the *Dependovirus*, *Parvovirus*, and *Densovirus* genera. { ˌpär·və'vir·əˌdī }

parvovirus [VIROL] The equivalent name for picodnavirus. { ¦pär·vō'vī·rəs }

PAS *See para*-aminosalicylic acid.

Paschen bodies [PATH] Accumulations of the elementary bodies of smallpox. { 'päsh·ən ˌbäd·ēz }

passage number [MICROBIO] The number of times that a culture has been subcultured. { 'pas·ij ˌnəm·bər }

Passalidae [INV ZOO] The peg beetles, a family of tropical coleopteran insects in the superfamily Scarabaeoidea. { pə'sal·əˌdē }

passenger [CELL MOL] A deoxyribonucleic acid segment that will be spliced into a plasmid or bacteriophage for subsequent cloning. { 'pas·ən·jər }

Passeres [VERT ZOO] The equivalent name for Oscines. { 'pas·əˌrēz }

Passeriformes [VERT ZOO] A large order of perching birds comprising two major divisions: Suboscines and Oscines. { ˌpas·ə·rə'fȯrˌmēz }

Passifloraceae [BOT] A family of dicotyledonous, often climbing plants in the order Violales; flowers are polypetalous and hypogynous with a corona, and seeds are arillate with an oily endosperm. { ˌpas·ə·flə'rās·ēˌē }

passive anaphylaxis [IMMUNOL] Anaphylaxis elicited by temporary sensitization with antibodies followed by injection of the corresponding sensitizing antigen. { 'pas·iv ˌan·ə·fə'lak·səs }

passive congestion [MED] An increased content of blood in an organ or other body part due to impaired return of venous blood. { 'pas·iv kən'jes·chən }

passive cutaneous anaphylaxis [IMMUNOL] The vascular reaction at the site of intradermally injected antibody when, 3 hours later, the specific antigen, usually mixed with Evans blue dye, is injected intravenously. { 'pas·iv kyü'tā·nē·əs ˌan·ə·fə'lak·səs }

passive immunity [IMMUNOL] **1.** Immunity acquired by injection of antibodies in another individual or in an animal. **2.** Immunity acquired by the fetus by the transfer of maternal antibodies through the placenta. { 'pas·iv i'myün·əd·ē }

passive immuno-gene therapy [IMMUNOL] Immuno-gene therapy in which autologous T cells are repeatedly sensitized in culture by exposure to cytokines or by transfer of cytokine genes and then are transferred back into the patient. { ¦pas·iv ¦im·yə·nō'jēn ˌther·ə·pē }

passive immunotherapy [IMMUNOL] Immunotherapy involving the transfer of antibodies to tumor-bearing recipients. { ¦pas·iv ˌim·yə'ther·ə·pē }

Pasteur-Chamberland filter [MICROBIO] A porcelain filter with small pores used in filtration sterilization. { pa'stər 'chām·bər·lənd ˌfil·tər }

Pasteur effect [MICROBIO] Inhibition of fermentation by supplying an abundance of oxygen to replace anerobic conditions. { pa'stər iˌfekt }

Pasteurella [MICROBIO] A genus of gram-negative, nonmotile, nonsporulating, facultatively anaerobic coccobacillary to rod-shaped bacteria which are parasitic and often pathogens in many species of mammals, birds, and reptiles, it was named to honor Louis Pasteur in 1887. { ˌpas·chə'rel·ə }

pasteurellosis *See* hemorrhagic septicemia. { ˌpa·stə·rə'lō·səs }

Pasteuriaceae [MICROBIO] Formerly a family of stalked bacteria in the order Hyphomicrobiales. { ˌpa·stə·rēˌās·ēˌē }

patagium [VERT ZOO] **1.** A membrane or fold of skin extending between the forelimbs and hindlimbs of flying squirrels, flying lizards, and other arboreal animals. **2.** A membrane or fold of skin on a bird's wing anterior to the humeral and radioulnar bones. { pə'tā·jē·əm }

patch budding [BOT] Budding in which a small rectangular patch of bark bearing the scion (a bud) is fitted into a corresponding opening in the bark of the stock. { 'pach ˌbəd·iŋ }

patch test [IMMUNOL] A test in which material is applied and left in contact with intact skin surfaces for 48 hours in order to demonstrate tissue sensitivity. { 'pach ˌtest }

patella [ANAT] A sesamoid bone in front of the knee, developed in the tendon of the quadriceps femoris muscle. Also known as kneecap. { pə 'tel·ə }

Patellidae [INV ZOO] The true limpets, a family of gastropod mollusks in the order Archeogastropoda. { pə'tel·əˌdē }

patent [MED] Open; exposed. { 'pat·ənt }

patent foramen ovale [MED] Persistence, usually functional, of the fetal foramen ovale after birth. { 'pat·ənt fə'rā·mən 'ōv·yül }

patent medicine [PHARM] A medicine, generally trademarked, whose composition is incompletely disclosed. { 'pat·ənt 'med·ə·sən }

patent period [MED] The period of an infective disease during which the causative agent can be detected. { 'pat·ənt ˌpir·ē·əd }

paternity test [IMMUNOL] Identification of the blood groups of a mother, her child, and a putative father in order to establish the probability of paternity or nonpaternity; actually, only nonpaternity can be established. { pə'tər·nəd·ē ˌtest }

Paterson-Kelly syndrome *See* Plummer-Vinson syndrome. { 'pad·ər·sən 'kel·ē ˌsinˌdrōm }

path clamp [PHYSIO] An electrophysiologic recording technique for studying the function and regulation of ionic currents in cell membranes down to the level of single ion channels. { 'path ˌklamp }

pathergy [IMMUNOL] Either a subnormal response to an allergen or an unusually intense one in which the individual becomes sensitive not only to the specific substance but to others. { 'path·ər·jē }

pathogen [MED] A disease-producing agent; usually refers to living organisms. { 'path·ə·jən }

pathogenesis [MED] The origin and course of development of disease. { ¦path·ə'jen·ə·səs }

pathogenic [MED] **1.** Producing or capable of producing disease. **2.** Pertaining to pathogenesis. { ¦path·ə¦jen·ik }

pathogenicity [MED] The ability of an organism to enter a host and cause disease. { ˌpath·ə·jə'nis·əd·ē }

pathogenicity island [GEN] A deoxyribonucleic acid cluster commonly containing genes associated with pathogenesis. Also known as virulence cassette. { ˌpath·ə·jə'nis·əd·ē ˌī·lənd }

pathognomonic [MED] Characteristic of a given disease, enabling it to be distinguished from other diseases. { ¦path·əg·nō'män·ik }

pathologic anatomy [PATH] The study of structural changes caused by disease. Also known as morbid anatomy. { ¦path·ə¦läj·ik }

pathologist [MED] A person who specializes in the study and practice of pathology. { pə'thäl·ə·jəst }

pathology [MED] The study of the causes, nature, and effects of diseases and other abnormalities. { pə'thäl·ə·jē }

pathovar [MICROBIO] A pathological variant of a nonpathological bacterial species. Abbreviated pv. { 'path·əˌvär }

patroclinous inheritance [GEN] Inheritance in which the offspring more closely resembles the male parent than the female parent. { ¦pa·trəˌklī·nəs in'her·əd·əns }

pattern formation [EMBRYO] The embryogenic process in which the spatial differentiation of cells is specified in a structure that initially is largely homogeneous. { 'pad·ərn fərˌmā·shən }

pattern gene [GEN] A gene involved in the establishment of a particular pattern during cell differentiation. { 'pad·ərn ˌjēn }

patulin [PHARM] $C_7H_6O_4$ An antibiotic derived from several fungi (*Aspergillus*, *Penicillium* species); crystalline compound soluble in water and most organic solvents; melting point is 111°C; used as an antimicrobial agent; also appears to be a potent carcinogenic mycotoxin. Also known as penicidin. { 'pach·ə·lən }

paua *See* abalone. { paü·ə }

Paucituberculata [VERT ZOO] An order of marsupial mammals in some systems of classification, including the opossum, rats, and polydolopids. { ¦pȯs·ē·təˌbər·kyə'läd·ə }

Paul-Bunnell test *See* heterophile agglutination test. { 'pȯl 'bən·əl ˌtest }

Pauli exclusion principle *See* exclusion principle. { 'pȯl·ē ik'sklü·zhən ˌprin·sə·pəl }

paunch [ANAT] In colloquial usage, the cavity of the abdomen and its contents. [VERT ZOO] *See* rumen. { pȯnch }

paurometabolous metamorphosis [INV ZOO] A simple, gradual, direct metamorphosis in which immature forms resemble the adult except in size and are referred to as nymphs. { ˌpȯr·ē·mə'tab·ə·ləs ˌmed·ə'mȯr·fə·səs }

Pauropoda [INV ZOO] A class of the Myriapoda distinguished by bifurcate antennae, 12ny trunk segments with 9 pairs of functional legs, and the lack of eyes, spiracles, tracheae, and a circulatory system. { pȯ'räp·ə·də }

Paussidae [INV ZOO] The flat-horned beetles, a family of coleopteran insects in the suborder Adephaga. { 'pȯs·əˌdē }

Pavlov's pouch [PHYSIO] A small portion of stomach, completely separated from the main stomach, but retaining its vagal nerve branches, which communicates with the exterior; used in the long-term investigation of gastric secretion, and particularly in the study of conditioned reflexes. { 'pavˌläfs ˌpaüch }

paw [VERT ZOO] The foot of an animal, especially a quadruped having claws. { pȯ }

paxilla [INV ZOO] A pillarlike spine in certain starfishes that sometimes has a flattened summit covered with spinules. { pak'sil·ə }

Paxillosida [INV ZOO] An order of the Asteroidea in some systems of classification, equivalent to the Paxillosina. { ˌpak·sə'läs·əd·ə }

Paxillosina [INV ZOO] A suborder of the Phanerozonida with pointed tube feet which lack suckers, and with paxillae covering the upper body surface. { ˌpak·sə'läs·ə·nə }

PBI *See* protein-bound iodine.

PBI test *See* protein-bound iodine test. { ¦pē ¦bē'ī ˌtest }

P blood group [IMMUNOL] A system of immunologically distinct, genetically determined erythrocyte antigens first defined by their reaction with anti-P, and immune rabbit antiserum, and later broadened to include related antigens. { 'pē 'bləd ˌgrüp }

PC *See* phosphocreatine.

PCP *See* phencyclidine.

PCR *See* polymerase chain reaction.

PDGF *See* platelet-derived growth factor.

pea [BOT] **1.** *Pisum sativum*. The garden pea, an annual leafy leguminous vine cultivated for its smooth or wrinkled, round edible seeds which are borne in dehiscent pods. **2.** Any of several related or similar cultivated plants. { pē }

peach [BOT] *Prunus persica*. A low, spreading, freely branching tree of the order Rosales, cultivated in less rigorous parts of the temperate zone for its edible fruit, a juicy drupe with a single large seed, a pulpy yellow or white mesocarp, and a thin firm epicarp. { pēch }

peak expiratory flow rate [MED] A measurement of the amount of air that leaves the lungs on forced exhalation. Abbreviated PEFR. { ¦pēk ik,spī·rəˌtȯr·ē 'flō ˌrāt }

peak-flow monitor [IMMUNOL] A device that measures the amount of air that enters and leaves the lungs.

peanut [BOT] *Arachis hypogaea*. A low, branching, self-pollinated annual legume cultivated for its edible seed, which is a one-loculed legume formed beneath the soil in a pod. { 'pē·nət }

pear [BOT] Any of several tree species of the

genus *Pyrus* in the order Rosales, cultivated for their fruit, a pome that is wider at the apical end and has stone cells throughout the flesh. { per }

pearl [PATH] **1.** Rounded mass of concentrically arranged squamous epithelial cells, seen in some carcinomas. **2.** Mucous cast of the bronchi or bronchioles found in the sputum of asthmatic persons. { pərl }

pearl moss *See* carrageen. { 'pərl ,mȯs }

pearly tumor *See* cholesteatoma. { 'pər·lē 'tü·mər }

peat ball [ECOL] A lake ball containing an abundance of peaty fragments. { 'pēt ,bȯl }

peat flow [ECOL] A mudflow of peat produced in a peat bog by a bog burst. { 'pēt ,flō }

peat moss [ECOL] Moss, especially sphagnum moss, from which peat has been produced. { 'pēt ,mȯs }

pébrine [INV ZOO] A contagious protozoan disease of silkworms and other caterpillars caused by *Nosema bombycis*. { pā'brēn }

pecan [BOT] *Carya illinoensis*. A large deciduous hickory tree in the order Fagales which produces an edible, oblong, thin-shelled nut. { pi'kän }

peccary [VERT ZOO] Either of two species of small piglike mammals in the genus *Tayassu*, composing the family Tayassuidae. { 'pek·ə·rē }

pecking order [VERT ZOO] A hierarchy of social dominance within a flock of poultry where each bird is allowed to peck another lower in the scale and must submit to pecking by one of higher rank. { 'pek·iŋ ,ȯr·dər }

Pecora [VERT ZOO] An infraorder of the Artiodactyla; includes those ruminants with a reduced ulna and usually with antlers, horns, or deciduous horns. { 'pek·ə·rə }

pecten [ZOO] Any of various comblike structures possessed by animals. { 'pek·tən }

Pectenidae [INV ZOO] A family of bivalve mollusks in the order Anisomyaria; contains the scallops. { pek'ten·ə,dē }

pectic acid [BIOCHEM] A complex acid, partially demethylated, obtained from the pectin of fruits. { 'pek·tik 'as·əd }

pectin [BIOCHEM] A purified carbohydrate obtained from the inner portion of the rind of citrus fruits, or from apple pomace; consists chiefly of partially methoxylated polygalacturonic acids. { 'pek·tən }

Pectinariidae [INV ZOO] The cone worms, a family of polychaete annelids belonging to the Sedentaria. { ,pek·tə·nə'rī·ə,dē }

pectinase [BIOCHEM] An enzyme that catalyzes the transformation of pectin into sugars and galacturonic acid. { 'pek·tə,nās }

pectinesterase [BIOCHEM] An enzyme that catalyzes the hydrolytic breakdown of pectins to pectic acids. { ¦pek·tən'es·tə,rās }

pectineus [ANAT] A muscle arising from the pubis and inserted on the femur. { pek'tin·ē·əs }

Pectinibranchia [INV ZOO] An order of gastropod mollusks which contains many families of snails; respiration is by means of ctenidia, the nervous system is not concentrated, and sexes are separate. { ,pek·tə·nə'braŋ·kē·ə }

Pectobothridia [INV ZOO] A subclass of parasitic worms in the class Trematoda, characterized by caudal hooks or hard posterior suckers or both. { ,pek·tə·bä'thrid·ē·ə }

pectoral fin [VERT ZOO] One of the pair of fins of fishes corresponding to forelimbs of a quadruped. { 'pek·tə·rəl 'fin }

pectoral girdle [ANAT] The system of bones supporting the upper or anterior limbs in vertebrates. Also known as shoulder girdle. { 'pek·tə·rəl 'gərd·əl }

pectoralis major [ANAT] The large muscle connecting the anterior aspect of the chest with the shoulder and upper arm. { ,pek·tə'ral·əs 'mā·jər }

pectoralis minor [ANAT] The small, deep muscle connecting the third to fifth ribs with the scapula. { ,pek·tə'ral·əs 'mīn·ər }

pedal [BIOL] Of or pertaining to the foot. { 'ped·əl }

pedal disk [INV ZOO] The broad, flat base of many sea anemones, used for attachment to a substrate. { 'ped·əl ,disk }

pedal ganglion [INV ZOO] One of the paired ganglia supplying nerves to the foot muscles in most mollusks. { 'ped·əl 'gaŋ·glē·ən }

pedal gland *See* foot gland. { 'ped·əl ,gland }

pedate [BIOL] **1.** Having toelike parts. **2.** Having a foot. **3.** Having tube feet. { 'pe,dāt }

pediatrician [MED] A physician who specializes in pediatrics. { ,pēd·ē·ə'trish·ən }

pediatrics [MED] The branch of medicine that deals with the growth and development of the child through adolescence, and with the care, treatment, and prevention of diseases, injuries, and defects of children. { ,ped·ē'a·triks }

pedicel [BOT] **1.** The stem of a fruiting or spore-bearing organ. **2.** The stem of a single flower. [ZOO] A short stalk in an animal body. { 'ped·ə,sel }

pedicellaria [INV ZOO] In echinoids and starfishes, any of various small grasping organs in the form of a beak carried on a stalk. { ,ped·ə·sə'ler·ē·ə }

pedicellate [BIOL] Having a pedicel. { pe'dis·ə,lāt }

Pedicellinea [INV ZOO] The single order of the class Calyssozoa, including all entoproct bryozoans. { ,ped·ə·sə'lin·ē·ə }

pedicle [ANAT] A slender process acting as a foot or stalk (as the base of a tumor), or the basal portion of an organ that is continuous with other structures. { 'ped·ə·kəl }

pediculosis [MED] Infestation with lice, especially of the genus *Pediculus*. { pə,dik·yə'lō·səs }

pedigree [GEN] Diagrammatic representation of the ancestry of an individual. { 'ped·ə,grē }

pedilidae [INV ZOO] The false ant-loving flower beetles, a family of coleopteran insects in the superfamily Tenebrionoidea. { pə'dil·ə,dē }

Pedinidae [INV ZOO] The single family of the order Pedinoida. { pe'din·ə,dē }

Pedinoida [INV ZOO] An order of Diadematacea

making up those forms of echinoderms which possess solid spines and a rigid test. { ˌped·ən'ȯid·ə }

Pedionomidae [VERT ZOO] A family of quaillike birds in the order Gruiformes. { ˌped·ē·ə 'näm·əˌdē }

Pedipalpida [INV ZOO] Former order of the Arachnida; these animals are now placed in the orders Uropygi and Amblypygi. { ˌped·ə'pal·pəd·ə }

pedipalpus [INV ZOO] One of the second pair of appendages of an arachnid. { ¦ped·ə¦pal·pəs }

pedodontics [MED] The branch of dentistry concerned with the care of children's teeth. { ¦pēd·ō¦dän·tiks }

pedology [MED] The science of the study of the physiological as well as the psychological aspects of childhood. { pe'däl·ə·jē }

peduncle [ANAT] A band of white fibers joining different portions of the brain. [BOT] **1.** A flower-bearing stalk. **2.** A stalk supporting the fruiting body of certain thallophytes. [INV ZOO] The stalk supporting the whole or a large part of the body of certain crinoids, brachiopods, and barnacles. { 'pēˌdəŋ·kəl }

pedunculate [BIOL] **1.** Having or growing on a peduncle. **2.** Being attached to a peduncle. { pē'dəŋ·kyə·lət }

PEFR *See* peak expiratory flow rate.

Pegasidae [VERT ZOO] The single family of the order Pegasiformes. { pə'gas·əˌdē }

Pegasiformes [VERT ZOO] The sea moths or sea dragons, a small order of actinopterygian fishes; the anterior of the body is encased in bone, and the nasal bones are enlarged to form a rostrum that projects well forward of the mouth. { pəˌgas·ə'fȯrˌmēz }

peg graft [BOT] A graft made by driving a scion of leafless dormant wood with wedge-shaped base into an opening in the stock and sealing with wax or other material. { 'peg ˌgraft }

Peisidicidae [INV ZOO] A monogeneric family of polychaete annelids belonging to the Errantia. { ˌpī·sə'dis·əˌdē }

pelagism *See* seasickness. { 'pel·əˌjiz·əm }

pelargonidin [BIOCHEM] An anthocyanidin pigment obtained by hydrolysis of pelargonin in the form of its red-brown crystalline chloride, $C_{15}H_{11}ClO_5$. { ¦peˌlär'gän·əd·ən }

pelargonin [BIOCHEM] An anthocyanin obtained from the dried petals of red pelargoniums or blue cornflowers in the form of its red crystalline chloride, $C_{27}H_{31}ClO_{15}$. { ˌpeˌlär'gō·nən }

Pel-Ebstein fever [MED] A relapsing fever characteristic of Hodgkin's disease. { 'pel 'ebˌstīn ˌfē·vər }

Pelecanidae [VERT ZOO] The pelicans, a family of aquatic birds in the order Pelecaniformes. { ˌpel·ə'kan·əˌdē }

Pelecaniformes [VERT ZOO] An order of aquatic, fish-eating birds characterized by having all four toes joined by webs. { ˌpel·əˌkan·ə'fȯrˌmēz }

Pelecanoididae [VERT ZOO] The diving petrels, a family of oceanic birds in the order Procellariiformes. { ˌpel·ə·kə'nȯid·əˌdē }

Pelecinidae [INV ZOO] The pelecinid wasps, a monospecific family of hymenopteran insects in the superfamily Proctotrupoidea. { ˌpel·ə'sin·əˌdē }

Pelecypoda [INV ZOO] The equivalent name for Bivalvia. { ˌpel·ə'si·pə·də }

Pelger anomaly [MED] A hereditary anomaly of granulocytes in the peripheral blood with no more than one or two nuclear lobes and unusually coarse nuclear chromatin. { 'pel·gər əˌnäm·ə·lē }

pelican [VERT ZOO] Any of several species of birds composing the family Pelecanidae, distinguished by the extremely large bill which has a distensible pouch under the lower mandible. { 'pel·ə·kən }

pellagra [MED] A disease caused by nicotinic acid deficiency characterized by skin lesions, inflammation of the soft tissues of the mouth, diarrhea, and central nervous system disorders. { pə'lag·rə }

pellet [PHARM] A small pill. [VERT ZOO] A mass of undigestible material regurgitated by a carnivorous bird. { 'pel·ət }

pelletierine [PHARM] $C_5H_{10}N(CH_2)_2CHO$ An alkaloid obtained as a liquid from the root of the pomegranate; boiling point is 195°C; soluble in water, alcohol, and ether; used in medicine. { ˌpel·ə'tiˌrēn }

pellicle [CELL MOL] A plasma membrane. [INV ZOO] A thin protective membrane, as on certain protozoans. { 'pel·ə·kəl }

Pelmatozoa [INV ZOO] A division of the Echinodermata made up of those forms which are anchored to the substrate during at least part of their life history. { pelˌmad·ə'zō·ə }

Pelobatidae [VERT ZOO] A family of frogs in the suborder Anomocoela, including the spadefoot toads. { ˌpel·ō'bad·əˌdē }

Pelodytidae [VERT ZOO] A family of frogs in the suborder Anomocoela. { ˌpel·ə'did·əˌdē }

Pelogonidae [INV ZOO] The equivalent name for Ochteridae. { ˌpel·ə'gän·əˌdē }

Pelomedusidae [VERT ZOO] The side-necked or hidden-necked turtles, a family of the order Chelonia. { ˌpel·ō·mə'düs·əˌdē }

Pelonemataceae [MICROBIO] A family of gliding bacteria of uncertain affiliation; straight, flexuous, or spiral, unbranched filaments containing colorless, cylindrical cells. { ¦pel·ōˌnem·ə'tās·ēˌē }

Pelopidae [INV ZOO] A family of oribatid mites, order Sarcoptiformes. { pə'läp·əˌdē }

Peloplocaceae [MICROBIO] Formerly a family in the order Chlamydobacteriales; long, unbranched trichomes in a delicate sheath. { ˌpel·ō·plə'kās·ēˌē }

Peloridiidae [INV ZOO] The single family of the homopteran series Coleorrhyncha. { ˌpel·ō·rə'dī·əˌdē }

pelotherapy [MED] Therapeutic treatment with earth or mud. { ¦pel·ō'ther·ə·pē }

pelouse *See* meadow. { pə'lüz }

peltate [BOT] Of leaves, having the petiole attached to the lower surface instead of the base. { 'pel,tāt }

pelvic cavity *See* pelvis. { 'pel·vik ¦kav·əd·ē }

pelvic fin [VERT ZOO] One of the pair of fins of fishes corresponding to the hindlimbs of a quadruped. { 'pel·vik ¦fin }

pelvic girdle [ANAT] The system of bones supporting the lower limbs, or the hindlimbs, of vertebrates. { 'pel·vik ¦gərd·əl }

pelvic index [ANAT] The ratio of the anteroposterior diameter to the transverse diameter of the pelvis. { 'pel·vik ¦in,deks }

pelvis [ANAT] **1.** The main, basin-shaped cavity of the kidney into which urine is discharged by nephrons. **2.** The basin-shaped structure formed by the hipbones together with the sacrum and coccyx, or caudal vertebrae. **3.** The cavity of the bony pelvis. Also known as pelvic cavity. { 'pel·vəs }

pelviscope [MED] An endoscope for examination of the pelvic organs of the female. { 'pel·və,skōp }

pemphigoid [MED] An autoimmune skin disorder resembling, but histologically and clinically distinct from, pemphigus and characterized by bloody blisters, especially on the trunk and limbs. { 'pem·fə,gȯid }

pemphigus [MED] An acute or chronic disease of the skin characterized by the appearance of bullae, which develop in crops or in continuous succession. { 'pem·fə·gəs }

pemphigus contageosus [MED] A vesicular dermatitis endemic in tropical areas, chiefly affecting the armpits and groin. { 'pem·fə·gəs kən,tā·jē'ō·səs }

pen [INV ZOO] The inner horny, feather-shaped, chitnous shell of a squid. Also known as gladius. { pen }

Penaeidea [INV ZOO] A primitive section of the Decapoda in the suborder Natantia; in these forms, the pleurae of the first abdominal somite overlap those of the second, the third legs are chelate, and the gills are dendrobranchiate. { ,pen·ē'id·ē·ə }

penesaline [ECOL] Referring to an environment intermediate between normal marine and saline, characterized by evaporitic carbonates often interbedded with gypsum or anhydrite, and by a salinity high enough to be toxic to normal marine organisms. { ,pēn·ə'sā,lēn }

penetrance [GEN] The proportion of individuals carrying a dominant gene in the heterozygous condition or a recessive gene in the homozygous condition in which the specific phenotypic effect is apparent. Also known as gene penetrance. { 'pen·ə·trəns }

penetrant [INV ZOO] A large barbed nematocyst that pierces the body of the prey and injects a paralyzing agent. { 'pen·ə·trənt }

penetration gland [INV ZOO] A gland at the anterior end of certain cercariae that secretes a histolytic substance. { ,pen·ə'trā·shən ,gland }

penguin [VERT ZOO] Any member of the avian order Sphenisciformes; structurally modified wings do not fold and they function like flippers, the tail is short, feet are short and webbed, and the legs are set far back on the body. { 'peŋ·gwən }

penicidin *See* patulin. { ,pen·ə'sīd·ən }

penicillamine [PHARM] $C_5H_{11}NO_2S$ The most characteristic degradation product of penicillin-type antibiotics; used as a chelating agent, as a drug to reduce cystine excretion in cystinurea, and in the treatment of rheumatoid arthritis. { ,pen·ə'sil·ə,mēn }

penicillate [BIOL] Having a tuft of fine hairs. { ¦pen·ə¦sil·ət }

penicillin [MICROBIO] **1.** The collective name for salts of a series of antibiotic organic acids produced by a number of *Penicillium* and *Aspergillus* species; active against most gram-positive bacteria and some gram-negative cocci. **2.** *See* benzyl penicillin sodium. { ,pen·ə'sil·ən }

penicillinase [BIOCHEM] A bacterial enzyme that hydrolyzes and inactivates penicillin. { ,pen·ə'sil·ə,nās }

penicillin V *See* phenoxymethylpenicillin. { ,pen·ə'sil·ən 'fīv *or* 'vē }

Peniculina [INV ZOO] A suborder of the Hymenostomatida. { pe,nik·yə'lī·nə }

penis [ANAT] The male organ of copulation in humans and certain other vertebrates. Also known as phallus. { 'pē·nəs }

penna *See* contour feather. { 'pen·ə }

pennaceous [ZOO] Referring to the stiff, tightly bound portion of the feather vane on a bird. { pə'nā·shəs }

Pennales [BOT] An order of diatoms (Bacillariophyceae) in which the form is often circular, and the markings on the valves are radial. { pə'nā·lēz }

pennate [BIOL] **1.** Wing-shaped. **2.** Having wings. **3.** Having feathers. { 'pe,nāt }

Pennatulacea [INV ZOO] The sea pens, an order of the subclass Alcyonaria; individuals lack stolons and live with their bases embedded in the soft substratum of the sea. { pə,nach·ə'lā·shə }

Pennellidae [INV ZOO] A family of copepod crustaceans in the suborder Caligoida; skin-penetrating external parasites of various marine fishes and whales. { ,pen·ə'lī·ə,dē }

penniculus [ANAT] A tuft of arterioles in the spleen. [BIOL] A brush-shaped structure. { pə'nik·yə·ləs }

pentacrinoid [INV ZOO] The larva of a feather star. { pen'tak·rə,nȯid }

pentactinal [ZOO] Having five rays or branches. { pen'tak·tə·nəl }

pentacula [INV ZOO] The five-tentacled stage in the life history of echinoderms. { pen'tak·yə·lə }

pentacyanium bis(methyl sulfate) [PHARM] $C_{29}H_{45}N_3O_9S_2$ A water-soluble, crystalline compound obtained from ethanol solution, and melting at 173–175°C; used in medicine in the dichloride form as an antihypertensive agent. { ,pen·tə·sī'an·ē·əm ¦bis ¦meth·əl 'səl,fāt }

pentadactyl [VERT ZOO] Having five digits on the hand or foot. { ¦pen·tə¦dakt·əl }

pentadelphous [BOT] Having the stamens in five sets with the filaments more or less united within each set. { ¦pen·tə¦del·fəs }

pentahydroxyhexoic acid *See* galactonic acid. { ˌpen·tə·hī¦dräk·sē·hek'sō·ik 'as·əd }

pentalogy [MED] Five symptoms or defects which together characterize a disease or syndrome. { pen'tal·ə·jē }

pentalogy of Fallot [MED] A syndrome consisting of the tetralogy of Fallot plus an interatrial septal defect. { pen'tal·ə·jē əv fa'lō }

pentamerous [BOT] Having each whorl of the flower consisting of five members, or a multiple of five. { pen'tam·ə·rəs }

pentandrous [BOT] Having five stamens. { pen'tan·drəs }

pentasaccharide [BIOCHEM] A carbohydrate which, on hydrolysis, yields five molecules of monosaccharides. { ˌpen·tə'sak·əˌrīde }

Pentastomida [INV ZOO] A class of bloodsucking parasitic arthropods; the adult is vermiform, and there are two pairs of hooklike, retractile claws on the cephalothorax. { ˌpen·tə'stäm·ə·də }

Pentatomidae [INV ZOO] The true stink bugs, a family of hemipteran insects in the superfamily Pentatomoidea. { ˌpen·tə'täm·əˌdē }

Pentatomoidea [INV ZOO] A subfamily of the hemipteran group Pentatomorpha distinguished by marginal trichobothria and by antennae which are usually five-segmented. { ˌpen·tə·tə'mȯid·ē·ə }

Pentatomorpha [INV ZOO] A large group of hemipteran insects in the subdivision Geocorisae in which the eggs are nonoperculate, a median spermatheca is present, accessory salivary glands are tubular, and the abdomen has trichobothria. { ˌpen·tə·də'mȯr·fə }

pentobarbital sodium [PHARM] $C_{11}H_{17}N_2NaO_3$ A short- to intermediate-acting barbiturate; used as a hypnotic and sedative drug. Also known as sodium pentobarbitone. { ¦pen·tō'bär·bəˌtȯl 'sōd·ē·əm }

pentoglycan [BIOCHEM] A polysaccharide yielding a pentose sugar on hydrolysis. { ˌpent·ə'glīˌkan }

pentolinium tartrate [PHARM] $C_{23}H_{42}N_2O_{12}$ A white to cream-colored powder, soluble in water; used in medicine. { ˌpen·tə'lin·ē·əm 'tärˌtrāt }

pentosan [BIOCHEM] A hemicellulose present in cereal, straws, brans, and other woody plant tissues; yields five-carbon-atom sugars. { 'pen·təˌsan }

pentose [BIOCHEM] Any one of a class of carbohydrates containing five atoms of carbon. { 'penˌtōs }

pentose phosphate pathway [BIOCHEM] A pathway by which glucose is metabolized or transformed in plants and microorganisms; glucose-6-phosphate is oxidized to 6-phosphogluconic acid, which then undergoes oxidative decarboxylation to form ribulose-5-phosphate, which is ultimately transformed to fructose-6-phosphate. { 'penˌtōs 'fäsˌfāt 'pathˌwā }

pentosuria [MED] The presence of pentose in the urine. { ˌpen·təs'yu̇r·ē·ə }

pentylenetetrazol [PHARM] $C_6H_{10}N_4$ A central nervous system stimulant used as an analeptic. { ¦pent·əlˌēn'te·trəˌzȯl }

pepo [BOT] A fleshy indehiscent berry with many seeds and a hard rind; characteristic of the Cucurbitaceae. { 'pēˌpō }

pepper [BOT] Any of several warm-season perennials of the genus *Capsicum* in the order Polemoniales, especially *C. annum* which is cultivated for its fruit, a many-seeded berry with a thickened integument. { 'pep·ər }

peppermint [BOT] Any of various aromatic herbs of the genus *Mentha* in the family Labiatae, especially *M. piperita*. { 'pep·ərˌmint }

PeP reaction *See* proton-electron-proton reaction. { 'pēˌē'pē rēˌak·shən }

pepsin [BIOCHEM] A proteolytic enzyme found in the gastric juice of mammals, birds, reptiles, and fishes. { 'pep·sən }

pepsinogen [BIOCHEM] The precursor of pepsin, found in the stomach mucosa. { pep'sin·ə·jən }

peptic [PHYSIO] **1.** Of or pertaining to pepsin. **2.** Of or pertaining to digestion. { 'pep·tik }

peptic ulcer [MED] An ulcer involving the mucosa, submucosa, and muscular layer on the lower esophagus, stomach, or duodenum, due in part at least to the action of acid-pepsin gastric juice. { 'pep·tik 'əl·sər }

peptidase [BIOCHEM] An enzyme that catalyzes the hydrolysis of peptides to amino acids. { 'pep·təˌdās }

peptide [BIOCHEM] A compound of two or more amino acids joined by peptide bonds. { 'pepˌtīd }

peptide map [CELL MOL] A fragmentation pattern generated by digestion of a particular protein with proteolytic enzymes of known specificity; used in protein identification. { 'pepˌtīd ˌmap }

peptide synthetase [BIOCHEM] A ribosomal synthetase that catalyzes the formation of peptide bonds during protein synthesis. { 'pepˌtīd 'sin·thəˌtās }

Peptococcaceae [MICROBIO] A family of gram-positive cocci; organisms can use either amino acids or carbohydrates for growth and energy. { ˌpep·tə·käk'sās·ēˌē }

peptone [BIOCHEM] A water-soluble mixture of proteoses and amino acids derived from albumin, meat, or milk; used as a nutrient and to prepare nutrient media for bacteriology. { 'pepˌtōn }

Peracarida [INV ZOO] A superorder of the Eumalacostraca; these crustaceans have the first thoracic segment united with the head, the cephalothorax usually larger than the abdomen, and some thoracic segments free from the carapace. { ˌper·ə'kar·ə·də }

Peramelidae [VERT ZOO] The bandicoots, a family of insectivorous mammals in the order Marsupialia. { ˌpər·ə'mel·əˌdē }

perazine [PHARM] $C_{20}H_{25}N_3S$ A crystalline compound, melting at 51–53°C; used in medicine as a tranquilizer. { 'per·ə,zēn }

perception [PHYSIO] Recognition in response to sensory stimuli; the act or process by which the memory of certain qualities of an object is associated with other qualities impressing the senses, thereby making possible recognition of the object. { pər'sep·shən }

perceptual overload [NEUROSCI] Saturation of the nervous system by an input of excess sensory information, resulting in an absence of response. { pər¦sep·chə·wəl 'ō·vər,lōd }

perch [VERT ZOO] **1.** Any member of the family Percidae. **2.** The common name for a number of unrelated species of fish belonging to the Centrarchidae, Anabantoidei, and Percopsiformes. { pərch }

Percidae [VERT ZOO] A family of fresh-water actinopterygian fishes in the suborder Percoidei; comprises the true perches. { 'pər·sə,dē }

Perciformes [VERT ZOO] The typical spiny-rayed fishes, comprising the largest order of vertebrates; characterized by fin spines, a swim bladder without a duct, usually ctenoid scales, and 17 or fewer caudal fin rays. { ,pər·sə'fȯr,mēz }

Percoidei [VERT ZOO] A large, typical suborder of the order Perciformes; includes over 50% of the species in this order. { pər'kȯid·ē,ī }

Percomorphi [VERT ZOO] An equivalent, ordinal name for the Perciformes. { ,pər·kə'mȯr,fī }

Percopsidae [VERT ZOO] A family of fishes in the order Percopsiformes. { pər'käp·sə,dē }

Percopsiformes [VERT ZOO] A small order of actinopterygian fishes characterized by single, ray-supported dorsal and anal fins and a subabdominal pelvic fin with three to eight soft rays. { pər,käp·sə'fȯr,mēz }

percussion [MED] The act of striking or firmly tapping the surface of the body with a finger or a small hammer to elicit sounds, or vibratory sensations, of diagnostic value. { pər'kəsh·ən }

perennial [BOT] A plant that lives for an indefinite period, dying back seasonally and producing new growth from a perennating part. { pə'ren·ē·əl }

perfect flower [BOT] A flower having both stamens and pistils. { 'pər·fikt 'flau̇·ər }

perfoliate [BOT] Pertaining to the form of a leaf having its base united around the stem. [INV ZOO] Pertaining to the form of certain insect antennae having the terminal joints expanded and flattened to form plates which encircle the stalk. { pər'fōl·ē,āt }

perforatorium *See* acrosome. { ,pər·fə·rə'tȯr·ē·əm }

perforin [IMMUNOL] An enzyme that is secreted by natural killer cells and cytotoxic T cells and destroys foreign cells by puncturing their membranes, causing leakage of the cell contents. Also known as cytolysin; pore-forming protein. { 'pər·fər·ən }

performing arts medicine [MED] A subspecialty in medicine that deals with problems specific to the activities of dancers, musicians, and vocalists. { pər¦fȯrm·iŋ 'ärts ,med·ə·sən }

perfusion [PHYSIO] The pumping of a fluid through a tissue or organ by way of an artery. { pər'fyü·zhən }

Pergidae [INV ZOO] A small family of hymenopteran insects in the superfamily Tenthredinoidea. { 'pər·jə,dē }

perianal [ANAT] Situated or occurring around the anus. { ¦per·ē'ān·əl }

perianth [BOT] The calyx and corolla considered together. { 'per·ē,anth }

periapical cyst *See* radicular cyst. { ¦per·ē'ap·ə·kəl 'sist }

periappendicitis [MED] Inflammation of the tissue around the vermiform process, or of the serosal region of the vermiform appendix. { ¦per·ē·ə,pen·də'sīd·əs }

periarteritis [MED] Inflammation of the outer coat of an artery and of the periarterial tissues. { ¦per·ē,ärd·ə'rīd·əs }

periarteritis nodosa *See* polyarteritis nodosa. { ¦per·ē,ärd·ə'rīd·əs nō'dō·sə }

periblast [EMBRYO] The nucleated layer of cytoplasm that surrounds the blastodisk of an egg undergoing discoidal cleavage. { 'per·ə,blast }

periblastula [EMBRYO] The blastula of a centrolecithal egg, formed by superficial segmentation. { ¦per·ə'blas·tyə·lə }

periblem [BOT] A layer of primary meristem which produces the cortical cells. { 'per·ə,blem }

pericardial [ANAT] **1.** Of or pertaining to the pericardium. **2.** Located around the heart. { ,per·ə'kärd·ē·əl }

pericardial cavity [ANAT] A potential space between the inner layer of the pericardium and the epicardium of the heart. { ,per·ə'kärd·ē·əl 'kav·əd·ē }

pericardial fluid [PHYSIO] The fluid in the pericardial cavity. { ,per·ə'kärd·ē·əl 'flü·əd }

pericardial organ [INV ZOO] One of the neurohemal organs associated with the pericardial cavity in crustaceans. { ,per·ə'kärd·ē·əl 'ȯr·gən }

pericarditis [MED] Inflammation of the pericardium. { ,per·ə,kär'dīd·əs }

pericardium [ANAT] The membranous sac that envelops the heart; it contains 5–20 grams of clear serous fluid. { ,per·ə'kärd·ē·əm }

pericarp [BOT] The wall of a fruit, developed by ripening and modification of the ovarian wall. { 'per·ə,kärp }

pericentric inversion [GEN] An aberration resulting from a break in each arm of a chromosome and rejoining with the middle fragment containing the centromere now inverted with respect to the two terminal fragments. { ,per·ə'sen·trik in'vər·zhən }

pericholangitis [MED] Inflammation of the tissues around the bile ducts or the interlobular bile capillaries. { ¦per·ē,kō,lan'jīd·əs }

perichondrium [ANAT] The fibrous connective tissue covering cartilage, except at joints. { ,per·ə'kän·drē·əm }

periclinal [BOT] Pertaining to a cell layer that is parallel to the surface of a plant part. { ¦per·ə¦klīn·əl }

periclinal chimera [GEN] A plant composed of cells of two distinct species separated into distinctive zones. { ¦per·ə¦klīn·əl kī'mir·ə }

pericolitis [MED] Inflammation of the peritoneum or tissues around the colon. { ¦per·ə·kə'līd·əs }

pericoronitis [MED] Inflammation of the tissue surrounding the coronal portion of the tooth, usually a partially erupted third molar. { ¦per·ə‚kär·ə'nīd·əs }

pericranium [ANAT] The periosteum on the outer surface of the cranial bones. { ¦per·ə'krā·nē·əm }

pericycle [BOT] The outer boundary of the stele of plants; may not be present as a distinct layer of cells. { 'per·ə‚sī·kəl }

pericystium [ANAT] The tissues surrounding a bladder. [MED] The vascular wall of a cyst. { ‚per·ə'sis·tē·əm }

pericyte [HISTOL] A mesenchymal cell found around a capillary; it may or may not be contractile. { 'per·ə‚sīt }

periderm [BOT] A group of secondary tissues forming a protective layer which replaces the epidermis of many plant stems, roots, and other parts; composed of cork cambium, phelloderm, and cork. [EMBRYO] The superficial transient layer of epithelial cells of the embryonic epidermis. { 'per·ə‚dərm }

peridium [BOT] The outer investment of the sporophore of many fungi. { pə'rid·ē·əm }

peri-endothelial cell [HISTOL] A cell that resides next to and supports the endothelial cells. { ‚pe·rē ‚en·də‚thē·lē·əl 'sel }

perifollicular [HISTOL] Surrounding a follicle. { ¦per·ə·fə'lik·yə·lər }

perigonium [BOT] The perianth of a liverwort. [INV ZOO] The sac containing the generative bodies in the gonophore of a hydroid. { ‚per·ə'gō·nē·əm }

perigynium [BOT] A fleshy cup- or tubelike structure surrounding the archegonium of various bryophytes. { ‚per·ə'jin·ē·əm }

perigynous [BOT] Bearing the floral organs on the rim of an expanded saucer- or cup-shaped receptacle or hypanthium. { pə'rij·ə·nəs }

perihepatitis [MED] Inflammation of the peritoneum and tissues surrounding the liver. { ¦per·ē‚hep·ə'tīd·əs }

peri-infarction block *See* intraventricular heart block. { ¦per·ē·in'färk·shən ‚bläk }

perikaryon [CELL MOL] A cytoplasmic mass surrounding a nucleus. [NEUROSCI] The body of a nerve cell, containing the nucleus. { ¦per·ə'kar·ē‚än }

Perilampidae [INV ZOO] A family of hymenopteran insects in the superfamily Chalcidoidea. { ‚per·ə'lam·pə‚dē }

perilymph [PHYSIO] The fluid separating the membranous from the osseous labyrinth of the internal ear. { 'per·ə‚limpf }

perimetrium [ANAT] The serous covering of the uterus. { ‚per·ə'mē·trē·əm }

perimysium [ANAT] The connective tissue sheath enveloping a muscle or a bundle of muscle fibers. { ‚per·ə'mī·sē·əm }

perineum [ANAT] **1.** The portion of the body included in the outlet of the pelvis, bounded in front by the pubic arch, behind by the coccyx and sacrotuberous ligaments, and at the sides by the tuberosities of the ischium. **2.** The region between the anus and the scrotum in the male, between the anus and the posterior commissure of the vulva in the female. { ‚per·ə'nē·əm }

perineural [ANAT] Situated around nervous tissue or a nerve. { ‚per·ə'nu̇r·əl }

perineural fibroblastoma *See* neurofibroma. { ‚per·ə'nu̇r·əl ¦fī·brō·bla'stō·mə }

perineural fibroma *See* neurofibroma. { ‚per·ə'nu̇r·əl fī'brō·mə }

periocular [ANAT] Surrounding the eye. { ¦per·ē'äk·yə·lər }

periodic disease *See* familial Mediterranean fever. { ¦pir·ē¦äd·ik di‚zēz }

periodicity [CELL MOL] The number of base pairs in one turn of the deoxyribonucleic acid duplex. { ‚pir·ē·ə'dis·əd·ē }

periodic peritonitis *See* familial Mediterranean fever. { ¦pir·ē¦äd·ik ‚pər·ə·tə'nīd·əs }

periodontal [ANAT] **1.** Surrounding a tooth. **2.** Of or pertaining to the periodontium. { ¦per·ē·ō¦dänt·əl }

periodontitis [MED] Inflammation of the periodontium. { ¦per·ē·ō‚dän'tīd·əs }

periodontium [ANAT] The tissues surrounding a tooth. { ¦per·ē·ō'dan·chəm }

periodontoclasia [MED] Any periodontal disease that results in the destruction of the periodontium. { ¦per·ē·ō‚dänt·ə'klā·zhə }

periodontics [MED] The branch of dentistry devoted to diseases of the gingiva (gum tissue), alveolar bone, periodontal ligament, and cementum. { ‚per·ē·ə'dän·tiks }

periodontosis [MED] A degenerative disturbance of the periodontium, characterized by degeneration of connective-tissue elements of the periodontal ligament and by bone resorption. { ¦per·ē·ō‚dän'tō·səs }

perionychium [ANAT] The border of epidermis surrounding an entire nail. Also known as paronychium. { ‚per·ē·ō'nik·ē·əm }

periosteum [ANAT] The fibrous membrane enveloping bones, except at joints and the points of tendonous and ligamentous attachment. { ‚per·ē'äs·tē·əm }

periostitis [MED] Inflammation of the periosteum. { ‚per·ē‚ä'stīd·əs }

periostracum [INV ZOO] A protective layer of chitin covering the outer portion of the shell in many mollusks, especially fresh-water forms. { ‚per·ē'äs·trə·kəm }

periotic [ANAT] **1.** Situated about the ear. **2.** Of or pertaining to the parts immediately about the internal ear. { ¦per·ē¦äd·ik }

peripheral [ANAT] Pertaining to or located at or

near the surface of a body or an organ. { pə'rif·ə·rəl }

peripheral hemodynamics [PHYSIO] A division of hematology concerned with blood flow in regions of the body that are close to the surface, such as the extremities. { pə¦rif·ə·rəl ˌhē·mō·dī'nam·iks }

peripheral membrane protein [CELL MOL] A protein that is associated with the plasma membrane via electrostatic interactions with integral membrane proteins or membrane lipids. Also known as extrinsic protein. { pə¦rif·ə·rəl 'mem ˌbrān ˌprōˌtēn }

peripheral nervous system [NEUROSCI] The autonomic nervous system, the cranial nerves, and the spinal nerves including their associated sensory receptors. { pə'rif·ə·rəl 'nər·vəs ˌsis·təm }

peripheral vision [PHYSIO] The act of seeing images that fall upon parts of the retina outside the macula lutea. Also known as indirect vision. { pə'rif·ə·rəl 'vizh·ən }

periphlebitis [MED] Inflammation of the tissues around a vein or of the adventitia of a vein. { ¦per·ə·flə'bīd·əs }

periphyton [ECOL] Sessile biotal components of a fresh-water ecosystem. { pə'rif·əˌtän }

periplasm [CELL MOL] The region between the cytoplasmic and outer membranes of a cell. { 'per·əˌplaz·əm }

periplast [CELL MOL] **1.** A cell membrane. **2.** A pellicle covering ectoplasm. [HISTOL] The stroma of an animal organ. [INV ZOO] The ectoplasm of a flagellate. { 'per·əˌplast }

periproct [INV ZOO] The area surrounding the anus of echinoids. { 'per·əˌpräkt }

Peripylina [INV ZOO] An equivalent name for Porulosida. { ¦per·ə'plī·nə }

perisarc [INV ZOO] The outer integument of a hydroid. { 'per·əˌsärk }

Periscelidae [INV ZOO] A family of myodarian cyclorrhaphous dipteran insects in the subsection Acalypteratae. { ˌper·ə'sel·əˌdē }

Perischoechinoidea [INV ZOO] A subclass of principally extinct echinoderms belonging to the Echinoidea and lacking stability in the number of columns of plates that make up the ambulacra and interambulacra. { pə¦ris·kōˌek·ə'nȯid·ē·ə }

perisperm [BOT] In a seed, the nutritive tissue that is derived from the nucellus and deposited on the outside of the embryo sac. { 'per·əˌspərm }

Perissodactyla [VERT ZOO] An order of exclusively herbivorous mammals distinguished by an odd number of toes and mesaxonic feet, that is, with the axis going through the third toe. { pəˌris·ō'dak·tə·lə }

peristalsis [PHYSIO] The rhythmic progressive wave of muscular contraction in tubes, such as the intestine, provided with both longitudinal and transverse muscular fibers. { ˌper·ə'stäl·səs }

peristome [BOT] The fringe around the opening of a moss capsule. [INV ZOO] The area surrounding the mouth of various invertebrates. { 'per·əˌstōm }

perithecial ascomycetes *See* Pyrenomycetes. { ˌper·iˌthē·shəl ˌas·kə·mī'sēdˌēz }

perithecium [MYCOL] A spherical, cylindrical, or oval ascocarp which usually opens by a terminal slit or pore. { ˌper·ə'thē·shəm }

peritoneal cavity [ANAT] The potential space between the visceral and parietal layers of the peritoneum. { ¦per·ə·tə¦nē·əl 'kav·əd·ē }

peritoneoscope [MED] A long, slender endoscope equipped with sheath, obturator, biopsy forceps, a sphygmomanometer bulb and tubing, scissors, and a syringe; introduced through a small incision in the abdominal wall to permit visualization of the gas-inflated peritoneal cavity. { ˌper·ə·tə'nē·əˌskōp }

peritoneoscopy *See* laparoscopy. { ˌper·ə·tə ˌnē'äs·kə·pē }

peritoneum [ANAT] The serous membrane enveloping the abdominal viscera and lining the abdominal cavity. { ˌper·ə·tə'nē·əm }

peritonitis [MED] Inflammation of the peritoneum. { ˌper·ə·tə'nīd·əs }

peritonsillar abscess [MED] An abscess forming in acute tonsillitis around one or both tonsils. { ˌper·ə'täns·əl·ər 'abˌses }

Peritrichia [INV ZOO] A specialized subclass of the class Ciliatea comprising both sessile and mobile forms. { ˌper·ə'trik·ē·ə }

Peritrichida [INV ZOO] The single order of the protozoan subclass Peritrichia. { ˌper·ə'trik·ə·də }

peritrichous [INV ZOO] Of certain protozoans, having spirally arranged cilia around the oral disk. [MICROBIO] Of bacteria, having a uniform distribution of flagella on the body surface. { pə'ri·trə·kəs }

perityphlitis [MED] Inflammation of the peritoneum surrounding the cecum and vermiform appendix. { ˌper·ə'tif·ləd·əs }

perivitelline space [CELL MOL] In mammalian ova, the space formed between the ovum and the zona pellucida at the time of maturation, into which the polar bodies are given off. { ¦per·ə'vid·əlˌēn ˌspās }

periwinkle *See* Vinca rosea. { 'per·iˌwiŋ·kəl }

perlèche [MED] An inflammatory condition occurring at the angles of the mouth with resultant fissuring. { pər'lesh }

permanent teeth [ANAT] The second set of teeth of a mammal, following the milk teeth; in humans, the set of 32 teeth consists of 8 incisors, 4 canines, 8 premolars, and 12 molars. { 'pər·mə·nənt ¦tēth }

permease [BIOCHEM] Any of a group of enzymes which mediate the phenomenon of active transport. { 'pər·mēˌās }

permissive cell [VIROL] A cell that supports replication of a virus. { pər'mis·iv 'sel }

permissive condition [GEN] An environment under which a conditional lethal mutant can survive and show the wild phenotype. { pər'mis·iv kənˌdish·ən }

Permo-Triassic mass extinction [EVOL] A mass extinction event marking the division between the Permian Period and the Triassic Period as well as the border between the Paleozoic and Mesozoic eras. It is estimated to have triggered the extinction of 85% or more of all ocean species, approximately 70% of land vertebrates, and significant extinctions of plants and insects. { ¦pər·mō trī¦as·ik ¦mas ik'stiŋk·shən }

pernicious anemia [MED] A megaloblastic macrocytic anemia resulting from lack of vitamin B_{12}, secondary to gastric atrophy and loss of intrinsic factor necessary for vitamin B_{12} absorption, and accompanied by degeneration of the posterior and lateral columns of the spinal cord. { pər'nish·əs ə'nēm·ē·ə }

pernicious malaria *See* falciparum malaria. { pər'nish·əs mə'ler·ē·ə }

perniosis [MED] Any dermatitis resulting from chilblain. { ˌpər·nē'ō·səs }

Perognathinae [VERT ZOO] A subfamily of the rodent family Heteromyidae, including the pocket and kangaroo mice. { ˌper·əg'nath·əˌnē }

peronine [PHARM] $C_{17}H_{17}NO(OH)OC_7H_7 \cdot HCl$ A white, crystalline powder having a bitter taste; soluble in boiling water; used in medicine. { 'per·əˌnēn }

Peronosporales [MYCOL] An order of aquatic and terrestrial phycomycetous fungi with a hyphal thallus and zoospores with two flagella. { ˌper·əˌnäs·pə'rā·lēz }

Perothopidae [INV ZOO] A small family of coleopteran insects in the superfamily Elateroidea found only in the United States. { ˌper·ə 'thäp·əˌdē }

peroxidase [BIOCHEM] An enzyme that catalyzes reactions in which hydrogen peroxide is an electron acceptor. { pə'räk·səˌdās }

peroxisome [CELL MOL] Any of a subclass of microbodies that contain at least four enzymes involved in the metabolism of hydrogen peroxide. { pə'räk·səˌsōm }

perpetuation [VIROL] Maintenance of a viral genome within bacterial host cells without killing them due to weakening of viral virulence. { pərˌpech·ə'wā·shən }

Persian melon [BOT] A variety of muskmelon (*Cucumis melo*) in the order Violales; the fruit is globular and without sutures, and has dark-green skin, thin abundant netting, and firm, thick, orange flesh. { 'pər·zhən 'mel·ən }

persistence of vision [PHYSIO] The ability of the eye to retain the impression of an image for a short time after the image has disappeared. { pər'sis·təns əv 'vizh·ən }

persistent [BOT] Of a leaf, withering but remaining attached to the plant during the winter. { pər'sis·tənt }

persistent virus infection [VIROL] A covert viral infection in which a degree of equilibrium is established between the virus and the host's immune system, resulting in an infection of long duration. { pər'sis·tənt 'vī·rəs inˌfek·shən }

perspiration *See* sweat. { ˌpər·spə'rā·shən }

pertussis [MED] An infectious inflammatory bacterial disease of the air passages, caused by *Hemophilus pertussis* and characterized by explosive coughing ending in a whooping inspiration. Also known as whooping cough. { pər'təs·əs }

pessary [MED] An appliance of varied form placed in the vagina for uterine support or contraception. [PHARM] Any suppository or other form of medication placed in the vagina for therapeutic purposes. { 'pes·ə·rē }

pessulus [VERT ZOO] A bar composed of cartilage or bone that crosses the windpipe of a bird at its division into bronchi. { 'pes·yə·ləs }

pestilence [MED] **1.** Any epidemic contagious disease. **2.** Infection with the plague organism *Pasteurella pestis*. { 'pes·tə·ləns }

PET *See* positron emission tomography. { ¦pē¦ē'tē *or* pet }

petal [BOT] One of the sterile, leaf-shaped flower parts that make up the corolla. { 'ped·əl }

Petaluridae [INV ZOO] A family of dragonflies in the suborder Anisoptera. { ˌped·ə'lu̇r·əˌdē }

petasma [INV ZOO] A modified endopodite of the first abdominal appendage in a male decapod crustacean. { pə'taz·mə }

petechiae [MED] Hemorrhages the size of the head of a pin. { pə'tēk·ēˌē }

petiole [BOT] The stem which supports the blade of a leaf. { 'ped·ēˌōl }

petit mal [MED] A generalized epileptic seizure of the absence type, that is, characterized by different degrees of impaired consciousness. { pə¦tē¦mäl }

petrel [VERT ZOO] A sea bird of the families Procellariidae and Hydrobatidae, generally small to medium-sized with long wings and dark plumage with white areas near the rump. { 'pe·trəl }

petri dish [MICROBIO] A shallow glass or plastic dish with a loosely fitting overlapping cover used for bacterial plate cultures and plant and animal tissue cultures. { 'pē·trē ˌdish }

Petriidae [INV ZOO] A small family of coleopteran insects in the superfamily Tenebrionoidea. { pə'trī·əˌdē }

petroleum microbiology [MICROBIO] Those aspects of microbiological science and engineering of interest to the petroleum industry, including the role of microbes in petroleum formation, and the exploration, production, manufacturing, storage, and food synthesis from petroleum. { pə'trō·lē·əm ˌmī·krō·bī'äl·ə·jē }

Petromyzonida [VERT ZOO] The lampreys, an order of eellike, jawless vertebrates (Agnatha) distinguished by a single, dorsal nasal opening, and the mouth surrounded by an oral disk and provided with a rasping tongue. { ˌpe·trō·mī'zän·ə·də }

Petromyzontiformes [VERT ZOO] The equivalent name for Petromyzonida. { ˌpe·trō·mīˌzänt·ə'fȯrˌmēz }

petrosal nerve [NEUROSCI] Any of several small nerves passing through the petrous part of the temporal bone and usually attached to the geniculate ganglion. { pə'trō·səl ˌnərv }

petrosal process [ANAT] A sharp process of the sphenoid bone located below the notch for the passage of the abducens nerve, which articulates with the apex of the petrous portion of the temporal bone and forms the medial boundary of the foramen lacerum. { pə'trō·səl ˌprä·səs }

Petrosaviaceae [BOT] A small family of monocotyledonous plants in the order Triuridales characterized by perfect flowers, three carpels, and numerous seeds. { ˌpe·trōˌsav·ē'ās·ēˌē }

Peyer's patches [HISTOL] Aggregates of lymph nodules beneath the epithelium of the ileum. { 'pī·ərz ˌpach·əz }

Peyronie's disease [MED] Scarring of the shaft of the penis; may interfere with normal erections. { ¦pā·rə¦nēz diˌzēz }

Pfeiffer's disease *See* infectious mononucleosis. { 'fī·fərz diˌzēz }

p53 [GEN] A multifunctional tumor suppressor protein consisting of several domains that exhibit a number of biochemical activities and interact with a large variety of cellular and viral proteins; its encoding gene is the most commonly mutated gene in tumor cells. { ˌpēˌfif·tē'thrē }

PGA *See* folic acid.

phacella *See* gastric filament. { fə'sel·ə }

Phaenocephalidae [INV ZOO] A monospecific family of coleopteran insects in the superfamily Cucujoidea, found only in Japan. { ˌfē·nō·sə'fal·əˌdē }

Phaenothontidae [VERT ZOO] The tropic birds, a family of fish-eating aquatic forms in the order Pelecaniformes. { ˌfē·nō'thänt·əˌdē }

Phaeocoleosporae [MYCOL] A spore group of the Fungi Imperfecti with dark filiform spores. { fēˌkō·lē'äs·pəˌrē }

Phaeodictyosporae [MYCOL] A spore group of the Fungi Imperfecti with dark muriform spores. { ˌfē·ōˌdik·tē'äs·pəˌrē }

Phaeodidymae [MYCOL] A spore group of the Fungi Imperfecti with dark two-celled spores. { ˌfē·ō'did·əˌmē }

Phaeodorina [INV ZOO] The equivalent name for Tripylina. { ˌfē·ə·də'rī·nə }

Phaeohelicosporae [MYCOL] A spore group of the Fungi Imperfecti with dark, spirally coiled, septate spores. { ˌfē·ōˌhel·ə'käs·pəˌrē }

Phaeophragmiae [MYCOL] A spore group of the Fungi Imperfecti with dark three- to many-celled spores. { ˌfē·ō'frag·mēˌē }

Phaeophyta [BOT] The brown algae, constituting a division of plants; the plant body is multicellular, varying from a simple filamentous form to a complex, sometimes branched body having a basal attachment. { fē'äf·əd·ə }

Phaeosporae [MYCOL] A spore group of Fungi Imperfecti characterized by dark one-celled, nonfiliform spores. { fē'äs·pəˌrē }

Phaeostaurosporae [MYCOL] A spore group of the Fungi Imperfecti with dark star-shaped or forked spores. { ˌfē·ō·stȯ'räs·pəˌrē }

phage *See* bacteriophage. { fāj }

phage cross [VIROL] Multiple infection of a single bacterium by phages that differ at one or more genetic sites, leading to the production of recombinant progeny phage. { 'fāj ˌkrȯs }

phage induction [VIROL] Prophage stimulation by a variety of means that induce the vegetative state. { 'fāj inˌdək·shən }

phage restriction [VIROL] The inability of a phage to replicate due to an enzyme mechanism for degrading foreign deoxyribonucleic acid that enter the bacterial host cell. { 'fāj riˌstrik·shən }

phagocyte [CELL MOL] An ameboid cell that engulfs foreign material. { 'fag·əˌsīt }

phagocytic vacuole *See* food vacuole. { ¦fag·ə¦sid·ik 'vak·yəˌwōl }

phagocytin [BIOCHEM] A type of bactericidal agent present within phagocytic cells. { ¦fag·ə¦sīt·ən }

phagocytosis [CELL MOL] A specialized form of macropinocytosis in which cells engulf large solid objects such as bacteria and deliver the internalized objects to special digesting vacuoles; exists in certain cell types, such as macrophages and neutrophils. { ˌfag·əˌsī'tō·səs }

phagolysosome [CELL MOL] An intracellular vesicle formed by fusion of a lysosome with a phagosome. { ˌfag·ə'lī·səˌsōm }

phagosome [CELL MOL] A closed intracellular vesicle containing material captured by phagocytosis. { 'fag·əˌsōm }

phagotroph [INV ZOO] An organism that ingests nutrients by phagocytosis. { 'fag·əˌträf }

Phalacridae [INV ZOO] The shining flower beetles, a family of coleopteran insects in the superfamily Cucujoidea. { fə'lak·rəˌdē }

Phalacrocoracidae [VERT ZOO] The cormorants, a family of aquatic birds in the order Pelecaniformes. { ˌfal·əˌkrō·kə'ras·əˌdē }

Phalaenidae [INV ZOO] The equivalent name for Noctuidae. { fə'len·əˌdē }

Phalangeridae [VERT ZOO] A family of marsupial mammals in which the marsupium is well developed and opens anteriorly, the hindfeet are syndactylous, and the hallux is opposable and lacks a claw. { ˌfal·ən'jer·əˌdē }

Phalangida [INV ZOO] An order of the class Arachnida characterized by an unsegmented cephalothorax broadly joined to a segmented abdomen, paired chelate chelicerae, and paired palpi. { fə'lan·jə·də }

phalanx [ANAT] One of the bones of the fingers or toes. { 'fāˌlaŋks }

Phalaropodidae [VERT ZOO] The phalaropes, a family of migratory shore birds characterized by lobate toes and by reversal of the sex roles with respect to dimorphism and care of the young. { fəˌler·ə'päd·əˌde }

Phallostethidae [VERT ZOO] A family of actinopterygian fishes in the order Atheriniformes. { ˌfal·ō'steth·əˌdē }

Phallostethiformes [VERT ZOO] An equivalent name for Atheriniformes. { ˌfal·ō'steth·ə'fȯr ˌmēz }

phallotoxin [BIOCHEM] One of a group of toxic

peptides produced by the mushroom *Amanita phalloides*. { ¦fal·ō¦täk·sən }

phallus [ANAT] *See* penis. [EMBRYO] An undifferentiated embryonic structure derived from the genital tubercle that differentiates into the penis in males and the clitoris in females. { 'fal·əs }

phanerogam [BOT] A plant that produces seeds, for example, an angiosperm or gymnosperm. { 'fan·ə·rə,gam }

phanerophyte [ECOL] A perennial tree or shrub with dormant buds borne on aerial shoots. { 'fan·ə·rō,fīt }

Phanerozonida [INV ZOO] An order of the Asteroidea in which the body margins are defined by two conspicuous series of plates and in which pentamerous symmetry is generally constant. { ,fan·ə·rō'zän·ə·də }

Phanodermatidae [INV ZOO] A family of free-living nematodes in the superfamily Enoploidea. { ,fa·nō·dər'mad·ə,dē }

Pharetronida [INV ZOO] An order of calcareous sponges in the subclass Calcinea characterized by a leuconoid structure. { ¦far·ə'trän·ə·də }

pharmaceutical [PHARM] A chemical produced industrially (medicinal drug), which is useful in preventive or therapeutic treatment of a physical, mental, or behavioral condition. { ,fär·mə'süd·i·kəl }

pharmaceutical biotechnology [PHARM] A field that uses micro- and macroorganisms and hybridomas to create pharmaceuticals that are safer and more cost-effective than conventionally produced pharmaceuticals. { ,fär·mə,süd·i·kəl ,bī·ō·tek'näl·ə·jē }

pharmaceutics *See* pharmacy. { ,fär·mə'süd·iks }

pharmacodynamics [PHARM] The science that deals with the actions of drugs. { ¦fär·mə·kō·dī'nam·iks }

pharmacogenetics [GEN] The science of genetically determined variations in drug responses. { ¦fär·mə·kō·jə'ned·iks }

pharmacognosy [PHARM] A subfield of pharmacology which studies the biological and chemical components of medically useful substances that occur naturally (primarily those synthesized by plants). { ,fär·mə'käg·nə·sē }

pharmacokinetics [PHARM] The study of the way that drugs move through the body after they are swallowed or injected. { ¦fär·mə·kō·ki'ned·iks }

pharmacologic pyrogen [PHARM] A naturally occurring pharmacologic agent, such as serotonin or a catecholamine, that controls body temperature; it can cause fever when injected under experimental conditions. { ¦fär·mə·kə'läj·ik 'pī·rə·jən }

pharmacopoeia [PHARM] A book containing a selected list of medicinal substances and their dosage forms, providing also a description and the standards for purity and strength for each. { ,fär·mə·kə'pē·ə }

pharmacotherapy [MED] The treatment of disease by means of drugs. { ¦fär·mə·kō'ther·ə·pē }

pharyngeal aponeurosis [ANAT] The fibrous submucous layer of the pharynx. { fə'rin·jē·əl ,ap·ō·nü'rō·səs }

pharyngeal bursa [EMBRYO] A small pit caudal to the pharyngeal tonsil, resulting from the ingrowth of epithelium along the course of the degenerating tip of the notochord of the vertebrate embryo. { fə'rin·jē·əl 'bər·sə }

pharyngeal cleft [EMBRYO] One of the paired open clefts on the sides of the embryonic pharynx between successive visceral arches in vertebrates. { fə'rin·jē·əl 'kleft }

pharyngeal plexus [NEUROSCI] A plexus of veins situated at the side of the pharynx. [ANAT] A nerve plexus innervating the pharynx. { fə'rin·jē·əl 'plek·səs }

pharyngeal pouch [EMBRYO] One of the five paired sacculations in the lateral aspect of the pharynx in vertebrate embryos. Also known as visceral pouch. { fə'rin·jē·əl 'paüch }

pharyngeal tonsil *See* adenoid. { fə'rin·jē·əl 'tän·səl }

pharyngeal tooth [VERT ZOO] A tooth developed on the pharyngeal bone in many fishes. { fə'rin·jē·əl 'tüth }

pharyngitis [MED] Inflammation of the pharynx. { ,far·ən'jīd·əs }

Pharyngobdellae [INV ZOO] A family of leeches in the order Arhynchobdellae that is distinguished by the lack of jaws. { fə,riŋ,gäb'del·ə,dē }

pharyngology [MED] The science of the pharyngeal mechanism, functions, and diseases. { ,far·iŋ'gäl·ə·jē }

pharyngoscope [MED] An instrument for examining the pharynx. { fə'riŋ·gə,skōp }

pharyngo-tonsillar diphtheria [MED] A type of diphtheria that is characterized by a sore throat, difficulty in swallowing, and low-grade fever. { fə,riŋ·gō ,täns·əl·ər dif'thir·ē·ə }

pharynx [ANAT] A chamber at the oral end of the vertebrate alimentary canal, leading to the esophagus. { 'far·iŋks }

phase-shift mutation *See* frameshift mutation. { 'fāz ¦shift myü,tā·shən }

Phasianidae [VERT ZOO] A family of game birds in the order Galliformes; typically, members are ground feeders, have bare tarsi and copious plumage, and lack feathers around the nostrils. { ,fāz·ē'an·ə,dē }

phasmid [INV ZOO] One of a pair of lateral caudal pores which function as chemoreceptors in certain nematodes. { 'faz·məd }

Phasmidae [INV ZOO] A family of the insect order Orthoptera including the walking sticks and leaf insects. { 'faz·mə,dē }

Phasmidea [INV ZOO] An equivalent name for Secernentea. { faz'mid·ē·ə }

Phasmidia [INV ZOO] An equivalent name for Secernentea. { faz'mid·ē·ə }

pheasant [VERT ZOO] Any of various large sedentary game birds with long tails in the family Phasianidae; sexual dimorphism is typical of the group. { 'fez·ənt }

phellem [BOT] Cork; the outer tissue layer of the periderm. { 'fel·əm }

phelloderm [BOT] Layers of parenchymatous cells formed as inward derivatives of the phellogen. { 'fel·ə,dərm }

phellogen [BOT] The meristematic portion of the periderm, consisting of one layer of cells that initiate formation of the cork and secondary cortex tissue. { 'fel·ə·jən }

phenacaine hydrochloride [PHARM] White crystals with a slight bitter taste and a melting point of 190°C; soluble in boiling water, chloroform, and alcohol; used in medicine. { 'fen·ə,kān ,hī·drə'klȯr,īd }

phenacemide [PHARM] $C_6H_5CH_2CONHCONH_2$ A white to cream-colored, crystalline solid with a melting point of 212–216°C; soluble in chloroform, benzene, ether, and alcohol; used in medicine. { fə'nas·ə·məd }

phencyclidine [PHARM] $C_{17}H_{25}N$ An addicting drug originally used as an intravenous anesthetic that was subsequently removed from medical use in humans because it produces hallucinations and delusions. It has been illegally manufactured and sold on the street, sometimes causing serious adverse reactions. Also known as angel dust; PCP. { ,fen·ə'sī·klə,dēn }

Phengodidae [INV ZOO] The fire beetles, a New World family of coleopteran insects in the superfamily Cantharoidea. { fen'gäd·ə,dē }

phenindione [PHARM] $C_{15}H_{10}O_2$ Pale yellow crystals, soluble in alcohol, ether, acetone, and benzene; used as an anticoagulant. { ,fen·ən'dī,ōn }

pheniramine maleate [PHARM] $C_{16}H_{20}N_2 \cdot C_4H_4O_4$ A white, crystalline powder with a melting point of 104–108°C; soluble in alcohol and water; used as an antihistamine. { fə'nir·ə,mēn 'mal·ē,āt }

phenobarbital [PHARM] $C_{12}H_{12}N_2O_3$ A crystalline compound, 5-ethyl-5-phenylbarbituric acid, with a slightly bitter taste, melting at 174–178°C; soluble in water, alcohol, chloroform, and ether; used in medicine as a long-acting sedative, anticonvulsant, and hypnotic. { ¦fē·nō'bär·bə,tȯl }

phenocoll hydrochloride [PHARM] $C_2H_5OC_6H_4$-$NHCOCH_2NH_2 \cdot HCl$ A white, crystalline powder with a melting point of 95°C; soluble in warm alcohol and in water; used in medicine. { 'fēn·ə,käl ,hī·drə'klȯr,īd }

phenocopy [GEN] The nonhereditary alteration of a phenotype to a form imitating a mutant trait; caused by external conditions during development. { ¦fēn·ə¦käp·ē }

phenocritical period [EMBRYO] During development, that period during which a gene's effect can most readily be influenced by external factors. { ¦fē·nə,krid·ə·kəl 'pir·ē·əd }

phenogenetics [GEN] The study of the phenotypic effects of the genetic material. Also known as physiological genetics. { ¦fēn·ə·jə'ned·iks }

phenological shift [ECOL] A change in the timing of growth and breeding events in the life of an individual organism. { ,fēn·ə,läj·i·kəl 'shift }

phenotype [GEN] The observable characters of an organism, dependent upon genotype and environment. { 'fē·nə,tīp }

phenotypic lag [GEN] Delay in the expression of a newly acquired character. { ¦fē·nə¦tip·ik 'lag }

phenotypic masking [MICROBIO] Masking of the phenotype in strains of bacteria that are drug-dependent. { ¦fē·nə,tip·ik 'mask·iŋ }

phenotypic mixing [VIROL] The production of virus particles having different structural components in the protein coats, each synthesized under the direction of a different genome. { ¦fē·nə,tip·ik 'mik·siŋ }

phenotypic plasticity [GEN] The extent of genotype expression in different environments. { ¦fē·nə,tip·ik plas'tis·əd·ē }

phenotypic sex determination [BIOL] Control of the development of gonads by environmental stimuli, such as temperature. { ¦fē·nə,tip·ik 'seks di,tər·mə,nā·shən }

phenotypic suppression [BIOL] Prevention of mutant phenotype expression. { ¦fē·nə,tip·ik sə'presh·ən }

phenoxymethylpenicillin [PHARM] $C_{16}H_{18}N_2O_5S$ A white, crystalline powder, soluble in alcohol and acetone; used as an oral antibiotic. Also known as penicillin V. { fə¦näk·sē,meth·əl,pen·ə'sil·ən }

phenylalanine [BIOCHEM] $C_9H_{11}O_2N$ An essential amino acid, obtained in the levo form by hydrolysis of proteins (as lactalbumin); converted to tyrosine in the normal body. Also known as α-aminohydrocinnamic acid; α-amino-β-phenylpropionic acid; β-phenylalanine. { ¦fen·əl'al·ə,nēn }

β-phenylalanine *See* phenylalanine. { ¦bād·ə ¦fen·əl'al·ə,nēn }

phenylephrine [PHARM] $C_9H_{13}NO_2$ A sympathomimetic amine, used in its hydrochloride salt form as a vasoconstrictor. { ¦fen·əl'ef·rən }

phenylketonuria [MED] A hereditary disorder of metabolism, transmitted as an autosomal recessive, in which there is a lack of the enzyme phenylalanine hydroxylase, resulting in excess amounts of phenylalanine in the blood and of excess phenylpyruvic and other acids in the urine. Abbreviated PKU. Also known as phenylpyruvic oligophrenia. { ¦fen·əl,kēd·ə 'nyu̇r·ē·ə }

phenylpropanolamine hydrochloride [PHARM] $C_9H_{13}ON \cdot HCl$ A white, crystalline powder with a melting point of 198–199°C; soluble in alcohol and water; used in medicine. { ¦fen·əl,prō·pə'näl·ə,mēn ,hī·drə'klȯr,īd }

phenylpyruvic acid [BIOCHEM] $C_6H_5CH_2 \cdot CO \cdot COOH$ A keto acid, occurring as a metabolic product of phenylalanine. { ¦fen·əl·pī'rü·vik 'as·əd }

phenylpyruvic oligophrenia *See* phenylketonuria. { ¦fen·əl·pī'rü·vik ,äl·ə·gə'frē·nē·ə }

phenytoin [PHARM] $C_{15}H_{12}N_2O_2$ A powder with a melting point of 295–298°C; used as an anticonvulsant and antiepileptic in humans. { fə'nid·ə·wən }

pheochromoblast [HISTOL] A precursor of a pheochromocyte. { ˌfē·ō'krō·məˌblast }

pheochromocytoma [MED] A tumor of the sympathetic nervous system composed principally of chromaffin cells; found most often in the adrenal medulla. { ˌfē·ōˌkrō·mō·sī 'tō·mə }

pheoplast [CELL MOL] A plastid containing brown pigment and found in diatoms, dinoflagellates, and brown algae. { 'fē·əˌplast }

pheromone [PHYSIO] Any substance secreted by an animal which influences the behavior of other individuals of the same species. { 'fer·əˌmōn }

phialospore [MYCOL] One of a chain of spores produced successively on phialides. { 'fī·əˌlə,spȯr }

Philadelphia chromosome [PATH] An abnormally small G-group chromosome found in the hematopoietic cells of most patients with chronic granulocytic leukemia. { ¦fil·ə¦del·fyə 'krō·məˌskōp }

Philippine fowl disease *See* Newcastle disease. { 'fil·əˌpēn 'fau̇l diˌzēz }

Philomycidae [INV ZOO] A family of pulmonate gastropods composed of slugs. { ˌfil·ə'mīs·əˌdē }

philopatry [ECOL] A dispersal method in which reproductive particles remain near their point of origin. { ˌfī·lə'pa·trē }

Philopteridae [INV ZOO] A family of biting lice (Mallophaga) that are parasitic on most land birds and water birds. { ˌfil·əp'ter·əˌdē }

phimosis [MED] Elongation of the prepuce and constriction of the orifice, so that the foreskin cannot be retracted to uncover the glans penis. { fə'mō·səs }

phlebitis [MED] Inflammation of a vein. { flə·bīd·əs }

phlebography [MED] **1.** X-ray photography of a vein or veins following intravenous injection of a radiopaque substance. **2.** Recording of venous pulsations. { flə'bäg·rə·fē }

phlebolith [MED] A calculus in a vein. { 'fleb·əˌlith }

phlebosclerosis [MED] **1.** Sclerosis of a vein. **2.** Chronic phlebitis. { ¦fle·bō·sklə'rō·səs }

phlebotaxis [BIOL] Movement of a simple motile organism in response to the presence of blood. { 'fle·bəˌtak·səs }

phlebothrombosis [MED] A venous thrombus not associated with inflammation of the vein. { ¦fle·bō·thräm'bō·səs }

phlebotomus fever [MED] An acute viral infection, transmitted by the fly *Phlebotomus papatosii* and characterized by fever, pains in the head and eyes, inflammation of the conjunctiva, leukopenia, and general malaise. Also known as Chitral fever; pappataci fever; sandfly fever; three-day fever. { flə'bäd·ə·məs ˌfēv·ər }

phlebotomy [MED] The withdrawal of blood from a vein. { flə'bäd·ə·mē }

Phlebovirus [VIROL] A genus of the family Bunyaviridae that causes sandfly fever. { 'flē·bəˌvī·rəs }

phlegm [PHYSIO] A viscid, stringy mucus, secreted by the mucosa of the air passages. { flem }

phlegmasia alba dolens [MED] A painful swelling of the leg usually seen postpartum, due to femoral vein thrombophlebitis or lymphatic obstruction. Also known as milk leg. { fleg'mā·zhə 'al·bə 'dō·lənz }

phleomycin [MICROBIO] An antibacterial antibiotic produced by *Streptomyces verticillatus*; antitumor activity has also been demonstrated. { 'flē·əˌmīs·ən }

Phloeidae [INV ZOO] The bark bugs, a small neotropical family of hemipteran insects in the superfamily Pentatomoidea. { 'flē·əˌdē }

phloem [BOT] A complex, food-conducting vascular tissue in higher plants; principal conducting cells are sieve elements. Also known as bast; sieve tissue. { 'flō·əm }

Phocaenidae [VERT ZOO] The porpoises, a family of marine mammals in the order Cetacea. { fō'sē·nəˌdē }

Phocidae [VERT ZOO] The seals, a pinniped family of carnivoran mammals in the superfamily Canoidea. { 'fō·səˌdē }

phocomelia [MED] A congenital or inherited condition in which the proximal part of a limb is missing. { ˌfō·kə'mēl·yə }

Phodilidae [VERT ZOO] A family of birds in the order Strigiformes; the bay owl (*Pholidus bodius*) is the single species. { fō'dil·əˌdē }

Phoenicopteridae [VERT ZOO] The flamingos, a family of long-legged, long-necked birds in the order Ciconiiformes. { ˌfēn·ə·käp'ter·əˌdē }

Phoenicopteriformes [VERT ZOO] An order comprising the flamingos in some systems of classification. { ˌfē·nəˌkäp·tə·rə'fȯrˌmēz }

Phoeniculidae [VERT ZOO] The African wood hoopoes, a family of birds in the order Coraciiformes. { ˌfēn·ə'kyü·ləˌdē }

Pholadidae [INV ZOO] A family of bivalve mollusks in the subclass Eulamellibranchia; individuals may have one or more dorsal accessory plates, and the visceral mass is attached to the valves in the dorsal portion of the body. { fō 'lad·əˌdē }

Pholidota [VERT ZOO] An order of mammals comprising the living pangolins and their fossil predecessors; characterized by an elongate tubular skull with no teeth, a long protrusive tongue, strong legs, and five-toed feet with large claws. { ˌfäl·ə'dōd·ə }

Phomaceae [MYCOL] The equivalent name for Sphaerioidaceae. { fō'mās·ēˌē }

Phomales [MYCOL] The equivalent name for Sphaeropsidales. { fō'mā·lēz }

phonocardiograph [MED] An instrument that provides a graphic record of heart murmurs and other sounds. { ¦fō·nō'kärd·ē·əˌgraf }

phonoelectrocardioscope [MED] An electronic medical instrument that uses a double-beam cathode-ray oscilloscope to show simultaneously the waveforms of two different quantities related to the heart. { ¦fō·nō·iˌlek·trə'kärd·ē·əˌskōp }

phonoreception [PHYSIO] The perception of sound through specialized sense organs. { ¦fō·nō·ri'sep·shən }

phoresy [ECOL] A relationship between two different species of organisms in which the larger, or host, organism transports a smaller organism, the guest. { 'fȯr·ə·sē }

Phoridae [INV ZOO] The hump-backed flies, a family of cyclorrhaphous dipteran insects in the series Aschiza. { 'fȯr·ə,dē }

Phoronida [INV ZOO] A small, homogeneous group, or phylum, of animals having an elongate body, a crown of tentacles surrounding the mouth, and the anus occurring at the level of the mouth. { fə'rän·ə·də }

phosphatase [BIOCHEM] An enzyme that catalyzes the hydrolysis and synthesis of phosphoric acid esters and the transfer of phosphate groups from phosphoric acid to other compounds. { 'fäs·fə,tās }

phosphatide *See* phospholipid. { 'fäs·fə,tīd }

phosphaturia *See* hyperphosphaturia. { ,fäs·fə'tu̇r·ē·ə }

phosphocreatine [BIOCHEM] $C_4H_{10}N_3O_5P$ Creatine phosphate, a phosphoric acid derivative of creatine which contains an energy-rich phosphate bond; it is present in muscle and other tissues, and during the anaerobic phase of muscular contraction it hydrolyzes to creatine and phosphate and makes energy available. Abbreviated PC. { ¦fäs·fō'krē·ə,tēn }

phosphoenolpyruvic acid [BIOCHEM] $CH_2O(OPO_3H_2)COOH$ A high-energy phosphate formed by dehydration of 2-phosphoglyceric acid; it reacts with adenosine diphosphate to form adenosine triphosphate and enolpyruvic acid. { ¦fäs·fō¦ē,nȯl·pī'rü·vik 'as·əd }

phosphofructokinase [BIOCHEM] A glycolytic enzyme that functions in carbohydrate metabolism by catalyzing the phosphorylation of fructose phosphate. { ,fäs·fə,fru̇k,tō'kī,nās }

phosphoglucoisomerase [BIOCHEM] An enzyme that catalyzes the conversion of galactose-1-phosphate to glucose-1-phosphate. { ¦fäs·fō¦glü·kō·ī'säm·ə,rās }

phosphoglucomutase [BIOCHEM] An enzyme that catalyzes the conversion of glucose-1-phosphate to glucose-6-phosphate. { ¦fäs·fō¦glü·kō'myü,tās }

phosphoglycolate [BOT] A two-carbon by-product of photorespiration. { 'fäs·fə'glī·kə ,lāt }

phosphohexose isomerase [BIOCHEM] An enzyme found in muscle and yeast that catalyzes the interconversion of glucose-6-phosphate and fructose-6-phosphate. { ,fäs·fə,hek,sōs ī'säm·ə,rās }

phospholipase [BIOCHEM] An enzyme that catalyzes a hydrolysis of a phospholipid, especially a lecithinase that acts in this manner on a lecithin. { ¦fäs·fō'lī,pās }

phospholipase A_1 [BIOCHEM] A principal phospholipase that releases fatty acids from the first position (F1) of phospholipids. Abbreviated PLA1. { ,fäs·fə'lī,pās ¦ā'wən }

phospholipase A_2 [BIOCHEM] A principal phospholipase that cleaves fatty acids (principally arachidonic acid) from the second position (F2) of phospholipids. Abbreviated PLA2. { ,fäs·fə'lī,pās ¦ā'tü }

phospholipase C [BIOCHEM] A widely distributed enzyme of considerable physiological significance that hydrolyzes phospholipids at the phosphodiester bond to produce diacylglycerol and a phosphorylated head group, as found in phosphocholine or phosphoinositol. Abbreviated PLC. { ,fäs·fə'lī,pās ¦sē }

phospholipase D [BIOCHEM] An enzyme found in almost all mammalian cells that hydrolyzes phosphatidylcholine to produce the signaling molecule phosphatidic acid, which acts on many regulatory enzymes and other proteins in the cell. Abbreviated PLD. { ,fäs·fə'lī,pās 'dē }

phospholipid [BIOCHEM] Any of a class of esters of phosphoric acid containing one or two molecules of fatty acid, an alcohol, and a nitrogenous base. Also known as phosphatide. { ¦fäs·fō'lip·əd }

phosphomonoesterase [BIOCHEM] An enzyme catalyzing hydrolysis of phosphoric acid esters containing one ester linkage. { ¦fäs·fō,män·ō'es·tə,rās }

phosphorolysis [BIOCHEM] A reaction by which elements of phosphoric acid are incorporated into the molecule of a compound. { ,fäs·fə'räl·ə·səs }

phosphorylase [BIOCHEM] An enzyme that catalyzes the formation of glucose-1-phosphate (Cori ester) from glycogen and inorganic phosphate; it is widely distributed in animals, plants, and microorganisms. { fäs'fȯr·ə,lās }

phosphotransacetylase [BIOCHEM] An enzyme that catalyzes the reversible transfer of an acetyl group from acetyl coenzyme A to a phosphate, with formation of acetyl phosphate. { ¦fä·sfō ,tranz·ə'sed·əl,ās }

photic zone [ECOL] The uppermost layer of a body of water (approximately the upper 330 feet or 100 meters) that receives enough sunlight to permit the occurrence of photosynthesis. { 'fōd·ik }

Photidae [INV ZOO] A family of amphipod crustaceans in the suborder Gammaridea. { 'fäd·ə,dē }

photoautotroph [ECOL] An autotroph that uses energy from light to produce organic molecules. { ,fōd·ō'ȯd·ō,träf }

photoautotrophic [BIOL] Pertaining to organisms which derive energy from light and manufacture their own food. { ¦fōd·ō,ȯd·ō'träf·ik }

photobiont [ECOL] A photosynthetic partner of a symbiotic pair, such as the algal component of the fungal-algal association in lichens. { ¦fōd·ō'bī,änt }

photocoagulator [MED] An instrument that uses a xenon flash lamp and an associated train of optics to focus an intense beam of light on a detached retina for the purpose of inducing coagulation and a lesion that welds the retina back into position. { ¦fōd·ō·kō'ag·yə,lād·ər }

photoelectric plethysmograph [MED] A medical instrument for measuring and recording ear opacity by means of a tiny phototube and lamp clipped to the ear, as a measure of the state of fullness of blood vessels; also worn by aircraft pilots during high-altitude flights, as an alarm indicating the need for more oxygen. { ¦fōd·ō·i'lek·trik plə'thiz·mə,graf }

photoinhibition [BOT] Damage to the light-harvesting reactions of the photosynthetic apparatus caused by excess light energy trapped by the chloroplast. { ¦fōd·ō,in·ə'bish·ən }

photomorphogenesis [BOT] The control exerted by light over growth, development, and differentiation of plants that is independent of photosynthesis. { ¦fōd·ō,mȯr·fō'jen·ə·səs }

photoperiodism [PHYSIO] The physiological responses of an organism to the length of night or day or both. { ¦fōd·ō'pir·ē·ə,diz·əm }

photophilic [BIOL] Thriving in full light. { ¦fōd·ō¦fil·ik }

photophobic [BIOL] **1.** Avoiding light. **2.** Exhibiting negative phototropism. { ,fōd·ə'fō·bik }

photophore gland [VERT ZOO] A highly modified integumentary gland which develops into a luminous organ composed of a lens and a light-emitting gland; occurs in deep-sea teleosts and elasmobranchs. { 'fōd·ə,fȯr ,gland }

photophobic response [PHYSIO] A transient alteration in swimming direction or velocity when a motile organism is exposed to a sudden change in light intensity. { ,fōd·o,fō·bik ri'späns }

photophosphorylase [BIOCHEM] An enzyme that is associated with the surface of a thylakoid membrane and is involved in the final stages of adenosine triphosphate production by photosynthetic phosphorylation. { ¦fōd·ō,fä'sfȯr·ə,lās }

photophosphorylation [BIOCHEM] Phosphorylation that is induced by light energy in photosynthesis. { ¦fōd·ō,fä·sfə·rə'lā·shən }

photophygous [BIOL] Thriving in shade. { fə'täf·ə·gəs }

photopic vision *See* foveal vision. { fō'täp·ik 'vizh·ən }

photopigment [BIOCHEM] A pigment that is unstable in the presence of light of appropriate wavelengths, such as the chromophore pigment which combines with opsins to form rhodopsin in the rods and cones of the vertebrate eye. { ¦fōd·ō¦pig·mənt }

photoreactive chlorophyll [BIOCHEM] Chlorophyll molecules which receive light quanta from antenna chlorophyll and constitute a photoreaction center where light energy conversion occurs. { ¦fōd·ō·rē'ak·tiv 'klȯr·ə,fil }

photoreactivation *See* photoreversal. { ,fōd·ō,rē·ak·tə'vā·shən }

photoreception [PHYSIO] The process of absorption of light energy by plants and animals and its utilization for biological functions, such as photosynthesis and vision. { ¦fōd·ō·ri'sep·shən }

photoreceptor [PHYSIO] A highly specialized, light-sensitive cell or group of cells containing photopigments. { ¦fōd·ō·ri'sep·tər }

photorespiration [BIOCHEM] Respiratory activity taking place in plants during the light period; CO_2 is released and O_2 is taken up, but no useful form of energy, such as adenosinetriphosphate, is derived. { ¦fōd·ō,res·pə'rā·shən }

photoreversal [BIOPHYS] An enzymatic repair system that uses short-wavelength visible (violet and blue) or long-wavelength ultraviolet light to reconstitute the deoxyribonucleic acid of cells that have been irradiated by ultraviolet light. Also known as photoreactivation. { ,fōd·ō·ri'vər·səl }

photosynthesis [BIOCHEM] Synthesis of chemical compounds in light, especially the manufacture of organic compounds (primarily carbohydrates) from carbon dioxide and a hydrogen source (such as water), with simultaneous liberation of oxygen, by chlorophyll-containing plant cells. { ¦fōd·ō'sin·thə·səs }

photosystem I [BIOCHEM] One of two reaction sequences of the light phase of photosynthesis in green plants that involves a pigment system which is excited by wavelengths shorter than 700 nanometers and which transfers this energy to energy carriers such as NADPH that are subsequently utilized in carbon dioxide fixation. { 'fōd·ō,sis·təm 'wən }

photosystem II [BIOCHEM] One of two reaction sequences of the light phase of photosynthesis in green plants which involves a pigment system excited by wavelengths shorter than 685 nanometers and which is directly involved in the splitting or photolysis of water. { 'fōd·ō,sis·təm 'tü }

phototaxis [BIOL] Movement of a motile organism or free plant part in response to light stimulation. { ¦fōd·ə¦tak·səs }

phototroph [BIOL] An organism that utilizes light as a source of metabolic energy. { 'fōd·ə,träf }

phototrophic bacteria [MICROBIO] Primarily aquatic bacteria comprising two principal groups: purple bacteria and green sulfur bacteria; all contain bacteriochlorophylls. { ¦fōd·ə¦träf·ik bak'tir·ē·ə }

phototropism [BOT] A growth-mediated response of a plant to stimulation by visible light. { fō'tä·trə,piz·əm }

phototropy *See* phototropism. { fō'tä·trə·pē }

Phoxichilidiidae [INV ZOO] A family of marine arthropods in the subphylum Pycnogonida; typically, chelifores are present, palpi are lacking, and ovigers have five to nine joints in males only. { ,fäk·sə,kil·ə'dī·ə,dē }

Phoxocephalidae [INV ZOO] A family of amphipod crustaceans in the suborder Gammaridea. { ,fäk·sō·sə'fal·ə,dē }

Phractolaemidae [VERT ZOO] A family of tropical African fresh-water fishes in the order Gonorynchiformes. { ,frak·tə'lē·mə,dē }

phragmacone *See* phragmocone. { 'frag·mə,kōn }

Phragmobasidiomycetes [MYCOL] An equivalent name for Heterobasidiomycetidae. { ¦frag·mō·bə¦sid·ē·ō·mī'sēd·ēz }

phragmocone [INV ZOO] The siphuncular tube of the chambered part of the shell of certain mollusks. Also spelled phragmacone. { 'frag·mə,kōn }

phragmoid [BOT] Having septae perpendicular to the long axis, as the conidia of certain fungi. { 'frag,mȯid }

phragmoplast [CELL MOL] A thin barrier which is formed across the spindle equator in late cytokinesis in plant cells and within which the cell wall is laid down. { 'frag·mə,plast }

phragmosome [CELL MOL] A differentiated cytoplasmic partition in which the phragmoplast and cell plate develop during cell division in plant cells. { 'frag·mə,sōm }

Phragmosporae [MYCOL] A spore group of the Fungi Imperfecti with three- to many-celled spores. { frag'mäs·pə,rē }

Phreatoicidae [INV ZOO] A family of isopod crustaceans in the suborder Phreatoicoidea in which only the left mandible retains a lacinia mobilis. { frē,ad·ō'īs·ə,dē }

Phreatoicoidea [INV ZOO] A suborder of the Isopoda having a subcylindrical body that appears laterally compressed, antennules shorter than the antennae, and the first thoracic segment fused with the head. { frē,ad·ō·i'kȯid·ē·ə }

phreatophyte [ECOL] A plant with a deep root system which obtains water from the groundwater or the capillary fringe above the water table. { frē'ad·ə,fīt }

phrenectomy [MED] Resection of a section of a phrenic nerve or removal of an entire phrenic nerve. { frə'nek·tə·mē }

phrenic nerve [NEUROSCI] A nerve, arising from the third, fourth, and fifth cervical (cervical plexus) segments of the spinal cord; innervates the diaphragm. { 'fren·ik ¦nərv }

phrynoderma [MED] Dryness of the skin with follicular hyperkeratosis, caused by vitamin A deficiency. { ,frī·nə'dər·mə }

Phrynophiurida [INV ZOO] An order of the Ophiuroidea in which the vertebrae usually articulate by means of hourglass-shaped surfaces, and the arms are able to coil upward or downward in the vertical plane. { ,frī·nə'fyu̇r·ə·də }

Phycitinae [INV ZOO] A large subfamily of moths in the family Pyralididae in which the frenulum of the female is a simple spine rather than a bundle of bristles. { fī'sīt·ən,ē }

phycobilin [BIOCHEM] Any of various protein-bound pigments which are open-chain tetrapyrroles and occur in some groups of algae. { ,fī·kō'bī·lən }

phycobiliprotein [BIOCHEM] A water-soluble photosynthetic membrane protein that covalently binds with phycobilins (photosynthetic pigments) in some groups of algae. { ,fī·kō ,bil·ē'prō,tēn }

phycobilisome [BIOCHEM] A light-harvesting structure containing aggregates of photosynthetic accessory pigments that is located on the surface of thylakoid membranes in all cyanobacteria and red algae. { ,fī·kō'bil·ē,sōm }

phycobiliviolin [BIOCHEM] A yellow-light (575-nanometer) absorbing pigment found in all cryptophytes but in only a few cyanobacteria. Also known as cryptoviolin. { ,fī·kō,bil·ē'vī·ə·lən }

phycobiont [BOT] The algal component of a lichen, commonly the green unicell of the genus *Trebouxia*. { ,fī·kō'bī,änt }

phycocyanin [BIOCHEM] A blue phycobilin. { ¦fī·kō'sī·ə·nən }

phycocyanobilin [BIOCHEM] $C_{31}H_{38}O_2N_4$ Phycobilin with an ethylidene side chain ($=CH-CH_3$) and only one asymmetric carbon atom (C_1). { ¦fī·kō,sī·ə·nō'bī·lən }

phycoerythrin [BIOCHEM] A red phycobilin. { ¦fī·kō·ə'rith·rən }

phycoerythrobilin [BIOCHEM] $C_{31}H_{38}O_2N_4$ Phycobilin with seven conjugated double bonds, an ethylidine side chain ($=CH-CH_3$), and two asymmetric carbon atoms (C_1 and C_7). { ¦fī·kō·ə,rith·rə'bī·lən }

phycology *See* algology. { fī'käl·ə·jē }

Phycomycetes [MYCOL] A primitive class of true fungi belonging to the Eumycetes; they lack regularly spaced septa in the actively growing portions of the plant body, and have the sporangiospore, produced in the sporangium by cleavage, as the fundamental, asexual reproductive unit. { ¦fī·kō,mī'sēd·ēz }

Phycosecidae [INV ZOO] A small family of coleopteran insects of the superfamily Cucujoidea, including five species found in New Zealand, Australia, and Egypt. { ¦fī·kō'sē·sə,dē }

phycourobilin [BIOCHEM] A blue-green light (495-nanometer) absorbing pigment found in some cyanobacteria and red algae. { ,fī·kō·yu̇'rō·bi·lin }

Phylactolaemata [INV ZOO] A class of fresh-water ectoproct bryozoans; individuals have lophophores which are U-shaped in basal outline, and relatively short, wide zooecia. { fə,lak·tō'lē·məd·ə }

phyletic evolution [EVOL] The gradual evolution of population without separation into isolated parts. { fī'led·ik ,ev·ə'lü·shən }

phyletic gradualism *See* gradualism. { fī'led·ik 'gra·jə·wə,liz·əm }

phyllary [BOT] A bract of the involucre of a composite plant. { 'fil·ə·rē }

Phyllobothrioidea [INV ZOO] The equivalent name for Tetraphyllidea. { ,fil·ō,bäth·rē'ȯid·ē·ə }

phyllobranchiate gill [INV ZOO] A type of decapod crustacean gill with flattened branches, or lamellae usually arranged in two opposite series. { ,fil·ō'braŋ·kē·ət 'gil }

phylloclade [BOT] A flattened stem that fulfills the same functions as a leaf. { 'fil·ə,klād }

phyllode [BOT] A broad, flat petiole that replaces the blade of a foliage leaf. [INV ZOO] A petal-shaped group of ambulacra near the mouth of certain echinoderms. { 'fi,lōd }

Phyllodocidae [INV ZOO] A leaf-bearing family of errantian annelids in which the species are often brilliantly iridescent and are highly motile. { ,fil·ə'däs·ə,dē }

Phyllogoniaceae [BOT] A family of mosses in the order Isobryales in which the leaves are equitant. { ˌfil·əˌgō·nē'ās·ēˌē }

Phyllophoridae [INV ZOO] A family of dendrochirotacean holothurians in the order Dendrochirotida having a rather naked skin and a complex calcareous ring. { ˌfil·ə'fòr·əˌdē }

phyllosoma [INV ZOO] A flat, transparent, long-legged larval stage of various spiny lobsters. { ˌfil·ə'sō·mə }

phyllospondylous [VERT ZOO] Of vertebrae, having a hypocentrum but no pleurocentra; the neural arch extends ventrad to enclose the notochord and form transverse processes which articulate with the ribs. { ˌfil·ə'spän·də·ləs }

Phyllostictales [MYCOL] An equivalent name for Sphaeropsidales. { ˌfil·əˌstik'tā·lēz }

Phyllostomatidae [VERT ZOO] The New World leaf-nosed bats (Chiroptera), a large tropical and subtropical family of insect- and fruit-eating forms with narrow, pointed ears. { ˌfil·ə·stō 'mad·əˌdē }

phyllotaxy [BOT] The arrangement of leaves on a stem. { ¦fil·ə¦tak·sē }

Phylloxerinae [INV ZOO] A subfamily of homopteran insects in the family Chermidae in which the sexual forms lack mouthparts, and the parthenogenetic females have a beak but the digestive system is closed, and no honeydew is produced. { ˌfiˌläk'ser·əˌnē }

phylogenetic tree *See* evolutionary tree. { ˌfī·lō·jə¦ned·ik 'trē }

phylogeny [EVOL] The evolutionary or ancestral history of organisms. { fə'läj·ə·nē }

phylum [SYST] A major taxonomic category in classifying animals (and plants in some systems), composed of groups of related classes. { 'fī·ləm }

Phymatidae [INV ZOO] A family of carnivorous hemipteran insects characterized by strong, thick forelegs. { fī'mad·əˌdē }

Phymosomatidae [INV ZOO] A family of echinacean echinoderms in the order Phymosomatoida with imperforate crenulate tubercles; one surviving genus is known. { ˌfī·mə·sō'mad·əˌdē }

Phymosomatoida [INV ZOO] An order of Echinacea with a stirodont lantern and diademoid ambulacral plates. { ˌfī·mə·sə·mə'tòid·ē·ə }

physa [INV ZOO] The rounded basal portion of the body of certain sea anemones. { 'fī·sə }

Physalopteridae [INV ZOO] A family of parasitic nematodes in the superfamily Spiruroidea. { ˌfī·sə·läp'ter·əˌdē }

Physalopteroidea [INV ZOO] A superfamily of parasitic nematodes in the order Spirurida, characterized by two large lateral lips generally provided with teeth on their inner surfaces, reduced stoma, and a reduced or absent inner whorl of circumoral sensilla and an external circle with four fused sensilla. { 'fī·səˌläp·tə'ròid·ē·ə }

Physaraceae [MYCOL] A family of slime molds in the order Physarales. { ˌfī·sə'rās·ēˌē }

Physarales [MYCOL] An order of Myxomycetes in the subclass Myxogastromycetidae. { ˌfī·sə'rā·lēz }

physiatrics [MED] A branch of medicine dealing with treatment, prevention, and diagnosis of disorders and disabilities through the use of physical therapy, physical agents such as electricity, light, heat, and water, and mechanical apparatus. { ˌfiz·ē'a·triks }

physiatrist [MED] A physician specializing in physiatrics. { ˌfiz·ē'a·trəst }

physical medicine [MED] A consultative, diagnostic, and therapeutic medical specialty, coordinating and integrating the use of physical and occupational therapy and physical reconditioning in the professional management of the diseased and injured. { 'fiz·ə·kəl 'med·ə·sən }

physical therapy [MED] The treatment of disease and injury by physical means. { 'fiz·ə·kəl 'ther·ə·pē }

physician [MED] An individual authorized to practice medicine. { fi'zish·ən }

physiological biophysics [BIOPHYS] An area of biophysics concerned with the use of physical mechanisms to explain the behavior and the functioning of living organisms or parts thereof, and with the response of living organisms to physical forces. { ˌfiz·ē·ə'läj·ə·kəl ˌbī·ō'fiz·iks }

physiological dead space *See* dead space. { ˌfiz·ē·ə'läj·ə·kəl 'ded ˌspās }

physiological ecology [ECOL] The study of biophysical, biochemical, and physiological processes used by animals to cope with factors of their physical environment, or employed during ecological interactions with other organisms. { ˌfiz·ē·ə'läj·ə·kəl ē'käl·ə·jē }

physiological genetics *See* phenogenetics. { ˌfiz·ē·ə'läj·ə·kəl jə'ned·iks }

physiological homeostasis [PHYSIO] Maintenance of the body's natural resistance to disease. { ˌfiz·ē·əˌläj·i·kəl ˌhō·mē·ō'stās·əs }

physiological saline *See* normal saline. { ˌfiz·ē·ə'läj·ə·kəl 'sāˌlēn }

physiological salt solution *See* normal saline. { ˌfiz·ē·ə'läj·ə·kəl 'sòlt səˌlü·shən }

physiological sodium chloride solution *See* normal saline. { ˌfiz·ē·ə'läj·ə·kəl 'sōd·ē·əm 'klòrˌīd səˌlü·shən }

physiologic diplopia [PHYSIO] A normal phenomenon in which there is formation of images in noncorresponding retinal points, giving a perception of depth. Also known as introspective diplopia. { ˌfiz·ē·ə'läj·ik di'plō·pē·ə }

physiologic tremor [PHYSIO] A tremor in normal individuals, caused by fatigue, apprehension, or overexposure to cold. { ˌfiz·ē·ə'läj·ik 'trem·ər }

physiology [BIOL] The study of the basic activities that occur in cells and tissues of living organisms by using physical and chemical methods. { ˌfiz·ē'äl·ə·jē }

Physopoda [INV ZOO] The equivalent name for Thysanoptera. { fī'säp·ə·də }

Physosomata [INV ZOO] A superfamily of amphipod crustaceans in the suborder Hyperiidea; the eyes are small or rarely absent, and the inner plates of the maxillipeds are free at the apex. { ˌfī·sə'säm·əd·ə }

Phytalmiidae [INV ZOO] A family of myodarian cyclorrhaphous dipteran insects in the subsection Acalypteratae. { ˌfīd·əl'mī·əˌdē }

phytal zone [ECOL] The part of a lake bottom covered by water shallow enough to permit the growth of rooted plants. { 'fīd·əl ˌzōn }

Phytamastigophorea [INV ZOO] A class of the subphylum Sarcomastigophora, including green and colorless phytoflagellates. { ¦fīd·əˌma·stə¦gäf·ə'rē·ə }

phytase [BIOCHEM] An enzyme occurring in plants, especially cereals, which catalyzes hydrolysis of phytic acid to inositol and phosphoric acid. { 'fīˌtās }

phytoalexin [BIOCHEM] A natural substance that is toxic to fungi and is synthesized by a plant as a response to fungal infection. { 'fīd·ō·ə'lek·sən }

phytochemistry [BOT] The study of the chemistry of plants, plant products, and processes taking place within plants. { 'fīd·ōˌkem·ə·strē }

phytochorology *See* plant geography. { ¦fīd·ō·kō'räl·ə·jē }

phytochrome [BIOCHEM] A protein plant pigment which serves to direct the course of plant growth and development in response variously to the presence or absence of light, to photoperiod, and to light quality. { 'fīd·əˌkrōm }

phytocoenosis [ECOL] The entire plant population of a particular habitat. { ¦fīd·ō·sē'nō·səs }

phytogenic dam [ECOL] A natural dam consisting of plants and plant remains. { ¦fīd·ə¦jen·ik 'dam }

phytogenic dune [ECOL] Any dune in which the growth of vegetation influences the form of the dune, for example, by arresting the drifting of sand. { ¦fīd·ə¦jen·ik 'dün }

phytogeography *See* geobotany; plant geography. { ¦fīd·ō·jē'äg·rə·fē }

phytohemagglutinin *See* phytolectin. { ¦fīd·ōˌhē·mə'glüt·ən·ən }

phytohormone *See* plant hormone. { ¦fīd·ō 'hȯrˌmōn }

Phytolacca americana *See* pokeweed. { ˌfīd·ōˌlak·ə əˌmer·ə'kän·ə }

phytolectin [BIOCHEM] A lectin found in plants. Also known as phytohemagglutinin. { ˌfīd·ə 'lek·tən }

Phytomastigina [INV ZOO] The equivalent name for Phytamastigophorea. { ˌfīd·ō·mas·tə 'jī·nə }

Phytomonadida [INV ZOO] The equivalent name for Volvocida. { ˌfīd·ō·mō'näd·ə·də }

phytopathogen [ECOL] An organism that causes a disease in a plant. { ¦fīd·ō'path·ə·jən }

phytophagous [ZOO] Feeding on plants. { fī 'täf·ə·gəs }

Phytophthra citrophthora [MYCOL] A water mold that causes citrus gummosis. { fīˌdäf·thrə ˌsi·trəf'thȯr·ə }

phytoplankton [ECOL] Planktonic plant life. { ¦fīd·ə'plaŋk·tən }

Phytoseiidae [INV ZOO] A family of the suborder Mesostigmata. { ˌfīd·ō·sē'ī·əˌdē }

phytosociology [ECOL] A broad study of plants that includes the study of all phenomena affecting their lives as social units. { ¦fīd·ōˌsō·sē'äl·ə·jē }

phytosterol [BIOCHEM] Any of various sterols obtained from plants, including ergosterol and stigmasterol. { fī'täs·təˌrȯl }

phytotoxin [BIOCHEM] **1.** A substance toxic to plants. **2.** A toxin produced by plants. { ¦fīd·ə'täk·sən }

phytotron [BOT] A research tool used to study whole plants; contains a large number of individually controlled environments that provide the means of studying the effect of each environmental factor, such as temperature or light, at many levels simultaneously. { 'fīd·əˌträn }

pia arachnoid [VERT ZOO] The outer meninx of certain submammalian forms having two membranes covering the brain and spinal cord. { 'pī·ə ə'rakˌnȯid }

pia mater [ANAT] The vascular membrane covering the surface of the brain and spinal cord. { 'pē·ə ˌmād·ər }

piastrenemia *See* thrombocytosis. { pēˌas·trə'nē·mē·ə }

pica [MED] Craving for substances not normally used as food; an abnormal appetite; sometimes seen in hysterical patients or during pregnancy. { 'pī·kə }

Picidae [VERT ZOO] The woodpeckers, a large family of birds in the order Piciformes; adaptive modifications include a long tongue and hyoid mechanism, and stiffened tail feathers. { 'pis·əˌdē }

Piciformes [VERT ZOO] An order of birds characterized by the peculiar arrangement of the tendons of the toes. { ˌpis·ə'fȯrˌmēz }

Picinae [VERT ZOO] The true woodpeckers, a subfamily of the Picidae. { 'pis·əˌnē }

Pick's disease [MED] **1.** A form of presenile dementia characterized by severe atrophy of the frontal and temporal lobes of the cerebrum. **2.** A recurrent or progressive form of ascites with little or no edema. **3.** *See* constrictive pericarditis. **4.** *See* polyserositis. { 'piks diˌzēz }

Picornaviridae [VIROL] A viral family made up of the small (18–30 nanometers) ether-sensitive viruses that lack an envelope and have a Togaviridae genome, contains the genera *Enterovirus* (human polio), *Cardiovirus* (mengo), *Rhinovirus* (common cold), and *Aphtovirus* (foot-and-mouth disease). { pēˌkȯr·nə'vir·əˌdī }

picodnavirus [VIROL] A group of deoxyribonucleic acid-containing animal viruses including the adeno-satellite viruses. { pē'käd·nəˌvī·rəs }

picornavirus [VIROL] A viral group made up of small (18–30 nanometers), ether-sensitive viruses that lack an envelope and have a ribonucleic acid genome; among subgroups included are enteroviruses and rhinoviruses, both of human origin. { pē'kȯr·nəˌvī·rəs }

Picrodendraceae [BOT] A small family of dicotyledonous plants in the order Juglandales characterized by unisexual flowers borne in catkins, four apical ovules in a superior ovary, and trifoliate leaves. { ˌpik·rō·den'drās·ēˌē }

picrotoxin [BIOCHEM] $C_{30}H_{34}O_{13}$ A poisonous, crystalline plant alkaloid found primarily in *Cocculus indicus*; used as a stimulant and convulsant drug. Also known as cocculin. { ˌpik·rəˈtäk·sən }

Picumninae [VERT ZOO] The piculets, a subfamily of the avian family Picidae. { piˈkyüm·nəˌnē }

piebaldism [MED] A pigmentary disorder characterized by patterns of white spots on the skin and caused by an inherited absence of melanocytes. { ˈpīˌbȯlˌdiz·əm }

piece-root grafting [BOT] Grafting in which each piece of a cut seedling root is used as a stock. { ˈpēs ˌrüt ˌgraft·iŋ }

Pieridae [INV ZOO] A family of lepidopteran insects in the superfamily Papilionoidea including white, sulfur, and orange-tip butterflies; characterized by the lack of a prespiracular bar at the base of the abdomen. { pīˈer·əˌdē }

Piesmatidae [INV ZOO] The ash-gray leaf bugs, a family of hemipteran insects belonging to the Pentatomorpha. { pēzˈmad·əˌdē }

Piesmidae [INV ZOO] A small family of hemipteran insects in the superfamily Lygaeoidea. { ˈpez·məˌdē }

pig [VERT ZOO] Any wild or domestic mammal of the superfamily Suoidea in the order Artiodactyla; toes terminate in nails which are modified into hooves, the tail is short, and the body is covered sparsely with hair which is frequently bristlelike. { pig }

pigeon [VERT ZOO] Any of various stout-bodied birds in the family Columbidae having short legs, a bill with a horny tip, and a soft cere. { ˈpij·ən }

pigeon milk [PHYSIO] A milky glandular secretion of the crop of pigeons that is regurgitated to feed newly hatched young. { ˈpij·ən ˌmilk }

pigment [CELL MOL] Any coloring matter in plant or animal cells. { ˈpig·mənt }

pigmentation [PHYSIO] The normal color of the body and its organs, resulting from a summation of the natural color of the tissue, the pigments deposited therein, and the pigments carried through the blood bathing the tissue. { ˌpig·mənˈtā·shən }

pigment cell [CELL MOL] Any cell containing deposits of pigment. { ˈpig·mənt ˌsel }

pigment epithelium [HISTOL] A heavily pigmented layer of epithelial cells interposed between the photoreceptors of the vertebrate retina and their blood supply, the choroid. It forms a barrier regulating the exchange of fluids and substances between the blood and outer layer of the retina; it also plays a number of roles in support of photoreceptor structure and function in vertebrates. { ˌpig·mənt ˌep·əˈthē·lē·əm }

pika [VERT ZOO] Any member of the family Ochotonidae, which includes 14 species of lagomorphs resembling rabbits but having a vestigial tail and short, rounded ears. { ˈpī·kə }

pike [VERT ZOO] Any of about five species of predatory fish which compose the family Esocidae in the order Clupeiformes; the body is cylindrical and compressed, with cycloid scales that have deeply scalloped edges. { pīk }

Pilacraceae [MYCOL] A family of Basidiomycetes. { ˌpil·əˈkrās·ēˌē }

Pilargidae [INV ZOO] A family of small, short, depressed errantian polychaete annelids. { pəˈlär·jəˌdē }

pileum [VERT ZOO] The top of a bird's head, from the nape to the bill. { ˈpil·ē·əm }

pileus [BIOL] The umbrella-shaped upper cap of mushrooms and other basidiomycetous fungi. { ˈpil·ē·əs }

Pilidae [INV ZOO] A family of fresh-water snails in the order Pectinibranchia. { ˈpil·əˌdē }

Pilifera [VERT ZOO] Collective designation for animals with hair, that is, mammals. { pīˈlif·ə·rə }

pill [PHARM] A small, solid dosage form of a globular, ovoid, or lenticular shape, containing one or more medicinal substances. { pil }

pillotina [MICROBIO] A large spirochete that contains microtubules and lives symbiotically in the hind gut of termites. { ˌpil·əˈtē·nə }

pilomotor nerve [NEUROSCI] A nerve causing contraction of one of the arrectoris pilorum muscles. { ¦pī·lōˈmōd·ər ˌnərv }

pilomotor reflex [PHYSIO] Erection of the hairs of the skin (gooseflesh) in response to chilling or irritation of the skin or to an emotional stimulus. { ¦pī·lōˈmōd·ər ˈrēˌfleks }

pilosebaceous [ANAT] Pertaining to the hair follicles and sebaceous glands, as the pilosebaceous apparatus, comprising the hair follicle and its attached gland. { ¦pī·lō·səˈbā·shəs }

pilosis [MED] The abnormal or excessive development of hair. { pīˈlō·səs }

pilus [ANAT] A hair. [BIOL] A fine, slender, hairlike body. [MICROBIO] Any filamentous appendage other than flagella on certain gram-negative bacteria. Also known as fimbria. { ˈpī·ləs }

pimento [BOT] *Capsicum annuum*. A type of pepper in the order Polemoniales grown for its thick, sweet-fleshed red fruit. { pəˈment·ō }

pimple [MED] A small pustule or papule. { ˈpim·pəl }

piñ [BOT] A fiber obtained from the large leaves of the pineapple plant. Also known as pineapple fiber. { ˈpēn·yə }

pinacocyte [INV ZOO] A flattened polygonal cell occurring in the dermal epithelium of sponges, and lining the exhalant canals. { ˈpin·ə·kōˌsīt }

Pinales [BOT] An order of gymnospermous woody trees and shrubs in the class Pinopsida, including pine, spruce, fir, cypress, yew, and redwood; the largest plants are the conifers. { pīˈnā·lēz }

Pinatae *See* Pinopsida. { pīˈnäd·ē }

pincer [INV ZOO] A grasping apparatus, as on the anterior legs of a lobster, consisting of two hinged jaws. { ˈpin·sər }

pinch graft [MED] A small, full-thickness graft

lifted from the donor area by a needle, and cut free with a razor. { 'pinch ˌgraft }

pine [BOT] Any of the cone-bearing trees composing the genus *Pinus*; characterized by evergreen leaves (needles), usually in tight clusters of two to five. { 'pīn }

pineal body [ANAT] An unpaired, elongated, club-shaped, knoblike or threadlike organ attached by a stalk to the roof of the vertebrate forebrain. Also known as conarium; epiphysis. { 'pin·ē·əl ˌbäd·ē }

pineapple [BOT] *Ananas sativus*. A perennial plant of the order Bromeliales with long, swordlike, usually rough-edged leaves and a dense head of small abortive flowers; the fruit is a sorosis that develops from the fleshy inflorescence and ripens into a solid mass, covered by the persistent bracts and crowned by a tuft of leaves. { 'pīˌnap·əl }

pineapple fiber *See* piña. { 'pīˌnap·əl ˌfī·bər }

pine nut [BOT] The edible seed borne in the cone of various species of pine (*Pinus*), such as stone pine (*P. pinea*) and piñon pine (*P. cembroides* var. *edulis*). { 'pīn ˌnət }

pinfeather [VERT ZOO] A young, underdeveloped feather, especially one still enclosed in a cylindrical horny sheath which is afterward cast off. { 'pinˌfe<u>th</u>·ər }

pinguecula [ANAT] A small patch of yellowish-white connective tissue located on the conjunctiva, between the cornea and the canthus of the eye. { piŋ'gwek·yə·lə }

Pinicae [BOT] A large subdivision of the Pinophyta, comprising woody plants with a simple trunk and excurrent branches, simple, usually alternative, needlelike or scalelike leaves, and wood that lacks vessels and usually has resin canals. { 'pī·nəˌsē }

pinion [VERT ZOO] The distal portion of a bird's wing. { 'pin·yən }

pinkeye [MED] **1.** A contagious, mucopurulent conjunctivitis. **2.** *See* catarrhal conjunctivitis. { 'pinkˌī }

pink salmon [VERT ZOO] *Oncorhynchus*. Weighing less than 2 kilograms (5 pounds), it is the smallest but typically most abundant of the salmon. Also known as humpback salmon. { ˌpiŋk 'sam·ən }

pinna [ANAT] The cartilaginous, projecting flap of the external ear of vertebrates. Also known as auricle. { 'pin·ə }

pinnate [BOT] Having parts arranged like a feather, branching from a central axis. { 'piˌnāt }

pinnately compound leaf [BOT] A leaf with leaflets that are borne on the continuation of the petiole. { piˌnāt·lē ˌkämˌpau̇nd' 'lēf }

pinnate muscle [ANAT] A muscle having a central tendon onto which many short, diagonal muscle fibers attach at rather acute angles. { 'piˌnāt ¦məs·əl }

Pinnipedia [VERT ZOO] A suborder of aquatic mammals in the order Carnivora, including walruses and seals. { ˌpin·ə'pē·dē·ə }

Pinnotheridae [INV ZOO] The pea crabs, a family of decapod crustaceans belonging to the Brachygnatha. { ˌpin·ə'ther·əˌdē }

pinnulate [BIOL] Having pinnules. { 'pin·yə·lāt }

pinnule [BIOL] The secondary branch of a plumelike or pinnate organ. { 'pinˌyül }

pinocytosis [CELL MOL] Deprecated term formerly used to describe the process of uptake or internalization of particles, macromolecules, and fluid droplets by living cells; the process is now termed endocytosis. { ¦pin·ō·sī'tō·səs }

Pinophyta [BOT] The gymnosperms, a division of seed plants characterized as vascular plants with roots, stems, and leaves, and with seeds that are not enclosed in an ovary but are borne on cone scales or exposed at the end of a stalk. { pə'näf·əd·ə }

Pinopsida [BOT] A class of gymnospermous plants in the subdivision Pinicae characterized by entire-margined or slightly toothed, narrow leaves. { pə'näp·səd·ə }

pinosome [CELL MOL] A closed intracellular vesicle containing material captured by pinocytosis. { 'pin·əˌsōm }

pinta [MED] A disease of the skin seen most frequently in tropical America, characterized by dyschromic changes and hyperkeratosis in patches of skin; caused by the spirochete *Treponema carateum*. Also known as carate; mal de pinto; piquite; purupuru; quitiqua. { 'pēnt·ə }

pinulus [INV ZOO] A sponge spicule, usually with five rays, one of which develops numerous small spines. { 'pin·yə·ləs }

pinworm [INV ZOO] *Enterobius vermicularis*. A phasmid nematode of the superfamily Oxyuroidea; causes enterobiasis. Also known as human threadworm; seatworm. { 'pinˌwərm }

pioneer [ECOL] An organism that is able to establish itself in a barren area and begin an ecological cycle. { ˌpī·ə'nir }

Piophilidae [INV ZOO] The skipper flies, a family of myodarian cyclorrhaphous dipteran insects in the subsection Acalypteratae. { ˌpī·ə'fil·əˌdē }

Piperaceae [BOT] A family of dicotyledonous plants in the order Piperales characterized by alternate leaves, a solitary ovule, copious perisperm, and scanty endosperm. { ˌpip·ə'rās·ēˌē }

Piperales [BOT] An order of dicotyledonous herbaceous plants marked by ethereal oil cells, uniaperturate pollen, and reduced crowded flowers with orthotropous ovules. { 'pip·ə'rā·lēz }

piperoxan [PHARM] $C_{14}H_{19}NO_2$ An adrenergic blocking agent that has been used, as the hydrochloride salt, for diagnosis of pheochromocytoma. { ¦pī·pər'äk·sən }

Pipidea [VERT ZOO] A family of frogs sometimes included in the suborder Opisthocoela, but more commonly placed in its own suborder, Aglossa; a definitive tongue is lacking, and free ribs are present in the tadpole but they fuse to the vertebrae in the adult. { pə'pid·ē·ə }

Pipridae [VERT ZOO] The manakins, a family of

colorful, neotropical suboscine birds in the order Passeriformes. { 'pip·rə,dē }

piptoblast [INV ZOO] A statoblast that is free but has no float. { 'pip·tə,blast }

piquite *See* pinta. { pē'kēt }

piriformis [ANAT] A muscle arising from the front of the sacrum and inserted into the greater trochanter of the femur. { ,pir·ə'fȯr·məs }

Piroplasmea [INV ZOO] A class of parasitic protozoans in the superclass Sarcodina; includes the single genus *Babesia*. { ,pir·ə'plaz·mē·ə }

Pisces [VERT ZOO] The fish and fishlike vertebrates, including the classes Agnatha, Placodermi, Chondrichthyes, and Osteichthyes. { 'pī·sēz }

piscivorous [ZOO] Feeding on fishes. { pə'siv·ə·rəs }

Pisionidae [INV ZOO] A small family of errantian polychaete annelids; allies of the scale bearers. { ,pī·sē'än·ə,dē }

pistachio [BOT] *Pistacia vera*. A small, spreading dioecious evergreen tree with leaves that have three to five broad leaflets, and with large drupaceous fruit; the edible seed consists of a single green kernel covered by a brown coat and enclosed in a tough shell. { pə'stash·ē,ō }

pistil [BOT] The ovule-bearing organ of angiosperms; consists of an ovary, a style, and a stigma. { 'pist·əl }

pistillate [BOT] **1.** Having a pistil. **2.** Having pistils but no stamens. { 'pist·əl,āt }

pit [BOT] **1.** A cavity in the secondary wall of a plant cell, formed where secondary deposition has failed to occur, and the primary wall remains uncovered; two main types are simple pits and bordered pits. **2.** The stone of a drupaceous fruit. { pit }

pitch [CELL MOL] The distance between two adjacent turns of double-stranded deoxyribonucleic acid. { pich }

pitcher plant [BOT] Any of various insectivorous plants of the families Sarraceniaceae and Nepenthaceae; the leaves form deep pitchers in which water collects and insects are drowned and digested. { 'pich·ər ,plant }

pith [BOT] A central zone of parenchymatous tissue that occurs in most vascular plants and is surrounded by vascular tissue. { pith }

pith ray *See* medullary ray. { 'pith ,rā }

Pittidae [VERT ZOO] The pittas, a homogeneous family of brightly colored suboscine birds with an erectile crown of feathers, in the suborder Tyranni. { 'pid·ə,dē }

pitting [MED] **1.** The formation of pits; in the fingernails, a consequence and sign of psoriasis. **2.** The preservation for a short time of indentations on the skin made by pressing with the finger; seen in pitting edema. { 'pid·iŋ }

pitting edema [MED] Edema of such degree that the skin can be temporarily indented by pressure with the fingers. { 'pid·iŋ ə'dē·mə }

pituicyte [HISTOL] The characteristic cell of the neurohypophysis; these cells are pigmented and fusiform and are probably derived from neuroglial cells. { pə'tü·ə,sīt }

pituitary [ANAT] Of or pertaining to the hypophysis. [PHYSIO] Secreting phlegm or mucus (archaic usage). { pə'tü·ə,ter·ē }

pituitary dwarfism [MED] Stunted growth due to a deficiency of the primary growth hormone; characterized clinically by growth failure in early life, and in older persons by deficient subcutaneous fat with loose, wrinkled skin and precocious senility. { pə'tü·ə,ter·ē 'dwȯr,fiz·əm }

pituitary gland *See* hypophysis. { pə'tü·ə,ter·ē ,gland }

pityriasis [MED] A fine, branny desquamation of the skin. { ,pid·ə'rī·ə·səs }

pivot joint [ANAT] A diarthrosis that permits a rotation of one bone around another; an example is the articulation of the atlas with the axis. Also known as trochoid. { 'piv·ət ,jȯint }

PKA *See* cyclic AMP-dependent protein kinase. { ¦pē¦kā'ā }

PKC *See* protein kinase C.

PKU *See* phenylketonuria.

placebo [MED] A preparation, devoid of pharmacologic effect, given to patients for psychologic effect, or as a control in evaluating a medicinal believed to have a pharmacologic action. { plä'chā·bō *or* plə'sē·bō }

placenta [BOT] A plant surface bearing a sporangium. [EMBRYO] A vascular organ that unites the fetus to the wall of the uterus in all mammals except marsupials and monotremes. { plə'sent·ə }

placenta accreta [MED] A placenta that has partially grown into the myometrium of the uterus. { plə'sent·ə ə'krēd·ə }

placental barrier [EMBRYO] The tissues intervening between the maternal and the fetal blood of the placenta, which prevent or hinder certain substances or organisms from passing from mother to fetus. { plə'sent·əl 'bar·ē·ər }

placenta previa [MED] A pregnancy disorder in which the placenta is abnormally located near or over the cervix. { plə,sen·tə 'prē·vē·ə }

placentation [BOT] The attachment of ovules along the inner ovarian wall by means of the placenta. [EMBRYO] The formation and fusion of the placenta to the uterine wall. { plas·ən'tā·shən }

placode [EMBRYO] A platelike epithelial thickening, frequently marking, in the embryo, the anlage of an organ or part. { 'pla,kōd }

Placothuriidae [INV ZOO] A family of holothurian echinoderms in the order Dendrochirotida; individuals are invested in plates and have a complex calcareous ring mechanism. { ,pla·kə·thə'rī·ə,dē }

plagiocephaly [MED] A type of strongly asymmetric cranial deformation, in which the anterior portion of one side and the posterior portion of the opposite side of the skull are developed more than their counterparts so that the maximum length of the skull is not in the midline but on a diagonal. { ,plā·jē·ō'sef·ə·lē }

plagioclimax [ECOL] A plant community which is in equilibrium under present conditions, but which has not reached its natural climax, or has regressed from it, due to biotic factors such as human intervention. { ¦plā·jē·ō'klī,maks }

plagiodont [VERT ZOO] Of a snake, having obliquely set, or two converging series of, palatal teeth. { 'plā·jē·ə,dänt }

plagiogravitropism [BOT] A response of root and shoot branches to gravity where growth is at different angles from the vertical. { ¦plā·jē·ō,grav·ə'trō·piz·əm }

plagiosere [ECOL] A plant succession deflected from its normal course by biotic factors. { 'plā·jē·ə,sir }

plague [MED] **1.** An infectious bacterial disease of rodents and humans caused by *Pasteurella pestis*, transmitted to humans by the bite of an infected flea (*Xenopsylla cheopis*) or by inhalation. Also known as black death; bubonic plague. **2.** Any contagious, malignant, epidemic disease. { plāg }

plain arch [FOREN] A fingerprint pattern in which the ridges enter on one side of the impression and flow or tend to flow out the other side with a rise or wave in the center. { 'plān ,ärch }

plain whorl [FOREN] A whorl fingerprint pattern that has two deltas and at least one ridge making a complete circuit, which may be spiral, oval, circular, or any variant of a circle. An imaginary line drawn between the two deltas must touch or cross at least one of the recurving ridges within the inner pattern area. { 'plān ,wərl }

planaria [INV ZOO] Any flatworm of the turbellarian order Tricladida; the body is broad and dorsoventrally flattened, with anterior lateral projections, the auricles, and a pair of eyespots on the anterior dorsal surface. { plə'ner·ē·ə }

Planctomyces [MICROBIO] A genus of appendaged bacteria; spherical, oblong, or pear-shaped cells with long, slender stalks; reproduce by budding. { ,plaŋk·tə'mī·sēz }

planidium [INV ZOO] A first-stage legless larva of various insects in the orders Diptera and Hymenoptera. { plə'nid·ē·əm }

Planipennia [INV ZOO] A suborder of insects in the order Neuroptera in which the larval mandibles are modified for piercing and for sucking. { ,plan·ə'pen·ē·ə }

plankton [ECOL] Passively floating or weakly motile aquatic plants and animals. { 'plaŋk·tən }

planktonic [ECOL] Free-floating. { plaŋk'tän·ik }

planoblast [INV ZOO] The medusa form of a hydrozoan. { 'plan·ə,blast }

planomycin *See* fervenulin. { ,plan·ə'mīs·ən }

planospiral [INV ZOO] Having the shell coiled in one plane, used particularly of foraminiferans and mollusks. { ¦plā·nō'spī·rəl }

plant [BOT] Any organism belonging to the kingdom Plantae, generally distinguished by the presence of chlorophyll, a rigid cell wall, and abundant, persistent, active embryonic tissue, and by the absence of the power of locomotion. { plant }

Plantae [BOT] The plant kingdom. { 'plan,tē }

Plantaginaceae [BOT] The single family of the plant order Plantaginales. { ,plan·tə·jə'nās·ē,ē }

Plantaginales [BOT] An order of dicotyledonous herbaceous plants in the subclass Asteridae, marked by small hypogynous flowers with a persistent regular corolla and four petals. { ,plan·tə·jə'nā·lēz }

plantar [ANAT] Of or relating to the sole of the foot. { 'plan·tər }

plantaris [ANAT] A small muscle of the calf of the leg; origin is the lateral condyle of the femur, and insertion is the calcaneus; flexes the knee joint. { plan'tar·əs }

plantar reflex [PHYSIO] Flexion of the toes in response to stroking of the outer surface of the sole, from heel to little toe. { 'plan·tər 'rē ,fleks }

plant extract *See* crude drug. { 'plant 'ek,strakt }

plant fermentation [BIOCHEM] A form of plant metabolism in which carbohydrates are partially degraded without the consumption of molecular oxygen. { 'plant ,fər·mən'tā·shən }

plant geography [BOT] A major division of botany, concerned with all aspects of the spatial distribution of plants. Also known as geographical botany; phytochorology; phytogeography. { 'plant jē,äg·rə·fē }

plant hormone [BIOCHEM] An organic compound that is synthesized in minute quantities by one part of a plant and translocated to another part, where it influences physiological processes. Also known as phytohormone. { 'plant ,hȯr,mōn }

plantigrade [VERT ZOO] Pertaining to walking with the whole sole of the foot touching the ground. { 'plan·tə,grād }

plant key [BOT] An analytical guide to the identification of plants, based on the use of contrasting characters to subdivide a group under study into branches. { 'plant ,kē }

plant kingdom [BOT] The worldwide array of plant life constituting a major division of living organisms. { 'plant ,kiŋ·dəm }

plant pathology [BOT] The branch of botany concerned with diseases of plants. { 'plant pə'thäl·ə·jē }

plant physiology [BOT] The branch of botany concerned with the processes which occur in plants. { 'plant ,fiz·ē'äl·ə·jē }

plant propagation [BOT] The deliberate, directed reproduction of plants using seeds or spores (sexual propagation), or using vegetative cells, tissues, or organs (asexual reproduction). { 'plant ,präp·ə,gā·shən }

plant respiration [BOT] A biochemical process in plants whereby specific substrates are oxidized with a subsequent release of carbon dioxide, CO_2. { 'plant ,res·pə,rā·shən }

plant societies [ECOL] Assemblages of plants which constitute structural parts of plant communities. { 'plant sə,sī·əd·ēz }

plantula [INV ZOO] A small, cushionlike structure on the ventral surface of the segments of insect tarsi. { 'plan·chə·lə }

plant virus [VIROL] A virus that replicates only within plant cells. { 'plant ,vī·rəs }

planula [INV ZOO] The ciliated, free-swimming larva of coelenterates. { 'plan·yə·lə }

Planuloidea [INV ZOO] The equivalent name for Moruloidea. { ,plan·yə'lȯid·ē·ə }

plaque [MED] **1.** A patch, or an abnormal flat area on any internal or external body surface. **2.** A localized area of atherosclerosis. [VIROL] A clear area representing a colony of viruses on a plate culture formed by lysis of the host cell. { plak }

plasma [HISTOL] The fluid portion of blood or lymph. { 'plaz·mə }

plasma cell *See* plasmacyte. { 'plaz·mə ,sel }

plasmacyte [HISTOL] A fairly large, generally ovoid cell with a small, eccentrically placed nucleus; the chromatin material is adherent to the nuclear membrane and the cytoplasm is agranular and deeply basophilic everywhere except for a clear area adjacent to the nucleus in the area of the cytocentrum. Also known as plasma cell. { 'plaz·mə,sīt }

plasmagel [CELL MOL] The outer, gelated zone of protoplasm in a pseudopodium. { 'plaz·mə,jel }

plasmagene [CELL MOL] A cytoplasmic particle or substance, which may be present in bodies such as plastids or mitochondria, and which can reproduce and pass on inherited qualities to daughter cells. { 'plaz·mə,jēn }

plasmalemma *See* cell membrane. { ¦plaz·mə 'lem·ə }

plasmals [BIOCHEM] Aldehydic components of lipids which give positive color tests with reagents used for detecting aldehydes in tissues. { 'plaz·məlz }

plasmalogen [BIOCHEM] Any of a group of glycerol-based phospholipids in which a fatty acid group is replaced by a fatty aldehyde. { plaz 'mal·ə·jən }

plasma membrane *See* cell membrane. { 'plaz·mə 'mem,brān }

plasmapheresis [MED] The withdrawal of blood from a donor to obtain plasma, its components, or the nonerythrocytic formed elements of blood, followed by the return of the erythrocytes to the donor. { ¦plaz·mə·fə'rē·səs }

plasmasol [CELL MOL] The inner, solated zone of protoplasm in a pseudopodium. { 'plaz·mə,sȯl }

plasma thromboplastin antecedent *See* factor XI. { 'plaz·mə ¦thräm·bō¦plas·tən 'ant·i,sēd·ənt }

plasma thromboplastin component *See* Christmas factor. { 'plaz·mə ¦thräm·bō¦plas·tən kəm'pō·nənt }

Plasmaviridae [VIROL] A family of enveloped deoxyribonucleic acid (DNA) phages that are characterized by rounded pleomorphic particles with a small, densely stained center and a supercoiled, double-stranded, circular DNA genome. { ,plaz·mə'vir·ə,dī }

Plasmavirus [VIROL] The sole genus of the deoxyribonucleic acid-containing phage family Plasmaviridae. { 'plaz·mə,vī·rəs }

plasmid [GEN] An extrachromosomal genetic element found among various strains of *Escherichia coli* and other bacteria. { 'plaz·məd }

plasmid cloning vector [GEN] A plasmid that accepts foreign deoxyribonucleic acid (DNA) and is therefore used in recombinant DNA experiments. { 'plaz·mid ,klōn·iŋ ,vek·tər }

plasmid donation [GEN] The transfer of a nonconjugative plasmid from a donor cell to a recipient cell by way of a contact function provided by a conjugative plasmid. { 'plaz·mid dō,nā·shən }

plasmin [BIOCHEM] A proteolytic enzyme in plasma which can digest many proteins through the process of hydrolysis. Also known as fibrinolysin. { 'plaz·mən }

plasminogen [BIOCHEM] The inert precursor, or zymogen, of plasmin. Also known as profibrinolysin. { plaz'min·ə·jən }

plasmodesma [CELL MOL] An intercellular bridge, thought to be strands of cytoplasm connecting two cells. { ,plaz·mə'dez·mə }

plasmodesmata [BOT] Strands of cytoplasm connecting the protoplasts of two contiguous plant cells. { ,Plaz·mə·dez'mäd·ə }

Plasmodiidae [INV ZOO] A family of parasitic protozoans in the suborder Haemosporina inhabiting the erythrocytes of the vertebrate host. { ,plaz·mə'dī·ə,dē }

Plasmodiophorida [INV ZOO] An order of the protozoan subclass Mycetozoia occurring as endoparasites of plants. { ,plaz·mō,dī·ə'fȯr·ə·də }

Plasmodiophoromycetes [MYCOL] A class of the Fungi. { ,plaz·mō·dī,äf·ə·rō·mī'sē·dēz }

plasmoditrophoblast *See* syncytiotrophoblast. { ,plaz·mō·dī'träf·ə,blast }

plasmodium [MICROBIO] The noncellular, multinucleate, jellylike, ameboid, assimilative stage of the Myxomycetes. { plaz'mō·dē·əm }

Plasmodroma [INV ZOO] A subphylum of the Protozoa, including Mastigophora, Sarcodina, and Sporozoa, in some taxonomic systems. { plaz'mä·drə·mə }

plasmogamy [INV ZOO] Fusion of protoplasts, without nuclear fusion, to form a multinucleate mass; occurs in certain protozoans. { plaz'mäg·ə·mē }

plasmolysis [PHYSIO] Shrinking of the cytoplasm away from the cell wall due to exosmosis by immersion of a plant cell in a solution of higher osmotic activity. { plaz'mäl·ə·səs }

plasmon [GEN] The cytoplasmic genetic system in eukaryotes consisting primarily of mitochondrial deoxyribonucleic acid (DNA) and chloroplast DNA. { 'plaz,män }

plasmosome *See* nucleolus. { 'plaz·mə,sōm }

plasmotomy [INV ZOO] Subdivision of a plasmodium into two or more parts. { plas'mäd·ə·mē }

plastic surgery [MED] Surgical repair, replacement, or alteration of lost, injured, or deformed parts of the body by transfer of tissue. { 'plas·tik 'sər·jə·rē }

plastid [CELL MOL] One of the specialized cell

organelles containing pigments or protein materials, often serving as centers of special metabolic activities; examples are chloroplasts and leukoplasts. { 'plas·təd }

plastogene [CELL MOL] A cytoplasmic factor, controlled by or interacting with the nucleus, which determines differentiation of a plastid. { 'plas·tə,jēn }

plastoquinone [BIOCHEM] Any of a group of quinones that are involved in electron transport in chloroplasts during photosynthesis. { ,plas·tə·kwə'nōn }

plastron [INV ZOO] The ventral plate of the cephalothorax of spiders. [VERT ZOO] The ventral portion of the shell of tortoises and turtles. { 'plas,trän }

Platanaceae [BOT] A small family of monoecious dicotyledonous plants in which flowers have several carpels which are separate, three or four stamens, and more or less orthotropous ovules, and leaves are stipulate. { ,plat·ən'ās·ē,ē }

Plataspidae [INV ZOO] A family of shining, oval hemipteran insects in the superfamily Pentatomoidea. { plə'tas·pə,dē }

plate budding [BOT] Plant budding by inserting a rectangular scion with a bud under a flap of bark on the stock in such a manner that the exposed wood on the stock is covered. { 'plāt ,bəd·iŋ }

plate count [MICROBIO] The number of bacterial colonies that develop on a medium in a petri dish seeded with a known amount of inoculum. { 'plāt ,kaůnt }

platelet *See* thrombocyte. { 'plāt·lət }

platelet-activating factor [IMMUNOL] A phospholipid released by leukocytes that causes aggregation of platelets and other effects, such as an increase in vascular permeability and bronchoconstriction. Abbreviated PAF. { ¦plāt·lət 'ak·tə,vād·iŋ ,fak·tər }

platelet-derived growth factor [CELL MOL] A glycolytic protein released by platelets and other cells that stimulates growth of cells of mesenchymal origin, for example, bone cartilage, vascular tissue, and connective tissue. { ¦plāt·lət də,rīvd 'grōth ,fak·tər }

Platyasterida [INV ZOO] An order of Asteroidea in which traces of metapinnules persist, the ossicles of the arm skeleton being arranged in two growth gradient systems. { ¦plad·ē·a'ster·ə·də }

platycelous [VERT ZOO] Of a vertebra, having a flat or concave ventral surface and a convex dorsal surface. { ¦plad·ē¦sē·ləs }

Platycephalidae [VERT ZOO] The flatheads, a family of perciform fishes in the suborder Cottoidei. { ,plad·ē·sə'fal·ə,dē }

Platycopa [INV ZOO] A suborder of ostracod crustaceans in the order Podocopida including marine forms with two pairs of thoracic legs. { plə'tik·ə·pə }

Platycopina [INV ZOO] The equivalent name for Platycopa. { ,plad·ə'käp·ə·nə }

Platyctenea [INV ZOO] An order of the ctenophores whose members are sedentary or parasitic; adults often lack ribs and are flattened due to shortening of the main axis. { ,plad·ik'tē·nē·ə }

Platyctenida [ZOO] An order of the phylum Ctenophora comprising four families (Ctenoplanidae, Coeloplanidae, Tjalfiellidae, Savangiidae) and six genera. { ,plad·ē'ten·əd·ə }

Platygasteridae [INV ZOO] A family of hymenopteran insects in the superfamily Proctotrupoidea. { ,plad·ē·ga'ster·ə,dē }

Platyhelminthes [INV ZOO] A phylum of invertebrates composed of bilaterally symmetrical, nonsegmented, dorsoventrally flattened worms characterized by lack of coelom, anus, circulatory and respiratory systems, and skeleton. { ¦plad·ē·hel'min·thēz }

platymyarian [INV ZOO] In nematodes, pertaining to flat muscle cells with the fibrillar region limited to a basal zone. { ¦plad·ē,mī¦a·rē·ən }

Platypodidae [INV ZOO] The ambrosia beetles, a family of coleopteran insects in the superfamily Curculionoidea. { ,plad·ə'päd·ə,dē }

Platypsyllidae [INV ZOO] The equivalent name for Leptinidae. { ,plad·ə'sil·ə,dē }

platypus [VERT ZOO] *Ornithorhynchus anatinus*. A monotreme, making up the family Ornithorhynchidae, which lays and incubates eggs in a manner similar to birds, and retains some reptilian characters; the female lacks a marsupium. Also known as duckbill platypus. { 'plad·ə,pu̇s }

platysma [ANAT] A subcutaneous muscle of the neck, extending from the face to the clavicle; muscle of facial expression. { plə'tiz·mə }

platyspondylia [MED] A rare congenital skeletal defect marked by abnormally shaped vertebrae. { ,plad·ē·spän'dil·yə }

Platysternidae [VERT ZOO] The big-headed turtles, a family of Asiatic fresh-water Chelonia with a single species (*Platysternon megacephalum*), characterized by a large head, hooked mandibles, and a long tail. { ,plad·ē'stər·nə,dē }

Plaut-Vincent's infection *See* Vincent's infection. { 'plau̇t 'vin·səns in,fek·shən }

Plecoptera [INV ZOO] The stoneflies, an order of primitive insects in which adults differ only slightly from immature stages, except for wings and tracheal gills. { plə'käp·tə·rə }

plectane [INV ZOO] A cuticular plate supporting papillae in some nematodes. { 'plek,tān }

Plectascales [MYCOL] An equivalent name for Eurotiales. { ,plek·tə'skā·lēz }

Plectognathi [VERT ZOO] The equivalent name for Tetraodontiformes. { plek'täg·nə,thī }

Plectoidea [INV ZOO] A superfamily of small, free-living nematodes characterized by simple spiral amphids or variants thereof, elongate cylindroconoid stoma, and reflexed ovaries. { plek 'tȯid·ē·ə }

Plectomycetes [MYCOL] A class of the subdivision Ascomycotina, members produce a well-developed mycelium on which both sexual (asci) and asexual (conidia) states occur. { ˌplek·tō·mī'sēˌdēz }

plectostele [BOT] A protostele that has the xylem divided into plates. { 'plek·təˌstēl }

Pleidae [INV ZOO] A family of hemipteran insects in the superfamily Pleoidea. { 'plē·əˌdē }

pleiomorphism [GEN] The occurrence of variable phenotypes in a group of organisms with the same genotype. { ˌplē·ə'mȯrˌfiz·əm }

pleiotropic [GEN] Referring to a gene or mutation that has multiple effects. { ˌplī·ə'träp·ik }

pleiotropy [GEN] The quality of a gene having more than one phenotypic effect. { plī'ä·trə·pē }

pleocytosis [MED] Increase of cells in the cerebrospinal fluid. { ˌplē·ōˌsī'tō·səs }

pleodont [VERT ZOO] Having solid teeth. { 'plē·əˌdänt }

Pleoidea [INV ZOO] A superfamily of suboval hemipteran insects belonging to the subdivision Hydrocoriseae. { plē'ȯid·ē·ə }

pleomorphism [BIOL] The occurrence of more than one distinct form of an organism in a single life cycle. { ˌplē·ō'mȯrˌfiz·əm }

pleopod [INV ZOO] An abdominal appendage in certain crustaceans that is modified for swimming. { 'plē·əˌpäd }

Pleosporales [BOT] The equivalent name for the lichenized Pseudophaeriales. { ˌplē·ə·spə'rā·lēz }

plerocercoid [INV ZOO] The infective metacestode of certain cyclophyllidean tapeworms; distinguished by a solid body. { ˌplir·ə'sərˌkȯid }

plerome [BOT] Central core of primary meristem which gives rise to all cells of the stele from the pericycle inward. { 'pliˌrōm }

plesioaster [INV ZOO] A type of poriferan microscleric monaxonic spicule. { ˌplē·sē·ō'as·tər }

plesiomorph [EVOL] The original character of a branching phyletic lineage, found in the ancestral forms. { 'plē·sē·əˌmȯrf }

plesiotype [SYST] A specimen or specimens on which subsequent descriptions are based. { 'plē·sē·əˌtīp }

Plethodontidae [VERT ZOO] A large family of salamanders in the suborder Salamandroidea characterized by the absence of lungs and the presence of a fine groove from nostril to upper lip. { ˌpleth·ə'dänt·əˌdē }

plethora [MED] An excess of blood in an organ or the circulatory system. { 'pleth·ə·rə }

plethysmograph [MED] An instrument for measuring changes in the size of a part of the body by measuring changes in the amount of blood in that part. { plə'thiz·məˌgraf }

pleura [ANAT] The serous membrane covering the lung and lining the thoracic cavity. { plu̇r·ə }

pleural cavity [ANAT] The potential space included between the parietal and visceral layers of the pleura. { 'plu̇r·əl ¦kav·əd·ē }

pleural effusion [MED] Abnormal accumulation of fluid in the area between the membranes lining the lungs and the chest cavity (the pleural space). { ˌplu̇r·əl i'fyü·zhən }

pleural rib *See* ventral rib. { 'plu̇r·əl ¦rib }

pleurapophysis [ANAT] One of the lateral processes of a vertebra, corresponding morphologically to a rib. { ˌplu̇r·ə'päf·ə·səs }

pleurisy [MED] Inflammation of the pleura. Also known as pleuritis. { 'plu̇r·ə·sē }

pleuritis *See* pleurisy. { plu̇'rīd·əs }

pleurobranchia [INV ZOO] A gill that arises from the lateral wall of the thorax in certain arthropods. { ˌplu̇r·ə'braŋ·kē·ə }

pleurocarpous [BOT] Having the sporophyte in leaf axils along the side of the stem or on lateral branches; refers specifically to mosses. { ¦plu̇r·ə¦kär·pəs }

Pleuroceridae [INV ZOO] A family of fresh-water snails in the order Pectinibranchia. { ˌplu̇r·ə'ser·əˌdē }

Pleurocoelea [INV ZOO] An extinct superfamily of gastropod mollusks of the order Opisthobranchia in which the shell, mantle cavity, and gills were present. { ˌplu̇r·ə'sē·lē·ə }

Pleurodira [VERT ZOO] A suborder of turtles (Chelonia) distinguished by spines on the posterior cervical vertebrae so that the head is retractile laterally. { ˌplu̇r·ə'dī·rə }

pleurodontia [VERT ZOO] Attachment of the teeth to the inner surface of the jawbone. { ˌplu̇r·ə'dän·chə }

pleurodynia [MED] Severe paroxysmal pain and tenderness of the intercostal muscles. { ˌplu̇r·ə'dī·nē·ə }

pleurolophocercous cercaria [INV ZOO] A larval digenetic trematode distinguished by a long, powerful tail with a pair of fin folds, a protrusible oral sucker, and pigmented dorsal eyespots. { ¦plu̇r·ə¦läf·ə¦sər·kəs sər'kar·ē·ə }

pleuron [INV ZOO] The lateral portion of a single thoracic segment in arthropods. { 'plu̇rˌän }

Pleuronectiformes [VERT ZOO] The flatfishes, an order of actinopterygian fishes distinguished by the loss of bilateral symmetry. { ˌplu̇r·ōˌnek·tə'fȯr·mēz }

Pleuronematina [INV ZOO] A suborder of the Hymenostomatida. { ˌplu̇r·ōˌnem·ə'tī·nə }

pleuroperitoneal cavity [VERT ZOO] The body cavity containing both the lungs and the abdominal viscera in all pulmonate vertebrates except mammals. { ˌplu̇r·ō¦per·ə·tə¦nē·əl 'kav·əd·ē }

pleuropneumonia [MED] Combined pleurisy and pneumonia. [VET MED] An infectious disease of cattle producing pleural and lung inflammation, caused by *Mycoplasma* species. { ˌplu̇r·ō·nu̇'mō·nyə }

pleuropneumonialike organism [MICROBIO] A former classification for a poorly defined group of microorganisms classified in the order Mycoplasmatales, including the smallest organisms capable of independent life, and comparable in size to the large filterable viruses. Abbreviated PPLO. { ˌplu̇r·ō·nu̇¦mō·nyəˌlik 'ȯr·gəˌniz·əm }

Pleurostigmophora [INV ZOO] A subclass of the centipedes, in some taxonomic systems, distinguished by lateral spiracles. { ˌplu̇r·ə·stig'mäf·ə·rə }

plexus [ANAT] A network of interlacing nerves or anastomosing vessels. { 'plek·səs }

plica [BIOL] A fold, as of skin or a leaf. { 'plī·kə }

ploidy [GEN] Number of complete chromosome sets in a nucleus: haploid (N), diploid (2N), triploid (3N), tetraploid (4N), and so on. { 'plȯid·ē }

Plokiophilidae [INV ZOO] A small family of predacious hemipteran insects in the superfamily Cimicoidea; individuals live in the webs of spiders and embiids. { ˌpläk·ē·ō'fil·əˌdē }

Plotosidae [VERT ZOO] A family of Indo-Pacific salt-water catfishes (Siluriformes). { plə'täs·əˌdē }

plum [BOT] Any of various shrubs or small trees of the genus *Prunus* that bear smooth-skinned, globular to oval, drupaceous stone fruit. { pləm }

plumaceous [VERT ZOO] Referring to the portion of a feather vane near the base that lacks hooklets and is loosely bound. { plü'mā·shəs }

plumage [VERT ZOO] The entire covering of feathers of a bird. { 'plü·mij }

Plumatellina [INV ZOO] The single order of the ectoproct bryozoan class Phylactolaemata. { ˌplü·mə·tə'lī·nə }

Plumbaginaceae [BOT] The leadworts, the single family of the order Plumbaginales. { ˌpləm·bə·jə'nās·ēˌē }

Plumbaginales [BOT] An order of dicotyledonous plants in the subclass Caryophyllidae; flowers are pentamerous with fused petals, trinucleate pollen, and a compound ovary containing a single basal ovule. { ˌpləm·bə·jə'nā·lēz }

plumbism [MED] Lead poisoning. { 'pləm·biz·əm }

plumicome [BIOL] A spicule with plumelike tufts. { 'plü·məˌkōm }

Plummer-Vinson syndrome [MED] Dysphagia koilonychia, gastric achlorhydria, glossitis, and hypochromic microcytic anemia caused by iron deficiency. Also known as Paterson-Kelly syndrome; sideropenic dysphagia. { 'pləm·ər 'vin·sən ˌsinˌdrōm }

plumose [VERT ZOO] Having feathers or plumes. { 'plüˌmōs }

plumule [BOT] The primary bud of a plant embryo. [VERT ZOO] A down feather. { 'plü·myül }

plurilocular sporangium [BOT] A multicelled, compartmentalized sporangium, such as is found in some brown algae. { ˌplür·ə'läk·yə·lər spə'ran·jē·əm }

pluripotent cell [EMBRYO] A cell capable of differentiating into most cell types found in an organism but not capable of forming a functional organism. { 'plūr·əˌpōt·ent ˌsel }

plus strand [CELL MOL] A parental ribonucleic acid (RNA) strand in RNA bacteriophages that is used as a template for a complementary RNA strand (minus strand) produced with the formation of a double-stranded (double-helix) RNA. { 'pləs ˌstrand }

pluteus [INV ZOO] The free-swimming, bilaterally symmetrical, easel-shaped larva of ophiuroids and echinoids. { 'plüd·ē·əs }

plutonism [MED] A disease caused by exposure to plutonium, manifested in experimental animals by graying of the hair, liver degeneration, and tumor formation. { 'plüt·ənˌiz·əm }

pluviilignosa [ECOL] A tropical rain forest. { ¦plü·vēˌil·əg'nō·sə }

pneumatocele [MED] **1.** Herniation of the lung. **2.** A sac or tumor containing gas; especially the scrotum filled with gas. { 'nü·məd·ōˌsēl }

pneumatocodon [INV ZOO] Exumbrellar surface of the float or pneumatophore of siphonophorans. { ˌnü·məd·ō'kōˌdän }

pneumatophore [BOT] **1.** An air bladder in marsh plants. **2.** A submerged or exposed erect root that functions in the respiration of certain marsh plants. [INV ZOO] The air sac of a siphonophore. { 'nü·məd·əˌfȯr }

pneumatosaccus [INV ZOO] Subumbrellar surface of the float or pneumatophore of siphonophorans. { ˌnü·mə·dō'sak·əs }

pneumatosis [MED] The presence of air or gas in abnormal situations in the body. { ˌnü·mə'tō·səs }

pneumaturia [MED] The voiding of urine containing free gas. { ˌnü·mə'tür·ē·ə }

pneumoangiography [MED] The outline of the vessels of the lung by means of radiopaque material, for roentgenographic visualization. { ˌnü·mōˌan·jē'äg·rə·fē }

pneumobacillus *See* Klebsiella pneumoniae. { ¦nü·mō·bə'sil·əs }

pneumoconiosis [MED] Any lung disease caused by dust inhalation. { ¦nü·mōˌkō·nē'ō·səs }

Pneumocystis carinii pneumonia [MED] A lung infection in humans caused by the protozoan *Pneumocystis carinii*. Also known as interstitial plasma-cell pneumonia. { ¦nü·mə¦sis·təs kə 'rin·ēˌī nü'mō·nyə }

pneumoencephalography [MED] A method of visualizing the ventricular system and subarachnoid pathways of the brain by roentgenography after removal of spinal fluid followed by the injection of air or gas into the subarachnoid space. { ¦nü·mō·inˌsef·ə'läg·rə·fē }

pneumoenteritis [MED] Inflammation of the lungs and of the intestine. { ¦nü·mōˌent·ə'rīd·əs }

pneumography [MED] **1.** Roentgenography of the lung. **2.** The recording of the respiratory excursions. { nü'mäg·rə·fē }

pneumohemothorax [MED] The presence of air or gas and blood in the thoracic cavity. { ¦nü·mō¦hē·mō'thȯrˌaks }

pneumolithiasis [MED] The occurrence of calculi or concretions in a lung. { ¦nü·mō·li'thī·ə·səs }

pneumomycosis [MED] Any disease of the lungs caused by a fungus. { ¦nü·mō·mī'kō·səs }

pneumonectomy [MED] Surgical removal of an entire lung. { ˌnü·mə'nek·tə·mē }

pneumonia [MED] An acute or chronic inflammation of the lungs caused by numerous microbial, immunological, physical, or chemical agents, and associated with exudate in the alveolar lumens. { nu̇'mō·nyə }

pneumonic plague [MED] A virulent type of plague in humans, with lung involvement. { nu̇ 'män·ik 'plāg }

pneumonitis [MED] Inflammation of the lung. { ,nü·mə'nīd·əs }

pneumonolysis [MED] The loosening of any portion of lung adherent to the chest wall; a form of collapse therapy used in the treatment of pulmonary tuberculosis. { ,nü·mə'näl·ə·səs }

pneumopericardium [MED] The presence of air in the pericardial cavity. { ¦nü·mō,per·ə'kärd·ē·əm }

pneumoperitoneum [MED] **1.** The presence of air or gas in the peritoneal cavity. **2.** Injection of a gas into the peritoneal cavity as a diagnostic or therapeutic measure. { ¦nü·mō,per·ə·tə'nē·əm }

pneumostome [INV ZOO] The respiratory aperture of gastropod mollusks. { 'nü·mə,stōm }

pneumothorax [MED] The presence of air or gas in the pleural cavity. { ¦nü·mō'thȯr,aks }

Poaceae [BOT] The equivalent name for Gramineae. { pō'ās·ē,ē }

Poales [BOT] The equivalent name for Cyperales. { pō'ā·lēz }

pock [MED] A pustule of an eruptive fever, especially of smallpox. { päk }

pocket gopher *See* gopher. { 'päk·ət ,gō·fər }

pod [BOT] A dry dehiscent fruit; a legume. { päd }

Podargidae [VERT ZOO] The heavy-billed frogmouths, a family of Asian and Australian birds in the order Caprimulgiformes. { pə'där·jə,dē }

podiatrist *See* chiropodist. { pə'dī·ə·trəst }

Podicipedide [VERT ZOO] The single family of the avian order Podicipediformes. { ,päd·ə·sə'ped·ə,dē }

Podicipediformes [VERT ZOO] The grebes, an order of swimming and diving birds distinguished by dense, silky plumage, a rudimentary tail, and toes that are individually broadened and lobed. { ,päd·ə·sə,ped·ə'fȯr,mēz }

Podicipitiformes [VERT ZOO] The equivalent name for Podicipediformes. { ,päd·ə·sə,pid·ə'fȯr,mēz }

podite [INV ZOO] A segment of a limb of an arthropod. { 'pä,dīt }

podium [INV ZOO] The terminal portion of a body wall appendage in certain echinoderms. { 'pō·dē·əm }

podobranch [INV ZOO] A gill of a crustacean attached to the basal segment of a thoracic limb. { 'päd·ə,braŋk }

Podocopa [INV ZOO] A suborder of fresh-water ostracod crustaceans in the order Podocopida in which the inner lamella has a calcified rim joining the outer lamella along a chitinous zone of concrescence, and the two valves fit together firmly. { pə'däk·ə·pə }

Podocopida [INV ZOO] An order of the Ostracoda; contains all fresh-water ostracodes and is divided into the suborders Podocopa, Metacopina, and Platycopina. { ,päd·ə'käp·ə·də }

Podocopina [INV ZOO] The equivalent name for Podocopa. { ,päd·ə·kə'pī·nə }

podocyst [INV ZOO] A sinus in the foot of certain gastropod mollusks. { 'päd·ə,sist }

Podogona [INV ZOO] The equivalent name for Ricinuleida. { pə'däg·ə·nə }

podophyllin [PHARM] A light yellow to greenish-yellow or light brown powder with a bitter taste; soluble in alcohol, ether, and ammonium hydroxide; used in medicine. Also known as podophyllum resin. { ,päd·ə'fil·ən }

podophyllinic acid lactone *See* podophyllotoxin. { ,päd·ə·fə'lin·ik 'as·əd 'lak,tōn }

podophyllotoxin [PHARM] $C_{22}H_{22}O_8$ A crystalline compound with a melting point of 114–118°C; soluble in alcohol, chloroform, acetone, and warm benzene; used as an antineoplastic agent in medicine. Also known as podophyllinic acid lactone. { ,päd·ə,fil·ə'täk·sən }

podophyllum resin *See* podophyllin. { ,päd·ə'fil·əm 'rez·ən }

Podosphaera leucotricha [MYCOL] A fungal plant pathogen that causes apple powdery mildew. { ,päd·ə¦sfī·rə ,lü·kə'trik·ə }

Podostemaceae [BOT] The single family of the order Podostemales. { pə,däs·tə'mās·ē,ē }

Podostemales [BOT] An order of dicotyledonous plants in the subclass Rosidae; plants are submerged aquatics with modified, branching shoots and small, perfect flowers having a superior ovary and united carpels. { ,päd·ə·stə'mā·lēz }

Podoviridae [VIROL] A family of linear double-stranded deoxyribonucleic acid-containing bacterial viruses (bacteriophages) characterized by a short contractile tail and an icosahedral head, contains the T7 coliphage. { ,päd·ə'vir·ə,dī }

Poeciliidae [VERT ZOO] A family of fishes in the order Atheriniformes including the live-bearers, such as guppies, swordtails, and mollies. { ,pē·sə'lī·ə,dē }

Poecilosclerida [INV ZOO] An order of sponges of the class Demospongiae in which the skeleton includes two or more types of megascleres. { ,pē·sə·lō'skler·ə·də }

Poeobiidae [INV ZOO] A monotypic family of spioniform worms (*Poeobius meseres*) belonging to the Sedentaria and found in the North Pacific Ocean. { ,pē·ə'bī·ə,dē }

pogonochore [BOT] A type of plant that produces plumed disseminules. { pə'gän·ə,chȯr }

Pogonophora [INV ZOO] The single class of the phylum Brachiata; the elongate body consists of three segments, each with a separate coelom; there is no mouth, anus, or digestive canal, and sexes are separate. { ,pō·gə'näf·ə·rə }

poikilocytosis [MED] A condition in which erythrocytes are distorted in shape. { pȯi¦kil·ə·sī'tō·səs }

poikilotherm [ZOO] An animal, such as reptiles,

fishes, and invertebrates, whose body temperature varies with and is usually higher than the temperature of the environment; a cold-blooded animal. { pȯi'kil·ə,thərm }

point localization [PHYSIO] The ability to locate the point on the skin that has been touched. { 'pȯint ,lō·kə·lə'zā·shən }

point mutation [GEN] Mutation of a single gene due to addition, loss, replacement, or change of sequence in one or more base pairs of the deoxyribonucleic acid of that gene. { 'pȯint myü'tā·shən }

poison gland [VERT ZOO] Any of various specialized glands in certain fishes and amphibians which secrete poisonous mucuslike substances. { 'pȯiz·ən ,gland }

poison hemlock [BOT] *Conium maculatum*. A branching biennial poisonous herb that contains a volatile alkaloid, coniine, in its fruits and leaves. { 'pȯiz·ən 'hem,läk }

poison ivy [BOT] Any of several climbing, shrubby, or arborescent plants of the genus *Rhus* in the sumac family (Anacardiaceae); characterized by ternate leaves, greenish flowers, and white berries that produce an irritating oil. { 'pȯiz·ən 'ī·vē }

poison oak [BOT] Any of several bushy poison ivy plants or shrubby poison sumacs. { 'pȯiz·ən 'ōk }

poisonous plant [BOT] Any of about 400 species of vascular plants containing principles which initiate pathological conditions in man and animals. { 'pȯiz·ən·əs 'plant }

poison sumac [BOT] *Rhus vernix*. A tall bush of the sumac family (Anacardiaceae) bearing pinnately compound leaves with 7–13 entire leaflets, and drooping, axillary clusters of white fruits that produce an irritating oil. { 'pȯiz·ən 'sü,mak }

pokeweed [BOT] *Phytolacca americana*. A poisonous garden weed that is found throughout the United States but is native to the eastern and central areas. Symptoms of poisoning include an immediate burning bitter taste, salivation, vomiting, diarrhea, and possible shock. { 'pōk,wēd }

polar bear [VERT ZOO] *Thalarctos maritimus*. A large aquatic carnivore found in the polar regions of the Northern Hemisphere. { 'pō·lər ,ber }

polar body [CELL MOL] One of the small bodies cast off by the oocyte during maturation. { 'pō·lər ,bäd·ē }

polarity [CELL MOL] The orientation of a strand of polynucleotide with respect to its partner, expressed in terms of nucleotide linkages. { pə'lar·əd·ē }

polar mutations [GEN] A class of mutations in the genes of an operon that affect the expression not only of the gene in which the mutation resides, but also of the genes located to one side of the mutated gene. { 'pō·lər myü'tā·shənz }

polar nucleus [BOT] One of the two nuclei in the center of the embryo sac of a seed plant which fuse to form the endosperm nucleus. { 'pō·lər 'nü·klē·əs }

Polemoniaceae [BOT] A family of autotrophic dicotyledonous plants in the order Polemoniales distinguished by lack of internal phloem, corolla lobes that are convolute in the bud, three carpels, and axile placentation. { ,päl·ə,mō·nē'ās·ē,ē }

Polemoniales [BOT] An order of dicotyledonous plants in the subclass Asteridae, characterized by sympetalous flowers, a regular, usually five-lobed corolla, and stamens equal in number and alternate with the petals. { ,päl·ə,mō·nē'ā·lēz }

polian vesicle [INV ZOO] Interradial reservoirs connecting with the ring vessel in most asteroids and holothuroids. { 'pō·lē·ən 'ves·ə·kəl }

polioencephalomyelitis [MED] A disease in which both the gray matter of the brain and the spinal cord are inflamed. { ¦pō·lē·ō·in¦sef·ə·lō,mī·ə'līd·əs }

poliomyelitis [MED] An acute infectious viral disease which in its most serious form involves the central nervous system and, by destruction of motor neurons in the spinal cord, produces flaccid paralysis. Also known as Heine-Medin disease; infantile paralysis. { ¦pō·lē·ō,mī·ə'līd·əs }

poliovirus vaccine [IMMUNOL] A vaccine prepared from one or all three types of polioviruses in a live or attenuated state. { ¦pō·lē·ō'vī·rəs vak'sēn }

pollen [BOT] The small male reproductive bodies produced in pollen sacs of the seed plants. { 'päl·ən }

pollen count [BOT] The number of grains of pollen that collect on a specified area (often taken as 1 square centimeter) in a specified time. { 'päl·ən ,kaůnt }

pollen sac [BOT] In the anther of angiosperms and gymnosperms, a cavity that contains microspores. { 'päl·ən ,sak }

pollen tube [BOT] The tube produced by the wall of a pollen grain which enters the embryo sac and provides a passage through which the male nuclei reach the female nuclei. { 'päl·ən ,tüb }

pollination [BOT] The transfer of pollen from a stamen to a pistil; fertilization in flowering plants. { ,päl·ə'nā·shən }

pollinosis *See* hay fever. { ,päl·ə'nō·səs }

pollution [ECOL] Destruction or impairment of the purity of the environment. [PHYSIO] Emission of semen at times other than during coitus. { pə'lü·shən }

polyadelphous [BOT] Pertaining to stamens that are united by their filaments into several sets or bundles. { ¦päl·ē·ə¦del·fəs }

polyadenylation [CELL MOL] The addition of adenine nucleotides to the 3′ end of messenger ribonucleic acid molecules during posttranscriptional modification. { ,päl·ē·ə,den·ə'lā·shən }

Polyangiaceae [MICROBIO] A family of bacteria in the order Myxobacterales; vegetative cells and myxospores are cylindrical with blunt, rounded ends; the slime capsule is lacking; sporangia are sessile or stalked. { ,päl·ē,an·jē'ās·ē,ē }

polyarteritis [MED] Inflammation of several arteries simultaneously. Also known as panarteritis. { 'päl·ē,ärd·ə'rīd·əs }

polyarteritis nodosa [MED] A systemic disease characterized by widespread inflammation of small and medium-sized arteries in which some of the foci are nodular. Also known as disseminated necrotizing periarteritis; periarteritis nodosa. { 'päl·ē,ärd·ə'rīd·əs nō'dō·sə }

polyarthritis [MED] Inflammation of several joints simultaneously. { 'päl·ē·är'thrīd·əs }

polyaxon [INV ZOO] A spicule that is laid down along several axes. { ¦päl·ē'ak,sän }

Polybrachiidae [INV ZOO] A family of sedentary marine animals in the order Thecanephria. { ,päl·i·bra'kī·ə,dē }

Polychaeta [INV ZOO] The largest class of the phylum Annelida, distinguished by paired, lateral, fleshy appendages (parapodia) provided with setae, on most segments. { ,päl·i'kēd·ə }

Polycirrinae [INV ZOO] A subfamily of polychaete annelids in the family Terebellidae. { ,päl·i'sir·ə,nē }

polycistronic messenger [BIOCHEM] In ribonucleic acid viruses, messenger ribonucleic acid that contains the amino acid sequence for several proteins. { ¦päl·i·sis'trän·ik 'mes·ən·jər }

Polycladida [INV ZOO] A class of marine Turbellaria whose leaflike bodies have a central intestine with radiating branches, many eyes, and tentacles in most species. { ,päl·i'klad·əd·ə }

polyclimax [ECOL] A climax community under the controlling influence of many environmental factors, including soils, topography, fire, and animal interactions. { ,päl·i'klī,maks }

polyclonal [IMMUNOL] Pertaining to cells or molecules that arise from more than one clone. { ,päl·i'klō·nəl }

polyclonal mixed cryoglobulin [BIOCHEM] A cryoglobulin made of heterogeneous immunoglobin molecules belonging to two or more different classes, and sometimes additional serum proteins. { ¦päl·i¦klōn·əl 'mikst ¦krī·ə'gläb·yə·lən }

Polycopidae [INV ZOO] The single family of the suborder Cladocopa. { ,päl·i'käp·ə,dē }

Polyctenidae [INV ZOO] A family of hemipteran insects in the superfamily Cimicoidea; the individuals are bat ectoparasites which resemble bedbugs but lack eyes and have ctenidia and strong claws. { ,päl·ək'ten·ə,dē }

polycyesis [MED] Multiple pregnancy. { ¦päl·i·sī'ē·səs }

polycystic kidney [MED] A usually hereditary, congenital, and bilateral disease in which a large number of cysts are present on the kidney. { ¦päl·i'sis·tik 'kid·nē }

polycystic ovarian syndrome [MED] A disorder in which the ovaries are bilaterally enlarged with multiple follicular cysts due to abnormal regulation of the hypothalamic-pituitary-ovarian axis. Symptoms include amenorrhea, menstrual abnormalities, infertility, and hirsutism. { ,päl·ē,sis·tik ō,ver·ē·ən 'sin,drōm }

polycythemia [MED] A condition characterized by an increased number of erythrocytes in the circulation. { ,päl·i,sī'thē·mē·ə }

polycythemia vera [MED] An absolute increase in all blood cells derived from bone marrow, especially erythrocytes. { ,päl·i,sī'thē·mē·ə 'vir·ə }

polydactyly [MED] The condition of having supernumerary fingers or toes. { ,päl·i'dak·təl·ē }

polydipsia [MED] Excessive thirst. { ,päl·i 'dip·sē·ə }

polyembryony [ZOO] A form of sexual reproduction in which two or more offspring are derived from a single egg. { ¦päl·ē·im'brī·ə,nē }

polyestrous [PHYSIO] Having several periods of estrus in a year. { ¦päl·ē'es·trəs }

Polygalaceae [BOT] A family of dicotyledonous plants in the order Polygalales distinguished by having a bicarpellate pistil and monadelphous stamens. { ,päl·i·gə'lās·ē,ē }

polygalacturonase [BIOCHEM] An enzyme that catalyzes the hydrolysis of glycosidic linkage of polymerized galacturonic acids. { ,päl·i,ga ,lak'tür·ə,nās }

Polygalales [BOT] An order of dicotyledonous plants in the subclass Rosidae characterized by its simple leaves and usually irregular, hypogynous flowers. { ,päl·i·gə'lā·lēz }

polygamous [BOT] Having both perfect and imperfect flowers on the same plant. [VERT ZOO] Having more than one mate at one time. { pə'lig·ə·məs }

polygene [GEN] One of a group of nonallelic genes that collectively control a quantitative character. { 'päl·i,jēn }

polygenic inheritance [GEN] The phenotypic expression of a trait involving the interaction of many genes. { ,päl·i,jen·ik in'her·əd·əns }

polyglycolic acid polymer [MED] A synthetic biodegradable polymer material used as dissolvable sutures and tissue engineering scaffolds. { ,päl·ē·glī¦käl·ik ¦as·əd 'päl·i·mər }

Polygonaceae [BOT] The single family of the order Polygonales. { pə,lig·ə'nās·ē,ē }

Polygonales [BOT] An order of dicotyledonous plants in the subclass Caryophyllidae characterized by well-developed endosperm, a unilocular ovary, and often trimerous flowers. { pə,lig·ə'nā·lēz }

polyhedral disease *See* polyhedrosis. { ¦päl·i¦hē·drəl di'zēz }

polyhedrosis [INV ZOO] Any of several virus diseases of insect larvae characterized by the breakdown of tissues and presence of polyhedral granules. Also known as polyhedral disease. { ,päl·i·hē'drō·səs }

polyhidrosis *See* hyperhidrosis. { ¦päl·i·hī'drō·səs }

polyhydramnios [MED] An excessive volume of amniotic fluid. Also known as hydramnios. { ¦päl·i·hī'dram·nē,ōs }

polykaryocyte *See* syncytium. { ¦päl·i'kar·ē·ə,sīt }

polymenorrhea *See* metrorrhagia. { ¦päl·i,men·ə'rē·ə }

Polymera [INV ZOO] Formerly a subphylum of

the Vermes; equivalent to the phylum Annelida. { pə'lim·ə·rə }

polymerase [BIOCHEM] An enzyme that links nucleotides together to form polynucleotide chains. { pə'lim·ə,rās }

polymerase chain reaction [CELL MOL] A technique for copying and amplifying the complementary strands of a target deoxyribonucleic acid molecule. Abbreviated PCR. { pə¦lim·ə,rās 'chān rē¦ak·shən }

polymere *See* polygene. { 'päl·ə,mir }

polymorph [BIOL] An organism that exhibits polymorphism. [HISTOL] *See* granulocyte. { 'päl·i,mȯrf }

polymorphic modification *See* polymorph. { ¦päl·i¦mȯr·fik ,mäd·ə·fə'kā·shən }

polymorphism [BIOL] **1.** Occurrence of different forms of individual in a single species. **2.** Occurrence of different structural forms in a single individual at different periods in the life cycle. [GEN] The coexistence of genetically determined distinct forms in the same population, even the rarest of them being too common to be maintained solely by mutation; human blood groups are an example. { ,päl·i'mȯr,fiz·əm }

polymorphonuclear leukocyte *See* granulocyte. { ¦päl·i,mȯr·fō'nü·klē·ər 'lü·kə,sīt }

polymyarian [INV ZOO] Referring to the cross-sectional appearance of muscle cells in a nematode, having many cells in each quadrant. { ¦päl·i,mī¦ar·ē·ən }

polymyositis [MED] Inflammation of many muscles simultaneously. { ¦päl·i,mī·ə'sīd·əs }

polymyxin [MICROBIO] Any of the basic polypeptide antibiotics produced by certain strains of *Bacillus polymyxa*. { ,päl·i'mik·sən }

Polynemidae [VERT ZOO] A family of perciform shore fishes in the suborder Mugiloidei. { ,päl·i'nem·ə,dē }

polyneuritis [MED] Degenerative or inflammatory lesions of several nerves simultaneously, usually symmetrical. Also known as multiple neuritis. { ¦päl·i·nu̇'rīd·əs }

Polynoidae [INV ZOO] The largest family of polychaetes, included in the Errantia and having a body of varying size and shape that is covered with elytra. { ,päl·ə'nȯi,dē }

polynucleotide [BIOCHEM] A linear sequence of nucleotides. { ¦päl·ə'nü·klē·ə,tīd }

Polyodontidae [INV ZOO] A family of tubicolous, often large-bodied errantian polychaetes with characteristic cephalic and parapodial structures. { ,päl·ē·ō'dänt·ə,dē }

polyoma virus [VIROL] A small deoxyribonucleic acid virus normally causing inapparent infection in mice, but experimentally capable of producing parotid tumors and a wide variety of other tumors. { ,päl·ē'ō·mə 'vī·rəs }

polyopia [MED] A condition in which more than one image of an object is formed upon the retina. { ,päl·ē'ō·pē·ə }

Polyopisthocotylea [INV ZOO] An order of the trematode subclass Monogenea having a solid posterior holdfast bearing suckers or clamps. { ¦päl·ē·ō,pis·thə,kad·əl'ē·ə }

polyorrhymenitis *See* polyserositis. { ¦päl·ē·ȯ ,rim·ə'nīd·əs }

polyp [INV ZOO] A sessile cnidarian individual having a hollow, somewhat cylindrical body, attached at one end, with a mouth surrounded by tentacles at the free end; may be solitary (hydra) or colonial (coral). [MED] A smooth, rounded or oval mass projecting from a membrane-covered surface. { 'päl·əp }

polypectomy [MED] Surgical excision of a polyp. { ,päl·i'pek·tə·mē }

polypeptide [BIOCHEM] A chain of amino acids linked together by peptide bonds but with a lower molecular weight than a protein; obtained by synthesis, or by partial hydrolysis of protein. { ¦päl·i'pep,tīd }

polypetalous [BOT] Having distinct petals, in reference to a flower or a corolla. Also known as choripetalous. { ¦päl·i'ped·əl·əs }

Polyphaga [INV ZOO] A suborder of the order Coleoptera; members are distinguished by not having the hind coxae fused to the metasternum and by lacking notopleural sutures. { pə'lif·ə·gə }

polyphagous [ZOO] Feeding on many different kinds of plants or animals. { pə'lif·ə·gəs }

polyphenol oxidase [BIOCHEM] A copper-containing enzyme that catalyzes the oxidation of phenol derivatives to quinones. { ¦päl·i'fē,nȯl 'äk·sə,dās }

polyphyodont [VERT ZOO] Having teeth which may be constantly replaced. { ¦päl·i¦fī·ə,dänt }

polypide [INV ZOO] The internal contents of an ectoproct bryozoan zooid. { 'päl·i,pīd }

Polyplacophora [INV ZOO] The chitons, an order of mollusks in the class Amphineura distinguished by an elliptical body with a dorsal shell that comprises eight calcareous plates overlapping posteriorly. { ,päl·i·pla'käf·ə·rə }

polyploidy [GEN] The occurrence of related species possessing three, four, or larger multiples of the haploid set of chromosomes. { 'päl·i,plȯid·ē }

Polypodiales [BOT] The true ferns; the largest order of modern ferns, distinguished by being leptosporangiate and by having small sporangia with a definite number of spores. { ,päl·i,päd·ē'ā·lēz }

Polypodiatae *See* Polypodiopsida. { ,päl·i·pə'dī·ə,tē }

Polypodiophyta [BOT] The ferns, a division of the plant kingdom having well-developed roots, stems, and leaves that contain xylem and phloem and show well-developed alternation of generations. { ,päl·i,päd·ē'äf·əd·ē }

Polypodiopsida [BOT] A class of the division Polypodiophyta; stems of these ferns bear several large, spirally arranged, compound leaves with sporangia grouped in sori on their undermargins. { ,päl·i,päd·ē'äp·səd·ə }

polypore [MYCOL] Any member of the Basidiomycetes having basidia that line the numerous tubes or pores of the basidiocarp. { 'päl·ē,pȯr }

Polypteridae [VERT ZOO] The single family of the order Polypteriformes. { ,päl·əp'ter·ə,dē }

Polypteriformes [VERT ZOO] An ancient order of actinopterygian fishes distinguished by thick, rhombic, ganoid scales with an enamellike covering, a slitlike spiracle behind the eye, a symmetrical caudal fin, and a dorsal series of free, spinelike finlets. { ˌpäl·əpˌter·əˈfȯrˌmēz }

polyribosome *See* polysome. { ˌpäl·iˈrī·bəˌsōm }

polysaccharide [BIOCHEM] A carbohydrate composed of many monosaccharides. Also known as glycan. { ¦päl·iˈsak·əˌrīd }

polysaccharide vaccine [IMMUNOL] A noninfectious vaccine that contains the polysaccharide coats, or capsules, of encapsulated bacteria. { ˌpäl·ēˌsak·əˌrīd vakˈsēn }

polysaprobic [ECOL] Referring to a body of water in which organic matter is decomposing rapidly and free oxygen either is exhausted or is present in very low concentrations. { ¦päl·ə·səˈprō·bik }

polysepalous [BOT] Having separate sepals. Also known as chorisepalous. { ¦päl·əˈsep·ə·ləs }

polyserositis [MED] Widespread, chronic fibrosing inflammation of serous membranes, especially in the upper abdomen. Also known as chronic hyperplastic perihepatitis; Concato's disease; multiple serositis; Pick's disease; polyorrhymenitis. { ¦päl·iˌser·əˈsīd·əs }

polysome [CELL MOL] A complex of ribosomes bound together by a single messenger ribonucleic acid molecule. Also known as polyribosome. { ˈpäl·iˌsōm }

polysomy [GEN] The occurrence in a nucleus of an extra copy of one or more individual chromosomes. { päl·ēˌsō·mē }

polyspermy [PHYSIO] Penetration of the egg by more than one sperm. { ˈpäl·ēˌspər·mē }

polyspore [BOT] In certain red algae, an asexual spore, of which there are 12 to 16. { ˈpäl·ēˌspȯr }

polystele [BOT] A stele consisting of vascular units in the parenchyma. { ˈpäl·iˌstēl }

Polystomatoidea [INV ZOO] A superfamily of monogeneid trematodes characterized by strong suckers and hooks on the posterior end. { ˌpäl·iˌstō·məˈtȯid·ē·ə }

Polystylifera [INV ZOO] A suborder of the Hoplonemertini distinguished by many stylets. { ˌpäl·i·stəˈlif·ə·rə }

polytene chromosome [GEN] A giant, multistranded, chromosome produced by multiple sounds of endoreduplication and thus composed of many copies of the parental chromosome pair having their chromomeres in register. Also known as Balbiani chromosome. { ˈpäl·iˌtēn ˈkrō·məˌsōm }

Polytrichales [BOT] An order of ascocarpous perennial mosses; rigid, simple stems are highly developed and arise from a prostrate subterranean rhizome. { pəˌli·trəˈkā·lēz }

polytypic [SYST] A taxon that contains two or more taxa in the immediately subordinate category. { ¦päl·i¦tip·ik }

polyuria [MED] The passage of copious amounts of urine. { ˌpäl·ēˈyu̇r·ē·ə }

polyvalent [IMMUNOL] **1.** Of antigens, having many combining sites or determinants. **2.** Pertaining to vaccines composed of mixtures of different organisms, and to the resulting mixed antiserum. { ¦päl·iˈvā·lənt }

Polyzoa [INV ZOO] The equivalent name for Bryozoa. { ˌpäl·əˈzō·ə }

Pomacentridae [VERT ZOO] The damselfishes, a family of perciform fishes in the suborder Percoidei. { ˌpō·məˈsen·trəˌdē }

Pomadasyidae [VERT ZOO] The grunts and sweetlips, a family of perciform fishes in the suborder Percoidei. { ˌpō·mə·dəˈsī·əˌdē }

Pomatiasidae [INV ZOO] A family of land snails in the order Pectinibranchia. { ˌpō·mə·tīˈas·əˌdē }

Pomatomidae [VERT ZOO] A monotypic family of the Perciformes containing the bluefish (*Pomatomus saltatrix*). { ˌpō·məˈtäm·əˌdē }

pomegranate [BOT] *Punica granatum.* A small, deciduous ornamental tree of the order Myrtales cultivated for its fruit, which is a reddish, pomelike berry containing numerous seeds embedded in crimson pulp. { ˈpäm·əˌgran·ət }

Pompe disease [MED] A hereditary glycogen storage disease in humans arising from deficiency of a lysosomal enzyme and characterized by weakness, enlargement of the heart and cardiac failure, enlargement of the tongue, and moderate enlargement of the liver. { ˈpämp diˌzēz }

pompeii worm [INV ZOO] *Alvinella pompejana.* A polychaetous annelid that lives in sea-floor hydrothermal vent chimneys and may experience extreme thermal gradients between its anterior (80°C; 176°F) and posterior (22°C; 72°F) ends. { pämˈpā ˌwərm }

Pompilidae [INV ZOO] The spider wasps, the single family of the superfamily Pompiloidea. { pamˈpil·əˌdē }

Pompiloidea [INV ZOO] A monofamilial superfamily of hymenopteran insects in the suborder Apocrita with oval abdomen and strong spinose legs. { ˌpäm·pəˈlȯid·ē·ə }

ponderosa pine [BOT] *Pinus ponderosa.* A hard pine tree of western North America; attains a height of 150–225 feet (46–69 meters) and has long, dark-green leaves in bundles of two to five and tawny, yellowish bark. { ˌpän·dəˌrō·sə ˈpīn }

Ponerinae [INV ZOO] A subfamily of tropical carnivorous ants (Formicidae) in which pupae characteristically form in cocoons. { pōˈner·əˌnē }

Pongidae [VERT ZOO] A family of anthropoid primates in the superfamily Hominoidea; includes the chimpanzee, gorilla, and orangutan. { ˈpän·jəˌdē }

pons [ANAT] **1.** A process or bridge of tissue connecting two parts of an organ. **2.** A convex white eminence located at the base of the brain; consists of fibers receiving impulses from the cerebral cortex and sending fibers to the contralateral side of the cerebellum. { pänz }

pontine flexure [EMBRYO] A flexure in the embryonic brain concave dorsally, occurring in the region of the myelencephalon. { 'pän,tēn 'flek·shər }

Pontodoridae [INV ZOO] A monotypic family of pelagic polychaetes assigned to the Errantia. { ,pän·tə'dȯr·ə,dē }

poplar [BOT] Any tree of the genus *Populus*, family Salicaceae, marked by simple, alternate leaves, scaly buds, bitter bark, and flowers and fruits in catkins. { 'päp·lər }

popliteal artery [NEUROSCI] A continuation of the femoral artery in the posterior portion of the thigh above the popliteal space and below the buttock. { päp'lid·ē·əl 'ärd·ə·rē }

popliteal nerve [ANAT] Either of two branches of the sciatic nerve in the lower part of the thigh; the larger branch continues as the tibial nerve, and the smaller branch continues as the peroneal nerve. { päp'lid·ē·əl 'nərv }

popliteal space [ANAT] A diamond-shaped area behind the knee joint. { päp'lid·ē·əl 'spās }

popliteal vein [ANAT] A vein passing through the popliteal space, formed by merging of the tibial veins and continuing to become the femoral vein. { päp'lid·ē·əl 'vān }

popliteus [ANAT] **1.** The ham or hinder part of the knee joint. **2.** A muscle on the back of the knee joint. { päp'lid·ē·əs }

poppy [BOT] Any of various ornamental herbs of the genus *Papaver*, family Papaveraceae, with large, showy flowers; opium is obtained from the fruits of the opium poppy (*P. somniferum*). { 'päp·ē }

population [BIOL] A group of organisms occupying a specific geographic area or biome. { ,päp·yə'lā·shən }

population ecology [ECOL] The study of the vital statistics of populations, and the interactions within and between populations that influence survival and reproduction. { ,päp·yə,lā·shən ē'käl·ə·jē }

population bottleneck [EVOL] Genetic drift that occurs as a result of a drastic reduction in population by an event having little to do with the usual forces of natural selection. { ,päp·yə'lā·shən 'bäd·əl,nek }

population density [ECOL] The size of the population within a particular unit of space. { ,päp·yə'lā·shən 'den·səd·ē }

population dispersal [BIOL] The process by which groups of living organisms expand the space or range within which they live. { ,päp·yə'lā·shən di'spər·səl }

population dispersion [BIOL] The spatial distribution at any particular moment of the individuals of a species of plant or animal. { ,päp·yə'lā·shən di'spər·zhən }

population dynamics [BIOL] The aggregate of processes that determine the size and composition of any population. { ,päp·yə'lā·shən dī'nam·iks }

population genetics [GEN] The study of both experimental and theoretical consequences of Mendelian heredity on the population level; includes studies of gene frequencies, genotypes, phenotypes, mating systems, selection, and migration. { ,päp·yə'lā·shən jə'ned·iks }

Porcellanasteridae [INV ZOO] A family of essentially deep-water forms in the suborder Paxillosina. { pȯr,sel·ə·nə'ster·ə,dē }

Porcellanidae [INV ZOO] The rock sliders, a family of decapod crustaceans of the group Anomura which resemble true crabs but are distinguished by the reduced, chelate fifth pereiopods and the well-developed tail fan. { ,pȯr·sə'lan·ə,dē }

porcupine [VERT ZOO] Any of about 26 species of rodents in two families (Hystricidae and Erethizontidae) which have spines or quills in addition to regular hair. { 'pȯr·kyə,pīn }

pore [BIOL] Any minute opening by which matter passes through a wall or membrane. { pȯr }

pore-forming protein *See* perforin. { 'pȯr ,fȯrm·iŋ ,prō,tēn }

pore fungus [MYCOL] The common name for members of the families Boletaceae and Polyporaceae in the group Hymenomycetes; sporebearing surfaces are characteristically within tubes or pores. { 'pȯr ,fəŋ·gəs }

porencephaly [MED] A condition in which the cavity of a lateral ventricle extends to the surface of the cerebral hemisphere; may result from brain tissue destruction or maldevelopment. { ,pȯr·ən'sef·ə·lē }

Porifera [INV ZOO] The sponges, a phylum of the animal kingdom characterized by the presence of canal systems and chambers through which water is drawn in and released; tissues and organs are absent. { pə'rif·ə·rə }

porocyte [INV ZOO] One of the perforated, tubular cells which constitute the wall of the incurrent canals in certain Porifera. { 'pȯr·ə,sīt }

porogamy [BOT] Passage of the pollen tube through the micropyle of an ovule in a seed plant. { pȯ'räg·ə·mē }

porosis [MED] Condition characterized by increased porosity, as of bone. { pə'rō·səs }

porphin [BIOCHEM] A heterocyclic ring consisting of four pyrrole rings linked by methine (−CH=) bridges; the basic structure of chlorophyll, hemoglobin, the cytochromes, and certain other related substances. { 'pȯr·fən }

porphobilinogen [BIOCHEM] $C_{10}H_{14}O_4N_2$ Dicarboxylic acid derived from pyrrole; a product of hemoglobin breakdown that gives the urine a Burgundy-red color. { ¦pȯr·fō·bə'lin·ə·jən }

porphyria [MED] A usually hereditary, pathologic disorder of porphyrin metabolism characterized by porphyrinuria and photosensitivity. { pȯr'fir·ē·ə }

porphyrin [BIOCHEM] A class of red-pigmented compounds with a cyclic tetrapyrrolic structure in which the four pyrrole rings are joined through their α-carbon atoms by four methene bridges (=C−); the porphyrins form the active nucleus of chlorophylls and hemoglobin. { 'pȯr·fə·rən }

porphyrinuria [MED] The excretion of large

quantities of porphyrin in the urine. { ˌpȯr·fə·rə'nyu̇r·ē·ə }

porpoise [VERT ZOO] Any of several species of marine mammals of the family Phocaenidae which have small flippers, a highly developed sonar system, and smooth, thick, hairless skin. { 'pȯr·pəs }

porta hepatis [ANAT] The transverse fissure of the liver through which the portal vein and hepatic artery enter the liver and the hepatic ducts leave. { 'pȯrd·ə he'pad·əs }

portal [ANAT] **1.** Of or pertaining to the porta hepatis. **2.** Pertaining to the portal vein or system. { 'pȯrd·əl }

portal circulation [PHYSIO] The passage of venous blood through a portal system. { 'pȯrd·əl ˌsər·kyə'lā·shən }

portal cirrhosis [MED] Replacement of normal liver structure by abnormal lobules of liver cells, often hyperplastic, delimited by bands of fibrous tissue, giving the gross appearance of a finely nodular surface. Also known as Laennec's cirrhosis. { 'pȯrd·əl sə'rō·səs }

portal hypertension [MED] Portal venous pressure in excess of 20 mmHg (2666 pascals), resulting from intrahepatic or extrahepatic portal venous compression or occlusion. { 'pȯrd·əl ˌhī·pər'ten·shən }

portal system [ANAT] A system of veins that break into a capillary network before returning the blood to the heart. { 'pȯrd·əl ˌsis·təm }

portal vein [ANAT] Any vein that terminates in a network of capillaries. { 'pȯrd·əl ˌvān }

Portuguese man-of-war [INV ZOO] Any of several brilliantly colored tropical siphonophores in the genus *Physalia* which possess a large float and extremely long tentacles. { ˌpȯr·chə'gēz ¦man əv 'wȯr }

Portulacaceae [BOT] A family of dicotyledonous plants in the order Caryophyllales distinguished by a syncarpous gynoecium, few, cyclic tepals and stamens, two sepals, and two to many ovules. { ˌpȯr·chə·lə'kās·ēˌē }

Portunidae [INV ZOO] The swimming crabs, a family of the Brachyura having the last pereiopods modified as swimming paddles. { pȯr 'tü·nəˌdē }

port-wine stain [MED] A congenital hemangioma characterized by one or more red to purplish patches, usually on the face. { 'pȯrt ˌwīn ˌstān }

Porulosida [INV ZOO] An order of the protozoan subclass Radiolaria in which the central capsule shows many pores. { ˌpȯr·yə'lä·səd·ə }

positional cloning [GEN] The identification of a gene and its isolation as a cloned deoxyribonucleic acid fragment, starting from knowledge of its position on a genetic map or chromosome. { pə¦zish·ən·əl 'klon·iŋ }

position effect [GEN] **1.** Change in expressivity of a gene associated with changes in location on a chromosome, as from euchromatin to heterochromatin. **2.** Inherent gene expression as influenced by neighboring genes. { pə'zish·ən iˌfekt }

positive afterimage [PHYSIO] An afterimage persisting after the eyes are closed or turned toward a dark background, and of the same color as the stimulating light. { 'päz·əd·iv 'af·tər ˌim·ij }

positive allosteric control *See* allosteric activation. { ¦päz·əd·iv ˌal·əˌstir·ik kən'trōl }

positive gene control [CELL MOL] Enhancement of gene expression through binding of specific expressor molecules to promoter sites. { ¦päz·əd·iv 'jēn kənˌtrōl }

positive interference [GEN] The reduction of the likelihood of another crossover in the vicinity of a nearby crossover. { 'päz·əd·iv ˌin·tər'fir·əns }

positive photokinesis [PHYSIO] The faster movement of an organism upon entering an illuminated area. { ¦päz·əd·iv ˌfōd·ōkə'nē·səs }

positive phototaxis [PHYSIO] The orientation and movement of an organism toward the source of a light stimulus. { ¦päz·əd·iv ˌfōd·ō'tak·səs }

positron emission tomography [MED] A technique that uses measurements of the back-to-back emission of gamma rays from the annihilation of positrons emitted by radioactive tracers to map the distribution of these tracers in the human body. Abbreviated PET. { 'päz·əˌträn i¦mish·ən tō'mäg·rə·fē }

postcentral gyrus [ANAT] The cerebral convolution that lies immediately posterior to the central sulcus and extends from the longitudinal fissure above the posterior ramus of the lateral sulcus. { pōst'sen·trəl 'jī·rəs }

postcentral sulcus [ANAT] The first sulcus of the parietal lobe of the cerebrum, lying behind and roughly parallel to the central sulcus. { pōst'sen·trəl 'səl·kəs }

postencephalitic parkinsonism [MED] The parkinsonian syndrome occurring as a sequel to lethargic encephalitis within a variable period, from days to many years, after the acute process. { ˌpōst·inˌsef·ə'lid·ik 'pärk·ən·səˌniz·əm }

posterior [ZOO] **1.** The hind end of an organism. **2.** Toward the back, or hinder end, of the body. { pä'stir·ē·ər }

posterior chamber [ANAT] The space in the eye between the posterior surface of the iris and the ciliary body, and the lens. { pä'stir·ē·ər 'chām·bər }

posterior horn [NEUROSCI] The dorsal column of gray matter in the spinal cord containing the axons of sensory (afferent) neurons. { pä¦stir·ē·ər hȯrn }

posthitis [MED] Inflammation of the prepuce. { päs'thīd·əs }

postmeiotic fusion [CELL MOL] The union of two identical haploid nuclei produced by mitotic division of the egg nucleus, resulting in restoration of the diploid state. { ¦pōst·mēˌäd·ik 'fyü·zhən }

postmortem [MED] Occurring after death. { pōst'mȯrd·əm }

postnatal [MED] Subsequent to birth. { pōs 'nād·əl }

postnecrotic cirrhosis [MED] Cirrhosis, usually

due to toxic agents or viral hepatitis, characterized by necrosis of liver cells, regenerating nodules of hepatic tissue, and the presence of large bands of connective tissue. { ˌpōs·nə'kräd·ik sə'rō·səs }

postoperative hernia *See* incisional hernia. { pōst'äp·rəd·iv 'hər·nē·ə }

postpartum [MED] Following childbirth. { pōs 'pärd·əm }

postpartum pituitary necrosis *See* Sheehan's syndrome. { ˌpōst¦pärd·əm pəˌtü·əˌter·ē nə'krō·səs }

postprandial [MED] After a meal. { pōs'pran·dē·əl }

posttranslational modification [CELL MOL] Any polypeptide alteration that occurs after synthesis of the chain. { ˌpōs·tranz¦lā·shən·əl ˌmäd·ə·fə'kā·shən }

posttraumatic hernia *See* incisional hernia. { ¦pōs·trȯ'mad·ik 'hər·nē·ə }

postural edema [MED] A condition resulting from an increased hydrostatic pressure in the capillaries caused by fluid collection in the subcutaneous tissues of the feet and ankles that occurs when an individual has been standing motionless for a long period of time. { ˌpäs·chə·rəl ə'dē·mə }

Potamogalinae [VERT ZOO] An aberrant subfamily of West African tenrecs (Tenrecidae). { ˌpäd·ə·mō'gal·əˌnē }

Potamogetonaceae [BOT] A large family of monocotyledonous plants in the order Najadales characterized by a solitary, apical or lateral ovule, usually two or more carpels, flowers in spikes or racemes, and four each of tepals and stamens. { ˌpäd·ə·mōˌjēd·ə'nās·ēˌē }

Potamogetonales [BOT] The equivalent name for Najadales. { ˌpäd·ə·mōˌjēd·ə'nā·lēz }

Potamonidae [INV ZOO] A family of fresh-water crabs included in the Brachyura. { ˌpäd·ə 'män·əˌdē }

potamoplankton [BIOL] Plankton found in rivers. { ¦päd·ə·mō¦plaŋk·tən }

potato [BOT] *Solanum tuberosum.* An erect herbaceous annual that has a round or angular aerial stem, underground lateral stems, pinnately compound leaves, and white, pink, yellow, or purple flowers occurring in cymose inflorescences; produces an edible tuber which is a shortened, thickened underground stem having nodes (eyes) and internodes. Also known as Irish potato; white potato. { pə'tā·dō }

potato virus X group *See* Potexvirus. { pə¦tā·dō ˌvī·rəs 'eks ˌgrüp }

potato virus Y [VIROL] A species in the genus *Potyvirus* that is a common pathogen of potato plants. Symptoms of infection can include shortening of the stem internodes, spotting and severe malformation of the upper leaves, defoliation, and early plant death. Abbreviated PVY. { pə¦tā·dō ˌvī·rəs 'wī }

potentiator [PHARM] A drug that augments the response of a second drug so that the total response is greater than the predictible additive effect. { pə'ten·chēˌād·ər }

Potexvirus [VIROL] A genus of plant viruses characterized by flexuous helical rods containing one molecule of linear, positive-sense, single-stranded ribonucleic acid. Also known as potato virus X group. { pōt'eksˌvī·rəs }

Potter-Bucky grid [MED] An assembly of lead strips resembling an open venetian blind, placed between a patient being x-rayed and the screen or film, to reduce the effects of scattered radiation. Also known as Bucky diaphragm; grid. { 'päd·ər 'bək·ē ˌgrid }

Pottiales [BOT] An order of mosses distinguished by erect stems, lanceolate to broadly ovate or obovate leaves, a strong, mostly percurrent or excurrent costa, and a cucullate calyptra. { ˌpäd·ē'ā·lēz }

Pott's disease [MED] Abnormal backward curvature of the spine caused by tuberculous osteitis. { 'päts diˌzēz }

Potyvirus [VIROL] A genus of plant viruses characterized by virions that are flexuous helical rods containing one molecule of linear, positive-sense, single-stranded ribonucleic acid. Potato virus Y is the most common member. { pōt'wīˌvī·rəs }

poultice [MED] A soft mass of hot, moist material applied as an external counterirritant, analgesic, or antiseptic. { 'pōl·təs }

Poupart's ligament *See* inguinal ligament. { pü 'pärz ˌlig·ə·mənt }

pour-plate culture [MICROBIO] A technique for pure-culture isolation of bacteria; liquid, cooled agar in a test tube is inoculated with one loopful of bacterial suspension and mixed by rolling the tube between the hands; subsequent transfers are made from this to a second test tube, and from the second to a third; contents of each tube are poured into separate petri dishes; pure cultures can be isolated from isolated colonies appearing on the plates after incubation. { 'pȯr ˌplāt ˌkəl·chər }

Pourtalesiidae [INV ZOO] A family of exocyclic Euechinoidea in the order Holasteroida, including those forms with a bottle-shaped test. { ˌpȯrd·əl·ə'sī·əˌdē }

powdery mildew [MYCOL] A fungus characterized by production of abundant powdery conidia on the host; a member of the family Erysiphaceae or the genus *Oidium.* { 'pau̇d·ə·rē 'milˌdü }

pox [MED] A vesicular or pustular exanthematic disease that may leave pit scars. { päks }

Poxviridae [VIROL] A family of deoxyribonucleic acid-containing animal viruses that is characterized by its ability to replicate in the cell cytoplasm, includes the subfamilies Chordopoxviridae and Entomopoxviridae. { päks'vir·əˌdī }

poxvirus [VIROL] A deoxyribonucleic acid-containing animal virus group including the viruses of smallpox, molluscum contagiosum, and various animal pox and fibromas. { 'päksˌvī·rəs }

PPLO *See* pleuropneumonialike organism.

Prader-Willi syndrome [MED] A genetic disorder that is caused by defects on the paternally derived chromosome 15, causing mild mental retardation, neonatal hypotonia, hypogonadism, compulsive overeating, childhood onset obesity, and mild facial dysmorphism. { ¦präd·ər ¦wil·ē 'sin,drōm }

prairie dog [VERT ZOO] The common name for three species of stout, fossorial rodents belonging to the genus *Cynomys* in the family Sciuridae; all have a short, flat tail, small ears, and short limbs terminating in long claws. { 'prer·ē ,dȯg }

prairie wolf *See* coyote. { 'prer·ē ,wu̇lf }

Prasinovolvocales [BOT] An order of green algae in which there are lateral appendages in the flagellum. { ,prāz·ən·ō,väl·və'kā·lēz }

pratincolous [ECOL] **1.** Living in meadows. **2.** Living in low grass. { prə'tiŋ·kə·ləs }

preadaptation [EVOL] Possession by an organism or group of organisms, specialized to one mode of life, of characters which favor easy adaptation to a new environment. { ¦prē,ad·əp'tā·shən }

precancerous [PATH] Pertaining to any pathological condition of a tissue which is likely to develop into cancer. { prē'kan·sə·rəs }

precentral gyrus [ANAT] The cerebral convolution that lies between the precentral sulcus and the central sulcus and extends from the superomedial border of the hemisphere to the posterior ramus of the lateral sulcus. { 'prē¦sen·trəl 'jī·rəs }

precipitation [IMMUNOL] Aggregation of soluble antigen by an antibody. { prə,sip·ə'tā·shən }

precipitin [IMMUNOL] An antibody that chemically interacts with an antigen to form a precipitate. { prə'sip·ə·tən }

precipitin test [IMMUNOL] An immunologic test in which a specific reaction between antigen and antibody results in a visible precipitate. { prə'sip·ə·tən ,test }

precollagenous fiber *See* reticular fiber. { ,prē·kə'laj·ə·nəs 'fī·bər }

predation [BIOL] The killing and eating of an individual of one species by an individual of another species. { prə'dā·shən }

predator [ECOL] An animal that preys on other animals as a source of food. { 'pred·əd·ər }

prednisolone [BIOCHEM] $C_{21}H_{28}O_5$ A glucocorticoid that is a dehydrogenated analog of hydrocortisone. { pred'nis·ə,lōn }

prednisone [PHARM] $C_{21}H_{26}O_5$ An adrenocortical steroid drug, obtained in crystalline form, that is an analog of cortisone. { 'pred·nə,sōn }

preeclampsia [MED] A toxemia occurring in the latter half of pregnancy, characterized by an acute elevation of blood pressure and usually by edema and proteinuria, but without convulsions or coma. Also known as toxemia of pregnancy. { ¦prē·i'klamp·sē·ə }

preen gland *See* uropygial gland. { 'prēn ,gland }

prefoliation *See* vernation. { 'prē,fō·lē'ā·shən }

prefrontal [ANAT] Situated in the anterior part of the frontal lobe of the brain. [VERT ZOO] **1.** Of or pertaining to a bone of some vertebrate skulls, located anterior and lateral to the frontal bone. **2.** Of, pertaining to, or being a scale or plate in front of the frontal scale on the head of some reptiles and fishes. { prē'frənt·əl }

prefrontal lobotomy *See* lobotomy. { prē'frənt·əl lə'bäd·ə·mē }

pregnancy [MED] The state of being pregnant, from conception to childbirth. { 'preg·nən·sē }

pregnancy test [PATH] Any biologic or chemical procedure used to diagnose pregnancy. { 'preg·nən·sē ,test }

pregnanediol [BIOCHEM] $C_{21}H_{36}O_2$ A metabolite of progesterone, present in urine during the progestational phase of the menstrual cycle and also during pregnancy. { preg,nan·ə'dī,ȯl }

pregnenolone [BIOCHEM] $C_{21}H_{32}O_2$ A steroid ketone that is formed as an oxidation product of cholesterol, stigmasterol, and certain other steroids. { preg'nen·ə,lōn }

prehensile [VERT ZOO] Adapted for seizing, grasping, or plucking, especially by wrapping around some object. { prē'hen·səl }

prehension [PHYSIO] A movement that involves holding, seizing, or grasping. { prē'hen·shən }

premaxilla [ANAT] Either of two bones of the upper jaw of vertebrates located in front of and between the maxillae. { ¦prē·mak'sil·ə }

premessenger ribonucleic acid [CELL MOL] A giant molecule of ribonucleic acid that is transcribed from a cistron. { ,prē'mes·ən·jər ¦rī·bō¦nü¦klē·ik 'as·əd }

premitosis [CELL MOL] In certain protozoans, a kind of mitosis in which there is an intranuclear centriole and the mitotic apparatus is located within the nuclear membrane. { ,prē·mī'tō·səs }

premolar [ANAT] In each quadrant of the permanent human dentition, one of the two teeth between the canine and the first molar; a bicuspid. { prē'mō·lər }

premutation [GEN] A heritable change, such as a trinucleotide repeat expansion, that has no phenotypic effect but greatly increases the likelihood of a further change at the altered site, resulting in a characteristic phenotype. { ,prē·myü'tā·shən }

prenatal [MED] Existing or occurring before birth. { prē'nād·əl }

prepattern [EMBRYO] The organization in a developing organism before a definite organizational pattern is established. { 'prē,pad·ərn }

prepuce [ANAT] **1.** The foreskin of the penis, a fold of skin covering the glans penis. **2.** A similar fold over the glans clitoridis. { 'prēp·əs }

preputial gland *See* Tyson's gland. { prē'pyü·shəl ,gland }

presbycusis [MED] A condition of diminished auditory acuity associated with old age. { ,prez·bə'kyü·səs }

presbyopia [MED] Diminished ability to focus the eye on near objects due to gradual loss of elasticity of the crystalline lens with age. { ,prez·bē'ō·pē·ə }

prescutum [INV ZOO] The anterior part of the tergum of a segment of the thorax in insects. { prē'skyüd·əm }

preserve [ECOL] An area that is maintained for game or fish, especially for sport, and may have limited access requiring a permit for entry. { prə'zərv }

presoma [INV ZOO] The anterior portion of an invertebrate that lacks a definitive head structure. { prē'sō·mə }

pressure point [PHYSIO] A point of marked sensibility to pressure or weight, arranged like the temperature spots, and showing a specific end apparatus arranged in a punctate manner and connected with the pressure sense. { 'presh·ər pȯint }

pressure ulcer *See* decubitus ulcer. { 'presh·ər ,əl·sər }

pretransfer ribonucleic acid [CELL MOL] The primary transcription product of transfer ribonucleic acid-encoding genes. { ,prē¦tranz·fər ,rī·bō·nü¦klē·ik 'as·əd }

prevalence [GEN] The frequency with which a medical condition is found in specific population at a specific time. { 'prev·ə·ləns }

priapism [MED] Persistent erection of the penis, usually unaccompanied by sexual desire, as seen in certain pathologic conditions. { 'prī·ə,piz·əm }

Priapulida [INV ZOO] A minor phylum of wormlike marine animals; the body is made up of three distinct portions (proboscis, trunk, and caudal appendage) and is often covered with spines and tubercles, and the mouth is surrounded by concentric rows of teeth. { ,prī·ə'pyül·əd·ə }

Priapuloidea [INV ZOO] An equivalent name for Priapulida. { prī,ap·ə'lȯid·ē·ə }

Pribnow box [CELL MOL] In prokaryotes, a highly conserved sequence element located upstream from the transcriptional start site to which binds the sigma subunit of the ribonucleic acid polymerase. { 'prib,nō ,bäks }

pricker *See* needle. { 'prik·ər }

prickly heat *See* miliaria. { 'prik·lē 'hēt }

primaquine [PHARM] $C_{15}H_{21}N_3O$ An ether-soluble viscous liquid, used as the diphosphate salt in medicine to cure malaria. { 'prī·mə,kwēn }

primary [VERT ZOO] Of or pertaining to quills on the distal joint of a bird wing. { 'prī ,mer·ē }

primary atypical pneumonia *See* Eaton agent pneumonia. { 'prī,mer·ē ā'tip·ə·kəl nü'mō·nyə }

primary biliary cirrhosis [MED] A slowly progressive disease primarily of middle-aged women, caused by an autoimmune destruction of bile ducts that begins as inflammation in and around larger intrahepatic bile ducts and eventually results in liver cell damage. { ¦Prīm·ə·rē ,bil·ē·er·ē sə'rō·səs }

primary constriction *See* centromere. { 'prī,mer·ē kən'strik·shən }

primary consumer [ECOL] In an ecosystem, an animal that feeds on plants (producers) directly. Also known as a herbivore. { ¦Prī,mer·ē kən 'sü·mər }

primary culture [BIOL] A tissue culture started from cells, tissues, or organs taken directly from the organism. { ¦prī,mer·ē 'kəl·chər }

primary growth [BOT] Plant growth that originates in apical meristematic tissue of shoots and roots, giving rise to primary tissue. { 'prī,mer·ē 'grōth }

primary hypertension *See* essential hypertension. { 'prī,mer·ē ,hī·pər'ten·chən }

primary hypothermia [MED] A decrease in internal body temperature caused by environmental stress that overwhelms the body's thermoregulation capability. { ¦Prī·mərē ,hī·pōthər·mē·ə }

primary immune response [IMMUNOL] The activation and response of lymphocytes specific for a newly encountered antigen; generally slower and weaker than the secondary immune response. { ¦prī·mə·rē i'myün ri,späns }

primary lateral sclerosis [MED] A sclerotic disease of the crossed pyramidal tracts of the spinal cord, characterized by paralysis of the limbs, with rigidity, increased tendon reflexes, and absence of sensory and nutritive disorders. Also known as lateral sclerosis. { 'prī,mer·ē 'lad·ə·rəl sklə 'rō·səs }

primary lesion [MED] **1.** In syphilis, tuberculosis, and cowpox, a chancre. **2.** In dermatology, the earliest clinically recognizable manifestation of cutaneous disease, such as a macule, papule, vesicle, pustule, or wheal. { 'prī,mer·ē 'lē·zhən }

primary meristem [BOT] Meristem which is derived directly from embryonic tissue and which gives rise to epidermis, vascular tissue, and the cortex. { 'prī,mer·ē 'mer·ə,stem }

primary phloem [BOT] Phloem derived from apical meristem. { 'prī,mer·ē 'flō·əm }

primary producer [ECOL] In an ecosystem, an organism (primarily green photosynthetic plants) that utilizes the energy of the sun and inorganic molecules from the environment to synthesize organic molecules. { ¦Prī,mer·ē prə'dü·sər }

primary production [ECOL] The total amount of new organic matter produced by photosynthesis. { 'prī,mer·ē prə'dək·shən }

primary root [BOT] The first plant root to develop; derived from the radicle. { 'prī,mer·ē 'rüt }

primary structure [BIOCHEM] The sequence of amino acids in the molecule of a protein or a peptide. { 'prī,mer·ē 'strək·chər }

primary succession *See* prisere. { 'prī,mer·ē sək'sesh·ən }

primary syphilis [MED] The first stage of the venereal disease, characterized clinically by a painless ulcer, or chancre, at the point of infection and painless, discrete regional adenopathy. { 'prī,mer·ē 'sif·ə·ləs }

primary tissue [BOT] Plant tissue formed during primary growth. [HISTOL] Any of the four fundamental tissues composing the vertebrate body. { 'prī,mer·ē 'tish·ü }

primary transcript [CELL MOL] The initial transcription of a ribonucleic acid molecule from deoxyribonucleic acid. { 'prī,mer·ē 'tran ,skript }

primary xylem [BOT] Xylem derived from apical meristem. { 'prī,mer·ē 'zī·ləm }

Primates [VERT ZOO] The order of mammals to which man belongs; characterized in terms of evolutionary trends by retention of a generalized limb structure and dentition, increasing digital mobility, replacement of claws by flat nails, development of stereoscopic vision, and progressive development of the cerebral cortex. { prī'mād·ēz }

primed lymphocyte [IMMUNOL] A lymphocyte that is from an immunized individual or that has been exposed to antigen in cell culture and is therefore sensitized. { 'prīmd 'lim·fə,sīt }

prime mover [ANAT] A muscle that produces a specific motion or maintains a specific posture. { 'prīm 'müv·ər }

primer [CELL MOL] **1.** A short ribonucleic acid (RNA) sequence that is complementary to a sequence of deoxyribonucleic acid (DNA) and has a 3′-OH terminus at which a DNA polymerase begins synthesis of a DNA chain. **2.** A short sequence of DNA that is complementary to a messenger RNA (mRNA) sequence and enables reverse transcriptase to begin copying the neighboring sequences of mRNA. **3.** A transfer RNA whose elongation starts RNA-directed DNA synthesis in retroviruses. { 'prīm·ər }

primibrach [INV ZOO] In crinoids, the brachials of the unbranched arm. { 'prī·mə,brak }

primitive gut [EMBRYO] The tubular structure in embryos which differentiates into the alimentary canal. { 'prim·əd·iv 'gət }

primitive streak [EMBRYO] A dense, opaque band of ectoderm in the bilaminar blastoderm associated with the morphogenetic movements and proliferation of the mesoderm and notochord; indicates the first trace of the vertebrate embryo. { 'prim·əd·iv 'strēk }

primordial germ cell [EMBRYO] Embryonic cell giving rise to a germ cell from which a gamete (egg or sperm) develops. { prī¦mȯrd·ē·əl 'jərm ,sel }

primordial gut *See* archenteron. { prī'mȯrd·ē·əl 'gət }

primordium *See* anlage. { prī'mȯrd·ē·əm }

Primulaceae [BOT] A family of dicotyledonous plants in the order Primulales characterized by a herbaceous habit and capsular fruit with two to many seeds. { ,prim·yə'lās·ē,ē }

Primulales [BOT] An order of dicotyledonous plants in the subclass Dilleniidae distinguished by sympetalous flowers, stamens located opposite the corolla lobes, and a compound ovary with a single style. { ,prim·yə'lā·lēz }

prion [BIOCHEM] Any of a group of infectious proteins that cause fatal neurodegenerative diseases in humans and animals, including scrapie and bovine spongiform encephalopathy in animals and Creutzefeldt-Jakob disease and Gerstmann-Straussler-Scheinker disease in humans. { 'prī,än }

prion diseases [MED] A group of invariably fatal disorders affecting humans and animals that are clinically characterized by neurological and behavioral degeneration caused by the cerebral accumulation of an abnormal prion protein which is resistant to proteolytic enzymes and, in contrast to other infectious agents, does not require nucleic acid for replication. The diseases are transmissible either genetically (for example, Creutzfeldt-Jakob disease) or via infection (new variant Creutzfeldt-Jakob disease and mad cow disease) or can occur spontaneously (classical or sporadic Creutzfeldt-Jakob disease). Also known as spongiform encephalopathies. { 'prī,än di,zēz·əz }

prionodont [VERT ZOO] Having many simple, similar teeth set in a row like sawteeth. { prī'än·ə,dänt }

prisere [ECOL] The ecological succession of vegetation that occurs in passing from barren earth or water to a climax community. Also known as primary succession. { 'prī,sir }

Pristidae [VERT ZOO] The sawfishes, a family of modern sharks belonging to the batoid group. { 'pris·tə,dē }

Pristiophoridae [VERT ZOO] The saw sharks, a family of modern sharks often grouped with the squaloids which have a greatly extended rostrum with enlarged denticles along the margins. { ,pris·tē·ə'fȯr·ə,dē }

proaccelerin [BIOCHEM] A labile procoagulant in normal plasma but deficient in the blood of patients with parahemophilia; essential for rapid conversion of prothrombin to thrombin. Also known as factor V; labile factor. { ¦prō·ak'sel·ə·rən }

proamnion [EMBRYO] The part of the embryonic area at the sides and in front of the head of the developing amniote embryo, which remains without mesoderm for a considerable period. { prō'am·nē,än }

proband [GEN] The clinically affected individual through whom a family is found that can be used to study the genetics of a particular disorder. Also known as propositius. { 'prō,band }

probe [BIOL] A biochemical substance labeled with a radioactive isotope or tagged in some other way and used to identify or isolate a gene, a gene product, or a protein. { prōb }

probe-type microelectrode [NEUROSCI] A microelectrode consisting of one or more long thin shanks projecting from a larger carrier area that is designed to facilitate the investigation of processing in three-dimensional neural networks on the level of individual neurons. { ¦prōb,tīp ,mī·krō·i'lek,trōd }

problem space [PHYSIO] A mental representation of a problem that contains knowledge of the initial state and the goal state of the problem as well as possible intermediate states that must be searched in order to link up the beginning and the end of the task. { 'präb·ləm ,spās }

Proboscidea [VERT ZOO] An order of herbivorous placental mammals characterized by having a proboscis, incisors enlarged to become tusks, and pillarlike legs with five toes bound together on a broad pad. { ,prō·bə'sid·ē·ə }

proboscis [INV ZOO] A tubular organ of varying

form and function on a large number of invertebrates, such as insects, annelids, and tapeworms. [VERT ZOO] The flexible, elongated snout of certain mammals. { prə'bäs·kəs }

procambium [BOT] The part of the apical meristematic tissue from which primary vascular tissues are derived. { prō'käm·bē·əm }

Procampodeidae [INV ZOO] A family of the insect order Diplura. { prō,kam·pə'dē·ə,dē }

procarp [BOT] The reproductive structure of the female gametophyte that is found in certain red algae. { 'prō,kärp }

Procaviidae [VERT ZOO] A family of mammals in the order Hyracoidea including the hyraxes. { ,prō·kə'vī·ə,dē }

Procaviinae [VERT ZOO] A subfamily of ungulate mammals in the family Procaviidae. { ,prō·kə'vī·ə,nē }

Procellariidae [VERT ZOO] A family of birds in the order Procellariiformes comprising the petrels, fulmars, and shearwaters. { ,prō·sə·lə'rī·ə,dē }

Procellariiformes [VERT ZOO] An order of oceanic birds characterized by tubelike nostril openings, webbed feet, dense plumage, compound horny sheath of the bill, and, often, a peculiar musky odor. { ,prō·sə·lə,rī·ə'fȯr,mēz }

procephalon [INV ZOO] The part of an insect's head that lies anteriorly to the segment in which the mandibles are located. { prō'sef·ə,län }

procercoid [INV ZOO] The solid parasitic larva of certain eucestodes, such as pseudophyllideans, that develops in the body of the intermediate host. { prō'sər,kȯid }

process [ANAT] A projection from the central mass of an organism. { 'prä,ses }

procoagulant [BIOCHEM] Any of blood clotting factors V to VIII; accelerates the conversion of prothrombin to thrombin in the presence of thromboplastin and calcium. { ¦prō·kō'ag·yə·lənt }

Procoela [VERT ZOO] A suborder of the Anura characterized by a procoelous vertebral column and a free coccyx articulating with a double condyle. { prō'sēl·ə }

procoelous [VERT ZOO] The form of a vertebra that is concave anteriorly and convex posteriorly. { prō'sēl·əs }

procollagen [BIOCHEM] A high-molecular-weight form of collagen that is found in intracellular spaces and is believed to be the precursor of collagen. { prō'käl·ə·jən }

proctiger [INV ZOO] The cone-shaped, reduced terminal segment of the abdomen of an insect which contains the anus. { 'präk·tə·jər }

proctitis [MED] Inflammation of the anus or rectum. { präk'tīd·əs }

proctodone [INV ZOO] An insect hormone that causes diapause to end. { 'präk·tə,dōn }

proctology [MED] A branch of medicine concerned with the structure and disease of the anus, rectum, and sigmoid colon. { präk'täl·ə·jē }

proctoscope [MED] An instrument for inspecting the anal canal and rectum. { 'präk·tə,skōp }

proctosigmoidectomy [MED] The abdominoperineal excision of the anus and rectosigmoid, usually with the formation of an abdominal colostomy. { ¦präk·tō,sig,mȯi'dek·tə·mē }

Proctotrupidae [INV ZOO] A family of hymenopteran insects in the superfamily Proctotrupoidea. { ,präk·tə'trü·pə,dē }

Proctotrupoidea [INV ZOO] A superfamily of parasitic Hymenoptera in the suborder Apocrita. { ,präk·tə·trə'pȯid·ē·ə }

procumbent [BOT] Having stems that lie flat on the ground but do not root at the nodes. { prō'kəm·bənt }

Procyonidae [VERT ZOO] A family of carnivoran mammals in the superfamily Canoidea, including raccoons and their allies. { ,prō·sē'än·ə,dē }

prodrome [MED] **1.** An early or premonitory manifestation of impending disease before the specific symptoms begin. **2.** An aura. { 'prō,drōm }

prodrug [PHYSIO] An inactive precursor of a drug that is activated via the body's metabolism. { 'prō,drəg }

producer [ECOL] An autotrophic organism of the ecosystem; any of the green plants. { prə'dü·sər }

production ecology *See* ecological energetics. { prə¦duk·shən ē'käl·ə·jē }

productive cough [MED] A cough accompanied by expectoration. { prə'dək·tiv 'kȯf }

proenzyme *See* zymogen. { prō'en,zīm }

proerythroblast of Ferrata *See* pronormoblast. { ¦prō·ə'rith·rə,blast əv fe'räd·ə }

proestrus [PHYSIO] The beginning of the follicular phase of estrus. { prō'es·trəs }

profibrinolysin *See* plasminogen. { ¦prō,fī·brə'näl·ə·sən }

profunda [ANAT] Deep-seated; applied to certain arteries. { prō'fən·də }

profundal zone [ECOL] The region occurring below the limnetic zone and extending to the bottom in lakes deep enough to develop temperature stratification. { prō'fənd·əl ,zōn }

progenitor cell [CELL MOL] A precursor cell that completes a series of cell divisions to produce a distinct cell lineage. { prə'jen·əd·ər ,sel }

progeny [BIOL] Offspring; descendants. { 'präj·ə·nē }

progeny test [GEN] The assessment of parental genotype by study of its progeny under controlled conditions. { 'präj·ə·nē ,test }

progeria [MED] An abnormal childhood state of premature senescence, characterized by wrinkled skin, gray hair, lack of pubic or facial hair, development of atherosclerosis, and a short life span. Also known as Hutchinson-Gilford syndrome. { prō'jir·ē·ə }

progestational hormone [BIOCHEM] **1.** The natural hormone progesterone, which induces progestational changes of the uterine mucosa. **2.** Any derivative or modification or progesterone having similar actions. { ¦prō,je'stā·shən·əl 'hȯr,mōn }

progesterone [BIOCHEM] $C_{21}H_{30}O_2$ A steroid hormone produced in the corpus luteum, placenta, testes, and adrenals; plays an important physiological role in the luteal phase of the menstrual cycle and in the maintenance of pregnancy; it is an intermediate in the biosynthesis of androgens, estrogens, and the corticoids. { prō'jes·tə,rōn }

proglottid [INV ZOO] One of the segments of a tapeworm. { prō'gläd·əd }

prognosis [MED] A prediction as to the course and outcome of a disease, injury, or developmental abnormality. { präg'nō·səs }

progressive lateral sclerosis [MED] Amyotrophic lateral sclerosis with primary involvement of the pyramidal tracts. { prə'gres·iv 'lad·ə·rəl sklə'rō·səs }

progressive muscular dystrophy [MED] Chronic progressive dystrophy of the skeletal muscles. { prə'gres·iv 'məs·kyə·lər 'dis·trə·fē }

prohaptor [INV ZOO] The anterior attachment organ of a typical monogenetic trematode. { prō'hap·tər }

prohormone [BIOCHEM] The precursor of a hormone. { 'prō,hȯr,mōn }

Projapygidae [INV ZOO] A family of wingless insects in the order Diplura. { ¦prä·jə'pij·əd·ē }

projection area [ANAT] An area of the cortex connected with lower centers of the brain by projection fibers. { prə'jek·shən ,er·ē·ə }

projection fibers [ANAT] Fibers joining the cerebral cortex to lower centers of the brain, and vice versa. { prə'jek·shən ,fī·bərz }

Prokaryotae [BIOL] A superkingdom of predominantly unicellular microorganisms lacking a membrane-bound nucleus containing chromosomes and having asexual reproduction by binary fission; it includes the kingdom Monera, and viruses, which are acellular, are included by some. { ¦prō·kar·ē'ō,dē }

prokaryote [CELL MOL] **1.** A primitive nucleus, where the deoxyribonucleic acid-containing region lacks a limiting membrane. **2.** Any cell containing such a nucleus, such as the bacteria and the blue-green algae. { prō'kar·ē,ōt }

prolactin [BIOCHEM] A protein hormone produced by the adenohypophysis; stimulates lactation and promotes functional activity of the corpus luteum. Also known as lactogenic hormone; luteotropic hormone; mammary-stimulating hormone; mammogen; mammogenic hormone; mammotropin. { prō 'lak·tən }

prolamellar body [BOT] An accumulation of vesicles formed by the invagination of the proplastid membrane during etiolation. { ,prō·lə¦mel·ər 'bäd·ē }

prolamin [BIOCHEM] Any of the simple proteins, such as zein, found in plants; soluble in strong alcohol, insoluble in absolute alcohol and water. { prō'lam·ən }

prolapse [MED] The falling or sinking down of a part or organ. { 'prō,laps }

proliferative arthritis *See* rheumatoid arthritis. { prō'lif·ə,rād·iv ärth'rīd·əs }

proline [BIOCHEM] $C_5H_9O_2$ A heterocyclic amino acid occurring in essentially all proteins, and as a major constituent in collagen proteins. { 'prō,lēn }

prolinemia [MED] A rare hereditary disease caused by absence of the degradative enzymes that convert proline to glutamic acid, and characterized by a high content of proline in blood and urine with consequent mental retardation and renal malfunction. { ,prō·lə'nē·mē·ə }

prometaphase [CELL MOL] A stage between prophase and metaphase in mitosis in which the nuclear membrane disappears and the spindle forms. { prō'med·ə,fāz }

promoter [GEN] The site on deoxyribonucleic acid to which ribonucleic acid polymerase binds preparatory to initiating transcription of a gene or an operon. { prə'mōd·ər }

promyelocyte [HISTOL] The earliest myelocyte stage derived from the myeloblast. { prō'mī·ə·lə,sīt }

pronate [ANAT] **1.** To turn the forearm so that the palm of the hand is down or toward the back. **2.** To turn the sole of the foot outward with the lateral margin of the foot elevated; to evert. { 'prō,nāt }

pronator [PHYSIO] A muscle which pronates, as the muscles of the forearm attached to the ulna and radius. { 'prō,nād·ər }

pronephros [EMBRYO] One of the anterior pair of renal organs in higher vertebrate embryos; the pair initiates formation of the archinephric duct. { prō'ne·frəs }

pronghorn [VERT ZOO] *Antilocapra americana*. An antelopelike artiodactyl composing the family Antilocapridae; the only hollow-horned ungulate with branched horns present in both sexes. { 'präŋ,hȯrn }

pronormoblast [HISTOL] A nucleated erythrocyte precursor with scanty basophilic cytoplasm without hemoglobin. Also known as lymphoid hemoblast of Pappenheim; macroblast of Naegeli; megaloblast of Sabin; proerythroblast of Ferrata; prorubricyte; rubriblast; rubricyte. { prō'nȯr·mə,blast }

pronucleus [CELL MOL] One of the two nuclear bodies of a newly fertilized ovum, the male pronucleus and the female pronucleus, the fusion of which results in the formation of the germinal (cleavage) nucleus. { prō'nü·klē·əs }

proofreading [CELL MOL] Any mechanism for correcting errors in replication, transcription, or translation that involves monitoring of individual units after they have been added to the chain. Also known as editing. { 'prüf,rēd·iŋ }

propagule [BOT] **1.** A reproductive structure of brown algae. **2.** A propagable shoot. { 'präp·ə,gyül }

Propalticidae [INV ZOO] A family of coleopteran insects of the superfamily Cucujoidea found in Old World tropics and Pacific islands. { ,prō·pəl'sid·əd·ē }

properdin [IMMUNOL] A macroglobin of normal

plasma capable of killing various bacteria and viruses in the presence of complement and magnesium ions. { 'prō·pər·dən }

prophage [VIROL] Integrated unit formed by union of the provirus into the bacterial genome. { 'prō,fāj }

prophase [CELL MOL] The initial stage of mitotic or meiotic cell division in which chromosomes are condensed from the nuclear material and split logitudinally to form pairs. { 'prō,fāz }

prophylactic vaccination [IMMUNOL] Vaccination occurring before exposure to pathogens. { ,pro·fə,lak·tic ,vak·sə'nā·shən }

prophylaxis [MED] The prevention of disease. { ,prō·fə'lak·səs }

propiodal [PHARM] $C_5H_{13}ONI_2$ A white, crystalline solid with a melting point of 275°C; soluble in water; used in medicine for iodine therapy. { 'prō·pē·ə,dal }

Propionibacteriaceae [MICROBIO] A family of bacteria related to the actinomycetes; gram-positive, anaerobic to aerotolerant rods or filaments; ferment carbohydrates, with propionic acid as the principal product. { ¦prō·pē,än·ə,bak·tir·ē'ās·ē,ē }

proplastid [BOT] Precursor body of a cell plastid. { prō'plas·təd }

propleuron [INV ZOO] A pleuron of the prothorax in insects. { prō'plůr,än }

propodite [INV ZOO] The sixth leg joint of certain crustaceans. Also known as propodus. { 'präp·ə,dīt }

propodus *See* propodite. { 'präp·əd·əs }

propositus *See* proband. { prə'päz·əd·əs }

proprioception [PHYSIO] The reception of internal stimuli. { ,prō·prē·ə'sep·shən }

proprioceptor [PHYSIO] A sense receptor that signals spatial position and movements of the body and its members, as well as muscular tension. { ,prō·prē·ə'sep·tər }

prop root [BOT] A root that serves to support or brace the plant. Also known as brace root. { 'präp ,rüt }

propterygium [VERT ZOO] The anterior of the three principal basal cartilages forming a support of one of the paired fins of sharks, rays, and certain other fishes. { ,präp·tə'rij·ē·əm }

proptosis [MED] A falling downward or forward, especially of an eyeball. { präp'tō·səs }

propylhexedrine [PHARM] $C_{10}H_{21}N$ A clear, colorless liquid with a boiling point of 202–206°C; soluble in dilute acids; used in medicine. { ,prō·pəl,hek'sed·rən }

propylpiperidine *See* coniine. { ¦prō·pəl·pi'per·ə,dēn }

propylthiouracil [PHARM] $C_7H_{10}N_2OS$ White, crystalline powder with a melting point of 218–221°C; soluble in ammonia and alkali hydroxides; used in medicine. { ,prō·pəl¦thī·ō'yůr·ə,sil }

prorennin *See* renninogen. { prō'ren·ən }

Prorhynchidae [INV ZOO] A family of turbellarians in the order Alloeocoela. { prō'riŋ·kə,dē }

prorubricyte *See* pronormoblast. { prō'rü·brə ,sīt }

Prosobranchia [INV ZOO] The largest subclass of the Gastropoda; generally, respiration is by means of ctenidia, an operculum is present, there is one pair of tentacles, and the sexes are separate. { ,prä·sə'braŋ·kē·ə }

prosodus [INV ZOO] A canal leading from an incurrent canal to a flagellated chamber in Porifera. { 'präs·əd·əs }

prosoma [INV ZOO] The anterior part of the body of mollusks and other invertebrates; primitive segmentation is not apparent. { prō'sō·mə }

Prosopora [INV ZOO] An order of the class Oligochaeta comprising mesonephridiostomal forms in which there are male pores in the segment of the posterior testes. { prə'säp·ə·rə }

prosopyle [INV ZOO] The opening into a flagellated chamber from an inhalant canal in sponges. { 'präs·ə,pīl }

prostaglandin [BIOCHEM] Any of various physiologically active compounds containing 20 carbon atoms and formed from essential fatty acids; found in highest concentrations in normal human semen; activities affect the nervous system, circulation, female reproductive organs, and metabolism. { ,präs·tə'glan·dən }

prostate [ANAT] A glandular organ that surrounds the urethra at the neck of the urinary bladder in the male. { 'prä,stāt }

prostatectomy [MED] Surgical removal of all or part of the prostate. { ,präs·tə'tek·tə·mē }

prostate specific antigen [IMMUNOL] A glycoprotein with 240 amino acids that is expressed exclusively by human prostate epithelial cells and that exhibits protease activity; elevated levels in the blood are detected in individuals with prostate cancer. Abbreviated PSA. { ¦prä,stāt spə,sif·ik 'ant·i·jən }

prostatitis [MED] Inflammation of the prostate. { ,präs·tə'tīd·əs }

prosthecae [MICROBIO] Appendages that are part of the wall in bacteria in the genus *Caulobacter*. { präs'thē·sē }

prosthecate bacteria [MICROBIO] Single-celled microorganisms that differ from typical unicellular bacteria in having one or more appendages which extend from the cell surface; the best-known genus is *Caulobacter*. { 'präs·thə,kāt bak 'tir·ē·ə }

prosthesis [MED] An artificial substitute for a missing part of the body, such as a substitute hand, leg, eye, or denture. { präs'thē·səs }

prosthetic group [BIOCHEM] A characteristic nonamino acid substance that is strongly bound to a protein and necessary for the protein portion of an enzyme to function; often used to describe the function, as in hemeprotein for hemoglobin. { präs'thed·ik 'grüp }

prosthodontics [MED] The science and practice of replacement of missing dental and oral structures. { ¦präs·thə¦dän·tiks }

Prostigmata [INV ZOO] The equivalent name for Trombidiformes. { ,prō,stig'mäd·ə }

prostomium [INV ZOO] The portion of the head

anterior to the mouth in annelids and mollusks. { prō'stō·mē·əm }

protamine [BIOCHEM] Any of the simple proteins that are combined with nucleic acid in the sperm of certain fish, and that upon hydrolysis yield basic amino acids; used in medicine to control hemorrhage, and in the preparation of an insulin form to control diabetes. { 'prōd·ə,mēn }

protandry [PHYSIO] That condition in which an animal is first a male and then becomes a female. { prō'tan·drē }

protanomaly *See* protanopia. { ,prōd·ə'näm·ə·lē }

protanopia [MED] Partial color blindness in which there is defective red vision; green sightedness. Also known as protanomaly. { ,prōd·ə'nō·pē·ə }

Proteaceae [BOT] A large family of dicotyledonous plants in the order Proteales, notable for often having a large cluster of small or reduced flowers. { prō'tās·ē,ē }

Proteales [BOT] An order of dicotyledonous plants in the subclass Rosidae marked by its strongly perigynous flowers, a four-lobed, often corolla-like calyx, and reduced or absent true petals. { ,prōd·ē'ā·lēz }

protease [CELL MOL] An enzyme that digests proteins. { 'prōd·ē,ās }

proteasome [BIOCHEM] A large proteolytic particle found in the cytoplasm and nucleus of all eukaryotic cells that is the site for degradation of most intracellular proteins. { 'prōd·ē·ə,sōm }

protective coloration [ZOO] A color pattern that blends with the environment and increases the animal's probability of survival. { prə'tek·tiv ,kəl·ə'rā·shən }

Proteeae [MICROBIO] Formerly a tribe of the Enterobacteriaceae comprising the genus *Proteus*; included organisms which were characteristically motile, fermented dextrose with gas production, and produced urease. { ,prō'tē·ē,ē }

Proteida [VERT ZOO] A suborder coextensive with Proteidae in some classification systems. { prō'tē·əd·ə }

Proteidae [VERT ZOO] A family of the amphibian suborder Salamandroidea; includes the neotenic, aquatic *Necturus* and *Proteus* species. { prō'tē·ə,dē }

protein [BIOCHEM] Any of a class of high-molecular-weight polymer compounds composed of a variety of α-amino acids joined by peptide linkages. { 'prō,tēn }

proteinase [BIOCHEM] A type of protease which acts directly on native proteins in the first step of their conversion to simpler substances. { 'pròt·ən,ās }

protein-bound iodine [BIOCHEM] Iodine bound to blood protein. Abbreviated PBI. { 'prō,tēn ,baůnd 'ī·ə,dīn }

protein-bound iodine test [PATH] A test of thyroid function that reflects the level of circulating thyroid hormone by determination of the level of protein-bound iodine in the blood. Abbreviated PBI test. { 'prō,tēn ,baůnd 'ī·ə,dīn ,test }

protein coat *See* capsid. { 'prō,tēn ,kōt }

protein engineering [CELL MOL] The design and construction of new proteins or enzymes with novel or desired functions by modifying amino acid sequences by using recombinant deoxyribonucleic acid technology. { 'prō,tēn ,en·jə'nir·iŋ }

protein kinase [BIOCHEM] An enzyme that exerts regulatory effects on growth and malignant transformation by phosphorylating proteins. { ¦prō,tēn 'kī,nās }

protein kinase A *See* cyclic AMP-dependent protein kinase. { ¦prō,tēn ¦kī,nās 'ā }

protein kinase C [CELL MOL] A Ca^{2+}-dependent enzyme mediating the phosphorylation of certain cellular proteins, thereby regulating important physiological functions, such as cell growth, ion channel activity, secretion, and synaptic transmission. Abbreviated PKC. { ¦prō ,tēn ¦kī,nās 'sē }

protein translocator [CELL MOL] A protein that mediates the transport of other proteins across the membrane of an organelle such as the endoplasmic reticulum, nucleus, mitochondria, and chloroplast. { ¦prō,tēn tranz'lō,kād·ər }

proteinuria [MED] The presence of protein in the urine. { ,prōt·ən'ůr·ē·ə }

Proteocephalidae [INV ZOO] A family of tapeworms in the order Proteocephaloidea in which the reproductive organs are within the central mesenchyme of the segment. { ,prōd·ē·ō·sə 'fal·ə,dē }

Proteocephaloidea [INV ZOO] An order of tapeworms of the subclass Cestoda in which the holdfast organ bears four suckers and, frequently, a suckerlike apical organ. { ,prōd·ē·ō,sef·ə'lòid·ē·ə }

proteoglycan [BIOCHEM] A high-molecular-weight polyanionic substance covalently linked by numerous heteropolysaccharide side chains to a polypeptide chain backbone. { ,prōd·ē·ō'glī·kən }

proteolysin [BIOCHEM] A lysin that produces proteolysis. { ,prōd·ē'äl·ə·sən }

proteolysis [BIOCHEM] Fragmentation of a protein molecule by addition of water to the peptide bonds. { ,prōd·ē'äl·ə·səs }

proteolytic enzyme [BIOCHEM] Any enzyme that catalyzes the breakdown of protein. { ¦prōd·ē·ə¦lid·ik 'en,zīm }

proteome [CELL MOL] The complete set of proteins present in the various cells of an organism. { 'prōd·ē,ōm }

proteomics [CELL MOL] The separation, identification, and characterization of the complete set of proteins present in the various cells of an organism. { ,prōd·ē'äm·iks }

Proteomyxida [INV ZOO] The single order of the Proteomyxidia. { ¦prōd·ē·ə'mik·səd·ə }

Proteomyxidia [INV ZOO] A subclass of Actinopodea including protozoan organisms which lack protective coverings or skeletal elements and have reticulopodia, or filopodia. { ¦prōd·ē·ə·mik'sid·ē·ə }

proteoplast [CELL MOL] A type of cell plastid containing crystalline, fibrillar, or amorphous masses of protein. { 'prōd·ē·ə,plast }

proteose [BIOCHEM] One of a group of derived proteins intermediate between native proteins and peptones; soluble in water, not coagulable by heat, but precipitated by saturation with ammonium or zinc sulfate. { 'prōd·ē,ōs }

Proterostomia [ZOO] That part of the animal kingdom in which cleavage of the egg is of the determinate type; includes all bilateral phyla except Echinodermata, Chaetognatha, Pogonophora, Hemichordata, and Chordata. { ,präd·ə·rō'stō·mē·ə }

Proteutheria [VERT ZOO] A group of primatelike insectivores that contains the living tree shrews. { ,prōd·ē·yü'thir·ē·ə }

prothallium [BOT] The gametophyte of a pteridophyte in the form of a flat green thallus with thizoids. { prō'thal·ē·əm }

prothoracic gland [INV ZOO] One of the paired glands in the prothorax of insects which produce ecdysone. { ¦prō·thə'ras·ik 'gland }

prothorax [INV ZOO] The first thoracic segment of an insect; bears the first pair of legs. { prō'thȯr,aks }

prothrombin [BIOCHEM] An inactive plasma protein precursor of thrombin. Also known as factor II; thrombinogen. { prō'thräm·bən }

prothrombin factor *See* vitamin K. { prō'thräm·bən ,fak·tər }

prothrombin time [PATH] A one-stage clotting test based on the time required for clotting to occur after the addition of tissue thromboplastin and calcium to decalcified plasma. { prō 'thräm·bən ,tīm }

prothymocyte [IMMUNOL] A T-cell precursor. { prō'thī·mə,sīt }

proticity [BIOCHEM] In oxidative phosphorylation, the flowing of protons in the proton circuit from high to low protic potential. { prō'tis·əd·ē }

Protista [BIOL] A proposed kingdom to include all unicellular organisms lacking a definite cellular arrangement, such as bacteria, algae, diatoms, and fungi. { prə'tis·tə }

Protoariciinae [INV ZOO] A subfamily of polychaete annelids in the family Orbiniidae. { ,prōd·ō,ar·ə'sī·ə,nē }

Protobranchia [INV ZOO] A small and primitive order in the class Bivalvia; the hinge is taxodont in all but one family, there is a central ligament pit, and the anterior and posterior adductor muscles are nearly equal in size. { ,prōd·ō'braŋ·kē·ə }

Protochordata [INV ZOO] The equivalent name for Hemichordata. { ¦prōd·ō·kȯr'dad·ə }

Protococcaceae [BOT] A monogeneric family of green algae in the suborder Ulotrichineae in which reproduction is entirely vegetative. { ,prōd·ō·käk'sās·ē,ē }

Protococcida [INV ZOO] A small order of the protozoan subclass Coccidia; all are invertebrate parasites, and only sexual reproduction is known. { ,prōd·ō'käk·səd·ə }

Protocucujidae [INV ZOO] A small family of coleopteran insects in the superfamily Cucujoidea found in Chile and Australia. { ,prōd·ō·kə'kü·yə,dē }

protoderm *See* dermatogen. { 'prōd·ə,dərm }

Protodrilidae [INV ZOO] A family of annelids belonging to the Archiannelida. { ,prōd·ō'dril·ə,dē }

protogyny [PHYSIO] A condition in hermaphroditic or dioecious organisms in which the female reproductive structures mature before the male structures. { prō'täj·ə·nē }

Protomastigida [INV ZOO] The equivalent name for Kinetoplastida. { ,prōd·ō·ma'stij·ə·də }

protomer [BIOCHEM] One of the polypeptide chains composing an oligomeric protein. Also known as subunit. { 'prōd·ə·mər }

Protomonadina [INV ZOO] An order of flagellates, subclass Mastigophora, with one or two flagella, including many species showing protoplasmic collars ringing the base of the flagellum. { ,prōd·ō,män·ə'dī·nə }

Protomonida [INV ZOO] The equivalent name for Protomonadina. { ,prōd·ō'män·ə·də }

Protomyzostomidae [INV ZOO] A family of parasitic polychaetes belonging to the Myzostomaria and known for three species from Japan and the Murman Sea. { ,prōd·ō,mī·zə'stäm·ə,dē }

protonema [BOT] A green, filamentous structure that originates from an asexual spore of mosses and some liverworts and that gives rise by budding to a mature plant. { ,prōt·ən'ē·mə }

protonephridium [INV ZOO] **1.** A primitive excretory tube in many invertebrates. **2.** The duct of a flame cell. { ,prōd·ō·nə'frid·ē·əm }

proto-oncogene [GEN] The normal-functioning precursor of an oncogene; mutation of this gene to produce an oncogene causes excessive activity of the gene product, leading to cancer. { ¦prō·dō 'äŋ·kōjēn }

protophloem [BOT] The initial primary phloem developed from the procambium. { ¦prōd·ə'flō·əm }

Protophyta [BOT] A division of the plant kingdom, according to one system of classification, set up to include the bacteria, the blue-green algae, and the viruses. { prə'täf·əd·ə }

protoplasm [CELL MOL] The colloidal complex of protein that composes the living material of a cell. { 'prōd·ə,plaz·əm }

protoplast [CELL MOL] **1.** The living portion of a cell considered as a unit; includes the cytoplasm, the nucleus, and the plasma membrane. **2.** Plant, fungal, or bacterial cell that has had its cell wall removed. { 'prōd·ə,plast }

protoplast fusion [GEN] A technique by which two protoplasts are fused to form hybrid cells that can grow into mature hybrid organisms; usually performed on plants. { 'prōd·ə,plast ,fyü·zhən }

protopodite [INV ZOO] The basal segment of a

crustacean limb bearing an endopodite or exopodite, or both, at its distal extremity. { prə 'täp·ə,dīt }

Protospondyli [VERT ZOO] An equivalent name for Semionotiformes. { ,prōd·ō'spän·də,lī }

Protostomia [INV ZOO] A major division of bilateral animals; includes most worms, arthropods, and mollusks. { ,prōd·ə'stō·mē·ə }

Prototheria [VERT ZOO] A small subclass of Mammalia represented by a single order, the Monotremata. { ¦prōd·ō'thir·ē·ə }

prototroch [INV ZOO] The band of cilia characteristic of a trochophore larva. { 'prōd·ə,träk }

prototroph [MICROBIO] A microorganism that has the ability to synthesize all of its amino acids, nucleic acids, vitamins, and other cellular constituents from inorganic nutrients. { 'prōd·ə,träf }

prototrophic [MICROBIO] Pertaining to bacteria with the nutritional properties of the wild type, or the strains found in nature. { ¦prōd·ō¦träf·ik }

Prototrupoidea [INV ZOO] A superfamily of the Hymenoptera. { ,prōd·ō·trə'pöid·ē·ə }

protoxin [BIOCHEM] A chemical compound that is a precursor to a toxin. { ¦prō¦täk·sən }

protoxylem [BOT] The part of the primary xylem that differentiates from the procambium and is formed during elongation of an embryonic plant organ. { ¦prōd·ō'zī·ləm }

Protozoa [INV ZOO] A diverse phylum of eukaryotic microorganisms; the structure varies from a simple uninucleate protoplast to colonial forms, the body is either naked or covered by a test, locomotion is by means of pseudopodia or cilia or flagella, there is a tendency toward universal symmetry in floating species and radial symmetry in sessile types, and nutrition may be phagotrophic or autotrophic or saprozoic. { ¦prōd·ə¦zō·ə }

protozoology [INV ZOO] That branch of biology which deals with the Protozoa. { ¦prōd·ō·zō'äl·ə·jē }

Protrachaeta [INV ZOO] The equivalent name for Onychophora. { prō·trə'kēd·ə }

protrypsin *See* trypsinogen. { prō'trip·sən }

Protura [INV ZOO] An order of primitive wingless insects belonging to the subclass Apterygota; individuals are elongate and eyeless, lack antennae, and are from pale amber to white in color; anamorphosis is characteristic of the group. { prə'tyu̇r·ə }

proventriculus [INV ZOO] **1.** A sac anterior to the gizzard in earthworms. **2.** A dilation of the foregut anterior to the midgut of Mandibulata. [VERT ZOO] The true stomach of a bird, usually separated from the gizzard by a constriction. { ,prō·vən'trik·yə·ləs }

provirus [VIROL] The phage genome. { prō 'vī·rəs }

provitamin [BIOCHEM] A vitamin precursor; assumes vitamin activity upon activation or chemical change. { prō'vīd·ə·mən }

proximal [ANAT] Near the body or the median line of the body. { 'präk·sə·məl }

proximal convoluted tubule [ANAT] The convoluted portion of the vertebrate nephron lying between Bowman's capsule and the loop of Henle; functions in the resorption of sugar, sodium and chloride ions, and water. { 'präk·sə·məl ,kän·və'lüd·əd 'tüb,yül }

proximoceptor [PHYSIO] An exteroceptor involved in taste or cutaneous sensations. { ¦präk·sə·mō¦sep·tər }

pruritus [MED] Localized or generalized itch due to irritation of sensory nerve endings. { prü'rīd·əs }

PSA *See* prostate specific antigen.

psalterium *See* omasum. { söl'tir·ē·əm }

Psammettidae [INV ZOO] A family of Psamminida, with a strongly built test, haphazardly arranged xenophyae, no specialized surface layer, and no large openings in the test. { sə'med·ə,dē }

Psamminida [INV ZOO] An order of Xenophyophorea distinguished by the absence of linellae in the test and, in general, rigidity of the body. { ,sam·ə'nī·də }

Psamminidae [INV ZOO] A family of Psamminida, with a solid, sometimes fragile test and external xenophyae arranged in a distinct surface layer. { sə'min·ə,dē }

Psammodrilidae [INV ZOO] A small family of spioniform worms belonging to the Sedentaria. { ,sam·ə'dril·ə,dē }

psammoma [MED] A tumor, usually a meningioma, which contains psammoma bodies. { sa 'mäm·ə }

psammoma bodies [PATH] Laminated calcific spherical structures found in certain benign and malignant tumors. { sa'mäm·ə ,bäd·ēz }

psammomatus papilloma *See* serous cystadenoma. { sa'mäm·əd·əs ,pap·ə'lō·mə }

psammon [ECOL] **1.** In a body of fresh water, that part of the environment composed of a sandy beach and bottom lakeward from the water line. **2.** Organisms which inhabit the interstitial water in the sands on a lake shore. { 'sa,män }

psammophilic [ECOL] Pertaining to an organism found in sand. { ¦sam·ə¦fil·ik }

psammophyte [ECOL] Thriving (as a plant) on sandy soil. { 'sam·ə,fīt }

psammosere [ECOL] Stages in plant succession which begin in sandy soil. { 'sam·ə,sir }

Pselaphidae [INV ZOO] The ant-loving beetles, a large family of coleopteran insects in the superfamily Staphylinoidea. { sə'laf·ə,dē }

Psephenidae [INV ZOO] The water penny beetles, a small family of coleopteran insects in the superfamily Dryopoidea. { sə'fen·ə,dē }

Pseudaliidae [INV ZOO] A family of roundworms belonging to the Strongyloidea which occur as parasites of whales and porpoises. { ,süd·ə·də'lī·ə,dē }

pseudaposematic [ECOL] Pertaining to an imitation in coloration or form by an organism of another organism that possesses dangerous or disagreeable characteristics. { ¦süd·ə,pōz·ə 'mad·ik }

pseudoalleles [GEN] Closely linked genes of

similar function that can be separated from each other by crossing over. { ˌsü·dō·əˈlēlz }

pseudoautosomal [GEN] Pertaining to segments of the X and Y chromosomes that undergo obligatory crossing-over so that they show on autosomal pattern of inheritance instead of the typical x- or y-linked pattern. { ¦süd·ōˌȯd·ōˈsō·məl }

pseudocentrum [VERT ZOO] A centrum formed by fusion of the dorsal or dorsal and ventral arcualia, as in tailed amphibians. { ¦sü·dō ˈsen·trəm }

pseudocercus *See* urogomphus. { ¦sü·dōˈsər·kəs }

pseudocoele [INV ZOO] A space between the body wall and internal organs that is not formed by gastrulation and lacks a cellular lining. { ˈsüd·əˌsēl }

pseudocoelocyte [INV ZOO] In nematodes, a mesenchymal cell in the pseudocoelom. { ¦sü·dōˈsel·əˌsīt }

Pseudocoelomata [INV ZOO] A group comprising the animal phyla Entoprocta, Aschelminthes, and Acanthocephala; characterized by a pseudocoelom. { ¦sü·dōˌsē·ləˈmäd·ə }

pseudocolumella [INV ZOO] In anthozoans, a type of axial structure. { ¦sü·dōˌkal·yəˈmel·ə }

Pseudocycnidae [INV ZOO] A family of the Caligoida which comprises external parasites on the gills of various fishes. { ¦sü·dōˈsik·nəˌdē }

pseudocyesis [MED] A condition characterized by amenorrhea, enlargement of the abdomen, and other symptoms simulating gestation, due to an emotional disorder. { ¦sü·dō·sīˈē·səs }

Pseudodiadematidae [INV ZOO] A family of Jurassic and Cretaceous echinoderms in the order Phymosomatoida which had perforate crenulate tubercles. { ¦sü·dōˌdī·ə·dəˈmad·əˌdē }

pseudogamy *See* pseudomixis. { süˈdäg·ə·mē }

pseudogene [GEN] A sequence of deoxyribonucleic acid resembling but not functioning like a gene; usually produced by gene duplication followed by mutations that alter or abolish function. { ˈsü·dōˌjēn }

pseudoglanders [VET MED] An infectious bacterial lymphangitis of horses and other equines caused by *Corynebacterium pseudotuberculosis* and characterized by ulcerating nodules of the lymph nodes in the legs. { ¦sü·dōˈglan·dərz }

pseudohermaphroditism [PHYSIO] A condition in humans which simulates hermaphroditism, with gynandry in females and androgyny in males. { ¦sü·dō·hərˈmaf·rə·dəˌtiz·əm }

pseudomixis [BIOL] Formation of an embryo from the fusion of vegetative cells instead of gametes. Also known as pseudogamy; somatogamy. { ¦süd·ōˈmik·səs }

Pseudomonadaceae [MICROBIO] A family of gram-negative, aerobic, rod-shaped bacteria; cells are straight or curved and motile by polar flagella. { süˌdäm·ə·nəˈdās·ēˌē }

Pseudomonadales [MICROBIO] Formerly an order of ovoid, rod-shaped, comma-shaped, or spiral bacteria in the class Schizomycetes; cells characterized as rigid and motile by means of polar flagella. { süˌdäm·ə·nəˈdā·lēz }

Pseudomonadineae [MICROBIO] Formerly a suborder of bacteria of the order Pseudomonadales including those families whose cells lacked photosynthetic pigments. { süˌdäm·ə·nəˈdī·nēˌē }

Pseudomonas [MICROBIO] A genus of gram-negative, motile, non-spore-forming, rod-shaped bacteria that cause a variety of infectious diseases in animals and humans (such as glanders and melioidosis) and in plants. { ˌsüd·əˈmōn·əs }

Pseudomonas aeruginosa [MICROBIO] An opportunistic pathogen that is the most significant cause of hospital-acquired infections, particularly in predisposed patients with metabolic, hematologic, and malignant diseases. It produces toxic factors such as lipase, esterase, lecithinase, elastase, and endotoxin, some of which may contribute to its pathogenesis. { ˌsüd·əˌmōn·əs ˌar·ə·jəˈnō·sə }

Pseudomonas mallei [MICROBIO] A mammalian parasite that is the causative agent of glanders, an infectious disease of horses that is occasionally transmitted to humans by direct contact. { ˌsüd·əˌmōn·əs ˈmal·ēˌī }

Pseudomonas pseudomallei [MICROBIO] A bacteria that is the causative agent of melioidosis, an endemic glanders-like disease of humans and animals that occurs most frequently in southeastern Asia and northern Australia. { ˌsüd·əˌmōn·əs ˌsüd·əˈmal·ē·ī }

pseudomucinous cystadenoma [MED] A benign ovarian tumor composed of columnar mucin-producing cells lining multilocular cysts filled with mucinous material. { ˌsüd·əˈmyüs·ən·əs ¦sist·ad·ənˈō·mə }

pseudoparalysis [MED] An apparent motor paralysis that is caused by voluntary inhibition of motor impulses because of pain or other organic or psychic causes. { ¦sü·dō·pəˈral·ə·səs }

pseudoparasite [MED] Something in the blood that is mistaken for a parasite. { ¦sü·dōˈpar·əˌsīt }

pseudoparkinsonism [MED] A reversible syndrome resembling parkinsonism that may result from the dopamine-blocking action of antipsychotic drugs. Also known as drug-induced parkinsonism. { ˌsüd·ōˈpär·kən·səˌniz·əm }

Pseudophoracea [INV ZOO] An extinct superfamily of gastropod mollusks in the order Aspidobranchia. { ˌsüd·ə·fəˈrā·shə }

Pseudophyllidea [INV ZOO] An order of tapeworms of the subclass Cestoda, parasitic principally in the intestine of cold-blooded vertebrates. { ˌsüd·ə·fəˈlid·ē·ə }

pseudoplasmodium [INV ZOO] An aggregate of amebas resembling a plasmodium. { ¦sü·dō·plazˈmō·dē·əm }

pseudoplastic [BIOL] Referring to an organism which lacks the capacity for major modification or for evolutionary differentiation. { ¦sü·dōˈplas·tik }

pseudopodium [BOT] A slender, leafless branch

of the gametophyte in certain Bryatae. [CELL MOL] Temporary projection of the protoplast of ameboid cells in which cytoplasm streams actively during extension and withdrawal. [INV ZOO] Foot of a rotifer. { ˌsüd·ə'pōd·ē·əm }

Pseudoscorpionida [INV ZOO] An order of terrestrial Arachnida having the general appearance of miniature scorpions without the postabdomen and sting. { ¦sü·dōˌskȯr·pē'än·ə·də }

Pseudosphaeriales [BOT] An order of the class Ascolichenes, shared by the class Ascomycetes; the ascocarp is flask-shaped and lined with a layer of interwoven, branched pseudoparaphyses. { ¦sü·dōˌsfir·ē'ā·lēz }

Pseudosporidae [INV ZOO] A family of the protozoan subclass Proteomyxidia; flagellated stages invade Volvocidae and filamentous algae and become amebas. { ¦sü·dō'spȯr·əˌdē }

pseudostem [BOT] A false stem composed of concentric rolled or folded blades and sheaths that surround the growing point. { 'süd·ōˌstem }

pseudostratified epithelium [HISTOL] A type of epithelium in which all cells reach to the basement membrane but some extend toward the surface only part way, while others reach the surface. { ¦sü·dō'strad·əˌfīd ˌep·ə'thē·lē·əm }

Pseudothelphusidae [INV ZOO] A family of fresh-water crabs belonging to the Brachyura. { ¦sü·dō·thel'fyüz·əˌdē }

Pseudotriakidae [VERT ZOO] The false catsharks, a family of galeoids in the carcharinid line. { ¦sü·dō·trī'ak·əˌdē }

pseudotuberculosis [MED] A bacterial infection in humans and many animals caused by *Pasteurella pseudotuberculosis*; may be severe in humans with septicemia and symptoms resembling typhoid fever. { ¦sü·dō·təˌbər·kyə'lō·səs }

pseudo-wild type [GEN] A wild phenotype due to a second (suppressor) mutation of a mutant. { ¦sü·dō' wīld ˌtīp }

psilate [BOT] Lacking ornamentation; generally applied to pollen. { 'sīˌlāt }

Psilidae [INV ZOO] The rust flies, a family of myodarian cyclorrhaphous dipteran insects in the subsection Acalyptratae. { 'sil·əˌdē }

Psilopsida [BOT] A subdivision of the Tracheophyta. { sī'läp·səd·ə }

Psilorhynchidae [VERT ZOO] A small family of actinopterygian fishes belonging to the Cyprinoidei. { ˌsī·lō'riŋ·kəˌdē }

psilosis [MED] Falling out of the hair. { sī'lō·səs }

Psilotales [BOT] The equivalent name for Psilotophyta. { ˌsī·lō'tā·lēz }

Psilotatae [BOT] A class of the Psilotophyta. { sī'läd·əˌdē }

Psilotophyta [BOT] A division of the plant kingdom represented by three living species; the life cycle is typical of the vascular cryptogams. { ˌsī·lō'täf·əd·ə }

Psittacidae [VERT ZOO] The single family of the Psittaciformes. { sə'tas·əˌdē }

Psittaciformes [VERT ZOO] The parrots, a monofamilial order of birds that exhibit zygodactylism and have a strong hooked bill. { səˌtas·ə'fȯrˌmēz }

psittacosis [MED] Pneumonia and generalized infection of man and of birds caused by agents of the PLT-Bedsonia group; transmitted to humans by psittacine birds. { ˌsid·ə'kō·səs }

psoas [ANAT] Either of two muscles: psoas major which arises from the bodies and transverse processes of the lumbar vertebrae and is inserted into the lesser trochanter of the femur, and psoas minor which arises from the bodies and transverse processes of the lumbar vertebrae and is inserted on the pubis. { 'sō·əs }

Psocoptera [INV ZOO] An order of small insects in which wings may be present or absent, tarsi are two- or three-segmented, cerci are absent, and metamorphosis is gradual. { sō'käp·tə·rə }

Psolidae [INV ZOO] A family of echinoderms in the order Dendrochirotida characterized by a ventral adhesive sucker and a U-shaped gut, with the mouth and anus opening upward on the adoral surface. { 'säl·əˌdē }

Psophiidae [VERT ZOO] The trumpeters, a family of birds in the order Gruiformes. { sō'fil·əˌdē }

psoriasis [MED] A usually chronic, often acute inflammatory skin disease of unknown cause; characterized by dull red, well-defined lesions covered by silvery scales which when removed disclose tiny capillary bleeding points. { sə'rī·ə·səs }

psychiatric genetics [GEN] The study of the genetic causes and modes of inheritence that underlie the generally recognized mental illnesses. { ˌsī·kē'a·trik jə'ned·iks }

psychiatrist [MED] A person who specializes in psychiatry; a licensed physician trained in psychiatry. { sə'kī·ə·trəst }

psychiatry [MED] The medical science that deals with the origins, diagnosis, and treatment of mental and emotional disorders. { sə'kī·ə·trē }

Psychidae [INV ZOO] The bagworms, a family of lepidopteran insects in the superfamily Tineoidea; males are large, hairy moths, but females are degenerate, wingless, and legless and live in bag-shaped cases. { 'sī·kəˌdē }

Psychodidae [INV ZOO] The moth flies, a family of orthorrhaphous dipteran insects in the series Nematocera. { sī'käd·əˌdē }

psychogalvanic reflex [PHYSIO] A variation in the electric conductivity of the skin in response to emotional stimuli, which cause changes in blood circulation, secretion of sweat, and skin temperature. { ¦sī·kō·gal'van·ik 'rēˌfleks }

psychogenic [MED] Of psychic origin. { ¦sī·kō¦jen·ik }

psychology [BIOL] **1.** The science that deals with the functions of the mind and the behavior of an organism in relation to its environment. **2.** The mental activity characteristic of a person or a situation. { sī'käl·ə·jē }

psychosomatic [MED] Of or pertaining to the

interrelationship between mental processes and somatic functions. { ¦sī·kō·sə'mad·ik }

psychosurgery [MED] The branch of medicine that deals with the treatment of various psychoses, severe neuroses, and chronic painful conditions by means of operative procedures on the brain. { ¦sī·kō'sər·jə·rē }

psychrophile [BIOL] An organism that thrives at low temperatures. { 'sī·krə,fīl }

psychrophilic [PHYSIO] Relating to the ability to live and grow at low temperatures. { ,sī·krə'fil·ik }

psychrophobia [PHYSIO] Abnormal sensitivity to cold. { ,sī·krə'fō·bē·ə }

psychrophyte [ECOL] A plant adapted to the climatic conditions of the arctic or alpine regions. { 'sī·krə,fīt }

Psyllidae [INV ZOO] The jumping plant lice, a family of the Homoptera in the series Sternorrhyncha in which adults have a transverse head with protuberant eyes and three ocelli, 6- to 10-segmented antennae, and wings with reduced but conspicuous venation. { 'sil·ə,dē }

PTA *See* factor XI.

ptarmigan [VERT ZOO] Any of various birds of the genus *Lagopus* in the family Tetraonidae; during the winter, plumage is white and hairlike feathers cover the feet. { 'tar·mə·gən }

PTC *See* Christmas factor.

Pteraspidomorphi [VERT ZOO] The equivalent name for Diplorhina. { tə,ras·pə·də'mȯr,fī }

Pterasteridae [INV ZOO] A family of deep-water echinoderms in the order Spinulosida distinguished by having webbed spine fins. { ,ter·ə'ster·ə,dē }

Pteridophyta [BOT] The equivalent name for Polypodiophyta. { ,ter·ə'däf·əd·ə }

Pteriidae [INV ZOO] Pearl oysters, a family of bivalve mollusks which have nacreous shells. { tə'rī·ə,dē }

pterinophore [CELL MOL] A yellow to orange chromatophore that contains pterine pigment. { tə'rin·ə,fōr }

Pterobranchia [INV ZOO] A group of small or microscopic marine animals regarded as a class of the Hemichordata; all are sessile, tubicolous organisms with a U-shaped gut and three body segments. { ,ter·ə'braŋ·kē·ə }

pterochore [BOT] A type of plant that produces winged disseminules. { 'ter·ə,kȯr }

Pteroclidae [VERT ZOO] The sandgrouse, a family of gramnivorous birds in the order Columbiformes; mainly an Afro-Asian group resembling pigeons and characterized by cryptic coloration, usually corresponding with the soil color of the habitat. { tə'räk·lə,dē }

pteroic acid [BIOCHEM] $C_{14}H_{12}N_6O_3$ A crystalline amino acid formed by hydrolysis of folic acid or other pteroylglutamic acids. { tə'rō·ik 'as·əd }

Pteromalidae [INV ZOO] A family of hymenopteran insects in the superfamily Chalcidoidea. { ,ter·ə'mal·ə,dē }

Pteromedusae [INV ZOO] A suborder of hydrozoan cnidarians in the order Trachylina characterized by a modified, bipyramidal medusae. { ,ter·ə·mə'dü,sē }

Pterophoridae [INV ZOO] The plume moths, a family of the lepidopteran superfamily Pyralidoidea in which the wings are divided into featherlike plumes, maxillary palpi are lacking, and the legs are long. { ,ter·ə'fȯr·ə,dē }

Pteropidae [INV ZOO] The fruit bats, a large family of the Chiroptera found in Asia, Australia, and Africa. { tə'räp·ə,dē }

Pteropoda [INV ZOO] The sea butterflies, an order of pelagic gastropod mollusks in the subclass Opisthobranchia in which the foot is modified into a pair of large fins and the shell, when present, is thin and glasslike. { tə'räp·ə·də }

Pteropodidae [VERT ZOO] A family of fruit-eating bats in the suborder Megachiroptera, characterized by primitive ears and by shoulder joints. { ,ter·ə'päd·ə,dē }

Pteropsida [BOT] A large group of vascular plants characterized by having parenchymatous leaf gaps in the stele, and leaves which are thought to have originated in the distant past as branched stem systems. { tə'räp·səd·ə }

pterostigma [INV ZOO] Opaque thickened spot occurring on the costal margin of an insect wing. { ,ter·ə'stig·mə }

pteroylglutamic acid *See* folic acid. { ¦ter·ə·wəl·glü'tam·ik 'as·əd }

pterygium [MED] **1.** A triangular mass of mucous membrane growing on the conjunctiva, usually near the inner canthus. **2.** Overgrowth of the cuticle forward on the nail. [VERT ZOO] A generalized vertebrate limb. { tə'rij·ē·əm }

pterygoid bone [VERT ZOO] A rodlike bone or group of bones forming a portion of the palatoquadrate arch in lower vertebrates. { 'ter·ə,gȯid }

pterygopalatine fossa [ANAT] The gap between the pterygoid process of the sphenoid bone and the maxilla and palatine bone. { ¦ter·ə·gō'pal·ə,tēn 'fäs·ə }

pterygoquadrate [EMBRYO] Of, pertaining to, or being the first branchial arch in lower vertebrate embryos; gives rise to most of the upper jaw. { ¦ter·ə·gō'kwä,drāt }

Ptiliidae [INV ZOO] The feather-winged beetles, a family of coleopteran insects in the superfamily Staphylinoidea. { 'til·ə,dē }

Ptilodactylidae [INV ZOO] The toed-winged beetles, a family of the Coleoptera in the superfamily Dryopoidea. { ¦til·ō·dak'til·ə,dē }

Ptinidae [INV ZOO] The spider beetles, a family of coleopteran insects in the superfamily Bostrichoidea. { 'tin·ə,dē }

ptosis [MED] Prolapse, abnormal depression, or falling down of an organ or part; applied especially to drooping of the upper eyelid, from paralysis of the third cranial nerve. { 'tō·səs }

p27 [CELL MOL] A cyclin-dependent kinase inhibitor in normal and neoplastic cells. { ,pē ,twen·tē 'sev·ən }

ptyalase *See* ptyalin. { 'tī·ə,lās }

ptyalin [BIOCHEM] A diastatic enzyme found in saliva which catalyzes the hydrolysis of starch to dextrin, maltose, and glucose, and the hydrolysis of sucrose to glucose and fructose. Also known as ptyalase; salivary amylase; salivary diastase. { 'tī·ə·lən }

Ptychodactiaria [INV ZOO] An order of the zoantharian anthozoans of the phylum Cnidaria known only from two genera, *Ptychodactis* and *Dactylanthus*. { ¦tī·kō,dak·tē'ar·ē·ə }

Ptychomniaceae [BOT] A family of mosses in the order Isobryales distinguished by an eight-ribbed capsule. { tī,käm·nē'ās·ē,ē }

puberty [PHYSIO] The period at which the generative organs become capable of exercising the function of reproduction; signalized in the boy by a change of voice and discharge of semen, in the girl by the appearance of the menses. { 'pyü·bərd·ē }

puberulent [BOT] Having a surface covered with very fine downlike hairs. { pyü'ber·yə·lənt }

puberulic acid [BIOCHEM] $(HO)_3(C_7H_2O)COOH$ A keto acid formed as a metabolic product of certain species of *Penicillium*; has some germicidal activity against gram-positive bacteria. { pyü'bər·yə·lik 'as·əd }

pubic arch [ANAT] The arch formed by the conjoined rami of the pubis and ischium. { 'pyü·bik 'ärch }

pubic crest [ANAT] The crest extending from the pubic tubercle to the medial extremity of the pubis. { 'pyü·bik 'krest }

pubic symphysis [ANAT] The fibrocartilaginous union of the pubic bones. Also known as symphysis pubis. { 'pyü·bik 'sim·fə·səs }

pubis [ANAT] The pubic bone, the portion of the hipbone forming the front of the pelvis. { 'pyü·bəs }

public health [MED] **1.** The state of health of a community or of a population. **2.** The art and science dealing with the protection and improvement of community health. { 'pəb·lik 'helth }

public health dentistry [MED] The science of promoting dental health through community effort, it comprises research, education, prevention, diagnosis, prescription, and the evaluation of community dental care. { ,pəb·lik ,helth 'dent·ə·strē }

Puccinia asparagi [MYCOL] An autoecious fungus of the order Uredinales; the causative agent of asparagus rust. { pü,sin·ē·ə ə'spär·ə,gē }

Puccinia graminis [MYCOL] A macrocylic heteroecious fungus of the order Uredinales; the causative agent of black stem rust of cereal grains. { pü,sin·ē·ə 'gram·ə·nəs }

Puccinia malvacearum [MYCOL] A microcyclic fungus of the order Uredinales; the causative agent of hollyhock rust. { pü,sin·ē·ə ,mal·və'sē·ə·rəm }

puerperal sepsis [MED] A toxic condition caused by infection in the birth canal, occurring as a complication or sequel of pregnancy. { pyü'ər·prəl 'sep·səs }

puerperium [MED] **1.** The state of a woman during labor or immediately after delivery. **2.** The period from delivery to the time when the uterus returns to normal size, about 6 weeks. { ,pyü·ər'pir·ē·əm }

puff ball [BOT] A spherical basidiocarp that retains spores until fully mature and, when disturbed, releases them as puffs of fine dust. { 'pəf ,bȯl }

pullorum disease [VET MED] A highly fatal disease of chickens and other birds caused by *Salmonella pullorum*, characterized by weakness, lassitude, lack of appetite, and whitish or yellowish diarrhea. Also known as bacillary white diarrhea; white diarrhea. { pə'lȯr·əm di,zēz }

pulmonary anthrax [MED] A form of anthrax in humans caused by the inhalation of dust containing *Bacillus anthracis* spores. { 'pül·mə,ner·ē 'an,thraks }

pulmonary artery [ANAT] A large artery that conducts venous blood from the heart to the lungs of tetrapods. { 'pu̇l·mə,ner·ē 'ärd·ə·rē }

pulmonary circulation [PHYSIO] The circulation of blood through the lungs for the purpose of oxygenation and the release of carbon dioxide. Also known as lesser circulation. { 'pu̇l·mə,ner·ē ,sər·kyə'lā·shən }

pulmonary embolism [MED] Obstruction of the pulmonary artery or a branch of it by a free-floating blood clot (embolus) usually originating from a vein in the leg or pelvic area. { ,pu̇l·mə,ner·ē 'em·bə,liz·əm }

pulmonary edema [MED] An effusion of fluid into the alveoli and interstitial spaces of the lungs. { 'pu̇l·mə,ner·ē i'dē·mə }

pulmonary plexus [NEUROSCI] A nerve plexus composed chiefly of vagal fibers situated on the anterior and posterior aspects of the bronchi and accompanying them into the substance of the lung. { 'pu̇l·mə,ner·ē 'plek·səs }

pulmonary stenosis [MED] Narrowing of the orifice of the pulmonary artery. { 'pu̇l·mə,ner·ē stə'nō·səs }

pulmonary valve [ANAT] A valve consisting of three semilunar cusps situated between the right ventricle and the pulmonary trunk. { 'pu̇l·mə,ner·ē 'valv }

pulmonary vein [ANAT] A large vein that conducts oxygenated blood from the lungs to the heart in tetrapods. { 'pu̇l·mə,ner·ē 'vān }

pulmonary ventilation [PHYSIO] The volume of gas entering and exiting the lungs per unit time of respiration. { 'pu̇l·mə,ner·ē ,vent·əl'ā·shən }

Pulmonata [INV ZOO] A subclass of the gastropod mollusks which contains the "lung"-bearing snails; the gills have been lost and in their place the mantle cavity has become a pulmonary sac. { ,pu̇l·mə'näd·ə }

pulmonate [VERT ZOO] Possessing lungs or lunglike organs. { 'pu̇l·mə,nāt }

pulp [ANAT] A mass of soft spongy tissue in the interior of an organ. [BOT] The soft succulent portion of a fruit. { pəlp }

pulp cavity [ANAT] The space within the central part of a tooth containing the dermal pulp and made up of the pulp chamber and a root canal. { 'pəlp ˌkav·əd·ē }

pulp chamber [ANAT] The coronal portion of the central cavity of a tooth. { 'pəlp ˌchām·bər }

pulpotomy [MED] Surgical removal of the pulp of a tooth. { pəl'päd·ə·mē }

pulsation [PHYSIO] A beating or throbbing, usually rhythmic, as of the heart or an artery. { pəl'sā·shən }

pulse [PHYSIO] **1.** The regular, recurrent, palpable wave of arterial distention due to the pressure of the blood ejected with each contraction of the heart. **2.** A single wave. { pəls }

pulsed-gel electrophoresis [CELL MOL] A technique in which rare cutting restriction enzymes are used to generate very large fragments of deoxyribonucleic acid (up to 1 million base pairs long), which are separated in gels by applying alternating cycles of electric fields in different directions. { ¦pəlst,jel ¦ˌlek·trə·fə'rē·səs }

pulse pressure [PHYSIO] The difference between the systolic and diastolic blood pressure. { 'pəls ˌpresh·ər }

pulse rate [PHYSIO] The number of pulsations of an artery per minute. { 'pəls ˌrāt }

pulse wave [PHYSIO] A wave of increased pressure over the arterial system, started by contraction of septum and valves in the heart. { 'pəls ˌwāv }

pulvillus [INV ZOO] A small cushion or cushionlike pad, often covered with short hairs, on an insect's foot between the claws of the last segment. { ˌpəl'vil·əs }

pulvinus [BOT] A cushionlike enlargement of the base of a petiole which functions in turgor movements of leaves. { ˌpəl'vī·nəs }

puma [VERT ZOO] *Felis concolor.* A large, tawny brown wild cat (family Felidae) once widespread over most of the Americas. Also known as American lion; catamount; cougar; mountain lion. { 'pü·mə }

pumpkin [BOT] Any of several prickly vines with large lobed leaves and yellow flowers in the genus Cucurbita of the order Violales; the fruit is orange-colored and large, with a firm rind. { 'pəm·kən }

puna [ECOL] An alpine biological community in the central portion of the Andes Mountains of South America characterized by low-growing, widely spaced plants that lack much green color most of the year. { 'pü·nə }

punctate [BIOL] Dotted; full of minute points. { 'pəŋk,tāt }

punctuated equilibria [EVOL] In the fossil record, long periods of little change in lineages interspersed with brief periods of relatively rapid change.

punctuated evolution *See* punctuated equilibrium. { ¦pəŋk·chə,wād·əd ˌev·ə'lü·shən }

pupa [INV ZOO] The quiescent, intermediate form assumed by an insect that undergoes complete metamorphosis; it follows the larva and precedes the adult stages and is enclosed in a hardened cuticle or a cocoon. { 'pyü·pə }

pupate [INV ZOO] **1.** To develop into a pupa. **2.** To pass through a pupal stage. { 'pyü,pāt }

pupil [ANAT] The contractile opening in the iris of the vertebrate eye. { 'pyü·pəl }

pupillary reflex [PHYSIO] **1.** Contraction of the pupil in response to stimulation of the retina by light. Also known as Whytt's reflex. **2.** Contraction of the pupil on accommodation for close vision, and dilation of the pupil on accommodation for distant vision. **3.** Contraction of the pupil on attempted closure of the eye. Also known as Westphal-Pilcz reflex; Westphal's pupillary reflex. { 'pyü·pə,ler·ē 'rē,fleks }

Pupipara [INV ZOO] A section of cyclorrhaphous dipteran insects in the Schizophora series in which the young are born as mature maggots ready to become pupae. { pyü'pip·ə·rə }

pure culture [MICROBIO] A culture that contains cells of one kind, all progeny of a single cell. { 'pyu̇r ¦kəl·chər }

purging [MED] The condition in which there is rapid and continuous evacuation of the bowels. { 'pərj·iŋ }

purine [BIOCHEM] A heterocyclic compound containing fused pyrimidine and imidazole rings; adenine and guanine are the purine components of nucleic acids and coenzymes. { 'pyu̇,rēn }

Purkinje cell [HISTOL] Any of the cells of the cerebral cortex with large, flask-shaped bodies forming a single cell layer between the molecular and granular layers. { pər'kin·jē ˌsel }

Purkinje effect [PHYSIO] When illumination is reduced to a low level, slowly enough to allow adaptation by the eye, the sensation produced by the longer-wave stimuli (red, orange) decreases more rapidly than that produced by shorter-wave stimuli (violet, blue). Also spelled Parkinje effect. { pər'kin·jē i,fekt }

Purkinje fibers [HISTOL] Modified cardiac muscle fibers composing the terminal portion of the conducting system of the heart. { pər'kin·jē ˌfī·bərz }

puromycin [MICROBIO] $C_{22}H_{29}O_5N_7$ A colorless, crystalline broad-spectrum antibiotic produced by a strain of *Streptomyces.* { ¦pu̇r·ə¦mīs·ən }

purple bacteria [MICROBIO] Any of various photosynthetic bacteria that contain bacteriochlorophyll, distinguished by purplish or reddish-brown pigments. { 'pər·pəl bak'tir·ē·ə }

purple nonsulfur bacteria [MICROBIO] Any of various purple photosynthetic bacteria, especially members of the family Athiorhodaceae, that utilize organic hydrogen donor compounds. { 'pər·pəl ¦nän¦səl·fər bak'tir·ē·ə }

purple sulfur bacteria [MICROBIO] Any of various anaerobic photosynthetic purple bacteria, especially in the family Thiorhodaceae, that utilize H_2S and other inorganic sulfur compounds as a source of hydrogen, while the carbon source can be carbon monoxide. { 'pər·pəl 'səl·fər bak,tir·ē·ə }

purpura [MED] Spontaneous hemorrhages into tissues such as joints, skin, and mucosal surfaces. { 'pər·pyə·rə }

purulent [MED] Consisting of, containing, or forming pus. { 'pyu̇r·ə·lənt }

purupuru *See* pinta. { ¦pu̇r·ü¦pu̇r·ü }

pus [MED] A viscous, creamy, pale-yellow or yellow-green fluid produced by liquefaction necrosis in a neutrophil-rich exudate. { pəs }

pustule [MED] A small, circumscribed, pus-filled elevation on the skin. { 'pəs·chül }

pusule [INV ZOO] A noncontractile fluid-filled vacuole emptied by means of a duct; found in dinoflagellates. { 'pəs·yül }

putrefaction [BIOCHEM] Decomposition of organic matter, particularly the anaerobic breakdown of proteins by bacteria, with the production of foul-smelling compounds. { ˌpyü·trə 'fak·shən }

PVY *See* potato virus Y.

pyarthrosis [MED] Suppuration involving a joint. { ¦pī·är'thrō·səs }

pycnidiospore [MYCOL] A conidium produced by a pycnidium. { pik'nid·ē·ōˌspȯr }

pycnidium [MYCOL] A cavity that bears pycnidiospores in certain fungi. { pik'nid·ē·əm }

pycniospore [MYCOL] A haploid spore of a rust fungus that fuses with a haploid hypha of opposite sex to produce dikaryotic aeciospores. { 'pik·nē·əˌspȯr }

pycnium [MYCOL] A flask-shaped fruit body of a rust fungus formed in clusters just beneath the surface of a host tissue. { 'pik·nē·əm }

Pycnogonida [INV ZOO] The sea spiders, a subphylum of marine arthropods in which the body is reduced to a series of cylindrical trunk somites supporting the appendage. { ˌpik·nə'gän·əd·ə }

Pycnogonidae [INV ZOO] A family of the Pycnogonida lacking both chelifores and palpi and having six to nine jointed ovigers in the male only. { ˌpik·nə'gän·əˌdē }

pyelitis [MED] Inflammation of the renal pelvis. { ˌpī·ə'līd·əs }

pyelolithotomy [MED] Excision of a renal calculus through an incision in the renal pelvis. { ¦pī·ə·lō·li'thäd·ə·mē }

pyelonephritis [MED] The disease process resulting from the effects of infections of the parenchyma and the pelvis of the kidney. Also known as interstitial nephritis. { ¦pī·ə·lō·ne'frīd·əs }

pyemia [MED] A disease state due to the presence of pyogenic microorganisms in the blood and the formation, wherever these organisms lodge, of embolic or metastatic abscesses. { pī'ē·mē·ə }

pygidium [INV ZOO] **1.** A caudal shield on the abdomen of some Arthropoda. **2.** The terminal body segment of many invertebrates. { pī'jid·ē·əm }

Pygopodidae [VERT ZOO] The flap-footed lizards, a family of the suborder Sauria. { ˌpī·gə'päd·əˌdē }

pygostyle [VERT ZOO] A specialized bone in birds which is formed by a number of fused tail vertebrae. { 'pīg·əˌstīl }

pyknosis [PATH] The polymerization and contraction of the nuclear chromosomal components. { pik'nō·səs }

pylephlebitis [MED] Inflammation of the portal vein. { ¦pī·lə·flə'bīd·əs }

pylome [INV ZOO] An aperture for emission of pseudopodia and intake of food in some Sarcodina. { 'pīˌlōm }

pyloric caecum [INV ZOO] **1.** One of the tubular pouches that open into the ventriculus of an insect. **2.** One of the paired tubes having lateral glandular diverticula in each ray of a starfish. [VERT ZOO] One of the tubular pouches that open from the pyloric end of the stomach into the alimentary canal of most fishes. { pə'lȯr·ik 'sē·kəm }

pyloric sphincter [ANAT] The thickened ring of circular smooth muscle at the lower end of the pyloric canal of the stomach. { pə'lȯr·ik 'sfiŋk·tər }

pyloric stenosis [MED] Obstruction of the pyloric opening of the stomach due to hypertrophy of the pyloric sphincter. { pə'lȯr·ik stə'nō·səs }

pylorospasm [MED] Spasm of the pylorus. { pī'lȯr·əˌspaz·əm }

pylorus [ANAT] The orifice of the stomach communicating with the small intestine. { pə'lȯr·əs }

pyobacillosis [VET MED] A bacterial infection of sheep, swine, or rarely cattle caused by *Corynebacterium pyogenes*; usually marked by abscess formation, but in sheep takes the form of chronic purulent pneumonia. { ¦pī·ōˌbas·ə'lō·səs }

pyocyanin [MICROBIO] $C_{13}H_{10}N_{20}$ An antibiotic substance forming blue crystals, produced by *Pseudomonas aeruginosa*; active against many bacteria and fungi. { ˌpī·ō'sī·ə·nən }

pyoderma [MED] Any pus-producing skin lesion or lesions, used in reference to groups of furuncles, pustules, or even carbuncles. { ˌpī·ō'dər·mə }

pyogenesis [MED] The formation of pus. { ¦pī·ō'jen·ə·səs }

pyonephritis [MED] Suppurative inflammation of a kidney. { ¦pī·ō·ne'frīd·əs }

pyonephrosis [MED] Replacement of renal tissue by abscesses. { ¦pī·ō·nē'frō·səs }

pyorrhea [MED] A purulent discharge. { ˌpī·ə'rē·ə }

pyosalpinx [MED] A reproductive system disorder in which the Fallopian tubes are distended with pus. { ˌpī·ə'salˌpiŋks }

Pyralidae [INV ZOO] The equivalent name for Pyralididae. { pə'ral·əˌdē }

Pyralididae [INV ZOO] A large family of moths in the lepidopteran superfamily Pyralidoidea; the labial palpi are well developed, and the legs are usually long and slender. { ˌpir·ə'lid·əˌdē }

Pyralidinae [INV ZOO] A subfamily of the Pyralididae. { ˌpir·ə'lin·əˌnē }

Pyralidoidea [INV ZOO] A superfamily of the Lepidoptera belonging to the Heteroneura and including long-legged, slender-bodied moths with well-developed maxillary palpi. { ˌpir·ə·lə'dȯid·ē·ə }

pyramidal area *See* motor area. { ¦pir·ə¦mid·əl *or* pə'ram·ə·dəl 'er·ē·ə }

pyramidal system [ANAT] The corticospinal and corticobulbar tracts. { ¦pir·ə¦mid·əl 'sis·təm }

Pyramidellidae [INV ZOO] A family of gastropod mollusks in the order Tectibranchia; the operculum is present in this group. { pə,ram·ə'del·ə,dē }

pyramid of numbers [ECOL] The concept that an organism making up the base of a food chain is numerically abundant while each succeeding member of the chain is represented by successively fewer individuals; uses feeding relationship as a basis for the quantitative analysis of an ecological system. { 'pir·ə,mid əv 'nəm·bərz }

pyranose [BIOCHEM] A sugar whose cyclic or ring structure resembles that of pyran. { 'pī·rə,nōs }

pyranoside [BIOCHEM] A glycoside whose cyclic sugar component resembles that of pyran. { pī'ran·ə,sīd }

Pyraustinae [INV ZOO] A large subfamily of the Pyralididae containing relatively large, economically important moths. { pə'rós·tə,nē }

pyrazinamide [PHARM] $C_5H_5N_3O$ A crystalline compound with a melting point of 189–191°C; used as a drug in the treatment of tuberculosis. { ,pir·ə'zin·ə,mīd }

pyrenoid [BOT] A colorless body found within the chromatophore of certain algae; a center for starch formation and storage. { 'pir·ə,nòid }

Pyrenolichenes [BOT] The equivalent name for Pyrenulales. { ,pī·rə·nō·lī'kē·nēz }

Pyrenomycetes [MYCOL] The largest class in the subdivision Ascomycotina, distinguished by a single-walled ascus and the coiled branches that form on the hyphae to initiate ascocarp formation. Also known as perithecial ascomycetes. { pī,rē·nō·mī'sēd,ēz }

Pyrenulaceae [BOT] A family of the Pyrenulales; all species are crustose and most common on tree bark in the tropics. { pī,ren·yə'lās·ē,ē }

Pyrenulales [BOT] An order of the class Ascolichenes including only those lichens with perithecia that contain true paraphyses and unitunicate asci. { pī,ren·yə'lā·lēz }

pyrexia [MED] Elevation of temperature above the normal; fever. { pī'rek·sē·ə }

Pyrgotidae [INV ZOO] A family of myodarian cyclorrhaphous dipteran insects in the subsection Acalyptratae. { pər'gäd·ə,dē }

2,3-pyridinedicarboxylic acid [BIOCHEM] $C_7H_5NO_4$ An odorless, crystalline compound with a melting point of 190°C; soluble in water; inhibits glucose synthesis. Also known as quinolinic acid. { ¦tü ¦thrē ¦pir·ə,dēn,kär·bäk'sil·ik 'as·əd }

pyridostigmine bromide [PHARM] $C_9H_{13}O_2NBr$ A white, crystalline powder with a melting point of 154–157°C; soluble in water, alcohol, and chloroform; used in medicine. { ¦pir·ə·dō'stig,mēn 'brō,mīd }

pyridoxal hydrochloride *See* pyridoxine hydrochloride. { ,pir·ə'däk·səl ,hī·drə'klór,īd }

pyridoxal phosphate *See* codecarboxylase. { ,pir·ə'däk·səl 'fä,sfāt }

pyridoxine hydrochloride [BIOCHEM] $C_8H_{11}NO_3·HCl$ A crystalline compound, decomposing at about 208°C; used in medicine in vitamin therapy. Also known as pyridoxal hydrochloride; vitamin B_6 hydrochloride. { ,pir·ə'däk,sēn ,hī·drə'klór,īd }

pyrimidine [BIOCHEM] $C_4H_4N_2$ A heterocyclic organic compound containing nitrogen atoms at positions 1 and 3; naturally occurring derivatives are components of nucleic acids and coenzymes. { pə'rim·ə,dēn }

pyrogen [BIOCHEM] A group of substances thought to be polysaccharides of microbial origin that produce an increase in body temperature when injected into humans and some animals. { 'pī·rə,jən }

pyrophosphatase [BIOCHEM] An enzyme catalyzing hydrolysis of esters containing two or more molecules of phosphoric acid to form a simpler phosphate ester. { ,pī·rō'fä·sfə,tās }

Pyrosomida [INV ZOO] An order of pelagic tunicates in the class Thaliacea in which species form tubular swimming colonies and are often highly luminescent. { ,pī·rə'säm·əd·ə }

Pyrrhocoridae [INV ZOO] A family of hemipteran insects belonging to the superfamily Pyrrhocoroidea. { ,pir·ə'kór·ə,dē }

Pyrrhocoroidea [INV ZOO] A superfamily of the Pentatomorpha. { ,pir·ə·kə'ròid·ē·ə }

Pyrrhophyta [BOT] A small division of motile, generally unicellular flagellate algae characterized by the presence of yellowish-green to golden-brown plastids and by the general absence of cell walls. { pə'räf·əd·ə }

pyrrobutamine phosphate [PHARM] $C_{20}H_{22}NCl·2H_3PO_4$ A cream-colored, water-soluble powder with a melting range of 127–131°C; used in medicine. { ¦pi·rō'byüd·ə,mēn 'fä,sfāt }

pyruvate [BIOCHEM] Salt of pyruvic acid, such as sodium pyruvate, $NaOOCCOCH_3$. { pī'rü,vāt }

pyruvic acid [BIOCHEM] Important intermediate in protein and carbohydrate metabolism; liquid with acetic-acid aroma; melts at 11.8°C; miscible with alcohol, ether, and water; used in biochemical research. { pī'rü·vik 'as·əd }

pyruvic carboxylase [BIOCHEM] An enzyme found in yeast, bacteria, and plants that catalyzes the decarboxylation of pyruvate to acetaldehyde and carbon dioxide. { pī,rü·vik kär'bäk·sə,lās }

pyruvic decarboxylase [BIOCHEM] An enzyme found in yeast that, along with thiamine pyrophosphate and magnesium ions, catalyzes the removal of a carboxyl group from pyruvic acid to produce acetaldehyde and carbon dioxide. { pī,rü·vik ,dē·kär'bäk·sə,lās }

Pythidae [INV ZOO] An equivalent name for Salpingidae. { 'pith·ə,dē }

python [VERT ZOO] The common name for members of the reptilian subfamily Pythoninae. { 'pī,thän }

Pythoninae [VERT ZOO] A subfamily of the rep-

Q

Q fever [MED] An acute, febrile infectious disease of humans, characterized by sudden onset and patchy pneumonitis, and caused by a bacterialike organism, *Coxiella burneti*. { 'kyü ˌfē·vər }

QRS complex [MED] The electrocardiographic deflection representing ventricular depolarization; the initial downward deflection is termed a Q wave; the initial upward deflection, an R wave; and the downward deflection following the R wave, an S wave. Also known as ventricular depolarization complex. { ¦kyü¦är'es ˌkäm ˌpleks }

quadrant [ANAT] One of the four regions into which the abdomen may be divided for purposes of physical diagnosis. [PHYSIO] A sector of one-fourth of the field of vision of one or both eyes. { 'kwä·drənt }

quadrate bone [VERT ZOO] A small element forming part of the upper jaw joint on each side of the head in vertebrates below mammals. { 'kwä'drāt ˌbōn }

quadratojugal [VERT ZOO] A small bone connecting the quadrate and jugal bones on each side of the skull in many lower vertebrates. { kwä¦drā·dō'jü·gəl }

quadriceps [ANAT] Four-headed, as a muscle. { 'kwä·drəˌseps }

quadriceps femoris [ANAT] The large extensor muscle of the thigh, combining the rectus femoris and vastus muscles. { 'kwä·drəˌseps 'fem·ə·rəs }

quadrigeminal body *See* corpora quadrigemina. { ¦kwä·drə¦jem·ə·nəl 'bäd·ē }

quadrilocular [BIOL] Having four cells or cavities. { ˌkwä·drə'läk·yə·lər }

quadriplegia [MED] Paralysis affecting the four extremities of the body; may be spastic or flaccid. { ˌkwä·drə'plē·jə }

quadruped [VERT ZOO] An animal that has four legs. { 'kwä·drəˌped }

quagmire *See* bog. { 'kwägˌmīr }

quail [VERT ZOO] Any of several migratory game birds in the family Phasianidae. { kwāl }

quantasome [CELL MOL] One of the highly ordered array of units that has a "cobblestone" appearance in electron micrographs of the lamella of chloroplasts, and thought to be the most probable site of the light reaction in photosynthesis. { 'kwän·təˌsōm }

quantitative genetics [GEN] The study of continuously varying traits, such as height or milk yield. { 'kwän·ə·tād·iv jə'ned·iks }

quantitative inheritance [GEN] The acquisition of characteristics which show a quantitative and continuous type of variation. { 'kwän·ə·tād·iv in'her·əd·əns }

quantitative structure-activity relationships [BIOCHEM] The establishment of statistical correlations between the potencies of a series of structurally related compounds and one or more quantitative structural parameters, such as lipophilicity, polarity, and molecular size, by using multilinear regression analysis. { 'kwän·əˌtād·iv ¦strək·chər ak¦tiv·əd·ē ri'lā·shənˌships }

quantitative trait [GEN] A trait that is under the control of many factors, both genetic and environmental, each of which contributes only a small amount to the total variability of the trait. { ˌkwänt·əˌtād·iv 'trāt }

quantitative trait locus [GEN] The location of a gene that affects a quantitative trait. { ˌkwänt·əˌtād·iv ¦trāt 'lō·kəs }

quantum evolution [EVOL] A special but extreme case of phyletic evolution; the rapid evolution that takes place when relatively sudden and drastic change occurs in the environment or when organisms spread into new habitats where conditions differ from those to which they are adapted; the organisms must then adapt quickly to the new conditions if they are to survive. { 'kwän·təm ˌev·ə'lü·shən }

quarantine [MED] Limitation of freedom of movement of susceptible individuals who have been exposed to communicable disease, for a period of time equal to the incubation period of the disease. { 'kwär·ənˌtēn }

quebracho [BOT] Any of a number of South American trees in different genera in the order Sapindales, but all being a valuable source of wood, bark, and tannin. { kā'bra·chō }

quebracho bark [BOT] Bark of the white quebracho tree *Aspidosperma quebracho* of Chile and Argentina; main components are aspidospermine, tannin, and quebrachine; used in medicine and tanning. { kā'bra·chō ˌbärk }

queen [INV ZOO] A mature, fertile female in a colony of ants, bees, or termites, whose function is to lay eggs. { kwēn }

Queensland tick typhus [MED] A benign infectious disease of humans found in rural northeastern Australia, caused by the bacterialike microorganism *Rickettsia australis* and presumed to be carried by the tick *Ixodes holocyclus*. { 'kwēnz·lənd ¦tik 'tī·fəs }

quellung reaction [MICROBIO] Swelling of the capsule of a bacterial cell, caused by contact with serum containing antibodies capable of reacting with polysaccharide material in the capsule; applicable to *Pneumococcus*, *Klebsiella*, and *Hemophilus*. { 'kwel·əŋ rē,ak·shən }

quenching [IMMUNOL] An adaptation of immunofluorescence that uses two fluorochromes, one of which absorbs light emitted by the other; one fluorochrome labels that antigen, another the antibody, and the antigen-antibody complexes retain both; the initially emitted light is absorbed and so quenched by the second compound. { 'kwench·iŋ }

quercetin [BIOCHEM] $C_{15}H_5O_2(OH)_5$ A yellow, crystalline flavonol obtained from oak bark and Douglas-fir bark; used as an antioxidant and absorber of ultraviolet rays, and in rubber, plastics, and vegetable oils. { 'kwer·sə·tən }

quercite *See* quercitol. { 'kwer,sīt }

quercitol [PHARM] $C_6H_7(OH)_5$ Colorless, water-soluble, sweet-tasting crystals with a melting point of 234°C; used in medicine. Also known as acorn sugar; quercite. { 'kwer·sə,tȯl }

quiescence [CELL MOL] A period in which a cell is not increasing its mass or going through the cell cycle. { kwē'es·əns }

quiescent [MED] Inactive, latent, or dormant, referring to a disease or pathological process. { kwē'es·ənt }

quiescent center [BOT] A central region (usually hemispherical) of many root stems that consists of cells which rarely divide or synthesize deoxyribonucleic acid and have less ribonucleic acid and protein than adjacent cells. { kwē'es·ənt ¦sen·tər }

quill [VERT ZOO] The hollow, horny shaft of a large stiff wing or tail feather. { kwil }

quillwort [BOT] The common name for plants of the genus *Isoetes*. { 'kwil,wȯrt }

quinacrine [PHARM] $C_{23}H_{30}ClN_3O$ Formerly an important antimalarial drug but now used in the treatment of giardiasis, tapeworm infections, amebiasis, and a variety of other conditions. { 'kwin·ə·krən }

quinalbarbitone *See* secobarbital. { ,kwin·əl'bär·bə,tōn }

quinaphthol [PHARM] $C_{40}H_{40}N_2O_{10}S_2$ A yellow, crystalline powder with a melting point of 185–186°C; used in medicine. { kwi'naf,thȯl }

quince [BOT] *Cydonia oblonga*. A deciduous tree of the order Rosales characterized by crooked branching, leaves that are densely hairy on the underside and solitary white or pale-pink flowers; fruit is an edible pear- or apple-shaped tomentose pome. { kwins }

quinoa [BOT] *Chenopodium quinoa*. An annual herb of the family Chenopodiaceae grown at high altitudes in South America for the highly nutritious seeds. { kwi'nō·ə }

quinoprotein [BIOCHEM] A member of a class of proteins that uses pyrroloquinoline quinone as a cofactor. { ,kwin·ə'prō,tēn }

quinquefoliate [BOT] Of a leaf, having five leaflets. { ,kwin·kə'fō·lē,āt }

quintuplet [BIOL] One of five children who have been born at one birth. { kwin'təp·lət }

quitiqua *See* pinta. { 'kēt·ə,kä }

R

rabbit [VERT ZOO] Any of a large number of burrowing mammals in the family Leporidae. { 'rab·ət }

rabies [VET MED] An acute, encephalitic viral infection transmitted to humans by the bite of a rabid animal. Also known as hydrophobia. { 'rā·bēz }

raccoon [VERT ZOO] Any of 16 species of carnivorous nocturnal mammals belonging to the family Procyonidae; all are arboreal or semiarboreal and have a bushy, long ringed tail. { ra'kün }

race [BIOL] **1.** An infraspecific taxonomic group of organisms, such as subspecies or microspecies. **2.** A fixed variety or breed. { rās }

racemase [BIOCHEM] Any of group of enzymes that catalyze racemization reactions. { 'ras·ə,mās }

raceme [BOT] An inflorescence on which flowers are borne on stalks of equal length on an unbranched main stalk that continues to grow during flowering. { rā'sēm }

racemose [ANAT] Of a gland, compound and shaped like a bunch of grapes, with freely branching ducts that terminate in acini. [BOT] Bearing, or occurring in the form of, a raceme. { 'ras·ə,mōs }

rachiglossate radula [INV ZOO] A radula of certain gastropod mollusks which has one or three longitudinal series of teeth, each of which may bear many cusps. { ¦rā·kə'glä,sāt 'raj·ə·lə }

rachilla [BOT] The axis of a grass spikelet. { rə'kil·ə }

rachis [ANAT] The vertebral column. [BIOL] An axial structure such as the axis of an inflorescence, the central petiole of a compound leaf, or the central cord of an ovary in Nematoda. { 'rā·kəs }

radial artery [ANAT] A branch of the brachial artery in the forearm; principal branches are the radial recurrent and the main artery of the thumb. { 'rād·ē·əl 'ärd·ə·rē }

radial canal [INV ZOO] **1.** One of the numerous canals that radiate from the spongocoel in certain Porifera. **2.** Any of the canals extending from the coelenteron to the circular canal in the margin of the umbrella in jellyfishes. **3.** A canal radiating from the circumoral canal along each ambulacral area in many echinoderms. { 'rād·ē·əl kə'nal }

radial cleavage [EMBRYO] A cleavage pattern characterized by formation of a mass of cells that show radial symmetry. { 'rād·ē·əl 'klē·vij }

radial deviation [BIOPHYS] A position of the human hand in which the wrist is bent toward the thumb. { ¦rād·ē·əl ,dē,·vē'ā·shən }

radial loop [FOREN] A loop fingerprint pattern which flows in the direction of the radius bone. { ,rād·ē·əl 'lüp }

radial nerve [NEUROSCI] A large nerve that arises in the brachial plexus and branches to enervate the extensor muscles and skin of the posterior aspect of the arm, forearm, and hand. { 'rād·ē·əl 'nərv }

Radiata [INV ZOO] Members of the Eumetazoa which have a primary radial symmetry; includes the Cnidaria and Ctenophora. { ,rād·ē'äd·ə }

radiation biochemistry [BIOCHEM] The study of the response of the constituents of living matter to radiation. { ,rād·ē'ā·shən ¦bī·ō'kem·ə·strē }

radiation biology *See* radiobiology. { ,rād·ē'ā·shən bī'äl·ə·jē }

radiation biophysics [BIOPHYS] The study of the response of organisms to ionizing radiations and to ultraviolet light. { ,rād·ē'ā·shən ¦bī·ō'fiz·iks }

radiation burn [MED] A burn caused by overexposure to radiant energy. { ,rād·ē'ā·shən ,bərn }

radiation cataract *See* irradiation cataract. { ,rād·ē'ā·shən 'kad·ə,rakt }

radiation cytology [CELL MOL] An aspect of biology that deals with the effects of radiations on living cells. { ,rād·ē'ā·shən sī'täl·ə·jē }

radiation dermatitis *See* radiodermatitis. { ,rād·ē'ā·shən ,dər·mə'tīd·əs }

radiation effects [BIOL] The harmful effects of ionizing radiation on humans and other animals, such as production of cancers, cataracts, and radiation ulcers, loss of hair, reddening of skin, sterilization, nausea, vomiting, mucous or bloody diarrhea, purpura, epilation, and agranulocytic infections. { ,rād·ē'ā·shən i,feks }

radiation hazard [MED] Health hazard arising from exposure to ionizing radiation. { ,rād·ē'ā·shən ,haz·ərd }

radiation hybrid panel [GEN] A group of interspecific hybrid cell lines each containing a different set of very tiny fragments of the deoxyribonucleic acid of one parental species; those used in mapping human genes are produced by massive radiation of human cells before hybridizing them with rodent cells. { ˌrā·dē'ā·shən ¦hī·brəd ˌpan·əl }

radiation lobe *See* lobe. { ˌrād·ē'ā·shən ˌlōb }

radiation microbiology [MICROBIO] A field of basic and applied radiobiology concerned chiefly with the damaging effects of radiation on microorganisms. { ˌrād·ē'ā·shən ¦mī·krō·bī'äl·ə·jē }

radiation sickness [MED] **1.** Illness, usually manifested by nausea and vomiting, resulting from the effects of therapeutic doses of radiation. **2.** Radiation injury following exposure to excessive doses of radiation, such as the explosion of an atomic bomb. { ˌrād·ē'ā·shən ˌsik·nəs }

radiation therapy [MED] The use of ionizing radiation or radioactive substances to treat disease. Also known as actinotherapy; radiotherapy. { ˌrād·ē'ā·shən 'ther·ə·pē }

radical [BOT] **1.** Of, pertaining to, or proceeding from the root. **2.** Arising from the base of a stem or from an underground stem. { 'rad·ə·kəl }

radical mastectomy [MED] Surgical removal of the breast, subcutaneous fat, muscle, lymph glands, and a wide area of skin for cancer. { 'rad·ə·kəl ¦ma'stek·tə·mē }

radicle [BOT] The embryonic root of a flowering plant. { 'rad·ə·kəl }

radicotomy *See* rhizotomy. { ˌrad·ə'käd·ə·mē }

radicular cyst [MED] A cyst arising from a granuloma of the root of a tooth. Also known as periapical cyst. { rə'dik·yə·lər 'sist }

radiculitis [MED] Inflammation of a nerve root. { rə¦dik·yə'līd·əs }

radiculoneuropathy [MED] Disease of the peripheral spinal nerves and their roots. { ra¦dik·yə·lō·nu̇'räp·ə·thē }

radiobioassay [BIOL] The analysis of the kind, concentration, and location of radioactive material in the human body by direct measurement (in vivo counting) or the evaluation of materials removed (or excreted). { ˌrād·ē·ōˌbī·o'asˌā }

radiobiology [BIOL] Study of the scientific principles, mechanisms, and effects of the interaction of ionizing radiation with living matter. Also known as radiation biology. { 'rād·ē·ō·bī 'äl·ə·jē }

radiocardiogram [MED] An x-ray recording of the variation with time of the concentration of a radioisotope in a heart chamber; usually the radioactive material is injected intravenously. { ¦rad·ē·ō'kärd·ē·əˌgram }

radiodermatitis [MED] Degenerative changes in the skin following excessive exposure to ionizing radiation. Also known as radiation dermatitis. { ¦rad·ē·ōˌdər·mə'tīd·əs }

radioecology [ECOL] The interdisciplinary study of organisms, radionuclides, ionizing radiation, and the environment. { ¦rād·ē·ō·ē'käl·ə·jē }

radiohumeral joint [ANAT] The joint in the elbow between the radius and the humerus bones. { ¦rād·ē·ō'hyüm·ə·rəl 'jȯint }

radioimmunoassay [IMMUNOL] A sensitive method for determining the concentration of an antigenic substance in a sample by comparing its inhibitory effect on the binding of a radioactivity-labeled antigen to a limited amount of a specific antibody with the inhibitory effect of known standards. { ¦rād·ē·ō¦im·yə·nō'aˌsā }

radio knife [MED] A surgical knife that uses a high-frequency electric arc at its tip to cut tissue and simultaneously sterilize the edges of the wound. { 'rād·ē·ō ˌnīf }

Radiolaria [INV ZOO] A subclass of the protozoan class Actinopodea whose members are noted for their siliceous skeletons and characterized by a membranous capsule which separates the outer from the inner cytoplasm. { ˌrād·ē·ō 'lar·ē·ə }

radiole [INV ZOO] A spine on a sea urchin. { 'rād·ē·ōl }

radiologist [MED] A physician who specializes in the use of radiant energy in the diagnosis and treatment of disease. { ˌrād·ē'äl·ə·jəst }

radiology [MED] The medical science concerned with radioactive substances, x-rays, and other ionizing radiations, and the application of the principles of this science to diagnosis and treatment of disease. { ˌrād·ē'äl·ə·jē }

radiomimetic activity [BIOL] The radiationlike effects of certain chemicals, such as nitrogen mustard, urethane, and fluorinated pyrimidines. { ¦rād·ē·ō·mi'med·ik ak'tiv·əd·ē }

radiomutation [GEN] A chromosomal aberration which is the result of exposure of living tissue to ionizing radiation. { ¦rād·ē·ō·myü'tā·shən }

radionecrosis [PATH] Destruction of living tissue by radiation. { ¦rād·ē·ō·ne'krō·səs }

radiopaque agent [MED] A substance, such as barium sulfate, which is opaque to x-rays, administered orally to a patient to provide contrast in x-ray photographs of his gastrointestinal system. { ¦rād·ē·ō'pāk ˌā·jənt }

radiopharmaceutical [PHARM] A radioactive drug used for medicinal purposes, either diagnostic or therapeutic. { ˌrād·ē·ōˌfär·mə'süd·ə·kəl }

radioresistance [BIOL] The resistance of organisms or tissues to the harmful effects of various radiations. { ¦rād·ē·ō·ri'zis·təns }

radiotherapy *See* radiation therapy. { ¦rād·ē·ō'ther·ə·pē }

radiotoxicity [MED] A radioactive compound that is toxic to living cells or tissues, causing radiation sickness. { ˌrād·ē·ō·täk'sis·əd·ē }

radish [BOT] *Raphanus sativus*. **1.** An annual or biennial crucifer belonging to the order Capparales. **2.** The edible, thickened hypocotyl of the plant. { 'rad·ish }

radium therapy [MED] Radiotherapy using the radiations from radium. { 'rād·ē·əm ˌther·ə·pē }

radius [ANAT] The outer of the two bones of the human forearm or of the corresponding part in vertebrates other than fish. { 'rād·ē·əs }

radon breath analysis [MED] Examination of

exhaled air for the presence of radon to determine the presence and quantity of radium in the human body. { 'rā,dän 'breth ə'nal·ə·səs }

radula [INV ZOO] A filelike ribbon studded with horny or chitinous toothlike structures, found in the mouth of all classes of mollusks except Bivalvia. { 'raj·ə·lə }

raffinase [BIOCHEM] An enzyme that hydrolyzes raffinose, yielding fructose in the reaction. { 'raf·ə,nās }

raffinose [BIOCHEM] $C_{18}H_{32}O_{16}·5H_2O$ A white, crystalline trisaccharide found in sugarbeets, cottonseed meal, and molasses; yields glucose, fructose, and galactose on complete hydrolysis. Also known as gossypose; melitose; melitriose. { 'raf·ə,nōs }

Rafflesiales [BOT] A small order of dicotyledonous plants; members are highly specialized, nongreen, rootless parasites which grow from the roots of the host. { re,flē·zhē'ā·lēz }

Raillietiellidae [INV ZOO] A small family of parasitic arthropods in the order Cephalobaenida. { ,rāl·yə'tyel·ə,dē }

rain desert [ECOL] A desert in which rainfall is sufficient to maintain a sparse general vegetation. { 'rān ,dez·ərt }

Rainey's corpuscle [INV ZOO] The sickle-shaped spore of an encysted sarcosporidian. { 'rā·nēz ,kȯr·pə·səl }

rainforest [ECOL] A forest of broad-leaved, mainly evergreen, trees found in continually moist climates in the tropics, subtropics, and some parts of the temperate zones. { 'rān,fär·əst }

rain rot [VET MED] Weeping dermatitis accompanied by swelling of the skin and loss of hair in sheep exposed to prolonged periods of rain. { 'rān ,rät }

raised bog [ECOL] An area of acid, peaty soil, especially that developed from moss, in which the center is relatively higher than the margins. { 'rāzd 'bäg }

Rajidae [VERT ZOO] The skates, a family of elasmobranchs included in the batoid group. { 'raj·ə,dē }

Rajiformes [VERT ZOO] The equivalent name for Batoidea. { ,raj·ə'fȯr,mēz }

râle [MED] An abnormal sound accompanying the normal sounds of respiration within the air passages and heard on auscultation of the chest. { ral }

Rallidae [VERT ZOO] A large family of birds in the order Gruiformes comprising rails, gallinules, and coots. { 'ral·ə,dē }

ram [VERT ZOO] A male sheep or goat. { ram }

ramate [BIOL] Having branches. { 'ra,māt }

ramentum [BOT] A thin brownish scale consisting of a single layer of cells and occurring on the leaves and young shoots of many ferns. { rə'men·təm }

ramicolous [BOT] Living on twigs. { rə'mik·ə·ləs }

ramie [BOT] *Boehmeria nivea.* A shrub or half-shrub of the nettle family (Urticaceae) cultivated as a source of a tough, strong, durable, lustrous natural woody fiber resembling flax, obtained from the phloem of the plant; used for high-quality papers and fabrics. Also known as China grass; rhea. { 'rā·mē }

ramiform [BOT] Branched or comblike. { 'ram·ə,fȯrm }

Ramon flocculation test [IMMUNOL] A method of standardizing antitoxins; a toxin-antitoxin flocculation that is a precipitin reaction in which the end point is the zone of optimal proportion; that is, the zone in which there is no uncombined antigen or antibody. { rə'mōn ,fläk·yə'lā·shən ,test }

ramose [BIOL] Having lateral divisions or branches. { 'rā,mōs }

Ramphastidae [VERT ZOO] The toucans, a family of birds with large, often colorful bills in the order Piciformes. { ,ram'fas·tə,dē }

ramus [ANAT] A slender bone process branching from a large bone. [VERT ZOO] The barb of a feather. [ZOO] The branch of a structure such as a blood vessel, nerve, arthropod appendage, and so on. { 'rā·məs }

random mating [GEN] A mating system in which there is an equal opportunity for all male and female gametes to join in fertilization. { 'ran·dəm ,mād·iŋ }

range [ECOL] The area or region over which a species is distributed. { rānj }

range of motion [BIOPHYS] The degree of movement that can occur in a joint. { 'rānj əv 'mō·shən }

Ranidae [VERT ZOO] A family of frogs in the suborder Diplasiocoela including the large, widespread genus *Rana*. { 'ran·ə,dē }

RANK ligand [BIOCHEM] A local paracrine factor that originates from osteoblasts and mediates the effects of most, if not all, agents that are known to impact osteoclast development in bone. Also known as osteoprotegerin ligand. { ¦raŋk 'līg·ənd }

ranula [MED] A retention cyst of a salivary gland. { 'ran·yə·lə }

Ranunculaceae [BOT] A family of dicotyledonous herbs in the order Ranunculales distinguished by alternate leaves with net venation, two or more distinct carpels, and numerous stamens. { rə,nəŋ·kyə'lās·ē,ē }

Ranunculales [BOT] An order of dicotyledons in the subclass Magnoliidae characterized by its mostly separate carpels, triaperturate pollen, herbaceous or only secondarily woody habit, and frequently numerous stamens. { rə,nəŋ·kyə'lā·lēz }

rape [BOT] *Brassica napus.* A plant of the cabbage family in the order Capparales; the plant does not form a compact head, the leaves are bluish-green, deeply lobed, and curled, and the small flowers produce black seeds; grown for forage. { rāp }

raphania [MED] A disease thought to be due to chronic ingestion of the poison in seeds of the wild radish. { rə'fan·yə }

raphe [ANAT] A broad seamlike junction between two lateral halves of an organ or other body part. [BOT] **1.** The part of the funiculus attached along its full length to the integument of an anatropous ovule, between the chalaza and the attachment to the placenta. **2.** The longitudinal median line or slit on a diatom valve. { 'rā·fē }

Raphidae [VERT ZOO] A family of birds in the order Columbiformes that included the dodo (*Raphus calcullatus*); completely extirpated during the 17th and early 18th centuries. { 'raf·ə,dē }

raphide [BOT] One of the long, needle-shaped crystals, usually consisting of calcium oxalate, occurring as a metabolic by-product in certain plant cells. { 'rāf,īd }

rapid wasting syndrome [INV ZOO] A coral reef disease that is characterized by a rapid loss of tissue and destruction of the underlying skeleton. { ¦rap·əd 'wāst·iŋ ,sin,drōm }

raptorial [ZOO] **1.** Living on prey. **2.** Adapted for snatching or seizing prey, as birds of prey. { rap'tȯr·ē·əl }

rapture of the deep *See* nitrogen narcosis. { 'rap·chər əv thə 'dēp }

rash [MED] A lay term for nearly any skin eruption, but more commonly for acute inflammatory dermatoses. { rash }

rasorial [ZOO] Adapted for scratching for food; applied to birds. { rə'sȯr·ē·əl }

raspberry [BOT] Any of several species of upright shrubs of the genus *Rubus*, with perennial roots and prickly biennial stems, in the order Rosales; the edible black or red juicy berries are aggregate fruits, and when ripe they are easily separated from the fleshy receptacle. { 'raz ,ber·ē }

Ras protein [CELL MOL] A GTPase that is part of a signaling cascade that controls cell growth; it turns itself off by hydrolyzing guanosine triphosphate (GTP) to guanosine diphosphate (GDP); however, if left in its active (GTP-bound) form, it can lead to uncontrollable cell growth, or cancer. { ¦är¦ā'es ,prō,tēn }

rat [VERT ZOO] The name applied to over 650 species of mammals in several families of the order Rodentia; they differ from mice in being larger and in having teeth modified for gnawing. { rat }

rataria larva [INV ZOO] The second, hourglass-shaped, free-swimming larva of the siphonophore *Velella*. { rə'tar·ē·ə ,lär·və }

rat-bite fever [MED] Either of two diseases transmitted by the bite of a rat: spirillary rat-bite fever and streptobacillary fever. { 'rat ,bīt ,fē·vər }

ratfish [VERT ZOO] The common name for members of the chondrichthyan order Chimaeriformes. { 'rat,fish }

Rathke's pouch *See* craniobuccal pouch. { 'rät ,kēz ,pau̇ch }

ratites [VERT ZOO] A group of flightless, mostly large, running birds comprising several orders and including the emus, cassowaries, kiwis, and ostriches. { 'ra,tīts }

rattan [BOT] Any of several long-stemmed, climbing palms, especially of the genera *Calanius* and *Daemonothops*; stem material is used to make walking sticks, wickerwork, and cordage. { ra 'tan }

rattlesnake [VERT ZOO] Any of a number of species of the genera *Sistrurus* or *Crotalus* distinguished by the characteristic rattle on the end of the tail. { 'rad·əl,snāk }

rat typhus *See* murine typhus. { 'rat 'tī·fəs }

Raunkiaer system [BOT] A classification system for plant life-forms based on the position of perennating buds in relation to the soil surface. { 'rau̇n·kē·ir ,sis·təm }

Rauwolfia [BOT] A genus of mostly poisonous, tropical trees and shrubs of the dogbane family (Apocynaceae); certain species yield substances used as emetics and cathartics, while R. *serpentina* is a source of alkaloids used as tranquilizers. { rau̇'wu̇l·fē·ə }

ray [VERT ZOO] Any of about 350 species of the elasmobranch order Batoidea having flattened bodies with large pectoral fins attached to the side of the head, ventral gill slits, and long, spikelike tails. { rā }

rayfin fish [VERT ZOO] The common name for members of the Actinopterygii. { 'rā,fin ,fish }

ray flower [BOT] One of the small flowers with a strap-shaped corolla radiating from the margin of the head of a capitulum. { 'rā ,flau̇·ər }

ray initial [BOT] One of the cells of the cambium which divide to produce new phloem and xylem ray cells. { 'rā i,nish·əl }

Raynaud's disease [MED] A usually bilateral disease of blood vessels, especially of the extremities; excited by cold or emotion, characterized by intermittent pallor, cyanosis, and redness, and generally accompanied by pain. { rā 'nōz di,zēz }

ray shake [BOT] A radial crack in wood caused by wounds in a tree along the barrier zone. { 'rā ,shāk }

RDS *See* respiratory distress syndrome of newborn.

reaction time [PHYSIO] The interval between application of a stimulus and the beginning of the response. { rē'ak·shən ,tīm }

reaction wood [BOT] An abnormal development of a tree and therefore its wood as the result of unusual forces acting on it, such as an atypical gravitational pull. { rē'ak·shən ,wu̇d }

reading [CELL MOL] A linear process by which amino acid sequences are recognized by the protein-synthesizing system of a cell from messenger ribonucleic codes. { 'rēd·iŋ }

reading frame [CELL MOL] A nucleotide sequence that starts with an initiation codon, partitions the subsequent nucleotides into a series of amino acid-encoding triplets, and ends with a termination codon. { 'rēd·iŋ ,frām }

reading mistake [CELL MOL] The incorrect placement of one or more amino acid residues in a polypeptide chain during genetic translation. { 'rēd·iŋ mə'stāk }

readthrough [GEN] Transcription beyond a termination sequence due to failure of ribonucleic acid polymerase to recognize the termination codon. { 'rēd,thrü }

reagin [IMMUNOL] The class of immunoglobulin that mediates allergic reactions such as asthma and urticaria. Also known as IgE; reaginic antibody. { rē'ā·jən }

reaginic antibody *See* reagin. { ˌrē·əˌjin·ik 'an·təˌbäd·ē }

reassortant virus [VIROL] A virion containing deoxyribonucleic acid from one virus species and a protein coat from another. { ˌrē·ə¦sȯrt·ənt 'vī·rəs }

recapitulation theory [BIOL] The biological theory that an organism passes through developmental stages resembling various stages in the phylogeny of its group; ontogeny recapitulates phylogeny. Also known as biogenetic law; Haeckel's law. { ˌrē·kəˌpich·ə'lā·shən ˌthē·ə·rē }

recent [EVOL] Referring to taxa which still exist; the antonym of fossil. { 'rē·sənt }

receptacle [BOT] The pointed end of a pedicel or peduncle from which the flower parts grow. { ri'sep·tə·kəl }

receptive aphasia *See* sensory aphasia. { ri'sep·tiv ə'fā·zhə }

receptor [BIOCHEM] A site or structure in a cell which combines with a drug or other biological to produce a specific alteration of cell function. [PHYSIO] A sense organ. { ri'sep·tər }

recessive [GEN] **1.** An allele that is not expressed phenotypically when present in the heterozygous condition. **2.** An organism homozygous for a recessive gene. { ri'ses·iv }

reciprocal inhibition [PHYSIO] In muscular movement, the simultaneous relaxation of one muscle and the contraction of its antagonist. { ri'sip·rə·kəl ˌin·ə'bish·ən }

reciprocal recombination [GEN] In dihybrid gametes, the generation, by meiotic crossing over, of a new combination of alleles unlike those of the maternal and paternal homologues. { ri'sip·rə·kəl rēˌkäm·bə'nā·shən }

reciprocal translocation [CELL MOL] The special case of translocation in which two segments exchange positions. { ri'sip·rə·kəl ˌtranz·lō'kā·shən }

reclinate [BOT] Vernation in which the upper part of the leaf is bent down on the lower part. { 'rek·ləˌnāt }

recognition sequence [CELL MOL] A specific sequence of nucleotides at which a restriction endonuclease cleaves a deoxyribonucleic acid molecule. { ˌrek·ig'nish·ən ˌsē·kwəns }

recognition site [CELL MOL] The nucleotide sequence in duplex deoxyribonucleic acid (DNA) to which a restriction endonuclease binds initially and within which the endonuclease cuts the DNA. { ˌrek·ig'nish·ən ˌsīt }

recombinant [GEN] Any new cell, individual, or molecule that is produced in the laboratory by recombinant deoxyribonucleic acid technology or that arises naturally as a result of recombination. { rē'käm·bə·nənt }

recombinant DNA [CELL MOL] Deoxyribonucleic acid (DNA) that has been altered (caused to recombine) by rearrangement of its sequence, addition or deletion of DNA segments, or introduction of foreign DNA segments (for example, a gene from another organism introduced into an organism's genome). { ˌrē¦käm·bə·nənt ¦dē ¦en'ā }

recombinant technology [GEN] **1.** In genetic engineering, laboratory techniques used to join deoxyribonucleic acid (DNA) from different sources to produce novel DNA. Also known as gene splicing. **2.** In genetic engineering, laboratory techniques used to join ribonucleic acid (RNA) from different sources to produce novel RNA. { ri¦käm·bə·nənt tek'näl·ə·jē }

recombination [GEN] **1.** The occurrence of gene combinations in the progeny that differ from those of the parents as a result of independent assortment and crossing-over. **2.** The production of genetic information in which some elements of one line of descent are replaced or added to by those of another line. { ˌrēˌkäm·bə'nā·shən }

recombination frequency [GEN] The number of recombinants divided by the total number of progeny. { rē'käm·bə'nā·shən ˌfrē·kwən·sē }

recombination mosaic [GEN] A mosaic produced as the result of somatic crossing-over. { ˌrēˌkäm·bə'nā·shən mōˌzā·ik }

recombination repair [CELL MOL] A repair mechanism involving exchange of correct for incorrect segments between two damaged deoxyribonucleic acid molecules. { rēˌkäm·bə'nā·shən riˌper }

recombinator [CELL MOL] Any nucleotide sequence that stimulates genetic recombination at neighboring sites. { rē'käm·bəˌnād·ər }

recovery room [MED] A hospital room in which surgical patients are kept during the period immediately following an operation for care and recovery from anesthesia. { ri'kəv·ə·rē ˌrüm }

recruitment [PHYSIO] A serial discharge from neurons innervating groups of muscle fibers. { ri'krüt·mənt }

Recticornia [INV ZOO] A family of amphipod crustaceans in the superfamily Genuina containing forms in which the first antennae are straight, arise from the anterior margin of the head, and have few flagellar segments. { ˌrek·tə'kȯr·nē·ə }

rectocele [MED] Bulging or herniation of the rectum into the vagina. Also known as vaginal protocele. { 'rek·təˌsēl }

rectrix [VERT ZOO] One of the stiff tail feathers used by birds to control direction of flight. { 'rek·triks }

rectum [ANAT] The portion of the large intestine between the sigmoid flexure and the anus. { 'rek·təm }

rectus [ANAT] Having a straight course, as certain muscles. { 'rek·təs }

rectus abdominis [ANAT] The long flat muscle of the anterior abdominal wall which, as vertical fibers, arises from the pubic crest and symphysis, and is inserted into the cartilages of the fifth, sixth, and seventh ribs. { 'rek·təs ab'däm·ə·nəs }

rectus femoris [ANAT] A division of the quadriceps femoris inserting in the patella and ultimately into the tubercle of the tibia. { 'rek·təs 'fem·ə·rəs }

rectus oculi [ANAT] Any of four muscles (superior, inferior, lateral, and medial) of the eyeball, running forward from the optic foramen and inserted into the sclerotic coat. { 'rek·təs 'äk·yə·lī }

recumbent [BOT] Of or pertaining to a plant or plant part that tends to rest on the surface of the soil. { ri'kəm·bənt }

recurrent backcrossing [GEN] Repetitive sexual crossing of hybrids to one parent, used to eliminate all but the desired alleles and traits of the donor parent. { ri'kər·ənt 'bak,kròs·iŋ }

recurrent parent [GEN] In recurrent backcrossing, the parent that is crossed with the first and the subsequent generations. Also known as backcross parent. { ri'kər·ənt 'per·ənt }

recycling endosome [CELL MOL] An intracellular vesicle through which internalized receptors pass during their transport from early endosomes back to the plasma membrane for reuse. { rē¦sīk·liŋ 'en·də,sōmz }

red algae [BOT] The common name for members of the phylum Rhodophyta. { 'red 'al·jē }

red blood cell *See* erythrocyte. { 'red 'bləd ,sel }

red gum *See* eucalyptus gum. { 'red ¦gəm }

redia [INV ZOO] A larva produced within the miracidial sporocyst of certain digenetic trematodes which may give rise to daughter rediae or to cercariae. { 'rē·dē·ə }

redifferentiation [PHYSIO] The return to a position of greater specialization in actual and potential functions, or the developing of new characteristics. { rē,dif·ə,ren·chē'ā·shən }

red nucleus [HISTOL] A mass of reticular fibers in the gray matter of the tegmentum of the mesencephalon of higher vertebrates; it receives fibers from the cerebellum of the opposite side and gives rise to rubrospinal tract fibers of the opposite side. { 'red 'nü·klē·əs }

red tide [BIOL] A reddish discoloration of coastal surface waters due to concentrations of certain toxin-producing dinoflagellates. Also known as red water. { 'red 'tīd }

redtop grass [BOT] One of the bent grasses, *Agrostis alba* and its relatives, which grow on a wide variety of soils; it is a perennial, spreads slowly by rootstocks, and has top growth 2–3 feet (60–90 centimeters) tall. { 'red,täp ,gras }

reducer *See* decomposer. { ri'dü·sər }

redundancy [GEN] **1.** Repetition of a deoxyribonucleic acid sequence in a nucleus. **2.** Multiplicity of codons for individual amino acids. { ri'dən·dən·sē }

Reduviidae [INV ZOO] The single family of the hemipteran group Reduvioidea; nearly all have a stridulatory furrow on the prosternum, ocelli are generally present, and the beak is three-segmented. { ,rej·ə'vī·ə,dē }

Reduvioidea [INV ZOO] The assassin bugs or conenose bugs, a monofamilial group of hemipteran insects in the subdivision Geocorisae. { ,rej·ə,vē'òid·ē·ə }

red water [BIOL] *See* red tide. [VET MED] **1.** A babesiasis of cattle characterized by hematuria following release of hemoglobin by destruction of erythrocytes. **2.** A chronic disease of cattle attributed to oxalic acid in the forage; hematuria results from escape of blood from lesions in the bladder. **3.** An acute febrile septicemia of cattle, and sometimes horses, sheep, and swine, caused by the bacterium *Clostridium hemolyticum* and characterized by hemoglobinuria and sometimes intestinal hemorrhages. { 'red 'wòd·ər }

redwood [BOT] *Sequoia sempervirens*. An evergreen tree of the pine family; it is the tallest tree in the Americas, attaining 350 feet (107 meters); its soft heartwood is a valuable building material. { 'red,wu̇d }

reed [BOT] Any tall grass characterized by a slender jointed stem. { rēd }

Reed-Sternberg cell [PATH] An anaplastic reticuloendothelial cell constituting the characteristic microscopic feature of Hodgkin's disease. { 'rēd 'stərn,bərg ,sel }

reepithelialization [MED] **1.** Regrowth of epithelial tissue over a denuded surface. **2.** Surgical placement of a graft of epithelium over a denuded surface. { rē,ep·ə,thē·lē·ə·lə'zā·shən }

referred pain [MED] Pain felt in one area but originating in another area. { ri'fird 'pān }

refined lecithin *See* lecithin. { ri'fīnd 'les·ə·thən }

reflex [PHYSIO] An automatic response mediated by the nervous system. { 'rē,fleks }

reflex arc [NEUROSCI] A chain of neurons composing the anatomical substrate or pathway of the unconditioned reflex. { 'rē,fleks ,ärk }

reflex bladder [ANAT] A urinary bladder controlled only by the simple reflex arc through the sacral part of the spinal cord. { 'rē,fleks ,blad·ər }

reflexed [BOT] Turned abruptly backward. { 'rē,flekst }

reflex epilepsy [MED] Seizure brought on by specific sensory stimuli. { 'rē,fleks 'ep·ə,lep·sē }

reflex time [NEUROSCI] The time required for the nerve impulse to travel in a reflex action. { 'rē,fleks ,tīm }

reflux esophagitis [MED] An inflammation of the lower esophagus that results from excessive acid reflux or regurgitation from the stomach; symptoms include heartburn and indigestion. { ,rē,fləks i,säf·ə'jīd·əs }

refractory [MED] Not readily yielding to treatment. { ri'frak·trē }

refractory period [NEUROSCI] A brief period of time following the stimulation of a nerve during which the nerve will not respond to a second stimulus. { ri'frak·trē ,pir·ē·əd }

Refsum's disease [MED] A familial disorder characterized by visual disturbances, ataxia, neuritic changes, and cardiac damage, associated with high blood level of phytinic acid. { 'ref·səmz di,zēz }

refugium [ECOL] An area which has escaped the great changes which occurred in the region as a whole, often providing conditions in which relic colonies can survive; for example, a driftless area which has escaped the effects of glaciation because it projected above the ice. { rə'fyü·jē·əm }

regeneration [BIOL] The replacement by an organism of tissues or organs which have been lost or severely injured. { rē,jen·ə'rā·shən }

regeneration electrode [NEUROSCI] A silicon-based microelectrode with holes through which severed peripheral nerve fibers can regenerate and then reinnervate on the target tissue or organ. { rē,jen·ə'rā·shən i,lek,trōd }

regional anatomy [ANAT] The detailed study of the anatomy of a part or region of the body of an animal. { 'rēj·ən·əl ə'nad·ə·mē }

regional anesthesia *See* regional block anesthesia. { 'rēj·ən·əl ,an·əs'thē·zhə }

regional block anesthesia [MED] Anesthesia limited to a part or region of the body by injecting an anesthetic around the nerves that supply the area to block conduction from the area. Also known as regional anesthesia. { 'rēj·ən·əl ¦bläk ,an·əs'thē·zhə }

regional enteritis [MED] Inflammation of isolated segments of the small intestine, with involved parts becoming thick-walled and rigid. Also known as regional enterocolitis; regional ileitis. { 'rēj·ən·əl ,ent·ə'rīd·əs }

regional enterocolitis *See* regional enteritis. { 'rēj·ən·əl ,ent·ə·rō·kə'līd·əs }

regional ileitis *See* regional enteritis. { 'rēj·ən·əl ,il·ē'īd·əs }

regioselectivity [BIOCHEM] Control over the location of reaction in a complex molecule. { ,rē·jē·ō,si·lek'tiv·əd·ē }

regular [BOT] Having radial symmetry, referring to a flower. { 'reg·yə·lər }

regular connective tissue [HISTOL] Connective tissue in which the fibers are arranged in definite patterns. { 'reg·yə·lər kə'nek·tiv 'tish·ü }

Regularia [INV ZOO] An assemblage of echinoids in which the anus and periproct lie within the apical system; not considered a valid taxon. { ,reg·yə'lar·ē·ə }

regulative egg [EMBRYO] An egg in which unfertilized fragments develop as complete, normal individuals. { 'reg·yə,lād·iv 'eg }

regulator gene [GEN] A gene that controls the rate of transcription of one or more other genes. { 'reg·yə,lād·ər ,jēn }

regulatory sequence [CELL MOL] A sequence of deoxyribonucleic acid to which gene regulatory proteins bind to control the rate of transcription. { 'reg·yə·lə,tȯr·ē ,sē·kwəns }

regulatory site [BIOCHEM] A site on an enzyme, other than the active site, at which certain molecules bind, thereby affecting the enzyme's activity. { 'reg·yə·lə,tȯr·ē ,sit }

regulon [GEN] In bacteria, a system of genes, formed by one or more operons, which regulate enzyme induction and whose activity is controlled by a single repressor substance. { 'reg·yə,län }

regurgitation [MED] Reverse circulation of blood in the heart due to defective functioning of the valves. [PHYSIO] Bringing back into the mouth undigested food from the stomach. { ri,gər·jə'tā·shən }

rehabilitation [MED] The restoration to a disabled individual of maximum independence commensurate with his limitations by developing his residual capacity. { ,rē·ə,bil·ə'tā·shən }

Rehfuss tube [MED] A stomach tube designed for the removal of specimens of gastric contents for analysis after administration of a test meal. { 'rā·fu̇s ,tüb }

Reichert's cartilage [EMBRYO] The cartilage of the hyoid arch in a human embryo. { 'rī·kərts ,kärt·lij }

Reighardiidae [INV ZOO] A monotypic family of arthropods in the order Cephalobaenida; the posterior end of the organism is rounded, without lobes, and the cuticula is covered with minute spines. { ,rī·gär'dī·ə,dē }

reindeer [VERT ZOO] *Rangifer tarandus*. A migratory ruminant of the deer family (Cervidae) which inhabits the Arctic region and has a circumpolar distribution; characteristically, both sexes have antlers and are brown with yellow-white areas on the neck and chest. { 'rān,dir }

reinfection [MED] A second infection after recovery from an earlier infection with the same kind of organism. { ,rē·ən'fek·shən }

reishi mushroom *See* Ganoderma lucidum. { rā'ē·shē ,məsh,rüm }

Reissner's membrane [ANAT] The anterior wall of the cochlear duct, which separates the cochlear duct from the scala vestibuli. Also known as vestibular membrane of Reissner. { 'rīs·nərz ,mem,brān }

Reiter's syndrome [MED] The triad of idiopathic nongonococcal urethritis, conjunctivitis, and subacute or chronic polyarthritis. Also known as arthritis urethritica; idiopathic arthritis; infectious uroarthritis. { 'rīd·ərz ,sin,drōm }

rejection [IMMUNOL] Destruction of a graft by the immune system of the recipient. { ri'jek·shən }

relapsing fever [MED] An acute infectious disease caused by various species of the spirochete *Borrelia*, characterized by episodes of fever which subside spontaneously and recur over a period of weeks. { ri'laps·iŋ ,fē·vər }

relative erythrocytosis [MED] A form of erythrocytosis that occurs when the concentration of erythrocytes in the circulating blood increases through loss of plasma. { 'rel·əd·iv ə¦rith·rō,sī'tō·səs }

relative refractory period [NEUROSCI] A period of a few milliseconds following the absolute refractory period during which the excitation threshold of neural tissue is raised and a stronger-than-normal stimulus is required to initiate an action potential. { 'rel·əd·iv ri'frak·trē ˌpir·ē·əd }

relaxed circular deoxyribonucleic acid [CELL MOL] A form of circular deoxyribonucleic acid in which the circle of one strand is broken. Also known as open-circle deoxyribonucleic acid. { ri'lakst 'sər·kyə·lər dē¦äk·sē,rī·bō·nü'klē·ik 'as·əd }

relaxin [BIOCHEM] A hormone found in the serum of humans and certain other animals during pregnancy; probably acting with progesterone and estrogen, it causes relaxation of pelvic ligaments in the guinea pig. { ri'lak·sən }

release factor [CELL MOL] Any protein that responds to termination codons in messenger ribonucleic acid and causes the release of the finished polypeptide. { ri'lēs ˌfak·tər }

releaser stimulus [ZOO] A stimulus which affects an animal by initiating an instinctual behavior pattern. { ri'lēs·ər ˌstim·yə·ləs }

relict [BIOL] A persistent, isolated remnant of a once-abundant species. { 'rel·ikt }

Remak's ganglion [NEUROSCI] A ganglion near the junction of the coronary sinus and the right atrium. { 'rā,mäks ˌgaŋ·glē,än }

remex *See* flight feather. { 'rē,meks }

remiges [VERT ZOO] Wing feathers of a bird. { 'rem·ə,jēz }

remineralization [PHYSIO] The continual re-forming of tooth mineral that occurs at the surface of teeth, chiefly from constituents of saliva. { rē,min·rə·lə'zā·shən }

removable prosthodontics [MED] A subdivision of prosthodontics that focuses on the replacement of missing teeth by dentures. { ri ˌmüv·ə·bəl ˌpräs·thə'dän·tiks }

renal artery [ANAT] A branch of the abdominal or ventral aorta supplying the kidneys in vertebrates. { 'rēn·əl 'ärd·ə·rē }

renal calculus [MED] A concretion in the kidney. { 'rēn·əl 'kal·kyə·ləs }

renal-cell carcinoma [MED] A malignant renal tumor composed principally of large, often hyalin, polygonal cells. Also known as clear-cell carcinoma; Grawitz's tumor; hypernephroma. { 'rēn·əl ¦sel ˌkärs·ən'ō·mə }

renal clearance [PHYSIO] The volume of blood plasma completely cleared of a particular substrate by the kidneys per unit time, a measure of kidney function. { ¦rēn·əl klir·əns }

renal corpuscle [ANAT] The glomerulus together with its Bowman's capsule in the renal cortex. Also known as Malpighian corpuscle. { 'rēn·əl 'kȯr·pə·səl }

renal dwarfism [MED] Dwarfism due to any of various chronic diseases of the kidney in children. { 'rēn·əl 'dwȯr,fiz·əm }

renal failure [MED] Severe malfunction of the kidneys, producing uremia and the resulting constitutional symptoms. { 'rēn·əl 'fāl·yər }

renal papilla [ANAT] A fingerlike projection into the renal pelvis through which the collecting tubules discharge. { 'rēn·əl pə'pil·ə }

renal pyramid [ANAT] Any of the conical masses composing the medullary substance of the kidney. Also known as Malpighian pyramid. { 'rēn·əl 'pir·ə·məd }

renal rickets [MED] The metabolic bone disease due to increased bone resorption resulting from the acidosis and secondary hyperparathyroidism of renal insufficiency. { 'rēn·əl 'rik·əts }

renal threshold [PHYSIO] A concentration of a substance within the blood which, when reached, causes the substance to appear in the urine. { 'rēn·əl 'thresh,hōld }

renal tubular acidosis [MED] Defective hydrogen-ion excretion in the renal tubules, resulting in hyperchloremic acidosis and inadequate acidification of the urine. { 'rēn·əl 'tüb·yə·lər ˌas·ə'dō·səs }

renal tubule [ANAT] One of the glandular tubules which elaborate urine in the kidneys. { 'rēn·əl 'tü·byül }

renal vein [ANAT] A vein which returns blood from the kidney to the vena cava. { 'rēn·əl 'vān }

renaturation [BIOCHEM] The process of restoring denatured proteins to their original condition. { rē,nach·ə'rā·shən }

renette [INV ZOO] An excretory cell found in certain nematodes. { re'net }

renin [BIOCHEM] A proteolytic enzyme produced in the afferent glomerular arteriole which reacts with the plasma component hypertensinogen to produce angiotensin II. { 'rēn·ən }

rennet [VERT ZOO] The lining of the stomach of certain animals, especially the fourth stomach in ruminants. { 'ren·ət }

rennin [BIOCHEM] An enzyme found in the gastric juice of the fourth stomach of calfs; used for coagulating milk casein in cheesemaking. Also known as chymosin. { 'ren·ən }

renninogen [BIOCHEM] The zymogen of rennin. Also known as prorennin. { rə'nin·ə·jən }

Reoviridae [VIROL] A family of ribonucleic acid (RNA)-containing viruses characterized by an unenveloped icosahedral virion containing a genome of 10–12 segments (depending on the genus) of double-stranded RNA, hosts include vertebrates, invertebrates, and plants. { ˌrē·ə'vir·ə,dī }

reovirus [VIROL] A group of ribonucleic acid-containing animal viruses, including agents of encephalitis and phlebotomus fever. { 'rē·ō,vī·rəs }

repair synthesis [CELL MOL] Enzymatic excision and replacement of regions of damaged deoxyribonucleic acid, as in repair of thymine dimers by ultraviolet irradiation. { ri'per ˌsin·thə·səs }

repand [BOT] Having a margin that undulates slightly, referring to a leaf. { rə'pand }

repent [BOT] Of a stem, creeping along the ground and rooting at the nodes. { 'rē·pent }

repetition frequency [CELL MOL] The number of copies of a given nucleotide sequence present in the haploid genome. { ˌrep·ə'tish·ən ˌfrē·kwən·sē }

repetitious deoxyribonucleic acid [CELL MOL] Nucleotide sequences occurring repeatedly in chromosomal deoxyribonucleic acid. { rep·ə¦tish·əs dē,äk·sē¦rī·bō·nü,klē·ik 'as·əd }

replacement transfusion *See* exchange transfusion. { ri'plās·mənt tranz,fyü·zhən }

replica plating [MICROBIO] A method for the isolation of nutritional mutants in microorganisms; colonies are grown from a microorganism suspension previously exposed to a mutagenic agent, on a complete medium in a petri dish; a velour surface is used to transfer the impression of all these colonies to a petri dish containing a minimal medium; colonies that do not grow on the minimal medium are the mutants. { 'rep·lə·kə ,plād·iŋ }

replicating fork [CELL MOL] The Y-shaped region of a chromosome that is a growing point in the replication of deoxyribonucleic acid. { 'rep·lə,kād·iŋ ¦fȯrk }

replication [CELL MOL] Duplication, as of a nucleic acid, by copying from a molecular template. [VIROL] Multiplication of phage in a bacterial cell. { ,rep·lə'kā·shən }

replication bubble *See* replication eye. { ,rep·lə'kā·shən ,bəb·əl }

replication eye [CELL MOL] A replicated region of deoxyribonucleic acid contained within a longer, unreplicated region and presented in the shape of an eye. Also known as replication bubble. { ,rep·lə'kā·shən ī }

replicon [GEN] Each deoxyribonucleic segment that acts as a unit of replication. { 'rep·lə,kän }

replum [BOT] A thin wall separating the two valves or chambers of certain fruits. { 'rep·ləm }

reporter gene [GEN] A transfected gene that produces a signal, such as green fluorescence, when it is expressed; it is typically included in a larger cloned gene that is introduced into an organism to study its temporal and spatial pattern of expression. { ri'pȯrd·ər ,jēn }

repressing [BIOCHEM] The termination of enzyme synthesis when the products of the reaction catalyzed by the enzyme reach a critical concentration. [CELL MOL] Inhibition of transcription or translation due to binding of a repressor to an operator on a deoxyribonucleic acid molecule or to a specific messenger ribonucleic acid site. { ri 'pres·iŋ }

repression [BIOCHEM] The termination of enzyme synthesis when the products of the reaction catalyzed by the enzyme reach a critical concentration. [CELL MOL] Inhibition of transcription or translation due to binding of a repressor to an operator on a deoxyribonucleic acid molecule or to a specific messenger ribonucleic acid site. { ri'presh·ən }

repressor [BIOCHEM] An end product of metabolism which represses the synthesis of enzymes in the metabolic pathway. [GEN] The product of a regulator gene that acts to repress the transcription of another gene. { ri'pres·ər }

reproduction [BIOL] The mechanisms by which organisms give rise to other organisms of the same kind. { ¦rē·prə¦dək·shən }

reproductive behavior [ZOO] The behavior patterns in different types of animals by means of which the sperm is brought to the egg and the parental care of the resulting young insured. { ¦rē·prə¦dək·tiv bi'hā·vyər }

reproductive distribution [ECOL] The range of areas where conditions are favorable to maturation, spawning, and early development of marine animals. { ¦rē·prə¦dək·tiv ,dis·trə'byü·shən }

reproductive system [ANAT] The structures concerned with the production of sex cells and perpetuation of the species. { ¦rē·prə¦dək·tiv ,sis·təm }

reproductive technology [MED] Any procedure undertaken to aid in conception, intrauterine development, and birth when natural processes do not function normally. { ,rē·prə,dək·tiv tek'näl·ə·jē }

reproductive toxicity [MED] Adverse effects on the male and/or female reproductive systems caused by exposure to a toxic chemical. It may be expressed as alterations in sexual behavior, decreases in fertility, or fetal loss during pregnancy. Developmental toxicity may also be included. { ,rē·prə,dək·tiv täk'sis·əd·ē }

Reptantia [INV ZOO] A suborder of the crustacean order Decapoda including all decapods other than shrimp. { rep'tan·chē·ə }

reptile [VERT ZOO] Any member of the class Reptilia. { 'rep,tīl }

Reptilia [VERT ZOO] A class of terrestrial vertebrates composed of turtles, tuatara, lizards, snakes, and crocodileans; characteristically they lack hair, feathers, and mammary glands, the skin is covered with scales, they have a three-chambered heart, and the pleural and peritoneal cavities are continuous. { rep'til·yə }

RES *See* reticuloendothelial system.

rescinnamine [PHARM] $C_{35}H_{42}N_2O_2$ A white to buff-colored, crystalline powder with a melting point of 238°C (in vacuum); soluble in acetic acid and chloroform; an alkaloid extracted from a few species of *Rauwolfia* and used in medicine. { rə'sin·ə,mēn }

resection [MED] The surgical removal of a section or segment of an organ or other structure. { ri'sek·shən }

resectoscope [MED] A tubelike instrument containing an optical device and a sliding knife; permits excision of tissue in body cavities without an opening other than that made by the instrument itself. { ri'sek·tə,skōp }

Resedaceae [BOT] A family of dicotyledonous herbs in the order Capparales having irregular, hypogynous flowers. { ,res·ə'dās·ē,ē }

reserpine [PHARM] $C_{33}H_{40}N_2O_9$ An alkaloid extracted from certain species of *Rauwolfia* and used as a sedative and antihypertensive. { rə'sər·pēn }

reserve cell [HISTOL] **1.** One of the small, undifferentiated epithelial cells at the base of the stratified columnar lining of the bronchial tree. **2.** A chromophobe cell. { ri'zərv 'sel }

residual air *See* residual volume. { rə'zij·ə·wəl 'er }

residual volume [PHYSIO] Air remaining in the

lungs after the most complete expiration possible; it is elevated in diffuse obstructive emphysema and during an attack of asthma. Also known as residual air. { rə'zij·ə·wəl 'väl·yəm }

resin duct [BOT] A canal (intercellular space) lined with secretory cells that release resins into the canal; common in gymnosperms. { 'rez·ən ,dəkt }

resistance factor *See* R factor. { ri'zis·təns ,fak·tər }

resistance transfer factor [GEN] A carrier of genetic information in bacteria which is considered to control the ability of self-replication and conjugal transfer of R factors. Abbreviated RTF. { ri'zis·təns 'tranz·fər ,fak·tər }

respiration [PHYSIO] **1.** The processes by which tissues and organisms exchange gases with their environment. **2.** The act of breathing with the lungs, consisting of inspiration and expiration. { ,res·pə'rā·shən }

respiratory arrest [MED] Sudden cessation of spontaneous respiration due to failure of the respiratory center. { 'res·prə,tȯr·ē ə'rest }

respiratory center [PHYSIO] A large area of the brain involved in regulation of respiration. { 'res·prə,tȯr·ē ¦sen·tər }

respiratory dead space [PHYSIO] That part of the respiratory system which has no alveoli and in which little or no exchange of gas between air and blood takes place. { 'res·prə,tȯr·ē 'ded ,spās }

respiratory distress syndrome of newborn [MED] A disease occurring during the first days of life, characterized by respiratory distress and cyanosis; a hyaline membrane lines the alveoli when the disease persists for more than several hours. Abbreviated RDS. { 'res·prə,tȯr·ē di 'stres ,sin,drōm əv 'nü,bȯrn }

respiratory epithelium [HISTOL] The ciliated pseudostratified epithelium lining the respiratory tract. { 'res·prə,tȯr·ē ,ep·ə'thē·lē·əm }

respiratory minute volume [PHYSIO] The total amount of air which moves in and out of the lungs in a minute. { 'res·prə,tȯr·ē 'min·ət 'väl·yəm }

respiratory pigment [BIOCHEM] Any of various conjugated proteins that function in living organisms to transfer oxygen in cellular respiration. { 'res·prə,tȯr·ē ¦pig·mənt }

respiratory quotient [PHYSIO] The ratio of volumes of carbon dioxide evolved and oxygen consumed during a given period of respiration. Abbreviated RQ. { 'res·prə,tor·ē ¦kwō·shənt }

respiratory syncytial virus [VIROL] An enveloped, single-stranded RNA animal virus belonging to the Paramyxoviridae genus *Pneumovirus*; associated with a large proportion of respiratory illnesses in very young children, particularly bronchiolitis and pneumonia. { 'res·prə,tȯr·ē sin'sish·əl 'vī·rəs }

respiratory system [ANAT] The structures and passages involved with the intake, expulsion, and exchange of oxygen and carbon dioxide in the vertebrate body. { 'res·prə,tȯr·ē ¦sis·təm }

respiratory tree [ANAT] The trachea, bronchi, and bronchioles. [INV ZOO] Either of a pair of branched tubular appendages of the cloaca in certain holothurians that is thought to have a respiratory function. { 'res·prə,tor·ē ,trē }

restiform body *See* inferior cerebellar peduncle. { 'res·tə,fȯrm ,bäd·ē }

resting cell [CELL MOL] An interphase cell. { 'rest·iŋ ,sel }

resting metabolism [PHYSIO] The metabolism of a person at rest while seated or standing in a normal position. { 'rest·iŋ me,tab·ə,liz·əm }

resting potential [PHYSIO] The potential difference between the interior cytoplasm and the external aqueous medium of the living cell. { 'rest·iŋ pə,ten·chəl }

resting spore [BIOL] A spore that remains dormant for long periods before germination, withstanding adverse conditions; usually invested in a thickened cell wall. { 'rest·iŋ ,spȯr }

Restionaceae [BOT] A large family of monocotyledonous plants in the order Restionales characterized by unisexual flowers, wholly cauline leaves, unilocular anthers, and a more or less open inflorescence. { ,res·tē·ə'nās·ē,ē }

Restionales [BOT] An order of monocotyledonous plants in the subclass Commelinidae having reduced flowers and a single, pendulous, orthotropous ovule in each of the one to three locules of the ovary. { ,res·tē·ə'nā·lēz }

restless legs syndrome [MED] A condition that is characterized by intense disagreeable feelings in the legs at rest and repose with compulsion to move the legs to get relief from these symptoms, peak onset usually occurs during middle age, and the disorder tends to become more severe with age. { ,rest·ləs 'legz ,sin,drōm }

restoration [ECOL] A conservation measure involving the correction of past abuses that have impaired the productivity of the resources base. { ,res·tə'rā·shən }

restoration ecology [ECOL] The application of ecological principles and field methodologies to the successful restoration of damaged ecosystems. { ,res·tə,rā·shən ē'käl·ə·jē }

restorative dentistry [MED] A branch of dentistry that focuses on the preservation and restoration of decayed, defective, missing, and traumatized teeth. { rə,stir·əd·iv 'dent·ə·strē }

restorative immunotherapy [IMMUNOL] Immunotherapy that involves the direct and indirect restoration of deficient immunological function through any means other than the direct transfer of cells. { rə,stȯr·əd·iv ,im·yə·nō 'ther·ə·pē }

restriction [CELL MOL] The degradation of foreign deoxyribonucleic acid by restriction endonucleases capable of recognizing particular patterns of specificity. { ri¦strik·shən }

restriction endonuclease [CELL MOL] Any of the specific endonucleases that recognizes a short specific sequence within a deoxyribonucleic acid molecule and then catalyzes double-stranded cleavage of that molecule. Also known as endodeoxyribonuclease. { ri'strik·shən ¦en·dō'nü·klē,ās }

restriction endonuclease analysis [CELL MOL]

A technique in which deoxyribonucleic acid (DNA) fragments obtained from digestion with restriction enzymes are compared to construct a restriction map showing the position of specific sites along a sequence of DNA. { ri,strik·shən ,en·dō'nü·klē,ās ə,nal·ə·səs }

restriction fragment [CELL MOL] Any of the individual polynucleotide sequences produced by digestion of deoxyribonucleic acid with a restriction endonuclease. { ri'strik·shən ,frag·mənt }

restriction fragment length polymorphism [CELL MOL] Variations in the length of restriction fragments resulting from action by a specific endonuclease. { ri¦strik·shən ¦frag·mənt ,leŋkth ,päl·i'mȯr,fiz·əm }

restriction map [CELL MOL] A diagram of a deoxyribonucleic acid molecule showing sites at which restriction endonucleases produce cleavage. { ri'strik·shən ,map }

restriction point [CELL MOL] In the mammalian cell cycle, a time late in the G1 phase at which the cell commits to the replication of its deoxyribonucleic acid (DNA). { ri'strik·shən ,pȯint }

restriction site [CELL MOL] A sequence in a deoxyribonucleic acid molecule that can be cleaved with a specific restriction endonuclease. { ri 'strik·shən ,sīt }

restrictive condition [GEN] An environmental condition under which a conditional lethal mutant either cannot grow or shows the mutant phenotype. { ri¦strik·tiv kən¦dish·ən }

resupinate [BOT] Inverted, usually through 180°, so as to appear upside down or reversed. { rē'sü·pə,nāt }

resuscitation [MED] Restoration of consciousness or life functions after apparent death. { ri,səs·ə'tā·shən }

retardation [MED] Slow mental or physical functioning. { ,rē,tär'dā·shən }

rete cord [EMBRYO] One of the deep, anastomosing strands of cells of the medullary cords of the vertebrate embryo that form the rete testis or the rete ovarii. { 'rēd·ē ,kȯrd }

rete mirabile [VERT ZOO] A network of small blood vessels that are formed by the branching of a large vessel and that usually reunite into a single trunk; believed to have an oxygen-storing function in certain aquatic fauna. { 'rēd·ē mi'räb·ə·lē }

retention cyst [MED] A cyst caused by obstructed outflow of secretion from a gland. { ri 'ten·chən ,sist }

rete ovarii [ANAT] Vestigial tubules or cords of cells near the hilus of the ovary, corresponding with the rete testis, but not connected with the mesonephric duct. { 'rēd·ē ō'var·ē,ī }

rete testis [ANAT] The network of anastomosing tubules in the mediastinum testis. { 'rēd·ē 'tes·təs }

reticular *See* reticulate. { re'tik·yə·lər }

reticular cell *See* reticulocyte. { re'tik·yə·lər 'sel }

reticular degeneration [PATH] Rupture of epidermal cells with formation of multilocular bullae due to intracellular edema. { re'tik·yə·lər di- ,jen·ə'rā·shən }

reticular fiber [HISTOL] Any of the delicate, branching argentophile fibers conspicuous in the connective tissue of lymphatic tissue, myeloid tissue, the red pulp of the spleen, and most basement membranes. Also known as argentaffin fiber; argyrophil lattice fiber; precollagenous fiber. { re'tik·yə·lər 'fī·bər }

reticular formation [NEUROSCI] The portion of the central nervous system which consists of small islands of gray matter separated by fine bundles of nerve fibers running in every direction. { re'tik·yə·lər fȯr¦mā·shən }

Reticulariaceae [MYCOL] A family of plasmodial slime molds in the order Liceales. { rə¦tik·yə,lar·ē'ās·ē,ē }

reticular system *See* reticuloendothelial system. { re'tik·yə·lər ,sis·təm }

reticular tissue [HISTOL] Connective tissue having reticular fibers as the principal element. { re'tik·yə·lər ,tish·ü }

reticulate [BIOL] Having or resembling a network of fibers, veins, or lines. { rə'tik·yə·lət }

reticulated *See* reticulate. { rə'tik·yə,lād·əd }

reticulate venation [BOT] A branching vascular system with successively thinner veins diverging as branches from the thicker veins. { rə,tik·yə·lət ve'nā·shən }

reticulin [BIOCHEM] A protein isolated from reticular fibers. { rə'tik·yə·lən }

reticulocyte [HISTOL] Also known as reticular cell. **1.** A large, immature red blood cell, having a reticular appearance when stained due to retention of portions of the nucleus. **2.** A cell of reticular tissue. { rə'tik·yə·lə,sīt }

reticulocytopenia [MED] A decrease in the normal number of circulating reticulocytes. Also known as reticulopenia. { re,tik·yə·lō,sīd·ə'pē·nyə }

reticulocytosis [MED] An increase in the normal number of circulating reticulocytes. { rə,tik·yə·lō·sə'tō·səs }

reticuloendothelial granulomatosis [MED] A group of rare diseases characterized by generalized reticuloendothelial hyperplasia with or without intracellular lipid deposition. { rə¦tik·yə·lō,en·dō'thē·lē·əl ¦gran·yə·lō·mə'tō·səs }

reticuloendothelial system [ANAT] The macrophage system, including all phagocytic cells such as histiocytes, macrophages, reticular cells, monocytes, and microglia, except the granular white blood cells. Abbreviated RES. Also known as hematopoietic system; reticular system. { rə¦tik·yə·lō,en·dō'thē·lē·əl ,sis·təm }

reticuloendothelium [HISTOL] The cells making up the reticuloendothelial system. { rə,tik·yə·lō,en·də'thē·lē·əm }

reticulohistiocytoma [MED] A solitary skin nodule composed of large multinucleated histiocytes that contain glycolipids. { rə,tik·yə·lō,his·tē·ō·sə'tō·mə }

reticulopenia *See* reticulocytopenia. { re,tik·yə·lō'pē·nyə }

reticulopodia [INV ZOO] Pseudopodia in the form of a branching network. { rə¦tik·yə·lō'päd·ē·ə }

reticulosis [MED] An increase in the number of histiocytes, monocytes, or other reticuloendothelial elements.

reticulospinal tract [NEUROSCI] Nerve fibers descending from large cells of the reticular formation of the pons and medulla into the spinal cord. { rə¦tik·yə·lō'spīn·əl 'trakt }

reticulum [BIOL] A fine network. [VERT ZOO] The second stomach in ruminants. { rə'tik·yə·ləm }

retina [NEUROSCI] The photoreceptive layer and terminal expansion of the optic nerve in the dorsal aspect of the vertebrate eye. { 'ret·ən·ə }

retinaculum [INV ZOO] **1.** A clasp on the forewing of certain moths for retaining the frenulum of the hindwing. **2.** An appendage on the third abdominal somite of springtails that articulates with the furcula. { ˌret·ən'ak·yə·ləm }

retinal [BIOCHEM] A carotenoid, produced as an intermediate in the bleaching of rhodopsin and decomposition to vitamin A. Also known as vitamin A aldehyde. { 'ret·ən·əl }

retinal astigmatism [MED] Astigmatism due to changes in the localization of the fixation point. { 'ret·ən·əl ə'stig·məˌtiz·əm }

retinal detachment [MED] An eye disorder characterized by the separation of the sensory layers of the retina from their supporting foundations. { ˌret·ən·əl di'tach·mənt }

retinal pigment *See* rhodopsin. { 'ret·ən·əl 'pig·mənt }

retinal retinitis *See* vascular retinopathy. { 'ret·ən·əl ˌret·ən'īd·əs }

retinene [BIOCHEM] A pigment extracted from the retina, which turns yellow by the action of light; the chief carotenoid of the retina. { 'ret·ənˌēn }

retinitis [MED] Inflammation of the retina. { ˌret·ən'īd·əs }

retinitis pigmentosa [MED] A hereditary affection inherited as a sex-linked recessive and characterized by slowly progressing atrophy of the retinal nerve layers, and clumping of retinal pigment, followed by attenuation of the retinal arterioles and waxy atrophy of the optic disks. { ˌret·ən'īd·əs ˌpig·mən'tō·sə }

retinoblastoma [MED] A malignant tumor of the sensory layer of the retina. { ˌret·ən·ō·bla'stō·mə }

retinochoroiditis [MED] Inflammation of the retina and choroid. { ˌret·ən·ōˌkȯrˌȯi'dīd·əs }

retinoid [BIOCHEM] The set of molecules composing vitamin A and its synthetic analogs, such as retinal or retinyl acetate. { 'ret·ənˌȯid }

retinoid receptor [CELL MOL] Any member of a family of nuclear receptors that mediate the actions of natural and synthetic analogs of vitamin A (retinoids) by regulating the transcription of retinoid-responsive genes in the cell nucleus. { 'ret·ənˌȯid riˌsep·tər }

retinol *See* vitamin A. { 'ret·ənˌȯl }

retinopathy [MED] Any pathologic condition involving the retina. { ˌret·ən'äp·ə·thē }

retinoschisis [MED] **1.** Separation with hole formation of the layers composing the retina. **2.** A congenital anomaly characterized by cleavage of the retina. { ˌret·ən·ō'ski·səs }

retinula [INV ZOO] The receptor element at the inner end of the ommatidium in a compound eye. { rə'tin·yə·lə }

Retortamonadida [INV ZOO] An order of parasitic flagellate protozoans belonging to the class Zoomastigophorea, having two or four flagella and a complex blepharoplast-centrosome-axostyle apparatus. { ri'tȯr·də·mə'näd·ə·də }

retractor [ANAT] A muscle that draws a limb or other body part toward the body. [MED] A clawlike instrument for holding tissues away from the surgical field. { ri'trak·tər }

retroactive refit *See* retrofit. { ¦re·trō'ak·tiv 'rēˌfit }

retrocerebral gland [INV ZOO] Any of various endocrine glands located behind the brain in insects which function in postembryonic development and metamorphosis. { ¦re·trō·sə'rē·brəl 'gland }

retroflexion [ANAT] The state of being bent backward. [MED] A condition in which the uterus is bent backward on itself, producing a sharp angle in its longitudinal axis at the junction of the cervix and the fundus. { ¦re·trə'flek·shən }

retrograde amnesia [MED] Loss of memory for events occurring prior to, but not after, the onset of a current disease or trauma. { 're·trəˌgrād am'nē·zha }

retrogression [MED] Going backward, as in degeneration or atrophy of tissues. { ˌre·trə'gresh·ən }

retrolental fibroplasia [MED] An oxygen-induced disease of the retina in premature infants characterized by formation of an opaque membrane behind the lens of the eye. { ¦re·trə¦lent·əl ˌfī·brə'plā·zhə }

retroposon [GEN] A mobile genetic element that transposes to or from a chromosomal site by reverse transcription from a ribonucleic acid intermediate. { ˌre·trə'pōˌzän }

retrorse [BIOL] Bent downward or backward. { ri'trȯrs }

retrostalsis [PHYSIO] Reverse peristalsis. { ¦re·trō'stäl·səs }

retrotransposon [CELL MOL] A small, mobile deoxyribonucleic acid (DNA) sequence that can retrotranspose, that is, move from one genomic location to another by producing ribonucleic acid (RNA) that is transcribed by reverse transcriptase back into DNA which is then inserted at a new site. { ¦re·trō·tranz'pōˌzän }

retroversion [ANAT] A turning back. [MED] A condition in which the uterus is tilted backward without any change in the angle of its longitudinal axis. { ˌre·trə'vər·zhən }

Retroviridae [VIROL] A family of ribonucleic acid (RNA)-containing animal viruses characterized by spherical enveloped virions containing two single-stranded RNA molecules and reverse transcriptase, includes the subfamilies Oncovirinae, Spumavirinae, and Lentivirinae. { ˌre·trə'vir·əˌdī }

retrovirus [VIROL] A family of ribonucleic acid

viruses distinguished by virions which possess reverse transcriptase and which have two proteinaceous structures, a dense core, and an envelope that surrounds the core. { 're·trō,vī·rəs }

retuse [BOT] Having a rounded apex with a slight, central notch. { rə'tüs }

reverse genetics [CELL MOL] An experimental method in which information from cloned deoxyribonucleic acid (DNA) or protein sequences is used to find or to produce mutations that help identify the function of a gene or protein (in contrast to classical genetics in which a known function or trait is traced back to a particular gene. { ri¦vərs jə'ned·iks }

reverse graft [BOT] A plant graft made by inserting the scion in an inverted position. { ri'vərs 'graft }

reverse mutation [GEN] A mutation in a mutant allele which makes it capable of producing the nonmutant phenotype; may actually restore the original deoxyribonucleic acid sequence of the gene or produce a new one which has a similar effect. Also known as back mutation. { ri'vərs myü'tā·shən }

reverse passive anaphylaxis [IMMUNOL] Hypersensitivity produced when the antigen is injected first, then followed in several hours by the specific antibody, causing shock. { ri'vərs 'pas·iv ,an·ə·fə'lak·səs }

reverse pinocytosis *See* emiocytosis. { ri'vərs ¦pī·nō·sī'tō·səs }

reverse transcript [CELL MOL] A deoxyribonucleic acid sequence obtained from a ribonucleic acid sequence by means of reverse transcription. { ri¦vərs 'tran,skript }

reverse transcriptase [GEN] A polymerase that mediates deoxyribonucleic acid synthesis by using a ribonucleic acid template. { ri'vərs tran'skrip,tās }

revolute [BOT] Rolled backward and downward. { 'rev·ə,lüt }

Reye's syndrome [MED] An uncommon liver disorder primarily occurring in infants and young children; characterized by convulsions, hypoglycemia, and a liver showing diffuse microvacuolar fatty metamorphosis. { 'rīz ,sin,drōm }

R factor [GEN] A self-replicating, infectious agent that carries genetic information and transmits drug resistance from bacterium to bacterium by conjugation of cell. Also known as resistance factor. { 'är ,fak·tər }

Rhabdiasoidea [INV ZOO] An order or superfamily of parasitic nematodes. { ,rab·dē·ə'sȯid·ē·ə }

rhabdion [INV ZOO] One of the sclerotized segments lining the buccal cavity of nematodes. { 'rab·dē,än }

rhabdite [INV ZOO] A small rodlike or fusiform body secreted by epidermal or parenchymal cells of certain turbellarians and trematodes. { 'rab,dīt }

Rhabditia [INV ZOO] A subclass of nematodes in the class Secernentea. { rab'dish·ə }

Rhabditidia [INV ZOO] An order of nematodes in the subclass Rhabditia including parasites of humans and domestic animals. { ,rab·də'tid·ē·ə }

Rhabditoidea [INV ZOO] A superfamily of small to moderate-sized nematodes in the order Rhabditidia with small, porelike, anteriorly located amphids, and esophagus with corpus, isthmus, and valvulated basal bulb. { ,rab·də'tȯid·ē·ə }

Rhabdocoela [INV ZOO] Formerly an order of the Turbellaria, and now divided into three orders, Catenulida, Macrostomida, and Neorhabdocoela. { ,rab·də'sē·lə }

rhabdolith [BOT] A minute coccolith having a shield surmounted by a long stem and found at all depths in the ocean, from the surface to the bottom. { 'rab·də,lith }

rhabdome [INV ZOO] The central translucent cylinder in the retinula of a compound eye. { 'rab,dōm }

rhabdomyoblastoma *See* rhabdomyosarcoma. { ¦rab·dō,mī·ō·bla'stō·mə }

rhabdomyoma [MED] A benign tumor of skeletal muscle. { ¦rab·dō·mī'ō·mə }

rhabdomyosarcoma [MED] A malignant tumor of skeletal muscle in the extremities composed of anaplastic muscle cells. Also known as malignant rhabdomyoma; rhabdomyoblastoma. { ¦rab·dō,mī·ō·sär'kō·mə }

Rhabdophorina [INV ZOO] A suborder of ciliates in the order Gymnostomatida. { ,rab·dō·fə'rī·nə }

rhabdosome [INV ZOO] A colonial graptolite that develops from a single individual. { 'rab·də,sōm }

Rhabdoviridae [VIROL] A large family of negative-strand ribonucleic acid (RNA) viruses characterized by an enveloped bullet-shaped virion containing nonfragmented single-stranded RNA. They infect mammals, birds, fish, insects, and plants. The family includes the genera *Vesiculovirus* (vesicular stomatitis) and *Lyssavirus* (rabies). { ,rab·dō'vir·ə,dī }

rhabdovirus [VIROL] A group of ribonucleic acid-containing animal viruses, including rabies virus and certain infective agents of fish and insects. { ¦rab·dō'vī·rəs }

rhabdus [INV ZOO] A uniaxial sponge spicule. { 'rab·dəs }

rhachitomous [VERT ZOO] Being, having, or pertaining to vertebrae with centra whose parts do not fuse. { rə'kid·ə·məs }

Rhacophoridae [VERT ZOO] A family of arboreal frogs in the suborder Diplasiocoela. { ,rak·ō'fȯr·ə,dē }

Rhacopilaceae [BOT] A family of mosses in the order Isobryales generally having dimorphous leaves with smaller dorsal leaves and a capsule that is plicate when dry. { ,rak·ō·pə'lās·ē,ē }

Rhagionidae [INV ZOO] The snipe flies, a family of predatory orthorrhaphous dipteran insects in the series Brachycera that are brownish or gray with spotted wings. { ,rag·ē'än·ə,dē }

rhagon [INV ZOO] A pyramid-shaped, colonial sponge having an osculum at the apex and flagellated chambers in the upper wall only. { 'rā ,gän }

Rhamnaceae [BOT] A family of dicotyledonous plants in the order Rhamnales characterized by a solitary ovule in each locule, free stamens, simple leaves, and flowers that are hypogynous to perigynous or epigynous. { ram'nās·ē,ē }

Rhamnales [BOT] An order of dicotyledonous plants in the subclass Rosidae having a single set of stamens, opposite the petals, usually a well-developed intrastamenal disk, and two or more locules in the ovary. { ram'nā·lēz }

rhamnose [BIOCHEM] $C_6H_{12}O_5$ A deoxysugar occurring free in poison sumac, and in glycoside combination in many plants. Also known as isodulcitol. { 'ram,nōs }

rhamphoid [BIOL] Beak-shaped. { 'ram,fȯid }

rhampotheca [VERT ZOO] The horny sheath covering a bird's beak. { ,ram·fə'thē·kə }

Rh antigen *See* Rh factor. { ¦är¦āch 'ant·i·jən }

Rh blocking serum [IMMUNOL] A serum that reacts with Rh-positive blood without causing agglutination, but which blocks the action of anti-Rh serums that are subsequently introduced. { ¦är¦āch ¦bläk·iŋ ,sir·əm }

Rh blocking test [IMMUNOL] A test for the detection of Rh antibody in plasma wherein erythrocytes having the Rh antigen are incubated in the patient's serum so that the antibodies may be adsorbed on these cells, which are then employed in the antiglobulin test. Also known as indirect Coombs test; indirect developing test. { ¦är¦āch ¦bläk·iŋ ,test }

Rh blood group [IMMUNOL] The extensive, genetically determined system of red blood cell antigens defined by the immune serum of rabbits injected with rhesus monkey erythrocytes, or by human antisera. Also known as rhesus blood group. { ¦är¦āch 'bləd ,grüp }

rhea [BOT] *See* ramie. [VERT ZOO] The common name for members of the avian order Rheiformes. { 'rē·ə }

rhegmatogenous retinal detachment [MED] Retinal detachment that is due to a retinal hole or tear, it occurs spontaneously or following trauma. { ,reg·mə¦täj·ə·nəs ,ret·ən·əl di'tach·mənt }

Rheidae [VERT ZOO] The single family of the avian order Rheiformes. { 'rē·ə,dē }

Rheiformes [VERT ZOO] The rheas, an order of South American running birds; called American ostriches, they differ from the true ostrich in their smaller size, feathered head and neck, three-toed feet, and other features. { ,rē·ə'fȯr,mēz }

rheobase [PHYSIO] The intensity of the steady current just sufficient to excite a tissue when suddenly applied. { 'rē·ō,bās }

rheoelectroencephalograph [MED] Electroencephalograph for measuring differential blood flow in both sides of the brain and in any other part of the body. { ¦rē·ō·i,lek·trō·in'sef·ə·lə ,graf }

rheophile [ECOL] Living or thriving in running water. { 'rē·ə,fīl }

rheophilous bog [ECOL] A bog which draws its source of water from drainage. { rē'äf·ə·ləs 'bäg }

rheoplankton [ECOL] Plankton found in flowing water. { 'rē·ō,plaŋk·tən }

rheotaxis [BIOL] Movement of a motile cell or organism in response to the direction of water currents. { ¦rē·ə¦tak·səs }

rheotropism [BIOL] Orientation response of an organism to the stimulus of a flowing fluid, as water. { rē'ä·trə,piz·əm }

rhesus blood group *See* Rh blood group. { 'rē·səs 'bləd ,grüp }

rhesus factor *See* Rh factor. { 'rē·səs ,fak·tər }

rhesus macaque *See* rhesus monkey. { 'rē·səs mə'kak }

rhesus monkey [VERT ZOO] *Macaque mulatta*. An agile, gregarious primate found in southern Asia and having a short tail, short limbs of almost equal length, and a stocky build. Also known as rhesus macaque. { 'rē·səs ,məŋ·kē }

rheumatic arteritis [MED] A type of allergic arteritis associated with acute rheumatic fever. { rü'mad·ik ,ärd·ə'rīd·əs }

rheumatic carditis [MED] Inflammation of the heart resulting from rheumatic fever. { rü'mad·ik kär'dīd·əs }

rheumatic encephalopathy [MED] An inflammatory reaction of the brain and the smaller arteries of the cerebral cortex associated with rheumatic fever. { rü'mad·ik in,sef·ə'läp·ə·thē }

rheumatic endocarditis [MED] Inflammation of the endocardium in acute rheumatic fever, usually involving heart valves. { rü'mad·ik ,en·dō·kär'dīd·əs }

rheumatic fever [MED] A febrile disease occurring in childhood as a delayed sequel of infection by *Streptococcus hemolyticus*, group A; characterized by arthritis, carditis, nosebleeds, and chorea. { rü'mad·ik 'fē·vər }

rheumatic pneumonia [MED] Pneumonia associated with acute rheumatic fever. { rü'mad·ik nu̇'mō·nyə }

rheumatism [MED] Any combination of muscle or joint pain, stiffness, or discomfort arising from nonspecific disorders. { 'rü·mə,tiz·əm }

rheumatoid arthritis [MED] A chronic systemic inflammatory disease of connective tissue in which symptoms and changes predominate in articular and related structures. Also known as atrophic arthritis; chronic infectious arthritis; proliferative arthritis. { 'rü·mə,tȯid är'thrīd·əs }

rheumatoid factor [IMMUNOL] The immunoglobulin in the class IgM that is detected in the synovial fluid of individuals with rheumatoid arthritis. { 'rü·mə,tȯid ,fak·tər }

rheumatoid nodules [MED] Subcutaneous lateral foci of fibrinoid degeneration or necrosis surrounded by mononuclear cells in a regular palisade arrangement, occurring usually in association with rheumatoid arthritis or rheumatic fever. { 'rü·mə,tȯid 'näj·ülz }

rheumatoid spondylitis [MED] A chronic progressive arthritis of young men, affecting mainly the spine and sacroiliac joints, leading to fusion and deformity. Also known as Marie-Strümpell disease. { 'rü·mə,tȯid ,spän·də'līd·əs }

Rh factor [IMMUNOL] Any of several red blood cell antigens originally identified in the blood of rhesus monkeys. Also known as Rh antigen; rhesus factor. { ¦är'āch ˌfak·tər }

Rhigonematidae [INV ZOO] A family of nematodes in the superfamily Oxyuroidea. { ˌrig·ō·nə'mad·əˌdē }

Rhincodontidae [VERT ZOO] The whale sharks, a family of essentially tropical galeoid elasmobranchs in the isurid line. { ˌriŋ·kə'dänt·əˌdē }

Rh incompatibility [MED] A condition in which red blood cells of the fetus become coated with immunoglobulin G antibody (Rh antibody) of maternal origin which is directed against Rh-D antigen of paternal origin that is present on fetal cells. { ¦är¦āch ˌin·kəmˌpad·ə'bil·əd·ē }

rhinencephalon [ANAT] The anterior olfactory portion of the vertebrate brain. { ¦rīn·in'sef·əˌlän }

rhinitis [MED] Inflammation of the mucous membranes in the nose. { rī'nīd·əs }

Rhinobatidae [VERT ZOO] The guitarfishes, a family of elasmobranchs in the batoid group. { ˌrī·nō'bad·əˌdē }

Rhinoceratidae [VERT ZOO] A family of perissodactyl mammals in the superfamily Rhinoceratoidea, comprising the living rhinoceroses. { rīˌnäs·ə'räd·əˌdē }

Rhinoceratoidea [VERT ZOO] A superfamily of perissodactyl mammals in the suborder Ceratomorpha including living and extinct rhinoceroses. { rīˌnäs·ə·rə'tòid·ē·ə }

rhinocerebral mucormycosis [MED] A mold infection of the sinus that spreads rapidly to the eye and brain. { ˌrīn·ō·səˌrēb·rəl ˌmyü·kō·mī'kō·səs }

rhinoceros [VERT ZOO] The common name for the odd-toed ungulates composing the family Rhinoceratidae, characterized by massive, thick-skinned limbs and bodies, and one or two horns which are composed of a solid mass of hairs attached to the bony prominence of the skull. { rī'näs·ə·rəs }

Rhinochimaeridae [VERT ZOO] A family of ratfishes, order Chimaeriformes, distinguished by an extremely elongate rostrum. { ˌrīn·ə·ki'mer·əˌdē }

Rhinocryptidae [VERT ZOO] The tapaculos, a family of ground-inhabiting suboscine birds in the suborder Tyranni characterized by a large, movable flap which covers the nostrils. { ˌrīn·ə'krip·təˌdē }

rhinogenous [MED] Originating in the nose. { rī'näj·ə·nəs }

rhinolaryngology [MED] The science of the anatomy, physiology, and pathology of the nose and larynx. { ¦rī·nōˌlar·iŋ'gäl·ə·jē }

rhinolaryngoscope [MED] A scope containing a mirror and a light, used to examine the nose and larynx. { ¦rī·nō·lə'rin·jəˌskōp }

rhinology [MED] The science of the anatomy, functions, and diseases of the nose. { rī'näl·ə·jē }

Rhinolophidae [VERT ZOO] The horseshoe bats, a family of insect-eating chiropterans widely distributed in the Eastern Hemisphere and distinguished by extremely complex, horseshoe-shaped nose leaves. { ˌrīn·ə'läf·əˌdē }

rhinopharyngitis [MED] Inflammation of the nose and pharynx or of the nasopharynx. { ¦rī·nōˌfar·ən'jīd·əs }

rhinophore [INV ZOO] An olfactoreceptor of certain land mollusks, usually borne on a tentacle. { 'rīn·əˌfòr }

rhinoplasty [MED] A plastic operation on the nose. { 'rīn·əˌplas·tē }

Rhinopomatidae [VERT ZOO] The mouse-tailed bats, a small family of insectivorous chiropterans found chiefly in arid regions of northern Africa and southern Asia and characterized by long, wirelike tails and rudimentary nose leaves. { ˌrīn·ə·pə'mad·əˌdē }

Rhinopteridae [VERT ZOO] The cow-nosed rays, a family of batoid sharks having a fleshy pad at the front end of the head and a well-developed poison spine. { ˌrī·näp'ter·əˌdē }

rhinorrhea [MED] **1.** A mucous discharge from the nose. **2.** Escape of cerebrospinal fluid through the nose. { ˌrin·ə'rē·ə }

rhinoscleroma [MED] A chronic infectious bacterial disease caused by *Klebsiella rhinoscleromatis* and characterized by hard nodules and plaques of inflamed tissue in the nose and adjacent areas. { ˌrīn·ə·sklə'rō·mə }

rhinoscope [MED] An instrument for examining the nasal cavities. { 'rīn·əˌskōp }

Rhinotermitidae [INV ZOO] A family of lower termites of the order Isoptera. { ¦rī·nō·tər'mad·əˌdē }

rhinotheca [VERT ZOO] The horny sheath on the upper part of a bird's bill. { ¦rīn·ə'thē·kə }

rhinovirus [VIROL] A subgroup of the picornavirus group including small, ribonucleic acid-containing forms which are not inactivated by ether. { ¦rīn·ə'vī·rəs }

Rhipiceridae [INV ZOO] The cedar beetles, a family of coleopteran insects in the superfamily Elateroidea. { ˌrip·ə'ser·əˌdē }

Rhipidistia [VERT ZOO] The equivalent name for Osteolepiformes. { ˌrip·ə'dis·tē·ə }

rhipidium [BOT] A fan-shaped inflorescence with cymose branching in which branches lie in the same plane and are suppressed alternately on each side. { ri'pid·ē·əm }

Rhipiphoridae [INV ZOO] The wedge-shaped beetles, a family of coleopteran insects in the superfamily Meloidea. { ˌrip·ə'fòr·əˌdē }

rhizanthous [BOT] Producing flowers directly from the root. { rī'zan·thəs }

rhizautoicous [BOT] Of mosses, having the antheridial branch and the archegonial branch connected by rhizoids. { ¦rīzˌó¦tòi·kəs }

rhizine [BOT] The rhizoid of a lichen. { 'rīˌzēn }

Rhizobiaceae [MICROBIO] A family of gram-negative, motile, aerobic rods; utilize carbohydrates and produce slime on carbohydrate media. { rīˌzō·bē'ās·ēˌē }

Rhizobium [MICROBIO] A genus of rod-shaped,

gram-negative, aerobic, and nitrogen-fixing bacteria which form symbiotic nodules on the roots of leguminous plants, such as clover and beans. { rī'zō·bē·əm }

rhizocarpous [BOT] Pertaining to perennial herbs having perennating underground parts from which stems and foliage arise annually. { ¦rī·zō¦kär·pəs }

Rhizocephala [INV ZOO] An order of crustaceans which parasitize other crustaceans; adults have a thin-walled sac enclosing the visceral mass and show no trace of segmentation, appendages, or sense organs. { ˌrī·zō'sef·ə·lə }

Rhizochloridina [INV ZOO] A suborder of flagellate protozoans in the order Heterochlorida. { ˌrī·zō‚klȯr·ə'dī·nə }

rhizoid [BOT] A rootlike structure which helps to hold the plant to a substrate; found on fungi, liverworts, lichens, mosses, and ferns. { 'rī ˌzȯid }

Rhizomastigida [INV ZOO] An order of the protozoan class Zoomastigophorea; all species are microscopic and ameboid, and have one or two flagella. { ¦rī·zō·mas'tij·əd·ə }

Rhizomastigina [INV ZOO] The equivalent name for Rhizomastigida. { 'rī·zō·mas·tə'jī·nə }

rhizome [BOT] An underground horizontal stem, often thickened and tuber-shaped, and possessing buds, nodes, and scalelike leaves. { 'rī‚zōm }

rhizomorph [BOT] A rootlike structure, characteristic of many basidiomycetes, consisting of a mass of densely packed and intertwined hyphae. { 'rī·zə‚mȯrf }

Rhizophagidae [INV ZOO] The root-eating beetles, a family of minute coleopteran insects in the superfamily Cucujoidea. { ˌrī·zō'fā·jə‚dē }

Rhizophoraceae [BOT] A family of dicolyledonous plants in the order Cornales distinguished by opposite, stipulate leaves, two ovules per locule, folded or convolute bud petals, and a berry fruit. { rī‚zäf·ə'rās·ē‚ē }

Rhizophorales [BOT] An order of dicotyledonous flowering plants, class Magnoliopsida; mostly tanniferous trees and shrubs with leaves opposite, simple, and entire, and flowers regular, mostly perfect, and variously perigynous or epigynous. { ¦rīz·ə·fō'rā·lēz }

rhizophore [BOT] A leafless, downward-growing dichotomous *Selaginella* shoot that has tufts of adventitious roots at the apex. { 'rī·zə‚fȯr }

rhizoplast [CELL MOL] A delicate fiber or thread running between the nucleus and the blepharoplast in cells bearing flagella. { 'rī·zə‚plast }

rhizopod [INV ZOO] An anastomosing rootlike pseudopodium. { 'rī·zə‚päd }

Rhizopodea [INV ZOO] A class of the protozoan superclass Sarcodina in which pseudopodia may be filopodia, lobopodia, or reticulopodia, or may be absent. { ˌrī·zə'pō·dē·ə }

Rhizostomeae [INV ZOO] An order of the class Scyphozoa having the umbrella generally higher than it is wide with the margin divided into many lappets but not provided with tentacles. { ˌrī·zə'stō·mē‚ē }

rhizotomy [MED] Surgical division of any root, as of a nerve. Also known as radicotomy. { rī'zäd·ə·mē }

rhizotron [BOT] An underground laboratory system designed for examining plant root growth; contains enclosed columns of soil with transparent plastic windows which permit viewing, measuring, and photographing. { 'rīz·ə‚trän }

Rh negative [MED] Absence of the Rh-D antigen on the surface of red blood cells. { ¦är¦āch 'neg·əd·iv }

Rh null [MED] Total absence of all Rh antigens on the surface of red blood cells. { ¦är¦āch 'nəl }

L-rhodeose *See* L-fucose. { ¦el 'rōd·ē‚ōs }

Rhodesian trypanosomiasis [MED] A fulminating form of African sleeping sickness caused by *Trypanosoma rhodesiense*, transmitted by the tsetse fly, and characterized by parasitemia, edema, lymphadenitis, and myocarditis. Also known as East African sleeping sickness. { rō'dē·zhən trə‚pan·ə·sō'mī·ə·səs }

Rhodininae [INV ZOO] A subfamily of limivorous worms in the family Maldanidae. { rō 'din·ə‚nē }

Rhodobacteriineae [MICROBIO] Formerly a suborder of the order Pseudomonadales comprising all of the photosynthetic, or phototrophic, bacteria except those of the genus *Rhodomicrobium*. { ¦rō·dō·bak‚tir·ē'ī·nē‚ē }

rhodonea *See* rose. { ˌrōd·ən'ē·ə }

Rhodophyceae [BOT] A class of algae belonging to the division or subphylum Rhodophyta. { ˌrōd·ə'fīs·ē‚ē }

Rhodophyta [BOT] The red algae, a large diverse phylum or division of plants distinguished by having an abundance of the pigment phycoerythrin. { rō'däf·əd·ə }

rhodoplast [BOT] A reddish chromatophore occurring in red algae. { 'rōd·ə‚plast }

rhodopsin [BIOCHEM] A deep-red photosensitive pigment contained in the rods of the retina of marine fishes and most higher vertebrates. Also known as retinal pigment; visual purple. { rō'däp·sən }

Rhodospirillaceae [MICROBIO] A family of bacteria in the suborder Rhodospirillineae; cells are motile by flagella, multiplication is by budding or binary fission, and photosynthetic membranes are continuous with the cytoplasmic membrane. { ¦rō·dō‚spī·rə'lās·ē‚ē }

Rhodospirillales [MICROBIO] The single order of the phototrophic bacteria; cells are spherical, rod-shaped, spiral, or vibrio-shaped, and all contain bacteriochlorophylls and carotenoid pigments. { ¦rō·dō‚spī·rə'lā·lēz }

Rhodospirillineae [MICROBIO] The purple bacteria, a suborder of the order Rhodospirillales; contain bacteriochlorophyll *a* or *b*, located on internal membranes. { ¦rō·dō‚spī·rə'lin·ē‚ē }

rhodoxanthin [BIOCHEM] $C_{40}H_{50}O_2$ A xanthophyll carotenoid pigment. { ¦rō·dō'zan·thən }

rhohelos [ECOL] A stream-crossed, nonalluvial marsh typical of filled lake areas. { rō'hē‚lōs }

Rhoipteleaceae [BOT] A monotypic family of dicotyledonous plants in the order Juglandales having pinnately compound leaves, and flowers in triplets with four sepals and six stamens, and the lateral flowers female but sterile. { ˌrȯipˌtē·lē'ās·ēˌē }

rhombencephalon [EMBRYO] The most caudal of the primary brain vesicles in the vertebrate embryo. Also known as hindbrain. { ¦rämˌben 'sef·əˌlän }

rhombogen [INV ZOO] A form of reproductive individual of the mesozoan order Dicyemida found in the sexually mature host which arises from nematogens and gives rise to free-swimming infusorigens. { 'räm·bə·jən }

Rhopalidae [INV ZOO] A family of pentatomorphan hemipteran insects in the superfamily Coreoidea. { rō'pāl·əˌdē }

rhopalium [INV ZOO] A sense organ found on the margin of a discomedusan. { rō'pāl·ē·əm }

Rhopalocera [INV ZOO] Formerly a suborder of Lepidoptera comprising those forms with clubbed antennae. { ˌrō·pə'läs·ə·rə }

rhopalocercous cercaria [INV ZOO] A free-swimming digenetic trematode larva distinguished by a very wide tail. { ¦rō·pə·lō¦sər·kəs sər'kar·ē·ə }

Rhopalodinidae [INV ZOO] A family of holothurian echinoderms in the order Dactylochirotida in which the body is flask-shaped, the mouth and anus lying together. { ¦rō·pə·lō'din·əˌdē }

Rhopalosomatidae [INV ZOO] A family of hymenopteran insects in the superfamily Scolioidea. { ¦rō·pə·lō·sō'mad·əˌdē }

Rh positive [MED] Presence of the Rh-D antigen on the surface of red blood cells. { ¦är¦āch 'päz·əd·iv }

rhubarb [BOT] *Rheum rhaponticum*. A herbaceous perennial of the order Polygoniales grown for its thick, edible petioles. { 'rüˌbärb }

Rhynchobdellae [INV ZOO] An order of the class Hirudinea comprising leeches that possess an eversible proboscis and lack hemoglobin in the blood. { ˌriŋˌkäb'de·lē }

Rhynchocephalia [VERT ZOO] An order of lepidosaurian reptiles represented by a single living species, *Sphenodon punctatus*, and characterized by a diapsid skull, teeth fused to the edges of the jaws, and an overhanging beak formed by the upper jaw. { ˌriŋ·kō·sə'fāl·yə }

rhynchocoel [INV ZOO] A cavity that holds the inverted proboscis in nemertinean worms. { 'riŋ·kōˌsēl }

Rhynchocoela [INV ZOO] A phylum of bilaterally symmetrical, unsegmented, ribbonlike worms having an eversible proboscis and a complete digestive tract with an anus. { ˌriŋ·kō 'sē·lə }

rhynchodaeum [INV ZOO] The part of the proboscis lying anterior to the brain in nemertinean worms. { riŋ'kō·dē·əm }

Rhynchodina [INV ZOO] A suborder of ciliate protozoans in the order Thigmotrichida. { ˌriŋ·kə'dī·nə }

Rhynchonellida [INV ZOO] An order of articulate brachiopods; typical forms are dorsibiconvex, the posterior margin is curved, the dorsal interarea is absent, and the ventral one greatly reduced. { ˌriŋ·kə'nel·əd·ə }

rhynchophorous [ZOO] Having a beak. { riŋ 'käf·ə·rəs }

Rhynochetidae [VERT ZOO] A monotypic family of gruiform birds containing only the kagu of New Caledonia. { ˌrī·nə'kēd·əˌdē }

Rhysodidae [INV ZOO] The wrinkled bark beetles, a family of coleopteran insects in the suborder Adephaga. { rī'säd·əˌdē }

rib [ANAT] One of the long curved bones forming the wall of the thorax in vertebrates. [BOT] A primary vein in a leaf. { rib }

ribitol *See* adonitol. { 'rī·bəˌtȯl }

riboflavin [BIOCHEM] $C_{17}H_{20}N_4O_6$ A water-soluble, yellow orange fluorescent pigment that is essential to human nutrition as a component of the coenzymes flavin mononucleotide and flavin adenine dinucleotide. Also known as lactoflavin; vitamin B_2; vitamin G. { 'rī·bəˌflā·vən }

riboflavin 5′-phosphate [BIOCHEM] $C_{17}H_{21}N_4O_9P$ The phosphoric acid ester of riboflavin. Also known as flavin phosphate; flavin mononucleotide; FMN; isoalloxazine mononucleotide; vitamin B_2 phosphate. { 'rī·bəˌflā·vən ¦fīvˌprīm 'fäˌsfāt }

D-riboketose *See* ribulose. { ¦dē ¦rī·bō'kēˌtōs }

ribonuclease [BIOCHEM] $C_{587}H_{909}N_{171}O_{197}S_{12}$ An enzyme that catalyzes the depolymerization of ribonucleic acid. { ˌrī·bō'nü·klēˌās }

ribonucleic acid [BIOCHEM] A long-chain, usually single-stranded nucleic acid consisting of repeating nucleotide units containing four kinds of heterocyclic, organic bases: adenine, cytosine, quanine, and uracil; they are conjugated to the pentose sugar ribose and held in sequence by phosphodiester bonds; involved intracellularly in protein synthesis. Abbreviated RNA. { ¦rī·bō¦nü¦klē·ik 'as·əd }

ribonucleic acid polymerase [BIOCHEM] An enzyme that transcribes a ribonucleic acid (RNA) molecule from one strand of a deoxyribonucleic acid (DNA) molecule. Also known as transcriptase. { ¦rī·bō¦nü¦klē·ik 'as·əd pə'lim·əˌrās }

ribonucleoprotein [BIOCHEM] Any of a large group of conjugated proteins in which molecules of ribonucleic acid are closely associated with molecules of protein. { ¦rī·bō¦nü·klē·ō'prōˌtēn }

ribonucleotide [BIOCHEM] A ribose-containing nucleotide, the structural unit of ribonucleic acid. { ¦rī·bō'nü·klē·əˌtīd }

ribose [BIOCHEM] $C_5H_{10}O_5$ A pentose sugar occurring as a component of various nucleotides, including ribonucleic acid. { 'rīˌbōs }

riboside [BIOCHEM] Any glycoside containing ribose as the sugar component. { 'rī·bəˌsīd }

ribosomal ribonucleic acid [BIOCHEM] Any of three large types of ribonucleic acid found in ribosomes: 5S RNA, with molecular weight 40,000; 14–16S RNA, with molecular weight 600,000; and 18–22S RNA with molecular weight

1,200,000. Abbreviated r-RNA. { ¦rī·bə¦sō·məl ¦rī·bō¦nü¦klē·ik 'as·əd }

ribosome [CELL MOL] One of the small, complex particles composed of various proteins and three molecules of ribonucleic acid which synthesize proteins within the living cell. { 'rī·bə,sōm }

ribozyme [BIOCHEM] A ribonucleic acid molecule that can catalyze, or lower the activation energy for, specific biochemical reactions. { 'rīb·ə,zīm }

ribulose [BIOCHEM] $C_5H_{10}O_5$ A pentose sugar that exists only as a syrup; synthesized from arabinose by isomerization with pyridine; important in carbohydrate metabolism. Also known as D-erythropentose; D-riboketose. { 'rī·byə,lōs }

ribulose 1,5-bisphosphate carboxylase/oxygenase [BIOCHEM] A key enzyme in carbon fixation during photosynthesis by plant chloroplasts and their cyanobacterial relatives, as well as by photosynthetic proteobacteria. Also known as rubisco; RuBP carboxylase/oxygenase. { ¦rī·byə,lo ¦wən ¦fiv ,bis¦fäs,fāt kä¦bäk·sə,lās 'äk·sə·jə,nās }

ribulose diphosphate [BIOCHEM] $C_5H_{12}O_{11}P_2$ The phosphate ester of ribulose. { 'rī·byə,lōs dī'fä,sfāt }

rice [BOT] *Oryza sativa*. An annual cereal grass plant of the order Cyperales, cultivated as a source of human food for its carbohydrate-rich grain. { rīs }

Ricinidae [INV ZOO] A family of bird lice, order Mallophaga, which occur on numerous land and water birds. { rə'sin·ə,dē }

Ricinuleida [INV ZOO] An order of rare, ticklike arachnids in which the two anterior pairs of appendages are chelate, and the terminal segments of the third legs of the male are modified as copulatory structures. { ,ris·ən·yü'lē·ə·də }

rickets [MED] A disorder of calcium and phosphorus metabolism affecting bony structures, due to vitamin D deficiency. { 'rik·əts }

Rickettsiaceae [MICROBIO] A family of the order Rickettsiales; small, rod-shaped, coccoid, or diplococcoid cells often found in arthropods; includes human and animal parasites and pathogens. { ri,ket·sē'ās·ē,ē }

Rickettsiales [MICROBIO] An order of prokaryotic microorganisms; gram-negative, obligate, intracellular animal parasites (may be grown in tissue cultures); many cause disease in humans and animals. { ri,ket·sē'ā·lēz }

rickettsialpox [MED] An acute febrile disease caused by the organism *Rickettsia akari* and transmitted from the mouse to humans by the mite *Allodermanyssus sanguineus*; characterized by rash, a primary ulcer, and often swelling of glands. { ri¦ket·sē·əl¦päks }

Rickettsieae [MICROBIO] A tribe of the family Rickettsiaceae; cells are occasionally filamentous; infect arthropods and some vertebrates and are pathogenic for humans, most frequently an incidental host. { ri'ket·sē,ē }

rickettsiosis [MED] Any disease caused by rickettsiae. { ri,ket·sē'ō·səs }

rictus [VERT ZOO] The mouth aperture in birds. { 'rik·təs }

Riedel's disease [MED] A form of chronic thyroiditis with irregular localized areas of stony, hard fibrosis. { 'rēd·əlz di,zēz }

rifampicin [MICROBIO] An antibacterial and antiviral antibiotic; action depends upon its preferential inhibition of bacterial ribonucleic acid polymerase over animal-cell RNA polymerase. { rə'fam·pə·sən }

Rift Valley fever [MED] A toxic generalized febrile virus disease of humans and animals in South and East Africa, transmitted by a mosquito, and characterized by headache, photophobia, myalgia, and anorexia. { 'rift ¦val·ē 'fē·vər }

rigor [MED] **1.** Stiffness. **2.** A chill associated with muscular contraction and tremor. { 'rig·ər }

rigor mortis [PATH] Stiffening and rigidity of the musculature occurring after death, beginning within 5–10 hours, and disappearing after 3–4 days. { 'rig·ər 'mȯrd·əs }

Riley-Day syndrome *See* dysautonomia. { ¦rī·lē ¦dā 'sin,drōm }

rind [BOT] **1.** The bark of a tree. **2.** The thick outer covering of certain fruits. { rīnd }

rinderpest [VET MED] An acute, contagious, and often fatal virus disease of cattle, sheep, and goats which is characterized by fever and the appearance of ulcers on the mucous membranes of the intestinal tract. { 'rin·dər,pest }

ring canal [INV ZOO] In echinoderms, the circular tube of the water-vascular system that surrounds the esophagus. { 'riŋ kə,nal }

ring deoxyribonucleic acid *See* circular deoxyribonucleic acid. { 'riŋ dē¦äk·sē¦rī·bō¦nü¦klē·ik 'as·əd }

ringent [BOT] Having widely separated, gaping lips. [ZOO] Gaping irregularly. { 'rin·jənt }

ringtail *See* cacomistle. { 'riŋ,tāl }

ring test [IMMUNOL] The simplest of the precipitin tests for antigen-antibody reaction; the solution containing antigen is layered on a solution containing antibody; a white disk or precipitate forms at the point where the two solutions diffuse until optimum concentration for precipitation is reached. { 'riŋ ,test }

ring vessel [INV ZOO] A part of the water-vascular system in echinoderms; it is the circular canal around the mouth into which the stone canal empties, and from which a radial water vessel traverses to each of five radii. { 'riŋ ,ves·əl }

ringworm [MED] A fungus infection of skin, hair, or nails producing annular lesions with elevated margins. Also known as tinea. { 'riŋ,wərm }

Riodininae [INV ZOO] A subfamily of the lepidopteran family Lycaenidae in which prothoracic legs are nonfunctional in the male. { ,rī·ə'din·ə,dē }

riparian [BIOL] Living or located on a riverbank. { rə'per·ē·ən }

riparian zone [ECOL] The part of the watershed immediately adjacent to the stream channel. { rī'per·ē·ən ,zōn }

ripe [BOT] Of fruit, fully developed, having mature seed and so usable as food. { rīp }

river-delta marsh [EVOL] A brackish or freshwater marsh bordering the mouth of a distributary stream. { 'riv·ər ¦del·tə ,märsh }

rivulose [BOT] Marked by irregular, narrow lines. { 'riv·yə,lōs }

R loop hybridization [CELL MOL] A process by which ribonucleic acid is annealed with double-stranded deoxyribonucleic acid (DNA), thus displacing a single DNA strand. { ¦är ,lüp ,hī·brid·ə'zā·shən }

RNA *See* ribonucleic acid.

RNA editing [CELL MOL] Alteration in the nucleotide sequence of messenger RNA after it has been transcribed from DNA, changing the nature of its protein product from what was encoded; may entail insertion, deletion, or enzyme-catalyzed chemical alteration of nucleotides. { ¦är ¦en¦ā 'ed·əd·iŋ }

RNAi *See* RNA interference. { ¦är¦en¦ā'ī }

RNA interference [CELL MOL] The process by which foreign, double-stranded RNA is recognized and degraded by specialized protein complexes within many eukaryotic cells; believed to be an evolutionarily conserved defense mechanism against RNA viruses and transposable elements. Abbreviated RNAi. { ¦är¦en¦ā ,in·tər 'fir·əns }

RNA polymerase I [CELL MOL] An enzyme found in the nucleolus that copies ribosomal genes to produce ribosomal ribonucleic acid. { ¦är¦en¦ā pə¦lim·ə,rās 'wən }

RNA polymerase II [CELL MOL] An enzyme that transcribes messenger ribonucleic acid sequences in the chromatin. { ¦är¦en¦ā pə¦lim·ə,rās 'tü }

RNA polymerase III [CELL MOL] An enzyme that directs the synthesis of small ribonucleic acid (RNA) molecules such as the amino acyl transfer RNAs. { ¦är¦en¦ā pə¦lim·ə,rās 'thrē }

RNA primer [CELL MOL] A short strand of RNA that is synthesized along single-stranded DNA during replication, initiating DNA polymerase-catalyzed synthesis of the complementary strand. { ¦är¦en¦ā 'prī·mər }

roach [INV ZOO] An insect of the family Blattidae; the body is wide and flat, the anterior part of the thorax projects over the head, and antennae are long and filiform, with many segments. Also known as cockroach. { rōch }

Robertinacea [INV ZOO] A superfamily of marine, benthic foraminiferans in the suborder Rotaliina characterized by a trochospiral or modified test with a wall of radial aragonite, and having bilamellar septa. { rä,bərd·ə'nās·ē·ə }

Roccilaceae [BOT] A family of fruticose species of Hysteriales that grow profusely on trees and rocks along the coastlines of Portugal, California, and western South America. { ,räs·ə'lās·ē,ē }

rocket electrophoresis [IMMUNOL] A variant of crossed electrophoresis in which the medium contains only one antibody; test substances are driven directly into the medium that contains the antibody, forming rocket-shaped (inverted V) trails of precipitation. { 'räk·ət i,lek·trō·fə'rē·səs }

rock shell [INV ZOO] The common name for a large number of gastropod mollusks composing the family Muricidae and characterized by having conical shells with various sculpturing. { 'räk ,shel }

Rocky Mountain spotted fever [MED] An acute, infectious, typhuslike disease of man caused by the rickettsial organism *Rickettsia rickettsi* and transmitted by species of hard-shelled ticks; characterized by sudden onset of chills, headache, fever, and an exanthem on the extremities. Also known as American spotted fever; tick fever; tick typhus. { 'räk·ē 'maunt·ən 'späd·əd 'fē·vər }

rod [HISTOL] One of the rod-shaped sensory bodies in the retina which are sensitive to dim light. { räd }

rodent [VERT ZOO] The common name for members of the order Rodentia. { 'rōd·ənt }

Rodentia [VERT ZOO] An order of mammals characterized by a single pair of ever-growing upper and lower incisors, a maximum of five upper and four lower cheek teeth on each side, and free movement of the lower jaw in an anteroposterior direction. { rō'den·chə }

roentgenotherapy *See* x-ray therapy. { ,rent·gə·nō'ther·ə·pē }

roll-tube technique [MICROBIO] A pure-culture technique, employed chiefly in tissue culture, in which, during incubation, the test tubes are held in a wheellike instrument at an angle of about 15° from the horizontal and the wheel is rotated vertically about once every 2 minutes. { 'rōl ,tüb tek,nēk }

Romberg's sign [MED] **1.** A sign for obturator hernia in which there is pain radiating to the knee. **2.** A sign for loss of position sense in which the patient cannot maintain equilibrium when standing with feet together and eyes closed. { 'räm,bərgz ,sīn }

rookery [ZOO] A location used by birds for breeding and nesting. { 'rük·ə·rē }

rooster [VERT ZOO] An adult male of certain birds and fowl, such as pheasants and ptarmigans. { 'rüs·tər }

root [BOT] The absorbing and anchoring organ of a vascular plant; it bears neither leaves nor flowers and is usually subterranean. { rüt }

root canal [ANAT] The cavity within the root of a tooth, occupied by pulp, nerves, and vessels. { 'rüt kə,nal }

root cap [BOT] A thick, protective mass of parenchymal cells covering the meristematic tip of the root. { 'rüt ,kap }

root hair [BOT] One of the hairlike outgrowths of the root epidermis that function in absorption. { 'rüt ,her }

root-knot nematode [INV ZOO] A plant-parasitic nematode species that induces galls or knots to form on roots. { ,rüt ,nät 'nēm·ə,tōd }

rootstock [BOT] A root or part of a root used as the stock for grafting. { 'rüt,stäk }

rootworm [INV ZOO] **1.** An insect larva that

feeds on plant roots. **2.** A nematode that infests the roots of plants. { 'rüt,wərm }

Roproniidae [INV ZOO] A small family of hymenopteran insects in the superfamily Proctotrupoidea. { ,räp·rə'nī·ə,dē }

rosacea [MED] A chronic skin disorder of middle age characterized by redness, papules, and oiliness. { rō'zā·shē·ə }

Rosaceae [BOT] A family of dicotyledonous plants in the order Rosales typically having stipulate leaves and hypogynous, slightly perigynous, or epigynous flowers, numerous stamens, and several or many separate carpels. { rō'zās·ē,ē }

Rosales [BOT] A morphologically diffuse order of dicotyledonous plants in the subclass Rosidae. { rō'zā·lēz }

rose [BOT] A member of the genus *Rosa* in the rose family (Rosaceae); plants are erect, climbing, or trailing shrubs, generally prickly stemmed, and bear alternate, odd-pinnate single leaves. { rōz }

rose hip [BOT] The ripened false fruit of a rose plant. { 'rōz ,hip }

rosemary [BOT] *Rosmarinus officinalis.* A fragrant evergreen of the mint family from France, Spain, and Portugal; leaves have a pungent bitter taste and are used as an herb and in perfumes. { 'rōz,mer·ē }

Rosenmueller's organ *See* epoophoron. { 'rōz·ən,mül·ərz ,ȯr·gən }

roseola infantum *See* exanthem subitum. { ,rō·zē'ō·lə in'fan·təm }

roseola typhosa [MED] The rose-colored eruption characteristic of typhus or typhoid fever. { ,rō·zē'ō·ə tī'fō·sə }

rosette [BIOL] Any structure or marking resembling a rose. { rō'zet }

Rosidae [BOT] A large subclass of the class Magnoliatae; most have a well-developed corolla with petals separate from each other, binucleate pollen, and ovules usually with two integuments. { 'rōz·ə,dē }

rostellum [BIOL] The anterior, flattened region of the scolex of armed tapeworms. { rä'stel·əm }

Rostratulidae [VERT ZOO] A small family of birds in the order Charadriiformes containing the painted snipe; females are more brightly colored than males. { ,rä·strə'tyü·lə,dē }

rostrum [BIOL] A beak or beaklike process. { 'rä·strəm }

Rotaliacea [INV ZOO] A superfamily of foraminiferans in the suborder Rotaliina characterized by a planispiral or trochospiral test having apertural pores and composed of radial calcite, with secondarily bilamellar septa. { rō,tal·ē'ā·shə }

rotate [BOT] Of a sympetalous corolla, having a short tube and petals radiating like the spokes of a wheel. { 'rō,tāt }

rotation therapy [MED] Radiation therapy in which either the patient or the source of radiation is rotated, to permit a larger dose at the center of rotation within the patient's body than on any area of the skin. { rō'tā·shən ,ther·ə·pē }

rotator [ANAT] A muscle that partially rotates a part of the body on the part's axis. { 'rō,tād·ər }

Rotatoria [INV ZOO] The equivalent name for Rotifera. { ,rōd·ə'tȯr·ē·ə }

röteln *See* rubella. { 're,teln }

Rotifera [INV ZOO] A class of the phylum Aschelminthes distinguished by the corona, a retractile trochal disk provided with several groups of cilia and located on the head. { rō'tif·ə·rə }

rough colony [MICROBIO] A flattened, irregular, and wrinkled colony of bacteria indicative of decreased capsule formation and virulence. { 'rəf 'käl·ə·nē }

rouleau [PATH] A roll of erythrocytes resembling a stack of coins. { rü'lō }

round ligament [ANAT] **1.** A flattened band extending from the fovea on the head of the femur to attach on either side of the acetabular notch between which it blends with the transverse ligament. **2.** A fibrous cord running from the umbilicus to the notch in the anterior border of the liver; represents the remains of the obliterated umbilical vein. { 'raůnd 'lig·ə·mənt }

round window [ANAT] A membrane-covered opening between the middle and inner ears in amphibians and mammals through which energy is dissipated after traveling in the membranous labyrinth. { 'raůnd 'win·dō }

roundworm [INV ZOO] The name applied to nematodes. { 'raůnd,wərm }

Rous sarcoma [VET MED] A fibrosarcoma that can be produced in chickens, pheasants, and ducklings inoculated with the filterable, ribonucleic acid Rous virus. { 'raůs sär'kō·mə }

RQ *See* respiratory quotient.

r-RNA *See* ribosomal ribonucleic acid.

r selection [ECOL] Selection that favors rapid population growth (r represents the intrinsic rate of increase). { 'är si,lek·shən }

RTF *See* resistance transfer factor.

rubber tree [BOT] *Hevea brasiliensis.* A tall tree of the spurge family (Euphorbiaceae) from which latex is collected and coagulated to produce rubber. { 'rəb·ər ,trē }

rubella [MED] An infectious virus disease of humans characterized by coldlike symptoms, fever, and transient, generalized pale-pink rash; its occurrence in early pregnancy is associated with congenital abnormalities. Also known as epidemic roseola; French measles; German measles; röteln. { rü'bel·ə }

rubeola *See* measles. { ,rü·bē'ō·lə }

Rubiaceae [BOT] The single family of the plant order Rubiales. { ,rü·bē'ās·ē,ē }

Rubiales [BOT] An order of dicotyledonous plants marked by their inferior ovary, regular or nearly regular corolla, and opposite leaves with interpetiolar stipules or whorled leaves without stipules. { ,rü·bē'ā·lēz }

rubisco *See* ribulose-1,5-bisphosphate carboxylase/oxygenase. { rü'bis·kō }

RuBP carboxylase/oxygenase *See* ribulose 1,5-bisphosphate carboxylase/oxygenase. { ¦rü¦bē ¦pē kär¦bak·sə,lās 'äk·sə·jə,nās }

rubredoxin [BIOCHEM] A class of iron-sulfur proteins that contains one iron coordinated to the sulfur atom of four cysteine residues. { ¦rü·brə'däk·sən }

rubriblast *See* pronormoblast. { 'rü·brə,blast }

rubricyte *See* pronormoblast. { 'rü·brə,sīt }

ruddy turnstone [VERT ZOO] *Arenaria interpes*. A member of the avian order Charadriiformes that perform transpacific flights during their migration. { 'rəd·ē 'tərn,stōn }

ruderal [ECOL] **1.** Growing on rubbish, or waste or disturbed places. **2.** A plant that thrives in such a habitat. { 'rüd·ə·rəl }

Ruffini cylinder [NEUROSCI] A cutaneous nerve ending suspected as the mediator of warmth. { rü'fē·nē ,sil·ən·dər }

rufous [BOT] Having a reddish-brown color. { 'rü·fəs }

rugose [BIOL] Having a wrinkled surface. { 'rü,gōs }

rumen [VERT ZOO] The first chamber of the ruminant stomach. Also known as paunch. { 'rü·mən }

ruminant [PHYSIO] Characterized by the act of regurgitation and rechewing of food. [VERT ZOO] A mammal belonging to the Ruminantia. { 'rü·mə·nənt }

Ruminantia [VERT ZOO] A suborder of the Artiodactyla including sheep, goats, camels, and other forms which have a complex stomach and ruminate their food. { ,rü·mə'nan·chə }

rumination [MED] Voluntary regurgitation of food from the stomach, followed by remastication and swallowing in emotionally or mentally disturbed persons. Also known as merycism. [PHYSIO] Regurgitation and remastication of food in preparation for true digestion in ruminants. { ,rü·mə'nā·shən }

rumination disorder [MED] A childhood eating disorder that involves repeated regurgitation and rechewing of food. This behavior is not the result of a gastrointestinal or medical condition; the partially digested food comes back into the mouth without any observable nausea, disgust, or attempt to vomit. { ,rü·mə'nā·shən dis ,ȯrd·ər }

runcinate [BOT] Pinnately cut with downward-pointing lobes. { 'rən·sə·nət }

runner [BOT] A horizontally growing, sympodial stem system; adventitious roots form near the apex, and a new runner emerges from the axil of a reduced leaf. Also known as stolon. { 'rən·ər }

running bird [VERT ZOO] Any of the large, flightless, heavy birds usually categorized as ratites. { 'rən·iŋ ,bərd }

runoff desert [ECOL] An arid region in which local rain is insufficient to support any perennial vegetation except in drainage or runoff channels. { 'rən,ȯf ,dez·ərt }

rupicolous [ECOL] Living among or growing on rocks. { rü'pik·ə·ləs }

rupture *See* hernia. { 'rəp·chər }

Russell bodies [PATH] Hyaline eosinophilic globules 4–5 micrometers in diameter, thought to be particles of antibody globulin, occurring in the cytoplasm of plasma cells in chronic inflammatory exudates. { 'rəs·əl ,bäd·ēz }

Russell's viper *See* tic polonga. { 'rəs·əlz 'vī·pər }

rust fungi *See* Urediniomycetes. { 'rəst ,fən,jī }

rut [PHYSIO] The period during which the male animal has a heightened mating drive. { rət }

rutabaga [BOT] *Brassica napobrassica*. A biennial crucifer of the order Capparales probably resulting from the natural crossing of cabbage and turnip and characterized by a large, edible, yellowish fleshy root. { ¦rüd·ə¦bā·gə }

Rutaceae [BOT] A family of dicotyledonous plants in the order Sapindales distinguished by mostly free stamens and glandular-punctate leaves. { rü'tās·ē,ē }

rye [BOT] *Secale cereale*. A cereal plant of the order Cyperales cultivated for its grain, which contains the most desirable gluten, next to wheat. { rī }

rye buckwheat *See* tartary buckwheat. { 'rī 'bək,wēt }

Rynchopidae [VERT ZOO] The skimmers, a family of birds in the order Charadriiformes with a knifelike lower beak that is longer and narrower than the upper one. { ,riŋ'käp·ə,dē }

Rytiodinae [VERT ZOO] A subfamily of trichechiform sirenians in the family Dugongidae. { ,rid·ē'äd·ən,ē }

S

saba [BOT] A plant (*Musa sapientum* var. *compressa*) that is common in the Philippines; the fruit is a cooking banana. { sə'bä }

Sabellariidae [INV ZOO] The sand-cementing worms, a family of polychaete annelids belonging to the Sedentaria and characterized by a compact operculum formed of setae of the first several segments. { sə,bel·ə'rī·ə,dē }

Sabellidae [INV ZOO] A family of sedentary polychaete annelids often occurring in intertidal depths but descending to great abyssal depths; one of two families that make up the feather-duster worms. { sə'bel·ə,dē }

Sabellinae [INV ZOO] A subfamily of the Sabellidae including the most numerous and largest members. { sə'bel·ə,nē }

Sabin vaccine [IMMUNOL] A live-poliovirus vaccine that is administered orally. { 'sā·bən vak'sēn }

sable [VERT ZOO] *Martes zibellina.* A carnivore of the family Mustelidae; a valuable fur-bearing animal, quite similar to the American marten. { 'sā·bəl }

sablefish [VERT ZOO] *Anoplopoma fimbria.* An abundant black-skinned fish in the North Pacific. { 'sā·bəl,fish }

Sabouraud's agar [MICROBIO] A peptone-maltose agar used as a culture medium for pathogenic fungi, especially the dermatophytes. { sa·bü'rōz 'ag·ər }

sac [BIOL] A soft-walled cavity within a plant or animal, often containing a special fluid and usually having a narrow opening or none at all. { sak }

saccadic movement [PHYSIO] Rapid eye movement that transfers the gaze from one fixation point to another. { sə¦kad·ik 'müv·mənt }

saccate [BOT] Having a saclike or pouchlike form. { 'sa,kāt }

saccharase [BIOCHEM] An enzyme that catalyzes the hydrolysis of disaccharide to monosaccharides, specifically of sucrose to dextrose and levulose. Also known as invertase; invertin; sucrase. { 'sak·ə,rās }

saccharolytic [ZOO] Pertaining to an organism that metabolizes carbohydrates. { ,sak·ə·rə 'lid·ik }

Saccharomycetaceae [MYCOL] The single family of the order Saccharomycetales. { ¦sak·ə·rō,mī·sə'tās·ē,ē }

Saccharomycetales [MYCOL] An order of the subclass Hemiascomycetidae comprising typical yeasts, characterized by the presence of naked asci in which spores are formed by free cells. { ¦sak·ə·rō,mī·sə'tā·lēz }

Saccharomycetoideae [MYCOL] A subfamily of Saccharomycetacae in which spores may be hat-, sickle-, or kidney-shaped, or round or oval. { ¦sak·ə·rō,mī·sə'tȯid·ē,ē }

saccharopinuria [MED] An inborn error of amino acid metabolism characterized by abnormally high levels of saccharopine in the urine. { ¦sak·ə·rō·pi'nu̇r·ē·ə }

Saccoglossa [INV ZOO] An order of gastropod mollusks belonging to the Opisthobranchia. { ,sak·ə'gläs·ə }

Saccopharyngiformes [VERT ZOO] Formerly an order of actinopterygian fishes, the gulpers, now included in the Anguilliformes. { ,sak·ō·fə,rin·jə'fȯr,mēz }

Saccopharyngoidei [VERT ZOO] The gulpers, a suborder of actinopterygian fishes in the order Anguilliformes having degenerative adaptations, including loss of swim bladder, opercle, branchiostegal ray, caudal fin, scales, and ribs. { ,sak·ō,far·əŋ'gȯid·ē,ī }

saccular aneurysm [MED] A saclike arterial dilation communicating with the artery by a relatively small opening. { 'sak·yə·lər 'an·yə,riz·əm }

sacculus [ANAT] The smaller, lower saclike chamber of the membranous labyrinth of the vertebrate ear. { 'sak·yə·ləs }

saccus *See* vesicle. { 'sak·əs }

sac fungus [MYCOL] The common name for members of the class Ascomycetes. { 'sak ,fəŋ·gəs }

sacral block [MED] Anesthesia induced by injection of an anesthetic through the caudal hiatus. { 'sak·rəl 'bläk }

sacral nerve [NEUROSCI] Any of five pairs of spinal nerves in the sacral region which innervate muscles and skin of the lower back, lower extremities, and perineum, and branches to the hypogastric and pelvic plexuses. { 'sak·rəl 'nərv }

sacral vertebrae [ANAT] Three to five fused vertebrae that form the sacrum in most mammals; amphibians have one sacral vertebra, reptiles usually have two, and birds have 10–23 fused in the synsacrum. { 'sak·rəl 'vərd·ə,brā }

sacrococcygeus [ANAT] One of two inconstant thin muscles extending from the lower sacral vertebrae to the coccyx. { ¦sa·krō·käk'sij·ē·əs }
sacroiliac [ANAT] Pertaining to the sacrum and the ilium. { ¦sa·krō¦il·ē,ak }
sacrospinous [ANAT] Pertaining to the sacrum and the spine of the ischium. { ¦sa·krō'spī·nəs }
sacrum [ANAT] A triangular bone, consisting in humans of five fused vertebrae, located below the last lumbar vertebra, above the coccyx, and between the hipbones. { 'sak·rəm }
Saefftigen's pouch [INV ZOO] An elongated pouch inside the genital sheath in many acanthocephalans. { 'zef·ti·gənz ,pau̇ch }
safflower [BOT] *Carthamus tinctorius.* An annual thistlelike herb belonging to the composite family (Compositae); the leaves are edible, flowers yield dye, and seeds yield a cooking oil. { 'sa,flau̇·ər }
saffron [BOT] *Crocus sativus.* A crocus of the iris family (Iridaceae); the source of a yellow dye used for coloring food and medicine. { 'saf·rən }
Sagartiidae [INV ZOO] A family of zoantharians in the order Actiniaria. { ,sag·ər'tī·ə,dē }
sage [BOT] *Salvia officinalis.* A half-shrub of the mint family (Labiatae); the leaves are used as a spice. { sāj }
sagebrush [BOT] Any of various hoary undershrubs of the genus *Artemisia* found on the alkaline plains of the western United States. { 'sāj,brəsh }
sagitta [VERT ZOO] The larger of two otoliths in the ear of most fishes. { sə'jid·ə }
sagittal [ZOO] In the median longitudinal plane of the body, or parallel to it. { 'saj·əd·əl }
Sagittariidae [VERT ZOO] A family of birds in the order Falconiformes comprising a single species, the secretary bird, noted for its nuchal plumes resembling quill pens stuck behind an ear. { ,saj·ə·tə'rī·ə,dē }
sagittate [BOT] Shaped like an arrowhead, especially referring to leaves. { 'saj·ə,tāt }
sagittocyst [INV ZOO] A cyst in the epidermis of certain turbellarians containing a single spindle-shaped needle. { sə'jid·ə,sist }
sahel [ECOL] A region having characteristics of a savanna or a steppe and bordering on a desert. { sə'hel }
sailfish [VERT ZOO] Any of several large fishes of the genus *Istiophorus* characterized by a very large dorsal fin that is highest behind its middle. { 'sāl,fish }
Saint John's wort [PHARM] *Hypericum perforatum.* A herbacious perennial that has been used for millennia for its many medicinal properties, including wound healing and treatment of kidney and lung ailments, insomnia, and depression. { sānt 'jänz ,wȯrt }
Saint Louis encephalitis [MED] A mosquito-borne arbovirus infection of the central nervous system, occurring in the central and western United States and in Florida. { 'sānt 'lü·əs in ,sef·ə'līd·əs }
Saint Vitus dance [MED] Chorea associated with rheumatic fever. Also known as Sydenham's chorea. { 'sānt 'vīd·əs ,dans }
salamander [VERT ZOO] The common name for members of the order Urodela. { 'sal·ə,man·dər }
Salamandridae [VERT ZOO] A family of urodele amphibians in the suborder Salamandroidea characterized by a long row of prevomerine teeth. { ,sal·ə'man·drə,dē }
Salamandroidea [VERT ZOO] The largest suborder of the Urodela characterized by teeth on the roof of the mouth posterior to the openings of the nostrils. { ,sal·ə,man'drȯid·ē·ə }
Salangidae [VERT ZOO] A family of soft-rayed fishes, in the suborder Galaxioidei, which live in estuaries of eastern Asia. { sə'lan·jə,dē }
Saldidae [INV ZOO] The shore bugs, a family of predacious hemipteran insects in the superfamily Saldoidea. { 'sal·də,dē }
Saldoidea [INV ZOO] A superfamily of the hemipteran group Leptopodoidea. { sal'dȯi·dē·ə }
Saleniidae [INV ZOO] A family of echinoderms in the order Salenioida distinguished by imperforate tubercles. { ,sa·lə'nī·ə·dē }
Salenioida [INV ZOO] An order of the Echinacea in which the apical system includes one or several large angular plates covering the periproct. { sə,lē·nē'ȯi·də }
Salicaceae [BOT] The single family of the order Salicales. { ,sal·ə'kās·ē,ē }
Salicales [BOT] A monofamilial order of dicotyledonous plants in the subclass Dilleniidae; members are dioecious, woody plants, with alternate, simple, stipulate leaves and plumose-hairy mature seeds. { ,sal·ə'kā·lēz }
salicylanilide [PHARM] $C_{13}H_{11}NO_2$ A crystalline compound with a melting point of 135.8–136.2°C; freely soluble in alcohol, ether, chloroform, and benzene, slightly soluble in water; used as a topical antifungal agent in humans and animals and in medicine. { ¦sal·ə·səl'an·əl·əd }
salicylism [MED] A syndrome produced by excessive doses of salicylates; characterized by dizziness, headache, and nausea. { 'sal·ə·sə,liz·əm }
Salientia [VERT ZOO] The equivalent name for Anura. { ,sā·lē'en·chə }
saliva [PHYSIO] The opalescent, tasteless secretions of the oral glands. { sə'lī·və }
salivary amylase *See* ptyalin. { 'sal·ə,ver·ē 'am·ə,lās }
salivary diastase *See* ptyalin. { 'sal·ə,ver·ē 'dī·ə,stās }
salivary gland [PHYSIO] A gland that secretes saliva, such as the sublingual or parotid. { 'sal·ə,ver·ē ,gland }
salivary gland chromosomes [CELL MOL] Polytene chromosomes found in the interphase nuclei of salivary glands in the larvae of Diptera; chromosomes in the larva undergo complete somatic pairing to form two homologous polytene chromosomes fused side by side. { 'sal·ə,ver·ē ,gland 'krō·mə,sōmz }

salivation [MED] Mild mercury poisoning suffered by workers in amalgamation plants. [PHYSIO] Excessive secretion of saliva. { ˌsal·əˈvā·shən }

Salk vaccine [IMMUNOL] A killed-virus vaccine administered for active immunization against poliomyelitis. { ˈsȯk vakˌsēn }

salmon [VERT ZOO] The common name for a number of fish in the family Salmonidae which live in coastal waters of the North Atlantic and North Pacific and breed in rivers tributary to the oceans. { ˈsam·ən }

Salmonella [MICROBIO] A genus of gram-negative, facultatively anaerobic bacteria belonging to the family Enterobacteriaceae that cause enteric infections with or without blood invasion. Most species are motile, utilize citrate, decarboxylate ornithine, form gas from glucose, and produce hydrogen sulfide. Salmonellae do not ferment lactose, produce indole, or split urea; the Voges-Proskauer reaction is negative. { ˌsal·məˈnel·ə }

Salmonelleae [MICROBIO] Formerly a tribe of the Enterobacteriaceae comprising the pathogenic genera *Salmonella* and *Shigella*. { ˌsal·məˈnel·ēˌē }

salmonellosis [MED] Infection with any species of *Salmonella*. { ˌsal·mə·neˈlō·səs }

Salmonidae [VERT ZOO] A family of soft-rayed fishes in the suborder Salmonoidei including the trouts, salmons, whitefishes, and graylings. { salˈmän·əˌdē }

Salmoniformes [VERT ZOO] An order of soft-rayed fishes comprising salmon and their allies; the stem group from which most higher teleostean fishes evolved. { ˌsalˌmän·əˈfȯrˌmēz }

Salmonoidei [VERT ZOO] A suborder of the Salmoniformes comprising forms having an adipose fin. { sal·məˈnȯid·ēˌī }

Salmopercae [VERT ZOO] An equivalent name for Percopsiformes. { ˌsal·mōˈpərˌsē }

Salpida [INV ZOO] An order of tunicates in the class Thaliacea including transparent forms ringed by muscular bands. { ˈsal·pə·də }

Salpingidae [INV ZOO] The narrow-waisted bark beetles, a family of coleopteran insects in the superfamily Tenebrionoidea. { salˈpin·jəˌdē }

salpingitis [MED] **1.** Inflammation of the fallopian tube. **2.** Inflammation of the eustachian tube. { ˌsal·pənˈjīd·əs }

salpingo-oophoritis [MED] Inflammation of the fallopian tubes and ovaries. { salˌpiŋ·gōˌō·ə·fəˈrīd·əs }

salsoline [PHARM] $C_{11}H_{15}NO_2$ A compound that crystallizes from alcohol solution, melts at 221°C, soluble in hot alcohol and chloroform; used in medicine as an antihypertensive agent. { ˈsal·səˌlēn }

saltatorial [ZOO] Adapted for leaping. { ˌsal·təˌtȯr·ē·əl }

salt gland [VERT ZOO] A compound tubular gland, located around the eyes and nasal passages in certain marine turtles, snakes, and birds, which copiously secretes a watery fluid containing a high percentage of salt. { ˈsȯlt ˌgland }

Salticidea [INV ZOO] The jumping spiders, a family of predacious arachnids in the suborder Dipneumonomorphae having keen vision and rapid movements. { ˌsal·təˈsid·ē·ə }

saltmarsh [ECOL] A maritime habitat found in temperate regions, but typically associated with tropical and subtropical mangrove swamps, in which excess sodium chloride is the predominant environmental feature. { ˈsȯltˌmärsh }

saltmarsh plain [ECOL] A salt marsh that has been raised above the level of the highest tide and has become dry land. { ˈsȯltˌmärsh ˌplān }

salt-spray climax [ECOL] A climax community along exposed Atlantic and Gulf seacoasts composed of plants able to tolerate the harmful effects of salt picked up and carried by onshore winds from seawater. { ˈsȯlt ˌsprā ˈklīˌmaks }

Salviniales [BOT] A small order of heterosporous, leptosporangiate ferns (division Polypodiophyta) which float on the surface of the water. { ˌsalˌvin·ēˈāˌlēz }

samara [BOT] A dry, indehiscent, winged fruit usually containing a single seed, such as sugar maple (*Acer saccharum*). { səˈmar·ə }

Sambonidae [INV ZOO] A family of pentastomid arthropods in the suborder Porocephaloidea of the order Porocephalida. { samˈbän·əˌdē }

Samythinae [INV ZOO] A subfamily of sedentary polychaete annelids in the family Ampharetidae having a conspicuous dorsal membrane. { səˈmith·əˌnē }

Sandalidae [INV ZOO] The equivalent name for Rhipiceridae. { sanˈdal·əˌdē }

sandalwood [BOT] **1.** Any species of the genus *Santalum* of the sandalwood family (Santalaceae) characterized by a fragrant wood. **2.** *S. album*. A parasitic tree with hard, close-grained, aromatic heartwood used in ornamental carving and cabinetwork. { ˈsan·dəlˌwu̇d }

sandblow [ECOL] A patch of coarse, sandy soil denuded of vegetation by wind action. { ˈsanˌblō }

sand dollar [INV ZOO] The common name for the flat, disk-shaped echinoderms belonging to the order Clypeasteroida. { ˈsan ˌdäl·ər }

sandfly [INV ZOO] Any of various small biting Diptera, especially of the genus *Phlebotomus*, which are vectors for phlebotomus (sandfly) fever. { ˈsanˌflī }

sandfly fever *See* phlebotomus fever. { ˈsanˌflī ˌfē·vər }

sand hopper [INV ZOO] The common name for gammaridean crustaceans found on beaches. { ˈsan ˌhäp·ər }

sandpiper [VERT ZOO] Any of various small birds that are related to plovers and that frequent sandy and muddy shores in temperate latitudes; bill is moderately long with a soft, sensitive tip, legs and neck are moderately long, and plumage is streaked brown, gray, or black above and is white below. { ˈsanˌpī·pər }

sand shark [VERT ZOO] Any of various shallow-water predatory elasmobranchs of the family Carchariidae. Also known as tiger shark. { ˈsan ˌshark }

Sanfilippo's syndrome [MED] A hereditary metabolic disorder, transmitted as an autosomal recessive, characterized by excessive amounts of heparitin sulfate in the urine, and manifested by minor skeletal changes and slight hepatomegaly. { san·fə'lip·ōz ˌsinˌdrōm }

sanguineous [PHYSIO] Pertaining to or containing blood. { saŋ'gwin·ē·əs }

sanguivorous [ZOO] Feeding on blood. { saŋ 'gwiv·ə·rəs }

sanidaster [INV ZOO] A rod-shaped spicule having spines at intervals along its length. { ¦san·ə¦das·tər }

San Joaquin Valley fever *See* coccidioidomycosis. { ¦san wȯ¦kēn ¦val·ē 'fē·vər }

SA node *See* sinoauricular node. { ¦es¦ā 'nōd }

Santalaceae [BOT] A family of parasitic dicotyledonous plants in the order Santalales characterized by dry or fleshy indehiscent fruit, plants with chlorophyll, petals absent, and ovules without integument. { ˌsan·tə'lās·ēˌē }

Santalales [BOT] An order of dicotyledonous plants in the subclass Rosidae characterized by progressive adaptation to parasitism, accompanied by progressive simplification of the ovules. { ˌsan·tə'lā·lēz }

sap [BOT] The fluid part of a plant which circulates through the vascular system and is composed of water, gases, salts, and organic products of metabolism. { sap }

saphenous nerve [NEUROSCI] A somatic sensory nerve arising from the femoral nerve and innervating the skin of the medial aspect of the leg, foot, and knee joint. { sə'fē·nəs ˌnərv }

Sapindaceae [BOT] A family of dicotyledonous plants in the order Sapindales distinguished by mostly alternate leaves, usually one and less often two ovules per locule, and seeds lacking endosperm. { ˌsap·ən'dās·ēˌē }

Sapindales [BOT] An order of mostly woody dicotyledonous plants in the subclass Rosidae with compound or lobed leaves and polypetalous, hypogynous to perigynous flowers with one or two sets of stamens. { ˌsap·ən'dā·lēz }

sapling [BOT] A young tree with a trunk less than 4 inches (10 centimeters) in diameter at a point approximately 4 feet (1.2 meters) above the ground. { 'sap·liŋ }

Sapotaceae [BOT] A family of dicotyledonous plants in the order Ebenales characterized by a well-developed latex system. { ˌsap·ə'tās·ēˌē }

saprobe [ECOL] An organism that lives on decaying organic matter. { 'saˌprōb }

saprobic [BOT] Living on decaying organic matter; applied to plants and microorganisms. { sə'prō·bik }

saprogen [BIOL] An organism that lives on nonliving organic matter. { 'sap·rə·jən }

Saprolegniales [MYCOL] An order of aquatic fungi belonging to the class Phycomycetes, having a mostly hyphal thallus and zoospores with two flagella. { ˌsap·rəˌleg·nē'ā·lēz }

saprophage [BIOL] An organism that lives on decaying organic matter. { 'sap·rəˌfāj }

saprophyte [BOT] A plant that lives on decaying organic matter. { 'sap·rəˌfīt }

saprovore [ZOO] A detritus-eating animal. { 'sap·rəˌvȯr }

saprozoic [ZOO] Feeding on decaying organic matter; applied to animals. { ¦sap·rə¦zō·ik }

sapwood [BOT] The younger, softer, outer layers of a woody stem, between the cambium and heartwood. Also known as alburnum. { 'sapˌwu̇d }

Sapygidae [INV ZOO] A family of hymenopteran insects in the superfamily Scolioidea. { sə'pij·əˌdē }

Sarcina [MICROBIO] A genus of strictly anaerobic bacteria in the family Peptococcaceae; spherical cells occur in packets; ferment carbohydrates. { 'sär·sə·nə }

sarcochore [BOT] A plant dispersing minute, light disseminules. { 'sär·kəˌkȯr }

Sarcodina [INV ZOO] A superclass of Protozoa in the subphylum Sarcomastigophora in which movement involves protoplasmic flow, often with recognizable pseudopodia. { ˌsär·kə'dī·nə }

sarcoglia [CELL MOL] The protoplasm occurring at a myoneural junction. { sär'käg·lē·ə }

sarcoidosis [MED] A disease of unknown etiology characterized by granulomatous lesions, somewhat resembling true tubercles, but showing little or no necrosis, affecting the lymph nodes, skin, liver, spleen, heart, skeletal muscles, lungs, bones in distal parts of the extremities (osteitis cystica of Jüngling), and other structures, and sometimes by hyperglobulinemia, cutaneous anergy, and hypercalcinuria. { ˌsär ˌkȯi'dō·səs }

sarcolemma [HISTOL] The thin connective tissue sheath enveloping a muscle fiber. { ˌsär·kə'lem·ə }

sarcoleukemia *See* leukosarcoma. { ¦sär·kə·lü'kē·mē·ə }

sarcoma [MED] A malignant tumor arising in connective tissue and composed principally of anaplastic cells that resemble those of supportive tissues. { sär'kō·mə }

sarcoma botryoides [MED] A malignant mesenchymoma that forms grapelike structures; most common in the vagina of infants. { sär'kō·mə bä·trē'ȯi·dēz }

Sarcomastigophora [INV ZOO] A subphylum of Protozoa comprising forms that possess flagella or pseudopodia or both. { ˌsar·kəˌmas·tə'gäf·ə·rə }

sarcomere [HISTOL] One of the segments defined by Z disks in a skeletal muscle fibril. { 'sär·kəˌmir }

Sarcophagidae [INV ZOO] A family of the myodarian orthorrhaphous dipteran insects in the subsection Calypteratae comprising flesh flies, blowflies, and scavenger flies. { ˌsär·kə'fa·jəˌdē }

sarcoplasm [HISTOL] Hyaline, semifluid interfibrillar substance of striated muscle tissue. { 'sär·kəˌplaz·əm }

sarcoplasmic reticulum [CELL MOL] Collectively, the cysternae of a single muscle fiber. { ¦sar·kə¦plaz·mik rə'tik·yə·ləm }

Sarcopterygii [VERT ZOO] A subclass of Osteichthyes, including Crossopterygii and Dipnoi in some systems of classification. { sär,käp·tə 'rij·ē,ī }

Sarcoptiformes [INV ZOO] A suborder of the Acarina including minute globular mites without stigmata. { sär,käp·tə'fȯr,mēz }

sarcosoma [INV ZOO] The fleshy portion of an anthozoan. { ,sär·kə'sō·mə }

Sarcosporida [INV ZOO] An order of Protozoa of the class Haplosporea which comprises parasites in skeletal and cardiac muscle of vertebrates. { ,sär·kə'spȯr·əd·ə }

sarcosporidiosis [VET MED] A disease of mammals other than humans caused by muscle infestation by sporozoans of the order Sarcosporida. { ,sär·kō·spə,rid·ē'ō·səs }

sarcostyle [INV ZOO] A fibril or column of muscular tissue. { 'sär·kə,stīl }

sarcotubule [CELL MOL] A tubular invagination of a muscle fiber. { ¦sär·kō'tü,byül }

sardine [VERT ZOO] **1.** *Sardina pilchardus.* The young of the pilchard, a herringlike fish in the family Clupeidae found in the Atlantic along the European coasts. **2.** The young of any of various similar and related forms which are processed and eaten as sardines. { sär'dēn }

sarkomycin [MICROBIO] $C_7H_8O_3$ An antibiotic produced by an actinomycete which acts as a carcinolytic agent. { ,sär·kə'mīs·ən }

sarmentocymarin [BIOCHEM] A cardioactive, steroid glycoside from the seeds of *Strophanthus sarmentosus*; on hydrolysis it yields sarmentogenin and sarmentose. { sär¦men·tō'sī·mə·rən }

sarmentogenin [BIOCHEM] $C_{23}H_{34}O_5$ The steroid aglycon of sarmentocymarin; isometric with digitoxigenin, and characterized by a hydroxyl group at carbon number 11. { sär,men·tō'jen·ən }

sarmentose [BOT] Producing slender, prostrate stems or runners. { sär'men,tōs }

Sarothriidae [INV ZOO] The equivalent name for Jacobsoniidae. { ,sar·ə'thrī·ə,dē }

Sarraceniaceae [BOT] A small family of dicotyledonous plants in the order Sarraceniales in which leaves are modified to form pitchers, placentation is axile, and flowers are perfect with distinct filaments. { ,sar·ə,sē·nē'ās·ē,ē }

Sarraceniales [BOT] An order of dicotyledonous herbs or shrubs in the subclass Dilleniidae; plants characteristically have alternate, simple leaves that are modified for catching insects, and grow in waterlogged soils. { ,sar·ə,sē·nē'ā·lēz }

sarsaparilla [BOT] Any of various tropical American vines of the genus *Smilax* (family Liliaceae) found in dense, moist jungles; a flavoring material used in medicine and soft drinks is obtained from the dried roots of at least four species. { ,sas·pə'ril·ə }

sartorius [ANAT] A large muscle originating in the anterior superior iliac spine and inserting in the tibia; flexes the hip and knee joints, and rotates the femur laterally. { sär'tȯr·ē·əs }

sassafras [BOT] *Sassafras albidum.* A medium-sized tree of the order Magnoliales recognized by the bright-green color and aromatic odor of the leaves and twigs. { 'sas·ə,fras }

satellite [CELL MOL] A chromosome segment distant from but attached to the rest of the chromosome by an achromatic filament. { 'sad·əl,īt }

satellite band [CELL MOL] A fraction of the deoxyribonucleic acid (DNA) of an organism which has a different density from the rest and is therefore separable as a band in density gradient centrifugation; these bands are usually made up of highly repetitive sequences of DNA. { 'sad·əl,īt ,band }

satellite cell [HISTOL] One of the neurilemmal cells surrounding nerve cells in the peripheral nervous system. { 'sad·əl,īt ,sel }

satellite deoxyribonucleic acid [CELL MOL] Any fraction, usually highly repetitious, of chromosomal deoxyribonucleic acid that differs significantly in its base composition from the majority of other fractions. { ¦sad·ə,līt dē,äk·sē'rī·bō·nü,klē·ik 'as·əd }

satellitosis [MED] A condition, associated with inflammatory and degenerative diseases of the central nervous system, in which satellite cells increase around the nerve cells. { ,sad·əl,īd'ō·səs }

saturation diving [PHYSIO] Diving in which the tissues exposed to high pressure at great ocean depths for 24 hours become saturated with gases, especially inert gases, thereby reaching a new equilibrium state. { ,sach·ə'rā·shən 'dīv·iŋ }

Saturniidae [INV ZOO] A family of medium- to large-sized moths in the superfamily Saturnioidea including the giant silkworms, characterized by reduced, often vestigial, mouthparts and strongly bipectinate antennae. { ,sad·ər'nī·ə,dē }

Saturnioidea [INV ZOO] A superfamily of medium- to very-large-sized moths in the suborder Heteroneura having the frenulum reduced or absent, reduced mouthparts, no tympanum, and pectinate antennae. { ,sad·ər·nē'ȯid·ē·ə }

Satyrinae [INV ZOO] A large, cosmopolitan subfamily of lepidopterans in the family Nymphalidae, containing the wood nymphs, meadow browns, graylings, and arctics, characterized by bladderlike swellings at the bases of the forewing veins. { sə'tir·ə,nē }

Sauria [VERT ZOO] The lizards, a suborder of the Squamata, characterized generally by two or four limbs but sometimes none, movable eyelids, external ear openings, and a pectoral girdle. { 'sȯr·ē·ə }

Saururaceae [BOT] A family of dicotyledonous plants in the order Piperales distinguished by mostly alternate leaves, two to ten ovules per carpel, and carpels distinct or united into a compound ovary. { ,sȯ·rə'rās·ē,ē }

savane armée *See* thornbush. { sa'vän är'mā }

savane épineuse *See* thornbush. { sa'vän ā·pə'nüz }

savanna [ECOL] Any of a variety of physiognomically or environmentally similar vegetation types in tropical and extratropical regions; all contain grasses and one or more species of trees of the families Leguminosae, Bombacaceae, Bignoniaceae, or Dilleniaceae. { sə'van·ə }

savanna-woodland *See* tropical woodland. { sə 'van·ə 'wu̇d·lənd }

savory [BOT] A herb of the mint family in the genus *Satureia*; of the more than 100 species, only summer savory (*S. hortensis*) and winter savory (*S. montana*) are grown for flavoring purposes. { 'sav·ə·rē }

sawfish [VERT ZOO] Any of several elongate viviparous fishes of the family Pristidae distinguished by a dorsoventrally flattened elongated snout with stout toothlike projections along each edge. { 'sȯ,fish }

saxicolous [ECOL] Living or growing among rocks. { sak'sik·ə·ləs }

Saxifragaceae [BOT] A family of dicotyledonous plants in the order Rosales which are scarcely or not at all succulent and have two to five carpels usually more or less united, and leaves not modified into pitchers. { ,sak·sə·frə'gās·ē,ē }

saxitoxin [BIOCHEM] A nonprotein toxin produced by the dinoflagellate *Gonyaulax catenella*. { ¦sak·sə¦täk·sən }

SBMV *See* southern bean mosaic virus.

scab [MED] Crusty exudate covering a wound or ulcer during the healing process. { skab }

scabies [MED] A contagious skin disorder caused by the mite *Sarcoptes scabiei* burrowing beneath the skin, causing the formation of multiform lesions with intense itching. { 'skā·bēz }

scabrous [BIOL] Having a rough surface covered with stiff hairs or scales. { 'skab·rəs }

scaffold protein [CELL MOL] A protein that assembles interacting signaling proteins into multimolecular signaling complexes. { 'ska ,fōld ,prō,tēn }

scala media [ANAT] The middle channel of the cochlea, filled with endolymph and bounded above by Reissner's membrane and below by the basilar membrane. Also known as cochlear duct. { 'skā·lə 'mē·dē·ə }

scalariform [BIOL] Resembling a ladder; having transverse markings or bars. { skə'lar·ə,fȯrm }

scala tympani [ANAT] The lowest channel in the cochlea of the ear; filled with perilymph. { 'skā·lə tim'pan·ē }

scala vestibuli [ANAT] The uppermost channel of the cochlea; filled with perilymph. { 'skā·lə ve'stib·yə·lē }

scalded skin syndrome *See* toxic epidermal necrolysis. { ¦skȯl·dəd 'skin ,sin,drōm }

scale [BOT] The bract of a catkin. [VERT ZOO] A flat calcified or cornified platelike structure on the skin of most fishes and of some tetrapods. { skāl }

scale insect [INV ZOO] Any of various small, structurally degenerate homopteran insects in the superfamily Coccoidea which resemble scales on the surface of a host plant. { 'skāl 'in,sekt }

scalenus [ANAT] One of three muscles in the neck, arising from the transverse processes of the cervical vertebrae, and inserted on the first two ribs. { skā'lē·nəs }

scale scar [BOT] A mark left on a stem after bud scales have fallen off. { 'skāl ,skär }

Scalibregmidae [INV ZOO] A family of mud-swallowing worms belonging to the Sedentaria and found chiefly in sublittoral and great depths. { ,skal·ə'breg·mə,dē }

scaling [BIOL] **1.** The removing of scales from fishes. **2.** *See* root planing. { 'skāl·iŋ }

scallion *See* shallot. { 'skal·yən }

scallop [INV ZOO] Any of various bivalve mollusks in the family Pectinidae distinguished by radially ribbed valves with undulated margins. { 'skäl·əp }

Scalpellidae [INV ZOO] A primitive family of barnacles in the suborder Lepadomorpha having more than five plates. { skal'pel·ə,dē }

scaly leg [VET MED] A highly contagious disease of poultry caused by the sarcoptid mite *Knemidokoptes mutans*. { 'skā·lē ,leg }

scandent [BOT] Climbing by stem-roots or tendrils. { 'skan·dənt }

scansorial [BOT] Adapted for climbing. { skan 'sȯr·ē·əl }

Scapanorhychidae [VERT ZOO] The goblin sharks, a family of deep-sea galeoids in the isurid line having long, sharp teeth and a long, pointed rostrum. { ,skap·ə·nō'rik·ə,dē }

scapha [ANAT] The furrow of the auricle between the helix and the antihelix. { 'skaf·ə }

Scaphidiidae [INV ZOO] The shining fungus beetles, a family of coleopteran insects in the superfamily Staphylinoidea. { ,skaf·ə'dī·ə,dē }

scaphocephaly [MED] A condition of the skull characterized by elongation and narrowing, and a projecting, keellike sagittal suture, caused by its premature closure. { ,skaf·ə'sef·ə·lē }

scaphoid [ANAT] A boat-shaped bone of the carpus or of the tarsus. { 'skaf,ȯid }

Scaphopoda [INV ZOO] A class of the phylum Mollusca in which the soft body fits the external, curved and tapering, nonchambered, aragonitic shell which is open at both ends. { skə'fäp·əd·ə }

scapula [ANAT] The large, flat, triangular bone forming the back of the shoulder. Also known as shoulder blade. { 'skap·yə·lə }

scapulet [INV ZOO] In some medusae, fringed outgrowths on the outer surfaces of the arms near the bell. { 'skap·yə,let }

scapulus [INV ZOO] A modified submarginal region in some sea anemones. { 'skap·yə·ləs }

scapus [BIOL] The stem, shaft, or column of a structure. { 'skā·pəs }

scar [MED] A permanent mark on the skin or other tissue, formed from connective-tissue replacement of tissue destroyed by a wound or disease process. { skär }

Scarabaeidae [INV ZOO] The lamellicorn beetles, a large cosmopolitan family of coleopteran insects in the superfamily Scarabaeoidea including the Japanese beetle and other agricultural pests. { ˌskar·əˈbē·əˌdē }

Scarabaeoidea [INV ZOO] A superfamily of Coleoptera belonging to the suborder Polyphaga. { ˌskar·ə·bēˈȯid·ē·ə }

scarabiasis [MED] Invasion of the intestine by the dung beetle, characterized by anorexia, emaciation, and disturbance of the gastrointestinal tract. { ¦skar·əˈbī·ə·səs }

Scaridae [VERT ZOO] The parrotfishes, a family of perciform fishes in the suborder Percoidei which have the teeth of the jaw generally coalescent. { ˈskar·əˌdē }

scarification [MED] The operation of making numerous small, superficial incisions in skin or other tissue. { ˌskär·ə·fəˈkā·shən }

scarious [BOT] Having a thin, membranous texture. { ˈskar·ē·əs }

scarlet fever [MED] An acute, contagious bacterial disease caused by *Streptococcus hemolyticus*; characterized by a papular, or rough, bright-red rash over the body, with fever, sore throat, headache, and vomiting occurring 2–3 days after contact with a carrier. { ˈskär·lət ˈfē·vər }

scarlet fever streptococcus antitoxin [IMMUNOL] A sterile aqueous solution of antitoxins obtained from the blood of animals immunized against group A beta hemolytic streptococci toxin; formerly used in the treatment of, and to produce immunity against, scarlet fever. { ˈskär·lət ˈfē·vər ¦strep·tə¦käk·əs ˌant·iˈtäk·sən }

scarlet fever streptococcus toxin [IMMUNOL] Toxic filtrate of cultures of *Streptococcus pyogenes* responsible for the characteristic rash of scarlet fever; the toxin is used in the Dick test. { ˈskär·lət ˈfē·vər ¦strep·tə¦käk·əs ˈtäk·sən }

Scarpa's fascia [ANAT] The deep, membranous layer of the superficial fascia of the lower abdomen. { ˈskär·pəz ˌfā·shə }

scar tissue [MED] Contracted, dense connective tissue that is formed by the healing process of a wound or diseased tissue. { ˈskär ˌtish·ü }

Scatopsidae [INV ZOO] The minute black scavenger flies, a family of orthorrhaphous dipteran insects in the series Nematocera. { skəˈtäp·səˌdē }

scavenger [ECOL] An organism that feeds on carrion, refuse, and similar matter. { ˈskav·ən·jər }

Scelionidae [INV ZOO] A family of small, shining wasps in the superfamily Proctotrupoidea, characterized by elbowed, 11- or 12-segmented antennae. { ˌsel·ēˈän·əˌdē }

scent gland [VERT ZOO] A specialized skin gland of the tubuloalveolar or acinous variety which produces substances having peculiar odors; found in many mammals. { ˈsent ˌgland }

Schardinger dextrin *See* cycloamylose. { ˈshärd·ən·jər ˈdeks·trən }

Schatzki's ring [MED] A mucosal constriction at the junction of the esophagus and stomach in the presence of a hiatus hernia. Also known as lower esophageal ring. { ¦shät·skēz ˈriŋ }

Scheie's syndrome [MED] A hereditary disease transmitted as an autosomal recessive and characterized by high levels of chondroitin sulfate B in the urine, mild distortion of the facies, hypertrichosis, clouding of the cornea, and aortic valve disease. { ˈshī·əz ˌsinˌdrōm }

schemochrome [ZOO] A feather color that originates within the feather structures, through refraction of light independent of pigments. { ˈskē·məˌkrōm }

Schenck's disease *See* sporotrichosis. { ˈsheŋks diˌzēz }

Scheuermann's disease *See* osteochondrosis. { ˈshȯi·ərˌmänz diˌzēz }

Schick test [IMMUNOL] A skin test for determining susceptibility to diphtheria performed by the intradermal injection of diluted diphtheria toxin; a positive reaction, showing edema and scaling after 5 to 7 days, indicates lack of immunity. { ˈshik ˌtest }

Schilder's disease [MED] **1.** A retrogressive disease of the white matter in the central nervous system characterized by diffuse loss of myelin. **2.** Any of the progressive degenerative diseases of the white matter in the central nervous system. { ˈshil·dərz diˌzēz }

Schindleriidae [VERT ZOO] The single family of the order Schindlerioidei. { ˌshind·ləˈrī·əˌdē }

Schindlerioidei [VERT ZOO] A suborder of fishes in the order Perciformes composed of one monogeneric family comprising two tiny oceanic species that are transparent and neotenic. { ˌshind·lə·rēˈȯid·ēˌī }

schindylesis [ANAT] A synarthrosis in which a plate of one bone is fixed in a fissure of another. { ˌskin·dəˈlē·səs }

schistosome dermatitis [MED] A dermatitis caused by penetration of the skin by certain schistosome cercariae. Also known as swamp itch; swimmer's itch. { ˈshis·təˌsōm ˌdər·məˈtīd·əs }

schistosomiasis [MED] A disease in which humans are parasitized by any of three species of blood flukes: *Schistosoma mansoni*, *S. haematobium*, and *S. japonicum*; adult worms inhabit the blood vessels. Also known as bilharzias; snail fever. { ˌshis·tə·səˈmī·ə·səs }

Schistostegiales [BOT] A monospecific order of mosses; the small, slender, glaucous plants are distinguished by the luminous protonema. { ¦shis·təˌstej·ēˈā·lēz }

schizaxon [NEUROSCI] An axon that divides, in its course, into equal or nearly equal branches. { skizˈakˌsän }

schizocarp [BOT] A dry fruit that separates at maturity into single-seeded indehiscent carpels. { ˈskiz·əˌkärp }

Schizocoela [INV ZOO] A group of animal phyla, including Bryozoa, Brachiopoda, Phoronida, Sipunculoidea, Echiuroidea, Priapuloidea, Mollusca, Annelida, and Arthropoda, all characterized by the appearance of the coelom as a space in the embryonic mesoderm. { ¦skiz·ə¦sē·lə }

schizodont [INV ZOO] A multinucleate trophozoite that segments into merozoites. { 'skiz·ə,dänt }

schizogamy [BIOL] A form of reproduction involving division of an organism into a sexual and an asexual individual. { ski'zäg·ə·mē }

schizogenesis [BIOL] Reproduction by fission. { ,ski·zō'jen·ə·səs }

schizognathous [VERT ZOO] Descriptive of birds having a palate in which the vomer is small and pointed, the maxillopalatines are not united with each other or with the vomer, and the palatines articulate posteriorly with the rostrum. { ski'zäg·nə·thəs }

Schizogoniales [BOT] A small order of the Chlorophyta containing algae that are either submicroscopic filaments or macroscopic ribbons or sheets a few centimeters wide and attached by rhizoids to rocks. { ,skiz·ə,gō·nē'ā·lēz }

schizogony [INV ZOO] Asexual reproduction by multiple fission of a trophozoite; a characteristic of certain Sporozoa. { ski'zäg·ə·nē }

Schizomeridaceae [BOT] A family of green algae in the order Ulvales. { ¦skiz·ə,mer·ə'dās·ē,ē }

Schizomycetes [MICROBIO] Formerly a class of the division Protophyta which included the bacteria. { ,skiz·ə·mī'sēd·ēz }

Schizomycophyta [BOT] The designation for bacteria in those taxonomic systems that consider bacteria as plants. { ,skiz·ə·mī'käf·əd·ə }

schizont [INV ZOO] A multinucleate cell in certain members of the Sporozoa that is produced from a trophozoite in a cell of the host, and that segments into merozoites. { 'skī,zänt }

Schizopathidae [INV ZOO] A family of dimorphic zoantharians in the order Antipatharia. { ,skiz·ə'path·ə,dē }

schizopelmous [VERT ZOO] Having the two flexor tendons of the toes separate, as in certain birds. { ¦ski·zō¦pel·məs }

Schizophora [INV ZOO] A series of the dipteran suborder Cyclorrhapha in which adults possess a frontal suture through which a distensible sac, or ptilinum, is pushed to help the organism escape from its pupal case. { ski'zäf·ə·rə }

Schizophyceae [MICROBIO] The blue-green algae, a class of the division Protophyta. { ,skiz·ə'fī·sē,ē }

Schizophyta [BOT] The prokaryotes, a division of the plant subkingdom Thallobionta; includes the bacteria and blue-green algae. { ski'zäf·əd·ē }

schizopod [INV ZOO] **1.** Having the limbs split so that each has an endopodite and an exopodite, as in certain crustaceans. **2.** A biramous appendage. { 'skiz·ə,päd }

Schizopteridae [INV ZOO] A family of minute ground-inhabiting hemipterans in the group Dipsocoeoidea; individuals characteristically live in leaf mold. { ,ski·zäp'ter·ə,dē }

schizorhinal [VERT ZOO] Having a deep cleft on the posterior margin of the osseous external nares, as in certain birds. { ¦skiz·ə¦rīn·əl }

schizothecal [VERT ZOO] Having the horny envelope of the tarsus divided into scalelike plates; refers to most birds. { ¦skiz·ə'thē·kəl }

schizothoracic [INV ZOO] Having a prothorax that is large and loosely articulated with the thorax. { ¦ski·zō·thə'ras·ik }

Schlemm's canal [ANAT] A space or series of spaces at the junction of the sclera and cornea in the eye; drains aqueous humor from the anterior chamber. { 'shlemz kə,nal }

Schneiderian membrane [ANAT] The mucosa lining the nasal cavities and paranasal sinuses. { shnī'dir·ē·ən 'mem,brān }

Schneider's index [MED] A test of general physical and circulatory efficiency, consisting of pulse and blood pressure observations under standard conditions of rest and exercise. { 'shnī·dərz 'in,deks }

Schoenbiinae [INV ZOO] A subfamily of moths in the family Pyralididae, including the genus *Acentropus*, the most completely aquatic Lepidoptera. { shən'bī·ə,nē }

Schultz-Charlton test [IMMUNOL] A skin test for the diagnosis of scarlet fever, performed by the intradermal injection of human scarlet fever immune serum; a positive reaction consists of blanching of the rash in the area surrounding the point of injection. { 'shu̇lts 'chärlt·ən ,test }

Schultz-Dale reaction [IMMUNOL] A method for demonstrating anaphylactic hypersensitivity outside the body by suspending an excised intestinal loop or uterine strip from a sensitized animal in an oxygenated, physiological salt solution; addition of the proper allergen causes contraction of the smooth muscle. { 'shu̇lts 'dāl rē,ak·shən }

Schwann cell [HISTOL] One of the cells that surround peripheral axons forming sheaths of the neurilemma. { 'shwän ,sel }

schwannoma *See* neurilemmoma. { shwä'nō·mə }

Sciaenidae [VERT ZOO] A family of perciform fishes in the suborder Percoidei, which includes the drums. { sī'ēn·ə,dē }

sciatica [MED] Neuralgic pain in the lower extremities, hips, and back caused by inflammation or injury to the sciatic nerve. { sī'ad·ə·kə }

sciatic nerve [NEUROSCI] Either of a pair of long nerves that originate in the lower spinal cord and send fibers to the upper thigh muscles and the joints, skin, and muscles of the leg. { sī'ad·ik 'nərv }

SCIDS *See* severe combined immunodeficiency syndrome. { skidz *or* ¦es¦sē¦ī¦dē'es }

scillarenin [PHARM] $C_{24}H_{32}O_4$ A crystalline compound that forms prisms from a methanol solution and melts at 232–238°C; used in medicine for cardiac disease. { si'lar·ə·nən }

Scincidae [VERT ZOO] The skinks, a family of the reptilian suborder Sauria which have reduced limbs and snakelike bodies. { 'skiŋ·kə,dē }

Scinidae [INV ZOO] A family of bathypelagic, amphipod crustaceans in the suborder Hyperiidea. { 'skin·ə,dē }

scintillon [INV ZOO] An outpocketing of the cytoplasm in dinoflagellates which contains luciferase and luciferin-binding protein and is the source of bioluminescence. { 'sint·i,län }

Sciomyzidae [INV ZOO] A family of myodarian cyclorrhaphous dipteran insects in the subsection Acalypteratae. { ,sī·ə'miz·ə,dē }

scion [BOT] A section of a plant, usually a stem or bud, which is attached to the stock in grafting. { 'sī·ən }

sciophilous [ECOL] Capable of thriving in shade. { sī'äf·ə·ləs }

sciophyte [BOT] A plant that thrives at lowered light intensity. { 'sī·ə,fīt }

scirrhous carcinoma [MED] A hard, poorly differentiated adenocarcinoma in which the anaplastic cells are surrounded by dense bundles of collagenous fibers. { 'sir·əs ,kärs·ən'ō·mə }

Scitaminales [BOT] An equivalent name for Zingiberales. { ,sīd·ə·mə'nā·lēz }

Scitamineae [BOT] An equivalent name for Zingiberales. { ,sīd·ə'min·ē,ē }

Sciuridae [VERT ZOO] A family of rodents including squirrels, chipmunks, marmots, and related forms. { sī'yür·ə,dē }

Sciuromorpha [VERT ZOO] A suborder of Rodentia according to the classical arrangement of the order. { sī,yür·ə'mór·fə }

sclera [ANAT] The hard outer coat of the eye, continuous with the cornea in front and with the sheath of the optic nerve behind. { 'skler·ə }

Scleractinia [INV ZOO] An order of the subclass Zoantharia which comprises the true, or stony, corals; these are solitary or colonial anthozoans which attach to a firm substrate. { ,skler·ək'tin·ē·ə }

Scleraxonia [INV ZOO] A suborder of cnidarians in the order Gorgonacea in which the axial skeleton has calcareous spicules. { ,skleer·ək'sō·nē·ə }

sclereid [BOT] A thick-walled, lignified plant cell typically found in sclerenchyma. { 'sklir·ē·əd }

sclerema neonatorum [MED] A disease of the newborn, particularly the premature or the undernourished, dehydrated, and debilitated, characterized by waxy-white hardening of subcutaneous tissue. { sklə'rē·mə ,ne·ə·nə'tór·əm }

sclerenchyma [BOT] A supporting plant tissue composed principally of sclereids whose walls are often mineralized. { sklə'reŋ·kə·mə }

scleriasis *See* scleroderma. { sklə'rī·ə·səs }

sclerite [INV ZOO] One of the sclerotized plates of the integument of an arthropod. { 'skle,rīt }

scleroblast [INV ZOO] A spicule-secreting cell in Porifera. { 'skler·ə,blast }

scleroblastema [EMBRYO] Embryonic tissue from which bones are formed. { ¦skler·ō·bla'stē·mə }

sclerocaulous [BOT] Having a hard, dry stem because of exceptional development of sclerenchyma. { ¦skler·ō'kól·əs }

sclerochore [BOT] A plant that disperses disseminules without apparent morphological adaptations. { 'skler·ə,kór }

Sclerodactylidae [INV ZOO] A family of echinoderms in the order Dendrochirotida having a complex calcareous ring. { ,skler·ō,dak'til·ə,dē }

scleroderma [MED] An abnormal increase in collagenous connective tissue in the skin. Also known as chorionitis; dermatosclerosis; scleriasis. { ,skler·ō'dər·mə }

sclerodermatous [INV ZOO] Having a skeleton that is composed of scleroderm, as certain corals. [VERT ZOO] Having a hard outer covering, for example, hard plate or horny scale. { ¦sler·ō¦dər·məd·əs }

Sclerogibbidae [INV ZOO] A monospecific family of the hymenopteran superfamily Bethyloidea. { ,skler·ō'jib·ə,dē }

sclerophyllous [BOT] Characterized by thick, hard foliage due to well-developed sclerenchymatous tissue. { ¦skler·ə¦fil·əs }

scleroprotein [BIOCHEM] Any one of a class of proteins, such as keratin, fibroin, and the collagens, which occur in hard parts of the animal body and serve to support or protect. Also known as albuminoid. { ¦skler·ō'prō,tēn }

scleroseptum [INV ZOO] A calcareous radial septum in certain corals. { ¦skler·ō'sep·təm }

sclerosing adenomatosis [MED] A form of mammary dysplasia in which ductular structures are encased in fibrous tissue, simulating invading cancerous ductular structures. Also known as fibrosing adenomatosis. { sklə'rōs·iŋ ,ad·ən·ō·mə'tō·səs }

sclerosing hemangioma [MED] A type of benign histiocytoma marked by prominence of the capillary channels. { sklə'rōs·iŋ hi,mān¦jē'ō·mə }

sclerosis [PATH] Hardening of a tissue, especially by proliferation of fibrous connective tissue. { sklə'rō·səs }

sclerotesta [BOT] The middle hard layer of the testa in various seeds. { ,skler·ə'tes·tə }

sclerotic [ANAT] Pertaining to the sclera. [MED] **1.** Hard. **2.** Of or pertaining to sclerosis. { sklə'räd·ik }

sclerotium [MICROBIO] The hardened, resting or encysted condition of the plasmodium of Myxomycetes. [MYCOL] A hardened, resting mass of hyphae, usually black on the outside, from which fructifications may develop. { sklə'rō·shəm }

sclerotome [EMBRYO] The part of a mesodermal somite which enters into the formation of the vertebrae. [MED] A knife used in sclerotomy. [VERT ZOO] The fibrous tissue separating successive myotomes in certain lower vertebrates. { 'skler·ə,tōm }

sclerotomy [MED] Surgical incision of the sclera. { sklə'räd·ə·mē }

scobiform [BOT] Resembling sawdust. { 'skäb·ə,fórm }

Scolecosporae [MYCOL] A spore group of Fungi Imperfecti characterized by filiform spores. { ,skō·lə'käs·pə,rē }

scolex [INV ZOO] The head of certain tapeworms, typically having a muscular pad with hooks, and two pairs of lateral suckers. { 'skō,leks }

Scoliidae [INV ZOO] A family of the Hymenoptera in the superfamily Scolioidea. { skō'lī·ə,dē }

Scolioidea [INV ZOO] A superfamily of hymenopteran insects in the suborder Apocrita. { ,skō·lē'ȯid·ē·ə }

scoliosis [MED] Lateral curvature of the spine. { ,skō·lē'ō·səs }

scolop [INV ZOO] The thickened, distal tip of a vibration-sensitive organ in insects. { 'skäl·əp }

Scolopacidea [VERT ZOO] A large, cosmopolitan family of birds of the order Charadriiformes including snipes, sandpipers, curlews, and godwits. { ,skäl·ə·pə'sīd·ē·ə }

Scolopendridae [INV ZOO] A family of centipedes in the order Scolopendromorpha which characteristically possess eyes. { ,skäl·ə'pen·drə,dē }

Scolopendromorpha [INV ZOO] An order of the chilopod subclass Pleurostigmophora containing the dominant tropical forms, and also the largest of the centipedes. { ¦skäl·ə,pen·drə'mȯr·fə }

scolophore *See* scolopophore. { 'skäl·ə,fȯr }

scolopophore [INV ZOO] A spindle-shaped, bipolar nerve ending in the integument of insects, believed to be auditory in function. Also known as scolophore. { skə'läp·ə,fȯr }

Scolytidae [INV ZOO] The bark beetles, a large family of coleopteran insects in the superfamily Curculionoidea characterized by a short beak and clubbed antennae. { skə'lid·ə,dē }

Scombridae [VERT ZOO] A family of perciform fishes in the suborder Scombroidei including the mackerels and tunas. { 'skäm·brə,dē }

Scombroidei [VERT ZOO] A suborder of fishes in the order Perciformes; all are moderate- to large-sized shore and oceanic fishes having fixed premaxillae. { skäm'brȯi·dē,ī }

scopa [INV ZOO] A brushlike arrangement of short stiff hairs on the body surface of certain insects. { 'skō·pə }

Scopeumatidae [INV ZOO] The dung flies, a family of myodarian cyclorrhaphous dipteran insects in the subsection Calypteratae. { ,skä·pyü 'mad·ə,dē }

Scopidae [VERT ZOO] A family of birds in the order Ciconiiformes containing a single species, the hammerhead (*Scopus umbretta*) of tropical Africa. { 'skäp·ə,dē }

scopolamine [PHARM] $C_{17}H_{21}O_4N$ An alkaloid derivative of several plants in the family Solanaceae, used as an anticholinergic drug; its hydrobromide salt is used as a sedative. { skə'päl·ə,mēn }

scopula [ZOO] A tuft of hair, as on the feet and chelicerae of certain spiders. { 'skäp·yə·lə }

Scorpaenidae [VERT ZOO] The scorpion fishes, a family of Perciformes in the suborder Cottoidei, including many tropical shorefishes, some of which are venomous. { skȯr'pē·nə,dē }

Scorpaeniformes [VERT ZOO] An order of fishes coextensive with the perciform suborder Cottoidei in some systems of classification. { skȯr,pē·nə'fȯr,mēz }

scorpioid cyme [BOT] A cyme with a curved axis and flowers arising two-ranked on alternate sides of the axis. { 'skȯr·pē,ȯid 'sīm }

scorpion [INV ZOO] The common name for arachnids constituting the order Scorpionida. { 'skȯr·pē·ən }

Scorpionida [VERT ZOO] The scorpions, an order of arachnids characterized by a shieldlike carapace covering the cephalothorax and by large pedipalps armed with chelae. { ,skȯr·pē'än·əd·ə }

Scotch pine [BOT] *Pinus sylvestris.* A hard pine of North America having two short, bluish needles in a cluster. { 'skäch 'pīn }

scotochromogen [MICROBIO] **1.** Any microorganism which produces pigment when grown without light as well as with light. **2.** A member of group II of the atypical mycobacteria. { ,skäd·ə'krō·mə·jən }

scotodinia [MED] Dizziness and headache associated with the appearance of black spots before the eyes. { ,skäd·ə'din·ē·ə }

scotoma [MED] A blind spot or area of depressed vision in the visual field. { skə'tō·mə }

scotopic vision [PHYSIO] Vision that is due to the activity of the rods of the retina only; it is the type of vision that occurs at very low levels of illumination, and it can detect differences of brightness but not of hue. Also known as night vision. { skə'täp·ik 'vizh·ən }

scouring rush *See* horsetail. { 'skaůr·iŋ ,rəsh }

scrapie [VET MED] A transmissible, usually fatal, virus disease of sheep, characterized by degeneration of the central nervous system. { 'skrā·pē }

Scraptidae [INV ZOO] An equivalent name for Melandryidae. { 'skrap·tə,dē }

screw pine *See* pandanus. { 'skrü ,pīn }

scrod [VERT ZOO] A young fish, especially cod. { skräd }

scrofula [MED] Tuberculosis of cervical lymph nodes. { 'skräf·yə·lə }

Scrophulariaceae [BOT] A large family of dicotyledonous plants in the order Scrophulariales, characterized by a usually herbaceous habit, irregular flowers, axile placentation, and dry, dehiscent fruit. { ,skräf·yə,lar·ē'ās·ē,ē }

Scrophulariales [BOT] An order of flowering plants in the subclass Asteridae distinguished by a usually superior ovary and, generally, either by an irregular corolla or by fewer stamens than corolla lobes, or commonly both. { ,skräf·yə,lar·ē'ā·lēz }

scrotum [ANAT] The pouch containing the testes. { 'skrōd·əm }

scrub [ECOL] A tract of land covered with a generally thick growth of dwarf or stunted trees and shrubs and a poor soil. { skrəb }

scrub typhus *See* tsutsugamushi disease. { 'skrəb ,tī·fəs }

scruple [PHARM] A unit of mass in apothecaries' measure, equal to 20 grains or to 1.2959782 grams. { 'skrü·pəl }

sculpin [VERT ZOO] Any of several species of small fishes in the family Cottidae characterized by a large head that sometimes has spines, spiny fins, broad mouth, and smooth, scaleless skin. { 'skəl·pən }

scurf [MED] A branlike desquamation of the epidermis, especially from the scalp; dandruff. { skərf }

scurvy [MED] An acute or chronic nutritional disorder due to vitamin C deficiency; characterized by weakness, subcutaneous hemorrhages, and alterations of any tissue containing collagen, ground substance, dentine, intercellular cement, or osteoid. { 'skər·vē }

scute [INV ZOO] A cornified, epithelial, scalelike structure in lizards and snakes. { skyüt }

Scutechiniscidae [INV ZOO] A family of heterotardigrades in the suborder Echiniscoidea, with segmental and intersegmental thickenings of cuticle. { skü,tek·ə'nis·ə,dē }

Scutelleridae [INV ZOO] The shield bugs, a family of Hemiptera in the superfamily Pentatomoidea. { ,sküd·əl'er·ə,dē }

scutellum [BOT] **1.** A rounded apothecium with an elevated rim found in certain lichens. **2.** The flattened cotyledon of a monocotyledonous plant embryo, such as a grass. [INV ZOO] The third of four pieces forming the upper part of the thoracic segment in certain insects. [VERT ZOO] One of the scales on the tarsi and toes of birds. { skü'tel·əm }

Scutigeromorpha [INV ZOO] The single order of notostigmophorous centipedes; members are distinguished by a dorsal respiratory opening, compound-type eyes, long flagellate multisegmental antennae, and long thin legs with multisegmental tarsi. { skü,tij·ə·rə'mȯr·fə }

scutum [INV ZOO] **1.** A bony, horny, or chitinous plate. **2.** The second of four pieces forming the upper part of the thoracic segment in certain insects. **3.** One or two lower opercular valves in certain barnacles. { 'sküd·əm }

Scydmaenidae [INV ZOO] The antlike stone beetles, a large cosmopolitan family of the Coleoptera in the superfamily Staphylinoidea. { sid'mē·nə,dē }

Scyllaridae [INV ZOO] The Spanish, or shovel-nosed, lobsters, a family of the Scyllaridea. { si 'lar·ə,dē }

Scyllaridea [INV ZOO] A superfamily of decapod crustaceans in the section Macrura including the heavily armored spiny lobsters and the Spanish lobsters, distinguished by the absence of a rostrum and chelae. { ,sil·ə'rid·ē·ə }

Scylliorhinidae [VERT ZOO] The catsharks, a family of the cacharinid group of galeoids; members exhibit the most exotic color patterns of all sharks. { ,sil·ē·ō'rin·ə,dē }

scyphistoma [INV ZOO] A sessile, polyploid larva of many Scyphozoa which may produce either more scyphistomae or free-swimming medusae. { sī'fis·tə·mə }

scyphomedusa [INV ZOO] A medusa of the scyphozoans. { ¦sī·fō·mə'dü·sə }

Scyphomedusae [INV ZOO] A subclass of the class Scyphozoa characterized by reduced marginal tentacles, tetramerous medusae, and medusalike polyploids. { ¦sī·fō·mə'dü,sē }

scyphopolyp [INV ZOO] A polyp of the scyphozoans. { ¦sī·fō'päl·əp }

Scyphozoa [INV ZOO] A class of the phylum Cnidaria; all members are marine and are characterized by large, well-developed medusae and by small, fairly well-organized polyps. { ,sī·fə 'zō·ə }

SDS polyacrylamide gel electrophoresis [CELL MOL] An electrophoretic technique in which proteins are denatured by the negatively charged detergent sodium dodecyl sulfate, which masks their intrinsic electrical charge so that when the constituent polypeptide chains are run through a polyacrylamide gel, the protein molecules are separated according to size, not electrical charge. Abbreviated SDS-PAGE. { ¦es¦dē¦es ,päl·ē·ə¦kril·ə·mīd ,jel i,lek·trō·fə'rē·səs }

SDS-PAGE *See* SDS polyacrylamide gel electrophoresis. { ¦es¦dē¦es'pāj }

sea anemone [INV ZOO] Any of the 1000 marine cnidarians that constitute the order Actiniaria; the adult is a cylindrical polyp or hydroid stage with the free end bearing tentacles that surround the mouth. { 'sē ə,nem·ə·nē }

sea cucumber [INV ZOO] The common name for the echinoderms that make up the class Holothuroidea. { 'sē kyü,kəm·bər }

sea fan [INV ZOO] A form of horny coral that branches like a fan. { 'sē ,fan }

sea grass [BOT] Marine plants which are found in shallow brackish or marine waters, are more highly organized than algae, are seed-bearing, and attain lengths of up to 8 feet (2.4 meters). { 'sē ,gras }

sea horse [INV ZOO] Any of about 50 species of tropical and subtropical marine fishes constituting the genus *Hippocampus* in the family Syngnathidae; the body is compressed, the head is bent ventrally and has a tubiform snout, and the tail is tapering and prehensile. { 'sē ,hȯrs }

seal [VERT ZOO] Any of various carnivorous mammals of the suborder Pinnipedia, especially the families Phoridae, containing true seals, and Otariidae, containing the eared and fur seals. { sēl }

sea lily [INV ZOO] The common name for those crinoids in which the body is flower-shaped and is carried at the tip of an anchored stem. { 'sē ,lil·ē }

sea lion [VERT ZOO] Any of several large, eared seals of the Pacific Ocean; related to fur seals but lack a valuable coat. { 'sē ,lī·ən }

sea marsh [ECOL] A salt marsh periodically overflowed or flooded by the sea. Also known as sea meadow. { 'sē ,märsh }

sea meadow *See* sea marsh. { 'sē ,med·ō }

sea otter [VERT ZOO] *Enhydra lutris.* A large marine otter found close to the shoreline in the

North Pacific; these animals are diurnally active and live in herds. { 'sē ,äd·ər }

sea pen [INV ZOO] The common name for cnidarians constituting the order Pennatulacea. { 'sē ,pen }

sea-run [VERT ZOO] Having the habit of ascending a river from the sea, especially to spawn, as salmon and brook trout. { 'sē,rən }

seashell [INV ZOO] The shell of a marine invertebrate, especially a mollusk. { 'sē,shel }

seasickness [MED] Motion sickness occurring at sea. Also known as pelagism. { 'sē,sik·nəs }

sea slug [INV ZOO] The common name for the naked gastropods composing the suborder Nudibranchia. { 'sē ,sləg }

sea spider [INV ZOO] The common name for arthropods in the subphylum Pycnogonida. { 'sē ,spī·dər }

sea squirt [INV ZOO] A sessile, marine tunicate of the class Ascidiacea; it squirts water from two openings in the unattached end when touched or disturbed. { 'sē ,skwərt }

sea turtle [VERT ZOO] Any of various marine turtles, principally of the families Cheloniidae and Dermochelidae, having paddle-shaped feet. { 'sē ,tərd·əl }

seatworm *See* pinworm. { 'sēt,wərm }

sea urchin [INV ZOO] A marine echinoderm of the class Echinoidea; the soft internal organs are enclosed in and protected by a test or shell consisting of a number of close-fitting plates beneath the skin. { 'sē ,ər·chən }

seaweed [BOT] A marine plant, especially algae. { 'sē,wēd }

sebaceous gland [PHYSIO] A gland, arising in association with a hair follicle, which produces and liberates sebum. { si'bā·shəs 'gland }

Sebekidae [INV ZOO] A family of pentastomid arthropods in the suborder Porocephaloidea. { si'bek·ə,dē }

seborrheic dermatitis [MED] An acute inflammation of the skin, occurring usually on oily skin; characterized by scales, crusting yellowish patches, and itching. { ,seb·ə'rē·ik ,dər·mə'tīd·əs }

sebum [PHYSIO] The secretion of sebaceous glands, composed of fat, cellular debris, and keratin. { 'sē·bəm }

secalose [BIOCHEM] A polysaccharide consisting of fructose units; occurs in green rye and oats, and in rye flour. { 'sek·ə,lōs }

Secernentea [INV ZOO] A class of the phylum Nematoda in which the primary excretory system consists of intracellular tubular canals joined anteriorly and ventrally in an excretory sinus, into which two ventral excretory gland cells may also open. { ,se·sər'nen·chə }

secobarbital [PHARM] $C_{12}H_{18}N_2O_3$ 5-Allyl-5-(1-methylbutyl) barbituric acid, a short-acting barbiturate; white powder with bitterish taste; very soluble in alcohol and ether, slightly soluble in water; melts at 82°C; used as a sedative and hypnotic, frequently as the sodium derivative. Also known as quinolbarbitone; Seconal (trade name). { ,se·kō'bär·bə,tól }

secodont [VERT ZOO] Having teeth adapted for cutting. { 'sek·ə,dänt }

Seconal *See* secobarbital. { 'sek·ən,ól }

secondary amyloidosis [MED] Amyloidosis that usually follows chronic suppurative, inflammatory diseases, such as tuberculosis, osteomyelitis, and bronchiectasis. { 'sek·ən,der·ē ,am·ə,lói'dō·səs }

secondary biliary cirrhosis [MED] A type of cirrhosis caused by obstruction of the bile duct by calculi (stones). { ¦sek·ən,der·ē ,bil·ē,er·ē sə 'rō·səs }

secondary cambium [BOT] One of the tissue layers formed after the initial cambial layers in certain plant roots, and that produce a ring of tissue. { 'sek·ən,der·ē 'kam·bē·əm }

secondary consumer [ECOL] In an ecosystem, an animal that feeds on primary consumers. Also known as a carnivore. { ¦sek·ən,der·ē kən'sü·mər }

secondary hypothermia [MED] A decrease in core body temperature caused by an underlying pathology that prevents the body from generating enough core heat. { ,sek·ən,der·ē ,hī·pō 'thər·mē·ə }

secondary meristem [BOT] Meristem developed from differentiated living tissue. { 'sek·ən,der·ē 'mer·ə,stem }

secondary metabolite [BOT] A natural chemical product of plants not normally involved in primary metabolic processes such as photosynthesis and cell respiration. Also known as secondary plant product. { 'sek·ən,der·ē mə'tab·ə,līt }

secondary periderm [BOT] Any layer of the periderm except the first and outermost layer. { 'sek·ən,der·ē 'per·i,dərm }

secondary phloem [BOT] Phloem produced by the cambium, consisting of two interpenetrating systems, the vertical or axial and the horizontal or ray. { 'sek·ən,der·ē 'flō·əm }

secondary plant product *See* secondary metabolite. { 'sek·ən,der·ē 'plant ,präd·əkt }

secondary root [BOT] A root arising from a primary root. { 'sek·ən,der·ē 'rüt }

secondary structure [BIOCHEM] The conformation of a protein or peptide molecule with respect to nearest-neighbor amino acids. { 'sek·ən,der·ē 'strək·chər }

secondary succession [ECOL] Ecological succession that occurs in habitats where the previous community has been destroyed or severely disturbed, such as following forest fire, abandonment of agricultural fields, or epidemic disease or pest attack. { ¦sek·ən,der·ē sək'sesh·ən }

secondary tissue [BOT] Tissue that develops from the vascular cambium or from differentiated tissues. { 'sek·ən,der·ē 'tish·ü }

secondary tympanic membrane [ANAT] The membrane closing the fenestra cochleae. { 'sek·ən,der·ē tim'pan·ik 'mem,brān }

secondary wall [BOT] The portion of a plant cell wall produced internal to and following deposition of the primary wall; usually consists of several anisotropic layers, and often has prominent internal rings, spirals, bars, or reticulations. { 'sek·ən,der·ē 'wȯl }

secondary xylem [BOT] Xylem produced by cambium, composed of two interpenetrating systems, the horizontal (ray) and vertical (axial). { 'sek·ən,der·ē 'zī·ləm }

second-degree burn [MED] A burn that is more severe than a first-degree burn and is characterized by blistering as well as reddening of the skin, edema, and destruction of the superficial underlying tissues. { 'sek·ənd ¦di·grē 'bərn }

second generation *See* F_2. { 'sek·ənd ,jen·ə'rā·shən }

second messenger [CELL MOL] Any small molecule or ion that occurs in the cytoplasm of a cell, is generated in response to a hormone binding to a cell-surface receptor, and activates various kinases that regulate the activities of other enzymes. { 'sek·ənd 'mes·ən·jər }

secretin [BIOCHEM] A basic polypeptide hormone produced by the duodenum in response to the presence of acid; acts to excite the pancreas to activity. { si'krēt·ən }

secretion [PHYSIO] **1.** The act or process of producing a substance which is specialized to perform a certain function within the organism or is excreted from the body. **2.** The material produced by such a process. { si'krē·shən }

secretor gene [GEN] A dominant autosomal gene in humans which controls secretion of A and B antigenic material in saliva, urine, plasma, and other body fluids; it is not linked to the ABO locus. { si'krēd·ər ,jēn }

secretory diarrhea [MED] Diarrhea resulting from an increase in electrolyte secretion into the intestinal lumen by its epithelial cells, leading to an increased flow of water from the intestinal mucosa into the lumen, it persists during fasting. { si¦krēd·ə·rē dī·ə'rē·ə }

secretory granules [CELL MOL] Accumulations of material produced within a cell for secretion outside the cell. { si'krēd·ə·rē ,gran,yülz }

secretory structure [BOT] Plant cells or organizations of plant cells which produce a variety of secretions. { si'krēd·ə·rē ,strək·chər }

secund [BOT] Having lateral members arranged on one side only. { 'sē,kənd }

securinine [PHARM] $C_{13}H_{15}NO_2$ A crystalline compound that forms yellow crystals from a methanol solution and melts at 142–143°C; used to make the nitrate compound for cardiac insufficiency. { si'kyu̇r·ə,nēn }

sedation [MED] A state of lessened activity. { si'dā·shən }

sedative [PHARM] An agent or drug that has a quieting effect on the central nervous system. { 'sed·əd·iv }

Sedentaria [INV ZOO] A group of families of polychaete annelids in which the anterior, or cephalic, region is more or less completely concealed by overhanging peristomial structures, or the body is divided into an anterior thoracic and a posterior abdominal region. { ,sed·ən'tar·ē·ə }

sedimentation rate [PATH] The rate at which red blood cells settle out of anticoagulated blood. { ,sed·ə·mən'tā·shən ,rāt }

sedoheptulose [BIOCHEM] A seven-carbon ketose sugar widely distributed in plants of the Crassulaceae family; a significant intermediary compound in the cyclic regeneration of D-ribulose. { ¦sē·dō'hep·tə,lōs }

seed [BOT] A fertilized ovule containing an embryo which forms a new plant upon germination. { sēd }

seed coat [BOT] The envelope which encloses the seed except for a tiny pore, the micropyle. { 'sēd ,kōt }

seedling [BOT] **1.** A plant grown from seed. **2.** A tree younger and smaller than a sapling. **3.** A tree grown from a seed. { 'sēd·liŋ }

segmental reflex [NEUROSCI] A reflex arc having afferent inputs by way of the spinal dorsal roots, and efferent outputs over spinal ventral roots of the same or adjacent segments. { seg 'ment·əl 'rē,fleks }

segmentation *See* metamerism. { ,seg·mən'tā·shən }

segmentation cavity *See* blastocoele. { ,seg·mən'tā·shən ,kav·əd·ē }

segregation [GEN] The separation of homologous chromosomes, and thus the alleles they carry, during meiosis in the formation of gametes. { ,seg·rə'gā·shən }

segregation distorter [GEN] An abnormality of meiosis which produces a distortion of the 1:1 segregation ratio of alleles in a heterozygote. { ,seg·rə'gā·shən dis,tȯr·dər }

Seisonacea [INV ZOO] A monofamiliar order of the class Rotifera characterized by an elongated jointed body with a small head, a long slender neck region, a thick fusiform trunk, and an elongated foot terminating in a perforated disk. { ,sī·sə'nā·shə }

Seisonidea [INV ZOO] The equivalent name for Seisonacea. { ,sī·sə'nīd·ē·ə }

Seitz filter [MICROBIO] A bacterial filter made of asbestos and used to sterilize solutions without the use of heat. { 'zītz ,fil·tər }

seizure [MED] **1.** The sudden onset or recurrence of a disease or an attack. **2.** Specifically, an epileptic attack, fit, or convulsion. { 'sē·zhər }

Selachii [VERT ZOO] An order of elasmobranchs including all fossil sharks, except Cladoselachii and Pleuracanthodii. { sə'lāk·ē,ī }

Selaginellales [BOT] The plant order of small club mosses, containing one living genus, *Selaginella*; distinguished from other lycopods in being heterosporous and in having a ligule borne on the upper base of the leaf. { sə,laj·ə·nə'lā·lēz }

selectin [CELL MOL] Any of a family of carbohydrate-binding, calcium-dependent cell adhesion molecules that play an important role in leukocyte-endothelial binding. { sə'lek·tən }

selection [GEN] Any natural or artificial process which favors the survival and propagation of individuals of a given phenotype in a population. { si'lek·shən }

selection coefficient [GEN] A measure of the

rate of transmission through successive generations of a given allele compared to the rate of transmission of another (usually the wild-type) allele. { si'lek·shən ˌkō·iˌfish·ənt }

selection pressure [EVOL] Those factors that influence the direction of natural selection. { si'lek·shən ˌpresh·ər }

selective breeding [BIOL] Breeding of animals or plants having desirable characters. { si'lek·tiv 'brēd·iŋ }

selective medium [MICROBIO] A bacterial culture medium containing an individual organic compound as the sole source of carbon, nitrogen, or sulfur for growth of an organism. { si 'lek·tiv 'mē·dē·əm }

selenium disulfide *See* selenium sulfide. { sə'lē·nē·əm dī'səlˌfīd }

selenium sulfide [PHARM] SeS_2 A bright orange powder with a melting point of 100°C; used in medicine. Also known as selenium disulfide. { sə'lē·nē·əm 'səlˌfīd }

selenodont [VERT ZOO] **1.** Being or pertaining to molars having crescentic ridges on the crown. **2.** A mammal with selenodont dentition. { sə'lē·nəˌdänt }

selenosis [MED] Selenium poisoning. { ˌsel·ə 'nō·səs }

self-differentiation [PHYSIO] The differentiation of a tissue, even when isolated, solely as a result of intrinsic factors after determination. { ¦self ˌdif·əˌren·chē¦ā·shən }

self-incompatibility [BOT] Pertaining to an individual flower that cannot complete fertilization with its own pollen. { ¦self ˌin·kəmˌpad·ə'bil·əd·ē }

selfish deoxyribonucleic acid [CELL MOL] Any tandemly repeated or dispersed repetitive deoxyribonucleic acid sequence that has no obvious function but can spread and accumulate in the species because of its rapid replication. Also known as junk deoxyribonucleic acid. { ¦sel·fish dēˌäk·sē'rī·bō·nüˌklē·ik 'as·əd }

self-pollination [BOT] Transfer of pollen from the anther to the stigma of the same flower or of another flower on the same plant. { ¦self ˌpäl·ə¦nā·shən }

Seligeriales [BOT] An order of true mosses in the class Bryopsida; members grow on rocks and may be exceedingly small to moderate size and tufted; the double structure of the peristome is distinctive. { ˌsel·əˌjir·ē'ā·lēz }

sella turcica [ANAT] A depression in the upper surface of the sphenoid bone in which the pituitary gland rests in vertebrates. { ˌsel·ə 'tər·kə·kə }

selva *See* tropical rainforest. { 'sel·və }

Semaeostomeae [INV ZOO] An order of the class Scyphozoa including most of the common medusae, characterized by a flat, domelike umbrella whose margin is divided into many lappets. { səˌmē·ə'stō·mēˌē }

sematic *See* aposematic. { si'mat·ik }

semeiography [MED] Description of the signs and symptoms of a disease. { ˌsē·mē'äg·rə·fē }

semelparity [BIOL] Reproduction that occurs only one time during the life of an individual. { 'sem·əlˌpar·əd·ē }

semelparous [ZOO] Capable of breeding or reproducing only once. { ˌsem·əl'par·əs }

semen [PHYSIO] The fluid that carries the male germ cells. Also known as seminal fluid. { 'sē·mən }

semicircular canal [ANAT] Any of three loop-shaped tubular structures of the vertebrate labyrinth; they are arranged in three different spatial planes at right angles to each other, and function in the maintenance of body equilibrium. { ¦sem·i'sər·kyə·lər kə'nal }

semicoma [MED] A mildly or partially comatose state in which the patient can be roused and responds to strong stimuli with some purposeful movements. { ¦sem·i'kō·mə }

semiconservative replication [CELL MOL] Replication of deoxyribonucleic acid by longitudinal separation of the two complementary strands of the molecule, each being conserved and acting as a template for synthesis of a new complementary strand. { ¦sem·i·kən'sər·vəd·iv rep·li'kā·shən }

semidesert [ECOL] An area intermediate in character and often located between a desert and a grassland or woodland. { ¦sem·i'dez·ərt }

semidormancy [BOT] Decrease in plant growth rate; may be seasonal or associated with unfavorable environmental conditions. { ¦sem·i'dȯr·mən·sē }

semidouble [BOT] Pertaining to a flower that has more than the usual number of petals or disk florets while it retains some pollen-bearing stamens or some perfect disk florets. { ¦sem·i'dəb·əl }

semilate [BOT] Pertaining to a plant whose growing season is intermediate between midseason forms and late forms. { ¦sem·i'lāt }

semilethal gene [GEN] A mutant causing the death of some of the individuals of the relevant genotype, but never 100%. Also known as sublethal gene. { ¦sem·i'lēth·əl 'jēn }

semilunar cartilage [ANAT] One of the two interarticular knee cartilages. { ¦sem·i'lü·nər 'kärt·lij }

semilunar ganglion *See* Gasserian ganglion. { ¦sem·i'lü·nər 'gaŋ·glē·ən }

semilunar valve [ANAT] Either of two tricuspid valves in the heart, one at the orifice of the pulmonary artery and the other at the orifice of the aorta. { ¦sem·i'lü·nər 'valv }

semimembranosus [ANAT] One of the hamstring muscles, arising from the ischial tuber, and inserted into the tibia. { ¦sem·iˌmem·brə'nō·səs }

seminal bursa [INV ZOO] A sac which retains sperm for a period of time in turbellarians. { 'sem·ən·əl 'bər·sə }

seminal fluid *See* semen. { 'sem·ə·nəl ˌflü·əd }

seminal fructose [PHYSIO] Fructose that is normally produced in the seminal vesicles. { ¦sem·ən·əl 'frükˌtōs }

seminal groove [ZOO] A passage in many animals providing a pathway for sperm. { 'sem·ən·əl 'grüv }

seminal receptacle *See* spermatheca. { 'sem·ən·əl ri'sep·tə·kəl }

seminal vesicle [ANAT] A saclike, glandular diverticulum on each ductus deferens in male vertebrates; it is united with the excretory duct and serves for temporary storage of semen. { 'sem·ən·əl 'ves·i·kəl }

seminiferous tubule [ANAT] Any of the tubercles of the testes which produce spermatozoa. { ¦sem·ə¦nif·rəs 'tü,byül }

seminivorous [ZOO] Feeding on seeds. { ¦sem·ə¦niv·ə·rəs }

seminoma [MED] A malignant tumor of the testes composed of large, uniform cells with clear cytoplasm. { ,sem·ə'nō·mə }

semiochemical [PHYSIO] Any of a class of substances produced by organisms, especially insects, that participate in regulation of their behavior in such activities as aggregation of both sexes, sexual stimulation, and trail following. { ¦sem·ē·ə'kem·ə·kəl }

Semionotiformes [VERT ZOO] An order of actinopterygian fishes represented by the single living genus *Lepisosteus*, the gars; the body is characteristically encased in a heavy armor of interlocking ganoid scales. { ,sem·ē·ə,nō·də 'fór,mēz }

semipalmate [VERT ZOO] Having a web halfway down the toes. { ¦sem·i'päl,māt }

semiparasite *See* hemiparasite. { ¦sem·i'par·ə,sīt }

semispecies [SYST] **1.** The species that compose a superspecies. **2.** Populations that have acquired some attributes of species rank. **3.** Organisms that are borderline between species and subspecies. { ¦sem·i'spē·shēz }

semispinalis [ANAT] One of the deep longitudinal muscles of the back, attached to the vertebrae. { ,sem·i,spī'nal·əs }

semitendinosus [ANAT] One of the hamstring muscles, arising from the ischium and inserted into the tibia. { ,sem·ē,ten·də'nō·səs }

Semper's larva [INV ZOO] A cylindrical larva in the life history of certain zoanthid corals, characterized by a hole at each end and an annular or longitudinal band of long cilia. { 'sem·pərz ,lär·və }

senescence [BIOL] The study of the biological changes related to aging, with special emphasis on plant, animal, and clinical observations which may apply to humans. { si'nes·əns }

senescent arthritis *See* degenerative joint disease. { si'nes·ənt är'thrid·əs }

senile [MED] Of, pertaining to, or caused by the aging process or by the infirmities of old age. { 'sē,nīl }

senile cataract [MED] The most common type of cataract; it occurs with aging, and progressively blurred vision is the only symptom. { ,sē,nīl 'kad·ə,rakt }

senile eczema [MED] A form of eczema associated with aging and caused by factors such as dryness of skin, soap sensitivity, or poor hygiene or diet. { 'sē,nīl 'ek·sə·mə }

senile emphysema [MED] Degenerative changes in the lungs and thoracic cage associated with aging. { 'sē,nīl ,em·fə'sē·mə }

senile gangrene [MED] A form of tissue death caused by deterioration of the blood supply to the extremities in the elderly. { 'sē,nīl 'gaŋ ,grēn }

senile vaginitis [MED] Inflammation of the vagina occurring in elderly women following the chronic irritation of the thinned, atrophic mucosa. { 'sē,nīl ,vaj·ə'nīd·əs }

senility [MED] Old age and its characteristics. { si'nil·əd·ē }

senna [PHARM] The dried leaflets of plants of the *Cassia* genus; used in medicine as a cathartic. { 'sen·ə }

sensation [PHYSIO] The subjective experience that results from the stimulation of a sense organ. { sen'sā·shən }

sensation level *See* level above threshold. { sen'sā·shən ,lev·əl }

sense organ [PHYSIO] A structure which is a receptor for external or internal stimulation. { 'sens ,ór·gən }

sense strand [CELL MOL] The strand of a double-stranded deoxyribonucleic acid molecule that is complementary to the ribonucleic acid formed by transcription. Also known as coding strand. { 'sens ,strand }

sensillum [ZOO] A simple, epithelial sense organ composed of one cell or of a few cells. { sen'sil·əm }

sensitivity [PHYSIO] The capacity for receiving sensory impressions from the environment. { ,sen·sə'tiv·əd·ē }

sensitization [IMMUNOL] The alteration of a body's responsiveness to a foreign antigen, usually an allergen, such that upon subsequent exposures to the allergen there is a heightened immune response. { ,sen·səd·ə'zā·shən }

sensorineural deafness [MED] Deafness caused by an abnormality of the sense organ in the inner ear or the auditory nerve (cranial nerve VIII). { ¦sen·sə·rə'nůr·əl 'def·nəs }

sensorium [PHYSIO] **1.** A center, especially in the brain, for receiving and integrating sensations. **2.** The entire sensory apparatus of an individual. { sen'sór·ē·əm }

sensory aphasia [MED] A form of aphasia in which the perception of sounds as language is partially preserved, but the patient is unable to comprehend the meaning of words, and in speaking, words are evoked with difficulty, are used incorrectly, and do not convey ideas correctly, resulting frequently in other forms of language impairment, particularly in agrammatism. Also known as receptive aphasia. { 'sen·sə·rē a'fā·zhə }

sensory area [NEUROSCI] Any area of the cerebral cortex associated with the perception of sensations. { 'sen·sə·rē ,er·ē·ə }

sensory cell [NEUROSCI] A neuron having its terminal processes connected with sensory nerve endings. [PHYSIO] A modified epithelial or connective tissue cell adapted for the reception and transmission of sensations. { 'sen·sə·rē ˌsel }

sensory nerve [NEUROSCI] A nerve that conducts afferent impulses from the periphery to the central nervous system. { 'sen·sə·rē ˌnərv }

sensory-neural hearing loss [PHYSIO] A type of hearing loss resulting from damage to the receptive elements in the ear or to the auditory nerve itself. { ¦sen·sə·rē ˌnu̇r·əl 'hir·iŋ ˌlȯs }

sepal [BOT] One of the leaves composing the calyx. { sēp·əl }

separation disk [BOT] A layer of gelatinous material between two adjacent negative cells in some blue-green algae; associated with hormogonium formation. { ˌsep·ə'rā·shən ˌdisk }

separation layer [BOT] A structurally distinct layer of the abscission zone of a plant containing abundant starch and dense cytoplasm. { ˌsep·ə'rā·shən ˌlā·ər }

Sepioidea [INV ZOO] An order of the molluscan subclass Coleoidea having a well-developed eye, an internal shell, fins separated posteriorly, and chromatophores in the dermis. { ˌsē·pē'ȯid·ē·ə }

Sepsidae [INV ZOO] The spiny-legged flies, a family of myodarian cyclorrhaphous dipteran insects in the subsection Acalypteratae; development takes place in decaying organic matter. { 'sep·səˌdē }

sepsis [MED] **1.** Poisoning by products of putrefaction. **2.** The severe toxic, febrile state resulting from infection with pyogenic microorganisms, with or without associated septicemia. { 'sep·səs }

septal filament [INV ZOO] In anthozoans, the free edges of the septum containing gland cells and nematocysts. { 'sep·təl 'fil·ə·mənt }

septal ostium [INV ZOO] Any of the openings in septa of anthozoans. { 'sep·təl 'äs·tē·əm }

septate [BIOL] Having a septum. { 'sepˌtāt }

Septibranchia [INV ZOO] A small order of bivalve mollusks in which the anterior and posterior abductor muscles are about equal in size, the foot is long and slender, and the gills have been transformed into a muscular septum. { ˌsep·tə'braŋ·kē·ə }

septic [MED] Of or pertaining to sepsis. { 'sep·tik }

septic abortion [MED] An abortion complicated by acute infection of the endometrium. { 'sep·tik ə'bȯr·shən }

septic embolus [MED] An embolus formed by bacteria. { 'sep·tik 'em·bə·ləs }

septicemia [MED] A clinical syndrome in which infection is disseminated through the body in the bloodstream. Also known as blood poisoning. { ˌsep·tə'sē·mē·ə }

septicidal [BOT] A type of dehiscence exhibited by some fruit in which splitting of the pericarp occurs along the junction of component carpels. { ¦sep·tə¦sīd·əl }

septulum [ANAT] A small septum. { 'sep·tə·ləm }

septum [BIOL] A partition or dividing wall between two cavities. { 'sep·təm }

septum pellucidum [ANAT] A thin translucent septum forming the internal boundary of the lateral ventricles of the brain and enclosing between its two laminas the so-called fifth ventricle. { 'sep·təm pə'lü·səd·əm }

septum primum [EMBRYO] The first incomplete interatrial septum of the embryo. { 'sep·təm 'prē·məm }

septum secundum [EMBRYO] The second incomplete interatrial septum of the embryo, containing the foramen ovale; it develops to the right of the septum primum and fuses with it to form the adult interatrial septum. { 'sep·təm si'kən·dəm }

sequela [MED] The abnormal aftereffects or complications of an illness, infection, or injury. { si'kwel·ə }

sequestrum [MED] A piece of dead or detached bone within a cavity, abscess, or wound. { si'kwes·trəm }

Sequoia [BOT] A genus of conifers having overlapping, scalelike evergreen leaves and vertical grooves in the trunk; the giant sequoia (*Sequoia gigantea*) is the largest and oldest of all living things. { si'kwȯi·yə }

sere [ECOL] A temporary community which occurs during a successional sequence on a given site. { sir }

Sergestidae [INV ZOO] A family of decapod crustaceans including several species of prawns. { sər'jes·təˌdē }

serial homology [ZOO] The similarity between the members of a single series of structures, such as vertebrae, in an organism. { 'sir·ē·əl hə'mäl·ə·jē }

sericeous [BOT] Of, pertaining to, or consisting of silk. { sə'rish·əs }

serine [BIOCHEM] $C_3H_7O_3N$ An amino acid obtained by hydrolysis of many proteins; a biosynthetic precursor of several metabolites, including cysteine, glycine, and choline. { 'seˌrēn }

serine protease [CELL MOL] A family of endopeptidases whose proteolytic activity involves the hydroxy group of the serine residue; they play an important role in digestion, blood clotting, and the complement system. { ¦seˌrēn 'prōd·əˌās }

seritinous [ECOL] Of, pertaining to, or occurring during the latter, drier half of the summer. { ¦ser·ə'tī·nəs }

seroche *See* mountain sickness. { sə'rōsh }

serodiagnosis [MED] Diagnosis based upon the reaction of blood serum of a patient. { ¦si·rōˌdi·əg'nō·səs }

seroepidemiology [MED] The study of the distribution of serum antibodies. { ¦si·rōˌep·əˌdē·mē'äl·ə·jē }

serofibrinous [PHYSIO] Composed of serum and fibrin. { ¦si·rō'fī·brə·nəs }

Serolidae [INV ZOO] A family of isopod crustaceans which contains greatly flattened forms that live partially buried on sandy bottoms. { sə'räl·əˌdē }

serology [BIOL] The branch of science dealing with the properties and reactions of blood sera. { sə'räl·ə·jē }

seronegative [PATH] **1.** Having a negative serologic test for some condition. **2.** Specifically, having a negative serologic test for syphilis. { ¦si·rō'neg·əd·iv }

seropositive [PATH] **1.** Having a positive serologic test for some condition. **2.** Specifically, having a positive serologic test for syphilis. { ¦si·rō'päz·əd·iv }

seropurulent [MED] Composed of serum and pus, as a seropurulent exudate. { ¦si·rō'pyůr·ə·lənt }

seroresistance [PATH] Persistent positive serologic reaction for syphilis despite prolonged intensive treatment; the patient is said to be Wassermann-fast. { ¦si·rō·ri'zis·təns }

serosa [ANAT] The serous membrane lining the pleural, peritoneal, and pericardial cavities. [EMBRYO] The chorion of reptile and bird embryos. { sə'rō·sə }

serotherapy [MED] The treatment of disease by means of human or animal serum containing antibodies. Also known as immunotherapy. { ¦si·rō'ther·ə·pē }

serotinous [BOT] Plants which flower or develop late in a season. { sə'rät·ən·əs }

serotonin [BIOCHEM] $C_{10}H_{12}ON_2$ A compound derived from tryptophan which functions as a local vasoconstrictor, plays a role in neurotransmission, and has pharmacologic properties. Also known as 5-hydroxytryptamine. { ˌsir·ə'tō·nən }

serotype [MICROBIO] A serological type of intimately related microorganisms, distinguished on the basis of antigenic composition. { 'sir·əˌtīp }

serous cystadenoma [MED] A benign cystic tumor of the ovary, made up of cylindrical cells resembling those of the uterine tube; psammoma bodies often appear in the wall of the cyst. Also known as endosalpingioma; papillary adenoma of ovary; papillary cystadenoma of ovary; papillocystoma; psammomatus papilloma; serous cystoma. { 'sir·əs ˌsistˌad·ən'ō·mə }

serous cystoma *See* serous cystadenoma. { 'sir·əs si'stō·mə }

serous gland [PHYSIO] A structure that secretes a watery, albuminous fluid. { 'sir·əs ˌgland }

serous membrane [HISTOL] A delicate membrane covered with flat, mesothelial cells lining closed cavities of the body. { 'sir·əs 'mem ˌbrān }

serous plethora [MED] An increase in the watery part of the blood. { 'sir·əs 'pleth·ə·rə }

Serpentes [VERT ZOO] The snakes, a suborder of the Squamata characterized by the lack of limbs and pectoral girdle and external ear openings, immovable eyelids, and a braincase that is completely bony anteriorly. { sər'penˌtēz }

serpentine locomotion [VERT ZOO] The wavelike or undulating movements characteristic of snakes. { 'sər·pənˌtēn ˌlō·kə'mō·shən }

Serpulidae [INV ZOO] A family of polychaete annelids belonging to the Sedentaria including many of the feather-duster worms which construct calcareous tubes in the earth, sometimes in such abundance as to clog drains and waterways. { sər'pyů·ləˌdē }

Serranidae [VERT ZOO] A family of perciform fishes in the suborder Percoidei including the sea basses and groupers. { sə'ran·əˌdē }

serrate [BIOL] Possessing a notched or toothed edge. { 'seˌrāt }

Serratia marcescens [MICROBIO] A human pathogen that is intrinsically resistant to many antimicrobials (for example, cephalosporins, polymyxins, and nitrofurans) and occurs predominantly in hospitalized patients. { səˌrā·shē·ə mär'ses·əns }

Serratieae [MICROBIO] Formerly a tribe of the Enterobacteriaceae containing the genus *Serratia*, with soil and water forms characterized by the production of a bright-orange to deep-red pigment, prodigiosin. { sə'rāsh·ēˌē }

Serritermitidae [INV ZOO] A family of the Isoptera which contains the single monotypic genus *Serritermes*. { ˌser·ə·tər'mid·əˌdē }

Serropalpidae [INV ZOO] An equivalent name for Melandryidae. { ˌser·ə'pal·pəˌdē }

serrulate [BIOL] Finely serrate. { 'ser·ə·lət }

Sertoli cell [HISTOL] One of the sustentacular cells of the seminiferous tubules. { sər'tō·lē ˌsel }

serum [PHYSIO] The liquid portion that remains when blood clots spontaneously and the formed and clotting elements are removed by centrifugation; it differs from plasma by the absence of fibrinogen. { 'sir·əm }

serum accident [IMMUNOL] A serious allergic reaction which immediately follows the introduction of a foreign serum into a hypersensitive individual; dyspnea and flushing occur, soon followed by shock and occasionally by fatal termination. { 'sir·əm ˌak·sə·dənt }

serum albumin [BIOCHEM] The principal protein fraction of blood serum and serous fluids. { 'sir·əm al'byü·mən }

serum globulin [BIOCHEM] The globulin fraction of blood serum. { 'sir·əm 'glä·byə·lən }

serum hepatitis [MED] A form of viral hepatitis transmitted by parenteral injection of human blood or blood products contaminated with the type B virus. { 'sir·əm ˌhep·ə'tīd·əs }

serum shock [MED] An anaphylactic reaction following the injection of foreign serum into a sensitized individual. { 'sir·əm ˌshäk }

serum sickness [MED] A syndrome manifested in 8–12 days after the administration of serum by an urticarial rash, edema, enlargement of lymph nodes, arthralgia, and fever. { 'sir·əm ˌsik·nəs }

sesamoid bone [MED] A small bone developed in a tendon subjected to much pressure. { 'ses·əˌmȯid ˌbōn }

sessile [BOT] Attached directly to a branch or stem without an intervening stalk. [ZOO] Permanently attached to the substrate. { 'ses·əl }

Sessilina [INV ZOO] A suborder of ciliates in the order Peritrichida. { ˌses·əˈlī·nə }

sessoblast [INV ZOO] A statoblast that attaches to zooecial tubes or to the substratum. { ˈses·əˌblast }

seta [BIOL] **1.** A slender, usually rigid bristle or hair. Also known as chaeta. **2.** In mosses and liverworts, the stalk of the sporophyte supporting the capsule. { ˈsed·ə }

setigerous [INV ZOO] Referring to a segment with setae. { səˈtij·ə·rəs }

severe combined immunodeficiency syndrome [MED] A group of genetic disorders that result in a deficiency of B and T lymphocytes, causing great susceptibility to infection. Abbreviated SCIDS. { səˈvir kəmˈbīnd ˌim·yəˌnō·dəˈfish·ən·sē ˌsinˌdrōm }

sex [BIOL] **1.** The state of condition of an organism which comes to expression in the production of germ cells. **2.** To determine the sex of. { seks }

sex cell *See* gamete. { ˈseks ˌsel }

sex chromatin *See* Barr body. { ˈseks ˈkrō·mə·tən }

sex chromosome [GEN] Either member of a pair of chromosomes responsible for sex determination of an organism. { ˈseks ˌkrō·məˌsōm }

sex cords [EMBRYO] Cordlike masses of epithelial tissue that invaginate from germinal epithelium of the gonad and give rise to seminiferous tubules and rete testes in the male, and primary ovarian follicles and rete ovarii in the female. { ˈseks ˌkȯrdz }

sex determination [GEN] The mechanisms which determine whether the bipotential embryo will develop as male or female in a species. { ˈseks diˌtər·məˌnā·shən }

sex factor *See* fertility factor. { ˈseks ˌfak·tər }

sex hormone [BIOCHEM] Any hormone secreted by a gonad, but also found in other tissues. { ˈseks ˌhȯrˌmōn }

sex-influenced inheritance [GEN] That part of the inheritance pattern on which sex differences operate to promote character differences. { ˈseks inˈflü·ənst inˈher·ət·əns }

sex-limited inheritance [GEN] Expression of a phenotype in only one sex; may be due to either a sex-linked or autosomal gene. { ˈseks ˈlim·əd·əd inˈher·ət·əns }

sex-linked inheritance [GEN] The transmission to successive generations of traits that are due to alleles at gene loci on a sex chromosome. { ˈseks ˈliŋkt inˈher·ət·əns }

sex organs [ANAT] The organs pertaining entirely to the sex of an individual, both physiologically and anatomically. { ˈseks ˌȯr·gənz }

sex ratio [BIOL] The relative proportion of males and females in a population. { ˈseks ˌrā·shō }

sexual cycle [PHYSIO] A cycle of physiological and structural changes associated with sex; examples are the estrous cycle and the menstrual cycle. { ˈsek·shə·wəl ˈsī·kəl }

sexual dimorphism [BIOL] Diagnostic morphological differences between the sexes. { ˈsek·shə·wəl dīˈmȯrˌfiz·əm }

sexuality [BIOL] **1.** The sum of a person's sexual attributes, behavior, and tendencies. **2.** The psychological and physiological sexual impulses whose satisfaction affords pleasure. { ˌsek·shəˈwal·əd·ē }

sexually transmitted disease [MED] An infection acquired and transmitted primarily by sexual contact. Abbreviated STD. { ˈsek·shə·lē transˈmid·əd diˈzēz }

sexual reproduction [BIOL] Reproduction involving the paired union of special cells (gametes) from two individuals. { ˈsek·shə·wəl ˌrē·prəˈdək·shən }

sexual selection [EVOL] A special form of natural selection responsible for the evolution of traits that promote success in competition for mates. { ˈseksh·ə·wəl siˈlek·shən }

sexual spore [BIOL] A spore resulting from conjugation of gametes or nuclei of opposite sex. { ˈsek·shə·wəl ˈspȯr }

Sezary syndrome [MED] Exfoliative erythroderma with a cutaneous infiltrate of atypical mononuclear cells; similar cells are also present in the peripheral blood. { səˈzar·ē ˌsinˌdrōm }

shadscale [BOT] *Atriplex confertifolia*. A small shiny shrub found in the Great Basin Desert. { ˈshadˌskāl }

shake culture [MICROBIO] **1.** A method for isolating anaerobic bacteria by shaking a deep liquid culture of an agar or gelatin to distribute the inoculum before solidification of the medium. **2.** A liquid medium in a flask that has been inoculated with an aerobic microorganism and placed on a shaking machine; action of the machine continually aerates the culture. { ˈshāk ˌkəl·chər }

shallot [BOT] *Allium ascalonicum*. A bulbous onionlike herb. Also known as scallion. { ˈshal·ət }

shark [VERT ZOO] Any of about 225 species of carnivorous elasmobranchs which occur principally in tropical and subtropical oceans; the body is fusiform with a heterocercal tail and a tough, usually gray, skin roughened by tubercles, and the snout extends beyond the mouth. { shärk }

shearwater [VERT ZOO] Any of various species of oceanic birds of the genus *Puffinus* having tubular nostrils and long wings. { ˈshirˌwȯd·ər }

sheathed bacteria [MICROBIO] Chains of bacterial cells, usually rod-shaped, enclosed in a hyaline envelope or sheath. { ˈshēthd bakˈtir·ē·ə }

Sheehan's syndrome [MED] Pituitary necrosis caused by insufficient blood supply to the gland due to copious blood loss or intravascular clotting and hemorrhage associated with premature separation of the placenta during childbirth. Also known as postpartum pituitary necrosis. { ˈshē·ənz ˌsinˌdrōm }

sheep [VERT ZOO] Any of various mammals of the genus *Ovis* in the family Bovidae characterized by a stocky build and horns, when present, which tend to curl in a spiral. { shēp }

shell [ZOO] **1.** A hard, usually calcareous, outer covering on an animal body, as of bivalves and turtles. **2.** The hard covering of an egg. **3.** Chitinous exoskeleton of certain arthropods. { shel }

shellfish [INV ZOO] An aquatic invertebrate, such as a mollusk or crustacean, that has a shell or exoskeleton. { 'shel,fish }

shell gland [INV ZOO] An organ that secretes the embryonic shell in many mollusks. [VERT ZOO] A specialized structure attached to the oviduct in certain animals that secretes the eggshell material. { 'shel ,gland }

shell membrane [CELL MOL] Either of a pair of membranes lining the inner surface of an egg shell; they allow free entry of oxygen but prevent rapid evaporation of moisture. { 'shel 'mem ,brān }

shelterbelt [ECOL] A natural or planned barrier of trees or shrubs to reduce erosion and provide shelter from wind and storm activity. { 'shel·tər,belt }

Shetland sheep [VERT ZOO] A breed of sheep raised in the Shetland Isles of Scotland. { 'shet·lənd 'shēp }

shiitake mushroom *See* Letinula edodes. { ,shē·ē,tä·kē 'məsh,rüm }

shikimic acid [BIOCHEM] $C_7H_{10}O_5$ A crystalline acid that is a plant constituent, and an intermediate in the biochemical pathway from phosphoenolpyruvic acid to tyrosine. { shə'kim·ik 'as·əd }

shingles *See* herpes zoster. { 'shiŋ·gəlz }

shin splints [MED] An injury and an inflammation of the lower leg muscles and bones of the lower and middle third of the tibia and fibula, seen in athletes such as runners or basketball and football players. { 'shin ,splins }

shipping fever [VET MED] An acute, occasionally subacute, septicemic disease in cattle and sheep, probably caused by a combination of virus and *Pasteurella multocida* or *P. hemolytica*. { 'ship·iŋ ,fē·vər }

shipworm [INV ZOO] Any of several bivalve mollusk species belonging to the family Teredinidae and which superficially resemble earthworms because the two valves are reduced to a pair of plates at the anterior of the animal or are used for boring into wood. { 'ship,wərm }

shipyard eye *See* keratoconjunctivitis. { 'ship ,yärd 'ī }

shock [MED] Clinical manifestations of circulatory insufficiency, including hypotension, weak pulse, tachycardia, pallor, and diminished urinary output. { 'shäk }

shock organ [IMMUNOL] The organ or tissue that exhibits the most marked response to the antigen-antibody interaction in hypersensitivity, as the lungs in allergic asthma or the skin in allergic contact dermatitis. { 'shäk ,ȯr·gən }

shoot [BOT] **1.** The aerial portion of a plant, including stem, branches, and leaves. **2.** A new, immature growth on a plant. { shüt }

Shope papilloma [VET MED] A transmissible, virus-induced papilloma occurring naturally on the skin of rabbits. { 'shōp ,pap·ə'lō·mə }

shop typhus *See* murine typhus. { 'shäp ,tī·fəs }

shore bird [VERT ZOO] A general term applied to a large number of birds in 12 families of the suborder Charadrii which are always found near water, although the habitat and morphology is varied. Also known as wader. { 'shȯr ,bərd }

shore crab *See* Carcinus maenas. { 'shȯr ,krab }

short-day response [PHYSIO] A photoperiodic response to decreasing days and increasing nights. { ,shȯrt ,dā ri'späns }

short interspersed elements [GEN] Families of short deoxyribonucleic acid sequences that are individually inserted abundantly throughout the genome in mammals and other taxons; the 300-base-pair Alu short interspersed elements make up about 5% of the human genome. Abbreviated SINES. { 'shȯrt ,in·tər,spərst 'el·ə·məns }

short-term exposure limit [PHYSIO] The maximum amount of harmful gas or dust to which a person may be exposed for a brief period (usually 15 minutes) without being physically harmed. Abbreviated STEL. { ¦shȯrt ¦tərm ik'spō·zhər ,lim·ət }

shoulder [ANAT] **1.** The area of union between the upper limb and the trunk in humans. **2.** The corresponding region in other vertebrates. { 'shōl·dər }

shoulder blade *See* scapula. { 'shōl·dər ,blād }

shoulder girdle *See* pectoral girdle. { 'shōl·dər ,gərd·əl }

shoulder-hand syndrome [MED] A syndrome characterized by severe constant intractable pain in the shoulder and arm, limited joint motion, diffuse swelling of the distal part of the upper extremity, fibrosis and atrophy of muscles, and decalcification of underlying bones; the cause is not well understood; it is similar to, or may be a form of, causalgia. Also known as hand-shoulder syndrome. { 'shōl·dər 'hand 'sin ,drōm }

shrew [VERT ZOO] Any of more than 250 species of insectivorous mammals of the family Soricidae; individuals are small with a moderately long tail, minute eyes, a sharp-pointed snout, and small ears. { shrü }

shrimp [INV ZOO] The common name for a number of crustaceans, principally in the decapod suborder Natantia, characterized by having well-developed pleopods and by having the abdomen sharply bent in most species, producing a humped appearance. { shrimp }

shrub [BOT] A low woody plant with several stems. { shrəb }

shunt [MED] A vascular passage by which blood is diverted from its normal circulatory path; frequently it is a surgical passage created between two blood vessels, but it may also be an anatomical feature. { shənt }

shuttle vector [CELL MOL] A deoxyribonucleic acid vector able to replicate in two different organisms, and therefore able to shuttle foreign nucleic acids between two different hosts. Also known as bifunctional vector. { 'shəd·əl ,vek·tər }

Shwartzman phenomenon [IMMUNOL] A type of local tissue reactivity in the skin in which a preparatory injection of the endotoxin is followed by an intravenous injection of the same or another endotoxin 24 hours later, producing immediate neutropenia and thrombopenia with the development of leukocyte-platelet thrombi with subsequent hemorrhage. { 'shwȯrts·mənfə,näm·ə,nän }

sialadenitis [MED] Inflammation of a salivary gland. { sī¦al·ə·də'nīd·əs }

sialagogue [PHARM] A drug producing a flow of saliva. { sī'al·ə,gäg }

sialic acid [BIOCHEM] Any of a family of amino sugars, containing nine or more carbon atoms, that are nitrogen- and oxygen-substituted acyl derivatives of neuraminic acid; as components of lipids, polysaccharides, and mucoproteins, they are widely distributed in bacteria and in animal tissues. { sī'al·ik 'as·əd }

sialogram [MED] Roentgenogram of a salivary duct system after the injection of an opaque medium. { sī'al·ə,gram }

sialography [MED] Radiographic examination of a salivary gland following injection of an opaque substance into its duct. { ,sī·ə'läg·rə·fē }

sialolith [PATH] A salivary calculus. { sī'al·ə,lith }

sialolithiasis [MED] The presence of salivary calculi. { sī¦al·ə·li'thī·ə·səs }

sialomucin [BIOCHEM] An acid mucopolysaccharide containing sialic acid as the acid component. { sī¦al·ə'myüs·ən }

Siamese twins [MED] Viable conjoined twins. { 'sī·ə,mēz 'twinz }

Siberian tick typhus [MED] A relatively benign, rash- and eschar-producing spotted-fever-like disease in northern Asia, caused by *Rickettsia siberica*; transmitted by four species of *Dermacentor* and two of *Haemaphysalis*. { sī'bir·ē·ən ¦tik ,tī·fəs }

Siboglinidae [INV ZOO] A family of pogonophores in the order Athecanephria. { ,sī·bə'glī·nə,dē }

sickle-cell anemia [MED] A chronic, hereditary hemolytic and thrombotic disorder in which hypoxia causes the erythrocyte to assume a sickle shape; occurs in individuals homozygous for sickle-cell hemoglobin trait. Also known as sickle-cell disease. { 'sik·əl ¦sel ə'nē·mē·ə }

sickle-cell disease *See* sickle-cell anemia. { 'sik·əl ¦sel di,zēz }

sickle-cell hemoglobin [PATH] The hemoglobin found in sickle-cell anemia, differing in electrophoretic mobility and other physiochemical properties from normal adult hemoglobin. Also known as hemoglobin S. { 'sik·əl ¦sel 'hē·mə,glō·bən }

sicula [INV ZOO] The cone-shaped chitinous skeleton of the first zooid of a graptolite colony. { 'sik·yə·lə }

Siderocapsaceae [MICROBIO] A family of gram-negative, chemolithotrophic bacteria; characterized by the ability to deposit iron or manganese compounds on or around the cells. { ,sid·ə·rə,kap'sās·ē,ē }

siderocyte [CELL MOL] An erythrocyte which contains granules staining blue with the Prussian blue reaction. { 'sid·ə·rə,sīt }

siderofibrosis [MED] Fibrosis associated with deposits of iron-bearing pigment. { ,sid·ə·rə·fī'brō·səs }

sideropenic dysphagia *See* Plummer-Vinson syndrome. { ¦sid·ə·rə¦pē·nik dis'fā·jə }

siderophore [BIOCHEM] A molecular receptor that binds and transports iron. { 'sid·ə·rə,fȯr }

siderosilicosis [MED] A pneumoconiosis resulting from prolonged inhalation of silica and iron dusts. { ,sid·ə·rə,sil·ə'kō·səs }

siderosis [MED] Pneumoconiosis due to prolonged inhalation of dust containing iron salt. Also known as arc-welder's disease. [PATH] The presence or concentration of stainable iron pigment in a tissue or organ. { ,sid·ə'rō·səs }

SIDS *See* sudden infant death syndrome. { sidz }

Sierolomorphidae [INV ZOO] A small family of hymenopteran insects in the superfamily Scolioidea. { sē¦er·ə·lō'mȯr·fə,dē }

sieve area [BOT] An area in the wall of a sieve-tube element, sieve cell, or parenchyma cell characterized by clusters of pores through which strands of cytoplasm pass to adjoining cells. { 'siv ,er·ē·ə }

sieve cells [BOT] The food-conducting cells of phloem tissue in seedless vascular plants and gymnosperms; unlike sieve-tube members in angiosperms, they do not have sieve plates. { 'siv ,selz }

sieve elements [BOT] The food-conducting cells of phloem tissue; may be either sieve cells (in seedless vascular plants and gymnosperms) or sieve-tube members (in angiosperms). { 'siv ,el·ə·məns }

sieve plate [BOT] A perforated section of the wall of a component member of a sieve tube. { 'siv ,plāt }

sieve tissue *See* phloem. { 'siv ,tish·ü }

sieve tray *See* sieve plate. { 'siv ,trā }

sieve tube [BOT] A phloem element consisting of a series of thin-walled cells arranged end to end, in which some sieve areas are more specialized than others. { 'siv ,tüb }

sieve-tube member [BOT] The food-conducting cells of phloem tissue in angiosperms; the individual cells have a sieve plate at each end and are arranged end to end to form a sieve tube. { 'siv ,tüb ,mem·bər }

Sigalionidae [INV ZOO] A family of scale-bearing polychaete annelids belonging to the Errantia. { ,sig·ə·lē'än·ə,dē }

Siganidae [VERT ZOO] A small family of herbivorous perciform fishes in the suborder Acanthuroidei having minute concealed scales embedded in the skin and strong, sharp fin spines. { si 'gan·ə,dē }

sight *See* vision. { sīt }

sigma [INV ZOO] A C-shaped spicule. { 'sig·mə }

sigma factor [CELL MOL] A regulatory protein

in prokaryotes that combines with the ribonucleic acid (RNA) polymerase enzyme to facilitate the initiation of RNA synthesis. { 'sig·mə ˌfak·tər }

sigmaspire [INV ZOO] An S-shaped sponge spicule. { 'sig·məˌspīr }

sigmoid [BIOL] S-shaped. { 'sigˌmȯid }

sigmoid colon [ANAT] The S-shaped portion of the colon between the descending colon and the rectum. { 'sigˌmȯid 'kō·lən }

sigmoiditis [MED] Inflammation of the sigmoid flexure of the colon. { ˌsig·mȯi'dīd·əs }

sigmoidoscope [MED] An appliance for the inspection, by artificial light, of the sigmoid colon; it differs from the proctoscope in its greater length and diameter. { sig'mȯid·əˌskōp }

signaling cell [PHYSIO] A cell whose products induce a specific response in target cells. { 'sig·nə·liŋ ˌsel }

signal molecule [BIOCHEM] A molecule produced by a signaling cell. { 'sig·nəl ˌmäl·ə ˌkyül }

single nucleotide polymorphism [GEN] A single base-pair difference between two copies of a deoxyribonucleic acid sequence from two individuals. Abbreviated SNP. { ¦siŋ·gəl ¦nü·klē·əˌtīd ˌpäl·ē'mȯr·fiz·əm }

signal-recognition particle [CELL MOL] A ribonucleoprotein consisting of a ribonucleic acid (RNA) molecule and six distinct peptide chains that recognizes the signal sequence of a partially synthesized protein and guides it along with its ribosome to a signal recognition particle receptor in the endoplasmic reticulum. Abbreviated SRP. { ¦sig·nəl rek·ig'nish·ən ˌpärd·i·kəl }

signal sequence [CELL MOL] A discrete sequence of amino acids in a protein that serves to identify it to transport mechanisms within a cell so as to guide the protein to its destination. { 'sig·nəl ˌsē·kwəns }

signal transduction [CELL MOL] The relaying of molecular signals (for example, as contained in a hormone) or physical signals (for example, sensory stimuli) from a cell's exterior to its intracellular response mechanisms. { 'sig·nəl tranz ˌdək·shən }

signet-ring cell [HISTOL] A cell with a large fat- or carbohydrate-filled vacuole that pushes the nucleus against the cell membrane. { 'sig·nət ˌriŋ 'sel }

silent mutation [GEN] A mutation that does not result in amino acid sequence change. { 'sī·lənt myü'tā·shən }

silicle [BOT] A many-seeded capsule formed from two united carpels, usually of equal length and width, and divided on the inside by a replum. { 'sil·ə·kəl }

silicoblast [INV ZOO] Poriferan amebocytes involved in formation of siliceous spicules. { 'sil·ə·kəˌblast }

Silicoflagellata [BOT] A class of unicellular flagellates of the plant division Chrysophyta represented by a single living genus, *Dictyocha*. { ¦sil·ə·kōˌflaj·ə'läd·ə }

Silicoflagellida [INV ZOO] An order of marine flagellates in the class Phytamastigophorea which have an internal, siliceous, tubular skeleton, numerous yellow chromatophores, and a single flagellum. { ¦sil·ə·kō·flə'jel·əd·ə }

silicosiderosis [MED] Pneumoconiosis caused by inhalation of silicate-and iron-containing dust. { ¦sil·ə·kōˌsid·ə'rō·səs }

silicosis [MED] Pneumoconiosis due to the inhalation of silica (SiO_2) particles. { ˌsil·ə'kō·səs }

silique [BOT] A silicle-like capsule, but usually at least four times as long as it is wide, which opens by sutures at either margin and has parietal placentation. { si'lēk }

silk [INV ZOO] A continuous protein fiber consisting principally of fibroin and secreted by various insects and arachnids, especially the silkworm, for use in spinning cocoons, webs, egg cases, and other structures. { silk }

silk cotton *See* kapok. { 'silk ¦kät·ən }

silk cotton tree *See* kapok tree. { 'silk ¦kät·ən ˌtrē }

silk gland [INV ZOO] A gland in certain insects which secretes a viscous fluid in the form of filaments known as silk; it is a salivary gland in insects and an abdominal gland in spiders. { 'silk ˌgland }

silkworm [INV ZOO] The larva of various moths, especially *Bombyx mori*, that produces a large amount of silk for building its cocoon. { 'silkˌwərm }

Silphidae [INV ZOO] The carrion beetles, a family of coleopteran insects in the superfamily Staphylinoidea. { 'sil·fəˌdē }

Siluridae [VERT ZOO] A family of European catfish in the suborder Siluroidei in which the adipose dorsal fin is rudimentary or lacking. { si 'lu̇r·əˌdē }

Siluriformes [VERT ZOO] The catfishes, a distinctive order of actinopterygian fishes in the superorder Ostariophysi, distinguished by a complex Weberian apparatus that involves the fifth vertebrae and one to four pair of barbels. { siˌlu̇r·ə'fȯrˌmēz }

Siluroidei [VERT ZOO] A suborder of the Siluriformes. { ˌsil·yə'rȯid·ēˌī }

Silvanidae [INV ZOO] An equivalent name for Cucujidae. { sil'van·əˌdē }

silver arsphenamine [PHARM] A brownish-black powder soluble in water; used in medicine. { 'sil·vər är'sfen·əˌmēn }

silverfish [INV ZOO] Any of over 350 species of insects of the order Thysanura; they are small, wingless forms with biting mouthparts. { 'sil·vərˌfish }

silverline system [INV ZOO] A series of superficial argentophilic lines in many protozoans, especially ciliates. { 'sil·vərˌlīn ˌsis·təm }

similarity coefficient [SYST] In numerical taxonomy, a factor S used to calculate the similarity between organisms, according to the formula $S = n_s/(n_s + n_d)$, where n_s represents the number of positive features shared by two strains, and n_d represents the number of features positive for one strain and negative for the other. { ˌsim·ə'lar·əd·ē ˌkō·iˌfish·ənt }

Simmonds' disease [MED] Hypopituitarism with marked insufficiency of the target glands and profound cachexia. Also known as hypophyseal cachexia; hypopituitary cachexia. { 'sim·ənz di,zēz }

Simonsiellaceae [MICROBIO] A family of bacteria in the order Cytophagales; cells are arranged to form flat filaments capable of gliding motility when the flat surface is in contact with the substrate. { sə,män·sē·ə'lās·ē,ē }

simple [BIOL] **1.** Made up of one piece. **2.** Unbranched. **3.** Consisting of identical units, as a simple tissue. { 'sim·pəl }

simple branched tubular gland [ANAT] A structure consisting of two or more unbranched, tubular secreting units joining a common outlet duct. { 'sim·pəl 'brancht 'tüb·yə·lər 'gland }

simple fruit [BOT] A fruit that has developed from a single carpel or several united carpels. { 'sim·pəl 'früt }

simple gland [ANAT] A gland having a single duct. { 'sim·pəl 'gland }

simple goiter [MED] Diffuse enlargement of the thyroid gland, usually not associated with constitutional features. { 'sim·pəl 'gȯid·ər }

simple layering [BOT] A plant propagation technique in which one portion of the stem, still attached to the plant, is buried. { ,sim·pəl 'lā·ə·riŋ }

simple leaf [BOT] A leaf having one blade, or a lobed leaf in which the separate parts do not reach down to the midrib. { 'sim·pəl 'lēf }

simple pistil [BOT] A pistil that consists of a single carpel. { 'sim·pəl 'pis·təl }

simple pit [BOT] A pit that lacks a border. { 'sim·pəl 'pit }

simple protein [BIOCHEM] One of a group of proteins which, upon hydrolysis, yield exclusively amino acids; included are globulins, glutelins, histones, prolamines, and protamines. { 'sim·pəl 'prō,tēn }

simple stomach [ANAT] A stomach consisting of a single dilation of the alimentary canal, as found in humans, dogs, and many higher and lower vertebrates. { 'sim·pəl 'stəm·ək }

simple tubular gland [ANAT] A gland consisting of a single, tubular secreting unit. { 'sim·pəl 'tü·byə·lər 'gland }

simplex uterus [ANAT] A uterus consisting of a single cavity, representing the greatest degree of fusion of the Müllerian ducts; found in humans and apes. { 'sim,pleks ¦yüd·ə·rəs }

Simuliidae [INV ZOO] The black flies, a family of orthorrhaphous dipteran insects in the series Nematocera. { ,sim·yə'lī·ə,dē }

single-photon-emission computed tomography [MED] A technique which measures the emission of single photons of a given energy from radioactive tracers to construct images of the distribution of the tracers in the human body. Abbreviated SPECT. { 'siŋ·gəl ¦fō,tän i¦mish·ən kəm,pyüd·əd tō'mäg·rə·fē }

single radial immunodiffusion [IMMUNOL] A technique for quantitating soluble proteins that involves placing the solution to be measured into a well cut into an agar or agarose gel containing antiserum specific for the protein. As the solution to be measured diffuses out of the well, it complexes with the antiserum and forms a ring, the size of which is proportional to the quantity of soluble protein in the well. Abbreviated SRID. Also known as Mancini method. { ¦siŋ·gəl ¦rād·ē·əl ,im·yə·nō·də'fyü·zhən }

singultus [MED] A repeated involuntary spasmodic contraction of the diaphragm followed by sudden closures of the glottis. Also known as hiccup. { 'siŋ,gəl·təs }

sinistrorse [BIOL] Twisting or coiling counterclockwise. { ¦sin·ə¦strȯrs }

sinoatrial node [ANAT] A bundle of Purkinje fibers located near the junction of the superior vena cava with the right atrium which acts as a pacemaker for cardiac excitation. Abbreviated SA node. Also known as sinoauricular node. { ¦sī·nō'ā·trē·əl 'nōd }

sinoauricular node *See* sinoatrial node. { ¦sī·nō·ȯ'rik·yə·lər 'nōd }

sinuate [BOT] Having a wavy margin with strong indentations. { 'sin·yə,wət }

sinus [BIOL] A cavity, recess, or depression in an organ, tissue, or other part of an animal body. { 'sī·nəs }

sinus gland [INV ZOO] An endocrine gland in higher crustaceans, lying in the eyestalk in most stalk-eyed species, which is the site of storage and release of a molt-inhibiting hormone. { 'sī·nəs ,gland }

sinus hairs *See* vibrissae. { 'sī·nəs ,herz }

sinusitis [MED] Inflammation of a paranasal sinus. { ,sī·nə'sīd·əs }

sinus of Morgagni [ANAT] The space between the upper border of the levator veli palatini muscle and the base of the skull. { 'sī·nəs əv mȯr 'gän·yē }

sinusoid [ANAT] Any of the relatively large spaces comprising part of the venous circulation in certain organs, such as the liver. { 'sī·nə ,sȯid }

sinus venosus [EMBRYO] The vessel in the transverse septum of the embryonic mammalian heart into which open the vitelline, allantoic, and common cardinal veins. [VERT ZOO] The chamber of the lower vertebrate heart to which the veins return blood from the body. { 'sī·nəs və'nō·səs }

Siphinodentallidae [INV ZOO] A family of mollusks in the class Scaphopoda characterized by a subterminal epipodial ridge which is not slit dorsally and which terminates with a crenulated disk. { ¦sī·fə·nō·den'tal·ə,dē }

siphon [BOT] A tubular element in various algae. [INV ZOO] **1.** A tubular structure for intake or output of water in bivalves and other mollusks. **2.** The sucking-type of proboscis in many arthropods. { 'sī·fən }

Siphonales [BOT] A large order of green algae (Chlorophyta) which are coenocytic, nonseptate, and mostly marine. Also known as Bryopsidales; Caulerpales. { ,sī·fə'nā·lez }

Siphonaptera [INV ZOO] The fleas, an order of

insects characterized by a small, laterally compressed, oval body armed with spines and setae, three pairs of legs modified for jumping, and sucking mouthparts. { ˌsī·fə'näp·trə }

Siphonocladaceae [BOT] A family of green algae in the order Siphonocladales. { ¦sī·fə·nō·klə'dās·ē,ē }

Siphonocladales [BOT] An order of green algae in the division Chlorophyta including marine, mostly tropical forms. { ¦sī·fə·nō·klə'dā·lēz }

siphonogamous [BOT] In plants, especially seed plants, the accomplishment of fertilization by means of a pollen tube. { ˌsī·fə'näg·ə·məs }

siphonoglyph [INV ZOO] A ciliated groove leading from the mouth to the gullet in certain anthozoans. { sī'fän·əˌglif }

Siphonolaimidae [INV ZOO] A family of nematodes in the superfamily Monhysteroidea in which the stoma is modified into a narrow, elongate, hollow, spearlike apparatus. { ¦sī·fə·nō ˌlā'im·əˌdē }

Siphonolaimoidea [INV ZOO] A superfamily of marine nematodes in the order Monhysterida, having a stoma in the form of a very narrow tube or a spear, and very large amphids. { ¦sī·fə·nō·lə'mȯid·ē·ə }

Siphonophora [INV ZOO] An order of the cnidarian class Hydrozoa characterized by the complex organization of components which may be connected by a stemlike region or may be more closely united into a compact organism. { ˌsī·fə'näf·rə }

siphonosome [INV ZOO] The lower part of a siphonophore colony, bearing the nutritive and reproductive zooids. { sī'fän·əˌsōm }

siphonostele [BOT] A type of stele consisting of pith surrounded by xylem and phloem. { sī'fän·əˌstēl }

siphonozooid [INV ZOO] A zooid of certain alcyonarians that lacks tentacles and gonads. { ¦sī·fə·nə'zōˌȯid }

Siphoviridae [VIROL] A family of linear double-stranded deoxyribonucleic acid-containing bacterial viruses (bacteriophages) characterized by a long noncontractile tail. Formerly known as Styloviridae. { ˌsī·fə'vir·əˌdī }

siphuncle [INV ZOO] **1.** A honeydew-secreting tube (cornicle) in aphids. **2.** A tubular extension of the mantle extending through all the chambers to the apex of a shelled cephalopod. { 'sīˌfəŋ·kəl }

Siphunculata [INV ZOO] The equivalent name for Anoplura. { siˌfəŋ·kyə'läd·ə }

Sipunculida [INV ZOO] A phylum of marine worms which dwell in burrows, secreted tubes, or adopted shells; the mouth and anus occur close together at one end of the elongated body, and the jawless mouth, surrounded by tentacles, is situated in an eversible proboscis. { ˌsīˌpəŋ'kyü·lə·də }

Sipunculoidea [INV ZOO] An equivalent name for Sipunculida. { sīˌpəŋ·kyə'lȯid·ē·ə }

Sirenia [VERT ZOO] An order of aquatic placental mammals which include the living manatees and dugongs; these are nearly hairless, thick-skinned mammals without hindlimbs and with paddlelike forelimbs. { sī'rē·nē·ə }

Siricidae [INV ZOO] The horntails, a family of the Hymenoptera in the superfamily Siricoidea; females use a stout, hornlike ovipositor to deposit eggs in wood. { sə'ris·əˌdē }

Siricoidea [INV ZOO] A superfamily of wasps of the suborder Symphala in the order Hymenoptera. { ˌsir·ə'kȯid·ē·ə }

sisal [BOT] *Agave sisalina*. An agave of the family Amaryllidaceae indigenous to Mexico and Central America; a coarse, stiff yellow fiber produced from the leaves is used for making twine and brush bristles. { 'sī·səl }

sister chromatids [CELL MOL] The two daughter strands of a chromosome after it has duplicated. { 'sis·tər 'krō·məˌtədz }

situs inversus [ANAT] Complete mirror-image inversion of the internal organs. { 'sīd·əs in'vər·səs }

situs solitus [ANAT] The normal pattern of internal organ placement. { ˌsīd·əs 'säl·ə·təs }

Sjögren's syndrome [IMMUNOL] An autoimmune disease characterized by the destruction of salivary and lacrimal glands, damage by T lymphocytes within the glands may be accompanied by damage by immune complexes throughout the body. { 'shō·grənz ˌsinˌdrōm }

skate [VERT ZOO] Any of various batoid elasmobranchs in the family Rajidae which have flat bodies with winglike pectoral fins and a slender tail with two small dorsal fins. { skāt }

skeletal muscle [ANAT] A striated, voluntary muscle attached to a bone and concerned with body movements. { 'skel·əd·əl ˌməs·əl }

skeletal system [ANAT] Structures composed of bone or cartilage or a combination of both which provide a framework for the vertebrate body and serve as attachment for muscles. Also known as skeleton. { 'skel·əd·əl ˌsis·təm }

skeleton *See* skeletal system. { 'skel·ət·ən }

skimmer [VERT ZOO] Any of various ternlike birds, members of the Rhynchopidae, that inhabit tropical waters around the world and are unique in having the knifelike lower mandible substantially longer than the wider upper mandible. { 'skim·ər }

skin [ANAT] The external covering of the vertebrate body, consisting of two layers, the outer epidermis and the inner dermis. { skin }

skink [VERT ZOO] Any of numerous small- to medium-sized lizards comprising the family Scincidae with a cylindrical body; short, sometimes vestigial, legs; cores of bone in the body scales; and pleurodont dentition. { skiŋk }

skin test [IMMUNOL] A procedure for evaluating immunity status involving the introduction of a reagent into or under the skin. { 'skin ˌtest }

skipjack *See* bluefish. { 'skipˌjak }

skull [ANAT] The bones and cartilages of the vertebrate head which form the cranium and the face. { skəl }

skunk [VERT ZOO] Any one of a group of carnivores in the family Mustelidae characterized by a glossy black and white coat and two musk glands at the base of the tail. { skəŋk }

slant culture [MICROBIO] A method for maintaining bacteria in which the inoculum is streaked on the surface of agar that has solidified in inclined glass tubes. { 'slant ˌkəl·chər }

slavery [INV ZOO] An interspecific association among ants in which members of one species bring pupae of another species to their nest, which, when adult, become slave workers in the colony. { 'slav·ə·rē }

sleep [PHYSIO] A state of rest in which consciousness and activity are diminished. { 'slēp }

sleeping sickness *See* African sleeping sickness; encephalitis lethargica. { 'slēp·iŋ ˌsik·nəs }

sleep paralysis [MED] Transient paralysis with spontaneous recovery occurring on falling asleep or on awakening. { 'slēp pəˌral·ə·səs }

sleep therapy *See* narcosis therapy. { 'slēp ˌther·ə·pē }

slime bacteria [MICROBIO] The common name for bacteria in the order Myxobacterales, so named for the layer of slime deposited behind cells as they glide on a surface. { 'slīm bakˌtir·ē·ə }

slime fungus *See* slime mold. { 'slīm ˌfəŋ·gəs }

slime gland [ZOO] A glandular structure in many animals producing a mucous material. { 'slīm ˌgland }

slime mold [MYCOL] The common name for members of the Myxomycetes. Also known as slime fungus. { 'slīm ˌmōld }

slipped disk *See* herniated disk. { 'slipt 'disk }

slop culture [BOT] A method of growing plants in which surplus nutrient fluid is allowed to run through the sand or other medium in which the plants are growing. { 'släp ˌkəl·chər }

sloth [VERT ZOO] Any of several edentate mammals in the family Bradypodidae found exclusively in Central and South America; all are slow-moving, arboreal species that cling to branches upside down by means of long, curved claws. { slāth }

slough [MED] A necrotic mass of tissue in, or separating from, healthy tissue. { slaü }

slow-reacting substance of anaphylaxis [BIOCHEM] A group of leukotrienes released by mast cells during anaphylaxis that induces prolonged bronchoconstriction. { ¦slō rēˌak·tiŋ ¦səb·stəns əv ˌan·ə·fə'lak·səs }

slow virus [VIROL] Any member of a group of animal viruses characterized by prolonged periods of incubation and an extended clinical course lasting months or years. { 'slō 'vī·rəs }

slow virus infection [VIROL] A persistent viral infection characterized by a long preclinical period extending for months or years from the time of exposure. { 'slō ¦vir·əs inˌfek·shən }

sludge barrel *See* calyx. { 'sləj ˌbar·əl }

sludge bucket *See* calyx. { 'sləj ˌbək·ət }

sludged blood [MED] The intracapillary aggregation of erythrocytes associated with decreased blood flow in the involved capillary bed. { 'sləjd 'bləd }

slug [INV ZOO] Any of a number of pulmonate gastropods which have a rudimentary shell and the body elevated toward the middle and front end where the mantle covers the lung region. { sləg }

small intestine [ANAT] The anterior portion of the intestine in humans and other mammals; it is divided into three parts, the duodenum, the jejunum, and the ileum. { 'smȯl in'tes·tən }

small nucleolar ribonucleic acid [CELL MOL] A type of antisense ribonucleic acid (RNA) that mediates site-specific cleavage and modification of ribosomal RNA precursors. Abbreviated snoRNA. { ˌsmȯl nüˌklē·ə·lər ˌrī·bō·nüˌklē·ik 'as·əd }

smallpox [MED] An acute, infectious, viral disease characterized by severe systemic involvement and a single crop of skin lesions which proceeds through macular, papular, vesicular, and pustular stages. Also known as variola. { 'smȯlˌpäks }

smallpox vaccine [IMMUNOL] A vaccine prepared from a glycerinated suspension of the exudate from cowpox vesicles obtained from healthy vaccinated calves or sheep. Also known as antismallpox vaccine; glycerinated vaccine virus; Jennerian vaccine; virus vaccinium. { 'smȯlˌpäks vakˌsēn }

smear [BIOL] A preparation for microscopic examination made by spreading a drop of fluid, such as blood, across a slide and using the edge of another slide to leave a uniform film. { smir }

smegma [PHYSIO] The sebaceous secretion that accumulates around the glans penis and the clitoris. { 'smeg·mə }

smell [PHYSIO] To perceive by olfaction. { smel }

smelling salts *See* aromatic spirits of ammonia. { 'smel·iŋ ˌsȯls }

Smilacaceae [BOT] A family of monocotyledonous plants in the order Liliales; members are usually climbing, leafy-stemmed plants with tendrils, trimerous flowers, and a superior ovary. { ˌsmī·lə'kās·ēˌē }

Sminthuridae [INV ZOO] A family of insects in the order Collembola which have simple tracheal systems. { smin'thyu̇r·əˌdē }

Smith-Petersen nail [MED] A three-flanged nail used to fix fractures of the neck of the femur; it is inserted from just below the greater trochanter, through the neck, and into the head of the femur. { 'smith 'pēd·ər·sən ˌnāl }

Smith-Recklinghausen's disease *See* neurofibromatosis. { 'smith 'rek·liŋˌhau̇z·ənz diˌzēz }

smooth muscle [ANAT] The involuntary muscle tissue found in the walls of viscera and blood vessels, consisting of smooth muscle fibers. { 'smüth 'məs·əl }

smooth muscle fiber [HISTOL] Any of the elongated, nucleated, spindle-shaped cells comprising smooth muscles. Also known as involuntary fiber; nonstriated fiber; unstriated fiber. { 'smüth 'məs·əl ˌfī·bər }

smut fungus [MYCOL] The common name for

members of the Ustilaginales. { 'smət ˌfəŋ·gəs }

snail [INV ZOO] Any of a large number of gastropod mollusks distinguished by a spiral shell that encloses the body, a head, a foot, and a mantle. { snāl }

snail fever *See* schistosomiasis. { 'snāl ˌfē·vər }

snake [VERT ZOO] Any of about 3000 species of reptiles which belong to the 13 living families composing the suborder Serpentes in the order Squamata. { snāk }

snapping finger *See* trigger finger. { 'snap·iŋ ˌfiŋ·gər }

sneeze [PHYSIO] A sudden, noisy, spasmodic expiration through the mouth and nose. { snēz }

Snellen test [MED] A test for visual acuity presenting letters, numbers, or letter E's in various positions, with the symbols varying in size; if the smallest are read at a distance of 20 feet (6 meters), vision is recorded as 20/20, or normal. { 'snel·ən ˌtest }

snout [VERT ZOO] The elongated nose of various mammals. { snau̇t }

snow blindness [MED] A transient visual impairment and actinic keratoconjunctivitis caused by exposure of the eyes to ultraviolet rays reflected from snow. Also known as solar photophthalmia. { 'snō ˌblīnd·nəs }

snow fungus *See* Tremella fuciformis. { 'snō ˌfəŋ·gəs }

SNP *See* single nucleotide polymorphism.

snuffles [MED] Discharge from the nasal mucosa in congenital syphilis in infants. { 'snəf·əlz }

Sobemovirus [VIROL] A genus of plant viruses with nonenveloped icosahedral virions containing one molecule of linear, single-stranded, positive-sense ribonucleic acid, the type species is Southern bean mosaic virus. Also known as Southern bean mosaic virus group. { sə'bē·məˌvī·rəs }

sobole [BOT] An underground creeping stem. { 'sä·bəˌlē }

social animal [ZOO] An animal that exhibits social behavior. { 'sō·shəl 'an·ə·məl }

social behavior [ZOO] Any behavior on the part of an organism stimulated by, or acting upon, another member of the same species. { 'sō·shəl bi'hā·vyər }

social hierarchy [VERT ZOO] The establishment of a dominance-subordination relationship among higher animal societies. { 'sō·shəl 'hī·ərˌär·kē }

social insects [ECOL] Insect species in which individuals share resources and reproduce cooperatively. { ¦sō·shəl 'inˌseks }

social parasitism [VERT ZOO] An aberrant type of parasitism occurring in some birds, in which the female of one species lays her eggs in the nests of other species and permits the foster parents to raise the young. { 'sō·shəl 'par·əˌsəˌdiz·əm }

social releaser [ZOO] A releaser stimulus which an animal receives from a member of its species. { 'sō·shəl ri'lē·sər }

society [ECOL] A secondary or minor plant community forming part of a community. [ZOO] An organization of individuals of the same species in which there are divisions of resources and of labor as well as mutual dependence. { sə'sī·əd·ē }

sockeye salmon [VERT ZOO] The species *Oncorhynchus nerka*, which is generally smaller and is uniquely adapted to rearing in interior lakes rather than streams or rivers. Also known as red salmon. { ˌsäkˌī 'sam·ən }

sodium *para*-aminobenzoate [PHARM] $NH_2C_6H_4COONa$ Water-soluble crystals, used in medicine. Also known as PABA sodium. { 'sōd·ē·əm ¦par·ə ¦am·ə·nō'ben·zə·wāt }

sodium aurothiomalate *See* gold sodium thiomalate. { 'sōd·ē·əm ˌȯr·əˌthī·ō'maˌlāt }

sodium barbital [PHARM] $C_8H_{11}N_2NaO_3$ A bitter, white powder, soluble in water; used in medicine. Also known as soluble barbital. { 'sōd·ē·əm 'bär·bə·təl }

sodium chloride solution *See* normal saline. { 'sōd·ē·əm 'klȯrˌīd səˌlü·shən }

sodium glucosulfone [PHARM] $C_{24}H_{34}N_2Na_2O_{18}S_3$ A leprostatic drug, and suppressant for dermatitis herpetiformis. Also known as glucosulfone sodium. { 'sōd·ē·əm ¦glü·kō'səlˌfōn }

sodium pentobarbitone *See* pentobarbital sodium. { 'sōd·ē·əm ˌpen·tə'bär·bəˌtōn }

sodium/potassium pump [CELL MOL] The cell membrane protein channel through which active transport of sodium ions out of the cell and potassium ions into the cell against their electrochemical gradients takes place. { ¦sōd·ē·əm pə'tas·ē·əm ˌpəmp }

sodium thiopental [PHARM] $C_{11}H_{17}N_2O_2SNa$ A yellowish-white powder, soluble in water and alcohol; used in medicine as a barbiturate. { 'sōd·ē·əm ˌthī·ə'pent·əl }

soft cataract [MED] A cataract, affecting the cortex of the lens of the eye, which is of soft consistency and has a milky appearance. { 'sȯft 'kad·əˌrakt }

soft chancre *See* chancroid. { 'sȯft 'shəŋ·kər }

soft coral [INV ZOO] The common name for cnidarians composing the order Alcyonacea; the colony is supple and leathery. { 'sȯft 'kär·əl }

soft palate [ANAT] The posterior part of the palate which consists of an aggregation of muscles, the tensor veli palatini, levator veli palatini, azygos uvulae, palatoglossus, and palatopharyngeus, and their covering mucous membrane. { 'sȯft 'pal·ət }

soft-shell disease [INV ZOO] A disease of lobsters caused by a chitinous bacterium which extracts chitin from the exoskeleton. { 'sȯft ¦shel diˌzēz }

soil conservation [ECOL] Management of soil to prevent or reduce soil erosion and depletion by wind and water. { ¦sȯil ˌkän·sərˌvā·shən }

soil ecology [ECOL] The study of interactions among soil organisms and interactions between biotic and abiotic aspects of the soil environment. { 'sȯil i,käl·ə·jē }

soil microbiology [MICROBIO] A study of the microorganisms in soil, their functions, and the effect of their activities on the character of the soil and the growth and health of plant life. { ¦sȯil ¦mī·krə·bī'äl·ə·jē }

Solanaceae [BOT] A family of dicotyledonous plants in the order Polemoniales having internal phloem, mostly numerous ovules and seeds on axile placentae, and mostly cellular endosperm. { ,sō·lə'nās·ē,ē }

solanine [BIOCHEM] A bitter poisonous alkaloid derived from potato sprouts (*Solanum tuberosum*), tomatoes, and nightshade. { 'sō·lə,nēn }

solar dermatitis [MED] Any of various skin eruptions caused by exposure to the sun, excluding sunburn. { 'sō·lər ,dər·mə'tīd·əs }

solar photophthalmia *See* snow blindness. { 'sō·lər ¦fōd·ō'thal·mē·ə }

solar propagation [BOT] A method of rooting plant cuttings involving the use of a modified hotbed; bottom heat is provided by radiation of stored solar heat from bricks or stones in the bottom of the hotbed frame. { 'sō·lər ,präp·ə'gā·shən }

Solasteridae [INV ZOO] The sun stars, a family of asteroid echinoderms in the order Spinulosida. { ,säl·ə'ster·ə,dē }

Solemyidae [INV ZOO] A family of bivalve mollusks in the order Protobranchia. { ,säl·ə'mī·ə,dē }

Solenichthyes [VERT ZOO] An equivalent name for Gasterosteiformes. { ,säl·ə'nik·thē,ēz }

solenium [INV ZOO] A diverticulum of the enteron in certain hydroids. { sō'lē·nē·əm }

solenocyte [INV ZOO] Any of various hollow, flagellated cells in the nephridia of the larvae of certain annelids, mollusks, rotifers, and lancelets. { sō'lē·nə,sīt }

solenodon [VERT ZOO] Either of two species of insectivorous mammals comprising the family Solenodontidae; the almique (*Atopogale cubana*) is found only in Cuba, while the white agouta (*Solenodon paradoxus*) is confined to Haiti. { sō'lē·nə,dän }

Solenodontidae [VERT ZOO] The solenodons, a family of insectivores belonging to the group Lipotyphla. { sō,lē·nə'dänt·ə,dē }

Solenogastres [INV ZOO] The equivalent name for Aplacophora. { sō,lē·nə'ga,strēz }

soleus [ANAT] A flat muscle of the calf; origin is the fibula, popliteal fascia, and tibia, and insertion is the calcaneus; plantarflexes the foot. { 'sō·lē·əs }

solfataric [ECOL] Relating to a volcanic area. { ,sōl·fə'tär·ik }

Solpugida [INV ZOO] The sun spiders, an order of nonvenomous, spiderlike, predatory arachnids having large chelicerae for holding and crushing prey. { säl'pyü·jəd·ə }

soluble barbital *See* sodium barbital. { 'säl·yə·bəl 'bär·bə,tal }

solute compartmentation [BOT] The sequestering of a plant cell's salt in a vacuole so that the salt does not poison the cell. { 'säl·yüt kəm,pärt·mən'tā·shən }

soma [BIOL] The whole of the body of an individual, excluding the germ tract. { 'sō·mə }

Somasteroidea [INV ZOO] A subclass of Asterozoa comprising sea stars of generalized structure, the jaws often only partly developed, and the skeletal elements of the arm arranged in a double series of transverse rows termed metapinnules. { ,sō·mə·stə'rȯid·ē·ə }

somatic aneuploidy [CELL MOL] An irregular variation in number of one or more individual chromosomes in the cells of a tissue. { sō'mad·ik 'a·nyü,plȯid·ē }

somatic cell [BIOL] Any cell of the body of an organism except the germ cells. { sō'mad·ik 'sel }

somatic copulation [MYCOL] A form of reproduction in ascomycetes and basidiomycetes involving sexual fusion of undifferentiated vegetative cells. { sō'mad·ik ,käp·yə'lā·shən }

somatic crossing-over [CELL MOL] Crossing-over during mitosis in somatic or vegetative cells. { sō'mad·ik ¦krȯs·iŋ 'ō·vər }

somatic death [BIOL] The cessation of characteristic life functions. { sō'mad·ik 'deth }

somatic embryogenesis [BOT] The production of embryoids from sporophytic or somatic plant cells. { sə¦mad·ik ,em·brē·ə'jen·ə·səs }

somatic mesoderm [EMBRYO] The external layer of the lateral mesoderm associated with the ectoderm after formation of the coelom. { sō'mad·ik 'mez·ə,dərm }

somatic mutation [GEN] A genetic change limited to a somatic cell lineage; a major cause of cancer in humans. { sō¦mad·ik myü'tā·shən }

somatic nervous system [PHYSIO] The portion of the nervous system concerned with the control of voluntary muscle and relating the organism with its environment. { sō'mad·ik 'nər·vəs ,sis·təm }

somatic pairing [CELL MOL] The pairing of homologous chromosomes at mitosis in somatic cells; occurs in Diptera. { sō'mad·ik 'per·iŋ }

somatic reflex system [PHYSIO] An involuntary control system characterized by a control loop which includes skeletal muscles. { sō'mad·ik 'rē,fleks ,sis·təm }

somatoblast [INV ZOO] **1.** An undifferentiated cleavage cell that gives rise to somatic cells in annelids. **2.** The outer cell layer of the nematogen in Dicyemida. { 'sō·məd·ə,blast }

somatochrome [NEUROSCI] A nerve cell possessing a well-defined body completely surrounding the nucleus on all sides, the cytoplasm having a distinct contour, and readily taking a stain. { 'sō·məd·ə,krōm }

somatoclonal variation [GEN] The appearance of new traits in plants that regenerate from a callus in tissue culture. { sə,mad·ə¦klōn·əl ,ver·ē'ā·shən }

somatocyst [INV ZOO] A cavity filled with air in the float of certain Siphonophora. { 'sō·məd·ə,sist }

somatogamy *See* pseudomixis. { ˌsō·mə'täg·ə·mē }

somatomedin [BIOCHEM] A growth factor similar to insulin that may be produced by a variety of cell types. Also known as insulin-like growth factor. { səˌmad·ə'med·ən }

somatometry [ANAT] Measurement of the human body with the soft parts intact. { ˌsō·mə'täm·ə·trē }

somatophyte [BOT] A plant composed of distinct somatic cells that develop especially into mature or adult tissue. { 'sō·məd·əˌfīt }

somatopleure [EMBRYO] A complex layer of tissue consisting of the somatic layer of the mesoblast together with the epiblast, forming the body wall in craniate vertebrates and the amnion and chorion in amniotes. { 'sō·məd·əˌplùr }

somatopsychic [MED] Pertaining to both the body and mind. { ¦sō·məd·ə'sī·kik }

somatostatin [BIOCHEM] A peptide secreted by the hypothalamus which acts primarily to inhibit the release of growth hormone from the anterior pituitary. { ¦sō·məd·ə'stat·ən }

somatotropin [BIOCHEM] The growth hormone of the pituitary gland. { ˌsō·mə'tä·trə·pən }

somesthesis [PHYSIO] The general name for all systems of sensitivity present in the skin, muscles and their attachments, visceral organs, and nonauditory labyrinth of the ear. { ¦sōm·es 'thē·səs }

somite *See* metamere. { 'sōˌmīt }

somnambulism [PHYSIO] **1.** Sleepwalking. **2.** The performance of any fairly complex act while in a sleeplike state or trance. { säm'näm·byəˌliz·əm }

somnificant *See* hypnotic. { säm'nif·ə·kənt }

Sonne dysentery [MED] An intestinal bacterial infection caused by *Shigella sonnei*. { 'zȯn·ə 'dis·ənˌter·ē }

sonoencephalograph *See* echoencephalograph. { ¦sän·ō·in'sef·ə·ləˌgraf }

sooty mold [MYCOL] Ascomycetous fungi of the family Capnodiaceae, with dark mycelium and conidia. { 'sùd·ē 'mōld }

soporific *See* hypnotic. { ¦säp·ə¦rif·ik }

sorbin *See* sorbose. { 'sȯr·bən }

sorbose [BIOCHEM] $C_6H_{12}O_6$ A carbohydrate prepared by fermentation; produced as water-soluble crystals that melt at 165°C; used in the production of vitamin C. Also known as sorbin. { 'sȯrˌbōs }

soredium [BOT] A structure comprising algal cells wrapped in the hyphal tissue of lichens, as in certain Lecanorales; when separated from the thallus, it grows into a new thallus. { sȯ'rē·dē·əm }

sorghum [BOT] Any of a variety of widely cultivated grasses, especially *Sorghum bicolor* in the United States, grown for grain and herbage; growth habit and stem form are similar to Indian corn, but leaf margins are serrate and spikelets occur in pairs on a hairy rachis. { 'sȯr·gəm }

Soricidae [VERT ZOO] The shrews, a family of insectivorous mammals belonging to the Lipotyphla. { sə'ris·əˌdē }

sorrel tree *See* sourwood. { 'sär·əl ˌtrē }

sorus [BOT] **1.** A cluster of sporangia on the lower surface of a fertile fern leaf. **2.** A clump of reproductive bodies or spores in lower plants. { 'sȯr·əs }

SOS response [GEN] A bacterial deoxyribonucleic acid (DNA) repair system in which cell division and DNA replication are blocked, and DNA repair, recombination, and mutation genes are induced. { ¦es¦ō'es riˌspäns }

souma [VET MED] A disease caused by *Trypanosoma vivax* in domestic and wild animals; the insect vectors are the tsetse fly and the stable fly. { 'sü·mə }

sourwood [BOT] *Oxydendrum arboreum*. A deciduous tree of the heath family (Ericaceae) indigenous along the Alleghenies and having long, simple, finely toothed, long-pointed leaves that have an acid taste, and white, urn-shaped flowers. Also known as sorrel tree. { 'saùrˌwùd }

South African tick-bite fever [MED] An infectious tick-borne rickettsial disease of humans which is similar to fièvre boutonneuse. { 'saùth 'af·ri·kən 'tik ˌbīt ˌfē·vər }

South American blastomycosis [MED] An infectious, yeastlike fungus disease of humans seen primarily in Brazil; caused by *Blastomyces brasiliensis* and characterized by massive enlargement of the cervical lymph nodes. Also known as paracoccidioidomycosis. { 'saùth ə'mer·ə·kən ˌblas·tōˌmī'kō·səs }

South American leishmaniasis *See* American mucocutaneous leishmaniasis. { 'saùth ə'mer·ə·kən ˌlēsh·mə'nī·ə·səs }

South American trypanosomiasis *See* Chagas' disease. { 'saùth ə'mer·ə·kən tripˌan·ə·sə'mī·ə·səs }

South Australian faunal region [ECOL] A marine littoral region along the southwestern coast of Australia. { 'saùth ȯ'strāl·yən 'fȯn·əl ˌrē·jən }

southern bean mosaic virus [VIROL] The type species of the plant-virus genus *Sobemovirus*. It is transmitted mechanically or via seed or the bean leaf beetle. Symptoms include crinkled leaves expressing a mild mosaic. Abbreviated SBMV. { ¦sət͟h·ərn ˌbēn mō'zā·ik ˌvī·rəs }

southern bean mosaic virus group *See* Sobemovirus. { ¦sət͟h·ərn ˌbēn mō'zā·ik ˌvī·rəs ˌgrüp }

Southern blotting [CELL MOL] A technique for the detection of specific sequences among deoxyribonucleic acid (DNA) fragments whereby the fragments are separated by gel electrophoresis and then blotted onto a sheet of nitrocellulose for detection with radioactively labeled nucleic acid probes. It was first described by E. M. Southern in 1975. { 'sə·t͟h·ərn ˌbläd·iŋ }

sow [VERT ZOO] An adult female swine. { saù }

soybean [BOT] *Glycine max*. An erect annual legume native to China and Manchuria and widely cultivated for forage and for its seed. { 'sȯiˌbēn }

soybean lecithin *See* lecithin. { 'sȯiˌbēn 'les·ə·thən }

soy lecithin *See* lecithin. { 'sȯi 'les·ə·thən }

space biology [BIOL] A term for the various biological sciences and disciplines that are concerned with the study of living things in the space environment. { 'spās bī,äl·ə·jē }

space medicine [MED] A branch of medicine that deals with the physiologic disturbances and diseases produced in man by high-velocity projection through and beyond the earth's atmosphere, flight through interplanetary space, and return to earth. { 'spās ,med·ə·sən }

space perception [PHYSIO] The awareness of the spatial properties and relations of an object, or of one's own body, in space; especially, the sensory appreciation of position, size, form, distance, and direction of an object, or of the observer himself, in space. { 'spās pər,sep·shən }

spacer deoxyribonucleic acid [CELL MOL] Untranscribed deoxyribonucleic acid (DNA) segments, usually containing repetitious DNA, of eukaryotic and some viral genomes flanking functional genetic regions (cistrons). { ¦spās·ər dē¦äk·sē,rī·bō·nü,klē·ik 'as·əd }

spadix [BOT] A fleshy spike that is enclosed in a leaflike spathe and is the characteristic inflorescence of palms and arums. [INV ZOO] A cone-shaped structure in male Nautiloidea formed of four modified tentacles, and believed to be homologous with the hectocotylus in male squids. { 'spā·diks }

Spanish collar *See* paraphimosis. { 'span·ish 'käl·ər }

Sparganiaceae [BOT] A family of monocotyledonous plants in the order Typhales distinguished by the inflorescence of globose heads, a vestigial perianth, and achenes that are sessile or nearly sessile. { spär,gā·nē'ās·ē,ē }

sparganosis [VET MED] An infection by the plerocercoid larva, or sparganum, of certain species of *Spirometra*; the adult form normally occurs in the intestine of dogs and cats. { ,spär·gə'nō·səs }

sparganum [INV ZOO] The plerocercoid larva of a tapeworm. { 'spär·gə·nəm }

Sparidae [VERT ZOO] A family of perciform fishes in the suborder Percoidei, including the porgies. { 'spar·ə,dē }

spasm [MED] An involuntary and abnormal contraction of isolated bundles of muscle or groups of muscles resulting from a chemical imbalance due to fatigue, ischemia, or trauma. { 'spaz·əm }

spasmophilia [MED] A morbid tendency to convulsions, and to tonic spasms, such as those observed in tetany, infantile spasms, or spasmus nutans. { ,spaz·mə'fil·ē·ə }

spastic colon *See* irritable colon. { 'spas·tik 'kō·lən }

spastic diplegia [MED] **1.** Spastic paralysis of the arms and legs caused by diffuse lesions of the cerebral cortex. **2.** A form of cerebral palsy, possibly due to prenatal or perinatal hypoxia or other injuries resulting in atrophic lobar sclerosis, or to congenital or developmental abnormalities. { 'spas·tik dī'plē·jə }

spastic ileus [MED] A form of ileus in which temporary obstruction is due to segmental intestinal spasm. Also known as dynamic ileus; hyperdynamic ileus. { 'spas·tik 'il·ē·əs }

spasticity *See* spastic paralysis. { spas'tis·əd·ē }

spastic paralysis [MED] A condition in which a group of muscles manifest increased tone, exaggerated tendon reflexes, depressed or absent superficial reflexes, and sometimes clonus, due to an upper motor neuron lesion. Also known as spasticity. { 'spas·tik pə'ral·ə·səs }

spastic paraplegia [MED] Paralysis of the lower limbs with increased muscular tone and hyperactive tendon reflexes; commonly seen in diseases and injuries involving pyramidal tracts of the spinal cord. { 'spas·tik ,par·ə'plē·jə }

spastic strabismus [MED] A squint resulting from the contraction of an ocular muscle. { 'spas·tik strə'biz·məs }

Spatangoida [INV ZOO] An order of exocyclic Euechinoidea in which the posterior ambulacral plates form a shield-shaped area behind the mouth. { ,spat·ən'gȯid·ə }

spathe [BOT] A large, usually colored bract or pair of bracts enclosing an inflorescence, especially a spadix, on the same axis. { spāth }

spatulate [BIOL] Shaped like a spoon. { 'spach·ə·lət }

spawn [BOT] Mycelium used for initiating mushroom propagation. [ZOO] **1.** The collection of eggs deposited by aquatic animals, such as fish. **2.** To produce or deposit eggs or discharge sperm; applied to aquatic animals. { spȯn }

spay [VET MED] To remove the ovaries. { spā }

spearmint [BOT] *Mentha spicata.* An aromatic plant of the mint family, Labiatae; the leaves are used as a flavoring in foods. { 'spir,mint }

speciation [EVOL] The evolution of species. { ,spē·sē'ā·shən }

species [SYST] A taxonomic category ranking immediately below a genus and including closely related, morphologically similar individuals which actually or potentially interbreed. { 'spē·shēz }

species concept [EVOL] The idea that the diversity of nature is divisible into a finite number of definable species. { 'spē·shēz ,kän,sept }

species population [ECOL] A group of similar organisms residing in a defined space at a certain time. { 'spē·shēz ,päp·yə'lā·shən }

specific active immunotherapy [IMMUNOL] Active immunotherapy that attempts to stimulate specific antitumor responses with tumor-associated antigens as the immunizing materials. { spə¦sif·ik ¦ak·tiv ,im·yə·nō'ther·ə·pē }

specific locus test [GEN] A technique used to detect recessive induced mutations in diploid organisms; a strain which carries several known recessive mutants in a homozygous condition is crossed with a nonmutant strain treated to induce mutations in its germ cells; induced recessive mutations allelic with those of the test strain will be expressed in the progeny. { spə'sif·ik 'lō·kəs ,test }

SPECT *See* single-photon-emission computed tomography. { spekt }

spectacle [ZOO] A colored marking in the form of rings around the eyes, as in certain birds, reptiles, and mammals (as the raccoon). { 'spek·tə·kəl }

speculum [MED] A tubular instrument for inserting into a passage or cavity of the body to facilitate visual inspection or medication. { 'spek·yə·ləm }

speech [PHYSIO] A complex process in which the eating and breathing mechanisms are used to generate patterns of sounds that form words and sentences to express thoughts. { spēch }

Spelaeogriphacea [INV ZOO] A peracaridan order of the Malacostraca comprised of the single species *Spelaeogriphus lepidops*, a small, blind, transparent, shrimplike crustacean with a short carapace that coalesces dorsally with the first thoracic somite. { ¦spē·lē·ō·gri'fās·ē·ə }

spelter shakes *See* metal fume fever. { 'spel·tər ˌshāks }

sperm *See* spermatozoon. { spərm }

spermatheca [ZOO] A sac in the female for receiving and storing sperm until fertilization; found in many invertebrates and certain vertebrates. Also known as seminal receptacle. { ¦spər·mə'thē·kə }

spermatic cord [ANAT] The cord consisting of the ductus deferens, epididymal and testicular nerves and blood vessels, and connective tissue that extends from the testis to the deep inguinal ring. { spər'mad·ik 'kȯrd }

spermatid [HISTOL] A male germ cell immediately before assuming its final typical form. { 'spər·məd·əd }

spermatin [BIOCHEM] An albuminoid material occurring in semen. { 'spər·məd·ən }

spermatocele [MED] A cystic dilation of a duct in the head of the epididymis or in the rete testis. { spər'mad·əˌsēl }

spermatocyte [HISTOL] A cell of the last or next to the last generation of male germ cells which differentiates to form spermatozoa. { spər 'mad·əˌsīt }

spermatogenesis [PHYSIO] The process by which spermatogonia undergo meiosis and transform into spermatozoa. { spərˌmad·ə'jen·ə·səs }

spermatogonium [HISTOL] A primitive male germ cell, the last generation of which gives rise to spermatocytes. { spərˌmad·ə'gō·nē·əm }

spermatophore [ZOO] A bundle or packet of sperm produced by certain animals, such as annelids, arthropods, and some vertebrates. { spər'mad·əˌfȯr }

spermatophyte [BOT] Any one of the seed-bearing vascular plants. { spər'mad·əˌfīt }

spermatorrhea [MED] Involuntary discharge of semen without orgasm. { spərˌmad·ə'rē·ə }

spermatozoon [HISTOL] A mature male germ cell. Also known as sperm. { spərˌmad·ə'zō·ən }

spermaturia [MED] The presence of sperm in the urine. { ˌspər·mə'tu̇r·ē·ə }

spermidine [BIOCHEM] $H_2N(CH_2)_3NH(CH_2)_4NH_2$ The triamine found in semen and other animal tissues. { 'spər·məˌdēn }

spermine [BIOCHEM] $C_{10}H_{26}N_4$ A tetramine found in semen, blood serum, and other body tissues. { 'spərˌmēn }

spermiogenesis [CELL MOL] Nuclear and cytoplasmic transformation of spermatids into spermatozoa. { ˌspər·mē·ō'jen·ə·səs }

sperm nucleus [BOT] One of the two nuclei in a pollen grain that function in double fertilization in seed plants. { 'spərm ˌnü·klē·əs }

sperm whale [VERT ZOO] *Physeter catadon*. An aggressive toothed whale belonging to the group Odontoceti of the order Cetacea; it produces ambergris and contains a mixture of spermaceti and oil in a cavity of the nasal passage. { 'spərm ˌwāl }

Sphaeriales [MYCOL] An order of fungi in the subclass Euascomycetes characterized by hard, dark perithecia with definite ostioles. { ˌsfir·ē'ā·lēz }

Sphaeriidae [INV ZOO] The minute bog beetles, a small family of coleopteran insects in the suborder Myxophaga. { sfə'rī·əˌdē }

Sphaerioidaceae [MYCOL] A family of fungi of the order Sphaeropsidales in which the pycnidia are black or dark-colored and are flask-, cone-, or lens-shaped with thin walls and a round, relatively small pore. { ˌsfir·ēˌȯi'dās·ēˌē }

Sphaerocarpales [BOT] An order of liverworts in the subclass Marchantiidae, characterized by an envelope surrounding each antheridium and archegonium, absence of elaters, poor development of seta, and absence of thickenings in the unilayered wall of an indehiscent capsule. { ˌsfir·ō·kär'pā·lēz }

Sphaeroceridae [INV ZOO] A family of myodarian cyclorrhaphous dipteran insects in the subsection Acalypteratae. { ˌsfir·ō'ser·əˌdē }

Sphaerodoridae [INV ZOO] A family of polychaete annelids belonging to the Errantia in which species are characterized by small bodies, and are usually papillated. { ˌsfir·ō'dȯr·əˌdē }

Sphaerolaimidae [INV ZOO] A family of free-living nematodes in the superfamily Monhysteroidea characterized by a spacious and deep stoma. { ˌsfir·ō'lī·məˌdē }

Sphaeromatidae [INV ZOO] A family of isopod crustaceans in the suborder Flabellifera in which the body is broad and oval and the inner branch of the uropod is immovable. { ˌsfir·ō'mad·əˌdē }

Sphaerophoraceae [BOT] A family of the Ascolichenes in the order Caliciales which are fruticose with a solid thallus. { sfəˌräf·ə'rās·ēˌē }

Sphaeropleineae [BOT] A suborder of green algae in the order Ulotrichales distinguished by long, coenocytic cells, numerous bandlike chloroplasts, and heterogametes produced in undifferentiated vegetative cells. { ˌsfir·ō'plān·ēˌē }

Sphaeropsidaceae [MYCOL] An equivalent name for Sphaerioidaceae. { sfəˌräp·sə'dās·ēˌē }

Sphaeropsidales [MYCOL] An order of fungi of

the class Fungi Imperfecti in which asexual spores are formed in pycnidia, which may be separate or joined to vegetative hyphae, conidiophores are short or absent, and conidia are always slime spores. { sfə,räp·sə'dā·lēz }

Sphagnaceae [BOT] The single monogeneric family of the order Sphagnales. { sfag'nās·ē,ē }

Sphagnales [BOT] The single order of mosses in the subclass Sphagnobrya containing the single family Sphagnaceae. { sfag'nā·lēz }

Sphagnobrya [BOT] A subclass of the Bryopsida; plants are grayish-green with numerous, spirally arranged branches and grow in deep tufts or mats, commonly in bogs and in other wet habitats. { sfag'näb·rē·ə }

sphagnum bog [ECOL] A bog composed principally of mosses of the genus *Sphagnum* (Sphagnales) but also of other plants, especially acid-tolerant species, which tend to form peat. { 'sfag·nəm 'bäg }

Sphecidae [INV ZOO] A large family of hymenopteran insects in the superfamily Sphecoidea. { 'sfes·ə,dē }

Sphecoidea [INV ZOO] A superfamily of wasps belonging to the suborder Apocrita. { sfə'kȯid·ē·ə }

sphenethmoid [VERT ZOO] A bone that surrounds the anterior portion of the brain in many amphibians. { sfēn·'eth,mȯid }

Spheniscidae [VERT ZOO] The single family of the avian order Sphenisciformes. { sfə'nis·ə,dē }

Sphenisciformes [VERT ZOO] The penguins, an order of aquatic birds found only in the Southern Hemisphere and characterized by paddlelike wings, erect posture, and scalelike feathers. { sfə,nis·ə'fȯr,mēz }

Sphenodontidae [VERT ZOO] A family of lepidosaurian reptiles in the order Rhynchocephalia represented by a single living species, *Sphenodon punctatus*, a lizardlike form distinguished by lack of a penis. { ,sfē·nə'dänt·ə,dē }

sphenoid bone [ANAT] The butterfly-shaped bone forming the anterior part of the base of the skull and portions of the cranial, orbital, and nasal cavities. { 'sfē,nȯid ¦bōn }

sphenoid sinus [ANAT] Either of a pair of paranasal sinuses located centrally between and behind the eyes, below the ethymoid sinus. { 'sfē,nȯid ¦sī·nəs }

sphenopalatine [ANAT] Of or pertaining to the region of or surrounding the sphenoid and palatine bones. { ,sfē·nō'pal·ə,tēn }

sphenopalatine foramen [ANAT] The space between the sphenoid and orbital processes of the palatine bone; it opens into the nasal cavity and gives passage to branches from the pterygopalatine ganglion and the sphenopalatine branch of the maxillary artery. { ,sfē·nō'pal·ə,tēn fə'rā·mən }

Sphenopsida [BOT] A group of vascular cryptogams characterized by whorled, often very small leaves and by the absence of true leaf gaps in the stele; essentially equivalent to the division Equisetophyta. { sfə'näp·səd·ə }

spherocyte [PATH] A spherical red blood cell. { 'sfir·ə,sīt }

spherocytosis [MED] Preponderance of spherocytes in the blood. { ,sfir·ə,sī'tō·səs }

spheroplast [CELL MOL] A plant cell which possesses only a partial or modified cell wall. [MICROBIO] A bacterial cell that assumes a spherical shape due to partial or complete absence of the wall. { 'sfir·ə,plast }

sphincter [ANAT] A muscle that surrounds and functions to close an orifice. { 'sfiŋk·tər }

sphincter of Oddi [ANAT] Sphincter of the hepatopancreatic ampulla. { 'sfiŋk·tər əv 'äd·ē }

Sphindidae [INV ZOO] The dry fungus beetles, a family of coleopteran insects in the superfamily Cucujoidea. { 'sfin·də,dē }

Sphingidae [INV ZOO] The single family of the lepidopteran superfamily Sphingoidea. { 'sfin·jə,dē }

Sphingoidea [INV ZOO] A superfamily of Lepidoptera in the suborder Heteroneura consisting of the sphinx, hawk, or hummingbird moths; these are heavy-bodied forms with antennae that are thickened with a pointed apex, a well-developed proboscis, and narrow wings. { sfiŋ'gȯid·ē·ə }

sphingolipid [BIOCHEM] Any lipid, such as a sphingomyelin, that yields sphingosine or one of its derivatives as a product of hydrolysis. { ¦sfiŋ·gō'lip·əd }

sphingolipidosis [MED] Any of a group of hereditary metabolic disorders characterized by excessive accumulations of certain glycolipids and phospholipids in various tissues of the body. { ¦sfiŋ·gō,lip·ə'dō·səs }

sphingomyelin [BIOCHEM] A phospholipid consisting of choline, sphingosine, phosphoric acid, and a fatty acid. { ¦sfiŋ·gō'mī·ə·lən }

sphingosine [BIOCHEM] $C_{18}H_{37}O_2N$ A moiety of sphingomyelin, cerebrosides, and certain other phosphatides. { 'sfiŋ·gə,sēn }

sphygmomanometer [MED] An instrument for measuring the arterial blood pressure. { ¦sfig·mō·mə'näm·əd·ər }

sphygmophone [MED] A microphone attached to the wrist to pick up the sounds of the pulse. { 'sfig·mə,fōn }

Sphyraenidae [VERT ZOO] A family of shore fishes in the suborder Mugiloidei of the order Perciformes comprising the barracudas. { sfə'rē·nə,dē }

Sphyriidae [INV ZOO] A family of ectoparasitic Crustacea belonging to the group Lernaeopodoida; the parasite embeds its head and part of its thorax into the host. { sfə'rī·ə,dē }

spicule [BOT] An empty diatom shell. [INV ZOO] A calcareous or siliceous, usually spikelike supporting structure in many invertebrates, particularly in sponges and alcyonarians. { 'spik·yül }

spiculin [BIOCHEM] An organic material making up a portion of a spicule. { 'spik·yə·lən }

spiculum [INV ZOO] A bristlelike copulatory organ in certain nematodes. Also known as copulatory spicule. { 'spik·yə·ləm }

spider [INV ZOO] The common name for arachnids comprising the order Araneida. { 'spīd·ər }

spider nevus [MED] A type of telangiectasis characterized by a central, elevated, tiny red dot, pinhead in size, from which blood vessels radiate like strands of a spider's web. Also known as stellar nevus. { 'spīd·ər 'nē·vəs }

Spiegler's test [PATH] A test for the presence of protein in urine performed by overlaying clear acidulated urine on Spiegler's reagent (mercuric chloride, tartaric acid, glycerin, distilled water); opalescence at the fluid junction indicates protein. { 'spēg·lərz ,test }

spike [BOT] An indeterminate inflorescence with sessile flowers. { spīk }

spikelet [BOT] The compound inflorescence of a grass consisting of one or several bracteate spikes. { 'spīk·lət }

spina bifida [MED] A congenital anomaly characterized by defective closure of the vertebral canal with herniation of the spinal cord meninges. { ¦spī·nə 'bī·fəd·ə }

spina bifida occulta [MED] An asymptomatic congenital anomaly consisting of incomplete fusion of the posterior arch of the vertebral canal without hernial protrusion of the meninges. { ¦spī·nə 'bī·fəd·ə ə'kəl·tə }

spinacene *See* squalene. { 'spin·ə,sēn }

spinach [BOT] *Spinacia oleracea*. An annual potherb of Asiatic origin belonging to the order Caryophyllales and grown for its edible foliage. { 'spin·ich }

spinal anesthesia [MED] **1.** Anesthesia due to a lesion of the spinal cord. **2.** Anesthesia produced by the injection of an anesthetic into the spinal subarachnoid space. { 'spīn·əl ,an·əs 'thē·zhə }

spinal column *See* spine. { 'spīn·əl ,käl·əm }

spinal cord [NEUROSCI] The cordlike posterior portion of the central nervous system contained within the spinal canal of the vertebral column of all vertebrates. { 'spīn·əl ,kȯrd }

spinal foramen [ANAT] Central canal of the spinal cord. { 'spīn·əl fə'rā·mən }

spinal ganglion [NEUROSCI] Any one of the sensory ganglions, each associated with the dorsal root of a spinal nerve. { 'spīn·əl 'gaŋ·glē·ən }

spinal nerve [NEUROSCI] Any of the paired nerves arising from the spinal cord. { 'spīn·əl ¦nərv }

spinal reflex [NEUROSCI] A reflex mediated through the spinal cord without the participation of the more cephalad structures of the brain or spinal cord. { 'spīn·əl 'rē,fleks }

spindle [CELL MOL] A structure formed of fiberlike elements just before metaphase that extends between the poles of the achromatic figure and is attached to the centromeric regions of the chromatid pairs. { 'spin·dəl }

spindle fiber [CELL MOL] One of the fiberlike elements of the spindle; an aggregation of microtubules resulting from the polymerization of a series of small protein fibrils by primary —S—S— linkages. { 'spin·dəl ,fī·bər }

spine [ANAT] An articulated series of vertebrae forming the axial skeleton of the trunk and tail, and being a characteristic structure of vertebrates. Also known as backbone; spinal column; vertebral column. [BOT] A rigid sharp-pointed process in plants; many are modified leaves. [INV ZOO] One of the processes covering the surface of a sea urchin. [VERT ZOO] **1.** One of the spiny rays supporting the fins of most fishes. **2.** A sharp-pointed modified hair on certain mammals, such as the porcupine. { spīn }

spiniger [INV ZOO] Seta that tapers to a fine point, most frequently used in connection with compound seta (thus, compound spiniger). { 'spin·ə·jər }

spinneret [INV ZOO] An organ that spins fiber from the secretion of silk glands. { ,spin·ə'ret }

spinney [ECOL] A small grove of trees or a thicket with undergrowth. { 'spin·ē }

spinoblast [INV ZOO] A statoblast having a float of air cells and barbs or hooks on the surface. { 'spī·nə,blast }

spinochrome [BIOCHEM] A type of echinochrome; an organic pigment that is known only from sea urchins and certain homopteran insects. { 'spī·nə,krōm }

spinous process [ANAT] Any slender, sharp-pointed projection on a bone. { 'spī·nəs ,prä·səs }

Spintheridae [INV ZOO] An amphinomorphan family of small polychaete annelids included in the Errantia. { spin'ther·ə,dē }

Spinulosida [INV ZOO] An order of Asteroidea in which pedicellariae rarely occur, marginal plates bounding the arms and disk are small and inconspicuous, and spines occur in groups on the upper surface. { ,spin·yə'läs·əd·ē }

spiny-rayed fish [VERT ZOO] The common designation for actinopterygian fishes, so named for the presence of stiff, unbranched, pointed fin rays, known as spiny rays. { 'spī·nē 'rād 'fish }

Spionidae [INV ZOO] A family of spioniform annelid worms belonging to the Sedentaria. { spī'än·ə,dē }

spioniform worm [INV ZOO] A polychaete annelid characterized by the presence of a pair of short to long, grooved palpi near the mouth. { spī'än·ə,fȯrm 'wərm }

spiracle [INV ZOO] An external breathing orifice of the tracheal system in insects and certain arachnids. [VERT ZOO] **1.** The external respiratory orifice in cetaceous and amphibian larvae. **2.** The first visceral cleft in fishes. { 'spir·ə·kəl }

spiral cleavage [EMBRYO] A cleavage pattern characterized by formation of a cell mass showing spiral symmetry; occurs in mollusks. { 'spī·rəl 'klē·vij }

spiral ganglion [NEUROSCI] The ganglion of the cochlear part of the vestibulocochlear nerve embedded in the spiral canal of the modiolus. { 'spī·rəl 'gaŋ·glē·ən }

spiral ligament [ANAT] The reticular connective tissue connecting the basilar membrane to the outer cochlear wall in the ear of mammals. { 'spī·rəl 'lig·ə·mənt }

spiral organ *See* organ of Corti. { 'spī·rəl ¦ȯr·gən }

spiral valve [VERT ZOO] A spiral fold of mucous membrane in the small intestine of elasmobranchs and some primitive fishes which increases the surface area for absorption. { 'spī·rəl ¦valv }

spiramycin [MICROBIO] A complex of related antibiotics, which resemble erythromycin structurally and in antibacterial spectrum, produced by *Streptomyces ambofaciens*. { ¦spī·rə'mīs·ən }

spiraster [INV ZOO] A spiral spicule bearing rays in Porifera. { spī'ras·tər }

spire [BOT] A narrow, tapering blade or stalk. { spīr }

spiricle [BOT] Any of the coiled threads in certain seed coats which uncoil when moistened. { 'spir·ə·kəl }

Spirillaceae [MICROBIO] A family of bacteria; motile, helically curved rods that move with a characteristic corkscrewlike motion. { 'spī·rə 'lās·ē,ē }

Spirillinacea [INV ZOO] A superfamily of foraminiferan protozoans in the suborder Rotaliina characterized by a planispiral or low conical test with a wall composed of radial calcite. { spə,ril·ə'nās·ē·ə }

Spirobrachiidae [INV ZOO] A family of the Brachiata in the order Thecanephria. { ,spī·rə·brə'kī·ə,dē }

Spirochaetaceae [MICROBIO] The single family of the order Spirochaetales. { ,spī·rə·kē'tās·ē,ē }

Spirochaetales [MICROBIO] An order of bacteria characterized by slender, helically coiled cells sometimes occurring in chains. { ,spī·rə·kē 'tā·lēz }

spirochetal jaundice *See* Weil's disease. { ¦spī·rə¦kēd·əl 'jȯn·dəs }

spirochete [MICROBIO] The common name for any member of the order Spirochaetales. { 'spī·rə,kēt }

spirochetemia [MED] The presence of spirochetes in the blood. { ,spī·rə,kēd'ē·mē·ə }

spirocyst [INV ZOO] A thin-walled capsule that contains a long, unarmed, eversible, spirally coiled thread of uniform diameter; found in cnidarians. { 'spī·rə,sist }

spirograph [MED] An instrument for registering respiration. { 'spī·rə,graf }

spirometry [PHYSIO] The measurement, by a form of gas meter (spirometer), of volumes of air that can be moved in or out of the lungs. { spī'räm·ə·trē }

spironolactone [PHARM] $C_{24}H_{32}O_4S$ A steroid having a lactone ring attached at carbon-17; used as a diuretic. { 'spī·rə·nō'lak,tōn }

Spirotrichia [INV ZOO] A subclass of the protozoan class Ciliatea which contains those ciliates characterized by conspicuous, compound ciliary structures, known as cirri, and buccal organelles. { ,spī·rō'trik·ē·ə }

Spirulidae [INV ZOO] A family of cephalopod mollusks containing several species of squids. { spī'rül·ə,dē }

Spiruria [INV ZOO] A subclass of nematodes in the class Secernentea. { spī'ru̇r·ē·ə }

Spirurida [INV ZOO] An order of phasmid nematodes in the subclass Spiruria. { spī'ru̇r·əd·ə }

Spiruroidea [INV ZOO] A superfamily of spirurid nematodes which are parasitic in the respiratory and digestive systems of vertebrates. { ,spī·rə'rȯid·ē·ə }

splanchnic mesoderm [EMBRYO] The internal layer of the lateral mesoderm that is associated with the entoderm after the formation of the coelom. { 'splaŋk·nik 'mez·ə,dərm }

splanchnic nerve [NEUROSCI] A nerve carrying nerve fibers from the lower thoracic paravertebral ganglions to the collateral ganglions. { 'splaŋk·nik ¦nərv }

splanchnocranium [ANAT] Portions of the skull derived from the primitive skeleton of the gill apparatus. { ¦splaŋk·nə'krā·nē·əm }

splanchnopleure [EMBRYO] The inner layer of the mesoblast from which part of the wall of the alimentary canal and portions of the visceral organs are derived in coelomates. { 'splaŋk·nə,plu̇r }

spleen [ANAT] A blood-forming lymphoid organ of the circulatory system, present in most vertebrates. { splēn }

splenectomy [MED] Surgical removal of the spleen. { splə'nek·tə·mē }

splenic fever *See* anthrax. { 'splen·ik 'fē·vər }

splenic flexure [ANAT] An abrupt turn of the colon beneath the lower end of the spleen, connecting the descending with the transverse colon. { 'splen·ik 'flek·shər }

splenium [ANAT] The rounded posterior extremity of the corpus callosum. [MED] A bandage. { 'splē·nē·əm }

splenomegaly [MED] Enlargement of the spleen. { ,splē·nə'meg·ə·lē }

splint [MED] A stiff or flexible material applied to an anatomical part in order to protect it, immobilize it, or restrict its motion. { splint }

split gene [GEN] A eukaryotic gene in which the coding sequence is divided into two or more exons that are interrupted by a number of noncoding intervening sequences (introns). Also known as interrupted gene. { 'split 'jēn }

splitter [SYST] A taxonomist who divides taxa very finely. { 'splid·ər }

spondylitis [MED] Inflammation of the vertebrae. Also known as ankylosing spondylitis. { ,spän·də'līd·əs }

spondyloarthropathy [MED] Any of a group of arthritis disorders characterized by involvement of the axial (central) skeleton and association with the histocompatibility system. { ,spän·də·lō·är'thräp·ə·thē }

spondylolisthesis [MED] Forward displacement of a vertebra upon the one below as a result of a bilateral defect in the vertebral arch, or erosion of the articular surface of the posterior facets due to degenerative joint disease. { ¦span·də·lō,lis'thē·səs }

sponge [INV ZOO] The common name for members of the phylum Porifera. { spənj }

spongiform encephalopathy *See* prion disease. { ˌspən·jəˌfȯrm inˌsef·ə'läp·ə·thē }

Spongiidae [INV ZOO] A family of sponges of the order Dictyoceratida; members are encrusting, massive, or branching in form and have small spherical flagellated chambers which characteristically join the exhalant canals by way of narrow channels. { spən'jī·əˌdē }

Spongillidae [INV ZOO] A family of fresh- and brackish water sponges in the order Haplosclerida which are chiefly gray, brown, or white in color, and encrusting, massive, or branching in form. { spən'jil·əˌdē }

spongin [BIOCHEM] A scleroprotein, occurring as the principal component of skeletal fibers in many sponges. { 'spən·jən }

spongioblast [EMBRYO] A primordial cell arising from the ectoderm of the embryonic neural tube which differentiates to form the neuroglia, the ependymal cells, the neurolemma sheath cells, the satellite cells of ganglions, and Müller's fibers of the retina. { 'spən·jē·ōˌblast }

spongiocyte [HISTOL] **1.** A neuroglia cell. **2.** A cell of the adrenal cortex which has a spongy appearance due to the solution of lipids during tissue preparation for microscopical examination. { 'spən·jē·ōˌsīt }

spongocoel [INV ZOO] The branching, internal cavity of a sponge, connected to the outside by way of the osculum. { 'späŋ·gəˌsēl }

spongy mesophyll [BOT] A system of loosely and irregularly arranged parenchymal cells with numerous intercellular spaces found near the lower surface in well-differentiated broad leaves. Also known as spongy parenchyma. { 'spən·jē 'mē·zōˌfil }

spongy parenchyma *See* spongy mesophyll. { 'spən·jē pə'reŋ·kə·mə }

spontaneous abortion [MED] An unexpected, premature expulsion of the fetus. Also known as miscarriage. { spän'tā·nē·əs ə'bȯr·shən }

spontaneous amputation [MED] **1.** Congenital amputation. **2.** Amputation not caused by external trauma or injury, as in ainhum. { spän'tā·nē·əs ˌam·pyə'tā·shən }

spontaneous generation *See* abiogenesis. { spän'tā·nē·əs ˌjen·ə'rā·shən }

spontaneous mutation [GEN] A mutation that occurs spontaneously, that is, in an individual not specifically exposed to a known mutagen. { spän'tā·nē·əs myü'tā·shən }

spontaneous pneumothorax [MED] Air that has entered the chest cavity and caused lung collapse without being the result of injury or deliberate introduction of air; caused by an overexpansion of air in the lungs. { spän'tā·nē·əs ˌnü·mō'thȯrˌaks }

spoon nail *See* koilonychia. { 'spün ˌnāl }

sporangiolum [MYCOL] A small sporangium that has few sporangiospores. { spəˌran·jē'ōl·əm }

sporangiophore [BOT] A stalk or filament on which sporangia are borne. { spə'ran·jē·əˌfȯr }

sporangiospore [BOT] A spore that forms in a sporangium. { spə'ran·jē·əˌspȯr }

sporangium [BOT] A case in which asexual spores are formed and borne. { spə'ran·jē·əm }

spore [BIOL] A uni- or multicellular, asexual, reproductive or resting body that is resistant to unfavorable environmental conditions and produces a new vegetative individual when the environment is favorable. { spȯr }

spore mother cell [BOT] One of the cells of the archespore of a spore-bearing plant from which a spore, but usually a tetrad of spores, is produced. Also known as sporocyte. { 'spȯr 'mət͟h·ər ˌsel }

spore stain [MICROBIO] A type of staining technique that utilizes the fact that the spore does not take up dyes readily but, once stained, it resists decolorization. The differentiating agent used may be a dilute solution of an organic acid, an acid dye, or another basic dye. { 'spȯr ˌstān }

sporidium [MYCOL] A small spore, especially one formed on a promycelium. { spə'rid·ē·əm }

sporoblast [INV ZOO] A sporozoan cell from which sporozoites arise. { 'spȯr·əˌblast }

Sporobolomycetaceae [MYCOL] The single family of the order Sporobolomycetales. { spəˌräb·ə·lōˌmī·sə'tās·ēˌē }

Sporobolomycetales [MYCOL] An order of yeastlike and moldlike fungi assigned to the class Basidiomycetes characterized by the formation of sterigmata, upon which the asexual ballistospores are formed. { spəˌräb·ə·lōˌmī·sə'tā·lēz }

sporocarp [BOT] Any multicellular structure in or on which spores are formed. { 'spȯr·əˌkärp }

sporocyst [BOT] A unicellular resting body from which asexual spores arise. [INV ZOO] **1.** A resistant envelope containing an encysted sporozoan. **2.** An encysted sporozoan. **3.** The first reproductive form of a digenetic trematode in which rediae develop. { 'spȯr·əˌsist }

sporocyte *See* spore mother cell. { 'spȯr·əˌsīt }

sporogenesis [BIOL] **1.** Reproduction by means of spores. **2.** Formation of spores. { ¦spȯr·ə'jen·ə·səs }

sporogony [BIOL] Reproduction by means of spores. [INV ZOO] Propagative reproduction involving formation, by sexual processes, and subsequent division of a zygote. { spə'räg·ə·nē }

sporont [INV ZOO] A stage in the life history of sporozoans which gives rise to spores. { 'spȯrˌänt }

sporophore [MYCOL] A structure on the thallus of fungi which produces spores. { 'spȯr·əˌfȯr }

sporophyll [BOT] A modified leaf that develops sporangia. { 'spȯr·əˌfil }

sporophyte [BOT] **1.** An individual of the spore-bearing generation in plants exhibiting alternation of generation. **2.** The spore-producing generation. **3.** The diplophase in a plant life cycle. { 'spȯr·əˌfīt }

sporopollenin [BIOCHEM] A substance related to suberin and cutin but more resistant to decay that is found in the exine of pollen grains. { ¦spȯr·ō'päl·ə·nən }

sporosac [INV ZOO] A degenerate gonophore in certain hydroid cnidarians. { 'spȯr·ə,sak }

sporotrichosis [MED] A granulomatous fungus disease caused by *Sporotrichum schenckii*, with cutaneous lesions along the lymph channels and occasionally involving the internal organs. Also known as de Beurmann-Gougerot disease; Schenk's disease. { ,spȯr·ə·tri'kō·səs }

Sporozoa [INV ZOO] A subphylum of parasitic Protozoa, typically producing spores during the asexual stages of the life cycle. { ,spȯr·ə'zō·ə }

sporozoite [INV ZOO] A motile, infective stage of certain sporozoans, which is the result of sexual reproduction and which gives rise to an asexual cycle in the new host. { ,spȯr·ə'zō,īt }

sports medicine [MED] A branch of medicine concerned with the effects of exercise and sports on the human body, including treatment of injuries. { 'spȯrts ,med·ə·sən }

sporulation [BIOL] The act and process of spore formation. { ,spȯr·yə'lā·shən }

spot film [MED] A small, highly collimated radiograph of an anatomic part, usually obtained in conjunction with fluoroscopy. { 'spät ,film }

spout hole [VERT ZOO] **1.** A blowhole of a cetacean mammal. **2.** A nostril of a walrus or seal. { 'spau̇t ,hōl }

sprain [MED] A wrenching of a joint, producing a stretching or laceration of the ligaments. { sprān }

sprain fracture [MED] An injury in which a tendon or ligament, together with a shell of bone, is torn from its attachment; occurs most commonly at the ankle. { 'sprān ,frak·chər }

spreading factor *See* hyaluronidase. { 'spred·iŋ ,fak·tər }

spreading of inactivation [GEN] A type of position effect, as in the inactivation of nearby autosomal genes in an X-autosome translocation chromosome in mammalian females heterozygous for the translocation. { ¦spred·iŋ əv in,ak·tə'vā·shən }

Sprigginidae [INV ZOO] An extinct family of annelid worms distinguished by a horseshoe-shaped head. { spri'gin·ə,dē }

springwood [BOT] The portion of an annual ring that is formed principally during the growing season; it is softer, more porous, and lighter than summerwood because of its higher proportion of large, thin-walled cells. { 'spriŋ,wu̇d }

spruce [BOT] An evergreen tree belonging to the genus *Picea* characterized by single, four-sided needles borne on small peglike projections, pendulous cones, and resinous wood. { sprüs }

spruce budworm [INV ZOO] The larva of a common moth, *Choristoneura fumiferana*, that is a destructive pest primarily of spruce and balsam fir. { ,sprüs 'bəd,wərm }

sprue [MED] A syndrome characterized by impaired absorption of food, water, and minerals by the small intestine; symptoms are the result of nutritional deficiencies. { sprü }

Spumellina [INV ZOO] The equivalent name for Peripylina. { spyü'mel·ə·nə }

spur [BOT] **1.** A hollow process at the base of a petal or sepal. **2.** A short fruit-bearing tree branch. **3.** A short projecting root. [ZOO] A stiff, sharp outgrowth, as on the legs of certain birds and insects. { spər }

sputum [PHYSIO] Material discharged from the surface of the respiratory passages, mouth, or throat; may contain saliva, mucus, pus, microorganisms, blood, or inhaled particulate matter in any combination. { 'spyüd·əm }

squalene [BIOCHEM] $C_{30}H_{50}$A liquid triterpene which is found in large quantities in shark liver oil, and which appears to play a role in the biosynthesis of sterols and polycyclic terpenes; used as a bactericide and as an intermediate in the synthesis of pharmaceuticals. Also known as spinacene. { 'skwā,lēn }

Squalidae [VERT ZOO] The spiny dogfishes, a family of squaloid elasmobranchs recognized by their well-developed fin spines. { 'skwā·lə,dē }

Squamata [VERT ZOO] An order of reptiles, composed of the lizards and snakes, distinguished by a highly modified skull that has only a single temporal opening, or none, by the lack of shells or secondary palates, and by possession of paired penes on the males. { skwə'mäd·ə }

squamodisk [INV ZOO] In monogenetic trematodes, a disk bearing concentric circles of spines, scales, or ridges, and located on the opisthaptor. { 'skwā·mō,disk }

squamosal bone [ANAT] The part of the temporal bone in man corresponding with the squamosal bone in lower vertebrates. [VERT ZOO] A membrane bone lying external and dorsal to the auditory capsule of many vertebrate skulls. { skwə'mō·səl ,bōn }

squamous [BIOL] Covered with or composed of scales. { 'skwā·məs }

squamous-cell carcinoma [MED] A carcinoma composed principally of anaplastic, squamous epithelial cells. Also known as epidermoid carcinoma. { 'skwā·məs ¦sel ,kärs·ən'ō·mə }

squamous epithelium [HISTOL] A single-layered epithelium consisting of thin, flat cells. { 'skwā·məs ,ep·ə'thē·lē·əm }

squamulose [BIOL] Covered with or composed of minute scales. { 'skwäm·yə,lōs }

squarrose [BOT] Having stiff divergent bracts, or other processes. { 'skwä,rōs }

squarrulose [BOT] Mildly squarrose. { 'skwä·rə,lōs }

squash [BOT] Either of two plants of the genus *Cucurbita*, order Violales, cultivated for its fruit; some types are known as pumpkins. { skwäsh }

Squatinidae [VERT ZOO] A group of squaloid elasmobranchs of uncertain affinity characterized by a greatly extended rostrum with enlarged denticles along the margins; maximum length is under 4 feet (1.2 meters). { skwə'tin·ə,dē }

squid [INV ZOO] Any of a number of marine cephalopod mollusks characterized by a reduced internal shell, ten tentacles, an ink sac, and chromatophores. { skwid }

Squillidae [INV ZOO] The single family of the

eumalacostracan order Stomatopoda, the mantis shrimp. { 'skwil·ə,dē }

squint *See* strabismus. { skwint }

squirrel [VERT ZOO] Any of over 200 species of arboreal rodents of the families Sciuridae and Anomaluridae having a bushy tail and long, strong hind limbs. { 'skwərl }

SRID *See* single radial immunodiffusion. { ¦es ¦är¦ī'dē *or* 'es,rid }

SRP *See* signal-recognition particle. { ¦es¦är'pē }

SRS-A *See* slow-reacting substance of anaphylaxis.

SRY [GEN] The male sex-determining gene on the Y chromosome in mammals. { ¦es¦är'wī }

stab culture [MICROBIO] A culture of anaerobic bacteria made by piercing a solid agar medium in a test tube with an inoculating needle covered with the bacterial inoculum. { 'stab ,kəl·chər }

stable factor *See* factor VII. { 'stā·bəl 'fak·tər }

Stader's splint [MED] A metal bar with pins affixed at right angles; the pins are driven into the fragments of a fracture, and the bar maintains the alignment. { 'stād·ərz ,splint }

stage theory [NEUROSCI] A theory of color vision which proposes that there are three or more types of cone receptors whose responses are conducted to higher visual centers, and that interactions occur at some stage along the conducting paths so that strong activity in one type of response inhibits that of other response paths. Also known as zone theory. { 'stāj ,thē·ə·rē }

staggers [VET MED] Any of various diseases of livestock, sheep, and horses manifested by lack of coordination in movement and a staggering gait. { 'stag·ərz }

stalked barnacle [INV ZOO] The common name for crustaceans composing the suborder Lepadomorpha. { 'stȯkt 'bär·nə·kəl }

stallion [VERT ZOO] **1.** A mature male equine mammal. **2.** A male horse not castrated. { 'stal·yən }

stamen [BOT] The male reproductive structure of a flower, consisting of an anther and a filament. { 'stā·mən }

Stamey test [MED] A test of differential urinary excretion designed to detect unilateral renovascular disease. { 'stām·ē ,test }

staminate flower [BOT] A flower having stamens but lacking functional carpels. { 'stam·ə·nət 'flau̇·ər }

staminode [BOT] A stamen with no functional anther. { 'stā·mə,nōd }

stand [ECOL] A group of plants, distinguishable from adjacent vegetation, which is generally uniform in species composition, age, and condition. { stand }

standing crop [ECOL] The number of individuals or total biomass present in a community at one particular time. { 'stand·iŋ 'kräp }

stapedectomy [MED] Surgical reconstruction of the junction of the ossicular chain of the middle ear with the oval window of the inner ear, fixed in place by otosclerosis. { ,stā·pə'dek·tə·mē }

stapedius muscle [ANAT] The muscle which attaches to and controls the stapes in the inner ear. { stə'pēd·ē·əs ,məs·əl }

stapes [ANAT] The stirrup-shaped middle-ear ossicle, articulating with the incus and the oval window. Also known as columella. { 'stā·pēz }

Staphylinidae [INV ZOO] The rove beetles, a very large family of coleopteran insects in the superfamily Staphylinoidea. { ,staf·ə'lin·ə,dē }

Staphylinoidea [INV ZOO] A superfamily of Coleoptera in the suborder Polyphaga. { ,staf·ə·lə'nȯid·ē·ə }

staphylococcal pneumonia [MED] A severe form of pneumonia caused by *Staphylococcus aureus*. { ,staf·ə·lō'käk·əl nu̇'mō·nē·ə }

staphylomycin [MICROBIO] An antibiotic composed of three active components produced by a strain of *Actinomyces* that inhibits growth of gram-positive microorganisms and acid-fast bacilli. { ,staf·ə·lō'mīs·ən }

staphylotoxin [BIOCHEM] Any of the various toxins elaborated by strains of *Staphylococcus aureus*, including hemolysins, enterotoxins, and leukocidin. { ,staf·ə·lō'täk·sən }

starch [BIOCHEM] Any one of a group of carbohydrates or polysaccharides, of the general composition $(C_6H_{10}O_5)_n$, occurring as organized or structural granules of varying size and markings in many plant cells; it hydrolyzes to several forms of dextrin and glucose; its chemical structure is not completely known, but the granules consist of concentric shells containing at least two fractions: an inner portion called amylose, and an outer portion called amylopectin. { stärch }

starfish [INV ZOO] The common name for echinoderms belonging to the subclass Asteroidea. { 'stär,fish }

Starling's law of the heart [PHYSIO] The energy associated with cardiac contraction is proportional to the length of the myocardial fibers in diastole. { 'stär·liŋz 'lȯ əv t͟hə 'härt }

start codon *See* initiation codon. { ¦stärt 'kō,dän }

starter [MICROBIO] A culture of microorganisms, either pure or mixed, used to commence a process, for example, cheese manufacture. { 'stär·dər }

startle response [PHYSIO] The complex, involuntary, usually spasmodic psychophysiological response movement of an organism to a sudden unexpected stimulus. { 'stärd·əl ri,späns }

startpoint [CELL MOL] The deoxyribonucleic acid base pair that corresponds to the first nucleotide incorporated into the primary ribonucleic acid (RNA) transcript by RNA polymerase. { 'stärt,pȯint }

stasis [MED] A cessation of the normal flow of blood or other body fluids. { 'stā·səs }

stasis dermatitis [MED] Chronic inflammation of the skin of the legs, resulting from poor circulation. { 'stā·səs ,dər·mə'tīd·əs }

static contraction *See* isometric contraction. { 'stad·ik kən'trak·shən }

static reflex [PHYSIO] Any one of a series of reflexes which are involved in the establishment of muscular tone for postural purpose. { 'stad·ik 'rē,fleks }

static work *See* isometric work. { 'stad·ik 'wərk }

stationary phase [MICROBIO] The period following termination of exponential growth in a bacterial culture when the number of viable microorganisms remains relatively constant for a time. { 'stā·shə,ner·ē 'fāz }

stationary population [ECOL] A population containing a basically even distribution of age groups. { 'stā·shə,ner·ē ,päp·yə'lā·shən }

statoblast [INV ZOO] A chitin-encapsulated body which serves as a special means of asexual reproduction in the Phylactolaemata. { 'stad·ə,blast }

statocone [INV ZOO] One of the minute calcareous granules found in the statocyst of certain animals. { 'stad·ə,kōn }

statocyst [BOT] A cell containing statoliths in a fluid medium. Also known as statocyte. [INV ZOO] A sensory vesicle containing statoliths and which functions in the perception of the position of the body in space. { 'stad·ə,sist }

statocyte *See* statocyst. { 'stad·ə,sīt }

statokinetic [PHYSIO] Pertaining to the balance and posture of the body or its parts during movement, as in walking. { ¦stad·ō·ki'ned·ik }

statolith [BOT] A sand grain or other solid inclusion which moves readily in the fluid contents of a statocyst, comes to rest on the lower surface of the cell, and is believed to function in gravity perception. [INV ZOO] A secreted calcareous body, a sand grain, or other solid inclusion contained in a statocyst. { 'stad·ə,lith }

statoreceptor [PHYSIO] A sense organ concerned primarily with equilibrium. { ¦stād·ō·ri 'sep·tər }

statospore [BOT] In certain algae, an internally formed spore in its resting stage. { 'stad·ə,spȯr }

status asthmaticus [MED] Intractable asthma lasting from a few days to a week or longer. { 'stad·əs az'mad·ə·kəs }

status epilepticus [MED] Occurrence of prolonged, generalized epileptic seizures in rapid succession with brief intervals of coma. { 'stad·əs ,ep·ə'lep·tə·kəs }

stauractine [INV ZOO] A sponge spicule in which the four rays lie in one plane. { stȯ 'rak,tēn }

Stauromedusae [INV ZOO] An order of the class Scyphozoa in which the medusa is composed of a cuplike bell called a calyx and a stem that terminates in a pedal disk. { ¦stȯ·rō·mi'dü·sē }

Staurosporae [MYCOL] A spore group of the Fungi Imperfecti characterized by star-shaped or forked spores. { stȯ'räs·pə,rē }

STD *See* sexually transmitted disease.

steapsin [BIOCHEM] An enzyme in pancreatic juice that catalyzes the hydrolysis of fats. Also known as pancreatic lipase. { stē'ap·sən }

Steatornithidae [VERT ZOO] A family of birds in the order Caprimulgiformes which contains a single, South American species, the oilbird or guacharo (*Steatornis caripensis*). { 'stē·ə,tȯr'nith·ə,dē }

steatorrhea [MED] **1.** Fatty stool. **2.** Increased flow of sebum. { 'stē,ad·ə'rē·ə }

Steganopodes [VERT ZOO] Formerly, an order of birds that included the totipalmate swimming birds. { ,steg·ə'näp·ə,dēz }

Steiner's hypocycloid *See* deltoid. { ¦stīn·ərz ,hī·pə'sī,klȯid }

Stein-Leventhal syndrome [MED] A complex of symptoms characterized by amenorrhea or abnormal uterine bleeding or both, enlarged polycystic ovaries, frequently hirsutism, and occasionally retarded breast development. { 'stīn 'lev·ən,thȯl ,sin,drōm }

Steinmann pin [MED] A surgical nail inserted in distal portions of such bones as the femur or tibia for skeletal tractions. { 'stīn·mən ,pin }

STEL *See* short-term exposure limit. { stel *or* ¦es¦tē¦ē'el }

stele [BOT] The part of a plant stem including all tissues and regions of plants from the cortex inward, including the pericycle, phloem, cambium, xylem, and pith. { 'stēl }

Stelenchopidae [INV ZOO] A family of polychaete annelids belonging to the Myzostomaria, represented by a single species from Crozet Island in the Antarctic Ocean. { ,stel·ən'käp·ə,dē }

stellar nevus *See* spider nevus. { 'stel·ər 'nē·vəs }

stellate ganglion [NEUROSCI] The ganglion formed by the fusion of the inferior cervical and the first thoracic sympathetic ganglions. { 'ste ,lāt 'gaŋ·glē·ən }

stellate reticulum [HISTOL] The part of the epithelial dental organ of a developing tooth which lies between the inner and the outer dental epithelium; composed of stellate cells with long, anastomosing processes in a mucoid fluid in the interstitial spaces. { 'ste,lāt rə'tik·yə·ləm }

Stelleroidea [INV ZOO] The single class of echinoderms in the subphylum Asterozoa; characters coincide with those of the subphylum. { ,stel·ə'rȯid·ē·ə }

stem [BOT] The organ of vascular plants that usually develops branches and bears leaves and flowers. { stem }

stem cell [EMBRYO] A formative cell. [HISTOL] *See* hemocytoblast. { 'stem ,sel }

stem-cell leukemia [MED] A form of leukemia in which the predominant cell type is so poorly differentiated that its series cannot be identified. { 'stem ¦sel lü'kē·mē·ə }

Stemonitaceae [MYCOL] The single family of the order Stemonitales. { stē,män·ə'tās·ē,ē }

Stemonitales [MYCOL] An order of fungi in the subclass Myxogastromycetidae of the class Myxomycetes. { stē,män·ə'tā·lēz }

Stenetrioidea [INV ZOO] A group of isopod crustaceans in the suborder Asellota consisting mostly of tropical marine forms in which the first pleopods are fused. { stə,ne·trē'ȯid·ē·ə }

Stenocephalidae [INV ZOO] A family of Old World, neotropical Hemiptera included in the Pentatomorpha. { ,sten·ə·sə'fal·ə,dē }

stenocephaly [MED] Unusual narrowness of the head. { ˌsten·əˈsef·ə·lē }

Stenoglossa [INV ZOO] The equivalent name for Neogastropoda. { ˌsten·əˈgläs·ə }

stenohaline [ECOL] In marine organisms, indicating the ability to tolerate only a narrow range of salinities. { ¦sten·ə¦haˌlīn }

Stenolaemata [INV ZOO] A class of marine ectoproct bryozoans having lophophores which are circular in basal outline and zooecia which are long, slender, tubular or prismatic, and gradually tapering to their proximal ends. { ˌsten·ə·ləˈmäd·ə }

stenoplastic [BIOL] Relating to an organism which exhibits a limited capacity for modification or adaptation to a new environment. { ¦sten·ə¦plas·tik }

Stenopodidea [INV ZOO] A section of decapod crustaceans in the suborder Natantia which includes shrimps having the third pereiopods chelate and much longer and stouter than the first pair. { ˌsten·ə·pəˈdid·ē·ə }

stenosis [MED] Constriction or narrowing, as of the heart or blood vessels. { stəˈnō·səs }

Stenostomata [INV ZOO] The equivalent name for Cyclostomata. { ˌsten·ə·stəˈmäd·ə }

stenotherm [BIOL] An organism able to tolerate only a small variation of temperature in the environment. { ˈsten·əˌthərm }

stenothermic [BIOL] Indicating the ability to tolerate only a limited range of temperatures. { ¦sten·ə¦thər·mik }

Stenothoidae [INV ZOO] A family of amphipod crustaceans in the suborder Gammaridea containing semiparasitic and commensal species. { ˌsten·əˈthȯiˌdē }

stenotopic [ECOL] Referring to an organism with a restricted distribution. { ¦sten·ə¦täp·ik }

Stensen's duct *See* parotid duct. { ˈsten·sənz ˌdəkt }

step-down photophobic response [PHYSIO] A photophobic response elicited by a sudden decrease in light intensity. { ¦step ˌdau̇n ˌfōd·əˌfō·bik riˈspäns }

Stephanidae [INV ZOO] A small family of the Hymenoptera in the superfamily Ichneumonoidea characterized by many-segmented filamentous antennae. { stəˈfan·əˌdē }

stepping reflex [PHYSIO] A reflex response of the newborn and young infant, characterized by alternating stepping movements with the legs, as in walking, elicited when the infant is held upright so that both soles touch a flat surface while the infant is moved forward to accompany any step taken. { ˈstep·iŋ ˌrēˌfleks }

step-up photophobic response [PHYSIO] A photophobic response elicited by a sudden increase in light intensity. { ¦step ˌəp ˌfōd·əˌfō·bik riˈspäns }

stercobilin [BIOCHEM] Urobilin as a component of the brown fecal pigment. { ˌstər·kōˈbī·lən }

stercobilinogen [BIOCHEM] A colorless reduction product of stercobilin found in feces. { ˌstər·kō·bīˈlin·ə·jən }

Stercorariidae [VERT ZOO] A family of predatory birds of the order Charadriiformes including the skuas and jaegers. { ˌstər·kə·rəˈrī·əˌdē }

Sterculiaceae [BOT] A family of dicotyledonous trees and shrubs of the order Malvales distinguished by imbricate or contorted petals, bilocular anthers, and ten to numerous stamens arranged in two or more whorls. { stərˌkyü·lē ˈās·ē }

stereoblastula [EMBRYO] A blastula that lacks a cavity, making it unable to gastrulate. { ¦ster·ē·əˈblas·chə·lə }

stereocilia [CELL MOL] **1.** Nonmotile tufts of secretory microvilli on the free surface of cells of the male reproductive tract. **2.** Homogeneous cilia within simple membrane coverings; found on the free-surface hair cells. { ¦ster·ē·əˈsil·ē·ə }

stereogastrula [EMBRYO] A gastrula that lacks a cavity. { ¦ster·ē·əˈgas·trə·lə }

stereopsis *See* stereoscopy. { ˌster·ēˈäp·səs }

stereoscopic radius [PHYSIO] The greatest distance at which there is a sensation of depth in vision due to the fact that the two eyes do not perceive exactly the same view. { ¦ster·ē·ə¦skäp·ik ˈrād·ē·əs }

stereoscopic vision *See* stereoscopy. { ¦ster·ē·ə¦skäp·ik ˈvizh·ən }

stereoscopy [PHYSIO] The phenomenon of simultaneous vision with two eyes in which there is a vivid perception of the distances of objects from the viewer; it is present because the two eyes view objects in space from two points, so that the retinal image patterns of the same object are slightly different in the two eyes. Also known as stereopsis; stereoscopic vision. { ˌster·ēˈäs·kə·pē }

stereoselectivity [BIOCHEM] The selectivity of a reaction for forming one stereoisomer of a product in preference to another stereoisomer. { ¦ster·ē·ō·si·lekˈtiv·əd·ē }

stereotaxis [BIOL] An orientation movement in response to stimulation by contact with a solid body. Also known as thigmotaxis. { ˌster·ē·əˈtak·səs }

stereotropism [BIOL] Growth or orientation of a sessile organism or part of an organism in response to the stimulus of a solid body. Also known as thigmotropism. { ˌster·ēˈä·trəˌpiz·əm }

sterigma [BOT] A peg-shaped structure to which needles are attached in certain conifers. [MYCOL] A slender stalk arising from the basidium of some fungi, from the top of which basidiospores are formed by abstriction. { stəˈrig·mə }

sterile distribution [ECOL] A range of areas in which marine animals may live and spawn, but in which eggs do not hatch and larvae do not survive. { ˈster·əl ˌdis·trəˈbyü·shən }

sterility [PHYSIO] The inability to reproduce because of congenital or acquired reproductive system disorders involving lack of gamete production or production of abnormal gametes. { stəˈril·əd·ē }

sterilization [MICROBIO] An act or process of destroying all forms of microbial life on and in an object. { ˌster·ə·ləˈzā·shən }

Sternaspidae [INV ZOO] A monogeneric family of polychaete annelids belonging to the Sedentaria. { stər'nas·pə,dē }

sternebra [VERT ZOO] A segment of the sternum in vertebrates. { 'stər·nə·brə }

Sterninae [VERT ZOO] A subfamily of birds in the family Laridae, including the Arctic tern. { 'stər·nə,nē }

sternite [INV ZOO] **1.** The ventral part of an arthropod somite. **2.** The chitinous plate on the ventral surface of an abdominal segment of an insect. { 'stər,nīt }

sternocleidomastoid [ANAT] A muscle of the neck that flexes the head; origin is the manubrium of the sternum and clavicle, and insertion is the mastoid process. { ¦stər·nō¦klīd·ə'ma ,stȯid }

sternocostal [ANAT] Pertaining to the sternum and the ribs. { ¦stər·nə'käst·əl }

sternohyoid [ANAT] A muscle arising from the manubrium of the sternum and inserted into the hyoid bone. { ,stər·nō'hī,ȯid }

Sternorrhyncha [INV ZOO] A series of the insect order Homoptera in which the beak appears to arise either between or behind the fore coxae, and the antennae are long and filamentous with no well-differentiated terminal setae. { ¦stər·nə'riŋ·kə }

sternothyroid [ANAT] Pertaining to the sternum and thyroid cartilage. { ¦stər·nə'thī,rȯid }

Sternoxia [INV ZOO] The equivalent name for Elateroidea. { stər'näk·sē·ə }

sternum [ANAT] The bone, cartilage, or series of bony or cartilaginous segments in the median line of the anteroventral part of the body of vertebrates above fishes, connecting with the ribs or pectoral girdle. { 'stər·nəm }

steroid [BIOCHEM] A member of a group of compounds, occurring in plants and animals, that are considered to be derivatives of a fused, reduced ring system, cyclopenta[α]-phenanthrene, which consists of three fused cyclohexane rings in a nonlinear or phenanthrene arrangement. { 'sti,rȯid }

sterol [BIOCHEM] Any of the natural products derived from the steroid nucleus; all are waxy, colorless solids soluble in most organic solvents but not in water, and contain one alcohol functional group. { 'sti,rȯl }

sterol regulatory element binding protein [BIOCHEM] A transcription factor required for the active transcription of genes that encode the low-density lipoprotein receptor and enzymes in cholesterol synthesis. { ¦ste,rȯl ¦reg·yə·lə,tȯr·ē ,el·ə·mənt 'bīnd·iŋ ,prō,tēn }

sterone [BIOCHEM] A ketone derived from a steroid. { 'sti,rōn }

stethoscope [MED] An instrument for indirect auscultation for the detection and study of sounds arising within the body; sounds are conveyed to the ears of the examiner through rubber tubing connected to a funnel or disk-shaped endpiece. { 'steth·ə,skōp }

stibophen [PHARM] $C_{12}H_4Na_5O_{16}S_4Sb \cdot 7H_2O$ A crystalline compound that is soluble in water, insoluble in organic solvents; used in medicine for protozoan infections. { 'stib·ə,fen }

Stichaeidae [VERT ZOO] The pricklebacks, a family of perciform fishes in the suborder Blennioidei. { stə'kē·ə,dē }

Stichocotylidae [INV ZOO] A family of trematodes in the subclass Aspidogastrea in which adults are elongate and have a single row of alveoli. { ¦stik·ə·kə'til·ə,dē }

Stichopodidae [INV ZOO] A family of the echinoderm order Aspidochirotida characterized by tentacle ampullae and by left and right gonads. { ¦stik·ə'päd·ə,dē }

Stickland reaction [BIOCHEM] An amino acid fermentation involving the coupled decomposition of two or more substrates. { 'stik·lənd rē,ak·shən }

stickleback [VERT ZOO] Any fish which is a member of the family Gasterosteidae, so named for the variable number of free spines in front of the dorsal fin. { 'stik·əl,bak }

sticky end [BIOCHEM] Any of the single-stranded complementary ends of a deoxyribonucleic acid molecule. Also known as cohesive end. { 'stik·ē ,end }

stigma [BOT] The rough or sticky apical surface of the pistil for reception of the pollen. [INV ZOO] **1.** The eyespot of certain protozoans, such as *Euglena*. **2.** The spiracle of an insect or arthropod. **3.** A colored spot on many lepidopteran wings. { 'stig·mə }

stigmatism [PHYSIO] A condition of the refractive media of the eye in which rays of light from a point are accurately brought to a focus on the retina. { 'stig·mə,tiz·əm }

Stilbaceae [MYCOL] The equivalent name for Stilbellaceae. { stil'bās·ē,ē }

Stilbellaceae [MYCOL] A family of fungi of the order Moniliales in which conidiophores are aggregated in long bundles or fascicles, forming synnemata or coremia, generally having the conidia in a head at the top. { ,stil·bə'lās·ē,ē }

stilbesterol *See* diethylstilbesterol. { stil'bes·tə,rȯl }

stillbirth [MED] Birth of a dead infant. { 'stil ,bərth }

Still's disease [MED] Juvenile rheumatoid arthritis in which involvement of the viscera is prominent. Also known as Chauffard-Still disease. { 'stilz di,zēz }

stilt root [BOT] A prop root of a mangrove tree. { 'stilt ,rüt }

stimulant [PHARM] A drug or agent that temporarily acts on muscles, nerves, or a sensory end organ, producing an increase in its state of activity. { 'stim·yə·lənt }

stimulation deafness [MED] Deafness induced by noise; involves changes in the chemical interchange between the canals of the cochlea, as well as nerve destruction. { ,stim·yə'lā·shən 'def·nəs }

stimulator [MED] A neurosurgical device that supplies a controlled alternating-current voltage to two electrodes that are applied to a patient. { 'stim·yə,lād·ər }

stimulatory G protein [CELL MOL] A G protein (heterotrimetric GTP-binding protein) that activates adenylyl cyclase to produce cAMP in intracellular signaling pathways. Symbolized G∩. { ¦stim·yə·lə,tȯr·ē 'jē ,prō,tēn }

stimulon [GEN] A system of genes not physically linked together. { 'stim·yə,län }

stimulus [PHYSIO] An agent that produces a temporary change in physiological activity in an organism or in any of its parts. { 'stim·yə·ləs }

stinger [ZOO] A sharp piercing organ, as of a bee, stingray, or wasp, usually connected with a poison gland. { 'stiŋ·ər }

stinging cell *See* cnidoblast. { 'stiŋ·iŋ ,sel }

stingray [VERT ZOO] Any of various rays having a whiplike tail armed with a long serrated spine, at the base of which is a poison gland. { 'stiŋ,rā }

stipe [BOT] **1.** The petiole of a fern frond. **2.** The stemlike portion of the thallus in certain algae. [MYCOL] The short stalk or stem of the fruit body of a fungus, such as a mushroom. { stīp }

stipule [BOT] Either of a pair of appendages that are often present at the base of the petiole of a leaf. { 'stip·yül }

Stirodonta [INV ZOO] Formerly, an order of Euechinoidea that included forms with stirodont dentition. { ,stir·ə'dän·tə }

stirodont dentition [INV ZOO] In Echinoidea, the condition in which the teeth are keeled within and the foramen magnum is open. { 'stir·ə,dänt den'tish·ən }

Stokes-Adams syndrome [MED] Syncopic or convulsive attacks occurring in patients with complete heart block. { ¦stōks 'ad·əmz ,sin ,drōm }

Stokes stretcher [MED] A basket-type stretcher constructed of tubular steel and strong wire mesh, and which acts as a splint for the entire body. { 'stōks 'strech·ər }

stolon [BOT] *See* runner. [INV ZOO] An elongated projection of the body wall from which buds are formed giving rise to new zooids in Anthozoa, Hydrozoa, Bryozoa, and Ascidiacea. [MYCOL] A hypha produced above the surface and connecting a group of conidiophores. { 'stō·lən }

Stolonifera [INV ZOO] An order of the Alcyonaria, lacking a coenenchyme; they form either simple (*Clavularia*) or rather complex colonies (*Tubipora*). { ,stäl·ə'nif·rə }

stoma [BIOL] A small opening or pore in a surface. [BOT] One of the minute openings in the epidermis of higher plants which are regulated by guard cells and through which gases and water vapor are exchanged between internal spaces and the external atmosphere. { 'stō·mə }

stomach [ANAT] The tubular or saccular organ of the vertebrate digestive system located between the esophagus and the intestine and adapted for temporary food storage and for the preliminary stages of food breakdown. { 'stəm·ək }

stomatitis [MED] Inflammation of the soft tissues in the mouth. { ,stō·mə'tīd·əs }

stomatoblastula [INV ZOO] A blastula stage in some sponges capable of engulfing maternal amebocytes for nutrition. { ¦stō·məd·ō'blas·chə·lə }

stomatology [MED] The branch of medical science that concerns the anatomy, physiology, pathology, therapeutics, and hygiene of the oral cavity, of the tongue, teeth, and adjacent structures and tissues, and of the relationship of that field to the entire body. { ,stō·mə'täl·ə·jē }

Stomatopoda [INV ZOO] The single order of the Eumalacostraca in the superorder Hoplocarida distinguished by raptorial arms, especially the second pair of maxillipeds. { ,stō·mə'täp·əd·ə }

Stomiatoidei [VERT ZOO] A suborder of fishes of the order Salmoniformes including the lightfishes and allies, which are of small size and often grotesque form and are equipped with photophores. { ,stō·mē·ə'tȯid·ē,ī }

stomium [BOT] A region along a sporangium or pollen sac where dehiscence takes place. { 'stō·mē·əm }

stomocnidae nematocyst [INV ZOO] A nematocyst which has an open-ended thread. { stə'mäk·nə,dē nə'mad·ə,sist }

stomodaeum [EMBRYO] The anterior part of the embryonic alimentary tract formed as an invagination of the ectoderm. { ,stō·mə'dē·əm }

stone canal [INV ZOO] A canal in many echinoderms that has a more or less calcified wall and that leads from the madreporite to the ring vessel. { 'stōn kə,nal }

stone cell *See* brachysclereid. { 'stōn ,sel }

stone fruit *See* drupe. { 'stōn ,früt }

stonewort [BOT] The common name for algae comprising the class Charophyceae, so named because most species are lime-encrusted. { 'stōn,wȯrt }

Stonnomida [INV ZOO] An order of Xenophyophorea distinguished by the presence of linellae in the test and a flexible body. { stə'näm·əd·ə }

Stonnomidae [INV ZOO] A family coextensive with the order Stonnomida. { stə'näm·ə,dē }

stony coral [INV ZOO] Any coral characterized by a calcareous skeleton. { 'stō·nē 'kär·əl }

stop codon *See* terminator codon. { 'stäp 'kō ,dän }

storage disease [MED] Metabolic abnormality in which some substance (such as fats, proteins, or carbohydrates) accumulates in abnormal amounts in certain body tissues. { 'stȯr·ij di,zēz }

stork [VERT ZOO] Any of several species of long-legged wading birds in the family Ciconiidae. { stȯrk }

strabismus [MED] Incoordinate action of the extrinsic ocular muscles resulting in failure of the visual axes to meet at the desired objective point. Also known as cast; heterotropia; squint. { strə'biz·məs }

straight sinus [ANAT] A sinus of the dura mater running from the inferior sagittal sinus along the junction of the falx cerebri and tentorium to the transverse sinus. { 'strāt ¦sī·nəs }

strain [BIOL] An intraspecific group of organisms that possess only one or a few distinctive traits and are maintained as an artificial breeding group. [CELL MOL] A population of cells derived either from a primary culture or from a cell line by the selection or cloning of cells having specific properties or markers. { strān }

strain propagation [PHYSIO] Transmission of a response to external stress within the body by mechanical or biological processes. { 'strān ˌpräp·əˌgā·shən }

strangulated hernia [MED] A hernia involving the intestine in which circulation of the blood and fecal current are blocked. { 'straŋ·gyəˌlād·əd 'hər·nē·ə }

strangulation [MED] **1.** Asphyxiation due to obstruction of the air passages, as by external pressure on the neck. **2.** Constriction of a part producing arrest of the circulation, as strangulation of a hernia. { ˌstraŋ·gyə'lā·shən }

strategy [ECOL] A group of related traits that evolved under the influence of natural selection and solve particular problems encountered by organisms. { 'strad·ə·jē }

stratified squamous epithelium [HISTOL] A multiple-layered epithelium composed of thin, flat superficial cells and cuboidal and columnar deeper cells. { 'strad·əˌfīd 'skwā·məs ˌep·ə'thē·lē·əm }

Stratiomyidae [INV ZOO] The soldier flies, a family of orthorrhaphous dipteran insects in the series Brachycera. { ˌstrad·ē·ō'mī·əˌdē }

stratum corneum [HISTOL] The outer layer of flattened keratinized cells of the epidermis. { 'strad·əm 'kȯr·nē·əm }

stratum disjunctum [HISTOL] The outermost layer of desquamating keratinized cells of the stratum corneum. { 'strad·əm dis'jəŋk·təm }

stratum germinativum [HISTOL] The innermost germinative layer of the epidermis. { 'strad·əm ˌjər·mə·nə'tī·vəm }

stratum granulosum [HISTOL] A layer of granular cells interposed between the stratum corneum and the stratum germinativum in the thick skin of the palms and soles. { 'strad·əm ˌgran·yə'lō·səm }

stratum lucidum [HISTOL] A layer of irregular transparent epidermal cells with traces of nuclei interposed between the stratum corneum and stratum germinativum in the thick skin of the palms and soles. { 'strad·əm 'lü·səd·əm }

Strauss reaction [IMMUNOL] The exudative swelling of the scrotum in male hamsters and guinea pigs upon subcutaneous or intraperitoneal inoculation of *Pseudomonas mallei*, the causative agent of glanders. { 'strau̇s rēˌak·shən }

straw [BOT] A stem of grain, such as wheat or oats. { strȯ }

strawberry [BOT] A low-growing perennial of the genus *Fragaria*, order Rosales, that spreads by stolons; the juicy, usually red, edible fruit consists of a fleshy receptacle with numerous seeds in pits or nearly superficial on the receptacle. { 'strȯˌber·ē }

strawberry hemangioma [MED] A vascular birthmark characterized by a soft, raised, bright-red, lobular appearance. { 'strȯˌber·ē hiˌmān·jē'ō·mə }

streak plate [MICROBIO] A method of culturing aerobic bacteria by streaking the surface of a solid medium in a petri dish with an inoculating wire or glass rod in a continuous movement so that most of the surface is covered; used to isolate majority members of a mixed population. { 'strēk ˌplāt }

Streblidae [INV ZOO] The bat flies, a family of cyclorrhaphous dipteran insects in the section Pupipara; adults are ectoparasites on bats. { 'streb·ləˌdē }

Strelitziaceae [BOT] A family of monocotyledonous plants in the order Zingiberales distinguished by perfect flowers with five functional stamens and without an evident hypanthium, penniveined leaves, and symmetrical guard cells. { strəˌlit·sē'ās·ēˌē }

strepaster [INV ZOO] A short, spiny microscleric, monaxonic spicule. { stre'pas·tər }

strepogenin [BIOCHEM] A factor, possibly a peptide derivative of glutamic acid, reported to exist in certain proteins, acting as a growth stimulant in bacteria and mice in the presence of completely hydrolyzed protein. Also known as streptogenin. { strə'päj·ə·nən }

Strepsiptera [INV ZOO] An order of the Coleoptera that is coextensive with the family Stylopidae. { strep'sip·tə·rə }

streptobacillary fever *See* Haverhill fever. { ¦strep·tō·bə'sil·ə·rē 'fē·vər }

streptobiosamine [BIOCHEM] $C_{13}H_{23}NO_9$ A nitrogen-containing disaccharide, obtained when streptomycin undergoes acid hydrolysis; in the streptomycin molecule it is glycosidally linked to streptidine. { ¦strep·tō·bī'ä·səˌmēn }

Streptococcaceae [MICROBIO] A family of gram-positive cocci; chemoorganotrophs with fermentative metabolism. { ¦strep·tə·käk'sās·ēˌē }

Streptococceae [MICROBIO] Formerly a tribe of the family Lactobacillaceae including cocci that occur in pairs, short chains, or tetrads and which generally obtain energy by fermentation of carbohydrates or related compounds. { ¦strep·tə'käk·sēˌē }

streptogenin *See* strepogenin. { strep'tä·jə·nən }

streptokinase [BIOCHEM] An enzyme occurring as a component of fibrinolysin in cultures of certain hemolytic streptococci. { ¦strep·tō'kīˌnās }

streptolysin [BIOCHEM] Any of a group of hemolysins elaborated by *Streptococcus pyogenes*. { ¦strep·tō'līs·ən }

Streptomycetaceae [MICROBIO] A family of soil-inhabiting bacteria in the order Actinomycetales; branched mycelia are produced by vegetative hyphae; spores are produced on aerial hyphae. { ˌstrep·tōˌmī·sə'tās·ēˌē }

streptomycin [MICROBIO] $C_{21}H_{39}O_{12}N_7$ A water-soluble antibiotic obtained from *Streptomyces griseus* that is used principally in the treatment of tuberculosis. { ˌstrep·tə'mīs·ən }

streptothricin [MICROBIO] $C_{19}H_{34}O_7N_8$ An antibiotic produced by *Streptomyces lavendulae*; active against various gram-negative and gram-positive microorganisms. { ˌstrep·tə'thrīs·ən }

stress [BIOL] A stimulus or succession of stimuli of such magnitude as to tend to disrupt the homeostasis of the organism. { stres }

stress test [MED] A procedure involving continuous electrocardiographic monitoring during exercise, as running a treadmill, to test the circulatory response to physical stress. { 'stres ˌtest }

stretcher [MED] A litter usually made of canvas stretched on a frame for carrying injured, disabled, or dead persons. { 'strech·ər }

stretch reflex [PHYSIO] Contraction of a muscle in response to a sudden, brisk, longitudinal stretching of the same muscle. Also known as myostatic reflex. { 'strech ˌrēˌfleks }

stria [BIOL] A minute line, band, groove, or channel. { 'strī·ə }

striated muscle [HISTOL] Muscle tissue consisting of muscle fibers having cross striations. { 'strīˌād·əd 'məs·əl }

stricture [MED] Abnormal narrowing of a passage, such as a vessel, duct, or the intestine. Also known as constriction. { 'strik·chər }

stridor [MED] A peculiar, harsh, vibrating sound produced during respiration. { 'strīd·ər }

stridulation [INV ZOO] Creaking and other audible sounds made by certain insects, produced by rubbing various parts of the body together. { ˌstrij·ə'lā·shən }

Strigidae [VERT ZOO] A family of birds of the order Strigiformes containing the true owls. { 'strij·əˌdē }

Strigiformes [VERT ZOO] The order of birds containing the owls. { ˌstrij·ə'fȯrˌmēz }

strigose [BIOL] Covered with stiff, pointed, hairlike scales or bristles. { 'strīˌgōs }

Strigulaceae [BOT] A family of Ascolichenes in the order Pyrenulales comprising crustose species confined to tropical evergreens, and which form extensive crusts on or under the cuticle of leaves. { ˌstrig·yə'lās·ēˌē }

strobilation [INV ZOO] Asexual reproduction by segmentation of the body into zooids, proglottids, or separate individuals. { ˌstrāb·ə'lā·shən }

strobilocercus [INV ZOO] A larval tapeworm that has undergone strobilation. { ˌstrāb·ə·lō'sər·kəs }

strobilus [BOT] **1.** A conelike structure made up of sporophylls, or spore-bearing leaves, as in Equisetales. **2.** The cone of members of the Pinophyta. { 'strāb·ə·ləs }

stroke [MED] A sudden cerebrovascular accident. { strōk }

stroke volume [PHYSIO] The amount of blood pumped during each cardiac contraction; quantitatively, the diastolic volume of the left ventricle minus the volume of blood in the ventricle at the end of systole. { 'strōk ˌväl·yəm }

stroma [ANAT] The supporting tissues of an organ, including connective and nervous tissues and blood vessels. { 'strō·mə }

stromal endometriosis *See* interstitial endometriosis. { 'strō·məl ¦en·dōˌmē·trē'ō·səs }

stromal myosis *See* interstitial endometriosis. { 'strō·məl mī'ō·səs }

Stromateidae [INV ZOO] A family of perciform fishes in the suborder Stromateoidei containing the butterfishes. { 'strō·mə'tē·əˌdē }

Stromateoidei [VERT ZOO] A suborder of fishes of the order Perciformes in which most species have teeth in pockets behind the pharyngeal bone. { 'strō·mə·tē'ȯid·ēˌī }

stromatosis *See* interstitial endometrosis. { ˌstrō·mə'tō·səs }

Strombidae [INV ZOO] A family of gastopod mollusks comprising tropical conchs. { 'strām·bəˌdē }

strongyle [INV ZOO] A monaxonic spicule rounded at each end. { 'strānˌjīl }

Strongyloidea [INV ZOO] The hookworms, an order or superfamily of roundworms which, as adults, are endoparasites of most vertebrates, including humans. { ˌstrān·jə'lȯid·ē·ə }

strongyloidiasis [MED] An infestation of humans with one of the roundworms of the genus *Strongyloides*. { ˌstrān·jəˌlȯi'dī·ə·səs }

strongylote [INV ZOO] Rounded at one end, referring to sponge spicules. { 'strān·jəˌlōt }

strophanthin [PHARM] A glycoside or mixture of glycosides extracted from the plant *Strophanthus kombe*; used as a cardioactive drug in the treatment of various heart ailments. { strə'fan·thən }

strophiole [BOT] A crestlike excrescence around the hilum in certain seeds. { 'strä·fēˌōl }

structural gene *See* cistron. { 'strək·chə·rəl 'gēn }

Struthionidae [VERT ZOO] The single family of the avian order Struthioniformes. { ˌstrü·thē'än·əˌdē }

Struthioniformes [VERT ZOO] A monofamilial order of ratite birds containing the single living species of ostrich (*Struthio camelus*). { ˌstrü·thēˌän·ə'fȯrˌmēz }

strychninization [MED] The condition resulting from large doses of strychnine. { ˌstrik·nə·nə'zā·shən }

Strychnos [BOT] A genus of tropical trees and shrubs of the order Loganiaceae. { 'strikˌnōs }

Stuart factor [BIOCHEM] A procoagulant in normal plasma but deficient in the blood of patients with a hereditary bleeding disorder; may be closely related to prothrombin since both are formed in the liver by action of vitamin K. Also known as factor X; Stuart-Power factor. { 'stü·ərt ˌfak·tər }

Stuart-Power factor *See* Stuart factor. { 'stü·ərt 'pau̇·ər ˌfak·tər }

sturgeon [VERT ZOO] Any of 10 species of large bottom-living fish which comprise the family Acipenseridae; the body has five rows of bony plates, and the snout is elongate with four barbels on its lower surface. { 'stər·jən }

stutter [MED] A speech disorder marked by repetition of words, syllables, or sounds, or by hesitations in manner by the speaker. { 'stəd·ər }

sty *See* hordeolum. { stī }

Styginae [INV ZOO] A subfamily of butterflies in the family Lycaenidae in which the prothoracic legs in the male are nonfunctional. { 'stij·ə·nē }

Stygocaridacea [INV ZOO] An order of crustaceans in the superorder Syncarida characterized by having a furca. { ˌstig·əˌkar·ə'dās·ē·ə }

Stylasterina [INV ZOO] An order of the class Hydrozoa, including several brightly colored branching or encrusting corallike cnidarians of warm seas. { stəˌlas·tə'rī·nə }

style [BOT] The portion of a pistil connecting the stigma and ovary. [ZOO] A slender elongated process on an animal. { stīl }

stylet [INV ZOO] A slender, rigid, elongated appendage. [MED] **1.** A slender probe used for surgery. **2.** A thin wire inserted in a catheter to provide support or in a hollow needle to clear the passage. { 'stī·lət }

styloglossus [ANAT] A muscle arising from the styloid process of the temporal bone, and inserted into the tongue. { ˌstī·lō'gläs·əs }

stylohyoid [ANAT] Pertaining to the styloid process of the temporal bone and the hyoid bone. { ¦stī·lō¦hīˌȯid }

styloid [ZOO] Resembling a style. { 'stīˌlȯid }

stylomastoid [ANAT] Relating to the styloid and the mastoid processes of the temporal bone. { ¦stī·lō'maˌstȯid }

Stylommatophora [INV ZOO] A large order of the molluscan subclass Pulmonata characterized by having two pairs of retractile tentacles with eyes located on the tips of the large tentacles. { stiˌlam·ə'täf·ə·rə }

stylopodium [BOT] A conical or disk-shaped enlargement at the base of the style in plants of the family Umbelliferae. { ˌstī·lə'pōd·ē·əm }

Styloviridae *See* Siphoviridae. { ˌstil·ə'vir·əˌdī }

Stypocapitellidae [INV ZOO] A family of polychaete annelids belonging to the Sedentaria and consisting of a monotypic genus found in western Germany. { ˌstī·pōˌkap·ə'tel·əˌdē }

styramate [PHARM] $C_9H_{11}NO_3$ A compound that crystallizes from chloroform solution, and melts at 111–112°C; used in medicine as a muscle relaxant. { 'stī·rəˌmāt }

subacute bacterial endocarditis *See* bacterial endocarditis. { ¦səb·ə'kyüt bak'tir·ē·əl ¦en·dō·kär'dīd·əs }

subalpine *See* alpestrine. { ¦səb'alˌpīn }

subarachnoid hemorrhage [MED] Bleeding between the pia mater and the arachnoid of the brain. { ¦səb·ə'rakˌnȯid 'hem·rij }

subarachnoid space [ANAT] The space between the pia mater and the arachnoid of the brain. { ¦səb·ə'rakˌnȯid ¦spās }

subboreal [ECOL] A biogeographic zone whose climatic condition approaches that of the boreal. { ¦səb'bȯr·ē·əl }

subcardinal vein [VERT ZOO] Either of a pair of longitudinal veins of the mammalian embryo or the adult of some lower vertebrates which partly replace the postcardinals in the abdominal region, ventromedial to the mesonephros. { ¦səb'kärd·nəl 'vān }

subclavian artery [ANAT] The proximal part of the principal artery in the arm or forelimb. { 'səb¦klā·vē·ən 'ärd·ə·rē }

subclavian vein [ANAT] The proximal part of the principal vein in the arm or forelimb. { 'səb¦klā·vē·ən 'vān }

subclavius [ANAT] A small muscle attached to the clavicle and the first rib. { ˌsəb'klā·vē·əs }

subclimax [ECOL] A community immediately preceding a climax in an ecological succession. { ¦səb'klīˌmaks }

subcollateral [ANAT] Ventrad of the collateral sulcus of the brain. { ¦səb·kə'lad·ə·rəl }

subcutaneous connective tissue [HISTOL] The layer of loose connective tissue beneath the dermis. { ¦səb·kyü'tā·nē·əs kə'nek·tiv ˌtish·ü }

subcutaneous emphysema [MED] The presence of air in the tissues just under the skin; when seen in diving it usually involves the skin of the neck and nearby areas. { ¦səb·kyü'tā·nē·əs ˌem·fə'sē·mə }

subcutaneous mycosis [MED] Any of a wide spectrum of infections caused by a heterogeneous group of fungi characterized by the development of lesions (that usually remain localized or spread slowly by direct extension via the lymphatics) at sites of inoculation and initially involve the deeper layers of the dermis and subcutaneous tissues but eventually extend into the epidermis. { ˌsəb·kyüˌtān·ē·əs mī'kō·səs }

subdominant [ECOL] A species which may appear more abundant at particular times of the year than the true dominant in a climax; for example, in a savannah trees and shrubs are more conspicuous than the grasses, which are the true dominants. { ¦səb'däm·ə·nənt }

subdural hematoma [MED] A mass of blood between the arachnoid and the dura mater. { səb'dủr·əl ˌhē·mə'tō·mə }

subdural hemorrhage [MED] Bleeding between the dura mater and the arachnoid. { səb'dủr·əl 'hem·rij }

subendothelial layer [HISTOL] The middle layer of the tunica intima of veins and of medium and larger arteries, consisting of collagenous and elastic fibers and a few fibroblasts. { ¦səbˌen·də'thē·lē·əl 'lā·ər }

suberin [BIOCHEM] A fatty substance found in many plant cell walls, especially cork. { 'sü·bə·rən }

suberization [BOT] Infiltration of plant cell walls by suberin resulting in the formation of corky tissue that is impervious to water. { ˌsü·bə·rə'zā·shən }

suberose [BOT] Having a texture like cork due to or resembling that due to suberization. { 'sü·bəˌrōs }

sublethal gene *See* semilethal gene. { ¦səb'lē·thəl 'jēn }

subleukemic [MED] Less than leukemic; usually applied to states in which the peripheral blood manifestations of leukemia are temporarily suppressed. { ¦səb·lü'kē·mik }

sublimed sulfur *See* flowers of sulfur. { sə'blīmd 'səl·fər }

subliminal [PHYSIO] Below the threshold of responsiveness, consciousness, or sensation to a stimulus. { sə'blim·ə·nəl }

sublingual gland [ANAT] A complex of salivary glands located in the sublingual fold on each side of the floor of the mouth. { ¦səb'liŋ·gwəl 'gland }

subluxation [MED] An incomplete dislocation. { ˌsəb·lək'sā·shən }

submandibular duct [ANAT] The duct of the submandibular gland which empties into the mouth on the side of the frenulum of the tongue. { ¦səb·man'dib·yə·lər 'dəkt }

submandibular gland [ANAT] A large seromucous or mixed salivary gland located below the mandible on each side of the jaw. Also known as mandibular gland; submaxillary gland. { ¦səb·man'dib·yə·lər 'gland }

submaxillary gland *See* submandibular gland. { ¦səb'mak·səˌler·ē ˌgland }

submerged culture [MICROBIO] A method for growing pure cultures of aerobic bacteria in which microorganisms are incubated in a liquid medium subjected to continuous, vigorous agitation. { səb'mərjd 'kəl·chər }

submerged fermentation [MICROBIO] Industrial production of antibiotics, enzymes, and other substances by growing the microorganisms that produce the product in a submerged culture. { səb'mərjd ˌfər·mən'tā·shən }

submucosa [HISTOL] The layer of fibrous connective tissue that attaches a mucous membrane to its subadjacent parts. { ¦səb·myü'kō·sə }

submucous plexus [NEUROSCI] A visceral nerve network lying in the submucosa of the digestive tube. Also known as Meissner's plexus. { ¦səb'myü·kəs 'plek·səs }

Suboscines [VERT ZOO] A major division of the order Passeriformes, usually divided into the suborders Eurylaimi, Tyranni, and Memirae. { ¦səb'äs·əˌnēz }

subscapularis [ANAT] A muscle arising from the costal surface of the scapula and inserted on the lesser tubercle of the humerus. { ˌsəbˌskap·yə'lar·əs }

subsere [ECOL] A secondary community that succeeds an interrupted climax. { 'səbˌsir }

subspecies [SYST] A geographically defined grouping of local populations which differs taxonomically from similar subdivisions of species. { ¦səb'spē·shēz }

substance P [BIOCHEM] An undecapeptide widely distributed in the central nervous system and found in highest concentrations in superficial layers of the dorsal horn of the spinal cord, in the trigeminal nerve nucleus, and in the substantia nigra; acts as a neurotransmitter. { 'səb·stəns 'pē }

substrain [CELL MOL] A strain derived by isolation of a single cell or group of cells having properties or markers not shared by the other cells of the cell strain. { 'səb·ˌstrān }

substrate [BIOCHEM] The substance with which an enzyme reacts. [ECOL] The foundation to which a sessile organism is attached. { 'səb ˌstrāt }

subtend [BOT] To lie adjacent to and below another structure, often enclosing it. { səb'tend }

subtilin [MICROBIO] An antibiotic substance obtained from *Bacillus subtilis*, active against gram-positive bacteria. { 'səb·tə·lən }

Subtriquetridae [INV ZOO] A family of arthropods in the suborder Porocephaloidea. { ˌsəb·trə'ke·trəˌdē }

subtropical forest *See* temperate rainforest. { ˌsəb'träp·ə·kəl 'fär·əst }

subulate [BOT] Linear, delicate, and tapering to a sharp point. { 'səb·yə·lət }

Subuluridae [INV ZOO] The equivalent name for Heterakidae. { ¦səb·yə'lu̇r·əˌdē }

Subuluroidea [INV ZOO] A superfamily of parasitic nematodes in the order Ascaridida characterized by weakly developed lips with sensilla and a thick-walled stoma that is armed with three teeth. { ˌsəb·yə·lə'ro̊id·ē·ə }

subumbrella [INV ZOO] The concave undersurface of the body of a jellyfish. { ¦səb·əm'brel·ə }

subunit *See* protomer. { 'səbˌyü·nət }

subunit vaccine [IMMUNOL] A type of noninfectious vaccine that consists of immunogenic viral proteins stripped free from whole virus particles, then purified from other irrelevant components, thereby reducing the risk of adverse reactions and residual infectious virus. { ˌsəbˌyü·nit vak'sēn }

succession [ECOL] A gradual process brought about by the change in the number of individuals of each species of a community and by the establishment of new species populations which may gradually replace the original inhabitants. { sək'sesh·ən }

succinamide [BIOCHEM] $H_2NCOCH_2CONH_2$ The amide of succinic acid. { sək'sin·ə·mīd }

succinate dehydrogenase [BIOCHEM] A key enzyme in the citric acid cycle; it oxidizes succinate to fumarate. { ¦sək·səˌnāt ˌdē·hī'dräj·əˌnās }

succinic acid dehydrogenase [BIOCHEM] An enzyme that catalyzes the dehydrogenation of succinic acid to fumaric acid in the presence of a hydrogen acceptor. Also known as succinic dehydrogenase. { sək'sin·ik 'as·əd dē'hī·drə·jəˌnās }

succinic dehydrogenase *See* succinic acid dehydrogenase. { sək'sin·ik dē'hī·drə·jəˌnās }

succinoxidase [BIOCHEM] A complex enzyme system containing succinic dehydrogenase and cytochromes that catalyze the conversion of succinate ion and molecular oxygen to fumarate ion. { ˌsək·sən'äk·səˌdās }

succinyldicholine [PHARM] A drug that is used to produce relaxation of muscle during surgery, it is hydrolyzed by serum cholinesterase rather than by acetylcholinesterase, limiting the duration of paralysis. { sək,sin·əl·dī'kō,lēn }

succinylsulfathiazole [PHARM] $C_{13}H_{13}N_3O_5S_2$ A poorly absorbed sulfonamide used as an intestinal antibacterial agent in preoperative preparation of patients for abdominal surgery, and also postoperatively to maintain a low bacterial count. { ¦sək·sən·əl,səl·fə'thī·ə,zōl }

succulent [BOT] Describing a plant having juicy, fleshy tissue. { 'sək·yə·lənt }

succus entericus [PHYSIO] The intestinal juice secreted by the glands of the intestinal mucous membrane; it is thin, opalescent, alkaline, and has a specific gravity of 1.011. { 'sək·əs in'ter·ə·kəs }

sucker [BOT] A shoot that develops rapidly from the lower portion of a plant, and usually at the expense of the plant. [ZOO] A disk-shaped organ in various animals for adhering to or holding onto an individual, usually of another species. { 'sək·ər }

sucking louse [INV ZOO] The common name for insects of the order Anoplura, so named for the slender, tubular mouthparts. { 'sək·iŋ ,laủs }

sucrase *See* saccharase. { 'sü,krās }

Suctoria [INV ZOO] A small subclass of the protozoan class Ciliatea, distinguished by having tentacles which serve as mouths. { sək'tȯr·ē·ə }

Suctorida [INV ZOO] The single order of the protozoan subclass Suctoria. { sək'tȯr·əd·ə }

sudamen [MED] A skin disease in which sweat accumulates under the superficial horny layers of the epidermis to form small, clear, transparent vesicles. { sü'dā·mən }

sudatoria *See* hyperhidrosis. { ,süd·ə'tȯr·ē·ə }

sudden death syndrome *See* sudden infant death syndrome. { 'səd·ən 'deth ,sin,drōm }

sudden infant death syndrome [MED] The sudden and unexpected death of an apparently normal infant that remains unexplained after the performance of an adequate autopsy. Abbreviated SIDS. Also known as crib death; sudden death syndrome. { 'səd·ən 'in·fənt ¦deth 'sin ,drōm }

sudomotor [NEUROSCI] Pertaining to the efferent nerves that control the activity of sweat glands. { ¦süd·ə'mōd·ər }

suffrutescent [BOT] Of or pertaining to a stem intermediate between herbaceous and shrubby, becoming partly woody and perennial at the base. { ¦sə,frü'tes·ənt }

suffruticose [BOT] Low stems which are woody, grading into herbaceous at the top. { sə'früd·ə,kōs }

sugar [BIOCHEM] A generic term for a class of carbohydrates usually crystalline, sweet, and water soluble; examples are glucose and fructose. { 'shu̇g·ər }

sugarbeet [BOT] *Beta vulgaris*. A beet characterized by a white root and cultivated for the high sugar content of the roots. { 'shu̇g·ər,bēt }

sugarcane [BOT] *Saccharum officinarum*. A stout, perennial grass plant characterized by two-ranked leaves, and a many-jointed stalk with a terminal inflorescence in the form of a silky panicle; the source of more than 50% of the world's annual sugar production. { 'shu̇g·ər,kān }

sugar maple [BOT] *Acer saccharum*. A commercially important species of maple tree recognized by its gray furrowed bark, sharp-pointed scaly winter buds, and symmetrical oval outline of the crown. { 'shu̇g·ər 'mā·pəl }

suicide [IMMUNOL] Death of cells that have selectively taken up heavily radioactively labeled antigen. { 'sü·ə,sīd }

suicide inhibitor [BIOCHEM] A compound which resembles the normal substrate for an enzyme, but which interacts with the enzyme to form a covalent bond and thus inactivates the enzyme. { 'sü·ə,sīd in'hib·əd·ər }

Suidae [VERT ZOO] A family of paleodont artiodactyls in the superfamily Suoidae including wild and domestic pigs. { 'sü·ə,dē }

sulcate [ZOO] Having furrows or grooves on the surface. { 'səl,kāt }

sulculus [ZOO] A small sulcus. { 'səl·kyə·ləs }

sulcus [ZOO] A furrow or groove, especially one on the surface of the cerebrum. { 'səl·kəs }

sulfadiazine [PHARM] $C_{10}H_{10}O_2N_4S$ An antibacterial sulfonamide used in the treatment of a variety of infections. { ¦səl·fə'dī·ə,zēn }

sulfa drug [PHARM] Any of a family of drugs of the sulfonamide type with marked bacteriostatic properties. { 'səl·fə ,drəg }

sulfaguanidine [PHARM] $C_7H_{10}N_4O_2S$ An intestinal antibacterial sulfonamide proposed for treatment of dysentery and for sterilization of the colon prior to gastrointestinal tract surgery. { ¦səl·fə'gwän·ə,dēn }

sulfamerazine [PHARM] $C_{11}H_{12}N_4O_2S$ An antibacterial agent with uses similar to those of sulfadiazine, but generally used in combination with sulfadiazine and with sulfamethazine. { ¦səl·fə'mer·ə,zēn }

sulfanilamide [PHARM] $C_6H_8O_2N_2S$ White crystals slightly soluble in water, soluble in alcohol and most sulfa drugs, but less effective and more toxic than its derivatives. { ,səl·fə'nil·ə,mīd }

sulfapyridine [PHARM] $C_{11}H_{11}N_3O_2S$ A sulfonamide formerly used for the treatment of various infections but found to be too toxic for general use; now employed only as a suppressant for dermatitis herpetiformis. { ¦səl·fə'pir·ə,dēn }

sulfatase [BIOCHEM] Any of a group of esterases that catalyze the hydrolysis of sulfuric esters. { 'səl·fə,tās }

sulfathiazole [PHARM] $C_9H_9N_3O_2S_2$ A sulfa drug formerly widely used in the treatment of pneumococcal, staphylococcal, and urinary tract infections; it has been replaced by less toxic sulfonamides. { ¦səl·fə'thī·ə,zōl }

sulfatide lipidosis *See* metachromatic leukodystrophy. { 'səl·fə,tīd ,lip·ə'dō·səs }

sulf-heme protein [BIOCHEM] A heme protein that has reacted with sulfur to yield a new structure. { 'səlf ,hēm 'prō,tēn }

sulfhemoglobin [BIOCHEM] A greenish substance derived from hemoglobin by the action of hydrogen sulfide; it may appear in the blood following the ingestion of sulfanilamide and other substances. { ¦səlf'hē·mə,glō·bən }

sulfidogen [MICROBIO] A strict anaerobe that reduces sulfur to hydrogen sulfide. { səl'fīd·ə,jen }

sulfisoxazole [PHARM] $C_{11}H_{13}N_3O_3S$ A sulfonamide of general therapeutic utility; for parenteral administration the soluble salt sulfisoxazole diethanolamine is used; for pediatric use the tasteless derivative acetyl sulfisoxazole is given. { ,səl·fə'säk·sə,zōl }

sulfobromophthalein sodium [PHARM] $C_{20}H_8Br_4Na_2O_{10}S_2$ A hygroscopic, crystalline compound that has a bitter taste; soluble in water; used in humans and animals as a diagnostic and in liver function tests. { ¦səl·fō,brō·mō'thā,lēn 'sōd·ē·əm }

Sulfolobus [MICROBIO] A genus of bacteria that is gram-negative, coccoid, chemolithotrophic, and thermoacidophilic. It is found worldwide in sulfur-rich hot springs and oxidizes sulfur for energy production. Its cells are highly irregular in shape, often lobed, but occasionally spherical. { səl'fäl·ə·bəs }

4,4′-sulfonyldianiline [PHARM] $C_{12}H_{12}N_2O_2$ S A sulfone that precipitates as crystals from alcohol; melting point 175–176°C; used in the treatment of leprosy. Also known as dapsone. { ¦fór ,fór,prīm ¦səl·fə,nil·dī'an·ə·lən }

sulfur bacteria [MICROBIO] Any of various bacteria having the ability to oxidize sulfur compounds. { 'səl·fər bak,tir·ē·ə }

Sulidae [VERT ZOO] A family of aquatic birds in the order Pelecaniformes including the gannets and boobies. { 'sü·lə,dē }

summation wave [PHYSIO] A sustained contraction of muscles, caused by the rapid firing of nerve impulses. { sə'mā·shən ,wāv }

summerwood [BOT] The less porous, usually harder portion of an annual ring that forms in the latter part of the growing season. { 'səm·ər,wu̇d }

sunburn [MED] Skin inflammation due to overexposure to sunlight. { 'sən,bərn }

sundew [BOT] Any plant of the genus *Drosera* of the family Droseraceae; the genus comprises small, herbaceous, insectivorous plants that grow on all continents, especially Australia. { 'sən,dü }

sunfish [VERT ZOO] Any of several species of marine and freshwater fishes in the families Centrarchidae and Molidae characterized by brilliant metallic skin coloration. { 'sən,fish }

sunflower [BOT] *Helianthus annuus*. An annual plant native to the United States characterized by broad, ovate leaves growing from a single, usually long (3–20 feet or 1–6 meters) stem, and large, composite flowers with yellow petals. { 'sən,flau̇·ər }

sunstroke [MED] Heat stroke resulting from prolonged exposure to the sun, characterized by extreme pyrexia, prostration, convulsion, and coma. Also known as thermic fever. { 'sən ,strōk }

Suoidea [VERT ZOO] A superfamily of artiodactyls of the suborder Paleodonta which comprises the pigs and peccaries. { sü'òid·ē·ə }

supercilium [ANAT] The eyebrow. { ,sü·pər 'sil·ē·əm }

supercoiling [CELL MOL] Winding of the deoxyribonucleic acid duplex on itself so that it crosses its own axis; may be in the same (positive) direction as, or opposite (negative) direction to, the turns of the double helix. { 'sü·pər,kȯil·iŋ }

superfecundation [PHYSIO] Multiple, simultaneous fertilization by a number of sperm of many eggs released at ovulation. { ¦sü·pər,fē,kən 'dā·shən }

superfemale [GEN] In *Drosophila*, a female with three X chromosomes and two sets of autosomes resulting in sterility and generally early death. In humans, XXX females are fertile. { ¦sü·pər'fē,māl }

superficial cleavage [EMBRYO] Meroblastic cleavage restricted to the peripheral cytoplasm, as in the centrolecithal insect ovum. { ¦sü·pər¦fish·əl 'klē·vij }

superficial palmar arch [ANAT] The arterial anastomosis formed by the ulnar artery in the palm with a branch from the radial artery. Also known as palmar arch. { ¦sü·pər¦fish·əl 'päm·ər 'ärch }

supergene [GEN] A chromosomal segment protected from crossing over and therefore transmitted from generation to generation as if it were a single recon. { 'sü·pər,jēn }

superhelix [BIOCHEM] A macromolecular structure consisting of a number of alpha-helical polypeptide strands which are twisted together. { ¦sü·pər'hē,liks }

superinfection [VIROL] An attack on a bacterial cell by several phages due to the introduction of large numbers of viruses into the bacterial culture. { ¦sü·pər·in'fek·shən }

superior [BOT] **1.** Positioned above another organ or structure. **2.** Referring to a calyx that is attached to the ovary. **3.** Referring to an ovary that is above the insertion of the floral parts. { sə'pir·ē·ər }

superior alveolar canals [ANAT] The alveolar canals of the maxilla. { sə'pir·ē·ər al'vē·ə·lər kə'nalz }

superior fruit *See* true fruit. { sə,pir·ē·ər 'früt }

superior ganglion [NEUROSCI] **1.** The upper sensory ganglion of the glossopharyngeal nerve, located in the upper part of the jugular foramen; it is inconstant. **2.** The upper sensory ganglion of the vagus nerve, located in the jugular foramen. { sə'pir·ē·ər 'gaŋ·glē·ən }

superior mesenteric artery [ANAT] A major branch of the abdominal aorta with branches to the pancreas and intestine. { sə'pir·ē·ər ,mez·ən'ter·ik 'ärd·ə·rē }

superior vena cava [ANAT] The principal vein collecting blood from the head, chest wall, and upper extremities and draining into the right atrium. { sə'pir·ē·ər ¦vē·nə 'kā·və }

supermale [GEN] In *Drosophila*, a male with one

X chromosome and three or more sets of autosomes, resulting in sterility and generally early death. { 'sü·pər,māl }

supernatant [BIOCHEM] The overlying soluble liquid fraction of a sample that remains after precipitation of the insoluble solid component by centrifugation. { ,sü·pər'nāt·ənt }

supernumerary bud *See* accessory bud. { ¦sü·pər'nü·mə,rer·ē 'bəd }

supernumerary chromosome [CELL MOL] A chromosome present in addition to the normal chromosome complement. Also known as accessory chromosome. { ¦sü·pər'nü·mə,rer·ē 'krō·mə,sōm }

superposed [BOT] **1.** Growing vertically over another part. **2.** Of or pertaining to floral parts that are opposite each other. { ¦sü·pər'pōzd }

superposition eye [INV ZOO] A compound eye in which a given rhabdome receives light from a number of facets; visual acuity is reduced in this type of eye. { ,sü·pər·pə'zish·ən 'ī }

supination [ANAT] **1.** Turning the palm upward. **2.** Inversion of the foot. { ,sü·pə'nā·shən }

suppository [PHARM] A medicated solid body of varying weight and shape, intended for introduction into different orifices of the body, as the rectum, urethra, or vagina; usually suppositories melt or are softened at body temperature, though in some instances release of medication is effected through use of a hydrophilic vehicle; typical vehicles or bases are theobroma oil (cocoa butter), glycerinated gelatin, sodium stearate, and propylene glycol monostearate. { sə'päz·ə,tȯr·ē }

suppressor gene [GEN] A gene that reverses the effect of a mutation in another gene. { sə'pres·ər ,jēn }

suppressor mutation [GEN] A mutation that restores the function lost after a primary mutation at a different locus. { sə'pres·ər myü,tā·shən }

suppuration [MED] Pus formation. { ,səp·yə'rā·shən }

supracardinal veins [VERT ZOO] Paired longitudinal veins in the mammalian embryo and various adult lower vertebrates in the thoracic and abdominal regions, dorsolateral to and on the sides of the descending aorta; they replace the postcardinal and subcardinal veins. { ¦sü·prə'kärd·nəl ,vānz }

suprahyoid muscles [ANAT] The muscles attached to the upper margin of the hyoid bone. { ¦sü·prə'hī,ȯid ,məs·əlz }

supraliminal [PHYSIO] Above, or in excess of, a threshold. { ¦sü·prə'lim·ə·nəl }

supranuclear [ANAT] In the nervous system, central to a nucleus. { ¦sü·prə'nü·klē·ər }

supraoccipital [ANAT] Situated above the occipital bone. { ¦sü·prə·äk'sip·əd·əl }

supraoptic [ANAT] Situated above the optic tract. { ¦sü·prə¦äp·tik }

suprarenal gland *See* adrenal gland. { ¦sü·prə'rēn·əl ,gland }

suprascapula [ANAT] An anomalous bone sometimes found between the superior border of the scapula and the spines of the lower cervical or first thoracic vertebrae. { ¦sü·prə'skap·yə·lə }

suprasegmental reflex [PHYSIO] A reflex employing complex multineuronal channels to integrate the body and limb musculature with fixed positions or movements of the head. { ¦sü·prə·seg'ment·əl 'rē,fleks }

suprasternal notch [ANAT] Jugular notch of the sternum. { ¦sü·prə'stərn·əl 'näch }

suprasternal space [ANAT] The triangular space above the manubrium, enclosed by the layers of the deep cervical fascia which are attached to the front and back of this bone. { ¦sü·prə'stərn·əl 'spās }

supravital [BIOL] Pertaining to the staining of living cells after removal from a living animal or of still living cells from a recently killed animal. { ¦sü·prə'vīd·əl }

surgery [MED] The branch of medicine that deals with conditions requiring operative procedures. { 'sər·jə·rē }

surgical debridement [MED] The removal of foreign material and devitalized tissue using a scalpel or other sharp instrument. { ,sər·ji·kəl də'brēd·mənt }

surgical needle [MED] Any sewing needle used in a surgical operation. { 'sər·jə·kəl 'nēd·əl }

survival ratio [BIOL] The number of organisms surviving irradiation by ionizing radiation divided by the number of organisms before irradiation. { sər'vī·vəl ,rā·shō }

suspension feeder [ZOO] An animal that feeds on small particles suspended in water; particles may be minute living plants or animals, or products of excretion or decay from these or larger organisms. { sə'spen·shən ¦fēd·ər }

suspensor [BOT] A mass of cells in higher plants that pushes the embryo down into the embryo sac and into contact with the nutritive tissue. [MYCOL] A hypha which bears an apical gametangium in fungi of the Mucorales. { sə'spen·sər }

sustained yield [BIOL] In a biological resource such as timber or grain, the replacement of a harvest yield by growth or reproduction before another harvest occurs. { sə'stānd 'yēld }

sustentacular cell [HISTOL] One of the supporting cells of an epithelium as contrasted with other cells with special function, as the nonnervous cells of the olfactory epithelium or the Sertoli cells of the seminiferous tubules. { ¦səs·tən¦tak·yə·lər ,sel }

suture [BIOL] A distinguishable line of union between two closely united parts. [MED] A fine thread used to close a wound or surgical incision. { 'sü·chər }

swamp [ECOL] A waterlogged land supporting a natural vegetation predominantly of shrubs and trees. { swämp }

swamp fever *See* infectious anemia. { 'swämp ,fē·vər }

swamp itch *See* schistosome dermatitis. { 'swämp ,ich }

swan [VERT ZOO] Any of several species of large

waterfowl comprising the subfamily Anatinae; they are herbivorous stout-bodied forms with long necks and spatulate bills. { swän }

swarmer cell [MICROBIO] The daughter cell which separates from the stalked mother cell in bacteria in the genus *Caulobacter.* { 'swȯr·mər ˌsel }

swayback [MED] Increased lumbar lordosis with compensatory increased thoracic kyphosis. [VET MED] Sinking of the back, or lordosis. { 'swāˌbak }

sweat [PHYSIO] The secretion of the sweat glands. Also known as perspiration. { swet }

sweat gland [PHYSIO] A coiled tubular gland of the skin which secretes sweat. { 'swet ˌgland }

sweepstakes route [ECOL] A means that allows chance migration across a sea on natural rafts, so that oceanic islands can be colonized. { 'swēpˌstāks ˌrüt }

sweetgum [BOT] *Liquidambar styraciflua.* A deciduous tree of the order Hamamelidales found in the southeastern United States, and distinguished by its five-lobed, or star-shaped, leaves, and by the corky ridges developed on the twigs. { 'swētˌgəm }

sweet potato [BOT] *Ipomoea batatas.* A tropical vine having variously shaped leaves, purplish flowers, and a sweet, fleshy, edible tuberous root. { 'swēt pəˌtād·ō }

swim bladder [VERT ZOO] A gas-filled cavity found in the body cavities of most bony fishes; has various functions in different fishes, acting as a float, a lung, a hearing aid, and a sound-producing organ. { 'swim ˌblad·ər }

swimmeret [INV ZOO] Any of a series of paired biramous appendages under the abdomen of many crustaceans, used for swimming and egg carrying. { ˌswim·əˌret }

swimmer's conjunctivitis *See* inclusion conjunctivitis. { 'swim·ərz kənˌjəŋk·tə'vīd·əs }

swimmer's itch *See* schistosome dermatitis. { 'swim·ərz 'ich }

swimming bird [VERT ZOO] Any bird belonging to the orders Charadriiformes and Pelacaniformes. { 'swim·iŋ ˌbird }

swimming-pool conjunctivitis *See* inclusion conjunctivitis. { 'swim·iŋ ¦pül kənˌjəŋk·tə'vīd·əs }

swine [VERT ZOO] Any of various species comprising the Suidae. { swīn }

swine erysipelas [VET MED] An infectious bacterial disease of swine caused by *Erysipelothrix insidiosa* in which involvement of the skin is predominant. { 'swin ˌer·ə'sip·ə·ləs }

swine influenza [VET MED] A disease of swine caused by the associated effects of a filterable virus and *Hemophilus suis*, characterized by inflammation of the upper respiratory tract. { 'swin ˌin·flü'en·zə }

swine plague [VET MED] Hemorrhagic septicemia of swine caused by *Pasteurella suiseptica*, characterized by pleuropneumonia. { 'swin ¦plāg }

swine pox [VET MED] A benign infection of young hogs characterized by pox lesions on the body and inner surfaces of the legs. { 'swin ¦päks }

switch gene [GEN] A gene that causes the epigenotype to switch to a different developmental pathway. { 'swich ˌjēn }

sycamore [BOT] **1.** Any of several species of deciduous trees of the genus *Platanus*, especially *P. occidentalis* of eastern and central North America, distinguished by simple, large, three- to five-lobed leaves and spherical fruit heads. **2.** The Eurasian maple (*Acer pseudoplatanus*). { 'sik·əˌmȯr }

Sycettida [INV ZOO] An order of calcareous sponges of the subclass Calcaronea in which choanocytes occur in flagellated chambers, and the spongocoel is not lined with these cells. { sə'sed·əd·ə }

Sycettidae [INV ZOO] A family of sponges in the order Sycettida. { sə'sed·əˌdē }

sycon [INV ZOO] A canal system in sponges in which the flagellated layer is confined to outpocketings of the paragaster that are indirectly connected to the incurrent canals. { 'sīˌkän }

syconium [BOT] A fleshy fruit, as a fig, with an enlarged pulpy receptacle internally lined with minute flowers. { sī'kō·nē·əm }

sycosis [MED] An inflammatory disease affecting the hair follicles, particularly of the beard, and characterized by papules, pustules, and tubercles, perforated by hairs, together with infiltration of the skin and crusting. { sī'kō·səs }

Sydenham's chorea *See* Saint Vitus dance. { 'sīd·ənˌhamzkə'rē·ə }

Syllidae [INV ZOO] A large family of polychaete annelids belonging to the Errantia; identified by their long, linear, translucent bodies with articulated cirri; size ranges from minute *Exogone* to *Trypanosyllis*, which may be 4 inches (100 millimeters) long. { 'sil·əˌdē }

Syllinae [INV ZOO] A subfamily of polychaete annelids of the family Syllidae. { 'sil·əˌnē }

Sylonidae [INV ZOO] A family of parasitic crustaceans in the order Rhizocephala. { sə'län·əˌdē }

Sylopidae [INV ZOO] A family of coleopteran insects in the superfamily Meloidea in which the elytra in males are reduced to small leathery flaps while the hindwings are large and fan-shaped. { sə'läp·əˌdē }

sylvatic plague [VET MED] Plague occurring in rodents; may be transmitted to humans. Also known as endemic rural plague. { sil'vad·ik 'plāg }

Sylvicolidae [INV ZOO] A family of orthorrhaphous dipteran insects in the series Nematocera. { ˌsil·və'käl·əˌdē }

symbiont [ECOL] A member of a symbiotic pair. { 'sim·bēˌänt }

symbiosis [ECOL] **1.** An interrelationship between two different species. **2.** An interrelationship between two different organisms in which the effects of that relationship is expressed as being harmful or beneficial. Also known as consortism. { ˌsim·bē'ō·səs }

symblepharon [MED] Adhesion of the eyelids to the eyeball. { sim'blef·əˌrän }

symmetry [BIOL] The disposition of organs and

other constituent parts of the body of living organisms with respect to imaginary axes. { 'sim·ə,trē }

sympathetic nervous system [NEUROSCI] The portion of the autonomic nervous system, innervating smooth muscle and glands of the body, which upon stimulation produces a functional state of preparation for flight or combat. { ,sim·pə'thed·ik 'nər·vəs ,sis·təm }

sympathetic ophthalmia [MED] A granulomatous inflammation of the uveal tract following ocular injury or intraocular surgery. { ,sim·pə 'thed·ik äf'thal·mē·ə }

sympathicotropic cell [HISTOL] Any of various cells possessing special affinity for the sympathetic nervous system. { sim¦path·ə·kō¦träp·ik 'sel }

sympathochromaffin cell [NEUROSCI] One of the precursors of sympathetic and medullary cells in the adrenal medulla. { ¦sim·pə·thō·krō 'maf·ən ,sel }

sympatholitic [PHARM] Of or pertaining to an effect antagonistic to that of the sympathetic nervous system. { ¦sim·pə·thō¦lid·ik }

sympathomimetic [PHARM] Having the ability to produce physiologic changes similar to those caused by action of the sympathetic nervous system. { ¦sim·pə·thō·mə'med·ik }

sympatric [ECOL] Of a species, occupying the same range as another species but maintaining identity by not interbreeding. { sim'pa·trik }

sympatric speciation [EVOL] Speciation that occurs without geographic isolation of a population. { sim¦pa·trik ,spē·shē'ā·shən }

sympetalous *See* gamopetalous. { sim'ped·əl·əs }

symphile [ECOL] An organism, usually a beetle, living as a guest in the nest of a social insect, such as an ant, where it is reared and bred in exchange for its exudates. { 'sim,fīl }

Symphyla [INV ZOO] A class of the Myriapoda comprising tiny, pale, centipedelike creatures which inhabit humus or soil. { 'sim·fə·lə }

symphysis [ANAT] An immovable articulation of bones connected by fibrocartilaginous pads. { 'sim·fə·səs }

symphysis pubis *See* pubic symphysis. { 'sim·fə·səs 'pyü·bəs }

Symphyta [INV ZOO] A suborder of the Hymenoptera including the sawflies and horntails characterized by a broad base attaching the abdomen to the thorax. { 'sim·fəd·ə }

sympodium [BOT] A branching system in trees in which the main axis is composed of successive secondary branches, each representing the dominant fork of a dichotomy. { sim'pōd·ē·əm }

symporter [CELL MOL] A channel protein that simultaneously transports two different types of substrates (for example, sodium ion plus glucose) across a cell membrane, both in the same direction. { sim'pȯrd·ər }

symptom [MED] A phenomenon of physical or mental disorder or disturbance which leads to complaints on the part of the patient. { 'sim·təm }

symptomatology [MED] **1.** The science of symptoms. **2.** In common usage, the symptoms of disease taken together as a whole. { ,sim·tə·mə'täl·ə·jē }

Synallactidae [INV ZOO] A family of echinoderms of the order Aspidochirotida comprising mainly deep-sea forms which lack tentacle ampullae. { ,sin·ə'lak·tə,dē }

synandrous [BOT] Having several united stamens. { sə'nan·drəs }

synangium [BOT] A compound sorus made up of united sporangia. [VERT ZOO] In lower vertebrates, a peripheral arterial trunk from which branches arise. { sə'nan·jē·əm }

Synanthae [BOT] An equivalent name for Cyclanthales. { sə'nan,thē }

Synanthales [BOT] An equivalent name for Cyclanthales. { ,sin·ən'thā·lēz }

synapomorphy [SYST] A derived trait shared by two or more taxa that is believed to reflect their shared ancestry. { si'nap·ə,mȯr·fē }

synapse [NEUROSCI] A site where the axon of one neuron comes into contact with and influences the dendrites of another neuron or a cell body. { 'si,naps }

synapsis [CELL MOL] Pairing of homologous chromosomes during the zygotene stage of meiosis. { sə'nap·səs }

synaptic transmission [NEUROSCI] The mechanisms by which a presynaptic neuron influences the activity of an anatomically adjacent postsynaptic neuron. { sə'nap·tik tranz'mish·ən }

synapticulum [INV ZOO] A conical or cylindrical supporting process, as those extending between septa in some corals, or connecting gill bars in *Branchiostoma*. { ,sin·ap'tik·yə·ləm }

synaptic vesicle [NEUROSCI] A small membrane-bound structure in the axon terminals of nerve cells that contains neurotransmitters and releases them by exocytosis when an action potential reaches the terminal. { si¦nap·tik 'ves·ə·kəl }

Synaptidae [INV ZOO] A family of large sea cucumbers of the order Apodida lacking a respiratory tree and having a reduced water-vascular system. { sə'nap·tə,dē }

synaptinemal complex [CELL MOL] Ribbonlike structures that extend the length of synapsing chromosomes and are believed to function in exchange pairing. { sə,nap·tə'nē·məl 'käm ,pleks }

synarthrosis [ANAT] An articulation in which the connecting material (fibrous connective tissue) is continuous, immovably binding the bones. { ¦sin·är'thrō·səs }

Synbranchiformes [VERT ZOO] An order of eellike actinopterygian fishes that, unlike true eels, have the premaxillae present as distinct bones. { sin,braŋ·kə'fȯr,mēz }

Synbranchii [VERT ZOO] The equivalent name for Synbranchiformes. { sin'braŋ·kē,ī }

Syncarida [INV ZOO] A superorder of crustaceans of the subclass Melacostraca lacking a carapace and oostegites and having exopodites on all thoracic limbs. { siŋ'kar·əd·ə }

syncarp [BOT] A compound fleshy fruit. { 'sin,kärp }

syncarpous [BOT] Descriptive of a gynoecium having the carpels united in a compound ovary. { sin'kär·pəs }

synchondrosis [ANAT] A type of synarthrosis in which the bone surfaces are connected by cartilage. { ,sin·kän'drō·səs }

synchorology [ECOL] A study which involves the distribution ranges of plant communities, phytosociological zones, vegetation and geographical complexes, dissemination spectra, and current plant migration patterns. { ,sin·kə'räl·ə·jē }

synchronous growth [MICROBIO] A population of bacteria in which all cells divide at approximately the same time. { 'siŋ·krə·nəs 'grōth }

syncope [MED] Swooning or fainting; temporary suspension of consciousness. { 'siŋ·kə·pē }

syncytial trophoblast *See* syncytiotrophoblast. { sin'sish·əl 'träf·ə,blast }

syncytiotrophoblast [CELL MOL] An irregular sheet or net of deeply staining cytoplasm in which nuclei are irregularly scattered. Also known as plasmoditrophoblast; syncytial trophoblast. { sin¦sish·ē·ō'träf·ə,blast }

syncytium [CELL MOL] A mass of multinucleated cytoplasm without division into separate cells. Also known as polykaryocyte. [INV ZOO] Multinucleated cell or gland. { sin'sish·ē·əm }

syndactyly [ANAT] The condition characterized by union of two or more digits, as in certain birds and mammals; it is a familial anomaly in humans. { sin'dakt·əl·ē }

syndesmosis [ANAT] An articulation in which the bones are joined by collagen fibers. { ,sin ,dez'mō·səs }

syndrome [MED] A group of signs and symptoms which together characterize a disease. Also known as complex. { 'sin,drōm }

syndromic hearing loss [MED] Hearing loss that occurs in the presence of one or more other symptoms. { sin,drōm·ik 'hir·iŋ ,lȯs }

syndynamics [ECOL] The study of the causes of and trends in successional changes within a plant community. { ¦sin·dī'nam·iks }

synecology [ECOL] The study of environmental relations of groups of organisms, such as communities. { ¦sin·i'käl·ə·jē }

Synentognathi [VERT ZOO] The equivalent name for Beloniformes. { ,sin·ən'täg·nə·thē }

synergid [BOT] Either of two small cells lying in the embryo sac in seed plants adjacent to the egg cell toward the micropylar end. { sə'nər·jəd }

synergism [ECOL] An ecological association in which the physiological processes or behavior of an individual are enhanced by the nearby presence of another organism. { 'sin·ər,jiz·əm }

synergist [ANAT] A muscle that assists a prime mover muscle in performing a specific action. { 'sin·ər,jist }

synergy [PHARM] Suppression of a strain of infectious microbes by concentrations of two or more drugs which are not active singly. { 'sin·ər·jē }

Syngamidae [INV ZOO] A family of roundworms belonging to the Strongyloidea and including parasites of birds and mammals. { siŋ'gam·ə,dē }

syngamy [BIOL] Sexual reproduction involving union of gametes. { 'siŋ·gə·mē }

syngeneic *See* isogeneic. { ,sin·jə'nē·ik }

syngenesious [BOT] Pertaining to an aggregate of stamens fused at the anthers. { ¦sin·jə¦nē·zhəs }

Syngnathidae [VERT ZOO] A family of fishes in the order Gasterosteiformes including the seahorses and pipefishes. { siŋ'nath·ə,dē }

synkinesia [PHYSIO] Involuntary movement coincident with purposeful movements carried out by a distant part of the body, such as swinging the arms while walking. Also known as accessory movement; associated automatic movement. { ¦sin,kī¦nē·zhə }

synonym [SYST] A taxonomic name that is rejected as being incorrectly applied, or incorrect in form, or not representative of a natural genetic grouping. { 'sin·ə,nim }

synonymous substitution *See* isonymous substitution. { si·'nän·ə·məs ,səb·stə'tü·shən }

synopsis [SYST] In taxonomy, a brief summary of current knowledge about a taxon. { sə'näp·səs }

synorchidism [MED] Partial or complete fusion of the two testes within the abdomen or scrotum. { sə'nȯr·kə,diz·əm }

synostosis [ANAT] A type of synarthrosis in which the bones are continuous. [MED] Union of originally separate bones into a single bone structure. { si,nä'stō·səs }

synovia *See* synovial fluid. { sə'nō·vē·ə }

synovial fluid [PHYSIO] A transparent viscid fluid secreted by synovial membranes. Also known as synovia. { sə'nō·vē·əl 'flü·əd }

synovial membrane [HISTOL] A layer of connective tissue which lines sheaths of tendons at freely moving articulations, ligamentous surfaces of articular capsules, and bursae. { sə'nō·vē·əl 'mem,brān }

synovitis [MED] Inflammation of a synovial membrane. { ,sin·ə'vīd·əs }

synpelmous [VERT ZOO] Having the two main flexor tendons of the toes united beyond the branches that go to each digit, as in certain birds. { sin'pel·məs }

synphylogeny [ECOL] The study of the trends and changes in plant communities through historical and evolutionary perspectives. { ¦sin·fə'läj·ə·nē }

synphysiology [ECOL] The study of the metabolic processes of plant communities or species which constantly compete with each other, by investigating water needs, transpiration, assimilation and production or organic matter, physiological effects of light, temperature, root exudates, and various other ecological factors. { ¦sin,fiz·ē'äl·ə·jē }

synsepalous *See* gamosepalous. { sin'sep·ə·ləs }

Synteliidae [INV ZOO] The sap-flow beetles, a small family of coleopteran insects in the superfamily Histeroidea. { ,sint·əl'ī·ə,dē }

syntenic group [GEN] The loci on one chromosome in the complement, whether or not they show linkage in family studies (pedigree analysis). { sin'ten·ik ,grüp }

Syntexidae [INV ZOO] A family of the Hymenoptera in the superfamily Siricoidea. { sin'tek·sə,dē }

synthetase *See* ligase. { 'sin·thə,tās }

synthetic chromosome *See* artificial chromosome. { sin¦thed·ik 'krō·mə,sōm }

syntrophism [BIOL] Mutual dependence of cells for nutritional needs, especially between strains of bacteria. { 'sin·trə,fiz·əm }

syntrophoblast [EMBRYO] The outer synctial layer of the trophoblast that forms the outermost fetal element of the placenta. { sin'träf·ə,blast }

syntype [SYST] Any specimen of a series when no specimen is designated as the holotype. Also known as cotype. { 'sin,tīp }

synusia [ECOL] A structural unit of a community characterized by uniformity of life-form or of height. { sə'nü·zhə }

syphilis [MED] An infectious disease caused by the spirochete *Treponema pallidum*, transmitted principally by sexual intercourse. { 'sif·ə·ləs }

syphilitic meningoencephalitis *See* general paresis. { ¦sif·ə¦lid·ik mə¦niŋ·gō·in,sef·ə'līd·əs }

Syringamminidae [INV ZOO] A family of Psamminida, with a fragile test constructed of tubes of xenophyae tightly cemented together. { ,sir·iŋ·gə'min·ə,dē }

syringe [MED] **1.** An apparatus commonly made of glass or plastic, fitted snugly onto a hollow needle, used to aspirate or inject fluids for diagnostic or therapeutic purposes. Also known as hypodermic syringe. **2.** A large glass barrel with a fitted rubber bulb at one end and a nozzle at the other, used primarily for irrigation purposes. { sə'rinj }

syringobulbia [MED] The presence of cavities in the medulla oblongata similar to those found in syringomyelia. { sə,riŋ·gō'bəl·bē·ə }

syringocystadenoma *See* syringoma. { sə,riŋ·gō ¦sist,ad·ən'ō·mə }

syringocystoma *See* syringoma. { sə,riŋ·gō·si 'stō·mə }

syringoma [MED] A multiple nevoid tumor of sweat glands. Also known as syringocystadenoma; syringocystoma. { ,sir·əŋ'gō·mə }

syringomyelia [MED] A chronic disease characterized by the presence of cavities surrounded by gliosis near the canal of the spinal cord and often extending to the medulla. { sə,riŋ·gō,mī'ē·lē·ə }

syrinx [VERT ZOO] The vocal organ in birds. { 'sir·iŋks }

Syrphidae [INV ZOO] The flower flies, a family of cyclorrhaphous dipteran insects in the series Aschiza. { 'sər·fə,dē }

Systellomatophora [INV ZOO] An order of the subclass Pulmonata in which the eyes are contractile but stalks are not retractile, the body is sluglike, oval, or lengthened, and the lung is posterior. { ¦sis·tə·lō·mə'täf·ə·rə }

systematics [BIOL] The science of animal and plant classification. { ,sis·tə'mad·iks }

systemic circulation [PHYSIO] The general circulation, as distinct from the pulmonary circulation. { si'stem·ik ,sər·kyə'lā·shən }

systemic inflammatory response syndrome [MED] The spectrum of elicited pathophysiologic changes (including blood clotting and changes in metabolism, heart rate, and respiration) resulting from excess production of inflammatory mediators (for example, histamines and leukotrienes), which orchestrate the process of inflammation through various processes. { si¦stem·ik in,flam·ə,tȯr·ē ri'späns ,sin,drōm }

systemic lupus erythematosus [MED] An inflammatory, multisystem, usually chronic disorder in which tissue injury is mediated by immune complexes. { sis¦tem·ik ,lü·pəs ,er·ə,thē·mə'tō·səs }

systems ecology [ECOL] The combined approaches of systems analysis and the ecology of whole ecosystems and subsystems. { 'sis·təmz i'käl·ə·jē }

systole [PHYSIO] The contraction phase of the heart cycle. { 'sis·tə·lē }

syzygy [INV ZOO] End-to-end union of the sporonts of certain gregarine protozoans. { 'siz·ə·jē }

T

Tabanidae [INV ZOO] The deer and horse flies, a family of orthorrhaphous dipteran insects in the series Brachycera. { tə'ban·ə,dē }

tabes dorsalis [MED] A form of parenchymatous neurosyphilis in which there is demyelination and sclerosis of the posterior columns of the spinal cord. Also known as locomotor ataxia. { 'tā·bēz dȯr'sal·əs }

tabled whelk [INV ZOO] *Neptunea tabulata*. A marine gastropod mollusk about 5 inches (13 centimeters) in length and 2 inches (5 centimeters) in diameter, found at depths of 150–200 feet (45–60 meters), off the west coast of Canada and the United States. { 'tā·bəld 'welk }

TAB vaccine [IMMUNOL] A vaccine containing killed typhoid bacilli and the paratyphoid organisms (*Salmonella paratyphi* A and B) most frequently involved in paratyphoid fever. { 'tab vak'sēn }

tache noire [MED] The primary painless lesion of the tick-borne typhus fevers of Africa, manifested by a raised red area with a typical black necrotic center which appears at the site of the tick bite. { täsh 'nwär }

Tachinidae [INV ZOO] The tachina flies, a family of bristly, grayish or black Diptera whose larvae are parasitic in caterpillars and other insects. { tə'kin·ə,dē }

tachycardia [MED] Excessive rapidity of the heart's action. { ¦tak·ə¦kärd·ē·ə }

Tachyglossidae [VERT ZOO] A family of monotreme mammals having relatively large brains with convoluted cerebral hemispheres; comprises the echidnas or spiny anteaters. { ,tak·ə'gläs·ə,dē }

Tachyniscidae [INV ZOO] A family of myodarian cyclorrhaphous dipteran insects in the subsection Acalypteratae. { ,tak·ə'nis·ə,dē }

tachyphylaxis [IMMUNOL] Rapid desensitization against doses of organ extracts or serum by the previous inoculation of small subtoxic doses of the same preparation. { ,tak·ə·fə'lak·səs }

tachypnea [MED] An abnormally rapid rate of respiration. { tə'kip·nē·ə }

tachysterol [BIOCHEM] The precursor of calciferol in the irradiation of ergosterol; an isomer of ergosterol. { tə'kis·tə,rȯl }

tachytely [EVOL] Evolution at a rapid rate resulting in differential selection and fixation of new types. { ¦tak·ə¦tel·ē }

tactile [PHYSIO] Pertaining to the sense of touch. { 'tak·təl }

tactile agnosia *See* astereognosis. { 'tak·təl ag'nō·zhə }

tactile hairs *See* vibrissae. { 'tak·təl 'herz }

tactile receptor *See* tactoreceptor. { 'tak·təl ri 'sep·tər }

tactoid [BIOCHEM] A particle that appears as a spindle-shaped body under the polarizing microscope and occurs in mosaic virus, fibrin, and myosin. { 'tak,tȯid }

tactoreceptor [PHYSIO] A sense organ that responds to touch. Also known as tactile receptor. { ¦tak·tō·ri'sep·tər }

tadpole [VERT ZOO] The larva of a frog or toad; at hatching it has a rounded body with a long fin-bordered tail, and the gills are external but shortly become enclosed. { 'tad,pōl }

tadpole shrimp [INV ZOO] Any of the phyllopod crustaceans that are members of the genus *Lepidurus*. { 'tad,pōl ,shrimp }

taenia [ANAT] A ribbon-shaped band of nerve fibers or muscle. { 'tē·nē·ə }

Taeniodidea [INV ZOO] An equivalent name for Cyclophyllidea. { ,tē·nē·ə'did·ē·ə }

Taenioidea [INV ZOO] An equivalent name for Cyclophyllidea. { ,tē·nē'ȯid·ē·ə }

taenoglossate radula [INV ZOO] A long, narrow radula with seven teeth in each transverse row, found in certain pectinibranch bivalves. { ,tē·nə'glä,sāt 'raj·ə·lə }

tagma [INV ZOO] A compound body section of a metameric animal that results from embryonic fusion of two or more somites. { 'tag·mə }

tagmosis [INV ZOO] The formation of groups of metameres into body regions with functional differences. { tag'mō·səs }

tagua palm [BOT] *Phytelephas macrocarpa*. A palm tree of tropical America; the endosperm of the seed is used as an ivory substitute. { 'täg·wə ,päm }

taiga [ECOL] A zone of forest vegetation encircling the Northern Hemisphere between the arctic-subarctic tundras in the north and the steppes, hardwood forests, and prairies in the south. Also known as boreal forest. { 'tī·gə }

tail [VERT ZOO] **1.** The usually slender appendage that arises immediately above the anus in many vertebrates and contains the caudal vertebrae. **2.** The uropygium, and its feathers, of a bird. **3.** The caudal fin of a fish or aquatic mammal. { tāl }

talbutal [PHARM] $C_{11}H_{16}N_2O_3$ A crystalline compound that melts at 108–110°C, and is soluble in alcohol, chloroform, acetone, and ether; used in medicine as a short-acting hypnotic and sedative. { 'täl·byə,tal }

talcosis [MED] A lung disease caused by inhalation of talc dust and characterized by chronic induration and fibrosis. { tal'kō·səs }

talipes [MED] Any of several foot deformities, especially of congenital origin. { 'tal·ə,pēz }

talipes cavus [MED] A deformity of the foot marked by exaggeration of the longitudinal arch and by dorsal contractures of the toes. { 'tal·ə,pēz 'kā·vəs }

talipes equinovarus [MED] The most common form of clubfoot, characterized by an extreme turning down and under of the foot, it is seen more often in boys and tends to affect one foot only. { ¦tal·ə,pēz ,ek·wi·no'va·rəs }

Talitridae [INV ZOO] A family of terrestrial amphipod crustaceans in the suborder Gammaridea. { tə'li·trə,dē }

tall fescue toxicosis [VET MED] A group of several animal disorders caused by grazing on tall fescue infected with the endophytic symbiotic fungus *Acremonium coenophialum*. { ,tȯl ,fes·kyü ,tak·sə'kō·səs }

talon [VERT ZOO] A sharply hooked claw on the foot of a bird of prey. { 'tal·ən }

Talpidae [VERT ZOO] The moles, a family of insectivoran mammals; distinguished by the forelimbs which are adapted for digging, having powerful muscles and a spadelike bony structure. { 'tal·pə,dē }

talus *See* astragalus. { 'tal·əs }

tamarack [BOT] *Larix laricina*. A larch and a member of the pine family; it has an erect narrowly pyramidal habit, and grows in wet and moist soils in the northeastern United States, west to the Lake States and across Canada to Alaska; used for railroad ties, posts, sills, and boats. Also known as hackmarack. { 'tam·ə,rak }

tampon [MED] A plug of absorbent material, such as cotton or sponge, inserted into a cavity as packing. { 'tam,pän }

Tanaidacea [INV ZOO] An order of eumalacostracans of the crustacean superorder Peracarida; the body is linear, more or less cylindrical or dorsoventrally depressed, and the first and second thoracic segments are fused with the head, forming a carapace. { ,tan·ē·ə'dā·shə }

Tanaostigmatidae [INV ZOO] A small family of hymenopteran insects in the superfamily Chalcidoidea. { tə,nā·ō·stig'mad·ə,dē }

tandem duplication [CELL MOL] The occurrence of two identical sequences, one following the other, in a chromosome segment. { 'tan·dəm ,dü·plə'kā·shən }

tangelo [BOT] A tree that is hybrid between a tangerine or other mandarin and a grapefruit or shaddock; produces an edible fruit. { 'tan·jə·lō }

tangerine [BOT] Any of several trees of the species *Citrus reticulata*; the fruit is a loose-skinned mandarin with a deep-orange or scarlet rind. { 'tan·jə,rēn }

tangoreceptor [PHYSIO] A sense organ in the skin that responds to touch and pressure. { ¦taŋ·gō·ri'sep·tər }

tannase [BIOCHEM] An enzyme that catalyzes the hydrolysis of tannic acid to gallic acid; found in cultures of *Aspergillus* and *Penicillium*. { 'ta ,nās }

Tanyderidae [INV ZOO] The primitive crane flies, a family of orthorrhaphous dipteran insects in the series Nematocera. { ,tan·ə'der·ə,dē }

Tanypezidae [INV ZOO] A family of myodarian cyclorrhaphous dipteran insects in the subsection Acalyptreatae. { ,tan·ə'pez·ə,dē }

tape grass [BOT] *Vallisnerida spiralis*. An aquatic flowering plant belonging to the family Hydrocharitaceae. Also known as eel grass. { 'tāp ,gras }

tapetum [BOT] A layer of nutritive cells surrounding the spore mother cells in the sporangium in higher plants; it is broken down to provide nourishment for developing spores. [NEUROSCI] **1.** A reflecting layer in the choroid coat behind the neural retina, chiefly in the eyes of nocturnal mammals. **2.** A tract of nerve fibers forming part of the roof of each lateral ventricle in the vertebrate brain. { tə'pēd·əm }

tapeworm [INV ZOO] Any member of the class Cestoidea; all are vertebrate endoparasites, characterized by a ribbonlike body divided into proglottids, and the anterior end modified into a holdfast organ. { 'tāp,wərm }

Taphrina caerulescens [MYCOL] A fungal pathogen that is the cause of leaf blister of oaks. { ta,frī·nə ,kī·rə'les·ənz }

Taphrina deformans [MYCOL] A fungal pathogen that is the cause of leaf curl of peach and almond trees. { ta,frī·nə di'fȯr·mənz }

tapir [VERT ZOO] Any of several large odd-toed ungulates of the family Tapiridae that have a heavy, sparsely hairy body, stout legs, a prehensile muzzle, a short tail, and small eyes. { 'tā·pər }

Tapiridae [VERT ZOO] The tapirs, a family of perissodactyl mammals in the superfamily Tapiroidea. { tə'pir·ə,dē }

Tapiroidea [VERT ZOO] A superfamily of the mammalian order Perissodactyla in the suborder Ceratomorpha. { ,tap·ə'rȯid·ē·ə }

taproot [BOT] A root system in which the primary root forms a dominant central axis that penetrates vertically and rather deeply into the soil; it is generally larger in diameter than its branches. { 'tap,rüt }

tarantula [INV ZOO] **1.** Any of various large hairy spiders of the araneid suborder Mygalomorphae. **2.** Any of the wolf spiders comprising the family Lycosidae. { tə'ran·chə·lə }

Tardigrada [INV ZOO] A class of microscopic, bilaterally symmetrical invertebrates in the subphylum Malacopoda; the body consists of an anterior prostomium and five segments surrounded by a soft, nonchitinous cuticle, with four pairs of ventrolateral legs. { tär'dig·rə·də }

tardive dyskinesia [MED] A movement disorder marked by involuntary twitching of the mouth, lips, tongue, arms, legs, or trunk; frequently associated with the use of neuroleptic drugs. { 'tär·div ˌdis·kə'nē·zhə }

target cell [PHYSIO] A cell that has receptors for the product of a signaling cell. { 'tär·gət ˌsel }

tarpon [VERT ZOO] *Megalops atlantica*. A herringlike fish of the family Elopidae weighing up to 300 pounds (136 kilograms) and reaching a length of 8 feet (2.4 meters); it has a single soft, rayed dorsal fin, strong jaws, a bony plate under the mouth, numerous small teeth, and coarse, bony flesh covered with large scales. { 'tär·pən }

tarsal gland [ANAT] Any of the sebaceous glands in the tarsal plates of the eyelids. Also known as Meibomian gland. { 'tär·səl ˌgland }

tarsal tunnel syndrome [MED] A neurological foot disorder in which the posterior tibial nerve becomes compressed and damaged within the tarsal canal, symptoms include pain and burning that arises from behind the inside of the ankle and that may travel to the bottom of the foot. { ˌtärs·əl 'tən·əl ˌsinˌdrōm }

tarsier [VERT ZOO] Any of several species of primates comprising the genus *Tarsius* of the family Tarsiidae characterized by a round skull, a flattened face, and large eyes that are separated from the temporal fossae in the orbital depression, and by adhesive pads on the expanded ends of the fingers and toes. { 'tär·sēˌā }

Tarsiidae [VERT ZOO] The tarsiers, a family of prosimian primates distinguished by incomplete postorbital closure and a greatly elongated ankle region. { tär'sī·əˌdē }

Tarsonemidae [INV ZOO] A small family of phytophagous mites in the suborder Trombidiformes. { ˌtär·sə'nem·əˌdē }

tarsoplasty *See* blepharoplasty. { 'tär·səˌplas·tē }

tarsus [ANAT] **1.** The instep of the foot consisting of the calcaneus, talus, cuboid, navicular, medial, intermediate, and lateral cuneiform bones. **2.** The dense connective tissues supporting an eyelid. { 'tär·səs }

tartar *See* dental calculi. { 'tärd·ər }

tartary buckwheat [BOT] One of three buckwheat species grown commercially; the leaves are narrower than the other two species and arrow-shaped, and the flowers are smaller with inconspicuous greenish-white sepals. Also known as duck wheat; hulless buckwheat; rye buckwheat. { 'tärd·ə·rē 'bəkˌwēt }

tassel [BOT] The male inflorescence of corn and certain other plants. { 'tas·əl }

taste [PHYSIO] A chemical sense by which flavors are perceived depending on taste, tactile, and warm and cold receptors in the mouth, as well as smell receptors in the nose. { tāst }

taste bud [ANAT] An end organ consisting of goblet-shaped clusters of elongate cells with microvilli on the distal end to mediate the sense of taste. { 'tāst ˌbəd }

TATA box [GEN] In eukaryotes, a short sequence of base pairs that is rich in adenine (A) and thymidine (T) residues and located about 25–30 nucleotides upstream of the transcriptional initiation site. { 'täˌtä ˌbäks or ¦tē¦ā¦tē'ā ˌbäks }

tau protein [NEUROSCI] A protein found in the axons of healthy neurons, where it binds to other proteins called microtubules to form the cytoskeleton of the neuron and provide the tracks over which material can be transported from one part of the neuron to another. { 'taů ˌprōˌtēn }

taurocholic acid [BIOCHEM] $C_{26}H_{45}NO_7S$ A common bile acid with a five-carbon chain; it is the product of the conjugation of taurine with cholic acid; crystallizes from an alcohol ether solution, and decomposes at about 125°C. Also known as cholaic acid; cholytaurine. { ¦tȯr·ə¦kōl·ik 'as·əd }

taurodont [ANAT] Of teeth, having a large pulp cavity and reduced roots. { 'tȯr·əˌdänt }

Taxales [BOT] A small order of gymnosperms in the class Pinatae; members are trees or shrubs with evergreen, often needlelike leaves, with a well-developed fleshy covering surrounding the individual seeds, which are terminal or subterminal on short shoots. { tak'sā·lēz }

taxis [PHYSIO] A mechanism of orientation by means of which an animal moves in a direction related to a source of stimulation. { 'tak·səs }

Taxodonta [INV ZOO] A subclass of pelecypod mollusks in which the hinge is of the taxodontal type, that is, the dentition is a series of similar alternating teeth and sockets along the hinge margin. { ˌtak·sə'dänt·ə }

taxol [PHARM] $C_{47}H_{51}NO_{14}$ An alkaloid derived from the Pacific yew tree (*Taxus brevifolia*) that is used in the treatment of ovarian and breast cancer. { 'takˌsōl }

taxon [SYST] A taxonomic group or entity. { 'takˌsän }

taxonomic category [SYST] One of a hierarchy of levels in the biological classification of organisms; the seven major categories are kingdom, phylum, class, order, family, genus, species. { ¦tak·sə¦näm·ik 'kad·əˌgȯr·ē }

taxonomy [SYST] A study aimed at producing a hierarchical system of classification of organisms which best reflects the totality of similarities and differences. { tak'sän·ə·mē }

Taxus brevifolia [BOT] The Pacific yew tree, from which the anticancer compound taxol is derived. { ˌtak·səs ˌbrev·i'fō·lē·ə }

Tayassuidae [VERT ZOO] The peccaries, a family of artiodactyl mammals in the superfamily Suoidae. { ˌtā·yə'sü·əˌdē }

Tay-Sachs disease [MED] A form of sphingolipidosis, transmitted as an autosomal recessive, in which there is an accumulation in neuronal cells of the neuraminic fraction of gangliosides; manifested clinically within the first few months of life by hypotonia progressing to spasticity, convulsions, and visual loss accompanied by the appearance of a cherry-red spot at the macula lutea. Also known as infantile amaurotic familial idiocy. { ¦tā ¦saks di,zēz }

TBSV *See* tomato bushy stunt virus.

T cell [IMMUNOL] One of a heterogeneous population of thymus-derived lymphocytes which participates in the immune responses. Also known as T lymphocyte. { 'tē ,sel }

T-cell receptor [IMMUNOL] Protein on the surface of T lymphocytes that specifically recognizes molecules of the major histocompatibility complex, either alone or in association with foreign antigens. Abbreviated TCR. { 'tē ,sel ri,sep·tər }

TCR *See* T-cell receptor.

tea [BOT] *Thea sinensis.* A small tree of the family Theaceae having lanceolate leaves and fragrant white flowers; a caffeine beverage is made from the leaves of the plant. { tē }

tear gland *See* lacrimal gland. { 'tir ,gland }

technosphere [ECOL] The part of the physical environment affected through building or modification by humans. { 'tek·nə,sfir }

Tectibranchia [INV ZOO] An order of mollusks in the subclass Opisthobranchia containing the sea hares and the bubble shells; the shell may be present, rudimentary, or absent. { ,tek·tə'braŋ·kē·ə }

Tectiviridae [VIROL] A family of nontailed bacterial viruses (bacteriophages) characterized by a nonenveloped icosahedral particle containing a linear double-stranded deoxyribonucleic acid genome, includes the PRDI phage. { ,tek·tə'vir·ə,dī }

tectorial membrane [ANAT] **1.** A jellylike membrane covering the organ of Corti in the ear. **2.** A strong sheet of connective tissue running from the basilar part of the occipital bone to the dorsal surface of the bodies of the axis and third cervical vertebra. { tek'tȯr·ē·əl 'mem,brān }

tectum [ANAT] A rooflike structure of the body, especially the roof of the midbrain including the corpora quadrigemina. { 'tek·təm }

Teflon shakes *See* metal fume fever. { 'tef,län 'shāks }

tegmen [BIOL] An integument or covering. [BOT] The inner layer of a seed coat. [INV ZOO] A thickened forewing of Orthoptera, Coleoptera, and certain other insects. { 'teg·mən }

tegmentum [ANAT] A mass of white fibers with gray matter in the cerebral peduncles of higher vertebrates. [BOT] The outer layer, or scales, of a leaf bud. [INV ZOO] The upper layer of a shell plate in Amphineura. { teg'men·təm }

teichoic acid [BIOCHEM] A polymer of ribitol or glycerol phosphate with additional compounds such as glucose linked to the backbone of the polymer; found in the cell walls of some bacteria. { tā'kō·ik 'as·əd }

Teiidae [VERT ZOO] The tegus lizards, a diverse family of the suborder Sauria that is especially abundant and widespread in South America. { 'tē·ə,dē }

T-E index *See* temperature-efficiency index. { ¦tē'ē ,in,deks }

telamon [INV ZOO] A curved chitinous outgrowth of the cloacal wall in various male nematodes. { 'tel·ə,män }

telangiectasis [MED] Localized dilation of capillaries forming dark-red, wartlike elevations varying in size from about 1 to 7 millimeters. { tə¦lan·jē'ek·tə·səs }

teleceptor [PHYSIO] A sense receptor which transmits information about portions of the external environment which are not necessarily in direct contact with the organism, such as the receptors of the ear, eye, and nose. { 'tel·ə,sep·tər }

Telegeusidae [INV ZOO] The long-lipped beetles, a small family of colepteran insects in the superfamily Cantharoidea confined to the western United States. { ,tel·ə'gyüs·ə,dē }

telemedicine [MED] The use of teleconferencing in medical diagnosis and treatment, allowing rural health-care facilities to perform diagnosis and treatment that would otherwise be available only in metropolitan areas. { ,tel·ə'med·ə·sən }

telencephalon [EMBRYO] The anterior subdivision of the forebrain in a vertebrate embryo; gives rise to the olfactory lobes, cerebral cortex, and corpora striata. { ¦tel·en'sef·ə,län }

Teleostei [VERT ZOO] An infraclass of the subclass Actinopterygii, or rayfin fishes; distinguished by paired bracing bones in the supporting skeleton of the caudal fin, a homocercal caudal fin, thin cycloid scales, and a swim bladder with a hydrostatic function. { ,tel·ē'äs·tē,ī }

telephone theory *See* frequency theory. { 'tel·ə,fōn ,thē·ə·rē }

Telestacea [INV ZOO] An order of the subclass Alcyonaria comprising of individuals which form erect branching colonies by lateral budding from the body wall of an elongated axial polyp. { ,tel·ə'stā·shē·ə }

teletherapy [MED] Radiation treatment administered by using a source that is at a distance from the body, usually employing gamma-ray beams from radioisotope sources. { ¦tel·ə'ther·ə·pē }

teleutospore *See* teliospore. { tə'lüd·ə,spȯr }

teliospore [MYCOL] A thick-walled spore of the terminal stage of Uredinales and Ustilaginales which is a probosidium or a group of probosidia. Also known as teleutospore. { 'tē·lē·ə,spȯr }

telium [MYCOL] In rust fungi, a specialized fruiting structure that produces teliospores. { 'tēl·ē·əm }

teloblast [INV ZOO] A large cell that produces many smaller cells at the growing end of many embryos, especially in annelids and mollusks. { 'tel·ə,blast }

telocentric [CELL MOL] Pertaining to a chromosome with a terminal centromere. { ¦tel·ə¦sen·trik }

telocoel [EMBRYO] A cavity of the telencephalon. { 'tel·ə,sēl }

telodendrion [NEUROSCI] The terminal branching of an axon. Also known as telodendron. { ¦tel·ə'den·drē,än }

telodendron *See* telodendrion. { ¦tel·ə'den·drən }

telogen [PHYSIO] A quiescent phase in the cycle of hair growth when the hair is retained in the hair follicle as a dead or "club" hair. { 'tel·ə·jən }

telogen effluvium [MED] Hair loss that is due to increased shedding occurring after a metabolic disturbance. { ,tel·ə·jən e'flü·vē·əm }

telolecithal [CELL MOL] Of an ovum, having a large, evenly dispersed volume of yolk and a small amount of cytoplasm concentrated at one pole. { ¦tel·ō'les·ə·thəl }

telomerase [BIOCHEM] A deoxyribonucleic acid polymerase that elongates telomeres. { tə'läm·ə,rās }

telomere [CELL MOL] A centromere in the terminal position on a chromosome. { 'tel·ə,mir }

telophase [CELL MOL] The phase of meiosis or mitosis at which the chromosomes, having reached the poles, reorganize into interphase nuclei with the disappearance of the spindle and the reappearance of the nuclear membrane; in many organisms telophase does not occur at the end of the first meiotic division. { 'tel·ə,fāz }

Telosporea [INV ZOO] A class of the protozoan subphylum Sporozoa in which the spores lack a polar capsule and develop from an oocyst. { ,tel·ə'spȯr·ē·ə }

telotaxis [BIOL] Tactic movement of an organism by the orientation of one or the other of two bilaterally symmetrical receptors toward the stimulus source. { ¦tel·ō'tak·səs }

telotroch [INV ZOO] A preanal tuft of cilia in a trochophore larva. { 'tel·ə,träk }

telson [INV ZOO] The postabdominal segment in lobsters, amphipods, and certain other invertebrates. { 'tel·sən }

Temnocephalida [INV ZOO] A group of rhabdocoeles sometimes considered a distinct order but usually classified under the Neorhabdocoela; members are characterized by the possession of tentacles and adhesive organs. { ,tem·nō·sə'fal·əd·ə }

Temnochilidae [INV ZOO] The equivalent name for Ostomidae. { ,tem·nō'kil·ə,dē }

Temnopleuridae [INV ZOO] A family of echinoderms in the order Temnopleuroida whose tubercles are imperforate, though usually crenulate. { ,tem·nō'plu̇r·ə,dē }

Temnopleuroida [INV ZOO] An order of echinoderms in the superorder Echinacea with a camarodont lantern, smooth or sculptured test, imperforate or perforate tubercles, and bronchial slits which are usually shallow. { ,tem·nō·plə'rȯid·ē·ə }

temperate and cold savannah [ECOL] A regional vegetation zone, very extensively represented in North America and in Eurasia at high altitudes; consists of scattered or clumped trees (very often conifers and mostly needle-leaved evergreens) and a shrub layer of varying coverage; mosses and, even more abundantly, lichens form an almost continuous carpet. { ¦tem·prət ən ¦kōld sə'van·ə }

temperate and cold scrub [ECOL] Regional vegetation zone whose density and periodicity vary a good deal; requires a considerable amount of moisture in the soil, whether from mist, seasonal downpour, or snowmelt; shrubs may be evergreen or deciduous; and undergrowth of ferns and other large-leaved herbs are quite frequent, especially at subalpine level; wind shearing and very cold winters prevent tree growth. Also known as bosque; fourré; heath. { ¦tem·prət ən ¦kōld 'skrəb }

temperate mixed forest [ECOL] A forest of the North Temperate Zone containing a high proportion of conifers with a few broad-leafed species. { 'tem·prət 'mikst ¦fär·əst }

temperate phage [VIROL] A deoxyribonucleic acid phage, the genome of which can under certain circumstances become integrated with the genome of the host. { 'tem·prət 'fāj }

temperate rainforest [ECOL] A vegetation class in temperate areas of high and evenly distributed rainfall characterized by comparatively few species with large populations of each species; evergreens are somewhat short with small leaves, and there is an abundance of large tree ferns. Also known as cloud forest; laurel forest; laurisilva; moss forest; subtropical forest. { 'tem·prət 'rān,fär·əst }

temperate woodland [ECOL] A vegetation class similar to tropical woodland in spacing, height, and stratification, but it can be either deciduous or evergreen, broad-leaved or needle-leaved. Also known as parkland; woodland. { 'tem·prət 'wu̇d·lənd }

temperature-efficiency index [ECOL] For a given location, a measure of the long-range effectiveness of temperature (thermal efficiency) in promoting plant growth. Abbreviated T-E index. Also known as thermal-efficiency index. { 'tem·prə·chər i'fish·ən·sē ,in,deks }

temperature-efficiency ratio [ECOL] For a given location and month, a measure of thermal efficiency; it is equal to the departure, in degrees Fahrenheit, of the normal monthly temperature above 32°F (0°C) divided by 4: $(T - 32)/4$. Abbreviated T-E ratio. Also known as thermal-efficiency ratio. { 'tem·prə·chər i'fish·ən·sē ,rā·shō }

temperature-sensitive mutant [GEN] A mutant gene that is functional at high (low) temperature but is inactivated by lowering (elevating) the temperature. { 'tem·prə·chər ¦sen·səd·iv 'myüt·ənt }

template [CELL MOL] The macromolecular model for the synthesis of another macromolecule. { 'tem·plət }

temples [INV ZOO] The posterolateral angles of the head, in lice. { 'tem·pəlz }

temporal arteritis *See* giant-cell arteritis. { 'tem·prəl ,ärd·ə'rīd·əs }

temporal bone [ANAT] The bone forming a portion of the lateral aspect of the skull and part of the base of the cranium in vertebrates. { 'tem·prəl ,bōn }

temporal-lobe epilepsy [MED] Recurrent seizures originating in lesions of the temporal lobe of the brain. { 'tem·prəl ¦lōb 'ep·ə,lep·sē }

temporamandibular joint disease [MED] A condition characterized by a faulty bite or misalignment of the teeth. Abbreviated TMJ. { ,tem·pə·rə·man,dib·yə·lər 'jȯint di,zēz }

temporary plankton *See* meroplankton. { 'tem·pə,rer·ē 'plaŋk·tən }

tender plant [BOT] A plant that is incapable of resisting cold. { ¦ten·dər ¦plant }

Tendipedidae [INV ZOO] The midges, a family of orthorrhaphous dipteran insects in the series Nematocera whose larvae occupy intertidal wave-swept rocks on the seacoasts. { ,ten·də'ped·ə,dē }

tendon [ANAT] A white, glistening, fibrous cord which joins a muscle to some movable structure such as a bone or cartilage; tendons permit concentration of muscle force into a small area and allow the muscle to act at a distance. { 'ten·dən }

tendonitis [MED] Inflammation of a tendon. { ,ten·də'nīd·əs }

tendon sheath [ANAT] The synovial membrane surrounding a tendon. { 'ten·dən ,shēth }

tendril [BOT] A stem modification in the form of a slender coiling structure capable of twining about a support to which the plant is then attached. { 'ten·drəl }

Tenebrionidae [INV ZOO] The darkling beetles, a large cosmopolitan family of coleopteran insects in the superfamily Tenebrionoidea; members are common pests of grains, dried fruits, beans, and other food products. { tə,neb·rē'än·ə,dē }

Tenebrionoidea [INV ZOO] A superfamily of the Coleoptera in the suborder Polyphaga. { tə,neb·rē·ə'nȯid·ē·ə }

teniae coli [HISTOL] The three bands comprising the longitudinal layer of the tunica muscularis of the colon: the tenia libera, tenia mesocolica, and tenia omentalis. { 'tē·nē,ē 'kō·lī }

tenosynovitis [MED] Inflammation of a tendon and its sheath. { ¦ten·ō,sī·nə'vīd·əs }

tenrec [VERT ZOO] Any of about 30 species of unspecialized, insectivorous mammals which are indigenous to Madagascar and have poor vision and clawed digits. { 'ten,rek }

Tenrecidae [VERT ZOO] The tenrecs, a family of insectivores in the group Lipotyphla. { ten 'res·ə,dē }

tension headache *See* muscle-contraction headache. { 'ten·chən ,hed,āch }

tension wood [BOT] In some hardwood trees, wood characterized by the presence of gelatinous fibers and excessive longitudinal shrinkage; causes trees to lean. { 'ten·chən ,wu̇d }

tensor muscle [PHYSIO] A muscle that stretches a part or makes it tense. { 'ten·sər ,məs·əl }

tentacle [INV ZOO] Any of various elongate, flexible processes with tactile, prehensile, and sometimes other functions, and which are borne on the head or about the mouth of many animals. { 'ten·tə·kəl }

Tentaculata [INV ZOO] A class of the phylum Ctenophora whose members are characterized by having variously modified tentacles. { ten ,tak·yə'läd·ə }

tentaculocyst [INV ZOO] A sense organ located at the margin of the umbrella in some cnidarian medusoids, consisting of a modified tentacle with a cavity that often contains lithites. { ten 'tak·yə·lō,sist }

tentaculozoid [INV ZOO] A slender tentacular individual of a hydrozoan colony. { ten¦tak·yə·lō'zō,ȯid }

tented arch [FOREN] A fingerprint pattern which possesses either an angle, an upthrust, or two of the three basic characteristics of a loop. { 'ten·təd 'ärch }

Tenthredinidae [INV ZOO] A family of hymenopteran insects in the superfamily Tenthredinoidea including economically important species whose larvae are plant pests. { ,ten·thrə'din·ə,dē }

Tenthredinoidea [INV ZOO] A superfamily of Hymenoptera in the suborder Symphyla. { ,ten·thrə·də'nȯid·ē·ə }

tentillum [INV ZOO] A contractile branch of a tentacle containing many nematocysts in certain siphonophores. { ten'til·əm }

Tenuipalpidae [INV ZOO] A small family of mites in the suborder Trombidiformes. { ,ten·yə·wə'pal·pə,dē }

tepary bean [BOT] *Phaseolus acutifolius* var. *latifolius*. One of the four species of beans of greatest economic importance in the United States. { 'tep·ə·rē ,bēn }

Tephritidae [INV ZOO] The fruit flies, a family of myodarian cyclorrhaphous dipteran insects in the subsection Acalyptratae. { tə'frid·ə,dē }

T-E ratio *See* temperature-efficiency ratio. { ¦tē'ē 'rā·shō }

teratocarcinoma [MED] A teratoma with carcinomatous elements. { ¦ter·ə·tō,kärs·ən'ō·mə }

teratogen [MED] An agent causing formation of a congenital anomaly or monstrosity. { tə'rad·ə·jən }

teratogenesis [MED] The formation of a fetal monstrosity. { ,te·rə·tō'jen·ə·səs }

teratology [MED] The science of fetal malformations and monstrosities. { ,ter·ə'täl·ə·jē }

teratoma [MED] A true neoplasm composed of bizarre and chaotically arranged tissues that are foreign embryologically as well as histologically to the area in which the tumor is found. { ,ter·ə'tō·mə }

Terebellidae [INV ZOO] A family of polychaete annelids belonging to the Sedentaria which are chiefly large, thick-bodied, tubicolous forms with the anterior end covered by a matted mass of tentacular cirri. { ,ter·ə'bel·ə,dē }

Terebratulida [INV ZOO] An order of articulate brachiopods that has a punctate shell structure and is characterized by the possession of a loop extending anteriorly from the crural bases, providing some degree of support for the lophophore. { ˌter·ə·brə'tül·əd·ə }

Terebratulidina [INV ZOO] A suborder of articulate brachiopods in the order Terebratulida distinguished by a short V- or W-shaped loop. { ˌter·ə·brach·ə·lə'dīn·ə }

Teredinidae [INV ZOO] The pileworms or shipworms, a family of bivalve mollusks in the subclass Eulamellibranchia distinguished by having the two valves reduced to a pair of small plates at the anterior end of the animal. { ˌter·ə'din·əˌdē }

terete [BOT] Of a stem, cylindrical in section, but tapering at both ends. { tə'rēt }

tergite [INV ZOO] The dorsal plate covering a somite in arthropods and certain other articulate animals. { 'tərˌjīt }

tergum [INV ZOO] A dorsal plate of the operculum in barnacles. { 'tər·gəm }

terminal bar [CELL MOL] One of the structures formed in certain epithelial cells by the combination of local modifications of contiguous surfaces and intervening intercellular substances; they become visible with the light microscope after suitable staining and appear to close the spaces between the epithelial cells of the intestine at their free surfaces. { 'tər·mən·əl ˌbär }

terminal bud [BOT] A bud that develops at the apex of a stem. Also known as apical bud. { 'tər·mən·əl ˌbəd }

terminal disinfection [MICROBIO] The disinfection of sickrooms occupied by patients suffering from contagious disease. { ¦tərm·ən·əl ˌdis·in'fek·shən }

terminal endocarditis *See* verrucous endocarditis. { 'tər·mən·əl ¦en·dō·kär'dīd·əs }

terminal hair [ANAT] One of three types of hair in man based on hair size, time of appearance, and structural variations; the larger, coarser hair in the adult that replaces the vellus hair. { 'tər·mən·əl ¦her }

terminal nerve [NEUROSCI] Either of a pair of small cranial nerves that run from the nasal area to the forebrain, present in most vertebrates; the function is not known. { 'tər·mən·əl ¦nərv }

terminal sinus [EMBRYO] The vascular sinus bounding the area vasculosa of the blastoderm of a meroblastic ovum. Also known as marginal sinus. { 'tər·mən·əl ¦sī·nəs }

terminator codon [GEN] A UAA, UAG, or UGA trinucleotide in messenger ribonucleic acid (mRNA) that specifies termination of synthesis of the polypeptide (protein) product of the gene. Also known as stop codon. { 'tər·məˌnād·ər 'kōˌdän }

Termitaphididae [INV ZOO] The termite bugs, a small family of Hemiptera in the superfamily Aradoidea. { terˌmīd·ə'fid·əˌdē }

termitarium [INV ZOO] A termites' nest. { ˌtər·mə'ter·ē·əm }

termite [INV ZOO] A soft-bodied insect of the order Isoptera; individuals feed on cellulose and live in colonies with a caste system comprising three types of functional individuals: sterile workers and soldiers, and the reproductives. Also known as white ant. { 'tərˌmīt }

termiticole [ECOL] An organism that lives in a termites' nest. { tər'mīd·əˌkōl }

Termitidae [INV ZOO] A large family of the order Isoptera which contains the higher termites, representing 80% of the species. { tər'mid·əˌdē }

termitophile [ECOL] An organism that lives in a termites' nest in a symbiotic association with the termites. { tər'mīd·əˌfīl }

Termopsidae [INV ZOO] A family of insects in the order Isoptera composed of damp wood-dwelling forms. { tər'mäp·səˌdē }

ternate [BOT] Composed of three subdivisions, as a leaf with three leaflets. { 'tərˌnāt }

terrapin [VERT ZOO] Any of several North American tortoises in the family Testudinidae, especially the diamondback terrapin. { 'ter·ə·pən }

terrestrial ecosystem [ECOL] A community of organisms and their environment that occurs on the landmasses of continents and islands. { təˌres·trē·əl 'ek·ōˌsis·təm }

territoriality [ZOO] A pattern of behavior in which one or more animals occupy and defend a definite area or territory. { ter·əˌtòr·ē'al·əd·ē }

tertian malaria *See* vivax malaria. { 'tər·shən mə'ler·ē·ə }

tertiary structure [BIOCHEM] The characteristic three-dimensional folding of the polypeptide chains in a protein molecule. { 'tər·shēˌer·ē 'strək·chər }

Tessaratomidae [INV ZOO] A family of large tropical Hemiptera in the superfamily Pentatomoidea. { ˌtes·ə·rə'täm·əˌdē }

tessellate [BOT] Marked by a pattern of small squares resembling a tiled pavement. { 'tes·əˌlāt }

test [INV ZOO] A hard external covering or shell that is calcareous, siliceous, chitinous, fibrous, or membranous. { test }

testa [BOT] A seed coat. Also known as episperm. { 'tes·tə }

Testacellidae [INV ZOO] A family of pulmonate gastropods that includes some species of slugs. { ˌtes·tə'sel·əˌdē }

test cross [GEN] A cross between an individual homozygous at one or more loci and a test subject; the phenotype of the progeny will reveal the subject's genotype at these loci. { 'test ˌkròs }

testicular feminization [MED] A hereditary disorder in which affected individuals are chromosomally XY but have a feminine phenotype and are sterile. The X-linked testicular feminization (Tfm) gene results in absent or defective testosterone receptors, leaving cells unable to bind androgens and respond even though normal male levels of androgen are present. Also known as androgen insensitivity syndrome. { tes¦tik·yə·lər ˌfem·ə·nə'zā·shən }

testicular hormone [BIOCHEM] Any of various hormones secreted by the testes. { te'stik·yə·lər 'hòrˌmōn }

testis [ANAT] One of a pair of male reproductive glands in vertebrates; after sexual maturity, the source of sperm and hormones. { 'tes·təs }

testosterone [BIOCHEM] $C_{19}H_{28}O_2$ The principal androgenic hormone released by the human testis; may be synthesized from cholesterol and certain other sterols. { tes'täs·tə,rōn }

Testudinata [VERT ZOO] The equivalent name for Chelonia. { te,styüd·ən'äd·ə }

Testudinellidae [INV ZOO] A family of free-swimming rotifers in the suborder Flosculariacea. { te,styüd·ən'el·ə,dē }

Testudinidae [VERT ZOO] A family of tortoises in the suborder Cryptodira; there are about 30 species found on all continents except Australia. { ,test·yü'din·ə,dē }

tetanolysin [BIOCHEM] A hemolysin produced by *Clostridium tetani*. { ¦tet·ən·ō'līs·ən }

tetanospasmin [BIOCHEM] A neurotoxin elaborated by the bacterium *Clostridium tetani* and which is responsible for the manifestations of tetanus. { ,tet·ən·ō'spaz·mən }

tetanus [MED] An infectious disease of humans and animals caused by the toxin of *Clostridium tetani* and characterized by convulsive tonic contractions of voluntary muscles; infection commonly follows dirt contamination of deep wounds or other injured tissue. Also known as lockjaw. { 'tet·ən·əs }

tetanus antitoxin [IMMUNOL] A serum containing antibodies that neutralize tetanus toxin. { 'tet·ən·əs 'ant·i,täk·sən }

tetanus toxoid [IMMUNOL] Detoxified tetanus toxin used to produce active immunity against tetanus. { 'tet·ən·əs 'täk,sȯid }

tetany [MED] A state of increased neuromuscular irritability caused by a decrease of serum calcium, manifested by intermittent numbness and cramps or twitchings of the extremities, laryngospasm, bizarre behavior, loss of consciousness, and convulsions. { 'tet·ən·ē }

tetany of the newborn [MED] A temporary increase of neuromuscular irritability during the first two months of life, especially in infants who are premature, are delivered by cesarean section, or receive an exchange transfusion, or in twins. { 'tet·ən·ē əv t͟hə 'nü,bȯrn }

Tethinidae [INV ZOO] A family of myodarian cyclorrhaphous dipteran insects in the subsection Acalyptratae. { tə'thin·ə,dē }

Tetrabranchia [INV ZOO] A subclass of primitive mollusks of the class Cephalopoda; *Nautilus* is the only living form and is characterized by having four gills. { ,te·trə'braŋ·kē·ə }

Tetracentraceae [BOT] A family of dicotyledonous trees in the order Trochodendrales distinguished by possession of a perianth, four stamens, palmately veined leaves, and secretory idioblasts. { ,te·trə,sen'trās·ē,ē }

Tetractinomorpha [INV ZOO] A heterogeneous subclass of Porifera in the class Demospongiae. { tə¦trak·tə·nə'mȯr·fə }

tetracycline [MICROBIO] **1.** Any of a group of broad-spectrum antibiotics produced biosynthetically by fermentation with a strain of *Streptomyces aureofaciens* and certain other species or chemically by hydrogenolysis of chlortetracycline. **2.** $C_{22}H_{24}O_8N_2$ A broad-spectrum antibiotic belonging to the tetracycline group of antibiotics; useful because of broad antimicrobial action, with low toxicity, in the therapy of infections caused by gram-positive and gram-negative bacteria as well as rickettsiae and large viruses such as psittacosis-lymphogranuloma viruses. { ,te·trə'sī,klēn }

tetrad [CELL MOL] A group of four chromatids lying parallel to each other as a result of the longitudinal division of each of a pair of homologous chromosomes during the pachytene and later stages of the prophase of meiosis. { 'te,trad }

tetradactylous [VERT ZOO] Having four digits on a limb. { ¦te·trə¦dakt·əl·əs }

tetrad analysis [GEN] A method of genetic analysis possible in fungi, algae, bryophytes, and orchids in which the four products of an individual cell which has gone through meiosis are recovered as a group; it provides more direct and complete information regarding segregation and recombination mechanisms than is possible to obtain from meiotic products collected at random. { 'te,trad ə,nal·ə·səs }

tetrad of Fallot *See* tetralogy of Fallot. { 'te,trad əv fa'lō }

4,6,3′,4′-tetrahydroxyaurone *See* aureusidin. { ¦fȯr ¦siks ¦thrē,prīm ¦fȯr,prīm ¦te·trə·hī¦dräk·sē'ō,rōn }

Tetrahymenina [INV ZOO] A suborder of ciliated protozoans in the order Hymenostomatida. { ¦te·trə,hī·mə'nīn·ə }

tetralogy of Fallot [MED] A congenital abnormality of the heart consisting of pulmonary stenosis, defect of the interventricular septum, hypertrophy of the right ventricle, and overriding or dextroposition of the aorta. Also known as tetrad of Fallot. { tə'träl·ə·jē əv fa'lō }

tetramerous [BIOL] Characterized by or having four parts. { te'tram·ə·rəs }

Tetranychidae [INV ZOO] The spider mites, a family of phytophagous trombidiform mites. { ,te·trə'nik·ə,dē }

Tetraodontiformes [VERT ZOO] An order of specialized teleost fishes that includes the triggerfishes, puffers, trunkfishes, and ocean sunfishes. { ¦te·trə·ō,dänt·ə'fȯr,mēz }

Tetraonidae [VERT ZOO] The ptarmigans and grouse, a family of upland game birds in the order Galliformes characterized by rounded tails and wings and feathered nostrils. { ,te·trə'än·ə,dē }

Tetraphidaceae [BOT] The single family of the plant order Tetraphidales. { ,te·trə·fə'dās·ē,ē }

Tetraphidales [BOT] A monofamilial order of mosses distinguished by scalelike protonema and the peristomes of four rigid, nonsegmented teeth. { ,te·trə·fə'dā·lēz }

Tetraphyllidea [INV ZOO] An order of small tapeworms of the subclass Cestoda characterized by the variation in the structure of the scolex; all species are intestinal parasites of elasmobranch fishes. { ,te·trə·fə'līd·ē·ə }

tetraploidy [CELL MOL] The occurrence of related forms possessing in the somatic cells chromosome numbers four times the haploid number. { 'te·trə,plȯid·ē }

tetrapod [VERT ZOO] A four-footed animal. { 'te·trə,päd }

Tetrapoda [VERT ZOO] The superclass of the subphylum Vertebrata whose members typically possess four limbs; includes all forms above fishes. { te'träp·əd·ə }

Tetrarhynchoidea [INV ZOO] The equivalent name for Trypanorhyncha. { ,te·trə·riŋ'kȯid·ē·ə }

tetrasaccharide [BIOCHEM] A carbohydrate which, on hydrolysis, yields four molecules of monosaccharides. { ,te·trə'sak·ə,rīd }

Tetrasporales [BOT] A heterogeneous and artificial assemblage of colonial fresh-water and marine algae in the division Chlorophyta. { ,te·trə·spə'rā·lēz }

tetraspore [BOT] One of the haploid asexual spores of the red algae formed in groups of four. { 'te·trə,spȯr }

tetraxon [INV ZOO] A type of sponge spicule with four axes. { tə'trak,sän }

Tetrigidae [INV ZOO] The grouse locusts or pygmy grasshoppers in the order Orthoptera in which the front wings are reduced to small scalelike structures. { te'trij·ə,dē }

tetrodotoxin [BIOCHEM] $C_{11}H_{17}N_3O_8$ A toxin that blocks the action potential in the nerve impulse. { ¦te·trə·dō'täk·sən }

tetrose [BIOCHEM] Any of a group of monosaccharides that have a four-carbon chain; an example is erythrose, $CH_2OH\cdot(CHOH)_2\cdot CHO$. { 'te ,trōs }

Tettigoniidae [INV ZOO] A family of insects in the order Orthoptera which have long antennae, hindlegs fitted for jumping, and an elongate, vertically flattened ovipositor; consists of the longhorn or green grasshopper. { ,ted·ə·gə'nī·ə,dē }

Teuthidae [VERT ZOO] The rabbitfishes, a family of perciform fishes in the suborder Acanthuroidei. { tü'thid·ə,dē }

Teuthoidea [INV ZOO] An order of the molluscan subclass Coleoidea in which the rostrum is not developed, the proostracum is represented by the elongated pen or gladus, and ten arms are present. { tü'thȯid·ē·ə }

Texas fever [VET MED] A tick-borne infectious disease of cattle caused by the parasite *Babesia annulatus* which invades erythrocytes; characterized by fever, hemoglobinuria, and splenomegaly. { 'tek·səs ,fē·vər }

textile microbiology [MICROBIO] That branch of microbiology concerned with textile materials; deals with microorganisms that are harmful either to the fibers or to the consumer, and microorganisms that are useful, as in the retting process. { 'tek·stəl ¦mī·krō·bī'äl·ə·jē }

Textulariina [INV ZOO] A suborder of foraminiferan protozoans characterized by an agglutinated wall. { ,tek·styə·lə'rī·ən·ə }

thalamus [ANAT] Either one of two masses of gray matter located on the sides of the third ventricle and forming part of the lateral wall of that cavity. { 'thal·ə·məs }

thalassemia [MED] A hereditary form of hemolytic anemia resulting from a defective synthesis of hemoglobin: thalassemia major is the homozygous state accompanied by clinical illness, and thalassemia minor is the heterozygous state and may not have evident clinical manifestations. Also known as Mediterranean anemia. { ,thal·ə'sē·mē·ə }

Thalassinidea [INV ZOO] The mud shrimps, a group of thin-shelled, burrowing decapod crustaceans belonging to the Macrura; individuals have large chelate or subchelate first pereiopods, and no chelae on the third pereiopods. { thə,las·ə'nid·ē·ə }

Thaliacea [INV ZOO] A small class of pelagic Tunicata in which oral and atrial apertures occur at opposite ends of the body. { ,thā·lē'ā·shē·ə }

thalidomide [PHARM] $C_{13}H_{10}N_2O_4$ A drug used as a sedative and hypnotic; may produce teratogenic effects when administered during pregnancy. { thə'lid·ə,mīd }

Thallobionta [BOT] One of the two subkingdoms of plants, characterized by the absence of specialized tissues or organs and multicellular sex organs. { ¦thal·ō·bī'änt·ə }

Thallophyta [BOT] The equivalent name for Thallobionta. { thə'läf·əd·ə }

thallospore [BOT] A spore that develops by budding of hyphal cells. { 'thal·ə,spȯr }

thallotoxicosis [MED] Poisoning due to ingestion of thallium or its derivatives. { ¦thal·ō,täk·sə'kō·səs }

thallus [BOT] A plant body that is not differentiated into special tissue systems or organs and may vary from a single cell to a complex, branching multicellular structure. { 'thal·əs }

thanatology [MED] The study of the phenomenon of somatic death. { ,than·ə'täl·ə·jē }

Thaumaleidae [INV ZOO] A family of orthorrhaphous dipteran insects in the series Nematocera. { ,thȯ·mə'lē·ə,dē }

Thaumastellidae [INV ZOO] A monospecific family of the Hemiptera assigned to the Pentatomorpha found only in Ethiopia. { ,thȯ·mə'stel·ə,dē }

Thaumastocoridae [INV ZOO] The single family of the hemipteran superfamily Thaumastocoroidea. { thə,mas·tə'kȯr·ə,dē }

Thaumastocoroidea [INV ZOO] A monofamilial superfamily of the Hemiptera in the subdivision Geocorisae which occurs in Australia and the New World tropics. { thə,mas·tə·kə'rȯid·ē·ə }

Thaumatoxenidae [INV ZOO] A family of cyclorrhaphous dipteran insects in the series Aschiza. { ,thȯ·mə·täk'sen·ə,dē }

Theaceae [BOT] A family of dicotyledonous erect trees or shrubs in the order Theales characterized by alternate, exstipulate leaves, usually five petals, and mostly numerous stamens. { thē'ās·ē,ē }

Theales [BOT] An order of dicotyledonous mostly woody plants in the subclass Dilleniidae with simple or occasionally compound leaves, petals usually separate, numerous stamens, and the calyx arranged in a tight spiral. { thē'ā·lēz }

thebaine [PHARM] $C_{19}N_{21}NO_3$ A white, crystalline alkaloid derived from opium; melting point is 193°C; soluble in alcohol and ether; used in medicine. Also known as *para*-morphine. { 'thē,bān }

theca [ANAT] The sheath of dura mater which covers the spinal cord. [BOT] **1.** A moss capsule. **2.** A pollen sac. [HISTOL] The layer of stroma surrounding a Graafian follicle. [INV ZOO] The test of a testate protozoan or a rotifer. { 'thē·kə }

theca folliculi [HISTOL] The capsule surrounding a developing or mature Graafian follicle; consists of two layers, theca interna and theca externa. { 'thē·kə fə'lik·yə·lī }

Thecanephria [INV ZOO] An order of the phylum Brachiata containing a group of elongate, tube-dwelling tentaculate, deep-sea animals of bizarre structure. { ,thē·kə'nef·rē·ə }

thecate [BIOL] Having a theca. { 'thē,kāt }

Thelastomidae [INV ZOO] A family of nematode worms in the superfamily Oxyuroidea. { ,thel·ə'stäm·ə,dē }

thenar [ANAT] The ball of the thumb. { 'thē ,när }

Theophrastaceae [BOT] A family of tropical and subtropical dicotyledonous woody plants in the order Primulales characterized by flowers having staminodes alternate with the corolla lobes. { ,thē·ə·fras'tās·ē,ē }

theoretical community ecology [ECOL] A branch of theoretical ecology that is concerned with factors determining the species composition and functional organization of communities, with a particular emphasis on interspecific interactions such as competition, predation, and mutualism. { ,thē·ə¦red·i·kəl kə,myün·əd·ē ē'käl·ə·jē }

theoretical ecology [ECOL] The use of mathematical models and verbal reasoning to provide a conceptual framework for the analysis of ecological systems. { ,thē·ə,red·ə·kəl ē'käl·ə·jē }

therapeutic abortion [MED] Abortion performed when pregnancy jeopardizes the health or life of the mother. { ¦ther·ə¦pyüd·ik ə'bór·shən }

therapeutic vaccination [IMMUNOL] Vaccination strategies that attempt to induce immune responses after infection with the target disease. { ,ther·ə,pyüd·ik ,vak·sə'nā·shən }

Therevidae [INV ZOO] The stiletto flies, a family of orthorrhaphous dipteran insects in the series Brachycera. { thə'rev·ə,dē }

Theria [VERT ZOO] A subclass of the class Mammalia including all living mammals except the monotremes. { 'ther·ē·ə }

Theridiidae [INV ZOO] The comb-footed spiders, a family of the suborder Dipneumonomorphae. { ,ther·ə'dī·ə,de }

thermal belt [ECOL] Any one of several possible horizontal belts of a vegetation type found in mountainous terrain, resulting primarily from vertical temperature variation. Also known as thermal zone. { 'thər·məl ,belt }

thermal burn [MED] Tissue reaction to or injury resulting from application of heat. { 'thər·məl ¦bərn }

thermal ecology [ECOL] Study of the independent and interactive biotic and abiotic components of naturally heated environments. { 'thər·məl i'käl·ə·jē }

thermal-efficiency index *See* temperature-efficiency index. { 'thər·məl i¦fish·ən·sē ,in,deks }

thermal-efficiency ratio *See* temperature-efficiency ratio. { 'thər·məl i¦fish·ən·sē ,rā·shō }

thermal pollution [ECOL] The discharge of heated effluent into natural waters that causes a rise in temperature sufficient to upset the ecological balance of the waterway. { 'thər·məl pə'lü·shən }

thermal zone *See* thermal belt. { 'thər·məl 'zōn }

thermic fever *See* sunstroke. { 'thər·mik 'fē·vər }

thermoacidophile [BIOL] An organism that grows under extremely acidic conditions and at very high temperatures. { ¦thər·mō·a'sid·ə,fīl }

thermocoagulation [MED] Destruction of tissue by means of electrocautery or a high-frequency current. { ¦thər·mō·kō,ag·yə'lā·shən }

thermoduric bacteria [MICROBIO] Bacteria which survive pasteurization, but do not grow at temperatures used in a pasteurizing process. { ¦thər·mō'dúr·ik bak'tir·ē·ə }

thermogenesis [PHYSIO] The production of heat. { ,thər·mō'jen·ə·səs }

thermography [MED] A medical imaging technique based on detection of heat emitted by the body. { thər'mäg·rə·fē }

thermolysis [PHYSIO] The dissipation and dispersion of body heat by radiation or evaporation processes.

thermometer bird [VERT ZOO] The name applied to the brush turkey, native to Australia, because it lays its eggs in holes in mounds of earth and vegetation, with the heat from the decaying vegetation serving to incubate the eggs. { thər'mäm·əd·ər ,bərd }

thermoperiodicity [BOT] The totality of responses of a plant to appropriately fluctuating temperatures. { ¦thər·mō,pir·ē·ə'dis·əd·ē }

thermophile [BIOL] An organism that thrives at high temperatures. { 'thər·mə,fīl }

thermoreception [PHYSIO] The process by which environmental temperature affects specialized sense organs (thermoreceptors). { ¦thər·mō·ri'sep·shən }

thermoreceptor [PHYSIO] A sense receptor that responds to stimulation by heat and cold. { ¦thər·mō·ri'sep·tər }

thermoregulation [PHYSIO] A mechanism by which mammals and birds attempt to balance heat gain and heat loss in order to maintain a constant body temperature when exposed to variations in cooling power of the external medium. { ¦thər·mō,reg·yə'lā·shən }

Thermosbaenacea [INV ZOO] An order of small crustaceans in the superorder Pancarida. { ,thər·məs·bə'nās·ē·ə }

Thermosbaenidae [INV ZOO] A family of the crustacean order Thermosbaenacea. { ˌthər·məs'bē·nəˌdē }

thermotaxis [BIOL] Orientation movement of a motile organism in response to the stimulus of a temperature gradient. { ¦thər·mō¦tak·səs }

thermotherapy [MED] The treatment of disease by heat of any kind; involves the local or general application of heat to the body. { ¦thər·mō'ther·ə·pē }

therophyte [ECOL] An annual plant whose seed is the only overwintering structure. { 'ther·əˌfīt }

Thesium [BOT] The hymenium of the apothecium in lichens. { 'thē·sē·əm }

thesocyte [INV ZOO] An amebocyte in Porifera containing ergastic cytoplasmic inclusions. { 'thes·əˌsīt }

theta antigen [IMMUNOL] A cell membrane constituent which distinguishes T cells from other lymphocytes. { 'thād·ə 'ant·i·jən }

thiamine [BIOCHEM] $C_{12}H_{17}ClN_4OS$ A member of the vitamin B complex that occurs in many natural sources, frequently in the form of cocarboxylase. Also known as aneurine; vitamin B_1. { 'thī·ə·mən }

thiamine pyrophosphate [BIOCHEM] The coenzyme or prosthetic component of carboxylase; catalyzes decarboxylation of various α-keto acids. Also known as cocarboxylase. { 'thī·ə·mən ¦pī·rō'făˌsfāt }

Thiaridae [INV ZOO] A family of freshwater gastropod mollusks in the order Pectinibranchia. { ˌthī'ar·əˌdē }

thicket *See* tropical scrub. { 'thik·ət }

thickhead *See* bluetongue. { 'thikˌhed }

thick-tailed bushbaby [VERT ZOO] *Galago crassicaudatus*. A primate animal in the family Lorisidae; one of the six species of bushbaby, the thick-tailed bushbaby is more aggressive and solitary than the other species, and grows to over 1 foot (30 centimeters) in length with an equally long tail. Also known as great galago. { 'thik ¦tāld 'bu̇shˌbā·bē }

thigh [ANAT] The upper part of the leg, from the pelvis to the knee. { thī }

thigmotaxis *See* stereotaxis. { ¦thig·mə¦tak·səs }

Thigmotrichida [INV ZOO] An order of ciliated protozoans in the subclass Holotrichia. { ¦thig·mō'trik·əd·ə }

thigmotropism *See* stereotropism. { thig'mä·trəˌpiz·əm }

Thinocoridae [VERT ZOO] The seed snipes, family of South American birds in the order Charadriiformes. { ˌthin·ə'kȯr·əˌdē }

Thiobacillus ferrooxidans [MICROBIO] An aerobic rod-shaped microorganism that derives its energy from the oxidation of various sulfide minerals and soluble ferrous ion (Fe^{2+}); it thrives in acidic environments of pH 1–3, conditions that would be fatal to most other life-forms. { ˌthī·ō·bəˌsil·əs fə'räk·səˌdanz }

Thiobacteriaceae [MICROBIO] Formerly a family of nonfilamentous, gram-negative bacteria of the suborder Pseudomonadineae characterized by the ability to oxidize hydrogen sulfide, free sulfur, and inorganic sulfur compounds to sulfuric acid. { ¦thī·ō·bakˌtir·ē'ās·ēˌē }

thiobarbiturate [PHARM] A derivative of thiobarbituric acid that differs from the barbiturates in the replacement of one oxygen atom by sulfur but resembles the barbiturates in its effects. { ¦thī·ō·bär'bich·əˌrāt }

thiocarbamazine [PHARM] $C_{21}H_{17}AsN_2O_5S_2$ A white crystalline powder, freely soluble in dilute alkali; used in medicine as an amebicide. { ¦thī·ō·kär'bam·əˌzēn }

thiocarbarsone [PHARM] $C_{11}H_{13}AsN_2O_5S_2$ A white crystalline powder, freely soluble in dilute alkali; used in medicine as an amebicide. { ¦thī·ō'kär·bəˌsōn }

thiocresol [PHARM] $CH_3C_6H_4SH$ A compound with three isomers, of which only *meta*-thiocresol is a liquid; all three have a boiling point of 195°C and are soluble in alcohol or ether; used as an antiseptic. Also known as toluenethiol; tolylmercaptan. { ¦thī·ō'krēˌsōl }

Thiorhodaceae [MICROBIO] Formerly a family of bacteria in the suborder Rhodobacteriineae composed of the purple, red, orange, and brown sulfur bacteria; characterized as strict anaerobes which oxidize hydrogen sulfide and store sulfur globules internally. { ˌthī·ə·rō'dās·ēˌē }

thiosemicarbazone [PHARM] A class of chemical compounds used in treating tuberculosis; the most prominent member of the group is *para*-acetamidobenzaldehyde thiosemicarbazone. { ¦thī·ōˌsem·i'kär·bəˌzōn }

thiostreptone [MICROBIO] A polypeptide antibiotic produced by a species of *Streptomyces* that crystallizes from a chloroform methanol solution; used in veterinary medicine. { ¦thī·ō 'strepˌtōn }

thiouracil [PHARM] $C_4H_4N_2OS$ A bitter-tasting white powder; used as an antithyroid drug that acts by interfering with thyroxine synthesis. { 'thī·ō'yu̇r·əˌsil }

thirst [PHYSIO] A sensation, as of dryness in the mouth and throat, resulting from water deprivation. { 'thərst }

thistle [BOT] Any of the various prickly plants comprising the family Compositae. { 'this·əl }

thonzylamine hydrochloride [PHARM] $C_{16}H_{22}N_4O·HCl$ A white, crystalline powder with a melting point of 173–176°C; soluble in water, alcohol, and chloroform; used in medicine as an antihistamine. { thän'zil·əˌmēn ¦hī·drə'klȯrˌīd }

Thoracica [INV ZOO] An order of the subclass Cirripedia; individuals are permanently attached in the adult stage, the mantle is usually protected by calcareous plates, and six pairs of biramous thoracic appendages are present. { thə 'ras·ə·kə }

thoracic cavity *See* thorax. { thə'ras·ik 'kav·əd·ē }

thoracic duct [ANAT] The common lymph trunk beginning in the crura of the diaphragm at the level of the last thoracic vertebra, passing upward, and emptying into the left subclavian vein at its junction with the left internal jugular vein. { thə'ras·ik 'dəkt }

thoracic vertebrae [ANAT] The vertebrae associated with the chest and ribs in vertebrates; there are 12 in humans. { thə'ras·ik 'vərd·ə ˌbrā }

thoracoabdominal breathing [PHYSIO] The process of air breathing in reptiles, birds, and mammals that depends upon aspiration or sucking inspiration, and involves trunk musculature to supply pulmonary ventilation. { ¦thȯr·ə·kō·ab'däm·ə·nəl 'brēth·iŋ }

Thoracostomopsidae [INV ZOO] A family of marine nematodes in the superfamily Enoploidea, which have the stomatal armature modified to form a hollow tube. { ¦thȯr·ə·kō·stō'mäp·sə,dē }

thorax [ANAT] The chest; the cavity of the mammalian body between the neck and the diaphragm, containing the heart, lungs, and mediastinal structures. Also known as thoracic cavity. [INV ZOO] The middle of three principal divisions of the body of certain classes of arthropods. { 'thȯr,aks }

Thorictidae [INV ZOO] The ant blood beetles, a family of coleopteran insects in the superfamily Dermestoidea. { thə'rik·tə,dē }

thorn [BOT] A short, sharp, rigid, leafless branch on a plant. [ZOO] Any of various sharp spinose structures on an animal. { thȯrn }

thornback [VERT ZOO] *Raja clavata.* A ray found in European waters and characterized by spines on its back. { 'thȯrn,bak }

thornbush [ECOL] A vegetation class that is dominated by tall succulents and profusely branching smooth-barked deciduous hardwoods which vary in density from mesquite bush in the Caribbean to the open spurge thicket in Central Africa; the climate is that of a warm desert, except for a rather short intense rainy season. Also known as Dorngeholz; Dorngestrauch; dornveld; savane armée; savane épineuse; thorn scrub. { 'thȯrn,bu̇sh }

thorn forest [ECOL] A type of forest formation, mostly tropical and subtropical, intermediate between desert and steppe; dominated by small trees and shrubs, many armed with thorns and spines; leaves are absent, succulent, or deciduous during long dry periods, which may also be cool; an example is the caatinga of northeastern Brazil. { 'thȯrn 'fär·əst }

thorn scrub *See* thornbush. { 'thȯrn 'skrəb }

threadfin [VERT ZOO] Common name for any of the fishes in the family Polynemidae. { 'thred,fin }

three-day fever *See* phlebotomus fever. { 'thrē ¦dā 'fē·vər }

threonine [BIOCHEM] $CH_3CHOHCH(NH_2)COOH$ A crystalline α-amino acid considered essential for normal growth of animals; it is biosynthesized from aspartic acid and is a precursor of isoleucine in microorganisms. { 'thrē·ə,nēn }

thresher shark [VERT ZOO] Common name for fishes in the family Alopiidae; pelagic predacious sharks of generally wide distribution that have an extremely long, whiplike tail with which they thrash the water, destroying schools of small fishes. { 'thresh·ər ˌshärk }

threshold [PHYSIO] The minimum level of a stimulus that will evoke a response in an irritable tissue. { 'thresh,hōld }

threshold limit value [MED] The average concentration of toxic gas to which the normal person can be exposed without injury for 8 hours per day, 5 days per week for an unlimited period; differs slightly from maximum allowable concentration in that threshold limit value is an average concentration. Abbreviated TLV. { 'thresh ˌhōld 'lim·ət ˌval·yü }

threshold of audibility [PHYSIO] The minimum effective sound pressure of a specified signal that is capable of evoking an auditory sensation in a specified fraction of the trials; the threshold may be expressed in decibels relative to 0.0002 microbar (2×10^{-5} pascal) or 1 microbar (0.1 pascal). Also known as threshold of detectability; threshold of hearing. { 'thresh,hōld əv ˌȯd·ə'bil·əd·ē }

threshold of detectability *See* threshold of audibility. { 'thresh,hōld əv di,tek·tə'bil·əd·ē }

threshold of feelings [PHYSIO] The minimum effective sound pressure of a specified signal that, in a specified fraction of trials, will stimulate the ear to a point at which there is a sensation of feeling, discomfort, tickle, or pain; normally expressed in decibels relative to 0.0002 microbar (0.00002 pascal) or 1 microbar (0.1 pascal). { 'thresh,hōld əv 'fēl·iŋz }

threshold of hearing *See* threshold of audibility. { 'thresh,hōld əv 'hir·iŋ }

Threskiornithidae [VERT ZOO] The ibises, a family of long-legged birds in the order Ciconiiformes. { ˌthres·kē·ȯr'nith·ə,dē }

thrip [INV ZOO] A small, slender-bodied phytophagous insect of the order Thysanoptera with suctorial mouthparts, a stout proboscis, a vestigial right mandible, and a fully developed left mandible, while wings may be present or absent. { thrip }

Thripidae [INV ZOO] A large family of thrips, order Thysanoptera, which includes the most common species. { 'thrip·ə,dē }

throat [ANAT] The region of the vertebrate body that includes the pharynx, the larynx, and related structures. [BOT] The upper, spreading part of the tube of a gamopetalous calyx or corolla. { thrōt }

thrombin [BIOCHEM] An enzyme elaborated from prothrombin in shed blood which induces clotting by converting fibrinogen to fibrin. { 'thräm·bən }

thrombinogen *See* prothrombin. { thräm'bin·ə·jən }

thromboangitis obliterans [MED] Thrombosis with organization and a variable degree of associated inflammation in the arteries and veins of the extremities, occasionally of the viscera, progressing to fibrosis about these structures and associated nerves, and complicated by ischemic changes in the parts supplied. Also known as Buerger's disease. { ¦thräm·bō·an'jīd·əs ō'blid·ə,ranz }

thrombocyte [HISTOL] One of the minute protoplasmic disks found in vertebrate blood; thought to be fragments of megakaryocytes. Also known as blood platelet; platelet. { 'thräm·bə,sīt }

thrombocythemia *See* thrombocytosis. { ¦thräm·bō,sī'thē·mē·ə }

thrombocytopenia [MED] The condition of having an abnormally small number of platelets in the circulating blood. { ¦thräm·bō,sīd·ō'pē·nē·ə }

thrombocytopenic purpura [MED] Hemorrhages in the skin, mucous membranes, and elsewhere associated with a decreased number of thrombocytes per unit volume of blood. { ¦thräm·bō¦sīd·ō¦pē·nik 'pər·pə·rə }

thrombocytosis [MED] A condition characterized by an increase in the absolute number of thrombocytes in the circulation. Also known as piastrenemia; thrombocythemia. { ,thräm·bō,sī'tō·səs }

thromboembolectomy [MED] Surgical removal of an embolus that stems from a dislodged thrombus or part of a thrombus. { ¦thräm·bō ,em·bə'lek·tə·mē }

thromboembolism [MED] An embolism resulting from a dislodged thrombus or part of a thrombus. { ¦thräm·bō'em·bə,liz·əm }

thrombokinase [BIOCHEM] A proteolytic enzyme in blood plasma that, together with thromboplastin, calcium, and factor V, converts prothrombin to thrombin. { ¦thräm·bō'kī,nās }

thrombophlebitis [MED] Inflammation of a vein associated with thrombosis. { ¦thräm·bō·fle 'bīd·əs }

thromboplastin [BIOCHEM] Any of a group of lipid and protein complexes in blood that accelerate the conversion of prothrombin to thrombin. Also known as factor III; plasma thromboplastin component (PTC). { ,thräm·bō'plas·tən }

thromboplastinogen *See* antihemophilic factor. { ¦thräm·bō,plas'tin·ə·jən }

thrombosis [MED] Formation of a thrombus. { thräm'bō·səs }

thrombotic thrombocytopenic purpura [MED] Thrombi in blood vessels associated with deposits of hyaline substances in the walls and with thrombocytopenia. Also known as Moschcowitz's disease. { thräm'bäd·ik ¦thräm·bō¦sīd·ō¦pen·ik 'pər·pə·rə }

thromboxane [BIOCHEM] Any member of a group of 20-carbon fatty acids related to the prostaglandins and derived mainly from arachidonic acid. { ,thräm'bäk,sān }

thrombus [MED] A blood clot occurring on the wall of a blood vessel where the endothelium is damaged. { 'thräm·bəs }

Throscidae [INV ZOO] The false metallic woodboring beetles, a cosmopolitan family of the Coleoptera in the superfamily Elateroidea. { 'thräs·kə,dē }

thrush [MED] A form of candidiasis due to infection by *Candida albicans* and characterized by small whitish spots on the tip and sides of the tongue and the mucous membranes of the buccal cavity. Also known as mycotic stomatitis; parasitic stomatitis. [VET MED] A disease of the frog of a horse's foot accompanied by a fetid discharge. { thrəsh }

Thunburg technique [BIOCHEM] A technique used to study oxidation of a substrate occurring by dehydrogenation reactions; methylene blue, a reversibly oxidizable indicator, substitutes for molecular oxygen as the ultimate hydrogen acceptor (oxidant), becoming reduced to the colorless leuco form. { 'thən,bərg tek,nēk }

Thunnidae [VERT ZOO] The tunas, a family of perciform fishes; there are no scales on the posterior part of the body, and those on the anterior are fused to form an armored covering, the body is streamlined, and the tail is crescent-shaped. { 'thən·ə,dē }

thunniform motion [VERT ZOO] A type of locomotion in which a fish, such as a tuna, moves only the latter third of its body. { 'thən·ə,fórm ,mō·shən }

Thurniaceae [BOT] A small family of monocotyledonous plants in the order Juncales distinguished by an inflorescence of one or more dense heads, vascular bundles of the leaf in vertical pairs, and silica bodies in the leaf epidermis. { ,thər·nē'ās·ē,ē }

Thylacinidae [VERT ZOO] A family of Australian carnivorous marsupials in the superfamily Dasyuroidea. { ,thī·lə'sīn·ə,dē }

thylakoid [CELL MOL] An internal membrane system which occupies the main body of a plastid; particularly well developed in chloroplasts. { 'thī·lə,kóid }

thyme [BOT] A perennial mint plant of the genus *Thymus*; pungent aromatic herb is made from the leaves. { tīm }

thymectomy [MED] Surgical removal of the thymus gland. { thī'mek·tə·mē }

Thymelaeceae [BOT] A family of dicotyledonous woody plants in the order Myrtales characterized by a superior ovary with a solitary ovule, and petals, if present, are scalelike. { ,thī·mə·lē·ə'ās·ē,ē }

thymic aplasia [MED] Congenital absence of the thymus and of the parathyroids with deficient cellular immunity. Also known as Di George's syndrome. { 'thī·mik ə'plā·zhə }

thymic corpuscle [HISTOL] A characteristic, rounded, acidophil body in the medulla of the thymus; composed of hyalinized cells concentrically arranged about a core which is occasionally calcified. Also known as Hassal's body. { 'thī·mik 'kór·pə·səl }

thymidine [BIOCHEM] $C_{10}H_{14}N_2O_5$ A nucleoside derived from deoxyribonucleic acid; essential growth factor for certain microorganisms in mediums lacking vitamin B_{12} and folic acid. { 'thī·mə,dēn }

thymidylic acid [BIOCHEM] $C_{10}H_{15}N_2O_8P$ A mononucleotide component of deoxyribonucleic acid which yields thymine, D-ribose, and phosphoric acid on complete hydrolysis. { ¦thī·mə¦dil·ik 'as·əd }

thymine [BIOCHEM] $C_5H_6N_2O_2$ A pyrimidine component of nucleic acid, first isolated from the thymus. { 'thī,mēn }

thymocyte [HISTOL] A lymphocyte formed in the thymus. { 'thī·mə,sīt }

thymoma [MED] A usually benign primary tumor of the thymus composed principally of lymphocytic and epithelial cells in varying proportions. { thī'mō·mə }

thymopharyngeal duct [EMBRYO] The third pharyngobranchial duct; it may elongate between the pharynx and thymus. { ¦thī·mō·fə'rin·jē·əl 'dəkt }

thymosin [IMMUNOL] Any of a group of hormones secreted by the thymus gland that stimulate lymphocyte production within the thymus and confer on lymphocytes elsewhere in the body the capacity to respond to antigenic stimulation. { 'thī·mə·sən }

thymulin [IMMUNOL] A zinc-dependent thymic hormone that regulates the differentiation of the immature thymocyte subpopulation and the function of mature T and natural killer cells and also functions as a transmitter between the neuroendocrine and immune systems. { 'thī·myü·lən }

thymus gland [ANAT] A lymphoid organ in the neck or upper thorax of all vertebrates; it is most prominent in early life and is essential for normal development of the circulating pool of lymphocytes. { 'thī·məs ,gland }

Thyrididae [INV ZOO] The window-winged moths, a small tropical family of lepidopteran insects in the suborder Heteroneura. { thī'rid·ə,dē }

thyrocalcitonin *See* calcitonin. { ¦thī·rō,kal·sə'tō·nən }

thyroglobulin [BIOCHEM] An iodinated protein found as the storage form of the iodinated hormones in the thyroid follicular lumen and epithelial cells. { ¦thī·rō'gläb·yə·lən }

thyroglossal cyst [MED] A cyst formed from the remnants of the thyroglossal duct. { ¦thī·rō 'gläs·əl 'sist }

thyroglossal duct [EMBRYO] A narrow temporary channel connecting the anlage of the thyroid with the surface of the tongue. { thī·rō'gläs·əl 'dəkt }

thyroid [PHARM] Dried and powdered thyroid gland which contains about 0.2% iodine in combination, especially as thyroxine, and is used therapeutically in the treatment of thyroid deficiencies. { 'thī,rȯid }

thyroid cartilage [ANAT] The largest of the laryngeal cartilages in humans and most other mammals, located anterior to the cricoid; in humans, it forms the Adam's apple. { 'thī,rȯid ,kärt·lij }

thyroidectomy [MED] Surgical removal of the thyroid gland. { ,thī·rȯi'dek·tə·mē }

thyroid gland [ANAT] An endocrine gland found in all vertebrates that produces, stores, and secretes the thyroid hormones. { 'thī,rȯid ,gland }

thyroid hormone [BIOCHEM] Commonly, thyroxine or triiodothyronine, or both; a metabolically active compound formed and stored in the thyroid gland which functions to regulate the rate of metabolism. { 'thī,rȯid 'hȯr,mōn }

thyroiditis [MED] Inflammation of the thyroid gland. { ,thī,rȯi'dīd·əs }

thyroid-stimulating hormone *See* thyrotropic hormone. { 'thī,rȯid ¦stim·yə,lād·iŋ 'hȯr,mōn }

thyroprotein [BIOCHEM] A protein secreted in the thyroid gland, such as thyroxine. { ¦thī·rō'prō,tēn }

Thyropteridae [VERT ZOO] The New World disk-winged bats, a family of the Chiroptera found in Central and South America, characterized by a stalked sucking disk and a well-developed claw on the thumb. { ,thī,räp'ter·ə,dē }

thyrotoxic myopathy [MED] A chronic disease associated with hyperthyroidism resulting in muscular atrophy. { ¦thī·rō¦täk·sik mī'äp·ə·thē }

thyrotoxicosis *See* hyperthyroidism. { ¦thī·rō¦täk·sə'kō·səs }

thyrotropic hormone [BIOCHEM] A hormone produced by the adenohypophysis which regulates thyroid gland function. Also known as thyroid-stimulating hormone (TSH). { ¦thī·rə ¦träp·ik 'hȯr,mōn }

thyrotropin [BIOCHEM] A thyroid-stimulating hormone produced by the adenohypophysis. { thī'rä·trə·pən }

thyroxine [BIOCHEM] $C_{15}H_{11}I_4NO_4$ The active physiologic principle of the thyroid gland; used in the form of the sodium salt for replacement therapy in states of hypothyroidism or absent thyroid function. { thī'räk,sēn }

thyrse [BOT] An inflorescence with a racemose primary axis and cymose secondary and later axes. { thərs }

Thysanidae [INV ZOO] A family of hymenopteran insects in the superfamily Chalcidoidea. { thī'san·ə,dē }

Thysanoptera [INV ZOO] The thrips, an order of small, slender insects having exopterygote development, sucking mouthparts, and exceptionally narrow wings with few or no veins and bordered by long hairs. { ,thī·sə'näp·tə·rə }

Thysanura [INV ZOO] The silverfish, machilids, and allies, an order of primarily wingless insects with soft, fusiform bodies. { ,thī·sə'nu̇r·ə }

tibia [ANAT] The larger of the two leg bones, articulating with the femur, fibula, and talus. { 'tib·ē·ə }

tibialis [ANAT] **1.** A muscle of the leg arising from the proximal end of the tibia and inserted into the first cuneiform and first metatarsal bones. **2.** A deep muscle of the leg arising proximally from the tibia and fibula and inserted into the navicular and first cuneiform bones. { ,tib·ē'al·əs }

tic douloureux *See* trigeminal neuralgia. { ,tik ,dü·lə'rü }

tick [INV ZOO] Any arachnid comprising Ixodoidea; a bloodsucking parasite and important vector of various infectious diseases of humans and lower animals. { tik }

tick-bite paralysis [VET MED] A flaccid paralysis in animals, and occasionally in humans, caused by a feeding tick attached to the body. { 'tik ¦bīt pə'ral·ə·səs }

tick-borne typhus fever of Africa [MED] Any of several infections caused by *Rickettsia conori*, transmitted by ixodid ticks, and occurring in Africa and adjacent areas; includes boutonneuse fever, Marseilles fever, Kenya tick typhus fever, and South African tick-bite fever. { 'tik ¦bȯrn 'tī·fəs ,fē·vər əv 'af·ri·kə }

tick fever *See* Rocky Mountain spotted fever. { 'tik ,fē·vər }

tickle [PHYSIO] A tingling sensation of the skin or a mucous membrane following light, tactile stimulation. { 'tik·əl }

tick typhus *See* Rocky Mountain spotted fever. { 'tik ,ti·fəs }

tic polonga [VERT ZOO] *Vipera russellii*. A member of the Viperidae; one of the most deadly and most common snakes in India; it may reach a length of 5 feet (1.5 meters), is nocturnal in its habits, and pursues rodents into houses. Also known as Russell's viper. { ,tik pə'lȯŋ·gə }

tidal air [PHYSIO] That air which is inspired and expired during normal breathing. { 'tīd·əl 'er }

tidal stand *See* stand. { 'tīd·əl 'stand }

tidal volume [PHYSIO] The volume of air moved in and out of the lungs during a single normal respiratory cycle. { 'tīd·əl 'väl·yəm }

Tiedenmann's body [INV ZOO] One of the small glands opening into the ring vessel in many echinoderms in which amebocytes are produced. { 'tēd·ən,mänz ,bäd·ē }

tiger [VERT ZOO] *Felis tigris*. An Asiatic carnivorous mammal in the family Felidae characterized by a tawny coat with transverse black stripes and white underparts. { 'tī·gər }

tiger beetle [INV ZOO] The common name for any of the bright-colored beetles in the family Cicindelidae; there are about 1300 species distributed all over the world. { 'tī·gər ,bēd·əl }

tiger salamander [VERT ZOO] *Ambystoma tigrinum*. A salamander in the family Ambystomatidae, found in a variety of subspecific forms from Canada to Mexico and over most of the United States; lives in arid and humid regions, and is the only salamander in much of the Great Plains and Rocky Mountains. { 'tī·gər ¦sal·ə,man·dər }

tiger shark *See* sand shark. { 'tī·gər ,shärk }

tight junction [CELL MOL] An intercellular junction composed of a series of fusions of the junctional membrane, forming a continuous seal; serves as a selective barrier to small molecules and as a total barrier to large molecules. Also known as impermeable junction; occluding junction; zonula occludens. { ¦tīt 'jəŋk·shən }

tiller [BOT] A shoot that develops from an axillary or adventitious bud at the base of a stem. { 'til·ər }

Tilletiaceae [MYCOL] A family of fungi in the order Ustilaginales in which basidiospores form at the tip of the apibasidium. { tə,lē·shē'ās·ē,ē }

timberline [ECOL] The elevation or latitudinal limits for arboreal growth. Also known as tree line. { 'tim·bər,līn }

timothy [BOT] *Phleum pratense*. A perennial hay grass of the order Cyperales characterized by moderately leafy stems and a dense cylindrical inflorescence. { 'tim·ə·thē }

Tinamidae [VERT ZOO] The single family of the avian order Tinamiformes. { ti'nam·ə,dē }

Tinamiformes [VERT ZOO] The tinamous, an order of South and Central American birds which are superficially fowllike but have fully developed wings and are weak fliers. { ti,nam·ə'fȯr,mēz }

tincture of iodine [PHARM] A medicinal preparation used as an anti-infective containing 20 grams iodine and 24 grams sodium iodide in 1000 milliliters of alcohol. Also known as iodine solution; iodine tincture. { 'tiŋk·chər əv 'ī·ə,dīn }

tinea [MED] Group of skin diseases caused by various fungi, for example, tinea pedis (athlete's foot) and tinea capitis (ringworm infection of the scalp). { ,tin·ē·ə }

tinea favosa *See* favus. { 'tin·ē·ə fə'vō·sə }

Tineidae [INV ZOO] A family of small moths in the superfamily Tineoidea distinguished by an erect, bristling vestiture on the head. { ti'nē·ə,dē }

Tineoidea [INV ZOO] A superfamily of heteroneuran Lepidoptera which includes small moths that usually have well-developed maxillary palpi. { ,tin·ē'ȯid·ē·ə }

Tingidae [INV ZOO] The lace bugs, the single family of the hemipteran superfamily Tingoidea. { 'tin·jə,dē }

Tingoidea [INV ZOO] A superfamily of the Hemiptera in the subdivision Geocorisae characterized by the wings with many lacelike areolae. { tiŋ'gȯid·ē·ə }

tinnitus [MED] A ringing, roaring, or hissing sound in one or both ears. { 'tin·əd·əs }

Tintinnida [INV ZOO] An order of ciliated protozoans in the subclass Spirotrichia whose members are conical or trumpet-shaped pelagic forms bearing shells. { tin'tin·əd·ə }

Tiphiidae [INV ZOO] A family of the Hymenoptera in the superfamily Scolioidea. { tə'fī·ə,dē }

tip layering [BOT] A plant propagation technique in which only the stem tip is buried, used to reproduce trailing blackberries and black raspberries. { 'tip ,lā·ə·riŋ }

Tipulidae [INV ZOO] The crane flies, a family of orthorrhaphous dipteran insects in the series Nematocera. { tə'pyül·ə,dē }

tissue [HISTOL] An aggregation of cells more or less similar morphologically and functionally. { 'tish·ü }

tissue culture [CELL MOL] Growth of tissue cells in artificial media. { 'tish·ü ,kəl·chər }

tissue engineering [MED] The creation of tissues or organs to replace lost form or function. { 'tish·ü ,en·jə,nir·iŋ }

tissue plasminogen activator [MED] A proteolytic enzyme that can convert plasminogen to plasmin. { ¦tish·ü plaz,min·ə·jən 'ak·tə,vād·ər }

tissue typing [MED] A procedure involving a test or a series of tests to determine the compatibility of tissues from a prospective donor and a recipient prior to transplantation. { 'tish·ü ˌtīp·iŋ }

TLV *See* threshold limit value.

T lymphocyte *See* T cell. { ¦tē 'lim·fəˌsīt }

T method [BOT] A budding method in which a T-shaped cut is made through the bark at the internode of the stock, the bark of the scion is separated from the xylem along the cambium and removed, and the scion is forced into the incision on the stock. { 'tē ˌmeth·əd }

TMJ *See* temporamandibular joint disease.

TMV *See* tobacco mosaic virus.

toad [VERT ZOO] Any of several species of the amphibian order Anura, especially in the family Bufonidae; glandular structures in the skin secrete acrid, irritating substances of varying toxicity. { tōd }

toadstool [MYCOL] Any of various fleshy, poisonous or inedible fungi with a large umbrella-shaped fruiting body. { 'tōdˌstül }

tobacco [BOT] **1.** Any plant of the genus *Nicotinia* cultivated for its leaves, which contain 1–3% of the alkaloid nicotine. **2.** The dried leaves of the plant. { tə'bak·ō }

tobacco budworm [INV ZOO] The larva of a noctuid moth, *Heliothis virescens*, that damages the buds and young leaves of tobacco. { təˌbak·ō 'bədˌwərm }

tobacco mosaic virus [VIROL] The type species of the genus *Tobamovirus*, it infects tobacco, tomato, and other solanaceous plants, causing defoliation and/or mosaic symptoms on leaves, stems, and fruit. Abbreviated TMV. { tə¦bak·ō mō'zā·ik ˌvī·rəs }

tobacco mosaic virus group *See* Tobamovirus. { tə¦bak·ō mō'zā·ik ¦vī·rəs ˌgrüp }

tobacco rattle virus [VIROL] The type species of the genus *Tobravirus*, it infects a wide range of plants, usually via a nematode vector. Abbreviated TRV. { tə¦bak·ō 'rad·əl ˌvī·rəs }

tobacco rattle virus group *See* Tobravirus. { tə¦bak·ō 'rad·əl ¦vī·rəs ˌgrüp }

tobacco ring spot virus [VIROL] The type species of the genus *Nepovirus*, it has a wide host range and is transmitted via seeds and nematodes. Abbreviated TRSV. { tə¦bak·ō 'riŋ ˌspät ˌvī·rəs }

tobacco ring spot virus group *See* Nepovirus. { tə¦bak·ō 'riŋ ˌspät ¦vī·rəs ˌgrüp }

tobacco streak virus [VIROL] The type species of the genus *Ilarvirus*. { tə¦bak·ō 'stāk ˌvī·rəs }

tobacco streak virus group *See* Ilarvirus. { tə¦bak·ō 'stāk ¦vī·rəs ˌgrüp }

Tobamovirus [VIROL] A genus of plant viruses characterized by particles with a rigid helical rod containing linear single-stranded ribonucleic acid. Tobacco mosaic virus is the type species. Also known as tobacco mosaic virus group. { tə'bam·əˌvī·rəs }

Tobravirus [VIROL] A genus of plant viruses characterized by two types of rigid helical rods containing linear single-stranded ribonucleic acid. The type species is tobacco rattle virus. Also known as tobacco rattle virus group. { 'tō·brəˌvī·rəs }

Todidae [VERT ZOO] The todies, a family of birds in the order Coraciiformes found in the West Indies. { 'tō·dəˌdē }

toe [ANAT] One of the digits on the foot of humans and other vertebrates. { tō }

Togaviridae [VIROL] A family of positive-strand ribonucleic acid (RNA)-containing viruses characterized by spherical enveloped particles with an icosahedral nucleocapsid containing linear single-stranded RNA, it contains the genera *Alphavirus* (arbovirus A; prototype Sindbis virus), *Flavivirus* (arbovirus B; prototype yellow fever), Rubivirus (rubella virus), and Pestivirus (mucosal disease virus). { ˌtō·gə'vir·əˌdī }

tolbutamide [PHARM] $C_{12}H_{18}N_2O_3S$ A hypoglycemic drug effective when administered orally. { täl'byüd·əˌmīd }

toleragen [IMMUNOL] A substance which, in appropriate dosages, produces a state of specific immunological tolerance in humans or animals. { 'täl·ə·rə·jən }

tolerance [PHARM] **1.** The ability of enduring or being less responsive to the influence of a drug or poison, particularly when acquired by continued use of the substance. **2.** The allowable deviation from a standard, as the range of variation permitted for the content of a drug in one of its dosage forms. { 'täl·ə·rəns }

toluenethiol *See* thiocresol. { ¦täl·yəˌwēn'thīˌȯl }

tolylmercaptan *See* thiocresol. { ¦täˌlil·mər'kap ˌtan }

tomato [BOT] A plant of the genus *Lycopersicon*, especially *L. esculentum*, in the family Solanaceae cultivated for its fleshy edible fruit, which is red, pink, orange, yellow, white, or green, with fleshy placentas containing many small, oval seeds with short hairs and covered with a gelatinous matrix. { tə'mād·ō }

tomato bushy stunt virus [VIROL] The type species of the plant-virus genus *Tombusvirus*. Abbreviated TBSV. { təˌmād·ō 'bu̇sh·ē ˌstənt ˌvī·rəs }

tomato bushy stunt virus group *See* Tombusvirus. { təˌmād·ō 'bu̇sh·ē ˌstənt ˌvī·rəs ˌgrüp }

Tombusviridae [VIROL] A family of plant viruses characterized by nonenveloped icosahedral particles containing linear, positive-sense ribonucleic acid, genera include *Tombusvirus*, *Carmovirus*, *Necrovirus*, *Dianthovirus*, and *Machlomovirus*. { ˌtm·bəs'vir·əˌdī }

Tombusvirus [VIROL] A genus of the family Tombusviridae; tomato bushy stunt virus is the type species. Also known as Tomato bushy stunt virus group. { 'tam·bəsˌvī·rəs }

tomentose [BOT] Covered with densely matted hairs. { tə'menˌtōs }

tomentum [ANAT] The deep layer of the pia mater composed principally of minute blood vessels. [BIOL] Pubescence consisting of densely matted wooly hairs. { tə'men·təm }

tomium [VERT ZOO] The cutting edge of a bird's beak. { 'tō·mē·əm }

Tomopteridae [INV ZOO] The glass worms, a family of pelagic polychaete annelids belonging to the group Errantia. { tə,mäp'ter·ə,dē }

tongue [ANAT] A muscular organ located on the floor of the mouth in humans and most vertebrates which may serve various functions, such as taking and swallowing food or tasting or as a tactile organ or sometimes a prehensile organ. { təŋ }

tongue worm *See* acorn worm. { 'təŋ ,wərm }

tonic convulsion *See* tonic postural epilepsy; tonic spasm. { 'tän·ik kən'vəl·shən }

tonic labyrinthine reflexes [PHYSIO] Rotation or deviation of the head causes extension of the limbs on the same side as the chin, and flexion of the opposite extremities: dorsiflexion of the head produces increased extensor tonus of the upper extremities and relaxation of the lower limbs, while ventroflexion of the head produces the reverse; seen in the young infant and patients with a lesion at the midbrain level or above. { 'tän·ik ,lab·ə'rin,thēn 'rē,flek·səz }

tonic neck reflexes [PHYSIO] Reflexes in which rotation or deviation of the head causes extension of the limbs on the same side as the chin, and flexion of the opposite extremities; dorsiflexion of the head produces increased extension tonus of the upper extremities and relaxation of the lower limbs, and ventroflexion of the head, the reverse; seen normally in incomplete forms in the very young infant, and thereafter in patients with a lesion at the midbrain level or above. { 'tän·ik 'nek 'rē,flek·səz }

tonic postural epilepsy [MED] A form of epilepsy in which seizures are characterized by a rigid posture with the arms and legs extended, hands pronated, and feet held in plantar flexion. Also known as tonic convulsion. { 'tän·ik 'päs·chə·rəl 'ep·ə,lep·sē }

tonic spasm [MED] A spasm which persists for some time without relaxation. Also known as tonic convulsion. { 'tän·ik 'spaz·əm }

tonofibril [CELL MOL] Any of the fibrils converging on desmosomes in epithelial cells. { ¦tä·nō'fī·brəl }

tonometer [MED] An electronic instrument that measures hydrostatic pressure within the eye: when placed in position, a tiny movable plate is pressed against the eye, flattening a circular section of the cornea (no eyeball anesthesia is required); a current is then sent through a small electromagnet, of such value that it will just pull the plate away from the eye; the value of the current is proportional to eye pressure; a measurement can be made in about 1 second; used in diagnosis of glaucoma. Also known as electronic tonometer. { tō'näm·əd·ər }

tonoplast [BOT] The membrane surrounding a plant-cell vacuole. { 'tän·ə,plast }

tonsil [ANAT] **1.** Localized aggregation of diffuse and nodular lymphoid tissue found in the throat where the nasal and oral cavities open into the pharynx. **2.** *See* palatine tonsil. { 'tän·səl }

tonsillectomy [MED] Surgical removal of the palatine tonsil. { ,tan·sə'lek·tə·mē }

tonsillitis [MED] Inflammation of the tonsils. { ,tan·sə'līd·əs }

tonus [PHYSIO] The degree of muscular contraction when not undergoing shortening. Also known as muscle tone. { 'tō·nəs }

tooth [ANAT] One of the hard bony structures supported by the jaws in mammals and by other bones of the mouth and pharynx in lower vertebrates serving principally for prehension and mastication. [INV ZOO] Any of various sharp, horny, chitinous, or calcareous processes on or about any part of an invertebrate that functions like or resembles vertebrate jaws. { 'tüth }

toothache [MED] Pain in or about a tooth. Also known as odontalgia. { 'tü,thāk }

tooth decay [MED] Caries of the teeth. { 'tüth di,kā }

tooth shell [INV ZOO] A mollusk of the class Scaphopoda characterized by the elongate, tube-shaped, or cylindrical shell which is open at both ends and slightly curved. { 'tüth ,shel }

top grafting [BOT] Grafting a scion of one variety of tree onto the main branch of another. { ¦täp ,graft·iŋ }

tophus [MED] A localized swelling principally in cartilage and connective tissues in or adjacent to the small joints of the hands and feet; occurs specifically in gout. { 'tō·fəs }

topocline [ECOL] A graded series of characters exhibited by a species or other closely related organisms along a geographical axis. { 'täp·ə,klīn }

topogenesis *See* morphogenesis. { ¦täp·ə'jen·ə·səs }

topographic anatomy [ANAT] The use of bony and soft tissue landmarks on the surface of the body to indicate the known location of deeper structures. { ¦täp·ə¦graf·ik ə'nad·ə·mē }

topographic climax [ECOL] A climax plant community under a uniform macroclimate over which minor topographic features such as hills, rivers, valleys, or undrained depressions exert a controlling influence. { ¦täp·ə¦graf·ik 'klī,maks }

topoisomerases [BIOCHEM] Any of a group of enzymes capable of relaxing, unwinding, unpackaging, or changing the degree of supercoiling of deoxyribonucleic acid fiber. { ¦tä·pō·ī'säm·ə,rās·əz }

topotaxis *See* tropism. { ¦täp·ə¦tak·səs }

topotype [SYST] A specimen of a species not of the original type series collected at the type locality. { 'täp·ə,tīp }

top shell [INV ZOO] Any of the marine snails of the family Trochidae characterized by a spiral conical shell with a flat base. { ¦täp ,shel }

topwork [BOT] A procedure employed to propagate seedless varieties of fruit and hybrids, to change the variety of fruit, and to correct pollination problems, using any of three methods: root grafting, crown grafting, and top grafting. { 'täp,wərk }

tornaria [INV ZOO] The larva of some acorn worms (Enteropneusta) which is large and marked by complex bands of cilia. { tȯr'nar·ē·ə }

tornote [INV ZOO] A monaxon spicule in certain Porifera having both ends terminating abruptly in points. { 'tȯr,nōt }

torose [INV ZOO] **1.** Having knobby prominences on the surface. **2.** Cylindrical with alternate swellings and contractions. { 'tȯ,rōs }

Torpedinidae [VERT ZOO] The electric rays or torpedoes, a family of batoid sharks. { ,tȯr·pə'din·ə,dē }

torpor [PHYSIO] The condition in hibernating poikilotherms during winter when body temperature drops in a parallel relation to ambient environmental temperatures. { 'tȯr·pər }

Torridincolidae [INV ZOO] A small family of coleopteran insects in the suborder Myxophaga found only in Africa and Brazil. { tə,rid·ən'käl·ə,dē }

torticollis [MED] A deformity of the neck resulting from contraction of the cervical muscles or fascia. Also known as wryneck. { ,tȯrd·ə'käl·əs }

tortoise [VERT ZOO] Any of various large terrestrial reptiles in the order Chelonia, especially the family Testudinidae. { 'tȯrd·əs }

Tortricidae [INV ZOO] A family of phytophagous moths in the superfamily Tortricoidea which have a stout body, lightly fringed wings, and threadlike antennae. { tȯr'tris·ə,dē }

Tortricoidea [INV ZOO] A superfamily of small wide-winged moths in the suborder Heteroneura. { ,tȯr·trə'kȯid·ē·ə }

Torulopsidales [MYCOL] The equivalent name for Cryptococcales. { ,tȯr·ə,läp·sə'dā·lēz }

torulosis *See* cryptococcosis. { ,tȯr·ə'lō·səs }

torus [ANAT] A rounded protuberance occurring on a body part. [BOT] The thickened membrane closing a bordered pit. { 'tȯr·əs }

Torymidae [INV ZOO] A family of hymenopteran insects in the superfamily Chalcidoidea. { tȯ'rim·ə,dē }

total lung capacity [PHYSIO] The volume of gas contained within the lungs at the end of a maximum inspiration. { 'tōd·əl 'ləŋ kə,pas·əd·ē }

totipalmate [VERT ZOO] Having all four toes connected by webs, as in the Pelecaniformes. { ,tōd·ə'pä,māt }

totipotence [EMBRYO] Capacity of a blastomere to develop into a fully formed embryo. { tō'tip·əd·əns }

totipotent cell [EMBRYO] A cell capable of differentiating into every type of cell found in an organism, and of forming the entire organism. { 'tōd·ə,pōt·ent ,sel }

toucan [VERT ZOO] Any of numerous fruit-eating birds, of the family Ramphastidae, noted for their large and colorful bills. { 'tü,kan }

touch [PHYSIO] The array of sensations arising from pressure sensitivity of the skin. { təch }

Tourette's syndrome [MED] A syndrome characterized by repetitive tics, movement disorders, uncontrolled grunts, and occasionally verbal obscenities. Also known as Gilles de la Tourette syndrome. { tu̇'rets ,sin,drōm }

tourniquet [MED] An apparatus for controlling hemorrhage from, or circulation in, a limb or part of the body, where pressure can be brought upon the blood vessels by means of straps, cords, rubber tubes, or pads. { 'tu̇r·nə·kət }

toxa [INV ZOO] A curved sponge spicule. { 'täk·sə }

toxemia [MED] A condition in which the blood contains toxic substances, either of microbial origin or as by-products of abnormal protein metabolism. { täk'sē·mē·ə }

toxemia of pregnancy *See* preeclampsia. { täk'sē·mē·ə əv 'preg·nən·sē }

toxic [MED] Relating to a harmful effect by a poisonous substance on the human body by physical contact, ingestion, or inhalation. { 'täk·sik }

toxic amaurosis [MED] Blindness following the introduction of toxic substances into the body, such as ethyl and methyl alcohol, tobacco, lead, and metabolites of uremia and diabetes. { täk·sik ,a,mȯ'rō·səs }

toxic epidermal necrolysis [MED] Intraepidermal blistering and separation of the outer epidermis, giving the appearance and the management problems of a scald, caused by infection with *Staphylococcus aureus* strains producing one of the epidermolytic toxins, usually of phage group II. Also known as scalded skin syndrome. { ¦tak·sik ,ep·ə,dər·məl nə'kräl·ə·səs }

toxic goiter *See* hyperthyroidism. { 'täk·sik 'gȯid·ər }

toxic hepatitis [MED] Inflammation of the liver caused by chemical agents ingested or inhaled into the body, such as chlorinated hydrocarbons and some alkaloids. { 'täk·sik ,hep·ə'tīd·əs }

toxicity [PHARM] **1.** The quality of being toxic. **2.** The kind and amount of poison or toxin produced by a microorganism, or possessed by a chemical substance not of biological origin. { täk'sis·əd·ē }

toxicological study [MED] The study of how much poison must be present to produce an effect on animals or plant systems, may also include what type of effect is produced and how it is detected. { ,täk·sə·kə,läj·ə·kəl 'stəd·ē }

toxicology [PHARM] The study of poisons, including their nature, effects, and detection, and methods of treatment. { ,tak·sə'käl·ə·jē }

toxic psychosis [MED] A brain disorder due to a toxic agent such as lead or alcohol. { 'täk·sik sī'kō·səs }

toxic shock syndrome [MED] A serious, sometimes life-threatening disease usually caused by a toxin produced by some strains of the bacterium *Staphylococcus aureus*. The signs and symptoms are fever, abnormally low blood pressure, nausea and vomiting, diarrhea, muscle tenderness, and a reddish rash, followed by peeling of the skin. { ,täk·sik 'shäk ,sin,drōm }

toxicyst [INV ZOO] A type of trichocyst in Protozoa which may, upon contact, induce paralysis or lysis of the prey. { 'täk·sə,sist }

toxigenicity [MICROBIO] A microorganism's capability for producing toxic substances. { ,täk·sə·jə'nis·əd·ē }

toxin [BIOCHEM] Any of various poisonous substances produced by certain plant and animal cells, including bacterial toxins, phytotoxins, and zootoxins. { 'täk·sən }

Toxoglossa [INV ZOO] A group of carnivorous marine gastropod mollusks distinguished by a highly modified radula (toxoglossate). { ˌtäk·sə'gläs·ə }

toxoglossate radula [INV ZOO] A radula in certain carnivorous gastropods having elongated, spearlike teeth often perforated by the ducts of large poison glands. { ¦täk·sə¦glä,sāt 'raj·ə·lə }

toxoid [IMMUNOL] Detoxified toxin, but with antigenic properties intact; toxoids of tetanus and diphtheria are used for immunization. { 'täk,sȯid }

Toxoplasmea [INV ZOO] A class of the protozoan subphylum Sporozoa composed of small, crescent-shaped organisms that move by body flexion or gliding and are characterized by a two-layered pellicle with underlying microtubules, micropyle, paired organelles, and micronemes. { ¦täk·sə'plaz·mē·ə }

Toxoplasmida [INV ZOO] An order of the class Toxoplasmea; members are parasites of vertebrates. { ¦täk·sə'plaz· məd·ə }

toxoplasmin [BIOCHEM] The *Toxoplasma* antigen; used in a skin test to demonstrate delayed hypersensitivity to toxoplasmosis. { ¦täk·sə 'plaz·mən }

toxoplasmosis [MED] Infection by the protozoan *Toxoplasma gondii*, manifested clinically in severe cases by jaundice, hepatomegaly, and splenomegaly. { ˌtak·sə·plaz'mō·səs }

Toxopneustidae [INV ZOO] A family of Tertiary and extant echinoderms of the order Temnopleuroida where the branchial slits are deep and the test tends to be absent. { ˌtäk·sə'nyü·stə,dē }

Toxotidae [VERT ZOO] The archerfishes, a family of small fresh-water forms in the order Perciformes. { täk'säd·ə,dē }

T phage [VIROL] Any of a series (T1-T7) of deoxyribonucleic acid phages which lyse strains of the gram-negative bacterium *Escherichia coli* and its relatives. { 'tē ˌfāj }

TPN *See* triphosphopyridine nucleotide.

trabecula [ANAT] A band of fibrous or muscular tissue extending from the capsule or wall into the interior of an organ. { trə'bek·yə·lə }

trace element [BIOCHEM] A chemical element that is needed in minute quantities for the proper growth, development, and physiology of the organism. Also known as micronutrient. { 'trās ˌel·ə·mənt }

trachea [ANAT] The cartilaginous and membranous tube by which air passes to and from the lungs in humans and many vertebrates. [BOT] A xylem vessel resembling the trachea of vertebrates. [INV ZOO] One of the anastomosing air-conveying tubules composing the respiratory system in most insects. { 'trā·kē·ə }

tracheary [BOT] Water-conducting. { 'trā·kē ˌer·ē }

tracheid [BOT] An elongate, spindle-shaped xylem cell, lacking protoplasm at maturity, and having secondary walls laid in various thicknesses and patterns over the primary wall. { 'trā·kē·əd }

Tracheophyta [BOT] A large group of plants characterized by the presence of specialized conducting tissues (xylem and phloem) in the roots, stems, and leaves. { ˌtrā·kē'äf·əd·ē }

tracheophyte *See* Tracheophyta. { 'trā·kē·ə,fīt }

trachoma [MED] An infectious disease of the conjunctiva and cornea caused by *Chlamydia trachomatis* producing photophobia, pain, and excessive lacrimation. { trə'kō·mə }

trachomatous conjunctivitis [MED] Inflammation of the conjunctiva associated with trachoma, characterized by a subepithelial cellular infiltration with a follicular distribution. { trə'käm·əd·əs kən,jəŋk·tə'vīd·əs }

Trachylina [INV ZOO] An order of moderate-sized jellyfish of the class Hydrozoa distinguished by having balancing organs and either a small polyp stage or none. { ˌtrak·ə'līn·ə }

Trachymedusae [INV ZOO] A group of marine jellyfish, recognized as a separate order or as belonging to the order Trachylina whose tentacles have a solid core consisting of a single row of endodermal cells. { ¦tra·kē·mə'dü,sē }

Trachypsammiacea [INV ZOO] An order of colonial anthozoan cnidarians characterized by a dendroid skeleton. { ¦tra·kē,sam·ē'ās·ē·ə }

Trachystomata [VERT ZOO] The name given to the Meantes when the group is considered to be an order. { ¦tra·kē'stō·məd·ē }

traction alopecia [MED] Hair loss that is the result of prolonged, tightly pulled hairstyles. { ˌtrak·shən ˌal·ə'pē·shə }

traction diverticulum [MED] A circumscribed sacculation, usually of the esophagus, with bulging of the full thickness of the wall; caused by the pull of adhesions arising from adjacent organs. { 'trak·shən ˌdī·vər'tik·yə·ləm }

Tragulidae [VERT ZOO] The chevrotains, a family of pecoran ruminants in the superfamily Traguloidea. { trə'gyül·ə,dē }

Traguloidea [VERT ZOO] A superfamily of pecoran ruminants, composed of the most primitive forms with large canines; the chevrotain is the only extant member. { ˌtra·gyə'lȯid·ē·ə }

tragus [ANAT] **1.** The prominence in front of the opening of the external ear. **2.** One of the hairs in the external ear canal. { 'trā·gəs }

trail pheromone [PHYSIO] A type of pheromone used by social insects and some lepidopterans to recruit others of its species to a food source. { 'trāl ˌfer·ə,mōn }

trama [MYCOL] The loosely woven hyphal tissue between adjacent hymenia in basidiomycetes. { 'trā·mə }

Trametes versicolor [MYCOL] A brightly colored mushroom that appears to have antitumor properties, is a common inhabitant of the woods worldwide. Also known as Coriolus versicolor; turkey tail mushroom. { trə,mēd,ēz 'vər·sə ˌkəl·ər }

tranexamic acid [PHARM] $C_8H_{15}NO_2$ Crystals which soften at 270°C and are soluble in water;

used as a hemostatic agent. Abbreviated AMCHA. { ¦tran·ik¦sam·ik 'as·əd }

tranquilizer [PHARM] **1.** Any agent that brings about a state of relief from anxiety, or peace of mind. **2.** Any agent that produces a calming or sedative effect without inducing sleep. **3.** Any drug, such as chlorpromazine, used primarily for its calming and antipsychotic effects, or such as meprobamate, used for symptomatic treatment of common psychoneuroses and as an adjunct in somatic disorders complicated by anxiety and tension. { 'traŋ·kwə,līz·ər }

transaminase [BIOCHEM] One of a group of enzymes that catalyze the transfer of the amino group of an amino acid to a keto acid to form another amino acid. Also known as aminotransferase. { ¦tranz'am·ə,nās }

transcapsidation [VIROL] Change in the capsid of PARA (particle aiding replication of adenovirus) from one type of adenovirus to another. { ¦tranz,kap·sə'dā·shən }

transcranial magnetic stimulation [MED] A neurophysiologic technique that stimulates the human brain via magnetic pulses generated by a small electromagnet placed on the scalp. { tranz¦krā·nē·əl mag,ned·ik ,stim·yə'lā·shən }

transcriptase *See* ribonucleic acid polymerase. { tran'skrip,tās }

transcription [CELL MOL] The process by which ribonucleic acid is formed from deoxyribonucleic acid. { tranz'krip·shən }

transcription attenuation [CELL MOL] A form of gene expression regulation in bacteria whereby transcription is terminated shortly after it begins by preventing its continuation beyond the attenuator site. { tranz¦krip·shən a,ten·yə'wā·shən }

transcription unit [CELL MOL] The segment of deoxyribonucleic acid between the sites of initiation and termination of transcription by ribonucleic acid polymerase. { tranz'krip·shən ,yü·nət }

transcytosis [CELL MOL] A form of intracellular vesicular traffic in which endocytosed macromolecules are transferred across the cell and released (via exocytosis) at the opposite plasma membrane domain. { ,tranz,sī'tō·səs }

transduction [MICROBIO] Transfer of genetic material between bacterial cells by bacteriophages. { tranz'dək·shən }

transfection [GEN] Infection of a cell with viral deoxyribonucleic acid or ribonucleic acid. { trans'fek·shən }

transferase [BIOCHEM] Any of various enzymes that catalyze the transfer of a chemical group from one molecule to another. { 'tranz·fə,rās }

transfer immunity *See* adoptive immunity. { 'tranz,fər i,myün·əd·ē }

transfer ribonucleic acid [CELL MOL] The smallest ribonucleic acid molecule found in cells; its structure is complementary to messenger ribonucleic acid and it functions by transferring amino acids from the free state to the polymeric form of growing polypeptide chains. Abbreviated t-RNA. { 'tranz·fər ¦rī·bō·nü¦klē·ik 'as·əd }

transferrin [BIOCHEM] Any of various beta globulins in blood serum which bind and transport iron to the bone marrow and storage areas. { 'tranz'fer·ən }

transformation [GEN] **1.** Transfer and incorporation of foreign deoxyribonucleic acid (DNA) into a cell and subsequent recombination of part or all of that DNA into the cell's genome. Also known as bacterial transformation; genetic transformation. **2.** Conversion of a normal cell to a neoplastic cell by a cascade of events under the control of different classes of oncogenes. Also known as cellular transformation. [IMMUNOL] Change in a lymphocyte from a small, resting lymphocyte into a large lymphocyte following stimulation by antigens or lectin, or viral infection. Also known as lymphocyte transformation. { ,tranz·fər'mā·shən }

transforming principle [MICROBIO] Deoxyribonucleic acid which effects transformation in bacterial cells. { tranz'fórm·iŋ ,prin·sə·pəl }

transfusion [MED] The administration of blood, or one of its components, as a part of treatment. { tranz'fyü·zhən }

transgene [GEN] Genetic material from one organism that has been experimentally transferred to another, so that the host acquires the genetic traits of the transferred genes in its chromosomal composition. { 'tranz,jēn }

transgenic organism [GEN] An organism into whose genome a gene or genes from another organism has been experimentally transferred and can be expressed. { tranz'jen·ik 'ór·gə,niz·əm }

transient ischemic attack [MED] A brief loss of nerve function caused by a temporary lack of adequate blood flow and oxygen to the brain due to a rupture in the carotid arteries leading to the brain. { ,tranch·ənt i,skēm·ik ə'tak }

transition [CELL MOL] A mutation resulting from the substitution in deoxyribonucleic acid or ribonucleic acid of one purine or pyrimidine for another. { tran'zish·ən }

transitional epithelium [HISTOL] A form of stratified epithelium found in the urinary bladder; cells vary between squamous, when the tissue is stretched, and columnar, when not stretched. { tran'zish·ən·əl ,ep·ə'thē·lē·əm }

transketolase [BIOCHEM] An enzyme that cleaves a substrate at the position of the carbonyl carbon and transports a two-carbon fragment to an acceptor compound to form a new compound. { ¦tranz'kēd·ə,lās }

translation [CELL MOL] The process by which the linear sequence of nucleotides in a molecule of messenger ribonucleic acid directs the specific linear sequence of amino acids, as during protein synthesis. { tran'slā·shən }

translocation [BOT] Movement of water, mineral salts, and organic substances from one part of a plant to another. [CELL MOL] The transfer of a chromosome segment from its usual position to a new position in the same or in a different chromosome. { ¦tranz·lō'kā·shən }

transmethylase [BIOCHEM] A transferase enzyme involved in catalyzing chemical reactions in which methyl groups are transferred from a substrate to a new compound. { ¦tranz'meth·ə,lās }

transmethylation [BIOCHEM] A metabolic reaction in which a methyl group is transferred from one compound to another; methionine and choline are important donors of methyl groups. { ¦tranz,meth·ə'lā·shən }

transorbital lobotomy [MED] A lobotomy performed through the roof of the orbit. { tranz'ȯr·bəd·əl lə'bäd·ə·mē }

transpiration [BIOL] The passage of a gas or liquid (in the form of vapor) through the skin, a membrane, or other tissue. { ,tranz·pə'rā·shən }

transplantation [BIOL] **1.** The artificial removal of part of an organism and its replacement in the body of the same or of a different individual. **2.** To remove a plant from one location and replant it in another place. { ,tranz·plan'tā·shən }

transplantation antigen [IMMUNOL] An antigen in a cell which induces a histocompatibility reaction when the cell is transplanted into an organism not having that antigen. { ,tranz·plan'tā·shən 'ant·i·jən }

transplantation disease [MED] Disease ascribable to an immunological graft-versus-host reaction which occurs after transplantation of adult lymphoid cells to incompatible recipients who cannot reject them. { ,tranz·plan'tā·shən di ,zēz }

transposon [GEN] A genetic element which comprises large discrete segments of deoxyribonucleic acid and is capable of moving from one chromosomal site to another in the same organism or in a different organism. { ¦tranz'pō,zän }

transverse colon [ANAT] The portion of the colon between the right and left colic flexures, extending transversely across the upper abdomen. { trans¦vərs 'kō·lən }

transversion [CELL MOL] A mutation resulting from the substitution in deoxyribonucleic acid or ribonucleic acid of a purine for a pyrimidine or a pyrimidine for a purine. { trans'vər·zhən }

trapeziform [BIOL] Having the form of a trapezium. { trə'pē·zə,fȯrm }

trauma [MED] An injury caused by a mechanical or physical agent. { 'trau̇·mə }

traumatic pneumonosis [MED] The acute, noninflammatory pathologic pulmonary changes produced by a large momentary deceleration. { trə'mad·ik ,nü·mə'nō·səs }

traumatotropism [BIOL] Orientation response of an organ of a sessile organism in response to a wound. { ,trau̇·mə'tä·trə,piz·əm }

Trebidae [INV ZOO] A family of copepod crustaceans of the order Caligoida which are external parasites on selachians. { 'treb·ə,dē }

tree [BOT] A perennial woody plant at least 20 feet (6 meters) in height at maturity, having an erect stem or trunk and a well-developed crown or leaf canopy. { trē }

tree fern [BOT] The common name for plants belonging to the families Cyatheaceae and Dicksoniaceae; all are ferns that exhibit an arborescent habit. { 'trē ,fərn }

tree frog [VERT ZOO] Any of the arboreal frogs comprising the family Hylidae characterized by expanded digital adhesive disks. { 'trē ,fräg }

treeline [ECOL] The altitudinal or latitudinal limit beyond which conditions do not permit the growth of trees. { 'trē,līn }

Trematoda [INV ZOO] A loose grouping of acoelomate, parasitic flatworms of the phylum Platyhelminthes; they exhibit cephalization, bilateral symmetry, and well-developed holdfast structures. { trem·ə'tōd·ə }

trematodiasis [MED] Infection caused by a member of the Trematoda (trematodes). { ,trem·ə·tə'dī·ə·səs }

trembling ill *See* louping ill. { 'trem·bliŋ ¦il }

Tremella fuciformis [MYCOL] A mushroom that grows on deciduous trees in the southern United States and in warm climates worldwide. Once primarily grown for its medicinal properties (it boosts immunological function and stimulates leukocyte activity), it is now used mostly for food. Also known as snow fungus. { trə,mel·ə ,fyü·sə'fȯr·məs }

Tremellales [MYCOL] An order of basidiomycetous fungi in the subclass Heterobasidiomycetidae in which basidia have longitudinal walls. { ,trem·ə'lā·lēz }

tremor [MED] Involuntary, rhythmic trembling of voluntary muscles resulting from alternate contraction and relaxation of opposing muscle groups. { 'trem·ər }

trench fever [MED] A louse-borne infection that is caused by *Rickettsia quintana* and is characterized by headache, chills, rash, pain in the legs and back, and often by a relapsing fever. { 'trench ,fē·vər }

trench mouth *See* Vincent's infection. { 'trench ,mau̇th }

Trentepohliaceae [BOT] A family of green algae belonging to the Ulotrichales having thick walls, bandlike or reticulate chloroplasts, and zoospores or isogametes produced in enlarged, specialized cells. { ,tren·tə,pō·lē'ās·ē,ē }

Treponema pallidum [MICROBIO] A pathogenic spirochete that causes the sexually transmitted disease syphilis. { ,trep·ə,nē·mə 'pal·ə·dəm }

Treponemataceae [MICROBIO] Formerly a family of the bacterial order Spirochaetales including the spirochetes less than 20 micrometers long and less than 5 micrometers in diameter; most species are parasitic. { ,trep·ə,nē·mə'tās·ē,ē }

treponematosis [MED] Infection caused by any species of the genus *Treponema*. Also known as treponemiasis. { ,trep·ə,nē·mə'tō·səs }

treponemiasis *See* treponematosis. { ,trep·ə·nə'mī·ə·səs }

Treroninae [VERT ZOO] The fruit pigeons, a subfamily of the avian family Columbidae distinguished by the gaudy coloration of the feathers. { trə'rän·ə,nē }

Tretothoracidae [INV ZOO] A family of the Coleoptera in the superfamily Tenebrionoidea which contains a single species found in Queensland, Australia. { ˌtred·ə·thə'ras·əˌdē }

triacetyloleandomycin [MICROBIO] An antibiotic produced by *Streptomyces antibioticus* and used clinically in the treatment of pneumonia, osteomyelitis, furuncles, and carbuncles. { trī·əˌsēd·əlˌō·lēˌan·də'mīs ·ən }

triaene [INV ZOO] An elongated spicule in certain Porifera with three rays diverging from one end. { 'trīˌēn }

triage [MED] The process of determining which casualties (as from an accident, disaster, military battle, or explosion of nuclear weapons) need urgent treatment, which ones are well enough to go untreated, and which ones are beyond hope of benefit from treatment. { trē'äzh }

Triakidae [VERT ZOO] A family of galeoid sharks in the carcharinid line. { trī'ak·əˌdē }

triandrous [BOT] Possessing three stamens. { trī'an·drəs }

triangular ligament *See* urogenital diaphragm. { trī'aŋ·gyə·lər 'lig·ə·mənt }

Triatominae [INV ZOO] The kissing bugs, a subfamily of hemipteran insects in the family Reduviidae, distinguished by a long, slender rostrum. { trī·ə'täm·əˌnē }

triaxon [INV ZOO] A spicule in Porifera having three axes which cross each other at right angles. { trī'akˌsän }

tribromoethanol [PHARM] $C_2H_3Br_3O$ A white crystalline compound, melting at 79–82°C; used in medicine in anesthesia. Also known as tribromoethyl alcohol. { trīˌbrō·mō'eth·əˌnȯl }

tribromoethyl alcohol *See* tribromoethanol. { trīˌbrō·mō'eth·əl 'al·kəˌhȯl }

tricarboxylic acid cycle *See* Krebs cycle. { trīˌkär·bäk'sil·ik 'as·əd ˌsī·kəl }

Trichechidae [VERT ZOO] The manatees, a family of nocturnal, solitary sirenian mammals in the suborder Trichechiformes. { trə'kek·əˌdē }

Trichechiformes [VERT ZOO] A suborder of mammals in the order Sirenia which contains the manatees and dugongids. { trəˌkek·ə'fȯrˌmēz }

trichesthesia [PHYSIO] A form of tactile sensibility in hair-covered regions of the body. { ˌtrik·əs'thēzh·ə }

Trichiaceae [MYCOL] A family of slime molds in the order Trichiales. { ˌtrik·ē'ās·ēˌē }

Trichiales [MYCOL] An order of Myxomycetes in the subclass Myxogastromycetidae. { ˌtrik·ē'ā·lēz }

trichiasis [MED] An eye disorder characterized by the misdirected inward growth of the eyelashes, causing trauma to the cornea. { tri'kī·ə·səs }

trichinosis [MED] Infection by the nematode *Trichinella spiralis* following ingestion of encysted larvae in raw or partially cooked pork; characterized by eosinophilia, nausea, fever, diarrhea, stiffness and painful swelling of muscles, and facial edema. { ˌtrik·ə'nō·səs }

Trichiuridae [VERT ZOO] The cutlass-fishes, a family of the suborder Scombroidei. { ˌtrik·ē'yu̇r·əˌdē }

2,2,2-trichloroethanol [PHARM] CCl_3CH_2OH A hygroscopic liquid with a boiling point of 151–153°C; soluble in water and miscible with alcohol or ether; used in medicine as a hypnotic and anesthetic. Also known as trichloroethyl alcohol. { ¦tü ¦tü ¦tü trīˌklȯr·ō'eth·əˌnȯl }

trichloroethyl alcohol *See* 2,2,2-trichloroethanol. { trīˌklȯr·ō'eth·əl 'al·kəˌhȯl }

trichobezoar [MED] A ball of hair or similar concretion in the stomach or intestine. { ¦trik·ə'bē·zō·ər }

trichobothrium [INV ZOO] An erect, bristlelike sensory hair found on certain arthropods, insects, and other invertebrates. { ¦trik·ə'bäth·rē·əm }

trichobranchiate gill [INV ZOO] A gill with filamentous branches arranged in several series around the axis; found in some decapod crustaceans. { ¦trik·ə'braŋ·kēˌāt ¦gil }

Trichobranchidae [INV ZOO] A family of polychaete annelids belonging to the Sedentaria; most members are rare and live at great ocean depths. { ˌtrik·ə'braŋ·kəˌdē }

trichocercous cercaria [INV ZOO] A trematode larva distinguished by a spiny tail. { ¦trik·əˌsər·kəs sər'kar·ē·ə }

Trichocomaceae [MYCOL] A small tropical family of ascomycetous fungi in the order Eurotiales with ascocarps from which a tuft of capillitial threads extrudes, releasing the ascospores after dissolution of the asci. { ˌtrik·ə·kə'mās·ēˌē }

trichocyst [INV ZOO] A minute structure in the cortex of certain protozoans that releases filamentous or fibrillar threads when discharged. { 'trik·əˌsist }

Trichodactylidae [INV ZOO] A family of freshwater crabs in the section Brachyura, found mainly in tropical regions. { ˌtrik·ə·dak'til·əˌdē }

trichoepithelioma [MED] A benign tumor characterized by small, round, yellow, or flesh-colored papules, chiefly on the center of the face. { ˌtrik·ōˌep·əˌthē·lē'ō·mə }

Trichogrammatidae [INV ZOO] A family of the Hymenoptera in the superfamily Chalcidoidea whose larvae are parasitic in the eggs of other insects. { ˌtrik·ə·grə'mad·əˌdē }

trichogyne [BOT] A terminal portion of a procarp or archicarp which receives a spermatium. { 'trik·əˌjīn }

trichome [BOT] An appendage derived from the protoderm in plants, including hairs and scales. [INV ZOO] A brightly colored tuft of hairs on the body of a myrmecophile that releases an aromatic substance attractive to ants. { 'trī ˌkōm }

Trichomonadida [INV ZOO] An order of the protozoan class Zoomastigophorea which contains four families of uninucleate species. { ˌtrik·ə·mə'näd·əd·ə }

Trichomonadidae [INV ZOO] A family of flagellate protozoans in the order Trichomonadida. { ˌtrik·ə·mə'näd·əˌdē }

trichomoniasis [MED] An infection caused by a species of the genus *Trichomonas*. { ,trik·ə·mə'nī·ə·səs }

Trichomycetes [MYCOL] A class of true fungi, division Fungi. { ¦trik·ə,mī'sēd·ēs }

trichomycin [MICROBIO] An antibiotic produced by *Streptomyces hachijoensis* and *S. abikoensis*; a water-soluble yellow powder that inhibits yeasts and fungi. { ,trik·ə'mīs·ən }

Trichoniscidae [INV ZOO] A primitive family of isopod crustaceans in the suborder Oniscoidea found in damp littoral, halophilic, or riparian habitats. { ,trik·ə'nis·ə,dē }

Trichophilopteridae [INV ZOO] A family of lice in the order Mallophaga adapted to life upon the lemurs of Madagascar. { ,trik·ə,fil·əp'ter·ə,dē }

trichophytin [IMMUNOL] A group antigen obtained from filtrates of *Trichophyton mentagrophytes*; used in a skin test to ascertain past or present infection with the dermatophytes. { ,trik·ə'fīt·ən }

Trichoptera [INV ZOO] The caddis flies, an aquatic order of the class Insecta; larvae are wormlike and adults have two pairs of well-veined hairy wings, long antennae, and mouthparts capable of lapping only liquids. { trə'käp·tə·rə }

Trichopterygidae [INV ZOO] The equivalent name for Ptiliidae. { trə,käp·tə'rij·ə,dē }

trichosclereids [BOT] Sclereid cells that are long and slender, resembling fibers, with which they intergrade. { ,trī·kō'skler·ē·ədz }

Trichostomatida [INV ZOO] An order of ciliated protozoans in the subclass Holotrichia in which no true buccal ciliature is present but there is a vestibulum. { ,trik·ə·stō'mad·əd·ə }

Trichostrongylidae [INV ZOO] A family of parasitic roundworms belonging to the Strongyloidea; hosts are cattle, sheep, goats, swine, and cats. { ,trik·ə·strän'jil·ə,dē }

trichotillomania [MED] An obsessive-compulsive disorder characterized by artificial alopecia secondary to manipulation of the hair. { ,trik·ō,til·ə 'mānē·ə }

Trichuroidea [INV ZOO] A group of nematodes parasitic in various vertebrates and characterized by a slender body sometimes having a thickened posterior portion. { ,trik·yə'rȯid·ē·ə }

Tricladida [INV ZOO] The planarians, an order of the Turbellaria distinguished by diverticulated intestines with a single anterior branch and two posterior branches separated by the pharynx. { trī'klad·əd·ə }

triconodont [VERT ZOO] **1.** A tooth with three main conical cusps. **2.** Having such teeth. { trī'kän·ə,dänt }

Trictenotomidae [INV ZOO] A small family of Indian and Malaysian beetles in the superfamily Tenebrionoidea. { ,trik·tə·nō'täm·ə,dē }

tricuspid valve [ANAT] A valve consisting of three flaps located between the right atrium and right ventricle of the heart. { trī'kəs·pəd 'valv }

Tridacnidae [INV ZOO] A family of bivalve mollusks in the subclass Eulamellibranchia which contains the giant clams of the tropical Pacific. { trī'dak·nə,dē }

Tridactylidae [INV ZOO] The pygmy mole crickets, a family of insects in the order Orthoptera, highly specialized for fossorial existence. { ,trī·dak'til·ə,dē }

trifid [BIOL] Divided into three lobes separated by narrow sinuses partway to the base. { 'trī,fid }

trifoliate [BOT] Having three leaves or leaflets. { trī'fō·lē,āt }

trifoliosis [VET MED] An acute photosensitization characterized by superficial necrosis of white or light-skinned animals feeding on certain leguminous plants. { trī,fō·lē'ō·səs }

Trifolium hybridium *See* alsike clover. { trī,fō·lē·əm hī'brid·ē·əm }

trigeminal nerve [NEUROSCI] The fifth cranial nerve in vertebrates; either of a pair of composite nerves rising from the side of the medulla, and with three great branches: the ophthalmic, maxillary, and mandibular nerves. { trī'jem·ə·nəl 'nərv }

trigeminal neuralgia [MED] Sudden severe pains of unknown cause along the path of one or more branches of the trigeminal nerve. Also known as tic douloureux. { trī'jem·ə·nəl nu̇'ral·jə }

trigger finger [MED] A symptom of tenosynovitis manifested as a temporary partial obstruction in flexion or extension of a finger that is followed by a snapping into the final position; results from a thickening of a tendon or localized reduction in the tendon sheath. Also known as snapping finger. { 'trig·ər ,fiŋ·gər }

Triglidae [VERT ZOO] The searobins, a family of perciform fishes in the suborder Cottoidei. { 'trig·lə,dē }

Trigonalidae [INV ZOO] A small family of hymenopteran insects in the superfamily Proctotrupoidea. { ,trig·ə'nal·ə,dē }

trigone [ANAT] A triangular area inside the bladder limited by the openings of the ureters and urethra. [BOT] A thickening of plant cell walls formed when three or more cells adjoin. { 'trī,gōn }

trigonous [BIOL] **1.** Having three corners. **2.** Having a triangular cross section. { 'trig·ə·nəs }

Trilobitomorpha [INV ZOO] A subphylum of the Arthropoda including Trilobita. { trī,läb·əd·ə'mȯr·fə }

trilocular [BIOL] Having three cavities or cells. { trī'äk·yə·lər }

Trimenoponidae [INV ZOO] A family of lice in the order Mallophaga occurring as parasites on South American rodents. { ,trī·mə·nə'pän·ə,dē }

trimerous [BOT] Having parts in sets of three. [INV ZOO] In insects, having the tarsus divided or apparently divided into three segments. { 'trim·ə·rəs }

trimethadione [PHARM] C_6H_9NO White, crystalline granules with a melting point of 45–47°C; soluble in water, alcohol, chloroform, and ether;

used in medicine as an anticonvulsant. { trī,meth·ə'dī·ōn }

trimethylvinylammonium hydroxide *See* neurine. { trī¦meth·əl,vīn·əl·ə'mō·nē·əm hī'dräk,sīd }

trimorphous [BIOL] Characterized by occurring in three distinct forms, as an organ or whole organism. { trī'mȯr·fəs }

trinomial [SYST] A nomenclatural designation for an organism composed of three terms: genus, species, and subspecies or variety. { trī'nō·mē·əl }

trinomial nomenclature [SYST] The designation of subspecies by a three-word name. { trī'nō·mē·əl 'nō·mən,klā·chər }

trinucleotide repeat expansion [GEN] An increase in the number of copies of a trinucleotide that is normally already present in multiple adjacent copies. For example, the X-linked mental retardation 1 (XLMRI) locus in humans usually contains 6–50 tandem repeats of CCG, but this is expanded to 200–2000 copies in the fragile-X syndrome. { ,trī¦nü·klē·ə,tīd ri'pēt ik,span·shən }

Trionychidae [VERT ZOO] The soft-shelled turtles, a family of reptiles in the order Chelonia. { trī·ə'nik·ə,dē }

triose [BIOCHEM] A group of monsaccharide compounds that have a three-carbon chain length. { 'trī,ōs }

triphosphopyridine dinucleotide *See* triphosphopyridine nucleotide. { trī¦fä·sfō'pir·ə,dēn dī'nü·klē·ə,tīd }

triphosphopyridine nucleotide [BIOCHEM] $C_{12}H_{28}N_7O_{17}P_3$ A grayish-white powder, soluble in methanol and in water; a coenzyme and an important component of enzymatic systems concerned with biological oxidation-reduction systems. Abbreviated TPN. Also known as codehydrogenase II; coenzyme II; triphosphopyridine dinucleotide. { trī¦fä·sfō'pir·ə,dēn 'nü·klē·ə,tīd }

tripinnate [BIOL] Being bipinnate and having each division pinnate. { trī'pi,nāt }

triple response [PHYSIO] The three stages of vasomotor reaction consisting of reddening, flushing of adjacent skin, and development of wheals, when a pointed instrument is drawn heavily across the skin. { 'trip·əl ri'späns }

triplet [GEN] A three-base unit in deoxyribonucleic or ribonucleic acid which codes for a particular amino acid in a protein chain. { 'trip·lət }

triplet-state chlorophyll [BIOCHEM] The state of chlorophyll that occurs when two of the electrons in chlorophyll have their magnetic moments in parallel. { ,trip·lət ,stāt 'klȯr·ə,fil }

triploblastic [EMBRYO] Having three embryonic germ layers: an ectoderm, mesoderm, and endoderm. { ,trip·lə'blas·tik }

Triploblastica [ZOO] Animals that develop from three germ layers. { ,trip·ō'blas·tə·kə }

triploidy [CELL MOL] The occurrence of related forms possessing chromosome numbers three times the haploid number. { 'tri,plȯid·ē }

Tripylina [INV ZOO] A subdivision of the protozoan order Oculosida in which the major opening (astropyle) usually contains a perforated plate. { ,trip·ə'lī·nə }

trisaccate pollen [BOT] A three-pored pollen grain, often having a triangular outline in cross section. { trī'sa,kāt 'päl·ən }

trisaccharide [BIOCHEM] A carbohydrate which, on hydrolysis, yields three molecules of monosaccharides. { trī'sak·ə,rīd }

trisomic syndrome [MED] Any pathological condition characterized by the presence in triplicate of one of the chromosomes of a complement. { trī'sōm·ik 'sin,drōm }

trisomy [CELL MOL] The presence in triplicate of one of the chromosomes of the complement. { 'trī,sō·mē }

trisomy 13–15 *See* D_1 trisomy. { 'trī,sō·mē 'thir¦tēn 'fif¦tēn }

trisomy 18 syndrome [MED] A congenital disorder due to trisomy of all or a large part of chromosome 18, characterized by severe mental deficiency, hypertonicity with clenched hands, and anomalies of the hands, sternum, pelvis, and facies; most infants so afflicted fail to thrive. Also known as Edwards' syndrome; E trisomy. { 'trī,sō·mē 'ā¦tēn ,sin,drōm }

trisomy 21 syndrome *See* Down's syndrome. { 'trī,sō·mē 'twen·tē¦wən ,sin,drōm }

tritanopia [MED] A defect in a third constituent essential for color vision, as in violet blindness. { ,trīt·ən'ō·pē·ə }

tritonymph [INV ZOO] The third stage of development in certain acarids. { 'trīd·ō,nimf }

Triuridaceae [BOT] A family of monocotyledonous plants in the order Triuridales distinguished by unisexual flowers and several carpels with one seed per carpel. { trī,yu̇r·ə'dās·ē,ē }

Triuridales [BOT] A small order of terrestrial, mycotrophic monocots in the subclass Alismatidae without chlorophyll, and with separate carpels, trinucleate pollen, and a well-developed endosperm. { trī,yu̇r·ə'dā·lēz }

trivium [INV ZOO] The three rays opposite the madreporite in starfish. { 'triv·ē·əm }

t-RNA *See* transfer ribonucleic acid.

trochal disk [INV ZOO] A flat or funnel-shaped ciliated disk at the anterior end of a rotifer that functions in locomotion and food ingestion. { 'trō·kəl ,disk }

trochanter [ANAT] A process on the proximal end of the femur in many vertebrates, which serves for muscle attachment and, in birds, for articulation with the ilium. [INV ZOO] The second segment of an insect leg, counting from the base. { trō'kan·tər }

Trochidae [INV ZOO] A family of gastropod mollusks in the order Aspidobranchia, including many of the top shells. { 'träk·ə,dē }

Trochili [VERT ZOO] A suborder of the avian order Apodiformes. { 'träk·ə,lī }

Trochilidae [VERT ZOO] The hummingbirds, a tropical New World family of the suborder Trochili with tubular tongues modified for nectar feeding; slender bills and the ability to hover are further feeding adaptations. { trä'kil·ə,dē }

trochlea [ANAT] A pulleylike anatomical structure. { 'träk·lē·ə }

trochlear nerve [NEUROSCI] The fourth cranial nerve; either of a pair of somatic motor nerves which innervate the superior oblique muscle of the eyeball. { 'träk·lē·ər ˌnərv }

trochoblast [INV ZOO] A cell bearing cilia on a trochophore. { 'träk·əˌblast }

Trochodendraceae [BOT] A family of dicotyledonous trees in the order Trochodendrales distinguished by the absence of a perianth and stipules, numerous stamens, and pinnately veined leaves. { ˌträk·ō·den'drās·ēˌē }

Trochodendrales [BOT] An order of dicotyledonous trees in the subclass Hamamelidae characterized by primitively vesselless wood and unique, elongate, often branched idioblasts in the leaves. { ˌträk·ō·den'drā·lēz }

trochoid *See* pivot joint. { 'trōˌkȯid }

trochophore [INV ZOO] A generalized but distinct free-swimming larva found in several invertebrate groups, having a pear-shaped form with an external circlet of cilia, apical ciliary tufts, a complete functional digestive tract, and paired nephridia with excretory tubules. Also known as trochosphere. { 'träk·əˌfȯr }

trochosphere *See* trochophore. { 'träk·əˌsfir }

trochus [INV ZOO] The inner band of cilia on a trochal disk. { 'trō·kəs }

Troglodytidae [VERT ZOO] The wrens, a family of songbirds in the order Passeriformes. { ˌträg·lə'did·əˌdē }

Trogonidae [VERT ZOO] The trogons, the single, pantropical family of the avian order Trogoniformes. { trō'gän·əˌdē }

Trogoniformes [VERT ZOO] An order of brightly colored, slow-moving birds characterized by a unique foot structure with the first and second toes directed backward. { trōˌgän·ə'fȯrˌmēz }

Troland and Fletcher theories [PHYSIO] Theories of hearing according to which the time nature of a sound stimulation affects the sensation of pitch. { 'trō·lənd ən 'flech·ər ˌthē·ə·rēz }

Trombiculidae [INV ZOO] The chiggers, or red bugs, a family of mites in the suborder Trombidiformes whose larvae are parasites of vertebrates. { ˌträm·bə'kyül·əˌdē }

Trombidiformes [INV ZOO] The trombidiform mites, a suborder of the Acarina distinguished by the presence of a respiratory system opening at or near the base of the chelicerae. { ˌträm·bə·də'fȯrˌmēz }

Tropaeolaceae [BOT] A family of dicotyledonous plants in the order Geraniales characterized by strongly irregular flowers, simple peltate leaves, eight stamens, and schizocarpous fruit. { ˌtrō·pē·ō'lās·ēˌē }

trophallaxis [ECOL] Exchange of food between organisms, not only of the same species but between different species, especially among social insects. { ˌträf·ə'lak·səs }

trophic [BIOL] Pertaining to or functioning in nutrition. { 'träf·ik }

trophic ecology [ECOL] The study of the feeding relationships of organisms in communities and ecosystems. { ˌträf·ik ē'käl·ə·jē }

trophic level [ECOL] Any of the feeding levels through which the passage of energy through an ecosystem proceeds; examples are photosynthetic plants, herbivorous animals, and microorganisms of decay. { 'träf·ik ˌlev·əl }

trophobiosis [ECOL] A nutritional relationship associated only with certain species of ants in which alien insects supply food to the ants and are milked by the ants for their secretions. { ˌträf·ōˌbī'ō·səs }

trophoblast [EMBRYO] A layer of ectodermal epithelium covering the outer surface of the chorion and chorionic villi of many mammals. { 'träf·əˌblast }

trophocyte [INV ZOO] A nutritive cell of the ovary or testis of an insect. { 'träf·əˌsīt }

trophogenic [ECOL] Originating from nutritional differences rather than resulting from genetic determinants, such as various castes of social insects. { ¦träf·ə¦jen·ik }

tropholytic [ECOL] Pertaining to the deep zone in a lake where dissimilation of organic matter predominates. { ¦traf·ə¦lid·ik }

trophophase [MICROBIO] A period in culture production characterized by active microbial cell growth and the formation of primary metabolites. { 'träf·əˌfāz }

trophosome [INV ZOO] The nutritional zooids of a hydroid colony. { 'träf·əˌsōm }

trophotaeniae [VERT ZOO] Vascular rectal processes which establish placental relationships with the ovarian tissue in live-bearing fishes. { ˌträf·ō'tē·nēˌī }

trophozoite [INV ZOO] A vegetative protozoan; used especially of a parasite. { ˌträf·ə'zōˌīt }

trophus [INV ZOO] Masticatory apparatus in Rotifera. { 'träf·əs }

tropical life zone [ECOL] A subdivision of the eastern division of Merriam's life zones; an example is southern Florida, where the vegetation is the broadleaf evergreen forest; typical and important plants are palms and mangroves; typical and important animals are the armadillo and alligator; typical and important crops are citrus fruits, avocado, and banana. { 'träp·ə·kəl 'līf ˌzōn }

tropical myositis [MED] A severe purulent infection of muscle or a muscle group, usually in the leg, decompression drainage is frequently necessary. { ˌträp·ə·kəl ˌmī·ə'sīd·əs }

tropical rainforest [ECOL] A vegetation class consisting of tall, close-growing trees, their columnar trunks more or less unbranched in the lower two-thirds, and forming a spreading and frequently flat crown; occurs in areas of high temperature and high rainfall. Also known as hylaea; selva. { 'träp·ə·kəl 'rānˌfär·əst }

tropical savanna *See* tropical woodland. { 'träp·ə·kəlsə'van·ə }

tropical scrub [ECOL] A class of vegetation composed of low woody plants (shrubs), sometimes growing quite close together, but more often separated by large patches of bare ground, with clumps of herbs scattered throughout; an example is the Ghanaian evergreen coastal thicket. Also known as brush; bush; fourré; mallee; thicket. { 'träp·ə·kəl 'skrəb }

tropical sprue [MED] A malabsorption disease found in tropical and subtropical climates that resembles gluten enteropathy in its manifestations but does not improve with a gluten-free diet, it is thought that it may be an infectious diarrhea. { ,träp·ə·kəl 'sprü }

tropical woodland [ECOL] A vegetation class similar to a forest but with wider spacing between trees and sparse lower strata characterized by evergreen shrubs and seasonal graminoids; the climate is warm and moist. Also known as parkland; savanna-woodland; tropical savanna. { 'träp·ə·kəl 'wu̇d·lənd }

Tropiometridae [INV ZOO] A family of feather stars in the class Crinoidea which are bottom crawlers. { ,träp·ē·ō'me·trə,dē }

tropism [BIOL] Orientation movement of a sessile organism in response to a stimulus. Also known as topotaxis. { 'trō,piz·əm }

tropocollagen [BIOCHEM] The fundamental units of collagen fibrils. { ¦träp·ō'kal·ə·jən }

tropomyosin [BIOCHEM] A muscle protein similar to myosin and implicated as being part of the structure of the Z bands separating sarcomeres from each other. { ,träp·ə'mī·ə·sən }

troponin [BIOCHEM] A protein species located at specific stations every 36.5 nanometers on the actin helix in muscle sarcomere. { 'trō·pə·nən }

tropophytia [BOT] Plants that thrive in a climate that undergoes marked periodic changes. { ,träp·ə'fī·shə }

trout [VERT ZOO] Any of various edible freshwater fishes in the order Salmoniformes that are generally much smaller than the salmon. { 'trau̇t }

TRSV *See* tobacco ring spot virus.

true fruit [BOT] A fruit that is derived from only the ovary and its contents, usually derived from a superior (inserted above the other floral parts) ovary. Also known as superior fruit. { ¦trü 'früt }

truffle [BOT] The edible underground fruiting body of various European fungi in the family Tuberaceae, especially the genus *Tuber*. { 'trəf·əl }

trumpeter [VERT ZOO] A bird belonging to the Psophiidae, a family with three South American species; the common trumpeter (*Psophia crepitans*) is the size of a pheasant and resembles a long-legged guinea fowl. { 'trəm·pəd·ər }

truncate [BIOL] Abbreviated at an end, as if cut off. { 'trəŋ,kāt }

truncus arteriosis [EMBRYO] The embryonic arterial trunk between the bulbous arteriosis and the ventral aorta in anamniotes and early stages of amniotes. { ¦trəŋ·kəs är,tir·ē'ō·səs }

trunk [ANAT] The main mass of the human body, exclusive of the head, neck, and extremities; it is divided into thorax, abdomen, and pelvis. [BOT] The main stem of a tree. { trəŋk }

TRV *See* tobacco rattle virus.

Trypanorhyncha [INV ZOO] An order of tapeworms of the subclass Cestoda; all are parasites in the intestine of elasmobranch fishes. { trə,pan·ə'riŋ·kə }

Trypanosomatidae [INV ZOO] A family of Protozoa, order Kinetoplastida, containing flagellated parasites which exhibit polymorphism during their life cycle. { trə,pan·ə·sō'mad·ə,dē }

trypanosome [INV ZOO] A flagellated protozoan of the genus *Trypanosoma*. { trə'pan·ə ,sōm }

trypanosomiasis [MED] Any of many diseases of humans and animals caused by infection with species of *Trypanosoma* and transmitted by tsetse flies and other insects. { trə,pan·ə·sō'mī·ə·səs }

trypsin [BIOCHEM] A proteolytic enzyme which catalyzes the hydrolysis of peptide linkages in proteins and partially hydrolyzed proteins; derived from trypsinogen by the action of enterokinase in intestinal juice. { 'trip·sən }

trypsinogen [BIOCHEM] The zymogen of trypsin, secreted in the pancreatic juice. Also known as protrypsin. { trip'sin·ə·jən }

tryptophan [BIOCHEM] $C_{11}H_{12}O_2N_2$ An amino acid obtained from casein, fibrin, and certain other proteins; it is a precursor of indoleacetic acid, serotonin, and nicotinic acid. { 'trip·tə,fan }

tsetse fly [INV ZOO] Any of various South African muscoid flies of the genus *Glossina*; medically important as vectors of sleeping sickness or trypanosomiasis. { 'set,sē ,flī }

TSH *See* thyrotropic hormone.

tsutsugamushi disease [MED] A rickettsial disease of humans caused by *Rickettsia tsutsugamushi*, transmitted by larval mites, and characterized by headache, high fever, and a rash. Also known as scrub typhus. { ,tsü·sə·gə'mü·shē di,zēz }

tubal bladder [VERT ZOO] A urine reservoir organ that is an enlargement of the mesonephric ducts in most fish; there are four types: duplex, bilobed, simplex with ureters tied, and simplex with separate ureters. { 'tü·bəl 'blad·ər }

tubal ligation [MED] Surgical tying of the uterine tubes to prevent conception. { 'tü·bəl lī'gā·shən }

tube [BIOL] A narrow channel within the body of an animal or plant. { 'tüb }

tube cell [BOT] That nucleus of a pollen grain believed to influence the growth and development of the pollen tube. Also known as tube nucleus. { 'tüb ,sel }

tube foot [INV ZOO] One of the tentaclelike outpushings of the radial vessels of the water-vascular system in echinoderms; may be suctorial, or serve as stiltlike limbs or tentacles. { 'tüb ,fu̇t }

tube nucleus *See* tube cell. { 'tüb 'nü·klē·əs }

tuber [BOT] The enlarged end of a rhizome in which food accumulates, as in the potato. { 'tü·bər }

tuber cinereum [ANAT] An area of gray matter

extending from the optic chiasma to the mammillary bodies and forming part of the floor of the third ventricle. { 'tü·bər si'ner·ē·əm }

tubercle [BIOL] A small knoblike prominence. { 'tü·bər·kəl }

Tuberculariaceae [MYCOL] A family of fungi of the order Moniliales having short conidia that form cushion-shaped, often waxy or gelatinous aggregates (sporodochia). { tə‚bər·kyə‚la·rē 'ās·ē‚ē }

tuberculate [BIOL] Having or characterized by knoblike processes. { tə'bər·kyə·lət }

tuberculin [IMMUNOL] A preparation containing tuberculoproteins derived from *Mycobacterium tuberculosis* used in the tuberculin test to determine sensitization to tubercle bacilli. { tə'bər·kyə·lən }

tuberculin test [IMMUNOL] A test for past or present infection with tubercle bacilli based on a delayed hypersensitivity reaction at the site where tuberculin or purified protein derivative was introduced. { tə'bər·kyə·lən ‚test }

tuberculosis [MED] A chronic infectious disease of humans and animals primarily involving the lungs caused by the tubercle bacillus, *Mycobacterium tuberculosis*, or by M. *bovis*. Also known as consumption; phthisis. { tə‚bər·kyə'lō·səs }

tuberosity [ANAT] A large or obtuse prominence, especially as on bone for muscle attachment. { ‚tü·bə'räs·əd·ē }

tuberous organ [PHYSIO] An electroreceptor most sensitive to high-frequency electric signals, and distributed over the body surface of electric fish. { 'tü·bə·rəs 'ór·gən }

tuberous sclerosis [MED] A familial neurocutaneous syndrome characterized in its complete form by epilepsy, adenoma sebaceum, and mental deficiency, and pathologically by nodular sclerosis of the cerebral cortex. Also known as Bourneville's disease. { 'tü·bə·rəs sklə'rō·səs }

tubeworms [INV ZOO] Marine polychaete worms (particularly many species in the family Serpulidae) which construct permanent calcareous tubes on rocks, seaweeds, dock pilings, and ship bottoms. The individual tubes with hard walls of calcite-aragonite, ranging from 0.04 to 0.4 inch (1 millimeter to 1 centimeter) in diameter and from 0.16 to 4 inches (4 millimeters to 10 centimeters) in length, are firmly cemented to a hard substrate and to each other. { 'tüb‚wərmz }

Tubicola [INV ZOO] An order of sedentary polychaete annelids that surround themselves with a calcareous tube or one which is composed of agglutinated foreign particles. { tü'bik·ə·lə }

Tubulanidae [INV ZOO] A family of the order Palaeonemertini. { ‚tüb·yə'lan·ə‚dē }

tubular gland [ANAT] A secreting structure whose secretory endpieces are tubelike or cylindrical in shape. { 'tü·byə·lər 'gland }

tubule [ANAT] A slender, elongated microscopic tube in an anatomical structure. { 'tü‚byül }

Tubulidentata [VERT ZOO] An order of mammals which contains a single living genus, the aardvark (*Orycteropus*) of Africa. { ‚tü·byə·lə·den'täd·ə }

tubulin [BIOCHEM] A globular protein containing two subunits; 10–14 molecules are arranged to form a microtubule. { 'tü·byə·lən }

tubuloacinous gland *See* tubuloalveolar gland. { ¦tü·byə·lō'as·ə·nəs ‚gland }

tubuloalveolar gland [ANAT] A secreting structure having both tubular and alveolar secretory endpieces. Also known as acinotubular gland; tubuloacinous gland. { ¦tü·byə·lō·al'vē·ə·lər ‚gland }

tularemia [VET MED] A bacterial infection of wild rodents caused by *Pasteurella tularensis*; it may be generalized, or it may be localized in the eyes, skin, or lymph nodes, or in the respiratory tract or gastrointestinal tract; may be transmitted to humans and to some domesticated animals. { ‚tü·lə'rē·mē·ə }

tulip [BOT] Any of various plants with showy flowers constituting the genus *Tulipa* in the family Liliaceae; characterized by coated bulbs, lanceolate leaves, and a single flower with six equal perianth segments and six stamens. { 'tü·ləp }

tulip poplar *See* tulip tree. { 'tü·ləp 'päp·lər }

tulip tree [BOT] *Liriodendron tulipifera*. A tree belonging to the magnolia family (Magnoliaceae) distinguished by leaves which are squarish at the tip, true terminal buds, cone-shaped fruit, and large greenish-yellow and orange-colored flowers. Also known as tulip poplar. { 'tü·ləp ‚trē }

tumbleweed [BOT] Any of various plants that break loose from their roots in autumn and are driven by the wind in rolling masses over the ground. { 'təm·bəl‚wēd }

tumid [BIOL] Marked by swelling or inflation. { 'tü·məd }

tumor [MED] Any abnormal mass of cells resulting from excessive cellular multiplication. { 'tü·mər }

tumorigenic [MED] Tumor-forming. { ¦tü·mə·rə'jen·ik }

tumor necrosis factor [IMMUNOL] A monokine that induces leukocytosis, fever, weight loss, the acute-phase reaction, and necrosis of some tumors. { ¦tü·mər nə'krō·səs ‚fak·tər }

tumor nicotinamide adenine dinucleotide oxidase [BIOCHEM] A cancer-specific growth protein (an unregulated nicotinamide adenine dinucleotide oxidase form) whose inhibition may be the underlying mechanism of green and black tea catechins' slowing of cancer cell growth. Abbreviated tNOX. { ¦tü·mər ‚nik·ə‚tin·ə‚mīd ‚ad·ən‚ēn dī‚nü·klē·ə‚tīd 'äk·sə‚dās }

tumor suppressor gene [CELL MOL] A class of genes which, when mutated, predispose an individual to cancer by causing the loss of function of the particular tumor suppressor protein encoded by the gene. { ¦tüm·ər sə'pres·ər jēn }

tumor suppressor protein [CELL MOL] A protein that helps protect a normal cell from becoming malignant; for example, p53. { ¦tüm·ər sə'pres·ər prō‚tēn }

tuna [VERT ZOO] Any of the large, pelagic, cosmopolitan marine fishes which form the family Thunnidae including species that rank among the most valuable of food and game fish. { 'tü·nə }

tundra [ECOL] An area supporting some vegetation between the northern upper limit of trees and the lower limit of perennial snow on mountains, and on the fringes of the Antarctic continent and its neighboring islands. Also known as cold desert. { 'tən·drə }

tung nut [BOT] The seed of the tung tree (*Aleurites fordii*), which is the source of tung oil. { 'təŋ ˌnət }

tung tree [BOT] *Aleurites fordii.* A plant of the spurge family in the order Euphorbiales, native to central and western China and grown in the southern United States. { 'təŋ ˌtrē }

tunica [BIOL] A membrane or layer of tissue that covers or envelops an organ or other anatomical structure. { 'tü·nə·kə }

tunica adventitia *See* adventitia. { 'tü·nə·kə ˌad·vən'tish·ə }

tunica intima *See* intima. { 'tü·nə·kə 'in·tə·mə }

tunica mucosa *See* mucous membrane. { 'tü·nə·kə myü'kō·zə }

Tunicata [INV ZOO] A subphylum of the Chordata characterized by restriction of the notochord to the tail and posterior body of the larva, absence of mesodermal segmentation, and secretion of an outer covering or tunic about the body. { ˌtü·nə'käd·ə }

Tupaiidae [VERT ZOO] The tree shrews, a family of mammals in the order Insectivora. { tü'pī·əˌdē }

tupelo [BOT] Any of various trees belonging to the genus *Nyssa* of the sour gum family, Nyssaceae, distinguished by small, obovate, shiny leaves, a small blue-black drupaceous fruit, and branches growing at a wide angle from the axis. { 'tü·pəˌlō }

Turbellaria [INV ZOO] A class of the phylum Platyhelminthes having bodies that are elongate and flat to oval or circular in cross section. { ˌtər·bə'lar·ē·ə }

turbidostat [MICROBIO] A device in which a bacterial culture is maintained at a constant volume and cell density (turbidity) by adjusting the flow rate of fresh medium into the growth tube by means of a photocell and appropriate electrical connections. { tər'bid·əˌstat }

turbinate [BOT] Shaped like an inverted cone. [INV ZOO] Spiral with rapidly decreasing whorls from base to apex. { 'tər·bə·nət }

Turbinidae [INV ZOO] A family of gastropod mollusks including species of top shells. { tər'bin·əˌdē }

Turdidae [VERT ZOO] The thrushes, a family of passeriform birds in the suborder Oscines. { 'tər·dəˌdē }

turgid [MED] Swollen and congested. { 'tər·jəd }

turgor [BOT] Distension of a plant cell wall and membrane by the fluid contents. { 'tər·gər }

turgor movement [BOT] A reversible change in the position of plant parts due to a change in turgor pressure in certain specialized cells; movement of *Mimosa* leaves when touched is an example. { 'tər·gər ˌmüv·mənt }

turgor pressure [BOT] The actual pressure developed by the fluid content of a turgid plant cell. { 'tər·gər ˌpresh·ər }

turion [BOT] A scaly shoot, such as asparagus, developed from an underground bud. { 'tùr·ēˌän }

turkey [VERT ZOO] Either of two species of wild birds, and any of various derived domestic breeds, in the family Meleagrididae characterized by a bare head and neck, and in the male a large pendant wattle which hangs on one side from the base of the bill. { 'tər·kē }

turkey tail mushroom *See* Trametes versicolor. { 'tər·kē ˌtāl ˌməsh,rüm }

turmeric [BOT] *Curcuma longa.* An East Indian perennial of the ginger family (Zingiberaceae) with a short stem, tufted leaves, and short thick rhizomes; a spice with a pungent, bitter taste and a musky odor is derived from the rhizome. { 'tər·mər·ik }

Turner's syndrome [MED] A sex aberration in humans in which the chromosome complement includes only one sex chromosome, an X. { 'tər·nərz ˌsinˌdrōm }

Turnicidae [VERT ZOO] The button quails, a family of Old World birds in the order Gruiformes. { tər'nis·əˌdē }

turnip [BOT] *Brassica rapa* or *B. campestris* var. *rapa.* An annual crucifer of Asiatic origin belonging to the family Brassiaceae in the order Capparales and grown for its foliage and edible root. { 'tər·nəp }

turnip yellow mosaic virus [VIROL] The type species of the genus *Tymovirus*, it is transmitted mechanically and via beetles, causing chloroplast clumping. Abbreviated TYMV. { ¦tər·nəp ˌyel·ō mō'zā·ik ˌvī·rəs }

turnip yellow mosaic virus group *See* Tymovirus. { ¦tər·nəp ˌyel·ō mō'zā·ik ¦vī·rəs ˌgrüp }

turnover [CELL MOL] The number of substrate molecules transformed by a single molecule of enzyme per minute, when the enzyme is operating at maximum rate. { 'tərnˌō·vər }

turnover number [BIOCHEM] The number of molecules of a substrate acted upon in a period of 1 minute by a single enzyme molecule, with the enzyme working at a maximum rate. { 'tərn·ō·vər ˌnəm·bər }

turtle [VERT ZOO] Any of about 240 species of reptiles which constitute the order Chelonia distinguished by the two bony shells enclosing the body. { 'tərd·əl }

tussock [ECOL] A small hummock of generally solid ground in a bog or marsh, usually covered with and bound together by the roots of low vegetation such as grasses, sedges, or ericaceous shrubs. { 'təs·ək }

twin [BIOL] One of two individuals born at the same time. { 'twin }

twiner [BOT] A climbing stem that winds about its support, as pole beans or many tropical lianas. { 'twī·nər }

two-dimensional gel electrophoresis [CELL MOL] A type of gel electrophoresis in which proteins are first separated by charge and then by molecular weight, enabling the analysis of complex protein mixtures. { ¦tü·də,men·shən·əl ¦jel i,lek·trō·fə'rē·səs }

two-hit model [GEN] The hypothesis that a cancer arises in an individual heterozygous for a mutant gene for a dominantly inherited form of cancer only if an additional mutation occurs in a somatic cell; for example, in the normal allele at the same locus, as in human retinoblastoma. { 'tü ,hit ¦mäd·əl }

two-point threshold [PHYSIO] The distance on the skin separating two pointed stimulators that is required to experience two rather than one point of stimulation. { ,tü ,pȯint 'thresh,hōld }

two-step test [MED] Repeated ascents over two 9-inch (23-centimeter) steps as a simple exercise test of cardiovascular function. Also known as Master's two-step test. { 'tü ¦step 'test }

Tylenchida [INV ZOO] An order of soil-dwelling or phytoparasitic nematodes in the subclass Rhabdita. { tī'leŋ·kəd·ə }

Tylenchoidea [INV ZOO] A superfamily of mainly soil and insect-associated nematodes in the order Tylenchida with a stylet for piercing live cells and sucking the juices. { tī·lən'kȯid·ē·ə }

Tylopoda [VERT ZOO] An infraorder of artiodactyls in the suborder Ruminantia that contains the camels and extinct related forms. { tī'läp·əd·ə }

tylose [BOT] A mass of parenchymal cells appearing somewhat frothlike in the pores of some hardwood trees. { 'tī,lōs }

tylostyle [INV ZOO] A uniradiate spicule in Porifera with a point at one end and a knob at the other end. { 'tī·lə,stīl }

tylote [INV ZOO] A slender sponge spicule with a knob at each end. { 'tī,lōt }

Tymovirus [VIROL] A genus of plant viruses characterized by isometric particles containing one molecule of linear, positive-sense, single-stranded ribonucleic acid, turnip yellow mosaic virus is the type species. Also known as turnip yellow mosaic virus group. { 'tīm·ə,vī·rəs }

tympanic cavity [ANAT] The irregular, air-containing, mucous-membrane-lined space of the middle ear; contains the three auditory ossicles and communicates with the nasopharynx through the auditory tube. { tim'pan·ik 'kav·əd·ē }

tympanic membrane [ANAT] The membrane separating the external from the middle ear. Also known as eardrum; tympanum. { tim'pan·ik 'mem,brān }

tympanoplasty [MED] A surgical procedure performed to eradicate disease in the middle ear cavity or to reconstruct the conductive mechanism. { ¦tim·pə·nō¦plas·tē }

tympanum [ANAT] *See* tympanic membrane. [INV ZOO] A thin membrane covering an organ of hearing in insects. { 'tim·pə·nəm }

TYMV *See* turnip yellow mosaic virus.

type [SYST] A specimen on which a species or subspecies is based. { 'tīp }

type I of Cori *See* von Gierke's disease. { 'tīp 'wən əv 'kȯr·ē }

Typhaceae [BOT] A family of monocotyledonous plants in the order Typhales characterized by an inflorescence of dense, cylindrical spikes and absence of a perianth. { tī'fās·ē,ē }

Typhales [BOT] An order of marsh or aquatic monocotyledons in the subclass Commelinidae with emergent or floating stems and leaves and reduced, unisexual flowers having a single ovule in an ovary composed of a single carpel. { tī'fā·lēz }

Typhlopidae [VERT ZOO] A family of small, burrowing circumtropical snakes, suborder Serpentes, with vestigial eyes and toothless jaws. { ti'fläp·ə,dē }

Typhloscolecidae [INV ZOO] A family of pelagic polychaete annelids belonging to the Errantia. { ,tif·lō·skō'les·ə,dē }

typhlosole [INV ZOO] A dorsal longitudinal invagination of the intestinal wall in certain invertebrates serving to increase the absorptive surface. { 'tif·lə,sōl }

typhoid fever [MED] A highly infectious, septicemic disease of humans caused by *Salmonella typhi* which enters the body by the oral route through ingestion of food or water contaminated by contact with fecal matter. { 'tī,fȯid 'fē·vər }

typhoid vaccine [IMMUNOL] A type of killed vaccine used for active immunity production; made from killed typhoid bacillus (*Salmonella typhi*). { 'tī,fȯid vak'sēn }

typhus fever [MED] Any of three louse-borne human diseases caused by *Rickettsia prowazakii* characterized by fever, stupor, headaches, and a dark-red rash. { 'tī·fəs 'fē·vər }

tyramine [PHARM] $HOC_6H_4CH_2CH_2NH_2$ A crystalline compound with a melting point of 164–165°C; soluble in water and boiling alcohol; used in medicine as an adrenergic drug. Also known as tyrosamine. { 'tī·rə,mēn }

Tyranni [VERT ZOO] A suborder of suboscine Passeriformes containing birds with limited song power and having the tendon of the hind toe separate and the intrinsic muscles of the syrinx reduced to one pair. { tə'ra,nī }

Tyrannidae [VERT ZOO] The tyrant flycatchers, a family of passeriform birds in the suborder Tyranni confined to the Americas. { tə'ran·ə,dē }

Tyrannoidea [VERT ZOO] The flycatchers, a superfamily of suboscine birds in the suborder Tyranni. { tir·ə'nȯid·ē·ə }

tyrocidine [MICROBIO] A peptide antibiotic produced by *Bacillus brevis*; used to control fungi, bacteria, and protozoa. { ,tir·ə'sīd·ən }

tyrosamine *See* tyramine. { tə'räs·ə,mēn }

tyrosinase [BIOCHEM] An enzyme found in plants, molds, crustaceans, mollusks, and some bacteria which, in the presence of oxygen, catalyzes the oxidation of monophenols and polyphenols with the introduction of −OH groups and the formation of quinones. { 'tir·ə·sə,nās }

tyrosine [BIOCHEM] $C_9H_{11}NO_3$ A phenolic alpha amino acid found in many proteins; a precursor of the hormones epinephrine, norepinephrine, thyroxine, and triiodothyronine, and of the black pigment melanin. { 'tir·ə,sēn }

tyrosine hydroxylase [BIOCHEM] A specialized enzyme located only in catecholamine-containing nerve cells, where it serves as the primary regulatory or rate-limiting step in catecholamine biosynthesis. { ,tī·rə,sēn hī'dräk·sə,lās }

tyrosinemia [MED] An inborn metabolic disorder in which there is a deficiency of the enzyme *p*-hydroxyphenylpyruvic acid oxidase with abnormally high blood levels of tyrosine and sometimes methionine. { ,tir·ə'sē·mē·ə }

tyrosinosis [MED] Excretion of excessive amounts of tyrosine and its first oxidation products in the urine. { ,tir·ə·sə'nō·səs }

tyrothricin [MICROBIO] A polypeptide mixture produced by *Bacillus brevis* and consisting of the antibiotic substances gramicidin and tyrocidine; effective as an antibacterial applied locally in infections due to germ-positive organisms. { ,tir·ə'thrīs·ən }

Tyson's gland [ANAT] A small scent gland in the human male which secretes the smegma. Also known as preputial gland. { 'tī·sənz ,gland }

Tytonidae [VERT ZOO] The barn owls, a family of birds in the order Strigiformes distinguished by an unnotched sternum which is fused to large clavicles. { tī'tän·ə,dē }

tyvelose [BIOCHEM] A dideoxy sugar found in bacterial lipopolysaccharides. { 'tī·və,lōs }

U

ubiquitin [BIOCHEM] A small, 76-amino-acid, highly conserved protein present in the cytoplasm and nucleus of all eukaryotes (but not eubacteria and archaea). The covalent, ATP-dependent linkage of multiple ubiquitin molecules to proteins serves as a signal for their degradation by the 26S proteasome. { yü'bik·wə,tin }

udder [VERT ZOO] A pendulous organ consisting of several mammary glands enclosed in a single envelope; each gland has its own nipple; found in some mammals, such as the cow and goat. { 'əd·ər }

UDP *See* uridine diphosphate.

UDPG *See* uridine diphosphoglucose.

ulcer [MED] Localized interruption of the continuity of an epithelial surface, with an inflamed base. { 'əl·sər }

ulcerative colitis [MED] An idiopathic inflammatory disease of the mucosa and submucosa of the colon manifested clinically by pain, diarrhea, and rectal bleeding. { 'əl·sə,rād·iv kə'līd·əs }

ulcerative endocarditis [MED] Acute bacterial endocarditis. { 'əl·sə,rād·iv ¦en·dō·kär'dīd·əs }

ulcerative gingivitis *See* Vincent's infection. { 'əl·sə,rād·iv ,jin·jə'vīd·əs }

Ullrich-Turner syndrome [MED] A complex of symptoms including webbing of the neck, short stature, cubitus valgus, and hypogonadism in the male. Also known as male Turner's syndrome. { 'əl·rik 'tər·nər ,sin,drōm }

Ulmaceae [BOT] A family of dicotyledonous trees in the order Urticales distinguished by alternate stipulate leaves, two styles, a pendulous ovule, and lack of a latex system. { əl'mās·ē,ē }

ulna [ANAT] The larger of the two bones of the forearm or forelimb in vertebrates; articulates proximally with the humerus and radius and distally with the radius. { 'əl·nə }

ulnar deviation [BIOPHYS] A position of the human hand in which the wrist is bent toward the end finger. { ¦əl·nər ,dē·vē'ā·shən }

ulnar loop [FOREN] A loop fingerprint pattern which flows in the direction of the ulna bone, toward the little finger. { ,əl·nər 'lüp }

Ulotrichaceae [BOT] A family of green algae in the suborder Ulotrichineae; contains both attached and floating filamentous species with cells having parietal, platelike or bandlike chloroplasts. { yü,lä·trə'kās·ē,ē }

Ulotrichales [BOT] A large, artificial order of the Chlorophyta composed mostly of fresh-water, branched or unbranched filamentous species with mostly cylindrical, uninucleate cells having cellulose, but often mucilaginous walls. { yü,lä·trə'kā·lēz }

Ulotrichineae [BOT] A suborder of the Ulotrichales characterized by short cylindrical cells. { yü,lä·trə'kīn·ē,ē }

ultimobranchial body [EMBRYO] One of the small, endocrine structures which originate as terminal outpocketings from each side of the embryonic vertebrate pharynx; can produce the hormone calcitonin. { ¦əl·tə·mō'braŋ·kē·əl 'bäd·ē }

ultrasonic therapy *See* ultrasound diathermy. { ¦əl·trə'sän·ik 'ther·ə·pē }

ultrasound diathermy [MED] The application of high-frequency sound waves (0.7 to 1.0 megahertz) and the conversion of this mechanical energy into heat for local thermotherapy. Also known as ultrasonic therapy. { ¦əl·trə¦saund 'dī·ə,thər·mē }

ultrastructure [CELL MOL] The ultimate physiochemical organization of protoplasm. { ¦əl·trə'strək·chər }

Ulvaceae [BOT] A large family of green algae in the order Ulvales. { ,əl'vās·ē,ē }

Ulvales [BOT] An order of algae in the division Chlorophyta in which the thalli are macroscopic, attached tubes or sheets. { ,əl'vā·lēz }

umbel [BOT] An indeterminate inflorescence with the pedicels all arising at the top of the peduncle and radiating like umbrella ribs; there are two types, simple and compound. { 'əm·bəl }

Umbellales [BOT] An order of dicotyledonous herbs or woody plants in the subclass Rosidae with mostly compound or conspicuously lobed or dissected leaves, well-developed schizogenous secretory canals, separate petals, and an inferior ovary. { ,əm·bə'lā·lēz }

Umbelliferae [BOT] A large family of aromatic dicotyledonous herbs in the order Umbellales; flowers have an ovary of two carpels, ripening to form a dry fruit that splits into two halves, each containing a single seed. { ,əm·bə'lif·ə,rē }

umbilical artery [EMBRYO] Either of a pair of arteries passing through the umbilical cord to carry impure blood from the mammalian fetus to the placenta. { əm'bil·ə·kəl 'ärd·ə·rē }

umbilical cord [EMBRYO] The long, cylindrical structure containing the umbilical arteries and vein, and connecting the fetus with the placenta. { əm'bil·ə·kəl ˌkȯrd }

umbilical duct *See* vitelline duct. { əm'bil·ə·kəl ˌdəkt }

umbilical hernia [MED] Herniation through the umbilical ring. Also known as annular hernia. { əm'bil·ə·kəl 'hər·nē·ə }

umbilical vein [EMBRYO] A vein passing through the umbilical cord and conveying purified, nutrient-rich blood from placenta to fetus. { əm'bil·ə·kəl 'vān }

Umbilicariaceae [BOT] The rock tripes, a family of Asco-lichenes in the order Lecanorales having a large, circular, umbilicate thallus. { ˌəm·bə·ləˌkar·ē'ās·ēˌē }

umbilicus [ANAT] The navel; the round, depressed cicatrix in the median line of the abdomen, marking the site of the aperture through which passed the fetal umbilical vessels. { əm'bil·ə·kəs }

umbo [ANAT] A rounded elevation of the surface of the tympanic membrane. [INV ZOO] A prominence above the hinge of a bivalve mollusk shell. { 'əm·bō }

umbonate [BIOL] Having or forming an umbo. { 'əm·bə·nət }

UMP *See* uridylic acid.

uncinate trophus [INV ZOO] A trophus in rotifers characterized by a hooked or curved uncus. { 'ən·sə·nət 'träf·əs }

uncinus [INV ZOO] In annelids, a minute, pectiniform neuroseta. { ən'sī·nəs }

uncompetitive enzyme inhibition [BIOCHEM] The prevention of an enzymic process as a result of the interaction of an inhibitor with the enzyme-substrate complex or a subsequent intermediate form of the enzyme, but not with the free enzyme. { ¦ən·kəm'ped·əd·iv 'enˌzīm ˌin·ə'bish·ən }

unconscious [MED] Insensible; in a state lacking conscious awareness, with reflexes abolished. { ¦ən'kän·shəs }

underground stem [BOT] Any of the stems that grow underground and are often mistaken for roots; principal kinds are rhizomes, tubers, corms, bulbs, and rhizomorphic droppers. { ¦ən·dər¦graůnd 'stem }

underwing [INV ZOO] Either of a pair of posterior wings on certain insects, as the moth. { 'ən·dərˌwiŋ }

undulant fever *See* brucellosis. { 'ən·jə·lənt 'fē·vər }

unequal homologous recombination [GEN] Deoxyribonucleic acid exchange between identical chromosome regions that are not precisely paired. { ənˌē·kwəl hə¦mäl·ə·gəs rēˌkäm·bə'nā·shən }

ungula [VERT ZOO] A nail, hoof, or claw. { 'əŋ·gyə·lə }

ungulate [VERT ZOO] Referring to an animal that has hoofs. { 'əŋ·gyə·lət }

ungulicutate [VERT ZOO] Having claws or nails. { ¦əŋ·gyə'lik·əˌtāt }

unguligrade [VERT ZOO] Walking on hoofs. { 'əŋ·gyə·ləˌgrād }

unicellular [BIOL] Composed of a single cell. { ¦yü·nə'sel·yə·lər }

unicellular gland [ANAT] A gland consisting of a single cell. { ¦yü·nə'sel·yə·lər ˌgland }

unicuspid [ANAT] Having one cusp, as certain teeth. { ¦yü·nə'kəs·pəd }

unilateral hermaphroditism [ZOO] A type of hermaphroditism in which there is a combination of ovatestis on one side of the body with an ovary or testis on the other side. { ¦yü·nə'lad·ə·rəl hər'maf·rə·dīˌtiz·əm }

unilocular [BIOL] Having a single cavity. { ¦yü·nē'läk·yə·lər }

uninhibited bladder [MED] An abnormal urinary bladder that shows only a variable loss of cerebral inhibition over reflex bladder contractions, representing, of all neurogenic bladders, the least variance from normal. { ¦ən·in'hib·əd·əd 'blad·ər }

Unionidae [INV ZOO] The fresh-water mussels, a family of bivalve mollusks in the subclass Eulamellibranchia; the larvae, known as glochidia, are parasitic on fish. { ˌyü·nē'än·əˌdē }

uniparental disomy [GEN] Inheritance of both chromosomes or alleles of a homologous pair from one parent. { ¦yü·nə·pə¦rent·əl dī'sō·mē }

Unipolarina [INV ZOO] A suborder of the protozoan order Myxosporida characterized by spores with one to six (never five) polar capsules located at the anterior end. { ˌyü·nəˌpō·lə'rīn·ə }

uniporter [CELL MOL] A channel protein that transfers only one substrate at a time across the membrane. { 'yü·nəˌpȯrd·ər }

unitegmic [BOT] Referring to an ovule having a single integument. { ˌyü·nə'teg·mik }

universal donor [IMMUNOL] An individual of O blood group; can give blood to persons of all blood types. { ¦yü·nə¦vər·səl 'dō·nər }

universal recipient [IMMUNOL] An individual of AB blood group; can receive a blood transfusion of all blood types, A, AB, B, or O. { ¦yü·nə¦vər·səl ri'sip·ē·ənt }

unmyelinated [HISTOL] Lacking myelin, either as a normal condition or as the result of a disease. { ¦ən'mī·ə·ləˌnād·əd }

unproductive cough *See* dry cough. { ˌən·prə¦dək·tiv 'kȯf }

unresolved pneumonia *See* organizing pneumonia. { ¦ən·ri'zälvd nů'mō·nyə }

unstable colon *See* irritable colon. { ¦ən'stā·bəl 'kō·lən }

unstriated fiber *See* smooth muscle fiber. { ¦ən 'strīˌād·əd 'fī·bərz }

Upupidae [VERT ZOO] The Old World hoopoes, a family of birds in the order Coraciiformes whose young are hatched with sparse down. { yü'püp·əˌdē }

urachus [EMBRYO] A cord or tube of epithelium connecting the apex of the urinary bladder with the allantois; its connective tissue forms the median umbilical ligament. { 'yůr·ə·kəs }

uracil [BIOCHEM] $C_4H_4N_2O_2$ A pyrimidine base

important as a component of ribonucleic acid. { 'yu̇r·ə,sil }

Uraniidae [INV ZOO] A tropical family of moths in the superfamily Geometroidea including some slender-bodied, brilliantly colored diurnal insects which lack a frenulum and are often mistaken for butterflies. { ,yu̇r·ə'nī·ə,dē }

urate calculi [PATH] Kidney stones composed of uric acid salts and found particularly in people suffering from gout. { 'yu̇,rāt 'kal·kyə,lī }

urban typhus *See* murine typhus. { 'ər·bən 'tī·fəs }

urceolate [BIOL] Shaped like an urn. { ¦ər¦sē·ə·lət }

urease [BIOCHEM] An enzyme that catalyzes the degradation of urea to ammonia and carbon dioxide; obtained from the seed of jack bean. { 'yu̇r·ē,ās }

Urechinidae [INV ZOO] A family of echinoderms in the order Holasteroida which have an ovoid test lacking a marginal fasciole. { ,yu̇r·ə'kin·ə,dē }

Uredinales [MYCOL] An order of parasitic fungi of the subclass Heterobasidiomycetidae characterized by the teleutospore, a spore with one or more cells, each of which is a modified hypobasidium; members cause plant diseases known as rusts. { yə,red·ən'ā·lēz }

Urediniomycetes [MYCOL] A class of fungi in the subdivision Basidiomycotina that causes plant rust diseases, members have thick-walled teliospores, produced in the terminal state in pustules. Also known as rust fungi. { ,yu̇r·ə,din·ē·ō·mī'sē,dēz }

urediniospore [MYCOL] A spore produced by a uredinium. Also known as urediospore; uredospore. { ,yu̇r·ə'din·ē·ə,spȯr }

uredinium [MYCOL] A reproductive tissue of the rust fungus which gives rise to urediniospores. { ,yu̇r·ə'din·ē·əm }

urediospore *See* urediniospore. { yu̇'red·ē·ə ,spȯr }

uredospore *See* urediniospore. { yu̇'red·ə,spȯr }

uremia [MED] A condition resulting from kidney failure and characterized by azotemia, chronic acidosis, anemia, and a variety of systemic signs and symptoms. { yə'rē·mē·ə }

ureotelic [BIOL] Referring to animals that produce urea as their main nitrogenous excretion. { yə¦rē·ə¦tel·ik }

ureter [ANAT] A long tube conveying urine from the renal pelvis to the urinary bladder or cloaca in vertebrates. { 'yu̇r·əd·ər }

urethra [ANAT] The canal in most mammals through which urine is discharged from the urinary bladder to the outside. { yə'rē·thrə }

urethral gland [ANAT] One of the small, branched tubular mucous glands in the mucosa lining the urethra. { yə'rē·thrəl 'gland }

urethritis [MED] Inflammation of the urethra. { ,yu̇r·ə'thrīd·əs }

uric acid [BIOCHEM] $C_5H_4N_4O_3$ A white, crystalline compound, the excretory end product in amino acid metabolism by uricotelic species. { 'yu̇r·ik 'as·əd }

uricase [BIOCHEM] An enzyme present in the liver, spleen, and kidney of most mammals except humans; converts uric acid to allantoin in the presence of gaseous oxygen. { 'yu̇r·ə,kās }

uricotelism [PHYSIO] An adaptation of terrestrial reptiles and birds which effectively provides for detoxification of ammonia and also for efficient conservation of water due to a relatively low rate of glomerular filtration and active secretion of uric acid by the tubules to form a urine practically saturated with urate. { ,yu̇r·ə'käd·əl,iz·əm }

uridine [BIOCHEM] $C_9H_{12}N_2O_6$ A crystalline nucleoside composed of one molecule of uracil and one molecule of D-ribose; a component of ribonucleic acid. { 'yu̇r·ə,dīn }

uridine diphosphate [BIOCHEM] The chief transferring coenzyme in carbohydrate metabolism. Abbreviated UDP. { 'yu̇r·ə,dīn dī¦fä,sfāt }

uridine diphosphoglucose [BIOCHEM] A compound in which α-glucopyranose is esterified, at carbon atom 1, with the terminal phosphate group of uridine-5′-pyrophosphate (that is, uridine diphosphate); occurs in animal, plant, and microbial cells; functions as a key in the transformation of glucose to other sugars. Abbreviated UDPG. { 'yu̇r·ə,dīn dī¦fä'sfō'glü,kōs }

uridine monophosphate *See* uridylic acid. { 'yu̇r·ə,dīn ¦män·ə'fä,sfāt }

uridine phosphoric acid *See* uridylic acid. { 'yu̇r·ə,dīn fä'sfȯr·ik 'as·əd }

uridylic acid [BIOCHEM] $C_9H_{13}N_2O_9P$ Water-and alcohol-soluble crystals, melting at 202°C; used in biochemical research. Also known as uridine monophosphate (UMP); uridine phosphoric acid. { ¦yu̇r·ə¦dil·ik 'as·əd }

urinalysis [PATH] Analysis of the urine, involving chemical, physical, and microscopic tests. { ,yu̇r·ə'nal·ə·səs }

urinary bladder [ANAT] A hollow organ which serves as a reservoir for urine. { 'yu̇r·ə,ner·ē 'blad·ər }

urinary system [ANAT] The system which functions in the elaboration and excretion of urine in vertebrates; in humans and most mammals, consists of the kidneys, ureters, urinary bladder, and urethra. { 'yu̇r·ə,ner·ē ,sis·təm }

urinary tract infection [MED] An inflammatory process occurring in the kidney, ureter, bladder, or adjacent structures that occurs when microorganisms (usually *Escherichia coli*) enter through the urethra. { ,yu̇r·ə,ner·ē 'trakt in,fek·shən }

urination [PHYSIO] The discharge of urine from the bladder. Also known as micturition. { ,yu̇r·ə'nā·shən }

urine [PHYSIO] The fluid excreted by the kidneys. { 'yu̇r·ən }

uriniferous tubule [ANAT] One of the numerous winding tubules of the kidney. Also known as nephric tubule. { ¦yu̇r·ə¦nif·ə·rəs 'tü,byül }

urn [BOT] The theca of a moss. { ərn }

urobilin [BIOCHEM] A bile pigment produced by reduction of bilirubin by intestinal bacteria and excreted by the kidneys or removed by the liver. { ,yu̇r·ə'bī·lən }

urobilinogen [BIOCHEM] A chromogen, formed in feces and present in urine, from which urobilin is formed by oxidation. { ˌyu̇r·əˈbil·ə·jən }

urocanic acid [BIOCHEM] $C_6H_6N_2O_2$ A crystalline compound formed as an intermediate in the degradative pathway of histidine. { ¦yu̇r·ə¦kan·ik ˈas·əd }

Urochordata [INV ZOO] The equivalent name for Tunicata. { ¦yu̇r·ə·ko̊rˈdäd·ə }

urochrome [BIOCHEM] $C_{43}H_{51}O_{26}N$ Yellow pigment found in normal urine. { ˈyu̇r·əˌkrōm }

Urodela [VERT ZOO] The tailed amphibians or salamanders, an order of the class Amphibia distinguished superficially from frogs and toads by the possession of a tail, and from caecilians by the possession of limbs. { ˌyu̇r·əˈdē·lə }

urogenital diaphragm [ANAT] The sheet of tissue stretching across the pubic arch, formed by the deep transverse perineal and the sphincter urethrae muscles. Also known as triangular ligament. { ¦yu̇r·ə¦jen·əd·əl ˈdī·əˌfram }

urogenital system [ANAT] The combined urinary and genital system in vertebrates, which are intimately related embryologically and anatomically. Also known as genitourinary system. { ¦yu̇r·ə¦jen·əd·əl ˌsis·təm }

urogomphus [INV ZOO] In certain insect larvae, a process on the terminal segment. Also known as pseudocercus. { ˌyu̇r·əˈgäm·fəs }

urography [MED] Radiography of any portion of the urinary tract; most often follows the intravenous administration of iodinated contrast material. { yəˈräg·rə·fē }

urokinase [BIOCHEM] An enzyme, present in human urine, that catalyzes the conversion of plasminogen to plasmin. { ˌyu̇r·əˈkīˌnās }

urolithiasis [MED] **1.** Condition associated with the presence of urinary calculi. **2.** Formation or presence of urinary calculi. { ˌyu̇r·ə·ləˈthī·ə·səs }

urology [MED] The scientific study of urine and the diseases and abnormalities of the urinary and urogenital tracts. { yəˈräl·ə·jē }

uropepsin [BIOCHEM] The end product of the secretion of pepsinogen into the blood by gastric cells; occurs in urine. { ¦yu̇r·əˈpep·sən }

uropod [INV ZOO] One of the flattened abdominal appendages of various crustaceans that with the telson forms the tail fan. { ˈyu̇r·əˌpäd }

uroporphyrin [BIOCHEM] Any of several isomeric, metal-free porphyrins, occurring in small quantities in normal urine and feces; molecule has four acetic acid ($-CH_2COOH$) and four propionic acid ($-CH_2CH_2COOH$) groups. { ¦yu̇r·əˈpo̊r·fə·rən }

Uropygi [INV ZOO] The tailed whip scorpions, an order of arachnids characterized by an elongate, flattened body which bears in front a pair of thickened, raptorial pedipalps set with sharp spines and used to hold and crush insect prey. { ˌyu̇r·əˈpīˌjī }

uropygial gland [VERT ZOO] A relatively large, compact, bilobed, secretory organ located at the base of the tail (uropygium) of most birds having a keeled sternum. Also known as oil gland; preen gland. { ¦yu̇r·ə¦pij·ē·əl ˌgland }

urostyle [VERT ZOO] An unsegmented bone representing several fused vertebrae and forming the posterior part of the vertebral column in Anura. { ˈyu̇r·əˌstīl }

Urostylidae [INV ZOO] A family of hemipteran insects in the superfamily Pentatomoidea. { ˌyu̇r·əˈstīl·əˌdē }

Ursidae [VERT ZOO] A family of mammals in the order Carnivora including the bears and their allies. { ˈər·səˌdē }

ursolic acid [BIOCHEM] $C_{30}H_{48}O_3$ A pentacyclic terpene that crystallizes from absolute alcohol solution, found in leaves and berries of plants; used in pharmaceutical and food industries as an emulsifying agent. { ərˈsäl·ik ˈas·əd }

Urticaceae [BOT] A family of dicotyledonous herbs in the order Urticales characterized by a single unbranched style, a straight embryo, and the lack of milky juice (latex). { ˌərd·əˈkās·ēˌē }

Urticales [BOT] An order of dicotyledons in the subclass Hamamelidae; woody plants or herbs with simple, usually stipulate leaves, and reduced clustered flowers that usually have a vestigial perianth. { ˌərd·əˈkā·lēz }

urticaria [MED] Hives or nettle rash; a skin condition characterized by the appearance of intensely itching wheals or welts with elevated, usually white centers and a surrounding area of erythema. Also known as hives. { ˌərd·əˈkar·ē·ə }

use-dilution test [MICROBIO] A bioassay method for testing disinfectants for use on surfaces where a substantial reduction of bacterial contamination is not achieved by prior cleaning; the test organisms *Salmonella choleraesuis* and *Staphylococcus aureus* are deposited in stainless steel cylinders which are then exposed to the action of the test disinfectant. { ˈyüs də¦lü·shən ˌtest }

Usneaceae [BOT] The beard lichens, a family of Ascolichenes in the order Lecanorales distinguished by their conspicuous fruticose growth form. { ˌəs·nēˈās·ēˌē }

usnic acid [BIOCHEM] $C_{18}H_{16}O_7$ Yellow crystals, insoluble in water, slightly soluble in alcohol and ether, melts about 198°C; found in lichens; used as an antibiotic. Also known as usninic acid. { ˈəs·nik ˈas·əd }

usninic acid *See* usnic acid. { əsˈnin·ik ˈas·əd }

Ustilaginaceae [MYCOL] A family of fungi in the order Ustilaginales in which basidiospores bud from the sides of the septate epibasidium. { ˌəs·təˌlaj·əˈnās·ēˌē }

Ustilaginales [MYCOL] An order of the subclass Heterobasidiomycetidae comprising the smut fungi which parasitize plants and cause diseases known as smut or bunt. { ˌəs·təˌlaj·əˈnā·lēz }

uterus [ANAT] The organ of gestation in mammals which receives and retains the fertilized ovum, holds the fetus during development, and becomes the principal agent of its expulsion at term. { 'yüd·ə·rəs }

uterus bicornis [ANAT] A uterus divided into two horns or compartments; an abnormal condition in humans but normal in many mammals, such as carnivores. { 'yüd·ə·rəs bī'kȯr·nəs }

utricle *See* utriculus. { 'yü·trə·kəl }

utriculus [ANAT] **1.** That part of the membranous labyrinth of the ear into which the semicircular canals open. **2.** A small, blind pouch extending from the urethra into the prostate. Also known as utricle. { yü'trik·yə·ləs }

Uukuvirus [VIROL] A genus of the viral family Bunyaviridae that is transmitted via ticks to a wide range of vertebrate hosts. { yü'yük·ə,vī·rəs }

uvea [ANAT] The pigmented, vascular layer of the eye: the iris, ciliary body, and choroid. { 'yü·vē·ə }

uveitis *See* iridocyclochoroiditis. { ,yü·vē'īd·əs }

uvula [ANAT] **1.** A fingerlike projection in the midline of the posterior border of the soft palate. **2.** A lobe of the vermiform process of the lower surface of the cerebellum. { 'yü·vyə·lə }

V

vaccination [IMMUNOL] Inoculation of viral or bacterial organisms or antigens to produce immunity in the recipient. { ˌvak·səˈnā·shən }

vaccination encephalitis [MED] Encephalitis caused by vaccination with rabies vaccine. { ˌvak·səˈnā·shən inˌsef·əˈlīd·əs }

vaccine [IMMUNOL] A suspension of killed or attenuated bacteria or viruses or fractions thereof, injected to produce active immunity. { vakˈsēn }

vaccinia [VET MED] A contagious disease of cows which is characterized by vesicopustular lesions of the skin that are prone to appear on the teats and udder, and which is transmissible to humans by handling infected cows and by vaccination; confers immunity against smallpox. Also known as cowpox. { vakˈsin·ē·ə }

vacuole [CELL MOL] A membrane-bound cavity within a cell; may function in digestion, storage, secretion, or excretion. { ˈvak·yəˌwōl }

vagility [ECOL] The ability of organisms to disseminate. { vəˈjil·əd·ē }

vagina [ANAT] The canal from the vulvar opening to the cervix uteri. { vəˈjī·nə }

vagina fibrosa tendinis [ANAT] A fibrous sheath surrounding the tendon of a muscle and usually confining the tendon in a bony groove. { vəˈjī·nə fīˈbrō·sə ˈten·də·nəs }

vaginal protocele *See* rectocele. { ˈvaj·ən·əl ˈprōd·əˌsēl }

vaginate [BIOL] Invested in a sheath. { ˈvaj·əˌnāt }

vaginismus [MED] Painful vaginal spasm. { ˌvaj·əˈniz·məs }

vaginitis [MED] **1.** Inflammation of the vagina. **2.** Inflammation of a tendon sheath. { ˌvaj·əˈnīd·əs }

vagotonine [BIOCHEM] An endocrine substance which is thought to be elaborated by cells of the pancreas and which regulates autonomic tonus. { vəˈgäd·əˌnēn }

vagus [ANAT] The tenth cranial nerve; either of a pair of sensory and motor nerves forming an important part of the parasympathetic system in vertebrates. { ˈvā·gəs }

valence [BIOCHEM] The relative ability of a biological substance to react or combine. { ˈvā·ləns }

valid [SYST] Describing a taxon classified on the basis of distinctive characters of accepted importance. { ˈval·əd }

valine [BIOCHEM] $C_5H_{11}NO_2$ An amino acid considered essential for normal growth of animals, and biosynthesized from pyruvic acid. { ˈvaˌlēn }

vallate papilla [ANAT] One of the large, flat papillae, each surrounded by a trench, in a group anterior to the sulcus terminalis of the tongue. Also known as circumvallate papilla. { ˈvaˌlāt pəˈpil·ə }

Valoniaceae [BOT] A family of green algae in the order Siphonocladales consisting of plants that are essentially unicellular, coenocytic vesicles, spherical or clavate, and up to 2.4 inches (6 centimeters) in diameter. { vəˌlō·nēˈās·ēˌē }

valvate [BOT] Having valvelike parts, as those which meet edge to edge or which open as if by valves. { ˈvalˌvāt }

Valvatida [INV ZOO] An order of echinoderms in the subclass Asteroidea. { ˌval·vəˈtīd·ə }

Valvatina [INV ZOO] A suborder of echinoderms in the order Phanerozonida in which the upper marginals lie directly over, and not alternate with, the corresponding lower marginals. { ˌval·vəˈtīn·ə }

valve [ANAT] A flat of tissue, as in the veins or between the chambers in the heart, which permits movement of fluid in one direction only. [BOT] **1.** A segment of a dehiscing capsule or legume. **2.** The lidlike portion of certain anthers. [INV ZOO] **1.** One of the distinct, articulated pieces composing the shell of certain animals, such as barnacles and brachiopods. **2.** One of two shells encasing the body of a bivalve mollusk or a diatom. { valv }

Valvifera [INV ZOO] A suborder of isopod crustaceans distinguished by having a pair of flat, valvelike uropods which hinge laterally and fold inward beneath the rear part of the body. { valˈvif·ə·rə }

valvula [BIOL] A small valve. [INV ZOO] One of the small processes forming a sheath for the ovipositor in certain insects. { ˈval·vyə·lə }

valvulate [BIOL] Having valvules. { ˈval·vyə·lət }

vampire [VERT ZOO] The common name for bats making up the family Desmodontidae which have teeth specialized for cutting and which subsist on a blood diet. { ˈvamˌpīr }

Vampyrellidae [INV ZOO] A family of protozoans in the order Proteomyxida including species which invade filamentous algae and sometimes higher plants. { ,vam·pə'rel·ə,dē }

Vampyromorpha [INV ZOO] An order of dibranchiate cephalopod mollusks represented by *Vampyroteuthis infernalis*, an inhabitant of the deeper waters of tropical and temperate seas. { ¦vam·pə·rō'mȯr·fə }

vancomycin [MICROBIO] A complex antibiotic substance produced by *Streptomyces orientalis*; useful for treatment of severe staphylococcic infections. { ,vaŋ·kə'mīs·ən }

van Crevald-von Gierke's disease *See* von Gierke's disease. { van 'kre,väld fən 'gir·kēz di,zez }

Van den Bergh reaction [PATH] A liver function test in which diazotized serum or plasma is compared with a standard solution of diazotized bilirubin. { 'van dən ,bərg rē,ak·shən }

vane [VERT ZOO] The expanded web part of a feather. { vān }

vane feather *See* contour feather. { 'vān ,feth·ər }

Vaneyellidae [INV ZOO] A family of holothurian echinoderms in the order Dactylochirotida. { ,vā·nē'el·ə,dē }

Vanhorniidae [INV ZOO] A monospecific family of the Hymenoptera in the superfamily Proctotrupoidea. { ,van,hȯr'nī·ə,dē }

Varanidae [VERT ZOO] The monitors, a family of reptiles in the suborder Sauria found in the hot regions of Africa, Asia, Australia, and Malaya. { və'ran·ə,dē }

varicella *See* chickenpox. { ,var·ə'sel·ə }

varicocele [MED] Dilatation of the veins of the pampiniform plexus of the spermatic cord, forming a soft, elastic, often uncomfortable swelling. { 'var·ək·ə,sēl }

varicose [MED] Pertaining to blood vessels that are dilated, knotted, and tortuous. { 'var·ə ,kōs }

varicose vein [ANAT] An enlarged tortuous blood vessel that occurs chiefly in the superficial veins and their tributaries in the lower extremities. Also known as varicosity. { 'var·ə,kōs 'vān }

varicosity *See* varicose vein. { ,var·ə'käs·əd·ē }

variegate [BIOL] Having irregular patches of diverse colors. { 'ver·ē·ə,gāt }

variegated position effect [GEN] A phenomenon observed in some cases when a chromosome aberration causes a wild-type gene from euchromatin to be relocated adjacent to heterochromatin; the phenotypic expression of the wild-type allele will be unstable, producing patches of phenotypically mutant tissue that differ from the surrounding wild-type tissue. { 'ver·ē·ə,gād·əd pə¦zish·ən i,fekt }

variety [SYST] A taxonomic group or category inferior in rank to a subspecies. { və'rī·əd·ē }

variola *See* smallpox. { ,ver·ē'ō·lə }

varix [INV ZOO] A conspicuous ridge across each whorl of certain univalves marking the ancestral position of the outer lip of the aperture. [MED] A dilated and tortuous vein, artery, or lymphatic vessel. { 'var·iks }

varnish tree [BOT] *Rhus vernicifera*. A member of the sumac family (Anacardiaceae) cultivated in Japan; the cut bark exudes a juicy milk which darkens and thickens on exposure and is applied as a thin film to become a varnish of extreme hardness. Also known as lacquer tree. { 'vär·nish ,trē }

vasa vasorum [ANAT] The blood vessels supplying the walls of arteries and veins. { 'vā·zə va'sȯr·əm }

vascular [ANAT] Pertaining to blood vessels or other channels for the conveyance of a body fluid. { 'vas·kyə·lər }

vascular bundle [BOT] A strandlike part of the plant vascular system containing xylem and phloem. { 'vas·kyə·lər 'bənd·əl }

vascular cambium [BOT] The lateral meristem which produces secondary xylem and phloem. { 'vas·kyə·lər 'kam·bē·əm }

vascular endothelial growth factor [MED] A soluble factor that acts through specific cell-surface receptors on endothelial cells to critically regulate vasculogenesis. { ¦vas·kyə·lər ,en·dō¦thē·lē·əl 'grōth ,fak·tər }

vascularization [MED] Abnormal or excessive formation of blood vessels. [PHYSIO] The formation of new blood vessels within tissue. { ,vas·kyə·lə·rə'zā·shən }

vascular nevus [MED] A birthmark arising either as a developmental abnormality or as a postnatal benign neoplasm of a blood vessel. { 'vas·kyə·lər 'nē·vəs }

vascular ray [BOT] A ray derived from cambium and found in the stele of some vascular plants, often separating vascular bundles. { 'vas·kyə·lər 'rā }

vascular retinopathy [MED] Pathological changes in the retina associated with diseases such as arterial hypertension, chronic nephritis, eclampsia, and advanced arteriosclerosis. Also known as retinal retinitis. { 'vas·kyə·lər ,ret·ən'äp·ə·thē }

vascular tissue [BOT] The conducting tissue found in higher plants, consisting principally of xylem and phloem. { 'vas·kyə·lər 'tish·ü }

vasculitis [MED] Inflammation of a blood vessel or a lymph vessel. Also known as angiitis. { ,vas·kyə'līd·əs }

vasculogenesis [PHYSIO] The formation and differentiation of the vascular system. { ,vas·kyə·lə'jen·ə·səs }

vas deferens [ANAT] The portion of the excretory duct system of the testis which runs from the epididymal duct to the ejaculatory duct. Also known as ductus deferens. { 'vas 'def·ə·rənz }

vasectomy [MED] Cutting, or removing a section from, the ductus deferens. { va'sek·tə·mē }

vasoconstrictor [PHYSIO] A nerve or an agent that causes blood vessel constriction. { ¦vā·zō·kən'strik·tər }

vasodilator [PHYSIO] A nerve or an agent that causes blood vessel dilation. { ¦vā·zō'dī,lād·ər }

vasogenic shock [MED] Failure of peripheral

circulation due to vasodilation of arterioles and capillaries. { ¦vāz·ə¦jen·ik 'shäk }

vasography [MED] **1.** Radiography of blood vessels. **2.** Radiographic study of the vas deferens. { vā'zäg·rə·fē }

vasoinhibitor [PHARM] An agent that restricts or prevents the functioning of vasomotor nerves. { ˌvā·zō·in'hib·əd·ər }

vasoligation [MED] Surgical ligation of a vas deferens. { ˌvā·zō·lī'gā·shən }

vasomotion [PHYSIO] Change in the diameter of a blood vessel. Also known as angiokinesis. { 'vā·zəˌmō·shən }

vasomotor [PHYSIO] Pertaining to the regulation of the constriction or expansion of blood vessels. { 'vā·zəˌmōd·ər }

vasomotor center [PHYSIO] A large, diffuse area in the reticular formation of the lower brainstem; stimulation of different portions of this center causes either a rise in blood pressure and tachycardia (pressor area) or a fall in blood pressure and bradycardia (depressor area). { ¦vāz·ə'mōd·ər ˌsen·tər }

vasoneuropathy [MED] Any disease involving both blood vessels and nerves. { ˌvā·zō·nu̇'räp·ə·thē }

vasopressin [BIOCHEM] A peptide hormone which is elaborated by the posterior pituitary and which has a pressor effect; used medicinally as an antidiuretic. Also known as antidiuretic hormone (ADH). { ˌvā·zō'pres·ən }

vasospasm [MED] A spasm of the blood vessels. Also known as angiospasm. { 'vā·zō ˌspaz·əm }

vasotocin [BIOCHEM] A hormone from the neurosecretory cells of the posterior pituitary of lower vertebrates; increases permeability to water in amphibian skin and in bladder. { ˌvā·zə'tōs·ən }

vault [BIOL] An anatomical structure that is arched or dome-shaped. { vȯlt }

v-body *See* nucleosome. { 'vē ˌbäd·ē }

VD *See* venereal disease.

V(D)J recombination [GEN] A special type of chromatin diminution in which site-specific breakage within immunoglobulin heavy- and light-chain gene clusters and T-cell receptor gene clusters, elimination of part of each cluster, and rejoining of the ends of the remaining deoxyribonucleic acid fragments make possible an enormous number of different antibody and cell-surface antigen receptor specificities. { ¦vē¦dē¦jā rēˌkäm·bə'nā·shən }

vector [MED] An agent, such as an insect, capable of mechanically or biologically transferring a pathogen from one organism to another. { 'vek·tər }

vectorcardiogram [PHYSIO] The part of the pathway of instantaneous vectors during one cardiac cycle. Also known as monocardiogram. { ¦vek·tər'kärd·ē·əˌgram }

vectorcardiography [PHYSIO] A method of recording the magnitude and direction of the instantaneous cardiac vectors. { ¦vek·tərˌkärd·ē'äg·rə·fē }

vegetable [BOT] Resembling or relating to plants. { 'vej·tə·bəl }

vegetable diastase *See* diastase. { 'vej·tə·bəl 'dī·əsˌtās }

vegetation [BOT] The total mass of plant life that occupies a given area. { ˌvej·ə'tā·shən }

vegetational plant geography [ECOL] A field of study concerned with the mapping of vegetation regions and the interpretation of these in terms of environmental or ecological influences. { ˌvej·ə'tā·shən·əl 'plant jēˌäg·rə·fē }

vegetation and ecosystem mapping [BOT] An art and a science concerned with the drawing of maps which locate different kinds of plant cover in a geographic area. { ˌvej·ə¦tā·shən ən 'ek·ō¦sis·təm 'map·iŋ }

vegetation management [ECOL] The art and practice of manipulating vegetation such as timber, forage, crops, or wild life, so as to produce a desired part or aspect of that material in higher quantity or quality. { ˌvej·ə'tā·shən ˌman·ij·mənt }

vegetation zone [ECOL] **1.** An extensive, even transcontinental, band of physiognomically similar vegetation on the earth's surface. **2.** Plant communities assembled into regional patterns by the area's physiography, geological parent material, and history. { ˌvej·ə'tā·shən ˌzōn }

vegetative [BIOL] Having nutritive or growth functions, as opposed to reproductive. { 'vej·əˌtād·iv }

vegetative propagation [BOT] Production of a new plant from a portion of another plant, such as a stem or branch. { 'vej·əˌtād·iv ˌpräp·ə'gā·shən }

vegetative state [VIROL] The noninfective state during which the genome of a phage multiplies and directs host synthesis of substances needed for production of infective particles. { 'vej·ə ˌtād·iv ˌstāt }

veil *See* velum. { vāl }

Veillonella [MICROBIO] The type genus of the family Veillonellaceae; small cells occurring in pairs, chains, and clusters. { ˌvā·yō'nel·ə }

Veillonellaceae [MICROBIO] The single family of gram-negative, anaerobic cocci; characteristically occur in pairs with adjacent sides flattened; parasites of homotherms, including humans, rodents, and pigs. { ˌvā·yō·nə'lās·ēˌē }

vein [ANAT] A relatively thin-walled blood vessel that carries blood from capillaries to the heart in vertebrates. [BOT] One of the vascular bundles in a leaf. [INV ZOO] **1.** One of the thick, stiff ribs providing support for the wing of an insect. **2.** A venous sinus in invertebrates. { vān }

velamen [BOT] The corky epidermis covering the aerial roots of an epiphytic orchid. { və'lā·mən }

velarium [INV ZOO] The velum of certain scyphozoans and cubomedusans distinguished by the presence of canals lined with endoderm. { və'lar·ē·əm }

veld *See* veldt. { velt }

veldt [ECOL] Grasslands of eastern and southern Africa that are usually level and mixed with trees and shrubs. Also spelled veld. { velt }

veliger [INV ZOO] A mollusk larval stage following the trochophore, distinguished by an enlarged girdle of ciliated cells (velum). { 'vē·lə·jər }

Veliidae [INV ZOO] A family of the Hemiptera in the subdivision Amphibicorisae composed of small water striders which have short legs and a longitudinal groove between the eyes. { və'lī·ə,dē }

vellus [ANAT] Fine body hair that is present until puberty. { 'vel·əs }

Velocipedidae [INV ZOO] A tropical family of hemipteran insects in the superfamily Cimicoidea. { və,läs·ə'ped·ə,dē }

velum [BIOL] A veil- or curtainlike membrane. [INV ZOO] A swimming organ on the larva of certain marine gastropod mollusks that develops as a contractile ciliated collar-shaped ridge. Also known as veil. { 'vē·ləm }

vena cava [ANAT] One of two large veins which in air-breathing vertebrates conveys blood from the systemic circulation to the right atrium. { ,vē·nə 'kā·və }

venation [BOT] The system or pattern of veins in the tissues of a leaf. [INV ZOO] The arrangement of veins in an insect wing. { ve'nā·shən }

venereal bubo *See* lymphogranuloma venereum. { və'nir·ē·əl 'bü·bō }

venereal disease [MED] Any of several contagious diseases generally acquired during sexual intercourse; includes gonorrhea, syphilis, chancroid, granuloma inguinale, and lymphogranuloma venereum. Abbreviated VD. { və'nir·ē·əl di'zēz }

venereal wart [MED] A warty growth of the penis, frequent in some parts of the world, and probably acquired during sexual intercourse. { və'nir·ē·əl 'wȯrt }

venipuncture [MED] A surgical puncture of a vein, such as for withdrawing blood or injecting medication. { 'ven·ə,pəŋk·chər }

venom [PHYSIO] Any of various poisonous materials secreted by certain animals, such as snakes or bees. { 'ven·əm }

venous pressure [PHYSIO] Tension of the blood within the veins. { 'vē·nəs ,presh·ər }

vent [ZOO] The external opening of the cloaca or rectum, especially in fish, birds, and amphibians. { vent }

venter [ANAT] The abdomen, or other body cavity containing organs. [BOT] The thickened basal portion of an archegonium. [INV ZOO] **1.** The undersurface of an arthropod's abdomen. **2.** The outer, convex part of a curved or coiled gastropod or cephalopod shell. { 'ven·tər }

ventral [BOT] On the lower surface of a dorsiventral plant structure, such as a leaf. [ZOO] On or belonging to the lower or anterior surface of an animal, that is, on the side opposite the back. { 'ven·trəl }

ventral aorta [VERT ZOO] The arterial trunk or trunks between the heart and the first aortic arch in embryos or lower vertebrates. { 'ven·trəl ā'ȯrd·ə }

ventral hernia [MED] A hernia of the abdominal wall not involving the umbilical, femoral, or inguinal openings. Also known as abdominal hernia. { 'ven·trəl 'hər·nē·ə }

ventralia [INV ZOO] Paired sensory bristles on the ventral aspect of the head of gnathostomulids. { ven'tral·yə }

ventral light reflex [INV ZOO] A basic means of orientation in aquatic invertebrates, such as shrimp, which swim belly up toward the light. { 'ven·trəl 'līt ,rē,fleks }

ventral rib [VERT ZOO] Any of the ribs which lie in the septa dividing the trunk musculature into segments in fish. Also known as pleural rib. { 'ven·trəl 'rib }

ventricle [ANAT] **1.** A chamber, or one of two chambers, in the vertebrate heart which receives blood from the atrium and forces it into the arteries by contraction of the muscular wall. **2.** One of the interconnecting, fluid-filled chambers of the vertebrate brain that are continuous with the canal of the spinal cord. [ZOO] A cavity in a body part or organ. { 'ven·trə·kəl }

ventricose [BIOL] Swollen or distended, especially on one side. { ,ven·trə·kōs }

ventricular depolarization complex *See* QRS complex. { ven'trik·yə·lər di,pō·lə·rə'zā·shən ,käm ,pleks }

ventricular septum *See* interventricular septum. { ven'trik·yə·lər 'sep·təm }

ventriculus [ZOO] A ventricle that performs digestive functions, such as a stomach or a gizzard. { ven'trik·yə·ləs }

ventromedial nucleus [ANAT] A central nervous system nucleus in the hypothalamus that appears to be the satiation center; bilateral surgical damage to this nucleus results in overeating. { ¦ven·trō'mēd·ē·əl 'nü·klē·əs }

Venturia inaequalis [MYCOL] A fungal pathogen that causes apple scab disease. { ven,tu̇r·ē·ə ,in·ē'kwäl·əs }

venule [ANAT] A small vein. { 'ven·yül }

Venus' flytrap [BOT] *Dionaea muscipula*. An insectivorous plant (order Sarraceniales) of North and South Carolina; the two halves of a leaf blade can swing upward and inward as though hinged, thus trapping insects between the closing halves of the leaf blade. { 'vē·nəs 'flī,trap }

Verbenaceae [BOT] A family of variously woody or herbaceous dicotyledons in the order Lamiales characterized by opposite or whorled leaves and regular or irregular flowers, usually with four or two functional stamens. { ,vər·bə'nās·ē,ē }

Vermes [INV ZOO] An artificial taxon considered to be a phylum in some systems of classification, but variously defined as including all invertebrates except arthropods, or including all vermiform invertebrates. { 'vər·mēz }

vermiform [BIOL] Wormlike; resembling a worm. { 'vər·mə,fȯrm }

vermiform appendix [ANAT] A small, blind sac

projecting from the cecum. Also known as appendix. { 'vər·mə,fȯrm ə'pen,diks }

vermifuge [PHARM] An agent that expels worms or intestinal animal parasites. { 'vər·mə,fyüj }

Vermilingua [VERT ZOO] An infraorder of the mammalian order Edentata distinguished by lack of teeth and in having a vermiform tongue; includes the South American true anteaters. { ,vər·mə'liŋ·gwə }

vermis [ANAT] The median lobe of the cerebellum. { 'vər·məs }

vernalization [BOT] The induction in plants of the competence or ripeness to flower by the influence of cold, that is, at temperatures below the optimal temperature for growth. { ,vərn·əl·ə'zā·shən }

vernation [BOT] The characteristic arrangement of young leaves within the bud. Also known as prefoliation. { vər'nā·shən }

vernine *See* guanosine. { 'vər,nēn }

vernix caseosa [EMBRYO] A cheesy deposit on the surface of the fetus derived from the stratum corneum, sebaceous secretion, and remnants of the epitrichium. { 'vər·niks ,kā·sē'ō·sə }

veronal [PHARM] A white crystalline powder formerly used medicinally as a hypnotic drug to induce sleep. Also known as barbital. { 'ver·ə,nȯl }

verruca [BIOL] A wartlike elevation on the surface of a plant or animal. { və'rü·kə }

verruca peruana *See* verruca peruviana. { və'rü·kə ,per·ə'wä·nə }

verruca peruviana [MED] A benign eruptive form of bartonellosis with chronic cutaneous lesions. Also known as verruca peruana. { və'rü·kə pə,rü·vē'a·nə }

Verrucariaceae [BOT] A family of crustose lichens in the order Pyrenulales typically found on rocks, especially in intertidal or salt-spray zones along rocky coastlines. { və,rü·kə·rē'ās·ē,ē }

Verrucomorpha [INV ZOO] A suborder of the crustacean order Thoracica composed of sessile, asymmetrical barnacles. { və,rü·kə'mȯr·fə }

verrucose [BIOL] Having the surface covered with wartlike protuberances. { və'rü,kōs }

verrucous endocarditis [MED] Small thrombotic, nonbacterial, wartlike lesions on the heart valves and endocardium, occurring frequently in systemic lupus erythematosus. Also known as terminal endocarditis. { və'rü·kəs ¦en·dō·kär'dīd·əs }

versatile anther [BOT] An anther whose attachment is near its middle, thus enabling it to swing freely. { 'vər·səd·əl 'an·thər }

vertebra [ANAT] One of the bones that make up the spine in vertebrates. { 'vərd·ə·brə }

vertebral arch [ANAT] An arch formed by the paired pedicles and laminas of a vertebra; the posterior part of a vertebra which together with the anterior part, the body, encloses the vertebral foramen in which the spinal cord is lodged in vertebrates. Also known as neural arch. { 'vərd·ə·brəl 'ärch }

vertebral column *See* spine. { 'vərd·ə·brəl 'käl·əm }

Vertebrata [VERT ZOO] The major subphylum of the phylum Chordata including all animals with backbones, from fish to human. { ,vərd·ə'bräd·ə }

vertebrate zoology [ZOO] That branch of zoology concerned with the study of members of the Vertebrata. { 'vərd·ə·brət zō'äl·ə·jē }

vertical transmission [GEN] Passage of genetic information from one cell or individual organism to its progeny by conventional heredity mechanisms. { 'vərd·ə·kəl tranz'mish·ən }

verticillate [BOT] Whorled, in an arrangement resembling the spokes of a wheel. { ¦vərd·ə¦si,lāt }

vertigo [MED] The sensation that the outer world is revolving about the patient (objective vertigo) or that the patient is moving in space (subjective vertigo). { 'vərd·ə,gō }

vesicant [PHARM] An agent that causes blistering. { 'ves·ə·kənt }

vesication [MED] **1.** A blister. **2.** Formation of a blister. { ,ves·ə'kā·shən }

vesicle [BIOL] A small, thin-walled bladderlike cavity, usually filled with fluid. { 'ves·ə·kəl }

vesicular-arbuscular mycorrhizal fungi [MYCOL] Mycorrhizal fungi that grow into the root cortex of the host plant and penetrate root cells to form two kinds of specialized structures, arbuscules and vesicles. Also known as arbuscular mycorrhizae. { və¦sik·yə·lər är¦bəs·kyə·lər ,mī·kə,rīz·əl 'fən,jī }

vesicular stomatitis [VET MED] A viral disease, most often of horses, cattle, and pigs, characterized by fever and by vesicular and erosive lesions on the tongue, gums, lips, feet, and teats. { ve,sik·yə·lər ,stō·mə'tīd·əs }

vesicular stomatitis virus [VIROL] A virus in the genus *Vesiculovirus*, family Rhabdoviridae, the causative agent of vesicular stomatitis. { və¦sik·yə·lər ,stō·mə'tīd·əs ,vī·rəs }

Vesiculovirus [VIROL] A genus of the viral family Rhabdoviridae; type species is vesicular stomatitis virus. { və'sik·yə·lō,vī·rəs }

Vespertilionidae [VERT ZOO] The common bats, a large cosmopolitan family of the Chiroptera characterized by a long tail, extending to the edge of the uropatagium; almost all members are insect-eating. { ,ves·pər,til·ē'än·ə,dē }

vespertine [VERT ZOO] Active in the evening. { 'ves·pər,tīn }

Vespidae [INV ZOO] A widely distributed family of Hymenoptera in the superfamily Vespoidea including hornets, yellow jackets, and potter wasps. { 'ves·pə,dē }

Vespoidea [INV ZOO] A superfamily of wasps in the suborder Apocrita. { ve'spȯid·ē·ə }

vessel [BOT] A water-conducting tube or duct in the xylem. { 'ves·əl }

vessel segment [BOT] A single cell or unit of a plant vessel. { 'ves·əl ,seg·mənt }

vestibular apparatus [ANAT] The anatomical structures concerned with the vestibular portion of the eighth cranial nerve; includes the saccule, utricle, semicircular canals, vestibular nerve, and vestibular nuclei of the ear. { və'stib·yə·lər 'ap·ə,rad·əs }

vestibular membrane of Reissner *See* Reissner's membrane. { və'stib·yə·lər 'mem,brān əv 'rīs·nər }

vestibular nerve [NEUROSCI] A somatic sensory branch of the auditory nerve, which is distributed about the ampullae of the semicircular canals, macula sacculi, and macula utriculi. { və'stib·yə·lər 'nərv }

vestibular reflexes [PHYSIO] The responses of the vestibular apparatus to strong stimulation; responses include pallor, nausea, vomiting, and postural changes. { və'stib·yə·lər 'rē,flek·səz }

vestibule [ANAT] **1.** The central cavity of the bony labyrinth of the ear. **2.** The parts of the membranous labyrinth within the cavity of the bony labyrinth. **3.** The space between the labia minora. **4.** *See* buccal cavity. { 'ves·tə,byül }

vestibulocerebellar [NEUROSCI] Pertaining to the vestibular fibers and the cerebellum. { və¦stib·yə·lō,ser·ə'bel·ər }

vestibulocochlear nerve *See* auditory nerve. { və¦stib·yə·lə'käk·lē·ər ,nərv }

vestibulospinal tract [NEUROSCI] A tract of nerve fibers that originates principally from the lateral vestibular nucleus and descends in the anterior funiculus of the spinal cord. { və¦stib·yə·lə'spīn·əl ,trakt }

vestige [BIOL] A degenerate anatomical structure or organ that remains from one more fully developed and functional in an earlier phylogenetic form of the individual. { 'ves·tij }

vestigial [BIOL] Of, being, or resembling a vestige. { və'stij·ē·əl }

vetch [BOT] Any of a group of mostly annual legumes, especially of the genus *Vicia*, with weak viny stems terminating in tendrils and having compound leaves; some varieties are grown for their edible seed. { vech }

veterinary medicine [MED] The branch of medical practice which treats of the diseases and injuries of animals. { 'vet·ən,er·ē 'med·ə·sən }

V factor [MICROBIO] Phosphopyridine nucleotide, a growth factor required by the parasitic bacteria of the genus *Haemophilus*. { 'vē ,fak·tər }

viable [BIOL] Able to live and develop normally. { 'vī·ə·bəl }

Vianaidae [INV ZOO] A small family of South American Hemiptera in the superfamily Tingordea. { ,vē·ə'nā·ə,dē }

vibraculum [INV ZOO] A specially modified bryozoan zooid with a bristlelike seta that sweeps debris from the surface of the colony. { və'brak·yə·ləm }

vibratory saw [MED] A hand-operated saw, used in surgery, that can cut through hard materials such as bone or a cast, but leaves soft tissue such as skin or muscle untouched. { 'vī·brə,tȯr·ē 'sȯ }

Vibrionaceae [MICROBIO] A family of gram-negative, facultatively anaerobic rods; cells are straight or curved and usually motile by polar flagella; generally found in water. { ,vib·rē·ō'nās·ē,ē }

vibriosis [VET MED] An infectious bacterial disease, primarily of cattle, sheep, and goats, caused by *Vibrio fetus* and characterized by abortion, retained placenta, and metritis. { ,vib·rē'ō·səs }

vibrissae [VERT ZOO] Hairs with specialized erectile tissue; found on all mammals except humans. Also known as sinus hairs; tactile hairs; whiskers. { vī'bri,sē }

vicariants [ECOL] Two or more closely related taxa, presumably derived from one another or from a common immediate ancestor, that inhabit geographically distinct areas. { vī'kar·ē·əns }

vicuna [VERT ZOO] *Lama vicugna.* A rare, wild ruminant found in the Andes mountains; the fiber of the vicuna is strong, resilient, and elastic but is the softest and most delicate of animal fibers. { vī'kün·yə }

villous [BOT] Having a surface covered with long, soft, shaggy hairs. { 'vil·əs }

villous adenoma [MED] A slow-growing, potentially malignant neoplasm of the mucosa of the rectum; manifested by bleeding and mucoid diarrhea. { 'vil·əs ,ad·ən'ō·mə }

villous placenta *See* epitheliochorial placenta. { 'vil·əs plə'sen·tə }

villous tenosynovitis [MED] A chronic inflammatory reaction of a tendon sheath producing hypertrophy of the lining, with the formation of redundant folds and villi. { 'vil·əs ¦ten·ō,sin·ə'vīd·əs }

villus [ANAT] A fingerlike projection from the surface of a membrane. { 'vil·əs }

vinblastine [PHARM] $C_{46}H_{58}O_9N_4$ An alkaloid obtained from the periwinkle plant (*Vinca rosea*) and used, as the sulfate salt, as an antineoplastic drug. { vin'bla,stēn }

Vinca rosea [BOT] A low, creeping evergreen perennial. Also known as periwinkle. { ,viŋ·kə 'rō·zē·ə }

Vincent's angina *See* Vincent's infection. { 'vin·səns 'an·jə·nə }

Vincent's infection [MED] A noncontagious bacterial infection of the oral mucosa characterized by ulceration and formation of a gray pseudomembrane; caused by certain fusiform bacteria and spirochetes; formerly known as Vincent's angina. Also known as fusospirochetosis; Plaut-Vincent's infection; trench mouth; ulcerative gingivitis. { 'vin·səns in,fek·shən }

vincristine [PHARM] $C_{46}H_{56}O_{10}N_4$ An alkaloid extracted from the periwinkle plant (*Vinca rosea*) and used, as the sulfate salt, as an antineoplastic drug. Also known as leurocristine. { vin'kri ,stēn }

vine [BOT] A plant having a stem that is too flexible or weak to support itself. { vīn }

vinegar bacteria *See* Acetobacter. { 'vin·ə·gər bak,tir·ē·ə }

vinegar eel [INV ZOO] *Turbatrix aceti.* A very small nematode often found in large numbers in vinegar fermentation. Also known as vinegar worm. { 'vin·ə·gər ,ēl }

vinegar worm *See* vinegar eel. { 'vin·ə·gər ˌwərm }

Violaceae [BOT] A family of dicolyledonous plants in the order Violales characterized by polypetalous, mostly perfect, hypogynous flowers with a single style and five stamens. { ˌvī·ə'lās·ēˌē }

Violales [BOT] A heterogeneous order of dicotyledons in the subclass Dilleniidae distinguished by a unilocular, compound ovary and mostly parietal placentation. { ˌvī·ə'lā·lēz }

viomycin [MICROBIO] A polypeptide antibiotic or mixture of antibiotic substances produced by strains of *Streptomyces griseus* var. *purpureus* (*Streptomyces puniceus*); the sulfate salt is administered intramuscularly for treatment of tuberculosis resistant to other therapy. { ˌvī·ə'mīs·ən }

viper [VERT ZOO] The common name for reptiles of the family Viperidae; thick-bodied poisonous snakes having a pair of long fangs, present on the anterior part of the upper jaw, which fold against the roof of the mouth when the jaws are closed. { 'vī·pər }

Viperidae [VERT ZOO] A family of reptiles in the suborder Serpentes found in Eurasia and Africa; all species are proglyphodont. { vī'per·əˌdē }

viral encephalomyelitides [MED] A group of several encephalitis diseases caused by various viruses; includes epidemic encephalitis, equine encephalitides, and Japanese B encephalitis. { 'vī·rəl inˌsef·ə·lōˌmī·ə'lid·ə·dēz }

viral gastroenteritis [MED] An acute infectious gastroenteritis thought to be caused by various viruses and characterized by diarrhea, nausea, vomiting, and variable systemic symptoms. { 'vī·rəl ¦gas·trōˌent·ə'rīd·əs }

viral hepatitis [MED] A type of hepatitis caused by two distinct viruses, A and B; type A is also known as infectious hepatitis, type B as serum hepatitis. { 'vī·rəl ˌhep·ə'tīd·əs }

viral pneumonia [MED] A form of pneumonia caused by a virus of various types, in which the inflammatory reaction predominates in the septa, and the alveoli contain fibrin, edema fluid, and some inflammatory cells. { 'vī·rəl nu̇'mō·nyə }

viral shedding [VIROL] Excretion of virus from a specific site in the body or from a lesion. { 'vī·rəl 'shed·iŋ }

viremia [MED] Presence of viral particles in the blood. { vī'rē·mē·ə }

Vireonidae [VERT ZOO] The vireos, a family of New World passeriform birds in the suborder Oscines. { ˌvir·ē'än·əˌdē }

virgate [BOT] Banded. { 'vərˌgāt }

virgate trophus [INV ZOO] A piercing type of trophus in rotifers that is thin and slightly toothed. { 'vərˌgāt 'träf·əs }

virgo-forcipate trophus [INV ZOO] A type of muscular chamber in rotifers containing jaws of a cuticular material intermediate between a piercing and grasping type of structure. { 'vər·gō 'fȯr·sə·pət ˌträf·əs }

viridans streptococci [MICROBIO] A group of pathogenic and saprophytic streptococci including strains not causing beta hemolysis, although many cause alpha hemolysis, and none which elaborate a C substance. { 'vir·əˌdanz ¦strep·tə'käk·ē }

virilism *See* gynandry. { 'vir·əˌliz·əm }

virion [VIROL] The complete, mature virus particle. { 'vir·ēˌän }

viroid [MICROBIO] The smallest known agents of infectious disease, characterized by the absence of encapsidated proteins. { 'vīˌrȯid }

virology [MICROBIO] The study of submicroscopic organisms known as viruses. { vī'räl·ə·jē }

virotoxin [BIOCHEM] One of a group of toxins present in the mushroom *Amanita virosa*. { ¦vī·rə¦täk·sən }

virulence [MICROBIO] The disease-producing power of a microorganism; infectiousness. { 'vir·ə·ləns }

virulence cassette *See* pathogenicity island. { 'vir·ə·ləns kəˌset }

virus [VIROL] A large group of infectious agents ranging from 10 to 250 nanometers in diameter, composed of a protein sheath surrounding a nucleic acid core and capable of infecting all animals, plants, and bacteria; characterized by total dependence on living cells for reproduction and by lack of independent metabolism. { 'vī·rəs }

virus hepatitis *See* infectious hepatitis. { 'vī·rəs ˌhep·ə'tīd·əs }

virus interference [MICROBIO] A phenomenon which may be defined as protection of host cells against one virus, conferred as a result of prior infection with a different virus. { 'vī·rəs ˌin·tər'fir·əns }

viscera [ANAT] The organs within the cavities of the body of an organism. { 'vis·ə·rə }

visceral arch [ANAT] One of the series of mesodermal ridges covered by epithelium bounding the lateral wall of the oral and pharyngeal regions of vertebrates; embryonic in higher forms, they contribute to formation of the face and neck. [VERT ZOO] One of the first two arches of the series in gill-bearing forms. { 'vis·ə·rəl 'ärch }

visceral leishmaniasis [MED] A severe, generalized, and often fatal infection, caused by any of three pathogenic hemoflagellates of the genus *Leishmania*, affecting organs rich in endothelial cells; accompanied by fever, spleen and liver enlargement, anemia, leukopenia, skin pigmentation, and changes in plasma protein. { 'vis·ə·rəl ˌlēsh·mə'nī·ə·səs }

visceral peritoneum [ANAT] That portion of the peritoneum covering the organs of the abdominal cavity. { 'vis·ə·rəl ˌper·ə·tə'nē·əm }

visceral pouch *See* pharyngeal pouch. { 'vis·ə·rəl 'pau̇ch }

visceromegaly [MED] Enlargement of the organs in the abdomen, such as the liver, spleen, pancreas, stomach, or kidneys. { ˌvis·ə·rō'meg·ə·lē }

visceroptosis [MED] Prolapse of a viscus, especially the intestine; downward displacement of the intestine in the abdominal cavity. Also known as enteroptosis. { ˌvis·ə·räp'tō·səs }

viscid [BOT] Having a sticky surface, as certain leaves. { 'vis·əd }

viscus [ANAT] Singular of viscera. { 'vis·kəs }

vision [PHYSIO] The sense which perceives the form, color, size, movement, and distance of objects. Also known as sight. { 'vizh·ən }

visual acuity [PHYSIO] The ability to see fine details of an object; specifically, the ability to see an object whose angle subtended at the eye is 1 minute of arc. { 'vizh·ə·wəl ə'kyü·əd·ē }

visual pigment [BIOCHEM] Any of various photosensitive pigments of vertebrate and invertebrate photoreceptors. { 'vizh·ə·wəl 'pig·mənt }

visual projection area [NEUROSCI] The receptive center for visual images in the cortex of the brain, located in the walls and margins of the calcarine sulcus of the occipital lobe. Also known as Brodmann's area 17. { 'vizh·ə·wəl prə'jek·shən ˌer·ē·ə }

visual purple *See* rhodopsin. { 'vizh·ə·wəl 'pər·pəl }

visual receptive field [PHYSIO] That area of the retina within which stimulation with light or a light pattern causes a response in a particular receptor or neuron in the visual pathway. { ¦vizh·ə·wəl ri¦sep·tiv ˌfēld }

visual-righting reflex [PHYSIO] A reflex mechanism whereby righting of the head and body is caused by visual stimuli. Also known as optical-righting reflex. { 'vizh·ə·wəl 'rīd·iŋ 'rē ˌfleks }

visual seizure [MED] A form of epileptic seizure in which the patient experiences visual sensations in the form of light flashes, sometimes of varied colors. { 'vizh·ə·wəl 'sē·zhər }

visual yellow [BIOCHEM] An intermediary substance formed from rhodopsin in the retina after exposure to light; it is ultimately broken down to retinene and vitamin A. { 'vizh·ə·wəl 'yel·ō }

Vitaceae [BOT] A family of dicotyledonous plants in the order Rhamnales; mostly tendril-bearing climbers with compound or lobed leaves, as in grapes (*Vitis*). { vī'tās·ēˌē }

vital capacity [PHYSIO] The volume of air that can be forcibly expelled from the lungs after the deepest inspiration. { 'vīd·əl kə'pas·əd·ē }

vitalism [BIOL] The theory that the activities of a living organism are under the guidance of an agency which has none of the attributes of matter or energy. { 'vīd·əlˌiz·əm }

vitamer [BIOCHEM] One of several very similar chemical compounds that can perform a specific vitamin function. { 'vīd·ə·mər }

vitamin [BIOCHEM] An organic compound present in variable, minute quantities in natural foodstuffs and essential for the normal processes of growth and maintenance of the body; vitamins do not furnish energy, but are essential for energy transformation and regulation of metabolism. { 'vīd·ə·mən }

vitamin A [BIOCHEM] $C_{20}H_{29}OH$ A pale-yellow alcohol that is soluble in fat and insoluble in water; found in liver oils and carotenoids, and produced synthetically; it is a component of visual pigments and is essential for normal growth and maintenance of epithelial tissue. Also known as antiinfective vitamin; antixerophthalmic vitamin; retinol. { 'vīd·ə·mən ¦ā }

vitamin A aldehyde *See* retinal. { 'vīd·ə·mən ¦ā 'al·dəˌhīd }

vitamin B_1 *See* thiamine. { 'vīd·ə·mən ¦bē¦wən }

vitamin B_2 *See* riboflavin. { 'vīd·ə·mən ¦bē¦tü }

vitamin B_3 *See* pantothenic acid. { 'vīd·ə·mən ¦bē¦thrē }

vitamin B_6 [BIOCHEM] A vitamin which exists as three chemically related and water-soluble forms found in food: pyridoxine, pyridoxal, and pyridoxamine; dietary requirements and physiological activities are uncertain. { 'vīd·ə·mən ¦bē ¦siks }

vitamin B_{12} [BIOCHEM] A group of closely related polypyrrole compounds containing trivalent cobalt; the antipernicious anemia factor, essential for normal hemopoiesis. Also known as cobalamin; cyanocobalamin; extrinsic factor. { 'vīd·ə·mən ¦bē¦twelv }

vitamin B complex [BIOCHEM] A group of water-soluble vitamins that include thiamine, riboflavin, nicotinic acid, pyridoxine, panthothenic acid, inositol, *p*-aminobenzoic acid, biotin, folic acid, and vitamin B_{12}. { 'vīd·ə·mən ¦bē 'käm ˌpleks }

vitamin B_6 hydrochloride *See* pyridoxine hydrochloride. { 'vīd·ə·mən ¦bē¦siks ¦hī·drə'klórˌīd }

vitamin B_2 phosphate *See* riboflavin 5′-phosphate. { 'vīd·ə·mən ¦bē¦tü 'fäˌsfāt }

vitamin C *See* ascorbic acid. { 'vīd·ə·mən ¦sē }

vitamin D [BIOCHEM] Either of two fat-soluble, sterol-like compounds, calciferol or ergocalciferol (vitamin D_2) and cholecalciferol (vitamin D_3); occurs in fish liver oils and is essential for normal calcium and phosphorus deposition in bones and teeth. Also known as antirachitic vitamin. { 'vīd·ə·mən ¦dē }

vitamin D_3 *See* cholecalciferol. { 'vīd·ə·mən ¦dē ¦thrē }

vitamin E [BIOCHEM] Any of a series of eight related compounds called tocopherols, α-tocopherol having the highest biological activity; occurs in wheat germ and other naturally occurring oils and is believed to be needed in certain human physiological processes. { 'vīd·ə·mən ¦ē }

vitamin G *See* riboflavin. { 'vīd·ə·mən ¦jē }

vitamin K [BIOCHEM] Any of three yellowish oils which are fat-soluble, nonsteroid, and nonsaponifiable; it is essential for formation of prothrombin. Also known as antihemorrhagic vitamin; prothrombin factor. { 'vīd·ə·mən ¦kā }

vitamin P [BIOCHEM] A substance, such as citrin or one or more of its components, believed to be concerned with maintenance of the normal state of the walls of small blood vessels. { 'vīd·ə·mən ¦pē }

vitamin P complex *See* bioflavonoid. { 'vīd·ə·mən ¦pē ˌkämˌpleks }

vitellarium [INV ZOO] The part of the ovary in certain rotifers and flatworms that produces nutritive cells filled with yolk. Also known as yolk larva. { ˌvid·əl'ar·ē·əm }

vitelline artery [EMBRYO] An artery passing from the yolk sac to the primitive aorta in young vertebrate embryos. Also known as omphalomesenteric artery. { vī'tel,ēn 'ärd·ə·rē }

vitelline duct [EMBRYO] The constricted part of the yolk sac opening into the midgut region of the future ileum. Also known as omphalomesenteric duct; umbilical duct. { vī'tel,ēn 'dəkt }

vitelline membrane [CELL MOL] The cytoplasmic membrane on the surface of the mammalian ovum. { vī'tel,ēn 'mem,brān }

vitelline vein [EMBRYO] Any of the embryonic veins in vertebrates uniting the yolk sac and the sinus venosus; their proximal fused ends form the portal vein. Also known as omphalomesenteric vein. { vī'tel,ēn 'vān }

vitellogenesis [PHYSIO] The process by which yolk is formed in the ooplasm of an oocyte. { vī,tel·ə'jen·ə·səs }

vitiligo [MED] A skin disease characterized by an acquired ochromia in areas of various sizes and shapes. { ,vid·əl,ī·gō }

Vitreoscillaceae [MICROBIO] Formerly a family of bacteria in the order Beggiatoales; included organisms which have a filamentous habit and move by gliding, but never store sulfur, and rely on organic nutrients in their metabolism. { ,vi·trē,äs·ə'lās·ē,ē }

vitreous body *See* vitreous humor. { 'vi·trē·əs 'bäd·ē }

vitreous chamber [ANAT] A cavity of the eye posterior to the crystalline lens and anterior to the retina, which is filled with vitreous humor. { 'vi·trē·əs 'chām·bər }

vitreous humor [PHYSIO] The transparent gel-like substance filling the greater part of the globe of the eye, the vitreous chamber. Also known as vitreous body. { 'vi·trē·əs 'hyü·mər }

vitrification [MED] An experimental procedure for preserving human organs in which chemicals are added prior to cooling to prevent crystallization of water within and outside the cells, so that with coolings the molecules essentially become fixed in place. { ,vi·trə·fə'kā·shən }

vittate [BOT] **1.** Having longitudinal stripes. **2.** Bearing specialized oil tubes (vittae). { 'vi,tāt }

vivax malaria [MED] Malaria caused by *Plasmodium vivax* and characterized by typical paroxysms occurring every few days, commonly every 2 days. Also known as benign tertian malaria; tertian malaria. { 'vī,vaks mə'ler·ē·ə }

Viverridae [VERT ZOO] A family of carnivorous mammals in the superfamily Feloidea composed of the civets, genets, and mongooses. { vī 'ver·ə,dē }

Viviparidae [INV ZOO] A family of fresh-water gastropod mollusks in the order Pectinibranchia. { ,vī·və'par·ə,dē }

viviparous [PHYSIO] Bringing forth live young. { vī'vip·ə·rəs }

vocal cord *See* vocal fold. { 'vō·kəl ,kȯrd }

vocal fold [ANAT] Either of a pair of folds of tissue covered by mucous membrane in the larynx. Also known as vocal cord. { 'vō·kəl ,fōld }

vocal sac [VERT ZOO] An expansible pocket of skin beneath the chin or behind the jaws of certain frogs; may be inflated to a great volume and serves as a resonator. { 'vō·kəl ,sak }

Vochysiaceae [BOT] A family of dicotyledonous plants in the order Polygalales characterized by mostly three carpels, usually stipulate leaves, one fertile stamen, and capsular fruit. { vō,kizh·ē'ās·ē,ē }

Voges-Proskauer test [MICROBIO] One of the four tests of the IMVIC test; a qualitative test for the formation of acetyl methylcarbinol from glucose, in which solutions of α-naphthol, potassium hydroxide, and creatinine are added to an incubated culture of test organisms in a glucose broth, and a pink to rose color indicates a positive reaction. { 'fō·gə 'präs,kau̇·ər ,test }

volar [ANAT] Pertaining to, or on the same side as, the palm of the hand or the sole of the foot. { 'vō·lər }

vole [VERT ZOO] Any of about 79 species of rodent in the tribe Microtini of the family Cricetidae; individuals have a stout body with short legs, small ears, and a blunt nose. { vōl }

voluntary muscle [PHYSIO] A muscle directly under the control of the will of the organism. { 'väl·ən,ter·ē 'məs·əl }

Volutidae [INV ZOO] A family of gastropod mollusks in the order Neogastropoda. { və'lüd·ə,dē }

volutin [BIOCHEM] A basophilic substance, thought to be a nucleic acid, occurring as granules in the cytoplasm and vacuoles of algae and other microorganisms. { 'väl·yəd·ən }

volva [MYCOL] A cuplike membrane surrounding the base of the stipe in certain gill fungi. { 'väl·və }

volvent nematocyst [INV ZOO] A nematocyst in the form of an unarmed, coiled tube that is closed at the end. { 'väl·vənt nə'mad·ə,sist }

Volvocales [BOT] An order of one-celled or colonial green algae in the division Chlorophyta; individuals are motile with two, four, or rarely eight whiplike flagella. { ,väl·və'kā·lēz }

Volvocida [INV ZOO] An order of the protozoan class Phytamastigophorea; individuals are grass-green or colorless, have one, two, four, or eight flagella, and thick cell walls of cellulose. { väl 'väs·əd·ə }

volvulus [MED] A twisting of the bowel upon itself so as to occlude the lumen and, in severe cases, compromise its circulation. { 'väl·vyə·ləs }

Vombatidae [VERT ZOO] A family of marsupial mammals in the order Diprotodonta in some classification systems. { väm'bad·ə,dē }

vomer [ANAT] A skull bone below the ethmoid region constituting part of the nasal septum in most vertebrates. { 'vō·mər }

vomeronasal cartilage [ANAT] A strip of hyaline cartilage extending from the anterior nasal spine upward and backward on either side of the septal cartilage of the nose and attached to the anterior margin of the vomer. Also known as Jacobson's cartilage. { ¦väm·ə·rō'nāz·əl 'kärt·lij }

W

wader *See* shore bird. { 'wād·ər }

wading bird [VERT ZOO] Any of the long-legged, long-necked birds composing the order Ciconiiformes, including storks, herons, egrets, and ibises. { 'wād·iŋ ˌbərd }

Waldeyer's ring [ANAT] A circular arrangement of the lymphatic tissues formed by the palatine and pharyngeal tonsils and the lymphatic follicles at the base of the tongue and behind the posterior pillars of the fauces. { 'välˌdī·ərz ˌriŋ }

walking bird [VERT ZOO] Any bird of the order Columbiformes, including the pigeons, doves, and sandgrouse. { 'wȯk·iŋ ˌbərd }

walnut [BOT] The common name for about a dozen species of deciduous trees in the genus *Juglans* characterized by pinnately compound, aromatic leaves and chambered or laminate pith; the edible nut of the tree is distinguished by a deeply furrowed or sculptured shell. { 'wȯl·nət }

walrus [VERT ZOO] *Odobenus rosmarus*. The single species of the pinniped family Odobenidae distinguished by the upper canines in both sexes being prolonged as tusks. { 'wȯl·rəs }

warm-blooded *See* homoiothermal. { 'wȯrm ¦bləd·əd }

wart [MED] A papillomatous growth which occurs singly or in groups on the skin surface; thought to be caused by a viral agent. { wȯrt }

wasp [INV ZOO] The common name for members of 67 families of the order Hymenoptera; all are important as parasites or predators of injurious pests. { wäsp }

Wasserman test [IMMUNOL] A complement-fixation test for syphilis using sensitized lipid extracts of beef heart as antigen. { 'was·ər·mən ˌtest }

water-borne disease [MED] A disease transmitted by drinking water or by contact with potable or bathing water. { 'wȯd·ər ˌbȯrn di'zēz }

water brash [MED] Severe distress caused by regurgitation of gastric acid into the throat. { 'wȯd·ər ˌbrash }

water bug [INV ZOO] Any insect which lives in an aquatic habitat during all phases of its life history. { 'wȯd·ər ˌbəg }

water conservation [ECOL] The protection, development, and efficient management of water resources for beneficial purposes. { 'wȯd·ər ˌkän·sər'vā·shən }

watercress [BOT] *Nasturtium officinale*. A perennial cress generally grown in flooded soil beds and used for salads and food garnishing. { 'wȯd·ərˌkres }

waterfowl [VERT ZOO] Aquatic birds which constitute the order Anseriformes, including the swans, ducks, geese, and screamers. { 'wȯd·ərˌfau̇l }

Waterhouse-Frederikson syndrome [MED] The association of bacteremia, particularly acute meningococcemia, massive skin hemorrhage, shock, and acute adrenal hemorrhage and insufficiency. { 'wȯd·ərˌhau̇s 'fred·rik·sən ˌsin ˌdrōm }

watermelon [BOT] *Citrullus vulgaris*. An annual trailing vine with light-yellow flowers and leaves having five to seven deep lobes; the edible, oblong or roundish fruit has a smooth, hard, green rind filled with sweet, tender, juicy, pink to red tissue containing many seeds. { 'wȯd·ərˌmel·ən }

water microbiology [MICROBIO] An aspect of microbiology that deals with the normal and adventitious microflora of natural and artificial water bodies. { 'wȯd·ər ¦mī·krō·bī'äl·ə·jē }

water moccasin [VERT ZOO] *Agkistrodon piscivorus*. A semiaquatic venomous pit viper; skin is brownish or olive on the dorsal aspect, paler on the sides, and has indistinct black bars. Also known as cottonmouth. { 'wȯd·ər ˌmäk·ə·sən }

water pollution [ECOL] Contamination of water by materials such as sewage effluent, chemicals, detergents, and fertilizer runoff. { 'wȯd·ər pəˌlü·shən }

water potential [PHYSIO] The difference in free energy or chemical potential (per unit molal volume) between pure water and water in cells and solutions. { 'wȯd·ər pəˌten·chəl }

water retting [MICROBIO] A type of retting process in which the stalks of fiber plants are immersed in cold or warm, slowly renewed water, for 4 days to several weeks. The active organism is *Clostridium felsineum* and related types, which break down the pectin to a mixture of organic acids (chiefly acetic and butyric), alcohols (butanol, ethanol, and methanol), carbon dioxide (CO_2), and hydrogen (H2). { 'wȯd·ər ˌred·iŋ }

water vascular system [INV ZOO] An internal closed system of reservoirs and ducts containing a watery fluid, found only in echinoderms. { 'wȯd·ər 'vas·kyə·lər ˌsis·təm }

wax gland *See* ceruminous gland. { 'waks ˌgland }

W chromosome [GEN] The sex chromosome present only in females in species with female heterogamety. { 'dəb·əl,yü ˌkrō·mə,sōm }

weasel [VERT ZOO] The common name for at least 12 species of small, slim carnivores which belong to the family Mustelidae and which have a reddish-brown coat with whitish underparts; species in the northern regions have white fur during the winter and are called ermine. { 'wē·zəl }

web [VERT ZOO] The membrane between digits in many birds and amphibians. { web }

Weber-Christian disease [MED] Febrile, relapsing, nodular nonsuppurative panniculitis. { 'vā·bər 'kris·chən di,zēz }

Weberian apparatus [VERT ZOO] A series of bony ossicles which form a chain connecting the swim bladder with the inner ear in fishes of the superorder Ostariophysi. { vā'bir·ē·ən ˌap·ə ˌrad·əs }

Weberian ossicle [VERT ZOO] One of a chain of three or four small bones that make up the Weberian apparatus. { vā'bir·ē·ən 'äs·ə·kəl }

Weber's law [PHYSIO] The law that the stimulus increment which can barely be detected (the just noticeable difference) is a constant fraction of the initial magnitude of the stimulus; this is only an approximate rule of thumb. { 'vā·bərz ˌlȯ }

weed [BOT] A plant that is useless or of low economic value, especially one growing on cultivated land to the detriment of the crop. { wēd }

weevil [INV ZOO] Any of various snout beetles whose larvae destroy crops by eating the interior of the fruit or grain, or bore through the bark into the pith of many trees. { 'wē·vəl }

Wegener's granulomatosis [MED] A rare disease of unknown causation characterized by necrotizing granulomas in the air passages, necrotizing vasculitis, and glomerulitis. { 'vā·gə·nərz ˌgran·yə·lə·mə'tō·səs }

Weil-Felix test [IMMUNOL] An agglutination test for various rickettsial infections based on production of nonspecific agglutinins in the blood of infected patients, and using various strains of *Proteus vulgaris* as antigen. { 'vīl 'fā·liks ˌtest }

Weil's disease [MED] A severe form of leptospirosis characterized by jaundice, oliguria, circulatory collapse, and tendency to hemorrhage. Also known as icterohemorrhagic fever; leptospirosis icterohemorrhagia; spirochetal jaundice. { 'vīlz di,zēz }

Welwitschiales [BOT] An order of gymnosperms in the subdivision Geneticae represented by the single species *Welwitschia mirabilis* of southwestern Africa; distinguished by only two leaves and short, unbranched, cushion- or saucer-shaped woody main stem which tapers to a very long taproot. { wel,wich·ē'ā·lēz }

Werdnig-Hoffmann disease [MED] Infantile spinal muscular atrophy. { 'ver,nik 'hōf,män di,zēz }

Werner's syndrome [MED] A complex of symptoms, thought to be inherited as an autosomal recessive, including premature senescence, dwarfism, cataracts, scleroderma-like changes of the skin, osteoporosis, and multiglandular dysfunction. { 'ver·nərz ˌsin,drōm }

Wernicke's area [NEUROSCI] An area located in the left temporal lobe just posterior to the primary auditory complex that is involved with speech comprehension, injury to this area results in fluent aphasia. { 'ver·ni·kēz ˌer·ē·ə }

Wernicke's encephalopathy [MED] A disease due to thiamine deficiency, characterized by vomiting, ophthalmoplegia, ptosis, nystagmus, ataxia, weakness, dementia, and hemorrhaging of neurons around the third ventricle, cerebral aqueduct, and mammillary bodies. { 'ver·nə,kēz in,sef·ə'läp·ə·thē }

Western blotting [CELL MOL] A protein detection technique in which proteins are separated by one- or two-dimensional gel electrophoresis, transferred (blotted) to a nitrocellulose sheet, and then treated with radioactive antibodies (or antibodies coupled to a fluorescent dye or an easily detectable enzyme) that are specific to the protein of interest. { 'wes·tərn ˌbläd·iŋ }

Western equine encephalitis [MED] A type of equine encephalitis which occurs chiefly west of the Mississippi River; the chief vector is the culicine mosquito *Culex tarsalis*. { 'wes·tərn 'ē,kwīn in,sef·ə'līd·əs }

West Nile fever [MED] An acute, usually mild, mosquito-borne virus disease occurring in summer, chiefly in Egypt, Israel, Africa, India, and Korea; signs are fever and lymphadenopathy, sometimes with a rash. { 'west ¦nīl 'fē·vər }

Westphal-Pilcz reflex *See* pupillary reflex. { west ˌfōl 'pils ˌrē,fleks }

Westphal's pupillary reflex *See* pupillary reflex. { 'west,fōlz ¦pyü·pə,ler·ē 'rē,fleks }

wetlands [ECOL] An area characterized by a high content of soil moisture, such as a swamp or bog. { 'wet ˌlanz }

whale [VERT ZOO] A large marine mammal of the order Cetacea; the body is streamlined, the broad flat tail is used for propulsion, and the limbs are balancing structures. { wāl }

whalebone *See* baleen. { 'wāl,bōn }

Wharton's duct *See* submandibular duct. { 'wȯrt·ənz ˌdəkt }

wheat [BOT] A food grain crop of the genus *Triticum*; plants are self-pollinating; the inflorescence is a spike bearing sessile spikelets arranged alternately on a zigzag rachis. { wēt }

wheat germ [BOT] The embryo of a wheat grain. { 'wēt ˌjərm }

whelk [INV ZOO] A gastropod mollusk belonging to the order Neogastropoda; species are carnivorous but also scavenge. { welk }

whip grafting [BOT] A method of grafting by fitting a small tongue and notch cut in the base of the scion into corresponding cuts in the stock. { 'wip ˌgraft·iŋ }

Whipple's disease [MED] A disease characterized by infiltration of the intestinal wall and lymphatics by macrophages filled with glycoprotein. Also known as intestinal lipodystrophy. { 'wip·əlz di,zēz }

whipworm disease [MED] A chronic, wasting diarrhea produced by heavy parasitization of the large intestine by the nematode *Trichuris trichiura*, particularly in undernourished children in the tropics. { 'wip,wərm di,zēz }

whiskers *See* vibrissae. { 'wis·kərz }

white adipose tissue [HISTOL] The most common type of adipose tissue, representing stored food reserves and thermal and physical insulation. { ¦wīt ,ad·ə,pōs 'tish·ü }

white ant *See* termite. { 'wīt 'ant }

white band disease [INV ZOO] A coral reef disease that is typified by a loss of tissue that is visible as a band of bare white skeleton. { ,wīt 'band diz,ēz }

white blood cell *See* leukocyte. { 'wīt 'bləd ,sel }

white corpuscle *See* leukocyte. { 'wīt 'kȯr·pə·səl }

white diarrhea *See* pullorum disease. { 'wīt ,dī·ə'rē·ə }

whitefish [VERT ZOO] Any of various food fishes in the family Salmonidae, especially of the genus *Coregonus*, characterized by an adipose dorsal fin and nearly toothless mouth. { 'wīt,fish }

white infarct [MED] An infarct in which hemorrhage is slight, or that has been decolorized by removal of blood or its pigments. { 'wīt 'in ,färkt }

white muscardine [INV ZOO] A disease of the silkworm caused by the fungus *Beauveria bassiana*. { ,wīt 'məs·kər,dēn }

white potato *See* potato. { 'wīt pə'tād·ō }

whooping cough *See* pertussis. { 'hüp·iŋ ,kȯf }

whooping crane [VERT ZOO] *Grus americana*. A member of a rare North American migratory species of wading birds; the entire species forms a single population. { 'hüp·iŋ ,krān }

whorl [BOT] An arrangement of several identical anatomical parts, such as petals, in a circle around the same point. [FOREN] A fingerprint pattern in which at least two deltas are present with a recurve in front of each. { wərl }

Whytt's reflex *See* pupillary reflex. { 'wī·əts ,rē ,fleks }

Widal test [IMMUNOL] A macroscopic or microscopic agglutination test for the diagnosis of typhoid fever and other *Salmonella* infections by using killed or preserved bacteria as the antigen. { we'däl ,test }

wide cross [GEN] A mating between individuals of different genera. { 'wīd 'krȯs }

Wiesen *See* meadow. { 'vēz·ən }

wild boar [VERT ZOO] *Sus scrofa*. A wild hog with coarse, grizzled hair and enlarged tusks or canines on both jaws. Also known as boar. { 'wīld 'bȯr }

wild cinnamon *See* bayberry. { 'wīld 'sin·ə·mən }

wild type [GEN] The most prevalent allele or character in the wild organism. { 'wīld ,tīp }

willow [BOT] A deciduous tree and shrub of the genus *Salix*, order Salicales; twigs are often yellow-green and bear alternate leaves which are characteristically long, narrow, and pointed, usually with fine teeth along the margins. { 'wil·ō }

Wilms' tumor [MED] A malignant renal tumor composed principally of mesodermal tissues. Also known as nephroblastoma. { 'vilmz ,tü·mər }

Wilson's disease [MED] A hereditary disease of ceruloplasmin formation transmitted as an autosomal recessive and characterized by decreased serum ceruloplasmin and copper values, and increased excretion of copper in the urine. Also known as hepatolenticular degeneration. { 'wil·sənz di,zēz }

wilting point [BOT] A condition in which a plant begins to use water from its own tissues for transpiration because soil water has been exhausted. { 'wilt·iŋ ,pȯint }

windburn [BOT] Injury to plant foliage, caused by strong, hot, dry winds. [MED] A superficial inflammation of the skin, analogous to sunburn, caused by exposure to wind, especially a hot dry wind, inducing a dilation of the surface blood vessels. { 'win,bərn }

wing [ZOO] Any of the paired appendages serving as organs of flight on many animals. { wiŋ }

Winteraceae [BOT] A family of dicotyledonous plants in the order Magnoliales distinguished by hypogynous flowers, exstipulate leaves, air vessels absent, and stamens usually laminar. { ,win·tə'rās·ē,ē }

Wirsung's duct [ANAT] The adult pancreatic duct in man, sheep, ganoid fish, teleost fish, and frog. { 'vir,zùŋz ,dəkt }

Wiskott-Aldrich syndrome [MED] A hereditary immunodeficiency disease characterized by a low number of small platelets in the blood, eczema, and defective cell-mediated immunity. { ¦wis·kət 'ȯl,drich ,sin,drōm }

Wnt proteins [CELL MOL] A widely conserved family of secreted signaling molecules that regulate many processes during animal development but, when misregulated, can also contribute to several types of cancer. { ¦dəb·əl·yü¦en¦tē 'prō ,tēnz }

wobble pairing [CELL MOL] The ability of a transfer ribonucleic acid molecule to recognize more than one codon. { 'wäb·əl ,per·iŋ }

Wolbachieae [MICROBIO] A tribe of the family Rickettsiaceae; rickettsialike organisms found principally in arthropods. { wōl'bak·ē·ē }

wolf [VERT ZOO] Any of several wild species of the genus *Canis* in the family Canidae which are fierce and rapacious, sometimes attacking humans; includes the red wolf, gray wolf, and coyote. { wùlf }

Wolffian duct *See* mesonephric duct. { 'wùl·fē·ən 'dəkt }

wolfsbane *See* aconite. { 'wùlfs,bān }

wolverine [VERT ZOO] *Gulo gulo*. A carnivorous mammal which is the largest and most vicious member of the family Mustelidae. { ¦wùl·və¦rēn }

wood [BOT] The hard fibrous substance that makes up the trunks and large branches of trees beneath the bark. [ECOL] A dense growth of trees, more extensive than a grove and smaller than a forest. { wu̇d }

woodland *See* forest; temperate woodland. { 'wu̇d·lənd }

woodpecker [VERT ZOO] A bird of the family Picidae characterized by stiff tail feathers and zygodactyl feet which enable them to cling to a tree trunk while drilling into the bark for insects. { 'wu̇d,pek·ər }

wood ray [BOT] A vascular ray consisting of a radial row of parenchyma cells in secondary xylem. Also known as xylem ray. { 'wu̇d ,rā }

wood sugar *See* xylose. { 'wu̇d ,shu̇g·ər }

wool [VERT ZOO] The soft undercoat of various animals such as sheep, angora, goat, camel, alpaca, llama, and vicuna. { wu̇l }

wool-sorter's disease *See* anthrax. { 'wu̇l ¦sȯrd·ərz di,zēz }

worker [INV ZOO] One of the neuter, usually sterile individuals making up a caste of social insects, such as ants, termites, or bees, which labor for the colony. { 'wər·kər }

work metabolism [PHYSIO] Metabolism in excess of resting metabolism that can be related to the performance of a specific task. { 'wərk mə,tab·ə,liz·əm }

work strain [PHYSIO] The response of the human body to work stress experienced in the performance of a task. { 'wərk ,strān }

worm [INV ZOO] **1.** The common name for members of the Annelida. **2.** Any of various elongated, naked, soft-bodied animals resembling an earthworm. { wərm }

wound botulism [MED] Botulism that involves production of toxin by the organisms infecting or colonizing a wound. { ,wünd 'bäch·ə,liz·əm }

wound periderm [BOT] A protective tissue that develops within injured plant organs beneath wound surfaces. { 'wünd ,per·ə,dərm }

wound shock *See* hypovolemic shock. { 'wünd ,shäk }

wren [VERT ZOO] Any of the various small brown singing birds in the family Troglodytidae; they are insectivorous and tend to inhabit dense, low vegetation. { ren }

Wright's inbreeding coefficient *See* inbreeding coefficient. { ¦rīts 'in,brēd·iŋ ,kō·ə,fish·ənt }

wrist [ANAT] The part joining forearm and hand. { rist }

wrist socket *See* wrist. { 'rist ,säk·ət }

wryneck *See* torticollis. { 'rī,nek }

wuchereriasis [MED] Infection with worms of the genus *Wuchereria*. Also known as Bancroft's filariasis. { ,wu̇·kə·rē'rī·ə·səs }

xanthan [BIOCHEM] A polysaccharide produced by *Xanthomonas campestris* that is used in oil recovery to help improve water flooding and oil displacement. { 'zan·thən }

xanthelasma [MED] Raised yellow plaques occurring around the eyelids, resulting from lipid-filled cells in the dermis. { ˌzan·thə'laz·mə }

Xanthidae [INV ZOO] The mud crabs, a family of decapod crustaceans in the section Brachyura. { 'zan·thəˌdē }

xanthine oxidase [BIOCHEM] A flavoprotein enzyme catalyzing the oxidation of certain purines. { 'zanˌthēn 'äk·səˌdās }

xanthism [BIOL] A color variation in which an animal's normal coloring is largely replaced by yellow pigments. Also known as xanthochroism. { 'zanˌthiz·əm }

xanthochroism *See* xanthism. { ¦zan·thə'krōˌiz·əm }

xanthoma [MED] A yellowish mass of lipid-filled histocytes occurring in subcutaneous tissue, often around tendons. { zan'thō·mə }

xanthomatosis [MED] A condition marked by the deposit of a yellowish or orange lipoid material in the reticuloendothelial cells, the skin, and the internal organs. { ˌzanˌthō·mə'tō·səs }

Xanthomonas citri [MICROBIO] The bacterial pathogen that causes citrus canker. { ˌzan·thəˌmō·nəs 'siˌtrē }

xanthomycin [MICROBIO] An antibiotic produced by a strain of *Streptomyces* and composed of two varieties, A and B; active in low concentrations against a number of gram-positive microorganisms. { ˌzan·thə'mīs·ən }

xanthophore [CELL MOL] A yellow chromatophore. { 'zan·thəˌfȯr }

Xanthophyceae [BOT] A class of yellow-green to green flagellate organisms of the division Chrysophyta; zoologists classify these organisms in the order Heterochlorida. { ˌzan·thə'fīs·ēˌē }

xanthophyll [BIOCHEM] $C_{40}H_{56}O_2$ Any of a group of yellow, alcohol-soluble carotenoid pigments that are oxygen derivatives of the carotenes, and are found in certain flowers, fruits, and leaves. Also known as carotenol; lutein. { 'zan·thəˌfil }

xanthurenic acid [BIOCHEM] $C_{10}H_7NO_4$ Sulfur yellow crystals with a melting point of 286°C; soluble in aqueous alkali hydroxides and carbonates; excreted by pyridoxine-deficient animals after ingestion of tryptophan. { ¦zan·thyə¦ren·ik 'as·əd }

Xantusiidae [VERT ZOO] The night lizards, a family of reptiles in the suborder Sauria. { ˌzan·tə'sī·əˌdē }

X chromosome [GEN] The sex chromosome occurring in double dose in the homogametic female sex and in single dose in the heterogametic male sex in mammals, *Diosophila*, and many other organisms. { 'eks 'krō·məˌsōm }

Xenarthra [VERT ZOO] A suborder of mammals in the order Edentata including sloths, anteaters, and related forms; posterior vertebrae have extra articular facets and vertebrae in the hip, and shoulder regions tend to be fused. { zə'när·thrə }

xenobiotic [BIOCHEM] A chemical that is not normally found in the body, such as a drug. { ˌzēn·ə·bī'äd·ik }

xenogamy [BOT] Cross-fertilization between flowers on different plants. { zə'näg·ə·mē }

xenogeneic [IMMUNOL] Referring to cells, tissues, or organs used in transplantation that originate in a different species. { ¦zēn·ə·jə'nē·ik }

xenograft [IMMUNOL] A graft performed between members of different species. { 'zēn·əˌgraft }

xenophyae [INV ZOO] In xenophyophores, the inorganic portion of the test consisting of foreign elements, such as sponge spicules, foraminiferan or radiolarian tests, and mineral particles. { zə'näf·ēˌē }

Xenophyophorea [INV ZOO] A class of giant marine benthic Rhizopoda. { zə¦näf·ē·ə'fȯr·ē·ə }

Xenophyophorida [INV ZOO] An order of Protozoa in the subclass Granuloreticulosia; includes deep-sea forms that develop as discoid to fan-shaped branching forms which are multinucleate at maturity. { ¦zē·nōˌfī·ə'fȯr·əd·ə }

Xenopneusta [INV ZOO] A small order of worm-like animals belonging to the Echiurida. { ˌzēn·əp'nüs·tə }

Xenopterygii [VERT ZOO] The equivalent name for Gobiesociformes. { zəˌnäp·tə'rij·ēˌī }

Xenosauridae [VERT ZOO] A family of four rare species of lizards in the suborder Sauria; composed of the Chinese lizard (*Shinisaurus crocodilurus*) and three Central American species of the genus *Xenosaurus*. { ˌzēn·ə'sȯr·əˌdē }

xerarch succession [ECOL] A type of succession that originates in a dry habitat. { 'zer,ärk sək,sesh·ən }

xeric [ECOL] **1.** Of or pertaining to a habitat having a low or inadequate supply of moisture. **2.** Of or pertaining to an organism living in such an environment. { 'zer·ik }

xeroderma pigmentosum [MED] A genodermatosis characterized by premature degenerative changes in the form of keratoses, malignant epitheliomatosis, and hyper- and hypopigmentation. { ,zir·ə'dər·mə ,pig·mən'tō·səm }

xerodermosteosis *See* Sjögren's syndrome. { ,zir·ō·dər,mäs·tē'ō·səs }

xerophthalmia [MED] Dryness and thickening of the conjunctiva, sometimes following chronic conjunctivitis, disease of the lacrimal apparatus, or vitamin A deficiency. { ,zi,räf'thal·mē·ə }

xerophyte [ECOL] A plant adapted to life in areas where the water supply is limited. { 'zir·ə,fīt }

xerosere [ECOL] A temporary community in an ecological succession on dry, sterile ground such as rock, sand, or clay. { 'zir·ə,sir }

xerosis conjunctivae [MED] A condition marked by silver-gray, shiny, triangular spots on both sides of the cornea, within the region of the palpebral aperture, consisting of dried epithelium, flaky masses, and microorganisms. { zə'rō·səs kən'jəŋk·tə,vē }

xerotolerance [PHYSIO] The ability to grow in extremely dry habitats. { 'zer·ə,täl·ə·rəns }

xiphidio cercaria [INV ZOO] A digenetic trematode larva having a stylet in the oral sucker. { zə'fid·ē·ō sər'kar·ē·ə }

Xiphiidae [VERT ZOO] The swordfishes, a family of perciform fishes in the suborder Scombroidei characterized by a tremendously pronounced bill. { zə'fī·ə,dē }

xiphisternum [ANAT] The elongated posterior portion of the sternum. { ¦zif·ə'stər·nəm }

Xiphosura [INV ZOO] The equivalent name for Xiphosurida. { ,zif·ə'su̇r·ə }

Xiphosurida [INV ZOO] A subclass of primitive arthropods in the class Merostomata characterized by cephalothoracic appendages, ocelli, book lungs, a somewhat trilobed body, and freely articulating styliform telson. { ,zif·ə'su̇r·ə·də }

Xiphydriidae [INV ZOO] A family of the Hymenoptera in the superfamily Siricoidea. { ,zif·ə'drī·ə,dē }

XIST [GEN] The X-linked gene whose RNA transcript mediates X-chromosome inactivation in mammals.

X-linked adrenoleukodystrophy [MED] An inherited peroxisomal disorder in which a defective peroxisomal membrane protein results in impaired oxidation of very long chain fatty acids, which subsequently accumulate in the adrenal glands, nervous system, and testes and disrupt normal activity. { ¦eks ,liŋkt ə,drē·nō,lü·kə'dis·trə·fē }

X organ [INV ZOO] A cluster of neurosecretory cells of the medulla terminales, a portion of the brain lying in the eyestalk in stalk-eyed crustaceans. { 'eks ,ȯr·gən }

x-ray therapy [MED] Medical treatment by controlled application of x-rays; a type of radiotherapy. { 'eks ,rā ,ther·ə·pē }

Xyelidae [INV ZOO] A family of hymenopteran insects in the superfamily Megalodontoidea. { zī'el·ə,dē }

xylan [BIOCHEM] A polysaccharide composed of the pentose sugar D-xylose. { 'zī,lan }

xylem [BOT] The principal water-conducting tissue and the chief supporting tissue of higher plants; composed of tracheids, vessel members, fibers, and parenchyma. { 'zī·ləm }

Xylocopidae [INV ZOO] A family of hairy tropical bees in the superfamily Apoidea. { ,zīl·ə'käp·ə,dē }

Xylomyiidae [INV ZOO] A family of orthorrhaphous dipteran insects in the series Brachycera. { ,zīl·ə'mī·ə,dē }

xylophagous [BIOL] Referring to an organism which feeds on wood. { zī'läf·ə·gəs }

xylose [BIOCHEM] $C_5H_{10}O_5$ A pentose sugar found in many woody materials; combustible, white crystals with a sweet taste; soluble in water and alcohol; melts about 148°C; used as a nonnutritive sweetener and in dyeing and tanning. Also known as wood sugar. { 'zī,lōs }

Xyridaceae [BOT] A family of terrestrial monocotyledonous plants in the order Commelinales characterized by an open leaf sheath, three stamens, and a simple racemose head for the inflorescence. { ,zir·ə'dās·ē,ē }

Y

yak [VERT ZOO] *Poephagus grunniens.* A heavily built, long-haired mammal of the order Artiodactyla, with a shoulder hump; related to the bison, and resembles it in having 14 pairs of ribs. { yak }

yam [BOT] **1.** A plant of the genus *Dioscorea* grown for its edible fleshy root. **2.** An erroneous name for the Puerto Rico variety of sweet potato; the edible, starchy tuberous root of the plant. { yam }

yaws [MED] An infectious tropical disease of humans caused by the spirochete *Treponema pertenue;* manifested by a primary cutaneous lesion followed by a granulomatous skin eruption. { 'yȯz }

Y chromosome [GEN] The sex chromosome found only in the heterogametic male sex, as in mammals and *Drosophila.* { 'wī 'krō·mə,sōm }

yeast [MYCOL] A collective name for those fungi which possess, under normal conditions of growth, a vegetative body (thallus) consisting, at least in part, of simple, individual cells. { yēst }

yellow fat cell [HISTOL] A large, generally spherical fat cell with a thin shell of protoplasm and a single enlarged fat droplet which appears yellowish. { 'yel·ō 'fat ,sel }

yellow fever [MED] An acute, febrile, mosquito-borne viral disease characterized in severe cases by jaundice, albuminuria, and hemorrhage. { 'yel·ō 'fē·vər }

yellow-green algae [BOT] The common name for members of the class Xanthophyceae. { 'yel·ō ¦grēn 'al·jē }

Yersinia [MICROBIO] A genus of gram-negative, facultative, rod-shaped bacteria in the Enterobacteriaceae family that shares many physiological properties with related *Escherichia coli,* including metabolic processes and sensitivity to certain bacteriophages. { yər'sin·ē·ə }

yew [BOT] A genus of evergreen trees and shrubs, *Taxus,* with the fruit, an aril, containing a single seed surrounded by a scarlet, fleshy, cuplike envelope; the leaves are flat and acicular. { yü }

Y ligament *See* iliofemoral ligament. { 'wī ,lig·ə·mənt }

yohimbine [PHARM] $C_{21}H_{26}N_2O_3$ An alkaloid derived from plants that is a weak blocker of alpha-adrenergic receptors, used as a mydriatic and in the treatment of impotence. { yō'him,bīn }

yolk [BIOCHEM] **1.** Nutritive material stored in an ovum. **2.** The yellow spherical mass of food material that makes up the central portion of the egg of a bird or reptile. { yōk }

yolk larva *See* vitellarium. { 'yōk ,lär·və }

yolk sac [EMBRYO] A distended extraembryonic extension, heavy-laden with yolk, through the umbilicus of the midgut of the vertebrate embryo. { 'yōk ,sak }

Y organ [INV ZOO] Either of a pair of nonneural structures found in the anterior portion of the crustacean body; source of the molting hormone, ecdysone. { 'wī ,ȯr·gən }

Young-Helmholtz theory [NEUROSCI] A theory of color vision according to which there are three types of color receptors that respond to short, medium, and long waves respectively; primary colors are those that stimulate most successfully the three types of receptors. Also known as Helmholtz theory. { 'yəŋ 'helm,hōlts ,thē·ə·rē }

Yponomeutidae [INV ZOO] A heterogeneous family of small, often brightly colored moths in the superfamily Tineoidea; the head is usually smooth with reduced or absent ocelli. { ē,pän·ə'myüd·ə,dē }

Ypsilothuriidae [INV ZOO] A family of echinoderms in the order Dactylochirotida having 8–10 tentacles, a permanent spire on the plates of the test, and the body fusiform or U-shaped. { ,ip·sə·lō'thu̇r·ə,dē }

Z

Zanclidae [VERT ZOO] The Moorish idols, a family of Indo-Pacific perciform fishes in the suborder Acanthuroidei. { 'zaŋ·klə,dē }

Zapodidae [VERT ZOO] The Northern Hemisphere jumping mice, a family of the order Rodentia with long legs and large feet adapted for jumping. { zə'pȯid·ə,dē }

Z chromosome [GEN] The sex chromosome found in both sexes in species with female heterogamety. { 'zē ,krō·mə,sōm }

zebra [VERT ZOO] Any of three species of African mammals belonging to the family Equidae distinguished by a coat of black and white stripes. { 'zē·brə }

zebu [VERT ZOO] A domestic breed of cattle, indigenous to India, belonging to the family Bovidae, distinguished by long drooping ears, a dorsal hump between the shoulders, and a dewlap under the neck; known as the Brahman in the United States. { 'zē·bü }

Zeiformes [VERT ZOO] The dories, a small order of teleost fishes, distinguished by the absence of an orbitosphenoid bone, a spinous dorsal fin, and a pelvic fin with a spine and five to nine soft rays. { ,zē·ə'fȯr,mēz }

Zeitgeber [PHYSIO] A periodic environmental condition or event that acts to set or reset an innate biological rhythm of an organism. { 'tsīt,gā·bər }

zenodiagnosis [MED] A procedure of using a suitable arthropod to transfer an infectious agent from a patient to a susceptible laboratory animal. { ¦zē·nō,dī·əg'nō·səs }

Zeoidea [VERT ZOO] An equivalent name for Zeiformes. { zē'ȯid·ē·ə }

Zeomorphi [VERT ZOO] An equivalent name for Zeiformes. { ,zē·ə'mȯr·fī }

zero population growth [ECOL] A theory which advocates that there be no increase in population, that each person replace only oneself, and that birth control be practiced in all nations. { 'zir·ō 'päp·yə,lā·shən ,grōth }

zeugmatography [MED] A technique that combines nuclear magnetic resonance spectroscopy with methods of focusing and scanning radio waves to produce cross-sectional images of the human body that indicate proton density and relaxation times, and distinguish normal tissue from tumors. Also known as nuclear magnetic resonance tomography. { ,züg·mə'täg·rə·fē }

Ziehl-Neelsen stain [MICROBIO] A procedure for acid-fast staining of tubercle bacilli with carbol fuchsin. { 'zēl 'nēl·sən ,stān }

zinc chills *See* metal fume fever. { 'ziŋk ,chilz }

zinc finger [BIOCHEM] A small structural domain that is organized around a zinc ion and is found in many gene-regulatory proteins. { ¦ziŋk ¦fiŋ·gər }

Zingiberaceae [BOT] A family of aromatic monocotyledonous plants in the order Zingiberales characterized by one functional stamen with two pollen sacs, distichously arranged leaves and bracts, and abundant oil cells. { ,zin·jə·bə 'rās·ē,ē }

Zingiberales [BOT] An order of monocotyledonous herbs or scarcely branched shrubs in the subclass Commelinidae characterized by pinnately veined leaves and irregular flowers that have well-differentiated sepals and petals, an inferior ovary, and either one or five functional stamens. { ,zin·jə·bə'rā·lēz }

Z line [HISTOL] The line formed by attachment of the actin filaments between two sarcomeres. { 'zē ,līn }

Zoantharia [INV ZOO] A subclass of the class Anthozoa; individuals are monomorphic and most have retractile, simple, tubular tentacles. { ,zō·ən'thar·ē·ə }

Zoanthidea [INV ZOO] An order of anthozoans in the subclass Zoantharia; these are mostly colonial, sedentary, skeletonless, anemonelike animals that live in warm, shallow waters and coral reefs. { ,zō·ən'thid·ē·ə }

Zoarcidae [VERT ZOO] The eelpouts, a family of actinopterygian fishes in the order Gadiformes which inhabit cold northern and far southern seas. { zō'är·sə,dē }

zoea [INV ZOO] An early larval stage of decapod crustaceans distinguished by a relatively large cephalothorax, conspicuous eyes, and large, fringed antennae. { zō'ē·ə }

Zollinger-Ellison syndrome [MED] Gastric hypersecretion and hyperacidity, fulminating intractable atypical peptic ulceration, and hyperplasia of the islet cells of the pancreas. { 'zäl·ən·jər 'el·ə·sən ,sin,drōm }

zonal centrifuge [BIOL] A centrifuge that uses a rotating chamber of large capacity in which to separate cell organelles by density-gradient centrifugation. { 'zōn·əl 'sen·trə,fyüj }

zona fasciculata [ANAT] The middle tissue layer of the adrenal cortex where glucocorticoids, mainly cortisol and corticosterone, are synthesized and secreted. { ,zō·nə fə,sik·yə'läd·ə }

zona glomerulosa [ANAT] The outer tissue layer of the adrenal cortex where mineralocorticoids, primarily aldosterone, are synthesized and secreted. { ,zō·nə glə,mer·yə'lō·sə }

zona pellucida [HISTOL] The thick, solid, elastic envelope of the ovum. Also known as oolemma. { 'zō·nə pə'lüs·əd·ə }

zona reticularis [ANAT] The inner tissue layer of the adrenal cortex that is involved in the synthesis and secretion of sex steroid precursors and, to less extent, glucocorticoids. { ,zō·nə re,tik·yə'lär·əs }

zonation [ECOL] Arrangement of organisms in biogeographic zones. { zō'nā·shən }

zone of optimal proportion [IMMUNOL] One of three zones considered to appear when antigen and antibody are mixed; it is that zone in which there is no uncombined antigen or antibody. Also known as equivalence zone. { 'zōn əv 'äp·tə·məl prə'pȯr·shən }

zone of physiological zero [PHYSIO] The band of skin temperatures between about 86 and 97°F (30 and 36°C). { ,zōn əv ,fiz·ē·ə,läj·ə·kəl 'zir·ō }

zone theory *See* stage theory. { 'zōn ,thē·ə·rē }

zonite [INV ZOO] A body segment in Diplopoda. { 'zō,nīt }

zonula occludens *See* tight junction. { ¦zōn·yə·lə ə'klüd·ənz }

zoochlorellae [BIOL] Unicellular green algae which live as symbionts in the cytoplasm of certain protozoans, sponges, and other invertebrates. { ¦zō·ə·klə'rel,ē }

zoochory [BOT] Dispersal of plant disseminules by animals. { 'zō·ə,klȯr·ē }

zooecium [INV ZOO] The exoskeleton of a feeding zooid in bryozoans. { zō·'ē·shē·əm }

zoogeographic region [ECOL] A major unit of the earth's surface characterized by faunal homogeneity. { ¦zō·ə¦jē·ə¦graf·ik 'rē·jən }

zoogeography [BIOL] The science that attempts to describe and explain the distribution of animals in space and time. { ¦zō·ə,jē'äg·rə·fē }

zoogloea [MICROBIO] A gelatinous or mucilaginous mass characteristic of certain bacteria grown in organic-rich fluid media. { ,zō·ə 'glē·ə }

zooid [INV ZOO] A more or less independent individual of colonial animals such as bryozoans and coral. { 'zō,ȯid }

zoology [BIOL] The science that deals with knowledge of animal life. { zō'äl·ə·jē }

Zoomastigina [INV ZOO] The equivalent name for Zoomastigophorea. { ,zō·ə,mas·tə'jīn·ə }

Zoomastigophorea [INV ZOO] A class of flagellate protozoans in the subphylum Sarcomastigophora; some are simple, some are specialized, and all are colorless. { ,zō·ə,mas·tə·gə'fȯr·ē·ə }

zoonoses [BIOL] Diseases which are biologically adapted to and normally found in lower animals but which under some conditions also infect humans. { ,zō·ə'nō·sēz }

zooplankton [ECOL] Microscopic animals which move passively in aquatic ecosystems. { ¦zō·ə'plaŋk·tən }

zoosphere [ECOL] The world community of animals. { 'zō·ə,sfir }

zoosporangium [BOT] A spore case bearing zoospores. { ¦zō·ə·spə'ran·jē·əm }

zoospore [BIOL] An independently motile spore. { 'zō·ə,spȯr }

zooxanthellae [BIOL] Microscopic yellow-green algae which live symbiotically in certain radiolarians and marine invertebrates. { ,zō·ə·zan 'thē,lē }

Zoraptera [INV ZOO] An order of insects, related to termites and psocids, which live in decaying wood, sheltered from light; most individuals are wingless, pale in color, and blind. { zə'rap·tə·rə }

Zoroasteridae [INV ZOO] A family of deep-water asteroid echinoderms in the order Forcipulatida. { ,zȯr·ō·ə'ster·ə,dē }

Zorotypidae [INV ZOO] The single family, containing one genus, *Zorotypus*, in the order Zoraptera. { ,zȯr·ə'tīp·ə,dē }

zoster *See* herpes zoster. { 'zäs·tər }

Zosteraceae [BOT] A family of monocotyledonous plants in the order Najadales; the group is unique among flowering plants in that they grow submerged in shallow ocean waters near the shore. { ,zäs·tə'rās·ē,ē }

Zygaenidae [INV ZOO] A diverse family of small, often brightly colored African moths in the superfamily Zygaenoidea. { zī'jēn·ə,dē }

Zygaenoidea [INV ZOO] A superfamily of moths in the suborder Heteroneura characterized by complete venation, rudimentary palpi, and usually a rudimentary proboscis. { ,zī·jə'nȯid·ē·ə }

Zygnemataceae [BOT] A family of filamentous plants in the order Conjugales; they are differentiated into genera by chloroplast morphology, which may be spiral, bandlike, or cushionlike. { zig,nēm·ə'tās·ē,ē }

zygodactyl [VERT ZOO] Of birds, having a toe arrangement of two in front and two behind. { ¦zī·gō¦dak·təl }

zygomatic bone [ANAT] A bone of the side of the face below the eye; forms part of the zygomatic arch and part of the orbit in mammals. Also known as malar bone. { ¦zī·gə¦mad·ik 'bōn }

zygomorphic [BIOL] Bilaterally symmetrical. { ¦zī·gə¦mȯr·fik }

Zygomycetes [MYCOL] A class of fungi in the division Eumycetes. { ,zī·gō,mī'sēd·ēz }

Zygomycotina [MYCOL] A subdivision of fungi characterized by distinctive sexual and asexual reproductive stages. { ,zī·gə,mī·kə'tē·nə }

zygopophysis [ANAT] One of the articular processes of the neural arch of a vertebra. { ,zī·gə'päf·ə·səs }

Zygoptera [INV ZOO] The damsel flies, a suborder of insects in the order Odonata; individuals are slender, dainty creatures, often with bright-blue or orange coloring and usually with clear or transparent wings. { zī'gäp·tə·rə }

zygospore [BOT] A thick-walled cell or resting spore that results from the fusion of similar reproductive cells, especially in organisms that reproduce by conjugation. { 'zī·gə,spȯr }

zygote [EMBRYO] **1.** An organism produced by the union of two gametes. **2.** The fertilized ovum before cleavage. { 'zī,gōt }

zygotene [CELL MOL] The stage of meiotic prophase during which homologous chromosomes synapse; visible bodies in the nucleus are now bivalents. Also known as amphitene. { 'zī·gə,tēn }

zygotic induction [VIROL] Phage induction following conjugation of a lysogenic bacterium with a nonlysogenic one. { zī'gäd·ik in'dək·shən }

zymase [BIOCHEM] A complex of enzymes that catalyze glycosis. { 'zī,mās }

zymogen [BIOCHEM] The inactive precursor of an enzyme; liberates an active enzyme on reaction with an appropriate kinose. Also known as proenzyme. { 'zī·mə·jən }

zymogen granules [BIOCHEM] Granules of zymogen in gland cells, particularly those of the pancreatic acini and of the gastric chief cells. { 'zī·mə·jən 'gran,yülz }

zymogenic [MICROBIO] Obtaining energy by amylolitic processes. { ¦zī·mə¦jen·ik }

zymophore [BIOCHEM] The active portion of an enzyme. { 'zī·mə,fȯr }

zymosis *See* fermentation. { zī'mō·səs }

zymosterol [BIOCHEM] $C_{27}H_{43}OH$ An unsaturated sterol obtained from yeast fat; yields cholesterol on hydrogenation. { zī'mäs·tə,rȯl }

Zythiaceae [MYCOL] A family of fungi of the order Sphaeropsidales which contains many plant and insect pathogens. { ,zith·ē·'ās·ē,ē }

Appendix

Appendix

Equivalents of commonly used units for the U.S. Customary System and the metric system

1 inch = 2.5 centimeters (25 millimeters) 1 foot = 0.3 meter (30 centimeters) 1 yard = 0.9 meter 1 mile = 1.6 kilometers	1 centimeter = 0.4 inch 1 meter = 3.3 feet 1 meter = 1.1 yards 1 kilometer = 0.62 mile	1 inch = 0.083 foot 1 foot = 0.33 yard (12 inches) 1 yard = 3 feet (36 inches) 1 mile = 5280 feet (1760 yards)
1 acre = 0.4 hectare 1 acre = 4047 square meters	1 hectare = 2.47 acres 1 square meter = 0.00025 acre	
1 gallon = 3.8 liters 1 fluid ounce = 29.6 milliliters 32 fluid ounces = 946.4 milliliters	1 liter = 1.06 quarts = 0.26 gallon 1 milliliter = 0.034 fluid ounce	1 quart = 0.25 gallon (32 ounces; 2 pints) 1 pint = 0.125 gallon (16 ounces) 1 gallon = 4 quarts (8 pints)
1 quart = 0.95 liter 1 ounce = 28.35 grams 1 pound = 0.45 kilogram 1 ton = 907.18 kilograms	1 gram = 0.035 ounce 1 kilogram = 2.2 pounds 1 kilogram = 1.1×10^{-3} ton	1 ounce = 0.0625 pound 1 pound = 16 ounces 1 ton = 2000 pounds
°F = (1.8 × °C) + 32	°C = (°F − 32) ÷ 1.8	

Appendix

Conversion factors for the U.S. Customary System, metric system, and International System

A. Units of length

Units	*cm*	*m*	*in.*	*ft*	*yd*	*mi*
1 cm	= 1	0.01	0.3937008	0.03280840	0.01093613	6.213712×10^{-6}
1 m	= 100.	1	39.37008	3.280840	1.093613	6.213712×10^{-4}
1 in.	= 2.54	0.0254	1	0.08333333...	0.02777777...	1.578283×10^{-5}
1 ft	= 30.48	0.3048	12.	1	0.3333333...	$1.893939... \times 10^{-4}$
1 yd	= 91.44	0.9144	36.	3.	1	$5.681818... \times 10^{-4}$
1 mi	$= 1.609344 \times 10^{5}$	1.609344×10^{3}	6.336×10^{4}	5280.	1760.	1

B. Units of area

Units	*cm*^{2}	*m*^{2}	*in.*^{2}	*ft*^{2}	*yd*^{2}	*mi*^{2}
1 cm^2	= 1	10^{-4}	0.1550003	1.076391×10^{-3}	1.195990×10^{-4}	3.861022×10^{-11}
1 m^2	$= 10^{4}$	1	1550.003	10.76391	1.195990	3.861022×10^{-7}
1 in.^{2}	= 6.4516	6.4516×10^{-4}	1	$6.944444... \times 10^{-3}$	7.716049×10^{-4}	2.490977×10^{-10}
1 ft^2	= 929.0304	0.09290304	144.	1	0.1111111...	3.587007×10^{-8}
1 yd^2	= 8361.273	0.8361273	1296.	9.	1	3.228306×10^{-7}
1 mi^2	$= 2.589988 \times 10^{10}$	2.589988×10^{6}	4.014490×10^{9}	2.78784×10^{7}	3.0976×10^{6}	1

C. Units of volume

Units	m^3	cm^3	liter	$in.^3$	ft^3	qt	gal
1 m^3	= 1	10^6	10^3	6.102374×10^4	35.31467×10^{-3}	1.056688	264.1721
1 cm^3	= 10^{-6}	1	10^{-3}	0.06102374	3.531467×10^{-5}	1.056688×10^{-3}	2.641721×10^{-4}
1 liter	= 10^{-3}	1000.	1	61.02374	0.03531467	1.056688	0.2641721
1 $in.^3$	= 1.638706×10^{-5}	16.38706	0.01638706	1	5.787037×10^{-4}	0.01731602	4.329004×10^{-3}
1 ft^3	= 2.831685×10^{-2}	28316.85	28.31685	1728.	1	2.992208	7.480520
1 qt	= 9.46352×10^{-4}	946.359	0.946351	57.75	0.03342014	1	0.25
1 gal (U.S.)	= 3.785412×10^{-3}	3785.412	3.785412	231.	0.1336806	4.	1

D. Units of mass

Units	g	kg	oz	lb	metric ton	ton
1 g	= 1	10^{-3}	0.03527396	2.204623×10^{-3}	10^{-6}	1.102311×10^{-6}
1 kg	= 1000.	1	35.27396	2.204623	10^{-3}	1.102311×10^{-3}
1 oz (avdp)	= 28.34952	0.02834952	1	0.0625	2.834952×10^{-5}	3.125×10^{-5}
1 lb (avdp)	= 453.5924	0.4535924	16.	1	4.535924×10^{-4}	$5. \times 10^{-4}$
1 metric ton	= 10^8	1000.	35273.96	2204.623	1	1.102311
1 ton	= 907184.7	907.1847	32000.	2000.	0.9071847	1

Conversion factors for the U.S. Customary System, metric system, and International System (*cont.*)

E. Units of density

Units	$g \cdot cm^{-3}$	$g \cdot L^{-1}$, $kg \cdot m^{-3}$	$oz \cdot in.^{-3}$	$lb \cdot in.^{-3}$	$lb \cdot ft^{-3}$	$lb \cdot gal^{-1}$
1 $g \cdot cm^{-3}$	= 1	1000.	0.5780365	0.03612728	62.42795	8.345403
1 $g \cdot L^{-1}$, $kg \cdot m^{-3}$	= 10^{-3}	1	5.780365×10^{-4}	3.612728×10^{-5}	0.06242795	8.345403×10^{-3}
1 $oz \cdot in.^{-3}$	= 1.729994	1729.994	1	0.0625	108.	14.4375
1 $lb \cdot in.^{-3}$	= 27.67991	27679.91	16.	1	1728.	231.
1 $lb \cdot ft^{-3}$	= 0.01601847	16.01847	9.259259×10^{-3}	5.787037×10^{-4}	1	0.1336806
1 $lb \cdot gal^{-1}$	= 0.1198264	119.8264	4.749536×10^{-3}	4.329004×10^{-3}	7.480519	1

F. Units of pressure

Units	Pa, $N \cdot m^{-2}$	$dyn \cdot cm^{-2}$	*bar*	*atm*	$kgf \cdot cm^{-2}$	*mmHg (torr)*	*in.* Hg	$lbf \cdot in.^{-2}$
1 Pa, 1 $N \cdot m^{-2}$	= 1	10	10^{-5}	9.869233×10^{-6}	1.019716×10^{-5}	7.500617×10^{-3}	2.952999×10^{-4}	1.450377×10^{-4}
1 $dyn \cdot cm^{-2}$	= 0.1	1	10^{-6}	9.869233×10^{-7}	1.019716×10^{-6}	7.500617×10^{-4}	2.952999×10^{-5}	1.450377×10^{-5}
1 bar	= 10^5	10^6	1	0.9869233	1.019716	750.0617	29.52999	14.50377
1 atm	= 101325	101325.0	1.01325	1	1.033227	760.	29.92126	14.69595
1 $kgf \cdot cm^{-2}$	= 98066.5	980665	0.980665	0.9678411	1	735.5592	28.95903	14.22334
1 mmHg (torr)	= 133.3224	1333.224	1.333224×10^3	1.315789×10^{-3}	1.359510×10^{-3}	1	0.03937008	0.01933678
1 in. Hg	= 3386.388	33863.88	0.03386388	0.03342105	0.03453155	25.4	1	0.4911541
1 $lbf \cdot in.^{-2}$	= 6894.757	68947.57	0.06894757	0.06804596	0.07030696	51.71493	2.036021	1

G. Units of energy

Units	g mass (energy equiv.)	J	eV	cal	cal_{IT}	Btu_{IT}	kWh	hp-h	ft-lbf	$ft^3 \cdot lbf \cdot in.^{-2}$	liter-atm
1 g mass (energy equiv.)	= 1	8.987552×10^{13}	5.609589×10^{32}	2.148076×10^{3}	2.146640×10^{13}	8.518555×10^{10}	2.496542×10^{7}	3.347918×10^{7}	6.628878×10^{13}	4.603388×10^{11}	8.870024×10^{11}
1 J	$= 1.112650 \times 10^{-14}$	1	6.241510×10^{18}	0.2390057	0.2388459	9.478172×10^{-4}	$2.777777\ldots \times 10^{-7}$	3.725062	0.7375622	5.121960×10^{-3}	9.869233×10^{-3}
1 eV	$= 1.782662 \times 10^{-33}$	1.602176×10^{-19}	1	3.829293×10^{-20}	3.826733×10^{-20}	1.518570×10^{-22}	4.450490×10^{-26}	5.968206×10^{-26}	1.181705×10^{-19}	8.206283×10^{-22}	1.581225×10^{-21}
1 cal	$= 4.655328 \times 10^{-14}$	4.184	2.611448×10^{19}	1	0.9993312	3.965667×10^{-3}	$1.1622222\ldots \times 10^{-6}$	1.558562×10^{-6}	3.085960	2.143028×10^{-2}	0.04129287
1 cal_{IT}	$= 4.658443 \times 10^{-14}$	4.1868	2.613195×10^{19}	1.000669	1	3.968321×10^{-3}	1.163×10^{-6}	1.559609×10^{-6}	3.088025	2.144462×10^{-2}	0.04132050
1 Btu_{IT}	$= 1.173908 \times 10^{-11}$	1055.056	6.585141×10^{21}	252.1644	251.9958	1	2.930711×10^{-4}	3.930148×10^{-4}	778.1693	5.403953	10.41259
1 kWh	$= 4.005540 \times 10^{-8}$	3600000.	2.246944×10^{25}	860420.7	859845.2	3412.142	1	1.341022	2655224.	18349.06	35529.24
1 hp-h	$= 2.986931 \times 10^{-8}$	2384519.	1.675545×10^{25}	641615.6	641186.5	2544.33	0.7456998	1	1980000.	13750.	26494.15
1 ft-lbf	$= 1.508551 \times 10^{-14}$	1.355818	8.462351×10^{18}	0.3240483	0.3238315	1.285067×10^{-3}	3.766161×10^{-7}	$5.050505\ldots \times 10^{-7}$	1	$6.944444\ldots \times 10^{-3}$	0.01338088
1 ft^3 lbf · $in.^{-2}$	$= 2.172313 \times 10^{-12}$	195.2378	1.218579×10^{21}	46.66295.	46.63174	0.1850497	5.423272×10^{-5}	$7.272727\ldots \times 10^{-5}$	144.	1	1.926847
1 liter-atm	$= 1.127393 \times 10^{-12}$	101.325	6.324210×10^{20}	24.21726	24.20106	0.09603757	2.814583×10^{-5}	3.774419×10^{-5}	74.73349	0.5189825	1

Appendix

ABO blood group system

Blood group	RBC (*red blood cell*) *antigens*	*Possible genotypes*	*Plasma antibody*
A	A	AA or A/O	Anti-B
B	B	B/B or B/O	Anti-A
O	—	O/O	Anti-A and anti-B
AB	A and B	A/B	—

Amino acid abbreviations

Ala	Alanine	Leu	Leucine
Arg	Arginine	Lys	Lysine
Asn	Asparagine	Met	Methionine
Asp	Aspartic acid	Phe	Phenylalanine
Cys	Cysteine	Pro	Proline
Gln	Glutamine	Ser	Serine
Glu	Glutamic acid	Thr	Threonine
Gly	Glycine	Trp	Tryptophan
His	Histidine	Tyr	Tyrosine
Ile	Isoleucine	Val	Valine

Appendix

Universal (standard) genetic code*

RNA BASES:	U		C		A		G	
U	UUU UUC	Phe	UCU UCC UCA UCG	Ser	UAU UAC	Tyr	UGU UGC	Cys
	UUA UUG	Leu			UAA UAG	Stop	UGA	Stop
							UGG	Trp
C	CUU CUC CUA CUG	Leu	CCU CCC CCA CCG	Pro	CAU CAC	His	CGU CGC CGA CGG	Arg
					CAA CAG	Gln		
A	AUU AUC AUA	Ile	ACU ACC ACA ACG	Thr	AAU AAC	Asn	AGU AGC	Ser
	AUG	Met			AAA AAG	Lys	AGA AGG	Arg
G	GUU GUC GUA GUG	Val	GCU GCC GCA GCG	Ala	GAU GAC	Asp	GGU GGC GGA GGG	Gly
					GAA GAG	Glu		

*Each of the 64 codons found in mRNA specifies an amino acid (indicated by the common three-letter abbreviation) or the end of the protein chain (stop). U, uracil; C, cytosine; A, adenine; G, guanine.

Appendix

Some functions of essential vitamins

Vitamins	Functions	Best sources	Deficiency	*Daily recommended dietary allowance* (RDA) *for adults*	
				Men	*Women*
FAT-SOLUBLE					
Vitamin A (retinoids, carotenes)	Maintenance of vision in dim light, growth, reproduction	Fish liver oils; liver; dairy products; yellow, orange, and green plants; carrots; sweet potatoes	Poor growth and night vision; blindness	900 μg or 3000 IU	700 μg or 2330 IU
Vitamin D (cholecalciferol)	Rickets-preventive factor, calcification of bones, calcium and phosphorus metabolism	Fish oils, fortified dairy products	Rickets, osteomalacia	5 μg or 200 IU* (ages 19–50) 10 μg or 400 IU* (ages 50–70) 15 μg or 600 IU* (ages 70+)	5 μg or 200 IU* (ages 19–50) 10 μg or 400 IU* (ages 50–70) 15 μg or 600 IU* (ages 70+)
Vitamin E (tocopherols)	Antioxidant, membrane integrity and metabolism, heme synthesis	Grains and vegetable oils	Neuropathy	15 mg or 22 IU	15 mg or 22 IU
Vitamin K	Blood-clotting factor	Green vegetables	Bleeding	120 μg	90 μg

WATER-SOLUBLE					
Ascorbic acid (vitamin C)	Antiscorbutic (scurvy-preventive) factor, collagen formation, neurotransmitter synthesis	Citrus fruits, fresh vegetables, potatoes	Scurvy	90 mg	75 mg
Thiamine (vitamin B-1)	Antiberiberi factor, energy utilization, particularly from carbohydrates	Pork, liver, whole grains	Beriberi	1.2 mg	1.1 mg
Riboflavin (vitamin B-2)	Energy utilization, protein metabolism	Milk, egg white, liver, leafy vegetables	Cheilosis, glossitis	1.3 mg	1.1 mg
Niacin (vitamin B-3)	Antipellagra factor, energy release from carbohydrate, fat, and protein	Yeast, wheat germ, meats	Pellagra	16 mg	14 mg
Vitamin B-6 (pyridoxine, pyridoxal, pyridoxamine)	Coenzyme for protein metabolism	Whole grains, yeast, egg yolk, liver	Skin disorders; convulsion in infants	1.3 mg (ages 19–50) 1.7 mg (ages 50+)	1.3 mg (ages 19–50) 1.5 mg (ages 50+)
Pantothenic acid	Metabolism of protein, carbohydrate, fat	Liver, kidney, green vegetables, egg yolk		5 mg*	5 mg*
Folate	Transfer of one-carbon units in metabolism	Liver, deep-green leafy vegetables	Macrocytic anemias	400 μg	400 μg
Vitamin B-12 (cobalamin)	Blood formation, nervous tissue metabolism	Liver, kidney, yeast	Skin disorders	2.4 μg	2.4 μg
Biotin	Synthesis and oxidation of fatty acids and carbohydrates	Liver, meats	Pernicious anemia	30 μg*	30 μg*

*Adequate intake (AI); used when there is insufficient data to establish an RDA.

SOURCES: *Facts About Dietary Supplements*, NIH Clinical Center, National Institutes of Health, http://www.cc.nih.gov/ccc/supplements/intro.html; Dietary Reference Intakes: Vitamins, Food and Nutrition Board, National Academy of Science, www4.nationalacademies.org/iom/iomhome.nsf/WFiles/webtablevitamins/$file/webtablevitamins.pdf.

Appendix

Major groups of viruses*

Group	Host	Morphology	Examples of viruses
Class I viruses: double-stranded DNA genomes			
Myoviridae	Bacteria	Complex	T4
Siphoviridae	Bacteria	Complex	λ
Podoviridae	Bacteria	Complex	T7
Papovaviridae	Animal	Naked icosahedral	Polyomavirus, SV40
Adenoviridae	Animal	Naked icosahedral	Adenovirus
Herpesviridae	Animal	Enveloped icosahedral	Herpes simplex, varicella-zoster
Poxviridae	Animal	Complex	Smallpox, vaccinia
Hepadnaviridae	Animal	Enveloped icosahedral	Hepatitis B
Caulimoviruses	Plant	Naked icosahedral	Cauliflower mosaic
Class II viruses: single-stranded DNA genomes			
Microviridae	Bacteria	Naked icosahedral	ϕX174
Parvoviridae	Animal	Naked icosahedral	Parvovirus, adeno-associated virus
Geminiviruses	Plant	Fused-pair icosahedral	Maize streak
Class III viruses: double-stranded RNA genomes			
Reoviridae	Animal	Naked icosahedral	Reovirus, rotavirus
Class IV viruses: positive-strand RNA genomes			
Leviviridae	Bacteria	Naked icosahedral	MS2, Qβ
Picornaviridae	Animal	Naked icosahedral	Poliovirus, rhinovirus, hepatitis A, coxsackievirus
Togaviridae	Animal	Enveloped icosahedral	Sindbis
Coronaviridae	Animal	Enveloped helical	Murine hepatitis
Potyvirus	Plant	Naked helical	Potato Y
Tymovirus	Plant	Naked icosahedral	Turnip yellow mosaic
Tobamovirus	Plant	Naked helical	Tobacco mosaic
Comovirus	Plant	Naked icosahedral	Cowpea mosaic
Class V viruses: negative-strand RNA genomes			
Rhabdoviridae	Animal and plant	Enveloped helical	Rabies, vesicular stomatitis
Paramyxoviridae	Animal	Enveloped helical	Mumps, measles, parainfluenza
Filoviridae	Animal	Enveloped helical	Ebola
Orthomyxoviridae	Animal	Enveloped helical	Influenza A, B
Bunyaviridae	Animal	Enveloped helical	Phlebovirus
Arenaviridae	Animal	Enveloped helical	Lassa
Class VI viruses: retroviruses			
Retroviridae	Animal	Enveloped icosahedral	Human immunodeficiency virus (HIV)

*Reproduced with permission from B.A. Voyles, *The Biology of Viruses*, 2d ed., McGraw-Hill, 2002.

Appendix

Cranial nerves of vertebrates

Number	*Name*	*Fiber types*	*Peripheral origin or destination*	*Vertebrates possessing this nerve*
—	Terminal	Somatic sensory	Anterior nasal epithelium	Almost all
I	Olfactory	Special sensory	Olfactory mucosa	All
—	Vomeronasal	Special sensory	Vomeronasal mucosa	Almost all
II	Optic	Special sensory	Retina of eye	All
III	Oculomotor	Somatic motor	Four extrinsic eye muscles	All
IV	Trochlear	Somatic motor	One extrinsic eye muscle	All
V	Trigeminal	Special visceral motor	Muscles of mandibular arch derivative	All
		Somatic sensory	Most of head	All
VI	Abducens	Somatic motor	One extrinsic eye muscle	All
—	Anterior lateral line	Special sensory	Lateral line organs of head	Fish and larval amphibians
VII	Facial	Special visceral motor	Muscles of hyoid arch derivative	All
		General visceral motor	Salivary glands	All
		Somatic sensory	Small part of head	All
		Visceral sensory	Anterior pharynx	All
		Special sensory	Taste, anterior tongue	All
VIII	Vestibulocochlear	Special sensory	Inner ear	All
	Posterior lateral line	Special sensory	Lateral line organs of trunk	Fish and larval amphibians
IX	Glossopharyngeal	Special visceral motor	Muscles of third branchial arch	All
		General visceral motor	Salivary gland	All
		Somatic sensory	Skin near ear	All
		Visceral sensory	Part of pharynx	All
		Special sensory	Taste, posterior tongue	All
X	Vagus	Special visceral motor	Muscles of arches 4–6	All
		General visceral motor	Most viscera of entire trunk	All
		Visceral sensory	Larynx and part of pharynx	All
		Special sensory	Taste, pharynx	All
XI	Spinal accessory	Special visceral motor	Some muscles of arches 4–6	Reptiles, birds, mammals
XII	Hypoglossal	Somatic motor	Muscles of tongue and anterior throat	Reptiles, birds, mammals

Appendix

Classification of living organisms

- Domain Archaea[a]
  - Phylum Crenarchaeota
    - Class Thermoprotei
      - Order Thermoproteales
      - Order Desulfurococcales
      - Order Sulfolobales
  - Phylum Euryarchaeota
    - Class Methanobacteria
      - Order Methanobacteriales
    - Class Methanococci
      - Order Methanococcales
      - Order Methanomicrobiales
      - Order Methanosarcinales
    - Class Halobacteria
      - Order Halobacteriales
    - Class Thermoplasmata
      - Order Thermoplasmatales
    - Class Thermococci
      - Order Thermococcales
    - Class Archaeoglobi
    - Class Methanopyri
      - Order Methanopyrales

- Domain Bacteria
  - Phylum Aquificae
    - Class Aquificae
      - Order Aquificales
  - Phylum Thermotogae
    - Class Thermotogae
      - Order Thermotogales
  - Phylum Thermodesulfobacteria
    - Class Thermodesulfobacteria
      - Order Thermodesulfobacteriales
  - Phylum Deinococcus-Thermus
    - Class Deinococci
      - Order Deinococcales
      - Order Thermales
  - Phylum Chrysiogenetes
    - Class Chrysiogenetes
      - Order Chrysiogenales
  - Phylum Chloroflexi
    - Class Chloroflexi
      - Order Chloroflexales
      - Order Herpetosiphonales
  - Phylum Thermomicrobia
    - Class Thermomicrobia
      - Order Thermomicrobiales
  - Phylum Nitrospira
    - Class Nitrospira
      - Order Nitrospirales
  - Phylum Deferribacteres
    - Class Deferribacteres
      - Order Deferribacterales
  - Phylum Cyanobacteria
    - Class Cyanobacteria
  - Phylum Chlorobi
    - Class Chlorobia
      - Order Chlorobiales
  - Phylum Proteobacteria
    - Class Alphaproteobacteria
      - Order Rhodospirillales
      - Order Rickettsiales
      - Order Rhodobacterales
      - Order Sphingomonadales
      - Order Caulobacterales
      - Order Rhizobiales
    - Class Betaproteobacteria
      - Order Burkholderiales
      - Order Hydrogenophilales
      - Order Methylophilales
      - Order Neisseriales
      - Order Nitrosomonadales
      - Order Rhodocyclales
    - Class Gammaproteobacteria
      - Order Chromatiales
      - Order Acidithiobacillales
      - Order Xanthomonadales
      - Order Cardiobacteriales
      - Order Thiotrichales
      - Order Legionellales
      - Order Methylococcales
      - Order Oceanospirillales
      - Order Pseudomonadales
      - Order Alteromonadales
      - Order Vibrionales
      - Order Aeromonadales
      - Order Enterobacteriales
      - Order Pasteurellales
    - Class Deltaproteobacteria
      - Order Desulfurellales
      - Order Desulfovibrionales
      - Order Desulfobacterales
      - Order Desulfuromonadales
      - Order Syntrophobacterales
      - Order Bdellovibrionales
      - Order Myxococcales
    - Class Epsilonproteobacteria
      - Order Campylobacterales
  - Phylum Firmicutes
    - Class Clostridia
      - Order Clostridiales
      - Order Thermoanaerobacteriales
      - Order Haloanaerobiales
    - Class Mollicutes
      - Order Mycoplasmatales
      - Order Entomoplasmatales
      - Order Acholeplasmatales
      - Order Anaeroplasmatales
    - Class Bacilli
      - Order Bacillales
      - Order Lactobacillales
  - Phylum Actinobacteria
    - Class Actinobacteria
    - Subclass Acidimicrobidae
      - Order Acidimicrobiales
      - Suborder Acidimicrobineae
    - Subclass Rubrobacteridae
      - Order Rubrobacterales
      - Suborder Rubrobacterineae
    - Subclass Coriobacteridae
      - Order Coriobacteriales
      - Suborder Coriobacterineae
    - Subclass Sphaerobacteridae
      - Order Sphaeriobacterales
      - Suborder Sphaerobacterineae
    - Subclass Actinobacteridae
      - Order Actinomycetales
      - Suborder Actinomycineae
      - Suborder Micrococcineae
      - Suborder Corynebacterineae
      - Suborder Micromonosporineae
      - Suborder Propionibacterineae
      - Suborder Pseudonocardineae
      - Suborder Streptomycineae
      - Suborder Streptosporangineae
      - Suborder Frankineae
      - Suborder Glycomycineae
      - Order Bifidobacteriales
  - Phylum Planctomycetes

Classification of living organisms (*cont.*)

Class Planctomycetacia
Order Planctomycetales
Phylum Chlamydiae
Class Chlamydiae
Order Chlamydiales
Phylum Spirochaetes
Class Spirochaetes
Order Spirochaetales
Phylum Fibrobacteres
Class Fibrobacteres
Order Fibrobacterales
Phylum Acidobacteria
Class Acidobacteria
Order Acidobacteriales
Phylum Bacteroidetes
Class Bacteroidetes
Order Bacteroidales
Class Flavobacteria
Order Flavobacteriales
Class Sphingobacteria
Order Sphingobacteriales
Phylum Fusobacteria
Class Fusobacteria
Order Fusobacteriales
Phylum Verrucomicrobia
Class Verrucomicrobiae
Order Verrucomicrobiales
Phylum Dictyoglomus
Class Dictyoglomi
Order Dictyoglomales

Domain Eukarya[b]

Kingdom Protista
Phylum Metamonada
Phylum Trichozoa
Subphylum Parabasala
Class Trichomonadea
Class Hypermastigotea

Subkingdom Neozoa
Phylum Choanozoa
Phylum Amoebozoa
Subphylum Lobosa
Subphylum Conosa
Class Archamoebae
Class Mycetozoa
Phylum Foraminifera
Phylum Percolozoa
Phylum Euglenozoa
Class Euglenoidea
Class Saccostomae
Phylum Sporozoa
Subphylum Gregarinae
Subphylum Coccidiomorpha
Subphylum Perkinsida
Subphylum Manubrispora
Phylum Ciliophora
Phylum Radiozoa
Phylum Heliozoa
Phylum Rhodophyta
Class Rhodophyceae
Subclass Bangiophycidae
Order Bangiales
Order Compsopogonales
Order Porphyridiales
Order Rhodochaetales
Subclass Florideophycidae
Order Acrochaetiales
Order Ahnfeltiales
Order Balbianiales
Order Balliales
Order Batrachospermales
Order Bonnemaisoniales
Order Ceramiales
Order Colaconematales
Order Corallinales
Order Gelidiales
Order Gigartinales
Order Gracilariales
Order Halymeniales
Order Hildenbrandiales
Order Nemaliales
Order Palmariales
Order Plocamiales
Order Rhodogorgonales
Order Rhodymeniales
Order Thoreales
Phylum Chrysophyta
Class Bacillariophyceae
Subclass Bacillariophycidae
Order Achnanthales
Order Bacillariales
Order Cymbellales
Order Dictyoneidales
Order Lyrellales
Order Mastogloiales
Order Naviculales
Order Rhopalodiales
Order Surirellales
Order Thallassiophysales
Subclass Biddulphiophycidae
Order Anaulales
Order Biddulphiales
Order Hemiaulales
Order Triceratiales
Subclass Chaetocerotophycidae
Order Chaetocerotales
Order Leptocylindrales
Subclass Corethrophycidae
Order Cymatosirales
Subclass Coscinodiscophycidae
Order Arachnoidiscales
Order Asterolamprales
Order Aulacoseirales
Order Chrysaanthemodiscales
Order Coscinodiscales
Order Ethmodiscales
Order Melosirales
Order Orthoseirales
Order Paraliales
Order Stictocyclales
Order Stictodiscales
Subclass Cymatosirophycidae
Order Cymatosirales
Subclass Eunotiophycidae
Order Eunotiales
Subclass Fragilariophycidae
Order Ardissoneales
Order Cyclophorales
Order Climacospheniales
Order Fragilariales
Order Licmorphorales
Order Protoraphidales
Order Rhabdonematales
Order Rhaphoneidales
Order Striatellales
Order Tabellariales
Order Thalassionematales
Order Toxariales
Subclass Lithodesmiophycidae
Order Lithodesmiales
Subclass Rhizosoleniophycidae
Order Rhizosoleniales
Subclass Thalassiosirophycidae

Classification of living organisms (*cont.*)

Order Thalassiosirales
Class Bolidophyceae
Order Bolidomonadales
Class Chrysomerophyceae
Order Chrysomeridales *nom. nud.*
Class Chrysophyceae
Order Chromulinales
Order Hibberdiales
Class Dictyochophyceae
Order Dictyochales
Order Pedinellales
Order Rhizochromulinales
Class Eustigmatophyceae
Order Eustigmatales
Class Pelagophyceae
Order Pelagomonadales
Order Sarcinochrysidales
Class Phaeophyceae
Order Ascoseirales
Order Chordariales
Order Cutleriales
Order Desmarestiales
Order Dictyosiphonales
Order Dictyotales
Order Durvillaeales
Order Ectocarpales
Order Fucales
Order Laminariales
Order Scytosiphonales
Order Sphacelariales
Order Sporochnales
Order Tilopteridiales
Class Phaeothamniophyceae
Order Phaeothamniales
Order Pleurochloridellales
Class Pinguiophyceae
Order Pinguiochrysidales
Class Raphidophyceae
Order Rhaphidomonadales
Class Synurophyceae
Order Synurales
Class Xanthophyceae (=Tribophyceae)
Order Botrydiales
Order Chloramoebales
Order Heterogloeales
Order Mischococcales
Order Rhizochloridales
Order Tribonematales
Order Vaucheriales
Phylum Cryptophyta
Class Cryptophyceae
Order Cryptomonadales
Order Cryptococcales
Phylum Glaucocystophyta
Class Glaucocystophyceae
Order Cyanophorales
Order Glaucocystales
Order Gloeochaetales
Phylum Prymnesiophyta (=Haptophyta)
Class Pavlovophyceae
Order Pavlovales
Class Prymnesiophyceae
Order Coccolithales
Order Isochrysidales
Order Phaeocystales
Order Prymnesiales
Phylum Dinophyta
Class Dinophyceae
Order Actiniscales
Order Blastodiniales
Order Chytriodiniales
Order Desmocapsales
Order Desmomonadales
Order Dinophysales
Order Gonyaulacales
Order Gymnodiniales
Order Kolkwitziellales
Order Nannoceratopsiales
Order Noctilucales
Order Oxyrrhinales
Order Peridiniales
Order Phytodiniales
Order Prorocentrales
Order Ptychodiscales
Order Pyrocysales
Order Suessiales
Order Syndiniales
Order Thoracosphaerales
Phylum Chlorophyta
Class Charophyceae
Order Charales
Order Chlorokybales
Order Coleochaetales
Order Klebsormidiales
Order Zygnematales
Class Chlorophyceae
Order Chaetophorales
Order Chlorococcales
Order Cladophorales
Order Odeogoniales
Order Sphaeropleales
Order Volvocales
Order Pleurastrales
Class Prasinophyceae
Order Chlorodendrales
Order Mamiellales
Order Pseudoscourfeldiales
Order Pyramimonidales
Class Trebouxiophyceae
Order Trebouxiales
Class Ulvophyceae
Order Bryopsidales
Order Caulerpales
Order Codiolales
Order Dasycladales
Order Halimedales
Order Prasiolales
Order Siphonocladales
Order Trentepohliales
Order Ulotrichales
Order Ulvales
Phylum Euglenophyta
Class Euglenophyceae
Order Euglenales
Order Euglenamorphales
Order Eutreptiales
Order Heteronematales
Order Rhabdomonadales
Order Sphenomonadales
Phylum Acrasiomycota
Class Acrasiomycetes
Order Acrasiales
Phylum Dictyosteliomycota
Class Dictyosteliomycetes
Order Dictyosteliales
Phylum Myxomycota
Class Myxomycetes
Order Liceales
Order Echinosteliales
Order Trichiales
Order Physarales
Order Stemonitales
Order Ceratiomyxales
Class Protosteliomycetes
Order Protosteliales
Phylum Plasmodiophoromycota
Class Plasmodiophoromycetes
Order Plasmodiophorales
Phylum Oomycota
Class Oomycetes
Order Saprolegniales
Order Salilagenidiales
Order Leptomitales
OrderMyzocytiopsidales
Order Rhipidiales
Order Pythiales
Order Peronosporales
Phylum Hyphochytriomycota

Classification of living organisms (*cont.*)

Class Hyphochytriomycetes
Order Hyphochytriales
Phylum Labyrinthulomycota
Class Labyrinthulomycetes
Order Labyrinthulales
Phylum Chytridiomycota
Class Chytridiomycetes
Order Blastocladiales
Order Chytridiales
Order Monoblepharidales
Order Neocallimastigales
Order Spizellomycetales
Phylum Zygomycota
Class Trichomycetes
Order Amoebidiales
Order Asellariales
Order Eccrinales
Order Harpellales
Class Zygomycetes
Order Mucorales
Order Dimargaritales
Order Kickxellales
Order Endogonales
Order Glomales
Order Entomophthorales
Order Zoopagales
Phylum Ascomycota
Class Archiascomycetes
Order Taphrinales
Order Schizosaccharomycetales
Class Saccharomycetes
Order Saccharomycetales
Class Plectomycetes
Order Eurotiales
Order Ascosphaerales
Order Onygenales
Class Laboulbeniomycetes
Order Laboulbeniales
Order Spathulosporales
Class Pyrenomycetes
Order Hypocreales
Order Melanosporales
Order Microascales
Order Phylachorales
Order Ophiostomatales
Order Diaporthales
Order Calosphaceriales
Order Xylariales
Order Sordariales
Order Meliolales
Order Halosphaeriales
Class Discomycetes
Order Medeolariales
Order Rhytismatales
Order Ostropales
Order Cyttariales
Order Helotiales
Order Neolectales
Order Gyalectales
Order Lecanorales
Order Lichinales
Order Peltigerales
Order Pertusariales
Order Teloschistales
Order Caliciales
Order Pezizales
Class Loculoascomycetes
Order Coryneliales
Order Dothideales
Order Myriangiales
Order Arthoniales
Order Pyrenulales
Order Asterinales
Order Capnodiales
Order Chaetothyriales
Order Patellariales
Order Pleosporales
Order Melanommatales
Order Trichotheliales
Order Verrucariales
Phylum Basidiomycota
Class Basidiomycetes
Subclass Heterobasidiomycetes
Order Agaricostilbales
Order Atractiellales
Order Auriculariales
Order Heterogastridiales
Order Tremellales
Subclass Homobasidiomycetes
Order Agaricales
Order Boletales
Order Bondarzewiales
Order Cantharellales
Order Ceratobasidiales
Order Cortinariales
Order Dacrymycetales
Order Fistulinales
Order Ganodermatales
Order Gautieriales
Order Gomphales
Order Hericiales
Order Hymenochaetales
Order Hymenogastrales
Order Lachnocladiales
Order Lycoperdales
Order Melanogastrales
Order Nidulariales
Order Phallales
Order Poriales
Order Russulales
Order Schizophyllales
Order Sclerodermatales
Order Stereales
Order Thelephorales
Order Tulasnellales
Order Tulostomatales
Class Ustomycetes
Order Cryptobasidiales
Order Cryptomycocolacales
Order Exobasidiales
Order Graphiolales
Order Platyglocales
Order Sporidiales
Order Ustilaginales
Class Teliomycetes
Order Septobasidiales
Order Uredinales
Phylum Deuteromycetes (Asexual Ascomycetes and Basidiomycetes)
Class Hyphomycetes
Order Hyphomycetales
Order Stilbellales
Order Tuberculariales
Class Agonomycetes
Order Agonomycetales
Class Coelomycetes
Order Melanconiales
Order Sphaeropsidales
Order Pycnothyriales

Kingdom Plantae

Subkingdom Embryobionta
Division Hepaticophyta
Class Junermanniopsida
Order Calobryales
Order Jungermanniales
Order Metzgeriales

Classification of living organisms (*cont.*)

Class Marchantiopsida
Order Sphaerocarpales
Order Monocleales
Order Marchantiales
Division Anthocerotophyta
Class Anthocerotopsida
Order Anthocerotales
Division Bryophyta
Class Sphagnicopsida
Order Sphagnicales
Class Andreaeopsida
Order Andreaeles
Class Bryopsida
Order Archidiales
Order Bryales
Order Buxbaumiales
Order Dicranales
Order Encalyptales
Order Fissidentales
Order Funariales
Order Grimmiales
Order Hookeriales
Order Hypnobryales
Order Isobryales
Order Orthotrichales
Order Pottiales
Order Orthotrichales
Order Seligerales
Order Splachnales
Division Lycophyta
Class Lycopsida
Order Isoetales
Order Lycopodiales
Order Selaginellales
Division Polypodiophyta
Class Polypodopsida
Order Equisetales
Order Marattiales
Order Ophioglossales
Order Polypodiales
Order Psilotales
Division Pinophyta
Class Ginkgopsida
Order Ginkgoales
Class Cycadopsida
Order Cycadales
Class Pinopsida
Order Pinales
Order Podocarpales
Order Gnetales
Division Magnoliophyta
[unplaced orders]
Order Ceratophyllales
Order Chloranthales
Class Amborellopsida
Order Amborellales
Class Austrobaileyales
Order Austrobaileyales
Class Liliopsida
Order Acorales
Order Alismatales
Order Arecales
Order Asparagales
Order Commelinales
Order Dioscoreales
Order Liliales
Order Pandanales
Order Poales
Order Zingiberales
Class Magnoliopsida
Order Magnoliales
Order Laurales
Order Piperales
Order Canellales
Class Nymphaeopsida
Order Nymphaeales
Class Rosopsida
[unplaced orders]
Order Berberidopsidales
Order Buxales
Order Gunnerales
Order Proteales
Order Saxifragales
Order Santalales
Order Trochodendrales
Subclass Caryophyllidae
Order Caryophyllales
Order Dilleniales
Subclass Ranunculidae
Order Ranunculales
Subclass Rosidae
[unplaced orders]
Order Crossosomatales
Order Geraniales
Order Myrtales
Order Vitales
Superorder Rosanae
Order Celastrales
Order Cucurbitales
Order Fabales
Order Fagales
Order Malpighiales
Order Oxalidales
Order Rosales
Order Zygophyllales
Superorder Malvanae
Order Brassicales
Order Malvales
Order Sapindales
Subclass Asteridae
[unplaced order]
Order Boraginales
Superorder Cornanae
Order Cornales
Superorder Ericanae
Order Ericales
Superorder Lamianae
Order Garryales
Order Gentianales
Order Lamiales
Order Solanales
Superorder Asteranae
Order Apiales
Order Aquifoliales
Order Asterales
Order Dipsacales

Kingdom Animalia

Subkingdom Parazoa
Phylum Porifera
Subphylum Cellularia
Class Demospongiae
Class Calcarea
Subphylum Symplasma
Class Hexactinellida
Phylum Placozoa

Subkingdom Eumetazoa
Phylum Cnidaria
(=Coelenterata)
Class Scyphozoa
Order Stauromedusae
Order Coronatae
Order Semaeostomeae
Order Rhizostomeae
Class Cubozoa
Order Cubomedusae
Class Hydrozoa
Order Hydroida
Order Milleporina
Order Stylasterina
Order Trachylina

Classification of living organisms (*cont.*)

Order Siphonophora
Order Chondrophora
Order Actinulida
Class Anthozoa
Subclass Alcyonaria (=Octocorallia)
Order Stolonifera
Order Gorgonacea
Order Alcyonacea
Order Pennatulacea
Subclass Zoantharia (=Hexacorallia)
Order Actinaria
Order Corallimorpharia
Order Scleractinia
Order Zoanthinaria (=Zoanthidea)
Order Ceriantharia
Order Ptychodactiaria
Order Antipatharia
Phylum Ctenophora
Class Tentaculata
Order Cydippida
Order Platyctenida
Order Lobata
Order Cestida
Order Ganeshida
Order Thalassocalycida
Class Nuda
Order Beroida
Phylum Platyhelminthes
Class Turbellaria
Order Acoela
Order Rhabdocoela
Order Catenulida
Order Macrostomida
Order Nemertodermatida
Order Lecithoepitheliata
Order Polycladida
Order Prolecithophora (=Holocoela)
Order Proseriata
Order Tricladida
Order Neorhabdocoela
Class Cestoda
Subclass Cestodaria
Subclass Eucestoda
Order Caryophyllidea
Order Spathebothriidea
Order Trypanorhyncha
Order Pseudophyllidea
Order Tetraphyllidea
Order Cyclophyllidea
Class Monogenea
Class Trematoda
Subclass Digenea
Order Strigeidida
Order Azygiida
Order Echinostomida
Order Plagiorchiida
Order Opisthorchiida
Subclass Aspidogastrea (=Aspidobothrea)
Phylum Mesozoa
Class Orthonectida
Class Rhombozoa
Order Dicyemida
Order Heterocyemida
Phylum Myxozoa (=Myxospora)
Phylum Nemertea (=Rhynchocoela, Nemertinea)
Class Anopla
Order Palaeonemertea (=Palaeonemertini)
Order Heteronemertea
Class Enopla
Order Hoplonemertea (=Hoplonemertini)
Order Bdellonemertea
Phylum Gnathostomuilda
Order Filospermoidea
Order Bursovaginoidea
Phylum Gastrotricha
Order Chaetonotida
Order Macrodasyida
Phylum Cycliophora
Phylum Rotifera
Class Monogononta
Order Ploima
Order Flosculariaceae
Order Collothecaceae
Class Bdelloidea
Class Seisonidea
Phylum Acanthocephala
Class Archiacanthocephala
Class Eoacanthocephala
Class Palaeacanthocephala
Phylum Nematoda (=Nemata)
Class Adenophorea
Subclass Enoplia
Order Enoplida
Order Dorylaimida
Order Trichocephalida
Order Mermithida
Subclass Chromadoria
Class Secernentea
Subclass Rhabditia
Order Rhabditida
Order Ascaridida
Order Strongylida
Subclass Spiruria
Order Spirurida
Order Camallanida
Subclass Diplogasteria
Phylum Nematomorpha
Class Nectonematoida
Class Gordioida
Phylum Priapulida
Phylum Kinorhyncha (=Echinoderida)
Class Cyclorhagida
Class Homalorhagida
Phylum Loricifera
Phylum Mollusca
Subphylum Aculifera
Class Polyplacophora
Class Aplacophora
Subclass Neomeniophora (=Solenogastres)
Subclass Chaetodermo-morpha (=Caudofoveata)
Subphylum Conchifera
Class Monoplacophora
Class Gastropoda
Subclass Prosobranchia
Order Archaeogastropoda
Order Mesogastropoda (=Taenioglossa)
Order Neogastropoda
Subclass Opisthobranchia
Order Cephalaspidea
Order Runcinoidea
Order Acochlidioidea
Order Sacoglossa (=Ascoglossa)
Order Anaspidea (=Aplysiacea)
Order Notaspidea
Order Thecosomata
Order Gymnosomata

Appendix

Classification of living organisms (*cont.*)

Order Nudibranchia
Subclass Pulmonata
Order Archaeopulmonata
Order Basommatophora
Order Stylommatophora
Order Systellommato-phora
Class Bivalvia (=Pelecypoda)
Subclass Protobranchia (=Palaeotaxodonta, Cryptodonta)
Subclass Pteriomorphia
Subclass Paleoheterodonta
Subclass Heterodonta
Subclass Anomalodesmata
Class Scaphopoda
Class Cephalopoda
Subclass Nautiloidea
Subclass Coleoidea (=Dibranchiata)
Order Sepioidea
Order Teuthoidea (=Decapoda)
Order Vampyromorpha
Order Octopoda
Phylum Annelida
Class Polychaeta
Order Phyllodocida
Order Spintherida
Order Eunicida
Order Spionida
Order Chaetopterida
Order Magelonida
Order Psammodrilida
Order Cirratulida
Order Flabelligerida
Order Opheliida
Order Capitellida
Order Oweniida
Order Terebellida
Order Sabellida
Order Protodrilida
Order Myzostomida
Class Clitellata
Subclass Oligochaeta
Order Lumbriculida
Order Haplotaxida
Subclass Hirudinea
Order Rhynchobdellae
Order Arhynchobdellae
Order Branchiobdellida
Order Acanthobdellida
Class Pogonophora (=Siboglinidae)
Subclass Perviata (=Frenulata)
Subclass Obturata (=Vestimentifera)
Class Echiura
Order Echiura
Order Xenopneusta
Order Heteromyota
Phylum Sipuncula
Phylum Arthropoda
Subphylum Chelicerata
Class Merostomata
Order Xiphosura
Class Arachnida
Order Scorpiones
Order Uropygi
Order Amblypygi
Order Araneae
Order Ricinulei
Order Pseudoscorpiones
Order Solifugae (=Solpugida)
Order Opiliones
Order Acari
Class Pycnogonida (=Pantopoda)
Subphylum Mandibulata
Class Myriapoda
Order Chilopoda
Order Diplopoda
Order Symphyla
Order Pauropoda
Class Insecta (=Hexapoda)
Subclass Apterygota
Order Thysanura
Order Collembola
Subclass Pterygota
Superorder Hemime-tabola
Order Ephemeroptera
Order Odonata
Order Blattaria
Order Mantodea
Order Isoptera
Order Grylloblattaria
Order Orthoptera
Order Phasmida (=Phasmatoptera)
Order Dermaptera
Order Embiidina
Order Plecoptera
Order Psocoptera
Order Anoplura
Order Mallophaga
Order Thysanoptera
Order Hemiptera
Order Homoptera
Superorder Holometabola
Order Neuroptera
Order Coleoptera
Order Strepsiptera
Order Mecoptera
Order Siphonaptera
Order Diptera
Order Trichoptera
Order Lepidoptera
Order Hymenoptera
Class Crustacea
Subclass Cephalocarida
Subclass Malacostraca
Superorder Syncarida
Superorder Hoplocarida
Order Stomatopoda
Superorder Peracarida
Order Thermosbaenacea
Order Mysidacea
Order Cumacea
Order Tanaidacea
Order Isopoda
Order Amphipoda
Superorder Eucarida
Order Euphausiacea
Order Decapoda
Subclass Branchiopoda
Order Notostraca
Order Cladocera
Order Conchostraca
Order Anostraca
Subclass Ostracoda
Order Myodocopa
Order Podocopa
Subclass Mystacocarida
Subclass Copepoda
Order Calanoida
Order Harpacticoida
Order Cyclopoida
Order Monstrilloida
Order Siphonostomatoida